IMPORTANT!

HERE IS YOUR REGISTRATION CODE TO AC⟨
YOUR PREMIUM McGRAW-HILL ONLINE RE⟨

D0189018

For key premium online resources you need THIS CODE to gain access. Once the code is entered, you will be able to use the Web resources for the length of your course.

If your course is using **WebCT** or **Blackboard**, you'll be able to use this code to access the McGraw-Hill content within your instructor's online course.

Access is provided if you have purchased a new book. If the registration code is missing from this book, the registration screen on our Website, and within your WebCT or Blackboard course, will tell you how to obtain your new code.

Registering for McGraw-Hill Online Resources

TO gain access to your McGraw-Hill web resources simply follow the steps below:

1. USE YOUR WEB BROWSER TO GO TO: **www.mhhe.com/prescott6e**
2. CLICK ON **FIRST TIME USER**.
3. ENTER THE REGISTRATION CODE* PRINTED ON THE TEAR-OFF BOOKMARK ON THE RIGHT.
4. AFTER YOU HAVE ENTERED YOUR REGISTRATION CODE, CLICK **REGISTER**.
5. FOLLOW THE INSTRUCTIONS TO SET-UP YOUR PERSONAL UserID AND PASSWORD.
6. WRITE YOUR UserID AND PASSWORD DOWN FOR FUTURE REFERENCE. KEEP IT IN A SAFE PLACE.

TO GAIN ACCESS to the McGraw-Hill content in your instructor's **WebCT** or **Blackboard** course simply log in to the course with the UserID and Password provided by your instructor. Enter the registration code exactly as it appears in the box to the right when prompted by the system. You will only need to use the code the first time you click on McGraw-Hill content.

Thank you, and welcome to your McGraw-Hill online Resources!

Mc Graw Hill **Higher Education**

* YOUR REGISTRATION CODE CAN BE USED ONLY ONCE TO ESTABLISH ACCESS. IT IS NOT TRANSFERABLE.

0-07-294742-X T/A PRESCOTT/HARLEY: MICROBIOLOGY, 6E

REGISTRATION CODE

0PSE-2HK6-OJYT-YKVR-HP41

MCGRAW-HILL
⟨URCES

Mc Graw Hill **Higher Education**

Microbiology

Sixth Edition

Lansing M. Prescott
Professor Emeritus
Augustana College

John P. Harley
Eastern Kentucky University

Donald A. Klein
Colorado State University

McGraw Hill **Higher Education**

Boston   Burr Ridge, IL   Dubuque, IA   Madison, WI   New York   San Francisco   St. Louis
Bangkok   Bogotá   Caracas   Kuala Lumpur   Lisbon   London   Madrid   Mexico City
Milan   Montreal   New Delhi   Santiago   Seoul   Singapore   Sydney   Taipei   Toronto

Higher Education

MICROBIOLOGY, SIXTH EDITION

Published by McGraw-Hill, a business unit of The McGraw-Hill Companies, Inc., 1221 Avenue of the Americas, New York, NY 10020.

Some ancillaries, including electronic and print components, may not be available to customers outside the United States.

 This book is printed on recycled, acid-free paper containing 10% postconsumer waste.

1 2 3 4 5 6 7 8 9 0 VNH/VNH 0 9 8 7 6 5 4

ISBN 0–07–255678–1

Publisher: *Martin J. Lange*
Senior sponsoring editor: *Patrick E. Reidy*
Senior developmental editor: *Jean Sims Fornango*
Managing developmental editor: *Patricia Hesse*
Director of development: *Kristine Tibbetts*
Marketing manager: *Tami Petsche*
Senior project manager: *Jayne Klein*
Lead production supervisor: *Sandy Ludovissy*
Senior media project manager: *Jodi K. Banowetz*
Senior media technology producer: *John J. Theodald*
Designer: *Rick D. Noel*
Cover/interior designer: *Rokusek Design*
Cover image: *©Dennis Kunkel Microscopy, Inc., Penicillium Roqueforti (92386A)*
Lead photo research coordinator: *Carrie K. Burger*
Photo research: *Karen Pugliano*
Supplement producer: *Brenda A. Ernzen*
Compositor: *Carlisle Communications, Ltd.*
Typeface: *10/12 Times Roman*
Printer: *Van Hoffmann Corporation*

The credits section for this book begins on page C-1 and is considered an extension of the copyright page.

Library of Congress Cataloging-in-Publication Data

Prescott, Lansing M.
 Microbiology / Lansing M. Prescott, John P. Harley, Donald A. Klein. — 6th ed.
  p. cm.
 Includes bibliographical references and index.
 ISBN 0–07–255678–1 (hbk. : alk. paper) — ISBN 0–07–111216–2 (ISE)
 1. Microbiology. I. Harley, John P. II. Klein, Donald A., 1935–. III. Title.

QR41.2.P74 2005
579—dc22

2003025747
CIP

www.mhhe.com

Brief Contents

Contents

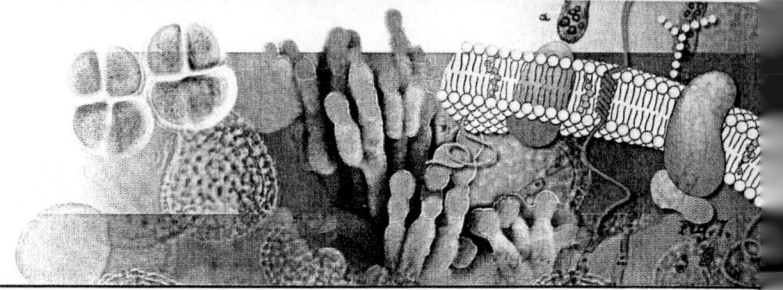

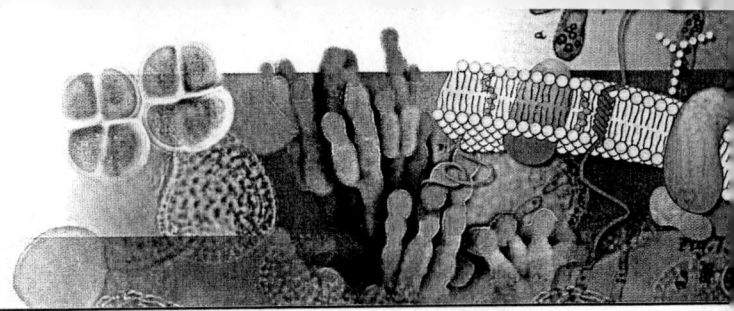

Preface

Because microbiology is an exceptionally broad discipline encompassing specialties as diverse as biochemistry, cell biology, genetics, taxonomy, pathogenic bacteriology, food and industrial microbiology, and ecology, our goal has been to provide a balanced introduction to the discipline. A microbiologist must be acquainted with many biological disciplines and with all major groups of microorganisms: viruses, bacteria, fungi, algae, and protozoa. Students new to the subject need an introduction to the whole before concentrating on specialized areas that might be of more interest to them. *Microbiology,* sixth edition, provides a balanced introduction to all major areas of microbiology for a variety of students. Because of this balance, the book is suitable for courses with orientations ranging from basic microbiology to medical and applied microbiology. Students preparing for careers in medicine, dentistry, nursing, and allied health professions will find the text just as useful as those aiming for careers in research, teaching, and industry. While two quarters/semesters each of biology and chemistry are assumed, we provide a strong overview of the relevant chemistry in appendix I.

OUR STRENGTHS

Main Themes

Seven themes permeate the text. They recur regularly and help integrate the specific information in an orderly manner. The seven themes are:

1. Development of microbiology as a science
2. Nature and importance of the techniques used to isolate, culture, observe, and identify microorganisms
3. Control of microorganisms and reduction of their detrimental effects
4. Importance of molecular biology and biochemistry for microbiology
5. Medical significance of microbiology
6. Ways in which microorganisms interact with their environments and the practical consequences of these interactions
7. Influences that microorganisms and microbiological applications have on everyday life

These themes help unify the text and enhance continuity. The student should get a feeling for what microbiologists do and for how their activities affect society.

Strong Biochemical Presentations

Despite the great variety in microbial structure and function, microorganisms share a biochemical unity that is basic to all life processes. Furthermore, specialized functions of individual microbial cells can only be described in biochemical terms. Thus it is not possible to understand microbiology in any fundamental sense without a consideration of biochemical mechanisms and the metabolic pathways common to all life. We provide biochemical background in two ways:

- First, you will find many illustrations that clarify the molecular processes being discussed in every chapter (for example, figures 9.19, 10.28, 11.17, and 31.19). The biochemical links are highlighted and described throughout.
- Second, two illustrated appendices (I and II) present graphic presentations of the Chemistry of Biological Molecules and Common Metabolic Pathways. This makes them easy to locate for reference purposes, and accessible for study and review.

Organizational Flexibility

Our flexible organization allows every instructor to sequence chapters and topics to suit their own syllabus. Each chapter is as self-contained as possible to promote this flexibility. For example, chapter 17, "The Viruses: Bacteriophages," contains all information critical to understanding the structure and function of bacteriophages. Students do not need to hunt through several chapters to assemble the information. They can return to chapter 17 to refer to specific details easily, making review a natural part of their study activities.

Readability

Because a student can not learn from a text they can not read, careful attention has been paid to the presentation of information in *Microbiology,* sixth edition. Comprehension is facilitated by a relatively simple, direct writing style. Information is broken up with numerous section headings and organized in an outline format within each chapter. The American Society for Microbiology's *ASM Style Manual* conventions for nomenclature and abbreviations were followed as consistently as possible. To help students with the many new terms they will encounter in the study of microbiology, new terminology is boldfaced when first used and clearly defined. Every term in the extensive glossary includes a page reference.

Study-Friendly Features

All students need help organizing their study time to maximize success. We have reorganized several key features to help them with this critical task. For example, the usefulness of the chapter summaries has been improved by organizing the summary statements under the appropriate chapter section number and title. A student can now go directly in the summary to a specific chapter section, rather than having to search for the desired statements. References in appendix V also are organized by numbered section headings within each chapter to facilitate the location of supplemental readings for specific topics.

In addition, no other text on the market today presents the reference resources that are part of *Microbiology*, sixth edition. These rich resources make it possible for students to expand their study, extend their reading, and use their text as a reference for many semesters to come. They include:

- A Review of the Chemistry of Biological Molecules (appendix I) is a visual reference on the chemistry of organic molecules. Definitions and line art provide a review or an introduction depending on the student's needs.
- Common Metabolic Pathways (appendix II) provides illustrations of nine critical biochemical pathways in one location, making review and reference more convenient for the student than if they were embedded in the text.
- Classification of Procaryotes (appendix III) summarizes the latest classification as reflected in the second edition of *Bergey's Manual of Systematic Bacteriology*.
- Classification of Viruses (appendix IV) provides a visual directory to a selected group of common viruses. The physical characteristics, family, and genera for each are provided.
- The Recommended Reading material (appendix V) is organized by chapter and provides direction to additional information for interested students.

NEW TO THIS EDITION!

Our New Look . . . Design

The interior of *Microbiology*, sixth edition, has been completely redesigned. Students today are very sensitive to visual presentations and our new design presents information within the framework of a bright, clean, modern-looking environment. We believe this appealing new look will help students move into the content and focus on the important topics. New icons call attention to the numbered main heads, and colorful headings help the students recognize shifts in focus. All of the boxed essays have been organized around five main themes and identified by category (Historical Highlights, Techniques & Applications, Disease, Microbial Diversity & Ecology, and Microbial Tidbits).

And . . . Illustrations

Tied in to this bright new look is our continuing improvement of our art program. New illustrations have been added to most chapters, and many older figures have been revised to improve their usefulness. Particular attention has been paid to consistency in the use of color. We have also tried to employ colors in such a way that the figures are easier to understand.

Not Just a New Look . . . New Content

Due to the fast pace of discoveries in the life sciences, substantial changes and updates have been made to keep the adopters of the sixth edition at the cutting edge of information. A summary of important new material by parts includes:

Parts One–Six (chapters 1–18) introduce the foundations of microbiology: the development of the field, the structure of microorganisms, microbial growth and control, metabolism, molecular biology and genetics, DNA technology and genomics, and the nature of viruses.

New and Significantly Updated Topics

Chapter 3—Protein secretion in procaryotes; fimbriae and bacterial movement

Chapter 6—Thermophile survival in high-temperature environments and the effect of salt on microbial growth

Chapter 11—Antiparallel nature of DNA

Chapter 12—Atomic structures of RNA polymerase and ribosomes; regulation by sRNA

Chapter 15—Thoroughly updated information on completed genomes

Chapter 18—Construction of the poliovirus from its genome sequence; mechanism of prion action; virus entry into host cells

Part Seven, The Diversity of the Microbial World (chapters 19–27) contains a survey of the procaryotes that closely follows the general organization of the second edition of *Bergey's Manual of Systematic Bacteriology*. Although principal attention is devoted to bacteria, the fungi, algae, and protozoa receive more than usual coverage.

New and Significantly Updated Topics

Chapter 19—Use of signature sequences in phylogenetic analysis; updated discussion of the classification system in the second edition of *Bergey's Manual of Systematic Bacteriology*

Chapter 20—Methane-consuming archaea and protein secretion in the archaea

Part Eight, Ecology and Symbiosis (chapters 28–30) focuses on the relationships of microorganisms to other organisms and the environment (microbial ecology). Aquatic and terrestrial microbiology are introduced here.

New and Significantly Updated Topics

Chapter 28—Methods of microbial ecology; discussion of lichens as controlled parasitic relationships; genomic reduction resulting from endosymbiosis; coevolution of gut microorganisms; inclusion of latest information on the hyperthermophile, *Geogemma barossii*

Chapter 29—Addition of *Cryptosporidium* to U.S. drinking water standards; removal of nitrogen and phosphorus by on-site water treatment processes; Canadian geese as a reservoir for *Giardia* and *Cryptosporidium*

Chapter 30—Occurrence of polyprosthecate bacteria such as *Verrucomicrobium*; the role of the oxidative burst in plant-microbe interactions; mycorrhizal interactions with achlorophyllous plants

Parts Nine and Ten, Nonspecific (Innate) Resistance and the Immune Response; Microbial Diseases and Their Control (chapters 31–33 in Part Nine and 34–40 in Part Ten) are concerned with pathogenicity, resistance, and disease. The disease survey is organized taxonomically on the chapter level; within each chapter diseases are covered according to mode of transmission. This provides flexibility and allows the student to easily locate information on a disease of interest.

New and Significantly Updated Topics

Chapter 31—Cathelicidin antimicrobial peptides; pattern-recognition receptors on macrophages, and Toll-like receptors in nonphagocytic host defense

Chapter 33—Vaccine table includes the latest recommendations approved for use in the United States including five new vaccines

Chapter 34—Use of actin-based motility by bacterial pathogens to spread within the host

Chapter 35—The Etest for antibiotic sensitivity; expansion of information on drug inactivation by chemical modification; discussion of antibiotic resistance genes on genetic elements other than plasmids

Chapter 37—New essays on the first recorded incidence of biological warfare and the SARS epidemic

Chapter 38—New or expanded discussion of smallpox, West Nile virus, and hepatitis G virus

Chapter 39—Weaponization of anthrax and expanded information on anthrax

Part Eleven, Food and Industrial Microbiology (chapters 41 and 42) concludes the text with an introduction to these fields.

New and Significantly Updated Topics

Chapter 41—Norwalk-like viruses in food and water; malo-lactic fermentation in wine production; use of probiotic *Lactobacillus* in feed to reduce the occurrence of *E. coli* in beef cattle

Chapter 42—Discussion of newest approaches for recovery from nature of previously "unculturable" microorganisms

SUPPLEMENTARY AND MEDIA MATERIALS

For the Student

- A **Student Study Guide** by Linda Sherwood of Montana State University is a valuable resource that provides learning objectives, study outlines, learning activities, and self-testing material to help students master course content.
- The *Microbiology*, **sixth edition, Online Learning Center** (www.mhhe.com/prescott6) provides self-quizzes, terminology exercises, study tips, web resources, etc., to aid students in mastering and integrating content.
- The sixth edition of *Laboratory Exercises in Microbiology* by John P. Harley has been prepared to accompany the text. Like the text, the laboratory manual provides a balanced introduction to laboratory techniques and principles that are important in each area of microbiology. The class-tested exercises are modular and short so that an instructor can eas-

ily choose only those exercises that fit his or her course. The sixth edition contains recipes for all reagents and media. New exercises in biotechnology have been added to this edition. A new appendix provides practice in solving dilution problems.

Dynamic Media

- *Microbes in Motion*, third edition, is an interactive, easy-to-use general microbiology CD-ROM that helps students actively explore and understand microbial structure and function through audio, video, animations, illustrations, slide shows, and text. Eighteen books cover topics from microbial genetics to vaccines.
- *HyperClinic*, second edition, CD-ROM allows students to evaluate realistic case studies that include patient histories and descriptions of signs and symptoms. Students can either analyze the results of physician-ordered clinical tests to reach a diagnosis, or evaluate a case study scenario and decide which clinical samples should be taken and which diagnostic tests should be run. More than 200 pathogens are profiled, 105 case studies presented, and 46 diagnostic tests covered.

For the Instructor

- The *Digital Content Manager* **CD-ROM** is the image resource for course presentations. The DCM contains virtually all of the line art, photos, and tables from *Microbiology*, sixth edition, as well as animations, videos, active-art, and a PowerPoint Lecture set for each chapter. See page xx for further details!
- **Instructor Testing and Resource CD-ROM** is offered free on request to adopters of the text. This cross-platform CD provides a database of over 2,500 objective questions for preparing exams and a grade-recording program.
- A set of 250 full-color acetate **transparencies** is available to supplement classroom lectures. These have been enhanced for projection and are available to adopters of the sixth edition.
- The **Online Learning Center** (www.mhhe.com/prescott6) provides multiple resources for course enhancement. Moreover, all the McGraw-Hill media resources are easily loaded into course management systems such as WebCT or Blackboard.

ACKNOWLEDGMENTS

The authors wish to thank the reviewers, who provided detailed criticism and analysis. Their suggestions greatly improved the final product.

REVIEWERS FOR THE SIXTH EDITION

Philip Achey, *University of Florida*

Susan Bagley, *Michigan Technological University*

Matthew Buechner, *University of Kansas*

Georganne Buescher, *Thomas Jefferson University*

Richard Ellis, *Bucknell University*

Harvey Friedman, *University of Missouri–St. Louis*

Bernard Frye, *University of Texas, Arlington*

Donald Glassman, *Des Moines Area Community College*

Tina Henkin, *Ohio State University*

Joan Henson, *Montana State University*

Judith Kandel, *California State University–Fullerton*

William A. Kuziel, *University of Texas*

Donald Lehman, *University of Delaware*

Elizabeth Machunis-Masouka, *University of Virginia*

Mark Maloney, *Spelman College*

Tamara Marsh, *Elmhurst College*

William McCleary, *Brigham Young University*

Kathleen McGuire, *San Diego State University*

Ruslan Medzhitov, *Yale University*

Rita Moyes, *Texas A&M University*

Michelle Nishiguchi, *New Mexico State University*

Saxena Pratibha, *University of Texas–Austin*

Sabine Rech, *San Jose State University*

Joanne Roehrs, *Owens Community College*

Wendy Schluchter, *University of New Orleans*

Richard Shippee, *Vincennes University*

Alan Spindler, *Brevard Community College*

Erica Suchman, *Colorado State University*

Marcello Tolmaskey, *California State University–Fullerton*

Jennifer Turco, *Valdosta State University*

Ruth Wrightsman, *University of California–Irvine*

Anne Zayaitz, *Kutztown University*

Lori Zeringue, *Louisiana State University*

Reviewers for the Fifth Edition

Stephen Aley, *University of Texas at El Paso*

Susan Bagley, *Michigan Technological University*

Robert Benoit, *Virginia Polytechnic Institute and State University*

Dennis Bazylinski, *Iowa State University*

Richard Bernstein, *San Francisco State University*

Paul Blum, *University of Nebraska*

Matthew Buechner, *University of Kansas*

Mary Burke, *Oregon State University*

James Champine, *Southeast Missouri State University*

John Clausz, *Carroll College*

James Cooper, *University of California at Santa Barbara*

Daniel DiMaio, *Yale University*

Leanne Field, *University of Texas*

Philip Johnson, *Grande Prairie Regional College*

Duncan Krause, *University of Georgia*

Diane Lavett, *Georgia Institute of Technology*

Ed Leadbetter, *University of Connecticut*

Donald Lehman, *University of Delaware*

Mark Maloney, *Spelman College*

Maura Meade-Callahan, *Allegheny College*

Ruslan Medzhitov, *Yale University School of Medicine*

Al Mikell, *University of Mississippi*

Craig Moyer, *Western Washington University*

Rita Moyes, *Texas A&M University*

David Mullin, *Tulane University*

Richard Myers, *Southwest Missouri State University*

Anthony Newsome, *Middle Tennessee State University*

Wade Nichols, *Illinois State University*

Ronald Porter, *Pennsylvania State University*

Sabine Rech, *San Jose State University*

Anna-Louise Reysenbach, *Portland State University*

Thomas Schmidt, *Michigan State University*

Linda Sherwood, *Montana State University*

Michele Shuster, *University of Pittsburgh*

Joan Slonczewski, *Kenyon College*

Daniel Smith, *Seattle University*

Kathleen C. Smith, *Emory University*

James Snyder, *University of Louisville School of Medicine*

William Staddon, *Eastern Kentucky University*

John Stolz, *DuQuesne University*

Thomas Terry, *University of Connecticut*

James VandenBosch, *Eastern Michigan University*

Reviewers for the Third and Fourth Editions

Laurie A. Achenbach, *Southern Illinois University*

Gary Armour, *MacMurray College*

Russell C. Baskett, *Germanna Community College*

George N. Bennett, *Rice University*

Prakash H. Bhuta, *Eastern Washington University*

James L. Botsford, *New Mexico State University*

Alfred E. Brown, *Auburn University*

Mary Burke, *Oregon State University*

David P. Clark, *Southern Illinois University*

William H. Coleman, *University of Hartford*

Donald C. Cox, *Miami University*

Phillip Cunningham, *Wayne State University*

Richard P. Cunningham, *SUNY at Albany*

James Daly, *Purchase College–SUNY*

Frank B. Dazzo, *Michigan State University*

Valdis A. Dzelzkalns, *Case Western Reserve University*

Richard J. Ellis, *Bucknell University*

Merrill Emmett, *University of Colorado at Denver*

Linda E. Fisher, *University of Michigan–Dearborn*

John Fitzgerald, *University of Georgia*

Harold F. Foerster, *Sam Houston State University*

B. G. Foster, *Texas A&M University*

Bernard Frye, *University of Texas at Arlington*

Katharine B. Gregg, *West Virginia Wesleyan College*

Eileen Gregory, *Rollins College*

Van H. Grosse, *Columbus College–Georgia*

Maria A. Guerrero, *Florida International University*

Robert Gunsalus, *UCLA*

Barbara B. Hemmingsen, *San Diego State University*

Joan Henson, *Montana State University*

William G. Hixon, *St. Ambrose University*

John G. Holt, *Michigan State University*

Ronald E. Hurlbert, *Washington State University*

Robert J. Kearns, *University of Dayton*

Henry Keil, *Brunel University*

Tim Knight, *Oachita Baptist University*

Robert Krasner, *Providence College*

Michael J. Lemke, *Kent State University*

Lynn O. Lewis, *Mary Washington College*

B. T. Lingappa, *College of the Holy Cross*

Vicky McKinley, *Roosevelt University*

Billie Jo Mello, *Mount Marty College*

James E. Miller, *Delaware Valley College*

David A. Mullin, *Tulane University*

Penelope J. Padgett, *Shippensburg University*

Richard A. Patrick, *Summit Editorial Group*

Bobbie Pettriess, *Wichita State University*

Thomas Punnett, *Temple University*

Jo Anne Quinlivan, *Holy Names College*

K. J. Reddy, *SUNY–Binghamton*

David C. Reff, *Middle Georgia College*

Jackie S. Reynolds, *Richland College*

Deborah Rochefort, *Shepherd College*

Allen C. Rogerson, *St. Lawrence University*

Michael J. San Francisco, *Texas Tech University*

Phillip Scheverman, *East Tennessee University*

Michael Shiaris, *University of Massachusetts at Boston*

Carl Sillman, *Penn State University*

Ann C. Smith, *University of Maryland*

David W. Smith, *University of Delaware*

Garriet W. Smith, *University of South Carolina at Aiken*

John Stolz, *Duquesne University*

Mary L. Taylor, *Portland State University*

Thomas M. Terry, *University of Connecticut*

Thomas M. Walker, *University of Central Arkansas*

Patrick M. Weir, *Felician College*

Jill M. Williams, *University of Glamorgan*

Heman Witmer, *University of Illinois at Chicago*

Elizabeth D. Wolfinger, *Meredith College*

Robert Zdor, *Andrews University*

Reviewers for the First and Second Editions

Richard J. Alperin, *Community College of Philadelphia*

Susan T. Bagley, *Michigan Technological University*

Dwight Baker, *Yale University*

R. A. Bender, *University of Michigan*

Hans P. Blaschek, *University of Illinois*

Dennis Bryant, *University of Illinois*

Douglas E. Caldwell, *University of Saskatchewan*

Arnold L. Demain, *Massachusetts Institute of Technology*

A. S. Dhaliwal, *Loyola University of Chicago*

Donald P. Durand, *Iowa State University*

John Hare, *Linfield College*

Robert B. Helling, *University of Michigan–Ann Arbor*

Barbara Bruff Hemmingsen, *San Diego State University*

R. D. Hinsdill, *University of Wisconsin–Madison*

John G. Holt, *Michigan State University*

Robert L. Jones, *Colorado State University*

Martha M. Kory, *University of Akron*

Robert I. Krasner, *Providence College*

Ron W. Leavitt, *Brigham Young University*

David Mardon, *Eastern Kentucky University*

Glendon R. Miller, *Wichita State University*

Richard L. Myers, *Southwest Missouri State University*

G. A. O'Donovan, *North Texas State University*

Pattle P. T. Pun, *Wheaton College*

Ralph J. Rascati, *Kennesaw State College*

Albert D. Robinson, *SUNY–Potsdam*

Ronald Wayne Roncadori, *University of Georgia–Athens*

Ivan Roth, *University of Georgia–Athens*

Thomas Santoro, *SUNY–New Paltz*

Ann C. Smith, *University of Maryland–College Park*

David W. Smith, *University of Delaware*

Paul Smith, *University of South Dakota*

James F. Steenbergen, *San Diego State University*

Henry O. Stone, Jr., *East Carolina University*

James E. Struble, *North Dakota State University*

Kathleen Talaro, *Pasadena City College*

Thomas M. Terry, *The University of Connecticut*

Michael J. Timmons, *Moraine Valley Community College*

John Tudor, *St. Joseph's University*

Robert Twarog, *University of North Carolina*

Blake Whitaker, *Bates College*

Oscar Will, *Augustana College*

Calvin Young, *California State University–Fullerton*

Publication of a textbook requires effort of many people besides the authors. We wish to express special appreciation to the editorial and production staffs of McGraw-Hill for their excellent work. In particular, we would like to thank Jean Fornango, our senior developmental editor, for her guidance, patience, prodding, and support. Our project manager, Jayne Klein, supervised production of this very complex project with commendable attention to detail. Beatrice Sussman, our copy editor for the second through fifth editions, once again corrected our errors in the sixth edition and contributed immensely to the text's clarity, consistency, and readability.

We wish to extend our appreciation to people who assisted us individually in completion of this project. Lansing Prescott wants to thank George M. Garrity, the editor-in-chief of *Bergey's Manual*, for his aid in the preparation of the fifth and sixth editions. Donald Klein wishes to acknowledge the aid of Jeffrey O. Dawson and Frank B. Dazzo.

Finally, but most importantly, we wish to extend appreciation to our families for their patience and encouragement, especially to our wives, Linda Prescott, Donna Dailey Harley, and Sandra Klein. To them, we dedicate this book.

Lansing M. Prescott
John P. Harley
Donald A. Klein

Visual Preview

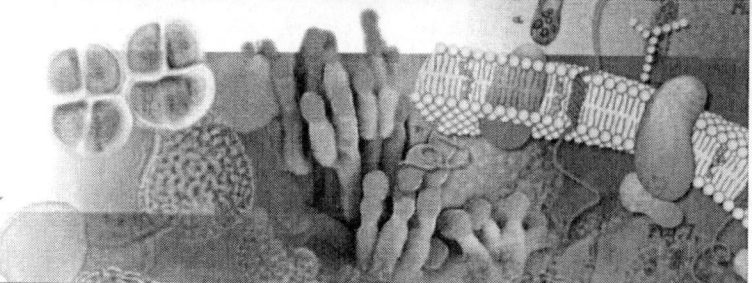

Rich tapestries reveal the grand scale of life.

Visual Program

The key to every biological problem must finally be sought in the cell.—E.B. Wilson

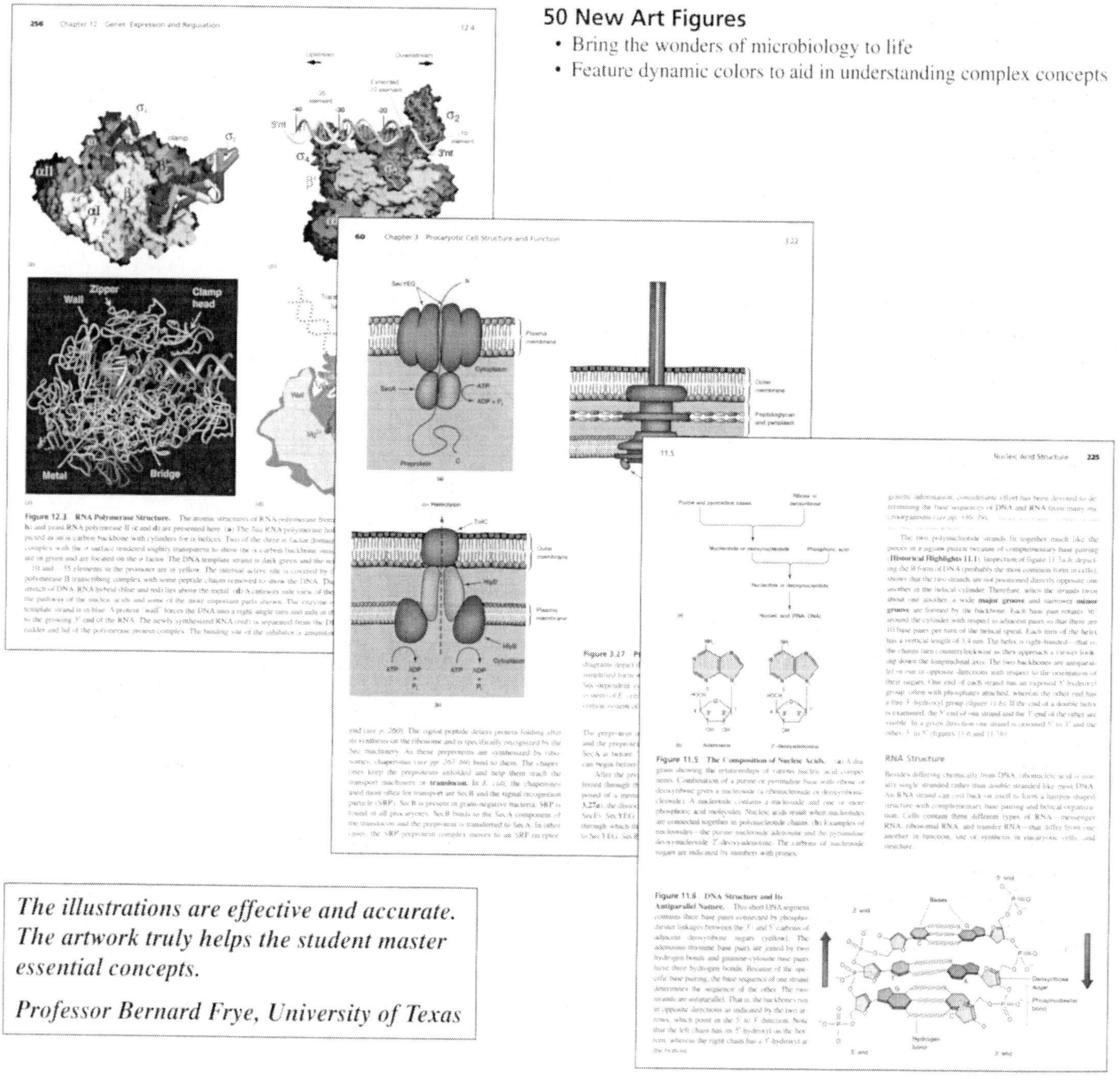

50 New Art Figures

- Bring the wonders of microbiology to life
- Feature dynamic colors to aid in understanding complex concepts

The illustrations are effective and accurate. The artwork truly helps the student master essential concepts.

Professor Bernard Frye, University of Texas

100 Revised Art Figures

- Present the unseen world in a consistent palette of color
- Inject new life into the study of microbiology

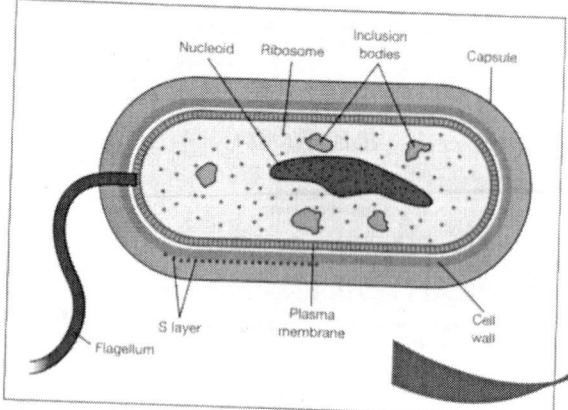

Figure 3.4

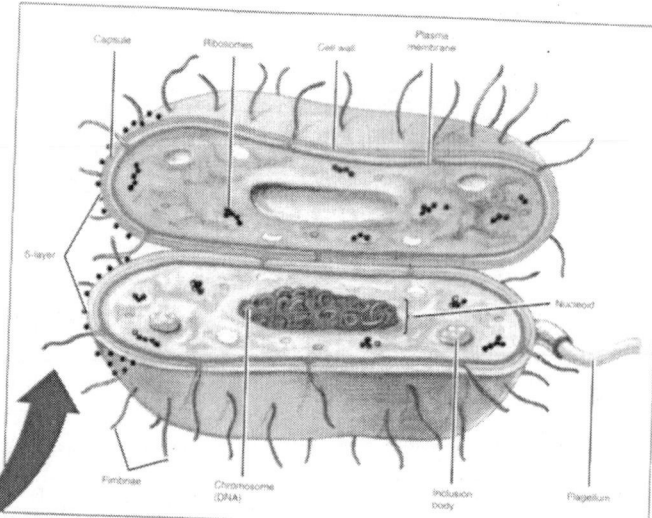

Figure 3.4

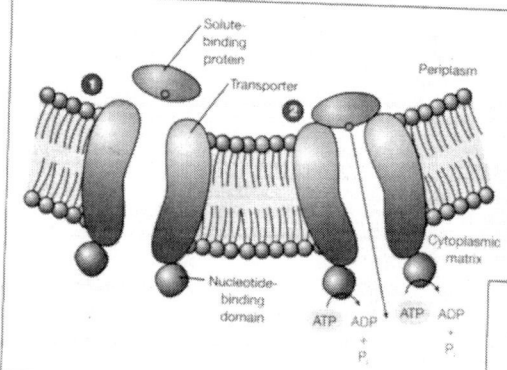

Figure 5.3

Art at Its Best

- Presents consistent color
- Aids in the mastery of complex concepts

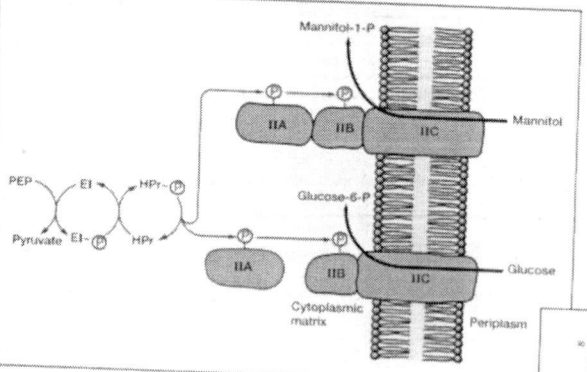

Figure 5.5

I was very impressed with the illustrations. Figure 6.3 detailed very nicely the difference between expression of cell number arithmetically vs. logarithmically.

Professor Richard Ellis, Bucknell University

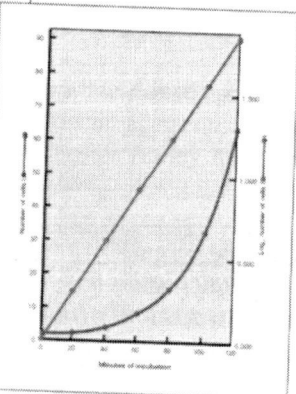

Figure 6.3

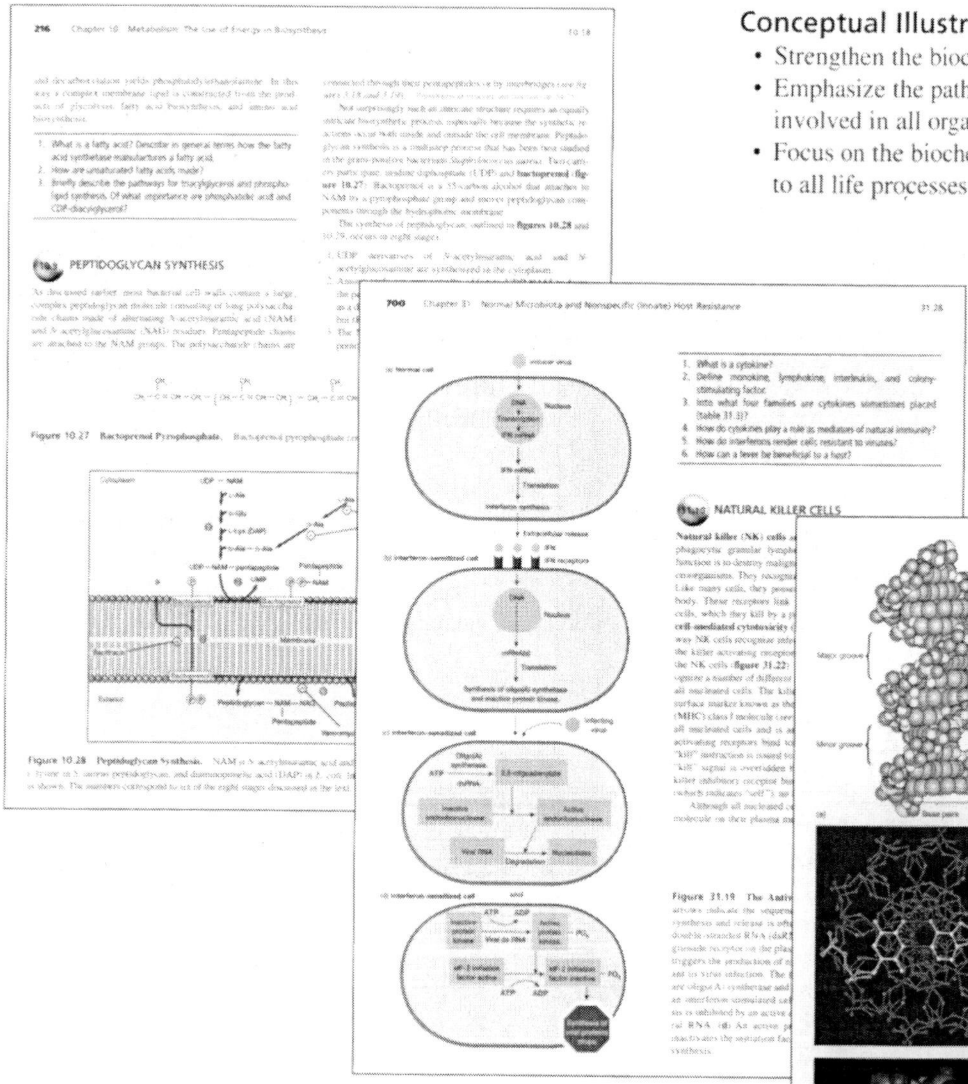

Conceptual Illustrations

- Strengthen the biochemical perspective
- Emphasize the pathways and interactions involved in all organism functions
- Focus on the biochemical unity that is basic to all life processes

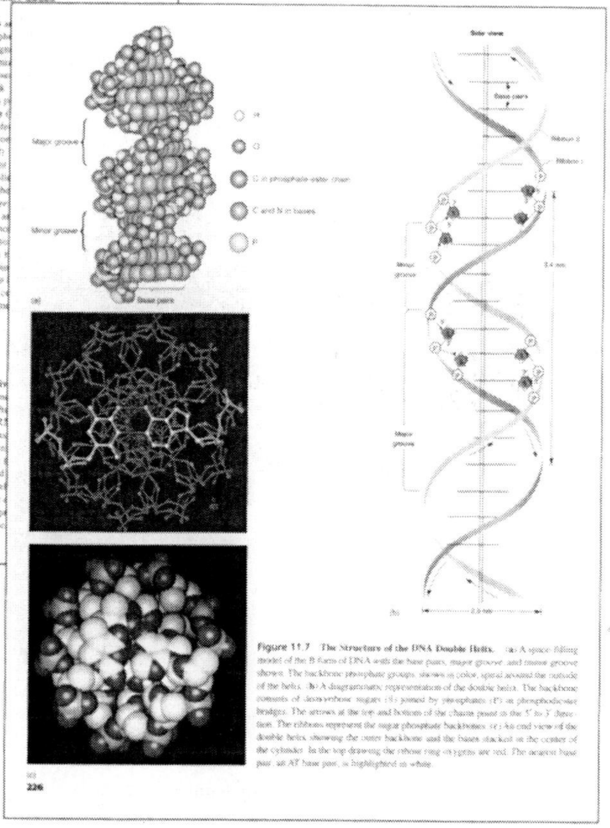

> *There is good use of specific examples to illustrate the application of general concepts.*
>
> *Professor Phil Achey, University of Florida*

Rich Photo Program

- Introduces students to a diverse microbial world

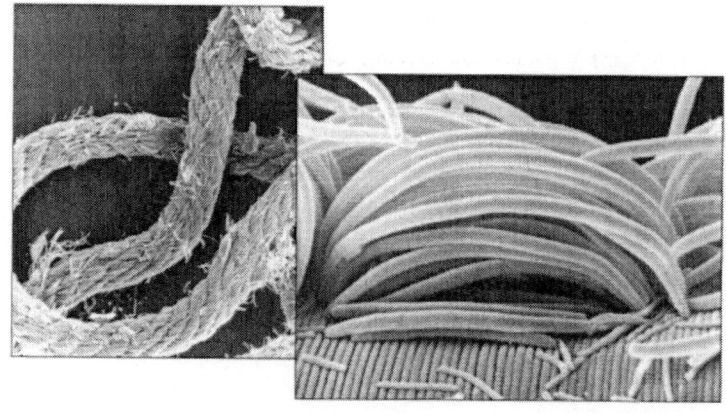

Content is poised on the cutting edge of life science discoveries.

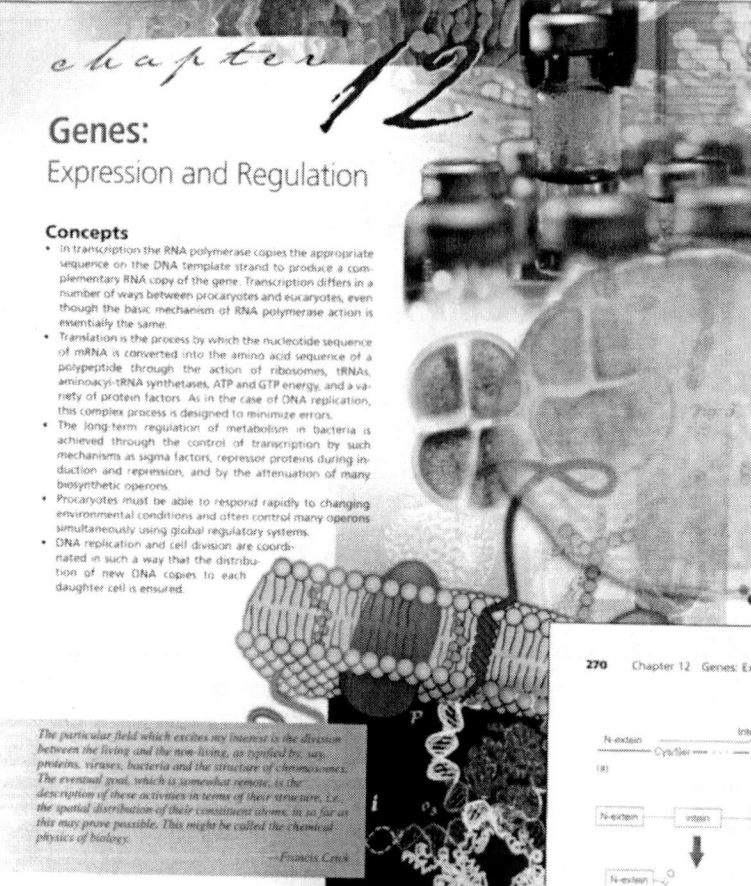

chapter 12

Genes:
Expression and Regulation

Concepts

- In transcription the RNA polymerase copies the appropriate sequence on the DNA template strand to produce a complementary RNA copy of the gene. Transcription differs in a number of ways between procaryotes and eucaryotes, even though the basic mechanism of RNA polymerase action is essentially the same.
- Translation is the process by which the nucleotide sequence of mRNA is converted into the amino acid sequence of a polypeptide through the action of ribosomes, tRNAs, aminoacyl-tRNA synthetases, ATP and GTP energy, and a variety of protein factors. As in the case of DNA replication, this complex process is designed to minimize errors.
- The long-term regulation of metabolism in bacteria is achieved through the control of transcription by such mechanisms as sigma factors, repressor proteins during induction and repression, and by the attenuation of many biosynthetic operons.
- Procaryotes must be able to respond rapidly to changing environmental conditions and often control many operons simultaneously using global regulatory systems.
- DNA replication and cell division are coordinated in such a way that the distribution of new DNA copies to each daughter cell is ensured.

The particular field which excites my interest is the division between the living and the non-living, as typified by, say, proteins, viruses, bacteria and the structure of chromosomes. The eventual goal, which is somewhat remote, is the description of these activities in terms of their structure, i.e. the spatial distribution of their constituent atoms, in so far as this may prove possible. This might be called the chemical physics of biology.

— Francis Crick

Lactose operon activity is under the control of a repressor protein. The lac repressor (violet) and catabolite activator protein (blue) are bound to the lac operon. The repressor blocks transcription when bound to the operator (red).

Chapter Concepts
- Briefly summarize important concepts

Cross-Referenced Notes
- Refer students to major topics that may require review in order to understand and integrate concepts

Review Questions Within Narrative
- Assist students in mastering section concepts before moving on to other topics

Numbered Headings
- Identify each major topic and are used for easy reference throughout the text

---

I believe the writing style is an easier read than my current text, particularly in the Genetics chapter.

Professor Donald Glassman, Des Moines Area Community College

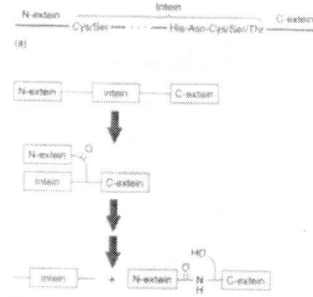

Figure 12.21   Protein Splicing. (a) A generalized illustration of intein structure. The amino acids that are commonly present at each end of the inteins are shown. Note that many are thiol- or hydroxyl-containing amino acids. (b) An overview of the proposed pattern or sequence of splicing. The precise mechanism is not yet known but presumably involves the hydroxyls or thiols located at each end of the intein.

polypeptide folds into its final shape. Self-splicing proteins begin as larger precursor proteins composed of one or more internal intervening sequences called **inteins** flanked by external sequences or **exteins**, the N-exteins and C-exteins (**figure 12.21a**). Inteins, which are between about 130 and 600 amino acids in length, are removed in an autocatalytic process involving a branched intermediate (**figure 12.21b**). Thus far, more than 130 inteins in 34 types of self-splicing protein have been discovered. Over 120 inteins have been found in bacteria and archaea. Some examples are an ATPase in the yeast *Saccharomyces cerevisiae*, the recA protein of *Mycobacterium tuberculosis*, and DNA polymerase in *Pyrococcus*. The presence of self-splicing proteins in all three domains may mean that these proteins are quite widespread and prevalent.

1. In which direction are polypeptides synthesized? What is a polyribosome and why is it useful?
2. Briefly describe the structure of transfer RNA and relate this to its function. How are amino acids activated for protein synthe-

Figure 12.22   The β-Galactosidase Reaction.

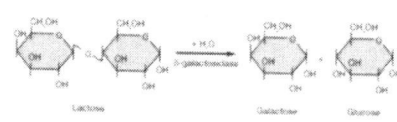

sis, and why is the specificity of the aminoacyl-tRNA synthetase reaction so important?
3. What are translational and exit domains? Contrast procaryotic and eucaryotic ribosomes in terms of structure. What roles does ribosomal RNA have?
4. Describe the nature and function of the following: fMet-tRNA, initiator codon, IF-3, IF-2, IF-1, elongation cycle, peptidyl and aminoacyl sites, EF-Tu, EF-Ts, transpeptidation reaction, peptidyl transferase, translocation, EF-G or translocase, nonsense codon, and release factors.
5. What are molecular chaperones and heat-shock proteins? Describe their functions.

12.3   REGULATION OF mRNA SYNTHESIS

The control of metabolism by regulation of enzyme activity is a fine-tuning mechanism; it acts rapidly to adjust metabolic activity from moment to moment. Microorganisms also are able to control the expression of their genome, although over longer intervals. For example, the *E. coli* chromosome can code for about 2,000 to 4,000 peptide chains, yet many fewer proteins are present in *E. coli* growing with glucose as its energy source. Regulation of gene expression serves to conserve energy and raw material, to maintain balance between the amounts of various cell proteins, and to adapt to long-term environmental change. Thus control of gene expression complements the regulation of enzyme activity.

Induction and Repression

The regulation of β-galactosidase synthesis has been intensively studied and serves as a primary example of how gene expression is controlled. This enzyme catalyzes the hydrolysis of the sugar lactose to glucose and galactose (**figure 12.22**). When *E. coli* grows with lactose as its carbon source, each cell contains about 3,000 β-galactosidase molecules, but has less than three molecules in the absence of lactose. The enzyme β-galactosidase is an **inducible enzyme**—that is, its level rises in the presence of a small molecule called an **inducer** (in this case the lactose derivative allolactose).

The genes for enzymes involved in the biosynthesis of amino acids and other substances often respond differently from genes coding for catabolic enzymes. An amino acid present in the surroundings may inhibit the formation of the enzymes responsible for its biosynthesis. This makes good sense because the microor-

Timely Topics
- Link the text topics to today's headlines

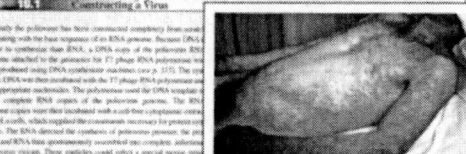

Special Interest Essays
Historical Highlights, Techniques & Applications, Microbial Diversity and Ecology, Disease, and Microbial Tidbits
- Provide additional perspective on the many facets of microbiology

Chapter Summaries
- Organized by numbered headings.
- Provide a snapshot of important chapter concepts

Key Terms
- Highlight chapter terminology and list term location in the chapter

Questions for Thought and Review
- Spur students to apply and integrate chapter content

Critical Thinking Questions
- Stimulate analytical problem solving skills

A Powerful Lecture Resource

This multimedia collection of visual resources allows instructors to utilize artwork from the text in multiple formats to create customized classroom presentations, visually-based tests and quizzes, dynamic course website content, or attractive printed support materials. The digital assets on this cross-platform CD-ROM are grouped within the following easy-to-use folders.

Art and Photo Library

Full-color digital files of all the tables and illustrations and most of the photos in *Microbiology,* sixth edition, can be readily incorporated into lecture presentations, exams, or custom-made classroom materials.

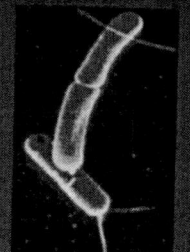

Information in DNA generates diversity

- Four bases – G (guanine), A (adenine), T (thymine), and C (cytosine) are the nucleotide building blocks of DNA

- DNA is a double stranded helix composed of A-T and G-C complementary bases

PowerPoint Lecture Outlines

Ready-made presentations that combine art and lecture notes cover each of the 42 chapters of the text. These lectures can be used as they are or can be tailored to reflect your preferred lecture topics and sequences.

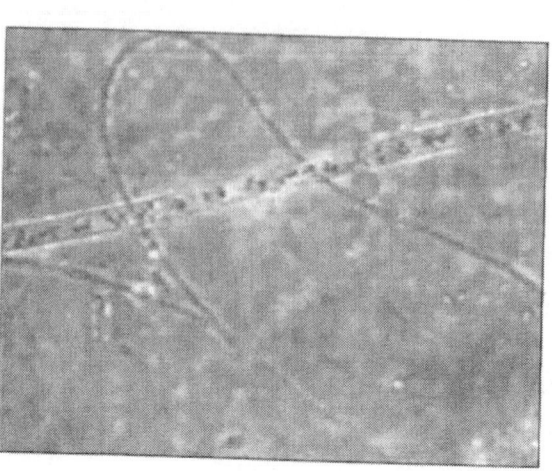

Animations and Videos

Instructive, full-color animations and videos are available to harness the visual impact of processes in motion. Import these animations into classroom presentations or online course materials.

Active Art Active Art allows you to customize key figures from the text to complement your lecture! Figures have been broken into smaller digestable parts so each piece can be used in any desired order or format. Using Microsoft PowerPoint's ungroupable art feature, instructors can also move, resize, or change the color of any or all objects in a piece of art.

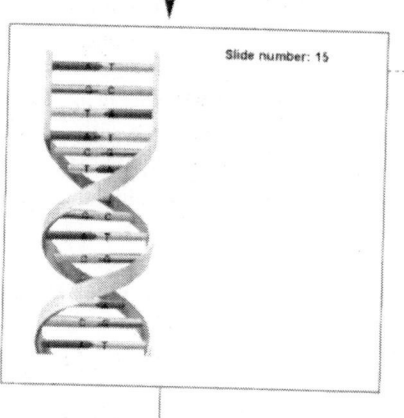 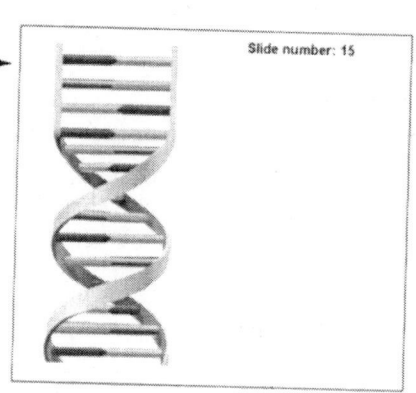

Remove Labels Labels and leader lines can be easily repositioned, edited, or removed. Your customized images can then be used for course assignments or additional quizzing for your students.

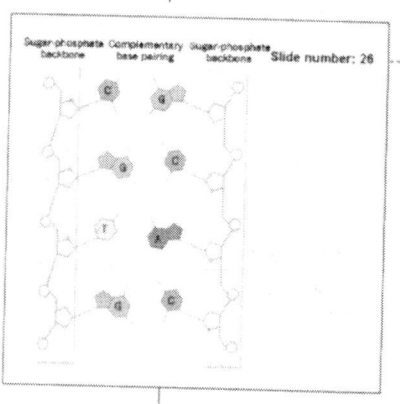 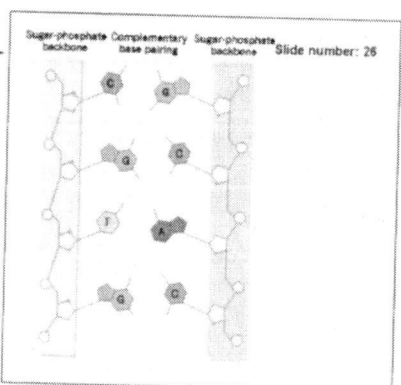

Change Colors Colors can be removed and/or changed from any Active Art slide or object. Black and white images can be created for use in lecture supplements or exams.

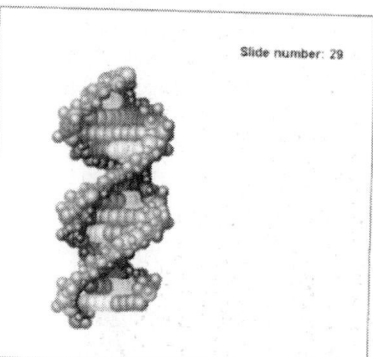

Resize Objects The entire image or parts of the image can be made larger or smaller depending on what you choose to emphasize.

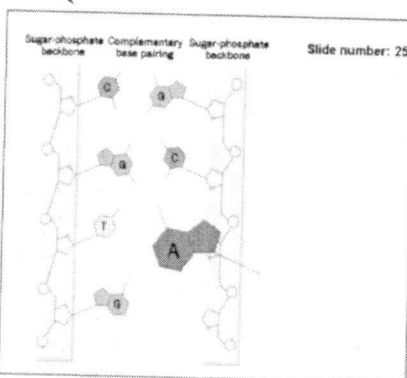

Lansing Prescott served as Professor of Biology and chair of the department at Augustana College in Sioux Falls, South Dakota, until May 1999. Dr. Prescott received his B.A. and M.A. in biology from Rice University and his Ph.D. in biochemistry from Brandeis University. He was a visiting lecturer at the University of Georgia in 1980. Dr. Prescott's research interests are the properties of bacterial aspartate transcarbamylases (particularly those from *Bacillus stearothermophilus* and *B. psychrophilus*) and the effect of toxicants on diatom morphology and physiology. As is the case in small, liberal arts colleges, one of Dr. Prescott's primary responsibilities was teaching undergraduates. He has taught courses in introductory microbiology for nursing and allied health students, general microbiology for majors, cell biology, biological chemistry, immunology, human physiology, and parasitology. In 1989, he received a faculty achievement award for excellence in teaching. Presently living just outside Austin, Texas, Dr. Prescott enjoys listening to music, playing golf and chess, and reading both fiction and nonfiction when he is not engaged in academic pursuits. Dr. Prescott's commitment to writing is long-standing. Besides his involvement in *Microbiology* and *Laboratory Exercises in Microbiology*, now in their sixth editions, he was a contributing author for a general biology textbook, L. G. Johnson's (1983) *Biology*. Dubuque, IA: Wm. C. Brown, and has been a Choice book reviewer for many years. Dr. Prescott can be reached at lansing_prescott@augie.edu.

John Harley is a professor at Eastern Kentucky University. He received his B.A. degree in biology and chemistry from Youngstown State University in 1964, an M.A. degree in parasitology and microbiology from Kent State University in 1966, and his Ph.D. in cardiovascular physiology from Kent State University in 1969. Dr. Harley did postdoctoral work at Baylor College of Medicine, Argonne National Lab, and Vanderbilt University. In 1972 he accepted a faculty position at Eastern Kentucky University, where he rose through the ranks to full professor and in 1990 was named a Foundation Professor by the EKU Alumni Association and Board of Reagents. He also holds full graduate status at the University of Kentucky where he teaches a pathophysiology course in the graduate program. Dr. Harley's research interests are in parasitology (abnormal and normal host relationships, biochemistry, life cycle studies, and pharmacology) and the effects of parasites on normal host physiology. He has published over 80 research papers and publications and was advisor to 16 graduate students. Together with Stephen Miller, he wrote the sixth edition of *Zoology*, published by McGraw-Hill. Dr. Harley teaches general zoology, general biology, human anatomy, microbiology, general physiology, human physiology, and pathophysiology. In addition to his academic pursuits, Dr. Harley raises and breeds Cavalier King Charles Spaniels, enjoys working on automobiles, traveling, gardening, reading science journals and fiction. Dr. Harley can be reached at john.harley@eku.edu.

Donald Klein is a Professor of Microbiology at Colorado State University, Fort Collins, Colorado. Dr. Klein received his B.S. and M.S. degrees in agriculture and agricultural microbiology from the University of Vermont. After predoctoral studies at the University of Tübingen, Germany, he received his Ph.D. in microbiology from the Pennsylvania State University. His research interests are in the area of environmental microbiology, with major emphases on plant-microbe relationships in the rhizosphere, plant community succession, and the fungal-bacterial structure of natural systems. In addition to work in these areas, Dr. Klein has had a long-standing interest in teaching, especially at the undergraduate level. He has taught courses in soil, aquatic and industrial microbiology, as well as courses in the areas of general microbiology, microbial diversity and microbial ecology. In addition to his contributions to the environmental, food, and industrial microbiology sections of *Microbiology*, Dr. Klein has edited books on microbial aspects of weather modification and soil reclamation. He is a member of the American Academy of Microbiology, and has served on the editorial boards of several scientific journals. In addition to his academic interests, Dr. Klein enjoys reading classic German literature, Jaguar XK150 restoration, sculpting mythical animals in stone, and sailing a beetlecat sailboat on Buzzards Bay, Massachusetts, in the summertime. Dr. Klein can be reached at dakspk@cape.com.

chapter 1

The History and Scope of Microbiology

Concepts

- Microbiology is the study of organisms that are usually too small to be seen by the unaided eye; it employs techniques—such as sterilization and the use of culture media—that are required to isolate and grow these microorganisms.
- Microorganisms are not spontaneously generated from inanimate matter but arise from other microorganisms.
- Many diseases result from viral, bacterial, fungal, or protozoan infections. Koch's postulates may be used to establish a causal link between the suspected microorganism and a disease.
- The development of microbiology as a scientific discipline has depended on the availability of the microscope and the ability to isolate and grow pure cultures of microorganisms.
- Microorganisms are responsible for many of the changes observed in organic and inorganic matter (e.g., fermentation and the carbon, nitrogen, and sulfur cycles that occur in nature).
- Microorganisms have two fundamentally different types of cells—procaryotic and eucaryotic—and are distributed among several kingdoms or domains.
- Microbiology is a large discipline, which has a great impact on other areas of biology and general human welfare.

Dans les champs de l'observation, le hasard ne favorise que les esprits préparés.
(In the field of observation, chance favors only prepared minds.)
—Louis Pasteur

Louis Pasteur, one of the greatest scientists of the nineteenth century, maintained that "Science knows no country, because knowledge belongs to humanity, and is a torch which illuminates the world."

*O*ne can't overemphasize the importance of microbiology. Society benefits from microorganisms in many ways. They are necessary for the production of bread, cheese, beer, antibiotics, vaccines, vitamins, enzymes, and many other important products. Indeed, modern biotechnology rests upon a microbiological foundation. Microorganisms are indispensable components of our ecosystem. They make possible the cycles of carbon, oxygen, nitrogen, and sulfur that take place in terrestrial and aquatic systems. They also are a source of nutrients at the base of all ecological food chains and webs.

Of course microorganisms also have harmed humans and disrupted society over the millennia. Microbial diseases undoubtedly played a major role in historical events such as the decline of the Roman Empire and the conquest of the New World. In 1347 plague or black death (*see chapter 39*) struck Europe with brutal force. By 1351, only four years later, the plague had killed 1/3 of the population (about 25 million people). Over the next 80 years, the disease struck again and again, eventually wiping out 75% of the European population. Some historians believe that this disaster changed European culture and prepared the way for the Renaissance. Today the struggle by microbiologists and others against killers like AIDS and malaria continues. The biology of AIDS and its impact (pp. 855–60)

In this introductory chapter the historical development of the science of microbiology is described, and its relationship to medicine and other areas of biology is considered. The nature of the microbial world is then surveyed to provide a general idea of the organisms and agents that microbiologists study. Finally, the scope, relevance, and future of modern microbiology are discussed.

Microbiology often has been defined as the study of organisms and agents too small to be seen clearly by the unaided eye—that is, the study of **microorganisms.** Because objects less than about one millimeter in diameter cannot be seen clearly and must be examined with a microscope, microbiology is concerned primarily with organisms and agents this small and smaller. Its subjects are viruses, bacteria, many algae and fungi, and protozoa (*see table 34.1*). Yet other members of these groups, particularly some of the algae and fungi, are larger and quite visible. For example, bread molds and filamentous algae are studied by microbiologists, yet are visible to the naked eye. Two bacteria that are visible without a microscope, *Thiomargarita* and *Epulopiscium*, also have been discovered (*see p. 43*). The difficulty in setting the boundaries of microbiology led Roger Stanier to suggest that the field be defined not only in terms of the size of its subjects but also in terms of its techniques. A microbiologist usually first isolates a specific microorganism from a population and then cultures it. Thus microbiology employs techniques—such as sterilization and the use of culture media—that are necessary for successful isolation and growth of microorganisms.

The development of microbiology as a science is described in sections 1.1 to 1.4. **Table 1.1** presents a summary of some of the major events in this process and their relationship to other historical landmarks.

1.1   THE DISCOVERY OF MICROORGANISMS

Even before microorganisms were seen, some investigators suspected their existence and responsibility for disease. Among others, the Roman philosopher Lucretius (about 98–55 B.C.) and the physician Girolamo Fracastoro (1478–1553) suggested that disease was caused by invisible living creatures. The earliest microscopic observations appear to have been made between 1625 and 1630 on bees and weevils by the Italian Francesco Stelluti, using a microscope probably supplied by Galileo. However, the first person to observe and describe microorganisms accurately was the amateur microscopist Antony van Leeuwenhoek (1632–1723) of Delft, Holland (**figure 1.1***a*). Leeuwenhoek earned his living as a draper and haberdasher (a dealer in men's clothing and accessories), but spent much of his spare time constructing simple microscopes composed of double convex glass lenses held between two silver plates (figure 1.1*b*). His microscopes could magnify around 50 to 300 times, and he may have illuminated his liquid specimens by placing them between two pieces of glass and shining light on them at a 45° angle to the specimen plane. This would have provided a form of dark-field illumination (*see chapter 2*) and made bacteria clearly visible (figure 1.1*c*). Beginning in 1673 Leeuwenhoek sent detailed letters describing his discoveries to the Royal Society of London. It is clear from his descriptions that he saw both bacteria and protozoa.

1.2   THE CONFLICT OVER SPONTANEOUS GENERATION

From earliest times, people had believed in **spontaneous generation**—that living organisms could develop from nonliving matter. Even the great Aristotle (384–322 B.C.) thought some of the simpler invertebrates could arise by spontaneous generation. This view finally was challenged by the Italian physician Francesco Redi (1626–1697), who carried out a series of experiments on decaying meat and its ability to produce maggots spontaneously. Redi placed meat in three containers. One was uncovered, a second was covered with paper, and the third was covered with a fine gauze that would exclude flies. Flies laid their eggs on the uncovered meat and maggots developed. The other two pieces of meat did not produce maggots spontaneously. However, flies were attracted to the gauze-covered container and laid their eggs on the gauze; these eggs produced maggots. Thus the generation of maggots by decaying meat resulted from the presence of fly eggs, and meat did not spontaneously generate maggots as previously believed. Similar experiments by others helped discredit the theory for larger organisms.

Leeuwenhoek's discovery of microorganisms renewed the controversy. Some proposed that microorganisms arose by spontaneous generation even though larger organisms did not. They

Table 1.1

Some Important Events in the Development of Microbiology

Date	Microbiological History	Other Historical Events
1546	Fracastoro suggests that invisible organisms cause disease	Publication of Copernicus's work on the heliocentric solar system (1543)
1590–1608	Jansen develops first useful compound microscope	Shakespeare's *Hamlet* (1600–1601)
1676	Leeuwenhoek discovers "animalcules"	J. S. Bach and Handel born (1685)
1688	Redi publishes work on spontaneous generation of maggots	Isaac Newton publishes the *Principia* (1687)
		Linnaeus's *Systema Naturae* (1735)
		Mozart born (1756)
1765–1776	Spallanzani attacks spontaneous generation	
1786	Müller produces first classification of bacteria	French Revolution (1789)
1798	Jenner introduces cowpox vaccination for smallpox	Beethoven's first symphony (1800)
		The battle of Waterloo and the defeat of Napoleon (1815)
		Faraday demonstrates the principle of an electric motor (1821)
1838–1839	Schwann and Schleiden proposed the Cell Theory	England issues first postage stamp (1840)
1835–1844	Bassi discovers that silkworm disease is caused by a fungus and proposes that many diseases are microbial in origin	
		Marx's *Communist Manifesto* (1848)
1847–1850	Semmelweis shows that childbed fever is transmitted by physicians and introduces the use of antiseptics to prevent the disease	Velocity of light first measured by Fizeau (1849)
		Clausius states the first and second laws of thermodynamics (1850)
1849	Snow studies the epidemiology of a cholera epidemic in London	Graham distinguishes between colloids and crystalloids (1850)
		Melville's *Moby Dick* (1851)
		Otis installs first safe elevator (1854)
		Bunsen introduces the use of the gas burner (1855)
1857	Pasteur shows that lactic acid fermentation is due to a microorganism	
1858	Virchow states that all cells come from cells	Darwin's *On the Origin of Species* (1859)
1861	Pasteur shows that microorganisms do not arise by spontaneous generation	American Civil War (1861–1865)
		Mendel publishes his genetics experiments (1865)
		Cross-Atlantic cable laid (1865)
1867	Lister publishes his work on antiseptic surgery	Dostoevski's *Crime and Punishment* (1866)
1869	Miescher discovers nucleic acids	Franco-German War (1870–1871)
1876–1877	Koch demonstrates that anthrax is caused by *Bacillus anthracis*	Bell invents telephone (1876)
		Edison's first light bulb (1879)
1880	Laveran discovers *Plasmodium*, the cause of malaria	
1881	Koch cultures bacteria on gelatin	Ives produces first color photograph (1881)
	Pasteur develops anthrax vaccine	
1882	Koch discovers tubercle bacillus, *Mycobacterium tuberculosis*	First central electric power station constructed by Edison (1882)
1884	Koch's postulates first published	Mark Twain's *The Adventures of Huckleberry Finn* (1884)
	Metchnikoff describes phagocytosis	
	Autoclave developed	
	Gram stain developed	
1885	Pasteur develops rabies vaccine	First motor vehicles developed by Daimler (1885–1886)
	Escherich discovers *Escherichia coli*, a cause of diarrhea	
1886	Fraenkel discovers *Streptococcus pneumoniae*, a cause of pneumonia	
1887	Petri dish (plate) developed by Richard Petri	
1887–1890	Winogradsky studies sulfur and nitrifying bacteria	Hertz discovers radio waves (1888)
1889	Beijerinck isolates root nodule bacteria	Eastman makes box camera (1888)
1890	Von Behring prepares antitoxins for diphtheria and tetanus	

Table 1.1   Continued

Date	Microbiological History	Other Historical Events
1892	Ivanowsky provides evidence for virus causation of tobacco mosaic disease	First zipper patented (1895)
1894	Kitasato and Yersin discover *Yersinia pestis*, the cause of plague	
1895	Bordet discovers complement	Röntgen discovers X rays (1895)
1896	Van Ermengem discovers *Clostridium botulinum*, the cause of botulism	
1897	Buchner prepares extract of yeast that ferments	Thomson discovers the electron (1897)
	Ross shows that malaria parasite is carried by the mosquito	Spanish-American War (1898)
1899	Beijerinck proves that a virus particle causes the tobacco mosaic disease	
1900	Reed proves that yellow fever is transmitted by the mosquito	Planck develops the quantum theory (1900)
1902	Landsteiner discovers blood groups	First electric typewriter (1901)
1903	Wright and others discover antibodies in the blood of immunized animals	First powered aircraft (1903)
1905	Schaudinn and Hoffmann show *Treponema pallidum* causes syphilis	Einstein's special theory of relativity (1905)
1906	Wassermann develops complement fixation test for syphilis	
1909	Ricketts shows that Rocky Mountain spotted fever is transmitted by ticks and caused by a microbe (*Rickettsia rickettsii*)	First model T Ford (1908)
		Peary and Hensen reach North Pole (1909)
1910	Ehrlich develops chemotherapeutic agent for syphilis	Rutherford presents his theory of the atom (1911)
1911	Rous discovers a virus that causes cancer in chickens	Picasso and cubism (1912)
		World War I begins (1914)
1915–1917	D'Herelle and Twort discover bacterial viruses	Einstein's general theory of relativity (1916)
		Russian Revolution (1917)
1921	Fleming discovers lysozyme	
1923	First edition of *Bergey's Manual*	Lindberg's transatlantic flight (1927)
1928	Griffith discovers bacterial transformation	
1929	Fleming discovers penicillin	Stock market crash (1929)
1931	Van Niel shows that photosynthetic bacteria use reduced compounds as electron donors without producing oxygen	
1933	Ruska develops first transmission electron microscope	Hitler becomes chancellor of Germany (1933)
1935	Stanley crystallizes the tobacco mosaic virus	
	Domagk discovers sulfa drugs	
1937	Chatton divides living organisms into procaryotes and eucaryotes	Krebs discovers the citric acid cycle (1937)
1941	Beadle and Tatum proposed the one-gene-one-enzyme hypothesis	World War II begins (1939)
1944	Avery shows that DNA carries information during transformation	The insecticide DDT introduced (1944)
	Waksman discovers streptomycin	
1946	Lederberg and Tatum describe bacterial conjugation	Atomic bombs dropped on Hiroshima and Nagasaki (1945)
		United Nations formed (1945)
		First electronic computer (1946)
1949	Enders, Weller, and Robbins grow poliovirus in human tissue cultures	
1950	Lwoff induces lysogenic bacteriophages	
1952	Hershey and Chase show that bacteriophages inject DNA into host cells	Korean War begins (1950)
		First hydrogen bomb exploded (1952)
	Zinder and Lederberg discover generalized transduction	Stalin dies (1952)
		First commercial transistorized product (1952)

Table 1.1 Continued

Date	Microbiological History	Other Historical Events
1953	Phase-contrast microscope developed	U.S. Supreme Court rules against segregated schools (1954)
	Medawar discovers immune tolerance	
	Watson and Crick propose the double helix structure for DNA	
1955	Jacob and Wollman discover the F factor is a plasmid	Montgomery bus boycott (1955)
	Jerne and Burnet propose the clonal selection theory	Sputnik launched by Soviet Union (1957)
1959	Yalow develops the radioimmunoassay technique	Birth control pill (1960)
1961	Jacob and Monod propose the operon model of gene regulation	First humans in space (1961)
1961–1966	Nirenberg, Khorana, and others elucidate the genetic code	Cuban missile crisis (1962)
1962	Porter proposes the basic structure for immunoglobulin G	Nuclear test ban treaty (1963)
	First quinolone antimicrobial (nalidixic acid) synthesized	Civil Rights March on Washington (1963)
		President Kennedy assassinated (1963)
		Arab-Israeli War (1967)
		Martin Luther King assassination (1968)
1970	Discovery of restriction endonucleases by Arber and Smith	Neil Armstrong walks on the moon (1969)
	Discovery of reverse transcriptase in retroviruses by Temin and Baltimore	
1973	Ames develops a bacterial assay for the detection of mutagens	Salt I Treaty (1972)
	Cohen, Boyer, Chang, and Helling use plasmid vectors to clone genes in bacteria	Vietnam War ends (1973)
1975	Kohler and Milstein develop technique for the production of monoclonal antibodies	President Nixon resigns because of Watergate cover-up (1974)
	Lyme disease discovered	
1977	Recognition of archaea as a distinct microbial group by Woese and Fox	Panama Canal Treaty (1977)
	Gilbert and Sanger develop techniques for DNA sequencing	
1979	Insulin synthesized using recombinant DNA techniques	Hostages seized in Iran (1978)
	Smallpox declared officially eliminated	Three Mile Island disaster (1979)
1980	Development of the scanning tunneling microscope	Home computers marketed (1980)
1982	Recombinant hepatitis B vaccine developed	AIDS first recognized (1981)
1982–1983	Discovery of catalytic RNA by Cech and Altman	First artificial heart implanted (1982)
1983–1984	The human immunodeficiency virus isolated and identified by Gallo and Montagnier	Meter redefined in terms of distance light travels (1983)
	The polymerase chain reaction developed by Mullis	
1986	First vaccine (hepatitis B vaccine) produced by genetic engineering approved for human use	Gorbachev becomes Communist party general secretary (1985)
1990	First human gene-therapy testing begun	Berlin Wall falls (1989)
		Persian Gulf War with Iraq begins (1990)
1992	First human trials of antisense therapy	Soviet Union collapse: Boris Yeltsin comes to power (1991)
1995	Chickenpox vaccine approved for U.S. use	
	Haemophilus influenzae genome sequenced	
1996	*Methanococcus jannaschii* genome sequenced	
	Yeast genome sequenced	
1997	Discovery of *Thiomargarita namibiensis*, the largest known bacterium	
	Escherichia coli genome sequenced	
2000	Discovery that *Vibrio cholerae* has two separate chromosomes	Water found on the moon (1998)
2002	Infectious poliovirus synthesis from basic chemical building blocks	First draft of human genome released (2001)
	Genome of malaria parasite, *Plasmodium falciparum*, sequenced	World Trade Center attack (2001)
		Second war with Iraq (2003)
		Successful completion of Human Genome Project (2003)

(a)

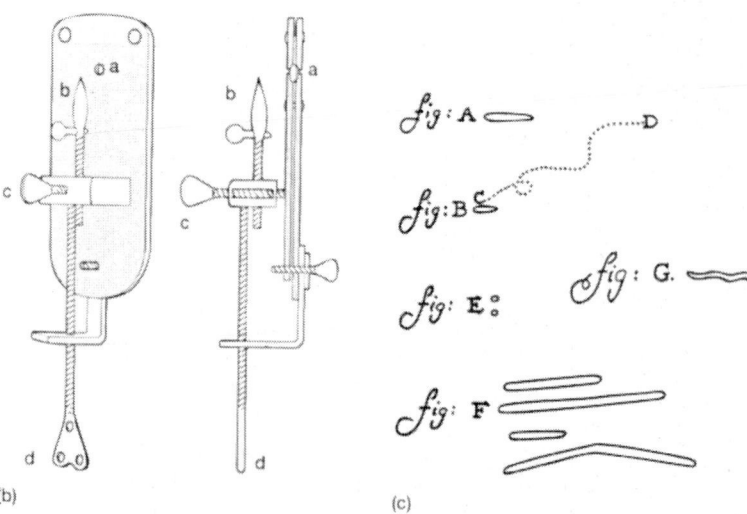

(b)

(c)

Figure 1.1   Antony van Leeuwenhoek. (a) Leeuwenhoek (1632–1723) and his microscopes. (b) A drawing of one of the microscopes showing the lens, *a;* mounting pin, *b;* and focusing screws, *c* and *d.* (c) Leeuwenhoek's drawings of bacteria from the human mouth.   *(b) Source: C. E. Dobell,* Antony van Leeuwenhoek and His Little Animals *(1932), Russell and Russell, 1958.*

pointed out that boiled extracts of hay or meat would give rise to microorganisms after sitting for a while. In 1748 the English priest John Needham (1713–1781) reported the results of his experiments on spontaneous generation. Needham boiled mutton broth and then tightly stoppered the flasks. Eventually many of the flasks became cloudy and contained microorganisms. He thought organic matter contained a vital force that could confer the properties of life on nonliving matter. A few years later the Italian priest and naturalist Lazzaro Spallanzani (1729–1799) improved on Needham's experimental design by first sealing glass flasks that contained water and seeds. If the sealed flasks were placed in boiling water for 3/4 of an hour, no growth took place as long as the flasks remained sealed. He proposed that air carried germs to the culture medium, but also commented that the external air might be required for growth of animals already in the medium. The supporters of spontaneous generation maintained that heating the air in sealed flasks destroyed its ability to support life.

Several investigators attempted to counter such arguments. Theodore Schwann (1810–1882) allowed air to enter a flask containing a sterile nutrient solution after the air had passed through a red-hot tube. The flask remained sterile. Subsequently Georg Friedrich Schroder and Theodor von Dusch allowed air to enter a flask of heat-sterilized medium after it had passed through sterile cotton wool. No growth occurred in the medium even though the air had not been heated. Despite these experiments the French naturalist Felix Pouchet claimed in 1859 to have carried out ex-

periments conclusively proving that microbial growth could occur without air contamination. This claim provoked Louis Pasteur (1822–1895) to settle the matter once and for all. Pasteur (**figure 1.2**) first filtered air through cotton and found that objects resembling plant spores had been trapped. If a piece of the cotton was placed in sterile medium after air had been filtered through it, microbial growth occurred. Next he placed nutrient solutions in flasks, heated their necks in a flame, and drew them out into a variety of curves, while keeping the ends of the necks open to the atmosphere (**figure 1.3**). Pasteur then boiled the solutions for a few minutes and allowed them to cool. No growth took place even though the contents of the flasks were exposed to the air. Pasteur pointed out that no growth occurred because dust and germs had been trapped on the walls of the curved necks. If the necks were broken, growth commenced immediately. Pasteur had not only resolved the controversy by 1861 but also had shown how to keep solutions sterile.

The English physicist John Tyndall (1820–1893) dealt a final blow to spontaneous generation in 1877 by demonstrating that dust did indeed carry germs and that if dust was absent, broth remained sterile even if directly exposed to air. During the course of his studies, Tyndall provided evidence for the existence of exceptionally heat-resistant forms of bacteria. Working independently, the German botanist Ferdinand Cohn (1828–1898) discovered the existence of heat-resistant bacterial endospores (*see chapter 3*).

Figure 1.2   Louis Pasteur. Pasteur (1822–1895) working in his laboratory.

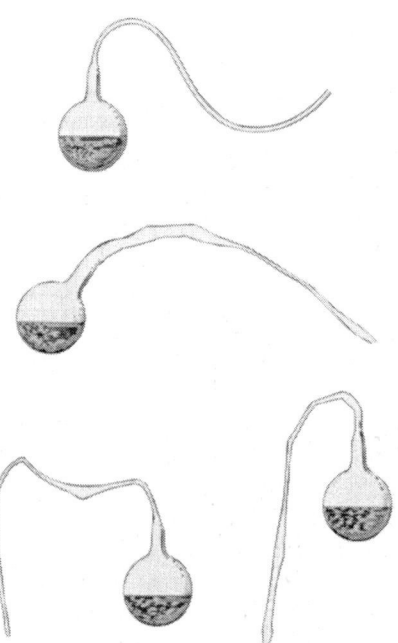

Figure 1.3   The Spontaneous Generation Experiment. Pasteur's swan neck flasks used in his experiments on the spontaneous generation of microorganisms. *Source: Annales Sciences Naturelle, 4th Series, Vol. 16, pp. 1–98, Pasteur, L., 1861, "Mémoire sur les Corpuscules Organisés Qui Existent Dans L'Atmosphère: Examen de la Doctrine des Générations Spontanées."*

became clear that disease could be caused by microbial infections, microbiologists began to examine the way in which hosts defended themselves against microorganisms and to ask how disease might be prevented. The field of immunology was born.

1. Describe the field of microbiology in terms of the size of its subject material and the nature of its techniques.
2. How did Pasteur and Tyndall finally settle the spontaneous generation controversy?

 1.3   THE ROLE OF MICROORGANISMS IN DISEASE

The importance of microorganisms in disease was not immediately obvious to people, and it took many years for scientists to establish the connection between microorganisms and illness. Recognition of the role of microorganisms depended greatly upon the development of new techniques for their study. Once it

Recognition of the Relationship between Microorganisms and Disease

Although Fracastoro and a few others had suggested that invisible organisms produced disease, most believed that disease was due to causes such as supernatural forces, poisonous vapors called miasmas, and imbalances between the four humors thought to be present in the body. The idea that an imbalance between the four humors (blood, phlegm, yellow bile [choler], and black bile [melancholy]) led to disease had been widely accepted since the time of the Greek physician Galen (129–199). Support for the germ theory of disease began to accumulate in the early nineteenth century. Agostino Bassi (1773–1856) first showed a microorganism could cause disease when he demonstrated in 1835 that a silkworm disease was due to a fungal infection. He also suggested that many diseases were due to microbial infections. In 1845 M. J. Berkeley proved that the great Potato Blight of Ireland was caused by a fungus, and in 1853 Heinrich de Bary showed that smut and rust fungi caused cereal crop diseases. Following his successes with the study of fermentation, Pasteur was asked by the French government to investigate the pèbrine disease of silkworms that was disrupting the silk industry (*see figure 27.5*). After several years of work, he showed that the disease was due to a protozoan parasite. The disease was controlled by raising caterpillars from eggs produced by healthy moths.

Indirect evidence that microorganisms were agents of human disease came from the work of the English surgeon Joseph Lister (1827–1912) on the prevention of wound infections. Lister, impressed with Pasteur's studies on the involvement of microorganisms in fermentation and putrefaction, developed a system of antiseptic surgery designed to prevent microorganisms from entering wounds. Instruments were heat sterilized, and phenol was used on surgical dressings and at times sprayed over the surgical area. The approach was remarkably successful and transformed surgery after Lister published his findings in 1867. It also provided strong indirect evidence for the role of microorganisms in disease because phenol, which killed bacteria, also prevented wound infections.

The first direct demonstration of the role of bacteria in causing disease came from the study of anthrax (*see chapter 39*) by the German physician Robert Koch (1843–1910). Koch (**figure 1.4**) used the criteria proposed by his former teacher, Jacob Henle (1809–1885), to establish the relationship between *Bacillus anthracis* and anthrax, and published his findings in 1876 (**Techniques & Applications 1.1** briefly discusses the scientific method). Koch injected healthy mice with material from diseased animals, and the mice became ill. After transferring anthrax by inoculation through a series of 20 mice, he incubated a piece of spleen containing the anthrax bacillus in beef serum. The bacilli grew, re-

produced, and produced spores. When the isolated bacilli or spores were injected into mice, anthrax developed. His criteria for proving the causal relationship between a microorganism and a specific disease are known as **Koch's postulates** and can be summarized as follows:

1. The microorganism must be present in every case of the disease but absent from healthy organisms.
2. The suspected microorganism must be isolated and grown in a pure culture.
3. The same disease must result when the isolated microorganism is inoculated into a healthy host.
4. The same microorganism must be isolated again from the diseased host.

Although Koch used the general approach described in the postulates during his anthrax studies, he did not outline them fully until his 1884 publication on the cause of tuberculosis (**Disease 1.2**).

Koch's proof that *Bacillus anthracis* caused anthrax was independently confirmed by Pasteur and his coworkers. They discovered that after burial of dead animals, anthrax spores survived and were brought to the surface by earthworms. Healthy animals then ingested the spores and became ill.

The Development of Techniques for Studying Microbial Pathogens

During Koch's studies on bacterial diseases, it became necessary to isolate suspected bacterial pathogens. At first he cultured bacteria on the sterile surfaces of cut, boiled potatoes. This was unsatisfactory because bacteria would not always grow well on potatoes. He then tried to solidify regular liquid media by adding gelatin. Separate bacterial colonies developed after the surface had been streaked with a bacterial sample. The sample could also be mixed with liquefied gelatin medium. When the gelatin medium hardened, individual bacteria produced separate colonies. Despite its advantages gelatin was not an ideal solidifying agent because it was digested by many bacteria and melted when the temperature rose above 28°C. A better alternative was provided by Fannie Eilshemius Hesse, the wife of Walther Hesse, one of Koch's assistants (**figure 1.5**). She suggested the use of agar as a solidifying agent—she had been using it successfully to make jellies for some time. Agar was not attacked by most bacteria and did not melt until reaching a temperature of 100°C. One of Koch's assistants, Richard Petri, developed the petri dish (plate), a container for solid culture media. These developments made possible the isolation of pure cultures that contained only one type of bacterium, and directly stimulated progress in all areas of bacteriology. Isolation of bacteria and pure culture techniques (pp. 104–7)

Koch also developed media suitable for growing bacteria isolated from the body. Because of their similarity to body fluids, meat extracts and protein digests were used as nutrient sources. The result was the development of nutrient broth and nutrient agar, media that are still in wide use today.

By 1882 Koch had used these techniques to isolate the bacillus that caused tuberculosis. There followed a golden age of about

Figure 1.4   Robert Koch. Koch (1843–1910) examining a specimen in his laboratory.

Techniques & Applications
1.1   The Scientific Method

Although biologists employ a variety of approaches in conducting research, microbiologists and other experimentally oriented biologists often use the general approach known as the scientific method. They first gather observations of the process to be studied and then develop a tentative **hypothesis**—an educated guess—to explain the observations (see **Box figure**). This step often is inductive and creative because there is no detailed, automatic technique for generating hypotheses. Next they decide what information is required to test the hypothesis and collect this information through observation or carefully designed experiments. After the information has been collected, they decide whether the hypothesis has been supported or falsified. If it has failed to pass the test, the hypothesis is rejected, and a new explanation or hypothesis is constructed. If the hypothesis passes the test, it is subjected to more severe testing. The procedure often is made more efficient by constructing and testing alternative hypotheses and then refining the hypothesis that survives testing. This general approach is often called the hypothetico-deductive method. One deduces predictions from the currently accepted hypothesis and tests them. In deduction the conclusion about specific cases follows logically from a general premise ("if . . ., then . . ." reasoning). Induction is the opposite. A general conclusion is reached after considering many specific examples. Both types of reasoning are used by scientists.

When carrying out an experiment, it is essential to use a control group as well as an experimental group. The control group is treated precisely the same as the experimental group except that the experimental manipulation is not performed on it. In this way one can be sure that any changes in the experimental group are due to the experimental manipulation rather than to some other factor not taken into account.

If a hypothesis continues to survive testing, it may be accepted as a valid theory. A **theory** is a set of propositions and concepts that provides a reliable, systematic, and rigorous account of an aspect of nature. It is important to note that hypotheses and theories are never absolutely proven. Scientists simply gain more and more confidence in their accuracy as they continue to survive testing, fit with new observations and experiments, and satisfactorily explain the observed phenomena.

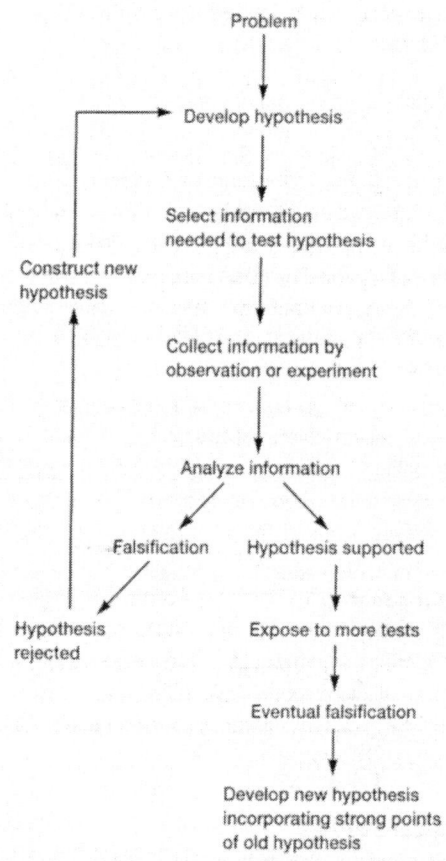

The Hypothetico-Deductive Method. This approach is most often used in scientific research.

Figure 1.5   Fannie Eilshemius (1850–1934) and Walther Hesse (1846–1911). Fannie Hesse suggested to her husband Walther (a physican and bacteriologist) that he should try using agar in his culture medium when more typical media failed to meet his needs.

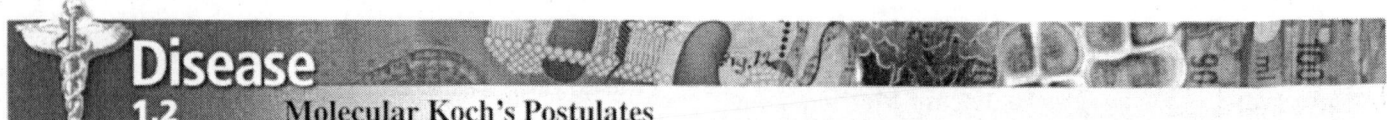

Disease

1.2   Molecular Koch's Postulates

Although the criteria that Koch developed for proving a causal relationship between a microorganism and a specific disease have been of great importance in medical microbiology, it is not always possible to apply them in studying human diseases. For example, some pathogens cannot be grown in pure culture outside the host; because other pathogens grow only in humans, their study would require experimentation on people. The identification, isolation, and cloning of genes responsible for pathogen virulence (*see p. 768*) have made possible a new molecular form of Koch's postulates that resolves some of these difficulties. The emphasis is on the virulence genes present in the infectious agent rather than on the agent itself. The molecular postulates can be briefly summarized as follows:

1. The virulence trait under study should be associated much more with pathogenic strains of the species than with nonpathogenic strains.

2. Inactivation of the gene or genes associated with the suspected virulence trait should substantially decrease pathogenicity.

3. Replacement of the mutated gene with the normal wild-type gene should fully restore pathogenicity.

4. The gene should be expressed at some point during the infection and disease process.

5. Antibodies or immune system cells directed against the gene products should protect the host.

The molecular approach cannot always be applied because of problems such as the lack of an appropriate animal system. It also is difficult to employ the molecular postulates when the pathogen is not well characterized genetically.

---

30 to 40 years in which most of the major bacterial pathogens were isolated (table 1.1).

The discovery of viruses and their role in disease was made possible when Charles Chamberland (1851–1908), one of Pasteur's associates, constructed a porcelain bacterial filter in 1884. The first viral pathogen to be studied was the tobacco mosaic disease virus (*see chapter 16*). The development of virology (pp. 352–53)

Immunological Studies

In this period progress also was made in determining how animals resisted disease and in developing techniques for protecting humans and livestock against pathogens. During studies on chicken cholera, Pasteur and Roux discovered that incubating their cultures for long intervals between transfers would attenuate the bacteria, which meant they had lost their ability to cause the disease. If the chickens were injected with these attenuated cultures, they remained healthy but developed the ability to resist the disease. He called the attenuated culture a vaccine [Latin *vacca*, cow] in honor of Edward Jenner because, many years earlier, Jenner had used vaccination with material from cowpox lesions to protect people against smallpox (*see section 16.1*). Shortly after this, Pasteur and Chamberland developed an attenuated anthrax vaccine in two ways: by treating cultures with potassium bichromate and by incubating the bacteria at 42 to 43°C. Vaccines and immunizations (pp. 740–44)

Pasteur next prepared rabies vaccine by a different approach. The pathogen was attenuated by growing it in an abnormal host, the rabbit. After infected rabbits had died, their brains and spinal cords were removed and dried. During the course of these studies, Joseph Meister, a nine-year-old boy who had been bitten by a rabid dog, was brought to Pasteur. Since the boy's death was certain in the absence of treatment, Pasteur agreed to try vaccination.

Joseph was injected 13 times over the next 10 days with increasingly virulent preparations of the attenuated virus (*see illustration, p. 739*). He survived.

In gratitude for Pasteur's development of vaccines, people from around the world contributed to the construction of the Pasteur Institute in Paris, France. One of the initial tasks of the Institute was vaccine production.

After the discovery that the diphtheria bacillus produced a toxin, Emil von Behring (1854–1917) and Shibasaburo Kitasato (1852–1931) injected inactivated toxin into rabbits, inducing them to produce an antitoxin, a substance in the blood that would inactivate the toxin and protect against the disease. A tetanus antitoxin was then prepared and both antitoxins were used in the treatment of people.

The antitoxin work provided evidence that immunity could result from soluble substances in the blood, now known to be antibodies (humoral immunity). It became clear that blood cells were also important in immunity (cellular immunity) when Elie Metchnikoff (1845–1916) discovered that some blood leukocytes could engulf disease-causing bacteria (**figure 1.6**). He called these cells phagocytes and the process phagocytosis [Greek *phagein*, eating].

---

1. Discuss the contributions of Lister, Pasteur, and Koch to the germ theory of disease and to the treatment or prevention of diseases.

2. What other contributions did Koch make to microbiology?

3. Describe Koch's postulates. What are the molecular Koch's postulates and why are they important?

4. How did von Behring and Metchnikoff contribute to the development of immunology?

Figure 1.6 Elie Metchnikoff. Metchnikoff (1845–1916) shown here at work in his laboratory.

1.4 INDUSTRIAL MICROBIOLOGY AND MICROBIAL ECOLOGY

Although Theodore Schwann and others had proposed in 1837 that yeast cells were responsible for the conversion of sugars to alcohol, a process they called alcoholic fermentation, the leading chemists of the time believed microorganisms were not involved. They were convinced that fermentation was due to a chemical instability that degraded the sugars to alcohol. Pasteur did not agree. It appears that early in his career Pasteur became interested in fermentation because of his research on the stereochemistry of molecules. He believed that fermentations were carried out by living organisms and produced asymmetric products such as amyl alcohol that had optical activity. There was an intimate connection between molecular asymmetry, optical activity, and life. Then in 1856 M. Bigo, an industrialist in Lille, France, where Pasteur worked, requested Pasteur's assistance. His business produced ethanol from the fermentation of beet sugars, and the alcohol yields had recently declined and the product had become sour. Pasteur discovered that the fermentation was failing because the yeast normally responsible for alcohol formation had been replaced by microorganisms producing lactic acid rather than

ethanol. In solving this practical problem, Pasteur demonstrated that all fermentations were due to the activities of specific yeasts and bacteria, and he published several papers on fermentation between 1857 and 1860. His success led to a study of wine diseases and the development of pasteurization (*see chapter 7*) to preserve wine during storage. Pasteur's studies on fermentation continued for almost 20 years. One of his most important discoveries was that some fermentative microorganisms were anaerobic and could live only in the absence of oxygen, whereas others were able to live either aerobically or anaerobically. Fermentation (pp. 174–77); The effect of oxygen on microorganisms (pp. 125–27).

A few of the early microbiologists chose to investigate the ecological role of microorganisms. In particular they studied microbial involvement in the carbon, nitrogen, and sulfur cycles taking place in soil and aquatic habitats. Two of the pioneers in this endeavor were Sergei N. Winogradsky (1856–1953) and Martinus W. Beijerinck (1851–1931). Biogeochemical cycles (pp. 593–600)

The Russian microbiologist Sergei N. Winogradsky made many contributions to soil microbiology. He discovered that soil bacteria could oxidize iron, sulfur, and ammonia to obtain energy, and that many bacteria could incorporate CO_2 into organic matter much like photosynthetic organisms do. Winogradsky also isolated anaerobic nitrogen-fixing soil bacteria and studied the decomposition of cellulose.

Martinus W. Beijerinck was one of the great general microbiologists who made fundamental contributions to microbial ecology and many other fields. He isolated the aerobic nitrogen-fixing bacterium *Azotobacter;* a root nodule bacterium also capable of fixing nitrogen (later named *Rhizobium*); and sulfate-reducing bacteria. Beijerinck and Winogradsky developed the enrichment-culture technique and the use of selective media (*see chapter 5*), which have been of such great importance in microbiology.

---

1. Briefly describe the work of Pasteur on microbial fermentations.
2. How did Winogradsky and Beijerinck contribute to the study of microbial ecology?

1.5 MEMBERS OF THE MICROBIAL WORLD

Although the kingdoms of organisms and the differences between procaryotic and eucaryotic cells are discussed in much more detail later, a brief introduction to the organisms a microbiologist studies is given here. Comparison of procaryotic and eucaryotic cells (pp. 90–91)

Two fundamentally different types of cells exist. **Procaryotic cells** [Greek *pro*, before, and *karyon,* nut or kernel; organism with a primordial nucleus] have a much simpler morphology than eucaryotic cells and lack a true membrane-delimited nucleus. All bacteria are procaryotic. In contrast, **eucaryotic cells** [Greek *eu*, true, and *karyon,* nut or kernel] have a membrane-enclosed nucleus; they are more complex morphologically and are usually larger than procaryotes. Algae, fungi, protozoa, higher plants, and animals are eucaryotic. Procaryotic and eucaryotic cells differ in many other ways as well (*see chapter 4*).

The early description of organisms as either plants or animals clearly is too simplified, and for many years biologists have divided organisms into five kingdoms: the *Monera, Protista, Fungi, Animalia,* and *Plantae* (*see chapter 19*). Microbiologists study primarily members of the first three kingdoms. Although they are not included in the five kingdoms, viruses are also studied by microbiologists.   Fungi (chapter 25); Algae (chapter 26); Protozoa (chapter 27); Introduction to the viruses (chapters 16–18)

In the last few decades great progress has been made in three areas that profoundly affect microbial classification. First, much has been learned about the detailed structure of microbial cells from the use of electron microscopy. Second, microbiologists have determined the biochemical and physiological characteristics of many different microorganisms. Third, the sequences of nucleic acids and proteins from a wide variety of organisms have been compared. It is now clear that there are two quite different groups of procaryotic organisms: Bacteria and Archaea. Furthermore, the protists are so diverse that it may be necessary to divide the kingdom *Protista* into three or more kingdoms. Thus many taxonomists have concluded that the five-kingdom system is too simple and have proposed a variety of alternatives (*see section 19.7*). The differences between Bacteria, Archaea, and the eucaryotes seem so great that many microbiologists have proposed that organisms should be divided among three domains: Bacteria (the true bacteria or eubacteria), Archaea,[1] and Eucarya (all eucaryotic organisms). This system, which we shall use here, and the results leading to it are discussed in chapter 19.

---

1. Describe and contrast procaryotic and eucaryotic cells.
2. Briefly describe the five-kingdom system and give the major characteristics of each kingdom.

1.6   THE SCOPE AND RELEVANCE OF MICROBIOLOGY

As the late scientist-writer Steven Jay Gould emphasized, we live in the Age of Bacteria. They were the first living organisms on our planet, live virtually everywhere life is possible, are more numerous than any other kind of organism, and probably constitute the largest component of the earth's biomass. The whole ecosystem depends on their activities, and they influence human society in countless ways. Thus modern microbiology is a large discipline with many different specialties; it has a great impact on fields such as medicine, agricultural and food sciences, ecology, genetics, biochemistry, and molecular biology.

---

[1]Although this will be discussed further in chapter 19, it should be noted here that several names have been used for the Archaea. The two most important are archaeobacteria and archaebacteria. In this text, we shall use only the name Archaea for sake of clarity and consistency.

For example, microbiology has been a major contributor to the rise of molecular biology, the branch of biology dealing with the physical and chemical aspects of living matter and its function. Microbiologists have been deeply involved in studies on the genetic code and the mechanisms of DNA, RNA, and protein synthesis. Microorganisms were used in many of the early studies on the regulation of gene expression and the control of enzyme activity (*see chapters 8 and 12*). In the 1970s new discoveries in microbiology led to the development of recombinant DNA technology and genetic engineering.   The mechanisms of DNA, RNA, and protein synthesis (chapters 11 and 12); Recombinant DNA and genetic engineering (chapter 14)

One indication of the importance of microbiology is the Nobel Prize given for work in physiology or medicine. About 1/3 of these have been awarded to scientists working on microbiological problems (*see inside front cover*).

Microbiology has both basic and applied aspects. Many microbiologists are interested primarily in the biology of the microorganisms themselves (**figure 1.7**). They may focus on a specific group of microorganisms and be called virologists (viruses), bacteriologists (bacteria), phycologists or algologists (algae), mycologists (fungi), or protozoologists (protozoa). Others are interested in microbial morphology or particular functional processes and work in fields such as microbial cytology, microbial physiology, microbial ecology, microbial genetics, molecular biology, and microbial taxonomy. Of course a person can be thought of in both ways (e.g., as a bacteriologist who works on taxonomic problems). Many microbiologists have a more applied orientation and work on practical problems in fields such as medical microbiology, food and dairy microbiology, and public health microbiology (basic research is also conducted in these fields). Because the various fields of microbiology are interrelated, an applied microbiologist must be familiar with basic microbiology. For example, a medical microbiologist must have a good understanding of microbial taxonomy, genetics, immunology, and physiology to identify and properly respond to the pathogen of concern.

What are some of the current occupations of professional microbiologists? One of the most active and important is medical microbiology, which deals with the diseases of humans and animals. Medical microbiologists identify the agent causing an infectious disease and plan measures to eliminate or control it. Frequently they are involved in tracking down new, unidentified pathogens such as the agent that causes variant Creutzfeldt-Jacob disease, the hantavirus, the West Nile virus, and the virus responsible for SARS. These microbiologists also study the ways in which microorganisms cause disease.   Hantavirus pulmonary syndrome (p. 853); West Nile fever (p. 854); SARS (p. 835)

Public health microbiology is closely related to medical microbiology. Public health microbiologists try to control the spread of communicable diseases. They often monitor community food establishments and water supplies in an attempt to keep them safe and free from infectious disease agents.

Immunology is concerned with how the immune system protects the body from pathogens and the response of infectious

(a) *Rita Colwell*

(b) *R. G. E. Murray*

(c) *Stanley Falkow*

(d) *Martha Howe*

(e) *Frederick Neidhardt*

(f) *Jean Brenchley*

Figure 1.7   Important Contibutors to Microbiology. (**a**) Rita R. Colwell has studied the genetics and ecology of marine bacteria such as *Vibrio cholerae* and helped establish the field of marine biotechnology. (**b**) R. G. E. Murray has contributed greatly to the understanding of bacterial cell envelopes and bacterial taxonomy. (**c**) Stanley Falkow has advanced our understanding of how bacterial pathogens cause disease. (**d**) Martha Howe has made fundamental contributions to our knowledge of the bacteriophage Mu. (**e**) Frederick C. Neidhardt has contributed to microbiology through his work on the regulation of *E. coli* physiology and metabolism, and by coauthoring advanced textbooks. (**f**) Jean E. Brenchley has studied the regulation of glutamate and glutamine metabolism, helped found the Pennsylvania State University Biotechnology Institute, and is now finding biotechnological uses for psychrophilic (cold-loving) microorganisms.

agents. It is one of the fastest growing areas in science; for example, techniques for the production and use of monoclonal antibodies have developed extremely rapidly. Immunology also deals with practical health problems such as the nature and treatment of allergies and autoimmune diseases like rheumatoid arthritis.   Monoclonal antibodies and their uses (section 32.3 and Techniques & Applications 36.2)

Many important areas of microbiology do not deal directly with human health and disease but certainly contribute to human welfare. Agricultural microbiology is concerned with the impact of microorganisms on agriculture. Agricultural microbiologists try to combat plant diseases that attack important food crops, work on methods to increase soil fertility and crop yields, and study the role of microorganisms living in the digestive tracts of ruminants such as cattle. Currently there is great interest in using bacterial and viral insect pathogens as substitutes for chemical pesticides.

The field of microbial ecology is concerned with the relationships between microorganisms and their living and nonliving habitats. Microbial ecologists study the contributions of microorganisms to the carbon, nitrogen, and sulfur cycles in soil and in freshwater. The study of pollution effects on microorganisms also is important because of the impact these organisms have on the environment. Microbial ecologists are employing microorganisms in bioremediation to reduce pollution effects.

Scientists working in food and dairy microbiology try to prevent microbial spoilage of food and the transmission of foodborne diseases such as botulism and salmonellosis (*see chapter 39*). They also use microorganisms to make foods such as cheeses, yogurts, pickles, and beer. In the future microorganisms themselves may become a more important nutrient source for livestock and humans.

In industrial microbiology microorganisms are used to make products such as antibiotics, vaccines, steroids, alcohols and other solvents, vitamins, amino acids, and enzymes. Microorganisms can even leach valuable minerals from low-grade ores.

Research on the biology of microorganisms occupies the time of many microbiologists and also has practical applications. Those working in microbial physiology and biochemistry study the synthesis of antibiotics and toxins, microbial energy production, the ways in which microorganisms survive harsh environmental conditions, microbial nitrogen fixation, the effects of chemical and physical agents on microbial growth and survival, and many other aspects.

Microbial genetics and molecular biology focus on the nature of genetic information and how it regulates the development and function of cells and organisms. The use of microorganisms has been very helpful in understanding gene function. Microbial geneticists play an important role in applied microbiology by producing new microbial strains that are more efficient in synthesizing useful products. Genetic techniques are used to test substances for their ability to cause cancer. More recently the field of genetic engineering (*see chapter 14*) has arisen from work in microbial genetics and molecular biology and will contribute substantially to microbiology, biology as a whole, and medicine. Engineered microorganisms are used to make hormones, antibiotics, vaccines, and other products (*see chapter 42*). New genes can be inserted into plants and animals; for example, it may be possible to give corn and wheat nitrogen-fixation genes so they will not require nitrogen fertilizers.

 1.7   THE FUTURE OF MICROBIOLOGY

As the preceding sections have shown, microbiology has had a profound influence on society. What of the future? Science writer

Bernard Dixon is very optimistic about microbiology's future for two reasons. First, microbiology has a clearer mission than do many other scientific disciplines. Second, it is confident of its value because of its practical significance. Dixon notes that microbiology is required both to face the threat of new and reemerging human infectious diseases and to develop industrial technologies that are more efficient and environmentally friendly.

What are some of the most promising areas for future microbiological research and their potential practical impacts? What kinds of challenges do microbiologists face? This brief list should give some idea of what the future may hold.

1. New infectious diseases are continually arising and old diseases are once again becoming widespread and destructive. AIDS, SARS, hemorrhagic fevers, and tuberculosis are excellent examples of new and reemerging infectious diseases. Microbiologists will have to respond to these threats, many of them presently unknown.

2. Microbiologists must find ways to stop the spread of established infectious diseases. Increases in antibiotic resistance will be a continuing problem, particularly the spread of multiple drug resistance that can render a pathogen resistant to current medical treatment. Microbiologists have to create new drugs and find ways to slow or prevent the spread of drug resistance. New vaccines must be developed to protect against diseases such as AIDS. It will be necessary to use techniques in molecular biology and recombinant DNA technology to solve these problems.

3. Research is needed on the association between infectious agents and chronic diseases such as autoimmune and cardiovascular diseases. It may be that some of these chronic afflictions partly result from infections.

4. We are only now beginning to understand how pathogens interact with host cells and the ways in which diseases arise. There also is much to learn about how the host resists pathogen invasions.

5. Microorganisms are increasingly important in industry and environmental control, and we must learn how to use them in a variety of new ways. For example, microorganisms can (a) serve as sources of high-quality food and other practical products such as enzymes for industrial applications, (b) degrade pollutants and toxic wastes, and (c) be used as vectors to treat diseases and enhance agricultural productivity. There also is a continuing need to protect food and crops from microbial damage.

6. Microbial diversity is another area requiring considerable research. Indeed, it is estimated that less than 1% of the earth's microbial population has been cultured. Greater efforts to grow previously uncultivated microorganisms will be required. We must develop new isolation techniques and an adequate classification of microorganisms, one which includes those microbes that cannot be cultivated in the laboratory. Much work needs to be done on microorganisms living in extreme environments. The discovery of new microorganisms may well lead to further advances in industrial processes and enhanced environmental control.

7. Microbial communities often live in biofilms, and these biofilms are of profound importance in both medicine and microbial ecology. Research on biofilms is in its infancy; it will be many years before we more fully understand their nature and are able to use our knowledge in practical ways. In general, microbe-microbe interactions have not yet been extensively explored.

8. The genomes of many microorganisms already have been sequenced, and many more will be determined in the coming years. These sequences are ideal for learning how the genome is related to cell structure and the minimum assortment of genes necessary for life. Analysis of the genome and its activity will require continuing advances in the field of bioinformatics and the use of computers to investigate biological problems.

9. Further research on unusual microorganisms and microbial ecology will lead to a better understanding of the interactions between microorganisms and the inanimate world. Among other things, this understanding should enable us to more effectively control pollution. Similarly, it has become clear that microorganisms are essential partners with higher organisms in symbiotic relationships. Greater knowledge of symbiotic relationships can help improve our appreciation of the living world. It also will lead to improvements in the health of plants, livestock, and humans.

10. Because of their relative simplicity, microorganisms are excellent subjects for the study of a variety of fundamental questions in biology. For example, how do complex cellular structures develop and how do cells communicate with one another and respond to the environment?

11. Finally, microbiologists will be challenged to carefully assess the implications of new discoveries and technological developments. They will need to communicate a balanced view of both the positive and negative long-term impacts of these events on society.

The future of microbiology is bright. The microbiologist René Dubos has summarized well the excitement and promise of microbiology:

> How extraordinary that, all over the world, microbiologists are now involved in activities as different as the study of gene structure, the control of disease, and the industrial processes based on the phenomenal ability of microorganisms to decompose and synthesize complex organic molecules. Microbiology is one of the most rewarding of professions because it gives its practitioners the opportunity to be in contact with all the other natural sciences and thus to contribute in many different ways to the betterment of human life.

Summary

Microbiology may be defined in terms of the size of the organisms studied and the techniques employed.

1.1 The Discovery of Microorganisms
a. Antony van Leeuwenhoek was the first person to describe microorganisms.

1.2 The Conflict over Spontaneous Generation
a. Experiments by Redi and others disproved the theory of spontaneous generation in regard to larger organisms.
b. The spontaneous generation of microorganisms was disproved by Spallanzani, Pasteur, Tyndall, and others.

1.3 The Role of Microorganisms in Disease
a. Support for the germ theory of disease came from the work of Bassi, Pasteur, Koch, and others. Lister provided indirect evidence with his development of antiseptic surgery.
b. Koch's postulates and molecular Koch's postulates are used to prove a direct relationship between a suspected pathogen and a disease.
c. Koch developed the techniques required to grow bacteria on solid media and to isolate pure cultures of pathogens.
d. Vaccines against anthrax and rabies were made by Pasteur; von Behring and Kitasato prepared antitoxins for diphtheria and tetanus.
e. Metchnikoff discovered some blood leukocytes could phagocytize and destroy bacterial pathogens.

1.4 Industrial Microbiology and Microbial Ecology
a. Pasteur showed that fermentations were caused by microorganisms and that some microorganisms could live in the absence of oxygen.
b. The role of microorganisms in carbon, nitrogen, and sulfur cycles was first studied by Winogradsky and Beijerinck.

1.5 Members of the Microbial World
a. Procaryotic cells differ from eucaryotic cells in lacking a membrane-delimited nucleus, and in other ways as well.
b. The Archaea are so different that many microbiologists divide organisms into three domains: *Bacteria, Archaea,* and *Eucarya.*

1.6 The Scope and Relevance of Microbiology
a. In the twentieth century microbiology contributed greatly to the fields of biochemistry and genetics. It also helped stimulate the rise of molecular biology.
b. There is a wide variety of fields in microbiology, and many have a great impact on society. These include the more applied disciplines such as medical, public health, industrial, food, and dairy microbiology. Microbial ecology, physiology, biochemistry, and genetics are examples of basic microbiological research fields.

1.7 The Future of Microbiology
a. Microbiologists will be faced with many exciting and important future challenges such as finding new ways to combat disease, reduce pollution, and feed the world's population.

Key Terms

eucaryotic cell   11
hypothesis   9

Koch's postulates   8
microbiology   2

microorganism   2
procaryotic cell   11

spontaneous generation   2
theory   9

Questions for Thought and Review

1. Why was the belief in spontaneous generation an obstacle to the development of microbiology as a scientific discipline?
2. Describe the major contributions of the following people to the development of microbiology: Leeuwenhoek, Spallanzani, Fracastoro, Pasteur, Tyndall, Cohn, Bassi, Lister, Koch, Chamberland, von Behring, Metchnikoff, Winogradsky, and Beijerinck.
3. Would microbiology have developed more slowly if Fannie Hesse had not suggested the use of agar? Give your reasoning. What is a pure culture?
4. Why do you think viruses are not included in the five-kingdom or three domain systems?
5. Why are microorganisms so useful to biologists as experimental models?
6. What do you think were the most important discoveries in the development of microbiology? Why?
7. List all the activities or businesses you can think of in your community that are directly dependent on microbiology.
8. Describe in your own words the scientific method. How does a theory differ from a hypothesis? Why is it important to have a control group?
9. What do you think are the five most important research areas to pursue in microbiology? Give reasons for your choices.

Critical Thinking Questions

1. Consider the impact of microbes on the course of world history. History is full of examples of instances or circumstances under which one group of people lost a struggle against another. In fact, when examined more closely, the "losers" often had the misfortune of being exposed to, more susceptible to, or unable to cope with an infectious agent. Thus weakened in physical strength or demoralized by the course of a devastating disease, they were easily overcome by human "conquerors."
   a. Choose an example of a battle or other human activity such as exploration of new territory and determine the impact of microorganisms, either indigenous or transported to the region, on that activity.
   b. Discuss the effect that the microbe(s) had on the outcome in your example.
   c. Suggest whether the advent of antibiotics, food storage and preparation technology, or sterilization technology would have made a difference in the outcome.
2. Vaccinations against various childhood diseases have contributed to the entry of women, particularly mothers, into the full-time workplace.
   a. Is this statement supported by data—comparing availability and extent of vaccination with employment statistics in different places or at different times?
   b. Before vaccinations for measles, mumps, and chickenpox, what was the incubation time and duration of these childhood diseases? What impact would such diseases have on mothers with several elementary schoolchildren at home if they had full-time jobs and lacked substantial child care support?
   c. What would be the consequence if an entire generation of children (or a group of children in one country) were not vaccinated against any diseases? What do you predict would happen if these children went to college and lived in a dormitory in close proximity with others who had received all of the recommended childhood vaccines?

For recommended readings on these and related topics, see Appendix V.

chapter 2

The Study of Microbial Structure:
Microscopy and Specimen Preparation

Concepts

- Light microscopes use glass lenses to bend and focus light rays and produce enlarged images of small objects. The resolution of a light microscope is determined by the numerical aperture of its lens system and by the wavelength of the light it employs; maximum resolution is about 0.2 μm.

- The most common types of light microscopes are the bright-field, dark-field, phase-contrast, and fluorescence microscopes. Each yields a distinctive image and may be used to observe different aspects of microbial morphology.

- Because most microorganisms are colorless and therefore not easily seen in the bright-field microscope, they are usually fixed and stained before observation. Either simple or differential staining can be used to enhance contrast. Specific bacterial structures such as capsules, endospores, and flagella also can be selectively stained.

- The transmission electron microscope achieves great resolution (about 0.5 nm) by using electron beams of very short wavelength rather than visible light. Although one can prepare microorganisms for observation in other ways, one normally views thin sections of plastic-embedded specimens treated with heavy metals to improve contrast.

- External features can be observed in great detail with the scanning electron microscope, which generates an image by scanning a fine electron beam over the surface of specimens rather than projecting electrons through them.

- New forms of microscopy are improving our ability to observe microorganisms and molecules. Two examples are the confocal scanning laser microscope and the scanning probe microscope.

There are more animals living in the scum on the teeth in a man's mouth than there are men in a whole kingdom.

—Antony van Leeuwenhoek

Clostridium botulinum is a rod-shaped bacterium that forms endospores and releases botulinum toxin, the cause of botulism food poisoning. In this phase-contrast micrograph, the endospores are the bright, oval objects located at the ends of the rods; some endospores have been released from the cells that formed them.

icrobiology usually is concerned with organisms so small they cannot be seen distinctly with the unaided eye. Because of the nature of this discipline, the microscope is of crucial importance. Thus it is important to understand how the microscope works and the way in which specimens are prepared for examination.

The chapter begins with a detailed treatment of the standard bright-field microscope and then describes other common types of light microscopes. Next preparation and staining of specimens for examination with the light microscope are discussed. This is followed by a description of transmission and scanning electron microscopes, both of which are used extensively in current microbiological research. The chapter closes with a brief introduction to two newer forms of microscopy: scanning probe microscopy and confocal microscopy.

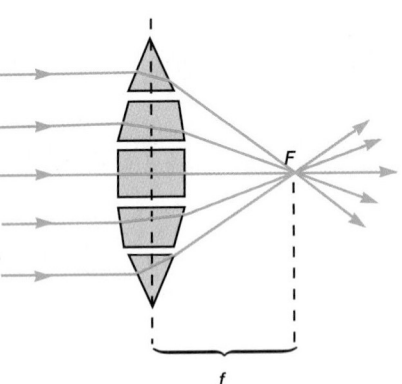

Figure 2.2   Lens Function. A lens functions somewhat like a collection of prisms. Light rays from a distant source are focused at the focal point F. The focal point lies a distance f, the focal length, from the lens center.

2.1   LENSES AND THE BENDING OF LIGHT

To understand how a light microscope operates, one must know something about the way in which lenses bend and focus light to form images. When a ray of light passes from one medium to another, **refraction** occurs—that is, the ray is bent at the interface. The **refractive index** is a measure of how greatly a substance slows the velocity of light; the direction and magnitude of bending is determined by the refractive indexes of the two media forming the interface. When light passes from air into glass, a medium with a greater refractive index, it is slowed and bent toward the normal, a line perpendicular to the surface (**figure 2.1**). As light leaves glass and returns to air, a medium with a lower refractive index, it accelerates and is bent away from the normal. Thus a prism bends light because glass has a different refractive index from air, and the light strikes its surface at an angle.

Lenses act like a collection of prisms operating as a unit. When the light source is distant so that parallel rays of light strike the lens, a convex lens will focus these rays at a specific point, the **fo-**

Table 2.1		
Common Units of Measurement		
Unit	Abbreviation	Value
1 centimeter	cm	10^{-2} meter or 0.394 inches
1 millimeter	mm	10^{-3} meter
1 micrometer	μm	10^{-6} meter
1 nanometer	nm	10^{-9} meter
1 Angstrom	Å	10^{-10} meter

cal point (F in **figure 2.2**). The distance between the center of the lens and the focal point is called the **focal length** (f in figure 2.2).

Our eyes cannot focus on objects nearer than about 25 cm or 10 inches (**table 2.1**). This limitation may be overcome by using a convex lens as a simple magnifier (or microscope) and holding it close to an object. A magnifying glass provides a clear image at much closer range, and the object appears larger. Lens strength is related to focal length; a lens with a short focal length will magnify an object more than a weaker lens having a longer focal length.

1. Define refraction, refractive index, focal point, and focal length.
2. Describe the path of a light ray through a prism or lens.
3. How is lens strength related to focal length?

2.2   THE LIGHT MICROSCOPE

Microbiologists currently employ a variety of light microscopes in their work; bright-field, dark-field, phase-contrast, and fluorescence microscopes are most commonly used. Modern microscopes are all compound microscopes. That is, the magnified image formed by the objective lens is further enlarged by one or more additional lenses.

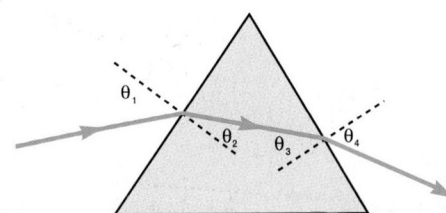

Figure 2.1   The Bending of Light by a Prism. Normals (lines perpendicular to the surface of the prism) are indicated by dashed lines. As light enters the glass, it is bent toward the first normal (angle θ_2 is less than θ_1). When light leaves the glass and returns to air, it is bent away from the second normal (θ_4 is greater than θ_3). As a result the prism bends light passing through it.

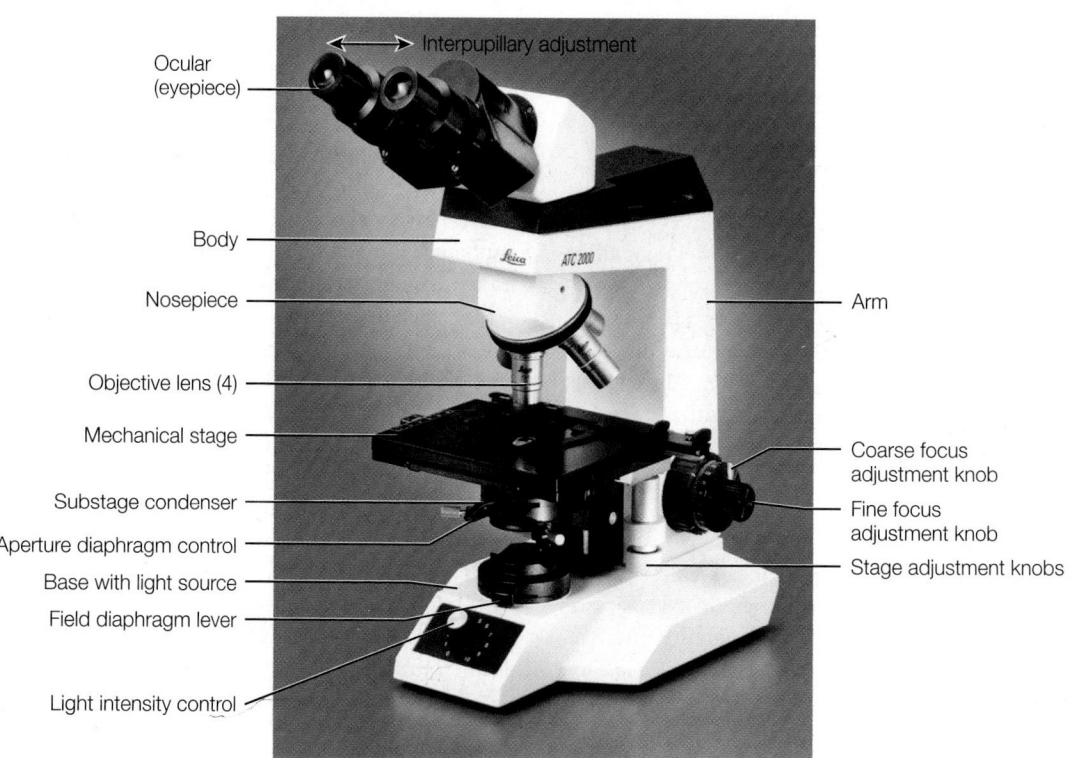

Interpupillary adjustment
Ocular (eyepiece)
Body
Nosepiece
Objective lens (4)
Mechanical stage
Substage condenser
Aperture diaphragm control
Base with light source
Field diaphragm lever
Light intensity control
Arm
Coarse focus adjustment knob
Fine focus adjustment knob
Stage adjustment knobs

Figure 2.3   A Bright-Field Microscope. The parts of a modern bright-field microscope. The microscope pictured is somewhat more sophisticated than those found in many student laboratories. For example, it is binocular (has two eyepieces) and has a mechanical stage, an adjustable substage condenser, and a built-in illuminator.

The Bright-Field Microscope

The ordinary microscope is called a **bright-field microscope** because it forms a dark image against a brighter background. The microscope consists of a sturdy metal body or stand composed of a base and an arm to which the remaining parts are attached (**figure 2.3**). A light source, either a mirror or an electric illuminator, is located in the base. Two focusing knobs, the fine and coarse adjustment knobs, are located on the arm and can move either the stage or the nosepiece to focus the image.

The stage is positioned about halfway up the arm and holds microscope slides by either simple slide clips or a mechanical stage clip. A mechanical stage allows the operator to move a slide around smoothly during viewing by use of stage control knobs. The **substage condenser** is mounted within or beneath the stage and focuses a cone of light on the slide. Its position often is fixed in simpler microscopes but can be adjusted vertically in more advanced models.

The curved upper part of the arm holds the body assembly, to which a nosepiece and one or more **eyepieces** or **oculars** are attached. More advanced microscopes have eyepieces for both eyes and are called binocular microscopes. The body assembly itself contains a series of mirrors and prisms so that the barrel holding the eyepiece may be tilted for ease in viewing (**figure 2.4**). The

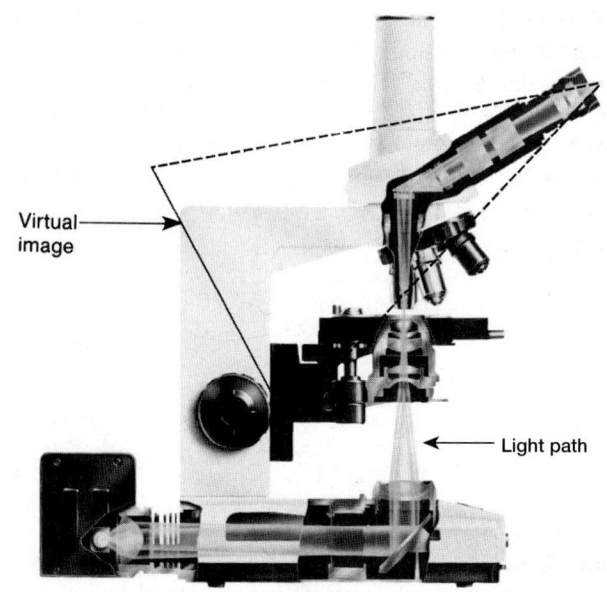

Virtual image
Light path

Figure 2.4   A Microscope's Light Path. The light path in an advanced bright-field microscope and the location of the virtual image are shown. (See also figure 2.23.)

nosepiece holds three to five **objectives** with lenses of differing magnifying power and can be rotated to position any objective beneath the body assembly. Ideally a microscope should be **parfocal**—that is, the image should remain in focus when objectives are changed.

The path of light through a bright-field microscope is shown in figure 2.4. The objective lens forms an enlarged real image within the microscope, and the eyepiece lens further magnifies this primary image. When one looks into a microscope, the enlarged specimen image, called the virtual image, appears to lie just beyond the stage about 25 cm away. The total magnification is calculated by multiplying the objective and eyepiece magnifications together. For example, if a 45× objective is used with a 10× eyepiece, the overall magnification of the specimen will be 450×.

Microscope Resolution

The most important part of the microscope is the objective, which must produce a clear image, not just a magnified one. Thus resolution is extremely important. **Resolution** is the ability of a lens to separate or distinguish between small objects that are close together. Much of the optical theory underlying microscope design was developed by the German physicist Ernst Abbé in the 1870s. The minimum distance (d) between two objects that reveals them as separate entities is given by the Abbé equation, in which lambda (λ) is the wavelength of light used to illuminate the specimen and $n \sin \theta$ is the numerical aperture (NA).

$$d = \frac{0.5\lambda}{n \sin \theta}$$

As d becomes smaller, the resolution increases, and finer detail can be discerned in a specimen.

The preceding equation indicates that a major factor in resolution is the wavelength of light used. The wavelength must be shorter than the distance between two objects or they will not be seen clearly. Thus the greatest resolution is obtained with light of the shortest wavelength, light at the blue end of the visible spectrum (in the range of 450 to 500 nm). The electromagnetic spectrum of radiation (p. 127)

The **numerical aperture** ($n \sin \theta$) is more difficult to understand. Theta is defined as ½ the angle of the cone of light entering an objective (**figure 2.5**). Light that strikes the microorganism after passing through a condenser is cone-shaped. When this cone has a narrow angle and tapers to a sharp point, it does not spread out much after leaving the slide and therefore does not adequately separate images of closely packed objects. The resolution is low. If the cone of light has a very wide angle and spreads out rapidly after passing through a specimen, closely packed objects appear widely separated and are resolved. The angle of the cone of light that can enter a lens depends on the refractive index (n) of the medium in which the lens works, as well as upon the objective itself. The refractive index for air is 1.00. Since $\sin \theta$ cannot be greater than 1 (the maximum θ is 90° and sin 90° is 1.00), no lens working in air can have a numerical aperture greater than 1.00. The only practical way to raise the numerical aperture above 1.00, and therefore achieve higher resolution, is to increase the refractive index with immersion oil, a colorless liquid with the same refractive index as glass (**table 2.2**). If air is replaced with

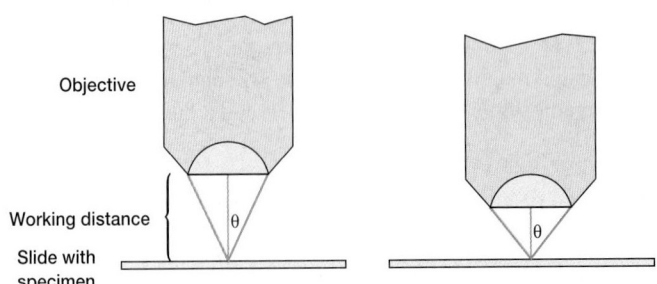

Figure 2.5 Numerical Aperture in Microscopy. The angular aperture θ is ½ the angle of the cone of light that enters a lens from a specimen, and the numerical aperture is $n \sin \theta$. In the right-hand illustration the lens has larger angular and numerical apertures; its resolution is greater and its working distance smaller.

Table 2.2				

The Properties of Microscope Objectives

	Objective			
Property	Scanning	Low Power	High Power	Oil Immersion
Magnification	4×	10×	40–45×	90–100×
Numerical aperture	0.10	0.25	0.55–0.65	1.25–1.4
Approximate focal length (f)	40 mm	16 mm	4 mm	1.8–2.0 mm
Working distance	17–20 mm	4–8 mm	0.5–0.7 mm	0.1 mm
Approximate resolving power with light of 450 nm (blue light)	2.3 μm	0.9 μm	0.35 μm	0.18 μm

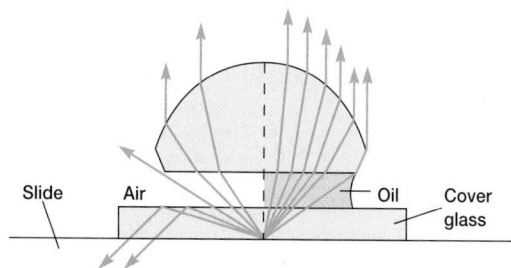

Figure 2.6   The Oil Immersion Objective. An oil immersion objective operating in air and with immersion oil.

immersion oil, many light rays that did not enter the objective due to reflection and refraction at the surfaces of the objective lens and slide will now do so (**figure 2.6**). An increase in numerical aperture and resolution results.

The resolution of a microscope depends upon the numerical aperture of its condenser as well as that of the objective. This is evident from the equation describing the resolution of the complete microscope.

$$d_{\text{microscope}} = \frac{\lambda}{(\text{NA}_{\text{objective}} + \text{NA}_{\text{condenser}})}$$

Most microscopes have a condenser with a numerical aperture between 1.2 and 1.4. However, the condenser numerical aperture will not be much above about 0.9 unless the top of the condenser is oiled to the bottom of the slide. During routine microscope operation, the condenser usually is not oiled and this limits the overall resolution, even with an oil immersion objective.

The limits set on the resolution of a light microscope can be calculated using the Abbé equation. The maximum theoretical resolving power of a microscope with an oil immersion objective (numerical aperture of 1.25) and blue-green light is approximately 0.2 μm.

$$d = \frac{(0.5)(5.30\text{nm})}{1.25} = 212 \text{ nm or } 0.2\mu\text{m}$$

At best, a bright-field microscope can distinguish between two dots around 0.2 μm apart (the same size as a very small bacterium).

Normally a microscope is equipped with three or four objectives ranging in magnifying power from 4× to 100× (table 2.2). The **working distance** of an objective is the distance between the front surface of the lens and the surface of the cover glass (if one is used) or the specimen when it is in sharp focus. Objectives with large numerical apertures and great resolving power have short working distances.

The largest useful magnification increases the size of the smallest resolvable object enough to be visible. Our eye can just detect a speck 0.2 mm in diameter, and consequently the useful limit of magnification is about 1,000 times the numerical aperture of the objective lens. Most standard microscopes come with 10× eyepieces and have an upper limit of about 1,000× with oil immersion. A 15× eyepiece may be used with good objectives to achieve a useful magnification of 1,500×. Any further magnification increase does not enable a person to see more detail. A light microscope can be built to yield a final magnification of 10,000×, but it would simply be magnifying a blur. Only the electron microscope provides sufficient resolution to make higher magnifications useful.

Proper specimen illumination also is extremely important in determining resolution. A microscope equipped with a concave mirror between the light source and the specimen illuminates the slide with a fairly narrow cone of light and has a small numerical aperture. Resolution can be improved with a substage condenser, a large light-gathering lens used to project a wide cone of light through the slide and into the objective lens, thus increasing the numerical aperture.

The Dark-Field Microscope

Living, unstained cells and organisms can be observed by simply changing the way in which they are illuminated. A hollow cone of light is focused on the specimen in such a way that unreflected and unrefracted rays do not enter the objective. Only light that has been reflected or refracted by the specimen forms an image (**figure 2.7**). The field surrounding a specimen appears black, while the object itself is brightly illuminated (**figure 2.8a,b**); because the background is dark, this type of microscopy is called **dark-field microscopy.** Considerable internal structure is often visible in larger eucaryotic microorganisms (figure 2.8b). The dark-field microscope is used to identify bacteria like the thin and distinctively shaped *Treponema pallidum* (figure 2.8a), the causative agent of syphilis.

The Phase-Contrast Microscope

Unpigmented living cells are not clearly visible in the bright-field microscope because there is little difference in contrast between the cells and water. Thus microorganisms often must be fixed and stained before observation to increase contrast and create variations in color between cell structures. A **phase-contrast microscope** converts slight differences in refractive index and cell density into easily detected variations in light intensity and is an excellent way to observe living cells (figure 2.8c–e).

The condenser of a phase-contrast microscope has an annular stop, an opaque disk with a thin transparent ring, which

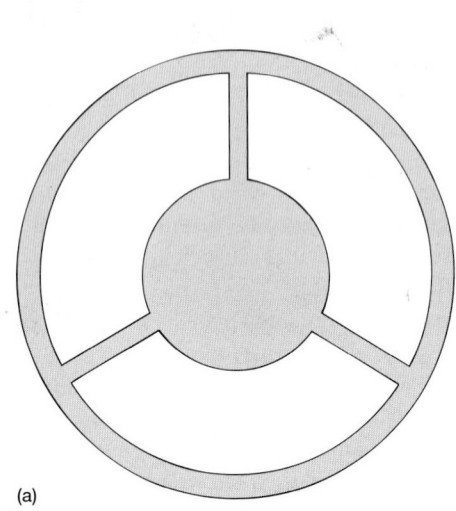

(a)

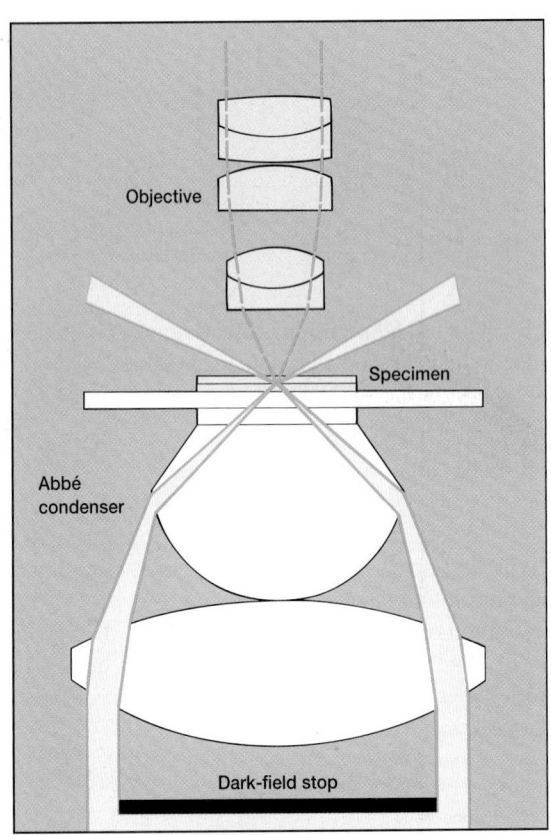

(b)

Figure 2.7   Dark-Field Microscopy. The simplest way to convert a microscope to dark-field microscopy is to place (**a**) a dark-field stop underneath (**b**) the condenser lens system. The condenser then produces a hollow cone of light so that the only light entering the objective comes from the specimen.

produces a hollow cone of light (**figure 2.9**). As this cone passes through a cell, some light rays are bent due to variations in density and refractive index within the specimen and are retarded by about $\frac{1}{4}$ wavelength. The deviated light is focused to form an image of the object. Undeviated light rays strike a phase ring in the phase plate, a special optical disk located in the objective, while the deviated rays miss the ring and pass through the rest of the plate. If the phase ring is constructed in such a way that the undeviated light passing through it is advanced by $\frac{1}{4}$ wavelength, the deviated and undeviated waves will be about $\frac{1}{2}$ wavelength out of phase and will cancel each other when they come together to form an image (**figure 2.10**). The background, formed by undeviated light, is bright, while the unstained object appears dark and well-defined. This type of microscopy is called **dark-phase-contrast microscopy.** Color filters often are used to improve the image (figure 2.8*c,d*).

Phase-contrast microscopy is especially useful for studying microbial motility, determining the shape of living cells, and detecting bacterial components such as endospores and inclusion bodies that contain poly-β-hydroxybutyrate, polymetaphosphate, sulfur, or other substances (*see chapter 3*). These are clearly visible (figure 2.8*d*) because they have refractive indexes markedly different from that of water. Phase-contrast microscopes also are widely used in studying eucaryotic cells.

The Differential Interference Contrast Microscope

The **differential interference contrast (DIC) microscope** is similar to the phase-contrast microscope in that it creates an image by detecting differences in refractive indices and thickness. Two beams of plane-polarized light at right angles to each other are generated by prisms. In one design, the object beam passes through the specimen, while the reference beam passes through a clear area of the slide. After passing through the specimen, the two beams are combined and interfere with each other to form an image. A live, unstained specimen appears brightly colored and three-dimensional (**figure 2.11**). Structures such as cell walls, endospores, granules, vacuoles, and eucaryotic nuclei are clearly visible.

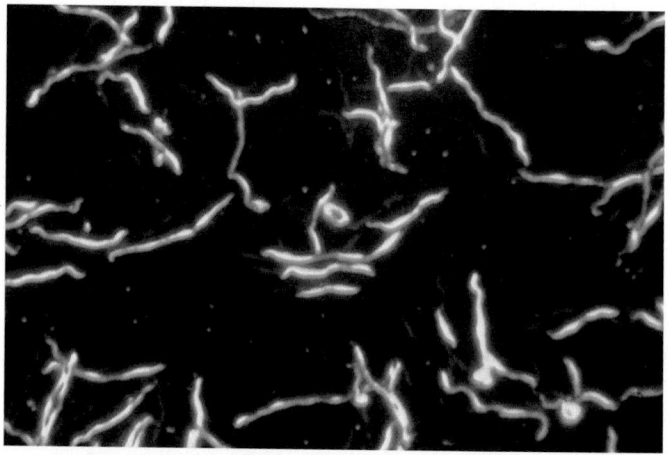

(a) *T. pallidum*

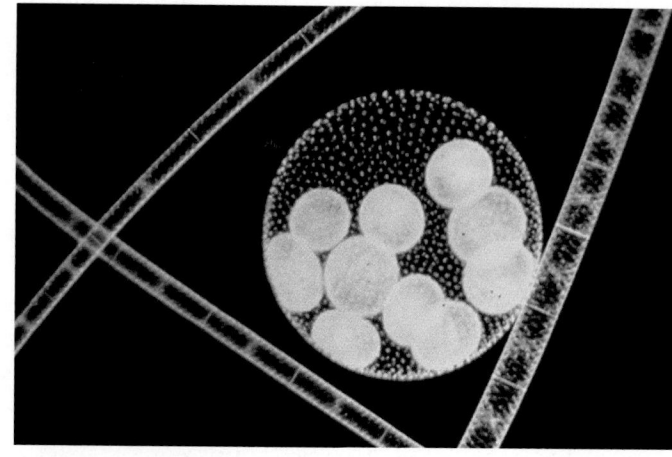

(b) *Volvox* and *Spirogyra*

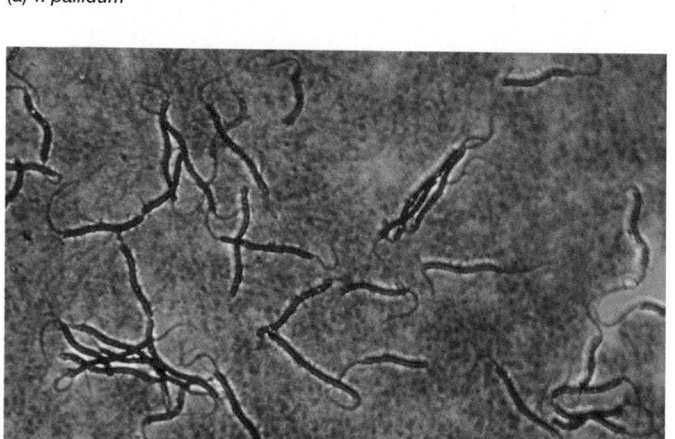

(c) *S. volutans*

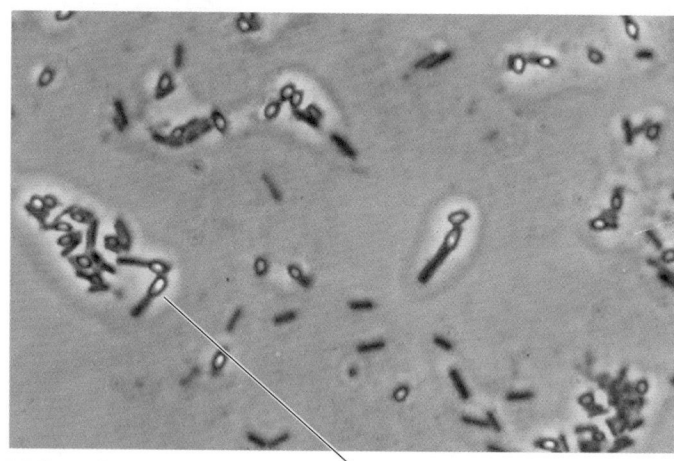

(d) *C. botulinum*   Endospore

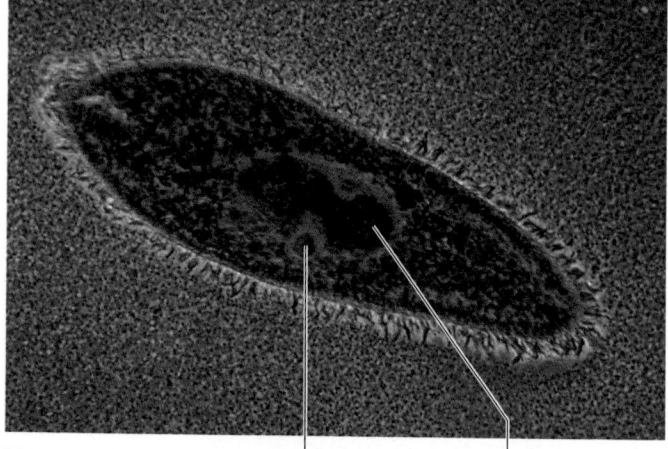

(e) *Paramecium*   Micronucleus   Macronucleus

Figure 2.8   Examples of Dark-Field and Phase-Contrast Microscopy. (**a**) *Treponema pallidum,* the spirochete that causes syphilis; dark-field microscopy (×500). (**b**) *Volvox* and *Spirogyra;* dark-field microscopy (×175). Note daughter colonies within the mature *Volvox* colony (center) and the spiral chloroplasts of *Spirogyra* (left and right). (**c**) *Spirillum volutans,* a very large bacterium with flagellar bundles; phase-contrast microscopy (×210). (**d**) *Clostridium botulinum,* the bacterium responsible for botulism, with subterminal oval endospores; phase-contrast microscopy (×600). (**e**) *Paramecium* stained to show a large central macronucleus with a small spherical micronucleus at its side; phase-contrast microscopy (×100).

Figure 2.9 Phase-Contrast Microscopy. The optics of a dark-phase-contrast microscope.

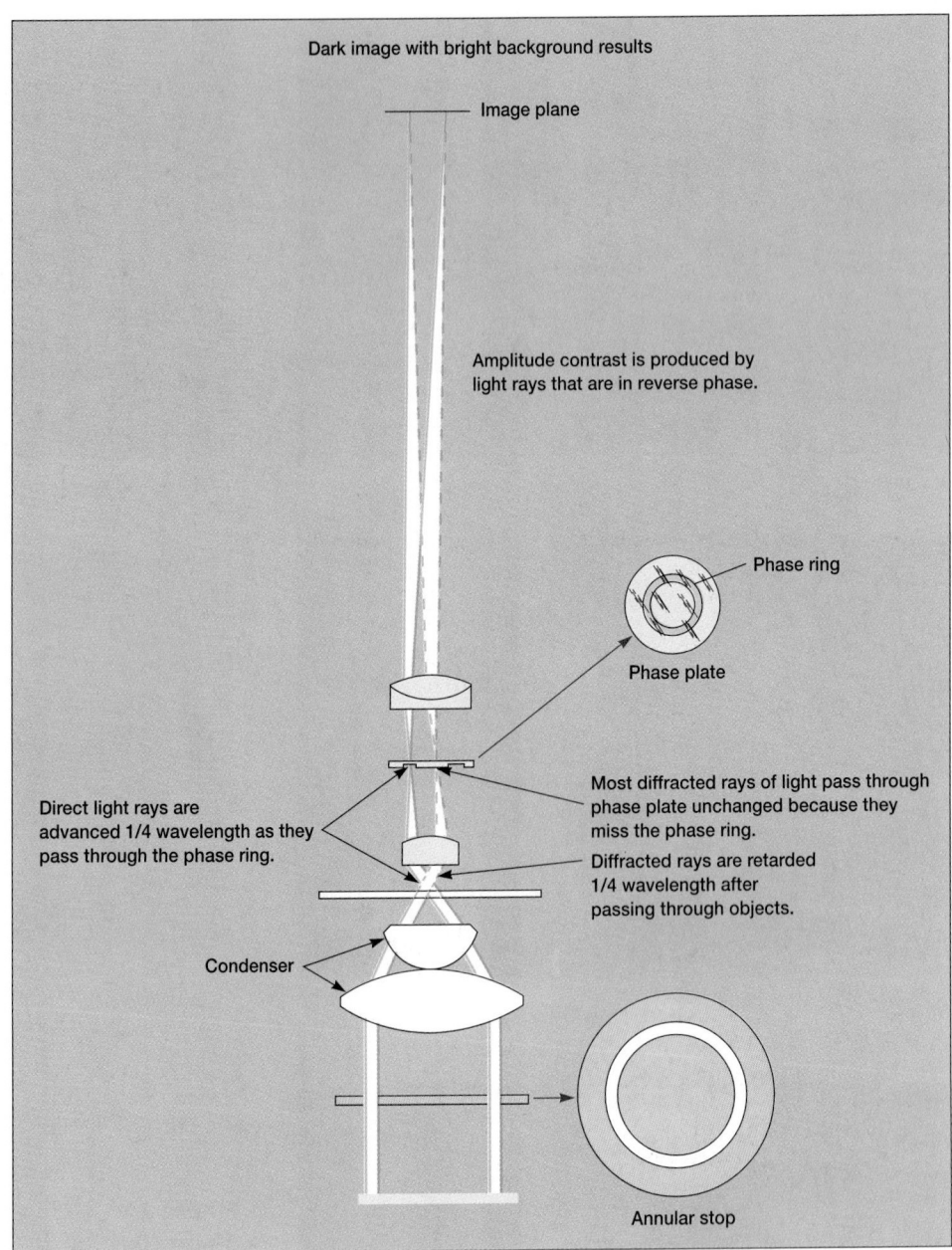

Dark image with bright background results

Image plane

Amplitude contrast is produced by light rays that are in reverse phase.

Phase ring

Phase plate

Most diffracted rays of light pass through phase plate unchanged because they miss the phase ring.

Direct light rays are advanced 1/4 wavelength as they pass through the phase ring.

Diffracted rays are retarded 1/4 wavelength after passing through objects.

Condenser

Annular stop

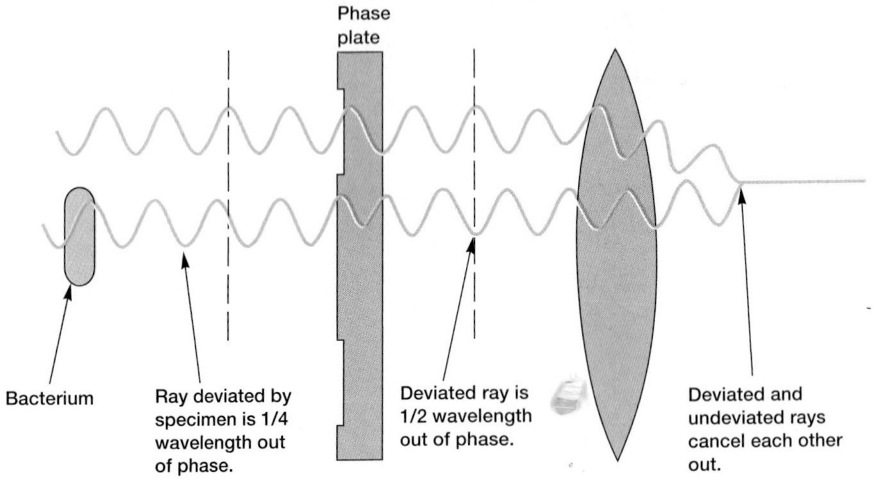

Figure 2.10 The Production of Contrast in Phase Microscopy. The behavior of deviated and undeviated or undiffracted light rays in the dark-phase-contrast microscope. Because the light rays tend to cancel each other out, the image of the specimen will be dark against a brighter background.

Phase plate

Bacterium

Ray deviated by specimen is 1/4 wavelength out of phase.

Deviated ray is 1/2 wavelength out of phase.

Deviated and undeviated rays cancel each other out.

The Fluorescence Microscope

The microscopes thus far considered produce an image from light that passes through a specimen. An object also can be seen because it actually emits light, and this is the basis of fluorescence microscopy. When some molecules absorb radiant energy, they become excited and later release much of their trapped energy as light. Any light emitted by an excited molecule will have a longer wavelength (or be of lower energy) than the radiation originally absorbed. **Fluorescent light** is emitted very quickly by the excited molecule as it gives up its trapped energy and returns to a more stable state.

The **fluorescence microscope (figure 2.12)** exposes a specimen to ultraviolet, violet, or blue light and forms an image of the object with the resulting fluorescent light. A mercury vapor arc lamp or other source produces an intense beam, and heat transfer is limited by a special infrared filter. The light passes through an exciter filter that transmits only the desired wavelength. A dark-field condenser provides a black background against which the fluorescent objects glow. Usually the specimens have been stained with dye molecules, called **fluorochromes,** that fluoresce brightly upon exposure to light of a specific wavelength, but some microorganisms are autofluorescing. The microscope forms an image of the fluorochrome-labeled microorganisms from the light emitted when they fluoresce **(figure 2.13)**. A barrier filter positioned after the objective lenses removes any remaining ultraviolet light, which could damage the viewer's eyes, or blue and violet light, which would reduce the image's contrast.

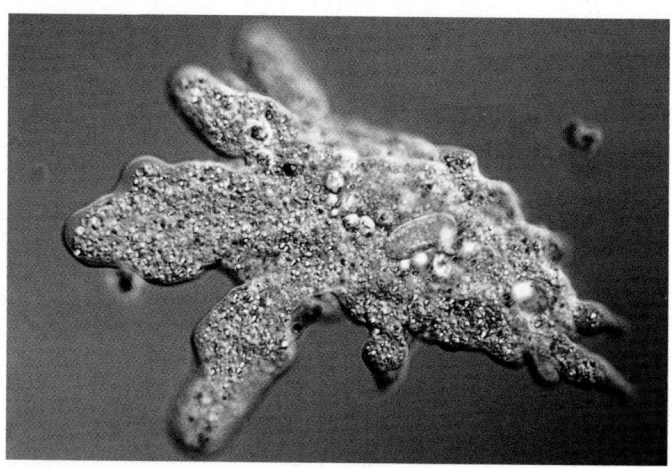

Figure 2.11   Differential Interference Contrast Microscopy. A micrograph of the protozoan *Amoeba proteus*. The three-dimensional image contains considerable detail and is artificially colored (×160).

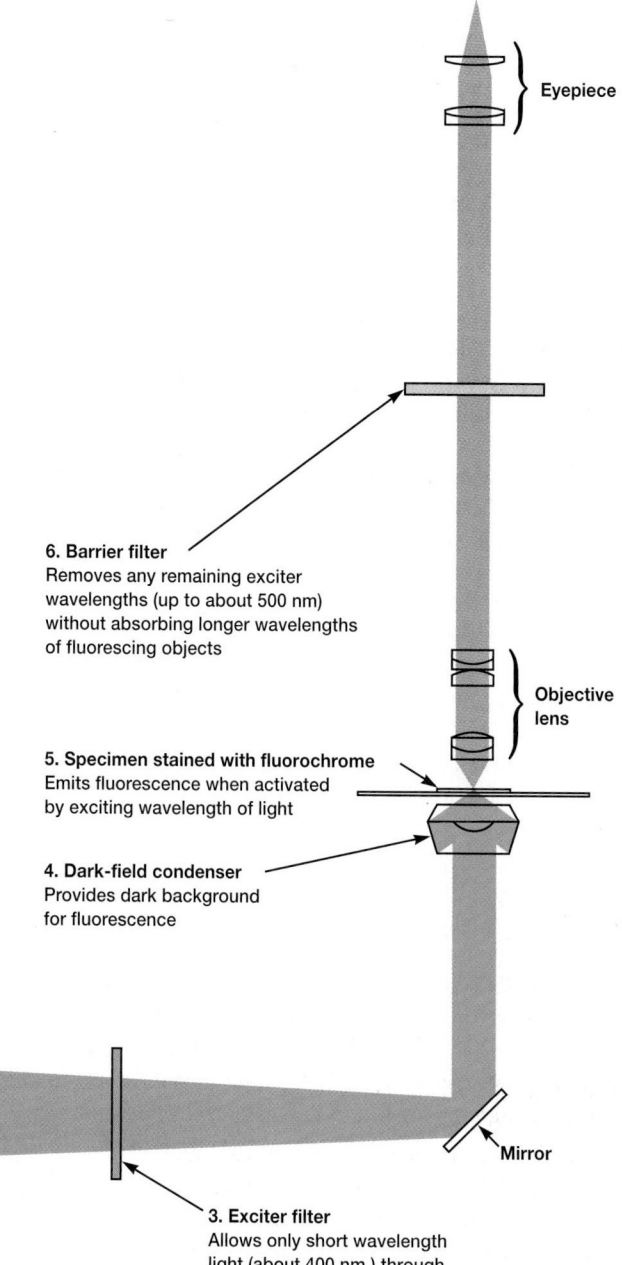

Eyepiece

6. Barrier filter
Removes any remaining exciter
wavelengths (up to about 500 nm)
without absorbing longer wavelengths
of fluorescing objects

Objective
lens

5. Specimen stained with fluorochrome
Emits fluorescence when activated
by exciting wavelength of light

4. Dark-field condenser
Provides dark background
for fluorescence

1. Mercury vapor arc lamp

Mirror

2. Heat filter

3. Exciter filter
Allows only short wavelength
light (about 400 nm) through

Figure 2.12   Fluorescence Microscopy.
The principles of operation of a fluorescence
microscope.

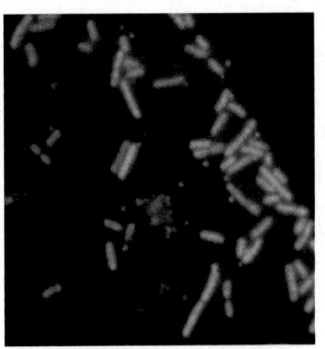

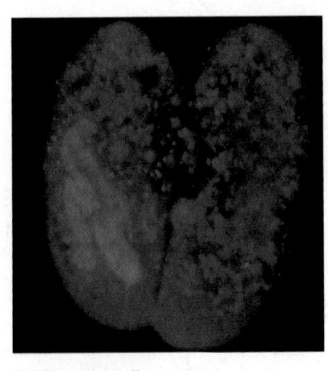

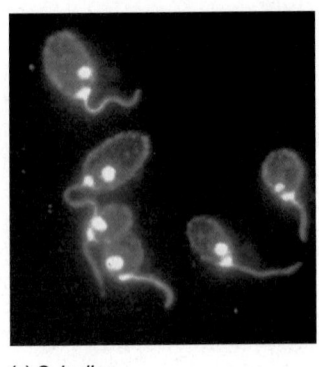

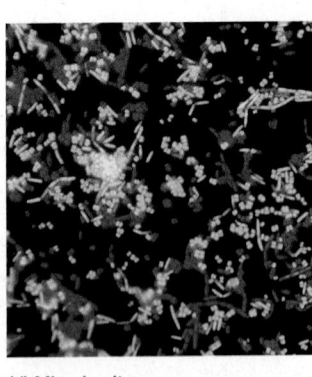

(a) *E. coli*

(b) *P. tetraurelia*

(c) *C. lucilae*

(d) Mixed culture

Figure 2.13 **Examples of Fluorescence Microscopy.** (**a**) *Escherichia coli* stained with fluorescent antibodies (×600). The green material is debris. (**b**) *Paramecium tetraurelia* conjugating; acridine-orange fluorescence (×125). (**c**) The flagellate protozoan *Crithidia luciliae* stained with fluorescent antibodies to show the kinetoplast (×1,000). (**d**) A mixture of *Micrococcus luteus* and *Bacillus cereus* (the rods). The live bacteria fluoresce green; dead bacteria are red.

The fluorescence microscope has become an essential tool in medical microbiology and microbial ecology. Bacterial pathogens (e.g., *Mycobacterium tuberculosis,* the cause of tuberculosis) can be identified after staining them with fluorochromes or specifically labeling them with fluorescent antibodies using immunofluorescence procedures. In ecological studies the fluorescence microscope is used to observe microorganisms stained with fluorochrome-labeled probes or fluorochromes such as acridine orange and DAPI (diamidino-2-phenylindole, a DNA-specific stain). The stained organisms will fluoresce orange or green and can be detected even in the midst of other particulate material. It is even possible to distinguish live bacteria from dead bacteria by the color they fluoresce after treatment with a special mixture of stains (figure 2.13*d*). Thus the microorganisms can be viewed and directly counted in a relatively undisturbed ecological niche. Immunofluorescence and diagnostic microbiology (pp. 755–56, 804–5)

1. List the parts of a light microscope and their functions.
2. Define resolution, numerical aperture, working distance, and fluorochrome.
3. How does resolution depend upon the wavelength of light, refractive index, and the numerical aperture? What are the functions of immersion oil and the substage condenser?
4. Briefly describe how dark-field, phase-contrast, differential interference contrast, and fluorescence microscopes work and the kind of image provided by each. Give a specific use for each type.

 2.3 PREPARATION AND STAINING OF SPECIMENS

Although living microorganisms can be directly examined with the light microscope, they often must be fixed and stained to increase visibility, accentuate specific morphological features, and preserve them for future study.

Fixation

The stained cells seen in a microscope should resemble living cells as closely as possible. **Fixation** is the process by which the internal and external structures of cells and microorganisms are preserved and fixed in position. It inactivates enzymes that might disrupt cell morphology and toughens cell structures so that they do not change during staining and observation. A microorganism usually is killed and attached firmly to the microscope slide during fixation.

There are two fundamentally different types of fixation. (1) Bacteriologists heat-fix bacterial smears by gently flame heating an air-dried film of bacteria. This adequately preserves overall morphology but not structures within cells. (2) Chemical fixation must be used to protect fine cellular substructure and the morphology of larger, more delicate microorganisms. Chemical fixatives penetrate cells and react with cellular components, usually proteins and lipids, to render them inactive, insoluble, and immobile. Common fixative mixtures contain such components as ethanol, acetic acid, mercuric chloride, formaldehyde, and glutaraldehyde.

Dyes and Simple Staining

The many types of dyes used to stain microorganisms have two features in common. (1) They have **chromophore groups,** groups with conjugated double bonds that give the dye its color. (2) They can bind with cells by ionic, covalent, or hydrophobic bonding. For example, a positively charged dye binds to negatively charged structures on the cell.

Ionizable dyes may be divided into two general classes based on the nature of their charged group.

1. **Basic dyes**—methylene blue, basic fuchsin, crystal violet, safranin, malachite green—have positively charged groups (usually some form of pentavalent nitrogen) and are generally sold as chloride salts. Basic dyes bind to negatively charged molecules like nucleic acids and many proteins. Because the surfaces of bacterial cells also are negatively charged, basic dyes are most often used in bacteriology.

2. **Acid dyes**—eosin, rose bengal, and acid fuchsin—possess negatively charged groups such as carboxyls (—COOH) and phenolic hydroxyls (—OH). Acid dyes, because of their negative charge, bind to positively charged cell structures.

The pH may alter staining effectiveness since the nature and degree of the charge on cell components change with pH. Thus anionic dyes stain best under acidic conditions when proteins and many other molecules carry a positive charge; basic dyes are most effective at higher pHs.

Although ionic interactions are probably the most common means of attachment, dyes also bind through covalent bonds or because of their solubility characteristics. For instance, DNA can be stained by the Feulgen procedure in which Schiff's reagent is covalently attached to its deoxyribose sugars after hydrochloric acid treatment. Sudan III (Sudan Black) selectively stains lipids because it is lipid soluble but will not dissolve in aqueous portions of the cell.

Microorganisms often can be stained very satisfactorily by **simple staining,** in which a single staining agent is used. Simple staining's value lies in its simplicity and ease of use. One covers the fixed smear with stain for the proper length of time, washes the excess stain off with water, and blots the slide dry. Basic dyes like crystal violet, methylene blue, and carbolfuchsin are frequently used to determine the size, shape, and arrangement of bacteria.

Differential Staining

Differential staining procedures divide bacteria into separate groups based on staining properties. The **Gram stain,** developed in 1884 by the Danish physician Christian Gram, is the most widely employed staining method in bacteriology. It is a differential staining procedure because it divides bacteria into two classes—gram negative and gram positive. Gram-positive and gram-negative bacteria (pp. 52–59, 428–29)

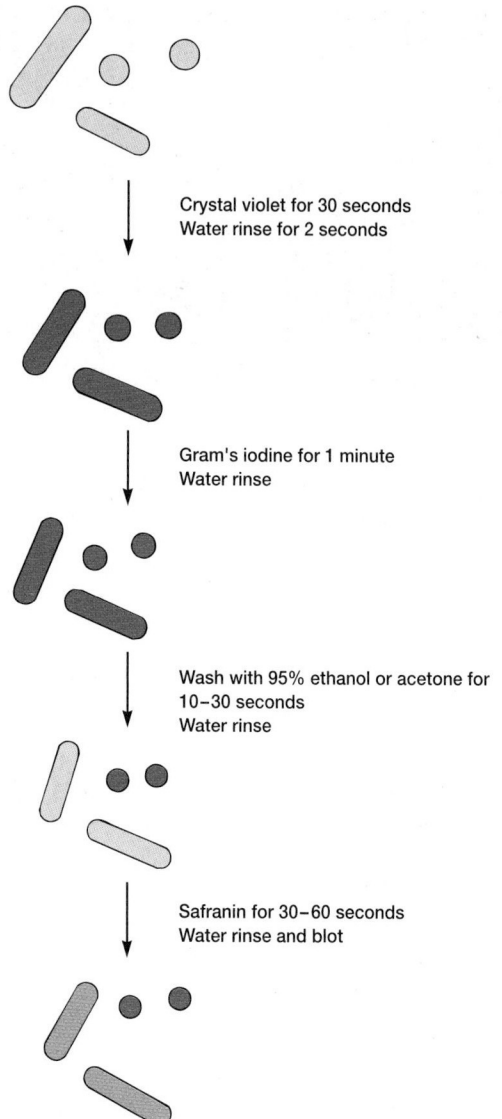

Crystal violet for 30 seconds
Water rinse for 2 seconds

Gram's iodine for 1 minute
Water rinse

Wash with 95% ethanol or acetone for 10–30 seconds
Water rinse

Safranin for 30–60 seconds
Water rinse and blot

Figure 2.14 The Gram-Staining Procedure. Note that decolorization with ethanol or acetone removes crystal violet from gram-negative cells but not from gram-positive cells. The gram-negative cells then turn pink to red when counterstained with safranin.

In the first step of the Gram-staining procedure (**figure 2.14**), the smear is stained with the basic dye crystal violet, the primary stain. It is followed by treatment with an iodine solution functioning as a **mordant.** That is, the iodine increases the interaction between the cell and the dye so that the cell is stained more strongly. The smear is next decolorized by washing with ethanol or acetone. This step generates the differential aspect of the Gram stain; gram-positive bacteria retain the crystal violet, whereas gram-negative bacteria lose their crystal violet and become colorless. Finally, the

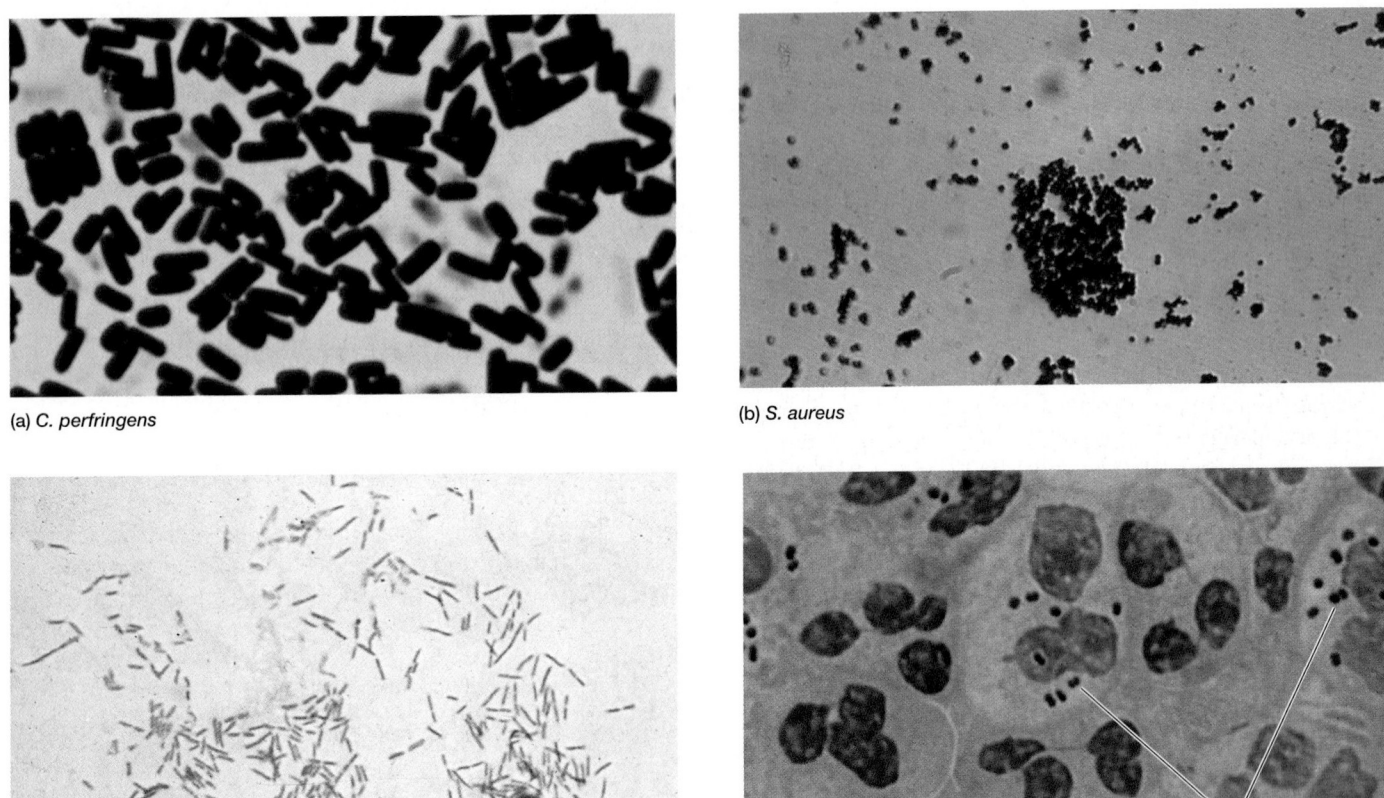

(a) *C. perfringens*

(b) *S. aureus*

(c) *E. coli*

(d) *N. gonorrhoeae*   Diplococci

Figure 2.15   Examples of Gram Staining. (**a**) Gram-positive *Clostridium perfringens* (×800). Some rods have stained pink rather than purple, as often happens when gram-positive cells age. (**b**) *Staphylococcus aureus.* Gram stain, bright-field microscopy (×1,000). The gram-positive cocci associate in grapelike clusters. (**c**) *Escherichia coli,* Gram stain (×500). (**d**) *Neisseria gonorrhoeae.* The diplococci are often within white blood cells (×1,000).

smear is counterstained with a simple, basic dye different in color from crystal violet. Safranin, the most common counterstain, colors gram-negative bacteria pink to red and leaves gram-positive bacteria dark purple (**figure 2.15**).   Cell wall structure and the mechanism of the Gram stain (pp. 58–59)

Acid-fast staining is another important differential staining procedure. A few species, particularly those in the genus *Mycobacterium* (*see chapter 24*) do not bind simple stains readily and must be stained by a harsher treatment: heating with a mixture of basic fuchsin and phenol (the Ziehl-Neelsen method). Once basic fuchsin has penetrated with the aid of heat and phenol, acid-fast cells are not easily decolorized by an acid-alcohol wash and hence remain red. This is due to the quite high lipid content of acid-fast cell walls; in particular, mycolic acid—a group of branched chain hydroxy lipids—appears responsible for acid-fastness. Non-acid-fast bacteria are decolorized by acid-alcohol and thus are stained blue by methylene blue counterstain. This method is used to identify *Mycobacterium tuberculosis* and *M. leprae* (**figure 2.16**), the pathogens responsible for tuberculosis and leprosy, respectively.

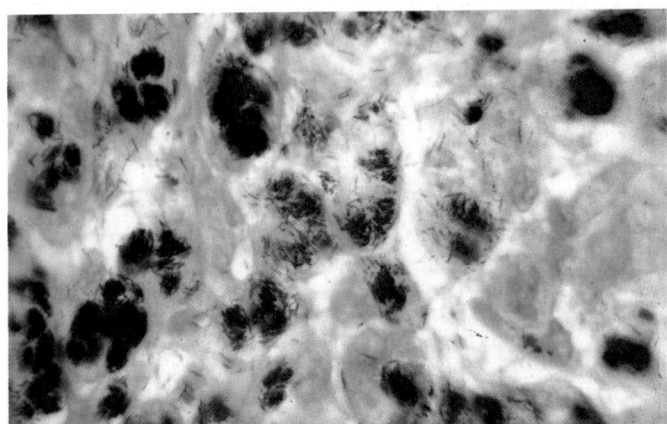

Figure 2.16   Acid-Fast Staining. *Mycobacterium leprae.* Acid-fast stain (×380). Note the masses of red bacteria within host cells.

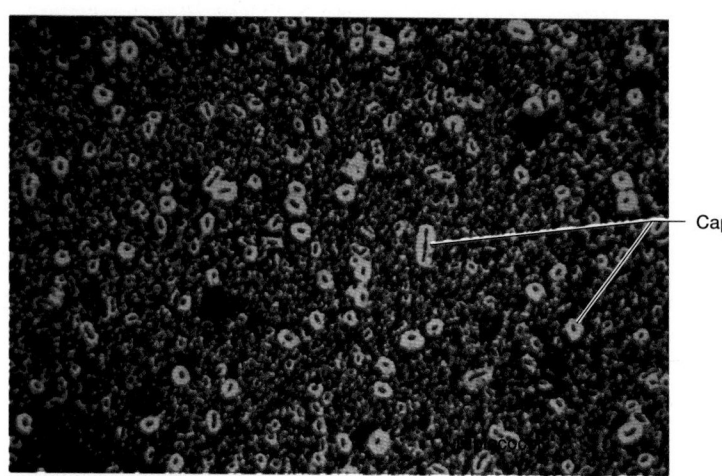

Figure 2.17 **Negative Staining.** *Klebsiella pneumoniae* negatively stained with India ink to show its capsules (×900).

Capsules

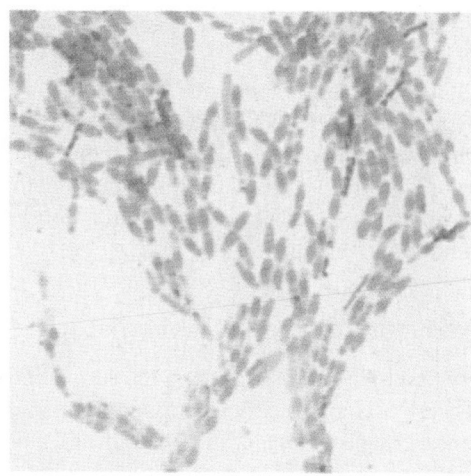

Figure 2.18 **Spore Staining.** *Bacillus cereus* stained with the Schaeffer-Fulton procedure. Note the central, elliptical blue to green spores within the red to purple cells (×1,000).

Staining Specific Structures

Many special staining procedures have been developed over the years to study specific bacterial structures with the light microscope. One of the simplest is **negative staining,** a technique that reveals the presence of the diffuse capsules surrounding many bacteria. Bacteria are mixed with India ink or nigrosin dye and spread out in a thin film on a slide. After air-drying, bacteria appear as lighter bodies in the midst of a blue-black background because ink and dye particles cannot penetrate either the bacterial cell or its capsule. The extent of the light region is determined by the size of the capsule and of the cell itself. There is little distortion of bacterial shape, and the cell can be counterstained for even greater visibility (**figure 2.17**). Capsules and slime layers (pp. 61–62)

Bacteria in the genera *Bacillus* and *Clostridium* (*see chapter 23*) form an exceptionally resistant structure capable of surviving for long periods in an unfavorable environment. This dormant structure is called an endospore since it develops within the cell. Endospore morphology and location vary with species and often are valuable in identification; endospores may be spherical to elliptical and either smaller or larger than the diameter of the parent bacterium. They can be observed with the phase-contrast microscope or negative staining. Endospores are not stained well by most dyes, but once stained, they strongly resist decolorization. This property is the basis of most **spore staining** methods (**figure 2.18**). In the Schaeffer-Fulton procedure, endospores are first stained by heating bacteria with malachite green, which is a very strong stain that can penetrate endospores. After malachite green treatment, the rest of the cell is washed free of dye with water and is counterstained with safranin. This technique yields a green endospore resting in a pink to red cell. Bacterial endospore structure (pp. 68–71)

Bacterial flagella are fine, threadlike organelles of locomotion that are so slender (about 10 to 30 nm in diameter) they can only be seen directly using the electron microscope. To observe

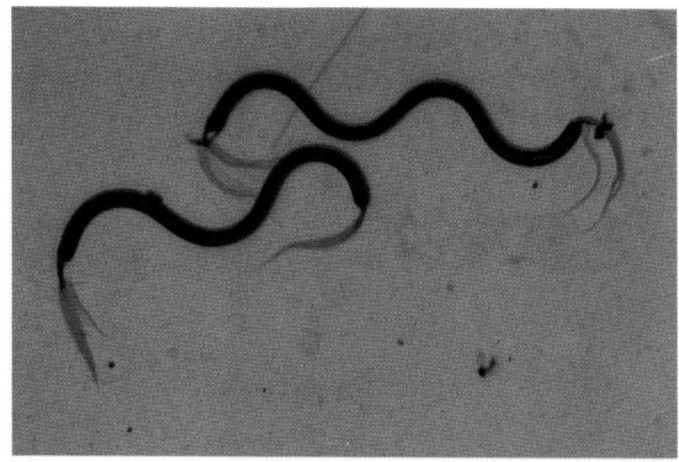

Figure 2.19 **Example of Flagella Staining.** *Spirillum volutans* with bipolar tufts of flagella (×400). (*See also figure 3.32.*)

them with the light microscope, the thickness of flagella is increased by coating them with mordants like tannic acid and potassium alum, and they are stained with pararosaniline (Leifson method) or basic fuchsin (Gray method). **Flagella staining** procedures provide taxonomically valuable information about the presence and distribution pattern of flagella (**figure 2.19;** *see also figure 3.32*). The bacterial flagellum (pp. 63–66)

1. Define fixation, dye, chromophore, basic dye, acid dye, simple staining, differential staining, mordant, negative staining, and acid-fast staining.
2. Describe the Gram-stain procedure and how it works.
3. How would you visualize capsules, endospores, and flagella?

2.4   ELECTRON MICROSCOPY

For centuries the light microscope has been the most important instrument for studying microorganisms. The electron microscope now has transformed microbiology and added immeasurably to our knowledge. The nature of the electron microscope and the ways in which specimens are prepared for observation are reviewed briefly in this section.

The Transmission Electron Microscope

The very best light microscope has a resolution limit of about 0.2 μm. Because bacteria usually are around 1 μm in diameter, only their general shape and major morphological features are visible in the light microscope. The detailed internal structure of larger microorganisms also cannot be effectively studied by light microscopy. These limitations arise from the nature of visible light waves, not from any inadequacy of the light microscope itself.

Recall that the resolution of a light microscope increases with a decrease in the wavelength of the light it uses for illumination. Electron beams behave like radiation and can be focused much as light is in a light microscope. If electrons illuminate the specimen, the microscope's resolution is enormously increased because the wavelength of the radiation is around 0.005 nm, approximately 100,000 times shorter than that of visible light. The transmission electron microscope has a practical resolution roughly 1,000 times better than the light microscope; with many electron microscopes, points closer than 5 Å or 0.5 nm can be distinguished, and the useful magnification is well over 100,000× (**figure 2.20**). The value of the electron microscope is evident on comparison of the photographs in **figure 2.21**; microbial morphology can now be studied in great detail.

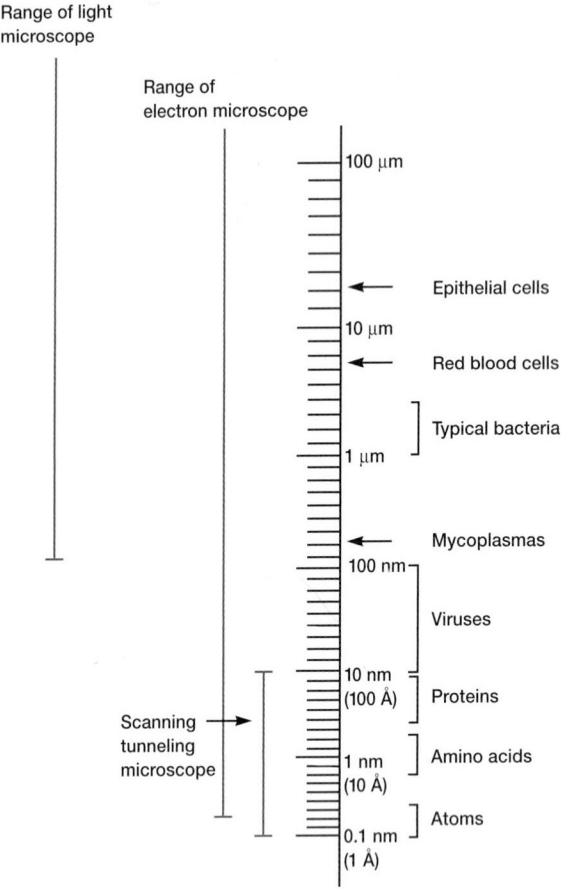

Figure 2.20   The Limits of Microscopic Resolution. Dimensions are indicated with a logarithmic scale (each major division represents a tenfold change in size). To the right side of the scale are the approximate sizes of cells, bacteria, viruses, molecules, and atoms.

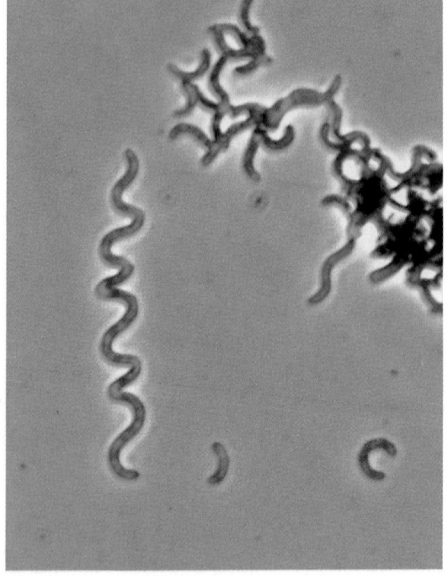

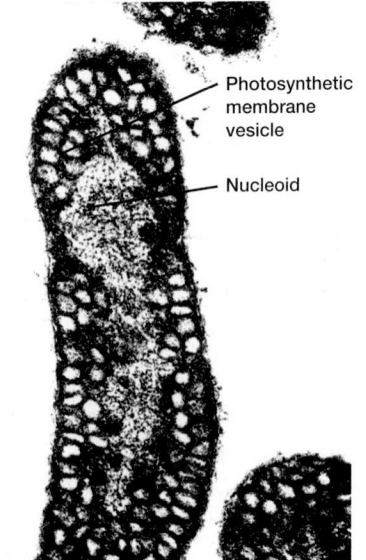

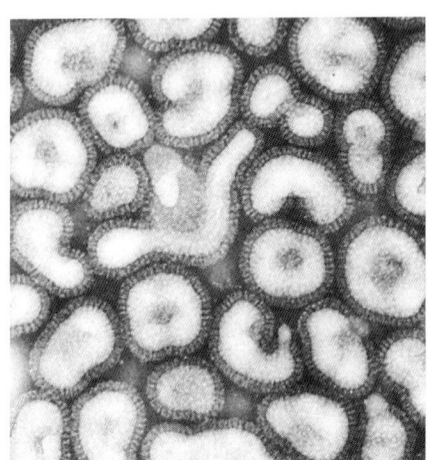

Figure 2.21   Light and Electron Microscopy. A comparison of light and electron microscopic resolution. (**a**) *Rhodospirillum rubrum* in phase-contrast light microscope (×600). (**b**) A thin section of *R. rubrum* in transmission electron microscope (×100,000). (**c**) A micrograph of human influenza viruses (×282,000). The particles are about 100 nm in diameter, much smaller than bacterial cells.

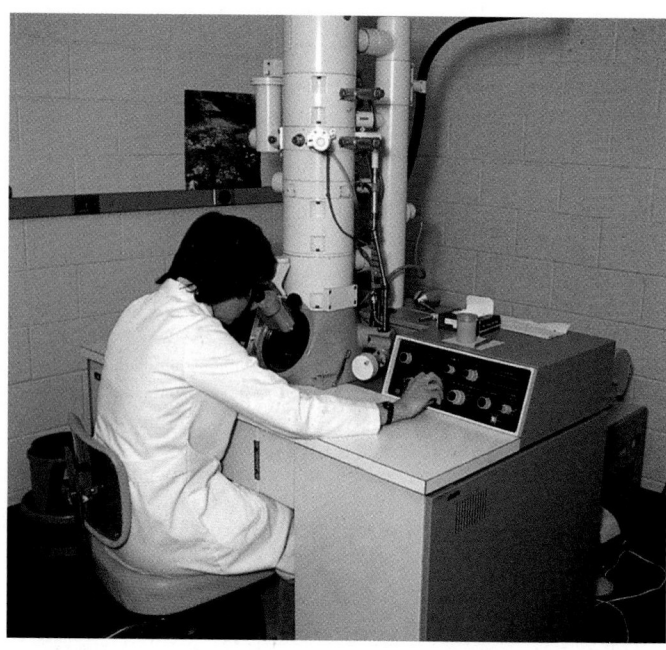

Figure 2.22   A Transmission Electron Microscope.
The electron gun is at the top of the central column, and the magnetic lenses are within the column. The image on the fluorescent screen may be viewed through a magnifier positioned over the viewing window. The camera is in a compartment below the screen.

A modern **transmission electron microscope (TEM)** is complex and sophisticated (**figure 2.22**), but the basic principles behind its operation can be readily understood. A heated tungsten filament in the electron gun generates a beam of electrons that is then focused on the specimen by the condenser (**figure 2.23**). Since electrons cannot pass through a glass lens, doughnut-shaped electromagnets called magnetic lenses are used to focus the beam. The column containing the lenses and specimen must be under high vacuum to obtain a clear image because electrons are deflected by collisions with air molecules. The specimen scatters electrons passing through it, and the beam is focused by magnetic lenses to form an enlarged, visible image of the specimen on a fluorescent screen. A denser region in the specimen scatters more electrons and therefore appears darker in the image since fewer electrons strike that area of the screen. In contrast, electron-transparent regions are brighter. The screen can also be moved aside and the image captured on photographic film as a permanent record.

Specimen Preparation

Table 2.3 compares some of the important features of light and electron microscopes. The distinctive features of the TEM place harsh restrictions on the nature of samples that can be viewed and the means by which those samples must be prepared. Since electrons are quite easily absorbed and scattered by solid matter, only

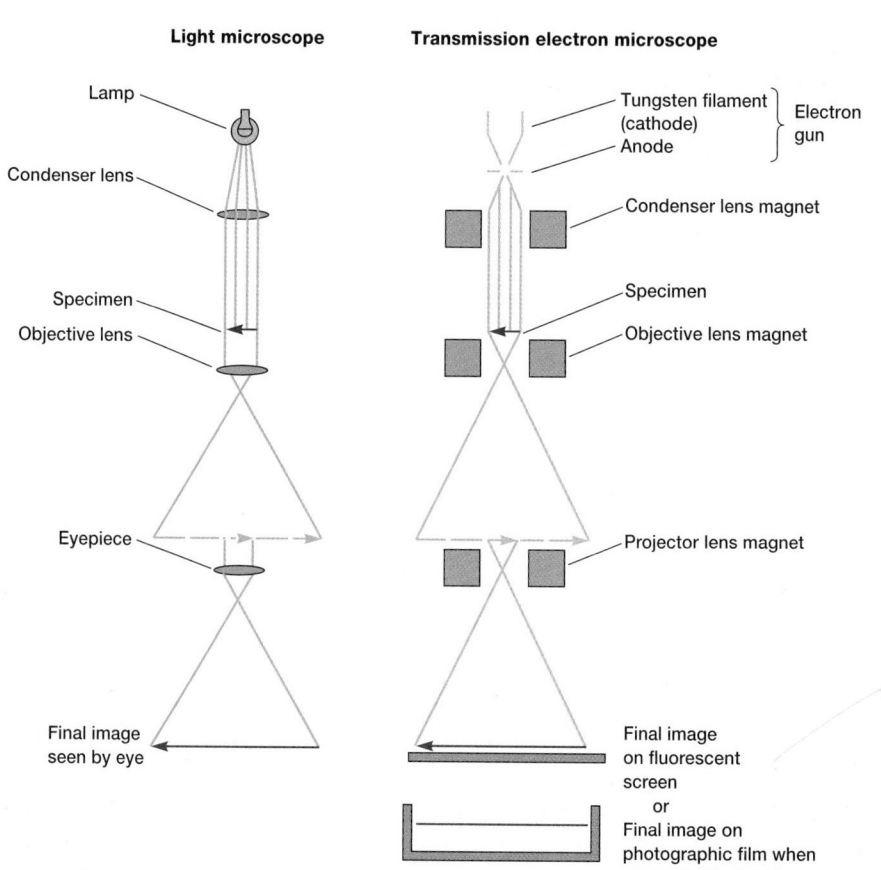

Figure 2.23   Transmission Electron Microscope Operation. An overview of TEM operation and a comparison of the operation of light and transmission electron microscopes.

Table 2.3

Characteristics of Light and Transmission Electron Microscopes

Feature	Light Microscope	Electron Microscope
Highest practical magnification	About 1,000–1,500	Over 100,000
Best resolution[a]	0.2 μm	0.5 nm
Radiation source	Visible light	Electron beam
Medium of travel	Air	High vacuum
Type of lens	Glass	Electromagnet
Source of contrast	Differential light absorption	Scattering of electrons
Focusing mechanism	Adjust lens position mechanically	Adjust current to the magnetic lens
Method of changing magnification	Switch the objective lens or eyepiece	Adjust current to the magnetic lens
Specimen mount	Glass slide	Metal grid (usually copper)

[a]The resolution limit of a human eye is about 0.2 mm.

extremely thin slices of a microbial specimen can be viewed in the average TEM. The specimen must be around 20 to 100 nm thick, about 1/50 to 1/10 the diameter of a typical bacterium, and able to maintain its structure when bombarded with electrons under a high vacuum! Such a thin slice cannot be cut unless the specimen has support of some kind; the necessary support is provided by plastic. After fixation with chemicals like glutaraldehyde or osmium tetroxide to stabilize cell structure, the specimen is dehydrated with organic solvents (e.g., acetone or ethanol). Complete dehydration is essential because most plastics used for embedding are not water soluble. Next the specimen is soaked in unpolymerized, liquid epoxy plastic until it is completely permeated, and then the plastic is hardened to form a solid block. Thin sections are cut from this block with a glass or diamond knife using a special instrument called an ultramicrotome.

Cells usually must be stained before they can be seen clearly in the bright-field microscope; the same is true for observations with the TEM. The probability of electron scattering is determined by the density (atomic number) of the specimen atoms. Biological molecules are composed primarily of atoms with low atomic numbers (H, C, N, and O), and electron scattering is fairly constant throughout the unstained cell. Therefore specimens are prepared for observation by soaking thin sections with solutions of heavy metal salts like lead citrate and uranyl acetate. The lead and uranium ions bind to cell structures and make them more electron opaque, thus increasing contrast in the material. Heavy osmium atoms from the osmium tetroxide fixative also "stain" cells and increase their contrast. The stained thin sections are then mounted on tiny copper grids and viewed.

Although specimens normally are embedded in plastic and thin sectioned to reveal the internal structure of the smallest cell, there are other ways in which microorganisms and smaller objects can be readied for viewing. One very useful technique is negative staining. The specimen is spread out in a thin film with either phosphotungstic acid or uranyl acetate. Just as in negative staining for light microscopy, heavy metals do not penetrate the specimen but render the background dark, whereas the specimen appears bright in photographs. Negative staining is an excellent

way to study the structure of viruses, bacterial gas vacuoles, and other similar objects. A microorganism also can be viewed after **shadowing** with metal. It is coated with a thin film of platinum or other heavy metal by evaporation at an angle of about 45° from horizontal so that the metal strikes the microorganism on only one side. The area coated with metal scatters electrons and appears light in photographs, whereas the uncoated side and the shadow region created by the object is dark (**figure 2.24**). The specimen

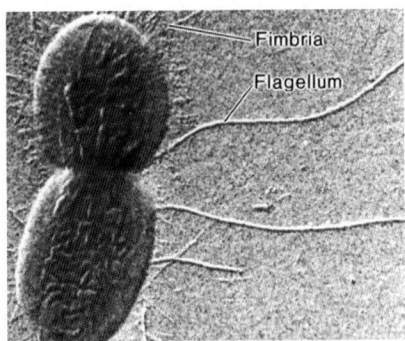

(a) *P. mirabilis*

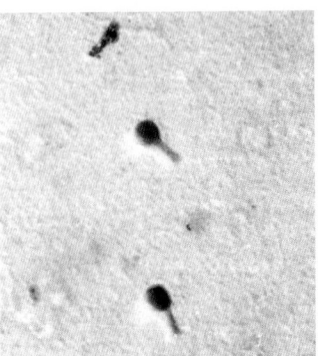

(b) *T4 coliphage*

Figure 2.24   Specimen Shadowing for the TEM. Examples of specimens viewed in the TEM after shadowing with uranium metal. (**a**) *Proteus mirabilis* (×42,750); note flagella and fimbriae. (**b**) T4 coliphage (×72,000).

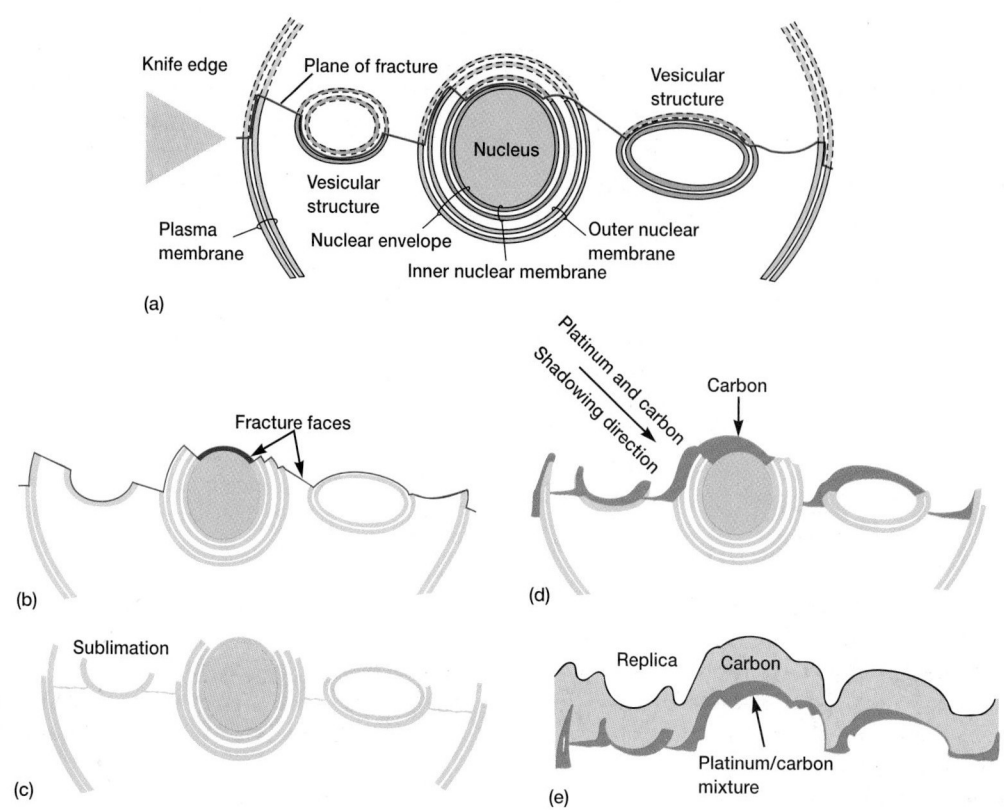

(a)

(b)

(c) Sublimation

(d)

(e)

Figure 2.25 **The Freeze-Etching Technique.** In steps (**a**) and (**b**), a frozen eucaryotic cell is fractured with a cold knife. Etching by sublimation is depicted in (**c**). Shadowing with platinum plus carbon and replica formation are shown in (**d**) and (**e**). See text for details.

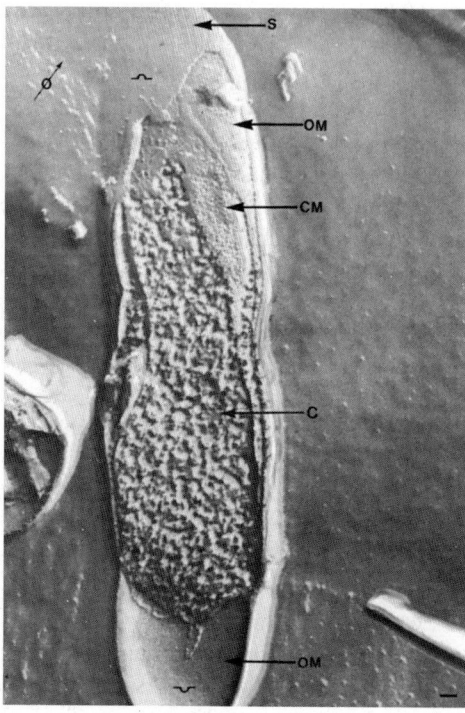

looks much as it would if light were shining on it to cast a shadow. This technique is particularly useful in studying virus morphology, bacterial flagella, and plasmids (*see chapter 13*).

The TEM will also disclose the shape of organelles within microorganisms if specimens are prepared by the **freeze-etching** procedure. Cells are rapidly frozen in liquid nitrogen and then warmed to −100°C in a vacuum chamber. Next a knife that has been precooled with liquid nitrogen (−196°C) fractures the frozen cells, which are very brittle and break along lines of greatest weakness, usually down the middle of internal membranes (**figure 2.25**). The specimen is left in the high vacuum for a minute or more so that some of the ice can sublimate away and uncover more structural detail (sometimes this etching step is eliminated). Finally, the exposed surfaces are shadowed and coated with layers of platinum and carbon to form a replica of the surface. After the specimen has been removed chemically, this replica is studied in the TEM and provides a detailed, three-dimensional view of intracellular structure (**figure 2.26**). An advantage of freeze-etching is that it minimizes the danger of artifacts because the cells are frozen quickly rather than being subjected to chemical fixation, dehydration, and plastic embedding.

The Scanning Electron Microscope

The previously described microscopes form an image from radiation that has passed through a specimen. More recently the

Figure 2.26 **Example of Freeze-Etching.** A freeze-etched preparation of the bacterium *Thiobacillus kabobis*. Note the differences in structure between the outer surface, S; the outer membrane of the cell wall, OM; the cytoplasmic membrane, CM; and the cytoplasm, C. Bar = 0.1 μm.

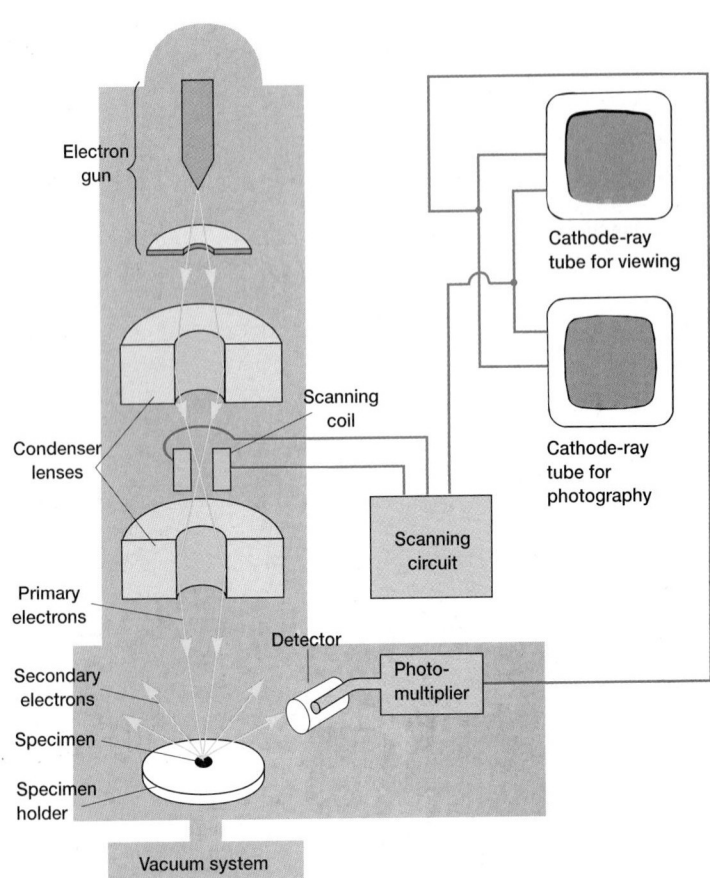

Figure 2.27   The Scanning Electron Microscope.
See text for explanation.

scanning electron microscope (SEM) has been used to examine the surfaces of microorganisms in great detail; many instruments have a resolution of 7 nm or less. The SEM differs from other electron microscopes in producing an image from electrons emitted by an object's surface rather than from transmitted electrons.

Specimen preparation is easy, and in some cases air-dried material can be examined directly. Most often, however, microorganisms must first be fixed, dehydrated, and dried to preserve surface structure and prevent collapse of the cells when they are exposed to the SEM's high vacuum. Before viewing, dried samples are mounted and coated with a thin layer of metal to prevent the buildup of an electrical charge on the surface and to give a better image.

The SEM scans a narrow, tapered electron beam back and forth over the specimen (**figure 2.27**). When the beam strikes a particular area, surface atoms discharge a tiny shower of electrons called secondary electrons, and these are trapped by a special detector. Secondary electrons entering the detector strike a scintillator causing it to emit light flashes that a photomultiplier converts to an electrical current and amplifies. The signal is sent to a cathode-ray tube and produces an image like a television picture, which can be viewed or photographed.

The number of secondary electrons reaching the detector depends on the nature of the specimen's surface. When the electron beam strikes a raised area, a large number of secondary

electrons enter the detector; in contrast, fewer electrons escape a depression in the surface and reach the detector. Thus raised areas appear lighter on the screen and depressions are darker. A realistic three-dimensional image of the microorganism's surface with great depth of focus results (**figure 2.28**). The actual in situ location of microorganisms in ecological niches such as the human skin and the lining of the gut also can be examined.

---

1. Why does the transmission electron microscope have much greater resolution than the light microscope? Describe in general terms how the TEM functions.
2. Describe how specimens are prepared for viewing in the TEM. How are sections usually stained to increase contrast? What are negative staining, shadowing, and freeze-etching?
3. How does the scanning electron microscope operate and in what way does its function differ from that of the TEM? The SEM is used to study which aspects of morphology?

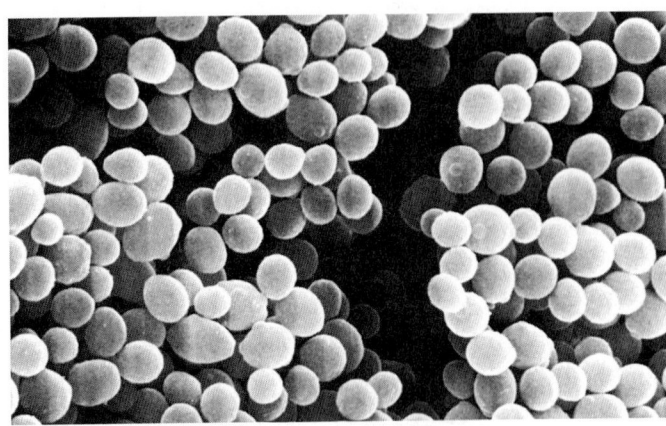

(a) *S. aureus*

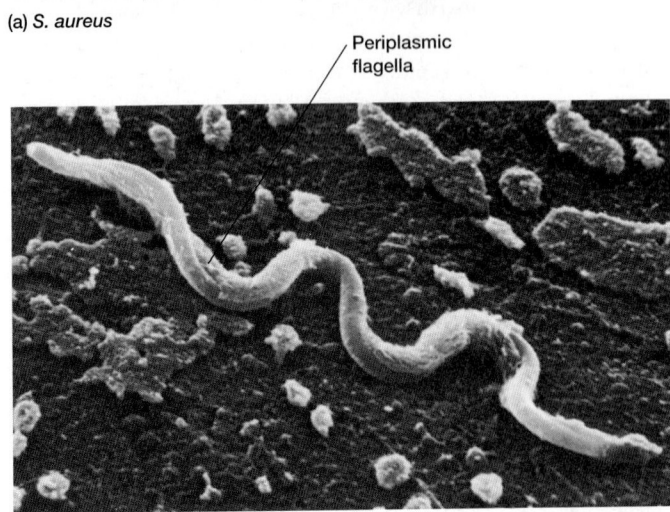

(b) *Cristispira*

Figure 2.28   Scanning Electron Micrographs of Bacteria.
(**a**) *Staphylococcus aureus* (×32,000). (**b**) *Cristispira,* a spirochete from the crystalline style of the oyster, *Ostrea virginica*. The axial fibrils or periplasmic flagella are visible around the protoplasmic cylinder (×6,000).

NEWER TECHNIQUES IN MICROSCOPY

Confocal Microscopy

A conventional light microscope, which uses a mixed wavelength light source and illuminates a large area of the specimen, will have a relatively great depth of field. Even if not in focus, images of bacteria from all levels within the field will be visible. These will include cells above, in, and below the plane of focus (**figure 2.29**). As a result the image can be murky, fuzzy, and crowded.

The solution to this problem is the **confocal scanning laser microscope (CSLM)** or confocal microscope. Fluorescently stained specimens are usually examined. A focused laser beam strikes a point in the specimen (**figure 2.30**). Light from the illuminated spot is focused by an objective lens onto a plane above the objective. An aperture above the objective lens blocks out stray light from parts of the specimen that lie above and below the plane of focus. The laser is scanned over a plane in the specimen (beam scanning) or the stage is moved (stage scanning) and a detector measures the illumination from each point to produce an image of the optical section. When many optical sections are scanned, a computer can combine them to form a three-

dimensional image from the digitized signals. This image can be measured and analyzed quantitatively.

The confocal microscope improves images in two ways. First, illumination of one spot at a time reduces interference from light scattering by the rest of the specimen. Second, the aperture above the objective lens blocks out stray light as previously mentioned. Consequently the image has excellent contrast and resolution. A depth of 1 μm or less in a thick preparation can be directly observed. Special computer software is used to create high-resolution, three-dimensional images of cell structures and complex specimens such as biofilms (**figure 2.31**).

Scanning Probe Microscopy

Although light and electron microscopes have become quite sophisticated and reached an advanced state of development, powerful new microscopes are still being created. A new class of microscopes, called **scanning probe microscopes,** measure surface features by moving a sharp probe over the object's surface. The **scanning tunneling microscope,** invented in 1980, is an excellent example of a scanning probe microscope. It can achieve magnifications of 100 million and allow scientists to view atoms on the surface of a solid. The electrons surrounding surface atoms tunnel or project out from the surface

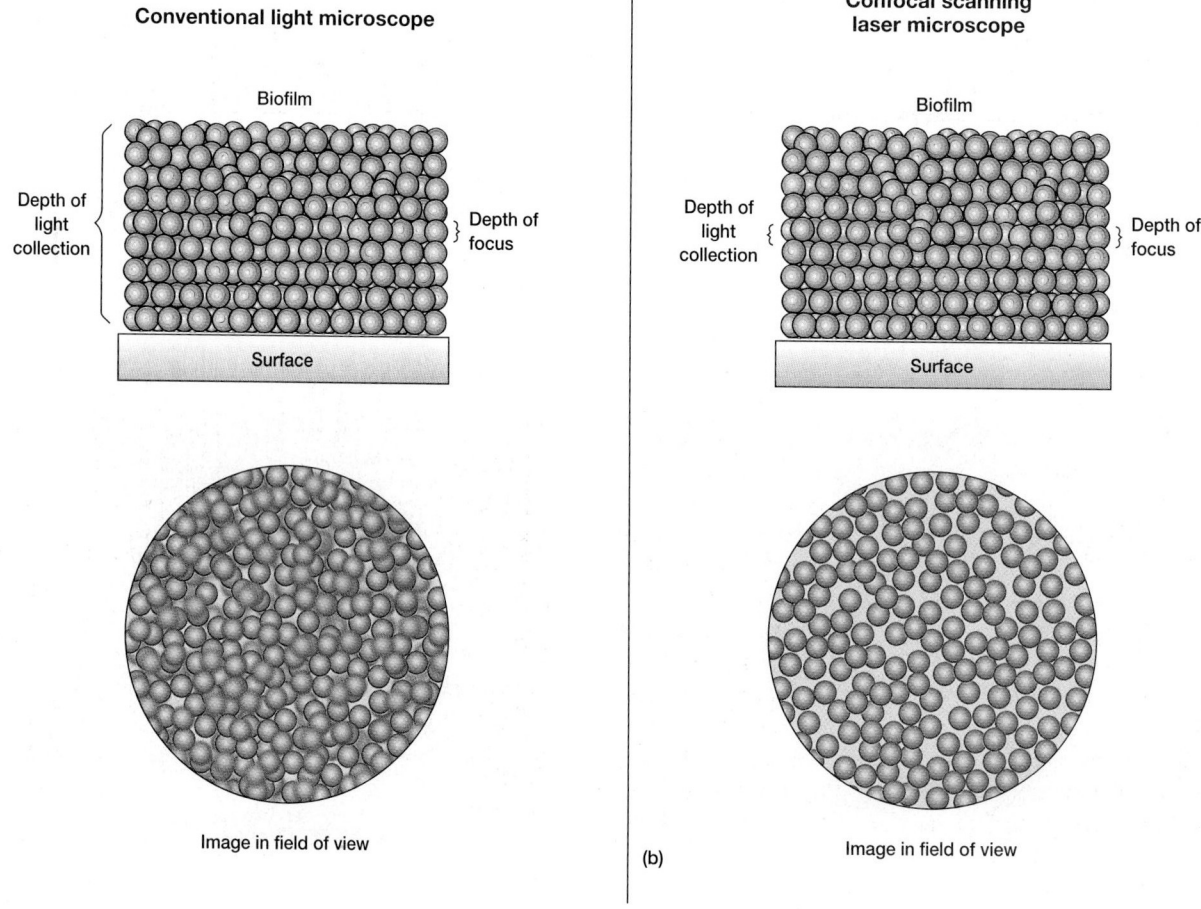

Figure 2.29   Confocal Scanning Laser Microscopy: Light Collection Depth and Image Clarity.
(**a**) Conventional light microscopic observation. (**b**) Confocal scanning laser microscopic observation.

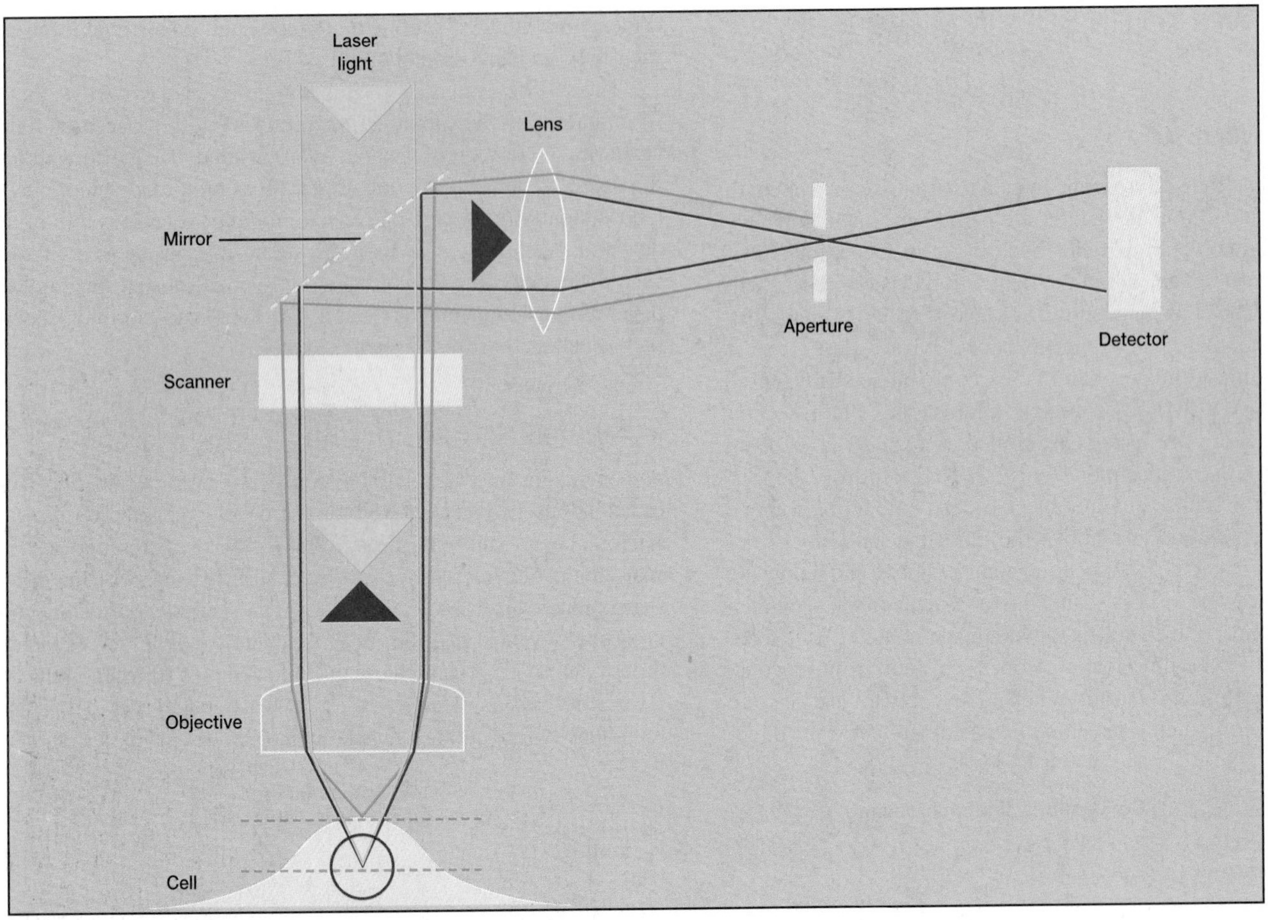

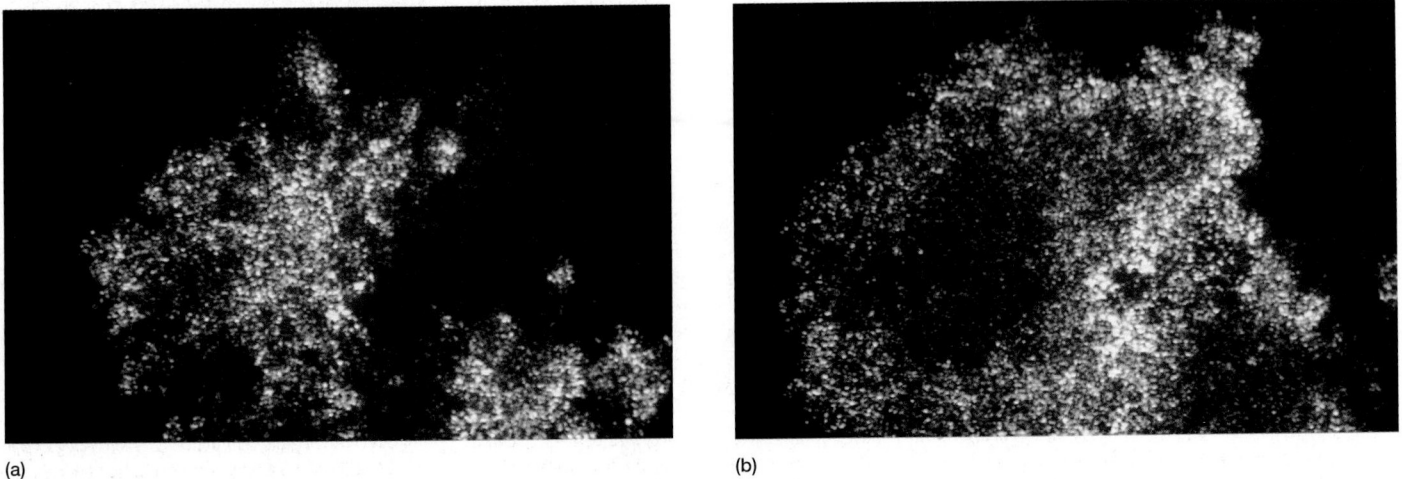

Figure 2.30   A Ray Diagram of a Confocal Laser Scanning Microscope.   The yellow lines represent laser light used for illumination. Red lines symbolize the light arising from the plane of focus, and the blue lines stand for light from parts of the specimen above and below the focal plane. See text for explanation.

(a)

(b)

Figure 2.31   Confocal Images at Various Depths below the Top of a Biofilm.   (**a**) 20 μm. (**b**) 40 μm. Each of these confocal images— which combines fluorescent and reflection images—has a depth of 2 μm and shows red-colored fluorescent tracer beads.

boundary a very short distance. The scanning tunneling microscope has a needlelike probe with a point so sharp that often there is only one atom at its tip. The probe is lowered toward the specimen surface until its electron cloud just touches that of the surface atoms. If a small voltage is applied between the tip and specimen, electrons flow through a narrow channel in the electron clouds. This tunneling current, as it is called, is extraordinarily sensitive to distance and will decrease about a thousandfold if the probe is moved away from the surface by a distance equivalent to the diameter of an atom.

The arrangement of atoms on the specimen surface is determined by moving the probe tip back and forth over the surface while keeping it at a constant height by adjusting the probe distance to maintain a steady tunneling current. As the tip moves up and down while following the surface contours, its motion is recorded and analyzed by a computer to create an accurate three-dimensional image of the surface atoms. The surface map can be displayed on a computer screen or plotted on paper. The resolution is so great that individual atoms are observed easily. The microscope's inventors, Gerd Binnig and Heinrich Rohrer, shared the 1986 Nobel Prize in Physics for their work, together with Ernst Ruska, the designer of the first transmission electron microscope.

The scanning tunneling microscope will likely have a major impact in biology. Recently it has been used to directly view DNA (**figure 2.32**). Since the microscope can examine objects when they are immersed in water, it may be particularly useful in studying biological molecules.

More recently a second type of scanning probe microscope has been developed. The **atomic force microscope** moves a sharp probe over the specimen surface while keeping the distance between the probe tip and the surface constant. It does this by exerting a very small amount of force on the tip, just enough to maintain a constant distance but not enough force to damage the surface. The vertical motion of the tip usually is followed by measuring the deflection of a laser beam that strikes the lever holding the probe. Unlike the scanning tunneling microscope, the atomic force microscope can be used to study surfaces that do not conduct electricity well. The atomic force microscope has been used to study the interactions between the *E. coli* GroES and GroEL chaperonin proteins, to map plasmids by locating restriction enzymes bound to specific sites, and to follow the behavior of living bacteria and other cells.

1. How does a confocal microscope operate and why does it provide better images of thick specimens than the standard compound microscope?
2. Briefly describe the scanning probe microscope and its most popular versions, the scanning tunneling microscope and the atomic force microscope. What are these microscopes used for?

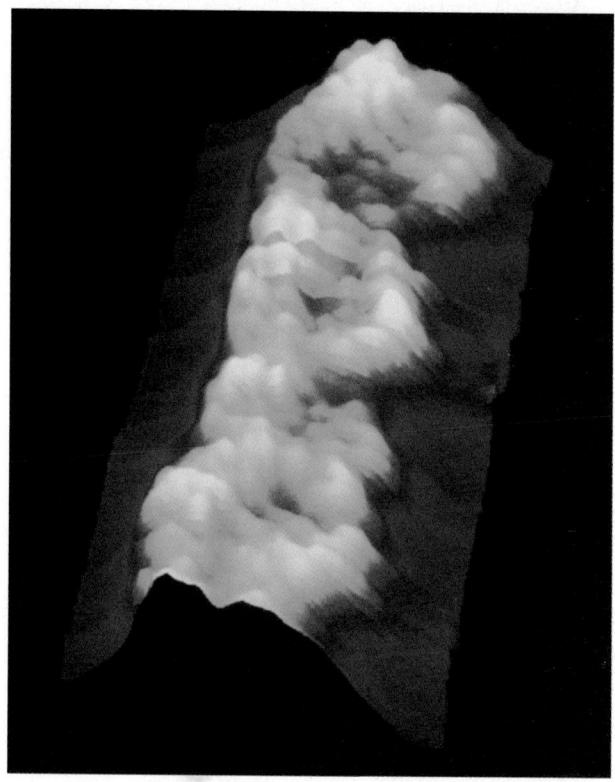

Figure 2.32   Scanning Tunneling Microscopy of DNA. The DNA double helix with approximately three turns shown (false color; ×2,000,000).

Summary

2.1 Lenses and the Bending of Light

a. A light ray moving from air to glass, or vice versa, is bent in a process known as refraction. Lenses focus light rays at a focal point and magnify images (**figure 2.2**).

2.2 The Light Microscope

a. In a compound microscope like the bright-field microscope, the primary image is formed by an objective lens and enlarged by the eyepiece or ocular lens to yield a virtual image (**figure 2.3**).
b. A substage condenser focuses a cone of light on the specimen.
c. Microscope resolution increases as the wavelength of radiation used to illuminate the specimen decreases. The maximum resolution of a light microscope is about 0.2 μm.
d. The dark-field microscope uses only refracted light to form an image (**figure 2.7**), and objects glow against a black background.
e. The phase-contrast microscope converts variations in the refractive index and density of cells into changes in light intensity and thus makes colorless, unstained cells visible (**figure 2.9**).
f. The differential interference contrast microscope uses two beams of light to create high-contrast, three-dimensional images of live specimens.
g. The fluorescence microscope illuminates a fluorochrome-labeled specimen and forms an image from its fluorescence (**figure 2.12**).

2. 3 Preparation and Staining of Specimens

a. Specimens usually must be fixed and stained before viewing them in the bright-field microscope.
b. Most dyes are either positively charged basic dyes or negative acid dyes and bind to ionized parts of cells.
c. In simple staining a single dye mixture is used to stain microorganisms.

d. Differential staining procedures like the Gram stain and acid-fast stain distinguish between microbial groups by staining them differently.

e. Some staining techniques are specific for particular structures like bacterial capsules, flagella, and endospores.

2.4 Electron Microscopy

a. The transmission electron microscope uses magnetic lenses to form an image from electrons that have passed through a very thin section of a specimen (**figure 2.23**). Resolution is high because the wavelength of electrons is very short.

b. Thin section contrast can be increased by treatment with solutions of heavy metals like osmium tetroxide, uranium, and lead.

c. Specimens are also prepared for the TEM by negative staining, shadowing with metal, or freeze-etching.

d. The scanning electron microscope (**figure 2.27**) is used to study external surface features of microorganisms.

2.5 Newer Techniques in Microscopy

a. The confocal scanning laser microscope (**figure 2.29**) is used to study thick, complex specimens. Scanning probe microscopes can visualize molecules and cells.

Key Terms

acid dyes 27
acid-fast staining 28
atomic force microscope 37
basic dyes 27
bright-field microscope 19
chromophore groups 27
confocal scanning laser microscope (CSLM) 35
dark-field microscopy 21
dark-phase-contrast microscopy 22

differential interference contrast (DIC) microscope 22
differential staining procedures 27
eyepieces 19
fixation 26
flagella staining 29
fluorescence microscope 25
fluorescent light 25
fluorochromes 25
focal length 18
focal point 18

freeze-etching 33
Gram stain 27
mordant 27
negative staining 29
numerical aperture 20
objectives 20
oculars 19
parfocal 20
phase-contrast microscope 21
refraction 18
refractive index 18

resolution 20
scanning electron microscope (SEM) 34
scanning probe microscope 35
scanning tunneling microscope 35
shadowing 32
simple staining 27
spore staining 29
substage condenser 19
transmission electron microscope (TEM) 31
working distance 21

Questions for Thought and Review

1. How are real and virtual images produced in a light microscope? Which one is a person actually seeing?

2. If a specimen is viewed using a 43× objective in a microscope with a 15× eyepiece, how many times has the image been magnified?

3. Why don't most light microscopes use 30× eyepieces for greater magnification?

4. Describe the two general types of fixation. Which would you normally use for bacteria? For protozoa?

5. Why would one expect basic dyes to be more effective under alkaline conditions?

6. What step in the Gram-stain procedure could be dropped without losing the ability to distinguish between gram-positive and gram-negative bacteria? Why?

7. Why must the TEM use a high vacuum and very thin sections?

8. Material is often embedded in paraffin before sectioning for light microscopy. Why can't this approach be used when preparing a specimen for the TEM?

9. Under what circumstances would it be desirable to prepare specimens for the TEM by use of negative staining? Shadowing? Freeze-etching?

10. Compare the microscopes described in this chapter—bright-field, dark-field, phase-contrast, DIC, fluorescence, TEM, SEM, confocal, and scanning probe—in terms of the images they provide and the purposes for which they are most often used.

11. Describe briefly how the scanning probe microscope operates. For what is it used? Distinguish between the two types of scanning probe microscopes with respect to their mechanism of operation.

12. Prepare a summary table showing the advantages of each type of microscope described in the chapter.

Critical Thinking Questions

1. If you prepared a sample of a specimen for light microscopy, stained with the Gram stain, and failed to see anything when you looked through your light microscope, list the things that you may have done incorrectly.

2. In a journal article, find an example of a light micrograph, a scanning or transmission electron micrograph, or a confocal image. Discuss why the figure was included in the article and why that particular type of microscopy was the method of choice for the research. What other figures would you like to see used in this study? Outline the steps that the investigators would take in order to obtain such photographs or figures.

For recommended readings on these and related topics, see Appendix V.

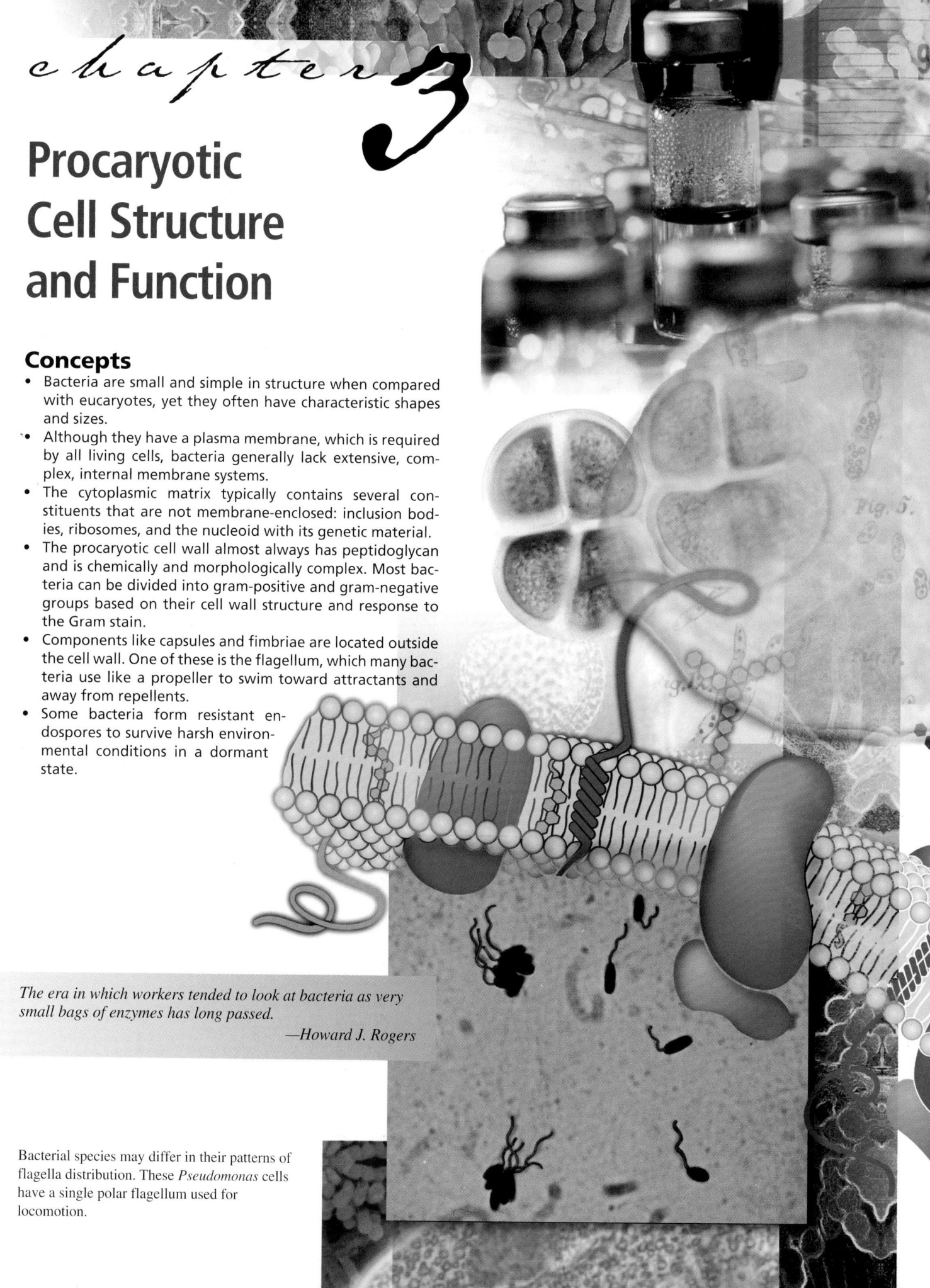

chapter 3

Procaryotic Cell Structure and Function

Concepts

- Bacteria are small and simple in structure when compared with eucaryotes, yet they often have characteristic shapes and sizes.
- Although they have a plasma membrane, which is required by all living cells, bacteria generally lack extensive, complex, internal membrane systems.
- The cytoplasmic matrix typically contains several constituents that are not membrane-enclosed: inclusion bodies, ribosomes, and the nucleoid with its genetic material.
- The procaryotic cell wall almost always has peptidoglycan and is chemically and morphologically complex. Most bacteria can be divided into gram-positive and gram-negative groups based on their cell wall structure and response to the Gram stain.
- Components like capsules and fimbriae are located outside the cell wall. One of these is the flagellum, which many bacteria use like a propeller to swim toward attractants and away from repellents.
- Some bacteria form resistant endospores to survive harsh environmental conditions in a dormant state.

The era in which workers tended to look at bacteria as very small bags of enzymes has long passed.

—*Howard J. Rogers*

Bacterial species may differ in their patterns of flagella distribution. These *Pseudomonas* cells have a single polar flagellum used for locomotion.

*E*ven a superficial examination of the microbial world shows that bacteria are one of the most important groups by any criterion: numbers of organisms, general ecological importance, or practical importance for humans. Indeed, much of our understanding of phenomena in biochemistry and molecular biology comes from research on bacteria. Although considerable space is devoted to eucaryotic microorganisms, the major focus is on procaryotes. Therefore the unit on microbial morphology begins with the structure of procaryotes. As mentioned in chapter 1 (*see p. 12*), there are two quite different groups of procaryotes: Bacteria and Archaea. This chapter focuses primarily on bacterial morphology; chapter 20 will discuss archaeal cell structure and composition. A comment about nomenclature is necessary to avoid confusion. The word procaryote will be used in a general sense to include both the bacteria and archaea; the term bacterium will refer specifically to bacteria.   Eucaryotes, procaryotes, and the composition of the microbial world (pp. 11–12; 90–91). The Archaea (pp. 437–52)

3.1 AN OVERVIEW OF PROCARYOTIC CELL STRUCTURE

Because much of this chapter is devoted to a discussion of individual cell components, a preliminary overview of the procaryotic cell as a whole is in order.

Size, Shape, and Arrangement

One might expect that small, relatively simple organisms like procaryotes would be uniform in shape and size. Although it is true that many procaryotes are similar in morphology, there is a remarkable amount of variation due to differences in genetics and ecology (**figures 3.1** and **3.2;** *see also figures 2.8 and 2.15*). Major morphological patterns are described here, and interesting variants are mentioned in the procaryotic survey (*see chapters 20–24*).

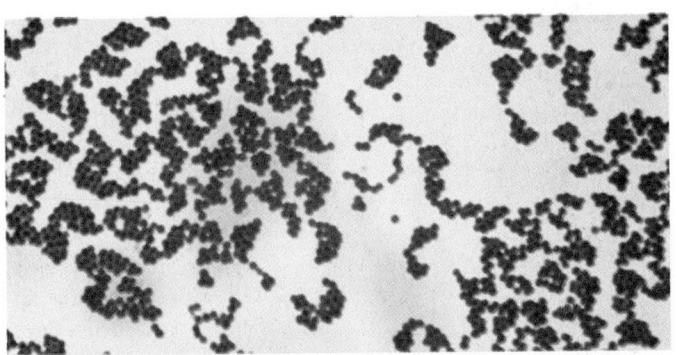

(a) *S. aureus*

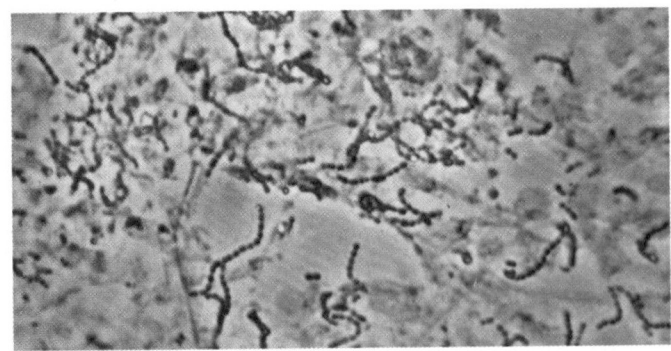

(b) *E. faecalis*

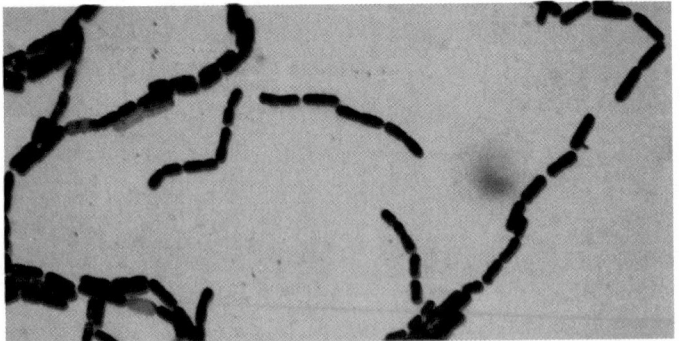

(c) *B. megaterium*

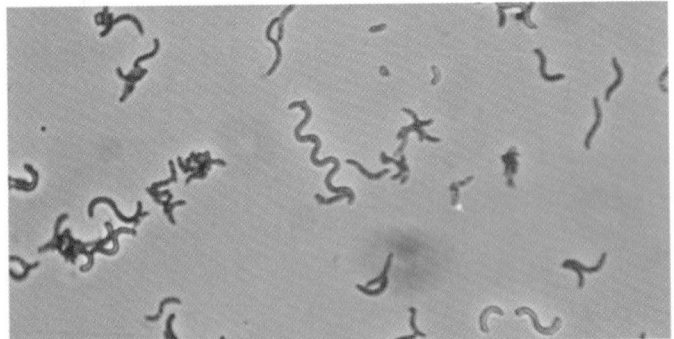

(d) *R. rubrum*

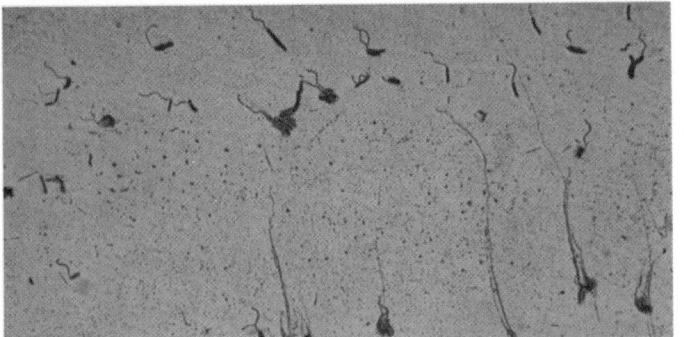

(e) *V. cholerae*

Figure 3.1   Representative Bacteria.   Stained bacterial cultures as seen in the light microscope. (**a**) *Staphylococcus aureus*. Note the gram-positive spheres in irregular clusters. Gram stain (×1,000). (**b**) *Enterococcus faecalis*. Note the chains of cocci; phase contrast (×200). (**c**) *Bacillus megaterium*, a rod-shaped bacterium in chains. Gram stain (×600). (**d**) *Rhodospirillum rubrum*. Phase contrast (×500). (**e**) *Vibrio cholerae*. Curved rods with polar flagella (×1,000).

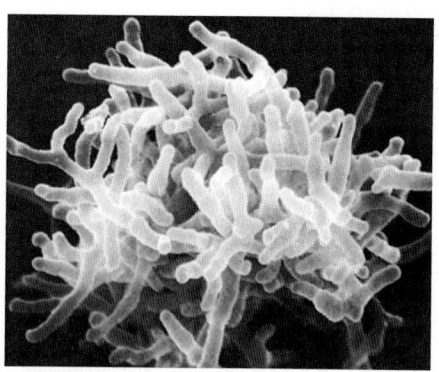

(a) *Actinomyces*

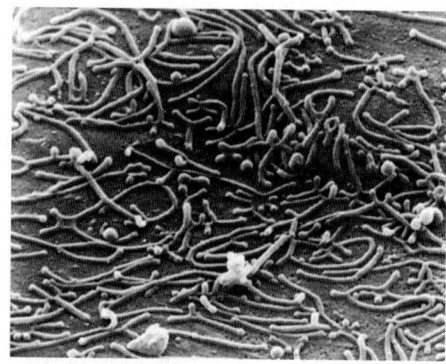

(b) *M. pneumoniae*

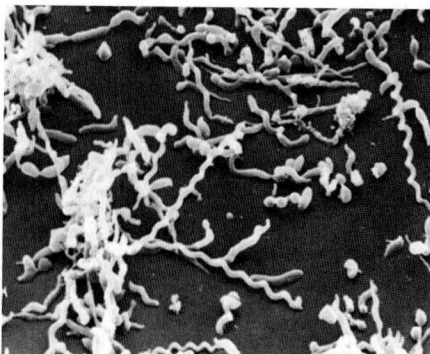

(c) *Spiroplasma*

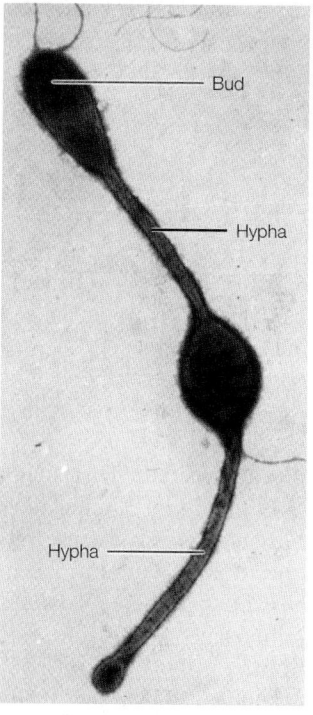

(d) *Hyphomicrobium*

(e) *Walsby's square bacterium*

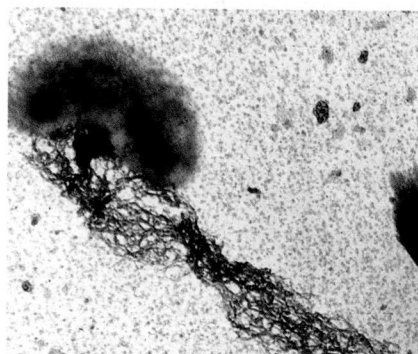

(f) *G. ferruginea*

Figure 3.2   Unusually Shaped Bacteria.
Examples of bacteria with shapes quite different from bacillus and coccus types. (**a**) *Actinomyces,* SEM (×21,000). (**b**) *Mycoplasma pneumoniae,* SEM (×62,000). (**c**) *Spiroplasma,* SEM (×13,000). (**d**) *Hyphomicrobium* with hyphae and bud, electron micrograph with negative staining. (**e**) Walsby's square bacterium. (**f**) *Gallionella ferruginea* with stalk.

Most commonly encountered bacteria have one of two shapes. **Cocci** (s., **coccus**) are roughly spherical cells. They can exist as individual cells, but also are associated in characteristic arrangements that are frequently useful in bacterial identification. **Diplococci** (s., **diplococcus**) arise when cocci divide and remain together to form pairs (*Neisseria; see figure 2.15*d). Long chains of cocci result when cells adhere after repeated divisions in one plane; this pattern is seen in the genera *Streptococcus, Enterococcus,* and *Lactococcus* (figure 3.1*b*). *Staphylococcus* divides in random planes to generate irregular grapelike clumps (figure 3.1*a*). Divisions in two or three planes can produce symmetrical clusters of cocci. Members of the genus *Micrococcus* often divide in two planes to form square groups of four cells called tetrads. In the genus *Sarcina,* cocci divide in three planes producing cubical packets of eight cells.

The other common bacterial shape is that of a **rod,** often called a **bacillus** (pl., **bacilli**). *Bacillus megaterium* is a typical example of a bacterium with a rod shape (figure 3.1*c; see also fig-*

*ure 2.15*a,c). Bacilli differ considerably in their length-to-width ratio, the coccobacilli being so short and wide that they resemble cocci. The shape of the rod's end often varies between species and may be flat, rounded, cigar-shaped, or bifurcated. Although many rods do occur singly, they may remain together after division to form pairs or chains (e.g., *Bacillus megaterium* is found in long chains). A few rod-shaped bacteria, the **vibrios,** are curved to form distinctive commas or incomplete spirals (figure 3.1*e*).

Bacteria can assume a great variety of shapes, although they often are simple spheres or rods. Actinomycetes characteristically form long filaments or hyphae that may branch to produce a network called a **mycelium** (figure 3.2*a*). Many bacteria are shaped like long rods twisted into spirals or helices; they are called **spirilla** if rigid and **spirochetes** when flexible (figures 3.1*d,* 3.2*c; see also figure* 2.8a,c). The oval- to pear-shaped *Hyphomicrobium* (figure 3.2*d*) produces a bud at the end of a long hypha. Other bacteria such as *Gallionella* produce nonliving stalks (figure 3.2*f*). A few bacteria actually are flat. For example, Anthony

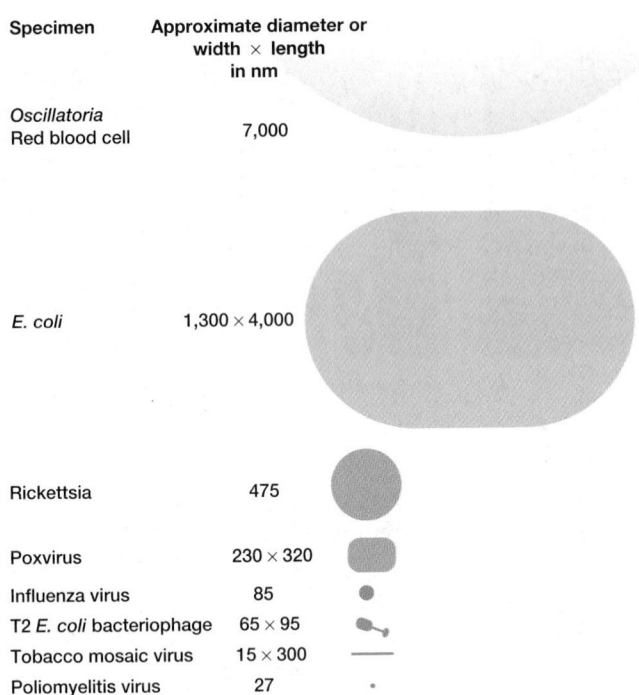

Specimen	Approximate diameter or width × length in nm
Oscillatoria Red blood cell	7,000
E. coli	1,300 × 4,000
Rickettsia	475
Poxvirus	230 × 320
Influenza virus	85
T2 E. coli bacteriophage	65 × 95
Tobacco mosaic virus	15 × 300
Poliomyelitis virus	27

Figure 3.3 Sizes of Bacteria and Viruses. The sizes of selected bacteria relative to a red blood cell and viruses.

Table 3.1

Functions of Procaryotic Structures

Plasma membrane	Selectively permeable barrier, mechanical boundary of cell, nutrient and waste transport, location of many metabolic processes (respiration, photosynthesis), detection of environmental cues for chemotaxis
Gas vacuole	Buoyancy for floating in aquatic environments
Ribosomes	Protein synthesis
Inclusion bodies	Storage of carbon, phosphate, and other substances
Nucleoid	Localization of genetic material (DNA)
Periplasmic space	Contains hydrolytic enzymes and binding proteins for nutrient processing and uptake
Cell wall	Gives bacteria shape and protection from lysis in dilute solutions
Capsules and slime layers	Resistance to phagocytosis, adherence to surfaces
Fimbriae and pili	Attachment to surfaces, bacterial mating
Flagella	Movement
Endospore	Survival under harsh environmental conditions

E. Walsby has discovered square bacteria living in salt ponds (figure 3.2e). These bacteria are shaped like flat, square-to-rectangular boxes about 2 μm by 2 to 4 μm, and only 0.25 μm thick. Finally, some bacteria are variable in shape and lack a single, characteristic form (figure 3.2b). These are called **pleomorphic** even though they may, like *Corynebacterium,* have a generally rodlike form.

Bacteria vary in size as much as in shape (**figure 3.3**). The smallest (e.g., some members of the genus *Mycoplasma*) are about 0.3 μm in diameter, approximately the size of the largest viruses (the poxviruses). Recently there have been reports of even smaller cells. Nanobacteria or ultramicrobacteria appear to range from around 0.2 μm to less than 0.05 μm in diameter. A few strains have been cultured, but most are simply very small bacteria-like objects only observed microscopically. It has been thought that the smallest possible cell is about 0.14 to 0.2 μm in diameter, but many nanobacteria are reported to be smaller. Some microbiologists think nanobacteria are artifacts, and more research will be required before the significance of these forms becomes clear. *Escherichia coli,* a bacillus of about average size, is 1.1 to 1.5 μm wide by 2.0 to 6.0 μm long. A few bacteria become fairly large; some spirochetes occasionally reach 500 μm in length, and the cyanobacterium *Oscillatoria* is about 7 μm in diameter (the same diameter as a red blood cell). A huge bacterium lives in the intestine of the brown surgeonfish, *Acanthurus nigrofuscus. Epulopiscium fishelsoni* grows as large as 600 by 80 μm, a little smaller than a printed hyphen. More recently an even larger bacterium, *Thiomargarita namibiensis,* has been discovered in ocean sedi-ment (**Microbial Diversity & Ecology 3.1**). Thus a few bacteria are much larger than the average eucaryotic cell (typical plant and animal cells are around 10 to 50 μm in diameter).

Procaryotic Cell Organization

A variety of structures is found in procaryotic cells. Their major functions are summarized in **table 3.1,** and **figure 3.4** illustrates many of them. Not all structures are found in every genus. Furthermore, gram-negative and gram-positive cells differ, particularly with respect to their cell walls. Despite these variations procaryotes are consistent in their fundamental structure and most important components.

Procaryotic cells almost always are bounded by a chemically complex cell wall. Inside this wall, and separated from it by a periplasmic space, lies the plasma membrane. This membrane can be invaginated to form simple internal membranous structures. Since the procaryotic cell does not contain internal membrane-bound organelles, its interior appears morphologically simple. The genetic material is localized in a discrete region, the nucleoid, and is not separated from the surrounding cytoplasm by membranes. Ribosomes and larger masses called inclusion bodies are scattered about in the cytoplasmic matrix. Both gram-positive and gram-negative cells can use flagella for locomotion. In addition, many cells are surrounded by a capsule or slime layer external to the cell wall.

Procaryotic cells are morphologically much simpler than eucaryotic cells. These two cell types are compared following the review of eucaryotic cell structure (*see pp. 90–91*).

1. What characteristic shapes can bacteria assume? Describe the ways in which bacterial cells cluster together.
2. Draw a bacterial cell and label all important structures.

Microbial Diversity & Ecology

3.1 Monstrous Microbes

Biologists often have distinguished between procaryotes and eucaryotes based in part on cell size. Generally, procaryotic cells are supposed to be smaller than eucaryotic cells. Procaryotes grow extremely rapidly compared to most eucaryotes and lack the complex vesicular transport systems of eucaryotic cells (*see chapter 4*). It has been assumed that they must be small because of the slowness of nutrient diffusion and the need for a large surface-to-volume ratio. Thus when Fishelson, Montgomery, and Myrberg discovered a large, cigar-shaped microorganism in the intestinal tract of the Red Sea brown surgeonfish, *Acanthurus nigrofuscus,* they suggested in their 1985 publication that it was a protist. It seemed too large to be anything else. In 1993 Esther Angert, Kendall Clemens, and Norman Pace used rRNA sequence comparisons (*see p. 420*) to identify the microorganism, now called *Epulopiscium fishelsoni,* as a procaryote related to the gram-positive genus *Clostridium.*

E. fishelsoni [Latin, *epulum,* a feast or banquet, and *piscium,* fish] can reach a size of 80 μm by 600 μm, and normally ranges from 200 to 500 μm in length (see **Box figure**). It is about a million times larger in volume than *Escherichia coli.* Despite its huge size the organism does have procaryotic cell structure. It is motile and swims at about two body lengths a second (approximately 2.4 cm/min) using the bacterial-type flagella that cover its surface. The cytoplasm contains large nucleoids and many ribosomes, as would be required for such a large cell. *Epulopiscium* appears to overcome the size limits set by diffusion by having an outer layer consisting of a highly convoluted plasma membrane. This increases the cell's surface area and aids in nutrient transport.

It appears that *Epulopiscium* is transmitted between hosts through fecal contamination of the fish's food. The bacterium can be eliminated by starving the surgeonfish for a few days. If juvenile fish that lack the bacterium are placed with infected hosts, they are reinoculated. Interestingly this does not work with uninfected adult surgeonfish.

In 1997, Heidi Schulz discovered an even larger procaryote in the ocean sediment off the coast of Namibia. *Thiomargarita namibiensis* is a spherical bacterium, between 100 and 750 μm in diameter, that often forms chains of cells enclosed in slime sheaths. It is over 100 times larger in volume than *E. fishelsoni.* A vacuole occupies about 98 percent of the cell and contains fluid rich in nitrate; it is surrounded by a 0.5 to 2.0 μm layer of cytoplasm filled with sulfur granules. The cytoplasmic layer is the same thickness as most bacteria and sufficiently thin for adequate diffusion rates. Nitrate is used as an electron acceptor for sulfur oxidation and energy production.

The discovery of these procaryotes greatly weakens the distinction between procaryotes and eucaryotes based on cell size. They are certainly larger than a normal eucaryotic cell. In addition, some eucaryotic cells have been discovered that are smaller than previously thought possible. The best example is *Nanochlorum eukaryotum. Nanochlorum* is only about 1 to 2 μm in diameter, yet is truly eucaryotic and has a nucleus, a chloroplast, and a mitochondrion. Our understanding of the factors limiting procaryotic cell size must be reevaluated. It is no longer safe to assume that large cells are eucaryotic and small cells are procaryotic.

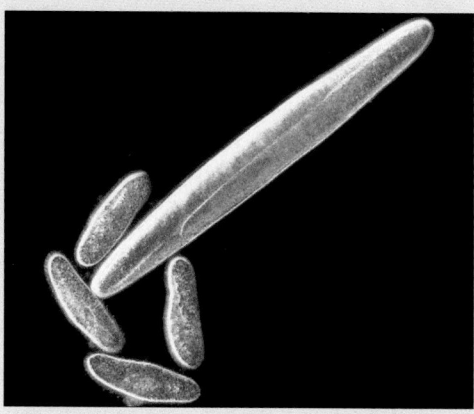

(a)

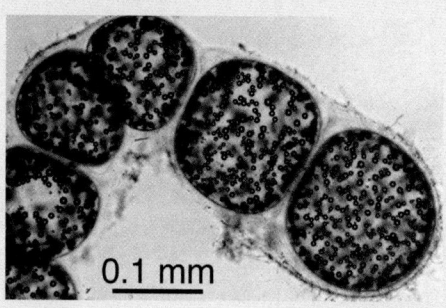

(b)

Giant Bacteria. (**a**) This photograph, taken with pseudo dark-field illumination, shows *Epulopiscium fishelsoni* at the top of the figure dwarfing the paramecia at the bottom (×200). (**b**) A chain of *Thiomargarita namibiensis* cells as viewed with the light microscope. Note the external mucous sheath and the internal sulfur globules.

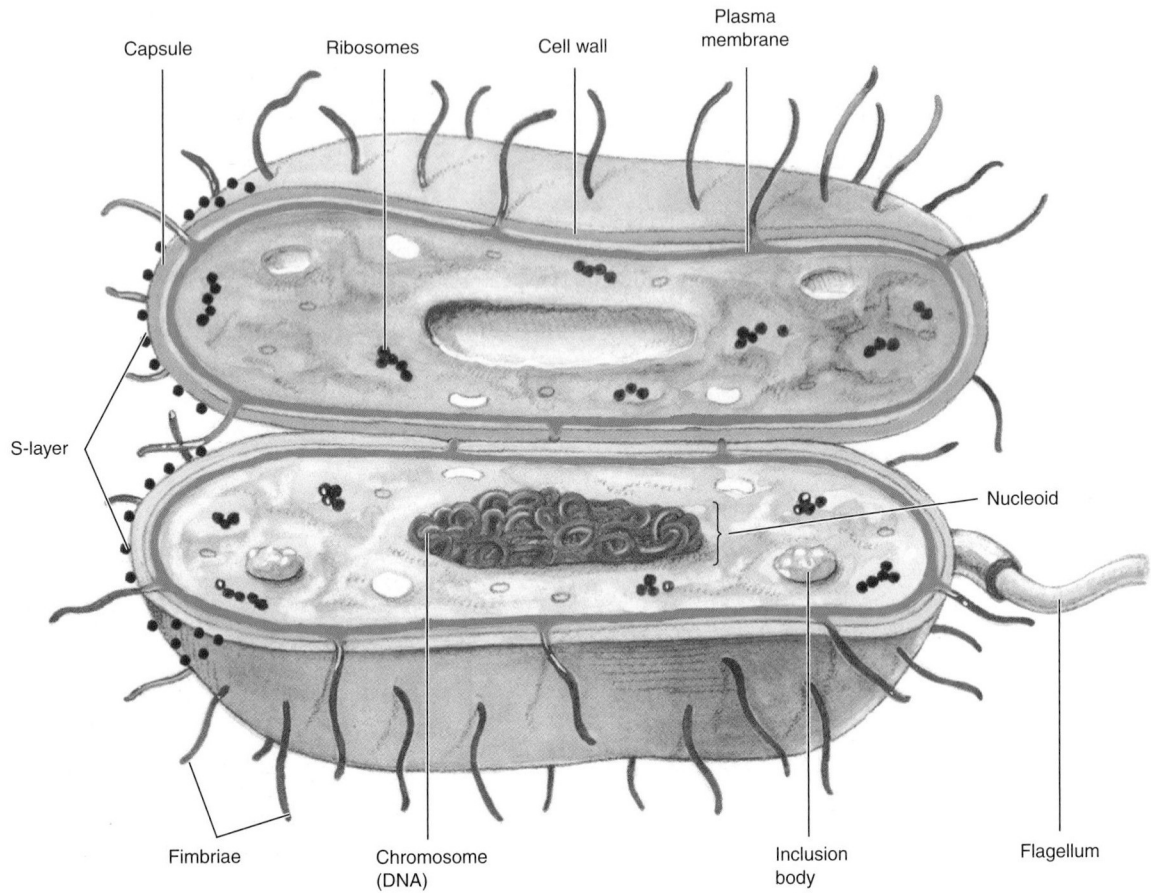

Figure 3.4 Morphology of a Procaryotic Cell.   Only a small area of surface proteins in the S-layer has been included to simplify the drawing; when present, these proteins cover the surface.

 PROCARYOTIC CELL MEMBRANES

Membranes are an absolute requirement for all living organisms. Cells must interact in a selective fashion with their environment, whether it is the internal environment of a multicellular organism or a less protected and more variable external environment. Cells must not only be able to acquire nutrients and eliminate wastes, but they also have to maintain their interior in a constant, highly organized state in the face of external changes. The **plasma membrane** encompasses the cytoplasm of both procaryotic and eucaryotic cells. This membrane is the chief point of contact with the cell's environment and thus is responsible for much of its relationship with the outside world. To understand membrane function, it is necessary to become familiar with membrane structure, and particularly with plasma membrane structure.

The Plasma Membrane

Membranes contain both proteins and lipids, although the exact proportions of protein and lipid vary widely. Bacterial plasma membranes usually have a higher proportion of protein than do eucaryotic membranes, presumably because they fulfill so many

different functions that are carried out by other organelle membranes in eucaryotes.

Most membrane-associated lipids are structurally asymmetric with polar and nonpolar ends (**figure 3.5**) and are called amphipathic. The polar ends interact with water and are **hydrophilic;** the nonpolar **hydrophobic** ends are insoluble in water and tend to associate with one another. This property of lipids enables them to form a bilayer in membranes. The outer surfaces are hydrophilic, whereas hydrophobic ends are buried in the interior away from the surrounding water. Many of these amphipathic lipids are phospholipids (figure 3.5). The lipid composition of bacterial membranes varies with environmental temperature in such a way that the membrane remains fluid during growth (*see pp. 123–24*). For example, bacteria growing at lower temperatures will have fatty acids with lower melting points in their membrane phospholipids. Bacterial membranes usually differ from eucaryotic membranes in lacking sterols such as cholesterol (**figure 3.6a**). However, many bacterial membranes do contain pentacyclic sterol-like molecules called hopanoids (figure 3.6b), and huge quantities of hopanoids are present in our ecosystem. It has been estimated that the total mass of hopanoids in sediments is around $10^{11–12}$ tons, about as much as the total mass of organic

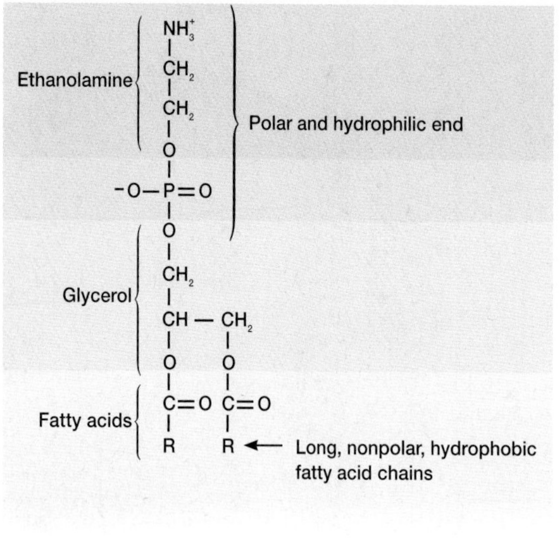

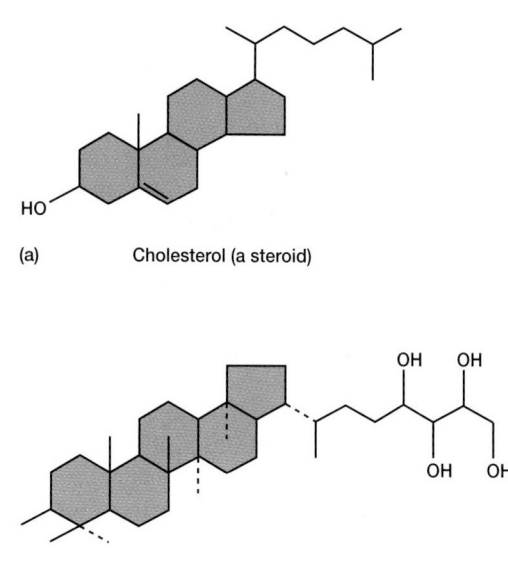

(a)   Cholesterol (a steroid)

(b)   A bacteriohopanetetrol (a hopanoid)

Figure 3.5 The Structure of a Polar Membrane Lipid.
Phosphatidylethanolamine, an amphipathic phospholipid often found in bacterial membranes. The R groups are long, nonpolar fatty acid chains.

Figure 3.6   Membrane Steroids and Hopanoids. Common examples.

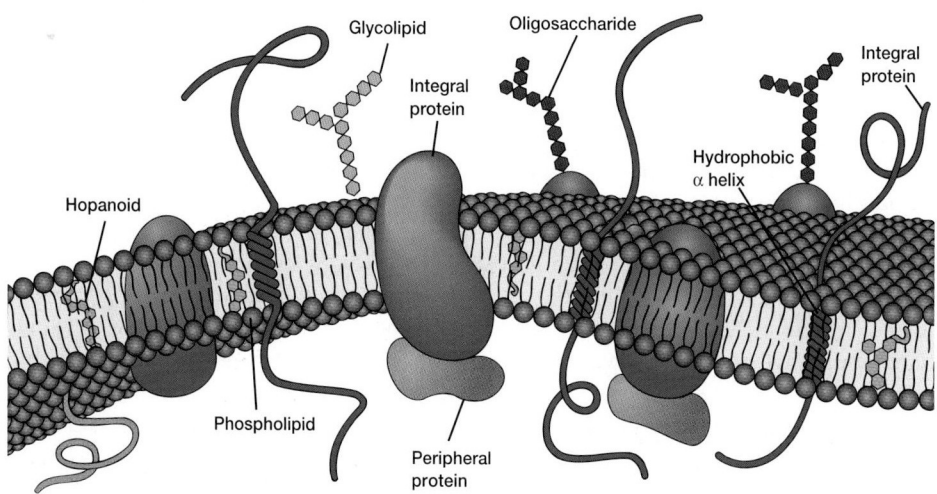

Figure 3.7   Plasma Membrane Structure.   This diagram of the fluid mosaic model of bacterial membrane structure shows the integral proteins (blue) floating in a lipid bilayer. Peripheral proteins (purple) are associated loosely with the inner membrane surface. Small spheres represent the hydrophilic ends of membrane phospholipids and wiggly tails, the hydrophobic fatty acid chains. Other membrane lipids such as hopanoids (pink) may be present. For the sake of clarity, phospholipids are shown in proportionately much larger size than in real membranes.

carbon in all living organisms (10^{12} tons). There is evidence that hopanoids have contributed significantly to the formation of petroleum. Hopanoids are synthesized from the same precursors as steroids. Like steroids in eucaryotes, they probably stabilize the bacterial membrane. The membrane lipid is organized in two layers, or sheets, of molecules arranged end-to-end (**figure 3.7**).

The distribution of lipids in eucaryotic membranes is asymmetric. Lipids in the outer monolayer or sheet differ from those of the inner monolayer. Although most lipids in individual mono-

layers mix freely with each other, there are microdomains that differ in lipid and protein composition. One such microdomain is the lipid raft, which is enriched in cholesterol and lipids with many saturated fatty acids such as some sphingolipids. The lipid raft spans the membrane bilayer, and lipids in the adjacent monolayers interact. These lipid rafts appear to participate in a variety of cellular processes (e.g., cell movement and signal transduction). They also may be involved in the entrance and assembly of some viruses (*see p. 392*).

Many archaeal membranes differ from bacterial membranes in having a monolayer with lipid molecules spanning the whole membrane.   Archaea (chapter 20)

Cell membranes are very thin structures, about 5 to 10 nm thick, and can only be seen with the electron microscope. The freeze-etching technique has been used to cleave membranes down the center of the lipid bilayer, splitting them in half and exposing the interior. In this way it has been discovered that many membranes, including the plasma membrane, have a complex internal structure. The small globular particles seen in these membranes are thought to be membrane proteins that lie within the membrane lipid bilayer (*see figure 2.26*).   Freeze-etching technique (p. 33)

The most widely accepted current model for membrane structure is the **fluid mosaic model** of S. Jonathan Singer and Garth Nicholson (figure 3.7). They distinguish between two types of membrane proteins. **Peripheral proteins** are loosely connected to the membrane and can be easily removed. They are soluble in aqueous solutions and make up about 20 to 30% of total membrane protein. About 70 to 80% of membrane proteins are **integral proteins.** These are not easily extracted from membranes and are insoluble in aqueous solutions when freed of lipids.   Protein and lipid chemistry (appendix I)

Integral proteins, like membrane lipids, are amphipathic; their hydrophobic regions are buried in the lipid while the hydrophilic portions project from the membrane surface (figure 3.7). Some of these proteins even extend all the way through the lipid layer. Integral proteins can diffuse laterally around the surface to new locations, but do not flip-flop or rotate through the lipid layer. Often carbohydrates are attached to the outer surface of plasma membrane proteins and seem to have important functions.

The emerging picture of the cell membrane is one of a highly organized and asymmetric system, which also is flexible and dynamic. Although membranes apparently have a common basic design, there are wide variations in both their structure and functional capacities. The differences are so large and characteristic that membrane chemistry can be used in bacterial identification.

The plasma membranes of procaryotic cells must fill an incredible variety of roles successfully. Many major plasma membrane functions are noted here even though they are discussed individually at later points in the text. The plasma membrane retains the cytoplasm, particularly in cells without cell walls, and separates it from the surroundings. The plasma membrane also serves as a selectively permeable barrier: it allows particular ions and molecules to pass, either into or out of the cell, while preventing the movement of others. Thus the membrane prevents the loss of essential components through leakage while allowing the movement of other molecules. Because many substances cannot cross the plasma membrane without assistance, it must aid such movement when necessary. Transport systems can be used for such tasks as nutrient uptake, waste excretion, and protein secretion. The procaryotic plasma membrane also is the location of a variety of crucial metabolic processes: respiration, photosynthesis, the synthesis of lipids and cell wall constituents, and probably chromosome segregation. Finally, the membrane contains

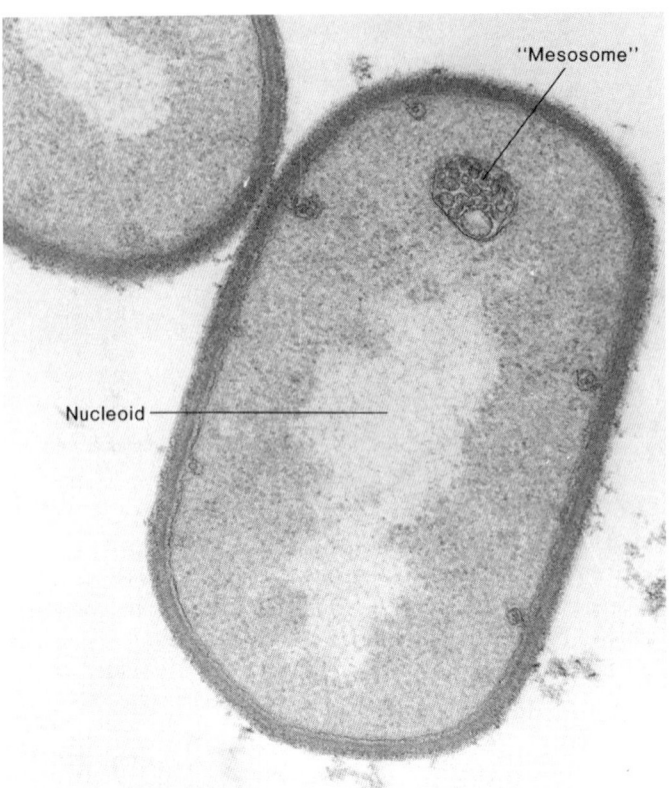

Figure 3.8 Mesosome Structure. *Bacillus fastidiosus* (×91,000). A large mesosome lies adjacent to the nucleoid.

special receptor molecules that help procaryotes detect and respond to chemicals in their surroundings. Clearly the plasma membrane is essential to the survival of microorganisms.   Osmosis (p. 59); Transport of substances across membranes (pp. 98–101)

Internal Membrane Systems

Although procaryotic cytoplasm does not contain complex membranous organelles like mitochondria or chloroplasts, membranous structures of several kinds can be observed. A commonly observed structure is the mesosome. Mesosomes are invaginations of the plasma membrane in the shape of vesicles, tubules, or lamellae (**figure 3.8** and figure 3.11). They are seen in both gram-positive and gram-negative bacteria, although they are generally more prominent in the former.

Mesosomes often are found next to septa or cross-walls in dividing bacteria and sometimes seem attached to the bacterial chromosome. Thus they may be involved in cell wall formation during division or play a role in chromosome replication and distribution to daughter cells.

Currently many bacteriologists believe that mesosomes are artifacts generated during the chemical fixation of bacteria for electron microscopy. Possibly they represent parts of the plasma

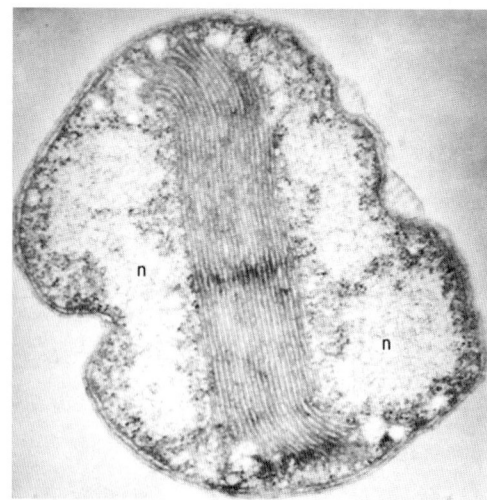

(a)

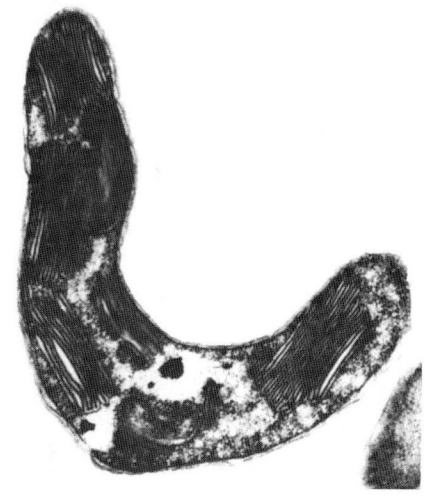

(b)

Figure 3.9 Internal Bacterial Membranes. Membranes of nitrifying and photosynthetic bacteria. (**a**) *Nitrocystis oceanus* with parallel membranes traversing the whole cell. Note nucleoplasm (n) with fibrillar structure. (**b**) *Ectothiorhodospira mobilis* with an extensive intracytoplasmic membrane system (×60,000).

membrane that are chemically different and more disrupted by fixatives.

Many bacteria have internal membrane systems quite different from the mesosome (**figure 3.9**). Plasma membrane infoldings can become extensive and complex in photosynthetic bacteria such as the cyanobacteria and purple bacteria or in bacteria with very high respiratory activity like the nitrifying bacteria (*see chapter 22*). They may be aggregates of spherical vesicles, flattened vesicles, or tubular membranes. Their function may be to provide a larger membrane surface for greater metabolic activity.

1. Describe with a labeled diagram and in words the fluid mosaic model for cell membranes.
2. List the functions of the plasma membrane.
3. Discuss the nature, structure, and possible functions of the mesosome.

 THE CYTOPLASMIC MATRIX

Procaryotic cytoplasm, unlike that of eucaryotes, lacks unit membrane-bound organelles. The **cytoplasmic matrix** is the substance lying between the plasma membrane and the nucleoid (pp. 50–52). The matrix is largely water (about 70% of bacterial mass is water). It is featureless in electron micrographs but often is packed with ribosomes and highly organized (**figure 3.10**). Specific proteins are positioned at particular sites such as the cell pole and the place where the bacterial cell will divide. Thus although bacteria may lack a true cytoskeleton, they do have a cytoskeleton-like system of proteins in their cytoplasmic matrix that may help determine cell shape. The plasma membrane and everything within is called the **protoplast;** thus the cytoplasmic matrix is a major part of the protoplast.

Inclusion Bodies

A variety of **inclusion bodies,** granules of organic or inorganic material that often are clearly visible in a light microscope, is present in the cytoplasmic matrix. These bodies usually are used for storage (e.g., carbon compounds, inorganic substances, and energy), and also reduce osmotic pressure by tying up molecules in particulate form. Some inclusion bodies are not bounded by a membrane and lie free in the cytoplasm—for example, polyphosphate granules, cyanophycin granules, and some glycogen granules. Other inclusion bodies are enclosed by a membrane about 2.0 to 4.0 nm thick, which is single-layered and not a typical bilayer membrane. Examples of membrane-enclosed inclusion bodies are poly-β-hydroxybutyrate granules, some glycogen and sulfur granules, carboxysomes, and gas vacuoles. Inclusion body membranes vary in composition. Some are protein in nature, whereas others contain lipid. Because inclusion bodies are used for storage, their quantity will vary with the nutritional status of the cell. For example, polyphosphate granules will be depleted in freshwater habitats that are phosphate limited. A brief description of several important inclusion bodies follows.

Organic inclusion bodies usually contain either glycogen or poly-β-hydroxybutyrate. **Glycogen** is a polymer of glucose units composed of long chains formed by $\alpha(1\rightarrow4)$ glycosidic bonds and branching chains connected to them by $\alpha(1\rightarrow6)$ glycosidic bonds (*see appendix I*). **Poly-β-hydroxybutyrate (PHB)** contains β-hydroxybutyrate molecules joined by ester bonds between the carboxyl and hydroxyl groups of adjacent molecules. Usually only one of these polymers is found in a species, but purple photosynthetic

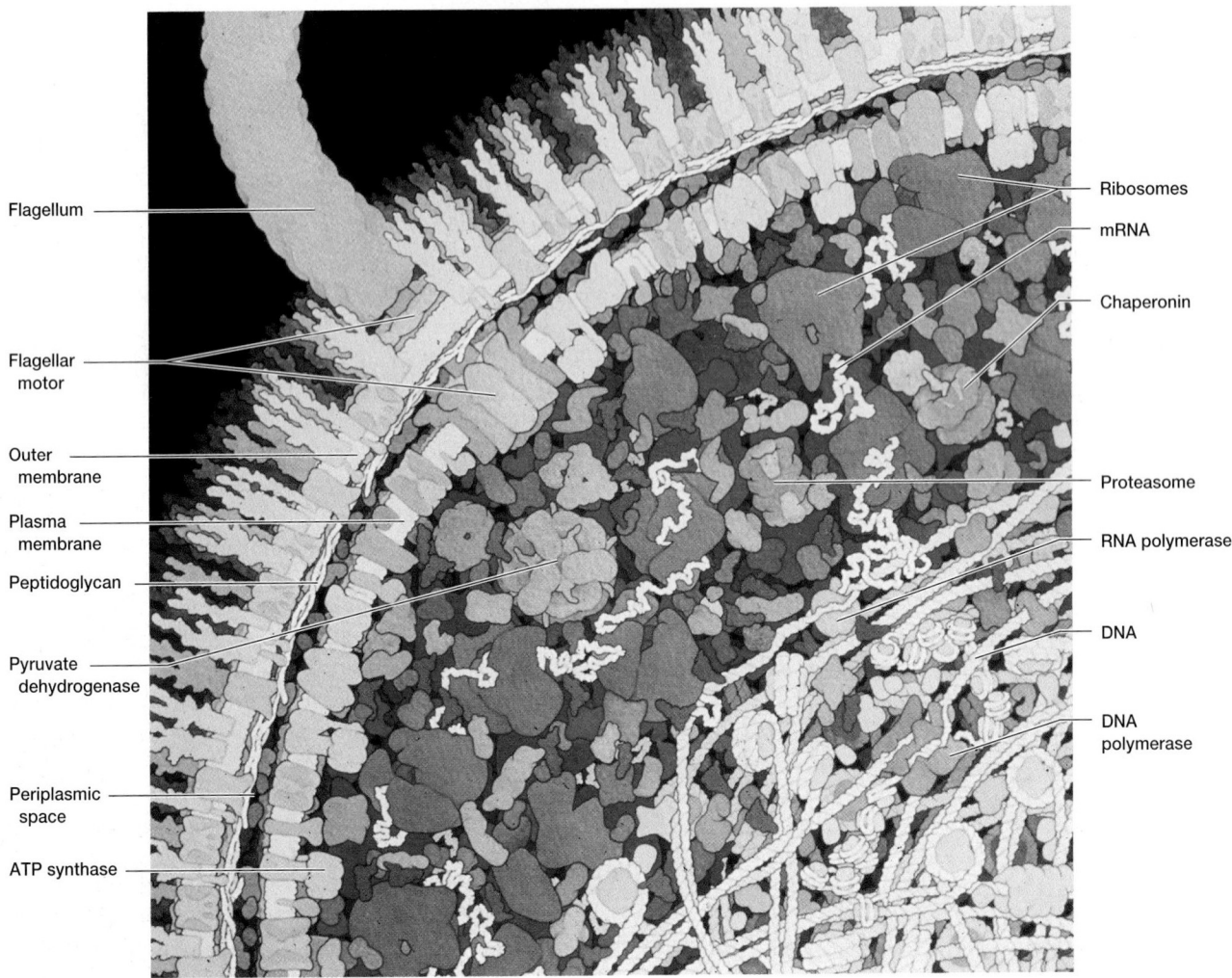

Figure 3.10 **A Cross Section of the Bacterium** *Escherichia coli* **Drawn at a Magnification of a Million Times.** The glycocalyx, flagellum, gram-negative cell wall, and plasma membrane are at the top. Ribosomes synthesizing proteins fill the underlying cytoplasmic matrix. At the bottom right is the nucleoid with its dense tangle of DNA and associated proteins. Note that the cytoplasmic matrix and nucleoid are highly organized and tightly packed.

bacteria have both. Poly-β-hydroxybutyrate accumulates in distinct bodies, around 0.2 to 0.7 μm in diameter, that are readily stained with Sudan black for light microscopy and are clearly visible in the electron microscope (**figure 3.11**). Glycogen is dispersed more evenly throughout the matrix as small granules (about 20 to 100 nm in diameter) and often can be seen only with the electron microscope. If cells contain a large amount of glycogen, staining with an iodine solution will turn them reddish-brown. Glycogen and PHB inclusion bodies are carbon storage reservoirs providing material for energy and biosynthesis. Many bacteria also store carbon as lipid droplets.

Cyanobacteria have two distinctive organic inclusion bodies. **Cyanophycin granules** (figure 3.13*a*) are composed of large polypeptides containing approximately equal amounts of the amino acids arginine and aspartic acid. The granules often are large enough to be visible in the light microscope and store extra

nitrogen for the bacteria. **Carboxysomes** are present in many cyanobacteria, nitrifying bacteria, and thiobacilli. They are polyhedral, about 100 nm in diameter, and contain the enzyme ribulose-1,5-bisphosphate carboxylase (*see p. 202*) in a paracrystalline arrangement. They serve as a reserve of this enzyme and may be a site of CO_2 fixation.

A most remarkable organic inclusion body, the **gas vacuole,** is present in many cyanobacteria (*see section 21.3*), purple and green photosynthetic bacteria, and a few other aquatic forms such as *Halobacterium* and *Thiothrix*. These bacteria float at or near the surface, because gas vacuoles give them buoyancy. This is vividly demonstrated by a simple but dramatic experiment. Cyanobacteria held in a full, tightly stoppered bottle will float, but if the stopper is struck with a hammer, the bacteria sink to the bottom. Examination of the bacteria at the beginning and end of the experiment shows that the sudden pressure in-

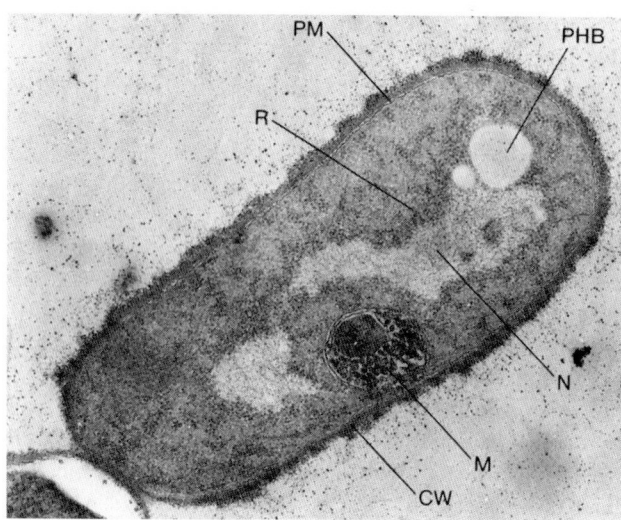

Figure 3.11 The Structure of a Typical Gram-Positive Cell.
Electron micrograph of *Bacillus megaterium* (×30,500). Note the
thick cell wall, CW; "mesosome," M; nucleoid, N; poly-β-
hydroxybutyrate inclusion body, PHB; plasma membrane, PM;
and ribosomes, R.

crease has collapsed the gas vacuoles and destroyed the microorganisms' buoyancy.

Gas vacuoles are aggregates of enormous numbers of small,
hollow, cylindrical structures called **gas vesicles** (**figure 3.12**).
Gas vesicle walls do not contain lipid and are composed entirely
of a single small protein. These protein subunits assemble to form
a rigid enclosed cylinder that is hollow and impermeable to water but freely permeable to atmospheric gases. Bacteria with gas
vacuoles can regulate their buoyancy to float at the depth necessary for proper light intensity, oxygen concentration, and nutrient
levels. They descend by simply collapsing vesicles and float upward when new ones are constructed.

Two major types of inorganic inclusion bodies are seen. Many
bacteria store phosphate as **polyphosphate granules** or **volutin
granules** (**figure 3.13a**). Polyphosphate is a linear polymer of orthophosphates joined by ester bonds. Thus volutin granules function as storage reservoirs for phosphate, an important component
of cell constituents such as nucleic acids. In some cells they act
as an energy reserve, and polyphosphate can serve as an energy
source in reactions. These granules are sometimes called
metachromatic granules because they show the metachromatic
effect; that is, they appear red or a different shade of blue when
stained with the blue dyes methylene blue or toluidine blue. Some
bacteria also store sulfur temporarily as sulfur granules, a second
type of inorganic inclusion body (figure 3.13b). For example,
purple photosynthetic bacteria can use hydrogen sulfide as a photosynthetic electron donor (*see section 9.11*) and accumulate the
resulting sulfur in either the periplasmic space or in special cytoplasmic globules.

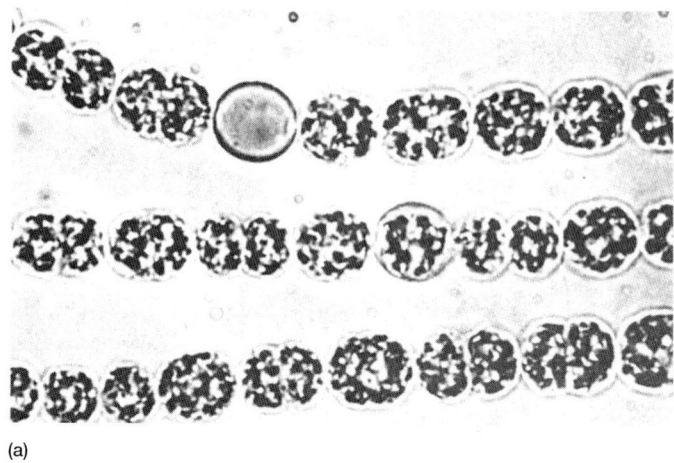

(a)

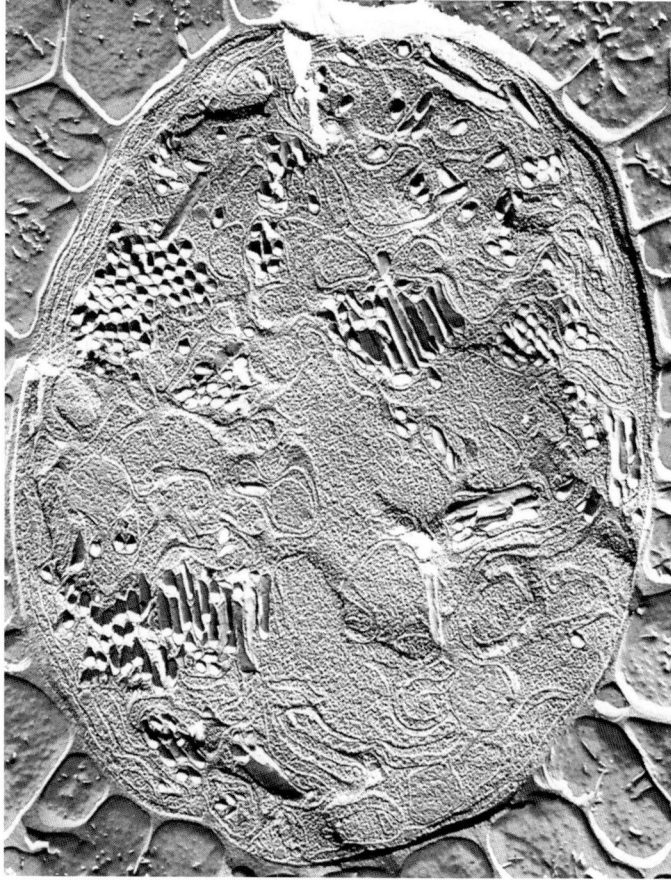

(b)

Figure 3.12 Gas Vesicles and Vacuoles. (a) Filaments of the
cyanobacterium *Anabaena flos-aquae* as seen in the light microscope. (b) A freeze-fracture preparation of *Anabaena flos-
aquae* (×89,000). Clusters of the cigar shaped vesicles form gas
vacuoles. Both longitudinal and cross-sectional views of gas
vesicles can be seen.

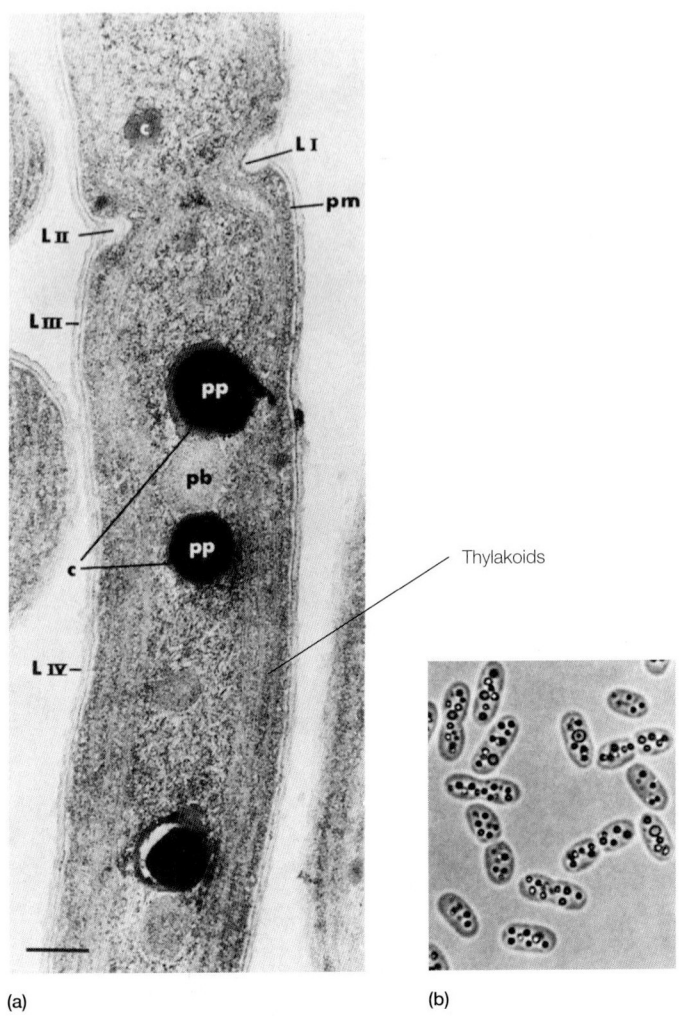

(a)   (b)

Thylakoids

Figure 3.13   Inclusion Bodies in Bacteria.   (a) Ultrastructure of the cyanobacterium *Anacystis nidulans.* The bacterium is dividing, and a septum is partially formed, LI and LII. Several structural features can be seen, including cell wall layers, LIII and LIV; the plasma membrane, pm; polyphosphate granules, pp; a polyhedral body, pb; and cyanophycin material, c. Thylakoids run along the length of the cell. Bar = 0.1 μm. **(b)** *Chromatium vinosum,* a purple sulfur bacterium, with intracellular sulfur granules, light field (×2,000).

Inorganic inclusion bodies can be used for purposes other than storage. An excellent example is the **magnetosome,** which is used by some bacteria to orient in the Earth's magnetic field. These inclusion bodies contain iron in the form of magnetite (**Microbial Diversity & Ecology 3.2**).

Ribosomes

As mentioned earlier, the cytoplasmic matrix often is packed with **ribosomes;** they also may be loosely attached to the plasma membrane. Ribosomes look like small, featureless particles at low magnification in electron micrographs (figure 3.11) but are actually very complex objects made of both protein and ribonucleic acid (RNA). They are the site of protein synthesis; matrix ribosomes synthesize proteins destined to remain within the cell, whereas the plasma membrane ribosomes make proteins for transport to the outside. The newly formed polypeptide folds into its final shape either as it is synthesized by the ribosome or shortly after completion of protein synthesis. The shape of each protein is determined by its amino acid sequence. Special proteins called molecular chaperones, or chaperones, aid the polypeptide in folding to its proper shape. Protein synthesis, including a detailed treatment of ribosomes and chaperones, is discussed at considerable length in chapter 12.

Note that procaryotic ribosomes are smaller than eucaryotic ribosomes. They commonly are called 70S ribosomes, have dimensions of about 14 to 15 nm by 20 nm, a molecular weight of approximately 2.7 million, and are constructed of a 50S and a 30S subunit. The S in 70S and similar values stands for **Svedberg unit.** This is the unit of the sedimentation coefficient, a measure of the sedimentation velocity in a centrifuge; the faster a particle travels when centrifuged, the greater its Svedberg value or sedimentation coefficient. The sedimentation coefficient is a function of a particle's molecular weight, volume, and shape (*see figure 16.7*). Heavier and more compact particles normally have larger Svedberg numbers or sediment faster. Ribosomes in the cytoplasmic matrix of eucaryotic cells are 80S ribosomes and about 22 nm in diameter. Despite their overall difference in size, both types of ribosomes are similarly composed of a large and a small subunit.

---

1. Briefly describe the nature and function of the cytoplasmic matrix and the ribosome. What is a protoplast?
2. What kinds of inclusion bodies do procaryotes have? What are their functions?
3. What is a gas vacuole? Relate its structure to its function.

 THE NUCLEOID

Probably the most striking difference between procaryotes and eucaryotes is the way in which their genetic material is packaged. Eucaryotic cells have two or more chromosomes contained within a membrane-delimited organelle, the nucleus. In contrast, procaryotes lack a membrane-delimited nucleus. The procaryotic chromosome is located in an irregularly shaped region called the **nucleoid** (other names are also used: the nuclear body, chromatin body, nuclear region). Usually procaryotes contain a single circle of double-stranded **deoxyribonucleic acid (DNA),** but some have a linear DNA chromosome. Recently it has been discovered that some bacteria such as *Vibrio cholerae* have more than one chromosome. Although nucleoid appearance varies with the method of fixation and staining, fibers often are seen in electron micrographs (figure 3.11 and **figure 3.14**) and are probably DNA.

Microbial Diversity & Ecology

3.2   Living Magnets

Bacteria can respond to environmental factors other than chemicals. A fascinating example is that of the aquatic magnetotactic bacteria that orient themselves in the Earth's magnetic field. Most of these bacteria have intracellular chains of magnetite (Fe_3O_4) particles or magnetosomes, around 40 to 100 nm in diameter and bounded by a membrane (see **Box figure**). Some species from sulfidic habitats have magnetosomes containing greigite (Fe_3S_4) and pyrite (FeS_2). Since each iron particle is a tiny magnet, the Northern Hemisphere bacteria use their magnetosome chain to determine northward and downward directions, and swim down to nutrient-rich sediments or locate the optimum depth in freshwater and marine habitats (*see pp. 616–20*). Magnetotactic bacteria in the Southern Hemisphere generally orient southward and downward, with the same result. Magnetosomes also are present in the heads of birds, tuna, dolphins, green turtles, and other animals, presumably to aid navigation. Animals and bacteria share more in common behaviorally than previously imagined.

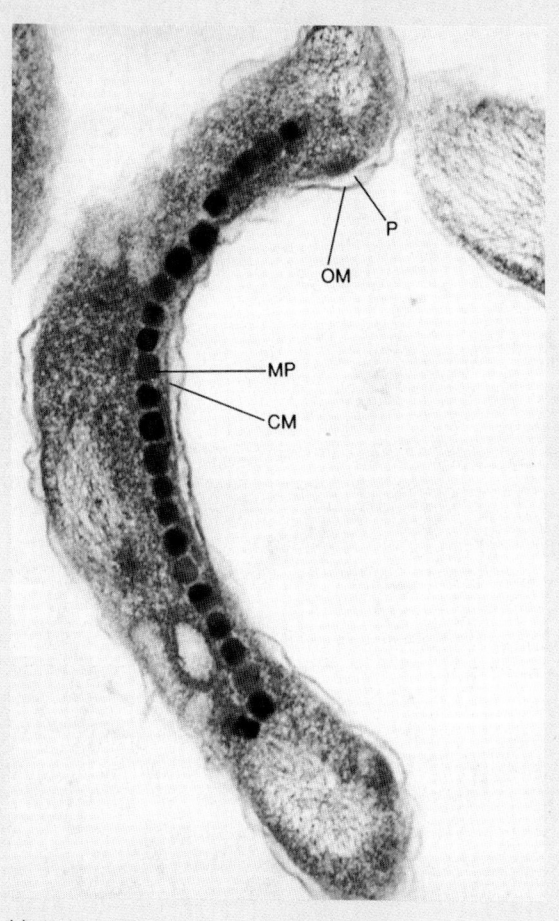

(a)

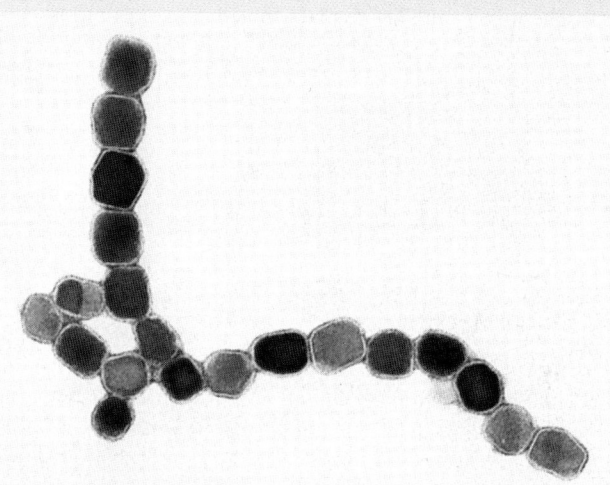

(b)

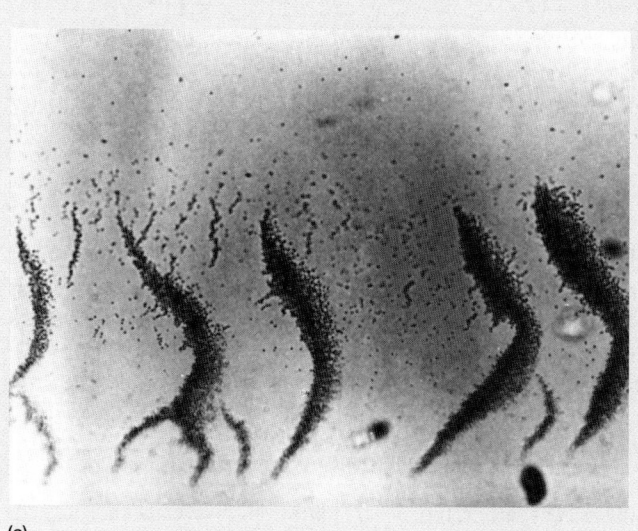

(c)

Magnetotactic Bacteria.   (**a**) Transmission electron micrograph of the magnetotactic bacterium *Aquaspirillum magnetotacticum* ($\times$123,000). Note the long chain of electron-dense magnetite particles, MP. Other structures: OM, outer membrane; P, periplasmic space; CM, cytoplasmic membrane. (**b**) Isolated magnetosomes ($\times$140,000). (**c**) Bacteria migrating in waves when exposed to a magnetic field.

The nucleoid also is visible in the light microscope after staining with the Feulgen stain, which specifically reacts with DNA. A cell can have more than one nucleoid when cell division occurs after the genetic material has been duplicated (figure 3.14*a*). In actively growing bacteria, the nucleoid has projections that extend into the cytoplasmic matrix (figure 3.14*b,c*). Presumably these projections contain DNA that is being actively transcribed to produce mRNA.

Careful electron microscopic studies often have shown the nucleoid in contact with either the mesosome or the plasma membrane. Membranes also are found attached to isolated nucleoids. Thus there is evidence that bacterial DNA is attached to cell membranes, and membranes may be involved in the separation of DNA into daughter cells during division.

Nucleoids have been isolated intact and free from membranes. Chemical analysis reveals that they are composed of about 60% DNA, 30% RNA, and 10% protein by weight. In *Escherichia coli*, a rod-shaped cell about 2 to 6 μm long, the closed DNA circle measures approximately 1,400 μm. Obviously it must be very efficiently packaged to fit within the nucleoid. The DNA is looped and coiled extensively (*see figure 11.9*), probably with the aid of RNA and nucleoid proteins (these proteins differ from the histone proteins present in eucaryotic nuclei).

There are a few exceptions to the above picture. Membrane-bound DNA-containing regions are present in two genera of planctomycetes. *Pirellula* has a single membrane that surrounds a region, the pirellulosome, which contains a fibrillar nucleoid and ribosome-like particles. The nuclear body of *Gemmata obscuriglobus* is bounded by two membranes (*see figure 21.12*). More work will be required to determine the functions of these membranes and how widespread this phenomenon is. The cell cycle and cell division (pp. 280–82). Procaryotic DNA and its function (chapters 11 and 12)

Many bacteria possess **plasmids** in addition to their chromosome. These are double-stranded DNA molecules, usually circular, that can exist and replicate independently of the chromosome or may be integrated with it; in either case they normally are inherited or passed on to the progeny. However, plasmids are not usually attached to the plasma membrane and sometimes are lost to one of the progeny cells during division. Plasmids are not required for host growth and reproduction, although they may carry genes that give their bacterial host a selective advantage. Plasmid genes can render bacteria drug-resistant, give them new metabolic abilities, make them pathogenic, or endow them with a number of other properties. Because plasmids often move between bacteria, properties such as drug resistance can spread throughout a population. Plasmids (pp. 288–91)

---

1. Characterize the nucleoid with respect to its structure and function.
2. What is a plasmid?

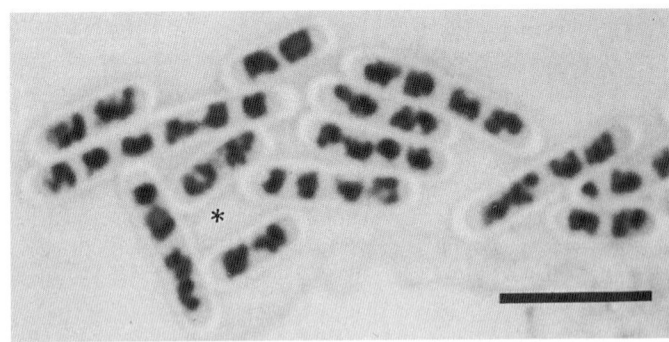

(a)

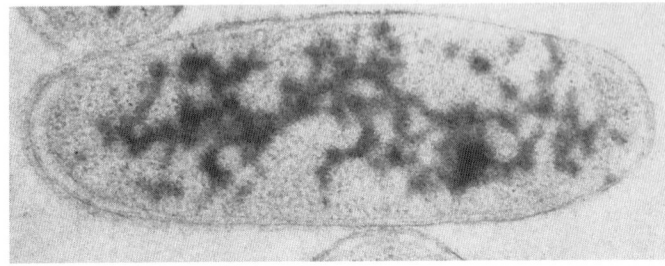

(b)

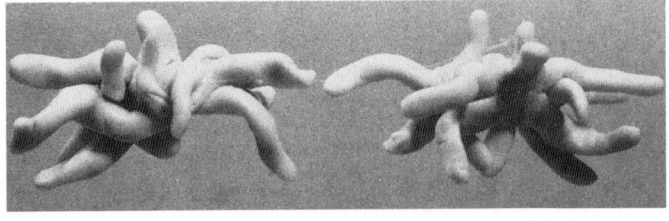

(c)

Figure 3.14   The Bacterial Nucleoid.   (a) Nucleoids in growing *Bacillus* cells stained using HCl-Giemsa stain and viewed with a light microscope (bar = 5 μm). **(b)** A section of actively growing *E. coli* immunostained specifically for DNA and examined in the transmission electron microscope. Coupled transcription and translation occur in parts of the nucleoid that extend out into the cytoplasm. **(c)** A model of two nucleoids in an actively growing *E. coli* cell. Note that a metabolically active nucleoid is not compact and spherical but has projections that extend into the cytoplasmic matrix.

3.5   THE PROCARYOTIC CELL WALL

The cell wall is the layer, usually fairly rigid, that lies just outside the plasma membrane. It is one of the most important parts of a procaryotic cell for several reasons. Except for the mycoplasmas (*see section 23.1*) and some Archaea (*see chapter 20*), most bacteria have strong walls that give them shape and protect them from osmotic lysis (p. 59); wall shape and strength is primarily due to peptidoglycan, as we will see shortly. The cell walls of

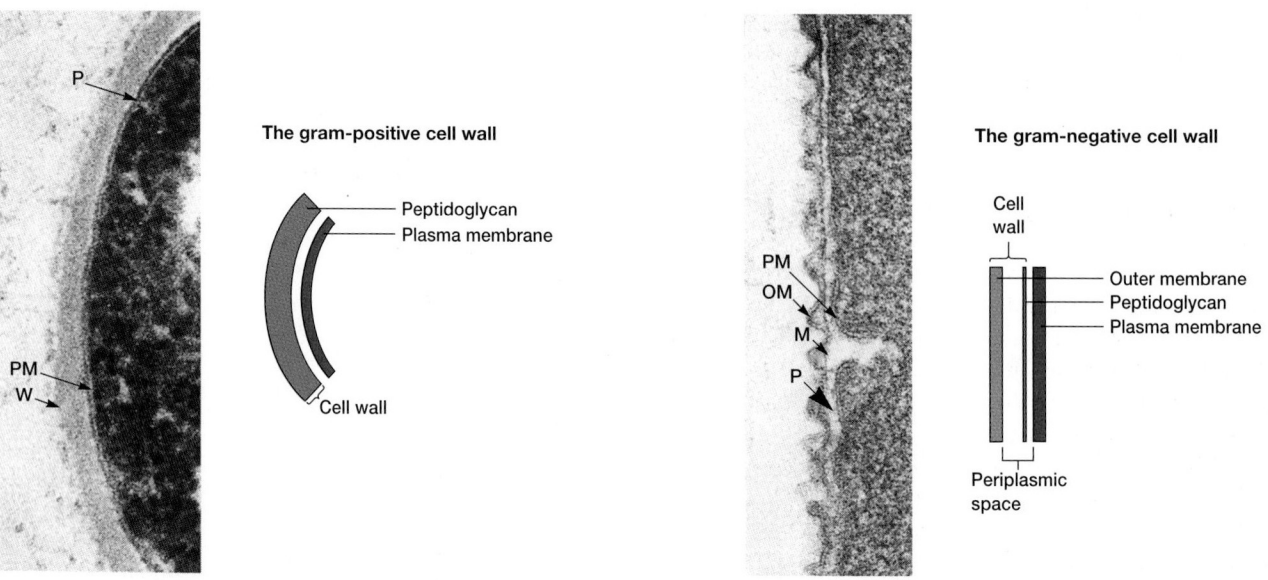

Figure 3.15   Gram-Positive and Gram-Negative Cell Walls.   The gram-positive envelope is from *Bacillus licheniformis* (left), and the gram-negative micrograph is of *Aquaspirillum serpens* (right). M; peptidoglycan or murein layer; OM, outer membrane; PM, plasma membrane; P, periplasmic space; W, gram-positive peptidoglycan wall.

many pathogens have components that contribute to their pathogenicity. The wall can protect a cell from toxic substances and is the site of action of several antibiotics.

After Christian Gram developed the Gram stain in 1884, it soon became evident that bacteria could be divided into two major groups based on their response to the Gram-stain procedure (*see table 19.9*). Gram-positive bacteria stained purple, whereas gram-negative bacteria were colored pink or red by the technique. The true structural difference between these two groups became clear with the advent of the transmission electron microscope. The gram-positive cell wall consists of a single 20 to 80 nm thick homogeneous **peptidoglycan** or **murein** layer lying outside the plasma membrane (**figure 3.15**). In contrast, the gram-negative cell wall is quite complex. It has a 2 to 7 nm peptidoglycan layer surrounded by a 7 to 8 nm thick **outer membrane.** Because of the thicker peptidoglycan layer, the walls of gram-positive cells are stronger than those of gram-negative bacteria. Microbiologists often call all the structures from the plasma membrane outward the **envelope** or cell envelope. This includes the wall and structures like capsules (p. 61) when present.   Gram-stain procedure (pp. 27–28)

Frequently a space is seen between the plasma membrane and the outer membrane in electron micrographs of gram-negative bacteria, and sometimes a similar but smaller gap may be observed between the plasma membrane and wall in gram-positive bacteria. This space is called the **periplasmic space.** Recent evidence indicates that the periplasmic space may be filled with a loose network of peptidoglycan. Possibly it is more a gel than a fluid-filled space. The substance that occupies the periplasmic space is the **periplasm.** Gram-positive cells may have periplasm even if they lack a discrete, obvious periplasmic space. Size esti-

mates of the periplasmic space in gram-negative bacteria range from 1 nm to as great as 71 nm. Some recent studies indicate that it may constitute about 20 to 40% of the total cell volume (around 30 to 70 nm), but more research is required to establish an accurate value. When cell walls are disrupted carefully or removed without disturbing the underlying plasma membrane, periplasmic enzymes and other proteins are released and may be easily studied. The periplasmic space of gram-negative bacteria contains many proteins that participate in nutrient acquisition—for example, hydrolytic enzymes attacking nucleic acids and phosphorylated molecules, and binding proteins involved in transport of materials into the cell. Denitrifying and chemolithoautotrophic bacteria (*see sections 9.6 and 9.10*) often have electron transport proteins in their periplasm. The periplasmic space also contains enzymes involved in peptidoglycan synthesis and the modification of toxic compounds that could harm the cell. Gram-positive bacteria may not have a visible periplasmic space and do not appear to have as many periplasmic proteins; rather, they secrete several enzymes that ordinarily would be periplasmic in gram-negative bacteria. Such secreted enzymes are often called **exoenzymes.** Some enzymes remain in the periplasm and are attached to the plasma membrane.

The Archaea differ from other procaryotes in many respects (*see chapter 20*). Although they may be either gram positive or gram negative, their cell walls are distinctive in structure and chemical composition. The walls lack peptidoglycan and are composed of proteins, glycoproteins, or polysaccharides.

Following this overview of the envelope, peptidoglycan structure and the organization of gram-positive and gram-negative cell walls are discussed in more detail.

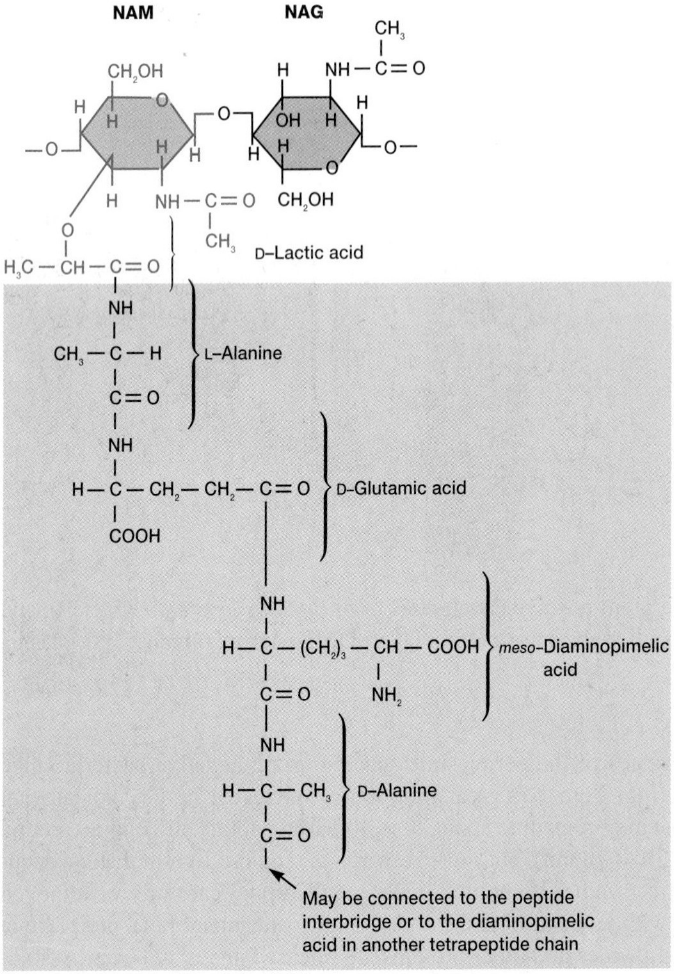

Figure 3.16   Peptidoglycan Subunit Composition.   The peptidoglycan subunit of *Escherichia coli,* most other gram-negative bacteria, and many gram-positive bacteria. NAG is *N*-acetylglucosamine. NAM is *N*-acetylmuramic acid (NAG with lactic acid attached by an ether linkage). The tetrapeptide side chain is composed of alternating D- and L-amino acids since *meso*-diaminopimelic acid is connected through its L-carbon. NAM and the tetrapeptide chain attached to it are shown in different shades of color for clarity.

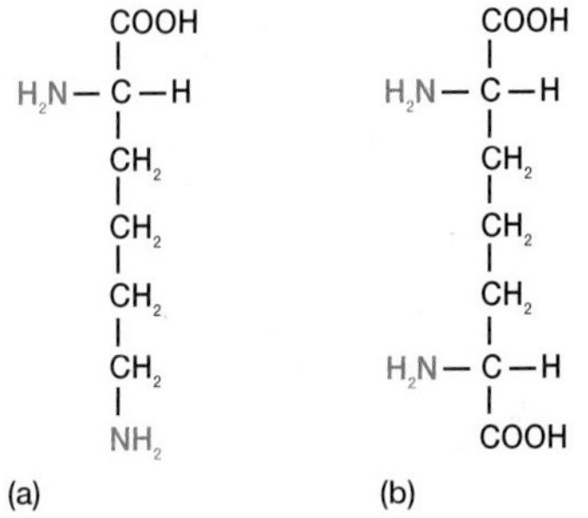

Peptidoglycan Structure

Peptidoglycan or murein is an enormous polymer composed of many identical subunits. The polymer contains two sugar derivatives, *N*-acetylglucosamine and *N*-acetylmuramic acid (the lactyl ether of *N*-acetylglucosamine), and several different amino acids, three of which—D-glutamic acid, D-alanine, and *meso*-diaminopimelic acid—are not found in proteins. The presence of D-amino acids protects against attack by most peptidases. The peptidoglycan subunit present in most gram-negative bacteria and many gram-positive ones is shown in **figure 3.16.** The backbone of this polymer is composed of alternating *N*-acetylglucosamine and *N*-acetylmuramic acid residues. A peptide

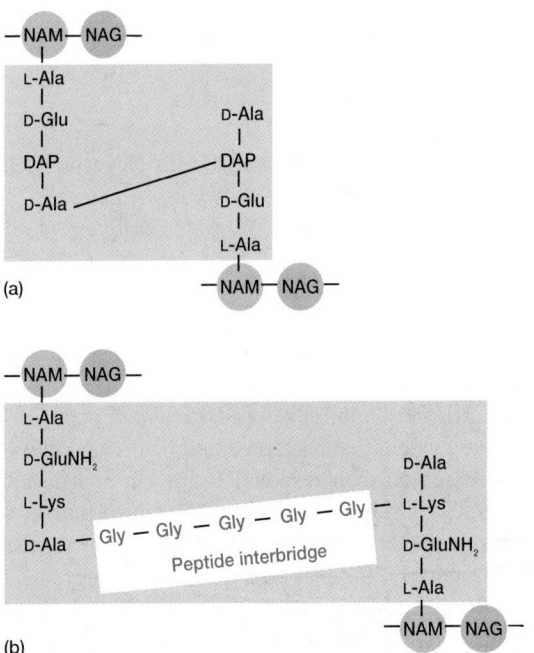

Figure 3.17   Diaminoacids Present in Peptidoglycan. (**a**) L-Lysine. (**b**) *meso*-Diaminopimelic acid.

Figure 3.18   Peptidoglycan Cross-Links.   (**a**) *E. coli* peptidoglycan with direct cross-linking, typical of many gram-negative bacteria. (**b**) *Staphylococcus aureus* peptidoglycan. *S. aureus* is a gram-positive bacterium. NAM is *N*-acetylmuramic acid. NAG is *N*-acetylglucosamine. Gly is glycine. Although the polysaccharide chains are drawn opposite each other for the sake of clarity, two chains lying side-by-side may be linked together (see figure 3.19).

chain of four alternating D- and L-amino acids is connected to the carboxyl group of *N*-acetylmuramic acid. Many bacteria substitute another diaminoacid, usually L-lysine, in the third position for *meso*-diaminopimelic acid (**figure 3.17**).   A review of the chemistry of biological molecules (appendix I); Peptidoglycan structural variations (pp. 507–8)

Chains of linked peptidoglycan subunits are joined by cross-links between the peptides. Often the carboxyl group of the terminal D-alanine is connected directly to the amino group of diaminopimelic acid, but a **peptide interbridge** may be used

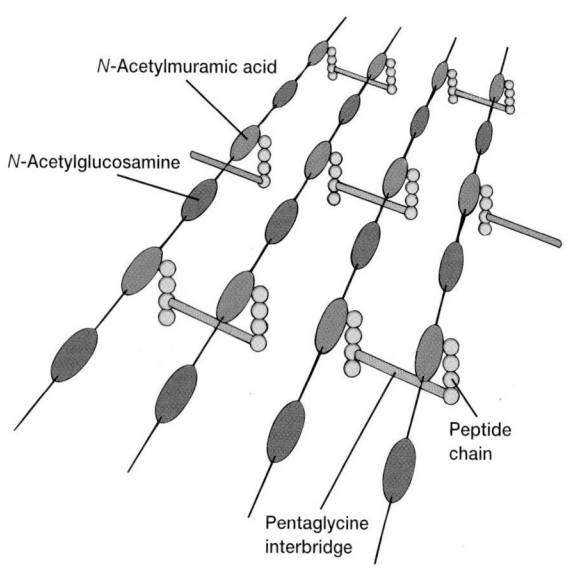

(a)   (b)

Figure 3.19   Peptidoglycan Structure. A peptidoglycan segment showing the polysaccharide chains, tetrapeptide side chains, and peptide interbridges. (**a**) A schematic diagram. (**b**) A space-filling model of gram-negative murein with four repeating peptidoglycan subunits in the plane of the paper. Two chains are arranged vertical to this direction.

Figure 3.20   Isolated Gram-Positive Cell Wall. The peptidoglycan wall from *Bacillus megaterium*, a gram-positive bacterium. The latex spheres have a diameter of 0.25 μm.

instead (**figure 3.18**). Most gram-negative cell wall peptidoglycan lacks the peptide interbridge. This cross-linking results in an enormous peptidoglycan sac that is actually one dense, interconnected network (**figure 3.19**). These sacs have been isolated from gram-positive bacteria and are strong enough to retain their shape and integrity (**figure 3.20**), yet they are elastic and somewhat stretchable, unlike cellulose. They also must be porous, as molecules can penetrate them.

Gram-Positive Cell Walls

Normally the thick, homogeneous cell wall of gram-positive bacteria is composed primarily of peptidoglycan, which often contains a peptide interbridge (figure 3.20 and **figure 3.21**). However, gram-positive cell walls usually also contain large amounts of **teichoic acids,** polymers of glycerol or ribitol joined by phosphate groups (figure 3.21 and **figure 3.22**). Amino acids such as D-alanine or sugars like glucose are attached to the glycerol and ribitol groups. The teichoic acids are connected to either the peptidoglycan itself by a covalent bond with the six hydroxyl of N-acetylmuramic acid or to plasma membrane lipids; in the latter case they are called lipoteichoic acids. Teichoic acids appear to extend to the surface of the peptidoglycan, and, because they are negatively charged, help give the gram-positive cell wall its negative charge. The functions of these molecules are still unclear, but they may be important in maintaining the structure of the wall. Teichoic acids are not present in gram-negative bacteria.

Staphylococci and most other gram-positive bacteria have a layer of proteins on the surface of their cell wall peptidoglycan. These proteins are involved in the interactions of the cell with its environment. Some are noncovalently attached by binding to the peptidoglycan, teichoic acids, or other receptors. For example, the S-layer proteins (p. 62) bind noncovalently to polymers scattered throughout the wall. Enzymes involved in peptidoglycan synthesis and turnover also seem to interact noncovalently with

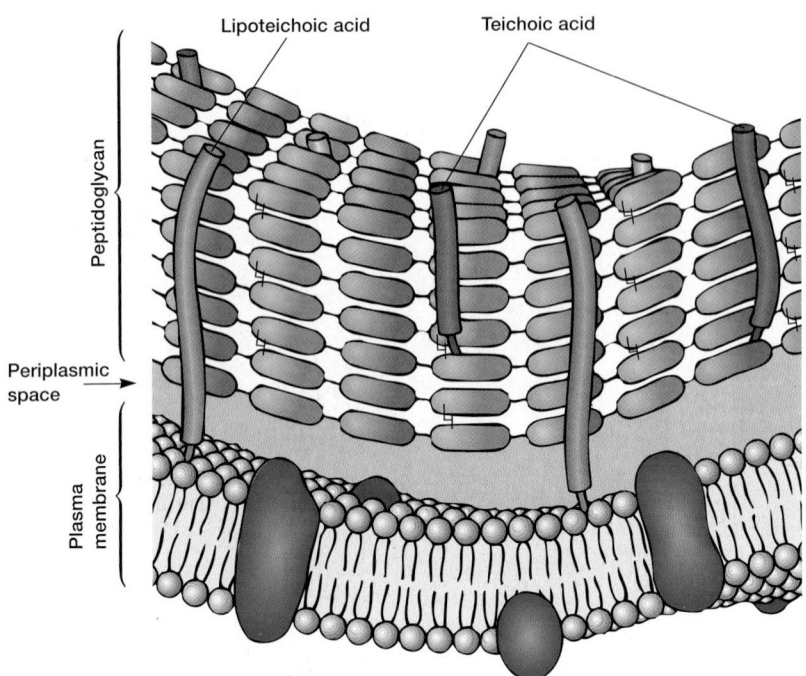

Figure 3.21   **The Gram-Positive Envelope.**

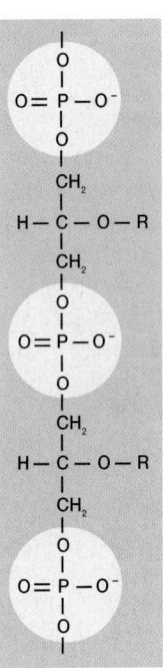

Figure 3.22   Teichoic Acid Structure.   The segment of a teichoic acid made of phosphate, glycerol, and a side chain, R. R may represent D-alanine, glucose, or other molecules.

the cell wall. Other surface proteins are covalently attached to the peptidoglycan. Many covalently attached proteins of gram-positive pathogens have roles such as aiding in adhesion to host tissues, preventing opsonization (*see p. 694*), and blocking phagocytosis. In staphylococci, these surface proteins are covalently joined to the pentaglycine bridge of the cell wall peptidoglycan. An enzyme called a sortase catalyzes the attachment of these surface proteins to the gram-positive peptidoglycan. Sortases are attached to the plasma membrane of the bacterial cell.

Gram-Negative Cell Walls

Even a brief inspection of figure 3.15 shows that gram-negative cell walls are much more complex than gram-positive walls. The thin peptidoglycan layer next to the plasma membrane may constitute not more than 5 to 10% of the wall weight. In *E. coli* it is about 2 nm thick and contains only one or two layers or sheets of peptidoglycan.

The outer membrane lies outside the thin peptidoglycan layer (**figures 3.23** and **3.24**). The most abundant membrane protein is Braun's lipoprotein, a small lipoprotein covalently joined to the underlying peptidoglycan and embedded in the outer membrane by its hydrophobic end. The outer membrane and peptidoglycan are so firmly linked by this lipoprotein that they can be isolated as one unit.

Another structure that may strengthen the gram-negative wall and hold the outer membrane in place is the adhesion site. The outer membrane and plasma membrane appear to be in direct contact at many locations in the gram-negative wall. In *E. coli* 20 to 100 nm areas of contact between the two membranes are seen in plasmolyzed cells. Adhesion sites may be regions of direct contact or possibly true membrane fusions. It has been proposed that substances can move into the cell through these adhesion sites rather than traveling through the periplasm.

Possibly the most unusual constituents of the outer membrane are its **lipopolysaccharides (LPSs)**. These large, complex molecules contain both lipid and carbohydrate, and consist of three parts: (1) lipid A, (2) the core polysaccharide, and (3) the O side chain. The LPS from *Salmonella typhimurium* has been studied most, and its general structure is described here (**figure 3.25**). The **lipid A** region contains two glucosamine sugar derivatives, each with three fatty acids and phosphate or pyrophosphate attached. It is buried in the outer membrane and the remainder of the LPS molecule projects from the surface. The **core polysaccharide** is joined to lipid A. In *Salmonella* it is constructed of 10 sugars, many of them unusual in structure. The **O side chain** or **O antigen** is a polysaccharide chain extending outward from the core. It has several peculiar sugars and varies in composition between bacterial strains. Although O side chains are readily recognized by host antibodies, gram-negative bacteria may thwart host

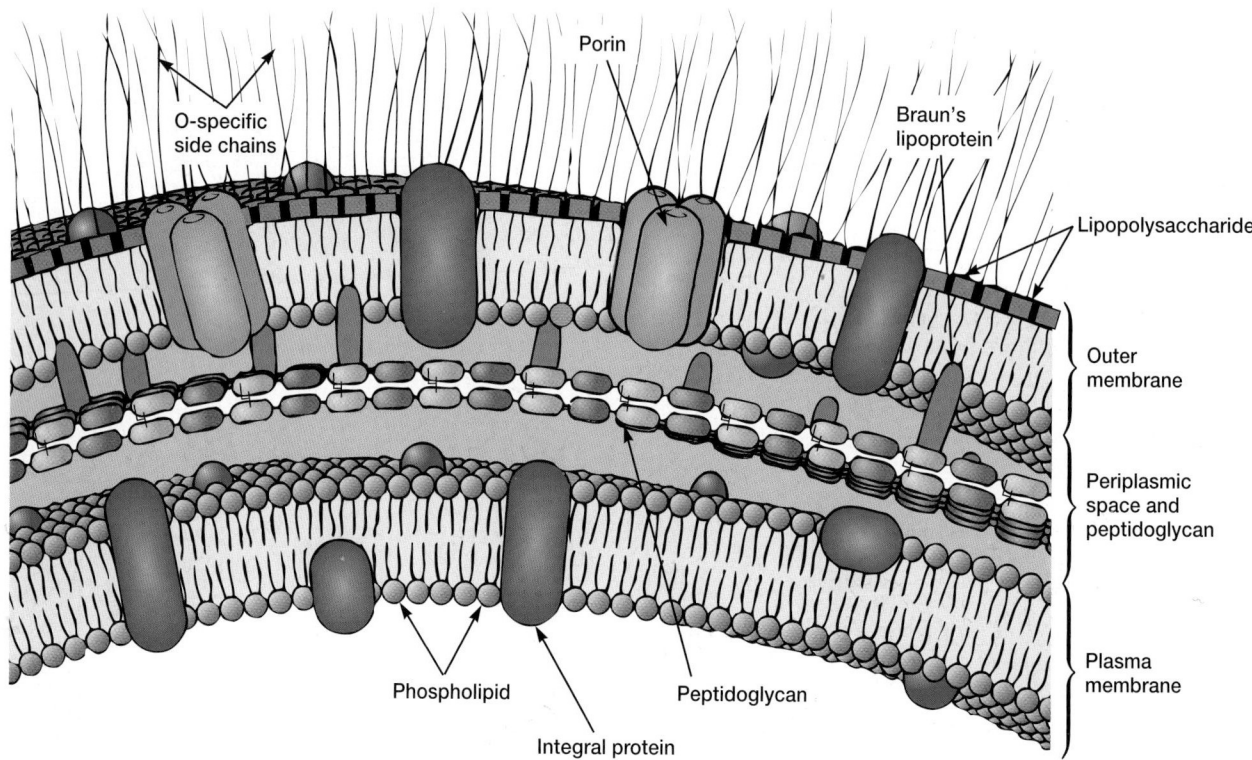

Figure 3.23   The Gram-Negative Envelope.

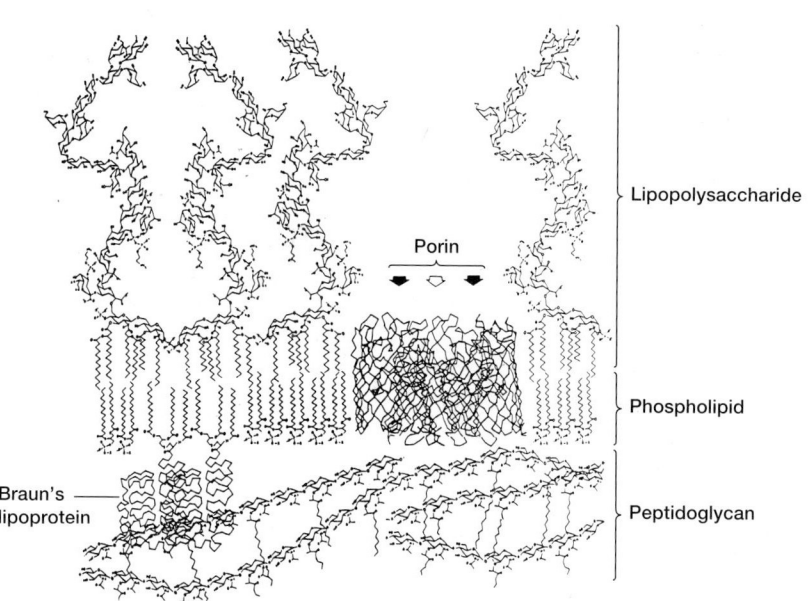

Figure 3.24   A Chemical Model of the *E. coli* Outer Membrane and Associated Structures.   This cross-section is to-scale. The porin OmpF has two channels in the front (solid arrows) and one channel in the back (open arrow) of the trimeric protein complex. LPS molecules can be longer than the ones shown here.

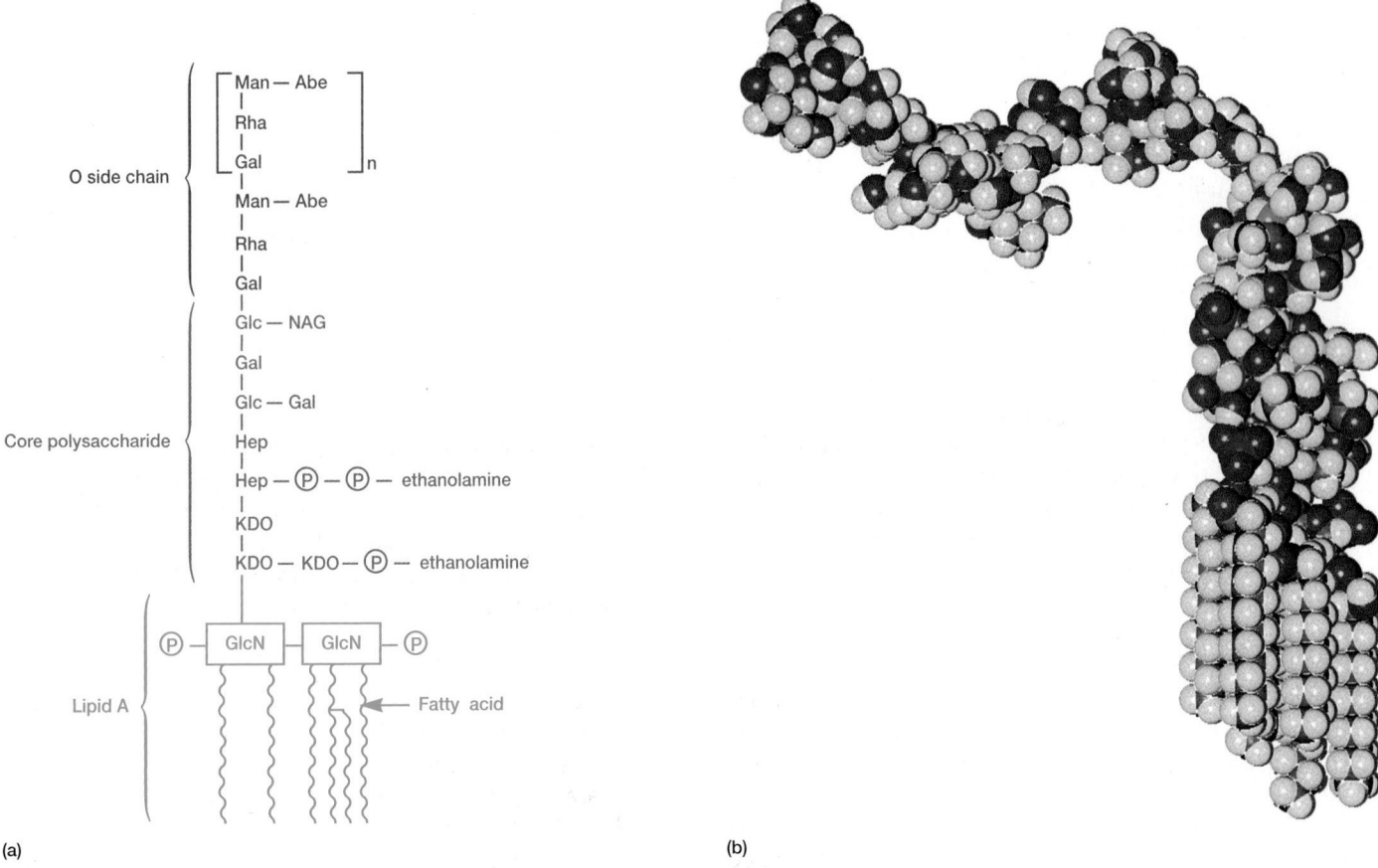

(a)   (b)

Figure 3.25   Lipopolysaccharide Structure.   (a) The lipopolysaccharide from *Salmonella*. This slightly simplified diagram illustrates one form of the LPS. Abbreviations: Abe, abequose; Gal, galactose; Glc, glucose; GlcN, glucosamine; Hep, heptulose; KDO, 2-keto-3-deoxyoctonate; Man, mannose; NAG, *N*-acetylglucosamine; P, phosphate; Rha, L-rhamnose. Lipid A is buried in the outer membrane. (b) Molecular model of an *Escherichia coli* lipopolysaccharide. The lipid A and core polysaccharide are straight; the O side chain is bent at an angle in this model.

defenses by rapidly changing the nature of their O side chains to avoid detection. Antibody interaction with the LPS before reaching the outer membrane proper may also protect the cell wall from direct attack.   Antibodies and antigens (chapters 32 and 33)

The LPS is important for several reasons other than the avoidance of host defenses. Since the core polysaccharide usually contains charged sugars and phosphate (figure 3.25), LPS contributes to the negative charge on the bacterial surface. Lipid A is a major constituent of the outer membrane, and the LPS helps stabilize membrane structure. Furthermore, lipid A often is toxic; as a result the LPS can act as an endotoxin (*see section 34.3*) and cause some of the symptoms that arise in gram-negative bacterial infections.

A most important outer membrane function is to serve as a protective barrier. It prevents or slows the entry of bile salts, antibiotics, and other toxic substances that might kill or injure the bacterium. Even so, the outer membrane is more permeable than the plasma membrane and permits the passage of small molecules like glucose and other monosaccharides. This is due to the presence of special **porin proteins** (figures 3.23 and 3.24). Three

porin molecules cluster together and span the outer membrane to form a narrow channel through which molecules smaller than about 600 to 700 daltons can pass. Larger molecules such as vitamin B_{12} must be transported across the outer membrane by specific carriers. The outer membrane also prevents the loss of constituents like periplasmic enzymes.

The Mechanism of Gram Staining

Although several explanations have been given for the Gram-stain reaction results, it seems likely that the difference between gram-positive and gram-negative bacteria is due to the physical nature of their cell walls. If the cell wall is removed from gram-positive bacteria, they become gram negative. The peptidoglycan itself is not stained; instead it seems to act as a permeability barrier preventing loss of crystal violet. During the procedure the bacteria are first stained with crystal violet and next treated with iodine to promote dye retention. When gram-positive bacteria then are decolorized with ethanol, the alcohol is thought to shrink the pores of the thick peptidoglycan. Thus the dye-iodine com-

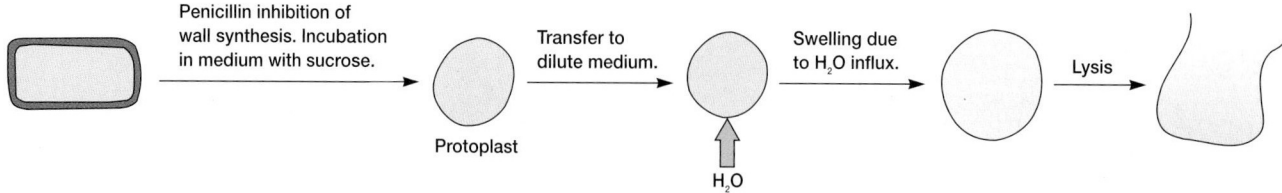

Figure 3.26   Protoplast Formation.   Protoplast formation induced by incubation with penicillin in an isotonic medium. Transfer to dilute medium will result in lysis.

plex is retained during the short decolorization step and the bacteria remain purple. In contrast, gram-negative peptidoglycan is very thin, not as highly cross-linked, and has larger pores. Alcohol treatment also may extract enough lipid from the gram-negative wall to increase its porosity further. For these reasons, alcohol more readily removes the purple crystal violet-iodine complex from gram-negative bacteria.

The Cell Wall and Osmotic Protection

The cell wall usually is required to protect bacteria against destruction by osmotic pressure. Solutes are much more concentrated in bacterial cytoplasm than in most microbial habitats, which are hypotonic. During **osmosis,** water moves across selectively permeable membranes such as the plasma membrane from dilute solutions (higher water concentration) to more concentrated solutions (lower water concentration). Thus water normally enters bacterial cells and the osmotic pressure may reach 20 atmospheres or 300 pounds/square inch. The plasma membrane cannot withstand such pressures and the cell will swell and be physically disrupted and destroyed, a process called **lysis,** without the wall that resists cell swelling and protects it. Solutes are more concentrated in hypertonic habitats than in the cell. Thus water flows outward, and the cytoplasm shrivels up and pulls away from the cell wall. This phenomenon is known as **plasmolysis** and is useful in food preservation because many microorganisms cannot grow in dried foods and jellies as they cannot avoid plasmolysis (*see pp. 118–21, chapter 41*).

The importance of the cell wall in protecting bacteria against osmotic lysis is demonstrated by treatment with lysozyme or penicillin. The enzyme **lysozyme** attacks peptidoglycan by hydrolyzing the bond that connects *N*-acetylmuramic acid with carbon four of *N*-acetylglucosamine. **Penicillin** inhibits peptidoglycan synthesis (*see section 35.6*). If bacteria are incubated with penicillin in an isotonic solution, gram-positive bacteria are converted to protoplasts that continue to grow normally when isotonicity is maintained even though they completely lack a wall. Gram-negative cells retain their outer membrane after penicillin treatment and are classified as **spheroplasts** because some of their cell wall remains. Protoplasts and spheroplasts are osmotically sensitive. If they are transferred to a dilute solution, they will lyse due to uncontrolled water influx (**figure 3.26**).

Although most bacteria require an intact cell wall for survival, some have none at all. For example, the mycoplasmas lack a cell wall and are osmotically sensitive, yet often can grow in dilute media or terrestrial environments because their plasma membranes are stronger than normal. The precise reason for this is not known, although the presence of sterols in the membranes of many species may provide added strength. Without a rigid cell wall, mycoplasmas tend to be pleomorphic or variable in shape.

---

1. Describe in some detail the composition and structure of peptidoglycan, gram-positive cell walls, and gram-negative cell walls. Include labeled diagrams in the answer.
2. Define or describe the following: outer membrane, periplasmic space, periplasm, envelope, teichoic acid, adhesion site, lipopolysaccharide, and porin protein.
3. Explain the role of the cell wall in protecting against lysis and how this role may be experimentally demonstrated. What are protoplasts and spheroplasts?

3.6   PROTEIN SECRETION IN PROCARYOTES

Section 3.5 has shown that the procaryotic cell wall is strong, complex, and often quite thick. Many important structures are located outside the wall as will be discussed in section 3.7. Furthermore, procaryotes often release a variety of proteins into their surroundings. Thus bacteria and archaea must be able to transport or translocate proteins from the cytoplasmic matrix through both the plasma membrane and the cell wall. Depending on the classification system, there are at least five or six different protein secretion pathways used by procaryotes. This section will briefly describe four of the most important routes of protein secretion: the Sec-dependent pathway, the type II pathway, the type I or ABC protein secretion pathway, and the type III pathway.

The Sec-Dependent Pathway

The **Sec-dependent pathway,** sometimes called the **general secretion pathway (GSP),** is the most commonly used and either translocates proteins across the plasma membrane or integrates them into the membrane itself. The presecretory protein or preprotein has a **signal peptide** at its amino-terminal or N-terminal

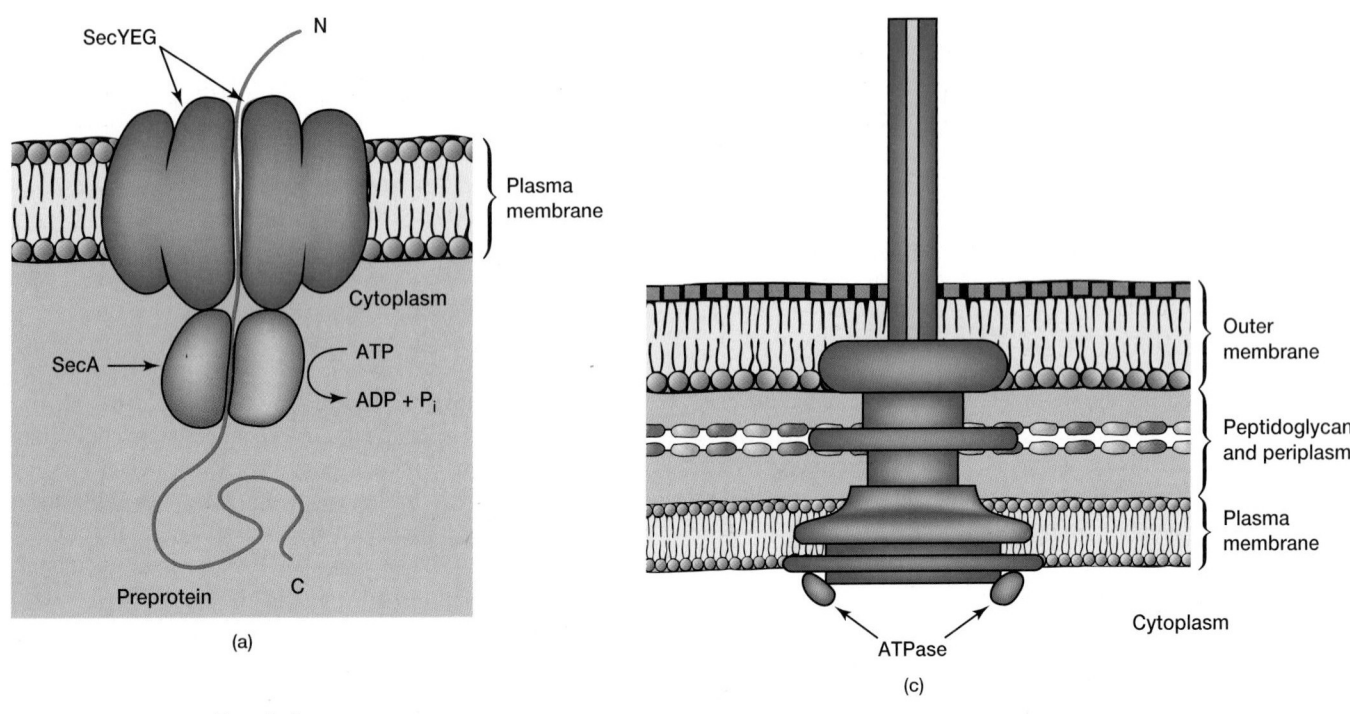

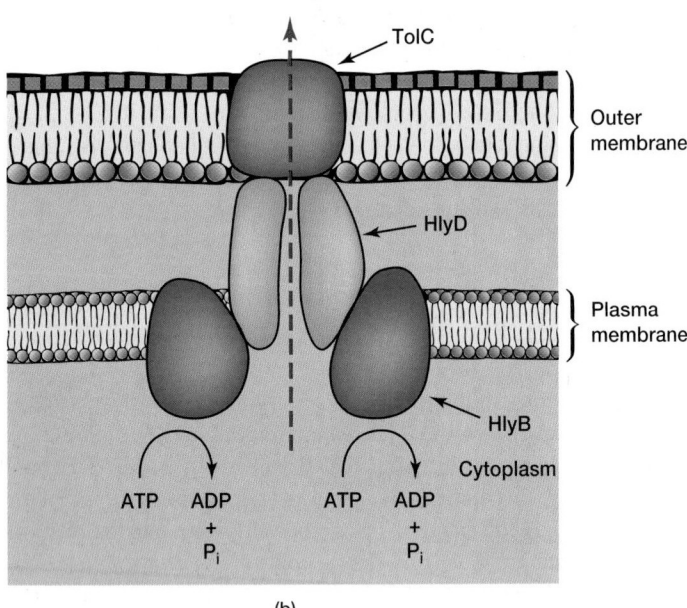

Figure 3.27   Protein Secretion Systems.   These schematic diagrams depict three major types of protein secretion systems in a simplified form with only some of the components shown: (**a**) the Sec-dependent system of *E. coli,* (**b**) the ABC transporter (type I system) of *E. coli* that secretes α-hemolysin, and (**c**) the type III secretion system of *Salmonella typhimurium.*

end (*see p. 260*). The signal peptide delays protein folding after its synthesis on the ribosome and is specifically recognized by the Sec machinery. As these preproteins are synthesized by ribosomes, chaperones (*see pp. 267–69*) bind to them. The chaperones keep the preproteins unfolded and help them reach the transport machinery or **translocon.** In *E. coli,* the chaperones used most often for transport are SecB and the signal recognition particle (SRP). SecB is present in gram-negative bacteria; SRP is found in all procaryotes. SecB binds to the SecA component of the translocon and the preprotein is transferred to SecA. In other cases, the SRP-preprotein complex moves to an SRP receptor.

The preprotein is then released with the aid of GTP hydrolysis and the preprotein associates with the translocon, probably with SecA as before. There is evidence that translocation of proteins can begin before the completion of their synthesis by ribosomes.

After the preprotein has bound to the translocon, it is transferred through the membrane. The bacterial translocon is composed of a membrane protein complex called SecYEG (**figure 3.27a**), the dissociable SecA, and other accessory proteins (SecD, SecF). SecYEG is thought to form a channel in the membrane through which the preprotein passes. SecA recognizes and binds to SecYEG, SecB, and the preprotein. It acts as a motor to translocate the preprotein through the membrane using ATP hydrolysis. Proton motive force (PMF, *see p. 182*) often stimulates the rate of translocation. Archaea lack SecA; the energetics and mechanism of their translocation system is not known in any detail.

When the preprotein emerges from the plasma membrane, the signal peptide is removed by the enzyme signal peptidase. The protein then folds into the proper shape, and disulfide bonds are formed when necessary. Some proteins are destined to be plasma membrane integral proteins. These lipoproteins probably have a different signal peptidase but are transported by the same translocon. Instead of moving outside the cell, they embed in the membrane or are anchored to it.

The Type II Protein Secretion Pathway

Some gram-negative bacteria have pathways to transport proteins across the outer membrane after they have reached the periplasm, usually through the activity of the Sec-dependent pathway. The **type II protein secretion pathway** is the most important of these routes and is present in the pathogens *Erwinia carotovora, Pseudomonas aeruginosa,* and *Vibrio cholerae.* It is responsible for the secretion of proteins such as cellulases, pectinases, proteases, lipases, cholera toxin, and pili proteins. Type II systems are quite complex and can contain as many as 12 to 14 proteins, most of which appear to be integral membrane proteins.

The Type I Protein Secretion Pathway (ABC Protein Secretion Pathway)

Many Sec-independent secretion pathways are members of the ATP-binding cassette (ABC) family of transport systems. The **type I protein secretion pathway** or **ABC protein secretion pathway** is involved in the secretion of proteins such as toxins (α-hemolysin), proteases, lipases, and specific peptides. Secreted proteins usually contain C-terminal secretion signals. These systems use ATP hydrolysis to drive translocation. They have two hydrophobic, membrane-embedded domains and two hydrophilic ATP-binding domains (figure 3.27*b*). The protein moves through a channel formed by the hydrophobic domains. Both ATP and proton motive force seem to be required. In gram-negative bacteria such as *E. coli,* a protein links the plasma membrane and outer membrane during transport. Another protein forms a channel in the outer membrane; for example, in *E. coli* the TolC trimer appears to have this role for at least some type I systems. Thus protein translocation occurs in one step across both membranes.

Type III Protein Secretion Pathway

Several gram-negative pathogens have the **type III protein secretion pathway** and use it to secrete virulence factors and inject them into plant and animal host cells. Important examples are *Salmonella, Yersinia, Shigella, E. coli, Bordetella, Pseudomonas aeruginosa,* and *Erwinia.* Contact between the bacteria and their host cells stimulates the process, and an activating signal such as a low calcium ion level is required for efficient secretion. About

20 different proteins are components of type III systems. The *Salmonella typhimurium* type III secretion machinery is shaped like a needle. The slender needlelike portion extends out from the cell surface; a cylindrical base is connected to both the outer membrane and the plasma membrane and looks somewhat like the flagellar basal body (figure 3.27*c*). It is thought that proteins may move through a translocation channel.

Type III systems secrete four different kinds of proteins: (1) those required for secretion (e.g., the needle proteins), (2) secretion regulators, (3) proteins that aid in the insertion of secreted proteins into target cells, and (4) effector proteins that alter host cell functions. Effector proteins are sometimes enzymes and can do many things. For example, they may act as cytotoxins, phagocytosis inhibitors, stimulators of cytoskeleton reorganization, or apoptosis promoters. The proteins to be secreted are normally associated with small chaperones, whose function is unclear. The chaperones may deliver proteins to the secretory system, maintain the proteins in the proper state for secretion, or prevent association with other proteins before they can be secreted. The participation of type III systems in bacterial virulence is further discussed in chapter 34 (*see p. 768*).

3.7 COMPONENTS EXTERNAL TO THE CELL WALL

Bacteria have a variety of structures outside the cell wall that can function in protection, attachment to objects, and cell movement. Several of these are discussed.

Capsules, Slime Layers, and S-Layers

Some bacteria have a layer of material lying outside the cell wall. When the layer is well organized and not easily washed off, it is called a **capsule.** A **slime layer** is a zone of diffuse, unorganized material that is removed easily. A **glycocalyx (figure 3.28)** is a network of polysaccharides extending from the surface of bacteria and other cells (in this sense it could encompass both capsules and slime layers). Capsules and slime layers usually are composed of polysaccharides, but they may be constructed of other materials. For example, *Bacillus anthracis* has a capsule of poly-D-glutamic acid. Capsules are clearly visible in the light microscope when negative stains or special capsule stains are employed (figure 3.28*a*); they also can be studied with the electron microscope (figure 3.28*b*).

Although capsules are not required for bacterial growth and reproduction in laboratory cultures, they do confer several advantages when bacteria grow in their normal habitats. They help bacteria resist phagocytosis by host phagocytes. *Streptococcus pneumoniae* provides a classic example. When it lacks a capsule, it is destroyed easily and does not cause disease, whereas the capsulated variant quickly kills mice. Capsules contain a great deal of water and can protect bacteria against desiccation. They exclude bacterial viruses and most hydrophobic toxic materials such as detergents. The glycocalyx also aids bacterial attachment to surfaces of solid objects in aquatic environments or to tissue

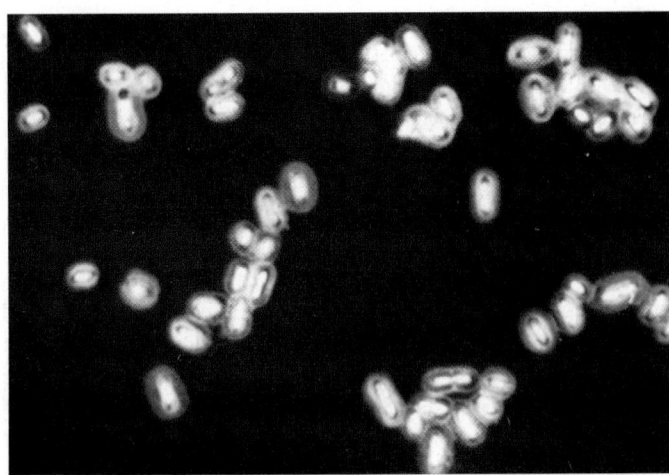

(a) *K. pneumoniae*

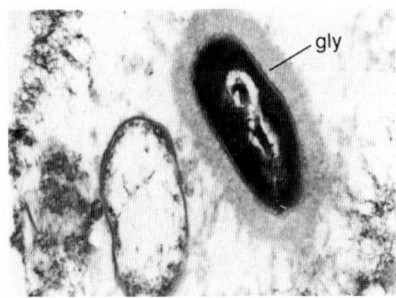

(b) *Bacteroides*

Figure 3.28 Bacterial Capsules. (**a**) *Klebsiella pneumoniae* with its capsule stained for observation in the light microscope (×1,500). (**b**) *Bacteroides* glycocalyx (gly), TEM (×71,250).

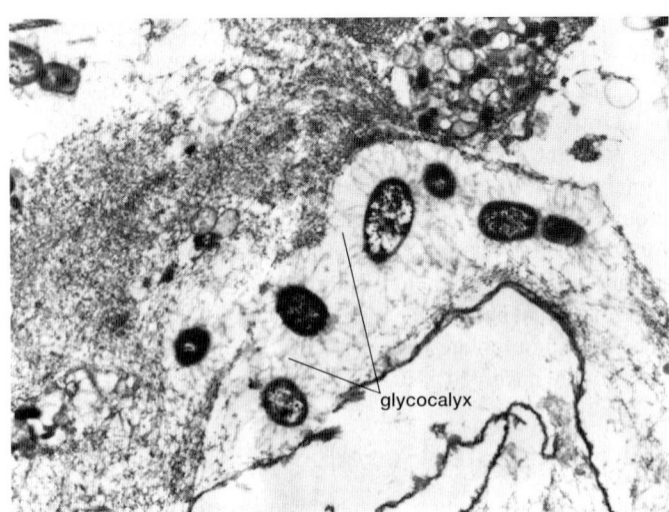

Figure 3.29 Bacterial Glycocalyx. Bacteria connected to each other and to the intestinal wall, by their glycocalyxes, the extensive networks of fibers extending from the cells (×17,500).

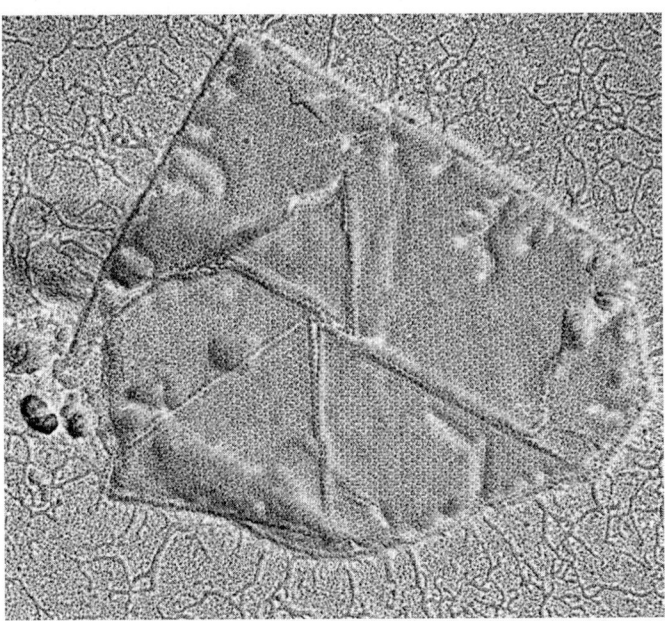

Figure 3.30 The S-Layer. An electron micrograph of the S-layer of *Deinococcus radiodurans* after shadowing.

surfaces in plant and animal hosts (**figure 3.29**). Gliding bacteria often produce slime, which presumably aids in their motility (*see Microbial Diversity & Ecology 21.1*). The relationship of surface polysaccharides to phagocytosis and host colonization (chapters 31 and 34)

Many gram-positive and gram-negative bacteria have a regularly structured layer called an **S-layer** on their surface. S-layers also are very common among Archaea, where they may be the only wall structure outside the plasma membrane. The S-layer has a pattern something like floor tiles and is composed of protein or glycoprotein (**figure 3.30**). In gram-negative bacteria the S-layer adheres directly to the outer membrane; it is associated with the peptidoglycan surface in gram-positive bacteria. It may protect the cell against ion and pH fluctuations, osmotic stress, enzymes, or the predacious bacterium *Bdellovibrio* (*see section 22.4*). The S-layer also helps maintain the shape and envelope rigidity of at least some bacterial cells. It can promote cell adhesion to surfaces. Finally, the layer seems to protect some pathogens against complement attack and phagocytosis, thus contributing to their virulence.

Pili and Fimbriae

Many gram-negative bacteria have short, fine, hairlike appendages that are thinner than flagella and not involved in motil-ity. These are usually called **fimbriae** (s., **fimbria**). Although many people use the terms fimbriae and pili interchangeably, we shall distinguish between fimbriae and sex pili. A cell may be covered with up to 1,000 fimbriae, but they are only visible in an electron microscope due to their small size (**figure 3.31**). They seem to be slender tubes composed of helically arranged protein subunits and are about 3 to 10 nm in diameter and up to several micrometers long. At least some types of fimbriae attach bacteria to solid surfaces such as rocks in streams and host tissues.

Fimbriae are responsible for more than bacterial attachment. Type IV fimbriae are present at one or both poles of a cell. They

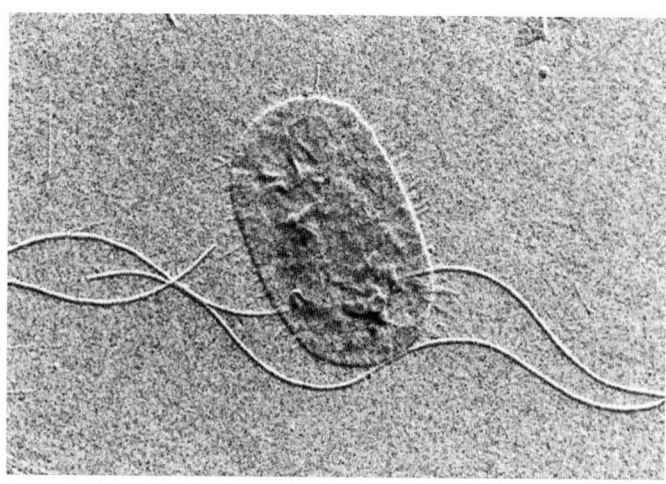

Figure 3.31 Flagella and Fimbriae. The long flagella and the numerous shorter fimbriae are very evident in this electron micrograph of *Proteus vulgaris* (×39,000).

can aid in attachment to objects, and also are required for the twitching motility that occurs in bacteria such as *P. aeruginosa, N. gonorrhoeae,* and some strains of *E. coli.* Movement is by short, intermittent jerky motions of up to several micrometers in length and normally is seen on very moist surfaces. There is evidence that the fimbriae actively retract to move these bacteria. Type IV fimbriae also seem to be involved in some sorts of gliding motility by myxobacteria (*see Microbial Diversity & Ecology 21.1*).

Sex pili (s., **pilus**) are similar appendages, about 1 to 10 per cell, that differ from fimbriae in the following ways. Pili often are larger than fimbriae (around 9 to 10 nm in diameter). They are genetically determined by sex factors or conjugative plasmids and are required for bacterial mating (*see chapter 13*). Some bacterial viruses attach specifically to receptors on sex pili at the start of their reproductive cycle.

Flagella and Motility

Most motile bacteria move by use of **flagella** (s., **flagellum**), threadlike locomotor appendages extending outward from the plasma membrane and cell wall. They are slender, rigid structures, about 20 nm across and up to 15 or 20 μm long. Flagella are so thin they cannot be observed directly with a bright-field microscope, but must be stained with special techniques designed to increase their thickness (*see chapter 2*). The detailed structure of a flagellum can only be seen in the electron microscope (figure 3.31).

Bacterial species often differ distinctively in their patterns of flagella distribution. **Monotrichous** bacteria (*trichous* means hair) have one flagellum; if it is located at an end, it is said to be a **polar flagellum** (**figure 3.32***a*). **Amphitrichous** bacteria (*amphi* means "on both sides") have a single flagellum at each pole. In contrast, **lophotrichous** bacteria (*lopho* means tuft) have a cluster of flagella at one or both ends (figure 3.32*b*). Flagella are spread fairly evenly over the whole surface of **peritrichous** (*peri* means "around") bacteria (figure 3.32*c*). Flagellation patterns are very useful in identifying bacteria.

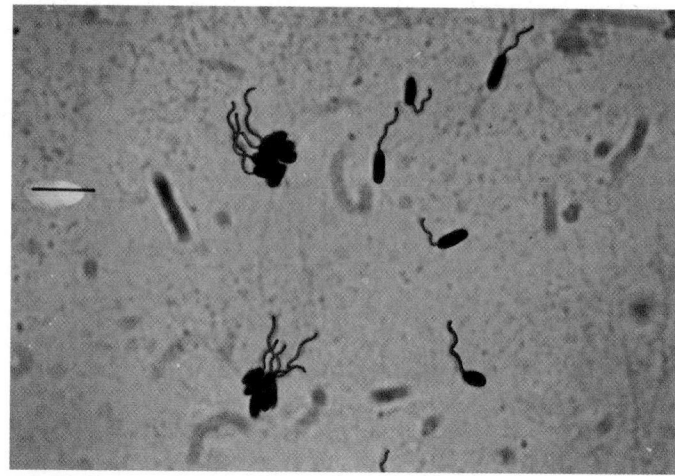

(a) *Pseudomonas*

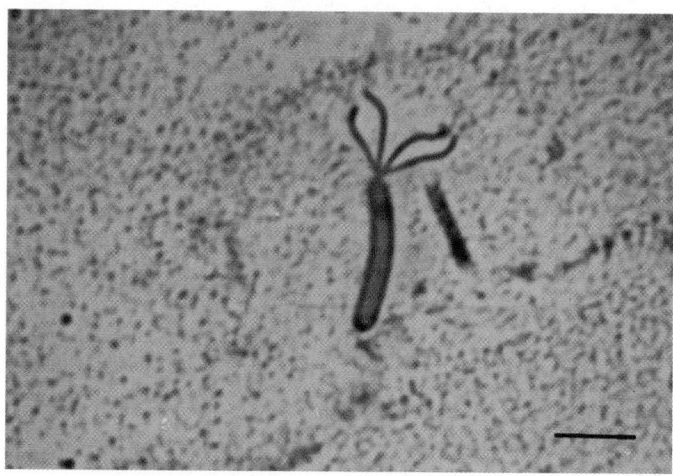

(b) *Spirillum*

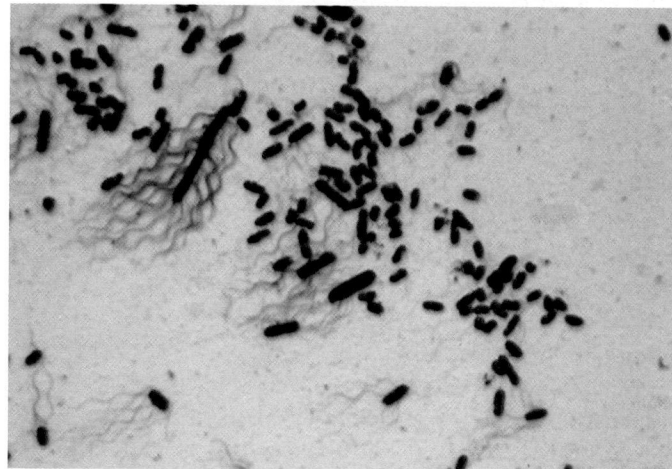

(c) *P. vulgaris*

Figure 3.32 Flagellar Distribution. Examples of various patterns of flagellation as seen in the light microscope. (**a**) Monotrichous polar (*Pseudomonas*). (**b**) Lophotrichous (*Spirillum*). (**c**) Peritrichous (*Proteus vulgaris,* ×600). Bars = 5 μm.

Figure 3.33   The Ultrastructure of Gram-Negative Flagella. (a) Negatively stained flagella from *Escherichia coli* (×66,000). Arrows indicate the location of curved hooks and basal bodies. (b) An enlarged view of the basal body of an *E. coli* flagellum (×485,000). All four rings (L, P, S, and M) can be clearly seen. The uppermost arrow is at the junction of the hook and filament. Bar = 30 nm.

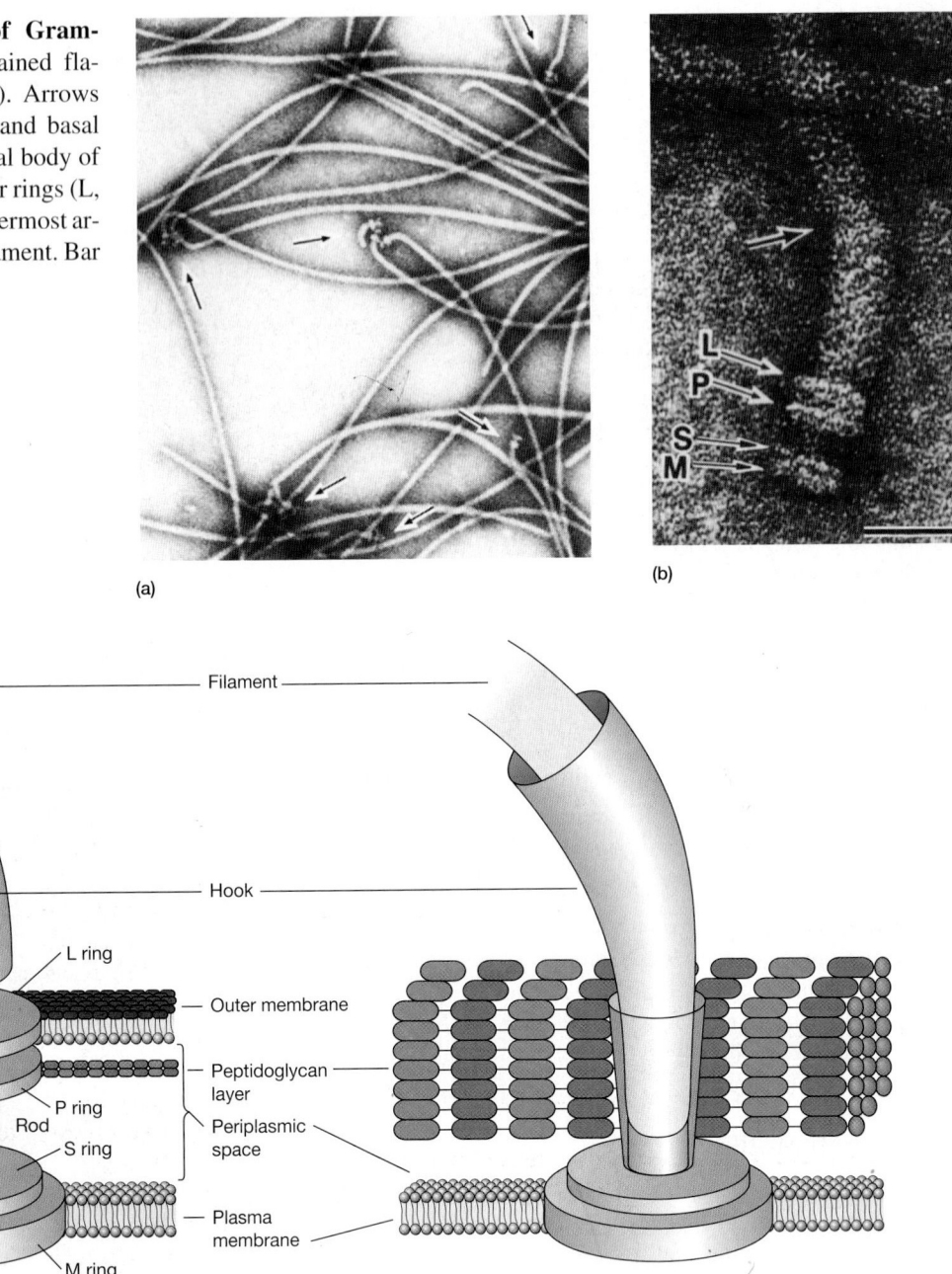

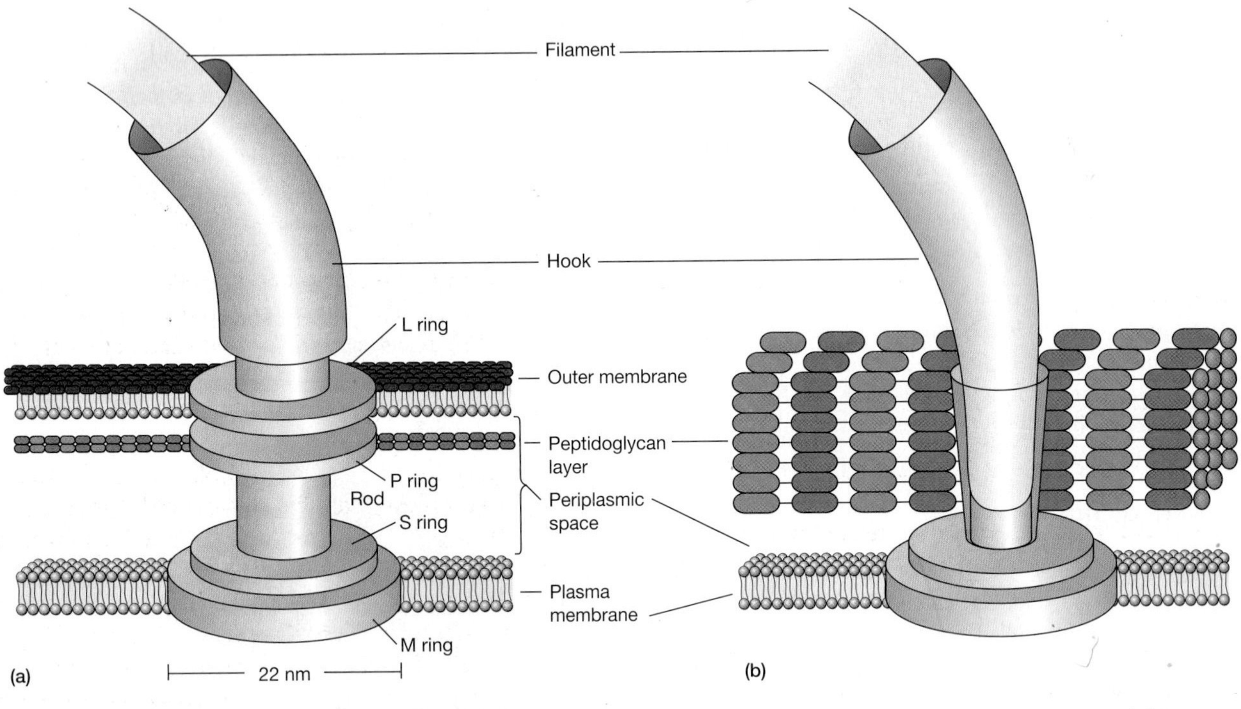

(a)

(b)

Figure 3.34   The Ultrastructure of Bacterial Flagella.   Flagellar basal bodies and hooks in (a) gram-negative and (b) gram-positive bacteria.

Flagellar Ultrastructure

Transmission electron microscope studies have shown that the bacterial flagellum is composed of three parts. (1) The longest and most obvious portion is the **filament,** which extends from the cell surface to the tip. (2) A **basal body** is embedded in the cell; and (3) a short, curved segment, the **hook,** links the filament to its basal body and acts as a flexible coupling. The filament is a hollow, rigid cylinder constructed of a single protein called **flagellin,** which ranges in molecular weight from 30,000 to 60,000. The filament ends with a capping protein. Some bacteria have sheaths surrounding their flagella. For example *Bdellovibrio* has a membranous structure surrounding the filament. *Vibrio cholerae* has a lipopolysaccharide sheath.

The hook and basal body are quite different from the filament (**figure 3.33**). Slightly wider than the filament, the hook is made of different protein subunits. The basal body is the most complex part of a flagellum (figure 3.33 and **figure 3.34**). In *E. coli* and most gram-negative bacteria, the body has four rings connected to a central rod. The outer L and P rings associate with the lipopolysaccharide and peptidoglycan layers, respectively. The

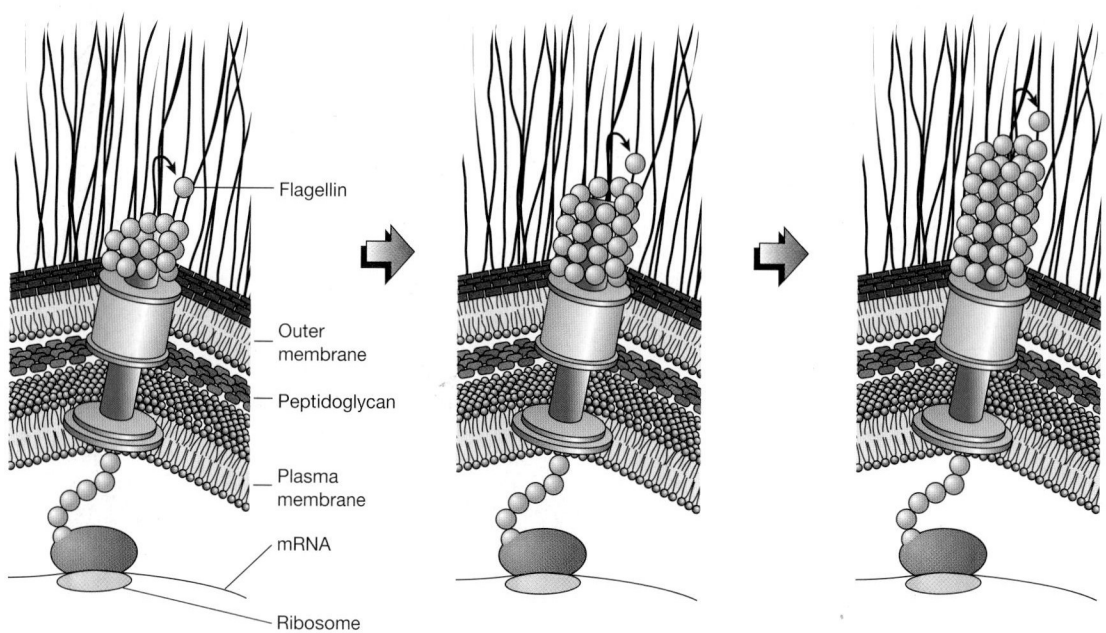

Figure 3.35   Growth of Flagellar Filaments.   Flagellin subunits travel through the flagellar core and attach to the growing tip.

inner M ring contacts the plasma membrane. Gram-positive bacteria have only two basal body rings, an inner ring connected to the plasma membrane and an outer one probably attached to the peptidoglycan.

Flagellar Synthesis

The synthesis of flagella is a complex process involving at least 20 to 30 genes. Besides the gene for flagellin, 10 or more genes code for hook and basal body proteins; other genes are concerned with the control of flagellar construction or function. It is not known how the cell regulates or determines the exact location of flagella.

Bacteria can be deflagellated, and the regeneration of the flagellar filament can then be studied. Export of flagellar components is carried out by an apparatus in the basal body that appears to be closely related to the type III protein secretion system used to secrete virulence factors (p. 61). It is believed that flagellin subunits are transported through the filament's hollow internal core. When they reach the tip, the subunits spontaneously aggregate under the direction of a special filament cap so that the filament grows at its tip rather than at the base (**figure 3.35**). Filament synthesis is an excellent example of **self-assembly.** Many structures form spontaneously through the association of their component parts without the aid of any special enzymes or other factors. The information required for filament construction is present in the structure of the flagellin subunit itself.

The Mechanism of Flagellar Movement

Procaryotic flagella operate differently from eucaryotic flagella. The filament is in the shape of a rigid helix, and the bacterium moves when this helix rotates. Considerable evidence shows that flagella act just like propellers on a boat. Bacterial mutants with straight flagella or abnormally long hook regions (polyhook mutants) cannot swim. When bacteria are tethered to a glass slide using antibodies to filament or hook proteins, the cell body rotates rapidly about the stationary flagellum. If polystyrene-latex beads are attached to flagella, the beads spin about the flagellar axis due to flagellar rotation. The flagellar motor can rotate very rapidly. The *E. coli* motor rotates 270 revolutions per second; *Vibrio alginolyticus* averages 1,100 rps.   Eucaryotic flagella and motility (pp. 88–90)

The direction of flagellar rotation determines the nature of bacterial movement. Monotrichous, polar flagella rotate counterclockwise (when viewed from outside the cell) during normal forward movement, whereas the cell itself rotates slowly clockwise. The rotating helical flagellar filament thrusts the cell forward in a run with the flagellum trailing behind (**figure 3.36**). Monotrichous bacteria stop and tumble randomly by reversing the direction of flagellar rotation. Peritrichously flagellated bacteria operate in a somewhat similar way. To move forward, the flagella rotate counterclockwise. As they do so, they bend at their hooks to form a rotating bundle that propels them forward. Clockwise rotation of the flagella disrupts the bundle and the cell tumbles.

Because bacteria swim through rotation of their rigid flagella, there must be some sort of motor at the base. A rod or shaft extends from the hook and ends in the M ring, which can rotate freely in the plasma membrane (**figure 3.37**). It is believed that the S ring is attached to the cell wall in gram-positive cells and does not rotate. The P and L rings of gram-negative bacteria would act as bearings for the rotating rod. There is some evidence that the basal body is a passive structure and rotates within a membrane-embedded protein

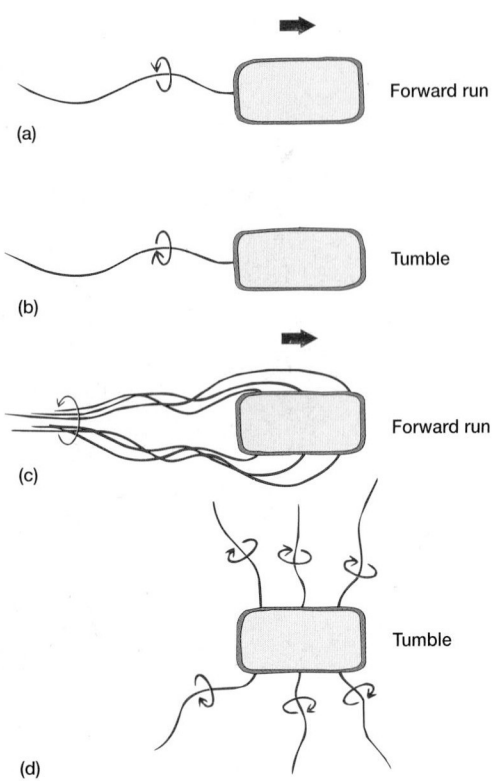

Figure 3.36 Flagellar Motility. The relationship of flagellar rotation to bacterial movement. Parts (**a**) and (**b**) describe the motion of monotrichous, polar bacteria. Parts (**c**) and (**d**) illustrate the movements of peritrichous organisms.

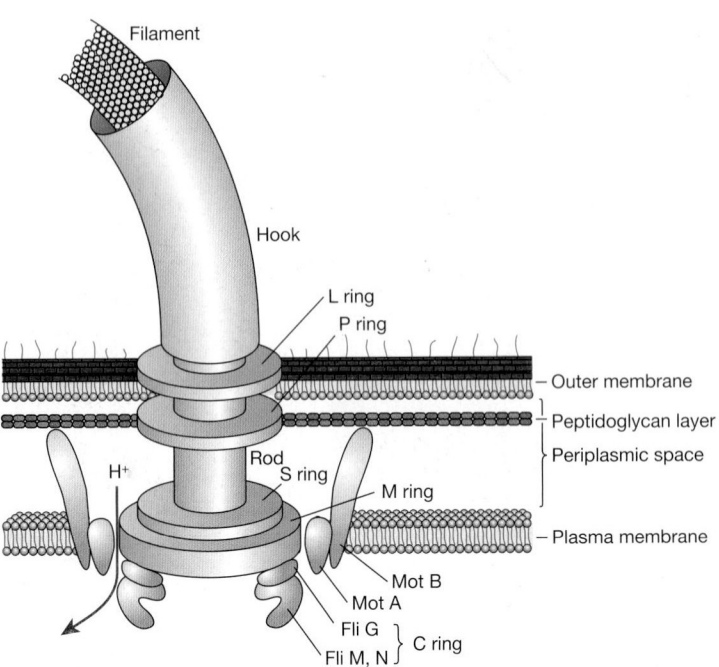

Figure 3.37 Mechanism of Flagellar Movement. This diagram of a gram-negative flagellum shows some of the more important components and the flow of protons that drives rotation. Five of the many flagellar proteins are labeled (Mot A, Mot B, Fli G, Fli M, Fli N).

complex much like the rotor of an electrical motor turns in the center of a ring of electromagnets (the stator).

The exact mechanism that drives basal body rotation still is not clear. Figure 3.37 provides a more detailed depiction of the basal body in gram-negative bacteria. The rotor portion of the motor seems to be made primarily of a rod, the M ring, and a C ring joined to it on the cytoplasmic side of the basal body. These two rings are made of several proteins; Fli G is particularly important in generating flagellar rotation. The two most important proteins in the stator part of the motor are Mot A and Mot B. These form a proton channel through the plasma membrane, and Mot B also anchors the Mot complex to cell wall peptidoglycan. There is some evidence that Mot A and Fli G directly interact during flagellar rotation. This rotation is driven by proton or sodium gradients in procaryotes, not directly by ATP as is the case with eucaryotic flagella.

The flagellum is a very effective swimming device. From the bacterium's point of view, swimming is quite a task because the surrounding water seems as thick and viscous as molasses. The cell must bore through the water with its helical or corkscrew-shaped flagella, and if flagellar activity ceases, it stops almost instantly. Despite such environmental resistance to movement, bacteria can swim from 20 to almost 90 μm/second. This is equivalent to traveling from 2 to over 100 cell lengths per second. In contrast, an exceptionally fast 6 ft human might be able to run around 5 body lengths per second.

Bacteria can move by mechanisms other than flagellar rotation. Spirochetes are helical bacteria that travel through viscous substances such as mucus or mud by flexing and spinning movements caused by a special **axial filament** composed of periplasmic flagella (*see section 21.6*). A very different type of motility, **gliding motility,** is employed by many bacteria: cyanobacteria (*see section 21.3*), myxobacteria (*see section 22.4*) and cytophagas (*see section 21.7*), and some mycoplasmas (*see section 23.1*). Although there are no visible external structures associated with gliding motility, these bacteria can coast along solid surfaces at rates up to 3 μm/second. The mechanism of gliding motility (Microbial Diversity & Ecology 21.1)

---

1. Give the major characteristics and functions of the four protein secretion pathways described in section 3.6. How do the Sec-dependent pathway and the type I protein secretion pathway operate?

2. Briefly describe capsules, slime layers, glycocalyxes, and S-layers. What are their functions?

3. Distinguish between fimbriae and sex pili, and give the function of each.

4. Be able to discuss the following: flagella distribution patterns, flagella structure and synthesis, and the way in which flagella operate to move a bacterium.

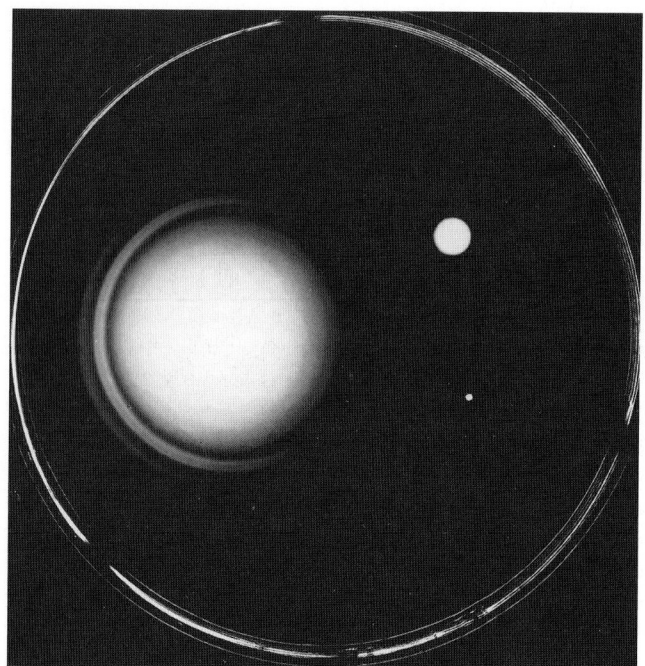

Figure 3.38 Positive Bacterial Chemotaxis. Chemotaxis can be demonstrated on an agar plate that contains various nutrients. Positive chemotaxis by *Escherichia coli* on the left. The outer ring is composed of bacteria consuming serine. The second ring was formed by *E. coli* consuming aspartate, a less powerful attractant. The upper right colony is composed of motile, but nonchemotactic mutants. The bottom right colony is formed by nonmotile bacteria.

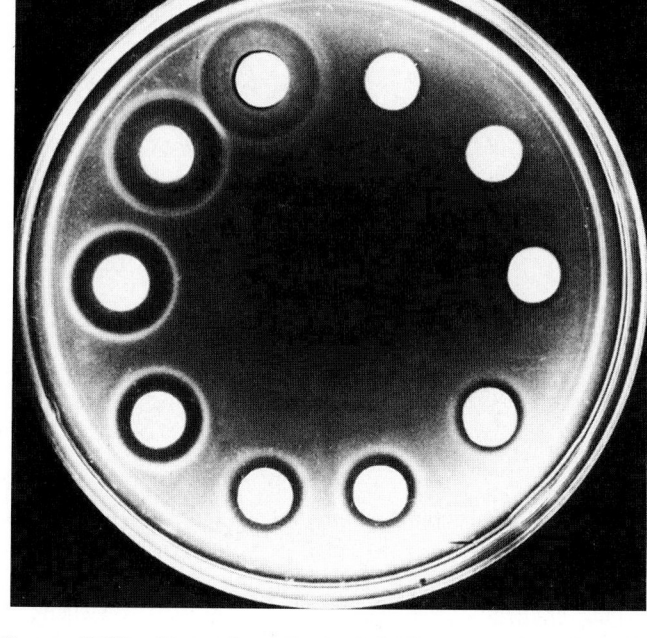

Figure 3.39 Negative Bacterial Chemotaxis. Negative chemotaxis by *E. coli* in response to the repellent acetate. The bright disks are plugs of concentrated agar containing acetate that have been placed in dilute agar inoculated with *E. coli*. Acetate concentration increases from zero at the top right to 3 M at top left. Note the increasing size of bacteria-free zones with increasing acetate. The bacteria have migrated for 30 minutes.

 3.8   CHEMOTAXIS

Bacteria do not always swim aimlessly but are attracted by such nutrients as sugars and amino acids, and are repelled by many harmful substances and bacterial waste products. (Bacteria also can respond to other environmental cues such as temperature, light, and gravity; Microbial Diversity & Ecology 3.2.) Movement toward chemical attractants and away from repellents is known as **chemotaxis.** Such behavior is of obvious advantage to bacteria.

Chemotaxis may be demonstrated by observing bacteria in the chemical gradient produced when a thin capillary tube is filled with an attractant and lowered into a bacterial suspension. As the attractant diffuses from the end of the capillary, bacteria collect and swim up the tube. The number of bacteria within the capillary after a short length of time reflects the strength of attraction and rate of chemotaxis. Positive and negative chemotaxis also can be studied with petri dish cultures (**figure 3.38**). If bacteria are placed in the center of a dish of agar containing an attractant, the bacteria will exhaust the local supply and then swim outward following the attractant gradient they have created. The result is an expanding ring of bacteria. When a disk of repellent is placed in a petri dish of semisolid agar and bacteria, the bacteria will swim away from the repellent, creating a clear zone around the disk (**figure 3.39**).

Bacteria can respond to very low levels of attractants (about 10^{-8} M for some sugars), the magnitude of their response increasing with attractant concentration. Usually they sense repellents only at higher concentrations. If an attractant and a repellent are present together, the bacterium will compare both signals and respond to the chemical with the most effective concentration.

Attractants and repellents are detected by **chemoreceptors,** special proteins that bind chemicals and transmit signals to the other components of the chemosensing system. About 20 attractant chemoreceptors and 10 chemoreceptors for repellents have been discovered thus far. These chemoreceptor proteins may be located in the periplasmic space or the plasma membrane. Some receptors participate in the initial stages of sugar transport into the cell.

The chemotactic behavior of bacteria has been studied using the tracking microscope, a microscope with a moving stage that automatically keeps an individual bacterium in view. In the absence of a chemical gradient, *E. coli* and other bacteria move randomly. A bacterium travels in a straight or slightly curved line, a **run,** for a few seconds; then it will stop and **tumble** or **twiddle** about. The tumble is followed by a run in a different direction

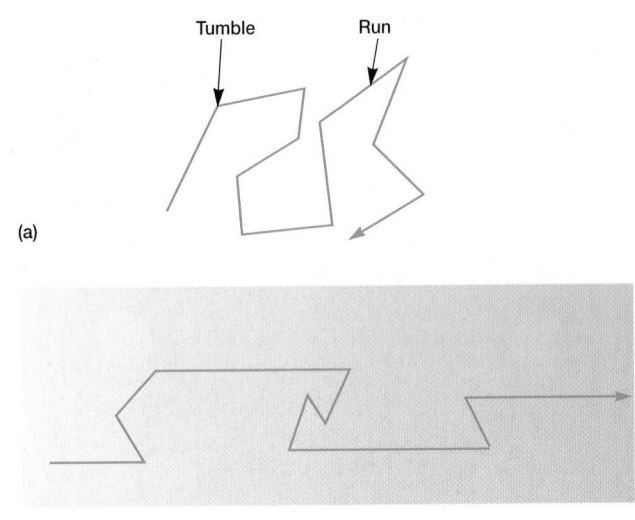

(a)

(b)

Figure 3.40 Directed Movement in Bacteria. (a) Random movement of a bacterium in the absence of a concentration gradient. Tumbling frequency is fairly constant. (b) Movement in an attractant gradient. Tumbling frequency is reduced when the bacterium is moving up the gradient. Therefore runs in the direction of increasing attractant are longer.

(figure 3.40). When the bacterium is exposed to an attractant gradient, it tumbles less frequently (or has longer runs) when traveling up the gradient, but tumbles at normal frequency if moving down the gradient. Consequently the bacterium moves up the gradient. Behavior is shaped by temporal changes in chemical concentration: the bacterium compares its current environment with that experienced a few moments previously; if the attractant concentration is higher, tumbling is suppressed and the run is longer. The opposite response occurs with a repellent gradient. Tumbling frequency decreases (the run lengthens) when the bacterium moves down the gradient away from the repellent.

Although bacterial chemotaxis appears to be deliberate, directed movement, it is important to keep in mind that this is not the case. When the environment is constant, bacteria tend to move in a random walk. That is, there is a random sequence of runs followed by tumbles. If a run is in the direction of improving conditions, tumbles are suppressed so that the cell tends to move in the preferred direction. This is said to be a biased random walk toward attractants and away from repellants. Individual cells do not choose a particular direction. Instead, they determine whether or not to continue in the same direction.

Much work has been done on the mechanism of chemotaxis in *Escherichia coli*. Recall that forward swimming is due to counterclockwise rotation of the flagellum, whereas tumbling results from clockwise rotation. The bacteria must be able to respond to gradients in such a way that they collect in nutrient-rich regions and at the proper oxygen level while avoiding toxic materials. *E. coli* has four different chemoreceptors that recognize serine, aspartate and maltose, ribose and galactose, and

dipeptides, respectively. These chemoreceptors often are called methyl-accepting chemotaxis proteins (MCPs). They seem to be localized in patches, often at the end of a rod-shaped cell like *E. coli*. The MCPs do not directly influence flagellar rotation but act through a series of proteins. The whole process is so efficient that a stimulus can trigger a motor response in less than 200 milliseconds.

The molecular mechanism underlying chemotaxis is quite complex. It involves conformational changes in proteins, protein methylation, and protein phosphorylation. When an attractant such as a nutrient molecule is not bound to an MCP, the CheA protein is phosphorylated using ATP. This phosphorylated protein can donate its phosphate to the CheY protein, which then interacts with the Fli M switch protein at the base of the flagellum to promote clockwise rotation and tumbling. An increase in nutrient binding will lead to dephosphorylated CheA, counterclockwise flagellar rotation, and a run. When no attractants or repellants are present, the system maintains intermediate levels of CheA phosphate and CheY phosphate. This produces a normal random walk pattern. In very general terms, the system has a sensory protein that can be phosphorylated and then phosphorylate another protein to cause a response. As we shall see later, this is called a two-component phosphorelay system. The molecular details of the chemotaxis system will be described in chapter 12 when two-component systems are discussed. It should be noted here that a similar mechanism is used to respond to other environmental factors such as oxygen (aerotaxis), light (phototaxis), temperature (thermotaxis), and osmotic pressure (osmotaxis). Two-component phosphorelay systems (pp. 279–80)

1. Define chemotaxis, run, and tumble or twiddle.
2. Explain in a general way how bacteria are attracted to substances like nutrients while being repelled by toxic materials.

 3.9 **THE BACTERIAL ENDOSPORE**

A number of gram-positive bacteria can form a special resistant, dormant structure called an **endospore.** Endospores develop within vegetative bacterial cells of several genera: *Bacillus* and *Clostridium* (rods), *Sporosarcina* (cocci), and others. These structures are extraordinarily resistant to environmental stresses such as heat, ultraviolet radiation, gamma radiation, chemical disinfectants, and desiccation. In fact, some endospores have remained viable for around 100,000 years, and actinomycete spores (which are not true endospores) have been recovered alive after burial in the mud for 7,500 years. Because of their resistance and the fact that several species of endospore-forming bacteria are dangerous pathogens, endospores are of great practical importance in food, industrial, and medical microbiology. This is because it is essential to be able to sterilize solutions and solid objects. Endospores often survive boiling for an hour or more; therefore autoclaves (*see chapter 7*) must be used to sterilize many materials. Endospores are also of considerable theoretical

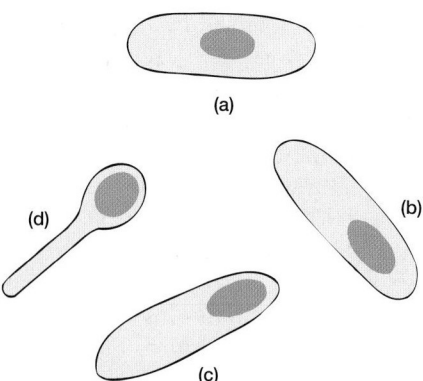

Figure 3.41   Examples of Endospore Location and Size.
(**a**) Central spore. (**b**) Subterminal spore. (**c**) Terminal spore.
(**d**) Terminal spore with swollen sporangium.

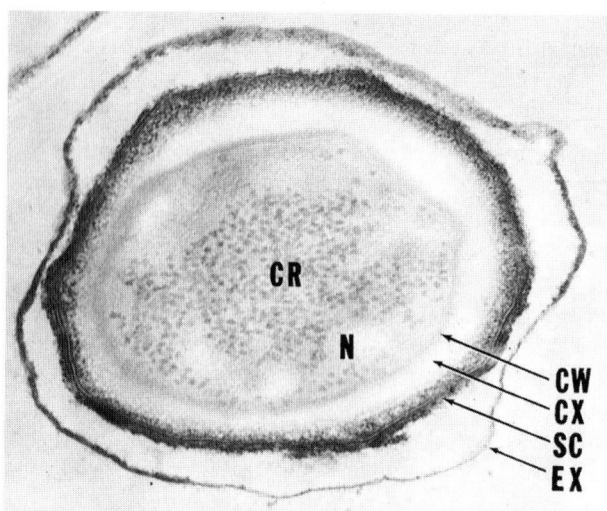

Figure 3.42   Endospore Structure. *Bacillus anthracis* endospore (×151,000). Note the following structures: exosporium, EX; spore coat, SC; cortex, CX; core wall, CW; and the protoplast or core with its nucleoid, N, and ribosomes, CR.

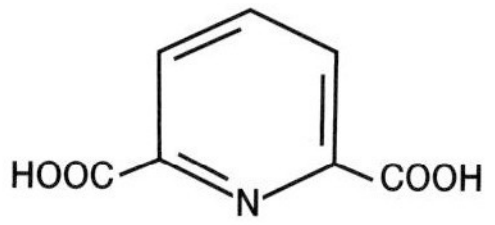

interest. Because bacteria manufacture these intricate entities in a very organized fashion over a period of a few hours, spore formation is well suited for research on the construction of complex biological structures. In the environment, endospores aid in survival when moisture or nutrients are scarce. Resistance of endospores to high temperature (chapter 7)

Endospores can be examined with both light and electron microscopes. Because spores are impermeable to most stains, they often are seen as colorless areas in bacteria treated with methylene blue and other simple stains; special spore stains are used to make them clearly visible (*see chapter 2*). Spore position in the mother cell or **sporangium** frequently differs among species, making it of considerable value in identification. Spores may be centrally located, close to one end (subterminal), or definitely terminal (**figure 3.41**). Sometimes a spore is so large that it swells the sporangium.

Electron micrographs show that endospore structure is complex (**figure 3.42**). The spore often is surrounded by a thin, delicate covering called the **exosporium.** A **spore coat** lies beneath the exosporium, is composed of several protein layers, and may be fairly thick. It is impermeable to many toxic molecules and responsible for the spore's resistance to chemicals. The coat also is thought to contain enzymes that are involved in germination. The **cortex,** which may occupy as much as half the spore volume, rests beneath the spore coat. It is made of a peptidoglycan that is less cross-linked than that in vegetative cells. The **spore cell wall** (or core wall) is inside the cortex and surrounds the protoplast or **core.** The core has the normal cell structures such as ribosomes and a nucleoid, but is metabolically inactive.

It is still not known precisely why the endospore is so resistant to heat and other lethal agents. As much as 15% of the spore's dry weight consists of **dipicolinic acid** complexed with calcium ions (**figure 3.43**), which is located in the core. It has long been thought that dipicolinic acid was directly involved in spore heat resistance, but heat-resistant mutants lacking dipicolinic acid now have been isolated. Calcium does aid in resistance to wet heat, oxidizing agents, and sometimes dry heat. It may be that calcium-dipicolinate often stabilizes spore nucleic acids. Re-

$$\text{HOOC}\underset{N}{\bigcirc}\text{COOH}$$

Figure 3.43   Dipicolinic Acid.

cently specialized small, acid-soluble DNA-binding proteins have been discovered in the endospore. They saturate spore DNA and protect it from heat, radiation, dessication, and chemicals. Dehydration of the protoplast appears to be very important in heat resistance. The cortex may osmotically remove water from the protoplast, thereby protecting it from both heat and radiation damage. The spore coat also seems to protect against enzymes and chemicals such as hydrogen peroxide. Finally, spores contain some DNA repair enzymes. DNA is repaired during germination and outgrowth after the core has become active once again. In summary, endospore heat resistance probably is due to several factors: calcium-dipicolinate and acid-soluble protein stabilization of DNA, protoplast dehydration, the spore coat, DNA repair, the greater stability of cell proteins in bacteria adapted to growth at high temperatures, and others.

Spore formation, **sporogenesis** or **sporulation,** normally commences when growth ceases due to lack of nutrients. It is a complex process and may be divided into seven stages (**figure 3.44**). An axial filament of nuclear material forms (stage I), followed by an inward folding of the cell membrane to enclose part of the DNA and produce the forespore septum (stage II). The membrane continues to grow and engulfs the immature spore in a second membrane (stage III). Next, cortex is laid down in the space between the two membranes, and both calcium and dipicolinic acid are accumulated (stage IV). Protein coats then are formed around the cortex (stage V), and maturation of the spore occurs (stage VI).

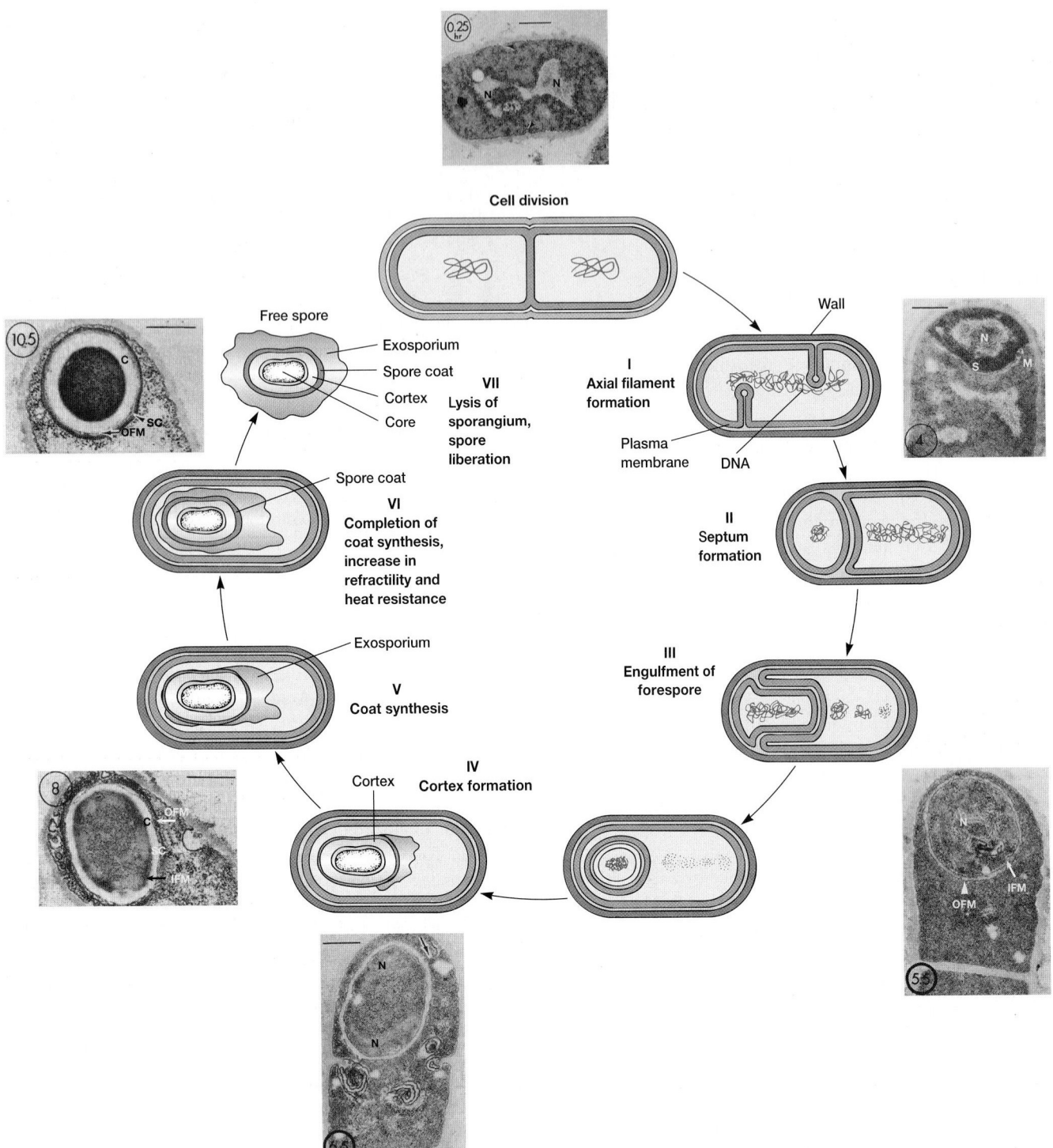

Figure 3.44   Endospore Formation: Life cycle of *Bacillus megaterium*.   The stages are indicated by Roman numerals. The circled numbers in the photographs refer to the hours from the end of the logarithmic phase of growth: 0.25 h—a typical vegetative cell; 4 h—stage II cell, septation; 5.5 h—stage III cell, engulfment; 6.5 h—stage IV cell, cortex formation; 8 h—stage V cell, coat formation; 10.5 h—stage VI cell, mature spore in sporangium. Abbreviations used: C, cortex; IFM and OFM, inner and outer forespore membranes; M, mesosome; N, nucleoid; S, septum; SC, spore coats. Bars = 0.5 μm.

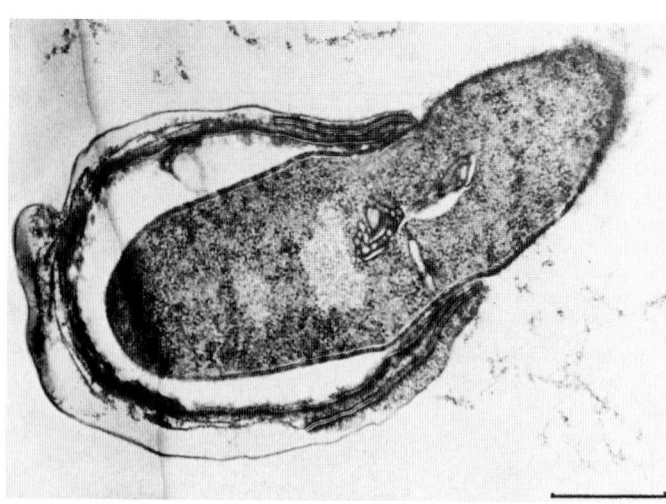

Figure 3.45 Endospore Germination. *Clostridium pectinovorum* emerging from the spore during germination. Bar = 0.5 μm.

Finally, lytic enzymes destroy the sporangium releasing the spore (stage VII). Sporulation requires only about 10 hours in *Bacillus megaterium.* Regulation of sporulation in *Bacillus* (pp. 277–78, 279)

The transformation of dormant spores into active vegetative cells seems almost as complex a process as sporogenesis. It occurs in three stages: (1) activation, (2) germination, and (3) outgrowth. Often an endospore will not germinate successfully, even in a nutrient-rich medium, unless it has been activated. Activation is a process that prepares spores for germination and usually results from treatments like heating. It is followed by **germination,** the breaking of the spore's dormant state. This process is characterized by spore swelling, rupture or absorption of the spore coat, loss of resistance to heat and other stresses, loss of refractility, release of spore components, and increase in metabolic activity. Many normal metabolites or nutrients (e.g., amino acids and sugars) can trigger germination after activation. Germination is followed by the third stage, outgrowth. The spore protoplast makes new components, emerges from the remains of the spore coat, and develops again into an active bacterium (**figure 3.45**).

1. Describe the structure of the bacterial endospore using a labeled diagram.
2. Briefly describe endospore formation and germination. What is the importance of the endospore? What might account for its heat resistance?

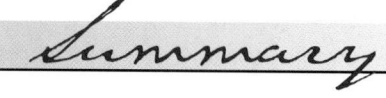

3.1 An Overview of Procaryotic Cell Structure

a. Bacteria may be spherical (cocci), rod-shaped (bacilli), spiral, or filamentous; form buds and stalks; or even have no characteristic shape at all (pleomorphic).
b. Bacterial cells can remain together after division to form pairs, chains, and clusters of various sizes and shapes.
c. All bacteria are procaryotes and much simpler structurally than eucaryotes. **Table 3.1** summarizes the major functions of bacterial cell structures.

3.2 Procaryotic Cell Membranes

a. The plasma membrane and most other membranes are composed of a lipid bilayer in which integral proteins are buried (**figure 3.7**). Peripheral proteins are more loosely attached to membranes.
b. The plasma membrane may invaginate to form some simple structures such as membrane systems containing photosynthetic and respiratory assemblies, and possibly mesosomes.

3.3 The Cytoplasmic Matrix

a. The cytoplasmic matrix contains inclusion bodies and ribosomes.

3.4 The Nucleoid

a. Procaryotic genetic material is located in an area called the nucleoid and is not usually enclosed by a membrane.

3.5 The Procaryotic Cell Wall

a. Most bacteria have a cell wall outside the plasma membrane to give them shape and protect them from osmotic stress.
b. Bacterial walls are chemically complex and usually contain peptidoglycan or murein (**figures 3.16–3.19**).
c. Bacteria often are classified as either gram positive or gram negative based on differences in cell wall structure and their response to Gram staining.
d. Gram-positive walls have thick, homogeneous layers of peptidoglycan and teichoic acids (**figure 3.21**). Gram-negative bacteria have a thin peptidoglycan layer surrounded by a complex outer membrane containing lipopolysaccharides (LPSs) and other components (**figure 3.23**).
e. Bacteria such as mycoplasmas lack a cell wall.

3.6 Protein Secretion in Procaryotes

a. Procaryotes can secrete proteins to the cell exterior using several different pathways.

3.7 Components External to the Cell Wall

a. Structures such as capsules, fimbriae, and sex pili are found outside the cell wall.
b. Many bacteria are motile, usually by means of threadlike locomotory organelles called flagella (**figure 3.33**).
c. Bacterial species differ in the number and distribution of their flagella.
d. The flagellar filament is a rigid helix that rotates like a propeller to push the bacterium through the water (**figures 3.36** and **3.37**).

3.8 Chemotaxis

a. Motile bacteria can respond to gradients of attractants and repellents, a phenomenon known as chemotaxis.

3.9 The Bacterial Endospore

a. Some bacteria survive adverse environmental conditions by forming endospores, dormant structures resistant to heat, desiccation, and many chemicals (**figure 3.42**).

Key Terms

ABC protein secretion
  pathway 61
amphitrichous 63
axial filament 66
bacillus 41
basal body 64
capsule 61
carboxysomes 48
chemoreceptors 67
chemotaxis 67
coccus 41
core 69
core polysaccharide 56
cortex 69
cyanophycin granules 48
cytoplasmic matrix 47
deoxyribonucleic acid (DNA)
  50
dipicolinic acid 69
diplococcus 41
endospore 68
envelope 53
exoenzyme 53
exosporium 69
filament 64

fimbriae 62
flagellin 64
flagellum 63
fluid mosaic model 46
gas vacuole 48
gas vesicles 49
general secretion pathway
  (GSP) 59
germination 71
gliding motility 66
glycocalyx 61
glycogen 47
hook 64
hydrophilic 44
hydrophobic 44
inclusion body 47
integral proteins 46
lipid A 56
lipopolysaccharides (LPSs) 56
lophotrichous 63
lysis 59
lysozyme 59
magnetosomes 50
metachromatic granules 49
monotrichous 63

murein 53
mycelium 41
nucleoid 50
O antigen 56
O side chain 56
osmosis 59
outer membrane 53
penicillin 59
peptide interbridge 54
peptidoglycan 53
peripheral proteins 46
periplasm 53
periplasmic space 53
peritrichous 63
plasma membrane 44
plasmid 52
plasmolysis 59
pleomorphic 42
polar flagellum 63
poly-β-hydroxybutyrate (PHB) 47
polyphosphate granules 49
porin proteins 58
protoplast 47
ribosome 50
rod 41

run 67
Sec-dependent pathway 59
self-assembly 65
sex pili 63
signal peptide 59
S-layer 62
slime layer 61
spheroplast 59
spirilla 41
spirochete 41
sporangium 69
spore cell wall 69
spore coat 69
sporogenesis 69
sporulation 69
Svedberg unit 50
teichoic acid 55
translocon 60
tumble 67
twiddle 67
type I protein secretion pathway 61
type II protein secretion pathway 61
type III protein secretion pathway 61
vibrio 41
volutin granules 49

Questions for Thought and Review

1. List all the major procaryotic structures discussed in this chapter and provide a brief description of the functions of each.
2. Some microbiologists believe that the plasma membrane is involved in DNA synthesis during bacterial reproduction. How might one prove that this is the case?
3. Discuss a possible mechanism of Gram staining in terms of differences in structure and chemistry between the walls of gram-positive and gram-negative bacteria.
4. What is self-assembly and why is it so important to cells?
5. How might one go about showing that a bacterium could form true endospores?

Critical Thinking Questions

1. Propose a model for the assembly of a flagellum in a gram-positive membrane. How would that model need to be modified for the assembly of a flagellum in a gram-negative membrane?
2. If you could not use a microscope, how would you determine whether a cell is procaryotic or eucaryotic? Assume the organism can be cultured easily in the laboratory.
3. The peptidoglycan of bacteria has been compared with the chain mail, worn beneath a medieval knight's suit of armor. It provides both protection and flexibility. Can you describe other structures in biology that have an analogous function? How are they replaced or modified to accommodate the growth of the inhabitant?

**For recommended readings on these
and related topics, see Appendix V.**

Eucaryotic Cell Structure and Function

Concepts

- Eucaryotic cells differ most obviously from procaryotic cells in having a variety of complex membranous organelles in the cytoplasmic matrix and the majority of their genetic material within membrane-delimited nuclei. Each organelle has a distinctive structure directly related to specific functions.
- A cytoskeleton composed of microtubules, microfilaments, and intermediate filaments helps give eucaryotic cells shape; microtubules and microfilaments are also involved in cell movements and intracellular transport.
- In eucaryotes, genetic material is distributed between cells by the highly organized, complex processes called mitosis and meiosis.
- Despite great differences between eucaryotes and procaryotes with respect to such things as morphology, they are similar on the biochemical level.

The key to every biological problem must finally be sought in the cell.

—*E. B. Wilson*

Often we emphasize procaryotes and viruses, but eucaryotic microorganisms also have major impacts on human welfare. For example, the protozoan parasite *Trypanosoma brucei gambiense* is a cause of African sleeping sickness. The organism invades the nervous system and the victim frequently dies after suffering several years from symptoms such as weakness, headache, apathy, emaciation, sleepiness, and coma.

In chapter 3 considerable attention is devoted to procaryotic cell structure and function because bacteria are immensely important in microbiology and have occupied a large portion of microbiologists' attention in the past. Nevertheless, algae, fungi, and protozoa also are microorganisms and have been extensively studied. These eucaryotes often are extraordinarily complex, interesting in their own right, and prominent members of the ecosystem (**figure 4.1**). In addition, fungi (and to some extent, algae) are exceptionally useful in industrial microbiology. Many fungi and protozoa are also major human pathogens; one only need think of either malaria or African sleeping sickness (see chapter opener) to appreciate the significance of eucaryotes in pathogenic microbiology. So although this text emphasizes bacteria, eucaryotic microorganisms are discussed at many points.

Chapter 4 focuses on eucaryotic cell structure and its relationship to cell function. Because many valuable studies on eucaryotic cell ultrastructure have used organisms other than microorganisms, some work on nonmicrobial cells is presented. At the end of the chapter, procaryotic and eucaryotic cells are compared in some depth.

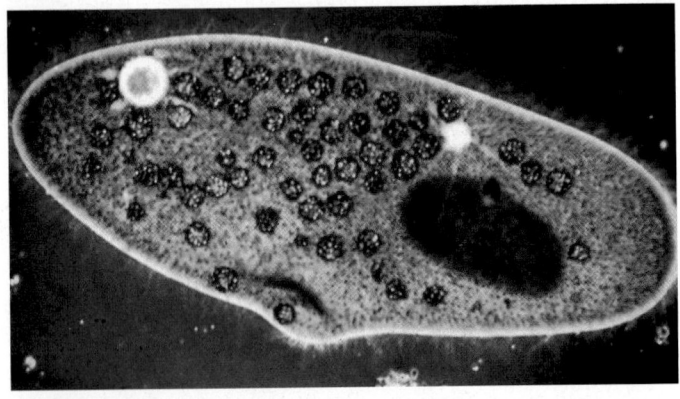

(a) *Paramecium*

(b) *Diatom frustules*

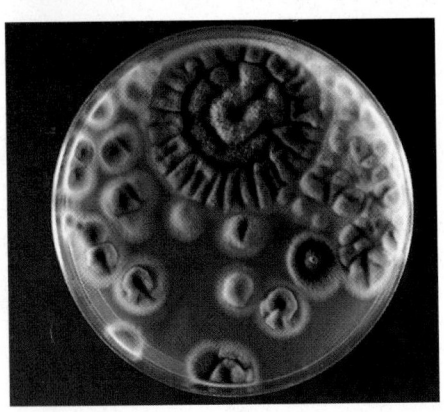

(c) *Penicillium*

(d) *Penicillium*

(e) *Stentor*

(f) *Amanita muscaria*

Figure 4.1   Representative Examples of Eucaryotic Microorganisms.   (a) *Paramecium* as seen with interference-contrast microscopy (×115). (b) Mixed diatom frustules (×100). (c) *Penicillium* colonies, and (d) a microscopic view of the mold's hyphae and conidia (×220). (e) *Stentor.* The ciliated protozoa are extended and actively feeding, dark-field microscopy (×100). (f) *Amanita muscaria,* a large poisonous mushroom (×5).

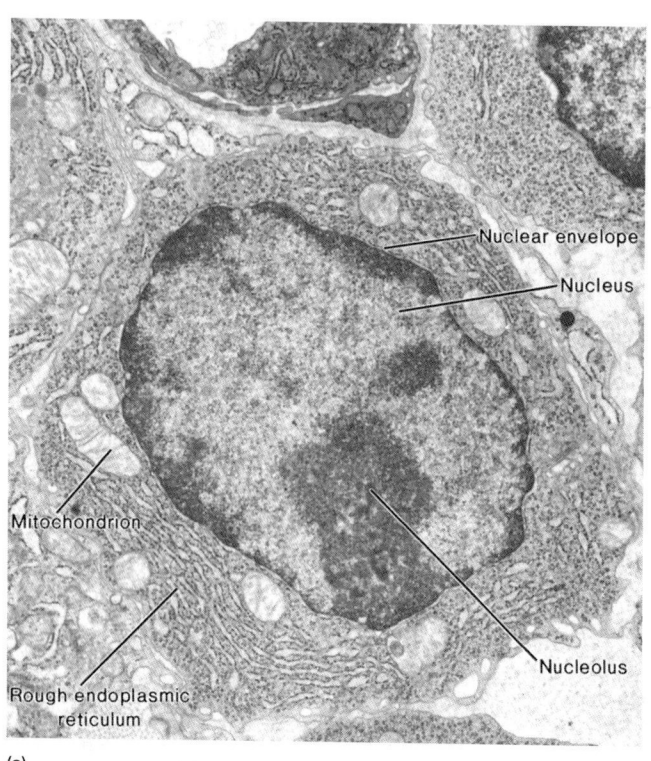

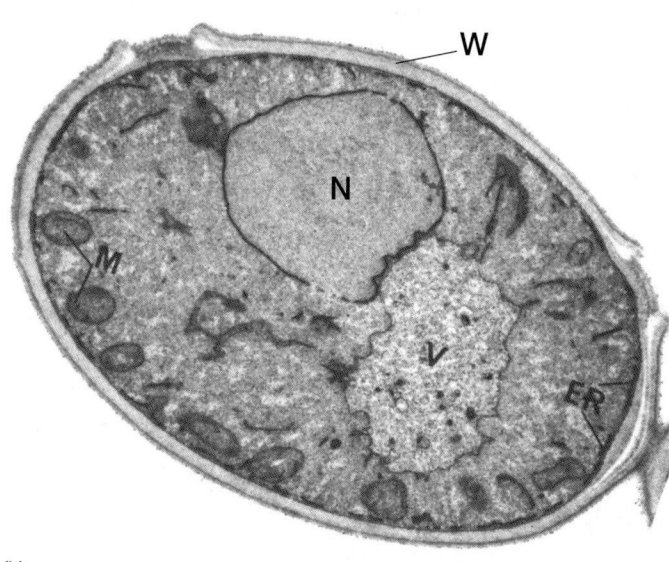

(a)

(b)

Figure 4.2 Eucaryotic Cell Ultrastructure. (a) A lymphoblast in the rat lymph node (×17,500). (b) The yeast *Saccharomyces* (×7,200). Note the nucleus (n), mitochondrion (m), vacuole (v), endoplasmic reticulum (er), and cell wall (w).

4.1 AN OVERVIEW OF EUCARYOTIC CELL STRUCTURE

The most obvious difference between eucaryotic and procaryotic cells is in their use of membranes. Eucaryotic cells have membrane-delimited nuclei, and membranes also play a prominent part in the structure of many other organelles (**figures 4.2 and 4.3**). **Organelles** are intracellular structures that perform specific functions in cells analogous to the functions of organs in the body. The name organelle (little organ) was coined because biologists saw a parallel between the relationship of organelles to a cell and that of organs to the whole body. It is not satisfactory to define organelles as membrane-bound structures because this would exclude such components as ribosomes and bacterial

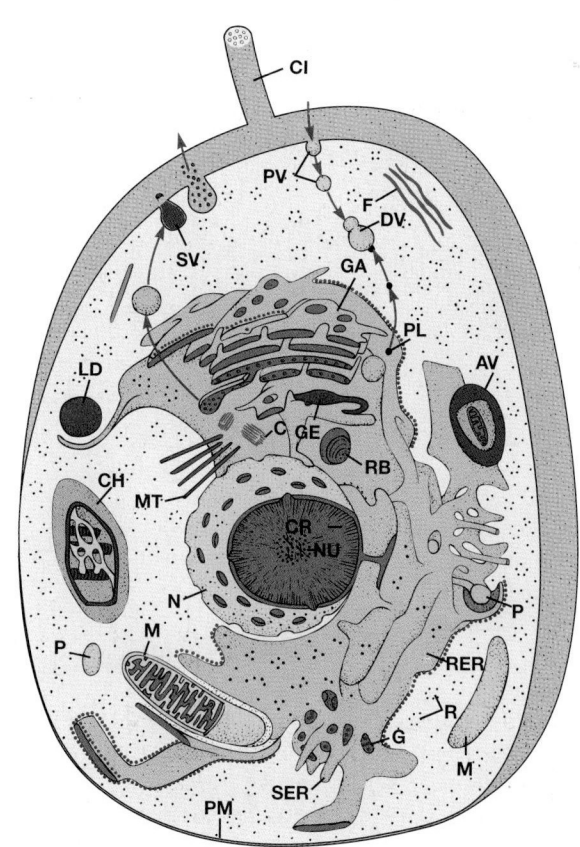

Figure 4.3 Eucaryotic Cell Ultrastructure. This is a schematic, three-dimensional diagram of a cell with the most important organelles identified in the illustration. AV, autophagic vacuole; C, centriole; CH, chloroplast; CI, cilium; CR, chromatin; DV, digestion vacuole; F, microfilaments; G, glycogen; GA, Golgi apparatus; GE, GERL; LD, lipid droplet; M, mitochondrion; MT, microtubules; N, nucleus; NU, nucleolus; P, peroxisome; PL, primary lysosome; PM, plasma membrane; PV, pinocytotic vesicle; R, ribosomes and polysomes; RB, residual body; RER, rough endoplasmic reticulum; SER, smooth endoplasmic reticulum; SV, secretion vacuole.

Table 4.1
Functions of Eucaryotic Organelles

Plasma membrane	Mechanical cell boundary, selectively permeable barrier with transport systems, mediates cell-cell interactions and adhesion to surfaces, secretion
Cytoplasmic matrix	Environment for other organelles, location of many metabolic processes
Microfilaments, intermediate filaments, and microtubules	Cell structure and movements, form the cytoskeleton
Endoplasmic reticulum	Transport of materials, protein and lipid synthesis
Ribosomes	Protein synthesis
Golgi apparatus	Packaging and secretion of materials for various purposes, lysosome formation
Lysosomes	Intracellular digestion
Mitochondria	Energy production through use of the tricarboxylic acid cycle, electron transport, oxidative phosphorylation, and other pathways
Chloroplasts	Photosynthesis—trapping light energy and formation of carbohydrate from CO_2 and water
Nucleus	Repository for genetic information, control center for cell
Nucleolus	Ribosomal RNA synthesis, ribosome construction
Cell wall and pellicle	Strengthen and give shape to the cell
Cilia and flagella	Cell movement
Vacuole	Temporary storage and transport, digestion (food vacuoles), water balance (contractile vacuole)

flagella. A comparison of figures 4.2 and 4.3 with figure 3.11 (*p. 49*) shows how much more structurally complex the eucaryotic cell is. This complexity is due chiefly to the use of internal membranes for several purposes. The partitioning of the eucaryotic cell interior by membranes makes possible the placement of different biochemical and physiological functions in separate compartments so that they can more easily take place simultaneously under independent control and proper coordination. Large membrane surfaces make possible greater respiratory and photosynthetic activity because these processes are located exclusively in membranes. The intracytoplasmic membrane complex also serves as a transport system to move materials between different cell locations. Thus abundant membrane systems probably are necessary in eucaryotic cells because of their large volume and the need for adequate regulation, metabolic activity, and transport.

Figures 4.2, 4.3, and 4.25*b* provide generalized views of eucaryotic cell structure and illustrate most of the organelles to be discussed. **Table 4.1** briefly summarizes the functions of the major eucaryotic organelles. Those organelles lying inside the plasma membrane are first described, and then components outside the membrane are discussed.

4.2 THE CYTOPLASMIC MATRIX, MICROFILAMENTS, INTERMEDIATE FILAMENTS, AND MICROTUBULES

When a eucaryotic cell is examined at low power with the electron microscope, its larger organelles are seen to lie in an apparently featureless, homogeneous substance called the **cytoplasmic matrix.** The matrix, although superficially uninteresting, is actually one of the most important and complex parts of the cell. It is the "environment" of the organelles and the location of many important biochemical processes. Several physical changes seen in cells—viscosity changes, cytoplasmic streaming, and others—also are due to matrix activity.

Water constitutes about 70 to 85% by weight of a eucaryotic cell. Thus a large part of the cytoplasmic matrix is water. Cellular water can exist in two different forms. Some of it is bulk or free water; this is normal, osmotically active water. Osmosis, water activity, and growth (pp. 59, 118–21)

Water also can exist as bound water or water of hydration. This water is bound to the surface of proteins and other macromolecules and is osmotically inactive and more ordered than bulk water. There is some evidence that bound water is the site of many metabolic processes. The protein content of cells is so high that the cytoplasmic matrix often may be semicrystalline. Usually matrix pH is around neutrality, about pH 6.8 to 7.1, but can vary widely. Protozoan digestive vacuoles may reach pHs as low as 3 to 4.

Probably all eucaryotic cells have **microfilaments,** minute protein filaments, 4 to 7 nm in diameter, which may be either scattered within the cytoplasmic matrix or organized into networks and parallel arrays. Microfilaments are involved in cell motion and shape changes. Some examples of cellular movements associated with microfilament activity are the motion of pigment granules, amoeboid movement, and protoplasmic streaming in slime molds (*see chapter 25*).

The participation of microfilaments in cell movement is suggested by electron microscopic studies showing that they frequently are found at locations appropriate for such a role. For example, they are concentrated at the interface between stationary and flowing cytoplasm in plant cells and slime molds. Experiments using the drug cytochalasin B have provided additional evidence. Cytochalasin B disrupts microfilament structure and often simultaneously inhibits cell movements. However, because the drug has additional effects in cells, a direct cause-and-effect interpretation of these experiments is sometimes difficult.

Microfilament protein has been isolated and analyzed chemically. It is an actin, very similar to the actin contractile protein of muscle tissue. This is further indirect evidence for microfilament involvement in cell movement.

Some pathogens such as *Listeria monocytogenes* make use of eucaryotic actin to move rapidly through the host cell. The ActA protein released by *Listeria* causes the polymerization of actin filaments at the end of the bacterium. A tail of actin is formed and

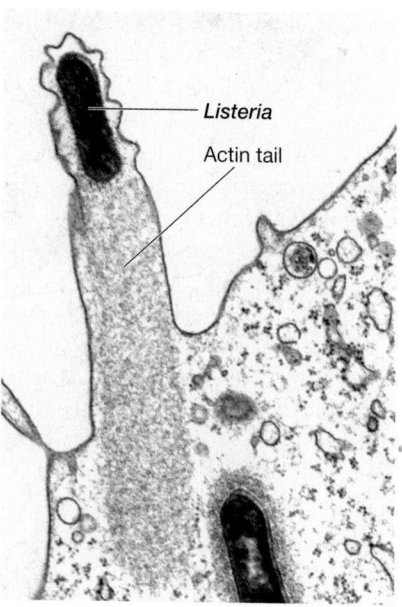

Figure 4.4 *Listeria* **Motility and Actin Filaments.** A *Listeria* cell is propelled through the cell surface by a bundle of actin filaments.

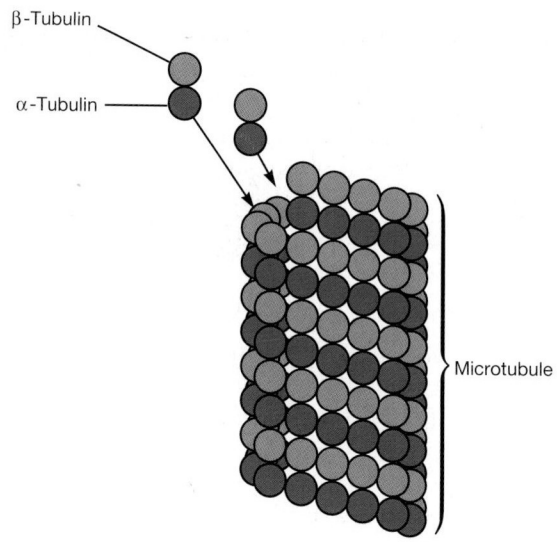

Figure 4.5   Microtubule Structure. The hollow cylinder, about 25 nm in diameter, is made of two kinds of protein subunits, α-tubulin and β-tubulin.

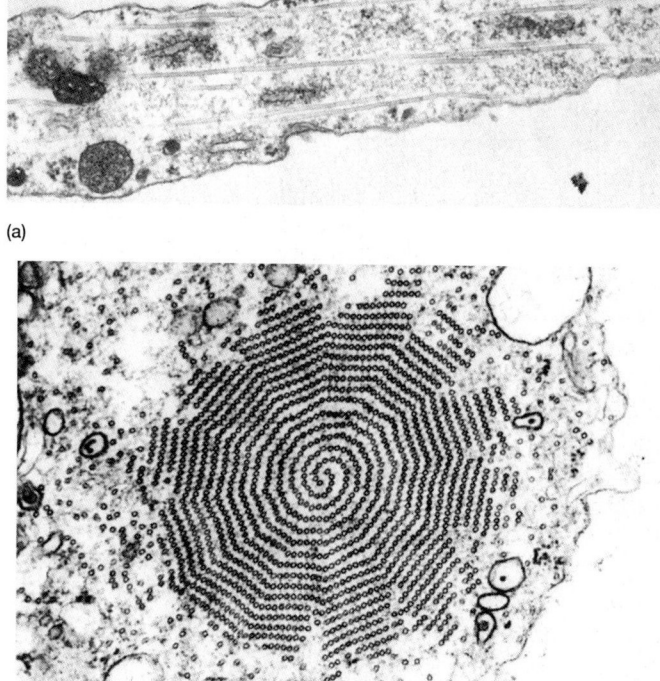

trapped in the host cytoskeleton. Its continued elongation pushes the bacterium along at rates up to 11 μm/minute. The bacterium can even be propelled through the cell surface and into neighboring cells (**figure 4.4**).

A second type of small filamentous organelle in the cytoplasmic matrix is shaped like a thin cylinder about 25 nm in diameter. Because of its tubular nature this organelle is called a **microtubule.** Microtubules are complex structures constructed of two slightly different spherical protein subunits named tubulins, each of which is approximately 4 to 5 nm in diameter. These subunits are assembled in a helical arrangement to form a cylinder with an average of 13 subunits in one turn or circumference (**figure 4.5**).

Microtubules serve at least three purposes: (1) they help maintain cell shape, (2) are involved with microfilaments in cell movements, and (3) participate in intracellular transport processes. Evidence for a structural role comes from their intracellular distribution and studies on the effects of the drug colchicine. Long, thin cell structures requiring support such as the axopodia (long, slender, rigid pseudopodia) of protozoa contain microtubules (**figure 4.6**). When migrating embryonic nerve and heart cells are exposed to colchicine, they simultaneously lose their microtubules and their characteristic shapes. The shapeless cells seem to wander aimlessly as if incapable of directed movement without their normal form. Their microfilaments are still intact, but due to the disruption of their microtubules by colchicine, they no longer behave normally.

Microtubules also are present in structures that participate in cell or organelle movements—the mitotic spindle, cilia, and flagella. For example, the mitotic spindle is constructed of microtubules;

Figure 4.6 Cytoplasmic Microtubules. Electron micrographs of pseudopodia with microtubules. (**a**) Microtubules in a pseudopodium from the protozoan *Reticulomyxa* (×65,000). (**b**) A transverse section of a heliozoan axopodium (×48,000). Note the parallel array of microtubules organized in a spiral pattern.

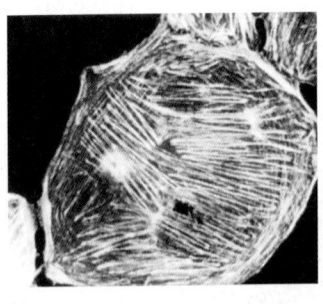
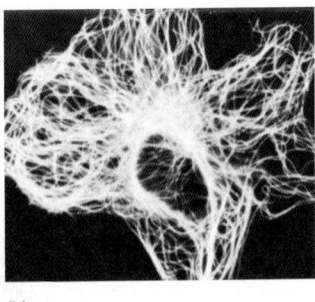

(a)   (b)

Figure 4.7   The Eucaryotic Cytoskeleton.   (a) Antibody-stained microfilament system in a mammal cell (×400). **(b)** Antibody-stained microtubule system in a mammal cell (×1,000).

when a dividing cell is treated with colchicine, the spindle is disrupted and chromosome separation blocked. Microtubules also are essential to the movement of eucaryotic cilia and flagella.

Other kinds of filamentous components also are present in the matrix, the most important of which are the **intermediate filaments** (about 8 to 10 nm in diameter). The microfilaments, microtubules, and intermediate filaments are major components of a vast, intricate network of interconnected filaments called the **cytoskeleton** (**figure 4.7**). As mentioned previously, the cytoskeleton plays a role in both cell shape and movement. Procaryotes lack a true, organized cytoskeleton and may not possess actinlike proteins.

1. What is an organelle?
2. Define cytoplasmic matrix, bulk or free water, bound water, microfilament, microtubule, and tubulin. Discuss the roles of microfilaments, intermediate filaments, and microtubules.
3. Describe the cytoskeleton. What are its functions?

4.3   THE ENDOPLASMIC RETICULUM

Besides the cytoskeleton, the cytoplasmic matrix is permeated with an irregular network of branching and fusing membranous tubules, around 40 to 70 nm in diameter, and many flattened sacs called **cisternae** (s., **cisterna**). This network of tubules and cisternae is the **endoplasmic reticulum (ER)** (figure 4.2*a* and **figure 4.8**). The nature of the ER varies with the functional and physiological status of the cell. In cells synthesizing a great deal of protein for purposes such as secretion, a large part of the ER is studded on its outer surface with ribosomes and is called **rough endoplasmic reticulum (RER)**. Other cells, such as those producing large quantities of lipids, have ER that lacks ribosomes. This is **smooth ER (SER)**.

The endoplasmic reticulum has many important functions. It transports proteins, lipids, and probably other materials through the cell. Lipids and proteins are synthesized by ER-associated enzymes and ribosomes. Polypeptide chains synthesized on RER-bound ribosomes may be inserted either into the ER membrane or into its lumen for transport elsewhere. The ER is also a major site of cell membrane synthesis.

New endoplasmic reticulum is produced through expansion of the old. Many biologists think the RER synthesizes new ER proteins and lipids. "Older" RER then loses its connected ribosomes and is modified to become SER. Not everyone agrees with this interpretation, and other mechanisms of growth of ER are possible.

4.4   THE GOLGI APPARATUS

The **Golgi apparatus** is a membranous organelle composed of flattened, saclike cisternae stacked on each other (**figure 4.9**). These membranes, like the smooth ER, lack bound ribosomes.

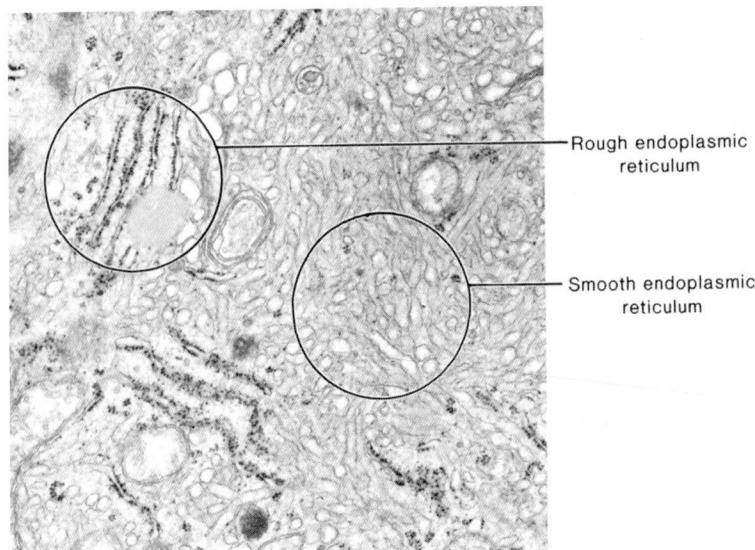

Rough endoplasmic reticulum

Smooth endoplasmic reticulum

Figure 4.8   The Endoplasmic Reticulum.   A transmission electron micrograph of the corpus luteum in a human ovary showing structural variations in eucaryotic endoplasmic reticulum. Note the presence of both rough endoplasmic reticulum lined with ribosomes and smooth endoplasmic reticulum without ribosomes (×26,500).

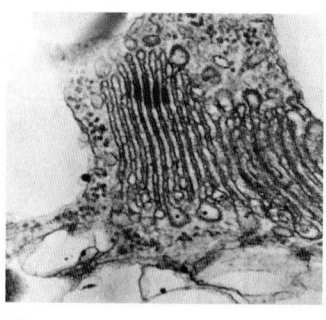

(a)

Figure 4.9   Golgi Apparatus Structure.
Golgi apparatus of *Euglena gracilis.*
Cisternal stacks are shown in the electron
micrograph (×165,000) in (**a**) and diagram-
matically in (**b**).

(b)

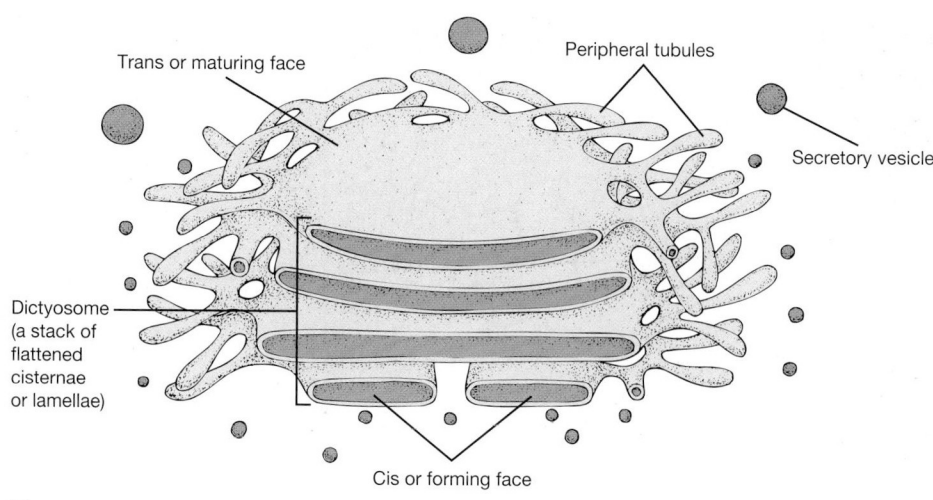

Trans or maturing face

Peripheral tubules

Secretory vesicle

Dictyosome
(a stack of
flattened
cisternae
or lamellae)

Cis or forming face

There are usually around 4 to 8 cisternae or sacs in a stack, although there may be many more. Each sac is 15 to 20 nm thick
and separated from other cisternae by 20 to 30 nm. A complex
network of tubules and vesicles (20 to 100 nm in diameter) is located at the edges of the cisternae. The stack of cisternae has a
definite polarity because there are two ends or faces that are quite
different from one another. The sacs on the cis or forming face often are associated with the ER and differ from the sacs on the
trans or maturing face in thickness, enzyme content, and degree
of vesicle formation. It appears that material is transported from
cis to trans cisternae by vesicles that bud off the cisternal edges
and move to the next sac.

The Golgi apparatus is present in most eucaryotic cells, but
many fungi and ciliate protozoa may lack a well-formed structure.
Sometimes it consists of a single stack of cisternae; however,
many cells may contain up to 20, and sometimes more, separate
stacks. These stacks of cisternae, often called **dictyosomes,** can be
clustered in one region or scattered about the cell.

The Golgi apparatus packages materials and prepares them
for secretion, the exact nature of its role varying with the organism. The surface scales of some flagellated algae and radiolarian
protozoa appear to be constructed within the Golgi apparatus and
then transported to the surface in vesicles. It often participates in
the development of cell membranes and in the packaging of cell
products. The growth of some fungal hyphae occurs when Golgi
vesicles contribute their contents to the wall at the hyphal tip.

In all these processes, materials move from the ER to the
Golgi apparatus. Most often vesicles bud off the ER, travel to the
Golgi apparatus, and fuse with the cis cisternae. Thus the Golgi
apparatus is closely related to the ER in both a structural and a
functional sense. Most proteins entering the Golgi apparatus from
the ER are glycoproteins containing short carbohydrate chains.
The Golgi apparatus frequently modifies proteins destined for
different fates by adding specific groups and then sends the proteins on their way to the proper location (e.g., lysosomal proteins
have phosphates added to their mannose sugars).

4.5   LYSOSOMES AND ENDOCYTOSIS

A very important function of the Golgi apparatus and endoplasmic reticulum is the synthesis of another organelle, the **lysosome.**
This organelle (or a structure very much like it) is found in a variety of microorganisms—protozoa, some algae, and fungi—as
well as in plants and animals. Lysosomes are roughly spherical
and enclosed in a single membrane; they average about 500 nm
in diameter, but range from 50 nm to several μm in size. They are
involved in intracellular digestion and contain the enzymes
needed to digest all types of macromolecules. These enzymes,
called hydrolases, catalyze the hydrolysis of molecules and function best under slightly acid conditions (usually around pH 3.5 to
5.0). Lysosomes maintain an acidic environment by pumping
protons into their interior. Digestive enzymes are manufactured
by the RER and packaged to form lysosomes by the Golgi apparatus. A segment of smooth ER near the Golgi apparatus also may
bud off lysosomes.

Lysosomes are particularly important in those cells that obtain nutrients through **endocytosis.** In this process a cell takes up
solutes or particles by enclosing them in vacuoles and vesicles
pinched off from its plasma membrane. Vacuoles and vesicles are
membrane-delimited cavities that contain fluid, and often solid
material. Larger cavities will be called vacuoles, and smaller cavities, vesicles. There are two major forms of endocytosis: phagocytosis and pinocytosis. During **phagocytosis** large particles and
even other microorganisms are enclosed in a phagocytic vacuole
or phagosome and engulfed (**figure 4.10***a*). In **pinocytosis** small
amounts of the surrounding liquid with its solute molecules are
pinched off as tiny pinocytotic vesicles (also called pinocytic
vesicles) or pinosomes. Often phagosomes and pinosomes are
collectively called **endosomes** because they are formed by endocytosis.

Eucaryotic microorganisms can employ several kinds of
pinocytosis. In **fluid-phase endocytosis,** a small portion of extracellular fluid is pinched off nonselectively. That is, there is no

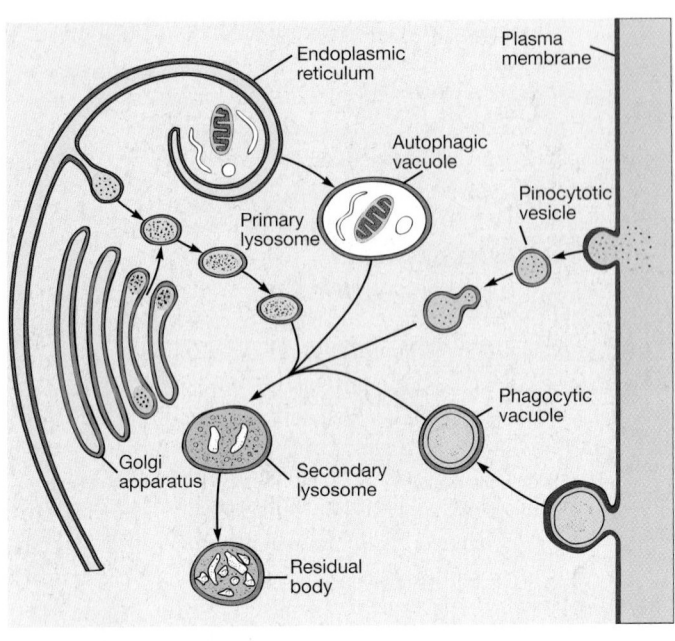

(a)

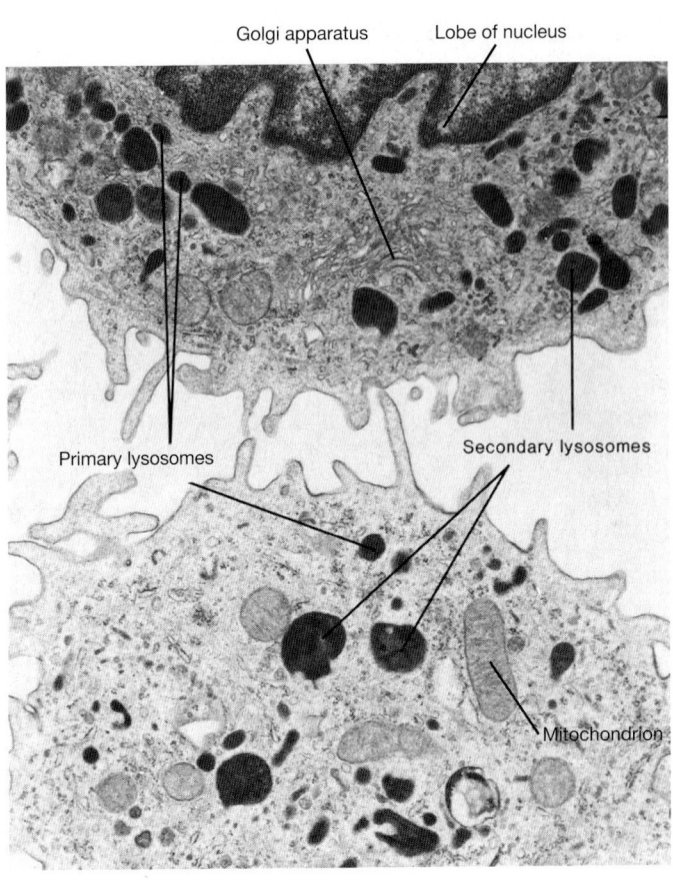

(b)

Figure 4.10   Lysosome Structure, Formation, and Function.
(**a**) A diagrammatic overview of lysosome formation and function.
(**b**) Lysosomes in macrophages from the lung. Secondary lysosomes contain partially digested material and are formed by fusion
of primary lysosomes and phagocytic vacuoles ($\times$14,137).

concentration of ingested material or selection of the contents, and
the ingested fluid is the same as the extracellular environment.
Fluid-phase endocytosis seems to go on continuously in many
cells and may be a way to internalize plasma membrane that has
been added during exocytosis. A second type of pinocytosis is
receptor-mediated endocytosis to form **coated vesicles.** It begins with coated pits, which are specialized membrane regions
coated on the cytoplasmic side with the protein **clathrin.** These
pits then pinch off to form coated vesicles. Coated pits have external receptors that specifically bind macromolecules; the
process is used to concentrate and ingest such things as hormones,
growth factors, antibodies, iron, and cholesterol. A third type of
pinocytosis forms what have been called **caveolae** ("little caves")
that look like tiny, flask-shaped invaginations of the plasma membrane (about 50 to 80 nm in diameter). Calveolae are enriched in
cholesterol and the membrane protein caveolin, and appear to be
closely related to lipid rafts (*see p. 45*). They may be involved in
signal transduction and in the active import of small molecules
such as folic acid and macromolecules. There is evidence that toxins such as the cholera toxin and many viruses, bacteria, and protozoa (e.g., simian virus 40, mycobacteria, *E. coli, Campylobacter
jejuni,* and *Plasmodium falciparum*) can enter the host cell using
caveolae. This route might favor pathogen survival because caveolae do not seem to fuse with endocytic vesicles and lysosomes.
Thus caveolar vesicle contents are not degraded.

Material in endosomes other than caveolar vesicles is digested with the aid of lysosomes. Newly formed lysosomes, or
primary lysosomes, fuse with phagocytic vacuoles to yield **secondary lysosomes** filled with material being digested (figure
4.10). These phagocytic vacuoles or secondary lysosomes often
are called food vacuoles. Digested nutrients then leave the secondary lysosome and enter the cytoplasm. When the lysosome
has accumulated large quantities of indigestible material, it is
known as a **residual body.**

Lysosomes join with phagosomes for defensive purposes as
well as to acquire nutrients. Invading bacteria, ingested by a
phagocytic cell, usually are destroyed when lysosomes fuse with
the phagosome. This is commonly seen in the leukocytes (white
blood cells) of vertebrates. Phagocytosis and resistance to pathogens
(pp. 693–97)

Cells can selectively digest portions of their own cytoplasm
in a type of secondary lysosome called an **autophagic vacuole**
(figure 4.10*a*). It is thought that these arise by lysosomal engulfment of a piece of cytoplasm (**figure 4.11**), or when the ER
pinches off cytoplasm to form a vesicle that subsequently fuses
with lysosomes. Autophagy probably plays a role in the normal
turnover or recycling of cell constituents. A cell also can survive
a period of starvation by selectively digesting portions of itself to
remain alive. Following cell death, lysosomes aid in digestion
and removal of cell debris.

Figure 4.11   Membrane Flow in the Vacuome.
The flow of material and membranes between organelles in a eucaryotic cell.

(1) Vesicles shuttling between the ER and Golgi apparatus.

(2) The Golgi–plasma membrane shuttle for secretion of materials.

(3) The Golgi-lysosome shuttle.

(4) The movement of material and membranes during endocytosis.

(5) Pathways of plasma membrane recovery from endosomes, lysosomes, and through the Golgi apparatus.

(6) Movement of vesicles from endosomes to lysosomes.

(7) Autophagy by a lysosome.

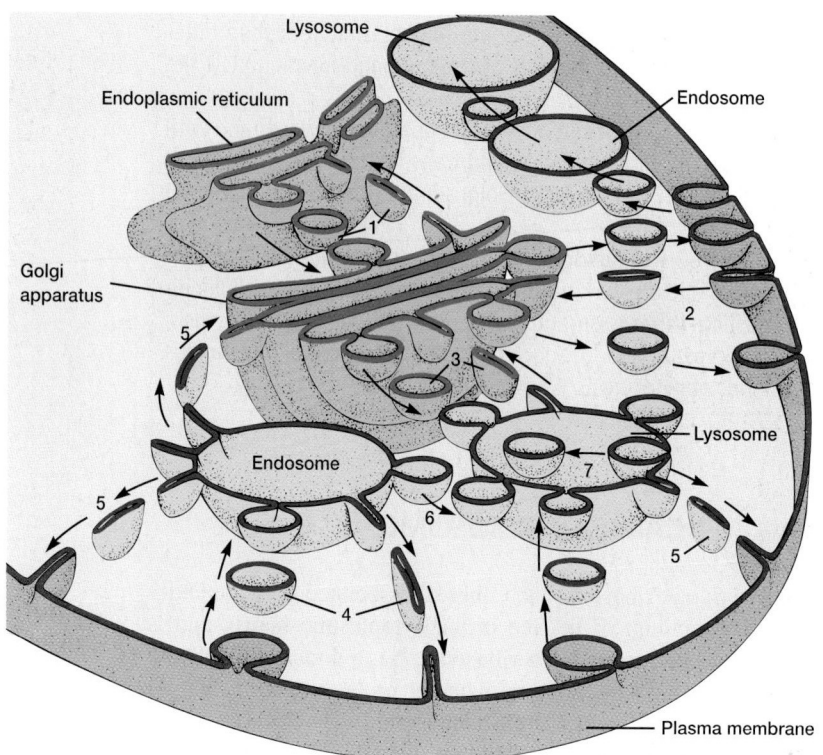

A most remarkable thing about lysosomes is that they accomplish all these tasks without releasing their digestive enzymes into the cytoplasmic matrix, a catastrophe that would destroy the cell. The lysosomal membrane retains digestive enzymes and other macromolecules while allowing small digestion products to leave.

The intricate complex of membranous organelles composed of the Golgi apparatus, lysosomes, endosomes, and associated structures seems to operate as a coordinated whole whose main function is the import and export of materials (figure 4.11). Christian de Duve (Nobel Prize, 1974) has suggested that this complex be called the vacuome in recognition of its functional unity. The ER manufactures secretory proteins and membrane, and contributes these to the Golgi apparatus. The Golgi apparatus then forms secretory vesicles that fuse with the plasma membrane and release material to the outside. It also produces lysosomes that fuse with endosomes to digest material acquired through phagocytosis and pinocytosis. Membrane movement in the region of the vacuome lying between the Golgi apparatus and the plasma membrane is two-way. Empty vesicles often are recycled and returned to the Golgi apparatus and plasma membrane rather than being destroyed. These exchanges in the vacuome occur without membrane rupture so that vesicle contents never escape directly into the cytoplasmic matrix.

More recently a nonlysosomal protein degradation system has been discovered in eucaryotic cells, a few bacteria, and many archaea. The majority of eucaryotic proteins may be de-

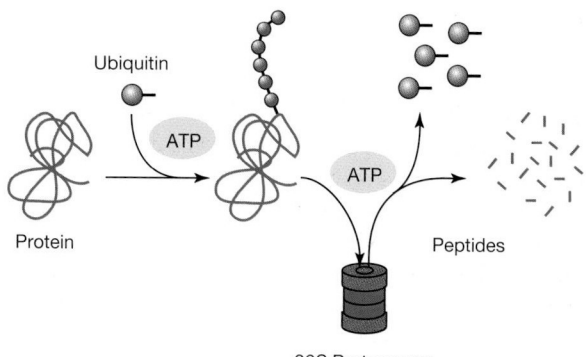

Figure 4.12   Proteasome Degradation of Proteins.   See text for details.

graded by this system. In eucaryotes, proteins are targeted for destruction by the attachment of several small ubiquitin polypeptides (**figure 4.12**). The marked protein then enters a huge cylindrical complex called a 26S **proteasome**, where it is degraded to peptides in an ATP-dependent process and the ubiquitins are released. The peptides may be hydrolyzed to amino acids. In this case the system is being used to recycle proteins. The proteasome also is involved in producing peptides for antigen presentation during many immunological responses (*see section 32.4*).

---

1. How do the rough and smooth endoplasmic reticulum differ from one another in terms of structure and function? List the processes in which the ER is involved.
2. Describe the structure of a Golgi apparatus in words and with a diagram. How do the cis and trans faces of the Golgi apparatus differ? List the major Golgi apparatus functions discussed in the text.
3. How are lysosomes formed? Describe the various forms of lysosomes and the way in which they participate in intracellular digestion. What is an autophagic vacuole? Define endocytosis, pinocytosis, and phagocytosis. Briefly describe the various kinds of pinocytosis. What is a proteasome?

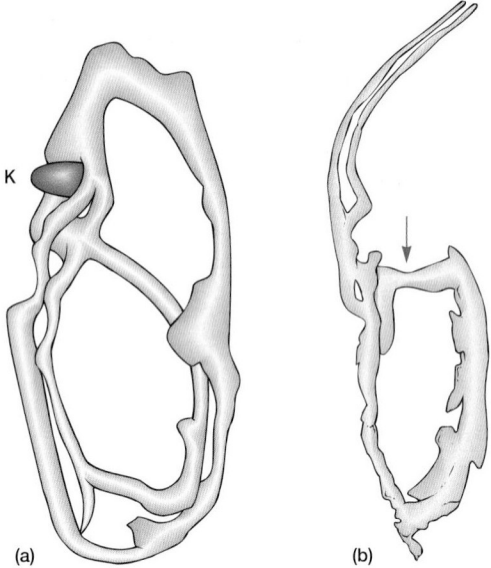

Figure 4.13 **Trypanosome Mitochondria.** The giant mitochondria from trypanosomes. (**a**) *Crithidia fasciculata* mitochondrion with kinetoplast, K. The kinetoplast contains DNA that codes for mitochondrial RNA and protein. (**b**) *Trypanosoma cruzi* mitochondrion with arrow indicating position of kinetoplast.

 EUCARYOTIC RIBOSOMES

The eucaryotic ribosome can either be associated with the endoplasmic reticulum or be free in the cytoplasmic matrix and is larger than the bacterial 70S ribosome. It is a dimer of a 60S and a 40S subunit, about 22 nm in diameter, and has a sedimentation coefficient of 80S and a molecular weight of 4 million. When bound to the endoplasmic reticulum to form rough ER, it is attached through its 60S subunit.

Both free and RER-bound ribosomes synthesize proteins. As mentioned earlier, proteins made on the ribosomes of the RER either enter its lumen for transport, and often for secretion, or are inserted into the ER membrane as integral membrane proteins. Free ribosomes are the sites of synthesis for nonsecretory and nonmembrane proteins. Some proteins synthesized by free ribosomes are inserted into organelles such as the nucleus, mitochondrion, and chloroplast. As discussed in chapters 3 and 12 (*see pp. 50, 267–69*), molecular chaperones aid the proper folding of proteins after synthesis. They also assist the transport of proteins into eucaryotic organelles such as mitochondria. Several ribosomes usually attach to a single messenger RNA (mRNA) and simultaneously translate its message into protein. These complexes of mRNA and ribosomes are called **polyribosomes** or **polysomes.** Ribosomal participation in protein synthesis is dealt with later. The role of ribosomes in protein synthesis (pp. 262–67)

As will be discussed later, eucaryotic mRNA synthesis occurs in the nucleus and is more complex than in procaryotes. The process often involves the removal of segments called introns and splicing of the RNA. The completed mRNA then complexes with cytoplasmic ribosomes and directs protein synthesis. Eucaryotic mRNA synthesis (pp. 257–60)

---

1. Describe the structure of the eucaryotic 80S ribosome and contrast it with the procaryotic ribosome.
2. How do free ribosomes and those bound to the ER differ in function?

 MITOCHONDRIA

Found in most eucaryotic cells, **mitochondria** (s., **mitochondrion**) frequently are called the "powerhouses" of the cell. Tricarboxylic acid cycle activity and the generation of ATP by electron transport and oxidative phosphorylation take place here. In the transmission electron microscope, mitochondria usually are cylindrical structures and measure approximately 0.3 to 1.0 μm by 5 to 10 μm. (In other words, they are about the same size as bacterial cells.) Although cells can possess as many as 1,000 or more mitochondria, at least a few cells (some yeasts, unicellular algae, and trypanosome protozoa) have a single giant tubular mitochondrion twisted into a continuous network permeating the cytoplasm (**figure 4.13**). The tricarboxylic acid cycle, electron transport, and oxidative phosphorylation (pp. 178–84)

The mitochondrion is bounded by two membranes, an outer mitochondrial membrane separated from an inner mitochondrial membrane by a 6 to 8 nm intermembrane space (**figure 4.14**). Special infoldings of the inner membrane, called **cristae** (s., **crista**), greatly increase its surface area. Their shape differs in mitochondria from various species. Fungi have platelike (laminar) cristae, whereas euglenoid flagellates may have cristae shaped like disks. Tubular cristae are found in a variety of eucaryotes; however, amoebae can possess mitochondria with cristae in the shape of vesicles (**figure 4.15**). The inner membrane encloses the mitochondrial matrix, a dense matrix containing ribosomes, DNA, and often large calcium phosphate granules. Mitochondrial ribosomes are smaller than cytoplasmic ribosomes and

Figure 4.14   Mitochondrial Structure.   (**a**) A diagram of mitochondrial structure. The insert shows F_1F_0 complexes lining the inner surface of the cristae. (**b**) Scanning electron micrograph ($\times 70,000$) of a freeze-fractured mitochondrion showing the cristae (arrows). The outer and inner mitochondrial membranes also are evident. (**c**) Transmission electron micrograph of a mitochondrion from a bat pancreas ($\times 85,000$). Note outer and inner mitochondrial membranes, cristae, and inclusions in the matrix. The mitochondrion is surrounded by rough endoplasmic reticulum.

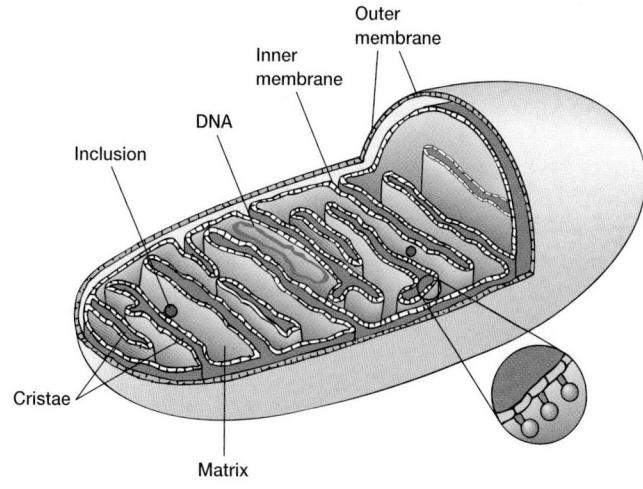

(a)

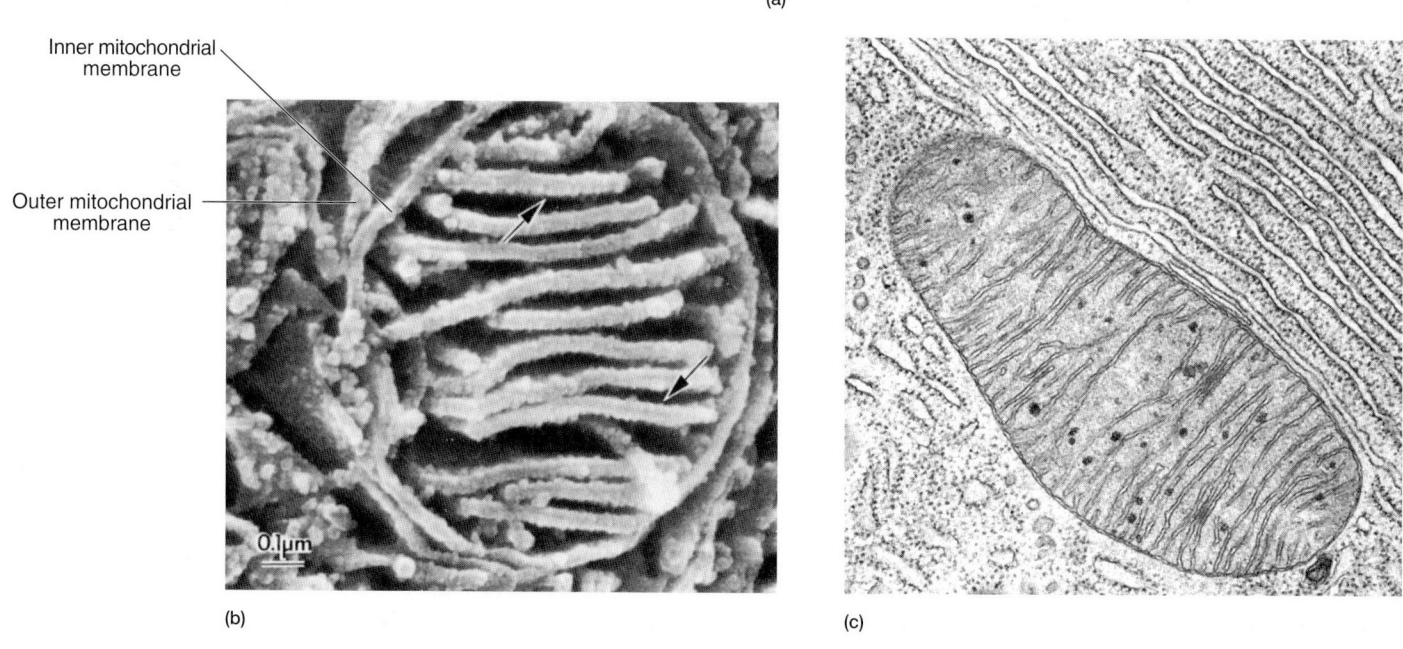

(b)

(c)

Figure 4.15   Mitochondrial Cristae. Mitochondria with a variety of cristae shapes. (**a**) Mitochondria from the protostelid slime mold *Schizoplasmodiopsis micropunctata.* Note the tubular cristae ($\times 49,500$). (**b**) The protozoan *Actinosphaerium* with vesicular cristae ($\times 75,000$).

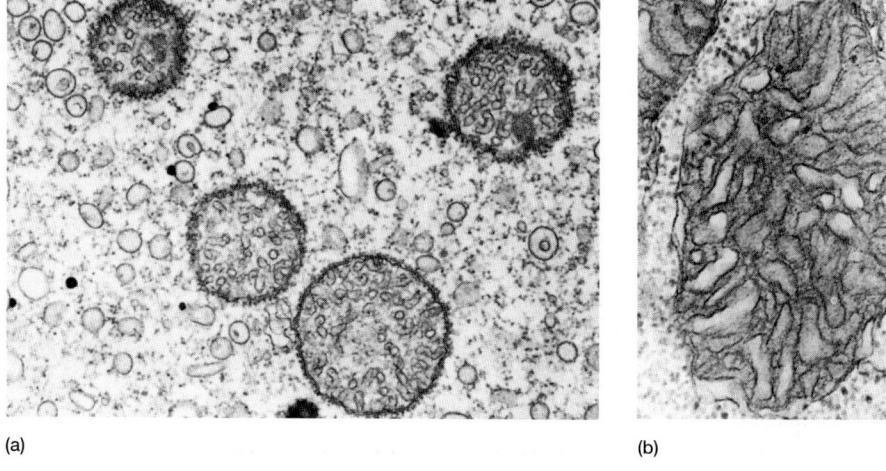

(a)

(b)

resemble those of bacteria in several ways, including their size and subunit composition. Mitochondrial DNA is a closed circle like bacterial DNA.

Each mitochondrial compartment is different from the others in chemical and enzymatic composition. The outer and inner mitochondrial membranes, for example, possess different lipids. Enzymes and electron carriers involved in electron transport and oxidative phosphorylation (the formation of ATP as a consequence of electron transport) are located only in the inner membrane. The enzymes of the tricarboxylic acid cycle and the β-oxidation pathway for fatty acids (*see chapter 9*) are located in the matrix.

The inner membrane of the mitochondrion has another distinctive structural feature related to its function. Many small spheres, about 8.5 nm diameter, are attached by stalks to its inner surface. The spheres are called **F₁ particles** and synthesize ATP during cellular respiration (*see pp. 182–84*).

The mitochondrion uses its DNA and ribosomes to synthesize some of its own proteins. In fact, mutations in mitochondrial DNA often lead to serious diseases in humans. Most mitochondrial proteins, however, are manufactured under the direction of the nucleus. Mitochondria reproduce by binary fission. Chloroplasts show similar partial independence and reproduction by binary fission. Because both organelles resemble bacteria to some ex-

tent, it has been suggested that these organelles arose from symbiotic associations between bacteria and larger cells (**Microbial Diversity & Ecology 4.1**).

 4.8   CHLOROPLASTS

Plastids are cytoplasmic organelles of algae and higher plants that often possess pigments such as chlorophylls and carotenoids, and are the sites of synthesis and storage of food reserves. The most important type of plastid is the chloroplast. **Chloroplasts** contain chlorophyll and use light energy to convert CO_2 and water to carbohydrates and O_2. That is, they are the site of photosynthesis.

Although chloroplasts are quite variable in size and shape, they share many structural features. Most often they are oval with dimensions of 2 to 4 μm by 5 to 10 μm, but some algae possess one huge chloroplast that fills much of the cell. Like mitochondria, chloroplasts are encompassed by two membranes (**figure 4.16**). A matrix, the **stroma,** lies within the inner membrane. It contains DNA, ribosomes, lipid droplets, starch granules, and a complex internal membrane system whose most prominent components are flattened, membrane-delimited sacs, the **thylakoids.** Clusters of two or more thylakoids are dispersed within the

Microbial Diversity & Ecology
4.1   The Origin of the Eucaryotic Cell

The profound differences between eucaryotic and procaryotic cells have stimulated much discussion about how the more complex eucaryotic cell arose. Some biologists believe the original "protoeucaryote" was a large aerobic archaean or bacterium that formed mitochondria, chloroplasts, and nuclei when its plasma membrane invaginated and enclosed genetic material in a double membrane. The organelles could then evolve independently. It also is possible that a large blue-green bacterium lost its cell wall and became phagocytic. Subsequently, primitive chloroplasts, mitochondria, and nuclei would be formed by the fusion of thylakoids and endoplasmic reticulum cisternae to enclose specific areas of cytoplasm.

By far the most popular theory for the origin of eucaryotic cells is the **endosymbiotic theory.** In brief, it is supposed that the ancestral procaryotic cell, which may have been an archaean, lost its cell wall and gained the ability to obtain nutrients by phagocytosing other procaryotes. When photosynthetic cyanobacteria arose, the environment slowly became aerobic. If an anaerobic, amoeboid, phagocytic procaryote—possibly already possessing a developed nucleus—engulfed an aerobic bacterial cell and established a permanent symbiotic relationship with it, the host would be better adapted to its increasingly aerobic environment. The endosymbiotic aerobic bacterium eventually would develop into the mitochondrion. Similarly, symbiotic associations with cyanobacteria could lead to the formation of chloroplasts and

photosynthetic eucaryotes. Some have speculated that cilia and flagella might have arisen from the attachment of spirochete bacteria (*see chapter 21*) to the surface of eucaryotic cells, much as spirochetes attach themselves to the surface of the motile protozoan *Myxotricha paradoxa* that grows in the digestive tract of termites.

There is evidence to support the endosymbiotic theory. Both mitochondria and chloroplasts resemble bacteria in size and appearance, contain DNA in the form of a closed circle like that of bacteria, and reproduce semiautonomously. Mitochondrial and chloroplast ribosomes resemble procaryotic ribosomes more closely than those in the eucaryotic cytoplasmic matrix. The sequences of the chloroplast and mitochondrial genes for ribosomal RNA and transfer RNA are more similar to bacterial gene sequences than to those of eucaryotic rRNA and tRNA nuclear genes. Finally, there are symbiotic associations that appear to be bacterial endosymbioses in which distinctive procaryotic characteristics are being lost. For example, the protozoan flagellate *Cyanophora paradoxa* has photosynthetic organelles called cyanellae with a structure similar to that of cyanobacteria and the remains of peptidoglycan in their walls. Their DNA is much smaller than that of cyanobacteria and resembles chloroplast DNA. Despite such evidence, the endosymbiotic theory still is somewhat speculative and the center of much continuing research and discussion.

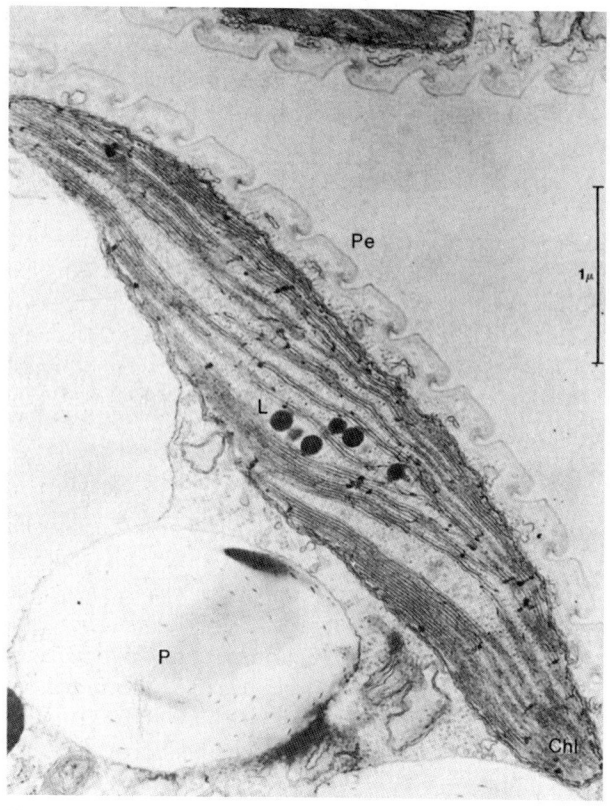

(a)

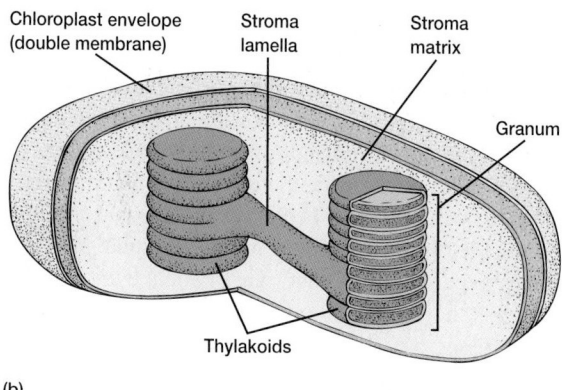

(b)

Figure 4.16   Chloroplast Structure.   (**a**) The chloroplast (Chl), of the euglenoid flagellate *Colacium cyclopicolum.* The chloroplast is bounded by a double membrane and has its thylakoids in groups of three or more. A paramylon granule (P), lipid droplets (L), and the pellicular strips (Pe), can be seen (×40,000). (**b**) A diagram of chloroplast structure.

stroma of most algal chloroplasts (figures 4.16 and 4.25*b*). In some groups of algae, several disklike thylakoids are stacked on each other like coins to form **grana** (s., **granum**).

Photosynthetic reactions are separated structurally in the chloroplast just as electron transport and the tricarboxylic acid cycle are in the mitochondrion. The formation of carbohydrate from CO_2 and water, the dark reaction, takes place in the stroma.

The trapping of light energy to generate ATP, NADPH, and O_2, the light reaction, is located in the thylakoid membranes, where chlorophyll and electron transport components are also found. Photosynthesis (pp. 190–96)

The chloroplasts of many algae contain a **pyrenoid** (figure 4.25*b*), a dense region of protein surrounded by starch or another polysaccharide. Pyrenoids participate in polysaccharide synthesis.

---

1. Describe in detail the structure of mitochondria and chloroplasts. Where are the different components of these organelles' energy trapping systems located?
2. Define F_1 particle, plastid, dark reaction, light reaction, and pyrenoid.
3. What is the role of mitochondrial DNA?

4.9 THE NUCLEUS AND CELL DIVISION

The cell **nucleus** is by far the most visually prominent organelle. It was discovered early in the study of cell structure and was shown by Robert Brown in 1831 to be a constant feature of eucaryotic cells. The nucleus is the repository for the cell's genetic information and is its control center.

Nuclear Structure

Nuclei are membrane-delimited spherical bodies about 5 to 7 μm in diameter (figures 4.2 and 4.25*b*). Dense fibrous material called **chromatin** can be seen within the nucleoplasm of the nucleus of a stained cell. This is the DNA-containing part of the nucleus. In nondividing cells, chromatin exists in a dispersed condition, but condenses during mitosis to become visible as **chromosomes.** Some nuclear chromatin, the euchromatin, is loosely organized and contains those genes that are expressing themselves actively. Heterochromatin is coiled more tightly, appears darker in the electron microscope, and is not genetically active most of the time.   Organization of DNA in eucaryotic nuclei (pp. 227–28)

The nucleus is bounded by the **nuclear envelope** (figures 4.2 and 4.25*b*), a complex structure consisting of inner and outer membranes separated by a 15 to 75 nm perinuclear space. The envelope is continuous with the ER at several points and its outer membrane is covered with ribosomes. A network of intermediate filaments, called the nuclear lamina, lies against the inner surface of the envelope and supports it. Chromatin usually is associated with the inner membrane.

Many **nuclear pores** penetrate the envelope (**figure 4.17**), each pore formed by a fusion of the outer and inner membranes. Pores are about 70 nm in diameter and collectively occupy about 10 to 25% of the nuclear surface. A complex ringlike arrangement of granular and fibrous material called the annulus is located at the edge of each pore.

The nuclear pores serve as a transport route between the nucleus and surrounding cytoplasm. Particles have been observed moving into the nucleus through the pores. Although the function

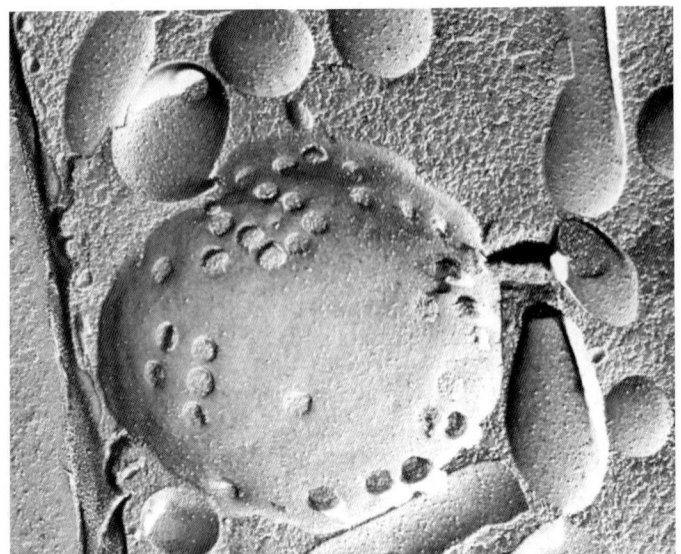

Figure 4.17   The Nucleus.   A freeze-etch preparation of the conidium of the fungus *Geotrichum candidum* (×44,600). Note the large convex nuclear surface with nuclear pores scattered over it.

of the annulus is not understood, it may either regulate or aid the movement of material through the pores. Substances also move directly through the nuclear envelope by unknown mechanisms.

The Nucleolus

Often the most noticeable structure within the nucleus is the **nucleolus** (figures 4.2 and 4.25*b*). A nucleus may contain from one to many nucleoli. Although the nucleolus is not membrane-enclosed, it is a complex organelle with separate granular and fibrillar regions. It is present in nondividing cells, but frequently disappears during mitosis. After mitosis the nucleolus reforms around the nucleolar organizer, a particular part of a specific chromosome.

The nucleolus plays a major role in ribosome synthesis. The nucleolar organizer DNA directs the production of ribosomal RNA (rRNA). This RNA is synthesized in a single long piece that then is cut to form the final rRNA molecules. The processed rRNAs next combine with ribosomal proteins (which have been synthesized in the cytoplasmic matrix) to form partially completed ribosomal subunits. The granules seen in the nucleolus are probably these subunits. Immature ribosomal subunits then leave the nucleus, presumably by way of the nuclear envelope pores and mature in the cytoplasm.   RNA splicing (pp. 257–60)

Mitosis and Meiosis

When a eucaryotic microorganism reproduces, its genetic material must be duplicated and then separated so that each new nucleus possesses a complete set of chromosomes. This process of nuclear division and chromosome distribution in eucaryotic cells

is called **mitosis.** Mitosis actually occupies only a small portion of a microorganism's life as can be seen by examining the **cell cycle** (**figure 4.18**). The cell cycle is the total sequence of events in the growth-division cycle between the end of one division and the end of the next. Cell growth takes place in the **interphase,** that portion of the cycle between periods of mitosis. Interphase is composed of three parts. The G_1 period (gap 1 period) is a time of active synthesis of RNA, ribosomes, and other cytoplasmic constituents accompanied by considerable cell growth. This is followed by the S period (synthesis period) in which DNA is replicated and doubles in quantity. Finally, there is a second gap, the G_2 period, when the cell prepares for mitosis, the M period, by activities such as the synthesis of special division proteins. The total length of the cycle differs considerably between microorganisms, usually due to variations in the length of G_1.

Mitotic events are summarized in figure 4.18. During mitosis, the genetic material duplicated during the S period is distributed equally to the two new nuclei so that each has a full set of genes. There are four phases in mitosis. In prophase, the chromosomes—each with two chromatids—become visible and move toward the equator of the cell. The mitotic spindle forms, the nucleolus disappears, and the nuclear envelope begins to dissolve. The chromosomes are arranged in the center of the spindle during metaphase and the nuclear envelope has disappeared. During anaphase the chromatids in each chromosome separate and move toward the opposite poles of the spindle. Finally during telophase the chromatids become less visible, the nucleolus reappears, and a nuclear envelope reassembles around each set of chromatids to form two new nuclei.

Mitosis in eucaryotic microorganisms can differ from that pictured in figure 4.18. For example, the nuclear envelope does not disappear in many fungi and some protozoa and algae (**figure 4.19**). Frequently cytokinesis, the division of the parental cell's cytoplasm to form new cells, begins during anaphase and finishes by the end of telophase. However, mitosis can take place without cytokinesis to generate multinucleate or coenocytic cells.

In mitosis the original number of chromosomes is the same after division and a diploid organism will remain diploid or 2N (i.e., it still has two copies of each chromosome). Frequently a microorganism reduces its chromosome number by half, from the diploid state to the haploid or 1N (a single copy of each chromosome). Haploid cells may immediately act as gametes and fuse to reform diploid organisms or may form gametes only after a considerable delay (**figure 4.20**). The process by which the number of chromosomes is reduced in half with each daughter cell receiving one complete set of chromosomes is called **meiosis.** Life cycles can be quite complex in eucaryotic microorganisms; a classic example is the life cycle of *Plasmodium,* the cause of malaria (*see pp. 929–31*).   Life cycles of eucaryotic microorganisms (chapters 25–27)

Meiosis is quite complex and involves two stages. The first stage differs markedly from mitosis. During prophase, homologous chromosomes come together and lie side-by-side, a process known as synapsis. Then the double-stranded chromosomes from each homologous pair move to opposite poles in anaphase. In

Figure 4.18 The Eucaryotic Cell Cycle.
The length of the M period has been increased disproportionately in order to show the phases of mitosis. G$_1$ period: synthesis of mRNA, tRNA, ribosomes, and cytoplasmic constituents. Nucleolus grows rapidly. S period: rapid synthesis and doubling of nuclear DNA and histones. G$_2$ period: preparation for mitosis and cell division. M period: mitosis (prophase, metaphase, anaphase, telophase) and cytokinesis.

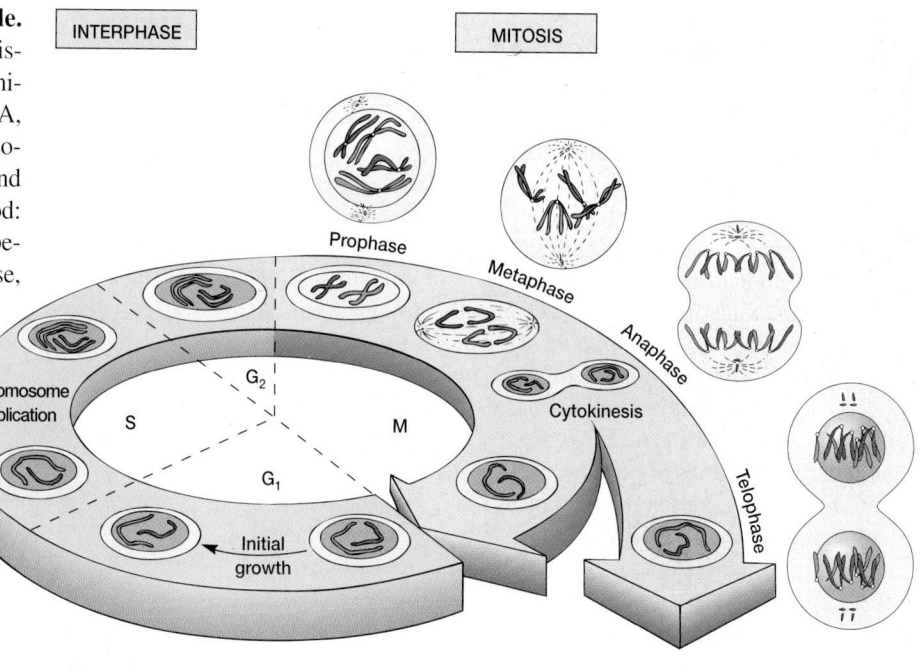

INTERPHASE

MITOSIS

Prophase

Metaphase

Anaphase

Chromosome replication

G$_2$

S

M

Cytokinesis

G$_1$

Telophase

Initial growth

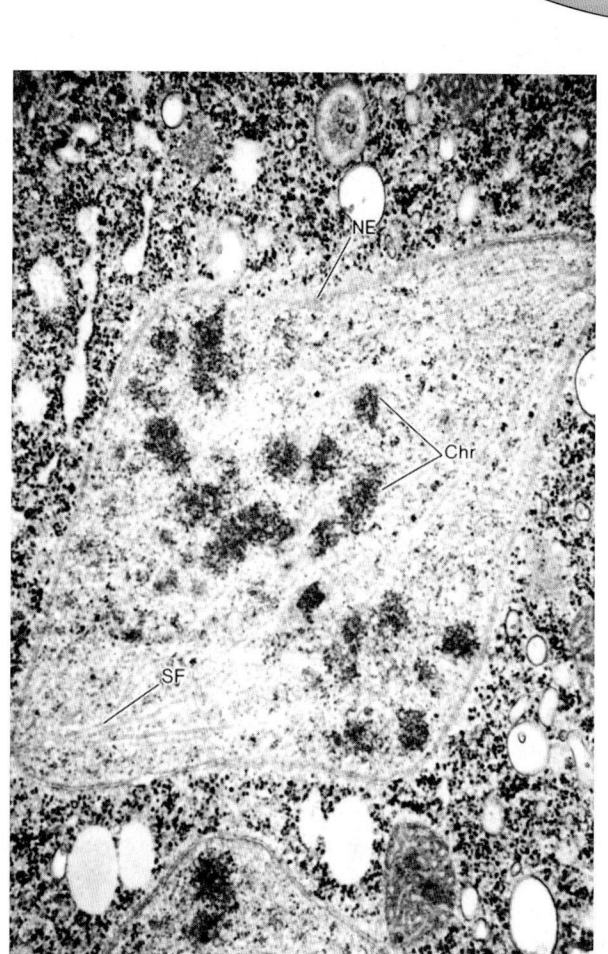

Figure 4.19 Mitosis with an Intact Nuclear Envelope.
Mitosis in the slime mold *Physarum flavicomum.* The nuclear envelope, NE, remains intact, and the spindle is intranuclear. The process is at metaphase with the chromosomes, Chr, aligned in the center and attached to spindle fibers, SF (×15,000).

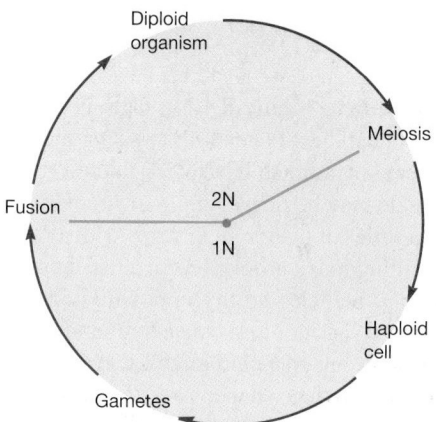

Diploid organism

Meiosis

Fusion

2N

1N

Haploid cell

Gametes

Figure 4.20 Generalized Eucaryotic Life Cycle.

contrast, during mitotic anaphase the two strands of each chromosome separate and move to opposite poles. Consequently the number of chromosomes is halved in meiosis but not in mitosis. The second stage of meiosis is similar to mitosis in terms of mechanics, and single-stranded chromosomes are separated. After completion of meiosis I and meiosis II, the original diploid cell has been transformed into four haploid cells.

---

1. Describe the structure of the nucleus. What are euchromatin and heterochromatin? What is the role of the pores in the nuclear envelope?
2. Briefly discuss the structure and function of the nucleolus. What is the nucleolar organizer?
3. Describe the eucaryotic cell cycle, its periods, and the process of mitosis. What is meiosis, how does it take place, and what is its role in the microbial life cycle?

4.10 EXTERNAL CELL COVERINGS

Eucaryotic microorganisms differ greatly from procaryotes in the supporting or protective structures they have external to the plasma membrane. In contrast with most bacteria, many eucaryotes lack an external cell wall. The amoeba is an excellent example. Eucaryotic cell membranes, unlike most procaryotic membranes, contain sterols such as cholesterol in their lipid bilayers, and this may make them mechanically stronger, thus reducing the need for external support. (However, as mentioned on page 44, many procaryotic membranes are strengthened by hopanoids.) Of course many eucaryotes do have a rigid external **cell wall.** Algal cell walls usually have a layered appearance and contain large quantities of polysaccharides such as cellulose and pectin. In addition, inorganic substances like silica (in diatoms) or calcium carbonate (some red algae) may be present. Fungal cell walls normally are rigid. Their exact composition varies with the organism; but usually, cellulose, chitin, or glucan (a glucose polymer different from cellulose) are present. Despite their nature the rigid materials in eucaryotic walls are chemically simpler than procaryotic peptidoglycan. Bacterial cell wall structure and chemistry (pp. 52–58)

Many protozoa and some algae have a different external structure, the **pellicle** (figure 4.16*a*). This is a relatively rigid layer of components just beneath the plasma membrane (sometimes the plasma membrane is also considered part of the pellicle). The pellicle may be fairly simple in structure. For example, *Euglena* has a series of overlapping strips with a ridge at the edge of each strip fitting into a groove on the adjacent one. In contrast, ciliate protozoan pellicles are exceptionally complex with two membranes and a variety of associated structures. Although pellicles are not as strong and rigid as cell walls, they do give their possessors a characteristic shape.

4.11 CILIA AND FLAGELLA

Cilia (s., **cilium**) and **flagella** (s., **flagellum**) are the most prominent organelles associated with motility. Although both are whiplike and beat to move the microorganism along, they differ from one another in two ways. First, cilia are typically only 5 to 20 μm in length, whereas flagella are 100 to 200 μm long. Second, their patterns of movement are usually distinctive (**figure 4.21**). Flagella move in an undulating fashion and generate planar or helical waves originating at either the base or the tip. If the wave moves from base to tip, the cell is pushed along; a beat traveling from the tip toward the base pulls the cell through the water. Sometimes the flagellum will have lateral hairs called flimmer filaments (thicker, stiffer hairs are called mastigonemes). These filaments change flagellar action so that a wave moving down the filament toward the tip pulls the cell along instead of pushing it. Such a flagellum often is called a tinsel flagellum, whereas the naked flagellum is referred to as a whiplash flagellum (**figure**

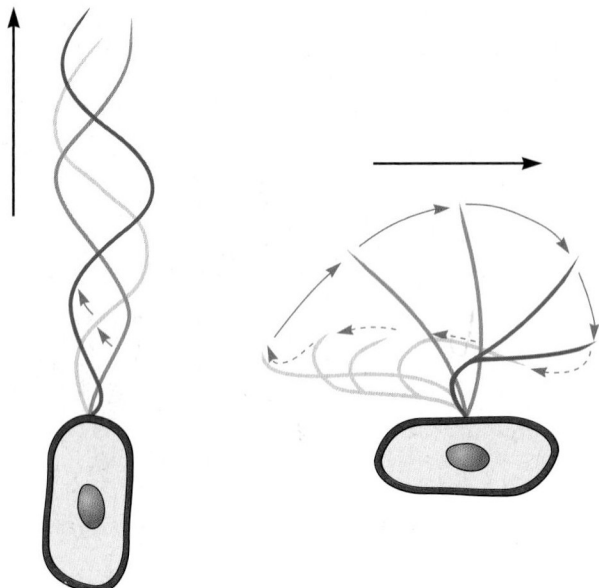

Figure 4.21 Patterns of Flagellar Movement. Flagellar movement (left illustration) often takes the form of waves that move either from the base of the flagellum to its tip or in the opposite direction. The motion of these waves propels the organism along. The beat of a cilium (right illustration) may be divided into two phases. In the effective stroke, the cilium remains fairly stiff as it swings through the water. This is followed by a recovery stroke in which the cilium bends and returns to its initial position. The black arrows indicate the direction of water movement in these examples.

4.22). Cilia, on the other hand, normally have a beat with two distinctive phases. In the effective stroke, the cilium strokes through the surrounding fluid like an oar, thereby propelling the organism along in the water. The cilium next bends along its length while it is pulled forward during the recovery stroke in preparation for another effective stroke. A ciliated microorganism actually coordinates the beats so that some of its cilia are in the recovery phase while others are carrying out their effective stroke (**figure 4.23**). This coordination allows the organism to move smoothly through the water.

Despite their differences, cilia and flagella are very similar in ultrastructure. They are membrane-bound cylinders about 0.2 μm in diameter. Located in the matrix of the organelle is a complex, the **axoneme,** consisting of nine pairs of microtubule doublets arranged in a circle around two central tubules (**figure 4.24**). This is called the 9 + 2 pattern of microtubules. Each doublet also has pairs of arms projecting from subtubule A (the complete microtubule) toward a neighboring doublet. A radial spoke extends from subtubule A toward the internal pair of microtubules with their central sheath. These microtubules are similar to those found in the cytoplasm. Each is constructed of two types of tubulin subunits, α- and β-tubulins, that resemble the contractile protein actin in their composition. Bacterial flagella and motility (pp. 63–66)

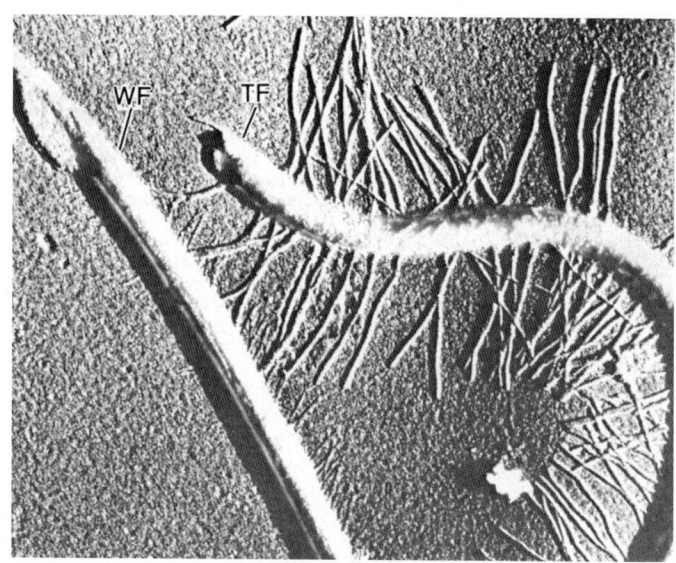

Figure 4.22   Whiplash and Tinsel Flagella.   Transmission electron micrograph of a shadowed whiplash flagellum, WF, and a tinsel flagellum, TF, with mastigonemes.

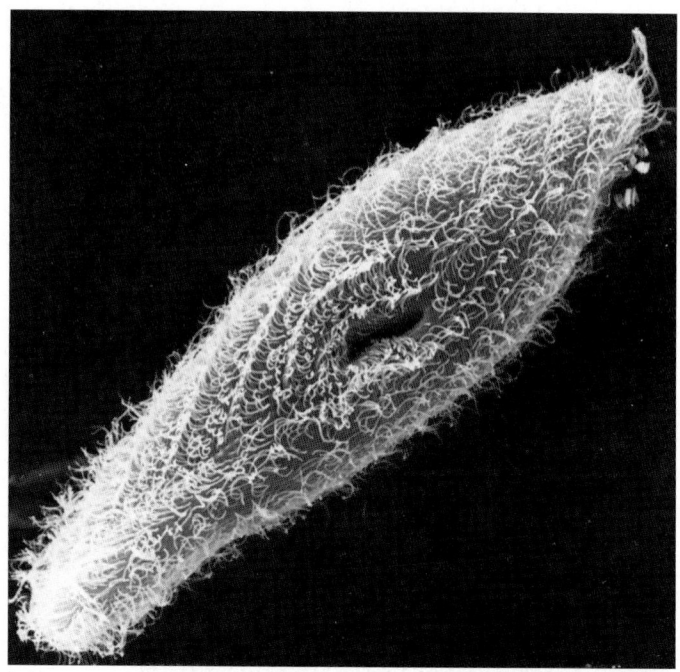

Figure 4.23   Coordination of Ciliary Activity.   A scanning electron micrograph of *Paramecium* showing cilia (×1,500). The ciliary beat is coordinated and moves in waves across the protozoan's surface, as can be seen in the photograph.

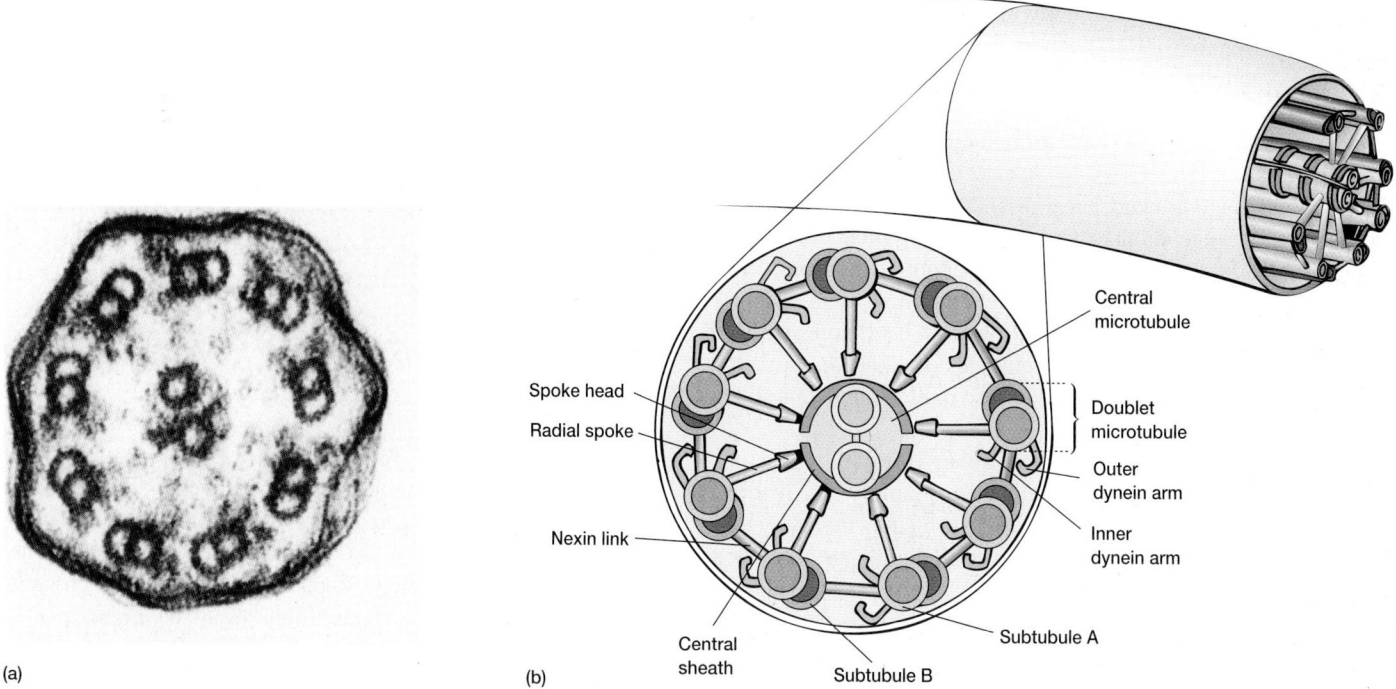

(a)   (b)

Figure 4.24   Cilia and Flagella Structure.   (**a**) An electron micrograph of a cilium cross section. Note the two central microtubules surrounded by nine microtubule doublets (×160,000). (**b**) A diagram of cilia and flagella structure with two doublets removed for sake of visibility.

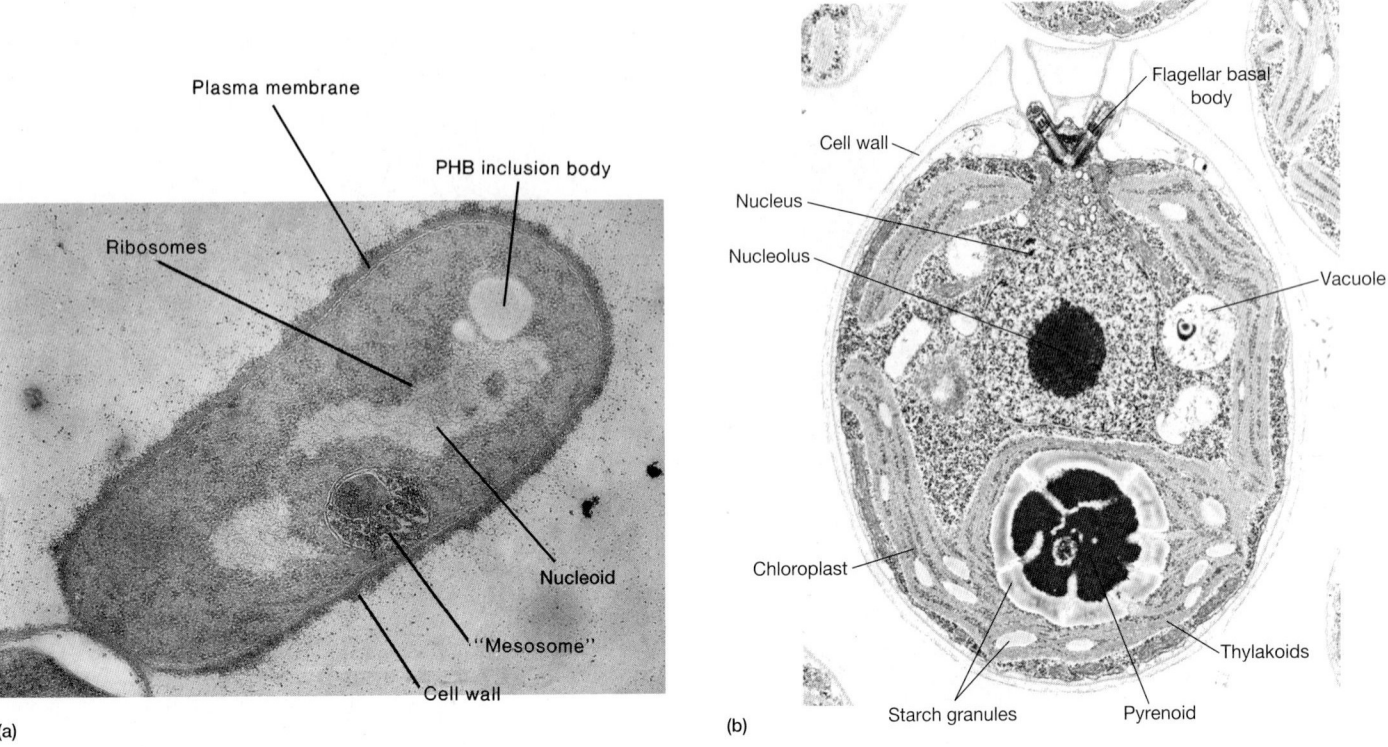

(a)   (b)

Figure 4.25   Comparison of Procaryotic and Eucaryotic Cell Structure. (a) The procaryote *Bacillus megaterium* (×30,500). (b) The eucaryotic alga *Chlamydomonas reinhardtii,* a deflagellated cell. Note the large chloroplast with its pyrenoid body (×30,000).

A **basal body** lies in the cytoplasm at the base of each cilium or flagellum. It is a short cylinder with nine microtubule triplets around its periphery (a 9 + 0 pattern) and is separated from the rest of the organelle by a basal plate. The basal body directs the construction of these organelles. Cilia and flagella appear to grow through the addition of preformed microtubule subunits at their tips.

Cilia and flagella bend because adjacent microtubule doublets slide along one another while maintaining their individual lengths. The doublet arms (figure 4.24), about 15 nm long, are made of the protein **dynein.** ATP powers the movement of cilia and flagella, and isolated dynein hydrolyzes ATP. It appears that dynein arms interact with the B subtubules of adjacent doublets to cause the sliding. The radial spokes also participate in this sliding motion.

Cilia and flagella beat at a rate of about 10 to 40 strokes or waves per second and propel microorganisms rapidly. The record holder is the flagellate *Monas stigmatica,* which swims at a rate of 260 μm/second (approximately 40 cell lengths per second); the common euglenoid flagellate, *Euglena gracilis,* travels at around 170 μm or 3 cell lengths per second. The ciliate protozoan *Paramecium caudatum* swims at about 2,700 μm/second (12 lengths per second). Such speeds are equivalent to or much faster than those seen in higher animals.

1. How do eucaryotic microorganisms differ from procaryotes with respect to supporting or protective structures external to the plasma membrane? Describe the pellicle and indicate which microorganisms have one.
2. Prepare and label a diagram showing the detailed structure of a cilium or flagellum. How do cilia and flagella move, and what is dynein's role in the process?

4.12   COMPARISON OF PROCARYOTIC AND EUCARYOTIC CELLS

A comparison of the cells in **figure 4.25** demonstrates that there are many fundamental differences between eucaryotic and procaryotic cells. **Eucaryotic cells** have a membrane-enclosed nucleus. In contrast, **procaryotic cells** lack a true, membrane-delimited nucleus. Bacteria and Archaea are procaryotes; all other organisms—algae, fungi, protozoa, higher plants, and animals—are eucaryotic. Most procaryotes are smaller than eucaryotic cells, often about the size of eucaryotic mitochondria and chloroplasts.

Table 4.2
Comparison of Procaryotic and Eucaryotic Cells

Property	Procaryotes	Eucaryotes
Organization of Genetic Material		
True membrane-bound nucleus	Absent	Present
DNA complexed with histones	No	Yes
Number of chromosomes	One[a]	More than one
Introns in genes	Rare	Common
Nucleolus	Absent	Present
Mitosis occurs	No	Yes
Genetic Recombination	Partial, unidirectional transfer of DNA	Meiosis and fusion of gametes
Mitochondria	Absent	Present
Chloroplasts	Absent	Present
Plasma Membrane with Sterols	Usually no[b]	Yes
Flagella	Submicrosopic in size; composed of one fiber	Microscopic in size; membrane bound; usually 20 microtubules in 9 + 2 pattern
Endoplasmic Reticulum	Absent	Present
Golgi Apparatus	Absent	Present
Cell Walls	Usually chemically complex with peptidoglycan[c]	Chemically simpler and lacking peptidoglycan
Differences in Simpler Organelles		
Ribosomes	70S	80S (except in mitochondria and chloroplasts)
Lysosomes and peroxisomes	Absent	Present
Microtubules	Absent or rare	Present
Cytoskeleton	May be absent	Present
Differentiation	Rudimentary	Tissues and organs

[a]Plasmids may provide additional genetic information.
[b]Only the mycoplasmas and methanotrophs (methane utilizers) contain sterols. The mycoplasmas cannot synthesize sterols and require them preformed. Many procaryotes contain hopanoids.
[c]The mycoplasmas and Archaea do not have peptidoglycan cell walls.

The presence of the eucaryotic nucleus is the most obvious difference between these two cell types, but several other major distinctions should be noted. It is clear from **table 4.2** that procaryotic cells are much simpler structurally. In particular, an extensive and diverse collection of membrane-delimited organelles is missing. Furthermore, procaryotes are simpler functionally in several ways. They lack mitosis and meiosis, and have a simpler genetic organization. Many complex eucaryotic processes are absent in procaryotes: phagocytosis and pinocytosis, intracellular digestion, directed cytoplasmic streaming, ameboid movement, and others.

Despite the many significant differences between these two basic cell forms, they are remarkably similar on the biochemical level as will be discussed in succeeding chapters. Procaryotes and eucaryotes are composed of similar chemical constituents. With a few exceptions the genetic code is the same in both, as is the way in which the genetic information in DNA is expressed. The principles underlying metabolic processes and most of the more important metabolic pathways are identical. Thus beneath the profound structural and functional differences between procaryotes and eucaryotes, there is an even more fundamental unity: a molecular unity that is basic to all known life processes.

4.1 An Overview of Eucaryotic Cell Structure
a. The eucaryotic cell has a true, membrane-delimited nucleus and many membranous organelles (**table 4.1**).

4.2 The Cytoplasmic Matrix, Microfilaments, Intermediate Filaments, and Microtubules
a. The cytoplasmic matrix contains microfilaments, intermediate filaments, and microtubules, small organelles partly responsible for

cell structure and movement. These and other types of filaments are organized into a cytoskeleton.

4.3 The Endoplasmic Reticulum
a. The matrix is permeated by an irregular network of tubules and flattened sacs or cisternae known as the endoplasmic reticulum (ER). The ER may have attached ribosomes and be active in protein synthesis (rough endoplasmic reticulum) or lack ribosomes (smooth ER).

b. The ER can donate materials to the Golgi apparatus, an organelle composed of one or more stacks of cisternae (**figure 4.9**). This organelle prepares and packages cell products for secretion.

4.4 The Golgi Apparatus
a. The Golgi apparatus also forms lysosomes (**figures 4.10** and **4.11**). These organelles contain digestive enzymes and aid in intracellular digestion of materials.

4.5 Lysosomes and Endocytosis
a. Eucaryotes ingest materials using two major kinds of endocytosis: phagocytosis and pinocytosis. Three types of pinocytosis are fluid-phase endocytosis, receptor-mediated endocytosis, and the formation of caveolae.

4.6 Eucaryotic Ribosomes
a. Eucaryotic ribosomes found free in the cytoplasmic matrix or bound to the ER are 80S ribosomes. Several may be attached to the same messenger RNA forming polyribosomes or polysomes.

4.7 Mitochondria
a. Mitochondria are organelles bounded by two membranes, with the inner membrane folded into cristae, and are responsible for energy generation by the tricarboxylic acid cycle, electron transport, and oxidative phosphorylation (**figure 4.14**).

4.8 Chloroplasts
a. Chloroplasts are the site of photosynthesis. The trapping of light energy takes place in the thylakoid membranes, whereas CO_2 incorporation is located in the stroma (**figure 4.16**).

4.9 The Nucleus and Cell Division
a. The nucleus is a large organelle containing the cell's chromosomes. It is bounded by a complex, double-membrane envelope perforated by pores through which materials can move.
b. The nucleolus lies within the nucleus and participates in the synthesis of ribosomal RNA and ribosomal subunits.
c. Eucaryotic chromosomes are distributed to daughter cells during regular cell division by mitosis (**figure 4.18**). Meiosis is used to halve the chromosome number during sexual reproduction.

4.10 External Cell Coverings
a. When a cell wall is present, it is constructed from polysaccharides, like cellulose, that are chemically simpler than procaryotic peptidoglycan. Many protozoa have a pellicle rather than a cell wall.

4.11 Cilia and Flagella
a. Many eucaryotic cells are motile because of cilia and flagella, membrane-delimited organelles with nine microtubule doublets surrounding two central microtubules (**figure 4.24**). The doublets slide along each other to bend the cilium or flagellum.

4.12 Comparison of Procaryotic and Eucaryotic Cells
a. Despite the fact that eucaryotes and procaryotes differ in many ways (**table 4.2**), they are quite similar metabolically.

Key Terms

autophagic vacuole   80
axoneme   88
basal body   90
caveolae   80
cell cycle   86
cell wall   88
chloroplast   84
chromatin   85
chromosome   85
cilia   88
cisternae   78
clathrin   80
coated vesicles   80
cristae   82
cytoplasmic matrix   76

cytoskeleton   78
dictyosome   79
dynein   90
endocytosis   79
endoplasmic reticulum (ER)   78
endosome   79
endosymbiotic theory   84
eucaryotic cells   90
F_1 particle   84
flagella   88
fluid-phase endocytosis   79
Golgi apparatus   78
grana   85
intermediate filament   78
interphase   86

lysosome   79
meiosis   86
microfilament   76
microtubule   77
mitochondrion   82
mitosis   86
nuclear envelope   85
nuclear pores   85
nucleolus   86
nucleus   85
organelle   75
pellicle   88
phagocytosis   79
pinocytosis   79

plastid   84
polyribosomes   82
polysomes   82
primary lysosomes   80
procaryotic cells   90
proteasome   81
pyrenoid   85
receptor-mediated endocytosis   80
residual body   80
rough ER (RER)   78
secondary lysosomes   80
smooth ER (SER)   78
stroma   84
thylakoid   84

Questions for Thought and Review

1. Describe the structure and function of every eucaryotic organelle discussed in the chapter.
2. Discuss the statement: "The most obvious difference between eucaryotic and procaryotic cells is in their use of membranes." What general roles do membranes play in eucaryotic cells?

3. Describe how the Golgi apparatus distributes proteins it receives from the ER to different organelles.
4. Briefly discuss how the complex of membranous organelles that de Duve calls the "vacuome" functions as a coordinated whole. What is its function?

5. Describe and contrast the ways in which flagella and cilia propel microorganisms through the water.
6. Outline the major differences between procaryotes and eucaryotes. How are they similar?

For recommended readings on these and related topics, see Appendix V.

Microbial Nutrition

Concepts

- Microorganisms require about 10 elements in large quantities, in part because they are used to construct carbohydrates, lipids, proteins, and nucleic acids. Several other elements are needed in very small amounts and are parts of enzymes and cofactors.
- All microorganisms can be placed in one of a few nutritional categories on the basis of their requirements for carbon, energy, and hydrogen atoms or electrons.
- Nutrient molecules frequently cannot cross selectively permeable plasma membranes through passive diffusion. They must be transported by one of three major mechanisms involving the use of membrane carrier proteins. Eucaryotic microorganisms also employ endocytosis for nutrient uptake.
- Culture media are needed to grow microorganisms in the laboratory and to carry out specialized procedures like microbial identification, water and food analysis, and the isolation of specific microorganisms. Many different media are available for these and other purposes.
- Pure cultures can be obtained through the use of spread plates, streak plates, or pour plates and are required for the careful study of an individual microbial species.

The whole of nature, as has been said, is a conjugation of the verb to eat, in the active and passive.

—*William Ralph Inge*

Staphylococcus aureus forms large, golden colonies when growing on blood agar. This human pathogen causes diseases such as boils, abscesses, bacteremia, endocarditis, food poisoning, pharyngitis, and pneumonia.

To obtain energy and construct new cellular components, organisms must have a supply of raw materials or nutrients. **Nutrients** are substances used in biosynthesis and energy release and therefore are required for microbial growth. This chapter describes the nutritional requirements of microorganisms, how nutrients are acquired, and the cultivation of microorganisms.

Environmental factors such as temperature, oxygen levels, and the osmotic concentration of the medium are critical in the successful cultivation of microorganisms. These topics are discussed in chapter 6 after an introduction to microbial growth.

5.1 THE COMMON NUTRIENT REQUIREMENTS

Analysis of microbial cell composition shows that over 95% of cell dry weight is made up of a few major elements: carbon, oxygen, hydrogen, nitrogen, sulfur, phosphorus, potassium, calcium, magnesium, and iron. These are called **macroelements** or macronutrients because they are required by microorganisms in relatively large amounts. The first six (C, O, H, N, S, and P) are components of carbohydrates, lipids, proteins, and nucleic acids. The remaining four macroelements exist in the cell as cations and play a variety of roles. For example, potassium (K^+) is required for activity by a number of enzymes, including some of those involved in protein synthesis. Calcium (Ca^{2+}), among other functions, contributes to the heat resistance of bacterial endospores. Magnesium (Mg^{2+}) serves as a cofactor for many enzymes, complexes with ATP, and stabilizes ribosomes and cell membranes. Iron (Fe^{2+} and Fe^{3+}) is a part of cytochromes and a cofactor for enzymes and electron-carrying proteins.

All organisms, including microorganisms, require several **micronutrients** or **trace elements** besides macroelements. The micronutrients—manganese, zinc, cobalt, molybdenum, nickel, and copper—are needed by most cells. However, cells require such small amounts that contaminants from water, glassware, and regular media components often are adequate for growth. Therefore it is very difficult to demonstrate a micronutrient requirement. In nature, micronutrients are ubiquitous and probably do not usually limit growth. Micronutrients are normally a part of enzymes and cofactors, and they aid in the catalysis of reactions and maintenance of protein structure. For example, zinc (Zn^{2+}) is present at the active site of some enzymes but is also involved in the association of regulatory and catalytic subunits in *E. coli* aspartate carbamoyltransferase (*see section 8.9*). Manganese (Mn^{2+}) aids many enzymes catalyzing the transfer of phosphate groups. Molybdenum (Mo^{2+}) is required for nitrogen fixation, and cobalt (Co^{2+}) is a component of vitamin B_{12}. Electron carriers and enzymes (pp. 155–60)

Besides the common macroelements and trace elements, microorganisms may have particular requirements that reflect the special nature of their morphology or environment. Diatoms (*see figure 26.6c,d*) need silicic acid (H_4SiO_4) to construct their beautiful cell walls of silica [$(SiO_2)_n$]. Although most bacteria do not require large amounts of sodium, many bacteria growing in saline lakes and oceans (*see pp. 120–21, 447–49*) depend on the presence of high concentrations of sodium ion (Na^+).

Finally, it must be emphasized that microorganisms require a balanced mixture of nutrients. If an essential nutrient is in short supply, microbial growth will be limited regardless of the concentrations of other nutrients.

5.2 REQUIREMENTS FOR CARBON, HYDROGEN, AND OXYGEN

The requirements for carbon, hydrogen, and oxygen often are satisfied together. Carbon is needed for the skeleton or backbone of all organic molecules, and molecules serving as carbon sources normally also contribute both oxygen and hydrogen atoms. They are the source of all three elements.

In addition to supplies of carbon, hydrogen, and oxygen, organisms require a source of electrons. Electron movement through electron transport chains and during other oxidation-reduction reactions can provide energy for use in work. Electrons also are needed to reduce molecules during biosynthesis (e.g., the reduction of CO_2 to form organic molecules). Because organic nutrients that supply carbon, hydrogen, and oxygen are almost always reduced, they have electrons to donate for use in energy release and biosynthesis. Indeed, the more reduced organic molecules are, the higher their energy content (e.g., lipids have a higher energy content than carbohydrates). This is because, as we shall see later, electron transfers release energy when the electrons move from reduced donors with more negative reduction potentials to oxidized electron acceptors with more positive potentials. Thus carbon sources frequently also serve as energy sources, although they don't have to. Oxidation-reduction reactions and energy (pp. 153–55)

One important carbon source that does not supply hydrogen or energy is carbon dioxide (CO_2). This is because CO_2 is oxidized and lacks hydrogen. Probably all microorganisms can fix CO_2—that is, reduce it and incorporate it into organic molecules. However, by definition, only **autotrophs** can use CO_2 as their sole or principal source of carbon. Many microorganisms are autotrophic, and most of these carry out photosynthesis and use light as their energy source. Some autotrophs oxidize inorganic molecules and derive energy from electron transfers. Photosynthetic carbon dioxide fixation (pp. 202–3)

The reduction of CO_2 is a very energy-expensive process. Thus many microorganisms cannot use CO_2 as their sole carbon source but must rely on the presence of more reduced, complex molecules such as glucose for a supply of carbon. Organisms that use reduced, preformed organic molecules as carbon sources are **heterotrophs** (these preformed molecules normally come from other organisms). As mentioned previously, most heterotrophs use reduced organic compounds as sources of both carbon and energy. For example, the glycolytic pathway produces carbon skeletons for use in biosynthesis and also releases energy as ATP and NADH. The glycolytic pathway (p. 171)

A most remarkable nutritional characteristic of microorganisms is their extraordinary flexibility with respect to carbon sources. Laboratory experiments indicate that there is no naturally occurring organic molecule that cannot be used by some microorganism. Actinomycetes will degrade amyl alcohol, paraffin, and even rubber. Some bacteria seem able to employ almost anything as a carbon source; for example, *Burkholderia cepacia* can use over 100 different carbon compounds. In contrast to these bacterial omnivores, some bacteria are exceedingly fastidious and catabolize only a few carbon compounds. Cultures of methylotrophic bacteria metabolize methane, methanol, carbon monoxide, formic acid, and related one-carbon molecules. Parasitic members of the genus *Leptospira* use only long-chain fatty acids as their major source of carbon and energy.

It appears that in natural environments complex populations of microorganisms often will metabolize even relatively indigestible human-made substances such as pesticides. Indigestible molecules sometimes are oxidized and degraded in the presence of a growth-promoting nutrient that is metabolized at the same time, a process called cometabolism. The products of this breakdown process can then be used as nutrients by other microorganisms. Degradation and microorganisms (pp. 979–84)

5.3 NUTRITIONAL TYPES OF MICROORGANISMS

In addition to the need for carbon, hydrogen, and oxygen, all organisms require sources of energy and electrons for growth to take place. Microorganisms can be grouped into nutritional classes based on how they satisfy all these requirements (**table 5.1**). We have already seen that microorganisms can be classified as either heterotrophs or autotrophs with respect to their preferred source of carbon. There are only two sources of energy available

to organisms: (1) light energy, and (2) the energy derived from oxidizing organic or inorganic molecules. **Phototrophs** use light as their energy source; **chemotrophs** obtain energy from the oxidation of chemical compounds (either organic or inorganic). Microorganisms also have only two sources for electrons. **Lithotrophs** (i.e., "rock-eaters") use reduced inorganic substances as their electron source, whereas **organotrophs** extract electrons from organic compounds. Photosynthesis light reactions (pp. 190–96); Oxidation of organic and inorganic molecules (pp. 171–90)

Despite the great metabolic diversity seen in microorganisms, most may be placed in one of four nutritional classes based on their primary sources of carbon, energy, and electrons (**table 5.2**). The large majority of microorganisms thus far studied are either photolithotrophic autotrophs or chemoorganotrophic heterotrophs.

Table 5.1
Sources of Carbon, Energy, and Electrons

Carbon Sources

Autotrophs	CO_2 sole or principal biosynthetic carbon source (*pp. 202–3*)[a]
Heterotrophs	Reduced, preformed, organic molecules from other organisms (*chapters 9 and 10*)

Energy Sources

Phototrophs	Light (*pp. 190–96*)
Chemotrophs	Oxidation of organic or inorganic compounds (*chapter 9*)

Electron Sources

Lithotrophs	Reduced inorganic molecules (*pp. 188–90*)
Organotrophs	Organic molecules (*chapter 9*)

[a]For each category, the location of material describing the participating metabolic pathways is given within the parentheses.

Table 5.2
Major Nutritional Types of Microorganisms

Major Nutritional Types[a]	Sources of Energy, Hydrogen/Electrons, and Carbon	Representative Microorganisms
Photolithotrophic autotrophy (Photolithoautotrophy)	Light energy Inorganic hydrogen/electron (H/e^-) donor CO_2 carbon source	Algae Purple and green sulfur bacteria Cyanobacteria
Photoorganotrophic heterotrophy (Photoorganoheterotrophy)	Light energy Organic H/e^- donor Organic carbon source (CO_2 may also be used)	Purple nonsulfur bacteria Green nonsulfur bacteria
Chemolithotrophic autotrophy (Chemolithoautotrophy)	Chemical energy source (inorganic) Inorganic H/e^- donor CO_2 carbon source	Sulfur-oxidizing bacteria Hydrogen bacteria Nitrifying bacteria Iron-oxidizing bacteria
Chemoorganotrophic heterotrophy (Chemoorganoheterotrophy)	Chemical energy source (organic) Organic H/e^- donor Organic carbon source	Protozoa Fungi Most nonphotosynthetic bacteria (including most pathogens)

[a]Bacteria in other nutritional categories have been found. The categories are defined in terms of energy, electron, and carbon sources. Condensed versions of these names are given in parentheses.

Photolithotrophic autotrophs (often called **photoautotrophs** or photolithoautotrophs) use light energy and have CO_2 as their carbon source. Eucaryotic algae and cyanobacteria employ water as the electron donor and release oxygen. Purple and green sulfur bacteria cannot oxidize water but extract electrons from inorganic donors like hydrogen, hydrogen sulfide, and elemental sulfur. **Chemoorganotrophic heterotrophs** (often called **chemoheterotrophs,** chemoorganoheterotrophs, or even heterotrophs) use organic compounds as sources of energy, hydrogen, electrons, and carbon. Frequently the same organic nutrient will satisfy all these requirements. It should be noted that essentially all pathogenic microorganisms are chemoheterotrophs.

The other two nutritional classes have fewer microorganisms but often are very important ecologically. Some purple and green bacteria are photosynthetic and use organic matter as their electron donor and carbon source. These **photoorganotrophic heterotrophs** (photoorganoheterotrophs) are common inhabitants of polluted lakes and streams. Some of these bacteria also can grow as photoautotrophs with molecular hydrogen as an electron donor. The fourth group, the **chemolithotrophic autotrophs** (chemolithoautotrophs), oxidizes reduced inorganic compounds such as iron, nitrogen, or sulfur molecules to derive both energy and electrons for biosynthesis. Carbon dioxide is the carbon source. A few chemolithotrophs can derive their carbon from organic sources and thus are heterotrophic. Chemolithotrophs contribute greatly to the chemical transformations of elements (e.g., the conversion of ammonia to nitrate or sulfur to sulfate) that continually occur in the ecosystem. Photosynthetic and chemolithotrophic bacteria (sections 21.3, 22.1, and 22.3)

Although a particular species usually belongs in only one of the four nutritional classes, some show great metabolic flexibility and alter their metabolic patterns in response to environmental changes. For example, many purple nonsulfur bacteria (*see section 22.1*) act as photoorganotrophic heterotrophs in the absence of oxygen but oxidize organic molecules and function chemotrophically at normal oxygen levels. When oxygen is low, photosynthesis and oxidative metabolism may function simultaneously. Another example is provided by bacteria such as *Beggiatoa* (*see pp. 486–88*) that rely on inorganic energy sources and organic (or sometimes CO_2) carbon sources. These microbes are sometimes called **mixotrophic** because they combine chemolithoautotrophic and heterotrophic metabolic processes. This sort of flexibility seems complex and confusing, yet it gives these microbes a definite advantage if environmental conditions frequently change.

---

1. What are nutrients, and on what basis are they divided into macroelements and micronutrients or trace elements? Describe some ways in which macroelements and micronutrients are used by an organism.
2. Define autotroph and heterotroph.
3. Discuss the ways in which microorganisms are classified based on their requirements for energy and electrons.
4. Describe the nutritional requirements of the four major nutritional groups and give some microbial examples of each. What is a mixotroph?

5.4 REQUIREMENTS FOR NITROGEN, PHOSPHORUS, AND SULFUR

To grow, a microorganism must be able to incorporate large quantities of nitrogen, phosphorus, and sulfur. Although these elements may be acquired from the same nutrients that supply carbon, microorganisms usually employ inorganic sources as well. Biochemical mechanisms for the incorporation of nitrogen, phosphorus, and sulfur (pp. 205–9)

Nitrogen is needed for the synthesis of amino acids, purines, pyrimidines, some carbohydrates and lipids, enzyme cofactors, and other substances. Many microorganisms can use the nitrogen in amino acids, and ammonia often is directly incorporated through the action of such enzymes as glutamate dehydrogenase or glutamine synthetase and glutamate synthase (*see section 10.4*). Most phototrophs and many nonphotosynthetic microorganisms reduce nitrate to ammonia and incorporate the ammonia in assimilatory nitrate reduction (*see p. 206*). A variety of bacteria (e.g., many cyanobacteria and the symbiotic bacterium *Rhizobium*) can reduce and assimilate atmospheric nitrogen using the nitrogenase system (*see section 10.4*).

Phosphorus is present in nucleic acids, phospholipids, nucleotides like ATP, several cofactors, some proteins, and other cell components. Almost all microorganisms use inorganic phosphate as their phosphorus source and incorporate it directly. Low phosphate levels actually limit microbial growth in many aquatic environments. Phosphate uptake by *E. coli* has been intensively studied. This bacterium can use both organic and inorganic phosphate. Some organophosphates such as hexose 6-phosphates can be taken up directly by transport proteins. Other organophosphates are often hydrolyzed in the periplasm by the enzyme alkaline phosphatase to produce inorganic phosphate, which then is transported across the plasma membrane. When inorganic phosphate is outside the bacterium, it crosses the outer membrane by the use of a porin protein channel. One of two transport systems subsequently moves the phosphate across the plasma membrane. At high phosphate concentrations, transport probably is due to the Pit system. When phosphate concentrations are low, the PST, (*p*hosphate-*s*pecific *t*ransport) system is more important. The PST system has higher affinity for phosphate; it is an ABC transporter (pp. 99–100) and uses a periplasmic binding protein.

Sulfur is needed for the synthesis of substances like the amino acids cysteine and methionine, some carbohydrates, biotin, and thiamine. Most microorganisms use sulfate as a source of sulfur and reduce it by assimilatory sulfate reduction (*see section 10.4*); a few require a reduced form of sulfur such as cysteine.

5.5 GROWTH FACTORS

Microorganisms often grow and reproduce when minerals and sources of energy, carbon, nitrogen, phosphorus, and sulfur are supplied. These organisms have the enzymes and pathways necessary to synthesize all cell components required for their well-

Table 5.3		
Functions of Some Common Vitamins in Microorganisms		

Vitamin	Functions	Examples of Microorganisms Requiring Vitamin[a]
Biotin	Carboxylation (CO_2 fixation)	*Leuconostoc mesenteroides* (B)
	One-carbon metabolism	*Saccharomyces cerevisiae* (F)
		Ochromonas malhamensis (A)
		Acanthamoeba castellanii (P)
Cyanocobalamin (B_{12})	Molecular rearrangements	*Lactobacillus* spp. (B)
	One-carbon metabolism—carries methyl groups	*Euglena gracilis* (A)
		Diatoms and many other algae (A)
		Acanthamoeba castellanii (P)
Folic acid	One-carbon metabolism	*Enterococcus faecalis* (B)
		Tetrahymena pyriformis (P)
Lipoic acid	Transfer of acyl groups	*Lactobacillus casei* (B)
		Tetrahymena spp. (P)
Pantothenic acid	Precursor of coenzyme A—carries acyl groups	*Proteus morganii* (B)
	(pyruvate oxidation, fatty acid metabolism)	*Hanseniaspora* spp. (F)
		Paramecium spp. (P)
Pyridoxine (B_6)	Amino acid metabolism (e.g., transamination)	*Lactobacillus* spp. (B)
		Tetrahymena pyriformis (P)
Niacin (nicotinic acid)	Precursor of NAD and NADP—carry electrons	*Brucella abortus, Haemophilus influenzae* (B)
	and hydrogen atoms	*Blastocladia pringsheimii* (F)
		Crithidia fasciculata (P)
Riboflavin (B_2)	Precursor of FAD and FMN—carry electrons	*Caulobacter vibrioides* (B)
	or hydrogen atoms	*Dictyostelium* spp. (F)
		Tetrahymena pyriformis (P)
Thiamine (B_1)	Aldehyde group transfer	*Bacillus anthracis* (B)
	(pyruvate decarboxylation, α-keto acid oxidation)	*Phycomyces blakesleeanus* (F)
		Ochromonas malhamensis (A)
		Colpidium campylum (P)

[a]The representative microorganisms are members of the following groups: bacteria (B), fungi (F), algae (A), and protozoa (P).

being. Many microorganisms, on the other hand, lack one or more essential enzymes. Therefore they cannot manufacture all indispensable constituents but must obtain them or their precursors from the environment. Organic compounds required because they are essential cell components or precursors of such components and cannot be synthesized by the organism are called **growth factors.** There are three major classes of growth factors: (1) amino acids, (2) purines and pyrimidines, and (3) vitamins. Amino acids are needed for protein synthesis, purines and pyrimidines for nucleic acid synthesis. **Vitamins** are small organic molecules that usually make up all or part of enzyme cofactors (*see section 8.6*), and only very small amounts sustain growth. The functions of selected vitamins, and examples of microorganisms requiring them, are given in **table 5.3.** Some microorganisms require many vitamins; for example, *Enterococcus faecalis* needs eight different vitamins for growth. Other growth factors are also seen; heme (from hemoglobin or cytochromes) is required by *Haemophilus influenzae,* and some mycoplasmas need cholesterol.

Knowledge of the specific growth factor requirements of many microorganisms makes possible quantitative growth-response assays for a variety of substances. For example, species from the bacterial genera *Lactobacillus* and *Streptococcus* can be used in microbiological assays of most vitamins and amino acids. The appropriate bacterium is grown in a series of culture vessels, each containing medium with an excess amount of all required components except the growth factor to be assayed. A different amount of growth factor is added to each vessel. The standard curve is prepared by plotting the growth factor quantity or concentration against the total extent of bacterial growth. Ideally the amount of growth resulting is directly proportional to the quantity of growth factor present; if the growth factor concentration doubles, the final extent of bacterial growth doubles. The quantity of the growth factor in a test sample is determined by comparing the extent of growth caused by the unknown sample with that resulting from the standards. Microbiological assays are specific, sensitive, and simple. They still are used in the assay of substances like vitamin B_{12} and biotin, despite advances in chemical assay techniques.

The observation that many microorganisms can synthesize large quantities of vitamins has led to their use in industry. Several water-soluble and fat-soluble vitamins are produced partly or completely using industrial fermentations. Good examples of such vitamins and the microorganisms that synthesize them are riboflavin (*Clostridium, Candida, Ashbya, Eremothecium*), coenzyme A (*Brevibacterium*), vitamin B_{12} (*Streptomyces, Propionibacterium,*

Pseudomonas), vitamin C (*Gluconobacter, Erwinia, Corynebac-terium*), β-carotene (*Dunaliella*), and vitamin D (*Saccharomyces*). Current research focuses on improving yields and finding microorganisms that can produce large quantities of other vitamins.

---

1. Briefly summarize the ways in which microorganisms obtain nitrogen, phosphorus, and sulfur from their environment.
2. What are growth factors? What are vitamins? How can microorganisms be used to determine the quantity of a specific substance in a sample?

 ## 5.6   UPTAKE OF NUTRIENTS BY THE CELL

The first step in nutrient use is uptake of the required nutrients by the microbial cell. Uptake mechanisms must be specific—that is, the necessary substances, and not others, must be acquired. It does a cell no good to take in a substance that it cannot use. Since microorganisms often live in nutrient-poor habitats, they must be able to transport nutrients from dilute solutions into the cell against a concentration gradient. Finally, nutrient molecules must pass through a selectively permeable plasma membrane that will not permit the free passage of most substances. In view of the enormous variety of nutrients and the complexity of the task, it is not surprising that microorganisms make use of several different transport mechanisms. The most important of these are facilitated diffusion, active transport, and group translocation. Eucaryotic microorganisms do not appear to employ group translocation but take up nutrients by the process of endocytosis (*see section 4.5*).

Plasma membrane structure and properties (pp. 44–46)

Passive Diffusion

A few substances, such as glycerol, can cross the plasma membrane by **passive diffusion.** Passive diffusion, often simply called diffusion, is the process in which molecules move from a region of higher concentration to one of lower concentration because of random thermal agitation. The rate of passive diffusion is dependent on the size of the concentration gradient between a cell's exterior and its interior (**figure 5.1**). A fairly large concentration gradient is required for adequate nutrient uptake by passive diffusion (i.e., the external nutrient concentration must be high), and the rate of uptake decreases as more nutrient is acquired unless it is used immediately. Very small molecules such as H_2O, O_2, and CO_2 often move across membranes by passive diffusion. Larger molecules, ions, and polar substances do not cross membranes by passive or simple diffusion.

Facilitated Diffusion

The rate of diffusion across selectively permeable membranes is greatly increased by using carrier proteins, sometimes called **permeases,** which are embedded in the plasma membrane. Be-

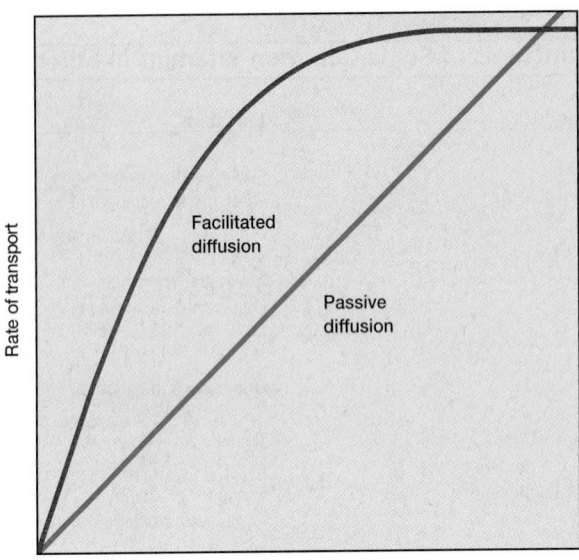

Figure 5.1   Passive and Facilitated Diffusion.   The dependence of diffusion rate on the size of the solute's concentration gradient (the ratio of the extracellular concentration to the intracellular concentration). Note the saturation effect or plateau above a specific gradient value when a facilitated diffusion carrier is operating. This saturation effect is seen whenever a carrier protein is involved in transport.

cause a carrier aids the diffusion process, it is called **facilitated diffusion.** The rate of facilitated diffusion increases with the concentration gradient much more rapidly and at lower concentrations of the diffusing molecule than that of passive diffusion (figure 5.1). Note that the diffusion rate levels off or reaches a plateau above a specific gradient value because the carrier is saturated—that is, the carrier protein is binding and transporting as many solute molecules as possible. The resulting curve resembles an enzyme-substrate curve (*see section 8.6*) and is different from the linear response seen with passive diffusion. Carrier proteins also resemble enzymes in their specificity for the substance to be transported; each carrier is selective and will transport only closely related solutes. Although a carrier protein is involved, facilitated diffusion is truly diffusion. A concentration gradient spanning the membrane drives the movement of molecules, and no metabolic energy input is required. If the concentration gradient disappears, net inward movement ceases. The gradient can be maintained by transforming the transported nutrient to another compound or by moving it to another membranous compartment in eucaryotes. Interestingly, some of these carriers are related to the major intrinsic protein of mammalian eye lenses and thus belong to the MIP family of proteins. The two most widespread MIP channels in bacteria are aquaporins that transport water and glycerol facilitators, which aid glycerol diffusion.

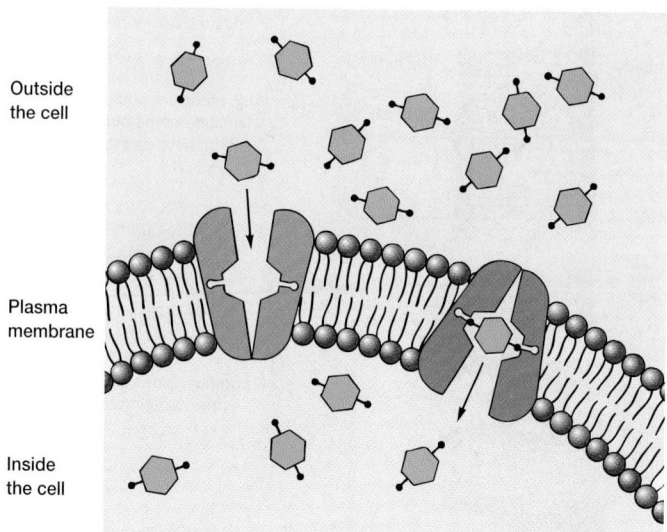

Outside
the cell

Plasma
membrane

Inside
the cell

Figure 5.2   A Model of Facilitated Diffusion. The membrane carrier can change conformation after binding an external molecule and subsequently releases the molecule on the cell interior. It then returns to the outward oriented position and is ready to bind another solute molecule. Because there is no energy input, molecules will continue to enter only as long as their concentration is greater on the outside.

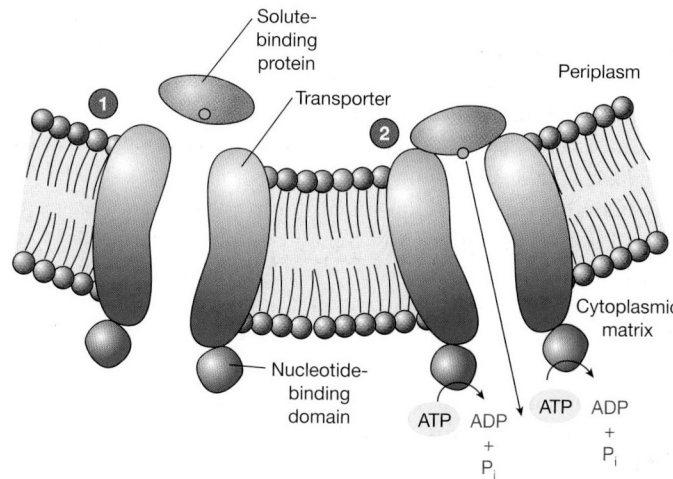

Solute-
binding
protein

Transporter

Periplasm

Cytoplasmic
matrix

Nucleotide-
binding
domain

ATP   ADP
+
P_i

ATP   ADP
+
P_i

Figure 5.3   ABC Transporter Function. (*1*) The solute binding protein binds the substrate to be transported and approaches the ABC transporter complex. (*2*) The solute binding protein attaches to the transporter and releases the substrate, which is moved across the membrane with the aid of ATP hydrolysis. See text for details.

Although much work has been done on the mechanism of facilitated diffusion, the process is not yet understood completely. It appears that the carrier protein complex spans the membrane (**figure 5.2**). After the solute molecule binds to the outside, the carrier may change conformation and release the molecule on the cell interior. The carrier would subsequently change back to its original shape and be ready to pick up another molecule. The net effect is that a hydrophilic molecule can enter the cell in response to its concentration gradient. Remember that the mechanism is driven by concentration gradients and therefore is reversible. If the solute's concentration is greater inside the cell, it will move outward. Because the cell metabolizes nutrients upon entry, influx is favored.

Facilitated diffusion does not seem to be important in procaryotes because nutrient concentrations often are lower outside the cell so that facilitated diffusion cannot be used in uptake. Glycerol is transported by facilitated diffusion in *E. coli, Salmonella typhimurium, Pseudomonas, Bacillus,* and many other bacteria. The process is much more prominent in eucaryotic cells where it is used to transport a variety of sugars and amino acids.

Active Transport

Although facilitated diffusion carriers can efficiently move molecules to the interior when the solute concentration is higher on the outside of the cell, they cannot take up solutes that are already

more concentrated within the cell (i.e., against a concentration gradient). Microorganisms often live in habitats characterized by very dilute nutrient sources, and, to flourish, they must be able to transport and concentrate these nutrients. Thus facilitated diffusion mechanisms are not always adequate, and other approaches must be used. The two most important transport processes in such situations are active transport and group translocation, both energy-dependent processes.

Active transport is the transport of solute molecules to higher concentrations, or against a concentration gradient, with the use of metabolic energy input. Because active transport involves protein carrier activity, it resembles facilitated diffusion in some ways. The carrier proteins or permeases bind particular solutes with great specificity for the molecules transported. Similar solute molecules can compete for the same carrier protein in both facilitated diffusion and active transport. Active transport is also characterized by the carrier saturation effect at high solute concentrations (figure 5.1). Nevertheless, active transport differs from facilitated diffusion in its use of metabolic energy and in its ability to concentrate substances. Metabolic inhibitors that block energy production will inhibit active transport but will not affect facilitated diffusion (at least for a short time).

Binding protein transport systems or **A**TP-***b**inding **c**assette* **transporters** (**ABC transporters**) are active in bacteria, archaea, and eucaryotes. Usually these transporters consist of two hydrophobic membrane-spanning domains associated on their cytoplasmic surfaces with two nucleotide-binding domains (**figure 5.3**). The membrane-spanning domains form a pore in the membrane and the nucleotide-binding domains bind and hydrolyze ATP to drive uptake. ABC transporters employ special substrate binding proteins, which are located in the periplasmic space of gram-negative bacteria (*see figure 3.23*) or are attached to membrane lipids on the external face of the gram-positive plasma membrane. These binding proteins, which also may participate in chemotaxis (*see pp. 67–68*), bind the molecule to be transported and then interact with the membrane transport proteins to move the solute molecule inside the cell. *E. coli* transports a variety of sugars (arabinose, maltose, galactose, ribose) and amino acids (glutamate, histidine, leucine) by this mechanism. It can also pump antibiotics out with a multidrug-resistance ABC transporter.

Substances entering gram-negative bacteria must pass through the outer membrane before ABC transporters and other active transport systems can take action. There are several ways in which this is accomplished. When the substance is small, a generalized porin protein (*see p. 58*) such as OmpF can be used; larger molecules require specialized porins. In some cases (e.g., for uptake of iron and vitamin B_{12}), specialized high-affinity outer membrane receptors and transporters are used.

It should be noted that eucaryotic ABC transporters are sometimes of great medical importance. Some tumor cells pump chemotherapeutic drugs out using these transporters. Cystic fibrosis results from a mutation that inactivates an ABC transporter that acts as a chloride ion channel in the lungs.

Bacteria also use proton gradients generated during electron transport to drive active transport. The membrane transport proteins responsible for this process lack special periplasmic solute-binding proteins. The lactose permease of *E. coli* is a well-studied example. The permease is a single protein having a molecular weight of about 30,000. It transports a lactose molecule inward as a proton simultaneously enters the cell (a higher concentration of protons is maintained outside the membrane by electron transport chain activity). Such linked transport of two substances in the same direction is called **symport.** Here, energy stored as a proton gradient drives solute transport. Although the mechanism of transport is not completely understood, recent X-ray diffraction studies show that the transport protein exists in outward- and inward-facing conformations. When lactose and a proton bind to separate sites on the outward-facing conformation, the protein changes to its inward-facing conformation. Then the sugar and proton are released into the cytoplasm. *E. coli* also uses proton symport to take up amino acids and organic acids like succinate and malate.   The chemiosmotic hypothesis (p. 182)

A proton gradient also can power active transport indirectly, often through the formation of a sodium ion gradient. For example, an *E. coli* sodium transport system pumps sodium outward in response to the inward movement of protons (**figure 5.4**). Such linked transport in which the transported substances move in opposite directions is termed **antiport.** The sodium gradient generated by this proton antiport system then drives the uptake of sugars and amino acids. A sodium ion could attach to a carrier protein, causing it to change shape. The carrier would then bind the sugar or amino acid tightly and orient its binding sites toward the cell interior. Because of the low intracellular sodium concentration, the sodium ion would dissociate from the carrier, and the

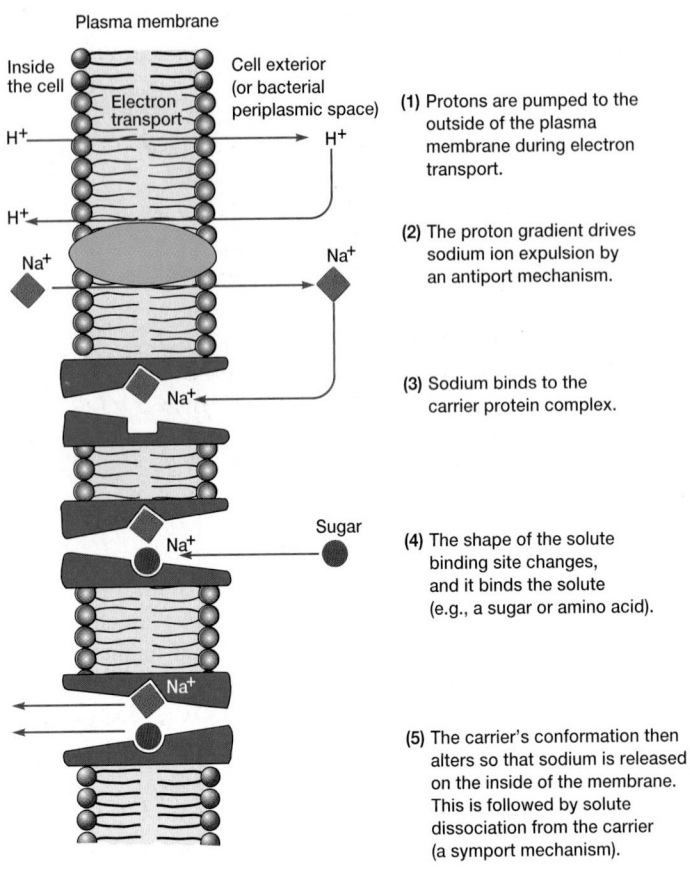

(1) Protons are pumped to the outside of the plasma membrane during electron transport.

(2) The proton gradient drives sodium ion expulsion by an antiport mechanism.

(3) Sodium binds to the carrier protein complex.

(4) The shape of the solute binding site changes, and it binds the solute (e.g., a sugar or amino acid).

(5) The carrier's conformation then alters so that sodium is released on the inside of the membrane. This is followed by solute dissociation from the carrier (a symport mechanism).

Figure 5.4   Active Transport Using Proton and Sodium Gradients.

other molecule would follow. *E. coli* transport proteins carry the sugar melibiose and the amino acid glutamate when sodium simultaneously moves inward.

Sodium symport or cotransport also is an important process in eucaryotic cells where it is used in sugar and amino acid uptake. ATP, rather than proton motive force, usually drives sodium transport in eucaryotic cells.

Often a microorganism has more than one transport system for each nutrient, as can be seen with *E. coli*. This bacterium has at least five transport systems for the sugar galactose, three systems each for the amino acids glutamate and leucine, and two potassium transport complexes. When there are several transport systems for the same substance, the systems differ in such properties as their energy source, their affinity for the solute transported, and the nature of their regulation. Presumably this diversity gives its possessor an added competitive advantage in a variable environment.

Group Translocation

In active transport, solute molecules move across a membrane without modification. Many procaryotes also take up molecules by **group translocation,** a process in which a molecule is trans-

Figure 5.5   Group Translocation: Bacterial PTS Transport.   Two examples of the phosphoenolpyruvate: sugar phosphotransferase system (PTS) are illustrated. The following components are involved in the system: phosphoenolpyruvate (PEP), enzyme I (EI), the low molecular weight heat-stable protein (HPr), and enzyme II (EII). The high-energy phosphate is transferred from HPr to the soluble EIIA. EIIA is attached to EIIB in the mannitol transport system and is separate from EIIB in the glucose system. In either case the phosphate moves from EIIA to EIIB, and then is transferred to the sugar during transport through the membrane. Other relationships between the EII components are possible. For example, IIA and IIB may form a soluble protein separate from the membrane complex; the phosphate still moves from IIA to IIB and then to the membrane domain(s).

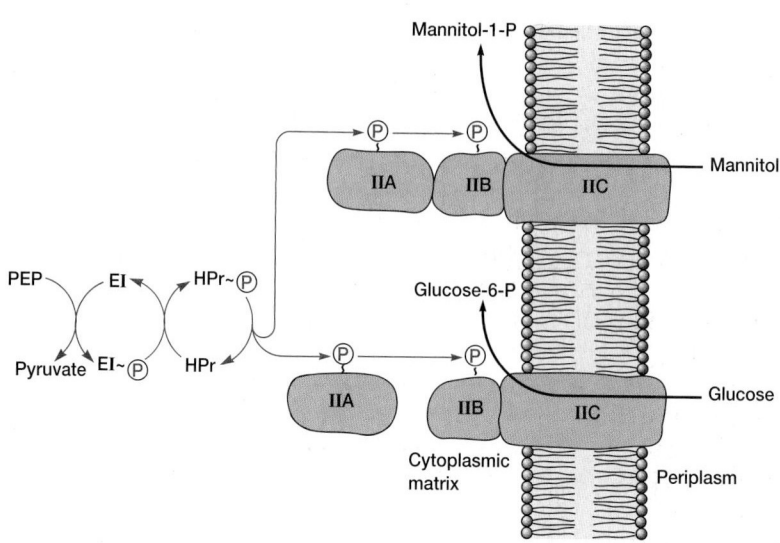

ported into the cell while being chemically altered (this can be classified as a type of energy-dependent transport because metabolic energy is used). The best-known group translocation system is the **phosphoenolpyruvate: sugar phosphotransferase system (PTS).** It transports a variety of sugars into procaryotic cells while phosphorylating them using phosphoenolpyruvate (PEP) as the phosphate donor.

PEP + sugar (outside) → pyruvate + sugar − P (inside)

The PTS is quite complex. In *E. coli* and *Salmonella typhimurium,* it consists of two enzymes and a low molecular weight heat-stable protein (HPr). HPr and enzyme I (EI) are cytoplasmic. Enzyme II (EII) is more variable in structure and often composed of three subunits or domains. EIIA (formerly called EIII) is cytoplasmic and soluble. EIIB also is hydrophilic but frequently is attached to EIIC, a hydrophobic protein that is embedded in the membrane. A high-energy phosphate is transferred from PEP to enzyme II with the aid of enzyme I and HPr (**figure 5.5**). Then, a sugar molecule is phosphorylated as it is carried across the membrane by enzyme II. Enzyme II transports only specific sugars and varies with PTS, whereas enzyme I and HPr are common to all PTSs.

PTSs are widely distributed in procaryotes. Except for some species of *Bacillus* that have both glycolysis and the phosphotransferase system, aerobic bacteria seem to lack PTSs. Members of the genera *Escherichia, Salmonella, Staphylococcus,* and other facultatively anaerobic bacteria (*see p. 125*) have phosphotransferase systems; some obligately anaerobic bacteria (e.g., *Clostridium*) also have PTSs. Many carbohydrates are transported by these systems. *E. coli* takes up glucose, fructose, mannitol, sucrose, *N*-acetylglucosamine, cellobiose, and other carbohydrates by group translocation. Besides their role in transport, PTS proteins can act as chemoreceptors for chemotaxis.

Iron Uptake

Almost all microorganisms require iron for use in cytochromes and many enzymes. Iron uptake is made difficult by the extreme insolubility of ferric iron (Fe^{3+}) and its derivatives, which leaves little free iron available for transport. Many bacteria and fungi have overcome this difficulty by secreting siderophores [Greek for iron bearers]. **Siderophores** are low molecular weight organic molecules that are able to complex with ferric iron and supply it to the cell. These iron-transport molecules are normally either hydroxamates or phenolates-catecholates. Ferrichrome is a hydroxamate produced by many fungi; enterobactin is the catecholate formed by *E. coli* (**figure 5.6a,b**). It appears that three siderophore groups complex with iron orbitals to form a six-coordinate, octahedral complex (figure 5.6c).

Microorganisms secrete siderophores when little iron is available in the medium. Once the iron-siderophore complex has reached the cell surface, it binds to a siderophore-receptor protein. Then the iron is either released to enter the cell directly or the whole iron-siderophore complex is transported inside by an ABC transporter. In *E. coli* the siderophore receptor is in the outer membrane of the cell envelope; when the iron reaches the periplasmic space, it moves through the plasma membrane with the aid of the transporter. After the iron has entered the cell, it is reduced to the ferrous form (Fe^{2+}). Iron is so crucial to microorganisms that they may use more than one route of iron uptake to ensure an adequate supply.

---

1. Describe facilitated diffusion, active transport, and group translocation in terms of their distinctive characteristics and mechanisms.
2. How do binding protein transport systems and membrane-bound transport systems differ with respect to energy sources? What are symport and antiport processes?
3. How are siderophores involved in iron transport?

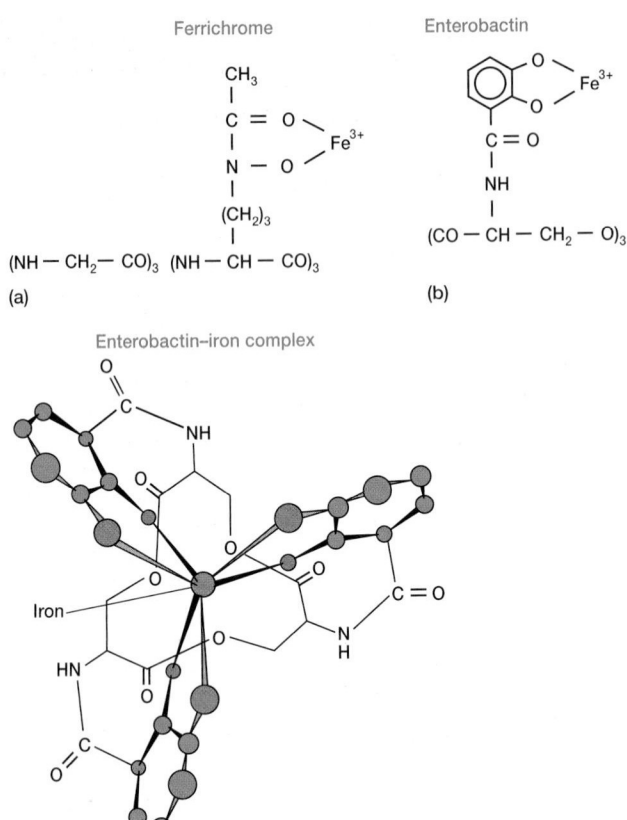

(a)

(b)

Enterobactin–iron complex

(c)

Figure 5.6 Siderophore Ferric Iron Complexes. (a) Ferrichrome is a cyclic hydroxamate [—CO—N(O⁻)—] molecule formed by many fungi. (b) *E. coli* produces the cyclic catecholate derivative, enterobactin. (c) Ferric iron probably complexes with three siderophore groups to form a six-coordinate, octahedral complex as shown in this illustration of the enterobactin-iron complex.

Table 5.4
Examples of Defined Media

BG–11 Medium for Cyanobacteria	Amount (g/liter)
NaNO$_3$	1.5
K$_2$HPO$_4$ · 3H$_2$O	0.04
MgSO$_4$ · 7H$_2$O	0.075
CaCl$_2$ · 2H$_2$O	0.036
Citric acid	0.006
Ferric ammonium citrate	0.006
EDTA (Na$_2$Mg salt)	0.001
Na$_2$CO$_3$	0.02
Trace metal solution[a]	1.0 ml/liter
Final pH 7.4	

Medium for *Escherichia coli*	Amount (g/liter)
Glucose	1.0
Na$_2$HPO$_4$	16.4
KH$_2$PO$_4$	1.5
(NH$_4$)$_2$SO$_4$	2.0
MgSO$_4$ · 7H$_2$O	200.0 mg
CaCl$_2$	10.0 mg
FeSO$_4$ · 7H$_2$O	0.5 mg
Final pH 6.8–7.0	

Sources: Data from Rippka, et al. *Journal of General Microbiology*, 111:1–61, 1979; and S. S. Cohen, and R. Arbogast, *Journal of Experimental Medicine*, 91:619, 1950.
[a]The trace metal solution contains H$_3$BO$_3$, MnCl$_2$·4H$_2$O, ZnSO$_4$·7H$_2$O, Na$_2$Mo$_4$·2H$_2$O, CuSO$_4$·5H$_2$O, and Co(NO$_3$)$_2$·6H$_2$O.

5.7 CULTURE MEDIA

Much of the study of microbiology depends on the ability to grow and maintain microorganisms in the laboratory, and this is possible only if suitable culture media are available. A culture medium is a solid or liquid preparation used to grow, transport, and store microorganisms. To be effective, the medium must contain all the nutrients the microorganism requires for growth. Specialized media are essential in the isolation and identification of microorganisms, the testing of antibiotic sensitivities, water and food analysis, industrial microbiology, and other activities. Although all microorganisms need sources of energy, carbon, nitrogen, phosphorus, sulfur, and various minerals, the precise composition of a satisfactory medium will depend on the species one is trying to cultivate because nutritional requirements vary so greatly. Knowledge of a microorganism's normal habitat often is useful in selecting an appropriate culture medium because its nutrient requirements reflect its natural surroundings. Frequently a medium is used to select and grow specific microorganisms or to help identify a particular species. In such cases the function of the medium also will determine its composition.

Synthetic or Defined Media

Some microorganisms, particularly photolithotrophic autotrophs such as cyanobacteria and eucaryotic algae, can be grown on relatively simple media containing CO$_2$ as a carbon source (often added as sodium carbonate or bicarbonate), nitrate or ammonia as a nitrogen source, sulfate, phosphate, and a variety of minerals (**table 5.4**). Such a medium in which all components are known is a **defined medium** or **synthetic medium.** Many chemoorganotrophic heterotrophs also can be grown in defined media with glucose as a carbon source and an ammonium salt as a nitrogen source. Not all defined media are as simple as the examples in table 5.4 but may be constructed from dozens of components. Defined media are used widely in research, as it is often desirable to know what the experimental microorganism is metabolizing.

Complex Media

Media that contain some ingredients of unknown chemical composition are **complex media.** Such media are very useful, as a single complex medium may be sufficiently rich and complete to meet the nutritional requirements of many different microorganisms. In addition, complex media often are needed because the

Historical Highlights
5.1   The Discovery of Agar as a Solidifying Agent and the Isolation of Pure Cultures

The earliest culture media were liquid, which made the isolation of bacteria to prepare pure cultures extremely difficult. In practice, a mixture of bacteria was diluted successively until only one organism, as an average, was present in a culture vessel. If everything went well, the individual bacterium thus isolated would reproduce to give a pure culture. This approach was tedious, gave variable results, and was plagued by contamination problems. Progress in isolating pathogenic bacteria understandably was slow.

The development of techniques for growing microorganisms on solid media and efficiently obtaining pure cultures was due to the efforts of the German bacteriologist Robert Koch and his associates. In 1881 Koch published an article describing the use of boiled potatoes, sliced with a flame-sterilized knife, in culturing bacteria. The surface of a sterile slice of potato was inoculated with bacteria from a needle tip, and then the bacteria were streaked out over the surface so that a few individual cells would be separated from the remainder. The slices were incubated beneath bell jars to prevent airborne contamination, and the isolated cells developed into pure colonies. Unfortunately many bacteria would not grow well on potato slices.

At about the same time, Frederick Loeffler, an associate of Koch's, developed a meat extract peptone medium for cultivating

pathogenic bacteria. Koch decided to try solidifying this medium. Koch was an amateur photographer—he was the first to take photomicrographs of bacteria—and was experienced in preparing his own photographic plates from silver salts and gelatin. Precisely the same approach was employed for preparing solid media. He spread a mixture of Loeffler's medium and gelatin over a glass plate, allowed it to harden, and inoculated the surface in the same way he had inoculated his sliced potatoes. The new solid medium worked well, but it could not be incubated at 37°C (the best temperature for most human bacterial pathogens) because the gelatin would melt. Furthermore, some bacteria digested the gelatin.

About a year later, in 1882, agar was first used as a solidifying agent. It had been discovered by a Japanese innkeeper, Minora Tarazaemon. The story goes that he threw out extra seaweed soup and discovered the next day that it had jelled during the cold winter night. Agar had been used by the East Indies Dutch to make jellies and jams. Fannie Eilshemius Hesse (*see figure 1.5*), the New Jersey–born wife of Walther Hesse, one of Koch's assistants, had learned of agar from a Dutch acquaintance and suggested its use when she heard of the difficulties with gelatin. Agar-solidified medium was an instant success and continues to be essential in all areas of microbiology.

Table 5.5	
Some Common Complex Media	

Nutrient Broth	**Amount (g/liter)**
Peptone (gelatin hydrolysate)	5
Beef extract	3
Tryptic Soy Broth	
Tryptone (pancreatic digest of casein)	17
Peptone (soybean digest)	3
Glucose	2.5
Sodium chloride	5
Dipotassium phosphate	2.5
MacConkey Agar	
Pancreatic digest of gelatin	17.0
Pancreatic digest of casein	1.5
Peptic digest of animal tissue	1.5
Lactose	10.0
Bile salts	1.5
Sodium chloride	5.0
Neutral red	0.03
Crystal violet	0.001
Agar	13.5

Complex media contain undefined components like peptones, meat extract, and yeast extract. **Peptones** are protein hydrolysates prepared by partial proteolytic digestion of meat, casein, soya meal, gelatin, and other protein sources. They serve as sources of carbon, energy, and nitrogen. Beef extract and yeast extract are aqueous extracts of lean beef and brewer's yeast, respectively. Beef extract contains amino acids, peptides, nucleotides, organic acids, vitamins, and minerals. Yeast extract is an excellent source of B vitamins as well as nitrogen and carbon compounds. Three commonly used complex media are (1) nutrient broth, (2) tryptic soy broth, and (3) MacConkey agar (**table 5.5**).

If a solid medium is needed for surface cultivation of microorganisms, liquid media can be solidified with the addition of 1.0 to 2.0% agar; most commonly 1.5% is used. **Agar** is a sulfated polymer composed mainly of D-galactose, 3,6-anhydro-L-galactose, and D-glucuronic acid (**Historical Highlights 5.1**). It usually is extracted from red algae (*see figure 26.8*). Agar is well suited as a solidifying agent because after it has been melted in boiling water, it can be cooled to about 40 to 42°C before hardening and will not melt again until the temperature rises to about 80 to 90°C. Agar is also an excellent hardening agent because most microorganisms cannot degrade it.

Other solidifying agents are sometimes employed. For example, silica gel is used to grow autotrophic bacteria on solid media in the absence of organic substances and to determine carbon sources for heterotrophic bacteria by supplementing the medium with various organic compounds.

nutritional requirements of a particular microorganism are unknown, and thus a defined medium cannot be constructed. This is the situation with many fastidious bacteria that have complex nutritional or cultural requirements; they may even require a medium containing blood or serum.

Types of Media

Media such as tryptic soy broth and tryptic soy agar are called general purpose media because they support the growth of many microorganisms. Blood and other special nutrients may be added to general purpose media to encourage the growth of fastidious heterotrophs. These specially fortified media (e.g., blood agar) are called enriched media.

Selective media favor the growth of particular microorganisms. Bile salts or dyes like basic fuchsin and crystal violet favor the growth of gram-negative bacteria by inhibiting the growth of gram-positive bacteria without affecting gram-negative organisms. Endo agar, eosin methylene blue agar, and MacConkey agar (table 5.5), three media widely used for the detection of *E. coli* and related bacteria in water supplies and elsewhere, contain dyes that suppress gram-positive bacterial growth. MacConkey agar also contains bile salts. Bacteria also may be selected by incubation with nutrients that they specifically can use. A medium containing only cellulose as a carbon and energy source is quite effective in the isolation of cellulose-digesting bacteria. The possibilities for selection are endless, and there are dozens of special selective media in use.

Differential media are media that distinguish between different groups of bacteria and even permit tentative identification of microorganisms based on their biological characteristics. Blood agar is both a differential medium and an enriched one. It distinguishes between hemolytic and nonhemolytic bacteria. Hemolytic bacteria (e.g., many streptococci and staphylococci isolated from throats) produce clear zones around their colonies because of red blood cell destruction. MacConkey agar is both differential and selective. Since it contains lactose and neutral red dye, lactose-fermenting colonies appear pink to red in color and are easily distinguished from colonies of nonfermenters.

---

1. Describe the following kinds of media and their uses: defined or synthetic media, complex media, general purpose media, enriched media, selective media, and differential media. Give an example of each kind.
2. What are peptones, yeast extract, beef extract, and agar? Why are they used in media?

5.8   ISOLATION OF PURE CULTURES

In natural habitats microorganisms usually grow in complex, mixed populations containing several species. This presents a problem for the microbiologist because a single type of microorganism cannot be studied adequately in a mixed culture. One needs a **pure culture,** a population of cells arising from a single cell, to characterize an individual species. Pure cultures are so important that the development of pure culture techniques by the German bacteriologist Robert Koch transformed microbiology. Within about 20 years after the development of pure culture techniques most pathogens responsible for the major human bacterial diseases

had been isolated (*see table 1.1*). There are several ways to prepare pure cultures; a few of the more common approaches are reviewed here.   A brief survey of some major milestones in microbiology (chapter 1)

The Spread Plate and Streak Plate

If a mixture of cells is spread out on an agar surface so that every cell grows into a completely separate **colony,** a macroscopically visible growth or cluster of microorganisms on a solid medium, each colony represents a pure culture. The **spread plate** is an easy, direct way of achieving this result. A small volume of dilute microbial mixture containing around 30 to 300 cells is transferred to the center of an agar plate and spread evenly over the surface with a sterile bent-glass rod (**figure 5.7**). The dispersed cells develop into isolated colonies. Because the number of colonies should equal the number of viable organisms in the sample, spread plates can be used to count the microbial population.

Pure colonies also can be obtained from **streak plates.** The microbial mixture is transferred to the edge of an agar plate with an inoculating loop or swab and then streaked out over the surface in one of several patterns (**figure 5.8**). At some point in the process, single cells drop from the loop as it is rubbed along the agar surface and develop into separate colonies (**see chapter-opening photo**). In both spread-plate and streak-plate techniques, successful isolation depends on spatial separation of single cells.

The Pour Plate

Extensively used with bacteria and fungi, a **pour plate** also can yield isolated colonies. The original sample is diluted several times to reduce the microbial population sufficiently to obtain separate colonies when plating (**figure 5.9**). Then small volumes of several diluted samples are mixed with liquid agar that has been cooled to about 45°C, and the mixtures are poured immediately into sterile culture dishes. Most bacteria and fungi are not killed by a brief exposure to the warm agar. After the agar has hardened, each cell is fixed in place and forms an individual colony. Plates containing between 30 and 300 colonies are counted. The total number of colonies equals the number of viable microorganisms in the diluted sample. Colonies growing on the surface also can be used to inoculate fresh medium and prepare pure cultures (**Techniques & Applications 5.2**).

The preceding techniques require the use of special culture dishes named **petri dishes** or plates after their inventor Julius Richard Petri, a member of Robert Koch's laboratory; Petri developed these dishes around 1887 and they immediately replaced agar-coated glass plates. They consist of two round halves, the top half overlapping the bottom (figure 5.8). Petri dishes are very easy to use, may be stacked on each other to save space, and are one of the most common items in microbiology laboratories.

Colony Morphology and Growth

Colony development on agar surfaces aids the microbiologist in identifying bacteria because individual species often form

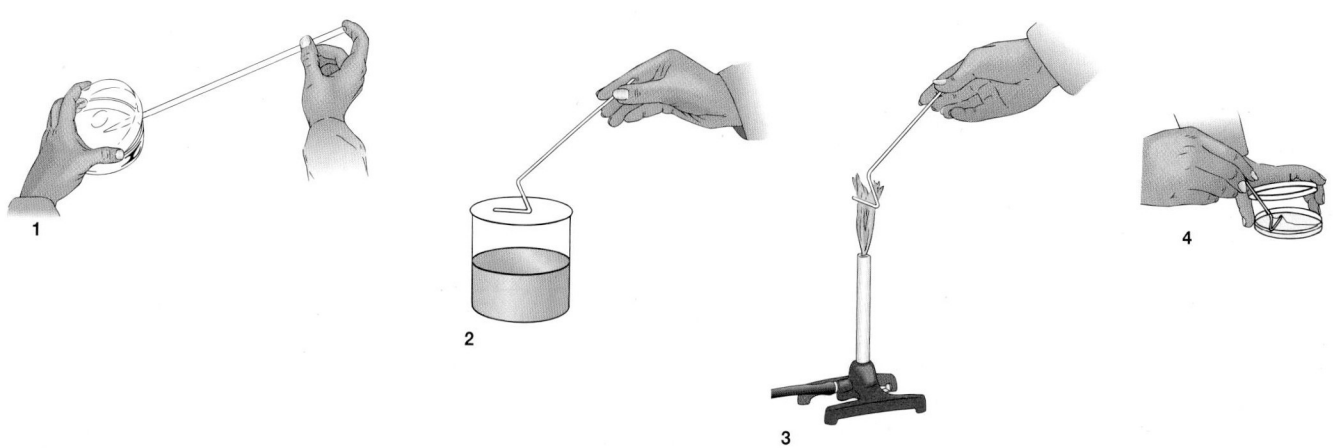

Figure 5.7 Spread-Plate Technique. The preparation of a spread plate. (*1*) Pipette a small sample onto the center of an agar medium plate. (*2*) Dip a glass spreader into a beaker of ethanol. (*3*) Briefly flame the ethanol-soaked spreader and allow it to cool. (*4*) Spread the sample evenly over the agar surface with the sterilized spreader. Incubate.

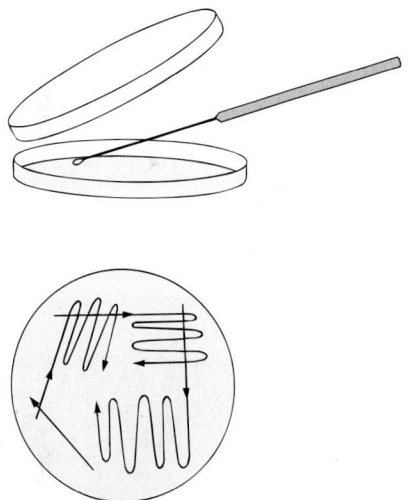

Figure 5.8 Streak-Plate Technique. Preparation of streak plates. The upper illustration shows a petri dish of agar being streaked with an inoculating loop. A commonly used streaking pattern is pictured at the bottom.

colonies of characteristic size and appearance (**figure 5.10**). When a mixed population has been plated properly, it sometimes is possible to identify the desired colony based on its overall appearance and use it to obtain a pure culture. The structure of bacterial colonies also has been examined with the scanning electron microscope. The microscopic structure of colonies is often as variable as their visible appearance (**figure 5.11**).

In nature bacteria and many other microorganisms often grow on surfaces in biofilms. However, sometimes they do form discrete colonies. Therefore an understanding of colony growth is important, and the growth of colonies on agar has been frequently studied. Generally the most rapid cell growth occurs at the colony edge. Growth is much slower in the center, and cell autolysis takes place in the older central portions of some colonies. These differences in growth appear due to gradients of oxygen, nutrients, and toxic products within the colony. At the colony edge, oxygen and nutrients are plentiful. The colony center, of course, is much thicker than the edge. Consequently oxygen and nutrients do not diffuse readily into the center, toxic metabolic products cannot be quickly eliminated, and growth in the colony center is

Techniques & Applications

5.2   The Enrichment and Isolation of Pure Cultures

A major practical problem is the preparation of pure cultures when microorganisms are present in very low numbers in a sample. Plating methods can be combined with the use of selective or differential media to enrich and isolate rare microorganisms. A good example is the isolation of bacteria that degrade the herbicide 2,4-dichlorophenoxyacetic acid (2,4-D). Bacteria able to metabolize 2,4-D can be obtained with a liquid medium containing 2,4-D as its sole carbon source and the required nitrogen, phosphorus, sulfur, and mineral components. When this medium is inoculated with soil, only bac-

teria able to use 2,4-D will grow. After incubation, a sample of the original culture is transferred to a fresh flask of selective medium for further enrichment of 2,4-D metabolizing bacteria. A mixed population of 2,4-D degrading bacteria will arise after several such transfers. Pure cultures can be obtained by plating this mixture on agar containing 2,4-D as the sole carbon source. Only bacteria able to grow on 2,4-D form visible colonies and can be subcultured. This same general approach is used to isolate and purify a variety of bacteria by selecting for specific physiological characteristics.

Figure 5.9 The Pour-Plate Technique. The original sample is diluted several times to thin out the population sufficiently. The most diluted samples are then mixed with warm agar and poured into petri dishes. Isolated cells grow into colonies and can be used to establish pure cultures. The surface colonies are circular; subsurface colonies would be lenticular or lens shaped.

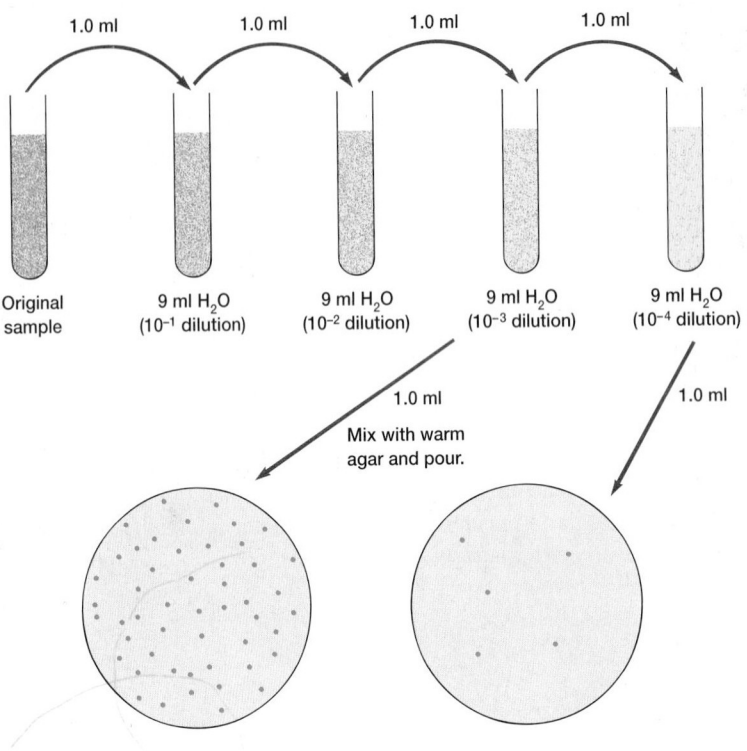

1.0 ml   1.0 ml   1.0 ml   1.0 ml

Original   9 ml H_2O 9 ml H_2O 9 ml H_2O 9 ml H_2O
sample   (10^{-1} dilution)   (10^{-2} dilution)   (10^{-3} dilution)   (10^{-4} dilution)

1.0 ml
Mix with warm
agar and pour.

1.0 ml

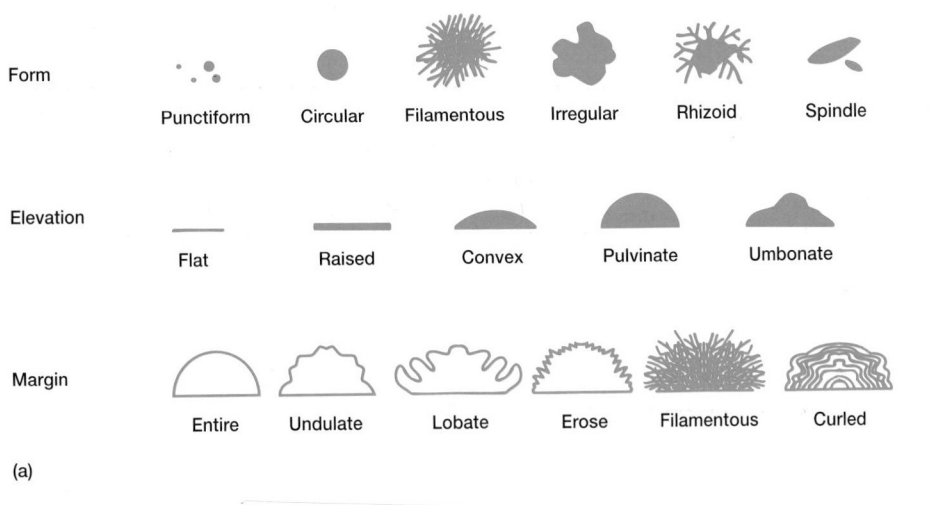

Form   Punctiform   Circular   Filamentous   Irregular   Rhizoid   Spindle

Elevation   Flat   Raised   Convex   Pulvinate   Umbonate

Margin   Entire   Undulate   Lobate   Erose   Filamentous   Curled

(a)

Figure 5.10 Bacterial Colony Morphology. (a) Variations in bacterial colony morphology seen with the naked eye. The general form of the colony and the shape of the edge or margin can be determined by looking down at the top of the colony. The nature of colony elevation is apparent when viewed from the side as the plate is held at eye level. (b) Colony morphology can vary dramatically with the medium on which the bacteria are growing. These beautiful snowflakelike colonies were formed by *Bacillus subtilis* growing on nutrient-poor agar. The bacteria apparently behave cooperatively when confronted with poor growth conditions, and often the result is an intricate structure that resembles the fractal patterns seen in nonliving systems.

(b)

Figure 5.11 Scanning Electron Micrographs of Bacterial Colonies. (**a**) *Micrococcus* on agar (×31,000). (**b**) *Clostridium* (×12,000). (**c**) *Mycoplasma pneumoniae* (×26,000). (**d**) *Escherichia coli* (×14,000).

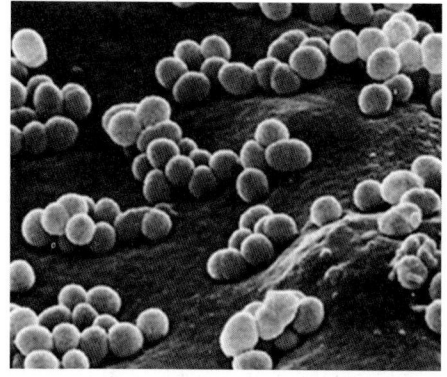

(a) *Micrococcus*

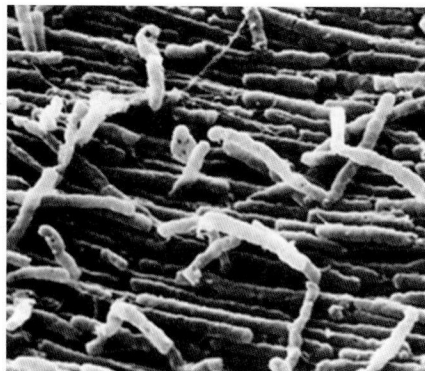

(b) *Clostridium*

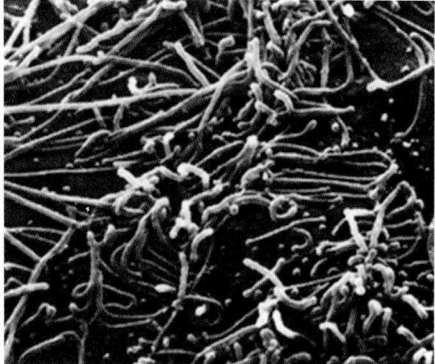

(c) *M. pneumoniae*

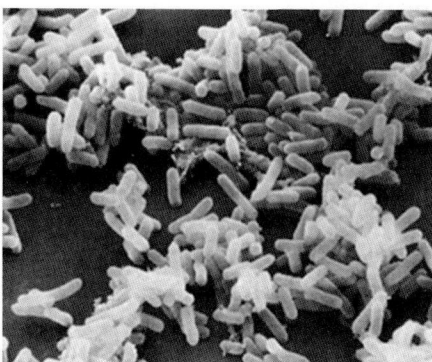

(d) *E. coli*

slowed or stopped. Because of these environmental variations within a colony, cells on the periphery can be growing at maximum rates while cells in the center are dying. Biofilms (pp. 602–3)

It is obvious from the colonies pictured in figure 5.10 that bacteria growing on solid surfaces such as agar can form quite complex and intricate colony shapes. These patterns vary with nutrient availability and the hardness of the agar surface. It is not yet clear how characteristic colony patterns develop. Nutrient diffusion and availability, bacterial chemotaxis, and the presence of liquid on the surface all appear to play a role in pattern formation. Undoubtedly cell-cell communication and quorum sensing (*see pp. 129–31*) is important as well. Much work will be required to understand the formation of bacterial colonies and biofilms.

1. What are pure cultures, and why are they important? How are spread plates, streak plates, and pour plates prepared?
2. In what way does microbial growth vary within a colony? What factors might cause these variations in growth?

Summary

Microorganisms require nutrients, materials that are used in biosynthesis and to make energy available.

5.1 The Common Nutrient Requirements

a. Macronutrients or macroelements (C, O, H, N, S, P, K, Ca, Mg, and Fe) are needed in relatively large quantities; micronutrients or trace elements (e.g., Mn, Zn, Co, Mo, Ni, and Cu) are used in very small amounts.

5.2 Requirements for Carbon, Hydrogen, and Oxygen

a. Autotrophs use CO_2 as their primary or sole carbon source; heterotrophs employ organic molecules.

5.3 Nutritional Types of Microorganisms

a. Microorganisms can be classified based on their energy and electron sources (**table 5.1**). Phototrophs use light energy, and chemotrophs obtain energy from the oxidation of chemical compounds. Electrons are extracted from reduced inorganic substances by lithotrophs and from organic compounds by organotrophs (**table 5.2**).

5.4 Requirements for Nitrogen, Phosphorus, and Sulfur

a. Nitrogen, phosphorus, and sulfur may be obtained from the same organic molecules that supply carbon, from the direct incorporation of ammonia and phosphate, and by the reduction and assimilation of oxidized inorganic molecules.

5.5 Growth Factors

a. Probably most microorganisms need growth factors. Growth factor requirements make microbiological assays possible.

5.6 Uptake of Nutrients by the Cell

a. Although some nutrients can enter cells by passive diffusion, a membrane carrier protein is usually required.
b. In facilitated diffusion the transport protein simply carries a molecule across the membrane in

the direction of decreasing concentration, and no metabolic energy is required (**figure 5.2**).
c. Active transport systems use metabolic energy and membrane carrier proteins to concentrate substances actively by transporting them against a gradient. ATP is used as an energy source by ABC transporters (**figure 5.3**). Gradients of protons and sodium ions also drive solute uptake across membranes (**figure 5.4**).
d. Bacteria also transport organic molecules while modifying them, a process known as group translocation. For example, many sugars are transported and phosphorylated simultaneously (**figure 5.5**).

e. Iron is accumulated by the secretion of siderophores, small molecules able to complex with ferric iron (**figure 5.6**). When the iron-siderophore complex reaches the cell surface, it is taken inside and the iron is reduced to the ferrous form.

5.7 Culture Media
a. Culture media can be constructed completely from chemically defined components (defined media or synthetic media) or may contain constituents like peptones and yeast extract whose precise composition is unknown (complex media).
b. Culture media can be solidified by the addition of agar, a complex polysaccharide from red algae.

c. Culture media are classified based on function and composition as general purpose media, enriched media, selective media, and differential media.

5.8 Isolation of Pure Cultures
a. Pure cultures usually are obtained by isolating individual cells with any of three plating techniques: the spread-plate, streak-plate, and pour-plate methods (**figures 5.7** and **5.8**).
b. Microorganisms growing on solid surfaces tend to form colonies with distinctive morphology. Colonies usually grow most rapidly at the edge where larger amounts of required resources are available.

Key Terms

active transport 99
agar 103
antiport 100
ATP-binding cassette transporters (ABC transporters) 99
autotrophs 94
chemoheterotrophs 96
chemolithotrophic autotrophs 96
chemoorganotrophic heterotrophs 96

chemotrophs 95
colony 104
complex medium 102
defined medium 102
differential media 104
facilitated diffusion 98
group translocation 100
growth factors 97
heterotrophs 94
lithotrophs 95
macroelements 94
micronutrients 94

mixotrophic 96
nutrient 94
organotrophs 95
passive diffusion 98
peptones 103
permease 98
petri dish 104
phosphoenolpyruvate: sugar phosphotransferase system (PTS) 101
photoautotrophs 96
photolithotrophic autotrophs 96

photoorganotrophic heterotrophs 96
phototrophs 95
pour plate 104
pure culture 104
selective media 104
siderophores 101
spread plate 104
streak plate 104
symport 100
synthetic medium 102
trace elements 94
vitamins 97

Questions for Thought and Review

1. Why is it so difficult to demonstrate the micronutrient requirements of microorganisms?
2. List some of the most important uses of the nitrogen, phosphorus, and sulfur that microorganisms obtain from their surroundings.
3. Why are amino acids, purines, and pyrimidines often growth factors, whereas glucose is usually not?

4. Why do microorganisms normally take up nutrients using transport proteins or permeases? What advantage does a microorganism gain by employing active transport rather than facilitated diffusion?
5. If you wished to obtain a pure culture of bacteria that could degrade benzene and use it as a carbon and energy source, how would you proceed?

6. Describe the nutritional requirements of a chemolithotrophic heterotroph. Where might you search for such a bacterium?
7. Suppose that you carry out a serial dilution of a 0.1 ml sample as shown in figure 5.9. The 10^{-3} plate gives 80 colonies and the 10^{-4} plate yields four colonies. Calculate the concentration (bacteria/ml) of the original, undiluted sample.

Critical Thinking Questions

1. Discuss the advantages and disadvantages of group translocation versus endocytosis for the host cell.

2. Explain why isolation of a pure culture on selective solid medium may not be successful.

For recommended readings on these and related topics, see Appendix V.

Microbial Growth

Concepts

- Growth is defined as an increase in cellular constituents and may result in an increase in a microorganism's size, population number, or both.

- When microorganisms are grown in a closed system, population growth remains exponential for only a few generations and then enters a stationary phase due to factors such as nutrient limitation and waste accumulation. In an open system with continual nutrient addition and waste removal, the exponential phase can be maintained for long periods.

- A wide variety of techniques can be used to study microbial growth by following changes in the total cell number, the population of viable microorganisms, or the cell mass.

- Water availability, pH, temperature, oxygen concentration, pressure, radiation, and a number of other environmental factors influence microbial growth. Yet many microorganisms, and particularly bacteria, have managed to adapt and flourish under environmental extremes that would destroy most higher organisms.

- In the natural environment, growth is often severely limited by available nutrient supplies and many other environmental factors.

- Bacteria can communicate with each other and behave cooperatively using population density–dependent signals.

The paramount evolutionary accomplishment of bacteria as a group is rapid, efficient cell growth in many environments.

—*J. L. Ingraham, O. Maaløe,*
and F. C. Neidhardt

Membrane filters are used in counting microorganisms. This membrane has been used to obtain a total bacterial count using an indicator to color colonies for easy counting.

hapter 5 emphasizes that microorganisms need access to a source of energy and the raw materials essential for the construction of cellular components. All organisms must have carbon, hydrogen, oxygen, nitrogen, sulfur, phosphorus, and a variety of minerals; many also require one or more special growth factors. The cell takes up these substances by membrane transport processes, the most important of which are facilitated diffusion, active transport, and group translocation. Eucaryotic cells also employ endocytosis.

Chapter 6 concentrates more directly on the growth. The nature of growth and the ways in which it can be measured are described first, followed by consideration of continuous culture techniques. An account of the influence of environmental factors on microbial growth completes the chapter.

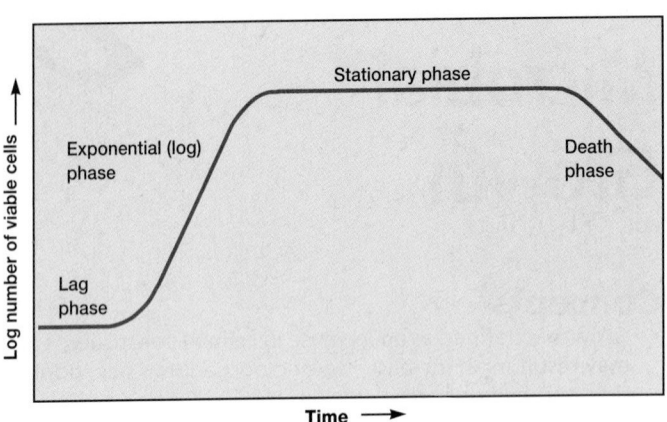

Figure 6.1   Microbial Growth Curve in a Closed System. The four phases of the growth curve are identified on the curve and discussed in the text.

Growth may be defined as an increase in cellular constituents. It leads to a rise in cell number when microorganisms reproduce by processes like budding or binary fission. In the latter, individual cells enlarge and divide to yield two progeny of approximately equal size. Growth also results when cells simply become longer or larger. If the microorganism is **coenocytic**—that is, a multinucleate organism in which nuclear divisions are not accompanied by cell divisions—growth results in an increase in cell size but not cell number. It is usually not convenient to investigate the growth and reproduction of individual microorganisms because of their small size. Therefore, when studying growth, microbiologists normally follow changes in the total population number.

The cell cycle (pp. 86; 280–82)

6.1   THE GROWTH CURVE

Population growth is studied by analyzing the growth curve of a microbial culture. When microorganisms are cultivated in liquid medium, they usually are grown in a **batch culture** or closed system—that is, they are incubated in a closed culture vessel with a single batch of medium. Because no fresh medium is provided during incubation, nutrient concentrations decline and concentrations of wastes increase. The growth of microorganisms reproducing by binary fission can be plotted as the logarithm of the number of viable cells versus the incubation time. The resulting curve has four distinct phases (**figure 6.1**).

Lag Phase

When microorganisms are introduced into fresh culture medium, usually no immediate increase in cell number occurs, and therefore this period is called the **lag phase.** Although cell division does not take place right away and there is no net increase in mass, the cell is synthesizing new components. A lag phase prior to the start of cell division can be necessary for a variety of reasons. The cells may be old and depleted of ATP, essential cofac-

tors, and ribosomes; these must be synthesized before growth can begin. The medium may be different from the one the microorganism was growing in previously. Here new enzymes would be needed to use different nutrients. Possibly the microorganisms have been injured and require time to recover. Whatever the causes, eventually the cells retool, replicate their DNA, begin to increase in mass, and finally divide.

The lag phase varies considerably in length with the condition of the microorganisms and the nature of the medium. This phase may be quite long if the inoculum is from an old culture or one that has been refrigerated. Inoculation of a culture into a chemically different medium also results in a longer lag phase. On the other hand, when a young, vigorously growing exponential phase culture is transferred to fresh medium of the same composition, the lag phase will be short or absent.

Exponential Phase

During the **exponential** or **log phase,** microorganisms are growing and dividing at the maximal rate possible given their genetic potential, the nature of the medium, and the conditions under which they are growing. Their rate of growth is constant during the exponential phase; that is, the microorganisms are dividing and doubling in number at regular intervals. Because each individual divides at a slightly different moment, the growth curve rises smoothly rather than in discrete jumps (figure 6.1). The population is most uniform in terms of chemical and physiological properties during this phase; therefore exponential phase cultures are usually used in biochemical and physiological studies.

Exponential growth is **balanced growth.** That is, all cellular constituents are manufactured at constant rates relative to each other. If nutrient levels or other environmental conditions change, **unbalanced growth** results. This is growth during which the rates of synthesis of cell components vary relative to one another until a new balanced state is reached. This response is readily observed in a shift-up experiment in which bacteria are transferred from a nutritionally poor medium to a richer one. There is a lag

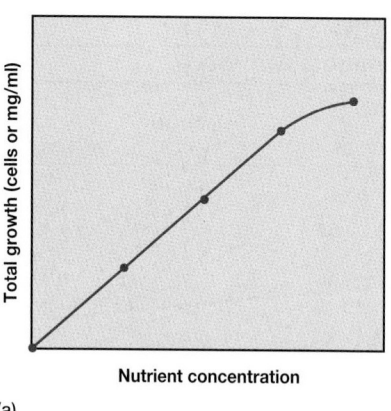

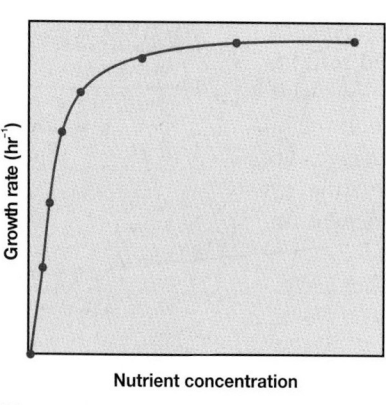

Figure 6.2   Nutrient Concentration and Growth.   (**a**) The effect of changes in limiting nutrient concentration on total microbial yield. At sufficiently high concentrations, total growth will plateau. (**b**) The effect on growth rate.

while the cells first construct new ribosomes to enhance their capacity for protein synthesis. This is followed by increases in protein and DNA synthesis. Finally, the expected rise in reproductive rate takes place.   Protein and DNA synthesis (sections 11.3 and 12.2)

Unbalanced growth also results when a bacterial population is shifted down from a rich medium to a poor one. The organisms may previously have been able to obtain many cell components directly from the medium. When shifted to a nutritionally inadequate medium, there is a lag in growth because they need time to make the enzymes required for the biosynthesis of unavailable nutrients. Consequently cell division and DNA replication continue after the shift-down, but net protein and RNA synthesis slow. The cells become smaller and reorganize themselves metabolically until they are able to grow again. Then balanced growth is resumed and the culture enters the exponential phase.   Regulation of nucleic acid synthesis (pp. 270–79)

These shift-up and shift-down experiments demonstrate that microbial growth is under precise, coordinated control and responds quickly to changes in environmental conditions.

When microbial growth is limited by the low concentration of a required nutrient, the final net growth or yield of cells increases with the initial amount of the limiting nutrient present (**figure 6.2a**). This is the basis of microbiological assays for vitamins and other growth factors. The rate of growth also increases with nutrient concentration (figure 6.2b), but in a hyperbolic manner much like that seen with many enzymes (*see figure 8.17*). The shape of the curve seems to reflect the rate of nutrient uptake by microbial transport proteins. At sufficiently high nutrient levels the transport systems are saturated, and the growth rate does not rise further with increasing nutrient concentration.   Microbiological assays (p. 97); Nutrient transport systems (pp. 98–101)

Stationary Phase

Eventually population growth ceases and the growth curve becomes horizontal (figure 6.1). This **stationary phase** usually is attained by bacteria at a population level of around 10^9 cells per ml. Other microorganisms normally do not reach such high population densities; protozoan and algal cultures often have maximum concentrations of about 10^6 cells per ml. Of course final population size depends on nutrient availability and other factors, as well as the type of microorganism being cultured. In the stationary phase the total number of viable microorganisms remains constant. This may result from a balance between cell division and cell death, or the population may simply cease to divide though remaining metabolically active.

Microbial populations enter the stationary phase for several reasons. One obvious factor is nutrient limitation; if an essential nutrient is severely depleted, population growth will slow. Aerobic organisms often are limited by O_2 availability. Oxygen is not very soluble and may be depleted so quickly that only the surface of a culture will have an O_2 concentration adequate for growth. The cells beneath the surface will not be able to grow unless the culture is shaken or aerated in another way. Population growth also may cease due to the accumulation of toxic waste products. This factor seems to limit the growth of many anaerobic cultures (cultures growing in the absence of O_2). For example, streptococci can produce so much lactic acid and other organic acids from sugar fermentation that their medium becomes acidic and growth is inhibited. Streptococcal cultures also can enter the stationary phase due to depletion of their sugar supply. Finally, there is some evidence that growth may cease when a critical population level is reached. Thus entrance into the stationary phase may result from several factors operating in concert.

As we have seen, bacteria in a batch culture may enter stationary phase in response to starvation. This probably often occurs in nature as well because many environments have quite low nutrient levels. Starvation can be a positive experience for bacteria. Many do not respond with obvious morphological changes such as endospore formation, but only decrease somewhat in overall size, often accompanied by protoplast shrinkage and nucleoid condensation. The more important changes are in gene expression and physiology. Starving bacteria frequently produce a variety of **starvation proteins,** which make the cell much more resistant to damage in a variety of ways. They increase peptidoglycan cross-linking and cell wall strength. The Dps (*D*NA-binding *p*rotein from *s*tarved cells) protein protects DNA. Chaperones prevent protein denaturation and renature damaged proteins. As a result of these and many other mechanisms, the starved cells become harder to kill and more resistant to starvation itself, damaging temperature changes, oxidative and osmotic damage, and toxic chemicals such as chlorine. These changes are so effective that some bacteria can survive starvation for years. Clearly, these considerations are of great practical importance in medical and industrial microbiology. There is even evidence that *Salmonella typhimurium* and some other bacterial pathogens become more virulent when starved.

Death Phase

Detrimental environmental changes like nutrient deprivation and the buildup of toxic wastes lead to the decline in the number of viable cells characteristic of the **death phase.** The death of a microbial population, like its growth during the exponential phase, may be logarithmic (that is, a constant proportion of cells dies every hour). This pattern in viable cell count holds even when the total cell number remains constant because the cells simply fail to lyse after dying. Often the only way of deciding whether a bacterial cell is viable is by incubating it in fresh medium; if it does not grow and reproduce, it is assumed to be dead. That is, death is defined to be the irreversible loss of the ability to reproduce. It should be noted that the rate of population death sometimes is not logarithmic and can vary with both environmental conditions such as the growth medium employed and the microorganisms involved. For example, some bacteria respond to harsh conditions by forming endospores.

Although most of a microbial population usually dies in a logarithmic fashion, the death rate may decrease after the population has been drastically reduced. This is due to the extended survival of particularly resistant cells. For this and other reasons, the death phase curve may be complex.

The Mathematics of Growth

Knowledge of microbial growth rates during the exponential phase is indispensable to microbiologists. Growth rate studies contribute to basic physiological and ecological research and the solution of applied problems in industry. Thus the quantitative aspects of exponential phase growth will be discussed. The analysis applies to microorganisms dividing by binary fission (*see* p. 280–82, 476).

During the exponential phase each microorganism is dividing at constant intervals. Thus the population will double in number during a specific length of time called the **generation time** or **doubling time.** This situation can be illustrated with a simple example. Suppose that a culture tube is inoculated with one cell that divides every 20 minutes (**table 6.1**). The population will be 2 cells after 20 minutes, 4 cells after 40 minutes, and so forth. Because the population is doubling every generation, the increase in population is always 2^n where n is the number of generations. The resulting population increase is exponential or logarithmic (**figure 6.3**).

Table 6.1
An Example of Exponential Growth

Time[a]	Division Number	2^n	Population $(N_o \times 2^n)$	$\log_{10}N_t$
0	0	$2^0 = 1$	1	0.000
20	1	$2^1 = 2$	2	0.301
40	2	$2^2 = 4$	4	0.602
60	3	$2^3 = 8$	8	0.903
80	4	$2^4 = 16$	16	1.204
100	5	$2^5 = 32$	32	1.505
120	6	$2^6 = 64$	64	1.806

[a]The hypothetical culture begins with one cell having a 20-minute generation time.

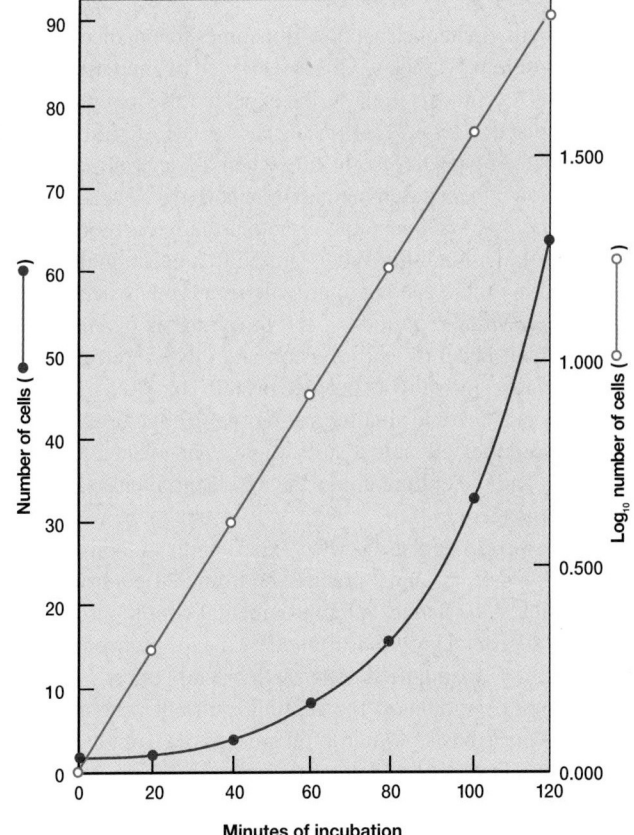

Figure 6.3   Exponential Microbial Growth.   The data from table 6.1 for six generations of growth are plotted directly (•–•) and in the logarithmic form (○–○). The growth curve is exponential as shown by the linearity of the log plot.

These observations can be expressed as equations for the generation time.

Let N_0 = the initial population number

N_t = the population at time t

n = the number of generations in time t

Then inspection of the results in table 6.1 will show that

$$N_t = N_0 \times 2^n.$$

Solving for n, the number of generations, where all logarithms are to the base 10,

$$\log N_t = \log N_0 + n \cdot \log 2, \text{ and}$$

$$n = \frac{\log N_t - \log N_0}{\log 2} = \frac{\log N_t - \log N_0}{0.301}$$

The rate of growth during the exponential phase in a batch culture can be expressed in terms of the **mean growth rate constant** *(k)*. This is the number of generations per unit time, often expressed as the generations per hour.

$$k = \frac{n}{t} = \frac{\log N_t - \log N_0}{0.301t}$$

The time it takes a population to double in size—that is, the **mean generation time** or mean doubling time (*g*), can now be calculated. If the population doubles ($t = g$), then

$$N_t = 2N_0.$$

Substitute $2N_0$ into the mean growth rate equation and solve for k.

$$k = \frac{\log (2N_0) - \log N_0}{0.301g} = \frac{\log 2 + \log N_0 - \log N_0}{0.301g}$$

$$k = \frac{1}{g}$$

The mean generation time is the reciprocal of the mean growth rate constant.

$$g = \frac{1}{k}$$

The mean generation time (*g*) can be determined directly from a semilogarithmic plot of the growth data (**figure 6.4**) and the growth rate constant calculated from the *g* value. The generation time also may be calculated directly from the previous equations. For example, suppose that a bacterial population increases from 10^3 cells to 10^9 cells in 10 hours.

$$k = \frac{\log 10^9 - \log 10^3}{(0.301)(10 \text{ hr})} = \frac{9 - 3}{3.01 \text{ hr}} = 2.0 \text{ generations/hr}$$

$$g = \frac{1}{2.0 \text{ gen./hr}} = 0.5 \text{ hr/gen. or } 30 \text{ min/gen.}$$

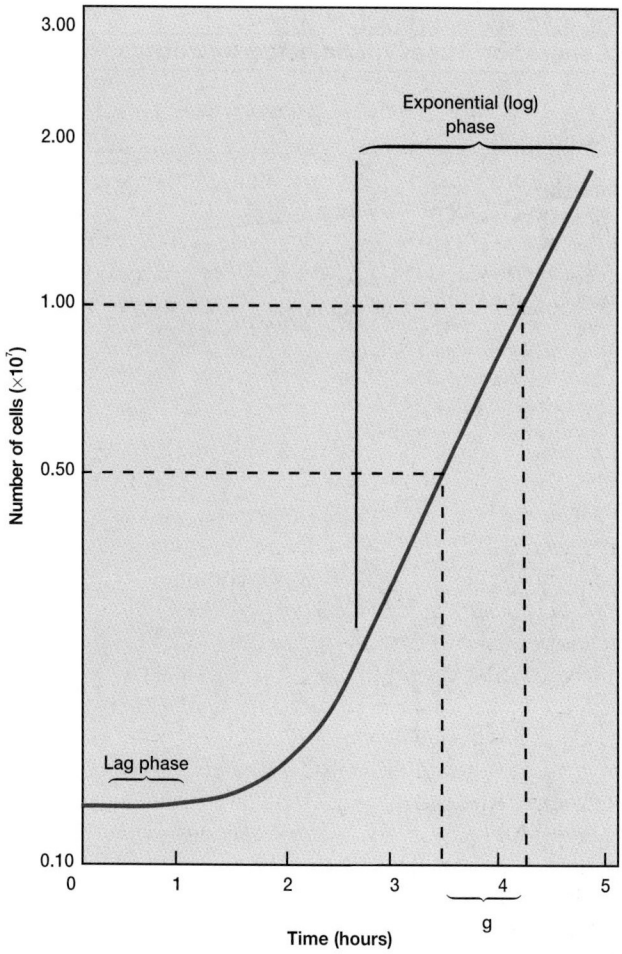

Figure 6.4   Generation Time Determination. The generation time can be determined from a microbial growth curve. The population data are plotted with the logarithmic axis used for the number of cells. The time to double the population number is then read directly from the plot. The log of the population number can also be plotted against time on regular axes.

Generation times vary markedly with the species of microorganism and environmental conditions. They range from less than 10 minutes (0.17 hours) for a few bacteria to several days with some eucaryotic microorganisms (**table 6.2**). Generation times in nature are usually much longer than in culture.

1. Define growth. Describe the four phases of the growth curve in a closed system and discuss the causes of each.
2. Define balanced growth, unbalanced growth, shift-up experiment, and shift-down experiment.
3. What effect does increasing a limiting nutrient have on the yield of cells and the growth rate?
4. What are the generation or doubling time and the mean growth rate constant? How can they be determined from growth data?

Table 6.2
Generation Times for Selected Microorganisms

Microorganism	Temperature (°C)	Generation Time (Hours)
Bacteria		
Beneckea natriegens	37	0.16
Escherichia coli	40	0.35
Bacillus subtilis	40	0.43
Staphylococcus aureus	37	0.47
Pseudomonas aeruginosa	37	0.58
Clostridium botulinum	37	0.58
Rhodospirillum rubrum	25	4.6–5.3
Anabaena cylindrica	25	10.6
Mycobacterium tuberculosis	37	≈12
Treponema pallidum	37	33
Algae		
Scenedesmus quadricauda	25	5.9
Chlorella pyrenoidosa	25	7.75
Asterionella formosa	20	9.6
Euglena gracilis	25	10.9
Ceratium tripos	20	82.8
Protozoa		
Tetrahymena geleii	24	2.2–4.2
Leishmania donovani	26	10–12
Paramecium caudatum	26	10.4
Acanthamoeba castellanii	30	11–12
Giardia lamblia	37	18
Fungi		
Saccharomyces cerevisiae	30	2
Monilinia fructicola	25	30

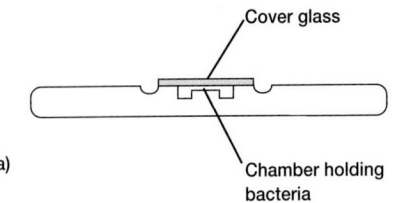

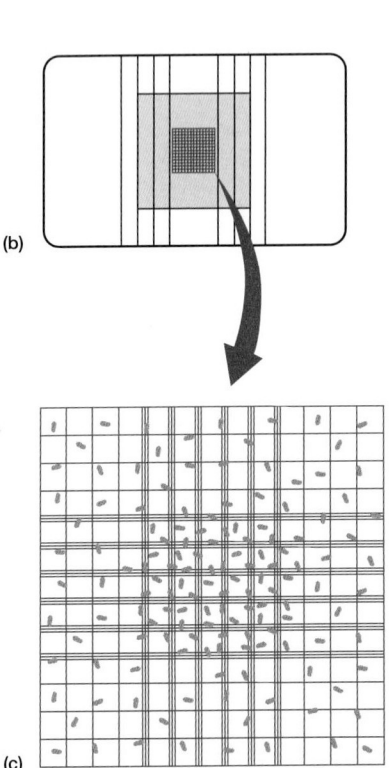

(a)

(b)

(c)

Figure 6.5 The Petroff-Hausser Counting Chamber.
(**a**) Side view of the chamber showing the cover glass and the space beneath it that holds a bacterial suspension. (**b**) A top view of the chamber. The grid is located in the center of the slide. (**c**) An enlarged view of the grid. The bacteria in several of the central squares are counted, usually at ×400 to ×500 magnification. The average number of bacteria in these squares is used to calculate the concentration of cells in the original sample. Since there are 25 squares covering an area of 1 mm^2, the total number of bacteria in 1 mm^2 of the chamber is (number/square)(25 squares). The chamber is 0.02 mm deep and therefore,

$$\text{bacteria/mm}^3 = (\text{bacteria/square})(25 \text{ squares})(50).$$

The number of bacteria per cm^3 is 10^3 times this value. For example, suppose the average count per square is 28 bacteria:

$$\text{bacteria/cm}^3 = (28 \text{ bacteria})(25 \text{ squares})(50)(10^3) = 3.5 \times 10^7.$$

 6.2 MEASUREMENT OF MICROBIAL GROWTH

There are many ways to measure microbial growth to determine growth rates and generation times. Either population mass or number may be followed because growth leads to increases in both. The most commonly employed techniques for growth measurement are examined briefly and the advantages and disadvantages of each noted. No single technique is always best; the most appropriate approach will depend on the experimental situation.

Measurement of Cell Numbers

The most obvious way to determine microbial numbers is through direct counting. Using a counting chamber is easy, inexpensive, and relatively quick; it also gives information about the size and morphology of microorganisms. Petroff-Hausser counting chambers can be used for counting procaryotes; hemocytometers can be used for both procaryotes and eucaryotes. Procaryotes are more easily counted in these chambers if they are stained, or when a phase-contrast or a fluorescence microscope is employed. These specially designed slides have chambers of known depth with an etched grid on the chamber bottom (**figure 6.5**). The number of mi-

croorganisms in a sample can be calculated by taking into account the chamber's volume and any sample dilutions required. There are some disadvantages to the technique. The microbial population must be fairly large for accuracy because such a small volume is

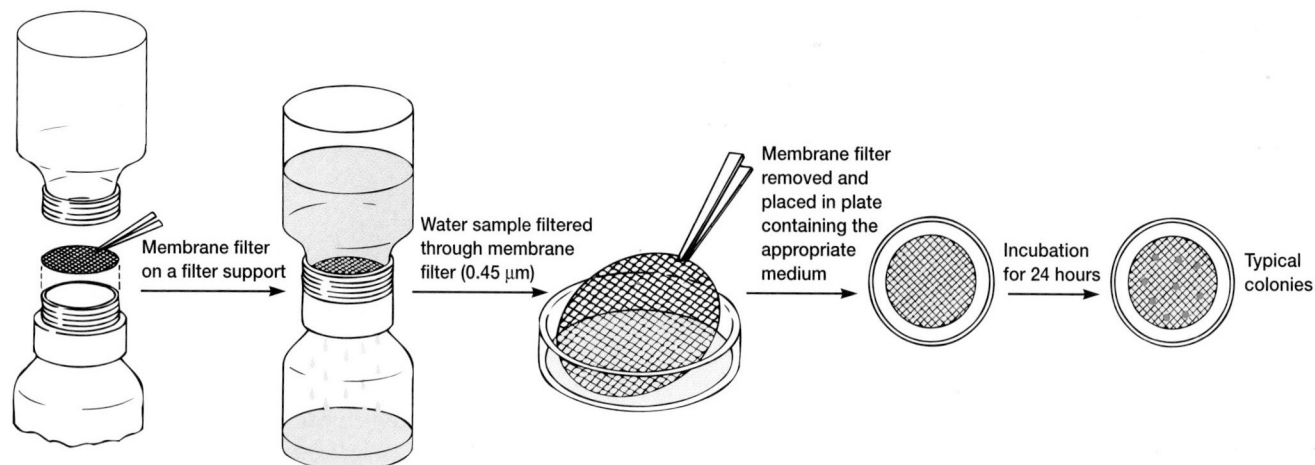

Figure 6.6   The Membrane Filtration Procedure.   Membranes with different pore sizes are used to trap different microorganisms. Incubation times for membranes also vary with the medium and microorganism.

sampled. It is also difficult to distinguish between living and dead cells in counting chambers without special techniques.

Larger microorganisms such as protozoa, algae, and nonfilamentous yeasts can be directly counted with electronic counters such as the Coulter Counter. The microbial suspension is forced through a small hole or orifice. An electrical current flows through the hole, and electrodes placed on both sides of the orifice measure its electrical resistance. Every time a microbial cell passes through the orifice, electrical resistance increases (or the conductivity drops) and the cell is counted. The Coulter Counter gives accurate results with larger cells and is extensively used in hospital laboratories to count red and white blood cells. It is not as useful in counting bacteria because of interference by small debris particles, the formation of filaments, and other problems.

Counting chambers and electronic counters yield counts of all cells, whether alive or dead. There are also several viable counting techniques, procedures specific for cells able to grow and reproduce. In most viable counting procedures, a diluted sample of bacteria or other microorganisms is dispersed over a solid agar surface. Each microorganism or group of microorganisms develops into a distinct colony. The original number of viable microorganisms in the sample can be calculated from the number of colonies formed and the sample dilution. For example, if 1.0 ml of a 1×10^{-6} dilution yielded 150 colonies, the original sample contained around 1.5×10^8 cells per ml. Usually the count is made more accurate by use of a special colony counter. In this way the spread-plate and pour-plate techniques may be used to find the number of microorganisms in a sample.

Plating techniques are simple, sensitive, and widely used for viable counts of bacteria and other microorganisms in samples of food, water, and soil. Several problems, however, can lead to inaccurate counts. Low counts will result if clumps of cells are not broken up and the microorganisms well dispersed. Because it is not possible to be absolutely certain that each colony arose from an individual cell, the results are often expressed in terms of **colony forming units (CFU)** rather than the number of microor-

ganisms. The samples should yield between 30 and 300 colonies for best results. Of course the counts will also be low if the agar medium employed cannot support growth of all the viable microorganisms present. The hot agar used in the pour-plate technique may injure or kill sensitive cells; thus spread plates sometimes give higher counts than pour plates.   Spread-plate and pour-plate techniques (p. 104)

Microbial numbers are frequently determined from counts of colonies growing on special membrane filters having pores small enough to trap bacteria. In the membrane filter technique, a sample is drawn through a special **membrane filter (figure 6.6)**. The filter is then placed on an agar medium or on a pad soaked with liquid media and incubated until each cell forms a separate colony. A colony count gives the number of microorganisms in the filtered sample, and special media can be used to select for specific microorganisms (**figure 6.7**). This technique is especially useful in analyzing aquatic samples.   Analysis of water purity (pp. 633–36)

Membrane filters also are used to count bacteria directly. The sample is first filtered through a black polycarbonate membrane filter to provide a good background for observing fluorescent objects. The bacteria then are stained with a fluorescent dye such as acridine orange or the DNA stain DAPI and observed microscopically. Acridine orange-stained microorganisms glow orange or green and are easily counted with an epifluorescence microscope (*see section 2.2*). Usually the counts obtained with this approach are much higher than those from culture techniques because some of the bacteria are dead. Commercial kits that use fluorescent reagents to stain live and dead cells differently are now available. This makes it possible to directly count the number of live and dead microorganisms in a sample (*see figures 2.13d and 28.32*).

Measurement of Cell Mass

Increases in the total cell mass, as well as in cell numbers, accompany population growth. Therefore techniques for measuring changes in cell mass can be used in following growth. The most

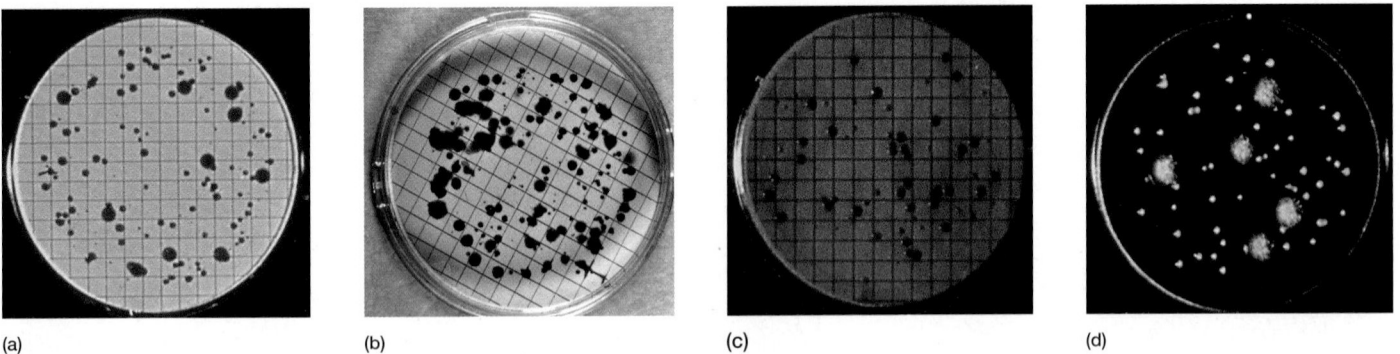

(a) (b) (c) (d)

Figure 6.7 Colonies on Membrane Filters. Membrane-filtered samples grown on a variety of media. (**a**) Standard nutrient media for a total bacterial count. An indicator colors colonies red for easy counting. (**b**) Fecal coliform medium for detecting fecal coliforms that form blue colonies. (**c**) m-Endo agar for detecting *E. coli* and other coliforms that produce colonies with a green sheen. (**d**) Wort agar for the culture of yeasts and molds.

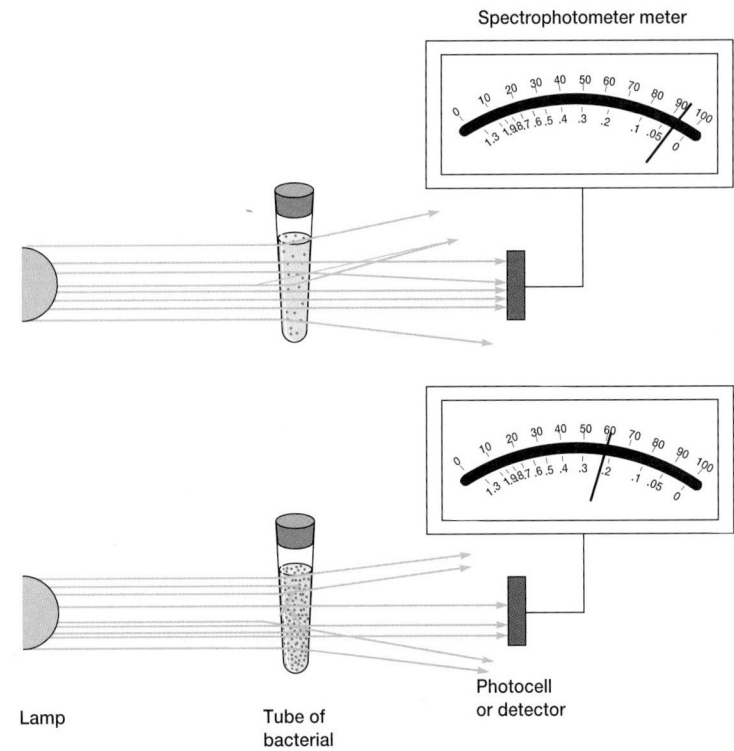

Figure 6.8 Turbidity and Microbial Mass Measurement. Determination of microbial mass by measurement of light absorption. As the population and turbidity increase, more light is scattered and the absorbance reading given by the spectrophotometer increases. The spectrophotometer meter has two scales. The bottom scale displays absorbance and the top scale, percent transmittance. Absorbance increases as percent transmittance decreases.

direct approach is the determination of microbial dry weight. Cells growing in liquid medium are collected by centrifugation, washed, dried in an oven, and weighed. This is an especially useful technique for measuring the growth of fungi. It is time consuming, however, and not very sensitive. Because bacteria weigh so little, it may be necessary to centrifuge several hundred milliliters of culture to collect a sufficient quantity.

More rapid, sensitive techniques depend on the fact that microbial cells scatter light striking them. Because microbial cells in a population are of roughly constant size, the amount of scattering is directly proportional to the biomass of cells present and indirectly related to cell number. When the concentration of bacteria reaches about 10 million cells (10^7) per ml, the medium appears slightly cloudy or turbid. Further increases in concentration result in greater turbidity and less light is transmitted through the medium. The extent of light scattering can be measured by a spectrophotometer and is almost linearly related to bacterial concentration at low absorbance levels (**figure 6.8**). Thus population growth can be easily measured spectrophotometrically as long as the population is high enough to give detectable turbidity.

If the amount of a substance in each cell is constant, the total quantity of that cell constituent is directly related to the total microbial cell mass. For example, a sample of washed cells collected from a known volume of medium can be analyzed for total

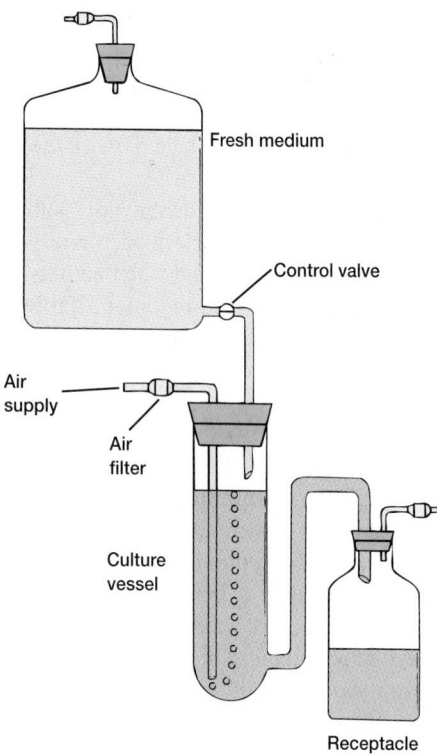

Figure 6.9   A Continuous Culture System: The Chemostat.
Schematic diagram of the system. The fresh medium contains a
limiting amount of an essential nutrient. Growth rate is determined by the rate of flow of medium through the culture vessel.

protein or nitrogen. An increase in the microbial population will
be reflected in higher total protein levels. Similarly, chlorophyll
determinations can be used to measure algal and cyanobacterial
populations, and the quantity of ATP can be used to estimate the
amount of living microbial mass.

---

1. Briefly describe each technique by which microbial population
   numbers may be determined and give its advantages and disadvantages.
2. Why are plate count results often expressed as colony forming
   units?

6.3   THE CONTINUOUS CULTURE OF MICROORGANISMS

Up to this point the focus has been on closed systems called batch
cultures in which nutrient supplies are not renewed nor wastes removed. Exponential growth lasts for only a few generations and
soon the stationary phase is reached. However, it is possible to
grow microorganisms in an open system, a system with constant
environmental conditions maintained through continual provision of nutrients and removal of wastes. These conditions are met
in the laboratory by a **continuous culture system**. A microbial

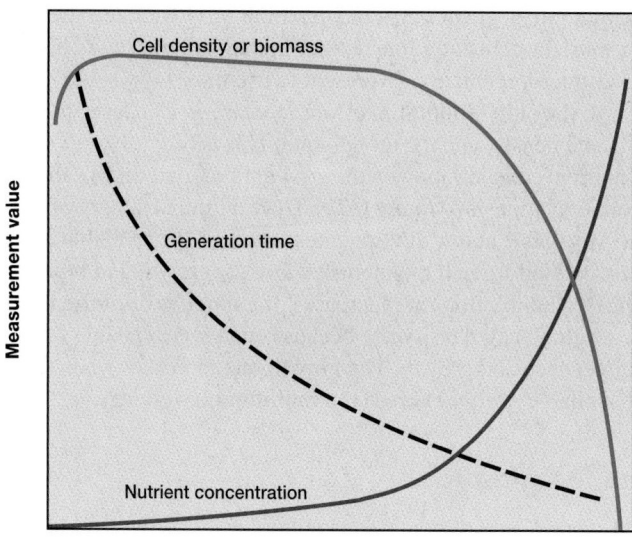

**Figure 6.10   Chemostat Dilution Rate and Microbial
Growth.**   The effects of changing the dilution rate in a chemostat.

population can be maintained in the exponential growth phase
and at a constant biomass concentration for extended periods in a
continuous culture system.

The Chemostat

Two major types of continuous culture systems commonly are
used: (1) chemostats and (2) turbidostats. A **chemostat** is constructed so that sterile medium is fed into the culture vessel at the
same rate as the media containing microorganisms is removed
(**figure 6.9**). The culture medium for a chemostat possesses an essential nutrient (e.g., an amino acid) in limiting quantities. Because of the presence of a limiting nutrient, the growth rate is
determined by the rate at which new medium is fed into the
growth chamber, and the final cell density depends on the concentration of the limiting nutrient. The rate of nutrient exchange
is expressed as the dilution rate (D), the rate at which medium
flows through the culture vessel relative to the vessel volume,
where f is the flow rate (ml/hr) and V is the vessel volume (ml).

$$D = f/V$$

For example, if f is 30 ml/hr and V is 100 ml, the dilution rate is
0.30 hr^{-1}.

    Both the microbial population level and the generation time
are related to the dilution rate (**figure 6.10**). The microbial population density remains unchanged over a wide range of dilution
rates. The generation time decreases (i.e., the growth rate rises) as
the dilution rate increases. The limiting nutrient will be almost
completely depleted under these balanced conditions. If the dilution rate rises too high, the microorganisms can actually be
washed out of the culture vessel before reproducing because the

dilution rate is greater than the maximum growth rate. The limiting nutrient concentration rises at higher dilution rates because fewer microorganisms are present to use it.

At very low dilution rates, an increase in D causes a rise in both cell density and the growth rate. This is because of the effect of nutrient concentration on the growth rate, sometimes called the Monod relationship (figure 6.2b). Only a limited supply of nutrient is available at low dilution rates. Much of the available energy must be used for cell maintenance, not for growth and reproduction. As the dilution rate increases, the amount of nutrients and the resulting cell density rise because energy is available for both maintenance and growth. The growth rate increases when the total available energy exceeds the **maintenance energy.**

The Turbidostat

The second type of continuous culture system, the **turbidostat,** has a photocell that measures the absorbance or turbidity of the culture in the growth vessel. The flow rate of media through the vessel is automatically regulated to maintain a predetermined turbidity or cell density. The turbidostat differs from the chemostat in several ways. The dilution rate in a turbidostat varies rather than remaining constant, and its culture medium lacks a limiting nutrient. The turbidostat operates best at high dilution rates; the chemostat is most stable and effective at lower dilution rates.

Continuous culture systems are very useful because they provide a constant supply of cells in exponential phase and growing at a known rate. They make possible the study of microbial growth at very low nutrient levels, concentrations close to those present in natural environments. These systems are essential for research in many areas—for example, in studies on interactions between microbial species under environmental conditions resembling those in a freshwater lake or pond. Continuous systems also are used in food and industrial microbiology.

---

1. How does an open system differ from a closed culture system or batch culture?
2. Describe how the two different kinds of continuous culture systems, the chemostat and turbidostat, operate.
3. What is the dilution rate? What is maintenance energy?

 6.4 **THE INFLUENCE OF ENVIRONMENTAL FACTORS ON GROWTH**

As we have seen (pp. 110–12), microorganisms must be able to respond to variations in nutrient levels, and particularly to nutrient limitation. The growth of microorganisms also is greatly affected by the chemical and physical nature of their surroundings. An understanding of environmental influences aids in the control of microbial growth and the study of the ecological distribution of microorganisms.

The ability of some microorganisms to adapt to extreme and inhospitable environments is truly remarkable. Procaryotes are present anywhere life can exist. Many habitats in which procaryotes thrive would kill most other organisms. Procaryotes such as *Bacillus infernus* even seem able to live over 1.5 miles below the Earth's surface, without oxygen and at temperatures above 60°C. Microorganisms that grow in such harsh conditions are often called **extremophiles.**

In this section we shall briefly review how some of the most important environmental factors affect microbial growth. Major emphasis will be given to solutes and water activity, pH, temperature, oxygen level, pressure, and radiation. **Table 6.3** summarizes the way in which microorganisms are categorized in terms of their response to these factors.

Solutes and Water Activity

Because a selectively permeable plasma membrane separates microorganisms from their environment, they can be affected by changes in the osmotic concentration of their surroundings. If a microorganism is placed in a hypotonic solution (one with a lower osmotic concentration), water will enter the cell and cause it to burst unless something is done to prevent the influx. The osmotic concentration of the cytoplasm can be reduced by use of inclusion bodies (*see pp. 47–50*). Bacteria and Archaea also can have mechanosensitive (MS) channels in their plasma membrane. In a hypotonic environment, the membrane stretches due to an increase in hydrostatic pressure and cellular swelling. MS channels then open and allow solutes to leave. Thus they can act as escape valves to protect cells from bursting.

Most bacteria, algae, and fungi have rigid cell walls that maintain the shape and integrity of the cell. When microorganisms with rigid cell walls are placed in a hypertonic environment, water leaves and the plasma membrane shrinks away from the wall, a process known as plasmolysis. This dehydrates the cell and may damage the plasma membrane; the cell usually becomes metabolically inactive and ceases to grow.

Many microorganisms keep the osmotic concentration of their protoplasm somewhat above that of the habitat by the use of compatible solutes, so that the plasma membrane is always pressed firmly against their cell wall. **Compatible solutes** are solutes that are compatible with metabolism and growth when at high intracellular concentrations. Most procaryotes increase their internal osmotic concentration in a hypertonic environment through the synthesis or uptake of choline, betaine, proline, glutamic acid, and other amino acids; elevated levels of potassium ions are also involved to some extent. Algae and fungi employ sucrose and polyols—for example, arabitol, glycerol, and mannitol—for the same purpose. Polyols and amino acids are ideal solutes for this function because they normally do not disrupt enzyme structure and function. A few procaryotes like *Halobacterium salinarium* raise their osmotic concentration with potassium ions (sodium ions are also elevated but not as much as potassium). *Halobacterium*'s enzymes have been altered so that they actually require high salt concentrations for normal activity (*see section 20.3*). Since protozoa do not have a cell wall, they must use contractile vacuoles (*see figure 27.3*) to eliminate excess

Table 6.3
Microbial Responses to Environmental Factors

Descriptive Term	Definition	Representative Microorganisms
Solute and Water Activity		
Osmotolerant	Able to grow over wide ranges of water activity or osmotic concentration	*Staphylococcus aureus, Saccharomyces rouxii*
Halophile	Requires high levels of sodium chloride, usually above about 0.2 M, to grow	*Halobacterium, Dunaliella, Ectothiorhodospira*
pH		
Acidophile	Growth optimum between pH 0 and 5.5	*Sulfolobus, Picrophilus, Ferroplasma, Acontium, Cyanidium caldarium*
Neutrophile	Growth optimum between pH 5.5 and 8.0	*Escherichia, Euglena, Paramecium*
Alkalophile	Growth optimum between pH 8.5 and 11.5	*Bacillus alcalophilus, Natronobacterium*
Temperature		
Psychrophile	Grows well at 0°C and has an optimum growth temperature of 15°C or lower	*Bacillus psychrophilus, Chlamydomonas nivalis*
Psychrotroph	Can grow at 0–7°C; has an optimum between 20 and 30°C and a maximum around 35°C	*Listeria monocytogenes, Pseudomonas fluorescens*
Mesophile	Has growth optimum around 20–45°C	*Escherichia coli, Neisseria gonorrhoeae, Trichomonas vaginalis*
Thermophile	Can grow at 55°C or higher; optimum often between 55 and 65°C	*Bacillus stearothermophilus, Thermus aquaticus, Cyanidium caldarium, Chaetomium thermophile*
Hyperthermophile	Has an optimum between 80 and about 113°C	*Sulfolobus, Pyrococcus, Pyrodictium*
Oxygen Concentration		
Obligate aerobe	Completely dependent on atmospheric O_2 for growth	*Micrococcus luteus, Pseudomonas, Mycobacterium;* most algae, fungi, and protozoa
Facultative anaerobe	Does not require O_2 for growth, but grows better in its presence	*Escherichia, Enterococcus, Saccharomyces cerevisiae*
Aerotolerant anaerobe	Grows equally well in presence or absence of O_2	*Streptococcus pyogenes*
Obligate anaerobe	Does not tolerate O_2 and dies in its presence	*Clostridium, Bacteroides, Methanobacterium, Trepomonas agilis*
Microaerophile	Requires O_2 levels below 2–10% for growth and is damaged by atmospheric O_2 (20%)	*Campylobacter, Spirillum volutans, Treponema pallidum*
Pressure		
Barophilic	Growth more rapid at high hydrostatic pressures	*Photobacterium profundum, Shewanella benthica, Methanococcus jannaschii*

water when living in hypotonic environments.   Osmosis and the protective function of the cell wall (p. 59)

The amount of water available to microorganisms can be reduced by interaction with solute molecules (the osmotic effect) or by adsorption to the surfaces of solids (the matric effect). Because the osmotic concentration of a habitat has such profound effects on microorganisms, it is useful to be able to express quantitatively the degree of water availability. Microbiologists generally use **water activity** (**a_w**) for this purpose (water availability also may be expressed as water potential, which is related to a_w). The water activity of a solution is 1/100 the relative humidity of the solution (when expressed as a percent). It is also equivalent to the ratio of the solution's vapor pressure (P_{soln}) to that of pure water (P_{water}).

$$a_w = \frac{P_{soln}}{P_{water}}$$

The water activity of a solution or solid can be determined by sealing it in a chamber and measuring the relative humidity after the system has come to equilibrium. Suppose after a sample is treated in this way, the air above it is 95% saturated—that is, the air contains 95% of the moisture it would have when equilibrated at the same temperature with a sample of pure water. The relative humidity would be 95% and the sample's water activity, 0.95. Water activity is inversely related to osmotic pressure; if a solution has high osmotic pressure, its a_w is low.

Microorganisms differ greatly in their ability to adapt to habitats with low water activity (**table 6.4**). A microorganism must expend extra effort to grow in a habitat with a low a_w value because it must maintain a high internal solute concentration to retain water. Some microorganisms can do this and are **osmotolerant;** they will grow over wide ranges of water activity or osmotic concentration. For example, *Staphylococcus aureus* is halotolerant (**figure 6.11**) and can be cultured in media containing any sodium

Table 6.4

Approximate Lower a_w Limits for Microbial Growth

Water Activity	Environment	Bacteria	Fungi	Algae
1.00—Pure water	Blood {Vegetables, Plant wilt {meat, fruit	Most gram-negative nonhalophiles		
	Seawater			
0.95	Bread	Most gram-positive rods	*Basidiomycetes*	Most algae
0.90	Ham	Most cocci, *Bacillus*	*Fusarium* *Mucor, Rhizopus* Ascomycetous yeasts	
0.85	Salami	*Staphylococcus*	*Saccharomyces rouxii* (in salt)	
0.80	Preserves		*Penicillium*	
0.75	Salt lakes Salted fish	*Halobacterium* *Actinospora*	*Aspergillus*	*Dunaliella*
0.70	Cereals, candy, dried fruit		*Aspergillus*	
0.60	Chocolate Honey Dried milk		*Saccharomyces rouxii* (in sugars) *Xeromyces bisporus*	
0.55—DNA disordered				

Adapted from A. D. Brown, "Microbial Water Stress," in *Bacteriological Reviews*, 40(4):803–846 1976. Copyright © 1976 by the American Society for Microbiology. Reprinted by permission.

chloride concentration up to about 3 M. It is well adapted for growth on the skin. The yeast *Saccharomyces rouxii* will grow in sugar solutions with a_w values as low as 0.6. The alga *Dunaliella viridis* tolerates sodium chloride concentrations from 1.7 M to a saturated solution.

Although a few microorganisms are truly osmotolerant, most only grow well at water activities around 0.98 (the approximate a_w for seawater) or higher. This is why drying food or adding large quantities of salt and sugar is so effective in preventing food spoilage. As table 6.4 shows, many fungi are osmotolerant and thus particularly important in the spoilage of salted or dried foods.   Food spoilage (pp. 940–43)

Halophiles grow optimally in the presence of NaCl or other salts at a concentration above about 0.2 M (figure 6.11). Extreme halophiles have adapted so completely to hypertonic, saline conditions that they require high levels of sodium chloride to grow, concentrations between about 2 M and saturation (about 6.2 M). The archaeon *Halobacterium* can be isolated from the Dead Sea (a salt lake between Israel and Jordan and the lowest lake in the world), the Great Salt Lake in Utah, and other aquatic habitats with salt concentrations approaching saturation. *Halobacterium* and other extremely halophilic bacteria have significantly modified the structure of their proteins and membranes rather than simply increasing the intracellular concentrations of solutes, the approach used by most osmotolerant microorganisms. These extreme halophiles accumulate enormous quantities of potassium in order to remain hypertonic to their environment; the internal potassium concentration may reach 4 to 7 M. The enzymes, ribosomes, and transport proteins of these bacteria require high potas-

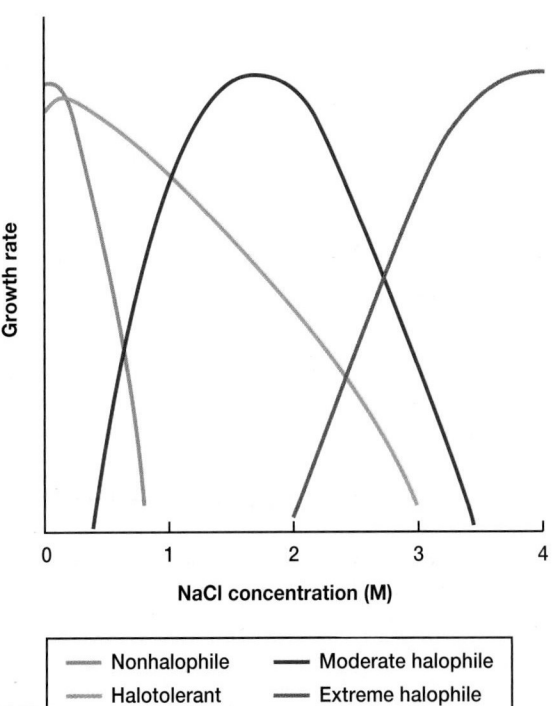

Figure 6.11   The Effects of Sodium Chloride on Microbial Growth.   Four different patterns of microbial dependence on NaCl concentration are depicted. The curves are only illustrative and not meant to provide precise shapes or salt concentrations required for growth.

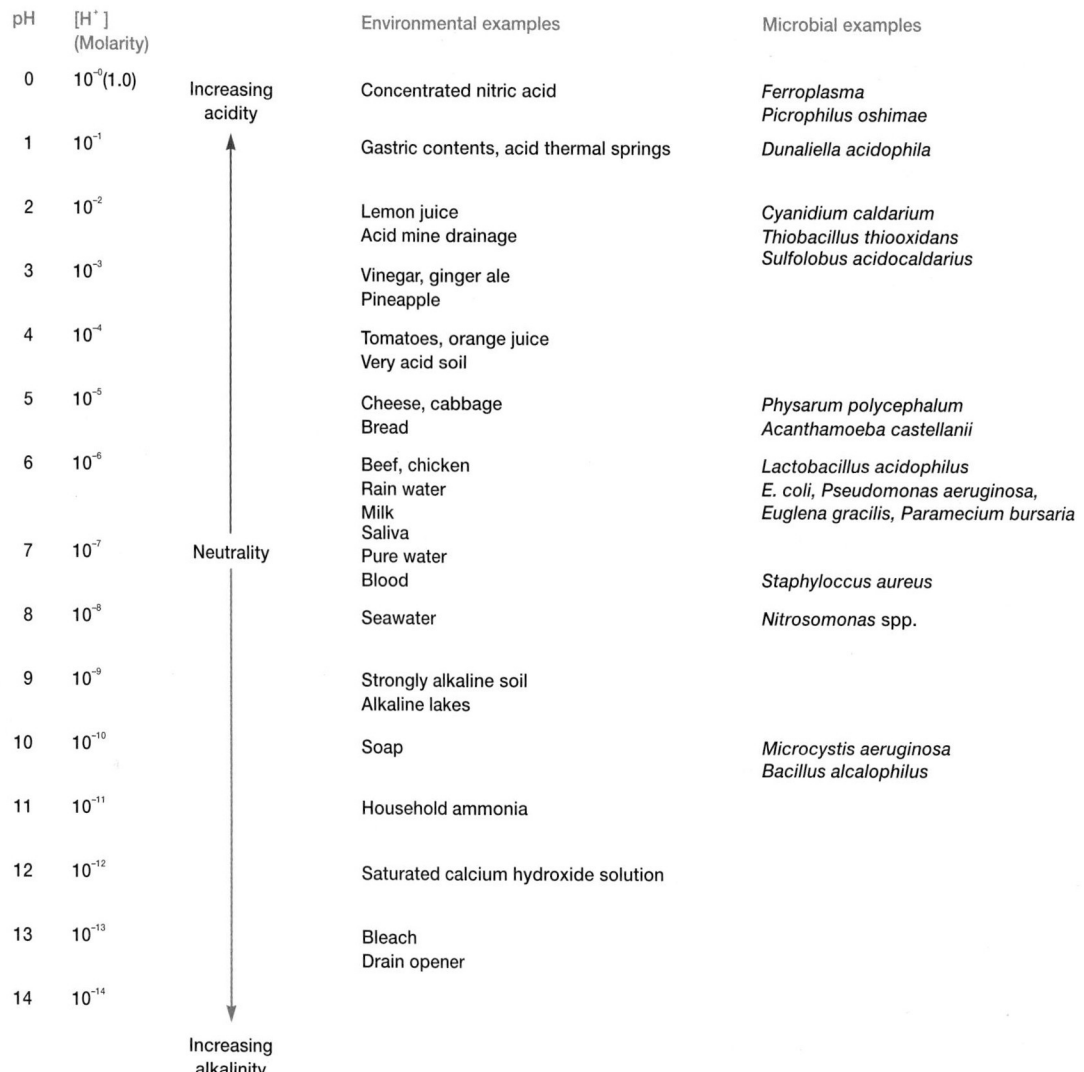

pH	$[H^+]$ (Molarity)		Environmental examples	Microbial examples
0	10^{-0}(1.0)	Increasing acidity ↑	Concentrated nitric acid	*Ferroplasma* *Picrophilus oshimae*
1	10^{-1}		Gastric contents, acid thermal springs	*Dunaliella acidophila*
2	10^{-2}		Lemon juice Acid mine drainage	*Cyanidium caldarium* *Thiobacillus thiooxidans* *Sulfolobus acidocaldarius*
3	10^{-3}		Vinegar, ginger ale Pineapple	
4	10^{-4}		Tomatoes, orange juice Very acid soil	
5	10^{-5}		Cheese, cabbage Bread	*Physarum polycephalum* *Acanthamoeba castellanii*
6	10^{-6}		Beef, chicken Rain water Milk Saliva	*Lactobacillus acidophilus* *E. coli, Pseudomonas aeruginosa,* *Euglena gracilis, Paramecium bursaria*
7	10^{-7}	Neutrality	Pure water Blood	*Staphyloccus aureus*
8	10^{-8}		Seawater	*Nitrosomonas* spp.
9	10^{-9}		Strongly alkaline soil Alkaline lakes	
10	10^{-10}		Soap	*Microcystis aeruginosa* *Bacillus alcalophilus*
11	10^{-11}		Household ammonia	
12	10^{-12}		Saturated calcium hydroxide solution	
13	10^{-13}		Bleach Drain opener	
14	10^{-14}	↓ Increasing alkalinity		

Figure 6.12 The pH Scale. The pH scale and examples of substances with different pH values. The microorganisms are placed at their growth optima.

sium levels for stability and activity. In addition, the plasma membrane and cell wall of *Halobacterium* are stabilized by high concentrations of sodium ion. If the sodium concentration decreases too much, the wall and plasma membrane literally disintegrate. Extreme halophilic bacteria have successfully adapted to environmental conditions that would destroy most organisms. In the process they have become so specialized that they have lost ecological flexibility and can prosper only in a few extreme habitats. The halobacteria (section 20.3)

1. How do microorganisms adapt to hypotonic and hypertonic environments? What is plasmolysis?
2. Define water activity and briefly describe how it can be determined.
3. Why is it difficult for microorganisms to grow at low a_w values?
4. What are halophiles and why does *Halobacterium* require sodium and potassium ions?

pH

pH is a measure of the hydrogen ion activity of a solution and is defined as the negative logarithm of the hydrogen ion concentration (expressed in terms of molarity).

$$pH = -\log [H^+] = \log(1/[H^+])$$

The pH scale extends from pH 0.0 (1.0 M H^+) to pH 14.0 (1.0×10^{-14} M H^+), and each pH unit represents a tenfold change in hydrogen ion concentration. **Figure 6.12** shows that the habitats in which microorganisms grow vary widely—from pH 0 to 2 at the acid end to alkaline lakes and soil that may have pH values between 9 and 10.

It is not surprising that pH dramatically affects microbial growth. Each species has a definite pH growth range and pH growth optimum. **Acidophiles** have their growth optimum between pH 0 and 5.5; **neutrophiles,** between pH 5.5 and 8.0; and

alkalophiles prefer the pH range of 8.5 to 11.5. Extreme alkalophiles have growth optima at pH 10 or higher. In general, different microbial groups have characteristic pH preferences. Most bacteria and protozoa are neutrophiles. Most fungi prefer slightly acid surroundings, about pH 4 to 6; algae also seem to favor slight acidity. There are many exceptions to these generalizations. For example, the alga *Cyanidium caldarium* and the archaeon *Sulfolobus acidocaldarius* are common inhabitants of acidic hot springs; both grow well around pH 1 to 3 and at high temperatures. The Archaea *Ferroplasma acidarmanus* and *Picrophilus oshimae* can actually grow at pH 0, or very close to it.

Although microorganisms will often grow over wide ranges of pH and far from their optima, there are limits to their tolerance. Drastic variations in cytoplasmic pH can harm microorganisms by disrupting the plasma membrane or inhibiting the activity of enzymes and membrane transport proteins. Procaryotes die if the internal pH drops much below 5.0 to 5.5. Changes in the external pH also might alter the ionization of nutrient molecules and thus reduce their availability to the organism.

Several mechanisms for the maintenance of a neutral cytoplasmic pH have been proposed. The plasma membrane may be relatively impermeable to protons. Neutrophiles appear to exchange potassium for protons using an antiport transport system (see *p. 100*). Extreme alkalophiles like *Bacillus alcalophilus* maintain their internal pH closer to neutrality by exchanging internal sodium ions for external protons. Internal buffering also may contribute to pH homeostasis.

Microorganisms often must adapt to environmental pH changes to survive. In bacteria, potassium/proton and sodium/proton antiport systems probably correct small variations in pH. If the pH becomes too acidic, other mechanisms come into play. When the pH drops below about 5.5 to 6.0, *Salmonella typhimurium* and *E. coli* synthesize an array of new proteins as part of what has been called their acidic tolerance response. A proton-translocating ATPase contributes to this protective response, either by making more ATP or by pumping protons out of the cell. If the external pH decreases to 4.5 or lower, chaperones such as acid shock proteins and heat shock proteins are synthesized (*see pp. 267–69*). Presumably these prevent the acid denaturation of proteins and aid in the refolding of denatured proteins.

Microorganisms frequently change the pH of their own habitat by producing acidic or basic metabolic waste products. Fermentative microorganisms form organic acids from carbohydrates, whereas chemolithotrophs like *Thiobacillus* oxidize reduced sulfur components to sulfuric acid. Other microorganisms make their environment more alkaline by generating ammonia through amino acid degradation. Microbial fermentations (pp. 174–77); Sulfur-oxidizing bacteria (pp. 483–85)

Buffers often are included in media to prevent growth inhibition by large pH changes. Phosphate is a commonly used buffer and a good example of buffering by a weak acid ($H_2PO_4^-$) and its conjugate base (HPO_4^{2-}).

$$H^+ + HPO_4^{2-} \longrightarrow H_2PO_4^-$$

$$OH^- + H_2PO_4^- \longrightarrow HPO_4^{2-} + HOH$$

If protons are added to the mixture, they combine with the salt form to yield a weak acid. An increase in alkalinity is resisted because the weak acid will neutralize hydroxyl ions through proton donation to give water. Peptides and amino acids in complex media also have a strong buffering effect.

Temperature

Environmental temperature profoundly affects microorganisms, like all other organisms. Indeed, microorganisms are particularly susceptible because they are usually unicellular and their temperature varies with that of the external environment. For these reasons, microbial cell temperature directly reflects that of the cell's surroundings. A most important factor influencing the effect of temperature on growth is the temperature sensitivity of enzyme-catalyzed reactions. At low temperatures a temperature rise increases the growth rate because the velocity of an enzyme-catalyzed reaction, like that of any chemical reaction, will roughly double for every 10°C rise in temperature. Because the rate of each reaction increases, metabolism as a whole is more active at higher temperatures, and the microorganism grows faster. Beyond a certain point further increases actually slow growth, and sufficiently high temperatures are lethal. High temperatures damage microorganisms by denaturing enzymes, transport carriers, and other proteins. Microbial membranes are also disrupted by temperature extremes; the lipid bilayer simply melts and disintegrates. Thus although functional enzymes operate more rapidly at higher temperatures, the microorganism may be damaged to such an extent that growth is inhibited because the damage cannot be repaired. At very low temperatures, membranes solidify and enzymes don't work rapidly. In summary, when organisms are above their optimum temperature, both function and cell structure are affected. If temperatures are very low, function is affected but not necessarily cell chemical composition and structure. The temperature dependence of enzyme activity (p. 159)

Because of these opposing temperature influences, microbial growth has a fairly characteristic temperature dependence with distinct **cardinal temperatures**—minimum, optimum, and maximum growth temperatures (**figure 6.13**). Although the shape of the temperature dependence curve can vary, the temperature optimum is always closer to the maximum than to the minimum. The cardinal temperatures for a particular species are not rigidly fixed but often depend to some extent on other environmental factors such as pH and the available nutrients. For example, *Crithidia fasciculata,* a flagellated protozoan living in the gut of mosquitos, will grow in a simple medium at 22 to 27°C. However, it cannot be cultured at 33 to 34°C without the addition of extra metals, amino acids, vitamins, and lipids.

The cardinal temperatures vary greatly between microorganisms (**table 6.5**). Optima usually range from 0°C to as high as 75°C, whereas microbial growth occurs at temperatures extending from −20°C to over 120°C. Recently the archaen *Geogemma barossii* that grows anaerobically at 121°C (250°F), the temperature normally used in autoclaves, has been isolated from a marine hydrothermal vent. The major factor determining this growth range seems to be water. Even at the most extreme temperatures,

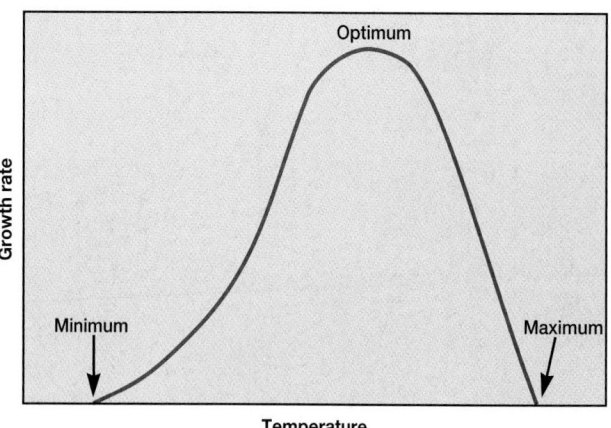

Figure 6.13 Temperature and Growth. The effect of temperature on growth rate.

Table 6.5
Temperature Ranges for Microbial Growth

Microorganism	Cardinal Temperatures (°C)		
	Minimum	Optimum	Maximum
Nonphotosynthetic Procaryotes			
Bacillus psychrophilus	−10	23–24	28–30
Micrococcus cryophilus	−4	10	24
Pseudomonas fluorescens	4	25–30	40
Staphylococcus aureus	6.5	30–37	46
Enterococcus faecalis	0	37	44
Escherichia coli	10	37	45
Neisseria gonorrhoeae	30	35–36	38
Thermoplasma acidophilum	45	59	62
Bacillus stearothermophilus	30	60–65	75
Thermus aquaticus	40	70–72	79
Sulfolobus acidocaldarius	60	80	85
Pyrococcus abyssi	67	96	102
Pyrodictium occultum	82	105	110
Pyrolobus fumarii	90	106	113
Photosynthetic Bacteria			
Rhodospirillum rubrum	ND[a]	30–35	ND
Anabaena variabilis	ND	35	ND
Oscillatoria tenuis	ND	ND	45–47
Synechococcus eximius	70	79	84
Eucaryotic Algae			
Chlamydomonas nivalis	−36	0	4
Fragilaria sublinearis	−2	5–6	8–9
Chlorella pyrenoidosa	ND	25–26	29
Euglena gracilis	ND	23	ND
Skeletonema costatum	6	16–26	>28
Cyanidium caldarium	30–34	45–50	56
Fungi			
Candida scottii	0	4–15	15
Saccharomyces cerevisiae	1–3	28	40
Mucor pusillus	21–23	45–50	50–58
Protozoa			
Amoeba proteus	4–6	22	35
Naegleria fowleri	20–25	35	40
Trichomonas vaginalis	25	32–39	42
Paramecium caudatum		25	28–30
Tetrahymena pyriformis	6–7	20–25	33
Cyclidium citrullus	18	43	47

[a]ND, no data.

microorganisms need liquid water to grow. The growth temperature range for a particular microorganism usually spans about 30 degrees. Some species (e.g., *Neisseria gonorrhoeae*) have a small range; others, like *Enterococcus faecalis,* will grow over a wide range of temperatures. The major microbial groups differ from one another regarding their maximum growth temperature. The upper limit for protozoa is around 50°C. Some algae and fungi can grow at temperatures as high as 55 to 60°C. Procaryotes have been found growing at or close to 100°C, the boiling point of water at sea level (*see figure 20.8*). Recently strains growing at even higher temperatures have been discovered (**Microbial Diversity & Ecology 6.1**). Clearly, procaryotic organisms can grow at much higher temperatures than eucaryotes. It has been suggested that eucaryotes are not able to manufacture organellar membranes that are stable and functional at temperatures above 60°C. The photosynthetic apparatus also appears to be relatively unstable because photosynthetic organisms are not found growing at very high temperatures.

Microorganisms such as those in table 6.5 can be placed in one of five classes based on their temperature ranges for growth (**figure 6.14**).

1. **Psychrophiles** grow well at 0°C and have an optimum growth temperature of 15°C or lower; the maximum is around 20°C. They are readily isolated from Arctic and Antarctic habitats; because 90% of the ocean is 5°C or colder (*see chapter 29*), it constitutes an enormous habitat for psychrophiles. The psychrophilic alga *Chlamydomonas nivalis* can actually turn a snowfield or glacier pink with its bright red spores. Psychrophiles are widespread among bacterial taxa and found in such genera as *Pseudomonas, Vibrio, Alcaligenes, Bacillus, Arthrobacter, Moritella, Photobacterium,* and *Shewanella*. The psychrophilic archaeon *Methanogenium* has recently been isolated from Ace Lake in Antarctica. Psychrophilic microorganisms have adapted to their environment in several ways. Their enzymes, transport systems, and protein synthetic mechanisms function well at low temperatures. The cell membranes of psychrophilic microorganisms have

high levels of unsaturated fatty acids and remain semifluid when cold. Indeed, many psychrophiles begin to leak cellular constituents at temperatures higher than 20°C because of cell membrane disruption.

2. Many species can grow at 0 to 7°C even though they have optima between 20 and 30°C, and maxima at about 35°C. These are called **psychrotrophs** or **facultative psychrophiles.** Psychrotrophic bacteria and fungi are major factors in the spoilage of refrigerated foods (*see chapter 41*).

Microbial Diversity & Ecology
6.1   Life above 100°C

Until recently the highest reported temperature for procaryotic growth was 105°C. It seemed that the upper temperature limit for life was about 100°C, the boiling point of water. Now thermophilic procaryotes have been reported growing in sulfide chimneys or "black smokers," located along rifts and ridges on the ocean floor, that spew sulfide-rich super-heated vent water with temperatures above 350°C (see **Box figure**). Evidence has been presented that these microbes can grow and repro-duce at or above 113°C. The pressure present in their habitat is sufficient to keep water liquid (at 265 atm; seawater doesn't boil until 460°C).

The implications of this discovery are many. The proteins, mem-branes, and nucleic acids of these procaryotes are remarkably tem-perature stable and provide ideal subjects for studying the ways in which macromolecules and membranes are stabilized. In the future it may be possible to design enzymes that can operate at very high tem-peratures. Some thermostable enzymes from these organisms have important industrial and scientific uses. For example, the Taq poly-merase from the thermophile *Thermus aquaticus* is used extensively in the polymerase chain reaction (*see pp. 316–19*).

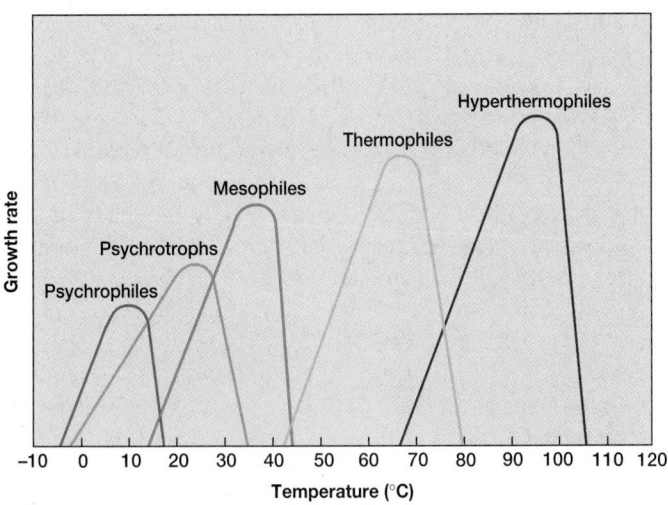

Figure 6.14   Temperature Ranges for Microbial Growth. Microorganisms can be placed in different classes based on their temperature ranges for growth. They are ranked in order of increas-ing growth temperature range as psychrophiles, psychrotrophs, mesophiles, thermophiles, and hyperthermophiles. Representative ranges and optima for these five types are illustrated here.

3. **Mesophiles** are microorganisms with growth optima around 20 to 45°C; they often have a temperature minimum of 15 to 20°C. Their maximum is about 45°C or lower. Most microor-ganisms probably fall within this category. Almost all human pathogens are mesophiles, as might be expected since their environment is a fairly constant 37°C.

4. Some microorganisms are **thermophiles;** they can grow at temperatures of 55°C or higher. Their growth minimum is usually around 45°C and they often have optima between 55 and 65°C. The vast majority are procaryotes although a few algae and fungi are thermophilic (table 6.5). These organ-isms flourish in many habitats including composts, self-heating hay stacks, hot water lines, and hot springs.

Thermophiles differ from mesophiles in many ways. They have more heat-stable enzymes and protein synthesis systems, which function properly at high temperatures. These proteins are stable for a variety of reasons. Heat-stable proteins have highly organized, hydrophobic interiors; more hydrogen bonds and other noncovalent bonds strengthen the structure. Larger quantities of amino acids such as proline also make the polypeptide chain less flexible (*see appendix I*). In addition, the proteins are stabilized and aided in fold-ing by special chaperones (*see pp. 267–69*). There is evi-dence that thermophile DNA is stabilized by special histonelike proteins. The membrane lipids of thermophiles are also quite temperature stable. They tend to be more satu-rated, more branched, and of higher molecular weight. This increases the melting points of membrane lipids. Archaeal thermophiles have membrane lipids with ether linkages, which protect the lipids from hydrolysis at high tempera-tures. Sometimes archaeal lipids actually span the membrane to form a rigid, stable monolayer (*see pp. 439–40*).

5. As mentioned previously, a few thermophiles can grow at 90°C or above and some have maxima above 100°C. Procary-otes that have growth optima between 80°C and about 113°C are called **hyperthermophiles.** They usually do not grow well

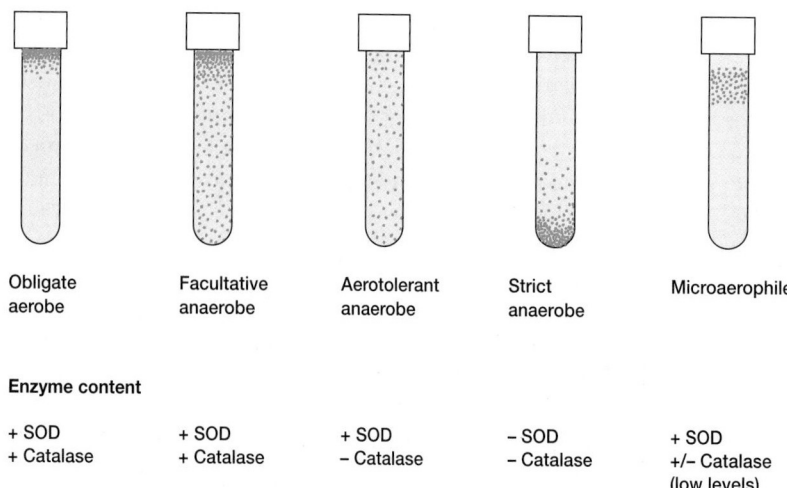

| Obligate aerobe | Facultative anaerobe | Aerotolerant anaerobe | Strict anaerobe | Microaerophile |

Enzyme content

| + SOD | + SOD | + SOD | − SOD | + SOD |
| + Catalase | + Catalase | − Catalase | − Catalase | +/− Catalase (low levels) |

Figure 6.15   Oxygen and Bacterial Growth. Each dot represents an individual bacterial colony within the agar or on its surface. The surface, which is directly exposed to atmospheric oxygen, will be aerobic. The oxygen content of the medium decreases with depth until the medium becomes anaerobic toward the bottom of the tube. The presence and absence of the enzymes superoxide dismutase (SOD) and catalase for each type are shown.

below 55°C. *Pyrococcus abyssi* and *Pyrodictium occultum* are examples of marine hyperthermophiles found in hot areas of the seafloor.

---

1. Define pH, acidophile, neutrophile, and alkalophile. How can microorganisms change the pH of their environment, and how does the microbiologist minimize this effect?
2. What are cardinal temperatures?
3. Why does the growth rate rise with increasing temperature and then fall again at higher temperatures?
4. Define psychrophile, psychrotroph, mesophile, thermophile, and hyperthermophile.

Oxygen Concentration

An organism able to grow in the presence of atmospheric O_2 is an **aerobe,** whereas one that can grow in its absence is an **anaerobe.** Almost all multicellular organisms are completely dependent on atmospheric O_2 for growth—that is, they are **obligate aerobes** (table 6.3). Oxygen serves as the terminal electron acceptor for the electron-transport chain in aerobic respiration. In addition, aerobic eucaryotes employ O_2 in the synthesis of sterols and unsaturated fatty acids. **Facultative anaerobes** do not require O_2 for growth but do grow better in its presence. In the presence of oxygen they will use aerobic respiration. **Aerotolerant anaerobes** such as *Enterococcus faecalis* simply ignore O_2 and grow equally well whether it is present or not. In contrast, **strict** or **obligate anaerobes** (e.g., *Bacteroides, Fusobacterium, Clostridium pasteurianum, Methanococcus, Neocallimastix*) do not tolerate O_2 at all and die in its presence. Aerotolerant and strict anaerobes cannot generate energy through respiration and must employ fermentation or anaerobic respiration pathways for this purpose. Finally, there are aerobes such as *Campylobacter,* called **microaerophiles,** that are damaged by the normal atmospheric level of O_2 (20%) and require O_2 levels below the range of 2 to 10% for growth. The nature of bacterial O_2 responses can be readily determined by growing the bacteria in culture tubes filled with a solid culture medium or a special medium like thioglycollate broth, which contains a reducing agent to lower O_2 levels (**figure 6.15**). Electron transport and aerobic respiration (pp. 179–84); Fermentation (pp. 174–77); Anaerobic respiration (pp. 185–86)

A microbial group may show more than one type of relationship to O_2. All five types are found among the procaryotes and protozoa. Fungi are normally aerobic, but a number of species—particularly among the yeasts—are facultative anaerobes. Algae are almost always obligate aerobes. It should be noted that the ability to grow in both aerobic and anaerobic environments provides considerable flexibility and is an ecological advantage.

Although strict anaerobes are killed by O_2, they may be recovered from habitats that appear to be aerobic. In such cases they associate with facultative anaerobes that use up the available O_2 and thus make the growth of strict anaerobes possible. For example, the strict anaerobe *Bacteroides gingivalis* lives in the mouth where it grows in the anaerobic crevices around the teeth.

These different relationships with O_2 appear due to several factors, including the inactivation of proteins and the effect of toxic O_2 derivatives. Enzymes can be inactivated when sensitive groups like sulfhydryls are oxidized. A notable example is the nitrogen-fixation enzyme nitrogenase (*see section 10.4*), which is very oxygen sensitive.

Oxygen accepts electrons and is readily reduced because its two outer orbital electrons are unpaired. Flavoproteins (*see section 8.5*), several other cell constituents, and radiation (pp. 127–28) promote oxygen reduction. The result is usually some combination of the reduction products **superoxide radical, hydrogen peroxide,** and **hydroxyl radical.**

$$O_2 + e^- \rightarrow O_2 \cdot^- \text{ (superoxide radical)}$$

$$O_2 \cdot^- + e^- + 2H^+ \rightarrow H_2O_2 \text{ (hydrogen peroxide)}$$

$$H_2O_2 + e^- + H^+ \rightarrow H_2O + OH \cdot \text{ (hydroxyl radical)}$$

These products of oxygen reduction are extremely toxic because they are powerful oxidizing agents and rapidly destroy cellular constituents. A microorganism must be able to protect itself against such oxygen products or it will be killed. Neutrophils and macrophages use these toxic oxygen products to destroy invading pathogens. Oxygen-dependent killing of pathogens (pp. 696–97)

Many microorganisms possess enzymes that afford protection against toxic O_2 products. Obligate aerobes and facultative anaerobes usually contain the enzymes **superoxide dismutase (SOD)** and **catalase,** which catalyze the destruction of superoxide radical and hydrogen peroxide, respectively. Peroxidase also can be used to destroy hydrogen peroxide.

$$2O_2^{-} + 2H^{+} \xrightarrow{\text{superoxide dismutase}} O_2 + H_2O_2$$

$$2H_2O_2 \xrightarrow{\text{catalase}} 2H_2O + O_2$$

$$H_2O_2 + NADH + H^{+} \xrightarrow{\text{peroxidase}} 2H_2O + NAD^{+}$$

Aerotolerant microorganisms may lack catalase but almost always have superoxide dismutase. The aerotolerant *Lactobacillus plantarum* uses manganous ions instead of superoxide dismutase to destroy the superoxide radical. All strict anaerobes lack both enzymes or have them in very low concentrations and therefore cannot tolerate O_2.

Because aerobes need O_2 and anaerobes are killed by it, radically different approaches must be used when growing the two types of microorganisms. When large volumes of aerobic microorganisms are cultured, either the culture vessel is shaken to aerate the medium or sterile air must be pumped through the culture vessel. Precisely the opposite problem arises with anaerobes; all O_2 must be excluded. This can be accomplished in several ways. (1) Special anaerobic media containing reducing agents such as thioglycollate or cysteine may be used. The medium is boiled during preparation to dissolve its components; boiling also drives off oxygen very effectively. The reducing agents will eliminate any dissolved O_2 remaining within the medium so that anaerobes can grow beneath its surface. (2) Oxygen also may be eliminated from an anaerobic system by removing air with a vac-

uum pump and flushing out residual O_2 with nitrogen gas (**figure 6.16**). Often CO_2 as well as nitrogen is added to the chamber since many anaerobes require a small amount of CO_2 for best growth. (3) One of the most popular ways of culturing small numbers of anaerobes is by use of a GasPak jar (**figure 6.17**). In this procedure the environment is made anaerobic by using hydrogen and a palladium catalyst to remove O_2 through the formation of water. The reducing agents in anaerobic agar also remove oxygen, as mentioned previously. (4) Plastic bags or pouches make conven-

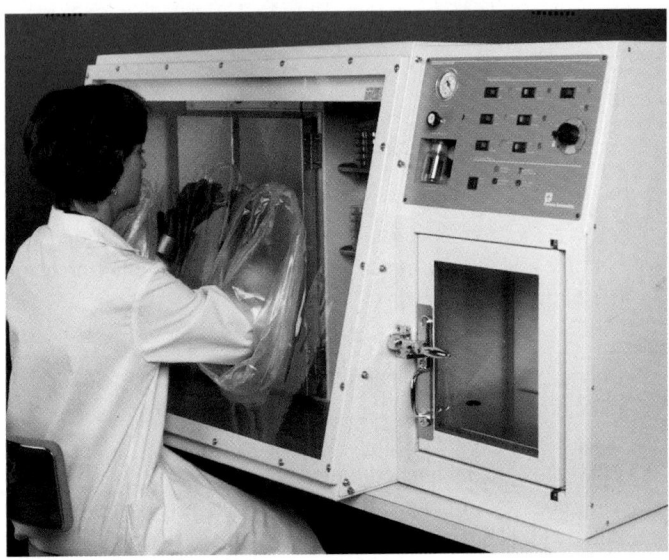

Figure 6.16   An Anaerobic Work Chamber and Incubator. This anaerobic system contains an oxygen-free work area and an incubator. The interchange compartment on the right of the work area allows materials to be transferred inside without exposing the interior to oxygen. The anaerobic atmosphere is maintained largely with a vacuum pump and nitrogen purges. The remaining oxygen is removed by a palladium catalyst and hydrogen. The oxygen reacts with hydrogen to form water, which is absorbed by desiccant.

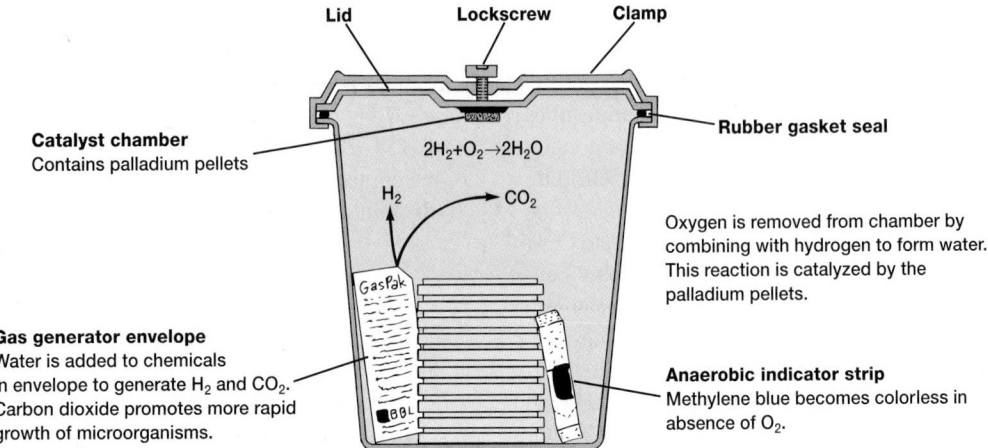

Figure 6.17   The GasPak Anaerobic System.   Hydrogen and carbon dioxide are generated by a GasPak envelope. The palladium catalyst in the chamber lid catalyzes the formation of water from hydrogen and oxygen, thereby removing oxygen from the sealed chamber.

ient containers when only a few samples are to be incubated anaerobically. These have a catalyst and calcium carbonate to produce an anaerobic, carbon-dioxide-rich atmosphere. A special solution is added to the pouch's reagent compartment; petri dishes or other containers are placed in the pouch; it then is clamped shut and placed in an incubator. A laboratory may make use of all these techniques since each is best suited for different purposes.

1. Describe the five types of O₂ relationships seen in microorganisms.
2. For what do aerobes use O₂? Why is O₂ toxic to many microorganisms and how do they protect themselves?
3. Describe four ways in which anaerobes may be cultured.

Pressure

Most organisms spend their lives on land or on the surface of water, always subjected to a pressure of 1 atmosphere (atm), and are never affected significantly by pressure. Yet the deep sea (ocean of 1,000 m or more in depth) is 75% of the total ocean volume. The hydrostatic pressure can reach 600 to 1,100 atm in the deep sea, while the temperature is about 2 to 3°C. Despite these extremes, bacteria survive and adapt. Many are **barotolerant:** increased pressure does adversely affect them but not as much as it does nontolerant bacteria. Some bacteria in the gut of deep-sea invertebrates such as amphipods and holothurians are truly **barophilic**—they grow more rapidly at high pressures. These gut bacteria may play an important role in nutrient recycling in the deep sea. One barophile has been recovered from the Mariana trench near the Philippines (depth about 10,500 m) that is actually unable to grow at pressures below about 400 to 500 atm when incubated at 2°C. Thus far, barophiles have been found among several bacterial genera (e.g., *Photobacterium, Shewanella, Colwellia*). Some members of the Archaea are thermobarophiles (e.g., *Pyrococcus* spp., *Methanococcus jannaschii*). The marine environment (pp. 624–27)

Radiation

Our world is bombarded with electromagnetic radiation of various types (**figure 6.18**). This radiation often behaves as if it were composed of waves moving through space like waves traveling on the surface of water. The distance between two wave crests or troughs is the wavelength. As the wavelength of electromagnetic radiation decreases, the energy of the radiation increases—gamma rays and X rays are much more energetic than visible light or infrared waves. Electromagnetic radiation also acts like a stream of energy packets called photons, each photon having a quantum of energy whose value will depend on the wavelength of the radiation.

Sunlight is the major source of radiation on the Earth. It includes visible light, ultraviolet (UV) radiation, infrared rays, and radio waves. Visible light is a most conspicuous and important aspect of our environment: most life is dependent on the ability of photosynthetic organisms to trap the light energy of the sun. Almost 60% of the sun's radiation is in the infrared region rather

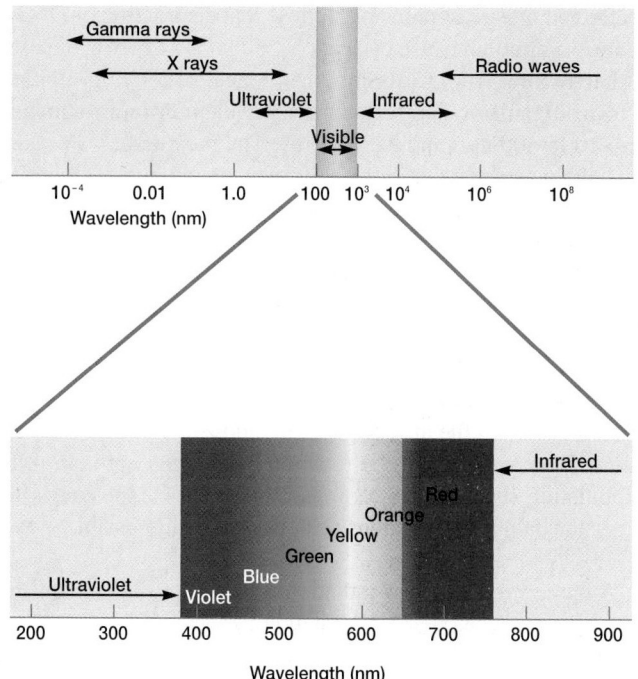

Figure 6.18   The Electromagnetic Spectrum. The visible portion of the spectrum is expanded at the bottom of the figure.

than the visible portion of the spectrum. Infrared is the major source of the Earth's heat. At sea level, one finds very little ultraviolet radiation below about 290 to 300 nm. UV radiation of wavelengths shorter than 287 nm is absorbed by O₂ in the Earth's atmosphere; this process forms a layer of ozone between 25 and 30 miles above the Earth's surface. The ozone layer then absorbs somewhat longer UV rays and reforms O₂. This elimination of UV radiation is crucial because it is quite damaging to living systems (*see section 11.6*). The fairly even distribution of sunlight throughout the visible spectrum accounts for the fact that sunlight is generally "white." Microbial photosynthesis (pp. 190–96)

Many forms of electromagnetic radiation are very harmful to microorganisms. This is particularly true of **ionizing radiation,** radiation of very short wavelength or high energy, which can cause atoms to lose electrons or ionize. Two major forms of ionizing radiation are (1) X rays, which are artificially produced, and (2) gamma rays, which are emitted during radioisotope decay. Low levels of ionizing radiation will produce mutations and may indirectly result in death, whereas higher levels are directly lethal. Although microorganisms are more resistant to ionizing radiation than larger organisms, they will still be destroyed by a sufficiently large dose. Ionizing radiation can be used to sterilize items. Some procaryotes (e.g., *Deinococcus radiodurans*) and bacterial endospores can survive large doses of ionizing radiation. Use of radiation in destroying microorganisms (p. 141); *Deinococcus* (pp. 455–56)

A variety of changes in cells are due to ionizing radiation; it breaks hydrogen bonds, oxidizes double bonds, destroys ring structures, and polymerizes some molecules. Oxygen enhances these destructive effects, probably through the generation of hydroxyl radicals (OH·). Although many types of constituents can be affected, it is reasonable to suppose that destruction of DNA is the most important cause of death.

Ultraviolet (UV) radiation, mentioned earlier, kills all kinds of microorganisms due to its short wavelength (approximately from 10 to 400 nm) and high energy. The most lethal UV radiation has a wavelength of 260 nm, the wavelength most effectively absorbed by DNA. The primary mechanism of UV damage is the formation of thymine dimers in DNA (*see pp. 241–42*). Two adjacent thymines in a DNA strand are covalently joined to inhibit DNA replication and function. This damage is repaired in several ways. In **photoreactivation,** blue light is used by a photoreactivating enzyme to split thymine dimers. A short sequence containing the thymine dimer can also be excised and replaced. This process occurs in the absence of light and is called **dark reactivation.** Damage also can be repaired by the recA protein in recombination repair and SOS repair. When UV exposure is too heavy, the damage is so extensive that repair is impossible. DNA repair mechanisms (pp. 248–49)

Although very little UV radiation below 290 to 300 nm reaches the Earth's surface, near-UV radiation between 325 and 400 nm can harm microorganisms. Exposure to near-UV radiation induces tryptophan breakdown to toxic photoproducts. It appears that these toxic tryptophan photoproducts plus the near-UV radiation itself produce breaks in DNA strands. The precise mechanism is not known, although it is different from that seen with 260 nm UV.

Visible light is immensely beneficial because it is the source of energy for photosynthesis. Yet even visible light, when present in sufficient intensity, can damage or kill microbial cells. Usually pigments called photosensitizers and O_2 are required. All microorganisms possess pigments like chlorophyll, bacteriochlorophyll, cytochromes, and flavins, which can absorb light energy, become excited or activated, and act as photosensitizers. The excited photosensitizer (P) transfers its energy to O_2 generating **singlet oxygen** ($^1O_2$).

$$P \xrightarrow{\text{light}} P \text{ (activated)}$$

$$P \text{ (activated)} + O_2 \longrightarrow P + {}^1O_2$$

Singlet oxygen is a very reactive, powerful oxidizing agent that will quickly destroy a cell. It is probably the major agent employed by phagocytes to destroy engulfed bacteria (*see section 31.8*).

Many microorganisms that are airborne or live on exposed surfaces use carotenoid pigments for protection against photooxidation. Carotenoids effectively quench singlet oxygen—that is, they absorb energy from singlet oxygen and convert it back into the unexcited ground state. Both photosynthetic and nonphotosynthetic microorganisms employ pigments in this way.

---

1. What are barotolerant and barophilic bacteria? Where would you expect to find them?
2. List the types of electromagnetic radiation in the order of decreasing energy or increasing wavelength.
3. Why is it so important that the Earth receives an adequate supply of sunlight? What is the importance of ozone formation?

---

4. How do ionizing radiation, ultraviolet radiation, and visible light harm microorganisms? How do microorganisms protect themselves against damage from UV and visible light?

 ## 6.5 MICROBIAL GROWTH IN NATURAL ENVIRONMENTS

Section 6.4 surveyed the effects on microbial growth of individual environmental factors such as water availability, pH, and temperature. Although microbial ecology will be introduced in more detail at a later point, we will now briefly consider the effect of the environment as a whole on microbial growth. Microbial ecology (chapters 28–30)

Growth Limitation by Environmental Factors

The microbial environment is complex and constantly changing. Characteristically microorganisms in a particular location are exposed to many overlapping gradients of nutrients and various other environmental factors. This is particularly true of microorganisms growing in biofilms. Microorganisms will grow in "microenvironments" until an environmental or nutritional factor limits growth. **Liebig's law of the minimum** states that the total biomass of an organism will be determined by the nutrient present in the lowest concentration relative to the organism's requirements. This law applies in both the laboratory (figure 6.2) and in terrestrial and aquatic environments. An increase in a limiting essential nutrient such as phosphate will result in an increase in the microbial population until some other nutrient becomes limiting. If a specific nutrient is limiting, changes in other nutrients will have no effect. The situation may be even more complex than this. Multiple limiting factors can influence a population over time. Furthermore, as we have seen, factors such as temperature, pH, light, and salinity influence microbial populations and limit growth. **Shelford's law of tolerance** states that there are limits to environmental factors below and above which a microorganism cannot survive and grow, regardless of the nutrient supply. This can readily be seen for temperature in figure 6.14. Each microorganism has a specific temperature range in which it can grow. The same rule applies to other factors such as pH, oxygen level, and hydrostatic pressure in the marine environment. The growth of a microorganism depends on both the nutrient supply and its tolerance of the environmental conditions. Biofilms (pp. 602–3)

Most microorganisms are confronted with deficiencies that limit their activities except when excess nutrients allow unlimited growth. Such rapid growth will quickly deplete nutrients and possibly result in the release of toxic waste products, which will limit further growth.

In response to low nutrient levels (**oligotrophic environments**) and intense competition, many microorganisms become more competitive in nutrient capture and exploitation of available resources. Often the organism's morphology will change in order to increase its surface area and ability to absorb nutrients. This can involve con-

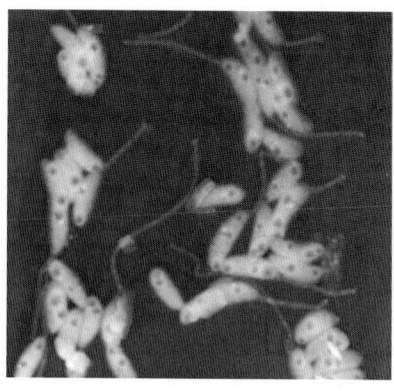
(a)

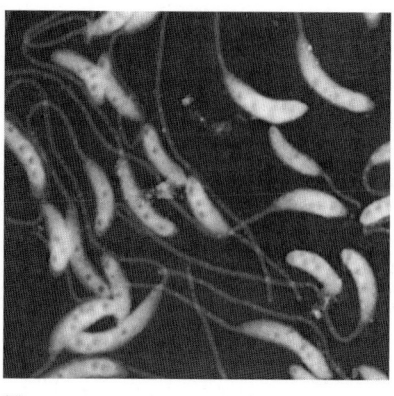
(b)

Figure 6.19   Morphology and Nutrient Absorption. Microorganisms can change their morphology in response to starvation and different limiting factors to improve their ability to survive. (**a**) *Caulobacter* has relatively short stalks when nitrogen is limiting. (**b**) The stalks are extremely long under phosphorus-limited conditions.

version of rod-shaped procaryotes to "mini" and "ultramicro" cells or changes in the morphology of prosthecate (*see pp. 476–79*) procaryotes (**figure 6.19**), in response to starvation. Nutrient deprivation induces many other changes as discussed previously. For example, microorganisms can undergo a step-by-step shutdown of metabolism except for housekeeping maintenance genes.

Many factors can alter nutrient levels in oligotrophic environments. Microorganisms may sequester critical limiting nutrients, such as iron, making them less available to competitors. The atmosphere can contribute essential nutrients and support microbial growth. This is seen in the laboratory as well as natural environments. Airborne organic substances have been found to stimulate microbial growth in dilute media, and enrichment of growth media by airborne organic matter can allow significant populations of microorganisms to develop. Even distilled water, which already contains traces of organic matter, can absorb one-carbon compounds from the atmosphere and grow microorganisms. The presence of such airborne nutrients and microbial growth, if not detected, can affect experiments in biochemistry and molecular biology, as well as studies of microorganisms growing in oligotrophic environments.

Natural substances also can directly inhibit microbial growth and reproduction in low-nutrient environments. These agents include phenolics, tannins, ammonia, ethylene, and volatile sulfur compounds. This may be a means by which microorganisms avoid expending limited energy reserves until an adequate supply of nutrients becomes available. Such chemicals are also important in plant pathology and may aid in controlling soil-borne microbial diseases.

Counting Viable But Nonculturable Vegetative Procaryotes

In order to study the growth of natural procaryotic populations outside the laboratory, it is essential to determine the number of viable microorganisms present. For most of microbiology's history, a viable microorganism has been defined as one that is able to grow actively, resulting in the formation of a colony or visible turbidity in a liquid medium. John R. Postgate of the University of Sussex in England was one of the first to note that microorganisms stressed by survival in natural habitats—or in many se-

lective laboratory media—were particularly sensitive to secondary stresses. Such stresses can produce viable microorganisms that have lost the ability to grow on media normally used for their cultivation. To determine the growth potential of such microorganisms, Postgate developed what is now called the Postgate Microviability Assay, in which microorganisms are cultured in a thin agar film under a coverslip. The ability of a cell to change its morphology, even if it does not grow beyond the single-cell stage, indicates that the microorganism does show "life signs."

Since that time many workers have developed additional sensitive microscopic and isotopic procedures to evaluate the presence and significance of these viable but nonculturable bacteria in both lab and field. For example, levels of fluorescent antibody and acridine orange–stained cells often are compared with population counts obtained by the most probable number (MPN) method (*see pp. 633–34*) and plate counts using selective and nonselective media. The release of radioactive-labeled cell materials also is used to monitor stress effects on microorganisms. Despite these advances, the estimation of substrate-responsive viable cells by Postgate's method is still important. These studies show that even when bacteria such as *Escherichia coli, Vibrio cholerae, Klebsiella pneumoniae, Enterobacter aerogenes,* and *Enterococcus faecalis* have lost their ability to grow on conventional laboratory media using standard cultural techniques, they still might be able to play a role in infectious disease.

The situation in natural environments with mixed populations is much more complex. Here, where often only 1 to 10% of observable cells are able to form colonies, the microbiologist is attempting to grow microorganisms that perhaps never have been cultured or characterized. In the future it is possible that media or proper environmental conditions for their growth will be developed. At present, molecular techniques involving PCR amplification and small subunit ribosomal RNA analysis are increasingly used to analyze the diversity of uncultured microbial populations (*see pp. 607–10*).

Quorum Sensing and Microbial Populations

For decades microbiologists tended to think of bacterial populations as collections of individuals growing and behaving independently. More recently it has become clear that many bacteria can communicate with one another and behave cooperatively. A major way in which this cooperation is accomplished is by a process known as **quorum sensing** or autoinduction. This is a phenomenon in which bacteria monitor their own population density through sensing the levels of signal molecules, sometimes called autoinducers because they can stimulate the cell that releases them. The concentration of these signal molecules increases along with the bacterial population until it rises to a specific threshold and signals the bacteria that the population density has reached a critical level or quorum. The bacteria then begin expressing sets of quorum-dependent genes. Quorum sensing has been found among both gram-negative and gram-positive bacteria. Regulation of gene expression (pp. 270–79)

Quorum sensing makes good practical sense. Take the production and release of extracellular enzymes as an example. If such enzymes were released by only a few bacteria, they would diffuse away and be rendered ineffective because of dilution. With control by quorum sensing, the bacteria reach a high population density before they release enzymes, and as a consequence enzyme levels are concentrated enough to have significant effects. This is an advantage within a host's body as well as in the soil or an aquatic habitat. If a pathogen can reach high levels at a particular site before producing virulence factors and escaping into surrounding host tissues, it has a much better chance of counteracting host defenses and successfully spreading throughout the host's body. This explains another pattern in quorum sensing. It seems to be very important in many bacteria that establish symbiotic or parasitic relationships with hosts.

An alternative function for the production and release of autoinducers has been recently proposed. It could be that autoinducer secretion is a way to determine the extent of diffusion and mixing in a cell's immediate environment. When there is too much diffusion and mixing, it would not make sense to release molecules such as proteases, siderophores, and antibiotics. If the autoinducer concentration rapidly dropped after release, it could signal the cell that other molecules should not be secreted. In fact, autoinducers often do regulate the production of secreted substances. Perhaps autoinducer release serves both functions: to sense population densities and diffusion rates. Only further experimentation can determine the role of this phenomenon in various procaryotes.

Quorum sensing was first discovered in gram-negative bacteria and is best understood in these microorganisms. The most common signals in gram-negative bacteria are acyl *homoserine lactones* (HSLs). These are small molecules composed of a 4- to 14-carbon acyl chain attached by an amide bond to homoserine lactone (**figure 6.20a**). The acyl chain may have a keto group or hydroxyl group on its third carbon. Acyl HSLs diffuse into the target cell (figure 6.20b). Once they reach a sufficiently high level, acyl HSLs bind to special receptor proteins and trigger a conformational change. Usually the activated complexes act as inducers—that is, they bind to target sites on the DNA and stimulate transcription of quorum-sensitive genes. The gene needed to synthesize acyl HSL is also produced frequently, thus amplifying the effect by the production and release of more autoinducer molecules.

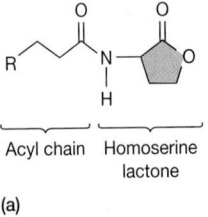

Acyl chain Homoserine lactone

(a)

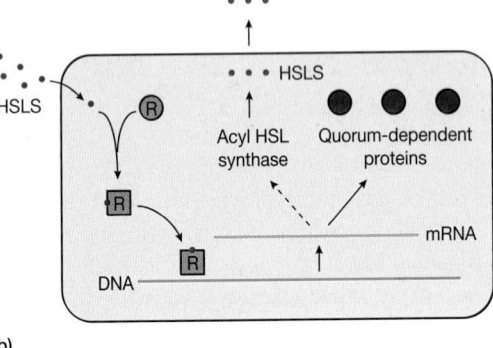

(b)

Figure 6.20 Quorum Sensing in Gram-Negative Bacteria. (**a**) A generalized structure for acyl homoserine lactone, the best-known quorum sensing signal or autoinducer. (**b**) A schematic diagram giving an overview of the way in which quorum sensing functions in many gram-negative bacteria. The receptor protein that acts as an inducer is labeled R. The dashed lines indicate that acyl HSL synthase is not always made in response to the autoinducer. See text for more details.

Many different processes are sensitive to acyl HSL signals and quorum sensing in gram-negative bacteria. Some well-studied examples are (1) bioluminescence production by *Vibrio fischeri*, (2) *Pseudomonas aeruginosa* synthesis and release of virulence factors, (3) conjugal transfer of genetic material by *Agrobacterium tumefaciens*, and (4) antibiotic production by *Erwinia carotovora* and *Pseudomonas aureofaciens*.

Gram-positive bacteria also regulate activities by quorum sensing, often using an oligopeptide pheromone signal. Good examples are mating in *Enterococcus faecalis*, competence induction in *Streptococcus pneumoniae*, stimulation of sporulation by *Bacillus subtilis*, and production of many toxins and other virulence factors by *Staphylococcus aureus*. Quorum sensing even stimulates the development of aerial mycelia and the production of streptomycin by *Streptomyces griseus*. In this case, the signal seems to be γ-butyrolactone rather than an oligopeptide.

An interesting and important function of quorum sensing is to promote the formation of mature biofilms by the pathogen *Pseudomonas aeruginosa*, and it may play a role in cystic fibrosis. Biofilm formation makes sense for the pathogen because biofilms protect against antibiotics and detergents. Quorum sensing should be very effective within biofilms because there will be less dilution and acyl HSL levels will increase rapidly. Under such circumstances, two different bacteria might stimulate each other by releasing similar signals; this appears to be the case in biofilms containing the pathogens *P. aeruginosa* and *Burkholderia cepacia*.

Quorum sensing is an example of what might be called multicellular behavior in that many individual cells communicate and coordinate their activities to act as a unit. Other examples of such complex behavior is pattern formation in colonies (*see pp. 104–7*) and fruiting body formation in the myxobacteria (*see pp. 497–99*).

1. How are Liebig's law of the minimum and Shelford's law of tolerance related?
2. Describe how microorganisms respond to oligotrophic environments.
3. Briefly discuss the Postgate microviability assay and other ways in which viable but nonculturable microorganisms can be counted or studied.
4. What is quorum sensing? Describe how it occurs and briefly discuss its importance to microorganisms.

Summary

Growth is an increase in cellular constituents and results in an increase in cell size, cell number, or both.

6.1 The Growth Curve

a. When microorganisms are grown in a closed system or batch culture, the resulting growth curve usually has four phases: the lag, exponential or log, stationary, and death phases (**figure 6.1**).

b. In the exponential phase, the population number of cells undergoing binary fission doubles at a constant interval called the doubling or generation time (**figure 6.3**). The mean growth rate constant (k) is the reciprocal of the generation time.

c. Exponential growth is balanced growth, cell components are synthesized at constant rates relative to one another. Changes in culture conditions (e.g., in shift-up and shift-down experiments) lead to unbalanced growth. A portion of the available nutrients is used to supply maintenance energy.

6.2 Measurement of Microbial Growth

a. Microbial populations can be counted directly with counting chambers, electronic counters, or fluorescence microscopy. Viable counting techniques such as the spread plate, the pour plate, or the membrane filter can be employed.

b. Population changes also can be followed by determining variations in microbial mass through the measurement of dry weight, turbidity, or the amount of a cell component.

6.3 The Continuous Culture of Microorganisms

a. Microorganisms can be grown in an open system in which nutrients are constantly provided and wastes removed.

b. A continuous culture system is an open system that can maintain a microbial population in the log phase. There are two types of these systems: chemostats and turbidostats.

6.4 The Influence of Environmental Factors on Growth

a. Most bacteria, algae, and fungi have rigid cell walls and are hypertonic to the habitat because of solutes such as amino acids, polyols, and potassium ions. The amount of water actually available to microorganisms is expressed in terms of the water activity (a_w).

b. Although most microorganisms will not grow well at water activities below 0.98 due to plasmolysis and associated effects, osmotolerant organisms survive and even flourish at low a_w values. Halophiles actually require high sodium chloride concentrations for growth (**figure 6.11** and **table 6.3**).

c. Each species of microorganism has an optimum pH for growth and can be classified as an acidophile, neutrophile, or alkalophile.

d. Microorganisms can alter the pH of their surroundings, and most culture media must be buffered to stabilize the pH.

e. Microorganisms have distinct temperature ranges for growth with minima, maxima, and optima—the cardinal temperatures. These ranges are determined by the effects of temperature on the rates of catalysis, protein denaturation, and membrane disruption.

f. There are five major classes of microorganisms with respect to temperature preferences: (1) psychrophiles, (2) facultative psychrophiles or psychrotrophs, (3) mesophiles, (4) thermophiles, and (5) hyperthermophiles (**figure 6.14 and table 6.3**).

g. Microorganisms can be placed into at least five different categories based on their response to the presence of O_2: obligate aerobes, facultative anaerobes, aerotolerant anaerobes, strict or obligate anaerobes, and microaerophiles (**figure 6.15 and table 6.3**).

h. Oxygen can become toxic because of the production of hydrogen peroxide, superoxide radical, and hydroxyl radical. These are destroyed by the enzymes superoxide dismutase, catalase, and peroxidase.

i. Most deep-sea microorganisms are barotolerant, but some are barophilic and require high pressure for optimal growth.

j. High-energy or short-wavelength radiation harms organisms in several ways. Ionizing radiation—X rays and gamma rays—ionizes molecules and destroys DNA and other cell components.

k. Ultraviolet (UV) radiation induces the formation of thymine dimers and strand breaks in DNA. Such damage can be repaired by photoreactivation or dark reactivation mechanisms.

l. Visible light can provide energy for the formation of reactive singlet oxygen, which will destroy cells.

6.5 Microbial Growth in Natural Environments

a. Microbial growth in natural environments is profoundly affected by nutrient limitations and other adverse factors. Some microorganisms can be viable but unculturable and must be studied with special techniques.

b. Often, bacteria will communicate with one another in a density-dependent way and carry out a particular activity only when a certain population density is reached. This phenomenon is called quorum sensing.

Key Terms

acidophile   121
aerobe   125
aerotolerant anaerobe   125

alkalophile   122
anaerobe   125
balanced growth   110

barophilic   127
barotolerant   127
batch culture   110

cardinal temperatures   122
catalase   126
chemostat   117

Questions for Thought and Review

1. Discuss the reasons why a culture might have a long lag phase after inoculation.
2. Why can't one always tell when a culture enters the death phase by the use of total cell counts?
3. Calculate the mean growth rate and generation time of a culture that increases in the exponential phase from 5×10^2 to 1×10^8 cells in 12 hours.
4. If the generation time is 90 minutes and the initial population contains 10^3 cells, how many bacteria will there be after 8 hours of exponential growth?
5. Why are continuous culture systems so useful to microbiologists?
6. How do bacterial populations respond in shift-up and shift-down experiments? Account for their behavior in molecular terms.
7. Does the internal pH remain constant despite changes in the external pH? How might this be achieved? Explain how extreme pH values might harm microorganisms.
8. What metabolic and structural adaptations for extreme temperatures have psychrophiles and thermophiles made?
9. Why are generation times in nature usually much longer than in culture?

Critical Thinking Questions

1. As an alternative to diffusable signals, suggest another mechanism by which bacteria can quorum sense.
2. Design an "enrichment" culture medium and a protocol for the isolation and purification of a soil bacterium (e.g., *Bacillus subtilis*) from a sample of soil. Note possible contaminants and competitors. How will you adjust conditions of growth and what conditions will be adjusted to differentially enhance the growth of the *Bacillus?*

**For recommended readings on these
and related topics, see Appendix V.**

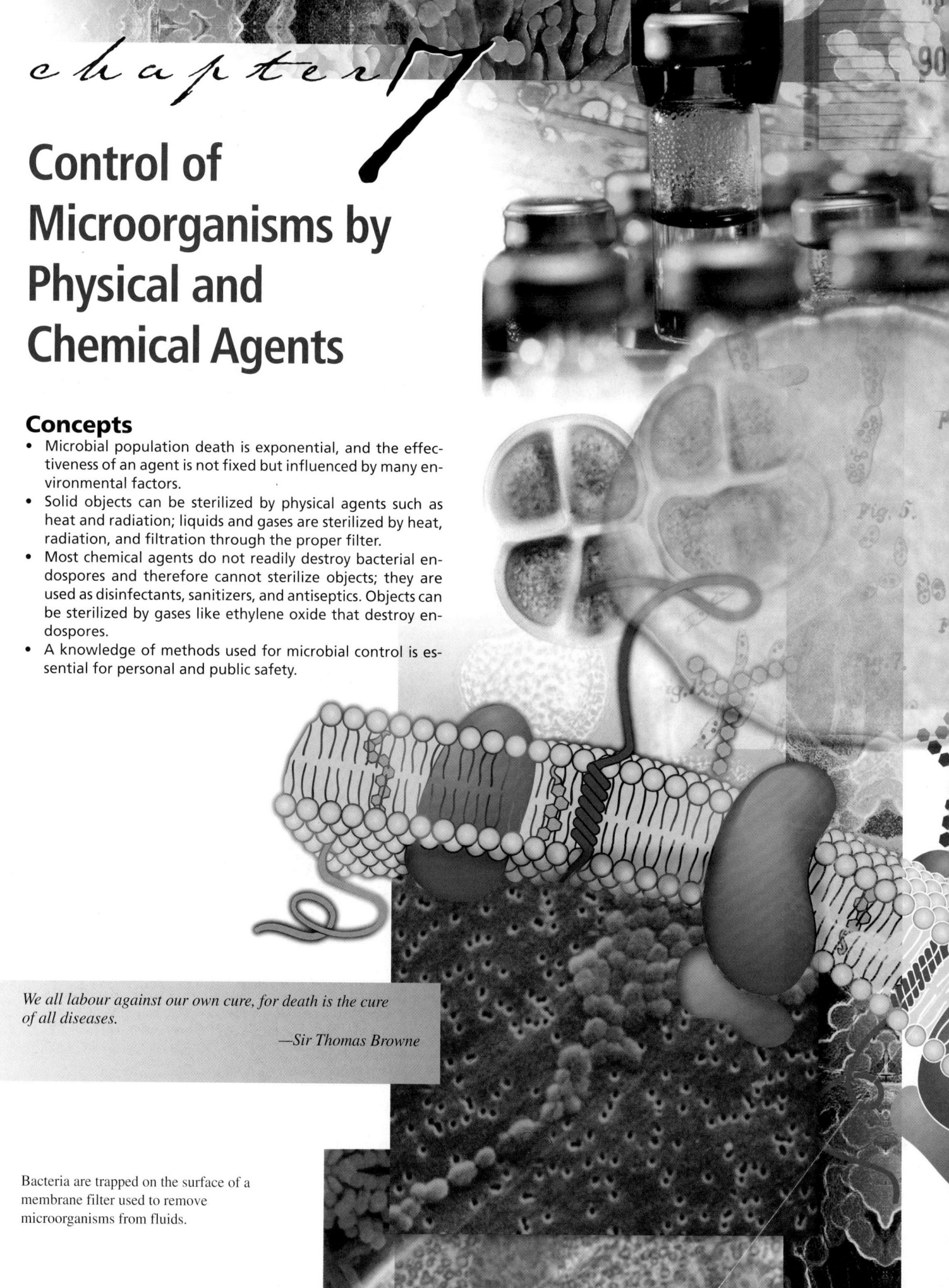

chapter 7

Control of Microorganisms by Physical and Chemical Agents

Concepts

- Microbial population death is exponential, and the effectiveness of an agent is not fixed but influenced by many environmental factors.
- Solid objects can be sterilized by physical agents such as heat and radiation; liquids and gases are sterilized by heat, radiation, and filtration through the proper filter.
- Most chemical agents do not readily destroy bacterial endospores and therefore cannot sterilize objects; they are used as disinfectants, sanitizers, and antiseptics. Objects can be sterilized by gases like ethylene oxide that destroy endospores.
- A knowledge of methods used for microbial control is essential for personal and public safety.

We all labour against our own cure, for death is the cure of all diseases.

—*Sir Thomas Browne*

Bacteria are trapped on the surface of a membrane filter used to remove microorganisms from fluids.

hapters 5 and 6 were concerned with microbial nutrition and growth. This chapter addresses the subject of the nonspecific control and destruction of microorganisms, a topic of immense practical importance. Although many microorganisms are beneficial and necessary for human well-being, microbial activities may have undesirable consequences, such as food spoilage and disease. Therefore it is essential to be able to kill a wide variety of microorganisms or inhibit their growth to minimize their destructive effects. The goal is twofold: (1) to destroy pathogens and prevent their transmission, and (2) to reduce or eliminate microorganisms responsible for the contamination of water, food, and other substances.

This chapter focuses on the control of microorganisms by nonspecific physical and chemical agents. Chapter 35 introduces the use of antimicrobial chemotherapy to control microbial disease.

From the beginning of recorded history, people have practiced disinfection and sterilization, even though the existence of microorganisms was long unsuspected. The Egyptians used fire to sterilize infectious material and disinfectants to embalm bodies, and the Greeks burned sulfur to fumigate buildings. Mosaic law commanded the Hebrews to burn any clothing suspected of being contaminated with the leprosy bacterium. Today the ability to destroy microorganisms is no less important: it makes possible the aseptic techniques used in microbiological research, the preservation of food, and the prevention of disease. The techniques described in this chapter are also essential to personal safety in both the laboratory and hospital (**Techniques & Applications 7.1**).

There are several ways to control microbial growth that have not been included in this chapter, but they should be considered for a more complete picture of how microorganisms are controlled. Chapter 6 describes the effects of osmotic activity, pH, temperature, O_2, and radiation on microbial growth and survival (*see pp. 118–28*). Chapter 41 discusses the use of physical and chemical agents in food preservation (*see pp. 943–46*).

7.1 DEFINITION OF FREQUENTLY USED TERMS

Terminology is especially important when the control of microorganisms is discussed because words like disinfectant and antiseptic often are used loosely. The situation is even more confusing because a particular treatment can either inhibit growth or kill depending on the conditions.

The ability to control microbial populations on inanimate objects, like eating utensils and surgical instruments, is of considerable practical importance. Sometimes it is necessary to eliminate all microorganisms from an object, whereas only partial destruction of the microbial population may be required in other situations. **Sterilization** [Latin *sterilis*, unable to produce offspring or barren] is the process by which all living cells, viable spores, viruses, and viroids (*see chapter 18*) are either destroyed or removed from an object or habitat. A sterile object is totally free of viable microorganisms, spores, and other infectious agents. When sterilization is achieved by a chemical agent, the chemical is called a sterilant. In contrast, **disinfection** is the killing, inhibition, or removal of microorganisms

Techniques & Applications
7.1   Safety in the Microbiology Laboratory

Personnel safety should be of major concern in all microbiology laboratories. It has been estimated that thousands of infections have been acquired in the laboratory, and many persons have died because of such infections. The two most common laboratory-acquired bacterial diseases are typhoid fever and brucellosis. Most deaths have come from typhoid fever (20 deaths) and Rocky Mountain spotted fever (13 deaths). Infections by fungi (histoplasmosis) and viruses (Venezuelan equine encephalitis and hepatitis B virus from monkeys) are also not uncommon. Hepatitis is the most frequently reported laboratory-acquired viral infection, especially in people working in clinical laboratories and with blood. In a survey of 426 U.S. hospital workers, 40% of those in clinical chemistry and 21% in microbiology had antibodies to the hepatitis B virus, indicating their previous exposure (though only about 19% of these had disease symptoms).

Efforts have been made to determine the causes of these infections in order to enhance the development of better preventive measures. Although often it is not possible to determine the direct cause of infection, some major potential hazards are clear. One of the most frequent causes of disease is the inhalation of an infectious aerosol. An aerosol is a gaseous suspension of liquid or solid particles that may be generated by accidents and laboratory operations such as spills, centrifuge accidents, removal of closures from shaken culture tubes, and plunging of contaminated loops into a flame. Accidents with hypodermic syringes and needles, such as self-inoculation and spraying solutions from the needle, also are common. Hypodermics should be employed only when necessary and then with care. Pipette accidents involving the mouth are another major source of infection; pipettes should be filled with the use of pipette aids and operated in such a way as to avoid creating aerosols.

People must exercise care and common sense when working with microorganisms. Operations that might generate infectious aerosols should be carried out in a biological safety cabinet. Bench tops and incubators should be disinfected regularly. Autoclaves must be maintained and operated properly to ensure adequate sterilization. Laboratory personnel should wash their hands thoroughly before and after finishing work.

that may cause disease. The primary goal is to destroy potential pathogens, but disinfection also substantially reduces the total microbial population. **Disinfectants** are agents, usually chemical, used to carry out disinfection and are normally used only on inanimate objects. A disinfectant does not necessarily sterilize an object because viable spores and a few microorganisms may remain. **Sanitization** is closely related to disinfection. In sanitization, the microbial population is reduced to levels that are considered safe by public health standards. The inanimate object is usually cleaned as well as partially disinfected. For example, sanitizers are used to clean eating utensils in restaurants.

It is frequently necessary to control microorganisms on living tissue with chemical agents. **Antisepsis** [Greek *anti,* against, and *sepsis,* putrefaction] is the prevention of infection or sepsis and is accomplished with **antiseptics.** These are chemical agents applied to tissue to prevent infection by killing or inhibiting pathogen growth; they also reduce the total microbial population. Because they must not destroy too much host tissue, antiseptics are generally not as toxic as disinfectants.

A suffix can be employed to denote the type of antimicrobial agent. Substances that kill organisms often have the suffix -cide [Latin *cida,* to kill]: a **germicide** kills pathogens (and many nonpathogens) but not necessarily endospores. A disinfectant or antiseptic can be particularly effective against a specific group, in which case it may be called a **bactericide, fungicide, algicide,** or **viricide.** Other chemicals do not kill, but they do prevent growth. If these agents are removed, growth will resume. Their names end in -static [Greek *statikos,* causing to stand or stopping]—for example, **bacteriostatic** and **fungistatic.**

Although these agents have been described in terms of their effects on pathogens, it should be noted that they also kill or inhibit the growth of nonpathogens as well. Their ability to reduce the total microbial population, not just to affect pathogen levels, is quite important in many situations.

---

1. Define the following terms: sterilization, sterilant, disinfection, disinfectant, sanitization, antisepsis, antiseptic, germicide, bactericide, bacteriostatic.

 ## 7.2 THE PATTERN OF MICROBIAL DEATH

A microbial population is not killed instantly when exposed to a lethal agent. Population death, like population growth, is generally exponential or logarithmic—that is, the population will be reduced by the same fraction at constant intervals (**table 7.1**). If the logarithm of the population number remaining is plotted against the time of exposure of the microorganism to the agent, a straight line plot will result (compare **figure 7.1** with figure 6.2). When the population has been greatly reduced, the rate of killing may slow due to the survival of a more resistant strain of the microorganism.

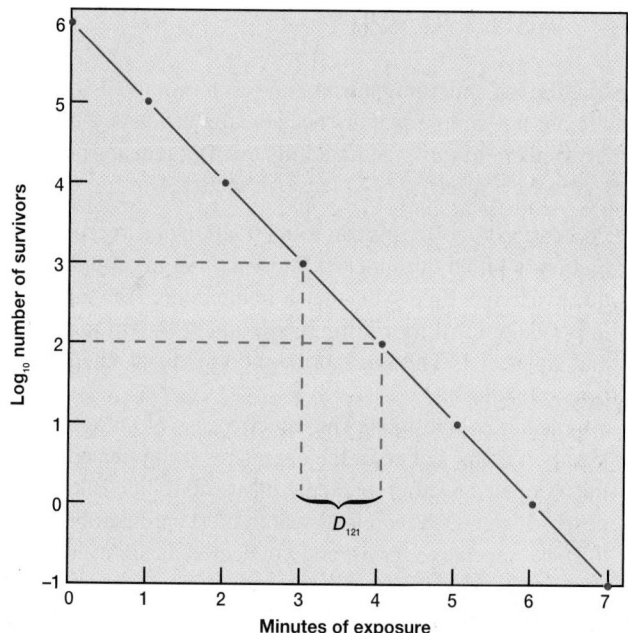

Figure 7.1   The Pattern of Microbial Death.   An exponential plot of the survivors versus the minutes of exposure to heating at 121°C. In this example the D_{121} value is 1 minute. The data are from table 7.1.

Table 7.1				
A Theoretical Microbial Heat-Killing Experiment				
Minute	**Microbial Number at Start of Minute**[a]	**Microorganisms Killed in 1 Minute (90% of total)**[a]	**Microorganisms at End of 1 Minute**	**Log$_{10}$ of Survivors**
1	10^6	9×10^5	10^5	5
2	10^5	9×10^4	10^4	4
3	10^4	9×10^3	10^3	3
4	10^3	9×10^2	10^2	2
5	10^2	9×10^1	10	1
6	10^1	9	1	0
7	1	0.9	0.1	−1

[a]Assume that the initial sample contains 10^6 vegetative microorganisms per ml and that 90% of the organisms are killed during each minute of exposure. The temperature is 121°C.

To study the effectiveness of a lethal agent, one must be able to decide when microorganisms are dead, a task by no means as easy as with macroorganisms. It is hardly possible to take a bacterium's pulse. A bacterium is defined as dead if it does not grow and reproduce when inoculated into culture medium that would normally support its growth. In like manner an inactive virus cannot infect a suitable host.

---

1. Describe the pattern of microbial death and how one decides whether microorganisms are actually dead.

7.3 CONDITIONS INFLUENCING THE EFFECTIVENESS OF ANTIMICROBIAL AGENT ACTIVITY

Destruction of microorganisms and inhibition of microbial growth are not simple matters because the efficiency of an **antimicrobial agent** (an agent that kills microorganisms or inhibits their growth) is affected by at least six factors.

1. Population size. Because an equal fraction of a microbial population is killed during each interval, a larger population requires a longer time to die than a smaller one. This can be seen in the theoretical heat-killing experiment shown in table 7.1 and figure 7.1. The same principle applies to chemical antimicrobial agents.

2. Population composition. The effectiveness of an agent varies greatly with the nature of the organisms being treated because microorganisms differ markedly in susceptibility. Bacterial endospores are much more resistant to most antimicrobial agents than are vegetative forms, and younger cells are usually more readily destroyed than mature organisms. Some species are able to withstand adverse conditions better than others. *Mycobacterium tuberculosis,* which causes tuberculosis, is much more resistant to antimicrobial agents than most other bacteria.

3. Concentration or intensity of an antimicrobial agent. Often, but not always, the more concentrated a chemical agent or intense a physical agent, the more rapidly microorganisms are destroyed. However, agent effectiveness usually is not directly related to concentration or intensity. Over a short range a small increase in concentration leads to an exponential rise in effectiveness; beyond a certain point, increases may not raise the killing rate much at all. Sometimes an agent is more effective at lower concentrations. For example, 70% ethanol is more effective than 95% ethanol because its activity is enhanced by the presence of water.

4. Duration of exposure. The longer a population is exposed to a microbicidal agent, the more organisms are killed (figure 7.1). To achieve sterilization, an exposure duration sufficient to reduce the probability of survival to 10^{-6} or less should be used.

5. Temperature. An increase in the temperature at which a chemical acts often enhances its activity. Frequently a lower con-centration of disinfectant or sterilizing agent can be used at a higher temperature.

6. Local environment. The population to be controlled is not isolated but surrounded by environmental factors that may either offer protection or aid in its destruction. For example, because heat kills more readily at an acid pH, acid foods and beverages such as fruits and tomatoes are easier to pasteurize than foods with higher pHs like milk. A second important environmental factor is organic matter that can protect microorganisms against heating and chemical disinfectants. Biofilms are a good example. The organic matter in a surface biofilm will protect the biofilm's microorganisms; furthermore, the biofilm and its microbes often will be hard to remove. It may be necessary to clean an object before it is disinfected or sterilized. Syringes and medical or dental equipment should be cleaned before sterilization because the presence of too much organic matter could protect pathogens and increase the risk of infection. The same care must be taken when pathogens are destroyed during the preparation of drinking water. When a city's water supply has a high content of organic material, more chlorine must be added to disinfect the water.

---

1. Briefly explain how the effectiveness of antimicrobial agents varies with population size, population composition, concentration or intensity of the agent, treatment duration, temperature, and local environmental conditions.

7.4 THE USE OF PHYSICAL METHODS IN CONTROL

Heat and other physical agents are normally used to control microbial growth and sterilize objects, as can be seen from the continual operation of the autoclave in every microbiology laboratory. The four most frequently employed physical agents are heat, low temperatures, filtration, and radiation.

Heat

Fire and boiling water have been used for sterilization and disinfection since the time of the Greeks, and heating is still one of the most popular ways to destroy microorganisms. Either moist or dry heat may be applied.

Moist heat readily kills viruses, bacteria, and fungi (**table 7.2**). Exposure to boiling water for 10 minutes is sufficient to destroy vegetative cells and eucaryotic spores. Unfortunately the temperature of boiling water (100°C or 212°F at sea level) is not high enough to destroy bacterial endospores that may survive hours of boiling. Therefore boiling can be used for disinfection of drinking water and objects not harmed by water, but boiling does not sterilize.

Because heat is so useful in controlling microorganisms, it is essential to have a precise measure of the heat-killing efficiency. Initially effectiveness was expressed in terms of thermal death

Table 7.2

Approximate Conditions for Moist Heat Killing

Organism	Vegetative Cells	Spores
Yeasts	5 minutes at 50–60°C	5 minutes at 70–80°C
Molds	30 minutes at 62°C	30 minutes at 80°C
Bacteria[a]	10 minutes at 60–70°C	2 to over 800 minutes at 100°C
		0.5–12 minutes at 121°C
Viruses	30 minutes at 60°C	

[a]Conditions for mesophilic bacteria.

point (TDP), the lowest temperature at which a microbial suspension is killed in 10 minutes. Because TDP implies that a certain temperature is immediately lethal despite the conditions, **thermal death time (TDT)** is now more commonly used. This is the shortest time needed to kill all organisms in a microbial suspension at a specific temperature and under defined conditions. However, such destruction is logarithmic, and it is theoretically not possible to "completely destroy" microorganisms in a sample, even with extended heating. Therefore an even more precise figure, the **decimal reduction time (D)** or **D value** has gained wide acceptance. The decimal reduction time is the time required to kill 90% of the microorganisms or spores in a sample at a specified temperature. In a semilogarithmic plot of the population remaining versus the time of heating (figure 7.1), the D value is the time required for the line to drop by one log cycle or tenfold. The D value is usually written with a subscript, indicating the temperature for which it applies. D values are used to estimate the relative resistance of a microorganism to different temperatures through calculation of the z **value.** The z value is the increase in temperature required to reduce D to 1/10 its value or to reduce it by one log cycle when log D is plotted against temperature (**figure 7.2**). Another way to describe heating effectiveness is with the F value. The **F value** is the time in minutes at a specific temperature (usually 250°F or 121.1°C) needed to kill a population of cells or spores.

The food processing industry makes extensive use of D and z values. After a food has been canned, it must be heated to eliminate the risk of botulism arising from *Clostridium botulinum* spores. Heat treatment is carried out long enough to reduce a population of 10^{12} *C. botulinum* spores to 10^0 (one spore); thus there is a very small chance of any can having a viable spore. The D value for these spores at 121°C is 0.204 minutes. Therefore it would take 12D or 2.5 minutes to reduce 10^{12} spores to one spore by heating at 121°C. The z value for *C. botulinum* spores is 10°C—that is, it takes a 10°C change in temperature to alter the D value tenfold. If the cans were to be processed at 111°C rather than at 121°C, the D value would increase by tenfold to 2.04 minutes and the 12D value to 24.5 minutes. D values and z values for some common food-borne pathogens are given in **table 7.3.** Three D values are included for *Staphylococcus aureus* to illustrate the variation of killing rate with environment and the protective effect of organic material. Food processing (pp. 943–46); Botulism (pp. 905–7)

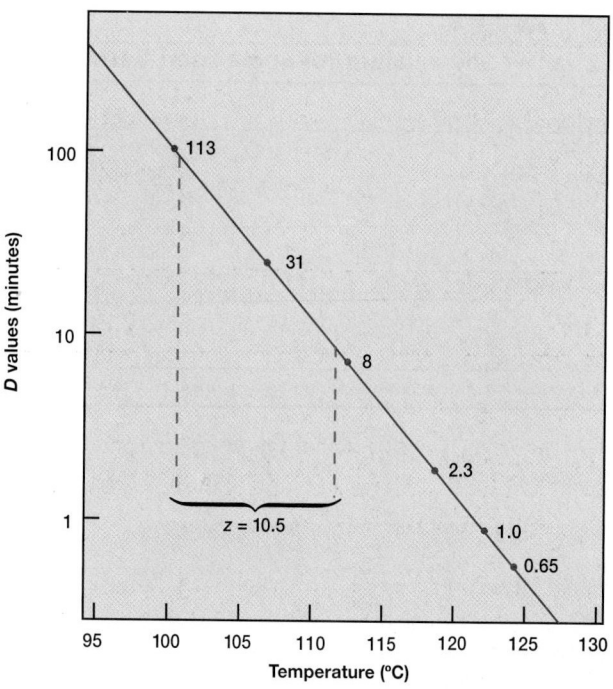

Figure 7.2 z Value Calculation. The z value used in calculation of time-temperature relationships for survival of a test microorganism, based on D value responses at various temperatures. The z value is the increase in temperature needed to reduce the decimal reduction time (D) to 10% of the original value. For this homogeneous sample of a test microorganism the z value is 10.5°. The D values are plotted on a logarithmic scale.

Moist heat sterilization must be carried out at temperatures above 100°C in order to destroy bacterial endospores, and this requires the use of saturated steam under pressure. Steam sterilization is carried out with an **autoclave (figure 7.3),** a device somewhat like a fancy pressure cooker. The development of the autoclave by Chamberland in 1884 tremendously stimulated the growth of microbiology. Water is boiled to produce steam, which is released through the jacket and into the autoclave's chamber. The air initially present in the chamber is forced out until the chamber is filled with saturated steam and the outlets are closed. Hot, saturated steam continues to enter until the chamber reaches the desired temperature and pressure, usually 121°C and 15 pounds of pressure. At this temperature saturated steam destroys all vegetative cells and endospores in a small volume of liquid within 10 to 12 minutes. Treatment is continued for about 15 minutes to provide a margin of safety. Of course, larger containers of liquid such as flasks and carboys will require much longer treatment times.

Moist heat is thought to kill so effectively by degrading nucleic acids and by denaturing enzymes and other essential proteins. It also may disrupt cell membranes.

Autoclaving must be carried out properly or the processed materials will not be sterile. If all air has not been flushed out of the chamber, it will not reach 121°C even though it may reach a

Table 7.3

D Values and z Values for Some Food-Borne Pathogens

Organism	Substrate	D Value (°C) in Minutes	z Value (°C)
Clostridium botulinum	Phosphate buffer	$D_{121} = 0.204$	10
Clostridium perfringens (heat-resistant strain)	Culture media	$D_{90} = 3–5$	6–8
Salmonella	Chicken à la king	$D_{60} = 0.39–0.40$	4.9–5.1
Staphylococcus aureus	Chicken à la king	$D_{60} = 5.17–5.37$	5.2–5.8
	Turkey stuffing	$D_{60} = 15.4$	6.8
	0.5% NaCl	$D_{60} = 2.0–2.5$	5.6

Values taken from F. L. Bryan, 1979, "Processes That Affect Survival and Growth of Microorganisms," *Time-Temperature Control of Foodborne Pathogens*, 1979. Atlanta: Centers for Disease Control and Prevention, Atlanta, GA.

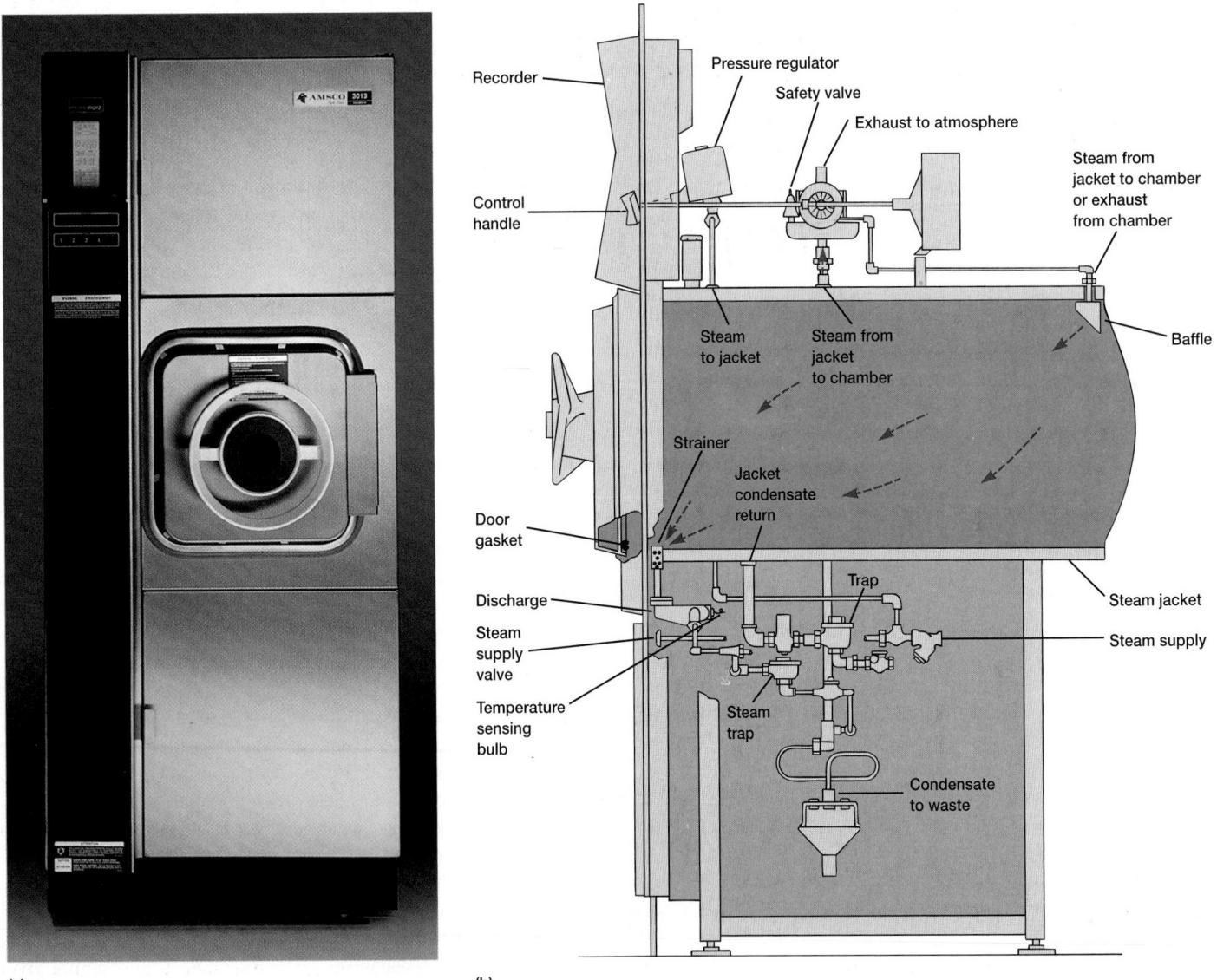

(a)

(b)

Figure 7.3   The Autoclave or Steam Sterilizer.   (**a**) A modern, automatically controlled autoclave or sterilizer. (**b**) Longitudinal cross section of a typical autoclave showing some of its parts and the pathway of steam.   *From John J. Perkins,* Principles and Methods of Sterilization in Health Science, *2nd edition, 1969. Courtesy of Charles C. Thomas, Publisher, Springfield, Illinois.*

pressure of 15 pounds. The chamber should not be packed too tightly because the steam needs to circulate freely and contact everything in the autoclave. Bacterial endospores will be killed only if they are kept at 121°C for 10 to 12 minutes. When a large volume of liquid must be sterilized, an extended sterilization time will be needed because it will take longer for the center of the liquid to reach 121°C; 5 liters of liquid may require about 70 minutes. In view of these potential difficulties, a biological indicator is often autoclaved along with other material. This indicator commonly consists of a culture tube containing a sterile ampule of medium and a paper strip covered with spores of *Bacillus stearothermophilus* or *Clostridium* PA3679. After autoclaving, the ampule is aseptically broken and the culture incubated for several days. If the test bacterium does not grow in the medium, the sterilization run has been successful. Sometimes either special tape that spells out the word *sterile* or a paper indicator strip that changes color upon sufficient heating is autoclaved with a load of material. If the word appears on the tape or if the color changes after autoclaving, the material is supposed to be sterile. These approaches are convenient and save time but are not as reliable as the use of bacterial endospores.

Many substances, such as milk, are treated with controlled heating at temperatures well below boiling, a process known as **pasteurization** in honor of its developer Louis Pasteur. In the 1860s the French wine industry was plagued by the problem of wine spoilage, which made wine storage and shipping difficult. Pasteur examined spoiled wine under the microscope and detected microorganisms that looked like the bacteria responsible for lactic acid and acetic acid fermentations. He then discovered that a brief heating at 55 to 60°C would destroy these microorganisms and preserve wine for long periods. In 1886 the German chemists V. H. and F. Soxhlet adapted the technique for preserving milk and reducing milk-transmissible diseases. Milk pasteurization was introduced into the United States in 1889. Milk, beer, and many other beverages are now pasteurized. Pasteurization does not sterilize a beverage, but it does kill any pathogens present and drastically slows spoilage by reducing the level of nonpathogenic spoilage microorganisms.

Milk can be pasteurized in two ways. In the older method the milk is held at 63°C for 30 minutes. Large quantities of milk are now usually subjected to **flash pasteurization** or high-temperature short-term (HTST) pasteurization, which consists of quick heating to about 72°C for 15 seconds, then rapid cooling. The dairy industry also sometimes uses **ultrahigh-temperature (UHT) sterilization.** Milk and milk products are heated at 140 to 150°C for 1 to 3 seconds. UHT-processed milk does not require refrigeration and can be stored at room temperature for about 2 months without flavor changes. The small coffee creamer portions provided by restaurants often are prepared using UHT sterilization. Pasteurization and the dairy industry (pp. 944–45)

Many objects are best sterilized in the absence of water by **dry heat sterilization.** The items to be sterilized are placed in an oven at 160 to 170°C for 2 to 3 hours. Microbial death apparently results from the oxidation of cell constituents and denaturation of proteins. Although dry air heat is less effective than moist heat—

Clostridium botulinum spores are killed in 5 minutes at 121°C by moist heat but only after 2 hours at 160°C with dry heat—it has some definite advantages. Dry heat does not corrode glassware and metal instruments as moist heat does, and it can be used to sterilize powders, oils, and similar items. Most laboratories sterilize glass petri dishes and pipettes with dry heat. Despite these advantages, dry heat sterilization is slow and not suitable for heat-sensitive materials like many plastic and rubber items.

Low Temperatures

Although our emphasis is on the destruction of microorganisms, often the most convenient control technique is to inhibit their growth and reproduction by the use of either freezing or refrigeration. This approach is particularly important in food microbiology (*see p. 944*). Freezing items at −20°C or lower stops microbial growth because of the low temperature and the absence of liquid water. Some microorganisms will be killed by ice crystal disruption of cell membranes, but freezing does not destroy contaminating microbes. In fact, freezing is a very good method for long-term storage of microbial samples when carried out properly, and many laboratories have a low-temperature freezer for culture storage at −30 or −70°C. Because frozen food can contain many microorganisms, it should be prepared and consumed promptly after thawing in order to avoid spoilage and pathogen growth. Effect of temperature on microbial growth (pp. 122–25)

Refrigeration greatly slows microbial growth and reproduction, but does not halt it completely. Fortunately most pathogens are mesophilic and do not grow well at temperatures around 4°C. Refrigerated items may be ruined by growth of psychrophilic and psychrotrophic microorganisms, particularly if water is present. Thus refrigeration is a good technique only for shorter-term storage of food and other items.

Filtration

Filtration is an excellent way to reduce the microbial population in solutions of heat-sensitive material, and sometimes it can be used to sterilize solutions. Rather than directly destroying contaminating microorganisms, the filter simply removes them. There are two types of filters. **Depth filters** consist of fibrous or granular materials that have been bonded into a thick layer filled with twisting channels of small diameter. The solution containing microorganisms is sucked through this layer under vacuum, and microbial cells are removed by physical screening or entrapment and also by adsorption to the surface of the filter material. Depth filters are made of diatomaceous earth (Berkefeld filters), unglazed porcelain (Chamberlain filters), asbestos, or other similar materials.

Membrane filters have replaced depth filters for many purposes. These circular filters are porous membranes, a little over 0.1 mm thick, made of cellulose acetate, cellulose nitrate, polycarbonate, polyvinylidene fluoride, or other synthetic materials. Although a wide variety of pore sizes are available, membranes with pores about 0.2 μm in diameter are used to remove most

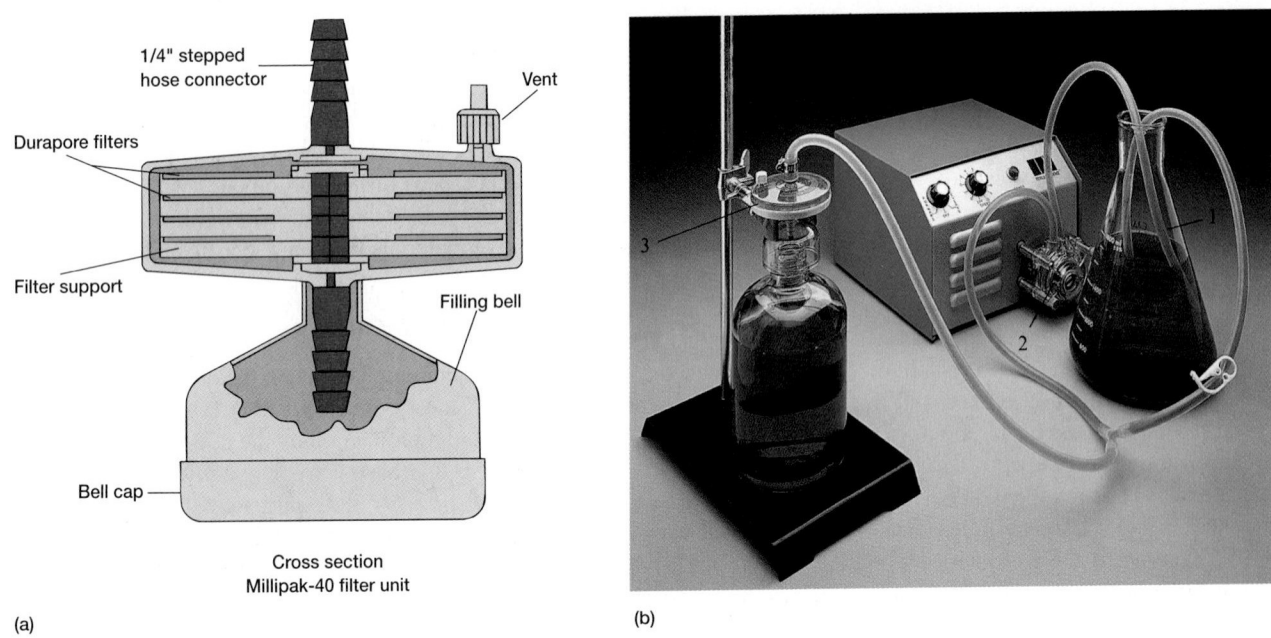

(a)

(b)

Figure 7.4 Membrane Filter Sterilization. A membrane filter outfit for sterilizing medium volumes of solution. **(a)** Cross section of the membrane filtering unit. Several membranes are used to increase capacity. **(b)** A complete filtering setup. The solution to be sterilized is kept in the Erlenmeyer flask, *1,* and forced through the filter by a peristaltic pump, *2.* The solution is sterilized by flowing through a membrane filter unit, *3,* and into a sterile container. A wide variety of other kinds of filtering outfits are also available.

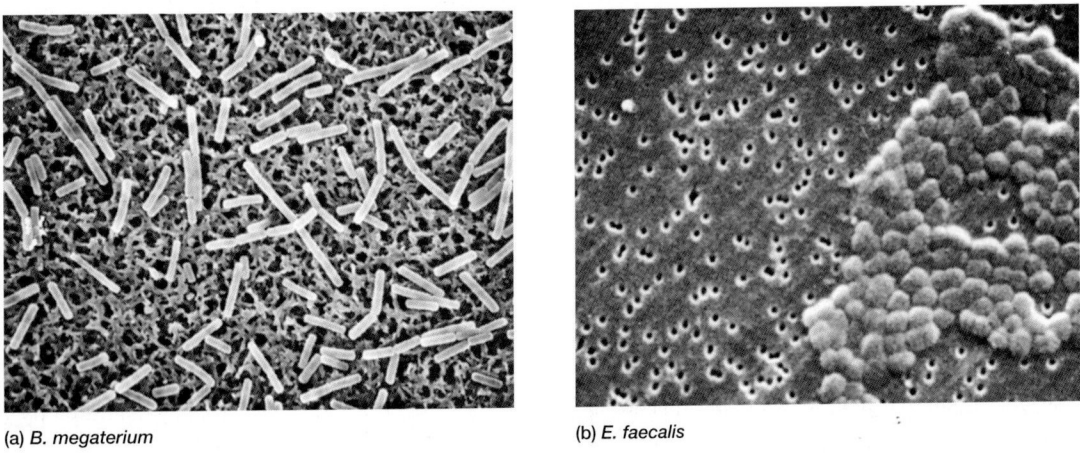

(a) *B. megaterium*

(b) *E. faecalis*

Figure 7.5 Membrane Filter Types. **(a)** *Bacillus megaterium* on an Ultipor nylon membrane with a bacterial removal rating of 0.2 μm (×2,000). **(b)** *Enterococcus faecalis* resting on a polycarbonate membrane filter with 0.4 μm pores (×5,900).

vegetative cells, but not viruses, from solutions ranging in volume from 1 ml to many liters. The membranes are held in special holders (**figure 7.4**) and often preceded by depth filters made of glass fibers to remove larger particles that might clog the membrane filter. The solution is pulled or forced through the filter with a vacuum or with pressure from a syringe, peristaltic pump, or nitrogen gas bottle, and collected in previously sterilized containers. Membrane filters remove microorganisms by screening them out much as a sieve separates large sand particles from small ones (**figure 7.5**). These filters are used to sterilize pharmaceuticals, ophthalmic solutions, culture media, oils, antibiotics, and other heat-sensitive solutions. The use of membrane filters in microbial counting (p. 115)

Air also can be sterilized by filtration. Two common examples are surgical masks and cotton plugs on culture vessels that let air in but keep microorganisms out. **Laminar flow biological safety cabinets** employing **high-efficiency particulate air (HEPA) filters,** which remove 99.97% of 0.3 μm particles, are one of the most important air filtration systems. Laminar flow biological safety cabinets force air through HEPA filters, then project a vertical curtain of sterile air across the cabinet opening. This protects a worker from microorganisms being handled within the

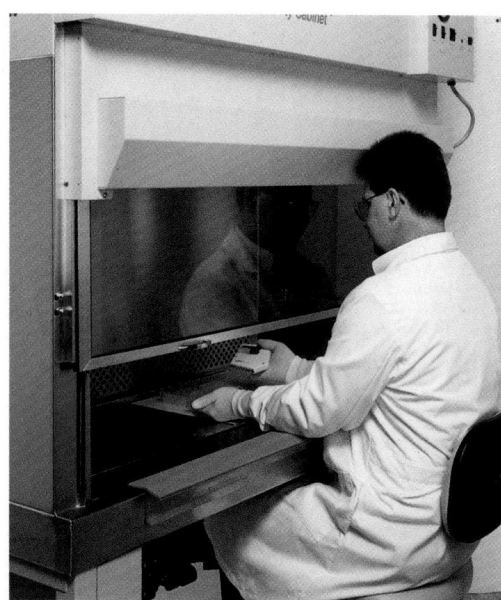

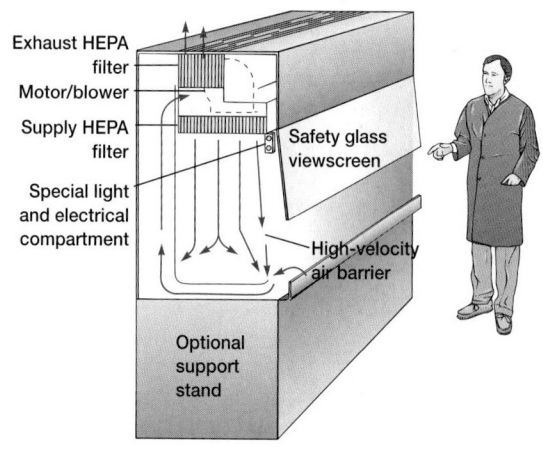

Figure 7.6   **A Laminar Flow Biological Safety Cabinet.**   (**a**) A technician pipetting potentially hazardous material in a safety cabinet. (**b**) A schematic diagram showing the airflow pattern.

cabinet and prevents contamination of the room (**figure 7.6**). A person uses these cabinets when working with dangerous agents such as *Mycobacterium tuberculosis,* tumor viruses, and recombinant DNA. They are also employed in research labs and industries, such as the pharmaceutical industry, when a sterile working surface is needed for conducting assays, preparing media, examining tissue cultures, and the like.

Radiation

The types of radiation and the ways in which radiation damages or destroys microorganisms have already been discussed. The practical uses of ultraviolet and ionizing radiation in sterilizing objects are briefly described next.   Radiation and its effects on microorganisms (pp. 127–28)

    Ultraviolet (UV) radiation around 260 nm (*see figure 6.18*) is quite lethal but does not penetrate glass, dirt films, water, and other substances very effectively. Because of this disadvantage, UV radiation is used as a sterilizing agent only in a few specific situations. UV lamps are sometimes placed on the ceilings of rooms or in biological safety cabinets to sterilize the air and any exposed surfaces. Because UV radiation burns the skin and damages eyes, people working in such areas must be certain the UV lamps are off when the areas are in use. Commercial UV units are available for water treatment. Pathogens and other microorganisms are destroyed when a thin layer of water is passed under the lamps.

    Ionizing radiation is an excellent sterilizing agent and penetrates deep into objects. It will destroy bacterial endospores and vegetative cells, both procaryotic and eucaryotic; however, ionizing radiation is not always as effective against viruses. Gamma radiation from a cobalt 60 source is used in the cold sterilization of antibiotics, hormones, sutures, and plastic disposable supplies

such as syringes. Gamma radiation has also been used to sterilize and "pasteurize" meat and other food. Irradiation can eliminate the threat of such pathogens as *Escherichia coli* O157:H7, *Staphylococcus aureus,* and *Campylobacter jejuni.* Both the Food and Drug Administration and the World Health Organization have approved food irradiation and declared it safe. A commercial irradiation plant operates near Tampa, Florida. However, this process has not yet been widely employed in the United States because of the cost and concerns about the effects of gamma radiation on food. The U.S. government currently approves the use of radiation to treat poultry, beef, pork, veal, lamb, fruits, vegetables, and spices. It will probably be more extensively employed in the future.

---

1. Define thermal death point (TDP), thermal death time (TDT), decimal reduction time (*D*) or *D* value, *z* value, and the *F* value.
2. Describe how an autoclave works. What conditions are required for sterilization by moist heat, and what three things must one do when operating an autoclave to help ensure success?
3. How are pasteurization, flash pasteurization, ultrahigh-temperature sterilization, and dry heat sterilization carried out? Give some practical applications for each of these procedures.
4. How can low temperature be used to control microorganisms?
5. What are depth filters and membrane filters, and how are they used to sterilize liquids? Describe the operation of a biological safety cabinet.
6. Give the advantages and disadvantages of ultraviolet light and ionizing radiation as sterilizing agents. Provide a few examples of how each is used for this purpose.

Techniques & Applications
7.2 Universal Precautions for Microbiology Laboratories

Blood and other body fluids from all patients should be considered infective.

1. All specimens of blood and body fluids should be put in a well-constructed container with a secure lid to prevent leaking during transport. Care should be taken when collecting each specimen to avoid contaminating the outside of the container and of the laboratory form accompanying the specimen.
2. All persons processing blood and body-fluid specimens should wear gloves. Masks and protective eyewear should be worn if mucous membrane contact with blood or body fluids is anticipated. Gloves should be changed and hands washed after completion of specimen processing.
3. For routine procedures, such as histologic and pathological studies or microbiologic culturing, a biological safety cabinet is not necessary. However, biological safety cabinets should be used whenever procedures are conducted that have a high potential for generating droplets. These include activities such as blending, sonicating, and vigorous mixing.
4. Mechanical pipetting devices should be used for manipulating all liquids in the laboratory. Mouth pipetting must not be done.

5. Use of needles and syringes should be limited to situations in which there is no alternative, and the recommendations for preventing injuries with needles outlined under universal precautions (*see Techniques & Applications 36.1, p. 802*) should be followed.
6. Laboratory work surfaces should be decontaminated with an appropriate chemical germicide after a spill of blood or other body fluids and when work activities are completed.
7. Contaminated materials used in laboratory tests should be decontaminated before reprocessing or be placed in bags and disposed of in accordance with institutional policies for disposal of infective waste.
8. Scientific equipment that has been contaminated with blood or other body fluids should be decontaminated and cleaned before being repaired in the laboratory or transported to the manufacturer.
9. All persons should wash their hands after completing laboratory activities and should remove protective clothing before leaving the laboratory.
10. There should be no eating, drinking, or smoking in the work area.

Source: Adapted from *Morbidity and Mortality Weekly Report*, 36 (Suppl. 2S) 5S–10S, 1987, the Centers for Disease Control and Prevention Guidelines.

7.5 THE USE OF CHEMICAL AGENTS IN CONTROL

Although objects are sometimes disinfected with physical agents, chemicals are more often employed in disinfection and antisepsis. Many factors influence the effectiveness of chemical disinfectants and antiseptics as previously discussed. Factors such as the kinds of microorganisms potentially present, the concentration and nature of the disinfectant to be used, and the length of treatment should be considered. Dirty surfaces must be cleaned before a disinfectant or antiseptic is applied. The proper use of chemical agents is essential to laboratory and hospital safety (**Techniques & Applications 7.2;** *see also Techniques & Applications 36.1*). It should be noted that chemicals also are employed to prevent microbial growth in food. This is discussed in the chapter on food microbiology (*see p. 945*).

Many different chemicals are available for use as disinfectants, and each has its own advantages and disadvantages. In selecting an agent, it is important to keep in mind the characteristics of a desirable disinfectant. Ideally the disinfectant must be effective against a wide variety of infectious agents (gram-positive and gram-negative bacteria, acid-fast bacteria, bacterial endospores, fungi, and viruses) at high dilutions and in the presence of organic matter. Although the chemical must be toxic for infectious agents, it should not be toxic to people or corrosive for common materials. In practice, this balance between effectiveness and low toxicity for animals is hard to achieve. Some chemicals are used despite their low effectiveness because they are relatively non-

toxic. The disinfectant should be stable upon storage, odorless or with a pleasant odor, soluble in water and lipids for penetration into microorganisms, and have a low surface tension so that it can enter cracks in surfaces. If possible the disinfectant should be relatively inexpensive.

One potentially serious problem is the overuse of the antiseptic triclosan and other germicides. This antibacterial agent is now found in products such as deodorants, mouthwashes, soaps, cutting boards, and baby toys. Triclosan seems to be everywhere. Unfortunately we are already seeing the emergence of triclosan-resistant bacteria. *Pseudomonas aeruginosa* actively pumps the antiseptic out of the cell. Bacteria seem to be responding to antiseptic overuse in the same way as they reacted to antibiotic overuse (*see pp. 792–94*). There is now some evidence that extensive use of triclosan also increases the frequency of antibiotic resistance in bacteria. Thus overuse of antiseptics can have unintended harmful consequences.

The properties and uses of several groups of common disinfectants and antiseptics are surveyed next. Many of their characteristics are summarized in **tables 7.4** and **7.5.** Structures of some common agents are given in **figure 7.7.**

Phenolics

Phenol was the first widely used antiseptic and disinfectant. In 1867 Joseph Lister employed it to reduce the risk of infection during operations. Today phenol and phenolics (phenol derivatives) such as cresols, xylenols, and orthophenylphenol are used

Table 7.4
Activity Levels of Selected Germicides

Class	Use Concentration of Active Ingredient	Activity Level[a]
Gas		
Ethylene oxide	450–500 mg/liter[b]	High
Liquid		
Glutaraldehyde, aqueous	2%	High to intermediate
Formaldehyde + alcohol	8 + 70%	High
Stabilized hydrogen peroxide	6–30%	High to intermediate
Formaldehyde, aqueous	6–8%	High to intermediate
Iodophors	750–5,000 mg/liter[c]	High to intermediate
Iodophors	75–150 mg/liter[c]	Intermediate to low
Iodine + alcohol	0.5 + 70%	Intermediate
Chlorine compounds	0.1–0.5%[d]	Intermediate
Phenolic compounds, aqueous	0.5–3%	Intermediate to low
Iodine, aqueous	1%	Intermediate
Alcohols (ethyl, isopropyl)	70%	Intermediate
Quaternary ammonium compounds	0.1–0.2% aqueous	Low
Chlorhexidine	0.75–4%	Low
Hexachlorophene	1–3%	Low
Mercurial compounds	0.1–0.2%	Low

Source: From Seymour S. Block, *Disinfection, Sterilization and Preservation.* Copyright © 1983 Lea & Febiger, Malvern, Pa. 1983. Reprinted by permission.
[a]High-level disinfectants destroy vegetative bacterial cells including *M. tuberculosis,* bacterial endospores, fungi, and viruses. Intermediate-level disinfectants destroy all of the above except endospores. Low-level agents kill bacterial vegetative cells except for *M. tuberculosis,* fungi, and medium-sized lipid-containing viruses (but not bacterial endospores or small, nonlipid viruses).
[b]In autoclave-type equipment at 55 to 60°C.
[c]Available iodine.
[d]Free chlorine.

Table 7.5
Relative Efficacy of Commonly Used Disinfectants and Antiseptics

Class	Disinfectant	Antiseptic	Comment
Gas			
Ethylene oxide	3–4[a]	0[a]	Sporicidal; toxic; good penetration; requires relative humidity of 30% or more; microbicidal activity varies with apparatus used; absorbed by porous material; dry spores highly resistant; moisture must be present, and presoaking is most desirable
Liquid			
Glutaraldehyde, aqueous	3	0	Sporicidal; active solution unstable; toxic
Stabilized hydrogen peroxide	3	0	Sporicidal; use solution stable up to 6 weeks; toxic orally and to eyes; mildly skin toxic; little inactivated by organic matter
Formaldehyde + alcohol	3	0	Sporicidal; noxious fumes; toxic; volatile
Formaldehyde, aqueous	1–2	0	Sporicidal; noxious fumes; toxic
Phenolic compounds	3	0	Stable; corrosive; little inactivation by organic matter; irritates skin
Chlorine compounds	1–2	0	Fast action; inactivation by organic matter; corrosive; irritates skin
Alcohol	1	3	Rapidly microbicidal except for bacterial spores and some viruses; volatile; flammable; dries and irritates skin
Iodine + alcohol	0	4	Corrosive; very rapidly microbicidal; causes staining; irritates skin; flammable
Iodophors	1–2	3	Somewhat unstable; relatively bland; staining temporary; corrosive
Iodine, aqueous	0	2	Rapidly microbicidal; corrosive; stains fabrics; stains and irritates skin
Quaternary ammonium compounds	1	0	Bland; inactivated by soap and anionics; compounds absorbed by fabrics; old or dilute solution can support growth of gram-negative bacteria
Hexachlorophene	0	2	Bland; insoluble in water, soluble in alcohol; not inactivated by soap; weakly bactericidal
Chlorhexidine	0	3	Bland; soluble in water and alcohol; weakly bactericidal
Mercurial compounds	0	±	Bland; much inactivated by organic matter; weakly bactericidal

Source: From Seymour S. Block, *Disinfection, Sterilization and Preservation.* Copyright © 1983 Lea & Febiger, Malvern, Pa. 1983. Reprinted by permission.
[a]Subjective ratings of practical usefulness in a hospital environment—4 is maximal usefulness; 0 is little or no usefulness; ± signifies that the substance is sometimes useful but not always.

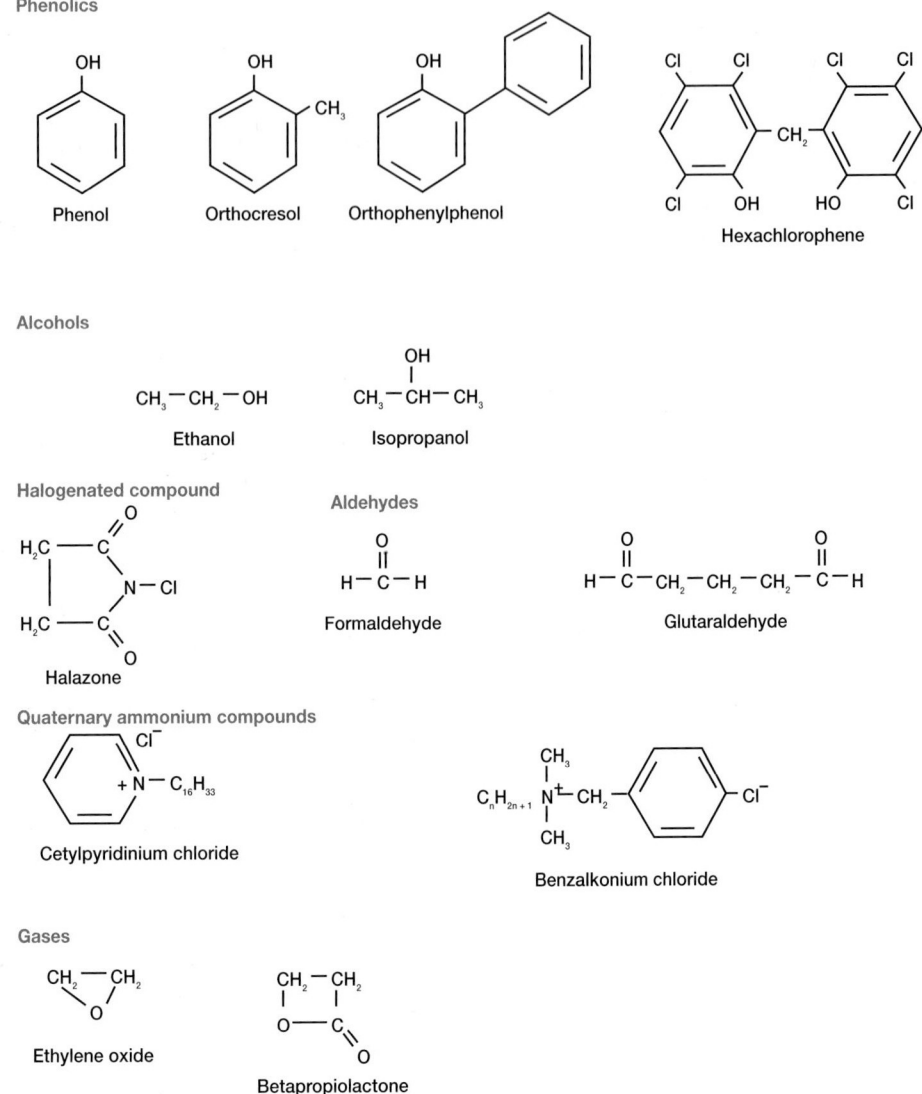

Figure 7.7 Disinfectants and Antiseptics. The structures of some frequently used disinfectants and antiseptics.

as disinfectants in laboratories and hospitals. The commercial disinfectant Lysol is made of a mixture of phenolics. Phenolics act by denaturing proteins and disrupting cell membranes. They have some real advantages as disinfectants: phenolics are tuberculocidal, effective in the presence of organic material, and remain active on surfaces long after application. However, they do have a disagreeable odor and can cause skin irritation.

Hexachlorophene (figure 7.7) has been one of the most popular antiseptics because once applied it persists on the skin and reduces skin bacteria for long periods. However, it can cause brain damage and is now used in hospital nurseries only in response to a staphylococcal outbreak.

Alcohols

Alcohols are among the most widely used disinfectants and antiseptics. They are bactericidal and fungicidal but not sporicidal; some lipid-containing viruses are also destroyed. The two most

popular alcohol germicides are ethanol and isopropanol, usually used in about 70 to 80% concentration. They act by denaturing proteins and possibly by dissolving membrane lipids. A 10 to 15 minute soaking is sufficient to disinfect thermometers and small instruments.

Halogens

A halogen is any of the five elements (fluorine, chlorine, bromine, iodine, and astatine) in group VIIA of the periodic table. They exist as diatomic molecules in the free state and form saltlike compounds with sodium and most other metals. The halogens iodine and chlorine are important antimicrobial agents. Iodine is used as a skin antiseptic and kills by oxidizing cell constituents and iodinating cell proteins. At higher concentrations, it may even kill some spores. Iodine often has been applied as tincture of iodine, 2% or more iodine in a water-ethanol solution of potassium iodide. Although it is an effective antiseptic, the skin may be damaged, a stain is left, and iodine allergies can result. More recently iodine has been complexed with an organic carrier to form an **iodophor.** Iodophors are water soluble, stable, and nonstaining, and release iodine slowly to minimize skin burns and irritation. They are used in hospitals for preoperative skin degerming and in hospitals and laboratories for disinfecting. Some popular brands are Wescodyne for skin and laboratory disinfection and Betadine for wounds.

Chlorine is the usual disinfectant for municipal water supplies and swimming pools and is also employed in the dairy and food industries. It may be applied as chlorine gas, sodium hypochlorite, or calcium hypochlorite, all of which yield hypochlorous acid (HClO) and then atomic oxygen. The result is oxidation of cellular materials and destruction of vegetative bacteria and fungi, although not spores.

$$Cl_2 + H_2O \longrightarrow HCl + HClO$$
$$Ca(OCl)_2 + 2H_2O \longrightarrow Ca(OH)_2 + 2HClO$$
$$HClO \longrightarrow HCl + O$$

Death of almost all microorganisms usually occurs within 30 minutes. Since organic material interferes with chlorine action by reacting with chlorine and its products, an excess of chlorine is added to ensure microbial destruction. One potential problem is that chlorine reacts with organic compounds to form carcinogenic trihalomethanes, which must be monitored in drinking water. Ozone sometimes has been used successfully as an alternative to chlorination in Europe and Canada. Municipal water purification (pp. 630–33)

Chlorine is also an excellent disinfectant for individual use because it is effective, inexpensive, and easy to employ. Small quantities of drinking water can be disinfected with halazone tablets. Halazone (parasulfone dichloramidobenzoic acid) slowly releases chloride when added to water and disinfects it in about a half hour. It is frequently used by campers lacking access to uncontaminated drinking water.

Chlorine solutions make very effective laboratory and household disinfectants. An excellent disinfectant-detergent combination can be prepared if a 1/100 dilution of household bleach (e.g., 1.3 fl oz of Clorox or Purex bleach in 1 gal or 10 ml/liter) is combined with sufficient nonionic detergent such as Palmolive Original (about 1 oz/gal or 7.8 ml/liter) to give a 0.8% detergent concentration. This mixture will remove both dirt and bacteria.

Heavy Metals

For many years the ions of heavy metals such as mercury, silver, arsenic, zinc, and copper were used as germicides. More recently these have been superseded by other less toxic and more effective germicides (many heavy metals are more bacteriostatic than bactericidal). There are a few exceptions. A 1% solution of silver nitrate is often added to the eyes of infants to prevent ophthalmic gonorrhea (in many hospitals, erythromycin is used instead of silver nitrate because it is effective against *Chlamydia* as well as *Neisseria*). Silver sulfadiazine is used on burns. Copper sulfate is an effective algicide in lakes and swimming pools.

Heavy metals combine with proteins, often with their sulfhydryl groups, and inactivate them. They may also precipitate cell proteins.

Quaternary Ammonium Compounds

Detergents [Latin *detergere,* to wipe off or away] are organic molecules that serve as wetting agents and emulsifiers because they have both polar hydrophilic and nonpolar hydrophobic ends. Due to their amphipathic nature (*see section 3.2*), detergents solubilize otherwise insoluble residues and are very effective cleansing agents. They are different than soaps, which are derived from fats.

Although anionic detergents have some antimicrobial properties, only cationic detergents are effective disinfectants. The most popular of these disinfectants are quaternary ammonium compounds characterized by a positively charged quaternary nitrogen and a long hydrophobic aliphatic chain (figure 7.7). They disrupt microbial membranes and may also denature proteins.

Cationic detergents like benzalkonium chloride and cetylpyridinium chloride kill most bacteria but not *M. tuberculosis* or endospores. They do have the advantages of being stable, nontoxic, and bland but they are inactivated by hard water and soap. Cationic detergents are often used as disinfectants for food utensils and small instruments and as skin antiseptics. Several brands are on the market. Zephiran contains benzalkonium chloride and Ceepryn, cetylpyridinium chloride.

Aldehydes

Both of the commonly used aldehydes, formaldehyde and glutaraldehyde, are highly reactive molecules that combine with nucleic acids and proteins and inactivate them, probably by cross-linking and alkylating molecules (figure 7.7). They are sporicidal and can be used as chemical sterilants. Formaldehyde is usually dissolved in water or alcohol before use. A 2% buffered solution of glutaraldehyde is an effective disinfectant. It is less irritating than formaldehyde and is used to disinfect hospital and laboratory equipment. Glutaraldehyde usually disinfects objects within about 10 minutes but may require as long as 12 hours to destroy all spores.

Sterilizing Gases

Many heat-sensitive items such as disposable plastic petri dishes and syringes, heart-lung machine components, sutures, and catheters are now sterilized with ethylene oxide gas (figure 7.7). Ethylene oxide (EtO) is both microbicidal and sporicidal and kills by combining with cell proteins. It is a particularly effective sterilizing agent because it rapidly penetrates packing materials, even plastic wraps.

Sterilization is carried out in a special ethylene oxide sterilizer, very much resembling an autoclave in appearance, that controls the EtO concentration, temperature, and humidity. Because pure EtO is explosive, it is usually supplied in a 10 to 20% concentration mixed with either CO_2 or dichlorodifluoromethane.

The ethylene oxide concentration, humidity, and temperature influence the rate of sterilization. A clean object can be sterilized if treated for 5 to 8 hours at 38°C or 3 to 4 hours at 54°C when the relative humidity is maintained at 40 to 50% and the EtO concentration at 700 mg/liter. Extensive aeration of the sterilized materials is necessary to remove residual EtO because it is so toxic.

Betapropiolactone (BPL) is occasionally employed as a sterilizing gas. In the liquid form it has been used to sterilize vaccines and sera. BPL decomposes to an inactive form after several hours and is therefore not as difficult to eliminate as EtO. It also destroys microorganisms more readily than ethylene oxide but does not penetrate materials well and may be carcinogenic. For these reasons, BPL has not been used as extensively as EtO.

Recently vapor-phase hydrogen peroxide has been used to decontaminate biological safety cabinets.

---

1. Why are most antimicrobial chemical agents disinfectants rather than sterilants? What general characteristics should one look for in a disinfectant?
2. Describe each of the following agents in terms of its chemical nature, mechanism of action, mode of application, common uses and effectiveness, and advantages and disadvantages: phenolics, alcohols, halogens (iodine and chlorine), heavy metals, quaternary ammonium compounds, aldehydes, and ethylene oxide.

 7.6 EVALUATION OF ANTIMICROBIAL AGENT EFFECTIVENESS

Testing of antimicrobial agents is a complex process regulated by two different federal agencies. The U.S. Environmental Protection Agency regulates disinfectants, whereas agents used on humans and animals are under the control of the Food and Drug Administration. Testing of antimicrobial agents often begins with an initial screening test to see if they are effective and at what concentrations. This may be followed by more realistic in-use testing.

The best-known disinfectant screening test is the **phenol coefficient test** in which the potency of a disinfectant is compared with that of phenol. A series of dilutions of phenol and the experimental disinfectant are inoculated with the test bacteria *Salmonella typhi* and *Staphylococcus aureus,* then placed in a 20 or 37°C water bath. These inoculated disinfectant tubes are next subcultured to regular fresh medium at 5 minute intervals, and the subcultures are incubated for two or more days. The highest dilutions that kill the bacteria after a 10 minute exposure, but not after 5 minutes, are used to calculate the phenol coefficient. The reciprocal of the appropriate test disinfectant dilution is divided by that for phenol to obtain the coefficient. Suppose that the phenol dilution was 1/90 and maximum effective dilution for disin-

Table 7.6
Phenol Coefficients for Some Disinfectants

Disinfectant	Phenol Coefficients[a]	
	Salmonella typhi	Staphylococcus aureus
Phenol	1	1
Cetylpyridinium chloride	228	337
O-phenylphenol	5.6 (20°C)	4.0
p-cresol	2.0–2.3	2.3
Hexachlorophene	5–15	15–40
Merthiolate	600	62.5
Mercurochrome	2.7	5.3
Lysol	1.9	3.5
Isopropyl alcohol	0.6	0.5
Ethanol	0.04	0.04
2% I$_2$ solution in EtOH	4.1–5.2 (20°C)	4.1–5.2 (20°C)

[a]All values were determined at 37°C except where indicated.

fectant X was 1/450. The phenol coefficient of X would be 5. The higher the phenol coefficient value, the more effective the disinfectant under these test conditions. A value greater than 1 means that the disinfectant is more effective than phenol. A few representative phenol coefficient values are given in **table 7.6.**

The phenol coefficient test is a useful initial screening procedure, but the phenol coefficient can be misleading if taken as a direct indication of disinfectant potency during normal use. This is because the phenol coefficient is determined under carefully controlled conditions with pure bacterial strains, whereas disinfectants are normally used on complex populations in the presence of organic matter and with significant variations in environmental factors like pH, temperature, and presence of salts.

To more realistically estimate disinfectant effectiveness, other tests are often used. The rates at which selected bacteria are destroyed with various chemical agents may be experimentally determined and compared. A **use dilution test** can also be carried out. Stainless steel cylinders are contaminated with specific bacterial species under carefully controlled conditions. The cylinders are dried briefly, immersed in the test disinfectants for 10 minutes, transferred to culture media, and incubated for two days. The disinfectant concentration that kills the organisms in the sample with a 95% level of confidence under these conditions is determined. Disinfectants also can be tested under conditions designed to simulate normal in-use situations. In-use testing techniques allow a more accurate determination of the proper disinfectant concentration for a particular situation.

---

1. Briefly describe the phenol coefficient test.
2. Why might it be necessary to employ procedures like the use dilution and in-use tests?

Summary

7.1 Definition of Frequently Used Terms

a. Sterilization is the process by which all living cells, viable spores, viruses, and viroids are either destroyed or removed from an object or habitat. Disinfection is the killing, inhibition, or removal of microorganisms (but not necessarily endospores) that can cause disease.

b. Many terms are used to define how microorganisms are controlled: sterilant, disinfectant, sanitization, antisepsis, and antiseptic.

c. Antimicrobial agents that kill organisms often have the suffix -cide, whereas agents that prevent growth and reproduction have the suffix -static.

7.2 The Pattern of Microbial Death

a. Microbial death is usually exponential or logarithmic (**figure 7.1**).

7.3 Conditions Influencing the Effectiveness of Antimicrobial Agent Activity

a. The effectiveness of a disinfectant or sterilizing agent is influenced by population size, population composition, concentration or intensity of the agent, exposure duration, temperature, and nature of the local environment.

7.4 The Use of Physical Methods in Control

a. The efficiency of heat killing is often indicated by the thermal death time or the decimal reduction time.

b. Although treatment with boiling water for 10 minutes kills vegetative forms, an autoclave must be used to destroy endospores by heating at 121°C and 15 pounds of pressure (**figure 7.3**).

c. Moist heat kills by degrading nucleic acids, denaturing enzymes and other proteins, and disrupting cell membranes.

d. Heat-sensitive liquids can be pasteurized by heating at 63°C for 30 minutes or at 72°C for 15 seconds (flash pasteurization). Heating at 140 to 150°C for 1 to 3 seconds (ultrahigh-temperature sterilization) may be used.

e. Glassware and other heat-stable items may be sterilized by dry heat at 160 to 170°C for 2 to 3 hours.

f. Refrigeration and freezing can be used to control microbial growth and reproduction.

g. Microorganisms can be efficiently removed by filtration with either depth filters or membrane filters.

h. Biological safety cabinets with high-efficiency particulate filters sterilize air by filtration (**figure 7.6**).

i. Radiation of short wavelength or high-energy ultraviolet and ionizing radiation can be used to sterilize objects.

7.5 The Use of Chemical Agents in Control

a. Chemical agents usually act as disinfectants because they cannot readily destroy bacterial endospores. Disinfectant effectiveness depends on concentration, treatment duration, temperature, and presence of organic material (**tables 7.4** and **7.5**).

b. Phenolics and alcohols are popular disinfectants that act by denaturing proteins and disrupting cell membranes (**figure 7.7**).

c. Halogens (iodine and chlorine) kill by oxidizing cellular constituents; cell proteins may also be iodinated. Iodine is applied as a tincture or iodophor. Chlorine may be added to water as a gas, hypochlorite, or an organic chlorine derivative.

d. Heavy metals tend to be bacteriostatic agents. They are employed in specialized situations such as the use of silver nitrate in the eyes of newborn infants and copper sulfate in lakes and pools.

e. Cationic detergents are often used as disinfectants and antiseptics; they disrupt membranes and denature proteins.

f. Aldehydes such as formaldehyde and glutaraldehyde can sterilize as well as disinfect because they kill spores.

g. Ethylene oxide gas penetrates plastic wrapping material and destroys all life forms by reacting with proteins. It is used to sterilize packaged, heat-sensitive materials.

7.6 Evaluation of Antimicrobial Agent Effectiveness

a. A variety of procedures can be used to determine the effectiveness of disinfectants, among them the following: phenol coefficient test, measurement of killing rates with germicides, use dilution testing, and in-use testing.

Key Terms

algicide   135
antimicrobial agent   136
antisepsis   135
antiseptics   135
autoclave   137
bactericide   135
bacteriostatic   135
decimal reduction time (D)   137
depth filters   139
detergent   145

disinfectant   135
disinfection   134
dry heat sterilization   139
D value   137
flash pasteurization   139
fungicide   135
fungistatic   135
F value   137
germicide   135

high-efficiency particulate air (HEPA) filters   140
iodophor   145
ionizing radiation   141
laminar flow biological safety cabinets   140
membrane filters   139
pasteurization   139
phenol coefficient test   146

sanitization   135
sterilization   134
thermal death time (TDT)   137
ultrahigh-temperature (UHT) sterilization   139
ultraviolet (UV) radiation   141
use dilution test   146
viricide   135
z value   137

Questions for Thought and Review

1. How can the D value be used to estimate the time required for sterilization? Suppose that you wanted to eliminate the risk of *Salmonella* poisoning by heating your food ($D_{60} = 0.40$ minutes, z value = 5.0). Calculate the 12D value at 60°C. How long would it take to achieve the same results by heating at 50, 55, and 65°C?

2. How can one alter treatment conditions to increase the effectiveness of a disinfectant?

3. How would the following be best sterilized: glass pipettes and petri plates, tryptic soy broth tubes, nutrient agar, antibiotic solution, interior of a biological safety cabinet, wrapped package of plastic petri plates?

4. Which disinfectants or antiseptics would be used to treat the following: oral thermometer, laboratory bench top, drinking water, patch of skin before surgery, small medical instruments (probes, forceps, etc.)? List all chemicals suitable for each task.

5. Until relatively recently, spoiled milk was responsible for a significant proportion of infant deaths.
    a. Why is untreated milk easily spoiled?

    b. Why is boiling milk over prolonged periods not desirable?

6. In table 7.3 why is the *D* value so different for the three conditions in which *S. aureus* might be found?

Critical Thinking Questions

1. Throughout history, spices have been used as preservatives and to cover up the smell/taste of food that is slightly spoiled. The success of some spices led to a magical, ritualized use of many of them and possession of spices was often limited to priests or other powerful members of the community.

    a. Choose a spice and trace its use geographically and historically. What is its common-day use today?

    b. Spices grow and tend to be used predominantly in warmer climates. Explain.

2. Design an experiment to determine whether an antimicrobial agent is acting as a cidal or static agent.

    How would you determine whether an agent is suitable for use as an antiseptic rather than as a disinfectant?

3. Suppose that you are testing the effectiveness of disinfectants with the phenol coefficient test and obtained the following results. What disinfectant can you safely say is the most effective? Can you determine its phenol coefficient from these results?

| | Bacterial Growth after Treatment | | |
Dilution	Disinfectant A	Disinfectant B	Disinfectant C
1/20	−	−	−
1/40	+	−	−
1/80	+	−	+
1/160	+	+	+
1/320	+	−	+

For recommended readings on these and related topics, see Appendix V.

chapter 8

Metabolism:
Energy, Enzymes, and Regulation

Concepts

- Energy is the capacity to do work. Living organisms can perform three major types of work: chemical work, transport work, and mechanical work.
- Most energy used by living organisms originally comes from sunlight trapped during photosynthesis by photoautotrophs. Chemoheterotrophs then consume autotroph-synthesized organic materials and use them as sources of energy and as building blocks.
- An energy currency is needed to connect energy-yielding exergonic reactions with energy-requiring endergonic reactions. The most commonly used currency is ATP.
- All living systems obey the laws of thermodynamics.
- When electrons are transferred from an electron donor with a more negative reduction potential to an acceptor with a more positive potential, energy is made available. A reversal of the direction of electron transfer—for example, during photosynthesis—requires energy input.
- Enzymes are protein catalysts that make life possible by increasing the rate of reactions at ambient temperatures. Enzymes do not change chemical equilibria or violate the laws of thermodynamics but accelerate reactions by lowering their activation energy.
- Metabolism is regulated in such a way that (a) cell components are maintained at the proper concentrations, even in the face of a changing environment, and (b) energy and material are conserved.
- The localization of enzymes and metabolites in separate compartments of a cell regulates and coordinates metabolic activity.
- The activity of regulatory enzymes may be changed through reversible binding of effectors to a regulatory site separate from the catalytic site or through covalent modification of the enzyme. Regulation of enzyme activity operates rapidly and serves as a fine-tuning mechanism to adjust metabolism from moment to moment.

This model shows *E. coli* aspartate carbamoyltransferase in the less active T state. The catalytic polypeptide chains are in blue and the regulatory chains are colored red.

• A pathway's activity is often controlled by its end products through feedback inhibition of regulatory enzymes located at the start of the sequence and at branch points.

Fresh oxygen flows
From the open stomata
The whole world inhales

—Crystal Cunningham

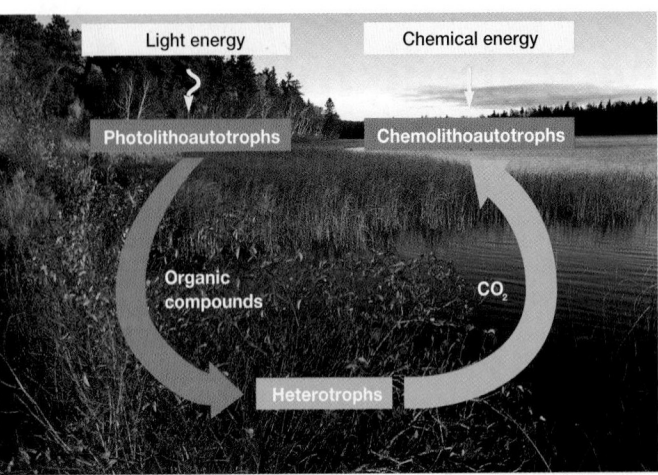

Figure 8.1 The Flow of Carbon and Energy in an Ecosystem. This diagram depicts the flow of energy and carbon in general terms. See text for discussion.

 hapters 3 and 4 contain many examples of an important principle: that a cell's structure is intimately related to its function. In each instance one can readily relate an organelle's construction to its function (and vice versa). A second unifying principle in biology is that life is sustained by the trapping and use of energy, a process made possible by the action of enzymes. Because this is so crucial to our understanding of microbial function, considerable attention is given to energy and enzymes in this chapter.

The organization of microbial metabolism will be briefly described in chapters 8 to 10. Metabolic pathways are treated as a sequence of enzymes functioning as a unit, with each enzyme using as its substrate a product of the preceding enzyme-catalyzed reaction. This picture of metabolic pathways is incomplete because we will usually ignore the regulation of pathway operation for the sake of space and simplicity. However, one should keep in mind that both regulation of the activity of individual pathways and coordination of the action of separate sequences are essential to the existence of life. Cells become disorganized and die without adequate control of metabolism, and regulation is just as important to life as is the efficient use of energy. Thus the last part of this chapter will be devoted to the regulation of metabolism as a foundation for the subsequent discussion of pathways.

This chapter begins with a brief survey of the nature of energy and the laws of thermodynamics. The participation of energy in metabolism and the role of ATP as an energy currency is considered next. An introduction to the nature and function of enzymes follows. The chapter ends with an overview of metabolic regulation, including an introduction to metabolic channeling and the regulation of the activity of critical enzymes.

8.1 ENERGY AND WORK

Energy may be most simply defined as the capacity to do work or to cause particular changes. Thus all physical and chemical processes are the result of the application or movement of energy. Living cells carry out three major types of work, and all are essential to life processes. **Chemical work** involves the synthesis of complex biological molecules required by cells from

much simpler precursors; energy is needed to increase the molecular complexity of a cell. Molecules and ions often must be transported across cell membranes against an electro-chemical gradient. For example, a molecule sometimes moves into a cell even though its concentration is higher internally. Similarly a solute may be expelled from the cell against a concentration gradient. This process is **transport work** and requires energy input in order to take up nutrients, eliminate wastes, and maintain ion balances. The third type of work is **mechanical work,** perhaps the most familiar of the three. Energy is required to change the physical location of organisms, cells, and structures within cells.

The ultimate source of most biological energy is the visible sunlight impinging on the Earth's surface. Light energy is trapped by phototrophs during **photosynthesis,** in which it is absorbed by chlorophyll and other pigments and converted to chemical energy. As noted in chapter 5, chemolithoautotrophs derive energy by oxidizing inorganic compounds rather than obtaining it from light absorption. Chemical energy from photosynthesis and chemolithotrophy can then be used by photolithoautotrophs and chemolithoautotrophs to transform CO_2 into biological molecules such as glucose (**figure 8.1**). Nutritional types (pp. 95–96)

The complex molecules manufactured by autotrophic organisms (both plant and microbial producers) serve as a carbon source for chemoheterotrophs and other consumers that use complex organic molecules as a source of material and energy for building their own cellular structures (it should be remembered that autotrophs also use complex organic molecules). Chemoheterotrophs often employ O_2 as an electron acceptor when oxidizing glucose and other organic molecules to CO_2. This process, in which O_2 acts as the final electron acceptor and is reduced to water, is called **aerobic respiration.** Much energy is released during this process. Thus in the ecosystem, energy is trapped by

Figure 8.2 **Adenosine Triphosphate and Adenosine Diphosphate.** (**a**) Structure of ATP and ADP. The two red bonds are more easily broken or have a high phosphate group transfer potential (see text). The pyrimidine ring atoms have been numbered. (**b**) A model of ATP. Carbon is in green; hydrogen in light blue; nitrogen in dark blue; oxygen in red; and phosphorus in orange.

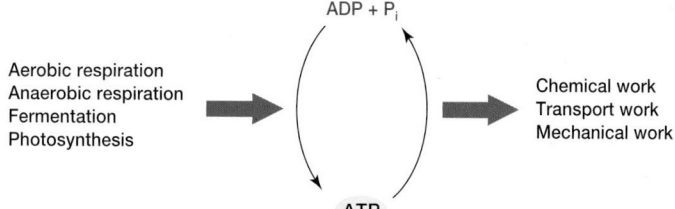

Figure 8.3 The Cell's Energy Cycle. ATP is formed from energy made available during aerobic respiration, anaerobic respiration, fermentation, and photosynthesis. Its breakdown to ADP and phosphate (P_i) makes chemical, transport, and mechanical work possible.

8.2 THE LAWS OF THERMODYNAMICS

To understand how energy is trapped or generated and how ATP functions as an energy currency, some knowledge of the basic principles of thermodynamics is required. The science of **thermodynamics** analyzes energy changes in a collection of matter (e.g., a cell or a plant) called a system. All other matter in the universe is called the surroundings. Thermodynamics focuses on the energy differences between the initial state and the final state of a system. It is not concerned with the rate of the process. For instance, if a pan of water is heated to boiling, only the condition of the water at the start and at boiling is important in thermodynamics, not how fast it is heated or on what kind of stove. Two important laws of thermodynamics must be understood. The **first law of thermodynamics** says that energy can be neither created nor destroyed. The total energy in the universe remains constant although it can be redistributed. For example, many energy exchanges do occur during chemical reactions (e.g., heat is given off by exothermic reactions and absorbed during endothermic reactions), but these heat exchanges do not violate the first law.

It is necessary to specify quantitatively the amount of energy used in or evolving from a particular process, and two types of energy units are employed. A **calorie** (cal) is the amount of heat energy needed to raise one gram of water from 14.5 to 15.5°C. The amount of energy also may be expressed in terms of **joules** (J), the units of work capable of being done. One cal of heat is equivalent to 4.1840 J of work. One thousand calories or a kilocalorie (kcal) is enough energy to boil 1.9 ml of water. A kilojoule is enough energy to boil about 0.44 ml of water, or enable a person weighing 70 kg to climb 35 steps. The joule is normally used by chemists and physicists. Because biologists most often speak of energy in terms of calories, this text will employ calories when discussing energy changes.

Although it is true that energy is conserved in the universe, the first law of thermodynamics does not account for many physical and chemical processes. A simple example may help make this clear. Suppose a full gas cylinder is connected to an empty one by a tube with a valve (**figure 8.4**). If the valve is opened, gas flows

photoautotrophs and chemolithoautotrophs; some of this energy subsequently flows to chemoheterotrophs when they use nutrients derived from autotrophs (figure 8.1; *see also figure 28.28*). The CO_2 produced during aerobic respiration can be incorporated again into complex organic molecules during photosynthesis and chemolithoautotrophy. Clearly the flows of carbon and energy in the ecosystem are intimately related.

Cells must efficiently transfer energy from their energy-generating or trapping apparatus to the systems actually carrying out work. That is, cells must have a practical form of energy currency. In living organisms the major currency is **adenosine 5′-triphosphate** (**ATP; figure 8.2**). When ATP breaks down to **adenosine diphosphate** (**ADP**) and orthophosphate (P_i), energy is made available for useful work. Later, energy from photosynthesis, aerobic respiration, anaerobic respiration, and fermentation is used to resynthesize ATP from ADP and P_i. An energy cycle is created in the cell (**figure 8.3**). Fermentation (pp. 174–77); Anaerobic respiration (pp. 185–86)

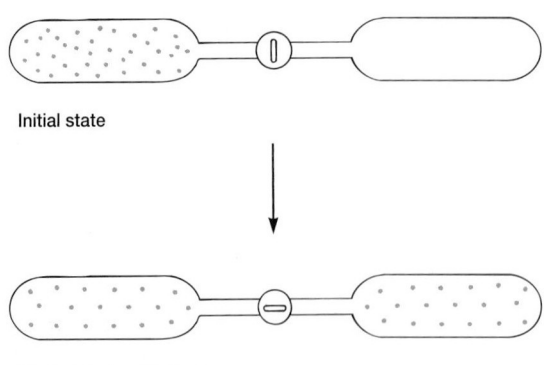

Initial state

Final state (equilibrium)

Figure 8.4 A Second Law Process.   The expansion of gas into an empty cylinder simply redistributes the gas molecules until equilibrium is reached. The total number of molecules remains unchanged.

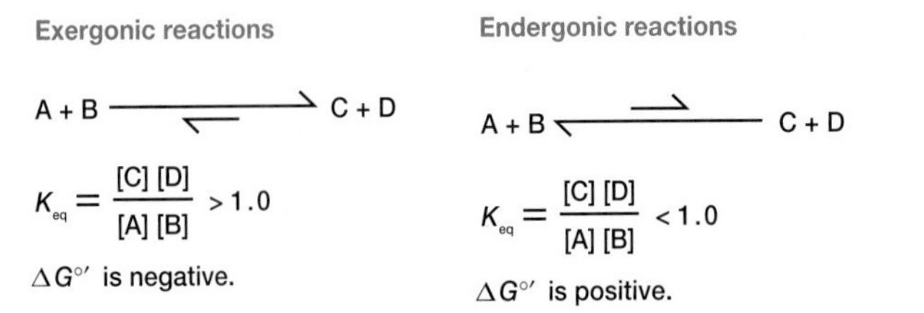

from the full to the empty cylinder until the gas pressure is equal on both sides. Energy has not only been redistributed but also conserved. The expansion of gas is explained by the **second law of thermodynamics** and a condition of matter called entropy. **Entropy** may be considered a measure of the randomness or disorder of a system. The greater the disorder of a system, the greater is its entropy. The second law states that physical and chemical processes proceed in such a way that the randomness or disorder of the universe (the system and its surroundings) increases to the maximum possible. Gas will always expand into an empty cylinder.

 FREE ENERGY AND REACTIONS

The first and second laws can be combined in a useful equation, relating the changes in energy that can occur in chemical reactions and other processes.

$$\Delta G = \Delta H - T \cdot \Delta S$$

ΔG is the change in free energy, ΔH is the change in enthalpy, T is the temperature in Kelvin (°C + 273), and ΔS is the change in entropy occurring during the reaction. The change in **enthalpy** is the change in heat content. Cellular reactions occur under conditions of constant pressure and volume. Thus the change in enthalpy is about the same as the change in total energy during the reaction. The **free energy change** is the amount of energy in a system available to do useful work at constant temperature and pressure. Therefore the change in entropy is a measure of the proportion of the total energy change that the system cannot use in performing work. Free energy and entropy changes do not depend on how the system gets from start to finish. A reaction will occur spontaneously at constant temperature and pressure if the free energy of the system decreases during the reaction or, in other words, if ΔG is negative. It follows from the equation that a reaction with a large positive change in entropy will normally tend to have a negative ΔG value and therefore occur sponta-

A + B ⟶ C + D   A + B ⟵ C + D

$$K_{eq} = \frac{[C]\,[D]}{[A]\,[B]} > 1.0 \qquad K_{eq} = \frac{[C]\,[D]}{[A]\,[B]} < 1.0$$

$\Delta G^{\circ\prime}$ is negative. $\Delta G^{\circ\prime}$ is positive.

Figure 8.5 $\Delta G^{\circ\prime}$ and Equilibrium.   The relationship of $\Delta G^{\circ\prime}$ to the equilibrium of reactions. Note the differences between exergonic and endergonic reactions.

neously. A decrease in entropy will tend to make ΔG more positive and the reaction less favorable.

The change in free energy has a definite, concrete relationship to the direction of chemical reactions. Consider this simple reaction.

$$A + B \rightleftharpoons C + D$$

If the molecules A and B are mixed, they will combine to form the products C and D. Eventually C and D will become concentrated enough to combine and produce A and B at the same rate as they are formed from A and B. The reaction is now at **equilibrium:** the rates in both directions are equal and no further net change occurs in the concentrations of reactants and products. This situation is described by the **equilibrium constant (K_{eq})**, relating the equilibrium concentrations of products and substrates to one another.

$$K_{eq} = \frac{[C][D]}{[A][B]}$$

If the equilibrium constant is greater than one, the products are in greater concentration than the reactants at equilibrium—that is, the reaction tends to go to completion as written.

The equilibrium constant of a reaction is directly related to its change in free energy. When the free energy change for a process is determined at carefully defined standard conditions of concentration, pressure, pH, and temperature, it is called the **standard free energy change (ΔG°)**. If the pH is set at 7.0 (which is close to the pH of living cells), the standard free energy change is indicated by the symbol $\Delta G^{\circ\prime}$. The change in standard free energy may be thought of as the maximum amount of energy available from the system for useful work under standard conditions. Using $\Delta G^{\circ\prime}$ values allows one to compare reactions without worrying about variations in the ΔG due to differences in environmental conditions. The relationship between $\Delta G^{\circ\prime}$ and K_{eq} is given by this equation.

$$\Delta G^{\circ\prime} = -2.303RT \cdot \log K_{eq}$$

R is the gas constant (1.9872 cal/mole-degree or 8.3145 J/mole-degree), and T is the absolute temperature. Inspection of this equation shows that when $\Delta G^{\circ\prime}$ is negative, the equilibrium constant is greater than one and the reaction goes to completion as written. It is said to be an **exergonic reaction (figure 8.5)**. In an **endergonic reaction** $\Delta G^{\circ\prime}$ is positive and the equilibrium constant is less than one. That is, the reaction is not favorable, and little product will be formed at equilibrium under standard conditions. Keep in mind that the $\Delta G^{o\prime}$ value shows only where the reaction lies at equilibrium, not how fast the reaction reaches equilibrium.

---

1. What is energy and what kinds of work are carried out in a cell? Describe the energy cycle and ATP's role in it.
2. What is thermodynamics? Summarize the first and second laws. Define free energy, entropy, and enthalpy.
3. How is the change in standard free energy related to the equilibrium constant for a reaction? What are exergonic and endergonic reactions?

 ## 8.4　THE ROLE OF ATP IN METABOLISM

Many reactions in the cell are endergonic and will not proceed far toward completion without outside assistance. One of ATP's major roles is to drive such endergonic reactions more to completion. ATP is a **high-energy molecule.** That is, it breaks down or hydrolyzes almost completely to the products ADP and P_i with a $\Delta G^{o\prime}$ of -7.3 kcal/mole.

$$ATP + H_2O \rightleftharpoons ADP + P_i$$

In reference to ATP the term high-energy molecule does not mean that there is a great deal of energy stored in a particular bond of ATP. It simply indicates that the removal of the terminal phosphate goes to completion with a large negative standard free energy change, or the reaction is strongly exergonic. In other words, ATP has a high **phosphate group transfer potential;** it readily transfers its phosphate to water. The phosphate group transfer potential is defined as the negative of $\Delta G^{o\prime}$ for the hydrolytic removal of phosphate. A molecule with a higher group transfer potential will donate phosphate to one with a lower potential.

Thus ATP is ideally suited for its role as an energy currency. It is formed in energy-trapping and -yielding processes such as photosynthesis, fermentation, and aerobic respiration. In the cell's economy, exergonic ATP breakdown is coupled with various endergonic reactions to promote their completion (**figure 8.6**). In other words, ATP links energy-yielding reactions, which liberate free energy, with energy-using reactions, which require free energy input to proceed toward completion. Facilitation of chemical work is the focus of the preceding example, but the same principles apply when ATP is coupled with endergonic processes involving transport work and mechanical work (figure 8.3).

 ## 8.5　OXIDATION-REDUCTION REACTIONS AND ELECTRON CARRIERS

Free energy changes are not only related to the equilibria of "regular" chemical reactions but also to the equilibria of oxidation-reduction reactions. The release of energy normally involves oxidation-reduction reactions. **Oxidation-reduction (redox) re-**

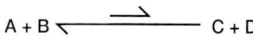

Figure 8.6　ATP as a Coupling Agent.　The use of ATP to make endergonic reactions more favorable. It is formed by exergonic reactions and then used to drive endergonic reactions.

actions are those in which electrons move from an **electron donor** to an **electron acceptor.**[1] By convention such a reaction is written with the donor to the right of the acceptor and the number (n) of electrons (e^-) transferred.

$$Acceptor + ne^- \rightleftharpoons donor$$

The acceptor and donor pair is referred to as a redox couple (table 8.1). When an acceptor accepts electrons, it becomes the donor of the couple. The equilibrium constant for the reaction is called the **standard reduction potential** (E_0) and is a measure of the tendency of the donor to lose electrons. The reference standard for reduction potentials is the hydrogen system with an E'_0 (the reduction potential at pH 7.0) of -0.42 volts or -420 millivolts.

$$2H^+ + 2e^- \rightleftharpoons H_2$$

In this reaction each hydrogen atom provides one proton (H^+) and one electron (e^-).

The reduction potential has a concrete meaning. Redox couples with more negative reduction potentials will donate electrons to couples with more positive potentials and greater affinity for electrons. Thus electrons will tend to move from donors at the top of the list in **table 8.1** to acceptors at the bottom because they have more positive potentials. This may be expressed visually in the form of an electron tower in which the most negative reduction potentials are at the top (**figure 8.7**). Electrons move from donors to acceptors down the potential gradient or fall down the tower to more positive potentials. Consider the case of the electron carrier **nicotinamide adenine dinucleotide (NAD⁺).** The $NAD^+/NADH$ couple has a very negative E'_0 and can therefore give electrons to many acceptors, including O_2.

$$NAD^+ + 2H^+ + 2e^- \rightleftharpoons NADH + H^+ \qquad E'_0 = -0.32 \text{ volts}$$
$$1/2 O_2 + 2H^+ + 2e^- \rightleftharpoons H_2O \qquad E'_0 = +0.82 \text{ volts}$$

---

[1]In an oxidation-reduction reaction, the electron donor is often called the reducing agent or reductant because it is donating electrons to the acceptor and thus reducing it. The electron acceptor is called the oxidizing agent or oxidant because it is removing electrons from the donor and oxidizing it.

Table 8.1

Selected Biologically Important Redox Couples

Redox Couple	E'_0 (Volts) [a]
$2H^+ + 2e^- \longrightarrow H_2$	-0.42
$Ferredoxin(Fe^{3+}) + e^- \longrightarrow ferredoxin\ (Fe^{2+})$	-0.42
$NAD(P)^+ + H^+ + 2e^- \longrightarrow NAD(P)H$	-0.32
$S + 2H^+ + 2e^- \longrightarrow H_2S$	-0.274
$Acetaldehyde + 2H^+ + 2e^- \longrightarrow ethanol$	-0.197
$Pyruvate^- + 2H^+ + 2e^- \longrightarrow lactate^{2-}$	-0.185
$FAD + 2H^+ + 2e^- \longrightarrow FADH_2$	-0.18 [b]
$Oxaloacetate^{2-} + 2H^+ + 2e^- \longrightarrow malate^{2-}$	-0.166
$Fumarate^{2-} + 2H^+ + 2e^- \longrightarrow succinate^{2-}$	0.031
$Cytochrome\ b\ (Fe^{3+}) + e^- \longrightarrow cytochrome\ b\ (Fe^{2+})$	0.075
$Ubiquinone + 2H^+ + 2e^- \longrightarrow ubiquinone\ H_2$	0.10
$Cytochrome\ c\ (Fe^{3+}) + e^- \longrightarrow cytochrome\ c\ (Fe^{2+})$	0.254
$Cytochrome\ a\ (Fe^{3+}) + e^- \longrightarrow cytochrome\ a\ (Fe^{2+})$	0.29
$Cytochrome\ a_3\ (Fe^{3+}) + e^- \longrightarrow cytochrome\ a_3\ (Fe^{2+})$	0.35
$NO_3^- + 2H^+ + 2e^- \longrightarrow NO_2^- + H_2O$	0.421
$NO_2^- + 8H^+ + 6e^- \longrightarrow NH_4^+ + 2H_2O$	0.44
$Fe^{3+} + e^- \longrightarrow Fe^{2+}$	0.771 [c]
$O_2 + 4H^+ + 4e^- \longrightarrow 2H_2O$	0.815

[a] E'_0 is the standard reduction potential at pH 7.0.
[b] The value for FADH/FADH$_2$ applies to the free cofactor because it can vary considerably when bound to an apoenzyme.
[c] The value for free Fe, not Fe complexed with proteins (e.g., cytochromes).

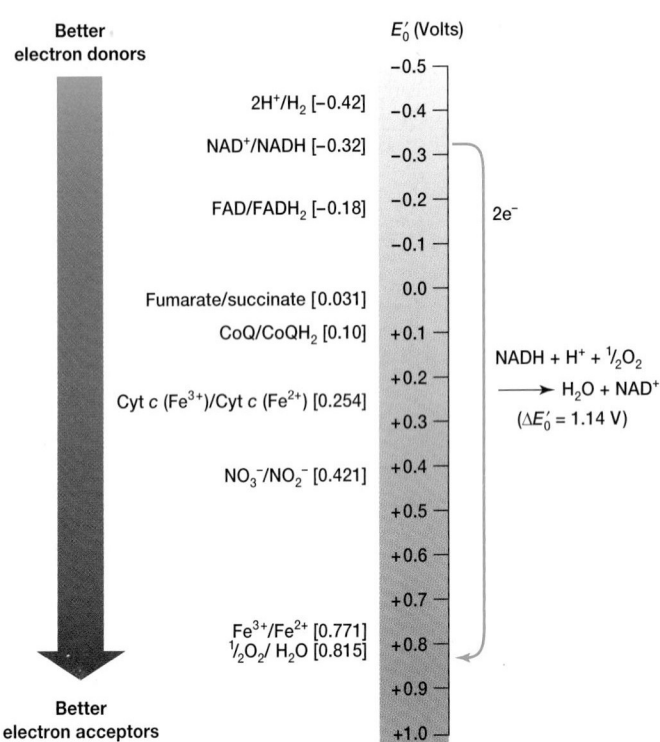

Figure 8.7   Electron Movement and Reduction Potentials.
The vertical electron tower in this illustration has the most negative reduction potentials at the top. Electrons will spontaneously move from donors higher on the tower (more negative potentials) to acceptors lower on the tower (more positive potentials). That is, the donor is always higher on the tower than the acceptor. For example, NADH will donate electrons to oxygen and form water in the process. Some typical donors and acceptors are shown on the left, and their redox potentials are given in brackets.

Because NAD^+/NADH is more negative than $1/2\ O_2$/H_2O, electrons will flow from NADH (the donor) to O_2 (the acceptor) as shown in figure 8.7.

$$NADH + H^+ + 1/2O_2 \rightarrow H_2O + NAD^+$$

When electrons move from a donor to an acceptor with a more positive redox potential, free energy is released. The $\Delta G^{o\prime}$ of the reaction is directly related to the magnitude of the difference between the reduction potentials of the two couples ($\Delta E'_0$). The larger the $\Delta E'_0$, the greater the amount of free energy made available, as is evident from the equation

$$\Delta G^{o\prime} = -nF \cdot \Delta E'_0$$

in which n is the number of electrons transferred and F is the Faraday constant (23,062 cal/mole-volt or 96,494 J/mole-volt). For every 0.1 volt change in $\Delta E'_0$, there is a corresponding 4.6 kcal change in $\Delta G^{o\prime}$ when a two-electron transfer takes place. This is similar to the relationship of $\Delta G^{o\prime}$ and K_{eq} in other chemical reactions—the larger the equilibrium constant, the greater the $\Delta G^{o\prime}$. The difference in reduction potentials between NAD^+/NADH and $1/2O_2$/H_2O is 1.14 volts, a large $\Delta E'_0$ value. When electrons move from NADH to O_2 during aerobic respiration, a large amount of free energy is made available to synthesize ATP (**figure 8.8**). If energy is released when electrons flow from negative to positive reduction potentials, then an input of energy is required to move electrons in the opposite direction, from more positive to more negative potentials. This is precisely what occurs during photosynthesis (figure 8.8). Light energy is trapped and used to move

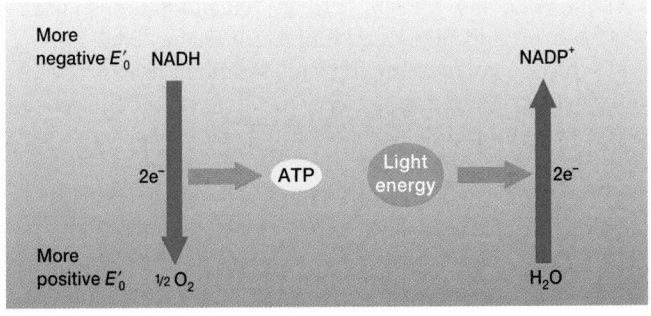

Figure 8.8   Energy Flow in Metabolism.   Examples of the relationship between electron flow and energy in metabolism. Oxygen and $NADP^+$ serve as electron acceptors for NADH and water, respectively.

electrons from water to the electron carrier **nicotinamide adenine dinucleotide phosphate ($NADP^+$)**.

The cycle of energy flow discussed earlier and illustrated in figure 8.1 can be understood from a different perspective, if the preceding concept is kept in mind. Photosynthetic organisms cap-

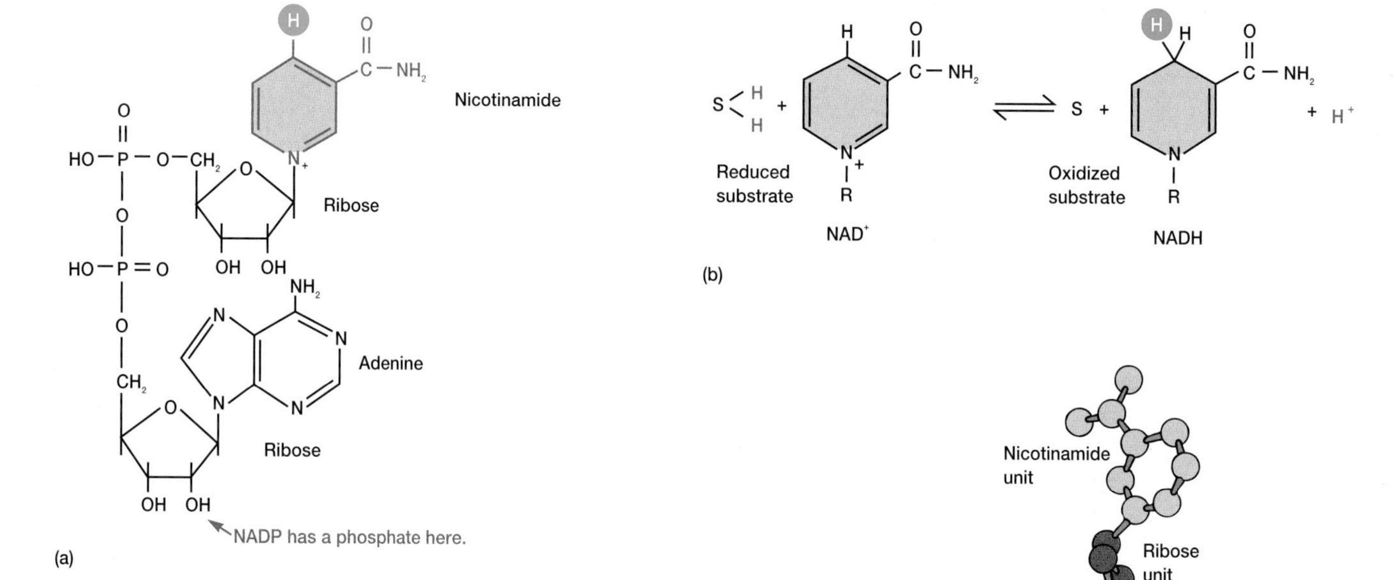

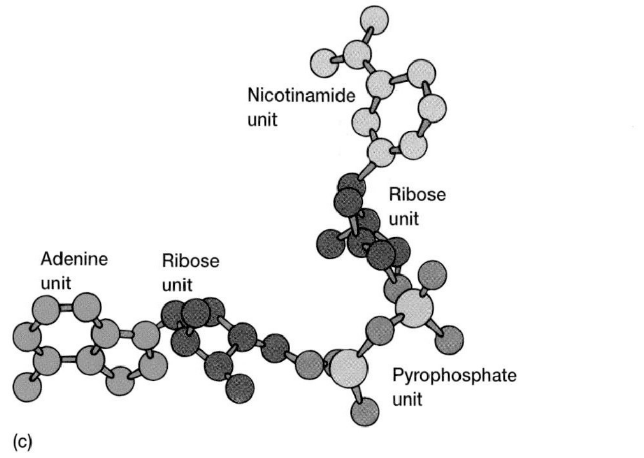

Figure 8.9   The Structure and Function of NAD. (**a**) The structure of NAD and NADP. NADP differs from NAD in having an extra phosphate on one of its ribose sugar units. (**b**) NAD can accept electrons and a hydrogen from a reduced substrate (SH_2). These are carried on the nicotinamide ring. (**c**) Model of NAD^+ when bound to the enzyme lactate dehydrogenase.

ture light energy and use it to move electrons from water (and other electron donors such as H_2S) to electron acceptors, such as $NADP^+$, that have more negative reduction potentials. These electrons can then flow back to more positive acceptors and provide energy for ATP production during photosynthesis. Photoautotrophs use ATP and NADPH to synthesize complex molecules from CO_2 (*see section 10.2*). Chemoheterotrophs also make use of energy released during the movement of electrons by oxidizing complex nutrients during respiration to produce NADH. NADH subsequently donates its electrons to O_2, and the energy released during electron transfer is trapped in the form of ATP. The energy from sunlight is made available to all living organisms because of this relationship between electron flow and energy.   Photosynthesis (pp. 190–96); Respiration and electron transport (pp. 179–84)

Electron transport is important in aerobic respiration, anaerobic respiration, chemolithotrophy, and photosynthesis. Electron movement in cells requires the participation of carriers such as NAD^+ and $NADP^+$, both of which can transport electrons between different locations. The nicotinamide ring of NAD^+ and $NADP^+$ (**figure 8.9**) accepts two electrons and one proton from a donor, while a second proton is released. There are several other electron carriers of importance in microbial metabolism (table 8.1), and they carry electrons in a variety of ways. **Flavin adenine dinucleotide (FAD)** and **flavin mononucleotide (FMN)** bear two electrons and two protons on the complex ring system shown in **figure 8.10**. Proteins bearing FAD and FMN are often called flavoproteins. **Coenzyme Q (CoQ)** or **ubiquinone** is a quinone that transports electrons and protons in many respiratory electron transport chains (**figure 8.11**). **Cytochromes** and several other

carriers use iron atoms to transport electrons by reversible oxidation and reduction reactions.

$$Fe^{3+}(\text{ferric iron}) + e^- \rightleftharpoons Fe^{2+}(\text{ferrous iron})$$

In the cytochromes these iron atoms are part of a heme group (**figure 8.12**) or other similar iron-porphyrin rings. Several different cytochromes, each of which consists of a protein and an iron-porphyrin ring, are a prominent part of respiratory electron transport chains. Some iron containing electron-carrying proteins lack a heme group and are called **nonheme iron proteins. Ferredoxin** is a nonheme iron protein active in photosynthetic electron transport and several other electron transport processes. Even though its iron atoms are not bound to a heme group, they still undergo reversible oxidation and reduction reactions. Although all the previously discussed molecules function in electron transport chains, some bear two electrons (NAD, FAD, and CoQ) while others carry only one electron at a time (cytochromes and nonheme iron proteins). This difference in the number of electrons carried is of great importance in the operation of electron transport chains (*see pp. 179–82*).

---

1. Why is ATP called a high-energy molecule? What is its role in the cell and how does it fulfill this role?
2. Write a generalized equation for a redox reaction. Define standard reduction potential.
3. How is the direction of electron flow between redox couples related to the standard reduction potential and the release of free energy? Name and briefly describe the major electron carriers found in cells.

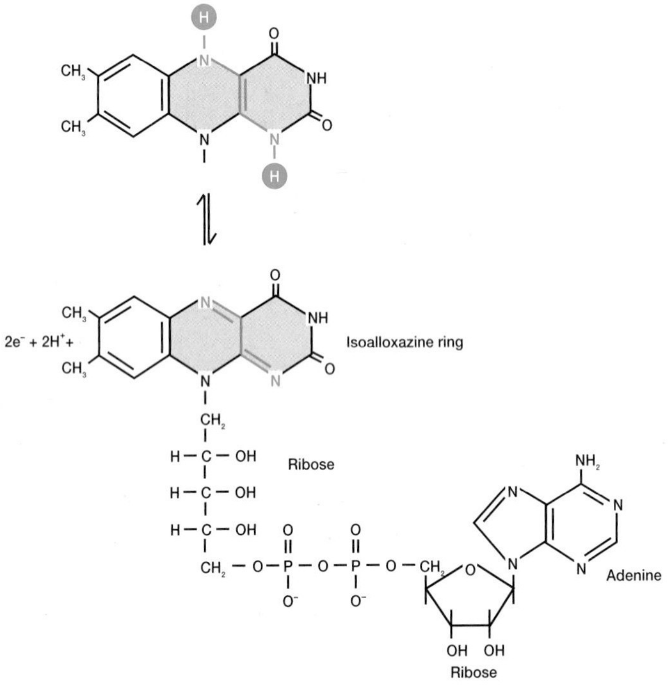

Figure 8.10   The Structure and Function of FAD. The vitamin riboflavin is composed of the isoalloxazine ring and its attached ribose sugar. FMN is riboflavin phosphate. The portion of the ring directly involved in oxidation-reduction reactions is in color.

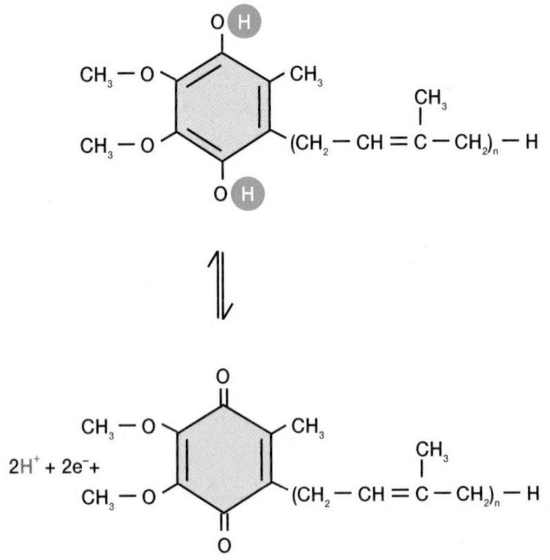

Figure 8.11   The Structure and Function of Coenzyme Q or Ubiquinone. The length of the side chain varies among organisms from n = 6 to n = 10.

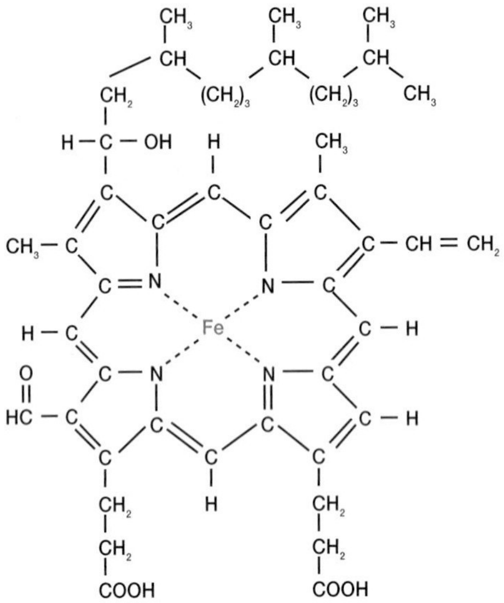

 **ENZYMES**

Recall that an exergonic reaction is one with a negative $\Delta G^{o\prime}$ and an equilibrium constant greater than one. An exergonic reaction will proceed to completion in the direction written (that is, toward the right of the equation). Nevertheless, one often can combine

Figure 8.12   The Structure of Heme. Heme is composed of a porphyrin ring and an attached iron atom. It is the nonprotein component of many cytochromes. The iron atom alternatively accepts and releases an electron.

the reactants for an exergonic reaction with no obvious result, even though products should be formed. It is precisely in these reactions that enzymes play their part.

Structure and Classification of Enzymes

Enzymes may be defined as protein catalysts that have great specificity for the reaction catalyzed and the molecules acted on. A **catalyst** is a substance that increases the rate of a chemical reaction without being permanently altered itself. Thus enzymes speed up cellular reactions. The reacting molecules are called **substrates,** and the substances formed are the **products.** Protein structure and properties (appendix I)

Many enzymes are indeed pure proteins. However, many enzymes consist of a protein, the **apoenzyme,** and also a nonprotein component, a **cofactor,** required for catalytic activity. The complete enzyme consisting of the apoenzyme and its cofactor is called the **holoenzyme.** If the cofactor is firmly attached to the apoenzyme it is a **prosthetic group.** Often the cofactor is loosely attached to the apoenzyme. It can even dissociate from the enzyme protein after products have been formed and carry one of these products to another enzyme (**figure 8.13**). Such a loosely bound cofactor is called a **coenzyme.** For example, NAD$^+$ is a coenzyme that carries electrons within the cell. Many vitamins that humans require serve as coenzymes or as their precursors. Niacin is incorporated into NAD$^+$ and riboflavin into FAD. Metal ions may also be bound to apoenzymes and act as cofactors.

Despite the large number and bewildering diversity of enzymes present in cells, they may be placed in one of six general classes (**table 8.2**). Enzymes usually are named in terms of the substrates they act on and the type of reaction catalyzed. For example, lactate dehydrogenase (LDH) removes hydrogens from lactate.

Table 8.2
Enzyme Classification

Type of Enzyme	Reaction Catalyzed by Enzyme	Example of Reaction
Oxidoreductase	Oxidation-reduction reactions	Lactate dehydrogenase: Pyruvate+NADH+H $\rightleftharpoons$ lactate+NAD$^+$
Transferase	Reactions involving the transfer of groups between molecules	Aspartate carbamoyltransferase: Aspartate+carbamoylphosphate $\rightleftharpoons$ carbamoylaspartate+phosphate
Hydrolase	Hydrolysis of molecules	Glucose-6-phosphatase: Glucose-6-phosphate + H_2O → glucose + P_i
Lyase	Removal of groups to form double bonds or addition of groups to double bonds	Fumarate hydratase: L-malate $\rightleftharpoons$ fumarate+H_2O
Isomerase	Reactions involving isomerizations	Alanine racemase: L-alanine $\rightleftharpoons$ D-alanine
Ligase	Joining of two molecules using ATP energy (or that of other nucleoside triphosphates)	Glutamine synthetase: Glutamate + NH_3 + ATP → glutamine + ATP + P_i

$$\begin{array}{cc} & x \quad y \\ & | \quad | \\ c=c+x-y \rightleftharpoons -c-c- \\ & | \quad | \end{array}$$

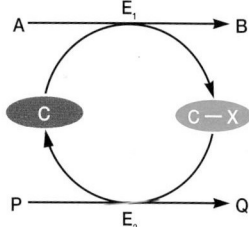

Figure 8.13 Coenzymes as Carriers. The function of a coenzyme in carrying substances around the cell. Coenzyme C participates with enzyme E_1 in the conversion of A to product B. During the reaction, it acquires X from the substrate A. The coenzyme can donate X to substrate P in a second reaction. This will convert it back to its original form, ready to accept another X. The coenzyme is not only participating in both reactions, but is also transporting X to various points in the cell.

$$\text{Lactate} + \text{NAD}^+ \overset{\text{LDH}}{\rightleftharpoons} \text{pyruvate} + \text{NADH} + \text{H}^+$$

Lactate dehydrogenase can also be given a more complete and detailed name, L-lactate:NAD oxidoreductase. This name describes the substrates and reaction type with even more precision.

The Mechanism of Enzyme Reactions

It is important to keep in mind that enzymes increase the rates of reactions but do not alter their equilibrium constants. If a reaction is endergonic, the presence of an enzyme will not shift its equilibrium so that more products can be formed. Enzymes simply speed up the rate at which a reaction proceeds toward its final equilibrium.

How do enzymes catalyze reactions? Although a complete answer would be long and complex, some understanding of the mechanism can be gained by considering the course of a normal exergonic chemical reaction.

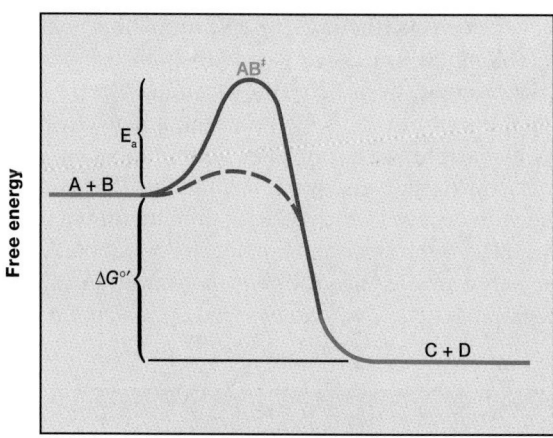

Progress of the reaction

Figure 8.14 Enzymes Lower the Energy of Activation. This figure traces the course of a chemical reaction in which A and B are converted to C and D. The transition-state complex is represented by AB‡, and the activation energy required to reach it, by E_a. The red line represents the course of the reaction in the presence of an enzyme. Note that the activation energy is much lower in the enzyme-catalyzed reaction.

$$\text{A} + \text{B} \rightleftharpoons \text{C} + \text{D}$$

When molecules A and B approach each other to react, they form a **transition-state complex,** which resembles both the substrates and the products (**figure 8.14**). The **activation energy** is required to bring the reacting molecules together in the correct way to reach the transition state. The transition-state complex can then decompose to yield the products C and D. The difference in free energy level between reactants and products is $\Delta G^{o\prime}$. Thus the equilibrium in our example will lie toward the products because $\Delta G^{o\prime}$ is negative (i.e., the products are at a lower energy level than the substrates).

Clearly A and B will not be converted to C and D in figure 8.14 if they are not supplied with an amount of energy equivalent to the activation energy. Enzymes accelerate reactions by lowering the activation energy; therefore more substrate molecules will have sufficient energy to come together and form products. Even though the equilibrium constant (or $\Delta G^{o\prime}$) is unchanged, equilibrium will be reached more rapidly in the presence of an enzyme because of this decrease in the activation energy.

Researchers have expended much effort in discovering how enzymes lower the activation energy of reactions, and the process is becoming clearer. Enzymes bring substrates together at a special place on their surface called the **active site** or **catalytic site** to form an **enzyme-substrate complex (figures 8.15, 8.16;** *see also appendix figure AI.19*). The enzyme can interact with a substrate in two general ways. It may be rigid and shaped to precisely fit the substrate so that the correct substrate binds specifically and is positioned properly for reaction. This mechanism is referred to as the lock-and-key model. An enzyme also may change shape when it binds the substrate so that the active site surrounds and precisely fits the substrate. This has been called the induced fit model and is used by hexokinase and many other enzymes (figure 8.16). The formation of an enzyme-substrate complex can lower the activation energy in many ways. For example, by bringing the substrates together at the active site, the enzyme is, in effect, concentrating them and speeding up the reaction. An enzyme does not simply concentrate its substrates, however. It also binds them so that they are correctly oriented with respect to each other in order to form a transition-state complex. Such an orientation lowers the amount of energy that the substrates require to reach the transition state. These and other catalytic site activities speed up a reaction hundreds

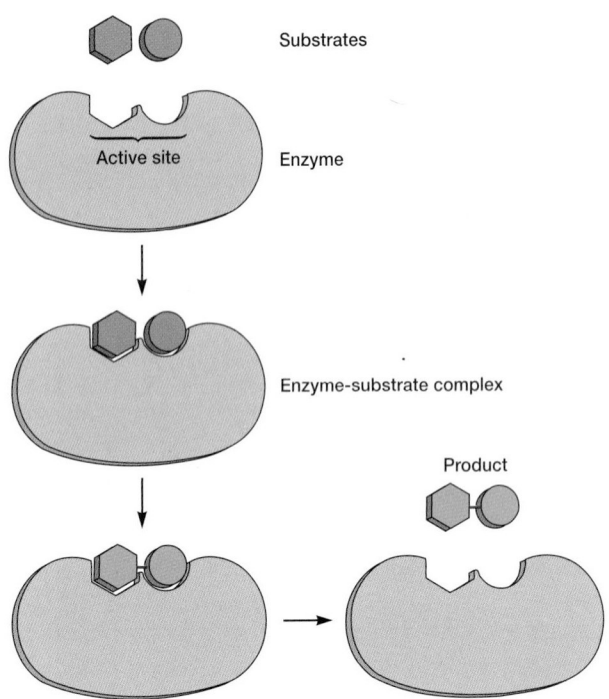

Figure 8.15   Enzyme Function.   The formation of the enzyme-substrate complex and its conversion to a product is shown.

of thousands of times, even though microorganisms grow between $-20°C$ and approximately $113°C$. These temperatures are not high enough to favor most organic reactions in the absence of enzyme catalysis, yet cells cannot survive at the high temperatures used by an organic chemist in routine organic syn-

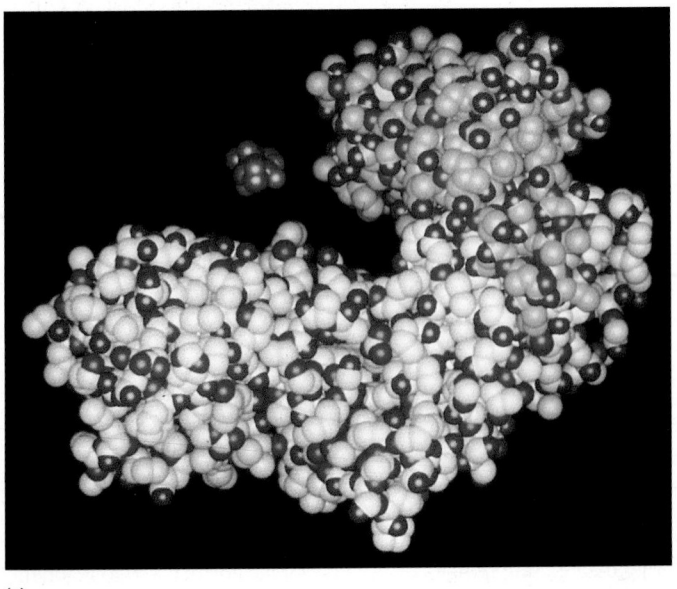

(a)

(b)

Figure 8.16   An Example of Enzyme-Substrate Complex Formation.   (**a**) A space-filling model of yeast hexokinase and its substrate glucose (purple). The active site is in the cleft formed by the enzyme's small lobe (green) and large lobe (gray). (**b**) When glucose binds to form the enzyme-substrate complex, hexokinase changes shape and surrounds the substrate.

theses. Enzymes make life possible by accelerating specific reactions at low temperatures.

The Effect of Environment on Enzyme Activity

Enzyme activity varies greatly with changes in environmental factors, one of the most important being the substrate concentration. As will be emphasized later, substrate concentrations are usually low within cells. At very low substrate concentrations, an enzyme makes product slowly because it seldom contacts a substrate molecule. If more substrate molecules are present, an enzyme binds substrate more often, and the reaction velocity (usually expressed in terms of the rate of product formation) is greater than at a lower substrate concentration. Thus the rate of an enzyme-catalyzed reaction increases with substrate concentration (**figure 8.17**). Eventually further increases in substrate concentration do not result in a greater reaction velocity because the available enzyme molecules are binding substrate and converting it to product as rapidly as possible. That is, the enzyme is saturated with substrate and operating at maximal velocity (V_{max}). The resulting substrate concentration curve is a hyperbola (figure 8.17). It is useful to know the substrate concentration an enzyme needs to function adequately. Usually the **Michaelis constant (K_m),** the substrate concentration required for the enzyme to achieve half maximal velocity, is used as a measure of the apparent affinity of an enzyme for its substrate. The lower the K_m value, the lower the substrate concentration at which an enzyme catalyzes its reaction.

Enzymes also change activity with alterations in pH and temperature (**figure 8.18**). Each enzyme functions most rapidly at a specific pH optimum. When the pH deviates too greatly from an enzyme's optimum, activity slows and the enzyme may be damaged. Enzymes likewise have temperature optima for maximum activity. If the temperature rises too much above the optimum, an enzyme's structure will be disrupted and its activity lost. This phenomenon, known as **denaturation,** may be caused by extremes of pH and temperature or by other factors. The pH and temperature optima of a microorganism's enzymes often reflect the pH and temperature of its habitat. Not surprisingly bacteria growing best at high temperatures often have enzymes with high temperature optima and great heat stability. Temperature and growth (pp. 122–25); Heat stable enzymes in biotechnology (p. 607)

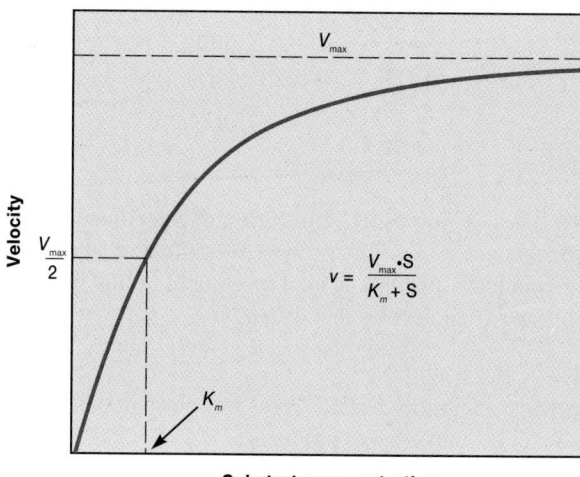

K_m = the substrate concentration required by the enzyme to operate at half its maximum velocity

V_{max} = the rate of product formation when the enzyme is saturated with substrate and operating as fast as possible

Figure 8.17   Michaelis-Menten Kinetics.   The dependence of enzyme activity upon substrate concentration. This substrate curve fits the Michaelis-Menten equation given in the figure, which relates reaction velocity (v) to the substrate concentration (S) using the maximum velocity and the Michaelis constant (K_m).

Enzyme Inhibition

Microorganisms can be poisoned by a variety of chemicals, and many of the most potent poisons are enzyme inhibitors. A **competitive inhibitor** directly competes with the substrate at an enzyme's catalytic site and prevents the enzyme from forming product. A classic example of this behavior is seen with the enzyme succinate dehydrogenase, which catalyzes the oxidation of succinate to fumarate in the tricarboxylic acid cycle (*see section 9.4*). Malonic acid is an effective competitive inhibitor of succinate dehydrogenase because it so closely resembles succinate, the normal substrate (**figure 8.19**). After malonate binds to the enzyme, it cannot be oxidized and the formation of fumarate is blocked.

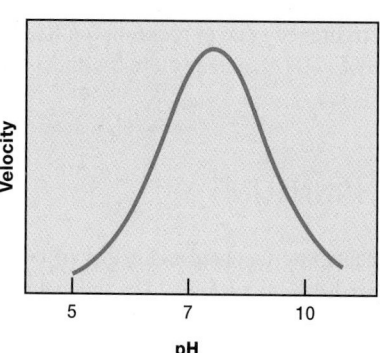

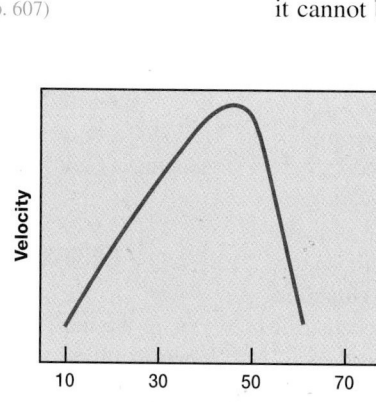

Figure 8.18   pH, Temperature, and Enzyme Activity.   The variation of enzyme activity with changes in pH and temperature. The ranges in pH and temperature are only representative. Enzymes differ from one another with respect to the location of their optima and the shape of their pH and temperature curves.

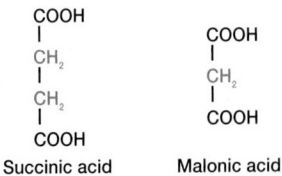

Figure 8.19 **Competitive Inhibition of Succinate Dehydrogenase.** A comparison of succinic acid and the competitive inhibitor, malonic acid. The colored atoms indicate the parts of the two molecules that differ.

Competitive inhibitors usually resemble normal substrates, but they cannot be converted to products.

Competitive inhibitors are important in the treatment of many microbial diseases. Sulfa drugs like sulfanilamide resemble *p*-aminobenzoate, a molecule used in the formation of the coenzyme folic acid. The drugs compete with *p*-aminobenzoate for the catalytic site of an enzyme involved in folic acid synthesis. This blocks the production of folic acid and inhibits bacterial growth (*see section 35.6*). Humans are not harmed because they cannot synthesize folic acid and must obtain it in their diet. Destruction of microorganisms by physical and chemical agents (chapter 7)

Inhibitors also can affect enzyme activity by binding to the enzyme at some location other than at the active site. This alters the enzyme's shape, rendering it inactive or less active. These inhibitors are often called **noncompetitive** because they do not directly compete with the substrate. Heavy metal poisons like mercury frequently are noncompetitive inhibitors of enzymes.

1. What is an enzyme and how does it speed up reactions? How are enzymes named? Define apoenzyme, holoenzyme, cofactor, coenzyme, prosthetic group, active or catalytic site, and activation energy.
2. How does enzyme activity change with substrate concentration, pH, and temperature? Define the terms Michaelis constant, maximum velocity, and denaturation.
3. What are competitive and noncompetitive inhibitors and how do they inhibit enzymes?

8.7 THE NATURE AND SIGNIFICANCE OF METABOLIC REGULATION

The task of the regulatory machinery is exceptionally complex and difficult. Pathways must be regulated and coordinated so effectively that all cell components are present in precisely the correct amounts. Furthermore, a microbial cell must be able to respond effectively to environmental changes by using those nutrients present at the moment and by switching on new catabolic pathways when different nutrients become available. Because all chemical components of a cell usually are not present in the surroundings, microorganisms also must synthesize unavailable components and alter biosynthetic activity in response to changes in nutrient availability. The chemical composition of a cell's surroundings is constantly changing, and these regulatory processes are dynamic and continuously responding to altered conditions.

Regulation is essential for the cell to conserve microbial energy and material and to maintain metabolic balance. If a particular energy source is unavailable, the enzymes required for its use are not needed and their further synthesis is a waste of carbon, nitrogen, and energy. Similarly it would be extremely wasteful for a microorganism to synthesize the enzymes required to manufacture a certain end product if that end product were already present in adequate amounts. Thus both catabolism and anabolism are regulated in such a way as to maximize efficiency of operation. Catabolism and anabolism (p. 168)

The drive to maintain balance and conserve energy and material is evident in the regulatory responses of a bacterium like *E. coli*. If the bacterium is grown in a very simple medium containing only glucose as a carbon and energy source, it will synthesize the required cell components in balanced amounts. Addition of a biosynthetic end product (the amino acid tryptophan, for example) to the medium will result in the immediate inhibition of the pathway synthesizing that end product; synthesis of the pathway's enzymes also will slow or cease. If *E. coli* is transferred to medium containing only the sugar lactose, it will synthesize the enzymes required for catabolism of this nutrient. In contrast, when *E. coli* grows in a medium possessing both glucose and lactose, glucose (the sugar supporting most rapid growth) is catabolized first. The culture will use lactose only after the glucose supply has been exhausted.

The flow of carbon through a pathway may be regulated in three major ways.

1. The localization of metabolites and enzymes in different parts of a cell, a phenomenon called **metabolic channeling,** influences pathway activity.
2. Critical enzymes often are directly stimulated or inhibited to alter pathway activity rapidly.
3. The number of enzyme molecules also may be controlled. The more catalyst molecules present, the greater the pathway's activity. In bacteria regulation is usually exerted at the level of transcription. Control of mRNA synthesis is slower than direct regulation of enzyme activity but does result in the saving of much energy and raw material because enzymes are not synthesized when not required.

Each of these mechanisms is described in detail. This chapter introduces the first two: metabolic channeling and direct control of enzyme activity. The discussion of gene expression regulation follows a description of DNA, RNA, and protein synthesis in chapters 11 and 12. Regulation of gene expression (pp. 270–79)

8.8 METABOLIC CHANNELING

One of the most common channeling mechanisms is that of **compartmentation,** the differential distribution of enzymes and metabolites among separate cell structures or organelles. Compartmentation is particularly important in eucaryotic microorganisms with their

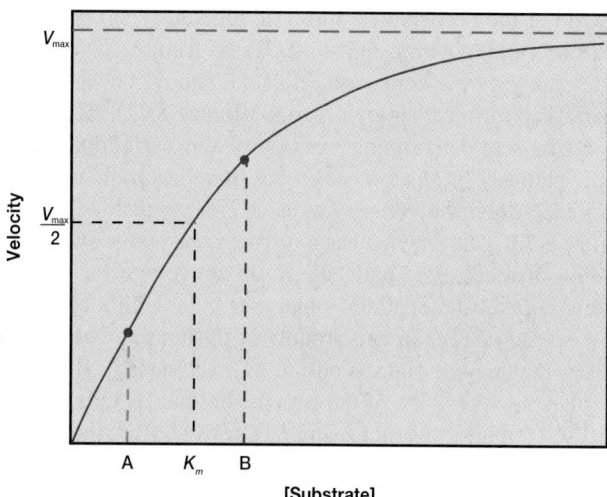

Figure 8.20   Control of Enzyme Activity by Substrate Concentration.   An enzyme-substrate saturation curve with the Michaelis constant (K_m) and the velocity equivalent to half the maximum velocity (V_{max}) indicated. The initial velocity of the reaction (v) is plotted against the substrate concentration [Substrate]. The maximum velocity is the greatest velocity attainable with a fixed amount of enzyme under defined conditions. When the substrate concentration is equal to or less than the K_m, the enzyme's activity will vary almost linearly with the substrate concentration. Suppose the substrate increases in concentration from level A to B. Because these concentrations are in the range of the K_m, a significant increase in enzyme activity results. A drop in concentration from B to A will lower the rate of product formation.

many membrane-bound organelles. For example, fatty acid oxidation is located within the mitochondrion, whereas fatty acid synthesis occurs in the cytoplasmic matrix. The periplasm in procaryotes can also be considered an example of compartmentation (*see p. 53*). Compartmentation makes possible the simultaneous, but separate, operation and regulation of similar pathways. Furthermore, pathway activities can be coordinated through regulation of the transport of metabolites and coenzymes between cell compartments. Suppose two pathways in different cell compartments require NAD for activity. The distribution of NAD between the two compartments will then determine the relative activity of these competing pathways, and the pathway with access to the most NAD will be favored.

Channeling also occurs within compartments such as the cytoplasmic matrix. The matrix is a structured dense material with many subcompartments. In eucaryotes it also is subdivided by the endoplasmic reticulum and cytoskeleton (*see chapter 4*). Metabolites and coenzymes do not diffuse rapidly in such an environment, and metabolite gradients will build up near localized enzymes or enzyme systems. This occurs because enzymes at a specific site convert their substrates to products, resulting in decreases in the concentration of one or more metabolites and increases in others. For example, product concentrations will be high near an enzyme and decrease with increasing distance from it.

Channeling can generate marked variations in metabolite concentrations and therefore directly affect enzyme activity. Substrate levels are generally around 10^{-3} moles/liter (M) to 10^{-6} M or even lower. Thus they may be in the same range as enzyme concentrations and equal to or less than the Michaelis constants (K_m) of many enzymes. Under these conditions the concentration of an enzyme's substrate may control its activity because the substrate concentration is in the rising portion of the hyperbolic substrate saturation curve (**figure 8.20**). As the substrate level increases, it is converted to product more rapidly; a decline in substrate concentration automatically leads to lower enzyme activity. If two enzymes in different pathways use the same metabolite, they may directly compete for it. The pathway winning this competition—the one with the enzyme having the lowest K_m value for the metabolite—will operate closer to full capacity. Thus channeling within a cell compartment can regulate and coordinate metabolism through variations in metabolite and coenzyme levels.   Enzyme kinetics and the substrate saturation curve (p. 159)

---

1. Give three ways in which the flow of carbon through a pathway may be regulated.
2. Define the terms metabolic channeling and compartmentation. How are they involved in the regulation of metabolism?

 CONTROL OF ENZYME ACTIVITY

Adjustment of the activity of regulatory enzymes controls the functioning of many metabolic pathways. This section describes these enzymes and discusses their role in regulating pathway activity.

Allosteric Regulation

Usually regulatory enzymes are **allosteric enzymes.** The activity of an allosteric enzyme is altered by a small molecule known as an **effector** or **modulator.** The effector binds reversibly by noncovalent forces to a **regulatory site** separate from the catalytic site and causes a change in the shape or conformation of the enzyme (**figure 8.21**). The activity of the catalytic site is altered as a result. A positive effector increases enzyme activity, whereas a negative effector decreases activity or inhibits the enzyme. These changes in activity often result from alterations in the apparent affinity of the enzyme for its substrate, but changes in maximum velocity also can occur.

The kinetic characteristics of nonregulatory enzymes show that the Michaelis constant (K_m) is the substrate concentration required for an enzyme to operate at half its maximal velocity. This constant applies only to hyperbolic substrate saturation curves, not to the sigmoidal curves often seen with allosteric enzymes (figure 8.23). The substrate concentration required for half maximal velocity with allosteric enzymes having sigmoidal substrate curves is called the $[S]_{0.5}$ or $K_{0.5}$ value.

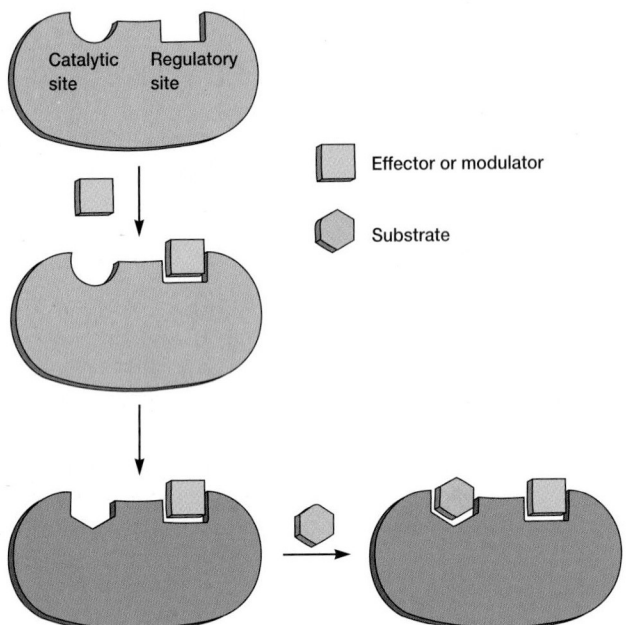

Figure 8.21 Allosteric Regulation. The structure and function of an allosteric enzyme. In this example the effector or modulator first binds to a separate regulatory site and causes a change in enzyme conformation that results in an alteration in the shape of the active site. The active site can now more effectively bind the substrate. This effector is a positive effector because it stimulates substrate binding and catalytic activity.

One of the best-studied allosteric regulatory enzymes is the aspartate carbamoyltransferase (ACTase) from *E. coli*. The enzyme catalyzes the condensation of carbamoyl phosphate with aspartate to form carbamoylaspartate (**figure 8.22**). ACTase catalyzes the rate-determining reaction of the pyrimidine biosynthetic pathway in *E. coli*. The substrate saturation curve is sigmoidal when the concentration of either substrate is varied (**figure 8.23**). The enzyme has more than one active site, and the binding of a substrate molecule to an active site increases the binding of substrate at the other sites. In addition, cytidine triphosphate (CTP), an end product of pyrimidine biosynthesis, inhibits the enzyme and the purine ATP activates it. Both effectors alter the $K_{0.5}$ value of the enzyme but not its maximum velocity. CTP inhibits by increasing $K_{0.5}$ or by shifting the substrate saturation curve to higher values. This allows the enzyme to operate more slowly at a particular substrate concentration when CTP is present. ATP activates by moving the curve to lower substrate concentration values so that the enzyme is maximally active over a wider substrate concentration range. Thus when the pathway is so active that the CTP concentration rises too high, ACTase activity decreases and the rate of end product formation slows. In contrast, when the purine end product ATP increases in concentration relative to CTP, it stimulates CTP synthesis through its effects on ACTase. Pyrimidine and purine biosynthesis (pp. 211–14)

E. coli aspartate carbamoyltransferase provides a clear example of separate regulatory and catalytic sites in allosteric enzymes. The enzyme is a large protein composed of two catalytic subunits and three regulatory subunits (**figure 8.24a**). The catalytic subunits contain only catalytic sites and are unaffected by CTP and ATP. Regulatory subunits do not catalyze the reaction

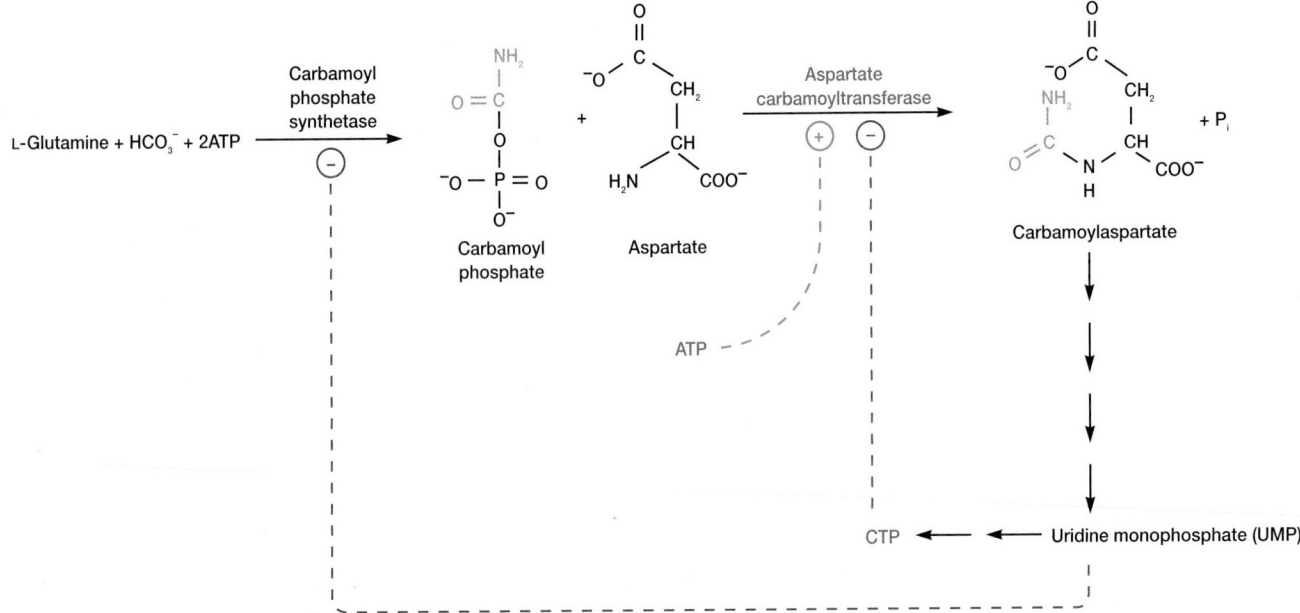

Figure 8.22 ACTase Regulation. The aspartate carbamoyltransferase reaction and its role in the regulation of pyrimidine biosynthesis. The end product CTP inhibits its activity (−) while ATP activates the enzyme (+). Carbamoyl phosphate synthetase is also inhibited by pathway end products such as UMP.

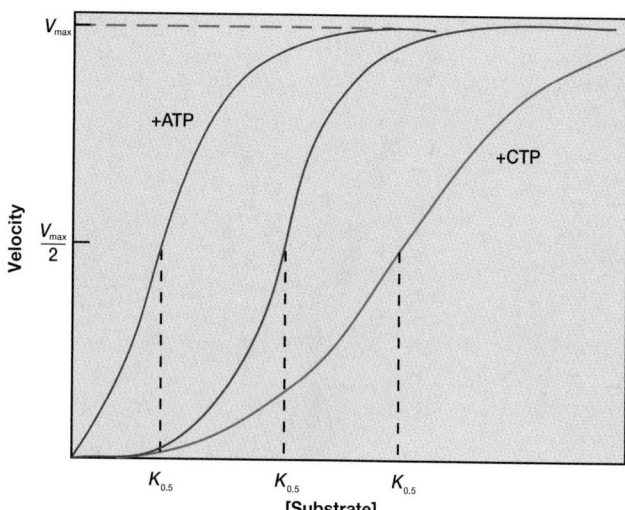

Figure 8.23   The Kinetics of *E. coli* Aspartate Carbamoyltransferase.   CTP, a negative effector, increases the $K_{0.5}$ value while ATP, a positive effector, lowers the $K_{0.5}$. The V_{max} remains constant.

but do possess regulatory sites to which CTP and ATP bind. When these effectors bind to the complete enzyme, they cause conformational changes in the regulatory subunits and subsequently in the catalytic subunits and their catalytic sites. The enzyme can change reversibly between a less active T form and a more active R form (figure 8.24*b,c*). Thus the regulatory site influences a catalytic site about 6.0 nm distant.

Covalent Modification of Enzymes

Regulatory enzymes also can be switched on and off by **reversible covalent modification.** Usually this occurs through the addition and removal of a particular group, one form of the enzyme being more active than the other. That is, the enzyme with an attached group can be either activated or inhibited. For example, glycogen

phosphorylase of the bread mold *Neurospora crassa* exists in phosphorylated and dephosphorylated forms called phosphorylase *a* and phosphorylase *b,* respectively (**figure 8.25**). Phosphorylase *b* is inactive because its required activator AMP is usually not present at sufficiently high levels. Phosphorylase *a,* the phosphorylated form, is active even without AMP. Glycogen phosphorylase is stimulated by phosphorylation of phosphorylase *b* to produce phosphorylase *a.* The attachment of phosphate changes the enzyme's conformation to an active form. Phosphorylation and dephosphorylation are catalyzed by separate enzymes, which also are regulated.   Phosphorylase and glycogen degradation (p. 187)

Enzymes can be regulated through the attachment of groups other than phosphate. One of the most intensively studied regulatory enzymes is *E. coli* glutamine synthetase, a large, complex enzyme existing in two forms (**figure 8.26**). When an adenylic acid residue is attached to each of its 12 subunits forming an adenylylated enzyme, glutamine synthetase is not very active. Removal of AMP groups produces more active deadenylylated glutamine synthetase, and glutamine is formed. The glutamine synthetase system differs from the phosphorylase system in two ways: (1) AMP is used as the modifying agent, and (2) the modified form of glutamine synthetase is less active. Glutamine synthetase also is allosterically regulated.   Glutamine synthetase and its role in nitrogen metabolism (pp. 206–7)

There are some advantages to using covalent modification for the regulation of enzyme activity. These interconvertible enzymes often are also allosteric. Because each form can respond differently to allosteric effectors, systems of covalently modified enzymes are able to respond to more stimuli in varied and sophisticated ways. Regulation can also be exerted on the enzymes that catalyze the covalent modifications, which adds a second level of regulation to the system.

Feedback Inhibition

The rate of many metabolic pathways is adjusted through control of the activity of the regulatory enzymes described in the preceding section. Every pathway has at least one **pacemaker enzyme**

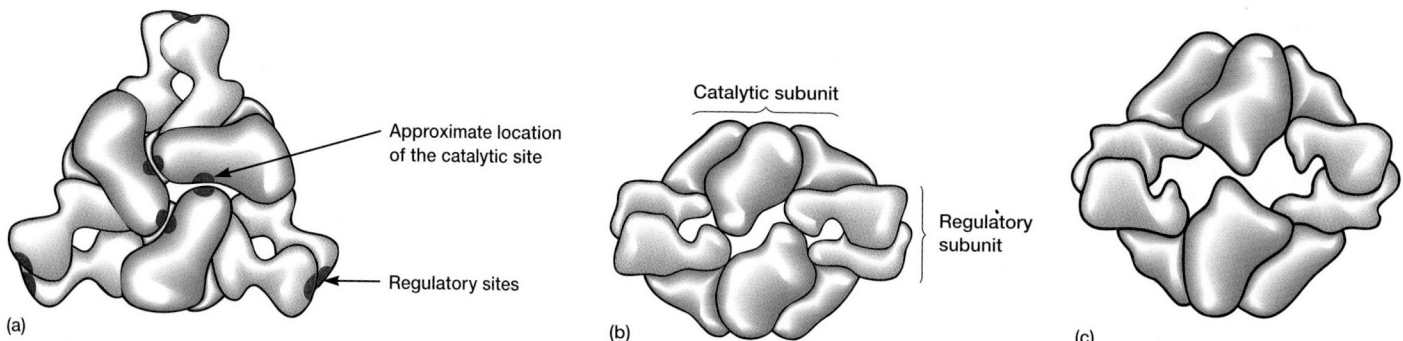

Figure 8.24   The Structure and Regulation of *E. coli* Aspartate Carbamoyltransferase.   (**a**) A schematic diagram of the enzyme showing the six catalytic polypeptide chains (blue), the six regulatory chains (orange), and the catalytic and regulatory sites. The enzyme is viewed from the top. Each catalytic subunit contains three catalytic chains, and each regulatory subunit has two chains. (**b**) The less active T state of ACTase viewed from the side. (**c**) The more active R state of ACTase. The regulatory subunits have rotated and pushed the catalytic subunits apart.

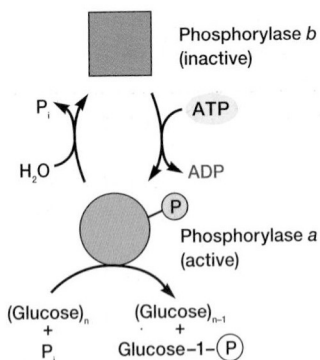

Figure 8.25 Reversible Covalent Modification of Glycogen Phosphorylase.   The active form, phosphorylase *a*, is produced by phosphorylation and is inactivated when the phosphate is removed hydrolytically to produce inactive phosphorylase *b*.

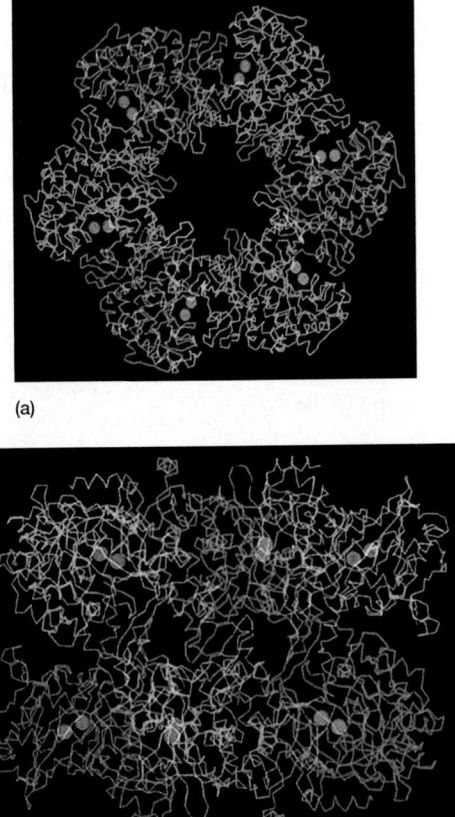

(a)

(b)

Figure 8.26   The Structure of *E. coli* Glutamine Synthetase. The enzyme contains 12 subunits in the shape of a hexagonal prism. For clarity the subunits are colored alternating green and blue. Each of the six catalytic sites has a pair of Mn^{2+} ions (red). The tyrosine residues to which adenyl groups can be attached are colored red. (**a**) Top view of molecule. (**b**) Side view showing the six nearest subunits.

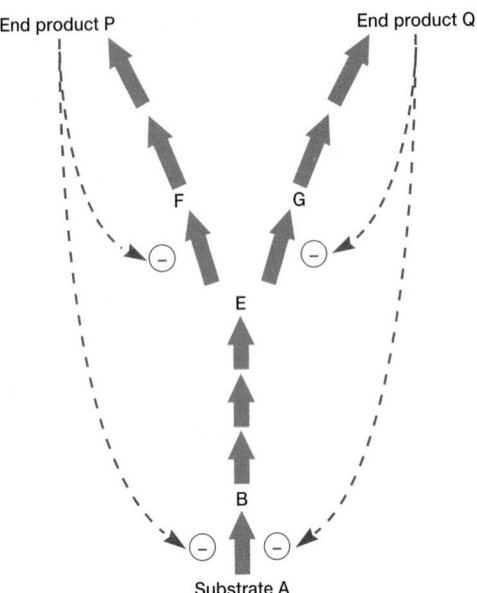

Figure 8.27   Feedback Inhibition.   Feedback inhibition in a branching pathway with two end products. The branch-point enzymes, those catalyzing the conversion of intermediate E to F and G, are regulated by feedback inhibition. Products P and Q also inhibit the initial reaction in the pathway. A colored line with a minus sign at one end indicates that an end product, P or Q, is inhibiting the enzyme catalyzing the step next to the minus. See text for further explanation.

that catalyzes the slowest or rate-limiting reaction in the pathway. Because other reactions proceed more rapidly than the pacemaker reaction, changes in the activity of this enzyme directly alter the speed with which a pathway operates. Usually the first step in a pathway is a pacemaker reaction catalyzed by a regulatory enzyme. The end product of the pathway often inhibits this regulatory enzyme, a process known as **feedback inhibition** or **end product inhibition.** Feedback inhibition ensures balanced production of a pathway end product. If the end product becomes too concentrated, it inhibits the regulatory enzyme and slows its own synthesis. As the end product concentration decreases, pathway activity again increases and more product is formed. In this way feedback inhibition automatically matches end product supply with the demand. The previously discussed *E. coli* aspartate carbamoyltransferase is an excellent example of end product or feedback inhibition.

Frequently a biosynthetic pathway branches to form more than one end product. In such a situation the synthesis of pathway end products must be coordinated precisely. It would not do to have one end product present in excess while another is lacking. Branching biosynthetic pathways usually achieve a balance between end products through the use of regulatory enzymes at branch points (**figure 8.27**). If an end product is present in excess, it often inhibits the branch-point enzyme on the sequence leading to its formation, in this way regulating its own formation without affecting the synthesis of other products. In figure 8.27 notice that both products also inhibit the initial enzyme in the pathway. An excess of one product slows the flow of carbon into the whole pathway while inhibiting the appropriate branch-point enzyme. Because less carbon is required when a branch is not functioning, feedback inhibition of the initial pacemaker enzyme helps match the supply with the demand in branching pathways. The regulation of multiple branched pathways is often made even more sophisticated by the presence of **isoenzymes,** different enzymes that catalyze the same reaction. The initial pacemaker step may be catalyzed by several isoenzymes, each under separate and independent control. In such a situation an excess of a single end product reduces pathway activity but does not completely block pathway function because some isoenzymes are still active.

---

1. Define the following: allosteric enzyme, effector or modulator, and $[S]_{0.5}$ or $K_{0.5}$.
2. How can regulatory enzymes be influenced by reversible covalent modification? What groups are used for this purpose with glycogen phosphorylase and glutamine synthetase, and which forms of these enzymes are active?
3. What is a pacemaker enzyme? Feedback inhibition? How does feedback inhibition automatically adjust the concentration of a pathway end product? What are isoenzymes and why are they important in pathway regulation?

Summary

8.1 Energy and Work

a. Energy is the capacity to do work. Living cells carry out three major kinds of work: chemical work of biosynthesis, transport work, and mechanical work.
b. The ultimate source of energy for most microbes is sunlight trapped by autotrophs and used to form organic material from CO_2. Photoautotrophs are then consumed by chemoheterotrophs.
c. ATP is the major energy currency and connects energy-yielding processes with energy-using processes (**figure 8.3**).

8.2 The Laws of Thermodynamics

a. The first law of thermodynamics states that energy is neither created nor destroyed.
b. The second law of thermodynamics states that changes occur in such a way that the randomness or disorder of the universe increases to the maximum possible. That is, entropy always increases during spontaneous processes.

8.3 Free Energy and Reactions

a. The first and second laws can be combined to determine the amount of energy made available for useful work.

$$\Delta G = \Delta H - T \cdot \Delta S$$

In this equation the change in free energy (ΔG) is the energy made available for useful work, the change in enthalpy (ΔH) is the change in heat content, and the change in entropy is ΔS.
b. The standard free energy change ($\Delta G^{o\prime}$) for a chemical reaction is directly related to the equilibrium constant.
c. In exergonic reactions $\Delta G^{o\prime}$ is negative and the equilibrium constant is greater than one; the reaction goes to completion as written. Endergonic reactions have a positive $\Delta G^{o\prime}$ and an equilibrium constant less than one (**figure 8.5**).

8.4 The Role of ATP in Metabolism

a. ATP is a high-energy molecule that serves as an energy currency; it links energy-trapping and -yielding processes with endergonic processes.

8.5 Oxidation-Reduction Reactions and Electron Carriers

a. In oxidation-reduction (redox) reactions, electrons move from an electron donor to an electron acceptor. The standard reduction potential measures the tendency of the donor to give up electrons.
b. Redox couples with more negative reduction potentials donate electrons to those with more positive potentials, and energy is made available during the transfer (**figure 8.7**).
c. Some most important electron carriers in cells are NAD^+, $NADP^+$, FAD, FMN, coenzyme Q, cytochromes, and the nonheme iron proteins.

8.6 Enzymes

a. Enzymes are protein catalysts that catalyze specific reactions.
b. Enzymes consist of a protein component, the apoenzyme, and often a nonprotein cofactor that may be a prosthetic group, a coenzyme, or a metal activator.
c. Enzymes speed reactions by binding substrates at their active sites and lowering the activation energy (**figure 8.14**).
d. The rate of an enzyme-catalyzed reaction increases with substrate concentration at low substrate levels and reaches a plateau (the maximum velocity) at saturating substrate concentrations. The Michaelis constant is the substrate concentration that the enzyme requires to achieve half maximal velocity (**figure 8.17**).
e. Enzymes have pH and temperature optima for activity.
f. Enzyme activity can be slowed by competitive and noncompetitive inhibitors.

8.7 The Nature and Significance of Metabolic Regulation

a. The regulation of metabolism keeps cell components in proper balance and conserves metabolic energy and material.

8.8 Metabolic Channeling

a. The localization of metabolites and enzymes in different parts of the cell, called metabolic channeling, influences pathway activity. A common channeling mechanism is compartmentation.

8.9 Control of Enzyme Activity

a. Regulatory enzymes are usually allosteric enzymes, enzymes in which an effector or modulator binds reversibly to a regulatory site separate from the catalytic site and causes a conformational change in the enzyme to alter its activity (**figure 8.21**).
b. Aspartate carbamoyltransferase is an allosteric enzyme that is inhibited by CTP and activated by ATP.
c. Enzyme activity also can be regulated by reversible covalent modification. Two examples of such regulation are glycogen phosphorylase (phosphate addition) and glutamine synthetase (AMP addition).
d. The first enzyme in a pathway and enzymes at branch points often are subject to feedback inhibition by one or more end products. Excess end product slows its own synthesis (**figure 8.27**).

Key Terms

activation energy 157
active site 158
adenosine diphosphate (ADP) 151
adenosine 5'-triphosphate (ATP) 151
aerobic respiration 150
allosteric enzymes 161
apoenzyme 156
calorie 151
catalyst 156
catalytic site 158
chemical work 150
coenzyme 156
coenzyme Q or CoQ (ubiquinone) 155
cofactor 156
compartmentation 160
competitive inhibitor 159
cytochrome 155

denaturation 159
effector or modulator 161
electron acceptor 153
electron donor 153
endergonic reaction 152
end product inhibition 164
energy 150
enthalpy 152
entropy 152
enzyme 156
enzyme-substrate complex 158
equilibrium 152
equilibrium constant (K_{eq}) 152
exergonic reaction 152
feedback inhibition 164
ferredoxin 155
first law of thermodynamics 151

flavin adenine dinucleotide
  (FAD) 155
flavin mononucleotide (FMN) 155
free energy change 152
high-energy molecule 153
holoenzyme 156
isoenzymes 165
joule 151
mechanical work 150
metabolic channeling 160
Michaelis constant (K_m) 159
nicotinamide adenine dinucleotide
  (NAD^+) 153
nicotinamide adenine dinucleotide
  phosphate ($NADP^+$) 154
noncompetitive inhibitor 160

nonheme iron protein 155
oxidation-reduction (redox) reaction 153
pacemaker enzyme 163
phosphate group transfer potential 153
photosynthesis 150
product 156
prosthetic group 156
regulatory site 161
reversible covalent modification 163
second law of thermodynamics 152
standard free energy change 152
standard reduction potential 153
substrate 156
thermodynamics 151
transition-state complex 157
transport work 150

Questions for Thought and Review

1. Describe in general terms how energy from sunlight is spread throughout the biosphere.
2. What sources of energy, other than sunlight, are used by microorganisms?
3. Under what conditions would it be possible to create more order in a system without violating the second law of thermodynamics?
4. Do living cells increase randomness or entropy within themselves? In the environment?
5. Suppose that a chemical reaction had a large negative $\Delta G^{o\prime}$ value. What would this indicate about its equilibrium constant? If displaced from equilibrium, would it proceed rapidly to com-

pletion? Would much or little free energy be made available?

6. Will electrons ordinarily move in an electron transport chain from cytochrome c ($E'_0 = +210\,\text{mV}$) to O_2 ($E'_0 = +820\,\text{mV}$) or in the opposite direction?
7. If a person had a niacin deficiency, what metabolic process might well be adversely affected? Why?
8. Draw a diagram showing how enzymes catalyze reactions by altering the activation energy. What is a transition-state complex? Use the diagram to discuss why enzymes do not change the equilibria of the reactions they catalyze.

9. What special properties might an enzyme isolated from a psychrophilic bacterium have? Will enzymes need to lower the activation energy more or less in thermophiles than in psychrophiles?
10. How might a substrate be able to regulate the activity of the enzyme using it?
11. Describe how E. coli aspartate carbamoyltransferase is regulated, both in terms of the effects of modulators and the mechanism by which they exert their influence.
12. What is the significance of the fact that regulatory enzymes often are located at pathway branch points?

Critical Thinking Questions

1. How could electron transport be driven in the opposite direction? Why would it be desirable to do this?

2. Take a look at the structures of macromolecules (appendix I). Which type has the most electrons to donate? Why are carbohydrates usually the primary source of electrons for nonautotrophic bacteria?

3. Most enzymes do not operate at their biochemical optima inside cells. Why not?

For recommended readings on these and related topics, see Appendix V.

chapter 9

Metabolism:
Energy Release and Conservation

Concepts

- Metabolism, the sum total of all chemical reactions occurring in the cell, can be divided into catabolism and anabolism. In catabolism, molecules are reduced in complexity and free energy is made available. Energy also can be derived from sunlight (photosynthesis) and the oxidation of inorganic nutrients. Anabolism involves the use of free energy to increase the complexity of molecules.

- During catabolism, nutrients are funneled into a few common pathways for more efficient use of enzymes (a few pathways process a wide variety of nutrients).

- The tricarboxylic acid cycle is the final pathway for the aerobic oxidation of nutrients to CO_2.

- The majority of energy released in catabolism is generated by the movement of electrons from electron transport carriers with more negative reduction potentials to ones with more positive reduction potentials. Thus aerobic respiration is much more efficient than anaerobic catabolism.

- A wide variety of electron acceptors can be used in catabolism: endogenous organic molecules (fermentation), O_2 (aerobic respiration), and oxidized exogenous inorganic and organic molecules other than O_2 (anaerobic respiration). Furthermore, reduced inorganic molecules as well as organic molecules can serve as electron donors for electron transport and ATP synthesis. Microbial catabolism is unique in the diversity of nutrients and mechanisms employed to make energy available.

- In photosynthesis trapped light energy boosts electrons to more negative reduction potentials or higher energy levels. These energized electrons are then used to make ATP and NADPH or NADH during electron transport.

- Proton motive force is generated by oxidative reactions and photosynthesis; it is used to power the production of ATP and other processes such as transport and bacterial motility.

The reaction center of the purple nonsulfur bacterium, *Rhodopseudomonas viridis*, with the bacteriochlorophylls and other prosthetic groups in yellow. These pigments trap light during photosynthesis.

It is in the fueling reactions that bacteria display their extraordinary metabolic diversity and versatility. Bacteria have evolved to thrive in almost all natural environments, regardless of the nature of available sources of carbon, energy, and reducing power. . . . The collective metabolic capacities of bacteria allow them to metabolize virtually every organic compound on this planet. . . .

—F. C. Neidhardt,
J. L. Ingraham, and M. Schaechter

hapter 8 introduces the basic principles of thermodynamics, the energy cycle and the use of ATP as an energy currency, the nature and function of enzymes, and regulation of enzyme activity. With this background material in hand, microbial metabolism can be discussed. **Metabolism** is the total of all chemical reactions occurring in the cell. The flow of energy and the participation of enzymes make metabolism possible.

This chapter begins with an overview of metabolism. An introduction to carbohydrate degradation and fermentation follows. Next the formation of ATP driven by aerobic and anaerobic respiration is described. Then the breakdown of organic substances other than carbohydrates (i.e., lipids, proteins, and amino acids) is briefly surveyed. The chapter concludes with sections on the oxidation of inorganic molecules and the trapping of light energy by photosynthetic light reactions.

9.1 AN OVERVIEW OF METABOLISM

Metabolism may be divided into two major parts. In **catabolism** [Greek *cata*, down, and *ballein*, to throw] larger and more complex molecules are broken down into smaller, simpler molecules with the release of energy. Some of this energy is trapped and made available for work; the remainder is released as heat. The trapped energy can then be used in anabolism, the second area of metabolism. **Anabolism** [Greek *ana*, up] is the synthesis of complex molecules from simpler ones with the input of energy. An anabolic process uses energy to increase the order of a system (*see chapter 8*).

Although the division of metabolism into two major parts is convenient and commonly employed, not all energy-yielding processes are comfortably encompassed by the previous definition of catabolism unless it is expanded to include processes that do not involve the degradation of complex organic molecules. In a broader sense, microorganisms usually use one of three sources of energy. Phototrophs capture radiant energy from the sun (**figure 9.1**). Chemoorganotrophs oxidize organic molecules to liber-

ate energy, while chemolithotrophs employ inorganic nutrients as energy sources. All three processes will be discussed in this chapter, beginning with chemoorganotrophy. Nutritional types of microorganisms (pp. 95–96)

Microorganisms vary not only in their energy sources, but also in the electron acceptors used by chemotrophs (**figure 9.2**). Three major kinds of acceptors are employed. In **fermentation** [Latin *fermentare*, to cause to rise or ferment] the energy substrate is oxidized and degraded without the participation of an exogenous or externally derived electron acceptor. Usually the catabolic pathway produces an intermediate such as pyruvate that acts as the electron acceptor. Fermentation normally occurs under anaerobic conditions, but also occurs sometimes when oxygen is present. Of course, energy-yielding metabolism can make use of exogenous or externally derived electron acceptors. This metabolic process is called **respiration** and may be divided into two different types. In **aerobic respiration,** the final electron acceptor is oxygen, whereas the acceptor in **anaerobic respiration** is a different exogenous acceptor. Most often the acceptor in anaerobic respiration is inorganic (e.g., NO_3^-, SO_4^{2-}, CO_2, Fe^{3+}, SeO_4^{2-}, and many others), but organic acceptors such as fumarate and humic acids may be used. Most respiration involves the activity of an electron transport chain. The amount of available energy is quite different for fermentation and respiration. The electron acceptor in fermentation is at the same oxidation state as the original nutrient and there is no overall net oxidation of the nutrient. Thus only a limited amount of energy is made available. The acceptor in respiratory processes has reduction potential much more positive that the substrate and thus considerably more energy will be released during respiration (*see pp. 153–55 and figure 8.7*). In both aerobic and anaerobic respiration, ATP is formed as a result of electron transport chain activity. Electrons for the chain can be obtained from inorganic nutrients, and it is possible to derive energy from the oxidation of inorganic molecules rather than from organic nutrients. This ability is confined to a small group of procaryotes called chemolithotrophs as mentioned previously (*see sections 22.1 and 22.2*).

It should be noted that these definitions of fermentation, aerobic respiration, and anaerobic respiration are slightly different from those often used by biologists and biochemists. Fermentation also may be defined as an energy-yielding process in which organic molecules serve as both electron donors and acceptors. Respiration is an energy-yielding process in which the acceptor is an inorganic molecule, either oxygen (aerobic respiration) or another inorganic acceptor (anaerobic respiration). Because microorganisms are so flexible and varied in their energy metabolism, the previous somewhat broader definitions will be used here (*see also table 42.7*).

Before learning about some of the more important catabolic pathways, it is best to look at the "lay of the land" and get our bearings. Albert Lehninger, a biochemist who worked at Johns Hopkins medical school, helped considerably in this task for chemoorganoheterotrophs by pointing out that aerobic metabolism may be divided into three stages (**figure 9.3**). In the first

Figure 9.1   Sources of Energy for Microorganisms. Most microorganisms employ one of three energy sources. Phototrophs trap radiant energy from the sun using pigments such as bacteriochlorophyll and chlorophyll. Chemotrophs oxidize reduced organic and inorganic nutrients to liberate and trap energy. The chemical energy derived from these three sources can then be used in work as discussed in chapter 8.

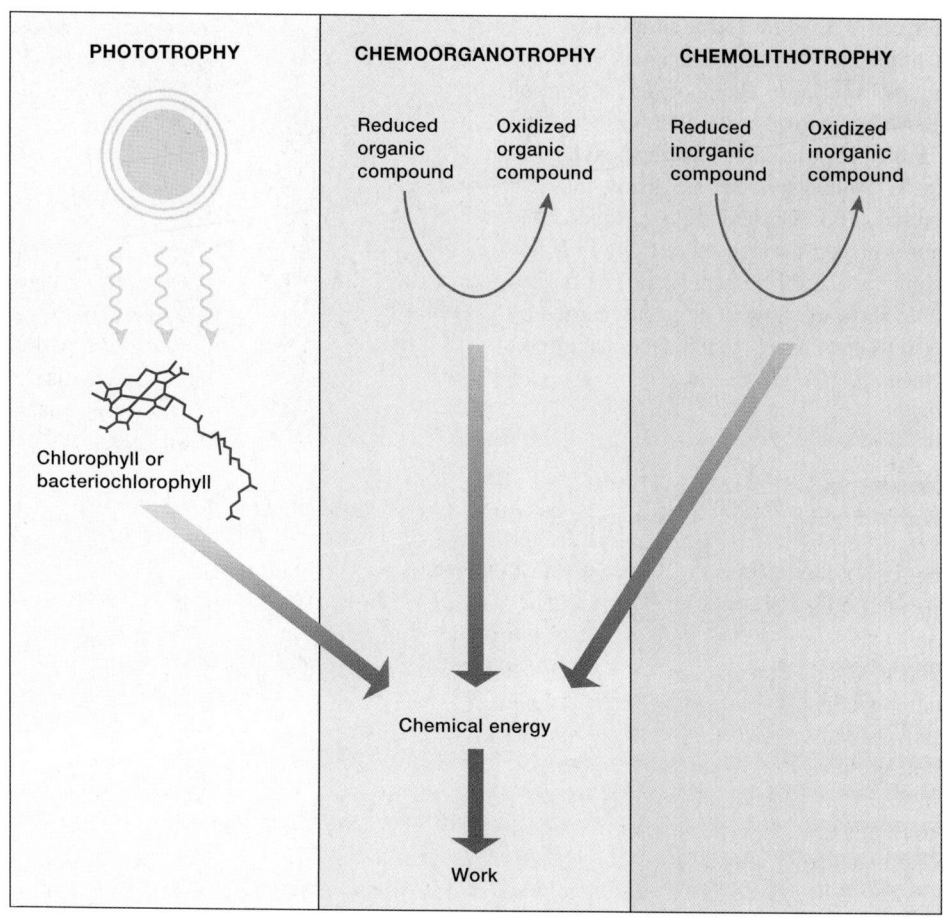

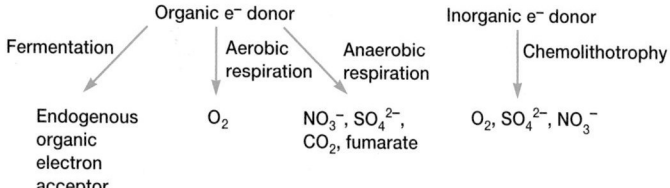

Figure 9.2   Patterns of Energy Release. Fermentation is the energy-yielding process in which an organic electron donor gives electrons to an endogenous acceptor, usually an intermediate derived from catabolism of the nutrient. In respiration, the electrons are donated to an exogenous acceptor, either oxygen (aerobic respiration) or some other acceptor such as nitrate or sulfate (anaerobic respiration). Reduced inorganic compounds also can serve as electron donors in energy production (chemolithotrophy).

stage of catabolism, larger nutrient molecules (proteins, polysaccharides, and lipids) are hydrolyzed or otherwise broken down into their constituent parts. The chemical reactions occurring during this stage do not release much energy. Amino acids, monosaccharides, fatty acids, glycerol, and other products of the first stage are degraded to a few simpler molecules in the second stage. Usually metabolites like acetyl coenzyme A, pyruvate, and

tricarboxylic acid cycle intermediates are formed. The second-stage process can operate either aerobically or anaerobically and often produces some ATP as well as NADH and/or $FADH_2$. Finally, nutrient carbon is fed into the tricarboxylic acid cycle during the third stage of catabolism, and molecules are oxidized completely to CO_2 with the production of ATP, NADH, and $FADH_2$. The cycle operates aerobically and is responsible for the release of much energy. Much of the ATP derived from the tricarboxylic acid cycle (and stage-two reactions) comes from the oxidation of NADH and $FADH_2$ by the electron transport chain. Oxygen, or sometimes another inorganic molecule, is the final electron acceptor.

Although this picture is somewhat oversimplified, it is useful in discerning the general pattern of catabolism. Notice that the microorganism begins with a wide variety of molecules and reduces their number and diversity at each stage. That is, nutrient molecules are funneled into ever fewer metabolic intermediates until they are finally fed into the tricarboxylic acid cycle. A common pathway often degrades many similar molecules (e.g., several different sugars). These metabolic pathways consist of enzyme-catalyzed reactions arranged so that the product of one reaction serves as a substrate for the next. The existence of a few common catabolic pathways, each degrading many nutrients, greatly increases metabolic efficiency by avoiding the need for a

Figure 9.3 The Three Stages of Catabolism. A general diagram of aerobic catabolism in a chemoorganoheterotroph showing the three stages in this process and the central position of the tricarboxylic acid cycle. Although there are many different proteins, polysaccharides, and lipids, they are degraded through the activity of a few common metabolic pathways. The dashed lines show the flow of electrons, carried by NADH and FADH$_2$, to the electron transport chain.

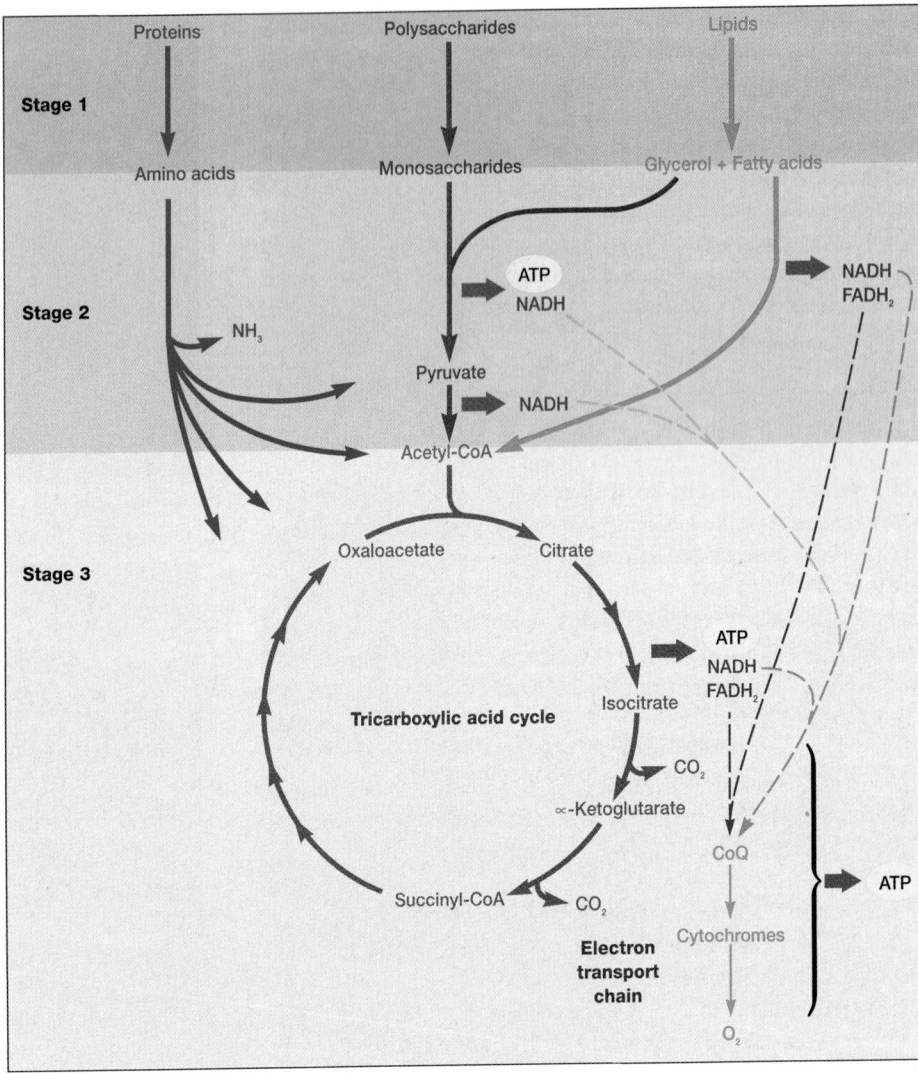

large number of less metabolically flexible pathways. It is in the catabolic phase that microorganisms exhibit their nutritional diversity. Most microbial biosynthetic pathways closely resemble their counterparts in higher organisms. The uniqueness of microbial metabolism lies in the diversity of sources from which it generates ATP and NADH (figures 9.1 and 9.2).

Carbohydrates and other nutrients serve two functions in the metabolism of heterotrophic microorganisms: (1) they are oxidized to release energy, and (2) they supply carbon or building blocks for the synthesis of new cell constituents. Although many anabolic pathways are separate from catabolic routes, there are **amphibolic pathways** [Greek *amphi*, on both sides] that function both catabolically and anabolically. Two of the most important are the glycolytic pathway and the tricarboxylic acid cycle. Most reactions in these two pathways are freely reversible and can be used to synthesize and degrade molecules. The few irreversible catabolic steps are bypassed in biosynthesis with special enzymes that

catalyze the reverse reaction (**figure 9.4**). For example, the enzyme fructose bisphosphatase reverses the phosphofructokinase step when glucose is synthesized from pyruvate (*see pp. 203–4*). The presence of two separate enzymes, one catalyzing the reversal of the other's reaction, permits independent regulation of the catabolic and anabolic functions of these amphibolic pathways.

1. Describe metabolism. How are catabolism and anabolism organized? What three major sources of energy are used by microorganisms? Define phototroph, chemoorganotroph, and chemolithotroph.
2. What kinds of electron acceptors do microorganisms use? Define fermentation, respiration, aerobic respiration, and anaerobic respiration.
3. Why are common pathways useful? What is an amphibolic pathway?

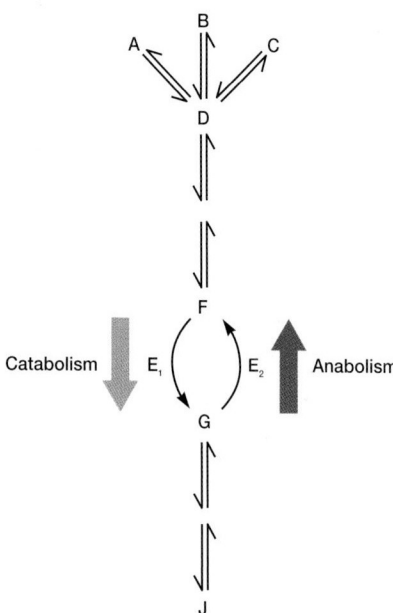

Figure 9.4   Amphibolic Pathway.   A simplified diagram of
an amphibolic pathway such as the glycolytic pathway. Note that
the interconversion of intermediates F and G is catalyzed by two
separate enzymes: E_1 operating in the catabolic direction and E_2
in the anabolic.

THE BREAKDOWN OF GLUCOSE TO PYRUVATE

Microorganisms employ several metabolic pathways to catabolize glucose and other sugars. Because of this metabolic diversity,
their metabolism is often confusing. To avoid confusion as much
as possible, the ways in which microorganisms degrade sugars to
pyruvate and similar intermediates are introduced by focusing on
only three routes: (1) glycolysis, (2) the pentose phosphate pathway, and (3) the Entner-Doudoroff pathway. Next the pathways
of aerobic and anaerobic pyruvate metabolism are described. For
the sake of simplicity, the chemical structures of metabolic intermediates are not used in pathway diagrams.   Diagrams of the glycolytic pathway and other major catabolic pathways with the structures of
intermediates and enzyme names (appendix II)

The Glycolytic Pathway

The **Embden-Meyerhof** or **glycolytic pathway** is undoubtedly
the most common pathway for glucose degradation to pyruvate in
stage two of catabolism. It is found in all major groups of microorganisms and functions in the presence or absence of O_2.
Glycolysis [Greek *glyco*, sweet, and *lysis*, a loosening] is located
in the cytoplasmic matrix of procaryotes and eucaryotes.

   The pathway as a whole may be divided into two parts (**figure 9.5** and *appendix II*). In the initial six-carbon stage, glucose
is phosphorylated twice and eventually converted to fructose 1,6-

bisphosphate. Other sugars are often fed into the pathway by conversion to glucose 6-phosphate or fructose 6-phosphate. This preliminary stage does not yield energy; in fact, two ATP molecules
are expended for each glucose. These initial steps "prime the
pump" by adding phosphates to each end of the sugar. The phosphates will soon be used to make ATP.

   The three-carbon stage of glycolysis begins when the enzyme
fructose 1,6-bisphosphate aldolase catalyzes the cleavage of fructose 1,6-bisphosphate into two halves, each with a phosphate
group. One of the products, glyceraldehyde 3-phosphate, is converted directly to pyruvate in a five-step process. Because the
other product, dihydroxyacetone phosphate, can be easily
changed to glyceraldehyde 3-phosphate, both halves of fructose
1,6-bisphosphate are used in the three-carbon stage. Glyceraldehyde 3-phosphate is first oxidized with NAD^+ as the electron acceptor, and a phosphate is simultaneously incorporated to give a
high-energy molecule called 1,3-bisphosphoglycerate. The high-
energy phosphate on carbon one is subsequently donated to ADP
to produce ATP. This synthesis of ATP is called **substrate-level
phosphorylation** because ADP phosphorylation is coupled with
the exergonic breakdown of a high-energy substrate molecule.
ATP as an energy currency (pp. 151, 153)

   A somewhat similar process generates a second ATP by
substrate-level phosphorylation. The phosphate group on 3-
phosphoglycerate shifts to carbon two, and 2-phosphoglycerate
is dehydrated to form a second high-energy molecule, phosphoenolpyruvate. This molecule donates its phosphate to ADP
forming a second ATP and pyruvate, the final product of the
pathway.

   The glycolytic pathway degrades one glucose to two pyruvates
by the sequence of reactions just outlined. ATP and NADH are also
produced. The yields of ATP and NADH may be calculated by considering the two stages separately. In the six-carbon stage two ATPs
are used to form fructose 1,6-bisphosphate. For each glyceraldehyde 3-phosphate transformed into pyruvate, one NADH and two
ATPs are formed. Because two glyceraldehyde 3-phosphates arise
from a single glucose (one by way of dihydroxyacetone phosphate), the three-carbon stage generates four ATPs and two
NADHs per glucose. Subtraction of the ATP used in the six-carbon
stage from that produced in the three-carbon stage gives a net yield
of two ATPs per glucose. Thus the catabolism of glucose to pyruvate in glycolysis can be represented by this simple equation.

$$\text{Glucose} + 2\text{ADP} + 2\text{P}_i + 2\text{NAD}^+ \longrightarrow$$
$$2 \text{ pyruvate} + 2\text{ATP} + 2\text{NADH} + 2\text{H}^+$$

The Pentose Phosphate Pathway

A second pathway, the **pentose phosphate** or **hexose monophosphate pathway** may be used at the same time as the glycolytic
pathway or the Entner-Doudoroff sequence. It can operate either
aerobically or anaerobically and is important in biosynthesis as
well as in catabolism.

   The pentose phosphate pathway begins with the oxidation
of glucose 6-phosphate to 6-phosphogluconate followed by

Figure 9.5 Glycolysis. The glycolytic pathway for the breakdown of glucose to pyruvate. The two stages of the pathway and their products are indicated.

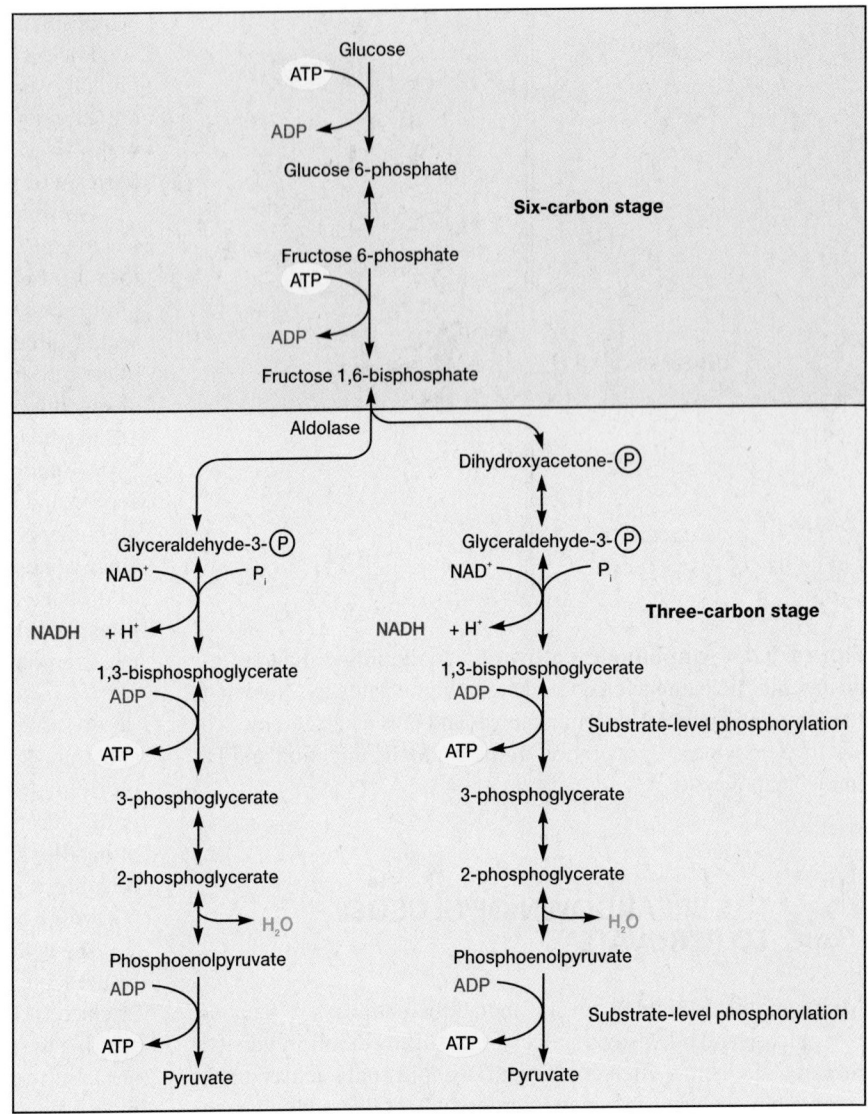

the oxidation of 6-phosphogluconate to the pentose ribulose 5-phosphate and CO_2 (**figure 9.6** and *appendix II*). NADPH is produced during these oxidations. Ribulose 5-phosphate is then converted to a mixture of three- through seven-carbon sugar phosphates. Two enzymes unique to this pathway play a central role in these transformations: (1) transketolase catalyzes the transfer of two-carbon ketol groups, and (2) transaldolase transfers a three-carbon group from sedoheptulose 7-phosphate to glyceraldehyde 3-phosphate (**figure 9.7**). The overall result is that three glucose 6-phosphates are converted to two fructose 6-phosphates, glyceraldehyde 3-phosphate, and three CO_2 molecules, as shown in this equation.

$$3 \text{ glucose 6-phosphate} + 6NADP^+ + 3H_2O \longrightarrow$$
$$2 \text{ fructose 6-phosphate} + \text{glyceraldehyde 3-phosphate} +$$
$$3CO_2 + 6NADPH + 6H^+$$

These intermediates are used in two ways. The fructose 6-phosphate can be changed back to glucose 6-phosphate while glyceraldehyde 3-phosphate is converted to pyruvate by glycolytic enzymes. The glyceraldehyde 3-phosphate also may be returned to the pentose phosphate pathway through glucose 6-phosphate formation. This results in the complete degradation of glucose 6-phosphate to CO_2 and the production of a great deal of NADPH.

$$\text{Glucose 6-phosphate} + 12NADP^+ + 7H_2O \longrightarrow$$
$$6CO_2 + 12NADPH + 12H^+ + P_i$$

The pentose phosphate pathway has several catabolic and anabolic functions that are summarized as follows:

1. NADPH from the pentose phosphate pathway serves as a source of electrons for the reduction of molecules during biosynthesis.

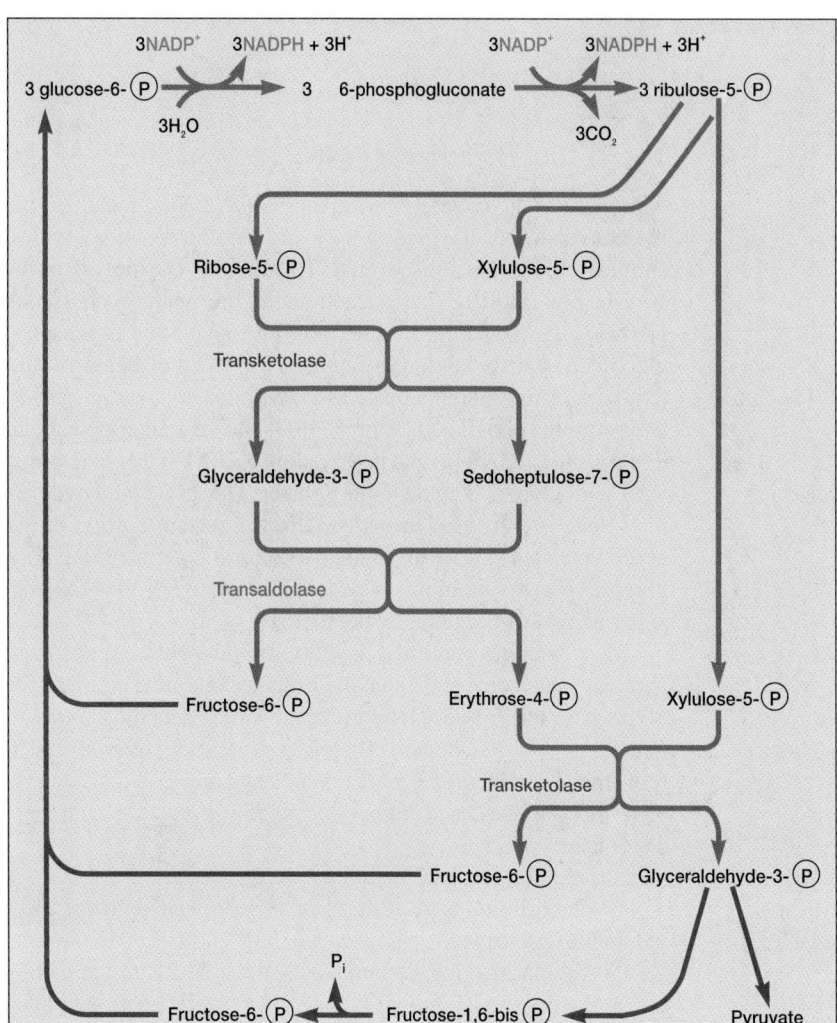

Figure 9.6   The Pentose Phosphate Pathway.
The conversion of three glucose 6-phosphate molecules to two fructose 6-phosphates and a glyceraldehyde 3-phosphate is traced. The fructose 6-phosphates are changed back to glucose 6-phosphate. The glyceraldehyde 3-phosphate can be converted to pyruvate or combined with a molecule of dihydroxyacetone phosphate (from the glyceraldehyde 3-phosphate formed by a second turn of the pathway) to yield fructose 6-phosphate.

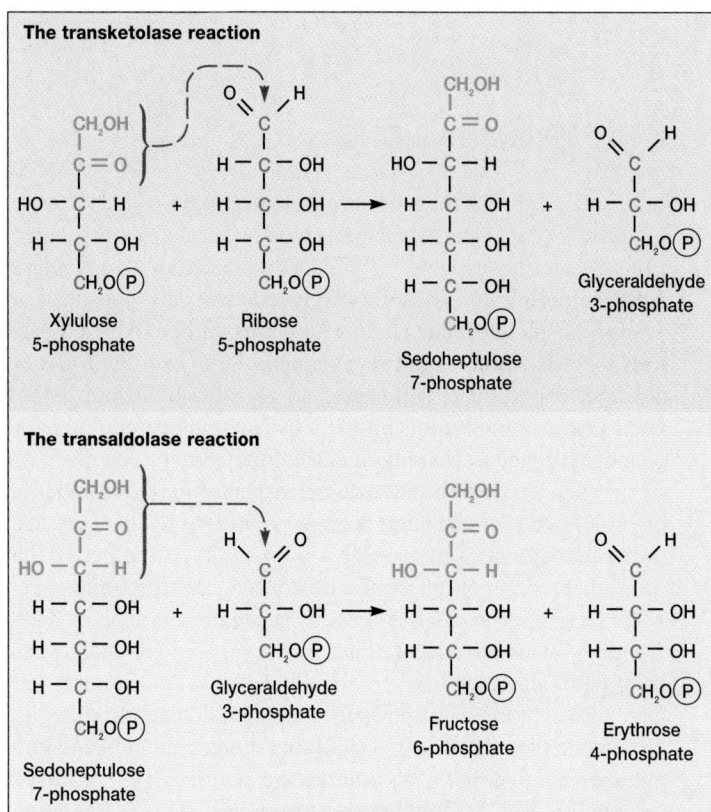

Figure 9.7   Transketolase and Transaldolase.
Examples of the transketolase and transaldolase reactions of the pentose phosphate pathway. The groups transferred in these reactions are in green.

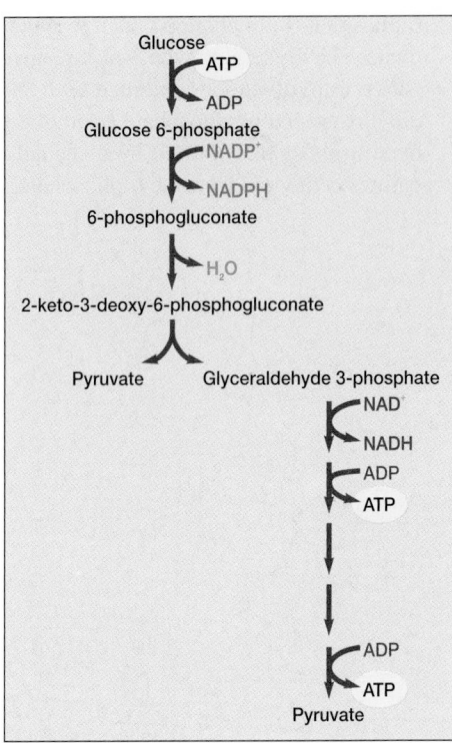

Figure 9.8   The Entner-Doudoroff Pathway.   The sequence leading from glyceraldehyde 3-phosphate to pyruvate is catalyzed by enzymes common to the glycolytic pathway.

2. The pathway synthesizes four- and five-carbon sugars for a variety of purposes. The four-carbon sugar erythrose 4-phosphate is used to synthesize aromatic amino acids and vitamin B_6 (pyridoxal). The pentose ribose 5-phosphate is a major component of nucleic acids, and ribulose 1,5-bisphosphate is the primary CO_2 acceptor in photosynthesis. Note that when a microorganism is growing on a pentose carbon source, the pathway also can supply carbon for hexose production (e.g., glucose is needed for peptidoglycan synthesis).

3. Intermediates in the pentose phosphate pathway may be used to produce ATP. Glyceraldehyde 3-phosphate from the pathway can enter the three-carbon stage of the glycolytic pathway and be converted to ATP and pyruvate. The latter may be oxidized in the tricarboxylic acid cycle to provide more energy. In addition, some NADPH can be converted to NADH, which yields ATP when it is oxidized by the electron transport chain. Because five-carbon sugars are intermediates in the pathway, the pentose phosphate pathway can be used to catabolize pentoses as well as hexoses.

Although the pentose phosphate pathway may be a source of energy in many microorganisms, it is more often of greater importance in biosynthesis. Several functions of the pentose phosphate pathway are mentioned again in chapter 10 when biosynthesis is considered more directly.

The Entner-Doudoroff Pathway

Although the glycolytic pathway is the most common route for the conversion of hexoses to pyruvate, another pathway with a similar role has been discovered. The **Entner-Doudoroff pathway** begins with the same reactions as the pentose phosphate pathway, the formation of glucose 6-phosphate and 6-phosphogluconate (**figure 9.8** and *appendix II*). Instead of being further oxidized, 6-phosphogluconate is dehydrated to form 2-keto-3-deoxy-6-phosphogluconate or KDPG, the key intermediate in this pathway. KDPG is then cleaved by KDPG aldolase to pyruvate and glyceraldehyde 3-phosphate. The glyceraldehyde 3-phosphate is converted to pyruvate in the bottom portion of the glycolytic pathway. If the Entner-Doudoroff pathway degrades glucose to pyruvate in this way, it yields one ATP, one NADPH, and one NADH per glucose metabolized.

Most bacteria have the glycolytic and pentose phosphate pathways, but some substitute the Entner-Doudoroff pathway for glycolysis. The Entner-Doudoroff pathway is generally found in *Pseudomonas, Rhizobium, Azotobacter, Agrobacterium,* and a few other gram-negative genera. Very few gram-positive bacteria have this pathway, with *Enterococcus faecalis* being a rare exception.

---

1. Summarize the major features of the glycolytic pathway, the pentose phosphate pathway, and the Entner-Doudoroff sequence. Include the starting points, the products of the pathways, the critical or unique enzymes, the ATP yields, and the metabolic roles each pathway has.
2. What is substrate-level phosphorylation?

 ## 9.3   FERMENTATIONS

In the absence of aerobic or anaerobic respiration, NADH is not oxidized by the electron transport chain because no external electron acceptor is available. Yet NADH produced in the glycolytic pathway during the oxidation of glyceraldehyde 3-phosphate to 1,3-bisphosphoglycerate (figure 9.5) must still be oxidized back to NAD^+. If NAD^+ is not regenerated, the oxidation of glyceraldehyde 3-phosphate will cease and glycolysis will stop. Many microorganisms solve this problem by slowing or stopping pyruvate dehydrogenase activity and using pyruvate or one of its derivatives as an electron and hydrogen acceptor for the reoxidation of NADH in a fermentation process (**figure 9.9**). This may lead to the production of more ATP. The process is so effective that some chemoorganoheterotrophs do not carry out respiration even when oxygen or another exogenous acceptor is available. There are many kinds of fermentations, and they often are characteristic of particular microbial groups (**figure 9.10**). A few more common fermentations are introduced here, and several others are discussed at later points. Two unifying themes should be kept in mind when microbial fermentations are examined: (1) NADH is oxidized to NAD^+, and (2) the electron acceptor is often either

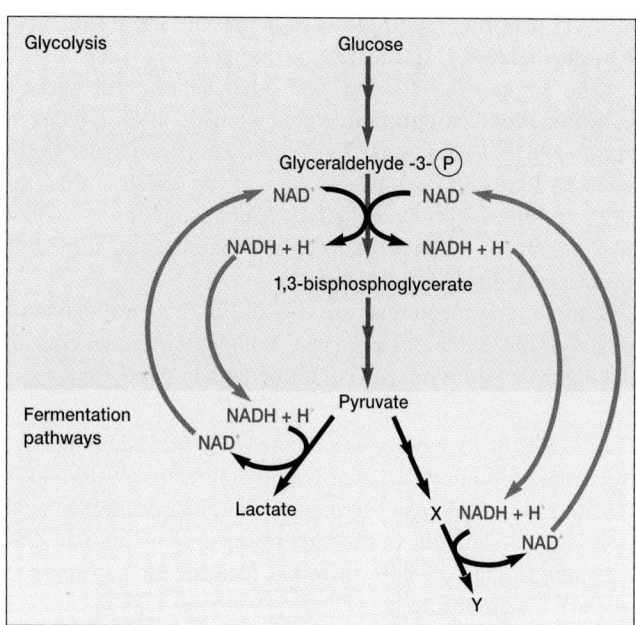

Figure 9.9   Reoxidation of NADH During Fermentation.   NADH from glycolysis is reoxidized by being used to reduce pyruvate or a pyruvate derivative (X). Either lactate or reduced product Y result.

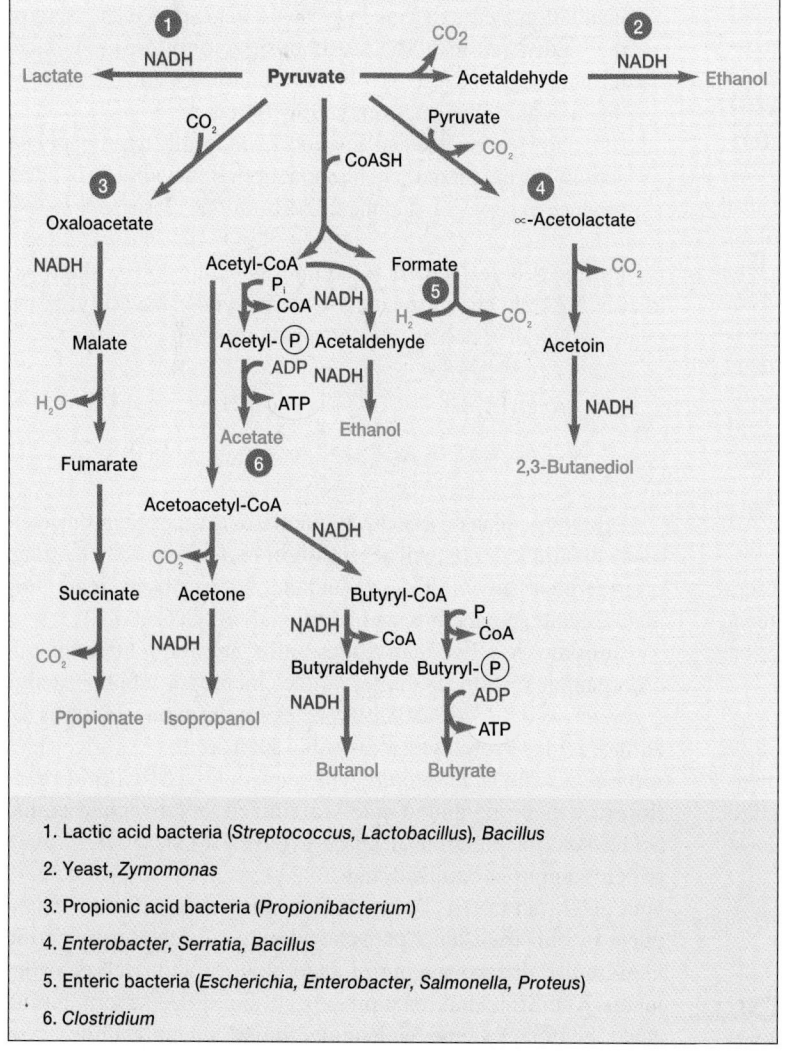

Figure 9.10   Some Common Microbial Fermentations.   Only pyruvate fermentations are shown for the sake of simplicity; many other organic molecules can be fermented. Most of these pathways have been simplified by deletion of one or more steps and intermediates. Pyruvate and major end products are shown in color.

1. Lactic acid bacteria (*Streptococcus, Lactobacillus*), *Bacillus*

2. Yeast, *Zymomonas*

3. Propionic acid bacteria (*Propionibacterium*)

4. *Enterobacter, Serratia, Bacillus*

5. Enteric bacteria (*Escherichia, Enterobacter, Salmonella, Proteus*)

6. *Clostridium*

Table 9.1

Mixed Acid Fermentation Products of
Escherichia coli

	Fermentation Balance (μM Product/100 μM Glucose)	
	Acid Growth (pH 6.0)	Alkaline Growth (pH 8.0)
Ethanol	50	50
Formic acid	2	86
Acetic acid	36	39
Lactic acid	80	70
Succinic acid	11	15
Carbon dioxide	88	2
Hydrogen gas	75	0.5
Butanediol	0	0

pyruvate or a pyruvate derivative. In fermentation the substrate is partially oxidized, ATP is formed by substrate-level phosphorylation only, and oxygen is not needed.

Many fungi and some bacteria, algae, and protozoa ferment sugars to ethanol and CO_2 in a process called **alcoholic fermentation.** Pyruvate is decarboxylated to acetaldehyde, which is then reduced to ethanol by alcohol dehydrogenase with NADH as the electron donor (figure 9.10, number 2). **Lactic acid fermentation,** the reduction of pyruvate to lactate (figure 9.10, number 1), is even more common. It is present in bacteria (lactic acid bacteria, *Bacillus*), algae (*Chlorella*), some water molds, protozoa, and even in animal skeletal muscle. Lactic acid fermenters can be separated into two groups. **Homolactic fermenters** use the glycolytic pathway and directly reduce almost all their pyruvate to lactate with the enzyme lactate dehydrogenase. **Heterolactic fermenters** form substantial amounts of products other than lactate; many produce lactate, ethanol, and CO_2 by way of the phosphoketolase pathway (*see section 23.4*).

The *Enterobacteriaceae* (pp. 492–93)

Alcoholic and lactic acid fermentations are quite useful. Alcoholic fermentation by yeasts produces alcoholic beverages; CO_2 from this fermentation causes bread to rise. Lactic acid fermentation can spoil foods, but also is used to make yogurt, sauerkraut, and pickles. The role of fermentations in food production is discussed in chapter 41.

Many bacteria, especially members of the family *Enterobacteriaceae,* can metabolize pyruvate to formic acid and other products in a process sometimes called the formic acid fermentation (figure 9.10, number 5). Formic acid may be converted to H_2 and CO_2 by formic hydrogenlyase (a combination of at least two enzymes).

$$HCOOH \longrightarrow CO_2 + H_2$$

There are two types of formic acid fermentation. **Mixed acid fermentation** results in the excretion of ethanol and a complex mixture of acids, particularly acetic, lactic, succinic, and formic acids

(table 9.1). If formic hydrogenlyase is present, the formic acid will be degraded to H_2 and CO_2. This pattern is seen in *Escherichia, Salmonella, Proteus,* and other genera. The second type, **butanediol fermentation,** is characteristic of *Enterobacter, Serratia, Erwinia,* and some species of *Bacillus* (figure 9.10, number 4). Pyruvate is converted to acetoin, which is then reduced to 2,3-butanediol with NADH. A large amount of ethanol is also produced, together with smaller amounts of the acids found in a mixed acid fermentation.

Formic acid fermentations are very useful in identification of members of the *Enterobacteriaceae.* Butanediol fermenters can be distinguished from mixed acid fermenters in three ways.

1. The Voges-Proskauer test is a colorimetric procedure that detects the acetoin precursor of butanediol (figure 9.10) and is positive with butanediol fermenters but not with mixed acid fermenters. The Voges-Proskauer test is used by both the Enterotube II and API 20E microbial identification systems to identify enteric bacteria.
2. Mixed acid fermenters produce four times more acidic products than neutral ones, whereas butanediol fermenters form mainly neutral products. Thus mixed acid fermenters acidify incubation media to a much greater extent. This is the basis of the methyl red test. The test is positive only for mixed acid fermentation because the pH drops below 4.4 and the color of the indicator changes from yellow to red.
3. CO_2 and H_2 arise in equal amounts from formic hydrogenlyase activity during mixed acid fermentation. Butanediol fermenters produce excess CO_2 and the CO_2/H_2 ratio is closer to 5:1.

Formic acid fermenters sometimes generate ATP while reoxidizing NADH. They use acetyl-CoA to synthesize acetyl phosphate, which then donates its phosphate to ADP.

$$\text{Acetyl-CoA} + P_i \longrightarrow \text{CoASH} + \text{acetyl-P}$$

$$\text{Acetyl-P} + \text{ADP} \longrightarrow \text{acetate} + \text{ATP}$$

Microorganisms carry out fermentations other than those already mentioned (**Historical Highlights 9.1**). Protozoa and fungi often ferment sugars to lactate, ethanol, glycerol, succinate, formate, acetate, butanediol, and additional products.

Substances other than sugars also are fermented by microorganisms. For example, some members of the genus *Clostridium* (*see pp. 508–9*) prefer to ferment mixtures of amino acids. Proteolytic clostridia such as the pathogens *C. sporogenes* and *C. botulinum* will carry out the **Stickland reaction** in which one amino acid is oxidized and a second amino acid acts as the electron acceptor. **Figure 9.11** shows the way in which alanine is oxidized and glycine reduced to produce acetate, CO_2, and NH_3. Some ATP is formed from acetyl phosphate by substrate-level phosphorylation, and the fermentation is quite useful for growing in anaerobic, protein-rich environments. The Stickland reaction is used to oxidize several amino acids: alanine, leucine, isoleucine, valine, phenylalanine, tryptophan, and histidine. Bacteria also ferment amino acids (e.g.,

Historical Highlights
9.1   Microbiology and World War I

The unique economic pressures of wartime sometimes provide incentive for scientific discovery. Two examples from the First World War involve the production of organic solvents by the microbial fermentation of readily available carbohydrates, such as starch or molasses.

The German side needed glycerol to make nitroglycerin. At one time the Germans had imported their glycerol, but such imports were prevented by the British naval blockade. The German scientist Carl Neuberg knew that trace levels of glycerol were usually produced during the alcoholic fermentation of sugar by *Saccharomyces cerevisiae.* He sought to develop a modified fermentation in which the yeasts would produce glycerol instead of ethanol. Normally acetaldehyde is reduced to ethanol by NADH and alcohol dehydrogenase (figure 9.10, pathway 2). Neuberg found that this reaction could be prevented by the addition of 3.5% sodium sulfite at pH 7.0. The bisulfite ions reacted with acetaldehyde and made it unavailable for reduction to ethanol. Because the yeast cells still had to regenerate their NAD^+ even though acetaldehyde was no longer available, Neuberg suspected that they would simply increase the rate of glycerol synthesis. Glycerol is normally produced by the reduction of dihydroxyacetone phosphate (a glycolytic intermediate) to glycerol phosphate with NADH, followed by the hydrolysis of glycerol phosphate to glycerol. Neuberg's hunch was correct, and German breweries were converted to glycerol manufacture by his procedure, eventually producing 1,000 tons of glycerol per month. Glycerol production by *S. cerevisiae* was not economically competitive under peacetime conditions and was ended. Today glycerol is produced microbially by the halophilic alga *Dunaliella salina,* in which high concentrations of intracellular glycerol accumulate to counterbalance the osmotic pressure from the high level of extracellular salt. *Dunaliella* grows in habitats such as the Great Salt Lake of Utah and seaside rock pools.

The British side needed the organic solvents acetone and butanol. Butanol was required for the production of artificial rubber, whereas acetone was used as a solvent from nitrocellulose in the manufacture of the smokeless explosive powder cordite. Prior to 1914 acetone was made by the dry heating (pyrolysis) of wood. Between 80 and 100 tons of birch, beech, or maple wood were required to make 1 ton of acetone. When war broke out, the demand for acetone quickly exceeded the existing world supply. However, by 1915 Chaim Weizmann, a young Jewish scientist working in Manchester, England, had developed a fermentation process by which the anaerobic bacterium *Clostridium acetobutylicum* converted 100 tons of molasses or grain into 12 tons of acetone and 24 tons of butanol (most clostridial fermentations stop at butyric acid).

$$2 \text{ pyruvate} \longrightarrow \text{acetoacetate} \longrightarrow \text{acetone} + CO_2$$

$$\text{Acetoacetate} \xrightarrow{\text{NADH}} \text{butyrate} \xrightarrow{\text{NADH}} \text{butanol}$$

This time the British and Canadian breweries were converted until new fermentation facilities could be constructed. Weizmann improved the process by finding a convenient way to select high-solvent producing strains of *C. acetobutylicum.* Because the strains most efficient in these fermentations also made the most heat-resistant spores, Weizmann merely isolated the survivors from repeated 100°C heat shocks. Acetone and butanol were made commercially by this fermentation process until it was replaced by much cheaper petrochemicals in the late 1940s and 1950s. In 1948 Chaim Weizmann became the first president of the State of Israel.

---

alanine, glycine, glutamate, threonine, and arginine) by other mechanisms. In addition to sugars and amino acids, organic acids such as acetate, lactate, propionate, and citrate are fermented. Some of these fermentations are of great practical importance. For example, citrate can be converted to diacetyl and give flavor to fermented milk (*see pp. 951–52*).

1. What are fermentations and why are they so useful to many microorganisms? Can ATP be produced during fermentation?
2. Briefly describe alcoholic, lactic acid, and formic acid fermentations. How do mixed acid fermenters and butanediol fermenters differ from each other?
3. What is the Stickland reaction?

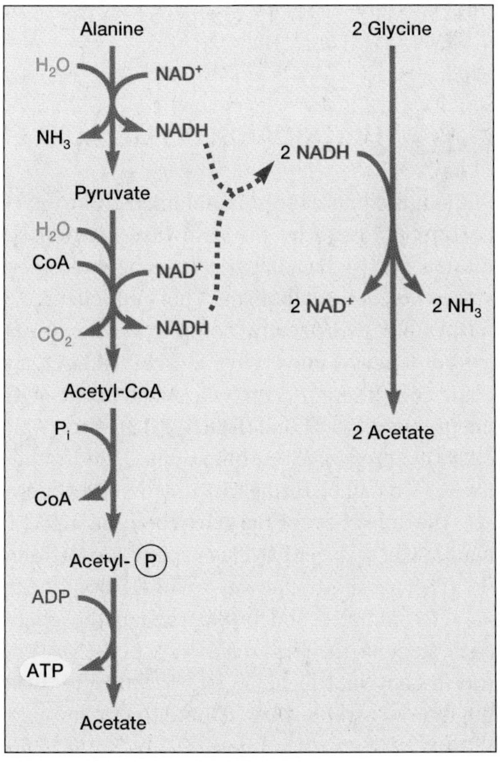

Figure 9.11   The Stickland Reaction.   Alanine is oxidized to acetate and glycine is used to reoxidize the NADH generated during alanine degradation. The fermentation also produces some ATP.

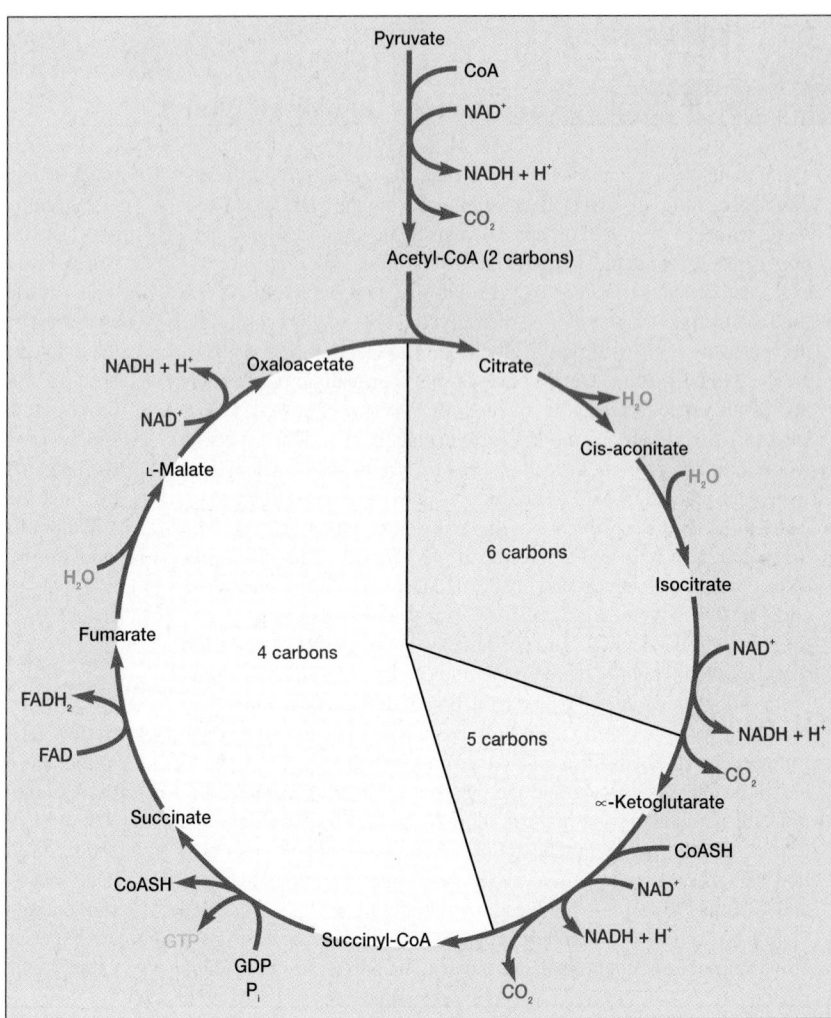

Figure 9.12   The Tricarboxylic Acid Cycle.
The cycle may be divided into three stages based on the size of its intermediates. The three stages are separated from one another by two decarboxylation reactions (reactions in which carboxyl groups are lost as CO_2). The pyruvate dehydrogenase complex forms acetyl-CoA through pyruvate oxidation.

 THE TRICARBOXYLIC ACID CYCLE

Although some energy is obtained from the breakdown of glucose to pyruvate by the pathways previously described, much more is released when pyruvate is degraded aerobically to CO_2 in stage three of catabolism. The multienzyme system called the pyruvate dehydrogenase complex first oxidizes pyruvate to form CO_2 and **acetyl coenzyme A (acetyl-CoA),** an energy-rich molecule composed of coenzyme A and acetic acid joined by a high-energy thiol ester bond (**figure 9.12**). Acetyl-CoA arises from the catabolism of many carbohydrates, lipids, and amino acids (figure 9.3). It can be further degraded in the tricarboxylic acid cycle.

The substrate for the **tricarboxylic acid (TCA) cycle, citric acid cycle,** or **Krebs cycle** is acetyl-CoA (figure 9.12 and *appendix II*). The traditional way to think about the cycle is in terms of its intermediates and products, and the chemistry involved in each step. In the first reaction acetyl-CoA is condensed with a four-carbon intermediate, oxaloacetate, to form citrate and to begin the six-carbon stage. Citrate (a tertiary alcohol) is rearranged

to give isocitrate, a more readily oxidized secondary alcohol. Isocitrate is subsequently oxidized and decarboxylated twice to yield α-ketoglutarate, then succinyl-CoA. At this point two NADHs are formed and two carbons are lost from the cycle as CO_2. Because two carbons were added as acetyl-CoA at the start, balance is maintained and no net carbon is lost. The cycle now enters the four-carbon stage during which two oxidation steps yield one $FADH_2$ and one NADH per acetyl-CoA. In addition, GTP (a high-energy molecule equivalent to ATP) is produced from succinyl-CoA by substrate-level phosphorylation. Eventually oxaloacetate is reformed and ready to join with another acetyl-CoA. Inspection of figure 9.12 shows that the TCA cycle generates two CO_2s, three NADHs, one $FADH_2$, and one GTP for each acetyl-CoA molecule oxidized.

Another way to think of the TCA cycle is in terms of its function as a pathway that oxidizes acetyl-CoA to CO_2. From this perspective, the first step is the attachment of an acetyl group to the acetyl carrier, oxaloacetate, to form citrate. The second stage begins with citrate and ends in the formation of succinyl-CoA.

Here, the acetyl carrier portion of citrate loses two carbons when it is oxidized to give two CO_2s. The third and last stage converts succinyl-CoA back to oxaloacetate, the acetyl carrier, so that it can pick up another acetyl group.

TCA cycle enzymes are widely distributed among microorganisms. The complete cycle appears to be functional in many aerobic bacteria, free-living protozoa, and most algae and fungi. This is not surprising because the cycle is such an important source of energy. However, the facultative anaerobe *E. coli* does not use the full TCA cycle under anaerobic conditions or when the glucose concentration is high but does at other times. Even those microorganisms that lack the complete TCA cycle usually have most of the cycle enzymes, because one of TCA cycle's major functions is to provide carbon skeletons for use in biosynthesis. The role of the tricarboxylic acid cycle in biosynthesis (pp. 209–11)

1. Give the substrate and products of the tricarboxylic acid cycle. Describe its organization in general terms. What are its two major functions?
2. What chemical intermediate links glycolysis to the TCA cycle?
3. In what eucaryotic organelle is the TCA cycle found? Where is the cycle located in procaryotes?

 ## 9.5 ELECTRON TRANSPORT AND OXIDATIVE PHOSPHORYLATION

Little ATP has been synthesized up to this point. Only the equivalent of four ATP molecules is directly synthesized when one glucose is oxidized to six CO_2 molecules by way of glycolysis and the TCA cycle. Most ATP generated comes from the oxidation of NADH and $FADH_2$ in the electron transport chain. The mitochondrial electron transport chain will be examined first because it has been so well studied. Then we will turn to bacterial chains, and finish with a discussion of ATP synthesis.

The Electron Transport Chain

The mitochondrial **electron transport chain** is composed of a series of electron carriers that operate together to transfer electrons from donors, like NADH and $FADH_2$, to acceptors, such as O_2 (refer to diagram of a mitochondrial electron transport chain in **figure 9.13**). The electrons flow from carriers with more negative reduction potentials to those with more positive potentials and eventually combine with O_2 and H^+ to form water. This pattern of electron flow is exactly the same as seen in the electron tower that was described in chapter 8 (*see figure 8.7*). The electrons move down this potential gradient much like water flowing down a series of rapids. The difference in reduction potentials between O_2 and NADH is large, about 1.14 volts, and makes possible the release of a great deal of energy. The potential changes at several points in the chain are large enough to provide sufficient energy for ATP production, much like the energy from waterfalls can be harnessed by waterwheels and used to generate electricity. The

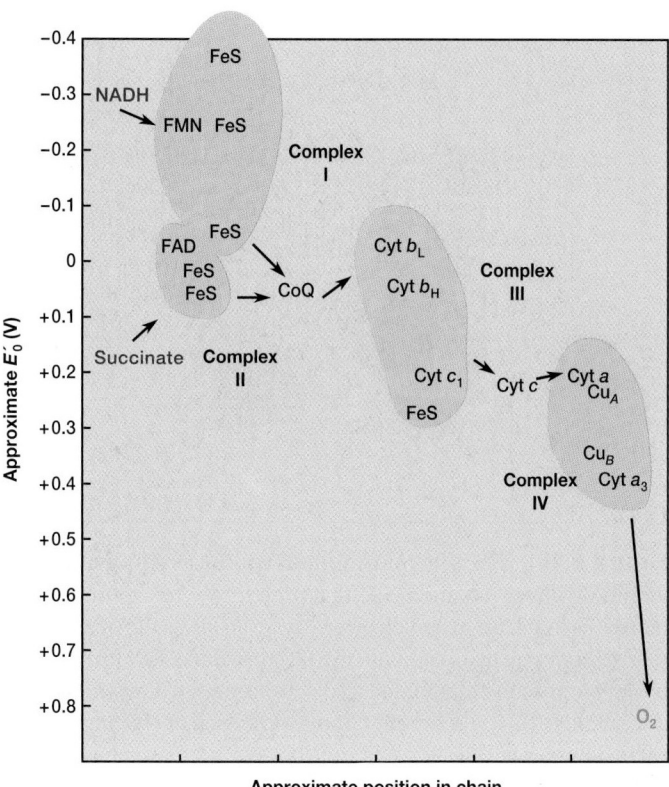

Figure 9.13 The Mitochondrial Electron Transport Chain. Many of the more important carriers are arranged at approximately the correct reduction potential and sequence. In the eucaryotic mitochondrion, they are organized into four complexes that are linked by coenzyme Q (CoQ) and cytochrome *c* (Cyt *c*). Electrons flow from NADH and succinate down the reduction potential gradient to oxygen. See text for details.

electron transport chain breaks up the large overall energy release into small steps. Some of the liberated energy is trapped in the form of ATP. As will be seen shortly, electron transport at these points may generate proton and electrical gradients. These gradients can then drive ATP synthesis.

The electron transport chain carriers reside within the inner membrane of the mitochondrion or in the bacterial plasma membrane. The mitochondrial system is arranged into four complexes of carriers, each capable of transporting electrons part of the way to O_2 (**figure 9.14**). Coenzyme Q and cytochrome *c* connect the complexes with each other.

The process by which energy from electron transport is used to make ATP is called **oxidative phosphorylation.** Thus as many as three ATP molecules may be synthesized from ADP and Pi when a pair of electrons pass from NADH to an atom of O_2. This is the same thing as saying that the phosphorus to oxygen (P/O) ratio is equal to 3. Because electrons from $FADH_2$ only pass two oxidative phosphorylation points, the maximum P/O ratio for $FADH_2$ is 2. The actual P/O ratios may be less than 3.0 and 2.0 in mitochondria.

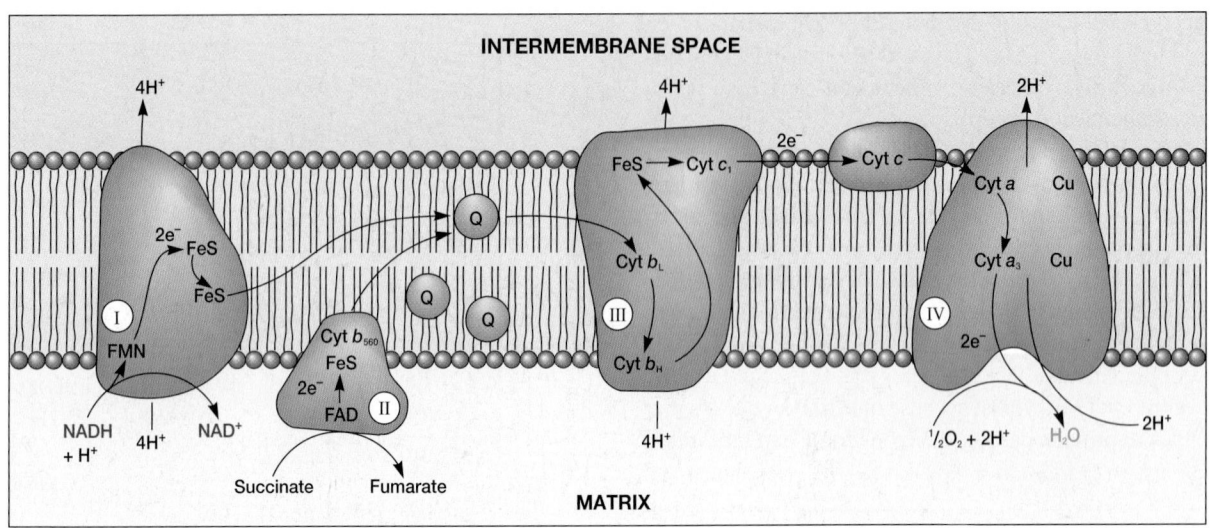

Figure 9.14 The Chemiosmotic Hypothesis Applied to Mitochondria. In this scheme the carriers are organized asymmetrically within the inner membrane so that protons are transported across as electrons move along the chain. Proton release into the intermembrane space occurs when electrons are transferred from carriers, such as FMN and coenzyme Q (Q), that carry both electrons and protons to components like nonheme iron proteins (FeS proteins) and cytochromes (Cyt) that transport only electrons. Complex IV pumps protons across the membrane as electrons pass from cytochrome a to oxygen. Coenzyme Q transports electrons from complexes I and II to complex III. Cytochrome c moves electrons between complexes III and IV. The number of protons moved across the membrane at each site per pair of electrons transported is still somewhat uncertain; the current consensus is that at least 10 protons must move outward during NADH oxidation.

The preceding discussion has focused on the eucaryotic mitochondrial electron transport chain. Although some bacterial chains resemble the mitochondrial chain, they are frequently very different. They vary in their electron carriers (e.g., in their cytochromes) and may be extensively branched. Electrons often can enter at several points and leave through several terminal oxidases. Bacterial chains also may be shorter and have lower P/O ratios than mitochondrial transport chains. Thus procaryotic and eucaryotic electron transport chains differ in details of construction although they operate using the same fundamental principles.

The electron transport chains of *Escherichia coli* and *Paracoccus denitrificans* will serve as examples of these differences. A simplified view of the *E. coli* transport chain is shown in **figure 9.15.** Although it transports electrons from NADH to acceptors and moves protons across the plasma membrane, the *E. coli* chain is quite different from the mitochondrial chain. For example, it is branched and contains a quite different array of cytochromes. Coenzyme Q or ubiquinol donates electrons to both branches, but they operate under different growth conditions. The cytochrome d branch has very high affinity for oxygen and functions at low oxygen levels. It is not as efficient as the cytochrome o branch because it does not actively pump protons. The cytochrome o branch has

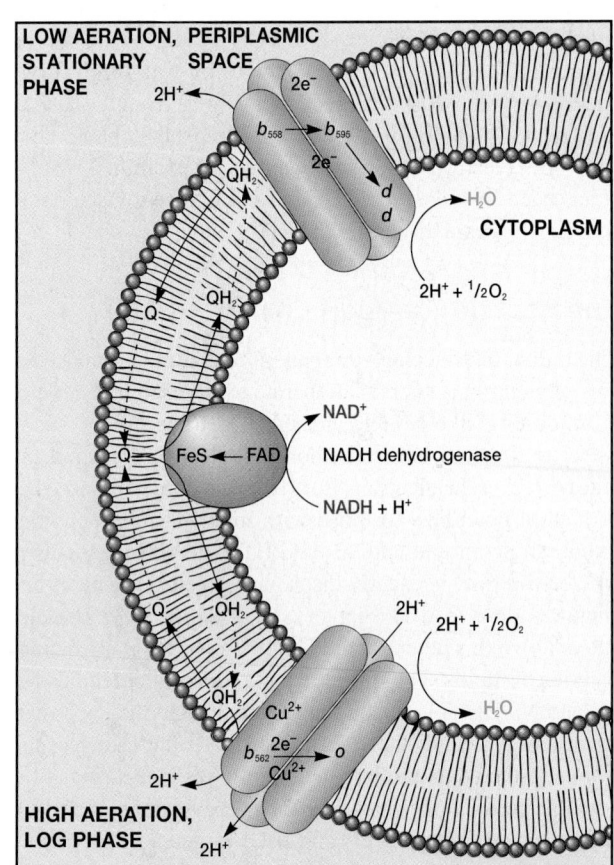

Figure 9.15 The Aerobic Respiratory System of *E. coli*. NADH is the electron source. Ubiquinone-8 (Q) connects the NADH dehydrogenase with two terminal oxidase systems. The upper branch operates when the bacterium is in stationary phase and there is little oxygen. At least five cytochromes are involved: b_{558}, b_{595}, b_{562}, d, and o. The lower branch functions when *E. coli* is growing rapidly with good aeration.

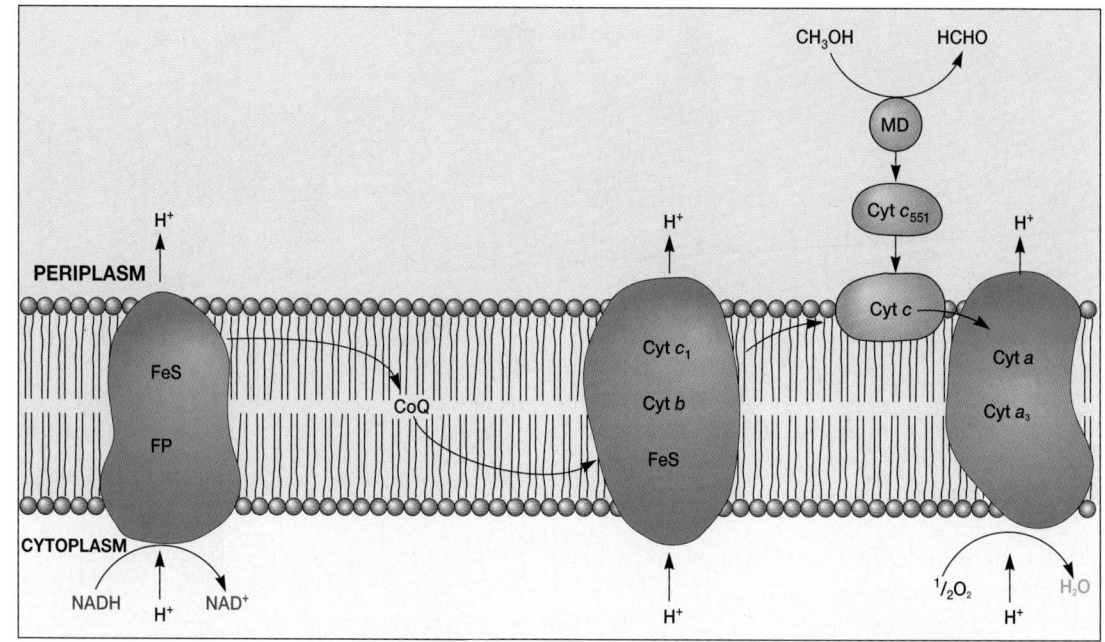

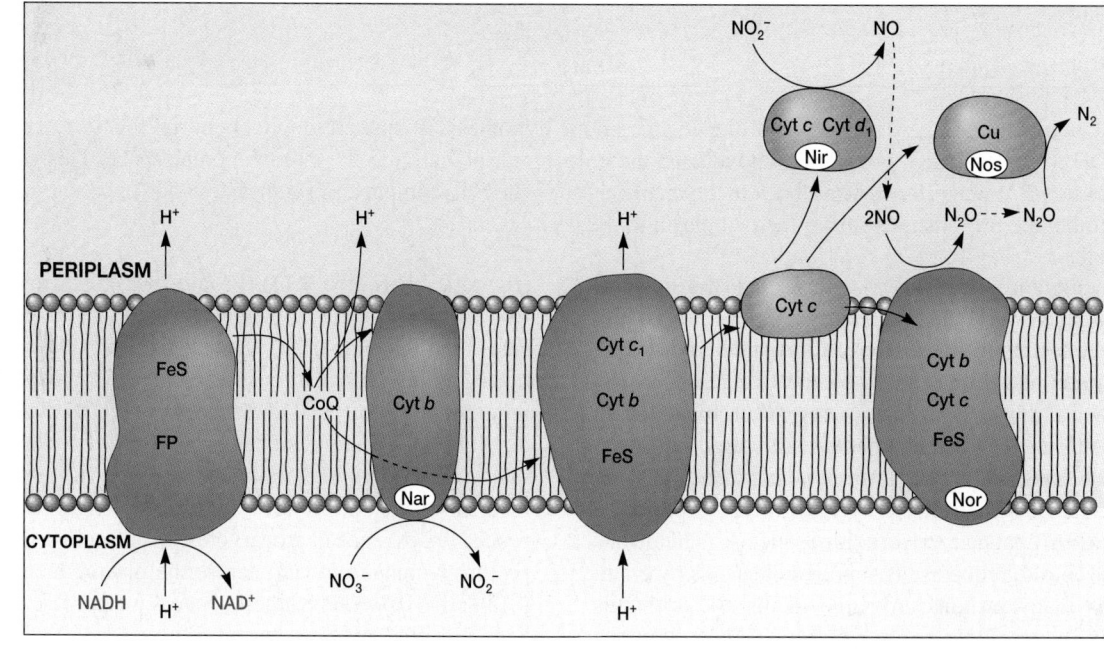

Figure 9.16   *Paracoccus denitrificans* **Electron Transport Chains.**   (**a**) The aerobic transport chain resembles a mitochondrial electron transport chain and uses oxygen as its acceptor. Methanol and methylamine can contribute electrons at the cytochrome *c* level. (**b**) The highly branched anaerobic chain is made of both membrane and periplasmic proteins. Nitrate is reduced to diatomic nitrogen by the collective action of four different reductases that receive electrons from CoQ and cytochrome *c*. Locations of proton movement are shown, but the number of protons involved has not been indicated. Abbreviations used: flavoprotein (FP), methanol dehydrogenase (MD), nitrate reductase (Nar), nitrite reductase (Nir), nitric oxide reductase (Nor), and nitrous oxide reductase (Nos).

moderately high affinity for oxygen, is a proton pump, and operates at higher oxygen concentrations.

Paracoccus denitrificans is a gram-negative, facultative anaerobic soil bacterium that can grow heterotrophically with a variety of nutrients or autotrophically on H_2 and CO_2 with NO_3^-

as the electron acceptor. The bacterium carries out either aerobic respiration or anaerobic respiration with nitrate as an acceptor. The aerobic electron transport chain has four complexes that correspond to the mitochondrial chain (**figure 9.16a**). In addition to donors such as NADH and succinate, *Paracoccus* oxidizes

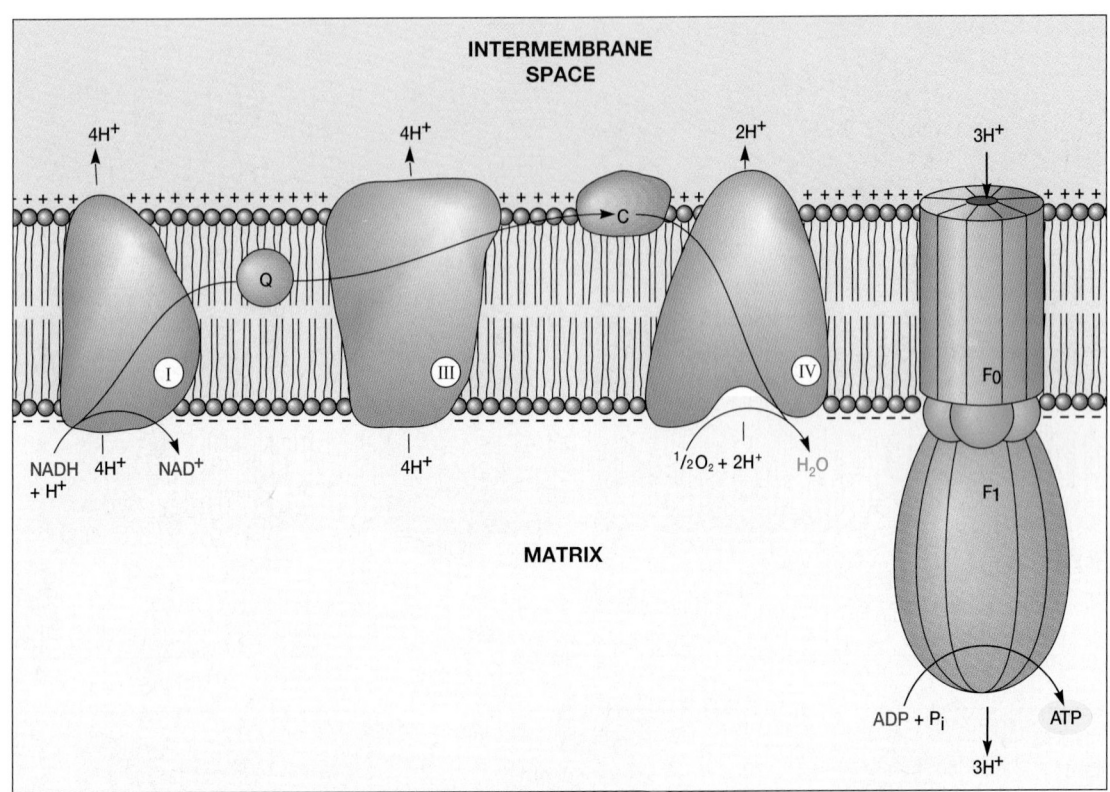

Figure 9.17 Chemiosmosis. An overview of the chemiosmotic hypothesis as applied to mitochondrial function. The flow of electrons from NADH to oxygen causes protons to move from the mitochondrial matrix to the intermembrane space. This generates proton and electrical gradients. When protons move back to the matrix through the F_1F_0 complex, F_1 synthesizes ATP. In procaryotes, the process is similar except that the protons move from the cytoplasm to the periplasm.

methanol and methylamine and grows with them as the sole carbon source. The electrons enter the transport chain at the cytochrome c level. Methanol is oxidized to formaldehyde, which is converted to CO_2 and incorporated by the Calvin cycle (*see pp. 202–3*). When the bacterium is growing anaerobically with nitrate as its electron acceptor, the chain is structured quite differently (figure 9.16b). The cytochrome aa_3 complex does not function. Rather electrons from the cytochrome c level of the chain move to nitrite reductase, nitric oxide reductase, and nitrous oxide reductase. Nitrate reductase is supplied electrons by coenzyme Q. Not as many protons cross the membrane with this arrangement, but it does allow anaerobic growth. We will return to this process in the context of anaerobic respiration and denitrification. Anaerobic respiration (pp. 185–86)

Oxidative Phosphorylation

The mechanism by which oxidative phosphorylation takes place has been studied intensively for years. Currently the most widely accepted hypothesis about how oxidative phosphorylation occurs is the chemiosmotic hypothesis. For the sake of clarity, we will confine our attention to this hypothesis.

According to the **chemiosmotic hypothesis,** first formulated in 1961 by the British biochemist Peter Mitchell, the electron transport chain is organized so that protons move outward from the mitochondrial matrix and electrons are transported inward

(figure 9.14; **figure 9.17**). Proton movement may result either from carrier loops, as shown in figure 9.14, or from the action of special proton pumps that derive their energy from electron transport. The result is **proton motive force (PMF),** composed of a gradient of protons and a membrane potential due to the unequal distribution of charges. When protons return to the mitochondrial matrix driven by the proton motive force, ATP is synthesized in a reversal of the ATP hydrolysis reaction (figure 9.17). A similar process takes place in procaryotes, with electron flow causing the protons to move outward across the plasma membrane (figures 9.15 and 9.16). ATP synthesis occurs when these protons diffuse back into the cell. The proton motive force also may drive the transport of molecules across membranes (*see section 5.6*) and the rotation of bacterial flagella (*see section 3.7*) and thus plays a central role in procaryotic physiology (**figure 9.18**). The chemiosmotic hypothesis is accepted by most microbiologists. There is considerable evidence for the generation of proton and charge gradients across membranes. However, the evidence for proton gradients as the direct driving force for oxidative phosphorylation is not yet conclusive. In some halophilic marine bacteria, sodium ions may be used to drive ATP synthesis.

Whatever the precise mechanism, ATP synthesis takes place at the F_1F_0 ATPase or **ATP synthase (figure 9.19).** The mitochondrial F_1 component appears as a spherical structure attached to the mitochondrial inner membrane surface by a stalk and the F_0 component, which is embedded in the membrane. The F_1F_0

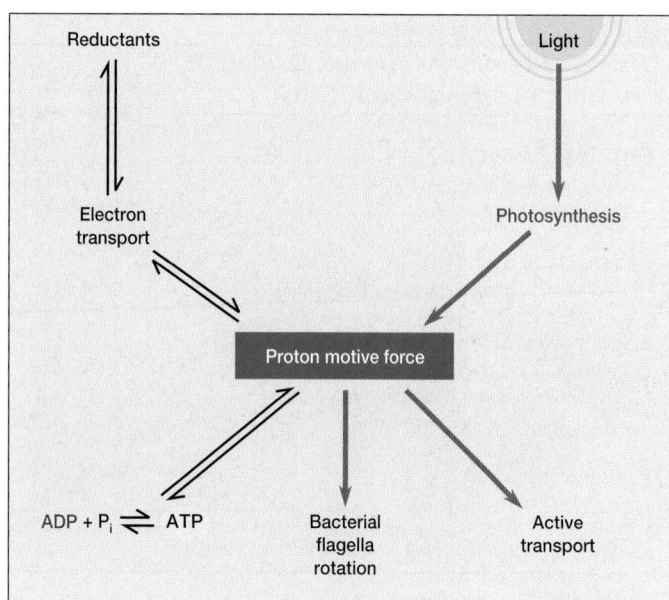

Figure 9.18 The Central Role of Proton Motive Force. It should be noted that active transport is not always driven by PMF.

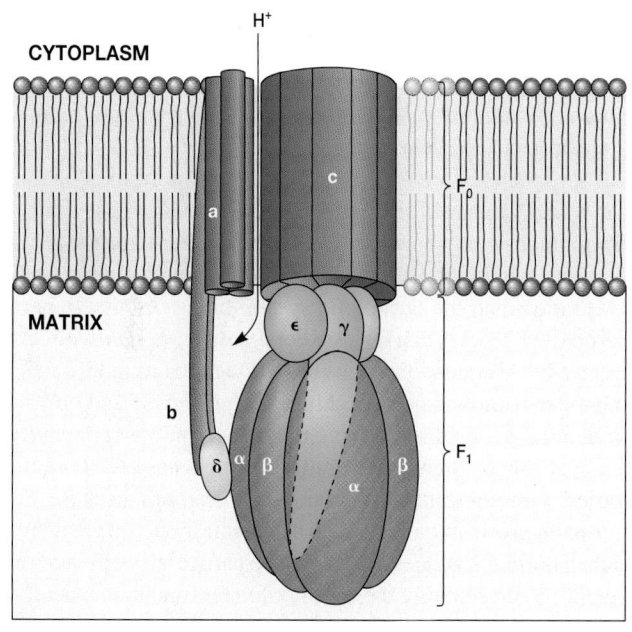

(a)

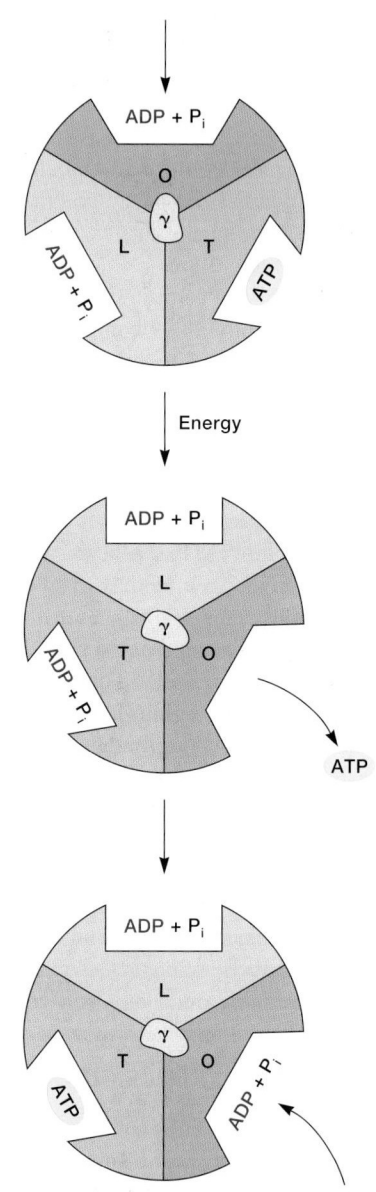

(b)

Figure 9.19 ATP Synthase Structure and Function. (a) The major structural features of ATP synthase deduced from X-ray crystallography and other studies. F_1 is a spherical structure composed largely of alternating α and β subunits; the three active sites are on the β subunits. The γ subunit extends upward through the center of the sphere and can rotate. The stalk (γ and ϵ subunits) connects the sphere to F_0, the membrane embedded complex that serves as a proton channel. F_0 contains one a subunit, two b subunits, and 9–12 c subunits. The stator arm is composed of subunit a, two b subunits, and the δ subunit; it is embedded in the membrane and attached to F_1. A ring of c subunits in F_0 is connected to the stalk and may act as a rotor and move past the a subunit of the stator. As the c subunit ring turns, it rotates the shaft ($\gamma\epsilon$ subunits). (b) The binding change mechanism of ATP synthesis is depicted in a simplified schematic diagram in which the F_1 sphere is viewed from the membrane side. The three active sites appear able to exist in three different conformations: an inactive open (O) conformation with low affinity for substrates, an inactive L conformation with fairly loose affinity for the substrates, and an active tight (T) conformation that has high affinity for substrates. In the first step, ADP and P_i bind to the O site. Then the γ subunit rotates $120°$ with the input of energy, presumably from proton flow through F_0. This rotation causes conformational changes in all three subunits resulting in a release of newly formed ATP and a conversion of the L site to an active T conformation. Finally, ATP is formed at the new T site while more ADP and P_i bind to the unoccupied O site and everything is ready for another energy-driven γ subunit rotation.

ATPase is on the inner surface of the plasma membrane in bacteria. F_0 participates in proton movement across the membrane, and this movement through a channel in F_0 is believed to drive oxidative phosphorylation. F_1 is a large complex in which three α subunits alternate with three β subunits. The γ subunit extends downward from the $\alpha_3\beta_3$ complex; it composes part of the stalk and interacts with F_0. The δ subunit also is located in the stalk. Much of the γ subunit is positioned in the center of F_1, surrounded by the α and β subunits. The γ subunit rotates rapidly in a counterclockwise direction within the $\alpha_3\beta_3$ complex much like a car's crankshaft and causes conformational changes that drive ATP synthesis at the active sites on the β subunits (figure 9.19b). Thus the ATP synthase is the smallest rotary motor known, much smaller than the bacterial flagellum.

Many chemicals inhibit the aerobic synthesis of ATP and can even kill cells at sufficiently high concentrations. These inhibitors generally fall into two categories. Some directly block the transport of electrons. The antibiotic piericidin competes with coenzyme Q; the antibiotic antimycin A blocks electron transport between cytochromes b and c; and both cyanide and azide stop the transfer of electrons between cytochrome a and O_2 because they are structural analogs of O_2. Another group of inhibitors known as **uncouplers** stops ATP synthesis without inhibiting electron transport itself. Indeed, they may even enhance the rate of electron flow. Normally electron transport is tightly coupled with oxidative phosphorylation so that the rate of ATP synthesis controls the rate of electron transport. The more rapidly ATP is synthesized during oxidative phosphorylation, the faster the electron transport chain operates to supply the required energy. Uncouplers disconnect oxidative phosphorylation from electron transport; therefore the energy released by the chain is given off as heat rather than as ATP. Many uncouplers like dinitrophenol and valinomycin may allow hydrogen ions, potassium ions, and other ions to cross the membrane without activating the F_1F_0 ATPase. In this way they destroy the pH and ion gradients. Valinomycin also may bind directly to the F_1F_0 ATPase and inhibit its activity.

The Yield of ATP in Glycolysis and Aerobic Respiration

The maximum ATP yield in eukaryotes from glycolysis, the TCA cycle, and electron transport can be readily calculated. The conversion of glucose to two pyruvate molecules during glycolysis gives a net gain of two ATPs and two NADHs. Because each cytoplasmic NADH can yield a maximum of two to three ATPs during electron transport and oxidative phosphorylation (a P/O ratio of 2 or 3), the total aerobic yield from the glycolytic pathway is six to eight ATP molecules (**table 9.2**).

When O_2 is present and the electron transport chain is operating, pyruvate is next oxidized to acetyl-CoA, the substrate for the TCA cycle. This reaction yields 2 NADHs because 2 pyruvates arise from a glucose; therefore 6 more ATPs are formed. Oxidation of each acetyl-CoA in the TCA cycle will yield 1 GTP (or ATP), 3 NADHs, and a single $FADH_2$ for a total of 2 GTPs

Table 9.2	
ATP Yield from the Aerobic Oxidation of Glucose by Eucaryotic Cells	
Glycolytic Pathway	
Substrate-level phosphorylation (ATP)	2 ATP[a]
Oxidative phosphorylation with 2 NADH	4–6 ATP[b]
2 Pyruvate to 2 Acetyl-CoA	
Oxidative phosphorylation with 2 NADH	6 ATP
Tricarboxylic Acid Cycle	
Substrate-level phosphorylation (GTP)	2 ATP
Oxidative phosphorylation with 6 NADH	18 ATP
Oxidative phosphorylation with 2 $FADH_2$	4 ATP
Total Aerobic Yield	**36–38 ATP**

[a]ATP yields are calculated with an assumed P/O ratio of 3.0 for NADH and 2.0 for $FADH_2$.
[b]The ATP yield depends on how NADH electrons enter the mitochondrion.

(ATPs), 6 NADHs, and 2 $FADH_2$s from two acetyl-CoA molecules. As table 9.2 shows, this amounts to 24 ATPs when NADH and $FADH_2$ from the cycle are oxidized in the electron transport chain. Thus the aerobic oxidation of glucose to 6 CO_2 molecules supplies a maximum of 38 ATPs. The calculations just summarized and presented in table 9.2 are theoretical and based on P/O ratios (the number of ATPs formed per oxygen atom reduced by 2 electrons in electron transport) of 3.0 for NADH oxidation and 2.0 for $FADH_2$. In fact, the P/O ratios are more likely about 2.5 for NADH and 1.5 for $FADH_2$. Thus the total ATP aerobic yield from glucose may be closer to 30 ATPs rather than 38.

Because bacterial electron transport systems often have lower P/O ratios than the eucaryotic system being discussed, bacterial aerobic ATP yields can be less. For example, *E. coli* with its truncated electron transport chains has a P/O ratio around 1.3 when using the cytochrome *bo* path at high oxygen levels and only a ratio of about 0.67 when employing the cytochrome *bd* branch (figure 9.15) at low oxygen concentrations. In this case ATP production varies with environmental conditions. Perhaps because *E. coli* normally grows in habitats such as the intestinal tract that are very rich in nutrients, it does not have to be particularly efficient in ATP synthesis. Presumably the transport chain functions when *E. coli* is in an aerobic freshwater environment between hosts.

Clearly, aerobic respiration is much more effective than anaerobic processes not involving electron transport and oxidative phosphorylation. For example, under anaerobic conditions when NADH is not oxidized by the electron transport chain, only two ATPS will be generated during the degradation of glucose to pyruvate. Many microorganisms, when moved from anaerobic to aerobic conditions, will drastically reduce their rate of sugar catabolism and switch to aerobic respiration, a regulatory phenomenon known as the **Pasteur effect.** This is of obvious advantage to the microorganism as less sugar must be degraded to obtain the same amount of ATP when the more efficient aerobic process can be employed.

Table 9.3			
Some Electron Acceptors Used in Respiration			
	Electron Acceptor	**Reduced Products**	**Examples of Microorganisms**
Aerobic	O_2	H_2O	All aerobic bacteria, fungi, protozoa, and algae
Anaerobic	NO_3^-	NO_2^-	Enteric bacteria
	NO_3^-	NO_2^-, N_2O, N_2	*Pseudomonas, Bacillus,* and *Paracoccus*
	SO_4^{2-}	H_2S	*Desulfovibrio* and *Desulfotomaculum*
	CO_2	CH_4	All methanogens
	S^0	H_2S	*Desulfuromonas* and *Thermoproteus*
	Fe^{3+}	Fe^{2+}	*Pseudomonas, Bacillus,* and *Geobacter*
	$HAsO_4^{2-}$	$HAsO_2$	*Bacillus, Desulfotomaculum, Sulfurospirillum*
	SeO_4^{2-}	$Se, HSeO_3^-$	*Aeromonas, Bacillus, Thauera*
	Fumarate	Succinate	*Wolinella*

---

1. Briefly describe the structure of the electron transport chain and its role in ATP formation. How do mitochondrial and bacterial chains differ?
2. By what mechanism might ATP be synthesized during oxidative phosphorylation? Briefly describe the structure of ATP synthase and how it functions. What is an uncoupler?
3. Calculate the ATP yield in eucaryotes for both glycolysis and the total aerobic oxidation of glucose. Explain your reasoning.

 9.6   ANAEROBIC RESPIRATION

Electrons derived from sugars and other organic molecules are usually donated either to endogenous organic electron acceptors or to molecular O_2 by way of an electron transport chain (figure 9.2). However, many bacteria have electron transport chains that can operate with exogenous electron acceptors other than O_2. As noted earlier, this energy-yielding process is called anaerobic respiration. The major electron acceptors are nitrate, sulfate, and CO_2, but metals and a few organic molecules can also be reduced (**table 9.3**).

Some bacteria can use nitrate as the electron acceptor at the end of their electron transport chain and still produce ATP. Often this process is called **dissimilatory nitrate reduction.** Nitrate may be reduced to nitrite by nitrate reductase, which replaces cytochrome oxidase.

$$NO_3^- + 2e^- + 2H^+ \longrightarrow NO_2^- + H_2O$$

However, reduction of nitrate to nitrite is not a particularly effective way of making ATP, because a large amount of nitrate is re-

quired for growth (a nitrate molecule will accept only two electrons). The nitrite formed is also quite toxic. Therefore nitrate often is further reduced all the way to nitrogen gas, a process known as **denitrification.** Each nitrate will then accept five electrons, and the product will be nontoxic.

$$2NO_3^- + 10e^- + 12H^+ \longrightarrow N_2 + 6H_2O$$

There is considerable evidence that denitrification is a multistep process with four enzymes participating: nitrate reductase, nitrite reductase, nitric oxide reductase, and nitrous oxide reductase.

$$NO_3^- \rightarrow NO_2^- \rightarrow NO \rightarrow N_2O \rightarrow N_2$$

Interestingly, one of the intermediates is nitric oxide (NO). In mammals this molecule acts as a neurotransmitter, helps regulate blood pressure, and is used by macrophages to destroy bacteria and tumor cells (*see p. 697*). Two types of bacterial nitrite reductases catalyze the formation of NO in bacteria. One contains cytochromes c and d_1 (e.g., *Paracoccus* and *Pseudomonas aeruginosa*), and the other is a copper protein (e.g., *Alcaligenes*). Nitrite reductase seems to be periplasmic in gram-negative bacteria. Nitric oxide reductase catalyzes the formation of nitrous oxide from NO and is a membrane-bound cytochrome bc complex. A well-studied example of denitrification is the gram-negative soil bacterium *Paracoccus denitrificans,* which reduces nitrate to N_2 anaerobically. Its chain contains membrane-bound nitrate reductase and nitric oxide reductase, whereas nitrite reductase and nitrous oxide reductase are periplasmic (figure 9.16b). The four enzymes use electrons from coenzyme Q and c-type cytochromes to reduce nitrate and generate PMF. *Paracoccus denitrificans* electron transport (pp. 181–82)

Denitrification is carried out by some members of the genera *Pseudomonas, Paracoccus,* and *Bacillus*. They use this route as an alternative to normal aerobic respiration and may be considered facultative anaerobes. If O_2 is present, these bacteria use aerobic respiration (the synthesis of nitrate reductase is repressed by O_2). Denitrification in anaerobic soil results in the loss of soil nitrogen and adversely affects soil fertility. The nitrogen cycle and denitrification (p. 598)

Two other major groups of bacteria employing anaerobic respiration are obligate anaerobes. Those using CO_2 or carbonate as a terminal electron acceptor are called methanogens because they reduce CO_2 to methane (*see chapter 20*). Sulfate also can act as the final acceptor in bacteria such as *Desulfovibrio*. It is reduced to sulfide (S^{2-} or H_2S), and eight electrons are accepted.

$$SO_4^{2-} + 8e^- + 8H^+ \longrightarrow S^{2-} + 4H_2O$$

Anaerobic respiration is not as efficient in ATP synthesis as aerobic respiration—that is, not as much ATP is produced by oxidative phosphorylation with nitrate, sulfate, or CO_2 as the terminal acceptors. Reduction in ATP yield arises from the fact that these alternate electron acceptors have less positive reduction potentials than O_2 (*see table 8.1*). The reduction potential difference between a donor like NADH and nitrate is smaller than the difference between NADH and O_2. Because energy yield is directly related to the magnitude of the reduction potential difference, less energy is available to make ATP in anaerobic respiration. Nevertheless, anaerobic respiration is useful because it is more efficient than fermentation and allows ATP synthesis by electron transport and oxidative phosphorylation in the absence of O_2. Anaerobic respiration is prevalent in oxygen-depleted soils and sediments.

Often one will see a succession of microorganisms in an environment when several electron acceptors are present. For example, if O_2, nitrate, manganese ion, ferric ion, sulfate, and CO_2 are available in a particular environment, a predictable sequence of electron acceptor use takes place when an oxidizable substrate is available to the microbial population. Oxygen is employed as an electron acceptor first because it inhibits nitrate use by microorganisms capable of respiration with either O_2 or nitrate. While O_2 is available, sulfate reducers and methanogens are inhibited because these groups are obligate anaerobes.

Once the O_2 and nitrate are exhausted and fermentation products, including hydrogen, have accumulated, competition for use of other electron acceptors begins. Manganese and iron will be used first, followed by competition between sulfate reducers and methanogens. This competition is influenced by the greater energy yield obtained with sulfate as an electron acceptor. Differences in enzymatic affinity for hydrogen, an important substrate used by both groups (section 9.10), also are important. The sulfate reducer *Desulfovibrio* grows rapidly and uses the available hydrogen at a faster rate than *Methanobacterium*. When the sulfate is exhausted, *Desulfovibrio* no longer oxidizes hydrogen, and the hydrogen concentration rises. The methanogens finally dominate the habitat and reduce CO_2 to methane.

---

1. Describe the process of anaerobic respiration. Is as much ATP produced in anaerobic respiration as in aerobic respiration? Why or why not?
2. What is denitrification?

 9.7 CATABOLISM OF CARBOHYDRATES AND INTRACELLULAR RESERVE POLYMERS

Microorganisms can catabolize many carbohydrates besides glucose. These carbohydrates may come either from outside the cell or from internal sources. Often the initial steps in the degradation of external carbohydrate polymers differ from those employed with internal reserves.

Carbohydrates

Figure 9.20 outlines some catabolic pathways for the monosaccharides (single sugars) glucose, fructose, mannose, and galactose. The first three are phosphorylated using ATP and easily enter the glycolytic pathway. In contrast, galactose must be converted to uridine diphosphate galactose (*see p. 204*) after initial

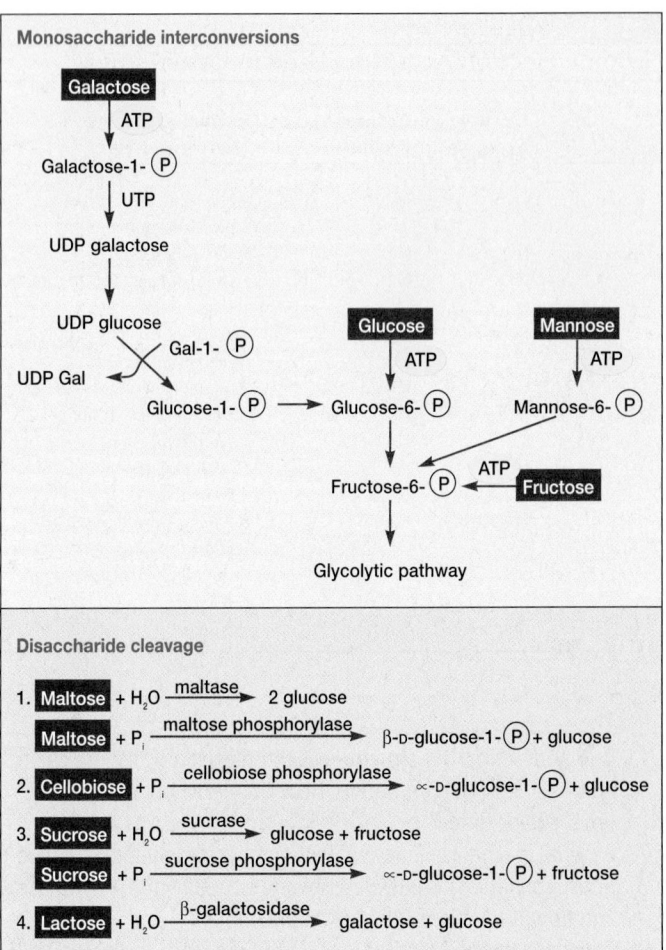

Figure 9.20 Carbohydrate Catabolism. Examples of enzymes and pathways used in disaccharide and monosaccharide catabolism. UDP is an abbreviation for uridine diphosphate.

phosphorylation, then changed into glucose 6-phosphate in a three-step process (figure 9.20).

The common disaccharides are cleaved to monosaccharides by at least two mechanisms (figure 9.20). Maltose, sucrose, and lactose can be directly hydrolyzed to their constituent sugars. Many disaccharides (e.g., maltose, cellobiose, and sucrose) are also split by a phosphate attack on the bond joining the two sugars, a process called phosphorolysis.

Polysaccharides, like disaccharides, are cleaved by both hydrolysis and phosphorolysis. Bacteria and fungi degrade external polysaccharides by secreting hydrolytic enzymes that cleave polysaccharides into smaller molecules, which can then be assimilated. Starch and glycogen are hydrolyzed by amylases to glucose, maltose, and other products. Cellulose is more difficult to digest; many fungi and a few bacteria (some gliding bacteria, clostridia, and actinomycetes) produce cellulases that hydrolyze cellulose to cellobiose and glucose. Some members of the bacterial genus *Cytophaga*, isolated from marine habitats, excrete an agarase that degrades agar. Many soil bacteria and bacterial plant pathogens degrade pectin, a polymer of galacturonic acid (a galactose derivative) that is an important constituent of plant cell walls and tissues.

In the context of compounds that are recalcitrant or difficult to digest, it should be noted that microorganisms also can degrade xenobiotic compounds (foreign substances not formed by natural biosynthetic processes) such as pesticides and various aromatic compounds. They transform these molecules to normal metabolic intermediates by use of special enzymes and pathways, then continue catabolism in the usual way. Biodegradation and bioremediation are discussed in chapter 42; the fungus *Phanerochaete chrysosporium* is an extraordinary example of the ability to degrade xenobiotics (*see Microbial Diversity & Ecology 42.3*).

Reserve Polymers

Microorganisms often survive for long periods in the absence of exogenous nutrients. Under such circumstances they catabolize intracellular stores of glycogen, starch, poly-β-hydroxybutyrate, and other energy reserves. Glycogen and starch are degraded by phosphorylases. Phosphorylases catalyze a phosphorolysis reaction that shortens the polysaccharide chain by one glucose and yields glucose 1-phosphate.

$$(\text{Glucose})_n + P_i \longrightarrow (\text{glucose})_{n-1} + \text{glucose-1-P}$$

Glucose 1-phosphate can enter the glycolytic pathway by way of glucose 6-phosphate (figure 9.20).

Poly-β-hydroxybutyrate (PHB) is an important, wide-spread reserve material. Its catabolism has been studied most thoroughly in *Azotobacter*. This bacterium hydrolyzes PHB to 3-hydroxybutyrate, then oxidizes the hydroxybutyrate to acetoacetate. Acetoacetate is converted to acetyl-CoA, which can be oxidized in the TCA cycle.

9.8 LIPID CATABOLISM

Microorganisms frequently use lipids as energy sources. Triglycerides or triacylglycerols, esters of glycerol and fatty acids (**figure 9.21**), are common energy sources and will serve as our examples. They can be hydrolyzed to glycerol and fatty acids by microbial lipases. The glycerol is then phosphorylated, oxidized to dihydroxyacetone phosphate, and catabolized in the glycolytic pathway (figure 9.5).

Fatty acids from triacylglycerols and other lipids are often oxidized in the **β-oxidation pathway** after conversion to coenzyme A esters (**figure 9.22**). In this cyclic pathway fatty acids are degraded to acetyl-CoA, which can be fed into the TCA cycle or used in biosynthesis (*see section 10.8*). One turn of the cycle produces acetyl-CoA, NADH, and FADH$_2$; NADH and FADH$_2$ can be oxidized by the electron transport chain to provide more ATP. The fatty acyl-CoA, shortened by two carbons, is ready for another turn of the cycle. Lipid fatty acids are a rich source of energy for microbial growth. In a similar fashion some microorganisms grow well on petroleum hydrocarbons under aerobic conditions.

Figure 9.21 A Triacylglycerol or Triglyceride. The R groups represent the fatty acid side chains.

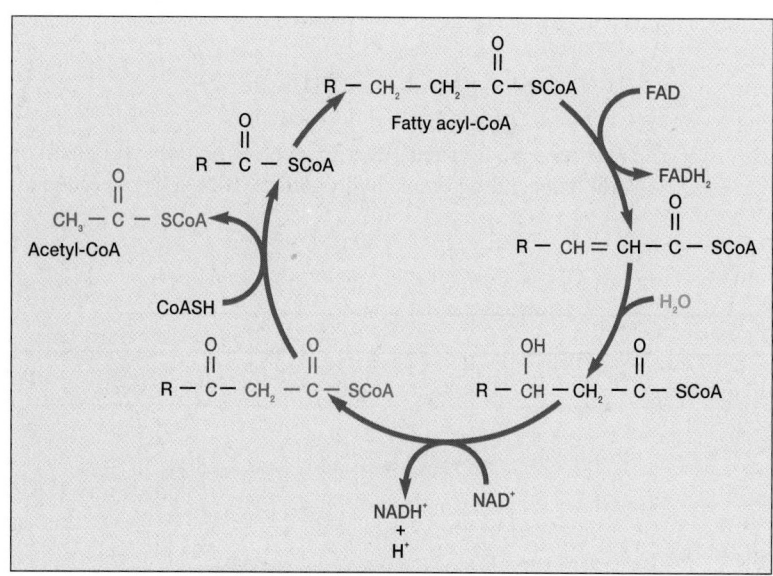

Figure 9.22 Fatty Acid β-Oxidation. The portions of the fatty acid being modified are shown in red.

9.9 PROTEIN AND AMINO ACID CATABOLISM

Some bacteria and fungi—particularly pathogenic, food spoilage, and soil microorganisms—can use proteins as their source of carbon and energy. They secrete **protease** enzymes that hydrolyze proteins and polypeptides to amino acids, which are transported into the cell and catabolized.

The first step in amino acid use is **deamination,** the removal of the amino group from an amino acid. This is often accomplished by **transamination.** The amino group is transferred from an amino acid to an α-keto acid acceptor (**figure 9.23**). The organic acid resulting from deamination can be converted to pyruvate, acetyl-CoA, or a TCA cycle intermediate and eventually oxidized in the TCA cycle to release energy. It also can be used as a source of carbon for the synthesis of cell constituents. Excess nitrogen from deamination may be excreted as ammonium ion, thus making the medium alkaline.

---

1. Briefly discuss the ways in which microorganisms degrade and use common monosaccharides, disaccharides, and polysaccharides from both external and internal sources.
2. Describe how a microorganism might derive carbon and energy from the lipids and proteins in its diet. What is β-oxidation? Transamination?

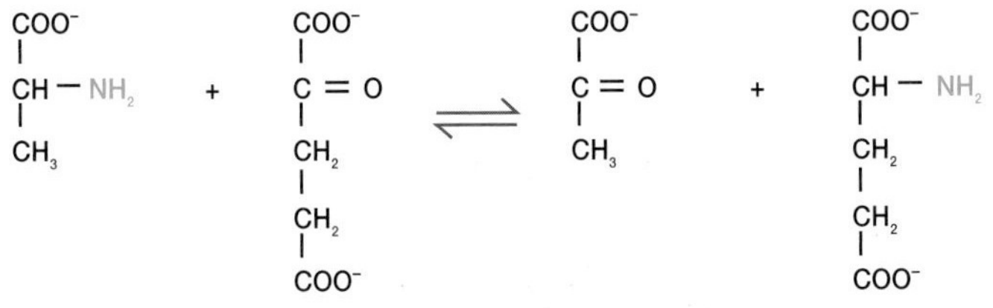

 9.10 OXIDATION OF INORGANIC MOLECULES

As we have seen, microorganisms can oxidize organic molecules such as carbohydrates, lipids, and proteins and synthesize ATP with the energy liberated. The electron acceptor is (1) another more oxidized endogenous organic molecule in fermentation, (2) O_2 in aerobic respiration, or (3) an oxidized exogenous molecule other than O_2 in anaerobic respiration (figure 9.2). In both aerobic and anaerobic respiration, ATP is formed as a result of electron transport chain activity. Electrons for the chain can be obtained from inorganic nutrients, and it is possible to derive energy from the oxidation of inorganic molecules rather than from organic nutrients. This ability is confined to a group of bacteria called **chemolithotrophs** (*see sections 22.1 and 22.2*). Each species is rather specific in its preferences for electron donors and acceptors (**table 9.4**). The acceptor is usually O_2, but sulfate and nitrate are also used. The most common electron donors are hydrogen, reduced nitrogen compounds, reduced sulfur compounds, and ferrous iron (Fe^{2+}).

Chemolithotrophic bacteria are usually autotrophic and employ the Calvin cycle to fix CO_2 as their carbon source (*see section 10.2*). However, some chemolithotrophs can function as heterotrophs if reduced organic compounds are available. Considerable energy is required to reduce CO_2 to carbohydrate; incorporation of one CO_2 in the Calvin cycle requires three ATPs and two NADPHs. Moreover, much less energy is available from the oxidation of inorganic molecules (**table 9.5**) than from the complete oxidation of glucose to CO_2, which is accompanied by a standard free energy change of -686 kcal/mole. The P/O ratios for oxidative phosphorylation in chemolithotrophs are probably around 1.0 (in the oxidation of hydrogen it is significantly higher). Because the yield of ATP is so low, chemolithotrophs must oxidize a large quantity of inorganic material to grow and reproduce, and this magnifies their ecological impact.

Several bacterial genera (table 9.4) can oxidize hydrogen gas to produce energy because they possess a hydrogenase enzyme that catalyzes the oxidation of hydrogen.

$$H_2 \longrightarrow 2H^+ + 2e^-$$

The electrons are donated either to an electron transport chain or to NAD^+, depending on the hydrogenase. If NADH is produced,

Figure 9.23 Transamination.

A common example of this process. The α-amino group (blue) of alanine is transferred to the acceptor α-ketoglutarate forming pyruvate and glutamate. The pyruvate can be catabolized in the tricarboxylic acid cycle or used in biosynthesis.

Alanine	α-Ketoglutarate	Pyruvate	Glutamate

Table 9.4			
Representative Chemolithotrophs and Their Energy Sources			
Bacteria	**Electron Donor**	**Electron Acceptor**	**Products**
Alcaligenes, Hydrogenophaga, and *Pseudomonas* spp.	H_2	O_2	H_2O
Nitrobacter	NO_2^-	O_2	NO_3^-, H_2O
Nitrosomonas	NH_4^+	O_2	NO_2^-, H_2O
Thiobacillus denitrificans	S^0, H_2S	NO_3^-	SO_4^{2-}, N_2
Thiobacillus ferrooxidans	Fe^{2+}, S^0, H_2S	O_2	Fe^{3+}, H_2O, H_2SO_4

Table 9.5
Energy Yields from Oxidations Used by Chemolithotrophs

Reaction	$\Delta G^{o\prime}$ (kcal/mole)[a]
$H_2 + \frac{1}{2} O_2 \longrightarrow H_2O$	-56.6
$NO_2^- + \frac{1}{2} O_2 \longrightarrow NO_3^-$	-17.4
$NH_4^+ + 1\frac{1}{2} O_2 \longrightarrow NO_2^- + H_2O + 2H^+$	-65.0
$S^0 + 1\frac{1}{2} O_2 + H_2O \longrightarrow H_2SO_4$	-118.5
$S_2O_3^{2-} + 2O_2 + H_2O \longrightarrow 2SO_4^{2-} + 2H^+$	-223.7
$2Fe^{2+} + 2H^+ + \frac{1}{2} O_2 \longrightarrow 2Fe^{3+} + H_2O$	-11.2

[a]The $\Delta G^{o\prime}$ for complete oxidation of glucose to CO_2 is -686 kcal/mole. A kcal is equivalent to 4.184kJ.

it can be used in ATP synthesis by electron transport and oxidative phosphorylation, with O_2 as the terminal electron acceptor. These hydrogen-oxidizing microorganisms often will use organic compounds as energy sources when such nutrients are available.

The best-studied nitrogen-oxidizing chemolithotrophs are the **nitrifying bacteria** (*see section 22.1*). These are soil and aquatic bacteria of considerable ecological significance. Ammonia oxidation to nitrate depends on the activity of at least two different genera. For example, *Nitrosomonas* and *Nitrosospira* oxidize ammonia to nitrite.

$$NH_4^+ + 1\frac{1}{2} O_2 \rightarrow NO_2^- + H_2O + 2H^+$$

The nitrite can then be further oxidized by *Nitrobacter* and *Nitrococcus* to yield nitrate.

$$NO_2^- + \frac{1}{2} O_2 \rightarrow NO_3^-$$

When two genera work together, ammonia in the soil is oxidized to nitrate in a process called **nitrification.** The role of chemolithotrophs in soil and aquatic ecosystems (pp. 593–600)

Energy released upon the oxidation of both ammonia and nitrite is used to make ATP by oxidative phosphorylation. However, microorganisms need a source of electrons (reducing power) as well as a source of ATP in order to reduce CO_2 and other molecules. Since molecules like ammonia and nitrite have more positive reduction potentials than NAD^+, they cannot directly donate their electrons to form the required NADH and NADPH. This is because electrons spontaneously move only from donors with more negative reduction potentials to acceptors with more positive potentials (*see section 8.5 and figure 8.7*). Sulfur-oxidizing bacteria face the same difficulty. Both types of chemolithotrophs solve this problem by using proton motive force to reverse the flow of electrons in their electron transport chains and reduce NAD^+ with electrons from nitrogen and sulfur donors (**figure 9.24**). Because energy is used to generate NADH as well as ATP, the net yield of ATP is fairly low. Chemolithotrophs can afford this inefficiency as they have no serious competitors for their unique energy sources.

The sulfur-oxidizing bacteria are the third major group of chemolithotrophs. The metabolism of *Thiobacillus* has been best studied. These bacteria oxidize sulfur (S^0), hydrogen sulfide (H_2S), thiosulfate ($S_2O_3^{2-}$), and other reduced sulfur compounds to sulfuric acid; therefore they have a significant ecological impact (**Microbial Diversity & Ecology 9.2**). Interestingly they generate ATP by both oxidative phosphorylation and substrate-level phosphorylation involving **adenosine 5′-phosphosulfate (APS).** APS is a high-energy molecule formed from sulfite and adenosine monophosphate (**figure 9.25**).

Some of these procaryotes are extraordinarily flexible metabolically. For example, *Sulfolobus brierleyi* and a few other bacteria can

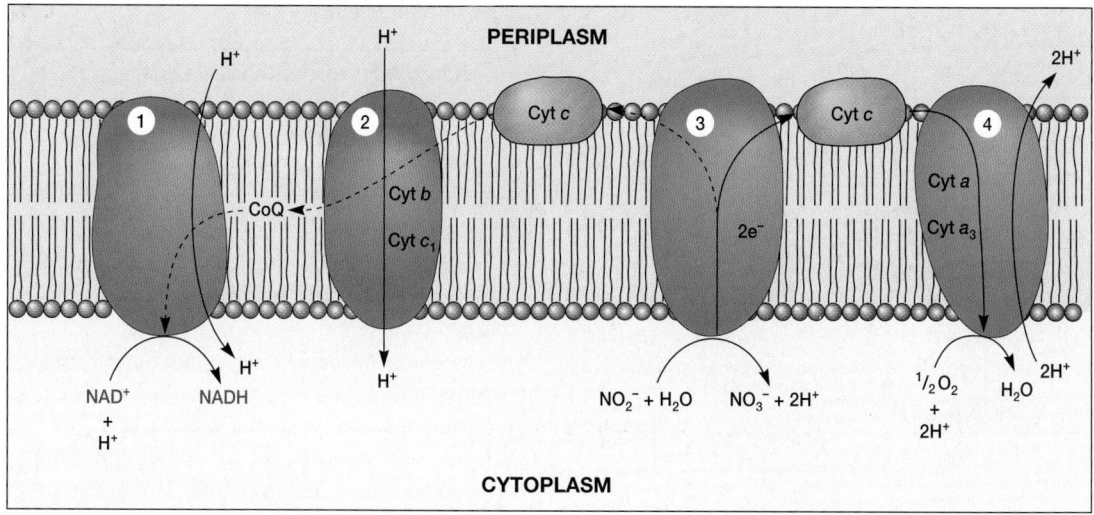

Figure 9.24   Electron Flow in *Nitrobacter* Electron Transport Chain.   *Nitrobacter* oxidizes nitrite and carries out normal electron transport to generate proton motive force for ATP synthesis. This is the right-hand branch of the diagram. Some of the proton motive force also is used to force electrons to flow up the reduction potential gradient from nitrite to NAD^+ (left-hand branch). Cytochrome c and four complexes are involved: NADH-ubiquinone oxidoreductase (1), ubiquinol-cytochrome c oxidoreductase (2), nitrite oxidase (3), and cytochrome aa_3 oxidase (4).

Microbial Diversity & Ecology

9.2   Acid Mine Drainage

Each year millions of tons of sulfuric acid flow to the Ohio River from the Appalachian Mountains. This sulfuric acid is of microbial origin and leaches enough metals from the mines to make the river reddish and acidic. The primary culprit is *Thiobacillus ferrooxidans,* a chemolithotrophic bacterium that derives its energy from oxidizing ferrous ion to ferric ion and sulfide ion to sulfate ion. The combination of these two energy sources is important because of the solubility properties of iron. Ferrous ion is somewhat soluble and can be formed at pH values of 3.0 or less in moderately reducing environments. However, when the pH is greater than 4.0 to 5.0, ferrous ion is spontaneously oxidized to ferric ion by O_2 in the water and precipitates as a hydroxide. If the pH drops below about 2.0 to 3.0 because of sulfuric acid production by spontaneous oxidation of sulfur or sulfur oxidation by thiobacilli and other bacteria, the ferrous ion remains reduced, soluble, and available as an energy source. Remarkably, *T. ferrooxidans* grows well at such acid pHs and actively oxidizes ferrous ion to an insoluble ferric precipitate. The water is rendered toxic for most aquatic life and unfit for human consumption.

The ecological consequences of this metabolic life-style arise from the common presence of pyrite (FeS_2) in coal mines. The bacteria oxidize both elemental components of pyrite for their growth and in the process form sulfuric acid, which leaches the remaining minerals.

Autoxidation or bacterial action
$$2FeS_2 + 7O_2 + 2H_2O \longrightarrow 2Fe^{2+} + 4SO_4^{2-} + 4H^+$$

T. ferrooxidans
$$2Fe^{2+} + \tfrac{1}{2}O_2 + 2H^+ \longrightarrow 2Fe^{3+} + H_2O$$

Pyrite oxidation is further accelerated because the ferric ion generated by bacterial activity readily oxidizes more pyrite to sulfuric acid and ferrous ion. In turn the ferrous ion supports further bacterial growth. It is difficult to prevent *T. ferrooxidans* growth as it requires only pyrite and common inorganic salts. Because *T. ferrooxidans* gets its O_2 and CO_2 from the air, the only feasible method of preventing its damaging growth is to seal the mines to render the habitat anaerobic.

(a) **Direct oxidation of sulfite**

$$SO_3^{2-} \xrightarrow{\text{sulfite oxidase}} SO_4^{2-} + 2e^-$$

(b) **Formation of adenosine 5′-phosphosulfate**

$$2SO_3^{2-} + 2AMP \longrightarrow 2APS + 4e^-$$
$$2APS + 2P_i \longrightarrow 2ADP + 2SO_4^{2-}$$
$$2ADP \longrightarrow AMP + ATP$$

$$2SO_3^{2-} + AMP + 2P_i \longrightarrow 2SO_4^{2-} + ATP + 4e^-$$

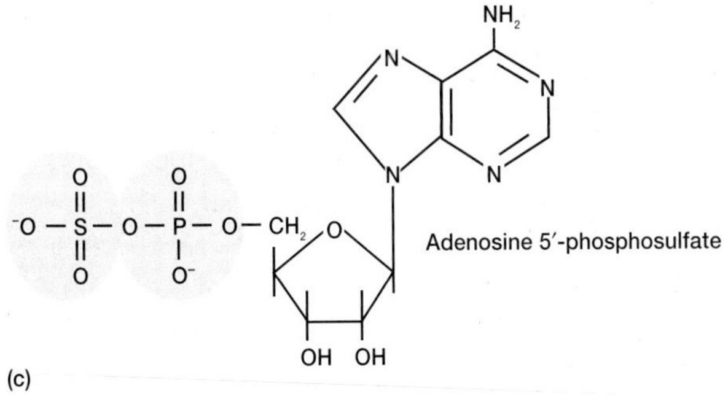

Adenosine 5′-phosphosulfate

(c)

Figure 9.25   Energy Generation by Sulfur Oxidation.
(**a**) Sulfite can be directly oxidized to provide electrons for electron transport and oxidative phosphorylation. (**b**) Sulfite can also be oxidized and converted to APS. This route produces electrons for use in electron transport and ATP by substrate-level phosphorylation with APS. (**c**) The structure of adenosine 5′-phosphosulfate.

grow aerobically by oxidizing sulfur with oxygen as the electron acceptor; in the absence of O_2, they carry out anaerobic respiration and oxidize organic material with sulfur as the electron acceptor.

Sulfur-oxidizing bacteria, like other chemolithotrophs, can use CO_2 as their carbon source. Many will grow heterotrophically if they are supplied with reduced organic carbon sources like glucose or amino acids.

---

1. How do chemolithotrophs obtain their ATP and NADH? What is their source of carbon?
2. Describe energy production by hydrogen-oxidizing bacteria, nitrifying bacteria, and sulfur-oxidizing bacteria.

9.11   PHOTOSYNTHESIS

Microorganisms derive energy not only from the oxidation of inorganic and organic compounds, but many can capture light energy and use it to synthesize ATP and NADH or NADPH (figure 9.1; *see also figure 8.8*). This process in which light energy is trapped and converted to chemical energy is called **photosynthesis.** Usually a photosynthetic organism reduces and incorporates CO_2, reactions that are also considered part of this process. Photosynthesis is one of the most significant metabolic processes on Earth because almost all our energy is ultimately derived from solar energy. It provides photosynthetic organisms with the ATP and NADPH necessary to synthesize the organic material required for

Table 9.6	
Diversity of Photosynthetic Organisms	
Eucaryotic Organisms	**Procaryotic Organisms**
Higher plants	Cyanobacteria (blue-green algae)
Multicellular green, brown, and red algae	Green sulfur bacteria
Unicellular algae (e.g., euglenoids, dinoflagellates, diatoms)	Green nonsulfur bacteria
	Purple sulfur bacteria
	Purple nonsulfur bacteria
	Prochloron

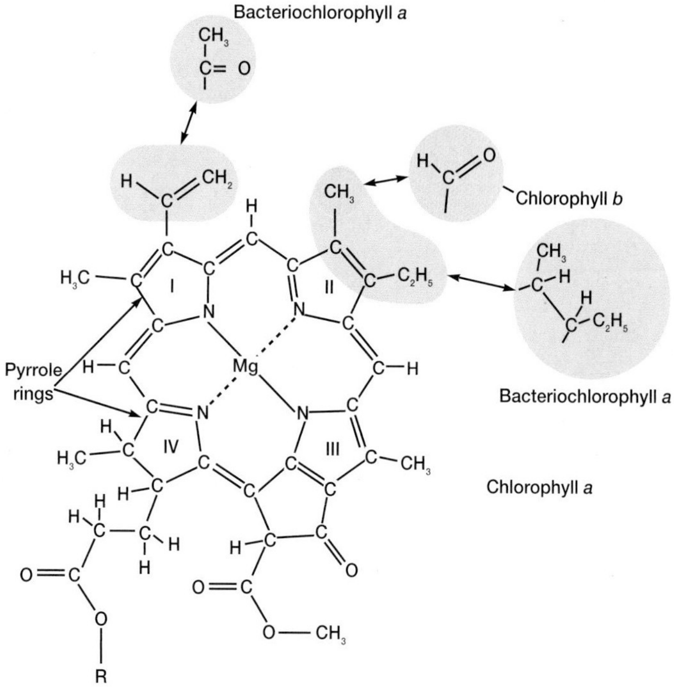

Figure 9.26 Chlorophyll Structure. The structures of chlorophyll *a*, chlorophyll *b*, and bacteriochlorophyll *a*. The complete structure of chlorophyll *a* is given. Only one group is altered to produce chlorophyll *b*, and two modifications in the ring system are required to change chlorophyll *a* to bacteriochlorophyll *a*. The side chain (R) of bacteriochlorophyll *a* may be either phytyl (a 20-carbon chain also found in chlorophylls *a* and *b*) or geranylgeranyl (a 20-carbon side chain similar to phytyl, but with three more double bonds).

growth. In turn these organisms serve as the base of most food chains in the biosphere. Photosynthesis is also responsible for replenishing our supply of O_2, a remarkable process carried out by a variety of organisms, both eucaryotic and procaryotic (**table 9.6**). Although most people associate photosynthesis with the more obvious higher plants, over half the photosynthesis on Earth is carried out by microorganisms.

Photosynthesis as a whole is divided into two parts. In the **light reactions** light energy is trapped and converted to chemical energy. This energy is then used to reduce or fix CO_2 and synthesize cell constituents in the **dark reactions.** In this section the nature of the light reactions is discussed; the dark reactions are reviewed in chapter 10. The photosynthetic dark reactions (pp. 202–3)

The Light Reaction in Eucaryotes and Cyanobacteria

All photosynthetic organisms have pigments for the absorption of light. The most important of these pigments are the **chlorophylls.** Chlorophylls are large planar rings composed of four substituted pyrrole rings with a magnesium atom coordinated to the four central nitrogen atoms (**figure 9.26**). Several chlorophylls are found in eucaryotes, the two most important of which are chlorophyll *a* and chlorophyll *b* (figure 9.26). These two molecules differ slightly in their structure and spectral properties. When dissolved in acetone, chlorophyll *a* has a light absorption peak at 665 nm; the corresponding peak for chlorophyll *b* is at 645 nm. In addition to absorbing red light, chlorophylls also absorb blue light strongly (the second absorption peak for chlorophyll *a* is at 430 nm). Because chlorophylls absorb primarily in the red and blue ranges, green light is transmitted. Consequently many photosynthetic organisms are green in color. The long hydrophobic tail attached to the chlorophyll ring aids in its attachment to membranes, the site of the light reactions.

Other photosynthetic pigments also trap light energy. The most widespread of these are the **carotenoids,** long molecules, usually yellowish in color, that possess an extensive conjugated double bond system (**figure 9.27**). β-Carotene is present in *Prochloron* and most divisions of algae; fucoxanthin is found in diatoms, dinoflagellates, and brown algae (*Phaeophyta*). Red algae and cyanobacteria have photosynthetic pigments called **phy-cobiliproteins,** consisting of a protein with a tetrapyrrole attached (figure 9.27). **Phycoerythrin** is a red pigment with a maximum absorption around 550 nm, and **phycocyanin** is blue (maximum absorption at 620 to 640 nm).

Carotenoids and phycobiliproteins are often called **accessory pigments** because of their role in photosynthesis. Although chlorophylls cannot absorb light energy effectively in the blue-green through yellow range (about 470 to 630 nm), accessory pigments do absorb light in this region and transfer the trapped energy to chlorophyll. In this way they make photosynthesis more efficient over a broader range of wavelengths. Accessory pigments also protect microorganisms from intense sunlight, which could oxidize and damage the photosynthetic apparatus in their absence.

Chlorophylls and accessory pigments are assembled in highly organized arrays called **antennas,** whose purpose is to create a large surface area to trap as many photons as possible. An antenna has about 300 chlorophyll molecules. Light energy is captured in an antenna and transferred from chlorophyll to chlorophyll until it reaches a special **reaction-center chlorophyll** directly involved in photosynthetic electron transport (**figure 9.28**). In eucaryotic cells and cyanobacteria, there are two kinds of antennas associated

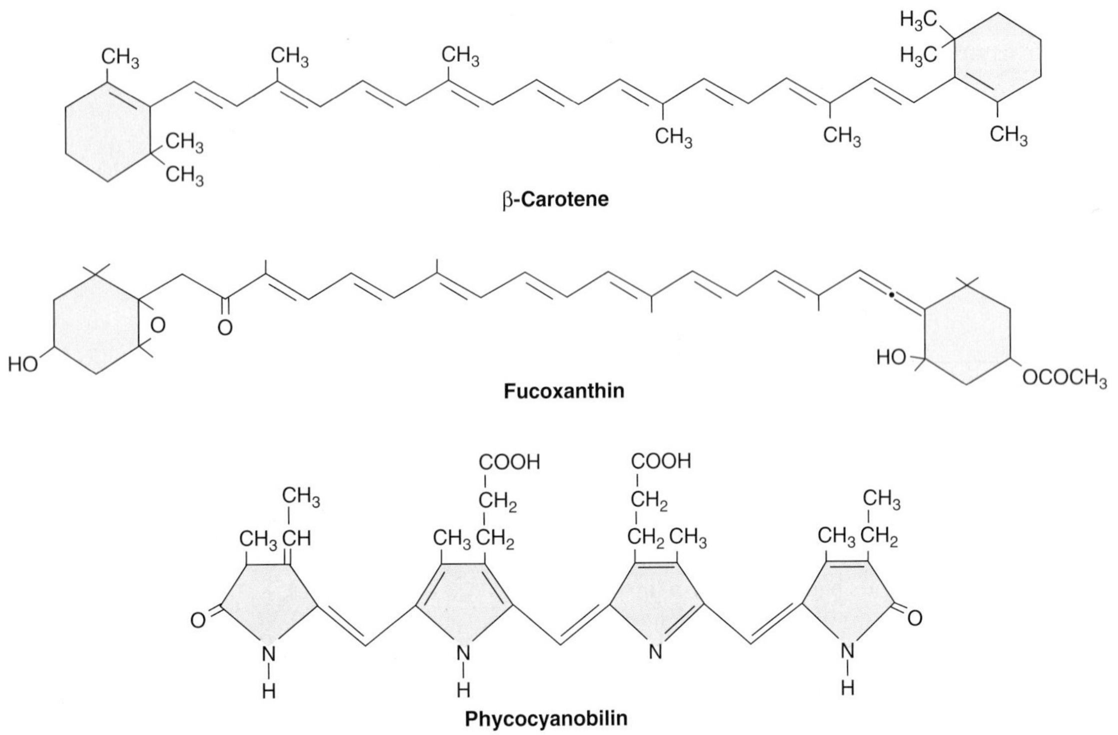

β-**Carotene**

Fucoxanthin

Phycocyanobilin

Figure 9.27 Representative Accessory Pigments. Beta-carotene is a carotenoid found in algae and higher plants. Note that it has a long chain of alternating double and single bonds called conjugated double bonds. Fucoxanthin is a carotenoid accessory pigment in several divisions of algae (the dot in the structure represents a carbon atom). Phycocyanobilin is an example of a linear tetrapyrrole that is attached to a protein to form a phycobiliprotein.

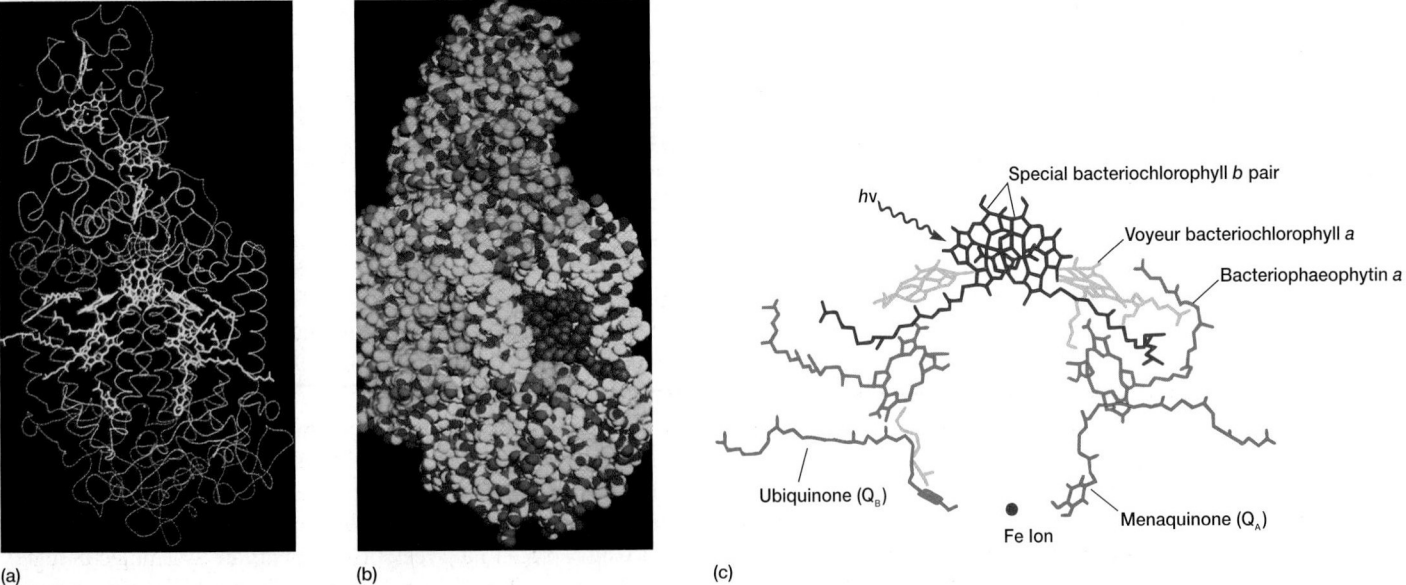

(a) (b) (c)

Figure 9.28 A Photosynthetic Reaction Chain. The reaction center of the purple nonsulfur bacterium, *Rhodopseudomonas viridis*. (**a**) The structure of the C_α backbone of the center's polypeptide chains with the bacteriochlorophylls and other prosthetic groups in yellow. (**b**) A space-filling model of the reaction center. The nitrogen atoms are blue; oxygen atoms, red; and sulfur atoms, yellow. The exposed prosthetic group atoms are reddish-brown. (**c**) A close-up view of the reaction center prosthetic groups. A photon is first absorbed by the "special pair" of bacteriochlorophyll *a* molecules, thus exciting them. An excited electron then moves to the bacteriopheophytin molecule in the right arm of the system.

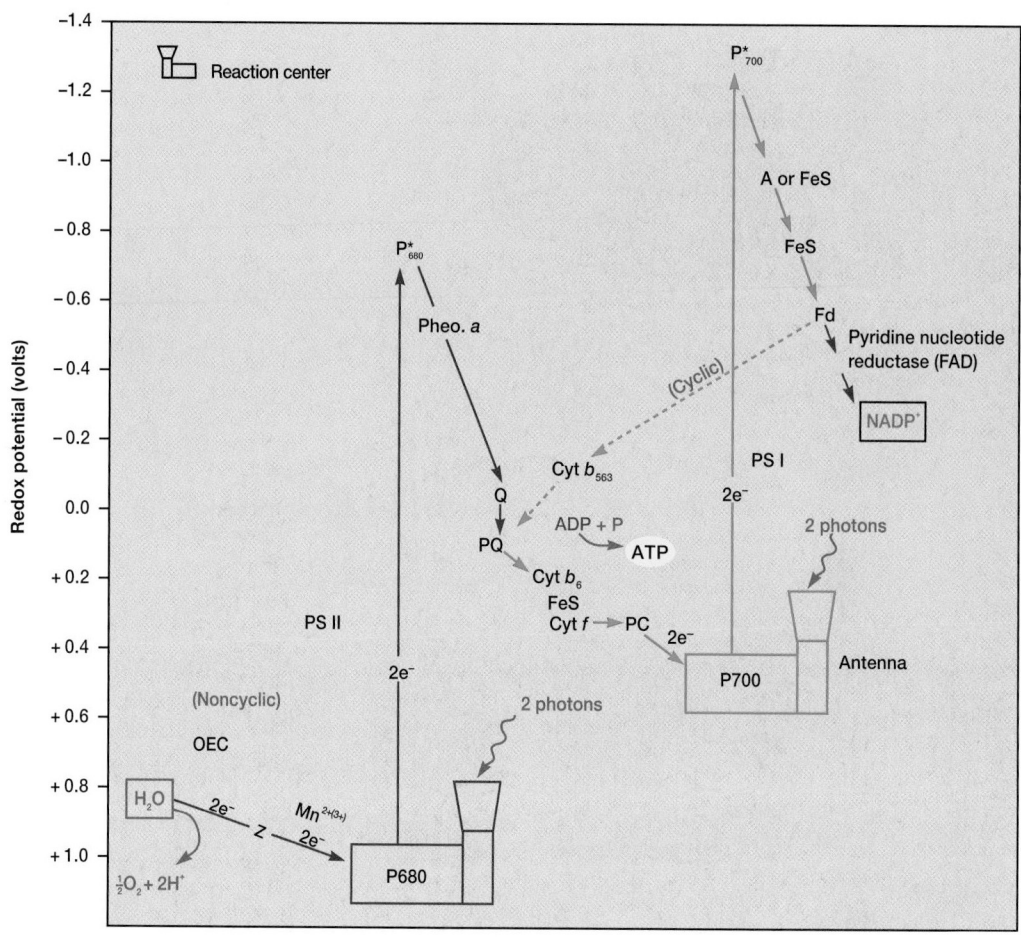

Figure 9.29 Green Plant Photosynthesis. Electron flow during photosynthesis in higher plants. Cyanobacteria and eucaryotic algae are similar in having two photosystems, although they may differ in some details. The carriers involved in electron transport are ferredoxin (Fd) and other FeS proteins; cytochromes b_6, b_{563}, and f; plastoquinone (PQ); copper containing plastocyanin (PC); pheophytin a (Pheo. a); possibly chlorophyll a (A); and the unknown quinone Q, which is probably a plastoquinone. Both photosystem I (PS I) and photosystem II (PS II) are involved in noncyclic photophosphorylation; only PS I participates in cyclic photophosphorylation. The oxygen evolving complex (OEC) that extracts electrons from water contains manganese ions and the substance Z, which transfers electrons to the PS II reaction center. See the text for further details.

with two different photosystems. **Photosystem I** absorbs longer wavelength light (≥680 nm) and funnels the energy to a special chlorophyll a molecule called P700. The term P700 signifies that this molecule most effectively absorbs light at a wavelength of 700 nm. **Photosystem II** traps light at shorter wavelengths (≤680 nm) and transfers its energy to the special chlorophyll P680.

When the photosystem I antenna transfers light energy to the reaction-center P700 chlorophyll, P700 absorbs the energy and is excited; its reduction potential becomes very negative. It then donates its excited or high-energy electron to a specific acceptor, probably a special chlorophyll a molecule (A) or an iron-sulfur protein (**figure 9.29**). The electron is eventually transferred to ferredoxin and can then travel in either of two directions. In the cyclic pathway (the dashed lines in figure 9.29), the electron moves in a cyclic route through a series of electron carriers and back to the oxidized P700. The pathway is termed cyclic because

the electron from P700 returns to P700 after traveling through the photosynthetic electron transport chain. PMF (p. 182) is formed during cyclic electron transport in the region of cytochrome b_6 and used to synthesize ATP. This process is called **cyclic photophosphorylation** because electrons travel in a cyclic pathway and ATP is formed. Only photosystem I participates.

Electrons also can travel in a noncyclic pathway involving both photosystems. P700 is excited and donates electrons to ferredoxin as before. In the noncyclic route, however, reduced ferredoxin reduces NADP$^+$ to NADPH (figure 9.29). Because the electrons contributed to NADP$^+$ cannot be used to reduce oxidized P700, photosystem II participation is required. It donates electrons to oxidized P700 and generates ATP in the process. The photosystem II antenna absorbs light energy and excites P680, which then reduces pheophytin a. Pheophytin a is chlorophyll a in which two hydrogen atoms have replaced the central magnesium.

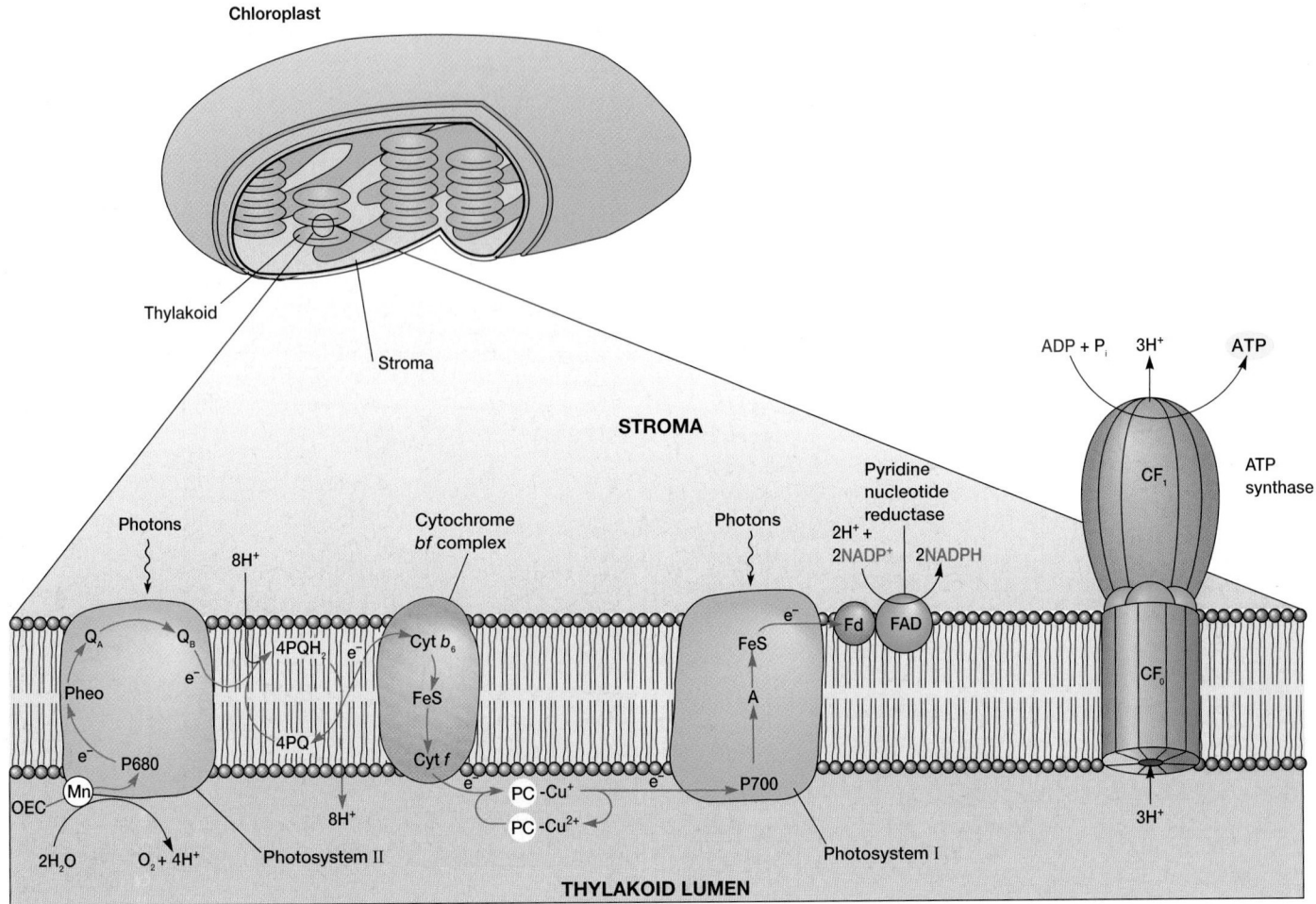

Figure 9.30   The Mechanism of Photosynthesis.   An illustration of the chloroplast thylakoid membrane showing photosynthetic electron transport chain function and noncyclic photophosphorylation. The chain is composed of three complexes: PS I, the cytochrome *bf* complex, and PS II. Two diffusible electron carriers connect the three complexes. Plastoquinone (PQ) connects PS I with the cytochrome *bf* complex, and plastocyanin (PC) connects the cytochrome *bf* complex with PS II. The light-driven electron flow pumps protons across the thylakoid membrane and generates an electrochemical gradient, which can then be used to make ATP. Water is the source of electrons and the oxygen-evolving complex (OEC) produces oxygen.

Electrons subsequently travel to Q (probably a plastoquinone) and down the electron transport chain to P700. Oxidized P680 then obtains an electron from the oxidation of water to O_2. Thus electrons flow from water all the way to $NADP^+$ with the aid of energy from two photosystems, and ATP is synthesized by **noncyclic photophosphorylation.** It appears that one ATP and one NADPH are formed when two electrons travel through the noncyclic pathway.

Just as is true of mitochondrial electron transport, photosynthetic electron transport takes place within a membrane. Chloroplast granal membranes contain both photosystems and their antennas. **Figure 9.30** shows a thylakoid membrane carrying out noncyclic photophosphorylation by the chemiosmotic mechanism. Protons move to the thylakoid interior during photosynthetic electron transport and return to the stroma when ATP is formed. It is believed that stromal lamellae possess only photosystem I and are involved in cyclic photophosphorylation alone. In cyanobacteria, photosynthetic light reactions are also located in membranes.

The dark reactions require three ATPs and two NADPHs to reduce one CO_2 and use it to synthesize carbohydrate (CH_2O).

$$CO_2 + 3ATP + 2NADPH + 2H^+ + H_2O \longrightarrow$$
$$(CH_2O) + 3ADP + 3P_i + 2NADP^+$$

The noncyclic system generates one NADPH and one ATP per pair of electrons; therefore four electrons passing through the system will produce two NADPHs and two ATPs. A total of 8 quanta of light energy (4 quanta for each photosystem) is needed to propel the four electrons from water to $NADP^+$. Because the ratio of ATP to NADPH required for CO_2 fixation is 3:2, at least one more ATP must be supplied. Cyclic photophosphorylation probably operates independently to generate the extra ATP. This requires absorption of another 2 to 4 quanta. It follows that around 10 to 12 quanta of light energy are needed to reduce and incorporate one molecule of CO_2 during photosynthesis.

Table 9.7

Properties of Microbial Photosynthetic Systems

Property	Eucaryotes	Cyanobacteria	Green and Purple Bacteria
Photosynthetic pigment	Chlorophyll *a*	Chlorophyll *a*	Bacteriochlorophyll
Photosystem II	Present	Present	Absent
Photosynthetic electron donors	H_2O	H_2O	H_2, H_2S, S, organic matter
O_2 production pattern	Oxygenic	Oxygenic[a]	Anoxygenic
Primary products of energy conversion	ATP + NADPH	ATP + NADPH	ATP
Carbon source	CO_2	CO_2	Organic and/or CO_2

[a]Some cyanobacteria can function anoxygenically under certain conditions. For example, *Oscillatoria* can use H_2S as an electron donor instead of H_2O.

The Light Reaction in Green and Purple Bacteria

Green and purple photosynthetic bacteria differ from cyanobacteria and eucaryotic photosynthesizers in several fundamental ways (**table 9.7**). In particular, green and purple bacteria do not use water as an electron source or produce O_2 photosynthetically—that is, they are **anoxygenic.** In contrast, cyanobacteria and eucaryotic photosynthesizers are almost always **oxygenic** (some cyanobacteria can carry out anoxygenic photosynthesis [*see p. 460*]). NADPH is not directly produced in the photosynthetic light reaction of purple bacteria. Green bacteria can reduce NAD^+ directly during the light reaction. To synthesize NADH and NADPH, green and purple bacteria must use electron donors like hydrogen, hydrogen sulfide, elemental sulfur, and organic compounds that have more negative reduction potentials than water and are therefore easier to oxidize (better electron donors). Finally, green and purple bacteria possess slightly different photosynthetic pigments, **bacteriochlorophylls** (figure 9.26), many with absorption maxima at longer wavelengths. Bacteriochlorophylls *a* and *b* have maxima in ether at 775 and 790 nm, respectively. In vivo maxima are about 830 to 890 nm (bacteriochlorophyll *a*) and 1,020 to 1,040 nm (Bchl *b*). This shift of absorption maxima into the infrared region better adapts these bacteria to their ecological niches (*see figure 21.3*).

There are four groups of green and purple photosynthetic bacteria, each containing several genera: green sulfur bacteria (*Chlorobium*), green nonsulfur bacteria (*Chloroflexus*), purple sulfur bacteria (*Chromatium*), and purple nonsulfur bacteria (*Rhodospirillum, Rhodopseudomonas*). The biology of these groups will be discussed later in much more detail.   The biology of photosynthetic bacteria (pp. 456–64, 475, 485–86)

Many differences found in green and purple bacteria are due to their lack of photosystem II; they cannot use water as an electron donor in noncyclic electron transport. Without photosystem II they cannot produce O_2 from H_2O photosynthetically and are restricted to cyclic photophosphorylation. Indeed, almost all purple and green sulfur bacteria are strict anaerobes. A tentative scheme for the photosynthetic electron transport chain of a purple nonsulfur bacterium is given in **figure 9.31.** When the special reaction-center chlorophyll P870 is excited, it donates an electron to bacteriopheophytin. Electrons then flow to quinones and through an electron transport chain back to P870 while driving ATP synthesis. Note that although both green and purple bacteria

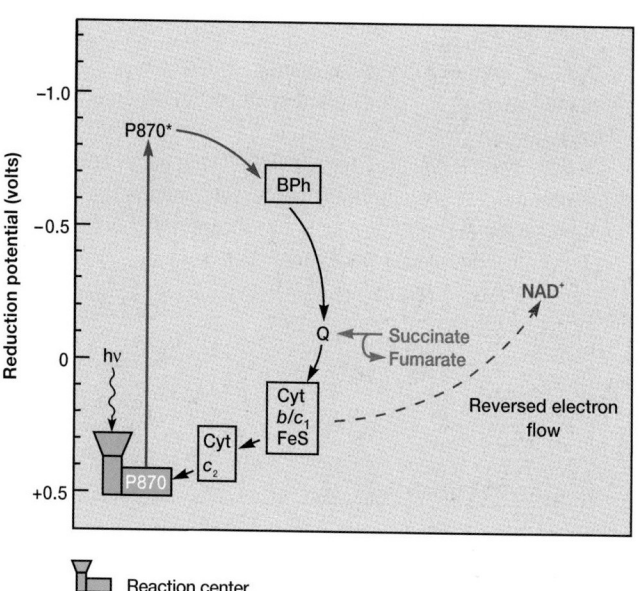

Figure 9.31   Purple Nonsulfur Bacterial Photosynthesis. The photosynthetic electron transport system in the purple nonsulfur bacterium, *Rhodobacter sphaeroides*. This scheme is incomplete and tentative. Ubiquinone (Q) is very similar to coenzyme Q. BPh stands for bacteriopheophytin. NAD^+ and the electron source succinate are in blue.

lack two photosystems, the purple bacteria have a photosynthetic apparatus similar to photosystem II, whereas the green sulfur bacteria have a system similar to photosystem I.

Green and purple bacteria face a further problem because they also require NADH or NADPH for CO_2 incorporation. They may synthesize NADH in at least three ways. If they are growing in the presence of hydrogen gas, which has a reduction potential more negative than that of NAD^+, the hydrogen can be used directly to produce NADH. Like chemolithotrophs, many photosynthetic purple bacteria use proton motive force to reverse the flow of electrons in an electron transport chain and move them from inorganic or organic donors to NAD^+ (figure 9.31 and **figure 9.32**). Green sulfur bacteria such as *Chlorobium* appear to carry out a simple form of noncyclic photosynthetic electron flow to reduce NAD^+ (**figure 9.33**).

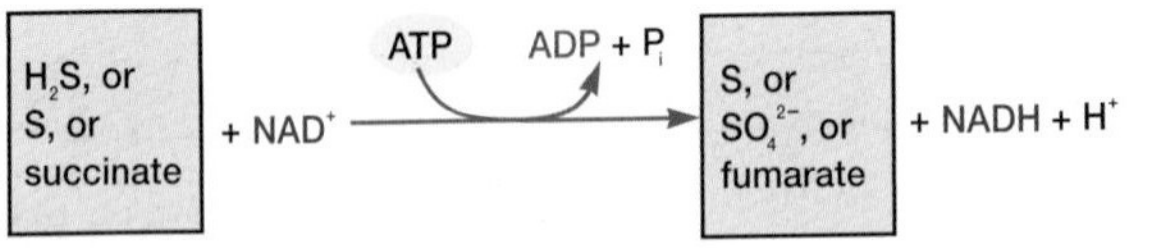

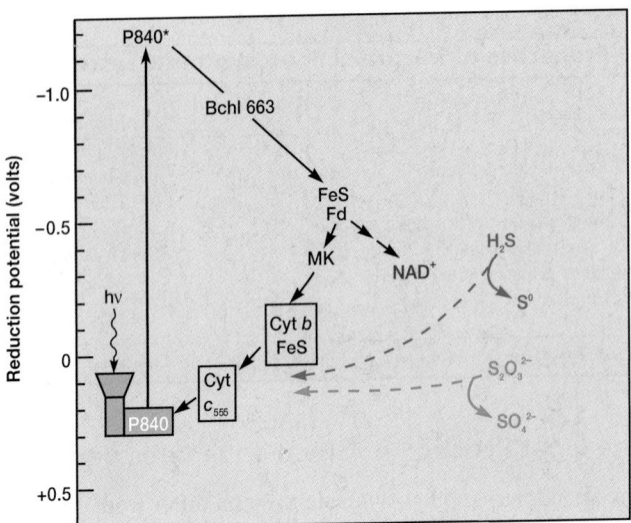

Figure 9.32 NAD Reduction in Green and Purple Bacteria. The use of reversed electron flow to reduce NAD$^+$. The arrow in this diagram represents an electron transport chain that is being driven in reverse by proton motive force or ATP. That is, electrons are moving from donors with more positive reduction potentials to an acceptor (NAD$^+$) with a more negative potential.

1. Describe photosynthesis as carried out by eucaryotes and cyanobacteria. How does photosynthesis in green and purple bacteria differ?
2. Define the following terms: light reaction, chlorophyll, carotenoid, phycobiliprotein, accessory pigment, antenna, photosystems I and II, cyclic photophosphorylation, noncyclic photophosphorylation, anoxygenic and oxygenic photosynthesis, and bacteriochlorophyll.

Figure 9.33 Green Sulfur Bacterial Photosynthesis. The photosynthetic electron transport system in the green sulfur bacterium, *Chlorobium limicola*. Light energy is used to make ATP by cyclic photophosphorylation and move electrons from sulfur donors (green and blue) to NAD$^+$. The electron transport chain has a quinone called menaquinone (MK).

Summary

Metabolism is the total of all chemical reactions in a cell and may be divided into catabolism and anabolism.

9.1 An Overview of Metabolism

a. Chemotrophic microorganisms can use three kinds of electron acceptors during energy metabolism (**figure 9.2**). The nutrient may be oxidized with an endogenous electron acceptor (fermentation), with oxygen as an exogenous electron acceptor (aerobic respiration), or with another external electron acceptor (anaerobic respiration).

b. Amphibolic pathways like glycolysis and the tricarboxylic acid cycle have both catabolic and anabolic functions.

9.2 The Breakdown of Glucose to Pyruvate

a. The glycolytic or Embden-Meyerhof pathway (**figure 9.5**) occurs by way of fructose 1,6-bisphosphate with the net production of two NADHs and two ATPs, the latter being produced by substrate-level phosphorylation.

b. In the pentose phosphate pathway, glucose 6-phosphate is oxidized twice and converted to pentoses and other sugars. It is a source of NADPH, ATP, pentoses, and tetroses.

c. In the Entner-Doudoroff pathway, glucose is oxidized to 6-phosphogluconate, which is then dehydrated and cleaved to pyruvate and glyceraldehyde 3-phosphate. The latter product can be oxidized by glycolytic enzymes to provide ATP and NADH.

9.3 Fermentations

a. In the absence of O_2 a microorganism often uses an oxidized endogenous organic molecule as an electron acceptor to reoxidize the NADH formed during glycolysis (fermentation).

9.4 The Tricarboxylic Acid Cycle

a. The tricarboxylic acid cycle is the final stage of catabolism in most aerobic cells (**figure 9.12**). It oxidizes acetyl-CoA to CO_2 and forms one GTP, three NADHs, and one $FADH_2$ per acetyl-CoA.

9.5 Electron Transport and Oxidative Phosphorylation

a. The NADH and $FADH_2$ produced from the oxidation of carbohydrates, fatty acids, and other nutrients can be oxidized in the electron transport chain. Electrons flow from carriers with more negative reduction potentials to those with more positive potentials (**figure 9.13;** *see also*

figure 8.7), and free energy is released for ATP synthesis by oxidative phosphorylation.

b. Bacterial electron transport chains are often different from eucaryotic chains with respect to such aspects as carriers and branching. In eucaryotes the P/O ratio for NADH is about 3 and that for $FADH_2$ is around 2; P/O ratios are usually much lower in bacteria.

c. Eucaryotic ATP synthesis takes place on small protein spheres, the F_1F_0 ATPases or ATP synthases, located on the inner surface of the inner mitochondrial membrane. Bacterial ATP synthase is on the inner surface of the plasma membrane.

d. The most widely accepted mechanism of oxidative phosphorylation is the chemiosmotic hypothesis in which proton motive force drives ATP synthesis (**figure 9.17**).

e. When the glycolytic pathway operates anaerobically, only 2 ATPs per glucose are formed, whereas aerobic respiration in eucaryotes can yield a maximum of 38 ATPs.

9.6 Anaerobic Respiration

a. Anaerobic respiration is the process of ATP production by electron transport in which the terminal electron acceptor is an exogenous, oxidized inorganic molecule other than O_2. The most common acceptors are nitrate, sulfate, and CO_2.

9.7 Catabolism of Carbohydrates and Intracellular Reserve Polymers

a. Microorganisms catabolize many extracellular carbohydrates. Monosaccharides are taken in and phosphorylated; disaccharides may be cleaved to monosaccharides by either hydrolysis or phosphorolysis.

b. External polysaccharides are degraded by hydrolysis and the products are absorbed. Intracellular glycogen and starch are converted to glucose 1-phosphate by phosphorolysis.

9.8 Lipid Catabolism

a. Fatty acids from lipid catabolism are usually oxidized to acetyl-CoA in the β-oxidation pathway.

9.9 Protein and Amino Acid Catabolism

a. Proteins are hydrolyzed to amino acids that are then deaminated; their carbon skeletons feed into the TCA cycle.

9.10 Oxidation of Inorganic Molecules

a. Chemolithotrophs or chemoautotrophs synthesize ATP by oxidizing reduced inorganic compounds—usually hydrogen, reduced nitrogen and sulfur compounds, or ferrous iron—with an electron transport chain and O_2 as the electron acceptor.

b. Chemolithotrophs usually incorporate CO_2 through the Calvin cycle and produce NADH by reversing electron transport.

9.11 Photosynthesis

a. In photosynthesis, eucaryotes and cyanobacteria trap light energy with chlorophyll and accessory pigments and move electrons through photosystems I and II to make ATP and NADPH (the light reactions).

b. Cyclic photophosphorylation involves the activity of photosystem I alone. In noncyclic photophosphorylation photosystems I and II operate together to move electrons from water to $NADP^+$ producing ATP, NADPH, and O_2 (**figure 9.29**).

c. Photosynthetic green and purple bacteria differ from eucaryotes and cyanobacteria in possessing bacteriochlorophyll and lacking photosystem II (**figures 9.31** and **9.33**). Thus they cannot use the noncyclic pathway to form NADPH and O_2; they are anoxygenic.

Key Terms

accessory pigments 191
acetyl coenzyme A (acetyl-CoA) 178
adenosine 5′-phosphosulfate (APS) 189
aerobic respiration 168
alcoholic fermentation 176
amphibolic pathways 170
anabolism 168
anaerobic respiration 168
anoxygenic photosynthesis 195
antenna 191
ATP synthase 182
bacteriochlorophyll 195
β-oxidation pathway 187
butanediol fermentation 176
carotenoids 191

catabolism 168
chemiosmotic hypothesis 182
chemolithotroph 188
chlorophylls 191
citric acid cycle 178
cyclic photophosphorylation 193
dark reactions 191
deamination 188
denitrification 185
dissimilatory nitrate reduction 185
electron transport chain 179
Embden-Meyerhof pathway 171
Entner-Doudoroff pathway 174
fermentation 168
glycolysis 171
glycolytic pathway 171

heterolactic fermenters 176
hexose monophosphate pathway 171
homolactic fermenters 176
Krebs cycle 178
lactic acid fermentation 176
light reactions 191
metabolism 168
mixed acid fermentation 176
nitrification 189
nitrifying bacteria 189
noncyclic photophosphorylation 194
oxidative phosphorylation 179
oxygenic photosynthesis 195
Pasteur effect 184
pentose phosphate pathway 171

photosynthesis 190
photosystem I 193
photosystem II 193
phycobiliproteins 191
phycocyanin 191
phycoerythrin 191
protease 188
proton motive force (PMF) 182
reaction-center chlorophyll 191
respiration 168
Stickland reaction 176
substrate-level phosphorylation 171
transamination 188
tricarboxylic acid (TCA) cycle 178
uncouplers 184

Questions for Thought and Review

1. Why might it be desirable for a microorganism with the Embden-Meyerhof pathway and the TCA cycle also to have the pentose phosphate pathway?

2. Of what advantage would it be for a microorganism to possess an electron transport chain and oxidative phosphorylation?

3. Describe two ways in which a microorganism can continue to produce energy when O_2 is absent.

4. Why would it be wasteful for anaerobic microorganisms to operate the complete TCA cycle?

5. How do substrate-level phosphorylation and oxidative phosphorylation differ from one another?

6. Can fermentation products be used in identifying bacteria? Give some examples if the answer is yes.

7. Describe what would happen to microbial metabolism if the enzyme lactate dehydrogenase was inhibited in a homolactic fermenter growing anaerobically on a medium containing glucose as the carbon source. What effects would an inhibitor of the ATP synthase have on an aerobically respiring microorganism? An uncoupler?

8. How would you isolate a thermophilic chemolithotroph that uses sulfur compounds

as a source of electrons? What changes in the incubation system would be needed to isolate bacteria using sulfur compounds in anaerobic respiration? How can one tell which process is taking place through an analysis of the sulfur molecules present in the medium?

9. Suppose that you isolated a bacterial strain that carried out oxygenic photosynthesis. What photosystems would it possess and what group of bacteria would it most likely belong to?

Critical Thinking Questions

1. Without looking in chapter 21, predict some characteristics that would describe niches occupied by green and purple photosynthetic bacteria.

2. From an evolutionary perspective, discuss why most microorganisms use aerobic respiration to generate ATP.

For recommended readings on these and related topics, see Appendix V.

chapter 10

Metabolism:
The Use of Energy in Biosynthesis

Concepts

- In anabolism or biosynthesis, cells use free energy to construct more complex molecules and structures from smaller, simpler precursors.
- Biosynthetic pathways are organized to optimize efficiency by conserving biosynthetic raw materials and energy.
- Autotrophs use ATP and NADPH from photosynthesis or from oxidation of inorganic molecules to reduce and incorporate CO_2 into organic material.
- Catabolic and anabolic pathways may differ in enzymes, regulation, intracellular location, and use of cofactors and nucleoside diphosphate carriers. Although many enzymes of amphibolic pathways participate in both catabolism and anabolism, some pathway enzymes are involved only in one of the two processes.
- Phosphorus, in the form of phosphate, can be directly assimilated, whereas inorganic sulfur and nitrogen compounds must often be reduced before incorporation into organic material.
- The tricarboxylic acid (TCA) cycle acts as an amphibolic pathway and requires anaplerotic reactions to maintain adequate levels of cycle intermediates.
- Most glycolytic enzymes participate in both the synthesis and catabolism of glucose. In contrast, fatty acids are synthesized from acetyl-CoA and malonyl-CoA by a pathway quite different from fatty acid β-oxidation.
- Peptidoglycan synthesis is a complex, multistep process that is begun in the cytoplasm and completed at the cell wall after the peptidoglycan repeat unit has been transported across the plasma membrane.

Biological structures are almost always constructed in a hierarchical manner, with subassemblies acting as important intermediates en route from simple starting molecules to the end products of organelles, cells, and organisms.
—W. M. Becker and D. W. Deamer

The nitrogenase Fe protein's subunits are arranged like a pair of butterfly wings. Nitrogenase consists of the Fe protein and the MoFe protein; it catalyzes the reduction of atmospheric nitrogen during nitrogen fixation.

s chapter 9 makes clear, microorganisms can obtain energy in many ways. Much of this energy is used in biosynthesis or anabolism. During biosynthesis a microorganism begins with simple precursors, such as inorganic molecules and monomers, and constructs ever more complex molecules until new organelles and cells arise (**figure 10.1**). A microbial cell must manufacture many different kinds of molecules; however, it is possible to discuss the synthesis of only the most important types of cell constituents.

This chapter begins with a general introduction to anabolism, then focuses on the synthesis of carbohydrates, amino acids, purines and pyrimidines, and lipids. It also describes the assimilation of CO_2, phosphorus, sulfur, and nitrogen. The chapter ends with a section on the synthesis of peptidoglycan and bacterial cell walls. Protein and nucleic acid synthesis is so significant and complex that it is described separately in chapters 11 and 12.

Because anabolism is the creation of order and a cell is highly ordered and immensely complex, much energy is required for biosynthesis. This is readily apparent from estimates of the biosynthetic capacity of rapidly growing *Escherichia coli* (**table 10.1**). Although most ATP dedicated to biosynthesis is employed in protein synthesis, ATP is also used to make other cell constituents.

Free energy is required for biosynthesis in mature cells of constant size because cellular molecules are continuously being degraded and resynthesized, a process known as **turnover.** Cells are never the same from instant to instant. Despite the continuous turnover of cell constituents, metabolism is carefully regulated so that the rate of biosynthesis is approximately balanced by that of catabolism. In addition to the energy expended in the turnover of molecules, many nongrowing cells also use energy to synthesize enzymes and other substances for release into their surroundings.

Regulation of metabolism (chapters 8 and 12)

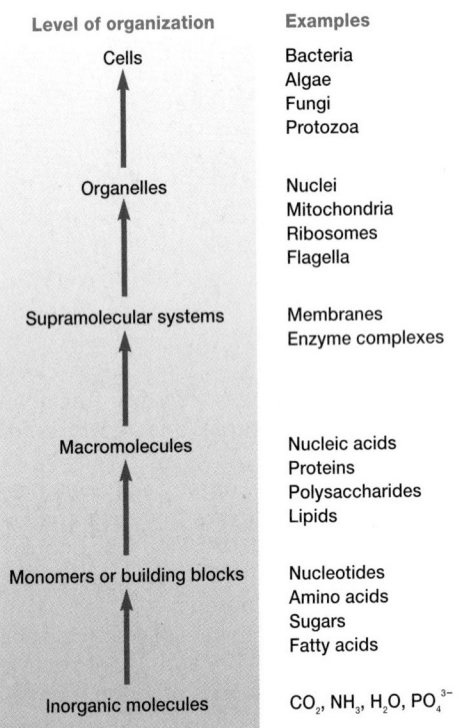

Level of organization	Examples
Cells	Bacteria Algae Fungi Protozoa
Organelles	Nuclei Mitochondria Ribosomes Flagella
Supramolecular systems	Membranes Enzyme complexes
Macromolecules	Nucleic acids Proteins Polysaccharides Lipids
Monomers or building blocks	Nucleotides Amino acids Sugars Fatty acids
Inorganic molecules	CO_2, NH_3, H_2O, PO_4^{3-}

Figure 10.1   The Construction of Cells.   The biosynthesis of procaryotic and eucaryotic cell constituents. Biosynthesis is organized in levels of ever greater complexity.

 PRINCIPLES GOVERNING BIOSYNTHESIS

Biosynthetic metabolism seems to follow certain patterns or be shaped by a few general principles. Six of these are now briefly discussed.

1. A microbial cell contains large quantities of proteins, nucleic acids, and polysaccharides, all of which are **macromole-**

Table 10.1			
Biosynthesis in *Escherichia coli*			

Cell Constituent	Number of Molecules per Cell[a]	Molecules Synthesized per Second	Molecules of ATP Required per Second for Synthesis
DNA	1[b]	0.00083	60,000
RNA	15,000	12.5	75,000
Polysaccharides	39,000	32.5	65,000
Lipids	15,000,000	12,500.0	87,000
Proteins	1,700,000	1,400.0	2,120,000

From *Bioenergetics* by Albert Lehninger. Copyright © 1971 by the Benjamin/Cummings Publishing Company. Reprinted by permission.
[a]Estimates for a cell with a volume of 2.25 μm^3, a total weight of 1×10^{-12}g, a dry weight of 2.5×10^{-13}g, and a 20 minute cell division cycle.
[b]It should be noted that bacteria can contain multiple copies of their genomic DNA.

cules or very large molecules that are polymers of smaller units joined together. The construction of large, complex molecules from a few simple structural units or **monomers** saves much genetic storage capacity, biosynthetic raw material, and energy. A consideration of protein synthesis clarifies this. Proteins—whatever size, shape, or function—are made of only 20 common amino acids joined by peptide bonds (*see appendix I*). Different proteins simply have different amino acid sequences but not new and dissimilar amino acids. Suppose that proteins were composed of 40 different amino acids instead of 20. The cell would then need the enzymes to manufacture twice as many amino acids (or would have to obtain the extra amino acids in its diet). Genes would be required for the extra enzymes, and the cell would have to invest raw materials and energy in the synthesis of these additional genes, enzymes, and amino acids. Clearly the use of a few monomers linked together by a single type of covalent bond makes the synthesis of macromolecules a highly efficient process. Almost all cell structures are built mainly of about 30 small precursors.

2. The cell often saves additional materials and energy by using many of the same enzymes for both catabolism and anabolism. For example, most glycolytic enzymes are involved in the synthesis and the degradation of glucose.

3. Although many enzymes in amphibolic pathways (*see section 9.1*) participate in both catabolic and anabolic activities, some steps are catalyzed by two different enzymes. One enzyme catalyzes the reaction in the catabolic direction, the other reverses this conversion (**figure 10.2**). Thus catabolic and anabolic pathways are never identical although many enzymes are shared. Use of separate enzymes for the two directions of a single step permits independent regulation of catabolism and anabolism. Although this has been discussed in more detail in sections 8.7 through 8.9, note that the regulation of anabolism is somewhat different from that of catabolism. Both types of pathways can be regulated by their end products as well as by the concentrations of ATP, ADP, AMP, and NAD^+. Nevertheless, end product regulation generally assumes more importance in anabolic pathways.

4. To synthesize molecules efficiently, anabolic pathways must operate irreversibly in the direction of biosynthesis. Cells can achieve this by connecting some biosynthetic reactions to the breakdown of ATP and other nucleoside triphosphates. When these two processes are coupled, the free energy made available during nucleoside triphosphate breakdown drives the biosynthetic reaction to completion (*see sections 8.3 and 8.4*).

5. In eucaryotic microorganisms biosynthetic pathways are frequently located in different cellular compartments from their corresponding catabolic pathways (**Techniques & Applications 10.1**). For example, fatty acid biosynthesis occurs in the cytoplasmic matrix, whereas fatty acid oxidation takes place within the mitochondrion. Compartmentation makes it easier for the pathways to operate simultaneously but independently.

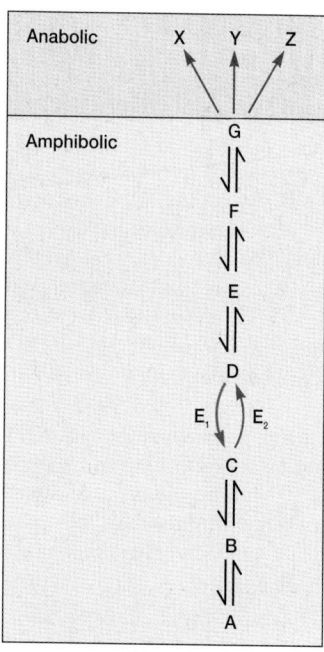

Figure 10.2 A Hypothetical Biosynthetic Pathway. The routes connecting G with X, Y, and Z are purely anabolic because they are used only for synthesis of the end products. The pathway from A to G is amphibolic—that is, it has both catabolic and anabolic functions. Most reactions are used in both roles; however, the interconversion of C and D is catalyzed by two separate enzymes, E_1 (catabolic) and E_2 (anabolic).

6. Finally, anabolic and catabolic pathways often use different cofactors. Usually catabolic oxidations produce NADH, a substrate for electron transport. In contrast, when an electron donor is needed during biosynthesis, NADPH rather than NADH normally serves as the donor. Fatty acid metabolism provides a second example. Fatty acyl-CoA molecules are oxidized to generate energy, whereas fatty acid synthesis involves acyl carrier protein thioesters (p. 214).

After macromolecules have been constructed from simpler precursors, they are assembled into larger, more complex structures such as supramolecular systems and organelles (figure 10.1). Macromolecules normally contain the necessary information to form spontaneously in a process known as **self-assembly.** For example, ribosomes are large assemblages of many proteins and ribonucleic acid molecules, yet they arise by the self-assembly of their components without the involvement of extra factors.

1. Define biosynthesis or anabolism and turnover.
2. List six principles by which biosynthetic pathways are organized.

Techniques & Applications
10.1   The Identification of Anabolic Pathways

There are three approaches to the study of pathway organization: (1) study of the pathway in vitro, (2) use of nutritional mutants, and (3) incubation of cells with precursors labeled radioisotopically. In vitro [Latin, in glass] studies employ cell-free extracts to search for enzymes and metabolic intermediates that might belong to a pathway. Although this direct approach was used to work out the organization of many catabolic pathways, progress in research on biosynthesis was slow until the other two techniques were developed in the early to middle 1940s.

Techniques using nutritional mutants were developed during Beadle and Tatum's work on the genetics of the red bread mold *Neurospora*. This approach is best illustrated with a hypothetical example. Suppose that a pathway for the synthesis of end product Z is organized with E_1, E_2, and so on, representing pathway enzymes.

$$A \xrightarrow{E_1} B \xrightarrow{E_2} C \xrightarrow{E_3} Z$$

The prototroph (*see section 11.6*), which will grow in medium lacking Z, can be treated with mutagenic agents such as ultraviolet light, X rays, or chemical mutagens. Some resulting mutants will be auxotrophs that require the presence of Z for growth because one of their biosynthetic enzymes is now inactive. When E_3 is inactive, the microorganism will grow only in the presence of Z, even though it can make C from the precursor A. When grown in the presence of a small amount of Z, intermediate C (the intermediate just before the blocked step) will accumulate in the medium. In this way a variety of mutants can be used to establish the identity of pathway intermediates. The order of intermediates can be determined by cross-feeding experiments. If E_2 has been inactivated by a mutation, the mutant will grow only when either C or Z is supplied. Because the medium in which the E_3 mutant has been cultured contains intermediate C, it will support growth of the E_2 mutant (other mutants would not produce enough C to support growth). If cross-feeding experiments are conducted with mutants of each step in the pathway, the steps can be placed in the correct order. Application of this technique quickly led to the elucidation of the pathways for the synthesis of tryptophan, folic acid, and other molecules.

Radioisotopes such as ^{14}C are used in the third approach to studying pathway organization. Potential biosynthetic precursors are synthesized in the laboratory with specific atoms made radioactive. The microorganism then is incubated with culture medium containing the radioactive molecule, and the biosynthetic end product is isolated and analyzed. If the molecule truly is a precursor of the end product, the latter should be radioactive. The location of the radioactive atom will determine what part of the product is contributed by the radioactively labeled precursor. Precisely the same approach can be employed with nonradioactive atoms like ^{15}N. This technique provided some of the first information about the nature of the purine biosynthetic pathway.

10.2 THE PHOTOSYNTHETIC FIXATION OF CO_2

Although most microorganisms can incorporate or fix CO_2, at least in anaplerotic reactions (pp. 209–11), only autotrophs use CO_2 as their sole or principal carbon source. The reduction and incorporation of CO_2 requires much energy. Usually autotrophs obtain energy by trapping light during photosynthesis, but some derive energy from the oxidation of reduced inorganic electron donors. Autotrophic CO_2 fixation is crucial to life on Earth because it provides the organic matter on which heterotrophs depend.   Photosynthetic light reactions and chemolithotrophy (pp. 188–96)

Microorganisms can fix CO_2 or convert this inorganic molecule to organic carbon and assimilate it in three major ways. Almost all microbial autotrophs incorporate CO_2 by a special metabolic pathway called by several names: the **Calvin cycle,** Calvin-Benson cycle, or reductive pentose phosphate cycle. Although the Calvin cycle is found in photosynthetic eucaryotes and most photosynthetic procaryotes, it is absent in the Archaea, some obligately anaerobic bacteria, and some microaerophilic bacteria. These microorganisms usually employ one of two other pathways. A reductive tricarboxylic acid pathway (*see figure 20.6a*) is used by some archaea (*Thermoproteus, Sulfolobus*) and by bacteria such as *Chlorobium* and *Desulfobacter*. The acetyl-CoA pathway (*see figure 20.6b*) is found in methanogens, sulfate reducers, and bacteria that can form acetate from CO_2 during fermentation (acetogens). Because of its importance, we will focus on the Calvin cycle here.   Alternate CO_2 fixation pathways (pp. 441–42)

The Calvin cycle is found in the chloroplast stroma of eucaryotic microbial autotrophs. Cyanobacteria, some nitrifying bacteria, and thiobacilli possess **carboxysomes.** These are polyhedral inclusion bodies that contain the enzyme ribulose-1,5-bisphosphate carboxylase (see following section). They may be the site of CO_2 fixation or may store the carboxylase and other proteins. Understanding the cycle is easiest if it is divided into three phases: carboxylation, reduction, and regeneration. An overview of the cycle is given in figure 10.4 and the details are presented in appendix II.

The Carboxylation Phase

Carbon dioxide fixation is accomplished by the enzyme **ribulose-1,5-bisphosphate carboxylase** or ribulosebisphosphate carboxylase/ oxygenase (rubisco) (**figure 10.3**), which catalyzes the addition of CO_2 to ribulose 1,5-bisphosphate (RuBP), forming two molecules of 3-phosphoglycerate (PGA).

The Reduction Phase

After PGA is formed by carboxylation, it is reduced to glyceraldehyde 3-phosphate. The reduction, carried out by two enzymes, is essentially a reversal of a portion of the glycolytic

Figure 10.3   The Ribulose-1,5-Bisphosphate Carboxylase Reaction.   This enzyme catalyzes the addition of carbon dioxide to ribulose 1,5-bisphosphate, forming an unstable intermediate, which then breaks down to two molecules of 3-phosphoglycerate.

pathway, although the glyceraldehyde 3-phosphate dehydrogenase differs from the glycolytic enzyme in using $NADP^+$ rather than NAD^+ (**figure 10.4**).

The Regeneration Phase

The third phase of the Calvin cycle regenerates RuBP and produces carbohydrates such as glyceraldehyde 3-phosphate, fructose, and glucose (figure 10.4). This portion of the cycle is similar to the pentose phosphate pathway and involves the transketolase and transaldolase reactions. The cycle is completed when phosphoribulokinase reforms RuBP.

To synthesize fructose 6-phosphate or glucose 6-phosphate from CO_2, the cycle must operate six times to yield the desired hexose and reform the six RuBP molecules.

$$6RuBP + 6CO_2 \longrightarrow 12PGA \longrightarrow 6RuBP + \text{fructose 6-P}$$

The incorporation of one CO_2 into organic material requires three ATPs and two NADPHs. The formation of glucose from CO_2 may be summarized by the following equation.

$$6CO_2 + 18ATP + 12NADPH + 12H^+ + 12H_2O \longrightarrow$$
$$\text{glucose} + 18ADP + 18P_i + 12NADP^+$$

ATP and NADPH are provided by photosynthetic light reactions or by oxidation of inorganic molecules in chemoautotrophs. Sugars formed in the Calvin cycle can then be used to synthesize other essential molecules.

10.3   SYNTHESIS OF SUGARS AND POLYSACCHARIDES

Many microorganisms cannot carry out photosynthesis and are heterotrophs that must synthesize sugars from reduced organic molecules rather than from CO_2. The synthesis of glucose from

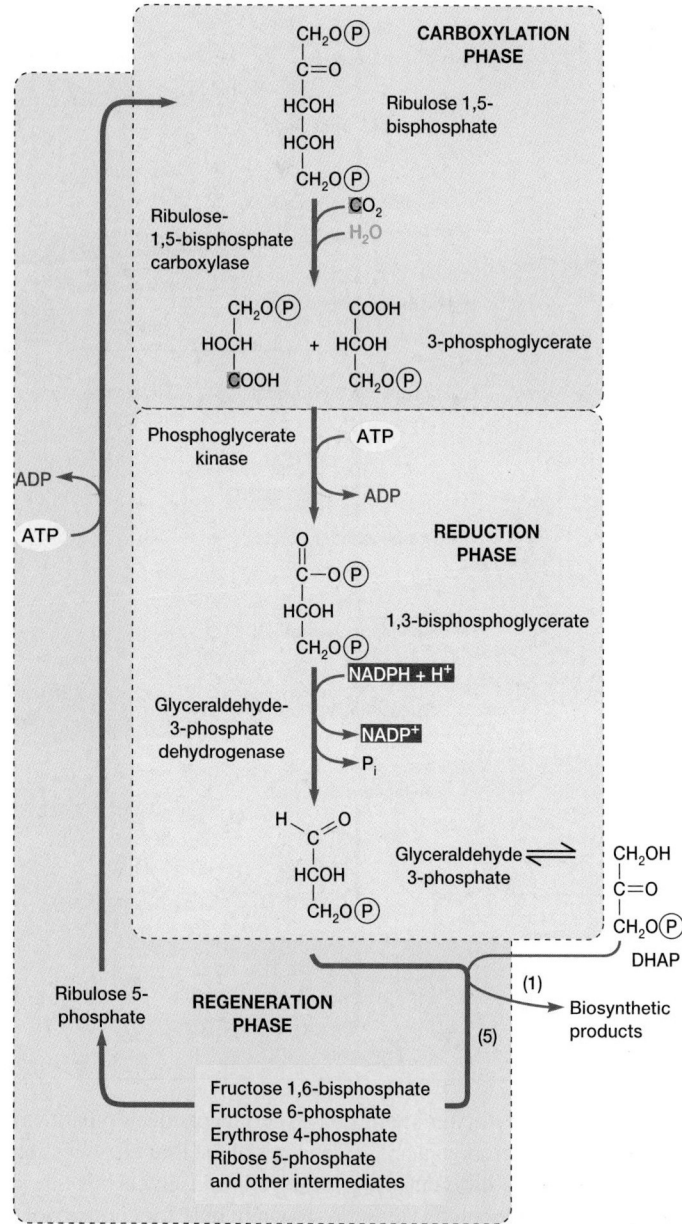

Figure 10.4   The Calvin Cycle.   This is an overview of the cycle with only the carboxylation and reduction phases in detail. Three ribulose 1,5-bisphosphates are carboxylated to give six 3-phosphoglycerates in the carboxylation phase. These are converted to six glyceraldehyde 3-phosphates, which can be converted to dihydroxyacetone phosphate (DHAP). Five of the six trioses (glyceraldehyde phosphate and dihydroxyacetone phosphate) are used to reform three ribulose 1,5-bisphosphates in the regeneration phase. The remaining triose is used in biosynthesis. The numbers in parentheses at the lower right indicate this carbon flow.

noncarbohydrate precursors is called **gluconeogenesis.** Although the gluconeogenic pathway is not identical with the glycolytic pathway, they do share seven enzymes (**figure 10.5**). Three glycolytic steps are irreversible in the cell: (1) the conversion of phosphoenolpyruvate to pyruvate, (2) the formation of fructose

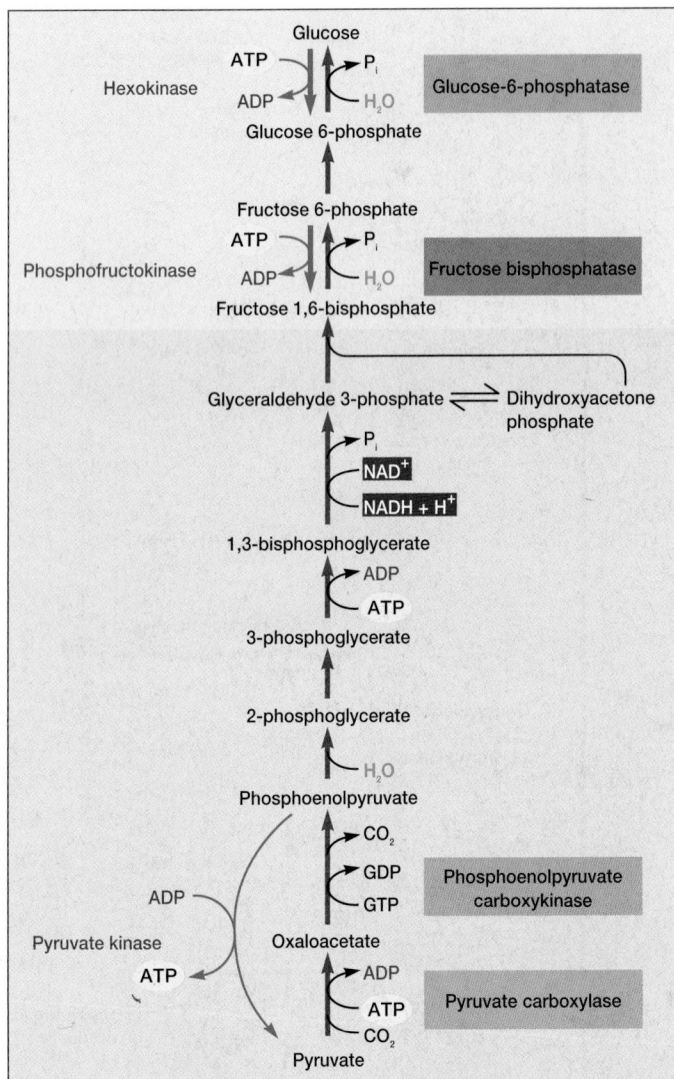

Figure 10.5   Gluconeogenesis.   The gluconeogenic pathway used in many microorganisms. The names of the four enzymes catalyzing reactions different from those found in glycolysis are in shaded boxes. Glycolytic steps are shown in blue for comparison.

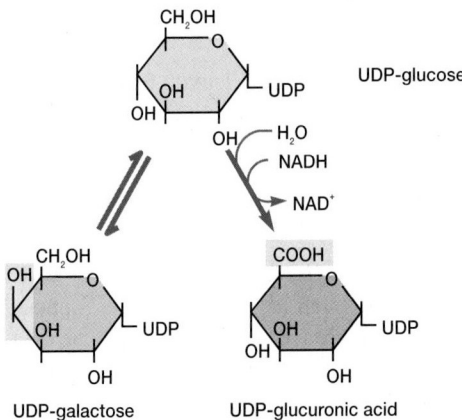

Figure 10.6   Uridine Diphosphate Glucose.   Glucose is in color.

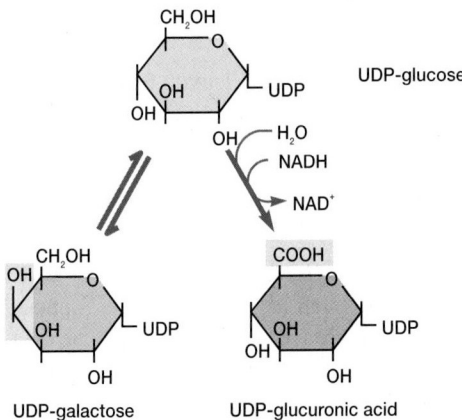

Figure 10.7   Uridine Diphosphate Galactose and Glucuronate Synthesis.   The synthesis of UDP-galactose and UDP-glucuronic acid from UDP-glucose. Structural changes are indicated by blue boxes.

1,6-bisphosphate from fructose 6-phosphate, and (3) the phosphorylation of glucose. These must be bypassed when the pathway is operating biosynthetically. For example, the formation of fructose 1,6-bisphosphate by phosphofructokinase is reversed by the enzyme, fructose bisphosphatase, which hydrolytically removes a phosphate from fructose bisphosphate. Usually at least two enzymes are involved in the conversion of pyruvate to phosphoenolpyruvate (the reversal of the pyruvate kinase step).

As can be seen in figure 10.5, the pathway synthesizes fructose as well as glucose. Once glucose and fructose have been formed, other common sugars can be manufactured. For example, mannose comes directly from fructose by a simple rearrangement.

$$\text{Fructose 6-phosphate} \rightleftharpoons \text{mannose 6-phosphate}$$

Several sugars are synthesized while attached to a nucleoside diphosphate. The most important nucleoside diphosphate sugar is **uridine diphosphate glucose (UDPG)**. Glucose is activated by attachment to the pyrophosphate of uridine diphosphate through a reaction with uridine triphosphate (**figure 10.6**). The UDP portion of UDPG is recognized by enzymes and carries glucose around the cell for participation in enzyme reactions much like ADP bears phosphate in the form of ATP. UDP-galactose is synthesized from UDPG through a rearrangement of one hydroxyl group. A different enzyme catalyzes the synthesis of UDP-glucuronic acid through the oxidation of UDPG (**figure 10.7**).

Nucleoside diphosphate sugars also play a central role in the synthesis of polysaccharides such as starch and glycogen. Again, biosynthesis is not simply a direct reversal of catabolism. Glycogen and starch catabolism (*see section 9.7*) proceeds either by hydrolysis to form free sugars or by the addition of phosphate to these polymers with the production of glucose 1-phosphate. Nucleoside diphosphate sugars are not involved. In contrast, during the synthesis of glycogen and starch in bacteria and algae, adenosine diphosphate glucose is formed from glucose 1-phosphate and then donates glucose to the end of growing glycogen and starch chains.

$$ATP + \text{glucose 1-phosphate} \longrightarrow \text{ADP-glucose} + PP_i$$

$$(Glucose)_n + \text{ADP-glucose} \longrightarrow (glucose)_{n+1} + ADP$$

Nucleoside diphosphate sugars also participate in the synthesis of complex molecules such as bacterial cell walls (pp. 216–17).

---

1. Briefly describe the three stages of the Calvin cycle.
2. What is gluconeogenesis and how does it usually occur? Describe the formation of mannose, galactose, starch, and glycogen. Why are nucleoside diphosphate sugars important?

10.4 THE ASSIMILATION OF INORGANIC PHOSPHORUS, SULFUR, AND NITROGEN

Besides carbon and oxygen, microorganisms also require large quantities of phosphorus, sulfur, and nitrogen for biosynthesis. Each of these is assimilated, or incorporated into organic molecules, by different routes. Microbial nutrition (chapter 5); Microbial participation in biogeochemical cycles (section 28.3)

Phosphorus Assimilation

Phosphorus is found in nucleic acids, proteins, phospholipids, ATP, and coenzymes like NADP. The most common phosphorus sources are inorganic phosphate and organic phosphate esters. Inorganic phosphate is incorporated through the formation of ATP in one of three ways: by (1) photophosphorylation (*see pp. 191–94*), (2) oxidative phosphorylation (*see pp. 182–84*), and (3) substrate-level phosphorylation. Glycolysis provides an example of the third process. Phosphate is joined with glyceraldehyde 3-phosphate to give 1,3-bisphosphoglycerate, which is next used in ATP synthesis.

$$\text{Glyceraldehyde 3-P} + P_i + NAD^+ \longrightarrow$$
$$\text{1,3-bisphosphoglycerate} + NADH + H^+$$

$$\text{1,3-bisphosphoglycerate} + ADP \longrightarrow$$
$$\text{3-phosphoglycerate} + ATP$$

Microorganisms may obtain organic phosphates from their surroundings in dissolved or particulate form. **Phosphatases** very often hydrolyze organic phosphate esters to release inorganic phosphate. Gram-negative bacteria have phosphatases in the periplasmic space between their cell wall and the plasma membrane, which allows phosphate to be taken up immediately after release. On the other hand, protozoa can directly use organic phosphates after ingestion or hydrolyze them in lysosomes and incorporate the phosphate.

Figure 10.8 **Phosphoadenosine 5′-phosphosulfate (PAPS).** The sulfate group is in color.

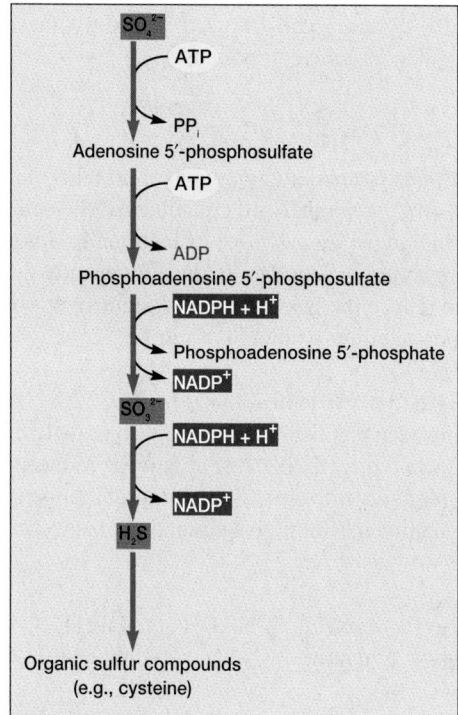

Figure 10.9 **The Sulfate Reduction Pathway.**

Sulfur Assimilation

Sulfur is needed for the synthesis of amino acids (cysteine and methionine) and several coenzymes (e.g., coenzyme A and biotin) and may be obtained from two sources. Many microorganisms use cysteine and methionine, obtained from either external sources or intracellular amino acid reserves. In addition, sulfate can provide sulfur for biosynthesis. The sulfur atom in sulfate is more oxidized than it is in cysteine and other organic molecules; thus sulfate must be reduced before it can be assimilated. This process is known as **assimilatory sulfate reduction** to distinguish it from the **dissimilatory sulfate reduction** that takes place when sulfate acts as an electron acceptor during anaerobic respiration (*see figure 28.20*). Anaerobic respiration (pp. 185–86)

Assimilatory sulfate reduction involves sulfate activation through the formation of **phosphoadenosine 5′-phosphosulfate** (**figure 10.8**), followed by reduction of the sulfate. The process is a complex one (**figure 10.9**) in which sulfate is first reduced to sulfite

(SO_3^{2-}), then to hydrogen sulfide. Cysteine can be synthesized from hydrogen sulfide in two ways. Fungi appear to combine hydrogen sulfide with serine to form cysteine (process 1), whereas many bacteria join hydrogen sulfide with O-acetylserine instead (process 2).

(1) H_2S + serine ⟶ cysteine + H_2O

(2)   Serine $\xrightarrow{\text{acetyl-CoA CoA}}$

O-acetylserine $\xrightarrow{H_2S \quad \text{acetate}}$ cysteine

Once formed, cysteine can be used in the synthesis of other sulfur-containing organic compounds.

Nitrogen Assimilation

Because nitrogen is a major component of proteins, nucleic acids, coenzymes, and many other cell constituents, the cell's ability to assimilate inorganic nitrogen is exceptionally important. Although nitrogen gas is abundant in the atmosphere, few microorganisms can reduce the gas and use it as a nitrogen source. Most must incorporate either ammonia or nitrate.

Ammonia Incorporation

Ammonia nitrogen can be incorporated into organic material relatively easily and directly because it is more reduced than other forms of inorganic nitrogen. Some microorganisms form the amino acid alanine in a reductive amination reaction catalyzed by alanine dehydrogenase.

$$\text{Pyruvate} + NH_4^+ + \text{NADH (NADPH)} + H^+$$
$$\rightleftharpoons \text{L-alanine} + NAD^+(NADP^+) + H_2O$$

The major route for ammonia incorporation often is the formation of glutamate from α-ketoglutarate (a TCA cycle intermediate). Many bacteria and fungi employ **glutamate dehydrogenase,** at least when the ammonia concentration is high.

$$\text{α-ketoglutarate} + NH_4^+ + \text{NADPH (NADH)} + H^+$$
$$\rightleftharpoons \text{glutamate} + NADP^+(NAD^+) + H_2O$$

Different species vary in their ability to use NADPH and NADH as the reducing agent in glutamate synthesis.

Once either alanine or glutamate has been synthesized, the newly formed α-amino group can be transferred to other carbon skeletons by transamination reactions (*see section 9.9*) to form different amino acids. **Transaminases** possess the coenzyme pyridoxal phosphate, which is responsible for the amino group transfer. Microorganisms have a number of transaminases, each of which catalyzes the formation of several amino acids using the same amino acid as an amino group donor. When glutamate dehydrogenase works in cooperation with transaminases, ammonia can be incorporated into a variety of amino acids (**figure 10.10**).

A second route of ammonia incorporation involves two enzymes acting in sequence, **glutamine synthetase** and **glutamate**

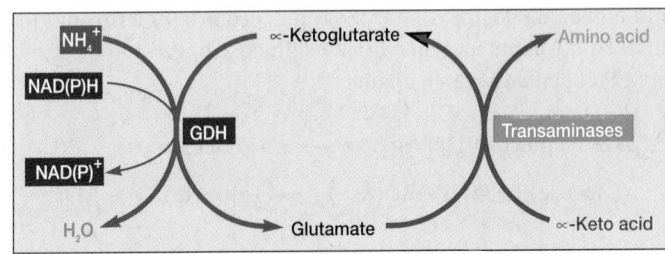

Figure 10.10   The Ammonia Assimilation Pathway.
Ammonia assimilation by use of glutamate dehydrogenase (GDH) and transaminases. Either NADP- or NAD-dependent glutamate dehydrogenases may be involved. This route is most active at high ammonia concentrations.

synthase (**figure 10.11**). Ammonia is used to synthesize glutamine from glutamate, then the amide nitrogen of glutamine is transferred to α-ketoglutarate to generate a new glutamate molecule. Because glutamate acts as an amino donor in transaminase reactions, ammonia may be used to synthesize all common amino acids when suitable transaminases are present (**figure 10.12**). Both ATP and a source of electrons, such as NADPH or reduced ferredoxin, are required. This route is present in *Escherichia coli, Bacillus megaterium,* and other bacteria. The two enzymes acting in sequence operate very effectively at low ammonia concentrations, unlike the glutamate dehydrogenase pathway. As we saw earlier, glutamine synthetase is tightly regulated by reversible covalent modification and allosteric effectors (*see p. 163*).

Assimilatory Nitrate Reduction

The nitrogen in nitrate (NO_3^-) is much more oxidized than that in ammonia. Nitrate must first be reduced to ammonia before the nitrogen can be converted to an organic form. This reduction of nitrate is called **assimilatory nitrate reduction,** which is not the same as that occurring during anaerobic respiration and dissimilatory nitrate reduction (*see sections 9.6 and 28.3*). In assimilatory nitrate reduction, nitrate is incorporated into organic material and does not participate in energy generation. The process is widespread among bacteria, fungi, and algae.

Assimilatory nitrate reduction takes place in the cytoplasm in bacteria. The first step in nitrate assimilation is its reduction to nitrite by **nitrate reductase,** an enzyme that contains both FAD and molybdenum (**figure 10.13**). NADPH is the electron source.

$$NO_3^- + \text{NADPH} + H^+ \longrightarrow NO_2^- + NADP^+ + H_2O$$

Nitrite is next reduced to ammonia with a series of two electron additions catalyzed by **nitrite reductase** and possibly other enzymes. Hydroxylamine may be an intermediate. The ammonia is then incorporated into amino acids by the routes already described.

Nitrogen Fixation

The reduction of atmospheric gaseous nitrogen to ammonia is called **nitrogen fixation.** Because ammonia and nitrate levels often are low and only a few procaryotes can carry out nitrogen

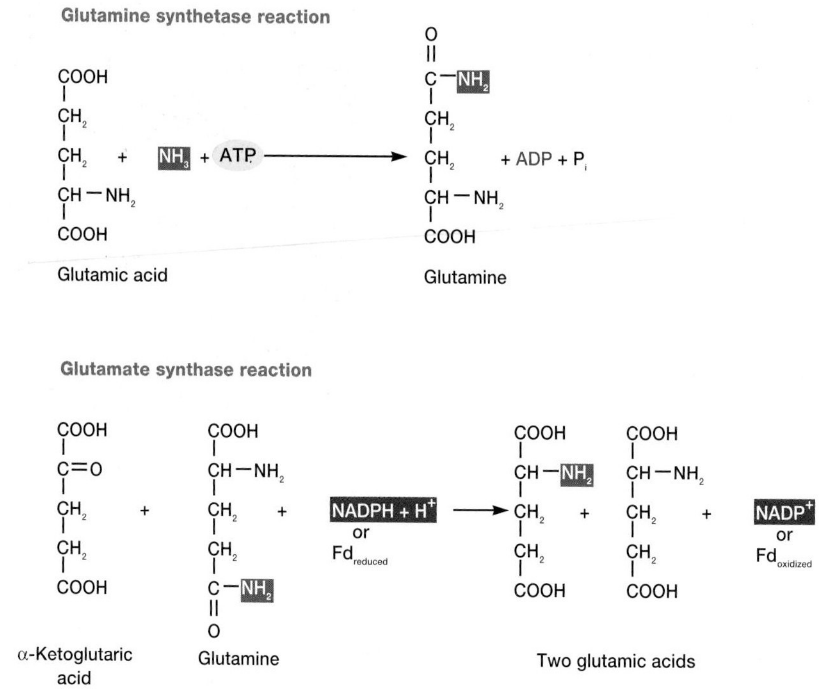

Glutamine synthetase reaction

$$\text{Glutamic acid} + NH_3 + ATP \longrightarrow \text{Glutamine} + ADP + P_i$$

Glutamic acid

Glutamine

Glutamate synthase reaction

$$\alpha\text{-Ketoglutaric acid} + \text{Glutamine} + NADPH + H^+ \text{ or } Fd_{reduced} \longrightarrow \text{Two glutamic acids} + NADP^+ \text{ or } Fd_{oxidized}$$

α-Ketoglutaric acid

Glutamine

Two glutamic acids

Figure 10.11   Glutamine Synthetase and Glutamate Synthase.   The glutamine synthetase and glutamate synthase reactions involved in ammonia assimilation. Some glutamine synthases use NADPH as an electron source; others use reduced ferredoxin (Fd). The nitrogen being incorporated and transferred is shown in turquoise.

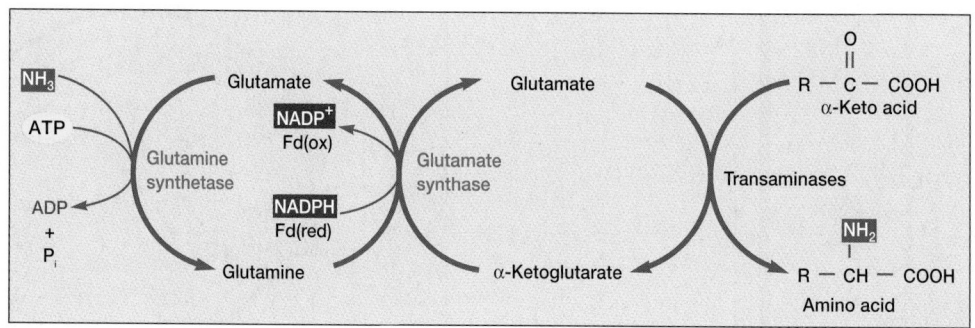

Figure 10.12   Ammonia Incorporation Using Glutamine Synthetase and Glutamate Synthase.   This route is effective at low ammonia concentrations.

fixation (eucaryotic cells completely lack this ability), the rate of this process limits plant growth in many situations. Nitrogen fixation occurs in (1) free-living bacteria (e.g., *Azotobacter, Klebsiella, Clostridium,* and *Methanococcus*), (2) bacteria living in symbiotic association with plants such as legumes (*Rhizobium*), and (3) cyanobacteria (*Nostoc* and *Anabaena*). The biological aspects of nitrogen fixation are discussed in chapter 30. The biochemistry of nitrogen fixation is the focus of this section.

The biology of nitrogen-fixing microorganisms (pp. 479–80, 598, 652–55)

The reduction of nitrogen to ammonia is catalyzed by the enzyme **nitrogenase.** Although the enzyme-bound intermediates in this process are still unknown, it is believed that nitrogen is reduced by two-electron additions in a way similar to that illustrated in **figure 10.14.** The reduction of molecular nitrogen to

ammonia is quite exergonic, but the reaction has a high activation energy because molecular nitrogen is an unreactive gas with a triple bond between the two nitrogen atoms. Therefore nitrogen reduction is expensive and requires a large ATP expenditure. At least 8 electrons and 16 ATP molecules, 4 ATPs per pair of electrons, are required.

$$N_2 + 8H^+ + 8e^- + 16ATP \longrightarrow$$
$$2NH_3 + H_2 + 16ADP + 16P_i$$

The electrons come from ferredoxin that has been reduced in a variety of ways: by photosynthesis in cyanobacteria, respiratory processes in aerobic nitrogen fixers, or fermentations in anaerobic bacteria. For example, *Clostridium pasteurianum* (an anaerobic

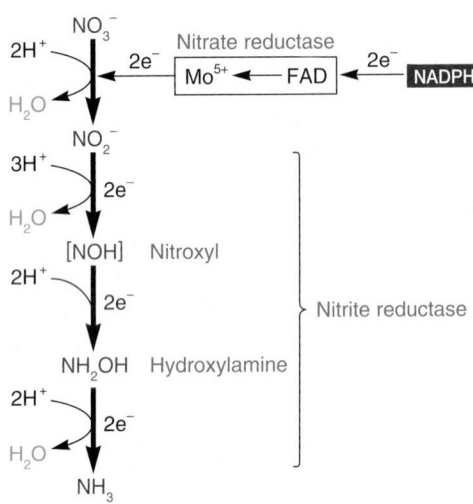

Figure 10.13   Assimilatory Nitrate Reduction. This sequence is thought to operate in bacteria that can reduce and assimilate nitrate nitrogen. See text for details.

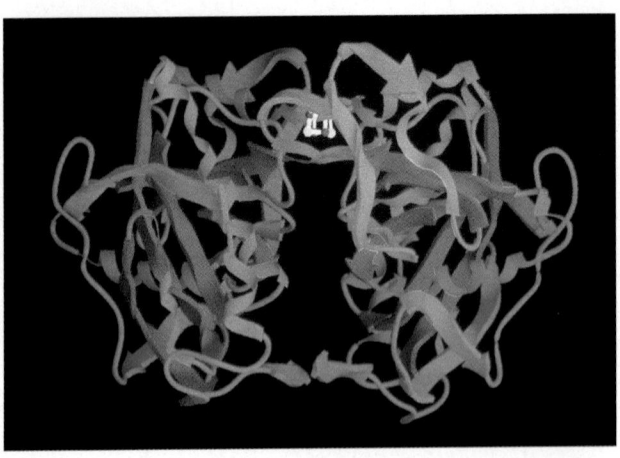

Figure 10.15   Structure of the Nitrogenase Fe Protein. The Fe protein's two subunits are arranged like a pair of butterfly wings with the iron sulfur cluster between the wings and at the "head" of the butterfly. The iron sulfur cluster is very exposed, which helps account for nitrogenase's sensitivity to oxygen. The oxygen can readily attack the exposed iron atoms.

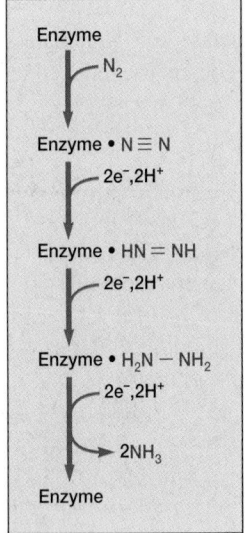

Figure 10.14   Nitrogen Reduction. A hypothetical sequence of nitrogen reduction by nitrogenase.

bacterium) reduces ferredoxin during pyruvate oxidation, whereas the aerobic *Azotobacter* uses electrons from NADPH to reduce ferredoxin.

Nitrogenase is a complex system consisting of two major protein components, a MoFe protein (MW 220,000) joined with one or two Fe proteins (MW 64,000). The MoFe protein contains 2 atoms of molybdenum and 28 to 32 atoms of iron; the Fe protein has 4 iron atoms (**figure 10.15**). Fe protein is first reduced by ferredoxin, then it binds ATP (**figure 10.16**). ATP binding changes the conformation of the Fe protein and lowers its reduction potential, enabling it to reduce the MoFe protein. ATP is hydrolyzed when this electron transfer occurs. Finally, reduced

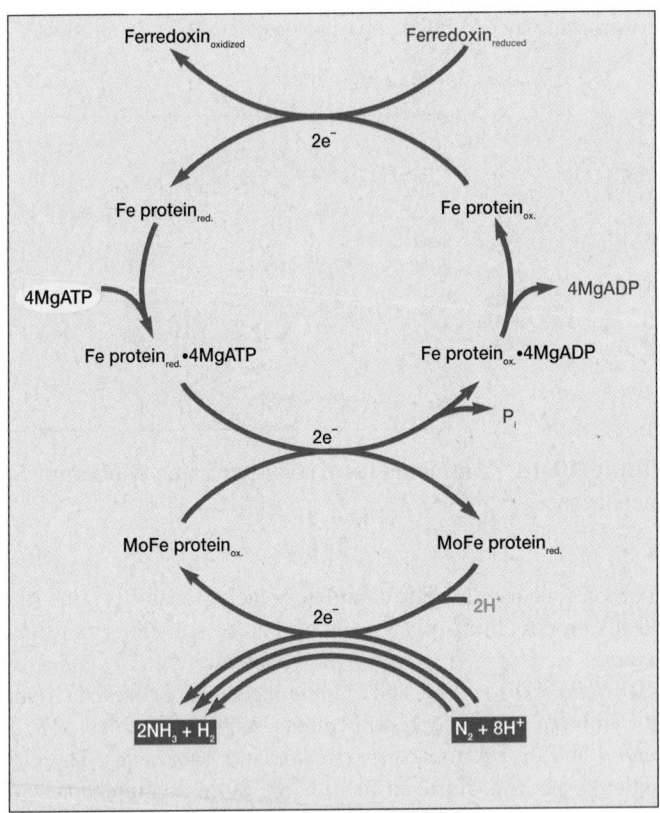

Figure 10.16   Mechanism of Nitrogenase Action. The flow of two electrons from ferredoxin to nitrogen is outlined. This process is repeated three times in order to reduce N_2 to two molecules of ammonia. The stoichiometry at the bottom includes proton reduction to H_2. See the text for a more detailed explanation.

MoFe protein donates electrons to atomic nitrogen. Nitrogenase is quite sensitive to O_2 and must be protected from O_2 inactivation within the cell. In many cyanobacteria, this protection against oxygen is provided by a special structure called the heterocyst (*see pp. 459–60*).

The reduction of N_2 to NH_3 occurs in three steps, each of which requires an electron pair (figures 10.14 and 10.16). Six electron transfers take place, and this requires a total 12 ATPs per N_2 reduced. The overall process actually requires at least 8 electrons and 16 ATPs because nitrogenase also reduces protons to H_2. The H_2 reacts with diimine (HN = NH) to form N_2 and H_2. This futile cycle produces some N_2 even under favorable conditions and makes nitrogen fixation even more expensive. Symbiotic nitrogen-fixing bacteria can consume almost 20% of the ATP produced by the host plant.

Nitrogenase can reduce a variety of molecules containing triple bonds (e.g., acetylene, cyanide, and azide).

$$HC \equiv CH + 2H^+ + 2e^- \longrightarrow H_2C = CH_2$$

The rate of reduction of acetylene to ethylene is even used to estimate nitrogenase activity.

Once molecular nitrogen has been reduced to ammonia, the ammonia can be incorporated into organic compounds. In the symbiotic nitrogen fixer *Rhizobium,* it appears that ammonia diffuses out of the bacterial cell and is assimilated in the surrounding legume cell. The primary route of ammonia assimilation seems to be the synthesis of glutamine by the glutamine synthetase–glutamate synthase system (figure 10.11). However, substances such as the purine derivatives allantoin and allantoic acid also are synthesized and used for the transport of nitrogen to other parts of the plant.

 ## 10.5   THE SYNTHESIS OF AMINO ACIDS

Microorganisms vary with respect to the type of nitrogen source they employ, but most can assimilate some form of inorganic nitrogen by the routes just described. Amino acid synthesis also requires construction of the proper carbon skeletons, and this is often a complex process involving many steps. Because of the need to conserve nitrogen, carbon, and energy, amino acid synthetic pathways are usually tightly regulated by allosteric and feedback mechanisms (*see section 8.9*). Although individual amino acid biosynthetic pathways are not described in detail, a survey of the general pattern of amino acid biosynthesis is worthwhile. Further details of amino acid biosynthesis may be found in introductory biochemistry textbooks.

The relationship of amino acid biosynthetic pathways to amphibolic routes is shown in **figure 10.17.** Amino acid skeletons are derived from acetyl-CoA and from intermediates of the TCA cycle, glycolysis, and the pentose phosphate pathway. To maximize efficiency and economy, the precursors for amino acid biosynthesis are provided by a few major amphibolic pathways. Sequences leading to individual amino acids branch off from

these central routes. Alanine, aspartate, and glutamate are made by transamination directly from pyruvate, oxaloacetate, and α-ketoglutarate, respectively. Most biosynthetic pathways are more complex, and common intermediates often are used in the synthesis of families of related amino acids for the sake of further economy. For example, the amino acids lysine, threonine, isoleucine, and methionine are synthesized from oxaloacetate by such a branching anabolic route (**figure 10.18**). The biosynthetic pathways for the aromatic amino acids phenylalanine, tyrosine, and tryptophan also share many intermediates (**figure 10.19**).

---

1. How do microorganisms assimilate sulfur and phosphorus?
2. Describe the roles of glutamate dehydrogenase, glutamine synthetase, glutamate synthase, and transaminases in ammonia assimilation. How is nitrate incorporated by assimilatory nitrate reduction?
3. What is nitrogen fixation? Briefly describe the structure and mechanism of action of nitrogenase.
4. Summarize in general terms the organization of amino acid biosynthesis.

 ## 10.6   ANAPLEROTIC REACTIONS

Inspection of figure 10.17 will show that TCA cycle intermediates are used in the synthesis of pyrimidines and a wide variety of amino acids. In fact, the biosynthetic functions of this pathway are so essential that most of it must operate anaerobically to supply biosynthetic precursors, even though NADH is not required for electron transport and oxidative phosphorylation in the absence of oxygen. Thus there is a heavy demand upon the TCA cycle to supply carbon for biosynthesis, and cycle intermediates could be depleted if nothing were done to maintain their levels. However, microorganisms have reactions that replenish cycle intermediates so that the TCA cycle can continue to function when active biosynthesis is taking place. Reactions that replace cycle intermediates are called **anaplerotic reactions** [Greek *anaplerotic,* filling up].

Most microorganisms can replace TCA cycle intermediates by CO_2 **fixation,** in which inorganic CO_2 is converted to organic carbon and assimilated. It should be emphasized that anaplerotic reactions do not serve the same function as the CO_2 fixation pathway that supplies the carbon required by autotrophs. In autotrophs CO_2 fixation provides most or all of the carbon required for growth. Anaplerotic CO_2 fixation reactions simply replace TCA cycle intermediates and maintain metabolic balance. Usually CO_2 is added to an acceptor molecule, either pyruvate or phosphoenolpyruvate, to form the cycle intermediate oxaloacetate (figure 10.17). Some microorganisms (e.g., *Arthrobacter globiformis,* yeasts) use pyruvate carboxylase in this role.

$$\text{Pyruvate} + CO_2 + ATP + H_2O \xrightarrow{\text{biotin}}$$
$$\text{oxaloacetate} + ADP + P_i$$

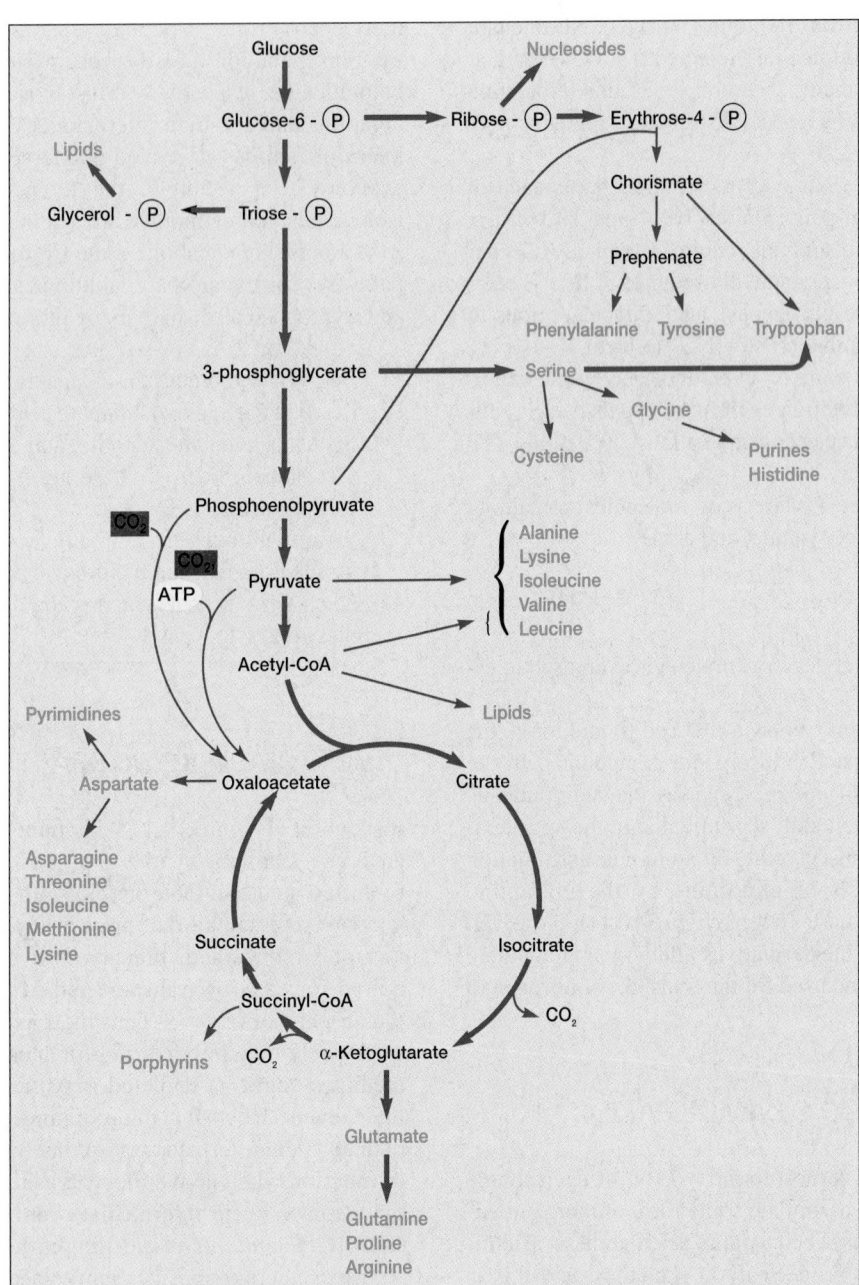

Figure 10.17   The Organization of Anabolism.   Biosynthetic products (in blue) are derived from intermediates of amphibolic pathways. Two major anaplerotic CO_2 fixation reactions are shown in red.

This enzyme requires the cofactor biotin and uses ATP energy to join CO_2 and pyruvate. Biotin is often the cofactor for enzymes catalyzing carboxylation reactions. Because of its importance, biotin is a required growth factor for many species. Other microorganisms, such as the bacteria *Escherichia coli* and *Salmonella typhimurium,* have the enzyme phosphoenolpyruvate carboxylase, which catalyzes the following reaction.

$$\text{Phosphoenolpyruvate} + CO_2 \longrightarrow \text{oxaloacetate} + P_i$$

Some bacteria, algae, fungi, and protozoa can grow with acetate as the sole carbon source by using it to synthesize TCA cy-

cle intermediates in the **glyoxylate cycle** (**figure 10.20**). This cycle is made possible by two unique enzymes, isocitrate lyase and malate synthase, that catalyze the following reactions.

$$\text{Isocitrate} \xrightarrow{\text{isocitrate lyase}} \text{succinate} + \text{glyoxylate}$$

$$\text{Glyoxylate} + \text{acetyl-CoA} \xrightarrow{\text{malate synthase}} \text{malate} + \text{CoA}$$

The glyoxylate cycle is actually a modified TCA cycle. The two decarboxylations of the latter pathway (the isocitrate dehydrogenase and α-ketoglutarate dehydrogenase steps) are bypassed,

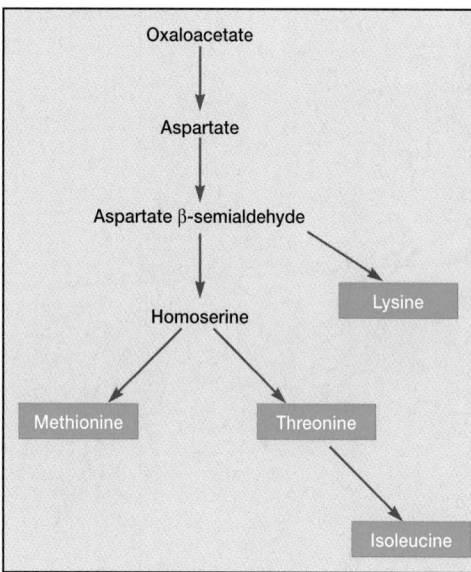

Figure 10.18   A Branching Pathway of Amino Acid Synthesis.   The pathways to methionine, threonine, isoleucine, and lysine. Although some arrows represent one step, most interconversions require the participation of several enzymes.

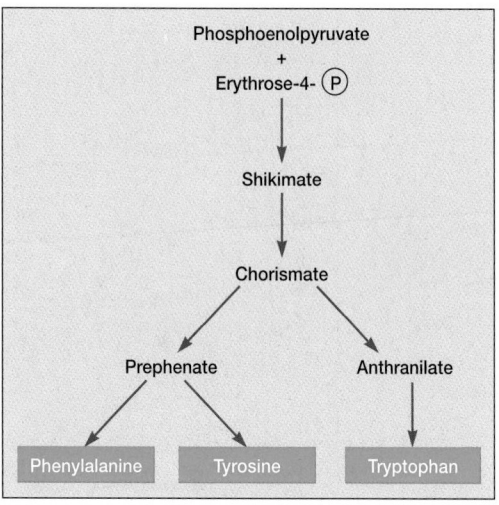

Figure 10.19   Aromatic Amino Acid Synthesis.   The synthesis of the aromatic amino acids phenylalanine, tyrosine, and tryptophan. Most arrows represent more than one enzyme reaction.

making possible the conversion of acetyl-CoA to form oxaloacetate without loss of acetyl-CoA carbon as CO_2. In this fashion acetate and any molecules that give rise to it can contribute carbon to the cycle and support microbial growth.   The TCA cycle (pp. 178–79)

---

1. Define an anaplerotic reaction and give an example.
2. How does the glyoxylate cycle convert acetyl-CoA to oxaloacetate, and what special enzymes are used?

 **THE SYNTHESIS OF PURINES, PYRIMIDINES, AND NUCLEOTIDES**

Purine and pyrimidine biosynthesis is critical for all cells because these molecules are used in the synthesis of ATP, several cofactors, ribonucleic acid (RNA), deoxyribonucleic acid (DNA), and other important cell components. Nearly all microorganisms can synthesize their own purines and pyrimidines as these are so crucial to cell function.   DNA and RNA synthesis (pp. 229–33, 254–60)

Purines and **pyrimidines** are cyclic nitrogenous bases with several double bonds and pronounced aromatic properties. Purines consist of two joined rings, whereas pyrimidines have only one (**figure 10.21** and figure 10.23). The purines **adenine** and **guanine** and the pyrimidines **uracil, cytosine,** and **thymine** are commonly found in microorganisms. A purine or pyrimidine base joined with a pentose sugar, either ribose or deoxyribose, is a **nucleoside.** A **nucleotide** is a nucleoside with one or more phosphate groups attached to the sugar.

Purine Biosynthesis

The biosynthetic pathway for purines is a complex, 11-step sequence (*see appendix II*) in which seven different molecules contribute parts to the final purine skeleton (figure 10.21). Because the pathway begins with ribose 5-phosphate and the purine skeleton is constructed on this sugar, the first purine product of the pathway is the nucleotide inosinic acid, not a free purine base. The cofactor folic acid is very important in purine biosynthesis. Folic acid derivatives contribute carbons two and eight to the purine skeleton. In fact, the drug sulfonamide inhibits bacterial growth by blocking folic acid synthesis. This interferes with purine biosynthesis and other processes that require folic acid.

Once inosinic acid has been formed, relatively short pathways synthesize adenosine monophosphate and guanosine monophosphate (**figure 10.22**) and produce nucleoside diphosphates and triphosphates by phosphate transfers from ATP. DNA contains deoxyribonucleotides (the ribose lacks a hydroxyl group on carbon two) instead of the ribonucleotides found in RNA. Deoxyribonucleotides arise from the reduction of nucleoside diphosphates or nucleoside triphosphates by two different routes. Some microorganisms reduce the triphosphates with a system requiring vitamin B_{12} as a cofactor. Others, such as *E. coli*, reduce the ribose in nucleoside diphosphates. Both systems employ a small sulfur-containing protein called thioredoxin as their reducing agent.

Pyrimidine Biosynthesis

Pyrimidine biosynthesis begins with aspartic acid and carbamoyl phosphate, a high-energy molecule synthesized from CO_2 and ammonia (**figure 10.23**). Aspartate carbamoyltransferase catalyzes the

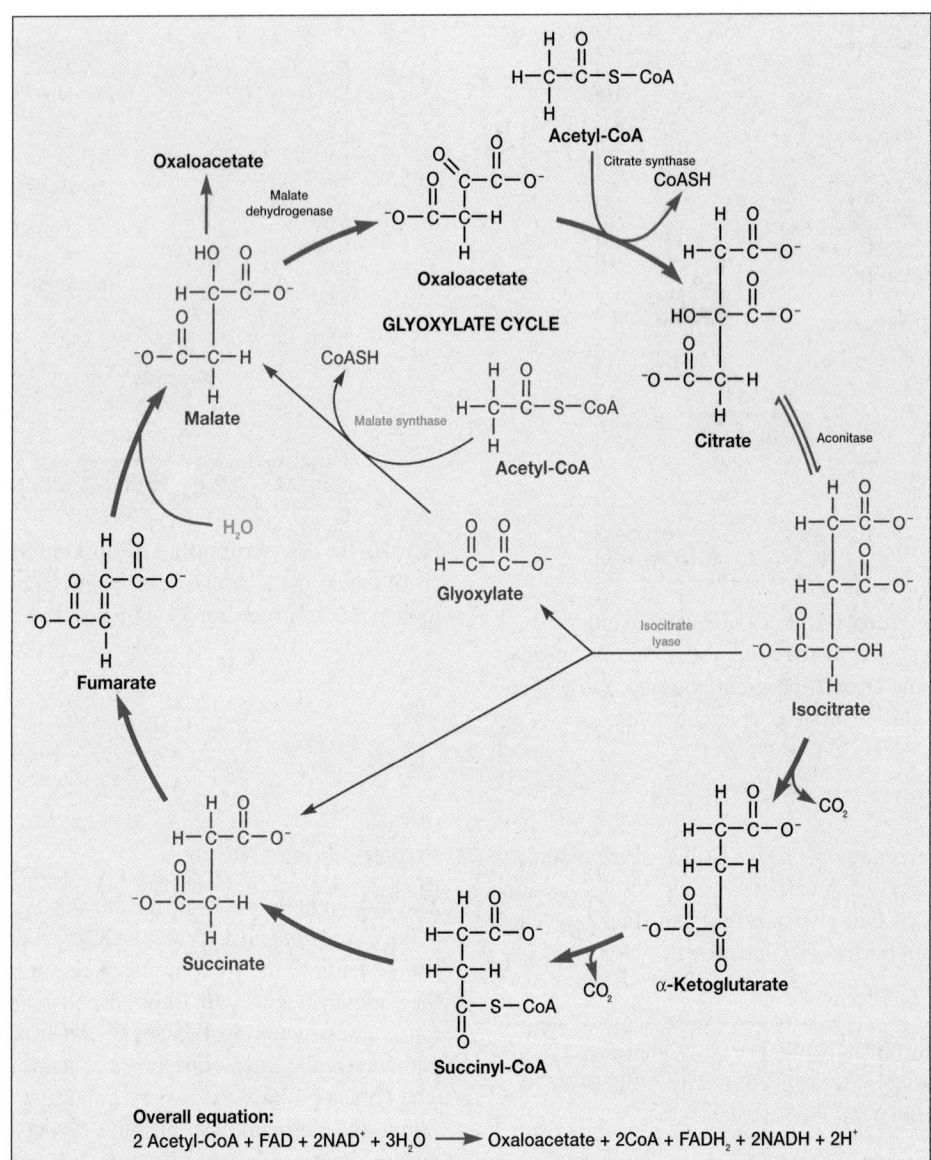

Figure 10.20 The Glyoxylate Cycle. The reactions and enzymes unique to the cycle are shown in red. The tricarboxylic acid cycle enzymes that have been bypassed are at the bottom.

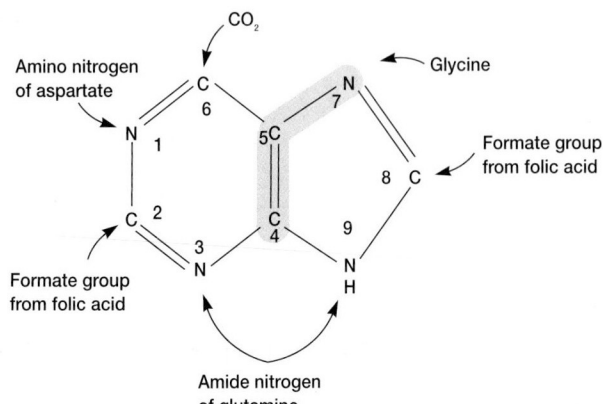

Figure 10.21 Purine Biosynthesis. The sources of purine skeleton nitrogen and carbon are indicated. The contribution of glycine is shaded in blue.

condensation of these two substrates to form carbamoylaspartate, which is then converted to the initial pyrimidine product, orotic acid. The regulation of aspartate carbamoyltransferase activity (pp. 162–63)

After synthesis of the pyrimidine skeleton, a nucleotide is produced by the ribose 5-phosphate addition using the high-energy intermediate 5-phosphoribosyl 1-pyrophosphate. Thus construction of the pyrimidine ring is completed before ribose is added, in contrast with purine ring synthesis, which begins with ribose 5-phosphate. Decarboxylation of orotidine monophosphate yields uridine monophosphate and eventually uridine triphosphate and cytidine triphosphate.

The third common pyrimidine is thymine, a constituent of DNA. The ribose in pyrimidine nucleotides is reduced in the same way as it is in purine nucleotides. Then deoxyuridine monophosphate is methylated with a folic acid derivative to form deoxythymidine monophosphate (**figure 10.24**).

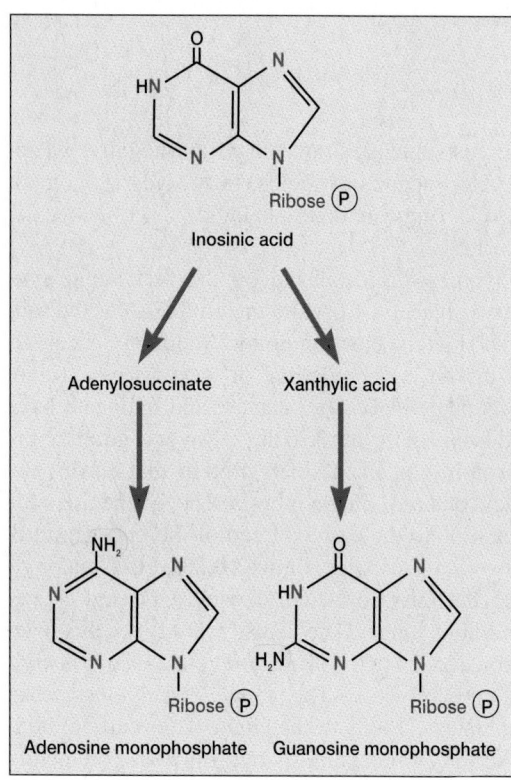

Inosinic acid

Adenylosuccinate   Xanthylic acid

Adenosine monophosphate   Guanosine monophosphate

Figure 10.22   Synthesis of Adenosine Monophosphate and Guanosine Monophosphate.   The shaded groups are the ones differing from those in inosinic acid.

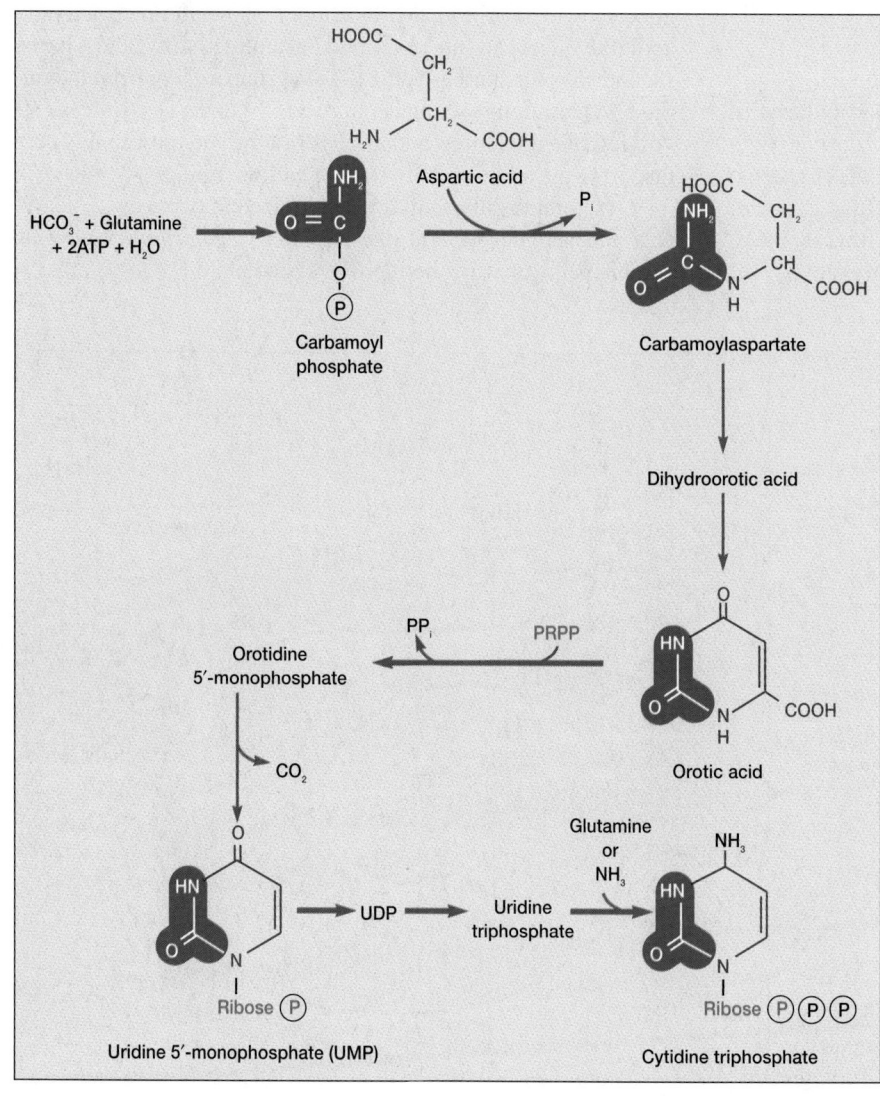

Aspartic acid

$HCO_3^- + Glutamine + 2ATP + H_2O$ P_i

Carbamoyl phosphate

Carbamoylaspartate

Dihydroorotic acid

PP_i PRPP

Orotidine 5'-monophosphate

Orotic acid

CO_2

Glutamine or NH_3

UDP   Uridine triphosphate

Uridine 5'-monophosphate (UMP)   Cytidine triphosphate

Figure 10.23   Pyrimidine Synthesis.   PRPP stands for 5-phosphoribose 1-pyrophosphoric acid, which provides the ribose 5-phosphate chain. The part derived from carbamoyl phosphate is shaded in turqoise.

Figure 10.24   Deoxythymidine Monophosphate Synthesis.
Deoxythymidine differs from deoxyuridine in having the shaded methyl group.

---

1. Define purine, pyrimidine, nucleoside, and nucleotide.
2. Outline the way in which purines and pyrimidines are synthesized. How is the deoxyribose component of deoxyribonucleotides made?

10.8   LIPID SYNTHESIS

A variety of lipids are found in microorganisms, particularly in cell membranes. Most contain **fatty acids** or their derivatives. Fatty acids are monocarboxylic acids with long alkyl chains that usually have an even number of carbons (the average length is 18 carbons). Some may be unsaturated—that is, have one or more double bonds. Most microbial fatty acids are straight chained, but some are branched. Gram-negative bacteria often have cyclopropane fatty acids (fatty acids with one or more cyclopropane rings in their chains). Lipid structure and nomenclature (appendix I)

Fatty acid synthesis is catalyzed by the **fatty acid synthetase** complex with acetyl-CoA and malonyl-CoA as the substrates and NADPH as the electron donor. Malonyl-CoA arises from the ATP-driven carboxylation of acetyl-CoA (figure 10.25). Synthesis takes place after acetate and malonate have been transferred from coenzyme A to the sulfhydryl group of the **acyl carrier protein (ACP),** a small protein that carries the growing fatty acid chain during synthesis. The synthetase adds two carbons at a time to the carboxyl end of the growing fatty acid chain in a two-stage process (**figure 10.25**). First, malonyl-ACP reacts with the fatty acyl-ACP to yield CO_2 and a fatty acyl-ACP two carbons longer. The loss of CO_2 drives this reaction to completion. Notice that ATP is used to add CO_2 to acetyl-CoA, forming malonyl-CoA. The same CO_2 is lost when malonyl-ACP donates carbons to the chain. Thus carbon dioxide is essential to fatty acid synthesis but it is not permanently incorporated. Indeed, some microorganisms require CO_2 for good growth, but they can do without it in the presence of a fatty acid like oleic acid (an 18-carbon unsaturated fatty acid). In the second stage of synthesis, the β-keto group arising from the initial condensation reaction is removed in a three-step process involving two reductions and a dehydration. The fatty acid is then ready for the addition of two more carbon atoms.

Unsaturated fatty acids are synthesized in two ways. Eucaryotes and aerobic bacteria like *Bacillus megaterium* employ an aerobic pathway using both NADPH and O_2.

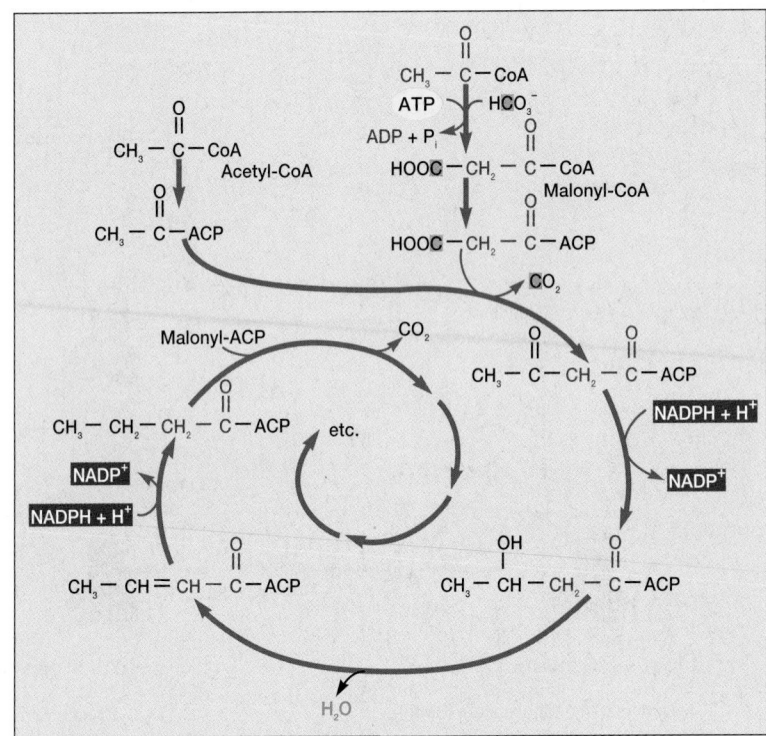

Figure 10.25   Fatty Acid Synthesis.   The cycle is repeated until the proper chain length has been reached. Carbon dioxide carbon and the remainder of malonyl-CoA are shown in red. ACP stands for acyl carrier protein.

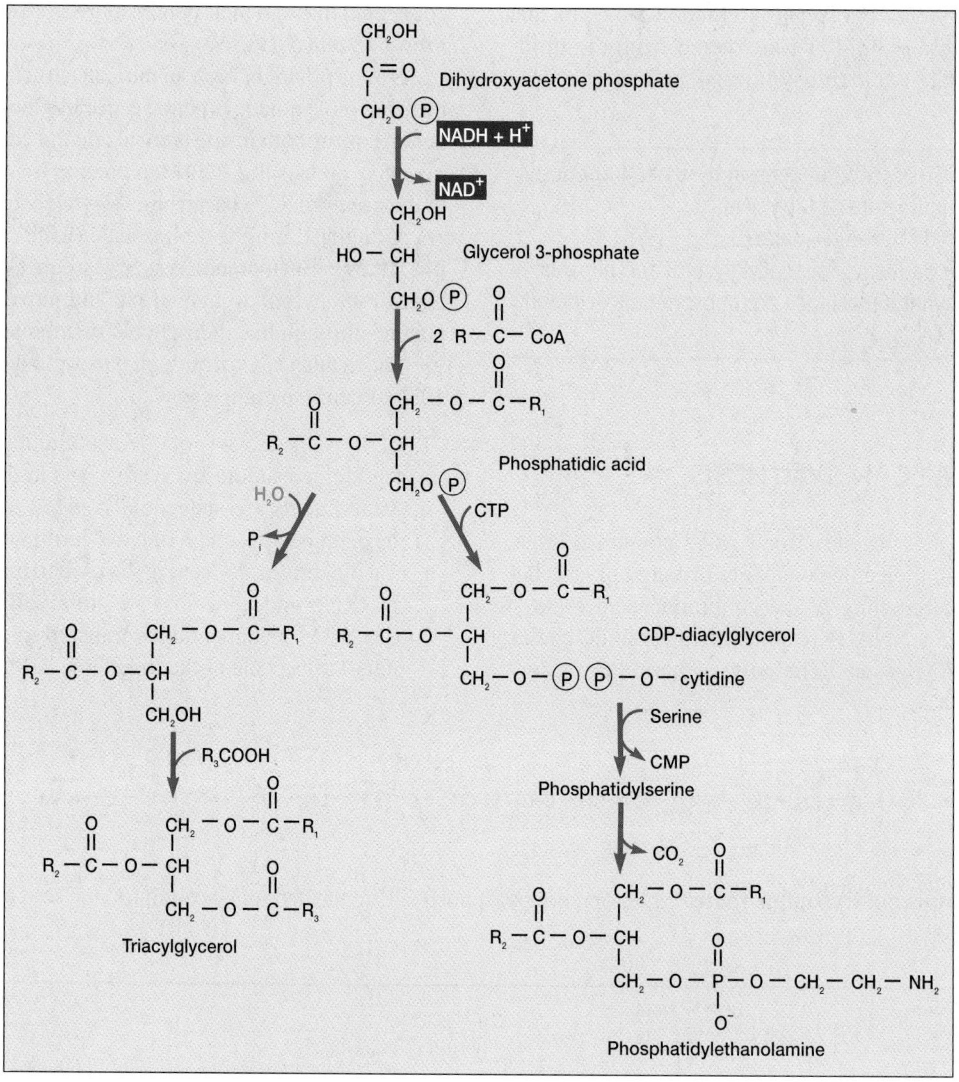

Figure 10.26 Triacylglycerol and Phospholipid Synthesis.

$$R — (CH_2)_9 — \overset{\overset{\textstyle O}{\|}}{C} — SCoA + NADPH + H^+ + O_2 \longrightarrow$$

$$R — CH = CH — (CH_2)_7 — \overset{\overset{\textstyle O}{\|}}{C} — SCoA + NADPH^+ + 2H_2O$$

A double bond is formed between carbons nine and ten, and O_2 is reduced to water with electrons supplied by both the fatty acid and NADPH. Anaerobic bacteria and some aerobes create double bonds during fatty acid synthesis by dehydrating hydroxy fatty acids. Oxygen is not required for double bond synthesis by this pathway. The anaerobic pathway is present in a number of common gram-negative bacteria (e.g., *Escherichia coli* and *Salmonella typhimurium*), gram-positive bacteria (e.g., *Lactobacillus plantarum* and *Clostridium pasteurianum*), and cyanobacteria.

Eucaryotic microorganisms frequently store carbon and energy as **triacylglycerol,** glycerol esterified to three fatty acids. Glycerol arises from the reduction of the glycolytic intermediate dihydroxyacetone phosphate to glycerol 3-phosphate, which is then esterified with two fatty acids to give **phosphatidic acid** (**figure 10.26**). Phosphate is hydrolyzed from phosphatidic acid giving a diacylglycerol, and the third fatty acid is attached to yield a triacylglycerol.

Phospholipids are major components of eucaryotic and most procaryotic cell membranes. Their synthesis also usually proceeds by way of phosphatidic acid. A special cytidine diphosphate (CDP) carrier plays a role similar to that of uridine and adenosine diphosphate carriers in carbohydrate biosynthesis. For example, bacteria synthesize phosphatidylethanolamine, a major cell membrane component, through the initial formation of CDP-diacylglycerol (figure 10.26). This CDP derivative then reacts with serine to form the phospholipid phosphatidylserine, and decarboxylation yields phosphatidylethanolamine. In this way a complex membrane lipid is constructed from the products of glycolysis, fatty acid biosynthesis, and amino acid biosynthesis.

---

1. What is a fatty acid? Describe in general terms how the fatty acid synthetase manufactures a fatty acid.
2. How are unsaturated fatty acids made?
3. Briefly describe the pathways for triacylglycerol and phospholipid synthesis. Of what importance are phosphatidic acid and CDP-diacylglycerol?

 10.9 PEPTIDOGLYCAN SYNTHESIS

As discussed earlier, most bacterial cell walls contain a large, complex peptidoglycan molecule consisting of long polysaccharide chains made of alternating *N*-acetylmuramic acid (NAM) and *N*-acetylglucosamine (NAG) residues. Pentapeptide chains are attached to the NAM groups. The polysaccharide chains are

connected through their pentapeptides or by interbridges (*see figures 3.18 and 3.19*). Peptidoglycan structure and function (pp. 54-55)

Not surprisingly such an intricate structure requires an equally intricate biosynthetic process, especially because the synthetic reactions occur both inside and outside the cell membrane. Peptidoglycan synthesis is a multistep process that has been best studied in the gram-positive bacterium *Staphylococcus aureus*. Two carriers participate: uridine diphosphate (UDP) and **bactoprenol (figure 10.27)**. Bactoprenol is a 55-carbon alcohol that attaches to NAM by a pyrophosphate group and moves peptidoglycan components through the hydrophobic membrane.

The synthesis of peptidoglycan, outlined in **figures 10.28** and 10.29, occurs in eight stages.

1. UDP derivatives of *N*-acetylmuramic acid and *N*-acetylglucosamine are synthesized in the cytoplasm.
2. Amino acids are sequentially added to UDP-NAM to form the pentapeptide chain (the two terminal D-alanines are added as a dipeptide). ATP energy is used to make the peptide bonds, but tRNA and ribosomes are not involved.
3. The NAM-pentapeptide is transferred from UDP to a bactoprenol phosphate at the membrane surface.

Figure 10.27 Bactoprenol Pyrophosphate.   Bactoprenol pyrophosphate connected to *N*-acetylmuramic acid (NAM).

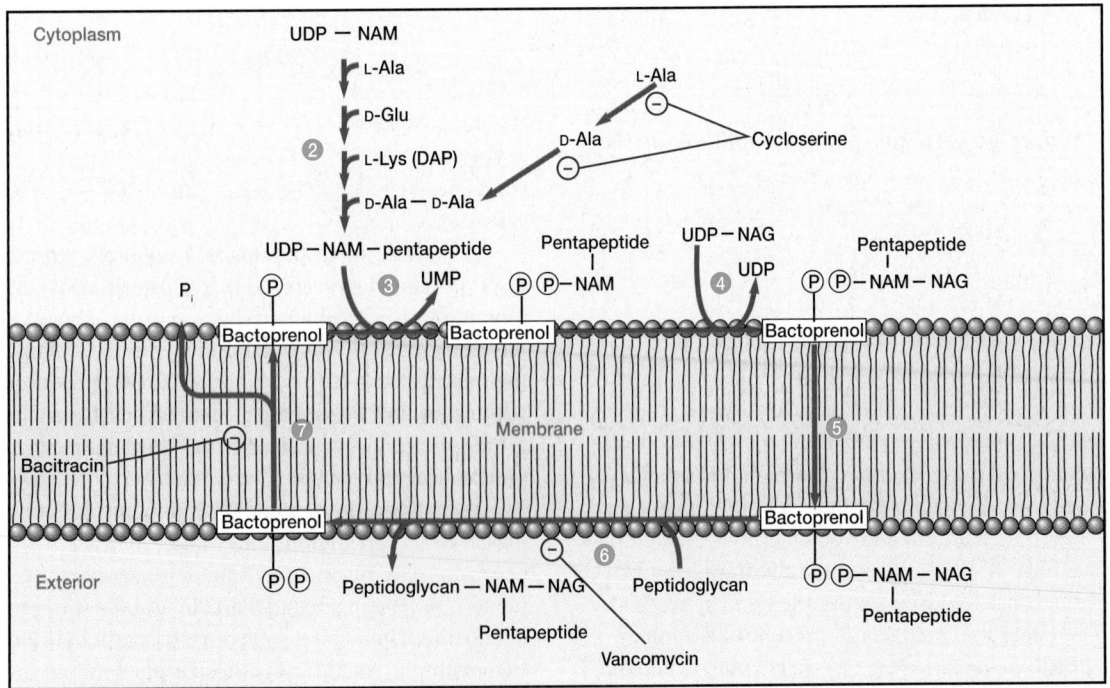

Figure 10.28 Peptidoglycan Synthesis.   NAM is *N*-acetylmuramic acid and NAG is *N*-acetylglucosamine. The pentapeptide contains L-lysine in *S. aureus* peptidoglycan, and diaminopimelic acid (DAP) in *E. coli*. Inhibition by bacitracin, cycloserine, and vancomycin also is shown. The numbers correspond to six of the eight stages discussed in the text. Stage eight is depicted in figure 10.29.

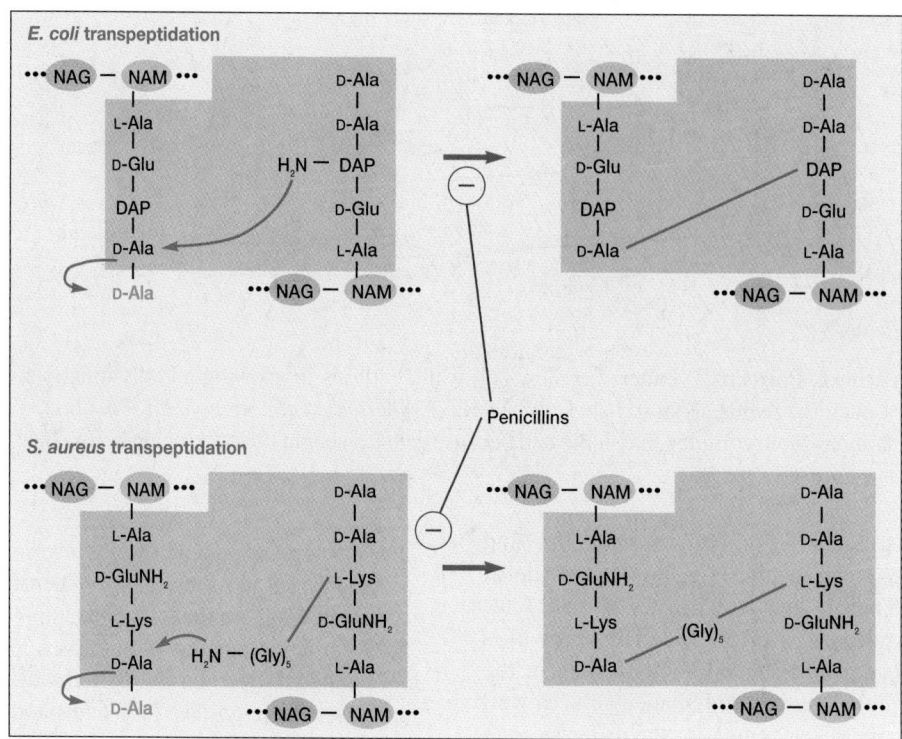

Figure 10.29   Transpeptidation.   The transpeptidation reactions in the formation of the peptidoglycans of *Escherichia coli* and *Staphylococcus aureus.*

4. UDP-NAG adds NAG to the NAM-pentapeptide to form the peptidoglycan repeat unit. If a pentaglycine interbridge is required, the glycines are added using special glycyl-tRNA molecules, not ribosomes.

5. The completed NAM-NAG peptidoglycan repeat unit is transported across the membrane to its outer surface by the bactoprenol pyrophosphate carrier.

6. The peptidoglycan unit is attached to the growing end of a peptidoglycan chain to lengthen it by one repeat unit.

7. The bactoprenol carrier returns to the inside of the membrane. A phosphate is released during this process to give bactoprenol phosphate, which can now accept another NAM-pentapeptide.

8. Finally, peptide cross-links between the peptidoglycan chains are formed by **transpeptidation (figure 10.29).** In *E. coli* the free amino group of diaminopimelic acid attacks the subterminal D-alanine, releasing the terminal D-alanine residue. ATP is used to form the terminal peptide bond inside the membrane. No more ATP energy is required when transpeptidation takes place on the outside. The same process occurs when an interbridge is involved; only the group reacting with the subterminal D-alanine differs.

Peptidoglycan synthesis is particularly vulnerable to disruption by antimicrobial agents. Inhibition of any stage of synthesis weakens the cell wall and can lead to osmotic lysis. Many antibiotics interfere with peptidoglycan synthesis. For example, penicillin inhibits the transpeptidation reaction (figure 10.29), and

bacitracin blocks the dephosphorylation of bactoprenol pyrophosphate (figure 10.28). Antibiotic effects on cell wall synthesis (pp. 788–90, 791)

10.10   PATTERNS OF CELL WALL FORMATION

To grow and divide efficiently, a bacterial cell must add new peptidoglycan to its cell wall in a precise and well-regulated way while maintaining wall shape and integrity in the presence of high osmotic pressure. Because the cell wall peptidoglycan is essentially a single enormous network, the growing bacterium must be able to degrade it just enough to provide acceptor ends for the incorporation of new peptidoglycan units. It must also reorganize peptidoglycan structure when necessary. This limited peptidoglycan digestion is accomplished by enzymes known as **autolysins,** some of which attack the polysaccharide chains, while others hydrolyze the peptide cross-links. Autolysin inhibitors keep the activity of these enzymes under tight control. Control of cell division (pp. 280–82)

Although the location and distribution of cell wall synthetic activity varies with species, there seem to be two general patterns (**figure 10.30**). Many gram-positive cocci (e.g., *Enterococcus faecalis* and *Streptococcus pyogenes*) have only one to a few zones of growth. The principal growth zone is usually at the site of septum formation, and new cell halves are synthesized back-to-back. The second pattern of synthesis occurs in

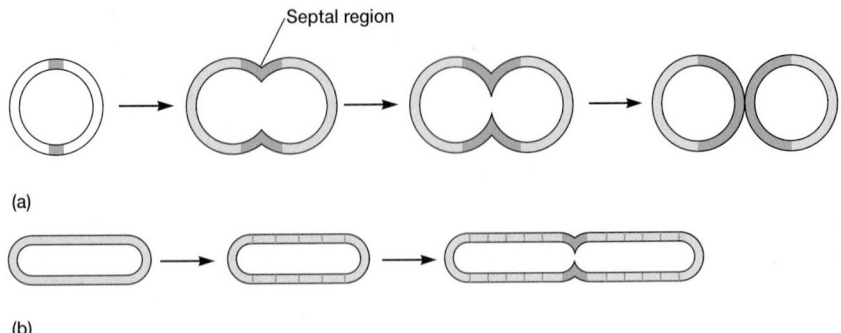

Septal region

(a)

(b)

Figure 10.30 Wall Synthesis Patterns. Patterns of new cell wall synthesis in growing and dividing bacteria. (**a**) Streptococci and some other gram-positive cocci. (**b**) Synthesis in rod-shaped bacteria (*Escherichia coli, Salmonella, Bacillus*). The zones of growth are in blue-green. The actual situation is more complex than indicated because cells can begin to divide again before the first division is completed.

the rod-shaped bacteria *Escherichia coli, Salmonella,* and *Bacillus.* Active peptidoglycan synthesis occurs at the site of septum formation just as before, but growth sites also are scattered along the cylindrical portion of the rod. Thus growth is distributed more diffusely in rod-shaped bacteria than in the streptococci. Synthesis must lengthen rod-shaped cells as well as divide them. Presumably this accounts for the differences in wall growth pattern.

1. Outline in a diagram the steps involved in the synthesis of peptidoglycan and show their relationship to the plasma membrane. What are the roles of bactoprenol and UDP?
2. What is the function of autolysins in cell wall peptidoglycan synthesis? Describe the patterns of peptidoglycan synthesis seen in gram-positive cocci and in rod-shaped bacteria such as *E. coli.*

In biosynthesis or anabolism, cells use energy to construct complex molecules from smaller, simpler precursors.

10.1 Principles Governing Biosynthesis
a. Many important cell constituents are macromolecules, large polymers constructed of simple monomers.
b. Although many catabolic and anabolic pathways share enzymes for the sake of efficiency, some of their enzymes are separate and independently regulated.
c. Macromolecular components often undergo self-assembly to form the final molecule or complex.

10.2 The Photosynthetic Fixation of CO_2
a. Photosynthetic CO_2 fixation is carried out by the Calvin cycle and may be divided into three phases: the carboxylation phase, the reduction phase, and the regeneration phase (**figure 10.4**). Three ATPs and two NADPHs are used during the incorporation of one CO_2.

10.3 Synthesis of Sugars and Polysaccharides
a. Gluconeogenesis is the synthesis of glucose and related sugars from nonglucose precursors.
b. Glucose, fructose, and mannose are gluconeogenic intermediates or made directly from

them; galactose is synthesized with nucleoside diphosphate derivatives. Bacteria and algae synthesize glycogen and starch from adenosine diphosphate glucose.

10.4 The Assimilation of Inorganic Phosphorus, Sulfur, and Nitrogen
a. Phosphorus is obtained from inorganic or organic phosphate.
b. Microorganisms can use cysteine, methionine, and inorganic sulfate as sulfur sources. Sulfate is reduced to sulfide during assimilatory sulfate reduction.
c. Ammonia nitrogen can be directly assimilated by the activity of transaminases and either glutamate dehydrogenase or the glutamine synthetase–glutamate synthase system (**figures 10.10–10.12**).
d. Nitrate is incorporated through assimilatory nitrate reduction catalyzed by the enzymes nitrate reductase and nitrite reductase.
e. Nitrogen fixation is catalyzed by the nitrogenase complex. Atmospheric molecular nitrogen is reduced to ammonia, which is then incorporated into amino acids (**figures 10.14** and **10.16**).

10.5 The Synthesis of Amino Acids
a. Amino acid biosynthetic pathways branch off from the central amphibolic pathways (**figure 10.17**).

10.6 Anaplerotic Reactions
a. Anaplerotic reactions replace TCA cycle intermediates to keep the cycle in balance while it supplies biosynthetic precursors. Many anaplerotic enzymes catalyze CO_2 fixation reactions. The glyoxylate cycle is also anaplerotic.

10.7 The Synthesis of Purines, Pyrimidines, and Nucleotides
a. Purines and pyrimidines are nitrogenous bases found in DNA, RNA, and other molecules. The purine skeleton is synthesized beginning with ribose 5-phosphate and initially produces inosinic acid. Pyrimidine biosynthesis starts with carbamoyl phosphate and aspartate, and ribose is added after the skeleton has been constructed.

10.8 Lipid Synthesis
a. Fatty acids are synthesized from acetyl-CoA, malonyl-CoA, and NADPH by the fatty acid synthetase system. During synthesis the intermediates are attached to the acyl carrier protein. Double bonds can be added in two different ways.
b. Triacylglycerols are made from fatty acids and glycerol phosphate. Phosphatidic acid is an important intermediate in this pathway.
c. Phospholipids like phosphatidylethanolamine can be synthesized from phosphatidic acid by forming CDP-diacylglycerol, then adding an amino acid.

10.9 Peptidoglycan Synthesis

a. Peptidoglycan synthesis is a complex process involving both UDP derivatives and the lipid carrier bactoprenol, which transports NAM-NAG-pentapeptide units across the cell membrane. Cross-links are formed by transpeptidation (**figures 10.28** and **10.29**).

10.10 Patterns of Cell Wall Formation

a. Peptidoglycan synthesis occurs in discrete zones in the cell wall. Existing peptidoglycan is selectively degraded by autolysins so new material can be added.

Key Terms

acyl carrier protein (ACP)   214
adenine   211
anaplerotic reactions   209
assimilatory nitrate reduction   206
assimilatory sulfate reduction   205
autolysins   217
bactoprenol   216
Calvin cycle   202
carboxysomes   202
CO_2 fixation   209
cytosine   211
dissimilatory sulfate reduction   205

fatty acid   214
fatty acid synthetase   214
gluconeogenesis   203
glutamate dehydrogenase   206
glutamate synthase   206
glutamine synthetase   206
glyoxylate cycle   210
guanine   211
macromolecule   200
monomers   201
nitrate reductase   206

nitrite reductase   206
nitrogenase   207
nitrogen fixation   206
nucleoside   211
nucleotide   211
phosphatase   205
phosphatidic acid   215
phosphoadenosine 5'-phosphosulfate   205
purine   211
pyrimidine   211

ribulose-1,5-bisphosphate carboxylase   202
self-assembly   201
thymine   211
transaminases   206
transpeptidation   217
triacylglycerol   215
turnover   200
uracil   211
uridine diphosphate glucose (UDPG)   204

Questions for Thought and Review

1. Discuss the relationship between catabolism and anabolism. How does anabolism depend on catabolism?
2. Suppose that a microorganism is growing on a medium that contains amino acids but no sugars. In general terms, how would it synthesize the pentoses and hexoses it requires?
3. Activated carriers participate in carbohydrate, lipid, and peptidoglycan synthesis. Briefly describe these carriers and their roles.
4. Which two enzymes discussed in the chapter appear to be specific to the Calvin cycle?
5. Why can phosphorus be directly incorporated into cell constituents whereas sulfur and nitrogen often cannot?
6. What is unusual about the synthesis of peptides that takes place during peptidoglycan construction?

Critical Thinking Questions

1. In metabolism important intermediates are covalently attached to carriers, as if to mark these as important so the cell does not lose track of them. Think about a hotel placing your room key on a very large ring. List a few examples of these carriers and indicate whether they are involved primarily in anabolism or catabolism.
2. Intermediary carriers are in a limited supply—when they cannot be recycled because of a metabolic block, serious consequences ensue. Think of some examples of these consequences.

For recommended readings on these and related topics, see Appendix V.

Genes:
Structure, Replication, and Mutation

Concepts

- The two kinds of nucleic acid, deoxyribonucleic acid (DNA) and ribonucleic acid (RNA), differ from one another in chemical composition and structure. In procaryotic and eucaryotic cells, DNA serves as the repository for genetic information.
- DNA is associated with basic proteins in the cell. In eucaryotes these are special histone proteins, whereas in procaryotes nonhistone proteins are complexed with DNA.
- The flow of genetic information usually proceeds from DNA through RNA to protein. A protein's amino acid sequence reflects the nucleotide sequence of its mRNA. This messenger is a complementary copy of a portion of the DNA genome.
- DNA replication is a very complex process involving a variety of proteins and a number of steps. It is designed to operate rapidly while minimizing errors and correcting those that arise when the DNA sequence is copied.
- Genetic information is contained in the nucleotide sequence of DNA (and sometimes RNA). When a structural gene directs the synthesis of a polypeptide, each amino acid is specified by a triplet codon.
- A gene is a nucleotide sequence that codes for a polypeptide, tRNA, or rRNA.
- Most bacterial genes have at least four major parts, each with different functions: promoters, leaders, coding regions, and trailers.
- Mutations are stable, heritable alterations in the gene sequence and usually, but not always, produce phenotypic changes. Nucleic acids are altered in several different ways, and these mutations may be either spontaneous or induced by chemical mutagens or radiation.
- It is extremely important to keep the nucleotide sequence constant, and microorganisms have several repair mechanisms designed to detect alterations in the genetic material and restore it to its original state. Often more than one repair system can correct a particular type of mutation. Despite these efforts some alterations remain uncorrected and provide material and opportunity for evolutionary change.

This model illustrates double-stranded DNA. DNA is the genetic material for procaryotes and eucaryotes. Genetic information is contained in the sequence of base pairs that lie in the center of the helix.

But the most important qualification of bacteria for genetic studies is their extremely rapid rate of growth. . . . a single E. coli cell will grow overnight into a visible colony containing millions of cells, even under relatively poor growth conditions. Thus, genetic experiments on E. coli usually last one day, whereas experiments on corn, for example, take months. It is no wonder that we know so much more about the genetics of E. coli than about the genetics of corn, even though we have been studying corn much longer.

—R. F. Weaver and P. W. Hedrick

hapters 8 through 10 have introduced the essentials of microbial metabolism. We now turn to microbial genetics and molecular biology. This chapter reviews some of the most basic concepts of molecular genetics: how genetic information is stored and organized in the DNA molecule, the way in which DNA is replicated, the nature of the genetic code, gene structure, mutagenesis, and DNA repair. In addition, the use of microorganisms to identify potentially dangerous mutagenic agents in the fight against cancer is described. Much of this information will be familiar to those who have taken an introductory genetics course. Because of the importance of procaryotes, primary emphasis is placed on their genetics.

Based on the foundation provided by this chapter, chapter 12 will focus on gene expression and its regulation. Chapter 13 contains information on plasmids and the nature of genetic recombination in microorganisms. These three chapters provide the background needed for understanding the material on recombinant DNA technology (chapter 14) and microbial genomics (chapter 15).

Geneticists, including microbial geneticists, use a specialized vocabulary because of the complexities of their discipline. Some knowledge of basic terminology is necessary at the beginning of this survey of general principles. The experimental material of the microbial geneticist is the **clone.** A clone is a population of cells that are derived asexually from a parental cell and are genetically identical. Sometimes a clone is called a pure culture. The term **genome** refers to all the genes present in a cell or virus. Procaryotes normally have one set of genes. That is, they are haploid (1N). Eucaryotic microorganisms usually have two sets of genes, or are diploid (2N). The genotype of an organism is the specific set of genes it possesses. In contrast, the phenotype is the collection of characteristics that are observable by the investigator. All genes are not expressed at the same time, and the environment profoundly influences phenotypic expression. Much genetics research has focused on the relationship between an organism's genotype and phenotype, and gene expression will be the focus of chapter 12.

Although genetic analysis began with the rediscovery of the work of Gregor Mendel in the early part of the twentieth century, subsequent elegant experimentation involving both bacteria and bacteriophages actually elucidated the nature of genetic informa-

tion, gene structure, the genetic code, and mutations. We will first review a few of these early experiments and then summarize the view of DNA, RNA and protein relationships—sometimes called the Central Dogma—that has guided much of modern research.

11.1 DNA AS GENETIC MATERIAL

The early work of Fred Griffith in 1928 on the transfer of virulence in the pathogen *Streptococcus pneumoniae* (**figure 11.1**) set the stage for the research that first showed that DNA was the genetic material. Griffith found that if he boiled virulent bacteria and injected them into mice, the mice were not affected and no pneumococci could be recovered from the animals. When he injected a combination of killed virulent bacteria and a living nonvirulent strain, the mice died; moreover, he could recover living virulent bacteria from the dead mice. Griffith called this change of nonvirulent bacteria into virulent pathogens **transformation.**

Oswald T. Avery and his colleagues then set out to discover which constituent in the heat-killed virulent pneumococci was responsible for Griffith's transformation. These investigators selectively destroyed constituents in purified extracts of virulent pneumococci, using enzymes that would hydrolyze DNA, RNA, or protein. They then exposed nonvirulent pneumococcal strains to the treated extracts. Transformation of the nonvirulent bacteria was blocked only if the DNA was destroyed, suggesting that DNA was carrying the information required for transformation (**figure 11.2**). The publication of these studies by O. T. Avery, C. M. MacLeod, and M. J. McCarty in 1944 provided the first evidence that Griffith's transforming principle was DNA and therefore that DNA carried genetic information.

Some years later (1952), Alfred D. Hershey and Martha Chase performed several experiments that indicated that DNA was the genetic material in the T2 bacteriophage. Some luck was involved in their discovery, for the genetic material of many viruses is RNA and the researchers happened to select a DNA virus for their studies. Imagine the confusion if T2 had been an RNA virus! The controversy surrounding the nature of genetic information might have lasted considerably longer than it did. Hershey and Chase either made the virus DNA radioactive with ^{32}P or labeled the viral protein coat with ^{35}S. They mixed radioactive bacteriophage with *E. coli* and incubated the mixture for a few minutes. The suspension was then agitated violently in a Waring blender to shear off any adsorbed bacteriophage particles (**figure 11.3**). After centrifugation, radioactivity in the supernatant and the bacterial pellet was determined. They found that most radioactive protein was released into the supernatant, whereas ^{32}P DNA remained within the bacteria. Since genetic material was injected and T2 progeny were produced, DNA must have been carrying the genetic information for T2.   The biology of bacteriophages (chapter 17)

Subsequent studies on the genetics of viruses and bacteria were largely responsible for the rapid development of molecular genetics. Furthermore, much of the new recombinant DNA tech-

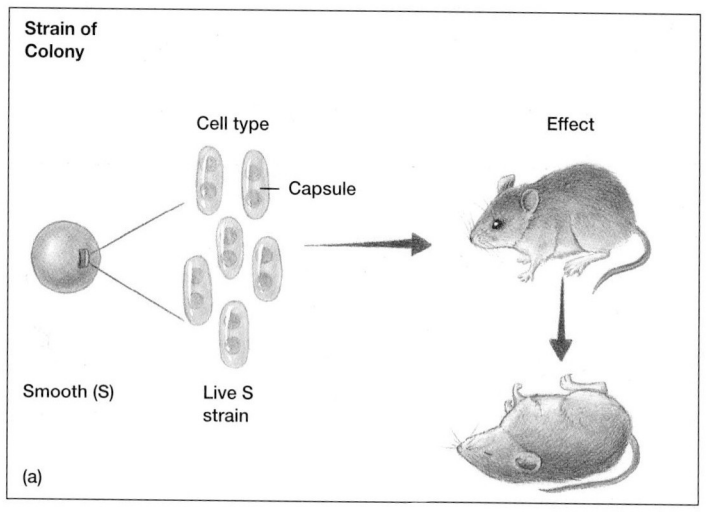

(a)

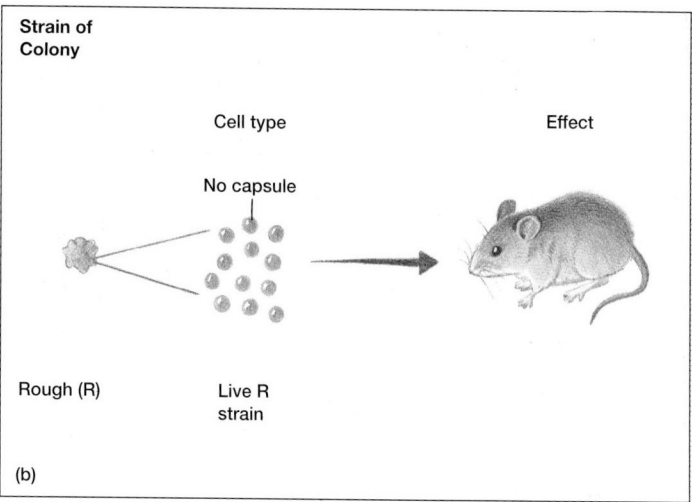

(b)

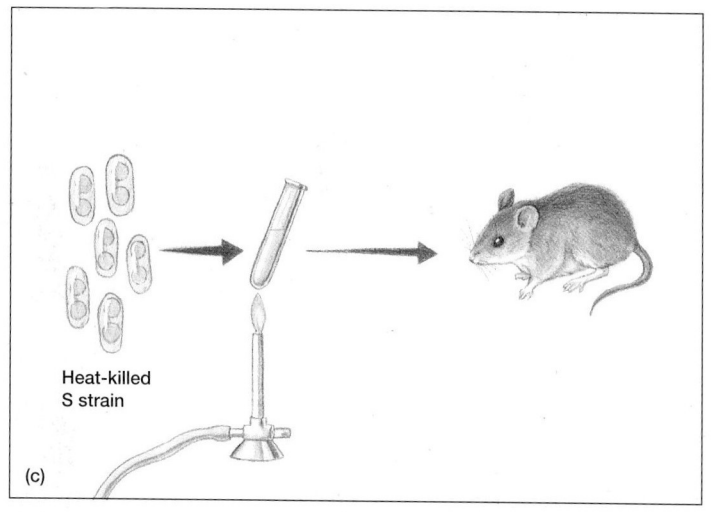

(c)

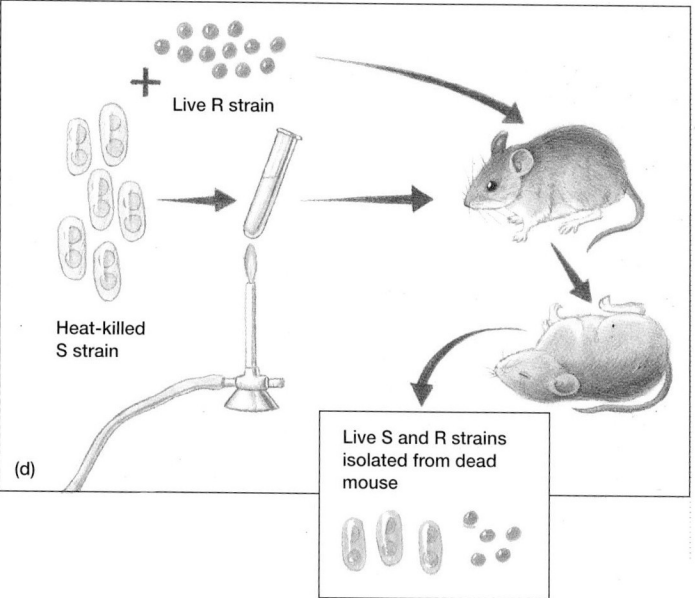

(d)

Figure 11.1 Griffith's Transformation Experiments. (a) Mice died of pneumonia when injected with pathogenic strains of S pneumococci, which have a capsule and form smooth-looking colonies. (b) Mice survived when injected with a nonpathogenic strain of R pneumococci, which lacks a capsule and forms rough colonies. (c) Injection with heat-killed strains of S pneumococci had no effect. (d) Injection with a live R strain and a heat-killed S strain gave the mice pneumonia, and live S strain pneumococci could be isolated from the dead mice.

nology (*see chapter 14*) has arisen from recent progress in bacterial and viral genetics. Research in microbial genetics has had a profound impact on biology as a science and on the technology that affects everyday life.

Biologists have long recognized a relationship between DNA, RNA, and protein (**figure 11.4**), and this recognition has guided a vast amount of research over the past decades. DNA is carefully copied during its synthesis or **replication.** The expression of the information encoded in the base sequence of DNA begins with the synthesis of an RNA copy of the DNA sequence making up a gene. A gene is a DNA segment or sequence that codes for a polypeptide, an rRNA, or a tRNA. Although DNA has two complementary strands, only the template strand is copied at any particular point on DNA. If both strands of DNA were transcribed, two different mRNAs

R cells + purified S cell polysaccharide	⟶ R colonies
R cells + purified S cell protein	⟶ R colonies
R cells + purified S cell RNA	⟶ R colonies
R cells + purified S cell DNA	⟶ S colonies
S cell extract + protease + R cells	⟶ S colonies
S cell extract + RNase + R cells	⟶ S colonies

Figure 11.2 Experiments on the Transforming Principle. Summary of the experiments of Avery, MacLeod, and McCarty on the transforming principle. DNA alone changed R to S cells, and this effect was lost when the extract was treated with deoxyribonuclease. Thus DNA carried the genetic information required for the R to S conversion or transformation.

Figure 11.3   The Hershey-Chase Experiment.
(**a**) When *E. coli* was infected with a T2 phage containing ^{35}S protein, most of the radioactivity remained outside the host cell. (**b**) When a T2 phage containing ^{32}P DNA was mixed with the host bacterium, the radioactive DNA was injected into the cell and phages were produced. Thus DNA was carrying the virus's genetic information.

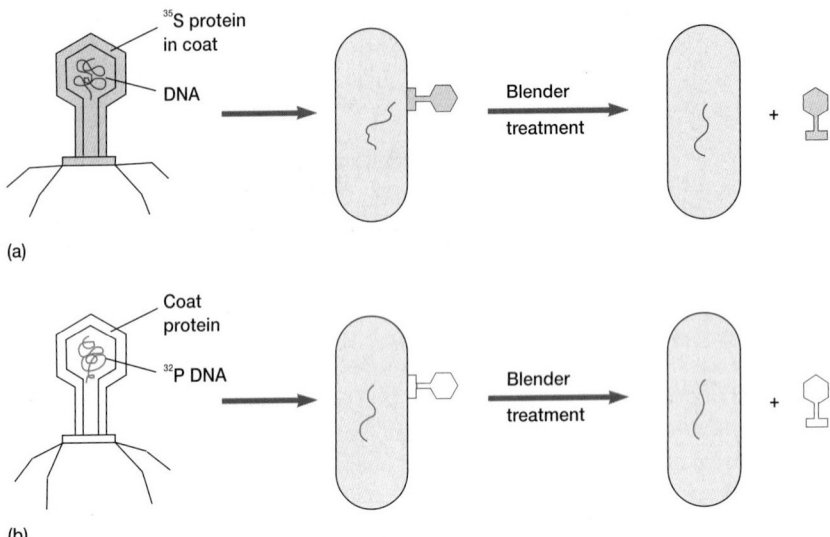

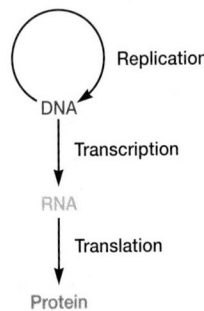

Figure 11.4   Relationships between DNA, RNA, and Protein Synthesis.   This conceptual framework is sometimes called the Central Dogma.

would result and cause genetic confusion. Thus the sequence corresponding to a gene is located only on one of the two complementary DNA strands. Different genes may be encoded on opposite strands. This process of DNA-directed RNA synthesis is called **transcription** because the DNA base sequence is being written into an RNA base sequence. The RNA that carries information from DNA and directs protein synthesis is **messenger RNA (mRNA).** The last phase of gene expression is **translation** or protein synthesis. The genetic information in the form of an mRNA nucleotide sequence is translated and governs the synthesis of protein. Thus the amino acid sequence of a protein is a direct reflection of the base sequence in mRNA. In turn the mRNA nucleotide sequence is a complementary copy of a portion of the DNA genome.

1. Define clone, genome, genotype, and phenotype.
2. Briefly summarize the experiments of Griffith; Avery, MacLeod, and McCarty; and Hershey and Chase. What did each show, and why were these experiments important to the development of microbial genetics?
3. Describe the general relationship between DNA, RNA, and protein.

 11.2   NUCLEIC ACID STRUCTURE

The structure and synthesis of purine and pyrimidine nucleotides are introduced in chapter 10 (*see pp. 211–14*). These nucleotides can be combined to form nucleic acids of two kinds (**figure 11.5***a*). **Deoxyribonucleic acid (DNA)** contains the 2′-deoxyribonucleosides (figure 11.5*b*) of adenine, guanine, cytosine, and thymine. **Ribonucleic acid (RNA)** is composed of the ribonucleosides of adenine, guanine, cytosine, and uracil (instead of thymine). In both DNA and RNA, nucleosides are joined by phosphate groups to form long polynucleotide chains (**figure 11.6**). The differences in chemical composition between the chains reside in their sugar and pyrimidine bases: DNA has deoxyribose and thymine; RNA has ribose and uracil in place of thymine.

DNA Structure

Deoxyribonucleic acids are very large molecules, usually composed of two polynucleotide chains coiled together to form a double helix 2.0 nm in diameter (figure 11.6 and **figure 11.7**). Each chain contains purine and pyrimidine deoxyribonucleosides joined by phosphodiester bridges (figure 11.6). That is, two adjacent deoxyribose sugars are connected by a phosphoric acid molecule esterified to a 3′-hydroxyl of one sugar and a 5′-hydroxyl of the other. Purine and pyrimidine bases are attached to the 1′-carbon of the deoxyribose sugars and extend toward the middle of the cylinder formed by the two chains. They are stacked on top of each other in the center, one base pair every 0.34 nm. The purine adenine (A) is always paired with the pyrimidine thymine (T) by two hydrogen bonds. The purine guanine (G) pairs with cytosine (C) by three hydrogen bonds (**figure 11.8**). This AT and GC base pairing means that the two strands in a DNA double helix are **complementary.** That is, the bases in one strand match up with those of the other according to the base pairing rules. Because the sequences of bases in these strands encode

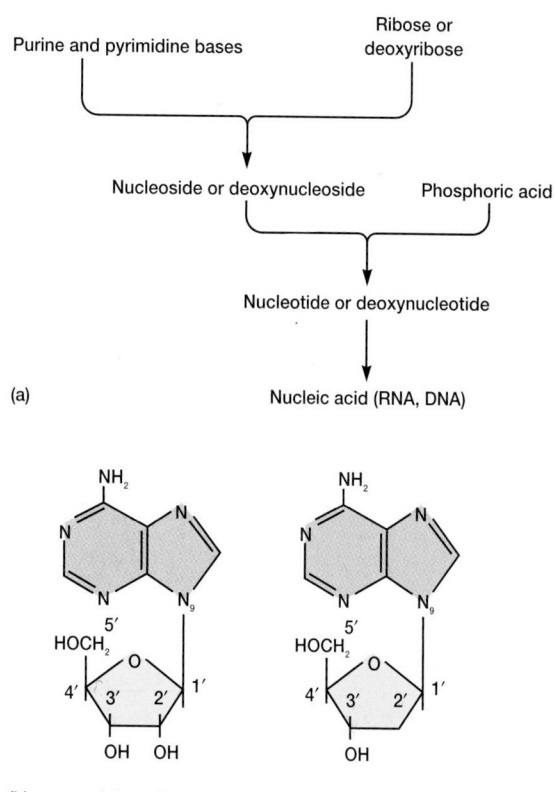

(b)   Adenosine   2′-deoxyadenosine

Figure 11.5   The Composition of Nucleic Acids. **(a)** A diagram showing the relationships of various nucleic acid components. Combination of a purine or pyrimidine base with ribose or deoxyribose gives a nucleoside (a ribonucleoside or deoxyribonucleoside). A nucleotide contains a nucleoside and one or more phosphoric acid molecules. Nucleic acids result when nucleotides are connected together in polynucleotide chains. **(b)** Examples of nucleosides—the purine nucleoside adenosine and the pyrimidine deoxynucleoside 2′-deoxyadenosine. The carbons of nucleoside sugars are indicated by numbers with primes.

genetic information, considerable effort has been devoted to determining the base sequences of DNA and RNA from many microorganisms (*see pp. 336–39*). *Nucleic acid sequence comparison and microbial taxonomy (chapter 19)*

The two polynucleotide strands fit together much like the pieces in a jigsaw puzzle because of complementary base pairing (**Historical Highlights 11.1**). Inspection of figure 11.7a,b, depicting the B form of DNA (probably the most common form in cells), shows that the two strands are not positioned directly opposite one another in the helical cylinder. Therefore, when the strands twist about one another, a wide **major groove** and narrower **minor groove** are formed by the backbone. Each base pair rotates 36° around the cylinder with respect to adjacent pairs so that there are 10 base pairs per turn of the helical spiral. Each turn of the helix has a vertical length of 3.4 nm. The helix is right-handed—that is, the chains turn counterclockwise as they approach a viewer looking down the longitudinal axis. The two backbones are antiparallel or run in opposite directions with respect to the orientation of their sugars. One end of each strand has an exposed 5′-hydroxyl group, often with phosphates attached, whereas the other end has a free 3′-hydroxyl group (figure 11.6). If the end of a double helix is examined, the 5′ end of one strand and the 3′ end of the other are visible. In a given direction one strand is oriented 5′ to 3′ and the other, 3′ to 5′ (figures 11.6 and 11.7b).

RNA Structure

Besides differing chemically from DNA, ribonucleic acid is usually single stranded rather than double stranded like most DNA. An RNA strand can coil back on itself to form a hairpin-shaped structure with complementary base pairing and helical organization. Cells contain three different types of RNA—messenger RNA, ribosomal RNA, and transfer RNA—that differ from one another in function, site of synthesis in eucaryotic cells, and structure.

Figure 11.6   DNA Structure and Its Antiparallel Nature.   This short DNA segment contains three base pairs connected by phosphodiester linkages between the 3′- and 5′-carbons of adjacent deoxyribose sugars (yellow). The adenosine-thymine base pairs are joined by two hydrogen bonds and guanine-cytosine base pairs have three hydrogen bonds. Because of the specific base pairing, the base sequence of one strand determines the sequence of the other. The two strands are antiparallel. That is, the backbones run in opposite directions as indicated by the two arrows, which point in the 5′ to 3′ direction. Note that the left chain has its 5′-hydroxyl on the bottom, whereas the right chain has a 3′-hydroxyl at the bottom.

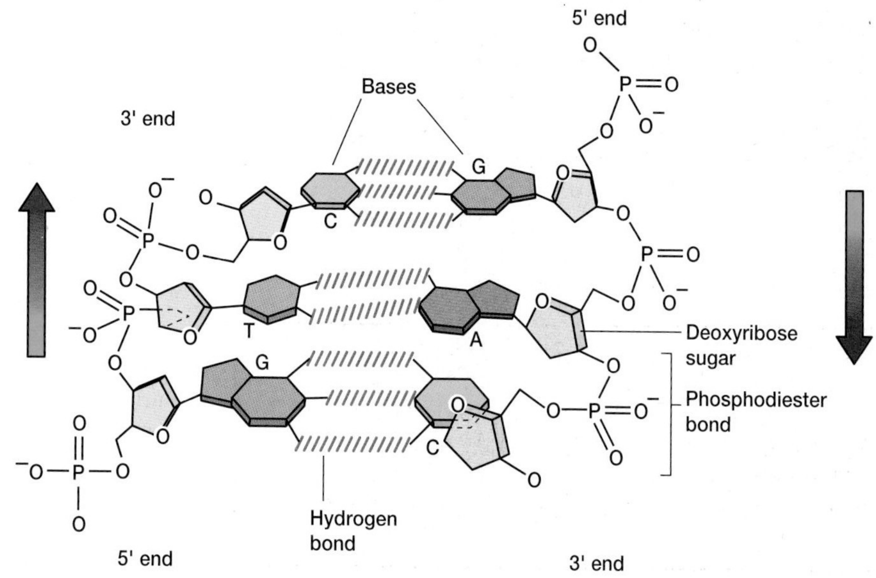

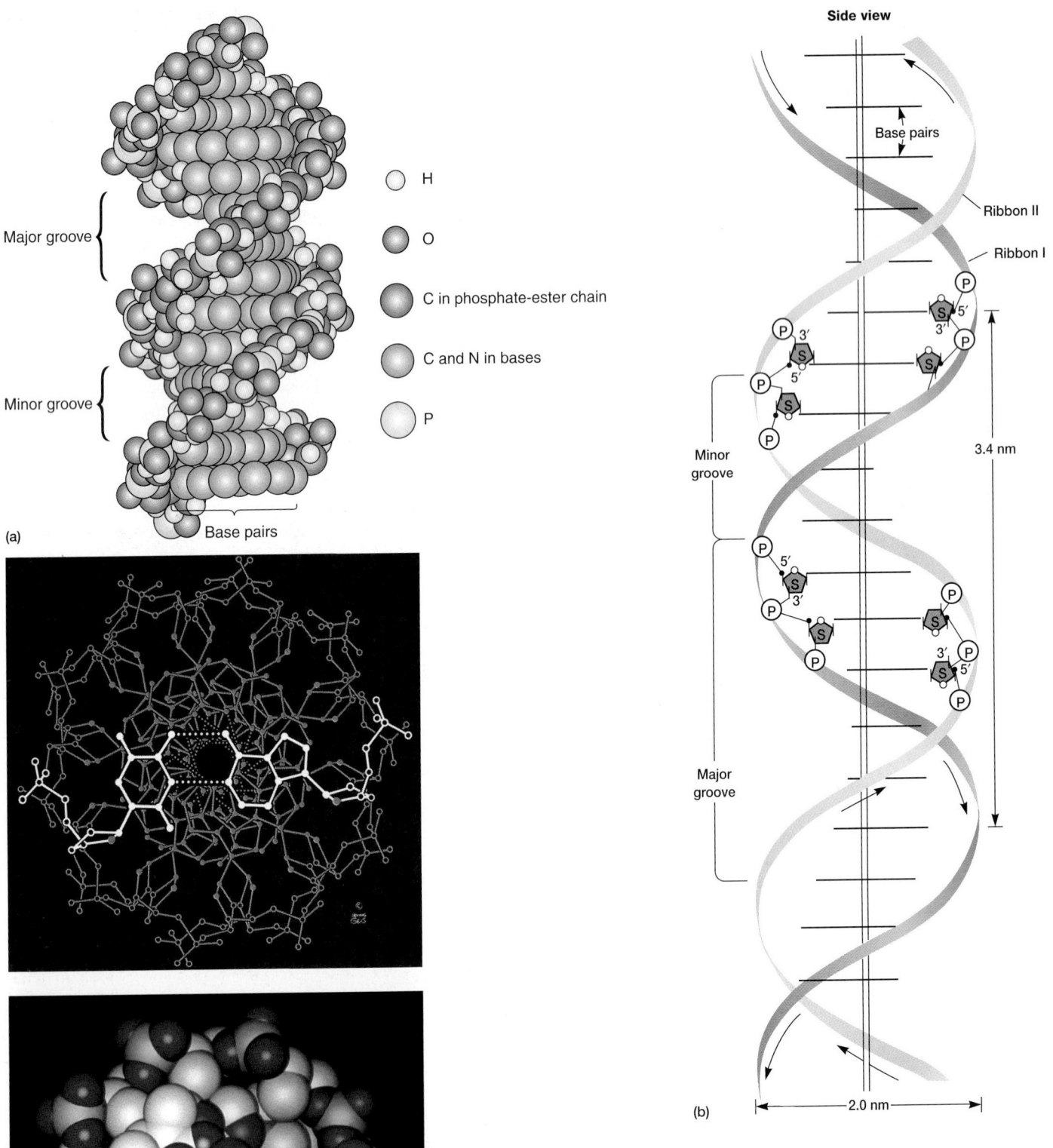

(a)

Major groove

Minor groove

Base pairs

H

O

C in phosphate-ester chain

C and N in bases

P

(c)

Side view

Base pairs

Ribbon II

Ribbon I

Minor groove

Major groove

3.4 nm

2.0 nm

(b)

Figure 11.7   The Structure of the DNA Double Helix.   (**a**) A space-filling model of the B form of DNA with the base pairs, major groove, and minor groove shown. The backbone phosphate groups, shown in color, spiral around the outside of the helix. (**b**) A diagrammatic representation of the double helix. The backbone consists of deoxyribose sugars (S) joined by phosphates (P) in phosphodiester bridges. The arrows at the top and bottom of the chains point in the 5′ to 3′ direction. The ribbons represent the sugar phosphate backbones. (**c**) An end view of the double helix showing the outer backbone and the bases stacked in the center of the cylinder. In the top drawing the ribose ring oxygens are red. The nearest base pair, an AT base pair, is highlighted in white.

Historical Highlights
11.1   The Elucidation of DNA Structure

The basic chemical composition of nucleic acids was elucidated in the 1920s through the efforts of P. A. Levene. Despite his major contributions to nucleic acid chemistry, Levene mistakenly believed that DNA was a very small molecule, probably only four nucleotides long, composed of equal amounts of the four different nucleotides arranged in a fixed sequence. Partly because of his influence, biologists believed for many years that nucleic acids were too simple in structure to carry complex genetic information. They concluded that genetic information must be encoded in proteins because proteins were large molecules with complex amino sequences that could vary among different proteins.

As so often happens, further advances in our understanding of DNA structure awaited the development of significant new analytical techniques in chemistry. One development was the invention of paper chromatography by Archer Martin and Richard Synge between 1941 and 1944. By 1948 the chemist Erwin Chargaff had begun using paper chromatography to analyze the base composition of DNA from a number of species. He soon found that the base composition of DNA from genetic material did indeed vary among species just as he expected. Furthermore, the total amount of purines always equaled the total amount of pyrimidines; and the adenine/thymine and guanine/cytosine ratios were always 1. These findings, known as Chargaff's rules, were a key to the understanding of DNA structure.

Another turning point in research on DNA structure was reached in 1951 when Rosalind Franklin arrived at King's College, London, and joined Maurice Wilkins in his efforts to prepare highly oriented DNA fibers and study them by X-ray crystallography. By the winter of 1952–1953, Franklin had obtained an excellent X-ray diffraction photograph of DNA.

The same year that Franklin began work at King's College, the American biologist James Watson went to Cambridge University and met Francis Crick. Although Crick was a physicist, he was very interested in the structure and function of DNA, and the two soon began to work on its structure. Their attempts were unsuccessful until Franklin's data provided them with the necessary clues. Her photograph of fibrous DNA contained a crossing pattern of dark spots, which showed that the molecule was helical. The dark regions at the top and bottom of the photograph showed that the purine and pyrimidine bases were stacked on top of each other and separated by 0.34 nm. Franklin had already concluded that the phosphate groups lay to the outside of the cylinder. Finally, the X-ray data and her determination of the density of DNA indicated that the helix contained two strands, not three or more as some had proposed.

Without actually doing any experiments themselves, Watson and Crick constructed their model by combining Chargaff's rules on base composition with Franklin's X-ray data and their predictions about how genetic material should behave. By building models, they found that a smooth, two-stranded helix of constant diameter could be constructed only when an adenine hydrogen bonded with thymine and when a guanine bonded with cytosine in the center of the helix. They immediately realized that the double helical structure provided a mechanism by which genetic material might be replicated. The two parental strands could unwind and direct the synthesis of complementary strands, thus forming two new identical DNA molecules (figure 11.11). Watson, Crick, and Wilkins received the Nobel Prize in 1962 for their discoveries. Franklin could not be considered for the prize because she had died of cancer in 1958 at the age of thirty-seven.

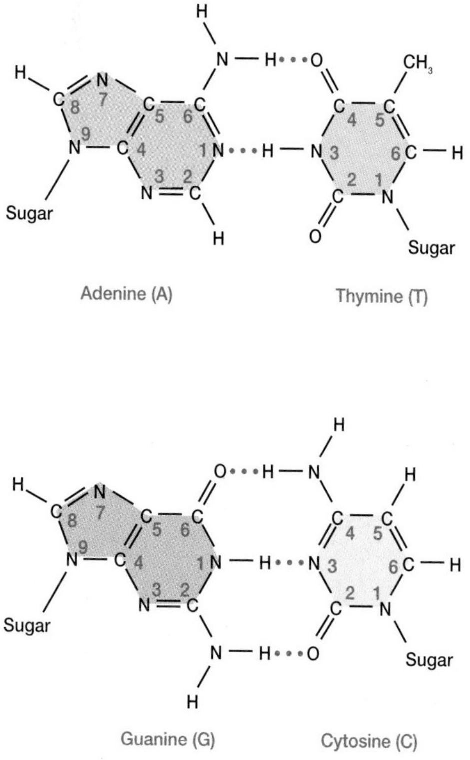

Adenine (A)   Thymine (T)

The Organization of DNA in Cells

Although DNA exists as a double helix in both procaryotic and eucaryotic cells, its organization differs in the two cell types (*see table 4.2*). DNA is organized in the form of a closed circle in almost all procaryotes (the chromosome of *Borrelia* is a linear DNA molecule). This circular double helix is further twisted into supercoiled DNA (**figure 11.9**) and is associated with basic proteins but not with the histones found complexed with almost all eucaryotic DNA. These histonelike proteins do appear to help organize bacterial DNA into a coiled chromatinlike structure. The structure of the bacterial nucleoid (pp. 50–52).

DNA is much more highly organized in eucaryotic chromatin (*see section 4.9*) and is associated with a variety of proteins, the most prominent of which are **histones.** These are small, basic proteins rich in the amino acids lysine and/or arginine. There are five types of histones in almost all eucaryotic cells studied: H1, H2A, H2B, H3, and H4. Eight histone molecules (two each of H2A, H2B, H3, and H4) form an ellipsoid about 11 nm long and 6.5 to

Guanine (G)   Cytosine (C)

Figure 11.8   DNA Base Pairs. DNA complementary base pairing showing the hydrogen bonds (. . .).

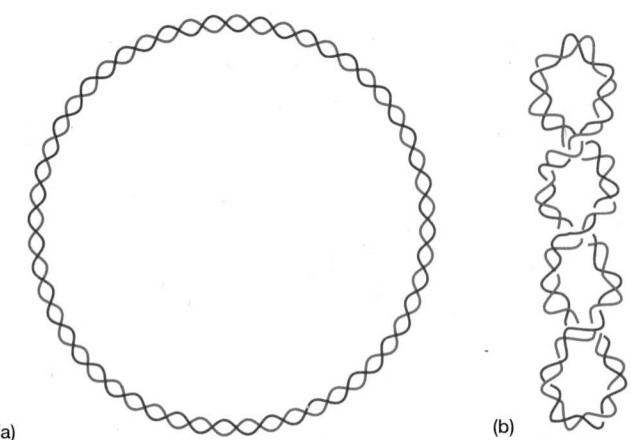

(a)   (b)

Figure 11.9   DNA Forms.   (a) The DNA double helix of almost all bacteria is in the shape of a closed circle. **(b)** The circular
DNA strands, already coiled in a double helix, are twisted a second
time to produce supercoils.

7 nm in diameter (**figure 11.10a**). DNA coils around the surface
of the ellipsoid approximately 1 3/4 turns or 166 base pairs before
proceeding on to the next. This complex of histones plus DNA is
called a **nucleosome.** Thus DNA gently isolated from chromatin
looks like a string of beads. The stretch of DNA between the beads
or nucleosomes, the linker region, varies in length from 14 to over
100 base pairs. Histone H1 appears to associate with the linker regions to aid the folding of DNA into more complex chromatin
structures (figure 11.10b). When folding reaches a maximum, the
chromatin takes the shape of the visible chromosomes seen in eucaryotic cells during mitosis and meiosis (*see figure 4.19*).

1. What are nucleic acids? How do DNA and RNA differ in structure?
2. Describe in some detail the structure of the DNA double helix.
   What does it mean to say that the two strands are complementary and antiparallel?
3. What are histones and nucleosomes? Describe the way in
   which DNA is organized in the chromosomes of procaryotes
   and eucaryotes.

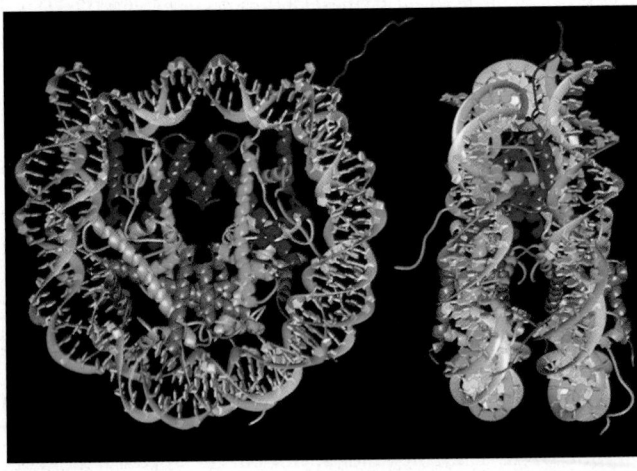

(a)

**Figure 11.10   Nucleosome Internal Organization
and Function.   (a)** The nucleosome core particle is a histone octamer surrounded by the 146 base pair DNA helix (brown and
turquoise). The octamer is a disk-shaped structure composed of
two H2A-H2B dimers and two H3-H4 dimers. The eight histone
proteins are colored differently: blue, H3; green, H4; yellow, H2A;
and red, H2B. Histone proteins interact with the backbone of the
DNA minor groove. The DNA double helix circles the histone octamer in a left-handed helical path. **(b)** An illustration of how a
string of nucleosomes, each associated with a histone H1, might be
organized to form a highly supercoiled chromatin fiber. The nucleosomes are drawn as cylinders.

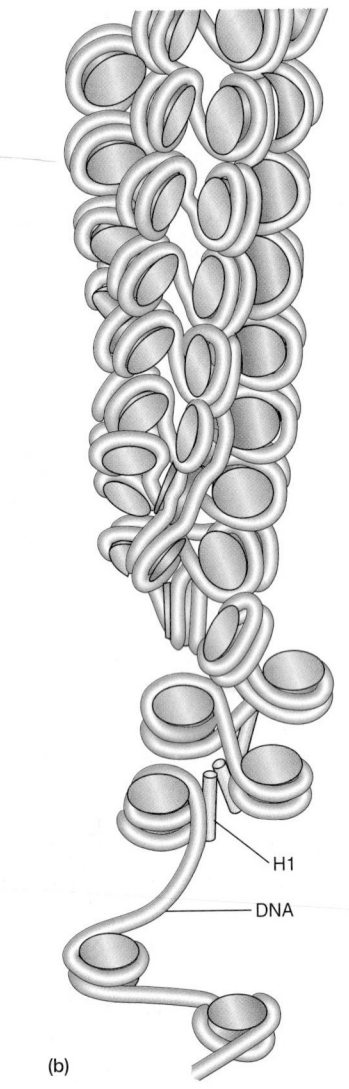

H1

DNA

(b)

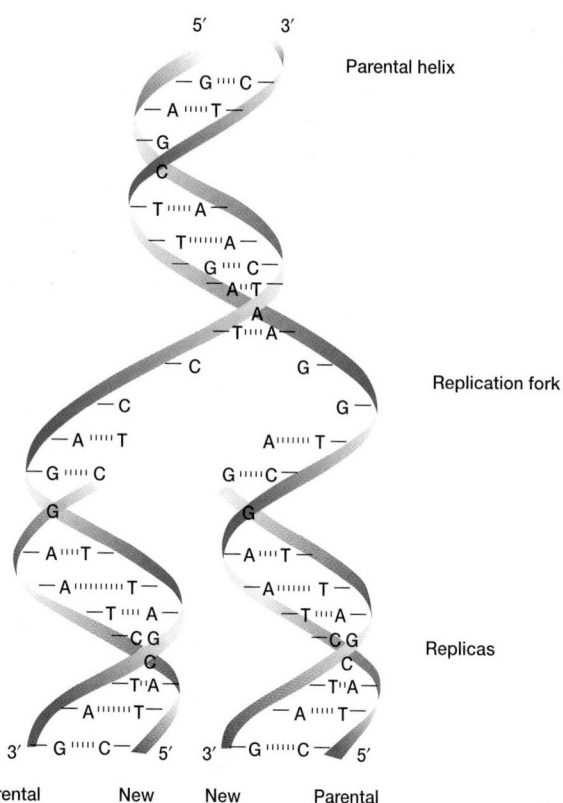

5′   3′

Parental helix

Replication fork

Replicas

Parental   New   New   Parental

Figure 11.11   Semiconservative DNA Replication. The replication fork of DNA showing the synthesis of two progeny strands. Newly synthesized strands are in maroon. Each copy contains one new and one old strand. This process is called semiconservative replication.

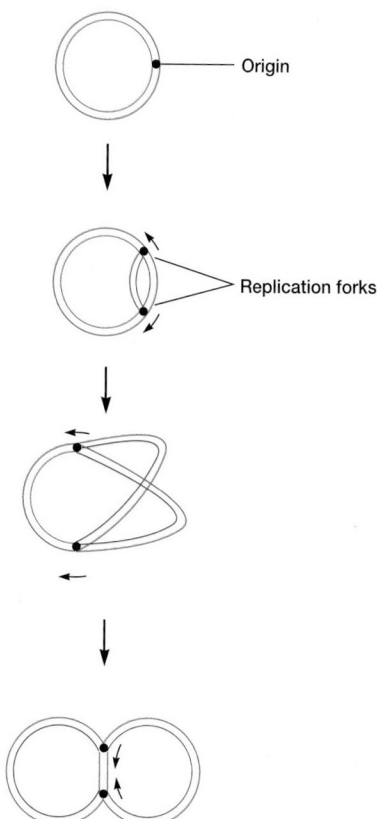

Origin

Replication forks

Figure 11.12   Bidirectional Replication. The replication of a circular bacterial genome. Two replication forks move around the DNA forming theta-shaped intermediates. Newly replicated DNA double helix is in red.

11.3   DNA REPLICATION

The replication of DNA is an extraordinarily important and complex process, one upon which all life depends. We shall first discuss the overall pattern of DNA synthesis and then examine the mechanism of DNA replication in greater depth.

Patterns of DNA Synthesis

Watson and Crick published their description of DNA structure in April 1953. Almost exactly one month later, a second paper appeared in which they suggested how DNA might be replicated. They hypothesized that the two strands of the double helix unwind from one another and separate (**figure 11.11**). Free deoxynucleoside triphosphates now line up along the two parental strands through complementary base pairing—A with T, G with C (figure 11.8). When these nucleotides are linked together by one or more enzymes, two replicas result, each containing a parental DNA strand and a newly formed strand. Research in subsequent years has proved Watson and Crick's hypothesis correct.

Replication patterns are somewhat different in procaryotes and eucaryotes. For example, when the circular DNA chromo-

some of *E. coli* is copied, replication begins at a single point, the origin. Synthesis occurs at the **replication fork,** the place at which the DNA helix is unwound and individual strands are replicated. Two replication forks move outward from the origin until they have copied the whole **replicon,** that portion of the genome that contains an origin and is replicated as a unit. When the replication forks move around the circle, a structure shaped like the Greek letter theta (θ) is formed (**figure 11.12**). Finally, since the bacterial chromosome is a single replicon, the forks meet on the other side and two separate chromosomes are released.

A different pattern of DNA replication occurs during *E. coli* conjugation (*see section 13.4*) and the reproduction of viruses, such as phage lambda (*see section 17.5*). In the **rolling-circle mechanism (figure 11.13),** one strand is nicked and the free 3′-hydroxyl end is extended by replication enzymes. As the 3′ end is lengthened while the growing point rolls around the circular template, the 5′ end of the strand is displaced and forms an ever-lengthening tail. The single-stranded tail may be converted to the double-stranded form by complementary strand synthesis. This mechanism is particularly useful to viruses (*see p. 379*) because it allows the rapid, continuous production of many genome copies from a single initiation event.

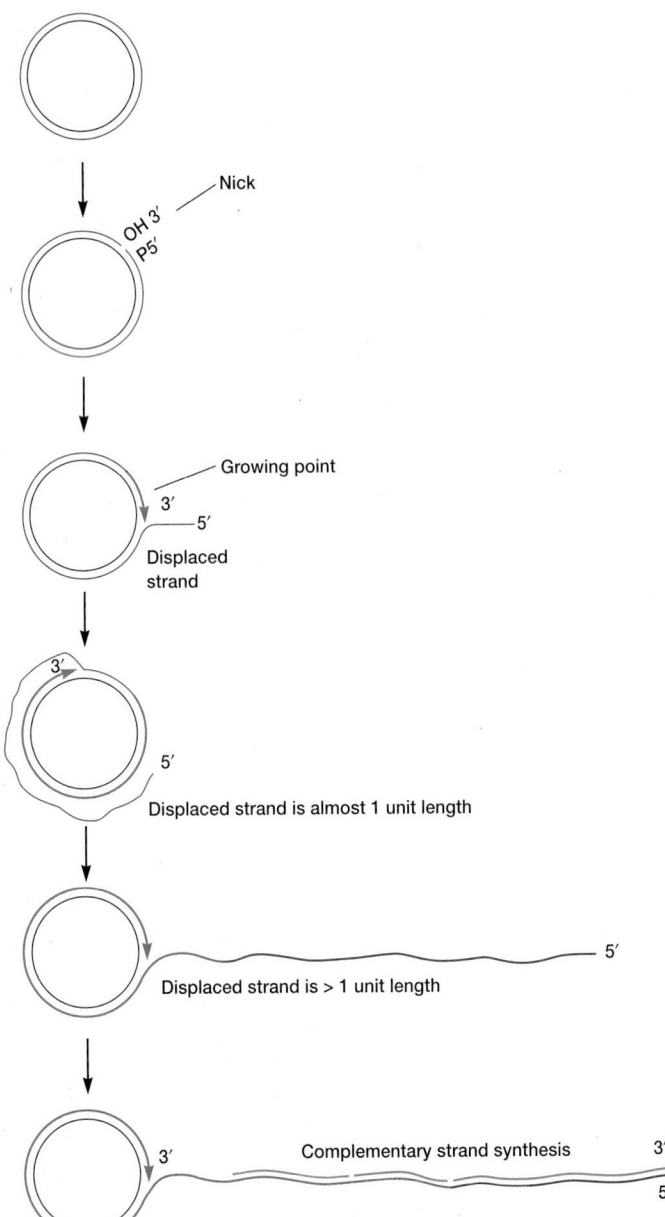

Figure 11.13   The Rolling-Circle Pattern of Replication.
A single-stranded tail, often composed of more than one genome copy, is generated and can be converted to the double-stranded form by synthesis of a complementary strand. The "free end" of the rolling-circle strand is probably bound to the primosome.

Eucaryotic DNA is linear and much longer than procaryotic DNA; *E. coli* DNA is about 1,300 μm in length, whereas the 46 chromosomes in the human nucleus have a total length of 1.8 m (almost 1,400 times longer). Clearly many replication forks must copy eucaryotic DNA simultaneously so that the molecule can be duplicated in a relatively short period, and so many replicons are present that there is an origin about every 10 to 100 μm along the DNA. Replication forks move outward from these sites and even-

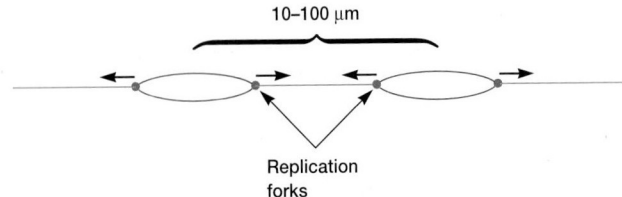

Figure 11.14   The Replication of Eucaryotic DNA.
Replication is initiated every 10 to 100 μm and the replication forks travel away from the origin. Newly copied DNA is in red.

tually meet forks that have been copying the adjacent DNA stretch (**figure 11.14**). In this fashion a large molecule is copied quickly.

Mechanism of DNA Replication

Because DNA replication is so essential to organisms, a great deal of effort has been devoted to understanding its mechanism. The replication of *E. coli* DNA is probably best understood and is the focus of attention in this section. The process in eucaryotic cells is thought to be similar.

DNA replication is initiated at the *oriC* locus. The DnaA protein binds to *oriC* while hydrolyzing ATP. This leads to the initial unwinding of double-stranded DNA at the initiation site. Further unwinding occurs through the activity of the DnaB protein, a helicase (see below).

E. coli has three different **DNA polymerase** enzymes, each of which catalyzes the synthesis of DNA in the 5′ to 3′ direction while reading the DNA template in the 3′ to 5′ direction (**figures 11.15** and **11.16**). The polymerases require deoxyribonucleoside triphosphates (dATP, dGTP, dCTP, and dTTP) as substrates and a DNA template to copy. Nucleotides are added to the 3′ end of the growing chain when the free 3′-hydroxyl group on the deoxyribose attacks the first or alpha phosphate group of the substrate to release pyrophosphate (figure 11.15). DNA polymerase III plays the major role in replication, although it is probably assisted by polymerase I. It is thought that polymerases I and II participate in the repair of damaged DNA (p. 248).

During replication the DNA double helix must be unwound to generate separate single strands. Unwinding occurs very quickly; the fork may rotate as rapidly as 75 to 100 revolutions per second. **Helicases** are responsible for DNA unwinding. These enzymes use energy from ATP to unwind short stretches of helix just ahead of the replication fork. Once the strands have separated, they are kept single through specific binding with **single-stranded DNA binding proteins (SSBs)** as shown in figure 11.16. Rapid unwinding can lead to tension and formation of supercoils or supertwists in the helix, just as rapid separation of two strands of a rope can lead to knotting or coiling of the rope. The tension generated by unwinding is relieved, and the unwinding process is promoted by enzymes known as **topoisomerases.** These enzymes change the structure of DNA by transiently breaking one or two strands in such a way that it remains unaltered as its shape is changed

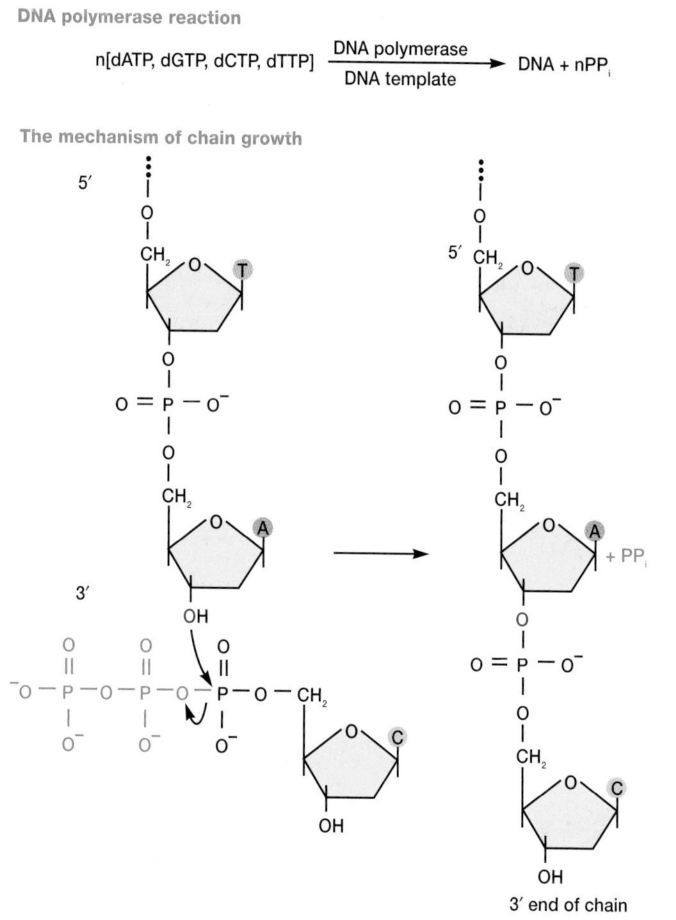

DNA polymerase reaction

$$n[dATP, dGTP, dCTP, dTTP] \xrightarrow[\text{DNA template}]{\text{DNA polymerase}} \text{DNA} + n\text{PP}_i$$

The mechanism of chain growth

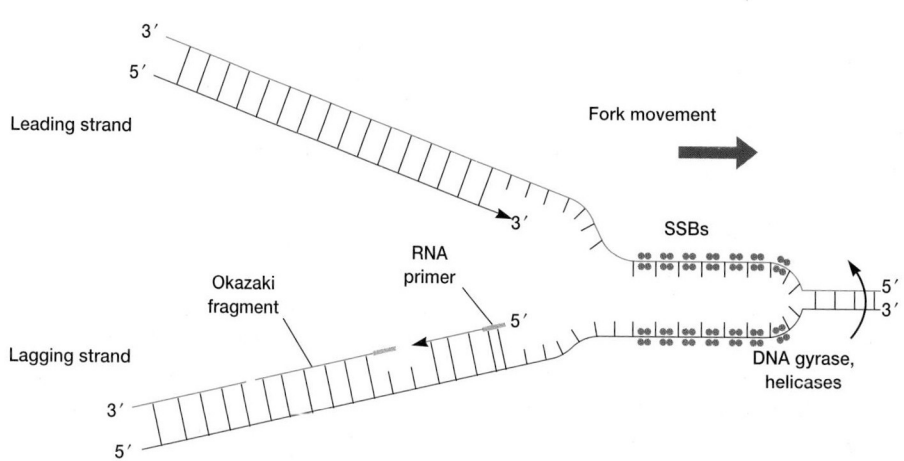

Figure 11.15   The DNA Polymerase Reaction and Its Mechanism.   The mechanism involves a nucleophilic attack by the hydroxyl of the 3′ terminal deoxyribose on the alpha phosphate group of the nucleotide substrate (in this example, adenosine attacks cytidine triphosphate).

(e.g., a topoisomerase might tie or untie a knot in a DNA strand). **DNA gyrase** is an *E. coli* topoisomerase that removes the supertwists produced during replication (*see figure 35.7*).

After the double helix has been unwound, successful replication requires the solution of two problems. First, DNA polymerase only synthesizes a new copy of DNA while moving in the 5′ to 3′ direction. Inspection of figure 11.16 shows that synthesis of the leading strand copy is relatively simple because the new strand can be extended continuously at its 3′ end as the DNA unwinds. In contrast, the lagging strand cannot be extended in the same direction because this would require 3′ to 5′ synthesis, which is not possible. As a result, the lagging strand copy is synthesized discontinuously in the 5′ to 3′ direction as a series of fragments; then the fragments are joined to form a complete copy. The second problem arises because DNA polymerase cannot start a new copy from scratch, but must build on an already existing strand. In figure 11.16, the leading strand copy already exists; however, the lagging strand fragments must be synthesized without a DNA strand to build upon. In this case, a special RNA primer is first synthesized and then a DNA copy can be built on the primer.

The details of DNA replication are outlined in a diagram of the replication fork (**figure 11.17**). The replication process takes place in four stages.

1. Helicases unwind the helix with the aid of topoisomerases like the DNA gyrase (figure 11.17, step 1). It appears that the DnaB protein is the helicase most actively involved in replication, but the n′ protein also may participate in unwinding. The single strands are kept separate by the DNA binding proteins (SSBs).

2. DNA is probably replicated continuously by DNA polymerase III when the leading strand is copied. Lagging strand replication is discontinuous, and the fragments are synthesized in the 5′ to 3′ direction just as in leading strand synthesis. First, a special RNA polymerase called a **primase** synthesizes a short RNA primer, usually around 10 nucleotides long, complementary to the DNA (figure 11.17, step 2). It appears that the primase

Figure 11.16   Bacterial DNA Replication.   A general diagram of the synthesis of DNA in *E. coli* at the replication fork. Bases and base pairs are represented by lines extending outward from the strands. The RNA primer is in gold. See text for details.

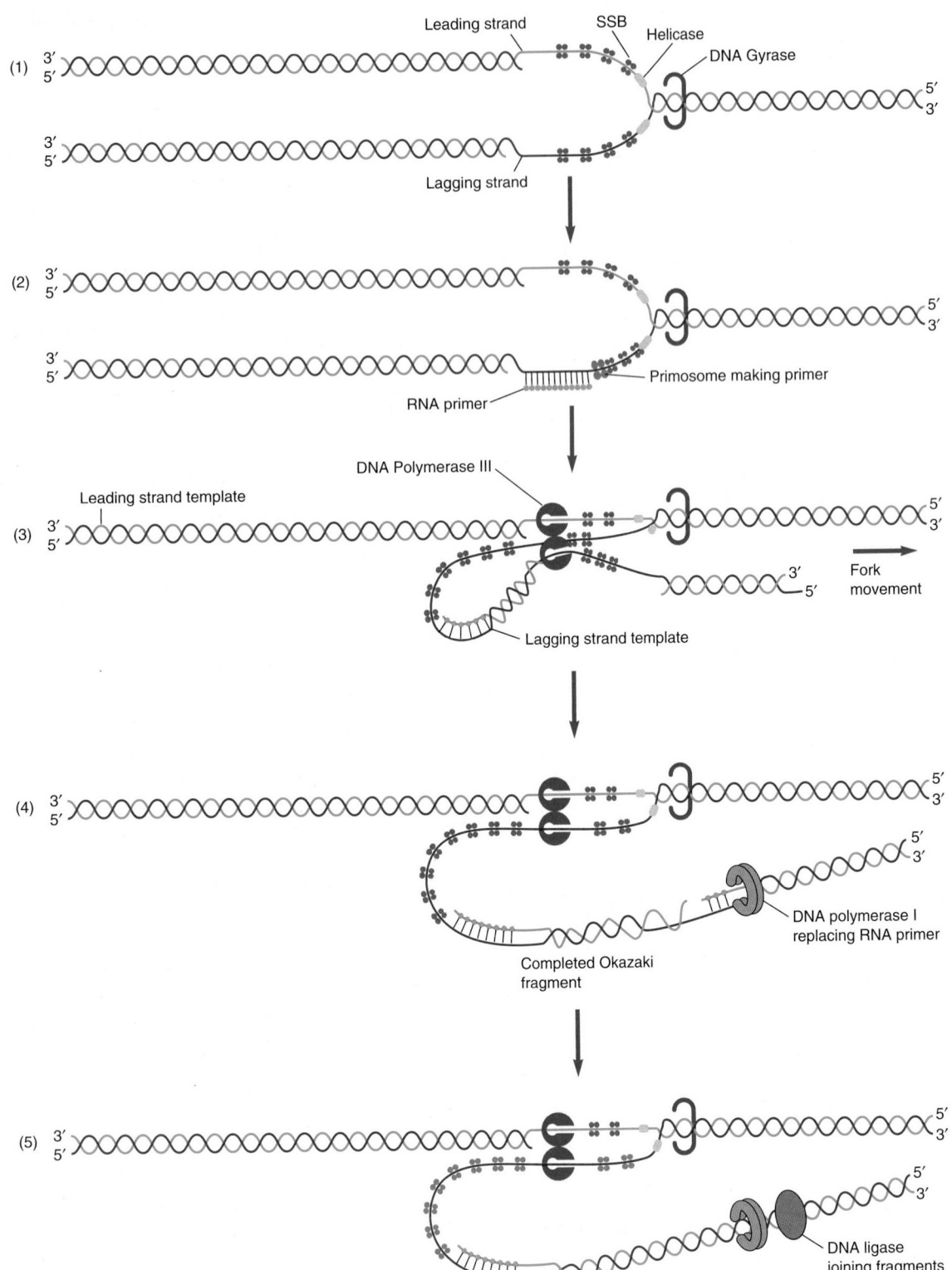

Figure 11.17 A Hypothetical Model for Activity at the Replication Fork.   The overall process is pictured in five stages with only one cycle of replication shown for sake of clarity. In practice, all these enzymes are functioning simultaneously and more than one round of replication can occur simultaneously; for example, new primer RNA can be synthesized at the same time as DNA is being replicated. (*1*) DNA gyrase, helicases, and single-stranded DNA binding proteins (SSBs) unwind DNA to produce a single-stranded stretch. (*2*) The primosome synthesizes an RNA primer. (*3*) The replisome has two DNA polymerase III complexes. One polymerase continuously copies the leading strand. The lagging strand loops around the other polymerase so that both strands can be replicated simultaneously. When DNA polymerase III encounters a completed Okazaki fragment, it releases the lagging strand. (*4*) DNA polymerase I removes the RNA primer and fills in the gap with complementary DNA. (*5*) DNA ligase seals the nick and joins the two fragments.

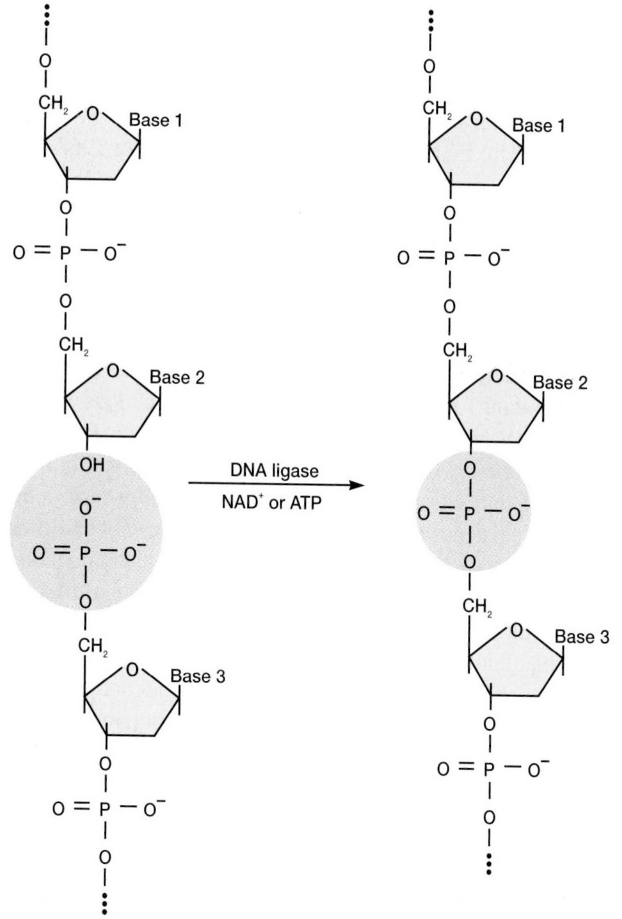

Figure 11.18   The DNA Ligase Reaction.   The groups being altered are shaded in blue.

requires the assistance of several other proteins, and the complex of the primase with its accessory proteins is called the **primosome.** DNA polymerase III holoenzyme then synthesizes complementary DNA beginning at the 3′ end of the RNA primer. Both leading and lagging strand synthesis probably occur concurrently on a single multiprotein complex with two catalytic sites. If this is the case, the lagging strand template must be looped around the complex (figure 11.17, step 3). The final fragments are around 1,000 to 2,000 nucleotides long in bacteria and approximately 100 nucleotides long in eucaryotic cells. They are called **Okazaki fragments** after their discoverer, Reiji Okazaki.

3. After most of the lagging strand has been duplicated by the formation of Okazaki fragments, DNA polymerase I or RNase H removes the RNA primer. Polymerase I synthesizes complementary DNA to fill the gap resulting from RNA deletion (figure 11.17, step 4). The polymerase appears to remove one primer nucleotide at a time and replace it with the appropriate complementary deoxyribonucleotide. Polymerase III holoenzyme also may be able to fill in the gap.

4. Finally, the fragments are joined by the enzyme **DNA ligase,** which forms a phosphodiester bond between the 3′-hydroxyl

of the growing strand and the 5′-phosphate of an Okazaki fragment (figure 11.17, step 5, and **figure 11.18**). Bacterial ligases use the pyrophosphate bond of NAD+ as an energy source; many other ligases employ ATP.

DNA polymerase III holoenzyme, the enzyme complex that synthesizes most of the DNA copy, is a very large entity containing DNA polymerase III and several other proteins. The γδ complex and β subunits of the holoenzyme bind it to the DNA template and primer. The α subunit carries out the actual polymerization reaction. It appears that most or all of the replication proteins form a huge complex or replication factory, sometimes called a **replisome,** that is relatively stationary and probably bound to the plasma membrane. The DNA moves through this factory and is copied, emerging as two daughter chromosomes. In slowly growing bacteria there seem to be two factories located at or close to the center of the cell. Rapidly growing cells might have four or more factories.

DNA replication stops when the polymerase complex reaches a termination site on the DNA in *E. coli*. The Tus protein binds to these *Ter* sites and halts replication. In many procaryotes, replication stops randomly when the forks meet.

DNA replication is an extraordinarily complex process. At least 30 proteins are required to replicate the *E. coli* chromosome. Presumably much of the complexity is necessary for accuracy in copying DNA. It would be very dangerous for any organism to make many errors during replication because a large number of mutations would certainly be lethal. In fact, *E. coli* makes errors with a frequency of only 10^{-9} or 10^{-10} per base pair replicated (or about 10^{-6} per gene per generation). Part of this precision results from the low error rate of the copying process itself. However, DNA polymerase III (and DNA polymerase I) also can proofread the newly formed DNA. As polymerase III moves along synthesizing a new DNA strand, it recognizes any errors resulting in improper base pairing and hydrolytically removes the wrong nucleotide through a special 3′ to 5′ exonuclease activity (which is found in the ε subunit). The enzyme then backs up and adds the proper nucleotide in its place. Polymerases delete errors by acting much like correcting typewriters.   DNA repair (pp. 248–49)

Despite its complexity and accuracy, replication occurs very rapidly. In procaryotes replication rates approach 750 to 1,000 base pairs per second. Eucaryotic replication is much slower, about 50 to 100 base pairs per second. This is not surprising because eucaryotic replication also involves operations like unwinding the DNA from nucleosomes.

1. Define the following terms: replication, transcription, messenger RNA, translation, replicon, replication fork, primosome, and replisome.
2. Be familiar with the nature and functions of the following replication components and intermediates: DNA polymerases I and III, topoisomerase, DNA gyrase, helicase, single-stranded DNA binding protein, Okazaki fragment, DNA ligase, leading strand, and lagging strand.

11.4   THE GENETIC CODE

The realization that DNA is the genetic material triggered efforts to understand how genetic instructions are stored and organized in the DNA molecule. Early studies on the nature of the genetic code showed that the DNA base sequence corresponds to the amino acid sequence of the polypeptide specified by the gene. That is, the nucleotide and amino acid sequences are colinear. It also became evident that many mutations are the result of changes of single amino acids in a polypeptide chain. However, the exact nature of the code was still unclear.

Establishment of the Genetic Code

Since only 20 amino acids normally are present in proteins, there must be at least 20 different code words in a linear, single strand of DNA. The code must be contained in some sequence of the four nucleotides commonly found in the linear DNA sequence. There are only 16 possible combinations (4^2) of the four nucleotides if only nucleotide pairs are considered, not enough to code for all 20 amino acids. Therefore a code word, or **codon,** must involve at least nucleotide triplets even though this would give 64 possible combinations (4^3), many more than the minimum of 20 needed to specify the common amino acids.

The actual codons were discovered in the early 1960s through the experiments carried out by Marshall Nirenberg, Heinrich Matthaei, Philip Leder, and Har Gobind Khorana. In 1968 Nirenberg and Khorana shared the Nobel Prize with Robert W. Holley, the first person to sequence a nucleic acid (phenylalanyl-tRNA).

Organization of the Code

The genetic code, presented in RNA form, is summarized in **table 11.1.** Note that there is **code degeneracy.** That is, there are up to six different codons for a given amino acid. Only 61 codons, the **sense codons,** direct amino acid incorporation into protein. The remaining three codons (UGA, UAG, and UAA) are involved in the termination of translation and are called **stop** or **nonsense codons.** Despite the existence of 61 sense codons, there are not 61 different tRNAs, one for each codon. The 5′ nucleotide in the anticodon can vary, but generally, if the nucleotides in the second and third anticodon positions complement the first two bases of the mRNA codon, an aminoacyl-tRNA with the proper amino acid will bind to the mRNA-ribosome complex. This pattern is evident on inspection of changes in the amino acid specified with variation in the third po-

Table 11.1
The Genetic Code

		Second Position				
		U	**C**	**A**	**G**	Third Position (3′ End)
First Position (5′ End)[a]	**U**	UUU ⎫ Phe UUC ⎭ UUA ⎫ Leu UUG ⎭	UCU ⎫ UCC UCA ⎬ Ser UCG ⎭	UAU ⎫ Tyr UAC ⎭ UAA ⎫ STOP UAG ⎭	UGU ⎫ Cys UGC ⎭ UGA STOP UGG Trp	U C A G
	C	CUU ⎫ CUC CUA ⎬ Leu CUG ⎭	CCU ⎫ CCC CCA ⎬ Pro CCG ⎭	CAU ⎫ His CAC ⎭ CAA ⎫ Gln CAG ⎭	CGU ⎫ CGC CGA ⎬ Arg CGG ⎭	U C A G
	A	AUU ⎫ AUC ⎬ Ile AUA ⎭ AUG Met	ACU ⎫ ACC ACA ⎬ Thr ACG ⎭	AAU ⎫ Asn AAC ⎭ AAA ⎫ Lys AAG ⎭	AGU ⎫ Ser AGC ⎭ AGA ⎫ Arg AGG ⎭	U C A G
	G	GUU ⎫ GUC GUA ⎬ Val GUG ⎭	GCU ⎫ GCC GCA ⎬ Ala GCG ⎭	GAU ⎫ Asp GAC ⎭ GAA ⎫ Glu GAG ⎭	GGU ⎫ GGC GGA ⎬ Gly GGG ⎭	U C A G

[a]The code is presented in the RNA form. Codons run in the 5′ to 3′ direction. See text for details.

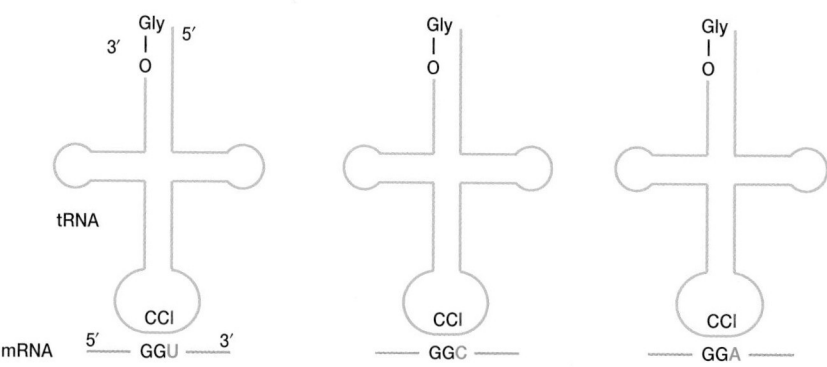

(a) Base pairing of one glycine tRNA with three codons due to wobble

(b) Glycine codons and anticodons (written in the 5′ ⟶ 3′ direction)

Glycine mRNA codons: GGU, GGC, GGA, GGG

Glycine tRNA anticodons: ICC, CCC

Figure 11.19 Wobble and Coding.   The use of wobble in coding for the amino acid glycine. **(a)** Inosine (I) is a wobble nucleoside that can base pair with uracil (U), cytosine (C), or adenine (A). Thus ICC base pairs with GGU, GGC, and GGA in the mRNA. **(b)** Because of the wobble produced by inosine, two tRNA anticodons can recognize the four glycine (Gly) codons. ICC recognizes GGU, GGC, and GGA; CCC recognizes GGG.

sition (table 11.1). This somewhat loose base pairing is known as **wobble** and relieves cells of the need to synthesize so many tRNAs (**figure 11.19**). Wobble also decreases the effects of DNA mutations.   The mechanism of protein synthesis and tRNA function (pp. 260–70)

---

1. Why must a codon contain at least three nucleotides?
2. Define the following: code degeneracy, sense codon, stop or nonsense codon, and wobble.

 GENE STRUCTURE

The **gene** has been defined in several ways. Initially geneticists considered it to be the entity responsible for conferring traits on the organism and the entity that could undergo recombination. Recombination involves exchange of DNA from one source with that from another (*see section 13.1*) and is responsible for generating much of the genetic variability found in viruses and living organisms. Genes were typically named for some mutant or altered phenotype. With the discovery and characterization of DNA, the gene was defined more precisely as a linear sequence of nucleotides or codons (this term can be used for RNA as well as DNA) with fixed start and end points.

At first, it was thought that a gene contained information for the synthesis of one enzyme, the one gene–one enzyme hypothesis. This has been modified to the one gene–one polypeptide hypothesis because of the existence of enzymes and other proteins composed of two or more different polypeptide chains coded for by separate genes. The segment that codes for a single polypeptide is sometimes also called a **cistron.** More recent results show that even this description is oversimplified. Not all genes are involved in protein synthesis; some code instead for rRNA and tRNA. Thus a gene might be defined as a polynucleotide sequence that codes for a functional product (i.e., a polypeptide,

tRNA, or rRNA). Some geneticists think of it as a segment of nucleic acid that is transcribed to give an RNA product. Most genes consist of discrete sequences of codons that are "read" only one way to produce a single product. That is, the code is not overlapping and there is a single starting point with one **reading frame** or way in which nucleotides are grouped into codons (**figure 11.20**). Chromosomes therefore usually consist of gene sequences that do not overlap one another (**figure 11.21a**). However, there are exceptions to the rule. Some viruses such as the phage ϕX174 do have overlapping genes (figure 11.21*b*), and parts of genes overlap in some bacterial genomes.

Procaryotic and viral gene structure differs greatly from that of eucaryotes. In procaryotic and viral systems, the coding information within a cistron normally is continuous (some bacterial genes do contain introns); however, in eucaryotic organisms, many genes contain coding information (exons) interrupted periodically by noncoding sequences (introns). An interesting exception to this rule is eucaryotic histone genes, which lack introns. Because procaryotic and viral systems are the best characterized, the more detailed description of gene structure that follows will focus on *E. coli* genes.   Exons and introns in eucaryotic genes (p. 257)

Genes That Code for Proteins

Recall from the discussion of transcription that although DNA is double stranded, only one strand contains coded information and directs RNA synthesis. This strand is called the **template strand,** and the complementing strand is known as the nontemplate strand (**figure 11.22**). Because the mRNA is made from the 5′ to the 3′ end, the polarity of the DNA template strand is therefore 3′ to 5′. Therefore the beginning of the gene is at the 3′ end of the template strand (also the 5′ end of the nontemplate strand). An RNA polymerase recognition/binding and regulatory site known as the **promoter** is located at the start of the gene.   The mechanism of transcription (pp. 254–60)

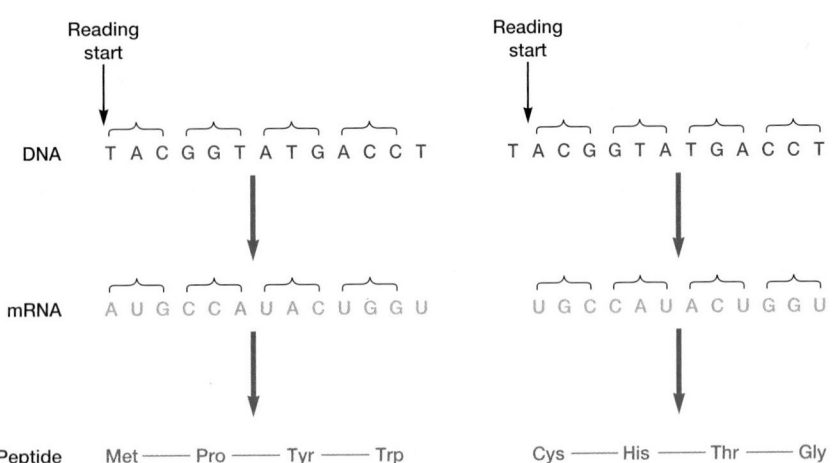

Figure 11.20 Reading Frames and Their Importance. The place at which DNA sequence reading begins determines the way nucleotides are grouped together in clusters of three (outlined with brackets), and this specifies the mRNA codons and the peptide product. In the example, a change in the reading frame by one nucleotide yields a quite different mRNA and final peptide.

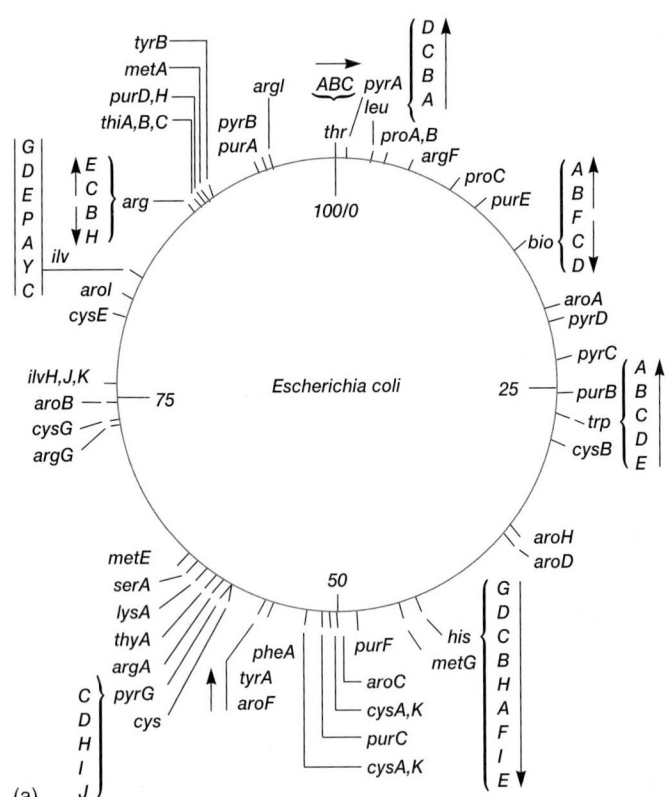

(a)

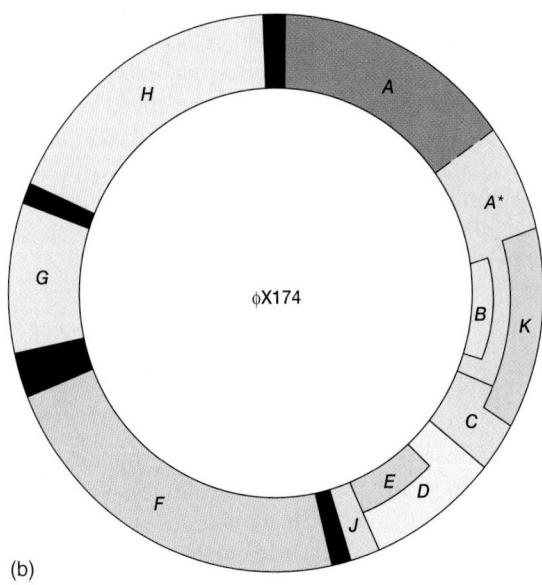

(b)

Figure 11.21 Chromosomal Organization in Bacteria and Viruses. (a) Simplified genetic map of *E. coli*. The *E. coli* map is divided into 100 minutes. (b) The map of phage φX174 shows the overlap of gene *B* with *A*, *K* with *A* and *C*, and *E* with *D*. The solid regions are spaces lying between genes. Protein A* consists of the last part of protein A and arises from reinitiation of transcription within gene A.

Promoters are sequences of DNA that are usually upstream from the actual coding or transcribed region (figure 11.22)—that is, the promoter is located before or upstream from the coding region in relationship to the direction of transcription (the direction of transcription is referred to as downstream). Different genes have different promoters, and promoters also vary in sequence between bacteria. In *E. coli* the promoter has two important functions, and these relate to two specific segments within the promoter (figure 11.22 and **figure 11.23**). Although these two segments do vary slightly in sequence between bacterial strains and different genes, they are fairly constant and may be represented by consensus sequences. These are idealized sequences composed of the bases most often found at each position when the sequences from different bacteria are compared. The RNA polymerase recognition site, with a consensus sequence of 5′TTGACA3′ on the nontemplate strand in *E. coli*, is centered about 35 base pairs before (the −35 region) the transcriptional start point (labeled as +1) of RNA synthesis. This sequence seems to be the site of the initial association

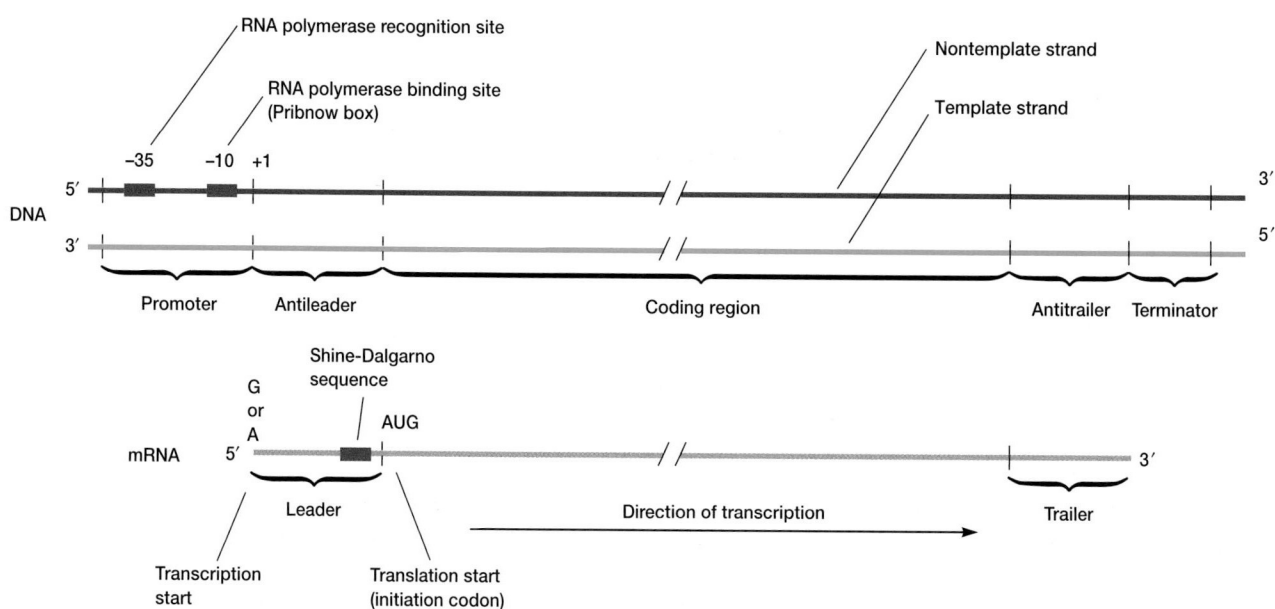

Figure 11.22 A Bacterial Structural Gene. The organization of a typical structural gene in bacteria. Leader and trailer sequences are included even though some genes lack one or both. Transcription begins at the +1 position in DNA, and the first nucleotide incorporated into mRNA is usually GTP or ATP. Translation of the mRNA begins with the AUG initiation codon. Regulatory sites are not shown.

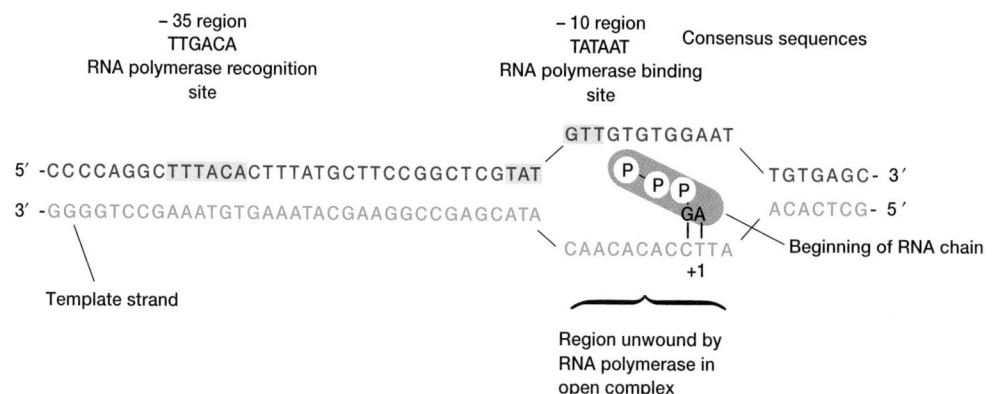

Figure 11.23 A Bacterial Promoter. The lactose operon promoter and its consensus sequences. The start point for RNA synthesis is labeled +1. The region around −35 is the site at which the RNA polymerase first attaches to the promoter. RNA polymerase binds and begins to unwind the DNA helix at the Pribnow box or RNA polymerase binding site, which is located in the −10 region.

of the RNA polymerase with the DNA. The **RNA polymerase binding site,** also known as the **Pribnow box,** is centered at the −10 region and has a consensus sequence 5′TATAAT3′ in *E. coli* (a sequence that favors the localized unwinding of DNA). This is where the RNA polymerase begins to unwind the DNA for eventual transcription. The initially transcribed portion of the gene is not necessarily coding material. Rather, a **leader sequence** may be synthesized first. The leader is usually a nontranslated sequence that is important in the initiation of translation and sometimes is involved in regulation of transcription.

The leader (figure 11.22) in procaryotes generally contains a consensus sequence known as the **Shine-Dalgarno sequence,** 5′AGGA3′, the transcript of which complements a sequence on the

16S rRNA in the small subunit of the ribosome. The binding of mRNA leader with 16S rRNA properly orients the mRNA on the ribosome. The leader also sometimes regulates transcription by attenuation (*see section 12.4*). Downstream and next to the leader is the most important part of the structural gene, the coding region.

The **coding region** (figure 11.22) of genes that direct the synthesis of proteins typically begins with the template DNA sequence 3′TAC5′. This produces the RNA translation initiation codon 5′AUG3′, which codes for *N*-formylmethionine. This modified form of methionine is the first amino acid incorporated in most procaryotic proteins. The remainder of the gene coding region consists of a sequence of codons that specifies the sequence of amino acids for that particular protein. Transcription does not

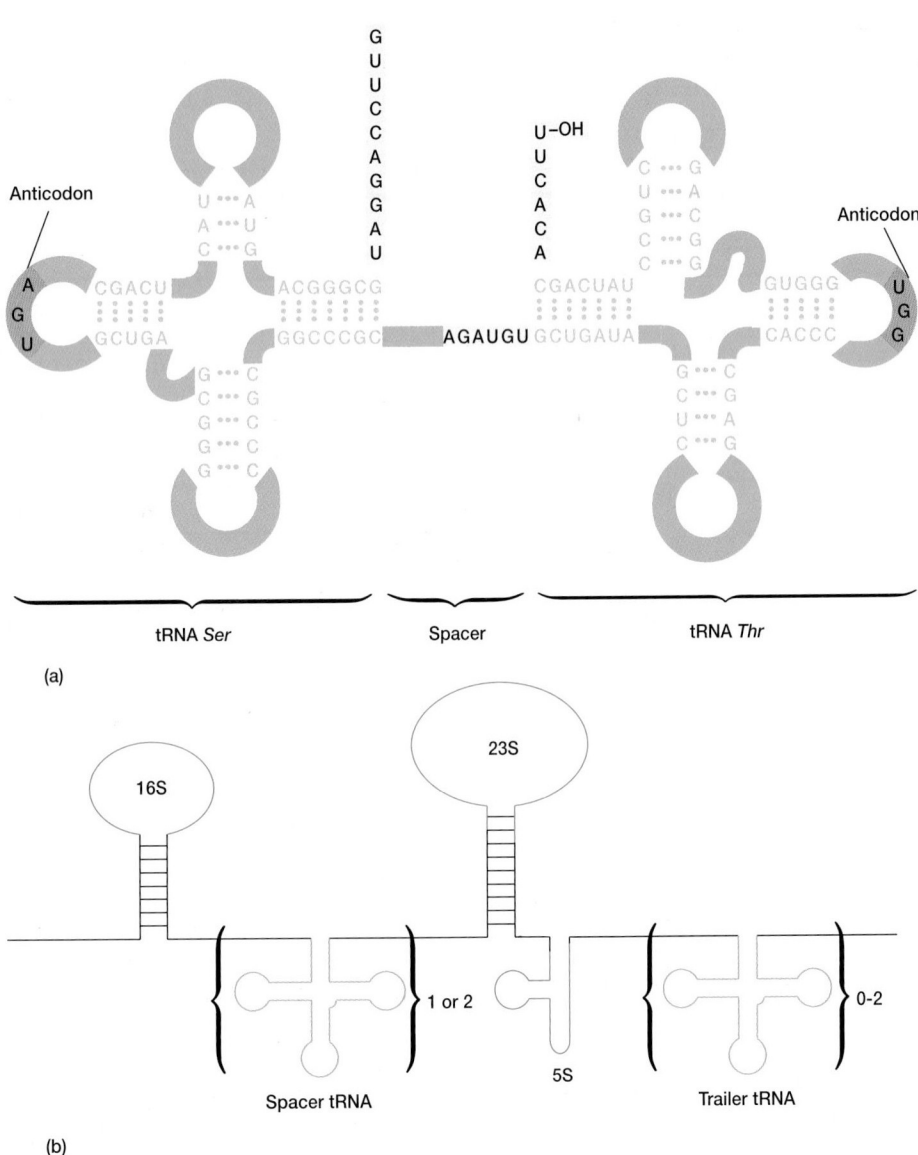

Figure 11.24 tRNA and rRNA Genes. (a) A tRNA precursor from *E. coli* that contains two tRNA molecules. The spacer and extra nucleotides at both ends are removed during processing. (b) The *E. coli* ribosomal RNA gene codes for a large transcription product that is cleaved into three rRNAs and one to three tRNAs. The 16S, 23S, and 5S rRNA segments are represented by blue lines, and tRNA sequences are placed in brackets. The seven copies of this gene vary in the number and kind of tRNA sequences.

stop at the translation stop codon but rather at a **terminator sequence** (*see section 12.1*). The terminator often lies after a nontranslated **trailer sequence** located downstream from the coding region (figure 11.22). The trailer sequence, like the leader, is needed for the proper expression of the coding region of the gene.

Besides the basic components described previously—the promoter, leader, coding region, trailer, and terminator—many procaryotic genes have a variety of regulatory sites. These are locations where DNA-recognizing regulatory proteins bind to stimulate or prevent gene expression. Regulatory sites often are associated with promoter function, and some consider them to be parts of special promoters. Two such sites, the operator and the CAP binding site, are discussed in *section 12.3*. Certainly every-

thing is not known about genes and their structure. With the ready availability of purified cloned genes and DNA sequencing technology, major discoveries continue to be made in this area. The operon and transcription regulation (pp. 270–74)

Genes That Code for tRNA and rRNA

The DNA segments that code for tRNA and rRNA also are considered genes, although they give rise to structurally important RNA rather than protein. In *E. coli* the genes for tRNA are fairly typical, consisting of a promoter and transcribed leader and trailer sequences that are removed during the maturation process (**figure 11.24a**). The precise function of the leader is not clear;

however, the trailer is required for termination. Genes coding for tRNA may code for more than a single tRNA molecule or type of tRNA (figure 11.24*a*). The segments coding for tRNAs are separated by short spacer sequences that are removed after transcription by special ribonucleases, at least one of which contains catalytic RNA. As mentioned in chapter 12 (*see p. 260*), mature tRNAs contain unusual nucleosides. Modified nucleosides such as inosine, ribothymidine, and pseudouridine almost always are formed after the tRNA has been synthesized. Special tRNA-modifying enzymes are responsible.   RNA splicing and ribozymes (pp. 257–60)

The genes for rRNA also are similar in organization to genes coding for proteins and have promoters, trailers, and terminators (figure 11.24*b*). Interestingly all the rRNAs are transcribed as a single, large precursor molecule that is cut up by ribonucleases after transcription to yield the final rRNA products. *E. coli* pre-rRNA spacer and trailer regions even contain tRNA genes. Thus the synthesis of tRNA and rRNA involve posttranscriptional modification, a relatively rare process in procaryotes.

---

1. Define or describe the following: gene, template and nontemplate strands, promoter, consensus sequence, RNA polymerase recognition and binding sites, Pribnow box, leader, Shine-Dalgarno sequence, coding region, reading frame, trailer, and terminator.
2. How do the genes of procaryotes and eucaryotes usually differ from each other?
3. Briefly discuss the general organization of tRNA and rRNA genes. How does their expression differ from that of structural genes with respect to posttranscriptional modification of the gene product?

 ## 11.6 MUTATIONS AND THEIR CHEMICAL BASIS

Considerable information is embedded in the precise order of nucleotides in DNA. For life to exist with stability, it is essential that the nucleotide sequence of genes is not disturbed to any great extent. However, sequence changes do occur and often result in altered phenotypes. These changes are largely detrimental but are important in generating new variability and contribute to the process of evolution. Microbial mutation rates also can be increased, and these genetic changes have been put to many important uses in the laboratory and industry.

Mutations [Latin *mutare,* to change] were initially characterized as altered phenotypes or phenotypic expressions. Long before the existence of direct proof that a mutation is a stable, heritable change in the nucleotide sequence of DNA, geneticists predicted that several basic types of transmitted mutations could exist. They believed that mutations could arise from the alteration of single pairs of nucleotides and from the addition or deletion of one or two nucleotide pairs in the coding regions of a gene. Clearly, mutations may be characterized according to either the kind of genotypic change that has occurred or their phenotypic consequences. In this section the molecular basis of mutations

and mutagenesis is first considered. Then the phenotypic effects of mutations, the detection of mutations, and the use of mutations in carcinogenicity testing are discussed.

Mutations and Mutagenesis

Mutations can alter the phenotype of a microorganism in several different ways. Morphological mutations change the microorganism's colonial or cellular morphology. Lethal mutations, when expressed, result in the death of the microorganism. Since the microorganism must be able to grow in order to be isolated and studied, lethal mutations are recovered only if they are recessive in diploid organisms or conditional (see the following) in haploid organisms.

Conditional mutations are those that are expressed only under certain environmental conditions. For example, a conditional lethal mutation in *E. coli* might not be expressed under permissive conditions such as low temperature but would be expressed under restrictive conditions such as high temperature. Thus the hypothetical mutant would grow normally at the permissive temperature but would die at high temperatures.

Biochemical mutations are those causing a change in the biochemistry of the cell. Since these mutations often inactivate a biosynthetic pathway, they frequently make a microorganism unable to grow on a medium lacking an adequate supply of the pathway's end product. That is, the mutant cannot grow on minimal medium and requires nutrient supplements. Such mutants are called **auxotrophs,** whereas microbial strains that can grow on minimal medium are **prototrophs.** Analysis of auxotrophy has been quite important in microbial genetics due to the ease of auxotroph selection and the relative abundance of this mutational type.   Mutant detection and replica plating (pp. 245–46); Nutrient requirements and nutritional types (pp. 94–98)

A resistant mutant is a particular type of biochemical mutant that acquires resistance to some pathogen, chemical, or antibiotic. Such mutants also are easy to select for and very useful in microbial genetics.   Mechanisms of drug resistance (pp. 792–94)

Mutations occur in one of two ways. (1) Spontaneous mutations arise occasionally in all cells and develop in the absence of any added agent. (2) Induced mutations, on the other hand, are the result of exposure of the organism to some physical or chemical agent called a **mutagen.**

Although most geneticists believe that spontaneous mutations occur randomly in the absence of an external agent and are then selected, observations by some microbiologists have led to a new and controversial hypothesis. John Cairns and his collaborators have reported that a mutant *E. coli* strain, which is unable to use lactose as a carbon and energy source, regains the ability to do so more rapidly when lactose is added to the culture medium as the only carbon source. Lactose appears to induce mutations that allow *E. coli* to use the sugar again. It has been claimed that these and similar observations on different mutations are examples of **directed** or **adaptive mutation**—that is, some bacteria seem able to choose which mutations occur so that they can better adapt to their surroundings. Many explanations

Figure 11.25   Transition and Transversion Mutations.
Errors in replication due to base tautomerization. (**a**) Normally AT and GC pairs are formed when keto groups participate in hydrogen bonds. In contrast, enol tautomers produce AC and GT base pairs. (**b**) Mutation as a consequence of tautomerization during DNA replication. The temporary enolization of guanine leads to the formation of an AT base pair in the mutant, and a GC to AT transition mutation occurs. The process requires two replication cycles. The mutation only occurs if the abnormal first-generation GT base pair is missed by repair mechanisms.

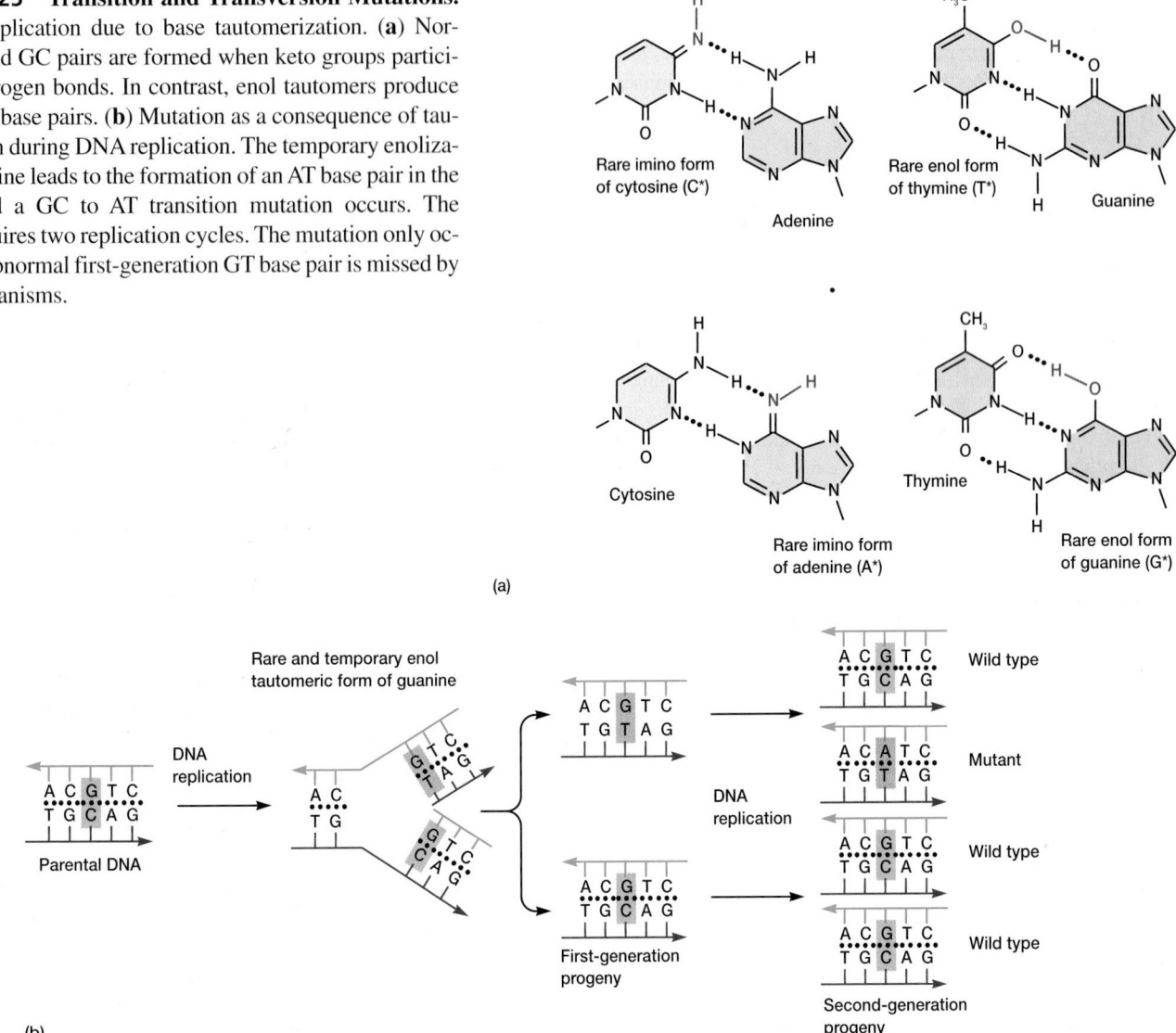

have been offered to account for this phenomenon without depending on bacterial selection of particular mutations. One of the most interesting is the proposal that **hypermutation** can produce such results. Some starving bacteria might rapidly generate multiple mutations through activation of special mutator genes. This would produce many mutant bacterial cells. In such a random process, the rate of production of favorable mutants would increase, with many of these mutants surviving to be counted. There would appear to be directed or adaptive mutation because many of the unfavorable mutants would die. There is support for this hypothesis. Mutator genes have been discovered and do cause hypermutation under nutritional stress. Even if the directed mutation hypothesis is incorrect, it has stimulated much valuable research and led to the discovery of new phenomena. Some of these issues will be discussed in chapter 42 in the context of natural genetic engineering.

Spontaneous Mutations

Spontaneous mutations arise without exposure to external agents. This class of mutations may result from errors in DNA replication, or even from the action of transposons (*see section 13.3*). A few of the more prevalent mechanisms are described here.

Generally replication errors occur when the base of a template nucleotide takes on a rare tautomeric form. Tautomerism is the relationship between two structural isomers that are in chemical equilibrium and readily change into one another. Bases typically exist in the keto form. However, they can at times take on either an imino or enol form (**figure 11.25a**). These tautomeric shifts change the hydrogen-bonding characteristics of the bases, allowing purine for purine or pyrimidine for pyrimidine substitutions that can eventually lead to a stable alteration of the nucleotide sequence (figure 11.25b). Such substitutions are known

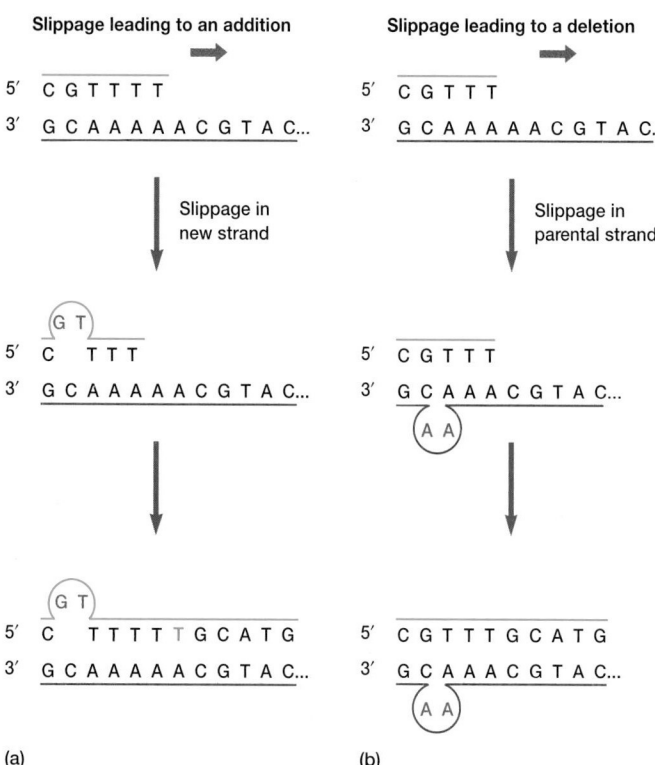

Slippage leading to an addition

Slippage leading to a deletion

(a)   (b)

Figure 11.26 Additions and Deletions. A hypothetical mechanism for the generation of additions and deletions during replication. The direction of replication is indicated by the large blue arrow. In each case there is strand slippage resulting in the formation of a small loop that is stabilized by the hydrogen bonding in the repetitive sequence, the AT stretch in this example. DNA synthesis proceeds to the right in this figure. (**a**) If the new strand slips, an addition of one T results. (**b**) Slippage of the parental strand yields a deletion (in this case, a loss of two Ts).

as **transition mutations** and are relatively common, although most of them are repaired by various proofreading functions (*see pp. 233 and 248*). In **transversion mutations,** a purine is substituted for a pyrimidine, or a pyrimidine for a purine. These mutations are rarer due to the steric problems of pairing purines with purines and pyrimidines with pyrimidines.

Spontaneous mutations also arise from **frameshifts,** usually caused by the deletion of DNA segments resulting in an altered codon reading frame. These mutations generally occur where there is a short stretch of the same nucleotide. In such a location, the pairing of template and new strand can be displaced by the distance of the repeated sequence leading to additions or deletions of bases in the new strand (**figure 11.26**).

Spontaneous mutations originate from lesions in DNA as well as from replication errors. For example, it is possible for purine nucleotides to be depurinated—that is, to lose their base. This results in the formation of an **apurinic site,** which will not base pair normally and may cause a transition type mutation after the next round of replication. Cytosine can be deaminated to uracil, which is then removed to form an **apyrimidinic site.** Reactive forms of

oxygen such as oxygen free radicals and peroxides are produced by aerobic metabolism (*see p. 125*). These may alter DNA bases and cause mutations. For example, guanine can be converted to 8-oxo-7,8-dihydrodeoxyguanine, which often pairs with adenine rather than cytosine during replication.

Finally, spontaneous mutations can result from the insertion of DNA segments into genes. This results from the movement of insertion sequences and transposons (*see pp. 291–95*), and usually inactivates the gene. Insertion mutations are very frequent in *E. coli* and many other bacteria.

Induced Mutations

Virtually any agent that directly damages DNA, alters its chemistry, or interferes with repair mechanisms (pp. 248–50) will induce mutations. Mutagens can be conveniently classified according to their mechanism of action. Four common modes of mutagen action are incorporation of base analogs, specific mispairing, intercalation, and bypass of replication.

Base analogs are structurally similar to normal nitrogenous bases and can be incorporated into the growing polynucleotide chain during replication. Once in place, these compounds typically exhibit base pairing properties different from the bases they replace and can eventually cause a stable mutation. A widely used base analog is 5-bromouracil (5-BU), an analog of thymine. It undergoes a tautomeric shift from the normal keto form to an enol much more frequently than does a normal base. The enol forms hydrogen bonds like cytosine and directs the incorporation of guanine rather than adenine (**figure 11.27**). The mechanism of action of other base analogs is similar to that of 5-bromouracil.

Specific mispairing is caused when a mutagen changes a base's structure and therefore alters its base pairing characteristics. Some mutagens in this category are fairly selective; they preferentially react with some bases and produce a specific kind of DNA damage. An example of this type of mutagen is methylnitrosoguanidine, an alkylating agent that adds methyl groups to guanine, causing it to mispair with thymine (**figure 11.28**). A subsequent round of replication could then result in a GC-AT transition. DNA damage also stimulates error-prone repair mechanisms. Other examples of mutagens with this mode of action are the alkylating agents ethylmethanesulfonate and hydroxylamine. Hydroxylamine hydroxylates the C-4 nitrogen of cytosine, causing it to base pair like thymine. There are many other DNA modifying agents that can cause mispairing.

Intercalating agents distort DNA to induce single nucleotide pair insertions and deletions. These mutagens are planar and insert themselves (intercalate) between the stacked bases of the helix. This results in a mutation, possibly through the formation of a loop in DNA. Intercalating agents include acridines such as proflavin and acridine orange.

Many mutagens, and indeed many carcinogens, directly damage bases so severely that hydrogen bonding between base pairs is impaired or prevented and the damaged DNA can no longer act as a template. For instance, UV radiation generates cyclobutane type dimers, usually thymine dimers, between adjacent pyrimidines

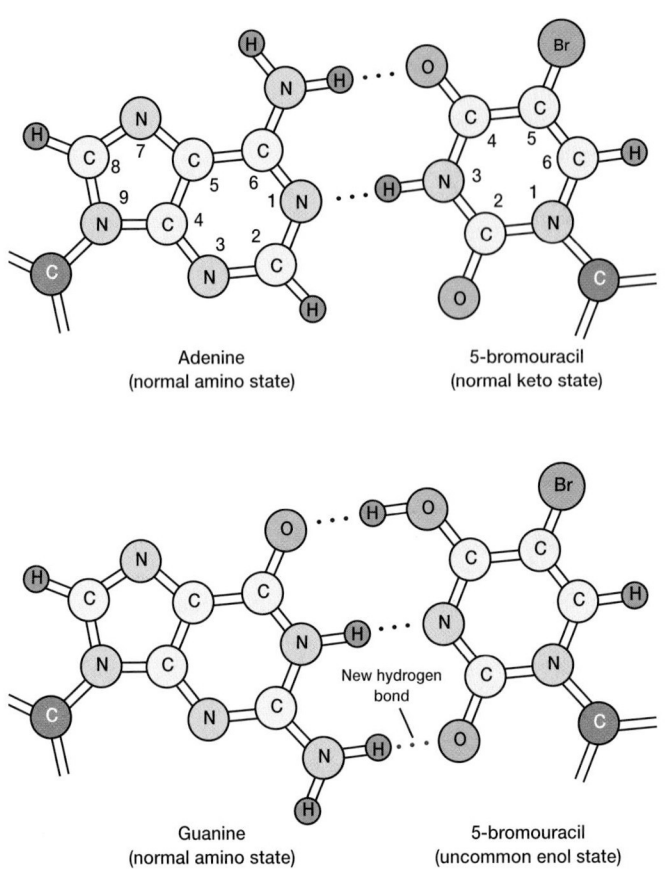

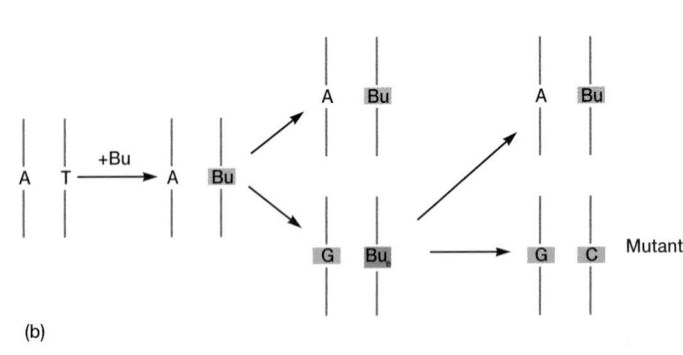

(b)

Figure 11.27 **Mutagenesis by the Base Analog 5-Bromouracil.**
(a) Base pairing of the normal keto form of 5-BU is shown in the top illustration. The enol form of 5-BU (bottom illustration) base pairs with guanine rather than with adenine as might be expected for a thymine analog. (b) If the keto form of 5-BU is incorporated in place of thymine, its occasional tautomerization to the enol form (BU_e) will produce an AT to GC transition mutation.

(**figure 11.29**). Other examples are ionizing radiation and carcinogens such as aflatoxin B1 and other benzo(a)pyrene derivatives. Such damage to DNA would generally be lethal but may trigger a repair mechanism that restores much of the damaged genetic material, although with considerable error incorporation (pp. 248–50).

Retention of proper base pairing is essential in the prevention of mutations. Often the damage can be repaired before a mutation is permanently established. If a complete DNA replication cycle takes place before the initial lesion is repaired, the mutation frequently becomes stable and inheritable.

The Expression of Mutations

The expression of a mutation will only be readily noticed if it produces a detectable, altered phenotype. A mutation from the most prevalent gene form, the **wild type,** to a mutant form is called a **forward mutation.** Later, a second mutation may make the mutant appear to be a wild-type organism again. Such a mutation is called a **reversion mutation** because the organism seems to have reverted back to its original phenotype. A true **back mutation** converts the mutant nucleotide sequence back to the wild-type sequence. The wild-type phenotype also can be regained by a second mutation in a different gene, a **suppressor mutation,** which

overcomes the effect of the first mutation (**table 11.2**). If the second mutation is within the same gene, the change may be called a second site reversion or intragenic suppression. Thus although revertant phenotypes appear to be wild types, the original DNA sequence may not be restored. In practice, a mutation is visibly expressed when a protein that is in some way responsible for the phenotype is altered sufficiently to produce a new phenotype. However, mutations may occur and not alter the phenotype for a variety of reasons.

Although very large deletion and insertion mutations exist, most mutations affect only one base pair in a given location and therefore are called **point mutations.** There are several types of point mutations (table 11.2).

One kind of point mutation that could not be detected until the advent of nucleic acid sequencing techniques is the **silent mutation.** If a mutation is an alteration of the nucleotide sequence of DNA, mutations can occur and have no visible effect because of code degeneracy. When there is more than one codon for a given amino acid, a single base substitution could result in the formation of a new codon for the same amino acid. For example, if the codon CGU were changed to CGC, it usually would still code for arginine even though a mutation had occurred. The expression of this mutation often would not be detected except at the level of the DNA or mRNA. When there is no change in

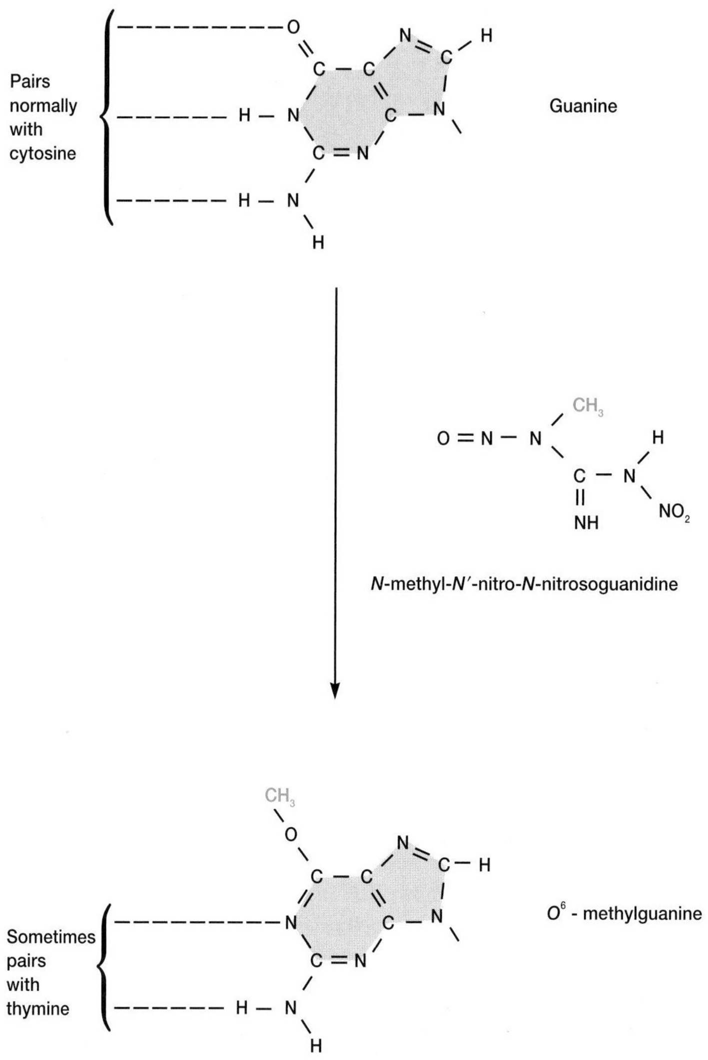

Pairs
normally
with
cytosine   Guanine

N-methyl-*N'*-nitro-*N*-nitrosoguanidine

O^6 - methylguanine

Sometimes
pairs
with
thymine

Figure 11.28   Methyl-Nitrosoguanidine Mutagenesis.   Mutagenesis by methyl-nitrosoguanidine due to the methylation of guanine.

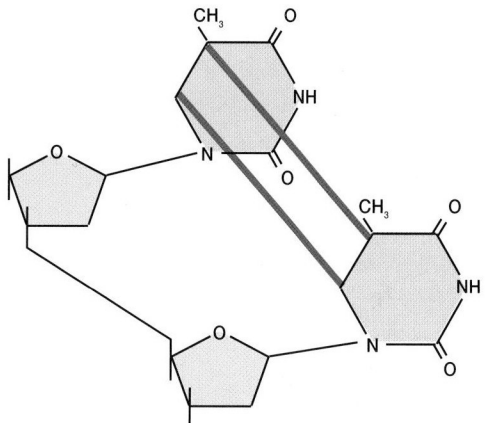

Figure 11.29   Thymine Dimer.   Thymine dimers are formed by ultraviolet radiation. The enzyme photolyase cleaves the two blue bonds during photoreactivation.

the protein or its concentration, there will be no change in the phenotype of the organism.

A second type of point mutation is the **missense mutation.** This mutation involves a single base substitution in the DNA that changes a codon for one amino acid into a codon for another. For example, the codon GAG, which specifies glutamic acid, could be changed to GUG, which codes for valine. The expression of missense mutations can vary. Certainly the mutation is expressed at the level of protein structure. However, at the level of protein function, the effect may range from complete loss of activity to no change at all.

Many proteins are still functional after the substitution of a single amino acid, but this depends on the type and location of the amino acid. For instance, replacement of a nonpolar amino acid in the protein's interior with a polar amino acid probably will drastically alter the protein's three-dimensional structure and therefore its function. Similarly the replacement of a critical amino acid at the active site of an enzyme will destroy its activity. However, the

Table 11.2	
Summary of Some Molecular Changes from Gene Mutations	

Type of Mutation	Result and Example
Forward Mutations	
Single Nucleotide-Pair (Base-Pair) Substitutions	
At DNA Level	
Transition	Purine replaced by a different purine, or pyrimidine replaced by a different pyrimidine (e.g., AT $\longrightarrow$ GC).
Transversion	Purine replaced by a pyrimidine, or pyrimidine replaced by a purine (e.g., AT $\longrightarrow$ CG).
At Protein Level	
Silent mutation	Triplet codes for same amino acid: AGG $\longrightarrow$ **CGG** both code for Arg
Neutral mutation	Triplet codes for different but functionally equivalent amino acid: AAA (Lys) $\longrightarrow$ A**G**A (Arg)
Missense mutation	Triplet codes for a different amino acid.
Nonsense mutation	Triplet codes for chain termination: CAG (Gln) $\longrightarrow$ **U**AG (stop)
Single Nucleotide-Pair Addition or Deletion: Frameshift Mutation	Any addition or deletion of base pairs that is not a multiple of three results in a frameshift in reading the DNA segments that code for proteins.
Intragenic Addition or Deletion of Several to Many Nucleotide Pairs	
Reverse Mutations	
True Reversion	AAA (Lys) $\xrightarrow{\text{forward}}$ GAA (Glu) $\xrightarrow{\text{reverse}}$ AAA (Lys) wild type mutant wild type
Equivalent Reversion	UCC (Ser) $\xrightarrow{\text{forward}}$ UGC (Cys) $\xrightarrow{\text{reverse}}$ AGC (Ser) wild type mutant wild type
	CGC (Arg, basic) $\xrightarrow{\text{forward}}$ CCC (Pro, not basic) $\xrightarrow{\text{reverse}}$ CAC (His, basic) wild type mutant pseudo-wild type
Suppressor Mutations	
Intragenic Suppressor Mutations	CATCATCATCATCATCAT
Frameshift of opposite sign at site within gene. Addition of X to the base sequence shifts the reading frame from the CAT codon to XCA followed by TCA codons. The subsequent deletion of a C base shifts the reading frame back to CAT.	(+) (−) $\downarrow$ $\downarrow$ CATXCATATCATCATCAT y x z y y y
Extragenic Suppressor Mutations	
Nonsense suppressors	Gene (e.g., for tyrosine tRNA) undergoes mutational event in its anticodon region that enables it to recognize and align with a mutant nonsense codon (e.g., UAG) to insert an amino acid (tyrosine) and permit completion of the translation.
Physiological suppressors	A defect in one chemical pathway is circumvented by another mutation—for example, one that opens up another chemical pathway to the same product, or one that permits more efficient uptake of a compound produced in small quantities because of the original mutation.

replacement of one polar amino acid with another at the protein surface may have little or no effect. Missense mutations may actually play a very important role in providing new variability to drive evolution because they often are not lethal and therefore remain in the gene pool.   Protein structure (appendix I)

A third type of point mutation causes the early termination of translation and therefore results in a shortened polypeptide. Such mutations are called **nonsense mutations** because they involve the conversion of a sense codon to a nonsense or stop codon. Depending on the relative location of the mutation, the phenotypic expression may be more or less severely affected. Most proteins retain some function if they are shortened by only one or two amino acids; complete loss of normal function will almost certainly result if the mutation occurs closer to the middle of the gene.

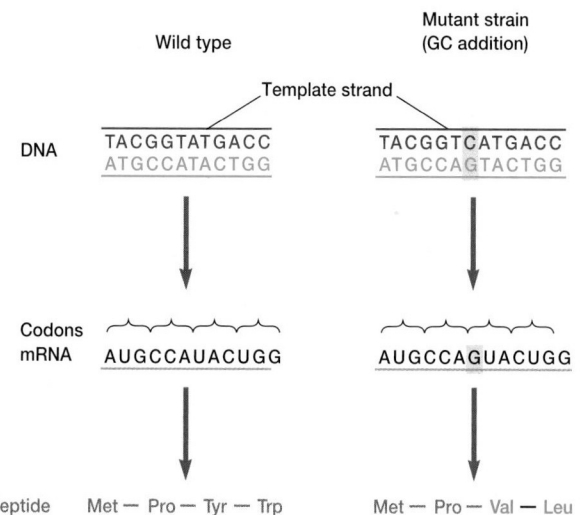

Figure 11.30   Frameshift Mutation.   A frameshift mutation resulting from the insertion of a GC base pair. The reading frameshift produces a different peptide after the addition.

The **frameshift mutation** is a fourth type of point mutation and was briefly mentioned earlier. Frameshift mutations arise from the insertion or deletion of one or two base pairs within the coding region of the gene. Since the code consists of a precise sequence of triplet codons, the addition or deletion of fewer than three base pairs will cause the reading frame to be shifted for all codons downstream (figure 11.20). **Figure 11.30** shows the effect of a frameshift mutation on a short section of mRNA and the amino acid sequence it codes for.

Frameshift mutations usually are very deleterious and yield mutant phenotypes resulting from the synthesis of nonfunctional proteins. The reading frameshift often eventually produces a nonsense or stop codon so that the peptide product is shorter as well as different in sequence. Of course if the frameshift occurred near the end of the gene, or if there were a second frameshift shortly downstream from the first that restored the reading frame, the phenotypic effect might not be as drastic. A second nearby frameshift that restores the proper reading frame is a good example of an intragenic suppressor mutation (table 11.2).

Mutations also occur in the regulatory sequences responsible for the control of gene expression and in other noncoding portions of structural genes. Constitutive lactose operon mutants in *E. coli* are excellent examples. These mutations map in the operator site and produce altered operator sequences that are not recognized by the repressor protein, and therefore the operon is continuously active in transcription. If a mutation renders the promoter sequence nonfunctional, the coding region of the structural gene will be completely normal, but a mutant phenotype will result due to the absence of a product. RNA polymerase rarely transcribes a gene correctly without a fully functional promoter.   The *lac* operon and gene regulation (pp. 270–74)

Mutations also occur in rRNA and tRNA genes and can alter the phenotype through disruption of protein synthesis. In fact,

these mutants often are initially identified because of their slow growth. One type of suppressor mutation is a base substitution in the anticodon region of a tRNA that allows the insertion of the correct amino acid at a mutant codon (table 11.2).

---

1. Define or describe the following: mutation, conditional mutation, auxotroph and prototroph, spontaneous and induced mutations, mutagen, transition and transversion mutations, frameshift, apurinic site, base analog, specific mispairing, intercalating agent, thymine dimer, wild type, forward and reverse mutations, suppressor mutation, point mutation, silent mutation, missense and nonsense mutations, directed or adaptive mutation and hypermutation, and frameshift mutation.
2. Give four ways in which spontaneous mutations might arise.
3. How do the mutagens 5-bromouracil, methyl-nitrosoguanidine, proflavin, and UV radiation induce mutations?
4. Give examples of intragenic and extragenic suppressor mutations.

 **DETECTION AND ISOLATION OF MUTANTS**

In order to study microbial mutants, one must be able to detect them readily, even when there are few, and then efficiently isolate them from the parent organism and other mutants. Fortunately, this often is easy to do. This section describes some techniques used in mutant detection, selection, and isolation.

Mutant Detection

When collecting mutants of a particular organism, one must know the normal or wild-type characteristics so as to recognize an altered phenotype. A suitable detection system for the mutant phenotype under study also is needed. Since mutations are generally rare, about one per 10^7 to 10^{11} cells, it is important to have a very sensitive detection system so that these events will not be missed. Geneticists often induce mutations to increase the probability of obtaining specific changes at high frequency (about one in 10^3 to 10^6); even so, mutations are rare.

Detection systems in bacteria and other haploid organisms are straightforward because any new allele should be seen immediately, even if it is a recessive mutation. Sometimes detection of mutants is direct. If albino mutants of a normally pigmented bacterium are being studied, detection simply requires visual observation of colony color. Other direct detection systems are more complex. For example, the **replica plating** technique is used to detect auxotrophic mutants. It distinguishes between mutants and the wild-type strain based on their ability to grow in the absence of a particular biosynthetic end product (**figure 11.31**). A lysine auxotroph, for instance, will grow on lysine-supplemented media but not on a medium lacking an adequate supply of lysine because it cannot synthesize this amino acid.

Once a detection method is established, mutants are collected. Since a specific mutation is a rare event, it is necessary to look at

Figure 11.31 Replica Plating. The use of replica plating in isolating a lysine auxotroph. Mutants are generated by treating a culture with a mutagen. The culture containing wild type and auxotrophs is plated on complete medium. After the colonies have developed, a piece of sterile velveteen is pressed on the plate surface to pick up bacteria from each colony. Then the velvet is pressed to the surface of other plates and organisms are transferred to the same position as on the master plate. After determining the location of Lys⁻ colonies growing on the replica with complete medium, the auxotrophs can be isolated and cultured.

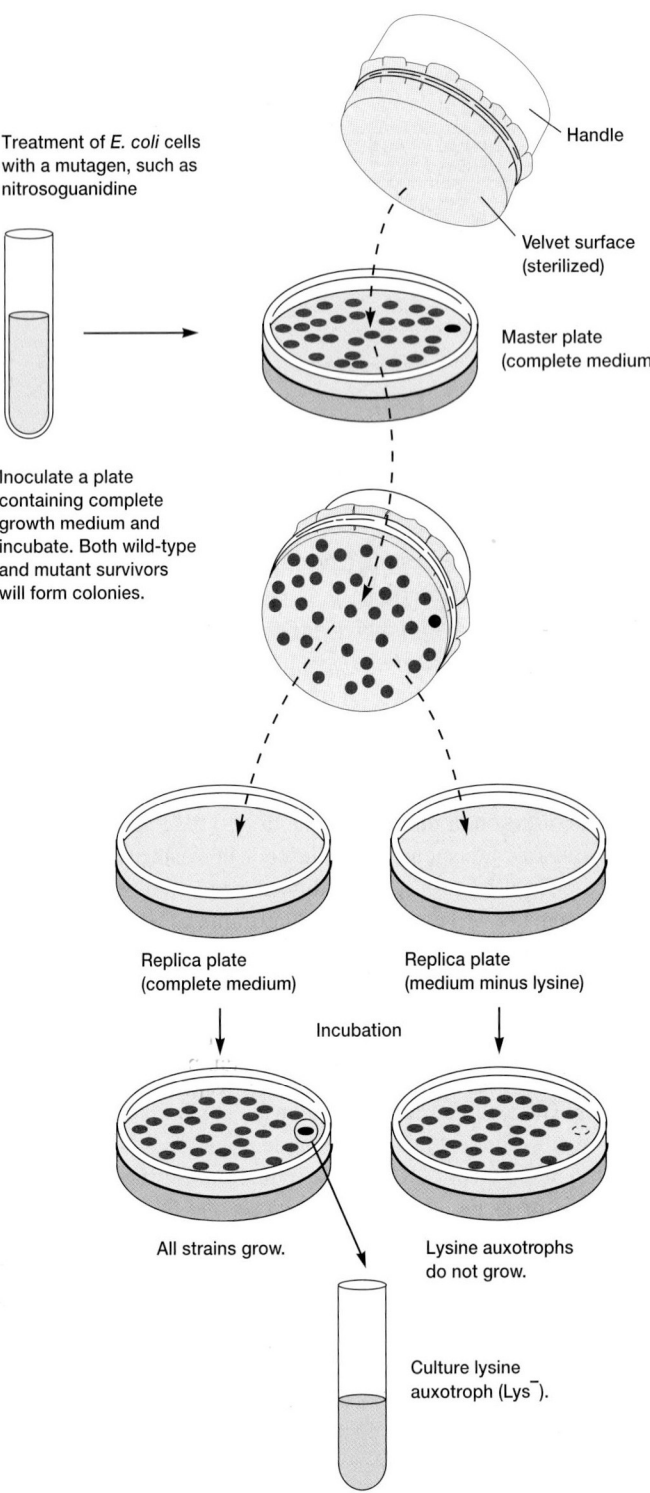

perhaps thousands to millions of colonies or clones. Using direct detection methods, this could become quite a task, even with microorganisms. Consider a search for the albino mutants mentioned previously. If the mutation rate were around one in a million, on the average a million or more organisms would have to be tested to find one albino mutant. This probably would require several thousand plates. The task of isolating auxotrophic mutants in this way would be even more taxing with the added labor of replica plating. This difficulty can be partly overcome by using mutagens to increase the mutation rate, thus reducing the number of colonies to be examined. However, it is more efficient to use a selection system employing some environmental factor to separate mutants from wild-type microorganisms.

Mutant Selection

An effective selection technique uses incubation conditions under which the mutant will grow, because of properties given it by the mutation, whereas the wild type will not. Selection methods often involve reversion mutations or the development of resistance to an environmental stress. For example, if the intent is to isolate revertants from a lysine auxotroph (Lys⁻), the approach is quite easy. A large population of lysine auxotrophs is plated on minimal medium lacking lysine, incubated, and examined for colony formation. Only cells that have mutated to restore the ability to manufacture lysine will grow on minimal medium (**figure 11.32**). Thus several million cells can be plated on a single petri dish, and many cells can be tested for mutations by scanning a few petri dishes for growth. This is because the auxotrophs will not grow on minimal medium and confuse the results; only the phenotypic revertants will form colonies. This method has proven very useful in determining the relative mutagenicity of many substances.

Resistance selection methods follow a similar approach. Often wild-type cells are not resistant to virus attack or antibiotic treatment, so it is possible to grow the bacterium in the presence of the agent and look for surviving organisms. Consider the example of a phage-sensitive wild-type bacterium. When the organism is cultured in medium lacking the virus and then plated out on selective medium containing phages, any colonies that form will be resistant to phage attack and very likely will be mutants in this regard. Resistance selection can be used together with virtually any environmental parameter; resistance to bacteriophages, antibiotics, or temperature are most commonly employed.

Substrate utilization mutations also are employed in bacterial selection. Many bacteria use only a few primary carbon sources. With such bacteria, it is possible to select mutants by a method similar to that employed in resistance selection. The culture is plated

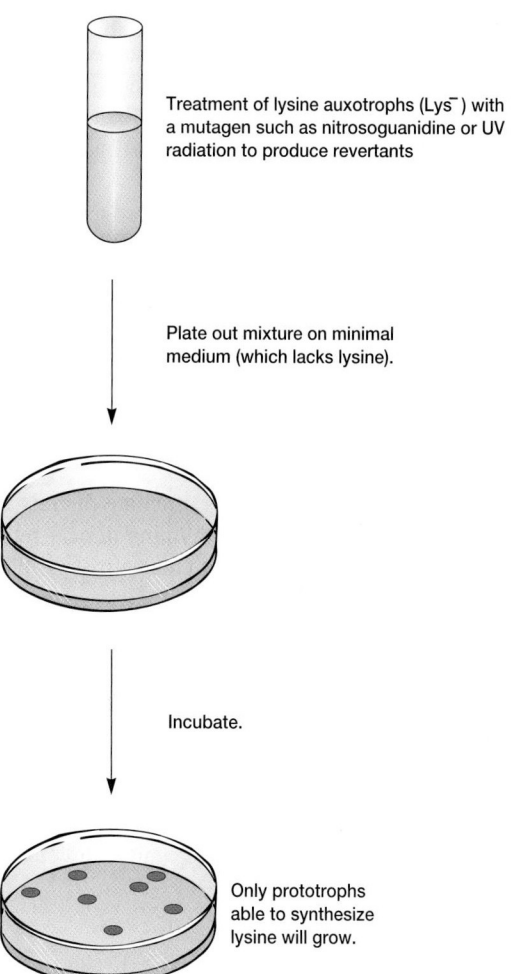

Figure 11.32 Mutant Selection.   The production and direct selection of auxotroph revertants. In this example, lysine revertants will be selected after treatment of a lysine auxotroph culture because the agar contains minimal medium that will not support auxotroph growth.

on medium containing an alternate carbon source. Any colonies that appear can use the substrate and are probably mutants.

Mutant detection and selection methods are used for purposes other than understanding more about the nature of genes or the biochemistry of a particular microorganism. One very important role of mutant selection and detection techniques is in the study of carcinogens. The next section briefly describes one of the first and perhaps best known of the carcinogen testing systems.

Carcinogenicity Testing

An increased understanding of the mechanisms of mutation and cancer induction has stimulated efforts to identify environmental carcinogens so that they can be avoided. The observation that many carcinogenic agents also are mutagenic is the basis for detecting potential carcinogens by testing for mutagenicity while taking advantage of bacterial selection techniques and short

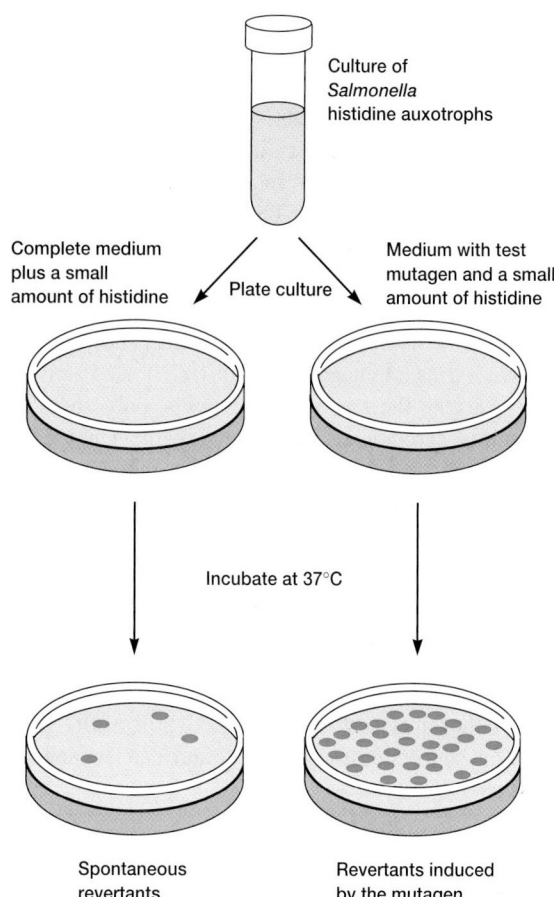

Figure 11.33   The Ames Test for Mutagenicity.   See text for details.

generation times. The **Ames test,** developed by Bruce Ames in the 1970s, has been widely used to test for carcinogens. The Ames test is a mutational reversion assay employing several special strains of *Salmonella typhimurium,* each of which has a different mutation in the histidine biosynthesis operon. The bacteria also have mutational alterations of their cell walls that make them more permeable to test substances. To further increase assay sensitivity, the strains are defective in the ability to carry out repair of DNA and have plasmid genes that enhance error-prone DNA repair.

In the Ames test these special tester strains of *Salmonella* are plated with the substance being tested and the appearance of visible colonies followed (**figure 11.33**). To ensure that DNA replication can take place in the presence of the potential mutagen, the bacteria and test substance are mixed in dilute molten top agar to which a trace of histidine has been added. This molten mix is then poured on top of minimal agar plates and incubated for 2 to 3 days at 37°C. All of the histidine auxotrophs will grow for the first few hours in the presence of the test compound until the histidine is depleted. Once the histidine supply is exhausted, only revertants that have mutationally regained the ability to synthesize histidine will grow. The visible colonies need only be counted and compared to controls in order to estimate the relative mutagenicity of the compound: the more colonies, the greater the mutagenicity.

A mammalian liver extract is also often added to the molten top agar prior to plating. The extract converts potential carcinogens into electrophilic derivatives that will readily react with DNA. This process occurs naturally when foreign substances are metabolized in the liver. Since bacteria do not have this activation system, liver extract often is added to the test system to promote the transformations that occur in mammals. Many potential carcinogens, such as the aflatoxins (*see p. 942*), are not actually carcinogenic until they are modified in the liver. The addition of extract shows which compounds have intrinsic mutagenicity and which need activation after uptake. Despite the use of liver extracts, only about half the potential animal carcinogens are detected by the Ames test.

---

1. Describe how replica plating is used to detect and isolate auxotrophic mutants.
2. Why are mutant selection techniques generally preferable to the direct detection and isolation of mutants?
3. Briefly discuss how reversion mutations, resistance to an environmental factor, and the ability to use a particular nutrient can be employed in mutant selection.
4. What is the Ames test and how is it carried out? What assumption concerning mutagenicity and carcinogenicity is it based upon?

 11.8 DNA REPAIR

Since replication errors and a variety of mutagens can alter the nucleotide sequence, a microorganism must be able to repair changes in the sequence that might be fatal. DNA is repaired by several different mechanisms besides **proofreading** by replication enzymes (DNA polymerases can remove an incorrect nucleotide immediately after its addition to the growing end of the chain). Repair in *E. coli* is best understood and is briefly described in this section. DNA replication and proofreading (pp. 229–33)

Excision Repair

Excision repair is a general repair system that corrects damage which causes distortions in the double helix. A repair endonuclease or uvrABC endonuclease removes the damaged bases along with some bases on either side of the lesion (**figure 11.34**). The resulting single-stranded gap, about 12 nucleotides long, is filled by DNA polymerase I, and DNA ligase joins the fragments (p. 233). This system can remove thymine dimers (figure 11.29) and repair almost any other injury that produces a detectable distortion in DNA.

Besides this general excision repair system, specialized versions of the system excise specific sites on the DNA where the sugar phosphate backbone is intact but the bases have been removed to form apurinic or apyrimidinic sites (AP sites). Special endonucleases called AP endonucleases recognize these locations and nick the backbone at the site. Excision repair then commences, beginning with the excision of a short stretch of nucleotides.

Another type of excision repair employs DNA glycosylases. These enzymes remove damaged or unnatural bases yielding AP sites that are then repaired as just described. Not all types of damaged bases are repaired in this way, but new glycosylases are being discovered and the process may be of more general importance than first thought.

Removal of Lesions

Thymine dimers and alkylated bases often are directly repaired. **Photoreactivation** is the repair of thymine dimers by splitting them apart into separate thymines with the help of visible light in a photochemical reaction catalyzed by the enzyme photolyase. Because this repair mechanism does not remove and replace nucleotides, it is error free.

Sometimes damage caused by alkylation is repaired directly as well. Methyls and some other alkyl groups that have been added to the O–6 position of guanine can be removed with the help of an enzyme known as alkyltransferase or methylguanine methyltransferase. Thus damage to guanine from mutagens such as methylnitrosoguanidine (figure 11.28) can be repaired directly.

Postreplication Repair

Despite the accuracy of DNA polymerase action and continual proofreading, errors still are made during DNA replication. Remaining mismatched bases and other errors are usually detected and repaired by the **mismatch repair system** in *E. coli*. The mismatch correction enzyme scans the newly replicated DNA for mismatched pairs and removes a stretch of newly synthesized DNA around the mismatch. A DNA polymerase then replaces the excised nucleotides, and the resulting nick is sealed with a ligase. Postreplication repair is a type of excision repair.

Successful postreplication repair depends on the ability of enzymes to distinguish between old and newly replicated DNA strands. This distinction is possible because newly replicated DNA strands lack methyl groups on their bases, whereas older DNA has methyl groups on the bases of both strands. **DNA methylation** is catalyzed by DNA methyltransferases and results in three different products: *N*6-methyladenine, 5-methylcytosine, and *N*4-methylcytosine. After strand synthesis, the *E. coli* DNA adenine methyltransferase (DAM) methylates adenine bases in d(GATC) sequences to form *N*6-methyladenine. For a short time after the replication fork has passed, the new strand lacks methyl groups while the template strand is methylated. The repair system cuts out the mismatch from the unmethylated strand.

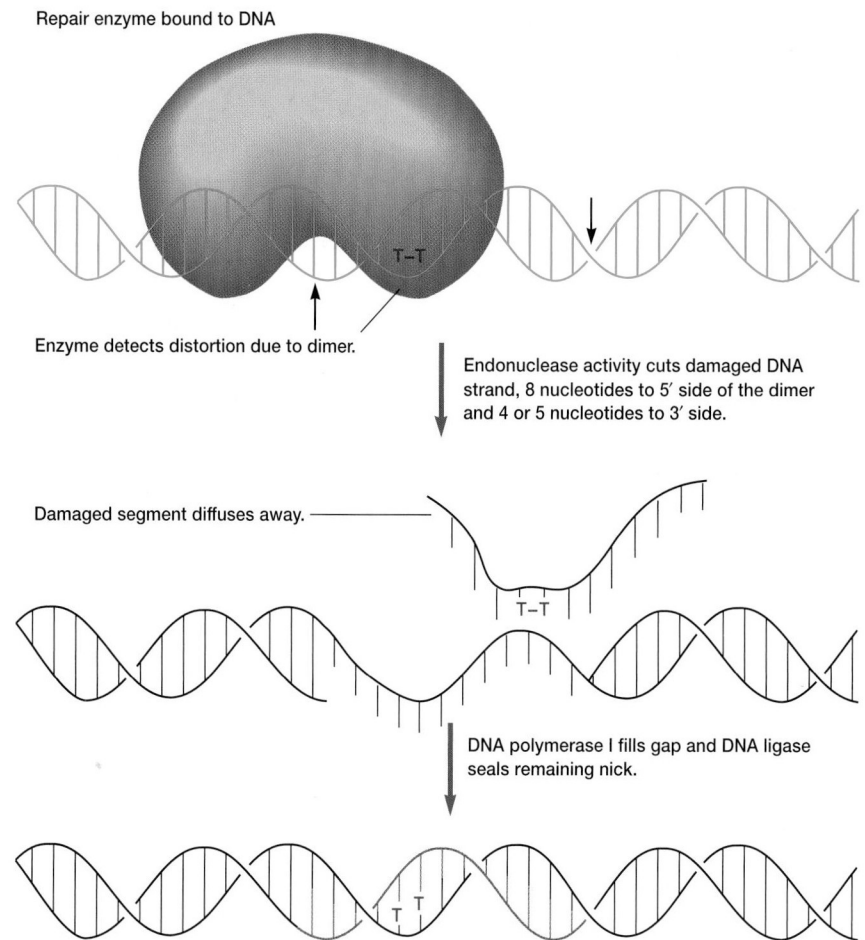

Repair enzyme bound to DNA

T–T

Enzyme detects distortion due to dimer.

Endonuclease activity cuts damaged DNA strand, 8 nucleotides to 5′ side of the dimer and 4 or 5 nucleotides to 3′ side.

Damaged segment diffuses away.

T–T

DNA polymerase I fills gap and DNA ligase seals remaining nick.

T T

Figure 11.34   Excision Repair.   Excision repair of a thymine dimer that has distorted the double helix. The repair endonuclease or uvrABC endonuclease is coded for by the *uvrA, B,* and *C* genes.

Recombination Repair

In **recombination repair,** damaged DNA for which there is no remaining template is restored. This situation arises if both bases of a pair are missing or damaged, or if there is a gap opposite a lesion. In this type of repair the **recA protein** cuts a piece of template DNA from a sister molecule and puts it into the gap or uses it to replace a damaged strand. Although bacteria are haploid, another copy of the damaged segment often is available because either it has recently been replicated or the cell is growing rapidly and has more than one copy of its chromosome. Once the template is in place, the remaining damage can be corrected by another repair system.   Bacterial recombination (pp. 286–88)

The recA protein also participates in a type of inducible repair known as **SOS repair.** In this instance the DNA damage is so great that synthesis stops completely, leaving many large gaps. RecA will bind to the gaps and initiate strand exchange. Si-

multaneously it takes on a proteolytic function that destroys the lexA repressor protein, which regulates the function of many genes involved in DNA repair and synthesis (**figure 11.35**). As a result many more copies of these enzymes are produced, accelerating the replication and repair processes. The system can quickly repair extensive damage caused by agents such as UV radiation, but it is error prone and does produce mutations. However, it is certainly better to have a few mutations than no DNA replication at all.

---

1. Define the following: proofreading, excision repair, photoreactivation, methylguanine methyltransferase, mismatch repair, DNA methylation, recombination repair, recA protein, SOS repair, and lexA repressor.
2. Describe in general terms the mechanisms of the following repair processes: excision repair, recombination repair, and SOS repair.

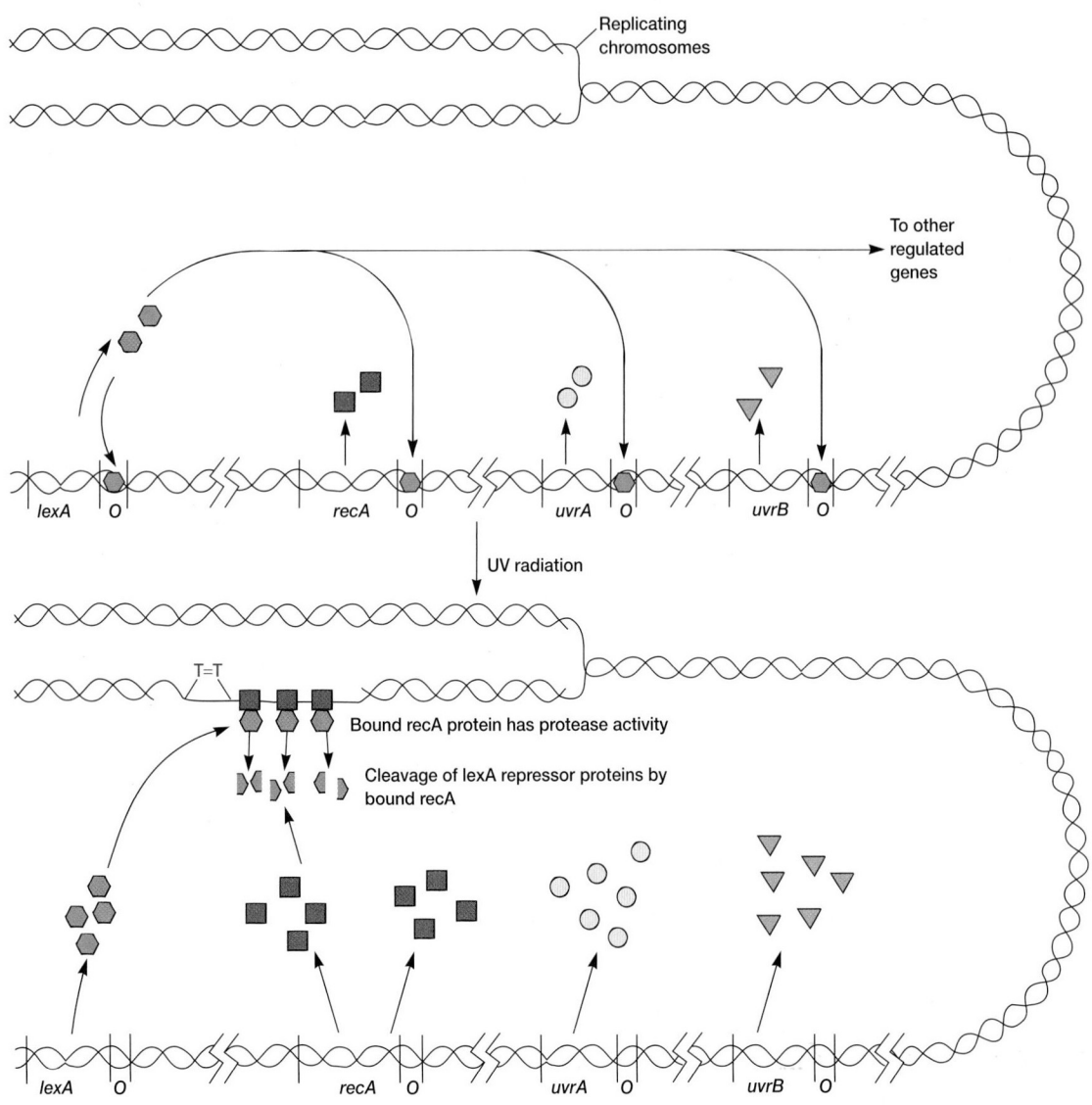

Figure 11.35   The SOS Repair Process.   In the absence of damage, repair genes are expressed in *E. coli* at low levels due to binding of the lexA repressor protein at their operators (*O*). When the recA protein binds to a damaged region—for example, a thymine dimer created by UV radiation—it destroys lexA and the repair genes are expressed more actively. The *uvr* genes code for the repair endonuclease or uvrABC endonuclease responsible for excision repair.

Summary

11.1 DNA as Genetic Material

a. The knowledge that DNA is the genetic material for cells came from studies on transformation by Griffith and Avery and from experiments on T2 phage reproduction by Hershey and Chase.

11.2 Nucleic Acid Structure

a. DNA differs in composition from RNA in having deoxyribose and thymine rather than ribose and uracil.

b. DNA is double stranded, with complementary AT and GC base pairing between the strands. The strands run antiparallel and are twisted into a right-handed double helix (**figures 11.6** and **11.7**).

c. RNA is normally single stranded, although it can coil upon itself and base pair to form hairpin structures.

d. In almost all procaryotes DNA exists as a closed circle that is twisted into supercoils and associated with histonelike proteins.

e. Eucaryotic DNA is associated with five types of histone proteins. Eight histones associate to form ellipsoidal octamers around which the DNA is coiled to produce the nucleosome (**figure 11.10**).

f. DNA synthesis is called replication. Transcription is the synthesis of an RNA copy of DNA and produces three types of RNA: messenger RNA (mRNA), transfer RNA (tRNA), and ribosomal RNA (rRNA).

g. The synthesis of protein under the direction of mRNA is called translation.

11.3 DNA Replication

a. Most circular procaryotic DNAs are copied by two replication forks moving around the circle to form a theta-shaped (θ) figure (**figure 11.12**). Sometimes a rolling-circle mechanism is employed instead (**figure 11.13**).

b. Eucaryotic DNA has many replicons and replication origins located every 10 to 100 μm along the DNA.

c. DNA polymerase enzymes catalyze the synthesis of DNA in the 5′ to 3′ direction while reading the DNA template in the 3′ to 5′ direction.

d. The double helix is unwound by helicases with the aid of topoisomerases like the DNA gyrase. DNA binding proteins keep the strands separate.

e. DNA polymerase III holoenzyme synthesizes a complementary DNA copy beginning with a short RNA primer made by a primase enzyme.

f. The leading strand is probably replicated continuously, whereas DNA synthesis on the lagging strand is discontinuous and forms Okazaki fragments (**figures 11.16** and **11.17**).

g. DNA polymerase I excises the RNA primer and fills in the resulting gap. DNA ligase then joins the fragments together.

11.4 The Genetic Code

a. Genetic information is carried in the form of 64 nucleotide triplets called codons (**table 11.1**); sense codons direct amino acid incorporation, and stop or nonsense codons terminate translation.

b. The code is degenerate—that is, there is more than one codon for most amino acids.

11.5 Gene Structure

a. A gene may be defined as the nucleic acid sequence that codes for a polypeptide, tRNA, or rRNA.

b. The template strand of DNA carries genetic information and directs the synthesis of the RNA transcript.

c. RNA polymerase binds to the promoter region, which contains RNA polymerase recognition and RNA polymerase binding sites (**figure 11.23**).

d. The gene also contains a coding region and a terminator; it may have a leader and a trailer (**figure 11.22**). Regulatory segments such as operators may be present.

e. The genes for tRNA and rRNA often code for a precursor that is subsequently processed to yield several products.

11.6 Mutations and Their Chemical Basis

a. A mutation is a stable, heritable change in the nucleotide sequence of the genetic material, usually DNA.

b. Mutations can be divided into many categories based on their effects on the phenotype, some major types are morphological, lethal, conditional, biochemical, and resistance mutations.

c. Spontaneous mutations can arise from replication errors (transitions, transversions, and frameshifts), from DNA lesions (apurinic sites, apyrimidinic sites, oxidations), and from insertions (**figures 11.25** and **11.26**).

d. Induced mutations are caused by mutagens. Mutations may result from the incorporation of base analogs, specific mispairing due to alteration of a base, the presence of intercalating agents, and a bypass of replication because of severe damage. Starvation and environmental stresses may stimulate mutator genes and lead to hypermutation.

e. The mutant phenotype can be restored to wild type by either a true reverse mutation or a suppressor mutation (**table 11.2**).

f. There are four important types of point mutations: silent mutations, missense mutations, nonsense mutations, and frameshift mutations (**table 11.2**).

11.7 Detection and Isolation of Mutants

a. It is essential to have a sensitive and specific detection technique to isolate mutants; an example is replica plating (**figure 11.31**) for the detection of auxotrophs (a direct detection system).

b. One of the most effective isolation techniques is to adjust environmental conditions so that the mutant will grow while the wild-type organism does not.

c. Because many carcinogens are also mutagenic, one can test for mutagenicity with the Ames test (**figure 11.33**) and use the results as an indirect indication of carcinogenicity.

11.8 DNA Repair

a. Mutations and DNA damage are repaired in several ways; for example: proofreading by replication enzymes, excision repair, removal of lesions (e.g., photoreactivation), postreplication repair (mismatch repair), and recombination repair.

Key Terms

Questions for Thought and Review

1. How do replication patterns differ between procaryotes and eucaryotes? Describe the operation of replication forks in the generation of theta-shaped intermediates and in the rolling-circle mechanism.
2. Outline the steps involved in DNA synthesis at the replication fork. How do DNA polymerases correct their mistakes?
3. Currently a gene is described in several ways. Which definition do you prefer and why?
4. How could one use small deletion mutations to show that codons are triplet (i.e., that the nucleotide sequence is read three bases at a time rather than two or four)?
5. Sometimes a point mutation does not change the phenotype. List all the reasons you can why this is so.
6. Why might a mutation leading to an amino acid change at a protein's surface not result in a phenotypic change while the substitution of an internal amino acid will?
7. Describe how you would isolate a mutant that required histidine for growth and was resistant to penicillin.
8. How would the following DNA alterations and replication errors be corrected (there may be *more* than one way): base addition errors by DNA polymerase III during replication, thymine dimers, AP sites, methylated guanines, and gaps produced during replication?

Critical Thinking Questions

1. Mutations are often considered harmful. Give an example of a mutation that would be beneficial to a microorganism. What gene would bear the mutation? How would the mutation alter the gene's role in the cell, and what conditions would select for this mutant allele?
2. Mistakes made during transcription affect the cell, but are not considered "mutations." Why not?
3. Given what you know about the difference between procaryotic and eucaryotic cells, give two reasons why the Ames test detects only about half of potential carcinogens, even when liver extracts are used.
4. Suppose that you have isolated a microorganism from a soil sample. Describe how you would go about determining the nature of its genetic material.

For recommended readings on these and related topics, see Appendix V.

chapter 12

Genes:
Expression and Regulation

Concepts

- In transcription the RNA polymerase copies the appropriate sequence on the DNA template strand to produce a complementary RNA copy of the gene. Transcription differs in a number of ways between procaryotes and eucaryotes, even though the basic mechanism of RNA polymerase action is essentially the same.
- Translation is the process by which the nucleotide sequence of mRNA is converted into the amino acid sequence of a polypeptide through the action of ribosomes, tRNAs, aminoacyl-tRNA synthetases, ATP and GTP energy, and a variety of protein factors. As in the case of DNA replication, this complex process is designed to minimize errors.
- The long-term regulation of metabolism in bacteria is achieved through the control of transcription by such mechanisms as sigma factors, repressor proteins during induction and repression, and by the attenuation of many biosynthetic operons.
- Procaryotes must be able to respond rapidly to changing environmental conditions and often control many operons simultaneously using global regulatory systems.
- DNA replication and cell division are coordinated in such a way that the distribution of new DNA copies to each daughter cell is ensured.

The particular field which excites my interest is the division between the living and the non-living, as typified by, say, proteins, viruses, bacteria and the structure of chromosomes. The eventual goal, which is somewhat remote, is the description of these activities in terms of their structure, i.e., the spatial distribution of their constituent atoms, in so far as this may prove possible. This might be called the chemical physics of biology.

—*Francis Crick*

Lactose operon activity is under the control of a repressor protein. The *lac* repressor (violet) and catabolite activator protein (blue) are bound to the *lac* operon. The repressor blocks transcription when bound to the operators (red).

hapter 11 is concerned with the genetic material of microorganisms. Its focus is on the structure and replication of DNA, the nature of the genetic code and genes, and the way in which genes change by mutation. This provides background for understanding chapters 12 through 15.

Chapter 12 is devoted to genetic expression and its regulation. An adequate knowledge of transcription and protein synthesis is essential to understanding molecular biology and microbial genetics. Thus the first two sections describe DNA transcription and protein synthesis in some detail. Because the role of these processes in the life of microorganisms depends on their proper regulation, the second half of the chapter is devoted to regulation. The control of mRNA synthesis and attenuation are discussed first. Then more complex levels of regulation are described in sections on global regulatory systems, two-component phospho-relay systems, and control of the bacterial cell cycle.

12.1 DNA TRANSCRIPTION OR RNA SYNTHESIS

As mentioned earlier, synthesis of RNA under the direction of DNA is called transcription. The RNA product has a sequence complementary to the DNA template directing its synthesis (**table 12.1**). Thymine is not normally found in mRNA and rRNA. Although adenine directs the incorporation of thymine during DNA replication, it usually codes for uracil during RNA synthesis. Transcription generates three kinds of RNA. Messenger RNA (mRNA) bears the message for protein synthesis. **Transfer RNA (tRNA)** carries amino acids during protein synthesis, and **ribosomal RNA (rRNA)** molecules are components of ribosomes. The structure and synthesis of procaryotic mRNA is described first.

Transcription in Procaryotes

Procaryotic mRNA is a single-stranded RNA of variable length containing directions for the synthesis of one to many polypeptides. Messenger RNA molecules also have sequences that do not code for polypeptides (**figure 12.1**). There is a nontranslated **leader sequence** of 25 to 150 bases at the 5′ end preceding the initiation codon. In addition, polygenic mRNAs (those directing the synthesis of more than one polypeptide) have spacer regions sep-

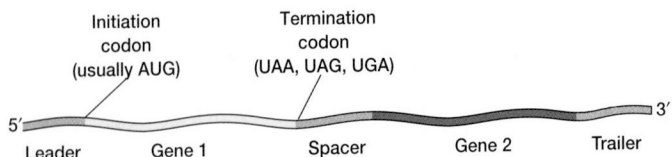

Figure 12.1   A Polygenic Bacterial Messenger RNA. See text for details.

arating the segments coding for individual polypeptides. Polygenic messenger polypeptides usually function together in some way (e.g., as part of the same metabolic pathway). At the 3′ end, following the last termination codon, is a nontranslated trailer.

Messenger RNA is synthesized under the direction of DNA by the enzyme **RNA polymerase.** An *E. coli* cell can have as many as 7,000 RNA polymerase molecules; only 2,000 to 5,000 polymerases may be active at any one time. The reaction is quite similar to that catalyzed by DNA polymerase. ATP, GTP, CTP, and UTP are used to produce an RNA copy of the DNA sequence. As mentioned earlier, these nucleotides contain ribose rather than deoxyribose.

$$n[ATP, GTP, CTP, UTP] \xrightarrow[\text{DNA template}]{\text{RNA polymerase}} RNA + nPP_i$$

RNA synthesis, like DNA synthesis, proceeds in a 5′ to 3′ direction with new nucleotides being added to the 3′ end of the growing chain at a rate of about 40 nucleotides per second at 37°C (**figure 12.2**). The RNA polymerase opens or unwinds the double helix to form a transcription bubble, about 12 to 20 base pairs in length, and transcribes the template strand to produce an RNA transcript that is complementary and antiparallel to the DNA template. It should be noted that pyrophosphate is produced in both DNA and RNA polymerase reactions. Pyrophosphate is then removed by hydrolysis to orthophosphate in a reaction catalyzed by the pyrophosphatase enzyme. Removal of the pyrophosphate product makes DNA and RNA synthesis irreversible. If the pyrophosphate level were too high, DNA and RNA would be degraded by a reversal of the polymerase reactions.

The RNA polymerase of *E. coli* is a very large molecule (about 480,000 daltons) containing four types of polypeptide chains: α, β, β′, and σ. The **core enzyme** is composed of four chains (α₂, β, β′) and catalyzes RNA synthesis. The **sigma factor** (σ) has no catalytic activity but helps the core enzyme recognize the start of genes. Once RNA synthesis begins, the sigma factor dissociates from the core enzyme–DNA complex and is available to aid another core enzyme. There are several different sigma factors in *E. coli*; σ⁷⁰ (about 70,000 molecular weight) is most often involved in transcription initiation. The precise functions of the α, β, and β′ polypeptides are not yet clear. The α subunit seems to be involved in the assembly of the core enzyme, recognition of promoters (see below), and interaction with some regulatory factors. The binding site for DNA is on β′, and the β subunit binds ribonucleotide substrates. Rifampin, an RNA polymerase inhibitor, binds to the β subunit.

The region to which RNA polymerase binds with the aid of the sigma factor is called the **promoter.** The procaryotic promoter se-

Table 12.1	
RNA Bases Coded for by DNA	
DNA Base	**Purine or Pyrimidine Incorporated into RNA**
Adenine	Uracil
Guanine	Cytosine
Cytosine	Guanine
Thymine	Adenine

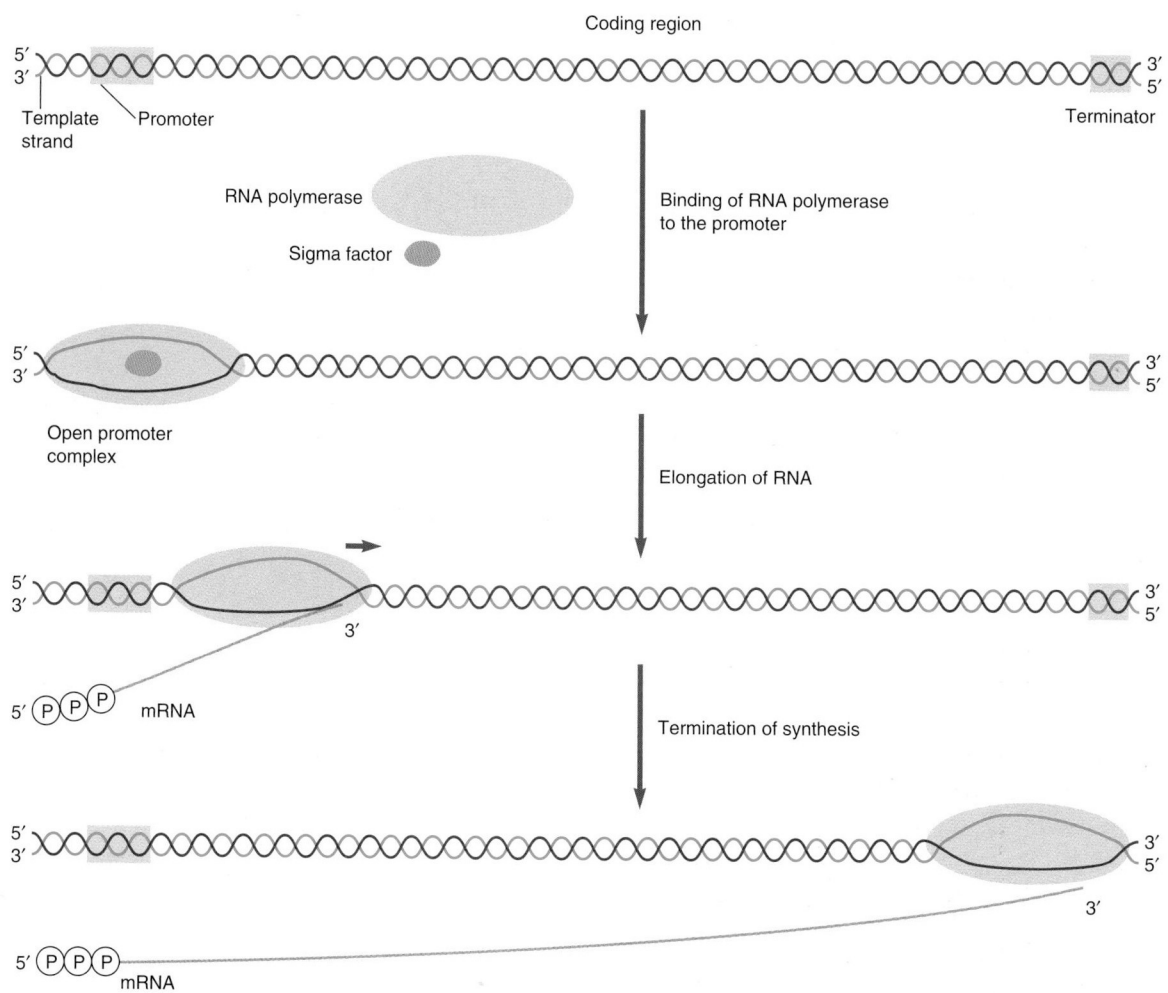

Figure 12.2   mRNA Transcription From DNA.   The lower DNA strand directs mRNA synthesis, and the RNA polymerase moves from left to right. Transcription has been simplified for clarity and divided into three general phases: binding of RNA polymerase to the promoter with the aid of a sigma factor, synthesis of RNA by the polymerase under the direction of the template strand, and termination of the process coupled with release of the RNA product. Upon binding, the RNA polymerase unwinds DNA to form a transcription bubble or open complex so that it can copy the template strand. See text for details.

quence is not transcribed. A 6 base sequence (usually TTGACA), approximately 35 base pairs before the transcription starting point, is present in *E. coli* promoters. A TATAAT sequence or **Pribnow box** lies within the promoter about 10 base pairs before the starting point of transcription or around 16 to 18 base pairs from the first hexamer sequence. The RNA polymerase recognizes these sequences, binds to the promoter, and unwinds a short segment of DNA beginning around the Pribnow box. Transcription starts 6 or 7 base pairs away from the 3′ end of the promoter.

Recently the atomic structure of RNA polymerase from *Thermus aquaticus* has been determined (**figure 12.3a,b**). The core enzyme is composed of four different subunits (α_2, β, β', and ω) and is complexed with the sigma factor (σ). The enzyme is claw-shaped with a clamp domain that closes on an internal channel, which contains an essential magnesium and the active site. Sigma interacts extensively with the core enzyme and specifically binds to the -10 and -35 elements of the promoter. It also may widen

the channel so that DNA can enter the interior of the polymerase complex. The RNA polymerase remains at the promoter while it constructs a chain about 9 nucleotides long, then it begins to move down the template strand. The first base used in RNA synthesis is usually a purine, either ATP or GTP. Since these phosphates are not removed during transcription, the 5′ end of procaryotic mRNA has a triphosphate attached to the ribose. Although only one polymerase can occupy the promoter at a time, more than one RNA polymerase can transcribe the same gene simultaneously. There is some evidence that a trailing polymerase may speed up transcription by "pushing" the leading polymerase and preventing it from backtracking.   Promoter structure and function (pp. 235–38)

There also must be stop signals to mark the end of a gene or sequence of genes and stop transcription by the RNA polymerase. Procaryotic **terminators** often contain a sequence coding for an RNA stretch that can hydrogen bond to form a hairpin-shaped loop and stem structure (**figure 12.4**). This structure appears to cause

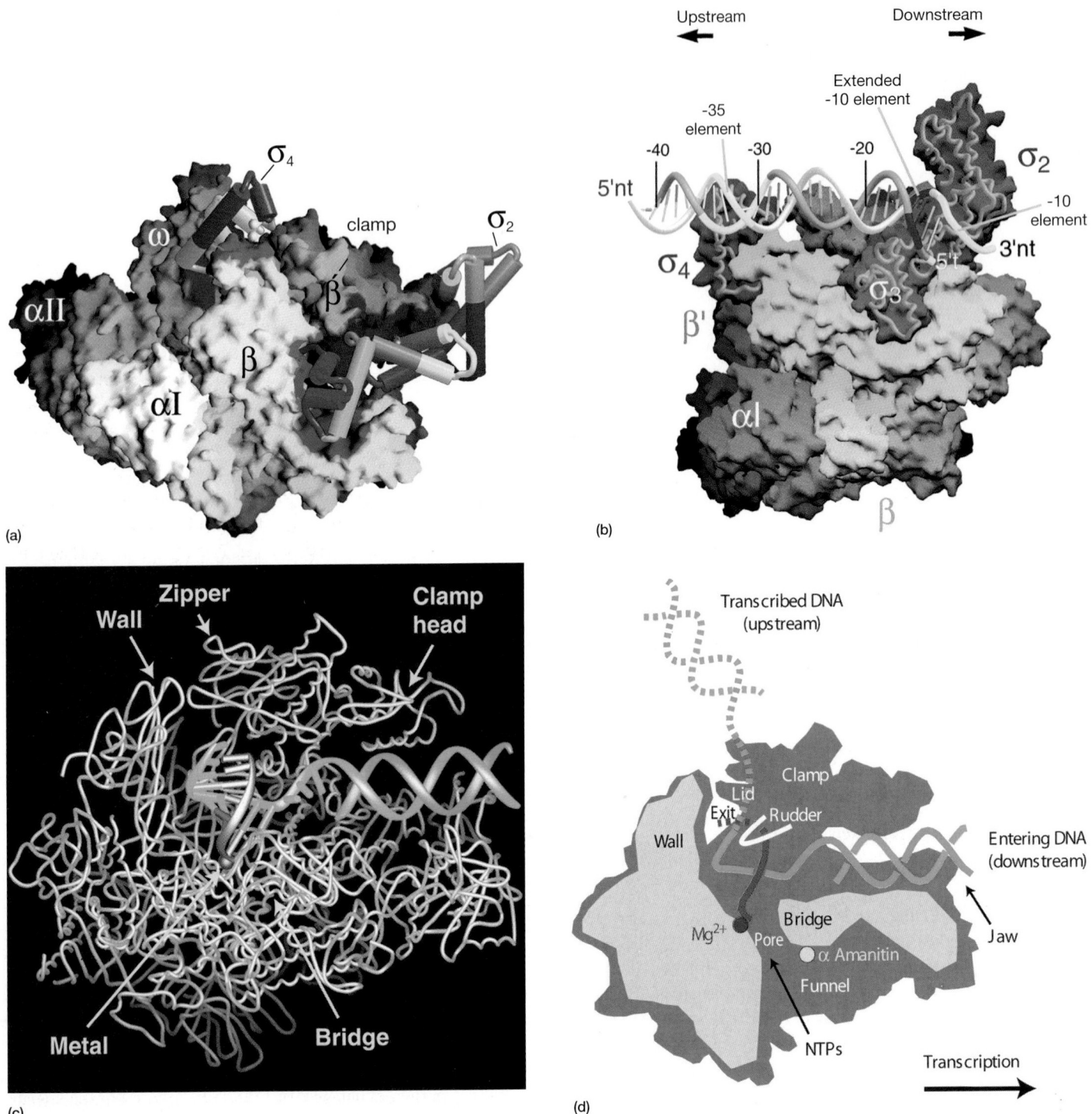

Figure 12.3   RNA Polymerase Structure.   The atomic structures of RNA polymerase from the procaryote *Thermus aquaticus* (**a** and **b**) and yeast RNA polymerase II (**c** and **d**) are presented here. (**a**) The *Taq* RNA polymerase holoenzyme is shown with the σ subunit depicted as an α-carbon backbone with cylinders for α-helices. Two of the three σ factor domains are labeled. (**b**) The holoenzyme-DNA complex with the σ surface rendered slightly transparent to show the α-carbon backbone inside. Protein surfaces that contact the DNA are in green and are located on the σ factor. The DNA template strand is dark green and the nontemplate strand (nt) is lighter green; the −10 and −35 elements in the promoter are in yellow. The internal active site is covered by the β subunit in this view. (**c**) Yeast RNA polymerase II transcribing complex with some peptide chains removed to show the DNA. The active site metal is a red sphere. A short stretch of DNA-RNA hybrid (blue and red) lies above the metal. (**d**) A cutaway side view of the polymerase II transcribing complex with the pathway of the nucleic acids and some of the more important parts shown. The enzyme is moving from left to right and the DNA template strand is in blue. A protein "wall" forces the DNA into a right-angle turn and aids in the attachment of nucleoside triphosphates to the growing 3′ end of the RNA. The newly synthesized RNA (red) is separated from the DNA template strand and exits beneath the rudder and lid of the polymerase protein complex. The binding site of the inhibitor α-amanitin also is shown.

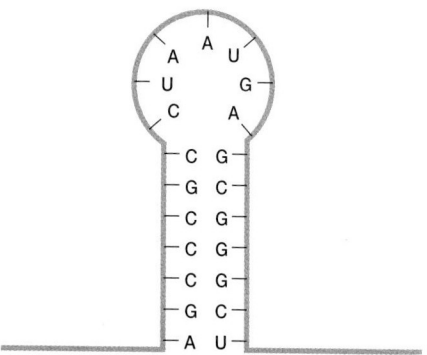

Figure 12.4 Procaryotic Terminators. An example of a hairpin structure formed by an mRNA terminator sequence.

the RNA polymerase to pause or stop transcribing DNA. There are two kinds of stop signals or terminators. The first type contains a stretch of about six uridine residues following the mRNA hairpin and causes the polymerase to stop transcription and release the mRNA without the aid of any accessory factors. The second kind of terminator lacks a poly-U region, and often the hairpin; it requires the aid of a special protein, the **rho factor** (ρ). It is thought that rho binds to mRNA and moves along the molecule until it reaches the RNA polymerase that has halted at a terminator. The rho factor then causes the polymerase to dissociate from the mRNA, probably by unwinding the mRNA-DNA complex.

Transcription in Eucaryotes

Transcriptional processes in eucaryotic microorganisms (and in other eucaryotic cells) differ in several ways from procaryotic transcription. There are three major RNA polymerases, not one as in procaryotes. RNA polymerase II, associated with chromatin in the nuclear matrix, is responsible for mRNA synthesis. Polymerases I and III synthesize rRNA and tRNA, respectively (**table 12.2**). The eucaryotic RNA polymerase II is a large aggregate, at least 500,000 daltons in size, with about 10 or more subunits. It is inhibited by the octapeptide α-amanitin. The atomic structure of the 10-subunit yeast DNA polymerase II associated with DNA and RNA has been determined (figure 12.3c,d). The entering DNA is held in a clamp that closes down on it. A magnesium ion is located at the active site and the 9 base pair DNA-RNA hybrid in the transcription bubble is bound in a cleft formed by the two large polymerase subunits. The newly synthesized RNA exits the polymerase beneath the rudder and lid regions. The substrate nucleoside triphosphates probably reach the active site through a pore in the complex. Unlike the bacterial polymerase, polymerase II requires extra transcription factors to recognize its promoters. The polymerase binds near the start point; the transcription factors bind to the rest of the promoter. Eucaryotic promoters also differ from those in procaryotes. They have combinations of several elements. Three of the most common are the TATA box (located about 30 base pairs before the start point or upstream), the CAAT box (about 75 base pairs upstream), and the GC box (90 base pairs upstream). Recently it has been shown that the TATA-

binding protein sharply bends the DNA on attachment. This makes the DNA more accessible to other initiation factors. A variety of general transcription factors, promoter specific factors, and promoter elements have been discovered in different eucaryotic cells. Each eucaryotic gene seems to be regulated differently, and more research will be required to understand the regulation of eucaryotic gene transcription.

Eucaryotic mRNA arises from **posttranscriptional modification** of large RNA precursors, about 5,000 to 50,000 nucleotides long, called **heterogeneous nuclear RNA (hnRNA)** molecules. These are the products of RNA polymerase II activity (**figure 12.5**). After hnRNA synthesis, the precursor RNA is cleaved by an endonuclease to yield the proper 3'-OH group. The enzyme polyadenylate polymerase then catalyzes the addition of adenylic acid to the 3' end of hnRNA to produce a poly-A sequence about 200 nucleotides long. The hnRNA finally is cleaved to generate the functional mRNA. Usually eucaryotic mRNA also differs in having a 5' cap consisting of 7-methylguanosine attached to the 5'-hydroxyl by a triphosphate linkage (**figure 12.6**). The adjacent nucleotide also may be methylated.

Eucaryotic mRNAs have 5' caps, unlike procaryotic mRNAs. Both types of cells can have mRNA with 3' poly-A, but procaryotes have poly-A stretches much less often and they are shorter. In addition, eucaryotic mRNA normally is monogenic in contrast to procaryotic mRNA, which often contains transcripts of two or more genes. The functions of poly-A and capping still are not completely clear. Poly-A protects mRNA from rapid enzymatic degradation. The poly-A tail must be shortened to about 10 nucleotides before mRNA can be degraded. Poly-A also seems to aid in mRNA translation. The 5' cap on eucaryotic messengers may promote the initial binding of ribosomes to the messenger. The cap also may protect the messenger from enzymatic attack.

Many eucaryotic genes differ from procaryotic genes in being split or interrupted, which leads to another type of posttranscriptional processing. **Split** or **interrupted genes** have **exons** (*ex*pressed sequences), regions coding for RNA that end up in the final RNA product (e.g., mRNA). Exons are separated from one another by **introns** (*int*ervening sequences), sequences coding for RNA that is missing from the final product (figure 12.5b). The initial RNA transcript has the intron sequences present in the interrupted gene. Genes coding for rRNA and tRNA may also be interrupted. Except for cyanobacteria and Archaea (*see chapters 20 and 21*), interrupted genes have not been found in procaryotes.

Introns are removed from the initial RNA transcript by a process called **RNA splicing** (figure 12.5b). The intron's borders must be clearly marked for accurate removal, and this is the case.

Figure 12.5 Eucaryotic mRNA Synthesis. (**a**) The production of eucaryotic messenger RNA. The addition of poly-A to the 3′ end of mRNA is included, but not the capping of the 5′ end. The poly-A sequence is in red and introns are purple. (**b**) The splicing of interrupted genes to produce mRNA. Exons are in red. The excised intron (blue) is in the shape of a circle or lariat.

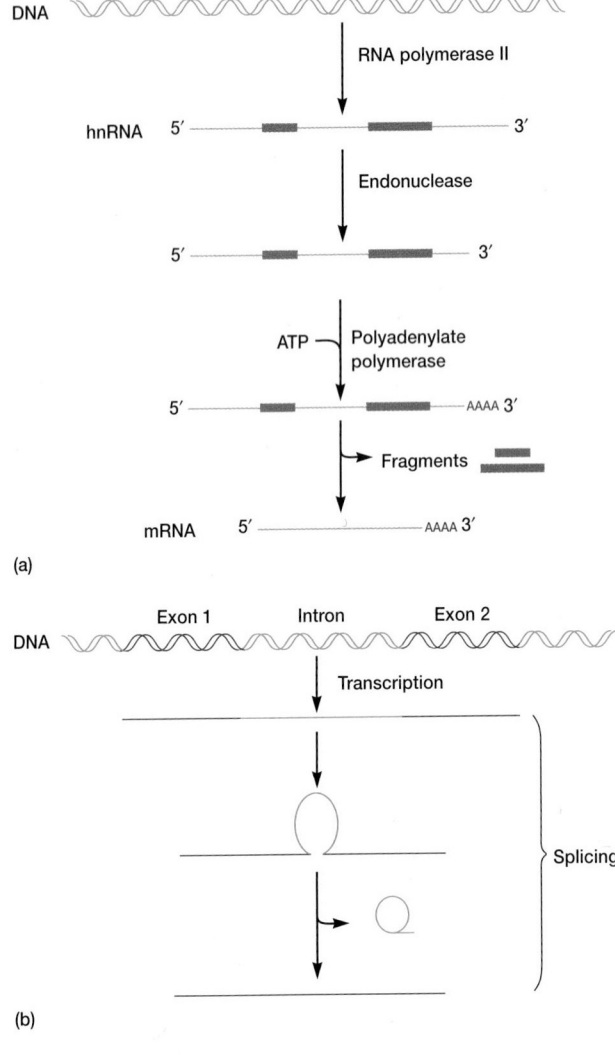

Figure 12.6 The 5′ Cap of Eucaryotic mRNA. Methyl groups are in blue.

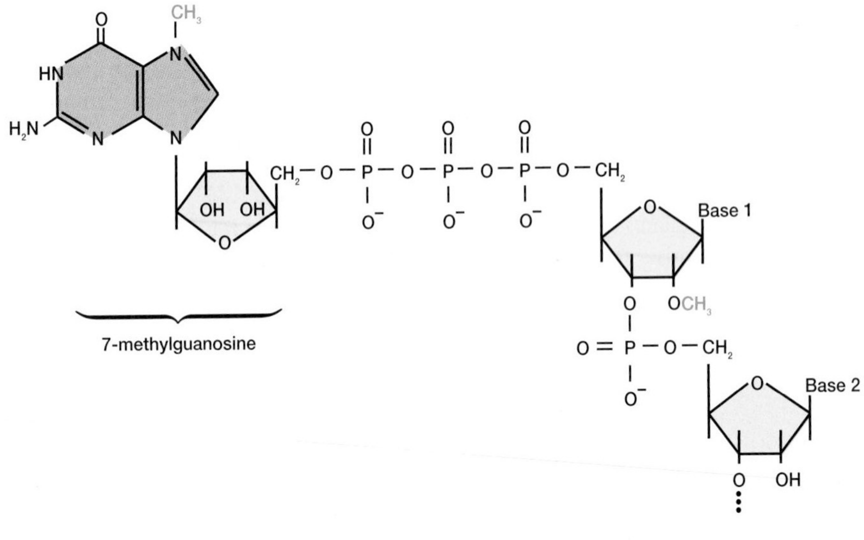

Microbial Tidbits

12.1   Catalytic RNA (Ribozymes)

Until recently biologists believed that all cellular reactions were catalyzed by proteins called enzymes (*see section 8.6*). The discovery during 1981–1984 by Thomas Cech and Sidney Altman that RNA also can sometimes catalyze reactions has transformed our way of thinking about topics as diverse as catalysis and the origin of life. It is now clear that some RNA molecules, called ribozymes, catalyze reactions that alter either their own structure or that of other RNAs.

This discovery has stimulated scientists to hypothesize that the early Earth was an "RNA world" in which RNA acted as both the genetic material and a reaction catalyst. Experiments showing that introns from *Tetrahymena thermophila* can catalyze the formation of polycytidylic acid under certain circumstances have further encouraged such speculations. Some have suggested that RNA viruses are "living fossils" of the original RNA world.

The best-studied ribozyme activity is the self-splicing of RNA. This process is widespread and occurs in *Tetrahymena* pre-rRNA; the mitochondrial rRNA and mRNA of yeast and other fungi; chloroplast tRNA, rRNA, and mRNA; in mRNA from some bacteriophages (e.g., the T4 phage of *E. coli*); and in the hepatitis delta virus. The 413-nucleotide rRNA intron of *T. thermophila* provides a good example of the self-splicing reaction. The reaction occurs in three steps and requires the presence of guanosine (see **Box figure**). First, the 3′-OH group of guanosine attacks the intron's 5′-phosphate group and cleaves the phosphodiester bond. Second, the new 3′-hydroxyl on the left exon attacks the 5′-phosphate of the right exon. This joins the two exons and releases the intron. Finally, the intron's 3′-hydroxyl attacks the phosphate bond of the nucleotide 15 residues from its end. This releases a terminal fragment and cyclizes the intron. Self-splicing of this rRNA occurs about 10 billion times faster than spontaneous RNA hydrolysis. Just as with enzyme proteins, the RNA's shape is essential to catalytic efficiency. The ribozyme even has Michaelis-Menten kinetics (*see p. 159*). The ribozyme from the hepatitis delta virus catalyzes RNA cleavage that is involved in virus replication. It is unusual in that the same RNA can fold into two shapes with quite different catalytic activities: the regular RNA cleavage activity and an RNA ligation reaction.

The discovery of ribozymes has many potentially important practical consequences. Ribozymes act as "molecular scissors" and will enable researchers to manipulate RNA easily in laboratory experiments. It also might be possible to protect hosts by specifically removing RNA from pathogenic viruses, bacteria, and fungi. For example, ribozymes are already being tested against the AIDS, herpes, and tobacco mosaic viruses.

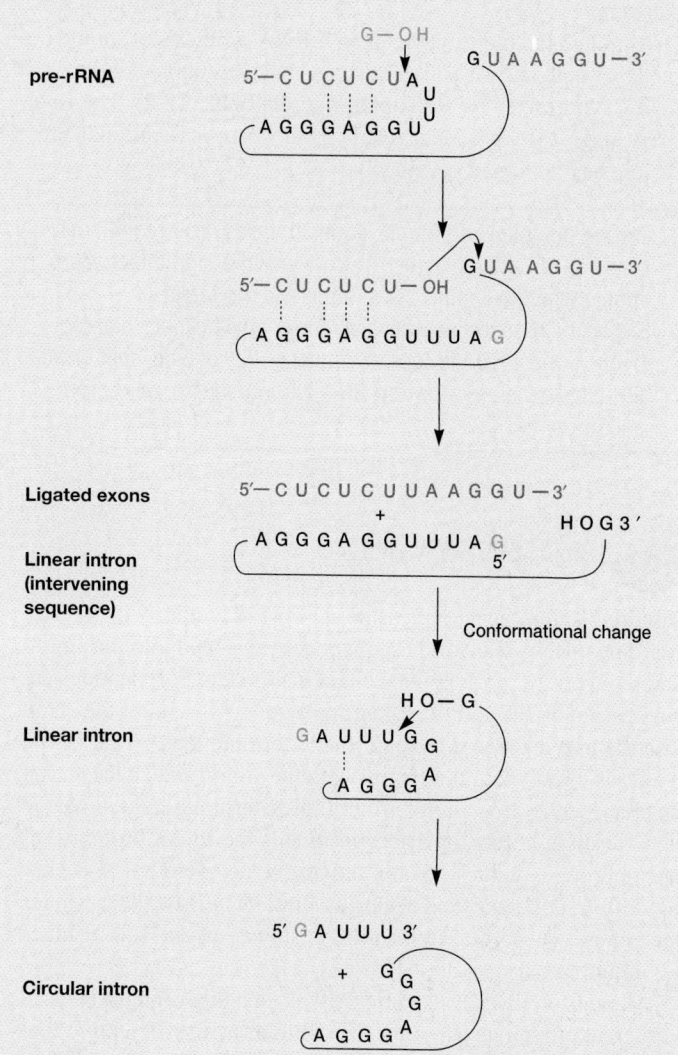

Ribozyme Action.   The mechanism of *Tetrahymena thermophila* pre-rRNA self-splicing. See text for details.

Exon-intron junctions have a GU sequence at the intron's 5′ boundary and an AG sequence at its 3′ end. These two sequences define the splice junctions and are recognized by special RNA molecules. The nucleus contains several **small nuclear RNA (snRNA)** molecules, about 60 to 300 nucleotides long. These complex with proteins to form *small nuclear ribonucleo*protein particles called snRNPs or snurps. Some of the snRNPs recognize splice junctions and ensure splicing accuracy. For example, U1-snRNP recognizes the 5′ splice junction, and U5-snRNP recog-

nizes the 3′ junction. Splicing of pre-mRNA occurs in a large complex called a **spliceosome** that contains the pre-mRNA, at least five kinds of snRNPs, and non-snRNP splicing factors.

As just mentioned, a few rRNA genes also have introns. Some of these pre-rRNA molecules are self-splicing. The RNA actually catalyzes the splicing reaction and now is called a **ribozyme** (**Microbial Tidbits 12.1**). Thomas Cech first discovered that pre-rRNA from the ciliate protozoan *Tetrahymena thermophila* is self-splicing. Sidney Altman then showed that ribonuclease P, which cleaves a fragment from one end of pre-tRNA, contains a piece of RNA that catalyzes the reaction. Several other self-splicing rRNA introns have since been discovered. Cech and Altman received the 1989 Nobel Prize in chemistry for these discoveries.

Although the focus has been on mRNA synthesis, it should be noted that both rRNAs and tRNAs also begin as parts of large precursors synthesized by RNA polymerases (table 12.2). The final rRNA and tRNA products result from posttranscriptional processing, as mentioned previously.

1. Define the following terms: leader, trailer, spacer region, polygenic mRNA, RNA polymerase core enzyme, sigma factor, promoter, template strand, terminator, and rho factor.
2. Define or describe posttranscriptional modification, heterogeneous nuclear RNA, 3′ poly-A sequence, 5′ capping, split or interrupted genes, exon, intron, RNA splicing, snRNA, spliceosome, and ribozyme.

12.2 PROTEIN SYNTHESIS

The final step in gene expression is protein synthesis or translation. The mRNA nucleotide sequence is translated into the amino acid sequence of a polypeptide chain in this step. Polypeptides are synthesized by the addition of amino acids to the end of the chain with the free α-carboxyl group (the C-terminal end). That is, the synthesis of polypeptides begins with the amino acid at the end of the chain with a free amino group (the N-terminal) and moves in the C-terminal direction. The ribosome is the site of protein synthesis. Protein synthesis is not only quite accurate but also very rapid. In *E. coli* synthesis occurs at a rate of at least 900 residues per minute; eucaryotic translation is slower, about 100 residues per minute. *Polypeptide and protein structure (appendix I)*

Many bacteria grow so quickly that each mRNA must be used with great efficiency to synthesize proteins at a sufficiently rapid rate. Ribosomal subunits are free in the cytoplasm if protein is not being synthesized. They come together to form the complete ribosome only when translation occurs. Frequently bacterial mRNAs are simultaneously complexed with several ribosomes, each ribosome reading the mRNA message and synthesizing a polypeptide. At maximal rates of mRNA use, there may be a ribosome every 80 nucleotides along the messenger or as many as 20 ribosomes simultaneously reading an mRNA that codes for a 50,000 dalton polypeptide. A complex of mRNA with several ribosomes is called a **polyribosome** or polysome. Polysomes are present in both procaryotes and eucaryotes. Bacteria can further increase the efficiency of gene expression through coupled transcription and translation (**figure 12.7**). While RNA polymerase is synthesizing an mRNA, ribosomes can already be attached to the messenger and involved in polypeptide synthesis. Coupled transcription and translation is possible in procaryotes because a nuclear envelope does not separate the translation machinery from DNA as it does in eucaryotes (*see figure 3.14*).

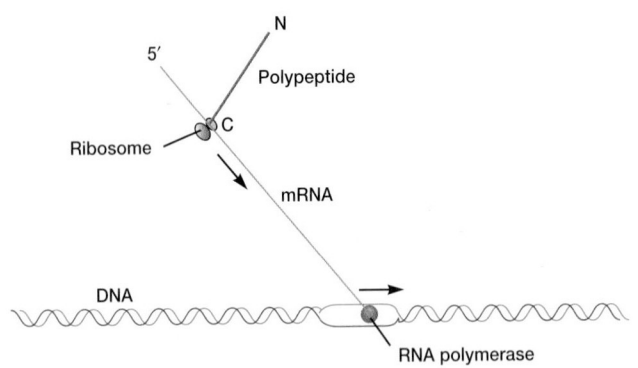

Figure 12.7 Coupled Transcription and Translation in Bacteria. mRNA is synthesized 5′ to 3′ by RNA polymerase while ribosomes are attaching to the newly formed 5′ end of mRNA and translating the message even before it is completed. Polypeptides are synthesized in the N-terminal to C-terminal direction.

Transfer RNA and Amino Acid Activation

The first stage of protein synthesis is **amino acid activation,** a process in which amino acids are attached to transfer RNA molecules. These RNA molecules are normally between 73 and 93 nucleotides in length and possess several characteristic structural features. The structure of tRNA becomes clearer when its chain is folded in such a way to maximize the number of normal base pairs, which results in a cloverleaf conformation of five arms or loops (**figure 12.8**). The acceptor or amino acid stem holds the activated amino acid on the 3′ end of the tRNA. The 3′ end of all tRNAs has the same —C—C—A sequence; the amino acid is attached to the terminal adenylic acid. At the other end of the cloverleaf is the anticodon arm, which contains the **anticodon triplet** complementary to the mRNA codon triplet. There are two other large arms: the D or DHU arm has the unusual pyrimidine nucleoside dihydrouridine; and the T or TΨC arm has ribothymidine (T) and pseudouridine (Ψ), both of which are unique to tRNA. Finally, the cloverleaf has a variable arm whose length changes with the overall length of the tRNA; the other arms are fairly constant in size.

Transfer RNA molecules are folded into an L-shaped structure (**figure 12.9**). The amino acid is held on one end of the L, the anticodon is positioned on the opposite end, and the corner of the L is formed by the D and T loops. Because there must be at least one tRNA for each of the 20 amino acids incorporated into proteins, at least 20 different tRNA molecules are needed. Actually more tRNA species exist (*see pp. 234–35*).

Amino acids are activated for protein synthesis through a reaction catalyzed by **aminoacyl-tRNA synthetases (figure 12.10).**

$$\text{Amino acid} + \text{tRNA} + \text{ATP} \xrightarrow{\text{Mg}^{2+}}$$
$$\text{aminoacyl-tRNA} + \text{AMP} + \text{PP}_i$$

Just as is true of DNA and RNA synthesis, the reaction is driven to completion when the pyrophosphate product is hydrolyzed to two

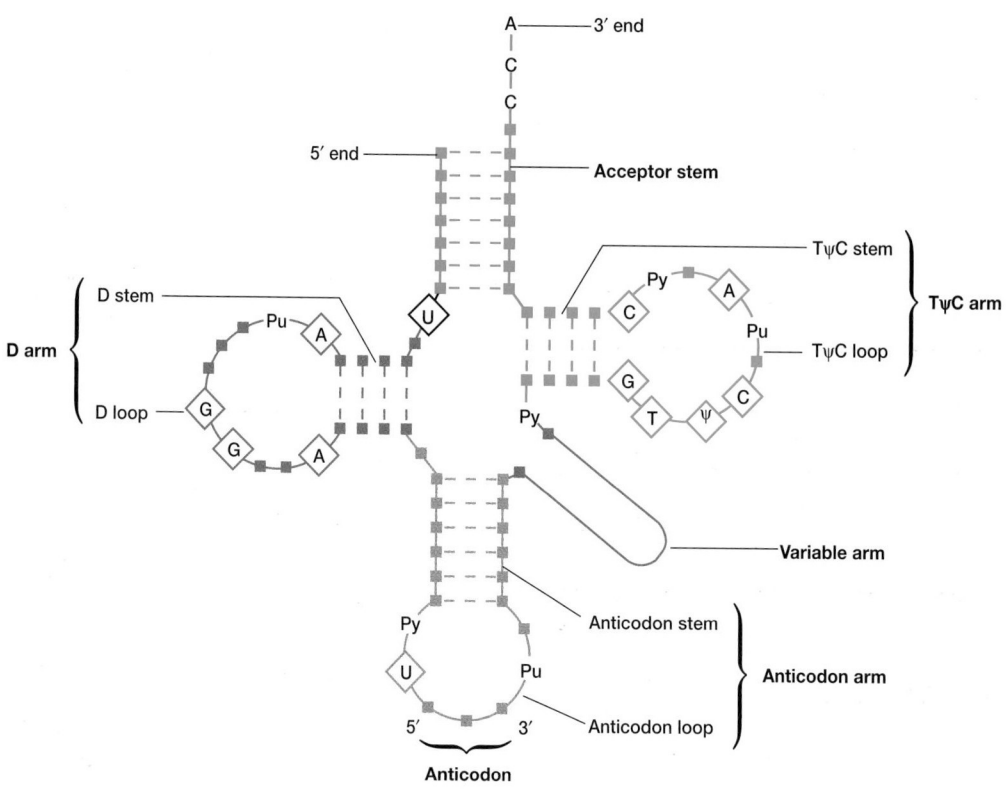

Figure 12.8 tRNA Structure. The cloverleaf structure for tRNA in procaryotes and eucaryotes. Bases found in all tRNAs are in diamonds; purine and pyrimidine positions in all tRNAs are labeled Pu and Py respectively.

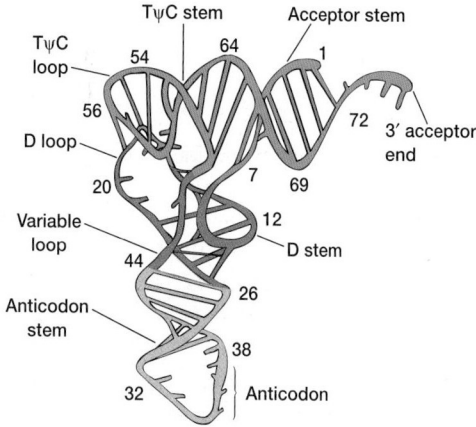

Figure 12.9 Transfer RNA Conformation. The three-dimensional structure of tRNA. The various regions are distinguished with different colors.

orthophosphates. The amino acid is attached to the 3′-hydroxyl of the terminal adenylic acid on the tRNA by a high-energy bond (**figure 12.11**), and is readily transferred to the end of a growing peptide chain. This is why the amino acid is called activated.

There are at least 20 aminoacyl-tRNA synthetases, each specific for a single amino acid and for all the tRNAs (cognate tRNAs)

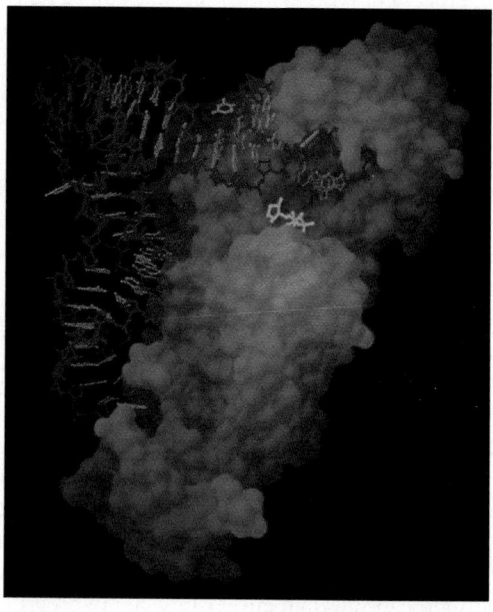

Figure 12.10 An Aminoacyl-tRNA Synthetase. A model of *E. coli* glutamyl-tRNA synthetase complexed with its tRNA and ATP. The enzyme is in blue, the tRNA in red and yellow, and ATP in green.

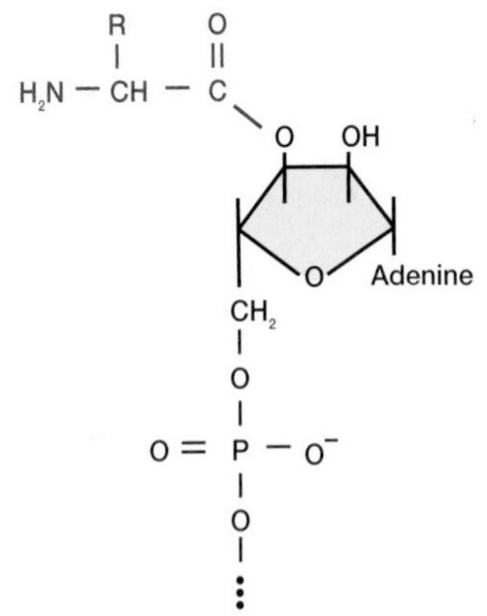

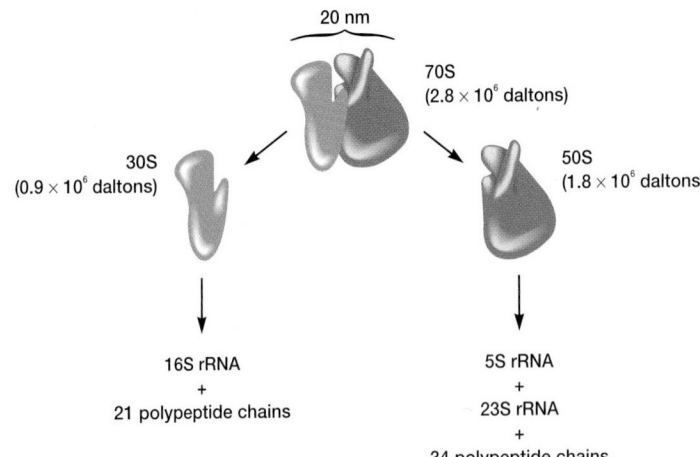

Figure 12.11 Aminoacyl-tRNA. The 3′ end of an aminoacyl-tRNA. The amino acid is attached to the 3′-hydroxyl of adenylic acid by a high-energy bond (red).

Figure 12.12 The 70S Ribosome.

to which each may be properly attached. This specificity is critical because once an incorrect acid is attached to a tRNA, it will be incorporated into a polypeptide in place of the correct amino acid. The protein synthetic machinery recognizes only the anticodon of the aminoacyl-tRNA and cannot tell whether the correct amino acid is attached. Some aminoacyl-tRNA synthetases will even proofread just like DNA polymerases do. If the wrong aminoacyl-tRNA is formed, aminoacyl-tRNA synthetases will hydrolyze the amino acid from the tRNA rather than release the incorrect product.

The Ribosome

The actual process of protein synthesis takes place on ribosomes that serve as workbenches, with mRNA acting as the blueprint. Procaryotic ribosomes have a sedimentation value of 70S and a mass of 2.8 million daltons. A rapidly growing *E. coli* cell may have as many as 15,000 to 20,000 ribosomes, about 15% of the cell mass. Introduction to ribosomal function and the Svedberg unit (p. 50)

The procaryotic ribosome is an extraordinarily complex organelle made of a 30S and a 50S subunit (**figure 12.12**). Each subunit is constructed from one or two rRNA molecules and many polypeptides. The shape of ribosomal subunits and their association to form the 70S ribosome are depicted in **figure 12.13**. The region of the ribosome directly responsible for translation is called the translational domain (figure 12.13*d*). Both subunits contribute to this domain, located in the upper half of the small subunit and in the associated areas of the large subunit. For example, the peptidyl transferase (p. 265) is found on the central protuberance of the large subunit. The growing peptide chain emerges from the large subunit at the exit domain. This is located on the side of the subunit opposite the central protuberance in both procaryotes and eucaryotes. X-ray diffraction studies have now confirmed this general picture of ribosome structure (figure 12.13*e*–*g*).

Eucaryotic cytoplasmic ribosomes are 80S, with a mass of 4 million daltons, and are composed of two subunits, 40S and 60S. Many of these ribosomes are found free in the cytoplasmic

matrix, whereas others are attached to membranes of the endoplasmic reticulum by their 60S subunit at a site next to the exit domain. The ribosomes of eucaryotic mitochondria and chloroplasts are smaller than cytoplasmic ribosomes and resemble the procaryotic organelle.

Ribosomal RNA is thought to have three roles. It obviously contributes to ribosome structure. The 16S rRNA of the 30S subunit also may aid in the initiation of protein synthesis in procaryotes. There is evidence that the 3′ end of the 16S rRNA complexes with an initiating signal site on the mRNA and helps position the mRNA on the ribosome. It also binds initiation factor 3 (pp. 263–65) and the 3′ CCA end of aminoacyl-tRNA. Finally, it appears that ribosomal RNA has a catalytic role in protein synthesis. The use of 16S rRNA sequences in the study of phylogeny (pp. 421–23)

Initiation of Protein Synthesis

Protein synthesis proper may be divided into three stages: initiation, elongation, and termination.

In the initiation stage *E. coli* and most bacteria begin protein synthesis with a specially modified aminoacyl-tRNA, *N*-formylmethionyl-tRNA^{fMet} (**figure 12.14**). Because the α-amino is blocked by a formyl group, this aminoacyl-tRNA can be used only for initiation. When methionine is to be added to a growing polypeptide chain, a normal methionyl-tRNA^{Met} is employed. Eucaryotic protein synthesis (except in the mitochondrion and chloroplast) and archaeal protein synthesis begin with a special initiator methionyl-tRNA^{Met}. Although most bacteria start protein synthesis with formylmethionine, the formyl group does not remain but is hydrolytically removed. In fact, one to three amino acids may be removed from the amino terminal end of the polypeptide after synthesis.

Figure 12.15 shows the initiation process in procaryotes. The initiator *N*-formylmethionyl-tRNA^{fMet} (fMet-tRNA) binds to the free 30S subunit first. Next mRNA attaches to the 30S subunit and is positioned properly through interactions with both the 3′ end of the 16S rRNA and the anticodon of fMet-tRNA. Messengers have a special **initiator codon** (AUG or sometimes

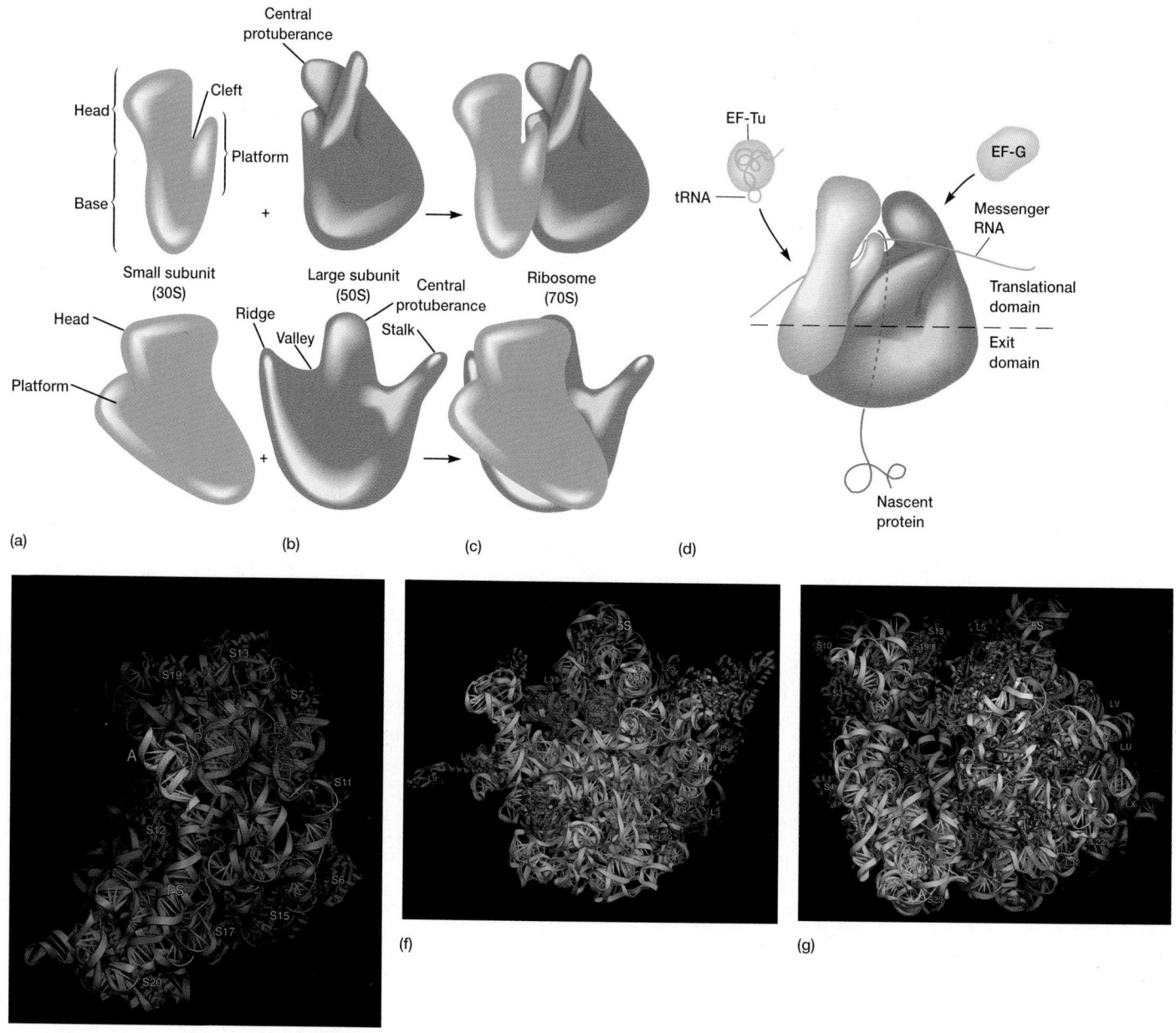

Figure 12.13   Procaryotic Ribosome Structure.   Parts (**a**)-(**d**) illustrate *E. coli* ribosome organization; parts (**e**)-(**g**) show the molecular structure of the *Thermus thermophilus* ribosome. (**a**) The 30S subunit. (**b**) The 50S subunit. (**c**) The complete 70S ribosome. (**d**) A diagram of ribosomal structure showing the translational and exit domains. The locations of elongation factor and mRNA binding are indicated. The growing peptide chain probably remains unfolded and extended until it leaves the large subunit. (**e**) Interior interface view of the 30S subunit of the *T. thermophilus* 70S ribosome showing the positions of the A, P, and E site tRNAs. (**f**) Interior interface view of the *T. thermophilus* 50S subunit and portions of its three tRNAs. (**g**) The complete *T. thermophilus* 70S ribosome viewed from the right-hand side with the 30S subunit on the left and the 50S subunit on the right. The anticodon arm of the A site tRNA is visible in the interface cavity. The components in figures (**e**)–(**g**) are colored as follows: 16S rRNA, cyan; 23S rRNA, gray; 5S rRNA, light blue; 30S proteins, dark blue; 50S proteins, magenta; and A, P, and E site tRNAs (gold, orange, and red, respectively).

GUG) that specifically binds with the fMet-tRNA anticodon (*see section 11.4*). Finally, the 50S subunit binds to the 30S subunit-mRNA forming an active ribosome-mRNA complex. The fMet-tRNA is positioned at the peptidyl or P site (see description of the elongation cycle). There is some uncertainty about the exact initiation sequence, and mRNA may bind before fMet-tRNA in pro-

caryotes. Eucaryotic initiation appears to begin with the binding of a special initiator Met-tRNA to the small subunit, followed by attachment of the mRNA.

In procaryotes three protein **initiation factors** are required (figure 12.15). Initiation factor 3 (IF-3) prevents 30S subunit binding to the 50S subunit and promotes the proper mRNA binding to

Figure 12.14 Procaryotic Initiator tRNA. The initiator aminoacyl-tRNA, *N*-formylmethionyl-tRNA^{fMet}, is used by bacteria. The formyl group is in color. Archaea use methionyl-tRNA for initiation.

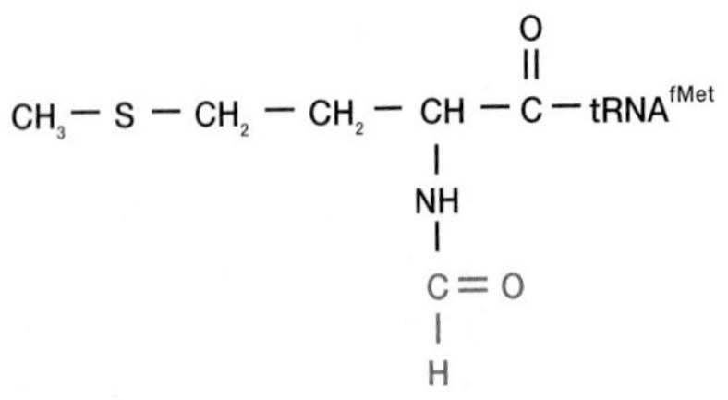

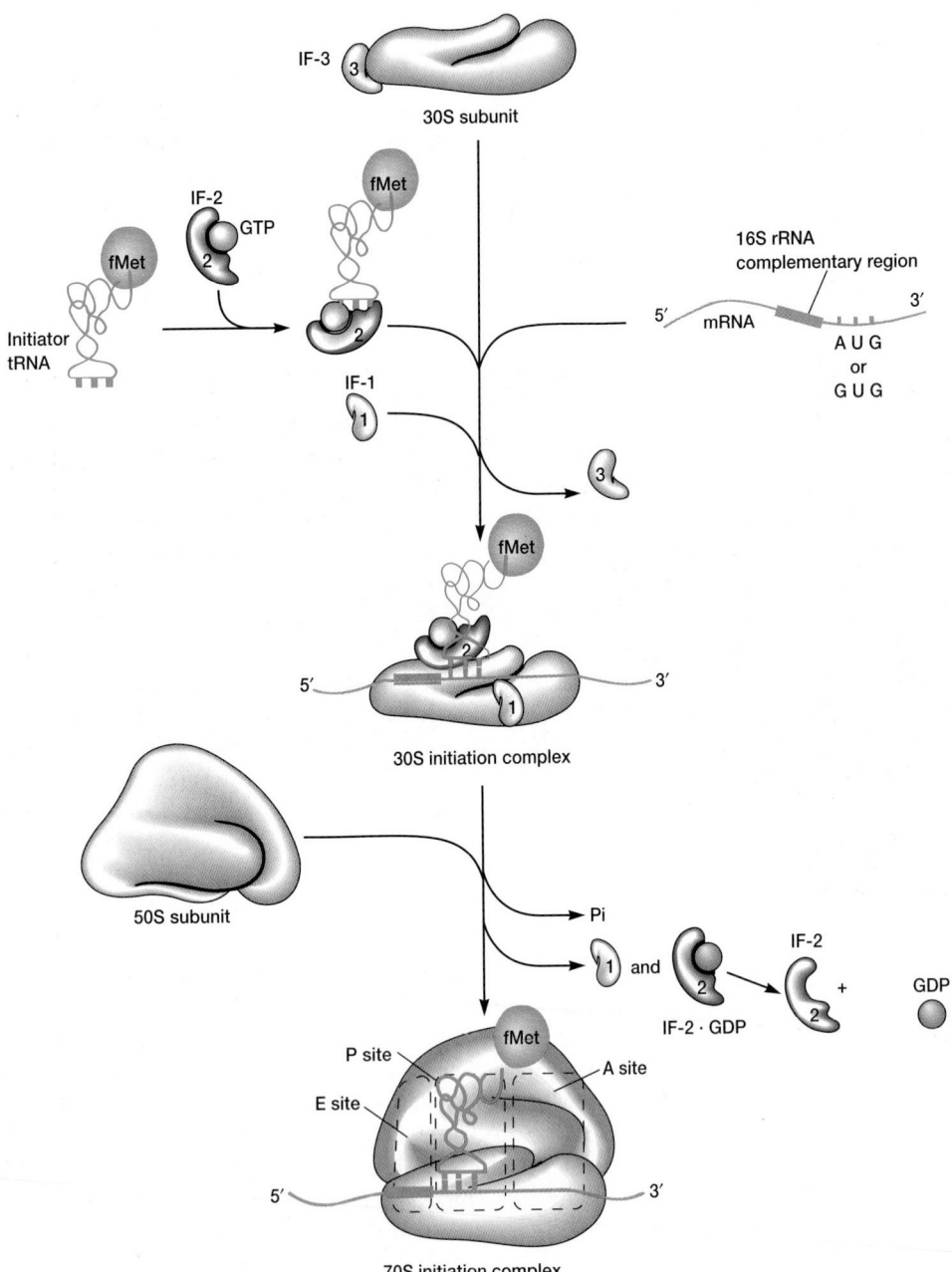

Figure 12.15 Initiation of Protein Synthesis. The initiation of protein synthesis in procaryotes. The following abbreviations are employed: IF-1, IF-2, and IF-3 stand for initiation factors 1, 2, and 3; initiator tRNA is *N*-formylmethionyl-tRNA^{fMet}. The ribosomal locations of initiation factors are depicted for illustration purposes only. They do not represent the actual initiation factor binding sites. See text for further discussion.

the 30S subunit. IF-2, the second initiation factor, binds GTP and fMet-tRNA and directs the attachment of fMet-tRNA to the P site of the 30S subunit. GTP is hydrolyzed during association of the 50S and 30S subunits. The third initiation factor, IF-1, appears to be needed for release of IF-2 and GDP from the completed 70S ribosome. IF-1 may aid in the binding of the 50S subunit to the 30S subunit. It also blocks tRNA binding to the A site. Eucaryotes require more initiation factors; otherwise the process is quite similar to that of procaryotes.

The initiation of protein synthesis is very elaborate. Apparently the complexity is necessary to ensure that the ribosome does not start synthesizing a polypeptide chain in the middle of a gene—a disastrous error.

Elongation of the Polypeptide Chain

Every amino acid addition to a growing polypeptide chain is the result of an **elongation cycle** composed of three phases: aminoacyl-tRNA binding, the transpeptidation reaction, and translocation. The process is aided by special protein **elongation factors** (just as with the initiation of protein synthesis). In each turn of the cycle, an amino acid corresponding to the proper mRNA codon is added to the C-terminal end of the polypeptide chain. The procaryotic elongation cycle is described next.

The ribosome has three sites for binding tRNAs: (1) the **peptidyl** or **donor site** (the **P site**), (2) the **aminoacyl** or **acceptor site** (the **A site**), and (3) the **exit site** (the **E site**). At the beginning of an elongation cycle, the peptidyl site is filled with either *N*-formylmethionyl-tRNA^{fMet} or peptidyl-tRNA and the aminoacyl and exit sites are empty (**figure 12.16**). Messenger RNA is bound to the ribosome in such a way that the proper codon interacts with the P site tRNA (e.g., an AUG codon for fMet-tRNA). The next codon (green) is located within the A site and is ready to direct the binding of an aminoacyl-tRNA.

The first phase of the elongation cycle is the aminoacyl-tRNA binding phase. The aminoacyl-tRNA corresponding to the green codon is inserted into the A site. GTP and the elongation factor EF-Tu, which donates the aminoacyl-tRNA to the ribosome, are required for this insertion. When GTP is bound to EF-Tu, the protein is in its active state and delivers aminoacyl-tRNA to the A site. This is followed by GTP hydrolysis, and the EF-Tu·GDP complex leaves the ribosome. EF-Tu·GDP is converted to EF-Tu·GTP with the aid of a second elongation factor, EF-Ts. Subsequently another aminoacyl-tRNA binds to EF-Tu·GTP (figure 12.16).

Aminoacyl-tRNA binding to the A site initiates the second phase of the elongation cycle, the **transpeptidation reaction** (figure 12.16 and **figure 12.17**). This is catalyzed by the **peptidyl transferase,** located on the 50S subunit. The α-amino group of the A site amino acid nucleophilically attacks the α-carboxyl group of the C-terminal amino acid on the P site tRNA in this reaction (figure 12.17). The peptide chain grows by one amino acid and is transferred to the A site tRNA. No extra energy source is required for peptide bond formation because the bond linking an amino acid to tRNA is high in energy. Recent evidence strongly suggests that 23S rRNA contains the peptidyl transferase func-

tion. Almost all protein can be removed from the 50S subunit, leaving the 23S rRNA and protein fragments. The remaining complex still has peptidyl transferase activity. The high-resolution structure of the large subunit has now been obtained by X-ray crystallography. There is no protein in the active site region. A specific adenine base seems to participate in catalyzing peptide bond formation. Thus the 23S rRNA appears to be the major component of the peptidyl transferase and contributes to both A and P site functions.

The final phase in the elongation cycle is **translocation.** Three things happen simultaneously: (1) the peptidyl-tRNA moves about 20 Å from the A site to the P site; (2) the ribosome moves one codon along mRNA so that a new codon is positioned in the A site; and (3) the empty tRNA leaves the P site. Instead of immediately being ejected from the ribosome, the empty tRNA moves from the P site to the E site and then leaves the ribosome. Ribosomal proteins are involved in these tRNA movements. The intricate process also requires the participation of the EF-G or translocase protein and GTP hydrolysis. The ribosome changes shape as it moves down the mRNA in the 5′ to 3′ direction.

Termination of Protein Synthesis

Protein synthesis stops when the ribosome reaches one of three special **nonsense codons**—UAA, UAG, and UGA (**figure 12.18**). Three **release factors** (RF-1, RF-2, and RF-3) aid the ribosome in recognizing these codons. After the ribosome has stopped, peptidyl transferase hydrolyzes the peptide free from its tRNA, and the empty tRNA is released. GTP hydrolysis seems to be required during this sequence, although it may not be needed for termination in procaryotes. Next the ribosome dissociates from its mRNA and separates into 30S and 50S subunits. IF-3 binds to the 30S subunit and prevents it from reassociating with the 50S subunit until the proper stage in initiation is reached. Thus ribosomal subunits associate during protein synthesis and separate afterward. The termination of eucaryotic protein synthesis is similar except that only one release factor appears to be active.

Protein synthesis is a very expensive process. Three GTP molecules probably are used during the elongation cycle, and two ATP high-energy bonds are required for amino acid activation (ATP is converted to AMP rather than to ADP). Therefore five high-energy bonds are required to add one amino acid to a growing polypeptide chain. GTP also is used in initiation and termination of protein synthesis (figures 12.15 and 12.18). Presumably this large energy expenditure is required to ensure the fidelity of protein synthesis. Very few mistakes can be tolerated.

Although the mechanism of protein synthesis is similar in procaryotes and eucaryotes, procaryotic ribosomes differ substantially from those in eucaryotes. This explains the effectiveness of many important chemotherapeutic agents. Either the 30S or the 50S subunit may be affected. For example, streptomycin binding to the 30S ribosomal subunit inhibits protein synthesis and causes mRNA misreading. Erythromycin binds to the 50S subunit and inhibits peptide chain elongation. The effect of antibiotics on protein synthesis (pp. 784–86,792)

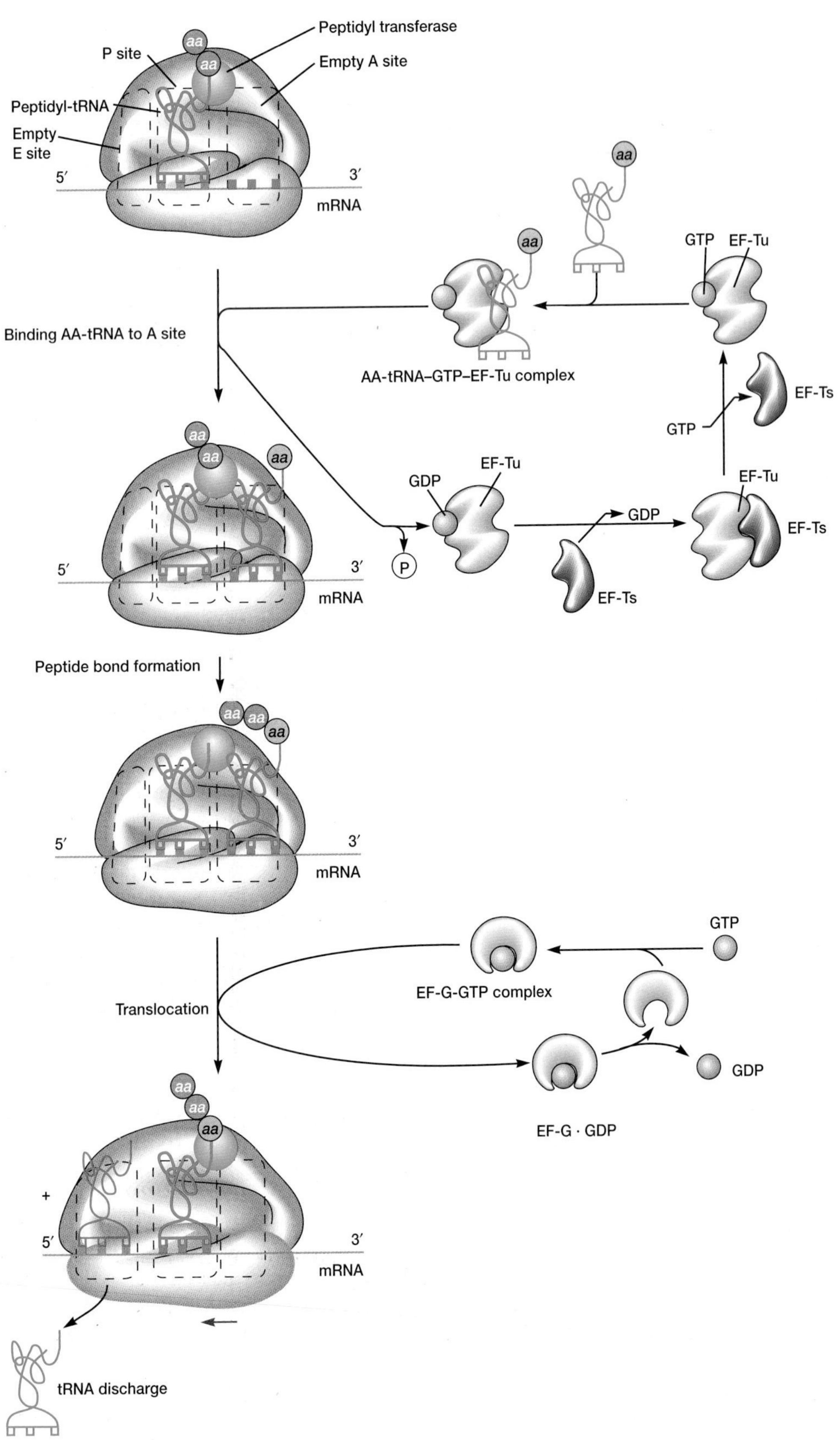

Figure 12.16   Elongation Cycle.   The elongation cycle of protein synthesis. The ribosome possesses three sites, a peptidyl or donor site (P site), an aminoacyl or acceptor site (A site), and an exit site (E site). The arrow below the ribosome in translocation step shows the direction of mRNA movement. See text for details.

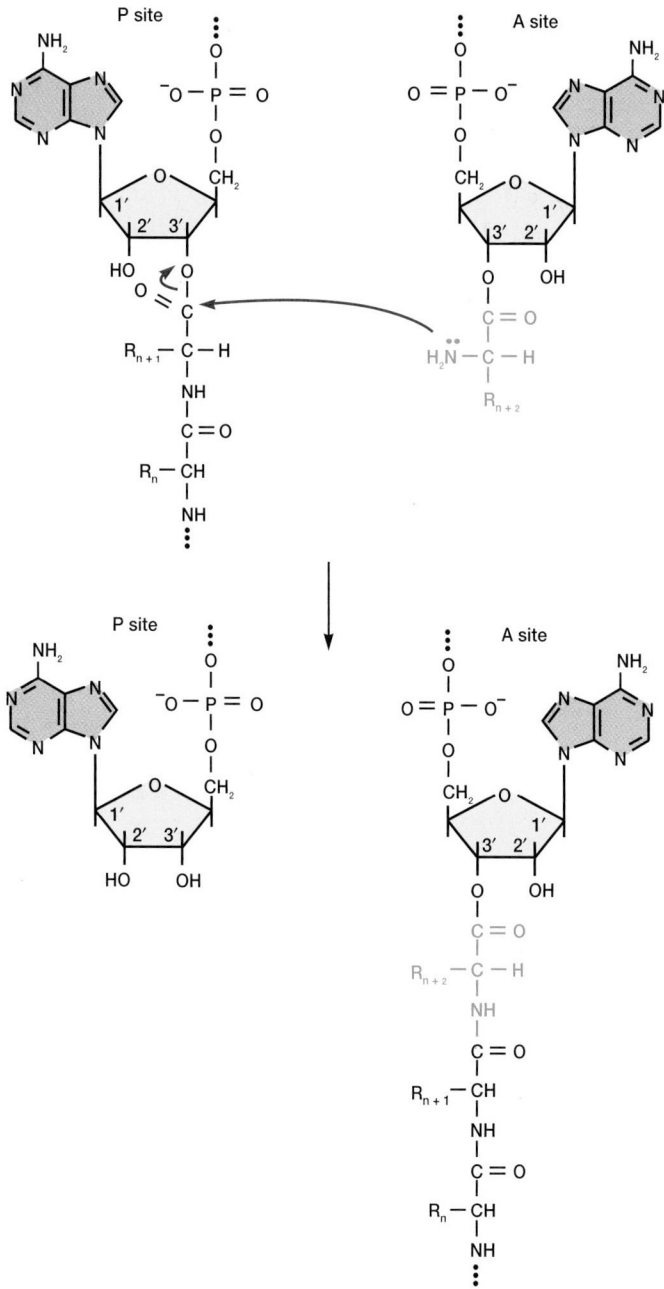

Figure 12.17 Transpeptidation. The peptidyl transferase reaction. The peptide grows by one amino acid and is transferred to the A site.

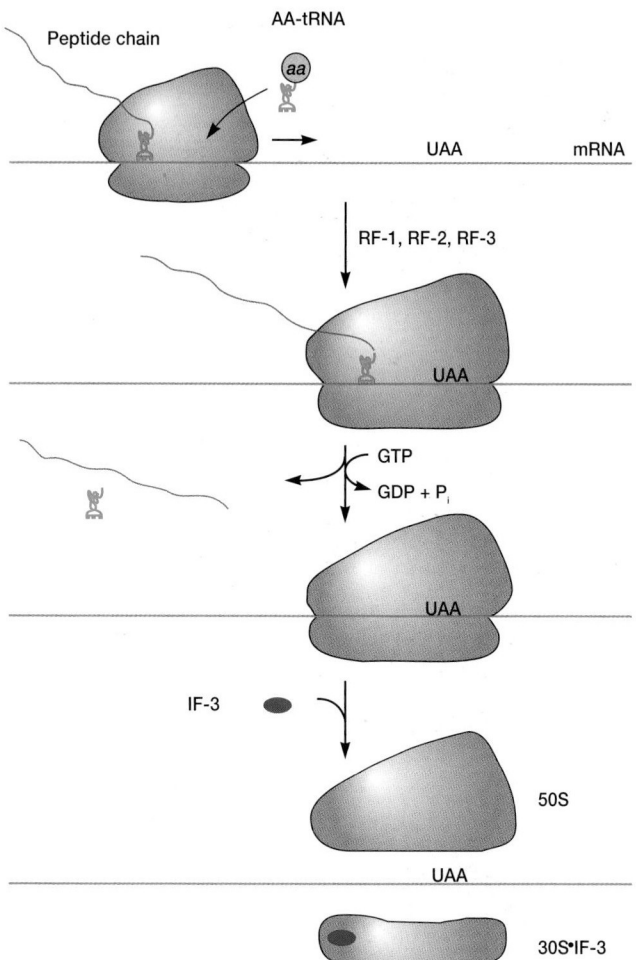

Figure 12.18 Termination of Protein Synthesis in Procaryotes. Although three different nonsense codons can terminate chain elongation, UAA is most often used for this purpose. Three release factors (RF) assist the ribosome in recognizing nonsense codons and terminating translation. GTP hydrolysis is probably involved in termination. Transfer RNAs are in green.

Protein Folding and Molecular Chaperones

For many years it was believed that polypeptides would spontaneously fold into their final native shape, either as they were synthesized by ribosomes or shortly after completion of protein synthesis. Although the amino acid sequence of a polypeptide does determine its final conformation, it is now clear that special helper proteins aid the newly formed or nascent polypeptide in folding to its proper shape. These proteins, called **molecular**

chaperones or chaperones, recognize only unfolded polypeptides or partly denatured proteins and do not bind to normal, functional proteins. Their role is essential because the cytoplasmic matrix is filled with nascent polypeptide chains and proteins. Under such conditions it is quite likely that new polypeptide chains often will fold improperly and aggregate to form nonfunctional complexes. Molecular chaperones suppress incorrect folding and may reverse any incorrect folding that has already taken place. They are so important that chaperones are present in all cells, procaryotic and eucaryotic.

Several chaperones and cooperating proteins aid proper protein folding in bacteria. The process has been most studied in *Escherichia coli* and involves at least four chaperones—DnaK, DnaJ, GroEL, and GroES—and the stress protein GrpE. After a sufficient length of nascent polypeptide extends from the ribosome,

Figure 12.19 Chaperones and Polypeptide Folding. The involvement of bacterial chaperones in the proper folding of a newly synthesized polypeptide chain is depicted in this diagram. Three possible outcomes of a chaperone reaction cycle are shown. A native protein may result, the partially folded polypeptide may bind again to DnaK and DnaJ, or the polypeptide may be transferred to GroEL and GroES. See text for details.

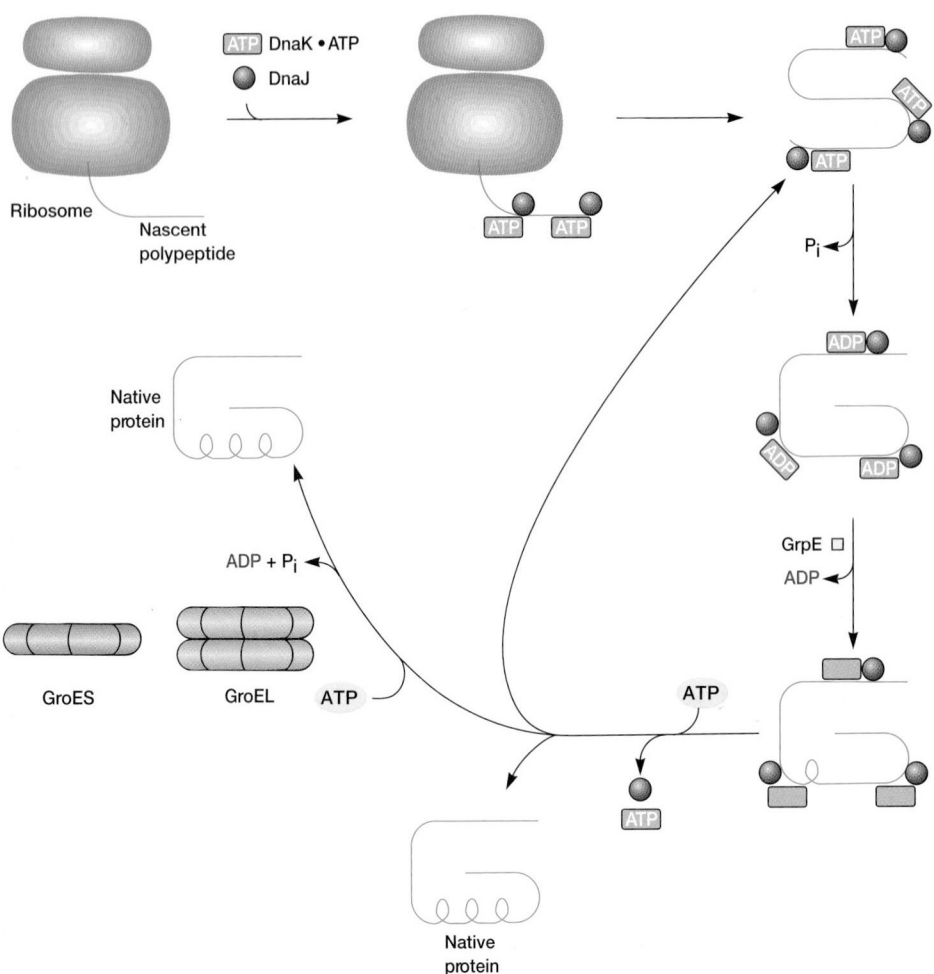

DnaJ binds to the unfolded chain (**figure 12.19**). DnaK, which is complexed with ATP, then attaches to the polypeptide. These two chaperones prevent the polypeptide from folding improperly as it is synthesized. The ATP is hydrolyzed to ADP after DnaK binding, and this increases the affinity of DnaK for the unfolded polypeptide. When the polypeptide has been synthesized, the GrpE protein binds to the chaperone-polypeptide complex and causes DnaK to release ADP. Then ATP binds to DnaK and both DnaK and DnaJ dissociate from the polypeptide. The polypeptide has been folding during this sequence of events and may have reached its final native conformation. If it is still only partially folded, it can bind DnaJ and DnaK again and repeat the process. Often DnaK and DnaJ will transfer the polypeptide to the chaperones GroEL and GroES, where the final folding takes place. GroEL is a large, hollow barrel-shaped complex of 14 subunits arranged in two stacked rings (**figure 12.20**). GroES exists as a single ring of seven subunits and can bind to one or both ends of the GroEL cylinder. As with DnaK, ATP binding to GroEL and ATP hydrolysis change the chaperone's affinity for the folding polypeptide and regulate polypeptide binding and release (polypeptide release is ATP-dependent). GroES binds to GroEL and assists in its binding and release of the refolding polypeptide.

Chaperones were first discovered because they dramatically increased in concentration when cells were exposed to high tem-

peratures, metabolic poisons, and other stressful conditions. Thus many chaperones often are called **heat-shock proteins** or stress proteins. When an *E. coli* culture is switched from 30 to 42°C, the concentrations of some 20 different heat-shock proteins increase greatly within about 5 minutes. If the cells are exposed to a lethal temperature, the heat-shock proteins are still synthesized but most proteins are not. Thus chaperones protect the cell from thermal damage and other stresses as well as promote the proper folding of new polypeptides. For example, DnaK protects *E. coli* RNA polymerase from thermal inactivation in vitro. In addition, DnaK reactivates thermally inactivated RNA polymerase, especially if ATP, DnaJ, and GrpE are present. GroEL and GroES also protect intracellular proteins from aggregation. As one would expect, large quantities of chaperones are present in hyperthermophiles such as *Pyrodictium occultum,* an archaeon that will grow at temperatures as high as 110°C. *Pyrodictium* has a chaperone similar to the GroEL of *E. coli.* The chaperone hydrolyzes ATP most rapidly at 100°C and makes up almost 3/4 of the cell's soluble protein when *P. occultum* grows at 108°C. Thermophilic and hyperthermophilic procaryotes (pp. 124–25)

Chaperones have other functions as well. They are particularly important in the transport of proteins across membranes. For example, in *E. coli* the chaperone SecB binds to the partially unfolded forms of many proteins and keeps them in an export-competent state

A

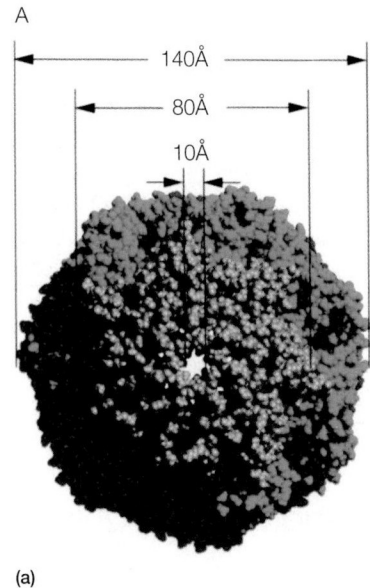

(a)

B

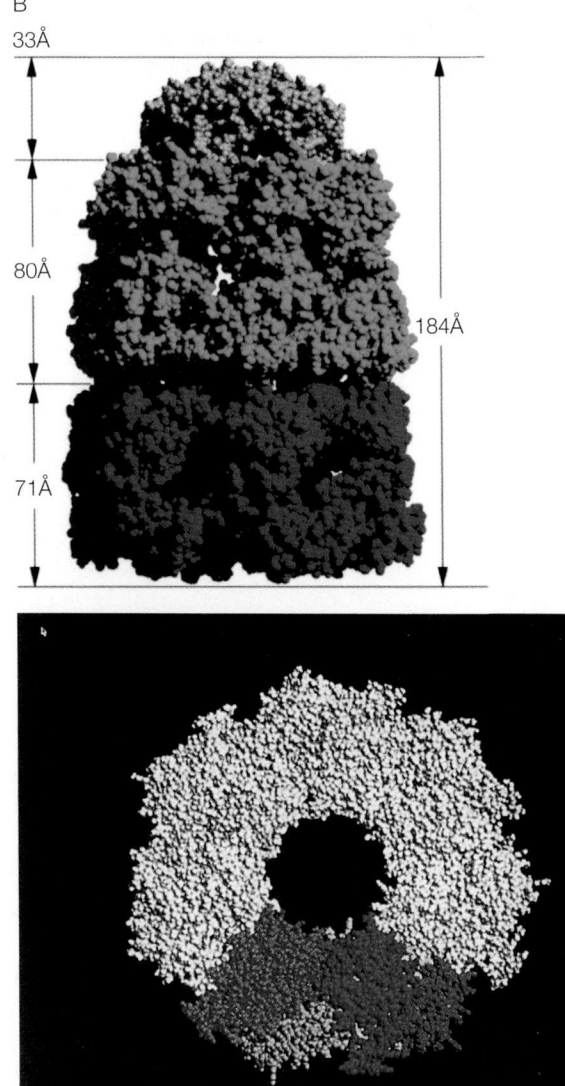

(b)

Figure 12.20 The GroEL-GroES Chaperone Complex. (**a**) Top and side views of the GroEL-GroES complex. The trans GroEL ring is red; the cis GroEL ring is green; GroES is gold. (**b**) A top view of the GroEL complex. Several domains have been colored to distinguish them from one another. Note the large central chamber in which a protein can fold.

until they are translocated across the plasma membrane. DnaK, DnaJ, and GroEL/GroES also can aid in protein translocation across membranes. Proteins destined for the periplasm or outer membrane (*see pp. 56–58*) are synthesized with the proper amino-terminal signal sequence. The signal sequence is a short stretch of amino acids that help direct the completed polypeptide to its final destination. Polypeptides associate with SecB and the chaperone then attaches to the membrane translocase. The polypeptides are transported through the membrane as ATP is hydrolyzed. When they enter the periplasm, the signal peptidase enzyme removes the signal sequence and the protein moves to its final location. Protein secretion in procaryotes (pp. 59–61)

As already noted, the polypeptide folds into its final shape after synthesis, often with the aid of molecular chaperones. This folding is possible because protein conformation is a direct function of the amino acid sequence (*see appendix I*). Recent research indicates that procaryotes and eucaryotes may differ with respect to the timing of protein folding. In terms of conformation, proteins are composed of compact, self-folding, structurally inde-

pendent regions. These regions, normally around 100 to 300 amino acids in length, are called **domains.** Larger proteins such as immunoglobulins (*see pp. 710–13*) may have two or more domains that are linked by less structured portions of the polypeptide chain. In eucaryotes, domains fold independently right after being synthesized by the ribosome. It appears that procaryotic polypeptides, in contrast, do not fold until after the complete chain has been synthesized. Only then do the individual domains fold. This difference in timing may account for the observation that chaperones seem to be more important in the folding of procaryotic proteins. Folding a whole polypeptide is more complex than folding one domain at a time and would require the aid of chaperones.

Protein Splicing

A further level of complexity in the formation of proteins has been discovered. Some microbial proteins are spliced after translation. In **protein splicing,** a part of the polypeptide is removed before the

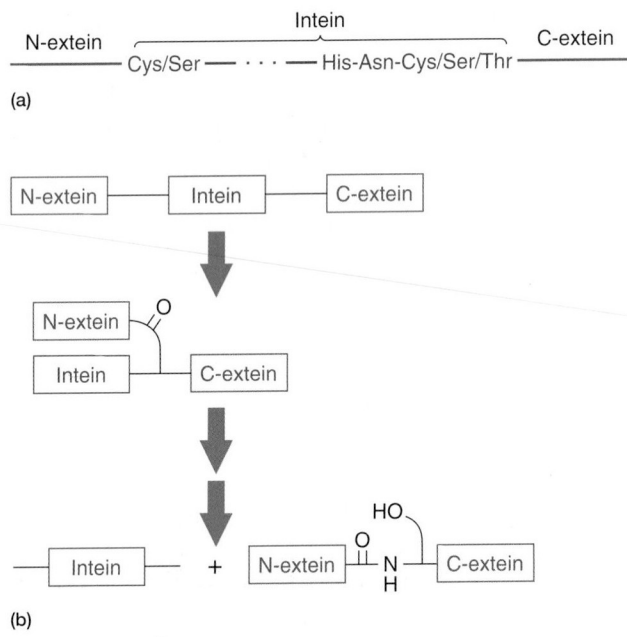

(a)

(b)

Figure 12.21   Protein Splicing.   (a) A generalized illustration of intein structure. The amino acids that are commonly present at each end of the inteins are shown. Note that many are thiol- or hydroxyl-containing amino acids. **(b)** An overview of the proposed pattern or sequence of splicing. The precise mechanism is not yet known but presumably involves the hydroxyls or thiols located at each end of the intein.

polypeptide folds into its final shape. Self-splicing proteins begin as larger precursor proteins composed of one or more internal intervening sequences called **inteins** flanked by external sequences or **exteins,** the N-exteins and C-exteins (**figure 12.21a**). Inteins, which are between about 130 and 600 amino acids in length, are removed in an autocatalytic process involving a branched intermediate (**figure 12.21b**). Thus far, more than 130 inteins in 34 types of self-splicing proteins have been discovered. Over 120 inteins have been found in bacteria and archaea. Some examples are an ATPase in the yeast *Saccharomyces cerevisiae,* the recA protein of *Mycobacterium tuberculosis,* and DNA polymerase in *Pyrococcus.* The presence of self-splicing proteins in all three domains may mean that these proteins are quite widespread and prevalent.

---

1. In which direction are polypeptides synthesized? What is a polyribosome and why is it useful?
2. Briefly describe the structure of transfer RNA and relate this to its function. How are amino acids activated for protein synthe-

sis, and why is the specificity of the aminoacyl-tRNA synthetase reaction so important?
3. What are translational and exit domains? Contrast procaryotic and eucaryotic ribosomes in terms of structure. What roles does ribosomal RNA have?
4. Describe the nature and function of the following: fMet-tRNA, initiator codon, IF-3, IF-2, IF-1, elongation cycle, peptidyl and aminoacyl sites, EF-Tu, EF-Ts, transpeptidation reaction, peptidyl transferase, translocation, EF-G or translocase, nonsense codon, and release factors.
5. What are molecular chaperones and heat-shock proteins? Describe their functions.

12.3   REGULATION OF mRNA SYNTHESIS

The control of metabolism by regulation of enzyme activity is a fine-tuning mechanism: it acts rapidly to adjust metabolic activity from moment to moment. Microorganisms also are able to control the expression of their genome, although over longer intervals. For example, the *E. coli* chromosome can code for about 2,000 to 4,000 peptide chains, yet many fewer proteins are present in *E. coli* growing with glucose as its energy source. Regulation of gene expression serves to conserve energy and raw material, to maintain balance between the amounts of various cell proteins, and to adapt to long-term environmental change. Thus control of gene expression complements the regulation of enzyme activity.   Regulation of enzyme activity (pp. 161–65)

Induction and Repression

The regulation of β-galactosidase synthesis has been intensively studied and serves as a primary example of how gene expression is controlled. This enzyme catalyzes the hydrolysis of the sugar lactose to glucose and galactose (**figure 12.22**). When *E. coli* grows with lactose as its carbon source, each cell contains about 3,000 β-galactosidase molecules, but has less than three molecules in the absence of lactose. The enzyme β-galactosidase is an **inducible enzyme**—that is, its level rises in the presence of a small molecule called an **inducer** (in this case the lactose derivative allolactose).

The genes for enzymes involved in the biosynthesis of amino acids and other substances often respond differently from genes coding for catabolic enzymes. An amino acid present in the surroundings may inhibit the formation of the enzymes responsible for its biosynthesis. This makes good sense because the microor-

Figure 12.22   The β-Galactosidase Reaction.

Lactose   + H_2O
                              β-galactosidase

Galactose   +   Glucose

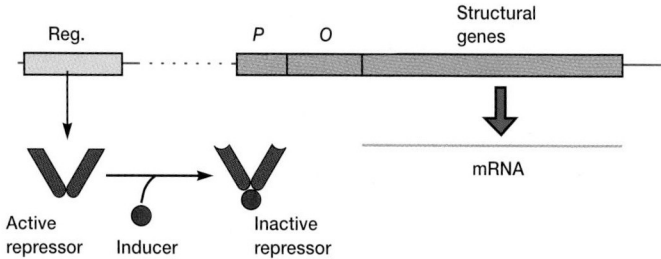

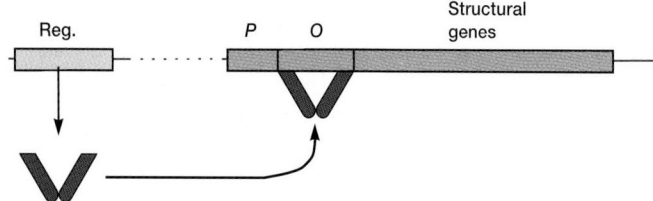

Figure 12.23   Gene Induction. The regulator gene, Reg., synthesizes an active repressor that binds to the operator, *O*, and blocks RNA polymerase binding to the promoter, *P*, unless the inducer inactivates it. In the presence of the inducer, the repressor protein is inactive and transcription occurs.

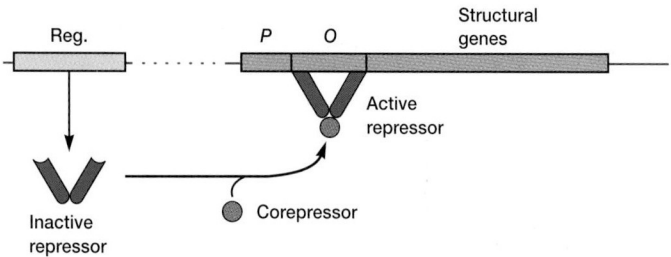

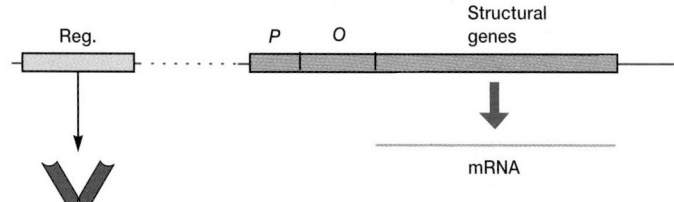

Figure 12.24   Gene Repression. The regulator gene, Reg., synthesizes an inactive repressor protein that must be activated by corepressor binding before it can bind to the operator, *O*, and block transcription. In the absence of the corepressor, the repressor is inactive and transcription occurs.

ganism will not need the biosynthetic enzymes for a particular substance if it is already available. Enzymes whose amount is reduced by the presence of an end product are **repressible enzymes,** and metabolites causing a decrease in the concentrations of repressible enzymes are **corepressors.** Generally, repressible enzymes are necessary for synthesis and always are present unless the end product of their pathway is available. Inducible enzymes, in contrast, are required only when their substrate is available; they are missing in the absence of the inducer.

Although variations in enzyme levels could be due to changes in the rates of enzyme degradation, most enzymes are relatively stable in growing bacteria. Induction and repression result principally from changes in the rate of transcription. When *E. coli* is growing in the absence of lactose, it often lacks mRNA molecules coding for the synthesis of β-galactosidase. In the presence of lactose, however, each cell has 35 to 50 β-galactosidase mRNA molecules. The synthesis of mRNA is dramatically influenced by the presence of lactose.   DNA transcription mechanism (pp. 254–60)

Negative Control

A controlling factor can either inhibit or activate transcription. Although the responses to the presence of metabolites are different, both induction and repression are forms of **negative control:** mRNA synthesis proceeds more rapidly in the absence of the active controlling factor.

The rate of mRNA synthesis is controlled by special **repressor proteins** that are synthesized under the direction of regulator genes. The repressor binds to a specific site on DNA called the **operator.** The importance of regulator genes and repressors is demonstrated by mutationally inactivating a regulator gene to form a **constitutive mutant.** A constitutive mutant produces the enzymes in question whether or not they are needed. Thus inactivation of repressor proteins blocks the regulation of transcription.   Gene structure (pp. 235–39)

Repressors must exist in both active and inactive forms because transcription would never occur if they were always active. In inducible systems the regulator gene directs the synthesis of an active repressor. The inducer stimulates transcription by reversibly binding to the repressor and causing it to change to an inactive shape (**figure 12.23**). Just the opposite takes place in a system controlled by repression (**figure 12.24**). The repressor protein initially is an inactive form called an **aporepressor** and becomes an active repressor only when the corepressor binds to it. The corepressor inhibits transcription by activating the aporepressor.

The synthesis of several proteins is often regulated by a single repressor. The **structural genes,** or genes coding for a polypeptide, are simply lined up together on the DNA, and a single mRNA carries all the messages. The sequence of bases coding for one or more polypeptides, together with the operator controlling its expression, is called an **operon.** This arrangement is of great advantage to the bacterium because coordinated

Historical Highlights
12.2   The Discovery of Gene Regulation

The ability of microorganisms to adapt to their environments by adjusting enzyme levels was first discovered by Emil Duclaux, a colleague of Louis Pasteur. He found that the fungus *Aspergillus niger* would produce the enzyme that hydrolyzes sucrose (invertase) only when grown in the presence of sucrose. In 1900 F. Dienert found that yeast contained the enzymes for galactose metabolism only when grown with lactose or galactose and would lose these enzymes upon transfer to a glucose medium. Such a response made sense because the yeast cells would not need enzymes for galactose metabolism when using glucose as its carbon and energy source. Further examples of adaptation were discovered and by the 1930s H. Karström could divide enzymes into two classes: (1) adaptive enzymes that are formed only in the presence of their substrates, and (2) constitutive enzymes that are always present. It was originally thought that enzymes might be formed from inactive precursors and that the presence of the substrate simply shifted the equilibrium between precursor and enzyme toward enzyme formation.

In 1942 Jacques Monod, working at the Pasteur Institute in Paris, began a study of adaptation in the bacterium *E. coli*. It was already known that the enzyme β-galactosidase, which hydrolyzes the sugar lactose to glucose and galactose, was present only when *E. coli* was grown in the presence of lactose. Monod discovered that nonmetabolizable analogues of β-galactosides, such as thiomethylgalactoside, also could induce enzyme production. This discovery made it possible to study induction in cells growing on carbon and energy sources other than lactose so that the growth rate and inducer concentration would not depend on the lactose supply. He next demonstrated that induction involved the synthesis of new enzyme, not just the conversion of already available precursor. Monod accomplished this by making *E. coli* proteins radioactive with ^{35}S, then transferring the labeled bacteria to nonradioactive medium and adding inducer. The

newly formed β-galactosidase was nonradioactive and must have been synthesized after addition of inducer.

A study of the genetics of lactose induction in *E. coli* was begun by Joshua Lederberg a few years after Monod had started his work. Lederberg isolated not only mutants lacking β-galactosidase but also a constitutive mutant in which synthesis of the enzyme proceeded in the absence of an inducer (LacI⁻). During bacterial conjugation (*see section 13.4*), genes from the donor bacterium enter the recipient to temporarily form an organism with two copies of those genes provided by the donor. When Arthur B. Pardee, François Jacob, and Monod transferred the gene for inducibility to a constitutive recipient not sensitive to inducers, the newly acquired gene made the recipient bacterium sensitive to inducer again. This functional gene was not a part of the recipient's chromosome. Thus the special gene directed the synthesis of a cytoplasmic product that inhibited the formation of β-galactosidase in the absence of the inducer. In 1961 Jacob and Monod named this special product the repressor and suggested that it was a protein. They further proposed that the repressor protein exerted its effects by binding to the operator, a special site next to the structural genes. They provided genetic evidence for their hypothesis. The name operon was given to the complex of the operator and the genes it controlled. Several years later in 1967, Walter Gilbert and Benno Müller-Hill managed to isolate the lac repressor and show that it was indeed a protein and did bind to a specific site in the lac operon.

The existence of repression was discovered by Monod and G. Cohen-Bazire in 1953 when they found that the presence of the amino acid tryptophan would repress the synthesis of tryptophan synthetase, the final enzyme in the pathway for tryptophan biosynthesis. Subsequent research in many laboratories showed that induction and repression were operating by quite similar mechanisms, each involving repressor proteins that bound to operators on the genome.

---

control of the synthesis of several metabolically related enzymes (or other proteins) can be achieved.

The Lactose Operon

The best-studied negative control system is the lactose operon of *E. coli*. The lactose or *lac* operon contains three structural genes and is controlled by the *lac* repressor (**figure 12.25**). One gene codes for β-galactosidase; a second gene directs the synthesis of β-galactoside permease, the protein responsible for lactose uptake. The third gene codes for the enzyme β-galactoside transacetylase, whose function still is uncertain. The presence of the first two genes in the same operon ensures that the rates of lactose uptake and breakdown will vary together (**Historical Highlights 12.2**).

The *lac* operon has three operators. The *lac* repressor protein finds an operator in a two-step process. First, the repressor binds to a DNA molecule, then rapidly slides along the DNA until it reaches an operator and stops. A portion of the repressor fits into the major groove of operator-site DNA by special N-terminal

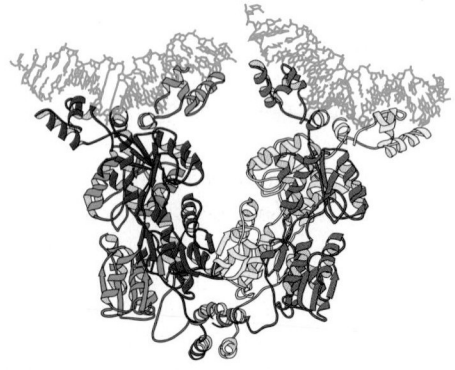

Figure 12.25   Lactose Repressor Binding to DNA.   The *lac* repressor-DNA complex is shown here. The repressor dimer binds to two stretches of DNA (blue) by specialized N-terminal headpiece subdomains that fit in the major groove.

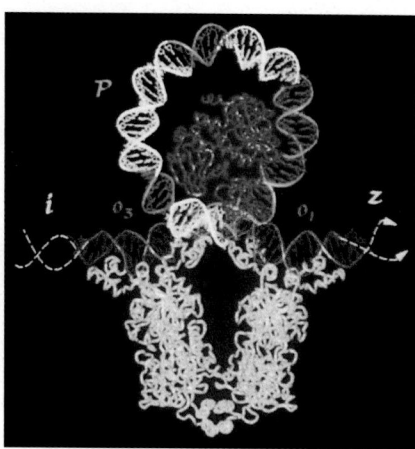

Figure 12.26 Repressor and CAP Bound to the *lac* Operon. The *lac* repressor is in violet, operators in red, promoter in green, and CAP (catabolite activator protein) in blue. RNA polymerase access to the promoter in the loop of DNA is hindered in this complex and transcription cannot begin.

subdomains. The shape of the repressor protein is ideally suited for specific binding to the DNA double helix.

How does the repressor inhibit transcription? The promoter to which RNA polymerase binds (*see pp. 235–38*) is located next to the operator. The repressor may bind simultaneously to more than one operator and bend the DNA segment that contains the promoter (**figure 12.26**). The bent promoter may not allow proper RNA polymerase binding or may not be able to initiate transcription after polymerase binding. Even if the polymerase is bound to the promoter, it is stored there and does not begin transcription until the repressor leaves the operator. A repressor does not affect the actual rate of transcription once it has begun.

Positive Control

The preceding section shows that operons can be under negative control, resulting in induction and repression. In contrast, some operons function only in the presence of a controlling factor— that is, they are under **positive operon control.** The *lac* operon is under positive control as well as negative control—that is, it is under dual control.

Lac operon function is regulated by the **catabolite activator protein (CAP)** or **cyclic AMP receptor protein (CRP)** and the small cyclic nucleotide **3′, 5′-cyclic adenosine monophosphate (cAMP; figure 12.27),** as well as by the *lac* repressor protein. The *lac* promoter contains a CAP site to which CAP must bind before RNA polymerase can attach to the promoter and begin transcription (**figure 12.28**). The catabolite activator protein is able to bind to the CAP site only when complexed with cAMP. Upon binding, CAP bends the DNA about 90° within two helical turns (figure 12.26 and **figure 12.29**). Interaction of CAP with RNA polymerase stimulates transcription. This positive control

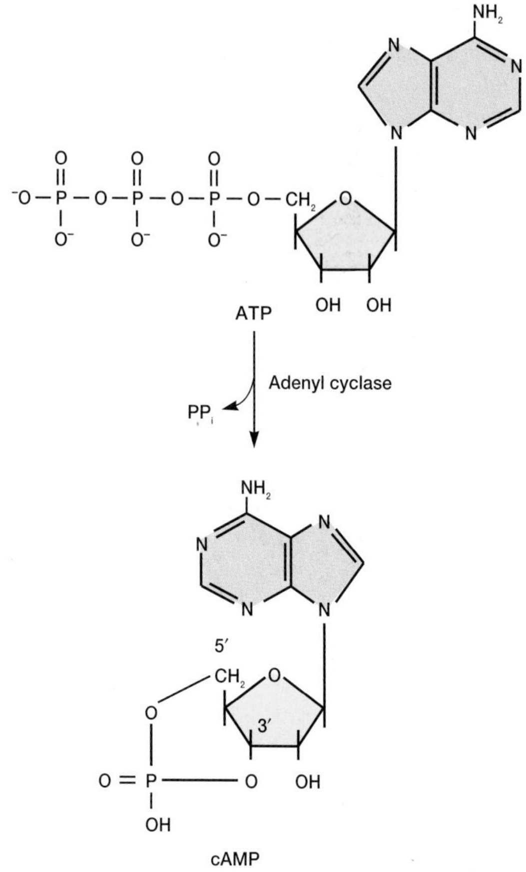

Figure 12.27 Cyclic Adenosine Monophosphate (cAMP). The phosphate group extends between the 3′ and 5′ hydroxyls of the ribose sugar. The enzyme adenyl cyclase forms cAMP from ATP.

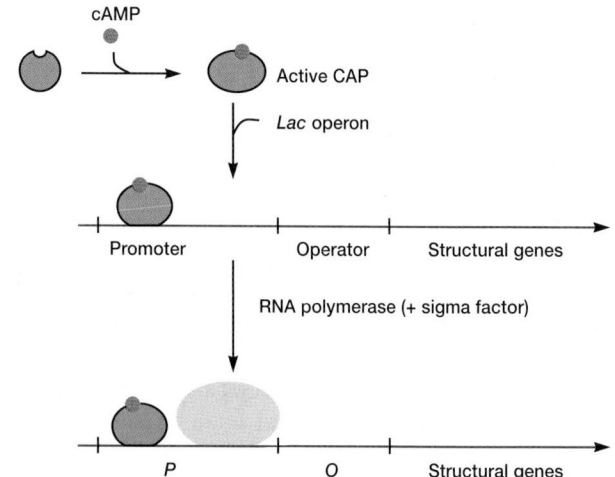

Figure 12.28 Positive Control of the *Lac* Operon. When cyclic AMP is absent or present at a low level, the CAP protein remains inactive and does not bind to the promoter. In this situation RNA polymerase also does not bind to the promoter and transcribe the operon's genes.

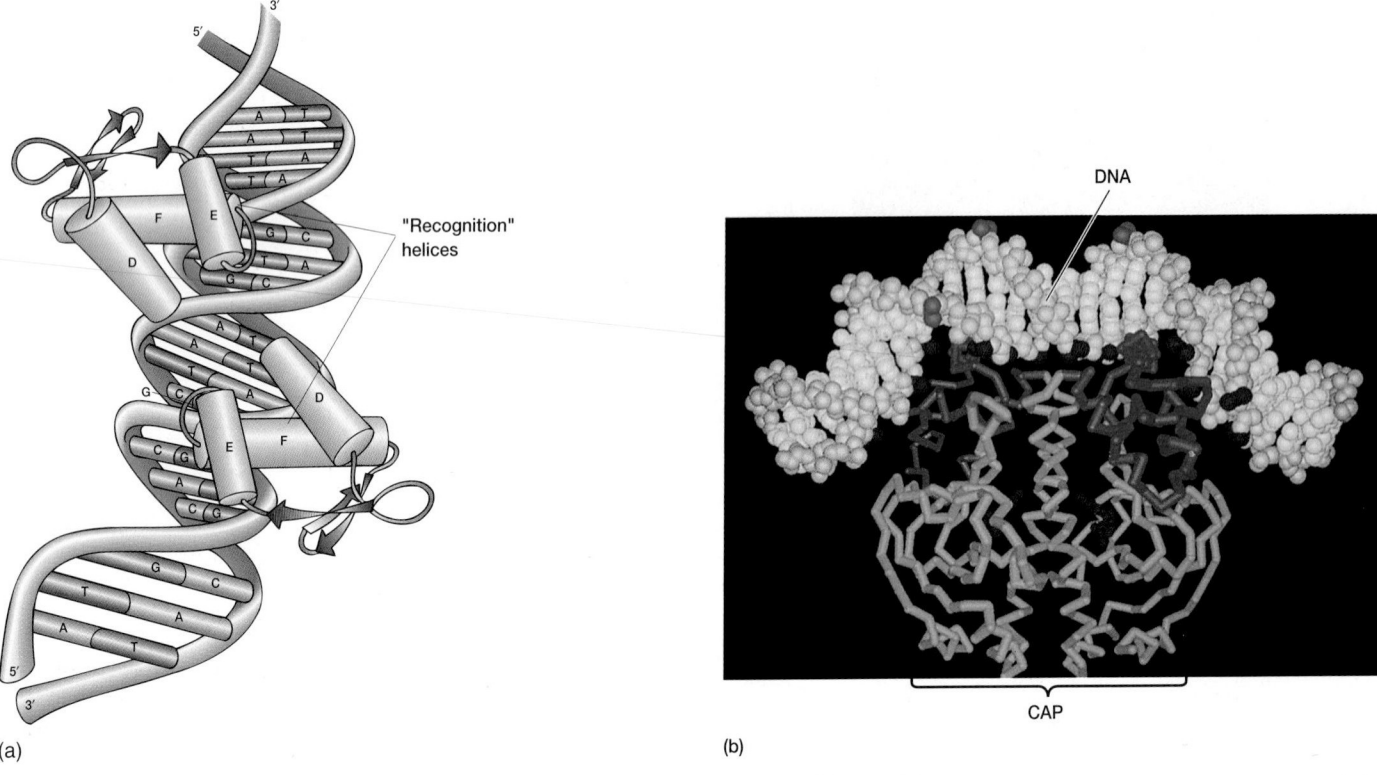

"Recognition" helices

DNA

CAP

(a)

(b)

Figure 12.29 CAP Structure and DNA Binding. (a) The CAP dimer binding to DNA at the lac operon promoter. The recognition helices fit into two adjacent major grooves on the double helix. (b) A model of the *E. coli* CAP-DNA complex derived from crystal structure studies. The cAMP binding domain is in blue and the DNA binding domain, in purple. The cAMP molecules bound to CAP are in red. Note that the DNA is bent by 90° when complexed with CAP.

system makes *lac* operon activity dependent on the presence of cAMP as well as on that of lactose.

 12.4 ATTENUATION

Bacteria can regulate transcription in other ways, as may be seen in the tryptophan operon of *E. coli*. The tryptophan operon contains structural genes for five enzymes in this amino acid's biosynthetic pathway. As might be expected, the operon is under the control of a repressor protein coded for by the *trpR* gene (trp stands for tryptophan), and excess tryptophan inhibits transcription of operon genes by acting as a corepressor and activating the repressor protein. Although the operon is regulated mainly by repression, the continuation of transcription also is controlled. That is, there are two decision points involved in transcriptional control, the initiation of transcription and the continuation of transcription past the attenuator region.

A **leader region** lies between the operator and the first structural gene in the operon, the *trpE* gene, and is responsible for controlling the continuation of transcription after the RNA polymerase has bound to the promoter (**figure 12.30a**). The leader region contains an **attenuator** and a sequence that codes for the synthesis of a leader peptide. The attenuator is a rho-

independent termination site (pp. 255–57) with a short GC-rich segment followed by a sequence of eight U residues. The four stretches marked off in figure 12.30a have complementary base sequences and can base pair with each other to form hairpin loops. In the absence of a ribosome, mRNA segments one and two pair to form a hairpin, while segments three and four generate a second loop next to the poly(U) sequence (figure 12.30b). The hairpin formed by segments three and four plus the poly(U) sequence will terminate transcription. If segment one is prevented from base pairing with segment two, segment two is free to associate with segment three. As a result segment four remains single stranded (figure 12.30c) and cannot serve as a terminator for transcription. It is important to note that the sequence coding for the leader peptide contains two adjacent codons that code for the amino acid tryptophan. Thus the complete peptide can be made only when there is an adequate supply of tryptophan. Since the leader peptide has not been detected, it must be degraded immediately after synthesis.

Ribosome behavior during translation of the mRNA regulates RNA polymerase activity as it transcribes the leader region. This is possible because translation and transcription are tightly coupled. When the active repressor is absent, RNA polymerase binds to the promoter and moves down the leader synthesizing mRNA. If there is no translation of the mRNA after the RNA polymerase

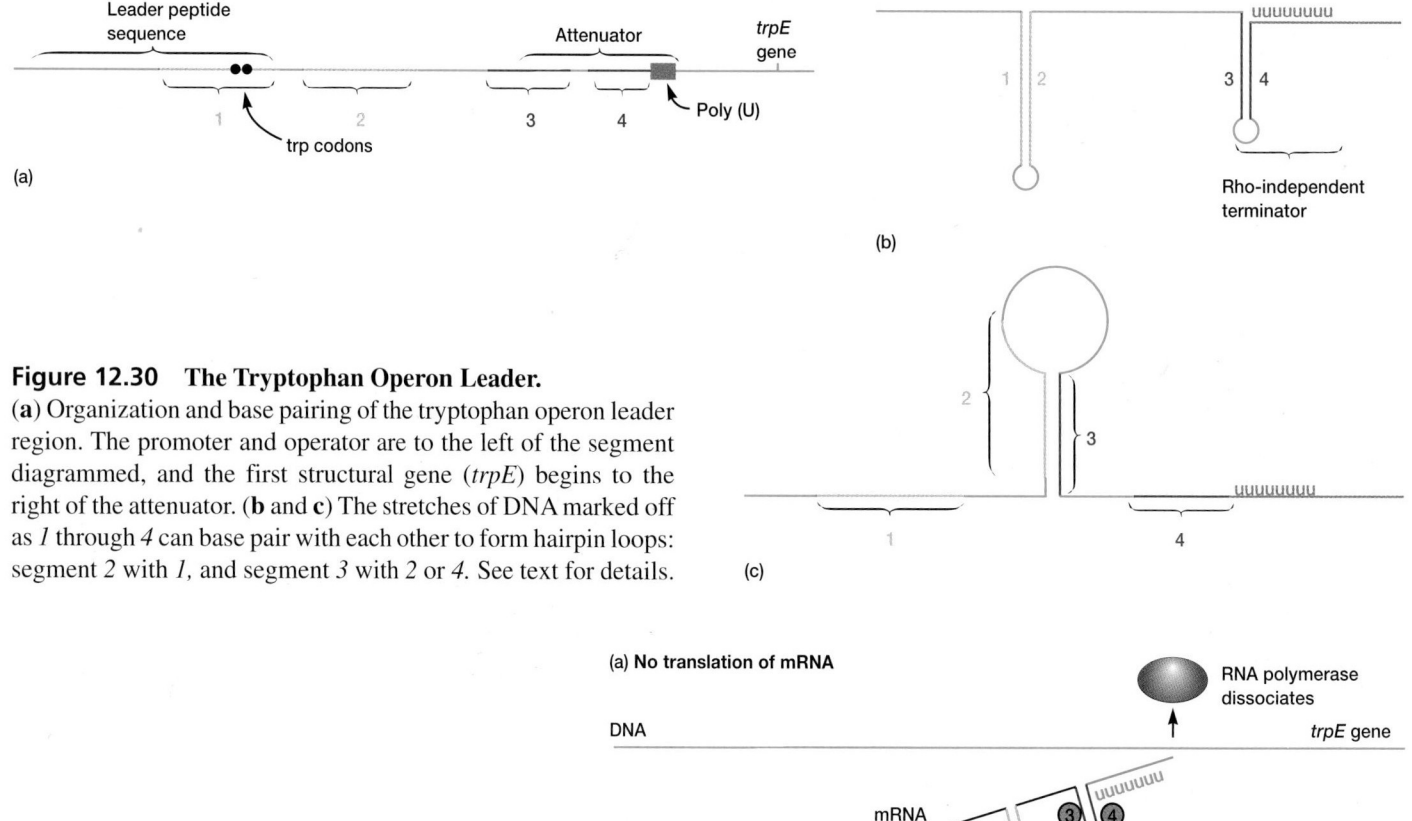

Figure 12.30 The Tryptophan Operon Leader.
(**a**) Organization and base pairing of the tryptophan operon leader region. The promoter and operator are to the left of the segment diagrammed, and the first structural gene (*trpE*) begins to the right of the attenuator. (**b** and **c**) The stretches of DNA marked off as *1* through *4* can base pair with each other to form hairpin loops: segment *2* with *1*, and segment *3* with *2* or *4*. See text for details.

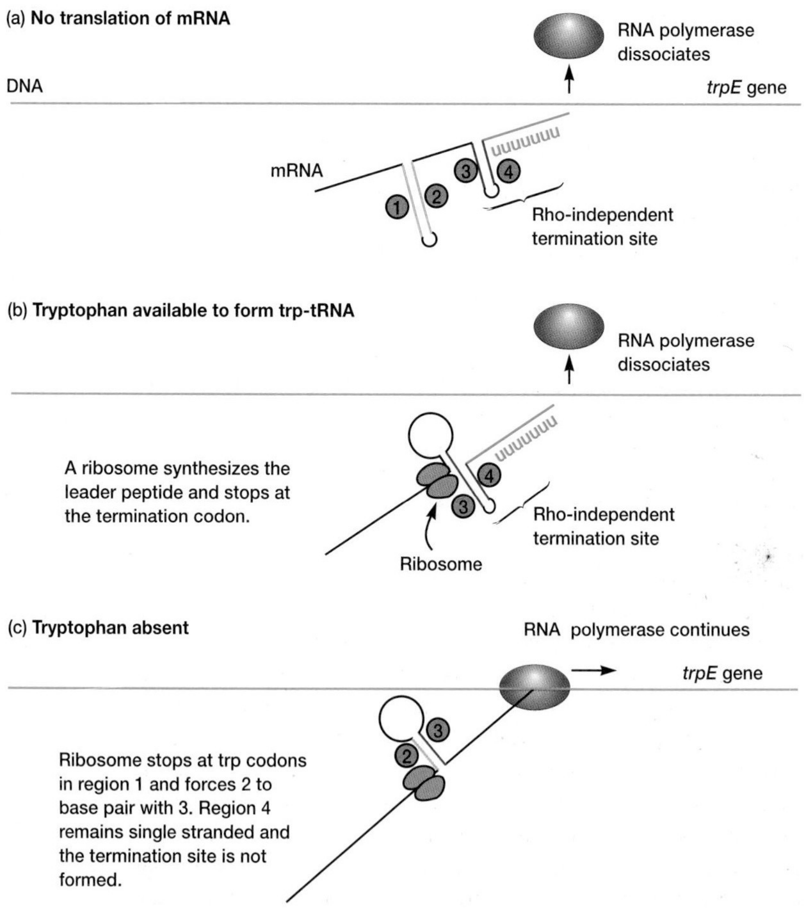

Figure 12.31 Attenuation Control. The control of tryptophan operon function by attenuation. See text for details.

has begun copying the leader region, segments three and four form a hairpin loop, and transcription terminates before the polymerase reaches the *trpE* gene (**figure 12.31a**). When tryptophan is present, there is sufficient tryptophanyl-tRNA for protein synthesis. Therefore the ribosome will synthesize the leader peptide and continue moving along the mRNA until it reaches a UGA stop codon (*see section 11.4*) lying between segments one and two. The ribosome halts at this codon and projects into segment two far enough to prevent it from pairing properly with segment three (figure 12.31b). Segments three and four form a hairpin loop, and the RNA polymerase terminates at the attenuator just as if no translation had taken place. If tryptophan is lacking, the ribosome will stop at the two adjacent tryptophan codons in the leader peptide sequence and prevent segment one from base pairing with segment two, because the tryptophan codons are located within segment one (figures 12.30a and 12.31c). If this happens while the RNA polymerase is still transcribing the leader region, segments two and three associate before segment four has been synthesized. Therefore segment four will remain single stranded and the terminator hairpin will not form. Consequently, when tryptophan is absent, the RNA polymerase continues on and transcribes tryptophan operon genes. Control of the continuation of transcription by a specific aminoacyl-tRNA is called **attenuation.**

Attenuation's usefulness is apparent. If the bacterium is deficient in an amino acid other than tryptophan, protein synthesis will slow and tryptophanyl-tRNA will accumulate. Transcription of the tryptophan operon will be inhibited by attenuation. When the bacterium begins to synthesize protein rapidly, tryptophan may be scarce and the concentration of tryptophanyl-tRNA may be low. This would reduce attenuation activity and stimulate operon transcription, resulting in larger quantities of the tryptophan biosynthetic enzymes. Acting together, repression and attenuation can coordinate the rate of synthesis of amino acid biosynthetic enzymes with the availability of amino acid end products and with the overall rate of protein synthesis. When tryptophan is present at high concentrations, any RNA polymerases not blocked by the activated repressor protein probably will not get past the attenuator sequence. Repression decreases transcription about seventyfold and attenuation slows it another eight- to tenfold; when both mechanisms operate together, transcription can be slowed about 600-fold.

Attenuation seems important in the regulation of several amino acid biosynthetic pathways. At least five other operons have leader peptide sequences that resemble the tryptophan system in organization. For example, the leader peptide sequence of the histidine operon codes for seven histidines in a row and is followed by an attenuator that is a terminator sequence.

12.5 GLOBAL REGULATORY SYSTEMS

Thus far, we have been considering the function of isolated operons. However, bacteria must respond rapidly to a wide variety of changing environmental conditions and be able to cope with such things as nutrient deprivation, dessication, and major temperature fluctuations. They also have to compete successfully with other organisms for scarce nutrients and use these nutrients efficiently. These challenges require a regulatory system that can rapidly control many operons at the same time. Such regulatory systems that affect many genes and pathways simultaneously are called **global regulatory systems.** There are many examples of these multigene global systems. Catabolite repression in enteric bacteria and sporulation in *Bacillus subtilis* will be discussed shortly. Two other previously discussed global systems are the SOS response (*see p. 249*) and the production of heat-shock proteins (p. 268).

Although it is usually possible to regulate all the genes of a metabolic pathway in a single operon, there are good reasons for more complex global systems. Some processes involve too many genes to be accommodated in a single operon. For example, the machinery required for protein synthesis is composed of 150 or more gene products, and coordination requires a regulatory network that controls many separate operons. Sometimes two levels of regulation are required because individual operons must be controlled independently and also cooperate with other operons. Regulation of sugar catabolism in *E. coli* is a good example. *E. coli* uses glucose when it is available; in such a case, operons for other catabolic pathways are repressed. If glucose is unavailable and another nutrient is present, the appropriate operon is activated.

Global regulation can be accomplished by several different mechanisms. A protein repressor or activator may affect several operons simultaneously. A sigma factor may cause RNA polymerase to recognize and transcribe an array of different operons with similar promoters. Sometimes a nonprotein regulator such as the nucleotide guanosine tetraphosphate controls several operons.

Global regulatory systems are so complex that a specialized nomenclature is used to describe the various kinds. Perhaps the most basic type is the **regulon.** A regulon is a collection of genes or operons that is controlled by a common regulatory protein. Usually the operons are associated with a single pathway or function (e.g., the production of heat-shock proteins or the catabolism of glycerol). A somewhat more complex situation is seen with a modulon. This is an operon network under the control of a common global regulatory protein, but whose constituent operons also are controlled separately by their own regulators. A good example of a modulon is catabolite repression. The most complex global systems are referred to as stimulons. A stimulon is a regulatory system in which all operons respond together in a coordinated way to an environmental stimulus. It may contain several regulons and modulons, and some of these may not share regulatory proteins. The genes involved in a response to phosphate limitation are scattered among several regulons and are part of one stimulon.

We will now briefly consider two examples of global regulation. First we will discuss catabolite repression and the use of positive operon control. Then an introduction to regulation by sigma factors and the induction of sporulation will follow.

Catabolite Repression

If *E. coli* grows in a medium containing both glucose and lactose, it uses glucose preferentially until the sugar is exhausted. Then after a short lag, growth resumes with lactose as the carbon source (**figure 12.32**). This biphasic growth pattern or response is called **diauxic growth.** The cause of diauxic growth or diauxie is complex and not completely understood, but **catabolite repression** or the glucose effect probably plays a part. The enzymes for glucose catabolism are constitutive and unaffected by CAP activity. When the bacterium is given glucose, the cAMP level drops, resulting in deactivation of the catabolite activator protein and inhibition of lac operon expression. The decrease in cAMP may be due to the effect of the phosphoenolpyruvate:phosphotransferase system (PTS) on the activity of adenyl cyclase, the enzyme that synthesizes cAMP. Enzyme III of the PTS donates a phosphate to

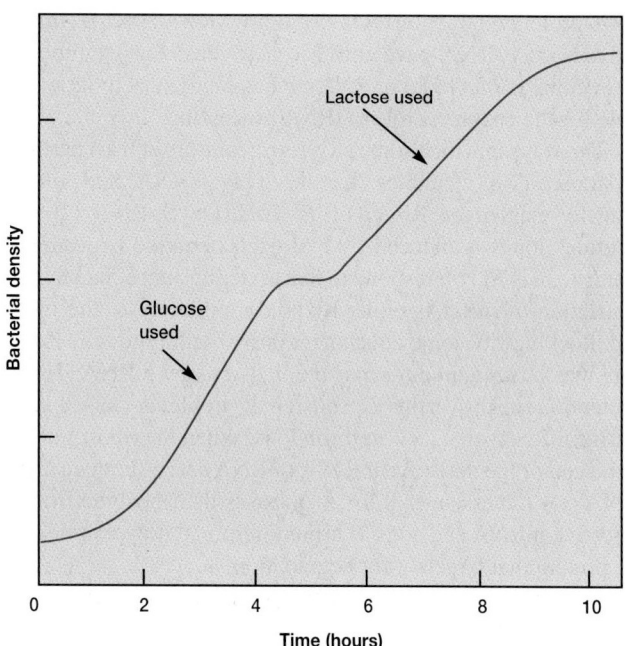

Figure 12.32   Diauxic Growth.   The diauxic growth curve of *E. coli* grown with a mixture of glucose and lactose. Glucose is first used, then lactose. A short lag in growth is present while the bacteria synthesize the enzymes needed for lactose use.

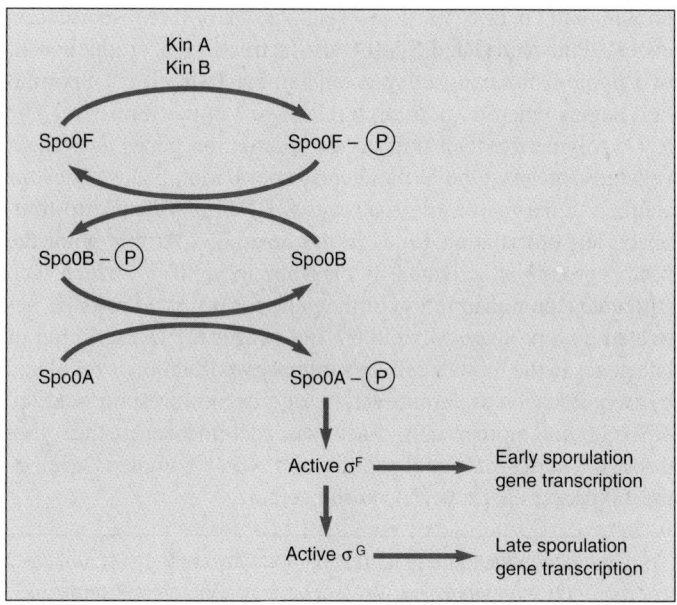

Figure 12.33   Initiation of Sporulation in *Bacillus subtilis*. A simplified diagram of the initial steps in triggering sporulation. The activation of kinases A and B begins a two-component phosphorelay system that activates the transcription regulator Spo0A. See text for more detail.

glucose during its transport; therefore, it enters the cell as glucose 6-phosphate. The phosphorylated form of enzyme III also activates adenyl cyclase. If glucose is being rapidly transported by PTS, the amount of phosphorylated enzyme III is low and the adenyl cyclase is less active, so the cAMP level drops. At least one other mechanism is involved in diauxic growth. When the PTS is actively transporting glucose into the cell, nonphosphorylated enzyme III is more prevalent. Nonphosphorylated enzyme III binds to the lactose permease and allosterically inhibits it, thus blocking lactose uptake.   The phosphoenolpyruvate: phosphotransferase system (pp. 100–101)

Whatever the precise mechanism, such control is of considerable advantage to the bacterium. It will use the most easily catabolized sugar (glucose) first rather than synthesize the enzymes necessary for another carbon and energy source. These control mechanisms are present in a variety of bacteria and metabolic pathways.

Regulation by Sigma Factors and Control of Sporulation

Although the RNA polymerase core enzyme can transcribe any gene to produce a messenger RNA copy, it needs the assistance of a sigma factor to bind the promoter and initiate transcription (*see p. 254*). This provides an excellent means of regulating gene expression. When a complex process requires a radical change in transcription, or the synthesis of several gene products in a precisely timed sequence, it may be regulated by a series of sigma factors. Each sigma factor enables the RNA polymerase core en-

zyme to recognize a specific set of promoters and transcribe only those genes. Substitution of the sigma factor immediately changes gene expression. Bacterial viruses often use sigma factors to control mRNA synthesis during their life cycle (*see chapter 17*). This regulatory mechanism also is common among both gram-negative and gram-positive bacteria. For example, *Escherichia coli* synthesizes several sigma factors. Under normal conditions the sigma factor σ^{70} directs RNA polymerase activity. (The superscript letter or number indicates the function or size of the sigma factor; 70 stands for 70,000 Da.) When flagella and chemotactic proteins are needed, *E. coli* produces σ^F (σ^{28}). If the temperature rises too high, σ^H (σ^{32}) appears and stimulates the formation of around 17 heat-shock proteins to protect the cell from thermal destruction. As would be expected, the promoters recognized by each sigma factor differ characteristically in sequence at the -10 and -35 positions (*see pp. 236–37*).

One of the best-studied examples of gene regulation by sigma factors is the control of sporulation in the gram-positive *Bacillus subtilis*. When *B. subtilis* is deprived of nutrients, it will form endospores in a complex developmental process lasting about 8 hours.   The bacterial endospore (pp. 68–71)

Normally the *B. subtilis* RNA polymerase uses sigma factor σ^A (σ^{43}) to recognize genes. Environmental signals such as nutrient deprivation stimulate a kinase (Kin A or Kin B) to catalyze the phosphorylation of the Spo0F protein (**figure 12.33**). Spo0F transfers the phosphate to Spo0B, which in turn phosphorylates Spo0A. Phosphorylated Spo0A has several effects. It binds to a promoter and represses the expression of the *abrB* gene. *abrB* codes for a protein that inhibits many genes not needed during growth with excess nutrients (e.g., at least three sporulation genes). Phosphorylated Spo0A also activates the production of two sigma factors: an active σ^F and an inactive pro-σ^E. Sporulation begins when σ^F partially replaces σ^A in the forespore. The RNA polymerase then transcribes sporulation genes as well as vegetative genes. One of these early sporulation genes codes for another sigma factor, σ^G, that causes RNA polymerase to transcribe late sporulation genes in the forespore. At this point the pro-σ^E protein is activated by cleavage to yield σ^E, which then stimulates the transcription of the gene for pro-σ^K. Pro-σ^K is activated by a protease to yield σ^K and trigger the transcription of late genes in the mother cell. In summary, sporulation is regulated by two cascades of sigma factors, one in the forespore and the other in the mother cell. Each cascade influences the other through a series of signals so that the whole complex developmental process is properly coordinated.

---

1. What are induction and repression and why are they useful? Define inducer, corepressor, repressor protein, aporepressor, regulator gene, negative control, constitutive mutant, operator, structural gene, and operon. Describe how the *lac* operon is regulated.
2. Define positive control, dual control, and catabolite activator protein. How is the *lac* operon controlled positively?
3. Define attenuation and describe how it works in terms of a labeled diagram, such as that provided in figure 12.31. What are the functions of the leader region and the attenuator in attenuation?
4. What are global regulatory systems and why are they necessary? Briefly describe regulons, modulons, and stimulons.
5. What is diauxic growth and how does it result from catabolite repression?
6. Briefly describe how sigma factors can be used to control gene expression. Describe the regulation of sporulation by sigma factors.

12.6 SMALL RNAs AND REGULATION

Microbiologists have known for many years that gene expression can be controlled by both regulatory proteins (e.g., repressor proteins and CAP) and aminoacyl-tRNA (attenuation). More recently it has been discovered that cellular processes can be controlled by special small regulatory RNA molecules. A large number of RNAs have been discovered that do not function as messenger RNAs or ribosomal RNAs. Microbiologists often refer to them as **small RNAs (sRNAs),** but they also have been called noncoding RNAs (ncRNAs). In *E. coli,* there may be as many as 50 to 200 sRNAs, which usually range around 50 to 400 nucleotides in length. It is thought that eucaryotes may have hundreds to thousands of these noncoding RNAs with lengths from 21 to over 10,000 nucleotides. Many procaryotic sRNAs regulate such activities as RNA processing, mRNA and protein stability, translation, and other cell processes. They appear to act by at least three different mechanisms. Some directly base pair with other RNAs (e.g., micF RNA), whereas others function by means of

RNA-protein interactions (OxyS). A few sRNAs such as tmRNA and RNase P RNA have intrinsic activities. For example, the RNA component (10sb) of RNase P has catalytic activity responsible for the enzyme's role in tRNA processing.

The ways in which some sRNAs operate are at least partly understood. Two examples are the OxyS RNA and tmRNA (transfer-messenger RNA) of *E. coli.* OxyS RNA (109 nucleotides long) is induced by hydrogen peroxide exposure and seems to help *E. coli* respond to this toxic substance. It can inhibit translation of mRNAs either by binding directly to the mRNA and blocking ribosome attachment or by binding to the Hfq protein, which is required for translation of a target mRNA. The 363 nucleotide tmRNA helps *E. coli* repair problems caused by defective mRNAs that lack stop condons. When the ribosome stalls at the end of the defective mRNA, tmRNA acts as both an alanyl-tRNA and a messenger RNA. It releases the ribosome from the defective mRNA and adds a carboxy-terminal polypeptide tag to the protein that targets it for degradation.

One kind of sRNA, called **antisense RNA,** has a base sequence complementary to a segment of another RNA molecule and specifically binds to the target RNA. Antisense RNA binding can block DNA replication, mRNA synthesis, or translation. The genes coding for these RNAs are sometimes called antisense genes.

This mode of regulation appears to be widespread among viruses and bacteria. Examples are the regulation of plasmid replication and Tn10 transposition, osmoregulation of porin protein expression, regulation of λ phage reproduction, and the autoregulation of cAMP-receptor protein synthesis. Antisense RNA regulation has not yet been demonstrated in eucaryotic cells, although there is evidence that it may exist. It is possible that antisense RNAs bind with some eucaryotic mRNAs and stimulate their degradation. Plasmids and transposons (pp. 288–95)

The regulation of *E. coli* outer membrane porin proteins provides an example of control by antisense RNA. The outer membrane contains channels made of porin proteins (*see p. 58*). The two most important porins in *E. coli* are the OmpF and OmpC proteins. OmpC pores are slightly smaller and are made when the bacterium grows at high osmotic pressures. It is the dominant porin in *E. coli* from the intestinal tract. This makes sense because the smaller pores would exclude many of the toxic molecules present in the intestine. The larger OmpF pores are favored when *E. coli* grows in a dilute environment, and they allow solutes to diffuse into the cell more readily.

The *ompF* and *ompC* genes are partly regulated by a special OmpR protein that represses the *ompF* gene and activates *ompC*. In addition, the *micF* gene produces a 174-nucleotide-long antisense micF RNA that blocks *ompF* action (mic stands for *m*RNA-interfering complementary RNA). The micF RNA is complementary to *ompF* at the translation initiation site. It complexes with *ompF* mRNA and represses translation. The *micF* gene is activated by conditions such as high osmotic pressure or the presence of some toxic materials that favor *ompC* expression. This helps ensure that OmpF protein is not produced at the same time as OmpC protein.

The fact that antisense RNA can bind specifically to mRNA and block its activity has great practical implications. Antisense

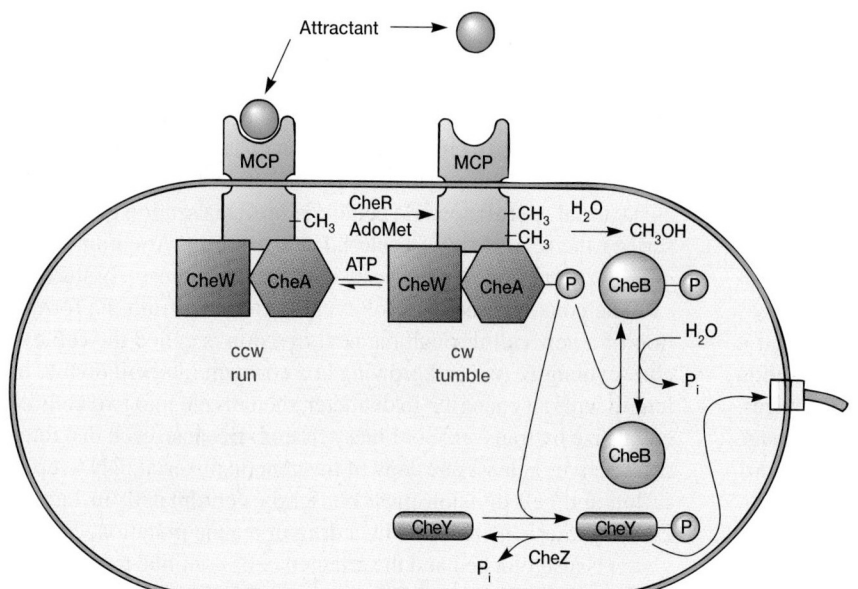

Figure 12.34   The Mechanism of Chemotaxis in *Escherichia coli*.   The chemotaxis system is designed to control counterclockwise (ccw) and clockwise (cw) flagellar rotation so that *E. coli* moves up an attractant gradient by a sequence of runs and tumbles. See the text for a description of the process.

RNA is already a valuable research tool. Suppose one desires to study the action of a particular gene. An antisense RNA that will bind to the gene's mRNA can be constructed and introduced into the cell, thus blocking gene expression. Changes in the cell are then observed. It also is possible to use the same approach with short strands of antisense DNA that bind to mRNA.

Antisense RNA and DNA may well be effective against a variety of cancers and infectious diseases. Promising preliminary results have been obtained using antisense oligonucleotides directed against *Trypanosoma brucei brucei* (the cause of African sleeping sickness), herpesviruses, the HIV virus, tumor viruses such as the RSV and polyoma viruses, ovarian cancer, cytomegalovirus infections, Crohn's disease, and chronic myelogenous leukemia. Although much further research is needed to determine the medical potential of these molecules, they may prove invaluable in the treatment of many diseases.

---

1. Describe small RNAs (noncoding RNAs) and how they are thought to function.
2. What is antisense RNA? How does it regulate gene expression?

 **TWO-COMPONENT PHOSPHORELAY SYSTEMS**

A **two-component phosphorelay system** is a signal transduction system that uses the transfer of phosphoryl groups to control gene transcription and protein activity. It has two major components: a sensor kinase and a response regulator. There are many phosphorelay systems; two good examples are the systems that control sporulation and chemotaxis.

In the sporulation regulation system (figure 12.33), kin A is a sensor kinase. It serves as a transmitter that phosphorylates itself

(autophosphorylation) on a special histidine residue in response to environmental signals. The Spo0F acts as a receiver and catalyzes the transfer of the phosphoryl group from kin A to a special aspartic acid residue on its surface; Spo0F then donates the phosphoryl group to a histidine on Spo0B. Spo0A is a response regulator. It has a receiver domain aspartate and picks up the phosphoryl group from Spo0B to become an active transcription regulator.   Control of sporulation by sigma factors (pp. 277–78)

Chemotaxis is controlled by a well-studied phosphorelay regulatory system. As we have seen previously, procaryotes sense various chemicals in their environment when these substances bind to chemoreceptors called **methyl-accepting chemotaxis proteins (MCPs).** The MCPs can influence flagellar rotation in such a way that the organisms swim toward attractants and away from repellants. This response is regulated by a complex system in which the CheA protein serves as a sensor kinase and the CheY protein is the response regulator.   Chemotaxis (pp. 67–68)

The MCP chemoreceptors are buried in the plasma membrane with major parts exposed on both sides. The periplasmic side of each MCP has a binding site for one or more attractant molecules and may also bind repellents. Although attractants often bind directly to the MCP, in some cases they may attach to special periplasmic binding proteins, which then interact with the MCPs. The cytoplasmic side of an MCP interacts with two proteins (**figure 12.34**). The CheW protein binds to MCPs and helps attach the CheA protein. The full complex is composed of an MCP dimer, two CheW monomers, and a CheA dimer. When the MCP is not bound to an attractant, it stimulates CheA to phosphorylate itself using ATP, a process called autophosphorylation. CheA autophosphorylation is inhibited when the attractant is bound to its MCP. Phosphorylated CheA can donate its phosphate to one of two receptor proteins, CheY or CheB. If CheY is phosphorylated by CheA, it changes to an active conformation, moves to the flagellum, and interacts with the switch protein (Fli M) at its base (*see* *figure 3.37*). This causes the flagellum to rotate clockwise. Thus a decrease in attractant level promotes clockwise rotation and tumbling. The phosphate is removed from CheY in about 10 seconds in a process aided by the protein CheZ. The short lifetime of phosphorylated CheY means that the bacterium is very responsive to changes in attractant concentration. It will not be stuck in the tumble mode for too long a time when the attractant level changes. When no attractants or repellents are present, the system maintains intermediate concentrations of CheA phosphate and CheY phosphate. This produces a normal run-tumble swimming pattern.

The *E. coli* cell must ignore past stimulus responses so that it can compare the most recent attractant or repellent concentration with the immediately previous one and respond to any changes. This means that it must be able to adapt to a concentration change in order to detect still further changes. That is, it must have a short-term memory with a retention time of only seconds. This adaptation is accomplished by methylation of the MCP receptors. The cytoplasmic portion or domain of MCP molecules usually has about four or five methylation sites containing special glutamic acid residues. Methyls can be added to these glutamic acid carboxyl groups using S-adenosylmethionine as the methylating agent. The reaction is catalyzed by the CheR protein and occurs at a fairly steady rate regardless of the attractant level. Methyl groups are hydrolytically removed from MCPs by the phosphorylated CheB protein, a methylesterase. These enzymes are part of a feedback circuit that stops motor responses a short time after they have commenced. The attractant-MCP complex is a good substrate for CheR and a poor substrate for CheB. When an attractant binds to the MCP, the levels of both CheY phosphate and CheB phosphate drop because the autophosphorylation of CheA is inhibited. This not only causes counterclockwise rotation and a run, but also lowers methylesterase activity so that MCP methylation increases. Increased methylation changes the conformation of the MCP so that it again supports an intermediate level of CheA autophosphorylation. CheY phosphate and CheB phosphate return to intermediate levels and restore the normal run-tumble behavior. Removal of the attractant causes the overmethylated MCP to stimulate CheA autophosphorylation and the levels of CheY phosphate and CheB phosphate increase. This induces tumbling and simultaneously promotes MCP demethylation so that the system returns to an intermediate level of CheA autophosphorylation.

The chemotactic response is a very complex one involving many different proteins and two forms of covalent protein regulation (*see section 8.9*). The actual response arises from a combination of (1) the control of CheA phosphorylation by attractant and repellent levels; (2) the clockwise rotation promoted by phosphorylated CheY; and (3) a feedback regulatory circuit involving CheR, phosphorylated CheB, and variations in MCP methylation.

---

1. What is a two-component phosphorelay system?
2. Explain in a general way how bacteria are attracted to substances like nutrients while being repelled by toxic materials.
3. Describe the molecular mechanism by which molecules attract *E. coli.*

12.8 CONTROL OF THE CELL CYCLE

Although much progress has been made in understanding the control of microbial enzyme activity and pathway function, much less is known about the regulation of more complex events such as bacterial sporulation and cell division. This section briefly describes the regulation of bacterial cell division. Attention is focused primarily on *E. coli* because it has been intensively studied.

The complete sequence of events extending from the formation of a new cell through the next division is called the **cell cycle.** A young *E. coli* cell growing at a constant rate will double in length without changing in diameter, then divide into two cells of equal size by transverse or binary fission. Because each daughter cell receives at least one copy of the genetic material, DNA replication and cell division must be tightly coordinated. In fact, if DNA synthesis is inhibited by a drug or a gene mutation, cell division is also blocked and the affected cells continue to elongate, forming long filaments. Termination of DNA replication also seems connected in some way with cell division. Although the growth rate of *E. coli* at 37°C may vary considerably, division usually takes place about 20 minutes after replication has finished. During this final interval the genetic material must be distributed between the daughter cells. The newly formed DNA copies are attached to adjacent sites on the plasma membrane at or close to the center of the cell, probably at their replication factories or replisomes (**figure 12.35a**). It is not yet clear how the two copies are separated. ParA, ParB, MukB and other proteins are involved in DNA partitioning. The ParA and ParB proteins are localized at the poles of late predivisional cells and may be part of a mitotic-like apparatus for bacterial division. In *Bacillus subtilis,* the RacA protein appears to attach chromosomes to the ends of the cell by their replication origin. It may aid in the movement of chromosomes to the developing endospore and the mother cell. There is some evidence for active separation of chromosomes by a force-generating mechanism. Possibly the chromosomes move apart because they are pushed by the replisomes and pulled by some sort of "mitotic" apparatus. DNA movement also may result from membrane growth and cell wall synthesis, but membrane growth is too slow to account for all the movement. After the chromosomes have been separated, a cross wall or septum forms between them. Patterns of cell wall formation (pp. 217–18)

Current evidence suggests that two sequences of events, operating in parallel but independently, control division and the cell cycle (**figure 12.36**). Like eucaryotic cells, procaryotes must reach a specific threshold size or initiation mass to trigger DNA replication. *E. coli* also has to reach a threshold length before it can partition its chromosomes and divide into two cells. Thus there seem to be two separate controls for the cell cycle, one sensitive to cell mass and the other responding to cell length. DNA replication takes about 40 minutes to complete.

Some of the *E. coli* cell cycle control mechanisms are becoming clearer, although much remains to be learned. The initiation of DNA replication requires binding of many copies of the DnaA protein to *oriC*, the replication origin site (*see pp. 229–33*). Active

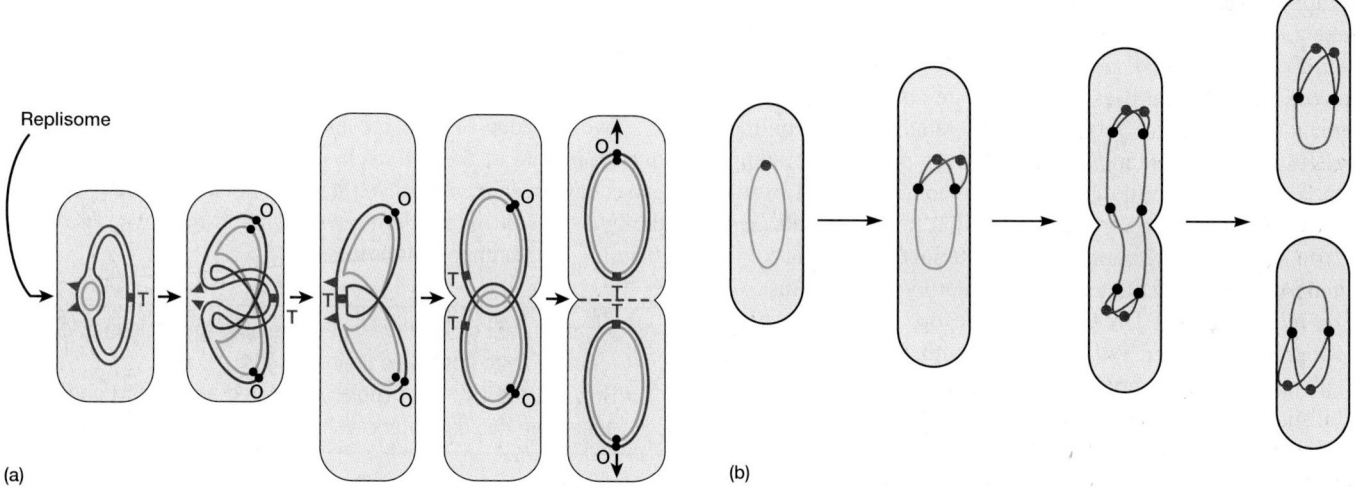

Figure 12.35   DNA Replication in Bacteria.   (**a**) In slowly growing bacteria, the chromosome is replicated once before division. In this simplified illustration, the replicating chromosomal DNA is spooled through a membrane-bound replisome or replication factory, the factory then divides into two replication foci, and eventually the duplicated chromosomes separate and move to opposite ends of the cell. O is the origin of replication and T is the termination region. (**b**) DNA replication in rapidly growing bacteria is more complex. A new round of DNA replication is initiated before the original cell divides so that the DNA in daughter cells is already partially replicated. The colored circles at the ends of the DNA loops are replication origins; the black circles along the sides are replication forks. Newly synthesized DNA is in color, with the red representing the most recently synthesized DNA. Membrane attachments are not shown for sake of simplicity.

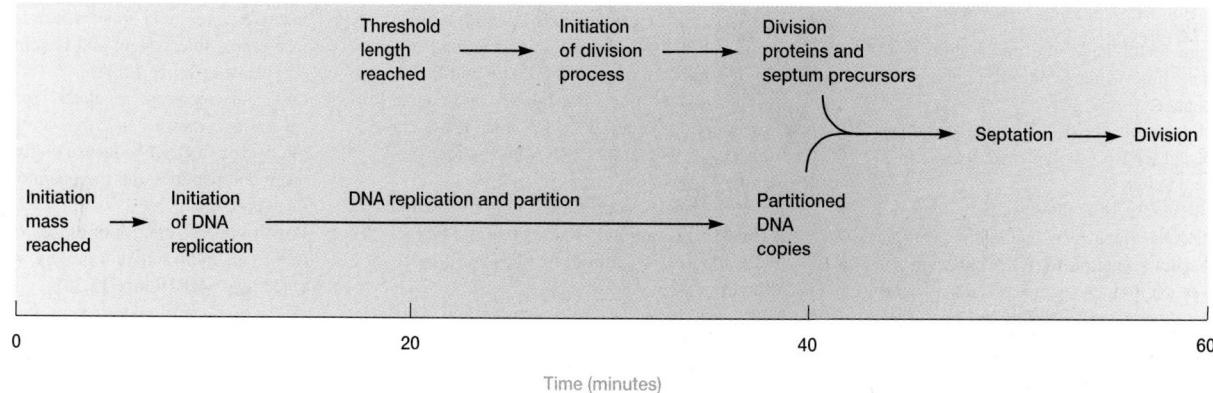

Figure 12.36   Control of the Cell Cycle in *E. coli*.   A 60-minute interval between divisions has been assumed for purposes of simplicity (the actual time between cell divisions may be shorter). *E. coli* requires about 40 minutes to replicate its DNA and 20 minutes after termination of replication to prepare for division. The position of events on the time line is approximate and meant to show the general pattern of occurrences.

DnaA protein has bound ATP, and the interconversion between DnaA-ATP and DnaA-ADP may help regulate initiation. Other factors also appear to participate in initiating DNA replication. After DNA replication is under way, another round does not immediately begin, partly because the parental DNA strand is methylated right after replication. The methylated replication origin binds to specific areas on the plasma membrane and is inactive.

The initiation of septation is equally complex and tightly regulated. Both termination of DNA replication and the attainment of threshold length are required to trigger septation and cell divi-

sion. This is at least partly due to the inhibition of septation by the proximity of chromosomes. The presence of DNA damage inhibits septation as well. A cell will complete chromosome replication, repair any DNA damage, and partition the chromosomes into opposite ends before it forms a septum and divides. There probably are one or more regulatory proteins that interact with various division proteins to promote septum formation and division. An adequate supply of peptidoglycan chains or precursors also must be available at the proper time. The FtsZ protein is a division protein essential in initiating septation. This protein is scattered throughout the cell between divisions. At the onset of septation, it forms a Z ring at the septation site, and the ring then becomes smaller. Because the FtsZ protein hydrolyzes GTP, the ring is a contractile structure that uses GTP energy; it also determines the placement of the septum. Several other proteins also are required for cell division. For example, the PBP3 or penicillin-binding protein 3 catalyzes peptidoglycan transglycosylation and peptidoglycan transpeptidation to help create the new cell wall (*see pp. 216–17*). Clearly regulation of the bacterial cell cycle is complex and involves several interacting regulatory mechanisms.

The relationship of DNA synthesis to the cell cycle varies with the growth rate. If *E. coli* is growing with a doubling time of about 60 minutes, DNA replication does not take place during the last 20 minutes—that is, replication is a discontinuous process when the doubling time is 60 minutes or longer. When the culture is growing with a doubling time of less than 60 minutes, a second

round of replication begins while the first round is still under way (figure 12.35b). The daughter cells may actually receive DNA with two or more replication forks, and replication is continuous because the cells are always copying their DNA.

Two decades of research have provided a fairly adequate overall picture of the cell cycle in *E. coli*. Several cell division genes have been identified. Yet it still is not known precisely how the cycle is controlled. Future work should improve our understanding of this important process.

1. What is a cell cycle? Briefly describe how the cycle in *E. coli* is regulated and how cycle timing results.
2. How are the two DNA copies separated and apportioned between the two daughter cells?

Summary

12.1 DNA Transcription or RNA Synthesis

a. Procaryotic mRNA has nontranslated leader and trailer sequences. Spacer regions exist between genes when mRNA is polygenic.

b. RNA is synthesized by RNA polymerase that copies the sequence of the DNA template strand (**figure 12.2**).

c. The sigma factor helps the procaryotic RNA polymerase bind to the promoter region at the start of a gene.

d. A terminator marks the end of a gene. A rho factor is needed for RNA polymerase release from some terminators.

e. RNA polymerase II synthesizes heterogeneous nuclear RNA, which then undergoes posttranscriptional modification by RNA cleavage and addition of a 3′ poly-A sequence and a 5′ cap to generate eucaryotic mRNA (**figure 12.6**).

f. Many eucaryotic genes are split or interrupted genes that have exons and introns. Exons are joined by RNA splicing. Splicing involves small nuclear RNA molecules, spliceosomes, and sometimes ribozymes.

12.2 Protein Synthesis

a. In translation, ribosomes attach to mRNA and synthesize a polypeptide beginning at the N-terminal end. A polysome or polyribosome is a complex of mRNA with several ribosomes.

b. Amino acids are activated for protein synthesis by attachment to the 3′ end of transfer RNAs. Activation requires ATP, and the reaction is catalyzed by aminoacyl-tRNA synthetases.

c. Ribosomes are large, complex organelles composed of rRNAs and many polypeptides. Amino acids are added to a growing peptide chain at the translational domain.

d. Protein synthesis begins with the binding of fMet-tRNA (procaryotes) or an initiator methionyl-tRNA^{Met} (eucaryotes) to an initiator

codon on mRNA and to the two ribosomal subunits. This involves the participation of protein initiation factors (**figure 12.15**).

e. In the elongation cycle the proper aminoacyl-tRNA binds to the A site with the aid of EF-Tu and GTP (**figure 12.16**). Then the transpeptidation reaction is catalyzed by peptidyl transferase. Finally, during translocation, the peptidyl-tRNA moves to the P site and the ribosome travels along the mRNA one codon. Translocation requires GTP and EF-G or translocase. The empty tRNA leaves the ribosome by way of the exit site.

f. Protein synthesis stops when a nonsense codon is reached. Procaryotes require three release factors for codon recognition and ribosome dissociation from the mRNA.

g. Molecular chaperones help proteins fold properly, protect cells against environmental stresses, and transport proteins across membranes.

h. Procaryotic proteins may not fold until completely synthesized, whereas eucaryotic protein domains fold as they leave the ribosome. Some proteins are self-splicing and excise portions of themselves before folding into their final shape.

12.3 Regulation of mRNA Synthesis

a. β-Galactosidase is an inducible enzyme whose concentration rises in the presence of its inducer.

b. Many biosynthetic enzymes are repressible enzymes whose levels are reduced in the presence of end products called corepressors.

c. Induction and repression result from regulation of the rate of transcription by repressor proteins coded for by regulator genes. This is an example of negative control. A regulator gene mutation can lead to a constitutive mutant, which continuously produces a metabolite.

d. The repressor inhibits transcription by binding to an operator and interfering with the binding of RNA polymerase to its promoter.

e. In inducible systems the newly synthesized repressor protein is active, and inducer binding inactivates it (**figure 12.23**). In contrast, an inactive repressor or aporepressor is synthesized in a repressible system and is activated by the corepressor (**figure 12.24**).

f. Often one repressor regulates the synthesis of several enzymes because they are part of a single operon, a DNA sequence coding for one or more polypeptides and the operator controlling its expression.

g. Positive operon control of the *lac* operon is due to the catabolite activator protein, which is activated by cAMP (**figure 12.28**).

12.4 Attenuation

a. In the tryptophan operon a leader region lies between the operator and the first structural gene (**figure 12.30**). It codes for the synthesis of a leader peptide and contains an attenuator, a rho-independent termination site.

b. The synthesis of the leader peptide by a ribosome while RNA polymerase is transcribing the leader region regulates transcription: therefore the tryptophan operon is expressed only when there is insufficient tryptophan available. This mechanism of transcription control is called attenuation (**figure 12.31**).

12.5 Global Regulatory Systems

a. Global regulatory systems can control many operons simultaneously and help procaryotes respond rapidly to a wide variety of environmental challenges.

b. Catabolite repression probably contributes to diauxic growth when *E. coli* is cultured in the presence of both glucose and lactose.

c. The transcription of genes can be regulated by altering the promoters to which RNA polymerase binds by changing the available sigma factors. A good example is the control of sporulation.

12.6 Small RNAs and Regulation

a. Small RNA molecules can regulate gene expression, RNA processing, translation, and other cell processes. For example, antisense RNAs help control porin protein levels.

12.7 Two-Component Phosphorelay Systems

a. Two-component phosphorelay systems are signal transduction systems that use phosphoryl group transfers in regulation. They have a sensor kinase and a response regulator.

b. Sporulation and chemotaxis regulatory systems are two-component phosphorelay systems.

12.8 Control of the Cell Cycle

a. The complete sequence of events extending from the formation of a new cell through the next division is called the cell cycle.

b. The end of DNA replication is tightly linked to cell division, so division in *E. coli* usually takes place about 20 minutes after replication is finished (**figure 12.35**). Special division and regulatory proteins are involved.

c. In very rapidly dividing bacterial cells, a new round of DNA replication begins before the cells divide.

Key Terms

amino acid activation   260
aminoacyl or acceptor site (A site)   265
aminoacyl-tRNA synthetases   260
anticodon triplet   260
antisense RNA   278
aporepressor   271
attenuation   276
attenuator   274
catabolite activator protein (CAP)   273
catabolite repression   276
cell cycle   280
constitutive mutant   271
core enzyme   254
corepressor   271
3′, 5′-cyclic adenosine monophosphate (cAMP)   273
cyclic AMP receptor protein (CRP)   273

diauxic growth   276
domains   269
elongation cycle   265
elongation factors   265
exit site (E site)   265
exon   257
exteins   270
global regulatory systems   276
heat-shock proteins   268
heterogeneous nuclear RNA (hnRNA)   257
inducer   270
inducible enzyme   270
initiation factors   263
initiator codon   262
inteins   270
intron   257
leader region   274
leader sequence   254

methyl-accepting chemotaxis protein (MCP)   279
molecular chaperones   267
negative control   271
nonsense codons   265
operator   271
operon   271
peptidyl or donor site (P site)   265
peptidyl transferase   265
polyribosome   260
positive operon control   273
posttranscriptional modification   257
Pribnow box   255
promoter   254
protein splicing   269
regulon   276
release factors   265
repressible enzyme   271

repressor proteins   271
rho factor   257
ribosomal RNA (rRNA)   254
ribozyme   259
RNA polymerase   254
RNA splicing   257
sigma factor   254
small nuclear RNA (snRNA)   259
small RNAs (sRNAs)   278
spliceosome   259
split or interrupted genes   257
structural gene   271
terminator   255
transfer RNA (tRNA)   254
translocation   265
transpeptidation reaction   265
two-component phosphorelay system   279

Questions for Thought and Review

1. Describe how RNA polymerase transcribes procaryotic DNA. How does the polymerase know where to begin and end transcription?

2. How do eucaryotic RNA polymerases and promoters differ from those in procaryotes? In what ways does eucaryotic mRNA differ from procaryotic mRNA with respect to synthesis and structure? How does eucaryotic synthesis of rRNA and tRNA resemble that of mRNA? How does it differ?

3. Draw diagrams summarizing the sequence of events in the three stages of protein synthesis (initiation, elongation, and termination) and accounting for the energy requirements of translation.

4. Describe in some detail the organization of the regulatory systems responsible for induction and repression, and the mechanism of their operation.

5. How is *E. coli* able to use glucose exclusively when presented with a mixture of glucose and lactose?

6. Of what practical importance is attenuation in coordinating the synthesis of amino acids and proteins? Describe how attenuation activity would vary when protein synthesis suddenly rapidly accelerated, then later suddenly decelerated.

7. How does the timing of DNA replication seem to differ between slow-growing and fast-growing cells? Be able to account for the fact that bacterial cells may contain more than a single copy of DNA.

Critical Thinking Questions

1. Attenuation affects anabolic pathways, whereas repression can affect either anabolic or catabolic pathways. Provide an explanation for this.

2. Many people say that RNA was the first of the information molecules (RNA, DNA, protein) to arise during evolution. Given the information in this chapter, what evidence is there to support this hypothesis?

3. Compare and contrast RNA and DNA synthesis.

For recommended readings on these and related topics, see Appendix V.

chapter 13

Microbial Recombination and Plasmids

Concepts

- Recombination is a one-way process in procaryotes: a piece of genetic material (the exogenote) is donated to the chromosome of a recipient cell (the endogenote) and integrated into it.
- The actual transfer of genetic material between bacteria usually takes place in one of three ways: direct transfer between two bacteria temporarily in physical contact (conjugation), transfer of a naked DNA fragment (transformation), or transport of bacterial DNA by bacteriophages (transduction).
- Plasmids and transposable elements can move genetic material between procaryotic chromosomes and within chromosomes to cause rapid changes in genomes and drastically alter phenotypes.
- The bacterial chromosome can be mapped with great precision, using Hfr conjugation in combination with transformational and transductional mapping techniques.
- Recombination of virus genomes occurs when two viruses with homologous chromosomes infect a bacterial host cell at the same time.

Deep in the cavern of the infant's breast
The father's nature lurks, and lives anew.

—Horace, Odes

The scanning electron micrograph shows *Streptococcus pneumoniae*, the bacterium first used to study transformation and obtain evidence that DNA is the genetic material of organisms.

hapter 12 introduces the fundamentals of molecular genetics: the way genetic information is organized and stored, the nature of mutations and techniques for their isolation and study, and DNA repair. This chapter focuses on genetic recombination in microorganisms, with primary emphasis placed on recombination in bacteria and viruses.

The chapter begins with a general overview of bacterial recombination and an introduction to both bacterial plasmids and transposable elements. Next, the three types of bacterial gene transfer—conjugation, transformation, and transduction—are discussed. Because an understanding of the techniques used to locate genes on chromosomes depends on knowledge of recombination mechanisms, mapping the bacterial genome is discussed after the introduction to recombination. The chapter ends with a description of recombination in viruses and a brief discussion of viral chromosome mapping.

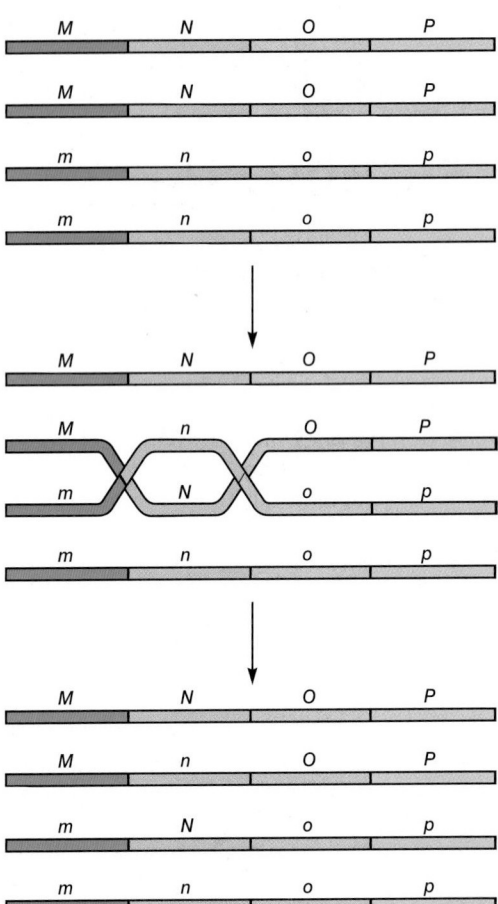

Figure 13.1 Crossing-Over. An example of recombination through crossing-over between homologous eucaryotic chromosomes. The *Nn* gene pair is exchanged. This process usually occurs during meiosis.

In a general sense, **recombination** is the process in which one or more nucleic acid molecules are rearranged or combined to produce a new nucleotide sequence. Usually genetic material from two parents is combined to produce a recombinant chromosome with a new, different genotype. Recombination results in a new arrangement of genes or parts of genes and normally is accompanied by a phenotypic change. Most eucaryotes exhibit a complete sexual life cycle, including meiosis, a process of extreme importance in generating new combinations of alleles (alternate forms of a particular gene) through recombination. These chromosome exchanges during meiosis result from **crossing-over** between homologous chromosomes, chromosomes containing identical sequences of genes (**figure 13.1**). Until about 1945 the primary focus in genetic analysis was on the recombination of genes in plants and animals. The early work on recombination in higher eucaryotes laid the foundations of classical genetics, but it was the development of bacterial and phage genetics between about 1945 and 1965 that really stimulated a rapid advance in our understanding of molecular genetics. Meiosis (pp. 86–87)

13.1 BACTERIAL RECOMBINATION: GENERAL PRINCIPLES

Microorganisms carry out several types of recombination. **General recombination,** the most common form, usually involves a reciprocal exchange between a pair of homologous DNA sequences. It can occur anyplace on the chromosome, and it results from DNA strand breakage and reunion leading to crossing-over (**figure 13.2**). General recombination is carried out by the products of *rec* genes such as the recA protein so important for DNA repair (*see pp. 248–49*). In bacterial transformation a nonreciprocal form of general recombination takes place (**figure 13.3**). A piece of genetic material is inserted into the chromosome through the incorporation of a single strand to form a stretch of **het-**

eroduplex DNA. A second type of recombination, one particularly important in the integration of virus genomes into bacterial chromosomes, is **site-specific recombination.** The genetic material is not homologous with the chromosome it joins, and generally the enzymes responsible for this event are specific for the particular virus and its host. A third kind of recombination, which may be considered a type of site-specific recombination, is called **replicative recombination.** It accompanies the replication of genetic material and does not depend on sequence homology. It is used by some genetic elements that move about the chromosome. DNA replication (pp. 229–33)

Although sexual reproduction with the formation of a zygote and subsequent meiosis is not present in bacteria, recombination can take place in several ways following **horizontal gene transfer.** In this process genes are transferred from one independent, mature organism to another. Horizontal gene transfer is quite different from the transmission of genes from parents to offspring (vertical gene transfer). In general, a piece of donor DNA, the **exogenote,** must enter the recipient cell and become a stable part of the recipient cell's genome, the **endogenote.** Two kinds of DNA

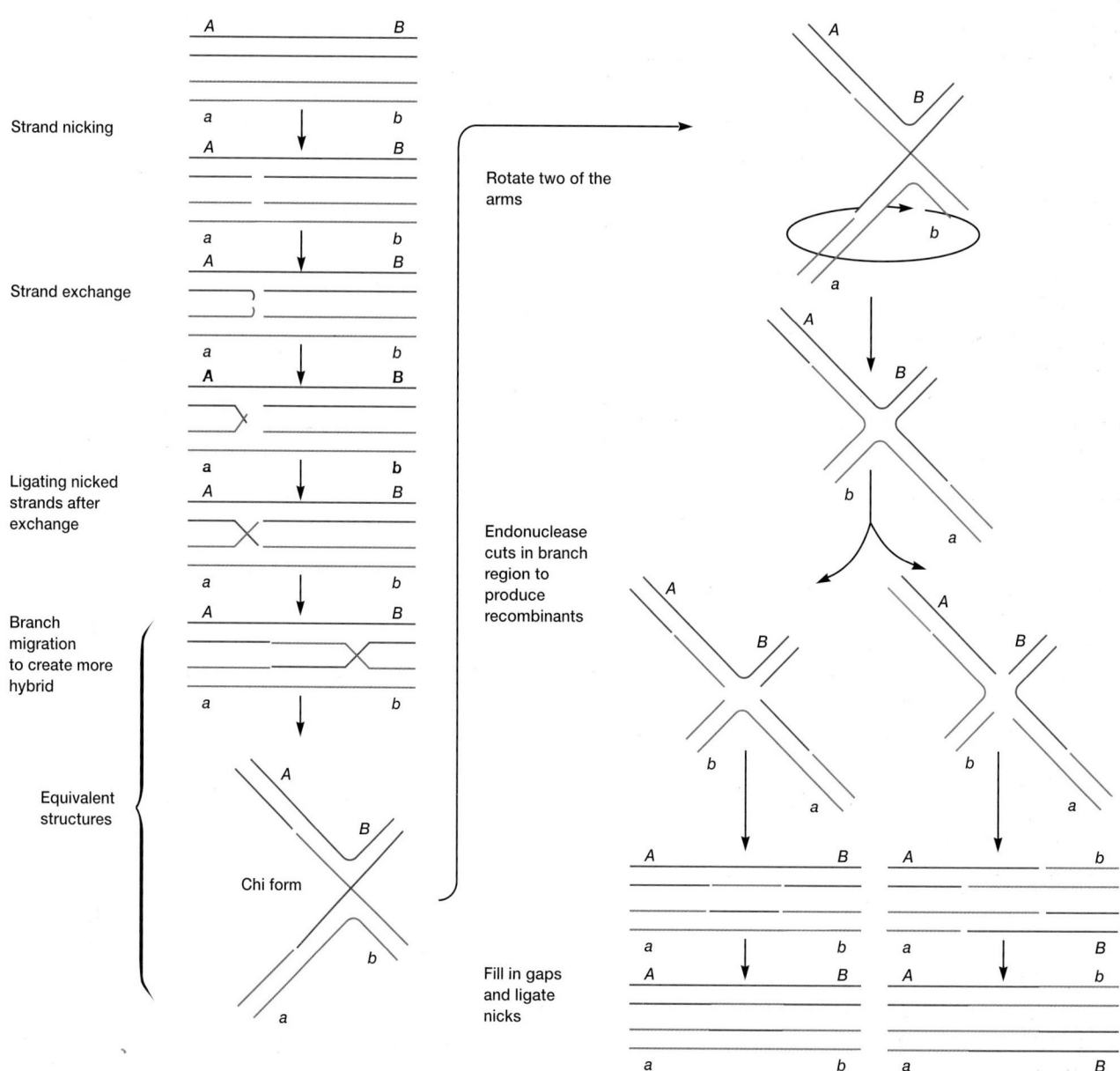

Figure 13.2   The Holliday Model for Reciprocal General Recombination.   Source: From H. Potter and D. Dressler, *Proceedings of the National Academy of Sciences,* 73:3000, 1976; after R. Holliday, *Genetics,* 78:273, 1974; and previous publications cited therein.

can move between bacteria. If a DNA fragment is the exchange vehicle, then the exogenote must get into the recipient cell and become incorporated into the endogenote as a replacement piece (or as an "extra" piece) without being destroyed by the host. During replacement of host genetic material, the recipient cell becomes temporarily diploid for a portion of the genome and is called a **merozygote (figure 13.4)**. Sometimes the DNA exists in a form that cannot be degraded by the recipient cell's endonucleases. In this case the DNA does not need to be integrated into the host genome but must only enter the recipient to confer its genetic information on the cell. Most linear DNA fragments are not stably maintained unless they have been integrated into the bacter-

ial genome. Resistant DNA, such as that in plasmids (see following), usually is circular and has sequences that allow it to maintain itself independent of the host chromosome. Recombination in bacteria is a one-way gene transfer from donor to recipient. Recombination in eucaryotes is reciprocal—that is, all of the DNA is conserved in the gametes that eventually arise from meiosis and recombination.

Movement of DNA from a donor bacterium to the recipient can take place in three ways: direct transfer between two bacteria temporarily in physical contact (conjugation), transfer of a naked DNA fragment (transformation), and transport of bacterial DNA by bacteriophages (transduction). Whatever the mode

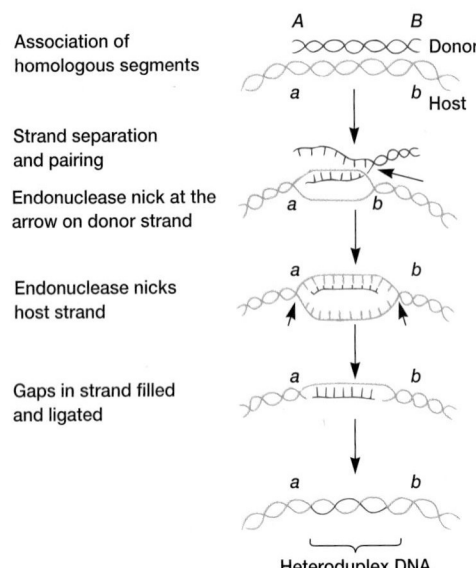

Figure 13.3 Nonreciprocal General Recombination.
The Fox model for nonreciprocal general recombination. This mechanism has been proposed for the recombination occurring during transformation in some bacteria.

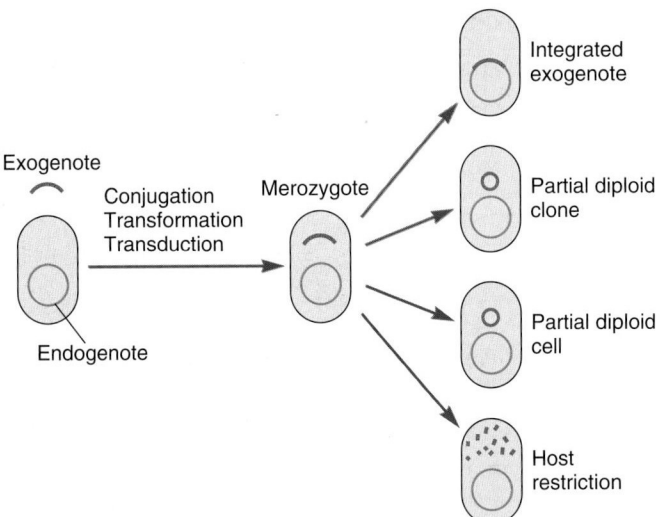

Figure 13.4 The Production and Fate of Merozygotes.
See text for discussion.

of transfer, the exogenote has only four possible fates in the recipient (figure 13.4). First, when the exogenote has a sequence homologous to that of the endogenote, integration may occur; that is, it may pair with the recipient DNA and be incorporated to yield a recombinant genome. Second, the foreign DNA sometimes persists outside the endogenote and replicates to produce a clone of partially diploid cells. Third, the exogenote may survive, but not replicate, so that only one cell is a partial diploid. Finally, host cell nucleases may degrade the exogenote, a process called **host restriction.**

1. Define the following terms: recombination, crossing-over, general recombination, site-specific recombination, replicative recombination, exogenote, endogenote, horizontal gene transfer, merozygote, and host restriction.
2. Distinguish among the three forms of recombination mentioned in this section.
3. What four fates can DNA have after entering a bacterium?

13.2 BACTERIAL PLASMIDS

Conjugation, the transfer of DNA between bacteria involving direct contact, depends on the presence of an "extra" piece of circular DNA known as a plasmid. Plasmids play many important roles in the lives of bacteria. They also have proved invaluable to microbiologists and molecular geneticists in constructing and transferring new genetic combinations and in cloning genes (*see chapter 14*). In this section the different types of bacterial plasmids are discussed.

Plasmids are small double-stranded DNA molecules, usually circular, that can exist independently of host chromosomes and are present in many bacteria (they are also present in some yeasts and other fungi). They have their own replication origins and are autonomously replicating and stably inherited. A **replicon** is a DNA molecule or sequence that has a replication origin and is capable of being replicated. Plasmids and bacterial chromosomes are separate replicons. Plasmids have relatively few genes, generally less than 30. Their genetic information is not essential to the host, and bacteria that lack them usually function normally. Single-copy plasmids produce only one copy per host cell. Multicopy plasmids may be present at concentrations of 40 or more per cell.

Characteristically, plasmids can be eliminated from host cells in a process known as **curing.** Curing may occur spontaneously or be induced by treatments that inhibit plasmid replication while not affecting host cell reproduction. The inhibited plasmids are slowly diluted out of the growing bacterial population. Some commonly used curing treatments are acridine mutagens, UV and ionizing radiation, thymine starvation, and growth above optimal temperatures.

Plasmids may be classified in terms of their mode of existence and spread. An **episome** is a plasmid that can exist either with or without being integrated into the host's chromosome. Some plasmids, **conjugative plasmids,** have genes for pili and can transfer copies of themselves to other bacteria during conjugation. A brief summary of the types of bacterial plasmids and their properties is given in **table 13.1.**

Fertility Factors

A plasmid called the fertility or **F factor** plays a major role in conjugation in *E. coli* and was the first to be described (**figure 13.5**). The F factor is about 100 kilobases long and bears genes responsible for cell attachment and plasmid transfer between specific bacterial strains during conjugation. Most of the information required for plasmid transfer is located in the *tra* operon, which

Table 13.1

Major Types of Bacterial Plasmids

Type	Representatives	Approximate Size (kbp)	Copy Number (Copies/Chromosome)	Hosts	Phenotypic Features[a]
Fertility Factor[b]	F factor	95–100	1–3	*E. coli, Salmonella, Citrobacter*	Sex pilus, conjugation
R Plasmids	RP4	54	1–3	*Pseudomonas* and many other gram-negative bacteria	Sex pilus, conjugation, resistance to Ap, Km, Nm, Tc
	R1	80	1–3	Gram-negative bacteria	Resistance to Ap, Km, Su, Cm, Sm
	R6	98	1–3	*E. coli, Proteus mirabilis*	Su, Sm, Cm, Tc, Km, Nm
	R100	90	1–3	*E. coli, Shigella, Salmonella, Proteus*	Cm, Sm, Su, Tc, Hg
	pSH6	21		*Staphylococcus aureus*	Gm, Tm, Km
	pSJ23a	36		*S. aureus*	Pn, Asa, Hg, Gm, Km, Nm, Em, etc.
	pAD2	25		*Enterococcus faecalis*	Em, Km, Sm
Col Plasmids	ColE1	9	10–30	*E. coli*	Colicin E1 production
	ColE2		10–15	*Shigella*	Colicin E2
	CloDF13			*Enterobacter cloacae*	Cloacin DF13
Virulence Plasmids	Ent (P307)	83		*E. coli*	Enterotoxin production
	K88 plasmid			*E. coli*	Adherence antigens
	ColV-K30	2		*E. coli*	Siderophore for iron uptake; resistance to immune mechanisms
	pZA10	56		*S. aureus*	Enterotoxin B
	Ti	200		*Agrobacter tumefaciens*	Tumor induction
Metabolic Plasmids	CAM	230		*Pseudomonas*	Camphor degradation
	SAL	56		*Pseudomonas*	Salicylate degradation
	TOL	75		*Pseudomonas putida*	Toluene degradation
	pJP4			*Pseudomonas*	2,4-dichlorophenoxyacetic acid degradation
				E. coli, Klebsiella, Salmonella	Lactose degradation
				Providencia	Urease
	sym			*Rhizobium*	Nitrogen fixation and symbiosis

[a]Abbreviations used for resistance to antibiotics and metals: Ap, ampicillin; Asa, arsenate; Cm, chloramphenicol; Em, erythromycin; Gm, gentamycin; Hg, mercury; Km, kanamycin; Nm, neomycin; Pn, penicillin; Sm, streptomycin; Su, sulfonamides; Tc, tetracycline.
[b]Many R plasmids, metabolic plasmids, and others are also conjugative.

contains at least 28 genes. Many of these direct the formation of sex pili that attach the F⁺ cell (the donor cell containing an F plasmid) to an F⁻ cell (**figure 13.6**). Other gene products aid DNA transfer. Sex pili (p. 63)

The F factor also has several segments called insertion sequences (p. 292) that assist plasmid integration into the host cell chromosome. Thus the F factor is an episome that can exist outside the bacterial chromosome or be integrated into it (**figure 13.7**).

Resistance Factors

Plasmids often confer antibiotic resistance on the bacteria that contain them. **R factors** or plasmids typically have genes that code for enzymes capable of destroying or modifying antibiotics. They are not usually integrated into the host chromosome. Genes coding for resistance to antibiotics such as ampicillin, chloramphenicol, and kanamycin have been found in plasmids. Some R plasmids have only a single resistance gene, whereas others can have as many as eight. Often the resistance genes are within a transposon (p. 291), and thus it is possible for bacterial strains to rapidly develop multiple resistance plasmids. R factors and antibiotic resistance (pp. 792–94)

Because many R factors also are conjugative plasmids, they can spread throughout a population, although not as rapidly as the F factor. Often, nonconjugative R factors also move between bacteria during plasmid promoted conjugation. Thus a whole population

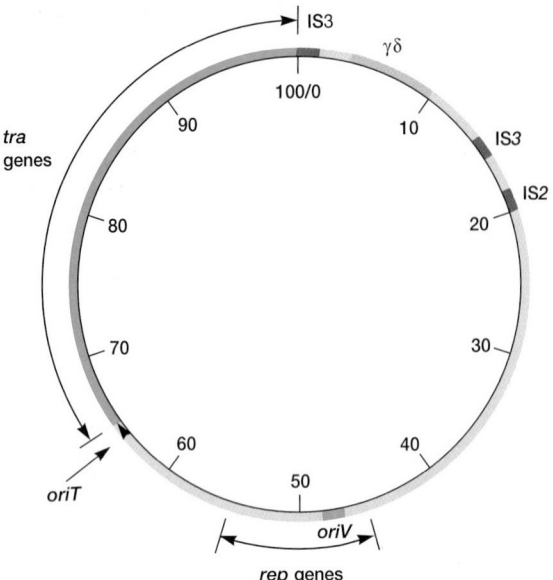

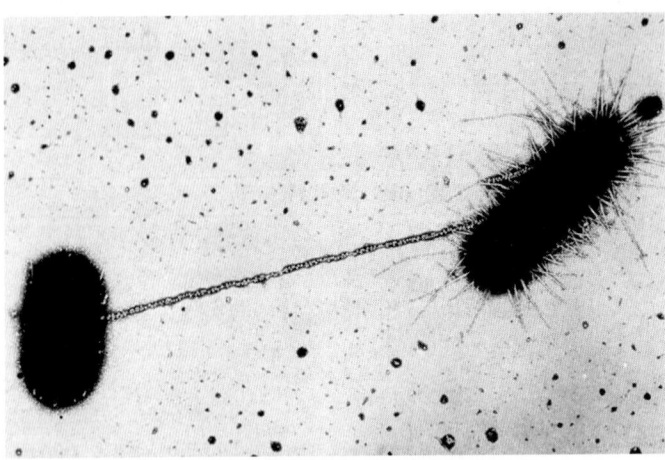

Figure 13.6 Bacterial Conjugation. An electron micrograph of two *E. coli* cells in an early stage of conjugation. The F⁺ cell to the right is covered with small pili or fimbriae, and a sex pilus connects the two cells.

Figure 13.5 The F plasmid. A map showing the size and general organization of the F plasmid. The plasmid contains several transposable elements. IS2 and IS3 are insertion sequences; γδ is also called transposon Tn*1000*. The *tra* genes code for proteins needed in pilus synthesis and conjugation. The *rep* genes code for proteins involved in DNA replication. *OriV* is the initiation site for circular DNA replication and *oriT,* the site for initiation of rolling circle replication and gene transfer during conjugation.

Figure 13.7 F Plasmid Integration. The reversible integration of an F plasmid or factor into a host bacterial chromosome. The process begins with association between plasmid and bacterial insertion sequences. The O arrowhead (blue-green) indicates the site at which the oriented transfer of chromosome to the recipient cell begins. A, B, 1, and 2 represent genetic markers.

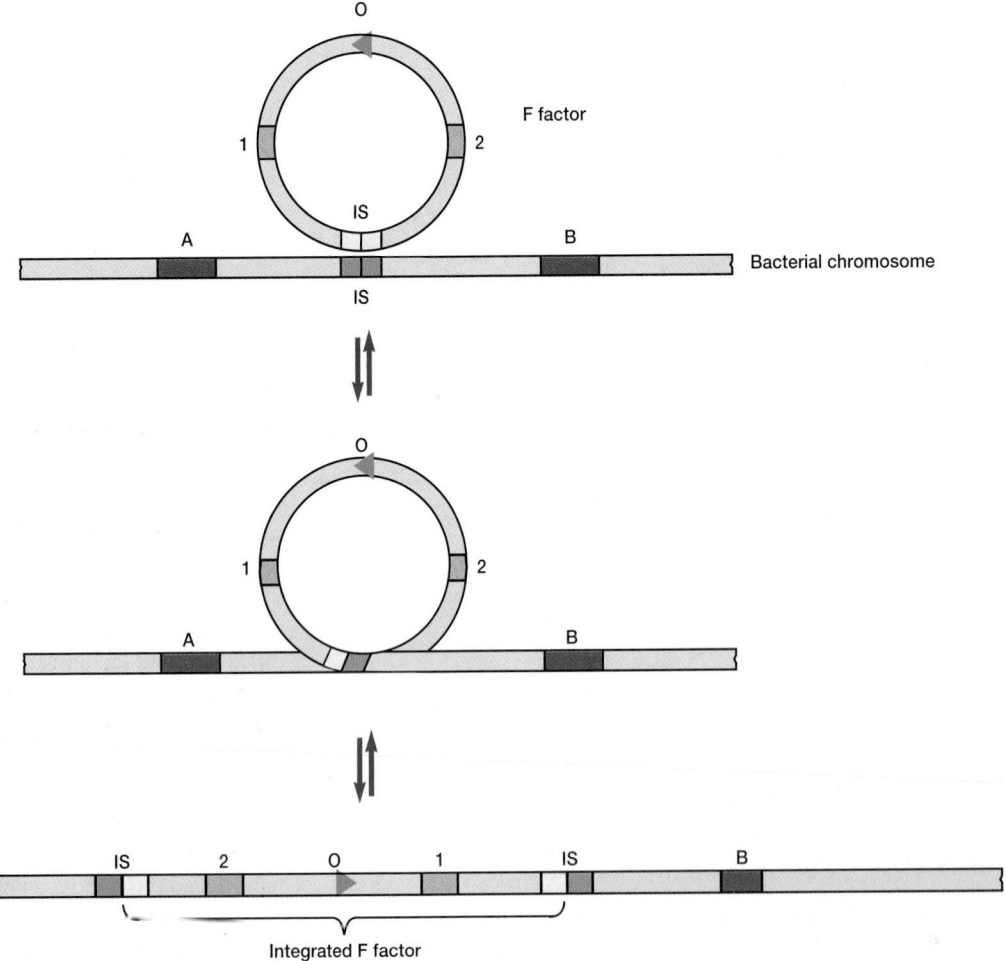

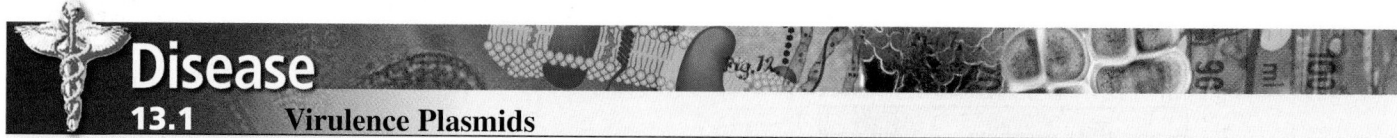

Disease
13.1   Virulence Plasmids

It is becoming increasingly evident that many bacteria are pathogenic because of their plasmids. These plasmids can carry genes for toxins, render the bacterium better able to establish itself in the host, or aid in resistance to host defenses. *E. coli* provides the best-studied example of virulence plasmids. Several strains of *E. coli* cause diarrhea. The enterotoxigenic strains responsible for traveler's diarrhea can produce two toxins: a heat-labile toxin (LT), which is a large protein very similar in structure and mechanism of action to cholera toxin (*see chapter 34*), and a heat-stable toxin (ST), a low molecular weight polypeptide. Both toxin genes are plasmid borne, and sometimes they are even carried by the same plasmid. The ST toxin gene is located on a transposon. Enterotoxigenic strains of *E. coli* also must be able to colonize the epithelium of the small intestine to cause diarrhea. This is made possible by the presence of special adhesive fimbriae encoded by genes on another plasmid. A second type of pathogenic *E. coli* invades the intestinal epithelium and causes a form of diarrhea very similar to the dysentery resulting from a *Shigella* infection. This *E. coli* strain and *Shigella* contain virulence plasmids that code for

special cell wall antigens and other factors enabling them to enter and destroy epithelial cells.

Some *E. coli* strains can invade the blood and organs of a host, causing a generalized infection. These pathogens often have ColV plasmids and produce colicin V. The ColV plasmid carries genes for two virulence determinants. One product increases bacterial resistance to host defense mechanisms involving complement (*see sections 31.7 and 32.7*). The other plasmid gene directs the synthesis of a hydroxamate that enables *E. coli* to accumulate iron more efficiently from its surroundings (*see section 5.6*). Since iron is not readily available in the animal host, but is essential for bacterial growth, this is an important factor in pathogenicity.

Several other pathogens carry virulence plasmids. Some *Staphylococcus aureus* strains produce an exfoliative toxin that is plasmid borne. The toxin causes the skin to loosen and often peel off in sheets, leading to the disease staphylococcal scalded skin syndrome (*see section 39.3*). Other plasmid-borne toxins are the tetanus toxin of *Clostridium tetani* and the anthrax toxin of *Bacillus anthracis*.

can become resistant to antibiotics. The fact that some of these plasmids are readily transferred between species further promotes the spread of resistance. When the host consumes large quantities of antibiotics, *E. coli* and other bacteria with R factors are selected for and become more prevalent. The R factors can then be transferred to more pathogenic genera such as *Salmonella* or *Shigella*, causing even greater public health problems (*see section 35.7*).

Col Plasmids

Bacteria also harbor plasmids with genes that may give them a competitive advantage in the microbial world. **Bacteriocins** are bacterial proteins that destroy other bacteria. They usually act only against closely related strains. Some bacteriocins kill cells by forming channels in the plasma membrane, thus increasing its permeability. They also may degrade DNA and RNA or attack peptidoglycan and weaken the cell wall. Col plasmids contain genes for the synthesis of bacteriocins known as colicins, which are directed against *E. coli*. Similar plasmids carry genes for bacteriocins against other species. For example, Col plasmids produce cloacins that kill *Enterobacter* species. Clearly the host is unaffected by the bacteriocin it produces. Some Col plasmids are conjugative and also can carry resistance genes. It should be noted that not all bacteriocin genes are on plasmids. For example, the bacteriocin genes of *Pseudomonas aeruginosa*, which code for proteins called pyocins, are located on the chromosome.   Bacteriocins and host defenses (p. 688)

Other Types of Plasmids

Several other important types of plasmids have been discovered. Some plasmids, called **virulence plasmids,** make their hosts more pathogenic because the bacterium is better able to resist

host defense or to produce toxins. For example, enterotoxigenic strains of *E. coli* cause traveler's diarrhea because of a plasmid that codes for an enterotoxin (**Disease 13.1**). **Metabolic plasmids** carry genes for enzymes that degrade substances such as aromatic compounds (toluene), pesticides (2,4-dichlorophenoxyacetic acid), and sugars (lactose). Metabolic plasmids even carry the genes required for some strains of *Rhizobium* to induce legume nodulation and carry out nitrogen fixation.

1. Give the major distinguishing features of a plasmid. What is an episome? A conjugative plasmid?
2. Describe each of the following plasmids and their importance: F factor, R factor, Col plasmid, virulence plasmid, and metabolic plasmid.

 13.3   TRANSPOSABLE ELEMENTS

The chromosomes of bacteria, viruses, and eucaryotic cells contain pieces of DNA that move around the genome. Such movement is called **transposition.** DNA segments that carry the genes required for this process and consequently move about chromosomes are **transposable elements** or **transposons.** Unlike other processes that reorganize DNA, transposition does not require extensive areas of homology between the transposon and its destination site. Transposons behave somewhat like lysogenic prophages (*see pp. 380–84*) except that they originate in one chromosomal location and can move to a different location in the same chromosome. Transposable elements differ from phages in lacking a virus life cycle and from plasmids in being unable to reproduce

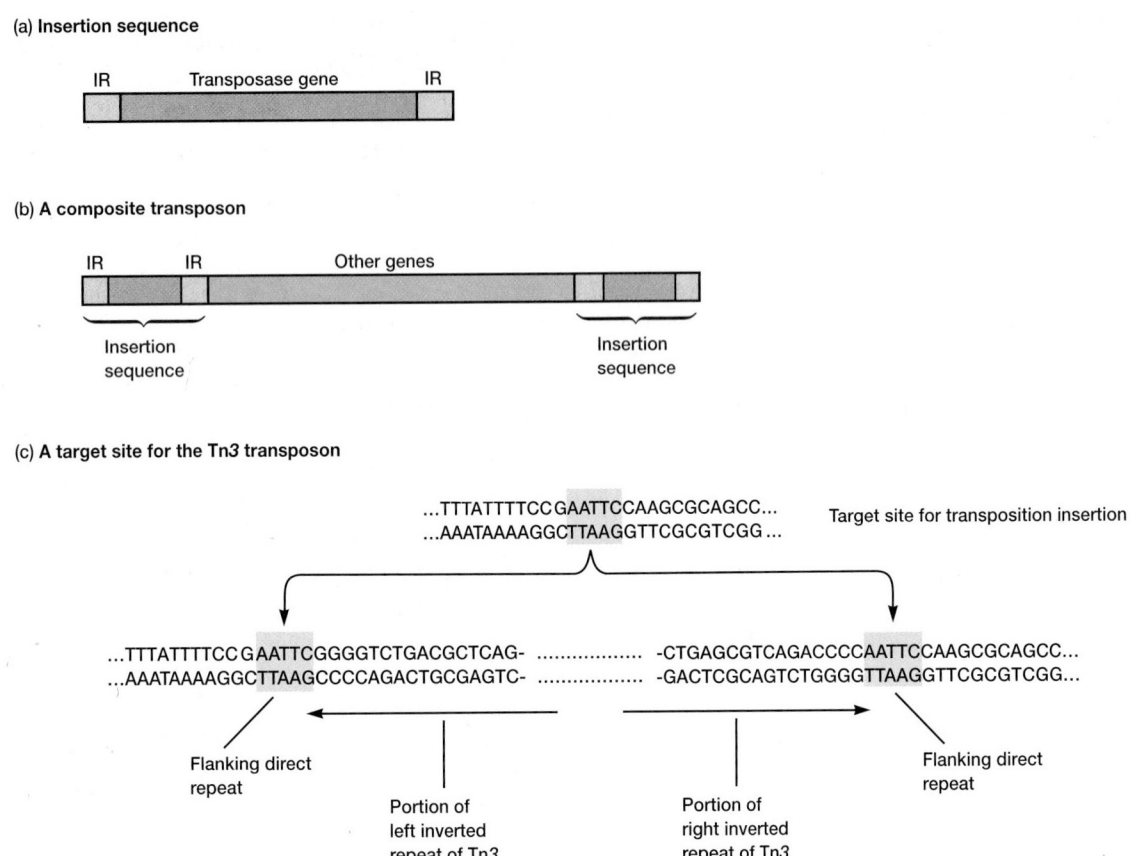

Figure 13.8 **Insertion Sequences and Transposons.** The structure of insertion sequences (**a**), composite transposons (**b**), and target sites (**c**). IR stands for inverted repeat. In (**c**), the highlighted five-base target site is duplicated during Tn*3* transposition to form flanking direct repeats. The remainder of Tn*3* lies between the inverted repeats.

autonomously and to exist apart from the chromosome. They were first discovered in the 1940s by Barbara McClintock during her studies on maize genetics (a discovery that won her the Nobel prize in 1983). They have been most intensely studied in bacteria.

The simplest transposable elements are **insertion sequences** or IS elements (**figure 13.8a**). An IS element is a short sequence of DNA (around 750 to 1,600 base pairs [bp] in length) containing only the genes for those enzymes required for its transposition and bounded at both ends by identical or very similar sequences of nucleotides in reversed orientation known as inverted repeats (figure 13.8c). Inverted repeats are usually about 15 to 25 base pairs long and vary among IS elements so that each type of IS has its own characteristic inverted repeats. Between the inverted repeats is a gene that codes for an enzyme called **transposase** (and sometimes a gene for another essential protein). This enzyme is required for transposition and accurately recognizes the ends of the IS. Each type of element is named by giving it the prefix IS followed by a number. In *E. coli* several copies of different IS elements have been observed; some of their properties are given in **table 13.2**.

Transposable elements also can contain genes other than those required for transposition (for example, antibiotic resist-

ance or toxin genes). These elements often are called **composite transposons** or elements. Complete agreement about the nomenclature of transposable elements has not yet been reached. Sometimes transposable elements are called transposons when they have extra genes, and insertion sequences when they lack these. Composite transposons often consist of a central region containing the extra genes, flanked on both sides by IS elements that are identical or very similar in sequence (figure 13.8b). Many composite transposons are simpler in organization. They are bounded by short inverted repeats, and the coding region contains both transposition genes and the extra genes. It is believed that composite transposons are formed when two IS elements associate with a central segment containing one or more genes. This association could arise if an IS element replicates and moves only a gene or two down the chromosome. Composite transposon names begin with the prefix Tn. Some properties of selected composites are given in **table 13.3**.

The process of transposition in procaryotes involves a series of events, including self-replication and recombinational processes. Typically in bacteria, the original transposon remains at the parental site on the chromosome, while a replicated copy inserts at the target DNA (figure 13.8c). This is called replicative

Table 13.2
The Properties of Selected Insertion Sequences

Insertion Sequence	Length (bp)	Inverted Repeat (Length in bp)	Target Site (Length in bp)	Number of Copies on *E. coli* Chromosome
IS*1*	768	23	9 or 8	6–10
IS*2*	1,327	41	5	4–13(1)[a]
IS*3*	1,400	38	3–4	5–6(2)
IS*4*	1,428	18	11 or 12	1–2
IS*5*	1,195	16	4	10–11

[a]The value in parentheses indicates the number of IS elements on the F factor.

Table 13.3
The Properties of Selected Composite Transposons

Transposon	Length (bp)	Terminal Repeat Length	Terminal Module	Genetic Markers[a]
Tn*3*	4,957	38		Ap
Tn*501*	8,200	38		Hg
Tn*951*	16,500	Unknown		Lactose utilization
Tn*5*	5,700		IS*50*	Km
Tn*9*	2,500		IS*1*	Cm
Tn*10*	9,300		IS*10*	Tc
Tn*903*	3,100		IS*903*	Km
Tn*1681*	2,061		IS*1*	Heat-stable enterotoxin
Tn*2901*	11,000		IS*1*	Arginine biosynthesis

[a]Abbreviations for antibiotics and metals same as in table 13.1.

transposition. Target sites are specific sequences about five to nine base pairs long. When a transposon inserts at a target site, the target sequence is duplicated so that short, direct-sequence repeats flank the transposon's terminal inverted repeats (**figure 13.9**). This can be seen in figure 13.8*c* where the five base pair target sequence moves to both ends of the transposon and retains the same orientation.

The transposition of the Tn*3* transposon is a well-studied example of replicative transposition. Its mechanism is outlined in **figure 13.10.** In the first stage the plasmid containing the Tn*3* transposon fuses with the target plasmid to form a cointegrate molecule (figure 13.10, steps 1 to 4). This process requires the Tn*3* transposase enzyme coded for by the *tnpA* gene (**figure 13.11**). Note that the cointegrate has two copies of the Tn*3* transposon. In the second stage the cointegrate is resolved to yield two plasmids, each with a copy of the transposon (figure 13.10, steps 5 and 6). Resolution involves a crossover at the two *res* sites and is catalyzed by a resolvase enzyme coded for by the *tnpR* gene (figure 13.11).

Transposable elements produce a variety of important effects. They can insert within a gene to cause a mutation or stimulate DNA rearrangement, leading to deletions of genetic material. If a transposon insertion produces an obvious pheno-

typic change, the gene can be tracked by following this altered phenotype. One can fragment the genome and isolate the mutated fragment, thereby partially purifying the gene. Thus transposons may be used to purify genes and study their functions. Because some transposons carry stop codons or termination sequences, they may block translation or transcription. Other elements carry promoters and thus activate genes near the point of insertion. Eucaryotic genes as well as procaryotic genes can be turned on and off by transposon movement. Transposons also are located within plasmids and participate in such processes as plasmid fusion and the insertion of F plasmids into the *E. coli* chromosome (figure 13.7).

In the previous discussion of plasmids, it was noted that an R plasmid can carry genes for resistance to several drugs. Transposons have antibiotic resistance genes and play a major role in generating these plasmids. Consequently the existence of these elements causes serious problems in the treatment of disease. Since plasmids can contain several different target sites, transposons will move between them; thus plasmids act as both the source and the target for transposons with resistance genes. In fact, multiple drug resistance plasmids probably often arise from transposon accumulation on a single plasmid (figure 13.11). Because transposons also move between plasmids and primary chromosomes,

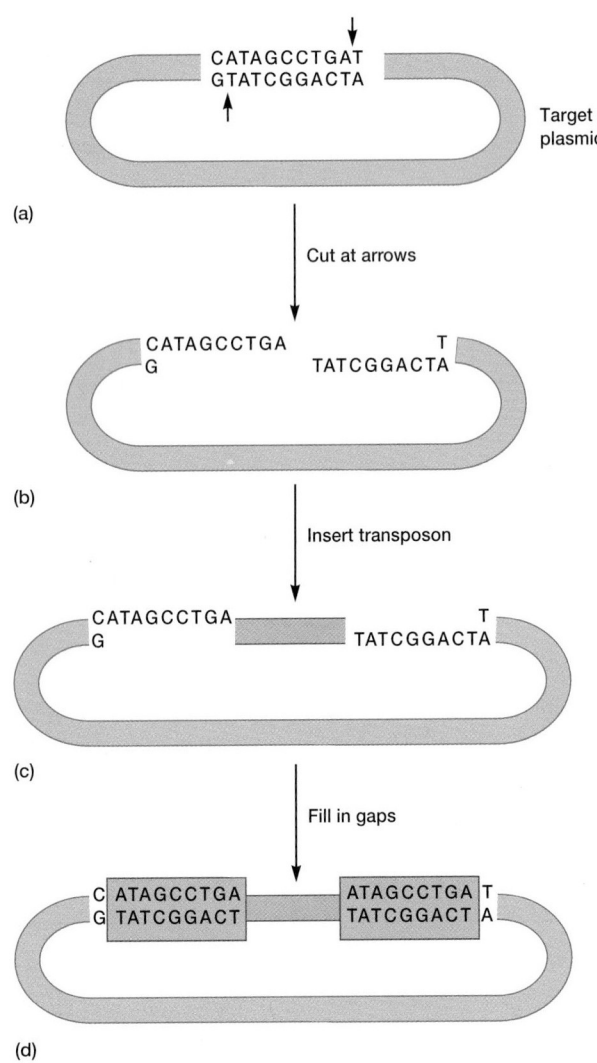

Figure 13.9 Generation of Direct Repeats in Host DNA Flanking a Transposon. (a) The arrows indicate where the two strands of host DNA will be cut in a staggered fashion, 9 base pairs apart. (b) After cutting. (c) The transposon (pink) has been ligated to one strand of host DNA at each end, leaving two 9 base gaps. (d) After the gaps are filled in, there are 9 base pair repeats of host DNA (purple boxes) at each end of the transposon.

Figure 13.10 Tn3 Transposition Mechanism. Step 1: The two plasmids are nicked to form the free ends labeled a–h. Step 2: Ends a and f are joined, as are g and d. This leaves b, c, e, and h free. Step 3: Two of these remaining free ends (b and c) serve as primers for DNA replication, which is shown in a blowup of the replicating region. Step 4: Replication continues until end b reaches e and end c reaches h. These ends are ligated to complete the cointegrate. Notice that the whole transposon has been replicated. The paired *res* sites are shown for the first time here, even though one *res* site existed in the previous steps. Steps 5 and 6: A crossover occurs between the two *res* sites in the two copies of the transposon, leaving two independent plasmids, each bearing a copy of the transposon.

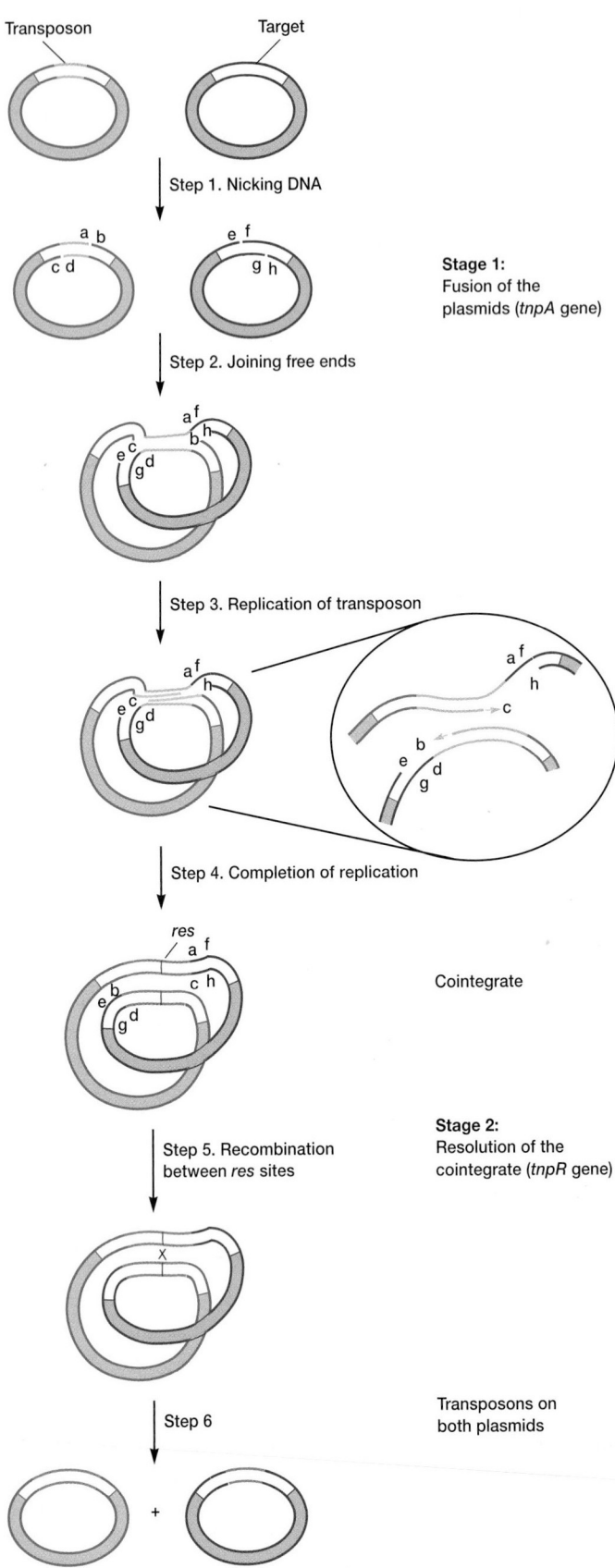

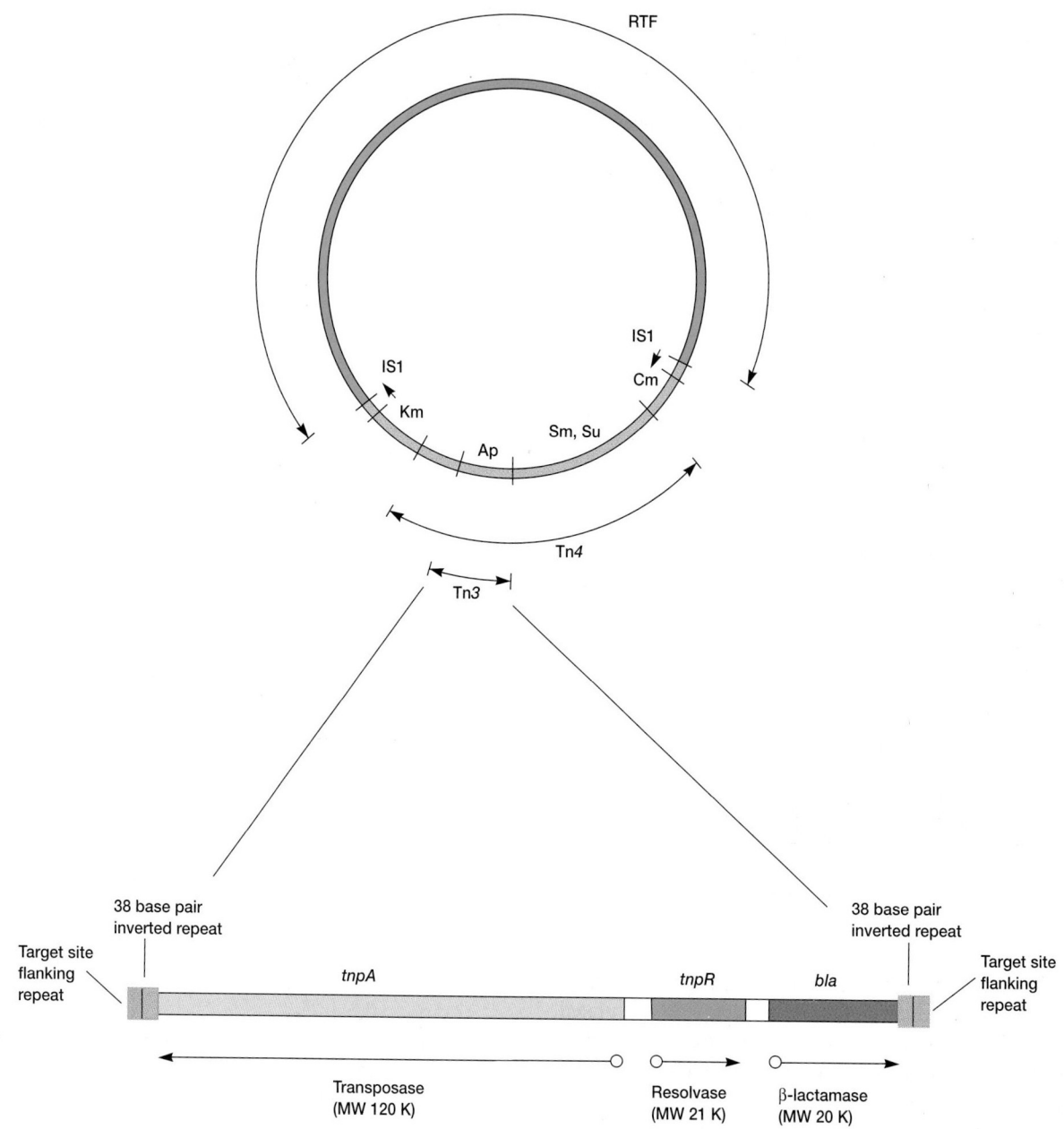

Figure 13.11 The Structure of R Plasmids and Transposons. The R1 plasmid carries resistance genes for five antibiotics: chloramphenicol (Cm), streptomycin (Sm), sulfonamide (Su), ampicillin (Ap), and kanamycin (Km). These are contained in the Tn3 and Tn4 transposons. The resistance transfer factor (RTF) codes for the proteins necessary for plasmid replication and transfer. The structure of Tn3 is shown in more detail. The arrows indicate the direction of gene transcription.

drug resistance genes can exchange between plasmids and chromosomes, resulting in the further spread of antibiotic resistance.

Some transposons bear transfer genes and can move between bacteria through the process of conjugation, as discussed in the next section. A well-studied example of such a **conjugative transposon** is Tn916 from *Enterococcus faecalis.* Although Tn916 cannot replicate autonomously, it will transfer itself from *E. faecalis* to a variety of recipients and integrate into their chro-

mosomes. Because it carries a gene for tetracycline resistance, this conjugative transposon also spreads drug resistance.

Transposable elements are widespread in nature. They are present in eucaryotes, bacteria, and the Archaea. For example, transposable elements have been found in yeast, maize, *Drosophila,* and humans. Clearly, transposable elements play an extremely important role in the generation and transfer of new gene combinations.

---

1. Define the following terms: transposition, transposable element or transposon, insertion sequence, transposase, conjugative transposon, and composite transposon. Be able to distinguish between an insertion sequence and a composite transposon.
2. How does transposition usually occur in bacteria, and what happens to the target site? What is replicative transposition?
3. Give several important effects transposable elements have on bacteria.

13.4 BACTERIAL CONJUGATION

The initial evidence for bacterial **conjugation,** the transfer of genetic information by direct cell to cell contact, came from an elegant experiment performed by Joshua Lederberg and Edward L. Tatum in 1946. They mixed two auxotrophic strains, incubated the culture for several hours in nutrient medium, and then plated it on minimal medium. To reduce the chance that their results were due to simple reversion, they used double and triple auxotrophs on the assumption that two or three reversions would not often occur simultaneously. For example, one strain required biotin (Bio⁻), phenylalanine (Phe⁻), and cysteine (Cys⁻) for growth, and another needed threonine (Thr⁻), leucine (Leu⁻), and thiamine (Thi⁻). Recombinant prototrophic colonies appeared on the minimal medium after incubation (**figure 13.12**). Thus the chromosomes of the two auxotrophs were able to associate and undergo recombination.

Lederberg and Tatum did not directly prove that physical contact of the cells was necessary for gene transfer. This evidence was provided by Bernard Davis (1950), who constructed a U tube consisting of two pieces of curved glass tubing fused at the base to form a U shape with a fritted glass filter between the halves. The filter allows the passage of media but not bacteria. The U tube was filled with nutrient medium and each side inoculated with a different auxotrophic strain of *E. coli* (**figure 13.13**). During incubation, the medium was pumped back and forth through the filter to ensure medium exchange between the halves. After a 4 hour incubation, the bacteria were plated on minimal medium. Davis discovered that when the two auxotrophic strains were separated from each other by the fine filter, gene transfer could not take place. Therefore direct contact was required for the recombination that Lederberg and Tatum had observed.

F⁺ × F⁻ Mating

In 1952 William Hayes demonstrated that the gene transfer observed by Lederberg and Tatum was polar. That is, there were definite donor (F⁺) and recipient (F⁻) strains, and gene transfer was nonreciprocal. He also found that in F⁺ × F⁻ mating the progeny were only rarely changed with regard to auxotrophy (that is, bacterial genes were not often transferred), but F⁻ strains frequently became F⁺.

These results are readily explained in terms of the F factor previously described (figure 13.5). The F⁺ strain contains an ex-

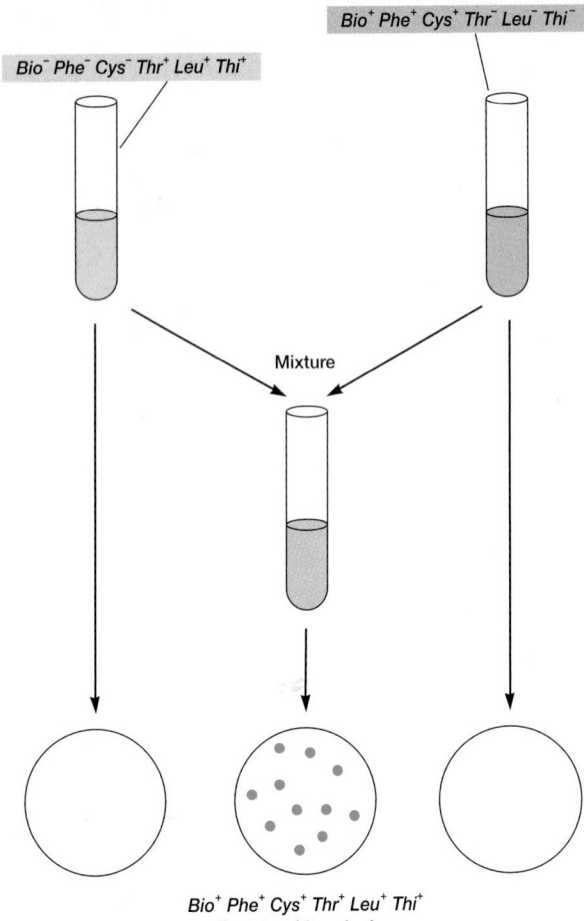

Figure 13.12 Evidence for Bacterial Conjugation. Lederberg and Tatum's demonstration of genetic recombination using triple auxotrophs. See text for details.

trachromosomal F factor carrying the genes for pilus formation and plasmid transfer. During F⁺ × F⁻ mating or conjugation, the F factor replicates by the rolling-circle mechanism, and a copy moves to the recipient (**figure 13.14a**). The entering strand is copied to produce double-stranded DNA. Because bacterial chromosome genes are rarely transferred with the independent F factor, the recombination frequency is low. It is still not completely clear how the plasmid moves between bacteria. The **sex pilus** or F pilus joins the donor and recipient and may contract to draw them together. The channel for DNA transfer could be either the hollow F pilus or a special conjugation bridge formed upon contact.   The rolling-circle mechanism of DNA replication (p. 229)

Although most research on plasmids and conjugation has been done using *E. coli* and other gram-negative bacteria, self-transmissible plasmids are present in gram-positive bacterial genera such as *Bacillus, Streptococcus, Enterococcus, Staphylococcus,* and *Streptomyces.* Much less is known about these systems. It appears that fewer transfer genes are involved, possibly because a sex pilus does not seem to be required for plasmid

In the figure: Bio⁻ Phe⁻ Cys⁻ Thr⁺ Leu⁺ Thi⁺ ; Bio⁺ Phe⁺ Cys⁺ Thr⁻ Leu⁻ Thi⁻ ; Mixture ; Bio⁺ Phe⁺ Cys⁺ Thr⁺ Leu⁺ Thi⁺ Prototrophic colonies

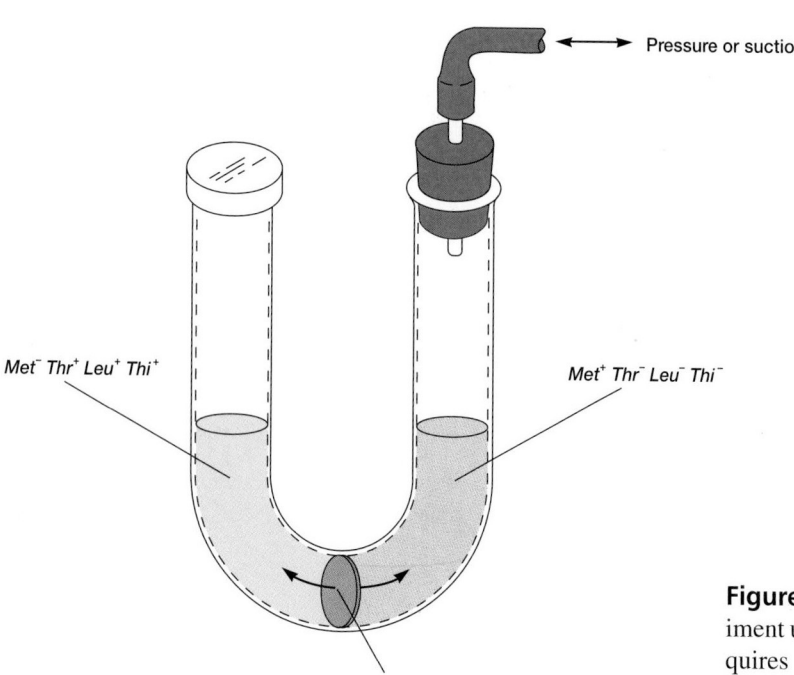

← Pressure or suction →

Met⁻ Thr⁺ Leu⁺ Thi⁺

Met⁺ Thr⁻ Leu⁻ Thi⁻

Fritted glass filter

Figure 13.13   The U-Tube Experiment.   The U-tube experiment used to show that genetic recombination by conjugation requires direct physical contact between bacteria. See text for details.

transfer. For example, *Enterococcus faecalis* recipient cells release short peptide chemical signals that activate transfer genes in donor cells containing the proper plasmid. Donor and recipient cells directly adhere to one another through special plasmid-encoded proteins released by the activated donor cell. Plasmid transfer then occurs.

Hfr Conjugation

Because certain donor strains transfer bacterial genes with great efficiency and do not usually change recipient bacteria to donors, a second type of conjugation must exist. The F factor is an episome and can integrate into the bacterial chromosome at several different locations by recombination between homologous insertion sequences present on both the plasmid and host chromosomes. When integrated, the F plasmid's *tra* operon is still functional; the plasmid can direct the synthesis of pili, carry out rolling-circle replication, and transfer genetic material to an F⁻ recipient cell. Such a donor is called an **Hfr strain** (for high frequency of recombination) because it exhibits a very high efficiency of chromosomal gene transfer in comparison with F⁺ cells. DNA transfer begins when the integrated F factor is nicked at its site of transfer origin (figure 13.14*b*). As it is replicated, the chromosome moves through the pilus or conjugation bridge connecting the donor and recipient. Because only part of the F factor is transferred at the start (the initial break is within the F plasmid), the F⁻ recipient does not become F⁺ unless the whole chromosome is transferred. Transfer is standardized at 100 minutes in *E. coli,* and the connection usually breaks before this process is finished. Thus a complete F factor usually is not transferred, and the recipient remains F⁻.

As mentioned earlier, when an Hfr strain participates in conjugation, bacterial genes are frequently transferred to the recipient. Gene transfer can be in either a clockwise or counterclockwise direction around the circular chromosome, depending on the orientation of the integrated F factor. After the replicated donor chromosome enters the recipient cell, it may be degraded or incorporated into the F⁻ genome by recombination.

F′ Conjugation

Because the F plasmid is an episome, it can leave the bacterial chromosome. Sometimes during this process the plasmid makes an error in excision and picks up a portion of the chromosomal material to form an **F′ plasmid** (**figure 13.15***a*). It is not unusual to observe the inclusion of one or more genes in excised F plasmids. The F′ cell retains all of its genes, although some of them are on the plasmid, and still mates only with an F⁻ recipient. F′ × F⁻ conjugation is virtually identical with F⁺ × F⁻ mating. Once again, the plasmid is transferred, but usually bacterial genes on the chromosome are not (figure 13.15*b*). Bacterial genes on the F′ plasmid are transferred with it and need not be incorporated into the recipient chromosome to be expressed. The recipient becomes F′ and is a partially diploid merozygote since it has two sets of the genes carried by the plasmid. In this way specific bacterial genes may spread rapidly throughout a bacterial population. Such transfer of bacterial genes is often called sexduction.

F′ conjugation is very important to the microbial geneticist. A partial diploid's behavior shows whether the allele carried by an F′ plasmid is dominant or recessive to the chromosomal gene. The formation of F′ plasmids also is useful in mapping the

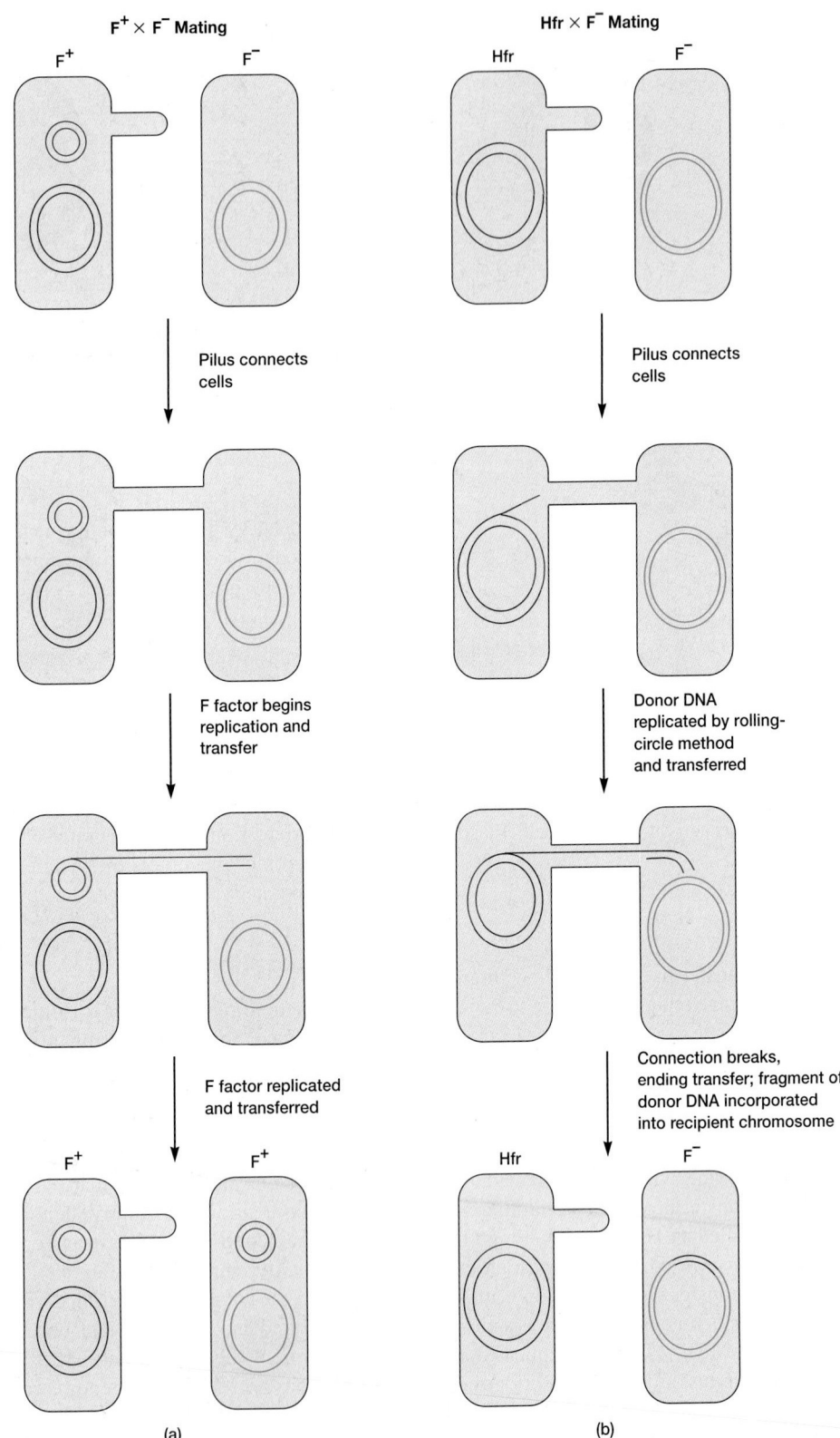

Figure 13.14 The Mechanism of Bacterial Conjugation. (a) F$^+$ × F$^-$ mating. (b) Hfr × F$^-$ mating (the integrated F factor is shown in red). See text for explanation.

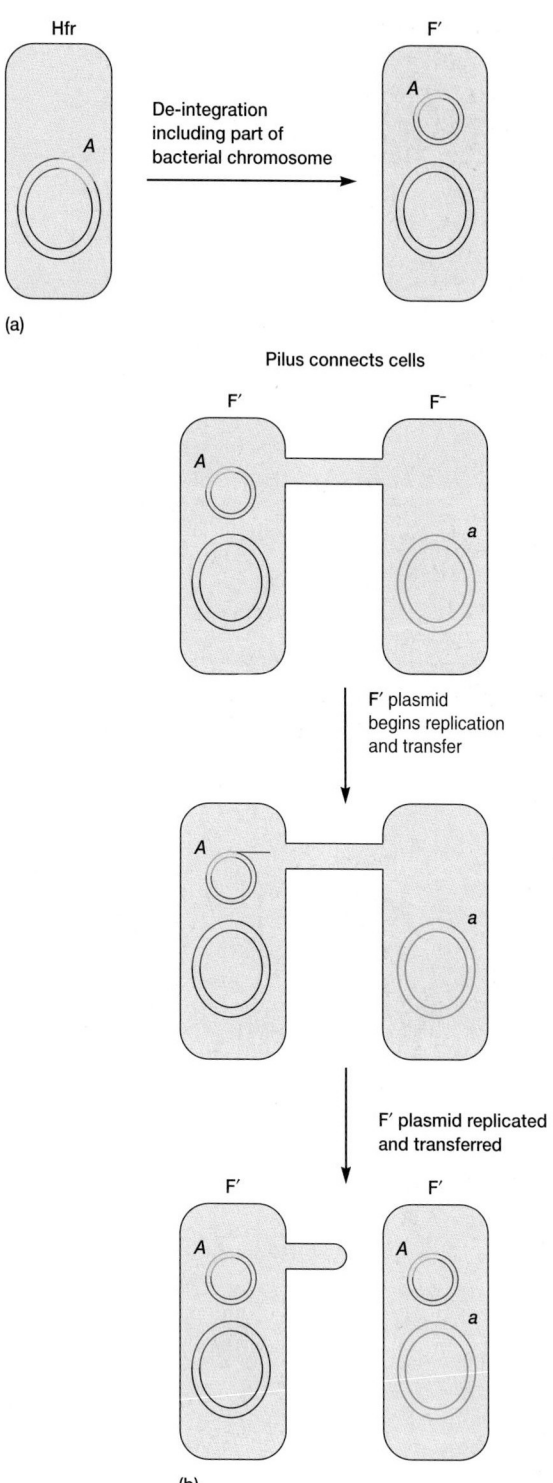

Figure 13.15   F′ Conjugation. (**a**) Due to an error in excision, the *A* gene of an Hfr cell is picked up by the F factor. (**b**) The *A* gene is then transferred to a recipient during conjugation. See text for explanation.

chromosome since if two genes are picked up by an F factor they must be neighbors.

1. What is bacterial conjugation and how was it discovered?
2. Distinguish between F⁺, Hfr, and F⁻ strains of *E. coli* with respect to their physical nature and role in conjugation.
3. Describe in some detail how F⁺ × F⁻ and Hfr conjugation processes proceed, and distinguish between the two in terms of mechanism and the final results.
4. What is F′ conjugation and why is it so useful to the microbial geneticist? How does the F′ plasmid differ from a regular F plasmid? What is sexduction?

13.5   DNA TRANSFORMATION

The second way in which DNA can move between bacteria is through transformation, discovered by Fred Griffith in 1928. **Transformation** is the uptake by a cell of a naked DNA molecule or fragment from the medium and the incorporation of this molecule into the recipient chromosome in a heritable form. In natural transformation the DNA comes from a donor bacterium. The process is random, and any portion of a genome may be transferred between bacteria. The discovery of transformation (pp. 222–23)

When bacteria lyse, they release considerable amounts of DNA into the surrounding environment. These fragments may be relatively large and contain several genes. If a fragment contacts a **competent** cell, one able to take up DNA and be transformed, it can be bound to the cell and taken inside (**figure 13.16a**). The transformation frequency of very competent cells is around 10^{-3} for most genera when an excess of DNA is used. That is, about one cell in every thousand will take up and integrate the gene. Competency is a complex phenomenon and is dependent on several conditions. Bacteria need to be in a certain stage of growth; for example, *S. pneumoniae* becomes competent during the exponential phase when the population reaches about 10^7 to 10^8 cells per ml. When a population becomes competent, bacteria such as pneumococci secrete a small protein called the competence factor that stimulates the production of 8 to 10 new proteins required for transformation. Natural transformation has been discovered so far only in certain gram-positive and gram-negative genera: *Streptococcus, Bacillus, Thermoactinomyces, Haemophilus, Neisseria, Moraxella, Acinetobacter, Azotobacter, Helicobacter,* and *Pseudomonas*. Other genera also may be capable of transformation. Gene transfer by this process occurs in soil and marine environments and may be an important route of genetic exchange in nature.

The mechanism of transformation has been intensively studied in *S. pneumoniae* (**figure 13.17a**). A competent cell binds a double-stranded DNA fragment if the fragment is moderately large; the process is random, and donor fragments compete with each other. The DNA then is cleaved by endonucleases to double-stranded fragments about 5 to 15 kilobases in size. DNA uptake requires energy expenditure. One strand is hydrolyzed by an

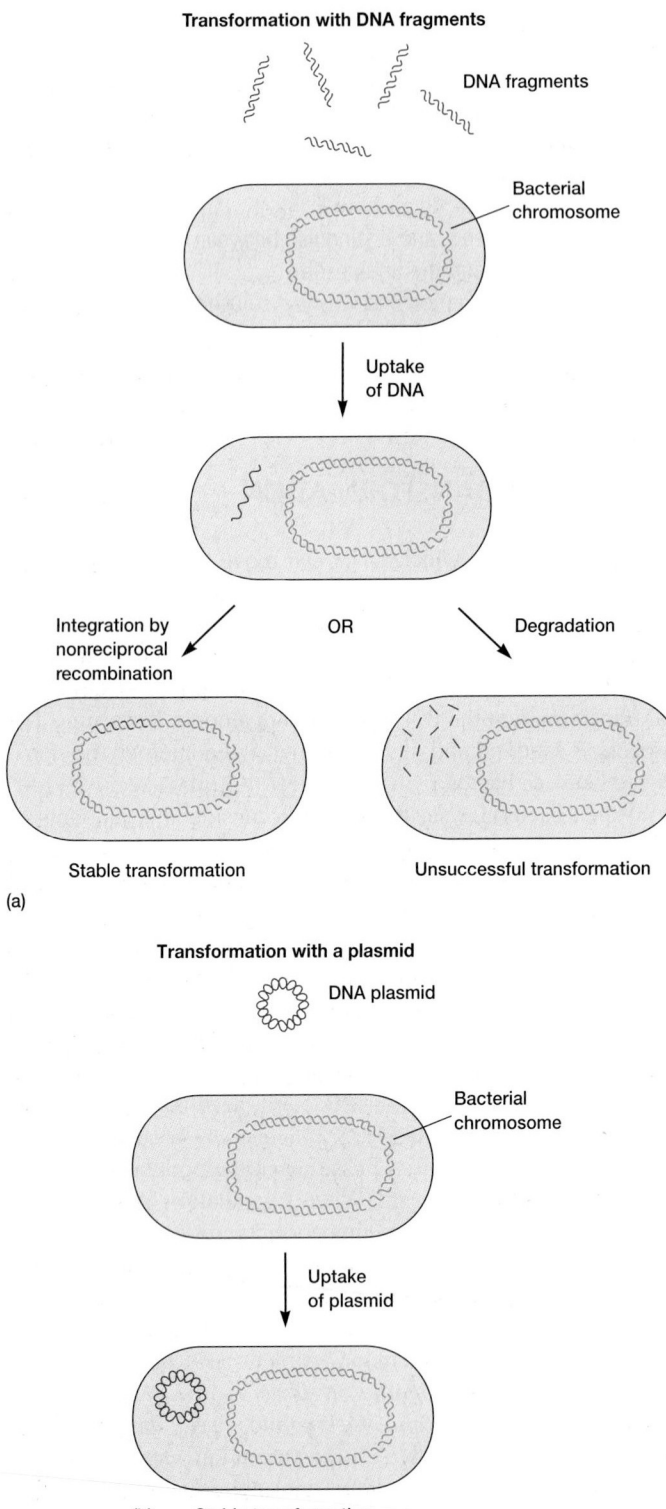

Transformation with DNA fragments

DNA fragments

Bacterial chromosome

Uptake of DNA

Integration by nonreciprocal recombination

OR

Degradation

Stable transformation

Unsuccessful transformation

(a)

Transformation with a plasmid

DNA plasmid

Bacterial chromosome

Uptake of plasmid

(b)   Stable transformation

Figure 13.16 Bacterial Transformation. Transformation with (**a**) DNA fragments and (**b**) plasmids. Transformation with a plasmid often is induced artificially in the laboratory. The transforming DNA is in red and integration is at a homologous region of the genome. See text for details.

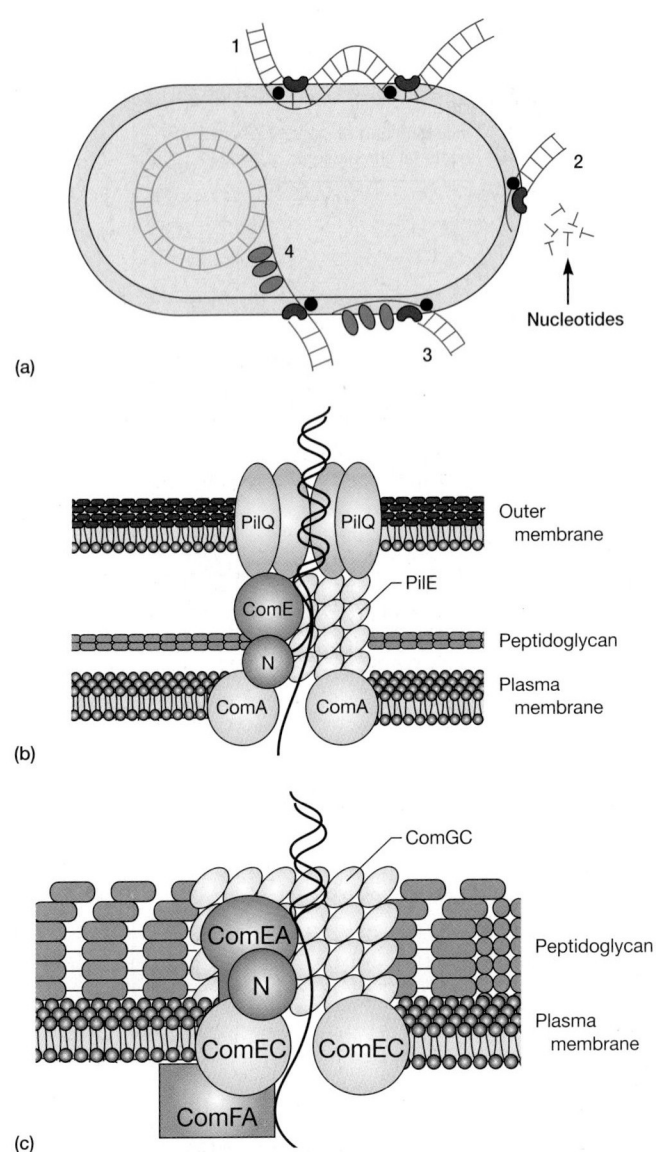

1

2

Nucleotides

4

3

(a)

PilQ   PilQ   Outer membrane

ComE   PilE

N   Peptidoglycan

ComA   ComA   Plasma membrane

(b)

ComGC

ComEA

N   Peptidoglycan

ComEC   ComEC   Plasma membrane

ComFA

(c)

Figure 13.17   The Mechanism of Transformation.
(**a**) An overview of the process (*1*) A long double-stranded DNA molecule binds to the surface with the aid of a DNA-binding protein (●) and is nicked by a nuclease (◖). (*2*) One strand is degraded by the nuclease. (*3*) The undegraded strand associates with a competence-specific protein (◖). (*4*) The single strand enters the cell and is integrated into the host chromosome in place of the homologous region of the host DNA. (**b**) DNA uptake machinery in *N. gonorrhoeae*. (**c**) Uptake machinery in *B. subtilis*. See text for details.

envelope-associated exonuclease during uptake; the other strand associates with small proteins and moves through the plasma membrane. The single-stranded fragment can then align with a homologous region of the genome and be integrated, probably by a mechanism similar to that depicted in figure 13.3.

Transformation in *Haemophilus influenzae*, a gram-negative bacterium, differs from that in *S. pneumoniae* in several respects.

Haemophilus does not produce a competence factor to stimulate the development of competence, and it takes up DNA from only closely related species (*S. pneumoniae* is less particular about the source of its DNA). Double-stranded DNA, complexed with proteins, is taken in by membrane vesicles. The specificity of *Haemophilus* transformation is due to a special 11 base pair sequence (5′AAGTGCG-GTCA3′) that is repeated over 1,400 times in *H. influenzae* DNA. DNA must have this sequence to be bound by a competent cell.

The protein complexes that take up free DNA must be able to move it through gram-negative and gram-positive walls, which may be both thick and complex. As expected, the machinery is quite large and complicated and appears related to protein secretion systems (*see section 3.6*). Figure 13.17*b* shows a schematic diagram of the complex used by the gram-negative bacterium *Neisseria gonorrhoeae*. PilQ aids in the movement across the outer membrane, and the pilin complex PilE moves the DNA through the periplasm and peptidoglycan. ComE is a DNA binding protein; N is the nuclease that degrades one strand before the DNA enters the cytoplasm through the transmembrane channel formed by ComA. The gram-positive machinery in *Bacillus subtilis* is depicted in figure 13.17*c*. Many of the components are similar: the pilin complex (ComGC), DNA binding protein (ComEA), nuclease (N), and channel protein (ComEC). ComFA is a DNA translocase that moves the DNA into the cytoplasm. A gram-negative equivalent of ComFA has not been identified yet in *N. gonorrhoeae*.

Artificial transformation is carried out in the laboratory by a variety of techniques, including treatment of the cells with calcium chloride, which renders their membranes more permeable to DNA. This approach succeeds even with species that are not naturally competent, such as *E. coli*. Relatively high concentrations of DNA, higher than would normally be present in nature, are used to increase transformation frequency. When linear DNA fragments are to be used in transformation, *E. coli* usually is rendered deficient in one or more exonuclease activities to protect the transforming fragments. It is even easier to transform bacteria with plasmid DNA since plasmids are not as easily degraded as linear fragments and can replicate within the host (figure 13.16*b*). This is a common method for introducing recombinant DNA into bacterial cells (*see sections 14.5 and 14.7*). DNA from any source can be introduced into bacteria by splicing it into a plasmid before transformation.

---

1. Define transformation and competence.
2. Describe how transformation occurs in *S. pneumoniae*. How does the process differ in *H. influenzae*?
3. Discuss two ways in which artificial transformation can be used to place functional genes within bacterial cells.

13.6 TRANSDUCTION

Bacterial viruses or bacteriophages participate in the third mode of bacterial gene transfer. These viruses have relatively simple structures in which virus genetic material is enclosed within an outer coat, composed mainly or solely of protein. The coat protects the genome and transmits it between host cells. The morphology and life cycle of bacteriophages is not discussed in detail until chapter 17. Nevertheless, it is necessary to briefly describe the life cycle here as background for a consideration of the bacteriophage's role in gene transfer. The lytic cycle (pp. 373–78); Lysogeny (pp. 380–84)

After infecting the host cell, a bacteriophage (phage for short) often takes control and forces the host to make many copies of the virus. Eventually the host bacterium bursts or lyses and releases new phages. This reproductive cycle is called a lytic cycle because it ends in lysis of the host. The cycle has four phases (**figure 13.18***a*). First, the virus particle attaches to a specific receptor site on the bacterial surface. The genetic material, which is often double-stranded DNA, then enters the cell. After adsorption and penetration, the virus chromosome forces the bacterium to make virus nucleic acids and proteins. The third stage begins after the synthesis of virus components. Phages are assembled from these components. The assembly process may be complex, but in all cases phage nucleic acid is packed within the virus's protein coat. Finally, the mature viruses are released by cell lysis.

Bacterial viruses that reproduce using a lytic cycle often are called virulent bacteriophages because they destroy the host cell. Many DNA phages, such as the lambda phage (*see p. 381*), are also capable of a different relationship with their host (figure 13.18*b*). After adsorption and penetration, the viral genome does not take control of its host and destroy it while producing new phages. Instead the genome remains within the host cell and is reproduced along with the bacterial chromosome. A clone of infected cells arises and may grow for long periods while appearing perfectly normal. Each of these infected bacteria can produce phages and lyse under appropriate environmental conditions. This relationship between the phage and its host is called **lysogeny.** Bacteria that can produce phage particles under some conditions are said to be **lysogens** or **lysogenic,** and phages able to establish this relationship are **temperate phages.** The latent form of the virus genome that remains within the host without destroying it is called the **prophage.** The prophage usually is integrated into the bacterial genome (figure 13.18*b*). Sometimes phage reproduction is triggered in a lysogenized culture by exposure to UV radiation or other factors. The lysogens are then destroyed and new phages released. This phenomenon is called induction.

Transduction is the transfer of bacterial genes by viruses. Bacterial genes are incorporated into a phage capsid because of errors made during the virus life cycle. The virus containing these genes then injects them into another bacterium, completing the transfer. Transduction may be the most common mechanism for gene exchange and recombination in bacteria. There are two very different kinds of transduction: generalized and specialized.

Generalized Transduction

Generalized transduction (**figure 13.19**) occurs during the lytic cycle of virulent and temperate phages and can transfer any part of the bacterial genome. During the assembly stage, when the viral chromosomes are packaged into protein capsids, random fragments of

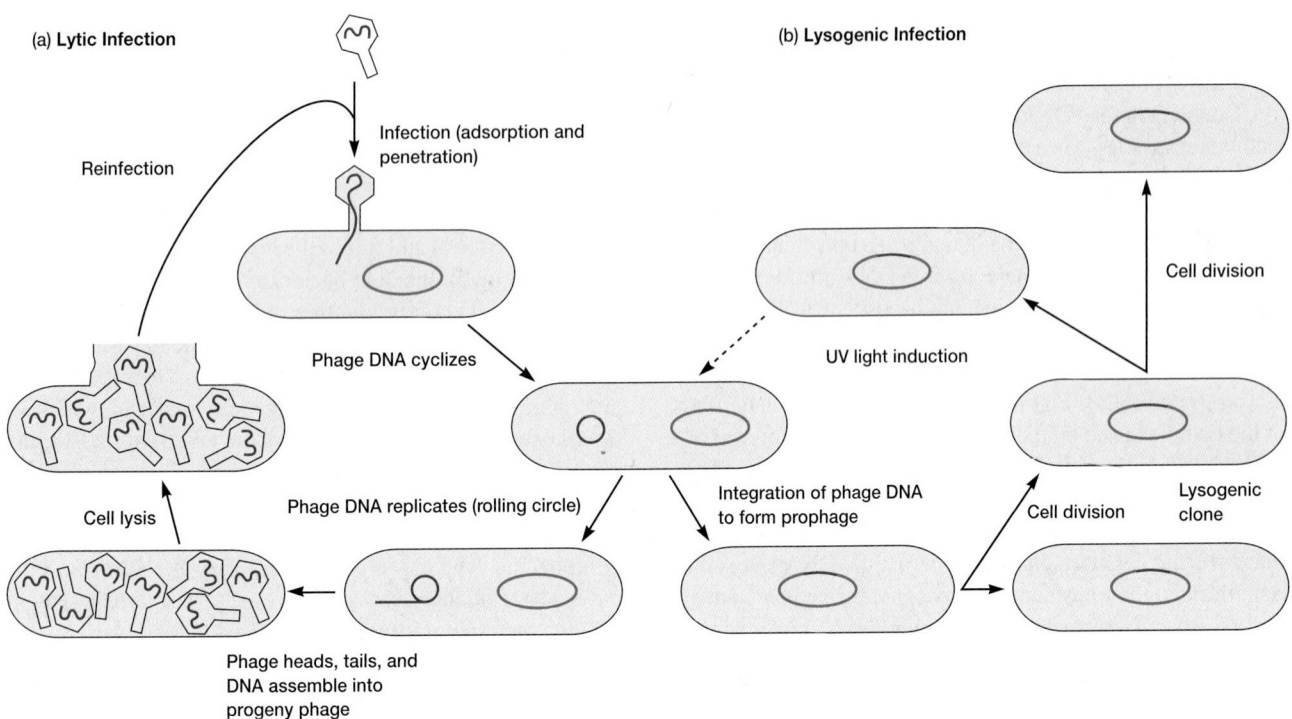

(a) Lytic Infection

Reinfection

Infection (adsorption and penetration)

Phage DNA cyclizes

Phage DNA replicates (rolling circle)

Cell lysis

Phage heads, tails, and DNA assemble into progeny phage

(b) Lysogenic Infection

Cell division

UV light induction

Integration of phage DNA to form prophage

Cell division

Lysogenic clone

Figure 13.18   Lytic versus Lysogenic Infection by Phage Lambda.   (**a**) Lytic infection. (**b**) Lysogenic infection. Viral and prophage DNA are in red.

the partially degraded bacterial chromosome also may be packaged by mistake. Because the capsid can contain only a limited quantity of DNA, the viral DNA is left behind. The quantity of bacterial DNA carried depends primarily on the size of the capsid. The P22 phage of *Salmonella typhimurium* usually carries about 1% of the bacterial genome; the P1 phage of *E. coli* and a variety of gram-negative bacteria carries about 2.0 to 2.5% of the genome. The resulting virus particle often injects the DNA into another bacterial cell but does not initiate a lytic cycle. This phage is known as a **generalized transducing particle** or phage and is simply a carrier of genetic information from the original bacterium to another cell. As in transformation, once the DNA has been injected, it must be incorporated into the recipient cell's chromosome to preserve the transferred genes. The DNA remains double stranded during transfer, and both strands are integrated into the endogenote's genome. About 70 to 90% of the transferred DNA is not integrated but often is able to survive and express itself. **Abortive transductants** are bacteria that contain this nonintegrated, transduced DNA and are partial diploids.

Generalized transduction was discovered in 1951 by Joshua Lederberg and Norton Zinder during an attempt to show that conjugation, discovered several years earlier in *E. coli*, could occur in other bacterial species. Lederberg and Zinder were repeating the earlier experiments with *Salmonella typhimurium*. They found that incubation of a mixture of two multiply auxotrophic strains yielded prototrophs at the level of about one in 10^5. This seemed like good evidence for bacterial recombination, and indeed it was, but their initial conclusion that the transfer resulted from conjugation was not borne out. When these investigators

performed the U-tube experiment (figure 13.13) with *Salmonella*, they still recovered prototrophs. The filter in the U tube had small enough pores to block the movement of bacteria between the two sides but allowed the phage P22 to pass. Lederberg and Zinder had intended to confirm that conjugation was present in another bacterial species and had instead discovered a completely new mechanism of bacterial gene transfer. The seemingly routine piece of research led to surprising and important results. A scientist must always keep an open mind about results and be prepared for the unexpected.

Specialized Transduction

In **specialized** or **restricted transduction,** the transducing particle carries only specific portions of the bacterial genome. Specialized transduction is made possible by an error in the lysogenic life cycle. When a prophage is induced to leave the host chromosome, excision is sometimes carried out improperly. The resulting phage genome contains portions of the bacterial chromosome (about 5 to 10% of the bacterial DNA) next to the integration site, much like the situation with F′ plasmids (**figure 13.20**). A transducing phage genome usually is defective and lacks some part of its attachment site. The transducing particle will inject bacterial genes into another bacterium, even though the defective phage cannot reproduce without assistance. The bacterial genes may become stably incorporated under the proper circumstances.

The best-studied example of specialized transduction is the lambda phage. The lambda genome inserts into the host chro-

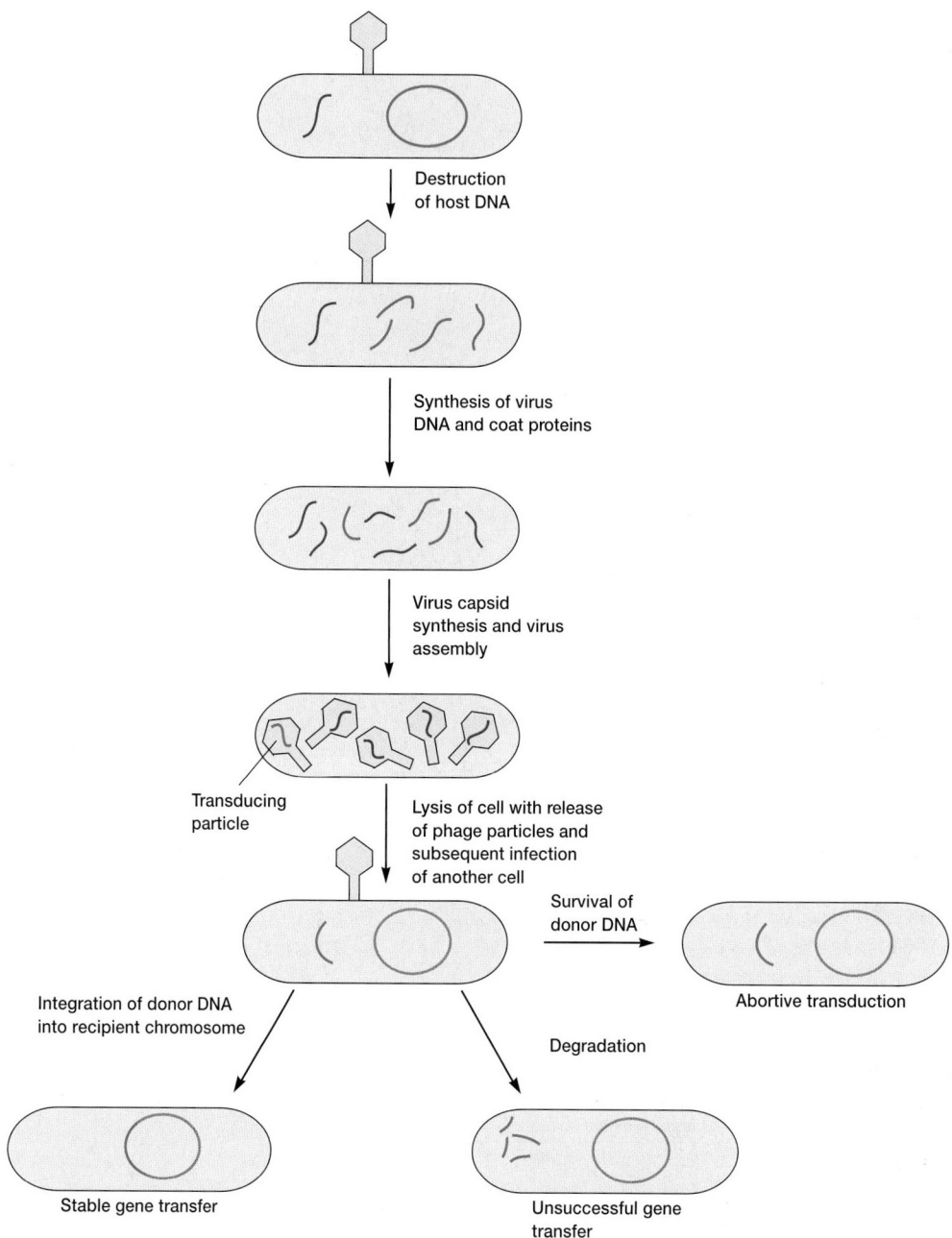

Destruction
of host DNA

Synthesis of virus
DNA and coat proteins

Virus capsid
synthesis and virus
assembly

Transducing
particle

Lysis of cell with release
of phage particles and
subsequent infection
of another cell

Survival of
donor DNA

Abortive transduction

Integration of donor DNA
into recipient chromosome

Degradation

Stable gene transfer

Unsuccessful gene
transfer

Figure 13.19   Generalized Transduction by Bacteriophages.   See text for details.

mosome at specific locations known as attachment or *att* sites (**figure 13.21;** *see also figures 17.17 and 17.21*). The phage *att* sites and bacterial *att* sites are similar and can complex with each other, although they are not identical. The *att* site for lambda is next to the *gal* and *bio* genes on the *E. coli* chromosome; consequently, specialized transducing lambda phages most often carry these bacterial genes. The lysate, or product of cell lysis, resulting from the induction of lysogenized *E. coli* contains normal phage and a few defective transducing particles. These particles are called lambda *dgal* because they carry

the galactose utilization genes (figure 13.21). Because these lysates contain only a few transducing particles, they often are called **low-frequency transduction lysates (LFT lysates).** Whereas the normal phage has a complete *att* site, defective transducing particles have a nonfunctional hybrid integration site that is part bacterial and part phage in origin. Integration of the defective phage chromosome does not readily take place. Transducing phages also may have lost some genes essential for reproduction. Stable transductants can arise from recombination between the phage and the bacterial chromosome because

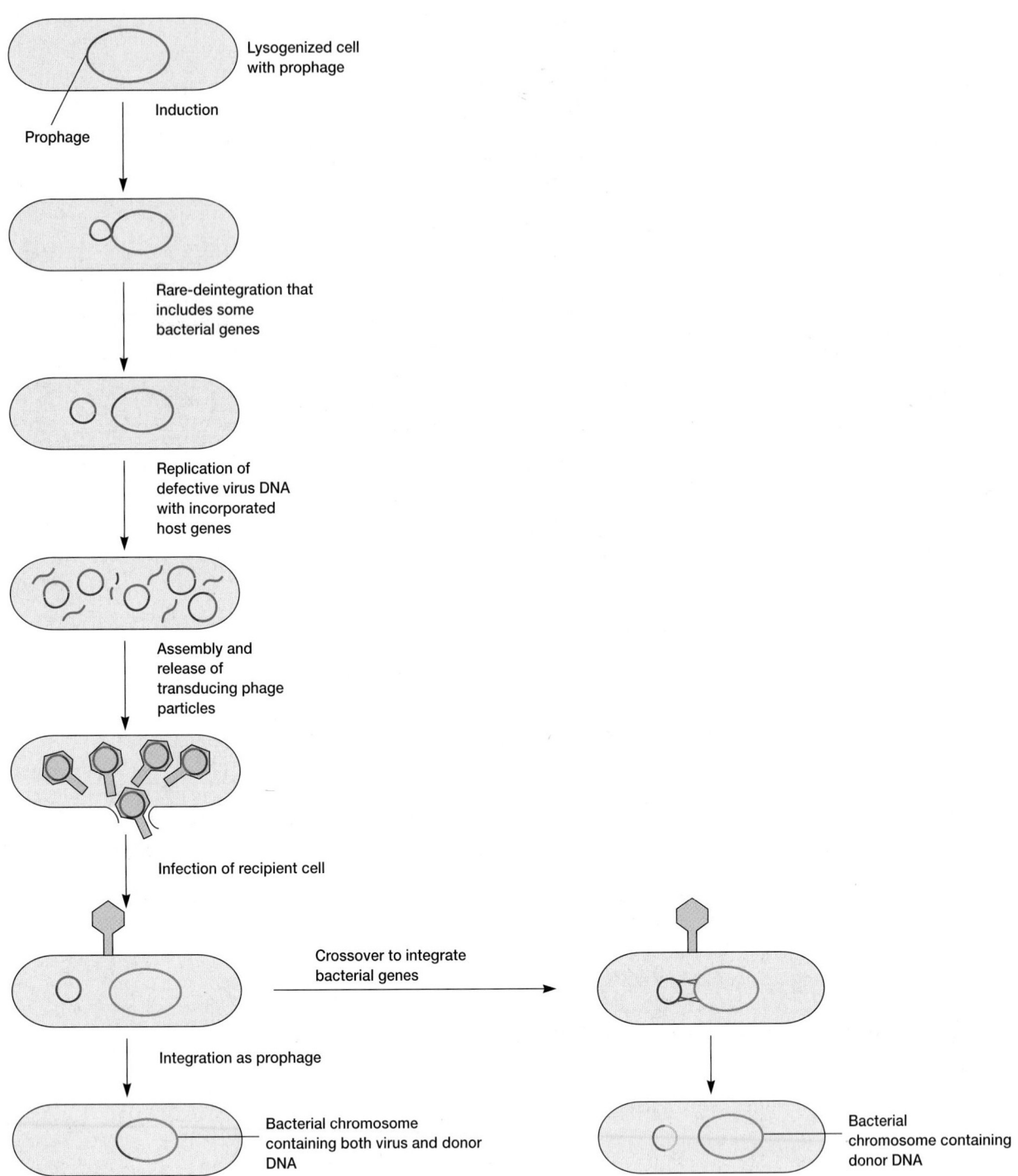

Lysogenized cell
with prophage

Prophage

Induction

Rare-deintegration that
includes some
bacterial genes

Replication of
defective virus DNA
with incorporated
host genes

Assembly and
release of
transducing phage
particles

Infection of recipient cell

Crossover to integrate
bacterial genes

Integration as prophage

Bacterial chromosome
containing both virus and donor
DNA

Bacterial
chromosome containing
donor DNA

Figure 13.20   Specialized Transduction by a Temperate Bacteriophage.   Recombination can produce two types of transductants. See text for details.

of crossovers on both sides of the *gal* site (figure 13.21).   The lysogenic cycle of lambda phage (pp. 381–85)

Defective lambda phages carrying the *gal* gene can integrate if there is a normal lambda phage in the same cell. The normal phage will integrate, yielding two bacterial/phage hybrid *att* sites

where the defective lambda *dgal* phage can insert (figure 13.21). It also supplies the genes missing in the defective phage. The normal phage in this instance is termed the **helper phage** because it aids integration and reproduction of the defective phage. These transductants are unstable because the prophages can be induced

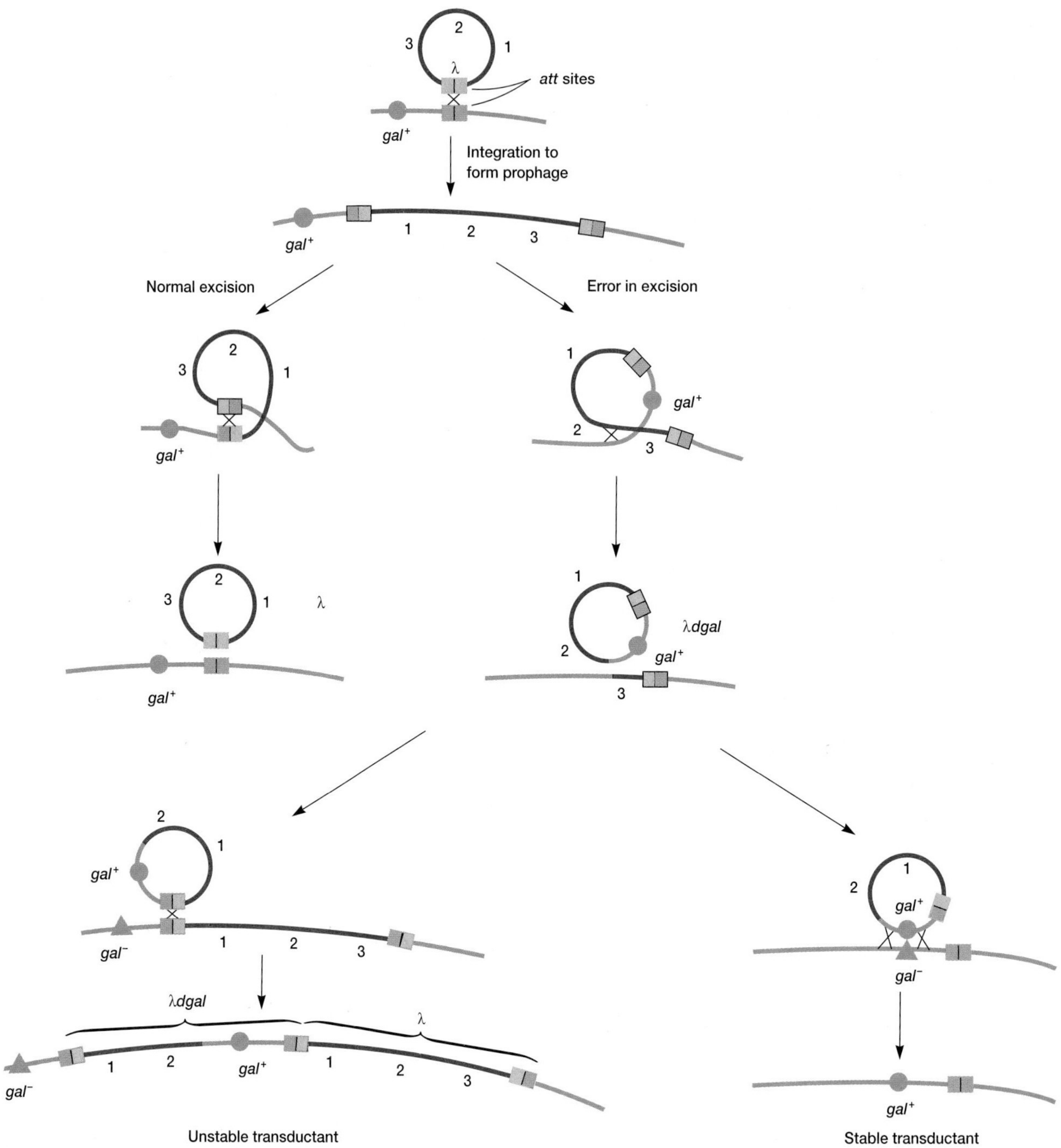

Figure 13.21 The Mechanism of Transduction for Phage Lambda and *E. coli.* Integrated lambda phage lies next to the *gal* genes. When it excises normally (top left), the new phage is complete and contains no bacterial genes. Rarely excision occurs asymmetrically (top right), and the *gal* genes are then picked up and some phage genes are lost. The result is a defective lambda phage that carries bacterial genes and can transfer them to a new recipient. See text for details.

to excise by agents such as UV radiation. Excision, however, produces a lysate containing a fairly equal mixture of defective lambda *dgal* phage and normal helper phage. Because it is very effective in transduction, the lysate is called a **high-frequency transduction lysate (HFT lysate).** Reinfection of bacteria with this mixture will result in the generation of considerably more transductants. LFT lysates and those produced by generalized transduction have one transducing particle in 10^5 or 10^6 phages; HFT lysates contain transducing particles with a frequency of about 0.1 to 0.5.

---

1. Briefly describe the lytic and lysogenic viral reproductive cycles. Define lysogeny, lysogen, temperate phage, prophage, and transduction.
2. Describe generalized transduction, how it occurs, and the way in which it was discovered. What is an abortive transductant?
3. What is specialized or restricted transduction and how does it come about? Be able to distinguish between LFT and HFT lysates and describe how they are formed.

 13.7   MAPPING THE GENOME

Finding the location of genes in any organism's genome is a very complex task. This section surveys approaches to mapping the bacterial genome, using *E. coli* as an example. All three modes of

gene transfer and recombination have been used in mapping. *Gene structure and the nature of mutations (pp. 235–47)*

Hfr conjugation is frequently used to map the relative location of bacterial genes. This technique rests on the observation that during conjugation the linear chromosome moves from donor to recipient at a constant rate. In an **interrupted mating experiment** the conjugation bridge is broken and Hfr × F⁻ mating is stopped at various intervals after the start of conjugation by mixing the culture vigorously in a blender (**figure 13.22*a***). The order and timing of gene transfer can be determined because they are a direct reflection of the order of genes on the bacterial chromosome (figure 13.22*b*). For example, extrapolation of the curves in figure 13.22*b* back to the x-axis will give the time at which each gene just began to enter the recipient. The result is a circular chromosome map with distances expressed in terms of the minutes elapsed until a gene is transferred. This technique can fairly precisely locate genes 3 min-

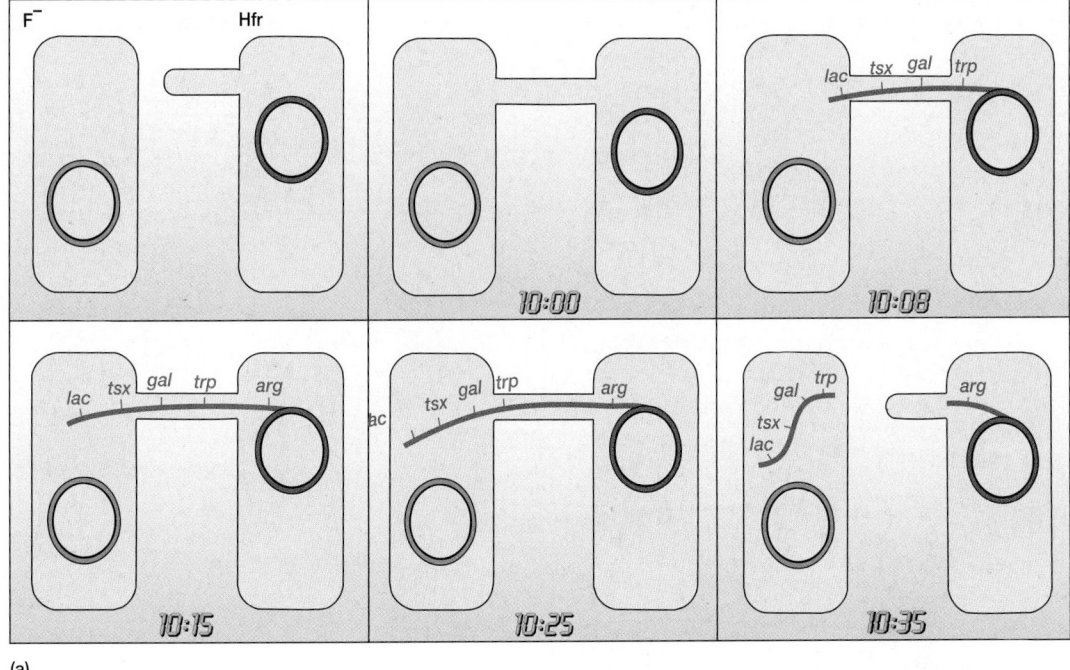

(a)

Figure 13.22   The Interrupted Mating Experiment.   An interrupted mating experiment on Hfr × F⁻ conjugation. (**a**) The linear transfer of genes is stopped by breaking the conjugation bridge to study the sequence of gene entry into the recipient cell. (**b**) An example of the results obtained by an interrupted mating experiment. The gene order is *lac-tsx-gal-trp*.

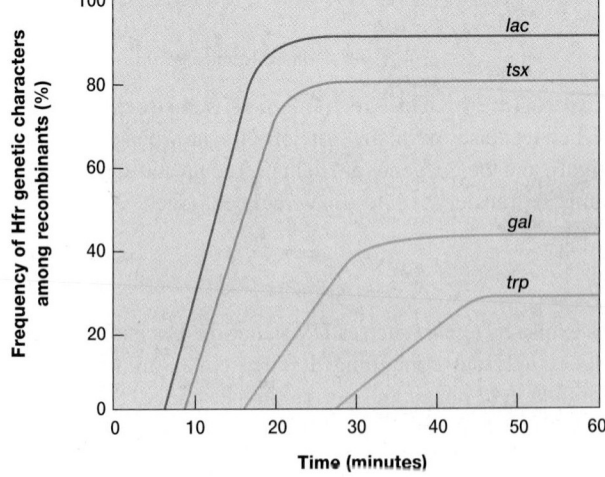

(b)

utes or more apart. The heights of the plateaus in figure 13.22*b* are lower for genes that are more distant from the F factor (the origin of transfer) because there is an ever-greater chance that the conjugation bridge will spontaneously break as transfer continues. Because of the relatively large size of the *E. coli* genome, it is not possible to generate a map from one Hfr strain. Therefore several Hfr strains with the F plasmid integrated at different locations must be used and their maps superimposed on one another. The overall map is adjusted to 100 minutes, although complete transfer may require somewhat more than 100 minutes. In a sense, minutes are an indication of map distance and not strictly a measure of time. Zero time is set at the threonine (*thr*) locus.

Gene linkage, or the proximity of two genes on a chromosome, also can be determined from transformation by measuring the frequency with which two or more genes simultaneously transform a recipient cell. Consider the case for cotransformation by two genes. In theory, a bacterium could simultaneously receive two genes, each carried on a separate DNA fragment. However, it is much more likely that the genes reside on the same fragment. If two genes are closely linked on the chromosome, then they should be able to cotransform. The closer the genes are together, the more often they will be carried on the same fragment and the higher will be the frequency of cotransformation. If genes are spaced a great distance apart, they will be carried on separate DNA fragments and the frequency of double transformants will equal the product of the individual transformation frequencies.

Generalized transduction can be used to obtain linkage information in much the same way as transformation. Linkages usually are expressed as cotransduction frequencies, using the argument that the closer two genes are to each other, the more likely they both will reside on the DNA fragment incorporated into a single phage capsid. The *E. coli* phage P1 is often used in such mapping because it can randomly transduce up to 1 to 2% of the genome.

Specialized transduction is used to find which phage attachment site is close to a specific gene. The relative locations of specific phage *att* sites are known from conjugational mapping, and the genes linked to each *att* site can be determined by means of specialized transduction. These data allow precise placement of genes on the chromosome.

A simplified genetic map of *E. coli* K12 is given in **figure 13.23**. Because conjugation data are not high resolution and cannot be used to position genes that are very close together, the whole map is developed using several mapping techniques. Usually, interrupted mating data are combined with those from cotransduction and cotransformation studies. Data from recombination studies also are used. Normally a new genetic marker in the *E. coli* genome is located within a relatively small region of the genome (10 to 15 minutes long) using a series of Hfr strains with F factor integration sites scattered throughout the genome. Once the genetic marker has been located with respect to several genes in the same region, its position relative to nearby neighbors is more accurately determined using transformation and transduction studies. Recent maps of the *E. coli* chromosome give the locations of more than a thousand genes. Remember that the genetic map only depicts physical reality in a relative sense.

A map unit in one region of the genome may not be the same physical distance as a unit in another part.

Genetic maps provide useful information in addition to the order of the genes. For example, there is considerable clustering of genes in *E. coli* K12 (figure 13.23). In the regions around 2, 17, and 27 minutes, there are many genes, whereas relatively few genetic markers are found in the 33 minute region. The areas apparently lacking genes may well have undiscovered genes, but perhaps their function is not primarily that of coding genetic information. One hypothesis is that the 33 minute region is involved in attachment of the *E. coli* chromosome to the plasma membrane during replication and cell division. It is interesting that this region is almost exactly opposite the origin of replication for the chromosome (*oriC*).

Of course a great deal could be learned by comparing a microorganism's genetic map with the actual nucleotide sequence of its genome. It is now possible to fairly rapidly sequence procaryotic genomes; genome sequencing and analysis will be discussed later in some detail.   Microbial genomics (chapter 15)

13.8   RECOMBINATION AND GENOME MAPPING IN VIRUSES

Bacteriophage genomes also undergo recombination, although the process is different from that in bacteria. Because phages themselves reproduce within cells and cannot recombine directly, crossing-over must occur inside a host cell. In principle, a virus recombination experiment is easy to carry out. If bacteria are mixed with enough phages, at least two virions will infect each cell on the average and genetic recombination should be observed. Phage progeny in the resulting lysate can be checked for alternate combinations of the initial parental genotypes.   Bacteriophage lytic cycle (pp. 373–78)

Alfred Hershey initially demonstrated recombination in the phage T2, using two strains with differing phenotypes. Two of the parental strains in Hershey's crosses were h^+r^+ and hr (**figure 13.24***a*). The gene *h* influences host range; when gene *h* changes, T2 infects different strains of *E. coli*. Phages with the r^+ gene have wild type plaque morphology, whereas T2 with the *r* genotype has a rapid lysis phenotype and produces larger than normal plaques with sharp edges (figures 13.24*b* and 13.24*c*). In one experiment Hershey infected *E. coli* with large quantities of the h^+r^+ and hr T2 strains (figure 13.24*a*). He then plated out the lysates with a mixture of two different host strains and was able to detect significant numbers of h^+r and hr^+ recombinants, as well as parental type plaques. As long as there are detectable phenotypes and methods for carrying out the crosses, it is possible to map phage genes in this way.   Plaque formation and morphology (p. 354)

Phage genomes are so small that often it is convenient to map them without determining recombination frequencies. Some techniques actually generate physical maps, which often are most useful in genetic engineering. Several of these methods require manipulation of the DNA with subsequent examination in the electron microscope. For example, one can directly compare wild-type

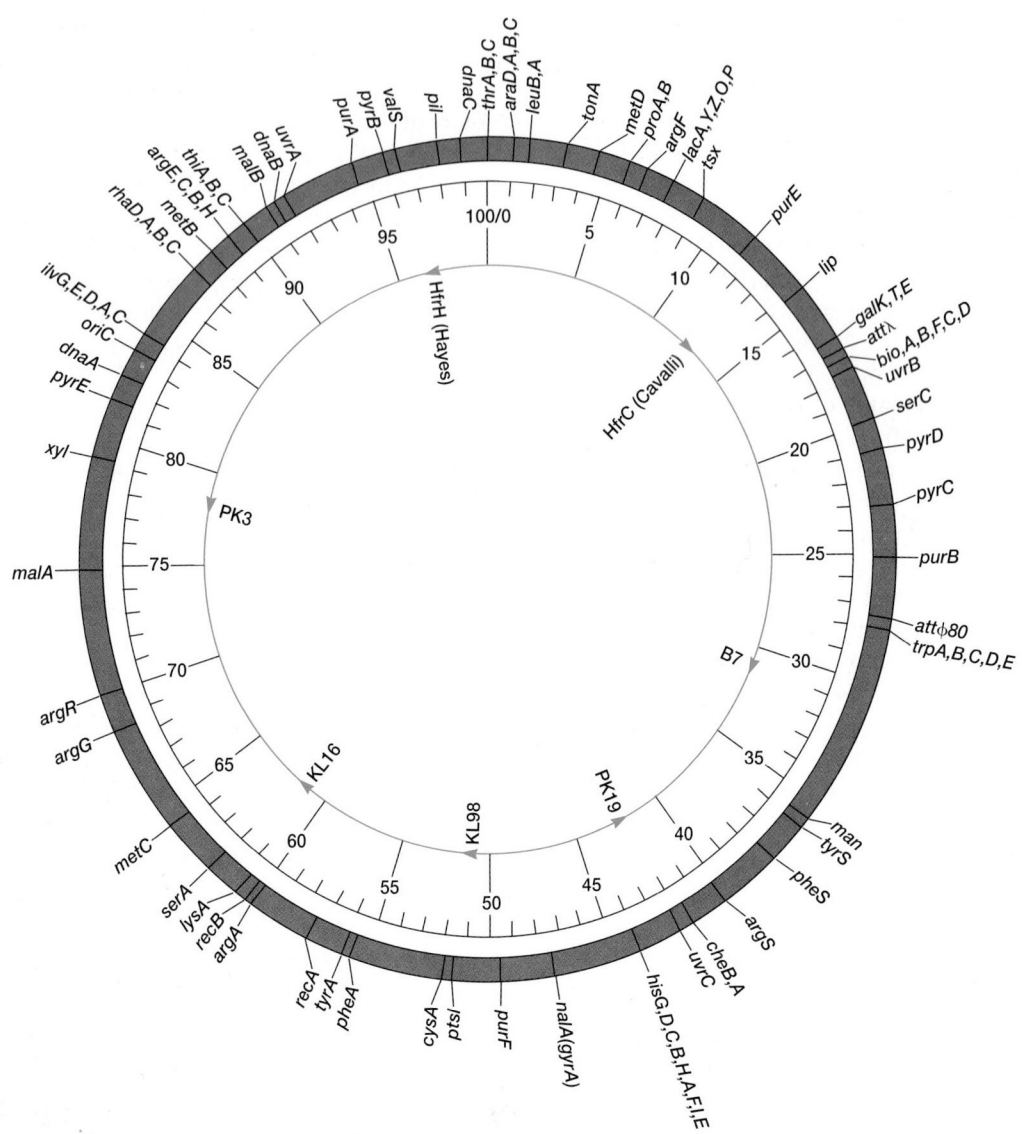

Figure 13.23 *E. coli* Genetic Map.   A circular genetic map of *E. coli* K12 with the location of selected genes. The inner circle shows the origin and direction of transfer of several Hfr strains. The map is divided into 100 minutes, the time required to transfer the chromosome from an Hfr cell to F⁻ at 37°C.

and mutant viral chromosomes. In heteroduplex mapping the two types of chromosomes are denatured, mixed, and allowed to rejoin or anneal. When joined, the homologous regions of the different DNA molecules form a regular double helix. In locations where the bases do not pair due to the presence of a mutation such as a deletion or insertion, bubbles will be visible in the electron microscope.

Several other direct techniques are used to map viral genomes or parts of them. Restriction endonucleases (*see section 14.1*) are employed together with electrophoresis to analyze DNA fragments and locate deletions and other mutations that affect electrophoretic mobility. Phage genomes also can be directly sequenced to locate particular mutations and analyze the changes that have taken place.

It should be noted that many of these physical mapping techniques also have been employed in the analysis of relatively small

portions of bacterial genomes. Furthermore, these methods are useful to the genetic engineer who is concerned with direct manipulation of DNA.

---

1. Describe how the bacterial genome can be mapped using Hfr conjugation, transformation, generalized transduction, and specialized transduction. Include both a description of each technique and any assumptions underlying its use.
2. Why is it necessary to use several different techniques in genome mapping? How is this done in practice?
3. How does recombination in viruses differ from that in bacteria? How did Hershey first demonstrate virus recombination?
4. Describe heteroduplex mapping.

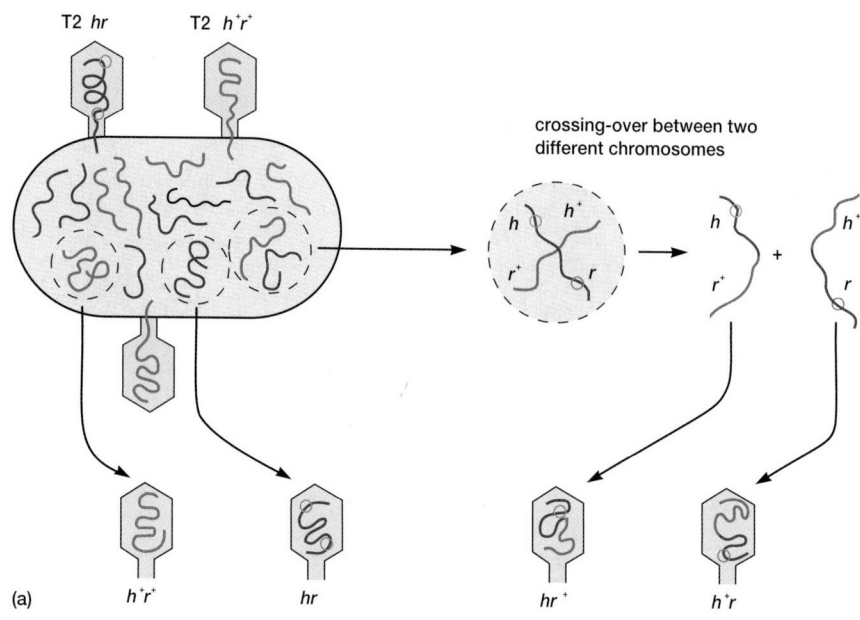

(a)

Figure 13.24 Genetic Recombination in Bacteriophages. (**a**) A summary of a genetic recombination experiment with the *hr* and *h*⁺r⁺ strains of the T2 phage. The *hr* chromosome is shown in color. (**b**) The types of plaques produced by this experiment on a lawn of *E. coli*. (**c**) A close-up of the four plaque types. See text for details.

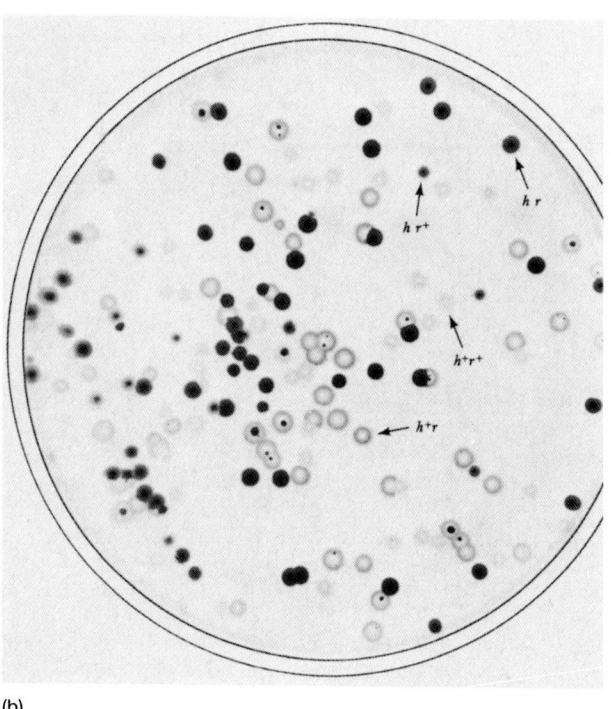

(b)

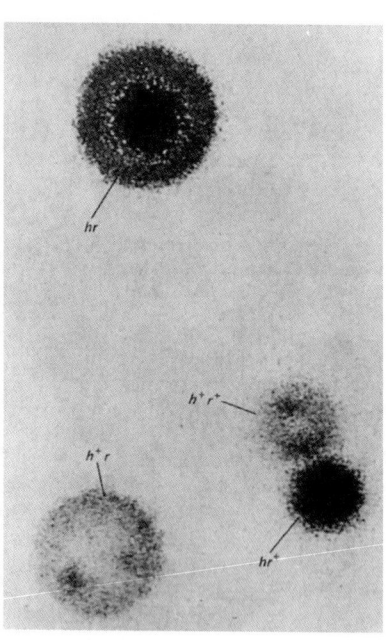

(c)

Summary

13.1 Bacterial Recombination: General Principles

a. In recombination, genetic material from two different chromosomes is combined to form a new, hybrid chromosome. There are three types of recombination: general recombination, site-specific recombination, and replicative recombination.

b. Bacterial recombination is a one-way process in which the exogenote is transferred from the donor to a recipient and integrated into the endogenote (**figure 13.4**).

13.2 Bacterial Plasmids

a. Plasmids are small, circular, autonomously replicating DNA molecules that can exist independent of the host chromosome. Their genes are not required for host survival.

b. Episomes are plasmids that can be reversibly integrated with the host chromosome.

c. Many important types of plasmids have been discovered: F factors, R factors, Col plasmids, virulence plasmids, and metabolic plasmids.

13.3 Transposable Elements

a. Transposons or transposable elements are DNA segments that move about the genome in a process known as transposition.

b. There are two types of transposable elements: insertion sequences and composite transposons.

c. Transposable elements cause mutations, block translation and transcription, turn genes on and off, aid F plasmid insertion, and carry antibiotic resistance genes.

13.4 Bacterial Conjugation

a. Conjugation is the transfer of genes between bacteria that depends upon direct cell-cell contact mediated by the F pilus.

b. In F⁺ × F⁻ mating the F factor remains independent of the chromosome and a copy is transferred to the F⁻ recipient; donor genes are not usually transferred (**figure 13.14**).

c. Hfr strains transfer bacterial genes to recipients because the F factor is integrated into the host chromosome. A complete copy of the F factor is not often transferred (**figure 13.14**).

d. When the F factor leaves an Hfr chromosome, it occasionally picks up some bacterial genes to become an F′ plasmid, which readily transfers these genes to other bacteria (**figure 13.15**).

13.5 DNA Transformation

a. Transformation is the uptake of naked DNA by a competent cell and its incorporation into the genome (**figure 13.16**).

13.6 Transduction

a. Bacterial viruses or bacteriophages can reproduce and destroy the host cell (lytic cycle) or become a latent prophage that remains within the host (lysogenic cycle) (**figure 13.18**).

b. Transduction is the transfer of bacterial genes by viruses.

c. In generalized transduction any host DNA fragment can be packaged in a virus capsid and transferred to a recipient (**figure 13.19**).

d. Temperate phages carry out specialized transduction by incorporating bacterial genes during prophage induction and then donating those genes to another bacterium (**figure 13.20**).

13.7 Mapping the Genome

a. The bacterial genome can be mapped by following the order of gene transfer during Hfr conjugation (**figure 13.22**); transformational and transductional mapping techniques also may be used.

13.8 Recombination and Genome Mapping in Viruses

a. When two viruses simultaneously enter a bacterial cell, their chromosomes can undergo recombination.

b. Virus genomes are mapped by recombination and heteroduplex mapping techniques.

Key Terms

abortive transductants 302	F′ plasmid 297	low-frequency transduction lysate	R factors 289
bacteriocin 291	generalized transducing particle 302	(LFT lysate) 303	sex pilus 296
competent 299	generalized transduction 301	lysogenic 301	site-specific recombination 286
composite transposons 292	general recombination 286	lysogens 301	specialized transduction 302
conjugation 296	helper phage 304	lysogeny 301	temperate phage 301
conjugative plasmid 288	heteroduplex DNA 286	merozygote 287	transduction 301
conjugative transposon 295	Hfr strain 297	metabolic plasmids 291	transformation 299
crossing-over 286	high-frequency transduction lysate	plasmid 288	transposable element 291
curing 288	(HFT lysate) 305	prophage 301	transposase 292
endogenote 286	horizontal gene transfer 286	recombination 286	transposition 291
episome 288	host restriction 288	replicative recombination 286	transposon 291
exogenote 286	insertion sequence 292	replicon 288	virulence plasmids 291
F factor 288	interrupted mating experiment 306	restricted transduction 302	

Questions for Thought and Review

1. How does recombination in procaryotes differ from that in most eucaryotes?

2. Distinguish between plasmids, transposons, and temperate phages.

3. How might one demonstrate the presence of a plasmid in a host cell?

4. What effect would you expect the existence of transposable elements and plasmids to have on the rate of microbial evolution? Give your reasoning.

5. How do multiple-drug-resistant plasmids often arise?

6. Suppose that you carried out a U-tube experiment with two auxotrophs and discovered that recombination was not blocked by the filter but was stopped by treatment with deoxyribonuclease. What gene transfer process is responsible? Why would it be best to use double or triple auxotrophs in this experiment?

7. List the similarities and differences between conjugation, transformation, and transduction.

8. How might one tell whether a recombination process was mediated by generalized or specialized transduction?

9. Why doesn't a cell lyse after successful transduction with a temperate phage?

10. Describe how you would precisely locate the *recA* gene and show that it was between 58 and 58.5 minutes on the *E. coli* chromosome.

Critical Thinking Questions

1. Diagram a double crossover event and a single crossover event. Which is more infrequent and why? Suggest experiments in which you would use one or the other event and what types of genetic markers you would employ. What kind of recognition features and catalytic capabilities would the recombination machinery need to possess?

2. Suppose that transduction took place when a U-tube experiment was conducted. How would you confirm that something like a virus was passed through the filter and transduced the recipient?

3. What would be the evolutionary advantage of having a period of natural "competence" in a bacterial life cycle? What would be possible disadvantages?

For recommended readings on these and related topics, see Appendix V.

chapter 14

Recombinant DNA Technology

Concepts

- Genetic engineering makes use of recombinant DNA technology to fuse genes with vectors and then clone them in host cells. In this way large quantities of isolated genes and their products can be synthesized.
- The production of recombinant DNA molecules depends on the ability of restriction endonucleases to cleave DNA at specific sites.
- Plasmids, bacteriophages and other viruses, and cosmids are used as vectors. They can replicate within a host cell while carrying foreign DNA and possess phenotypic traits that allow them to be detected.
- Genetic engineering is already making substantial contributions to biological research, medicine, industry, and agriculture. Future benefits are probably much greater.
- Genetic engineering also is accompanied by potential problems in such areas as safety, the ethics of its use with human subjects, environmental impact, and biological warfare.

The recombinant DNA breakthrough has provided us with a new and powerful approach to the questions that have intrigued and plagued man for centuries.

—Paul Berg

Four streaks of *E. coli* glow different colors because they contain different cloned luciferase genes.

hapters 12 and 13 introduce the essentials of microbial genetics. This chapter focuses on the practical applications of microbial genetics and the technology arising from it.

Although human beings have been altering the genetic makeup of organisms for centuries by selective breeding, only recently has the direct manipulation of DNA been possible. The deliberate modification of an organism's genetic information by directly changing its nucleic acid genome is called **genetic engineering** and is accomplished by a collection of methods known as **recombinant DNA technology.** First, the DNA responsible for a particular phenotype is identified and isolated. Once purified, the gene or genes are fused with other pieces of DNA to form recombinant DNA molecules. These are propagated (gene cloning) by insertion into an organism that need not even be in the same kingdom as the original gene donor. Recombinant DNA technology opens up totally new areas of research and applied biology. Thus it is an essential part of **biotechnology,** which is now experiencing a stage of exceptionally rapid growth and development. Although the term has several definitions, in this text biotechnology refers to those processes in which living organisms are manipulated, particularly at the molecular genetic level, to form useful products. The promise for medicine, agriculture, and industry is great; yet the potential risks of this technology are not completely known and may be considerable.   Biotechnology and industrial microbiology (chapter 42)

Recombinant DNA technology is very much the result of several key discoveries in microbial genetics. Section 14.1 briefly reviews some landmarks in the development of recombinant technology (**table 14.1**).

Table 14.1

Some Milestones in Biotechnology and Recombinant DNA Technology

1958	DNA polymerase purified
1970	A complete gene synthesized in vitro
	Discovery of the first sequence-specific restriction endonuclease and the enzyme reverse transcriptase
1972	First recombinant DNA molecules generated
1973	Use of plasmid vectors for gene cloning
1975	Southern blot technique for detecting specific DNA sequences
1976	First prenatal diagnosis using a gene-specific probe
1977	Methods for rapid DNA sequencing
	Discovery of "split genes" and somatostatin synthesized using recombinant DNA
1978	Human genomic library constructed
1979	Insulin synthesized using recombinant DNA
	First human viral antigen (hepatitis B) cloned
1981	Foot-and-mouth disease viral antigen cloned
	First monoclonal antibody-based diagnostic kit approved for use
1982	Commercial production by *E. coli* of genetically engineered human insulin
	Isolation, cloning, and characterization of a human cancer gene
	Transfer of gene for rat growth hormone into fertilized mouse eggs
1983	Engineered Ti plasmids used to transform plants
1985	Tobacco plants made resistant to the herbicide glyphosate through insertion of a cloned gene from *Salmonella*
	Development of the polymerase chain reaction technique
1987	Insertion of a functional gene into a fertilized mouse egg cures the shiverer mutation disease of mice, a normally fatal genetic disease
1988	The first successful production of a genetically engineered staple crop (soybeans)
	Development of the gene gun
1989	First field test of a genetically engineered virus (a baculovirus that kills cabbage looper caterpillars)
1990	Production of the first fertile corn transformed with a foreign gene (a gene for resistance to the herbicide bialaphos)
1991	Development of transgenic pigs and goats capable of manufacturing proteins such as human hemoglobin
	First test of gene therapy on human cancer patients
1994	The Flavr Savr tomato introduced, the first genetically engineered whole food approved for sale
	Fully human monoclonal antibodies produced in genetically engineered mice
1995	*Haemophilus influenzae* genome sequenced
1996	*Methanococcus jannaschii* and *Saccharomyces cerevisiae* genomes sequenced
1997	Human clinical trials of antisense drugs and DNA vaccines begun
	E. coli genome sequenced
1998	First cloned mammal (the sheep Dolly)
2002	*Plasmodium falciparum* genome sequenced
2003	Completion of the draft of the human genome

14.1   HISTORICAL PERSPECTIVES

Recombinant DNA is DNA with a new sequence formed by joining fragments from two or more different sources. One of the first breakthroughs leading to recombinant DNA (rDNA) technology was the discovery in the late 1960s by Werner Arber and Hamilton Smith of microbial enzymes that make cuts in double-stranded DNA. These enzymes recognize and cleave specific sequences about 4 to 8 base pairs long and are known as **restriction enzymes** or restriction endonucleases (**figure 14.1**). They normally protect the host cell by destroying phage DNA after its entrance. Cells protect their own DNA from restriction enzymes by methylating nucleotides in the sites that these enzymes recognize. Incoming foreign DNA is not methylated at the same sites and often is cleaved by host restriction enzymes. There are three general types of restriction enzymes. Types I and III cleave DNA away from recognition sites. Type II restriction endonucleases cleave DNA at specific recognition sites. The type II enzymes can be used to prepare DNA fragments containing specific genes or portions of genes. For example, the restriction

enzyme *Eco*RI, isolated by Herbert Boyer in 1969 from *E. coli*, cleaves the DNA between G and A in the base sequence GAATTC (**figure 14.2**). Note that in the double-stranded condition, the base sequence GAATTC will base pair with the same sequence running in the opposite direction. *Eco*RI therefore cleaves both DNA strands between the G and the A. When the two DNA fragments separate, they often contain single-stranded complementary ends, known as sticky ends or cohesive ends. There are hundreds of restriction enzymes that recognize many

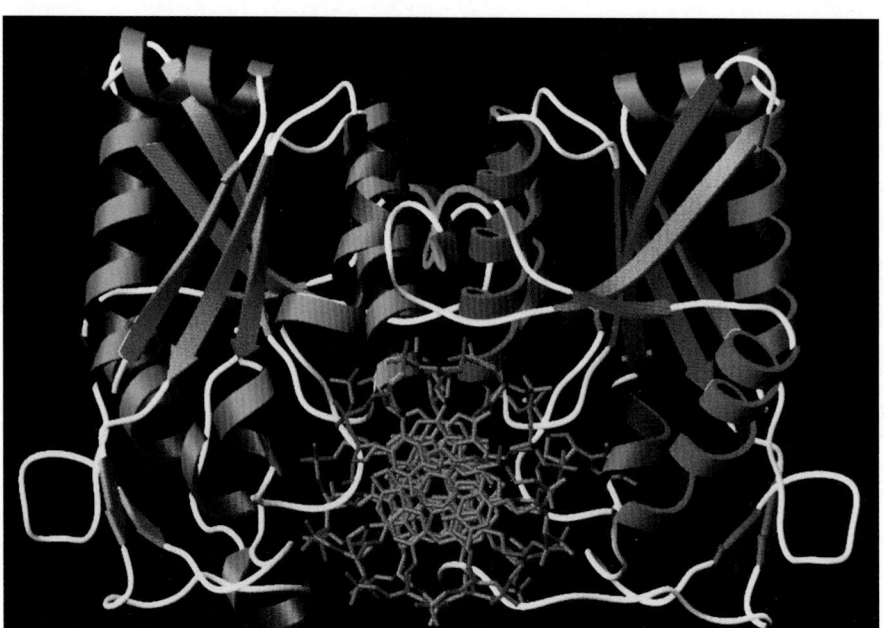

Figure 14.1 Restriction Endonuclease Binding to DNA. The structure of *Bam*HI binding to DNA viewed down the DNA axis. The enzyme's two subunits lie on each side of the DNA double helix. The α-helices are in green, the β conformations in purple to red, and DNA is in orange.

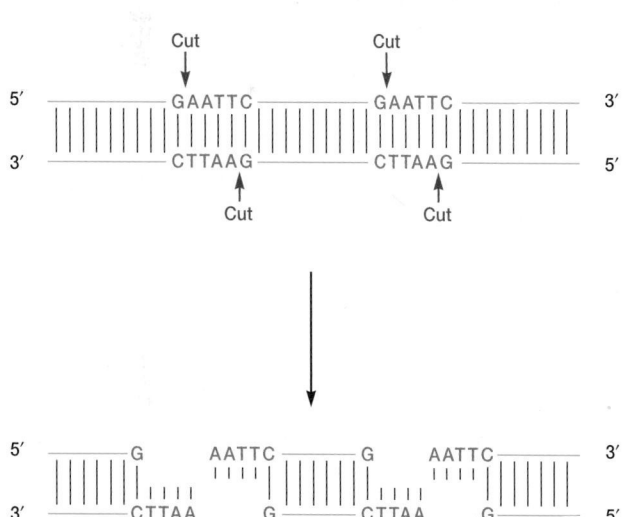

Figure 14.2 Restriction Endonuclease Action. The cleavage catalyzed by the restriction endonuclease *Eco*RI. The enzyme makes staggered cuts on the two DNA strands to form sticky ends.

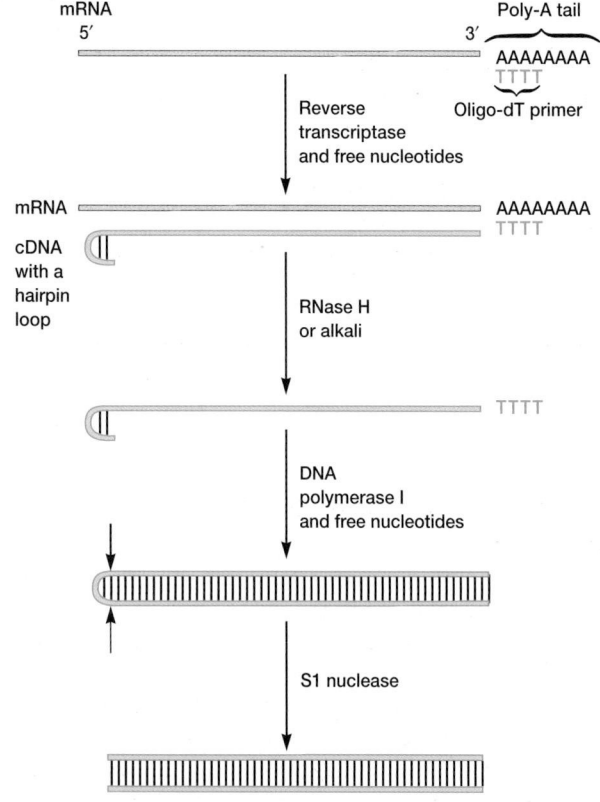

Figure 14.3 The Synthesis of Double-Stranded cDNA from mRNA. This figure briefly summarizes one procedure for synthesis of cDNA. Reverse transcriptase synthesizes a DNA copy of the mRNA using an oligo-dT primer. The RNA of the resulting RNA-DNA hybrid is degraded by RNase H to produce single-stranded DNA with a hairpin loop at one end. DNA polymerase I then converts this to a double-stranded form, and S1 nuclease nicks the DNA at the points indicated by the two arrows on the left to generate the final DNA copy.

different specific sequences (**table 14.2**). Each restriction enzyme name begins with three letters, indicating the bacterium producing it. For example, *Eco*RI is obtained from *E. coli*, whereas *Bam*HI comes from *Bacillus amyloliquefaciens H,* and *Sal*I from *Streptomyces albus.*   Restriction and phages (p. 375)

In 1970 Howard Temin and David Baltimore independently discovered the enzyme reverse transcriptase that retroviruses use to produce DNA copies of their RNA genome. This enzyme can be used to construct a DNA copy, called **complementary DNA (cDNA),** of any RNA (**figure 14.3**). Thus genes or major portions of genes can be synthesized from mRNA.   Reverse transcriptase and retroviruses (pp. 395–97)

Table 14.2			

Some Type II Restriction Endonucleases and Their Recognition Sequences

Enzyme	Microbial Source	Recognition Sequence[a]	End Produced[b]
*Alu*I	*Arthrobacter luteus*	↓ 5′—A—G—C—T—3′ 3′—T—C—G—A—5′ ↑	C—T—3′ G—A—5′
*Bam*HI	*Bacillus amyloliquefaciens* H	↓ 5′—G—G—A—T—C—C—3′ 3′—C—C—T—A—G—G—5′ ↑	G—A—T—C—C—3′ G—5′
*Eco*RI	*Escherichia coli*	↓ 5′—G—A—A—T—T—C—3′ 3′—C—T—T—A—A—G—5′ ↑	A—A—T—T—C—3′ G—5′
*Hae*III	*Haemophilus aegyptius*	↓ 5′—G—G—C—C—3′ 3′—C—C—G—G—5′ ↑	C—C—3′ G—G—5′
*Hin*dIII	*Haemophilus influenzae* b	↓ 5′—A—A—G—C—T—T—3′ 3′—T—T—C—G—A—A—5′ ↑	A—G—C—T—T—3′ A—5′
*Not*I	*Nocardia otitidis-caviarum*	↓ 5′—G—C—G—G—C—C—G—C—3′ 3′—C—G—C—C—G—G—C—G—5′ ↑	G—G—C—C—G—C—3′ C—G—5′
*Pst*I	*Providencia stuartii*	↓ 5′—C—T—G—C—A—G—3′ 3′—G—A—C—G—T—C—5′ ↑	G—3′ A—C—G—T—C—5′
*Sal*I	*Streptomyces albus*	↓ 5′—G—T—C—G—A—C—3′ 3′—C—A—G—C—T—G—5′ ↑	T—C—G—A—C—3′ G—5′

[a]The arrows indicate the sites of cleavage on each strand.
[b]Only the end of the right-hand fragment is shown.

The next advance came in 1972, when David Jackson, Robert Symons, and Paul Berg reported that they had successfully generated recombinant DNA molecules. They allowed the sticky ends of fragments to anneal—that is, to base pair with one another—and then covalently joined the fragments with the enzyme DNA ligase. Within a year, plasmid **vectors,** or carriers of foreign DNA fragments during gene cloning, had been developed and combined with foreign DNA (**figure 14.4**). The first such recombinant plasmid capable of being replicated within a bacterial host was the pSC101 plasmid constructed by Stanley Cohen and Herbert Boyer in 1973 (SC in the plasmid name stands for *S*tanley *C*ohen).

In 1975 Edwin M. Southern published a procedure for detecting specific DNA fragments so that a particular gene could be isolated from a complex DNA mixture. The **Southern blotting technique** depends on the specificity of base complementarity in nucleic acids (**figure 14.5**). DNA fragments are first separated by size with agarose gel electrophoresis. The fragments are then denatured (rendered single stranded) and transferred to a nitrocellu-

lose filter or nylon membrane so that each fragment is firmly bound to the filter at the same position as on the gel. Originally the transfer occurred when buffer flowed through the gel and the membrane as shown in figure 14.5. The negatively charged DNA fragments also can be electrophoresed from the gel onto the blotting membrane. The filter is bathed with solution containing a radioactive **probe,** a piece of labeled nucleic acid that hybridizes with complementary DNA fragments and is used to locate them. Those fragments complementary to the probe become radioactive and are readily detected by **autoradiography.** In this technique a sheet of photographic film is placed over the filter for several hours and then developed. The film is exposed and becomes dark everywhere a radioactive fragment is located because the energy released by the isotope causes the formation of dark-silver grains. Using Southern blotting, one can detect and isolate fragments with any desired sequence from a complex mixture.

More recently, nonradioactive probes to detect specific DNAs have been developed. In one approach the DNA probe is linked to

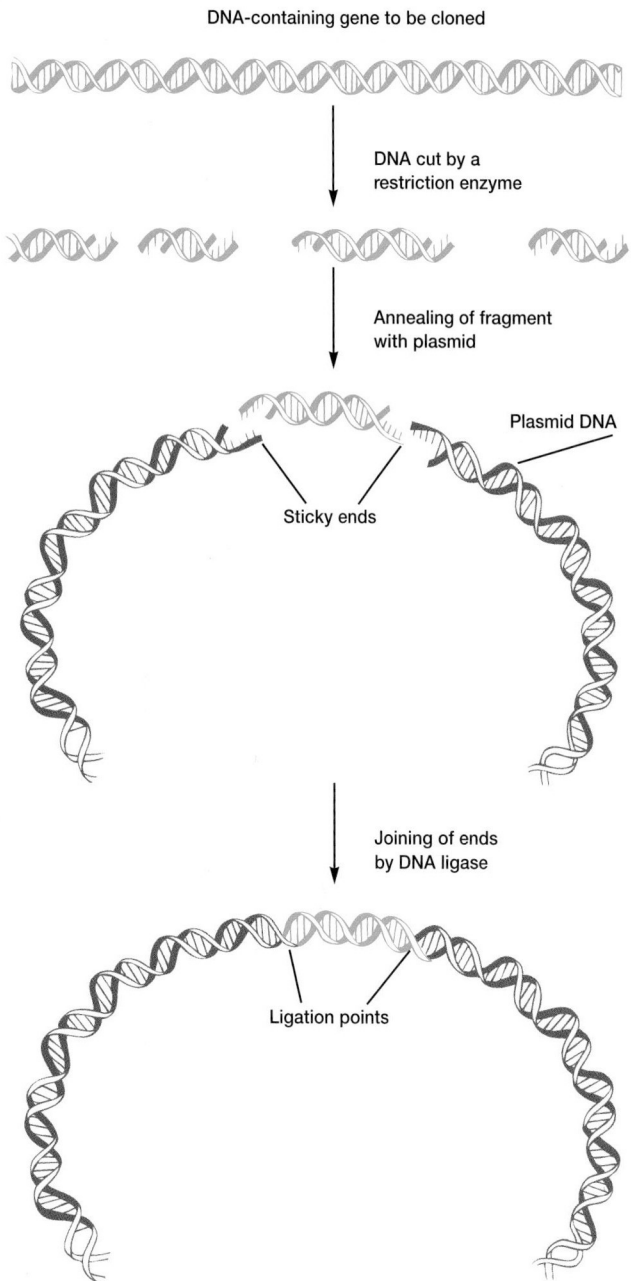

DNA-containing gene to be cloned

DNA cut by a
restriction enzyme

Annealing of fragment
with plasmid

Plasmid DNA

Sticky ends

Joining of ends
by DNA ligase

Ligation points

Figure 14.4 Recombinant Plasmid Construction. The general procedure used in constructing recombinant plasmid vectors is shown here.

an enzyme such as horseradish peroxidase. After the enzyme-DNA probe has bound to a DNA fragment on the filter, the substrate luminol that will emit light when acted on by the peroxidase is added. The chemiluminescent probe is detected by exposing the filter to photographic film for about 20 minutes. A second technique makes use of the vitamin biotin. A biotin-DNA probe is detected by incubating the filter with either the protein avidin or a similar bacterial protein, streptavidin (*see Techniques and Applications 42.4*). The protein specifically attaches to biotin, and is visualized with a spe-

cial reagent containing biotin complexed with the enzyme alkaline phosphatase (streptavidin also can be directly attached to the enzyme). The bands with the probe appear blue. These nonradioactive techniques are more rapid and safer than using radioisotopes but they may be less sensitive.

By the late 1970s techniques for easily sequencing DNA, synthesizing oligonucleotides, and expressing eucaryotic genes in procaryotes had also been developed. These techniques were then used to solve practical problems (table 14.1). Sections 14.2 through 14.8 describe how the previously discussed techniques and others are used in genetic engineering.

1. Define or describe restriction enzyme, sticky end, cDNA, vector, Southern blotting, probe, and autoradiography.

 SYNTHETIC DNA

Oligonucleotides [Greek *oligo,* few or scant] are short pieces of DNA or RNA between about 2 and 20 or 30 nucleotides long. The ability to synthesize DNA oligonucleotides of known sequence is extremely useful. For example, DNA probes can be synthesized and DNA fragments can be prepared for use in molecular techniques such as PCR (pp. 316–19). DNA structure (pp. 224–28)

DNA oligonucleotides are synthesized by a stepwise process in which single nucleotides are added to the end of the growing chain (**figure 14.6**). The 3′ end of the chain is attached to a solid support such as a silica gel particle. A DNA synthesizer or "gene machine" carries out the solid-phase synthesis. A specially activated nucleotide derivative is added to the 5′ end of the chain in a series of steps. At the end of an addition cycle, the growing chain is separated from the reaction mixture by filtration or centrifugation. The process is then repeated to attach another nucleotide. It takes about 40 minutes to add a nucleotide to the chain, and chains as large as 50 to 100 nucleotides can be synthesized.

Advances in DNA synthetic techniques have accelerated progress in the study of protein function. One of the most effective ways of studying the relationship of protein structure to function is by altering a specific part of the protein and observing functional changes. In the past this has been accomplished either by chemically modifying individual amino acids or by inducing mutations in the gene coding for the protein under study. There are problems with these two approaches. Chemical modification of a protein is not always specific; several amino acids may be altered, not just the one desired. It is not always possible to produce the proper mutation in the desired gene location. Recently these difficulties have been overcome with a technique called **site-directed mutagenesis.**

In site-directed mutagenesis an oligonucleotide of about 20 residues that contains the desired sequence change is synthesized. The altered oligonucleotide with its artificially mutated sequence is now allowed to bind to a single-stranded copy of the complete gene (**figure 14.7**). DNA polymerase is added to the gene-primer complex. The polymerase extends the primer and replicates the remainder of the target gene to produce a new gene copy with the

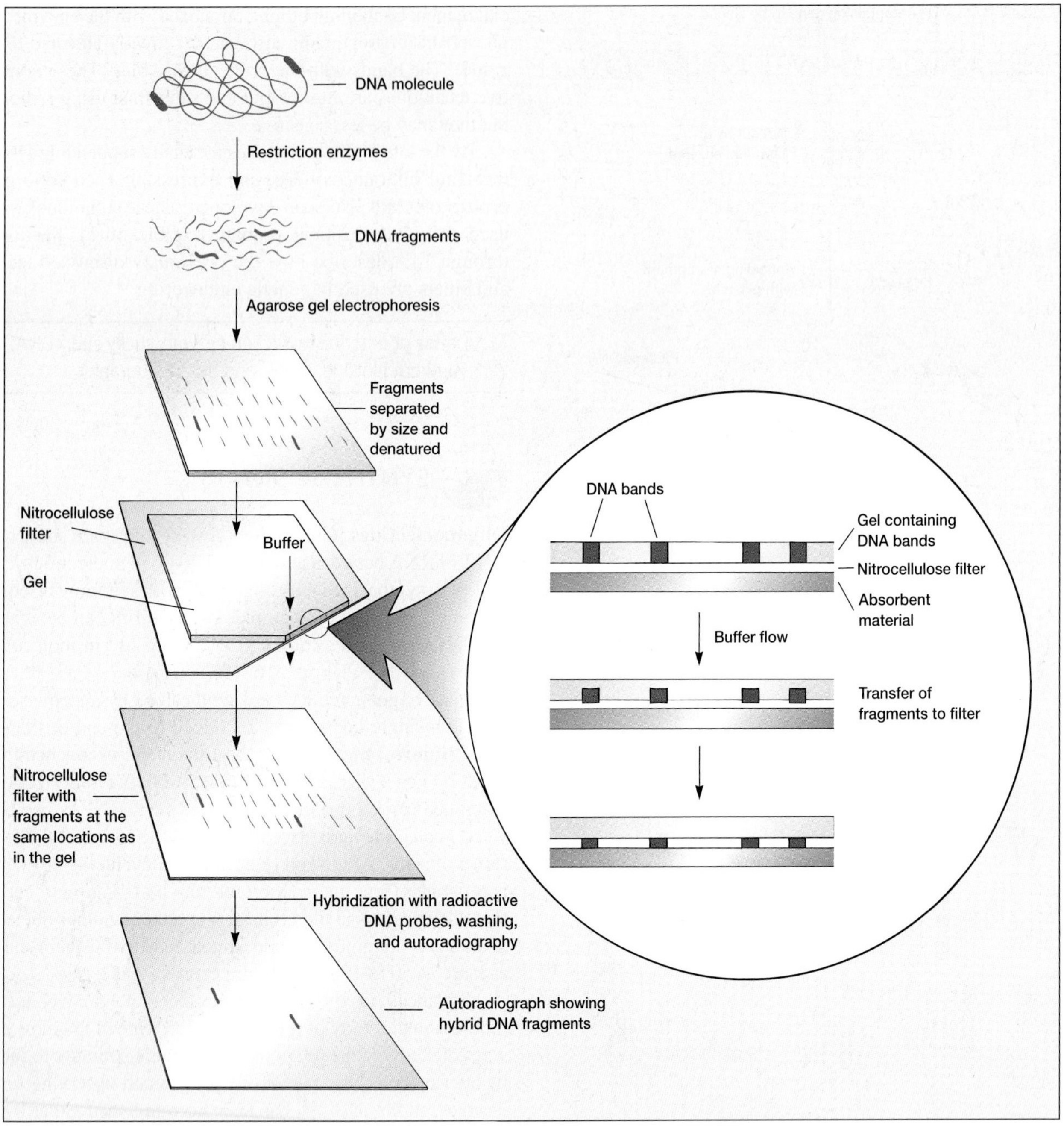

Figure 14.5   The Southern Blotting Technique.   The insert illustrates how buffer flow transfers DNA bands to the nitrocellulose filter. The fragments also can be transferred by electrophoresis. See text for further details.

desired mutation. If the gene is attached to a single-stranded DNA bacteriophage (*see pp. 362–64 and 378*) such as the M13 phage, it can be introduced into a host bacterium and cloned using the techniques to be described shortly. This will yield large quantities of the mutant protein for study of its function.

14.3   THE POLYMERASE CHAIN REACTION

Between 1983 and 1985 Kary Mullis developed a technique to synthesize large quantities of a DNA fragment without cloning it.

Although this chapter emphasizes recombinant DNA technology, the **polymerase chain reaction** or **PCR technique** will be introduced here because of its great practical importance and impact on biotechnology.

Figure 14.8 outlines how the PCR technique works. Suppose that one wishes to make large quantities of a particular DNA sequence. The first step is to synthesize fragments with sequences identical to those flanking the targeted sequence. This is readily accomplished with a DNA synthesizer machine as previously described. These synthetic oligonucleotides are usually about 20 nucleotides long and serve as primers for DNA synthesis. The

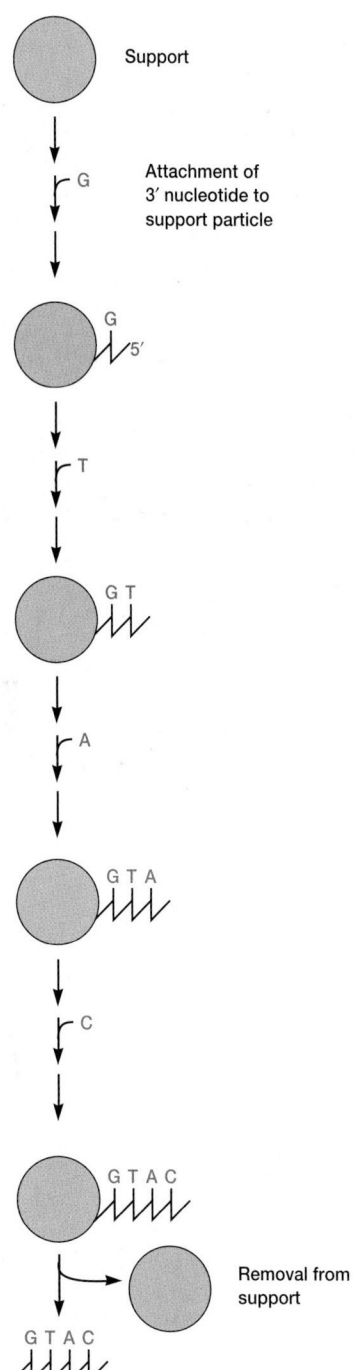

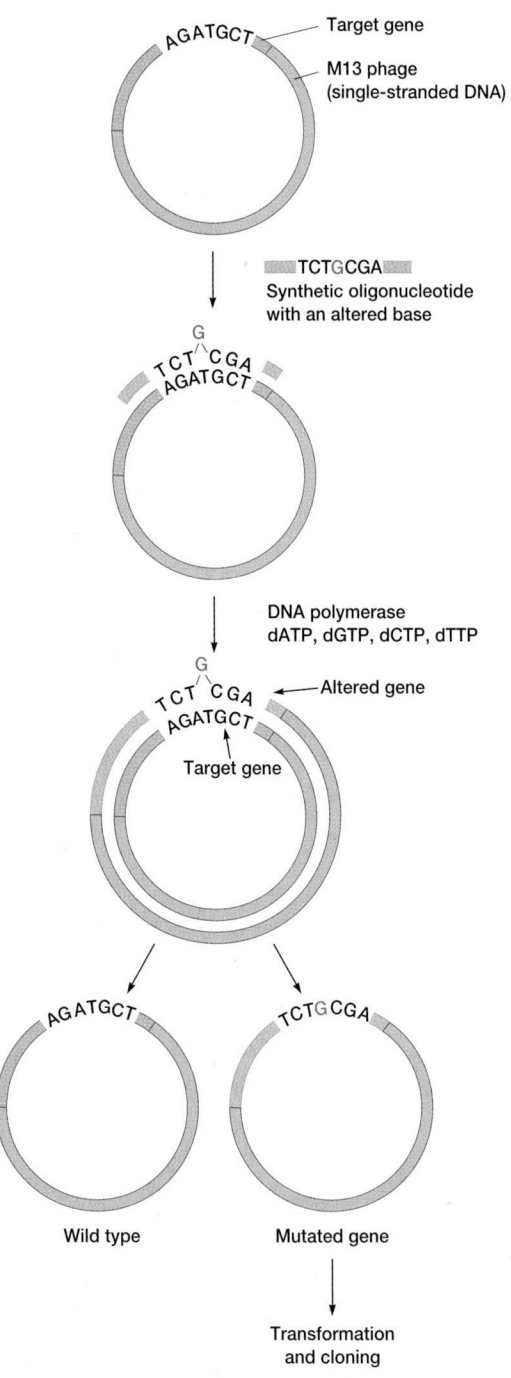

Figure 14.6 The Synthesis of a DNA Oligonucleotide.
During each cycle, the DNA synthesizer adds an activated nucleotide (A, T, G, or C) to the growing end of the chain. At the end of the process, the oligonucleotide is removed from its support.

Figure 14.7 Site-Directed Mutagenesis. A synthetic oligonucleotide is used to add a specific mutation to a gene. See text for details.

reaction mix contains the target DNA, a very large excess of the desired primers, a thermostable DNA polymerase, and four deoxyribonucleoside triphosphates. The PCR cycle itself takes place in three steps. First, the target DNA containing the sequence to be amplified is heat denatured to separate its complementary strands (step 1). Normally the target DNA is between 100 and 5,000 base pairs in length. Next, the temperature is lowered so that the

primers can hydrogen bond or anneal to the DNA on both sides of the target sequence (step 2). Because the primers are present in excess, the targeted DNA strands normally anneal to the primers rather than to each other. Finally, DNA polymerase extends the primers and synthesizes copies of the target DNA sequence using the deoxyribonucleoside triphosphates (step 3). Only polymerases able to function at the high temperatures employed in the PCR

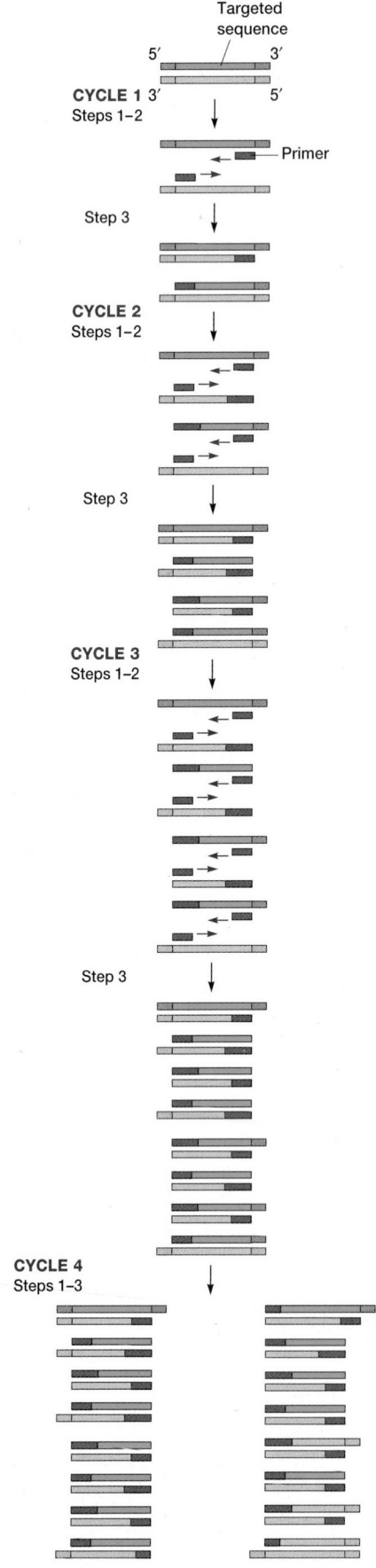

Figure 14.9 A Modern PCR Machine. PCR machines are now fully automated and microprocessor controlled. They can process up to 96 samples at a time.

technique can be used. Two popular enzymes are the Taq polymerase from the thermophilic bacterium *Thermus aquaticus* and the Vent polymerase from *Thermococcus litoralis*. At the end of one cycle, the targeted sequences on both strands have been copied. When the three-step cycle is repeated (figure 14.8), the four strands from the first cycle are copied to produce eight fragments. The third cycle yields 16 products. Theoretically, 20 cycles will produce about one million copies of the target DNA sequence; 30 cycles yield around one billion copies. Pieces ranging in size from less than 100 base pairs to several thousand base pairs in length can be amplified, and only 10 to 100 picomoles of primer are required. The concentration of target DNA can be as low as 10^{-20} to 10^{-15} M (or 1 to 10^5 DNA copies per 100 μl). The whole reaction mixture is often 100 μl or less in volume. DNA sequencing (p. 336); DNA replication (pp. 229–33)

The polymerase chain reaction technique has now been automated and is carried out by a specially designed machine (**figure 14.9**). Currently a PCR machine can carry out 25 cycles and amplify DNA 10^5 times in as little as 57 minutes. During a typical cycle the DNA is denatured at 94°C for 15 seconds, then the primers are annealed and extended (steps 2 and 3) at 68°C for 60 seconds.

Figure 14.8 The Polymerase Chain Reaction (PCR). In four cycles the targeted sequence has been amplified many times. See text for details.

PCR technology is improving continually. For example, RNA now can be efficiently used in PCR procedures. The *Tth* DNA polymerase, a recombinant *Thermus thermophilus* DNA polymerase, will transcribe RNA to DNA and then amplify the DNA. Cellular RNAs and RNA viruses may be studied even when the RNA is present in very small amounts (as few as 100 copies can be transcribed and amplified). PCR also can quantitate DNA products without the use of isotopes. This allows one to find the initial amount of target DNA in less than an hour using automated equipment. Quantitative PCR is quite valuable in virology and gene expression studies. As mentioned earlier, the target DNA to be amplified is normally less than about 5,000 base pairs in length. A "long PCR" technique has been developed that will amplify sequences up to 42 kilobases long. It depends on the use of error-correcting polymerases because Taq polymerase is error-prone.

The PCR technique has already proven exceptionally valuable in many areas of molecular biology, medicine, and biotechnology. It can be used to amplify very small quantities of a specific DNA and provide sufficient material for accurately sequencing the fragment or cloning it by standard techniques. PCR-based diagnostic tests for AIDS, Lyme disease, chlamydia, tuberculosis, hepatitis, the human papilloma virus, and other infectious agents and diseases are being developed. The tests are rapid, sensitive, and specific. PCR is particularly valuable in the detection of genetic diseases such as sickle cell anemia, phenylketonuria, and muscular dystrophy. The technique is already having an impact on forensic science where it is being used in criminal cases as a part of DNA fingerprinting technology. It is possible to exclude or incriminate suspects using extremely small samples of biological material discovered at the crime scene.

 ## 14.4   PREPARATION OF RECOMBINANT DNA

There are three ways to obtain adequate quantities of a DNA fragment. One can extract all the DNA from an organism, cleave the DNA into fragments, isolate the fragment of interest, and finally clone it. Alternatively, all of the fragments can be cloned by means of a suitable vector, and each clone (the population of identical molecules with a single ancestral molecule) can be tested for the desired gene. One also can directly synthesize the desired DNA fragment as described earlier, and then clone it.

Isolating and Cloning Fragments

Long, linear DNA molecules are fragile and easily sheared into fragments by passing the DNA suspension through a syringe needle several times. DNA also can be cut with restriction enzymes. The resulting fragments are either separated by electrophoresis or first inserted into vectors and cloned.

Agarose or polyacrylamide gels usually are used to separate DNA fragments electrophoretically. In **electrophoresis,** charged molecules are placed in an electrical field and allowed to migrate toward the positive and negative poles. The molecules separate because they move at different rates due to their differences in

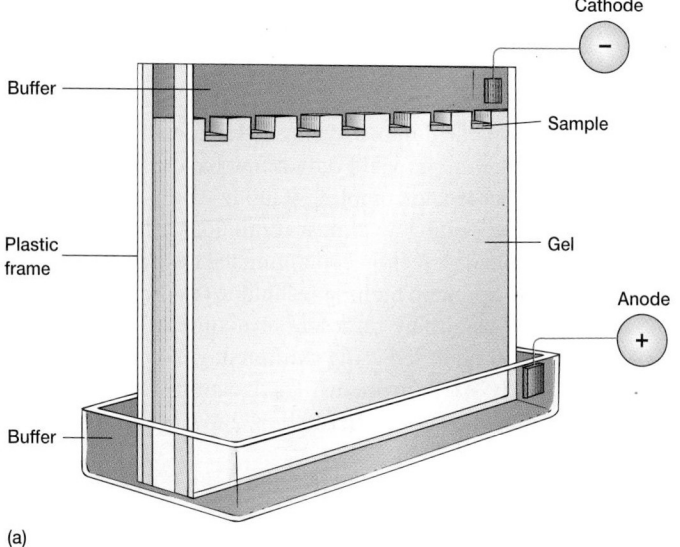

(a)

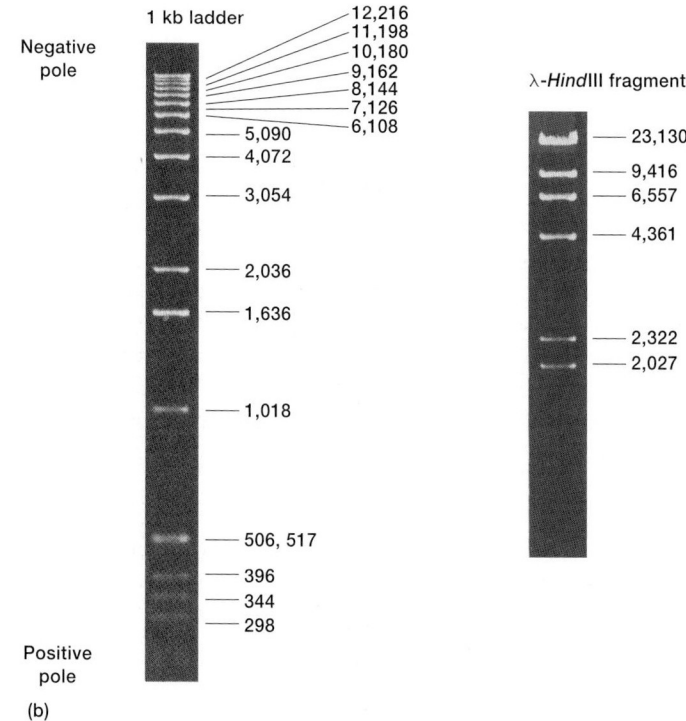

(b)

Figure 14.10   Gel Electrophoresis of DNA.   (**a**) A diagram of a vertical gel apparatus showing its parts. (**b**) The 1 kilobase ladder is an electrophoretic gel containing a series of DNA fragments of known size. The numbers indicate the number of base pairs each fragment contains. The smallest fragments have moved the farthest. The gel on the right shows many of the fragments that arise when lambda phage DNA is digested with the *Hind*III restriction enzyme.

charge and size. In practice, the fragment mixture is usually placed in wells molded within a sheet of gel (**figure 14.10**). The gel concentration varies with the size of DNA fragments to be separated. Usually 1 to 3% agarose gels or 3 to 20% polyacrylamide gels are used. When an electrical field is generated in the gel, the fragments move through the pores of the gel toward an electrode (negatively charged DNA fragments migrate toward the positive electrode or anode). Each fragment's migration rate is inversely proportional to the log of its molecular weight, and the fragments are separated into size classes (figure 14.10b). A simple DNA molecule might yield only a few bands. If the original DNA was very large and complex, staining of the gel would reveal a smear representing an almost continuous gradient in fragment size. The band or section containing the desired fragment is located with the Southern blotting technique (figure 14.5) and removed. Since a band may represent a mixture of several fragments, it is electrophoresed using a different gel concentration to separate similarly sized fragments. The location of the pure fragment is determined by Southern blotting, and it is extracted from the gel.

Once fragments have been isolated, they are ligated with an appropriate vector, such as a plasmid (*see section 13.2*), to form a recombinant molecule that can reproduce in a host cell. One of the easiest and most popular approaches is to cut the plasmid and donor DNA with the same restriction enzyme so that identical sticky ends are formed (**figure 14.11**). After a fragment has annealed with the plasmid through complementary base pairing, the breaks are joined by DNA ligase. A second method for creating recombinant molecules can be used with fragments and vectors lacking sticky ends. After cutting the plasmid and donor DNA, one can add poly(dA) to the 3′ ends of the plasmid DNA, using the enzyme terminal transferase (**figure 14.12**). Similarly, poly(dT) is added to the 3′ ends of the fragments. The ends will now base pair with each other and are joined by DNA ligase to form a recombinant plasmid. Some enzymes (e.g., the *Alu*I restriction enzyme from *Arthrobacter luteus*) cut DNA at the same position on both strands to form a blunt end. Fragments and vectors with blunt ends may be joined by T4 DNA ligase (blunt-end ligation).

The rDNA molecules are cloned by inserting them into bacteria, using transformation or phage injection (*see section 13.5*). Each strain reproduces to yield a population containing a single type of recombinant molecule. The overall process is outlined in **figure 14.13**. The same cloning techniques can be used with DNA fragments prepared using a DNA synthesizer machine.

Although selected fragments can be isolated and cloned as just described, it often is preferable to fragment the whole genome and clone all the fragments by using a vector. Then the desired clone can be identified. To be sure that the complete genome is represented in this collection of clones, called a **genomic library,** more than a thousand transformed bacterial strains must be maintained (the larger the genome, the more clones are needed). Libraries of cloned genes also can be generated using phage lambda as a vector and stored as phage lysates.

It is necessary to identify which clone in the library contains the desired gene. If the gene is expressed in the bacterium, it may be possible to assay each clone for a specific protein. However, a nucleic acid probe is normally employed in identification. The bacteria are replica plated on nitrocellulose paper and lysed in place with sodium hydroxide (**figure 14.14**). This yields a pattern of membrane-bound, denatured DNA corresponding to the colony pattern on the agar plate. The membrane is treated with a radioactive probe as in the original Southern blotting method. The radioactive spots identify colonies on the master plate that contain the desired DNA fragment. This approach also is used to analyze a library of cloned lambda phage (**figure 14.15**).

A cDNA library also can be constructed by using a suitable vector for cDNAs. Complementary DNAs are made from mRNAs, as described on page 313. The cDNA library differs from a genomic library in that it does not contain all the DNA fragments in the genome. Only DNA that is transcribed to yield mRNA is represented in the library. The advantage of the cDNA library is that it shows which genes are directing the synthesis of proteins in the cell.

Gene Probes

Success in isolating the desired recombinant clones depends on the availability of a suitable probe. Gene-specific probes are obtained in several ways. Frequently they are constructed with cDNA clones. If the gene of interest is expressed in a specific tissue or cell type, its mRNA is often relatively abundant. For example, reticulocyte mRNA may be enriched in globin mRNA, and pancreatic cells in insulin mRNA. Although mRNA is not available in sufficient quantity to serve as a probe, the desired mRNA species can be converted into cDNA by reverse transcription (figure 14.3). The cDNA copies are purified, spliced into appropriate vectors, and cloned to provide adequate amounts of the required probe.

Probes also can be generated if the gene codes for a protein of known amino acid sequence. Oligonucleotides, about 20 nucleotides or longer, that code for a characteristic amino acid sequence are synthesized. These often are satisfactory probes since they will specifically bind to the gene segment coding for the desired protein.

Sometimes previously cloned genes or portions of genes may be used as probes. This approach is effective when there is a reasonable amount of similarity between the nucleotide sequences of the two genes. Probes also can be generated by the polymerase chain reaction.

After construction, the probe is labeled to aid detection. Often ^{32}P is added to both DNA strands so that the radioactive strands can be located with autoradiography. Nonradioactively labeled probes may also be used.

Isolating and Purifying Cloned DNA

After the desired clone of recombinant bacteria or phages has been located with a probe, it can be picked from the master plate and propagated. The recombinant plasmid or phage DNA is then extracted and further purified when necessary. The DNA fragment is cut out of the plasmid or phage genome by means of restriction enzymes and separated from the remaining DNA by electrophoresis.

Clearly, DNA fragments can be isolated, purified, and cloned in several ways. Regardless of the exact approach, a key to successful cloning is choosing the right vector. Section 14.5 considers types of cloning vectors and their uses.

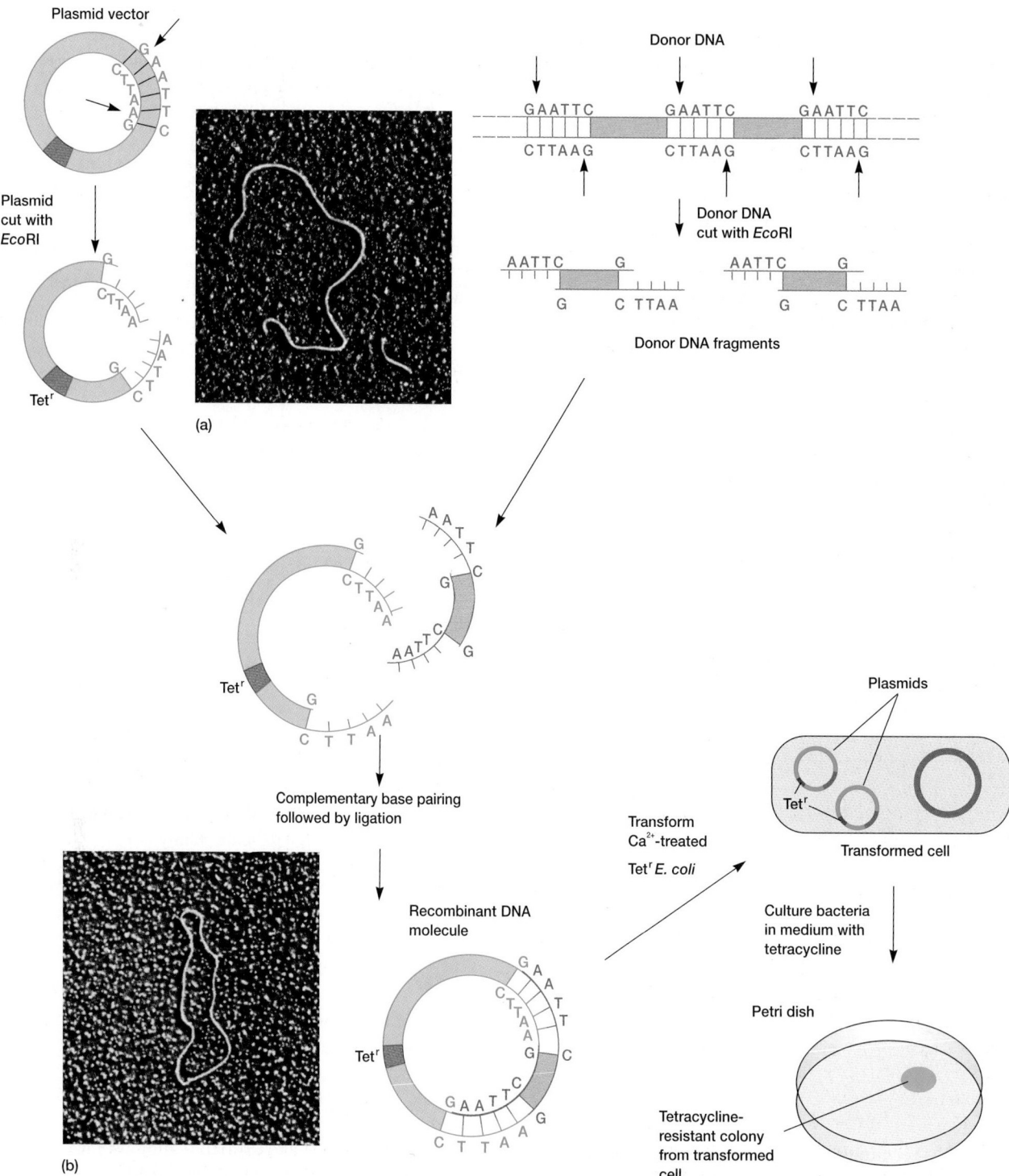

Figure 14.11   Recombinant Plasmid Construction and Cloning.   The construction and cloning of a recombinant plasmid vector using an antibiotic resistance gene to select for the presence of the plasmid. The scale of the sticky ends of the fragments and plasmid has been enlarged to illustrate complementary base pairing. (**a**) The electron micrograph shows a plasmid that has been cut by a restriction enzyme and a donor DNA fragment. (**b**) The micrograph shows a recombinant plasmid. See text for details.

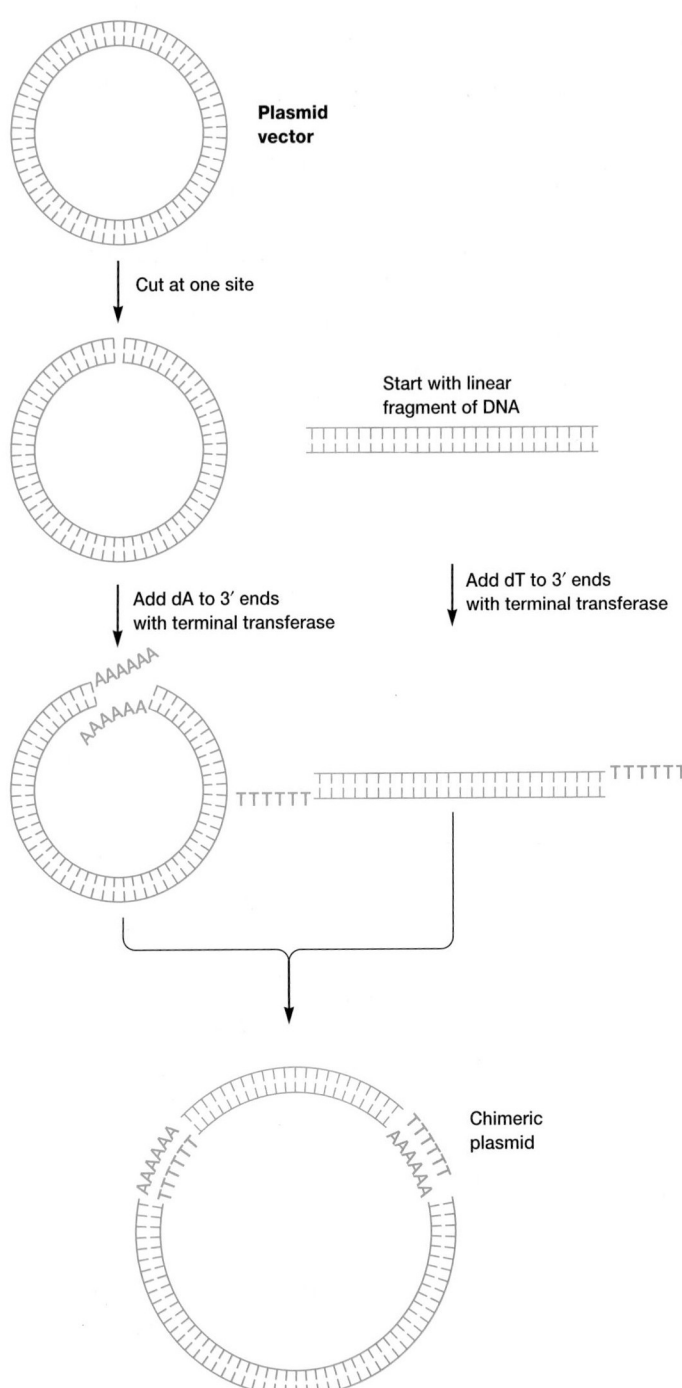

Figure 14.12 Terminal Transferase and the Construction of Recombinant Plasmids. The poly(dA-dT) tailing technique can be used to construct sticky ends on DNA and generate recombinant molecules.

---

1. How are oligonucleotides synthesized? What is site-directed mutagenesis?
2. Briefly describe the polymerase chain reaction technique. What is its importance?

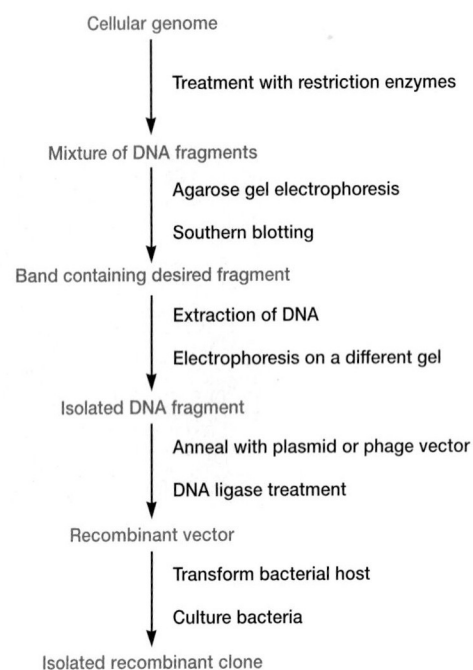

Figure 14.13 Cloning Cellular DNA Fragments. The preparation of a recombinant clone from previously isolated DNA fragments.

---

3. What is electrophoresis and how does it work?
4. Describe three ways in which a fragment can be covalently attached to vector DNA.
5. Outline in detail two different ways to isolate and clone a specific gene. What is a genomic library?
6. How are gene-specific probes obtained?

 CLONING VECTORS

There are four major types of vectors: plasmids, bacteriophages and other viruses, cosmids, and artificial chromosomes (**table 14.3**). Each type has its own advantages. Plasmids are the easiest to work with; rDNA phages and other viruses are more conveniently stored for long periods; larger pieces of DNA can be cloned with cosmids and artificial chromosomes. Besides these major types, there also are vectors designed for a specific function. For example, shuttle vectors are used in transferring genes between very different organisms and usually contain one replication origin for each host. The pYe type plasmids reproduce in both yeast and *E. coli* and can be used to clone yeast genes in *E. coli*.

All vectors share several common characteristics. They are typically small, well-characterized molecules of DNA. They contain at least one replication origin and can be replicated within the appropriate host, even when they contain "foreign" DNA. Finally, they code for a phenotypic trait that can be used to detect

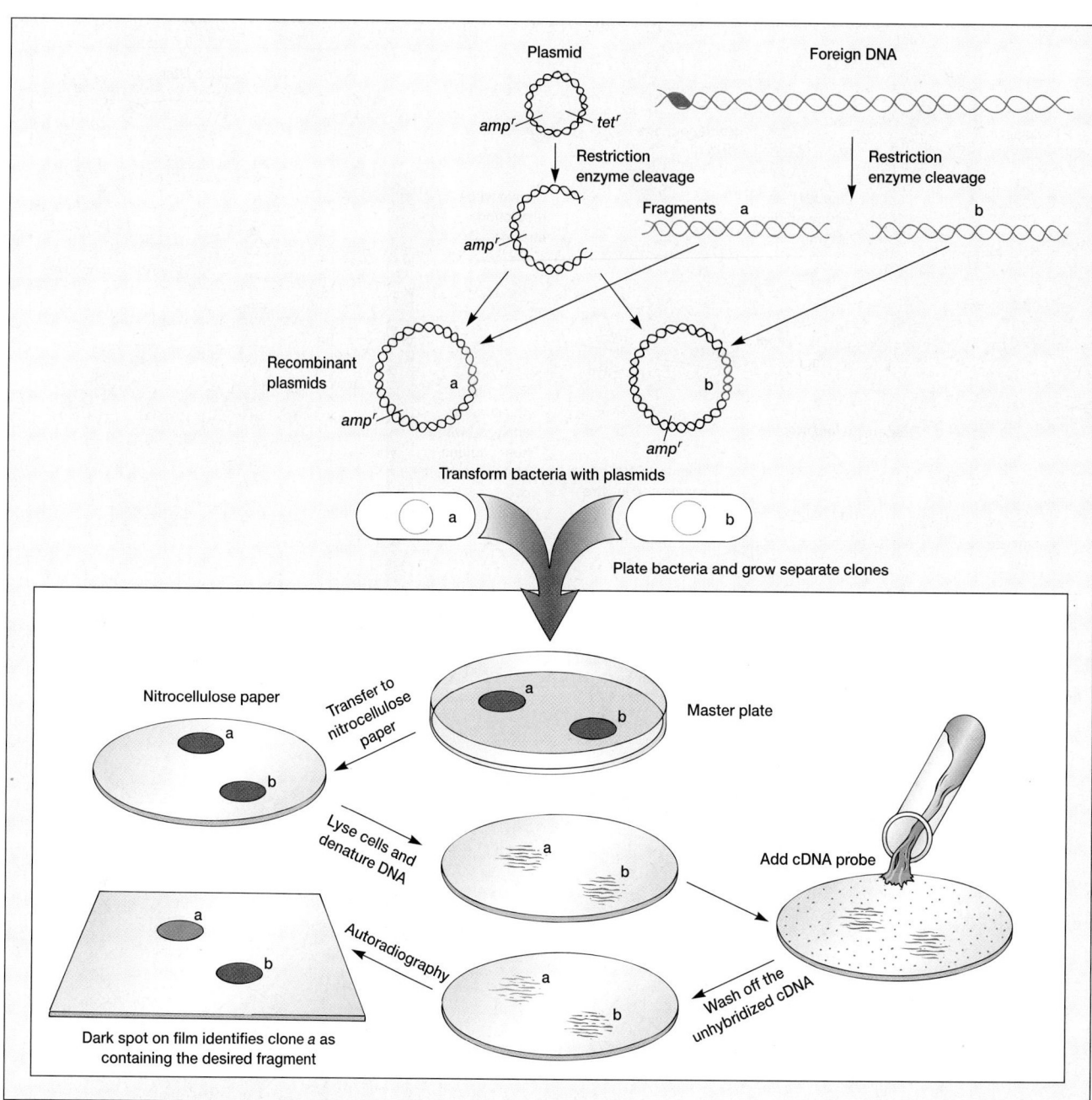

Figure 14.14   Cloning with Plasmid Vectors.   The use of plasmid vectors to clone a mixture of DNA fragments. The desired fragment is identified with a cDNA probe. See text for details.

their presence; frequently it is also possible to distinguish parental from recombinant vectors.

Plasmids

Plasmids were the first cloning vectors. They are easy to isolate and purify, and they can be reintroduced into a bacterium by transformation. Plasmids often bear antibiotic resistance genes, which are used to select their bacterial hosts. A recombinant plasmid containing foreign DNA often is called a **chimera,** after

the Greek mythological monster that had the head of a lion, the tail of a dragon, and the body of a goat. One of the most widely used plasmids is pBR322.   The biology of plasmids (pp. 288–91)

Plasmid pBR322 has both resistance genes for ampicillin and tetracycline and many restriction sites (**figure 14.16**). Several of these restriction sites occur only once on the plasmid and are located within an antibiotic resistance gene. This arrangement aids detection of recombinant plasmids after transformation (**figure 14.17**). For example, if foreign DNA is inserted into the ampicillin resistance gene, the plasmid will no

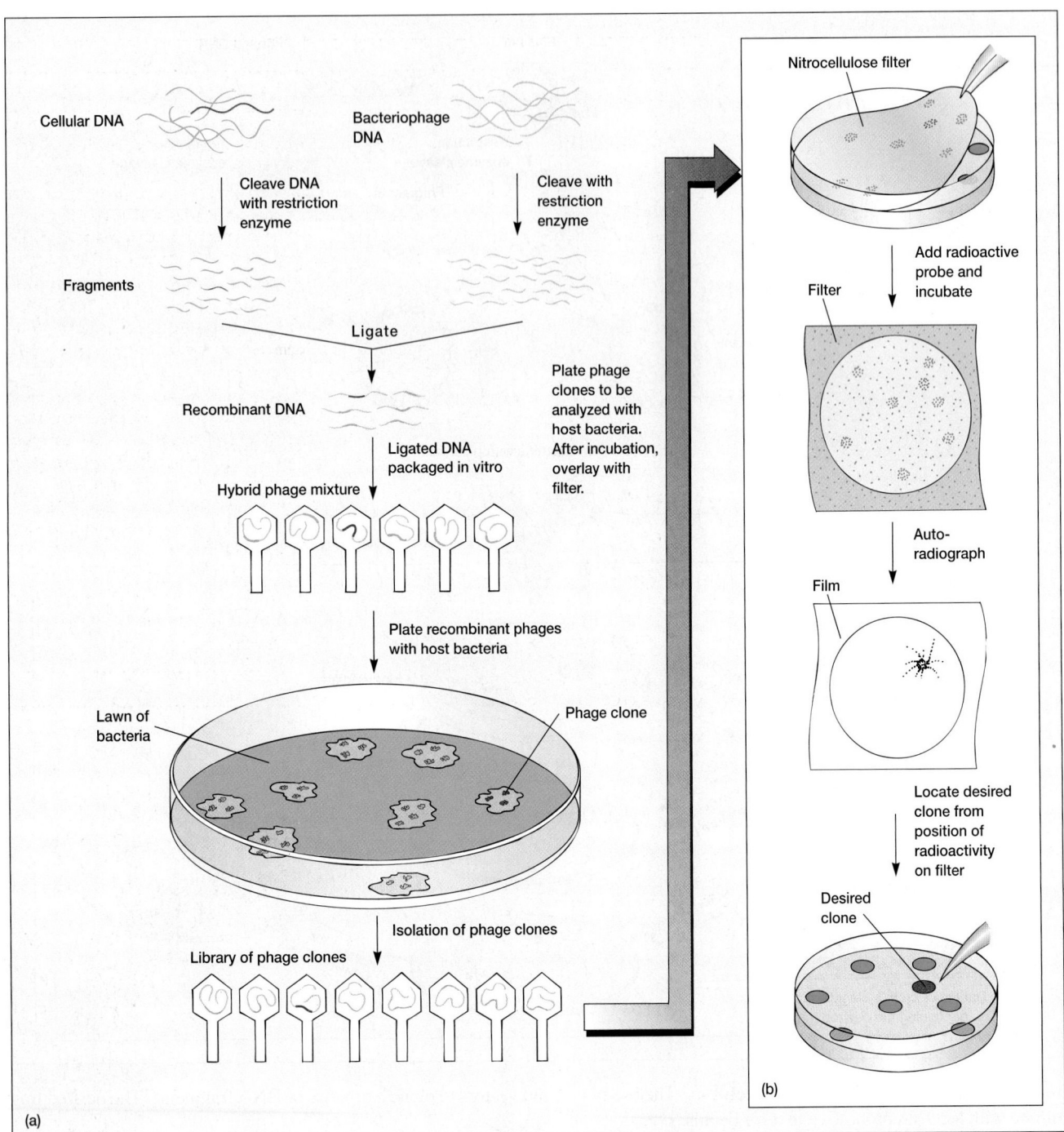

Figure 14.15 The Use of Lambda Phage as a Vector. (**a**) The preparation of a genomic library. Each colony on the bacterial lawn is a recombinant clone carrying a different DNA fragment. (**b**) Detection and cloning of the desired recombinant phage.

longer confer resistance to ampicillin. Thus tetracycline-resistant transformants that lack ampicillin resistance contain a chimeric plasmid.

Phage Vectors

Both single- and double-stranded phage vectors have been employed in recombinant DNA technology. For example, lambda

phage derivatives are very useful for cloning and can carry fragments up to about 45 kilobases in length. The genes for lysogeny and integration often are nonfunctional and may be deleted to make room for the foreign DNA. The modified phage genome also contains restriction sequences in areas that will not disrupt replication. After insertion of the foreign DNA into the modified lambda vector chromosome, the recombinant phage genome is packaged into viral capsids and can be used to infect host *E. coli* cells (figure 14.15).

Table 14.3
Some Recombinant DNA Cloning Vectors

Type	Vector	Restriction Sequences Present	Features
Plasmid (*E. coli*)	pBR322	*Bam*HI, *Eco*RI, *Hae*III, *Hin*dIII, *Pst*I, *Sal*I, *Xor*II	Carries genes for tetracycline and ampicillin resistance
Plasmid (yeast–*E. coli* hybrid)	pYe(CEN3)41	*Bam*HI, *Bgl*II, *Eco*RI, *Hin*dIII, *Pst*I, *Sal*I	Multiplies in *E. coli* or yeast cells
Cosmid (artificially constructed *E. coli* plasmid carrying lambda cos site)	pJC720	*Hin*dIII	Can be packaged in lambda phage particles for efficient introduction into bacteria; replicates as a plasmid; useful for cloning large DNA inserts
YAC (yeast artificial chromosome)	pYAC	*Sma*I, *Bam*HI	Carries gene for ampicillin resistance; multiplies in *Saccharomyces cerevisiae*
BAC (bacterial artificial chromosome)	pBAC108L	*Hin*dIII, *Bam*HI, *Not*I, *Sma*I, and others	Modified F plasmid that can carry 100–300 kb fragments; has a *cosN* site and a chloramphenicol resistance marker
Virus	Charon phage	*Eco*RI, *Hin*dIII, *Bam*HI, *Sst*I	Constructed using restriction enzymes and a ligase, having foreign DNA as its central portion, with lambda DNA at each end; carries β-galactosidase gene; packaged into lambda phage particles; useful for cloning large DNA inserts
Virus	Lambda 1059	*Bam*HI	Will carry large DNA fragments (8–21 kb); recombinant can grow on *E. coli* lysogenic for P2 phage, whereas vector cannot
Virus	M13	*Eco*RI	Single-stranded DNA virus; useful in studies employing single-stranded DNA insert and in producing DNA fragments for sequencing
Plasmid	Ti	*Sma*I, *Hpa*I	Maize plasmid

Adapted from G. D. Elseth and K. D. Baumgartner, *Genetics,* 1984 Benjamin/Cummings Publishing, Menlo Park, CA. Reprinted by permission of the author.

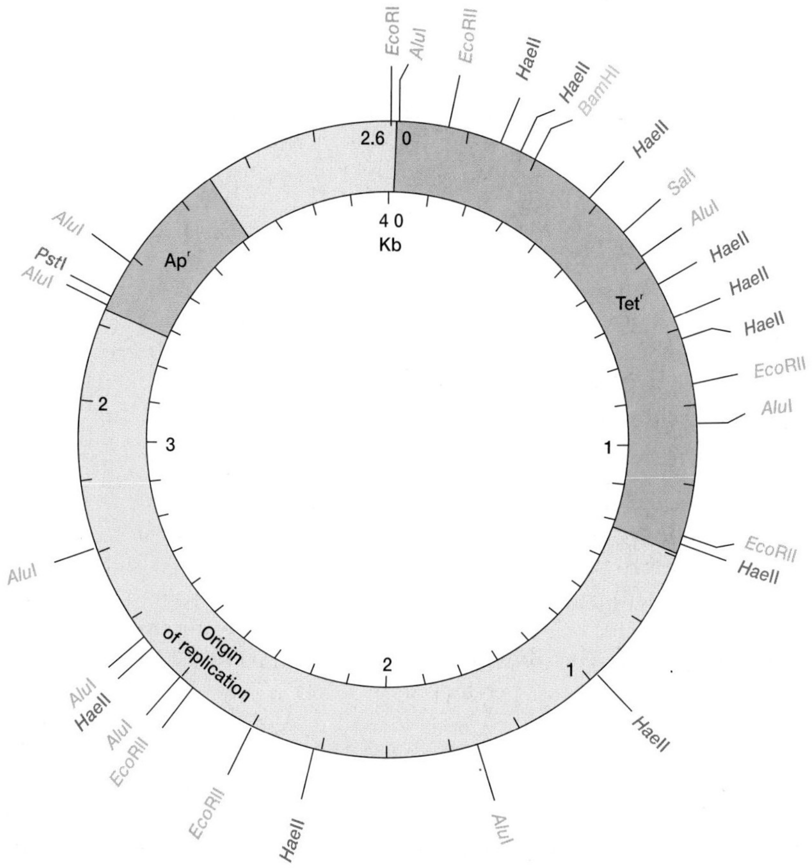

Figure 14.16 The pBR322 Plasmid. A map of the *E. coli* plasmid pBR322. The map is marked off in units of 1×10^5 daltons (outer circle) and 0.1 kilobases (inner circle). The locations of some restriction enzyme sites are indicated. The plasmid has resistance genes for ampicillin (Ap^r) and tetracycline (Tet^r).

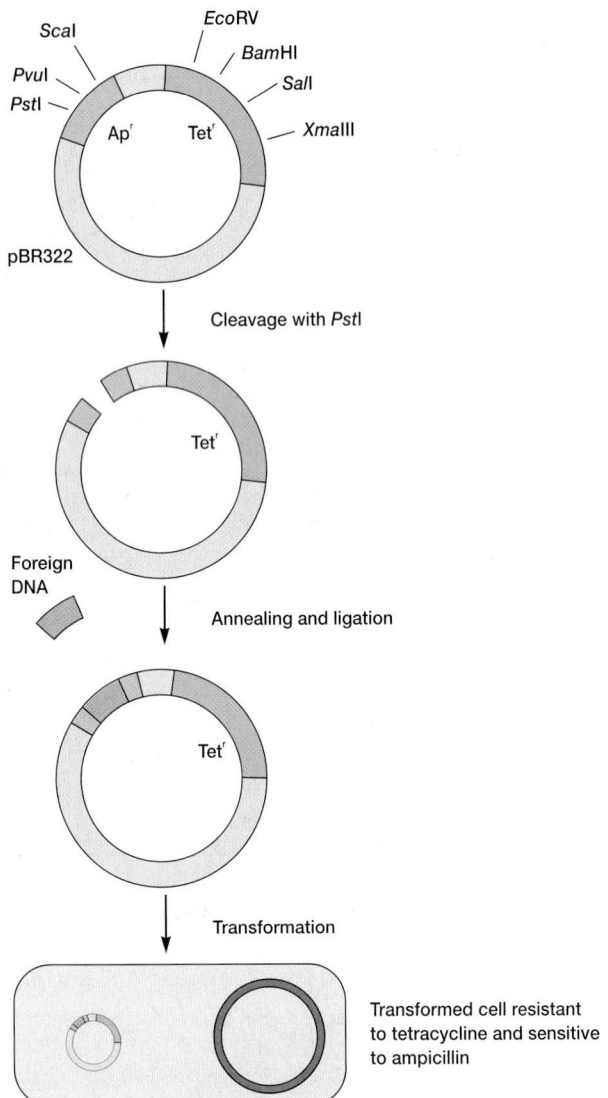

Figure 14.17 Detection of Recombinant Plasmids. The use of antibiotic resistance genes to detect the presence of recombinant plasmids. Because foreign DNA has been inserted into the ampicillin resistance gene, the recombinant host is only resistant to tetracycline. The restriction enzymes indicated at the top of the figure cleave only one site on the plasmid.

These vectors are often used to generate genomic libraries. *E. coli* also can be directly transformed with recombinant lambda DNA and produce phages. However, this approach is less efficient than the use of complete phage particles. The process is sometimes called **transfection.** Phages other than lambda also are used as vectors. For example, fragments as large as 95 kilobases can be carried by the P1 bacteriophage. The biology of bacteriophages (chapter 17)

Cosmids

Cosmids are plasmids that contain lambda phage cos sites and can be packaged into phage capsids. The lambda genome contains a recognition sequence called a cos site (or cohesive end) at each end. When the genome is to be packaged in a capsid, it is cleaved at one cos site and the linear DNA is inserted into the capsid until the second cos site has entered. Thus any DNA inserted between the cos sites is packaged. Cosmids typically contain several restriction sites and antibiotic resistance genes. They are packaged in lambda capsids for efficient injection into bacteria, but they also can exist as plasmids within a bacterial host. As much as 50 kilobases of DNA can be carried in this way.

Artificial Chromosomes

Artificial chromosomes are popular because they carry large amounts of genetic material. The **yeast artificial chromosome (YAC)** is one of the most widely used. YACs are stretches of DNA that contain all the elements required to propagate a chromosome in yeast: a replication origin, the centromere required to segregate chromatids into daughter cells, and two telomeres to mark the ends of the chromosome. They will also have restriction enzyme sites and genetic markers so that they can be traced and selected. Cleavage of a YAC with the proper restriction enzyme such as *Sma*I will open it up and allow the insertion of a piece of foreign DNA between the centromere and a telomere. In this way YACs containing DNA fragments between 100 and 2,000 kilobases in size can be placed in *Saccharomyces cerevisiae* cells and will be replicated along with the true chromosomes. The **bacterial artificial chromosome (BAC)** is an increasingly popular alternative cloning vector. BACs are cloning vectors based on the *E. coli* F-factor plasmid. They contain appropriate restriction enzyme sites and a marker such as chloramphenicol resistance. The modified plasmid is cleaved at a restriction site, and a foreign DNA fragment up to 300 kilobases in length is attached using DNA ligase. The BAC is reproduced in *E. coli* after insertion by electroporation. This vector is easy to reproduce and manipulate and does not undergo recombination as readily as YACs (which means that the insert is less likely to be rearranged). Because they can carry such large fragments of DNA, artificial chromosomes have been particularly useful in genome sequencing.

---

1. Define shuttle vector, chimera, cosmid, transfection, yeast, artificial chromosome, and bacterial artificial chromosome.
2. Describe how each of the major types of vectors is used in genetic engineering. Give an advantage of each.
3. How can the presence of two antibiotic resistance genes be used to detect recombinant plasmids?

14.6 INSERTING GENES INTO EUCARYOTIC CELLS

Because of its practical importance, much effort has been devoted to the development of techniques for inserting genes into eucaryotic cells. Some of these techniques have also been used successfully to transform bacterial cells. The most direct approach is the use of microinjection. Genetic material directly injected into animal cells such as fertilized eggs is sometimes stably incorporated into the host genome to produce a **transgenic animal,** one that has gained new genetic information from the acquisition of foreign DNA.

Another effective technique that works with mammalian cells and plant cell protoplasts is **electroporation.** If cells are mixed with a DNA preparation and then briefly exposed to pulses of high voltage (from about 250 to 4,000 V/cm for mammalian cells), the cells take up DNA through temporary holes in the plasma membrane. Some of these cells will be transformed.

One of the most effective techniques is to shoot microprojectiles coated with DNA into plant and animal cells. The **gene gun,** first developed at Cornell University, operates somewhat like a shotgun. A blast of compressed gas shoots a spray of DNA-coated metallic microprojectiles into the cells. The device has been used to transform corn and produce fertile corn plants bearing foreign genes. Other guns use electrical discharges to propel the DNA-coated projectiles. These guns are sometimes called biolistic devices, a name derived from biological and ballistic. They have been used to transform microorganisms (yeast, the mold *Aspergillus,* and the alga *Chlamydomonas*), mammalian cells, and a variety of plant cells (corn, cotton, tobacco, onion, and poplar).

Other techniques also are available. Plants can be transformed with *Agrobacterium* vectors as will be described shortly (p. 331). Viruses increasingly are used to insert desired genes into eucaryotic cells. For example, genes may be placed in a retrovirus (*see pp. 395–97*), which then infects the target cell and integrates a DNA copy of its RNA genome into the host chromosome. Adenoviruses also can transfer genes to animal cells. Recombinant baculoviruses will infect insect cells and promote the production of many proteins.

14.7 EXPRESSION OF FOREIGN GENES IN BACTERIA

After a suitable cloning vector has been constructed, rDNA enters the host cell, and a population of recombinant microorganisms develops. Most often the host is an *E. coli* strain that lacks restriction enzymes and is *recA⁻* to reduce the chances that the rDNA will undergo recombination with the host chromosome. *Bacillus subtilis* and the yeast *Saccharomyces cerevisiae* also may serve as hosts. Plasmid vectors enter *E. coli* cells by calcium chloride-induced transformation. Electroporation at 3 to 24 kV/cm is effective with both gram-positive and gram-negative bacteria.

A cloned gene is not always expressed in the host cell without further modification of the recombinant vector. To be transcribed, the recombinant gene must have a promoter that is recognized by the host RNA polymerase. Translation of its mRNA depends on the presence of leader sequences and mRNA modifications that allow proper ribosome binding. These are quite different in eucaryotes and procaryotes, and a procaryotic leader must be provided to synthesize eucaryotic proteins in a

bacterium. Finally, introns in eucaryotic genes must be removed because the procaryotic host will not excise them after transcription of mRNA; a eucaryotic protein is not functional without intron removal prior to translation.

The problems of expressing recombinant genes in host cells are largely overcome with the help of special cloning vectors called **expression vectors.** These vectors are often derivatives of plasmid pBR322 and contain the necessary transcription and translation start signals. They also have useful restriction sites next to these sequences so that foreign DNA can be inserted with relative ease. Some expression vectors contain portions of the lac operon and can effectively regulate the expression of the cloned genes in the same manner as the operon.

Somatostatin, the 14-residue hypothalamic polypeptide hormone that helps regulate human growth, provides an example of useful cloning and protein production. The gene for somatostatin was initially synthesized by chemical methods. Besides the 42 bases coding for somatostatin, the polynucleotide contained a codon for methionine at the 5′ end (the N-terminal end of the peptide) and two stop codons at the opposite end. To aid insertion into the plasmid vector, the 5′ ends of the synthetic gene were extended to form single-stranded sticky ends complementary to those formed by the *Eco*RI and *Bam*HI restriction enzymes. A modified pBR322 plasmid was cut with both *Eco*RI and *Bam*HI to remove a part of the plasmid DNA. The synthetic gene was then spliced into the vector by taking advantage of its cohesive ends (**figure 14.18**). Finally, a fragment containing the initial part of the lac operon (including the promoter, operator, ribosome binding site, and much of the β-galactosidase gene) was inserted next to the somatostatin gene. The plasmid now contained the somatostatin gene fused in the proper orientation to the remaining portion of the β-galactosidase gene.

After introduction of this chimeric plasmid into *E. coli,* the somatostatin gene was transcribed with the β-galactosidase gene fragment to generate an mRNA having both messages. Translation formed a protein consisting of the total hormone polypeptide attached to the β-galactosidase fragment by a methionine residue. Cyanogen bromide cleaves peptide bonds at methionine residues. Treatment of the fusion protein with cyanogen bromide broke the peptide chain at the methionine and released the hormone (**figure 14.19**). Once free, the polypeptide was able to fold properly and become active. Since production of the fusion protein was under the control of the *lac* operon, it could be easily regulated.

Many proteins have been produced since the synthesis of somatostatin. A similar approach was used to manufacture human insulin. Human growth hormone and some interferons also have been synthesized by rDNA techniques. The human growth hormone gene was too long to synthesize by chemical procedures and was prepared from mRNA as cDNA.

As mentioned earlier, introns in eucaryotic genes are not removed by bacteria and will render the final protein nonfunctional. The easiest solution is to prepare cDNA from processed mRNA that lacks introns and directly reflects the correct amino acid sequence of the protein product. In this instance it is particularly important to fuse the gene with an expression vector since a

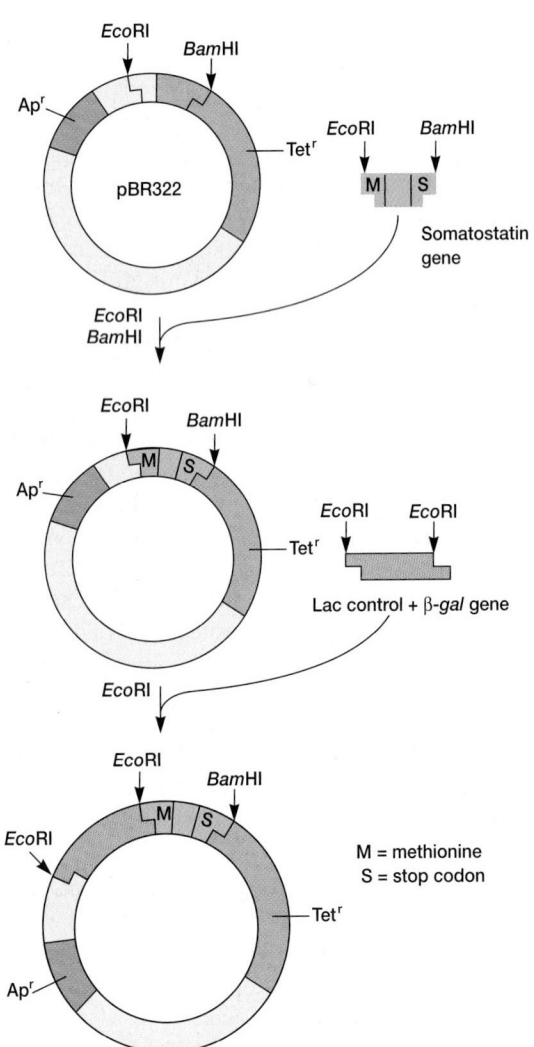

Figure 14.18 Cloning the Somatostatin Gene. An overview of the procedure used to synthesize a recombinant plasmid containing the somatostatin gene. See text for details.

promoter and other essential sequences will be missing in the cDNA. Eucaryotic mRNA processing (pp. 257–60)

If the mRNA is scarce, it may not be easy to obtain enough for cDNA synthesis. Often the sequence of the protein coded for by the gene is used to deduce the best DNA sequence for the specific polypeptide segment (a logical procedure called reverse translation). Then the DNA probe is synthesized and used to locate and isolate the desired mRNA after gel electrophoresis. Finally, the isolated mRNA is used to make cDNA.

1. What is a transgenic animal? Describe how electroporation and gene guns are used to insert foreign genes into eucaryotic cells. What other approaches may be used? How are bacteria transformed?

2. How can one prevent rDNA from undergoing recombination in a bacterial host cell?

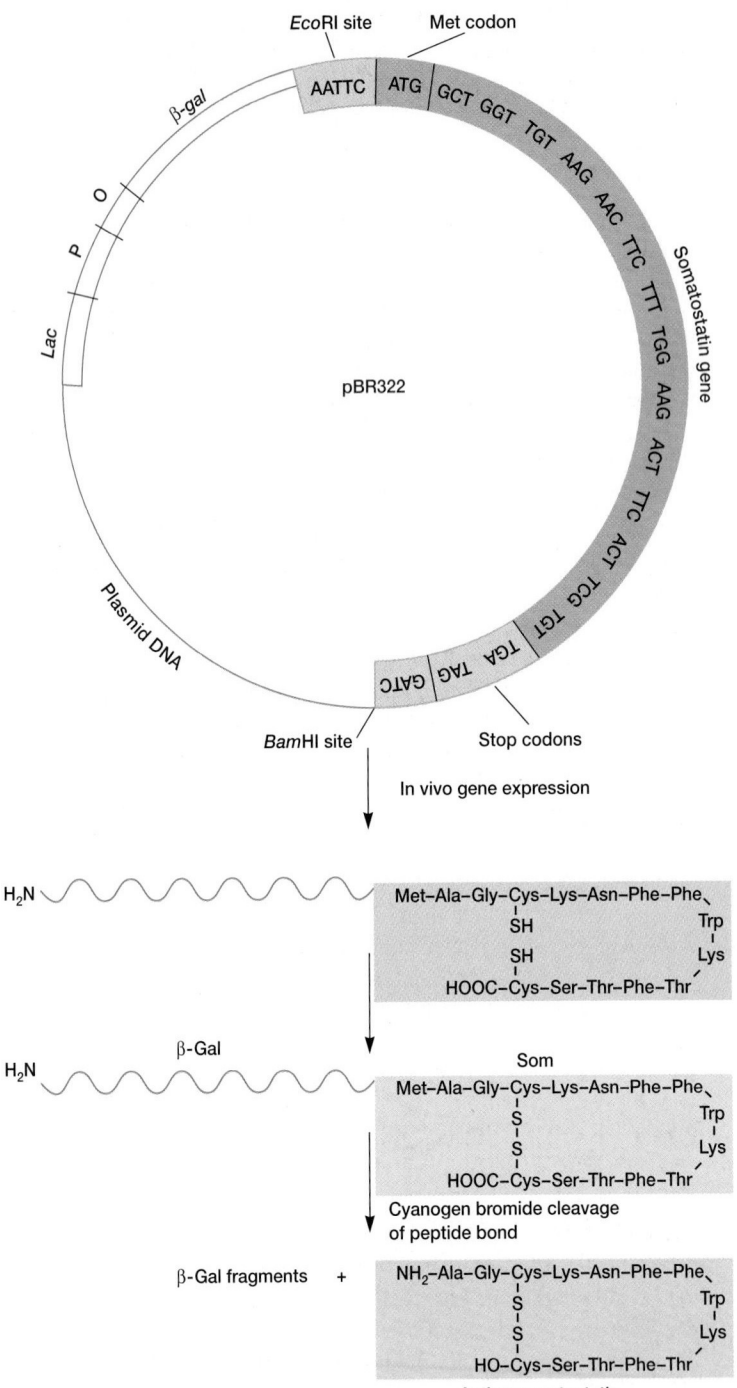

Figure 14.19 The Synthesis of Somatostatin by Recombinant *E. coli*. Cyanogen bromide cleavage at the methionine residue releases active hormone from the β-galactosidase fragment. The gene and associated sequences are shaded in color. Stop codons, the special methionine codon, and restriction enzyme sites are enclosed in boxes.

Techniques & Applications
14.1 Gene Expression and Kittyboo Colors

Genetic engineering techniques can yield unusual approaches to long-standing problems. To understand how the regulation of gene expression influences complex processes, the activity of more than one gene must be followed simultaneously. The use of genetic engineering techniques with luciferase genes has made this much easier.

A Jamaican beetle named the kittyboo has two light organs on its head and one on its abdomen. Recently four luciferase genes have been isolated from the beetle. Each luciferase enzyme produces a different colored light when it acts on the substrate luciferin. Keith V. Wood has cloned the genes and inserted them into *E. coli*. When the bacteria are exposed to luciferin, they glow (see **Box figure**).

These luciferase genes can be used to study gene regulation. Suppose that one wishes to follow the activity of the liver gene that codes for serum albumin. The albumin gene can be replaced with a luciferase gene, and the modified cell incubated with luciferin. Whenever the albumin gene is activated, the newly inserted luciferase gene will function and the cell glow. Measurement of light intensity is easy, rapid, and sensitive. By substituting a different luciferase gene for another liver gene, one can simultaneously follow the activity and coordination of two different liver genes. It is only necessary to measure light intensity at the two wavelengths characteristic of the luciferase genes.

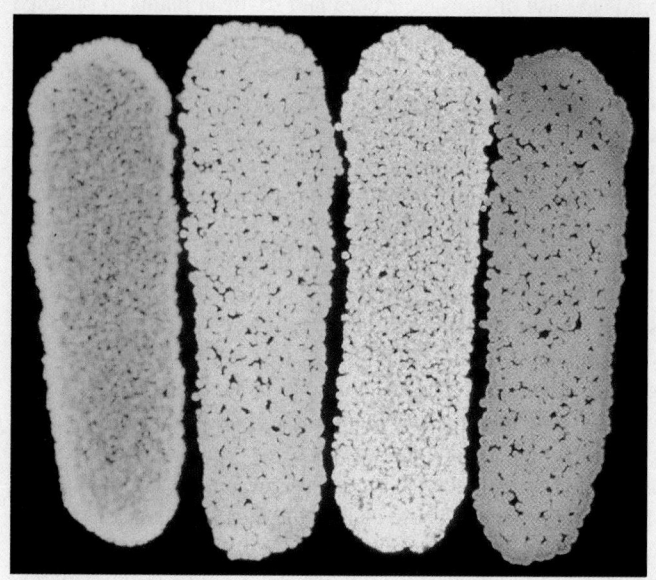

E. coli with Luciferase Genes. These four streaks of *E. coli* glow different colors because they contain four different luciferase genes cloned from the Jamaican click beetle or kittyboo, *Pyrophorus plagiophthalamus*.

Table 14.4
Some Human Peptides and Proteins Synthesized by Genetic Engineering

Peptide or Protein	Potential Use
α_1-antitrypsin	Treatment of emphysema
α-, β-, and γ-interferons	As antiviral, antitumor, and anti-inflammatory agents
Blood-clotting factor VIII	Treatment of hemophilia
Calcitonin	Treatment of osteomalacia
Epidermal growth factor	Treatment of wounds
Erythropoetin	Treatment of anemia
Growth hormone	Growth promotion
Insulin	Treatment of diabetes
Interleukins-1, 2, and 3	Treatment of immune disorders and tumors
Macrophage colony stimulating factor	Cancer treatment
Relaxin	Aid to childbirth
Serum albumin	Plasma supplement
Somatostatin	Treatment of acromegaly
Streptokinase	Anticoagulant
Tissue plasminogen activator	Anticoagulant
Tumor necrosis factor	Cancer treatment

3. List several reasons why a cloned gene might not be expressed in a host cell. What is an expression vector?
4. Briefly outline the procedure for somatostatin production.
5. Identify a way to eliminate eucaryotic introns during the synthesis of rDNA.

 14.8 APPLICATIONS OF GENETIC ENGINEERING

Genetic engineering and biotechnology will continue to contribute in the future to medicine, industry, and agriculture, as well as to basic research (**Techniques & Applications 14.1**). In this section some practical applications are briefly discussed.

Medical Applications

Certainly the production of medically useful proteins such as somatostatin, insulin, human growth hormone, and some interferons is of great practical importance (**table 14.4**). This is particularly true of substances that previously only could be obtained from human tissues. For example, in the past, human growth hormone for treatment of pituitary dwarfism was extracted from pituitaries obtained during autopsies and was available only in limited amounts. Interleukin-2 (a protein that helps regulate the immune response) and blood-clotting factor VIII have recently been cloned, and undoubtedly other important peptides and proteins will be produced in the future. A particularly interesting development is the use of transgenic corn and soybean plants to produce monoclonal antibodies (*see pp. 719–20*) for medical uses. It also is possible to use genetically engineered plants to produce oral vaccines. Genetically engineered mice now can produce fully human monoclonal antibodies. Synthetic vaccines—for instance, vaccines for malaria and rabies—are also being developed with recombinant techniques. A recombinant hepatitis B vaccine is already commercially available.

Other medical uses of genetic engineering are being investigated. Probes are now being used in the diagnosis of infectious disease. An individual could be screened for mutant genes with probes and hybridization techniques (even before birth when used together with amniocentesis). A type of genetic surgery called somatic cell gene therapy may be possible. For example, cells of an individual with a genetic disease could be removed, cultured, and transformed with cloned DNA containing a normal copy of the defective gene or genes. These cells could then be reintroduced into the individual; if they became established, the expression of the normal genes might cure the patient. An immune deficiency disease patient lacking the enzyme adenosine deaminase that destroys toxic metabolic by-products has been treated in this way. Some of the patient's lymphocytes (*see p. 681*) were removed, given the adenosine deaminase gene with the use of a modified retrovirus, and returned to the patient's body. It may be possible to use a defective retrovirus (*see section 18.2*) or another virus to directly insert the proper genes into host cells, perhaps even specifically targeted organs or tissues. Fusion toxins provide a third example of potentially important genetic engineering applications. The first to be developed are recombinant proteins in which the enzymatic and membrane translocation domains of the diphtheria toxin (*see pp. 771–72*) are combined with proteins that attach specifically to target cell surface receptors. Clinical trials are being carried out on a fusion toxin that contains interleukin-2. IL-2 binds to the cell, and the diphtheria toxin enzymatic domain enters and blocks protein synthesis. Encouraging results have been obtained in the treatment of leukemia, lymphomas, and rheumatoid arthritis. The use of probes and plasmid fingerprinting in clinical microbiology (pp. 816–17)

Research on gene therapy was given major setback in September 1999, when a teenage volunteer died during a clinical trial. Jesse Gelsinger apparently died because of an excessive immune response to the adenovirus being used to insert a gene for the liver enzyme ornithine transcarbamylase. This enzyme is needed to remove ammonia from the body and produce urea. Gelsinger's death has slowed clinical research and stimulated work on ways to improve virus vectors and also delivery genes without the use of viruses.

It now appears that livestock will also be important in medically oriented biotechnology through the use of an approach sometimes called molecular pharming. Pig embryos injected with human hemoglobin genes develop into transgenic pigs that synthesize human hemoglobin. Current plans are to purify the hemoglobin and use it as a blood substitute. A pig could yield 20 units of blood substitute a year. Somewhat similar techniques have produced transgenic goats whose milk contains up to 3 grams of human tissue plasminogen activator per liter. Tissue plasminogen activator (TPA) dissolves blood clots and is used to treat cardiac patients.

Recombinant DNA techniques are playing an increasingly important role in research on the molecular basis of disease. The transfer from research to practical application often is slow because it is not easy to treat a disease effectively unless its molecular mechanism is known. Thus genetic engineering may aid in the fight against a disease by providing new information about its nature, as well as by aiding in diagnosis and therapy.

Industrial Applications

Industrial applications of recombinant DNA technology include manufacturing protein products by using bacteria, fungi, and cultured mammalian cells as factories; strain improvement for existing bioprocesses; and the development of new strains for additional bioprocesses. As mentioned earlier, the pharmaceutical industry is already producing several medically important polypeptides using this technology. In addition, there is interest in making expensive, industrially important enzymes with recombinant bacteria. Bacteria that metabolize petroleum and other toxic materials have been developed. These bacteria can be constructed by assembling the necessary catabolic genes on a single plasmid and then transforming the appropriate organism. There also are many potential applications in the chemical and food industries. Food microbiology (chapter 41); Industrial microbiology and biotechnology (chapter 42)

Agricultural Applications

It is also possible to bypass the traditional methods of selective breeding and directly transfer desirable traits to agriculturally important animals and plants. Potential exists for increasing growth rates and overall protein yields of farm animals. The growth hormone gene has already been transferred from rats to mice, and both in vitro fertilization and embryo implantation methods are fairly well developed. Recently, recombinant bovine growth hormone has been used to increase milk production by at least 10%. Perhaps farm animals' disease resistance and tolerance to environmental extremes also can be improved.

Cloned genes can be inserted into plant as well as animal cells. Presently a popular way to insert genes into plants is with a recombinant **Ti plasmid** (tumor-inducing plasmid) obtained from the bacterium *Agrobacterium tumefaciens* (**Techniques & Applications 14.2**; *see also section 30.4*). It also is possible to donate genes by forming plant cell protoplasts, making them permeable to DNA, and then adding the desired rDNA. The advent of the gene gun (p. 327) will greatly aid the production of transgenic plants.

Much effort has been devoted to the transfer of the nitrogen-fixing abilities of bacteria associated with legumes to other crop plants. The genes largely responsible for the process have been cloned and transferred to the genome of plant cells; however, the

A plasmid from the plant pathogenic bacterium *Agrobacterium tumefaciens* is responsible for much success in the genetic engineering of plants. Infection of normal cells by the bacterium transforms them into tumor cells, and the crown gall disease develops in dicotyledonous plants such as grapes and ornamental plants. Normally, the gall or tumor is located near the junction of the plant's root and stem. The tumor forms because of the insertion of genes into the plant cell genome, and only strains of *A. tumefaciens* possessing a large conjugative plasmid called the Ti plasmid are pathogenic (*see figure 30.18*). The Ti plasmid carries genes for virulence and the synthesis of substances involved in the regulation of plant growth. The genes that induce tumor formation reside between two 23 base pair direct-repeat sequences. This region is known as T-DNA and is very similar to a transposon. T-DNA contains genes for the synthesis of plant growth hormones (an auxin and a cytokinin) and an amino acid derivative called opine that serves as a nutrient source for the invading bacteria. In diseased plant cells, T-DNA is inserted into the chromosomes at various sites and is stably maintained in the cell nucleus.

When the molecular nature of crown gall disease was recognized, it became clear that the Ti plasmid and its T-DNA had great potential as a vector for the insertion of rDNA into plant chromosomes. In one early experiment the yeast alcohol dehydrogenase gene was added to the T-DNA region of the Ti plasmid. Subsequent infection of cultured plant cells resulted in the transfer of the yeast gene. Since then, many modifications of the Ti plasmid have been made to improve its characteristics as a vector. Usually one or more antibiotic resistance genes are added, and the nonessential T-DNA, including the tumor inducing genes, is deleted. Those genes required for the actual infection of the plant cell by the plasmid are left. T-DNA also has been inserted into the *E. coli* pBR322 plasmid and other plasmids to produce cloning vectors that can move between bacteria and plants (see **Box figure**). The gene or genes of interest are spliced into the T-DNA region between the direct repeats. Then the plasmid is returned to *A. tumefaciens,* plant culture cells are infected with the bacterium, and transformants are selected by screening for antibiotic resistance (or another trait in the T-DNA). Finally, whole plants are regenerated from the transformed cells. In this way several plants have been made herbicide resistant (p. 332).

Unfortunately *A. tumefaciens* does not produce crown gall disease in monocotyledonous plants such as corn, wheat, and other grains; and it has been used only to modify plants such as potato, tomato, celery, lettuce, and alfalfa. However, there is evidence that T-DNA is transferred to monocotyledonous plants and expressed, although it does not produce tumors. This discovery plus the creation of new procedures for inserting DNA into plant cells may well lead to the use of rDNA techniques with many important crop plants.

T-DNA

Cloning vector

Ti plasmid sequence

Gene insertion into T-DNA region

Firefly luciferase gene and promoter

Recombinant plasmid

Transform *Agrobacterium* and infect tobacco plant

Plant genome

T-DNA and luciferase gene

Plant genome

(a)

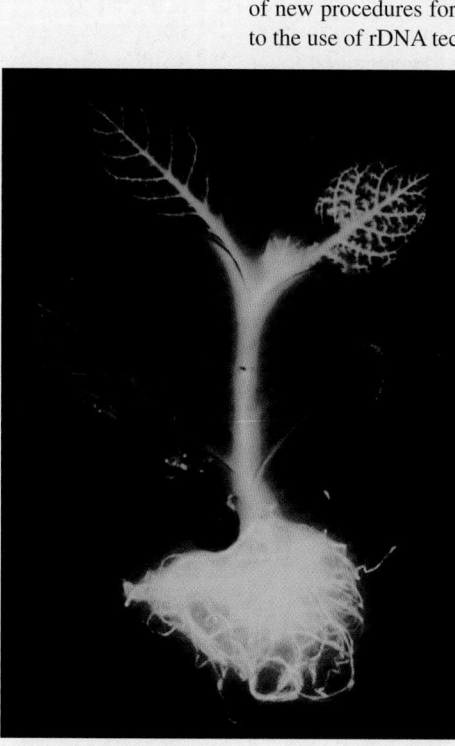

(b)

The Use of the Ti Plasmid Vector to Produce Transgenic Plants. (**a**) The formation of a cloning vector and its use in transformation. See text for details. (**b**) A tobacco plant, *Nicotiana tabacum,* that has been made bioluminescent by transfection with a special Ti plasmid vector containing the firefly luciferase gene. When the plant is watered with a solution of luciferin, the substrate for the luciferase enzyme, it glows. The picture was made by exposing the transgenic plant to Ektachrome film for 24 hours.

recipient cells have not been able to fix nitrogen. If successful, the potential benefit for crop plants such as corn is great. Nevertheless, there is some concern that new nitrogen-fixing varieties might spread indiscriminantly like weeds or disturb the soil nitrogen cycle (*see sections 28.3 and 30.4*).

Attempts at making plants resistant to environmental stresses have been more successful. For example, the genes for detoxification of glyphosate herbicides were isolated from *Salmonella,* cloned, and introduced into tobacco cells using the Ti plasmid. Plants regenerated from the recombinant cells were resistant to the herbicide. Herbicide-resistant varieties of cotton and fertile, transgenic corn also have been developed. This is of considerable importance because many crop plants suffer stress when treated with herbicides. Resistant crop plants would not be stressed by the chemicals being used to control weeds, and yields would presumably be much greater.

U.S. farmers grow substantial amounts of genetically modified (GM) crops. About a third of the corn, half of the soybeans, and a significant fraction of cotton crops are genetically modified. Cotton and corn are resistant to herbicides and insects. Soybeans have herbicide resistance and lowered saturated fat content. Other examples of genetically engineered commercial crops are canola, potato, squash, and tomato.

Many new agricultural applications are being explored. A good example is an altered strain of *Pseudomonas syringae* that protects against plant frost damage because it cannot produce the protein that induces ice-crystal formation. Recently, a salt-tolerant tomato plant has been developed. It may provide a cash crop for farmers with salinized land. Much effort is being devoted to defending plants against pests without the use of chemical pesticides. A strain of *Pseudomonas fluorescens* carrying the gene for the *Bacillus thuringiensis* toxin (*see pp. 988–90*) is under testing. This toxin destroys many insect pests such as the cabbage looper and the European corn borer. A variety of corn with the *B. thuringiensis* toxin gene has been developed. Unfortunately, some insects seem able to develop resistance to the toxin. There is considerable interest in insect-killing viruses and particularly in the baculoviruses. A scorpion toxin gene has been inserted into the autographa californica multicapsid nuclear polyhedrosis virus (AcMNPV). The engineered AcMNPV kills cabbage looper more rapidly than the normal virus and reduces crop damage significantly. Finally, virus-resistant strains of soybeans, potatoes, squash, rice, and other plants are under development.

---

1. List several important present or future applications of genetic engineering in medicine, industry, and agriculture.
2. What is the Ti plasmid and why is it so important?

14.9 SOCIAL IMPACT OF RECOMBINANT DNA TECHNOLOGY

Despite the positive social impact of rDNA technology, dangers may be associated with rDNA work and gene cloning. The potential to alter an organism genetically raises serious scientific and philosophical questions, many of which have not yet been adequately addressed. In this section some of the debate is briefly reviewed.

The initial concern raised by the scientific community was that recombinant *E. coli* and other genetically engineered microogranisms (GEMs) carrying dangerous genes might escape and cause widespread infections. Because of these worries, the federal government established guidelines to limit and regulate the locations and types of potentially dangerous experiments. Physical containment was to be practiced—that is, rDNA research was to be carried out in specially designed laboratories with extra safety precautions. In addition, rDNA researchers were also to practice biological containment. Only weakened microbial hosts unable to survive in a natural environment and non-conjugating bacterial strains were to be used to avoid the spread of the vector or rDNA itself. Further experiments suggested that the dangers were not as extreme as initially conceived. Yet little is known about what would happen if recombinant organisms escaped. For example, could dangerous genes such as oncogenes (*see section 18.5*) move from a weakened strain to a hardy one that would then spread? Will there be increased risk when extremely large quantities of recombinant microorganisms are grown in industrial fermenters? Some people worry because the use of rDNA techniques has become so widespread and the guidelines on their safe use have been relaxed to a considerable extent. Biomedical rDNA research has been regulated by the Recombinant DNA Advisory Committee (RAC) of the National Institutes of Health. More recently, the Food and Drug Administration (FDA) has assumed principal responsibility for oversight of gene therapy research. The Environmental Protection Agency and state governments have jurisdiction over agriculturally related field experiments. Industrial rDNA research can proceed without such close regulation, and this has also caused some anxiety about potential safety hazards.

Aside from these concerns over safety, the use of recombinant technology on human beings raises ethical and moral questions. These problems are not extreme in cases of somatic cell gene therapy to cure serious disease. There is much greater concern about attempts to correct defects or bestow traits considered to be desirable or cosmetic by introducing genes into human eggs and embryos. In 1983 two Nobel Prize winners, several other prominent scientists, and about 40 religious leaders signed a statement to Congress requesting that such alterations of human eggs or sperm not be attempted. There is a temptation to "improve" ourselves or reengineer the human body. Some might wish to try to change their children's intelligence, size, or physical attractiveness through genetic modifications. Others argue that the good arising from such interventions more than justifies the risks of misuse and that there is nothing inherently wrong in modifying the human gene pool to reduce the incidence of genetic disease. This discussion extends to other organisms as well. Some argue that we have no right to create new life-forms and further disrupt the genetic diversity that human activities have already severely reduced. United States courts have decided that researchers and companies can patent "nonnaturally occurring" living organisms, whether microbial, plant, or animal.

One of the major current efforts in biotechnology, the human genome project, has determined the sequences of all human chromosomes. Success in this project and further technical advances in biotechnology will make genetic screening very effective. Physicians will be able to detect critical flaws in DNA long before the resulting genetic disease becomes manifest. In some cases this may lead to immediate treatment. But often nothing can be done about the damage and all sorts of dilemmas arise. Should people be told about the situation? Do they even wish to know about defects that cannot be corrected? What happens if their employers and insurance companies learn of the problem? Employees may lose their jobs and insurance. The issue of privacy is crucial in a practical sense, especially if genetic data banks are established. There may be increasing pressure on parents to abort fetuses with genetic defects in order not to stress the medical and social welfare systems. Clearly modern biomedical technology holds both threat and promise. Society has not yet faced its implications despite rapid progress in the development of genetic screening technology.

Another area of considerable controversy involves the use of recombinant organisms in agriculture. Many ecologists are worried that the release of unusual, recombinant organisms without careful prior risk assessment may severely disrupt the ecosystem. They point to the many examples of ecosystem disruption from the introduction of foreign organisms. A major concern is the exchange of genes between transgenic plants and viruses or other plants. Viral nucleic acids inserted into plants to make them virus resistant might combine with the genome of an invading virus to make it even more virulent. Weeds might acquire resistance to viruses, pests, and herbicides from transgenic crop plants. Other potential problems trouble critics. For example, insect-killing viruses might destroy many more insect species than expected. Toxin genes inserted into one virus might move to other viruses. Genetically modified foods might trigger damaging allergic responses in consumers. Potential risks have been vigorously discussed because several recombinant organisms are already on the market. Thus far, no obvious ecological or health effects have been observed. Nevertheless, there have been calls for stricter regulation of GM foods. In Europe, GM crops are not commonly produced by farmers or purchased by consumers. Some large U.S. food producers have responded to public concern and quit using GM crops.

As with any technology, the potential for abuse exists. A case in point is the use of genetic engineering in biological warfare and terrorism. Although there are international agreements that limit the research in this area to defense against other biological weapons, the knowledge obtained in such research can easily be used in offensive biological warfare. Effective vaccines constructed using rDNA technology can protect the attacker's troops and civilian population. Because it is relatively easy and inexpensive to prepare bacteria capable of producing massive quantities of toxins or to develop particularly virulent strains of viral and bacterial pathogens, even small countries and terrorist organizations might acquire biological weapons. Many scientists and nonscientists also are concerned about increases in the military rDNA research carried out by the major powers. The emerging threat of bioterrorism (pp. 837–38)

Recombinant DNA technology has greatly enhanced our knowledge of genes and how they function, and it promises to improve our lives in many ways. Yet, as this brief discussion shows, problems and concerns remain to be resolved. Past scientific advances have sometimes led to unanticipated and unfortunate consequences, such as environmental pollution and nuclear weapons. With prudence and forethought we may be able to avoid past mistakes in the use of this new technology.

---

1. Describe four major areas of concern about the application of genetic engineering. In each case give both the arguments for and against the use of genetic engineering.

Summary

14.1 Historical Perspectives
a. Genetic engineering became possible after the discovery of restriction enzymes and reverse transcriptase, and the development of essential methods in nucleic acid chemistry such as the Southern blotting technique (**table 14.1**).

14.2 Synthetic DNA
a. Oligonucleotides of any desired sequence can be synthesized by a DNA synthesizer machine. This has made possible site-directed mutagenesis.

14.3 The Polymerase Chain Reaction
a. The polymerase chain reaction allows small amounts of DNA to be increased in concentration thousands of times (**figure 14.8**).

14.4 Preparation of Recombinant DNA
a. Agarose gel electrophoresis is used to separate DNA fragments according to size differences.
b. Fragments can be isolated and identified, then joined with plasmids or phage genomes and cloned; one also can first clone all fragments and subsequently locate the desired clone. (**figures 14.11, 14.14, and 14.15**).
c. Three techniques that can be used to join DNA fragments are the creation of similar sticky ends with a single restriction enzyme, the addition of poly(dA) and poly(dT) to create sticky ends, and blunt-end ligation with T4 DNA ligase.
d. Probes for the detection of recombinant clones are made in several ways and usually labeled with ^{32}P. Nonradioactive probes are also used.

14.5 Cloning Vectors
a. Several types of vectors may be used, each with different advantages: plasmids, phages and other viruses, cosmids, artificial chromosomes, and shuttle vectors (**table 14.3**).

14.6 Inserting Genes into Eucaryotic Cells
a. Genes can be inserted into eucaryotic cells by techniques such as microinjection, electroporation, and the use of a gene gun.

14.7 Expression of Foreign Genes in Bacteria
a. The recombinant vector often must be modified by the addition of promoters, leaders, and other elements. Eucaryotic gene introns also must be removed. An expression vector has the

necessary features to express any recombinant gene it carries.

b. Many useful products, such as the hormone somatostatin, have been synthesized using recombinant DNA technology (**figure 14.19**).

14.8 Applications of Genetic Engineering

a. Recombinant DNA technology will provide many benefits in medicine, industry, and agriculture.

14.9 Social Impact of Recombinant DNA Technology

a. Despite the great promise of genetic engineering, it also brings with it potential problems in areas of safety, human experimentation, potential ecological disruption, and biological warfare.

Key Terms

autoradiography 314
bacterial artificial chromosome (BAC) 326
biotechnology 312
chimera 323
complementary DNA (cDNA) 313
cosmid 326

electrophoresis 319
electroporation 327
expression vector 327
gene gun 327
genetic engineering 312
genomic library 320
oligonucleotides 315

polymerase chain reaction (PCR technique) 316
probe 314
recombinant DNA technology 312
restriction enzymes 312
site-directed mutagenesis 315
Southern blotting technique 314

Ti plasmid 330
transfection 326
transgenic animal 327
vectors 314
yeast artificial chromosome (YAC) 326

Questions for Thought and Review

1. Could the Southern blotting technique be applied to RNA and proteins? How might this be done?

2. Why could a band on an electrophoresis gel still contain more than one kind of DNA fragment?

3. What advantage might there be in creating a genomic library first rather than directly isolating the desired DNA fragment?

4. In what areas do you think genetic engineering will have the greatest positive impact in the future? Why?

5. What do you consider to be the greatest potential dangers of genetic engineering? Are there ethical problems with any of its potential applications?

Critical Thinking Questions

1. Initial attempts to perform PCR were carried out using the DNA polymerase from *E. coli.* What was the major difficulty?

2. Why can't PCR be carried out with RNA polymerase?

3. Write a one-page explanation for a reader with a high school education that explains gene therapy. Assume the reader has a son or daughter with a life-threatening disease for which the only treatment course is gene therapy.

4. Suppose that one inserted a simple plasmid (one containing an antibiotic resistance gene and a separate restriction site) carrying a human interferon gene into *E. coli,* but none of the transformed bacteria produced interferon. Give as many plausible reasons as possible for this result.

For recommended readings on these and related topics, see Appendix V.

chapter 15

Microbial Genomics

Concepts

- Genomics is the study of the molecular organization of genomes, their information content, and the gene products they encode. It may be divided into structural genomics, functional genomics, and comparative genomics.
- Individual pieces of DNA can be sequenced using the Sanger method. The easiest way to analyze microbial genomes is by whole-genome shotgun sequencing in which randomly produced fragments are sequenced individually and then aligned by computers to give the complete genome.
- Because of the mass of data to be analyzed, the use of sophisticated programs on high-speed computers is essential to genomics.
- Many bacterial genomes have already been sequenced and compared. The results are telling us much about such subjects as genome structure, microbial physiology, microbial phylogeny, and how pathogens cause disease. They will undoubtedly help in preparing new vaccines and drugs for the treatment of infectious disease.
- Genome function can be analyzed by annotation, the use of DNA chips to study mRNA synthesis, and the study of the organism's protein content (proteome) and its changes.

A prerequisite to understanding the complete biology of an organism is the determination of its entire genome sequence.
— *J. Craig Venter, et al.*

A DNA chip can be used to follow gene expression by a microorganism's complete genome. This chip contains probes for the over 4,200 known open reading frames in the *E. coli* genome.

hapter 13 provided a brief introduction to microbial recombination and plasmids, including the use of conjugation and other techniques in mapping the chromosome. Chapter 14 described the development and impact of recombinant DNA technology. This chapter will carry these themes further with the discussion of the current revolution in genome sequencing. We will begin with a general overview of the topic, followed by an introduction to the DNA sequencing technique. Next, the whole-genome shotgun sequencing method will be briefly described. This is followed by a comparison of selected microbial genomes and a discussion of what has been learned from them. After we have considered genome structure, we will turn to genome function and the array of transcripts and proteins produced by genomes. The focus will be on annotation, DNA chips, and the use of two-dimensional electrophoresis to study the proteome. The chapter concludes with a brief consideration of future challenges and opportunities in genomics.

15.1 INTRODUCTION

Genomics is the study of the molecular organization of genomes, their information content, and the gene products they encode. It is a broad discipline, which may be divided into at least three general areas. **Structural genomics** is the study of the physical nature of genomes. Its primary goal is to determine and analyze the DNA sequence of the genome. **Functional genomics** is concerned with the way in which the genome functions. That is, it examines the transcripts produced by the genome and the array of proteins they encode. The third area of study is **comparative genomics,** in which genomes from different organisms are compared to look for significant differences and similarities. This helps identify important, conserved portions of the genome and discern patterns in function and regulation. The data also provide much information about microbial evolution, particularly with respect to phenomena such as horizontal gene transfer.

It should be emphasized at the beginning that whole-genome sequence information provides an entirely new starting point for biological research. In the future, microbiologists will not have to spend as much time cloning genes because they will be able to generate new questions and hypotheses from computer analyses of genome data. Then they can test their hypotheses in the laboratory.

15.2 DETERMINING DNA SEQUENCES

The most widely used sequencing technique is that developed by Frederick Sanger in 1975. This approach uses dideoxynucleoside triphosphates (ddNTPs) in DNA synthesis. These molecules resemble normal nucleotides except that they lack a 3′-hydroxyl group (**figure 15.1**). They are added to the growing end of the chain, but terminate the synthesis catalyzed by DNA polymerase

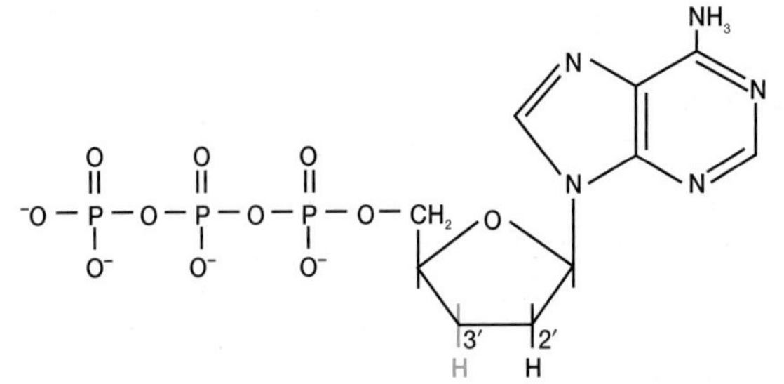

Figure 15.1   Dideoxyadenosine triphosphate (ddATP). Note the lack of a hydroxyl group on the 3′ carbon, which prevents further chain elongation by DNA polymerase.

because more nucleotides cannot be attached to further extend the chain. In the manual sequencing method, a single strand of the DNA to be sequenced is mixed with a primer, DNA polymerase I, four deoxynucleoside triphosphate substrates (one of which is radiolabeled), and a small amount of one of the dideoxynucleotides. DNA synthesis begins with the primer and terminates when a ddNTP is incorporated in place of a regular deoxynucleoside triphosphate. The result is a series of fragments of varying lengths. Four reactions are run, each with a different ddNTP. The mix with ddATP produces fragments with an A terminus; the mix with ddCTP produces fragments with C terminals, and so forth (**figure 15.2**). The radioactive fragments are removed from the DNA template and electrophoresed on a polyacrylamide gel to separate them from one another based on size. Four lanes are electrophoresed, one for each reaction mix, and the gel is autoradiographed (*see p. 314*). A DNA sequence is read directly from the gel, beginning with the smallest fragment or fastest-moving band and moving to the largest fragment or slowest band (figure 15.2a). More than 800 residues can be read from a single gel.

In automated systems dideoxynucleotides that have been labeled with fluorescent dyes are used (each ddNTP is labeled with a dye of a different color). The products from the four reactions are mixed and electrophoresed together. Because each ddNTP fluoresces with a different color, a detector can scan the gel and rapidly determine the sequence from the order of colors in the bands (figure 15.2b,c).

Fully automated capillary electrophoresis DNA analyzers are available. These sequencing machines are much faster than earlier systems and can run for 24 hours without operator attention. As many as 96 samples can be sequenced simultaneously; it is possible to generate over 1,000 kilobases of sequences a day. Current systems can sequence DNA strands around 700 bases long in about 1.5 hours.

15.3 WHOLE-GENOME SHOTGUN SEQUENCING

Although several virus genomes have been sequenced, in the past it has not been possible to sequence the genomes of bacteria. Prior to 1995, whole-genome approaches to sequencing

Figure 15.2   The Sanger Method for DNA Sequencing. (**a**) A sequencing gel with four separate lanes. The sequence begins, reading from the bottom, CAAAAAACGGACCGGGT-GTAC. (**b**) An example of sequencing by use of fluorescent dideoxynucleoside triphosphates. See text for details. (**c**) Part of an automated DNA sequencing run. Bases 493 to 499 were used as the example in (*b*).

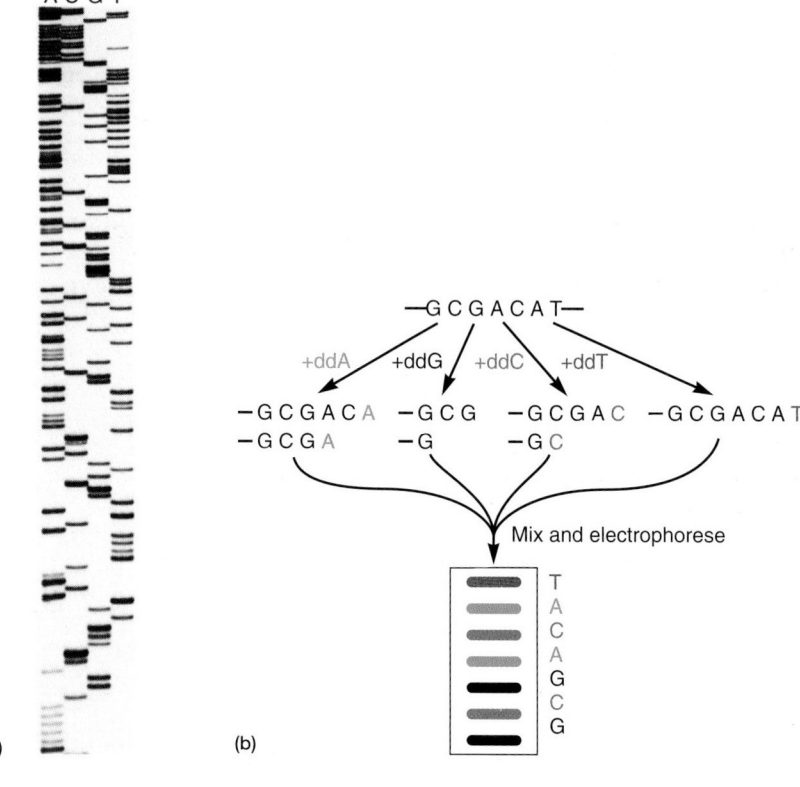

(a)

(b)

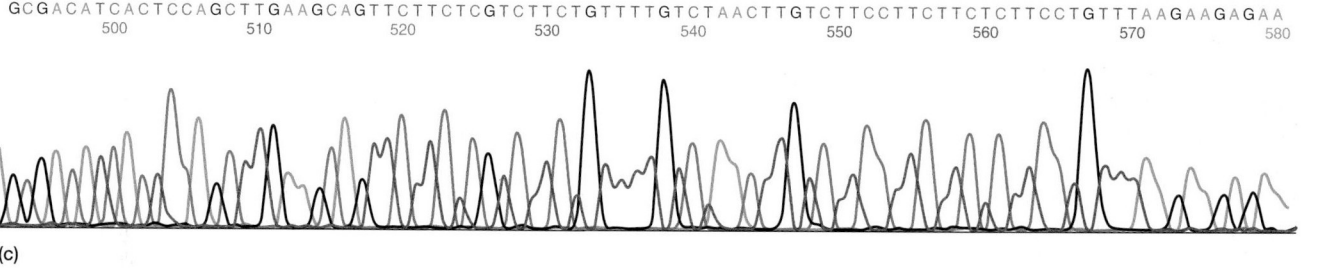

(c)

were not possible because available computational power was insufficient for assembling a genome from thousands of DNA fragments. J. Craig Venter, Hamilton Smith, and their collaborators initially sequenced the genomes of two free-living bacteria, *Haemophilus influenzae* and *Mycoplasma genitalium*. The genome of *H. influenzae,* the first to be sequenced, contains about 1,743 genes in 1,830,137 base pairs and is much larger than a virus genome.

Venter and Smith developed an approach called **whole-genome shotgun sequencing.** The process is fairly complex when considered in detail, and there are many procedures to ensure the accuracy of the results, but the following summary gives a general idea of the approach originally employed by The Institute of Genomic Research (TIGR). For simplicity, this approach may be broken into four stages: library construction, random sequencing, fragment alignment and gap closure, and editing.

1. *Library construction.* The large bacterial chromosomes were randomly broken into fairly small fragments, about the size of

a gene or less, using ultrasonic waves; the fragments were then purified (**figure 15.3**). These fragments were attached to plasmid vectors (*see pp. 323–24*), and plasmids with a single insert were isolated. Special *E. coli* strains lacking restriction enzymes were transformed with the plasmids to produce a library of the plasmid clones.

2. *Random sequencing.* After the clones were prepared and the DNA purified, thousands of bacterial DNA fragments were sequenced with automated sequencers, employing special dye-labeled primers. Thousands of templates were used, normally with universal primers that recognized the plasmid DNA sequences just next to the bacterial DNA insert. The nature of the process is such that almost all stretches of genome are sequenced several times, and this increases the accuracy of the final results.

3. *Fragment alignment and gap closure.* Using special computer programs, the sequenced DNA fragments were clustered and assembled into longer stretches of sequence by comparing nucleotide sequence overlaps between fragments.

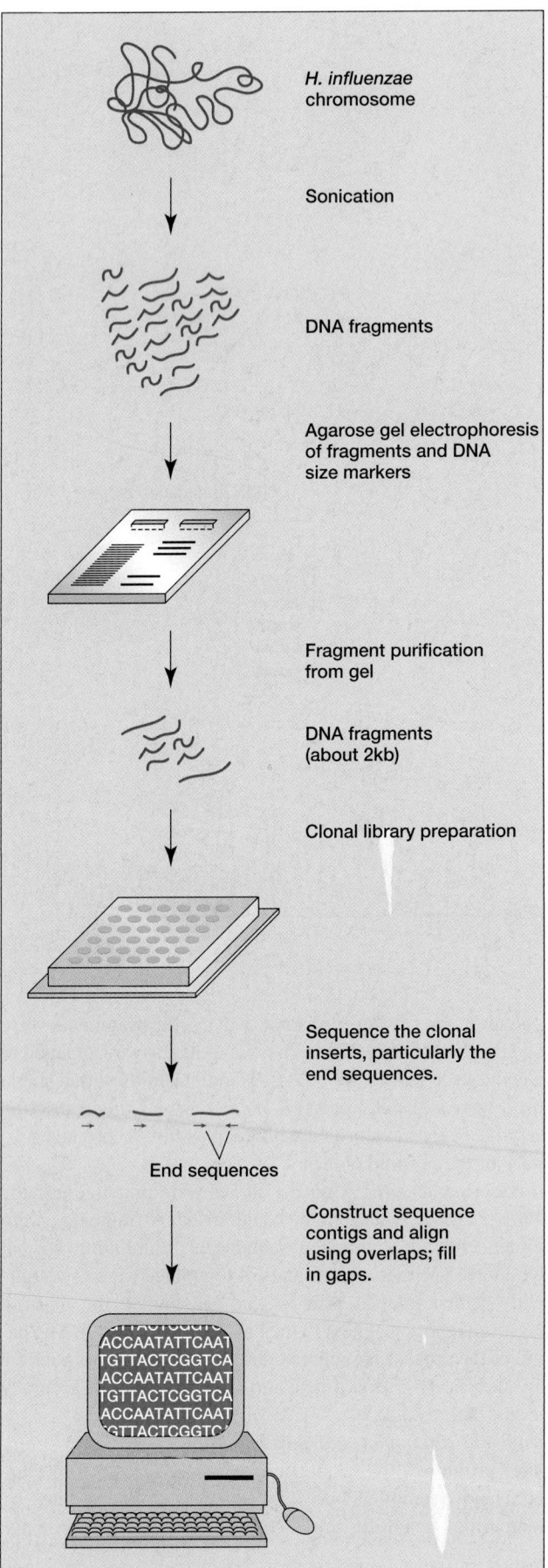

H. influenzae
chromosome

Sonication

DNA fragments

Agarose gel electrophoresis
of fragments and DNA
size markers

Fragment purification
from gel

DNA fragments
(about 2kb)

Clonal library preparation

Sequence the clonal
inserts, particularly the
end sequences.

End sequences

Construct sequence
contigs and align
using overlaps; fill
in gaps.

Figure 15.3 Whole-Genome Shotgun Sequencing. This general overview shows how the *Haemophilus influenzae* genome was sequenced. See text for details.

Two fragments were joined together to form a larger stretch of DNA if the sequences at their ends overlapped and matched (i.e., were the same). This overlap comparison process resulted in a set of larger contiguous nucleotide sequences or contigs.

Finally, the contigs were aligned in the proper order to form the completed genome sequence. If gaps existed between two contigs, sometimes fragment samples with their ends in the two adjacent contigs were available. These fragments could be analyzed and the gaps filled in with their sequences. When this approach was not possible, a variety of other techniques were used to align contigs and fill in gaps. For example, λ phage libraries containing large bacterial DNA fragments were constructed (*see pp. 320, 324–26*). The large fragments in these libraries overlapped the previously sequenced contigs. These fragments were then combined with oligonucleotide probes that matched the ends of the contigs to be aligned. If the probes bound to a λ library fragment, it could be used to prepare a stretch of DNA that represented the gap region. Overlaps in the sequence of this new fragment with two contigs would allow them to be placed side-by-side and fill in the gap between them.

4. *Editing.* The sequence was then carefully proofread in order to resolve any ambiguities in the sequence. Also the sequence was checked for unwanted frameshift mutations and corrected if necessary.

The approach worked so well that it took less than 4 months to sequence the *M. genitalium* genome (about 500,000 base pairs in size). The shotgun technique also has been used successfully by Celera Genomics in the Human Genome Project and to sequence the *Drosophila* genome.

Once the genome sequence has been established, the process of **annotation** begins. The goal of annotation is to determine the location of specific genes in the genome map. Every **open reading frame (ORF)**—a reading frame sequence (*see p. 235*) not interrupted by a stop codon—larger than 100 codons is considered to be a potential protein coding sequence. Computer programs are used to compare the sequence of the predicted ORF against large databases containing nucleotide and amino acid sequences of known enzymes and other proteins. If a bacterial sequence matches one in the database, it is assumed to code for the same protein. Although this comparison process is not without errors, it can provide tentative function assignments for about 40 to 50% of the presumed coding regions. It also gives some information about transposable elements, operons, repeat sequences, the presence of various metabolic pathways, and other genome features. Sequence analysis can be combined with other approaches to identify genes. For example, one can search for RNA or protein expression, or try to mutate the gene and inactivate its product. The results of genome sequencing and annotation for *Mycoplasma genitalium* and *Haemophilus influenzae* are shown in figures 15.5 and 15.6. Often the results of annotation are expressed in a diagram that summarizes the known metabolism and physiology of the organism. An example of this is given in figure 15.7.

---

1. Define genomics. What are the three general areas into which it can be divided?
2. Describe the Sanger method for DNA sequencing.
3. Outline the whole-genome shotgun sequencing method. What is annotation and how is it carried out?

 15.4 BIOINFORMATICS

DNA sequencing techniques have developed so rapidly that an enormous amount of data have already accumulated and genomes are being sequenced at an ever-increasing pace. The only way to organize and analyze all these data is through the use of computers, and this has led to the development of a new interdisciplinary field that combines biology, mathematics, and computer science. **Bioinformatics** is the field concerned with the management and analysis of biological data using computers. In the context of genomics, it focuses on DNA and protein sequences. The annotation process just described is one aspect of bioinformatics. DNA sequence data is stored in large databases. One of the largest genome databases is the International Nucleic Acid Sequence Data Library, often referred to as GenBank. Databases can be searched with special computer programs to find homologous sequences, DNA sequences that are similar to the one being studied. Protein coding regions also can be translated into amino acid sequences and then compared. These sequence comparisons can suggest functions of the newly discovered genes and proteins. The gene under study often will have a function similar to that of genes with homologous DNA or amino acid sequences.

 15.5 GENERAL CHARACTERISTICS OF MICROBIAL GENOMES

The development of shotgun sequencing and other genome sequencing techniques has led to the characterization of many procaryotic genomes in a very short time. Almost 100 genome sequences of microorganisms have been completed and published, and some of these are given in **table 15.1.** These procaryotes represent great phylogenetic diversity (**figure 15.4**). Over 100 more procaryotes, many of them major human pathogens, are being sequenced at present. Comparison of the genomes from different procaryotes will contribute significantly to the understanding of procaryotic evolution and help deduce which genes are responsible for various cellular processes. Genome sequences will aid in our understanding of genetic regulation and genome organization.

Currently published genome sequences already have provided new and important insights into genome organization and

Table 15.1
Examples of Complete Published Microbial Genomes

Genome	Domain[a]	Size (Mb)	% G + C
Agrobacterium tumefaciens	B	5.67[b]	57–63
Aquifex aeolicus	B	1.50	43
Archaeoglobus fulgidus	A	2.18	48
Bacillus subtilis	B	4.20	43
Borrelia burgdorferi	B	1.44	28
Campylobacter jejuni	B	1.64	31
Caulobacter crescentus	B	4.02	62–67
Chlamydia pneumoniae	B	1.23	40
Chlamydia trachomatis	B	1.05	41
Chlorobium tepidum	B	2.15	57
Clostridium perfringens	B	3.03	29
Corynebacterium glutamicum	B	3.3	55–58
Deinococcus radiodurans	B	3.28	67
Escherichia coli	B	4.60	50
Haemophilus influenzae Rd	B	1.83	39
Halobacterium sp. NRC-1	A	2.57	68
Helicobacter pylori	B	1.66	39
Listeria monocytogenes	B	2.9	37–39
Methanobacterium thermoautotrophicum	A	1.75	49
Methanococcus jannaschii	A	1.66	31
Mycobacterium leprae	B	3.27	58
Mycobacterium tuberculosis	B	4.40	65
Mycoplasma genitalium	B	0.58	31
Mycoplasma pneumoniae	B	0.81	40
Neisseria meningitidis	B	2.27	51
Nostoc sp. PCC 7120	B	6.41	41
Pseudomonas aeruginosa	B	6.3	67
Pyrococcus horiksohii	A	1.80	42
Rickettsia prowazekii	B	1.10	29
Saccharomyces cerevisiae	E	13	38
Salmonella typhimurium	B	4.86	50–53
Staphylococcus aureus	B	2.8	33
Streptococcus mutans	B	2.03	37
Streptococcus pneumoniae	B	2.16	40
Streptococcus pyogenes	B	1.9	39
Streptomyces coelicolor	B	8.67	72
Sulfolobus tokodaii	A	2.69	33
Synechocystis sp.	B	3.57	47
Thermoplasma acidophilum	A	1.56	46
Thermotoga maritima	B	1.80	46
Treponema pallidum	B	1.14	52
Vibrio cholerae	B	4.0	48
Yersinia pestis	B	4.65	48

[a]The following abbreviations are used: A, Archaea; B, Bacteria; E, Eucarya.
[b]The genome contains a circular chromosome, a linear chromosome, and two plasmids (AT and Ti).

function. *Mycoplasma genitalium* grows in human genital and respiratory tracts and has a genome of only 580 kilobases in size, one of the smallest genomes of any free-living organism (**figure 15.5**). Thus the sequence data are of great interest because they help establish the minimal set of genes needed for a free-living

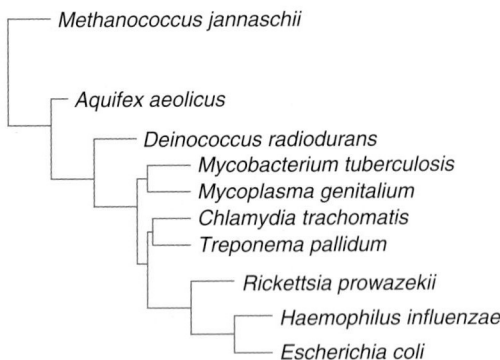

Figure 15.4 Phylogenetic Relationships of Some Procaryotes with Sequenced Genomes. These procaryotes are discussed in the text. *Methanococcus jannaschii* is in the domain *Archaea,* the rest are members of the domain *Bacteria.* Genomes from a broad diversity of procaryotes have been sequenced and compared. *Source: The Ribosomal Database Project.*

existence. There appear to be approximately 517 genes (480 protein-encoding genes and 37 genes for RNA species). About 90 proteins are involved in translation, and only around 29 proteins for DNA replication. Interestingly, 140 genes, or 29% of those in the genome, code for membrane proteins, and up to 4.5% of the genes seem to be involved in evasion of host immune responses. Only 5 genes have regulatory functions. Even in this smallest genome, 22% of the genes do not match any known protein sequence. Comparison with the *M. pneumoniae* genome and studies of gene inactivation by transposon insertion suggests that about 108 to 121 *M. genitalium* genes may not be essential for survival. Thus the minimum gene set required for laboratory growth conditions seems to be approximately 265 to 350 genes; about 100 of these have unknown functions. *Haemophilus influenzae* has a much larger genome, 1.8 megabases and 1,743 genes (**figure 15.6**). More than 40% of the genes have unknown functions. It has already been found that the bacterium lacks three Krebs cycle genes and thus a functional cycle. It does devote many more genes (64 genes) to regulatory functions than does *M. genitalium. Haemophilus influenzae* is a species capable of transformation (*see pp. 299–301*). The process must be very important to this bacterium because it contains 1,465 copies of the recognition sequence used in DNA uptake during transformation. *Methanococcus jannaschii,* a member of the Archaea, also has been sequenced. Only 44% of its 1,738 genes match those of other organisms, an indication of how different this archaeon is from bacteria and eucaryotes. Despite this profound difference, many of its genes for DNA replication, transcription, and translation are similar to eucaryotic genes and quite different from bacterial genes. However, the metabolism of *M. jannaschii* is more similar to that of bacteria than to that of eucaryotes. More recently the sequence of the 4.6 megabase *Escherichia coli* K12 genome has been published. About 5 to 6% of the genes code for proteins involved in cell and membrane structure; 12 to 14% for

transport proteins; 10% for the enzymes of energy and central intermediary metabolic pathways; 4% for regulatory genes; and 8% for replication, transcription, and translation proteins. The genome contains about 4,288 predicted genes, almost 2,500 of which do not resemble known genes. The large number of unknown genes in *Escherichia coli, Haemophilus influenzae,* and other procaryotes has great significance. It shows how little we know about microbial biology. Clearly there is much more to learn about the genetics, physiology, and metabolism of procaryotes, even of those that have been intensively studied.

Comparison of these and other genome sequences shows large differences. Not surprisingly, *E. coli* is most similar to *H. influenzae,* which is also a member of the γ-proteobacteria (1,130 similar genes). It differs more from the cyanobacterium *Synechocystis* sp. PCC6803 (675 similarities) and *Mycoplasma genitalium* (468 similarities). These four bacteria have only 111 proteins in common. *Escherichia coli* is even more unlike the archaeon *M. jannaschii* (231 similar genes) and the eucaryotic *Saccharomyces cerevisiae* (254 similar genes). Only 16 proteins, mostly translation proteins such as ribosomal proteins and aminoacyl synthetases, are essentially the same in all six organisms. There have been many gene losses and changes during the course of evolution.

Many of the genomes already sequenced belong to procaryotes that are either major human pathogens or of particular biological interest. Some recent sequences have yielded interesting discoveries and will be briefly discussed as examples of the kind of important information that can be obtained from genomics. Because of their practical importance, our focus will be primarily on human pathogens. As we will see, the results often pose more new questions than they answer old ones and open up many new areas of research.

The deinococci are soil bacteria of great interest because of their ability to survive a dose of radiation thousands of times greater than the amount needed to kill humans. They survive by stitching together their splintered chromosomes after radiation exposure. The genome consists of two circular chromosomes of different size (2.6 Mb and 0.4 Mb), a megaplasmid (177,466 bp), and a regular plasmid (45,704 bp). One would think that this genome should have some quite different DNA repair genes. However, despite its remarkable resistance to radiation, *Deinococcus radiodurans* has the same array of DNA repair mechanisms as other bacteria. It differs in simply having more of them. For example, most organisms have one MutT gene, which is involved in disposing of oxidized nucleotides; *Deinococcus* has 20 MutT-like genes. The genome also possesses many repeat sequences, which may be important in the repair process. It should be emphasized that many of the bacterium's genes have unknown functions, and some of these genes may aid in its unusually great resistance to radiation. The deinococci (pp. 455–56)

Rickettsia prowazekii is a member of the α-proteobacteria that is an obligate intracellular parasite of lice and humans. It is the causative agent of epidemic typhus that killed millions during and following the First and Second World Wars. Many microbiologists think that mitochondria may have arisen when a member

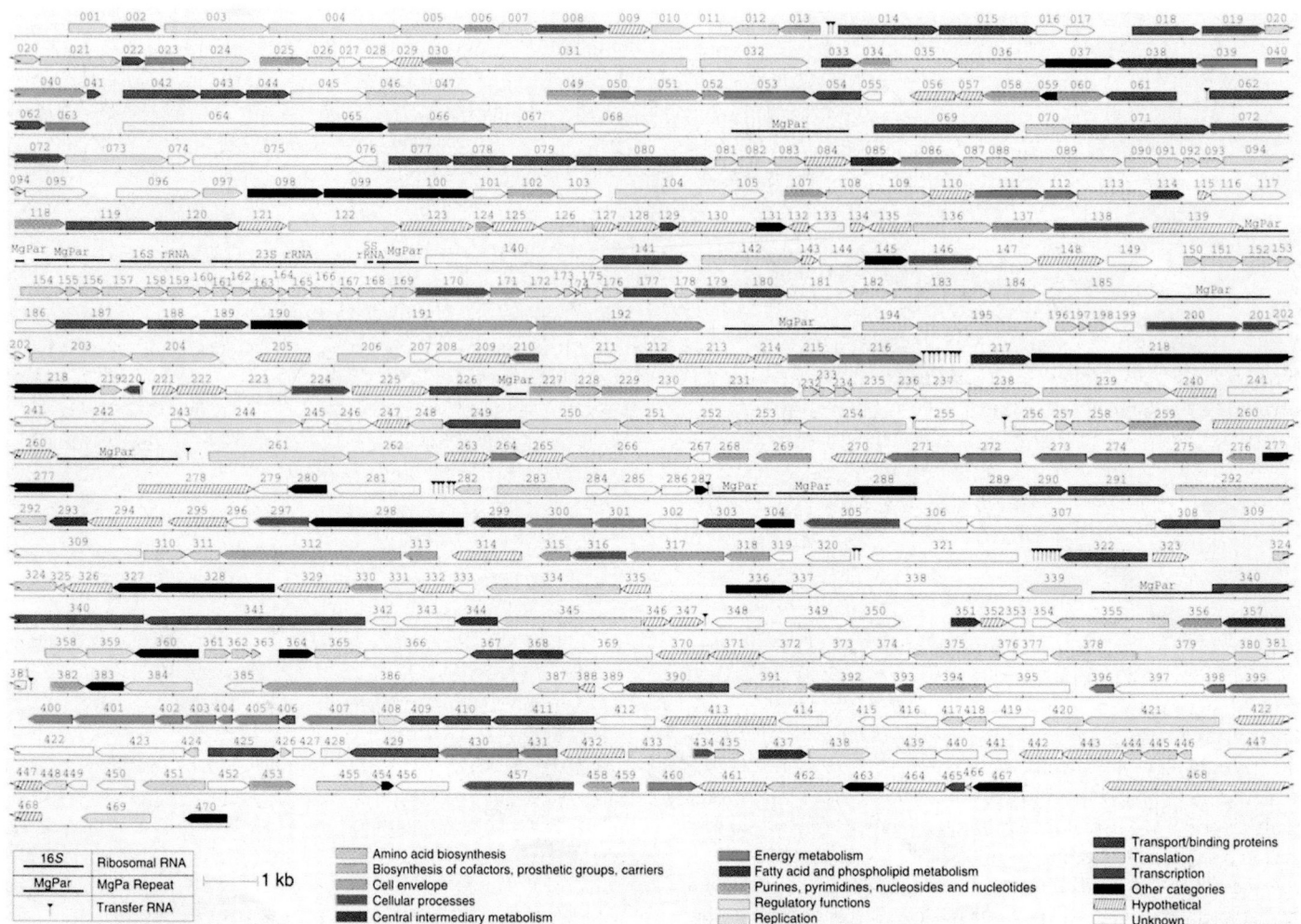

Figure 15.5   Map of the *Mycoplasma genitalium* genome.   The predicted coding regions are shown with the direction of transcription indicated by arrows. The genes are color coded by their functional role. The rRNA operon, tRNA genes, and adhesin protein operons (MgPa) are indicated.   *Reprinted with permission from Fraser, C. M., et al. Copyright 1995. The minimal gene complement of* Mycoplasma genitalium. Science *270:397–403. Figure 1, page 398 and The Institute for Genomic Research.*

of the α-proteobacteria established an endosymbiotic relationship with the ancestral eucaryotic cell (*see p. 412*). The sequenced genome of *R. prowazekii* is consistent with this hypothesis. About 25% of the 1.1 Mb genome is noncoding. Its protein-encoding genes show similarities to mitochondrial genes. Glycolysis is absent, but genes for the TCA cycle and electron transport are present, and ATP synthesis is similar to that in mitochondria. Both *Rickettsia* and the mitochondrion lack many genes for the biosynthesis of amino acids and nucleosides, in contrast with the situation in free-living α-proteobacteria. Thus aerobic respiration in eucaryotes may have arisen from an ancestor of *Rickettsia*. *Rickettsia* biology and clinical aspects (pp. 475–76, 885–86)

Chlamydiae are nonmotile, coccoid, gram-negative bacteria that reproduce only within cytoplasmic vesicles of eucaryotic cells by a unique life cycle. *Chlamydia trachomatis* infects humans and causes the sexually transmitted disease, nongonococcal

urethritis, probably the most commonly transmitted sexual disease in the United States. It also is the leading cause of preventable blindness in the world. The sequencing of its genome has revealed several surprises. The bacteria's life cycle is so unusual (*see pp. 464–66*) that one would expect its genome to be somewhat atypical. This has turned out not to be the case; the genome is similar to that of many other bacteria. Microbiologists have called *Chlamydia* an "energy parasite" and believed that it obtained all its ATP from the host cell. The genome results show that *Chlamydia* has the genes to make at least some ATP on its own, although it also has genes for ATP transporters. Another surprise is the presence of enzymes for the synthesis of peptidoglycan. Chlamydial cell walls lack peptidoglycan and microbiologists have been unable to explain why the antibiotic penicillin, which disrupts peptidoglycan synthesis, is able to inhibit chlamydial growth. The presence of peptidoglycan biosynthetic enzymes

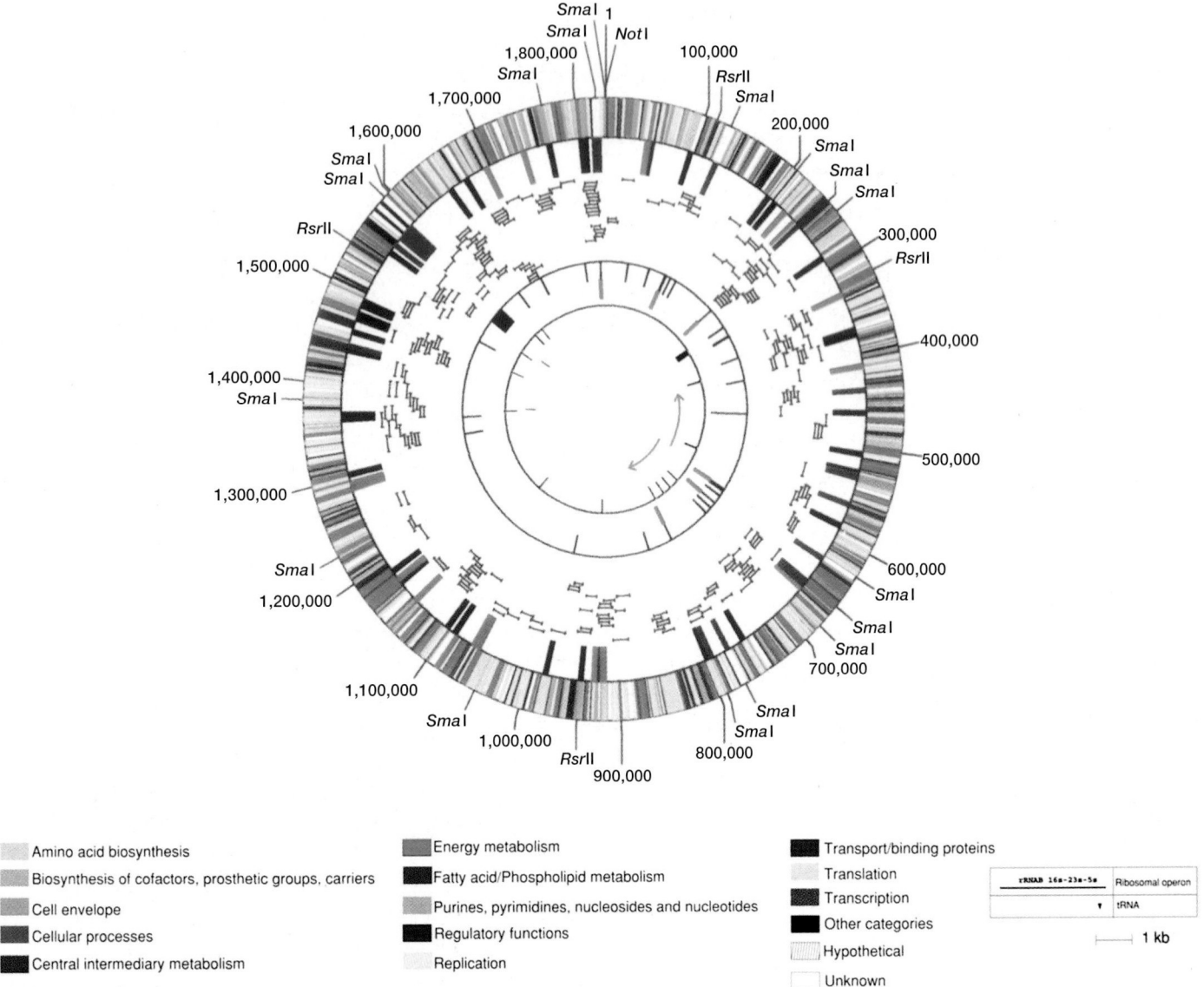

Figure 15.6 legend (keys):

- Amino acid biosynthesis
- Biosynthesis of cofactors, prosthetic groups, carriers
- Cell envelope
- Cellular processes
- Central intermediary metabolism
- Energy metabolism
- Fatty acid/Phospholipid metabolism
- Purines, pyrimidines, nucleosides and nucleotides
- Regulatory functions
- Replication
- Transport/binding proteins
- Translation
- Transcription
- Other categories
- Hypothetical
- Unknown

rRNA 16s-23s-5s Ribosomal operon
▼ tRNA
├─────┤ 1 kb

Figure 15.6 Map of the *Haemophilus influenzae* genome. The predicted coding regions in the outer concentric circle are indicated with colors representing their functional roles. The outer perimeter shows the *NotI, RsrII,* and *SmaI* restriction sites. The inner concentric circle shows regions of high G + C content (red and blue) and high A + T content (black and green). The third circle shows the coverage by λ clones (blue). The fourth circle shows the locations of rRNA operons (green), tRNAs (black), and the mu-like prophage (blue). The fifth circle shows simple tandem repeats and the probable origin of replication (outward pointing green arrows). The red lines are potential termination sequences.   *Reprinted with permission from Fleischman, R. D., et al. 1995. Whole-genome random sequencing and assembly of* Haemophilus influenzae *Rd.* Science *269:496–512. Figure 1, page 507 and The Institute of Genomic Research.*

helps account for the penicillin effect, but no one knows the purpose of peptidoglycan synthesis in this bacterium. Another major surprise is the absence of the *FtsZ* gene, which is thought to be required by all bacteria and archaea for septum formation during cell division (*see pp. 281–82*). The absence of this supposedly essential gene makes one wonder how *Chlamydia* divides. It may be that some of the genes with unknown functions play a major role in cell division. Perhaps *Chlamydia* employs a mechanism of

cell division different from that of other procaryotes. Finally, the genome contains at least 20 genes that have been obtained from eucaryotic host cells (most bacteria have no more than 3 or 4 such genes). Some of these genes are plantlike; originally *Chlamydia* may have infected a plantlike host and then moved to animals. *Chlamydia* (pp. 464–66; section 39.3)

One of the most difficult human pathogens to study has been the causative agent of syphilis, *Treponema pallidum*. This is be-

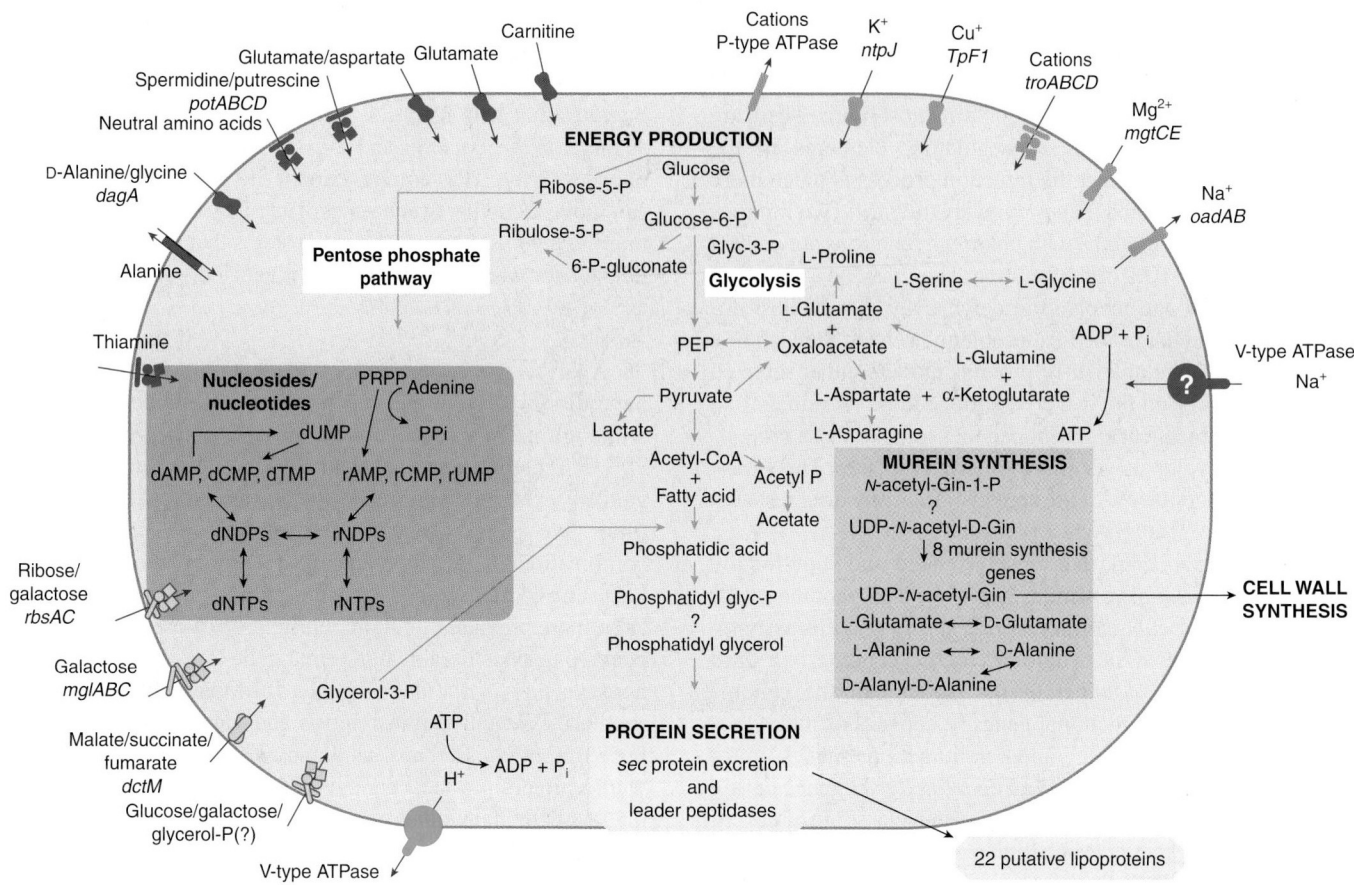

Figure 15.7 Metabolic Pathways and Transport Systems of *Treponema pallidum*. This depicts *T. pallidum* metabolism as deduced from genome annotation. Note the limited biosynthetic capabilities and extensive array of transporters. Although glycolysis is present, the TCA cycle and respiratory electron transport are lacking. Question marks indicate where uncertainties exist or expected activities have not been found.

cause it has not been possible to grow *Treponema* outside the human body. We know little about its metabolism or the way it avoids host defenses, and no vaccine for *Treponema* has yet been developed. Naturally the sequencing of the *Treponema pallidum* genome has generated considerable excitement and hope. It turns out that *Treponema* is metabolically crippled. It can use carbohydrates as an energy source, but lacks the TCA cycle and oxidative phosphorylation (**figure 15.7**). *Treponema* also lacks many biosynthetic pathways (e.g., for enzyme cofactors, fatty acids, nucleotides, and some electron transport proteins) and must rely on molecules supplied by its host. In fact, about 5% of its genes code for transport proteins. Given the lack of several critical pathways, it is not surprising that the pathogen has not been cultured successfully. The genes for surface proteins are of particular interest. *Treponema* has a family of surface protein genes characterized by many repetitive sequences. Some have speculated that these genes might undergo recombination in order to generate new surface proteins and allow the organism to avoid attack by the immune system, but this is not certain. However, it may be possible to develop a vaccine for syphilis using some of the newly

discovered surface proteins. We also may be able to identify strains of *Treponema* using these surface proteins, which would be of great importance in syphilis epidemiology. The genome results have not provided much of a clue about how *Treponema* causes syphilis. About 40% of the genes have unknown functions. Possibly some of them are responsible for avoiding host defenses and for the production of toxins and other virulence factors. *Treponema* and syphilis (pp. 466–68, 889–902)

For centuries, tuberculosis has been one of the major scourges of humankind and still kills about 3 million people annually. Furthermore, because of the spread of AIDS and noncompliance in drug treatment, *Mycobacterium tuberculosis* is increasing in frequency once again and is becoming ever more drug resistant. Anything that can be learned from genome studies could be of great importance in the fight to control the renewed spread of tuberculosis. The *Mycobacterium tuberculosis* genome is one of the largest yet found (4.40 Mb), exceeded only by *E. coli* (4.60 Mb genome) and *Pseudomonas aeruginosa* (6.26 Mb), and contains around 4,000 genes. Only about 40% of the genes have been given precise functions and 16% of its genes resemble no known proteins; presumably they are responsible for specific mycobacterial functions. More than 250 genes are devoted to lipid metabolism (*E. coli* has only about 50 such genes), and *M. tuberculosis* may obtain much of its energy by degrading host lipids. There are a surprisingly large number of regulatory elements in the genome. This may mean that the infection process is much more complex and sophisticated than previously thought. Two families of novel glycine-rich proteins with unknown functions are present and represent about 10% of the genome. They may be a source of antigenic variation and involved in defense against the host immune system. One major medical problem has been the lack of a good vaccine. A large number of proteins that are either secreted by the bacterium or on the bacterial surface have been identified from the genome sequence. It is hoped that some of these proteins can be used to develop newer, more effective vaccines. This is particularly important in view of the spread of multiply drug resistant *M. tuberculosis*.   *Mycobacterium tuberculosis* (pp. 528, 882–84)

It is tempting to think that closely related and superficially similar bacteria must have similar genomes. The genome of the leprosy bacillus, *Mycobacterium leprae*, shows that this assumption can be mistaken. The whole *M. leprae* genome is a third smaller than that of *M. tuberculosis*. About half the genome seems to be devoid of functional genes; it consists of junk DNA that represents over 1,000 degraded, nonfunctional genes. In total, *M. leprae* seems to have lost as many as 2,000 genes during its career as an intracellular parasite. It even lacks some of the enzymes required for energy production and DNA replication. This might explain why the bacterium has such a long doubling time, about two weeks in mice. One hope from genomics studies is that critical surface proteins can be discovered and used to develop a sensitive test for early detection of leprosy. This would allow immediate treatment of the disease before nerve damage occurs. Leprosy (pp. 893–94)

Recently the genomes of several major pathogenic gram-positive cocci have been sequenced; the results shed new light on their ability to cause disease. *Staphylococcus aureus* causes a wide variety of diseases ranging from food poisoning to abscesses. Unfortunately it is becoming resistant to all currently used antibiotics. The *S. aureus* chromosome is circular, 2.8 Mb in size, and contains around 2,600 protein coding regions. Analysis of its genome shows that several of the antibiotic resistance genes reside on plasmids and transposons. Furthermore, there are multiple copies of toxin genes and other virulence genes located on genomic islands. Three different strains of *Streptococcus pyogenes* also have been sequenced. Strain M1 is associated with pharyngitis and invasive infections, M3 causes highly invasive infections, and M18 is responsible for acute rheumatic fever. The chromosome is about 1.9 Mb in size. The three strains vary in the type of surface M protein they bear and their content of prophages (*see p. 380*), which contain genes for various virulence factors. Thus genes carried by bacteriophages appear to be at least partly responsible for the differences in pathogenicity between these strains of *S. pyogenes*. This is further evidence that bacteriophages can help produce new pathogen strains. Finally, the sequence of the 2.16 million base pair chromosome of *Streptococcus pneumoniae*, which causes such diseases as pneumonia, meningitis, and otitis media, has been published. The pathogen has genes for cell-surface enzymes that degrade carbohydrates that help hold cell membranes together. This provides sugars for the bacteria and aids in their spread. Insertion sequences (*see p. 292*) compose about 5% of the chromosome and can move around between genes. They may contribute to genome rearrangements. Several surface proteins that might be good candidates for vaccines have been identified. *Staphylococcus aureus* (pp. 896–99); *Streptococcus pyogenes* (pp. 879–81); *Streptococcus pneumoniae* (pp. 881–82)

Analysis and comparison of the genomes already sequenced have disclosed some general patterns in genome organization. Although protein sequences are usually conserved (i.e., about 70% of proteins contain ancient conserved regions), genome organization is quite variable in the Bacteria and Archaea. Sometimes two genes can fuse to form a new gene that has a combination of the functions possessed by the two separate genes. Less often, a gene can split or undergo fission; this seems to be more prevalent in thermophilic procaryotes. There also appears to be considerable horizontal gene transfer, particularly of housekeeping or operational genes. Informational genes, primarily those essential to transcription and translation, are not transferred as often. Perhaps, as James Lake has proposed, genes whose protein products are parts of large, complex systems and interact with many other molecules are not often transferred successfully. About 18% of the genes in *E. coli* seem to have been acquired by horizontal transfer after its divergence from *Salmonella*. There also is gene transfer between domains. The bacterium *Aquifex aeolicus* probably received about 16% of its genes from the Archaea, and 24% of the genes in *Thermotoga maritima* are similar to archaeal sequences. Some microbiologists have proposed that new species are created by the acquisition of genes that allow exploitation of a new ecological niche. For example, *E. coli* may have acquired the lactose operon and thus become able to metabolize the milk sugar lactose. This capacity would aid in colonization of the mammalian colon. Existence of extensive gene transfer between species and domains may require reevaluation of the bacterial taxonomic schemes that are based only on rRNA sequences (*see sections 19.6 and 19.7*). The comparison of many more genome sequences may clarify these phylogenetic relationships.   Horizontal gene transfer (section 13.1)

---

1. What sorts of general insights have been provided by the analysis of the genomes of *M. genitalium, H. influenzae, M. jannaschii,* and *E. coli?*
2. The genomes of *D. radiodurans, R. prowazekii, C. trachomatis, T. pallidum, M. tuberculosis, M. leprae* and the gram-positive cocci (*S. aureus, S. pyogenes, S. pneumoniae*) have been briefly discussed. Give one or two surprises or interesting insights that have arisen from each genome sequence.
3. Discuss what has been learned about horizontal gene transfer from genome comparisons.

Table 15.2									
Estimated Number of Genes Involved in Various Cell Functions[a]									
Gene Function	*Escherichia coli* K12	*Bacillus subtilis*	*Mycoplasma genitalium*	*Treponema pallidum*	*Rickettsia prowazekii*	*Chlamydia trachomatis*	*Mycobacterium tuberculosis*	*Methanococcus jannaschii*	*Pyrococcus abyssi*
Approximate total number of genes[b]	2,933	2,232	477	757	523	847	2,095	1,271	1,345
Cellular processes[c]	179	123	6	77	27	43	65	26	44
Cell envelope components	146	86	29	53	36	42	50	25	25
Transport and binding proteins	304	223	33	59	18	57	87	56	67
DNA metabolism	97	80	29	51	39	53	57	53	33
Transcription	38	45	13	25	23	23	26	21	19
Protein synthesis	121	105	90	97	87	100	90	117	99
Regulatory functions	159	163	5	22	6	15	77	18	19
Energy metabolism[d]	351	230	33	54	48	61	211	158	116
Central intermediary metabolism[e]	64	61	7	6	6	12	57	18	25
Amino acid biosynthesis	89	97	0	7	9	13	72	64	51
Fatty acid and phospholipid metabolism	67	53	8	11	11	25	78	9	8
Purines, pyrimidines, nucleosides, and nucleotides	75	68	19	21	12	15	48	37	40
Biosynthesis of cofactors and prosthetic groups	97	79	4	15	17	31	84	49	31

[a]Data adapted from TIGR (The Institute for Genomic Research) databases.
[b]The number of genes with known or hypothetical functions.
[c]Genes involved in cell division, chemotaxis and motility, detoxification, transformation, toxin production and resistance, pathogenesis, adaptations to atypical conditions, etc.
[d]Genes involved in amino acid and sugar catabolism, polysaccharide degradation and biosynthesis, electron transport and oxidative phosphorylation, fermentation, glycolysis/gluconeogenesis, pentose phosphate pathway, Entner-Doudoroff, pyruvate dehydrogenase, TCA cycle, photosynthesis, chemoautotrophy, etc.
[e]Amino sugars, phosphorus compounds, polyamine biosynthesis, sulfur metabolism, nitrogen fixation, nitrogen metabolism, etc.

 ## 15.6 FUNCTIONAL GENOMICS

Clearly, determination of genome sequences is only the start of genome research. It will take years to learn how the genome actually functions in a cell or organism (if that is completely possible) and to apply this knowledge in practical ways such as the conquest of disease and increased crop production. Sometimes the study of genome function and the practical application of this knowledge is referred to as postgenomics because it builds upon genome sequencing data. Functional genomics is a major postgenomics discipline. As mentioned earlier, functional genomics is concerned with learning how the genome operates. We will consider a few of the many approaches used to study genome function. First we will discuss annotation, which has already been introduced in the context of genome sequencing. Then techniques for the study of RNA- and protein-level expression will be described.

Genome Annotation

After sequencing, annotation can be used to tentatively identify many genes and this allows analysis of the kinds of genes and functions present in the microorganism (figure 15.7). **Table 15.2**

summarizes some of the data for several important procaryotic genomes, seven bacterial and two archaeal (*Methanococcus* and *Pyrococcus*). Even with these few examples, patterns can be seen. Genes responsible for essential informational functions (DNA metabolism, transcription, and protein synthesis) do not vary in number as much as other genes. There seems to be a minimum number of these essential genes necessary for life. Second, complex free-living bacteria such as *E. coli* and *B. subtilis* have many more operational or housekeeping genes than do most of the parasitic forms, which depend on the host for a variety of nutrients. Generally, larger genomes show more metabolic diversity. Parasitic bacteria derive many nutrients from their hosts and can shed genes for unnecessary pathways; thus they have smaller genomes.

Evaluation of RNA-Level Gene Expression

One of the best ways to evaluate gene expression is through the use of **DNA microarrays (DNA chips).** These are solid supports, typically of glass or silicon and about the size of a microscope slide, that have DNA attached in highly organized arrays. The chips can be constructed in several ways. In one approach, a programmable robotic machine delivers hundreds to thousands of microscopic

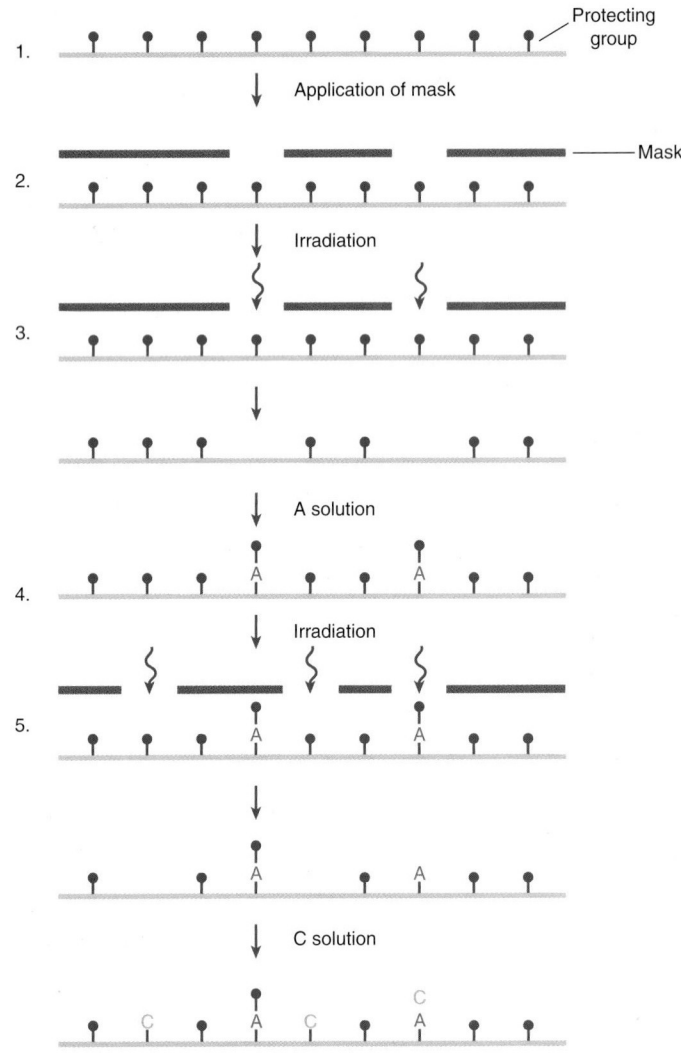

Figure 15.8 Construction of a DNA Chip with Attached Oligonucleotide Sequences. Only two cycles of synthesis are shown. See text for description of the steps.

droplets of DNA samples to specific positions on a chip using tiny pins to apply the solution (*see figure 42.21, p. 988.*) The spots are then dried and treated in order to bind the DNA tightly to the surface. Any DNA fragment can be attached in this way; often cDNA (*see p. 313*) about 500 to 5,000 bases long is used. A second procedure involves the synthesis of oligonucleotides directly on the chip in the following way (**figure 15.8**):

1. Coat the glass support with light-sensitive protecting groups that prevent random nucleoside attachment.
2. Cover the surface with a mask that has holes corresponding to the sites for attachment of the desired nucleosides.
3. Shine laser light through the mask holes to remove the exposed protecting groups.
4. Bathe the chip in a solution containing the first nucleoside to be attached. The nucleoside will chemically couple to the

light-activated sites. Each nucleoside has a light-removable protecting group to prevent addition of another nucleoside until the appropriate time.

5. Repeat steps 2 through 4 with a new mask each time to add nucleosides until all sequences on the chip have been completed.

This procedure can be used to construct any sequence. The commercial chip contains oligonucleotide probes that are 25 bases long. It is about 1.3 cm on a side and can have over 200,000 addressable positions (**figure 15.9**). The probes are often expressed sequence tags. An ***expressed sequence tag* (EST)** is a partial gene sequence unique to the gene in question that can be used to identify and position the gene during genomic analysis. It is derived from cDNA molecules. There are now chips that have probes for every expressed gene or open reading frame in the genome of *E. coli* (about 4,200 open reading frames) and the yeast *Saccharomyces cerevisiae* (approximately 6,100 open reading frames).

The nucleic acids to be analyzed, often called the targets, are isolated and labeled with fluorescent reporter groups. Nucleic acid targets may be mRNA or cDNA produced from mRNA by reverse transcription (*see figure 14.3*). The chip is incubated with the target mixture long enough to ensure proper binding to probes with complementary sequences. The unbound target is washed off and the chip is then scanned with laser beams. Fluorescence at a location indicates that the probe is bound to that particular sequence. Analysis of the hybridization pattern shows which genes are being expressed. Target samples from two experiments can be labeled with different fluorescent groups and compared using the same chip. Figure 15.9 and the chapter opening figure provide examples of fluorescently labeled DNA microarrays.

DNA chip results allow one to observe the characteristic expression of whole sets of genes during differentiation or in response to environmental changes. In some cases, many genes change expression in response to a single change in conditions. Patterns of gene expression can be detected and functions can be tentatively assigned based on expression. If an unknown gene is expressed under the same conditions as genes of known function, it is coregulated and quite likely shares the same general function. DNA chips also can be used to study regulatory genes directly by perturbing a regulatory gene and observing the effect on genome activity. Of course, only mRNAs that are currently expressed can be detected. If a gene is transiently expressed, its activity may be missed by a DNA chip analysis.

Evaluation of Protein-Level Gene Expression

Genome function can be studied at the translation level as well as the transcription level. The entire collection of proteins that an organism produces is called its **proteome.** Thus **proteomics** is the study of the proteome or the array of proteins an organism can produce. It is an essential discipline because proteomics provides information about genome function that mRNA studies cannot. There is not always a direct correlation between mRNA and protein levels because of the posttranslational modification of proteins and protein turnover. Measurement of mRNA levels can

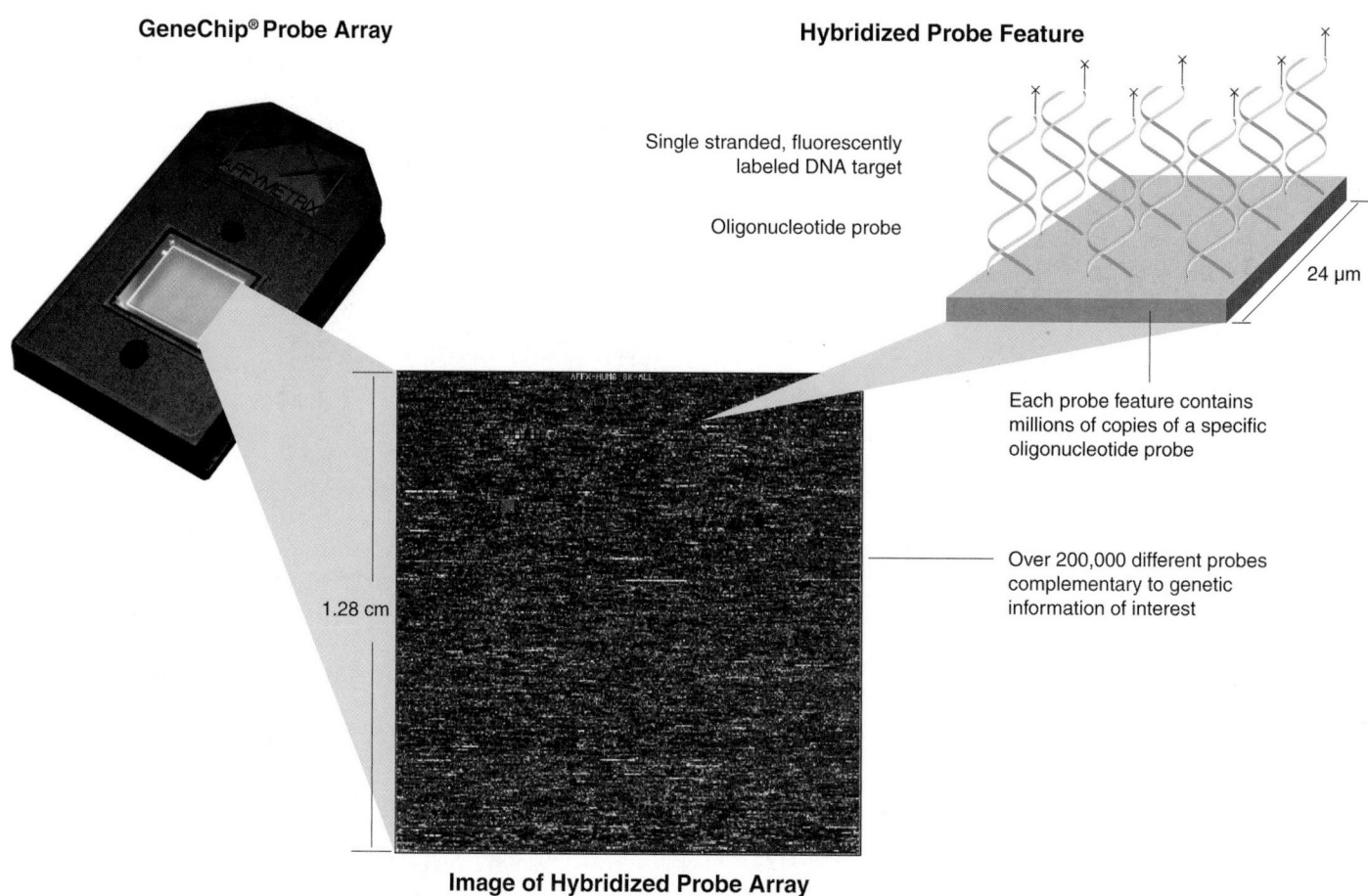

GeneChip® Probe Array

Hybridized Probe Feature

Single stranded, fluorescently labeled DNA target

Oligonucleotide probe

24 μm

Each probe feature contains millions of copies of a specific oligonucleotide probe

Over 200,000 different probes complementary to genetic information of interest

1.28 cm

Image of Hybridized Probe Array

Figure 15.9   The GeneChip Expression Probe Array.   The DNA chip manufactured by Affymetrix, Inc. contains probes designed to represent thousands or tens of thousands of genes.

show the dynamics of gene expression and tell what might occur in the cell, whereas proteomics discovers what is actually happening. Much research in this area is referred to as **functional proteomics.** It is focused on determining such things as the function of different cellular proteins, how they interact with one another, and the ways in which they are regulated.

Although new techniques in proteomics are currently being developed, we will focus briefly only on the traditional approach. A mixture of proteins is separated using two-dimensional electrophoresis. The first dimension makes use of isoelectric focusing, in which proteins move electrophoretically through a pH gradient (e.g., pH 3 to 10 or 4 to 7). The protein mixture is applied to a strip with an immobilized pH gradient and electrophoresed. Each protein moves along the strip until the pH on the strip equals its isoelectric point. At this point, the protein's net charge is zero and the protein stops moving. Thus the technique separates the proteins based on their content of ionizable amino acids. The second dimension is SDS *polyacrylamide gel electrophoresis* (SDS-PAGE). SDS (sodium dodecyl sulfate) is an anionic detergent that denatures proteins and coats the polypeptides with a negative charge. After the first electrophoretic run has been completed, the pH gradient strip is soaked in SDS buffer and then placed at the edge of an SDS-PAGE gel sheet. Then a voltage is applied at right angles to the strip at the edge of the sheet. Under these circumstances, polypeptides migrate through the polyacrylamide gel at a rate inversely proportional to their masses. That is, the smallest polypeptide will move the farthest in a particular length of time. This two-dimensional technique is very effective at separating proteins and can resolve thousands of proteins in a mixture (**figure 15.10**). If radiolabeled substrates are used, newly synthesized proteins can be distinguished and their rates of synthesis determined.   Gel electrophoresis (pp. 319–20)

Two-dimensional electrophoresis is even more powerful when coupled with mass spectrometry. The unknown protein spot is cut from the gel and cleaved into fragments by trypsin digestion. Then fragments are analyzed by a mass spectrometer and the mass of the fragments is plotted. This mass fingerprint can be used to estimate the probable amino acid composition of each fragment and tentatively identify the protein. The mass spectrometer can even be used to determine the sequence of a protein from the electrophoresis gel. The fragments from a protein, which may be further purified by liquid chromatography, are injected into a mass spectrometer and separated on the basis of their masses and charges. Computer analysis of the results often will give the

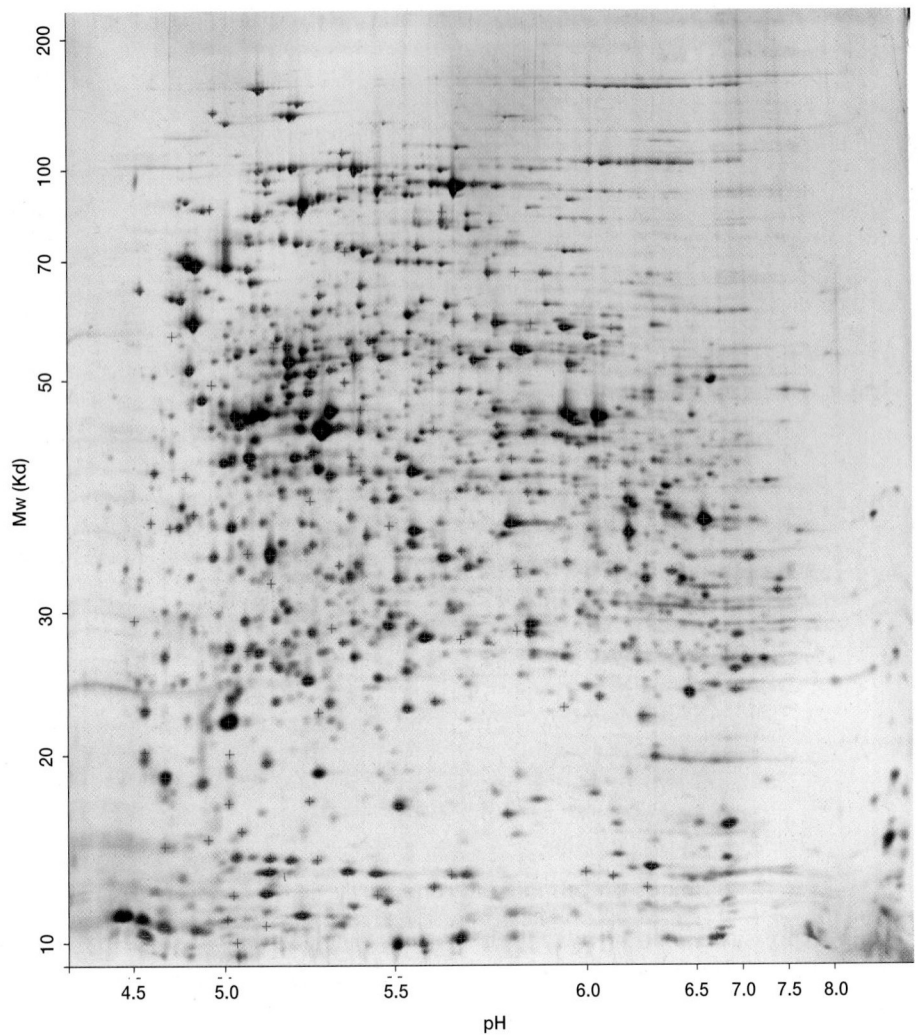

Figure 15.10 Two-Dimensional Electrophoresis of Proteins. The SWISS-2D PAGE map of *E. coli* K12 proteins. The first dimension pH gradient ran from pH 3 to 10. The second dimension comprised an 8 to 18% acrylamide gel for separation based on molecular weight. Identified proteins are indicated by red crosses.

sequence of each fragment. Sometimes nondigested complete proteins or collections of fragments are run through two mass spectrometers in sequence. The first spectrometer separates proteins and fragments, which are further fragmented. The second spectrometer then determines the sequence of each fragment produced in the first stage. The sequence of a whole protein often can be determined by analysis of this fragment sequence data. When electrophoresis and mass spectronomy are employed together, the proteome and its changes can be studied very effectively.

Proteomics has been used to study the physiology of *E. coli*. Some areas of research have been the effect of phosphate limitation, proteome changes under anaerobic conditions, heat-shock protein production, and the response to the toxicant 2,4-dinitrophenol. One particularly useful approach in studying genome function is to inactivate a specific gene and then look for changes in protein expression. Because changes in the whole proteome are followed, gene inactivation can tell much about gene function

and the large-scale effects of gene activity. A gene-protein database for *E. coli* has been established and provides information about the conditions under which each protein is expressed and where it is located in the cell.

The preceding discussion of functional genomics has emphasized areas of investigation with a record of success and a bright future. However it should be noted that many problems remain to be solved, and there may be limits to how much genomics can tell us about the living cell for a variety of reasons. For example, sequence information does not specify the nature and timing of gene regulation. Regulation of protein activity in living cells is extraordinarily complex and involves regulatory networks, which we do not yet understand completely. Functional assignments from annotation and other approaches sometimes may be inadequate because the function of a gene product often depends on its cellular context. Cells are extremely complex structural entities permeated by various physical compartments in which many processes are restricted to surfaces of membranes and macromolecular complexes (*see pp. 160–61*). Thus localization of proteins also affects function, and genomics cannot account for this. These and other problems should be kept in mind when thinking about future progress in functional genomics.

A second branch of proteomics is called **structural proteomics.** Here the focus is on directly determining the three-dimensional structures of many proteins and using these to predict the structures of other proteins and protein complexes. The assumption is that proteins fold into a limited number of shapes, and that proteins can be grouped into families of similar structures. If enough protein structures are determined for each family, all these natural shapes and the patterns of protein structure organization or protein-folding rules will be known. Then computers can use this information and the amino acid sequence of a newly discovered protein to predict its final shape. Such knowledge of proteome components will be useful in many ways—for example, in the study of microbial physiology and the design of new antimicrobial drugs.

---

1. What general lessons about genome function have been learned from the annotation results?
2. How are DNA microarrays or chips constructed and used to analyze gene expression? What sorts of things can be learned by this approach?
3. Describe two-dimensional electrophoresis and how it is used in the study of proteomics. What kinds of studies can be carried out with this technique? What role does mass spectrometry play in proteomic analysis? Define proteome, functional proteomics, and structural proteomics.

 ## 15.7 THE FUTURE OF GENOMICS

Although much has been accomplished in the past few years, the field of genomics is just beginning to mature. There are challenges ahead and many ways in which genomics can advance our knowledge of microorganisms and their practical uses. A few of these challenges and opportunities are outlined here.

1. We need to develop new methods for the large-scale analysis of genes and proteins so that more organisms can be studied.
2. All the new information about DNA and protein sequences, variations in mRNA and protein levels, and protein interactions must be integrated in order to understand genome organization and the workings of a living cell. One goal would be to have sufficient knowledge to model a cell on a computer and make predictions about how it would respond to environmental changes.
3. Genomics can be used to provide insights into pathogenicity and suggest treatments for infectious disease. Possible virulence genes can be identified, and the expression of genes during infection can be studied. Host responses to pathogens can be examined. More sensitive diagnostic tests, new antibiotics, and different vaccines may come from genomic studies of pathogens.
4. The field of pharmacogenomics should produce many new drugs to treat disease. Databases of human gene sequences can be searched both for proteins that might have therapeutic value and for new drug targets. Genomics also can be used to study variations in drug-metabolizing enzymes and individual responses to medication.
5. The nature of horizontal gene transfer and the process of microbial evolution can be studied by comparing a wide variety of genomes. Comparative genomics will aid in the study of microbial biodiversity.
6. The industrial applications are numerous. For example, genomics can be used to identify novel enzymes with industrial potential, enhance the bioremediation of hazardous wastes, and improve techniques for the microbial production of methane and other fuels.
7. Genomics will profoundly impact agriculture. It can be used to find new biopesticides and to improve sustainable agricultural practices through enhancements of processes such as nitrogen fixation.

It is clear that the possibilities are great and genomics will profoundly impact many areas of microbiology. Advances in our understanding of microorganisms also will aid in the genomic study of more complex eucaryotic organisms.

 # Summary

15.1 Introduction
a. Genomics is the study of the molecular organization of genomes, their information content, and the gene products they encode. It may be divided into three broad areas: structural genomics, functional genomics, and comparative genomics.

15.2 Determining DNA Sequences
a. DNA fragments are normally sequenced using dideoxynucleotides and the Sanger technique (**figure 15.2**).

15.3 Whole-Genome Shotgun Sequencing
a. Most often microbial genomes are sequenced using the whole-genome shotgun technique of Venter, Smith, and collaborators. Four stages are involved: library construction, sequencing of randomly produced fragments, fragment alignment and gap closure, and editing the final sequence (**figure 15.3**).
b. After the sequence has been determined, it is annotated. That is, computer analysis is used to identify genes and their functions by comparing them with gene sequences in databases.

15.4 Bioinformatics
a. Analysis of vast amounts of genome data requires sophisticated computers and programs; these analytical procedures are a part of the discipline of bioinformatics.

15.5 General Characteristics of Microbial Genomes
a. Many microbial genomes have already been sequenced (**table 15.1**) and about 100 more procaryotic genomes currently are being sequenced.

b. The genome of *Mycoplasma genitalium* is one of the smallest of any free-living organism. Analysis of this genome and others indicates that only about 265 to 350 genes are required for growth in the laboratory.

c. *Haemophilus influenzae* lacks a complete set of Krebs cycle genes and has 1,465 copies of the recognition sequence used in DNA uptake during transformation.

d. The archaeon *Methanococcus jannaschii* is quite different genetically from bacteria and eucaryotes; only around 44% of its genes match those of the bacteria and eucaryotes it was compared against. Its informational genes (replication, transcription, and translation) are more similar to eucaryotic genes, whereas its metabolic genes resemble those of bacteria more closely.

e. Even in the case of *E. coli*, perhaps the best-studied bacterium, about 58% of the predicted genes do not resemble known genes.

f. The genome sequence of *Rickettsia prowazekii* is very similar to that of mitochondria. Aerobic respiration in eucaryotes may have arisen from an ancestor of *Rickettsia*.

g. The genome of *Chlamydia trachomatis* has provided many surprises. For example, it appears able to make at least some ATP and peptidoglycan, despite the fact that it seems to obtain most ATP from the host and does not have a cell wall with peptidoglycan. The presence of plantlike genes indicates that it might have infected plantlike hosts before moving to animals.

h. *Treponema pallidum,* the causative agent of syphilis, has lost many of its metabolic genes, which may explain why it hasn't been cultivated outside a host.

i. *Mycobacterium tuberculosis* contains more than 250 genes for lipid metabolism and may obtain much of its energy from host lipids. Surface and secretory proteins have been identified and may help vaccine development.

j. The sequences of the gram-positive cocci *Staphylococcus aureus, Streptococcus pyogenes,* and *S. pneumoniae* show that often genes for antibiotic resistence and other virulence factors are carried by plasmids and bacteriophages.

k. There has been a great deal of horizontal gene transfer between genomes in both Bacteria and Archaea. This is particularly the case for housekeeping or operational genes.

15.6 Functional Genomics

a. Annotation of genomes can be used to identify many genes and their functions. There seem to be patterns in gene distribution. For example, parasitic forms tend to lose genes and obtain nutrients from their hosts.

b. DNA microarrays (DNA chips) can be used to follow gene expression and mRNA production (**figures 15.9** and **15.10**).

c. The entire collection of proteins that an organism can produce is the proteome, and its study is called proteomics. The proteome often is analyzed by two-dimensional electrophoresis followed in many cases by mass spectrometry. Proteomic experiments sometimes provide more evidence about gene expression than the use of DNA chips.

d. Despite great success in sequencing genomes, many problems still need to be resolved before the data can be interpreted adequately and applied to the understanding of organisms.

15.7 The Future of Genomics

a. In the future genomics will positively impact many areas of microbiology.

Key Terms

annotation 338
bioinformatics 339
comparative genomics 336
DNA microarrays (DNA chips) 345

expressed sequence tag (EST) 346
functional genomics 336
functional proteomics 347
genomics 336

open reading frame (ORF) 338
proteome 346
proteomics 346
structural genomics 336

structural proteomics 349
whole-genome shotgun
    sequencing 337

Questions for Thought and Review

1. What impact might genome comparisons have on the current phylogenetic schemes for Bacteria and Archaea that are discussed in chapter 19?

2. How would you use genomics data to develop new vaccines and antimicrobial drugs?

3. Why are proteomic studies necessary when one can use DNA chips to follow mRNA synthesis?

4. Discuss the importance of bioinformatics for genomics and the information it can supply.

5. Contrast informational and housekeeping or operational genes with respect to function and variation in quantity between genomes. How do free-living and parasitic microorganisms differ with respect to these genes?

6. Discuss some of the more important problems for postgenomic studies of microorganisms. What areas of microbiology do you think will be most positively impacted by genomics?

Critical Thinking Questions

1. Propose an experiment that can be done easily with a microchip that would have required years before this new technology.

2. What are the pitfalls of searches for homologous genes and proteins?

For recommended readings on these and related topics, see Appendix V.

chapter 16

The Viruses:
Introduction and General Characteristics

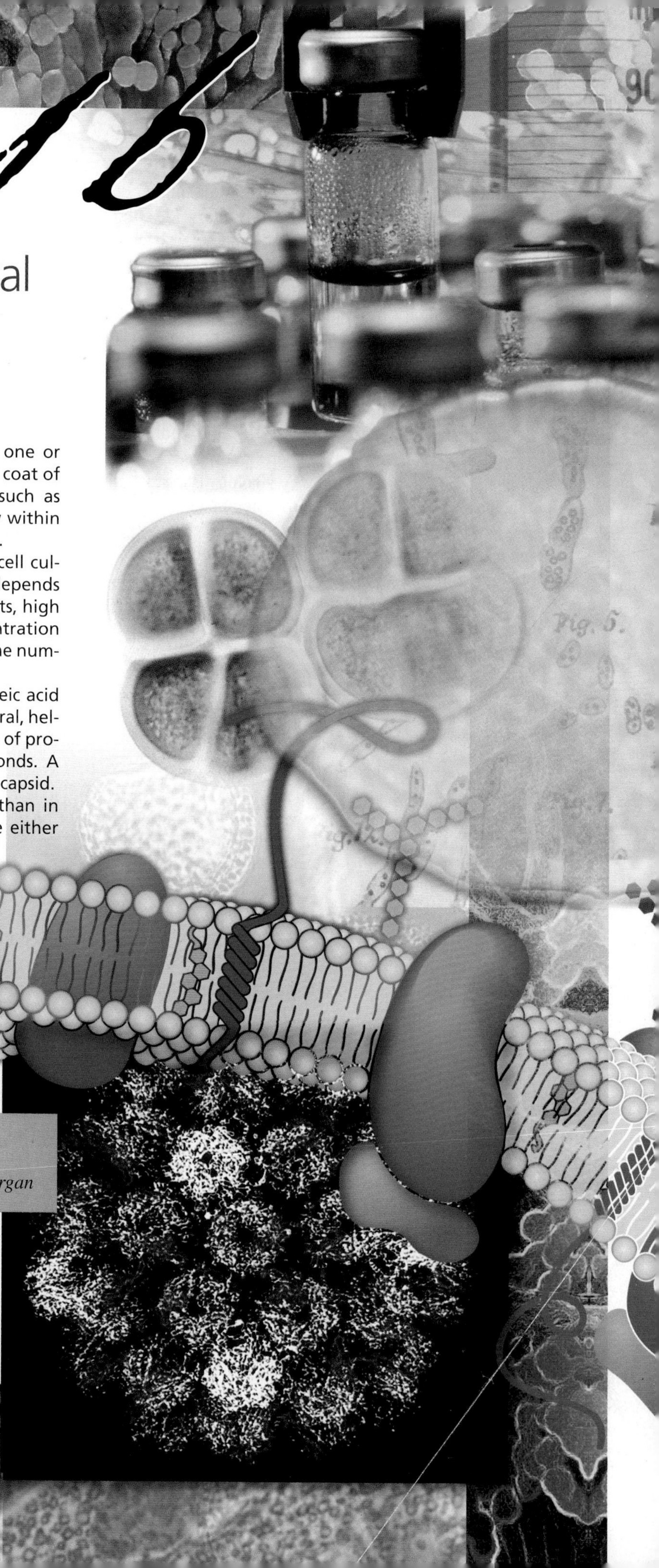

Concepts

- Viruses are simple, acellular entities consisting of one or more molecules of either DNA or RNA enclosed in a coat of protein (and sometimes, in addition, substances such as lipids and carbohydrates). They can reproduce only within living cells and are obligately intracellular parasites.
- Viruses are cultured by inoculating living hosts or cell cultures with a virion preparation. Purification depends mainly on their large size relative to cell components, high protein content, and great stability. The virus concentration may be determined from the virion count or from the number of infectious units.
- All viruses have a nucleocapsid composed of a nucleic acid surrounded by a protein capsid that may be icosahedral, helical, or complex in structure. Capsids are constructed of protomers that self-assemble through noncovalent bonds. A membranous envelope often lies outside the nucleocapsid.
- More variety is found in the genomes of viruses than in those of procaryotes and eucaryotes; they may be either single-stranded or double-stranded DNA or RNA. The nucleic acid strands can be linear, closed circle, or able to assume either shape.
- Viruses are classified on the basis of their nucleic acid's characteristics, capsid symmetry, the presence or absence of an envelope, their host, the diseases caused by animal and plant viruses, and other properties.

> *Great fleas have little fleas upon their backs to bite 'em*
> *And little fleas have lesser fleas, and so on ad infinitum.*
> —*Augustus De Morgan*

The simian virus 40 (SV-40) capsid shown here differs from most icosahedral capsids in containing only pentameric capsomers (pp. 360–62). SV-40 is a small double-stranded DNA polyomavirus with 72 capsomers. It may cause a central nervous system disease in rhesus monkeys and can produce tumors in hamsters. SV-40 was first discovered in cultures of monkey kidney cells during preparation of the poliovirus vaccine.

hapters 16, 17, and 18 focus on the viruses. These are infectious agents with fairly simple, acellular organization. They possess only one type of nucleic acid, either DNA or RNA, and only reproduce within living cells. Clearly viruses are quite different from procaryotic and eucaryotic microorganisms, and are studied by **virologists.**

Despite their simplicity in comparison with cellular organisms, viruses are extremely important and deserving of close attention. The study of viruses has contributed significantly to the discipline of molecular biology. Many human viral diseases are already known and more are discovered or arise every year, as demonstrated by the recent appearance of SARS. The whole field of genetic engineering is based in large part upon discoveries in virology. Thus it is easy to understand why **virology** (the study of viruses) is such a significant part of microbiology.

This chapter focuses on the broader aspects of virology: its development as a scientific discipline, the general properties and structure of viruses, the ways in which viruses are cultured and studied, and viral taxonomy. Chapter 17 is concerned with the bacteriophages, and chapter 18 is devoted to the viruses of eucaryotes.

Viruses have had enormous impact on humans and other organisms, yet very little was known about their nature until fairly recently. A brief history of their discovery and recognition as uniquely different infectious agents can help clarify their nature.

16.1 EARLY DEVELOPMENT OF VIROLOGY

Although the ancients did not understand the nature of their illnesses, they were acquainted with diseases, such as rabies, that are now known to be viral in origin. In fact, there is some evidence that the great epidemics of A.D. 165 to 180 and A.D. 251 to 266, which severely weakened the Roman Empire and aided its decline, may have been caused by measles and smallpox viruses. Smallpox had an equally profound impact on the New World. Hernán Cortés's conquest of the Aztec Empire in Mexico was made possible by an epidemic that ravaged Mexico City. The virus was probably brought to Mexico in 1520 by the relief expedition sent to join Cortés. Before the smallpox epidemic subsided, it had killed the Aztec King Cuitlahuac (the nephew and son-in-law of the slain emperor, Montezuma II) and possibly 1/3 of the population. Since the Spaniards were not similarly afflicted, it appeared that God's wrath was reserved for the Native Americans, and this disaster was viewed as divine support for the Spanish conquest (**Historical Highlights 16.1**).

The first progress in preventing viral diseases came years before the discovery of viruses. Early in the eighteenth century, Lady Wortley Montagu, wife of the English ambassador to Turkey, observed that Turkish women inoculated their children against smallpox. The children came down with a mild case and

subsequently were immune. Lady Montagu tried to educate the English public about the procedure but without great success. Later in the century an English country doctor, Edward Jenner, stimulated by a girl's claim that she could not catch smallpox because she had had cowpox, began inoculating humans with material from cowpox lesions. He published the results of 23 successful vaccinations in 1798. Although Jenner did not understand the nature of smallpox, he did manage to successfully protect his patients from the dread disease through exposure to the cowpox virus.

Until well into the nineteenth century, harmful agents were often grouped together and sometimes called viruses [Latin *virus*, poison or venom]. Even Louis Pasteur used the term virus for any living infectious disease agent. The development in 1884 of the porcelain bacterial filter by Charles Chamberland, one of Pasteur's collaborators and inventor of the autoclave, made possible the discovery of what are now called viruses. Tobacco mosaic disease was the first to be studied with Chamberland's filter. In 1892 Dimitri Ivanowski published studies showing that leaf extracts from infected plants would induce tobacco mosaic disease even after filtration to remove bacteria. He attributed this to the presence of a toxin. Martinus W. Beijerinck, working independently of Ivanowski, published the results of extensive studies on tobacco mosaic disease in 1898 and 1900. Because the filtered sap of diseased plants was still infectious, he proposed that the disease was caused by an entity different from bacteria, a filterable virus. He observed that the virus would multiply only in living plant cells, but could survive for long periods in a dried state. At the same time Friedrich Loeffler and Paul Frosch in Germany found that the hoof-and-mouth disease of cattle was also caused by a filterable virus rather than by a toxin. In 1900 Walter Reed began his study of the yellow fever disease whose incidence had been increasing in Cuba. Reed showed that this human disease was due to a filterable virus that was transmitted by mosquitoes. Mosquito control shortly reduced the severity of the yellow fever problem. Thus by the beginning of this century, it had been established that filterable viruses were different from bacteria and could cause diseases in plants, livestock, and humans.

Shortly after the turn of the century, Vilhelm Ellermann and Oluf Bang in Copenhagen reported that leukemia could be transmitted between chickens by cell-free filtrates and was probably caused by a virus. Three years later in 1911, Peyton Rous from the Rockefeller Institute in New York City reported that a virus was responsible for a malignant muscle tumor in chickens. These studies established that at least some malignancies were caused by viruses.

It was soon discovered that bacteria themselves also could be attacked by viruses. The first published observation suggesting that this might be the case was made in 1915 by Frederick W. Twort. Twort isolated bacterial viruses that could attack and destroy micrococci and intestinal bacilli. Although he speculated that his preparations might contain viruses, Twort did not follow up on these observations. It remained for Felix d'Herelle to establish decisively the existence of bacterial viruses. D'Herelle isolated bacterial viruses from patients with dysentery, probably

Historical Highlights
16.1   Disease and the Early Colonization of America

Although the case is somewhat speculative, there is considerable evidence that disease, and particularly smallpox, played a major role in reducing Indian resistance to the European colonization of North America. It has been estimated that Indian populations in Mexico declined about 90% within 100 years of initial contact with the Spanish. Smallpox and other diseases were a major factor in this decline, and there is no reason to suppose that North America was any different. As many as 10 to 12 million Indians may have lived north of the Rio Grande before contact with Europeans. In New England alone, there may have been over 72,000 in A.D. 1600; yet only around 8,600 remained in New England by A.D. 1674, and the decline continued in subsequent years.

Such an incredible catastrophe can be accounted for by consideration of the situation at the time of European contact with the Native Americans. The Europeans, having already suffered major epidemics in the preceding centuries, were relatively immune to the diseases they carried. On the other hand, the Native Americans had never been exposed to diseases like smallpox and were decimated by epidemics. In the sixteenth century, before any permanent English colonies had been established, many contacts were made by mission-

aries and explorers who undoubtedly brought disease with them and infected the natives. Indeed, the English noted at the end of the century that Indian populations had declined greatly but attributed it to armed conflict rather than to disease.

Establishment of colonies simply provided further opportunities for infection and outbreak of epidemics. For example, the Huron Indians decreased from a minimum of 32,000 people to 10,000 in 10 years. Between the time of initial English colonization and 1674, the Narraganset Indians declined from around 5,000 warriors to 1,000, and the Massachusetts Indians, from 3,000 to 300. Similar stories can be seen in other parts of the colonies. Some colonists interpreted these plagues as a sign of God's punishment of Indian resistance: the "Lord put an end to this quarrel by smiting them with smallpox. . . . Thus did the Lord allay their quarrelsome spirit and make room for the following part of his army."

It seems clear that epidemics of European diseases like smallpox decimated Native American populations and prepared the way for colonization of the North American continent. Many American cities—for example, Boston, Philadelphia, and Plymouth—grew upon sites of previous Indian villages.

---

caused by *Shigella dysenteriae*. He noted that when a virus suspension was spread on a layer of bacteria growing on agar, clear circular areas containing viruses and lysed cells developed. A count of these clear zones allowed d'Herelle to estimate the number of viruses present (plaque assay, p. 358). D'Herelle demonstrated that these viruses could reproduce only in live bacteria; therefore he named them bacteriophages because they could eat holes in bacterial "lawns."

The chemical nature of viruses was established when Wendell M. Stanley announced in 1935 that he had crystallized the tobacco mosaic virus (TMV) and found it to be largely or completely protein. A short time later Frederick C. Bawden and Norman W. Pirie managed to separate the TMV virus particles into protein and nucleic acid. Thus by the late 1930s it was becoming clear that viruses were complexes of nucleic acids and proteins able to reproduce only in living cells.

---

1. Describe the major technical advances and discoveries important in the early development of virology.
2. Give the contribution to virology made by each scientist mentioned in this section.

16.2   GENERAL PROPERTIES OF VIRUSES

Viruses are a unique group of infectious agents whose distinctiveness resides in their simple, acellular organization and pattern of reproduction. A complete virus particle or **virion**

consists of one or more molecules of DNA or RNA enclosed in a coat of protein, and sometimes also in other layers. These additional layers may be very complex and contain carbohydrates, lipids, and additional proteins. Viruses can exist in two phases: extracellular and intracellular. Virions, the extracellular phase, possess few if any enzymes and cannot reproduce independent of living cells. In the intracellular phase, viruses exist primarily as replicating nucleic acids that induce host metabolism to synthesize virion components; eventually complete virus particles or virions are released.

In summary, viruses differ from living cells in at least three ways: (1) their simple, acellular organization; (2) the presence of either DNA or RNA, but not both, in almost all virions (human cytomegalovirus has a DNA genome and four mRNAs); and (3) their inability to reproduce independent of cells and carry out cell division as procaryotes and eucaryotes do. Although some bacteria such as chlamydia and rickettsia (*see sections 21.5 and 22.1*) are obligately intracellular parasites like viruses, they do not meet the first two criteria.

16.3   THE CULTIVATION OF VIRUSES

Because they are unable to reproduce independent of living cells, viruses cannot be cultured in the same way as procaryotes and eucaryotic microorganisms. For many years researchers have cultivated animal viruses by inoculating suitable host animals or embryonated eggs—fertilized chicken eggs incubated about 6 to 8 days after laying (**figure 16.1**). To

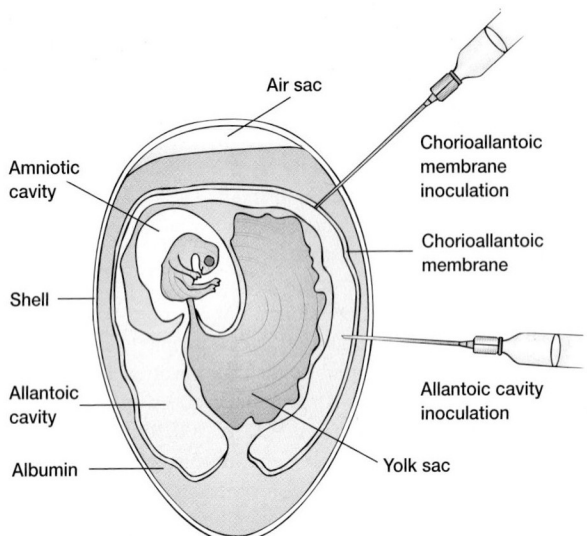

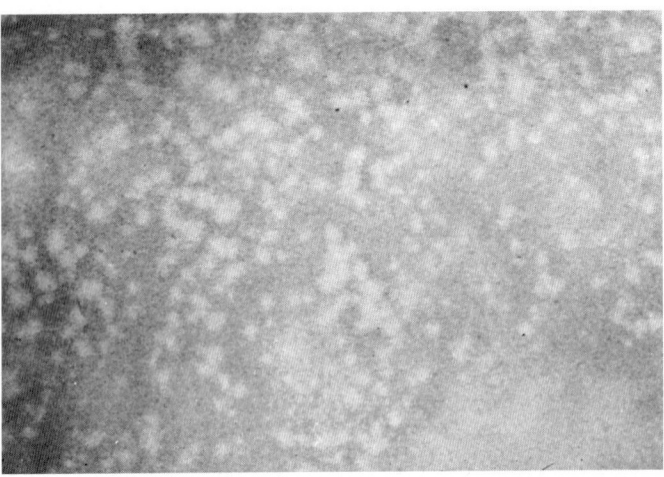

Figure 16.1 Virus Cultivation Using an Embryonated Egg. Two sites that are often used to grow animal viruses are the chorioallantoic membrane and the allantoic cavity. The diagram shows a 9 day chicken embryo.

Figure 16.2 Virus Plaques. Poliovirus plaques in a monkey kidney cell culture.

prepare the egg for virus cultivation, the shell surface is first disinfected with iodine and penetrated with a small sterile drill. After inoculation, the drill hole is sealed with gelatin and the egg incubated. Viruses may be able to reproduce only in certain parts of the embryo; consequently they must be injected into the proper region. For example, the myxoma virus grows well on the chorioallantoic membrane, whereas the mumps virus prefers the allantoic cavity. The infection may produce a local tissue lesion known as a pock, whose appearance often is characteristic of the virus.

More recently animal viruses have been grown in tissue (cell) culture on monolayers of animal cells. This technique is made possible by the development of growth media for animal cells and by the advent of antibiotics that can prevent bacterial and fungal contamination. A layer of animal cells in a specially prepared petri dish is covered with a virus inoculum, and the viruses are allowed time to settle and attach to the cells. The cells are then covered with a thin layer of agar to limit virion spread so that only adjacent cells are infected by newly produced virions. As a result localized areas of cellular destruction and lysis called **plaques** often are formed (**figure 16.2**) and may be detected if stained with dyes, such as neutral red or trypan blue, that can distinguish living from dead cells. Viral growth does not always result in the lysis of cells to form a plaque. Animal viruses, in particular, can cause microscopic or macroscopic degenerative changes or abnormalities in host cells and in tissues called **cytopathic effects** (**figure 16.3**). Cytopathic effects may be lethal, but plaque formation from cell lysis does not always occur.

Bacterial viruses or **bacteriophages** (**phages** for short) are cultivated in either broth or agar cultures of young, actively growing bacterial cells. So many host cells are destroyed that turbid bacterial cultures may clear rapidly because of cell lysis. Agar cultures are prepared by mixing the bacteriophage sample with cool, liquid agar and a suitable bacterial culture. The mixture is quickly poured into a petri dish containing a bottom layer of sterile agar. After hardening, bacteria in the layer of top agar grow and reproduce, forming a continuous, opaque layer or "lawn." Wherever a virion comes to rest in the top agar, the virus infects an adjacent cell and reproduces. Eventually, bacterial lysis generates a plaque or clearing in the lawn (**figure 16.4**). As can be seen in figure 16.4, plaque appearance often is characteristic of the phage being cultivated.

Plant viruses are cultivated in a variety of ways. Plant tissue cultures, cultures of separated cells, or cultures of protoplasts (*see p. 47*) may be used. Viruses also can be grown in whole plants. Leaves are mechanically inoculated when rubbed with a mixture of viruses and an abrasive such as carborundum. When the cell walls are broken by the abrasive, the viruses directly contact the plasma membrane and infect the exposed host cells. (The role of the abrasive is frequently filled by insects that suck or crush plant leaves and thus transmit viruses.) A localized **necrotic lesion** often develops due to the rapid death of cells in the infected area (**figure 16.5**). Even when lesions do not occur, the infected plant may show symptoms such as changes in pigmentation or leaf shape. Some plant viruses can be transmitted only if a diseased part is grafted onto a healthy plant.

1. What is a virus particle or virion, and how is it different from living organisms?
2. Discuss the ways in which viruses may be cultivated. Define the terms pock, plaque, cytopathic effect, bacteriophage, and necrotic lesion.

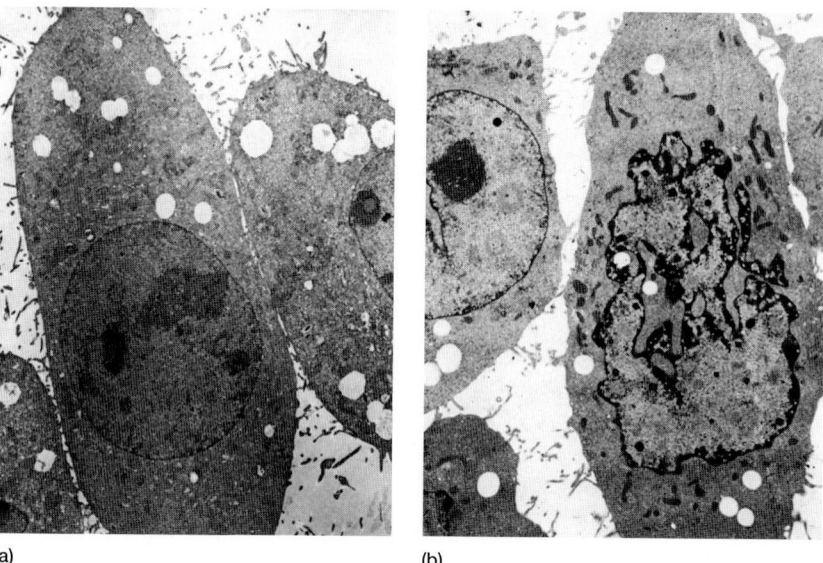

Figure 16.3 The Cytopathic Effects of Viruses.
(a) Normal mammalian cells in tissue culture. (b) Appearance of tissue culture cells 18 hours after infection with adenovirus. Transmission electron microscope photomicrographs ($\times 11,000$).

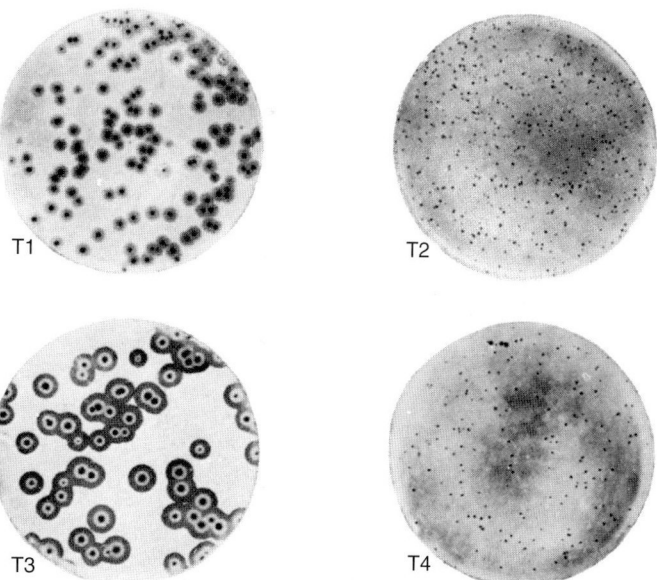

Figure 16.4 Phage Plaques.   Plaques produced on a lawn of *E. coli* by some of the T coliphages. Note the large differences in plaque appearance. The photographs are about 1/3 full size.

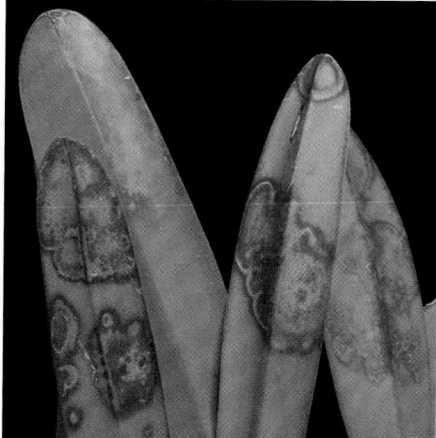

Figure 16.5 Necrotic Lesions on Plant Leaves.   (a) Tobacco mosaic virus on *Nicotiana glutinosa*. (b) Tobacco mosaic virus infection of an orchid showing leaf color changes.

16.4 VIRUS PURIFICATION AND ASSAYS

Virologists must be able to purify viruses and accurately determine their concentrations in order to study virus structure, reproduction, and other aspects of their biology. These methods are so important that the growth of virology as a modern discipline depended on their development.

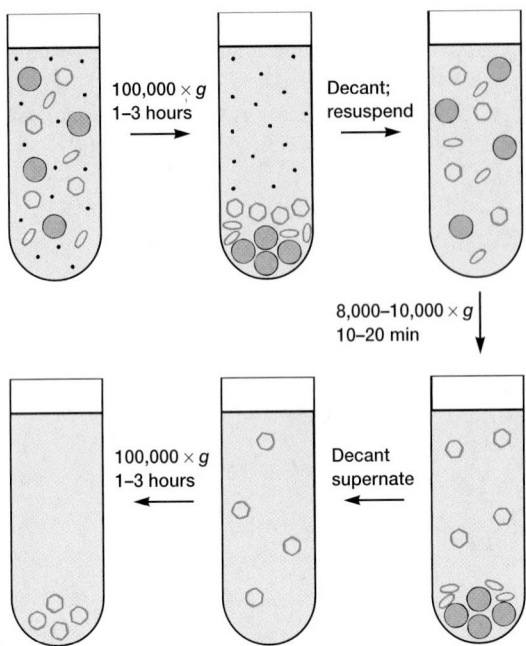

Figure 16.6   The Use of Differential Centrifugation to Purify a Virus.   At the beginning the centrifuge tube contains homogenate and icosahedral viruses (in green). First, the viruses and heavier cell organelles are removed from smaller molecules. After resuspension, the mixture is centrifuged just fast enough to sediment cell organelles while leaving the smaller virus particles in suspension; the purified viruses are then collected. This process can be repeated several times to further purify the virions.

Virus Purification

Purification makes use of several virus properties. Virions are very large relative to proteins, are often more stable than normal cell components, and have surface proteins. Because of these characteristics, many techniques useful for the isolation of proteins and organelles can be employed in virus isolation. Four of the most widely used approaches are (1) differential and density gradient centrifugation, (2) precipitation of viruses, (3) denaturation of contaminants, and (4) enzymatic digestion of cell constituents.

1. Host cells in later stages of infection that contain mature virions are used as the source of material. Infected cells are first disrupted in a buffer to produce an aqueous suspension or homogenate consisting of cell components and viruses. Viruses can then be isolated by **differential centrifugation,** the centrifugation of a suspension at various speeds to separate particles of different sizes (**figure 16.6**). Usually the homogenate is first centrifuged at high speed to sediment viruses and other large cellular particles, and the supernatant, which contains the homogenate's soluble molecules, is discarded. The pellet is next resuspended and centrifuged at a low speed to remove substances heavier than viruses. Higher speed centrifugation

then sediments the viruses. This process may be repeated to purify the virus particles further.

   Viruses also can be purified based on their size and density by use of **gradient centrifugation (figure 16.7)**. A sucrose solution is poured into a centrifuge tube so that its concentration smoothly and linearly increases between the top and the bottom of the tube. The virus preparation, often after purification by differential centrifugation, is layered on top of the gradient and centrifuged. As shown in figure 16.7a, the particles settle under centrifugal force until they come to rest at the level where the gradient's density equals theirs (isopycnic gradient centrifugation). Viruses can be separated from other particles only slightly different in density. Gradients also can separate viruses based on differences in their sedimentation rate (rate zonal gradient centrifugation). When this is done, particles are separated on the basis of both size and density; usually the largest virus will move most rapidly down the gradient. Figure 16.7b shows that viruses differ from one another and cell components with respect to either density (grams per milliliter) or sedimentation coefficient(s). Thus these two types of gradient centrifugation are very effective in virus purification.

2. Viruses, like many proteins, can be purified through precipitation with concentrated ammonium sulfate. Initially, sufficient ammonium sulfate is added to raise its concentration to a level just below that which will precipitate the virus. After any precipitated contaminants are removed, more ammonium sulfate is added and the precipitated viruses are collected by centrifugation. Viruses sensitive to ammonium sulfate often are purified by precipitation with polyethylene glycol.

3. Viruses frequently are less easily denatured than many normal cell constituents. Contaminants may be denatured and precipitated with heat or a change in pH to purify viruses. Because some viruses also tolerate treatment with organic solvents like butanol and chloroform, solvent treatment can be used to both denature protein contaminants and extract any lipids in the preparation. The solvent is thoroughly mixed with the virus preparation, then allowed to stand and separate into organic and aqueous layers. The unaltered virus remains suspended in the aqueous phase while lipids dissolve in the organic phase. Substances denatured by organic solvents collect at the interface between the aqueous and organic phases.
   Denaturation of enzymes (p. 159)

4. Cellular proteins and nucleic acids can be removed from many virus preparations through enzymatic degradation because viruses usually are more resistant to attack by nucleases and proteases than are free nucleic acids and proteins. For example, ribonuclease and trypsin often degrade cellular ribonucleic acids and proteins while leaving virions unaltered.

Virus Assays

The quantity of viruses in a sample can be determined either by counting particle numbers or by measurement of the infectious unit concentration. Although most normal virions are probably

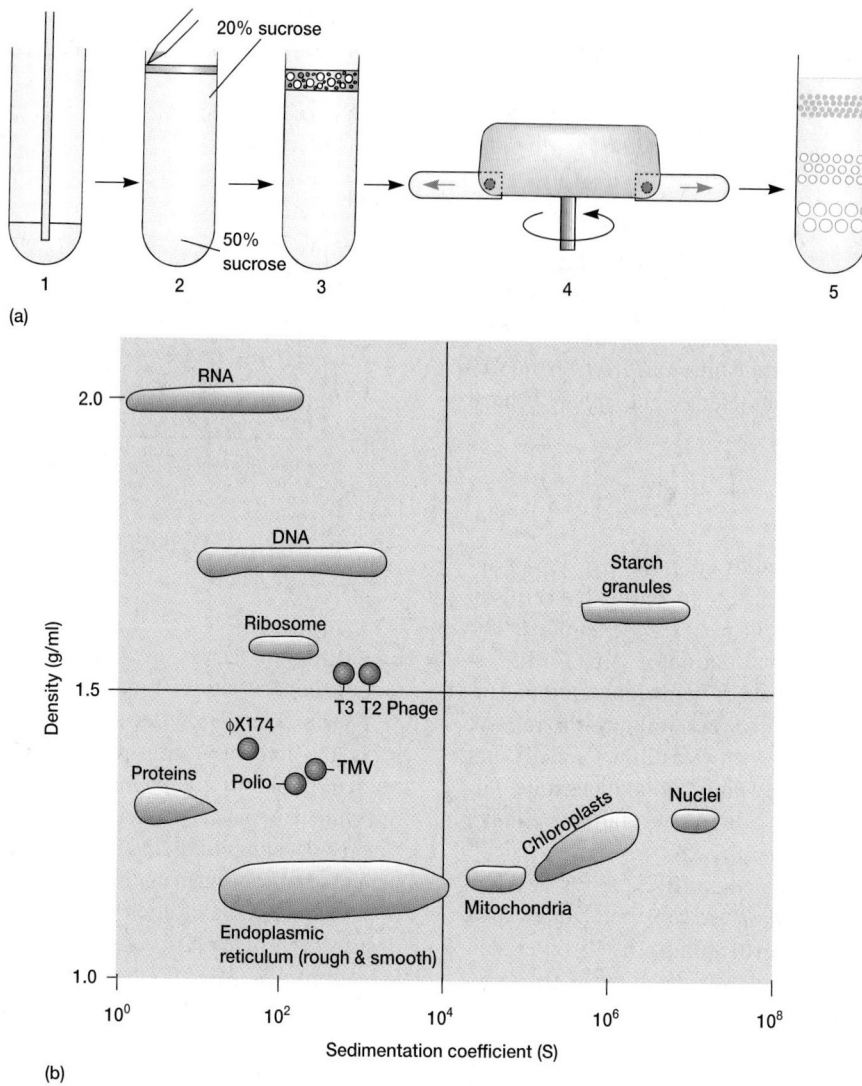

Figure 16.7 **Gradient Centrifugation.** (a) A linear sucrose gradient is prepared, *1,* and the particle mixture is layered on top, *2* and *3*. Centrifugation, *4,* separates the particles on the basis of their density and sedimentation coefficient, (the arrows in the centrifuge tubes indicate the direction of centrifugal force). *5*. In isopycnic gradient centrifugation, the bottom of the gradient is denser than any particle, and each particle comes to rest at a point in the gradient equal to its density. Rate zonal centrifugation separates particles based on their sedimentation coefficient, a function of both size and density, because the bottom of the gradient is less dense than the densest particles and centrifugation is carried out for a shorter time so that particles do not come to rest. The largest, most dense particles travel fastest. (b) The densities and sedimentation coefficients of representative viruses (shown in color) and other biological substances.

potentially infective, many will not infect host cells because they do not contact the proper surface site. Thus the total particle count may be from 2 to 1 million times the infectious unit number depending on the nature of the virion and the experimental conditions. Despite this, both approaches are of value.

Virus particles can be counted directly with the electron microscope. In one procedure the virus sample is mixed with a known concentration of small latex beads and sprayed on a coated specimen grid. The beads and virions are counted; the virus concentration is calculated from these counts and from the

bead concentration (**figure 16.8**). This technique often works well with concentrated preparations of viruses of known morphology. Viruses can be concentrated by centrifugation before counting if the preparation is too dilute. However, if the beads and viruses are not evenly distributed (as sometimes happens), the final count will be inaccurate.

The most popular indirect method of counting virus particles is the **hemagglutination assay.** Many viruses can bind to the surface of red blood cells (*see figure 33.9*). If the ratio of viruses to cells is large enough, virus particles will join the red blood cells

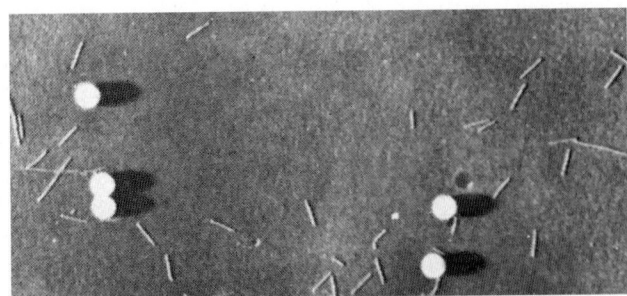

Figure 16.8 Tobacco Mosaic Virus. A tobacco mosaic virus preparation viewed in the transmission electron microscope. Latex beads 264 mm in diameter (white spheres) have been added.

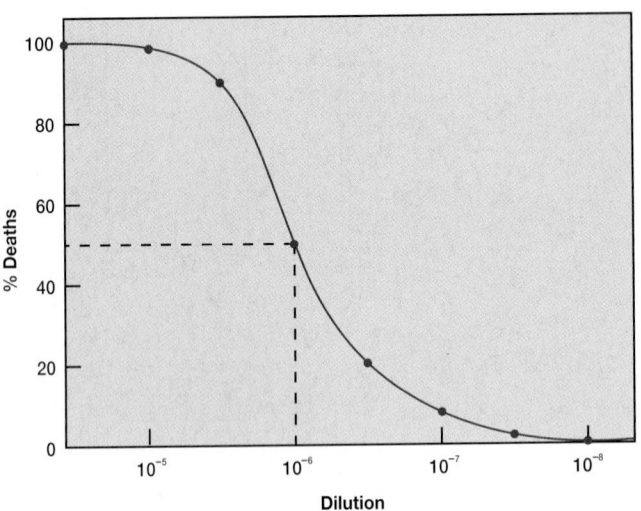

Figure 16.9 A Hypothetical Dose-Response Curve. The LD_{50} is indicated by the dashed line.

together, forming a network that settles out of suspension or agglutinates. In practice, red blood cells are mixed with a series of virus preparation dilutions and each mixture is examined. The hemagglutination titer is the highest dilution of virus (or the reciprocal of the dilution) that still causes hemagglutination. This assay is an accurate, rapid method for determining the relative quantity of viruses such as the influenza virus. If the actual number of viruses needed to cause hemagglutination is determined by another technique, the assay can be used to ascertain the number of virus particles present in a sample.

A variety of assays analyze virus numbers in terms of infectivity, and many of these are based on the same techniques used for virus cultivation. For example, in the **plaque assay** several dilutions of bacterial or animal viruses are plated out with appropriate host cells. When the number of viruses plated out are much lower than the number of host cells available for infection and when the viruses are distributed evenly, each plaque in a layer of bacterial or animal cells is assumed to have arisen from the reproduction of a single virus particle. Therefore a count of the plaques produced at a particular dilution will give the number of infectious virions or **plaque-forming units (PFU),** and the concentration of infectious units in the original sample can be easily calculated. Suppose that 0.10 ml of a 10^{-6} dilution of the virus preparation yields 75 plaques. The original concentration of plaque-forming units is

$$\text{PFU/ml} = (75 \text{ PFU}/0.10 \text{ ml})(10^6) = 7.5 \times 10^8.$$

Viruses producing different plaque morphology types on the same plate may be counted separately. Although the number of PFU does not equal the number of virus particles, their ratios are proportional: a preparation with twice as many viruses will have twice the plaque-forming units.

The same approach employed in the plaque assay may be used with embryos and plants. Chicken embryos can be inoculated with a diluted preparation or plant leaves rubbed with a mixture of diluted virus and abrasive. The number of pocks on embryonic membranes or necrotic lesions on leaves is multiplied

by the dilution factor and divided by the inoculum volume to obtain the concentration of infectious units.

When biological effects are not readily quantified in these ways, the amount of virus required to cause disease or death can be determined by the endpoint method. Organisms or cell cultures are inoculated with serial dilutions of a virus suspension. The results are used to find the endpoint dilution at which 50% of the host cells or organisms are destroyed (**figure 16.9**). The **lethal dose (LD_{50})** is the dilution that contains a dose large enough to destroy 50% of the host cells or organisms. In a similar sense, the **infectious dose (ID_{50})** is the dose which, when given to a number of test systems or hosts, causes an infection of 50% of the systems or hosts under the conditions employed.

---

1. Give the four major approaches by which viruses may be purified, and describe how each works. Distinguish between differential and density gradient centrifugation in terms of how they are carried out.
2. How can one find the virus concentration, both directly and indirectly, by particle counts and measurement of infectious unit concentration? Define plaque-forming units, lethal dose, and infectious dose.

 THE STRUCTURE OF VIRUSES

Virus morphology has been intensely studied over the past decades because of the importance of viruses and the realization that virus structure was simple enough to be understood. Progress has come from the use of several different techniques: electron microscopy, X-ray diffraction, biochemical analysis, and immunology. Although our knowledge is incomplete due to the large number of different viruses, the general nature of virus structure is becoming clear.

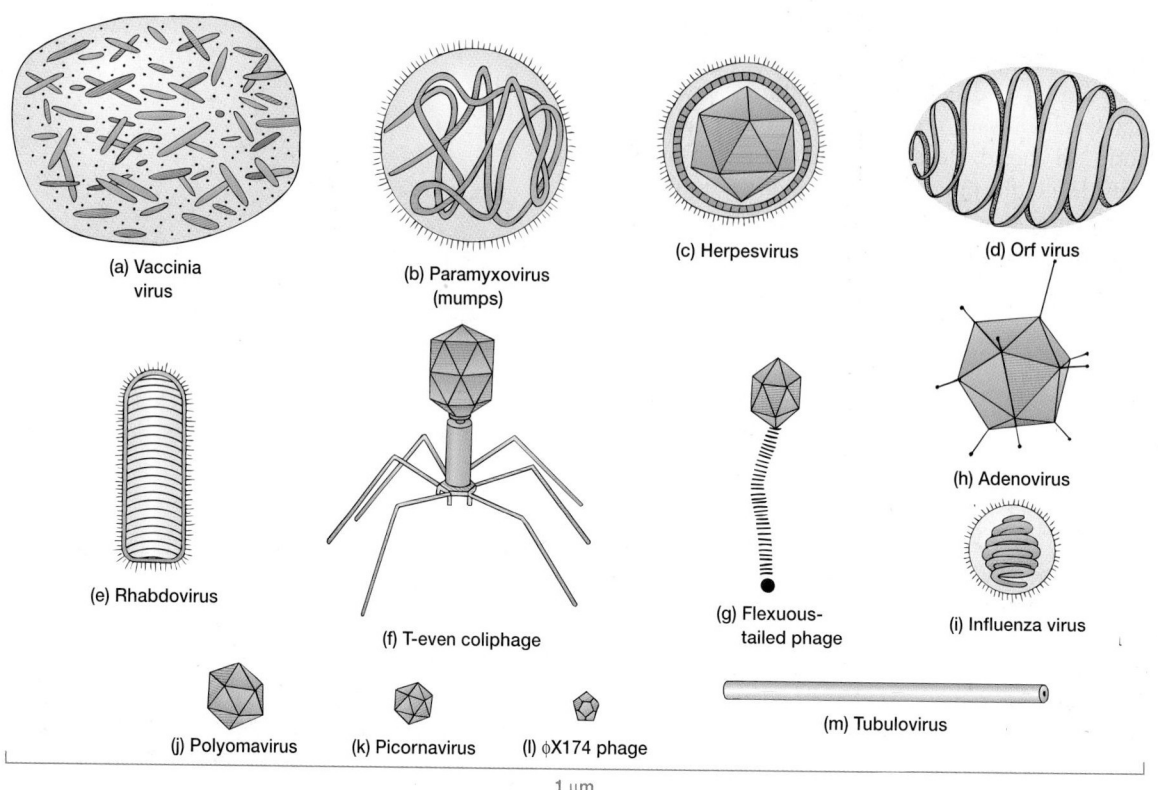

Figure 16.10   **The Size and Morphology of Selected Viruses.**   The viruses are drawn to scale. A 1 µm line is provided at the bottom of the figure.

Virion Size

Virions range in size from about 10 to 300 or 400 nm in diameter (**figure 16.10**). The smallest viruses are a little larger than ribosomes, whereas the poxviruses, like vaccinia, are about the same size as the smallest bacteria and can be seen in the light microscope. Most viruses, however, are too small to be visible in the light microscope and must be viewed with the scanning and transmission electron microscopes (*see section 2.4*).

General Structural Properties

All virions, even if they possess other constituents, are constructed around a **nucleocapsid** core (indeed, some viruses consist only of a nucleocapsid). The nucleocapsid is composed of a nucleic acid, either DNA or RNA, held within a protein coat called the **capsid**, which protects viral genetic material and aids in its transfer between host cells.

There are four general morphological types of capsids and virion structure.

1. Some capsids are **icosahedral** in shape. An icosahedron is a regular polyhedron with 20 equilateral triangular faces and 12 vertices (figure 16.10*h,j–l*). These capsids appear spherical when viewed at low power in the electron microscope.
2. Other capsids are **helical** and shaped like hollow protein cylinders, which may be either rigid or flexible (figure 16.10*m*).

3. Many viruses have an **envelope,** an outer membranous layer surrounding the nucleocapsid. Enveloped viruses have a roughly spherical but somewhat variable shape even though their nucleocapsid can be either icosahedral or helical (figure 16.10*b,c,i*).
4. **Complex viruses** have capsid symmetry that is neither purely icosahedral nor helical (figure 16.10*a,d,f,g*). They may possess tails and other structures (e.g., many bacteriophages) or have complex, multilayered walls surrounding the nucleic acid (e.g., poxviruses such as vaccinia).

Both helical and icosahedral capsids are large macromolecular structures constructed from many copies of one or a few types of protein subunits or **protomers.** Probably the most important advantage of this design strategy is that the information stored in viral genetic material is used with maximum efficiency. For example, the tobacco mosaic virus (TMV) capsid contains a single type of small subunit possessing 158 amino acids. Only about 474 nucleotides out of 6,000 in the virus RNA are required to code for coat protein amino acids. Unless the same protein is used many times in capsid construction, a large nucleic acid, such as the TMV RNA, cannot be enclosed in a protein coat without using much or all of the available genetic material to code for capsid proteins. If the TMV capsid were composed of six different protomers of the same size as the TMV subunit, about 2,900 of the 6,000 nucleotides would be required for its construction, and much less genetic material would be available for other purposes. The genetic code and translation (pp. 234–35)

Once formed and exposed to the proper conditions, protomers usually interact specifically with each other and spontaneously associate to form the capsid. Because the capsid is constructed without any outside aid, the process is called self-assembly (*see* p. 65). Some more complex viruses possess genes for special factors that are not incorporated into the virion but are required for its assembly.

Helical Capsids

Helical capsids are shaped much like hollow tubes with protein walls. The tobacco mosaic virus provides a well-studied example of helical capsid structure (**figure 16.11**). A single type of protomer associates together in a helical or spiral arrangement to produce a long, rigid tube, 15 to 18 nm in diameter by 300 nm long. The RNA genetic material is wound in a spiral and positioned toward the inside of the capsid where it lies within a groove formed by the protein subunits. Not all helical capsids are as rigid as the TMV capsid. Influenza virus RNAs are enclosed in thin, flexible helical capsids folded within an envelope (figures 16.10*i* and 16.12*a,b*).

The size of a helical capsid is influenced by both its protomers and the nucleic acid enclosed within the capsid. The diameter of the capsid is a function of the size, shape, and interactions of the protomers. The nucleic acid determines helical capsid length because the capsid does not seem to extend much beyond the end of the DNA or RNA.

Icosahedral Capsids

The icosahedron is one of nature's favorite shapes (the helix is probably most popular). Viruses employ the icosahedral shape because it is the most efficient way to enclose a space. A few genes, sometimes only one, can code for proteins that self-assemble to form the capsid. In this way a small number of linear genes can specify a large three-dimensional structure. Certain requirements must be met to construct an icosahedron. Hexagons pack together in planes and cannot enclose a space, and therefore pentagons must also be used.

When icosahedral viruses are negatively stained and viewed in the transmission electron microscope, a complex icosahedral capsid structure is revealed (**figure 16.12**). The capsids are constructed from ring- or knob-shaped units called **capsomers,** each usually made of five or six protomers. **Pentamers (pentons)** have five subunits; **hexamers (hexons)** possess six. Pentamers are at the vertices of the icosahedron, whereas hexamers form its edges and triangular faces (**figure 16.13**). The icosahedron in figure 16.13 is constructed of 42 capsomers; larger icosahedra are made if more hexamers are used to form the edges and faces (adenoviruses have a capsid with 252 capsomers as shown in figure 16.12*g,h*). In many plant and bacterial RNA viruses, both the pentamers and hexamers of a capsid are constructed with only one type of subunit, whereas adenovirus pentamers are composed

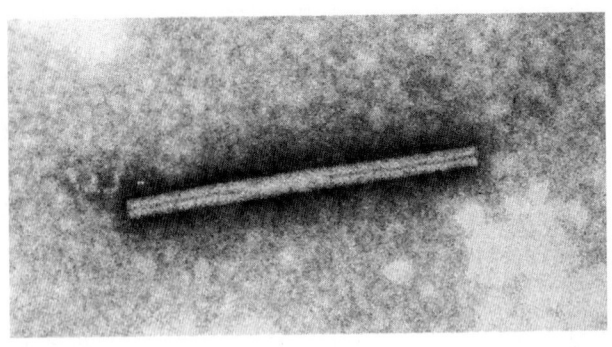

(a)

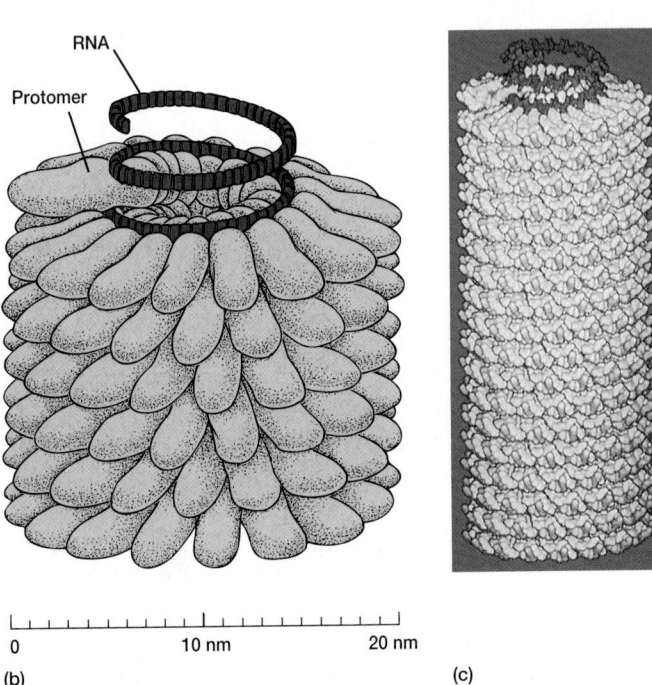

(b)   (c)

Figure 16.11   Tobacco Mosaic Virus Structure.   (a) An electron micrograph of the negatively stained helical capsid (×400,000). (**b**) Illustration of TMV structure. Note that the nucleocapsid is composed of a helical array of protomers with the RNA spiraling on the inside. (**c**) A model of TMV.

of different proteins than are adenovirus hexamers. Transmission electron microscopy and negative staining (pp. 30–32)

Protomers join to form capsomers through noncovalent bonding. The bonds between proteins within pentamers and hexamers are stronger than those between separate capsomers. Empty capsids can even dissociate into separate capsomers.

Recently it has been discovered that there is more than one way to build an icosahedral capsid. Although most icosahedral capsids appear to contain both pentamers and hexamers, simian virus 40 (SV-40), a small, double-stranded DNA polyomavirus, has only pentamers (**figure 16.14*a***). The virus is constructed of 72 cylindrical pentamers with hollow centers. Five flexible arms extend from the edge of each pentamer (figure 16.14*b*). Twelve pentamers occupy the icosahedron's vertices and associate with

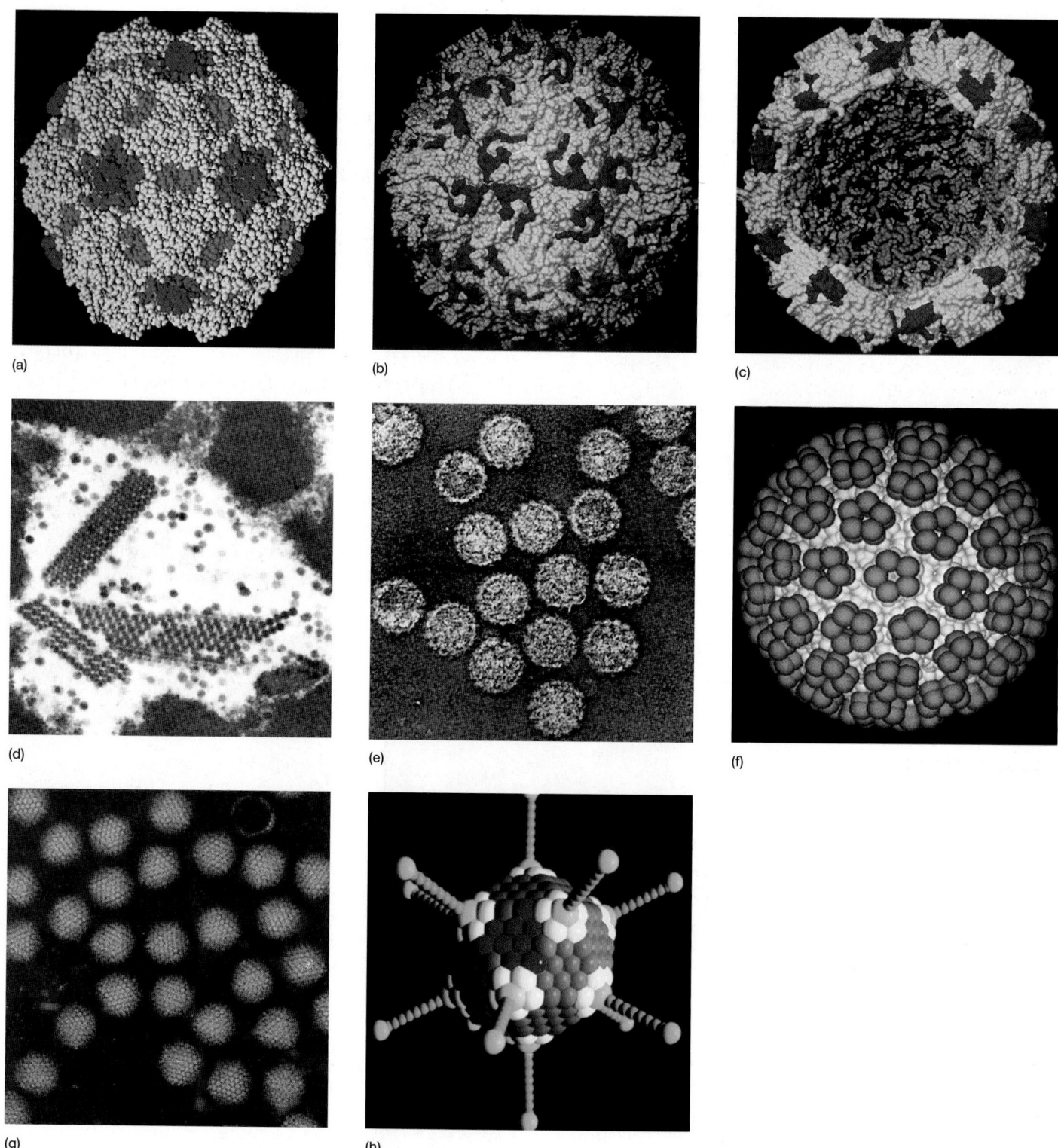

Figure 16.12 **Examples of Icosahedral Capsids.** (**a**) Canine parvovirus model, 12 capsomers, with the four parts of each capsid polypeptide given different colors. (**b**) and (**c**) Poliovirus model, 32 capsomers, with the four capsid proteins in different colors. The capsid surface is depicted in (*b*) and a cross section in (*c*). (**d**) Clusters of the human papilloma virus, 72 capsomers ($\times80,000$). (**e**) Simian virus 40 (SV-40), 72 capsomers ($\times340,000$). (**f**) Computer-simulated image of the polyomavirus, 72 capsomers, that causes a rare demyelinating disease of the central nervous system. (**g**) Adenovirus, 252 capsomers ($\times171,000$). (**h**) Computer-simulated model of adenovirus.

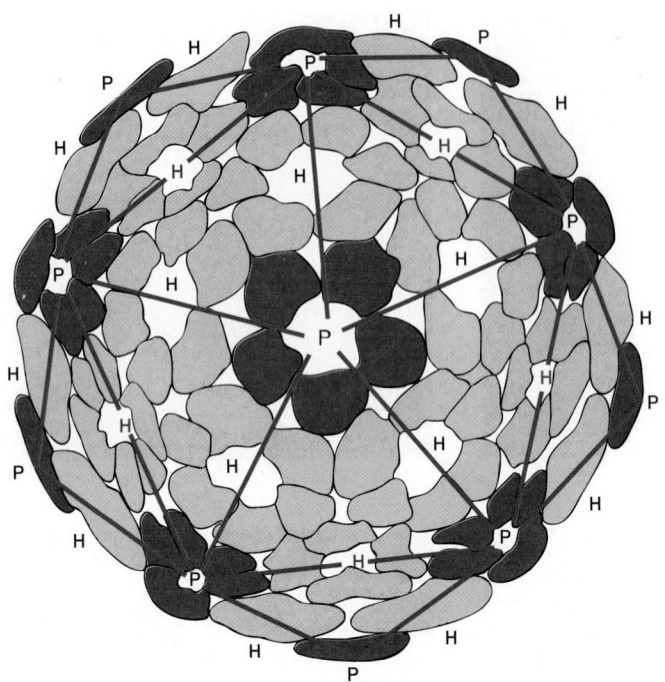

Figure 16.13 **The Structure of an Icosahedral Capsid.**
Pentons (P) are located at the 12 vertices. Hexons (H) form the edges and faces of the icosahedron. This capsid contains 42 capsomers; all protomers are identical.

five neighbors, just as they do when hexamers also are present. Each of the 60 nonvertex pentamers associates with its six adjacent neighbors as shown in figure 16.14c. An arm extends toward the adjacent vertex pentamer (pentamer 1) and twists around one of its arms. Three more arms interact in the same way with arms of other nonvertex pentamers (pentamers 3 to 5). The fifth arm binds directly to an adjacent nonvertex pentamer (pentamer 6) but does not attach to one of its arms. An arm does not extend from the central pentamer to pentamer 2; other arms hold pentamer 2 in place. Thus an icosahedral capsid is assembled without hexamers by using flexible arms as ropes to tie the pentamers together.

Nucleic Acids

Viruses are exceptionally flexible with respect to the nature of their genetic material. They employ all four possible nucleic acid types: single-stranded DNA, double-stranded DNA, single-stranded RNA, and double-stranded RNA. All four types are found in animal viruses. Plant viruses most often have single-stranded RNA genomes. Although phages may have single-stranded DNA or single-stranded RNA, bacterial viruses usually contain double-stranded DNA. **Table 16.1** summarizes many variations seen in viral nucleic acids. The size of viral genetic material also varies greatly (see table 16.2). The smallest genomes (those of the MS2

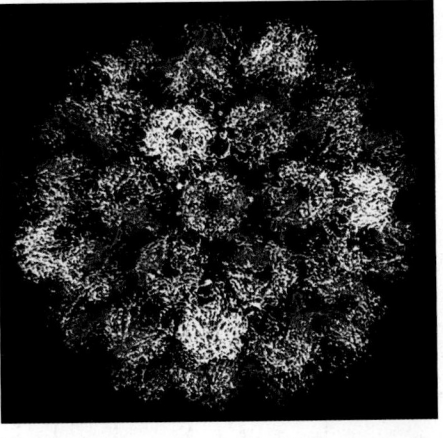

(a)

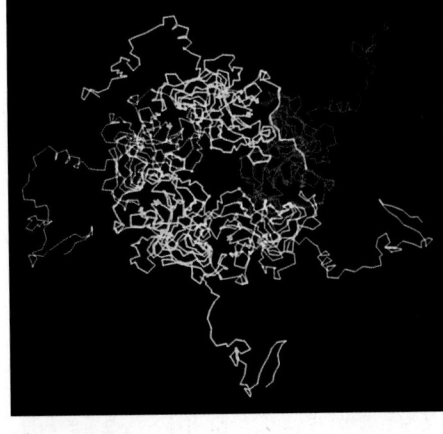

(b)

Figure 16.14 **An Icosahedral Capsid Constructed of Pentamers.** (**a**) The simian virus 40 capsid. The 12 pentamers at the icosahedron vertices are in white. The nonvertex pentamers are shown with each polypeptide chain in a different color. (**b**) A pentamer with extended arms. (**c**) A schematic diagram of the surface structure depicted in part **c.** The body of each pentamer is represented by a five-petaled flower design. Each arm is shown as a line or a line and cylinder (α-helix) with the same color as the rest of its protomer. The outer protomers are numbered clockwise beginning with the one at the vertex.

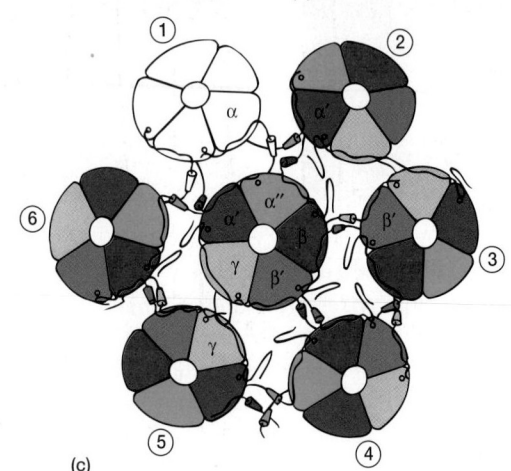

(c)

Table 16.1

Types of Viral Nucleic Acids

Nucleic Acid Type	Nucleic Acid Structure	Virus Examples
DNA		
Single Stranded	Linear single-stranded DNA	Parvoviruses
	Circular single-stranded DNA	φX174, M13, fd phages
Double Stranded	Linear double-stranded DNA	Herpesviruses (herpes simplex viruses, cytomegalovirus, Epstein-Barr virus), adenoviruses, T coliphages, lambda phage, and other bacteriophages
	Linear double-stranded DNA with single chain breaks	T5 coliphage
	Double-stranded DNA with cross-linked ends	Vaccinia, smallpox
	Closed circular double-stranded DNA	Polyomaviruses (SV-40), papillomaviruses, PM2 phage, cauliflower mosaic
RNA		
Single Stranded	Linear, single-stranded, positive-strand RNA	Picornaviruses (polio, rhinoviruses), togaviruses, RNA bacteriophages, TMV, and most plant viruses
	Linear, single-stranded, negative-strand RNA	Rhabdoviruses (rabies), paramyxoviruses (mumps, measles)
	Linear, single-stranded, segmented, positive-strand RNA	Brome mosaic virus (individual segments in separate virions)
	Linear, single-stranded, segmented, diploid (two identical single strands), positive-strand RNA	Retroviruses (Rous sarcoma virus, human immunodeficiency virus)
	Linear, single-stranded, segmented, negative-strand RNA	Paramyxoviruses, orthomyxoviruses (influenza)
Double Stranded	Linear, double-stranded, segmented RNA	Reoviruses, wound-tumor virus of plants, cytoplasmic polyhedrosis virus of insects, phage φ6, many mycoviruses

Modified from S. E. Luria, et al., *General Virology,* 3d edition, 1983. John Wiley & Sons, Inc., New York, NY.

and Qβ viruses) are around 1×10^6 daltons, just large enough to code for three to four proteins. MS2, Qβ, and some other viruses even save space by using overlapping genes (*see section 11.5*). At the other extreme, T-even bacteriophages, herpesvirus, and vaccinia virus have genomes of 1.0 to 1.6×10^8 daltons and may be able to direct the synthesis of over 100 proteins. In the following paragraphs the nature of each nucleic acid type is briefly summarized. Nucleic acid structure (pp. 224–27)

Tiny DNA viruses like φX174 and M13 bacteriophages or the parvoviruses possess single-stranded DNA (ssDNA) genomes (table 16.1). Some of these viruses have linear pieces of DNA, whereas others use a single, closed circle of DNA for their genome (**figure 16.15**).

Most DNA viruses use double-stranded DNA (dsDNA) as their genetic material. Linear dsDNA, variously modified, is found in many viruses; others have circular dsDNA. The lambda phage has linear dsDNA with cohesive ends—single-stranded complementary segments 12 nucleotides long—that enable it to cyclize when they base pair with each other (**figure 16.16**).

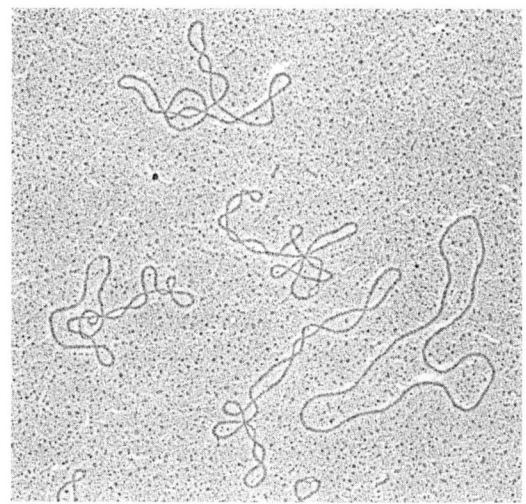

Figure 16.15   Circular Phage DNA.   The closed circular DNA of the phage PM2 (×93,000). Note both the relaxed and highly twisted or supercoiled forms.

Figure 16.16   Circularization of Lambda DNA. The linear DNA of the lambda phage can be reversibly circularized. This is made possible by cohesive ends (in blue) that have complementary sequences and can base pair with each other.

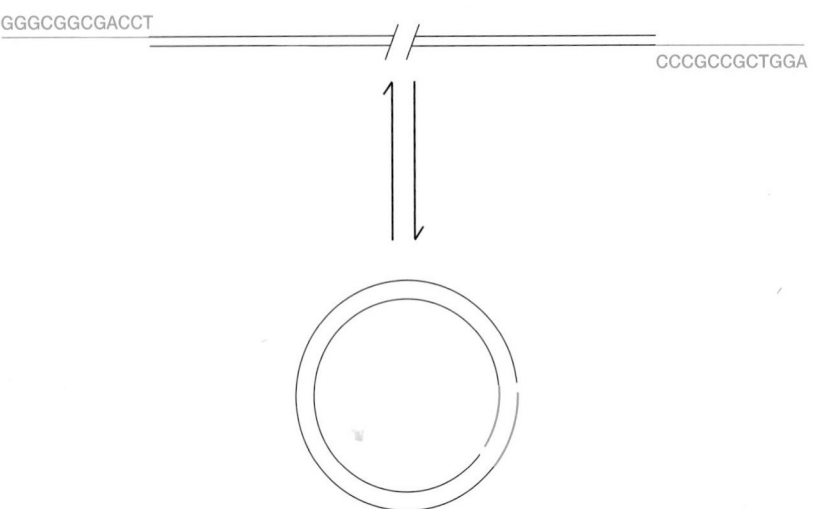

GGGCGGCGACCT

CCCGCCGCTGGA

Besides the normal nucleotides found in DNA, many virus DNAs contain unusual bases. For example, the T-even phages of *E. coli (see chapter 17)* have 5-hydroxymethylcytosine (*see figure 17.8*) instead of cytosine. Glucose is usually attached to the hydroxymethyl group.

Most RNA viruses employ single-stranded RNA (ssRNA) as their genetic material. The RNA base sequence may be identical with that of viral mRNA, in which case the RNA strand is called the **plus strand** or **positive strand** (viral mRNA is defined as plus or positive). However, the viral RNA genome may instead be complementary to viral mRNA, and then it is called a **minus** or **negative strand.** Polio, tobacco mosaic, brome mosaic, and Rous sarcoma viruses are all positive strand RNA viruses; rabies, mumps, measles, and influenza viruses are examples of negative strand RNA viruses. Many of these RNA genomes are **segmented genomes**—that is, they are divided into separate parts. It is believed that each fragment or segment codes for one protein. Usually all segments are probably enclosed in the same capsid even though some virus genomes may be composed of as many as 10 to 12 segments. However, it is not necessary that all segments be located in the same virion for successful reproduction. The brome mosaic virus genome is composed of four segments distributed among three different virus particles. All three of the largest segments are required for infectivity. Despite this complex and seemingly inefficient arrangement, the different brome mosaic virions manage to successfully infect the same host.

Plus strand viral RNA often resembles mRNA in more than the equivalence of its nucleotide sequence. Just as eucaryotic mRNA usually has a 5′ cap of 7-methylguanosine, many plant and animal viral RNA genomes are capped. In addition, most or all plus strand RNA animal viruses also have a poly-A sequence at the 3′ end of their genome, and thus closely resemble eucaryotic mRNA with respect to the structure of both ends. In fact, plus strand RNAs can direct protein synthesis immediately after entering the cell. Strangely enough, a number of single-stranded plant viral RNAs have 3′ ends that resemble eucaryotic transfer RNA, and the genomes of tobacco mosaic virus will actually ac-

cept amino acids. Capping is not seen in the RNA bacteriophages.
Eucaryotic mRNA structure and function (pp. 257–60)

A few viruses have double-stranded RNA (dsRNA) genomes. All appear to be segmented; some, such as the reoviruses, have 10 to 12 segments. These dsRNA viruses are known to infect animals, plants, fungi, and even one bacterial species.

Viral Envelopes and Enzymes

Many animal viruses, some plant viruses, and at least one bacterial virus are bounded by an outer membranous layer called an envelope (**figure 16.17**). Animal virus envelopes usually arise from host cell nuclear or plasma membranes; their lipids and carbohydrates are normal host constituents. In contrast, envelope proteins are coded for by virus genes and may even project from the envelope surface as **spikes** or **peplomers** (figure 16.17a,b,f). These spikes may be involved in virus attachment to the host cell surface. Since they differ among viruses, they also can be used to identify some viruses. Because the envelope is a flexible, membranous structure, enveloped viruses frequently have a somewhat variable shape and are called pleomorphic. However, the envelopes of viruses like the bullet-shaped rabies virus are firmly attached to the underlying nucleocapsid and endow the virion with a constant, characteristic shape (figure 16.17c). In some viruses the envelope is disrupted by solvents like ether to such an extent that lipid-mediated activities are blocked or envelope proteins are denatured and rendered inactive. The virus is then said to be "ether sensitive."

Influenza virus (figure 16.17a,b) is a well-studied example of an enveloped virus. Spikes project about 10 nm from the surface at 7 to 8 nm intervals. Some spikes possess the enzyme neuraminidase, which may aid the virus in penetrating mucous layers of the respiratory epithelium to reach host cells. Other spikes have hemagglutinin proteins, so named because they can bind the virions to red blood cell membranes and cause hemagglutination (*see figure 33.9*). Hemagglutinins participate in virion attachment to host cells. Proteins, like the spike proteins that are exposed on the outer envelope surface, are generally glycoproteins—that is,

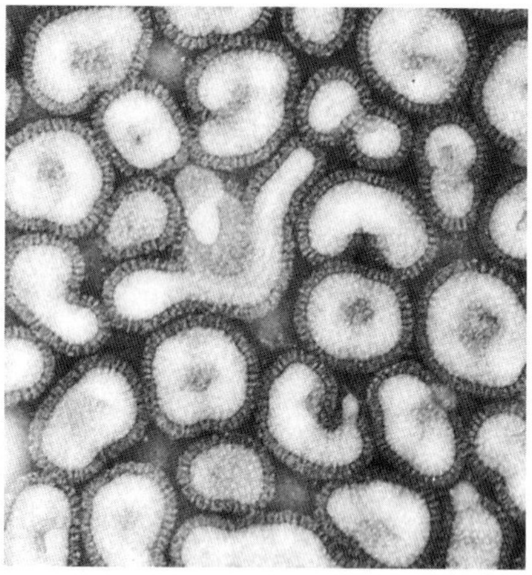

(a) Human influenza virus

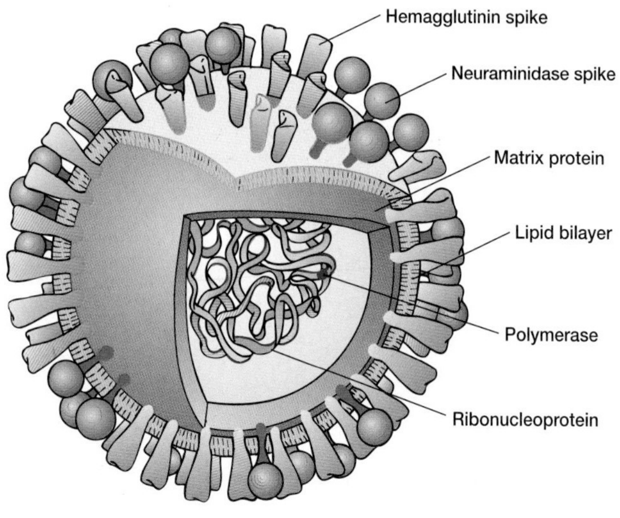

Hemagglutinin spike

Neuraminidase spike

Matrix protein

Lipid bilayer

Polymerase

Ribonucleoprotein

(b)   ◄──── 50 nm ────►

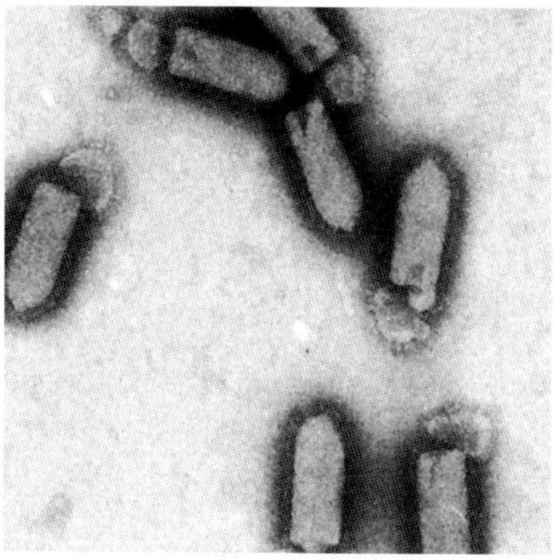

(c) Rhabdovirus particles

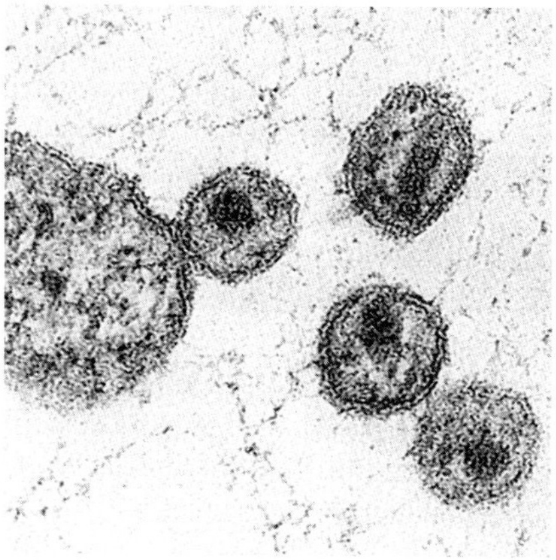

(d) HIV

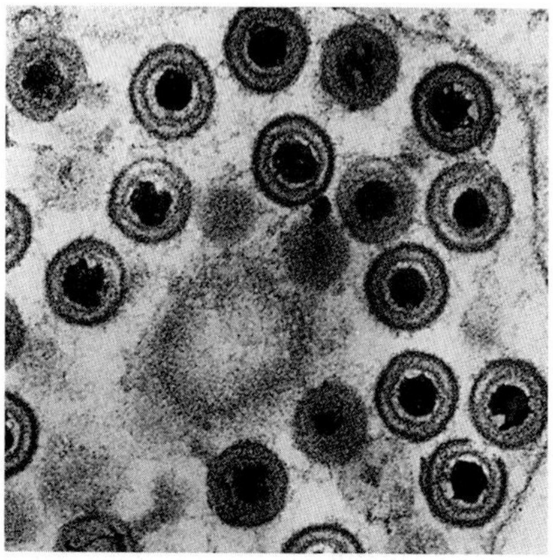

(e) Herpes viruses

Figure 16.17   Examples of Enveloped Viruses.   (a) Human influenza virus. Note the flexibility of the envelope and the spikes projecting from its surface (×282,000). **(b)** Diagram of the influenza virion. **(c)** Rhabdovirus particles (×250,000). This is the vesicular stomatitis virus, a relative of the rabies virus, which is similar in appearance. **(d)** Human immunodeficiency viruses (×33,000). **(e)** Herpesviruses (×100,000). **(f)** Computer image of the Semliki Forest virus, a virus that occasionally causes encephalitis in humans.

(f) Semliki Forest virus

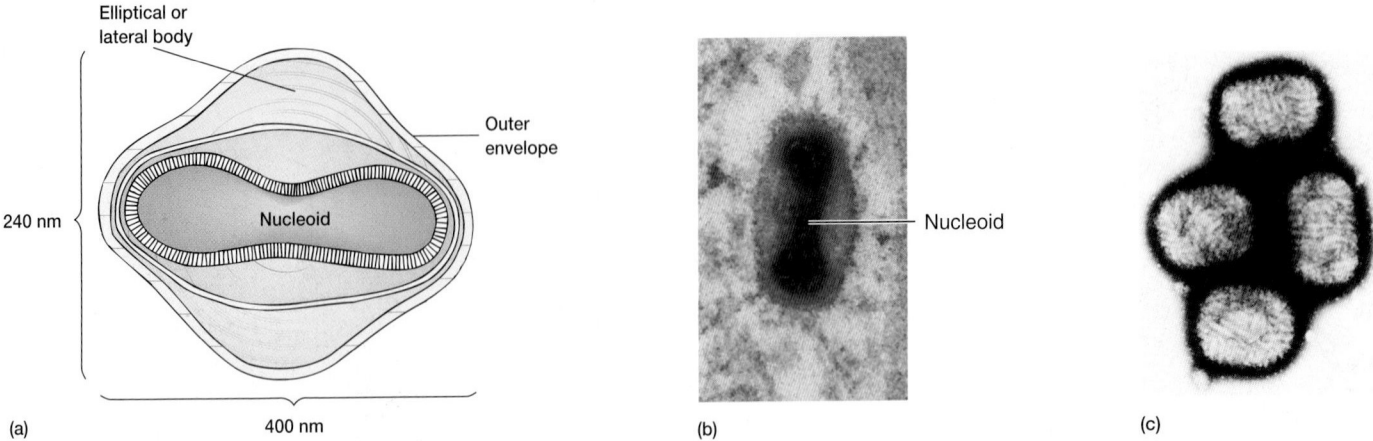

Figure 16.18   Vaccinia Virus Morphology.   (**a**) Diagram of vaccinia structure. (**b**) Micrograph of the virion clearly showing the nucleoid (×200,000). (**c**) Vaccinia surface structure. An electron micrograph of four virions showing the thick array of surface fibers (×150,000).

the proteins have carbohydrate attached to them. A nonglycosylated protein, the M or matrix protein, is found on the inner surface of the envelope and helps stabilize it.

Although it was originally thought that virions had only structural capsid proteins and lacked enzymes, this has proven not to be the case. In some instances, enzymes are associated with the envelope or capsid (e.g., influenza neuraminidase). Most viral enzymes are probably located within the capsid. Many of these are involved in nucleic acid replication. For example, the influenza virus uses RNA as its genetic material and carries an RNA-dependent RNA polymerase that acts both as a replicase and as an RNA transcriptase that synthesizes mRNA under the direction of its RNA genome. The polymerase is associated with ribonucleoprotein (figure 16.17*b*). Although viruses lack true metabolism and cannot reproduce independently of living cells, they may carry one or more enzymes essential to the completion of their life cycles.   Nucleic acid replication and transcription (sections 11.3 and 12.1); Animal virus reproduction (pp. 388–98)

Viruses with Capsids of Complex Symmetry

Although most viruses have either icosahedral or helical capsids, many viruses do not fit into either category. The poxviruses and large bacteriophages are two important examples.

The poxviruses are the largest of the animal viruses (about 400 × 240 × 200 nm in size) and can even be seen with a phase-contrast microscope or in stained preparations. They possess an exceptionally complex internal structure with an ovoid- to brick-shaped exterior. The double-stranded DNA is associated with proteins and contained in the nucleoid, a central structure shaped like a biconcave disk and surrounded by a membrane (**figure 16.18**). Two elliptical or lateral bodies lie between the nucleoid and its outer envelope, a membrane and a thick layer covered by an array of tubules or fibers.

Some large bacteriophages are even more elaborate than the poxviruses. The T2, T4, and T6 phages that infect *E. coli* have been intensely studied. Their head resembles an icosahedron

elongated by one or two rows of hexamers in the middle (**figure 16.19**) and contains the DNA genome. The tail is composed of a collar joining it to the head, a central hollow tube, a sheath surrounding the tube, and a complex baseplate. The sheath is made of 144 copies of the gp18 protein arranged in 24 rings, each containing six copies. In T-even phages, the baseplate is hexagonal and has a pin and a jointed tail fiber at each corner. The tail fibers are responsible for virus attachment to the proper site on the bacterial surface (*see section 17.2*).

There is considerable variation in structure among the large bacteriophages, even those infecting a single host. In contrast with the T-even phages, many coliphages have true icosahedral heads. T1, T5, and lambda phages have sheathless tails that lack a baseplate and terminate in rudimentary tail fibers. Coliphages T3 and T7 have short, noncontractile tails without tail fibers. Clearly these viruses can complete their reproductive cycles using a variety of tail structures.

Complex bacterial viruses with both heads and tails are said to have **binal symmetry** because they possess a combination of icosahedral (the head) and helical (the tail) symmetry.

---

1. Define the following terms: nucleocapsid, capsid, icosahedral capsid, helical capsid, complex virus, protomer, self-assembly, capsomer, pentamer or penton, and hexamer or hexon. How do pentamers and hexamers associate to form a complete icosahedron; what determines helical capsid length and diameter?

2. All four nucleic acid forms can serve as virus genomes. Describe each, the types of virion possessing it, and any distinctive physical characteristics the nucleic acid can have. What are the following: plus strand, minus strand, and segmented genome?

3. What is an envelope? What are spikes or peplomers? Why are some enveloped viruses pleomorphic? Give two functions spikes might serve in the virus life cycle, and the proteins that the influenza virus uses in these processes.

4. What is a complex virus? Binal symmetry? Briefly describe the structure of poxviruses and T-even bacteriophages.

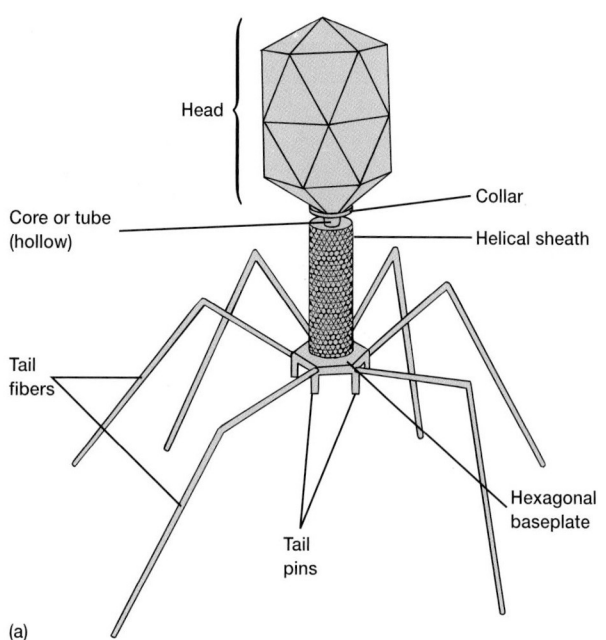

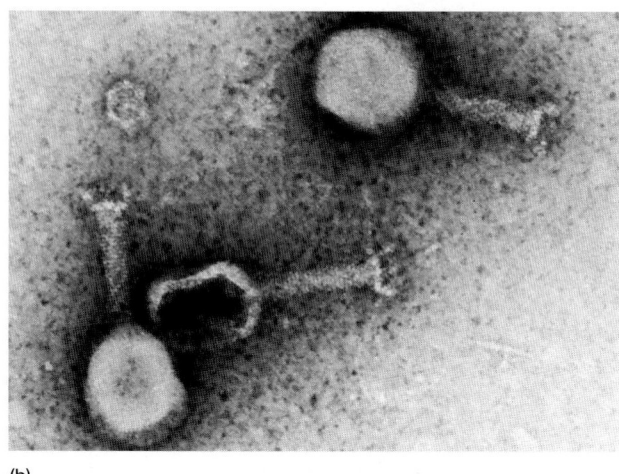

(a)

(b)

Figure 16.19   T-Even Coliphages.   (**a**) The structure of the T4 bacteriophage. (**b**) The micrograph shows the phage before injection of its DNA.

 PRINCIPLES OF VIRUS TAXONOMY

The classification of viruses is in a much less satisfactory state than that of either bacteria or eucaryotic microorganisms. In part, this is due to a lack of knowledge of their origin and evolutionary history (**Microbial Tidbits 16.2**). Usually viruses are separated into several large groups based on their host preferences: animal viruses, plant viruses, bacterial viruses, bacteriophages, and so forth. In the past virologists working with these groups were unable to agree on a uniform system of classification and nomenclature. Beginning with its 1971 report, the International Committee for Taxonomy of Viruses has developed a uniform classification system. The number of viruses and taxonomic categories continues to expand; at present over 1,500 virus species are placed within around 65 families, 9 subfamilies, and 200 to 300 genera. The committee places greatest weight on a few properties to define families: nucleic acid type, nucleic acid strandedness, the sense (positive or negative) of ssRNA genomes, presence or absence of an envelope, and the host. Virus family names end in *viridae;* subfamily names, in *virinae;* and genus (and species) names, in *virus.* For example, the poxviruses are in the family *Poxviridae;* the subfamily *Chorodopoxvirinae* contains poxviruses of vertebrates. Within the subfamily are several genera that are distinguished on the basis of immunologic characteristics and host specificity. The genus *Orthopoxvirus* contains several species, among them variola major (the cause of smallpox), vaccinia, and cowpox.

Viruses are divided into different taxonomic groups based on characteristics that are related to the type of host used, virion structure and composition, mode of reproduction, and the nature of any diseases caused. Some of the more important characteristics are:

1. Nature of the host—animal, plant, bacterial, insect, fungal
2. Nucleic acid characteristics—DNA or RNA, single or double stranded, molecular weight, segmentation and number of pieces of nucleic acid (RNA viruses), the sense of the strand in ssRNA viruses
3. Capsid symmetry—icosahedral, helical, binal
4. Presence of an envelope and ether sensitivity
5. Diameter of the virion or nucleocapsid
6. Number of capsomers in icosahedral viruses
7. Immunologic properties
8. Gene number and genomic map
9. Intracellular location of viral replication
10. The presence or absence of a DNA intermediate (ssRNA viruses), and the presence of reverse transcriptase
11. Type of virus release
12. Disease caused and/or special clinical features, method of transmission

In addition to these characteristics, genome sequences are available in public databases and are now being used in the construction of new families and genera. **Table 16.2** illustrates the use of some of these properties to describe a few common virus groups. Virus classification is further discussed when bacterial, animal, and plant viruses are considered more specifically, and a fairly complete summary of virus taxonomy is presented in appendix IV.

---

1. List some characteristics used in classifying viruses. Which seem to be the most important?
2. What are the endings for virus families, subfamilies, and genera or species?

Table 16.2

Some Common Virus Groups and Their Characteristics

Virus Group	Genome Size (kbp or kb)	Nucleic Acid	Strandedness	Capsid Symmetry[a]	Number of Capsomers	Presence of Envelope	Size of Capsid (nm)[b]	Host Range[c]
Picornaviridae	7–8	RNA	Single	I	32	−	22–30	A
Togaviridae	10–12		Single	I	32	+	40–70(e)	A
Retroviridae	7–12		Single	I?		+	100(e)	A
Orthomyxoviridae	10–15		Single	H		+	9(h), 80–120(e)	A
Paramyxoviridae	15		Single	H		+	18(h), 125–250(e)	A
Coronaviridae	27–31		Single	H		+	14–16(h), 80–160(e)	A
Rhabdoviridae	11–15		Single Single	H		+	18(h), 70–80 × 130–240 (bullet shaped)	A
Bromoviridae	8–9		Single	I,B		−	26–35; 18–26 × 30–85	P
Tobamovirus	7		Single	H		−	18 × 300	P
Leviviridae [Qβ]	3–4		Single	I	32	−	26–27	B
Reoviridae	19–32	RNA	Double	I	92	−	70–80	A,P
Cystoviridae	13		Double	I		+	100(e)	B
Parvoviridae	4–6	DNA	Single	I	12	−	20–25	A
Geminiviridae	3–6		Single	I		−	18 × 30 (paired particles)	P
Microviridae	4–6		Single	I		−	25–35	B
Inoviridae	7–9		Single	H		−	6 × 900–1,900	B
Polyomaviridae	5	DNA	Double	I	72	−	40	A
Papillomaviridae	7–8		Double	I	72	−	55	A
Adenoviridae	28–45		Double	I	252	−	60–90	A
Iridoviridae	140–383		Double	I		−	130–180	A
Herpesviridae	125–240		Double	I	162	+	100, 180–200(e)	A
Poxviridae	130–375		Double	C		+	200–260 × 250–290(e)	A
Baculoviridae	80–180		Double	H		+	40 × 300(e)	A
Hepadnaviridae	3		Double	C	42	+	28 (core), 42(e)	A
Caulimoviridae	8		Double	I,B		−	50; 30 × 60–900	P
Corticoviridae	9		Double	I		−	60	B
Myoviridae	39–169		Double	Bi		−	80 × 110, 110[d]	B, Arch
Lipothrixviridae	16		Double	H		+	38 × 410	Arch

[a]Types of symmetry: I, icosahedral; H, helical; C, complex; Bi, binal; B, bacilliform.
[b]Diameter of helical capsid (h); diameter of enveloped virion (e).
[c]Host range: A, animal; P, plant; B, bacterium; Arch, archaeon.
[d]The first number is the head diameter; the second number, the tail length.

Microbial Tidbits
16.2   The Origin of Viruses

The origin and subsequent evolution of viruses are shrouded in mystery, in part because of the lack of a fossil record. However, recent advances in the understanding of virus structure and reproduction have made possible more informed speculation on virus origins. At present there are two major hypotheses entertained by virologists. It has been proposed that at least some of the more complex enveloped viruses, such as the poxviruses and herpesviruses, arose from small cells, probably procaryotic, that parasitized larger, more complex cells. These parasitic cells would become ever simpler and more dependent on their hosts, much like multicellular parasites have done, in a process known as retrograde evolution. There are several problems with this hypothesis. Viruses are radically different from procaryotes, and it is difficult to envision the mechanisms by which such a transformation might have occurred or the selective pressures leading to it. In addition, one would expect to find some forms intermediate between procaryotes and at least the more complex enveloped viruses, but such forms have not been detected.

The second hypothesis is that viruses represent cellular nucleic acids that have become partially independent of the cell. Possibly a few mutations could convert nucleic acids, which are only synthesized at specific times, into infectious nucleic acids whose replication could not be controlled. This conjecture is supported by the observation that the nucleic acids of retroviruses (*see section 18.2*) and a number of other virions do contain sequences quite similar to those of normal cells, plasmids, and transposons (*see chapter 13*). The small, infectious RNAs called viroids (*see section 18.9*) have base sequences complementary to transposons, the regions around the boundary of mRNA introns (*see section 12.1*), and portions of host DNA. This has led to speculation that they have arisen from introns or transposons. It has been proposed that cellular proteins spontaneously assembled into icosahedra around infectious nucleic acids to produce primitive virions.

It is possible that viruses have arisen by way of both mechanisms. Because viruses differ so greatly from one another, it seems likely that they have originated independently many times during the course of evolution. Probably many viruses have evolved from other viruses just as cellular organisms have arisen from specific predecessors. The question of virus origins is complex and quite speculative; future progress in understanding virus structure and reproduction may clarify this question.

Summary

16.1 Early Development of Virology

a. Europeans were first protected from a viral disease when Edward Jenner developed a smallpox vaccine in 1798.

b. Chamberland's invention of a porcelain filter that could remove bacteria from virus samples enabled microbiologists to show that viruses were different from bacteria.

c. In the late 1930s Stanley, Bawden, and Pirie crystallized the tobacco mosaic virus and demonstrated that it was composed only of protein and nucleic acid.

16.2 General Properties of Viruses

a. A virion is composed of either DNA or RNA enclosed in a coat of protein (and sometimes other substances as well). It cannot reproduce independently of living cells.

16.3 The Cultivation of Viruses

a. Viruses are cultivated using tissue cultures, embryonated eggs, bacterial cultures, and other living hosts.

b. Sites of animal viral infection may be characterized by cytopathic effects such as pocks and plaques. Phages produce plaques in bacterial lawns. Plant viruses can cause localized necrotic lesions in plant tissues.

16.4 Virus Purification and Assays

a. Viruses can be purified by techniques such as differential and gradient centrifugation, precipitation, and denaturation or digestion of contaminants.

b. Virus particles can be counted directly with the transmission electron microscope or indirectly by the hemagglutination assay.

c. Infectivity assays can be used to estimate virus numbers in terms of plaque-forming units, lethal dose (LD_{50}), or infectious dose (ID_{50}).

16.5 The Structure of Viruses

a. All virions have a nucleocapsid composed of a nucleic acid, either DNA or RNA, held within a protein capsid made of one or more types of protein subunits called protomers.

b. There are four types of viral morphology: naked icosahedral, naked helical, enveloped icosahedral and helical, and complex.

c. Helical capsids resemble long hollow protein tubes and may be either rigid or quite flexible. The nucleic acid is coiled in a spiral on the inside of the cylinder (**figure 16.11***b*).

d. Icosahedral capsids are usually constructed from two types of capsomers: pentamers (pentons) at the vertices and hexamers (hexons) on the edges and faces of the icosahedron (**figure 16.13**).

e. Viral nucleic acids can be either single stranded or double stranded, DNA or RNA. Most DNA viruses have double-stranded DNA genomes that may be linear or closed circles (**table 16.1**).

f. RNA viruses usually have ssRNA that may be either plus (positive) or minus (negative) when compared with mRNA (positive). Many RNA genomes are segmented.

g. Viruses can have a membranous envelope surrounding their nucleocapsid. The envelope lipids usually come from the host cell; in contrast, many envelope proteins are viral and may project from the envelope surface as spikes or peplomers.

h. Although viruses lack true metabolism, some contain a few enzymes necessary for their reproduction.

i. Complex viruses (e.g., poxviruses and large phages) have complicated morphology not characterized by icosahedral and helical symmetry. Large phages often have binal symmetry: their heads are icosahedral and their tails, helical (**figure 16.19***a*).

16.6 Principles of Virus Taxonomy

a. Currently viruses are classified with a taxonomic system placing primary emphasis on the host, type and strandedness of viral nucleic acids, and on the presence or absence of an envelope.

Key Terms

Questions for Thought and Review

1. In what ways do viruses resemble living organisms?
2. Why might virology have developed much more slowly without the use of Chamberland's filter?
3. What advantage would an RNA virus gain by having its genome resemble eucaryotic mRNA?
4. A number of characteristics useful in virus taxonomy are listed on page 367. Can you think of any other properties that might be of considerable importance in future studies on virus taxonomy?

Critical Thinking Questions

1. Many classification schemes are used to identify bacteria. These start with Gram staining, progress to morphology/ arrangement characteristics, and include a battery of metabolic tests. Build an analogous scheme that could be used to identify viruses. You might start by considering the host, or you might start with viruses found in a particular environment, such as a marine filtrate.
2. Consider the different perspectives on the origin of viruses in Microbial Tidbits 16.2. Discuss whether you think viruses evolved before the first procaryote, or whether they have coevolved, and are perhaps still coevolving with their hosts.

For recommended readings on these and related topics, see Appendix V.

chapter 17

The Viruses:
Bacteriophages

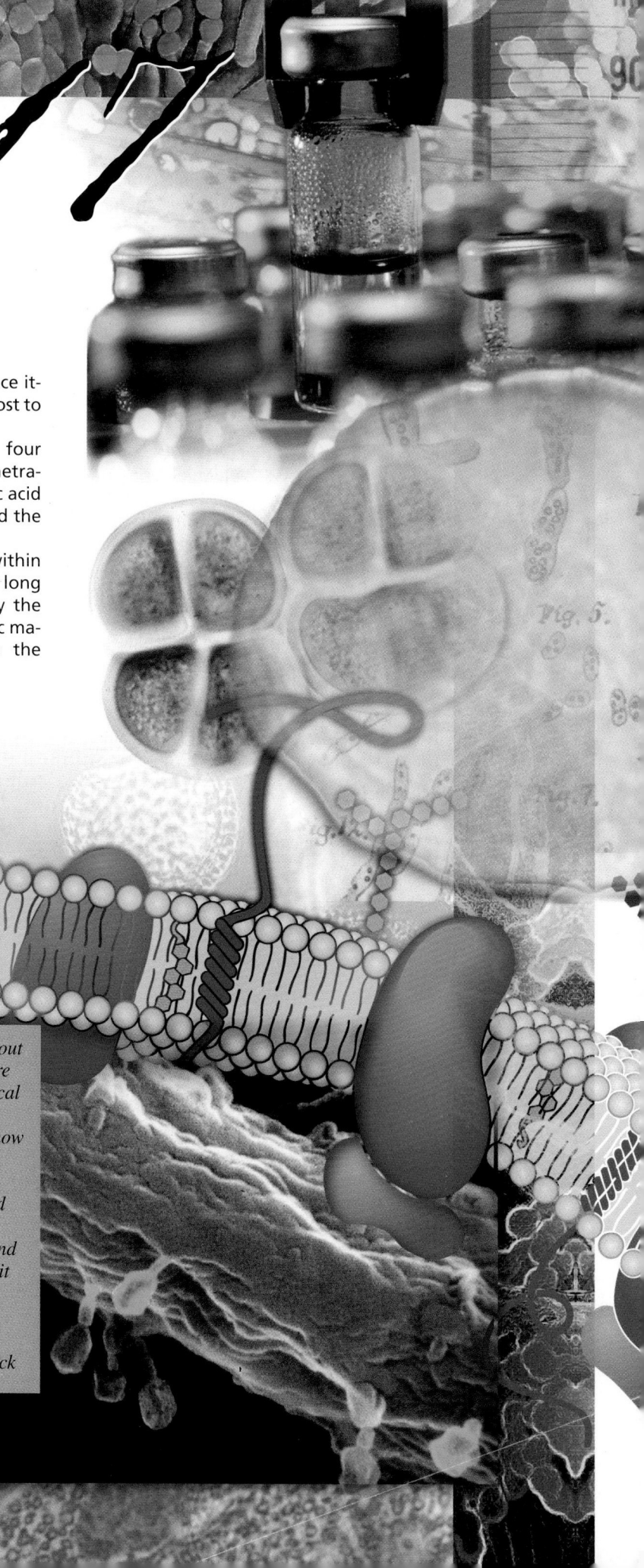

Concepts

- Since a bacteriophage cannot independently reproduce itself, the phage takes over its host cell and forces the host to reproduce it.
- The lytic bacteriophage life cycle is composed of four phases: adsorption of the phage to the host and penetration of virus genetic material, synthesis of virus nucleic acid and capsid proteins, assembly of complete virions, and the release of phage particles from the host.
- Temperate virus genetic material is able to remain within host cells and reproduce in synchrony with the host for long periods in a relationship known as lysogeny. Usually the virus genome is found integrated into the host genetic material as a prophage. A repressor protein keeps the prophage dormant and prevents virus reproduction.

You might wonder how such naive outsiders get to know about the existence of bacterial viruses. Quite by accident, I assure you. Let me illustrate by reference to an imaginary theoretical physicist, who knew little about biology in general, and nothing about bacterial viruses in particular. . . . Suppose now that our imaginary physicist, the student of Niels Bohr, is shown an experiment in which a virus particle enters a bacterial cell and 20 minutes later the bacterial cell is lysed and 100 virus particles are liberated. He will say: "How come, one particle has become 100 particles of the same kind in 20 minutes? That is very interesting. Let us find out how it happens!. . . Is this multiplying a trick of organic chemistry which the organic chemists have not yet discovered? Let us find out."

—Max Delbrück

A scanning electron micrograph of T-even bacteriophages infecting *E. coli*. The phages are colored blue.

hapter 16 introduces many of the facts and concepts underlying the field of virology, including information about the nature of viruses, their structure and taxonomy, and how they are cultivated and studied. Clearly the viruses are a complex, diverse, and fascinating group, the study of which has done much to advance disciplines such as genetics and molecular biology.

Chapters 17 and 18 focus on virus diversity. This chapter is concerned with bacterial viruses or **bacteriophages;** the next surveys animal, plant, and insect viruses. The taxonomy, morphology, and reproduction of each group are covered. Where appropriate, the biological and practical importance of viruses is emphasized, even though viral diseases are examined in chapter 38. Since the bacteriophages (or simply phages) have been the most intensely studied viruses and are best understood in a molecular sense, this chapter is devoted to them.

Because of their crucial role in the history of genetics and molecular biology, it is tempting to think of bacteriophages as useful only for laboratory research in these disciplines. However, phages are essential components of the ecosystem (**Microbial Diversity & Ecology 17.1**) and are important in both industry and medicine. For example, many phages destroy the gram-positive lactic acid bacteria (*see pp. 515–18*) that are critical to the production of fermented milk products such as yogurt and cheese. Bacteriophages also can carry a variety of virulence factors that make their bacterial hosts potent pathogens. This seems to be the case for major human pathogens such as *Streptococcus pyogenes, Staphylococcus aureus, Corynebacterium diphtheriae, Vibrio cholerae, E. coli* O157:H7, and *Salmonella enterica.* On the positive side, Russian physicians have used bacteriophages for years to treat bacterial diseases. Recent research indicates that phages may indeed be effective in treating bacterial infections, including those caused by antibiotic-resistant bacteria.

17.1 CLASSIFICATION OF BACTERIOPHAGES

Although properties such as host range and immunologic relationships are used in classifying phages, the most important are phage morphology and nucleic acid properties (**figure 17.1**). The genetic material may be either DNA or RNA; most known bacteriophages have double-stranded DNA. Most can be placed in one of a few morphological groups: tailless icosahedral phages, viruses with contractile tails, viruses with noncontractile tails, and filamentous phages. There are even a few phages with envelopes. The most complex forms are the phages with contractile tails, for example, the T-even phages of *E. coli.*

Although most phages have been isolated from bacteria, members of the families *Myoviridae* and *Siphoviridae* infect both bacteria and archaea. Viruses from the *Lipothrixviridae, Rudiviridae,* and *Fuselloviridae* have only been found in the Archaea thus far.

---

1. Briefly describe in general terms the morphology and nucleic acids of the major phage types.

Microbial Diversity & Ecology
17.1   An Ocean of Viruses

Microbiologists have previously searched without success for viruses in marine habitats. Thus it has been assumed the oceans probably did not contain many viruses. Recent discoveries have changed this view radically. Several groups have either centrifuged seawater at high speeds or passed it through an ultrafilter and then examined the sediment or suspension in an electron microscope. They have found that marine viruses are about 10 times more plentiful than marine bacteria. Between 10^6 and 10^9 virus particles per milliliter are present at the ocean's surface. It has been estimated that the top one millimeter of the world's oceans could contain a total of over 3×10^{30} virus particles!

Although little detailed work has been done on marine viruses, it appears that many contain double-stranded DNA. Most are probably bacteriophages and can infect both marine heterotrophs and cyanobacteria. Up to 70% of marine procaryotes may be infected by phages. Viruses that infect diatoms and other major algal components of the marine phytoplankton also have been detected.

Marine viruses may be very important ecologically. Viruses may control marine algal blooms such as red tides (p. 562), and bacterio-phages could account for 1/3 or more of the total aquatic bacterial mortality or turnover. If true, this is of major ecological significance because the reproduction of marine bacteria far exceeds marine protozoan grazing capacity. Virus-induced lysis of procaryotic and algal cells may well contribute greatly to carbon and nitrogen cycling in marine food webs. It could reduce the level of marine primary productivity in some situations.

Bacteriophages also may greatly accelerate the flow of genes between marine bacteria. Virus-induced bacterial lysis could generate most of the free DNA present in seawater. Gene transfer between aquatic bacteria by transformation (*see pp. 299–301*) does occur, and bacterial lysis by phages would increase its probability. Furthermore, such high phage concentrations can stimulate gene exchange by transduction (*see pp. 301–5*). These genetic exchanges could have both positive and negative consequences. Genes that enable marine bacteria to degrade toxic pollutants such as those in oil spills could spread through the native population. On the other hand, antibiotic resistance genes in bacteria from raw sewage released into the ocean also might be dispersed (*see section 35.7*).

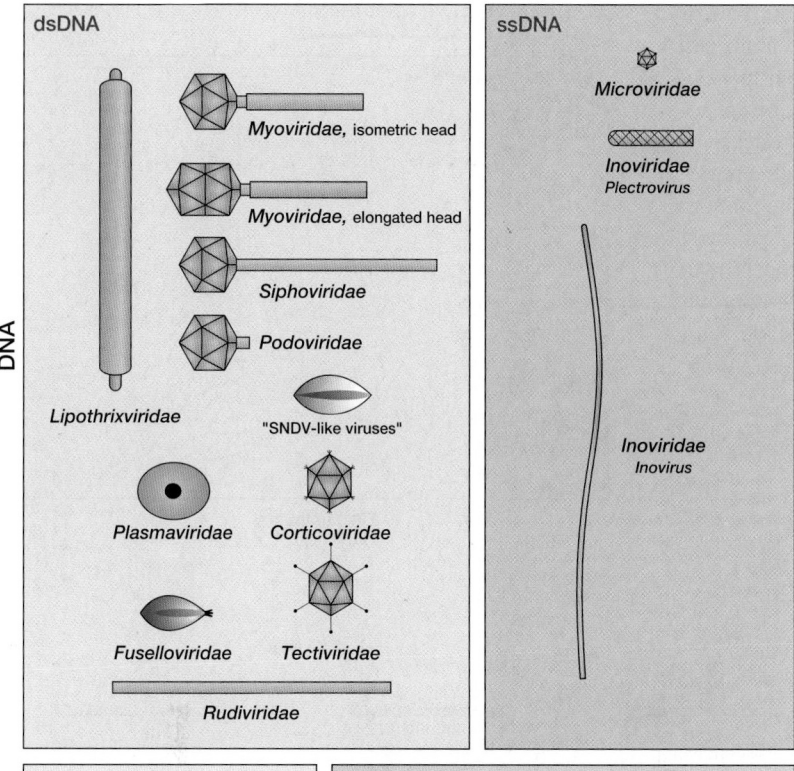

dsDNA

Myoviridae, isometric head

Myoviridae, elongated head

Siphoviridae

Podoviridae

Lipothrixviridae

"SNDV-like viruses"

Plasmaviridae

Corticoviridae

Fuselloviridae

Tectiviridae

Rudiviridae

DNA

ssDNA

Microviridae

Inoviridae
Plectrovirus

Inoviridae
Inovirus

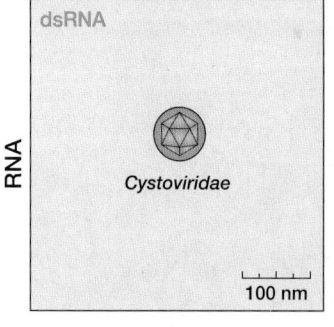

dsRNA

Cystoviridae

RNA

ssRNA

Leviviridae

100 nm

Figure 17.1 Major Bacteriophage Families and Genera. The *Myoviridae* are the only family with contractile tails. *Plasmaviridae* are pleomorphic. *Tectiviridae* have distinctive double capsids, whereas the *Corticoviridae* have complex capsids containing lipid.

bacteriophage research. In a **one-step growth experiment,** the reproduction of a large phage population is synchronized so that the molecular events occurring during reproduction can be followed. A culture of susceptible bacteria such as *E. coli* is mixed with bacteriophage particles, and the phages are allowed a short interval to attach to their host cells. The culture is then greatly diluted so that any virus particles released upon host cell lysis will not immediately infect new cells. This strategy works because phages lack a means of seeking out host cells and must contact them during random movement through the solution. Thus phages are less likely to contact host cells in a dilute mixture. The number of infective phage particles released from bacteria is subsequently determined at various intervals by a plaque count (*see section 16.4*).

A plot of the bacteriophages released from host cells versus time shows several distinct phases (**figure 17.2**). During the **latent period,** which immediately follows phage addition, there is no release of virions. This is followed by the **rise period** or **burst,** when the host cells rapidly lyse and release infective phages. Finally, a plateau is reached and no more viruses are liberated. The total number of phages released can be used to calculate the **burst size,** the number of viruses produced per infected cell.

The latent period is the shortest time required for virus reproduction and release. During the first part of this phase, host bacteria do not contain any complete, infective virions. This can be shown by lysing them with chloroform. This initial segment of the latent period is called the **eclipse period** because the virions were detectable before infection but are now concealed or eclipsed. The number of completed, infective phages within the host increases after the end of the eclipse period, and the host cell is prepared for lysis.

The one-step growth experiment with *E. coli* and phage T2 provides a well-studied example of this process. When the experiment is carried out with actively growing cells in rich medium at 37°C, the growth curve plateau is reached in approximately 30 minutes. Bacteriophage reproduction is an exceptionally rapid process, much faster than animal virus reproduction, which may take hours.

17.2 REPRODUCTION OF DOUBLE-STRANDED DNA PHAGES: THE LYTIC CYCLE

After DNA bacteriophages have reproduced within the host cell, many of them are released when the cell is destroyed by lysis. A phage life cycle that culminates with the host cell bursting and releasing virions is called a **lytic cycle.** The events taking place during the lytic cycle will be reviewed in this section, with the primary focus on the T-even phages of *E. coli.* These are double-stranded DNA bacteriophages with complex contractile tails and are placed in the family *Myoviridae.* They are some of the most complex viruses known. The structure of T-even coliphages (p. 366)

The One-Step Growth Experiment

The development of the one-step growth experiment in 1939 by Max Delbrück and Emory Ellis marks the beginning of modern

Adsorption to the Host Cell and Penetration

Bacteriophages do not randomly attach to the surface of a host cell; rather, they fasten to specific surface structures called **receptor sites.** The nature of these receptors varies with the phage; cell wall lipopolysaccharides and proteins, teichoic acids, flagella, and pili

Figure 17.2   The One-Step Growth Curve.   In the initial part of the latent period, the eclipse period, the host cells do not contain any complete, infective virions. During the remainder of the latent period, an increasing number of infective virions are present, but none are released. The latent period ends with host cell lysis and rapid release of virions during the rise period or burst. In this figure the blue line represents the total number of complete virions. The red line is the number of free viruses (the unadsorbed virions plus those released from host cells). When *E. coli* is infected with T2 phage at 37°C, the growth plateau is reached in about 30 minutes and the burst size is approximately 100 or more virions per cell. The eclipse period is 11–12 minutes, and the latent period is around 21–22 minutes in length.

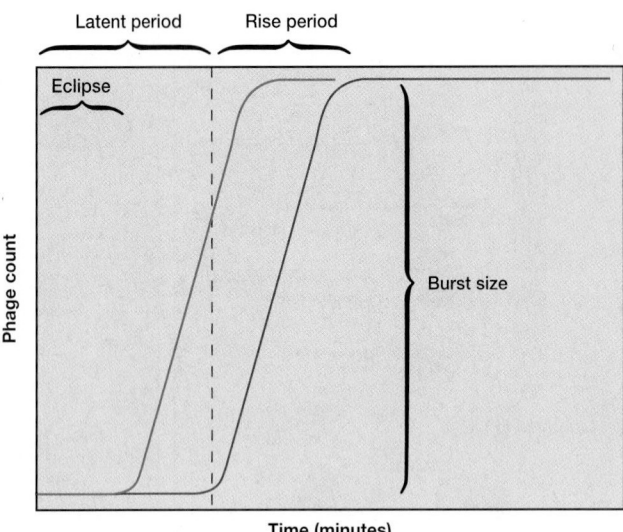

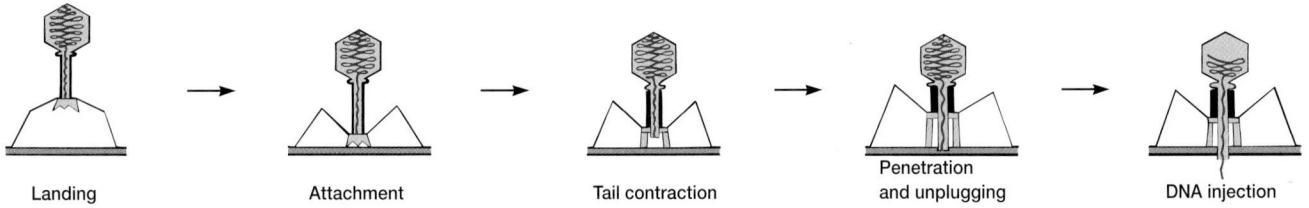

Figure 17.3   T4 Phage Adsorption and DNA Injection.   See text for details.

can serve as receptors. The T-even phages of *E. coli* use cell wall lipopolysaccharides or proteins as receptors. Variation in receptor properties is at least partly responsible for phage host preferences.
The structure of cell walls, flagella, and pili (pp. 52–59, 62–66)

T-even phage adsorption involves several tail structures (*see figure 16.19*). Phage attachment begins when a tail fiber contacts the appropriate receptor site. As more tail fibers make contact, the baseplate settles down on the surface (**figures 17.3** and **17.4**). Binding is probably due to electrostatic interactions and is influenced by pH and the presence of ions such as Mg^{2+} and Ca^{2+}. After the baseplate is seated firmly on the cell surface, conformational changes occur in the baseplate and sheath, and the tail sheath reorganizes so that it shortens from a cylinder 24 rings long (*see p. 366*) to one of 12 rings. That is, the sheath becomes shorter and wider, and the central tube or core is pushed through the bacterial wall. The baseplate contains the protein gp5 with lysozyme activity that may aid in the penetration of the tube through the peptidoglycan layer. Finally, the DNA is extruded from the head, through the tail tube, and into the host cell. The tube may interact with the plasma membrane to form a pore through which DNA passes.

The penetration mechanisms of other bacteriophages often appear different from that of the T-even phages, but most have not been studied in much detail. An exception is the PRD1 phage of the family *Tectiviridae* (figure 17.1), which has a membrane under its icosahedral capsid. It infects pseudomonads and members of the

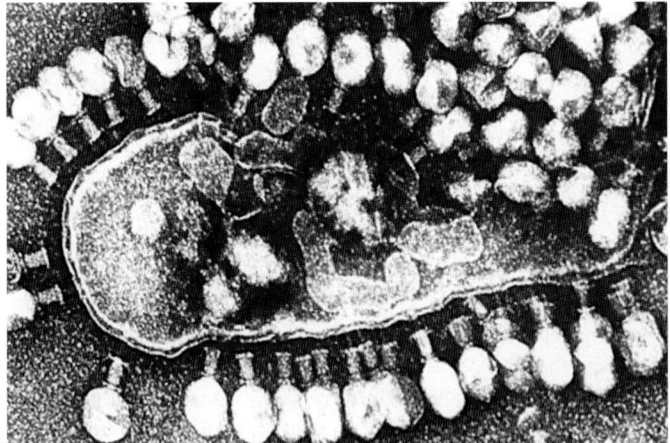

Figure 17.4   Electron Micrograph of *E. coli* Infected with Phage T4.   Baseplates, contracted sheaths, and tail tubes can be seen (×36,500).

Enterobacteriaceae. The PRD1 phage attaches to a surface receptor by a spike structure at one of its capsid vertexes (**figure 17.5**). This causes a conformational change in the capsid proteins. The attached spike-penton complex of the capsid dissociates from the virion, and the underlying membrane forms a tubular structure that penetrates the bacterial envelope. Virus DNA is then injected

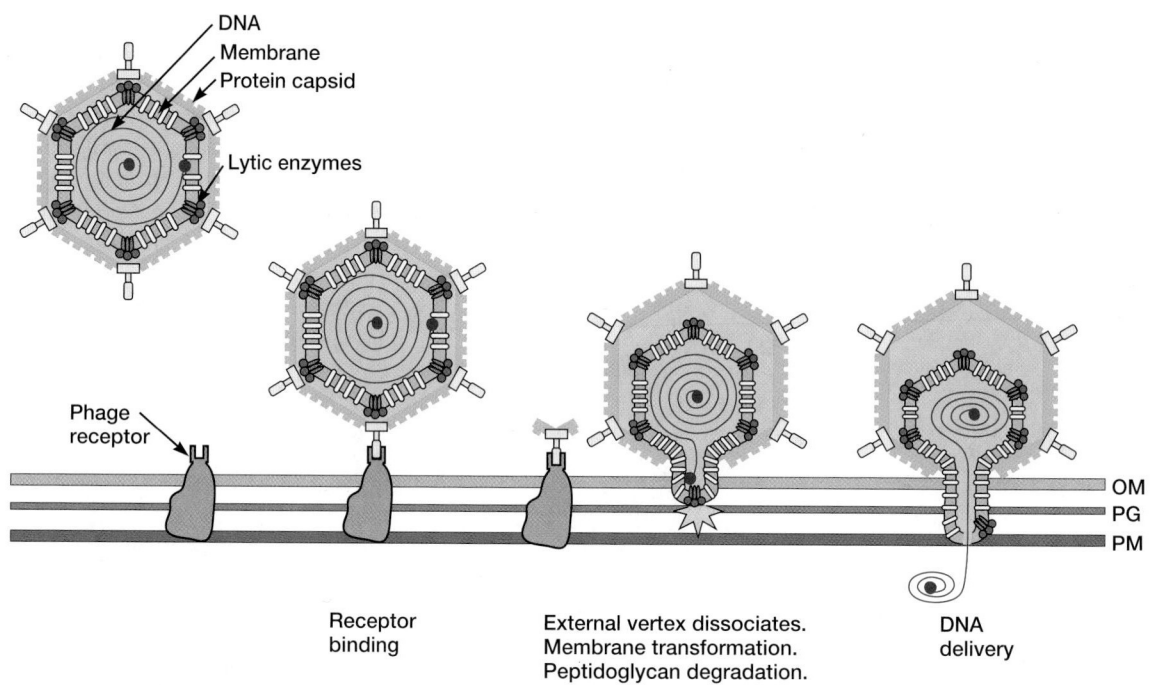

Figure 17.5 **Attachment and Penetration of Host Cell by Phage PRD1.** The gram-negative wall layers are indicated by OM (outer membrane), PG (peptidoglycan), and PM (plasma membrane). See text for more details.

through the membrane tube into the bacterium. Penetration of the membrane tube is made possible partly by the membrane enzyme P7, which breaks the same glycosidic bond that is attacked by lysozyme (*see figure 31.9*).

Synthesis of Phage Nucleic Acids and Proteins

Since the T4 phage of *E. coli* has been intensely studied, its reproduction will be used as our example (**figure 17.6**). Soon after phage DNA injection, the synthesis of host DNA, RNA, and protein is halted, and the cell is forced to make viral constituents. *E. coli* RNA polymerase (*see section 12.1*) starts synthesizing phage mRNA within 2 minutes. This mRNA and all other **early mRNA** (mRNA transcribed before phage DNA is made) direct the synthesis of the protein factors and enzymes required to take over the host cell and manufacture viral nucleic acids. Some early virus-specific enzymes degrade host DNA to nucleotides, thereby simultaneously halting host gene expression and providing raw material for virus DNA synthesis. Within 5 minutes, virus DNA synthesis commences. Promoters and transcription (pp. 254–57)

Virus gene expression follows an orderly sequence because of modifications of the RNA polymerase and changes in sigma factors. Initially T4 genes are transcribed by the regular host RNA polymerase and the sigma factor σ^{70}. After a short interval, a virus enzyme catalyzes the transfer of an ADP-ribosyl group from NAD to an α-subunit of RNA polymerase. This helps inhibit the transcription of host genes and promotes virus gene expression. Then the second α-subunit receives an ADP-ribosyl group and

this turns off some of the early T4 genes. The product of one early gene, *motA*, stimulates transcription of somewhat later genes, one of which produces the sigma factor gp55. This sigma factor helps RNA polymerase bind to late promoters and transcribe late genes, which become active around 10 to 12 minutes after infection.

It is clear from the sophisticated control of RNA polymerase and the precise order in which events occur in the reproductive cycle that the expression of T4 genes is tightly regulated. Even the organization of the genome appears suited for efficient control of the life cycle. As can be seen in **figure 17.7**, genes with related functions—such as the genes for phage head or tail fiber construction—are usually clustered together. Early and late genes also are clustered separately on the genome; they are even transcribed in different directions—early genes in the counterclockwise direction and late genes, clockwise. Since transcription always proceeds in the 5′ to 3′ direction, the early and late genes are located on different DNA strands (*see sections 11.5 and 12.1*).

Considerable preparation is required for synthesis of T4 DNA because it contains **hydroxymethylcytosine (HMC)** instead of cytosine (**figure 17.8**). HMC must be synthesized by two phage-encoded enzymes before DNA replication can begin. After T4 DNA has been synthesized, it is glucosylated by the addition of glucose to the HMC residues. Glucosylated HMC residues protect T4 DNA from attack by *E. coli* endonucleases called **restriction enzymes,** which would otherwise cleave the viral DNA at specific points and destroy it. This bacterial defense mechanism is called **restriction.** Other groups also can be used to modify phage DNA and protect it against restriction

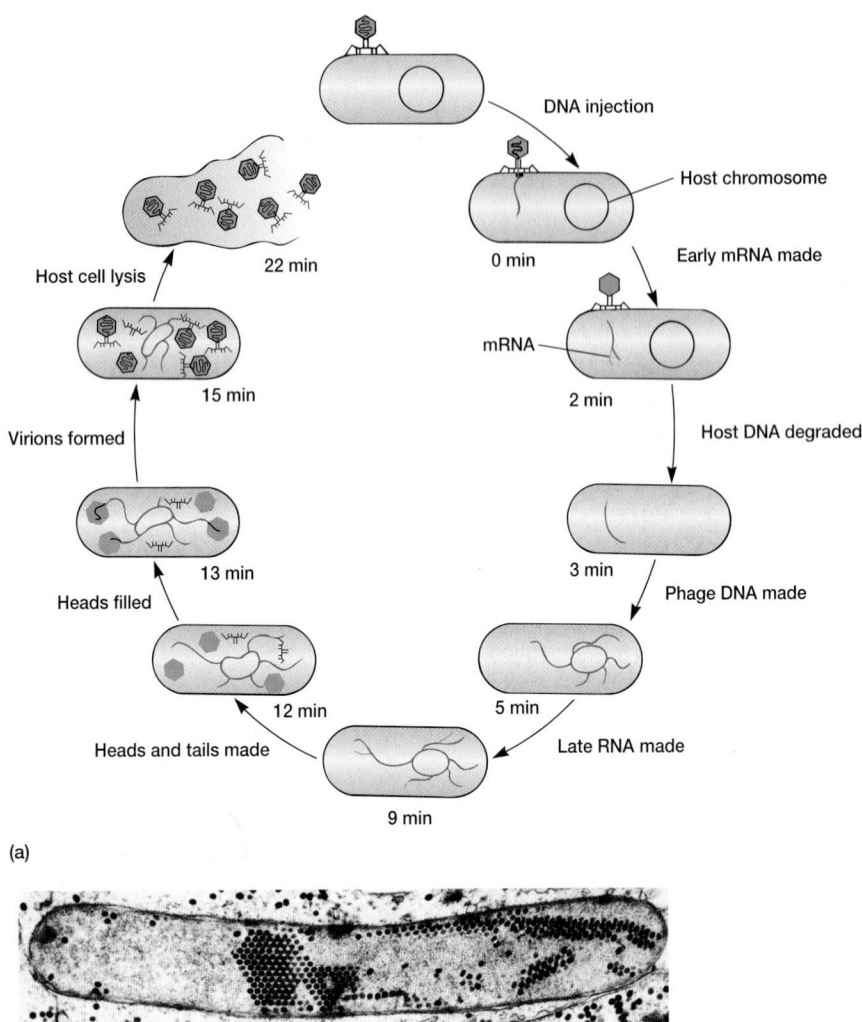

(a)

(b2)

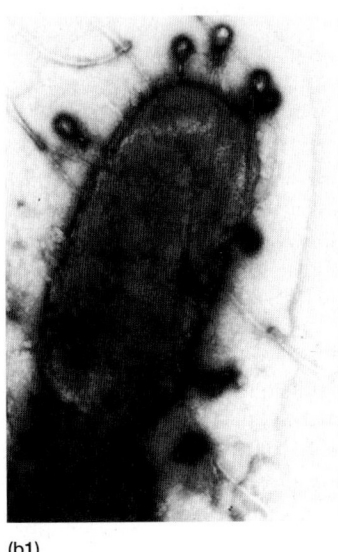

(b1)

Figure 17.6 The Life Cycle of Bacteriophage T4. (**a**) A schematic diagram depicting the life cycle with the minutes after DNA injection given beneath each stage. mRNA is drawn in only at the stage during which its synthesis begins. (**b**) Electron micrographs show the development of T2 bacteriophages in *E. coli.* (*b1*) Several phages are near the bacterium, and some are attached and probably injecting their DNA. (*b2*) By about 30 minutes after infection, the bacterium contains numerous completed phages.

enzymes. For example, methyl groups are added to the amino groups of adenine and cytosine in lambda phage DNA for the same reason. The replication of T4 DNA is an extremely complex process requiring at least seven phage proteins. Its mechanism resembles that described in chapter 11. Restriction enzymes and genetic engineering (pp. 312–13)

T4 DNA shows what is called terminal redundancy; that is, a base sequence is repeated at both ends of the molecule (**figure 17.9**). When many DNA copies have been made, about 6 to 10 copies are joined by their terminally redundant ends with the aid of several enzymes (figure 17.8). These very long DNA strands composed of several units linked together with the same orientation are called **concatemers.** During assembly, concatemers are cleaved in such a way that the genome is slightly longer than the T4 gene set. The genetic map is therefore drawn circular (figure 17.6) because T4 DNA is circularly permuted (**figure 17.10**). The sequence of genes in each T4 virus of a population is the same but starts with a different gene at the 5′ end. If all the linear pieces of DNA were coiled into circles, the DNA circles would have identical gene sequences.

The Assembly of Phage Particles

The assembly of the T4 phage is an exceptionally complex self-assembly process. **Late mRNA,** or that produced after DNA replication, directs the synthesis of three kinds of proteins: (1) phage structural proteins, (2) proteins that help with phage assembly without becoming part of the virion structure, and (3) proteins involved in cell lysis and phage release. Late mRNA transcription begins about 9 minutes after T4 DNA injection into *E. coli.* All the proteins required for phage assembly are synthesized simultaneously and then used in four fairly independent subassembly lines (**figure 17.11**). The baseplate is constructed of 15 gene products. After the baseplate is finished, the tail tube is built on it and the sheath is assembled around the tube. The phage prohead or procapsid is constructed separately of more than 10 proteins and then spontaneously combines with the tail assembly. The procapsid is assembled with the aid of **scaffolding proteins** that are degraded or removed after construction is completed. A special portal protein is located at the base of the procapsid where it connects to the tail. The portal protein is part of the DNA

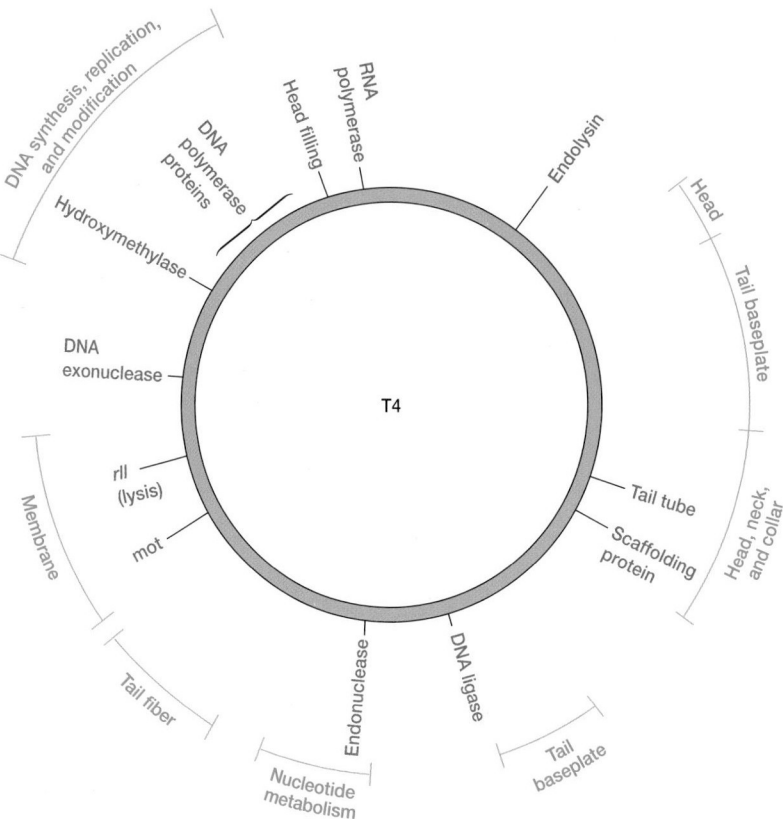

Figure 17.7   A Map of the T4 Genome.   Some of its genes and their functions are shown. Genes with related functions tend to be clustered together.

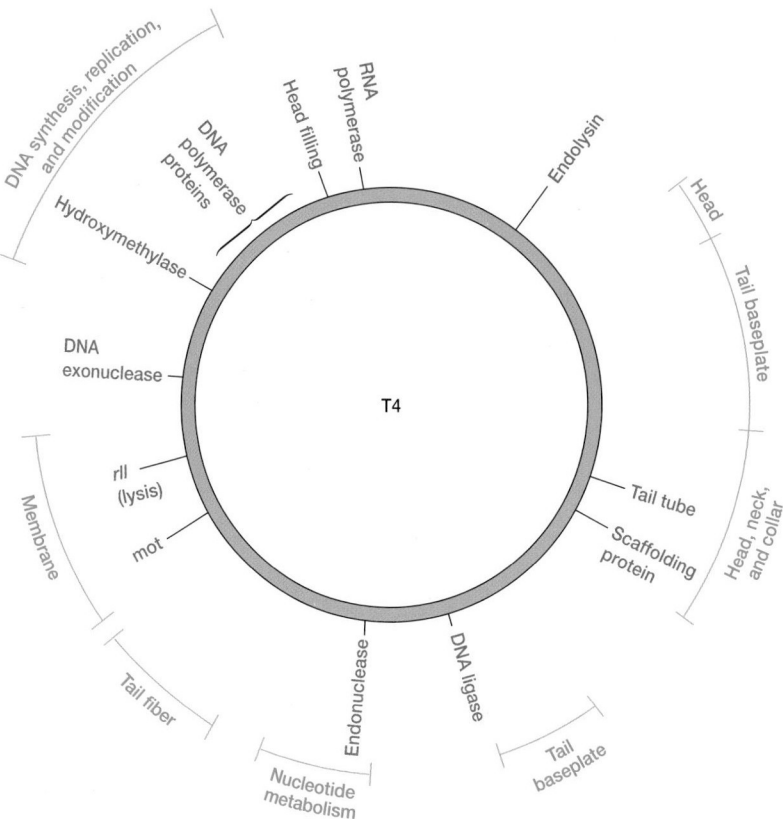

Figure 17.8   5-Hydroxymethlcytosine (HMC).   In T4 DNA, the HMC often has glucose attached to its hydroxyl.

translocating vertex, a structure that helps initiate head assembly and aids in DNA movement into and out of the head. Tail fibers attach to the baseplate after the head and tail have come together. Although many of these steps occur spontaneously, some require special virus proteins or host cell factors.

DNA packaging within the T4 head is still a somewhat mysterious process. In some way the DNA is drawn into the completed shell so efficiently that about 500 μm of DNA are packed into a cavity less than 0.1 μm across! It is thought that a long DNA concatemer enters the procapsid in an ATP-dependent process until it is packed full and contains about 2% more DNA than is needed for the full T4 genome. The concatemer is then cut, and T4 assembly is finished. The first complete T4 particles appear in *E. coli* at 37°C about 15 minutes after infection.

Release of Phage Particles

Many phages lyse their host cells at the end of the intracellular phase. The lysis of *E. coli* takes place after about 22 minutes at 37°C, and approximately 300 T4 particles are released. Several T4 genes are involved in this process. One directs the synthesis of an endolysin that attacks the cell wall peptidoglycan. Another phage protein called a holin produces a plasma membrane lesion that stops respiration and allows the endolysin to attack the peptidoglycan. Presumably it forms holes in the membrane.

Unlike the case with T4 and many other double-stranded DNA phages, lytic phages with smaller genomes often gain release by interfering with the enzymes that synthesize host peptidoglycan and thus disrupting wall construction. The bacterium then lyses because of osmotic pressure acting on a weakened cell wall. Good examples of this approach are the bacteriophages φX174 and Qβ, which will be described in sections 17.3 and 17.4. The phage φX174 produces the lytic enzyme E that blocks the activity of the bacterial protein MraY, which catalyzes the transfer of murein precursors to lipid carriers (*see figure 10.28*). Phage Qβ attacks an earlier step in peptidoglycan synthesis. It produces the A_2 maturation protein that interferes with the MurA enzyme. MurA participates in the synthesis of UDP-NAM, a precursor for peptidoglycan synthesis (*figure 10.28*).

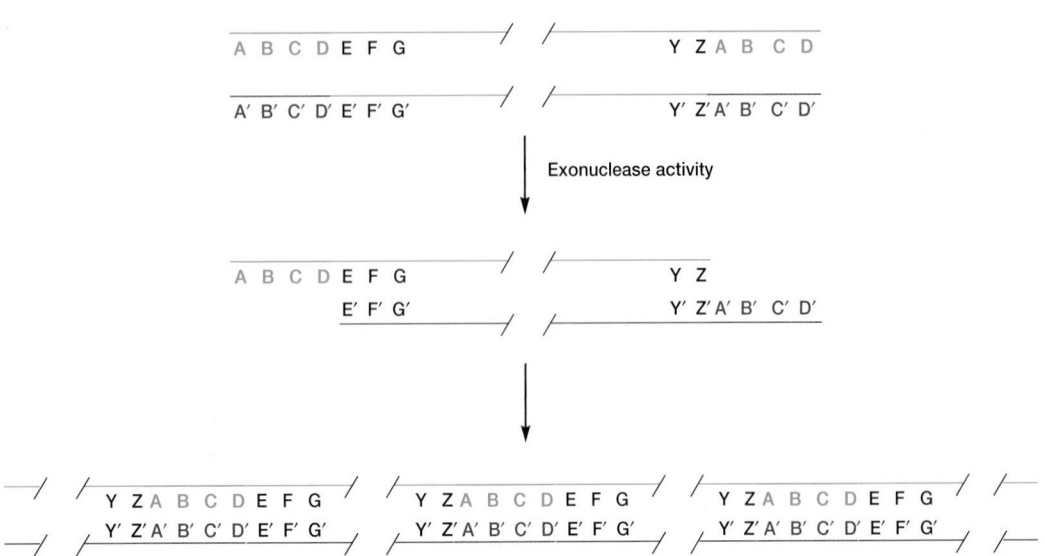

Figure 17.9   An Example of Terminal Redundancy.   The gene sequences in color are terminally redundant; they are repeated at each end of the DNA molecule. This makes it possible to join units together by their redundant ends forming a concatemer. For example, if the 3′ ends of each unit were partially digested by an exonuclease, the complementary 5′ ends would be exposed and could base pair to generate a long chain of repeated units. The breaks between terminal sequences indicate that the DNA molecules are longer than shown here.

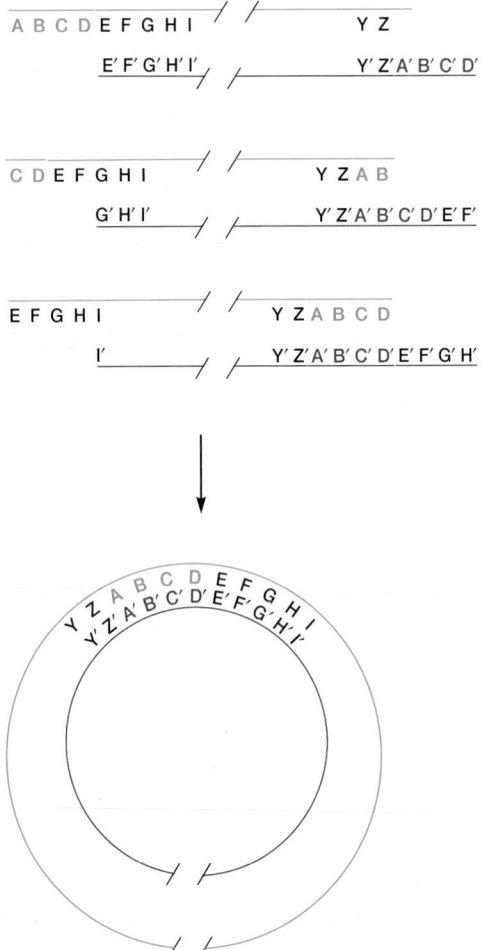

17.3   REPRODUCTION OF SINGLE-STRANDED DNA PHAGES

Thus far, only double-stranded DNA phage reproduction has been discussed, with the lytic phage T4 as an example. The reproduction of single-stranded DNA phages now will be briefly reviewed. The phage of φX174, family *Microviridae,* is a small ssDNA phage using *E. coli* as its host. Its DNA base sequence is the same as that of the viral mRNA (except that thymine is substituted for uracil) and is therefore positive; the genome contains overlapping genes (*see figure 11.20*b). The phage DNA must be converted to a double-stranded form before either replication or transcription can occur. When φX174 DNA enters the host, it is immediately copied by the bacterial DNA polymerase to form a double-stranded DNA, the **replicative form** or **RF** (**figure 17.12**). The replicative form then directs the synthesis of more RF copies, mRNA, and copies of the +DNA genome. The phage is released by host lysis through a different mechanism than used by the T4 phage.

The filamentous ssDNA bacteriophages behave quite differently in many respects from φX174 and other ssDNA phages.

Figure 17.10   Circularly Permuted Genomes Cut from a Concatemer.   The concatemer formed in figure 17.9 can be cut at any point into pieces of equal length that contain a complete complement of genes, even though different genes are found at their ends. If each piece has single-stranded cohesive ends as in figure 17.9, it will coil into a circle with the same gene order as the circles produced by other pieces.

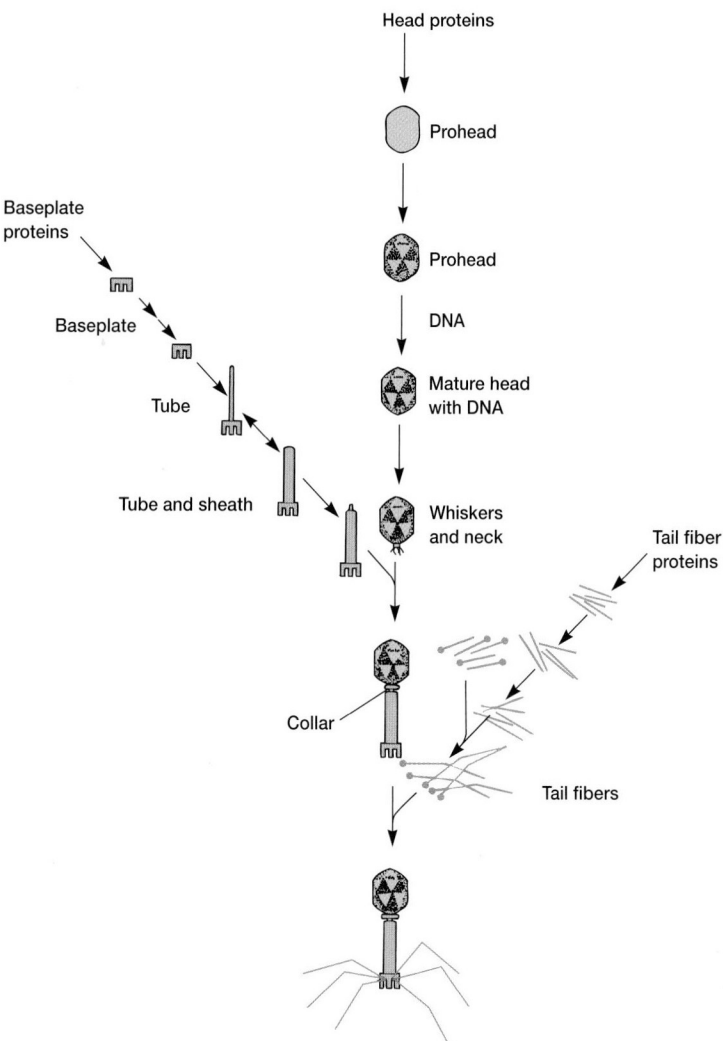

Figure 17.11   The Assembly of T4 Bacteriophage. Note the subassembly lines for the baseplate, tail tube and sheath, tail fibers, and head.

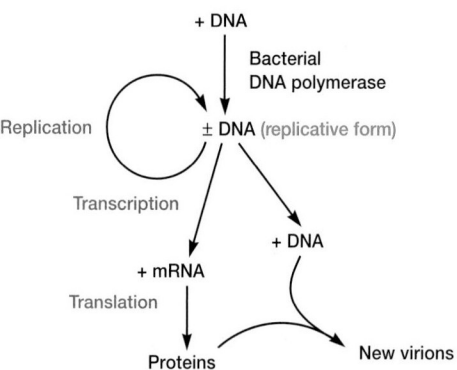

Figure 17.12   The Reproduction of φX174, a + Strand DNA Phage.   See text for details.

plasma membrane (**figure 17.13**). The host bacteria grow and divide at a slightly reduced rate.

---

1. How is a one-step growth experiment carried out? Summarize what occurs in each phase of the resulting growth curve. Define latent period, eclipse period, rise period or burst, and burst size.
2. Be able to describe in some detail what is occurring in each phase of the lytic dsDNA phage life cycle: adsorption and penetration, nucleic acid and protein synthesis, phage assembly, and phage release. Define the following terms: lytic cycle, receptor site, early mRNA, hydroxymethylcytosine, restriction, restriction enzymes, concatemers, replicative form, late mRNA, and scaffolding proteins.
3. How does the reproduction of the ssDNA phages φX174 and fd differ from each other and from the dsDNA T4 phage?

 REPRODUCTION OF RNA PHAGES

Many bacteriophages carry their genetic information as single-stranded RNA that can act as a messenger RNA and direct the synthesis of phage proteins. One of the first enzymes synthesized is a viral **RNA replicase,** an RNA-dependent RNA polymerase (**figure 17.14**). The replicase then copies the original RNA (a plus strand) to produce a double-stranded intermediate (±RNA), which is called the replicative form and is analogous to the ±DNA seen in the reproduction of ssDNA phages. The same replicase next uses this replicative form to synthesize thousands of copies of +RNA. Some of these plus strands are used to make more ±RNA in order to accelerate +RNA synthesis. Other +RNA acts as mRNA and directs the synthesis of phage proteins. Finally, +RNA strands are incorporated into maturing virus particles. The genome of these RNA phages serves as both a template for its own replication and an mRNA.

MS2 and Qβ, family *Leviviridae,* are small, tailless, icosahedral ssRNA phages of *E. coli,* which have been intensely studied (figure 17.1). They attach to the F-pili of their host and enter by

The fd phage, family *Inoviridae,* is one of the best studied and is shaped like a long fiber about 6 nm in diameter by 900 to 1,900 nm in length (figure 17.1). The circular ssDNA lies in the center of the filament and is surrounded by a tube made of a small coat protein organized in a helical arrangement. The virus infects male *E. coli* cells by attaching to the tip of the pilus; the DNA enters the host along or possibly through the pilus with the aid of a special adsorption protein. A replicative form is first synthesized and then transcribed. A phage-coded protein then aids in replication of the phage DNA by use of the rolling-circle method (*see section 11.3*).

The filamentous fd phages do not kill their host cell but establish a symbiotic relationship in which new virions are continually released by a secretory process. Filamentous phage coat proteins are first inserted into the membrane. The coat then assembles around the viral DNA as it is secreted through the host

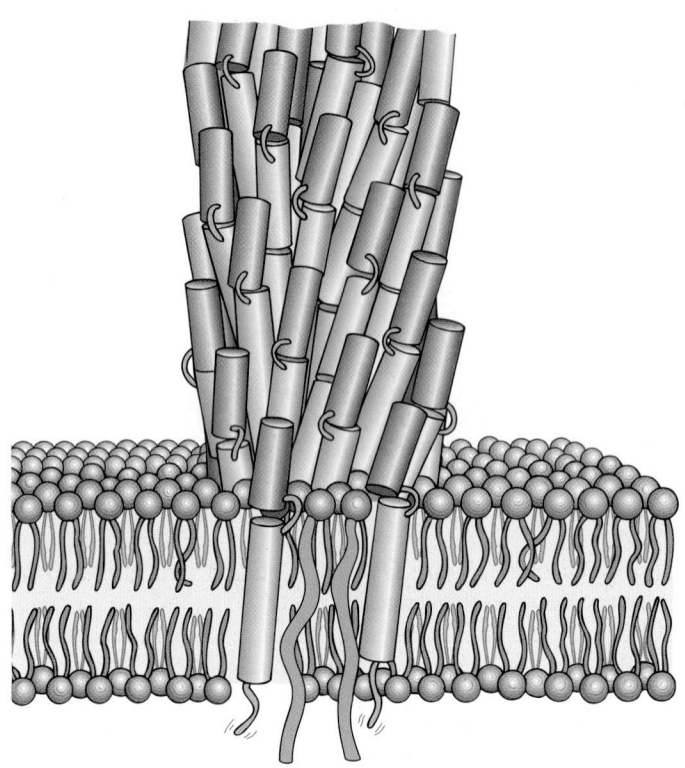

Figure 17.13 Release of Pf1 Phage. The Pf1 phage is a filamentous bacteriophage that is released from *Pseudomonas aeruginosa* without lysis. In this illustration the blue cylinders are hydrophobic α-helices that span the plasma membrane, and the red cylinders are amphipathic helices that lie on the membrane surface before virus assembly. In each protomer the two helices are connected by a short, flexible peptide loop (yellow). It is thought that the blue helix binds with circular, single-stranded viral DNA (green) as it is extruded through the membrane. The red helix simultaneously attaches to the growing virus coat that projects from the membrane surface. Eventually the blue helix leaves the membrane and also becomes part of the capsid.

an unknown mechanism. These phages have only three or four genes and are genetically the simplest phages known. In MS2, one protein is involved in phage adsorption to the host cell (and possibly also in virion construction or maturation). The other three genes code for a coat protein, an RNA replicase, and a protein needed for cell lysis.

Only one dsRNA phage has been discovered, the bacteriophage φ6 of *Pseudomonas phaseolicola* (figure 17.1). It is also unusual in possessing a membranous envelope. The icosahedral capsid within its envelope contains an RNA polymerase and three dsRNA segments, each of which directs the synthesis of an mRNA. It is not yet known how the dsRNAs are replicated.

---

1. How are ssRNA phages reproduced, and what role does RNA replicase play in the process?
2. What is peculiar about the structure of phage φ6?

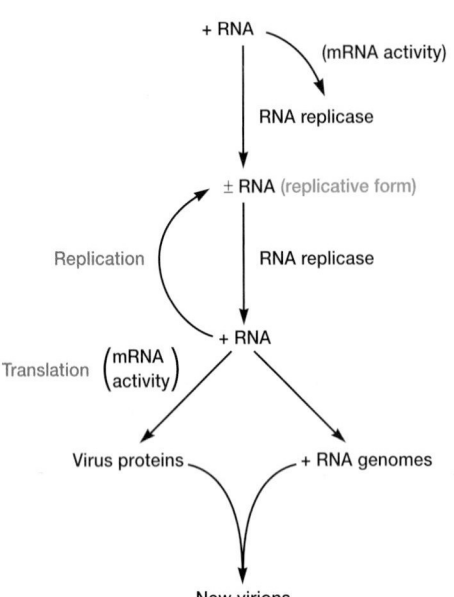

Figure 17.14 The Reproduction of Single-Stranded RNA Bacteriophages.

17.5 TEMPERATE BACTERIOPHAGES AND LYSOGENY

Up to this point many of the viruses we have discussed are **virulent bacteriophages;** these are phages that lyse their host cells during the reproductive cycle. Many DNA phages also can establish a different relationship with their host. After adsorption and penetration, the viral genome does not take control of its host and destroy it while producing new phages. Instead the viral genome remains within the host cell and replicates with the bacterial genome to generate a clone of infected cells that may grow and divide for long periods while appearing perfectly normal (*see figure 13.18*). Each of these infected bacteria can produce phages and lyse under appropriate environmental conditions. They cannot, for reasons that will become clear later, be reinfected by the same virus—that is, they have immunity to superinfection. This relationship between the phage and its host is called **lysogeny.** Bacteria having the potential to produce phage particles under some conditions are said to be **lysogens** or **lysogenic,** and phages able to enter into this relationship are **temperate phages.** The latent form of the virus genome that remains within the host but does not destroy it is called the **prophage.** The prophage usually is integrated into the bacterial genome but sometimes exists independently. **Induction** is the process by which phage reproduction is initiated in a lysogenized culture. It leads to the destruction of infected cells and the release of new phages—that is, induction of the lytic cycle. Lysogeny was briefly described earlier in the context of transduction and genetic recombination, but will be discussed in more detail here.  Generalized and specialized transduction (pp. 301–5)

Most bacteriophages that have been studied are temperate, and it appears that there are advantages in being able to lysogenize bacteria. Consider a phage-infected culture that is becoming dormant due to nutrient deprivation. Before bacteria enter dormancy, they degrade their own mRNA and protein. Thus the phage is faced with two problems: it can only reproduce in actively metabolizing bacteria, and phage reproduction is usually permanently interrupted by the mRNA and protein degradation. This predicament can be avoided if the phage becomes dormant (lysogenic) at the same time as its host; in fact, nutrient deprivation does favor lysogeny. Temperate phages also have an advantage in situations where many viruses per cell initiate an infection—that is, where there is a high multiplicity of infection (MOI). When every cell is infected, the last round of replication will destroy all host cells. Thus there is a risk that the phages may be left without a host and directly exposed to environmental hazards for months or years. This prospect is avoided if lysogeny is favored by a high MOI; some bacteria will survive, carry the virus genome, and synthesize new copies as they reproduce. Not surprisingly a high MOI does stimulate lysogeny.

A temperate phage may induce a change in the phenotype of its host cell that is not directly related to completion of its life cycle. Such a change is called a **lysogenic conversion** or a conversion and often involves alterations in bacterial surface characteristics or pathogenic properties. For example, when *Salmonella* is infected by an epsilon phage, the structure of its outer lipopolysaccharide layer (*see pp. 56–58*) may be modified. The phage changes the activities of several enzymes involved in construction of the lipopolysaccharide carbohydrate component and thus alters the antigenic properties of the host. These epsilon-induced changes appear to eliminate surface phage receptors and prevent infection of the lysogen by another epsilon phage. Another example is the temperate phage β of *Corynebacterium diphtheriae*, the cause of diphtheria. Only *C. diphtheriae* that is lysogenized with phage β will produce diphtheria toxin (*see sections 34.3 and 39.1*) because the phage, not the bacterium, carries the toxin gene.

The lambda phage, family *Siphoviridae*, that uses the K12 strain of *E. coli* as its host is the best-understood temperate phage and will serve as our example of lysogeny. Lambda is a double-stranded DNA phage possessing an icosahedral head 55 nm in diameter and a noncontractile tail with a thin tail fiber at its end (**figure 17.15**). The DNA is a linear molecule with cohesive ends—single-stranded stretches, 12 nucleotides long, that have complementary base sequences and can base pair with each other. Because of these cohesive ends, the linear genome cyclizes immediately upon infection (**figure 17.16**). *E. coli* DNA ligase then seals the breaks, forming a closed circle. The lambda genome has been carefully mapped, and over 40 genes have been located (**figure 17.17**). Most genes are clustered according to their function, with separate groups involved in head synthesis, tail synthesis, lysogeny and its regulation, DNA replication, and cell lysis.   DNA ligase (p. 233)

Lambda phage can reproduce using a normal lytic cycle. Immediately after lambda DNA enters *E. coli*, it is converted to a covalent circle, and transcription by the host RNA polymerase is initiated. As shown in figure 17.17, the polymerase binds to both a rightward promoter (PR) and a leftward promoter (PL) and begins to transcribe in both directions, copying different DNA strands. The first genes that are transcribed code for regulatory

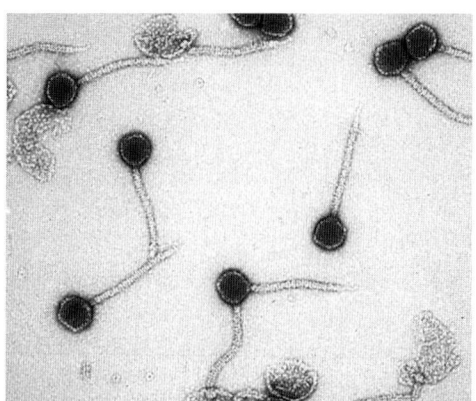
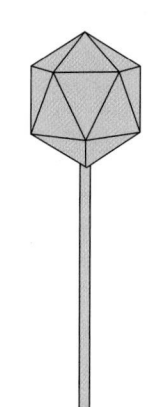

Figure 17.15   Bacteriophage Lambda.

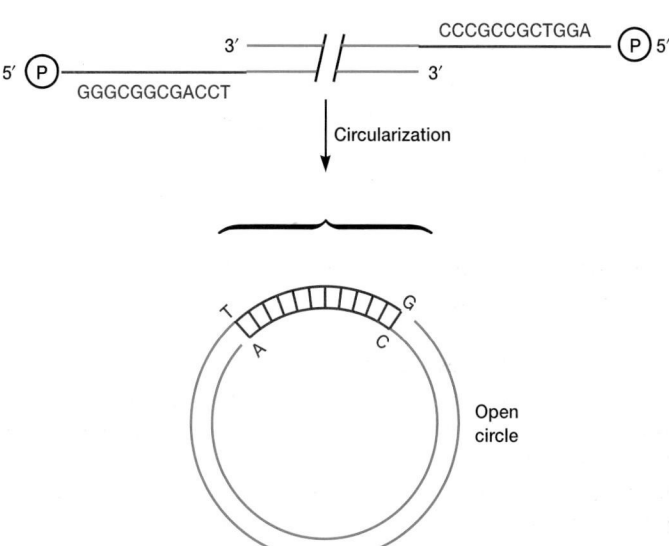

Figure 17.16   Lambda Phage DNA.   A diagram of lambda phage DNA showing its 12 base, single-stranded cohesive ends (printed in red) and the circularization their complementary base sequences make possible.

proteins that control the lytic cycle: leftward gene *N* and rightward genes *cro* and *cII* (figure 17.17). These and other regulatory genes ensure that virus proteins will be synthesized in an orderly time sequence and will be manufactured only when needed during the life cycle.   Regulation of transcription (pp. 270–74)

Lambda DNA replication and virion assembly are similar to the same processes already described for the T4 phage. One significant difference should be noted. Although initially bidirectional DNA replication is used and theta-shaped intermediates are seen (*see section 11.3*), lambda DNA is primarily synthesized by way of the rolling-circle mechanism to form long concatemers that are finally cleaved to give complete genomes (*see figure 11.13*).

The establishment of lysogeny and the earlier-mentioned immunity of lysogens to superinfection can be accounted for by the presence of the **lambda repressor** coded for by the *cI* gene. The

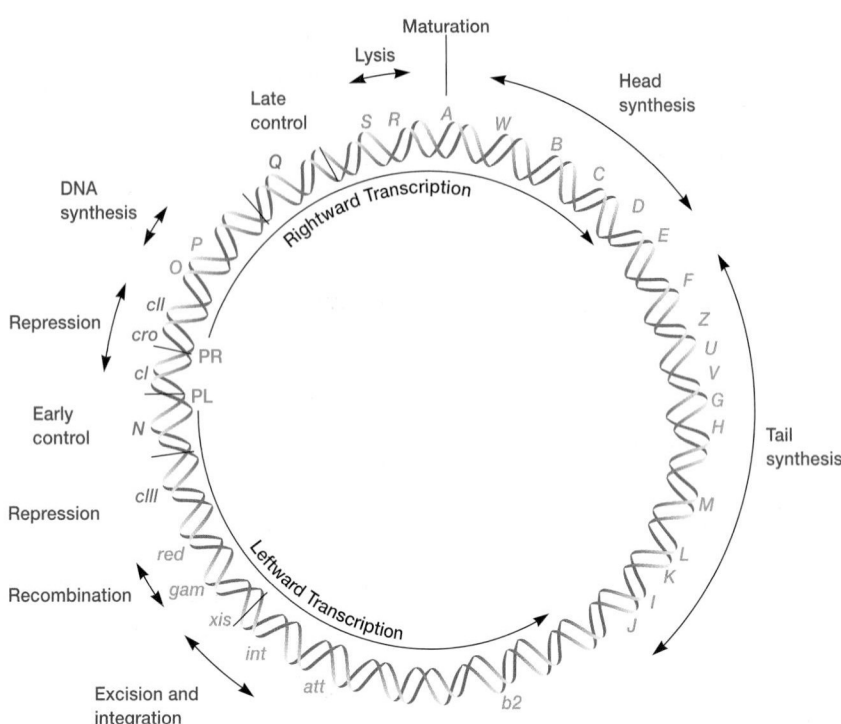

Figure 17.17 The Lambda Phage Genome. The direction of transcription and location of leftward and rightward promoters (PL and PR) are indicated on the inside of the map. The positions of major regulatory sites are shown by lines on the map and regulatory genes are in blue. Lambda DNA is double stranded, and transcription proceeds in opposite directions on opposite strands.

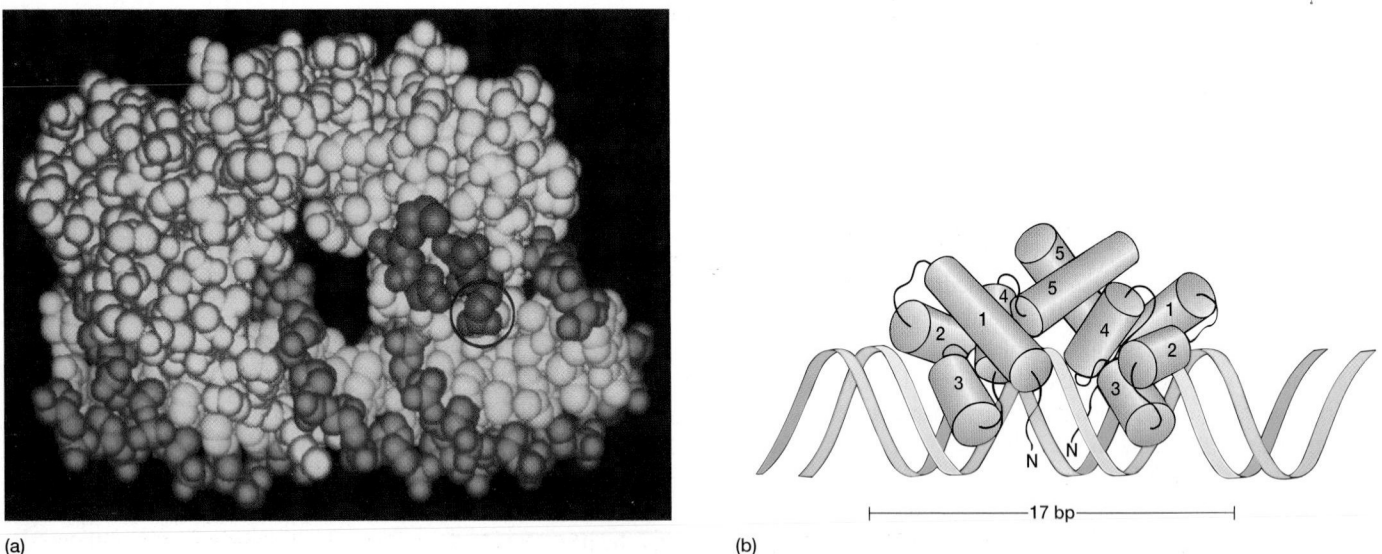

(a)   (b)

Figure 17.18 Lambda Repressor Binding. (a) A computer model of lambda repressor binding to the lambda operator. The lambda repressor dimer (brown and tan) is bound to DNA (blue and light blue). The arms of the dimer wrap around the major grooves of the double helix. (b) A diagram of the lambda repressor-DNA complex. The repressor binds to a 17 bp stretch of the operator. The α3-helices make closest contact with the major grooves of the operator (the helices are labeled in order, beginning at the N terminal of the chain).

repressor protein chain is 236 amino acids long and folds into a dumbbell shape with globular domains at each end (**figure 17.18**). One domain is concerned with binding to DNA, while the other binds with another repressor molecule to generate a dimer (the most active form of the lambda repressor). In a lysogen the repressor is synthesized continuously and binds to the operators O_L and O_R, thereby blocking RNA polymerase activity (figure 17.19c). If another lambda phage tries to infect the cell, its mRNA synthesis also will be inhibited. It should be noted that immunity always involves repressor activity. A potential host cell might remain uninfected due

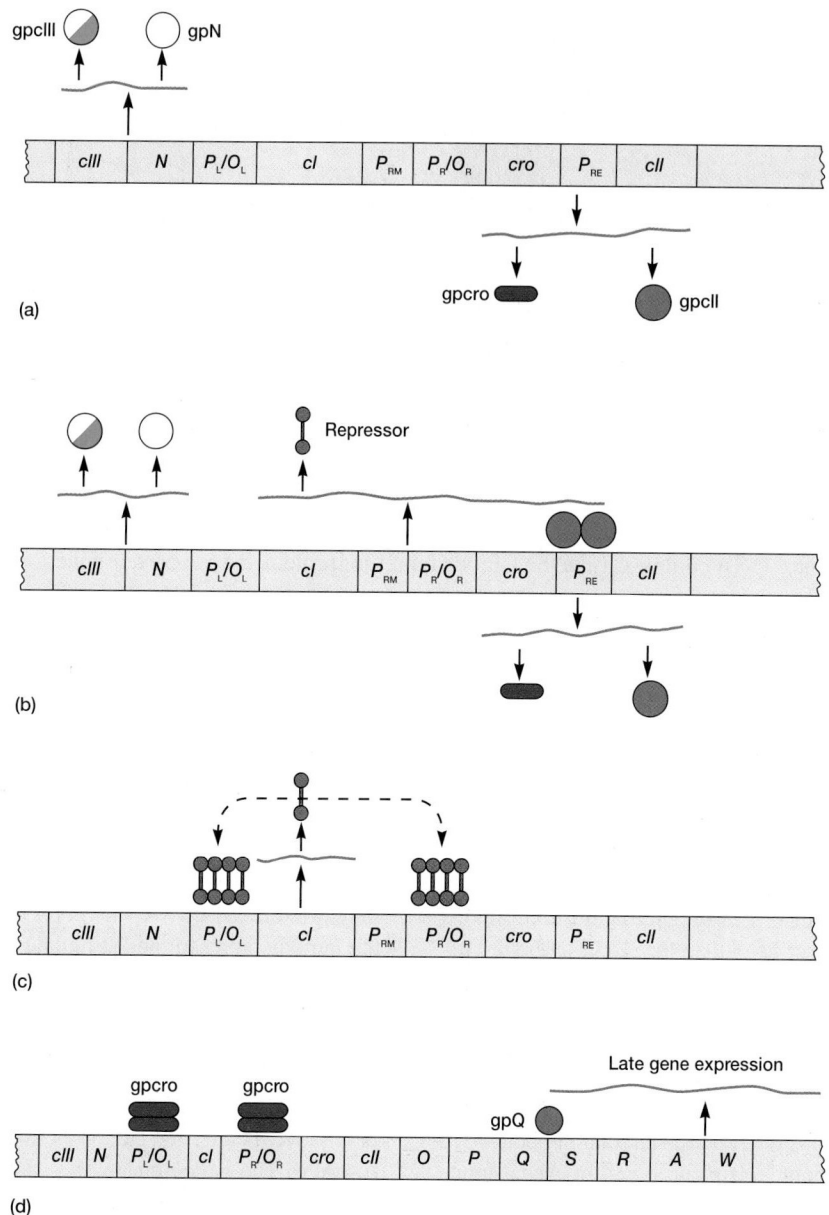

Figure 17.19   Choice Between Lysogeny and Lysis.   Events involved in the choice between the establishment of lysogeny and continuation of the lytic cycle. The action of gpN is ignored for the sake of simplicity, and the scale in part (*d*) differs from that in parts (*a*) to (*c*). The abbreviation gp stands for gene product (gpcro is the product of the *cro* gene). (**a**) and (**b**) illustrate the initial steps leading to lambda repressor synthesis. (**c**) represents the situation when repressor production overcomes cro synthesis and establishes lysogeny. In part (**d**) the cro protein has accumulated more rapidly than lambda repressor and cro protein dimers (the active form) have bound to O_L and O_R. This blocks both *cI* and *cro* gene function, but not late gene expression, since gpQ has already accumulated and promoted late mRNA synthesis. See text for further details.

to a mutation that alters its phage receptor site. In such an instance it is said to be resistant, not immune, to the phage.

   The sequence of events leading to the initial synthesis of repressor and the establishment of lysogeny is well known. Immediately after lambda DNA has been circularized and transcription has commenced, the cII and cIII proteins accumulate (**figure 17.19a**). The cII protein binds next to the promoter for the *cII* gene (P_{RE}, RE stands for repressor establishment), and stimulates RNA polymerase binding (figure 17.19b). The cIII protein

protects cII from degradation by a host enzyme, the HflA protease. Lambda repressor (gpcI) is rapidly synthesized and binds to O_R and O_L, thus turning off mRNA synthesis and the production of the cII and cIII proteins (figure 17.19c). The *cI* gene continues to be transcribed at a low rate because of the activity of a second promoter (P_{RM}, RM stands for repressor maintenance) that is activated by the repressor itself. This control circuit in which lambda repressor stimulates its own synthesis ensures that lysogeny will normally be stable when once established.

(a)

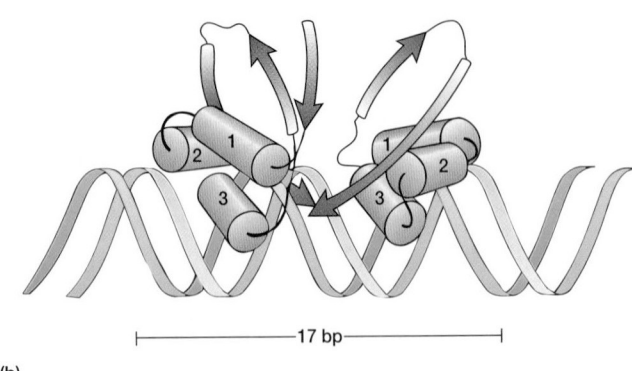

|-------17 bp-------|

(b)

Figure 17.20   Cro Protein Binding.   (a) A space-filling model of the cro protein–DNA complex. The cro protein is in yellow. (b) A diagram of the cro protein dimer–DNA complex. Like the lambda repressor protein, the cro protein functions as a dimer and binds to two adjacent DNA major grooves.

One might expect that lysogeny would be established every time but this is not the case. During this period the **cro protein** (gpcro) has also been accumulating. The cro protein binds to O_R and O_L, turns off the transcription of the repressor gene (as well as inhibiting the expression of other early genes), and represses P_{RM} function (figure 17.19*d* and **figure 17.20**). Because the lambda repressor can block *cro* transcription, there is a race between the production of lambda repressor and that of the cro protein. Although cro protein synthesis begins before that of the lambda repressor, gpcro binds to O_R more weakly and must rise to a higher level than the repressor before repressor synthesis is blocked and the lytic cycle started (figure 17.19*d*). The details of this competition are not yet completely clear, but it has been shown that a number of environmental factors influence the outcome of the race and the choice between the lytic and lysogenic pathways.

If the lambda repressor wins the race, the circular lambda DNA is inserted into the *E. coli* genome as first proposed by Alan Campbell. **Integration** or insertion is possible because the cII protein stimulates transcription of the *int* gene at the same time as that of the *cI* gene. The *int* gene codes for the synthesis of an **integrase** enzyme, and this protein becomes plentiful before lambda repressor turns off transcription. Lambda DNA has a phage attachment site (the *att* site) that can base pair with a bacterial attachment site located between the galactose or *gal* operon and the biotin operon on the *E. coli* chromosome. After these two sites match up, the integrase enzyme, with the aid of a special host protein, catalyzes the physical exchange of viral and bacterial DNA strands (**figure 17.21**). The circular lambda DNA is integrated into the *E. coli* DNA as a linear region next to the *gal* operon and is called a prophage. As can be seen in figure 17.21, the linear order of phage genes has been changed or permuted during integration.

The lambda prophage will leave the *E. coli* genome and begin the production of new phages when the host is unable to survive. The process is known as induction and is triggered by a drop in lambda repressor levels. Occasionally the repressor will spontaneously decline and the lytic cycle commence. However, induction usually is in response to environmental factors such as UV light or chemical mutagens that damage host DNA. This damage causes the recA protein, which normally plays a role in genetic recombination in *E. coli* (*see section 11.8*), to act as a protease and cleave the repressor chain between the two domains. The separated domains cannot assemble to form the normal active repressor dimer, and the lytic cycle genes become active again. There is some recent evidence that activated recA protein may not directly cleave the repressor. RecA may instead bind to the lambda repressor and stimulate it to proteolytically cleave itself. An early gene located next to the *int* gene, the *xis* gene, codes for the synthesis of an **excisionase** protein that binds to the integrase and enables it to reverse the integration process and free the prophage (figure 17.21). The lytic cycle then proceeds normally.

Most temperate phages exist as integrated prophages in the lysogen. Nevertheless, integration is not an absolute requirement for lysogeny. The *E. coli* phage P1 is similar to lambda in that it circularizes after infection and begins to manufacture repressor. However, it remains as an independent circular DNA molecule in the lysogen and is replicated at the same time as the host chromosome. When *E. coli* divides, P1 DNA is apportioned between the daughter cells so that all lysogens contain one or two copies of the phage genome.

---

1. Define virulent phage, lysogeny, temperate phage, lysogen, prophage, immunity, and induction.
2. What advantages might a phage gain by being capable of lysogeny?
3. Describe lysogenic conversion and its significance.
4. Precisely how, in molecular terms, is a bacterial cell made lysogenic by a temperate phage like lambda?
5. How is a prophage induced to become active again?
6. Be able to describe the roles of the lambda repressor, cro protein, the recA protein, integrase, and excisionase in lysogeny and induction.
7. How does the temperate phage P1 differ from lambda phage?

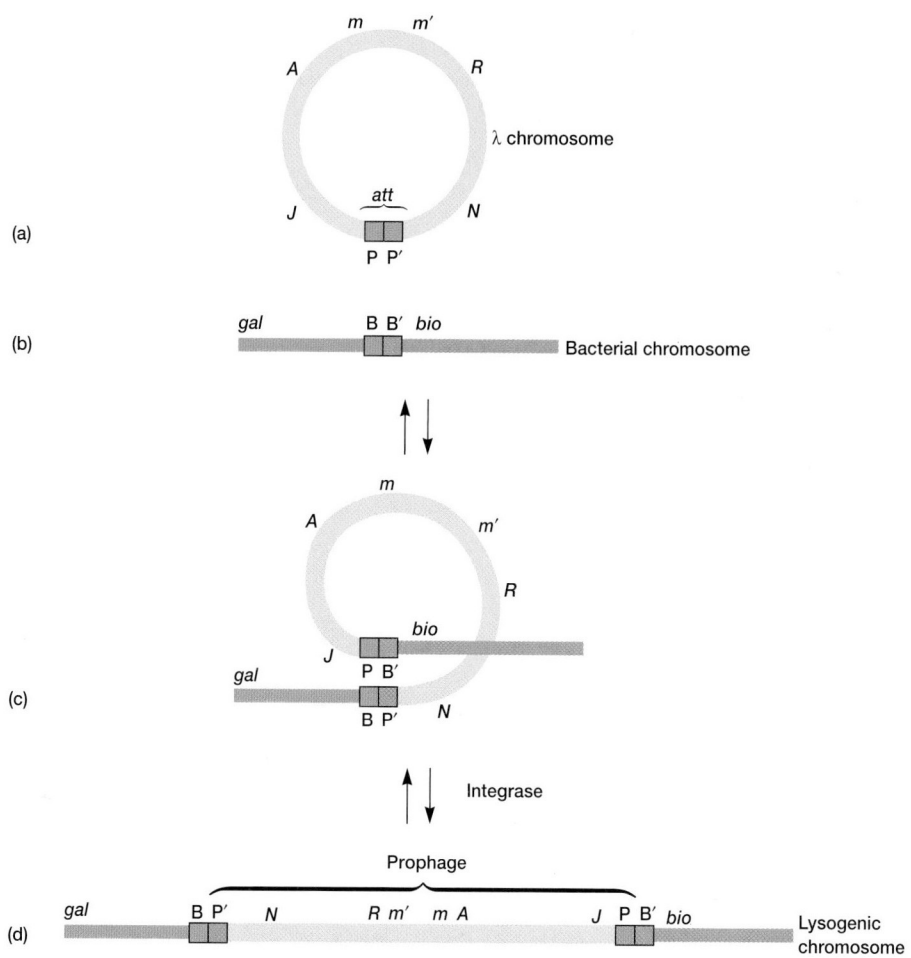

Figure 17.21   Reversible Insertion and Excision of Lambda Phage.   After circularization, the *att* site P, P′ (**a**) lines up with a corresponding bacterial sequence B, B′ (**b**) and is integrated between the *gal* and *bio* operons to form the prophage, (**c**) and (**d**). If the process is reversed, the circular lambda chromosome will be restored and can then reproduce.

17.1 Classification of Bacteriophages

a. There are four major morphological groups of phages: tailless icosahedral phages, phages with contractile tails, those with noncontractile tails, and filamentous phages (**figure 17.1**).

17.2 Reproduction of Double-Stranded DNA Phages: The Lytic Cycle

a. The lytic cycle of virulent bacteriophages is a life cycle that ends with host cell lysis and virion release.

b. The phage life cycle can be studied with a one-step growth experiment that is divided into an initial eclipse period within the latent period, and a rise period or burst (**figure 17.2**).

c. The life cycle of the dsDNA T4 phage of *E. coli* is composed of several phases. In the adsorption phase the phage attaches to a specific receptor site on the bacterial surface.

This is followed by penetration of the cell wall and insertion of the viral nucleic acid into the cell (**figure 17.3**).

d. Transcription of T4 DNA first produces early mRNA, which directs the synthesis of the protein factors and enzymes required to take control of the host and manufacture phage nucleic acids (**figure 17.6**).

e. T4 DNA contains hydroxymethylcytosine (HMC) in place of cytosine, and glucose is often added to the HMC to protect the phage DNA from attack by host restriction enzymes.

f. T4 DNA replication produces concatemers, long strands of several genome copies linked together.

g. Late mRNA is produced after DNA replication and directs the synthesis of capsid proteins, proteins involved in phage assembly, and those required for cell lysis and phage release.

h. Complete virions are assembled immediately after the separate components have been constructed. This is a self-assembly process, but does require participation of the bacterial membrane and a few extra proteins.

17.3 Reproduction of Single-Stranded DNA Phages

a. T4, φX174, and many other phages are released upon lysis of the host cell.

b. The replication of ssDNA phages proceeds through the formation of a double-stranded replicative form (RF) (**figure 17.12**). The filamentous ssDNA phages are continually released without host cell lysis.

17.4 Reproduction of RNA Phages

a. When ssRNA bacteriophage RNA enters a bacterial cell, it acts as a messenger and directs the synthesis of RNA replicase, which then produces double-stranded replicative forms and, subsequently, many +RNA copies (**figure 17.14**).

b. The φ6 phage is the only dsRNA phage known. It is also unusual in having a membranous envelope.

17.5 Temperate Bacteriophages and Lysogeny

a. Temperate phages, unlike virulent phages, often reproduce in synchrony with the host genome to yield a clone of virus-infected cells. This relationship is lysogeny, and the infected cell is called a lysogen. The latent form of the phage genome within the lysogen is the prophage (**figure 13.18**).

b. Lysogeny is reversible, and the prophage can be induced to become active again and lyse its host.

c. A temperate phage may induce a change in the phenotype of its host cell that is not directly related to the completion of its life cycle. Such a change is called a conversion.

d. Two of the first proteins to appear after infection with lambda are the lambda repressor and the cro protein. The lambda repressor blocks the transcription of both cro protein and those proteins required for the lytic cycle, while the cro protein inhibits transcription of the lambda repressor gene (**figure 17.19**).

e. There is a race between synthesis of lambda repressor and that of the cro protein. If the cro protein level rises high enough in time, lambda repressor synthesis is blocked and the lytic cycle initiated; otherwise, all genes other than the lambda repressor gene are repressed and the cell becomes a lysogen.

f. The final step in prophage formation is the insertion or integration of the lambda genome into the *E. coli* chromosome; this is catalyzed by a special integrase enzyme (**figure 17.21**).

g. Several environmental factors can lower repressor levels and trigger induction. The prophage becomes active and makes an excisionase protein that causes the integrase to reverse integration, free the prophage, and initiate a lytic cycle.

Key Terms

bacteriophages 372
burst size 373
concatemer 376
cro protein 384
early mRNA 375
eclipse period 373
excisionase 384
hydroxymethylcytosine (HMC) 375

induction 380
integrase 384
integration 384
lambda repressor 381
late mRNA 376
latent period 373
lysogen 380
lysogenic 380

lysogenic conversion 381
lysogeny 380
lytic cycle 373
one-step growth experiment 373
prophage 380
receptor sites 373
replicative form (RF) 378

restriction 375
restriction enzyme 375
rise period or burst 373
RNA replicase 379
scaffolding proteins 376
temperate phage 380
virulent bacteriophage 380

Questions for Thought and Review

1. Explain why the T4 phage genome is circularly permuted.
2. Can you think of a way to simplify further the genomes of the ssRNA phages MS2 and Qβ?

Would it be possible to eliminate one of their genes? If so, which one?

3. No temperate RNA phages have yet been discovered. How might this absence be explained?

4. How might a bacterial cell resist phage infections? Give those mechanisms mentioned in the chapter and speculate on other possible strategies.

Critical Thinking Questions

1. The choice between lysogeny and lysis is influenced by many factors. How would external conditions such as starvation or crowding be "sensed" and communicated to the transcriptional machinery and influence this choice?

2. If you were a doctor charged with curing a bacterium of its viral infection, what target would you choose for "chemotherapy" and why?

3. We don't know exactly how double-stranded RNA is replicated. Propose two possible models, and design experiments that would distin-

guish between them.

4. The most straightforward explanation as to why the endolysin of T4 is expressed so late in infection is that its promoter is recognized by the gp55 alternative sigma factor. Propose a different explanation.

For recommended readings on these and related topics, see Appendix V.

chapter 18

The Viruses:
Viruses of Eucaryotes

Concepts

- Although the details differ, animal virus reproduction is similar to that of the bacteriophages in having the same series of phases: adsorption, penetration and uncoating, replication of virus nucleic acids, synthesis and assembly of capsids, and virus release.
- Viruses may harm their host cells in a variety of ways, ranging from direct inhibition of DNA, RNA, and protein synthesis to the alteration of plasma membranes and formation of inclusion bodies.
- Not all animal virus infections have a rapid onset and relatively short duration. Some viruses establish long-term chronic infections; others are dormant for a while and then become active again. Slow virus infections may take years to develop.
- Cancer can be caused by a number of factors, including viruses. Viruses may bring oncogenes into a cell, carry promoters that stimulate a cellular oncogene, or in other ways transform cells into tumor cells.
- Plant viruses are responsible for many important diseases but have not been intensely studied due to technical difficulties. Most are RNA viruses. Insects are the most important transmission agents, and some plant viruses can even multiply in insect tissues before being inoculated into another plant.
- Members of at least seven virus families infect insects; the most important belong to the *Baculoviridae*, *Reoviridae*, or *Iridoviridae*. Many insect infections are accompanied by the formation of characteristic inclusion bodies. A number of these viruses show promise as biological control agents for insect pests.
- Infectious agents simpler than viruses also exist. Viroids are short strands of infectious RNA responsible for several plant diseases. Prions are somewhat mysterious proteinaceous particles associated with certain degenerative neurological diseases in humans and livestock.

A model of the ribonuclease H component of reverse transcriptase that is complexed with an RNA-DNA hybrid (protein in yellow, DNA backbone in lavender, RNA in pink, bases in blue).

The Virus
Observe this virus: think how small
Its arsenal, and yet how loud its call;
It took my cell, now takes your cell,
And when it leaves will take our genes as well.
Genes that are master keys to growth
That turn it on, or turn it off, or both;
Should it return to me or you
It will own the skeleton keys to do
A number on our tumblers; stage a coup.

—*Michael Newman*

 n chapter 17 the bacteriophages are introduced in some detail because they are very important to the fields of molecular biology and genetics, as well as to virology. The present chapter focuses on viruses that use eucaryotic organisms as hosts. Although plant and insect viruses are discussed, particular emphasis is placed on animal viruses since they are very well studied and the causative agents of so many important human diseases. The chapter closes with a brief summary of what is known about infectious agents that are even simpler in construction than viruses: the viroids and prions.

The chapter begins with a discussion of animal viruses. These viruses are not only of great practical importance but are the best studied of the virus groups to be described in this chapter.

18.1 CLASSIFICATION OF ANIMAL VIRUSES

When microbiologists first began to classify animal viruses, they naturally thought in terms of features such as the host preferences of each virus. Unfortunately not all criteria are equally useful. For example, many viruses will infect a variety of animals, and a particular animal can be invaded by several dissimilar viruses. Thus virus host preferences lack the specificity to distinguish precisely among different viruses. Modern classifications are primarily based on virus morphology, the physical and chemical nature of virion constituents, and genetic relatedness.

Morphology is probably the most important characteristic in virus classification. Animal viruses can be studied with the transmission electron microscope while still in the host cell or after release. As mentioned in chapter 16, the nature of virus nucleic acids is also extremely important. Nucleic acid properties such as the general type (DNA or RNA), strandedness, size, and segmentation are all useful. Genetic relatedness can be estimated by techniques such as nucleic acid hybridization, nucleic acid and protein sequencing, and by determining the ability to undergo recombination.

A brief diagrammatic description of DNA and RNA animal virus classification is presented in **figures 18.1** and **18.2. Figure 18.3** summarizes pictorially much of the same material.

1. List the most important characteristics used in identifying animal viruses (figures 18.1, 18.2, and 18.3). What other properties can be used to establish the relatedness of different viruses?

18.2 REPRODUCTION OF ANIMAL VIRUSES

The reproduction of animal viruses is very similar in many ways to that of phages. Animal virus reproduction may be divided into several stages: adsorption, penetration and uncoating, replication of virus nucleic acids, synthesis and assembly of virus capsids, and release of mature viruses. Each of these stages will be briefly described.

Adsorption of Virions

The first step in the animal virus cycle is adsorption to the host cell surface. This occurs through a random collision of the virion with a plasma membrane receptor site. The receptor is most often a protein, and frequently a glycoprotein. Several glycoprotein receptors have the sugar derivative sialic acid attached, which the virus recognizes. Sometimes a complex carbohydrate such as heparan sulfate is the receptor. Because the capacity of a virus to infect a cell depends greatly on its ability to bind to the cell, the distribution of these receptors plays a crucial role in the tissue and host specificity of animal viruses. For example, poliovirus receptors are found only in the human nasopharynx, gut, and spinal cord anterior horn cells; in contrast, measles virus receptors are present in most tissues. The dissimilarity in the distribution of receptors for these two viruses helps explain the difference in the nature of polio and measles.

The specific host cell receptors to which viruses attach vary greatly but they are always surface molecules necessary to the cell. Receptor distribution often varies over the surface of target cell plasma membranes. Cells have microdomains called lipid rafts that seem to be involved in both virion entrance and assembly. For example, the receptors for enveloped viruses such as the HIV and Ebola viruses are concentrated in the rafts. As will be discussed shortly, viruses often enter cells by endocytosis. They trick the host cell by attaching to surface molecules that are normally taken up by endocytosis. Thus they are passively carried into the cell. These host cell surface proteins usually are receptors that bind hormones and other important molecules essential to the cell's function and role in the body (**table 18.1**). Many host receptors are members of the immunoglobulin superfamily, a group of molecules that contain immunoglobulin domains (*see p. 712*). Most members of the immunoglobulin superfamily are surface proteins that are involved in the immune response and cell-cell interactions. Examples are the HIV CD4 receptor, the rhinovirus receptor, and the poliovirus ICAM (*inter*cellular *a*dhesion *m*olecule) receptor. In some cases, it appears that two or more host cell receptors are involved in attachment. Herpes simplex virus interacts

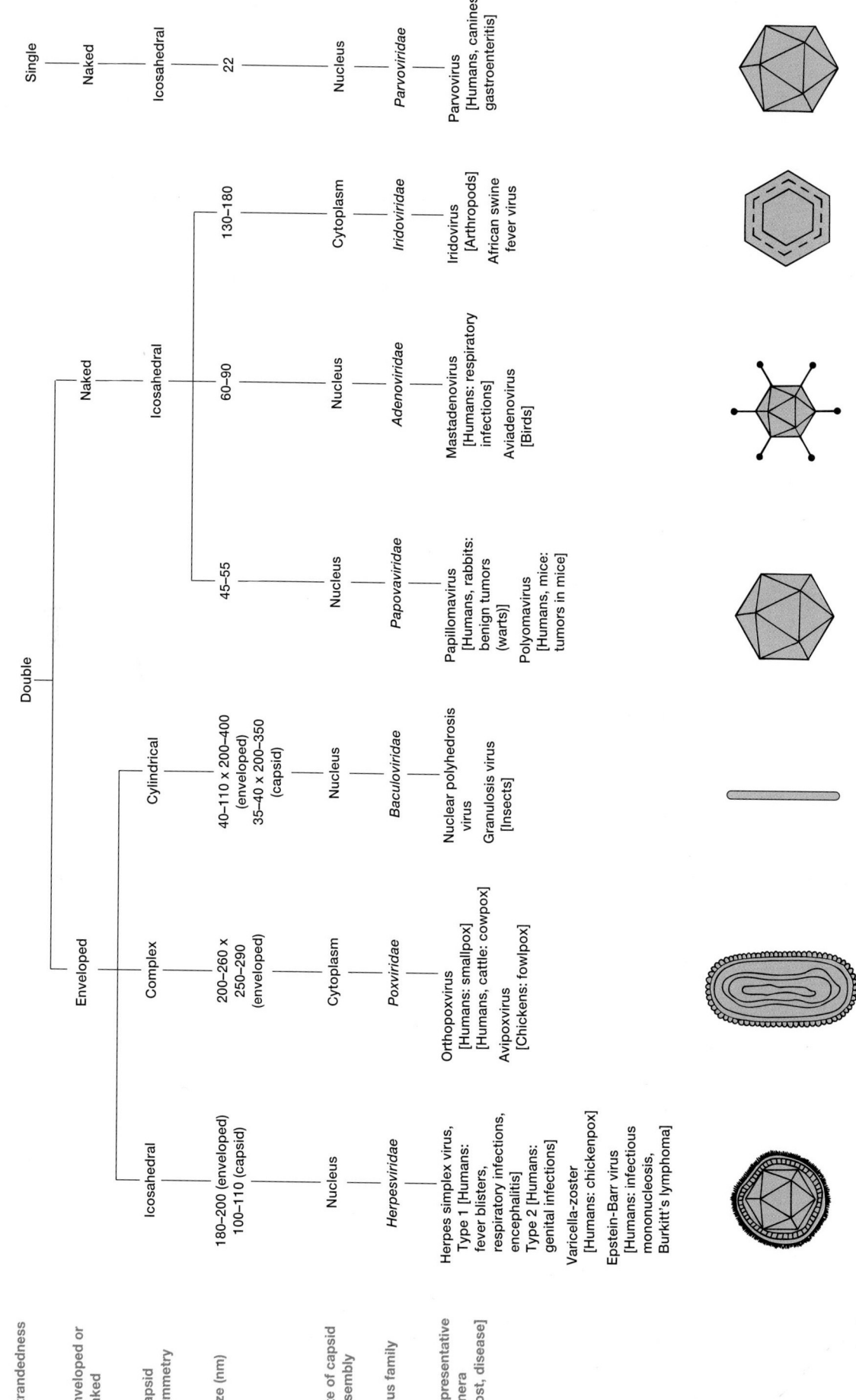

Figure 18.1 The Taxonomy of DNA Animal Viruses.

Strandedness	Double	Single									
Single-strand sense		Positive					Negative				
						DNA intermediate during replication					
Enveloped or naked	Naked	Enveloped	Enveloped	Naked	Naked	Enveloped	Enveloped	Enveloped	Enveloped	Enveloped	Enveloped
Capsid symmetry	Icosahedral	Icosahedral	Helical	Icosahedral	Icosahedral	Icosahedral	Helical	Helical	Helical	?	Helical
Size (nm)	75–80	40–75 (enveloped) 25–35 (capsid)	80–160 (enveloped) 14–16 (capsid diameter)	22–30	35–40	80–100 (enveloped)	80–120 (enveloped) 9 (capsid diam.)	70–80 x 130–240 (bullet-shaped)	90–100 (enveloped)	50–300	125–250 (enveloped) 18 (capsid diam.)
Site of capsid assembly	Cytoplasm	Cytoplasm	Cytoplasm	Cytoplasm	Cytoplasm	Cytoplasm	Cytoplasm	Cytoplasm	Cytoplasm	Cytoplasm	Cytoplasm
Virus family	*Reoviridae*	*Togaviridae*	*Coronaviridae*	*Picornaviridae*	*Calciviridae*	*Retroviridae*	*Orthomyxoviridae*	*Rhabdoviridae*	*Bunyaviridae*	*Arenaviridae*	*Paramyxoviridae*
Representative genera and groups [Host, disease]	Orbivirus [Humans: encephalitis] Rotavirus [Humans: diarrhea] Cypovirus [Insects: cytoplasmic polyhedrosis virus]	Alphavirus [Humans: encephalitis] Flavivirus [Humans: yellow fever, dengue] Rubivirus [Humans: rubella] Pestivirus [Pigs: hog cholera]	Infectious bronchitis virus [Humans: upper respiratory infection]	Enterovirus [Humans: polio] Rhinovirus [Humans: common cold] Hepatovirus [Humans: hepatitis A]	Vesicular exanthema of swine Norwalk virus Hepatitis E virus	Oncornavirus C [Birds, mice: sarcomas, leukemias] Human T-cell leukemia virus Human immunodeficiency virus	Influenza virus [Humans, swine]	Lyssavirus [Warm-blooded animals: rabies]	California encephalitis virus [Humans]	Lassa virus [Humans: hemorrhagic fever]	Paramyxovirus [Humans: colds, respiratory infections, mumps] Pneumovirus [Humans: pneumonia, common cold] Morbillivirus [Humans: measles]

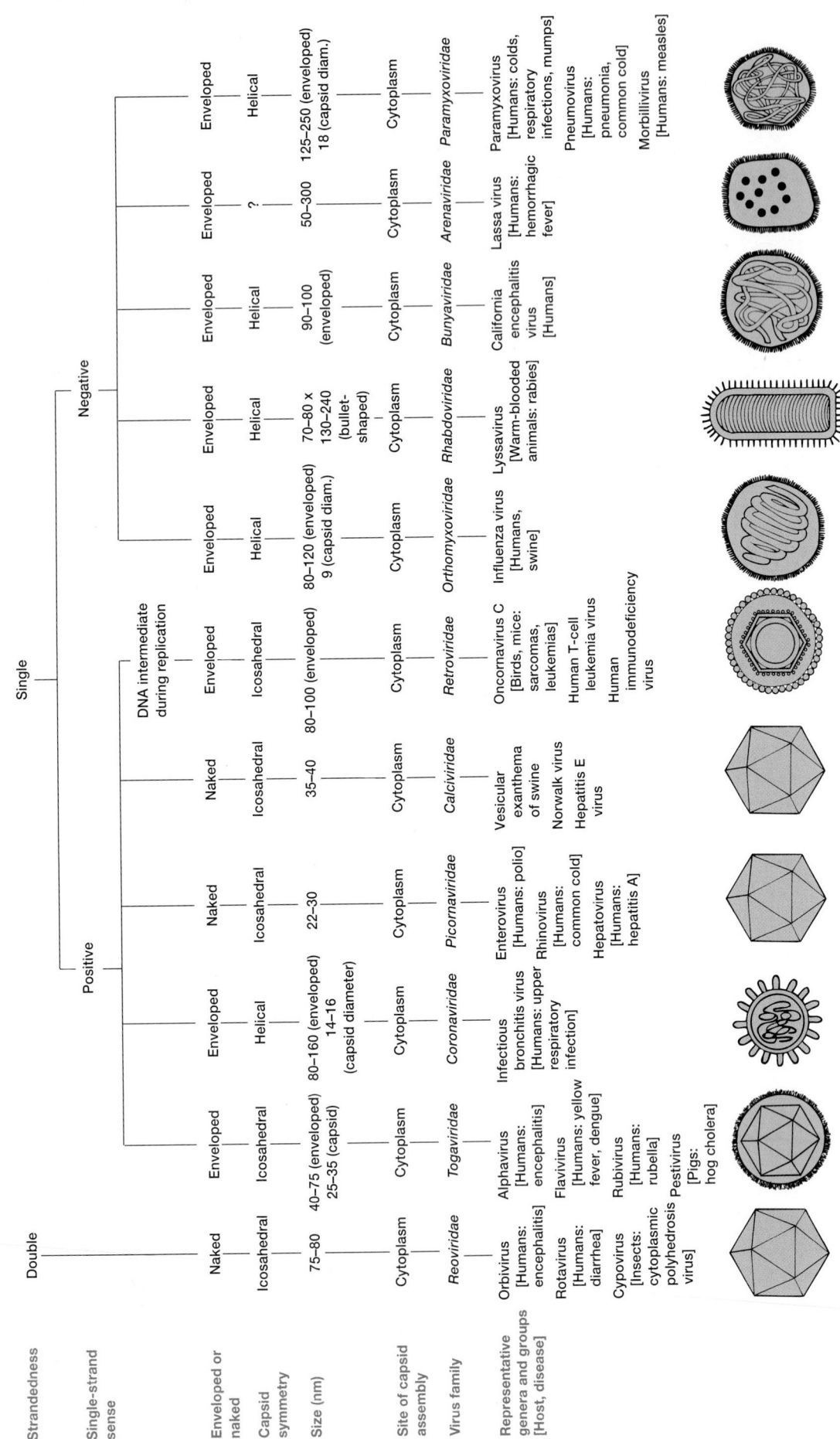

Figure 18.2 The Taxonomy of RNA Animal Viruses.

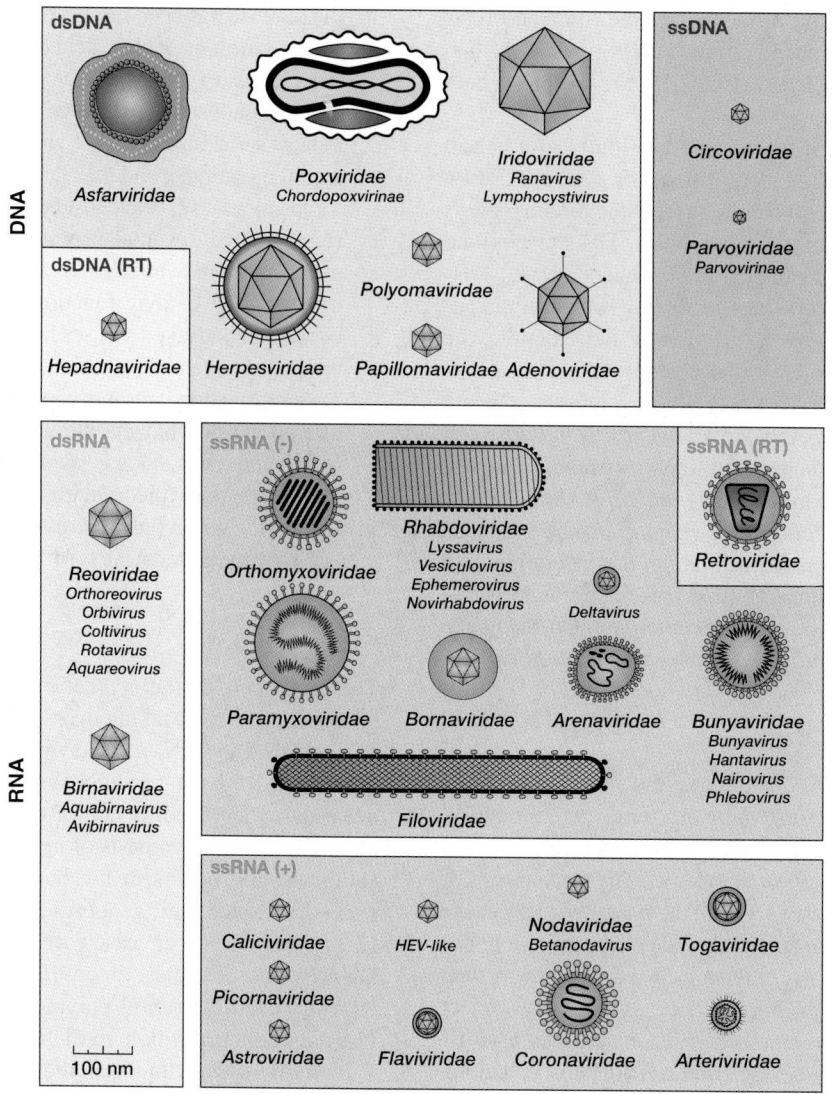

Figure 18.3   **A Diagrammatic Description of the Families and Genera of Viruses That Infect Vertebrates.**   RT stands for reverse transcriptase.

Table 18.1

Examples of Host Cell Surface Proteins That Serve as Virus Receptors

Virus	Cell Surface Protein
Adenovirus	Coxsackie adenovirus receptor (CAR) protein
Epstein-Barr virus	Receptor for the C3d complement protein on human B lymphocytes
Hepatitis A virus	Alpha 2-macroglobulin
Herpes simplex virus, type 1	Heparan sulfate
Human immunodeficiency virus	CD4 protein on T-helper cells, macrophages, and monocytes; CXCR-4 or the CCR5 receptor
Influenza A virus	Sialic acid–containing glycoprotein
Measles virus	CD46 complement regulator protein
Poliovirus	Immunoglobulin superfamily
Rabies virus	Acetylcholine receptor on neurons
Rhinovirus	Intercellular adhesion molecules (ICAMs) on the surface of respiratory epithelial cells
Reovirus, type 3	β-adrenergic receptor
Rotavirus	Acetylated sialic acid on glycoprotein
Vaccinia virus	Epidermal growth factor receptor

with a glycosaminoglycan and a member of the tumor necrosis factor/nerve growth factor receptor family. HIV uses CD4 and CXCR-4 (fusin) or the CCR5 (CC-CKR-5) receptor, both chemokine receptors.   Lipid rafts (p. 45)

The surface site on the virus can consist simply of a capsid structural protein or an array of such proteins. In some viruses—for example, the poliovirus and rhinoviruses—the binding site is at the bottom of a surface depression or valley. The site can bind to host cell surface projections but cannot be reached by host antibodies. Envelope glycoproteins also can be involved in the adsorption and penetration of enveloped viruses. For example, the herpes simplex virus has two envelope glycoproteins that are required for attachment, and at least four glycoproteins participate in penetration. In other cases the virus attaches to the host cell through special projections such as the fibers extending from the corners of adenovirus icosahedrons (*see figure 16.12*h) or the spikes of enveloped viruses. As mentioned in chapter 16, the influenza virus has two kinds of spikes: hemagglutinin and neuraminidase (*see figure 16.17*a,b). The hemagglutinin spikes appear to be involved in attachment to the host cell receptor site and recognize sialic acid (*N*-acetylneuraminic acid).   Human virus diseases (chapter 38)

Penetration and Uncoating

Viruses penetrate the plasma membrane and enter a host cell shortly after adsorption. Virus uncoating, the removal of the capsid and release of viral nucleic acid, occurs during or shortly after penetration. Both events will be reviewed together. The mechanisms of penetration and uncoating must vary with the type of virus because viruses differ so greatly in structure and mode of reproduction. For example, enveloped viruses may enter cells in a different way than naked or unenveloped virions. Furthermore, some viruses inject only their nucleic acid, whereas others must ensure that a virus-associated RNA or DNA polymerase also enters the host cell along with the virus genome. The entire process from adsorption to final uncoating may take from minutes to several hours.

The detailed mechanisms of penetration and uncoating are still often unclear; it appears that one of two different modes of entry are employed by most viruses (**figure 18.4**).

1. The envelopes of paramyxoviruses, members of the *Retroviridae,* and possibly some other viruses, fuse directly with the host cell plasma membrane. Fusion may involve special envelope fusion glycoproteins that bind to plasma membrane proteins. After attachment of paramyxoviruses, several things happen: membrane lipids rearrange, the adjacent halves of the contacting membranes merge, and a proteinaceous fusion pore forms. Finally the nucleocapsid enters the host cell cytoplasmic matrix, where uncoating is completed. A virus polymerase, associated with the nucleocapsid, begins transcribing the virus RNA while it is still within the capsid.
2. Nonenveloped viruses and other enveloped viruses enter cells by endocytosis. They may be engulfed by receptor-mediated endocytosis to form coated vesicles. The virions attach to coated pits, and the pits then pinch off to form coated vesicles filled with viruses. These vesicles fuse with lysosomes after

the clathrin has been removed; depending on the virus, escape of the virion or nucleic acid may occur either before or after vesicle fusion. Lysosomal enzymes can aid in virus uncoating and low endosomal pHs often trigger the uncoating process. In at least some instances, the virus envelope fuses with the lysosomal membrane, and the nucleocapsid is released into the cytoplasmic matrix (the capsid proteins may have been partially removed by lysosomal enzymes). Once in the cytoplasm, viral nucleic acid may be released from the capsid upon completion of uncoating or may function while still attached to capsid components. Naked viruses lack an envelope and thus cannot employ the membrane fusion mechanism. It appears that vesicle acidification causes a capsid conformational change. The altered capsid contacts the vesicle membrane and either releases the virus nucleic acid into the cytoplasm through a membrane pore (picornaviruses) or ruptures the membrane to release the virion (adenovirus). Viruses may also enter the host cell by way of caveolae formation.   Endocytosis (pp. 79–80)

Replication and Transcription in DNA Viruses

The early part of the synthetic phase, governed by the **early genes,** is devoted to taking over the host cell and to the synthesis of viral DNA and RNA. Some virulent animal viruses inhibit host cell DNA, RNA, and protein synthesis, though cellular DNA is not usually degraded. In contrast, nonvirulent viruses may actually stimulate the synthesis of host macromolecules. DNA replication usually occurs in the host cell nucleus; poxviruses are exceptions since their genomes are replicated in the cytoplasm. Messenger RNA—at least early mRNA—is transcribed from DNA by host enzymes, except for poxvirus early mRNA, which is synthesized by a viral polymerase. Some examples of DNA virus reproduction will help illustrate these generalizations.

Parvoviruses, with a genome composed of one small, single-stranded DNA molecule about 4,800 bases in size, are the simplest of the DNA viruses (*see figure 16.12*a). The genome is so small that it directs the synthesis of only three polypeptides, all capsid components. Even so, the virus must resort to the use of overlapping genes to fit three genes into such a small molecule. That is, the base sequences that code for the three polypeptide chains overlap each other and are read with different reading frames (*see section 11.5*). Since the genome does not code for any enzymes, the virus must use host cell enzymes for all biosynthetic processes. Thus viral DNA can only be replicated in the nucleus during the S period of the cell cycle, when the cell replicates its own DNA.   The genetic code and gene structure (pp. 234–39)

Herpesviruses are a large group of icosahedral, enveloped, double-stranded DNA viruses responsible for many important human and animal diseases (*see figure 16.17*e). The genome is a linear piece of DNA about 160,000 base pairs in size and contains at least 50 to 100 genes. Immediately upon uncoating, the DNA is transcribed by host RNA polymerase to form mRNAs directing the synthesis of several early proteins, principally regulatory proteins and the enzymes required for virus DNA replication (**figure 18.5, steps 1 and 2**). The DNA circularizes and replication with a virus-specific DNA polymerase begins in the cell nucleus within 4 hours

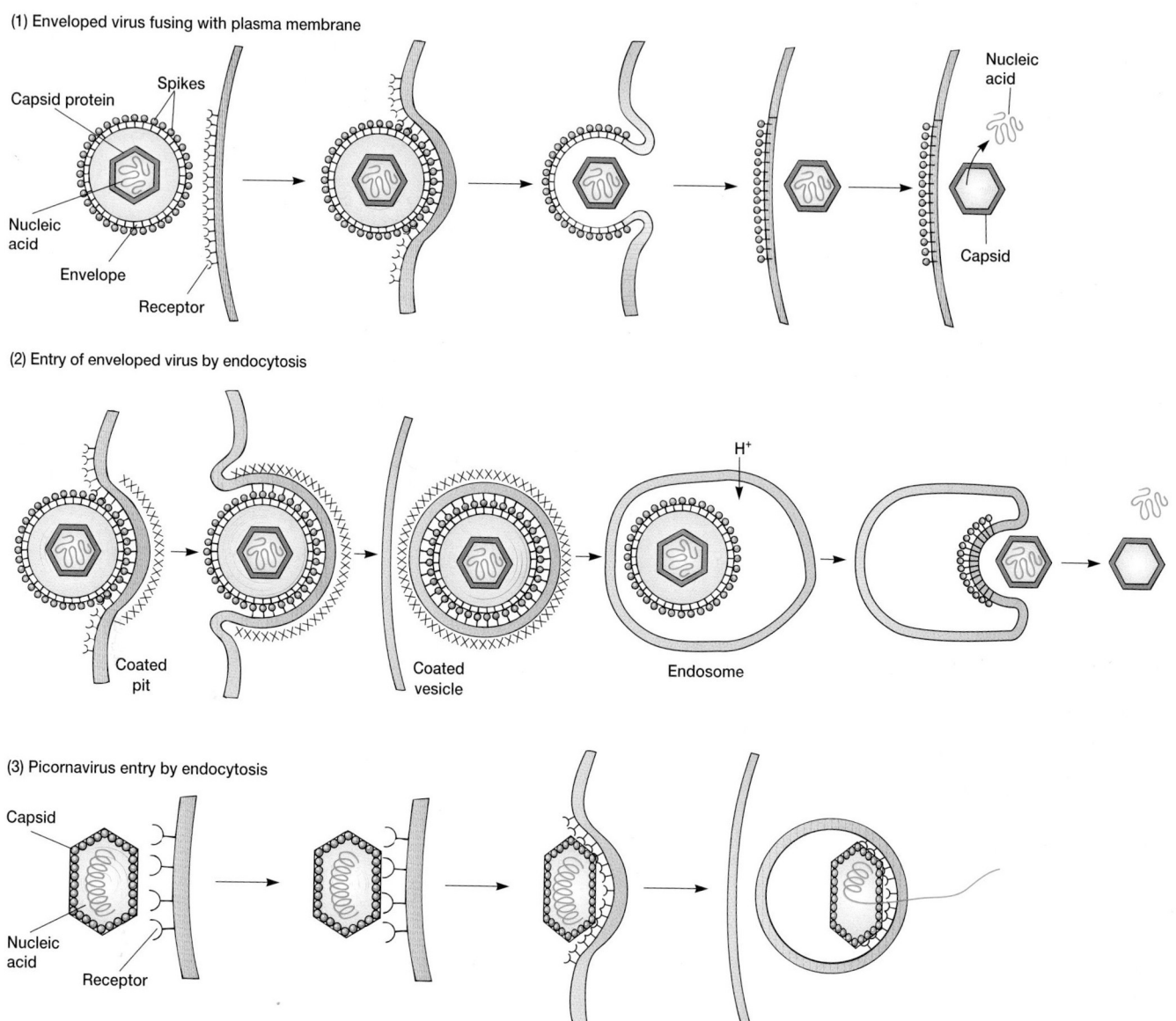

(1) Enveloped virus fusing with plasma membrane

Capsid protein
Spikes
Nucleic acid
Envelope
Receptor
Nucleic acid
Capsid

(2) Entry of enveloped virus by endocytosis

Coated pit
Coated vesicle
H^+
Endosome

(3) Picornavirus entry by endocytosis

Capsid
Nucleic acid
Receptor

Figure 18.4   **Animal Virus Entry.**   Examples of animal virus attachment and entry into host cells. Enveloped viruses can (*1*) enter after fusion of the envelope with the plasma membrane, or (*2*) escape from the vesicle after endocytosis. (*3*) Naked viruses such as poliovirus, a picornavirus, may be taken up by endocytosis and then insert their nucleic acid into the cytoplasm through the vesicle membrane. It also is possible that they insert the nucleic acid directly through the plasma membrane within a coated pit. See text for description of the entry modes.

after infection (step 3). Host DNA synthesis gradually slows during a lethal virus infection (not all herpes infections result in immediate cell death).   Herpesviruses and disease (pp. 846–47, 862–65)

Poxviruses such as the vaccinia virus are the largest viruses known and are morphologically complex (*see figure 16.18*). Their double-stranded DNA possesses over 200 genes. The virus enters through receptor-mediated endocytosis in coated vesicles; the central core escapes from the lysosome and enters the cytoplasmic matrix. The core contains both DNA and a DNA-dependent RNA polymerase that synthesizes early mRNAs, one of which directs the production of an enzyme that completes virus uncoating. DNA polymerase and other enzymes needed for DNA

replication are also synthesized early in the reproductive cycle, and replication begins about 1.5 hours after infection. About the time DNA replication commences, late mRNA transcription is initiated. Many late proteins are structural proteins used in capsid construction. The complete reproductive cycle in poxviruses takes about 24 hours.   Poxvirus diseases (pp. 851–52)

The hepadnaviruses such as hepatitis B virus are quite different from other DNA viruses with respect to genome replication. They have circular dsDNA genomes but replicate their DNA using the enzyme reverse transcriptase (p. 397). After infecting a cell, their DNA is released in the nucleus. Transcription occurs in the nucleus using host RNA polymerase and yields several mRNAs, including

Figure 18.5 Generalized Life Cycle for Herpes Simplex Virus Type I. Only one possible pathway for release of the virions is shown. See text for details.

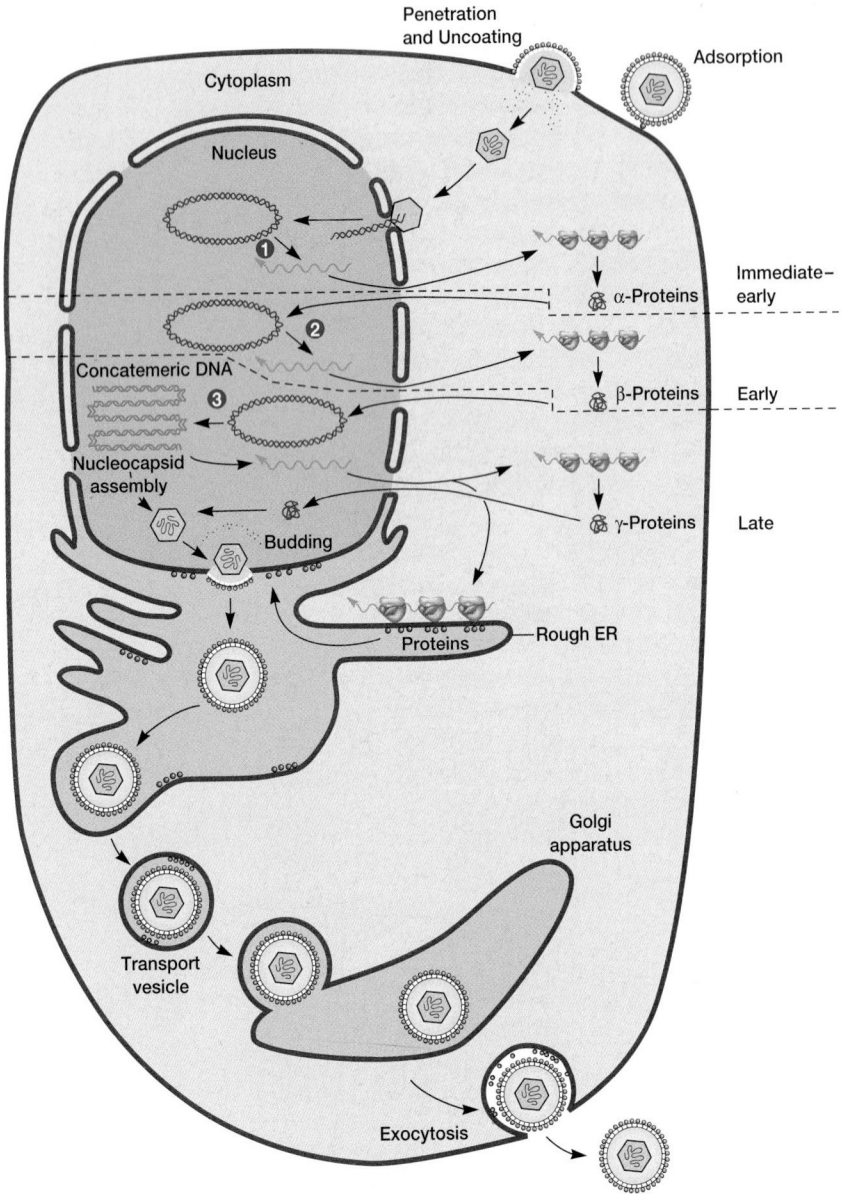

a large 3.4 kilobase RNA known as the pregenome. The RNAs move to the cytoplasm and are translated to produce virus proteins such as core proteins and a polymerase having three activities (DNA polymerase, reverse transcriptase, RNase H). Then the RNA pregenome associates with DNA polymerase and core protein to form an immature core particle. Reverse transcriptase subsequently transcribes the RNA using a protein primer to form a −DNA copy of the pregenome +RNA. After almost all the pregenome RNA has been degraded by RNase H, the remaining RNA fragment serves as a primer for DNA polymerase to copy the minus DNA and form a dsDNA genome. Finally the nucleocapsid is completed.   Hepatitis B virus and disease (pp. 866–67)

Replication and Transcription in RNA Viruses

The RNA viruses are much more diverse in their reproductive strategies than are the DNA viruses. Most RNA viruses can be placed in one of four general groups based on their modes of replication and transcription, and their relationship to the host cell genome. **Figure 18.6** summarizes the reproductive cycles characteristic of these groups. Transcription patterns are discussed first, and then mechanisms of RNA replication.

Transcription in RNA viruses other than the retroviruses (retroviruses are considered shortly) varies with the nature of the virus genome. The picornaviruses such as poliovirus are the best studied positive strand ssRNA viruses (**Techniques & Applications 18.1**). They use their RNA genome as a giant mRNA, and host ribosomes synthesize an enormous peptide that is then cleaved or processed by both host and viral encoded enzymes to form the proper polypeptides (figure 18.6a). In contrast, because their genome is complementary to the mRNA base sequence, negative strand ssRNA viruses (e.g., the orthomyxoviruses and paramyxoviruses) must employ a virus-associated RNA-dependent RNA polymerase or **transcriptase** to synthesize mRNA (figure 18.6c and **figure**

(a) Positive single-stranded RNA viruses (picornaviruses)

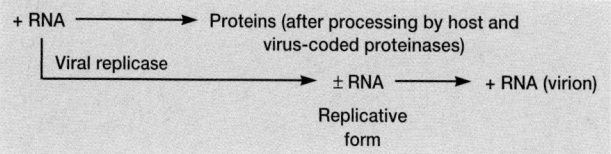

(b) Double-stranded RNA viruses (reoviruses)

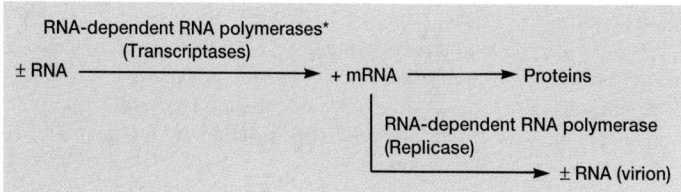

(c) Negative single-stranded RNA viruses (paramyxoviruses—mumps and measles;
orthomyxoviruses—influenza)

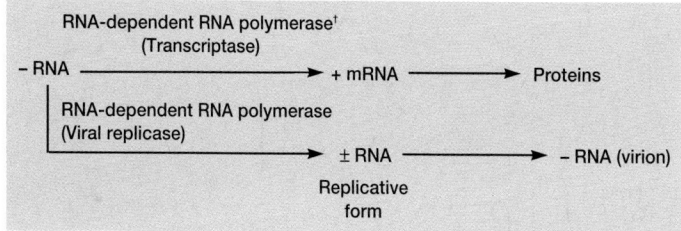

(d) Retroviruses (Rous sarcoma virus, HIV)

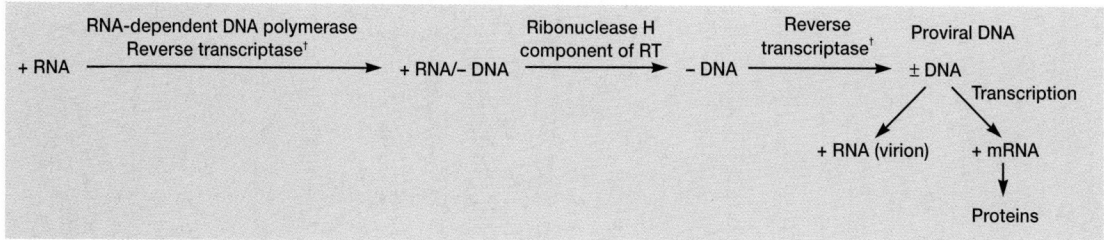

* Virus associated and newly synthesized enzymes

† Virus associated

Figure 18.6   RNA Animal Virus Reproductive Strategies.   Examples of viruses using the four strategies are given in parentheses.

18.7, step 1). The dsRNA reoviruses carry a virus-associated RNA-dependent RNA polymerase that copies the negative strand of their genome to generate mRNA. Later a new, virus-encoded polymerase continues transcription.

The nature of RNA replication also varies logically with the type of virus genetic material. Viral RNA is replicated in the host cell cytoplasmic matrix. Single-stranded RNA viruses, except retroviruses, use a viral **replicase** that converts the ssRNA into a double-stranded RNA called the **replicative form** (figure 18.6a,c). The appropriate strand of this intermediate then directs the synthesis of new viral RNA genomes (figure 18.7, step 2). Another way of looking at this process is that replication is governed by the principle of complementarity. The parental genome strand directs the synthesis of a complementary strand, which then serves as a template for the synthesis of new progeny virus

genomes. For example, picornaviruses have a +RNA genome and must make an intermediate −RNA complementary copy in order to produce new +RNA genomes. Reoviruses differ significantly from this pattern (figure 18.6b). The virion contains 10 to 13 different dsRNAs, each coding for an mRNA. Late in the reproductive cycle, a copy of each mRNA associates with the other mRNAs and special proteins to form a large complex. The RNAs in this complex are then copied by the viral replicase to form a double-stranded genome that is incorporated into a new virion. Reoviruses, orthomyxoviruses, and paramyxoviruses normally use a single polymerase for replication and transcription. Its mode of action depends on associated proteins and other factors.

Retroviruses such as the human immunodeficiency virus (*see figure 16.17*d) possess ssRNA genomes but differ from other RNA viruses in that they synthesize mRNA and replicate their

Recently the poliovirus has been constructed completely from scratch beginning with the base sequence of its RNA genome. Because DNA is easier to synthesize than RNA, a DNA copy of the poliovirus RNA genome attached to the promoter for T7 phage RNA polymerase was first produced using DNA synthesizer machines (*see p. 315*). The synthetic DNA was then incubated with the T7 phage RNA polymerase and the appropriate nucleotides. The polymerase used the DNA template to form complete RNA copies of the poliovirus genome. The RNA genome copies were then incubated with a cell-free cytoplasmic extract of HeLa cells, which supplied the constituents necessary for protein synthesis. The RNA directed the synthesis of poliovirus proteins; the proteins and RNA then spontaneously assembled into complete, infectious poliovirus virions. These particles could infect a special mouse strain and cause a disease that resembled human poliomyelitis.

Some scientists believe that it will be possible to make almost any virus once the genome sequence is known. This could be a particular threat if bioterrorists managed to create smallpox. Others believe that it will not be that easy to create a virulent pathogen like smallpox. The smallpox genome is much larger than that of poliovirus and would be harder to synthesize. Unlike the case with poliovirus, smallpox DNA is not infectious by itself and requires the presence of virion proteins such as polymerases. Thus one would have to find a way to supply these enzymes, perhaps by using helper viruses. A potentially greater threat is the synthesis of special hybrid strains that are very infectious and also quite lethal. On a more positive note, it may be possible to construct substantially weakened strains of viruses such as poliovirus for the production of more effective vaccines.

Figure 18.7   Simplified Life Cycle of Influenza Virus.
Several steps have been eliminated for simplicity and clarity. Transcription (step 1) requires a process called cap snatching in which host RNAs are cleaved about 10 to 13 nucleotides from the 5′ end and the capped fragments are used as primers for the synthesis of viral mRNA. Abbreviations: PB1, RNA polymerase; NP, nucleocapsid protein; HA, hemagglutinin; and NA, neuraminidase.

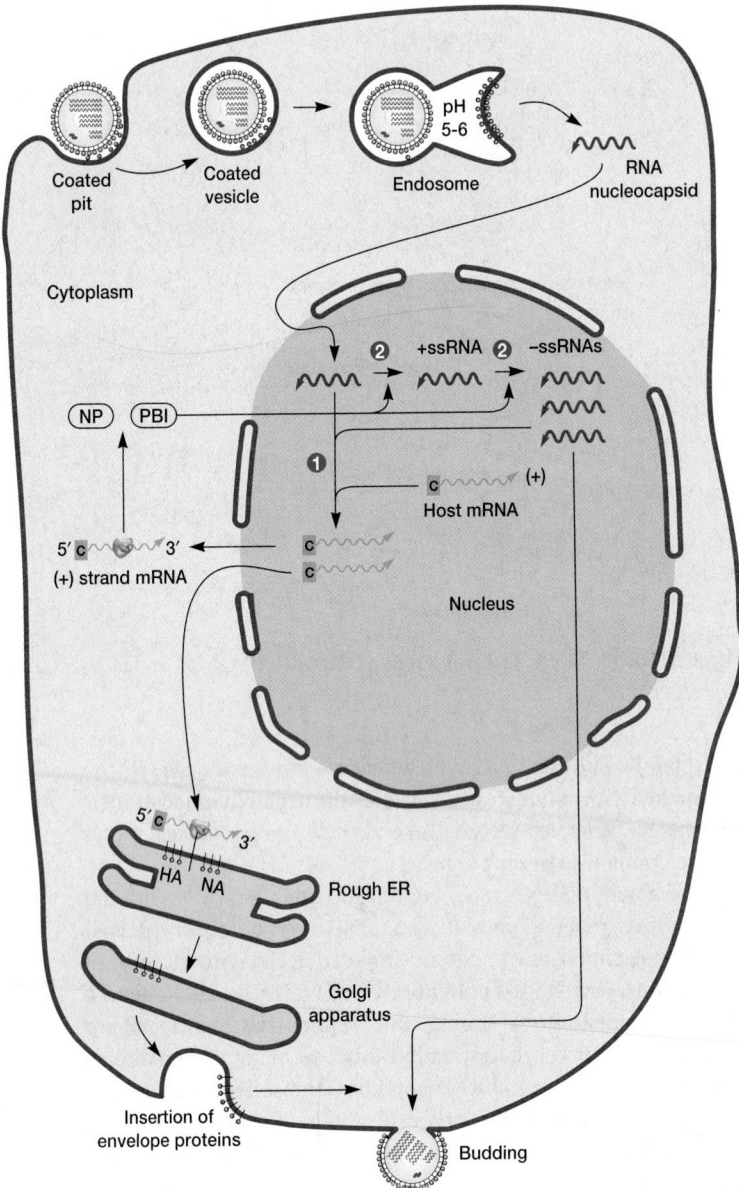

Table 18.2

Intracellular Sites of Animal Virus Reproduction

Virus	Nucleic Acid Replication	Capsid Assembly	Membrane Used in Budding
DNA Viruses			
Adenoviruses	Nucleus	Nucleus	
Hepadnaviruses	Cytoplasm	Cytoplasm	Endoplasmic reticulum
Herpesviruses	Nucleus	At nuclear membrane	Nucleus
Papillomaviruses	Nucleus	Nucleus	
Parvoviruses	Nucleus	Nucleus	
Polyomaviruses	Nucleus	Nucleus	
Poxviruses	Cytoplasm	Cytoplasm	
RNA Viruses			
Coronaviruses	Cytoplasm	Cytoplasm	Golgi apparatus and endoplasmic reticulum
Orthomyxoviruses	Nucleus	Cytoplasm	Plasma membrane
Paramyxoviruses	Cytoplasm	Cytoplasm	Plasma membrane
Picornaviruses	Cytoplasm	Cytoplasm	
Reoviruses	Cytoplasm	Cytoplasm	
Retroviruses	Cytoplasm and nucleus	At plasma membrane	Plasma membrane
Rhabdoviruses	Cytoplasm	Cytoplasm	Plasma membrane, intracytoplasmic membranes
Togaviruses	Cytoplasm	Cytoplasm	Plasma membrane, intracytoplasmic membranes

genome by means of DNA intermediates. The virus has an RNA-dependent DNA polymerase or **reverse transcriptase (RT)** that copies the +RNA genome to form a −DNA copy (chapter opening figure and figure 18.6*d; see also figure 38.9*). Interestingly, transfer RNA is carried by the virus and serves as the primer required for nucleic acid synthesis (*see section 11.3*). The transformation of RNA into DNA takes place in two steps. First, reverse transcriptase copies the +RNA to form a RNA-DNA hybrid. Then the **ribonuclease H** component of reverse transcriptase degrades the +RNA strand to leave −DNA. After synthesizing −DNA, the reverse transcriptase copies this strand to produce a double-stranded DNA called **proviral DNA,** which can direct the synthesis of mRNA and new +RNA virion genome copies. Notice that during this process genetic information is transferred from RNA to DNA rather than in the normal direction.

The reproduction of retroviruses is remarkable in other ways as well. After proviral DNA has been manufactured, it is converted to a circular form and incorporated or integrated into the host cell chromosome. Virus products are only formed after integration. Sometimes these integrated viruses can change host cells into tumor cells.   Biology of the HIV virus and AIDS (pp. 855–60)

Synthesis and Assembly of Virus Capsids

Some **late genes** direct the synthesis of capsid proteins, and these spontaneously self-assemble to form the capsid just as in bacteriophage morphogenesis. Recently the self-assembly process has been dramatically demonstrated (Techniques & Applications 18.1). It appears that during icosahedral virus assembly empty **procapsids** are first formed; then the nucleic acid is inserted in some unknown way. The site of morphogenesis varies with the virus (**table 18.2**). Large paracrystalline clusters of complete viri-

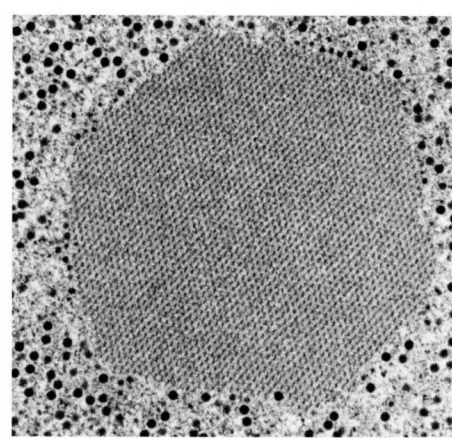

Figure 18.8 Paracrystalline Clusters. A crystalline array of adenoviruses within the cell nucleus (×35,000).

ons or procapsids are often seen at the site of virus maturation (**figure 18.8**). The assembly of enveloped virus capsids is generally similar to that of naked virions, except for poxviruses. These are assembled in the cytoplasm by a lengthy, complex process that begins with the enclosure of a portion of the cytoplasmic matrix through construction of a new membrane. Then newly synthesized DNA condenses, passes through the membrane, and moves to the center of the immature virus. Nucleoid and elliptical body construction takes place within the membrane.

Virion Release

Mechanisms of virion release differ between naked and enveloped viruses. Naked virions appear to be released most often by host cell

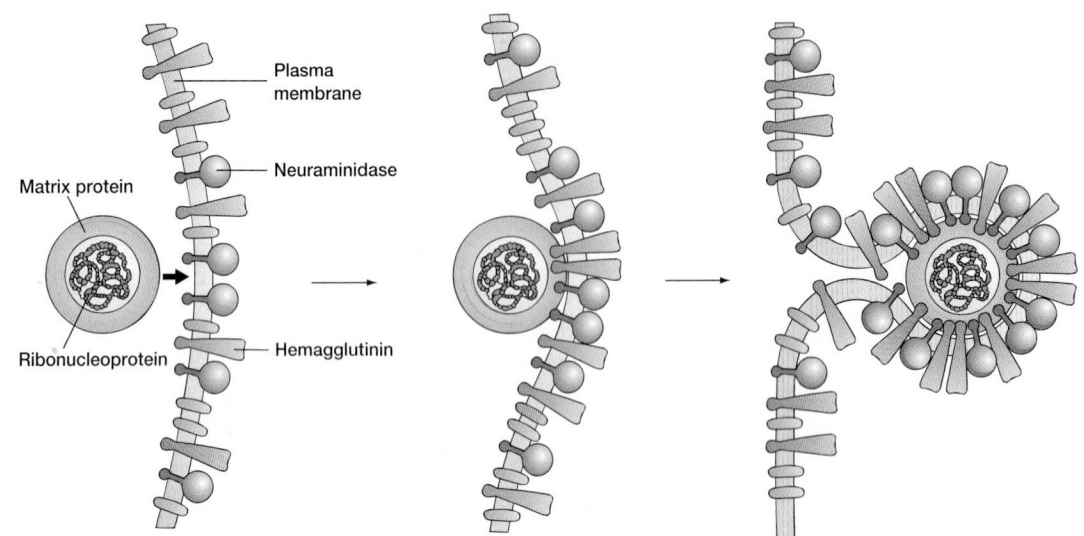

Figure 18.9   Release of Influenza Virus by Plasma Membrane Budding.   First, viral envelope proteins (hemagglutinin and neuraminidase) are inserted into the host plasma membrane. Then the nucleocapsid approaches the inner surface of the membrane and binds to it. At the same time viral proteins collect at the site and host membrane proteins are excluded. Finally, the plasma membrane buds to simultaneously form the viral envelope and release the mature virion.

lysis. In contrast, the formation of envelopes and the release of enveloped viruses are usually concurrent processes, and the host cell may continue virion release for some time. First, virus-encoded proteins are incorporated into the plasma membrane. Then the nucleocapsid is simultaneously released and the envelope formed by membrane budding (**figures 18.9** and **18.10**). In several virus families, a special M protein or matrix protein attaches to the plasma membrane and aids in budding. Although most envelopes arise from an altered plasma membrane, in herpesviruses, budding and envelope formation usually involves the nuclear envelope (table 18.2). The endoplasmic reticulum, Golgi apparatus, and other internal membranes also can be used to form envelopes.

Interestingly, it has been discovered that actin filaments can aid in virion release. Many viruses alter the actin microfilaments of the host cell cytoskeleton (*see p. 78*). For example, vaccinia virus appears to form long actin tails and use them to move intracellularly at up to 2.8 μm per minute. The actin filaments also propel vaccinia through the plasma membrane. In this way the virion escapes without destroying the host cell and infects adjacent cells.

1. Describe in detail each stage in animal virus reproduction and contrast it with the same stage in the bacteriophage life cycle (*see chapter 17*).
2. What probably plays the most important role in determining the tissue and host specificity of animal viruses?
3. Discuss how animal viruses can enter host cells.
4. Describe in general terms the replication and transcription processes that take place during animal virus reproduction.
5. Summarize the ways in which picornaviruses, -ssRNA viruses, and reoviruses carry out transcription and replication.

6. Outline in detail the retrovirus life cycle. What is proviral DNA?
7. In what ways are animal viruses released from their host cells?
8. Define the following terms: uncoating, coated vesicles, overlapping genes, replicative form, reverse transcriptase, and late genes.

18.3   CYTOCIDAL INFECTIONS AND CELL DAMAGE

An infection that results in cell death is a **cytocidal infection.** Animal viruses can harm their host cells in many ways; often this leads to cell death. As mentioned earlier (*see section 16.3*), microscopic or macroscopic degenerative changes or abnormalities in host cells and tissues are referred to as cytopathic effects (CPEs), and these often result from virus infections. Seven possible mechanisms of host cell damage are briefly described here.

1. Many viruses can inhibit host DNA, RNA, and protein synthesis. Cytocidal viruses (e.g., picornaviruses, herpesviruses, and adenoviruses) are particularly active in this regard. The mechanisms of inhibition are not yet clear.
2. Cell lysosomes may be damaged, resulting in the release of hydrolytic enzymes and cell destruction.
3. Virus infection can drastically alter plasma membranes through the insertion of virus-specific proteins so that the infected cells are attacked by the immune system. When in-

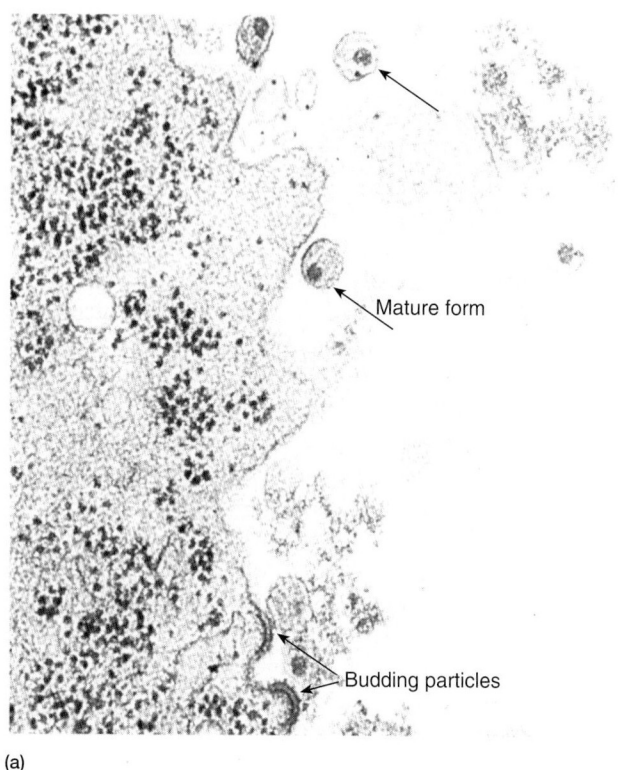

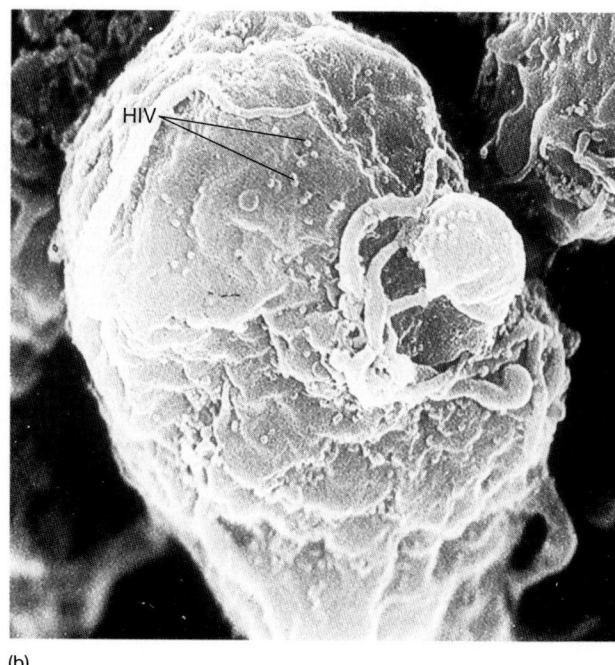

(a)   (b)

Figure 18.10   Human Immunodeficiency Virus (HIV) Release by Plasma Membrane Budding.   (**a**) A transmission electron micrograph of HIV particles beginning to bud, as well as some mature particles. (**b**) A scanning electron micrograph view of HIV particles budding from a lymphocyte.

fected by viruses such as herpesviruses and measles virus, as many as 50 to 100 cells may fuse into one abnormal, giant, multinucleate cell or syncytium called a polykaryocyte. The HIV virus, which causes AIDS (*see chapter 38*), appears to destroy CD4$^+$ T-helper cells at least partly through its effects on the plasma membrane.

4. High concentrations of proteins from several viruses (e.g., mumps virus and influenza virus) can have a direct toxic effect on cells and organisms.

5. Intracellular structures called **inclusion bodies** are formed during many virus infections. These may result from the clustering of subunits or virions within the nucleus or cytoplasm (e.g., the Negri bodies in rabies infections); they also may contain cell components such as ribosomes (arenavirus infections) or chromatin (herpesviruses). Regardless of their composition, these inclusion bodies can directly disrupt cell structure.

6. Chromosomal disruptions result from infections by herpesviruses and others.

7. Finally, the host cell may not be directly destroyed but transformed into a malignant cell. This will be discussed later in the chapter.

It should be emphasized that more than one of these mechanisms may be involved in a cytopathic effect.

18.4   PERSISTENT, LATENT, AND SLOW VIRUS INFECTIONS

Many virus infections (e.g., influenza) are **acute infections**—that is, they have a fairly rapid onset and last for a relatively short time. However, some viruses can establish **persistent infections** lasting many years. There are several kinds of persistent infections. In **chronic virus infections,** the virus is almost always detectable and clinical symptoms may be either mild or absent for long periods. Examples are hepatitis B virus (serum hepatitis) and human immunodeficiency virus. In **latent virus infections** the virus stops reproducing and remains dormant for a period before becoming active again. During latency, no symptoms, antibodies, or viruses are detectable. Examples are herpes simplex virus, varicella-zoster virus, cytomegalovirus, and Epstein-Barr virus. Herpes simplex type 1 virus often infects children and then becomes dormant within the nervous system ganglia; years later it can be activated to cause cold sores. The varicella-zoster virus causes chicken pox in children and then, after years of inactivity, may produce the skin disease shingles (initial adult infections result in chicken pox). These and other examples of such infections are discussed in more detail in chapter 38.

The causes of persistence and latency are probably multiple, although the precise mechanisms are still unclear. The virus genome may be integrated into the host genome. Viruses may become less antigenic and thus less susceptible to attack by the immune system. They may mutate to less virulent and slower reproducing forms. Sometimes a deletion mutation (*see section 11.6*) produces a **defective interfering (DI) particle** that cannot reproduce but slows normal virus reproduction, thereby reducing host damage and establishing a chronic infection.

A small group of viruses causes extremely slowly developing infections, often called **slow virus diseases** or slow infections, in which symptoms may take years to emerge. Measles virus occasionally produces a slow infection. A child may have a normal case of measles, then 5 to 12 years later develop a degenerative brain disease called subacute sclerosing panencephalitis (SSPE). Lentiviruses such as the human immunodeficiency virus also cause slow diseases. Many slow viruses may not be normal viruses at all. Several neurological diseases of animals and humans develop slowly. The best studied of these diseases are scrapie, a disease of sheep and goats, the human diseases kuru, fatal familial insomnia, Creutzfeldt-Jakob disease, and the new variant Creutzfeldt-Jakob disease. All are thought to be caused by simple, nonviral agents called prions, which are discussed later in the chapter.

---

1. Outline the ways in which viruses can damage host cells during cytocidal infections.
2. Define the following: acute infection, persistent infection, chronic infection, latent virus infection, and slow virus disease.
3. Why might an infection be chronic or latent?

 ## 18.5 VIRUSES AND CANCER

Cancer [Latin *cancer,* crab] is one of our most serious medical problems and the focus of an immense amount of interest and research. A **tumor** [Latin *tumere,* to swell] is a growth or lump of tissue resulting from **neoplasia** or abnormal new cell growth and reproduction due to a loss of regulation. Tumor cells have aberrant shapes and altered plasma membranes that contain distinctive tumor antigens. They may invade surrounding tissues to form unorganized cell masses. They often lose the specialized metabolic activities characteristic of differentiated tissue cells and rely greatly upon anaerobic metabolism. This reversion to a more primitive or less differentiated state is called **anaplasia.**

There are two major types of tumors with respect to overall form or growth pattern. If the tumor cells remain in place to form a compact mass, the tumor is benign. In contrast, malignant or cancerous tumor cells can actively spread throughout the body in a process known as **metastasis,** often by floating in the blood and establishing secondary tumors. Some cancers are not solid, but cell suspensions. For example, leukemias are composed of malignant white blood cells that circulate throughout the body. Indeed, dozens of kinds of cancers arise from a variety of cell types and afflict all kinds of organisms.

As one might expect from the wide diversity of cancers, there are many causes of cancer, only some of which are directly related to viruses. Possibly as many as 30 to 60% of cancers may be related to diet. Many chemicals in our surroundings are carcinogenic and may cause cancer by inducing gene mutations or interfering with normal cell differentiation. The Ames test for carcinogens (pp. 247–48).

Carcinogenesis is a complex, multistep process. It can be initiated by a chemical, usually a mutagen, but a cancer does not appear to develop until at least one more triggering event (possibly exposure to another chemical carcinogen or a virus) takes place. Often, as is discussed later, cancer-causing genes, or **oncogenes,** are directly involved and may come from the cell itself or be contributed by a virus. Many of these oncogenes seem to be involved in the regulation of cell growth and differentiation; for example, some code for all or part of growth factors that regulate cell growth. It may be that various cancers arise through different combinations of causes. Not surprisingly the chances of developing cancer rise with age because an older person will have been exposed to carcinogens and other causative factors for a longer time. Immune surveillance and destruction of cancer cells (*see chapter 32*) also may be less effective in older people.

Although viruses are known to cause many animal cancers, it is very difficult to prove that this is the case with human cancers since indirect methods of study must be used and Koch's postulates can't be applied completely. One tries to find virus particles and components within tumor cells, using techniques such as electron microscopy, immunologic tests, and enzyme assays. Attempts are also made to isolate suspected cancer viruses by cultivation in tissue culture or other animals. Sometimes a good correlation between the presence of a virus and cancer can be detected.

At present, viruses have been implicated in the genesis of at least eight human cancers:

1. The Epstein-Barr virus (EBV) is one of the best-studied human cancer viruses. EBV is a herpesvirus and the cause of two cancers. Burkitt's lymphoma is a malignant tumor of the jaw and abdomen found in children of central and western Africa. EBV also causes nasopharyngeal carcinoma. Both the virus particles and the EBV genome have been found within tumor cells; Burkitt's lymphoma patients also have high blood levels of antibodies to EBV. Interestingly there is some reason to believe that a person also must have had malaria to develop Burkitt's lymphoma. Environmental factors must play a role, because EBV does not cause much cancer in the United States despite its prevalence. Possibly this is due to a low incidence of malaria in the United States.
2. Hepatitis B virus appears to be associated with one form of liver cancer (hepatocellular carcinoma) and can be integrated into the human genome.
3. Hepatitis C virus causes cirrhosis of the liver, which can lead to liver cancer.
4. Human herpesvirus 8 is associated with the development of Kaposi's sarcoma.
5. Some strains of human papillomaviruses have been linked to cervical cancer.

6. At least two retroviruses, the human T-cell lymphotropic virus I (HTLV-1) and HTLV-2, seem able to cause cancer, adult T-cell leukemia, and hairy-cell leukemia, respectively (*see figure 38.17*). Other retrovirus-associated cancers may well be discovered in the future.

It appears that viruses can cause cancer in several ways. They may bring oncogenes into a cell and insert them into its genome. Rous sarcoma virus (a retrovirus) carries an *src* gene that codes for tyrosine kinase. This enzyme is located mainly in the plasma membrane and phosphorylates the amino acid tyrosine in several cellular proteins. It is thought that this alters cell growth and behavior. Since the activity of many proteins is regulated by phosphorylation and several other oncogenes also code for protein kinases, many cancers may result at least partly from altered cell regulation due to changes in kinase activity. The human T-cell lymphotropic viruses, HTLV-1 and HTLV-2, seem to transform T cells (*see chapter 32*) by producing a regulatory protein that sometimes activates genes involved in cell division as well as stimulating virus reproduction. Some oncogenic viruses carry one or more very effective promoters or enhancers (*see section 11.5*). If these viruses integrate themselves next to a cellular oncogene, the promoter or enhancer will stimulate its transcription, leading to cancer. In this case the oncogene might be necessary for normal cell growth (possibly it codes for a regulatory protein) and only causes cancer when it functions too rapidly or at the wrong time. For example, some chicken retroviruses induce lymphomas when they are integrated next to the *c-myc* cellular oncogene, which codes for a protein that appears to be involved in the induction of either DNA or RNA synthesis. With the possible exception of HTLV-1, it is not yet known how the viruses associated with human cancers actually aid in cancer development.

---

1. What are the major characteristics of cancer?
2. How might viruses cause cancer? Are there other ways in which a malignancy might develop?
3. Define the following terms: tumor, neoplasia, anaplasia, metastasis, and oncogene.

 18.6 PLANT VIRUSES

Although it has long been recognized that viruses can infect plants and cause a variety of diseases (*see figure 16.5*), plant viruses generally have not been as well studied as bacteriophages and animal viruses. This is mainly because they are often difficult to cultivate and purify. Some viruses, such as tobacco mosaic virus (TMV), can be grown in isolated protoplasts of plant cells just as phages and some animal viruses are cultivated in cell suspensions. However, many cannot grow in protoplast cultures and must be inoculated into whole plants or tissue preparations. Many plant viruses require insect vectors for transmission; some of these can be grown in monolayers of cell cultures derived from aphids, leafhoppers, or other insects.

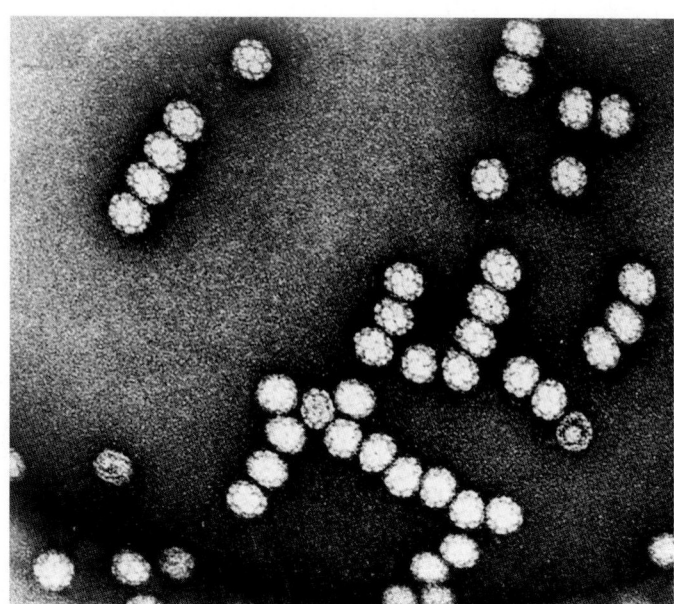

Figure 18.11 Turnip Yellow Mosaic Virus (TYMV). An RNA plant virus with icosahedral symmetry.

Virion Morphology

The essentials of capsid morphology are outlined in chapter 16 and apply to plant viruses since they do not differ significantly in construction from their animal virus and phage relatives. Many have either rigid or flexible helical capsids (tobacco mosaic virus, *see figure 16.11*). Others are icosahedral or have modified the icosahedral pattern with the addition of extra capsomers (turnip yellow mosaic virus, **figure 18.11**). Most capsids seem composed of one type of protein; no specialized attachment proteins have been detected. Almost all plant viruses are RNA viruses, either single stranded or double stranded (*see tables 16.1 and 16.2*). Caulimoviruses and geminiviruses with their DNA genomes are exceptions to this rule.

Plant Virus Taxonomy

The shape, size, and nucleic acid content of many plant virus groups are summarized in **figure 18.12.** Like other types of viruses, they are classified according to properties such as nucleic acid type and strandedness, capsid symmetry and size, and envelope presence (*see table 16.2*).

Plant Virus Reproduction

Since tobacco mosaic virus (TMV) has been studied the most extensively, its reproduction will be briefly described. The replication of virus RNA is an essential part of the reproduction process. Most plants contain RNA-dependent RNA polymerases, and it is possible that these normal constituents replicate the virus RNA. However, some plant virus genomes (e.g., turnip yellow virus and cowpea mosaic virus) appear to be copied by a virus-specific RNA replicase. Possibly TMV RNA is also replicated by a viral

Figure 18.12   A Diagrammatic Description of Families and Genera of Viruses That Infect Plants.   RT stands for reverse transcriptase.

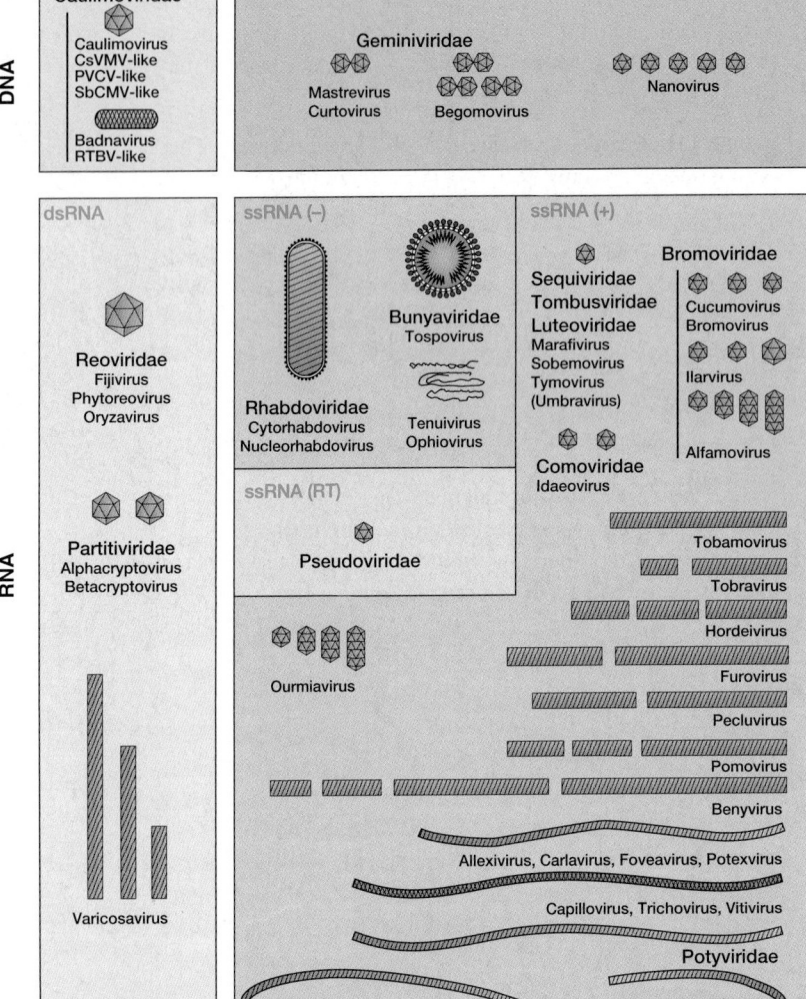

RNA polymerase, but the evidence is not clear on this matter. Four TMV-specific proteins, one of them the coat protein, are known to be made. Although TMV RNA is plus single-stranded RNA and could directly serve as mRNA, the production of messenger is complex. The coat protein mRNA has the same sequence as the 3′ end of the TMV genome and arises from it by some sort of intracellular processing.

After the coat protein and RNA genome have been synthesized, they spontaneously assemble into complete TMV virions in a highly organized process (**figure 18.13**). The protomers (*see section 16.5*) come together to form disks composed of two layers of protomers arranged in a helical spiral. Association of coat protein with TMV RNA begins at a special assembly initiation site close to the 3′ end of the genome. The helical capsid grows by the addition of protomers, probably as disks, to the end of the rod. As the rod lengthens, the RNA passes through a channel in its center and forms a loop at the growing end. In this way the RNA can easily fit as a spiral into the interior of the helical capsid.

Reproduction within the host depends on the virus's ability to spread throughout the plant. Viruses can move long distances through the plant vasculature; usually they travel in the phloem. The spread of plant viruses in nonvascular tissue is hindered by the presence of tough cell walls. Nevertheless, a virus such as TMV does spread slowly, about 1 mm/day or less. It moves from cell to cell through the plasmodesmata. These are slender cytoplasmic strands extending through holes in adjacent cell walls that join plant cells by narrow bridges. Special viral "movement proteins" are required for movement between cells. The TMV movement protein accumulates in the plasmodesmata, but the way in which it promotes virus movement is not understood.

Several cytological changes can take place in TMV-infected cells. Plant virus infections often produce microscopically visible intracellular inclusions, usually composed of virion aggregates, and hexagonal crystals of almost pure TMV virions sometimes do develop in TMV-infected cells (**figure 18.14**). The host cell

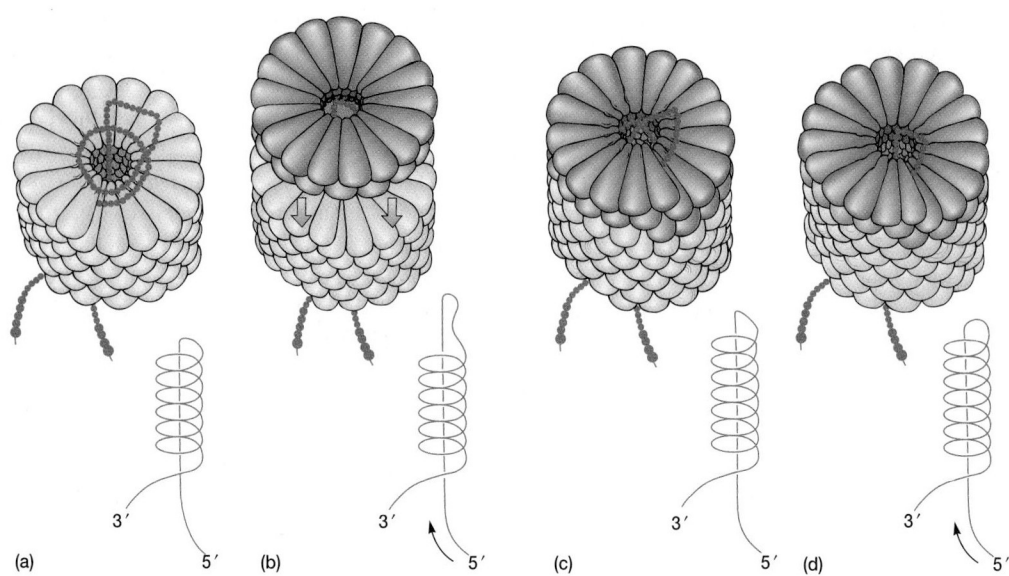

Figure 18.13   TMV Assembly.   The elongation phase of tobacco mosaic virus nucleocapsid construction. The lengthening of the helical capsid through the addition of a protein disk to its end is shown in a sequence of four illustrations; line drawings depicting RNA behavior are included. The RNA genome inserts itself through the hole of an approaching disk and then binds to the groove in the disk as it locks into place at the end of the cylinder.

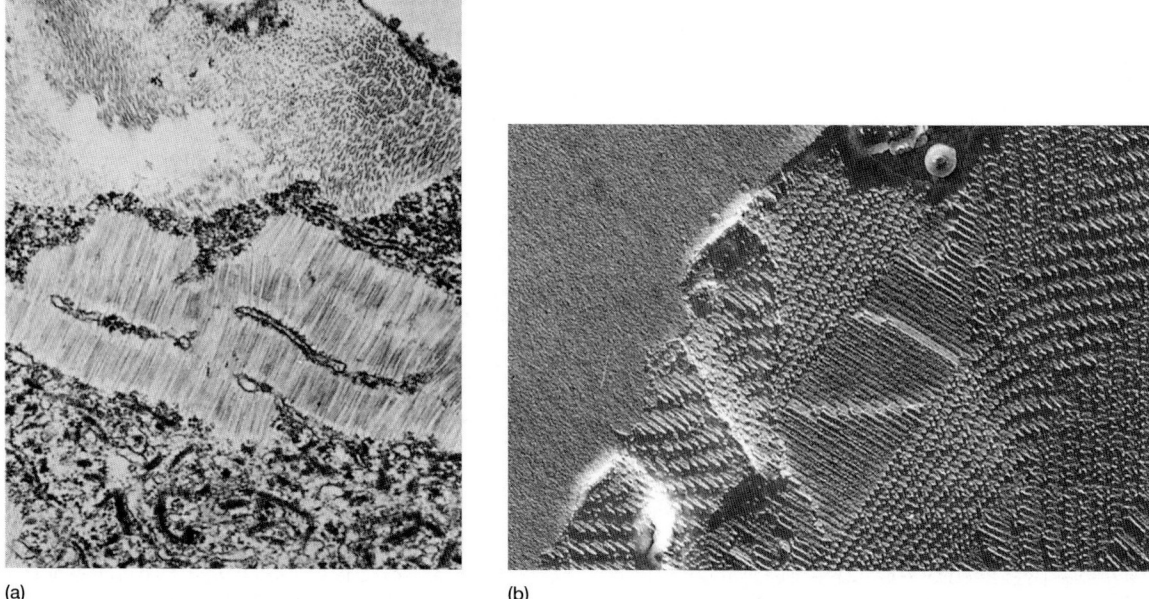

Figure 18.14   Intracellular TMV.   (a) A crystalline mass of tobacco mosaic virions from a 10-day-old lesion in a *Chenopodium amaranticolor* leaf. **(b)** Freeze-fracture view of a crystalline mass of tobacco mosaic virions in an infected leaf cell. In both views the particles can be seen longitudinally and in cross section.

chloroplasts become abnormal and often degenerate, while new chloroplast synthesis is inhibited.

Transmission of Plant Viruses

Since plant cells are protected by cell walls, plant viruses have a considerable obstacle to overcome when trying to establish themselves in a host. TMV and a few other viruses may be carried by the wind or animals and then enter when leaves are mechanically damaged. Some plant viruses are transmitted through contaminated seeds, tubers, or pollen. Soil nematodes can transmit viruses (e.g., the tobacco ringspot virus) while feeding on roots. Tobacco necrosis virus is transmitted by parasitic fungi. However, the most important agents of transmission are insects that feed on plants, particularly sucking insects such as aphids and leafhoppers.

Insects transmit viruses in several ways. They may simply pick up viruses on their mouth parts while feeding on an infected plant, then directly transfer the viruses to the next plant they visit. Viruses may be stored in an aphid's foregut; the aphid will infect plants when regurgitating while it is feeding. Several plant viruses—for example, the wound tumor virus—can multiply in leafhopper tissues before reaching the salivary glands and being inoculated into plants (i.e., it uses both insects and plants as hosts).

1. Why have plant viruses not been as well studied as animal and bacterial viruses?
2. Describe in molecular terms the way in which TMV is reproduced.
3. How are plant viruses transmitted between hosts?

18.7 VIRUSES OF FUNGI, ALGAE, AND PROTOZOA

Most of the fungal viruses or mycoviruses isolated from higher fungi such as *Penicillium* and *Aspergillus* (*see chapter 25*) contain double-stranded RNA and have isometric capsids (that is, their capsids are roughly spherical polyhedra), which are approximately 25 to 50 nm in diameter. Many appear to be latent viruses. Some mycoviruses do induce disease symptoms in hosts such as the mushroom *Agaricus bisporus,* but cytopathic effects and toxic virus products have not yet been observed.

Much less is known about the viruses of lower fungi, and only a few have been examined in any detail. Both dsRNA and dsDNA genomes have been found; capsids usually are isometric or hexagonal and vary in size from 40 to over 200 nm. Unlike the situation in higher fungi, virus reproduction is accompanied by host cell destruction and lysis.

Viruses have been detected during ultrastructural studies of eucaryotic algae, but few have been isolated. Those that have been studied have dsDNA genomes and polyhedral capsids. One virus of the green alga *Uronema gigas* resembles many bacteriophages in having a tail.

The viruses of protozoa also have not been extensively studied. It is known that *Giardia lamblia* and *Leishmania* spp. are infected by members of the *Totiviridae,* a family of naked icosahedral viruses with dsRNA genomes. A member of the family *Phycodnaviridae* (naked icosahedral viruses with dsDNA) infects *Chorella* strains that can live within *Paramecium bursaria* as symbionts. Finally, a new giant dsDNA virus, tentatively named a Mimivirus, has been discovered in the amoeba *Acanthamoeba polyphaga.* The virus is 400 nm in diameter and has a genome about 800 kilobase pairs in size; thus its genome is larger than that of several bacteria. Mimivirus is distantly related to the *Poxviridae* and *Phycodnaviridae.*

1. Describe the major characteristics of the viruses that infect higher fungi, lower fungi, algae, and protozoa. In what ways do they seem to differ from one another?

18.8 INSECT VIRUSES

Members of at least seven virus families (*Baculoviridae, Iridoviridae, Poxviridae, Reoviridae, Parvoviridae, Picornaviridae,* and *Rhabdoviridae*) are known to infect insects and reproduce or even use them as their primary host (*see table 16.2*). Of these, probably the three most important are the *Baculoviridae, Reoviridae,* and *Iridoviridae.*

The *Iridoviridae* are icosahedral viruses with lipid in their capsids and a linear double-stranded DNA genome. They are responsible for the iridescent virus diseases of the crane fly and some beetles. The group's name comes from the observation that larvae of infected insects can have an iridescent coloration due to the presence of crystallized virions in their fat bodies.

Many insect virus infections are accompanied by the formation of inclusion bodies within the infected cells. Granulosis viruses form granular protein inclusions, usually in the cytoplasm. Nuclear polyhedrosis and cytoplasmic polyhedrosis virus infections produce polyhedral inclusion bodies in the nucleus or the cytoplasm of affected cells. Although all three types of viruses generate inclusion bodies, they belong to two distinctly different families. The cytoplasmic polyhedrosis viruses are reoviruses; they are icosahedral with double shells and have double-stranded RNA genomes. Nuclear polyhedrosis viruses and granulosis viruses are baculoviruses—rod-shaped, enveloped viruses of helical symmetry and with double-stranded DNA.

The inclusion bodies, both polyhedral and granular, are protein in nature and enclose one or more virions (**figure 18.15**). Insect larvae are infected when they feed on leaves contaminated with inclusion bodies. Polyhedral bodies protect the virions against heat, low pH, and many chemicals; the viruses can remain viable in the soil for years. However, when exposed to alkaline insect gut contents, the inclusion bodies dissolve to liberate the virions, which then infect midgut cells. Some viruses remain in the midgut while others spread throughout the insect. Just as with bacterial and vertebrate viruses, insect viruses can persist in a la-

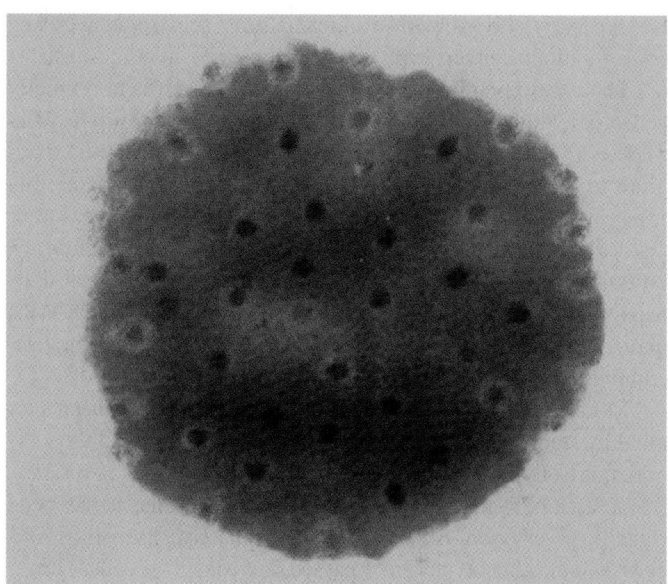

Figure 18.15   Inclusion Bodies.   A section of a cytoplasmic polyhedron from a gypsy moth *(Lymantria dispar)*. The occluded virus particles with dense cores are clearly visible (×50,000).

tent state within the host for generations while producing no disease symptoms. A reappearance of the disease may be induced by chemicals, thermal shock, or even a change in the insect's diet.

Much of the current interest in insect viruses arises from their promise as biological control agents for insect pests (*see chapter 42*). Many people hope that some of these viruses may partially replace the use of toxic chemical pesticides. Baculoviruses have received the most attention for at least three reasons. First, they attack only invertebrates and have considerable host specificity; this means that they should be fairly safe for nontarget organisms. Second, because they are encased in protective inclusion bodies, these viruses have a good shelf life and better viability when dispersed in the environment. Finally, they are well suited for commercial production since they often reach extremely high concentrations in larval tissue (as high as 10^{10} viruses per larva). The use of nuclear polyhedrosis viruses for the control of the cotton bollworm, Douglas fir tussock moth, gypsy moth, alfalfa looper, and European pine sawfly has either been approved by the Environmental Protection Agency or is being considered. The granulosis virus of the codling moth also is useful. Usually inclusion bodies are sprayed on foliage consumed by the target insects. More sensitive viruses are administered by releasing infected insects to spread the disease. As in the case of other pesticides, it is possible that resistance to these agents may develop in the future.

1. Summarize the nature of granulosis, nuclear polyhedrosis, and cytoplasmic polyhedrosis viruses and the way in which they are transmitted by inclusion bodies.
2. What are baculoviruses and why are they so promising as biological control agents for insect pests?

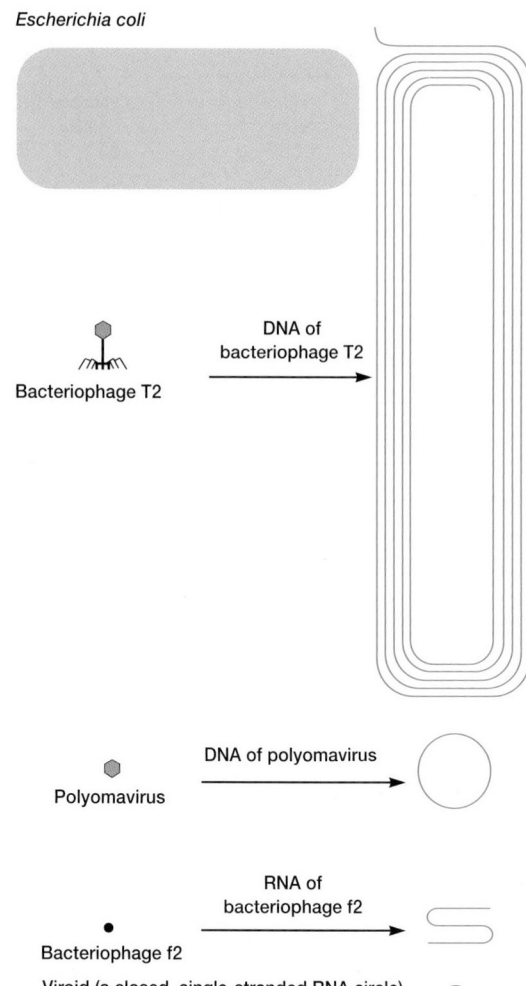

Escherichia coli

Bacteriophage T2

DNA of bacteriophage T2

Polyomavirus

DNA of polyomavirus

Bacteriophage f2

RNA of bacteriophage f2

Viroid (a closed, single-stranded RNA circle)

Figure 18.16   Viroids, Viruses, and Bacteria.   A comparison of *Escherichia coli,* several viruses, and the potato spindle-tuber viroid with respect to size and the amount of nucleic acid possessed. (All dimensions are enlarged approximately ×40,000.)

18.9   VIROIDS AND PRIONS

Although some viruses are exceedingly small and simple, even simpler infectious agents exist. Over 16 different plant diseases—for example, potato spindle-tuber disease, exocortis disease of citrus trees, and chrysanthemum stunt disease—are caused by a group of infectious agents called **viroids.** These are circular, single-stranded RNAs, about 250 to 370 nucleotides long (**figures 18.16** and **18.17**), that can be transmitted between plants through mechanical means or by way of pollen and ovules and are replicated in their hosts. The single-stranded RNA normally exists as a closed circle collapsed into a rodlike shape by intrastrand base pairing. Viroids are found principally in the nucleolus of infected cells; between 200 and 10,000 copies may be present. They do not act as mRNAs to direct protein synthesis, and it is not yet

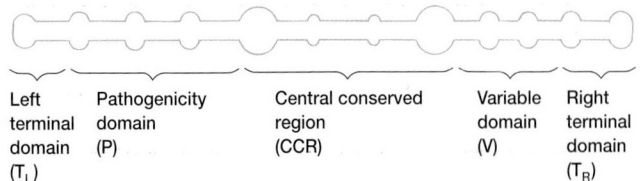

Figure 18.17   Viroid Structure. This schematic diagram shows the general organization of a viroid. The closed single-stranded RNA circle has extensive intrastrand base pairing and interspersed unpaired loops. Viroids have five domains. Most changes in viroid pathogenicity seem to arise from variations in the P and T$_L$ domains.

known how they cause disease symptoms. A plant may be infected without showing symptoms—that is, it may have a latent infection. The same viroid, when in another species, might well cause a severe disease. Although viroids could be replicated by an RNA-dependent RNA polymerase, they appear to be synthesized from RNA templates by a host RNA polymerase that mistakes them for a piece of DNA. A rolling-circle type mechanism seems to be involved (*see pp. 229–30*).

The potato spindle-tuber disease agent (PSTV) has been most intensely studied. Its RNA is about 130,000 daltons or 359 nucleotides in size, much smaller than any virus genome. Several PSTV strains have been isolated, ranging in virulence from ones that cause only mild symptoms to lethal varieties. All variations in pathogenicity are due to a few nucleotide changes in two short regions on the viroid. It is believed that these sequence changes alter the shape of the rod and thus its ability to cause disease.

There is evidence that an infectious agent different from both viruses and viroids can cause disease in livestock and humans. The agent has been called a **prion** (for proteinaceous infectious particle). The best studied of these prions causes a degenerative disorder of the central nervous system in sheep and goats; this disorder is named scrapie. Afflicted animals lose coordination of their movements, tend to scrape or rub their skin, and eventually cannot walk. No nucleic acid has yet been detected in the agent. It seems to be a 33 to 35 kDa hydrophobic membrane protein, often called PrP (for **pr**ion **p**rotein). The *PrP* gene is present in many normal vertebrates and invertebrates, and the prion protein is bound to the surface of neurons. Presumably an altered PrP is at least partly responsible for the disease.

Despite the isolation of PrP, the nature of prion diseases continues to stir controversy. Most researchers are convinced that prion diseases are transmitted by the PrP alone. They believe that the infective pathogen is an abnormal PrP (PrP^{Sc}; Sc, scrapie-associated), a protein that has either been changed in conformation or chemically modified. When PrP^{Sc} enters a normal brain, it might bind to the normal cellular PrP (PrP^C) and induce it to fold into the abnormal conformation. The newly produced PrP^{Sc} molecule could then convert more normal PrP^C proteins to the PrP^{Sc} form. Alternatively, PrP^{Sc} could activate enzymes that modify

PrP structure. Prions with different amino acid sequences and conformations convert PrP^C molecules to PrP^{Sc} in other hosts.

There is some recent evidence that PrP^{Sc} itself is not directly neurotoxic; rather a cytoplasmic PrP form is the neurotoxin. Misfolded PrP molecules are transported from the endoplasmic reticulum to the cytoplasm instead of to the plasma membrane, but sometimes they are not degraded by cytoplasmic proteasomes (*see p. 81*). These cytoplasmic PrP molecules or their cleavage products may be the true toxic species that kill the neuron. Perhaps they activate cell death pathways (*see pp. 725, 765*). PrP^{Sc} may also arise from these cytoplasmic PrP molecules if they accumulate sufficiently.

More support for the "protein-only" hypothesis has been supplied by studies on the yeast protein Sup35, which aids in the termination of protein synthesis in *Saccharomyces cerevisiae*. Because Sup35 acts much like mammalian prions, it has been called a yeast prion. Sup35 exists in an inactive form in [*PSI*$^+$] cells, and the phenotype can be passed on to daughter cells. The inactive, insoluble form of Sup35 aggregates and translation does not terminate properly. There is evidence that the [*PSI*$^+$] phenotype proliferates in yeast when the inactive prion form of Sup35 interacts with normal, soluble Sup35 and induces a self-propagating conformational change of active Sup35 to the inactive form.

Some scientists think that the "protein-only" hypothesis is inadequate. They are concerned about the existence of prion strains, which they believe are genetic, and about the problem of how genetic information can be transmitted between hosts by a protein. Doubters note that proteins have never been known to carry genetic information. It could be that prions somehow either directly aid an infectious agent such as an unknown virus or increase susceptibility to the agent. Possibly the real agent is a tiny nucleic acid that is coated with PrP and interacts with host cells to cause disease. This hypothesis is consistent with the finding that many strains of the scrapie agent have been isolated. There is also some evidence that a strain can change or mutate. Supporters of the "protein-only" hypothesis reply that strain characteristics are simply due to structural differences in the PrP molecule.

As mentioned earlier, some slow virus diseases may be due to prions. This is particularly true of certain neurological diseases of humans and animals. Bovine spongiform encephalopathy (BSE or "mad cow disease"), kuru, fatal familial insomnia, the Creutzfeldt-Jakob disease (CJD), and Gerstmann-Sträussler-Scheinker syndrome (GSS) appear to be prion diseases. They result in progressive degeneration of the brain and eventual death. At present, there is no effective treatment. Mad cow disease reached epidemic proportions in Great Britain and initially spread because cattle were fed bone meal made from cattle. Recently a small outbreak of BSE also occurred in Canada. It has now been shown that eating meat from cattle with BSE can cause a variant of Creutzfeldt-Jakob disease in humans. More than 90 people already have died in the United Kingdom and France from this source. CJD and GSS are rare and cosmopolitan in distribution among middle-aged people, while kuru has been found only in the Fore, an eastern New Guinea tribe. This tribe had a custom of consuming dead kinsmen, and women were given the honor of preparing the brain and practicing this ritual cannibalism. They and their children were infected by handling the diseased brain tissue. (Now that cannibalism has been eliminated, the incidence of kuru has decreased and is found only among the older adults.)

1. What are viroids and why are they of great interest?
2. How does a viroid differ from a virus?
3. What is a prion? In what way does a prion appear to differ fundamentally from viruses and viroids?

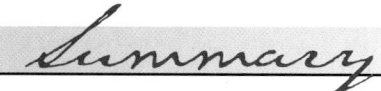

18.1 Classification of Animal Viruses

a. Animal viruses are classified according to many properties; the most important are their morphology and nucleic acids (**figures 18.1–18.3**).

18.2 Reproduction of Animal Viruses

a. The first step in the animal virus reproductive cycle is adsorption of the virus to a target cell receptor site; often special capsid structures are involved in this process.

b. Virus penetration of the host cell plasma membrane may be accompanied by capsid removal from the nucleic acid or uncoating. Most often penetration occurs through either endocytotic engulfment to form coated vesicles or fusion of the envelope with the plasma membrane (**figure 18.4**).

c. Early viral mRNA and proteins are involved in taking over the host cell and the synthesis of viral DNA and RNA. DNA replication usually takes place in the host nucleus and mRNA is initially manufactured by host enzymes.

d. Poxviruses differ from other DNA animal viruses in that DNA replication takes place in host cytoplasm and they carry an RNA polymerase. The parvoviruses are so small that they must conserve genome space by using overlapping genes and other similar mechanisms. Hepadnaviruses use reverse transcriptase to replicate their dsDNA genome.

e. The genome of positive ssRNA viruses can act as an mRNA, whereas negative ssRNA virus genomes direct the synthesis of mRNA by a virus-associated transcriptase. Double-stranded RNA reoviruses use both virus-associated and newly synthesized transcriptases to make mRNA (**figure 18.6**).

f. RNA virus genomes are replicated in the host cell cytoplasm. Most ssRNA viruses use a viral replicase to synthesize a dsRNA replicative form that then directs the formation of new genomes.

g. Retroviruses use reverse transcriptase to synthesize a DNA copy of their RNA genome. After the double-stranded proviral DNA has been synthesized, it is integrated into the host genome and directs the formation of virus RNA and protein.

h. Late genes code for proteins needed in (1) capsid construction by a self-assembly process and (2) virus release.

i. Usually, naked virions are released upon cell lysis. In enveloped virus reproduction, virus release and envelope formation normally occur simultaneously after modification of the host plasma membrane, and the cell is not lysed.

18.3 Cytocidal Infections and Cell Damage

a. Viruses can destroy host cells in many ways during cytocidal infections. These include such mechanisms as inhibition of host DNA, RNA, and protein synthesis; lysosomal damage; alteration of host cell membranes; and the formation of inclusion bodies.

18.4 Persistent, Latent, and Slow Virus Infections

a. Although most virus infections are acute or have a rapid onset and last for a relatively short time, some viruses can establish persistent infections lasting for years. Some infections are chronic. Viruses also can become dormant for a while and then resume activity in what is called a latent infection. Slow viruses may act so slowly that a disease develops over years.

18.5 Viruses and Cancer

a. Cancer is characterized by the formation of a malignant tumor that metastasizes or invades other tissues and can spread through the body. Carcinogenesis is a complex, multistep process involving many factors.

b. Viruses cause cancer in several ways. For example, they may bring a cancer-causing gene, or oncogene, into a cell, or the virion may insert a promoter next to a cellular oncogene and stimulate the gene to greater activity.

18.6 Plant Viruses

a. Most plant viruses have an RNA genome and may be either helical or icosahedral. Depending on the virus the RNA genome may be replicated by either a host RNA-dependent RNA polymerase or a virus-specific RNA replicase.

b. The TMV nucleocapsid forms spontaneously by self-assembly when disks of coat protein protomers complex with the RNA.

c. Plant viruses are transmitted in a variety of ways. Some enter through lesions in plant tissues, while others are transmitted by contaminated seeds, tubers, or pollen. Most are probably carried and inoculated by plant-feeding insects.

18.7 Viruses of Fungi, Algae, and Protozoa

a. Mycoviruses from higher fungi have isometric capsids and dsRNA, whereas the viruses of lower fungi may have either dsRNA or dsDNA genomes.

18.8 Insect Viruses

a. Members of several virus families—most importantly the *Baculoviridae*, *Reoviridae*, and *Iridoviridae*—infect insects, and many of these viruses produce inclusion bodies that aid in their transmission.

b. Baculoviruses and other viruses are finding uses as biological control agents for insect pests.

18.9 Viroids and Prions

a. Infectious agents simpler than viruses exist. For example, several plant diseases are caused by short strands of infectious RNA called viroids.

b. Prions are small proteinaceous agents associated with at least six degenerative nervous system disorders: scrapie, bovine spongiform encephalopathy, kuru, fatal familial insomnia, the Gerstmann-Sträussler-Scheinker syndrome, and Creutzfeldt-Jakob disease. The precise nature of prions is not yet clear.

acute infection   399
anaplasia   400
cancer   400
chronic virus infection   399
cytocidal infection   398
defective interfering (DI) particle   400
early genes   392

inclusion bodies   399
late genes   397
latent virus infection   399
metastasis   400
neoplasia   400
oncogene   400
persistent infection   399

prion   406
procapsids   397
proviral DNA   397
replicase   395
replicative form   395
retrovirus   395

reverse transcriptase (RT)   397
ribonuclease H   397
slow virus diseases   400
transcriptase   394
tumor   400
viroid   405

Questions for Thought and Review

1. Make a list of all the ways you can think of in which viruses differ from their eucaryotic host cells.

2. Compare animal and plant viruses in terms of entry into host cells. Why do they differ so greatly in this regard?

3. Compare lysogeny with the reproductive strategy of retroviruses. What advantage might there be in having one's genome incorporated into that of the host cell?

4. Inclusion bodies are mentioned more than once in the chapter. Where are they found, and what are their functions, if any?

5. Would it be advantageous for a virus to damage host cells? If it is not, why isn't damage to the host avoided? Is it possible that a virus might become less pathogenic when it has been associated with the host population for a longer time?

6. From what you know about cancer, is it likely that a single type of treatment can be used to cure it? What approaches might be effective in preventing cancer?

7. What is unusual about prions compared with viruses or bacterial pathogens that makes them so difficult to detect and treat?

Critical Thinking Questions

1. Consider each of the following viral reproduction steps: adsorption, penetration, replication, and transcription. Suggest a strategy by which one could pharmacologically inhibit or discourage entry and propagation of viruses in animal cells. Can you explain host range using some of the same rationale?

2. How does one prove that a virus is causing cancer? Try to think of approaches other than those discussed in the chapter. Give a major reason why it is so difficult to prove that a specific virus causes human cancer. Is it accurate to say that viruses cause cancer? Explain why or why not.

3. Propose some experiments that might be useful in determining what prions are and how they cause disease.

For recommended readings on these and related topics, see Appendix V.

chapter 19

Microbial Taxonomy and Phylogeny

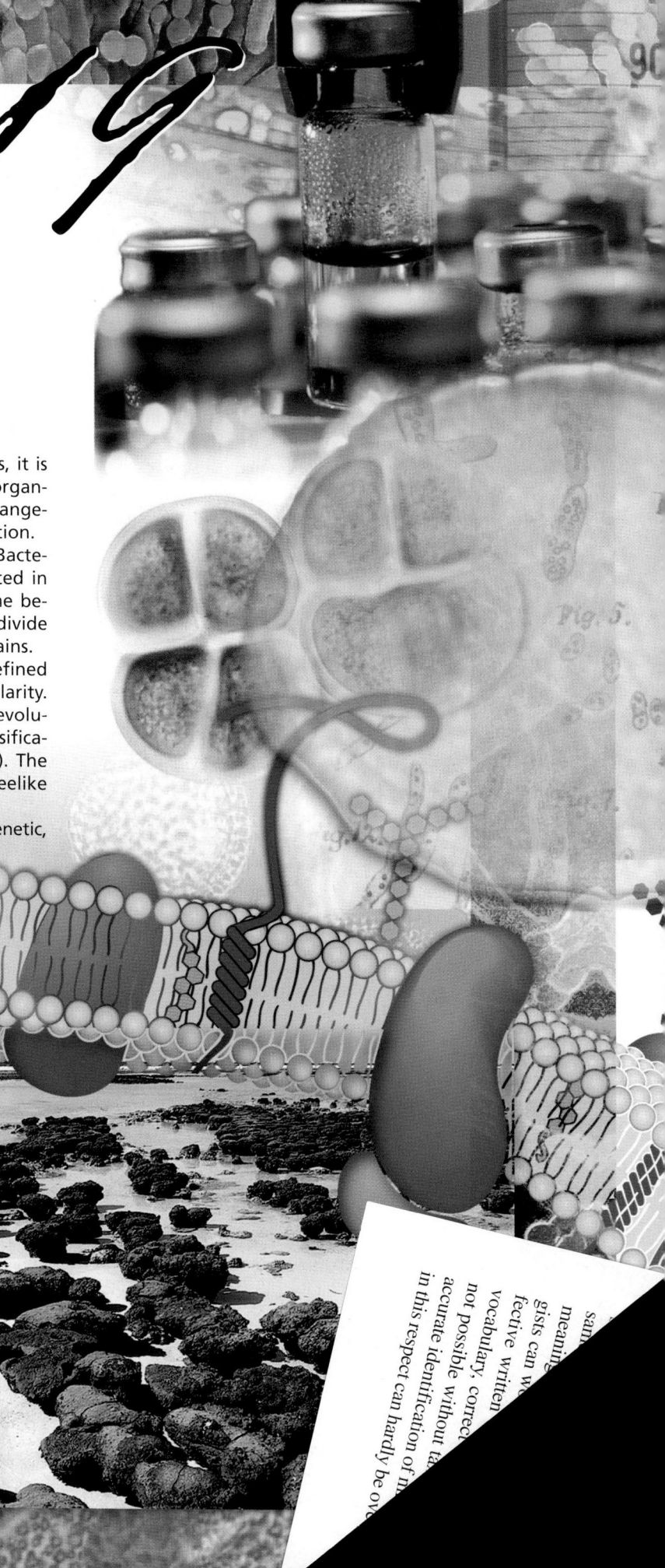

Concepts

- In order to make sense of the diversity of organisms, it is necessary to group similar organisms together and organize these groups in a nonoverlapping hierarchical arrangement. Taxonomy is the science of biological classification.
- It appears that the procaryotic groups (Archaea and Bacteria) first developed, then the eucaryotes. This resulted in three domains: *Bacteria, Archaea,* and *Eucarya.* Some believe that the living world is even more complex and divide it into five or more kingdoms rather than three domains.
- The basic taxonomic group is the species, which is defined in terms of either sexual reproduction or general similarity.
- Classifications are based on an analysis of possible evolutionary relationships (phylogenetic or phyletic classification) or on overall similarity (phenetic classification). The results of these analyses are often summarized in treelike diagrams called dendrograms.
- Morphological, physiological, metabolic, ecological, genetic, and molecular characteristics are all useful in taxonomy because they reflect the organization and activity of the genome. Nucleic acid sequences are probably the best indicators of microbial phylogeny and relatedness because nucleic acids are either the genetic material itself or the products of gene transcription.
- The first edition of *Bergey's Manual of Systematic Bacteriology* was largely phenotypic and divided procaryotes into groups based on easily determined characteristics such as shape, Gram-staining properties, oxygen relationships, and motility. The second edition, with its five volumes, is phylogenetically organized and distributes procaryotes among two domains and 25 phyla.
- Bacterial taxonomy is rapidly changing due to the acquisition of new data, particularly the use of molecular techniques such as the comparison of ribosomal RNA structure and chromosome sequences. This is leading to new phylogenetic classifications.

The stromatolites shown here are layered rocks formed by incorporation of minerals into microbial mats. Fossilized stromatolites indicate that microorganisms existed early in Earth's history.

What's in a name? that which we call a rose
By any other name would smell as sweet. . . .

—W. Shakespeare

One of the most fascinating and attractive aspects of the microbial world is its extraordinary diversity. It seems that almost every possible experiment in shape, size, physiology, and life-style has been tried. This section of the text focuses on microbial diversity. Chapter 19 introduces the general principles of microbial taxonomy. This is followed by a five-chapter (20–24) survey of the most important procaryotic groups. Our survey of microbial diversity ends with an extensive introduction to the major types of eucaryotic microorganisms: fungi, algae, and protozoa.

19.1 GENERAL INTRODUCTION AND OVERVIEW

Because of the bewildering diversity of living organisms, it is desirable to classify or arrange them into groups based on their mutual similarities. **Taxonomy** [Greek *taxis,* arrangement or order, and *nomos,* law, or *nemein,* to distribute or govern] is defined as the science of biological classification. In a broader sense it consists of three separate but interrelated parts: classification, nomenclature, and identification. **Classification** is the arrangement of organisms into groups or **taxa** (s., **taxon**) based on mutual similarity or evolutionary relatedness. **Nomenclature** is the branch of taxonomy concerned with the assignment of names to taxonomic groups in agreement with published rules. **Identification** is the practical side of taxonomy, the process of determining that a particular isolate belongs to a recognized taxon.

People often think of taxonomy as trivial and boring, simply a matter of splitting hairs over names of organisms. Actually, taxonomy is important for several reasons. First, it allows us to organize huge amounts of knowledge about organisms because all members of a particular group share many characteristics. In a sense it is something like a giant filing system or library catalogue that provides easy access to information. The more accurate the classification, the more information-rich and useful it is. Second, taxonomy allows us to make predictions and frame hypotheses for further research based on knowledge of similar organisms. If a relative has some property, the microorganism in question also may have the same characteristic. Third, taxonomy places microorganisms in meaningful, useful groups with precise names so that microbiologists can work with them and communicate efficiently. Just as effective communication is not possible without adequate vocabulary, spelling, and good grammar, microbiology is impossible without taxonomy. Fourth, taxonomy is essential for identifying microorganisms. Its practical importance cannot be overemphasized. For example, it is es-

sential to clinical microbiology (*see chapter 36*); treatment often is exceptionally difficult when the pathogen is unknown.

The term **systematics** often is used for taxonomy. However, many taxonomists define it in more general terms as "the scientific study of organisms with the ultimate object of characterizing and arranging them in an orderly manner." Any study of the nature of organisms, when the knowledge gained is used in taxonomy, is a part of systematics. Thus it encompasses disciplines such as morphology, ecology, epidemiology, biochemistry, molecular biology, and physiology.

Microbial taxonomy is too broad a subject for adequate coverage in a single chapter. Therefore this chapter emphasizes general principles and uses examples primarily from procaryotic taxonomy. The taxonomy of each major eucaryotic microbial group is reviewed where the group is introduced in subsequent chapters.

Currently microbial taxonomy is in ferment because of the use of new molecular techniques in classifying microorganisms. Even though these new advances have generated much excitement and are drastically changing microbial taxonomy, more traditional approaches still have value and also will be discussed. The chapter begins with a general overview of taxonomy and then briefly describes the major characteristics used in microbial taxonomy. Next, the assessment of microbial phylogeny is discussed, followed by a description of the major divisions of life and *Bergey's Manual of Systematic Bacteriology.* The chapter ends with a brief survey of procaryotic phylogeny and diversity.

19.2 MICROBIAL EVOLUTION AND DIVERSITY

It has been estimated that our planet is about 4.6 billion years old. What appear to be fossilized remains of procaryotic cells around 3.5 to 3.8 billion years old have been discovered in stromatolites and sedimentary rocks (**figure 19.1**), and microbial fossils are clearly present in rocks as old as 2 billion years. **Stromatolites** are layered or stratified rocks, often domed, that are formed by incorporation of mineral sediments into microbial mats (**figure 19.2**). Modern stromatolites are formed by cyanobacteria; presumably at least some fossilized stromatolites were formed in the same way. Thus procaryotic life arose very shortly after the Earth cooled. Very likely the earliest procaryotes were anaerobic. Cyanobacteria and oxygen-producing photosynthesis probably developed 2.5 to 3.0 billion or more years ago. Microbial diversity increased greatly as oxygen became more plentiful.

The studies of Carl Woese and his collaborators on rRNA sequences in procaryotic cells suggest that procaryotes divided into two distinct groups very early on. **Figure 19.3** depicts a universal phylogenetic tree that reflects these views. The tree is divided into three major branches representing the three primary groups: *Bacteria, Archaea,* and *Eucarya.* The archaea and bacteria first diverged, then the eucaryotes developed. These three primary groups are called **domains** and placed above the phylum and kingdom levels (the traditional kingdoms are distributed among these three domains). The domains differ markedly from one another. Eucaryotic organisms with primarily glycerol fatty acyl diester membrane lipids and eucaryotic rRNA belong to the

Figure 19.1 Fossilized Bacteria.
Several microfossils resembling bacteria are shown, some with interpretive drawings. **(a)** Thin sections of Archean Apex chert from Western Australia; what appear to be fossilized remains of procaryotes are about 3.5 billion years old. **(b)** *Gloeodiniopsis,* about 1.5 billion years old, from carbonaceous chert in the Satka Formation of the southern Ural Mountains. The arrow points to the enclosing sheath. **(c)** *Palaeolyngbya,* about 950 million years old, from carbonaceous shale of the Lakhanda Formation of the Khabarovsk region in eastern Siberia.

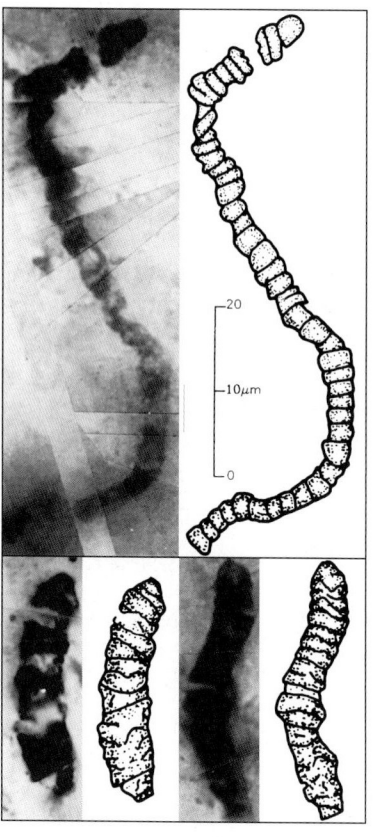

(a)

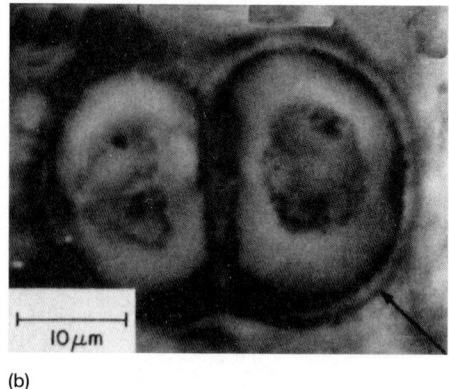

(b)

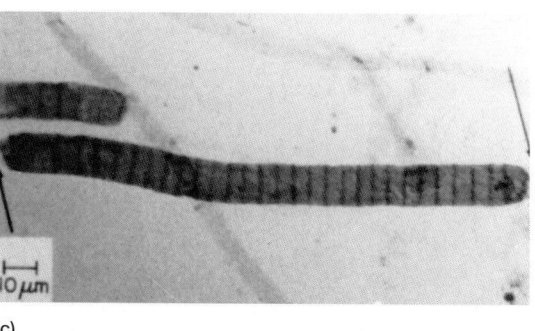

(c)

Eucarya. The domain *Bacteria* contains procaryotic cells with bacterial rRNA and membrane lipids that are primarily diacyl glycerol diesters. Procaryotes having isoprenoid glycerol diether or diglycerol tetraether lipids (*see pp. 439–40*) in their membranes and archaeal rRNA compose the third domain, *Archaea.*

It appears likely that modern eucaryotic cells arose from procaryotes about 1.4 billion years ago (some chemical evidence dates them as early as 2.7 billion years ago). There has been considerable speculation about how eucaryotes might have developed from procaryotic ancestors. It is not certain how this process occurred, and two hypotheses have been proposed. According to the first, nuclei, mitochondria, and chloroplasts arose by invagination of the plasma membrane to form double-membrane structures containing genetic material and capable of further development and specialization. The similarities between chloroplasts, mitochondria, and modern bacteria are due to conservation of primitive procaryotic features by the slowly changing organelles.

According to the more popular **endosymbiotic hypothesis,** the first event was nucleus formation in the pro-eucaryotic cell. The ancestral eucaryotic cell may have developed from a fusion of ancient bacteria and archaea. Possibly a gram-negative bacterial

Figure 19.2 Stromatolites. These are stromatolites at Shark Bay, Western Australia. Modern stromatolites are layered or stratified rocks formed by the incorporation of calcium sulfates, calcium carbonates, and other minerals into microbial mats. The mats are formed by cyanobacteria and other microorganisms.

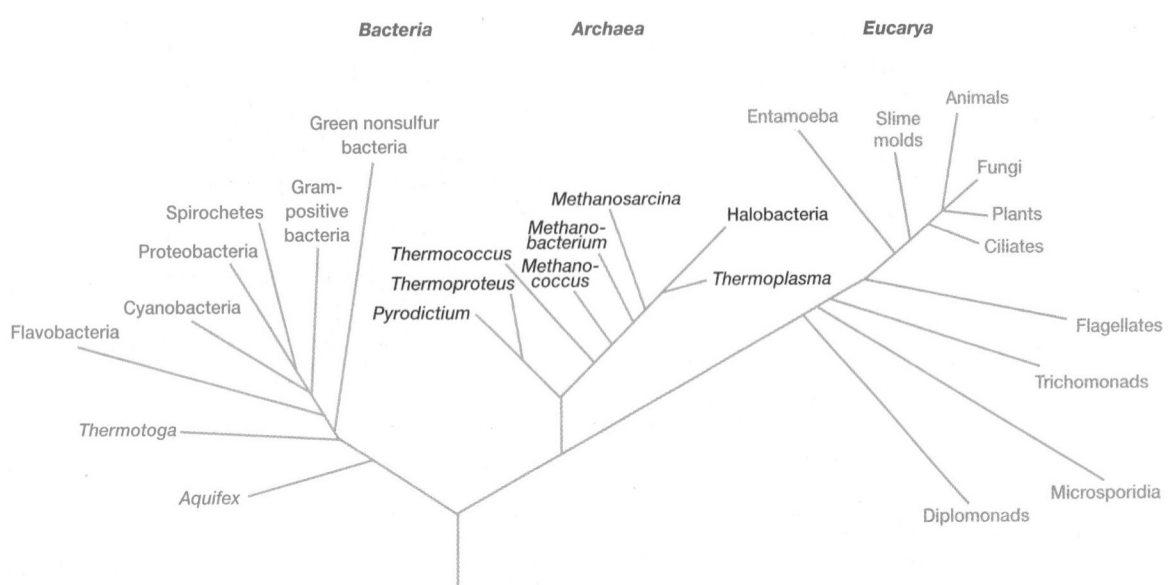

Figure 19.3 **Universal Phylogenetic Tree.** These relationships were determined from rRNA sequence comparisons. *Source: Adapted from G. J. Olsen and C. R. Woese. "Ribosomal RNA: A Key to Phylogeny" in* The FASEB Journal, *7:113–123, 1993.*

host cell that had lost its cell wall engulfed an archaeon to form an endosymbiotic association. The archaeon subsequently lost its wall and plasma membrane, while the host bacterium developed membrane infolds. Eventually the host genome was transferred to the original archaeon, and a nucleus and the endoplasmic reticulum was formed. Both bacterial and archaeal genes could be lost during formation of the eucaryotic genome. It should be noted that many believe the *Archaea* and *Eucarya* are more closely related than this hypothetical scenario implies. They propose that the eucaryotic line diverged from the *Archaea* and then the nucleus formed, possibly from the Golgi apparatus.

Mitochondria and chloroplasts appear to have developed later. The free-living, fermenting ancestral eucaryote with its nucleus established a permanent symbiotic relationship with photosynthetic bacteria, which then evolved into chloroplasts. Cyanobacteria have been considered the most likely ancestors of chloroplasts. More recently *Prochloron* has become the favorite candidate. *Prochloron* (*see pp. 460–63*) lives within marine invertebrates and resembles chloroplasts in containing both chlorophyll *a* and *b*, but not phycobilins. The existence of this bacterium suggests that chloroplasts arose from a common ancestor of prochlorophytes and cyanobacteria. Mitochondria arose from an endosymbiotic relationship between the free-living primitive eucaryote and bacteria with aerobic respiration (possibly an ancestor of three modern groups: *Agrobacterium, Rhizobium,* and *Rickettsia*). Some have proposed that aerobic respiration actually arose before oxygenic (oxygen-producing) photosynthesis and made use of small amounts of oxygen available at this early stage of planetary development. The exact sequence of development is still unclear.

The endosymbiotic hypothesis has received support from the discovery of an endosymbiotic cyanobacterium that inhabits the biflagellate protist *Cyanophora paradoxa* and acts as its chloroplast. This endosymbiont, called a cyanelle, resembles the cyanobacteria in its photosynthetic pigment system and fine structure and it is surrounded by a peptidoglycan layer. It differs from cyanobacteria in lacking the lipopolysaccharide outer membrane characteristic of gram-negative bacteria. The cyanelle may be a recently established endosymbiont that is evolving into a chloroplast. Further support is provided by rRNA trees, which locate chloroplast RNA within the cyanobacteria.

At present both hypotheses have supporters. It is possible that new data may help resolve the issue to everyone's satisfaction. However, these hypotheses concern processes that occurred in the distant past and cannot be directly observed. Thus a complete consensus on the matter may never be reached.

 TAXONOMIC RANKS

In preparing a classification scheme, one places the microorganism within a small, homogeneous group that is itself a member of larger groups in a nonoverlapping hierarchical arrangement. A category in any rank unites groups in the level below it based on shared properties (**figure 19.4**). In procaryotic taxonomy the most commonly used levels or ranks (in ascending order) are species, genera, families, orders, classes, and phyla. Microbial groups at each level or rank have names with endings or suffixes characteristic of that level (**table 19.1**). Microbiologists often use informal names in place of formal hierarchical ones. Typical examples of such names are purple bacteria, spirochetes, methane-oxidizing bacteria, sulfate-reducing bacteria, and lactic acid bacteria.

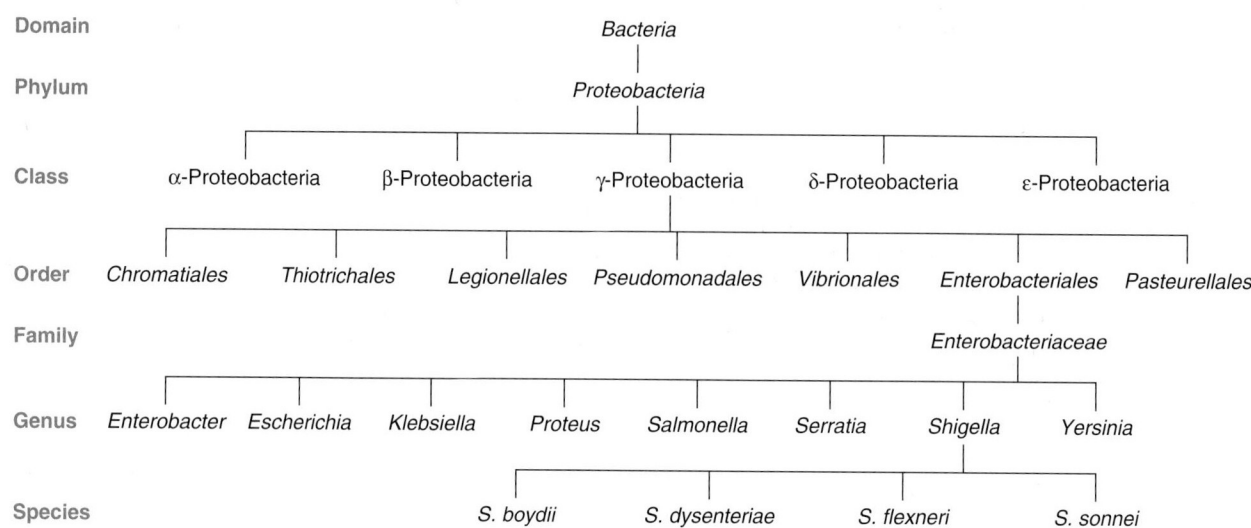

Figure 19.4 Hierarchical Arrangement in Taxonomy. In this example, members of the genus *Shigella* are placed within higher taxonomic ranks. Not all classification possibilies are given for each rank to simplify the diagram.

Table 19.1		
An Example of Taxonomic Ranks and Names		

Rank	Example
Domain	*Bacteria*
Phylum	*Proteobacteria*
Class	γ-*Proteobacteria*
Order	*Enterobacteriales*
Family	*Enterobacteriaceae*
Genus	*Shigella*
Species	*S. dysenteriae*

The basic taxonomic group in microbial taxonomy is the **species.** Taxonomists working with higher organisms define the term species differently than do microbiologists. Species of higher organisms are groups of interbreeding or potentially interbreeding natural populations that are reproductively isolated from other groups. This is a satisfactory definition for organisms capable of sexual reproduction but fails with many microorganisms because they do not reproduce sexually. Procaryotic species are characterized by phenotypic and genotypic (*see chapter 11*) differences. A **procaryotic species** is a collection of strains that share many stable properties and differ significantly from other groups of strains. This definition is very subjective and can be interpreted in many ways. The following more precise definition has been proposed by some bacterial taxonomists. A species (genomospecies) is a collection of strains that have a similar G + C composition and 70% or greater similarity as judged by DNA hybridization experiments (pp. 417–20). With an increasing amount of genome sequence data, some have argued that the definition of a procaryotic species needs further revision. Perhaps a species should be the collection

of organisms that share the same sequences in their core house-keeping genes (genes coding for products that are required by all cells and which are usually continually expressed). It will take much more work to resolve this complex issue. Whatever the definition, ideally a species also should be phenotypically distinguishable from other similar species. A **strain** is a population of organisms that is distinguishable from at least some other populations within a particular taxonomic category. It is considered to have descended from a single organism or pure culture isolate. Strains within a species may differ slightly from one another in many ways. **Biovars** are variant procaryotic strains characterized by biochemical or physiological differences, **morphovars** differ morphologically, and **serovars** have distinctive antigenic properties. One strain of a species is designated as the **type strain.** It is usually one of the first strains studied and often is more fully characterized than other strains; however, it does not have to be the most representative member. The type strain for the species is called the type species and is the nomenclatural type or the holder of the species name. A nomenclatural type is a device to ensure fixity of names when taxonomic rearrangements take place. For example, the type species must remain within the genus of which it is the nomenclatural type. Only those strains very similar to the type strain or type species are included in a species. Each species is assigned to a genus, the next rank in the taxonomic hierarchy. A **genus** is a well-defined group of one or more species that is clearly separate from other genera. In practice there is considerable subjectivity in assigning species to a genus, and taxonomists may disagree about the composition of genera.

Microbiologists name microorganisms by using the **binomial system** of the Swedish botanist Carl von Linné, or Carolus Linnaeus as he often is called. The Latinized, italicized name consists of two parts. The first part, which is capitalized, is the generic name, and the second is the uncapitalized specific epithet (e.g.,

Escherichia coli). The specific epithet is stable; the oldest epithet for a particular organism takes precedence and must be used. In contrast, a generic name can change if the organism is assigned to another genus because of new information. For example, the genus *Streptococcus* has been divided to form two new genera, *Enterococcus* and *Lactococcus* based on rRNA analysis and other characteristics. Thus *Streptococcus faecalis* is now *Enterococcus faecalis*. Often the name will be shortened by abbreviating the genus name with a single capital letter, for example *E. coli*. Approved lists of bacterial names were published in 1980 in the *International Journal of Systematic Bacteriology*, and new valid names are published periodically. *Bergey's Manual of Systematic Bacteriology* contains the currently accepted system of procaryotic taxonomy and will be discussed later in the chapter.

---

1. Define the following terms: taxonomy, classification, taxon, nomenclature, identification, systematics, species, strain, type strain, and binomial system.
2. Briefly describe the three domains into which living organisms may be divided.
3. How might the eucaryotic cell have arisen according to the endosymbiotic hypothesis?
4. How does the definition of species differ between organisms that reproduce sexually and those not able to do so?

19.4 CLASSIFICATION SYSTEMS

Once taxonomically relevant characteristics of microorganisms have been collected, they may be used to construct a classification system. The most desirable classification system, called a **natural classification,** arranges organisms into groups whose members share many characteristics and reflects as much as possible the biological nature of organisms. Linnaeus developed the first natural classification, based largely on anatomical characteristics, in the middle of the eighteenth century. It was a great improvement over previously employed artificial systems because knowledge of an organism's position in the scheme provided information about many of its properties. For example, classification of humans as mammals denotes that they have hair, self-regulating body temperature, and milk-producing mammary glands in the female.

There are two general ways in which classification systems can be constructed. Organisms can be grouped together based on overall similarity to form a phenetic system or they can be grouped based on probable evolutionary relationships to produce a phylogenetic system. Computers may be used to analyze data for the production of phenetic classifications. The process is called numerical taxonomy. This section briefly discusses phenetic and phylogenetic classifications, and describes numerical taxonomy.

Phenetic Classification

Many taxonomists maintain that the most natural classification is the one with the greatest information content or predictive value.

A good classification should bring order to biological diversity and may even clarify the function of a morphological structure. For example, if motility and flagella are always associated in particular microorganisms, it is reasonable to suppose that flagella are involved in at least some types of motility. When viewed in this way, the best natural classification system may be a **phenetic system,** one that groups organisms together based on the mutual similarity of their phenotypic characteristics. Although phenetic studies can reveal possible evolutionary relationships, they are not dependent on phylogenetic analysis. They compare many traits without assuming that any features are more phylogenetically important than others—that is, unweighted traits are employed in estimating general similarity. Obviously the best phenetic classification is one constructed by comparing as many attributes as possible. Organisms sharing many characteristics make up a single group or taxon.

Numerical Taxonomy

The development of computers has made possible the quantitative approach known as **numerical taxonomy.** Peter H. A. Sneath and Robert Sokal have defined numerical taxonomy as "the grouping by numerical methods of taxonomic units into taxa on the basis of their character states." Information about the properties of organisms is converted into a form suitable for numerical analysis and then compared by means of a computer. The resulting classification is based on general similarity as judged by comparison of many characteristics, each given equal weight. This approach was not feasible before the advent of computers because of the large number of calculations involved.

The process begins with a determination of the presence or absence of selected characters in the group of organisms under study. A character usually is defined as an attribute about which a single statement can be made. Many characters, at least 50 and preferably several hundred, should be compared for an accurate and reliable classification. It is best to include many different kinds of data: morphological, biochemical, and physiological.

After character analysis, an association coefficient, a function that measures the agreement between characters possessed by two organisms, is calculated for each pair of organisms in the group. The **simple matching coefficient (S_{SM}),** the most commonly used coefficient in bacteriology, is the proportion of characters that match regardless of whether the attribute is present or absent (**table 19.2**). Sometimes the **Jaccard coefficient (S_J)** is calculated by ignoring any characters that both organisms lack (table 19.2). Both coefficients increase linearly in value from 0.0 (no matches) to 1.0 (100% matches).

The simple matching coefficients, or other association coefficients, are then arranged to form a **similarity matrix.** This is a matrix in which the rows and columns represent organisms, and each value is an association coefficient measuring the similarity of two different organisms so that each organism is compared to every other one in the table (**figure 19.5a**). Organisms with great similarity are grouped together and separated from dissimilar or-

Table 19.2

The Calculation of Association Coefficients for Two Organisms

In this example, organisms A and B are compared in terms of the characters they do and do not share. The terms in the association coefficient equations are defined as follows:

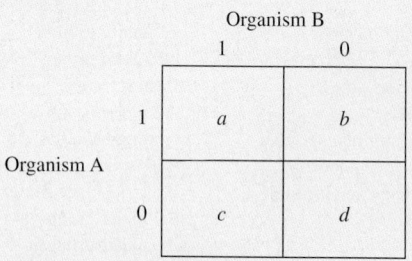

Organism B

	1	0
1	a	b
0	c	d

Organism A

a = number of characters coded as present (1) for both organisms
b and c = numbers of characters differing (1,0 or 0,1) between the two organisms
d = number of characters absent (0) in both organisms
Total number of characters compared = $a + b + c + d$

The simple matching coefficient $(S_{SM}) = \dfrac{a + d}{a + b + c + d}$

The Jaccard coefficient $(S_J) = \dfrac{a}{a + b + c}$

ganisms (figure 19.5*b*); such groups of organisms are called **phenons** (sometimes called phenoms).

The results of numerical taxonomic analysis are often summarized with a treelike diagram called a **dendrogram** (figure 19.5*c*). The diagram usually is placed on its side with the X-axis or abscissa graduated in units of similarity. Each branch point is at the similarity value relating the two branches. The organisms in the two branches share so many characteristics that the two groups are seen to be separate only after examination of association coefficients greater than the magnitude of the branch point value. Below the branch point value, the two groups appear to be one. The ordinate in such a dendrogram has no special significance, and the clusters may be arranged in any convenient order.

The significance of these clusters or phenons in traditional taxonomic terms is not always evident, and the similarity levels at which clusters are labeled species, genera, and so on, are a matter of judgment. Sometimes groups are simply called phenons and preceded by a number showing the similarity level above which they appear (e.g., a 70-phenon is a phenon with 70% or greater similarity among its constituents). Phenons formed at about 80% similarity often are equivalent to species.

Numerical taxonomy has already proved to be a powerful tool in microbial taxonomy. Although it often has simply reconfirmed already existing classification schemes, sometimes accepted classifications are found wanting. Numerical taxonomic methods also can be used to compare sequences of macromolecules such as RNA and proteins.

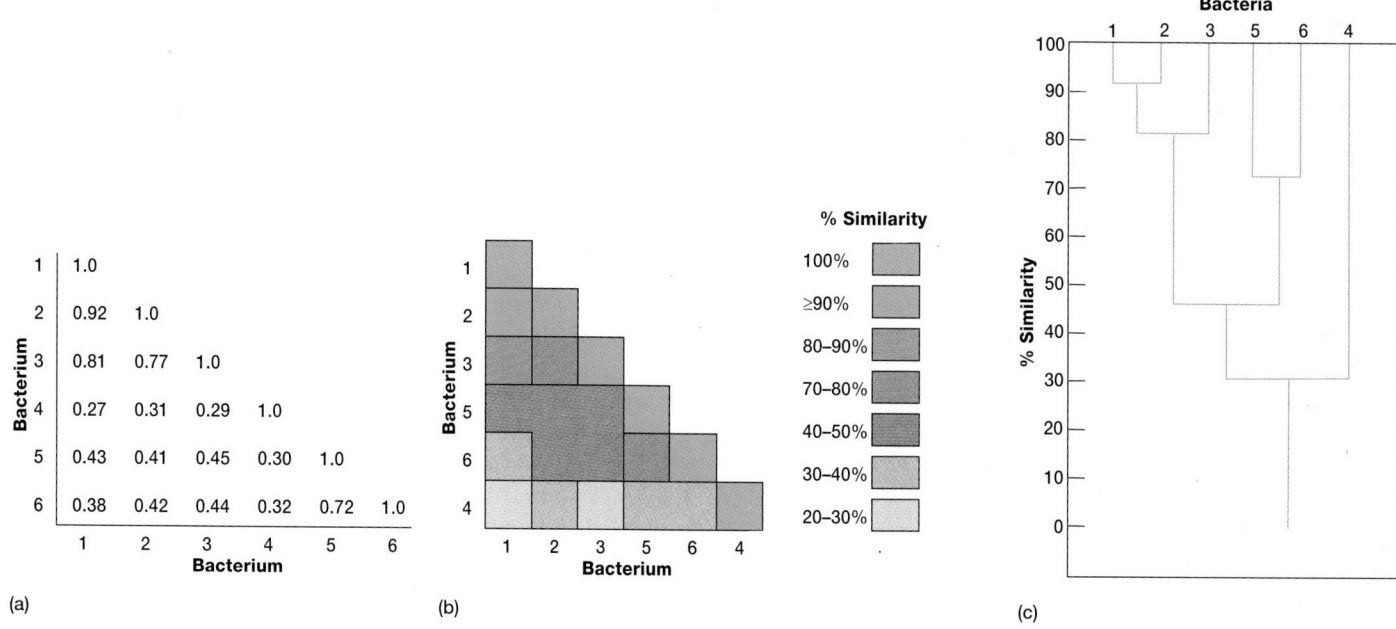

Figure 19.5   Clustering and Dendrograms in Numerical Taxonomy.   (**a**) A small similarity matrix that compares six strains of bacteria. The degree of similarity ranges from none (0.0) to complete similarity (1.0). (**b**) The bacteria have been rearranged and joined to form clusters of similar strains. For example, strains 1 and 2 are the most similar. The cluster of 1 plus 2 is fairly similar to strain 3, but not at all to strain 4. (**c**) A dendrogram showing the results of the analysis in part *b*. Strains 1 and 2 are members of a 90-phenon, and strains 1–3 form an 80-phenon. While strains 1–3 may be members of a single species, it is quite unlikely that strains 4–6 belong to the same species as 1–3.

Phylogenetic Classification

Following the publication in 1859 of Darwin's *On the Origin of Species,* biologists began trying to develop **phylogenetic** or **phyletic classification systems.** These are systems based on evolutionary relationships rather than general resemblance (the term **phylogeny** [Greek *phylon,* tribe or race, and *genesis,* generation or origin] refers to the evolutionary development of a species). This has proven difficult for procaryotes and other microorganisms, primarily because of the lack of a good fossil record. The direct comparison of genetic material and gene products such as RNA and proteins overcomes many of these problems.

1. What is a natural classification?
2. What are phylogenetic (phyletic) and phenetic classification systems? How do the two systems differ?
3. What is numerical taxonomy and why are computers so important to this approach?
4. Define the following terms: association coefficient, simple matching coefficient, Jaccard coefficient, similarity matrix, phenon, and dendrogram.
5. Which pair of species has more mutual similarity, a pair with an association coefficient of 0.9 or one with a coefficient of 0.6? Why?

19.5 MAJOR CHARACTERISTICS USED IN TAXONOMY

Many characteristics are used in classifying and identifying microorganisms. This section briefly reviews some of the most taxonomically important properties. For sake of clarity, characteristics have been divided into two groups: classical and molecular. Methods often employed in routine laboratory identification of bacteria are covered in the chapter on clinical microbiology (*see chapter 36*).

Classical Characteristics

Classical approaches to taxonomy make use of morphological, physiological, biochemical, ecological, and genetic characteristics. These characteristics have been employed in microbial taxonomy for many years. They are quite useful in routine identification and may provide phylogenetic information as well.

Morphological Characteristics

Morphological features are important in microbial taxonomy for many reasons. Morphology is easy to study and analyze, particularly in eucaryotic microorganisms and the more complex procaryotes. In addition, morphological comparisons are valuable because structural features depend on the expression of many genes, are usually genetically stable, and normally (at least in eucaryotes) these do not vary greatly with environmental changes. Thus morphological similarity often is a good indication of phylogenetic relatedness.

Table 19.3	
Some Morphological Features Used in Classification and Identification	
Feature	**Microbial Groups**
Cell shape	All major groups[a]
Cell size	All major groups
Colonial morphology	All major groups
Ultrastructural characteristics	All major groups
Staining behavior	Bacteria, some fungi
Cilia and flagella	All major groups
Mechanism of motility	Gliding bacteria, spirochetes
Endospore shape and location	Endospore-forming bacteria
Spore morphology and location	Bacteria, algae, fungi
Cellular inclusions	All major groups
Color	All major groups

[a]Used in classifying and identifying at least some bacteria, algae, fungi, and protozoa.

Many different morphological features are employed in the classification and identification of microorganisms (**table 19.3**). Although the light microscope has always been a very important tool, its resolution limit of about 0.2 µm (*see chapter 2*) reduces its usefulness in viewing smaller microorganisms and structures. The transmission and scanning electron microscopes, with their greater resolution, have immensely aided the study of all microbial groups.

Physiological and Metabolic Characteristics

Physiological and metabolic characteristics are very useful because they are directly related to the nature and activity of microbial enzymes and transport proteins. Since proteins are gene products, analysis of these characteristics provides an indirect comparison of microbial genomes. **Table 19.4** lists some of the most important of these properties.

Ecological Characteristics

Many properties are ecological in nature since they affect the relation of microorganisms to their environment. Often these are taxonomically valuable because even very closely related microorganisms can differ considerably with respect to ecological characteristics. Microorganisms living in various parts of the human body markedly differ from one another and from those growing in freshwater, terrestrial, and marine environments. Some examples of taxonomically important ecological properties are life cycle patterns; the nature of symbiotic relationships; the ability to cause disease in a particular host; and habitat preferences such as requirements for temperature, pH, oxygen, and osmotic concentration. Many growth requirements are also considered physiological characteristics (table 19.4).

Genetic Analysis

Because most eucaryotes are able to reproduce sexually, genetic analysis has been of considerable usefulness in the classification of these organisms. As mentioned earlier, the species is defined in

Table 19.4

Some Physiological and Metabolic Characteristics Used in Classification and Identification

Carbon and nitrogen sources
Cell wall constituents
Energy sources
Fermentation products
General nutritional type
Growth temperature optimum and range
Luminescence
Mechanisms of energy conversion
Motility
Osmotic tolerance
Oxygen relationships
pH optimum and growth range
Photosynthetic pigments
Salt requirements and tolerance
Secondary metabolites formed
Sensitivity to metabolic inhibitors and antibiotics
Storage inclusions

terms of sexual reproduction where possible. Although procaryotes do not reproduce sexually, the study of chromosomal gene exchange through transformation and conjugation is sometimes useful in their classification.

Transformation can occur between different procaryotic species but only rarely between genera. The demonstration of transformation between two strains provides evidence of a close relationship since transformation cannot occur unless the genomes are fairly similar. Transformation studies have been carried out with several genera: *Bacillus, Micrococcus, Haemophilus, Rhizobium,* and others. Despite transformation's usefulness, its results are sometimes hard to interpret because an absence of transformation may result from factors other than major differences in DNA sequence. Transformation (pp. 222, 299–301); Conjugation (pp. 296–99)

Conjugation studies also yield taxonomically useful data, particularly with the enteric bacteria (*see section 22.3*). For example, *Escherichia* can undergo conjugation with the genera *Salmonella* and *Shigella* but not with *Proteus* and *Enterobacter*. These observations fit with other data showing that the first three of these genera are more closely related to one another than to *Proteus* and *Enterobacter*.

Plasmids (*see section 13.2*) are undoubtedly important in taxonomy because they are present in most bacterial genera, and many carry genes coding for phenotypic traits. Because plasmids could have a significant effect on classification if they carried the gene for a trait of major importance in the classification scheme, it is best to base a classification on many characters. When the identification of a group is based on a few characteristics and some of these are coded for by plasmid genes, errors may result. For example, hydrogen sulfide production and lactose fermentation are very important in the taxonomy of the enteric bacteria, yet genes for both traits can be borne on plasmids as well as bacterial chromosomes. One must take care to avoid errors as a result of plasmid-borne traits.

Molecular Characteristics

Some of the most powerful approaches to taxonomy are through the study of proteins and nucleic acids. Because these are either direct gene products or the genes themselves, comparisons of proteins and nucleic acids yield considerable information about true relatedness. These more recent molecular approaches have become increasingly important in procaryotic taxonomy.

Comparison of Proteins

The amino acid sequences of proteins are direct reflections of mRNA sequences and therefore closely related to the structures of the genes coding for their synthesis. For this reason, comparisons of proteins from different microorganisms are very useful taxonomically. There are several ways to compare proteins. The most direct approach is to determine the amino acid sequence of proteins with the same function. The sequences of proteins with dissimilar functions often change at different rates; some sequences change quite rapidly, whereas others are very stable. Nevertheless, if the sequences of proteins with the same function are similar, the organisms possessing them are probably closely related. The sequences of cytochromes and other electron transport proteins, histones, heat-shock proteins, transcription and translation proteins, and a variety of metabolic enzymes have been used in taxonomic studies. Because protein sequencing is slow and expensive, more indirect methods of comparing proteins frequently have been employed. The electrophoretic mobility of proteins (*see pp. 319–20*) is useful in studying relationships at the species and subspecies levels. Antibodies can discriminate between very similar proteins, and immunologic techniques are used to compare proteins from different microorganisms. Antibody-antigen reactions in vitro (pp. 751–58)

The physical, kinetic, and regulatory properties of enzymes have been employed in taxonomic studies. Because enzyme behavior reflects amino acid sequence, this approach is useful in studying some microbial groups, and group-specific patterns of regulation have been found.

Nucleic Acid Base Composition

Microbial genomes can be directly compared, and taxonomic similarity can be estimated in many ways. The first, and possibly the simplest, technique to be employed is the determination of DNA base composition. DNA contains four purine and pyrimidine bases: adenine (A), guanine (G), cytosine (C), and thymine (T). In double-stranded DNA, A pairs with T, and G pairs with C. Thus the $(G + C)/(A + T)$ ratio or **G + C content,** the percent of $G + C$ in DNA, reflects the base sequence and varies with sequence changes as follows: DNA and RNA structure (pp 224–27)

$$\text{Mol\% G} + \text{C} = \frac{G + C}{G + C + A + T} \times 100$$

The base composition of DNA can be determined in several ways. Although the G + C content can be ascertained after hydrolysis of DNA and analysis of its bases with high-performance

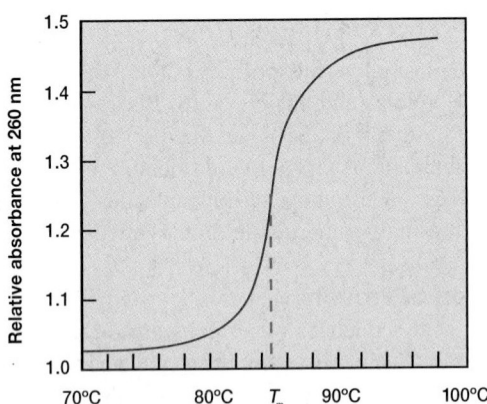

Figure 19.6 A DNA Melting Curve. The T_m is indicated.

liquid chromatography (HPLC), physical methods are easier and more often used. The G + C content often is determined from the **melting temperature (T_m)** of DNA. In double-stranded DNA three hydrogen bonds join GC base pairs, and two bonds connect AT base pairs (*see section 11.2*). As a result DNA with a greater G + C content will have more hydrogen bonds, and its strands will separate only at higher temperatures—that is, it will have a

higher melting point. DNA melting can be easily followed spectrophotometrically because the absorbance of 260 nm UV light by DNA increases during strand separation. When a DNA sample is slowly heated, the absorbance increases as hydrogen bonds are broken and reaches a plateau when all the DNA has become single stranded (**figure 19.6**). The midpoint of the rising curve gives the melting temperature, a direct measure of the G + C content. Since the density of DNA also increases linearly with G + C content, the percent G + C can be obtained by centrifuging DNA in a CsCl density gradient (*see chapter 16*).

The G + C content of many microorganisms has been determined (**table 19.5**). The G + C content of DNA from animals and higher plants averages around 40% and ranges between 30 and 50%. In contrast, the DNA of both eucaryotic and procaryotic microorganisms varies greatly in G + C content; procaryotic G + C content is the most variable, ranging from around 25 to almost 80%. Despite such a wide range of variation, the G + C content of strains within a particular species is constant. If two organisms differ in their G + C content by more than about 10%, their genomes have quite different base sequences. On the other hand, it is not safe to assume that organisms with very similar G + C contents also have similar DNA base sequences because two very different base sequences can be constructed from the same pro-

Table 19.5

Representative G + C Contents of Microorganisms

Organism	Percent G + C	Organism	Percent G + C	Organism	Percent G + C
Bacteria		*Salmonella*	50–53	*Plasmodium berghei*	41
Actinomyces	59–73	*Spirillum*	38	*Stentor polymorphus*	45
Anabaena	38–44	*Spirochaeta*	51–65	*Tetrahymena*	19–33
Bacillus	32–62	*Staphylococcus*	30–38	*Trichomonas*	29–34
Bacteroides	28–61	*Streptococcus*	33–44	*Trypanosoma*	45–59
Bdellovibrio	33–52	*Streptomyces*	69–73		
Caulobacter	63–67	*Sulfolobus*	31–37	**Slime Molds**	
Chlamydia	41–44	*Thermoplasma*	46	*Dictyostelium*	22–25
Chlorobium	49–58	*Thiobacillus*	52–68	*Lycogala*	42
Chromatium	48–70	*Treponema*	25–54	*Physarum polycephalum*	38–42
Clostridium	21–54				
Cytophaga	33–42	**Algae**		**Fungi**	
Deinococcus	62–70	*Acetabularia mediterranea*	37–53	*Agaricus bisporus*	44
Escherichia	48–52	*Chlamydomonas*	60–68	*Amanita muscaria*	57
Halobacterium	66–68	*Chlorella*	43–79	*Aspergillus niger*	52
Hyphomicrobium	59–67	*Cyclotella cryptica*	41	*Blastocladiella emersonii*	66
Methanobacterium	32–50	*Euglena gracilis*	46–55	*Candida albicans*	33–35
Micrococcus	64–75	*Nitella*	49	*Claviceps purpurea*	53
Mycobacterium	62–70	*Nitzschia angularis*	47	*Coprinus lagopus*	52–53
Mycoplasma	23–40	*Ochromonas danica*	48	*Fomes fraxineus*	56
Myxococcus	68–71	*Peridinium triquetrum*	53	*Mucor rouxii*	38
Neisseria	47–54	*Scenedesmus*	52–64	*Neurospora crassa*	52–54
Nitrobacter	60–62	*Spirogyra*	39	*Penicillium notatum*	52
Oscillatoria	40–50	*Volvox carteri*	50	*Polyporus palustris*	56
Prochloron	41			*Rhizopus nigricans*	47
Proteus	38–41	**Protozoa**		*Saccharomyces cerevisiae*	36–42
Pseudomonas	58–70	*Acanthamoeba castellanii*	56–58	*Saprolegnia parasitica*	61
Rhodospirillum	62–66	*Amoeba proteus*	66		
Rickettsia	29–33	*Paramecium* spp.	29–39		

portions of AT and GC base pairs. Only if two microorganisms also are alike phenotypically does their similar G + C content suggest close relatedness.

G + C content data are taxonomically valuable in at least two ways. First, they can confirm a taxonomic scheme developed using other data. If organisms in the same taxon are too dissimilar in G + C content, the taxon probably should be divided. Second, G + C content appears to be useful in characterizing procaryotic genera since the variation within a genus is usually less than 10% even though the content may vary greatly between genera. For example, *Staphylococcus* has a G + C content of 30 to 38%, whereas *Micrococcus* DNA has 64 to 75% G + C; yet these two genera of gram-positive cocci have many other features in common.

Nucleic Acid Hybridization

The similarity between genomes can be compared more directly by use of **nucleic acid hybridization** studies. If a mixture of single-stranded DNA formed by heating dsDNA is cooled and held at a temperature about 25°C below the T_m, strands with complementary base sequences will reassociate to form stable dsDNA, whereas noncomplementary strands will remain single (**figure 19.7**). Because strands with similar, but not identical, sequences associate to form less temperature stable dsDNA hybrids, incubation of the mixture at 30 to 50°C below the T_m will allow hybrids of more diverse ssDNAs to form. Incubation at 10 to 15°C below the T_m permits hybrid formation only with almost identical strands.

In one of the more widely used hybridization techniques, nitrocellulose filters with bound nonradioactive DNA strands are incubated at the appropriate temperature with single-stranded DNA fragments made radioactive with ^{32}P, ^{3}H, or ^{14}C. After radioactive fragments are allowed to hybridize with the membrane-bound ssDNA, the membrane is washed to remove any nonhybridized ssDNA and its radioactivity is measured. The quantity of radioactivity bound to the filter reflects the amount of hybridization and thus the similarity of the DNA sequences. The degree of similarity or homology is expressed as the percent of experimental DNA radioactivity retained on the filter compared with the percent of homologous DNA radioactivity bound under the same conditions (**table 19.6** provides examples). Two strains whose DNAs show at least 70% relatedness under optimal hybridization conditions and less than a 5% difference in T_m often are considered members of the same species.

If DNA molecules are very different in sequence, they will not form a stable, detectable hybrid. Therefore DNA-DNA hybridization is used to study only closely related microorganisms. More distantly related organisms are compared by carrying out DNA-RNA hybridization experiments using radioactive ribosomal or transfer RNA. Distant relationships can be detected because rRNA and tRNA genes represent only a small portion of the total DNA genome and have not evolved as rapidly as most other microbial genes. The technique is similar to that employed for DNA-DNA hybridization: membrane-bound DNA is incubated with radioactive rRNA, washed, and counted. An even more accurate measurement of homology is obtained by finding the temperature

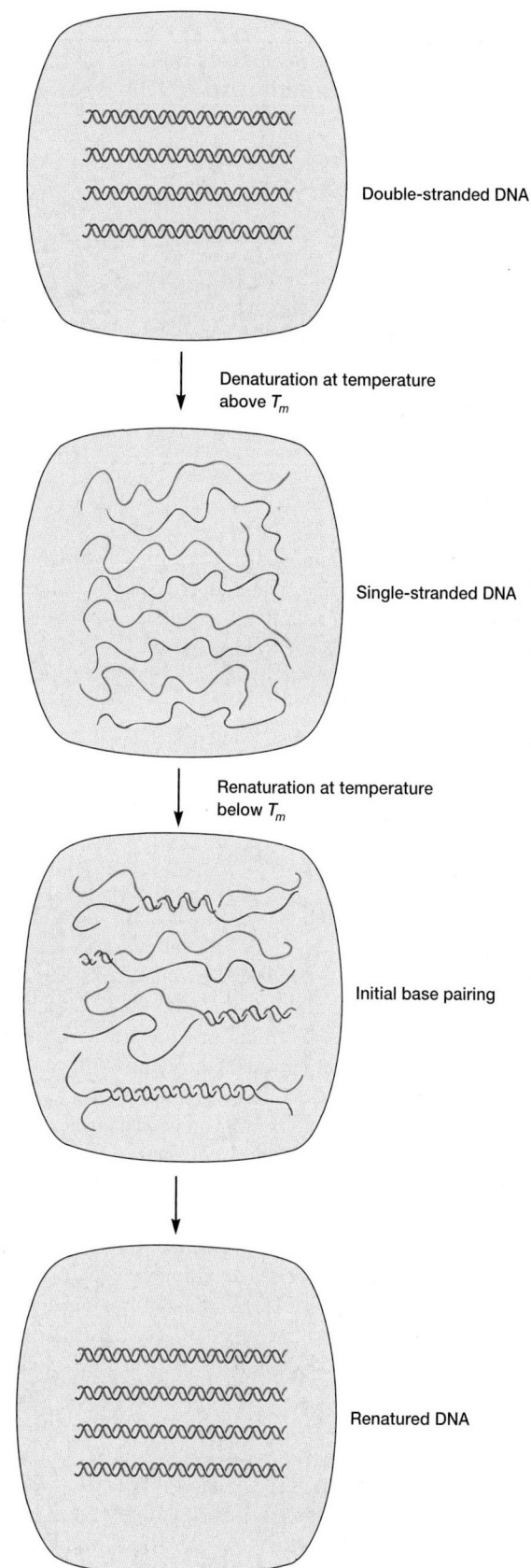

Double-stranded DNA

Denaturation at temperature above T_m

Single-stranded DNA

Renaturation at temperature below T_m

Initial base pairing

Renatured DNA

Figure 19.7   Nucleic Acid Melting and Hybridization. Complementary strands are shown in red and blue.

Table 19.6

Comparison of *Neisseria* Species by DNA Hybridization Experiments

Membrane-Attached DNA[a]	Percent Homology[b]
Neisseria meningitidis	100
N. gonorrhoeae	78
N. sicca	45
N. flava	35

Source: Data from T. E. Staley and R. R. Colwell, "Applications of Molecular Genetics and Numerical Taxonomy to the Classification of Bacteria" in *Annual Review of Ecology and Systematics*, 8: 282, 1973.
[a]The experimental membrane-attached nonradioactive DNA from each species was incubated with radioactive *N. meningitidis* DNA, and the amount of radioactivity bound to the membrane was measured. The more radioactivity bound, the greater the homology between DNA sequences.

[b] $\dfrac{N.meningitidis \text{ DNA bound to experimental DNA}}{\text{Amount bound to membrane attached } N.meningitidis \text{ DNA}} \times 100$

required to dissociate and remove half the radioactive rRNA from the membrane; the higher this temperature, the stronger the rRNA-DNA complex and the more similar the sequences. Ribosomes and ribosomal RNA (pp. 262–63); Transfer RNA (pp. 260–62)

Nucleic Acid Sequencing

Despite the usefulness of G + C content determination and nucleic acid hybridization studies, genome structures can be directly compared only by sequencing DNA and RNA. Techniques for rapidly sequencing both DNA and RNA are now available; thus far RNA sequencing has been used more extensively in microbial taxonomy.

Most attention has been given to sequences of the 5S and 16S rRNAs isolated from the 50S and 30S subunits, respectively, of procaryotic ribosomes (*see sections 3.3 and 12.2*). The rRNAs are almost ideal for studies of microbial evolution and relatedness since they are essential to a critical organelle found in all microorganisms. Their functional role is the same in all ribosomes. Furthermore, their structure changes very slowly with time, presumably because of their constant and critical role. Because rRNA contains variable and stable sequences, both closely related and very distantly related microorganisms can be compared. This is an important advantage as distantly related organisms can be studied only using sequences that change little with time.

Complete rRNAs can be sequenced using several different procedures. In one approach, RNA is first isolated and purified. Then, reverse transcriptase is used to make complementary DNA (cDNA) using primers that are complementary to conserved rRNA sequences. Next, the polymerase chain reaction amplifies the cDNA. Finally, the cDNA is sequenced and the rRNA sequence deduced from the results. Probably most often these days the rRNA genes are directly amplified and sequenced in the following way. The polymerase chain reaction is used to amplify the 16S rDNA from the bacterial genome. This process is not difficult because primers complementary to conserved sequences in the genes for small-subunit ribosomal RNA are readily available for use in the PCR reaction. The amplified rDNA is then sequenced either man-

ually or by an automated sequencer using the Sanger method. The polymerase chain reaction (pp. 316–19); DNA sequencing (pp. 336–39)

Recently complete procaryotic genomes have been sequenced (*see chapter 15*). Direct comparison of complete genome sequences undoubtedly will become important in procaryotic taxonomy.

---

1. Summarize the advantages of using each major group of characteristics (morphological, physiological/metabolic, ecological, genetic, and molecular) in classification and identification. How is each group related to the nature and expression of the genome? Give examples of each type of characteristic.
2. What two modes of genetic exchange in procaryotes have proved taxonomically useful? Why are plasmids of such importance in bacterial taxonomy?
3. Briefly describe some ways in which proteins from different organisms can be compared.
4. What is the G + C content of DNA, and how can it be determined through melting temperature studies and density gradient centrifugation?
5. Discuss the use of G + C content in taxonomy. Why is it not safe to assume that two microorganisms with the same G + C content belong to the same species? In what two ways are G + C content data taxonomically valuable?
6. Describe how nucleic acid hybridization studies are carried out using membrane-bound DNA. Why might one wish to vary the incubation temperature during hybridization? What is the advantage of conducting DNA-RNA hybridization studies?
7. How are rRNA sequencing studies conducted, and why is rRNA so suitable for determining relatedness?

 ### 19.6 ASSESSING MICROBIAL PHYLOGENY

Procaryotic taxonomy is changing rapidly. This is caused by ever-increasing knowledge of the biology of procaryotes and remarkable advances in computers and the use of molecular characteristics to determine phylogenetic relationships between procaryotic groups. This section briefly describes some of the ways in which phylogenetic relationships are assessed.

Molecular Chronometers

The sequences of nucleic acids and proteins change with time and are considered to be **molecular chronometers.** This concept, first suggested by Zuckerkandl and Pauling (1965), is important in the use of molecular sequences in determining phylogenetic relationships and is based on the assumption that there is an evolutionary clock. It is thought that the sequences of many rRNAs and proteins gradually change over time without destroying or severely altering their functions. One assumes that such changes are selectively neutral, occur fairly randomly, and increase linearly with time. When the sequences of similar molecules are quite different in two groups of organisms, the groups diverged from one

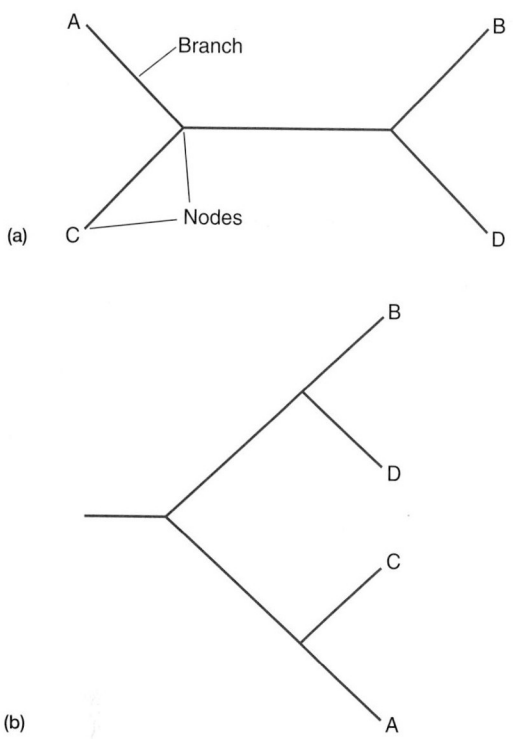

Figure 19.8   Examples of Phylogenetic Trees.   (**a**) Unrooted tree joining four taxonomic units. (**b**) Rooted tree. See text for details.

another a long time ago. Phylogenetic analysis using molecular chronometers is somewhat complex because the rate of sequence change can vary; some periods are characterized by especially rapid change. Furthermore, different molecules and various parts of the same molecule can change at different rates. Highly conserved molecules such as rRNAs are used to follow large-scale evolutionary changes, whereas rapidly changing molecules are employed in following speciation. Not everyone believes that molecular chronometers, and particularly protein clocks, are very accurate. Further studies will be required to establish their accuracy and usefulness.

Phylogenetic Trees

Phylogenetic relationships are illustrated in the form of branched diagrams or trees. A **phylogenetic tree** is a graph made of branches that connect nodes (**figure 19.8**). The nodes represent taxonomic units such as species or genes; the external nodes, those at the end of the branches, represent living organisms. The tree may have a time scale, or the length of the branches may represent the number of molecular changes that have taken place between the two nodes. Finally, a tree may be unrooted or rooted. An unrooted tree (figure 19.8a) simply represents phylogenetic relationships but does not provide an evolutionary path. Figure 19.8a shows that A is more closely related to C than it is to either B or D, but does not specify the common ancestor for the four

species or the direction of change. In contrast, the rooted tree (figure 19.8b) does give a node that serves as the common ancestor and shows the development of the four species from this root. It is much more difficult to develop a rooted tree. For example, there are 15 possible rooted trees that connect four species, but only three possible unrooted trees.

Phylogenetic trees are developed by comparing molecular sequences. To compare two molecules their sequences must first be aligned so that similar parts match up. The object is to align and compare homologous sequences, ones that are similar because they had a common origin in the past. This is not an easy task, and computers plus fairly complex mathematics must be employed to minimize the number of gaps and mismatches in the sequences being compared.

Once the molecules have been aligned, the number of positions that vary in the sequences can be determined. These data are used to calculate a measure of the difference between the sequences. Often the difference is expressed as the **evolutionary distance.** This is simply a quantitative indication of the number of positions that differ between two aligned macromolecules. Statistical adjustments can be made for back mutations and multiple substitutions that may have occurred. Organisms are then clustered together based on similarity in the sequences. The most similar organisms are clustered together, then compared with the remaining organisms to form a larger cluster associated together at a lower level of similarity or evolutionary distance. The process continues until all organisms are included in the tree.

Phylogenetic relationships also can be estimated by techniques such as parsimony analysis. In this approach, relationships are determined by estimating the minimum number of sequence changes required to give the final sequences being compared. It is presumed that evolutionary change occurs along the shortest pathway with the fewest changes or steps from an ancestor to the organism in question. The tree or pattern of relationships is favored that is simplest and requires the fewest assumptions.

rRNA, DNA, and Proteins as Indicators of Phylogeny

Although a variety of molecular techniques are used in estimating the phylogenetic relatedness of procaryotes, the comparison of 16S rRNAs isolated from thousands of strains is of particular importance (**figure 19.9**). Recall that either complete rRNA or rRNA fragments can be sequenced and compared (p. 420). The association coefficients or S_{ab} values from rRNA studies are assumed to be a true measure of relatedness; the higher the S_{ab} values obtained from comparing two organisms, the more closely the organisms are related to each other. If the sequences of the 16S rRNAs of two organisms are identical, the S_{ab} value is 1.0. S_{ab} values also are a measure of evolutionary time. A group of procaryotes that branched off from other procaryotes long ago will exhibit a large range of S_{ab} values because it has had more time to diversify than a group that developed more recently. That is, the narrower the range of S_{ab} values in a group of procaryotes, the more modern it is. After S_{ab} values have been determined, a computer calculates

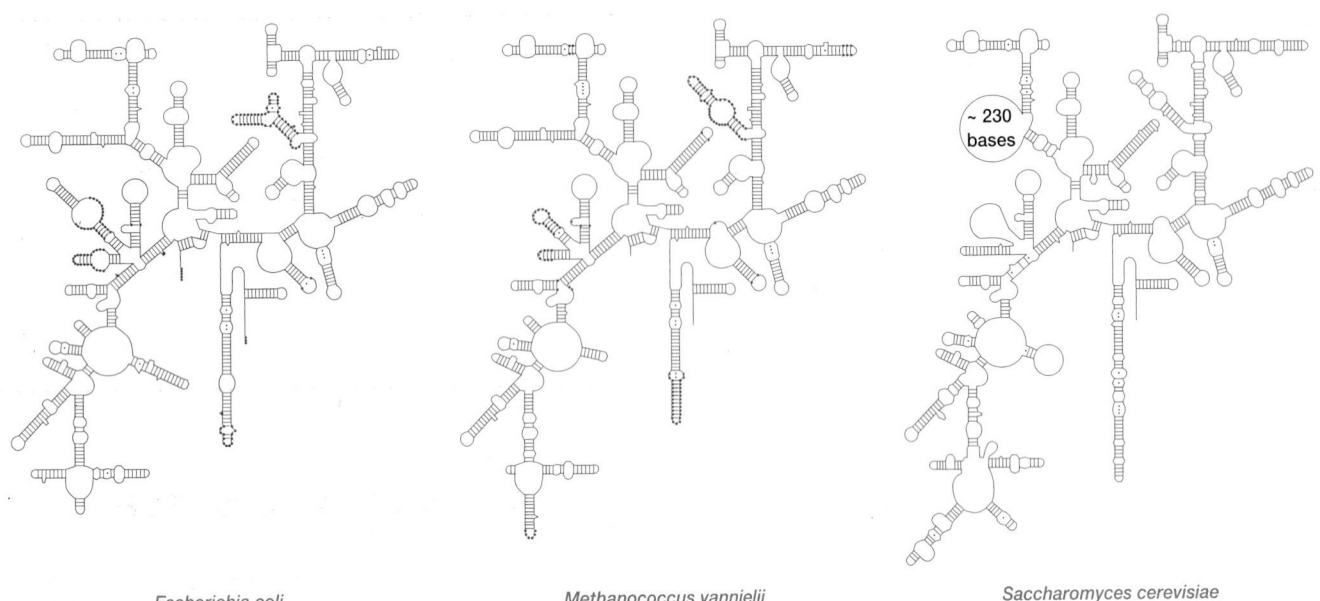

Escherichia coli   *Methanococcus vannielii*   *Saccharomyces cerevisiae*

Figure 19.9   Small Ribosomal Subunit RNA. Representative examples of rRNA secondary structures from the three primary domains: bacteria (*Escherichia coli*), archaea (*Methanococcus vannielii*), and eucarya (*Saccharomyces cerevisiae*). The red dots mark positions where bacteria and archaea normally differ.   *Source: Data from C. P. Woese.* Microbiological Reviews, *51(2):221–227, 1987.*

the relatedness of the organisms and summarizes their relationships in a tree or dendrogram (figures 19.5 and 19.8).

Ribosomal RNA sequence studies have uncovered a feature of great practical importance. The 16S rRNA of most major phylogenetic groups has one or more characteristic nucleotide sequences called oligonucleotide signatures. **Oligonucleotide signature sequences** are specific oligonucleotide sequences that occur in most or all members of a particular phylogenetic group. They are rarely or never present in other groups, even closely related ones. Thus signature sequences can be used to place microorganisms in the proper group. Signature sequences have been identified for bacteria, archaea, eucaryotes, and many major procaryotic groups (**table 19.7**).

Although rRNA comparisons are useful above the species level, DNA similarity studies sometimes are more effective in categorizing individual species and genera. These comparisons can be carried out using G + C content or hybridization studies, as already discussed. Techniques such as direct sequence analysis and analysis of DNA restriction fragment patterns also can be used. There are advantages to DNA comparisons. As with rRNA, the DNA composition of a cell does not change with growth conditions. DNA comparisons also are based on the complete genomes rather than a fraction, and make it easier to precisely define a species based on the 70% relatedness criterion. Full sequences of genomes now are being published and will make it easier to study the impact on phylogenetic schemes of such processes as lateral or horizontal gene transfer as will be discussed later.

A comparison of many different genomes and proteins has shown that signature sequences are present in genes other than

Table 19.7

Selected 16S rRNA Signature Sequences for Some Bacterial Groups[a]

Position in rRNA	Consensus Composition	γ-Proteobacteria	Cyanobacteria	Spirochetes	Bacteroides	Green Sulfur	Green Nonsulfur	Deinococcus	Gram Positive (Low GC)	Gram Positive (High GC)	Planctomyces
47	C	+	+	U	+	+	+	+	+	+	G
53	A	+	+	G	+	+	G	+	+	+	G
570	G	+	+	+	U	+	+	+	+	+	U
812	G	c	+	+	+	+	+	C	+	+	+
906	G	Ag	+	+	+	+	A	+	+	A	+
955	U	+	+	+	+	+	+	+	+	AC	C
1,207	G	+	C	+	+	+	+	+	C	C	+
1,234	C	+	+	a	U	A	+	+	+	+	+

[a]A plus sign in a column means that the group has the same base as the consensus sequence. If the letter is given in upper case, it is changed in more than 90% of the cases. A lowercase letter signifies a minor occurrence base (<15% of the cases).

those encoding ribosomal RNA. Many genes have inserts or deletions of specific lengths and sequences at fixed positions, and a particular insert or deletion may be found exclusively among all members of one or more phyla. R. S. Gupta refers to these as "conserved indels." These signature sequences are particularly useful in phylogenetic studies if they are flanked by conserved regions. In such cases, observed changes in the signature sequence will not be due to sequence misalignments. The signature sequences located in highly conserved housekeeping genes do not appear greatly affected by horizontal gene transfers (p. 425) and can be very effectively employed in phylogenetic analysis. In many cases the results are similar to those obtained from rRNA analyses.

Many protein sequences are currently used to develop phylogenetic trees. This approach does have some advantages over rRNA comparisons. A sequence of 20 amino acids has more information per site than a sequence of four nucleotides. Protein sequences are less affected by organism-specific differences in G + C content than are DNA and RNA sequences. Finally, protein sequence alignment is easier because it is not dependent on secondary structure as is an rRNA sequence. Proteins evolve at different rates, as might be expected. Indispensable proteins with constant functions do not change as rapidly (e.g., histones and heat-shock proteins), whereas proteins such as immunoglobulins evolve quite rapidly. Thus not all proteins are suitable for studying large-scale changes that occur over long periods. As mentioned earlier, there is a question about the adequacy of protein-based clocks.

It is clear that sequences of all three macromolecules can provide valuable phylogenetic information. However, different sequences sometimes produce different trees, and it may be difficult to decide which result is most accurate. Presumably more molecular data plus further study of phenotypic properties will help resolve uncertainties.

Polyphasic Taxonomy

Because phylogenetic results vary with the data used in analysis, many taxonomists believe that all possible valid data should be employed in determining phylogeny. In the approach called **polyphasic taxonomy**, taxonomic schemes are developed using a wide range of phenotypic and genotypic information ranging from molecular properties to ecological characteristics. The techniques that are appropriate for grouping organisms depend on the level of taxonomic resolution needed. For example, serological techniques can be used to identify strains, but not genera or species. Protein electrophoretic patterns are useful in determining species, but not genera or families. DNA hybridization and the analysis of % G + C can be used in studying species and genera. Characteristics such as chemical composition, DNA probe results, rRNA sequences, and DNA sequences can be used to define species, genera, and families. Where possible, as many properties as possible are used to get more stable and reliable results. Successful polyphasic approaches often will help one select techniques for rapid identification of the microorganism.

Because rRNA sequences have been used so extensively, we will focus mainly on the phylogenetic trees derived from rRNA studies.

1. What are molecular chronometers and upon what assumptions are they based?
2. Define phylogenetic tree and evolutionary distance. What is the difference between an unrooted and a rooted tree?
3. Discuss the use of S_{ab} values in determining the relatedness of procaryotes and the evolutionary age of taxonomic groups. What are oligonucleotide signature sequences?
4. Why might one choose to use DNA or protein sequences in phylogenetic studies?
5. Describe polyphasic taxonomy and discuss some of its advantages.

 ## 19.7 THE MAJOR DIVISIONS OF LIFE

Since the beginning of biology, organisms have been classified as either plants or animals. However, discoveries in microbiology over the past century have shown that the two-kingdom system is oversimplified. Although not all biologists would agree, most microbiologists now believe that living forms can be divided into three distinctly different groups. We will first review this system in more detail, then turn to alternate views.

Domains

As mentioned earlier and illustrated in figure 19.3 (p. 412), Carl Woese and his collaborators have used rRNA studies to group all living organisms into three domains: *Archaea, Bacteria,* and *Eucarya.* Thus there are two quite different groups of procaryotes, the bacteria and the archaea. The bacteria comprise the vast majority of procaryotes. Among other properties, bacteria either have cell wall peptidoglycan containing muramic acid or are related to bacteria with such cell walls, and have membrane lipids with ester-linked, straight-chained fatty acids that resemble eucaryotic membrane lipids (**table 19.8**). The second group, the archaea, differ from bacteria in many respects and resemble eucaryotes in some ways (table 19.8). Although the archaea are described in more detail at a later point, it should be noted that they differ from bacteria in lacking muramic acid in their cell walls and in possessing (1) membrane lipids with ether-linked branched aliphatic chains, (2) transfer RNAs without thymidine in the T or TψC arm (*see section 12.2*), (3) distinctive RNA polymerase enzymes, and (4) ribosomes of different composition and shape. Thus although archaea resemble bacteria in their procaryotic cell structure, they vary considerably on the molecular level. Both groups differ from eucaryotes in their cell ultrastructure (*see pp. 90–91*) and many other properties. However, inspection of table 19.8 shows that both bacteria and archaea do share some biochemical properties with eucaryotic cells. For example, bacteria and eucaryotes have ester-linked membrane lipids; archaea and eucaryotes are similar with respect to some components of the RNA and protein synthetic systems. The archaea (chapter 20)

Table 19.8
Comparison of *Bacteria, Archaea,* and *Eucarya*

Property	*Bacteria*	*Archaea*	*Eucarya*
Membrane-Enclosed Nucleus with Nucleolus	Absent	Absent	Present
Complex Internal Membranous Organelles	Absent	Absent	Present
Cell Wall	Almost always have peptidoglycan containing muramic acid	Variety of types, no muramic acid	No muramic acid
Membrane Lipid	Have ester-linked, straight-chained fatty acids	Have ether-linked, branched aliphatic chains	Have ester-linked, straight-chained fatty acids
Gas Vesicles	Present	Present	Absent
Transfer RNA	Thymine present in most tRNAs N-formylmethionine carried by initiator tRNA	No thymine in T or TψC arm of tRNA Methionine carried by initiator tRNA	Thymine present Methionine carried by initiator tRNA
Polycistronic mRNA	Present	Present	Absent
mRNA Introns	Absent	Absent	Present
mRNA Splicing, Capping, and Poly A Tailing	Absent	Absent	Present
Ribosomes			
Size	70S	70S	80S (cytoplasmic ribosomes)
Elongation factor 2	Does not react with diphtheria toxin	Reacts	Reacts
Sensitivity to chloramphenicol and kanamycin	Sensitive	Insensitive	Insensitive
Sensitivity to anisomycin	Insensitive	Sensitive	Sensitive
DNA-Dependent RNA Polymerase			
Number of enzymes	One	Several	Three
Structure	Simple subunit pattern (4 subunits)	Complex subunit pattern similar to eucaryotic enzymes (8–12 subunits)	Complex subunit pattern (12–14 subunits)
Rifampicin sensitivity	Sensitive	Insensitive	Insensitive
Polymerase II Type Promoters	Absent	Present	Present
Metabolism			
Similar ATPase	No	Yes	Yes
Methanogenesis	Absent	Present	Absent
Nitrogen fixation	Present	Present	Absent
Chlorophyll-based photosynthesis	Present	Absent	Present[a]
Chemolithotrophy	Present	Present	Absent

[a]Present in chloroplasts (of bacterial origin).

Although the preceding view is the most widely accepted, other phylogenetic trees have been proposed. Six or more different trees relating the major domains have been proposed. **Figure 19.10** provides a simplified view of some of these. The first (figure 19.10*a*) indicates that the three groups are about equi- distant from one another and fits with the early rRNA data. Fig- ure 19.10*b* represents the currently most popular tree in which ar- chaea and eucaryotes have a common ancestor; organisms like the bacteria may have existed before the other domains. The third tree, called the eocyte tree (figure 19.10*c*), is based on the pro-

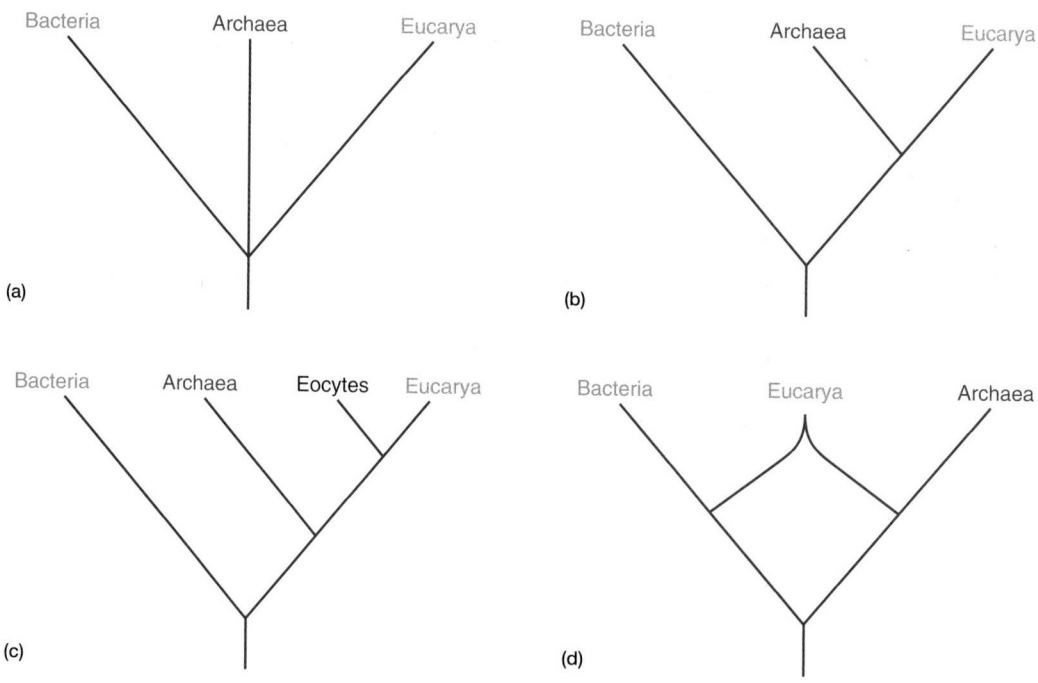

Figure 19.10 **Variations in the Design of the "Tree of Life."** These four alternative phylogenetic trees are discussed in the text.

posal that sulfur-dependent, extremely thermophilic procaryotes called eocytes (dawn + cell) are a separate group and more closely related to eucaryotes than are the archaea. Finally, some have proposed that eucaryotic cells are chimeric and arose from the fusion of a bacterium and archaeon (possibly a bacterium lacking a cell wall engulfed an eocytelike archaeon) (figure 19.10*d*).

Clearly the situation is confused and more than one model has been proposed, though most microbiologists favor the three-domain tree in figure 19.10*b*. When some protein sequences are used to construct phylogenetic trees, one does not even get a three-domain pattern. Many factors may account for these problems. There could be unrecognized gene duplications that occurred before the domains formed, leading to confusing patterns. Unequal rates of evolution could distort the trees. Phylogenetically important information may have been lost in some molecular sequences. There may be significant sequence variation between the same molecules from different strains of the same species. Unless several strains are analyzed, false conclusions may be drawn. Thus inaccurate universal trees may result when only the sequences from a few molecules are employed (as is usually the case).

One of the most important difficulties in constructing a satisfactory tree is widespread, frequent horizontal or lateral gene transfer. Recent genome sequence studies have shown that there is extensive horizontal gene transfer within and between domains (*see p. 344*). Eucaryotes possess genes from both bacteria and archaea, and there has been frequent gene swapping between the two procaryotic domains. It appears that at least some bacteria

even have acquired eucaryotic genes. Although a variety of mechanisms may be responsible, it has been suggested that much of this gene movement occurs by way of virus-mediated transfers. Clearly the pattern of microbial evolution is not as linear and treelike as previously thought. The trees shown in figures 19.3 and 19.10 are undoubtedly oversimplified. **Figure 19.11** depicts a more realistic reticulated tree in which horizontal gene transfer plays a major role. This tree resembles a web or network with many lateral branches linking various trunks, each branch representing the transfer of one or a few genes. Instead of having a single main trunk or common ancestor at its base, this tree has several trunks or groups of primitive cells that contribute to the original gene pool. Although there is extensive gene transfer between the two procaryotic domains throughout their development, the eucaryotic domain seldom participates in horizontal gene transfer after the formation of fungi, plants, and animals. It is possible that eucaryotic cells originated in a complex process involving many gene transfers from both bacteria and archaea. This hypothesis still allows for the formation of mitochondria and chloroplasts by endosymbiosis with α-proteobacteria and cyanobacteria, respectively. Presumably the three domains remain separate because there are many more gene transfers within each than between domains.

This brief discussion of the problems in developing a true universal phylogenetic tree is intended to show the difficulty in determining phylogenetic relationships. The best results will be obtained when all possible data, both molecular and phenotypic, are used in the analysis (for example, in polyphasic taxonomy).

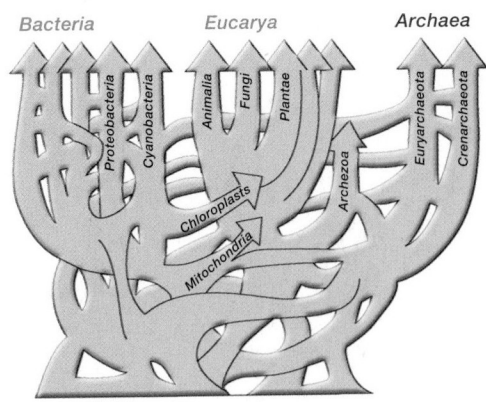

Figure 19.11 Universal Phylogenetic Tree with Frequent Horizontal or Lateral Gene Transfers. See text for discussion.

We will usually employ trees derived from 16S rRNA sequences because these data are most extensive and are used by most microbiologists. Keep in mind that such trees may well change as further data are collected and analyzed.

Kingdoms

While most bacteriologists favor the three-domain system, many protozoologists, botanists, and zoologists still think in terms of five or more kingdoms. This section briefly summarizes the nature of some of these classification systems.

The first classification system to have gained popularity in the last few decades is the five-kingdom system first suggested by Robert H. Whittaker in the 1960s. An overview of Whittaker's five-kingdom system is presented in **figure 19.12a.** Organisms are placed into five kingdoms based on at least three major criteria: (1) cell type—procaryotic or eucaryotic, (2) level of organization—solitary and colonial unicellular organization or multicellular, and (3) nutritional type. In this system the kingdom *Animalia* contains multicellular animals with wall-less eucaryotic cells and primarily ingestive nutrition, whereas the kingdom *Plantae* is composed of multicellular plants with walled eucaryotic cells and primarily photoautotrophic nutrition. Microbiologists study members of the other three kingdoms. The kingdom *Monera* or *Procaryotae* contains all procaryotic organisms. The kingdom *Protista* is the least homogeneous and hardest to define. **Protists** are eucaryotes with unicellular organization, either in the form of solitary cells or colonies of cells lacking true tissues. They may have ingestive, absorptive, or photoautotrophic nutrition, and they include most of the microorganisms known as algae, protozoa, and many of the simpler fungi. The kingdom *Fungi* contains eucaryotic and predominately multinucleate organisms, with nuclei dispersed in a walled and often septate mycelium (*see chapter 25*); their nutrition is absorptive. The taxonomy of the major protist and fungal phyla is discussed in more detail in chapters 25 to 27.

The five-kingdom system is not accepted by many biologists. A major problem is its lack of distinction between archaea and bacteria. The kingdom *Protista* also may be too diverse to be taxonomically useful. In addition, the boundaries between the kingdoms *Protista, Plantae,* and *Fungi* are ill-defined. For example, the brown algae are probably not closely related to the plants even though the five-kingdom system places them in the *Plantae.*

Because of such problems with the five-kingdom system, various alternatives have been suggested. The six-kingdom system is the simplest option; it divides the kingdom *Monera* or *Procaryotae* into two kingdoms, the *Eubacteria* and *Archaeobacteria* (figure 19.12b). Many attempts have been made to divide the protists into several better-defined kingdoms. The eight-kingdom system of Cavalier-Smith is a good example (figure 19.12c). Cavalier-Smith believes that differences in cellular structure and genetic organization are exceptionally important in determining phylogeny; thus he has used ultrastructural characteristics as well as rRNA sequences and other molecular data in developing his classification. He divides all organisms into two empires and eight kingdoms. The empire *Bacteria* contains two kingdoms, the *Eubacteria* and the *Archaeobacteria.* The second empire, the *Eucaryota,* contains six kingdoms of eucaryotic organisms. There are two new kingdoms of eucaryotes. The *Archezoa* are primitive eucaryotic unicellular organisms such as *Giardia* that have 70S ribosomes and lack Golgi apparatuses, mitochondria, chloroplasts, and peroxisomes. The kingdom *Chromista* contains mainly photosynthetic organisms that have their chloroplasts within the lumen of the rough endoplasmic reticulum rather than in the cytoplasmic matrix (as is the case in the kingdom *Plantae*). Diatoms, brown algae, cryptomonads, and oomycetes are all placed in the *Chromista.* The boundaries of the remaining four kingdoms—*Plantae, Fungi, Animalia,* and *Protozoa*—have been adjusted to better define each kingdom and distinguish it from the others. Sogin and his coworkers do not cluster the eucaryotes into a few major divisions, but rather consider them to be a single domain or empire composed of a collection of independently evolved lineages (figure 19.12d). In this scheme the protists do not comprise a separate kingdom, but simply represent a level of organization with many separate lineages and tremendous diversity.

1. How does Woese divide organisms into domains or empires in his universal phylogenetic tree? Describe several of the major differentiating characteristics of each domain.
2. Describe the two major alternatives to Woese's tree that are depicted in figure 19.10c and 19.10d. Why have there been difficulties in developing an accurate tree? Discuss the effect of frequent horizontal gene transfer on phylogenetic trees.
3. With what three major criteria did Whittaker divide organisms into five kingdoms?
4. Give the names and main distinguishing characteristics of the five kingdoms.
5. Briefly describe the six- and eight-kingdom systems. How do they differ from the five-kingdom system?

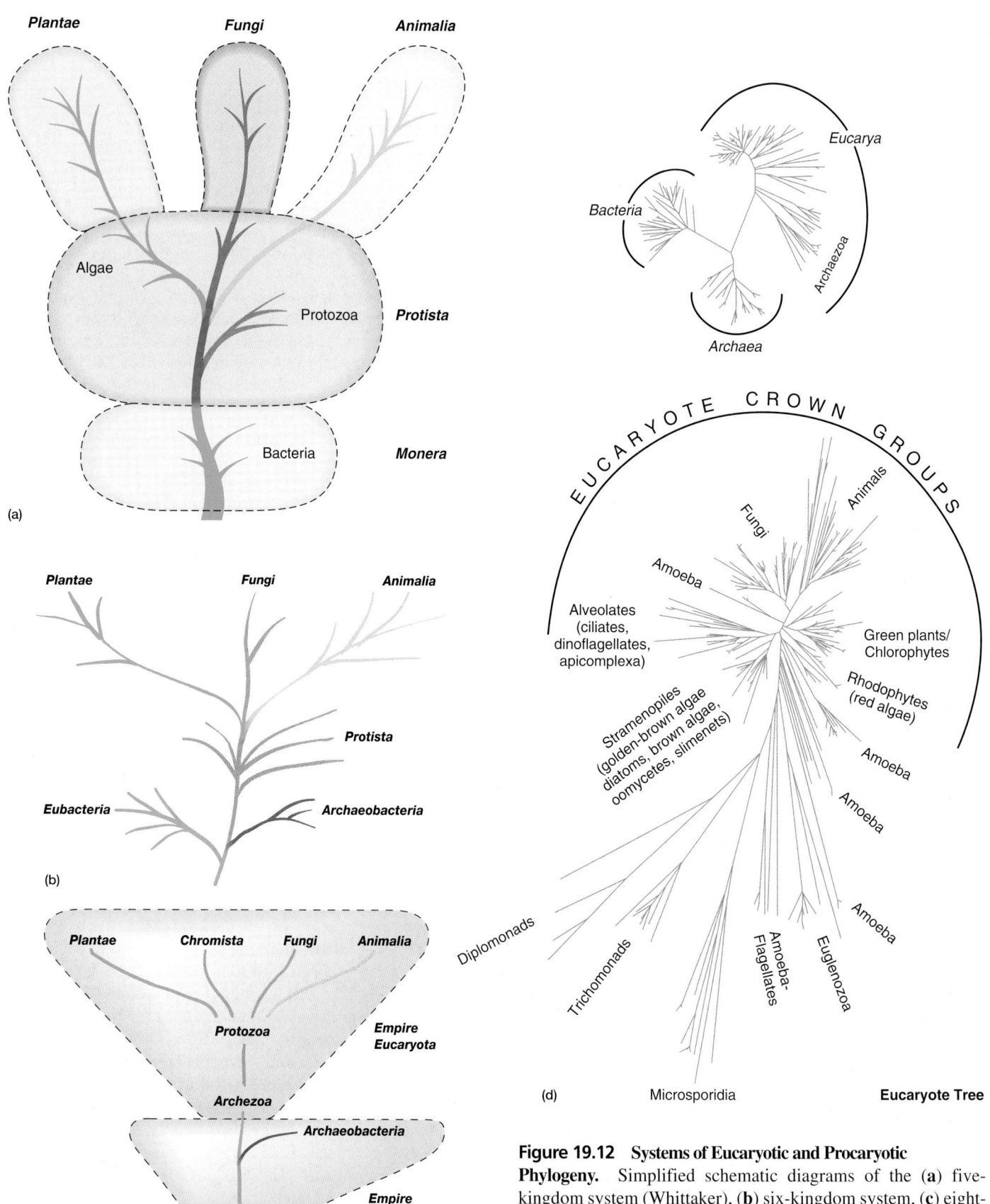

Figure 19.12 Systems of Eucaryotic and Procaryotic Phylogeny. Simplified schematic diagrams of the (**a**) five-kingdom system (Whittaker), (**b**) six-kingdom system, (**c**) eight-kingdom system (Cavalier-Smith), and (**d**) the universal and eucaryotic trees according to Sogin.

Microbial Diversity & Ecology
19.1   "Official" Nomenclature Lists—A Letter from Bergey's*

On a number of occasions lately, the impression has been given that the status of a bacterial taxon in *Bergey's Manual of Systematic Bacteriology* or *Bergey's Manual of Determinative Bacteriology* is in some sense official. Similar impressions are frequently given about the status of names in the *Approved List of Bacterial Names* and in the Validation Lists of newly proposed names that appear regularly in the *International Journal of Systematic Bacteriology*. It is therefore important to clarify these matters.

There is no such thing as an official classification. *Bergey's Manual* is not "official"—it is merely the best consensus at the time, and although great care has always been taken to obtain a sound and balanced view, there are also always regions in which data are lacking or confusing, resulting in differing opinions and taxonomic instability. When *Bergey's Manual* disavows that it is an official classification, many bacteriologists may feel that the solid earth is trembling. But many areas are in fact reasonably well established. Yet taxonomy is partly a matter of judgment and opinion, as is all science, and until new information is available, different bacteriologists may legitimately hold different views. They cannot be forced to agree to any "official classification." It must be remembered that, as yet, we know only a small percentage of the bacterial species in nature. Advances in technique also reveal new lights on bacterial relationships. Thus we must expect that existing boundaries of groups will have to be redrawn in the future, and it is expected that molecular biology, in particular, will imply a good deal of change over the next few decades.

The position with the Approved Lists and the Validation Lists is rather similar. When bacteriologists agreed to make a new start in bacteriological nomenclature, they were faced with tens of thousands of names in the literature of the past. The great majority were useless, because, except for about 2,500 names, it was impossible to tell exactly what bacteria they referred to. These 2,500 were therefore retained in the Approved Lists. The names are only approved in the sense that they were approved for retention in the new bacteriological nomenclature. The remainder lost standing in the nomenclature, which means they do not have to be considered when proposing new bacterial names (although names can be individually revived for good cause under special provisions).

The new International Code of Nomenclature of Bacteria requires all new names to be validly published to gain standing in the nomenclature, either by being published in papers in the *International Journal of Systematic Bacteriology* or, if published elsewhere, by being announced in the Validation Lists. The names in the Validation Lists are therefore valid only in the sense of being validly published (and therefore they must be taken account of in bacterial nomenclature). The names do not have to be adopted in all circumstances; if users believe the scientific case for the new taxa and validly published names is not strong enough, they need not adopt the names. For example, *Helicobacter pylori* was immediately accepted as a replacement for *Campylobacter pylori* by the scientific community, whereas *Tatlockia micdadei* had not generally been accepted as a replacement for *Legionella micdadei*. Taxonomy remains a matter of scientific judgment and general agreement.

*From P. H. A. Sneath and D. J. Brenner, "Official Nomenclature Lists in *ASM News*, 58(4):175, 1992. Copyright © by the American Society for Microbiology. Reprinted by permission.

19.8   BERGEY'S MANUAL OF SYSTEMATIC BACTERIOLOGY

In 1923, David Bergey, professor of bacteriology at the University of Pennsylvania, and four colleagues published a classification of bacteria that could be used for identification of bacterial species, the *Bergey's Manual of Determinative Bacteriology*. This manual is now in its ninth edition. The first edition of *Bergey's Manual of Systematic Bacteriology,* a more detailed work that contains descriptions of all procaryotic species currently identified, also is available (**Microbial Diversity & Ecology 19.1**). The first volume of the second edition was published in 2001. This section briefly describes the current edition of *Bergey's Manual of Systematic Bacteriology* (or *Bergey's Manual*) and then discusses at more length the new second edition.

The First Edition of *Bergey's Manual of Systematic Bacteriology*

Because it has not been possible in the past to classify procaryotes satisfactorily based on phylogenetic relationships, the system given in the first edition of *Bergey's Manual of Systematic Bacteriology* is primarily phenetic. Each of the 33 sections in the four volumes contains procaryotes that share a few easily determined characteristics and bears a title that either describes these properties or provides the vernacular names of the procaryotes included. The characteristics used to define sections are normally features such as general shape and morphology, Gram-staining properties, oxygen relationship, motility, the presence of endospores, the mode of energy production, and so forth. Procaryotic groups are divided among the four volumes in the following manner: (1) gram-negative bacteria of general, medical, or industrial importance; (2) gram-positive bacteria other than actinomycetes; (3) gram-negative bacteria with distinctive properties, cyanobacteria, and archaea; and (4) actinomycetes (gram-positive filamentous bacteria).

Gram-staining properties play a singularly important role in this phenetic classification; they even determine the volume into which a species is placed. There are good reasons for this significance. As noted in chapter 3, Gram staining usually reflects fundamental differences in bacterial wall structure. Gram-staining properties also are correlated with many other

Table 19.9			
Some Characteristic Differences between Gram-Negative and Gram-Positive Bacteria			
Property	**Gram-Negative Bacteria**	**Gram-Positive Bacteria**	**Mycoplasmas**
Cell wall	Gram-negative type wall with inner 2–7 nm peptidoglycan layer and outer membrane (7–8 nm thick) of lipid, protein, and lipopolysaccharide. (There may be a third outermost layer of protein.)	Gram-positive type wall with a homogeneous, thick cell wall (20–80 nm) composed mainly of peptidoglycan. Other polysaccharides and teichoic acids may be present.	Lack a cell wall and peptidoglycan precursors; enclosed by a plasma membrane
Cell shape	Spheres, ovals, straight or curved rods, helices or filaments; some have sheaths or capsules.	Spheres, rods, or filaments; may show true branching	Pleomorphic in shape; may be filamentous, can form branches
Reproduction	Binary fission, sometimes budding	Binary fission	Budding, fragmentation, and/or binary fission
Metabolism	Phototrophic, chemolithoautotrophic, or chemoorganoheterotrophic	Usually chemoorganoheterotrophic	Chemoorganoheterotrophic; most require cholesterol and long-chain fatty acids for growth.
Motility	Motile or nonmotile. Flagellation can be varied—polar, lophotrichous, peritrichous. Motility may also result from the use of axial filaments (spirochetes) or gliding motility.	Most often nonmotile; have peritrichous flagellation when motile	Usually nonmotile
Appendages	Can produce several types of appendages—pili and fimbriae, prosthecae, stalks	Usually lack appendages (may have spores on hyphae)	Lack appendages
Endospores	Cannot form endospores	Some groups can form endospores.	Cannot form endospores

properties of bacteria. Typical gram-negative bacteria, gram-positive bacteria, and mycoplasmas (bacteria lacking walls) differ in many characteristics, as can be seen in **table 19.9.** For these and other reasons, bacteria traditionally have been classified as gram positive or gram negative. This approach is retained to some extent in more phylogenetic classifications and is a useful way to think about bacterial diversity.   The procaryotic cell wall (pp. 52–59)

The Second Edition of *Bergey's Manual of Systematic Bacteriology*

There has been enormous progress in procaryotic taxonomy since 1984, the year the first volume of *Bergey's Manual of Systematic Bacteriology* was published. In particular, the sequencing of rRNA, DNA, and proteins has made phylogenetic analysis of procaryotes feasible. As a consequence, the second edition of *Bergey's Manual* is largely phylogenetic rather than phenetic and thus quite different from the first edition. Although the new edition will not be completed for some time, it is so important that its general features will be described here. Undoubtedly the details will change as work progresses, but the general organization of the new *Bergey's Manual* can be summarized.

The second edition will be published in five volumes. It will have more ecological information about individual taxa. The second edition will not group all the clinically important procaryotes together as the first edition did. Instead, pathogenic species will

be placed phylogenetically and thus scattered throughout the following five volumes.

> Volume 1—*The Archaea, and the Deeply Branching and Phototrophic Bacteria*
> Volume 2—*The Proteobacteria*
> Volume 3—*The Low G + C Gram-Positive Bacteria*
> Volume 4—*The High G + C Gram-Positive Bacteria*
> Volume 5—*The Planctomycetes, Spirochaetes, Fibrobacteres, Bacteroidetes,* and *Fusobacteria* (Volume 5 also will contain a section that updates descriptions and phylogenetic arrangements that have been revised since publication of volume 1.)

Volume 1 was published in 2001, and Volume 2 is scheduled for release in 2003. **Table 19.10** summarizes the planned organization of the second edition and indicates where the discussion of a particular group may be found in this textbook.   Bacterial classification according to the second edition of *Bergey's Manual of Systematic Bacteriology* (appendix III)

1. What characteristics are used to place procaryotes in different sections of the first edition of *Bergey's Manual?*
2. What are the major ways in which gram-negative and gram-positive bacteria differ? Distinguish mycoplasmas from other bacteria.
3. How in general does the second edition of *Bergey's Manual* differ from the first?

Table 19.10

Organization of *Bergey's Manual of Systematic Bacteriology*

Taxonomic Rank	Representative Genera	Textbook Coverage
Volume 1. *The Archaea and the Deeply Branching and Phototrophic Bacteria*		
Domain *Archaea*		
Phylum *Crenarchaeota*	*Thermoproteus, Pyrodictium, Sulfolobus*	pp. 442–44
Phylum *Euryarchaeota*		
Class I. *Methanobacteria*	*Methanobacterium*	pp. 444–47
Class II. *Methanococci*	*Methanococcus*	
Class III. *Halobacteria*	*Halobacterium, Halococcus*	pp. 447–49
Class IV. *Thermoplasmata*	*Thermoplasma, Picrophilus, Ferroplasma*	p. 449
Class V. *Thermococci*	*Thermococcus, Pyrococcus*	p. 449
Class VI. *Archaeoglobi*	*Archaeoglobus*	p. 451
Class VII. *Methanopyri*	*Methanopyrus*	p. 445
Domain *Bacteria*		
Phylum *Aquificae*	*Aquifex, Hydrogenobacter*	p. 454
Phylum *Thermotogae*	*Thermotoga, Geotoga*	pp. 454–55
Phylum *Thermodesulfobacteria*	*Thermodesulfobacterium*	
Phylum "*Deinococcus-Thermus*"	*Deinococcus, Thermus*	pp. 455–56
Phylum *Chrysiogenetes*	*Chrysogenes*	
Phylum *Chloroflexi*	*Chloroflexus, Herpetosiphon*	pp. 457–58
Phylum *Thermomicrobia*	*Thermomicrobium*	
Phylum *Nitrospira*	*Nitrospira*	
Phylum *Deferribacteres*	*Geovibrio*	
Phylum *Cyanobacteria*	*Prochloron, Synechococcus, Pleurocapsa, Oscillatoria, Anabaena, Nostoc, Stigonema*	pp. 458–64
Phylum *Chlorobi*	*Chlorobium, Pelodictyon*	p. 458
Volume 2. *The Proteobacteria*		
Phylum *Proteobacteria*		
Class I. Alphaproteobacteria	*Rhodospirillum, Rickettsia, Caulobacter, Rhizobium, Brucella, Nitrobacter, Methylobacterium, Beijerinckia, Hyphomicrobium*	pp. 474–81
Class II. Betaproteobacteria	*Neisseria, Burkholderia, Alcaligenes, Comamonas, Nitrosomonas, Methylophilus, Thiobacillus*	pp. 482–85
Class III. Gammaproteobacteria	*Chromatium, Leucothrix, Legionella, Pseudomonas, Azotobacter, Vibrio, Escherichia, Klebsiella, Proteus, Salmonella, Shigella, Yersinia, Haemophilus*	pp. 485–95
Class IV. Deltaproteobacteria	*Desulfovibrio, Bdellovibrio, Myxococcus, Polyangium*	pp. 495–99
Class V. Epsilonproteobacteria	*Campylobacter, Helicobacter*	p. 500
Volume 3. *The Low G + C Gram-Positive Bacteria*		
Phylum *Firmicutes*		
Class I. Clostridia	*Clostridium, Peptostreptococcus, Eubacterium, Desulfotomaculum, Heliobacterium, Veillonella*	pp. 508–11
Class II. *Mollicutes*	*Mycoplasma, Ureaplasma, Spiroplasma, Acholeplasma*	pp. 504–7
Class III. Bacilli	*Bacillus, Caryophanon, Paenibacillus, Thermoactinomyces, Lactobacillus, Streptococcus, Enterococcus, Listeria, Leuconostoc, Staphylococcus*	pp. 511–18
Volume 4. *The High G + C Gram-Positive Bacteria*		
Phylum *Actinobacteria*		
Class *Actinobacteria*	*Actinomyces, Micrococcus, Arthrobacter, Corynebacterium, Mycobacterium, Nocardia, Actinoplanes, Propionibacterium, Streptomyces, Thermomonospora, Frankia, Actinomadura, Bifidobacterium*	pp. 524–34
Volume 5. *The Planctomycetes, Spirochaetes, Fibrobacteres, Bacteriodetes, and Fusobacteria*		
Phylum *Planctomycetes*	*Planctomyces, Gemmata*	p. 464
Phylum *Chlamydiae*	*Chlamydia*	pp. 464–66
Phylum *Spirochaetes*	*Spirochaeta, Borrelia, Treponema, Leptospira*	pp. 466–69
Phylum *Fibrobacteres*	*Fibrobacter*	
Phylum *Acidobacteria*	*Acidobacterium*	
Phylum *Bacteroidetes*	*Bacteroides, Porphyromonas, Prevotella, Flavobacterium, Sphingobacterium, Flexibacter, Cytophaga*	pp. 469–71
Phylum *Fusobacteria*	*Fusobacterium, Streptobacillus*	
Phylum *Verrucomicrobia*	*Verrucomicrobium*	
Phylum *Dictyoglomus*	*Dictyoglomus*	

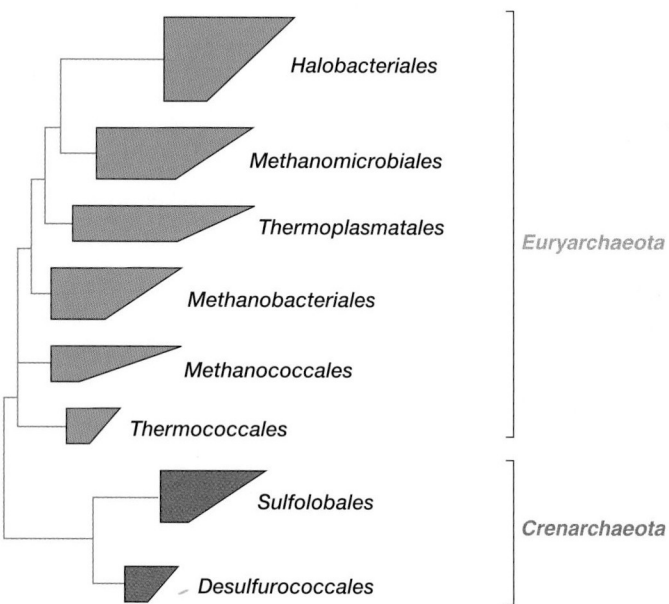

Figure 19.13 Phylogeny of the *Archaea*.   The tree is based on 16S rRNA data and shows relationships between the better-studied orders. Each tetrahedron represents a group of related organisms; its horizontal edges indicate the shortest and longest branches in the group. See text for discussion.

19.9 A SURVEY OF PROCARYOTIC PHYLOGENY AND DIVERSITY

Before beginning a detailed introduction to procaryotic diversity, it might be best to very briefly survey the major groups in the order they are discussed in the second edition of *Bergey's Manual.* This overview is meant only as a general survey of procaryotic diversity. The second edition places procaryotes into 25 phyla, only some of which will be mentioned here. Many of these groups will be discussed in much more detail in chapters 20 through 24. Recall that all organisms may be placed in one of three domains or empires as depicted in the universal phylogenetic tree (figure 19.3). We are concerned only with the procaryotic domains, *Archaea* and *Bacteria,* in this survey.

Volume 1 contains a wide diversity of procaryotes in two domains: the *Archaea* and the *Bacteria.* The *Archaea* differ from *Bacteria* in many ways as summarized in table 19.8. At present, they are divided into two phyla based on rRNA sequences (**figure 19.13**). The phylum *Crenarchaeota* contains thermophilic and hyperthermophylic sulfur-metabolizing organisms of the orders *Thermoproteales, Desulfurococcales,* and *Sulfolobales.* However, recently many other *Crenarchaeota* have been discovered. Some are inhibited by sulfur; others grow in the oceans at low temperatures as picoplankton. The phylum clearly is more diverse than first thought. The second phylum, the *Euryarchaeota,* contains primarily methanogenic procaryotes and halophilic procaryotes; thermophilic, sulfur-reducing organisms

(the thermoplasmas and thermococci) also are in this phylum. The two phyla are divided into eight classes and 12 orders.

The bacteria are an extraordinarily diverse assemblage of procaryotes that have been divided into 23 phyla (**figure 19.14**). In volume 1 are placed deeply branching bacterial groups and phototrophic bacteria. The more important phyla are described in these sections.

1. Phylum *Aquificae.* The phylum *Aquificae* contains autotrophic bacteria such as *Aquifex* and *Hydrogenobacter* that can use hydrogen for energy production. *Aquifex* (meaning "water maker") actually produces water by using hydrogen to reduce oxygen. This group contains some of the most thermophilic organisms known and is the deepest or earliest branch of the bacteria.

2. Phylum *Thermotogae.* This phylum is composed of one class and five genera. *Thermotoga* and other members of the class *Thermotogae* are anaerobic, thermophilic, fermentative, gram-negative bacteria that have unusual fatty acids and resemble *Aquifex* with respect to their ether-linked lipids.

3. Phylum "Deinococcus Thermus." The order *Deinococcales* contains bacteria that are extraordinarily radiation resistant. The genus *Deinococcus* is gram positive. It has high concentrations of carotenoid pigments, which may protect it from radiation, and unique lipids.

4. Phylum *Chloroflexi.* The phylum *Chloroflexi* has one class and two orders. Many members of this gram-negative group are called green nonsulfur bacteria. *Chloroflexus* carries out anoxygenic photosynthesis and is a gliding bacterium; in contrast, *Herpetosiphon* is a nonphotosynthetic, respiratory gliding bacterium. Both genera have unusual peptidoglycans and lack lipopolysaccharides in their outer membranes.

5. Phylum *Cyanobacteria.* The oxygenic photosynthetic bacteria are placed in the phylum *Cyanobacteria,* which contains the class *Cyanobacteria* and five subsections. Cyanobacteria have chlorophyll *a* and almost all species possess phycobilins. These bacteria can be unicellular or filamentous, either branched or unbranched. The cyanobacteria in the subsections differ from each other in general morphological characteristics and reproduction. Cyanobacteria incorporate CO_2 photosynthetically through use of the Calvin cycle just like plants and many purple photosynthetic bacteria.

6. Phylum *Chlorobi.* The phylum *Chlorobi* contains anoxygenic photosynthetic bacteria known as the green sulfur bacteria. They can incorporate CO_2 through the reductive tricarboxylic acid cycle rather than the Calvin cycle and oxidize sulfide to sulfur granules, which accumulate outside the cell.

Volume 2 of the second edition is devoted completely to the gram-negative proteobacteria, often called the purple bacteria. The phylum *Proteobacteria* is a large and extremely complex group that currently contains over 1,300 species in over 400 genera. Even though they are all related, the group is quite diverse in morphology, physiology, and life-style. All major nutritional types are represented: phototrophy, heterotrophy, and chemolithotrophy of several varieties. Many species important in medicine, industry, and biological research are proteobacteria. Obvious examples are the genera *Escherichia, Neisseria, Pseudomonas, Rhizobium,*

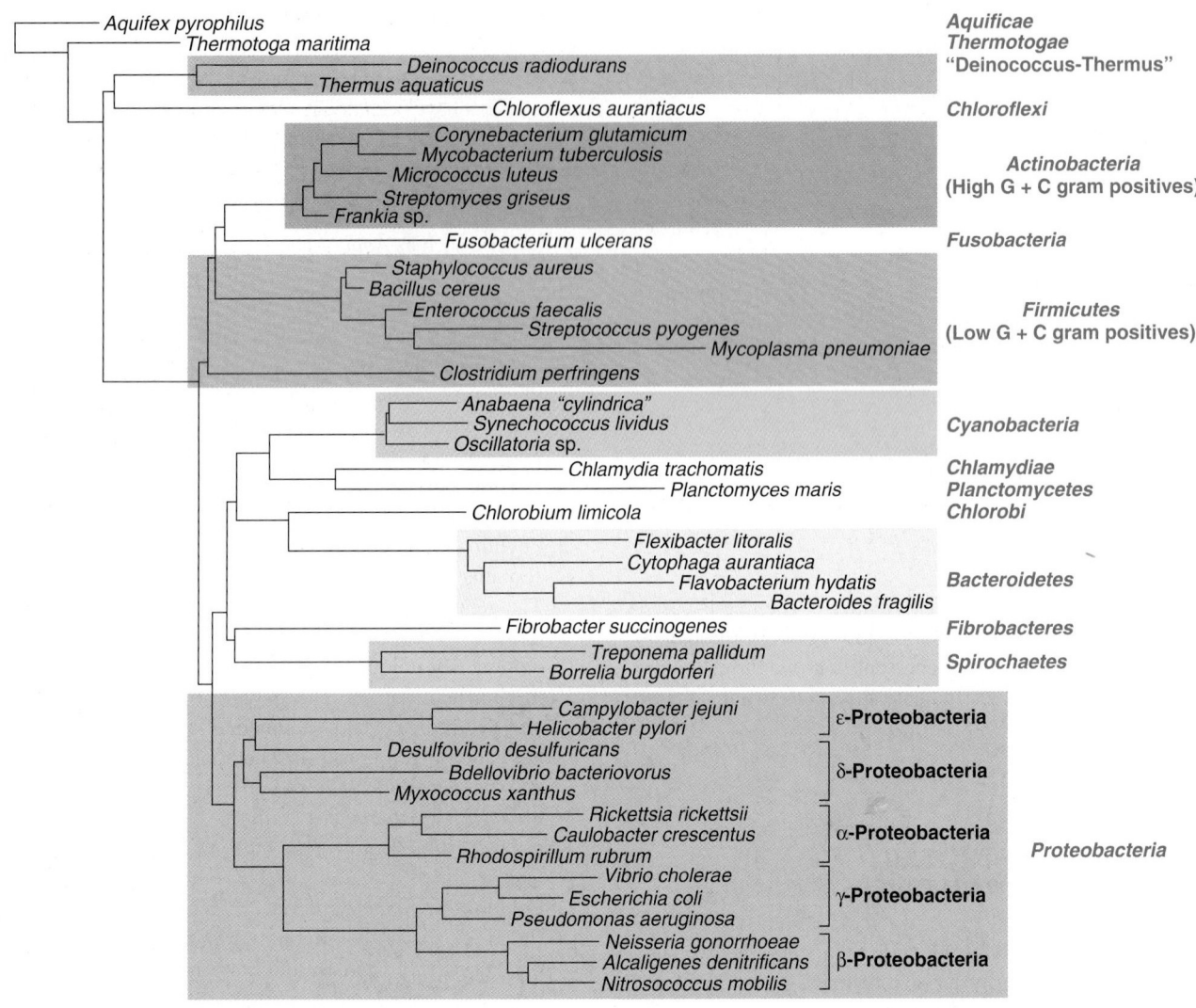

Figure 19.14 Phylogeny of the Bacteria. The tree is based on 16S rRNA comparisons. See text for discussion.
Source: The Ribosomal Database Project.

Rickettsia, Salmonella, and *Vibrio.* The phylum is divided into five classes based on rRNA data. Because photosynthetic bacteria are found in the α, β, and γ classes of the proteobacteria, many believe that the whole phylum arose from a photosynthetic ancestor. Presumably many strains lost photosynthesis when adapting metabolically to new ecological niches.

1. Class I—*Alphaproteobacteria.* The α-proteobacteria include most of the oligotrophic forms (those capable of growing at low nutrient levels). *Rhodospirillum* and other purple nonsulfur bacteria are photosynthetic. Some genera have unusual metabolic modes: methylotrophy (e.g., *Methylobacterium*), chemolithotrophy (*Nitrobacter*), and nitrogen fixation (*Rhizobium*). *Rickettsia* and *Brucella* are important pathogens. About half of the microbes in this group have distinctive morphology such as prosthecae (*Caulobacter, Hyphomicrobium*).

2. Class II—*Betaproteobacteria.* The β-proteobacteria overlap the α subdivision metabolically. However the β-proteobacteria tend to use substances that diffuse from organic decomposition in the anaerobic zone of habitats. Some of these bacteria use such substances as hydrogen (*Alcaligenes*), ammonia (*Nitrosomonas*), methane (*Methylobacillus*), or volatile fatty acids (*Burkholderia*).

3. Class III—*Gammaproteobacteria.* The γ-proteobacteria compose a large and complex group of 14 orders and 25 families. They often are chemoorganotrophic, facultatively anaerobic, and fermentative. However, there is considerable diversity among the γ-proteobacteria with respect to energy metabolism. Some important families such as *Enterobacteriaceae, Vibrionaceae,* and *Pasteurellaceae* use the Embden-Meyerhof pathway and the pentose phosphate pathway. Others such as the *Pseudomonadaceae* and *Azotobacteriaceae* are aerobes and have the Entner-Doudoroff and pentose phosphate pathways. A few are photosynthetic (e.g., *Chro-*

matium and *Ectothiorhodospira*), methylotrophic (*Methylococcus*), or sulfur-oxidizing (*Beggiatoa*).

4. Class IV—*Deltaproteobacteria*. The δ-proteobacteria contain seven orders, and 18 families. Many of these bacteria can be placed in one of three groups. Some are predators on other bacteria as the class name implies (e.g., *Bdellovibrio*). The order *Myxococcales* contains the fruiting myxobacteria such as *Myxococcus, Stigmatella,* and *Polyangium*. The myxobacteria often also prey on other bacteria. Finally, the class has a variety of anaerobes that generate sulfide from sulfate and sulfur while oxidizing organic nutrients (*Desulfovibrio*).

5. Class V—*Epsilonproteobacteria*. This section is composed of only one order, *Campylobacterales,* and three families. Despite its small size two important pathogenic genera are ε-proteobacteria: *Campylobacter* and *Helicobacter.*

Volume 3 of *Bergey's Manual* surveys the gram-positive bacteria with low G + C content in their DNA, which are members of the phylum *Firmicutes*. The dividing line is about 50% G + C; bacteria with a mol% lower than this value are in volume 3. Most of these bacteria are gram positive and heterotrophic. However, because of their close relationship to low G + C gram-positive bacteria, the mycoplasmas are placed here even though they lack cell walls and stain gram negative. There is considerable variation in morphology: some are rods, others are cocci, and mycoplasmas are pleomorphic. Endospores may be present. The phylum contains three classes.

1. Class I—Clostridia. This class contains three orders and 11 families. Although they vary in morphology and size, the members tend to be anaerobic. Genera such as *Clostridium, Desulfotomaculum,* and *Sporohalobacter* form true bacterial endospores; many others do not. *Clostridium* is one of the largest bacterial genera.

2. Class II—*Mollicutes*. The class *Mollicutes* contains five orders and six families. Members of the class often are called mycoplasmas. These bacteria lack cell walls and cannot make peptidoglycan or its precursors. Because mycoplasmas are bounded by the plasma membrane, they are pleomorphic and vary in shape from cocci to helical or branched filaments. They are normally nonmotile and stain gram negative because of the absence of a cell wall. In contrast with almost all other bacteria, most species require sterols for growth. The genera *Mycoplasma* and *Spiroplasma* contain several important animal and plant pathogens.

3. Class III—Bacilli. This large class comprises a wide variety of gram positive, aerobic or facultatively anaerobic, rods and cocci. The class Bacilli has two orders, *Bacillales* and *Lactobacillales,* and 17 families. As with the members of the class Clostridia, some genera (e.g., *Bacillus, Sporosarcina, Paenibacillus,* and *Sporolactobacillus*) form true endospores. The section contains many medically and industrially important genera: *Bacillus, Lactobacillus, Streptococcus, Lactococcus, Enterococcus, Listeria,* and *Staphylococcus.*

Volume 4 is devoted to the high G + C gram positives, those bacteria with mol% values above 50 to 55%. All bacteria in this volume are placed in the phylum *Actinobacteria* and class *Actinobacteria*. There is enormous morphological variety among these procaryotes. Some are cocci, others are regular or irregular rods. High G + C gram positives called actinomycetes often form complex branching hyphae. Although none of these bacteria produce true endospores, many genera do form a variety of asexual spores and some have complex life cycles. There is considerable variety in cell wall chemistry among the high G + C gram positives. For example, the composition of peptidoglycan varies greatly. Mycobacteria produce large mycolic acids that distinguish their walls from those of other bacteria.

The taxonomy of these bacteria is very complex. There are five subclasses, six orders, 14 suborders, and 44 families. Genera such as *Actinomyces, Arthrobacter, Corynebacterium, Micrococcus, Mycobacterium,* and *Propionibacterium* were placed in volume 2 of the first edition. They are now in the new volume 4 within the suborders *Actinomycineae, Micrococcineae, Corynebacterineae,* and *Propionibacterineae* because rRNA studies have shown them to be actinobacteria. The largest and most complex genus is *Streptomyces,* which contains over 500 species.

Volume 5 describes an assortment of nine phyla that are located here for convenience. The inclusion of these groups in volume 5 does not imply that they are directly related. Although they are all gram-negative bacteria, there is considerable variation in morphology, physiology, and life cycle pattern. Several genera are of considerable biological or medical importance. We will briefly consider four of the nine phyla.

1. Phylum *Planctomycetes*. The planctomycetes are related to the chlamydias according to their rRNA sequences. The phylum contains only one order, one family, and four genera. Planctomycetes are coccoid to ovoid or pear-shaped cells that lack peptidoglycan. Some have a membrane-enclosed nucleoid. Although they are normally unicellular, the genus *Isosphaera* will form chains. They divide by budding and may produce nonprosthecate appendages called stalks. Planctomycetes grow in aquatic habitats, and many move by flagella or gliding motility.

2. Phylum *Chlamydiae*. This small phylum contains one class, one order, and four families. The genus *Chlamydia* is by far the most important genus. *Chlamydia* is an obligately intracellular parasite with a unique life cycle involving two distinctive stages: elementary bodies and reticulate bodies. These bacteria do resemble planctomycetes in lacking peptidoglycan. They are small coccoid organisms with no appendages. Chlamydias are important pathogens and cause many human diseases.

3. Phylum *Spirochaetes*. This phylum contains helically shaped, motile, gram-negative bacteria characterized by a unique morphology and motility mechanism. The exterior boundary is a special outer membrane that surrounds the protoplasmic cylinder, which contains the cytoplasm and nucleoid. Periplasmic flagella lie between the protoplasmic cylinder and the outer membrane. The flagella rotate and move the cell even though they do not directly contact the external environment. These chemoheterotrophs can be free living, symbiotic, or parasitic. For example, the genera *Treponema* and *Borrelia*

contain several important human pathogens. The phylum has one class, *Spirochaetes,* three families, and 13 genera.

4. Phylum *Bacteroidetes.* This phylum has three classes (*Bacteroides, Flavobacteria,* and *Sphingobacteria*), three orders, and 12 families. Some of the better-known genera are *Bacteroides, Flavobacterium, Flexibacter,* and *Cytophaga.* The gliding bacteria *Flexibacter* and *Cytophaga* are ecologically significant and will be discussed later.

Because *Bergey's Manual* is the principal resource in procaryotic taxonomy and is used by microbiologists around the world, we will follow *Bergey's Manual* in organizing the survey of procaryotic diversity, Chapters 20 through 24. In so far as possible, the organization of the second edition of *Bergey's Manual* will be employed. Chapter 20 is devoted to the *Archaea.* Chapter 21 covers the bacteria of volumes one and five except the *Archaea.* Chapter 22 is devoted to the proteobacteria. Chapters 23 and 24 deal with the low G + C and high G + C gram-positive bacteria, respectively. Chapter contents will follow the overall phylogenetic scheme of *Bergey's Manual.* Phylogenetic and organizational details may well change somewhat before publication of each volume, but the general picture should adequately reflect the second edition. Appendix III describes the current classification scheme of the second edition of *Bergey's Manual.*

Finally, it must be emphasized that procaryotic nomenclature is as much in flux as classification. The names of families and genera are fairly well established and stable in the new system (at least in the absence of future discoveries); in fact, many family and genus names remain unchanged in the second edition of *Bergey's Manual.* In contrast, the names of orders and higher taxa are not always completely settled in the second edition. The cautions expressed in Microbial Diversity & Ecology 19.1 must always be kept in mind. Because the names of kingdoms, classes, and orders are still changing, their use will be kept to the minimum necessary for consistency with the use of taxa by the first edition, ease of communication, and student comprehension. Some higher taxonomic names may well change in the next few years, but we will still employ them here for the reasons just discussed.

1. Briefly summarize the contents of each of the five volumes in the second edition.
2. Give some ways in which the five classes of proteobacteria differ from each other.
3. In what phyla (and classes of *Proteobacteria* and *Firmicutes*) are the following placed: cyanobacteria, green nonsulfur bacteria, *Rickettsia,* the *Enterobacteriaceae, Campylobacter, Clostridium,* the mycoplasmas, *Bacillus, Streptomyces* and *Mycobacterium, Chlamydia, Treponema,* and *Cytophaga?*

Summary

19.1 General Introduction and Overview

a. Taxonomy, the science of biological classification, is composed of three parts: classification, nomenclature, and identification.

19.2 Microbial Evolution and Diversity

a. Living organisms can be divided into three domains: the *Eucarya,* the *Bacteria* and the *Archaea* (**figure 19.3**). The eucaryotic cell may have arisen from procaryotic cells by endosymbiotic events.

19.3 Taxonomic Ranks

a. The definition of species is different for sexually and asexually reproducing organisms. A bacterial species is a collection of strains that have many stable properties in common and differ significantly from other groups of strains.

b. Microorganisms are named according to the binomial system.

19.4 Classification Systems

a. The two major types of natural classification are phylogenetic systems and phenetic systems.

b. Classifications may be constructed by means of numerical taxonomy, in which the general similarity of organisms is determined using a computer to calculate and analyze association coefficients.

19.5 Major Characteristics Used in Taxonomy

a. Morphological, physiological, metabolic, and ecological characteristics are widely used in microbial taxonomy.

b. The study of transformation and conjugation in bacteria is sometimes taxonomically useful. Plasmid-borne traits can cause errors in bacterial taxonomy if care is not taken.

c. Proteins are direct reflections of mRNA sequences and may be used to compare genomes from different organisms.

d. The G + C content of DNA is easily determined and taxonomically valuable because it is an indirect reflection of the base sequence.

e. Nucleic acid hybridization studies are used to compare DNA or RNA sequences and thus determine genetic relatedness (**figure 19.7**).

f. Nucleic acid sequencing is the most powerful and direct method for comparing genomes. The sequences of 16S and 5S rRNA are used most often in phylogenetic studies. Complete procaryotic chromosomes are now being sequenced and compared.

19.6 Assessing Microbial Phylogeny

a. Phylogenetic relationships often are shown in the form of branched diagrams called phylogenetic trees (**figure 19.8**). Trees may be either rooted or unrooted and are created in several different ways.

b. The sequences of rRNA, DNA, and proteins are used to produce phylogenetic trees. Often members of a group will have a unique characteristic rRNA sequence that distinguishes them from members of other taxonomic groups.

19.7 The Major Divisions of Life

a. Studies on the sequence of 16S rRNA and other molecular properties suggest that the procaryotes are divided into two very different groups: *Bacteria* and *Archaea.* Archaea differ from bacteria in cell wall composition, membrane lipids, tRNA structure, ribosomes, and many other properties (**table 19.8**).

b. Although probably most microbiologists favor the three-domain system, there are alternatives such as the five-, six-, and eight-kingdom systems (**figure 19.12**).

19.8 Bergey's Manual of Systematic Bacteriology

a. *Bergey's Manual of Systematic Bacteriology* gives the accepted system of procaryotic taxonomy.

b. The first edition of *Bergey's Manual of Systematic Bacteriology* provides a primarily phenetic classification, and many taxa are not phylogenetically homogeneous. Easily determined features such as cell shape, Gram-staining characteristics, oxygen relationships, motility, and the mode of energy production are used to classify procaryotes (**table 19.9**).

c. The second edition of *Bergey's Manual* is phylogenetic and distributes procaryotes between two domains and 25 phyla (**table 19.10,** and **figures 19.13** and **19.14**). Comparisons of nucleic acid sequences, particularly 16S rRNA sequences, are the foundation of this new classification.

19.9 A Survey of Procaryotic Phylogeny and Diversity

a. The second edition of *Bergey's Manual* will have five volumes. The general organization of the five volumes is summarized in table 19.10 and briefly outlined here.

(1) Volume 1: *The Archaea and the Deeply Branching and Phototrophic Bacteria.* This volume describes the archaea, cyanobacteria, green sulfur and nonsulfur bacteria, deinococci, and other deeply branching groups.

(2) Volume 2: *The Proteobacteria.* All of the proteobacteria (purple bacteria) are placed in this volume and are divided into five major groups based on rRNA sequences and other characteristics: α-proteobacteria, β-proteobacteria, γ-proteobacteria, δ-proteobacteria, and ε-proteobacteria.

(3) Volume 3: *The Low G + C Gram-Positive Bacteria.* This volume contains gram-positive bacteria with G + C content below about 50%. Some of the major groups are the clostridia, bacilli, streptococci, and staphylococci. Mycoplasmas also are placed here.

(4) Volume 4: *The High G + C Gram-Positive Bacteria.* Gram-positive bacteria with G + C content above around 50 to 55% are in this volume. Such groups as *Corynebacterium, Mycobacterium, Nocardia,* and the actinomycetes are located here.

(5) Volume 5: *The Planctomycetes, Spirochaetes, Fibrobacteres, Bacteroidetes, and Fusobacteria.* Volume 5 has a variety of different gram-negative bacterial groups. The most practically important examples are the chlamydias and the spirochetes.

Key Terms

Archaea 411	genus 413	oligonucleotide signature	protists 426
Bacteria 411	identification 410	sequences 422	serovar 413
binomial system 413	Jaccard coefficient (S_J) 414	phenetic system 414	similarity matrix 414
biovar 413	melting temperature (T_m) 418	phenons 415	simple matching coefficient (S_{SM}) 414
classification 410	molecular chronometers 420	phylogenetic or phyletic classification	species 413
dendrogram 415	morphovar 413	systems 416	strain 413
domains 410	natural classification 414	phylogenetic tree 421	stromatolites 410
endosymbiotic hypothesis 411	nomenclature 410	phylogeny 416	systematics 410
Eucarya 411	nucleic acid hybridization 419	polyphasic taxonomy 423	taxon 410
evolutionary distance 421	numerical taxonomy 414	procaryotic species 413	taxonomy 410
G + C content 417			type strain 413

Questions for Thought and Review

1. Why are size and shape often less useful in characterizing procaryotic species than eucaryotic microbial species?

2. Why might a phylogenetic approach to microbial classification be preferable to a phenetic approach? Give arguments for the use of a phenetic classification. Which do you think is preferable and why?

3. Would a numerical taxonomist favor the phylogenetic or phenetic approach to obtaining a natural classification? Why might it be best to use unweighted characteristics in classification? Are there reasons to give some properties more weight or importance than others?

4. In what way does the simple matching coefficient (S_{SM}) differ from the Jaccard coefficient (S_J)? Give a reason why the former coefficient might be preferable to the latter.

5. What genetic feature of procaryotes makes it advisable to use as many characteristics as convenient in classifying and identifying an organism?

6. What properties of the molecules should be considered when selecting RNAs or proteins to sequence for the purpose of determining the relatedness of microorganisms that are only distantly related?

7. Discuss the problems in developing an accurate phylogenetic tree. Is it possible to create a completely accurate universal phylogenetic tree?

8. Why is the current procaryotic classification system likely to change considerably? How would one select the best features to use in identification of unknown procaryotes and determination of relatedness?

Critical Thinking Questions

1. A potential candidate for a molecular chronometer is tRNA. Explain why it would be an appropriate choice. What are some of the drawbacks of this choice?

2. Form a team and collect a heterogeneous group of screws, nails, tacks, and other similar fasteners. Create a classification system that groups the items logically. Assign "genus and species" names to the items, attempting to indicate phylogenetic relationships. This exercise will acquaint you with the problems faced by taxonomists and will provide an opportunity to sharpen your powers of observation. Compare your phylogenetic models with those of other teams. How are they similar and what are the differences?

For recommended readings on these and related topics, see Appendix V.

chapter 20

The *Archaea*

Concepts

- *Archaea* differ in many ways from both bacteria and eucaryotes. These include differences in cell wall structure and chemistry, membrane lipid structure, molecular biology, and metabolism.
- *Archaea* usually grow in a few restricted or specialized habitats: anaerobic, hypersaline, and high temperature.
- *Bergey's Manual* currently divides the archaea into five major groups: methanogenic archaea, extreme halophiles, cell wall-less archaea, extremely thermophilic S^0-metabolizers, and sulfate reducers.
- The second edition of *Bergey's Manual* divides the archaea into two phyla, the *Crenarchaeota* and *Euryarchaeota,* each with several orders.
- Methanogenic and sulfate-reducing archaea have unique cofactors that participate in methanogenesis.
- *Archaea* have special structural, chemical, and metabolic adaptations that enable them to grow in extreme environments.

As is often the case, epoch-making ideas carry with them implicit, unanalyzed assumptions that ultimately impede scientific progress until they are recognized for what they are. So it is with the prokaryote-eukaryote distinction. Our failure to understand its true nature set the stage for the sudden shattering of the concept when a "third form of life" was discovered in the late 1970s, a discovery that actually left many biologists incredulous. Archaebacteria, as this third form has come to be known, have revolutionized our notion of the prokaryote, have altered and refined the way in which we think about the relationship between prokaryotes and eukaryotes . . . and will influence strongly the view we develop of the ancestor that gave rise to all extant life.

—*C. R. Woese and R. S. Wolfe*

Archaea are often found in extreme environments such as this geyser in Yellowstone National Park.

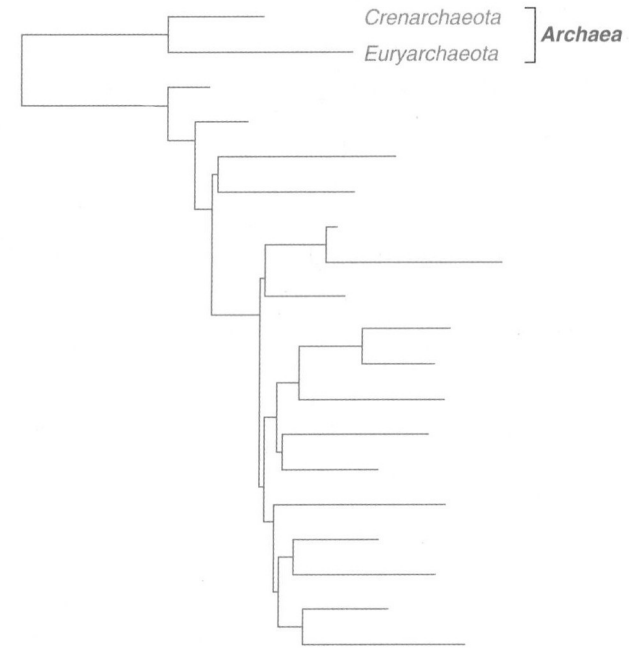

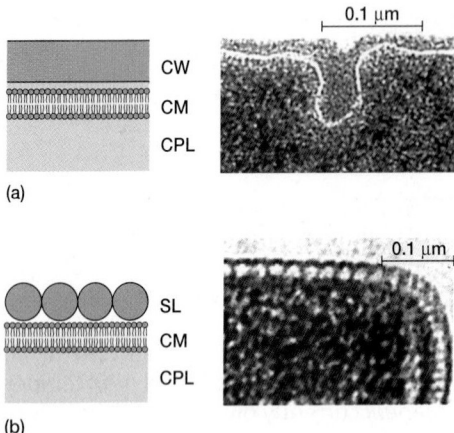

Figure 20.1 Cell Envelopes of *Archaea*. Schematic representations and electron micrographs of (**a**) *Methanobacterium formicicum,* a typical gram-positive organism, and (**b**) *Thermoproteus tenax,* a gram-negative archaeon. CW, cell wall; SL, surface layer; CM, cell membrane or plasma membrane; CPL, cytoplasm.

 hapters 20 through 24 survey the procaryotes described in *Bergey's Manual of Systematic Bacteriology.* Where possible, the order of presentation will follow that of the second edition of *Bergey's Manual.* Chapters 20 and 21 cover the material contained in volumes 1 and 5 of the second edition. Chapter 20 describes the *Archaea;* chapter 21 focuses on the remaining groups in volumes 1 and 5. Chapter 22 covers the proteobacteria, which are located in volume 2. Volume 3 is devoted to the low G + C gram-positive bacteria and will be discussed in chapter 23. Finally, chapter 24 deals with the high G + C bacteria of volume 4.

This chapter begins with a general introduction to the *Archaea.* Then it briefly discusses the biology of each major archaeal group.

Comparison of the sequences of rRNA from a great variety of organisms shows that organisms may be divided into three major groups: the bacteria, archaea, and eucaryotes (*see figure 19.3*). Some of the most important differences are summarized in table 19.8 (*see p. 424*). Because archaea are different from both bacteria and eucaryotes, their most distinctive properties will first be described in some detail and compared with those of the latter two groups. rRNA sequences and the archaea (pp. 410–11, 423–24)

20.1 INTRODUCTION TO THE *ARCHAEA*

As a group the *Archaea* [Greek *archaios,* ancient] are quite diverse, both in morphology and physiology. They can stain either gram positive or gram negative and may be spherical, rod-shaped, spiral, lobed, plate-shaped, irregularly shaped, or pleomorphic. Some are single cells, whereas others form filaments or aggregates. They range in diameter from 0.1 to over 15 μm, and some filaments can grow up to 200 μm in length. Multiplication may be by binary fission, budding, fragmentation, or other mechanisms. Archaea are just as diverse physiologically. They can be aerobic, facultatively anaerobic, or strictly anaerobic. Nutritionally they range from chemolithoautotrophs to organotrophs. Some are mesophiles; others are hyperthermophiles that can grow above 100°C.

Archaea often are found in extreme aquatic and terrestrial habitats. They are often present in anaerobic, hypersaline, or high-temperature environments. Recently archaea have been discovered in cold environments. It appears that they constitute up to 34% of the procaryotic biomass in coastal Antarctic surface waters (*see chapter 29*). A few are symbionts in animal digestive systems.

Archaeal Cell Walls

Although archaea can stain either gram positive or gram negative depending on the thickness and mass of the cell wall, their wall structure and chemistry differ from that of the bacteria. There is considerable variety in archaeal wall structure. Many gram-positive archaea have a wall with a single thick homogeneous layer resembling that in gram-positive bacteria and thus stain gram positive (**figure 20.1a**). Gram-negative archaea lack the outer membrane and complex peptidoglycan network or sacculus of gram-negative bacteria. Instead they usually have a surface

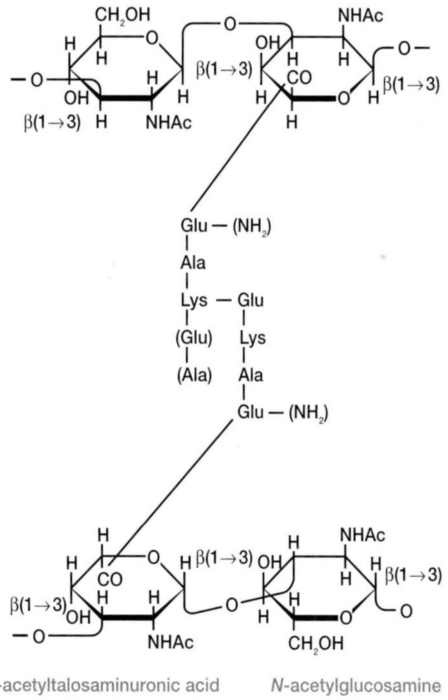

Figure 20.2   The Structure of Pseudomurein.   The components in parentheses are not always present. Ac represents the acetyl group.

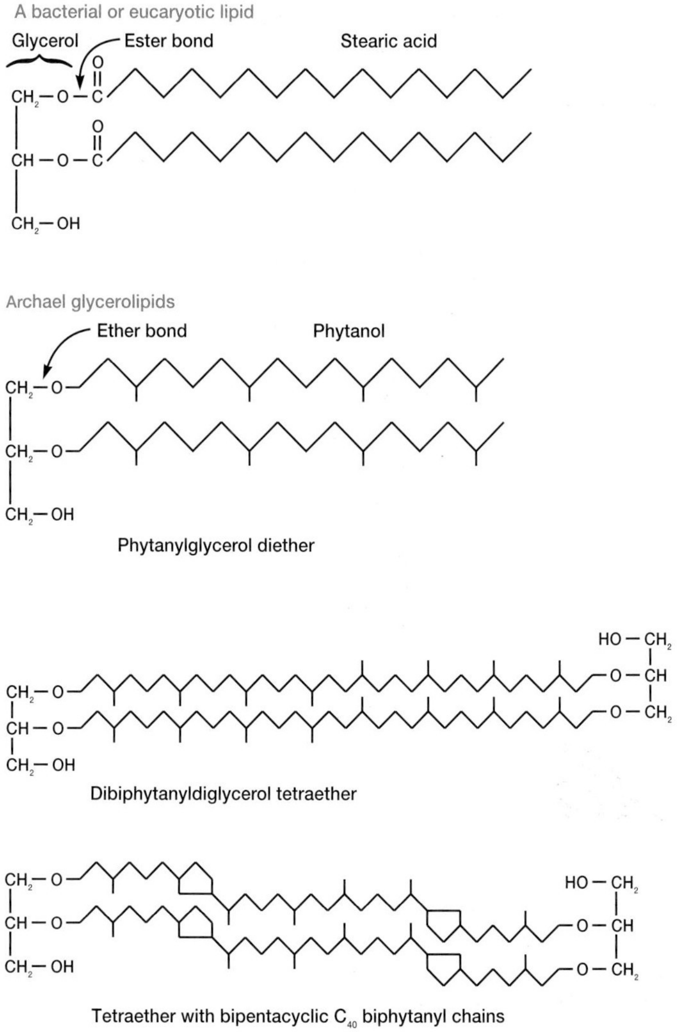

Figure 20.3   Archaeal Membrane Lipids.   An illustration of the difference between archaeal lipids and those of bacteria. Archaeal lipids are derivatives of isopranyl glycerol ethers rather than the glycerol fatty acid esters in bacteria. Three examples of common archaeal glycerolipids are given.

layer of protein or glycoprotein subunits (figure 20.1*b*).   Peptidoglycan structure and chemistry (pp. 54–55)

The chemistry of archaeal cell walls is also quite different from that of the bacteria. None have the muramic acid and D-amino acids characteristic of bacterial peptidoglycan. Not surprisingly, all archaea resist attack by lysozyme and β-lactam antibiotics such as penicillin. Gram-positive archaea can have a variety of complex polymers in their walls. *Methanobacterium* and some other methanogens have walls containing **pseudomurein,** a peptidoglycanlike polymer that has L-amino acids in its cross-links, *N*-acetyltalosaminuronic acid instead of *N*-acetylmuramic acid, and β(1→3) glycosidic bonds instead of β(1→4) glycosidic bonds (**figure 20.2**). *Methanosarcina* and *Halococcus* lack pseudomurein and contain complex polysaccharides similar to the chondroitin sulfate of animal connective tissue. Other heteropolysaccharides are also found in gram-positive walls.

Gram-negative archaea have a layer of protein or glycoprotein outside their plasma membrane. The layer may be as thick as 20 to 40 nm. Sometimes there are two layers, a sheath surrounding an electron-dense layer. The chemical content of these walls varies considerably. Some methanogens (*Methanolobus*), *Halobacterium,* and several extreme thermophiles (*Sulfolobus, Thermoproteus,* and *Pyrodictium*) have glycoproteins in their walls. In contrast, other methanogens (*Methanococcus, Methanomicrobium,* and *Methanogenium*) and the extreme thermophile *Desulfurococcus* have protein walls.

Archaeal Lipids and Membranes

As emphasized in table 19.8, a most distinctive feature of the archaea is the nature of their membrane lipids. They differ from both bacteria and eucaryotes in having branched chain hydrocarbons attached to glycerol by ether links rather than fatty acids connected by ester links (**figure 20.3**). Sometimes two glycerol groups are linked to form an extremely long tetraether. Usually the diether side chains are 20 carbons in length, and the tetraether chains contain 40 carbons. Cells can adjust the overall length of the tetraethers by cyclizing the chains to form pentacyclic rings (figure 20.3), and biphytanyl chains may contain from one to four cyclopentyl rings. Polar lipids are also present in archaeal membranes: phospholipids, sulfolipids, and glycolipids. From 7 to 30% of the membrane lipids

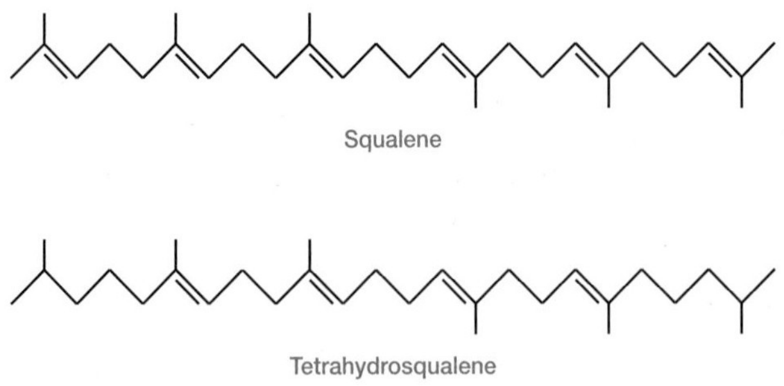

Squalene

Tetrahydrosqualene

Figure 20.4 Nonpolar Lipids of *Archaea*. Two examples of the most predominant nonpolar lipids are the C_{30} isoprenoid squalene and one of its hydroisoprenoid derivatives, tetrahydrosqualene.

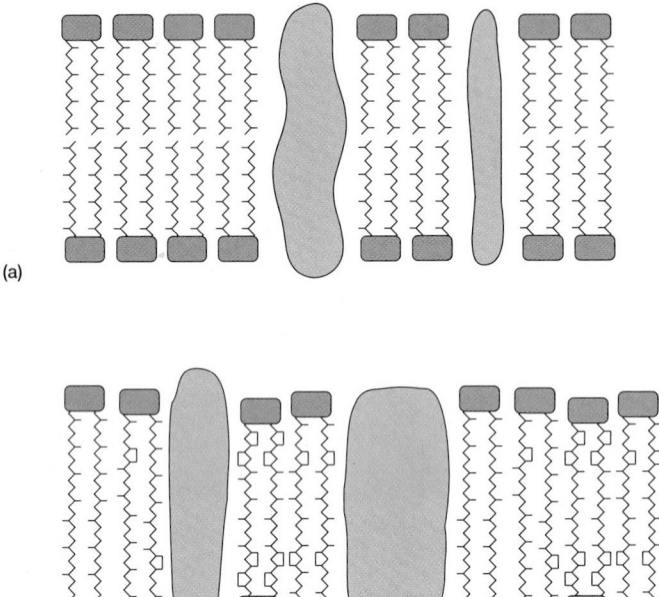

(a)

(b)

Figure 20.5 Examples of Archaeal Membranes. (a) A membrane composed of integral proteins and a bilayer of C_{20} diethers. (b) A rigid monolayer composed of integral proteins and C_{40} tetraethers.

are nonpolar lipids, which usually are derivatives of squalene (**figure 20.4**). These lipids can be combined in various ways to yield membranes of different rigidity and thickness. For example, the C_{20} diethers can be used to make a regular bilayer membrane (**figure 20.5a**). A much more rigid monolayer membrane may be constructed of C_{40} tetraether lipids (figure 20.5b). Of course archaeal membranes may contain a mix of diethers, tetraethers, and other lipids. As might be expected from their need for stability, the membranes of extreme thermophiles such as *Thermoplasma* and *Sulfolobus* are almost completely tetraether monolayers.

Genetics and Molecular Biology

Some features of archaeal genetics are similar to those in bacteria. The archaeal chromosomes that have been studied are single closed DNA circles. However, the genomes of some archaeons are significantly smaller than the normal bacterium. *E. coli* DNA has a size of about 2.5×10^9 daltons, whereas *Thermoplasma acidophilum* DNA is about 0.8×10^9 daltons and *Methanobacterium thermoautotrophicum* DNA is 1.1×10^9 daltons. The variation in G + C content is great, from about 21 to 68 mol%, and is another sign of archaeal diversity. Archaea have few plasmids. Recently the genome of the archaeon *Methanococcus jannaschii* was completely sequenced and compared with the genomes of other organisms. About 56% of its 1,738 genes are unlike those in bacteria and eucaryotes. If this degree of difference is characteristic of the domain *Archaea,* these organisms are as distinctive genotypically as they are in other respects.

Archaeal mRNA appears similar to that of bacteria rather than to eucaryotic mRNA. Polygenic mRNA has been discovered, and there is no evidence for mRNA splicing. Archaeal promoters are similar to those in bacteria.

Despite these and other similarities, there are also many differences between archaea and other organisms. Unlike both bacteria and eucaryotes, the TψC arm (*see p. 260*) of archaeal tRNA lacks thymine and contains pseudouridine or 1-methylpseudouridine. The archaeal initiator tRNA carries methionine as does the eucaryotic initiator tRNA. Although archaeal ribosomes are 70S, similar to bacterial ribosomes, electron microscopic studies show that their shape is quite variable and sometimes differs from that of both bacterial and eucaryotic ribosomes. They do resemble eucaryotic ribosomes in their sensitivity to anisomycin and insensitivity to chloramphenicol and kanamycin. Furthermore, their elongation factor 2 reacts with diphtheria toxin just as the eucaryotic EF–2 does. Some archaea, such as many methanogens in the phylum *Euryarchaeota,* differ from other procaryotes in having histone proteins that bind with DNA to form nucleosome-like structures. At least in *Pyrococcus,* another member of the *Euryarchaeota* (p. 449), there seems to be a single origin of bidirectional DNA replication just as in the bacteria. On the other hand, archaeal DNA replication proteins appear to be similar to those in eucaryotes. Finally, archaeal DNA-dependent RNA polymerases resemble the eucaryotic enzymes, not the bacterial RNA polymerase. They are large, complex enzymes and are insensitive to the drugs rifampin and streptolydigin. These and other differences distinguish the archaea from both bacteria and eucaryotes. Ribosomes and the mechanism of protein synthesis (pp. 260–67); DNA transcription (pp. 254–60)

Metabolism

Not surprisingly in view of the variety of their life-styles, archaeal metabolism varies greatly between the members of different groups. Some archaea are organotrophs; others are autotrophic. A few even carry out an unusual form of photosynthesis.

Archaeal carbohydrate metabolism is best understood. The enzyme 6-phosphofructokinase has not been found in archaea, and they do not appear to degrade glucose by way of the Embden-Meyerhof pathway. Extreme halophiles and thermophiles catabolize glucose using a modified form of the Entner-Doudoroff pathway (*see p. 174 and appendix II*) in which the initial intermediates are not phosphorylated. The halophiles have slightly different modifications of the pathway than do the extreme thermophiles but still produce pyruvate and NADH or NADPH. Methanogens do not catabolize glucose to any significant extent. In contrast with glucose degradation, gluconeogenesis proceeds by a reversal of the Embden-Meyerhof pathway in halophiles and methanogens. All archaea that have been studied can oxidize pyruvate to acetyl-CoA. They lack the pyruvate dehydrogenase complex present in eucaryotes and respiratory bacteria and use the enzyme pyruvate oxidoreductase for this purpose. Halophiles and the extreme thermophile *Thermoplasma* do seem to have a functional tricarboxylic acid cycle. No methanogen has yet been found with a complete tricarboxylic acid cycle. Evidence for functional respiratory chains has been obtained in halophiles and thermophiles. The Embden-Meyerhof pathway and tricarboxylic acid cycle (pp. 171–72, 178–79 and appendix II)

Very little is known in detail about biosynthetic pathways in the archaea. Preliminary data suggest that the synthetic pathways for amino acids, purines, and pyrimidines are similar to those in other organisms. Some methanogens can fix atmospheric dinitrogen. Not only do many archaea use a reversal of the Embden-Meyerhof pathway to synthesize glucose, but at least some methanogens and extreme thermophiles employ glycogen as their major reserve material.

Autotrophy is widespread among the methanogens and extreme thermophiles, and CO_2 fixation occurs in more than one way. *Thermoproteus* and possibly *Sulfolobus* incorporate CO_2 by the reductive tricarboxylic acid cycle (**figure 20.6a**). This pathway is also present in the green sulfur bacteria (*see p. 458*). Methanogenic archaea and probably most extreme thermophiles incorporate CO_2 by the reductive acetyl-CoA pathway (figure 20.6b). A similar pathway also is present in acetogenic bacteria and autotrophic sulfate-reducing bacteria.

As already noted, the mechanism of archaeal protein synthesis has both bacterial and eucaryotic features. The same thing may be said of its protein secretion system. All three domains have signal recognition particles (SRPs) that target new proteins to translocation sites (*see pp. 59–61, 268–69*), but the archaeal SRP differs from those in the other two domains. As in the bacteria, the archaeal SRP binds to the signal sequence of a preprotein and can direct it to the Sec-dependent protein secretion pathway for release outside the cell wall. The archaeal Sec-dependent pathway proteins, however, more closely resemble those of the eucaryotic pathway than the bacterial proteins. After the preprotein is moved across the membrane, its signal sequence is removed by a signal peptidase that resembles a subunit of the eucaryotic peptidase. Protein secretion in procaryotes (pp. 59–61)

Archaeal Taxonomy

As shown in figure 19.3 and in table 19.8 (*see p. 424*), the archaea are quite distinct from other living organisms. Within the domain, however, there is great diversity (*see figure 19.13*). The first edition of *Bergey's Manual* divided the archaea into five major groups based on physiological and morphological differences. **Table 20.1** summarizes some the characteristics of these five groups and gives representatives of each.

On the basis of rRNA data, the second edition of *Bergey's Manual* divides the archaea into the phyla *Euryarchaeota* [Greek *eurus*, wide, and Greek *archaios*, ancient or primitive] and *Crenarchaeota* [Greek *crene*, spring or fount, and *archaios*]. The euryarchaeotes are given this name because they occupy many different ecological niches and have a variety of metabolic patterns. The phylum *Euryarchaeota* is very diverse with seven classes (*Methanobacteria, Methanococci, Halobacteria, Thermoplasmata, Thermococci, Archaeglobi*, and *Methanopyri*), nine orders, and 16 families. The methanogens, extreme halophiles, sulfate reducers, and many extreme thermophiles with sulfur-dependent metabolism are located in the *Euryarchaeota*. Methanogens are the dominant physiological group.

The crenarchaeotes (**figure 20.7**) are thought to resemble the ancestor of the archaea, and almost all the well-characterized species are thermophiles or hyperthermophiles. The phylum *Crenarchaeota* is divided into one class, *Thermoprotei*, three orders and five families. The order *Thermoproteales* contains gram-negative anaerobic to facultative, hyperthermophilic rods. They often grow chemolithoautotrophically by reducing sulfur to hydrogen sulfide. Members of the order *Sulfolobales* are coccus-shaped thermoacidophiles. The order *Desulfurococcales* contains gram-negative coccoid or disk-shaped hyperthermophiles. They grow either chemolithotrophically by hydrogen oxidation or organotrophically by either fermentation or respiration with sulfur as the electron acceptor. The taxonomy of both phyla will undoubtedly undergo further revisions as more organisms are discovered. This is particularly the case with the crenarchaeotes because of the discovery of mesophilic forms in the ocean; these crenarchaeotes may constitute a significant fraction of the oceanic picoplankton.

---

1. What are the archaea? Briefly describe the major ways in which they differ from bacteria and eucaryotes.
2. How do archaeal cell walls differ from those of the bacteria? What is pseudomurein?
3. In what ways do archaeal membrane lipids differ from those of bacteria and eucaryotes? How might these differences lead to stronger membranes?
4. List the differences between archaea and other organisms with respect to tRNA, ribosome structure and behavior, EF-2 sensitivity, and RNA polymerases.
5. Briefly describe the way in which archaea degrade and synthesize glucose. In what two unusual ways do they incorporate CO_2?
6. Characterize the five different groups of archaea and distinguish between them. Distinguish between the phyla *Euryarchaeota* and *Crenarchaeota*.

Figure 20.6 Mechanisms of Autotrophic CO$_2$ Fixation. (**a**) The reductive tricarboxylic acid cycle. The cycle is reversed with ATP and reducing equivalents [H] to form acetyl-CoA from CO$_2$. The acetyl-CoA may be carboxylated to yield pyruvate, which can then be converted to glucose and other compounds. This sequence appears to function in *Thermoproteus neutrophilus.* (**b**) The synthesis of acetyl-CoA and pyruvate from CO$_2$ in *Methanobacterium thermoautotrophicum.* One carbon comes from the reduction of CO$_2$ to a methyl group, and the second is produced by reducing CO$_2$ to carbon monoxide through the action of the enzyme CO dehydrogenase (E$_1$). The two carbons are then combined to form an acetyl group. Corrin-E$_2$ represents the cobamide-containing enzyme involved in methyl transfers. Special methanogen coenzymes and enzymes are described in figures 20.11 and 20.12.

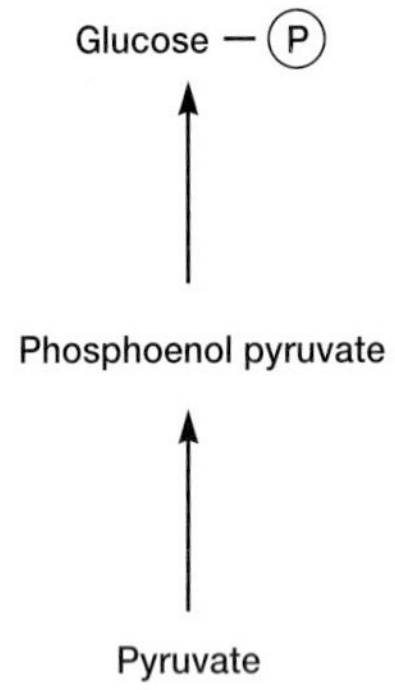

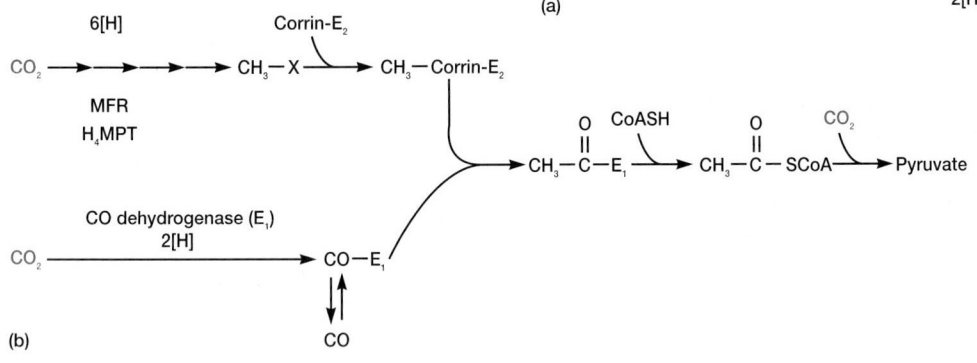

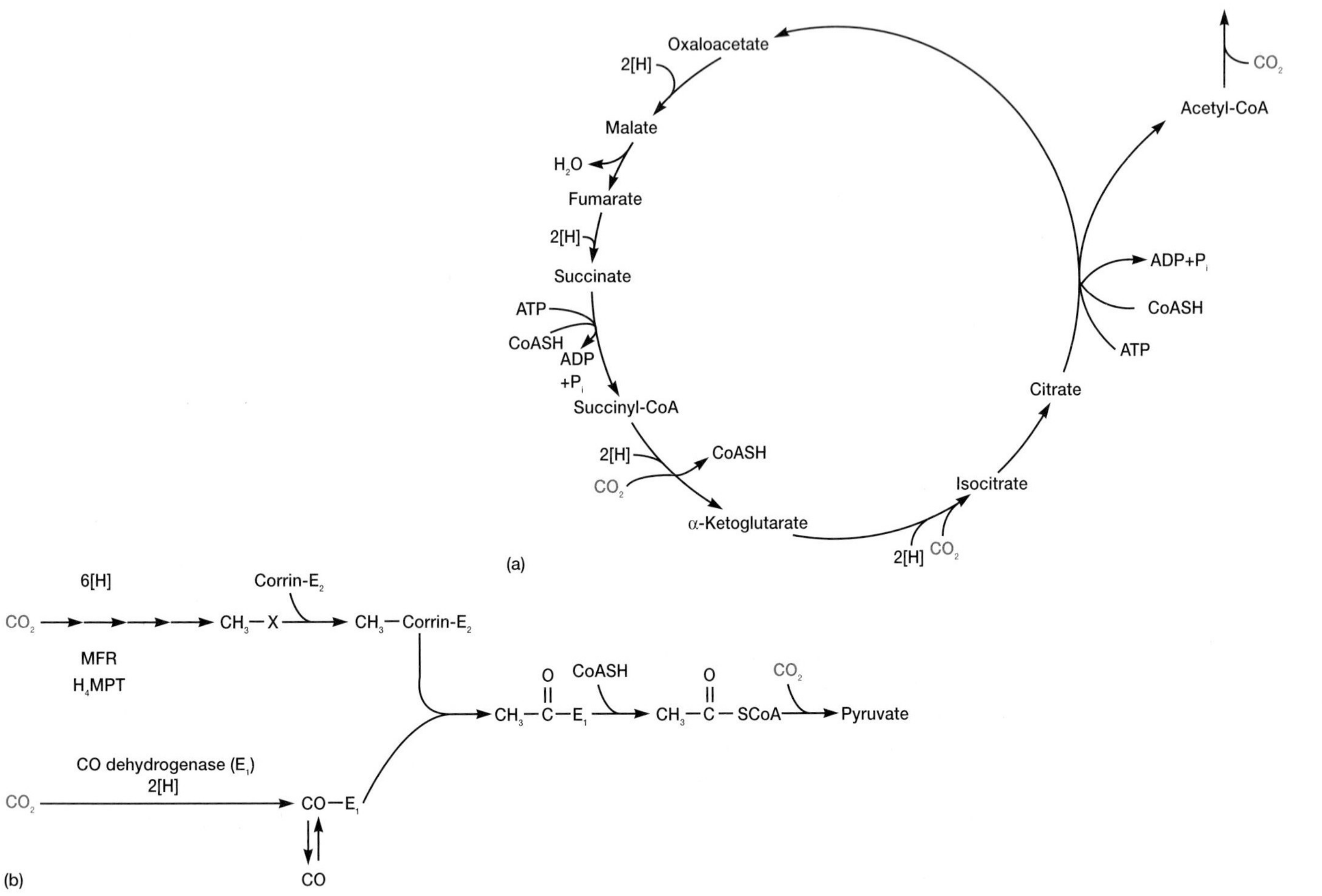

20.2 PHYLUM *CRENARCHAEOTA*

As mentioned previously, most of the crenarchaeotes that have been isolated are extremely thermophilic, and many are acidophiles and sulfur dependent. The sulfur may be used either as an electron acceptor in anaerobic respiration or as an electron source by lithotrophs. Almost all are strict anaerobes. They grow in geothermally heated water or soils that contain elemental sulfur. These environments are scattered all over the world. Exam-ples are the sulfur-rich hot springs in Yellowstone National Park, Wyoming, and the waters surrounding areas of submarine volcanic activity (**figure 20.8**). Such habitats are sometimes called solfatara. These archaea can be very thermophilic and often are classified as hyperthermophiles (*see pp. 124–25*). The most extreme example is *Pyrodictium,* an archaeon isolated from geothermally heated sea floors. *Pyrodictium* has a temperature minimum of 82°C, a growth optimum at 105°C, and a maximum at 110°C. Both organotrophic and lithotrophic growth occur in this group. Sulfur and H$_2$ are the most common electron sources

Table 20.1

Characteristics of the Major Archaeal Groups[a]

Group	General Characteristics	Representative Genera
Methanogenic archaea	Strict anaerobes. Methane is the major metabolic end product. S^0 may be reduced to H_2S without yielding energy production. Cells possess coenzyme M, factors 420 and 430, and methanopterin.	*Methanobacterium* *Methanococcus* *Methanomicrobium* *Methanosarcina*
Archaea sulfate reducers	Irregular gram-negative coccoid cells. H_2S formed from thiosulfate and sulfate. Autotrophic growth with thiosulfate and H_2. Can grow heterotrophically. Traces of methane also formed. Extremely thermophilic and strictly anaerobic. Possess factor 420 and methanopterin but not coenzyme M or factor 430.	*Archaeoglobus*
Extremely halophilic archaea	Coccoid or irregularly shaped rods. Gram-negative or gram-positive, primarily aerobic chemoorganotrophs. Require high sodium chloride concentrations for growth (≥ 1.5 M). Colonies are various shades of red. Neutrophilic or alkalophilic. Mesophilic or slightly thermophilic. Some species contain bacteriorhodopsin and use light for ATP synthesis.	*Halobacterium* *Halococcus* *Natronobacterium*
Cell wall-less archaea	Pleomorphic cells lacking a cell wall. Thermoacidophilic and chemoorganotrophic. Facultatively anaerobic. Plasma membrane contains a mannose-rich glycoprotein and a lipoglycan.	*Thermoplasma*
Extremely thermophilic S^0-metabolizers	Gram-negative rods, filaments, or cocci. Obligately thermophilic (optimum growth temperature between 70–110°C). Usually strict anaerobes but may be aerobic or facultative. Acidophilic or neutrophilic. Autotrophic or heterotrophic. Most are sulfur metabolizers. S^0 reduced to H_2S anaerobically; H_2S or S^0 oxidized to H_2SO_4 aerobically.	*Desulfurococcus* *Pyrodictium* *Pyrococcus* *Sulfolobus* *Thermococcus* *Thermoproteus*

a. Classification scheme employed by the first edition of *Bergey's Manual*.

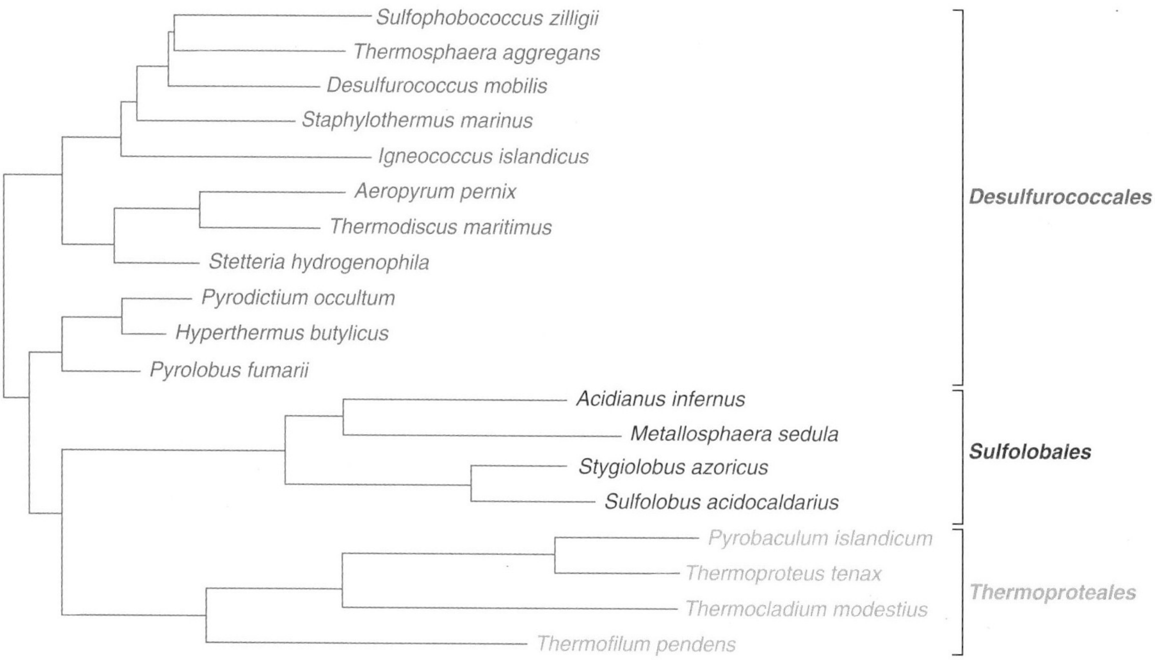

Figure 20.7   **The Phylum *Crenarchaeota*.**   A phylogenetic tree developed with 16S rRNA data for crenarchaeotal-type species. The three orders are indicated.

(a)

(b)

Figure 20.8 Habitats for Thermophilic Archaeans. (a) The pump geyser in Yellowstone National Park. The orange color is due to the carotenoid pigments of thermophilic bacteria. (b) The sulfur cauldron in Yellowstone National Park. The water is at its boiling point and very rich in sulfur. *Sulfolobus* grows well in such habitats.

for lithotrophs. At present, the phylum contains 24 genera; two of the better-studied genera are *Thermoproteus* and *Sulfolobus.*

Members of the genus *Sulfolobus* are gram-negative, aerobic, irregularly lobed spherical archaeons with a temperature optimum around 70 to 80°C and a pH optimum of 2 to 3 (**figure 20.9a,b**). For this reason, they are classed as **thermoacidophiles,** so called because they grow best at acid pH values and high temperatures. Their cell wall contains lipoprotein and carbohydrate but lacks peptidoglycan. They grow lithotrophically on sulfur granules in hot acid springs and soils while oxidizing the sulfur to sulfuric acid (figures 20.8*b* and 20.9*b*). Oxygen is the normal electron acceptor, but ferric iron may be used. Sugars and amino acids such as glutamate also serve as carbon and energy sources.

Thermoproteus is a long thin rod that can be bent or branched (figure 20.9*c*). Its cell wall is composed of glycoprotein. *Thermoproteus* is a strict anaerobe and grows at temperatures from 70 to 97°C and pH values between 2.5 and 6.5. It is found in hot springs and other hot aquatic habitats rich in sulfur. It can grow

organotrophically and oxidize glucose, amino acids, alcohols, and organic acids with elemental sulfur as the electron acceptor. That is, *Thermoproteus* can carry out anaerobic respiration. It will also grow chemolithotrophically using H_2 and S^0. Carbon monoxide or CO_2 can serve as the sole carbon source.

 20.3 PHYLUM *EURYARCHAEOTA*

As mentioned previously, the *Euryarchaeota* is a very diverse phylum with many classes, orders, and families. For the sake of clarity, the five major groups that comprise the euryarchaeotes will be briefly discussed with an emphasis on their physiology and ecology.

The Methanogens

Methanogens are strict anaerobes that obtain energy by converting CO_2, H_2, formate, methanol, acetate, and other compounds to

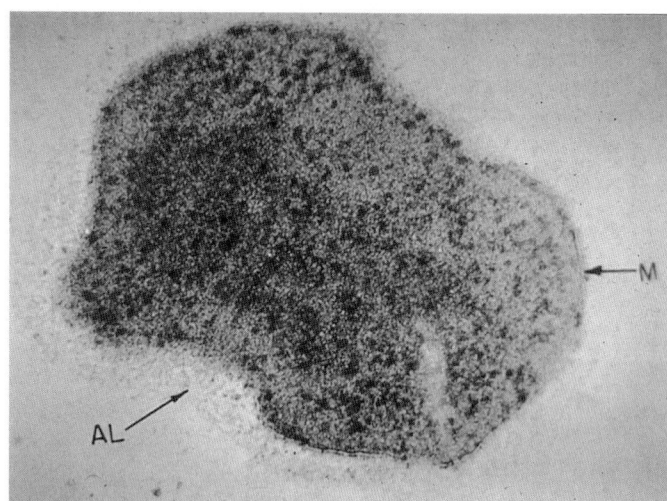

(a) *Sulfolobus brierleyi*

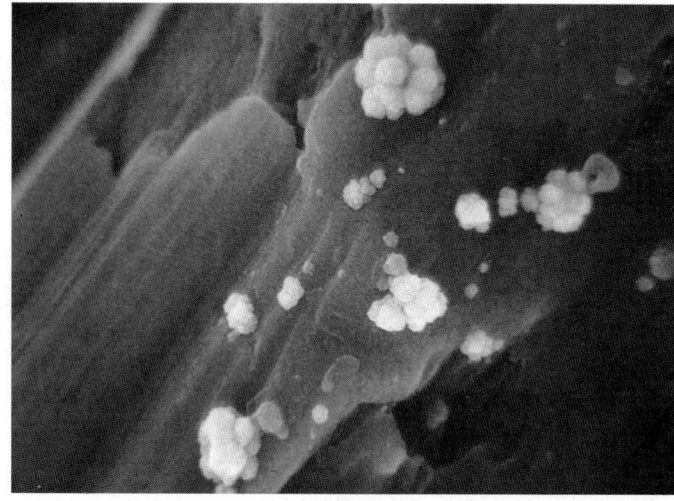

(b) *Sulfolobus colony*

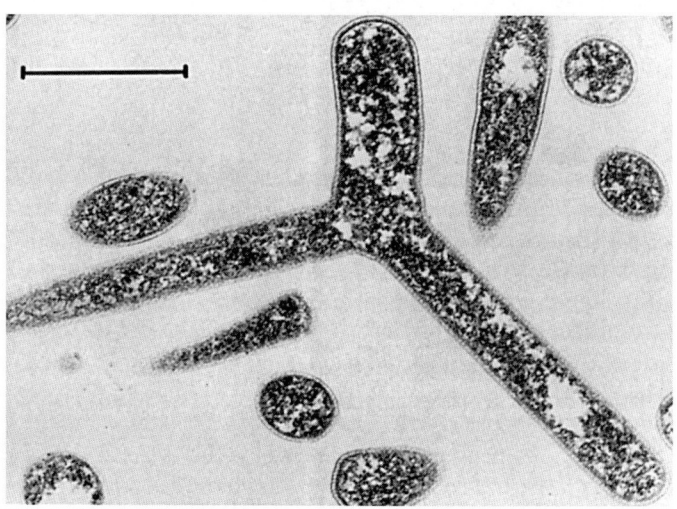

(c) *Thermoproteus tenax*

Figure 20.9 *Sulfolobus* **and** *Thermoproteus.* (**a**) A thin section of *Sulfolobus brierleyi*. The archaean, about 1 μm in diameter, is surrounded by an amorphous layer (AL) instead of a well-defined cell wall; the plasma membrane (M) is clearly visible. (**b**) A scanning electron micrograph of a colony of *Sulfolobus* growing on the mineral molybdenite (MoS_2) at 60°C. At pH 1.5–3, the organism oxidizes the sulfide component of the mineral to sulfate and solubilizes molybdenum. (**c**) Electron micrograph of *Thermoproteus tenax*. Bar = 1 μm.

either methane or methane and CO_2. They are autotrophic when growing on H_2 and CO_2. This is the largest group of archaea. There are five orders (*Methanobacteriales, Methanococcales, Methanomicrobiales, Methanosarcinales,* and *Methanopyrales*) and 27 genera, which differ greatly in overall shape, 16S rRNA sequence, cell wall chemistry and structure, membrane lipids, and other features. For example, methanogens construct three different types of cell walls. Several genera have walls with pseudomurein as mentioned earlier (figure 20.2). Other walls contain either proteins or heteropolysaccharides. The morphology of typical methanogens is shown in **figure 20.10,** and selected properties of representative genera are presented in **table 20.2**. It should be noted that although almost all archaea in these orders are methanogens, methanotrophs have recently been discovered in the *Methanosarcinales* (**Microbial Diversity & Ecology 20.1**).

One of the most unusual methanogenic groups is the class *Methanopyri*. It has one order, *Methanopyrales,* one family and a single genus, *Methanopyrus*. This extremely thermophilic rod-shaped methanogen has been isolated from a marine hydrothermal vent. *Methanopyrus kandleri* has a temperature minimum at 84°C and an optimum of 98°C; it will grow up to 110°C (above the boiling point of water). *Methanopyrus* occupies the deepest and most ancient branch of the euryarchaeotes. Perhaps methanogenic archaea were among the earliest organisms. They certainly seem well adapted to living under conditions similar to those presumed to have existed on a young Earth.

As might be inferred from the methanogens' ability to produce methane anaerobically, their metabolism is unusual. These procaryotes contain several unique cofactors: tetrahydromethanopterin (H_4MPT), methanofuran (MFR), coenzyme M (2-mercaptoethanesulfonic acid), coenzyme F_{420}, and coenzyme F_{430} (**figure 20.11**). The first three cofactors bear the C_1 unit when CO_2 is reduced to CH_4. F_{420} carries electrons and hydrogens, and F_{430} is a nickel tetrapyrrole serving as a cofactor for the enzyme methyl-CoM methylreductase. The pathway for methane synthesis is thought to function as shown in **figure 20.12**. It appears that

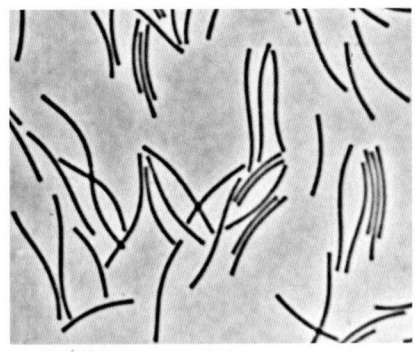

(a) *Methanospirillum hungatei*

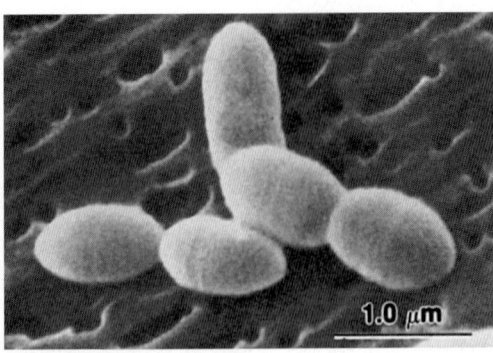

(b) *Methanobrevibacter smithii*

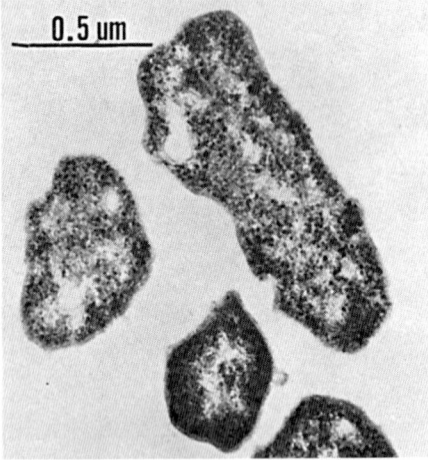

(c) *Methanogenium marisnigri*

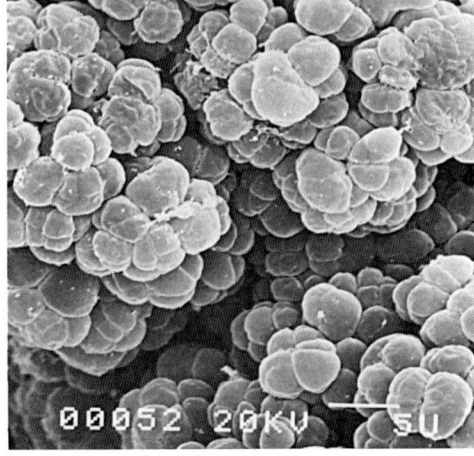

(d) *Methanosarcina mazei*

Figure 20.10   Selected Methanogens.   (a) *Methanospirillum hungatei;* phase contrast (×2,000). (b) *Methanobrevibacter smithii.* (c) *Methanogenium marisnigri;* electron micrograph (×45,000). (d) *Methanosarcina mazei;* SEM. Bar = 5 μm.

Table 20.2

Selected Characteristics of Representative Genera of Methanogens

Genus	Morphology	% G + C	Wall Composition	Gram Reaction	Motility	Methanogenic Substrates Used
Order *Methanobacteriales*						
Methanobacterium	Long rods or filaments	32–61	Pseudomurein	+ to variable	−	$H_2 + CO_2$, formate
Methanothermus	Straight to slightly curved rods	33	Pseudomurein with an outer protein S-layer	+	+	$H_2 + CO_2$
Order *Methanococcales*						
Methanococcus	Irregular cocci	29–34	Protein	−	+	$H_2 + CO_2$, formate
Order *Methanomicrobiales*						
Methanomicrobium	Short curved rods	45–49	Protein	−	+	$H_2 + CO_2$, formate
Methanogenium	Irregular cocci	52–61	Protein or glycoprotein	−	−	$H_2 + CO_2$, formate
Methanospirillum	Curved rods or spirilla	45–50	Protein	−	+	$H_2 + CO_2$, formate
Order *Methanosarcinales*						
Methanosarcina	Irregular cocci, packets	36–43	Heteropolysaccharide or protein	+ to variable	−	$H_2 + CO_2$, methanol, methylamines, acetate

Microbial Diversity & Ecology
20.1   Methanotrophic Archaea

The marine environment may contain as much as 10,000 billion tons of methane hydrate buried in the ocean floor (*see p. 625*), around twice the amount of all known fossil fuel reserves. Although some methane does rise toward the surface, it often is used before it escapes from the sediments in which it is buried. This is fortunate because methane is a much more powerful greenhouse gas than carbon dioxide. If the atmosphere were flooded with methane, the Earth could become too hot to support life as we know it. The reason for this disappearance of methane in sediments has been unclear until a recent discovery.

By using fluorescent probes for specific DNA sequences, a new assemblage of archaea and bacteria has been discovered in anoxic, methane-rich sediments. These clusters of procaryotes contain a core of about 100 archaea from the order *Methanosarcinales* surrounded by a layer of sulfate-reducing bacteria related to the *Desulfosarcina* (see **Box figure**). These two groups appear to cooperate metabolically in such a way that methane is anaerobically oxidized and sulfate reduced; perhaps the bacteria use waste products of methane oxidation to derive energy from sulfate reduction. Isotope studies show that the archaea feed on methane and the bacteria get much of their carbon from the archaea. These methanotrophs (*see pp. 488, 665*) may be crucial contributors to the Earth's carbon cycle because it is thought that they oxidize as much as 300 million tons of methane annually.

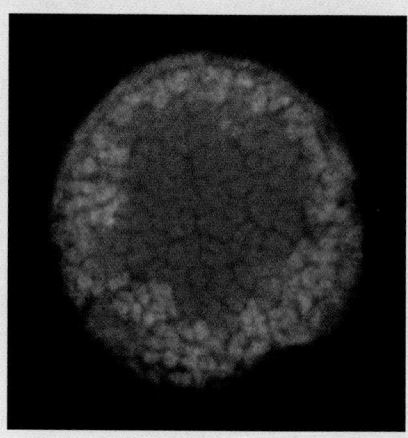

Methane-Consuming Archaea.   A cluster of methanotrophic archaea, stained red by a specific fluorescent probe, surrounded by a layer of bacteria, which are labeled by a green fluorescent probe.

ATP synthesis is linked with methanogenesis by electron transport, proton pumping, and a chemiosmotic mechanism (*see pp. 182–84*). Some methanogens can live autotrophically by forming acetyl-CoA from two molecules of CO_2 and then converting the acetyl-CoA to pyruvate and other products (figure 20.6).

Methanogens thrive in anaerobic environments rich in organic matter: the rumen and intestinal system of animals, freshwater and marine sediments, swamps and marshes, hot springs, anaerobic sludge digesters, and even within anaerobic protozoa. Methanogens often are of ecological significance. The rate of methane production can be so great that bubbles of methane will sometimes rise to the surface of a lake or pond. Rumen methanogens are so active that a cow can belch 200 to 400 liters of methane a day.

Methanogenic archaea are potentially of great practical importance since methane is a clean-burning fuel and an excellent energy source. For many years sewage treatment plants have been using the methane they produce as a source of energy for heat and electricity (*see figure 29.25*). Anaerobic digester microbes will degrade particulate wastes such as sewage sludge to H_2, CO_2, and acetate. CO_2-reducing methanogens form CH_4 from CO_2 and H_2, while aceticlastic methanogens cleave acetate to CO_2 and CH_4 (about 2/3 of the methane produced by an anaerobic digester comes from acetate). A kilogram of organic matter can yield up to 600 liters of methane. It is quite likely that future research will greatly increase the efficiency of methane production and make methanogenesis an important source of pollution-free energy.

Methanogenesis also can be an ecological problem (*see section 30.6*). Methane absorbs infrared radiation and thus is a greenhouse gas. There is evidence that atmospheric methane concentrations have been rising over the last 200 years. Methane production may significantly promote future global warming (*see Microbial Diversity & Ecology 30.4*). Recently it has been discovered that methanogens can oxidize Fe^0 and use it to produce methane and energy. This means that methanogens growing around buried or submerged iron pipes and other objects may contribute significantly to iron corrosion.

The Halobacteria

The **extreme halophiles** or **halobacteria,** class *Halobacteria,* are another major group of archaea, currently with 15 genera in one family, the *Halobacteriaceae* (**figure 20.13**). They are aerobic chemoheterotrophs with respiratory metabolism and require complex nutrients, usually proteins and amino acids, for growth. These procaryotes are either nonmotile or motile by lophotrichous flagella.

The most obvious distinguishing trait of this family is its absolute dependence on a high concentration of NaCl. These procaryotes require at least 1.5 M NaCl (about 8%, wt/vol), and usually have a growth optimum at about 3 to 4 M NaCl (17 to

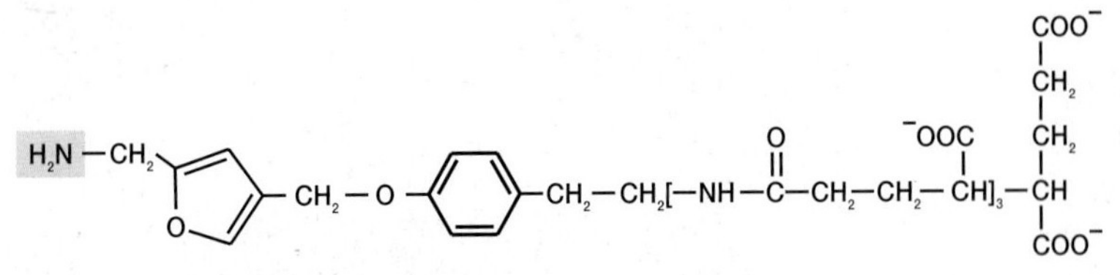

(a) Methanofuran (MFR)

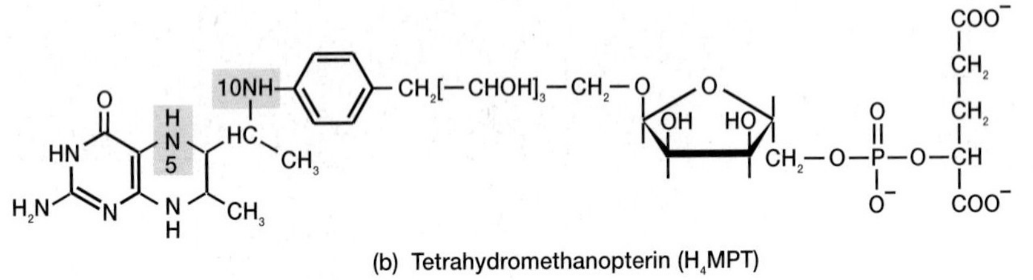

(b) Tetrahydromethanopterin (H₄MPT)

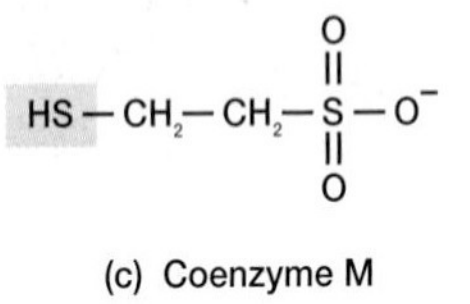

(c) Coenzyme M

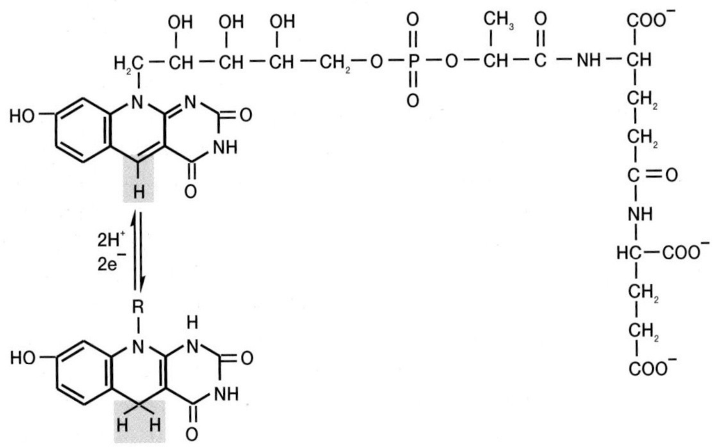

(d) Coenzyme F₄₂₀

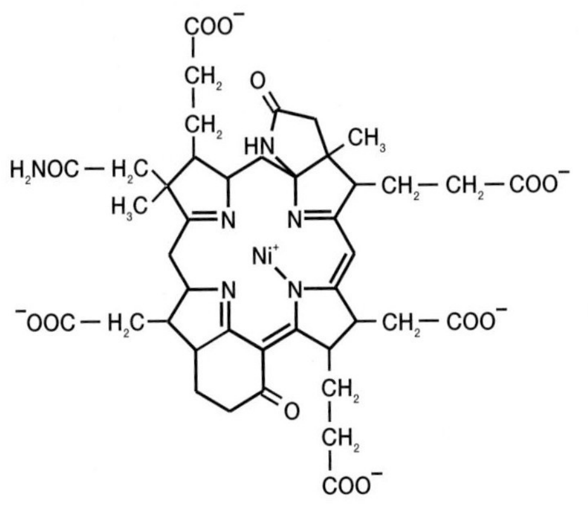

(e) Coenzyme F₄₃₀

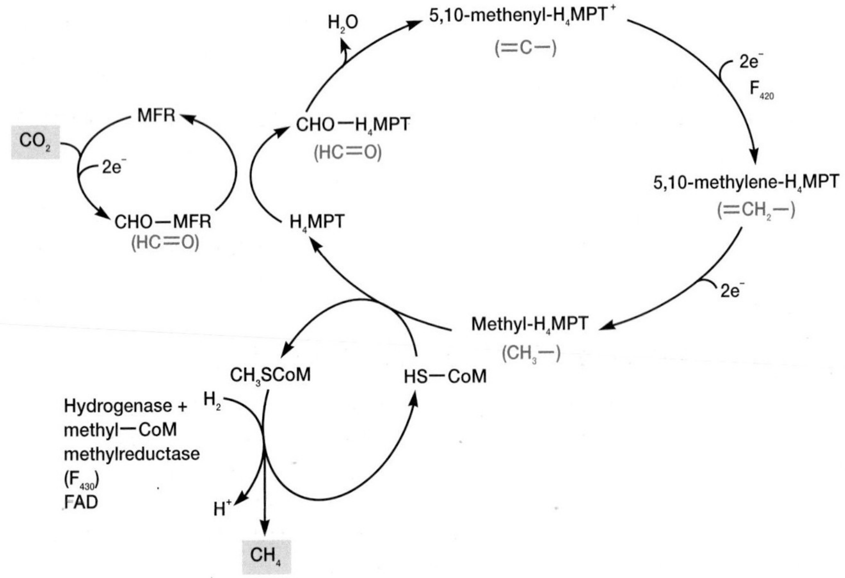

Figure 20.11 Methanogen Coenzymes. The portion of F_{420} (**d**) that is reversibly oxidized and reduced is shown in blue. MFR (**a**), H_4MPT (**b**), and coenzyme M (**c**) carry one-carbon units during methanogenesis (MFR and MPT also participate in the synthesis of acetyl-CoA). The places where the carbon units are attached are in blue. H_4MPT carries carbon units on nitrogens 5 and 10 in the same way as the coenzyme tetrahydrofolate. Coenzyme F_{430} (**e**) is a coenzyme for methyl-CoM methylreductase.

Figure 20.12 Methane Synthesis. Pathway for CH_4 synthesis from CO_2 in *M. thermoautotrophicum*. Cofactor abbreviations: methanopterin (MPT), methanofuran (MFR), and 2-mercaptoethanesulfonic acid or coenzyme M (CoM). The nature of the carbon-containing intermediates leading from CO_2 to CH_4 are indicated in parentheses. See text for further details.

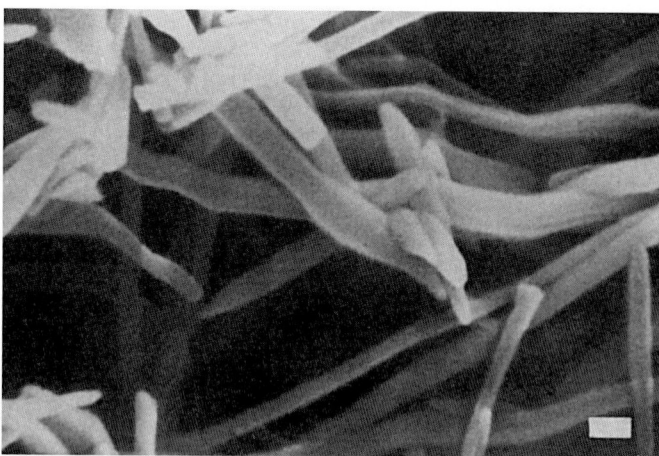

(a) *Halobacterium salinarium*

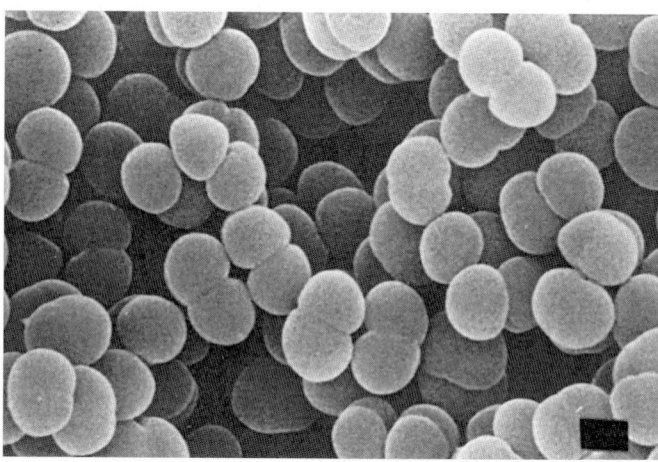

(b) *Halococcus morrhuae*

Figure 20.13 Examples of Halobacteria. (**a**) *Halobacterium salinarium.* A young culture that has formed long rods; SEM. Bar = 1 μm. (**b**) *Halococcus morrhuae;* SEM. Bar = 1 μm.

23%). They will grow at salt concentrations approaching saturation (about 36%). *Halobacterium's* cell wall is so dependent on the presence of NaCl that it disintegrates when the NaCl concentration drops to about 1.5 M. Thus halobacteria only grow in high-salinity habitats such as marine salterns (*see figure 28.29a*) and salt lakes such as the Dead Sea between Israel and Jordan, and the Great Salt Lake in Utah. They also can grow in food products such as salted fish and cause spoilage. Halobacteria often have red-to-yellow pigmentation from carotenoids that are probably used as protection against strong sunlight (*see section 6.4*). They can reach such high population levels that salt lakes, salterns, and salted fish actually turn red.   The effect of solutes on halophile growth (pp. 118–21)

Probably the best-studied member of the family is *Halobacterium salinarium (H. halobium).* This procaryote is unusual because it can trap light energy photosynthetically without the presence of chlorophyll. When exposed to low oxygen concentrations, some strains of *Halobacterium* synthesize a modified cell membrane called the **purple membrane,** which contains the protein **bacteriorhodopsin.** ATP is produced by a unique type of photosynthesis without the participation of bacteriochlorophyll or chlorophyll (**Microbial Tidbits 20.2**). *Halobacterium* actually has four rhodopsins, each with a different function. As already mentioned, bacteriorhodopsin drives outward proton transport for ATP synthesis. Halorhodopsin uses light energy to transport chloride ions into the cell and maintain a 4 to 5 M intracellular KCl concentration. Finally there are two rhodopsins that act as photoreceptors, one for red light and one for blue. They control flagellar activity to position the organism optimally in the water column. *Halobacterium* moves to a location of high light intensity, but one in which ultraviolet light is not sufficiently intense to be lethal.

The Thermoplasms

Procaryotes in the class *Thermoplasmata* are thermoacidophiles that lack cell walls. At present, three genera, *Thermoplasma Pi-*

crophilus, and *Ferroplasma* are known. They are sufficiently different from one another to be placed in separate families, *Thermoplasmataceae, Picrophilaceae,* and *Ferroplasmataceae.*

Thermoplasma grows in refuse piles of coal mines. These piles contain large amounts of iron pyrite (FeS), which is oxidized to sulfuric acid by chemolithotrophic bacteria. As a result the piles become very hot and acidic. This is an ideal habitat for *Thermoplasma* since it grows best at 55 to 59°C and pH 1 to 2. Although it lacks a cell wall, its plasma membrane is strengthened by large quantities of diglycerol tetraethers, lipopolysaccharides, and glycoproteins. The organism's DNA is stabilized by association with a special histonelike protein that condenses the DNA into particles resembling eucaryotic nucleosomes. At 59°C, *Thermoplasma* takes the form of an irregular filament, whereas at lower temperatures it is spherical (**figure 20.14**). The cells may be flagellated and motile.

Picrophilus is even more unusual than *Thermoplasma.* It originally was isolated from moderately hot solfataric fields in Japan. Although it lacks a regular cell wall, *Picrophilus* has an S-layer outside its plasma membrane (*see section 3.7*). The cells grow as irregularly shaped cocci, around 1 to 1.5 μm in diameter, and have large cytoplasmic cavities that are not membrane bounded. *Picrophilus* is aerobic and grows between 47 and 65°C with an optimum of 60°C. It is most remarkable in its pH requirements. The organism will grow only below pH 3.5 and has a growth optimum at pH 0.7. Growth actually occurs at about pH 0!

Extremely Thermophilic S⁰-Metabolizers

This physiological group contains the class *Thermococci,* with one order, *Thermococcales.* The *Thermococcales* are strictly anaerobic and can reduce sulfur to sulfide. They are motile by flagella and have optimum growth temperatures around 88 to 100°C. The order contains one family and three genera, *Thermococcus, Paleococcus,* and *Pyrococcus.*

The halophilic bacterium *Halobacterium salinarium* (formerly *H. halobium*) normally depends on respiration for the release of energy. However, under conditions of low oxygen and high light intensity, the bacteria synthesize a deep-purple pigment called bacteriorhodopsin, which closely resembles the sensory pigment rhodopsin from the rods and cones of vertebrate eyes. Bacteriorhodopsin's chromophore is the carotenoid derivative retinal (the aldehyde of vitamin A) that is covalently attached to the pigment protein by a Schiff base with the amino group of a lysine. The protein has seven membrane-spanning helices connected by loops on either side; its retinal rests in the center of the membrane. Individual bacteriorhodopsins aggregate in the membrane to form crystalline patches called the purple membrane.

Bacteriorhodopsin functions as a light-driven proton pump. When retinal absorbs light, the double bond between carbons 13 and 14 changes from a *trans* to a *cis* configuration and the Schiff base loses a proton. Protons are moved across the plasma membrane to the periplasmic space (*see section 3.5*) during these alterations, and the Schiff base changes are directly involved in this movement (see **Box figure**). The bacteriorhodopsin protein undergoes several conformational changes during the photocycle. These conformational changes also are involved in proton trans-

port. The light-driven proton pumping generates a pH gradient that can be used to power the synthesis of ATP by a chemiosmotic mechanism (*see section 9.5*).

This photosynthetic capacity is particularly useful to *Halobacterium* because oxygen is not very soluble in concentrated salt solutions and may decrease to an extremely low level in its habitat. When the surroundings become temporarily anaerobic, the bacterium uses light energy to synthesize sufficient ATP to remain alive until the oxygen level rises again. *Halobacterium* cannot grow anaerobically because it requires oxygen for continued retinal synthesis, but it can survive the stress of temporary oxygen limitation by means of photosynthesis.

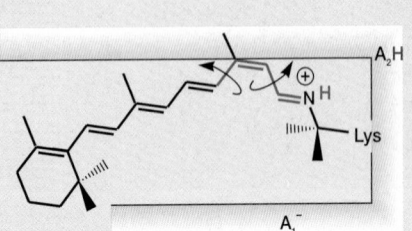

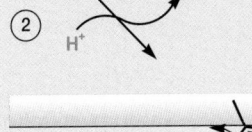

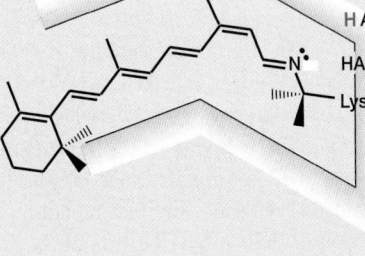

The Photocycle of Bacteriorhodopsin. In this hypothetical mechanism the retinal component of bacteriorhodopsin is buried in the membrane and retinal interacts with two amino acids, A_1 and A_2 (aspartates 96 and 85), that can reversibly accept and donate protons. A_2 is connected to the cell exterior, and A_1 is closer to the cell interior. The two amino acids may be special aspartic acid residues in bacteriorhodopsin. Light absorption by retinal in **step 1** triggers an isomerization from 13-*trans*-retinal to 13-*cis*-retinal. The retinal then donates a proton to A_2 in **steps 2 and 3,** while A_1 is picking up another proton from the interior and A_2 is releasing a proton to the outside. In **steps 4 and 5,** retinal obtains a proton from A_1 and isomerizes back to the 13-*trans* form. The cycle is then ready to begin again after **step 6.**

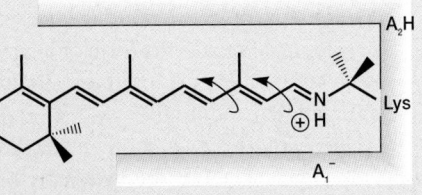

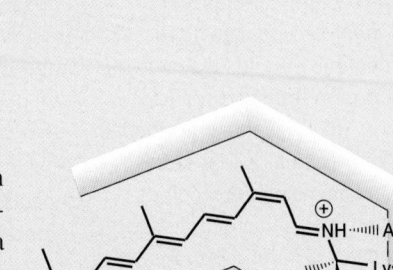

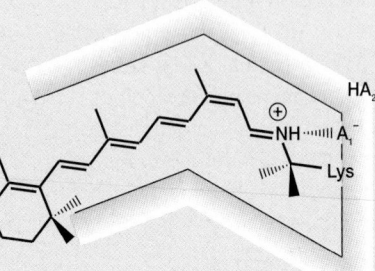

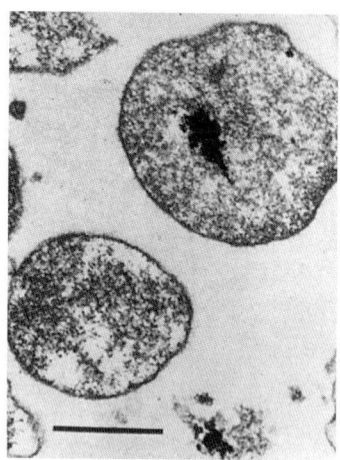

Figure 20.14 *Thermoplasma.*
Transmission electron micrograph. Bar = 0.5 μm.

Sulfate-Reducing Archaea

Archaeal sulfate reducers are found in the class *Archaeoglobi* and the order *Archaeoglobales*. This order has only one family and three genera. *Archaeoglobus* contains gram-negative, irregular coccoid cells with walls consisting of glycoprotein subunits. It can extract electrons from a variety of electron donors (e.g., H_2, lactate, glucose) and reduce sulfate, sulfite, or thiosulfate to sulfide. Ele-

mental sulfur is not used as an acceptor. *Archaeoglobus* is extremely thermophilic (the optimum is about 83°C) and can be isolated from marine hydrothermal vents. The organism is not only unusual in being able to reduce sulfate, unlike other archaea, but it also possesses the methanogen coenzymes F_{420} and methanopterin.

1. What are thermoacidophiles and where do they grow? In what ways do they use sulfur in their metabolism? Briefly describe *Sulfolobus* and *Thermoproteus*.
2. Generally characterize methanogenic archaea and distinguish them from other groups.
3. Briefly describe how methanogens produce methane and the roles of their unique cofactors in this process.
4. Where does one find methanogens? Discuss their ecological and practical importance.
5. Where are the extreme halophiles found and what is unusual about their cell walls and growth requirements?
6. Briefly describe how *Halobacterium* carries out photosynthesis. What is the purple membrane and what pigment does it contain?
7. How is *Thermoplasma* able to live in acidic, very hot coal refuse piles when it lacks a cell wall? How is its DNA stabilized? What is so remarkable about *Picrophilus*?
8. Characterize *Archaeoglobus*. In what way is it similar to the methanogens and how does it differ from other extreme thermophiles?

20.1 Introduction to the *Archaea*

a. Organisms may be divided into three major groups: *Archaea, Bacteria,* and *Eucarya* or eucaryotes.
b. The *Archaea* are highly diverse with respect to morphology, reproduction, physiology, and ecology. They grow in anaerobic, hypersaline, and high-temperature habitats.
c. Archaeal cell walls do not contain peptidoglycan and differ from bacterial walls in structure. They may be composed of pseudomurein, polysaccharides, or glycoproteins and other proteins.
d. The membrane lipids differ from those of other organisms in having branched chain hydrocarbons connected to glycerol by ether links. Bacterial and eucaryotic lipids have glycerol connected to fatty acids by ester bonds (**figure 20.3**).
e. Their tRNA, ribosomes, elongation factors, RNA polymerases, and other components distinguish archaea from bacteria and eucaryotes.
f. Although much of archaeal metabolism appears similar to that of other organisms, the archaea do differ with respect to glucose catabolism, pathways for CO_2 fixation, and the ability of some to synthesize methane.

g. Archaea may be divided into five groups: methanogenic archaea, sulfate reducers, extreme halophiles, cell wall-less archaea, and extremely thermophilic S^0-metabolizers (**table 20.1**).
h. The second edition of *Bergey's Manual* divides the *Archaea* into two phyla, the *Crenarchaeota* and *Euryarchaeota,* each with several orders (**figure 19.13**).

20.2 Phylum *Crenarchaeota*

a. The extremely thermophilic S^0-metabolizers in the phylum *Crenarchaeota* depend on sulfur for growth and are frequently acidophiles. The sulfur may be used as an electron acceptor in anaerobic respiration or as an electron donor by lithotrophs. They are almost always strict anaerobes and grow in geothermally heated soil and water that is rich in sulfur.

20.3 Phylum *Euryarchaeota*

a. The phylum *Euryarchaeota* contains five major groups: methanogens, halobacteria, the thermoplasms, extremely thermophilic S^0-metabolizers, and sulfate-reducing archaea.

b. Methanogenic archaea are strict anaerobes that can obtain energy through the synthesis of methane. They have several unusual cofactors that are involved in methanogenesis.
c. Extreme halophiles or halobacteria are aerobic chemoheterotrophs that require at least 1.5 M NaCl for growth. They are found in habitats such as salterns, salt lakes, and salted fish.
d. *Halobacterium salinarium* can carry out photosynthesis without chlorophyll or bacteriochlorophyll by using bacteriorhodopsin, which employs retinal to pump protons.
e. The thermophilic archaeon *Thermoplasma* grows in hot, acid coal refuse piles and survives despite the lack of a cell wall. Another thermoplasm, *Picrophilus,* can grow at pH 0.
f. The class *Thermococci* contains extremely thermophilic organisms that can reduce sulfur to sulfide.
g. Sulfate-reducing archaea are placed in the class *Archaeoglobi.* The extreme thermophile *Archaeoglobus* differs from other archaea in using a variety of electron donors to reduce sulfate. It also contains the methanogen cofactors F_{420} and methanopterin.

Key Terms

Archaea 438	extreme halophiles 447	methanogens 444	purple membrane 449
bacteriorhodopsin 449	halobacteria 447	pseudomurein 439	thermoacidophiles 444

Questions for Thought and Review

1. Discuss how the unusual properties of extreme halophiles and thermophiles reflect the habitats in which they grow.
2. Why do we classify *Thermoplasma* as an archaeon rather than as a mycoplasma?
3. Discuss what adaptations extremely thermophilic archaea require to grow at 100°C.
4. Do you think archaea should be separate from the bacteria although both groups are procaryotic? Give your reasoning and evidence.
5. Supposing you wished to isolate procaryotes from a hot spring in Yellowstone Park, how would you go about it?

Critical Thinking Questions

1. Explain why the fixation of CO_2 by *Thermoproteus* and possibly by *Sulfolobus* using a reductive reversal of the TCA cycle is *not* photosynthesis.
2. Often when the temperature increases, many procaryotes change their shapes from elongated rods into spheres. Suggest one reason for this change.
3. Why would ether linkages be more stable in membranes than ester lipids? How would the presence of tetraether linkages stabilize a thermophile's membrane?

For recommended readings on these and related topics, see Appendix V.

chapter 21

Bacteria:
The Deinococci and Nonproteobacteria Gram Negatives

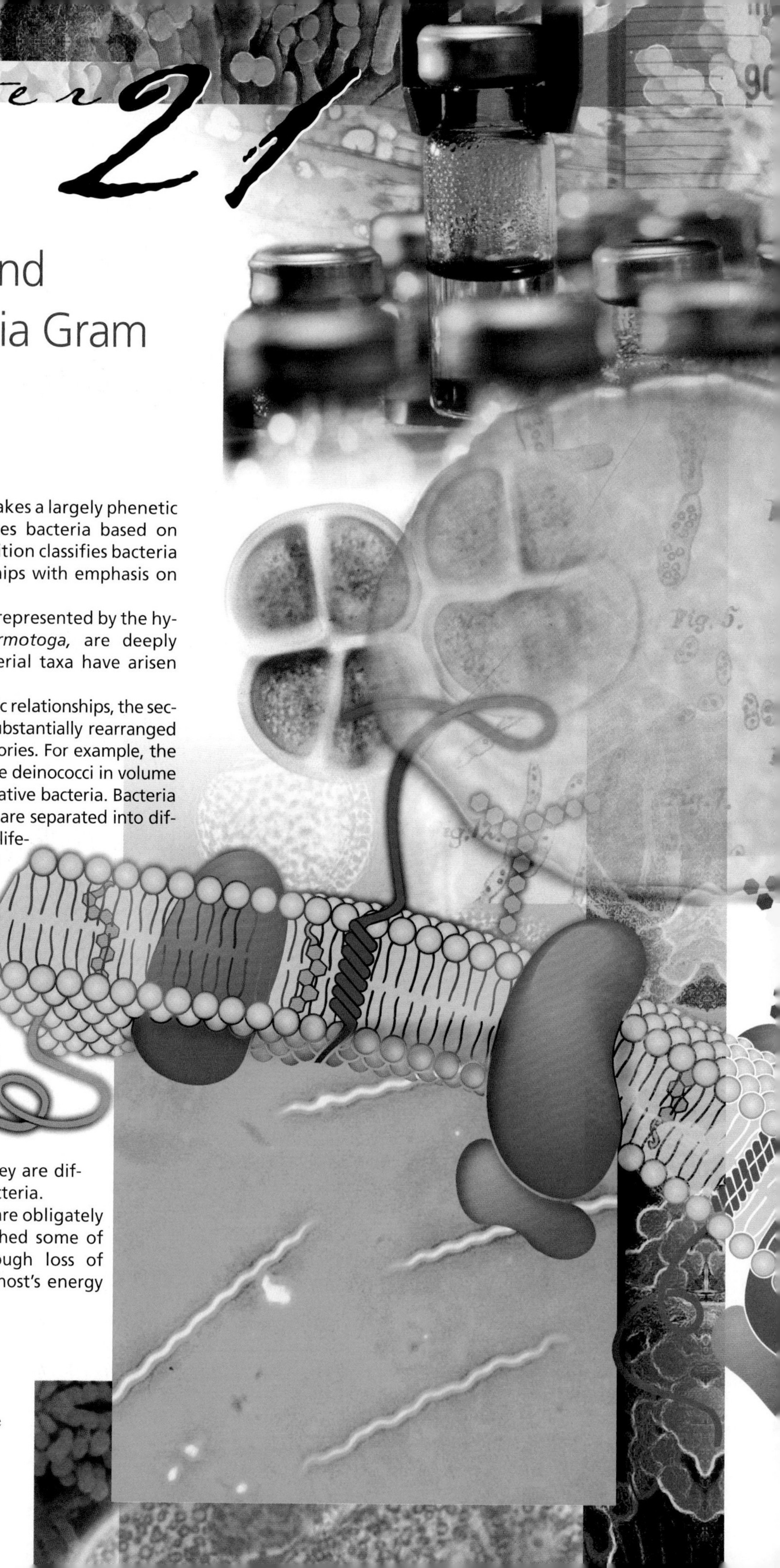

Concepts

- The first edition of *Bergey's Manual* takes a largely phenetic approach to classification and relates bacteria based on their overall similarity. The second edition classifies bacteria according to phylogenetic relationships with emphasis on 16S rRNA sequence comparisons.

- Some bacterial groups, such as those represented by the hyperthermophiles *Aquifex* and *Thermotoga,* are deeply branching and very old; other bacterial taxa have arisen more recently.

- Because of its emphasis on phylogenetic relationships, the second edition of *Bergey's Manual* has substantially rearranged bacterial groups and taxonomic categories. For example, the second edition places the gram-positive deinococci in volume 1, which otherwise contains gram-negative bacteria. Bacteria such as the rickettsias and chlamydiae are separated into different sections despite their similar lifestyles. The thermotogas and many other newly discovered groups have been added.

- Although most of the photosynthetic bacteria are located in volume 1 of the second edition, the purple bacteria have been moved to the proteobacteria in volume 2. The cyanobacteria are separated from other photosynthetic bacteria because they resemble eucaryotic phototrophs in having photosystem II and carrying out oxygenic photosynthesis. Their rRNA sequences also indicate that they are different from other photosynthetic bacteria.

- Bacteria such as the chlamydiae that are obligately intracellular parasites have relinquished some of their metabolic independence through loss of metabolic pathways. They use their host's energy supply and/or cell constituents.

Spirochetes are distinguished by their structure and mechanism of motility. *Treponema pallidum* shown here causes syphilis.

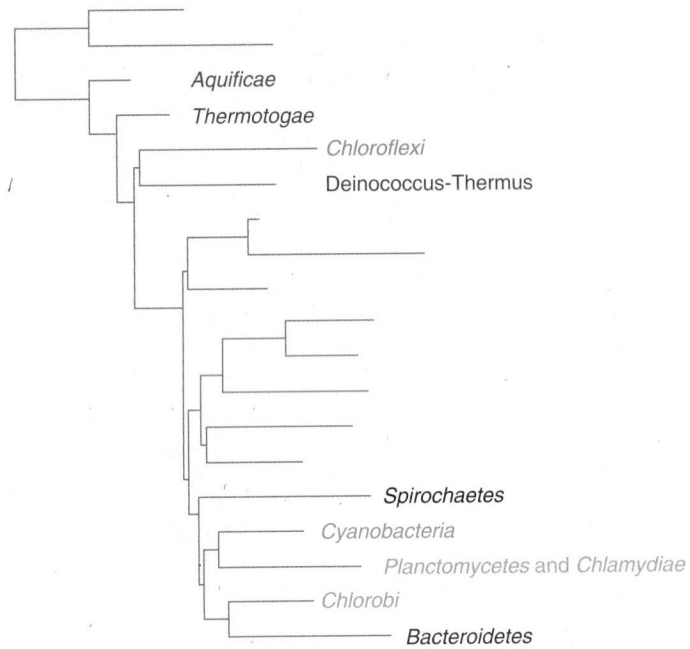

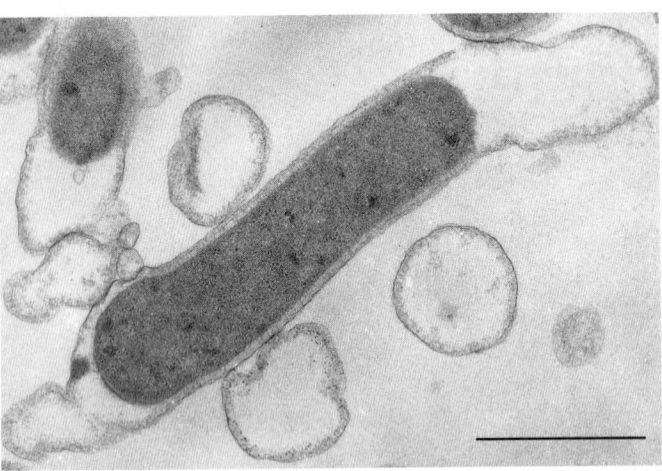

Figure 21.1 *Thermotoga maritima.* Note the loose sheath extending from each end of the cell. Bar = 1 μm.

• Gliding motility is widely distributed among bacteria and is very useful to organisms that digest insoluble nutrients or move over the surfaces of moist, solid substrata.

There are wide areas of the bacteriological landscape in which we have so far detected only some of the highest peaks, while the rest of the beautiful mountain range is still hidden in the clouds and the morning fogs of ignorance. The gold is still lying on the ground, but we have to bend down to grasp it.

—Preface to *The Prokaryotes*

hapter 20 surveyed the archaea, which are located in volume 1 of the second edition of *Bergey's Manual*. Volumes 1 and 5 of the new *Bergey's Manual* also will describe a wide variety of other procaryotic groups that are members of the second domain: Bacteria. Chapter 21 is devoted to 10 of the more interesting and important bacterial phyla from volumes 1 and 5. Their phylogenetic locations are depicted in the dendrogram on the top of this page. The general organization and perspective of the second edition of *Bergey's Manual* will be followed in most cases.

Although the treatment of each bacterial group varies somewhat from that of the others, usually an attempt is made to provide a brief review of the group's biology. Such aspects as distinguish-

ing characteristics, morphology, reproduction, physiology, metabolism, and ecology are included. The taxonomy of each major group is summarized, and representative species are discussed. This approach should help one appreciate bacteria as living organisms rather than simply as agents of disease with little interest or importance in other contexts.

21.1 AQUIFICAE AND THERMOTOGAE

The phylum *Aquificae*, which is thought to represent the deepest or oldest branch of bacteria, contains one class, one order, and five genera. Two of the best-studied genera are *Aquifex* and *Hydrogenobacter*. *Aquifex pyrophilus* is a gram-negative, microaerophilic rod. It is a hyperthermophile with a temperature optimum of 85°C and a maximum of 95°C. *Aquifex* is an autotroph and generates energy by oxidizing donors such as hydrogen, thiosulfate, and sulfur with oxygen as the acceptor. Because *Aquifex* and *Hydrogenobacter* are both thermophilic chemolithoautotrophs, it has been suggested that the bacterial ancestor was probably thermophilic and chemolithoautotrophic.

The second oldest or deepest branch is the phylum *Thermotogae*, which has one class, one order, and six genera. The members of the genus *Thermotoga* [Greek *therme*, heat; Latin *toga*, outer garment], like *Aquifex*, are hyperthermophiles with a growth optimum of 80°C and a maximum of 90°C. They are gram-negative rods with an outer sheathlike envelope (like a toga) that can extend or balloon out from the ends of the cell (**figure 21.1**). They grow in active geothermal areas, both marine hydrothermal systems and terrestrial solfataric springs. In contrast to *Aquifex*, *Thermotoga* is a chemoheterotroph with a functional glycolytic pathway that can grow anaerobically on carbohydrates and protein digests.

The genomes of *Aquifex aeolicus* and *Thermotoga maritima* have been sequenced. The *Aquifex* genome is about a third the

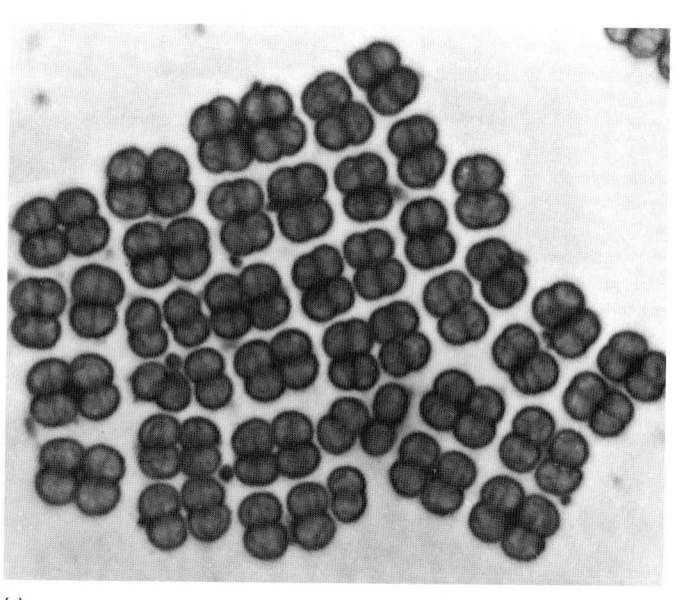

(a)

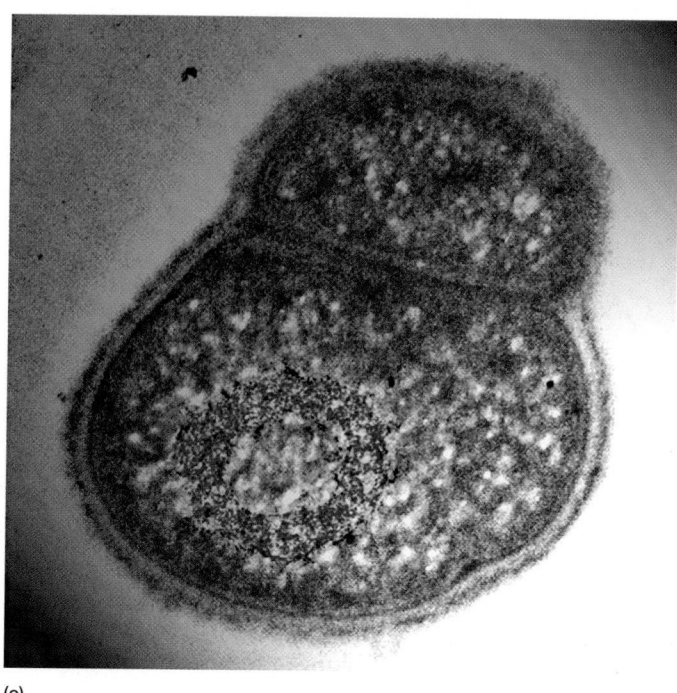

(c)

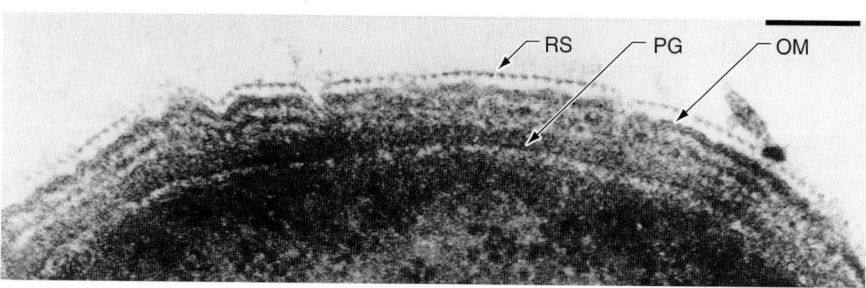

(b)

Figure 21.2 The Deinococci. (**a**) A *Deinococcus radiodurans* microcolony showing cocci arranged in tetrads (average cell diameter 2.5 μm). (**b**) The cell wall of *D. radiodurans* with a regular surface protein array (RS), a peptidoglycan layer (PG), and an outer membrane (OM). Bar = 100 nm. (**c**) A stained *D. radiodurans* cell showing a tightly packed, ringlike nucleoid stained purple.

size of the *E. coli* genome, but still contains the genes required for chemolithoautotrophy. The small genome does seem to reduce its metabolic flexibility; it will not grow on common organic substrates such as sugars and amino acids. The *Thermotoga* genome is somewhat larger and has genes for sugar degradation as already mentioned. About 24% of its coding sequences are similar to archaeal genes; this proportion is greater than that of other bacteria, including *Aquifex* (16% similarity) and may be due to horizontal gene transfer (*see pp. 425–26*).

 21.2   DEINOCOCCUS-THERMUS

The phylum Deinococcus-Thermus contains the class *Deinococci* and the orders *Deinococcales* and *Thermales*. There are only three genera in the phylum; the genus *Deinococcus* is best studied. Deinococci are spherical or rod-shaped with distinctively different 16S rRNA. They often are associated in pairs or tetrads (**figure 21.2a**) and are aerobic, mesophilic, and catalase positive; usually they can produce acid from only a few sugars. Although they stain gram positive, their cell wall is layered and

has an outer membrane like the gram-negative bacteria (figure 21.2b). They also differ from gram-positive cocci in having L-ornithine in their peptidoglycan, lacking teichoic acid, and having a plasma membrane with large amounts of palmitoleic acid rather than phosphatidylglycerol phospholipids. Almost all strains are extraordinarily resistant to both desiccation and radiation; they can survive as much as 3 to 5 million rad of radiation (an exposure of 100 rad can be lethal to humans).

Much remains to be discovered about the biology of these bacteria. Deinococci can be isolated from ground meat, feces, air, freshwater, and other sources, but their natural habitat is not yet known. Recent research indicates that their great resistance to desiccation and radiation may result from their ability to repair severely damaged chromosomes. The genome consists of two circular chromosomes, a megaplasmid, and a small plasmid. If the bacteria are exposed to enough radiation, their chromosomes are shattered into many fragments. Within 12 to 24 hours, they splice the fragments back together. A major reason for their ability to repair DNA is an unusually efficient RecA protein (*see p. 249*). However, it is thought that their chromosome structure also must be different for them to survive fragmentation. The

Table 21.1
Characteristics of the Major Groups of Gram-negative Photosynthetic Bacteria

	Anoxygenic Photosynthetic Bacteria				Oxygenic Photosynthetic Bacteria
Characteristic	**Green Sulfur[a]**	**Green Nonsulfur[b]**	**Purple Sulfur**	**Purple Nonsulfur**	**Cyanobacteria**
Major photosynthetic pigments	Bacteriochlorophylls a plus c, d, or e (the major pigment)	Bacteriochlorophylls a and c	Bacteriochlorophyll a or b	Bacteriochlorophyll a or b	Chlorophyll a plus phycobiliproteins
Morphology of photosynthetic membranes	Photosynthetic system partly in chlorosomes that are independent of the plasma membrane	Chlorosomes present when grown anaerobically	Photosynthetic system contained in spherical or lamellar membrane complexes that are continuous with the plasma membrane	Photosynthetic system contained in spherical or lamellar membrane complexes that are continuous with the plasma membrane	Membranes lined with phycobilisomes
Photosynthetic electron donors	H_2, H_2S, S	Photoheterotrophic donors—a variety of sugars, amino acids, and organic acids; photoautotrophic donors—H_2S, H_2	H_2, H_2S, S	Usually organic molecules: sometimes reduced sulfur compounds or H_2	H_2O
Sulfur deposition	Outside of the cell		Inside the cell[c]	Outside of the cell in a few cases	
Nature of photosynthesis	Anoxygenic	Anoxygenic	Anoxygenic	Anoxygenic	Oxygenic (sometimes facultatively anoxygenic)
General metabolic type	Obligately anaerobic photolithoautotrophs	Usually photoheterotrophic; sometimes photoautotrophic or chemoheterotrophic (when aerobic and in the dark)	Obligately anaerobic photolithoautotrophs	Usually anaerobic photoorganoheterotrophs; some facultative photolithoautotrophs (in dark, chemoorganoheterotrophs)	Aerobic photolithoautotrophs
Motility	Nonmotile; some have gas vesicles	Gliding	Motile with polar flagella; some are peritrichously flagellated	Motile with polar flagella or nonmotile; some have gas vesicles	Nonmotile or with gliding motility; some have gas vesicles
Percent G + C	48–58	53–55	45–70	61–72	35–71

[a]Characteristics of *Chlorobi*.
[b]Characteristics of *Chloroflexus*.
[c]With the exception of *Ectothiorhodospira*.

chromosome exists as a very compact ringlike structure (figure 21.2c). This may restrict DNA movement and maintain continuity even after radiation-induced breaks, thus facilitating repair. Recently the genome of *D. radiodurans* has been sequenced (*see* p. 340). The results show that the bacterium has an ample array of DNA repair mechanisms and many repeat sequences.

21.3 PHOTOSYNTHETIC BACTERIA

There are three groups of gram-negative photosynthetic bacteria: the purple bacteria, the green bacteria, and the cyanobacteria (**table 21.1**). The cyanobacteria differ most fundamentally from the green and purple photosynthetic bacteria in being able to carry out **oxygenic photosynthesis.** They use water as an electron donor and generate oxygen during photosynthesis. In contrast, purple and green bacteria use **anoxygenic photosynthesis.** Because they are unable to use water as an electron source, they employ reduced molecules such as hydrogen sulfide, sulfur, hydrogen, and organic matter as their electron source for the generation of NADH and NADPH. Consequently, purple and green bacteria do not produce oxygen but many form sulfur granules. Purple sulfur bacteria accumulate granules within their cells, whereas green sulfur bacteria deposit the sulfur granules outside their cells. The purple nonsulfur bacteria normally use organic molecules as an electron source. There also are differences in photosynthetic pigments, the organization of photosynthetic membranes, nutritional requirements, and oxygen relationships.

The mechanism of bacterial photosynthesis (pp. 190–96)

The connection between photosynthetic pigments, oxygen relationships, and ecological distribution should be noted here. As table 21.1 indicates, the purple and green bacteria differ

Table 21.2

Procaryotic Bacteriochlorophyll and Chlorophyll Absorption Maxima

	Long Wavelength Maxima (nm)	
Pigment	**In Ether or Acetone**	**Approximate Range of Values in Cells**
Chlorophyll *a*	665	680–685
Bacteriochlorophyll *a*	775	850–910 (purple bacteria)[a]
Bacteriochlorophyll *b*	790	1,020–1,035
Bacteriochlorophyll *c*	660	745–760
Bacteriochlorophyll *d*	650	725–745
Bacteriochlorophyll *e*	647	715–725

[a]The spectrum of bacteriochlorophyll *a* in green bacteria has a different maximum, 805–810 nm.

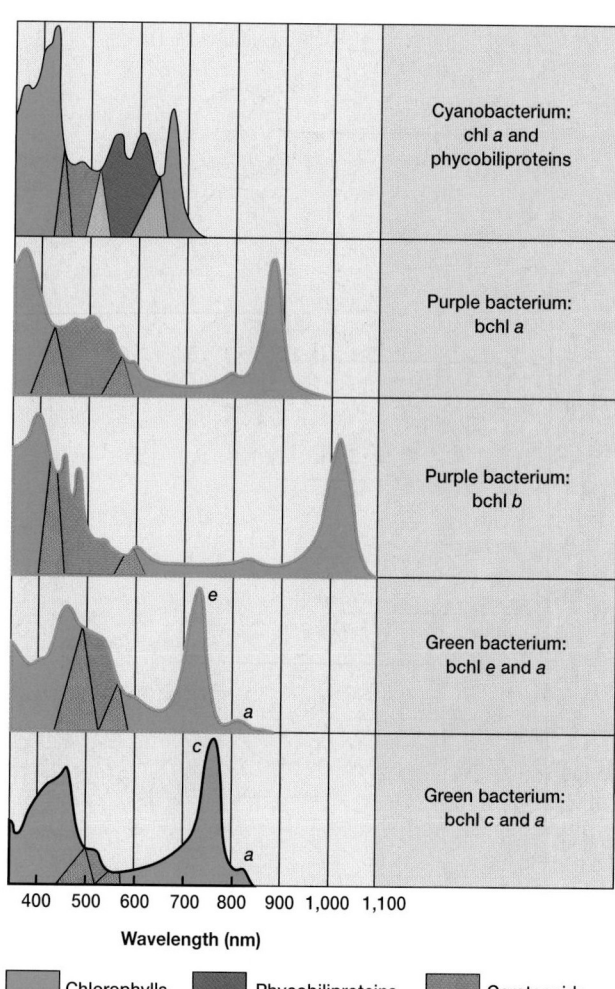

Figure 21.3 Photosynthetic Pigments. Absorption spectra of five photosynthetic bacteria showing the differences in absorption maxima and the contributions of various accessory pigments.

from the cyanobacteria in having bacteriochlorophylls rather than chlorophyll *a* (*see figure 9.26*). This proves to be quite useful because of the distinctive absorption spectra of the bacteriochlorophylls and accessory pigments. Normally green and purple bacteria are anaerobic and use H_2S and other reduced electron donors during photosynthesis (table 21.1). Because these bacteria grow best in deeper anaerobic zones of aquatic habitats, they cannot effectively use parts of the visible spectrum normally employed by photosynthetic organisms. There often is a dense surface layer of cyanobacteria and algae in lakes and ponds that absorbs a large amount of blue and red light. The bacteriochlorophyll pigments of purple and green bacteria absorb longer wavelength, far-red light (**table 21.2**) not used by other photosynthesizers (**figure 21.3**). In addition, the bacteriochlorophyll absorption peaks at about 350 to 550 nm enable them to grow at greater depths because shorter wavelength light can penetrate water farther. As a result, when the water is sufficiently clear, a layer of green and purple bacteria develops in the anaerobic, hydrogen sulfide–rich zone (**figure 21.4**). The microbial ecology of lakes (pp. 628–29)

The second edition of *Bergey's Manual* places photosynthetic bacteria into seven major groups distributed between five bacterial phyla. The phylum *Chloroflexi* contains the green nonsulfur bacteria, and the phylum *Chlorobi*, the green sulfur bacteria. The cyanobacteria are placed in their own phylum, *Cyanobacteria*. Purple bacteria are divided between three groups. Purple sulfur bacteria are placed in the γ-proteobacteria, families *Chromatiaceae* and *Ectothiorhodospiraceae*. The purple nonsulfur bacteria are distributed between the α-proteobacteria (five different families) and one family of the β-proteobacteria. Finally, the gram-positive heliobacteria in the phylum *Firmicutes* are also photosynthetic. There appears to have been considerable horizontal transfer of photosynthetic genes between the five phyla. About 50 genes related to photosynthesis were common to all five. The purple photosynthetic bacteria (pp. 475, 485–86); The heliobacteria (p. 509)

Phylum *Chloroflexi*

The phylum *Chloroflexi* has both photosynthetic and nonphotosynthetic members. *Chloroflexus* is the major representative of the photosynthetic **green nonsulfur bacteria.** It is a filamentous, gliding, thermophilic bacterium that often is isolated from neutral to alkaline hot springs where it grows in the form of orange-reddish mats, usually in association with cyanobacteria. Although it resembles the green bacteria in ultrastructure and photosynthetic pigments, its metabolism is more similar to that of the purple nonsulfur bacteria (*see p. 475*). *Chloroflexus* can carry out anoxygenic photosynthesis with organic compounds as carbon sources or grow aerobically as a chemoheterotroph. It doesn't appear closely related to any bacterial group based on 16S rRNA studies and is a deep and ancient branch of the bacterial tree. Nutritional types (section 5.3)

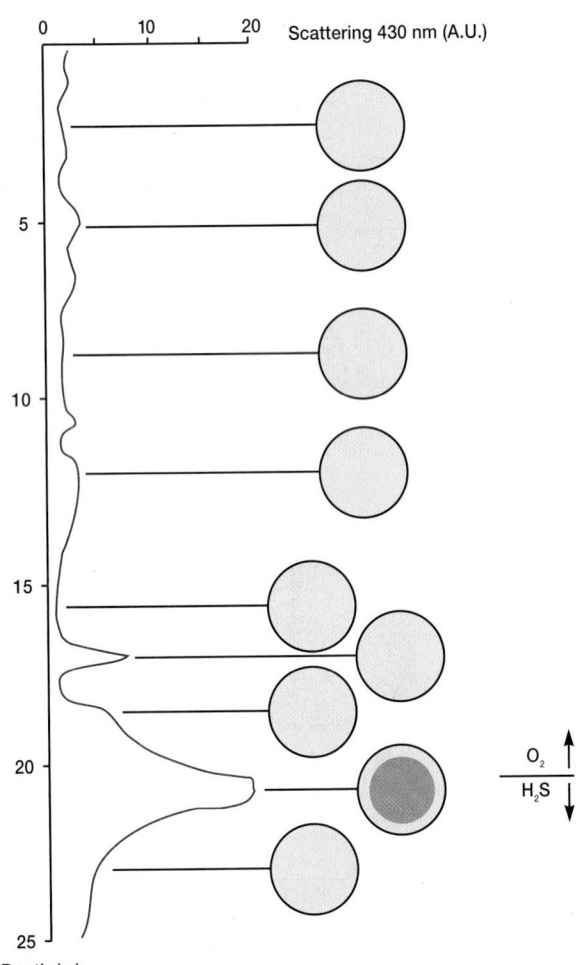

Figure 21.4 The Distribution of Photosynthetic Microorganisms in a Norwegian Fjord. Fifty-milliliter samples taken at various depths have been filtered and the filters mounted on a graph showing the light scattering (an indication of turbidity due to bacterial growth) as a function of depth. The peak at 20.8 m is caused by the purple bacteria *Chromatium* spp. growing at the interface of the aerobic zone and the anaerobic, sulfide-rich layer. The smaller peak at 17.1 m is caused by dinoflagellate growth. A.U. stands for absorbance units.

The nonphotosynthetic, gliding, rod-shaped or filamentous bacterium *Herpetosiphon* also is included in this phylum. *Herpetosiphon* is an aerobic chemoorganotroph with respiratory metabolism that uses oxygen as the electron acceptor. It can be isolated from freshwater and soil habitats.

Phylum *Chlorobi*

The phylum *Chlorobi* has only one class (*Chlorobia*), order (*Chlorobiales*), and family (*Chlorobiaceae*). The **green sulfur bacteria** are a small group of obligately anaerobic photolithoautotrophs that use hydrogen sulfide, elemental sulfur, and hydrogen as electron sources. The elemental sulfur produced by sulfide

oxidation is deposited outside the cell. Their photosynthetic pigments are located in ellipsoidal vesicles called **chlorosomes** or chlorobium vesicles, which are attached to the plasma membrane but are not continuous with it. The chlorosome membrane is not a normal lipid bilayer or unit membrane (*see section 3.2*). Chlorosomes contain accessory bacteriochlorophyll pigments, but the reaction center bacteriochlorophyll is located in the plasma membrane and must be able to obtain energy from chlorosome pigments. These bacteria flourish in the anaerobic, sulfide-rich zones of lakes. Although they lack flagella and are nonmotile, some species have gas vesicles (**figure 21.5a**) to adjust their depth for optimal light and hydrogen sulfide. Those forms without vesicles are found in sulfide-rich muds at the bottom of lakes and ponds.

The green sulfur bacteria are very diverse morphologically. They may be rods, cocci, or vibrios; some grow singly, and others form chains and clusters (figure 21.5b,c). They are either grass-green or chocolate-brown in color. Representative genera are *Chlorobium, Prosthecochloris,* and *Pelodictyon.*

---

1. Give the major distinguishing characteristics of *Aquifex, Thermotoga,* and the deinococci. What is thought to contribute to the dessication and radiation resistance of the deinococci?
2. How do oxygenic and anoxygenic photosynthesis differ from each other and why?
3. In general terms give the major characteristics of the following groups: purple sulfur bacteria, purple nonsulfur bacteria, and green sulfur bacteria. How do purple and green sulfur bacteria differ?
4. Compare the green nonsulfur (*Chloroflexi*) and green sulfur bacteria (*Chlorobi*).
5. What are chlorosomes or chlorobium vesicles?

Phylum *Cyanobacteria*

The **cyanobacteria** are the largest and most diverse group of photosynthetic bacteria. There is little agreement about the number of cyanobacterial species. Older classifications had as many as 2,000 or more species. In one recent system this has been reduced to 62 species and 24 genera. The second edition of *Bergey's Manual of Determinative Bacteriology* describes 57 genera in some detail. The G + C content of the group ranges from 35 to 71%. Although cyanobacteria are procaryotes, their photosynthetic system closely resembles that of the eucaryotes because they have chlorophyll *a* and photosystem II, and carry out oxygenic photosynthesis. Like the red algae, cyanobacteria use phycobiliproteins as accessory pigments. Photosynthetic pigments and electron transport chain components are located in thylakoid membranes lined with particles called **phycobilisomes (figure 21.6)**. These contain phycobilin pigments, particularly phycocyanin, and transfer energy to photosystem II. Carbon dioxide is assimilated through the Calvin cycle, and the reserve carbohydrate is glycogen. Sometimes they will store extra nitrogen as polymers of arginine or aspartic acid in cyanophycin granules. Since cyanobacteria lack the enzyme α-ketoglutarate dehydroge-

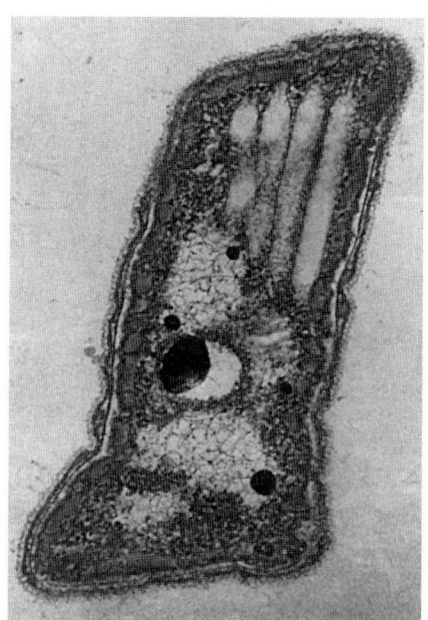

(a) *Pelodictyon clathratiforme*

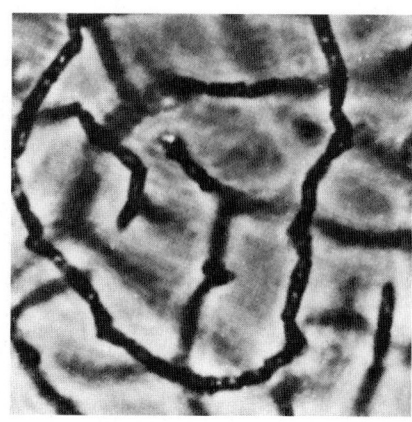

(b) *P. clathratiforme*

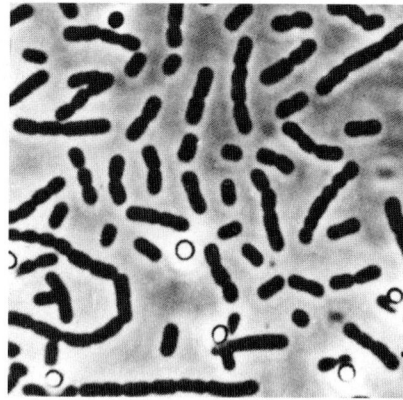

(c) *Chlorobium limicola*

Figure 21.5   Typical Green Sulfur Bacteria.   (a) An electron micrograph of *Pelodictyon clathratiforme* (×105,000). Note the chlorosomes (dark gray areas) and gas vesicles (light gray areas with pointed ends). **(b)** *P. clathratiforme.* Cells arranged in a netlike structure. **(c)** *Chlorobium limicola* with extracellular sulfur granules.

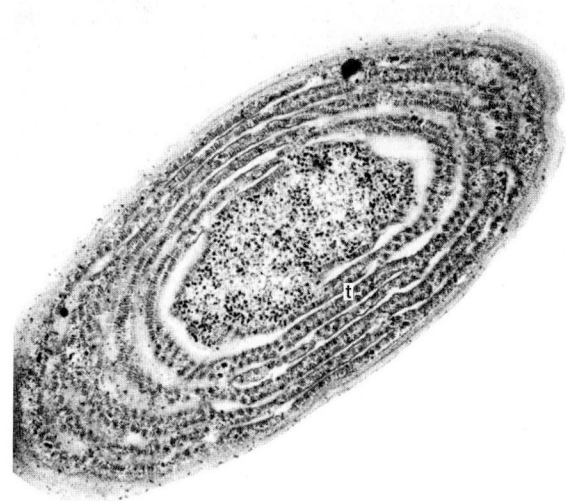

Figure 21.6   Cyanobacterial Thylakoids and Phycobilisomes. *Synechococcus lividus* with an extensive thylakoid system. The phycobilisomes lining these thylakoids are clearly visible as granules at location t (×85,000).

nase, they do not have a fully functional citric acid cycle. The pentose phosphate pathway plays a central role in their carbohydrate metabolism. Although many cyanobacteria are obligate photolithoautotrophs, some can grow slowly in the dark as chemoheterotrophs by oxidizing glucose and a few other sugars. Under anaerobic conditions *Oscillatoria limnetica* oxidizes hydrogen sulfide instead of water and carries out anoxygenic photosynthesis much like the green photosynthetic bacteria. As these

examples illustrate, cyanobacteria are capable of considerable metabolic flexibility.

Cyanobacteria also vary greatly in shape and appearance. They range in diameter from about 1 to 10 μm and may be unicellular, exist as colonies of many shapes, or form filaments called trichomes (**figure 21.7**; *see also figures 3.12 and 3.13*). A **trichome** is a row of bacterial cells that are in close contact with one another over a large area. In contrast, adjacent cells in a simple chain (such as those commonly found in the genus *Bacillus*) associate by only a small area of contact. Although most appear blue-green because of phycocyanin, a few are red or brown in color because of the red pigment phycoerythrin. Despite this variety, cyanobacteria have typical procaryotic cell structures and a normal gram-negative type cell wall (**figure 21.8**). They often use gas vesicles to move vertically in the water, and many filamentous species have gliding motility (**Microbial Diversity & Ecology 21.1**; *see also section 3.7*). Although cyanobacteria lack flagella, several strains of the marine genus *Synechococcus* are able to move at rates of up to 25 μm/second by means of an unknown mechanism.

Cyanobacteria show great diversity with respect to reproduction and employ a variety of mechanisms: binary fission, budding, fragmentation, and multiple fission. In the last process a cell enlarges and then divides several times to produce many smaller progeny, which are released upon the rupture of the parental cell. Fragmentation of filamentous cyanobacteria can generate small, motile filaments called **hormogonia.** Some species develop **akinetes,** specialized, dormant, thick-walled resting cells that are resistant to desiccation. Often these germinate to form new filaments.

Many filamentous cyanobacteria fix atmospheric nitrogen by means of special cells called **heterocysts** (**figure 21.9**). Around 5 to 10% of the cells can develop into heterocysts when

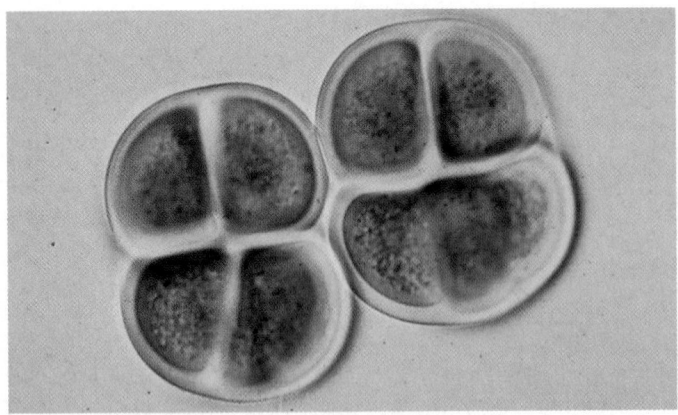

(a) *Chroococcus turgidus*

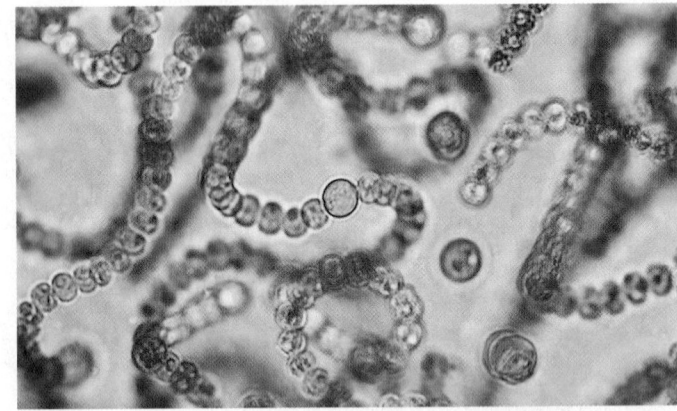

(b) *Nostoc*

(c) *Oscillatoria*

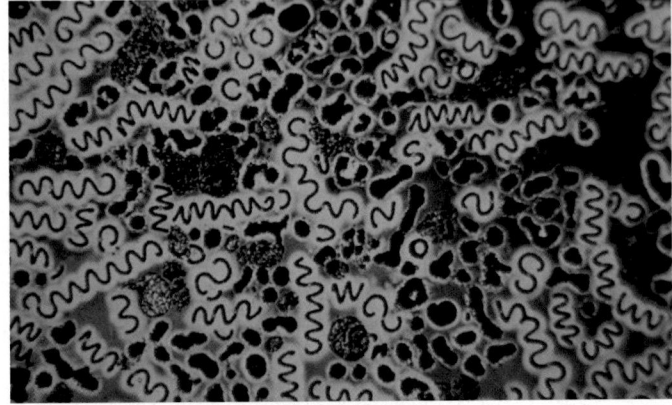

(d) *Anabaena spiroides* and *Microcystis aeruginosa*

Figure 21.7   Oxygenic Photosynthetic Bacteria.   Representative cyanobacteria. (**a**) *Chroococcus turgidus,* two colonies of four cells each (×600). (**b**) *Nostoc* with heterocysts (×550). (**c**) *Oscillatoria* trichomes seen with Nomarski interference-contrast optics (×250). (**d**) The cyanobacteria *Anabaena spiroides* and *Microcystis aeruginosa.* The spiral *A. spiroides* is covered with a thick gelatinous sheath (×1,000).

cyanobacteria are deprived of both nitrate and ammonia, their preferred nitrogen sources. When transforming themselves into heterocysts, cyanobacterial cells synthesize a very thick new wall, reorganize their photosynthetic membranes, discard their phycobiliproteins and photosystem II, and synthesize the nitrogen-fixing enzyme nitrogenase. Photosystem I is still functional and produces ATP, but no oxygen arises from noncyclic photophosphorylation because photosystem II is absent. This inability to generate O_2 is critical because the nitrogenase is extremely oxygen sensitive. The heterocyst wall slows or prevents O_2 diffusion into the cell, and any O_2 present is consumed during respiration. The structure and physiology of the heterocyst ensures that it will remain anaerobic; it is dedicated to nitrogen fixation. It obtains nutrients from adjacent vegetative cells and contributes fixed nitrogen in the form of the amino acid glutamine. It should be noted that nitrogen fixation also is carried out by cyanobacteria that lack heterocysts. Some fix nitrogen under dark, anoxic conditions in microbial mats. Planktonic forms such as *Trichodesmium* can fix nitrogen as well. The biochemistry of nitrogen fixation (pp. 206–9)

The classification of cyanobacteria is still in an unsettled state, partly due to the lack of pure cultures. At present all taxonomic schemes must be considered tentative. The second edition of *Bergey's Manual* divides the cyanobacteria into five subsections with 57 genera. **Table 21.3** briefly summarizes the major characteristics of the five subsections. These are distinguished using colony or trichome morphology and reproductive patterns. Some other properties important in cyanobacterial characterization are cell morphology, ultrastructure, genetic characteristics, physiology and biochemistry, and habitat/ecology (preferred habitat and growth habit). The authors consider genera as tentative and often do not provide species names.

Cyanobacterial taxonomy in the second edition of *Bergey's Manual* is very similar to that in the first edition. Probably the major difference is that the order *Prochlorales* is separate from the cyanobacteria in the first edition, while in the second edition the order is dropped and the prochlorophyte genera are placed within the phylum *Cyanobacteria.*

Prochlorophytes are oxygenic phototrophs that have both chlorophyll *a* and *b* but lack phycobilins. Thus although they

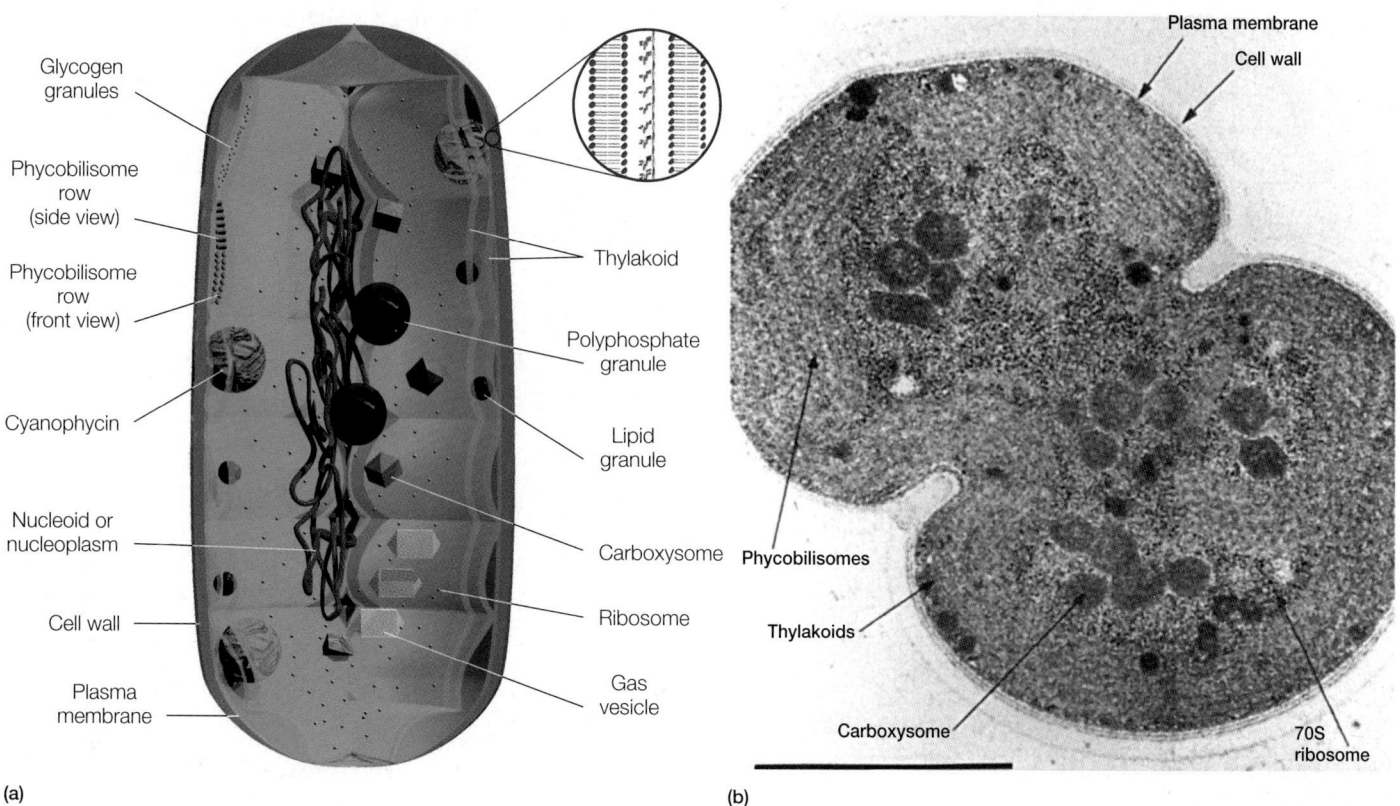

Figure 21.8 Cyanobacterial Cell Structure. (a) Schematic diagram of a vegetative cell. The insert shows an enlarged view of the envelope with its outer membrane and peptidoglycan. (b) Thin section of *Synechocystis* during division. Bar = 1 μm. Many structures are visible. (a) Illustration copyright © Hartwell T. Crim, 1998.

Microbial Diversity & Ecology

21.1   The Mechanism of Gliding Motility

Gliding motility varies greatly in rate (from about 2 μm per minute to over 600 μm per minute) and in the nature of the motion. Bacteria such as *Myxococcus* and *Flexibacter* glide along in a direction parallel to the longitudinal axis of their cells. Others (*Saprospira*) travel with a screwlike motion or even move in a direction perpendicular to the long axis of the cells in their trichome (*Simonsiella*). *Beggiatoa*, cyanobacteria, and some other bacteria rotate around their longitudinal axis while gliding, but this is not always seen. Many will flex or twitch as well as glide. Such diversity in gliding movement may indicate that more than one mechanism for motility exists. This conclusion is supported by the observation that some gliders (e.g., *Cytophaga, Flexibacter,* and *Flavobacterium*) move attached latex beads over their surface, whereas others such as *Myxococcus* either do not move beads or move them very slowly. (That is, not all gliding bacteria have rapidly moving cell-surface components.) Although slime is required for gliding, it does not appear to propel bacteria directly; rather, it probably attaches them to the substratum and lubricates the surface for more efficient movement.

A variety of mechanisms for gliding motility have been proposed. Cytoplasmic fibrils or filaments are associated with the envelope of many gliding bacteria. In *Oscillatoria* they seem to be contractile and may produce waves in the outer membrane, resulting in movement. Ringlike protein complexes or rotary assemblies resembling flagellar basal bodies are present in some envelopes. These assemblies may spin and move the bacterium along. Type IV fimbriae that can retract may be involved in several species. There is some evidence that differences in surface tension can propel *Myxococcus xanthus*. *Myxococcus* may secrete a surfactant at its posterior end (the end opposite the direction of movement) that lowers the surface tension at the rear end of the rod. The cell would be pulled forward by the greater surface tension exerted on its anterior end. Recently it has been found that slime emerges from bands of tiny nozzles encircling the ends of *Myxococcus*. Some have proposed that the gel swells in extracellular water and pushes the individual rods along. *Myxococcus* does use fimbriae extension and retraction when it is gliding as part of a moving group of cells.

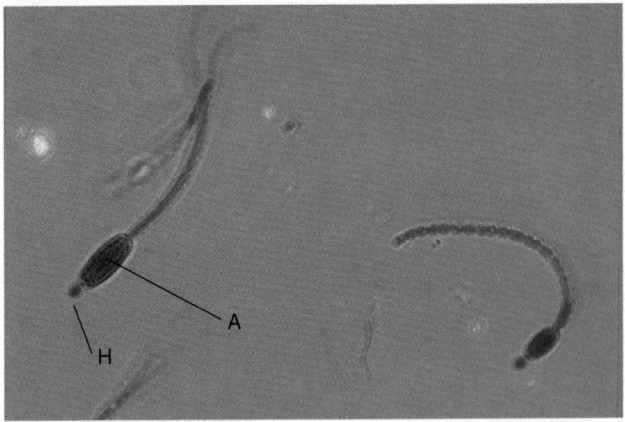

(a) *Cylindrospermum*

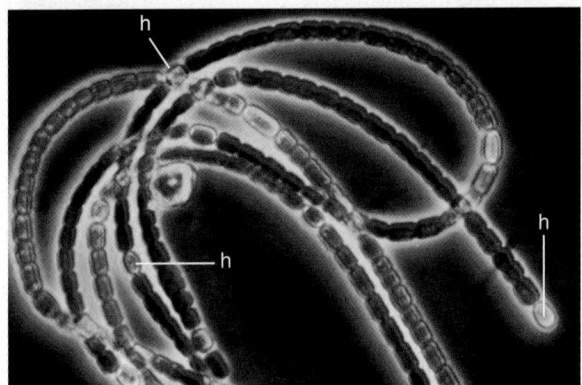

(b) *Anabaena*

Figure 21.9 Examples of Heterocysts and Akinetes.
(a) *Cylindrospermum* with terminal heterocysts (H) and subterminal akinetes (A) (×500). (b) *Anabaena,* with heterocysts. (c) An electron micrograph of an *Anabaena* heterocyst. Bar = 1 μm. Note the cell wall (W), additional outer walls (E), membrane system (M), and a pore channel to the adjacent cell (P).

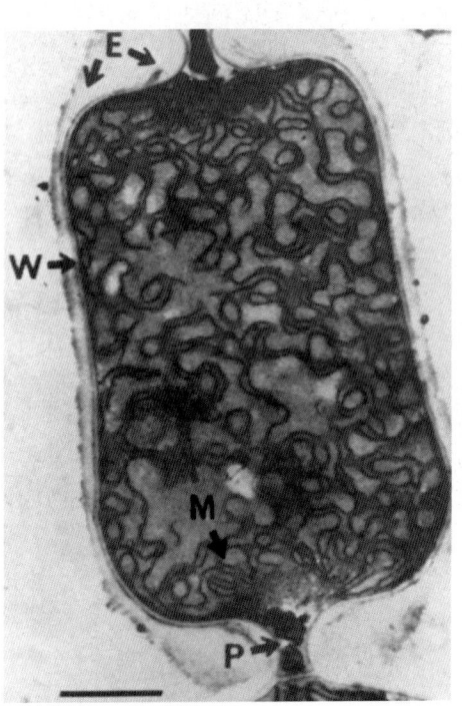

(c) *Anabaena* heterocyst

Table 21.3

Characteristics of the Cyanobacterial Subsections

Subsection	General Shape	Reproduction and Growth	Heterocysts	% G + C	Other Properties	Representative Genera
I	Unicellular rods or cocci; nonfilamentous aggregates	Binary fission, budding	−	31–71	Almost always nonmotile	*Chamaesiphon* *Chroococcus* *Gloeothece* *Gleocapsa* *Prochloron*
II	Unicellular rods or cocci; may be held together in aggregates	Multiple fission to form baeocytes	−	40–46	Only some baeocytes are motile	*Pleurocapsa* *Dermocarpa* *Chroococcidiopsis*
III	Filamentous, unbranched trichome with only vegetative cells	Binary fission in a single plane, fragmentation	−	34–67	Usually motile	*Lyngbya* *Oscillatoria* *Prochlorothrix* *Spirulina* *Pseudanabaena*
IV	Filamentous, unbranched trichome may contain specialized cells	Binary fission in a single plane, fragmentation to form hormogonia	+	38–47	Often motile, may produce akinetes	*Anabaena* *Cylindrospermum* *Aphanizomenon* *Nostoc* *Scytonema* *Calothrix*
V	Filamentous trichomes either with branches or composed of more than one row of cells	Binary fission in more than one plane, hormogonia formed	+	42–44	May produce akinetes; greatest morphological complexity and differentiation in cyanobacteria	*Fischerella* *Stigonema* *Geitleria*

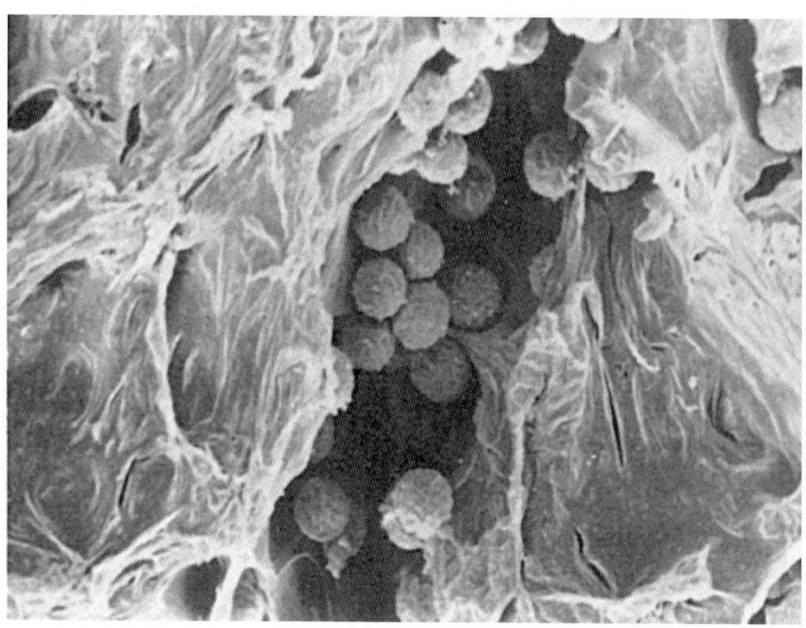

(a)

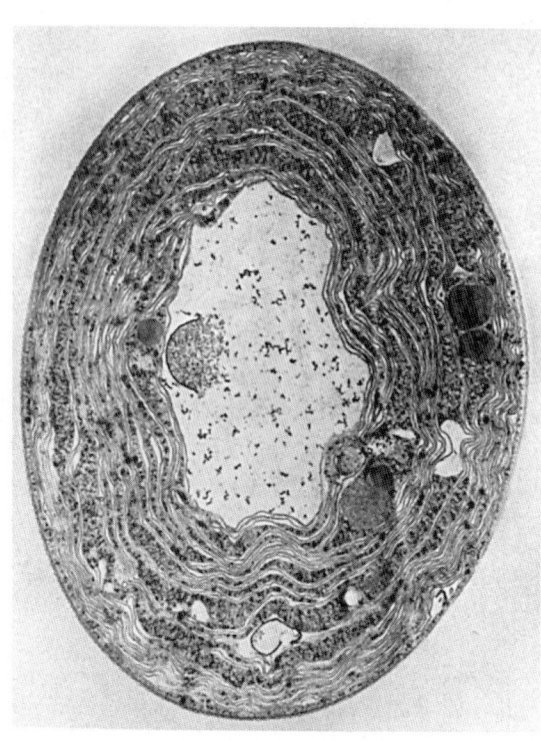

(b)

Figure 21.10   Prochloron.   (**a**) A scanning electron micrograph of *Prochloron* cells on the surface of a *Didemnum candidum* colony. (**b**) *Prochloron didemni* section; transmission electron micrograph (×23,500).

resemble other cyanobacteria with respect to chlorophyll *a,* they differ in also possessing chlorophyll *b,* the only procaryotes to do so. Because prochlorophytes lack phycobilin pigments and phycobilisomes, they are grass-green in color. They resemble chloroplasts in their pigments and thylakoid structure, but their 5S and 16S rRNAs place them within the cyanobacteria. The genera are located in subsections I and III.

The three recognized prochlorophyte genera are quite different from one another. *Prochloron* was first discovered as an extracellular symbiont growing either on the surface or within the cloacal cavity of marine colonial ascidian invertebrates (**figure 21.10**). These bacteria are single-celled, spherical, and from 8 to 30 μm in diameter. Their mol% of G + C is 31 to 41. *Prochlorothrix* is free living, has cylindrical cells that form filaments, and has been found in Dutch lakes. Its DNA has a higher G + C content (53 mol%). Unlike *Prochloron* it has been cultured in the laboratory.

Prochlorococcus marinus, less than 1 μm in diameter, has recently been discovered flourishing about 100 meters below the ocean surface. It differs from other prochlorophytes in having divinyl chlorophyll *a* and α-carotene instead of chlorophyll *a* and β-carotene. During the summer, it reaches concentrations of 5×10^5 cells per milliliter. It is one of the most numerous of the marine plankton and a significant component of the marine microbial food web.

As mentioned earlier, the five subsections differ in morphology and reproduction (table 21.3). Subsection I contains unicellular rods or cocci that are almost always nonmotile and reproduce by binary fission or budding. Organisms in subsection II are also

unicellular, though several individual cells may be held together in an aggregate by an outer wall. Members of this group reproduce by multiple fission to form spherical, very small, reproductive cells, often called **baeocytes,** which escape when the outer wall ruptures. Some baeocytes disperse through gliding motility. The other three subsections contain filamentous cyanobacteria. Usually the trichome is unbranched and often is surrounded by a sheath or slime layer. Cyanobacteria in subsection III form unbranched trichomes composed only of vegetative cells, whereas the other two subsections produce heterocysts in the absence of an adequate nitrogen source (they also may form akinetes). Heterocystous cyanobacteria are subdivided into those that form linear filaments (subsection IV) and cyanobacteria that divide in a second plane to produce branches or aggregates (subsection V).

Cyanobacteria are very tolerant of environmental extremes and are present in almost all waters and soils. Thermophilic species may grow at temperatures of up to 75°C in neutral to alkaline hot springs. Because these photoautotrophs are so hardy, they are primary colonizers of soils and surfaces that are devoid of plant growth. Some unicellular forms even grow in the fissures of desert rocks. In nutrient-rich warm ponds and lakes, surface cyanobacteria such as *Anacystis* and *Anabaena* can reproduce rapidly to form blooms (**figure 21.11**). The release of large amounts of organic matter upon the death of the bloom microorganisms stimulates the growth of chemoheterotrophic bacteria that subsequently depletes the available oxygen. This kills fish and other organisms (*see section 29.4*). Some species can produce toxins that kill livestock and other animals that drink the water.

Figure 21.11 Bloom of Cyanobacteria and Algae in a Eutrophic Pond.

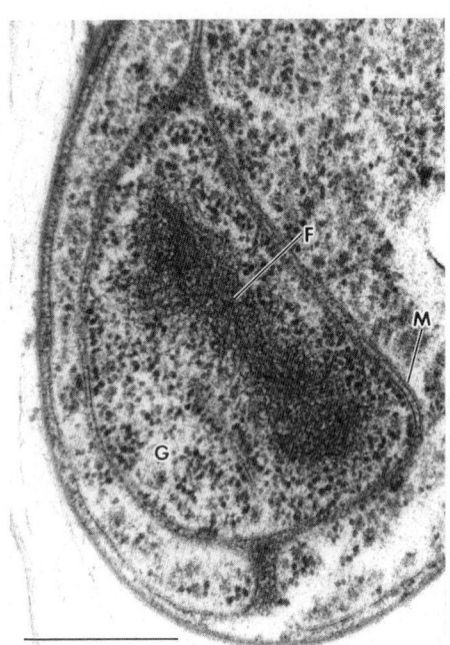

Figure 21.12 *Gemmata obscuriglobus* Nuclear Body. An electron micrograph showing the membrane-bounded nuclear region (Bar = 0.5 μm). Note the central fibrillar area (F), the outer granular area (G), and the outer double-membrane structure (M).

Other cyanobacteria, for example, *Oscillatoria,* are so pollution resistant and characteristic of freshwaters with high organic matter content that they are used as water pollution indicators.

Cyanobacteria are particularly successful in establishing symbiotic relationships with other organisms. They are the photosynthetic partner in most lichen associations. Cyanobacteria are symbionts with protozoa and fungi, and nitrogen-fixing species form associations with a variety of plants (liverworts, mosses, gymnosperms, and angiosperms). Types of symbiotic relationships (section 28.2); Lichens (p. 591)

1. Summarize the major characteristics of the cyanobacteria that distinguish them from other photosynthetic organisms.
2. Define or describe the following: phycobilisomes, hormogonia, akinetes, heterocysts, and baeocytes.
3. What is a trichome and how does it differ from a simple chain of cells?
4. Briefly discuss the ways in which cyanobacteria reproduce.
5. How are heterocysts modified to carry out nitrogen fixation? When do cyanobacteria develop heterocysts?
6. Give the features of the five major cyanobacterial groups.
7. Compare the prochlorophytes with other cyanobacteria and chloroplasts. Where does one find them?
8. List some important positive and negative impacts cyanobacteria have on humans and the environment.

21.4 PHYLUM *PLANCTOMYCETES*

The phylum *Planctomycetes* contains one class, one order, and four genera. Members of the phylum are spherical or oval, budding bacteria that lack peptidoglycan and have distinctive crateriform structures or pits in their walls. The nuclear body is membrane-bounded in the planctomycetes *Gemmata ob-*

scuriglobus and *Pirullela,* something not seen in other bacteria (**figure 21.12**). The genus *Planctomyces* attaches to surfaces through a stalk and holdfast; the other genera in the order lack stalks. Most of these bacteria have life cycles in which sessile cells bud to produce motile swarmer cells. The swarmer cells are flagellated and swim for a while before settling down to attach and begin reproduction.

21.5 PHYLUM *CHLAMYDIAE*

The first edition of *Bergey's Manual* places the chlamydiae with the rickettsias because both gram-negative groups are obligately intracellular parasites: they grow and reproduce only within host cells. Based on 16S rRNA data, the second edition moves the chlamydiae into the phylum *Chlamydiae* and the rickettsias to the α-proteobacteria. Although both chlamydiae and rickettsias are obligately intracellular parasites and not much larger than the poxviruses (*see p. 366*), they differ from viruses in having both DNA and RNA, a plasma membrane, functioning ribosomes, metabolic pathways, reproduction by binary fission, and other distinctive features.

The phylum *Chlamydiae* has one class, one order, four families, and only six genera. The genus *Chlamydia* is by far the most important and best studied; it will be the focus of our attention. **Chlamydiae** are nonmotile, coccoid, gram-negative bacteria, ranging in size from 0.2 to 1.5 μm. They can reproduce only within

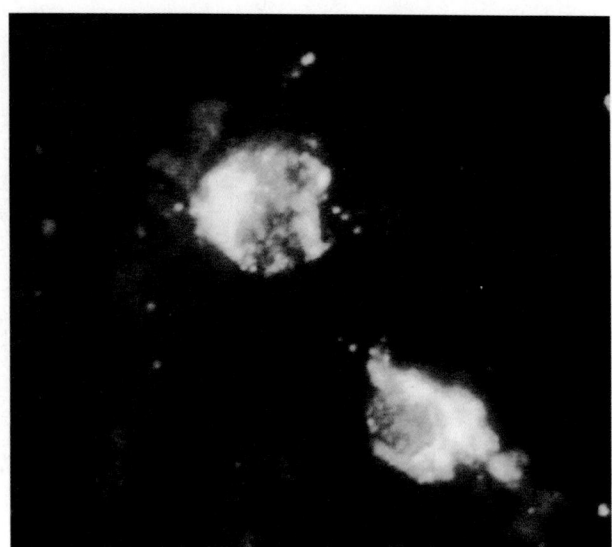

Figure 21.13   The Chlamydial Elementary Body. The infected cells glow a bright green because of the chlamydiae that are stained with fluorescently labeled monoclonal antibodies. The small yellow green dots are the chlamydiae.

cytoplasmic vesicles of host cells by a unique developmental cycle involving the formation of elementary bodies and reticulate bodies. Although their envelope resembles that of other gram-negative bacteria, the wall differs in lacking muramic acid and a peptidoglycan layer. The elementary bodies achieve osmotic stability by

cross-linking their outer membrane proteins, and possibly periplasmic proteins, with disulfide bonds. Chlamydiae are extremely limited metabolically and are obligately intracellular parasites of mammals and birds. (However, chlamydia-like bacteria have been isolated from spiders, clams, and freshwater invertebrates.) The size of their genome is 4 to 6×10^8 daltons, one of the smallest of all procaryotes, and the G + C content is 41 to 44%.

Chlamydial reproduction begins with the attachment of an **elementary body (EB)** to the cell surface (**figure 21.13**). Elementary bodies are 0.2 to 0.4 μm in diameter, contain electron-dense nuclear material and a rigid cell wall, and are infectious (**figure 21.14**). The host cell phagocytoses the EB, which then prevents the fusion of lysosomes with the phagosome and begins to reorganize itself to form a **reticulate body (RB)** or **initial body.** The RB is specialized for reproduction rather than infection. Reticulate bodies are 0.6 to 1.5 μm in diameter and have less dense nuclear material and more ribosomes than EBs; their walls are also more flexible. About 8 to 10 hours after infection, the reticulate body begins to divide and RB reproduction continues until the host cell dies. A chlamydia-filled vacuole or inclusion can become large enough to be seen in a light microscope and even fill the host cytoplasm. After 20 to 25 hours, RBs begin changing back into infectious EBs and continue this process until the host cell lyses and releases the chlamydiae 48 to 72 hours after infection.

Chlamydial metabolism is very different from that of other gram-negative bacteria. It has been thought that chlamydiae cannot catabolize carbohydrates or other substances and synthesize ATP. *Chlamydia psittaci,* one of the best-studied species, lacks

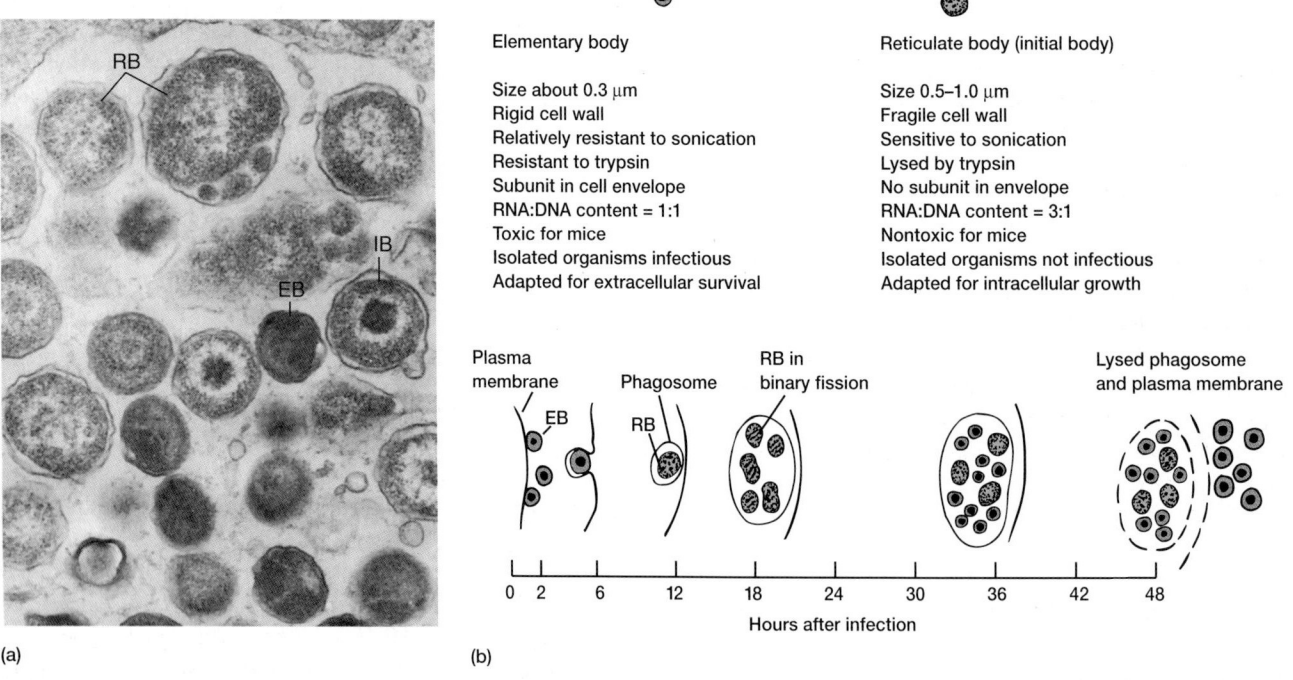

Elementary body

Size about 0.3 μm
Rigid cell wall
Relatively resistant to sonication
Resistant to trypsin
Subunit in cell envelope
RNA:DNA content = 1:1
Toxic for mice
Isolated organisms infectious
Adapted for extracellular survival

Reticulate body (initial body)

Size 0.5–1.0 μm
Fragile cell wall
Sensitive to sonication
Lysed by trypsin
No subunit in envelope
RNA:DNA content = 3:1
Nontoxic for mice
Isolated organisms not infectious
Adapted for intracellular growth

Plasma membrane   Phagosome   RB in binary fission   Lysed phagosome and plasma membrane
EB   RB

0   2   6   12   18   24   30   36   42   48
Hours after infection

(a)   (b)

Figure 21.14   The Chlamydial Life Cycle. (**a**) An electron micrograph of a microcolony of *Chlamydia trachomatis* in the cytoplasm of a host cell (×160,000). Three developmental stages are visible: the elementary body, EB; reticulate body, RB; and intermediate body, IB, a chlamydial cell intermediate in morphology between the first two forms. (**b**) A schematic representation of the infectious cycle of chlamydiae.

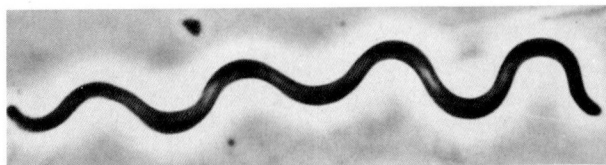

(a) *Cristispira*

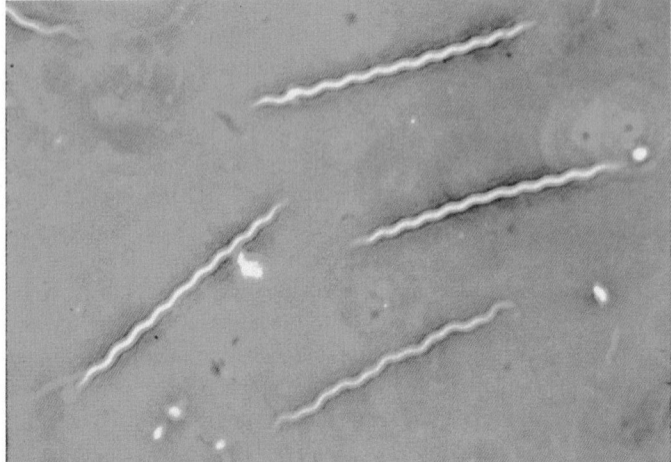

(b) *Treponema pallidum*

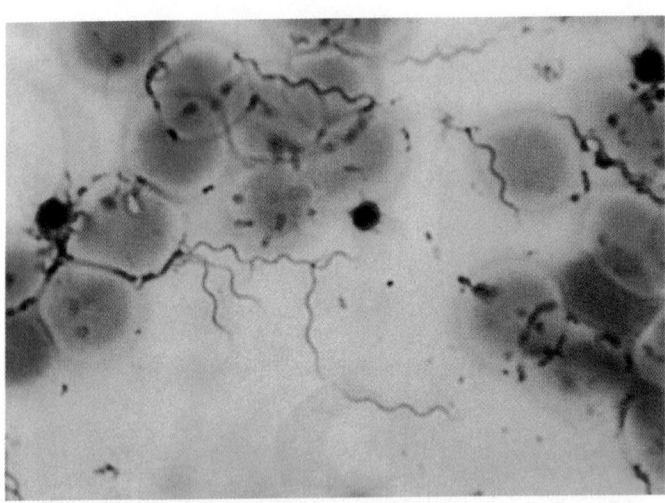

(c) *Borrelia duttonii*

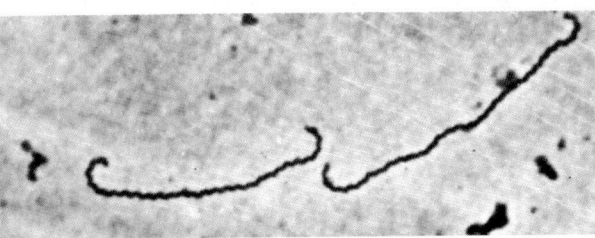

(d) *Leptospira interrogans*

Figure 21.15 The Spirochetes. Representative examples. (**a**) *Cristispira* sp. from the crystalline style of a clam; phase contrast (×2,200). (**b**) *Treponema pallidum* (×1,000). (**c**) *Borrelia duttonii* from human blood (×500). (**d**) *Leptospira interrogans* (×2,200).

both flavoprotein and cytochrome electron transport chain carriers, but does have a membrane translocase that acquires host ATP in exchange for ADP. Thus chlamydiae seem to be energy parasites that are completely dependent on their hosts for ATP. However, this might not be the complete story. The complete sequence of the *C. trachomatis* genome indicates that the bacterium may be able to synthesize at least some ATP. Although there are two genes for ATP/ADP translocases, there also are genes for substrate-level phosphorylation, electron transport, and oxidative phosphorylation. When supplied with precursors from the host, RBs can synthesize DNA, RNA, glycogen, lipids, and proteins. Presumably the RBs have porins and active membrane transport proteins, but little is known about these. They also can synthesize at least some amino acids and coenzymes. The EBs have minimal metabolic activity and cannot take in ATP or synthesize proteins. They seem to be dormant forms concerned exclusively with transmission and infection.   *C. trachomatis* genome sequence (pp. 341–42)

Three chlamydial species are important pathogens of humans and other warm-blooded animals. *C. trachomatis* infects humans and mice. In humans it causes trachoma, nongonococcal urethritis, and other diseases (*see section 39.3*). *C. psittaci* causes psittacosis in humans. However, unlike *C. trachomatis,* it also infects many other animals (e.g., parrots, turkeys, sheep, cattle, and cats) and invades the intestinal, respiratory, and gen-

ital tracts; the placenta and fetus; the eye; and the synovial fluid of joints. *Chlamydia pneumoniae* is a common cause of human pneumonia. There is now indirect evidence that infections by *C. pneumoniae* may be associated with the development of atherosclerosis and that chlamydial infections may cause severe heart inflammation and damage. Recently a fourth species, *C. pecorum,* has been recognized.

21.6 PHYLUM *SPIROCHAETES*

The phylum *Spirochaetes* [Greek *spira,* a coil, and *chaete,* hair] contains gram-negative, chemoheterotrophic bacteria distinguished by their structure and mechanism of motility. They are slender, long bacteria (0.1 to 3.0 μm by 5 to 250 μm) with a flexible, helical shape (**figure 21.15**). Many species are so slim that they are only clearly visible in a light microscope by means of phase-contrast or dark-field optics (*see section 2.2*). Spirochetes differ greatly from other bacteria with respect to motility and can move through very viscous solutions though they lack external rotating flagella. When in contact with a solid surface, they exhibit creeping or crawling movements. Their unique pattern of motility is due to an unusual morphological structure called the axial filament.

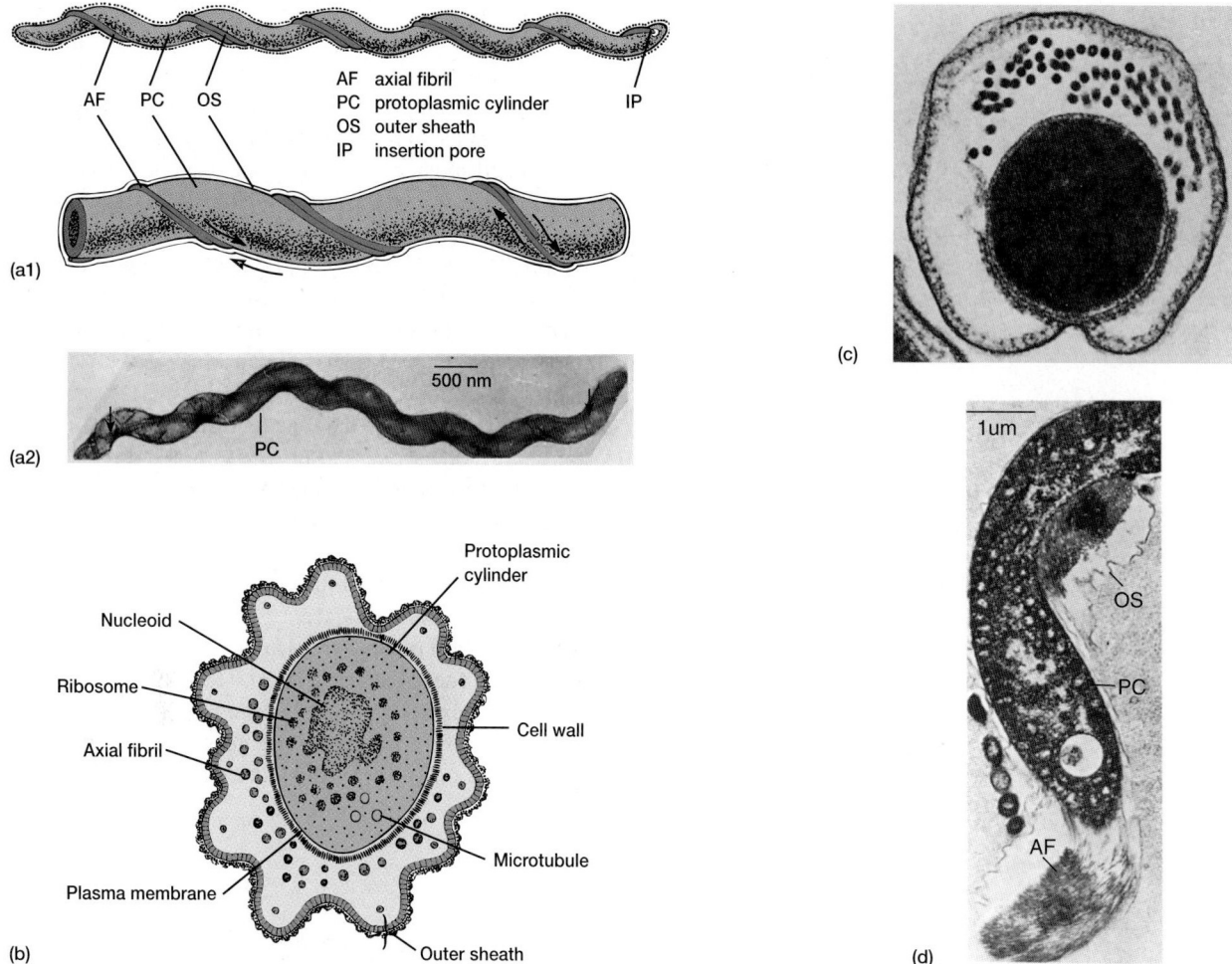

Figure 21.16 Spirochete Morphology. (**a1**) A surface view of spirochete structure as interpreted from electron micrographs. (**a2**) A longitudinal view of *T. zuelzerae* with axial fibrils extending most of the cell length. (**b**) A cross section of a typical spirochete showing morphological details. (**c**) Electron micrograph of a cross section of *Clevelandina* from the termite *Reticulitermes flavipes* showing the outer sheath, protoplasmic cylinder, and axial fibrils (×70,000). (**d**) Longitudinal section of *Cristispira* showing the outer sheath (OS), the protoplasmic cylinder (PC), and the axial fibrils (AF).

The distinctive features of spirochete morphology are evident in electron micrographs (**figure 21.16**). The central protoplasmic cylinder contains cytoplasm and the nucleoid, and is bounded by a plasma membrane and gram-negative type cell wall. It corresponds to the body of other gram-negative bacteria. Two to more than a hundred procaryotic flagella, called **axial fibrils, periplasmic flagella** or endoflagella, extend from both ends of the cylinder and often overlap one another in the center third of the cell (figure 21.16*c,d; see also figure 2.28b*). The whole complex of periplasmic flagella, the **axial filament,** lies inside a flexible outer sheath or outer membrane. The outer sheath contains lipid, protein, and carbohydrate and varies in structure between different genera. Its precise function is unknown, but the sheath is definitely important because spirochetes will die if it is damaged or removed. The outer sheath of *Treponema pallidum* has few pro-

teins exposed on its surface. This allows the syphilis spirochete to avoid attack by host antibodies.

Although the way in which periplasmic flagella propel the cell has not been established, they are responsible for motility because mutants with straight rather than curved flagella are nonmotile. Presumably the periplasmic flagella rotate like the external flagella of other bacteria. This could cause the corkscrew-shaped outer sheath to rotate and move the cell through the surrounding liquid (**figure 21.17**). Flagellar rotation could also flex or bend the cell and account for the crawling movement seen on solid surfaces.

Spirochetes can be anaerobic, facultatively anaerobic, or aerobic. Carbohydrates, amino acids, long-chain fatty acids, and long-chain fatty alcohols may serve as carbon and energy sources.

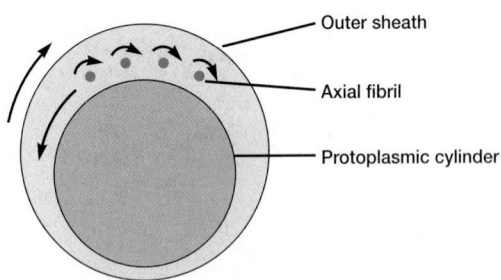

Outer sheath

Axial fibril

Protoplasmic cylinder

Figure 21.17 Spirochete Motility. A hypothetical mechanism for spirochete motility. See text for details.

The group is exceptionally diverse ecologically and grows in habitats ranging from mud to the human mouth. Members of the genus *Spirochaeta* are free-living and often grow in anaerobic and sulfide-rich freshwater and marine environments. Some species of the genus *Leptospira* grow in aerobic water and moist soil. In contrast, many spirochetes form symbiotic associations with other organisms and are found in a variety of locations: the hindguts of termites and wood-eating roaches, the digestive tracts of mollusks (*Cristispira*) and mammals, and the oral cavities of animals (*Treponema denticola, T. oralis*). Spirochetes from termite hindguts and freshwater sediments have nitrogenase and can

fix nitrogen. There is evidence that they contribute significantly to the nitrogen nutrition of termites. Spirochetes coat the surfaces of many protozoa from termite and wood-eating roach hindguts (**figure 21.18**). For example, the flagellate *Myxotricha paradoxa* is covered with slender spirochetes (0.15 by 10 μm in length) that are firmly attached and help move the protozoan. Some members of the genera *Treponema, Borrelia,* and *Leptospira* are important pathogens; for example, *Treponema pallidum* (*see figure 2.8a*) causes syphilis (*see section 39.3*), and *Borrelia burgdorferi* (*see figure 39.8*) is responsible for Lyme disease. The study of *Treponema* and its role in syphilis has been hindered by our inability to culture the spirochete outside its human host. The *T. pallidum* genome has now been sequenced (*see pp. 342–43*); the results show that this spirochete is metabolically crippled and quite dependent on its host. The *Borrelia burgdorferi* genome also has been sequenced. It contains a linear chromosome of 910,725 base pairs and at least 17 linear and circular plasmids, which constitute another 533,000 base pairs. The plasmids have some genes that are normally found on chromosomes, and plasmid proteins seem to be involved in bacterial virulence.

The second edition of *Bergey's Manual* divides the phylum *Spirochaetes* into one class, one order (*Spirochaetales*), and three families (*Spirochaetaceae, Serpulinaceae,* and *Leptospiraceae*). At present, there are 13 genera in the phylum. **Table 21.4** summarizes some of the more distinctive properties of selected genera.

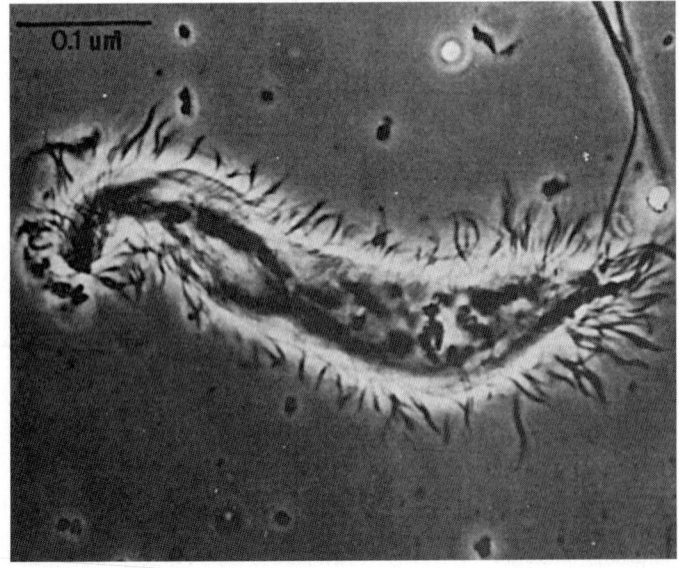

(a)

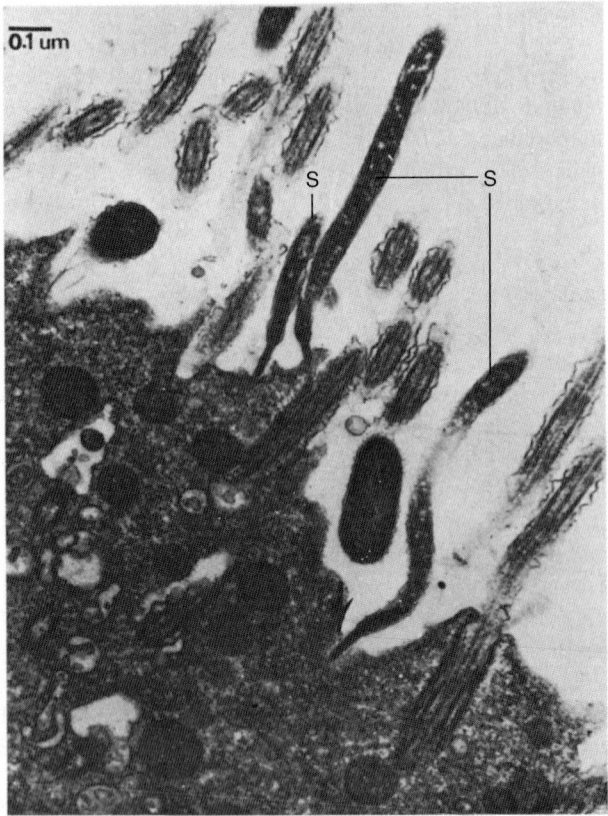

(b)

Figure 21.18 Spirochete-Protozoan Associations. The surface spirochetes serve as organs of motility for protozoa (see text). (**a**) The spirochete-*Personympha* association with the spirochetes projecting from the surface. (**b**) Electron micrograph of small spirochetes (S) attached to the membrane of the flagellate protozoan *Barbulanympha*.

Table 21.4

Characteristics of Spirochete Genera

Genus	Dimensions (μm) and Flagella	G + C Content (mol%)	Oxygen Relationship	Carbon + Energy Source	Habitats
Spirochaeta	0.2–0.75 × 5–250; 2–40 periplasmic flagella (almost always 2)	51–65	Facultatively anaerobic or anaerobic	Carbohydrates	Aquatic and free-living
Cristispira	0.5–3.0 × 30–180; ≥ 100 periplasmic flagella	N.A.[a]	Facultatively anaerobic?	N.A.[a]	Mollusk digestive tract
Treponema	0.1–0.4 × 5–20; 2–16 periplasmic flagella	25–53	Anaerobic or microaerophilic	Carbohydrates or amino acids	Mouth, intestinal tract, and genital areas of animals; some are pathogenic (syphilis, yaws)
Borrelia	0.2–0.5 × 3–20; 14–60 periplasmic flagella	27–32	Anaerobic or microaerophilic	Carbohydrates	Mammals and arthropods; pathogens (relapsing fever, Lyme disease)
Leptospira	0.1 × 6–24; 2 periplasmic flagella	35–49 (one strain is 53)	Aerobic	Fatty acids and alcohols	Free-living or pathogens of mammals, usually located in the kidney (leptospirosis)
Leptonema	0.1 × 6–20; 2 periplasmic flagella	51–53	Aerobic	Fatty acids	Mammals
Brachyspira	0.2 × 1.7–6.0; 8 periplasmic flagella	25–27	Anaerobic	N.A.[a]	Mammalian intestinal tract
Serpulina	0.3–0.4 × 7–9; 16–18 periplasmic flagella	25–26	Anaerobic	Carbohydrates and amino acids	Mammalian intestinal tract

[a]N.A., information not available

1. Describe the *Planctomycetes* and their more distinctive properties.
2. Compare chlamydiae with viruses.
3. Give the major characteristics of the phylum *Chlamydiae*. What are elementary and reticulate bodies? Briefly describe the steps in the chlamydial life cycle.
4. How does chlamydial metabolism differ from that of other bacteria?
5. List two or three chlamydial diseases of humans.
6. Give the most important characteristics of the spirochetes.
7. Define the following terms: protoplasmic cylinder, axial fibrils or periplasmic flagella, axial filament, outer sheath or outer membrane. Draw and label a diagram of spirochete morphology, locating these structures.
8. How might spirochetes use their axial filament to move?

21.7 PHYLUM *BACTEROIDETES*

The phylum *Bacteroidetes* is a new addition in the second edition of *Bergey's Manual*. It is a very diverse phylum and seems most closely related to the phylum *Chlorobi*. The phylum has three classes (*Bacteroides, Flavobacteria,* and *Sphingobacteria*), 12 families, and 60 genera.

The class *Bacteroides* contains anaerobic, gram-negative, nonsporing, motile or nonmotile rods of various shapes. These bacteria are chemoheterotrophic and usually produce a mixture of organic acids as fermentation end products, but they do not reduce sulfate or other sulfur compounds. The genera are identified using properties such as general shape, motility and flagellation pattern, and fermentation end products. These bacteria grow in habitats such as the oral cavity and intestinal tract of humans and other animals and the rumen (*see section 28.2*) of ruminants.

Although the difficulty of culturing these anaerobes has hindered the determination of their significance, they are clearly widespread and important. Often they benefit their host. *Bacteroides ruminicola* is a major component of the rumen flora; it ferments starch, pectin, and other carbohydrates. About 30% of the bacteria isolated from human feces are members of the genus *Bacteroides,* and these organisms may provide extra nutrition by degrading cellulose, pectins, and other complex carbohydrates. The family also is involved in human disease. Members of the genus *Bacteroides* are associated with diseases of major organ systems, ranging from the central nervous system to the skeletal system. *B. fragilis* is a particularly common anaerobic pathogen found in abdominal, pelvic, pulmonary, and blood infections.

Another important group in the *Bacteroidetes* is the class *Sphingobacteria*. Besides the similarity in their 16S rRNA sequences, sphingobacteria often have sphingolipids in their cell walls. Some genera in this class are *Sphingobacterium, Saprospira, Flexibacter, Cytophaga, Sporocytophaga,* and *Crenothrix.*

The genera *Cytophaga, Sporocytophaga,* and *Flexibacter* differ from each other in morphology, life cycle, and physiology.

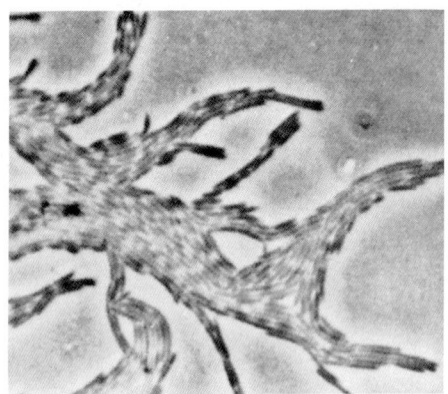

(a) *Cytophaga*

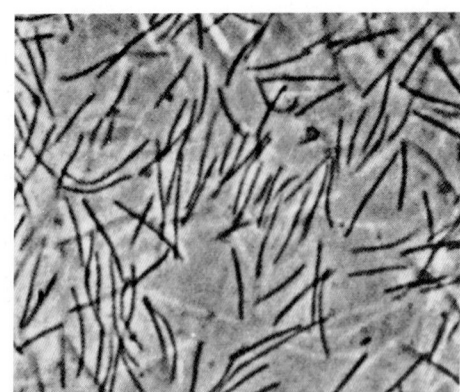

(b) *Sporocytophaga myxococcoides*

Figure 21.19   Nonphotosynthetic, Non-fruiting, Gliding Bacteria. Representative members of the order *Cytophagales.* (**a**) *Cytophaga* sp. (×1,150). (**b**) *Sporocytophaga myxococcoides,* vegetative cells on agar (×1,170). (**c**) *Sporocytophaga myxococcoides,* mature microcysts (×1,750). (**d**) Long thread cells of *Flexibacter elegans* (×1,100).

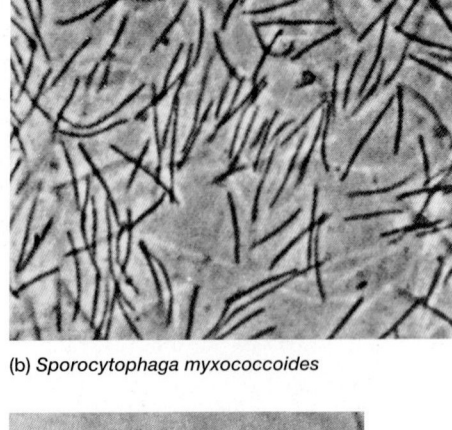

(c) *S. myxococcoides* microcysts

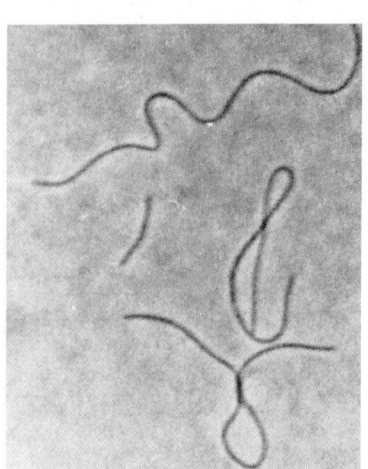

(d) *Flexibacter elegans*

Bacteria of the genus *Cytophaga* are slender rods, often with pointed ends (**figure 21.19***a*). They differ from the gliding, fruiting myxobacteria in lacking fruiting bodies and having a low G + C ratio. *Sporocytophaga* is similar to *Cytophaga* but forms spherical resting cells called microcysts (figure 21.19*b,c*). *Flexibacter* produces long, flexible threadlike cells when young (figure 21.19*d*) and is unable to use complex polysaccharides. Often colonies of these bacteria are colored yellow to orange because of carotenoid or flexirubin pigments. Some of the flexirubins are chlorinated, which is unusual for biological molecules.

Members of the genera *Cytophaga* and *Sporocytophaga* are aerobes that actively degrade complex polysaccharides. Soil cytophagas digest cellulose; both soil and marine forms attack chitin, pectin, and keratin. Some marine species even degrade agar. Cytophagas play a major role in the mineralization of organic matter and can cause great damage to exposed fishing gear and wooden structures. They also are a major component of the bacterial population in sewage treatment plants and presumably contribute significantly to this waste treatment process.

Although most cytophagas are free-living, some can be isolated from vertebrate hosts and are pathogenic. *Cytophaga columnaris* and others cause diseases such as columnaris disease, cold water disease, and fin rot in freshwater and marine fish.

The gliding motility so characteristic of these organisms is quite different from flagellar motility (*see section 3.7*). **Gliding motility** is present in a wide diversity of taxa: fruiting and nonfruiting aerobic chemoheterotrophs, cyanobacteria, green nonsulfur bacteria, and at least two gram-positive genera (*Heliobacterium* and *Desulfonema*). Gliding bacteria lack flagella and are stationary while suspended in liquid medium. When in contact with a surface, they glide along, leaving a slime trail; the gliding mechanism is unknown (Microbial Diversity & Ecology 21.1). Movement can be very rapid; some cytophagas travel 150 μm in a minute, whereas filamentous gliding bacteria may reach speeds of more than 600 μm/minute. Young organisms are the most motile, and motility often is lost with age. Low nutrient levels usually stimulate gliding.

Gliding motility gives a bacterium many advantages. Many aerobic chemoheterotrophic gliding bacteria actively digest insoluble macromolecular substrates such as cellulose and chitin, and gliding motility is ideal for searching these out. Because many of the digestive enzymes are cell bound, the bacteria must be in contact with insoluble nutrient sources; gliding motility makes this possible. Gliding movement is well adapted to drier habitats and to movement within solid masses such as soil, sediments, and rotting wood that are permeated by small channels. Finally, gliding bacteria, like flagellated bacteria, can position themselves at optimal conditions of light intensity, oxygen, hydrogen sulfide, temperature, and other factors that influence survival and growth.

1. Give the major properties of the class *Bacteroides*.
2. How do these bacteria benefit and harm their hosts?
3. Give three advantages of gliding motility.
4. Briefly describe the following genera: *Cytophaga*, *Sporocytophaga*, and *Flexibacter*.
5. Why are the cytophagas ecologically important?

Summary

21.1 *Aquificae* and *Thermotogae*

a. *Aquifex* and *Thermotoga* are hyperthermophilic gram-negative rods that represent the two deepest or oldest phylogenetic branches of the bacteria.

21.2 Deinococcus-Thermus

a. Members of the order *Deinococcales* are aerobic, gram-positive cocci and rods that are distinctive in their unusually great resistance to desiccation and radiation.

21.3 Photosynthetic Bacteria

a. Cyanobacteria carry out oxygenic photosynthesis; purple and green bacteria use anoxygenic photosynthesis.

b. The four most important groups of purple and green photosynthetic bacteria are the purple sulfur bacteria, the purple nonsulfur bacteria, the green sulfur bacteria, and the green nonsulfur bacteria (**table 21.1**).

c. The bacteriochlorophyll pigments of purple and green bacteria enable them to live in deeper, anaerobic zones of aquatic habitats.

d. Green nonsulfur bacteria such as *Chloroflexus* are placed in the phylum *Chloroflexi*. *Chloroflexus* is a filamentous, gliding thermophilic bacterium that is metabolically similar to the purple nonsulfur bacteria.

e. The phylum *Chlorobi* contains the green sulfur bacteria—obligately anaerobic photolithoautotrophs that use hydrogen sulfide, elemental sulfur, and hydrogen as electron sources.

f. Cyanobacteria carry out oxygenic photosynthesis by means of a photosynthetic apparatus similar to that of the eucaryotes. Like the red algae, they have phycobilisomes.

g. Cyanobacteria reproduce by binary fission, budding, multiple fission, and fragmentation by filaments to form hormogonia. Some will produce a dormant akinete.

h. Nitrogen-fixing cyanobacteria usually form heterocysts, specialized cells in which nitrogen fixation occurs.

i. The second edition of *Bergey's Manual* divides the cyanobacteria into five subsections and includes the prochlorophytes in the phylum *Cyanobacteria* (**table 21.3**).

21.4 Phylum *Planctomycetes*

a. The phyla *Planctomycetes* and *Chlamydiae* lack peptidoglycan in their walls.

21.5 Phylum *Chlamydiae*

a. Chlamydiae are nonmotile, coccoid, gram-negative bacteria that reproduce within the cytoplasmic vacuoles of host cells by a life cycle involving elementary bodies (EBs) and reticulate bodies (RBs) (**figure 21.14**). They are energy parasites.

21.6 Phylum *Spirochaetes*

a. The spirochetes are slender, long, helical, gram-negative bacteria that are motile because of the axial filament underlying an outer sheath or outer membrane (**figure 21.16**).

21.7 Phylum *Bacteroidetes*

a. Members of the class *Bacteroides* are obligately anaerobic, chemoheterotrophic, nonsporing, motile or nonmotile rods of various shapes. Some are important rumen and intestinal symbionts, others can cause disease.

b. Gliding motility is present in a diversity of bacteria, including the sphingobacteria.

c. Cytophagas degrade proteins and complex polysaccharides and are active in the mineralization of organic matter.

Key Terms

akinetes   459
anoxygenic photosynthesis   456
axial fibrils   467
axial filament   467
baeocytes   463

chlamydiae   464
chlorosomes   458
cyanobacteria   458
elementary body (EB)   465
gliding motility   470

green nonsulfur bacteria   457
green sulfur bacteria   458
heterocysts   459
hormogonia   459
initial body   465

oxygenic photosynthesis   456
periplasmic flagella   467
phycobilisomes   458
reticulate body (RB)   465
trichome   459

Questions for Thought and Review

1. Contrast the first and second editions of *Bergey's Manual* with respect to the treatment of the chlamydiae and cytophagas. What does this say about the change in approach between editions?
2. Suppose that you were a microbiologist working on the Mars probe project. How would you determine whether microorganisms are living on the planet, and why?
3. How do cyanobacteria differ from the purple and green photosynthetic bacteria?
4. Relate the physiology of each major group of photosynthetic bacteria to its preferred habitat.
5. Why is nitrogen fixation an oxygen-sensitive process? How are some cyanobacteria able to fix nitrogen when they also carry out oxygenic photosynthesis?
6. How might one try to culture a newly discovered rickettsia or chlamydia outside its host?

Critical Thinking Questions

1. Choose a bacterium discussed in this chapter. Write a short 1 to 2 page paper on where and how you would go about isolating the organism. What niche would you sample? What media and selective growth conditions would you use? Be sure to cite references.
2. Many types of movement are employed by bacteria in these phyla. Review them and propose mechanisms by which energy (ATP or proton gradients) might drive the locomotion.

For recommended readings on these and related topics, see Appendix V.

chapter 22

Bacteria:
The Proteobacteria

Concepts

- The proteobacteria of the second edition of *Bergey's Manual* come from volumes 1 and 3 of the first edition. In the first edition bacteria are placed in a particular section based on a few major phenotypic properties such as general shape, nutritional type, motility, oxygen relationships, and so forth. The second edition uses nucleic acid sequences, particularly 16S rRNA sequence comparisons, to place bacteria in phylogenetic groupings.
- Many of these gram-negative bacteria are of considerable importance, either as disease agents or because of their effects on the habitat. Others, such as *E. coli,* are major experimental organisms studied in many laboratories.
- Although many of these bacteria do not vary drastically in general appearance, they often are very diverse in their metabolism and life-styles, which range from obligately intracellular parasitism to a free-living existence in soil and aquatic habitats.
- Bacteria do not always have simple, unsophisticated morphology but may produce prosthecae, stalks, buds, sheaths, or complex fruiting bodies.
- Chemolithotrophic bacteria obtain energy and electrons by oxidizing inorganic compounds rather than the organic nutrients employed by most bacteria. They often have substantial ecological impact because of their ability to oxidize many forms of inorganic nitrogen and sulfur.
- Many bacteria that specialize in predatory or parasitic modes of existence, such as *Bdellovibrio* and the rickettsias, have relinquished some of their metabolic independence through the loss of metabolic pathways. They depend on the prey's or host's energy supply and/or cell constituents.

> *Microbes is a vigitable, an' ivry man is like a conservatory full iv millyons iv these potted plants.*
>
> —Finley Peter Dunne

Salmonella typhi, stained here with fluorescent acridine orange, is a significant human pathogen. It causes typhoid fever.

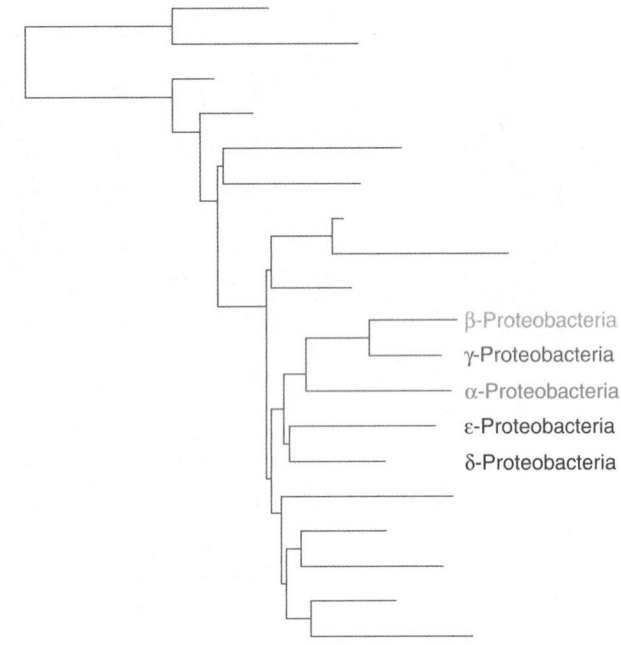

β-Proteobacteria
γ-Proteobacteria
α-Proteobacteria
ε-Proteobacteria
δ-Proteobacteria

 hapters 20 and 21 described many of the groups found in volumes 1 and 5 of the second edition of *Bergey's Manual.* Chapter 20 was devoted to the Archaea and chapter 21 surveyed the bacterial groups described in volumes 1 and 5. This chapter introduces the bacteria that are covered in volume 2 of the second edition of *Bergey's Manual.* These gram-negative bacteria are located in volumes 1 and 3 of the first edition. The general organization of the second edition will be followed here. This chapter depicts the major biological features of each group and a few selected representative forms of particular interest.

Volume 2 of the second edition of *Bergey's Manual* is devoted entirely to the **proteobacteria,** sometimes called the purple bacteria because of the purple photosynthetic bacteria scattered through several of its subgroups. The proteobacteria are the largest and most diverse group of bacteria; currently there are ver 450 genera. Although the 16S rRNA studies show that they hylogenetically related, proteobacteria vary markedly in espects. The morphology of these gram-negative bacteria om simple rods and cocci to genera with prosthecae, ven fruiting bodies. Physiologically these bacteria are e. Photoautotrophs, chemolithotrophs, and chemo-e all well represented. There is no obvious overall lism, morphology, or reproductive strategy of

Comparison of 16S rRNA sequences has revealed what appears to be five subgroups of proteobacteria (see the diagram at the left). The second edition of *Bergey's Manual* places all these procaryotes in the phylum *Proteobacteria,* which has five classes: *Alphaproteobacteria, Betaproteobacteria, Gammaproteobacteria, Deltaproteobacteria,* and *Epsilonproteobacteria.* Members of the purple photosynthetic bacteria are found among the α-, β-, and γ-proteobacteria. This has led to the proposal that the proteobacteria arose from a photosynthetic ancestor, presumably similar to the purple bacteria. Subsequently photosynthesis would have been lost by various lines, and new metabolic capacities would have been acquired as these bacteria adapted to different ecological niches.

22.1 CLASS *ALPHAPROTEOBACTERIA*

The **α-proteobacteria** include most of the oligotrophic proteobacteria (those capable of growing at low nutrient levels). Some have unusual metabolic modes such as methylotrophy (*Methylobacterium*), chemolithotrophy (*Nitrobacter*), and the ability to fix nitrogen (*Rhizobium*). Members of genera such as *Rickettsia* and *Brucella* are important pathogens; in fact, *Rickettsia* has become an obligately intracellular parasite. Many genera are characterized by distinctive morphology such as prosthecae.

The class *Alphaproteobacteria* has six orders and 18 families. **Figure 22.1** illustrates the phylogenetic relationships between some typical α-proteobacteria, and **table 22.1** summarizes the general characteristics of many of the bacteria discussed in the following sections.

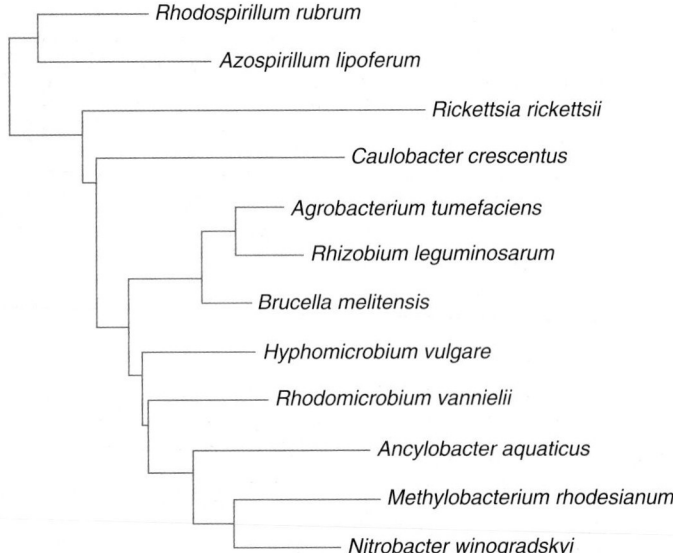

Rhodospirillum rubrum
Azospirillum lipoferum
Rickettsia rickettsii
Caulobacter crescentus
Agrobacterium tumefaciens
Rhizobium leguminosarum
Brucella melitensis
Hyphomicrobium vulgare
Rhodomicrobium vannielii
Ancylobacter aquaticus
Methylobacterium rhodesianum
Nitrobacter winogradskyi

Figure 22.1 Phylogenetic Relationships Among Selected α-Proteobacteria. The relationships of a few species based on 16S rRNA sequence data are shown here. *Source: The Ribosomal Database Project.*

Table 22.1

Characteristics of Selected α-Proteobacteria

Genus	Dimensions (μm) and Morphology	G + C Content (mol%)	Oxygen Requirement	Other Distinctive Characteristics
Agrobacterium	0.6–1.0 × 1.5–3.0; motile, nonsporing rods with peritrichous flagella	57–63	Aerobic	Chemoorganotroph that can invade plants and cause tumors
Caulobacter	0.4–0.6 × 1–2; rod- or vibrioid-shaped with a flagellum and prostheca and holdfast	62–67	Aerobic	Heterotrophic and oligotrophic; asymmetric cell division
Hyphomicrobium	0.3–1.2 × 1–3; rod-shaped or oval with polar prosthecae	59–65	Aerobic	Reproduces by budding; methylotrophic
Nitrobacter	0.5–0.8 × 1.0–2.0; rod- or pear-shaped, sometimes motile by flagella	60–62	Aerobic	Chemolithotroph, oxidizes nitrite to nitrate
Rhizobium	0.5–0.9 × 1.2–3.0; motile rods with flagella	59–64	Aerobic	Invades leguminous plants to produce nitrogen-fixing root nodules
Rhodospirillum	0.7–1.5 wide; spiral cells with polar flagella	62–64	Anaerobic, microaerobic, aerobic	Photoheterotroph under anaerobic conditions
Rickettsia	0.3–0.5 × 0.8–2.0; short nonmotile rods	29–33	Aerobic	Obligately intracellular parasite

The Purple Nonsulfur Bacteria

All the purple bacteria use anoxygenic photosynthesis, possess bacteriochlorophylls *a* or *b,* and have their photosynthetic apparatus in membrane systems that are continuous with the plasma membrane. Most are motile by polar flagella. With the exception of *Rhodocyclus* (β-proteobacteria), the purple nonsulfur bacteria are located among the α-proteobacteria. Photosynthetic Bacteria (pp. 456–64)

The **purple nonsulfur bacteria** are exceptionally flexible in their choice of an energy source. Normally they grow anaerobically as photoorganoheterotrophs; they trap light energy and employ organic molecules as both electron and carbon sources (*see table 21.1*). Although they are called nonsulfur bacteria, some species can oxidize very low, nontoxic levels of sulfide to sulfate, but they do not oxidize elemental sulfur to sulfate. In the absence of light, most purple nonsulfur bacteria can grow aerobically as chemoorganoheterotrophs, but some species carry out fermentations and grow anaerobically. Oxygen inhibits bacteriochlorophyll and carotenoid synthesis so that cultures growing aerobically in the dark are colorless.

Purple nonsulfur bacteria vary considerably in morphology (**figure 22.2**). They may be spirals (*Rhodospirillum*), rods (*Rhodopseudomonas*), half circles or circles (*Rhodocyclus*), or they may even form prosthecae and buds (*Rhodomicrobium*). Because of their metabolism, they are most prevalent in the mud and water of lakes and ponds with abundant organic matter and low sulfide levels. There also are marine species.

Rickettsia and Coxiella

In the second edition of *Bergey's Manual,* the genus *Rickettsia* is placed in the order *Rickettsiales* and family *Rickettsiaceae* of the α-proteobacteria, whereas *Coxiella* is in the order *Legionellales* and family *Coxiellaceae* of the γ-proteobacteria. The first edition places these two genera and other bacteria that grow intracellularly in a separate section, and we also will discuss *Rickettsia* and *Coxiella* together because of their similarity in life-style, despite their apparent phylogenetic distance.

These bacteria are rod-shaped, coccoid, or pleomorphic with typical gram-negative walls and no flagella. Although their size varies, they tend to be very small. For example, *Rickettsia* is 0.3 to 0.5 μm in diameter and 0.8 to 2.0 μm long; *Coxiella* is 0.2 to 0.4 μm by 0.4 to 1.0 μm. All species are parasitic or mutualistic. The parasitic forms grow in vertebrate erythrocytes, macrophages (*see figure 31.4*), and vascular endothelial cells. Often they also live in blood-sucking arthropods such as fleas, ticks, mites, or lice, which serve as vectors (*see section 37.6*) or primary hosts.

Because these genera contain important human pathogens, their reproduction and metabolism have been intensively studied. Rickettsias enter the host cell by inducing phagocytosis. Members of the genus *Rickettsia* immediately escape the phagosome and reproduce by binary fission in the cytoplasm (**figure 22.3**). In contrast, *Coxiella* remains within the phagosome after it has fused with a lysosome and actually reproduces within the phagolysosome. Eventually the host cell bursts, releasing new organisms. Besides incurring damage from cell lysis, the host is harmed by the toxic effects of rickettsial cell walls (wall toxicity appears related to the mechanism of penetration into host cells).

Rickettsias are very different from most other bacteria in physiology and metabolism. They lack the glycolytic pathway and do not use glucose as an energy source, but rather oxidize glutamate and tricarboxylic acid cycle intermediates such as succinate. The rickettsial plasma membrane has carrier-mediated transport systems, and host cell nutrients and coenzymes are absorbed and directly used. For example, rickettsias take up both NAD and uridine diphosphate glucose. Their membrane also has

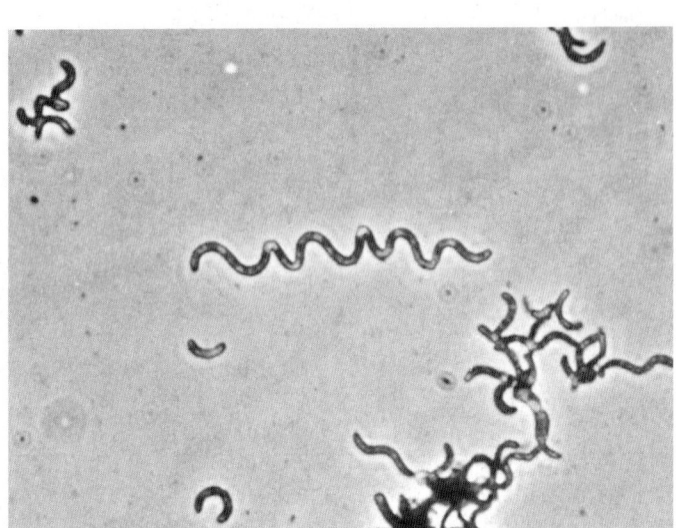

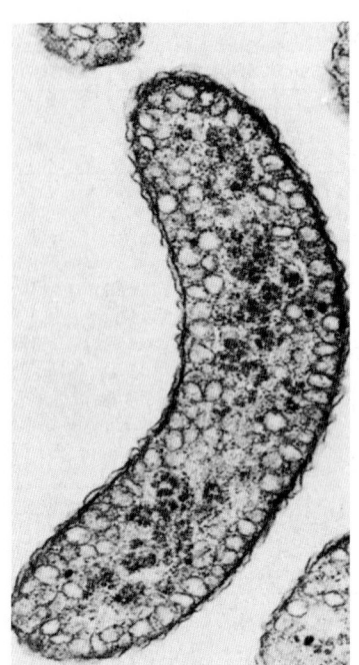

(a) *Rhodospirillum rubrum*

(b) *R. rubrum*

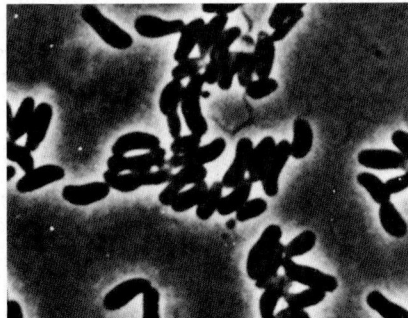

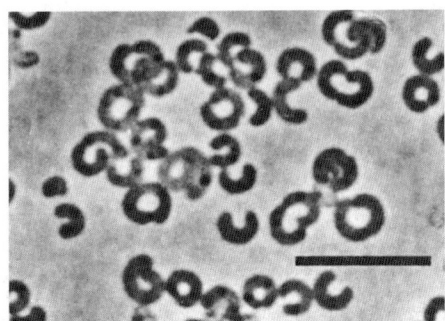

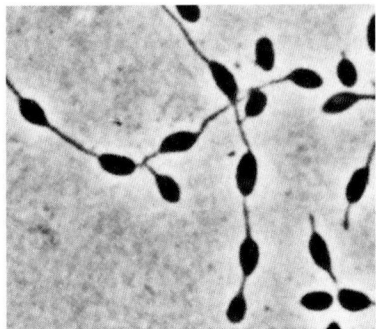

(c) *Rhodopseudomonas acidophila*

(d) *Rhodocyclus purpureus*

(e) *Rhodomicrobium vannielii*

Figure 22.2   Typical Purple Nonsulfur Bacteria.   (a) *Rhodospirillum rubrum;* phase contrast (×410). (b) *R. rubrum* grown anaerobically in the light. Note vesicular invaginations of the cytoplasmic membranes; transmission electron micrograph (×51,000). (c) *Rhodopseudomonas acidophila;* phase contrast. (d) *Rhodocyclus purpureus;* phase contrast. Bar = 10 μm. (e) *Rhodomicrobium vannielii* with vegetative cells and buds; phase contrast.

an adenylate exchange carrier that exchanges ADP for external ATP. Thus host ATP may provide much of the energy needed for growth. This metabolic dependence explains why many of these organisms must be cultivated in the yolk sacs of chick embryos or in tissue culture cells. Results from genome sequencing show that *R. prowazekii* is similar in many ways to mitochondria (*see pp. 340–41*). Possibly mitochondria arose from an endosymbiotic association with an ancestor of *Rickettsia.*

This order contains many important pathogens. *Rickettsia prowazekii* and *R. typhi* are associated with typhus fever, and *R. rickettsii,* with Rocky Mountain spotted fever. *Coxiella burnetii* causes Q fever in humans. These diseases are discussed later in some detail (*see chapter 39*). It should be noted that rickettsias are also important pathogens of domestic animals such as dogs, horses, sheep, and cattle.

The *Caulobacteraceae* and *Hyphomicrobiaceae*

A number of the proteobacteria are not simple rods or cocci but have some sort of appendage. These bacteria can have at least one of three different features: a prostheca, a stalk, or reproduction by budding. A **prostheca** (pl., prosthecae) is an extension of the cell, including the plasma membrane and cell wall, that is narrower than the mature cell. A **stalk** is a nonliving appendage produced by the cell and extending from it (*see figure 3.2f*). **Budding** is distinctly different from the **binary fission** normally used by bacteria (*see pp. 217–18*). The bud first appears as a small protrusion at a single point and enlarges to form a mature cell. Most or all of the bud's cell envelope is newly synthesized. In contrast, portions of the parental cell envelope are shared with the progeny cells during binary fission. Finally, the parental cell retains its identity

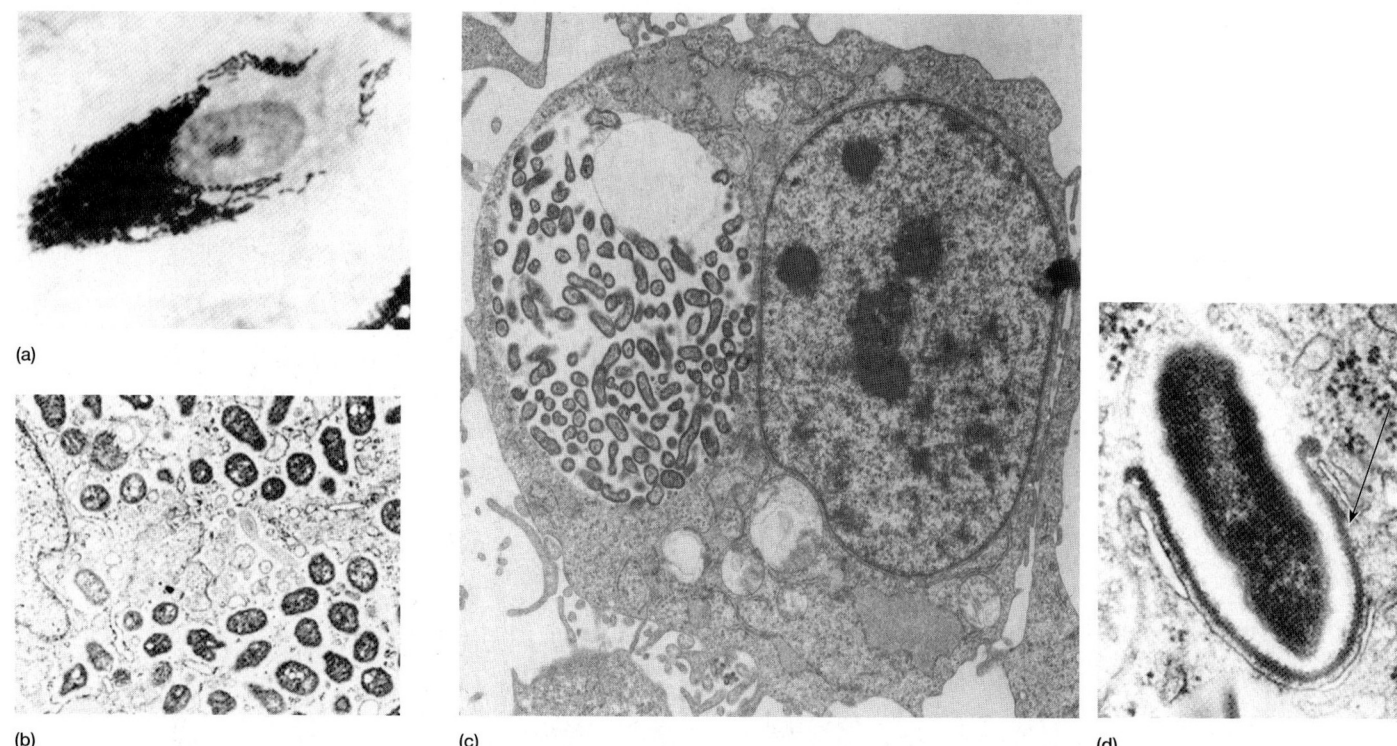

(a)

(b)   (c)   (d)

Figure 22.3 *Rickettsia* and *Coxiella.* Rickettsial morphology and reproduction. (**a**) A human fibroblast filled with *Rickettsia prowazekii* (×1,200). (**b**) A chicken embryo fibroblast late in infection with free cytoplasmic *R. prowazekii* (×13,600). (**c**) *Coxiella burnetti* growing within fibroblast vacuoles (×9,000). (**d**) *R. prowazekii* leaving a disrupted phagosome (arrow) and entering the cytoplasmic matrix (×46,000).

during budding, and the new cell is often smaller than its parent. In binary fission the parental cell disappears as it forms the progeny. The families *Caulobacteraceae* and *Hyphomicrobiaceae* of the α-proteobacteria contain two of the best studied of prosthecate genera: *Caulobacter* and *Hyphomicrobium.*

The genus *Hyphomicrobium* contains chemoheterotrophic, aerobic, budding bacteria that frequently attach to solid objects in freshwater, marine, and terrestrial environments. (They even grow in laboratory water baths.) The vegetative cell measures about 0.5 to 1.0 by 1 to 3 μm (**figure 22.4**). At the beginning of the reproductive cycle, the mature cell produces a hypha or prostheca, 0.2 to 0.3 μm in diameter, that grows to several μm in length (**figure 22.5**). The nucleoid divides, and a copy moves into the hypha while a bud forms at its end. As the bud matures, it produces one to three flagella, and a septum divides the bud from the hypha. The bud is finally released as an oval- to pear-shaped swarmer cell, which swims off, then settles down and begins budding. The mother cell may bud several times at the tip of its hypha.

Hyphomicrobium also has distinctive physiology and nutrition. Sugars and most amino acids do not support abundant growth; instead *Hyphomicrobium* grows on ethanol and acetate and flourishes with one-carbon compounds such as methanol, formate, and formaldehyde. That is, it is a facultative **methylotroph** and can derive both energy and carbon from reduced one-carbon compounds. It is so efficient at acquiring one-carbon

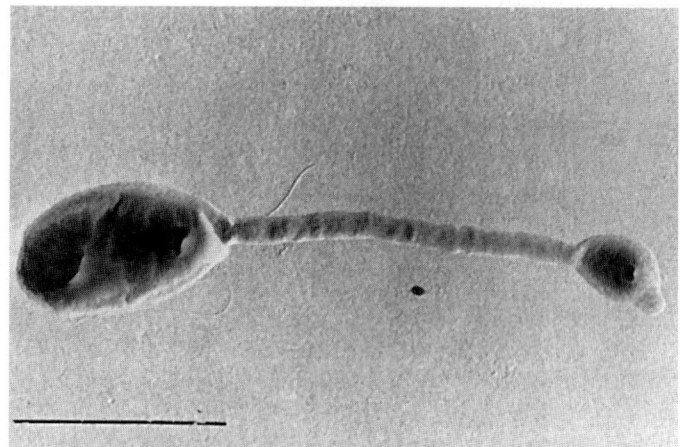

Figure 22.4 Prosthecate, Budding Bacteria. *Hyphomicrobium* morphology. *Hyphomicrobium facilis* with hypha and young bud. Bar = 1 μm.

molecules that it can grow in a medium without an added carbon source (presumably the medium absorbs sufficient atmospheric carbon compounds). *Hyphomicrobium* may comprise up to 25% of the total bacterial population in oligotrophic or nutrient-poor freshwater habitats.

Figure 22.5 **The Life Cycle of *Hyphomicrobium*.**
See text for details.

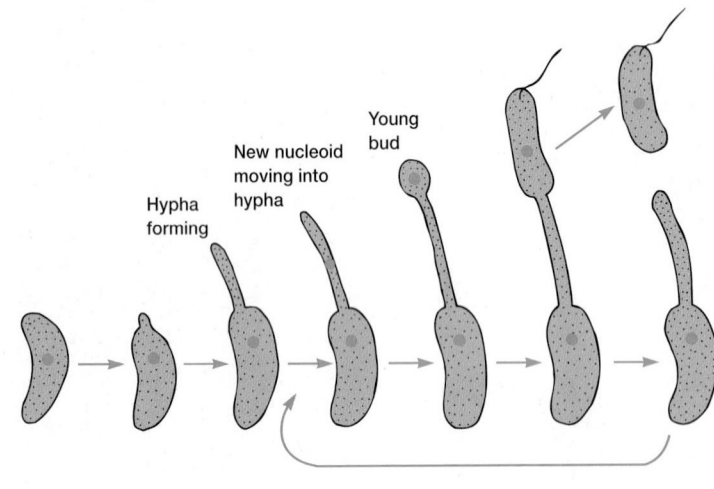

Figure 22.6 ***Caulobacter* Morphology and Reproduction.**
(**a**) Rosettes of cells adhering to each other by their prosthecae; phase contrast (×600). (**b**) A cell dividing to produce a swarmer (×6,030). Note prostheca and flagellum. (**c**) A cell with a flagellated swarmer (×6,030). (**d**) A rosette as seen in the electron microscope. Bar = 1 μm.

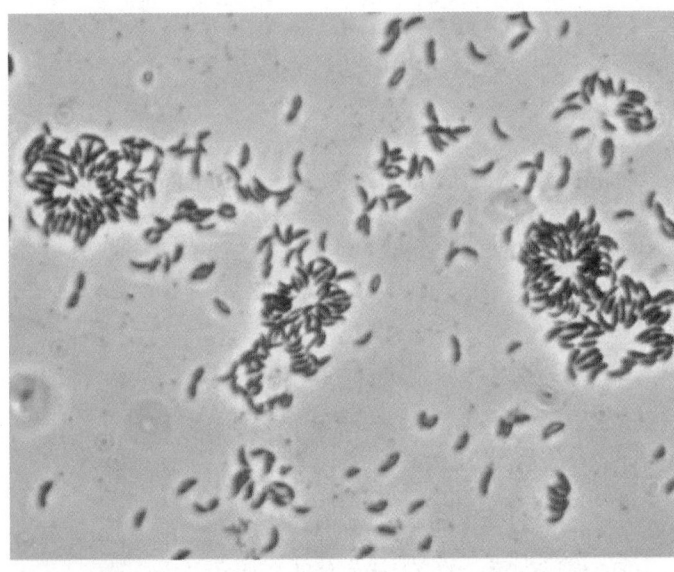

(a)

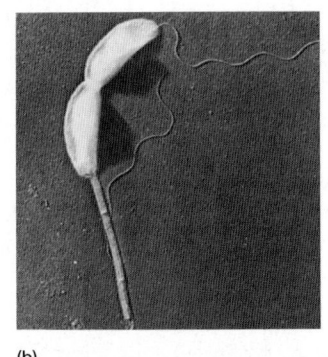

(b)

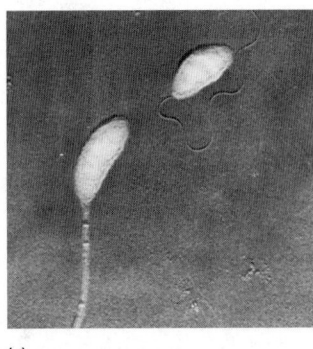

(c)

(d)

Bacteria in the genus *Caulobacter* may be polarly flagellated rods or possess a prostheca and **holdfast,** by which they attach to solid substrata (**figure 22.6**). Caulobacters are usually isolated from freshwater and marine habitats with low nutrient levels, but they also are present in the soil. They often adhere to bacteria, algae, and other microorganisms and may absorb nutrients released by their hosts. The prostheca differs from that of *Hyphomicrobium* in that it lacks cytoplasmic components and is composed almost totally of the plasma membrane and cell wall. It grows

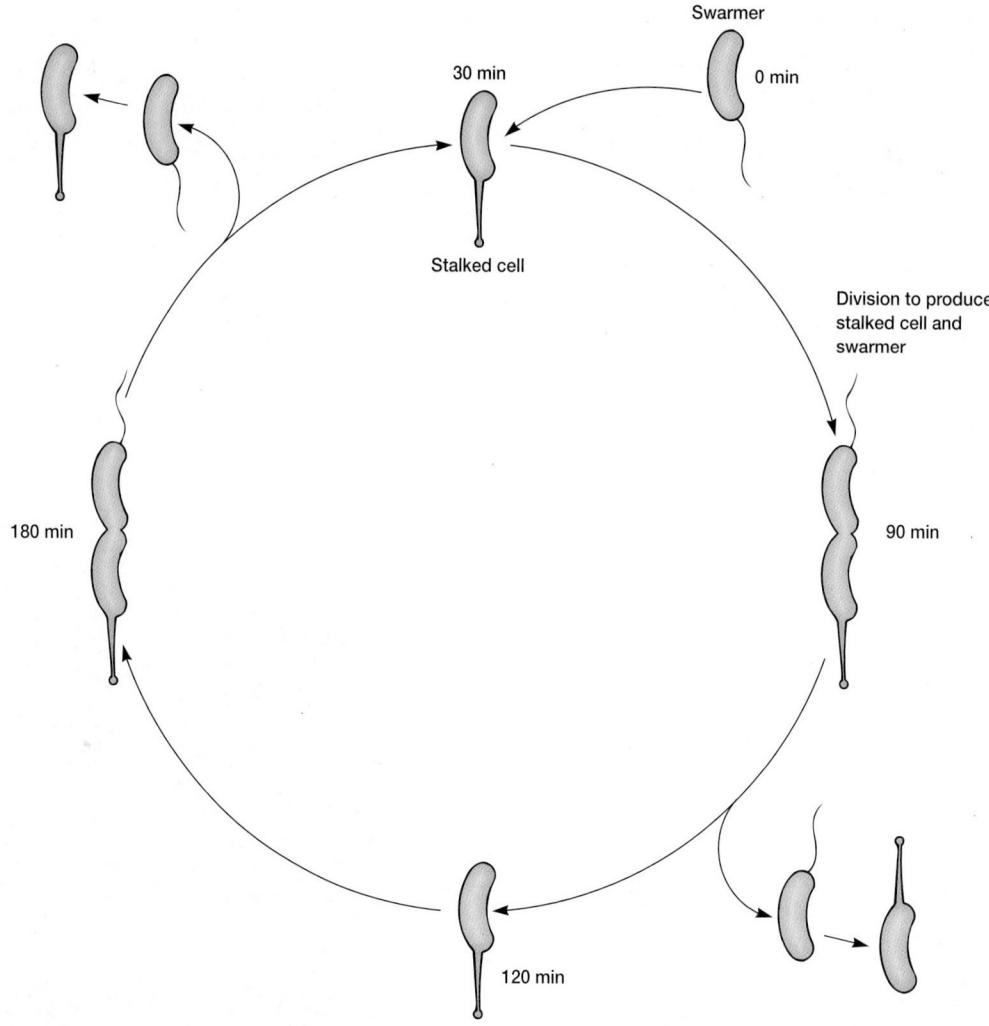

Figure 22.7 *Caulobacter* **Life Cycle.** A cycle starting with a swarmer cell at zero minutes is pictured. The cycle timing is for *Caulobacter* growing at 30°C in glucose minimal medium. See text for details.

longer in nutrient-poor media and can reach more than 10 times the length of the cell body. The prostheca may improve the efficiency of nutrient uptake from dilute habitats by increasing surface area; it also gives the cell extra buoyancy.

The life cycle of *Caulobacter* is unusual (**figure 22.7**). When ready to reproduce, the cell elongates and a single polar flagellum forms at the end opposite the prostheca. The cell then undergoes asymmetric transverse binary fission to produce a flagellated swarmer cell that swims off. The swarmer, which cannot reproduce, eventually comes to rest and forms a new prostheca on the flagellated end while its flagellum disappears. The new stalked form then begins to form swarmers. This cycle takes only about two hours for completion.

Family *Rhizobiaceae*

In the second edition of *Bergey's Manual,* the order *Rhizobiales* of the α-proteobacteria will contain 10 families with a great variety of phenotypes. The family *Hyphomicrobiaceae* has already

been discussed. The first family in this order is *Rhizobiaceae,* in which are located the gram-negative, aerobic genera *Rhizobium* and *Agrobacterium.*

Members of the genus *Rhizobium* are 0.5 to 0.9 by 1.2 to 3.0 μm motile rods, often containing poly-β-hydroxybutyrate granules, that become pleomorphic under adverse conditions (**figure 22.8**). They grow symbiotically within root nodule cells of legumes as nitrogen-fixing bacteroids (figure 22.8b). In contrast, *Azotobacter* is a free-living soil genus and fixes atmospheric nitrogen nonsymbiotically (p. 490). The biochemistry of nitrogen fixation (pp. 206–9); The *Rhizobium*-legume symbiosis (pp. 652–55)

The genus *Agrobacterium* is placed in the family *Rhizobiaceae* but differs from *Rhizobium* in not stimulating root nodule formation or fixing nitrogen. Instead agrobacteria invade the crown, roots, and stems of many plants and transform plant cells into autonomously proliferating tumor cells. Despite these differences, some believe that *Agrobacterium* and *Rhizobium* are actually members of the same species. Most of the distinguishing characteristics are carried on plasmids. The best-studied species

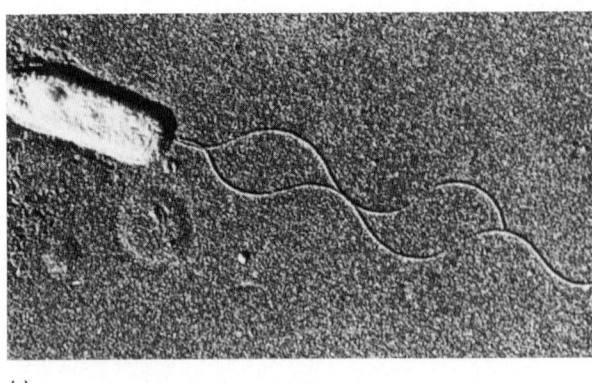

(a)　(b)

Figure 22.8 *Rhizobium.* (a) *Rhizobium leguminosarum* with two polar flagella (×14,000). (b) Scanning electron micrograph of bacteroids in alfalfa root (×640).

is *A. tumefaciens,* which enters many broad-leaved plants through wounds and causes crown gall disease (**figure 22.9**). The ability to produce tumors depends on the presence of a large Ti (for tumor-inducing) plasmid. Tumor production by *Agrobacterium* is discussed in greater detail in Techniques & Applications 14.2 and section 30.4. Plasmids (pp. 288–91)

Nitrifying Bacteria

The taxonomy of the aerobic chemolithotrophic bacteria, those bacteria that derive energy and electrons from reduced inorganic compounds, is quite complex. Normally these bacteria employ CO_2 as their carbon source and thus are chemolithoautotrophs, but some can function as chemolithoheterotrophs and use reduced organic carbon sources. In the second edition of *Bergey's Manual,* the chemolithotrophs are distributed between the α-, β-, and γ-proteobacteria. The nitrifying bacteria are α-, β-, and γ-proteobacteria; the sulfur-oxidizing bacteria are in the β and γ subgroups. Nitrifying bacteria will be discussed here. The colorless sulfur-oxidizing bacteria are covered in the section on β-proteobacteria. Chemolithotrophic metabolism is described in chapter 9, and the focus here is on the biology of these organisms. Their ecological impact is discussed further in the context of microbial ecology. The biochemistry of chemolithotrophy (pp. 188–90)

The **nitrifying bacteria** are a very diverse collection of bacteria. The second edition of *Bergey's Manual* places nitrifying genera in three classes and several families: *Nitrobacter* in the *Bradyrhizobiaceae,* α-proteobacteria; *Nitrosomonas* and *Nitrosospira* in the *Nitrosomonadaceae,* β-proteobacteria; *Nitrococcus* in the *Ectothiorhodospiraceae,* γ-proteobacteria; and *Nitrosococcus* in the *Chromatiaceae,* γ-proteobacteria. Although all are aerobic, gram-negative organisms without endospores and able to oxidize either ammonia or nitrite, they differ considerably in other properties (**table 22.2**). Nitrifiers may be rod-shaped, ellipsoidal, spherical, spirillar or lobate, and they may possess either polar or peritrichous flagella (**figure 22.10**). Often they have extensive membrane complexes lying in their cytoplasm. Identification is based on properties such as their preference for nitrite

Figure 22.9 *Agrobacterium.* Crown gall tumor of a tomato plant caused by *Agrobacterium tumefaciens.*

or ammonia, their general shape, and the nature of any cytomembranes present.

Nitrifying bacteria are very important ecologically and can be isolated from soil, sewage disposal systems, and freshwater and marine habitats. The genera *Nitrobacter* and *Nitrococcus* oxidize nitrite to nitrate; *Nitrosomonas, Nitrosospira,* and *Nitrosococcus* oxidize ammonia to nitrite. When two genera such as *Nitrobacter* and *Nitrosomonas* grow together in a niche, ammonia is converted to nitrate, a process called **nitrification** (*see section 9.10*). Nitrification occurs rapidly in soils treated with fertilizers containing ammonium salts. Nitrate nitrogen is readily used by plants, but it is also rapidly lost through leaching of the water soluble nitrate and by denitrification to nitrogen gas. Thus nitrification is a mixed blessing. The nitrogen cycle and microorganisms (pp. 597–98)

Table 22.2

Selected Characteristics of Representative Nitrifying Bacteria

Species	Cell Morphology and Size (μm)	Reproduction	Motility	Cytomembranes	G + C Content (mol%)	Habitat
Ammonia-Oxidizing Bacteria						
Nitrosomonas europaea	Rod; 0.8–1.0 × 1.0–2.0	Binary fission	+ or −; 1 or 2 subpolar flagella	Peripheral, lamellar	47.4–51.0	Soil, sewage, freshwater, marine
Nitrosococcus oceanus	Coccoid; 1.8–2.2 in diameter	Binary fission	+; 1 or more subpolar flagella	Centrally located parallel bundle, lamellar	50.5–51.0	Obligately marine
Nitrosospira briensis	Spiral; 0.3–0.4 in diameter	Binary fission	+ or −; 1 to 6 peritrichous flagella	Lacking	54.1 (1 strain)	Soil
Nitrite-Oxidizing Bacteria						
Nitrobacter winogradskyi	Rod; 0.6–0.8 × 1.0–2.0	Budding	+ or −; 1 polar flagellum	Polar cap of flattened vesicles in peripheral region of the cell	60.7–61.7	Soil, freshwater, marine
Nitrococcus mobilis	Coccoid; 1.5–1.8 in diameter	Binary fission	+; 1 or 2 subpolar flagella	Tubular cytomembranes randomly arranged in cytoplasm	61.2 (1 strain)	Marine

From S. W. Watson, et al., "The Family Nitrobacteraceae" in *The Prokaryotes,* Vol. 1, edited by M. P. Starr et al., 1981. Copyright © 1981 Springer-Verlag New York, Inc., New York, NY. Reprinted by permission.

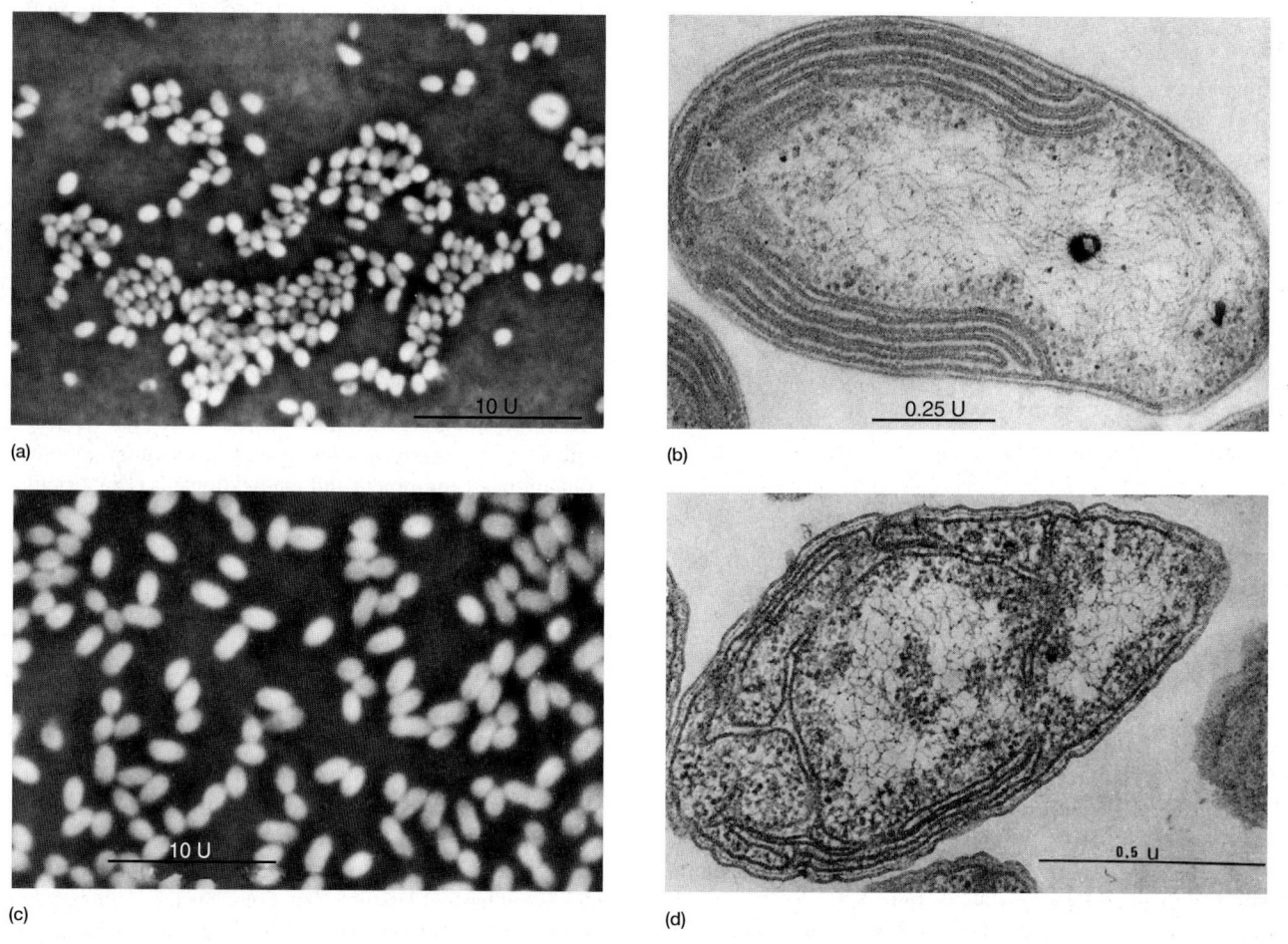

Figure 22.10 Representative Nitrifying Bacteria. (a) *Nitrobacter winogradskyi;* phase contrast (×2,500). (b) *N. winogradskyi.* Note the polar cap of cytomembranes (×213,000). (c) *Nitrosomonas europaea;* phase contrast (×2,500). (d) *N. europaea* with extensive cytoplasmic membranes (×81,700).

1. Describe the general properties of the α-proteobacteria.
2. Discuss the characteristics and physiology of the purple non-sulfur bacteria. Where would one expect to find them growing?
3. Briefly describe the characteristics and life cycle of the genus *Rickettsia*.
4. In what way does the physiology and metabolism of the rickettsias differ from that of other bacteria?
5. Name some important rickettsial diseases.
6. Define the following terms: prostheca, stalk, budding, swarmer cell, methylotroph, and holdfast.
7. Briefly describe the morphology and life cycles of *Hyphomicrobium* and *Caulobacter*.
8. What is unusual about the physiology of *Hyphomicrobium*? How does this influence its ecological distribution?
9. How do *Agrobacterium* and *Rhizobium* differ in life-style? What effect does *Agrobacterium* have on plant hosts?
10. What are chemolithotrophic bacteria?
11. Give the major characteristics of the nitrifying bacteria and discuss their ecological importance. How does the metabolism of *Nitrobacter* differ from that of *Nitrosomonas*?

22.2   CLASS *BETAPROTEOBACTERIA*

The **β-proteobacteria** overlap the α-proteobacteria metabolically but tend to use substances that diffuse from organic decomposition in the anaerobic zone of habitats. Some of these bacteria use hydrogen, ammonia, methane, volatile fatty acids, and similar substances. As with the α-proteobacteria, there is considerable metabolic diversity; the β-proteobacteria may be chemoheterotrophs, photolithotrophs, methylotrophs, and chemolithotrophs. The subgroup contains two genera with important human pathogens: *Neisseria* and *Bordetella*.

The class *Betaproteobacteria* has seven orders and 12 families. **Figure 22.11** shows the phylogenetic relationships between some species of β-proteobacteria, and **table 22.3** summarizes the general characteristics of many of the bacteria discussed in this section.

Order *Neisseriales*

The second edition places one family, *Neisseriaceae*, within the order and assigns 14 genera to it. The best-known and most intensely studied genus is *Neisseria*. Members of this genus are nonmotile, aerobic, gram-negative cocci that most often occur in pairs with adjacent sides flattened (*see figure 39.12*). They may have capsules and fimbriae. The genus is chemoorganotrophic, oxidase positive, and almost always catalase positive. They are inhabitants of the mucous membranes of mammals, and some are human pathogens. *Neisseria gonorrhoeae* is the causative agent of gonorrhea; *Neisseria meningitidis* is responsible for some cases of bacterial meningitis. Gonorrhea (pp. 892–93)

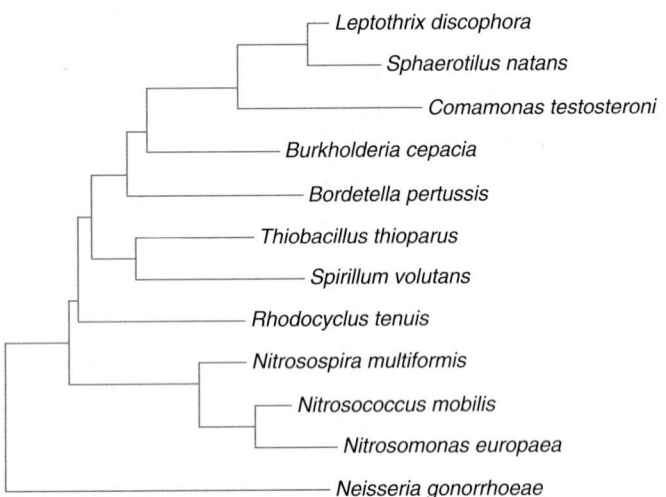

Figure 22.11   Phylogenetic Relationships Among Selected β-Proteobacteria. The relationships of a few species based on 16S rRNA sequence data are shown here. *Source: The Ribosomal Database Project.*

Order *Burkholderiales*

The order contains four families, three of them with well-known genera. The genus *Burkholderia* is placed in the family *Burkholderiaceae*. This genus was established when *Pseudomonas* was divided into at least seven new genera based on rRNA data: *Acidovorax, Aminobacter, Burkholderia, Comamonas, Deleya, Hydrogenophaga,* and *Methylobacterium*. Members of the genus *Burkholderia* are gram-negative, aerobic, nonfermentative, nonsporing, mesophilic straight rods. With the exception of one species, all are motile with a single polar flagellum or a tuft of polar flagella. Catalase is produced and they often are oxidase positive. Most species use poly-β-hydroxybutyrate as their carbon reserve. One of the most important species is *B. cepacia*, which will degrade over 100 different organic molecules and is very active in recycling organic materials in nature. This species also is a plant pathogen and causes disease in hospital patients due to contaminated equipment and medications. It is a particular problem for cystic fibrosis patients.

The family *Alcaligenaceae* contains the genus *Bordetella*. This genus is composed of gram-negative, aerobic coccobacilli, about 0.2 to 0.5 by 0.5 to 2.0 μm in size. *Bordetella* is a chemoorganotroph with respiratory metabolism that requires organic sulfur and nitrogen (amino acids) for growth. It is a mammalian parasite that multiplies in respiratory epithelial cells. *Bordetella pertussis* is a nonmotile, encapsulated species that causes whooping cough.

Some genera in the order (e.g., *Sphaerotilus* and *Leptothrix*) have a **sheath,** a hollow tubelike structure surrounding a chain of cells. Sheaths often are close fitting, but they are never in intimate contact with the cells they enclose and may contain ferric or manganic oxides. They have at least two functions. Sheaths help bacteria attach to solid surfaces and acquire nutrients from slowly running water as it flows past, even if it is nutrient-poor. Sheaths also protect against predators such as protozoa and *Bdellovibrio* (pp. 496–97).

Table 22.3

Characteristics of Selected β-Proteobacteria

Genus	Dimensions (μm) and Morphology	G + C Content (mol%)	Oxygen Requirement	Other Distinctive Characteristics
Bordetella	0.2–0.5 × 0.5–2.0; nonmotile coccobacillus	66–70	Aerobic	Requires organic sulfur and nitrogen; mammalian parasite
Burkholderia	0.3–1.0 × 1–5; straight rods with single flagella or a tuft at the pole	59–69.5	Aerobic	Poly-β-hydroxybutyrate as reserve; can be pathogenic
Leptothrix	0.6–1.4 × 1–12; straight rods in chains with sheath, free cells flagellated	69.5–71	Aerobic	Sheaths encrusted with iron and manganese oxides
Neisseria	0.6–1.0; cocci in pairs with flattened adjacent sides	46–54	Aerobic	Inhabitant of mucous membranes of mammals
Nitrosomonas	Size varies with strain; rod-shaped or ellipsoidal cells with intracytoplasmic membranes	45–54	Aerobic	Chemolithotroph that oxidizes ammonia to nitrite
Sphaerotilus	1.2–2.5 × 2–10; single chains of cells with sheaths, may have holdfasts	70	Aerobic	Sheaths not encrusted with iron and manganese oxides
Thiobacillus	0.5 × 1–4; rods, often with polar flagella	52–68	Aerobic	Chemolithotroph, oxidizes reduced sulfur compounds to sulfate

Two well-studied genera are *Sphaerotilus* and *Leptothrix* (**figures 22.12** and **22.13**). *Sphaerotilus* forms long sheathed chains of rods, 0.7 to 2.4 by 3 to 10 μm, attached to submerged plants, rocks, and other solid objects, often by a holdfast (figure 22.12). The sheaths are not usually encrusted by metal oxides. Single swarmer cells with a bundle of subpolar flagella escape the filament and form a new chain after attaching to a solid object at another site. *Sphaerotilus* prefers slowly running freshwater polluted with sewage or industrial waste. It grows so well in activated sewage sludge that it sometimes forms tangled masses of filaments and interferes with the proper settling of sludge (*see section 29.6*). *Leptothrix* characteristically deposits large amounts of iron and manganese oxides in its sheath. This seems to protect it and allow *Leptothrix* to grow in the presence of high concentrations of soluble iron compounds.

Order *Nitrosomonadales*

A number of chemolithotrophs are found in the order *Nitrosomonadales*. Two genera of nitrifying bacteria (*Nitrosomonas* and *Nitrosospira*) are located here in the family *Nitrosomonadaceae* but were discussed earlier along with other genera of nitrifying bacteria (p. 480). The stalked chemolithotroph *Gallionella* is in this order. The family *Spirillaceae* has one genus, *Spirillum* (**figure 22.14**).

Order *Hydrogenophilales*

This small order contains *Thiobacillus*, one of the best-studied chemolithotrophs and most prominent of the colorless sulfur bacteria. Like the nitrifying bacteria, **colorless sulfur bacteria** are a highly diverse group. Many are unicellular rod-shaped or spiral sulfur-oxidizing bacteria that are nonmotile or motile by flagella (**table 22.4**). The second edition disperses these bacteria between

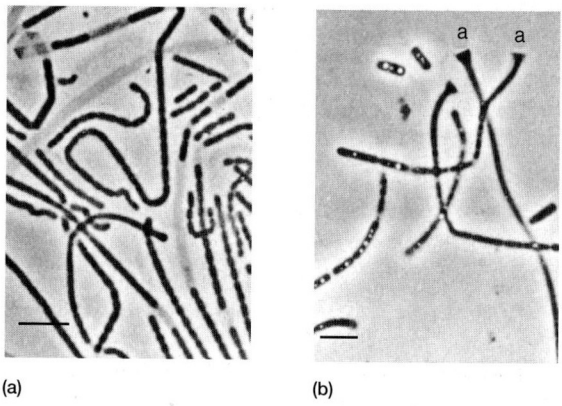

(a)   (b)

Figure 22.12   Sheathed Bacteria, *Sphaerotilus natans*. (a) Sheathed chains of cells and empty sheaths. (b) Chains with holdfasts (indicated by the letter a) and individual cells containing poly-β-hydroxybutyrate granules. Bars = 10 μm.

two classes; for example, *Thiobacillus* is in the β-proteobacteria, whereas *Thiomicrospira, Thiobacterium, Thiospira, Macromonas, Thiothrix, Beggiatoa,* and others are in the γ-proteobacteria. Only some of these bacteria have been isolated and studied in pure culture. Most is known about the genera *Thiobacillus* and *Thiomicrospira*. *Thiobacillus* is a gram-negative rod, and *Thiomicrospira* is a long spiral cell (**figure 22.15**); both are polarly flagellated. They differ from many of the nitrifying bacteria in that they lack extensive internal membrane systems.

The metabolism of *Thiobacillus* has been intensely studied (*see p. 189*). It grows aerobically by oxidizing a variety of inorganic sulfur compounds (elemental sulfur, hydrogen sulfide, thiosulfate) to sulfate. ATP is produced with a combination of oxidative phosphorylation and substrate-level phosphorylation by means of adenosine 5′-phosphosulfate (*see figure 9.25*).

Figure 22.13 Sheathed Bacteria, *Leptothrix* Morphology. (**a**) *L. lopholea* trichomes radiating from a collection of holdfasts. (**b**) *L. cholodnii* sheaths encrusted with MnO_2. Bars = 10 μm.

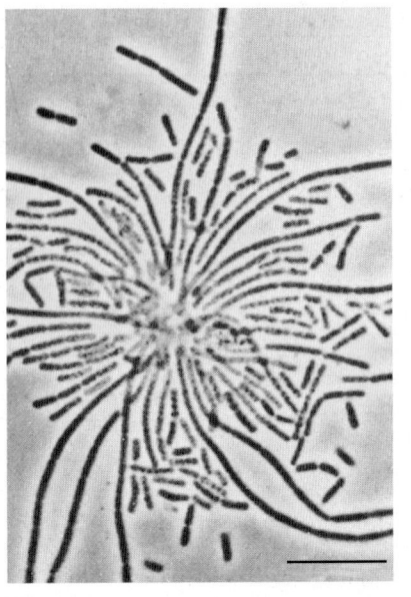

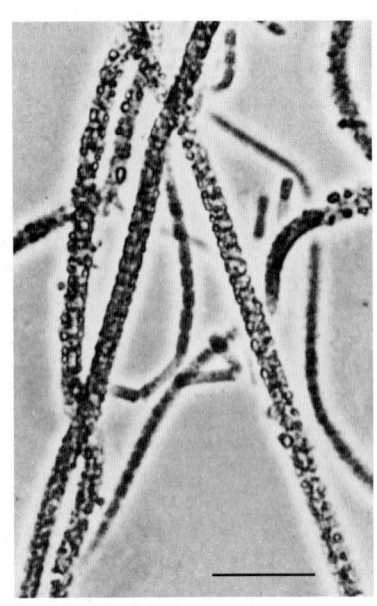

(a)   (b)

Figure 22.14 The Genus *Spirillum*. (**a**) *Spirillum volutans* with bipolar flagella visible (×450). (**b**) *Spirillum volutans;* phase contrast (×550).

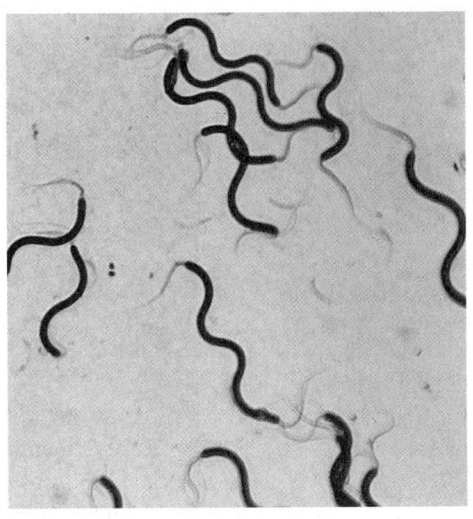

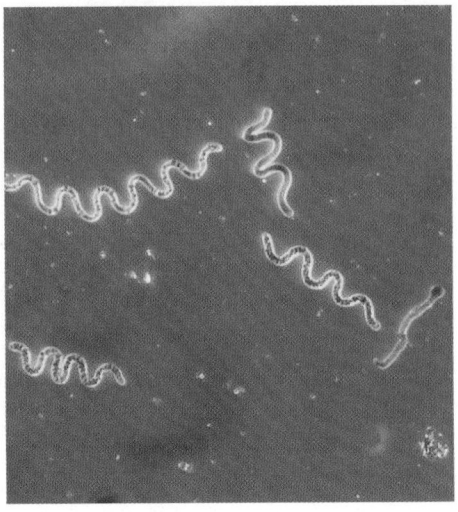

(a)   (b)

Table 22.4

Colorless Sulfur-Oxidizing Genera

Genus	Cell Shape	Motility, Flagella	G + C Content (mol%)	Location of Sulfur Deposit[a]	Nutritional Type
Thiobacillus	Rods	+; polar	52–68	Extracellular	Obligate or facultative chemolithotroph
Thiomicrospira	Spirals	+; polar	36–44	Extracellular	Obligate chemolithotroph
Thiobacterium	Rods embedded in gelatinous masses	–	N.A.[b]	Intracellular	Probably chemoorganoheterotroph
Thiospira	Spiral rods, usually with pointed ends	+; polar (single or in tufts)	N.A.	Intracellular	Unknown
Macromonas	Rods, cylindrical or bean shaped	+; polar	67	Intracellular	Probably chemoorganoheterotroph

[a]When hydrogen sulfide is oxidized to elemental sulfur.
[b]N.A., data not available.

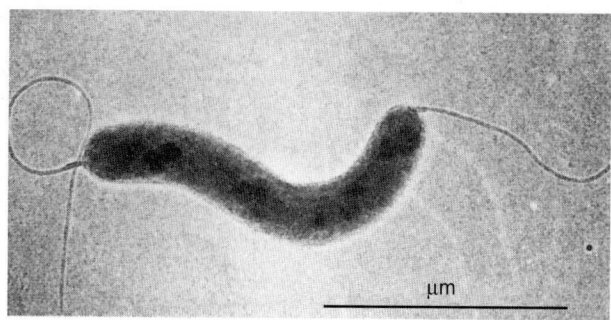

Figure 22.15 Colorless Sulfur Bacteria. *Thiomicrospira pelophila* with polar flagella. Bar = 1 μm.

Although *Thiobacillus* normally uses CO_2 as its major carbon source, *T. novellus* and a few other strains can grow heterotrophically. Some species are very flexible metabolically. For example, *Thiobacillus ferrooxidans* also uses ferrous iron as an electron donor and produces ferric iron as well as sulfuric acid. *T. denitrificans* even grows anaerobically by reducing nitrate to nitrogen gas. It should be noted that other sulfur-oxidizing bacteria such as *Thiobacterium* and *Macromonas* probably do not derive energy from sulfur oxidation. They may use the process to detoxify metabolically produced hydrogen peroxide.

Sulfur-oxidizing bacteria have a wide distribution and great practical importance. *Thiobacillus* grows in soil and aquatic habitats, both freshwater and marine. In marine habitats *Thiomicrospira* is more important than *Thiobacillus*. Because of their great acid tolerance (*T. thiooxidans* grows at pH 0.5 and cannot grow above pH 6), these bacteria prosper in habitats they have acidified by sulfuric acid production, even though most other organisms are dying. The production of large quantities of sulfuric acid and ferric iron by *T. ferrooxidans* corrodes concrete and pipe structures. Thiobacilli often cause extensive acid and metal pollution when they release metals from mine wastes. However, sulfur-oxidizing bacteria also are beneficial. They may increase soil fertility when they release elemental sulfur by oxidizing it to sulfate. Thiobacilli are used in processing low-grade metal ores because of their ability to leach metals from ore. The sulfur cycle and microorganisms (pp. 596–97)

---

1. Describe the general properties of the β-proteobacteria.
2. Briefly describe the following genera and their practical importance: *Neisseria, Burkholderia,* and *Bordetella.*
3. What is a sheath and of what advantage is it?
4. How does *Sphaerotilus* maintain its position in running water? How does it reproduce and disperse its progeny?
5. Characterize the colorless sulfur bacteria and discuss their placement in the second edition of *Bergey's Manual.*
6. How do colorless sulfur bacteria obtain energy by oxidizing sulfur compounds? What is adenosine 5'-phosphosulfate?
7. List several positive and negative impacts sulfur-oxidizing bacteria have on the environment and human activities.

22.3 CLASS *GAMMAPROTEOBACTERIA*

The γ-**proteobacteria** constitute the largest subgroup of proteobacteria with an extraordinary variety of physiological types. Many important genera are chemoorganotrophic and facultatively anaerobic. Other genera contain aerobic chemoorganotrophs, photolithotrophs, chemolithotrophs, or methylotrophs. According to some DNA-rRNA hybridization studies, the γ-proteobacteria are composed of several deeply branching groups. One consists of the purple sulfur bacteria; a second includes the intracellular parasites *Legionella* and *Coxiella.* The two largest groups contain a wide variety of nonphotosynthetic genera. Ribosomal RNA superfamily I is represented by the families *Vibrionaceae, Enterobacteriaceae,* and *Pasteurellaceae.* These bacteria use the glycolytic and pentose phosphate pathways to catabolize carbohydrates. Most are facultative anaerobes. Ribosomal RNA superfamily II contains mostly aerobes that often use the Entner-Doudoroff and pentose phosphate pathways to catabolize many different kinds of organic molecules. The genera *Pseudomonas, Azotobacter, Moraxella, Xanthomonas,* and *Acinetobacter* belong to this superfamily.

The exceptional diversity of these bacteria is evident from the fact that the second edition of *Bergey's Manual* divides the class *Gammaproteobacteria* into 14 orders and 25 families. **Figure 22.16** illustrates the phylogenetic relationships between major groups and selected γ-proteobacteria, and **table 22.5** outlines the general characteristics of some of the bacteria discussed in this section.

The Purple Sulfur Bacteria

As mentioned previously, the purple photosynthetic bacteria are distributed between three subgroups of the proteobacteria. Despite the diversity of these organisms, they do share some general characteristics, which were summarized in table 21.1 (*see p. 456*). Most of the purple nonsulfur bacteria are α-proteobacteria and were discussed earlier in this chapter (pp. 475–76). Because the purple sulfur bacteria are γ-proteobacteria, they will be described here.

Bergey's Manual divides the purple sulfur bacteria into two families: the *Chromatiaceae* and *Ectothiorhodospiraceae.* In the second edition these families are in the order *Chromatiales.* The family *Ectothiorhodospiraceae* contains eight genera. *Ectothiorhodospira* has red, spiral-shaped, polarly flagellated cells that deposit sulfur globules externally (**figure 22.17;** *see also figure 3.9*b). Internal photosynthetic membranes are organized as lamellar stacks. The typical purple sulfur bacteria are located in the family *Chromatiaceae,* which is much larger and contains 26 genera.

The **purple sulfur bacteria** are strict anaerobes and usually photolithoautotrophs. They oxidize hydrogen sulfide to sulfur and deposit it internally as sulfur granules (usually within invaginated pockets of the plasma membrane); often they eventually oxidize the sulfur to sulfate. Hydrogen also may serve as an

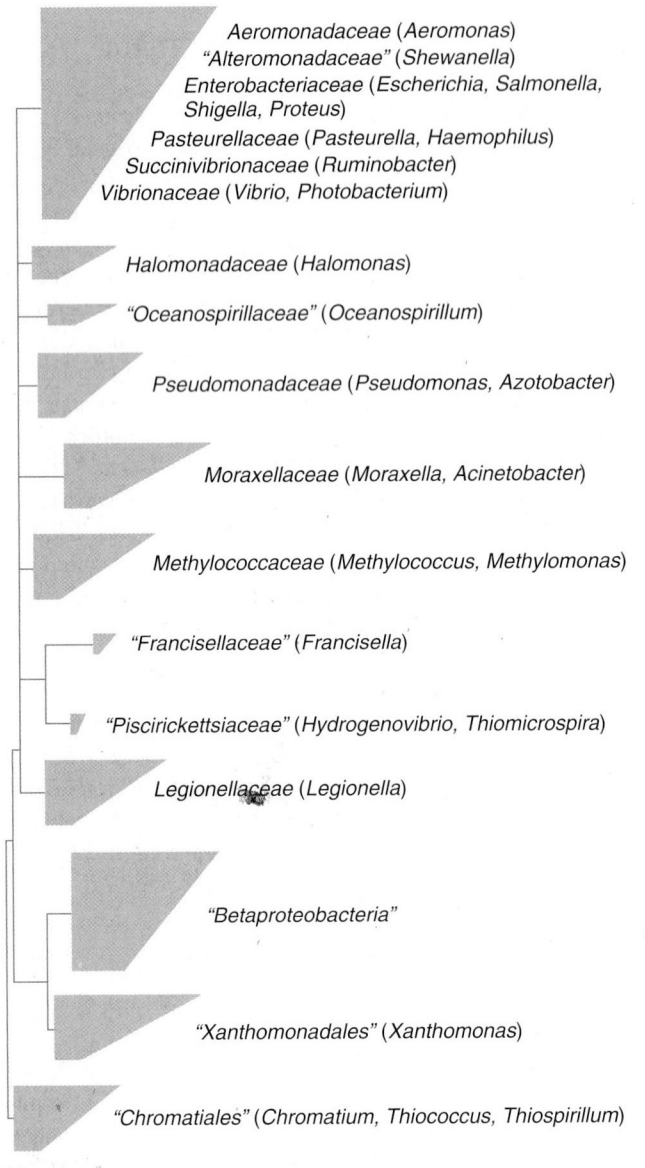

(a)

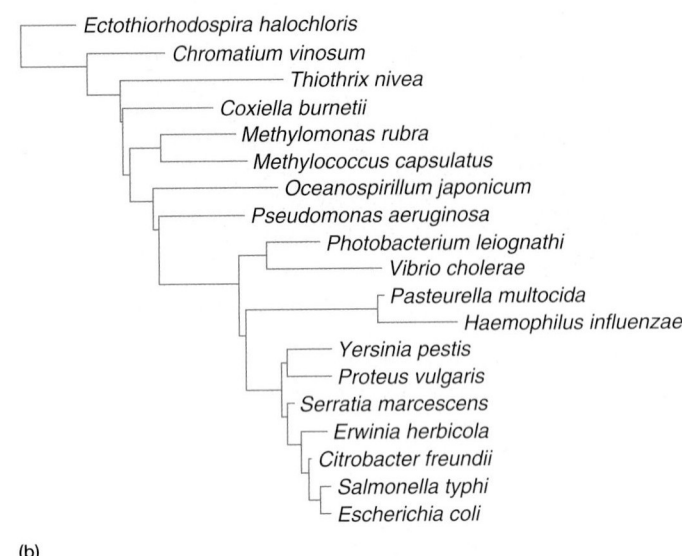

(b)

Figure 22.16   Phylogenetic Relationships Among γ-Proteobacteria.   (a) The major phylogenetic groups based on 16S rRNA sequence comparisons. Representative genera are given in parentheses. Each tetrahedron in the tree represents a group of related organisms; its horizontal edges show the shortest and longest branches in the group. Multiple branching at the same level indicates that the relative branching order of the groups cannot be determined from the data. The quotation marks around some names indicate that they are not formally approved taxonomic names. (b) The relationships of a few species based on 16S rRNA sequence data.   *Source: The Ribosomal Database Project.*

electron donor. *Thiospirillum, Thiocapsa,* and *Chromatium* are typical purple sulfur bacteria (**figure 22.18**). They are found in anaerobic, sulfide-rich zones of lakes (*see section 29.4*). Large blooms of purple sulfur bacteria occur in bogs and lagoons under the proper conditions (**figure 22.19**).

Order *Thiotrichales*

The order *Thiotrichales* contains three families, the largest of which is the family *Thiotrichaceae*. This family has several genera that oxidize sulfur compounds (see the colorless sulfur bacteria [p. 483] and chapter 9 for sulfur oxidation and chemolithotrophy). Morphologically both rods and filamentous forms are present.

Two of the best-studied gliding genera in this family are *Beggiatoa* and *Leucothrix* (**figures 22.20** and **22.21**). *Beggiatoa* is microaerophilic and grows in sulfide-rich habitats such as sulfur springs, freshwater with decaying plant material, rice paddies,

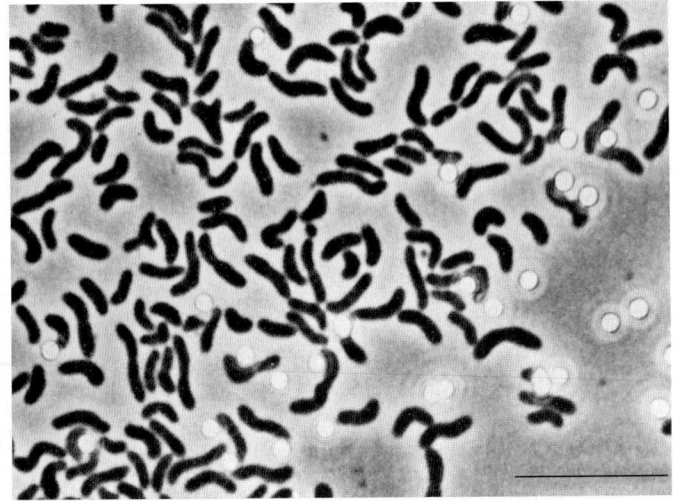

Figure 22.17   Purple Bacteria.   *Ectothiorhodospira mobilis;* light micrograph. Bar = 10 μm.

Table 22.5

Characteristics of Selected γ-Proteobacteria

Genus	Dimensions (μm) and Morphology	G + C Content (mol%)	Oxygen Requirement	Other Distinctive Characteristics
Azotobacter	1.5–2.0; ovoid cells, pleomorphic, peritrichous or nonmotile	63.2–67.5	Aerobic	Can form cysts; fix nitrogen nonsymbiotically
Beggiatoa	~1–50 × ~2–10; colorless cells form filaments, either single or in colonies	37–51	Aerobic or microaerophilic	Gliding motility; can form sulfur inclusions with hydrogen sulfide present
Chromatium	1–6 × 1.5–16; rod-shaped or ovoid, straight or slightly curved, polar flagella	48–70	Anaerobic	Photolithoautotroph that can use sulfide; sulfur stored within the cell
Ectothiorhodospira	0.5–1.5 in diameter; vibrioid- or rod-shaped, polar flagella	50.5–69.7	Anaerobic, some aerobic or microaerobic	Internal lamellar stacks of membranes; deposits sulfur granules outside cells
Escherichia	1.1–1.5 × 2–6; straight rods, peritrichous or nonmotile	48–52	Facultatively anaerobic	Mixed acid fermenter; formic acid converted to H_2 and CO_2, lactose fermented, citrate not used
Haemophilus	<1.0 in width; coccobacilli or rods, nonmotile	33–47	Facultative or aerobic	Fermentative; requires growth factors present in blood; parasites on mucous membranes
Leucothrix	Long filaments of short cylindrical cells, usually holdfast is present	46–51	Aerobic	Dispersal by gonidia, filaments don't glide; rosettes formed; heterotrophic
Methylococcus	1.0 in diameter; cocci with capsules, nonmotile	62–63	Aerobic	Can form a cyst; methane, methanol, and formaldehyde are sole carbon and energy sources
Photobacterium	0.8–1.3 × 1.8–2.4; straight, plump rods with polar flagella	40–44	Facultatively anaerobic	Two species can emit blue-green light; Na^+ needed for growth
Pseudomonas	0.5–1.0 × 1.5–5.0; straight or slightly curved rods, polar flagella	58–70	Aerobic	Respiratory metabolism with oxygen as acceptor; some are able to use H_2 or CO as energy source
Vibrio	0.5–0.8 × 1.4–2.6; straight or curved rods with sheathed polar flagella	38–51	Facultatively anaerobic	Fermentative or respiratory metabolism; sodium ions stimulate or are needed for growth; oxidase positive

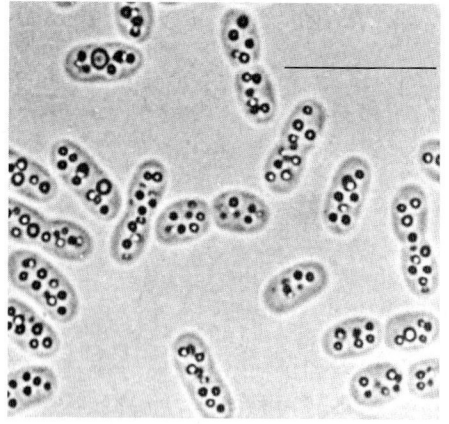

(a) *Chromatium vinosum*

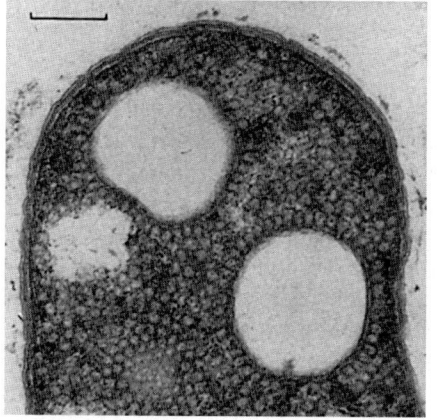

(b) *C. vinosum*

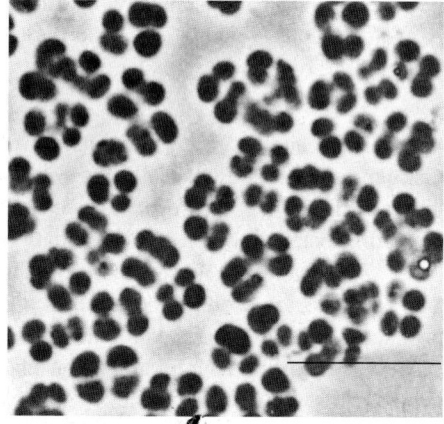

(c) *Thiocapsa roseopersicina*

Figure 22.18   Typical Purple Sulfur Bacteria. (**a**) *Chromatium vinosum* with intracellular sulfur granules. Bar = 10 μm. (**b**) Electron micrograph of *C. vinosum*. Note the intracytoplasmic vesicular membrane system. The large white areas are the former sites of sulfur globules. Bar = 0.3 μm. (**c**) *Thiocapsa roseopersicina*. Bar = 10 μm.

(a)

(b)

Figure 22.19   Purple Photosynthetic Sulfur Bacteria.   (a) Purple photosynthetic sulfur bacteria growing in a bog. (b) A sewage lagoon with a bloom of purple photosynthetic bacteria.

Figure 22.20   *Beggiatoa alba.*   A light micrograph showing part of a colony (×400). Note the dark sulfur granules within many of the filaments.

salt marshes, and marine sediments. Its filaments contain short, disklike cells and lack a sheath. *Beggiatoa* is very versatile metabolically. It oxidizes hydrogen sulfide to form large sulfur grains located in pockets formed by invaginations of the plasma membrane. *Beggiatoa* can subsequently oxidize the sulfur to sulfate. The electrons are used by the electron transport chain in energy production. Many strains also can grow heterotrophically with acetate as a carbon source, and some may incorporate CO_2 autotrophically.

Leucothrix (figure 22.21) is an aerobic chemoorganotroph that forms filaments or trichomes up to 400 μm long. It is usually marine and is attached to solid substrates by a holdfast. *Leucothrix* has a complex life cycle in which dispersal is by the formation of gonidia. Rosette formation often is seen in culture (figure 22.21*d*). *Thiothrix* is a related genus that forms sheathed

filaments and releases gonidia from the open end of the sheath (**figure 22.22**). In contrast with *Leucothrix*, *Thiothrix* is a chemolithotroph that oxidizes hydrogen sulfide and deposits sulfur granules internally. It also requires an organic compound for growth (i.e., it is a mixotroph). *Thiothrix* grows in sulfide-rich flowing water and activated sludge sewage systems.

Order *Methylococcales*

The family *Methylococcaceae* contains rods, vibrios, and cocci that use methane, methanol, and other reduced one-carbon compounds as their sole carbon and energy sources under aerobic or microaerobic (low oxygen) conditions. That is, they are methylotrophs (bacteria that use methane exclusively as their carbon and energy source often are called methanotrophs). The family contains seven genera, two of which are *Methylococcus* (spherical, nonmotile cells) and *Methylomonas* (straight, curved, or branched rods with a single, polar flagellum). When oxidizing methane, the bacteria contain complex arrays of intracellular membranes. Almost all can form some type of resting stage, often a cyst somewhat like that of the azotobacteria. Methylotrophic growth depends on the presence of methane and related compounds. Methanogenesis from substrates such as H_2 and CO_2 is widespread in anaerobic soil and water, and methylotrophic bacteria grow above anaerobic habitats all over the world.

Methane-oxidizing bacteria use methane as a source of both energy and carbon. Methane is first oxidized to methanol by the enzyme methane monooxygenase. The methanol is then oxidized to formaldehyde by methanol dehydrogenase, and the electrons from this oxidation are donated to an electron transport chain for ATP synthesis. Formaldehyde can be assimilated into cell material by the activity of either of two pathways, one involving the formation of the amino acid serine and the other proceeding through the synthesis of sugars such as fructose 6-phosphate and ribulose 5-phosphate.

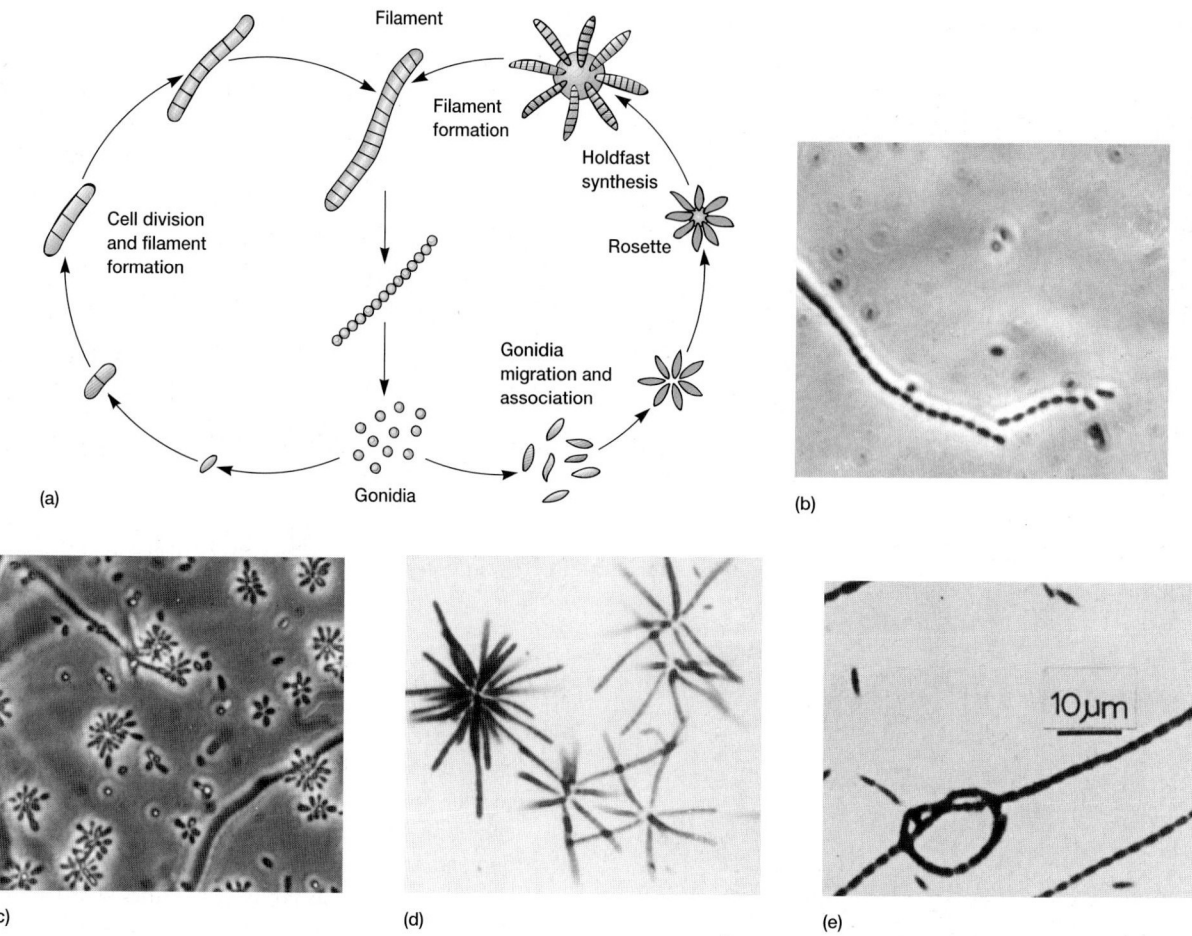

Figure 22.21   Morphology and Reproduction of *Leucothrix mucor*.   (a) Life cycle of *L. mucor*. **(b)** Separation of gonidia from the tip of mature filament; phase contrast (×1,400). **(c)** Gonidia aggregating to form rosettes; phase contrast (×950). **(d)** Young developing rosettes (×1,500). **(e)** A knot formed by a *Leucothrix* filament.

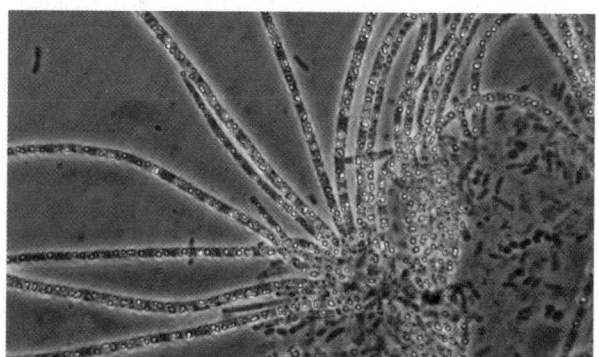

**Figure 22.22   *Thiothrix*.   A *Thiothrix* colony viewed with phase-contrast microscopy (×1,000).

Order *Pseudomonadales*

Pseudomonas is the most important genus in the order *Pseudomonadales,* family *Pseudomonadaceae.* The genus *Pseudomonas* contains straight or slightly gram-negative curved rods, 0.5 to 1.0 μm by 1.5 to 5.0 μm in length, that are motile by one or several polar flagella and lack prosthecae or sheaths (**figure 22.23;** *see figure 3.32*a). These chemoheterotrophs are aerobic and carry out respiratory metabolism with O_2 (and sometimes nitrate) as the electron acceptor. All pseudomonads have a functional tricarboxylic acid cycle and can oxidize substrates to CO_2. Most hexoses are degraded by the Entner-Doudoroff pathway rather than glycolytically.   The Entner-Doudoroff pathway and tricarboxylic acid cycle (pp. 174, 178–79, appendix II)

In the first edition, the genus is an exceptionally heterogeneous taxon composed of 70 or more species. Many can be placed in one of five rRNA homology groups. The three best-characterized groups, RNA groups I–III, are subdivided according to properties such as the presence of poly-β-hydroxybutyrate (PHB), the production of a fluorescent pigment, pathogenicity, the presence of arginine dihydrolase, and glucose utilization. For example, the fluorescent subgroup does not accumulate PHB and produces a diffusible, water-soluble, yellow-green pigment that fluoresces under UV radiation (**figure 22.24**). *Pseudomonas aeruginosa, P. fluorescens, P. putida,* and *P. syringae* are members of this group. The second edition of *Bergey's Manual* has reduced the size of the genus *Pseudomonas* drastically by

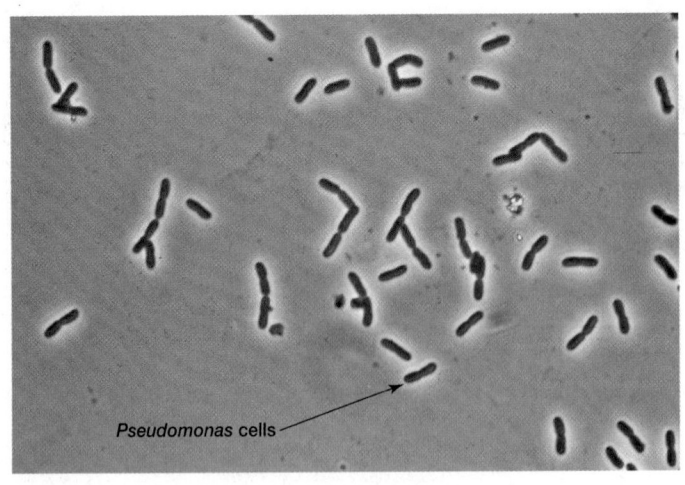

(a)

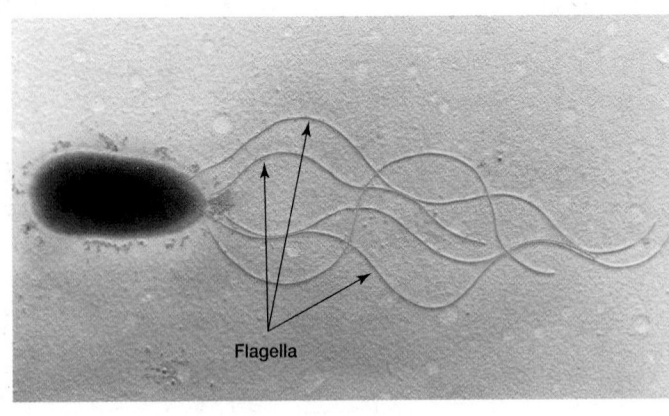

(b)

Figure 22.23 The Genus *Pseudomonas*. (a) A phase-contrast micrograph of *Pseudomonas* cells containing PHB (poly-β-hydroxybutyrate) granules. (b) A transmission electron micrograph of *Pseudomonas putida* with five polar flagella, each flagellum about 5–7 μm in length.

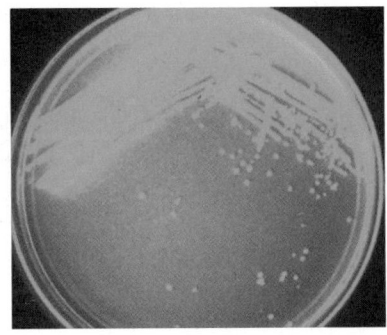

Figure 22.24 *Pseudomonas* Fluorescence. *Pseudomonas aeruginosa* colonies fluorescing under ultraviolet light.

transferring species to several newly named genera (e.g., *Burkholderia, Hydrogenophaga,* and *Methylobacterium*).

The pseudomonads have a great practical impact in several ways, including these:

1. Many can degrade an exceptionally wide variety of organic molecules. Thus they are very important in the **mineralization process** (the microbial breakdown of organic materials to inorganic substances) in nature and in sewage treatment. The fluorescent pseudomonads can use approximately 80 different substances as their carbon and energy sources; *Burkholderia cepacia* will degrade over 100 different organic molecules. Microbial decomposition of organic materials (pp. 593–96; 979–83)

2. Several species (e.g., *P. aeruginosa*) are important experimental subjects. Many advances in microbial physiology and biochemistry have come from their study. For example, the genome of *Pseudomonas aeruginosa* has been sequenced. It is very big (about 6.3 million base pairs) and much more com-

plex than the *E. coli* genome, although about 50% of its open reading frames are similar to those in *E. coli*. *P. aeruginosa* has an unusually large number of genes for catabolism, nutrient transport, the efflux of organic molecules, and metabolic regulation. This may explain its ability to grow in many environments and resist antibiotics.

3. Some pseudomonads are major animal and plant pathogens. *P. aeruginosa* infects people with low resistance such as cystic fibrosis patients and invades burn areas or causes urinary tract infections. *Burkholderia solanacearum* (*R. solanacearum*) causes wilts in many plants by producing pectinases, cellulases, and the plant hormones indoleacetic acid and ethylene. *P. syringae* and *B. cepacia* are also important plant pathogens.

4. Pseudomonads such as *P. fluorescens* are involved in the spoilage of refrigerated milk, meat, eggs, and seafood because they grow at 4°C and degrade lipids and proteins.

The genus *Azotobacter* also is in the family *Pseudomonadaceae*. The genus contains large, ovoid bacteria, 1.5 to 2.0 μm in diameter, that may be motile by peritrichous flagella. The cells are often pleomorphic, ranging from rods to coccoid shapes, and form cysts as the culture ages (**figure 22.25**). The genus is aerobic, catalase positive, and fixes nitrogen nonsymbiotically. *Azotobacter* is widespread in soil and water.

Order *Vibrionales*

Three closely related orders of the γ-proteobacteria contain a number of important bacterial genera. Each order has only one family of facultatively anaerobic gram-negative rods. **Table 22.6** summarizes the distinguishing properties of the families *Enterobacteriaceae, Vibrionaceae,* and *Pasteurellaceae*.

The order *Vibrionales* contains only one family, the *Vibrionaceae*. Members of the family *Vibrionaceae* are gram-

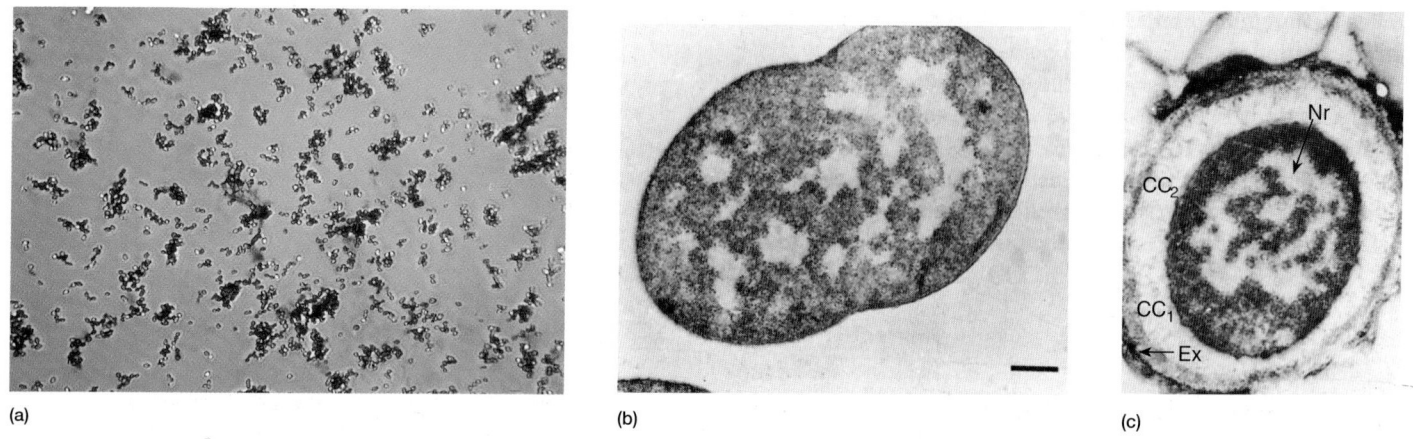

(a)   (b)   (c)

Figure 22.25 *Azotobacter.* (a) *A. chroococcum* (×270). (b) Electron micrograph of *A. chroococcum.* Bar = 0.2 μm. (c) *Azotobacter* cyst structure. The nuclear region (Nr), exine layers (CC₁ and CC₂), and exosporium (Ex) are visible.

Table 22.6			
Characteristics of Families of Facultatively Anaerobic Gram-Negative Rods			
Characteristics	*Enterobacteriaceae*	*Vibrionaceae*	*Pasteurellaceae*
Cell dimensions	0.3–1.0 × 1.0–6.0 μm	0.3–1.3 × 1.0–3.5 μm	0.2–0.3 × 0.3–2.0 μm
Morphology	Straight rods; peritrichous flagella or nonmotile	Straight or curved rods; polar flagella	Coccoid to rod-shaped cells, sometimes pleomorphic; nonmotile
Physiology	Oxidase negative	Oxidase positive; all can use D-glucose as sole or principal carbon source	Oxidase positive; heme and/or NAD often required for growth; organic nitrogen source required
G + C content	38–60%	38–63%	38–47%
Symbiotic relationships	Some parasitic on mammals and birds; some species plant pathogens	Most not pathogens (with a few exceptions)	Parasites of mammals and birds
Representative genera	*Escherichia, Shigella, Salmonella, Citrobacter, Klebsiella, Enterobacter, Erwinia, Serratia, Proteus, Yersinia*	*Vibrio, Photobacterium*	*Pasteurella, Haemophilus*

From J. G. Holt and N. R. Krieg (eds.), *Bergey's Manual of Systematic Bacteriology,* Vol. 1. Copyright © 1984 Williams and Wilkins Company, Baltimore, MD. Reprinted by permission.

negative, straight or curved rods with polar flagella (**figure 22.26**). Most are oxidase positive, and all use D-glucose as their sole or primary carbon and energy source (table 22.6). The majority are aquatic microorganisms, widespread in freshwater and the sea. There are six genera in the family: *Vibrio, Photobacterium, Enhydrobacter, Salinivibrio, Listonella,* and *Allomonas.*

Several vibrios are important pathogens. *V. cholerae* (*see figure 3.1*e) is the causative agent of cholera (*see section 39.4*), and *V. parahaemolyticus* sometimes causes gastroenteritis in humans following consumption of contaminated seafood. *V. anguillarum* and others are responsible for fish diseases.

The *Vibrio cholerae* genome has now been sequenced and found to contain 3,885 open reading frames distributed between two circular chromosomes, chromosome 1 (2.96 million base pairs) and chromosome 2 (1.07 million bp). The larger chromosome primarily has genes for essential cell functions such as DNA replication, transcription, and protein synthesis. It also has most of the virulence genes (e.g., the cholera toxin gene is located in an integrated CTXφ phage on chromosome 1). Chromosome

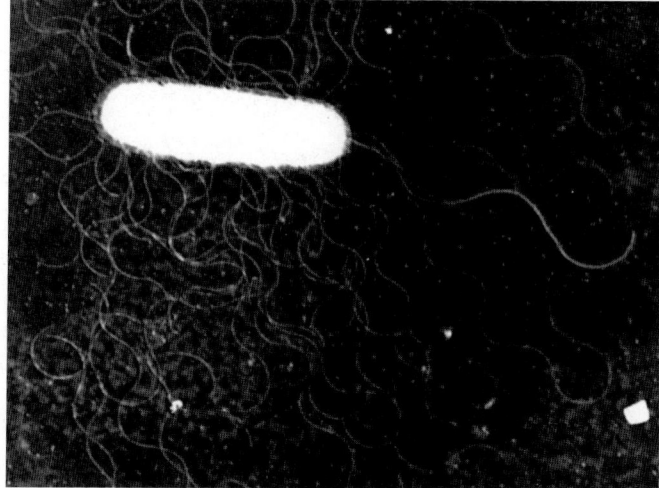

Figure 22.26 The *Vibrionaceae*. Electron micrograph of *Vibrio alginolyticus* grown on agar, showing a sheathed polar flagellum and unsheathed lateral flagella (×18,000).

Microbial Diversity & Ecology

22.1   Bacterial Bioluminescence

Several species in the genera *Vibrio* and *Photobacterium* can emit light of a blue-green color. The enzyme luciferase catalyzes the reaction and uses reduced flavin mononucleotide, molecular oxygen, and a long-chain aldehyde as substrates.

$$FMNH_2 + O_2 + RCHO \xrightarrow{luciferase} FMH$$
$$+ H_2O + RCOOH + light$$

The evidence suggests that an enzyme-bound, excited flavin intermediate is the direct source of luminescence. Because the electrons used in light generation are probably diverted from the electron transport chain and ATP synthesis, the bacteria expend considerable en-

ergy on luminescence. Luminescence activity is regulated and can be turned off or on under the proper conditions.

There is much speculation about the role of bacterial luminescence and its value to bacteria, particularly because it is such an energetically expensive process. Luminescent bacteria occupying the special luminous organs of fish do not emit light when they grow as free-living organisms in the seawater. Free-living luminescent bacteria can reproduce and infect young fish. Once settled in a fish's luminous organ, they begin emitting light, which the fish uses for its own purposes. Other luminescent bacteria growing on potential food items such as small crustacea may use light to attract fish to the food source. After ingestion, they could establish a symbiotic relationship in the host's gut.

(a)

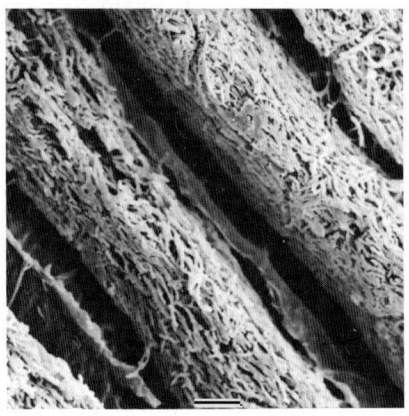
(b)

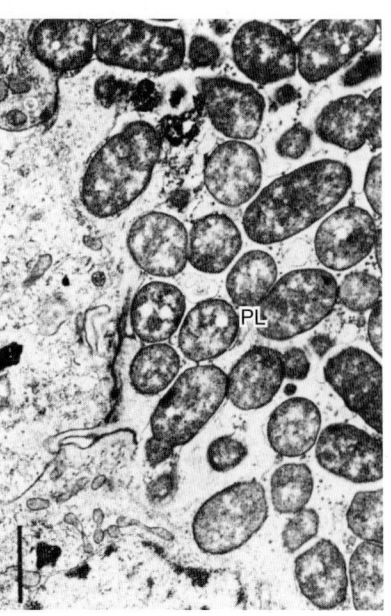
(c)

Figure 22.27   Bioluminescence. (a) A photograph of the Atlantic flashlight fish *Kryptophanaron alfredi.* The light area under the eye is the fish's luminous organ, which can be covered by a lid of tissue. (b) The masses of photobacteria in the SEM view are separated by thin epithelial cells. Line = 6.0 μm. (c) Ultrathin section of the luminous organ of a fish. *Equulities novaehollandiae,* with the bioluminescent bacterium *Photobacterium leiognathi,* PL. Bar = 2 μm.

2 also has essential genes such as transport genes and ribosomal protein genes. Copies of some genes are present on both chromosomes. Perhaps *V. cholerae* achieves faster genome duplication and cell division by distributing its genes between two chromosomes.

Several members of the family are unusual in being bioluminescent. *Vibrio fischeri* and at least two species of *Photobacterium* are among the few marine bacteria capable of **bioluminescence** and emit a blue-green light because of the activity of the enzyme luciferase (**Microbial Diversity & Ecology 22.1**). The peak emission of light is usually between 472 and 505 nm, but one strain of *V. fischeri* emits yellow light with a major peak at 545 nm. Although many of these bacteria are free-living, *V. fischeri, P. phosphoreum,* and *P. leiognathi* live symbiotically in the luminous

organs of fish (**figure 22.27**). Interestingly, *Vibrio fischeri* can degrade 3′, 5′-cyclic AMP and use it as a carbon, nitrogen, and phosphorus source for growth.

Order *Enterobacteriales*

The family *Enterobacteriaceae* is the largest of the families in table 22.6. It contains gram-negative, peritrichously flagellated or nonmotile, facultatively anaerobic, straight rods with simple nutritional requirements (*see figures 2.14c, 3.28a, 3.31,* and *3.32c*). In the second edition the order *Enterobacteriales* has only one family, *Enterobacteriaceae,* with over 40 genera. The relationship between *Enterobacteriales* and the orders *Vibrionales* and *Pasteurellales* can be seen by inspecting figure 22.16.

Figure 22.28   Identification of Enterobacterial Genera.   A dichotomous key to selected genera of enteric bacteria based on motility and biochemical characteristics. The following abbreviations are used: ONPG, *o*-nitrophenyl-β-D-galactopyranoside (a test for β-galactosidase); DNase, deoxyribonuclease; Gel. Liq., gelatin liquefaction; and VP, Voges-Proskauer (a test for acetoin).

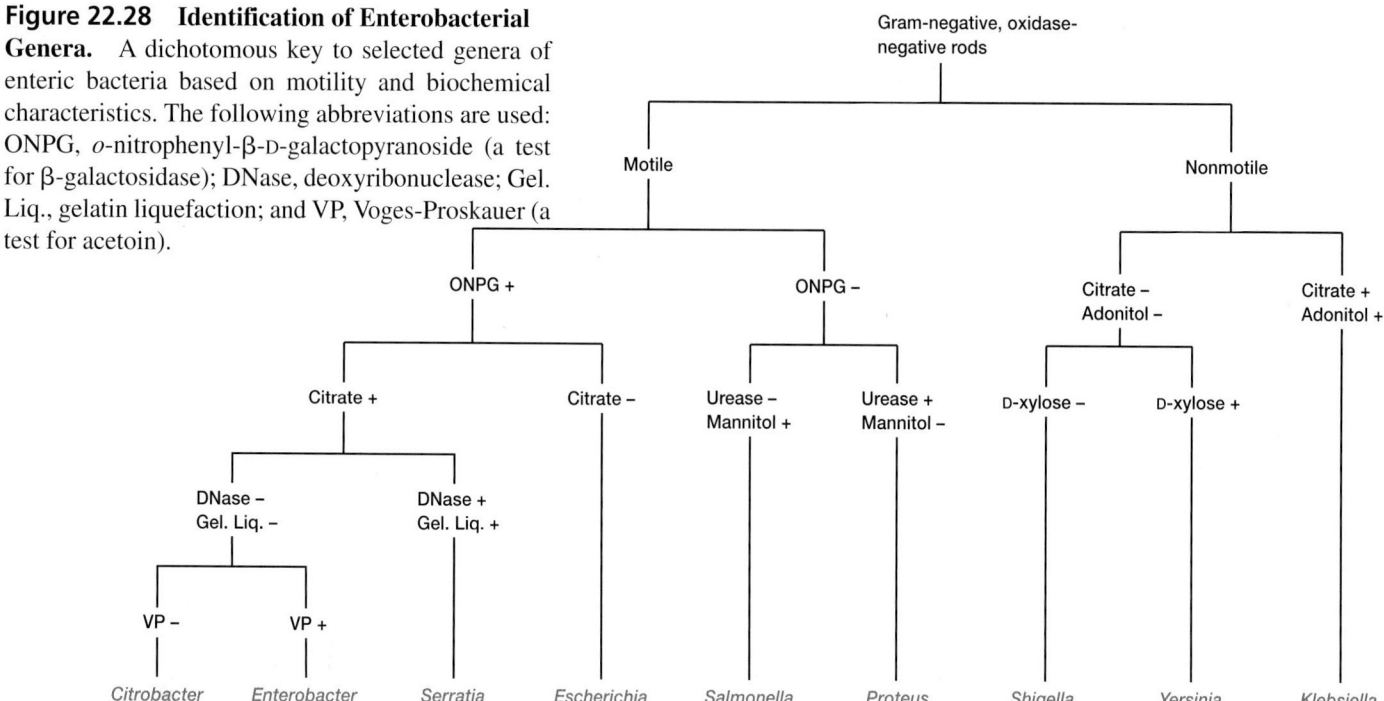

The metabolic properties of the *Enterobacteriaceae* are very useful in characterizing its constituent genera. Members of the family, often called **enterobacteria** or **enteric bacteria** [Greek *enterikos,* pertaining to the intestine], all degrade sugars by means of the Embden-Meyerhof pathway and cleave pyruvic acid to yield formic acid in formic acid fermentations. Those enteric bacteria that produce large amounts of gas during sugar fermentation, such as *Escherichia* spp., have the formic hydrogenlyase complex that degrades formic acid to H_2 and CO_2. The family can be divided into two groups based on their fermentation products. The majority (e.g., the genera *Escherichia, Proteus, Salmonella,* and *Shigella*) carry out mixed acid fermentation and produce mainly lactate, acetate, succinate, formate (or H_2 and CO_2), and ethanol. In butanediol fermentation the major products are butanediol, ethanol, and carbon dioxide. *Enterobacter, Serratia, Erwinia,* and *Klebsiella* are butanediol fermenters. As mentioned previously (*see section 9.3*), the two types of formic acid fermentation are distinguished by the methyl red and Voges-Proskauer tests. Formic acid fermentation and the family Enterobacteriaceae (p. 176)

Because the enteric bacteria are so similar in morphology, biochemical tests are normally used to identify them after a preliminary examination of their morphology, motility, and growth responses (**figure 22.28** provides a simple example). Some more commonly used tests are those for the type of formic acid fermentation, lactose and citrate utilization, indole production from tryptophan, urea hydrolysis, and hydrogen sulfide production. For example, lactose fermentation occurs in *Escherichia* and *Enterobacter* but not in *Shigella, Salmonella,* or *Proteus.* **Table 22.7** summarizes a few of the biochemical properties useful in distinguishing between genera of enteric bacteria. The mixed acid fermenters are located on the left in this table and the butanediol

fermenters on the right. The usefulness of biochemical tests in identifying enteric bacteria is shown by the popularity of commercial identification systems, such as the Enterotube and API 20-E systems, that are based on these tests. Commercial rapid identification systems (pp. 814–15); *E. coli* genome sequence (p. 340)

Members of the *Enterobacteriaceae* are so common, widespread, and important that they are probably more often seen in most laboratories than any other bacteria. *Escherichia coli* is undoubtedly the best-studied bacterium and the experimental organism of choice for many microbiologists. It is an inhabitant of the colon of humans and other warm-blooded animals, and it is quite useful in the analysis of water for fecal contamination (*see section 29.5*). Some strains cause gastroenteritis or urinary tract infections. Several enteric genera contain very important human pathogens responsible for a variety of diseases: *Salmonella* (**figure 22.29**), typhoid fever and gastroenteritis; *Shigella,* bacillary dysentery; *Klebsiella,* pneumonia; *Yersinia,* plague. Members of the genus *Erwinia* are major pathogens of crop plants and cause blights, wilts, and several other plant diseases. These and other members of the family are discussed in more detail at later points in the text (*see chapter 39*).

Order *Pasteurellales*

The second edition of *Bergey's Manual* places the family *Pasteurellaceae* in the order *Pasteurellales* and treats it much the same as the first edition does. The *Pasteurellaceae* differ from the other two families in several ways (table 22.6). Most notably, they are small (0.2 to 0.3 μm in diameter) and nonmotile, normally oxidase positive, have complex nutritional requirements of various kinds, and are parasitic in vertebrates. The family contains

Table 22.7

Some Characteristics of Selected Genera in the *Enterobacteriaceae*

Characteristics	*Escherichia*	*Shigella*	*Salmonella*	*Citrobacter*	*Proteus*
Methyl red	+	+	+	+	+
Voges-Proskauer	−	−	−	−	d
Indole production	(+)	d	−	d	d
Citrate use	−	−	(+)	+	d
H$_2$S production	−	−	(+)	d	(+)
Urease	−	−	−	(+)	+
β-galactosidase	(+)	d	d	+	−
Gas from glucose	+	−	(+)	+	+
Acid from lactose	+	−	(−)	d	−
Phenylalanine deaminase	−	−	−	−	+
Lysine decarboxylase	(+)	−	(+)	−	−
Ornithine decarboxylase	(+)	d	(+)	(+)	d
Motility	d	−	(+)	+	+
Gelatin liquifaction (22°C)	−	−	−	−	+
% G + C	48–52	49–53	50–53	50–52	38–41
Other characteristics	1.1–1.5 × 2.0–6.0 μm; peritrichous when motile	No gas from sugars	0.7–1.5 × 2–5 μm; peritrichous flagella	1.0 × 2.0–6.0 μm; peritrichous	0.4–0.8 × 1.0–3.0 μm; peritrichous

[a](+) usually present
[b](−) usually absent
[c]d, strains or species vary in possession of characteristic

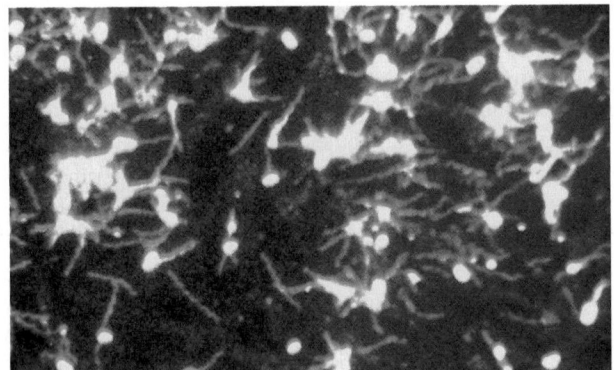

(a) *Salmonella enteritidis*

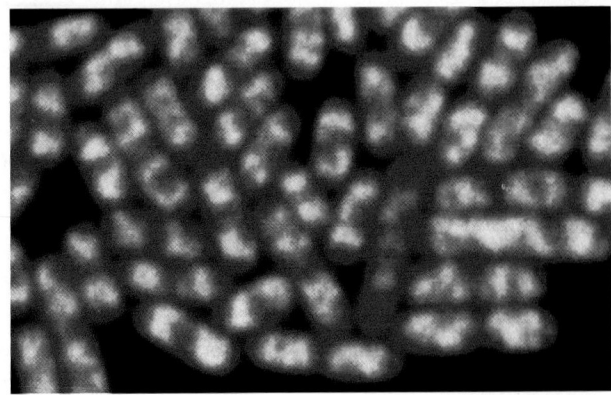

(b) *S. typhi*

six genera: *Pasteurella, Haemophilus, Actinobacillus, Lonepinella, Mannheimia,* and *Phocoenobacter.*

As might be expected, members of this family are best known for the diseases they cause in humans and many animals. *Pasteurella multilocida* and *P. haemolytica* are important animal pathogens. *P. multilocida* is responsible for fowl cholera, which destroys many chickens, turkeys, ducks, and geese each year. *P. haemolytica* is at least partly responsible for pneumonia in cattle, sheep, and goats (e.g., "shipping fever" in cattle). *H. influenzae* type b is a major human pathogen that causes a variety of diseases, including meningitis in children (*see section 39.1*). *H. influenzae* genome sequence (p. 340)

1. Describe the general properties of the γ-proteobacteria.
2. What are the major characteristics of the purple sulfur bacteria? Contrast the families *Chromatiaceae* and *Ectothiorhodospira.*
3. Describe the genera *Beggiatoa, Leucothrix,* and *Thiothrix.*
4. In what habitats would one expect to see the *Methylococcaceae* growing and why?

Figure 22.29   The *Enterobacteriaceae.* *Salmonella* treated with fluorescent stains. (**a**) *Salmonella enteritidis* with peritrichous flagella (×500). *S. enteritidis* is associated with gastroenteritis. (**b**) *Salmonella typhi* with acridine orange stain (×2,000). *S. typhi* causes typhoid fever.

Yersinia	*Klebsiella*	*Enterobacter*	*Erwinia*	*Serratia*
+	(+)[a]	(−)[b]	+	d[c]
− (37°C)	(+)	+	(+)	+
d	d	−	(−)	(−)
(−)	(+)	+	(+)	+
−	−	−	(+)	−
d	(+)	(−)	−	−
+	(+)	+	+	+
(−)	(+)	(+)	(−)	d
(−)	(+)	(+)	d	d
−	−	(−)	(−)	−
(−)	(+)	d	−	d
d	−	(+)	−	d
− (37°C)	−	+	+	+
(−)	−	d	d	(+)
46–50	53–58	52–60	50–58	53–59
0.5–0.8 × 1.0–3.0 μm; peritrichous when motile	0.3–1.0 × 0.6–6.0 μm; capsulated	0.6–1.0 × 1.2–3.0 μm; peritrichous	0.5–1.0 × 1.0–3.0μm; peritrichous; plant pathogens and saprophytes	0.5–0.8 × 0.9–2.0 μm; peritrichous; colonies often pigmented

5. What is a methylotroph? How do methane-oxidizing bacteria use methane as both an energy source and a carbon source?

6. Give the major distinctive properties of the genera *Pseudomonas* and *Azotobacter*. Briefly discuss the taxonomic changes that have occurred in the genus *Pseudomonas*.

7. Why are the pseudomonads such important bacteria? What is mineralization?

8. List the major distinguishing traits of the families *Vibrionaceae*, *Enterobacteriaceae*, and *Pasteurellaceae*.

9. What major human disease is associated with the *Vibrionaceae*, and what species causes it?

10. Briefly describe bioluminescence and the way it is produced.

11. Into what two groups can the enteric bacteria be placed based on their fermentation patterns?

12. Give two reasons why the enterobacteria are so important.

22.4 CLASS *DELTAPROTEOBACTERIA*

Although the **δ-proteobacteria** are not a large assemblage of genera, they show considerable morphological and physiological diversity. These bacteria can be divided into two general groups, all of them chemoorganotrophs. Some genera are predators such as the bdellovibrios and myxobacteria. Others are anaerobes that generate sulfide from sulfate and sulfur while oxidizing organic

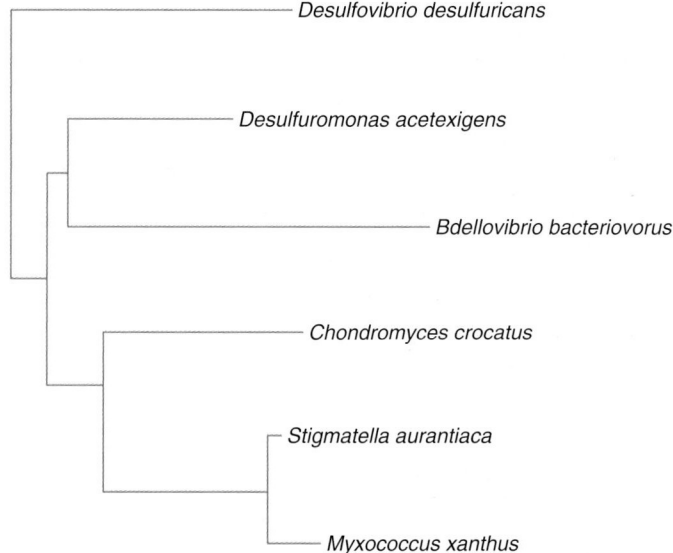

Figure 22.30   Phylogenetic Relationships Among Selected δ-Proteobacteria.   The relationships of a few species based on 16S rRNA sequence data are presented.   *Source: The Ribosomal Database Project.*

nutrients. The class has seven orders and 18 families. **Figure 22.30** illustrates the phylogenetic relationships between selected δ-proteobacteria, and **table 22.8** summarizes the general properties of some representative genera.

Table 22.8

Characteristics of Selected δ- and ε-Proteobacteria

Genus	Dimensions (μm) and Morphology	G + C Content (mol%)	Oxygen Requirement	Other Distinctive Characteristics
δ-Proteobacteria				
Bdellovibrio	0.2–0.5 × 0.5–1.4; comma-shaped rods with a sheathed polar flagellum	33.4–51.5	Aerobic	Preys on other gram-negative bacteria and grows in the periplasm, alternates between predatory and intracellular reproductive phases
Desulfovibrio	0.5–1.5 × 2.5–10; curved or sometimes straight rods, motile by polar flagella	46.1–61.2	Anaerobic	Oxidizes organic compounds to acetate and reduces sulfate or sulfur to H_2S
Desulfuromonas	0.4–0.9 × 1.0–4.0; straight or slightly curved or ovoid rods, lateral or subpolar flagella	50–63	Anaerobic	Reduces sulfur to H_2S, oxidizes acetate to CO_2, forms pink or peach-colored colonies
Myxococcus	0.4–0.7 × 2–10; slender rods with tapering ends, gliding motility	68–71	Aerobic	Forms fruiting bodies with microcysts not enclosed in a sporangium
Stigmatella	0.6–0.8 × 4–10; straight rods with tapered ends, gliding motility	68.5–68.7	Aerobic	Stalked fruiting bodies with sporangioles containing myxospores (0.9–1.2 × 2–4 μm)
ε-Proteobacteria				
Campylobacter	0.2–0.5 × 0.5–5; vibrioid cells with a single polar flagellum at one or both ends	30–38	Microaerophilic	Carbohydrates not fermented or oxidized; oxidase positive and urease negative; found in intestinal tract, reproductive organs, and oral cavity of animals
Helicobacter	0.5–1.0 × 2.5–5.0; helical, curved, or straight cells with rounded ends; multiple, sheathed flagella	33–42.5	Microaerophilic	Catalase and oxidase positive; urea rapidly hydrolyzed; found in the gastric mucosa of humans and other animals

Orders *Desulfovibrionales, Desulfobacterales, and Desulfuromonadales*

These sulfate- or sulfur-reducing bacteria are a diverse group united by their anaerobic nature and the ability to use elemental sulfur or sulfate and other oxidized sulfur compounds as electron acceptors during anaerobic respiration (**figure 22.31**). An electron transport chain generates ATP and reduces sulfur and sulfate to hydrogen sulfide. The best-studied sulfate-reducing genus is *Desulfovibrio; Desulfuromonas* uses only elemental sulfur as an acceptor. Anaerobic respiration (pp. 185–86)

These bacteria are very important in the cycling of sulfur within the ecosystem. Because significant amounts of sulfate are present in almost all aquatic and terrestrial habitats, sulfate-reducing bacteria are widespread and active in locations made anaerobic by microbial digestion of organic materials. *Desulfovibrio* and other sulfate-reducing bacteria thrive in habitats such as muds and sediments of polluted lakes and streams, sewage lagoons and digesters, and waterlogged soils. *Desulfuromonas* is most prevalent in anaerobic marine and estuarine sediments. It also can be isolated from methane digesters and anaerobic hydrogen sulfide-rich muds of freshwater habitats. It uses elemental sulfur, but not sulfate, as its electron acceptor. Often sulfate and sulfur reduction are apparent from the smell of hydrogen sulfide and the blackening of water and sediment by iron sulfide. Hydrogen sulfide production in waterlogged soils can kill animals, plants, and microorganisms. Sulfate-reducing bacteria

negatively impact industry because of their primary role in the anaerobic corrosion of iron in pipelines, heating systems, and other structures (*see section 42.4*). The sulfur cycle (pp. 596–97)

Order *Bdellovibrionales*

The order has only the family *Bdellovibrionaceae* and four genera. The genus *Bdellovibrio* [Greek *bdella*, leech] contains aerobic gram-negative, curved rods with polar flagella (**figure 22.32**). The flagellum is unusually thick due to the presence of a sheath that is continuous with the cell wall. *Bdellovibrio* has a distinctive life-style: it preys on other gram-negative bacteria and alternates between a nongrowing predatory phase and an intracellular reproductive phase.

The life cycle of *Bdellovibrio* is complex although it requires only 1 to 3 hours for completion (**figure 22.33**). The free bacterium swims along very rapidly (about 100 cell lengths per second) until it collides violently with its prey. It attaches to the bacterial surface, begins to rotate at rates as high as 100 revolutions per second, and bores a hole through the host cell wall in 5 to 20 minutes with the aid of several hydrolytic enzymes that it releases. Its flagellum is lost during penetration of the cell.

After entry, *Bdellovibrio* takes control of the host cell and grows in the space between the cell wall and plasma membrane while the host cell loses its shape and rounds up. The predator inhibits host DNA, RNA, and protein synthesis within minutes and disrupts the host's plasma membrane so that cytoplasmic con-

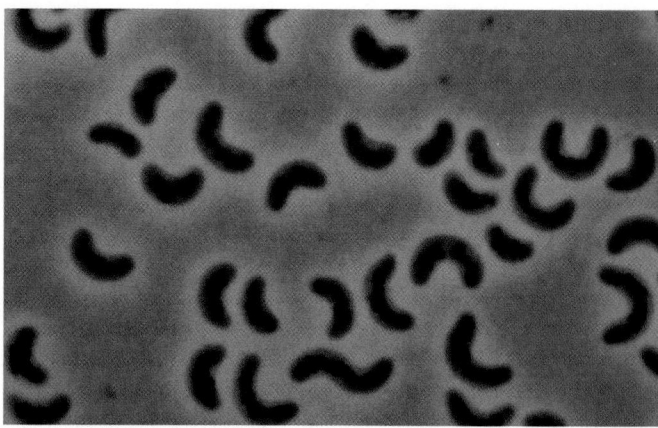

(a) *Desulfovibrio saprovorans*

(b) *Desulfovibrio gigas*

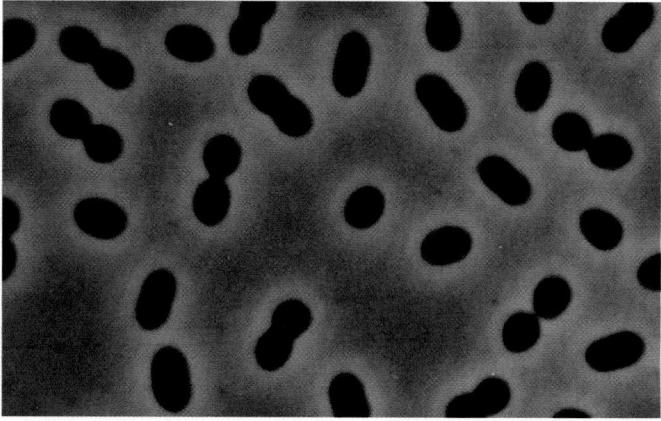

(c) *Desulfobacter postgatei*

Figure 22.31   The Dissimilatory Sulfate- or Sulfur-Reducing Bacteria.   Representative examples. (**a**) Phase-contrast micrograph of *Desulfovibrio saprovorans* with PHB inclusions (×2,000). (**b**) *Desulfovibrio gigas;* phase contrast (×2,000). (**c**) *Desulfobacter postgatei;* phase contrast (×2,000).

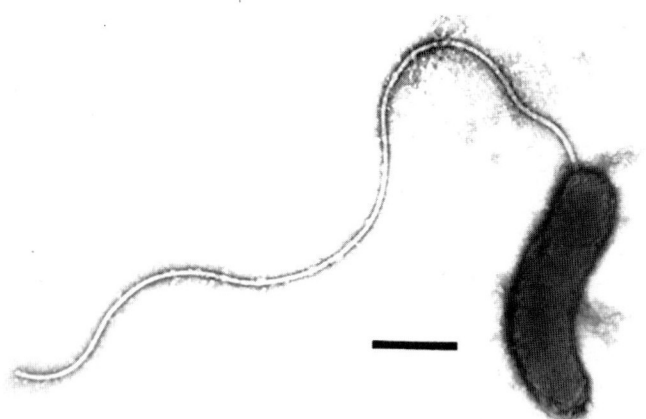

Figure 22.32   *Bdellovibrio* Morphology.   Negatively stained *Bdellovibrio bacteriovorus* with its sheathed polar flagellum. Bar = 0.2 μm.

stituents can leak out of the cell. The growing bacterium uses host amino acids as its carbon, nitrogen, and energy source. It employs fatty acids and nucleotides directly in biosynthesis, thus saving carbon and energy. The bacterium rapidly grows into a long filament under the cell wall and then divides into many smaller, flagellated progeny, which escape upon host cell lysis. Such multiple fission is rare in procaryotes.

The *Bdellovibrio* life cycle resembles that of bacteriophages in many ways. Not surprisingly, if a *Bdellovibrio* culture is plated out on agar with host bacteria, plaques will form in the bacterial lawn. This technique is used to isolate pure strains and count the number of viable organisms just as with phages.

Order *Myxococcales*

The **myxobacteria** are gram-negative, aerobic soil bacteria characterized by gliding motility (*see Microbial Diversity & Ecology 21.1*), a complex life cycle with the production of fruiting bodies, and the formation of dormant myxospores. In addition, their G + C content is around 67 to 71%, significantly higher than that of most gliding bacteria. Myxobacterial cells are rods, about 0.6 to 0.9 by 3 to 8 μm long, and may be either slender with tapered ends or stout with rounded, blunt ends (**figure 22.34**). The order *Myxobacteriales* is divided into five families based on the shape of vegetative cells, myxospores, and sporangia.

Most myxobacteria are micropredators or scavengers. They secrete an array of digestive enzymes that lyse bacteria and yeasts. Many myxobacteria also secrete antibiotics, which may kill their prey. The digestion products, primarily small peptides, are absorbed. Most myxobacteria use amino acids as their major source of carbon, nitrogen, and energy. All are chemoheterotrophs with respiratory metabolism.

The myxobacterial life cycle is quite distinctive and in many ways resembles that of the cellular slime molds (**figure 22.35**). In the presence of a food supply, myxobacteria migrate along a solid surface, feeding and leaving slime trails. During this stage the cells often form a swarm and move in a coordi-

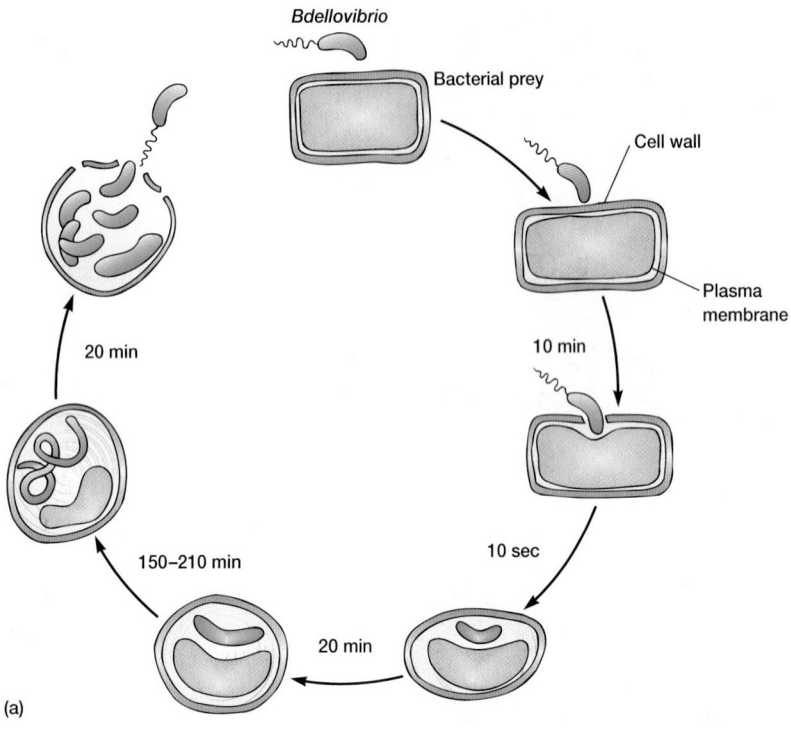

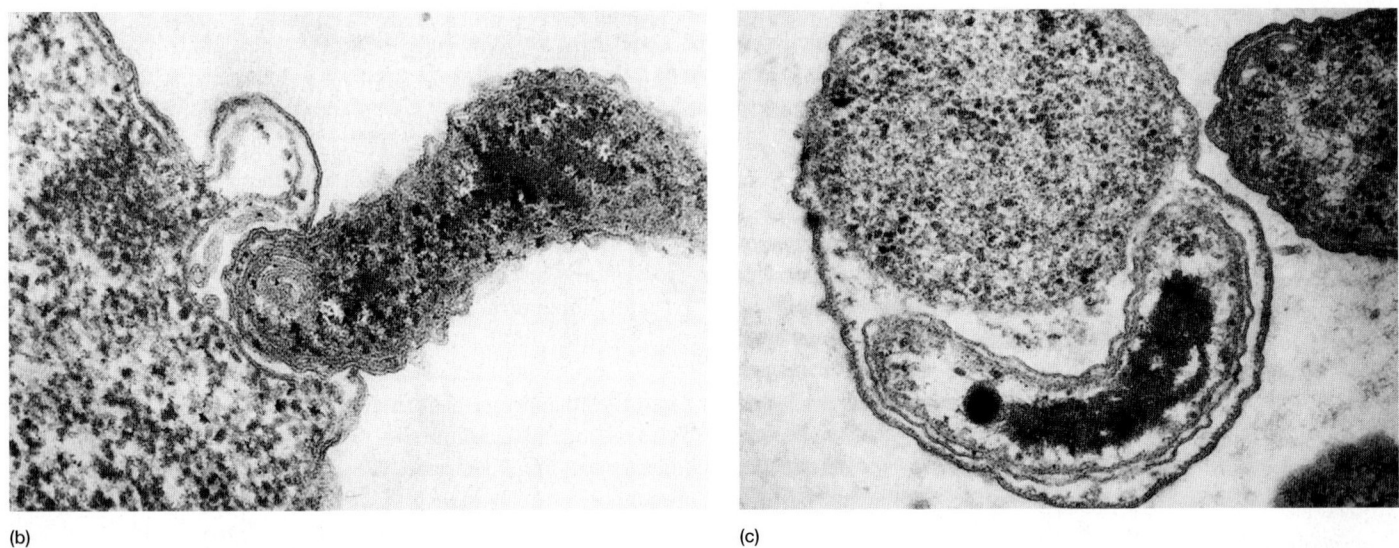

Figure 22.33   The Life Cycle of *Bdellovibrio.*   (a) A general diagram showing the complete life cycle (see text for details). **(b)** *Bdellovibrio bacteriovorus* penetrating the cell wall of *E. coli* (×55,000). **(c)** A *Bdellovibrio* encapsulated between the cell wall and plasma membrane of *E. coli* (×60,800).

nated fashion. Some species congregate to produce a sheet of cells that moves rhythmically to generate waves or ripples. When their nutrient supply is exhausted, the myxobacteria aggregate and differentiate into a **fruiting body.** This is a complex developmental process triggered by starvation and at least five different signals. Two of these signals have been characterized. Both the A factor, a mixture of peptides and amino

acids, and the protein C factor are released and help trigger the process. At least 15 new proteins are synthesized during fruiting body formation. Fruiting bodies range in height from 50 to 500 μm and often are attractively colored red, yellow, or brown by carotenoid pigments. They vary in complexity from simple globular objects made of about 100,000 cells (*Myxococcus*) to the elaborate, branching, treelike structures formed by *Stig-*

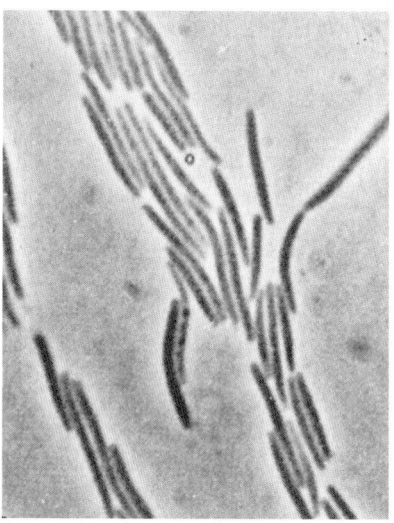
(a)

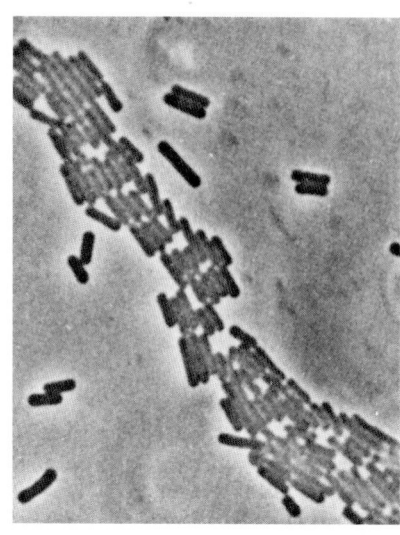
(b)

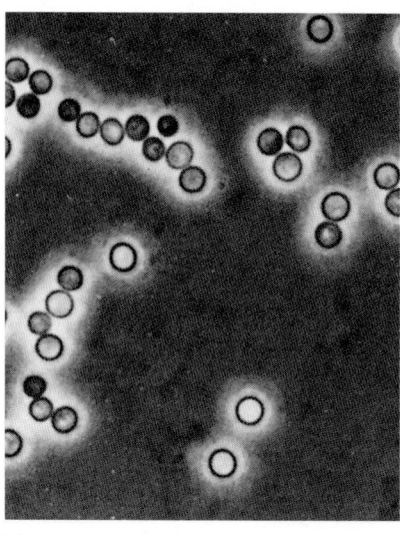
(c)

Figure 22.34 Gliding, Fruiting Bacteria (Myxobacteria). Myxobacterial cells and myxospores. (**a**) *Stigmatella aurantiaca* (×1,200). (**b**) *Chondromyces crocatus* (×950). (**c**) Myxospores of *Myxococcus xanthus* (×1,100). All photographs taken with a phase-contrast microscope.

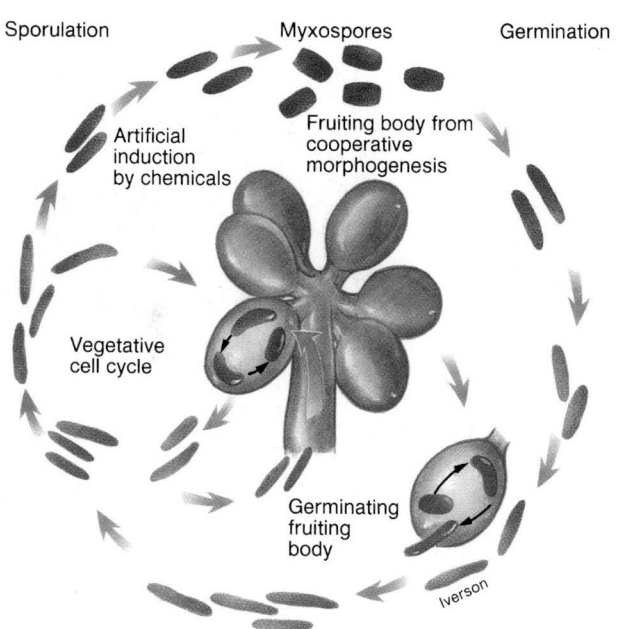

Figure 22.35 Myxobacterial Life Cycle. The outermost sequence depicts the chemical induction of myxospore formation followed by germination. Regular fruiting body production and myxospore germination are also shown.

matella and *Chondromyces* (**figure 22.36**). Some cells develop into dormant **myxospores** that frequently are enclosed in walled structures called sporangioles or sporangia. Each species forms a characteristic fruiting body.

Myxospores are not only dormant but desiccation-resistant, and they may survive up to 10 years under adverse conditions.

They enable myxobacteria to survive long periods of dryness and nutrient deprivation. The use of fruiting bodies provides further protection for the myxospores and assists in their dispersal. (The myxospores often are suspended above the soil surface.) Because myxospores are kept together within the fruiting body, a colony of myxobacteria automatically develops when the myxospores are released and germinate. This communal organization may be advantageous because myxobacteria obtain nutrients by secreting hydrolytic enzymes and absorbing soluble digestive products. A mass of myxobacteria can produce enzyme concentrations sufficient to digest their prey more easily than can an individual cell. Extracellular enzymes diffuse away from their source, and an individual cell will have more difficulty overcoming diffusional losses than a swarm of cells.

Myxobacteria are found in soils worldwide. They are most commonly isolated from neutral soils or decaying plant material such as leaves and tree bark, and from animal dung. Although they grow in habitats as diverse as tropical rain forests and the arctic tundra, they are most abundant in warm areas.

1. Briefly characterize the δ-proteobacteria.
2. Describe the metabolic specialization of the dissimilatory sulfate- or sulfur-reducing bacteria. Why are they important?
3. Characterize the genus *Bdellovibrio* and outline its life cycle in detail.
4. Give the major distinguishing characteristics of the myxobacteria. How do they obtain most of their nutrients?
5. Briefly describe the myxobacterial life cycle. What are fruiting bodies, myxospores, and sporangioles?

Figure 22.36 Myxobacterial Fruiting Bodies.
(**a**) An illustration of typical fruiting body structure.
(**b**) *Myxococcus fulvus.* Fruiting bodies are about
150–400 μm high. (**c**) *Myxococcus stipitatus.* The
stalk is as tall as 200 μm. (**d**) *Chondromyces crocatus* viewed with the SEM. The stalk may reach
700 μm or more in height.

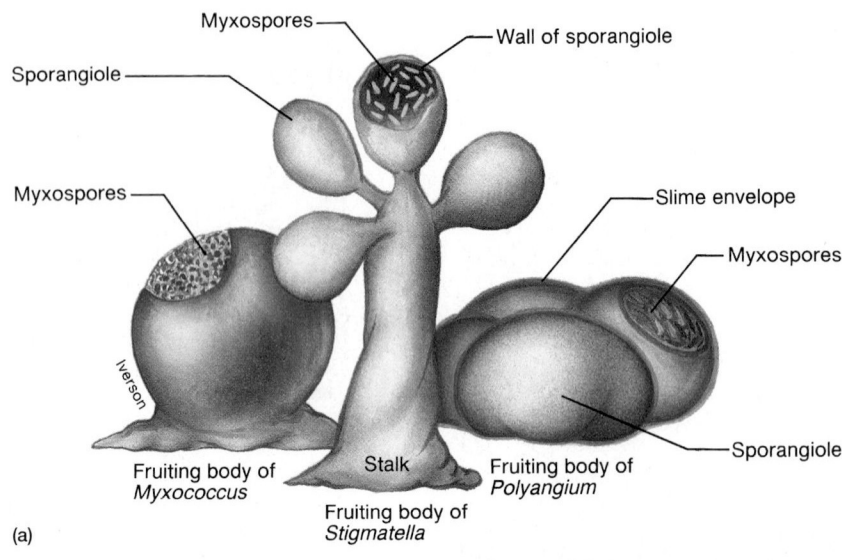

(a)

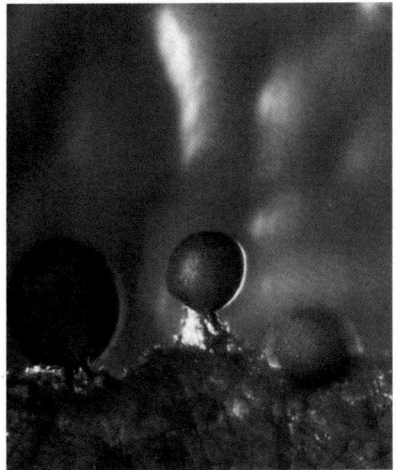 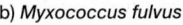

(b) *Myxococcus fulvus*

(c) *Myxococcus stipitatus*

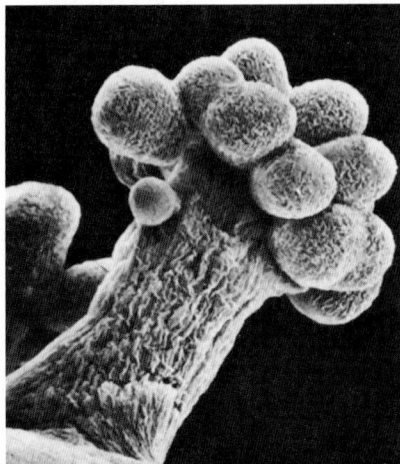

(d) *Chondromyces crocatus*

22.5 CLASS *EPSILONPROTEOBACTERIA*

The **ε-proteobacteria** are the smallest of the five proteobacterial classes. They all are slender gram-negative rods, which can
be straight, curved, or helical. The ε-proteobacteria have one
order, *Campylobacterales,* and three families. The two most
important genera, *Campylobacter* and *Helicobacter,* are microaerophilic, motile, helical or vibrioid, gram-negative rods.
Table 22.8 summarizes some of the characteristics of these two
genera.

The genus *Campylobacter* contains both nonpathogens and
species pathogenic for humans and other animals. *C. fetus* causes
reproductive disease and abortions in cattle and sheep. It is associated with a variety of conditions in humans ranging from **septicemia** (pathogens or their toxins in the blood) to **enteritis**
(inflammation of the intestinal tract). *C. jejuni* causes abortion in
sheep and enteritis diarrhea in humans.

There are at least 23 species of *Helicobacter,* all isolated from
the stomachs and upper intestines of humans, dogs, cats, and
other mammals (*see figure 39.16*). In developing countries 70 to
90% of the population is infected; developed countries range
from 25 to 50%. Most infections are probably acquired during
childhood, but the precise mode of transmission is unclear. The
major human pathogen is *Helicobacter pylori,* which is the cause
of gastritis and peptic ulcer disease (*see pp. 895–96*). *H. pylori*
produces large quantities of urease, and urea hydrolysis appears
to be associated with its virulence.

The genomes of *Campylobacter jejuni* and *Helicobacter pylori* (both about 1.6 million base pairs in size) have been sequenced. They are now being studied and compared in order to
understand the life styles and pathogenicity of these bacteria.

---

1. Briefly describe the properties of the ε-proteobacteria.
2. Give the general characteristics of *Campylobacter* and *Helicobacter.* What is their public health significance?

22.1 Class *Alphaproteobacteria*

The proteobacteria are the largest and most diverse group of bacteria. On the basis of rRNA sequence data, they are divided into five classes: the α-, β-, γ-, δ-, and ε-proteobacteria.

a. The purple nonsulfur bacteria can grow anaerobically as photoorganoheterotrophs and often aerobically as chemoorganoheterotrophs (**figure 22.2**). They are found in aquatic habitats with abundant organic matter and low sulfide levels.

b. Rickettsias are obligately intracellular parasites responsible for many diseases (**figure 22.3**). They have plasma membrane carriers and make extensive use of host cell nutrients, coenzymes, and ATP.

c. Many bacteria have prosthecae, stalks, or reproduction by budding. These are often placed among the α-proteobacteria.

d. Two examples of budding and/or appendaged bacteria are *Hyphomicrobium* (budding bacteria that produce swarmer cells) and *Caulobacter* (bacteria with prosthecae and holdfasts) (**figures 22.4–22.7**).

e. *Rhizobium* carries out nitrogen fixation, whereas *Agrobacterium* causes the development of plant tumors. Both are in the family *Rhizobiaceae* of the α-proteobacteria.

f. Chemolithotrophic bacteria derive energy and electrons from reduced inorganic compounds. Nitrifying bacteria are aerobes that oxidize either ammonia or nitrite to nitrate and are responsible for nitrification (**table 22.2**).

22.2 Class *Betaproteobacteria*

a. The genus *Neisseria* contains nonmotile, aerobic, gram-negative cocci that usually occur in pairs. They colonize mucous membranes and cause several human diseases.

b. Members of the genus *Burkholderia* are aerobic gram-negative rods that almost always have polar flagella. They degrade a wide variety of organic molecules and can cause disease.

c. *Sphaerotilus*, *Leptothrix*, and several other genera have sheaths, hollow tubelike structures that surround chains of cells without being in intimate contact (**figure 22.13**).

d. The colorless sulfur bacteria such as *Thiobacillus* oxidize elemental sulfur, hydrogen sulfide, and thiosulfate to sulfate while generating energy chemolithotrophically.

22.3 Class *Gammaproteobacteria*

a. The γ-proteobacteria are the largest subgroup of proteobacteria with great variety in physiological types (**table 22.5** and **figure 22.16**).

b. The purple sulfur bacteria are anaerobes and usually photolithoautotrophs. They oxidize hydrogen sulfide to sulfur and deposit the granules internally (**figure 22.18**).

c. Bacteria like *Beggiatoa* and *Leucothrix* grow in long filaments or trichomes (**figures 22.20** and **22.21**). Both genera have gliding motility. *Beggiatoa* is primarily a chemolithotroph and *Leucothrix*, a chemoorganotroph.

d. The *Methylococcaceae* are methylotrophs; they use methane, methanol, and other reduced one-carbon compounds as their sole carbon and energy sources.

e. The genus *Pseudomonas* contains straight or slightly curved, gram-negative, aerobic rods that are motile by one or several polar flagella and do not have prosthecae or sheaths (**figure 22.23**).

f. The pseudomonads participate in natural mineralization processes, are major experimental subjects, cause many diseases, and often spoil refrigerated food.

g. The most important facultatively anaerobic, gram-negative rods are found in three families: *Vibrionaceae*, *Enterobacteriaceae*, and *Pasteurellaceae* (**table 22.6**). These bacteria are in the γ-proteobacteria (orders *Vibrionales*, *Enterobacteriales*, and *Pasteurellales*).

h. The *Enterobacteriaceae*, often called enterobacteria or enteric bacteria, are gram-negative, peritrichously flagellated or nonmotile, facultatively anaerobic, straight rods with simple nutritional requirements.

i. The enteric bacteria are usually identified by a variety of physiological tests and are very important experimental organisms and pathogens of plants and animals (**table 22.7** and **figure 22.28**).

22.4 Class *Deltaproteobacteria*

a. The δ-proteobacteria contain gram-negative bacteria that are anaerobic and can use elemental sulfur and oxidized sulfur compounds as electron acceptors in anaerobic respiration (**table 22.8**). They are very important in sulfur cycling in the ecosystem.

b. *Bdellovibrio* is an aerobic curved rod with sheathed polar flagellum that preys on other gram-negative bacteria and grows within their periplasmic space (**figures 22.32** and **22.33**).

c. Myxobacteria are gram-negative, aerobic soil bacteria with gliding motility and a complex life cycle that leads to the production of dormant myxospores held within fruiting bodies (**figures 22.34–22.36**).

22.5 Class *Epsilonproteobacteria*

a. The ε-proteobacteria are the smallest of the proteobacterial classes and contain two important pathogenic genera: *Campylobacter* and *Helicobacter*. These are microaerophilic, motile, helical or vibrioid, gram-negative rods.

Key Terms

α-proteobacteria 474	enteric bacteria (enterobacteria) 493	mineralization process 490	proteobacteria 474
β-proteobacteria 482	enteritis 500	myxobacteria 497	purple nonsulfur bacteria 475
binary fission 476	ε-proteobacteria 500	myxospores 499	purple sulfur bacteria 485
bioluminescence 492	fruiting body 498	nitrification 480	septicemia 500
budding 476	γ-proteobacteria 485	nitrifying bacteria 480	sheath 482
colorless sulfur bacteria 483	holdfast 478	prostheca 476	stalk 476
δ-proteobacteria 495	methylotroph 477		

Questions for Thought and Review

1. Describe the general characteristics and organization of each of the five classes of proteobacteria.

2. Bacterial genera are distributed quite differently in the first and second editions of *Bergey's Manual*. Discuss the advantages and disadvantages of the phenetic and phylogenetic approaches to bacterial taxonomy shown by the first and second editions.

3. A number of bacterial groups are found in particular habitats to which they are well adapted. Give the habitat for each of the following groups and discuss the reasons for its preference: *Beggiatoa,* purple sulfur bacteria, *Hyphomicrobium,* and *Thiobacillus.*

4. Why might the ability to form dormant cysts be of great advantage to *Agrobacterium* but not as much to *Rhizobium?*

5. What advantage might a species gain by specializing in its nutrient requirements or habitat as methylotrophs do? Are there disadvantages to this strategy?

6. A pond with black water and a smell like rotten eggs might contain an excess of what kind of bacteria?

7. Why are biochemical tests of such importance in identifying enteric bacteria?

8. How does *Bdellovibrio* take advantage of its host's metabolism to grow rapidly and effi-

ciently? In what ways does its reproduction resemble that of bacteriophages?

9. Why are gliding and budding and/or appendaged bacteria distributed among so many different sections in *Bergey's Manual?*

10. Discuss the ways in which the distinctive myxobacterial life cycle may be of great advantage to these organisms.

Critical Thinking Questions

1. *Helicobacter pylori* produces large quantities of urease. Urease catalyzes the reaction:

$$H_2N-\overset{\overset{\textstyle O}{\|}}{C}-NH_2 \rightarrow CO_2 + 2NH_3$$

Suggest why this allows *H. pylori* to inhabit the acidity of the gastric mucosa.

2. Methylotrophs oxidize methane to methanol, then to formaldehyde, and finally into acetate. Suggest mechanisms by which the bacterium protects itself from the toxic effects of the intermediates, methanol and formaldehyde.

3. *Bdellovibrio* is an intracellular predator. Once it invades the periplasm, it manages to inhibit many aspects of host metabolism. Suggest a mechanism by which this inhibition could occur so rapidly.

For recommended readings on these and related topics, see Appendix V.

chapter 23

Bacteria:

The Low G + C Gram Positives

Concepts

- Volume 2 of the first edition of *Bergey's Manual* contains six sections covering all gram-positive bacteria except the actinomycetes. Bacteria are distributed among these sections on the basis of their shape, the ability to form endospores, acid fastness, oxygen relationships, the ability to temporarily form mycelia, and other properties.
- The second edition of *Bergey's Manual* groups the gram-positive bacteria phylogenetically into two major groups: the low G + C gram-positive bacteria and the high G + C gram-positive bacteria. The new classification is based primarily on nucleic acid sequences rather than phenotypic similarity.
- The low G + C gram positives contain (1) clostridia and relatives, (2) the mycoplasmas, and (3) the bacilli and lactobacilli. Endospore formers, cocci, and rods are found among the clostridia and bacilli groups rather than being placed in separate sections as in the first edition. Thus common possession of a complex structure such as an endospore does not necessarily indicate close relatedness between the genera.
- Peptidoglycan structure varies among different groups in ways that are often useful in identifying specific groups.
- Although most gram-positive bacteria are harmless free-living saprophytes, some species from most major groups are pathogens of humans, other animals, and plants. Other gram-positive bacteria are very important in the food and dairy industries.

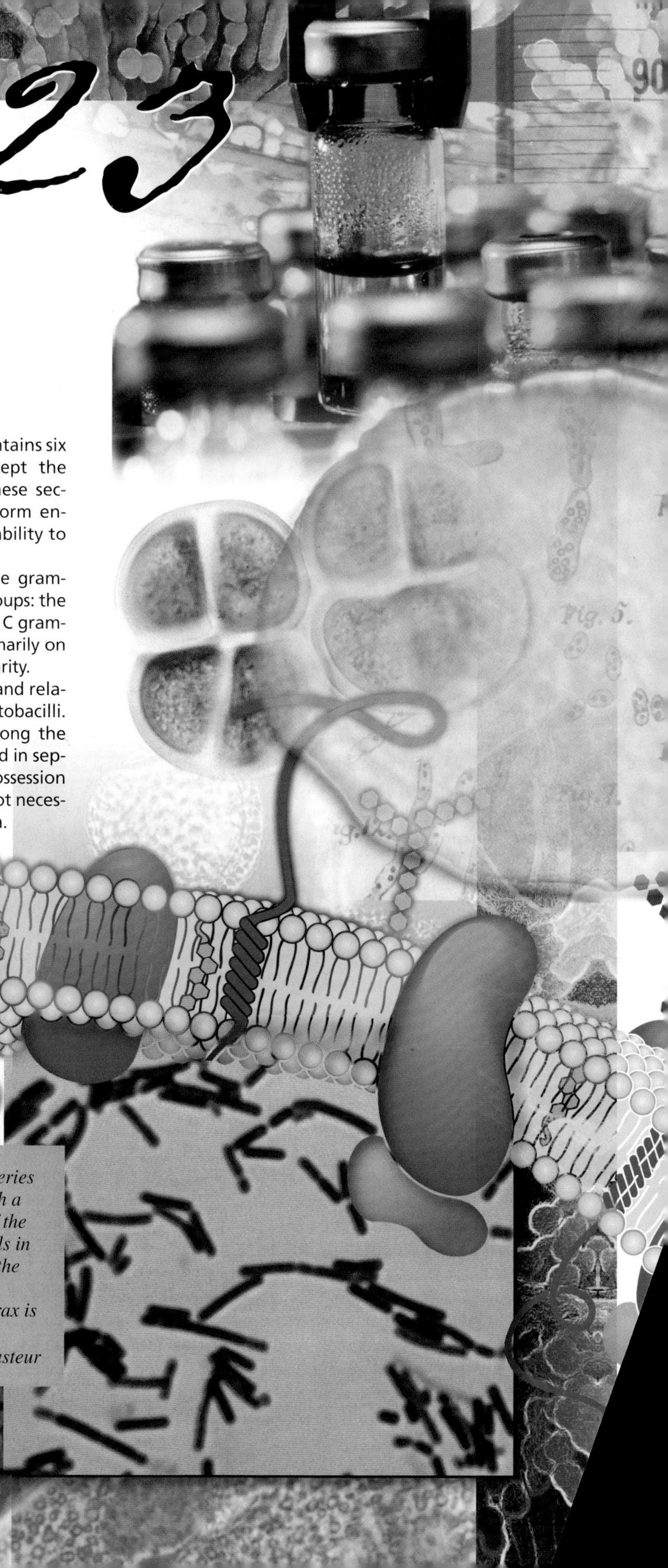

We noted, after having grown the bacterium through a series of such cultures, each fresh culture being inoculated with a droplet from the previous culture, that the last culture of the series was able to multiply and act in the body of animals in such a way that the animals developed anthrax with all the symptoms typical of this affection.

Such is the proof, which we consider flawless, that anthrax is caused by this bacterium.

—Louis Pasteur

Lactobacilli are indispensable to the food and dairy industry. They are not considered pathogens.

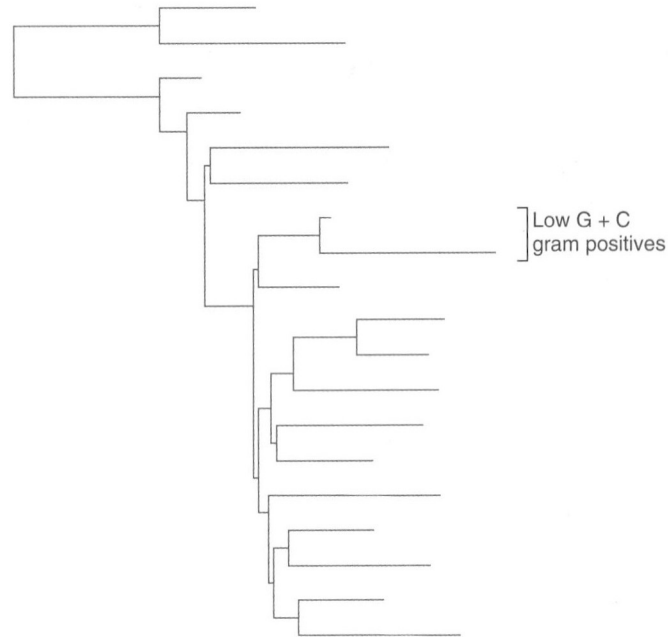

Low G + C
gram positives

Analysis of the phylogenetic relationships within the gram-positive bacteria by comparison of 16S rRNA sequences (*see figure 19.14*) shows that they are divided into a low G + C group (see bottom dendrogram) and high G + C or actinobacterial group. The distribution of genera within and between the groups shown in figure 19.14 differs markedly from the classification in the first edition of *Bergey's Manual*.  Bacterial phylogeny and diversity (pp. 431–34); The structure of the gram-positive cell wall (pp. 55–56)

The third volume of the second edition of *Bergey's Manual* describes the low G + C gram-positive bacteria. These are placed in the phylum *Firmicutes* and divided into three classes: *Clostridia, Mollicutes,* and *Bacilli.* The phylum *Firmicutes* is large and complex; it has 10 orders and 34 families. It differs most obviously from the classification system of the first edition in containing the class *Mollicutes.* The mycoplasmas, class *Mollicutes,* are now placed with the low G + C gram positives rather than with gram-negative bacteria. Ribosomal RNA data indicate that the mycoplasmas are closely related to the clostridia, despite their lack of cell walls. **Figure 23.1** shows the phylogenetic relationships between some of the bacteria in this chapter.

 hapter 23 surveys many of the bacteria found in volume 2 of the first edition of *Bergey's Manual of Systematic Bacteriology.* Volume 2 contains all the major groups of gram-positive bacteria except the actinomycetes, which are covered in volume 4. The second edition of *Bergey's Manual* will divide the gram-positive bacteria differently for reasons discussed a little later. Thus this chapter describes the general differences between the first and second edition, and then focuses on the mycoplasmas, *Clostridium* and its relatives, and the bacilli and lactobacilli. The remaining groups in volume 2 will be described in chapter 24 along with the actinomycetes.

23.2 CLASS *MOLLICUTES* (THE MYCOPLASMAS)

The class *Mollicutes* has five orders and six families. The best-studied genera are found in the orders *Mycoplasmatales* (*Mycoplasma, Ureaplasma*), *Entomoplasmatales* (*Entomoplasma, Mesoplasma, Spiroplasma*), *Acholeplasmatales* (*Acholeplasma*), and *Anaeroplasmatales* (*Anaeroplasma, Asteroleplasma*). **Table 23.1** summarizes some of the major characteristics of these genera.

Members of the class *Mollicutes* are commonly called **mycoplasmas.** These bacteria lack cell walls and cannot synthesize peptidoglycan precursors. Thus they are penicillin resistant but susceptible to lysis by osmotic shock and detergent treatment. Because they are bounded only by a plasma membrane, these procaryotes are pleomorphic and vary in shape from spherical or pear-shaped organisms, about 0.3 to 0.8 μm in diameter, to branched or helical filaments (**figure 23.2**). Some mycoplasmas (e.g., *M. genitalium*) have a specialized terminal structure that projects from the cell and gives them a flask or pear shape. This structure aids in attachment to eucaryotic cells. They are the smallest bacteria capable of self-reproduction. Although most are nonmotile, some can glide along liquid-covered surfaces. Most species differ from the vast majority of bacteria in requiring sterols for growth. They usually are facultative anaerobes, but a few are obligate anaerobes. When growing on agar, most species will form colonies with a "fried-egg" appearance because they grow into the agar surface at the center while spreading outward on the surface at the colony edges (**figure 23.3**). Their genome is one of the smallest found in procaryotes, about 5 to 10×10^8 daltons; the G + C content ranges from 23 to 41%. Recently the complete genome of *Mycoplasma genitalium,* a parasite of the human genital and respiratory tracts, has been sequenced. The

23.1 GENERAL INTRODUCTION

Volume 2 of the first edition of *Bergey's Manual* contains six sections that describe all gram-positive bacteria except the actinomycetes. Most of these bacteria are distributed among the first four sections on the basis of their general shape (whether they are ~ls, cocci, or irregular) and their ability to form endospores. The ~haped, acid-fast bacteria known as mycobacteria are placed ~arate section. The last section describes the nocardio- ~se are gram-positive filamentous bacteria that form a ~twork called a mycelium, which can break into rod- ~id forms.

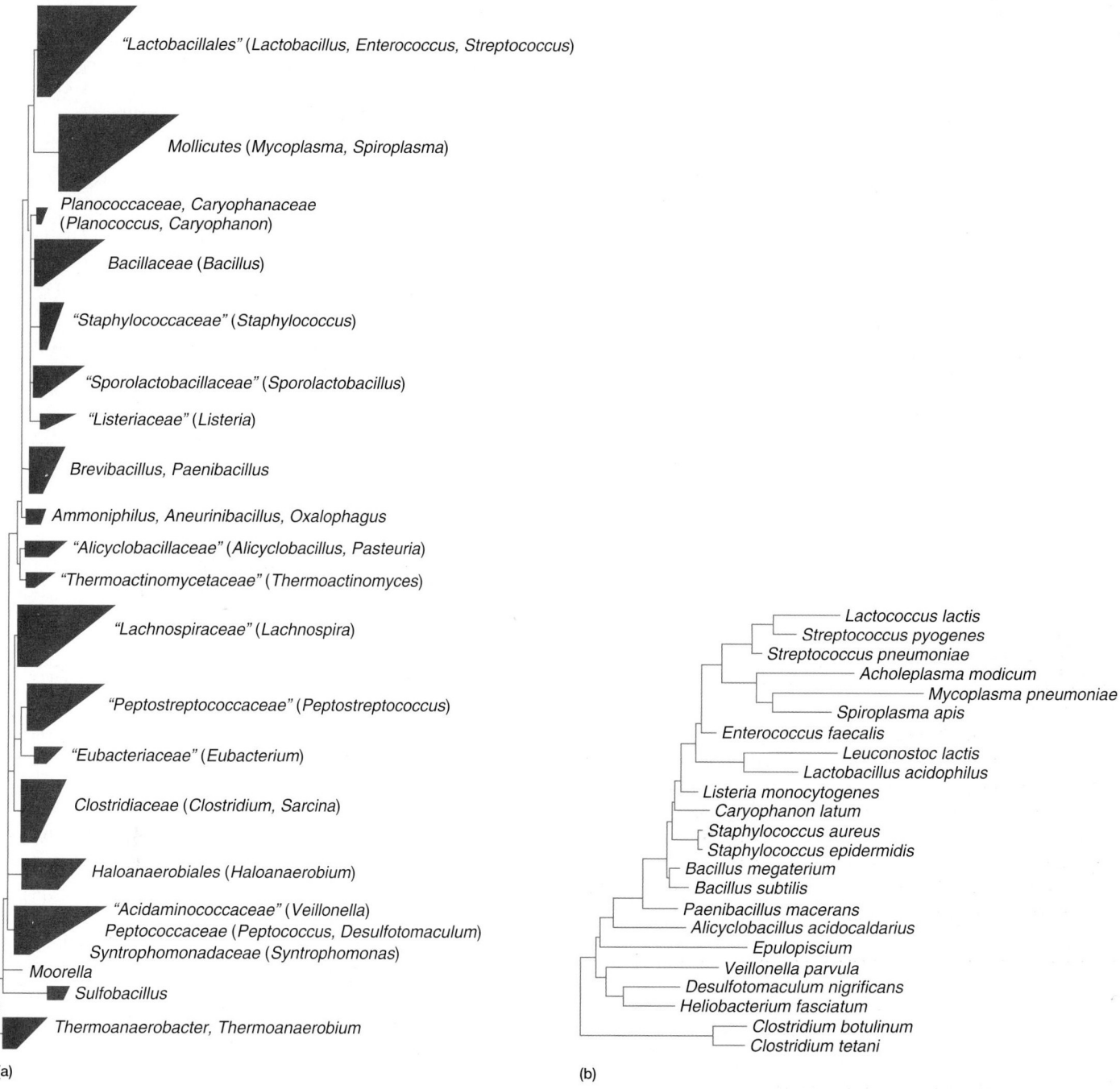

(a)

(b)

Figure 23.1   Phylogenetic Relationships in the Phylum *Firmicutes* (Low G + C Gram Positives).   (**a**) The major phylogenetic groups with representative genera in parentheses. Each tetrahedron in the tree represents a group of related organisms; its horizontal edges show the shortest and longest branches in the group. Multiple branching at the same level indicates that the relative branching order of the groups cannot be determined from the data. The quotation marks around some names indicate that they are not formally approved taxonomic names. (**b**) The relationships of a few species based on 16S rRNA sequence date.   *Source: The Ribosomal Database Project.*

Table 23.1

Properties of Some Members of the Class *Mollicutes*

Genus	No. of Recognized Species	G + C Content (mol%)	Genome Size (kpb)	Cholesterol Requirement	Habitat	Other Distinctive Features
Acholeplasma	13	26–36	1,500–1,650	No	Animals, some plants and insects	Optimum growth at 30–37°C
Anaeroplasma	4	29–34	1,500–1,600	Yes	Bovine or ovine rumen	Oxygen-sensitive anaerobes
Asteroleplasma	1	40	1,500	No	Bovine or ovine rumen	Oxygen-sensitive anaerobes
Entomoplasma	5	27–29	790–1,140	Yes	Insects, plants	Optimum growth, 30°C
Mesoplasma	12	27–30	870–1,100	No	Insects, plants	Optimum growth, 30°C; sustained growth in serum-free medium only with 0.04% Tween 80
Mycoplasma	104	23–40	600–1,350	Yes	Humans, animals	Optimum growth usually at 37°C
Spiroplasma	22	25–30	940–2,200	Yes	Insects, plants	Helical filaments; optimum growth at 30–37°C
Ureaplasma	6	27–30	760–1,170	Yes	Humans, animals	Urea hydrolysis

Adapted from J. G. Tully, et al., "Revised Taxonomy of the Class *Mollicutes*" in *International Journal of Systematic Bacteriology*, 43(2):378–85. Copyright © 1993 American Society for Microbiology, Washington, D.C. Reprinted by permission.

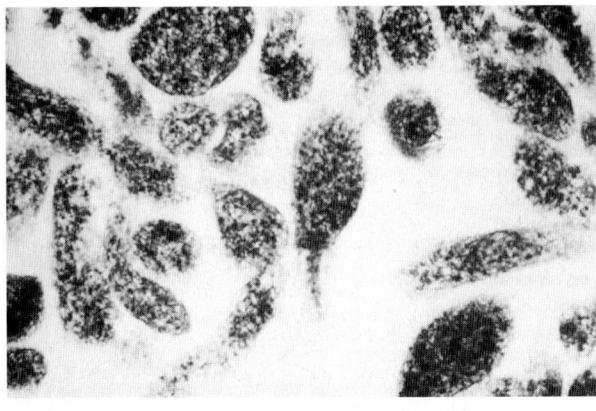

(a)

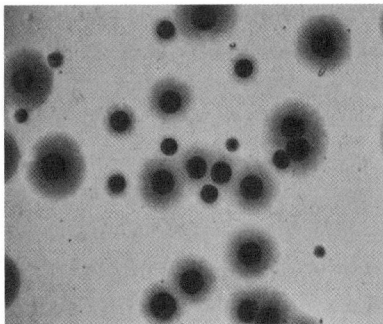

Figure 23.3 *Mycoplasma* **Colonies.**   Note the "fried-egg" appearance, colonies stained before viewing (×100).

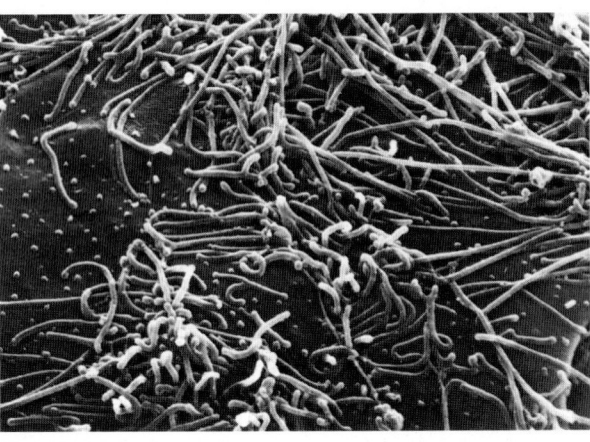

(b)

Figure 23.2   The Mycoplasmas.   Electron micrographs of *Mycoplasma pneumoniae* showing its pleomorphic nature. (**a**) A transmission electron micrograph of several cells (×47,880). The central cell appears flask or pear-shaped because of its terminal structure. (**b**) A scanning electron micrograph (×26,000).

M. genitalium genome is only 580 kilobases long and appears to have 482 genes; it seems that not many genes are required to sustain a free-living existence. Mycoplasmas can be saprophytes, commensals, or parasites, and many are pathogens of plants, animals, or insects.   *Mycoplasma genitalium* genome sequence (pp. 339–40)

The metabolism of mycoplasmas is not particularly unusual, although they are deficient in several biosynthetic sequences and often require sterols, fatty acids, vitamins, amino acids, purines, and pyrimidines. Those mycoplasmas needing sterols incorporate them into the plasma membrane. Mycoplasmas are usually more osmotically stable than bacterial protoplasts, and their membrane sterols may be a stabilizing factor. Some produce ATP by the Embden-Meyerhof pathway and lactic acid fermentation. Others catabolize arginine or urea to generate ATP. The pentose phosphate pathway seems functional in at least some mycoplasmas; none appear to have the complete tricarboxylic acid cycle.

Mycoplasmas are remarkably widespread and can be isolated from animals, plants, the soil, and even compost piles. Indeed, about 10% of the mammalian cell cultures in use are probably contaminated with mycoplasmas, which seriously interfere with tissue culture experiments and are difficult to detect and eliminate. In animals, mycoplasmas colonize mucous membranes and joints and often are associated with diseases of the respiratory and urogenital tracts. Mycoplasmas cause several major diseases in livestock, for example, contagious bovine pleuropneumonia in cattle (*M. mycoides*), chronic respiratory disease in chickens (*M. gallisepticum*), and pneumonia in swine (*M. hyopneumoniae*). *M. pneumoniae* causes primary atypical pneumonia in humans, and there is increasing evidence that *M. hominis* and *Ureaplasma urealyticum* also are human pathogens. The 0.8 million base pair genome of *M. pneumoniae* has been sequenced. This pathogen has lost many biosynthetic pathways (e.g., amino acid biosynthesis pathways), which accounts for its parasitic lifestyle. Spiroplasmas have been isolated from insects, ticks, and a variety of plants. They cause disease in citrus plants, cabbage, broccoli, corn, honey bees, and other hosts. Arthropods probably often act as vectors and carry the spiroplasmas between plants. Presumably many more pathogenic mollicutes will be discovered as techniques for their isolation and study improve.

---

1. What morphological feature distinguishes the mycoplasmas? In what class are they found? Why have they been placed with the low G + C gram-positive bacteria?
2. Give other distinguishing properties of the class *Mollicutes.*
3. What might mycoplasmas use sterols for?
4. Where are mycoplasmas found in animals? List several animal and human diseases caused by them. What kinds of organisms do spiroplasmas usually infect?

 ### 23.3 LOW G + C GRAM-POSITIVE BACTERIA IN *BERGEY'S MANUAL*

Because low G + C gram positives are treated so differently in the first and second editions of *Bergey's Manual,* it is necessary to compare the taxonomic approach used in the two editions before looking at individual groups of gram-positive bacteria. As we shall see, the fundamental difference in treatment is that of phenotypic versus phylogenetic approaches.

The first edition of *Bergey's Manual,* classifies the low G + C gram-positive bacteria largely on the basis of observable characteristics such as cell shape, the clustering and arrangement of cells, the presence or absence of endospores, oxygen relationships, fermentation patterns, peptidoglycan chemistry, and so forth. Because of the importance of peptidoglycan and endospores in the taxonomy of these bacteria, it is best to briefly discuss these two important components of gram-positive bacteria.

Peptidoglycan structure varies considerably among different gram-positive groups. Most gram-negative bacteria have a peptidoglycan structure in which *meso*-diaminopimelic acid in position 3 is directly linked through its free amino group with the free carboxyl of the terminal D-alanine of an adjacent peptide chain (**figure 23.4a;** *see also figure 3.18*). This same peptidoglycan

structure is present in many gram-positive genera, for example, *Bacillus, Clostridium, Lactobacillus, Corynebacterium, Mycobacterium,* and *Nocardia.* In other gram-positive bacteria, lysine is substituted for diaminopimelic acid in position 3, and the peptide subunits of the glycan chains are cross-linked by interpeptide bridges containing monocarboxylic L-amino acids or glycine, or both (figure 23.4b). Many genera including *Staphylococcus, Streptococcus, Micrococcus, Lactobacillus,* and *Leuconostoc* have this type of peptidoglycan. The genus *Streptomyces* and several other actinobacterial genera have replaced *meso*-diaminopimelic acid with L,L-diaminopimelic acid in position 3 and have one glycine residue as the interpeptide bridge. The plant pathogenic corynebacteria provide another example of peptidoglycan variation. In some of these bacteria, the interpeptide bridge connects positions 2 and 4 of the peptide subunits rather than 3 and 4 (figure 23.4c). Because the interpeptide bridge connects the carboxyl groups of glutamic acid and alanine, a diamino acid such as ornithine is used in the bridge. Many other variations in peptidoglycan structure are found, including other interbridge structures and large differences in the frequency of cross-linking between glycan chains. Bacilli and most gram-negative bacteria have fewer cross-links between chains than do gram-positive bacteria such as *Staphylococcus aureus* in which almost every muramic acid is cross-linked to another. These structural variants are often characteristic of particular groups and are therefore taxonomically useful.

Bacterial endospores have a complex structure with a spore coat, cortex, and inner spore membrane surrounding the protoplast (**figure 23.5**). They contain dipicolinic acid, are very heat resistant, and can remain dormant and viable for very long periods (**Microbial Tidbits 23.1**). In one well-documented experiment, endospores remained viable for about 70 years. A recent research article reports that viable endospores have been recovered from Dominican bees that were encased in 25- to 40-million-year-old amber. If this result is confirmed, endospores from an ancestor of *Bacillus sphaericus* have survived for more than 25 million years! Similar reports have been made subsequently; all these studies will have to be reconfirmed. Usually endospores are observed either in the light microscope after spore staining or by phase-contrast microscopy of unstained cells (*see sections 2.2 and 2.3*). They also can be detected by heating a culture at 70 to 80°C for 10 minutes followed by incubation in the proper growth medium. Because only endospores and some thermophiles would survive such heating, bacterial growth tentatively confirms their presence. Bacterial endospore properties and structure (pp. 68–71)

Although endospore-forming bacteria are distributed widely, they are primarily soil inhabitants. Soil conditions are often extremely variable, and endospores are an obvious advantage in surviving periods of dryness or nutrient deprivation.

The second edition takes a phylogenetic approach primarily based on 16S rRNA data rather than phenotypic similarity. The traditional low G + C gram-positive bacteria are divided into two classes: *Clostridia* (the clostridia and relatives), and *Bacilli* (the bacilli and lactobacilli) (figure 23.1). Each class has endospore-forming bacteria and both rods and cocci. Thus the

Figure 23.4 Representative Examples of Peptidoglycan Structure. (a) The peptidoglycan with a direct cross-linkage between positions 3 and 4 of the peptide subunits, which is present in most gram-negative and many gram-positive bacteria. (b) Peptidoglycan with lysine in position 3 and an interpeptide bridge. The bracket contains six typical bridges: (**1**) *Staphylococcus aureus*, (**2**) *S. epidermidis*, (**3**) *Micrococcus roseus* and *Streptococcus thermophilus*, (**4**) *Lactobacillus viridescens*, (**5**) *Streptococcus salvarius*, and (**6**) *Leuconostoc cremoris*. The arrows indicate the polarity of peptide bonds running in the C to N direction. (**c**) An example of the cross-bridge extending between positions 2 and 4 from *Corynebacterium poinsettiae*. The interbridge contains an L-diamino acid like ornithine, and L-homoserine (L-Hsr) is in position 3. The abbreviations and structures of amino acids in the figure are found in appendix I. See text for more detail.

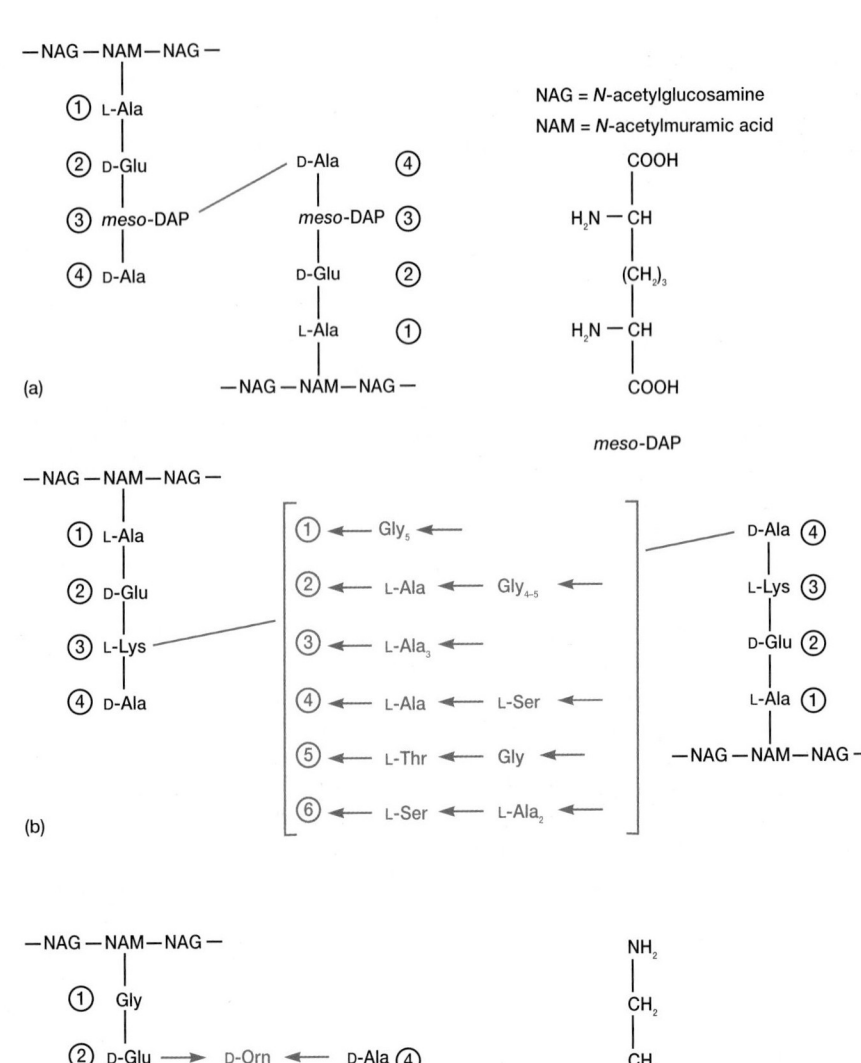

new groupings of gram-positive bacteria do not at all resemble those in the first edition. Phylogenetically coherent groups do not necessarily contain bacteria that closely resemble one another phenotypically. The following survey of low G + C gram-positive bacteria will follow the general organization of the second edition.

 23.4 CLASS *CLOSTRIDIA*

The class *Clostridia* has a very wide variety of gram-positive bacteria distributed into three orders and 11 families. The characteristics of some of the more important genera are summarized in **table 23.2.** Phylogenetic relationships are shown in figure 23.1.

Microbial Tidbits
23.1   Spores in Space

During the nineteenth-century argument over the question of the evolution of life, the panspermia hypothesis became popular. According to this hypothesis, life did not evolve from inorganic matter on Earth but arrived as viable bacterial spores that had escaped from another planet. More recently, the British astronomer Fred Hoyle has revived the hypothesis based on his study of the absorption of radiation by interstellar dust. Hoyle maintains that dust grains were initially viable bacterial cells that have been degraded, and that the beginning of life on Earth was due to the arrival of bacterial endospores that had survived their trip through space.

Even more recently Peter Weber and J. Mayo Greenberg from the University of Leiden in the Netherlands have studied the effect of very high vacuum, low temperature, and UV radiation on the survival of *Bacillus subtilis* endospores. Their data suggest that endospores within an interstellar molecular cloud might be able to survive between 4.5 to 45 million years. Molecular clouds move through space at speeds sufficient to transport spores between solar systems in this length of time. Although these results do not prove the panspermia hypothesis, they are consistent with the possibility that bacteria might be able to travel between planets capable of supporting life.

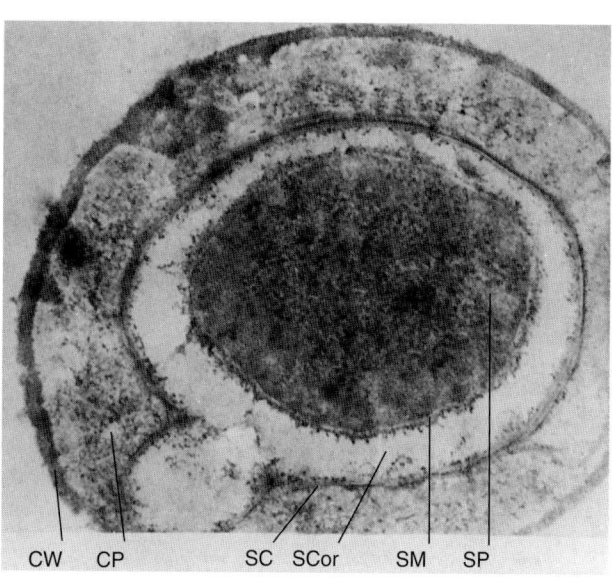

CW   CP   SC SCor   SM   SP

Figure 23.5   Bacterial Endospores.   A cross section of *Bacillus megaterium* and its endospore within the vegetative cell wall, CW; cell protoplast, CP; spore coat, SC; spore cortex, SCor; spore membrane, SM; and spore protoplast, SP (×120,000).

By far the largest genus is *Clostridium.* It includes obligately anaerobic, gram-positive bacteria that form endospores and do not carry out dissimilatory sulfate reduction. The genus contains well over 100 species in several distinct phylogenetic clusters. It is quite likely that the *Clostridium* will be subdivided into several genera in the future.

Members of the genus *Clostridium* have great practical impact. Because they are anaerobic and form heat-resistant endospores, they are responsible for many cases of food spoilage, even in canned foods. *C. botulinum* (**figure 23.6a;** *see also figure*

2.8d) is the causative agent of botulism (*see section 39.4*). Clostridia often can ferment amino acids to produce ATP by oxidizing one amino acid and using another as an electron acceptor in a process called the Stickland reaction (*see figure 9.11*).

This reaction generates ammonia, hydrogen sulfide, fatty acids, and amines during the anaerobic decomposition of proteins. These products are responsible for many unpleasant odors arising during putrefaction. Several clostridia produce toxins and are major disease agents. *C. tetani* (figure 23.6b) is the causative agent of tetanus, and *C. perfringens* (*see figure 2.15*a), of gas gangrene and food poisoning. Clostridia also are industrially valuable; for example, *C. acetobutylicum* is used to manufacture butanol in some countries. Microbiology of food (chapter 41)

In addition to the genus *Clostridium,* this section has several other genera of interest. *Desulfotomaculum* is an anaerobic, endospore-forming genus that reduces sulfate and sulfite to hydrogen sulfide during anaerobic respiration (**figure 23.7**). Although it stains gram negative, electron microscopic studies have shown that *Desulfotomaculum* has a gram-positive type cell wall despite its staining characteristics and really is a member of the low G + C gram positives.

An excellent example of the diversity in this section are the heliobacteria. The heliobacteria, genera *Heliobacterium* and *Heliophilum,* are a group of unusual anaerobic photosynthetic bacteria characterized by the presence of bacteriochlorophyll *g*. They have a photosystem I type reaction center like the green sulfur bacteria, but have no intracytoplasmic photosynthetic membranes; pigments are contained in the plasma membrane. Like *Desulfotomaculum,* they have a gram-positive type cell wall with lower than normal peptidoglycan content, and they stain gram negative. Some heliobacteria form endospores. Photosynthetic bacteria (pp. 456–64)

The genus *Veillonella* provides another good example of the difference between the first and second editions with respect to placement of genera. The second edition of *Bergey's*

Table 23.2

Characteristics of Clostridia and Relatives

Genus	Dimensions (μm) and Morphology	G + C Content (mol%)	Oxygen Relationship	Other Distinctive Characteristics
Clostridium	0.3–2.0 × 1.5–20; rod-shaped, often pleomorphic, nonmotile or peritrichous	22–55	Anaerobic	Does not carry out dissimilatory sulfate reduction; usually chemoorganotrophic, fermentative, and catalase negative; forms oval or spherical endospores
Desulfotomaculum	0.3–1.5 × 3–9; straight or curved rods, peritrichous or polar flagella	37–50	Anaerobic	Reduces sulfate to H₂S, forms subterminal to terminal endospores; stains gram negative but has gram-positive wall, catalase negative
Heliobacterium	1.0 × 4–10; rods that are frequently bent, gliding motility	52–55	Anaerobic	Photoheterotrophic with bacteriochlorophyll *g;* stains gram negative but has gram-positive wall, some form endospores
Veillonella	0.3–0.5; cocci in pairs, short chains, and masses; nonmotile	36–43	Anaerobic	Gram negative; pyruvate and lactate fermented, but not carbohydrates; acetate, propionate, CO₂, and H₂ produced from lactate; parasitic in mouths, intestines, and respiratory tracts of animals

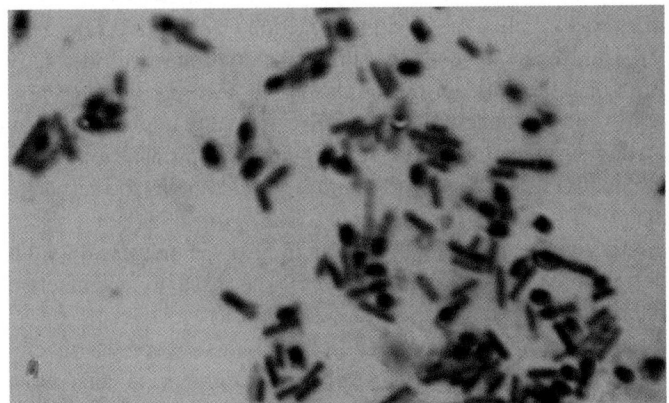

(a) *Clostridium botulinum*

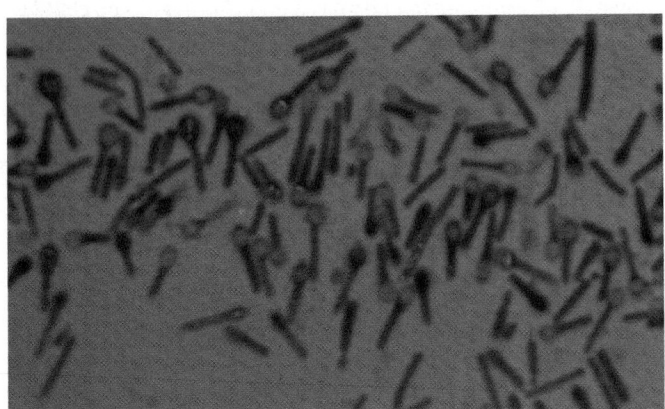

(b) *C. tetani*

Figure 23.6 The Clostridia. (**a**) *C. botulinum,* spores elliptical and subterminal, cells slightly swollen (×1,100). (**b**) *C. tetani,* spores round and terminal (×1,100).

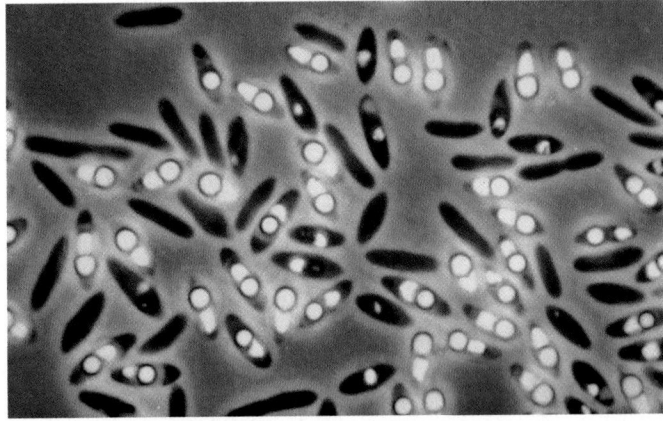

Figure 23.7 *Desulfotomaculum.* *Desulfotomaculum acetoxidans* with spores; phase contrast (×2,000).

Manual locates the genus in the low G + C gram positives. In the first edition *Veillonella* is placed with other anaerobic gram-negative cocci in section 8 and assigned to the family *Veillonellaceae.* The family *Veillonellaceae* contains anaerobic, chemoheterotrophic cocci ranging in diameter from about 0.3 to 2.5 μm. Usually they are diplococci (often with their adjacent sides flattened), but they may exist as single cells, clusters, or chains. All have complex nutritional requirements and ferment substances such as carbohydrates, lactate and other organic acids, and amino acids to produce gas (CO₂ and often H₂) plus a mixture of volatile fatty acids. They are parasites of homeothermic (warm-blooded) animals.

Like many groups of anaerobic bacteria, members of this family have not been thoroughly studied. Some species are part of the normal biota of the mouth, the gastrointestinal tract, and the urogenital tract of humans and other animals. For example, *Veil-*

lonella is plentiful on the tongue surface and dental plaque of humans (*see figure 39.25*) and it can be isolated about 10 to 20% of the time from the vagina. *Veillonella* is unusual in growing well on organic acids such as lactate, pyruvate, and malate while normally being unable to ferment glucose and other carbohydrates. It is well adapted to the oral environment because it can use the lactic acid produced from carbohydrates by the streptococci and other oral bacteria. Gram-negative, anaerobic cocci are found in infections of the head, lungs, and the female genital tract, but their precise role in such infections is unclear.

---

1. Briefly discuss the ways in which low G + C bacteria are treated differently in the first and second editions of *Bergey's Manual.*
2. Describe, in a diagram, the chemical composition and structure of the peptidoglycan found in gram-negative bacteria and many gram-positive genera.
3. Briefly discuss the ways in which three other peptidoglycan types differ from the gram-negative peptidoglycan.
4. How do bacilli and most gram-negative bacteria differ from gram-positive bacteria such as *S. aureus* with respect to cross-linking frequency?

---

5. What is a bacterial endospore? Give its most important properties and two ways to demonstrate its presence.
6. Give the general characteristics of *Clostridium, Desulfotomaculum,* the heliobacteria, and *Veillonella.* Briefly discuss why each is interesting or of practical importance.

 ## 23.5 CLASS *BACILLI*

The second edition of *Bergey's Manual* gathers a large variety of gram-positive bacteria into one class, *Bacilli,* and two orders, *Bacillales* and *Lactobacillales.* Within these orders are 17 families and over 70 gram-positive genera representing cocci, endospore-forming rods and cocci, and nonsporing rods. The biology of some members of the order *Bacillales* will be described first; then important representatives of the order *Lactobacillales* will be considered. The phylogenetic relationships between some of these organisms are pictured in figure 23.1, and the characteristics of selected genera are summarized in **table 23.3.**

Table 23.3

Characteristics of Members of the Class *Bacilli*

Genus	Dimensions (μm) and Morphology	G + C Content (mol%)	Oxygen Relationship	Other Distinctive Characteristics
Bacillus	0.5–2.5 × 1.2–10; straight rods, peritrichous	32–69	Aerobic or facultative	Forms endospores; catalase positive; chemoorganotrophic
Caryophanon	1.5–3.0 × 10–20; multicellular rods with rounded ends, peritrichous	41–46	Aerobic	Acetate only major carbon source; catalase positive; trichome cells have greater width than length, trichomes can be in short chains
Enterococcus	0.6–2.0 × 0.6–2.5; spherical or ovoid cells in pairs or short chains, nonsporing, sometimes motile	34–42	Facultative	Ferments carbohydrates to lactate with no gas; complex nutritional requirements; catalase negative; occurs widely, particularly in fecal material
Lactobacillus	0.5–1.2 × 1.0–10; usually long, regular rods, nonsporing, rarely motile	32–53	Facultative or microaerophilic	Fermentative, at least half the end-product is lactate; requires rich, complex media; catalase and cytochrome negative
Lactococcus	0.5–1.2 × 0.5–1.5; spherical or ovoid cells in pairs or short chains, nonsporing, nonmotile	38–40	Facultative	Chemoorganotrophic with fermentative metabolism; lactate and no gas produced; catalase negative; complex nutritional requirements; in dairy and plant products
Leuconostoc	0.5–0.7 × 0.7–1.2; cells spherical or ovoid, in pairs or chains; nonmotile and nonsporing	38–44	Facultative	Requires fermentable carbohydrate and nutritionally rich medium for growth; fermentation produces lactate, ethanol, and gas; catalase and cytochrome negative
Staphylococcus	0.9–1.3; spherical cells occurring singly and in irregular clusters, nonmotile and nonsporing	30–39	Facultative	Chemoorganotrophic with both respiratory and fermentative metabolism, usually catalase positive, associated with skin and mucous membranes of vertebrates
Streptococcus	0.5–2.0; spherical or ovoid cells in pairs or chains, nonmotile and nonsporing	34–46	Facultative	Fermentative, producing mainly lactate and no gas; catalase negative; commonly attack red blood cells (α- or β-hemolysis); complex nutritional requirements; commensals or parasites on animals
Thermoactinomyces	0.4–1.0 in diameter; branched, septate mycelium typical of actinomycetes	52.0–54.8	Aerobic	Usually thermophilic; true endospores form singly on hyphae; numerous in decaying hay, vegetable matter, and compost

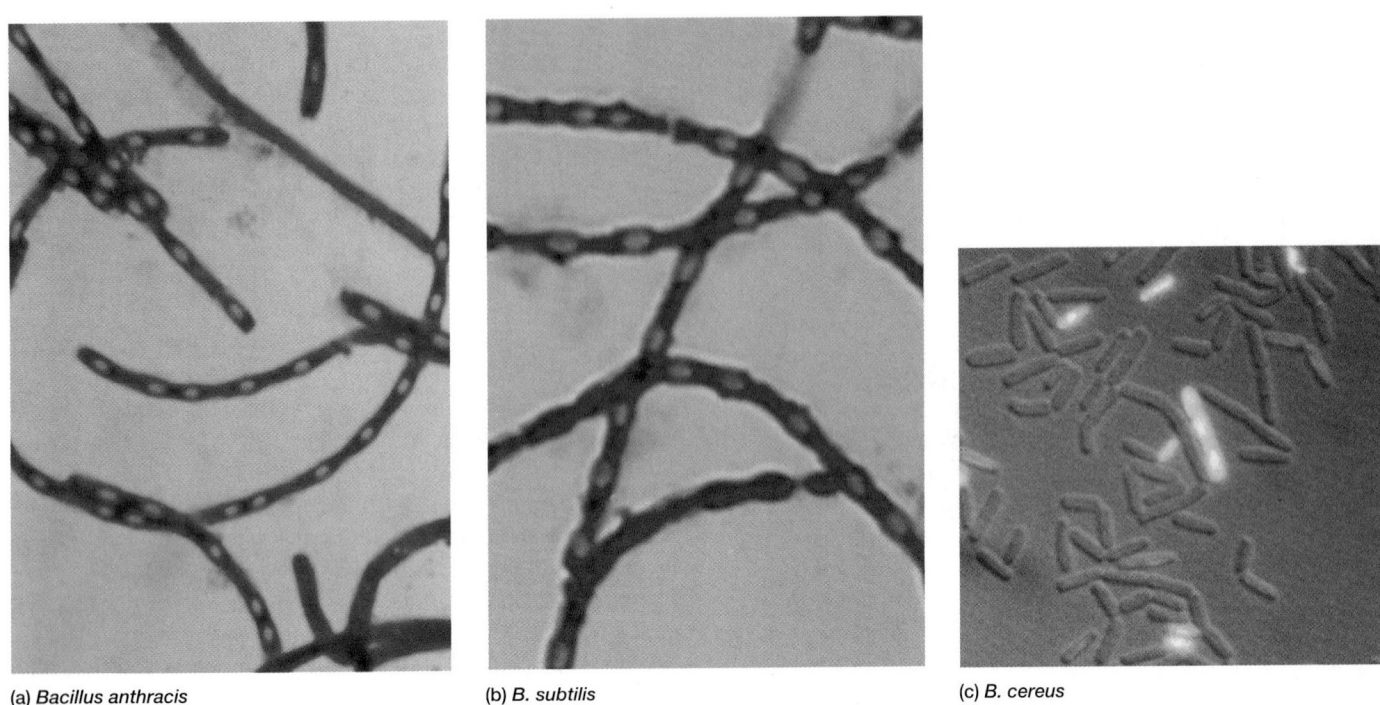

(a) *Bacillus anthracis*   (b) *B. subtilis*   (c) *B. cereus*

Figure 23.8   *Bacillus.* **(a)** *B. anthracis,* spores elliptical and central (×1,600). **(b)** *B. subtilis,* spores elliptical and central. **(c)** *B. cereus* stained with SYTOX Green nucleic acid stain and viewed by epifluorescence and differential interference contrast microscopy. The cells that glow green are dead.

Order Bacillales

The genus *Bacillus,* family *Bacillaceae,* is the largest in the order (**figure 23.8;** *see also figures 3.1c and 3.11*). The genus contains gram-positive, endospore-forming, chemoheterotrophic rods that are usually motile and peritrichously flagellated. It is aerobic, or sometimes facultative, and catalase positive. Many species have been placed in new families and genera based on rRNA sequence data. For example, the genus *Alicyclobacillus* contains acidophilic, sporing, gram-positive or gram-variable rods that have ω-alicyclic fatty acids with 6- or 7-carbon rings in their membranes. Members are aerobic or facultative and have a G + C content of about 51 to 60%. A second new genus is *Paenibacillus* [Latin *paene,* almost, and bacillus]. This genus contains gram-positive rods from rRNA group 3 that are facultative, motile by peritrichous flagella, have ellipsoidal endospores and swollen sporangia, produce acid and sometimes gas from glucose and various sugars, and have a G + C content of 40 to 54%. Some examples of organisms that were formerly in the genus *Bacillus* are *Paenibacillus alvei, P. macerans,* and *P. polymyxa.*

The 4.2 million base pair genome of *Bacillus subtilis,* the type species of the genus *Bacillus,* has been sequenced. Several families of genes have been expanded by gene duplication; the largest such family encodes ABC transporters (*see pp. 99–100*), which are the most frequent type of protein in *B. subtilis.* The genome contains genes for the catabolism of many diverse carbon sources, and for protein secretion and antibiotic synthesis. There are at least 10 integrated prophages or remnants of prophages.

Many species of *Bacillus* are of considerable importance. For example, members of the genus *Bacillus* produce the antibiotics bacitracin, gramicidin, and polymyxin. *B. cereus* (figure 23.8c) causes some forms of food poisoning and can infect humans. *B. anthracis* is the causative agent of the disease anthrax, which can affect both farm animals and humans (*see section 39.3*). Several species are used as insecticides. For example, *B. thuringiensis* and *B. sphaericus* form a solid protein crystal, the **parasporal body,** next to their spores during endospore formation (**figure 23.9**). The *B. thuringiensis* parasporal body contains protein toxins that will kill over 100 species of moths by dissolving in the alkaline gut contents of caterpillars and destroying their gut epithelium (*see section 42.5*). The solubilized toxin proteins are cleaved by midgut proteases to smaller toxic polypeptides that attack the epithelial cells. The alkaline gut contents escape into the blood, causing paralysis and death. One of these toxins has been isolated and shown to form pores in the plasma membrane. These channels allow monovalent cations such as potassium to pass through. Most of these *B. thuringiensis* toxin genes are carried on large plasmids. The *B. sphaericus* parasporal body contains proteins toxic for mosquito larvae and may be useful in controlling the mosquitos that carry the malaria parasite *Plasmodium.* Microbial insecticides (chapter 42)

The genus *Thermoactinomyces* has been moved from the actinomycetes to the family *Thermoactinomycetaceae* in the order *Bacillales.* This genus is thermophilic and grows between 45 and 60°C; it forms single spores on both its aerial and substrate mycelia (**figure 23.10**). Its G + C content is lower than that of the

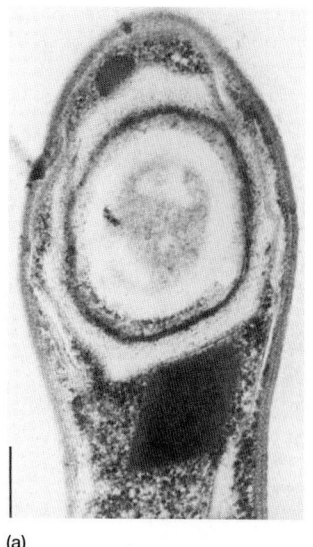

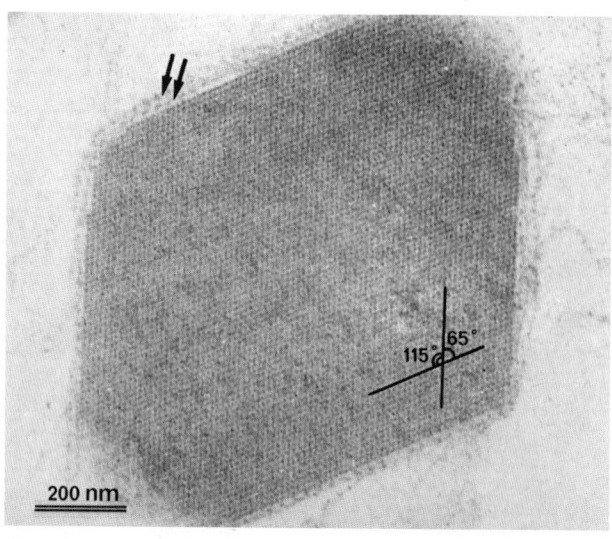

(a)   (b)

Figure 23.9 **The Parasporal Body.** (a) An electron micrograph of a *B. sphaericus* sporulating cell containing a parasporal body just beneath the endospore. Bar = 400 nm. (b) The crystalline parasporal body at a higher magnification. The crystal is surrounded by a two-layered envelope (arrows). Bar = 200 nm.

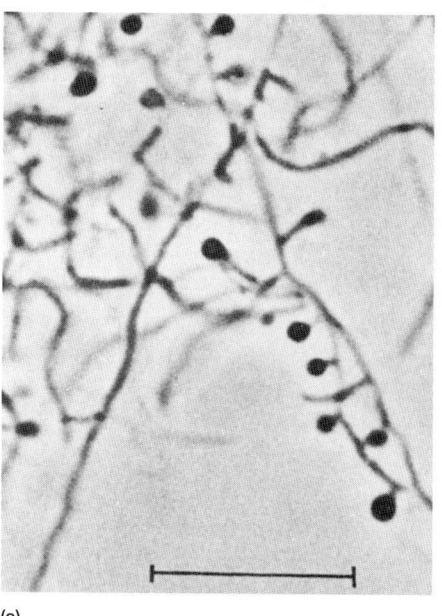

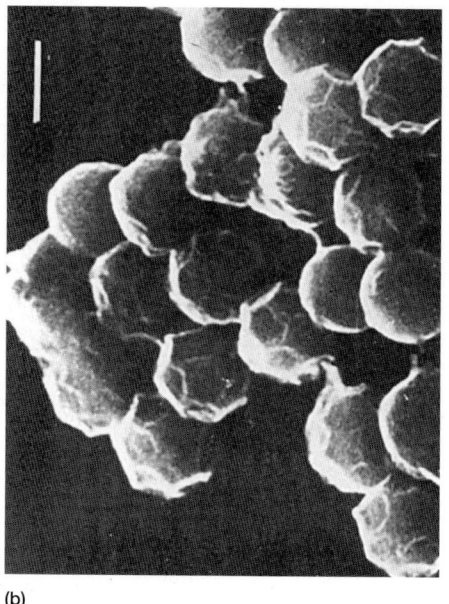

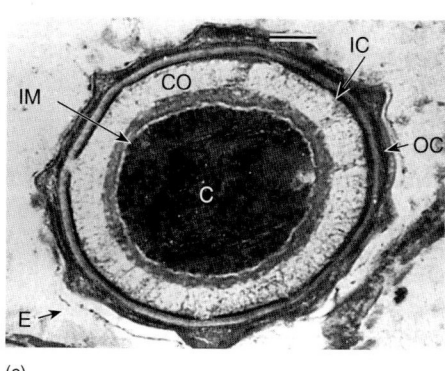

(a)   (b)   (c)

Figure 23.10 *Thermoactinomyces*. (a) *Thermoactinomyces vulgaris* aerial mycelium with developing endospores at tips of hyphae. Bar = 10 μm. (b) Scanning electron micrograph of *T. sacchari* spores. Bar = 1 μm. (c) Thin section of a *T. sacchari* endospore. Bar = 0.1 μm. E, exosporium; OC, outer spore coat; IC, inner spore coat; CO, cortex; IM, inner forespore membrane; C, core.

actinobacteria (52 to 55 mol%), and its 16S rRNA sequence suggests a relationship with the genus *Bacillus*. *Thermoactinomyces* is commonly found in damp haystacks, compost piles, and other high-temperature habitats. *Thermoactinomyces* spores (figure 23.10*b,c*) are true endospores and very heat resistant; they can survive at 90°C for 30 minutes. They are formed within hyphae and appear to have typical endospore structure, including the presence of calcium and dipicolinic acid. *Thermoactinomyces vulgaris* (figure 23.10*a*) from haystacks, grain storage silos, and compost piles is a causative agent of farmer's lung, an allergic disease of the respiratory system in agricultural workers. Recently spores from *Thermoactinomyces vulgaris* were recovered from the mud of a Minnesota lake and found to be viable after about 7,500 years of dormancy.

One of the more unusual genera in this order is *Caryophanon*. This gram-positive bacterium is strictly aerobic, catalase positive, and motile by peritrichous flagella. Its normal habitat is cow dung. *Caryophanon* morphology is distinctive. Individual cells are disk-shaped (1.5 to 2.0 μm wide by 0.5 to 1.0 μm long) and joined to form rods about 10 to 20 μm long (**figure 23.11**).

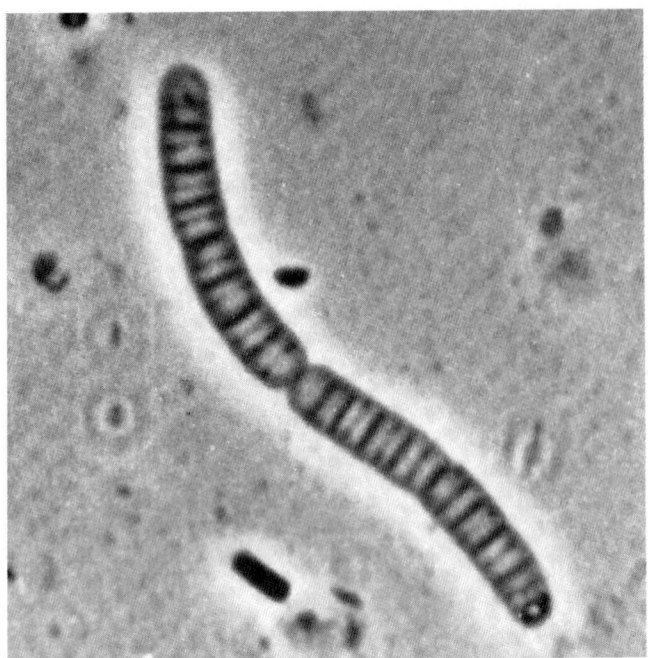

Figure 23.11   *Caryophanon* Morphology.   *Caryophanon la-tum* in a trichome chain. Note the disk-shaped cells stacked side by side; phase contrast (×3,450).

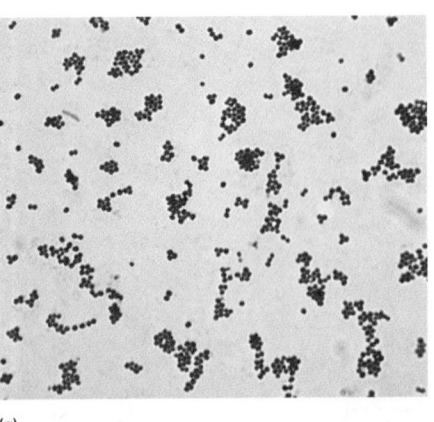

(a)

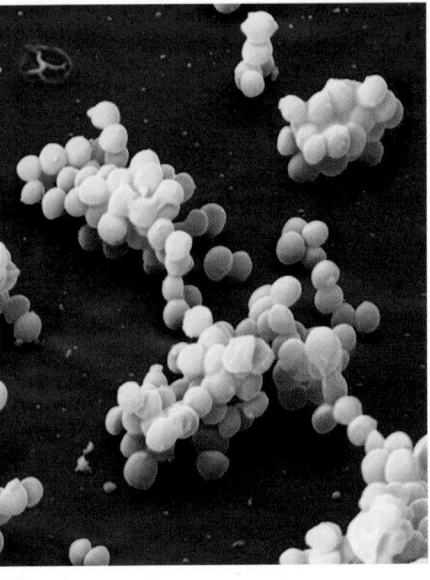

(b)

Figure 23.12   *Staphylococcus.*   (a) *Staphylococcus aureus,* Gram-stained smear (×1,500). (b) *Staphylococcus aureus* cocci arranged like clusters of grapes; color-enhanced scanning electron micrograph.

The family *Staphylococcaceae* contains four genera, the most important of which is the genus *Staphylococcus.* Members of this genus are facultatively anaerobic, nonmotile, gram-positive cocci that usually form irregular clusters (**figure 23.12,** *see also figure 3.1*a). They are catalase positive, oxidase negative, ferment glucose, and have teichoic acid in their cell walls. Staphylococci are normally associated with the skin, skin glands, and mucous membranes of warm-blooded animals.

Staphylococci are responsible for many human diseases. *S. epidermidis* is a common skin resident that is sometimes responsible for endocarditis and infections of patients with lowered resistance (e.g., wound infections, surgical infections, urinary tract infections). *S. aureus* is the most important human staphylococcal pathogen and causes boils, abscesses, wound infections, pneumonia, toxic shock syndrome, and other diseases. Recently strains of multiple drug resistant *S. aureus* have appeared and proven very difficult to treat (*see section 35.7*).

It also is a major cause of food poisoning as the following case illustrates. In March 1986 an outbreak of acute gastrointestinal illness occurred following a buffet supper served to 855 people at a New Mexico country club. At least 67 people became ill with diarrhea, nausea, or vomiting, and 24 required emergency medical treatment or hospitalization. The problem was *S. aureus* growing in the turkey and dressing served at the buffet. One or more of the food handlers carried *S. aureus* and had contaminated the food. Because the turkey was cooled for 3 hours after cooking, there was sufficient time for the bacterium to grow and produce toxins (*see sections 39.4 and 41.4*). This case is not uncommon. Turkey accounts for 10 to 21% of all bacterial food poisoning cases in which the source of poisoning is known, and it must be cooked and handled carefully.

Unlike other common staphylococci, *S. aureus* produces the enzyme **coagulase,** which causes blood plasma to clot. Growth and hemolysis patterns on blood agar are also useful in identifying these staphylococci (figure 23.17c). Recently the structure of staphylococcal α-hemolysin has been determined. The toxin lyses a cell by forming solvent-filled channels in its plasma membrane. Water-soluble toxin monomers bind to the cell surface and associate with each other to form pores. The hydrophilic channels then allow free passage of water, ions, and small molecules. *S. aureus* usually grows on the nasal membranes and skin; it also is found in the gastrointestinal and urinary tracts of warm-blooded animals.   Staphylococcal diseases (pp. 896–99)

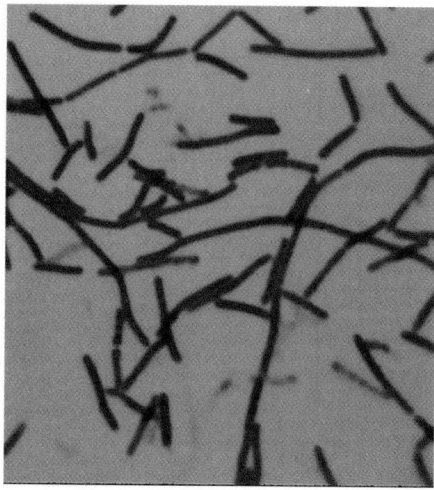

(a) *Lactobacillus acidophilus*

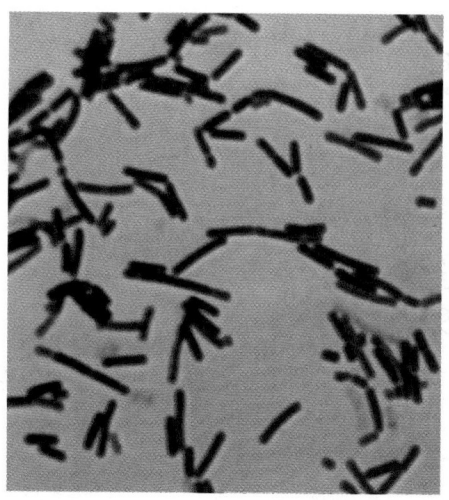

(b) *L. lactis*

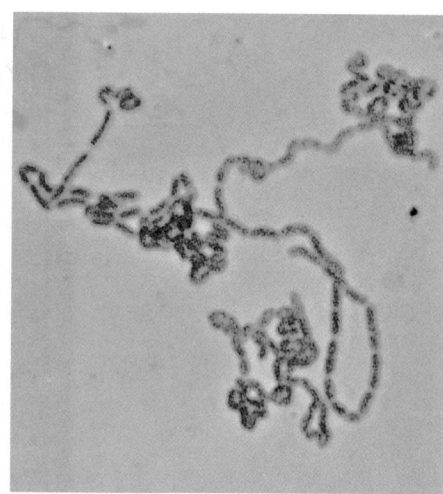

(c) *L. bulgaricus*

Figure 23.13 *Lactobacillus.* (**a**) *L. acidophilus* (×1,000). (**b**) *L. lactis.* Gram stain (×1,000). (**c**) *L. bulgaricus;* phase contrast (×600).

Listeria, family *Listeriaceae,* is another medically important genus in this order. The genus contains short rods that are aerobic or facultative, catalase positive, and motile by peritrichous flagella. It is widely distributed in nature, particularly in decaying matter. *Listeria monocytogenes* is a pathogen of humans and other animals and causes listeriosis, an important food infection (*see pp. 908–9*).

Order *Lactobacillales*

Many members of the order *Lactobacillales* produce lactic acid as their major or sole fermentation product and are sometimes collectively called **lactic acid bacteria.** *Streptococcus, Enterococcus, Lactococcus, Lactobacillus,* and *Leuconostoc* are all members of this group. Lactic acid bacteria are nonsporing and usually nonmotile. They lack cytochromes and obtain energy by substrate-level phosphorylation rather than by electron transport and oxidative phosphorylation. They normally depend on sugar fermentation for energy. Nutritionally they are fastidious and many vitamins, amino acids, purines, and pyrimidines must be supplied because of their limited biosynthetic capabilities. Lactic acid bacteria usually are categorized as facultative anaerobes, but some classify them as aerotolerant anaerobes. Oxygen relationships (pp. 125–27)

The largest genus in this order is *Lactobacillus* with around 100 species. *Lactobacillus* contains nonsporing rods and sometimes coccobacilli that lack catalase and cytochromes, are usually facultative or microaerophilic, produce lactic acid as their main or sole fermentation product, and have complex nutritional requirements (**figure 23.13**). Lactobacilli carry out either a homolactic fermentation using the Embden-Meyerhof pathway or a heterolactic fermentation with the pentose phosphate pathway (*see section 9.2*). They grow optimally under slightly acidic conditions, when the pH is between 4.5 to 6.4. The genus is found on plant surfaces and in dairy products, meat, water, sewage, beer, fruits, and many other materials. Lactobacilli also are part of the normal flora of the human body in the mouth, intestinal tract, and vagina. They usually are not pathogenic.

Lactobacillus is indispensable to the food and dairy industry (*see chapter 41*). Lactobacilli are used in the production of fermented vegetable foods (sauerkraut, pickles, silage), beverages (beer, wine, juices), sour dough bread, Swiss and other hard cheeses, yogurt, and sausage. Yogurt is probably the most popular fermented milk product in the United States and is produced commercially and by individuals using yogurt-making kits. In commercial production, nonfat or low-fat milk is pasteurized, cooled to 43°C or lower, inoculated with *Streptococcus thermophilus* and *Lactobacillus bulgaricus. S. thermophilus* grows more rapidly at first and renders the milk anaerobic and weakly acidic. *L. bulgaricus* then acidifies the milk even more. Acting together, the two species ferment almost all the lactose to lactic acid and flavor the yogurt with diacetyl (*S. thermophilus*) and acetaldehyde (*L. bulgaricus*). Fruits or fruit flavors to be added are pasteurized separately and then combined with the yogurt.

Lactobacilli also create problems. They sometimes are responsible for spoilage of beer, milk, and meat because their metabolic end products contribute undesirable flavors and odors.

Leuconostoc, family *Leuconostocaceae,* contains facultative gram-positive cocci, which may be elongated or elliptical and arranged in pairs or chains (**figure 23.14**). Leuconostocs lack catalase and cytochromes and carry out **heterolactic fermentation** (*see section 9.3*) by converting glucose to D-lactate and ethanol or acetic acid by means of the phosphoketolase pathway (**figure 23.15**). They can be isolated from plants, silage, and milk. The genus is used in wine production, in the fermentation of vegetables such as cabbage (sauerkraut; *see figure 41.18*) and cucumbers (pickles), and in the manufacture of buttermilk, butter, and cheese. *L. mesenteroides* synthesizes dextrans from sucrose and is important in industrial dextran production. *Leuconostoc* species are involved in food spoilage and tolerate high sugar concentrations so well that they grow in syrup and are a major problem in sugar refineries.

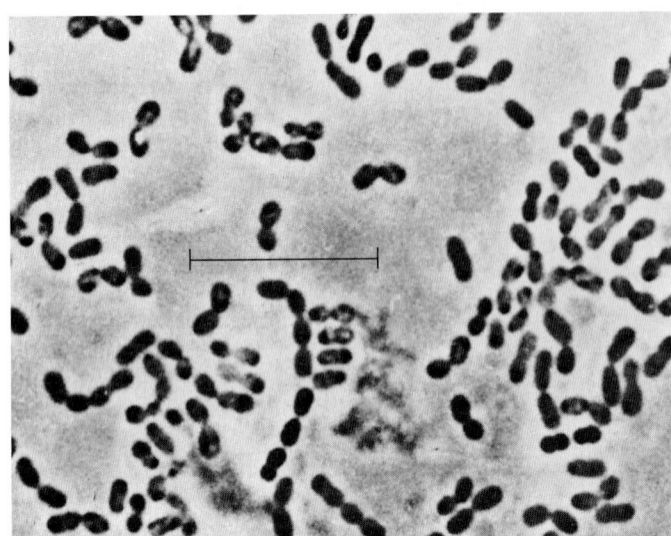

Figure 23.14 *Leuconostoc.* *Leuconostoc mesenteroides;* phase-contrast micrograph. Bar = 10 μm.

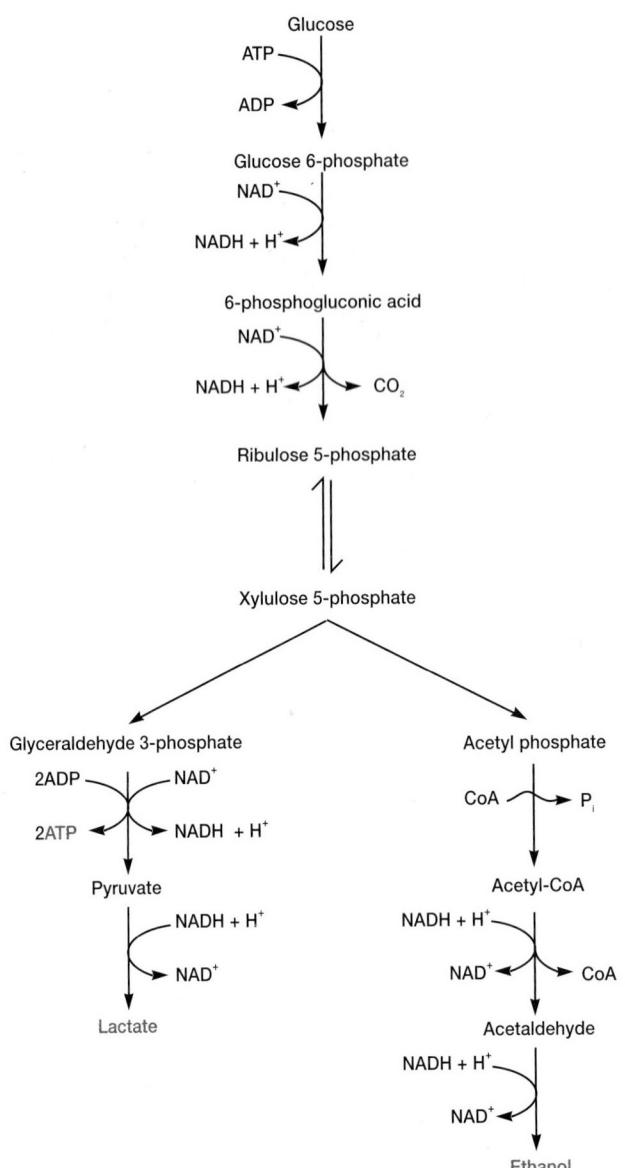

Figure 23.15 Heterolactic Fermentation and the Phosphoketolase Pathway. The phosphoketolase pathway converts glucose to lactate, ethanol, and CO_2.

Several important genera have chemoheterotrophic, mesophilic, nonsporing, gram-positive cocci and are placed in the families *Enterococcaceae (Enterococcus)* and *Streptococcaceae (Streptococcus, Lactococcus).* In practice, they are often distinguished primarily based on phenotypic properties such as oxygen relationships, cell arrangement, the presence of catalase and cytochromes, and peptidoglycan structure. The most important of these genera is *Streptococcus,* which is facultatively anaerobic and catalase negative. The streptococci and their close relatives, the enterococci and lactococci, occur in pairs or chains when grown in liquid media (**figure 23.16;** *see also figure 3.1*b), do not form endospores, and usually are nonmotile. They are all chemoheterotrophs that ferment sugars with lactic acid, but no gas, as the major product—that is, they carry out homolactic fermentation (*see section 9.3*). A few species are anaerobic rather that facultative.

The genus *Streptococcus* is large and complex. Often these bacteria have been clustered into four groups: pyogenic streptococci, oral streptococci, anaerobic streptococci, and other streptococci. Many bacteria that were placed within the genus have been moved to two new genera, *Enterococcus* and *Lactococcus.* Some major characteristics of these three closely related genera are summarized in **table 23.4. Table 23.5** lists a few properties of selected genera.

Many characteristics are used to identify these cocci. One of their most important taxonomic characteristics is the ability to lyse erythrocytes when growing on blood agar, an agar medium containing 5% sheep or horse blood (**figure 23.17**). In **α-hemolysis** a 1 to 3 mm greenish zone of incomplete hemolysis forms around the streptococcal colony; **β-hemolysis** is characterized by a zone of clearing or complete lysis without a marked color change. In addition, other hemolytic patterns are sometimes seen. Serological studies (*see chapters 33 and 36*) are also very important in identification because streptococci often have distinctive cell wall antigens. Polysaccharide and teichoic acid antigens found in the wall or between the wall and the plasma membrane are used to identify these cocci, particularly pathogenic β-hemolytic streptococci, by the **Lancefield grouping system** (*see Techniques & Applications 33.2*). Biochemical and physiological tests are essential in identification (e.g., growth temperature preferences, carbohydrate fermentation patterns, acetoin production, reduction of litmus milk, sodium chloride and bile salt tolerance, and the ability to hydrolyze arginine, esculin, hippurate, and starch). Sensitivity to bacitracin, sulfa drugs, and optochin (ethylhydrocuprein) also are used to identify particular species.

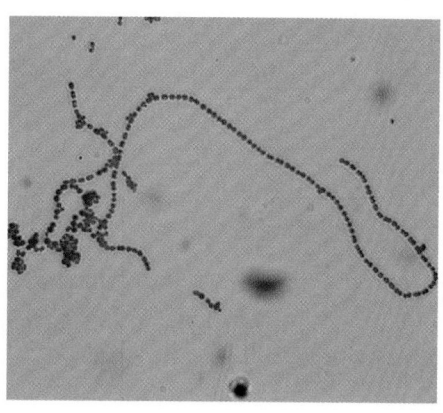

(a) *Streptococcus pyogenes*

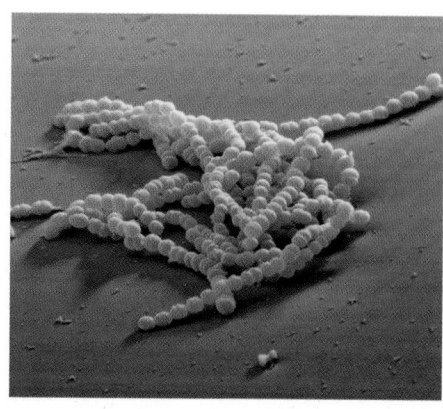

(b) *S. agalactiae*

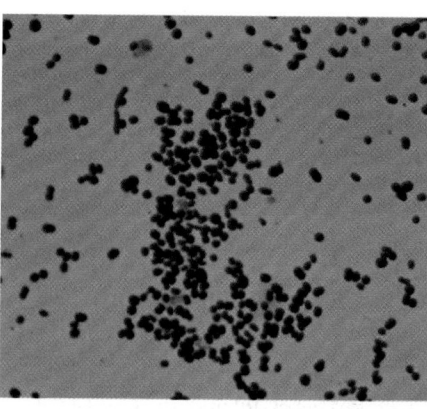

(c) *S. pneumoniae*

Figure 23.16   Streptococcus.   **(a)** *Streptococcus pyogenes* (×900). **(b)** *Streptococcus agalactiae,* the cause of Group B streptococcal infections. Note the long chains of cells; color-enhanced scanning electron micrograph (×4,800). **(c)** *Streptococcus pneumoniae* (×900).

Table 23.4

Classification of the Streptococci, Enterococci, and Lactococci

Characteristics	*Streptococcus*	*Enterococcus*	*Lactococcus*
Predominant arrangement (most common first)	Chains, pairs	Pairs, chains	Pairs, short chains
Capsule/slime layer	+	−	−
Habitat	Mouth, respiratory tract	Gastrointestinal tract	Dairy products
Growth at 45°C	Variable	+	−
Growth at 10°C	Variable	Usually +	+
Growth at 6.5% NaCl broth	Variable	+	−
Growth at pH 9.6	Variable	+	−
Hemolysis	Usually β (pyogenic) or α (oral)	α, β, −	Usually −
Serological group (Lancefield)	Variable (A–O)	Usually D	Usually N
Mol% G + C (normal range)	34–46	34–42	38–40
Representative species	Pyogenic streptococci	*E. faecalis*	*L. lactis*
	S. agalactiae	*E. faecium*	*L. raffinolactis*
	S. pyogenes	*E. avium*	*L. plantarum*
	S. equi	*E. durans*	
	S. dysgalactiae	*E. gallinarum*	
	Oral streptococci		
	S. gordonii		
	S. salvarius		
	S. sanguis		
	S. oralis		
	S. pneumoniae		
	S. mitis		
	S. mutans		
	Other streptococci		
	S. bovis		
	S. thermophilus		

Members of the three genera have considerable practical importance. Pyogenic streptococci usually are pathogens and associated with pus formation (pyogenic means pus producing). Most species produce β-hemolysis on blood agar and form chains of cells. The major human pathogen in this group is *S. pyogenes* (streptococcal sore throat, acute glomerulonephritis, rheumatic fever). The normal habitat of oral streptococci is the oral cavity and upper respiratory tract of humans and other animals. In other respects oral streptococci are not necessarily similar. *S. pneumoniae* is α-hemolytic and grows as pairs of cocci (figures 23.16*c* and 23.17*b*). It is associated with lobar pneumonia and otitis media (inflammation of the middle ear). *S. mutans* is associated with

Table 23.5

Properties of Selected Streptococci and Relatives

Characteristics	Pyogenic Streptococci S. pyogenes	Oral Streptococci S. pneumoniae	Oral Streptococci S. sanguis	Oral Streptococci S. mutans	Enterococci E. faecalis	Lactic Acid Streptococci L. lactis
Growth at 10°C	−[a]	−	−	−	+	+
Growth at 45°C	−	−	d	d	+	−
Growth at 6.5% NaCl	−	−	−	−	+	−
Growth at pH 9.6	−	−	−	−	+	−
Growth with 40% bile	−	−	d	d	+	+
α-hemolysis	−	+	+	−	−	d
β-hemolysis	+	−	−	−	+	−
Arginine hydrolysis	+	+	+	−	+	d
Hippurate hydrolysis	−	−	−	−	+	d
Mol% G + C of DNA	35–39	30–39	40–46	36–38	34–38	39

Modified from *Bergey's Manual of Systematic Bacteriology*, Vol. 2, edited by P. H. A. Sneath, et al. Copyright © 1986 Williams and Wilkins, Baltimore, MD. Reprinted by permission.
[a]Symbols: +, 90% or more of strains positive; −, 10% or less of strains positive; d, 11–89% of strains are positive.

Figure 23.17 Streptococcal and Staphylococcal Hemolytic Patterns. (**a**) *Streptococcus pyogenes* on blood agar, illustrating β-hemolysis. (**b**) *Streptococcus pneumoniae* on blood agar, illustrating α-hemolysis. (**c**) *Staphylococcus aureus* on blood agar, illustrating β-hemolysis. (**d**) *Staphylococcus epidermidis* on blood agar with no hemolysis.

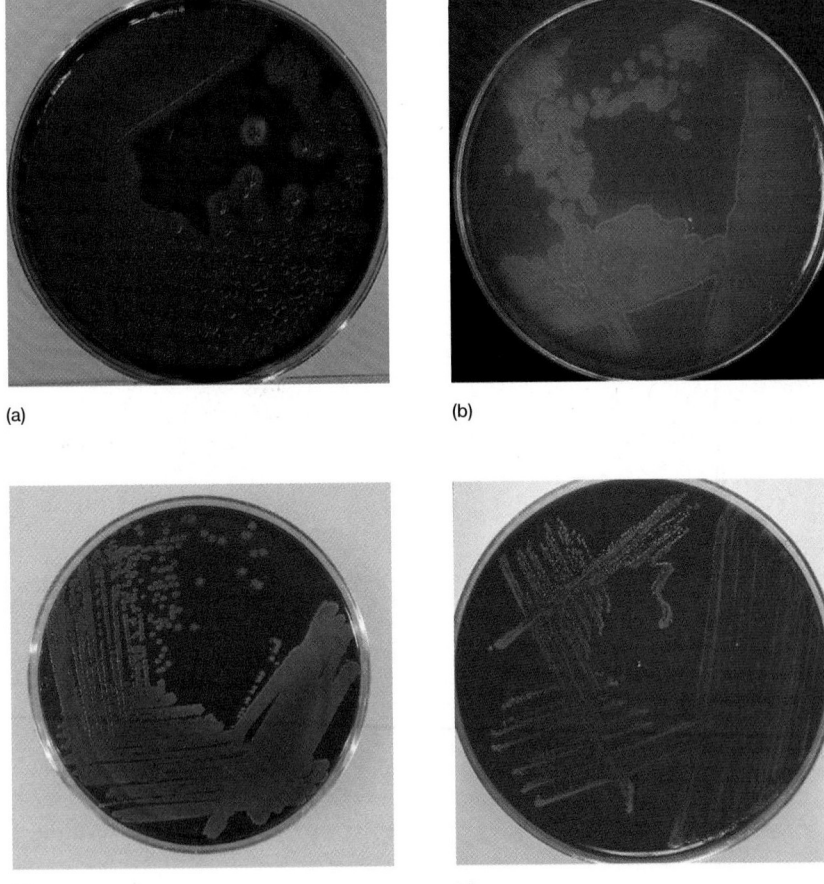

(a)

(b)

(c)

(d)

the formation of dental caries (*see section 39.6*). The enterococci such as *E. faecalis* are normal residents of the intestinal tracts of humans and most other animals. *E. faecalis* is an opportunistic pathogen that can cause urinary tract infections and endocarditis. Unlike the streptococci the enterococci will grow in 6.5% sodium chloride. The lactococci ferment sugars to lactic acid and can grow at 10°C but not at 45°C. *L. lactis* is widely used in the production of buttermilk and cheese (*see section 41.6*) because it can curdle milk and add flavor through the synthesis of diacetyl and other products. Streptococcal diseases (pp. 879–82)

1. List the major properties of the genus *Bacillus*. What practical impacts does it have on society? Define parasporal body and Strickland reaction.
2. Briefly describe the genus *Thermoactinomyces*, with particular emphasis on its unique features. What disease does it cause?
3. What is distinctive about the morphology of *Caryophanon*?
4. Describe the genus *Staphylococcus*. How does the pathogen *S. aureus* differ from the common skin resident *S. epidermidis*, and where is it normally found?
5. List the major properties of the genus *Lactobacillus*. Why is it important in the food and dairy industries?

6. Describe the major distinguishing characteristics of the following taxa: *Streptococcus, Enterococcus, Lactococcus,* and *Leuconostoc.*
7. Of what practical importance is *Leuconostoc*? What are lactic acid bacteria?
8. What are α-hemolysis, β-hemolysis, and the Lancefield grouping system?
9. Give a representative species and its importance for *Streptococcus, Enterococcus,* and *Lactococcus.* Distinguish between pyogenic and oral streptococci.

Summary

23.1 General Introduction
a. Bacteria are placed in sections of volume 2 of the first edition of *Bergey's Manual* according to characteristics such as general shape, the possession of endospores, and their response to acid-fast staining.
b. Based on 16S rRNA analysis, gram-positive bacteria are divided into low G + C and high G + C groups; this system is used in the second edition of *Bergey's Manual*.
c. The second edition of *Bergey's Manual* places the low G + C gram positives in the phylum *Firmicutes*, which contains three classes: *Clostridia, Mollicutes,* and *Bacilli* (**tables 23.1–23.3** and **figure 23.1**).

23.2 Class *Mollicutes* (the Mycoplasmas)
a. Mycoplasmas are gram-negative bacteria that lack cell walls and cannot synthesize peptidoglycan precursors. Many species require sterols for growth. They are the smallest bacteria capable of self-reproduction and usually grow on agar to give colonies a "fried-egg" appearance (**figure 23.3**).

23.3 Low G + C Gram-Positive Bacteria in *Bergey's Manual*
a. Peptidoglycan structure often differs between groups in taxonomically useful ways. Most variations are in amino acid 3 of the peptide subunit or in the interpeptide bridge (**figure 23.4**).

b. Endospores resist desiccation and heat; they are used by bacteria to survive harsh conditions, especially in the soil.

23.4 Class *Clostridia*
a. Members of the genus *Clostridium* are anaerobic gram-positive rods that form endospores and don't carry out dissimilatory sulfate reduction (**figure 23.6**). They are responsible for botulism, tetanus, food spoilage, and putrefaction.
b. *Desulfotomaculum* is an anaerobic, endospore-forming genus that reduces sulfate to sulfide during anaerobic respiration.
c. The heliobacteria are gram-positive, anaerobic, photosynthetic bacteria with bacteriochlorophyll g. Some form endospores.
d. The family *Veillonellaceae* contains anaerobic, gram-negative cocci, some of which are parasites of vertebrates.

23.5 Class *Bacilli*
a. The class *Bacilli* is divided into two orders: *Bacillales* and *Lactobacillales.*
b. The genus *Bacillus* contains aerobic and facultative, catalase-positive, endospore-forming, chemoheterotrophic, gram-positive rods that are usually motile and peritrichous (**figure 23.8**). *Bacillus* synthesizes antibiotics and insecticides, and causes food poisoning and anthrax.

c. *Thermoactinomyces* is a gram-positive thermophile that forms a mycelium and true endospores (**figure 23.10**). It causes allergic reactions and leads to farmer's lung.
d. Members of the genus *Staphylococcus* are facultatively anaerobic, nonmotile, gram-positive cocci that form irregular clusters (**figure 23.12**). They grow on the skin and mucous membranes of warm-blood animals, and some are important human pathogens.
e. Several important genera such as *Lactobacillus, Listeria,* and *Caryophanon* contain regular, nonsporing, gram-positive rods. *Lactobacillus* carries out lactic acid fermentation and is extensively used in the food and dairy industries.
f. *Leuconostoc* carries out heterolactic fermentation using the phosphoketolase pathway (**figure 23.15**) and is involved in the production of fermented vegetable products, buttermilk, butter, and cheese.
g. The genera *Streptococcus, Enterococcus,* and *Lactococcus* contain gram-positive cocci arranged in pairs and chains that are usually facultative and carry out homolactic fermentation (**tables 23.4** and **23.5**). Some important species are the pyogenic coccus *S. pyogenes,* the oral streptococci *S. pneumoniae* and *S. mutans,* the enterococcus *E. faecalis* and lactococcus *L. lactis* (**figure 23.16**).

Key Terms

α-hemolysis 516	coagulase 514	lactic acid bacteria 515	mycoplasmas 504
β-hemolysis 516	heterolactic fermentation 515	Lancefield grouping system 516	parasporal body 512

Questions for Thought and Review

1. On the basis of the treatment of gram-positive bacteria given in volume 2 of the first edition of *Bergey's Manual,* is it best to consider the volume as an identification guide, a description of phylogenetic relationships, or some of both? Explain your reasoning and conclusion.

2. Draw a diagram illustrating how gram-positive cocci might divide to produce the various clustering patterns observed (chains, tetrads, cubical packets, grapelike clusters).

3. Describe the characteristics most important in distinguishing between members of the following groups of genera: *Staphylococcus* and *Streptococcus, Bacillus* and *Clostridium.*

4. Which genera are noted for each of the following: coagulase, pathogenicity for insects, production of fermented foods and cheeses, use of the phosphoketolase pathway, causing allergic respiratory ailments, destruction of red blood cells, colonization of the tongue and teeth, and the generation of odors during putrefaction?

5. How might one go about determining whether the genome of *M. genitalium* is the smallest one compatible with a free-living existence?

6. Account for the ease with which anaerobic clostridia can be isolated from soil and other generally aerobic niches.

Critical Thinking Questions

1. Many low G + C bacteria are parasitic. The dependence on a host might be a consequence of the low G + C content. Elaborate on this concept.

2. The idea that *Bacillus* spores could be recovered from the intestines of bees preserved in amber is a bit *Jurassic Park*-like. Nevertheless, what types of controls would you wish to see performed in order to conclude that the spores were genuine and not contaminants from the current environment? (Assume the identification of the spore species was by PCR of rRNA sequences.)

For recommended readings on these and related topics, see Appendix V.

chapter 24

Bacteria:
The High G + C Gram Positives

Concepts

- Volume 4 of the first edition of *Bergey's Manual* contains aerobic, gram-positive bacteria—the actinomycetes—that form branching hyphae and asexual spores.
- The morphology and arrangement of spores, cell wall chemistry, and the types of sugars present in cell extracts are particularly important in actinomycete taxonomy and are used to divide these bacteria into different groups.
- The second edition of *Bergey's Manual* classifies the actinomycetes and other high G + C gram-positive bacteria using 16S rRNA data. They are placed in the phylum *Actinobacteria*, which is a large, complex grouping with one class and six orders.
- Actinomycetes have considerable practical impact because they play a major role in the mineralization of organic matter in the soil and are the primary source of most naturally synthesized antibiotics. The genera *Corynebacterium* and *Mycobacterium* contain important human pathogens.

Actinomycetes are very important from a medical point of view. . . . They may be a nuisance, as when they decompose rubber products, grow in aviation fuel, produce odorous substances that pollute water supplies, or grow in sewage-treatment plants where they form thick clogging foams. . . . In contrast, actinomycetes are the producers of most of the antibiotics.

—H. A. Lechevalier and M. P. Lechevalier

Frankia forms nonmotile spores and is symbiotic with a number of higher plants such as alder trees.

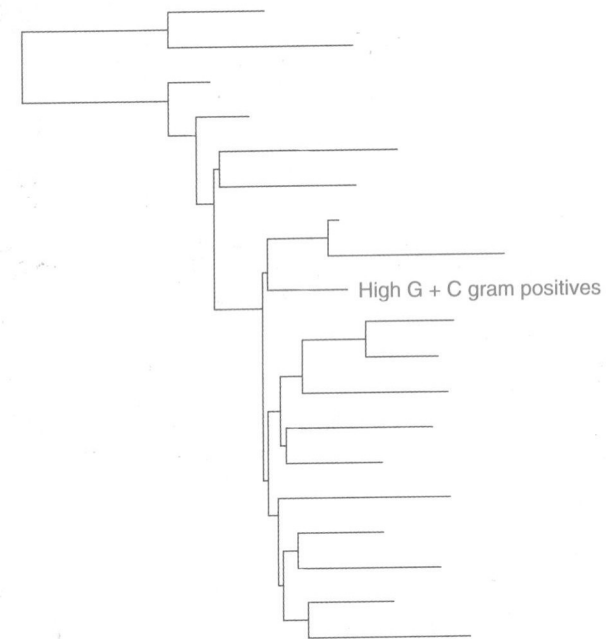

High G + C gram positives

general characteristics of the actinomycetes are summarized. Then the treatment of high G + C bacteria, essentially the actinomycetes and their relatives, by the two editions will be briefly compared. Finally, representatives are described, with emphasis on morphology, taxonomy, reproduction, and general importance. This survey of representative high G + C gram positives will follow the organization of the second edition. The **actinomycetes** [s., actinomycete] are aerobic, gram-positive bacteria that form branching filaments or hyphae and asexual spores. Although they are a diverse group, the actinomycetes do share many properties.

hapter 24, the last of the survey chapters on bacteria, describes the high G + C gram-positive bacteria (see dendrogram above). They will be in volume 4 of the second edition of *Bergey's Manual*. Many of these bacteria are called actinomycetes. Actinomycetes are gram-positive bacteria, but are distinctive because they have filamentous hyphae that do not normally undergo fragmentation and produce asexual spores. They closely resemble fungi in overall morphology. Presumably this resemblance results partly from adaptation to the same habitats. First, the

24.1 GENERAL PROPERTIES OF THE ACTINOMYCETES

When growing on a solid substratum such as agar, the branching network of hyphae developed by actinomycetes grows both on the surface of the substratum and into it to form a substrate mycelium. Septa usually divide the hyphae into long cells (20μm and longer) containing several nucleoids. Sometimes a tissuelike mass results and may be called a **thallus.** Many actinomycetes also have an aerial mycelium that extends above the substratum and forms asexual, thin-walled spores called **conidia** [s., conidium] or **conidiospores** on the ends of filaments (**figure 24.1**). If the spores are located in a sporangium, they are called **sporangiospores.** The spores can vary greatly in shape (**figure 24.2**). Actinomycete spores develop by septal formation at filament tips, usually in response to nutrient deprivation. Most are not par-

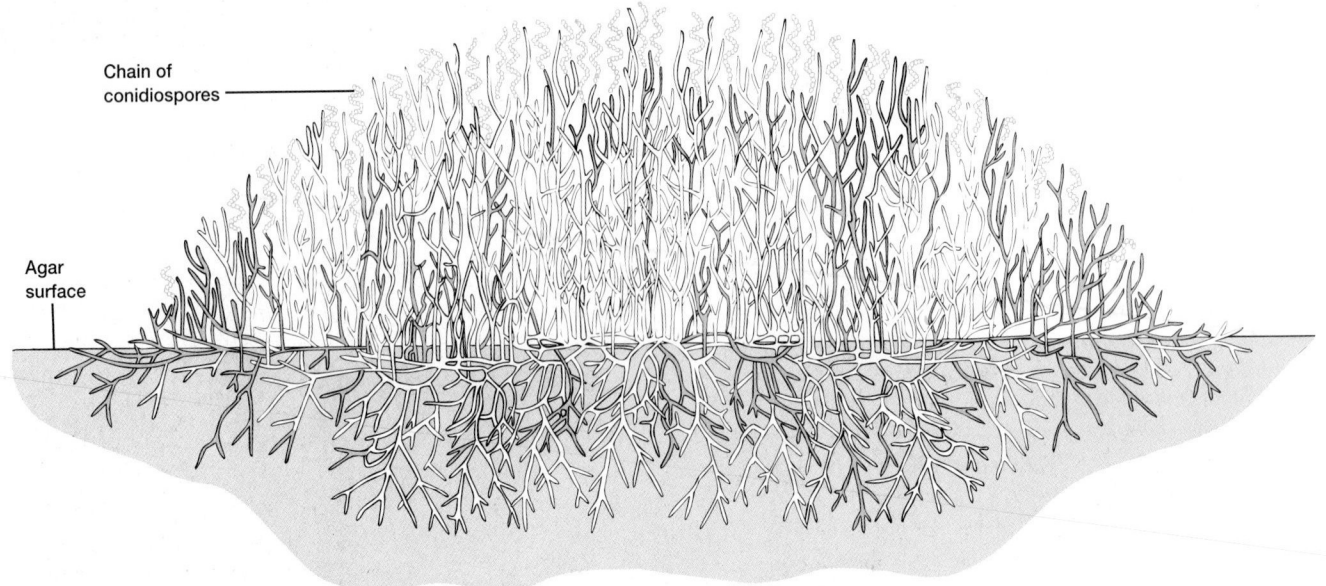

Chain of conidiospores

Agar surface

Figure 24.1 An Actinomycete Colony. The cross section of an actinomycete colony with living (blue-green) and dead (white) hyphae. The substrate mycelium and aerial mycelium with chains of conidiospores are shown.

Figure 24.2   Examples of Actinomycete Spores as Seen in the Scanning Electron Microscope.   (**a**) Sporulating *Saccharopolyspora* hyphae (×3,000). (**b**) Sporangia of *Pilimelia columellifera* on mouse hair (×520). (**c**) *Micromonospora echinospora.* Bar = 0.5 μm. (**d**) A chain of hairy streptomycete spores. Bar = 1.0 μm. (**e**) *Microbispora rosea,* paired spores on hyphae. (**f**) Aerial spores of *Kitasatosporia setae.* Bar = 5 μm.

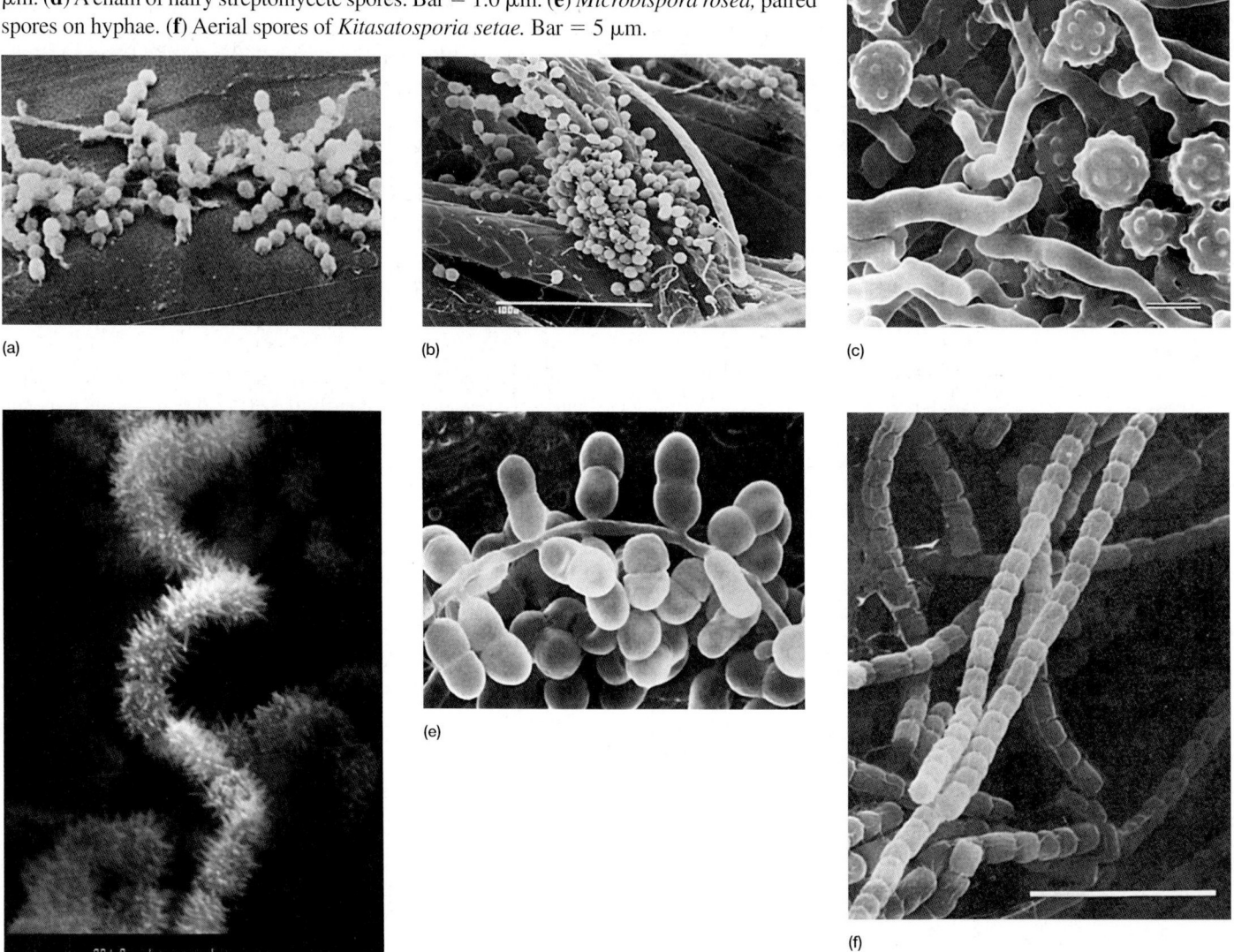

(a)

(b)

(c)

(d)

(e)

(f)

ticularly heat resistant but do withstand desiccation well and thus have considerable adaptive value.

Most actinomycetes are not motile. When motility is present, it is confined to flagellated spores.

Actinomycete cell wall composition varies greatly among different groups and is of considerable taxonomic importance. Four major cell wall types can be distinguished according to three features of peptidoglycan composition and structure: the amino acid in tetrapeptide side chain position 3, the presence of glycine in interpeptide bridges, and peptidoglycan sugar content (**table 24.1**). Cell extracts of actinomycetes with wall types II, III, and IV also contain characteristic sugars that are useful in identification (**table 24.2**). Some other taxonomically valuable properties are the morphology and color of mycelia and sporangia, the surface features and arrangement of conidiospores, the percent G + C in

DNA, the phospholipid composition of cell membranes, and spore heat resistance. Newer techniques are being applied to actinomycete taxonomy. Comparison of 16S rRNA sequences has proven valuable as we shall see. Another useful technique is the production of large DNA fragments by restriction enzyme digestion and their separation and comparison using pulsed-field electrophoresis. Gram-positive peptidoglycan structure and chemistry (pp. 507–8)

Actinomycetes have considerable practical significance. They are primarily soil inhabitants and are very widely distributed. They can degrade an enormous number and variety of organic compounds and are extremely important in the mineralization of organic matter. Actinomycetes produce most of the medically useful natural antibiotics. Although most actinomycetes are free-living microorganisms, a few are pathogens of humans, other animals, and some plants.

Table 24.1

Actinomycete Cell Wall Types

Cell Wall Type	Diaminopimelic Acid Isomer	Glycine in Interpeptide Bridge	Characteristic Sugars[a]	Representative Genera
I	L, L	+	NA	*Nocardioides, Streptomyces*
II	*meso*	+	NA	*Micromonospora, Pilimelia, Actinoplanes*
III	*meso*	−	NA	*Actinomadura, Frankia*
IV	*meso*	−	Arabinose, galactose	*Saccharomonospora, Nocardia*

[a]NA, either not applicable or no diagnostic sugars.

Table 24.2

Actinomycete Whole Cell Sugar Patterns

Sugar Pattern Types[a]	Characteristic Sugars	Representative Genera
A	Arabinose, galactose	*Nocardia, Rhodococcus, Saccharomonospora*
B	Madurose[b]	*Actinomadura, Streptosporangium, Dermatophilus*
C	None	*Thermomonospora, Actinosynnema, Geodermatophilus*
D	Arabinose, xylose	*Micromonospora, Actinoplanes*

[a]Characteristic sugar patterns are present only in wall types II–IV, those actinomycetes with *meso*-diaminopimelic acid.
[b]Madurose is 3-*O*-methyl-D-galactose.

24.2 HIGH G + C GRAM-POSITIVE BACTERIA IN *BERGEY'S MANUAL*

The first edition of *Bergey's Manual* divides the actinomycetes into seven sections, primarily based on properties such as cell wall type, conidia arrangement, and the presence or absence of a sporangium (**table 24.3**).

It has been clear for some time that the sections in volume 4 of the first edition are not homogeneous and do not always fit with 16S rRNA sequence data. The second edition of *Bergey's Manual* uses 16S rRNA sequences to classify the high G + C gram positives, gram-positive bacteria with a DNA base composition above approximately 50 mol% G + C. All these bacteria are placed in the phylum *Actinobacteria* and classified as shown in **figure 24.3**. The phylum is large and very complex; it contains one class (*Actinobacteria*), five subclasses, six orders, 14 suborders, and 44 families. In this system the **actinobacteria**

Table 24.3

Some Characteristics of Actinomycete Groups in the First Edition of *Bergey's Manual*

Group	Wall Type	Sugar Pattern	G + C Content (mol%)	Spore Arrangement	Presence of Sporangia	Selected Genera
Nocardioform actinomycetes[a]	I, IV, VI[b]	A	59–79	Varies	−	*Nocardia, Rhodococcus, Nocardioides, Saccharomonospora*
Actinomycetes with multilocular sporangia	III	B, C, D	57–75	Clusters of spores	+(−)[c]	*Geodermatophilus, Dermatophilus, Frankia*
Actinoplanetes	II	D	71–73	Varies	Usually +	*Actinoplanes, Pilimelia, Dactylosporangium, Micromonospora*
Streptomyces and related genera	I	Not of taxonomic value	69–78	Chains of 5 to more than 50 spores	−	*Streptomyces, Sporichthya*
Maduromycetes	III	B, C	64–74	Varies	+ or −	*Actinomadura, Microbispora, Planomonospora, Streptosporangium*
Thermomonospora and related genera	III	C (sometimes B)	64–73	Varies	−	*Thermomonospora, Actinosynnema, Nocardiopsis*

[a]Several genera have mycolic acid. Filaments readily fragment into rods and coccoid elements.
[b]Type VI cell wall has lysine instead of diaminopimelic acid.
[c]Members have clusters of spores that may not always be surrounded by a sporangial wall.

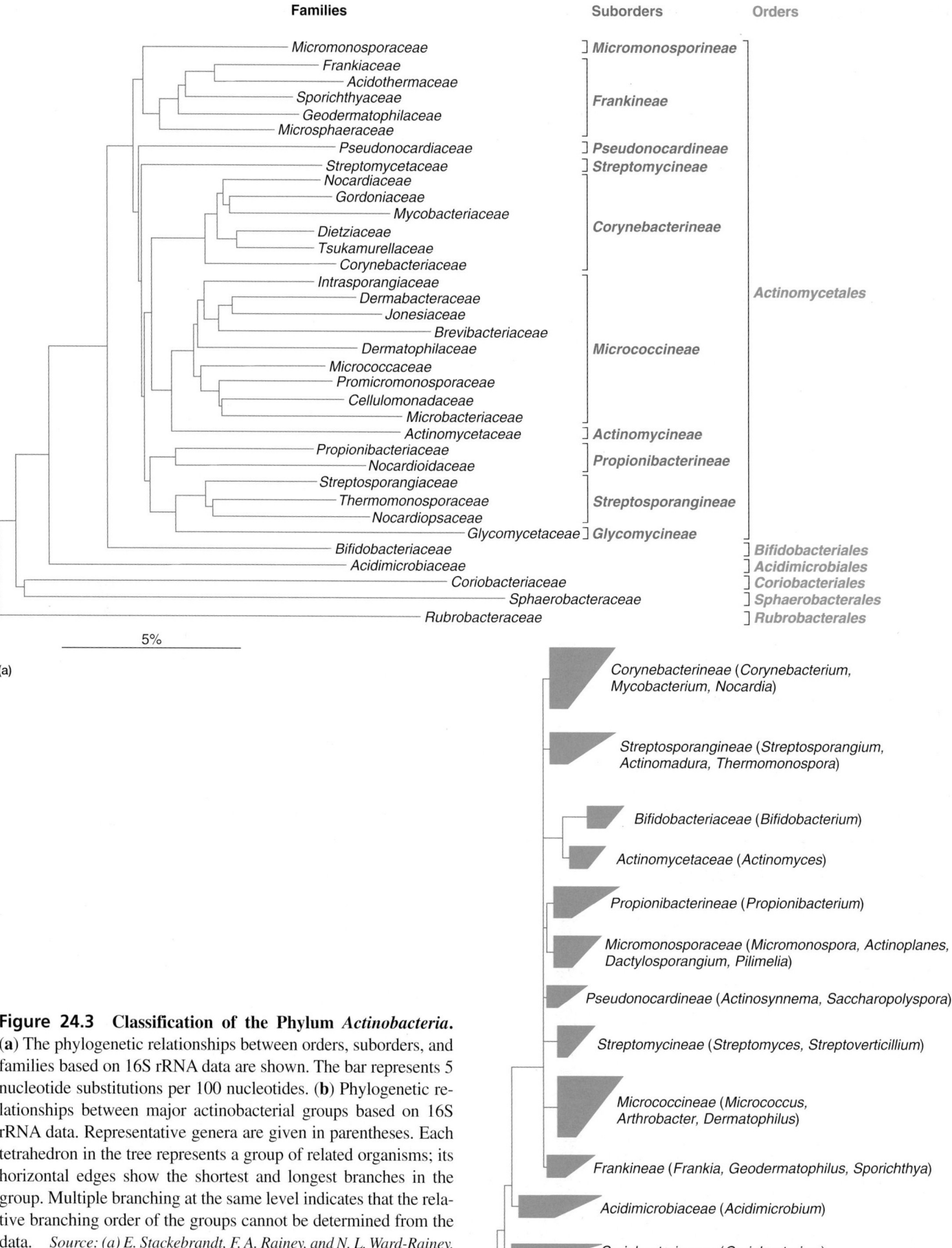

Families

Micromonosporaceae
Frankiaceae
Acidothermaceae
Sporichthyaceae
Geodermatophilaceae
Microsphaeraceae
Pseudonocardiaceae
Streptomycetaceae
Nocardiaceae
Gordoniaceae
Mycobacteriaceae
Dietziaceae
Tsukamurellaceae
Corynebacteriaceae
Intrasporangiaceae
Dermabacteraceae
Jonesiaceae
Brevibacteriaceae
Dermatophilaceae
Micrococcaceae
Promicromonosporaceae
Cellulomonadaceae
Microbacteriaceae
Actinomycetaceae
Propionibacteriaceae
Nocardioidaceae
Streptosporangiaceae
Thermomonosporaceae
Nocardiopsaceae
Glycomycetaceae
Bifidobacteriaceae
Acidimicrobiaceae
Coriobacteriaceae
Sphaerobacteraceae
Rubrobacteraceae

Suborders

Micromonosporineae
Frankineae
Pseudonocardineae
Streptomycineae
Corynebacterineae
Micrococcineae
Actinomycineae
Propionibacterineae
Streptosporangineae
Glycomycineae

Orders

Actinomycetales
Bifidobacteriales
Acidimicrobiales
Coriobacteriales
Sphaerobacterales
Rubrobacterales

5%

(a)

Figure 24.3 Classification of the Phylum *Actinobacteria*.
(**a**) The phylogenetic relationships between orders, suborders, and families based on 16S rRNA data are shown. The bar represents 5 nucleotide substitutions per 100 nucleotides. (**b**) Phylogenetic relationships between major actinobacterial groups based on 16S rRNA data. Representative genera are given in parentheses. Each tetrahedron in the tree represents a group of related organisms; its horizontal edges show the shortest and longest branches in the group. Multiple branching at the same level indicates that the relative branching order of the groups cannot be determined from the data.   *Source: (a) E. Stackebrandt, F. A. Rainey, and N. L. Ward-Rainey. Proposal for a new hierarchic classification system,* Actinobacteria, *classis nov.* Int. J. Syst. Bacteriol. *47(2):479–491, 1997. Figure 3, p. 482.*

Corynebacterineae (Corynebacterium, Mycobacterium, Nocardia)

Streptosporangineae (Streptosporangium, Actinomadura, Thermomonospora)

Bifidobacteriaceae (Bifidobacterium)

Actinomycetaceae (Actinomyces)

Propionibacterineae (Propionibacterium)

Micromonosporaceae (Micromonospora, Actinoplanes, Dactylosporangium, Pilimelia)

Pseudonocardineae (Actinosynnema, Saccharopolyspora)

Streptomycineae (Streptomyces, Streptoverticillium)

Micrococcineae (Micrococcus, Arthrobacter, Dermatophilus)

Frankineae (Frankia, Geodermatophilus, Sporichthya)

Acidimicrobiaceae (Acidimicrobium)

Coriobacteriaceae (Coriobacterium)

Rubrobacteraceae (Rubrobacter)

(b)

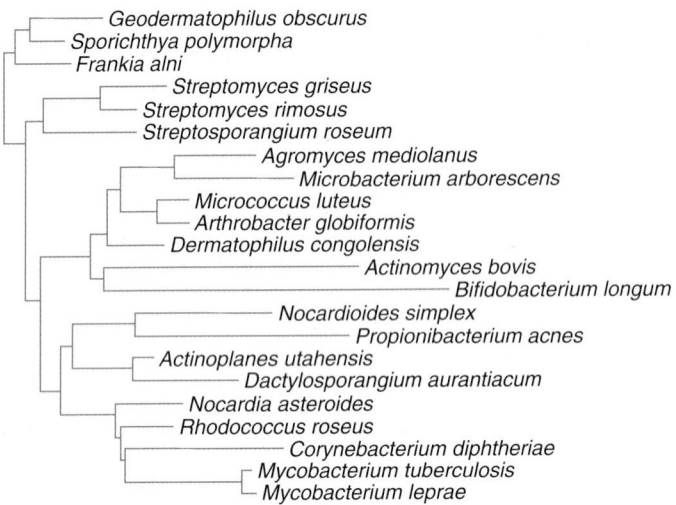

Figure 24.4 Phylogenetic Relationships Among Selected High G + C Gram-Positive Bacteria. The relationships of a few species based on 16S rRNA sequence data are shown. *Source: The Ribosomal Database Project.*

are composed of the actinomycetes and their high G + C relatives. The orders, suborders, and families are defined based on 16S rRNA sequences and distinctive signature nucleotides (*see* p. 422). **Figure 24.4** shows the phylogenetic relationships between selected representatives of the high G + C gram positives, and **table 24.4** summarizes the characteristics of some of the genera discussed in this chapter.

Most of the genera to be discussed in the following survey are in the subclass *Actinobacteridae* and order *Actinomycetales* that is divided into 10 suborders. The following survey will focus on several of these suborders. The order *Bifidobacteriales* also will be briefly described.

 SUBORDER *ACTINOMYCINEAE*

Most genera have irregularly shaped, nonsporing, gram-positive rods with aerobic or facultative metabolism. The rods may be straight or slightly curved and usually have swellings, club shapes, or other deviations from normal rod-shape morphology. *Actinomyces, Arcanobacterium,* and *Mobiluncus* are located in the suborder *Actinomycineae*.

Table 24.4
Characteristics of Actinobacteria

Genus	Dimensions (μm) and Morphology	G + C Content (mol%)	Oxygen Relationship	Other Distinctive Characteristics
Actinoplanes	Nonfragmenting, branching mycelium with little aerial growth; sporangia formed; motile spores with polar flagella	72–73	Aerobic	Hyphae often in palisade arrangement; highly colored; type II cell walls; found in soil and decaying plant material
Arthrobacter	0.8–1.2 × 1.0–8.0; young cells are irregular rods, older cells are small cocci; usually nonmotile	About 59–70	Aerobic	Have rod-coccus growth cycle; metabolism respiratory; catalase positive; mainly in soil
Bifidobacterium	0.5–1.3 × 1.5–8; rods of varied shape, usually curved; nonmotile	55–67	Anaerobic	Cells can be clubbed or branched, pairs often in V arrangement; ferment carbohydrates to acetate and lactate, but no CO_2; catalase negative
Corynebacterium	0.3–0.8 × 1.5–8.0; straight or slightly curved rods with tapered or clubbed ends; nonmotile	51–63	Facultatively anaerobic	Cells often arranged in a V formation or in palisades of parallel cells; catalase positive and fermentative; metachromatic granules
Frankia	0.5–2.0 in diameter; vegetative hyphae with limited to extensive branching and no aerial mycelium; multilocular sporangia formed	66–71	Aerobic to microaerophilic	Sporangiospores nonmotile; usually fixes nitrogen; type III cell walls; most strains are symbiotic with angiosperm plants and induce nodules
Micrococcus	0.5–2.0 diameter; cocci in pairs, tetrads, or irregular clusters; usually nonmotile	64–75	Aerobic	Colonies usually yellow or red; catalase positive with respiratory metabolism; primarily on mammalian skin and in soil
Mycobacterium	0.2–0.6 × 1.0–10; straight or slightly curved rods, sometimes branched; acid-fast; nonmotile and nonsporing	62–70	Aerobic	Catalase positive; can form filaments that are readily fragmented; walls have high lipid content; in soil and water; some parasitic
Nocardia	0.5–1.2 in diameter; rudimentary to extensive vegetative hyphae that can fragment into rod-shaped and coccoid forms	64–72	Aerobic	Aerial hyphae formed; catalase positive; type IV cell wall; widely distributed in soil
Propionibacterium	0.5–0.8 × 1–5; pleomorphic nonmotile rods, may be bifid or branched; nonsporing	53–67	Anaerobic to aerotolerant	Fermentation produces propionate and acetate, and often gas; catalase positive
Streptomyces	0.5–2.0 in diameter; vegetative mycelium extensively branched; aerial mycelium forms chains of three to many spores	69–78	Aerobic	Form discrete lichenoid, leathery, or butyrous colonies that often are pigmented; use many different organic compounds as nutrients; soil organisms

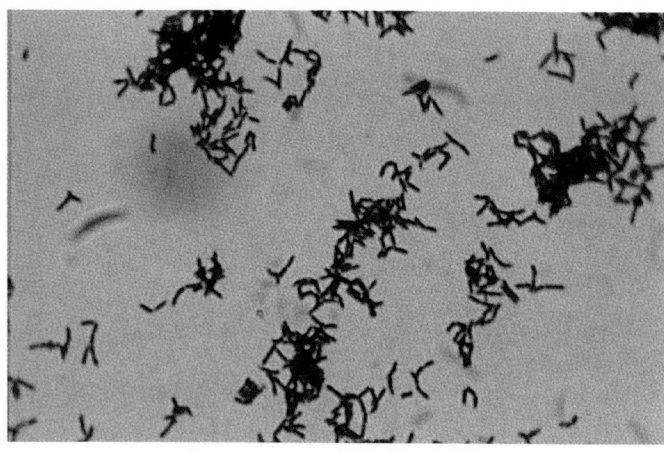

(a)

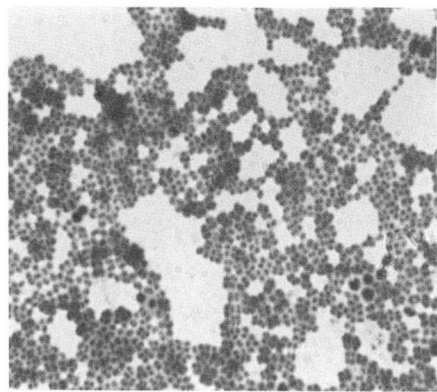

Figure 24.6 Micrococcus. *Micrococcus luteus,* methylene blue stain (×1,000).

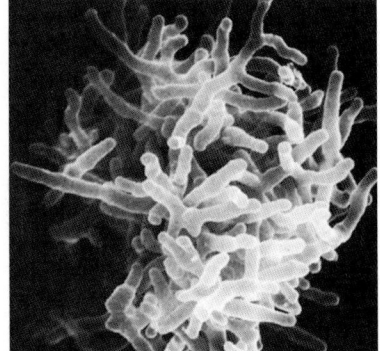

(b)

Figure 24.5 Representatives of the Genus *Actinomyces*. (**a**) *A. naeslundii;* Gram stain (×1,000). (**b**) *Actinomyces;* scanning electron micrograph (×18,000). Note filamentous nature of the colony.

Members of the genus *Actinomyces* are either straight or slightly curved rods that vary considerably in shape or slender filaments with true branching (**figure 24.5**). The rods and filaments may have swollen, clubbed, or clavate ends. They are either facultative or strict anaerobes that require CO_2 for best growth. The cell walls contain lysine but not diaminopimelic acid or glycine. *Actinomyces* species are normal inhabitants of mucosal surfaces of humans and other warm-blooded animals; the oral cavity is their preferred habitat. *A. bovis* causes lumpy jaw in cattle. *Actinomyces* is responsible for actinomycoses, ocular infections, and periodontal disease in humans. The most important human pathogen is *A. israelii*.

 24.4 SUBORDER *MICROCOCCINEAE*

The suborder *Micrococcineae* has 14 families and a wide variety of genera. Two of the best-known genera are *Micrococcus* and *Arthrobacter*.

The genus *Micrococcus* contains aerobic, catalase-positive cocci that occur mainly in pairs, tetrads, or irregular clusters and are usually nonmotile (**figure 24.6**). Micrococci often are yellow, orange, or red in color. They are widespread in soil, water, and on mammalian skin; the last habitat may be their normal one. Despite their frequent presence on the skin, micrococci do not seem particularly pathogenic.

The genus *Arthrobacter* contains aerobic, catalase-positive rods with respiratory metabolism and lysine in its peptidoglycan. Its most distinctive feature is a rod-coccus growth cycle (**figure 24.7**). When *Arthrobacter* grows in exponential phase, the bacteria are irregular, branched rods that may reproduce by a process called snapping division. As they enter stationary phase, the cells change to a coccoid form. Upon transfer to fresh medium, the coccoid cells produce outgrowths and again form actively growing rods. Although arthrobacters often are isolated from fish, sewage, and plant surfaces, their most important habitat is the soil, where they constitute a significant component of the microbial flora. They are well adapted to this niche because they are very resistant to desiccation and nutrient deprivation. This genus is unusually flexible nutritionally and can even degrade some herbicides and pesticides; it is probably important in the mineralization of complex organic molecules.

The mechanism of **snapping division** has been studied in *Arthrobacter*. These bacteria have a two-layered cell wall, and only the inner layer grows inward to generate a transverse wall dividing the new cells. The completed transverse wall or septum next thickens and puts tension on the outer wall layer, which still holds the two cells together. Eventually, increasing tension ruptures the outer layer at its weakest point, and a snapping movement tears the outer layer apart around most of its circumference. The new cells now rest at an angle to each other and are held together by the remaining portion of the outer layer that acts as a hinge.

A third genus in this suborder is *Dermatophilus*. *Dermatophilus* (cell wall type IIIB) also forms packets of motile spores with tufts of flagella, but it is a facultative anaerobe and a parasite of mammals responsible for the skin infection streptothrichosis.

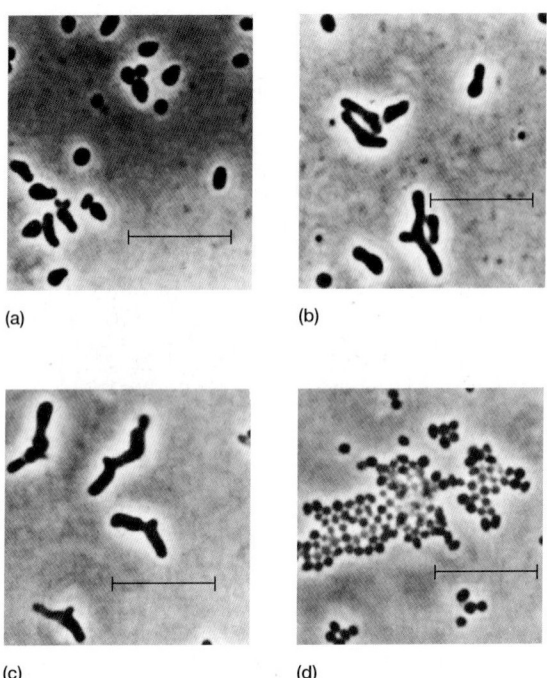

(a)　　　　(b)

(c)　　　　(d)

Figure 24.7 The Rod-Coccus Growth Cycle. The rod-coccus cycle of *Arthrobacter globiformis* when grown at 25°C. (**a**) Rods are outgrowing from cocci 6 hours after inoculation. (**b**) Rods after 12 hours of incubation.(**c**) Bacteria after 24 hours. (**d**) Cells after reaching stationary phase (3 days' incubation). The cells used for inoculation resembled these stationary-phase cocci. Bars = 10 μm.

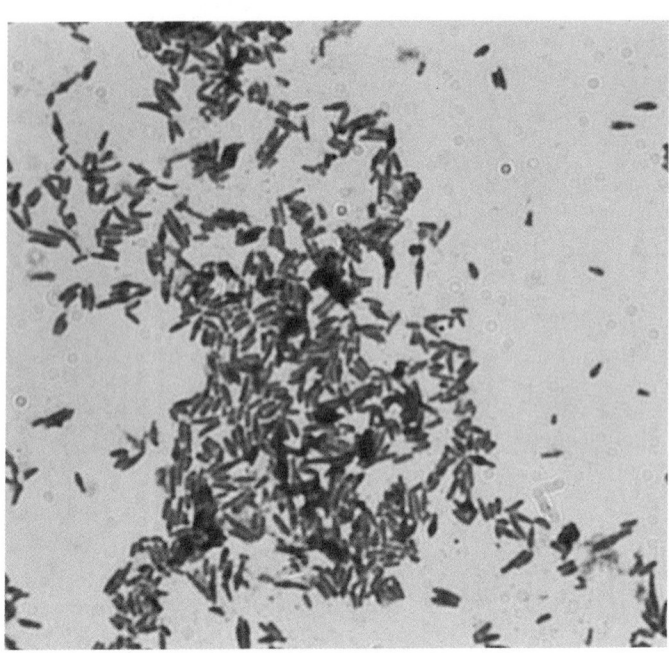

Figure 24.8 *Corynebacterium diphtheriae*. Note the irregular shapes of individual cells, the angular associations of pairs of cells, and palisade arrangements (×1,000). These gram-positive rods do not form endospores.

 24.5 SUBORDER *CORYNEBACTERINEAE*

This suborder contains seven families with several well-known genera. Three of the most important genera are *Corynebacterium, Mycobacterium,* and *Nocardia.*

The family *Corynebacteriaceae* has one genus, *Corynebacterium,* which contains aerobic and facultative, catalase-positive, straight to slightly curved rods, often with tapered ends. Club-shaped forms are also seen. The bacteria often remain partially attached after snapping division, resulting in angular arrangements of the cells, somewhat like Chinese letters, or a palisade arrangement in which rows of cells are lined up side by side (**figure 24.8**). Corynebacteria form metachromatic granules, and their walls have *meso*-diaminopimelic acid. Although some species are harmless saprophytes, many corynebacteria are plant or animal pathogens. For example, *C. diphtheriae* is the causative agent of diphtheria in humans (*see section 39.1*).

The family *Mycobacteriaceae* contains the genus *Mycobacterium,* which is composed of slightly curved or straight rods that sometimes branch or form filaments (**figure 24.9**). Mycobacterial filaments differ from those of actinomycetes in readily fragmenting into rods and coccoid bodies when distributed. They are aerobic and catalase positive. Mycobacteria grow very slowly

and must be incubated for 2 to 40 days after inoculation of a solidified complex medium to form a visible colony. Their cell walls have a very high lipid content and contain waxes with 60 to 90 carbon **mycolic acids.** These are complex fatty acids with a hydroxyl group on the β-carbon and an aliphatic chain attached to the α-carbon. The presence of mycolic acids and other lipids outside the peptidoglycan layer makes mycobacteria **acid-fast** (basic fuchsin dye cannot be removed from the cell by acid alcohol treatment). Extraction of wall lipid with alkaline ethanol destroys acid-fastness. Actinomycete morphology (pp. 522–23); Acid-fast staining (p. 28)

Although some mycobacteria are free-living saprophytes, they are best known as animal pathogens. *M. bovis* causes tuberculosis in cattle, other ruminants, and primates. Because this bacterium can produce tuberculosis in humans, dairy cattle are tested for the disease yearly; milk pasteurization kills the pathogen and affords further protection against disease transmission. Thus *M. tuberculosis* is the chief source of tuberculosis in humans. The other major mycobacterial human disease is leprosy, caused by *M. leprae.* Tuberculosis (pp. 882–84); Leprosy (pp. 893–94); The genome sequences of *M. tuberculosis* and *M. leprae* (pp. 343–44)

The family *Nocardiaceae* is composed of two genera, *Nocardia* and *Rhodococcus.* Because these and related genera resemble members of the genus *Nocardia* (named after Edmond Nocard [1850–1903], French bacteriologist and veterinary pathologist), they are collectively called **nocardioforms.** These bacteria de-

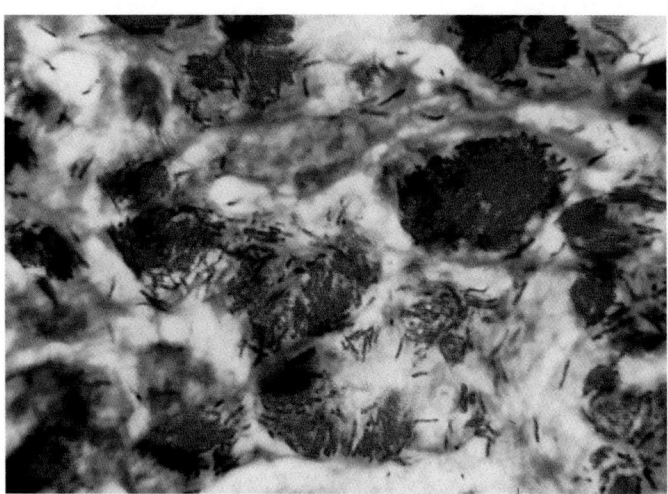

Figure 24.9 The Mycobacteria. *Mycobacterium leprae.* Acid-fast stain (×400). Note the masses of red mycobacteria within blue-green host cells.

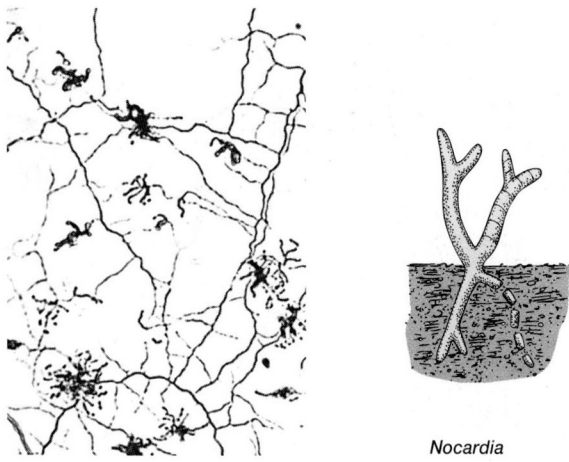

Nocardia

Figure 24.10 *Nocardia. Nocardia asteroides,* substrate mycelium and aerial mycelia with conidia illustration and light micrograph (×1,250).

velop a substrate mycelium that readily breaks into rods and coccoid elements (**figure 24.10**). Several genera also form an aerial mycelium that rises above the substratum and may produce conidia. All genera have a high G + C content like other actinomycetes, and almost all are strict aerobes. Most species have peptidoglycan with *meso*-diaminopimelic acid and no peptide interbridge. The wall usually contains a carbohydrate composed of arabinose and galactose; mycolic acids are present in *Nocardia* and *Rhodococcus.*

Nocardia is distributed worldwide in soil and also is found in aquatic habitats. Nocardiae are involved in the degradation of hydrocarbons and waxes and can contribute to the biodeterioration of rubber joints in water and sewage pipes. Although most are free-living saprophytes, some species, particularly *N. asteroides,* are opportunistic pathogens that cause nocardiosis in humans and other animals. People with low-resistance due to other health problems are most at risk. The lungs are most often infected, but the central nervous system and other organs may be invaded. *Rhodococcus* also is widely distributed in soils and aquatic habitats. It is of considerable interest because members of the genus can degrade an enormous variety of molecules such as petroleum hydrocarbons, detergents, benzene, polychlorinated biphenyls (PCBs), and various pesticides. It may be possible to use rhodococci to remove sulfur from fuels, thus reducing air pollution by sulfur oxide emissions.

1. Define actinomycete, thallus, substrate mycelium, aerial mycelium, conidium, conidiospore, and sporangiospore.
2. Describe how cell wall structure and sugar content are used to classify the actinomycetes. Include a brief description of the four major wall types.
3. Why are the actinomycetes of such practical interest?

4. How do the first and second editions of *Bergey's Manual* differ in their treatment of the high G + C gram-positive bacteria?
5. Give the major distinguishing characteristics of the actinomycetes. Describe the phylum *Actinobacteria* and its relationship to the actinomycetes.
6. Describe the major characteristics of the following genera: *Actinomyces, Micrococcus, Arthrobacter,* and *Corynebacterium.* Include comments on their normal habitat and importance.
7. What is snapping division? The rod-coccus growth cycle?
8. Give the distinctive properties of the genus *Mycobacterium,* and indicate how mycobacteria differ from actinomycetes.
9. Define mycolic acid and acid-fast.
10. List two human mycobacterial diseases and their causative agents. Which pathogen causes tuberculosis in cattle?
11. What is a nocardioform, and how can the group be distinguished from other actinomycetes?
12. Where is *Nocardia* found, and what problems may it cause? Consider both environmental and public health concerns.

24.6 SUBORDER *MICROMONOSPORINEAE*

The suborder *Micromonosporineae* contains only one family, *Micromonosporaceae.* Many of the genera in this family are placed among the actinoplanetes in the first edition of *Bergey's Manual.* Actinoplanetes [Greek *actinos,* a ray or beam, and *planes,* a wanderer] have an extensive substrate mycelium and are wall type IID. Often the hyphae are highly colored and diffusible pigments may be produced. Normally an aerial mycelium is absent or rudimentary. Conidiospores are usually formed within a sporangium raised above the surface of the substratum at the end of a special

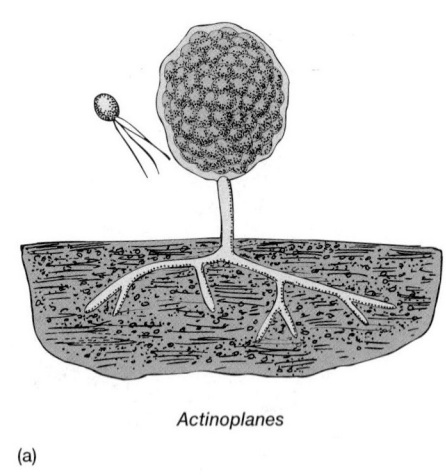

Actinoplanes

(a)

(b)

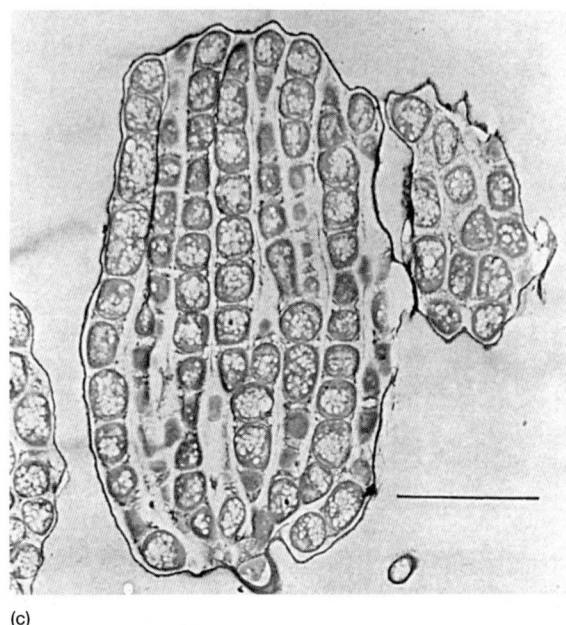

(c)

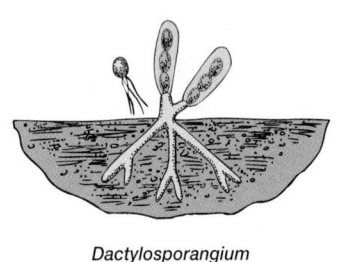

Dactylosporangium

(d)

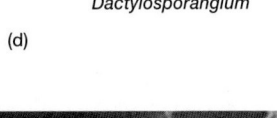

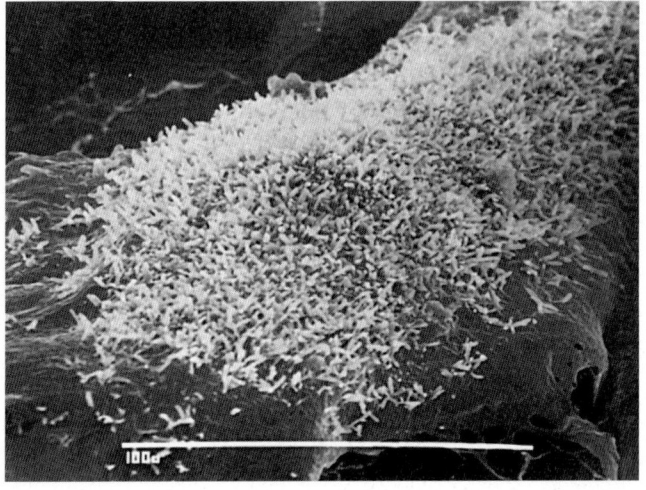

(e)

Figure 24.11 Family *Micromonosporaceae*. Actinoplanete morphology. (**a**) *Actinoplanes* structure. (**b**) A scanning electron micrograph of mature *Actinoplanes* sporangia. Bar = 5 μm. (**c**) A section of an *Actinoplanes rectilineatus* sporangium. Bar = 3 μm. (**d**) *Dactylosporangium* structure. (**e**) A *Dactylosporangium* colony covered with sporangia (×700). Bar = 100 μm.

hypha called a sporangiophore. The spores can be either motile or nonmotile. These bacteria vary in the arrangement and development of their spores. Some genera (*Actinoplanes, Pilimelia*) have spherical, cylindrical, or irregular sporangia with a few to several thousand spores per sporangium (figure 24.2*b* and **figure 24.11**). The sporangium develops above the substratum at the tip of a sporangiophore; the spores are arranged in coiled or parallel chains (**figure 24.12**). *Dactylosporangium* forms club-shaped, fingerlike, or pyriform sporangia with one to six spores (figure 24.11*d,e*). *Mi-*

cromonospora bears single spores, which often occur in branched clusters of sporophores (figure 24.2*c*).

Actinoplanetes [s., actinoplanete] grow in almost all soil habitats, ranging from forest litter to beach sand. They also flourish in freshwater, particularly in streams and rivers (probably because of abundant oxygen and plant debris). Some have been isolated from the ocean. The soil-dwelling species may have an important role in the decomposition of plant and animal material. *Pilmelia* grows in association with keratin. *Micromono-*

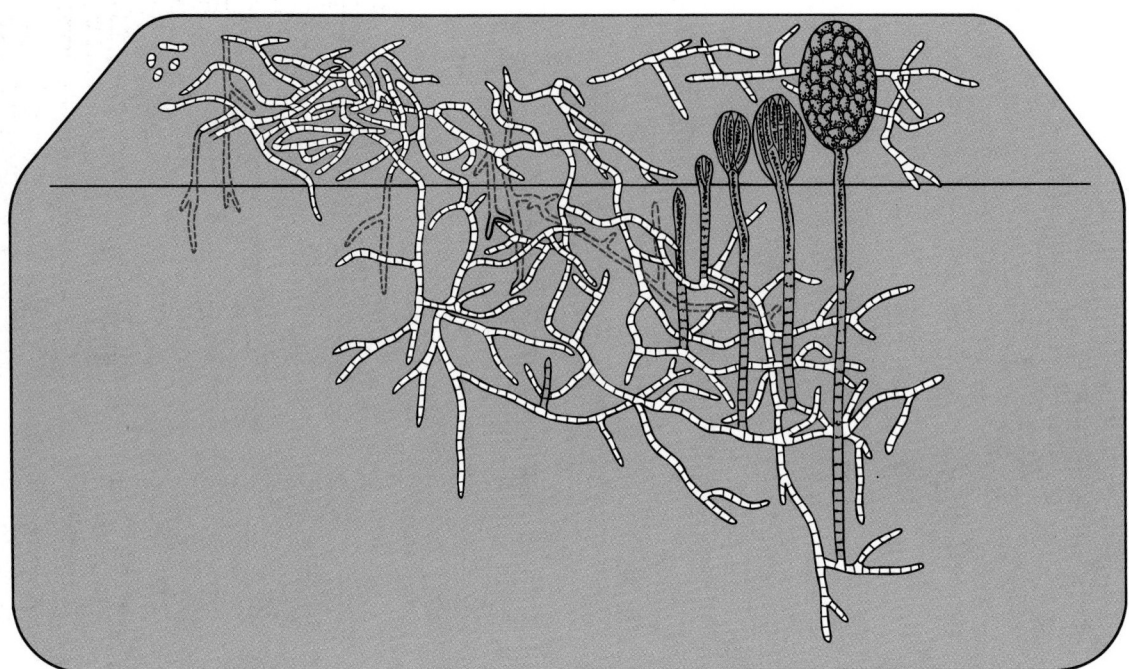

Figure 24.12   Sporangium Development in an Actinoplanete. The developing sporangium is shown in green; the more mature stages are to the right.

spora actively degrades chitin and cellulose, and it can produce antibiotics such as gentamicin.

24.7   SUBORDER *PROPIONIBACTERINEAE*

This suborder contains two families and 13 genera. The genus *Propionibacterium* will be placed in the family *Propionibacteriaceae* in the second edition. The genus contains pleomorphic, nonmotile, nonsporing rods that are often club-shaped with one end tapered and the other end rounded. Cells also may be coccoid or even branched. They can be single, in short chains, or in clumps. The genus is facultatively anaerobic or aerotolerant; lactate and sugars are fermented to produce large quantities of propionic and acetic acids, and often carbon dioxide. *Propionibacterium* is usually catalase positive. The genus is found growing on the skin and in the digestive tract of animals, and in dairy products such as cheese. *Propionibacterium* contributes substantially to the production of Swiss cheese (*see p. 953*). *P. acnes* is involved with the development of body odor and acne vulgaris (*see p. 677*).

24.8   SUBORDER *STREPTOMYCINEAE*

The suborder *Streptomycineae* has only one family, *Streptomycetaceae,* and three genera, the most important of which is *Streptomyces.* These bacteria have aerial hyphae that divide in a single plane to form chains of 3 to 50 or more nonmotile conidiospores with surface texture ranging from smooth to spiny and warty (fig-

ure 24.2*d;* **figures 24.13** and **24.14**). All have a type I cell wall and a G + C content of around 69 to 78%. The substrate mycelium, when present, does not undergo fragmentation. Members of this family and similar bacteria are often called **streptomycetes** [Greek *streptos,* bent or twisted, and *myces,* fungus].

Streptomyces is an enormous genus; there are around 500 species. Members of the genus are strict aerobes, are wall type I, and form chains of nonmotile spores within a thin, fibrous sheath (figure 24.14). The three to many conidia in each chain are often pigmented and can be smooth, hairy, or spiny in texture. *Streptomyces* species are determined by means of a mixture of morphological and physiological characteristics, including the following: the color of the aerial and substrate mycelia, spore arrangement, surface features of individual spores, carbohydrate use, antibiotic production, melanin synthesis, nitrate reduction, and the hydrolysis of urea and hippuric acid.

Streptomycetes are very important, both ecologically and medically. The natural habitat of most streptomycetes is the soil, where they may constitute from 1 to 20% of the culturable population. In fact, the odor of moist earth is largely the result of streptomycete production of volatile substances such as **geosmin.** Streptomycetes play a major role in mineralization. They are flexible nutritionally and can aerobically degrade resistant substances such as pectin, lignin, chitin, keratin, latex, and aromatic compounds. Streptomycetes are best known for their synthesis of a vast array of antibiotics, some of which are useful in medicine and biological research. Examples include amphotericin B, chloramphenicol, erythromycin, neomycin, nystatin, streptomycin, tetracycline, and so forth (**figure 24.15***a*). Although most streptomycetes are nonpathogenic saprophytes, a few are associated with plant and animal

Figure 24.13 *Streptomyces* **and Related Genera.** Strepto-mycete conidia arrangement. (**a**) An illustration of typical *Strepto-myces* morphology; a light micrograph of *S. carpinesis* spore chains. Bar = 5 μm. (**b**) *Streptoverticillium* (*Streptomyces*) morphology; a scanning electron micrograph of *Sv. salmonis* with developing spore chains. Bar = 2 μm.

Figure 24.14   Streptomycete Spores.
(**a**) Smooth spores of *S. niveus;* scanning electron micrograph. Bar = 0.25 μm. (**b**) Spiney spores of *S. viridochromogenes.* Bar = 0.5 μm. (**c**) Warty spores of *S. pulcher.* Bar = 0.25 μm.

diseases. *Streptomyces scabies* causes scab disease in potatoes and beets (figure 24.15*b*). *S. somaliensis* is the only streptomycete known to be pathogenic for humans. It is associated with **actino-mycetoma,** an infection of subcutaneous tissues that produces le-sions and leads to swelling, abscesses, and even bone destruction if

untreated. *S. albus* and other species have been isolated from pa-tients with various ailments and may be pathogenic.   Antibiotics and their properties (chapter 35)

Streptoverticillium, another genus in the family, has an aerial mycelium with a whorl of three to six short branches that are pro-

Figure 24.15 Streptomycetes of Practical Importance. (**a**) *Streptomyces griseus.* Colonies of the actinomycete that produces streptomycin. (**b**) *Streptomyces scabies,* an actinomycete growing on a potato.

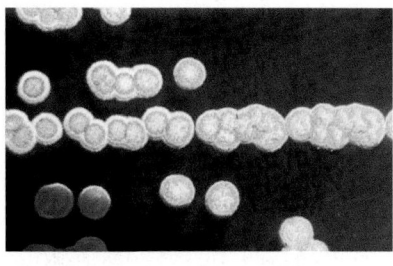

(a)

(b)

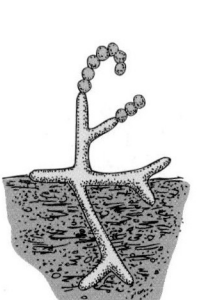

Actinomadura madurae

(a)

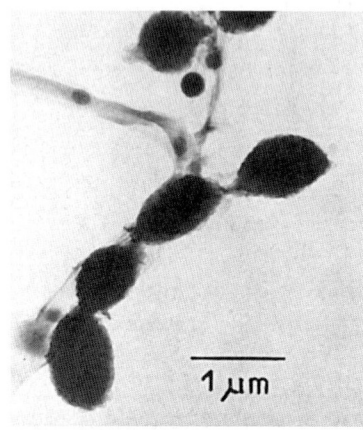

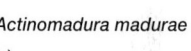

1 μm

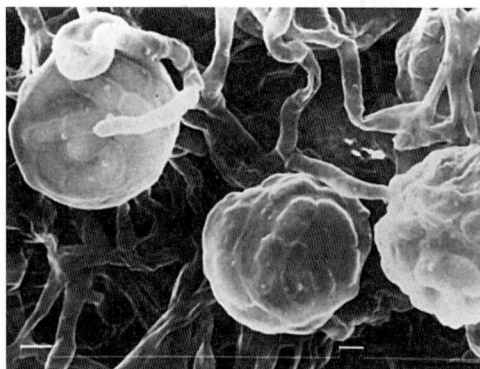

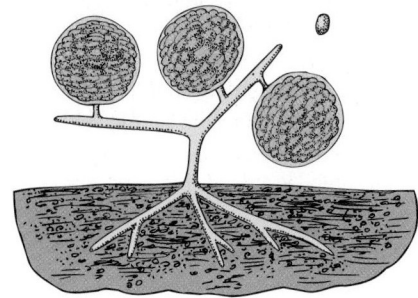

Streptosporangium

(b)

Figure 24.16 Maduromycetes. (**a**) *Actinomadura madurae* morphology; illustration and electron micrograph of a spore chain (×16,500). (**b**) *Streptosporangium* morphology; illustration and micrograph of *S. album* on oatmeal agar with sporangia and hyphae; SEM. Bar interval = 10 μm.

duced at fairly regular intervals. These branches have secondary branches bearing chains of spores (figure 24.13*b*).

24.9 SUBORDER *STREPTOSPORANGINEAE*

The suborder *Streptosporangineae* contains three families and 16 genera. The genera in this suborder are found in section 30 (maduromycetes) and section 31 (*Thermomonospora* and related genera) of the first edition of *Bergey's Manual.* All genera of the maduromycetes have type III cell walls and the sugar derivative **madurose** (3-*O*-methyl-D-galactose) in whole cell homogenates. Their G + C content is 64 to 74 mol%. Aerial mycelia bear pairs or short chains of spores, and the substrate mycelium is branched (figure 24.2*e* and **figure 24.16**). Some genera form sporangia; spores are not heat resistant. *Actinomadura* is another actinomycete associated with the disease actinomycetoma. *Thermomonospora* produces single spores on the aerial mycelium or on both the aerial and substrate mycelia. It has been isolated from high-temperature habitats such as compost piles and hay; it can grow at 40 to 48°C.

24.10 SUBORDER *FRANKINEAE*

Like the suborder *Streptosporangineae, Frankineae* contains genera from different parts of the first edition. *Frankia* and *Geodermatophilus* are actinomycetes with multilocular sporangia. Both genera form clusters of spores when a hypha divides both transversely and longitudinally. (Multilocular means having many cells or compartments.) They have type III cell walls (table 24.1), although the cell extract sugar patterns differ. The G + C content varies from 57 to 75 mol%. *Geodermatophilus* (type IIIC) has motile spores and is an aerobic soil organism. *Frankia* (type IIID) forms nonmotile sporangiospores in a sporogenous body (**figure 24.17**). It grows in symbiotic association with the roots of at least eight families of higher nonleguminous plants (e.g., alder trees) and is a microaerophile that can fix atmospheric nitrogen (*see section 30.4*).

The roots of infected plants develop nodules that fix nitrogen so efficiently that a plant such as an alder can grow in the absence of combined nitrogen when nodulated. Within the nodule cells, *Frankia* forms branching hyphae with globular vesicles at their ends (figure 24.17*c*). These vesicles may be the sites of nitrogen fixation. The nitrogen-fixation process resembles that of *Rhizobium* in that it is oxygen sensitive and requires molybdenum and cobalt.

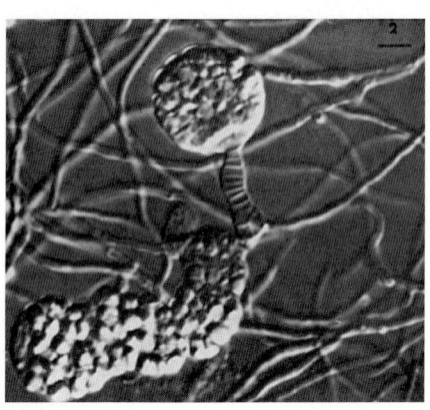
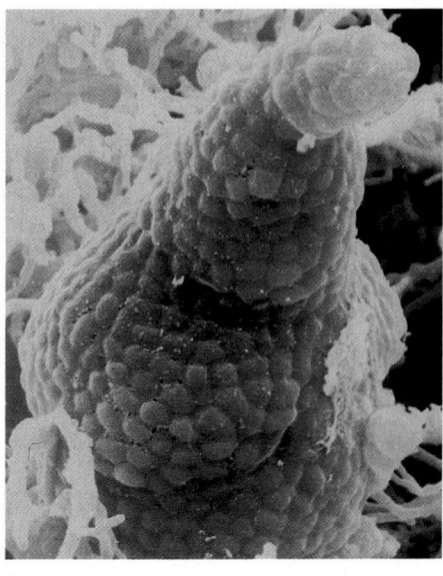
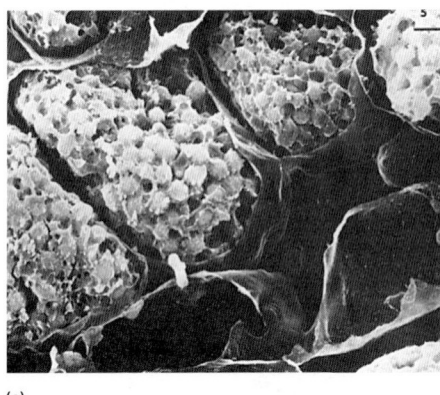

(a)   (b)   (c)

Figure 24.17 *Frankia.* (**a**) An interference contrast micrograph showing hyphae, multilocular sporangia, and spores. (**b**) A scanning electron micrograph of a sporangium surrounded by hyphae. (**c**) A nodule of the alder *Alnus rubra* showing cells filled with vesicles of *Frankia.* Scanning electron microscopy.

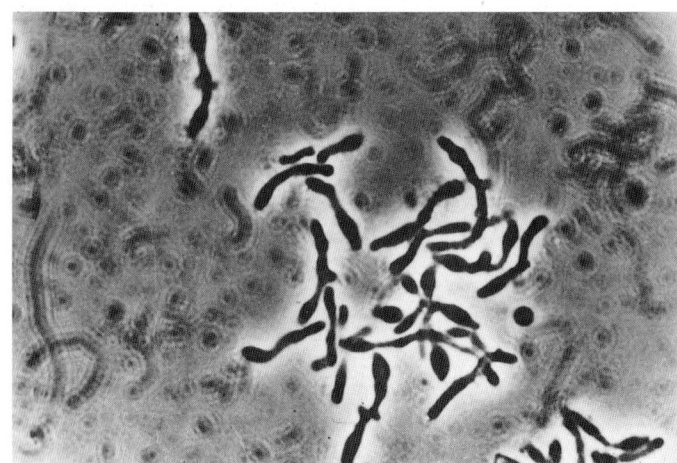

Figure 24.18 *Bifidobacterium. Bifidobacterium bifidum;* phase-contrast photomicrograph ($\times 1,500$).

Another genus in this suborder, *Sporichthya,* has been transferred from section 29 (streptomycetes and related genera) of the first edition. *Sporichthya* is one of the strangest of the actinomycetes. It lacks a substrate mycelium. The hyphae remain attached to the substratum by holdfasts and grow upward to form aerial mycelia that release motile, flagellate conidia in the presence of water.

 24.11 ORDER *BIFIDOBACTERIALES*

The order *Bifidobacteriales* has one true family, *Bifidobacteriaceae,* and ten genera (five of which have unknown affiliation). *Falcivibrio* and *Gardnerella* are found in the human genital/uri-nary tract; *Gardnerella* is thought to be a major cause of bacterial vaginitis. *Bifidobacterium* probably is the best-studied genus. Bifidobacteria are nonmotile, nonsporing, gram-positive rods of varied shapes that are slightly curved and clubbed; often they are branched (**figure 24.18**). The rods can be single or in clusters and V-shaped pairs. *Bifidobacterium* is anaerobic and actively ferments carbohydrates to produce acetic and lactic acids, but no carbon dioxide. It is found in the mouth and intestinal tract of warm-blooded vertebrates, in sewage, and in insects. *B. bifidus* is a pioneer colonizer of the human intestinal tract, particularly when babies are breast fed. A few *Bifidobacterium* infections have been reported in humans, but the genus does not appear to be a major cause of disease.

1. Give the distinguishing properties of the actinoplanetes.
2. Briefly describe the variations in sporangia and sporophore organization of the actinoplanetes.
3. Describe the genus *Propionibacterium* and comment on its practical importance.
4. What characteristics do streptomycetes have in common?
5. Describe the major properties of the genus *Streptomyces.*
6. Give three ways in which *Streptomyces* is of ecological and medical importance.
7. Briefly describe the genera of the suborder *Streptosporangineae.* What is madurose? Why is *Actinomadura* important?
8. Describe *Frankia* and discuss its importance.
9. Characterize the genus *Bifidobacterium.* Where is it found and why is it significant?

Summary

24.1 General Properties of the Actinomycetes

a. Actinomycetes are aerobic, gram-positive bacteria that form branching, usually non-fragmenting, hyphae and asexual spores (**figure 24.1**).

b. The asexual spores borne on aerial mycelia are called conidiospores or conidia if they are at the tip of hyphae and sporangiospores when they are within sporangia.

c. Actinomycetes have several distinctively different types of cell walls and often also vary in terms of the sugars present in cell extracts. Properties such as color and morphology are also taxonomically useful.

24.2 High G + C Gram-Positive Bacteria in *Bergey's Manual*

a. The first edition of *Bergey's Manual* classifies the actinomycetes based on such properties as conidial arrangement, the presence of a sporangium, cell wall type, and cell extract sugars (**tables 24.1–24.3**).

b. The second edition of *Bergey's Manual* classifies the high G + C bacteria phylogenetically using 16S rRNA data. The phylum *Actinobacteria* contains the actinomycetes and their high G + C relatives (**figure 24.3**).

24.3 Suborder *Actinomycineae*

a. The suborder *Actinomycineae* contains the genus *Actinomyces,* members of which are irregularly shaped, nonsporing rods that can cause disease in cattle and humans.

24.4 Suborder *Micrococcineae*

a. The suborder *Micrococcineae* has the genera *Micrococcus, Arthrobacter,* and *Dermatophilus. Arthrobacter* has an unusual rod-coccus growth cycle and carries out snapping division (**figure 24.7**)

24.5 Suborder *Corynebacterineae*

a. The genera *Corynebacterium, Mycobacterium,* and *Nocardia* will be placed in the suborder *Corynebacterineae.* Mycobacteria form either rods or filaments that readily fragment. Their cell walls have a high lipid content and mycolic acids; the presence of these lipids makes them acid-fast. The genera *Corynebacterium* and *Mycobacterium* contain several very important human pathogens.

b. Nocardioform actinomycetes have hyphae that readily fragment into rods and coccoid elements, and often form aerial mycelia with spores.

24.6 Suborder *Micromonosporineae*

a. The suborder *Micromonosporineae* has genera that are classified as actinoplanetes in section 28 of the first edition. These actinomycetes have an extensive substrate mycelium and form special aerial sporangia (**figure 24.12**). They are present in soil, fresh water, and the ocean. The soil forms are probably important in decomposition.

24.7 Suborder *Propionibacterineae*

a. The genus *Propionibacterium* has been placed in the suborder *Propionibacterineae* in the second edition. Members of this genus are common skin and intestinal inhabitants and are important in cheese manufacture and the development of acne vulgaris.

24.8 Suborder *Streptomycineae*

a. The suborder *Streptomycineae* of the second edition contain the genus *Streptomyces.* Members of this genus have type I cell walls and aerial hyphae bearing chains of 3 to 50 or more nonmotile conidiospores within a thin sheath (**figure 24.13**).

b. Streptomycetes are important in the degradation of more resistant organic material in the soil and produce many useful antibiotics. A few cause diseases in plants and animals.

24.9 Suborder *Streptosporangineae*

a. The suborder *Streptosporangineae* contains several genera that the first edition places in section 30 (the maduromycetes) and section 31 (*Thermomonospora* and related genera). Many of these organisms have the sugar derivative madurose and type III cell walls.

24.10 Suborder *Frankineae*

a. The genera *Frankia* and *Geodermatophilus* are placed in the suborder *Frankineae.* They produce clusters of spores at hyphal tips and have type III cell walls. *Frankia* grows in symbiotic association with nonleguminous plants and fixes nitrogen.

24.11 Order *Bifidobacteriales*

a. The genus *Bifidobacterium* is placed in the order *Bifidobacteriales* in the second edition. This irregular, anaerobic rod is one of the first colonizers of the intestinal tract in nursing babies.

Key Terms

acid-fast 528
actinobacteria 524
actinomycete 522
actinomycetoma 532

conidia 522
conidiospores 522
geosmin 531
madurose 533

mycolic acids 528
nocardioforms 528
snapping division 527

sporangiospores 522
streptomycetes 531
thallus 522

Questions for Thought and Review

1. Suppose that you discovered a nodulated plant that could fix atmospheric nitrogen. How might you show that a bacterial symbiont was involved and that *Frankia* rather than *Rhizobium* was responsible?

2. How could one decide whether a newly isolated spore-forming bacterium was producing conidiospores or true endospores?

3. List the properties that are most useful in actinomycete taxonomy and give some indication of their relative importance. How does the second edition of *Bergey's Manual* differ in its approach to classification of actinomycetes?

4. Which bacteria are noted for each of the following: the rod-coccus growth cycle, early colonization of the human intestine, metachromatic granules and snapping division, 60 to 90 carbon mycolic acids and acid fastness, the synthesis of antibiotics, the production of substrate mycelia that readily break into rods and cocci, and the production of large amounts of propionic and acetic acids.

5. Propose an explanation for the observation that mycobacteria are normally only weakly gram positive.

6. Which characteristics might be most important in distinguishing between the genera *Corynebacterium, Arthrobacter, Actinomyces, Propionibacterium, Mycobacterium,* and *Bifidobacterium?*

Critical Thinking Questions

1. Even though these are "high G + C" organisms, there are regions of the genome that must be more AT-rich. Suggest a few such regions and explain why they must be more AT-rich.

2. Choose two different species in the phylum *Actinobacteria* and investigate their physiology and ecology. Compare and contrast these two organisms. Can you determine why having a high G + C genomic content might confer an evolutionary advantage?

3. *Streptomyces coelicolor* has a developmental program for differentiation. Some of the genes involved in sporulation contain a rare codon, not used in vegetative genes. Suggest how *Streptomyces* might turn on sporulation, in response to starvation.

For recommended readings on these and related topics, see Appendix V.

chapter 25

The Fungi (Eumycota), Slime Molds, and Water Molds

Concepts

- Fungi are widely distributed and are found wherever moisture is present. They are of great importance to humans in both beneficial and harmful ways.
- Fungi exist primarily as filamentous hyphae. A mass of hyphae is called a mycelium.
- Like some bacteria, fungi digest insoluble organic matter by secreting exoenzymes, then absorbing the solubilized nutrients.
- Two reproductive structures occur in the fungi: (1) sporangia form asexual spores, and (2) gametangia form sexual gametes.
- The zygomycetes are characterized by resting structures called zygospores—cells in which zygotes are formed.
- The ascomycetes form zygotes within a characteristic saclike structure, the ascus. The ascus contains two or more ascospores.
- Yeasts are unicellular fungi—mainly ascomycetes.
- Basidiomycetes possess dikaryotic hyphae with two nuclei, one of each mating type. The hyphae divide uniquely, forming basidiocarps within which club-shaped basidia can be found. The basidia bear two or more basidiospores.
- The deuteromycetes (Fungi Imperfecti) have either lost the capacity for sexual reproduction, or it has never been observed.
- The chytrids are a group of terrestrial and aquatic fungi that reproduce by motile zoospores with single,

This is a scanning electron micrograph of the microscopic, unicellular yeast, *Saccharomyces cerevisiae* ($\times 21,000$). *S. cerevisiae* is the most thoroughly investigated eucaryotic microorganism in the world. This has led to a better understanding of the biology of the eucaryotic cell. Today it serves as a widely used biotechnological production organism as well as a eucaryotic model system.

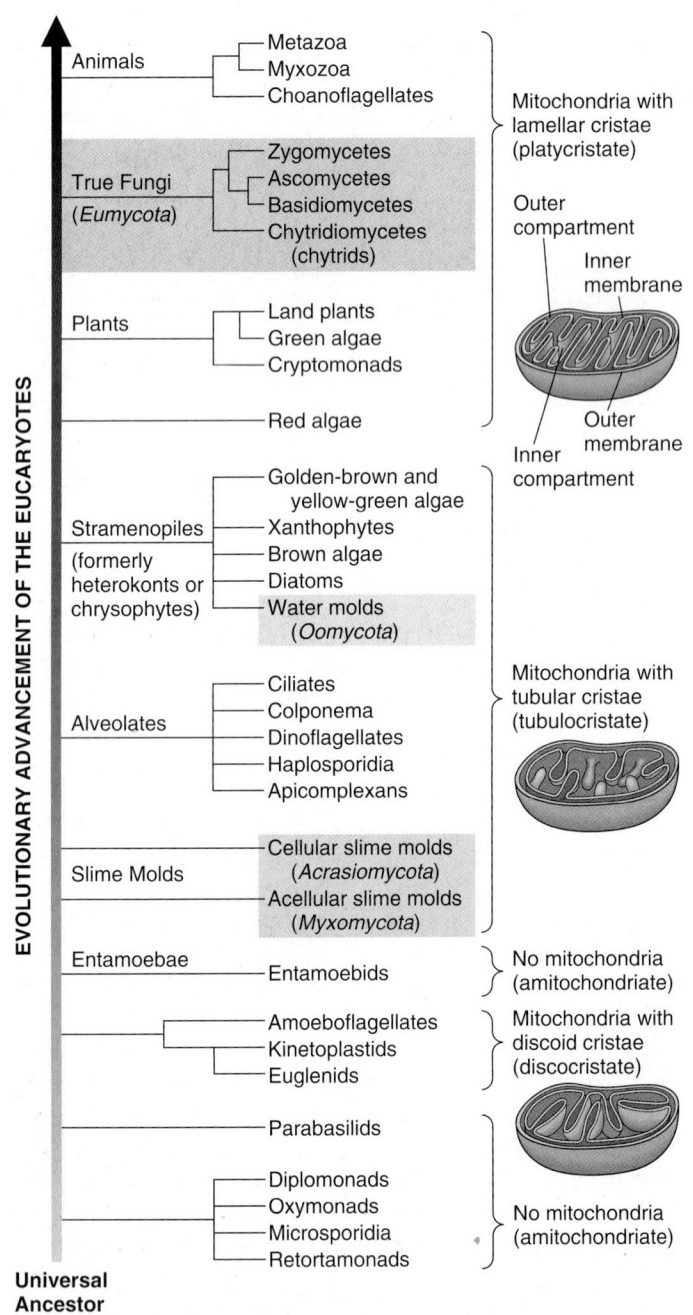

(see figure 19.12a)

Phylogenetic Diagram 25   Tentative Phylogeny of the True Fungi (*Eumycota*), Slime Molds, and Water Molds Based on 18S rRNA Sequence Comparisons.   Using molecular systematics, organisms are grouped together based on the molecular phylogeny of their nuclear SSU rRNA genes and the type of mitochondrial cristae present. According to this tentative phylogeny, the true fungi, slime molds, and water molds form three distinct monophyletic branches (highlighted by three different colors). Accordingly, the four major groups of fungi (Zygomycetes, Ascomycetes, Basidiomycetes, and Chytridiomycetes) form a phylogenetically coherent group. Molecular characters have been essential for phylogenetic analysis in cases when morphological characters are convergent, reduced, or missing. This is especially true for species that never reproduce sexually (e.g., the Fungi Imperfecti or *Deuteromycota*), because characters of sexual reproduction traditionally have been the basis for classifying fungi. The use of molecular characters allows the asexual fungi (*Deuteromycota*) to be placed among their closest relatives in the *Eumycota* and not in a separate division.

*T*his chapter is an introduction to the true fungi. It surveys their diversity, discusses their ecological and commercial impact, and presents some of their typical life cycles. A brief overview is also given of the eucaryotes that resemble the fungi—the slime molds and water molds.

Microbiologists use the term **fungus** [pl., fungi; Latin *fungus,* mushroom] to include eucaryotic, spore-bearing organisms with absorptive nutrition, no chlorophyll, and that reproduce sexually and asexually. Scientists who study fungi are **mycologists** [Greek *mykes,* mushroom, and *logos,* science], and the scientific discipline dealing with fungi is called **mycology.** The study of fungal toxins and their effects is called **mycotoxicology,** and the diseases caused by fungi in animals are known as **mycoses** [s., mycosis]. The five-kingdom system places the fungi in the kingdom *Fungi* (*see figure 19.12*a). According to the universal phylogenetic tree, fungi are members of the domain *Eucarya* (*see figure 19.3*). As presently deliminated, the kingdom *Fungi* is believed to constitute a monophyletic group (**Phylogenetic Diagram 25**) referred to as the true fungi or *Eumycota* (Greek *eu,* true, and *mykes,* fungus).

25.1 DISTRIBUTION

Fungi are primarily terrestrial organisms, although a few are freshwater or marine. Many are pathogenic and infect plants and animals. Fungi also form beneficial relationships with other organisms. For example, about three-fourths of all vascular plants

---

posterior, whiplash flagella, and represent a link between the true fungi and protists.

- The slime and water molds resemble the fungi only in appearance and life-style. In their cellular organization, reproduction, and life cycles, they are phylogenetically distinct.

Yeasts, molds, mushrooms, mildews, and the other fungi pervade our world. They work great good and terrible evil. Upon them, indeed, hangs the balance of life; for without their presence in the cycle of decay and regeneration, neither man nor any other living thing could survive.

—*Lucy Kavaler*

Table 25.1

Some Mycotoxicoses[a] Produced by Fungal Mycotoxins in Domestic Animals

Disease	Fungus	Mycotoxin	Contaminated Foodstuff	Animals Affected
Aflatoxicosis	*Aspergillus flavus*	Aflatoxins	Rice, corn, sorghum, cereals, peanuts, soybeans	Poultry, swine, cattle, sheep, dogs
Ergotism	*Claviceps purpurea*	Ergot alkaloids	Seedheads of many grasses, grains	Cattle, horses, swine, poultry
Mushroom poisoning	*Amanita verna*	Amanitins	Eaten from pastures	Cattle
Poultry hemorrhagic syndrome	*Aspergillus flavus* and others	Aflatoxins	Toxic grain and meal	Chickens
Slobbers	*Rhizoctonia*	Alkaloid slaframine	Red clover	Sheep, cattle
Tall fescue toxicosis	*Acremonium coenophialum* (an endophytic fungus)	Ergot alkaloids	Endophyte-infected tall fescue plants	Cattle, horses

[a]A mycotoxicosis [pl., mycotoxicoses] is a poisoning caused by a fungal toxin.

form associations (called mycorrhizae) between their roots and fungi. Fungi also are found in the upper portions of many plants. These endophytic fungi affect plant reproduction and palatability to herbivores. Lichens are associations of fungi and either algae or cyanobacteria. Mycorrhizae (pp. 656–58); Endophytic fungi (pp. 655–56); Lichens (p. 591)

25.2 IMPORTANCE

About 90,000 fungal species have been described; however, some estimates of total numbers suggest that 1.5 million species may exist.

Fungi are important to humans in both beneficial and harmful ways. With bacteria and a few other groups of heterotrophic organisms, fungi act as decomposers, a role of enormous significance. They degrade complex organic materials in the environment to simple organic compounds and inorganic molecules. In this way carbon, nitrogen, phosphorus, and other critical constituents of dead organisms are released and made available for living organisms (*see section 28.3*).

However, fungi are the major cause of plant diseases. Over 5,000 species attack economically valuable crops and garden plants and also many wild plants. In like manner many diseases of animals (**table 25.1**) and humans (*see section 40.1*) are caused by fungi.

Fungi, especially the yeasts, are essential to many industrial processes involving fermentation (*see chapter 41*). Examples include the making of bread, wine, and beer. Fungi also play a major role in the preparation of some cheeses, soy sauce, and sufu; in the commercial production of many organic acids (citric, gallic) and certain drugs (ergometrine, cortisone); and in the manufacture of many antibiotics (penicillin, griseofulvin) and the immunosuppressive drug cyclosporine.

In addition, fungi are important research tools in the study of fundamental biological processes. Cytologists, geneticists, biochemists, biophysicists, and microbiologists regularly use fungi

in their research. Based on this research the yeast *Saccharomyces cerevisiae* is the best understood eucaryotic cell.

25.3 STRUCTURE

The body or vegetative structure of a fungus is called a **thallus** [pl., thalli]. It varies in complexity and size, ranging from the single-cell microscopic yeasts to multicellular molds, macroscopic puffballs, and mushrooms (**figure 25.1**). The fungal cell usually is encased in a cell wall of **chitin.** Chitin is a strong but flexible nitrogen-containing polysaccharide consisting of *N*-acetylglucosamine residues.

A **yeast** is a unicellular fungus that has a single nucleus and reproduces either asexually by budding and transverse division or sexually through spore formation. Each bud that separates can grow into a new yeast, and some group together to form colonies. Generally yeast cells are larger than bacteria, vary considerably in size, and are commonly spherical to egg shaped. They have no flagella but do possess most of the other eucaryotic organelles (**figure 25.2**).

A **mold** (figure 25.1*a*) consists of long, branched, threadlike filaments of cells called **hyphae** [s., hypha; Greek *hyphe,* web] that form a **mycelium** [pl., mycelia], a tangled mass or tissuelike aggregation (**figure 25.3**). In some fungi, protoplasm streams through hyphae, uninterrupted by cross walls. These hyphae are called **coenocytic** (**figure 25.4***a*). The hyphae of other fungi (figure 25.4*b*) have cross walls called **septa** [s., septum] with either a single pore (figure 25.4*c*) or multiple pores (figure 25.4*d*) that permit cytoplasmic streaming. These hyphae are termed **septate.**

Hyphae are composed of an outer cell wall and an inner lumen, which contains the cytosol and organelles (**figure 25.5**). A plasma membrane surrounds the cytoplasm and lies next to the cell wall.

Many fungi, especially those that cause diseases in humans and animals, are dimorphic (**table 25.2**)—that is, they have two

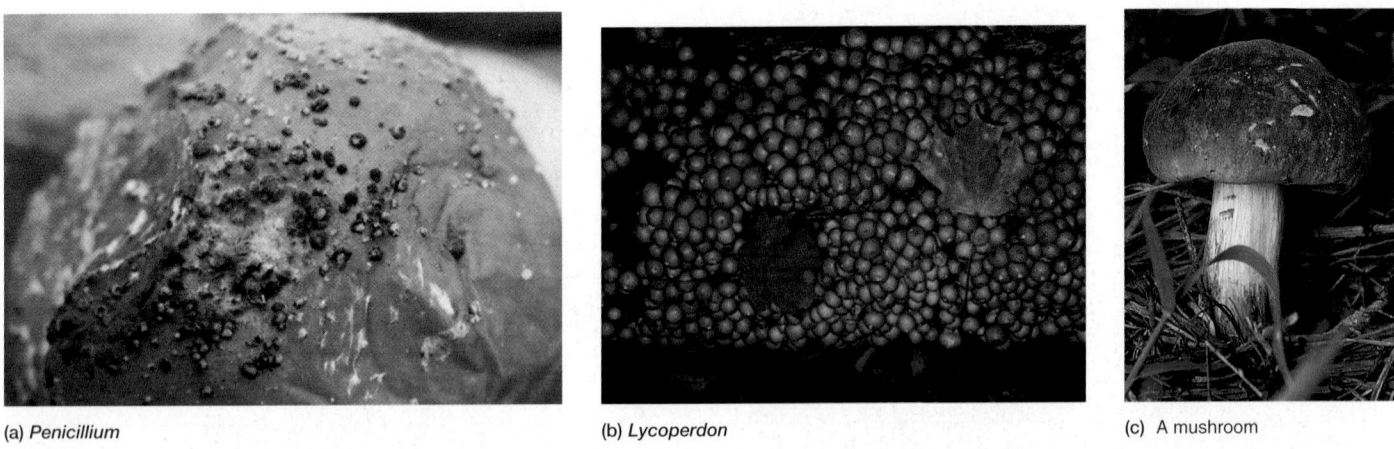

(a) *Penicillium*   (b) *Lycoperdon*   (c) A mushroom

Figure 25.1   Fungal Thalli.   (a) The multicellular common mold, *Penicillium,* growing on an apple. **(b)** A large group of puffballs, *Lycoperdon,* growing on a log. **(c)** A mushroom is made up of densely packed hyphae that form the mycelium or visible structure (thallus).

Figure 25.2   A Yeast.   Diagrammatic drawing of a yeast cell showing typical morphology. For clarity, the plasma membrane has been drawn separated from the cell wall. In a living cell the plasma membrane adheres tightly to the cell wall.

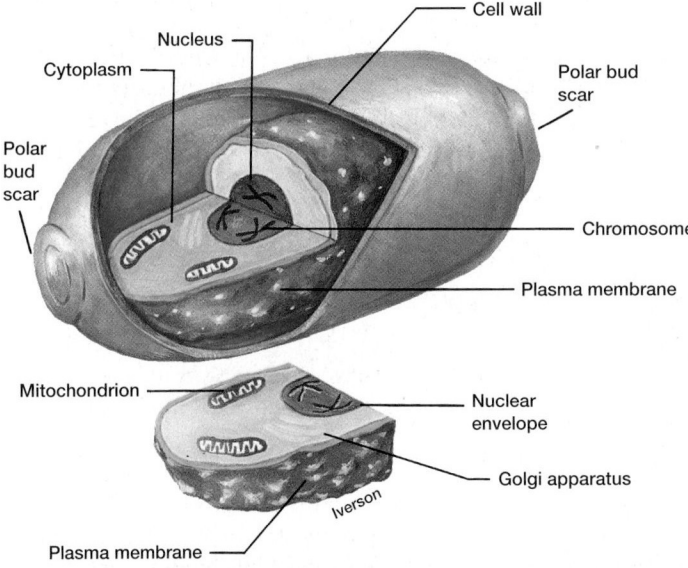

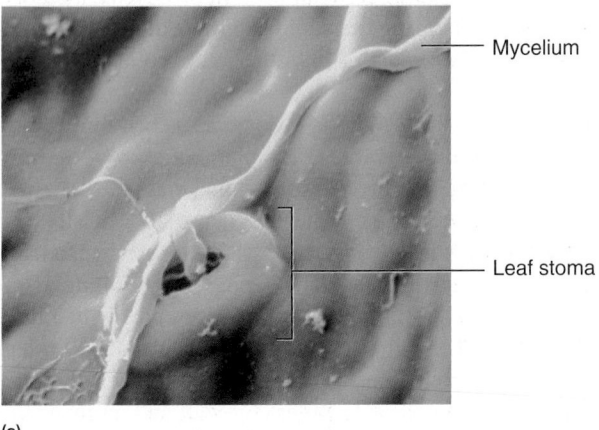

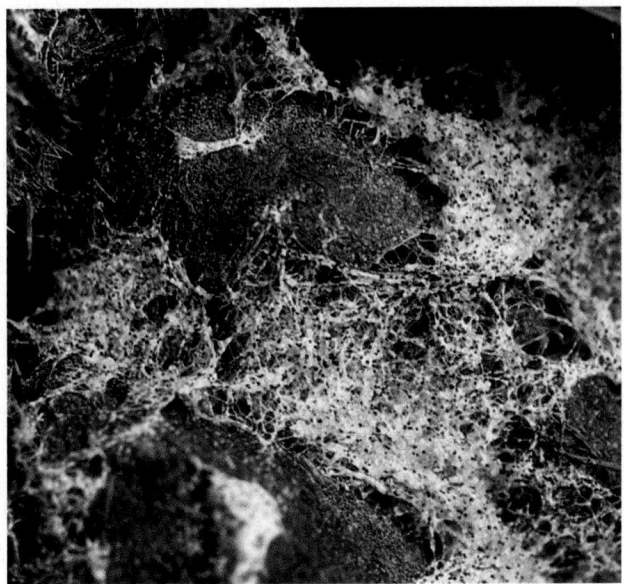

Figure 25.3   Mold Mycelia.   (a) Scanning electron micrograph of a young mycelial aggregate forming over a leaf stoma (×1,000). **(b)** A very large macroscopic mycelium of a basidiomycete growing on the soil.

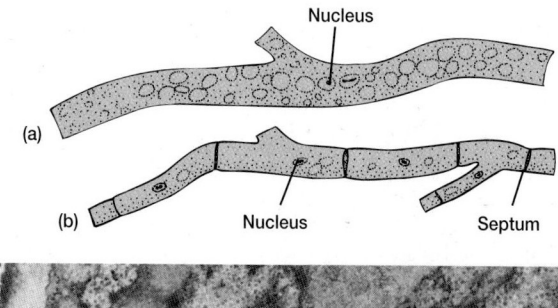

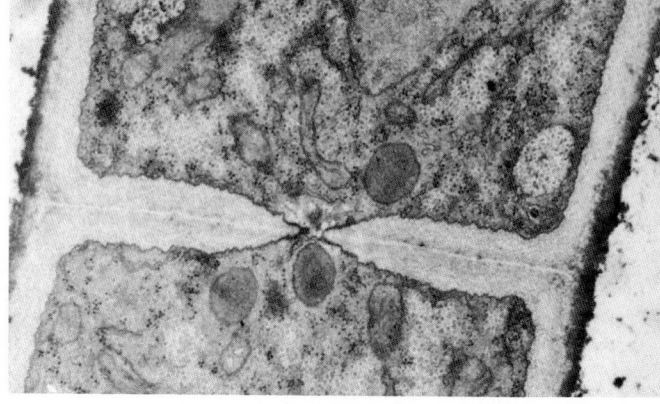

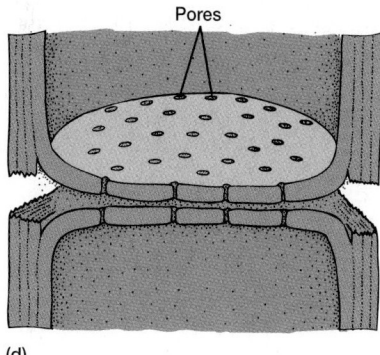

Figure 25.4 Hyphae. Drawings of (**a**) coenocytic hyphae and (**b**) hyphae divided into cells by septa. (**c**) Electron micrograph (×40,000) of a section of *Drechslera sorokiniana* showing wall differentiation and a single pore. (**d**) Drawing of a multiperforate septal structure.

Table 25.2
Some Medically Important Dimorphic Fungi

Fungus	Disease[a]
Blastomyces dermatitidis	Blastomycosis
Candida albicans	Candidiasis
Coccidioides capsulatum	Coccidioidomycosis
Histoplasma capsulatum	Histoplasmosis
Sporothrix schenckii	Sporotrichosis
Paracoccidioides brasiliensis	Paracoccidioidomycosis

[a]See chapter 40 for a discussion of each of these diseases.

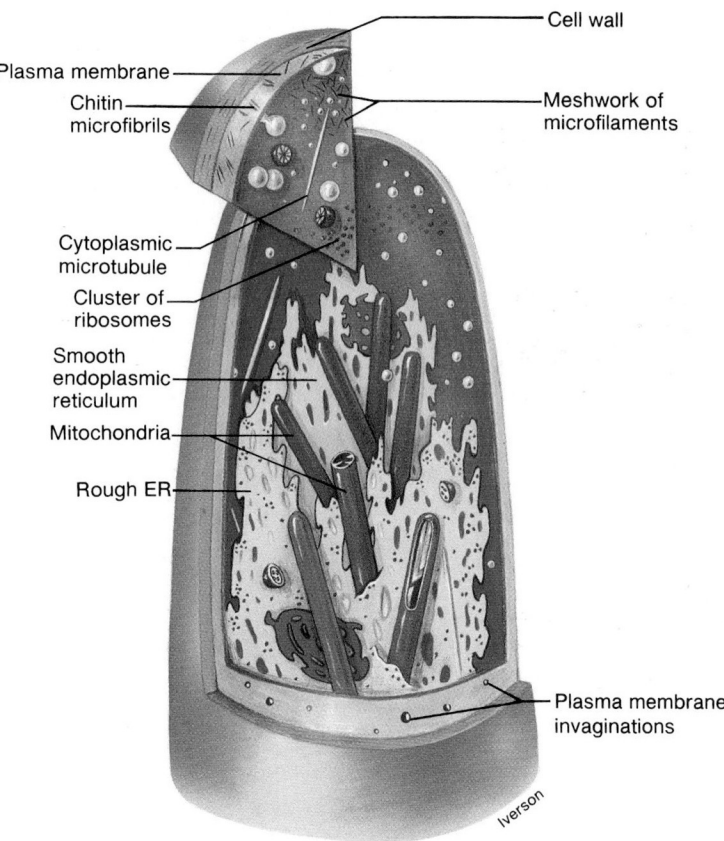

Figure 25.5 Hyphal Morphology. Diagrammatic representation of hyphal tip showing typical organelles and other structures.

forms. Dimorphic fungi can change from (1) the yeast (Y) form in the animal to (2) the mold or mycelial form (M) in the external environment in response to changes in various environmental factors (nutrients, CO_2 tension, oxidation-reduction potentials, temperature). This shift is called the **YM shift.** In plant-associated fungi the opposite type of dimorphism exists: the mycelial form occurs in the plant and the yeast form in the external environment.

1. How can a fungus be defined?
2. Where are fungi found?
3. Why are fungi important as decomposers?
4. What are some forms represented by different fungal thalli?
5. What organelles would you expect to find in the cytoplasm of a typical fungus?
6. Describe a typical yeast; a typical mold.

25.4 NUTRITION AND METABOLISM

Fungi grow best in dark, moist habitats, but they are found wherever organic material is available. Most fungi are **saprophytes,** securing their nutrients from dead organic material. Like many

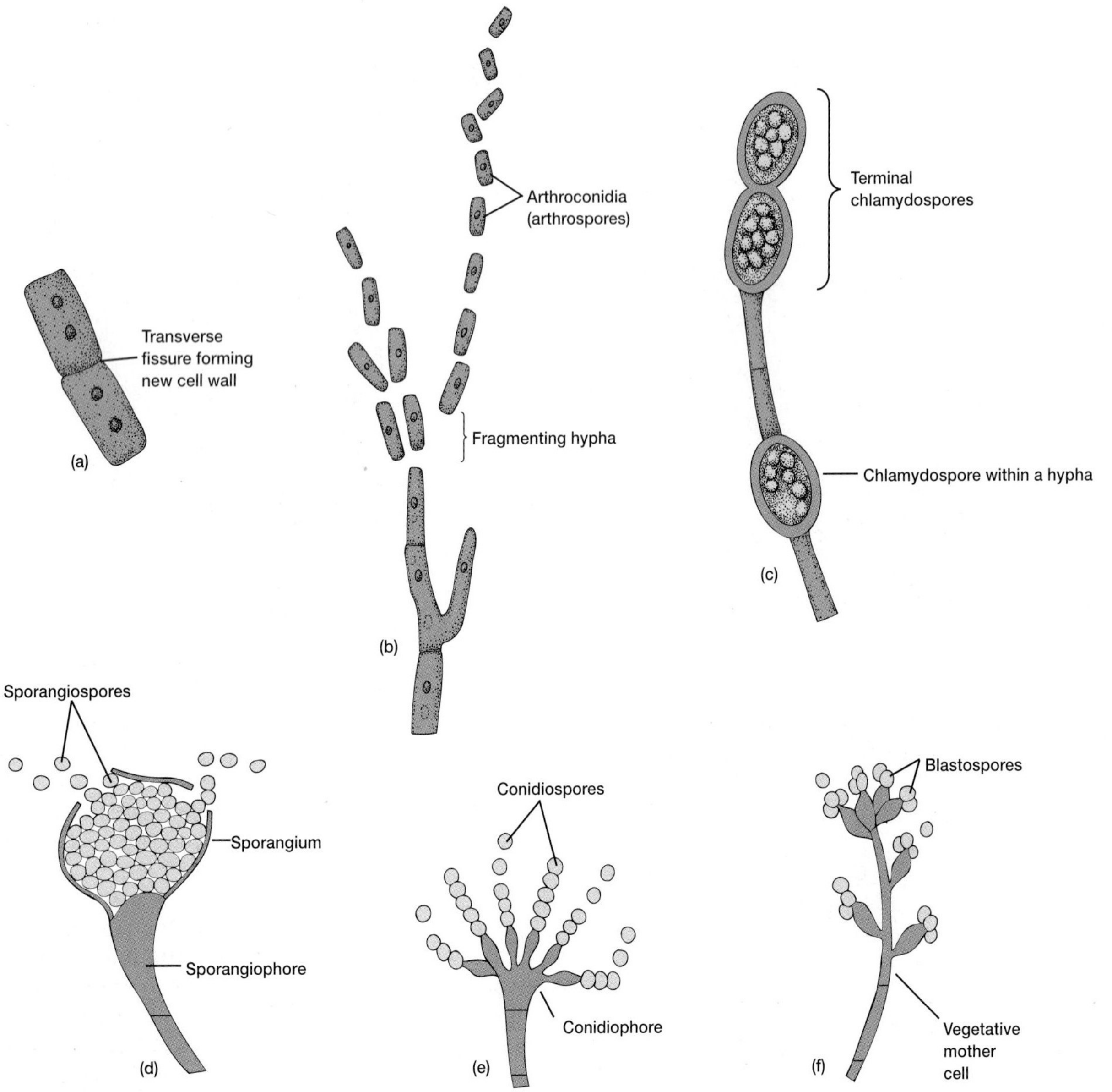

Arthroconidia
(arthrospores)

Transverse
fissure forming
new cell wall

(a)

Fragmenting hypha

(b)

Terminal
chlamydospores

Chlamydospore within a hypha

(c)

Sporangiospores

Sporangium

Sporangiophore

(d)

Conidiospores

Conidiophore

(e)

Blastospores

Vegetative
mother
cell

(f)

Figure 25.6   **Diagrammatic Representation of Asexual Reproduction in the Fungi and Some Representative Spores.**   (**a**) Transverse fission. (**b**) Hyphal fragmentation resulting in arthroconidia (arthrospores) and (**c**) chlamydospores. (**d**) Sporangiospores in a sporangium. (**e**) Conidiospores arranged in chains at the end of a conidiophore. (**f**) Blastospores are formed from buds off of the parent cell.

bacteria, fungi release hydrolytic exoenzymes that digest external substrates. They then absorb the soluble products. They are chemoorganoheterotrophs and use organic compounds as a source of carbon, electrons, and energy.

Glycogen is the primary storage polysaccharide in fungi. Most fungi use carbohydrates (preferably glucose or maltose) and nitrogenous compounds to synthesize their own amino acids and proteins.

Fungi usually are aerobic. Some yeasts, however, are facultatively anaerobic and can obtain energy by fermentation, such as in the production of ethyl alcohol from glucose. Obligately anaerobic fungi are found in the rumen of cattle.

25.5 REPRODUCTION

Reproduction in fungi can be either asexual or sexual. Asexual reproduction is accomplished in several ways:

1. A parent cell can divide into two daughter cells by central constriction and formation of a new cell wall (**figure 25.6a**).
2. Somatic vegetative cells may bud to produce new organisms. This is very common in the yeasts.
3. The most common method of asexual reproduction is spore production. Asexual spore formation occurs in an individual

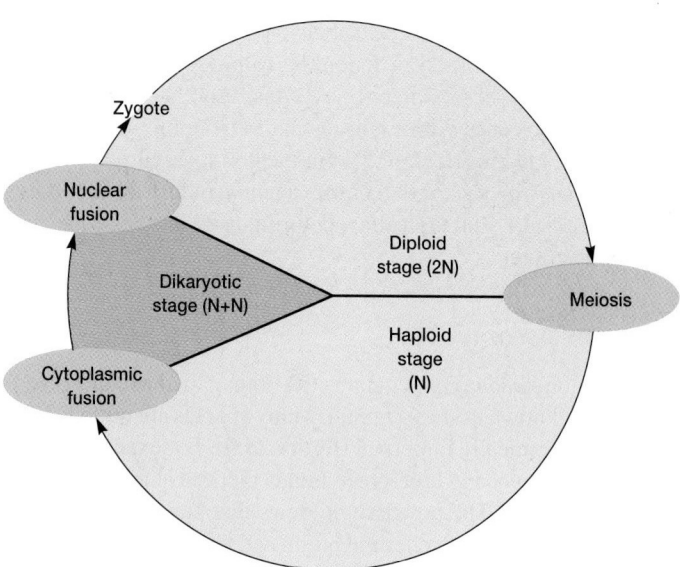

Figure 25.7 Reproduction in Fungi. A drawing of the generalized life cycle for fungi showing the alternation of haploid and diploid stages. Some fungal species do not pass through the dikaryotic stage indicated in this drawing. The asexual (haploid) stage is used to produce spores that aid in the dissemination of the species. The sexual (diploid) stage involves the formation of spores that survive adverse environmental conditions (e.g., cold, dryness, heat).

fungus through mitosis and subsequent cell division. There are several types of asexual spores:

a. A hypha can fragment (by the separation of hyphae through splitting of the cell wall or septum) to form cells that behave as spores. These cells are called **arthroconidia** or **arthrospores** (figure 25.6*b*).

b. If the cells are surrounded by a thick wall before separation, they are called **chlamydospores** (figure 25.6*c*).

c. If the spores develop within a sac [**sporangium;** pl., sporangia] at a hyphal tip, they are called **sporangiospores** (figure 25.6*d*).

d. If the spores are not enclosed in a sac but produced at the tips or sides of the hypha, they are termed **conidiospores** (figures 25.6*e* and 25.10).

e. Spores produced from a vegetative mother cell by budding (figure 25.6*f*) are called **blastospores.**

Sexual reproduction in fungi involves the union of compatible nuclei. Some fungal species are self-fertilizing and produce sexually compatible gametes on the same mycelium (homothallic). Other species require outcrossing between different but sexually compatible mycelia (heterothallic). Depending on the species, sexual fusion may occur between haploid gametes, gamete-producing bodies called **gametangia,** or hyphae. Sometimes both the cytoplasm and haploid nuclei fuse immediately to produce the diploid zygote. Usually, however, there is a delay between cytoplasmic and nuclear fusion. This produces a **dikaryotic stage** in which cells contain two separate haploid nuclei, one from each

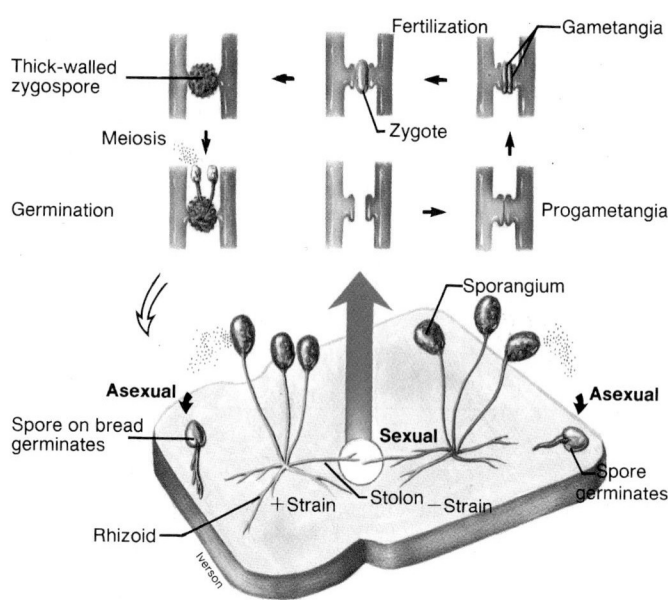

Figure 25.8 Division *Zygomycota*. Diagrammatic representation of the life cycle of *Rhizopus stolonifer*. Both the sexual and asexual phases are illustrated.

parent (**figure 25.7**). After a period of dikaryotic existence, the two nuclei fuse. This sexual reproduction yields spores. For example, in the zygomycetes the zygote develops into a **zygospore** (**figure 25.8**); in the ascomycetes, an **ascospore** (figure 25.11); and in the basidomycetes; a **basidiospore** (figure 25.13).

Fungal spores are important for several reasons. The size, shape, color, and number of spores are useful in the identification of fungal species. The spores are often small and light; they can remain suspended in air for long periods. Thus they frequently aid in fungal dissemination, a significant factor that explains the wide distribution of many fungi. Fungal spores often spread by adhering to the bodies of insects and other animals. The bright colors and fluffy textures of many molds often are due to their aerial hyphae and spores.

---

1. What are saprophytes? How do fungi usually obtain their nutrients?
2. How is asexual reproduction accomplished in the fungi? Sexual reproduction?
3. Describe each of the following types of asexual fungal spores: sporangiospore, conidiospore, and blastospore.
4. What generally governs the ecological distribution of fungi?

CHARACTERISTICS OF THE FUNGAL DIVISIONS
25.6

The traditional taxonomic scheme used by mycologists classifies the fungi into four divisions (**table 25.3**), based primarily on variations in sexual reproduction. (In mycology a division is equivalent

Table 25.3
Divisions of Fungi

Division	Common Name	Approximate Number of Species
Zygomycota	Zygomycetes	600
Ascomycota	Sac fungi	35,000
Basidiomycota	Club fungi	30,000
Deuteromycota[a]	Fungi Imperfecti	30,000

[a]According to the traditional system often followed by mycologists.

to a phylum in animal classification schemes.) Based on 18S rRNA studies, molecular microbiologists place the *Deuteromycota* (Fungi Imperfecti) among their closest relatives in either the *Zygomycota, Ascomycota,* or *Basidiomycota* and add the class *Chytridiomycetes* (Phylogenetic Diagram 25).

Division *Zygomycota*

The division *Zygomycota* contains the fungi called **zygomycetes.** Most live on decaying plant and animal matter in the soil; a few are parasites of plants, insects, animals, and humans. The hyphae of zygomycetes are coenocytic, with many haploid nuclei. Asexual spores, usually wind dispersed, develop in sporangia at the tips of aerial hyphae. Sexual reproduction produces tough, thick-walled zygotes called zygospores that can remain dormant when the environment is too harsh for growth of the fungus.

The bread mold, *Rhizopus stolonifer,* is a very common member of this division. This fungus grows on the surface of moist, carbohydrate-rich foods, such as breads, fruits, and vegetables. On breads, for example, *Rhizopus's* hyphae rapidly cover the surface. Special hyphae called rhizoids extend into the bread, and absorb nutrients (figure 25.8). Other hyphae (stolons) become erect, then arch back into the substratum forming new rhizoids. Still others remain erect and produce at their tips asexual sporangia filled with the black spores, giving the mold its characteristic color. Each spore, when liberated, can start a new mycelium.

Rhizopus usually reproduces asexually, but if food becomes scarce or environmental conditions unfavorable, it begins sexual reproduction. Sexual reproduction requires compatible strains of opposite mating types (figure 25.8). These have traditionally been labeled + and − strains because they are not morphologically distinguishable as male and female. When the two mating strains are close, hormones are produced that cause their hyphae to form projections called **progametangia** [Greek *pro,* before], and then mature gametangia. After fusion of the gametangia, the nuclei of the two gametes fuse, forming a zygote. The zygote develops a thick, rough, black coat and becomes a dormant zygospore. Meiosis often occurs at the time of germination; the zygospore then splits open and produces a hypha that bears an asexual sporangium and the cycle begins anew.

The zygomycetes also contribute to human welfare. For example, *Rhizopus* is used in Indonesia to produce a food called tempeh from boiled, skinless soybeans. Another zygomycete (*Mucor* spp.) is used with soybeans in the Orient to make a curd called sufu. Others are employed in the commercial preparation of some anesthetics, birth control agents, industrial alcohols, meat tenderizers, and the yellow coloring used in margarine and butter substitutes.

Division *Ascomycota*

The division *Ascomycota* contains the fungi called **ascomycetes,** commonly known as the sac fungi. Many species are quite familiar and economically important (**figure 25.9**). For example, most of the red, brown, and blue-green molds that cause food spoilage are ascomycetes. The powdery mildews that attack plant leaves and the fungi that cause chestnut blight and Dutch elm disease are ascomycetes. Many yeasts as well as edible morels and truffles are ascomycetes. The pink bread mold *Neurospora crassa,* also an ascomycete, has been an extremely important research tool in genetics and biochemistry.

Many ascomycetes are parasites on higher plants. *Claviceps purpurea* parasitizes rye and other grasses, causing the disease **ergot. Ergotism,** the toxic condition in humans and animals who eat grain infected with the fungus, is often accompanied by gangrene, psychotic delusions, nervous spasms, abortion, and convulsions. During the Middle Ages ergotism, then known as St. Anthony's fire, killed thousands of people. For example, over 40,000 deaths from ergot poisoning were recorded in France in the year 943. It has been suggested that the widespread accusations of witchcraft in Salem Village (now Danvers) and other Massachusetts communities in the late 1690s may have resulted from outbreaks of ergotism. The pharmacological activities of ergot are due to its active ingredient, lysergic acid diethylamide (LSD). In controlled dosages ergot can be used to induce labor, lower blood pressure, and ease migraine headaches.

The ascomycetes are named for their characteristic reproductive structure, the saclike **ascus** [pl., asci; Greek *askos,* sac]. The mycelium of the ascomycetes is composed of septate hyphae. Asexual reproduction is common in the ascomycetes and takes place by means of conidiospores (**figure 25.10**).

Sexual reproduction in the ascomycetes always involves the formation of an ascus containing two or more haploid ascospores (**figure 25.11a**). In the more complex ascomycetes, ascus formation is preceded by the development of special **ascogenous hyphae** into which pairs of nuclei migrate (figure 25.11b). One nucleus of each pair originates from a "male" mycelium (**antheridium**) or cell and the other from a "female" organ or cell (**ascogonium**) that has fused with it. As the ascogenous hyphae grow, the paired nuclei divide so that there is one pair of nuclei in each cell. After the ascogenous hyphae have matured, nuclear fusion occurs at the hyphal tips in the ascus mother cells. The diploid zygote nucleus then undergoes meiosis, and the resulting four haploid nuclei divide mitotically again to produce a row of eight nuclei in each developing ascus. These nuclei are walled off

(a) *Morchella esculenta*

(b) *Sarcoscypha coccinea*

(c) *Tuber brumale*

Figure 25.9   Division *Ascomycota.*   **(a)** The common morel, *Morchella esculenta,* is one of the choicest edible fungi. It fruits in the spring. **(b)** Scarlet cups, *Sarcoscypha coccinea,* with open ascocarps (apothecia). **(c)** The black truffle, *Tuber brumale,* is highly prized for its flavor by gourmet cooks. Technically truffles are mycorrhizal associations on oak trees.

Figure 25.10   Asexual Reproduction in Ascomycetes and Deuteromycetes.   Characteristic conidiospores of *Aspergillus* as viewed with the electron microscope ($\times 1,200$).

from one another. Thousands of asci may be packed together in a cup- or flask-shaped **ascocarp** (figure 25.9*b*). When the ascospores mature, they often are released from the asci with great force. If, perchance, the mature ascocarp is jarred, it may appear to belch puffs of "smoke" consisting of thousands of ascospores. Upon reaching a suitable environment, the ascospores germinate and start the cycle anew.

Although the term yeast is used in a general sense to refer to all unicellular fungi that reproduce asexually by either budding or bi-

nary fission (**figure 25.12*a,b***), many yeast genera are classified specifically within the ascomycetes because of their sexual reproduction (figure 25.12*c,d*). Yeasts are present in both terrestrial and aquatic habitats in which a suitable carbon source is available.

Division *Basidiomycota*

The division *Basidiomycota* contains the **basidiomycetes,** commonly known as the club fungi. Examples include the smuts, jelly fungi, rusts, shelf fungi, stinkhorns, puffballs, toadstools, mushrooms, and bird's nest fungi.

Basidiomycetes are named for their characteristic structure or cell, the **basidium,** that is involved in sexual reproduction (**figure 25.13**). A basidium [Greek *basidion,* small base] is produced at the tip of hyphae and normally is club shaped. Two or more basidiospores are produced by the basidium, and basidia may be held within fruiting bodies called **basidiocarps.**

The basidiomycetes affect humans in many ways. Most are saprophytes that decompose plant debris, especially cellulose and lignin. Many mushrooms are used as food throughout the world. The cultivation of *Agaricus campestris* is a multimillion-dollar business (*see figure 41.19*).

Many mushrooms produce specific alkaloids that act as either poisons or hallucinogens. One such example is the "destroying angel" mushroom, *Amanita phalloides.* Two toxins isolated from this species are phalloidin and α-amanitin. Phalloidin primarily attacks liver cells where it binds to plasma membranes, causing them to rupture and leak their contents. Alpha-amanitin attacks

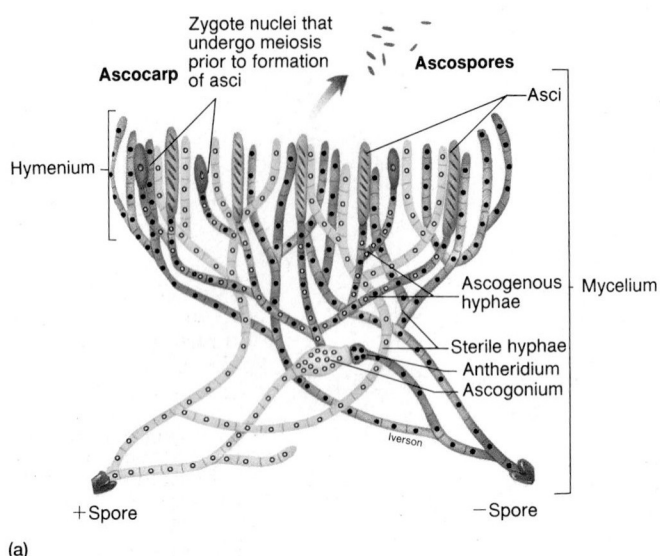

(a)

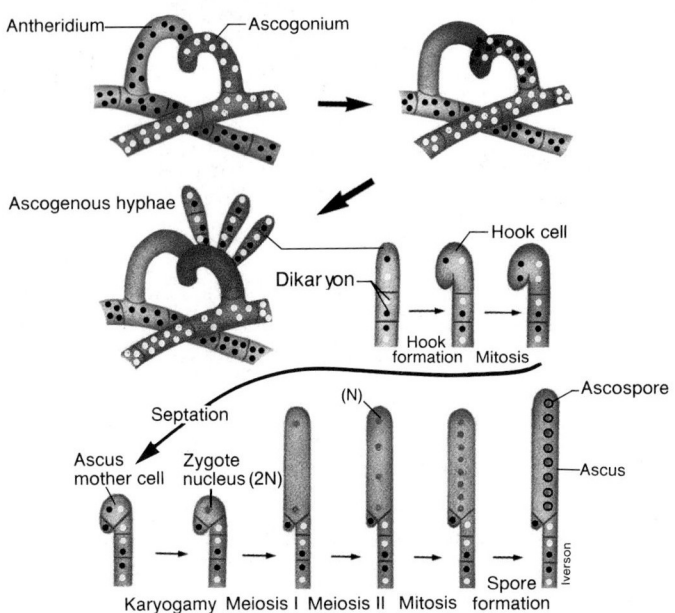

(b)

Figure 25.11 The Life Cycle of Ascomycetes. Sexual reproduction involves the formation of asci and ascospores. Within the ascus, karyogamy is followed by meiosis to produce the ascospores. (a) Sexual reproduction and ascocarp morphology of a cup fungus. (b) The details of sexual reproduction in ascogenous hyphae. The nuclei of the two mating types are represented by unfilled and filled circles. See text for details.

the cells lining the stomach and small intestine and is responsible for the severe gastrointestinal symptoms associated with mushroom poisoning.

The basidiomycete *Cryptococcus neoformans* is an important human pathogen. It produces the disease called **cryptococcosis,** a systemic infection primarily involving the lungs and central nervous system. Other basidiomycetes, the smuts and rusts, are

virulent plant pathogens that cause extensive damage to cereal crops; millions of dollars worth of crops are destroyed annually. In these fungi, large basidiocarps are not formed. Instead the small basidia arise from hyphae at the surface of the host plant. The mycelia grow either intra- or extracellularly in plant tissue. Fungal diseases (pp. 918–26)

Division *Deuteromycota*

To a large degree, classical fungal taxonomy is based on specific patterns of sexual reproduction. When a fungus lacks the sexual phase (perfect stage), or if this phase has not been observed, it is placed within the division *Deuteromycota,* commonly called the Fungi Imperfecti or **deuteromycetes** ("secondary fungi"). Most deuteromycetes reproduce by means of conidia (figures 25.6e and 25.10). Once a perfect stage is observed, the fungus is transferred to its proper division (often to the *Ascomycota*). Molecular systematics places the *Deuteromycota* among their closest relatives in the *Eumycota* (Phylogenetic Diagram 25).

Most Fungi Imperfecti are terrestrial, with only a few being reported from freshwater and marine habitats. The majority are either saprophytes or parasites of plants. A few are parasitic on other fungi.

Many imperfect fungi directly affect human welfare. Several are human pathogens, causing such diseases as athlete's foot, ringworm, and histoplasmosis (*see section 40.1*). The chemical activities of many Fungi Imperfecti are important industrially. For example, some species of *Penicillium* (*see figure 4.1c,d*) synthesize the well-known antibiotics penicillin and griseofulvin. Other species give characteristic aromas to cheeses such as Gorgonzola, Camembert, and Roquefort. Different species of *Aspergillus* are used to ferment soy sauce and to manufacture citric, gluconic, and gallic acids. *Aspergillus flavus* and *A. parasiticus* produce secondary metabolites, called aflatoxins, that are highly toxic and carcinogenic to animals and humans (*see section 41.2*). Another group of fungal toxins, the trichothecenes, are strong inhibitors of protein synthesis in eucaryotic cells.

Division *Chytridiomycota*

The simplest of the true fungi belong to the division *Chytridiomycota.* This division contains one class, *Chytridiomycetes,* and its members are known familiarly as the **chytrids.** These are simple terrestrial and aquatic fungi that reproduce asexually by forming motile zoospores with single, posterior, whiplash flagella. The entire organism is microscopic in size and may consist of a single cell, a small multinucleate mass, or a true mycelium. Usually chitin is the major constituent of chytrid cell walls. Chytrids are thought to have been derived from a protozoan ancestor having similar flagellation. They are likely to be ancestral to the remaining groups of true fungi (Phylogenetic Diagram, p. 538). Chytrids have a variety of life cycles. When sexual reproduction occurs, it results in a zygote that generally becomes a resting spore or sporangium. Some can grow saprophytically on dead organic matter; others are parasites of algae (*see figure 29.7*),

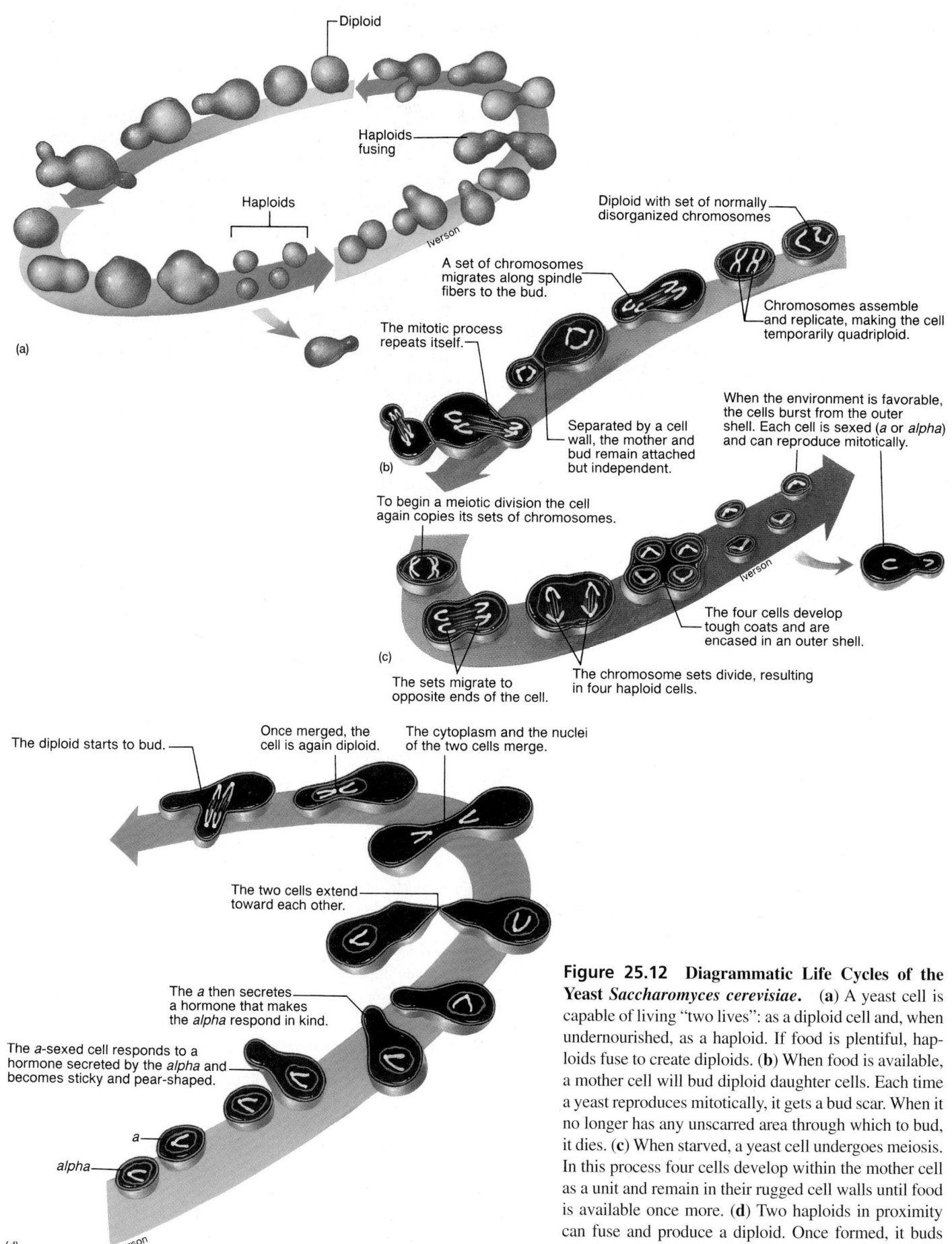

Diploid

Haploids
fusing

Haploids

(a)

Diploid with set of normally
disorganized chromosomes

A set of chromosomes
migrates along spindle
fibers to the bud.

Chromosomes assemble
and replicate, making the cell
temporarily quadriploid.

The mitotic process
repeats itself.

Separated by a cell
wall, the mother and
bud remain attached
but independent.

When the environment is favorable,
the cells burst from the outer
shell. Each cell is sexed (*a* or *alpha*)
and can reproduce mitotically.

(b)

To begin a meiotic division the cell
again copies its sets of chromosomes.

The four cells develop
tough coats and are
encased in an outer shell.

(c)

The sets migrate to
opposite ends of the cell.

The chromosome sets divide, resulting
in four haploid cells.

The diploid starts to bud.

Once merged, the
cell is again diploid.

The cytoplasm and the nuclei
of the two cells merge.

The two cells extend
toward each other.

The *a* then secretes
a hormone that makes
the *alpha* respond in kind.

The *a*-sexed cell responds to a
hormone secreted by the *alpha* and
becomes sticky and pear-shaped.

a

alpha

(d)

**Figure 25.12 Diagrammatic Life Cycles of the
Yeast *Saccharomyces cerevisiae*.** (**a**) A yeast cell is
capable of living "two lives": as a diploid cell and, when
undernourished, as a haploid. If food is plentiful, haploids fuse to create diploids. (**b**) When food is available,
a mother cell will bud diploid daughter cells. Each time
a yeast reproduces mitotically, it gets a bud scar. When it
no longer has any unscarred area through which to bud,
it dies. (**c**) When starved, a yeast cell undergoes meiosis.
In this process four cells develop within the mother cell
as a unit and remain in their rugged cell walls until food
is available once more. (**d**) Two haploids in proximity
can fuse and produce a diploid. Once formed, it buds
even before it attains its final shape. (*See section 4.9 and
figure 4.18 for a discussion of mitosis.*)

547

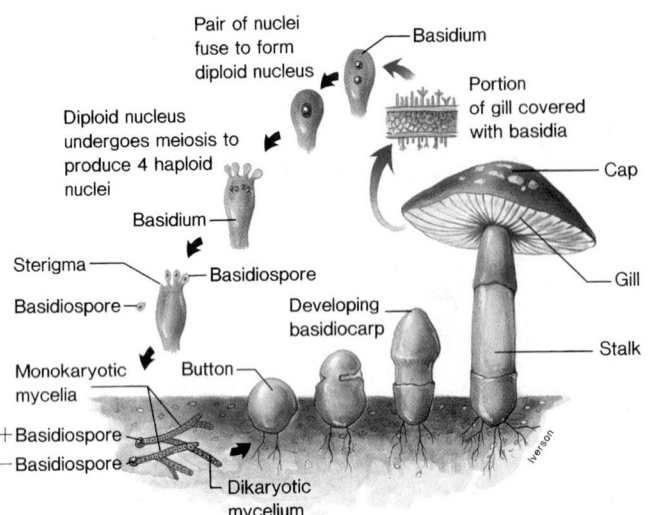

Figure 25.13   Division *Basidiomycota*.   The life cycle of a typical soil basidiomycete starts with a basidiospore germinating to produce a monokaryotic mycelium (one with a single nucleus in each septate cell). The mycelium quickly grows and spreads throughout the soil. When this primary mycelium meets another monokaryotic mycelium of a different mating type, the two fuse to initiate a new dikaryotic secondary mycelium. The secondary mycelium is divided by septa into cells, each of which contains two nuclei, one of each mating type. This dikaryotic mycelium is eventually stimulated to produce basidiocarps. A solid mass of hyphae forms a button that pushes through the soil, elongates, and develops a cap. The cap contains many platelike gills, each of which is coated with basidia. The two nuclei in the tip of each basidium fuse to form a diploid zygote nucleus, which immediately undergoes meiosis to form four haploid nuclei. These nuclei push their way into the developing basidiospores, which are then released at maturity.

other true fungi, and terrestrial and aquatic plants. Species such as *Allomyces* are used in the study of morphogenesis.

---

1. Describe how a typical zygomycete reproduces.
2. Give some beneficial uses for zygomycetes.
3. Describe the ascomycete life cycle. How are the ascomycetes important to humans?
4. How do yeasts reproduce sexually? Asexually?
5. Describe the life cycle of a typical basidiomycete.
6. How do some Fungi Imperfecti affect humans?
7. What are chytrids? Discuss their importance.

 25.7   **SLIME MOLDS AND WATER MOLDS**

The **slime molds** and **water molds** resemble fungi in only appearance and life-style. In their cellular organization, reproduction, and life cycles, they are phylogenetically distinct (Phylogenetic Diagram 25).

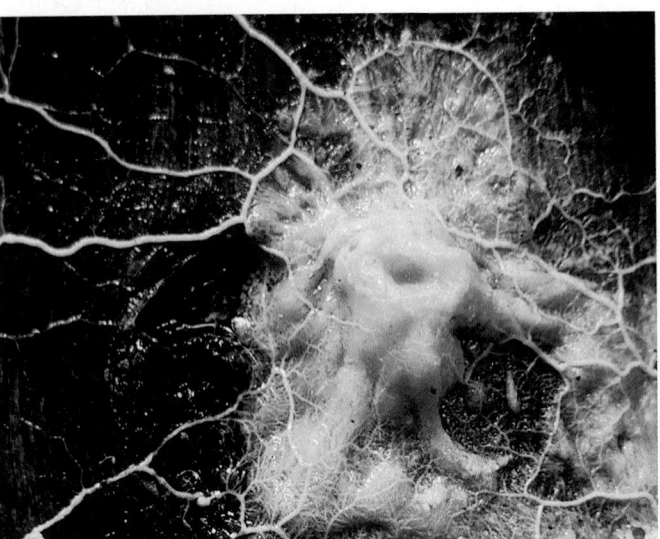

Figure 25.14   Slime Molds.   Plasmodium of the slime mold *Physarum;* light micrograph (×175).

Division *Myxomycota* (Acellular Slime Molds)

Under appropriate conditions **plasmodial (acellular) slime molds** exist as streaming masses of colorful protoplasm that creep along in an amoeboid fashion over moist, rotting logs, leaves, and other organic matter. Feeding is by phagocytosis. Because this streaming mass lacks cell walls, it is called a **plasmodium (figures 25.14** and **25.15*a***). The plasmodium contains many nuclei, and as the organism grows, the diploid nuclei divide repeatedly.

When the plasmodium matures or when food and/or moisture are scarce, it moves into a lighted area and develops delicate ornate fruiting bodies (figure 25.15*b, c*). As the fruiting bodies mature, they form spores with cellulose walls that are resistant to environmental extremes. The spores germinate in the presence of adequate moisture to release either nonflagellated amoeboid **myxamoebae** or flagellated **swarm cells.** Initially the myxamoebae or swarm cells feed and are haploid (figure 25.15*a*); eventually they fuse to form a diploid zygote. The zygote feeds, grows, and multiplies its nuclei through synchronous mitotic division to form the multinucleate plasmodium.

Division *Acrasiomycota* (Cellular Slime Molds)

The vegetative stage of **cellular slime molds** consists of individual amoeboid cells termed myxamoebae (**figure 25.16*a***). The myxamoebae feed phagocytically on bacteria and yeasts. When food is plentiful, they divide repeatedly by mitosis and cytokinesis, producing new daughter myxamoebae. As their food supply is exhausted, the myxamoebae begin to secrete cyclic adenosine monophosphate (cAMP). This attracts other myxamoebae that move toward the cAMP chemotactic source and in turn secrete more cAMP. When the individual myxamoebae aggregate (figure 25.16*b*), they form a sluglike **pseudoplasmodium**

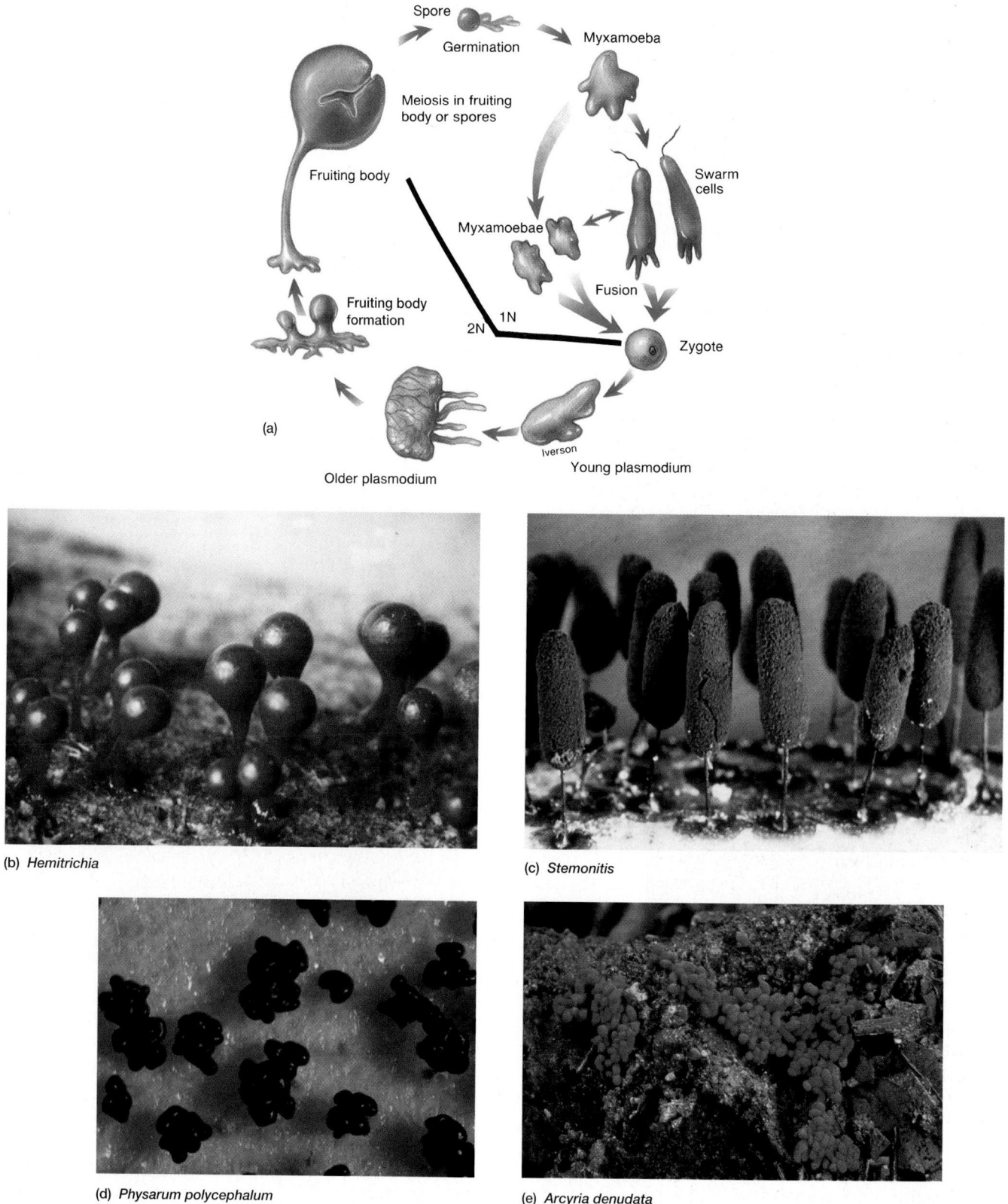

Figure 25.15   Reproduction in the *Myxomycota.* (**a**) The life cycle of a plasmodial slime mold. (Different parts of the life cycle are drawn at different magnifications.) Plasmodial slime-mold fruiting bodies: (**b**) *Hemitrichia* (×100), (**c**) *Stemonitis* (×100), (**d**) *Physarum polycephalum,* and (**e**) *Arcyria denudata.*

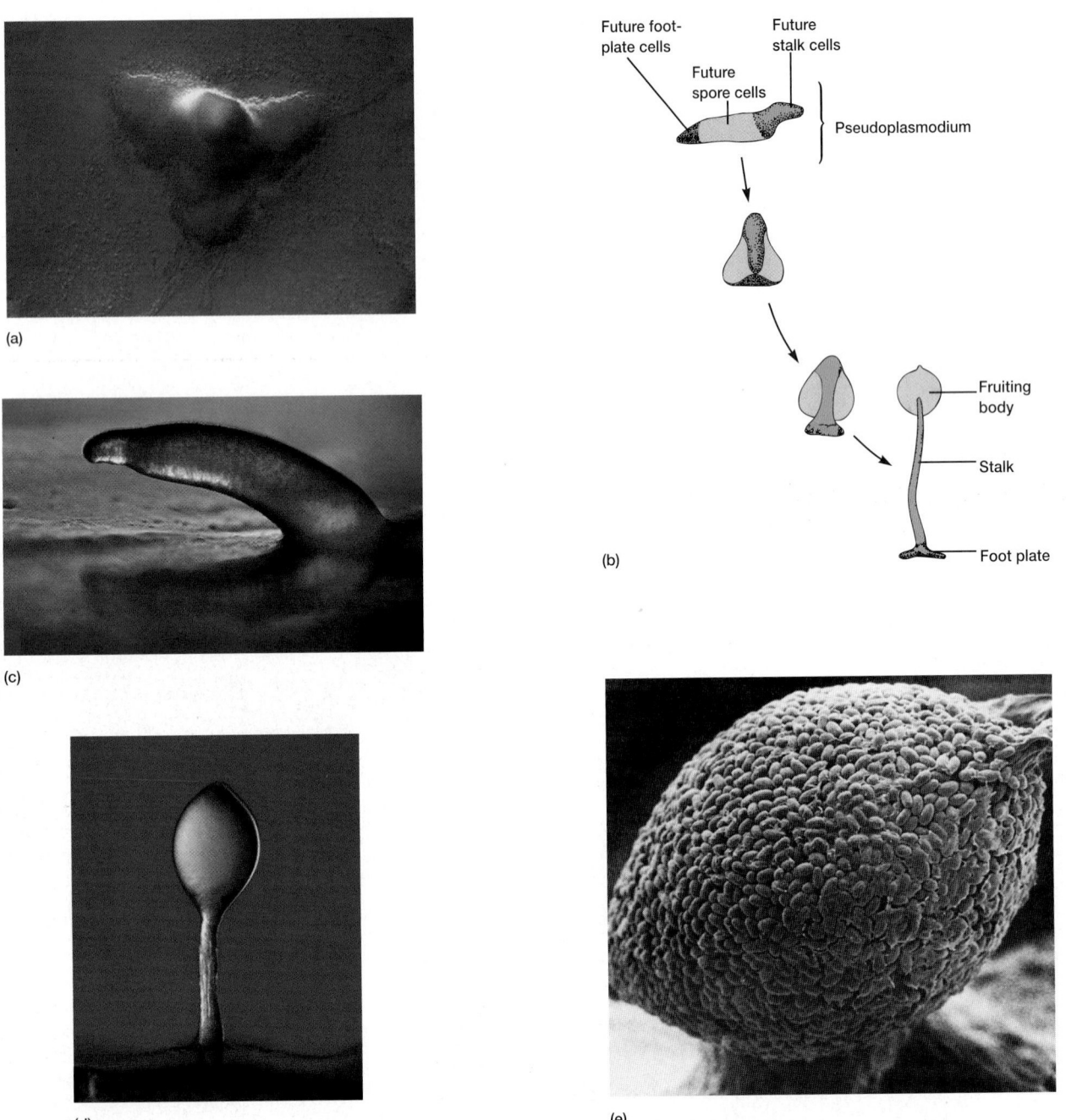

(a)

(b)

Future foot-
plate cells

Future spore cells

Future
stalk cells

Pseudoplasmodium

Fruiting
body

Stalk

Foot plate

(c)

(d)

(e)

Figure 25.16 Division *Acrasiomycota*. *Dictyostelium discoideum,* a cellular slime mold. In the free-living stage a myxamoeba resembles an irregularly shaped amoeba. (**a**) Aggregating myxamoebae become polar and begin to move in an oriented direction due to the influence of cAMP. (**b**) Diagrammatic drawing of cell migration involved in the formation of the sorocarp from (**c**) an initial pseudoplasmodium. (**d**) Light micrograph of a mature fruiting body (sorocarp). (**e**) Electron micrograph of a sorus showing spores ($\times 1,800$).

(figure 25.16c). The pseudoplasmodium may move around as a unit for a while, leaving a slime trail, but it eventually becomes sedentary. In the culmination of the asexual phase, pseudoplasmodial cells begin to differentiate into prestalk and prespore cells (figure 25.16b). A fruiting body called a **sorus** or **sorocarp** forms and matures (figure 25.16d) and then produces spores (figure 25.16e). The spores are eventually released, and when conditions

become favorable, they germinate to release haploid amoebae and repeat the cycle.

Division *Oomycota*

Members of the division *Oomycota* are collectively known as **oomycetes** or water molds. Oomycetes resemble true fungi only in appearance, consisting of finely branched filaments called hyphae. However, oomycetes have cell walls of cellulose, whereas the walls of most fungi are made of chitin. Oomycetes are also unlike the true fungi in that they have tubular mitochondrial cristae.

Oomycota means "egg fungi," a reference to the mode of sexual reproduction in water molds. A relatively large egg cell (oogonium) is fertilized by either a sperm cell or a smaller antheridium to produce a zygote. When the zygote germinates, it forms asexual zoospores that bear flagella.

Water molds such as *Saprolegnia* and *Achlya* are saprophytes that grow as cottony masses on dead algae and small animals, mainly in freshwater environments. They are important decomposers in aquatic ecosystems. Some water molds are parasitic on the gills of fish. The water mold *Peronospora hyoscyami* is currently responsible for the troublesome "blue mold" of tobacco plants throughout the world. In the United States alone, blue mold produces millions of dollars of damage yearly to tobacco crops. Other oomycetes cause late blight of potatoes (*Phytophthora infestans*) and grape downy mildew (*Plasmopara viticola*).

---

1. What is a plasmodium?
2. How do the plasmodial slime molds reproduce? Describe the cellular slime mold life cycle.
3. Where would you look for an oomycete?

25.1 Distribution

a. Fungi are widespread in the environment, being found wherever water, suitable organic nutrients, and an appropriate temperature occur. They secrete enzymes outside their body structure and absorb the digested food.

25.2 Importance

a. Fungi are important decomposers that break down organic matter; live as parasites on animals, humans, and plants; play a role in many industrial processes; and are used as research tools in the study of fundamental biological processes.

25.3 Structure

a. The body or vegetative structure of a fungus is called a thallus (**figure 25.1**). Fungi may be grouped into molds or yeasts based on the development of the thallus.

b. A fungus is a eucaryotic, spore-bearing organism that has absorptive nutrition and lacks chlorophyll; that reproduces asexually, sexually, or by both methods; and that normally has filamentous hyphae surrounded by cell walls, which usually contain chitin.

c. Yeasts are unicellular fungi that have a single nucleus and reproduce either asexually by budding and transverse division or sexually through spore formation (**figure 25.2**).

d. A mold consists of long, branched, threadlike filaments of cells, the hyphae, that form a tangled mass called a mycelium (**figure 25.3**). Hyphae may be either septate or coenocytic (nonseptate). The mycelium can produce reproductive structures.

e. Some fungi are dimorphic—they alternate between a yeast and a mold form (**table 25.2**).

25.4 Nutrition and Metabolism

a. Most fungi are saprophytes and grow best in moist, dark habitats. These chemoorganoheterotrophs are usually aerobic; some yeasts are fermentative.

25.5 Reproduction

a. Asexual reproduction often occurs in the fungi by the production of specific types of spores that are easily dispersed (**figure 25.6**).

b. Sexual reproduction is initiated in the fungi by the fusion of hyphae of different mating strains. In some fungi the nuclei in the fused hyphae immediately combine to form a zygote. In others the two genetically distinct nuclei remain separate, forming pairs that divide synchronously. Eventually some nuclei fuse (**figure 25.7**).

25.6 Characteristics of the Fungal Divisions

a. The zygomycetes are coenocytic. Most are saprophytic. One example is the common bread mold, *Rhizopus stolonifer*. Sexual reproduction occurs through a form of conjugation involving + and − strains (**figure 25.8**).

b. The ascomycetes are known as the sac fungi because they form a sac-shaped reproductive structure called an ascus (**figure 25.9**). In asexual reproduction they produce characteristic conidia (**figure 25.10**). Sexual reproduction involves + and − strains (**figure 25.11**).

c. The basidiomycetes are the club fungi. They are named after their basidium that produces basidiospores (**figure 25.13**).

d. The deuteromycetes (Fungi Imperfecti) are fungi with no known sexual (perfect) phase.

25.7 Slime Molds and Water Molds

a. The plasmodial (acellular) slime molds move about as a multinucleate plasmodium (**figure 25.14**). When food or moisture is scarce, these slime molds form sporangia within which spores are produced (**figure 25.15**).

b. The cellular slime molds consist of a vegetative stage called a myxamoeba (**figure 25.16**). The myxamoebae feed until their nutrients are exhausted, then the cells come together to form a moldlike multicellular structure called a sorus or sorocarp. The sorocarp produces haploid spores that germinate when conditions are favorable to form new myxamoebae.

c. The chytrids are a group of terrestrial and aquatic fungi that produce motile zoospores with single, posterior, whiplash flagella.

d. The *Oomycota* (water molds) are characterized by the production of motile spores (zoospores, gametes) and production of resistant sexual spores (oospores).

Key Terms

Questions for Thought and Review

1. Some fungi can reproduce both sexually and asexually. What are the advantages and disadvantages of each?
2. Why is nutrition in the fungi a property of membrane function?
3. Why are most fungi confined to a specific ecological niche?
4. Some authorities believe that the slime molds and water molds should be placed in the kingdom *Fungi*, whereas others believe that they should be placed in the kingdom *Protista*. What are the characteristics that these organisms share with both the fungi and protists that have led to this ambiguity?
5. At the present time those fungi that have no sexual reproduction cannot be classified with their sexually reproducing relatives. Why? Do you think that this is likely to change in the future?
6. Because asexual spores are such a rapid way of reproducing for some fungi, what adaptive "use" is there for an additional sexual phase?
7. Some fungi can be viewed as coenocytic organisms that exhibit little differentiation. When differentiation does occur, such as in the formation of reproductive structures, it is preceded by septum formation. Why does this occur?
8. The term mushrooming is a proverbial description for expanding rapidly. Why is this an accurate metaphor?
9. Both bacteria and fungi are major environmental decomposers. Obviously competition exists in a given environment, but fungi usually have an advantage. What is this advantage?
10. Very few antibiotics can control fungi, while there are many for bacterial control. Why is this so?

Critical Thinking Questions

1. What are some logical targets to exploit in treating animals or plants suffering from fungal infections? Are they different from the targets you would use when treating infections caused by bacteria? By viruses? Explain.
2. Fungi tend to reproduce asexually when nutrients are plentiful and conditions are favorable for growth, but reproduce sexually when environmental or nutrient conditions are not favorable. Why is this an evolutionarily important and successful strategy?
3. Compare and contrast the development of fruiting bodies by cellular slime molds and by myxobacteria.

For recommended readings on these and related topics, see Appendix V.

chapter 26

The Algae

Concepts

- Most algae are found in freshwater and marine environments; a few grow in terrestrial habitats.
- The algae are not a single, closely related taxonomic group but, instead, are a diverse, polyphyletic assemblage of unicellular, colonial, and multicellular eucaryotic organisms.
- Although algae can be autotrophic or heterotrophic, most are photoautotrophs. They store carbon in a variety of forms, including starch, oils, and various sugars.
- The body of an alga is called the thallus. Algal thalli range from small solitary cells to large, complex multicellular structures.
- Algae reproduce asexually and sexually.
- The following classical divisions of the algae are discussed: *Chlorophyta* (green algae), *Charophyta* (stoneworts/brittleworts), *Euglenophyta* (euglenoids), *Chrysophyta* (golden-brown and yellow-green algae; diatoms), *Phaeophyta* (brown algae), *Rhodophyta* (red algae), and *Pyrrhophyta* (dinoflagellates).

The term algae means different things to different people, and even the professional botanist and biologist find algae embarrassingly elusive of definition. Thus, laymen have given such names as "pond scums," "frog spittle," "water mosses," and "seaweeds," while some professionals shrink from defining them.

—Harold C. Bold and Michael J. Wynne

This photograph shows the giant kelp (*Macrocystis*) from along the coast of California. Kelps are brown algae; giant kelps can grow to more than 60 m long. This photograph shows the stipes, blades, and air bladders. Giant kelp are anchored to the ocean floor by holdfasts.

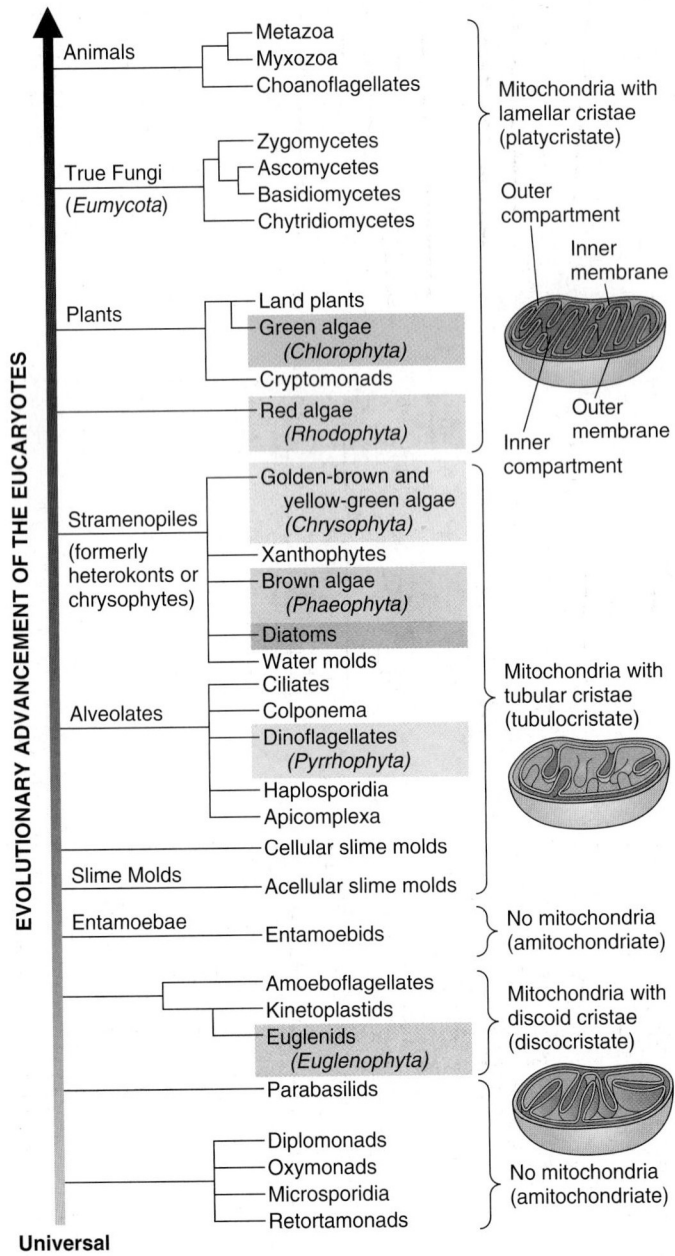

EVOLUTIONARY ADVANCEMENT OF THE EUCARYOTES

Animals — Metazoa, Myxozoa, Choanoflagellates

True Fungi (*Eumycota*) — Zygomycetes, Ascomycetes, Basidiomycetes, Chytridiomycetes

Mitochondria with lamellar cristae (platycristate)

Plants — Land plants, Green algae (*Chlorophyta*), Cryptomonads

Red algae (*Rhodophyta*)

Outer compartment — Inner membrane — Inner compartment — Outer membrane

Stramenopiles (formerly heterokonts or chrysophytes) — Golden-brown and yellow-green algae (*Chrysophyta*), Xanthophytes, Brown algae (*Phaeophyta*), Diatoms, Water molds

Alveolates — Ciliates, Colponema, Dinoflagellates (*Pyrrhophyta*), Haplosporidia, Apicomplexa

Mitochondria with tubular cristae (tubulocristate)

Slime Molds — Cellular slime molds, Acellular slime molds

Entamoebae — Entamoebids

No mitochondria (amitochondriate)

Amoeboflagellates, Kinetoplastids, Euglenids (*Euglenophyta*), Parabasilids

Mitochondria with discoid cristae (discocristate)

Diplomonads, Oxymonads, Microsporidia, Retortamonads

No mitochondria (amitochondriate)

Universal Ancestor

Phylogenetic Diagram 26   Tentative Phylogeny of the Algaelike Eucaryotes Based on 18S rRNA Sequence Comparisons. Using molecular systematics, organisms are grouped together based on the molecular phylogeny of their nuclear SSU rRNA genes and the type of mitochondrial cristae present. Accordingly, the algaelike eucaryotic organisms have arisen independently on five different occasions and are polyphyletic (highlighted by different colors).

Phycology or **algology** is the study of algae. The word phycology is derived from the Greek *phykos*, meaning seaweed. The term **algae** [s., alga] was originally used to define simple "aquatic plants." As just noted, it no longer has any formal significance in classification schemes. Instead the algae can be described as eucaryotic organisms that have chlorophyll *a* and carry out oxygen-producing photosynthesis. They differ from other photosynthetic eucaryotes in lacking a well-organized vascular conducting system and in having very simple reproductive structures. In sexual reproduction the whole organism may serve as a gamete; unicellular structures (gametangia) may produce gametes; or gametes can be formed by multicellular gametangia in which every cell is fertile. Unlike the case with plants, algal gametangia do not have nonfertile cells.

Table 26.2

Comparative Summary of Some Algal Characteristics

Division	Approximate Number of Species	Common Name and Representative
Chlorophyta	7,500	Green algae (*Chlamydomonas*)
Charophyta	250	Stoneworts or brittleworts (*Chara*)
Euglenophyta	700	Euglenoids (*Euglena*)
Chrysophyta	6,000	Golden-brown, yellow-green algae; diatoms, (*Cyclotella*)
Phaeophyta	1,500	Brown algae (*Sargassum*)
Rhodophyta	3,900	Red algae (*Corallina*)
Pyrrhophyta	1,100	Dinoflagellates (*Gymnodinium*)

[a] Refers specifically to the vegetative cells. Spores, akinetes, zygotes contain waxes, nonsaponifiable polymers, and phenolic substances.
[b] The following abbreviations are used: fresh water (fw), brackish water (bw), salt water (sw), terrestrial (t).

26.1 DISTRIBUTION OF ALGAE

Algae most commonly occur in water (fresh, marine, or brackish) in which they may be suspended (**planktonic**) or attached and living on the bottom (**benthic**). A few algae live at the water-atmosphere interface and are termed **neustonic.** **Plankton** [Greek *plankos,* wandering] consists of free-floating, mostly microscopic aquatic organisms. **Phytoplankton** is made up of algae and small plants, whereas **zooplankton** consists of animals and nonphotosynthetic protists. Some algae grow on moist rocks, wood, trees, and on the surface of moist soil. Algae also live as endosymbionts in various protozoa, mollusks, worms, and corals. Several algae grow as endosymbionts within plants, some are attached to the surface of various structures, and a few lead a parasitic existence. Algae also associate with fungi to form lichens.   Zooxanthella symbiosis (pp. 580–81); Lichen symbiosis (p. 591)

26.2 CLASSIFICATION OF ALGAE

According to the five-kingdom system of Whittaker, the algae belong to seven divisions distributed between two different kingdoms (**table 26.1**). This classical classification is based on cellular, not organismal, properties. Some more important properties include: (1) cell wall (if present) chemistry and morphology;

(2) form in which food or assimilatory products of photosynthesis are stored; (3) chlorophyll molecules and accessory pigments that contribute to photosynthesis; (4) flagella number and the location of their insertion in motile cells; (5) morphology of the cells and/or body (thallus); (6) habitat; (7) reproductive structures; and (8) life history patterns. Based on these properties the algae are arranged by divisions in **table 26.2,** which summarizes their more significant characteristics.

Table 26.1
Classical Classification of Algae[a]

Division (Common Name)	Kingdom
Chrysophyta (yellow-green and golden-brown algae; diatoms)	*Protista* (single cell or colonial; eucaryotic)
Euglenophyta (photosynthetic euglenoid flagellates)	*Protista*
Pyrrhophyta (dinoflagellates)	*Protista*
Charophyta (stoneworts)	*Protista*
Chlorophyta (green algae)	*Protista*
Phaeophyta (brown algae)	*Plantae* (multicellular; eucaryotic)
Rhodophyta (red algae)	*Plantae*

[a]Five-kingdom system.

	Pigments			Thylakoids per Stack in	Storage			
Chlorophylls	Phycobilins (Phycobiliproteins)	Carotenoids		Chloroplast	Products	Flagella	Cell Wall[a]	Habitat[b]
a, b	–	β-carotene, ± α-carotene, xanthophylls		3–6	Sugars, starch, fructosan	1, 2–8; equal, apical or subapical	Cellulose, mannan, protein, $CaCO_3$	fw, bw, sw, t
a, b	–	α-, β-, τ-carotene, xanthophylls		Many	Starch	2; subapical	Cellulose, $CaCO_3$	fw, bw
a, b	–	β-carotene, xanthophylls, ± τ-carotene		3	Paramylon, oils, sugars	1–3; slightly apical	Absent	fw, bw, sw, t
$a, c_1/c_2,$ rarely d	–	α-, β-, ε-carotene, fucoxanthin, xanthophylls		3	Chrysolaminarin, oils	1–2; equal or unequal, apical; or none	Cellulose, silica, $CaCO_3$, chitin, or absent	fw, bw, sw, t
a, c	–	β-carotene, fucoxanthin, xanthophylls		3	Laminarin, mannitol, oils	2; unequal, lateral	Cellulose, alginic acid, fucoidan	bw, sw
a, rarely d	C-phycocyanin, allophycocyanin, phycoerythrin	Xanthophylls (β-carotene, zeaxanthine, ± α-carotene)		1	Glycogenlike starch (floridean glycoside)	Absent	Cellulose, xylans, galactans, $CaCO_3$	fw, bw, sw
a, c_1, c_2	–	β-carotene, fucoxanthin, peridinin, dinoxanothin		3	Starch, glucan, oils	2; 1 trailing, 1 girdling	Cellulose, or absent	fw, bw, sw

Molecular systems (Phylogenetic Diagram 26) have placed some of the classical algae with plants (green algae); some as a separate lineage (red algae); some with the stramenopiles (golden-brown and yellow-green algae, brown algae, and diatoms); some with the alveolates (dinoflagellates); and still others with some protozoa (euglenoids). Two of these groups, the alveolates and stramenopiles, have been created recently as a result of rRNA comparisons and ultrastructural studies. The alveolates have mitochondria with tubular cristae and subsurface alveoli or sacs that abut the surface. Dinoflagellates, ciliate protozoa, and the apicomplexan protozoa are alveolates (Phylogenetic Diagram 26). The stramenopiles have mitochondria with tubular cristae and hollow hairs that give rise to a small number of fine hairs (tripartite tubular hairs). These hairs are usually on their flagella. Photosynthetic forms often have chlorophylls *a* and *c*. Some common stramenopiles are the opalinid protozoa, oomycetes, diatoms, brown algae or phaeophytes, chrysophytes, and xanthophytes. Although a few groups such as the diatoms have lost their hairs, they are still considered stramenopiles based on the rRNA data, mitochondrial characteristics, and other properties.

26.3 ULTRASTRUCTURE OF THE ALGAL CELL

The eucaryotic algal cell (**figure 26.1**) is surrounded by a thin, rigid cell wall. Some algae have an outer matrix lying outside the cell wall. This usually is flexible and gelatinous, similar to bacterial capsules. When present, the flagella are the locomotor organelles. The nucleus has a typical nuclear envelope with pores; within the nucleus are a nucleolus, chromatin, and karyolymph. The chloroplasts have membrane-bound sacs called thylakoids that carry out the light reactions of photosynthesis. These organelles are embedded in the stroma where the dark reactions of carbon dioxide fixation take place. A dense proteinaceous area, the **pyrenoid** that is associated with synthesis and storage of starch may be present in the chloroplasts.

Mitochondrial structure varies greatly in the algae. Some algae (euglenoids) have discoid cristae; some, lamellar cristae (green and red algae); and the remaining, have tubular cristae (golden-brown and yellow-green, brown, and diatoms).

---

1. Define the word algae.
2. To what two kingdoms do the algae belong? How do the classical and molecular classification systems differ?
3. What are some general characteristics of an algal cell?

26.4 ALGAL NUTRITION

Algae can be either autotrophic or heterotrophic. Most are photoautotrophic; they require only light and CO_2 as their principal source of energy and carbon. Chemoheterotrophic algae require external organic compounds as carbon and energy sources.

Types of microbial nutrition (pp. 95–96); Photosynthesis (pp. 190–94, 202–3)

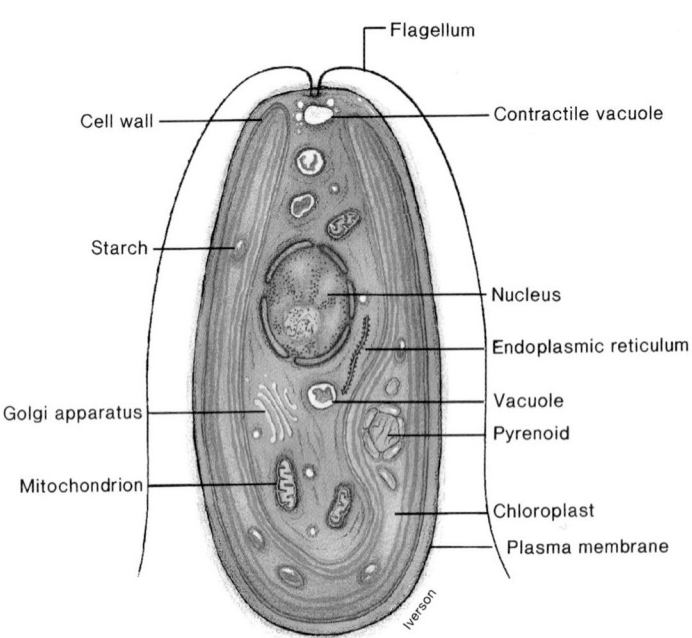

Figure 26.1   Algal Morphology.   Schematic drawing of a typical eucaryotic algal cell showing some of its organelles and other structures.

26.5 STRUCTURE OF THE ALGAL THALLUS (VEGETATIVE FORM)

The vegetative body of algae is called the **thallus** [pl., thalli]. It varies from the relative simplicity of a single cell to the more striking complexity of multicellular forms, such as the giant kelps. Single-celled algae may be as small as bacteria, whereas kelp can attain a size over 75 m in length. Algae are unicellular (**figure 26.2a,b,g**), colonial (figure 26.2c), filamentous (figure 26.2d), membranous and bladelike (figure 26.2e), or tubular (figure 26.2f).

26.6 ALGAL REPRODUCTION

Some unicellular algae reproduce asexually. In this kind of reproduction, gametes do not fuse to form a zygote. There are three basic types of asexual reproduction: fragmentation, spores, and binary fission. In **fragmentation** the thallus breaks up and each fragmented part grows to form a new thallus. **Spores** can be formed in ordinary vegetative cells or in specialized structures termed sporangia [s., sporangium; Greek *spora*, seed, and *angeion*, vessel]. Flagellated motile spores are called **zoospores.** Nonmotile spores produced by sporangia are termed **aplanospores.** In some unicellular algae **binary fission** occurs (nuclear division followed by division of the cytoplasm).

Other algae reproduce sexually. Eggs are formed within relatively unmodified vegetative cells called **oogonia** [s., oogonium] that function as female structures. Sperm are produced in

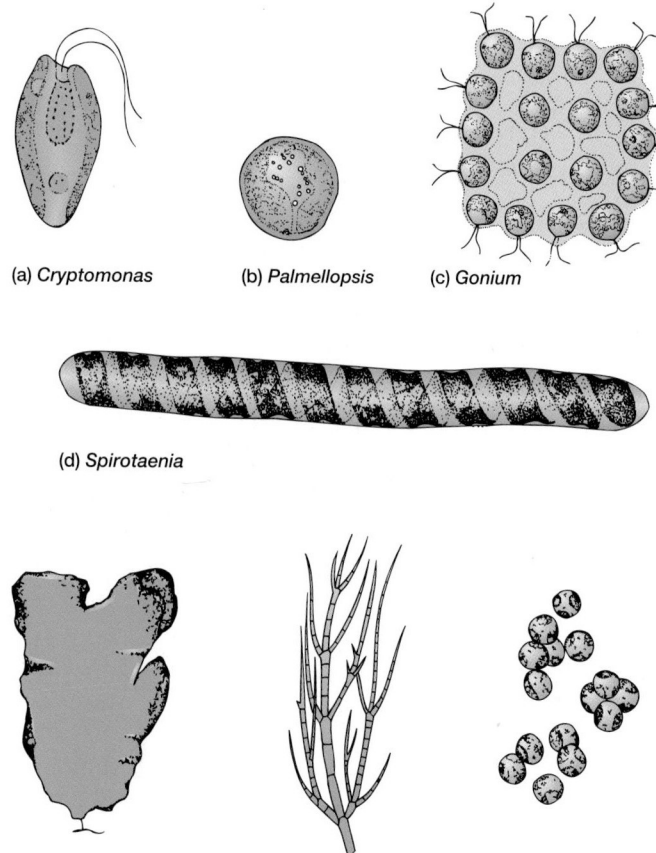

(a) *Cryptomonas* (b) *Palmellopsis* (c) *Gonium*

(d) *Spirotaenia*

(e) *Monostroma* (f) *Stigeoclonium* (g) *Chrysocapsa*

Figure 26.2 Diagrammatic Algal Bodies: (**a**) unicellular, motile, *Cryptomonas;* (**b**) unicellular, nonmotile, *Palmellopsis;* (**c**) colonial, *Gonium;* (**d**) filamentous, *Spirotaenia;* (**e**) bladelike kelp, *Monostroma;* (**f**) leafy tubular axis, branched tufts or plumes, *Stigeoclonium;* (**g**) unicellular, nonmotile, *Chrysocapsa.*

special male reproductive structures called **antheridia** [s., antheridium]. In sexual reproduction these gametes fuse to produce a diploid **zygote.**

---

1. How can algae be classified based on their mode of nutrition?
2. What are the different types of algal thalli?
3. How do algae reproduce asexually?
4. How do algae reproduce sexually?

 **26.7 CHARACTERISTICS OF THE ALGAL DIVISIONS**

Chlorophyta (Green Algae)

The *Chlorophyta* or green algae [Greek *chloros,* green] are an extremely varied division. They grow in fresh and salt water, in soil, on other organisms, and within other organisms. The *Chlorophyta* have chlorophylls *a* and *b* along with specific carotenoids, and

they store carbohydrates as starch. Many have cell walls of cellulose. They exhibit a wide diversity of body forms, ranging from unicellular to colonial, filamentous, membranous or sheetlike, and tubular types (**figure 26.3**). Some species have a holdfast structure that anchors them to the substratum. Both asexual and sexual reproduction occur in green algae. In molecular classification schemes, green algae are associated with the land plants and have mitochondria with lamellar cristae (Phylogenetic Diagram 26).

Chlamydomonas is a representative unicellular green alga (**figure 26.4**). Individuals have two flagella of equal length at the anterior end by which they move rapidly in water. Each cell has a single haploid nucleus, a large chloroplast, a conspicuous pyrenoid, and a **stigma (eyespot)** that aids the cell in phototactic responses. Two small contractile vacuoles at the base of the flagella function as osmoregulatory organelles that continuously remove water. *Chlamydomonas* reproduces asexually by producing zoospores through cell division. The alga also reproduces sexually when some products of cell division act as gametes and fuse to form a four-flagellated diploid zygote that ultimately loses its flagella and enters a resting phase. Meiosis occurs at the end of this resting phase and produces four haploid cells that give rise to adults.

From organisms like *Chlamydomonas,* several distinct lines of evolutionary specialization have evolved in the green algae. The first line contains nonmotile unicellular green algae, such as *Chlorella. Chlorella* (figure 26.3*a*) is widespread both in fresh and salt water and also in soil. It only reproduces asexually and lacks flagella, eyespots, and contractile vacuoles; the nucleus is very small.

Motile, colonial organisms such as *Volvox* represent a second major line of evolutionary specialization. A *Volvox* colony (figure 26.3*b; see also figure 2.8*b) is a hollow sphere made up of a single layer of 500 to 60,000 individual cells, each containing two flagella and resembling a *Chlamydomonas* cell. The flagella of all the cells beat in a coordinated way to rotate the colony in a clockwise direction as it moves through the water. Only a few cells are reproductive, and these are located at the posterior end of the colony. Some divide asexually and produce new colonies. Others produce gametes. After fertilization, the zygote divides to form a daughter colony. In both cases the daughter colonies stay within the parental colony until it ruptures.

A green alga, *Prototheca moriformis,* causes the disease **protothecosis** in humans and animals. *Prototheca* cells are fairly common in the soil, and it is from this site that most infections occur. Severe systemic infections, such as massive invasion of the bloodstream, have been reported in animals. More common in humans is the subcutaneous type of infection. It starts as a small lesion and spreads slowly through the lymph glands, covering large areas of the body.

---

1. What body forms do the green algae exhibit?
2. How do green algae reproduce?
3. Describe the structure of *Chlamydomonas, Chlorella,* and *Volvox.*
4. Why is the green alga, *Prototheca,* medically important?

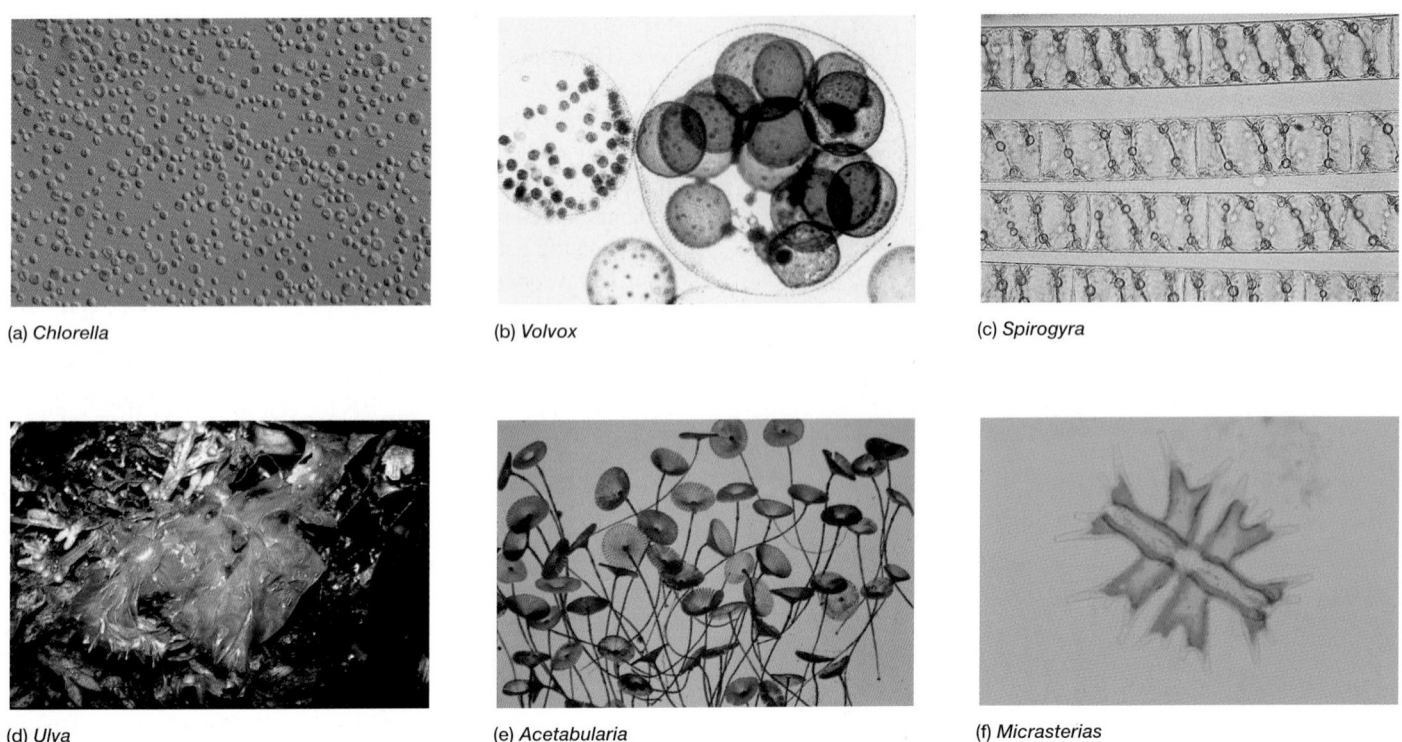

(a) *Chlorella*

(b) *Volvox*

(c) *Spirogyra*

(d) *Ulva*

(e) *Acetabularia*

(f) *Micrasterias*

Figure 26.3 *Chlorophyta* **(Green Algae); Light Micrographs.** (**a**) *Chlorella,* a unicellular nonmotile green alga (×160). (**b**) *Volvox,* a typical green algal colony (×450). (**c**) *Spirogyra,* a filamentous green alga (×100). Four filaments are shown. Note the ribbonlike, spiral chloroplasts within each filament. (**d**) *Ulva,* commonly called sea lettuce, has a leafy appearance. (**e**) *Acetabularia,* the mermaid's wine goblet. (**f**) *Micrasterias,* a large desmid (×150).

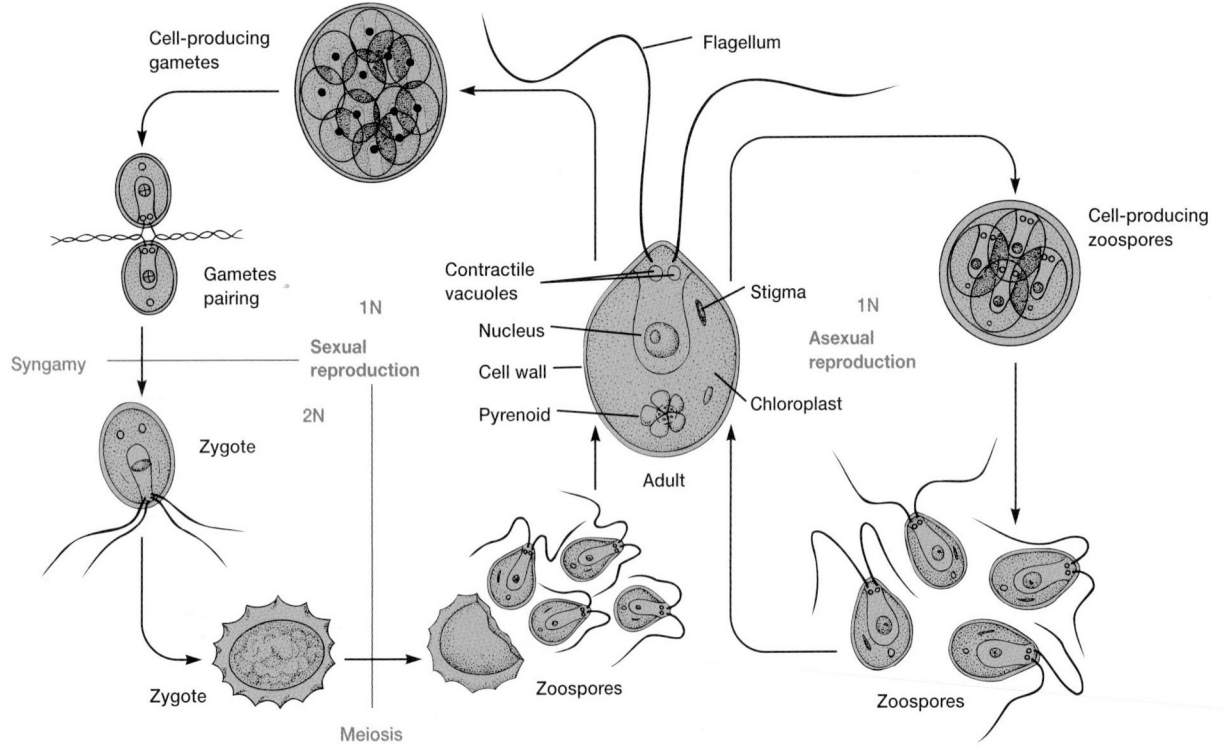

Figure 26.4 *Chlamydomonas:* **The Structure and Life Cycle of this Motile Green Alga.** During asexual reproduction, all structures are haploid; during sexual reproduction, only the zygote is diploid.

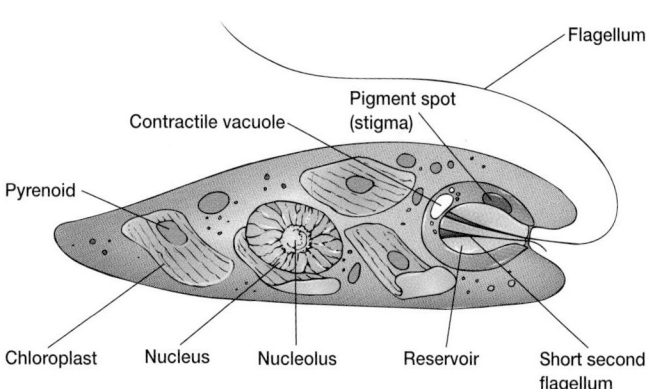

Figure 26.5 *Euglena.* **A Diagram Illustrating the Principal Structures Found in this Euglenoid.**   Notice that a short second flagellum does not emerge from the anterior invagination. In some euglenoids both flagella are emergent.

Charophyta (Stoneworts/Brittleworts)

The **stoneworts** are abundant in fresh to brackish waters and have a worldwide distribution. Often they appear as a dense covering on the bottom of shallow ponds. Some species precipitate calcium and magnesium carbonate from the water to form a limestone covering, thus giving the *Charophyta* their common names of stoneworts or brittleworts.

Euglenophyta (Euglenoids)

The **euglenoids** share with the *Chlorophyta* and *Charophyta* the presence of chlorophylls *a* and *b* in their chloroplasts. The primary storage product is paramylon (a polysaccharide composed of β-1,3 linked glucose molecules), which is unique to euglenoids. They occur in fresh, brackish, and marine waters and on moist soils; they often form water blooms in ponds and cattle water tanks. In molecular classification schemes, euglenoids are associated with the amoeboflagellates (flagellated protozoa) and kinetoplastids because all members have related rRNA sequences and mitochondria with discoid cristae at some stage in their life cycle (Phylogenetic Diagram 26).

The representative genus is *Euglena*. A typical *Euglena* cell (**figure 26.5**) is elongated and bounded by a plasma membrane. Inside the plasma membrane is a structure called the **pellicle,** which is composed of articulated proteinaceous strips lying side by side. The pellicle is elastic enough to enable turning and flexing of the cell, yet rigid enough to prevent excessive alterations in shape. The several chloroplasts contain chlorophylls *a* and *b* together with carotenoids. The large nucleus contains a prominent nucleolus. The stigma is located near an anterior reservoir. A large contractile vacuole near the reservoir continuously collects water from the cell and empties it into the reservoir, thus regulating the osmotic pressure within the organism. Two flagella arise from the base of the reservoir, although only one emerges from the canal and actively beats to move the cell. Reproduction in euglenoids is by longitudinal mitotic cell division.

Chrysophyta (Golden-Brown and Yellow-Green Algae; Diatoms)

The division *Chrysophyta* is quite diversified with respect to pigment composition, cell wall, and type of flagellated cells. In molecular classification schemes, these algae are associated with the stramenopiles and have mitochondria with tubular cristae (Phylogenetic Diagram 26). The division is divided into three major classes: golden-brown algae [Greek *chrysos,* gold], yellow-green algae, and diatoms. The major photosynthetic pigments are usually chlorophylls *a* and c_1/c_2, and the carotenoid fucoxanthin. When fucoxanthin is the dominant pigment, the cells have a golden-brown color. The major carbohydrate reserve in the *Chrysophyta* is **chrysolaminarin** (a polysaccharide storage product composed principally of β-1,3 linked glucose residues).

Some *Chrysophyta* lack cell walls; others have intricately patterned coverings external to the plasma membrane, such as **scales** (**figure 26.6a**), walls, and plates. Diatoms have a distinctive two-piece wall of silica, called a **frustule.** Two anteriorly attached flagella of unequal length are common among *Chrysophyta* (figure 26.6b), but some species have no flagella, and others have either one flagellum or two that are of equal length.

Most *Chrysophyta* are unicellular or colonial. Reproduction usually is asexual but occasionally sexual. Although some marine forms are known, most of the yellow-green and golden-brown algae live in fresh water. Blooms of some species produce unpleasant odors and tastes in drinking water.

The **diatoms** (figure 26.6c,d; see also figure 4.1b) are photosynthetic, circular or oblong chrysophyte cells with frustules composed of two halves or thecae that overlap like a petri dish [therefore their name is from the Greek *diatomsos,* cut in two]. The larger half is the **epitheca,** and the smaller half is the **hypotheca.** Diatoms grow in fresh water, salt water, and moist soil and comprise a large part of the phytoplankton (**Techniques & Applications 26.1**). The chloroplasts of these chrysophytes contain chlorophylls *a* and *c* as well as carotenoids. Some diatoms are facultative heterotrophs and can absorb carbon-containing molecules through the holes in their walls. The vegetative cells of diatoms are diploid; exist as unicellular, colonial, or filamentous shapes; lack flagella; and have a single large nucleus and smaller plastids. Reproduction consists of the organism dividing asexually, with each half then constructing a new theca within the old one. Because of this mode of reproduction, diatoms get smaller with each reproductive cycle. However, when they diminish to about 30% of their original size, sexual reproduction usually occurs. The diploid vegetative cells undergo meiosis to form gametes, which then fuse to produce a zygote. The zygote develops into an auxospore, which increases in size again and forms a new wall. The mature auxospore eventually divides mitotically to produce vegetative cells with normal frustules.

Diatom frustules are composed of crystallized silica [Si(OH)$_4$] with very fine markings (figure 26.6c,d). They have distinctive, and often exceptionally beautiful, patterns that are different for each species. Frustule morphology is very useful in diatom identification.

(a) *Mallomonas*

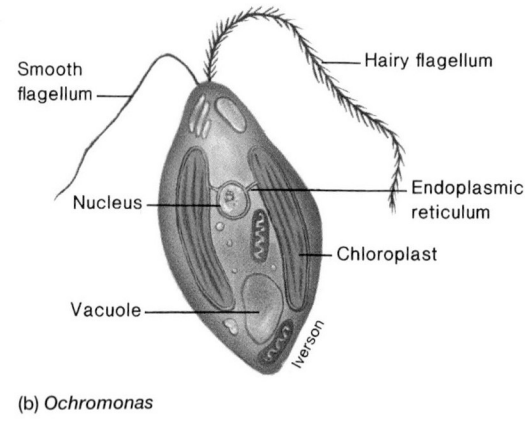

(b) *Ochromonas*

Smooth flagellum

Hairy flagellum

Nucleus

Endoplasmic reticulum

Chloroplast

Vacuole

Iverson

(c) *Cyclotella meneghiniana*

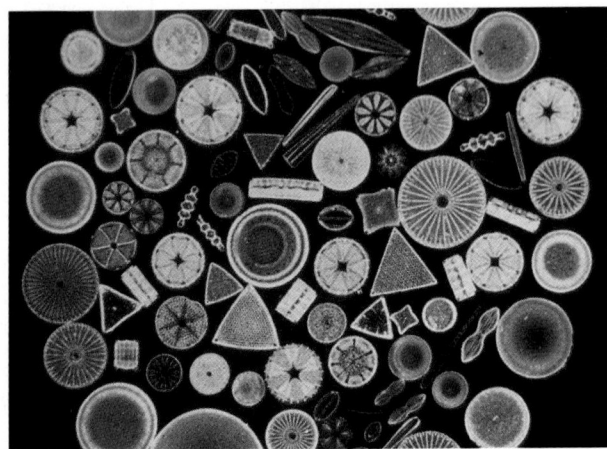

(d) Assorted diatoms

Figure 26.6 *Chrysophyta* **(Yellow-Green and Golden-Brown Algae; Diatoms).** (**a**) Scanning electron micrograph of *Mallomonas*, a chrysophyte, showing its silica scales. The scales are embedded in the pectin wall but synthesized within the Golgi apparatus and transported to the cell surface in vesicles ($\times$9,000). (**b**) *Ochromonas*, a unicellular chrysophyte. Diagram showing typical cell structure. (**c**) Scanning electron micrograph of a diatom, *Cyclotella meneghiniana* ($\times$750). (**d**) Assorted diatoms as arranged by a light microscopist ($\times$900).

Techniques & Applications

26.1 Practical Importance of Diatoms

Diatoms have both direct and indirect economic significance for humans. Because diatoms make up most of the phytoplankton of the cooler parts of the ocean, they are the most important ultimate source of food for fish and other marine animals in these regions. It is not unusual for 1 liter of seawater to contain almost a million diatoms. They are also extremely important for the biochemical cycling of silica and as contributors to global fixed carbon.

When diatoms die, their frustules sink to the bottom. Because the siliceous part of the frustule is not affected by the death of the cell, diatom frustules tend to accumulate at the bottom of aquatic environments. These form deposits of material called **diatomaceous earth.** This material is used as an active ingredient in many commercial preparations, including detergents, fine abrasive polishes, paint removers, decolorizing and deodorizing oils, and fertilizers. Diatomaceous earth also is used extensively as a filtering

agent, as a component in insulating (firebrick) and soundproofing products, and as an additive to paint to increase the night visibility of signs and license plates.

The use of diatoms as indicators of water quality and of pollution tolerance is becoming increasingly important. Specific tolerances for given species to various environmental parameters (concentrations of salts, pH, nutrients, nitrogen, temperature) have been compiled.

Recently a diatomaceous earth product to control insects, called INSECTO, has been introduced. Insects have their soft body parts exposed but covered by a waxy film to prevent dehydration. When they contact the diatoms in INSECTO, the silica frustules break the waxy film on the insects, causing them to dehydrate and die. INSECTO is a physical control product to which insects cannot build up resistance. Yet this product can be fed to poultry, livestock, and pets with no ill effects.

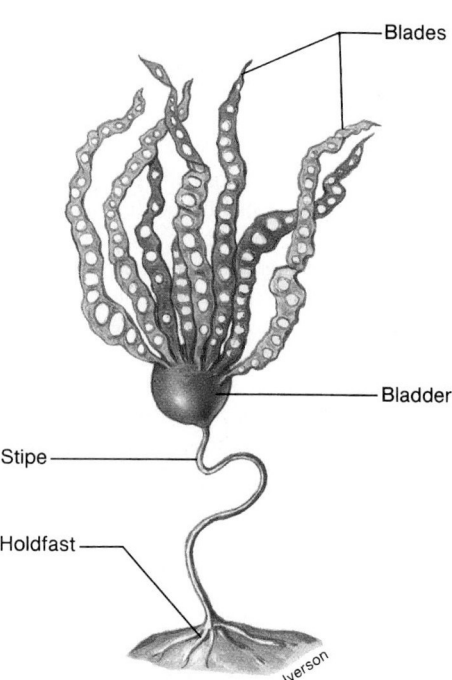

Figure 26.8 Rhodophyta (Red Algae). These algae (e.g., *Corallina gracilis*) are much smaller and more delicate than the brown algae. Most red algae have a filamentous, branched morphology as seen here.

Figure 26.7 *Phaeophyta* **(Brown Algae).** Diagram of the parts of the brown alga, *Nereocystis.* Due to the holdfast organ, the heaviest tidal action and surf seldom dislodge brown algae from their substratum. The stipe is a stalk that varies in length; the bladder is a gas-filled float.

---

1. Why are the *Charophyta* called stoneworts?
2. Briefly describe the major characteristics of the euglenoids, *Chrysophyta,* and diatoms.
3. How do euglenoids and diatoms reproduce?
4. How do the cells of diatoms differ from those of other organisms?

Phaeophyta (Brown Algae)

The *Phaeophyta* or brown algae [Greek *phaeo,* brown] consist of multicellular organisms that occur almost exclusively in the sea. Some species have the largest linear dimensions (length) known in the eucaryotic world (chapter opening figure). Since the brown algae have tubular cristae, they are associated with stramenopiles in molecular classification schemes (Phylogenetic Diagram 26). Most of the conspicuous seaweeds that are brown to olive green in color are assigned to this division. The simplest brown algae consist of small openly branched filaments; the larger, more advanced species have a complex arrangement. Some large **kelps** are conspicuously differentiated into flattened blades, stalks, and holdfast organs that anchor them to rocks (**figure 26.7**). Some, such as *Sargassum,* form huge floating masses that dominate the Sargasso Sea. The color of these algae reflects the presence of the brown pigment fucoxanthin, in addition to chlorophylls *a* and *c*, β-carotene, and violaxanthin. The main storage product is **laminarin,** which is quite similar in structure to chrysolaminarin.

Rhodophyta (Red Algae)

The division Rhodophyta, the red algae [Greek *rhodon,* rose], includes most of the seaweeds (**figure 26.8**). A few reds are unicellular but most are filamentous and multicellular. Some red algae are up to 1 m long. The stored food is the carbohydrate called floridean starch (composed of α-1,4 and α-1,6 linked glucose residues).

The red algae contain the red pigment phycoerythrin, one of the two types of phycobilins that they possess. The other accessory pigment is the blue pigment phycocyanin. The presence of these pigments explains how the red algae can live at depths of 100 m or more. The wavelengths of light (green, violet, and blue) that penetrate these depths are not absorbed by chlorophyll *a* but instead by these phycobilins. Not surprisingly the concentrations of these pigments often increases with depth as light intensity decreases. The phycobilins, after absorbing the light energy, pass it on to chlorophyll *a*. The algae appear decidedly red when phycoerythrin predominates over the other pigments. When phycoerythrin undergoes photodestruction in bright light, other pigments predominate and the algae take on shades of blue, brown, and dark green.

The cell walls of most red algae include a rigid inner part composed of microfibrils and a mucilaginous matrix. The matrix is composed of sulfated polymers of galactose called agar, funori, porphysan, and carrageenan. These four polymers give the red algae their flexible, slippery texture. Agar is used extensively in the laboratory as a culture medium component (*see section 5.7*). Many red algae also deposit calcium carbonate in their cell walls and play an important role in building coral reefs.

Pyrrhophyta (Dinoflagellates)

The *Pyrrhophyta* or **dinoflagellates** are unicellular, photosynthetic alveolate algae (Phylogenetic Diagram 26). Most dinoflagellates are marine, but some live in fresh water. Along with the

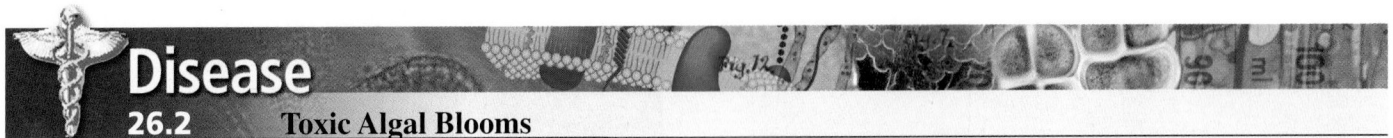

Disease

26.2   Toxic Algal Blooms

The Bible reports that the first plague Moses visited on the Egyptians was a blood-red tide that killed fish and fouled water. The Red Sea probably is named after these toxic algal blooms. Thousands of years later we still have problems with this plague.

The poisonous and destructive **red tides** that occur frequently in coastal areas often are associated with population explosions, or "blooms," of dinoflagellates. *Gymnodinium* and *Gonyaulax* species are the dinoflagellates most often involved. The pigments in the dinoflagellate cells are responsible for the red color of the water. Under these bloom conditions, the dinoflagellates produce a powerful neurotoxin called saxitoxin. The toxin paralyzes the striated respiratory muscles in many vertebrates by inhibiting sodium transport, which is essential to the function of their nerve cells. The toxin does not harm the shellfish that feed on the dinoflagellates. However, the shellfish do accumulate the toxin and are themselves highly poisonous to organisms, such as humans, who consume the shellfish, resulting in a condition known as **paralytic shellfish poisoning** or neurotoxic shellfish poisoning. Paralytic shellfish poisoning is characterized by numbness of the mouth, lips, face, and extremities. Duration of the illness ranges from a few hours to a few days and usually is not fatal.

Another type of poisoning in humans is called **ciguatera.** It results from eating marine fishes (e.g., grouper, snapper) that have consumed the dinoflagellate *Gambierdiscus toxicus.* The alga's toxin, called ciguatoxin, accumulates in the flesh of fish. This is one of the most powerful toxins known and remains in the flesh even after it has been cooked. Unfortunately it cannot be detected in the fishes and they are not visibly affected. In humans the toxin may cause gastrointestinal disturbances, profuse diarrhea, central nervous system involvement, and respiratory failure.

In 1988 a red tide that has long plagued the Gulf Coast of Florida spread northward to North Carolina. The dinoflagellates released a neurotoxin called brevetoxin, and this prompted state health authorities to shut down all shellfishing for three months. In 1987, in Prince Edward Island, Canada, several people died and hundreds became sick from eating mussels contaminated with domoic acid. The domoic acid was traced to a bloom of diatoms, golden-brown algae once thought to be innocent of all toxicity. The resulting disease, called **amnesic shellfish poisoning,** produces short-term memory loss in its victims. In 1991 pelicans eating anchovies off the California coast were found to be dying from domoic acid poisoning. Shellfish and crab fisheries from Washington to California were closed for several months, resulting in losses reaching hundreds of millions of dollars. In 1993 saxitoxin was found for the first time in crabs from Alaska. Unfortunately there are no treatments for the above types of poisonings. Supportive measures are the only therapy.

Overall, toxic algal blooms are on the rise. For example, in 1997 *Pfiesteria piscicida* (Latin for "fish killer") and other *Pfiesteria*-like dinoflagellates caused large fish kills along the coast of Maryland and Virginia. Similar fish kills have been occuring along the Atlantic coast at least since the 1980s. The flagellated form of the ambush-predator dinoflagellate swims toward the fish and attacks the prey, which it can then feed on (*see p. 627*). No one is certain why these toxic blooms are becoming more frequent, but most phycologists believe that the blooms are caused by the continuous pumping of nutrients such as nitrogen and phosphorus into coastal waters. Sewage and agricultural runoff are probably the major sources. Another possibility is world trade: oceangoing ships are unintentionally trafficking in harmful algae, giving the algae a free ride to foreign ports and new habitats in which they can flourish. With people eating more seafood, this toxic menace in the world's oceans will become increasingly more common in the future.

---

chrysophytes and diatoms, the dinoflagellates make up a large part of freshwater and marine plankton and are at the base of many food chains. Species of *Noctiluca, Pyrodinium, Gonyaulax,* and other genera can produce light and are responsible for much of the luminescence (phosphorescence) seen in ocean waters at night. Sometimes dinoflagellate populations reach such high levels that poisonous red tides result (**Disease 26.2**).

The flagella, protective coats (plates), and biochemistry of the dinoflagellates are distinctive. Many dinoflagellates are armored or clad in stiff, patterned, cellulose plates or thecae, which may become encrusted with silica (**figure 26.9**). Most have two flagella. In armored dinoflagellates these flagella beat in two grooves that girdle the cell—one transverse (the cingulum), the other longitudinal (the sulcus). The longitudinal flagellum extends posteriorly as a rudder; the flattened, ribbonlike transverse flagellum propels the cell forward while causing it to spin. Due to this spinning, the dinoflagellates received their name from the Greek *dinein,* "to whirl."

Most dinoflagellates have chlorophylls *a* and *c,* in addition to carotenoids and xanthophylls. As a result they usually have a yellowish-green to brown color. Their mitochondria have tubular cristae.

Some dinoflagellates can ingest other cells; others are colorless and heterotrophic. A few even occur as symbionts in many groups of jellyfish, sea anemones, mollusks, and corals. When dinoflagellates form symbiotic relationships, they lose their cellulose plates and flagella, become spherical golden-brown globules in the host cells, and are then termed **zooxanthellae.**

---

1. What is a seaweed?
2. Describe a kelp.
3. Why do red algae appear red?
4. What is unique about the flagellar arrangement in the dinoflagellates?

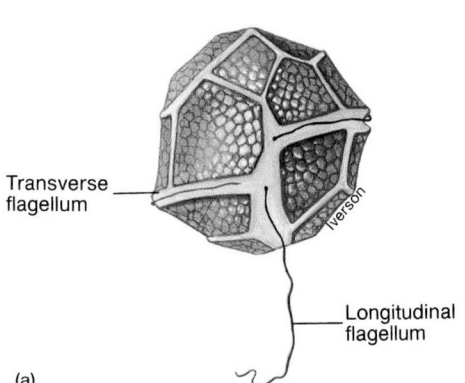

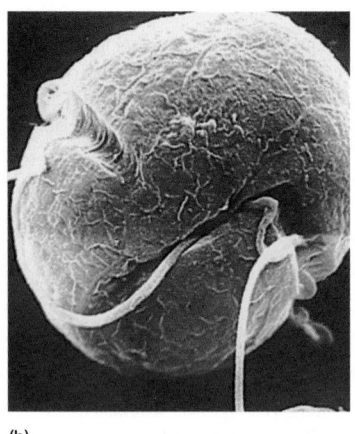

(a)

(b)

Figure 26.9 Dinoflagellates. (**a**) *Ceratium.* (**b**) Scanning electron micrograph of *Gymnodinium* (×4,000). Notice the plates of cellulose and the two flagella: one in the transverse groove and the other projecting outward.

Summary

Phycology (or algology) is the study of algae. Algae are eucaryotes that lack a well-developed vascular system and complex reproductive structures, but have chlorophyll and other pigments for carrying out oxygen-producing photosynthesis (or are closely related to photosynthetic species).

26.1 Distribution of Algae
a. Algae are found in almost every environmental niche—in fresh, marine, and brackish water; in certain terrestrial environments; and on moist inanimate objects. They may be endosymbionts, parasites, or components of lichens, and make up a large part of the phytoplankton on the earth.

26.2 Classification of Algae
a. Classical algal classification is based primarily on cellular properties (**table 26.1**). Examples include cell wall structure and chemistry, nutrient storage forms, types of chlorophyll molecules and accessory pigments, flagella, thallus morphology, habitat, reproductive structures, and life history patterns (**table 26.2**). Molecular classification systems have shown the algae to be polyphyletic.

26.3 Ultrastructure of the Algal Cell
a. The vegetative structure of the algae varies from the relative simplicity of a single cell (**figure 26.1**)

to the more striking complexity of organisms such as the giant kelps. Algae can be unicellular, colonial, filamentous, membranous, or tubular (**figure 26.2**).

26.4 Algal Nutrition
a. Most algae are photoautotrophic and require only light and CO_2.

26.5 Structure of the Algal Thallus (Vegetative Form)
a. The vegetative body of algae is called the thallus and varies from a single cell to large, multicellular forms.

26.6 Algal Reproduction
a. Both sexual and asexual reproduction occur in the algae. Asexual reproduction includes fragmentation, the production of spores, and binary fission. In sexual reproduction gametes fuse to form a zygote.

26.7 Characteristics of the Algal Divisions
a. Green algae (*Chlorophyta*) are a highly diverse group of organisms abundant in the sea, fresh water, and damp terrestrial habitats (**figure 26.3**).

b. Stoneworts/brittleworts (*Charophyta*) have more complex structures than the green algae. They contain whorls of short branches that arise regularly at their nodes. Their gametangia are complex and multicellular. Stoneworts are abundant in fresh to brackish water and are common as fossils.

c. Euglenoids (*Euglenophyta*) have chloroplasts that are biochemically similar to those of the green algae. They have a flexible proteinaceous pellicle inside the plasma membrane (**figure 26.5**).

d. The *Chrysophyta* contain golden-brown algae, yellow-green algae, and diatoms, and vary greatly in pigment composition, cell wall structure, and type of flagellated cells (**figure 26.6**).

e. Brown algae (*Phaeophyta*) are multicellular marine algae, some of which reach lengths of 75 m. The kelps—the largest of the brown algae—contribute greatly to the productivity of the sea as well as to many human needs (**figure 26.7**).

f. Red algae (*Rhodophyta*) have chlorophyll *a* and phycobilins and usually grow at great depths. Some produce agar (**figure 26.8**).

g. Dinoflagellates (*Pyrrhophyta*) are unicellular motile algae that play a role in the red tides that harm many forms of life (**figure 26.9**).

Key Terms

algae 554
algology 554
amnesic shellfish poisoning 562
antheridium 557
aplanospore 556

benthic 555
binary fission 556
chrysolaminarin 559
ciguatera 562
diatom 559

diatomaceous earth 560
dinoflagellate 561
epitheca 559
euglenoid 559
fragmentation 556

frustule 559
hypotheca 559
kelp 561
laminarin 561
neustonic 555

Questions for Thought and Review

1. How can algae be distinguished from the photosynthetic bacteria?
2. What characteristics are used in algal classification?
3. Although multicellularity must have originated many times in the different multicellular groups, how do present-day green algae provide an especially clear example of the probable origin of multicellularity?
4. What characteristics do euglenoids share with higher plants? With animals? Why are they considered protists?
5. Why do marine algae vary so much more in shape and size than those found in fresh water?
6. The phytoplankton of the open oceans (e.g., the Sargasso Sea) are predominantly algae, and these cells form the base of oceanic food chains. Which morphological characteristics of algae are adaptations to a floating (planktonic) habitat?
7. Freshwater algae are distributed worldwide. They rapidly colonize artificial lakes and water impoundments. How do algae accomplish such widespread dispersal?
8. What problem do diatoms have with continued asexual reproduction? How is this problem solved?
9. How are the red algae similar to the cyanobacteria?
10. What are some important characteristics of the green algae? The red algae? The brown algae? The dinoflagellates?
11. Describe the distinguishing characteristics of each major algal group discussed in this chapter.

Critical Thinking Questions

1. Why do algae have so many different types of pigments for photosynthesis? Do these pigments serve other purposes?
2. Why don't we know as much about the basic biology of algae as we do about the biology of fungi, viruses, and bacteria?

**For recommended readings on these
and related topics, see Appendix V.**

chapter 27

The Protozoa

Concepts

- Protozoa are protists exhibiting heterotrophic nutrition and various types of locomotion. They occupy a vast array of habitats and niches and have organelles similar to those found in other eucaryotic cells, and also specialized organelles.
- Current protozoan taxonomy divides the protozoa into seven phyla: *Sarcomastigophora*, *Labyrinthomorpha*, *Apicomplexa*, *Microspora*, *Ascetospora*, *Myxozoa*, and *Ciliophora*. These phyla represent four major groups: flagellates, amoebae, ciliates, and sporozoa. In molecular classification schemes, the protozoa are polyphyletic eucaryotes.
- Protozoa usually reproduce asexually by binary fission. Some have sexual cycles, involving meiosis and the fusion of gametes or gametic nuclei resulting in a diploid zygote. The zygote is often a thick-walled, resistant, and resting cell called a cyst. Some protozoa undergo conjugation in which nuclei are exchanged between cells.
- All protozoa have one or more nuclei; some have a macro- and micronucleus.
- Various protozoa feed by holophytic, holozoic, or saprozoic means; some are predatory or parasitic.

And a pleasant sight they are indeed. Their shapes range from teardrops to bells, barrels, cups, cornucopias, stars, snowflakes, and radiating suns, to the common amoebas, which have no real shape at all. Some live in baskets that look as if they were fashioned of exquisitely carved ivory filigree. Others use colored bits of silica to make themselves bright mosaic domes. Some even form graceful transparent containers shaped like vases or wine glasses of fine crystal in which they make their homes.

—Helena Curtis

This is a scanning electron micrograph (×2,160) of the protozoan *Naegleria fowleri*. Three *N. fowleri* from an axenic culture, attacking and beginning to devour or engulf a fourth, presumably dead amoeba, with their amoebastomes (suckerlike structures that function in phagocytosis). This amoeba is the major cause of the disease in humans called primary amebic meningoencephalitis.

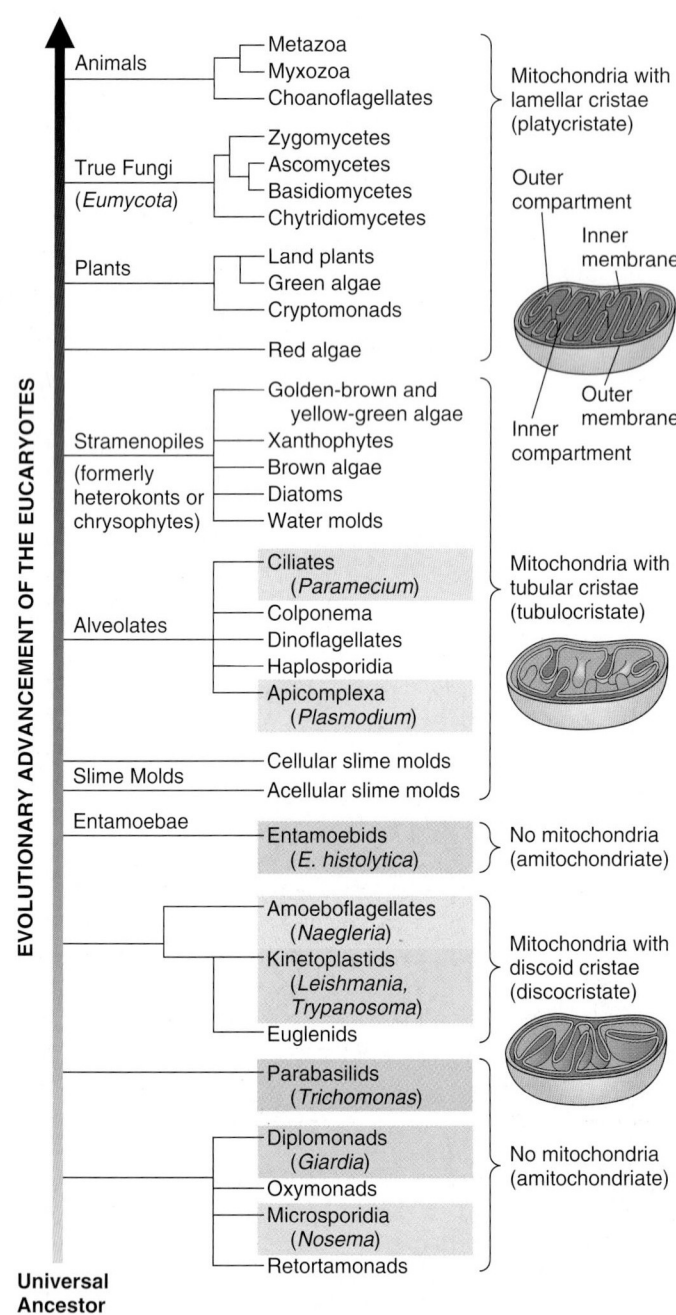

EVOLUTIONARY ADVANCEMENT OF THE EUCARYOTES

Animals
— Metazoa
— Myxozoa
— Choanoflagellates

True Fungi (*Eumycota*)
— Zygomycetes
— Ascomycetes
— Basidiomycetes
— Chytridiomycetes

Plants
— Land plants
— Green algae
— Cryptomonads

— Red algae

Stramenopiles (formerly heteronts or chrysophytes)
— Golden-brown and yellow-green algae
— Xanthophytes
— Brown algae
— Diatoms
— Water molds

Alveolates
— Ciliates (*Paramecium*)
— Colponema
— Dinoflagellates
— Haplosporidia
— Apicomplexa (*Plasmodium*)

Slime Molds
— Cellular slime molds
— Acellular slime molds

Entamoebae
— Entamoebids (*E. histolytica*)

— Amoeboflagellates (*Naegleria*)
— Kinetoplastids (*Leishmania, Trypanosoma*)
— Euglenids

— Parabasilids (*Trichomonas*)

— Diplomonads (*Giardia*)
— Oxymonads
— Microsporidia (*Nosema*)
— Retortamonads

Universal Ancestor

Mitochondria with lamellar cristae (platycristate)

Outer compartment

Inner membrane

Outer membrane

Inner compartment

Mitochondria with tubular cristae (tubulocristate)

No mitochondria (amitochondriate)

Mitochondria with discoid cristae (discocristate)

No mitochondria (amitochondriate)

Phylogenetic Diagram 27 Tentative Phylogeny of the Protozoan-Like Eucaryotes Based on 18S rRNA Sequence Comparisons. Recent molecular phylogeny of the nuclear SSU rRNA indicates that these eucaryotes are highly polyphyletic (protozoan groups are highlighted by different colors). Thus, like the algae, the protozoa do not represent a monophyletic group and the taxon "Protozoa" should not be used in classification schemes that seek to represent true molecular evolutionary histories. The word protozoa can still be used (as it is in this chapter) to denote a nonrelated polyphyletic group of eucaryotic organisms that share some morphological, reproductive, ecological, and biochemical characteristics.

The microorganisms called **protozoa** [s., protozoan; Greek *protos,* first, and *zoon,* animal] are studied in the discipline called **protozoology.** A protozoan can be defined as a usually motile eucaryotic unicellular protist. Protozoa are directly related only on the basis of a single negative characteristic—they are not multicellular. All, however, demonstrate the basic body plan of a single protistan eucaryotic cell.

27.1 DISTRIBUTION

Protozoa grow in a wide variety of moist habitats. Moisture is absolutely necessary for the existence of protozoa because they are susceptible to desiccation. Most protozoa are free living and inhabit freshwater or marine environments. Many terrestrial protozoa can be found in decaying organic matter, in soil, and even in beach sand; some are parasitic in plants or animals.

27.2 IMPORTANCE

Protozoa play a significant role in the economy of nature. For example, they make up a large part of **plankton**—small, free-floating organisms that are an important link in the many aquatic food chains and food webs of aquatic environments. A **food chain** is a series of organisms, each feeding on the preceding one. A **food web** is a complex interlocking series of food chains. Protozoa are also useful in biochemical and molecular biological studies. Many biochemical pathways used by protozoa are present in all eucaryotic cells. Finally, some of the most important diseases of humans (*see section 40.2*) and animals (**table 27.1**) are caused by protozoa. Microorganisms and ecosystems (pp. 603–4)

27.3 MORPHOLOGY

Because protozoa are eucaryotic cells, in many respects their morphology and physiology are the same as the cells of multicellular animals (*see figures 4.2 and 4.3*). However, because all of life's various functions must be performed within the individual

C hapter 27 presents the major biological features of the protists known as protozoa. The most important groups are the flagellates, amoebae, sporozoa, and ciliates. The protists demonstrate the great adaptive potential of the basic single eucaryotic cell, as evidenced by their many nonrelated polyphyletic origins (**Phylogenetic Diagram 27**).

Table 27.1				
Pathogenic Protozoa That Cause Major Diseases of Domestic Animals				
Protozoan Group[a]	**Genus**	**Host**	**Preferred Site of Infection**	**Disease**
Amoebae	*Entamoeba*	Mammals	Intestine	Amebiasis
	Iodamoeba	Swine	Intestine	Enteritis
Sporozoa	*Babesia*	Cattle	Blood cells	Babesiosis
	Theileria	Cattle, sheep, goats	Blood cells	Theileriasis
	Sarcocystis	Mammals, birds	Muscles	Sarcosporidiosis
	Toxoplasma	Cats	Intestine	Toxoplasmosis
	Isospora	Dogs	Intestine	Coccidiosis
	Eimeria	Cattle, cats, chickens, swine	Intestine	Coccidiosis
	Plasmodium	Many animals	Bloodstream, liver	Malaria
	Leucocytozoon	Birds	Spleen, lungs, blood	Leucocytozoonosis
	Cryptosporidium	Mammals	Intestine	Cryptosporidiosis
Ciliates	*Balantidium*	Swine	Large intestine	Balantidiasis
Flagellates	*Leishmania*	Dogs, cats, horses, sheep, cattle	Spleen, bone marrow, mucous membranes	Leishmaniasis
	Trypanosoma	Most animals	Blood	Trypanosomiasis
	Trichomonas	Horses, cattle	Genital tract	Trichomoniasis (abortion)
	Histomonas	Birds	Intestine	Blackhead disease
	Giardia	Mammals	Intestine	Giardiasis

[a]These groups are distinguished from one another largely by their mechanism of locomotion (see text).

protozoan, some morphological and physiological features are unique to protozoan cells. In some species the cytoplasm immediately under the plasma membrane is semisolid or gelatinous, giving some rigidity to the cell body. It is termed the **ectoplasm.** The bases of the flagella or cilia and their associated fibrillar structures are embedded in the ectoplasm. The plasma membrane and structures immediately beneath it are called the **pellicle.** Inside the ectoplasm is the area referred to as the **endoplasm,** which is more fluid and granular in composition and contains most of the organelles. Some protozoa have one nucleus, others have two or more identical nuclei. Still other protozoa have two distinct types of nuclei—a macronucleus and one or more micronuclei. The **macronucleus,** when present, is typically larger and associated with trophic activities and regeneration processes. The **micronucleus** is diploid and involved in both genetic recombination during reproduction and the regeneration of the macronucleus.

One or more vacuoles are usually present in the cytoplasm of protozoa. These are differentiated into contractile, secretory, and food vacuoles. **Contractile vacuoles** function as osmoregulatory organelles in those protozoa that live in a hypotonic environment, such as a freshwater lake. Osmotic balance is maintained by continuous water expulsion. Most marine protozoa and parasitic species are isotonic to their environment and lack such vacuoles. **Phagocytic vacuoles** are conspicuous in holozoic and parasitic species and are the sites of food digestion (*see figure 4.10*). **Secretory vacuoles** usually contain specific enzymes that perform various functions (such as excystation).

Most anaerobic protozoa (such as *Trichonympha,* which lives in the gut of termites; *see figure 28.2*) have no mitochondria, no

cytochromes, and an incomplete tricarboxylic acid cycle. However, some do have small, membrane-delimited organelles termed **hydrogenosomes.** These structures contain a unique electron transfer pathway in which hydrogenase transfers electrons to protons (which act as the terminal electron acceptors), and molecular hydrogen is formed. Other protozoa have mitochondria with discoid cristae (trypanosomes), tubular mitochondrial cristae (ciliates, sporozoa), and lamellar cristae (foraminiferans).

---

1. Describe a typical protozoan.
2. Where can protozoa be found?
3. What roles do protozoa play in the trophic structure of their communities and in the organisms with which they associate?
4. What is unique about the nuclei of some protozoa? How do macronuclei and micronuclei differ in structure and function?
5. What are the functions of contractile, phagocytic, and secretory vacuoles?

 27.4 NUTRITION

Most protozoa are chemoheterotrophic. Two types of heterotrophic nutrition are found in the protozoa: holozoic and saprozoic. In **holozoic nutrition,** solid nutrients such as bacteria are acquired by phagocytosis and the subsequent formation of phagocytic vacuoles. Some ciliates have a specialized structure for phagocytosis called the **cytostome** (cell mouth). In **saprozoic nutrition,** soluble nutrients such as amino acids and sugars cross the plasma membrane by pinocytosis, diffusion, or carrier-mediated transport (facilitated diffusion or active transport).

27.5 ENCYSTMENT AND EXCYSTMENT

Many protozoa are capable of **encystation.** They develop into a resting stage called a **cyst,** which is a dormant form marked by the presence of a wall and by very low metabolic activity. Cyst formation is particularly common among aquatic, free-living protozoa and parasitic forms. Cysts serve three major functions: (1) they protect against adverse changes in the environment, such as nutrient deficiency, desiccation, adverse pH, and low partial pressure of O_2; (2) they are sites for nuclear reorganization and cell division (reproductive cysts); and (3) they serve as a means of transfer between hosts in parasitic species.

Although the exact stimulus for **excystation** (escape from the cysts) is unknown, excystation generally is triggered by a return to favorable environmental conditions. For example, cysts of parasitic species excyst after ingestion by the host and form the vegetative form called the **trophozoite.**

27.6 LOCOMOTORY ORGANELLES

A few protozoa are nonmotile. Most, however, can move by one of three major types of locomotory organelles: pseudopodia, flagella, or cilia. **Pseudopodia** [s., pseudopodium; false feet] are cytoplasmic extensions found in the amoebae that are responsible for movement and food capture. There are many types of pseudopodia. Flagellates and ciliates move by flagella and cilia. Electron microscopy has shown that protozoan flagella and cilia are structurally the same and identical in function to those of other eucaryotic cells (*see figures 4.22–4.25*).

27.7 REPRODUCTION

Most protozoa reproduce asexually, and some also carry out sexual reproduction. The most common method of asexual reproduction is **binary fission.** During this process the nucleus first undergoes mitosis, then the cytoplasm divides by cytokinesis to form two identical individuals (**figure 27.1**).

The most common method of sexual reproduction is **conjugation.** In this process there is an exchange of gametes between paired protozoa of complementary mating types (**conjugants;** *see figure 2.13*b). Conjugation is most prevalent among ciliate protozoa. A well-studied example is *Paramecium caudatum* (**figure 27.2**). At the beginning of conjugation, two ciliates unite, fusing their pellicles at the contact point. The macronucleus in each is degraded. The individual micronuclei divide twice by meiosis to form four haploid pronuclei, three of which disintegrate. The remaining pronucleus divides again mitotically to form two gametic nuclei, a stationary one and a migratory one. The migratory nuclei pass into the respective conjugates. Then the cili-

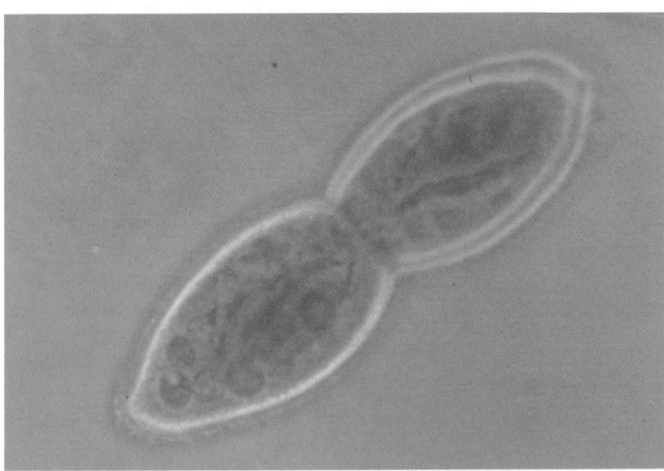

Figure 27.1 Protozoan Reproduction. Binary fission in *Paramecium caudatum* (×100).

ates separate, the gametic nuclei fuse, and the resulting diploid zygote nucleus undergoes three rounds of mitosis. The eight resulting nuclei have different fates: one nucleus is retained as a micronucleus; three others are destroyed; and the four remaining nuclei develop into macronuclei. Each separated conjugant now undergoes cell division. Eventually progeny with one macronucleus and one micronucleus are formed.

1. What specific nutritional types exist among protozoa?
2. What functions do cysts serve for a typical protozoan? What causes excystation to occur?
3. What is a pseudopodium?
4. How do protozoa reproduce asexually?
5. How do protozoa reproduce sexually? Describe the process of ciliate conjugation.

27.8 CLASSIFICATION

Many protozoan taxonomists regard the *Protozoa* as a subkingdom, which contains seven of the 14 phyla found within the kingdom *Protista* (**table 27.2**). The phylum *Sarcomastigophora* consists of flagellates and amoebae with a single type of nucleus. The phyla *Labyrinthomorpha, Apicomplexa, Microspora, Ascetospora,* and *Myxozoa* have either saprozoic or parasitic species. The phylum *Ciliophora* has ciliated protozoa with two types of nuclei. The classification of this subkingdom into phyla is based primarily on types of nuclei, mode of reproduction, and mechanism of locomotion.

More recent classifications are quite different. In 1993 T. Cavalier-Smith proposed that the protozoa be elevated to kingdom status with 18 phyla based on the structure of mitochondrial cristae and other characteristics (*see section 19.7*). The acceptance of this new classification by protozoologists, however, re-

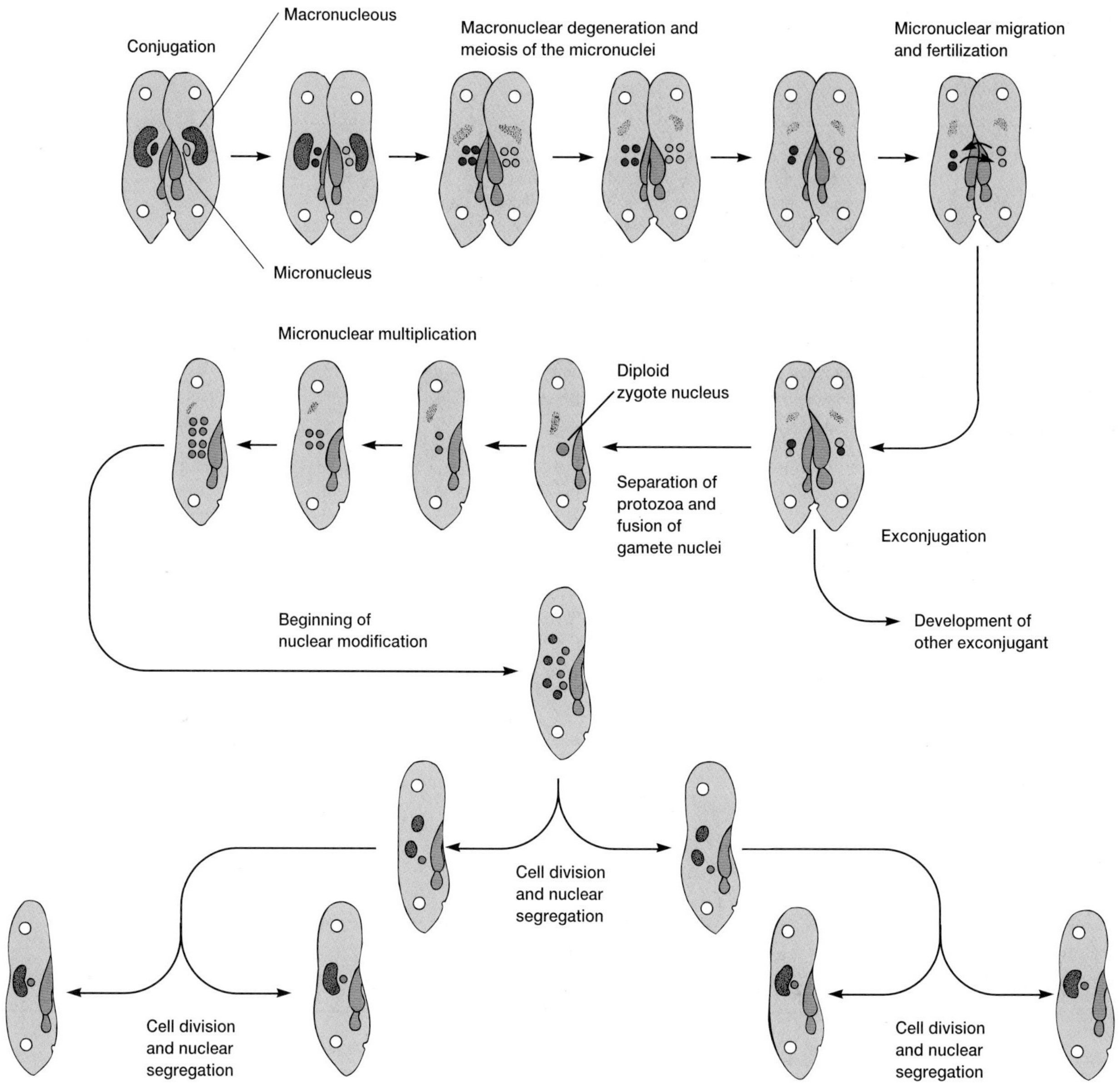

Figure 27.2   Conjugation in *Paramecium caudatum*, Schematic Drawing.   Follow the arrows. After the conjugants separate, only one of the exconjugants is followed; however, a total of eight new protozoa result from the conjugation.

mains to be determined. In recent molecular classification schemes, the protozoa do not exist as a discrete taxon. Protozoan-like eucaryotes are found at all evolutionary levels (Phylogenetic Diagram 27).

27.9   REPRESENTATIVE TYPES

This section describes some representatives of each group of protozoan protists to present an overview of protozoan diversity and to provide a basis for comparing different groups. For simplicity, the standard classification summarized in table 27.2 will be followed.

Phylum *Sarcomastigophora*

Protists that have a single type of nucleus and possess flagella (subphylum *Mastigophora*) or pseudopodia (subphylum *Sarcodina*) are placed in the phylum *Sarcomastigophora*. Both sexual and asexual reproduction are seen in this phylum.

The subphylum *Mastigophora* contains both phytoflagellates, chloroplast-bearing flagellates and close relatives, and **zooflagellates.** Zooflagellates do not have chlorophyll and are either holozoic, saprozoic, or symbiotic. Asexual reproduction occurs by longitudinal binary fission along the major body axis. Sexual reproduction is known for a few species, and encystment is common. Zooflagellates are characterized by the presence of

---

Table 27.2

Abbreviated Classification of the Subkingdom *Protozoa*[a]

Taxonomic Group	Characteristics	Examples
Phylum: *Sarcomastigophora*	Locomotion by flagella, pseudopodia, or both; when present, sexual reproduction is essentially syngamy (union of gametes external to the parents); single type of nucleus	
Subphylum: *Mastigophora*	One or more flagella; division by longitudinal binary fission; sexual reproduction in some groups	
Class: *Zoomastigophorea*	Chromatophores absent; one to many flagella; amoeboid forms, with or without flagella; sexuality known in some groups; mainly parasitic	*Giardia* *Leishmania* *Trichomonas* *Trichonympha* *Trypanosoma*
Subphylum: *Sarcodina*	Locomotion primarily by pseudopodia; shells (tests) often present; flagella restricted to reproductive stages when present; asexual reproduction by fission; mostly free living	
Superclass: *Rhizopoda*	Locomotion by pseudopodia or by protoplasmic flow with discrete pseudopodia; some contain tests	*Amoeba* *Coccodiscus* *Elphidium*
Phylum: *Labyrinthomorpha*	Spindle-shaped cells capable of producing mucous tracks; trophic stage as ectoplasmic network; nonamoeboid cells; saprozoic and parasitic on algae and seagrass	*Labyrinthula*
Phylum: *Apicomplexa*	All members have a spore-forming stage in their life cycle; contain an apical complex; sexuality by syngamy; all species parasitic; cysts often present; cilia absent; often called the Sporozoa	*Cryptosporidium* *Eimeria* *Plasmodium* *Toxoplasma*
Phylum: *Microspora*	Unicellular spores with spiroplasm containing polar filaments; obligatory intracellular parasites	*Nosema*
Phylum: *Ascetospora*	Spore with one or more spiroplasms; no polar capsules or polar filaments; all parasitic in invertebrates	*Haplosporidium*
Phylum: *Myxozoa*	Spores of multicellular origin; one or more polar capsules; all parasitic, especially in fish	*Myxosoma*
Phylum: *Ciliophora*	Simple cilia or compound ciliary organelles in at least one stage in the life cycle; two types of nuclei; contractile vacuole present; binary fission transverse; sexuality involving conjugation; most species free living, but many commensal, some parasitic	*Balantidium* *Didinium* *Entodinium* *Ichthyophthirius* *Nyctotherus* *Paramecium* *Stentor* *Tetrahymena* *Tokophrya* *Vorticella*

[a]Based on the 1980 Committee on Systematics and Evolution of the Society of Protozoologists.

---

one or more flagella. Most members are uninucleate. One major group, the kinetoplastids, has its mitochondrial DNA in a special region called the **kinetoplast** (**figure 27.3a**; *see also figures 2.13c and 4.13*).

Some zooflagellates are free living. The choanoflagellates are a distinctive example in that they have one flagellum, are solitary or colonial, and are on stalks. Other zooflagellates form symbiotic relationships. For example, *Trichonympha* species (*see figure 28.2*) are found in the intestine of termites and produce enzymes that the termite needs to digest the wood particles on which it feeds. The protozoan-termite relationship (pp. 580–81)

Many zooflagellates are important human parasites. *Giardia lamblia* (*see figure 40.17*) can be found in the human intestine where it may cause severe diarrhea. It is transmitted through water that has been contaminated with feces (*see section 40.2*). Trichomonads, such as *Trichomonas vaginalis,* live in the vagina and urethra of women and in the prostate, seminal vesicles, and urethra of men. They are transmitted primarily by sexual intercourse (*See table 39.4 on page 904 for a summary of all the sexually transmitted diseases covered in this textbook.*) Giardiasis (pp. 928–29). Trichomoniasis (p. 934)

The zooflagellates called **trypanosomes** are important blood pathogens of humans and animals in certain parts of the world. Because they live in the blood, they are also called hemoflagellates. These parasites (figure 27.3a) have a typical zooflagellate structure and appear to be the earliest diverging branch of protists with mitochondria and peroxisomes. A major human trypanosomal disease is African sleeping sickness caused by *Trypanosoma brucei rhodesiense* or *T. brucei gambiense* (*see section 40.2*).

The subphylum *Sarcodina* contains the amoeboid protists. They are found throughout the world in both fresh and salt water and are abundant in the soil. Several species are parasites of mammals. Simple amoebae [s., amoeba] move almost continually using their pseudopodia (**amoeboid movement**). Many have no definite shape, and their internal structures (figure 27.3b) occupy no particular position. The single nucleus, contractile and phagocytic vacuoles, and ecto- and endoplasm shift as the amoebae move. Amoebae engulf a variety of materials (small algae, bacteria, other protozoa) through

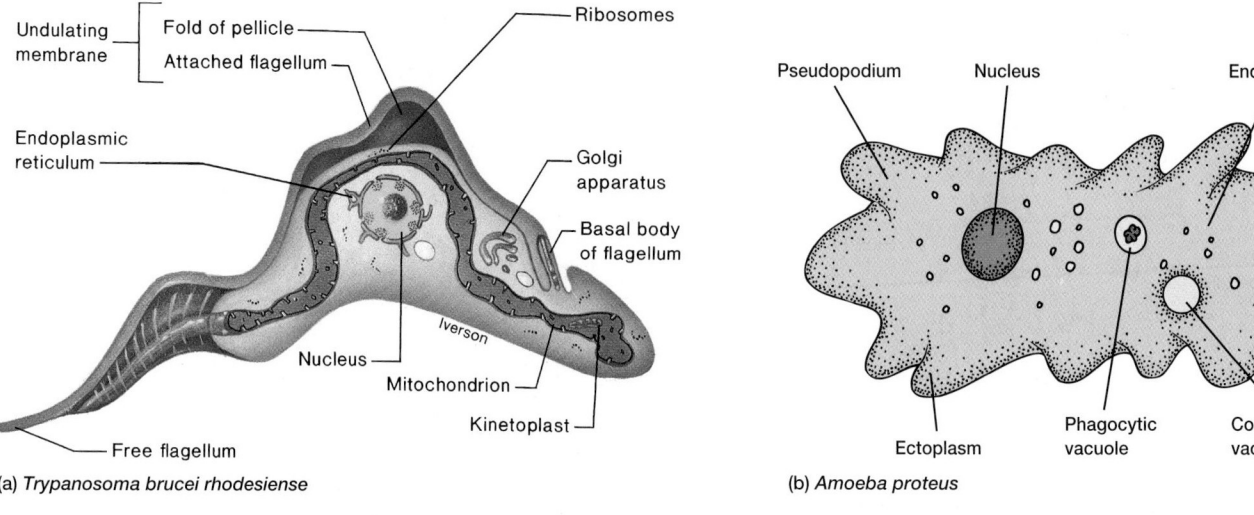

(a) *Trypanosoma brucei rhodesiense*

(b) *Amoeba proteus*

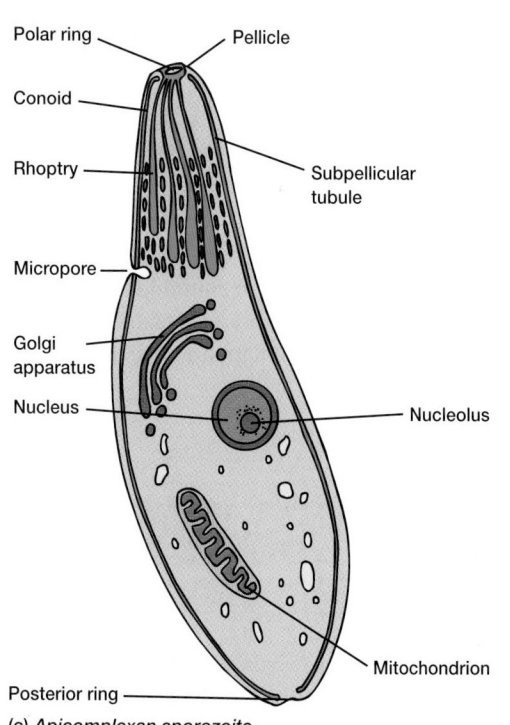

(c) *Apicomplexan sporozoite*

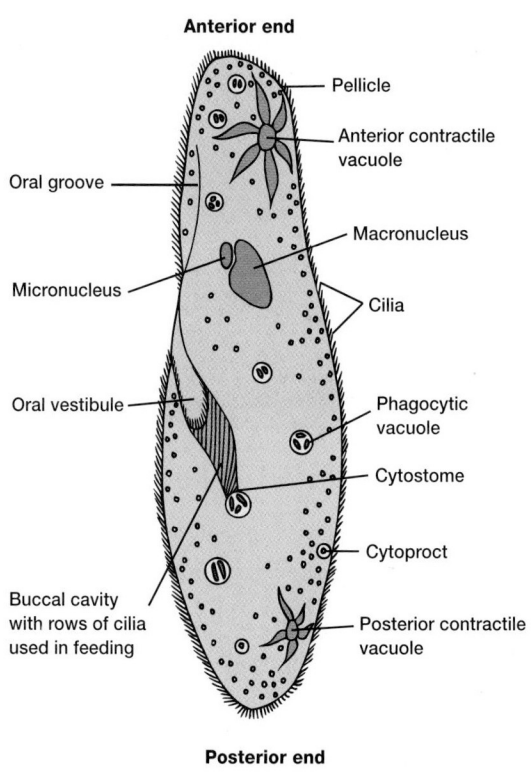

(d) *Paramecium caudatum*

Figure 27.3   Drawings of Some Representative Protozoa.   (a) Structure of the flagellate, *Trypanosoma brucei rhodesiense.* **(b)** The structure of the amoeboid protist, *Amoeba proteus.* **(c)** Structure of an apicomplexan sporozoite. **(d)** Structure of the ciliate *Paramecium caudatum.*

phagocytosis. Some material moves into and out of the plasma membrane by pinocytosis. Reproduction in the amoebae is by simple asexual binary fission. Some amoebae can form cysts.

Many free-living forms are more complex than simple amoebae. *Arcella* manufactures a loose-fitting shell or **test** for protection (**figure 27.4***a*). These amoebae extend their pseudopodia from the test aperture to either feed or creep along. The foraminiferans and radiolarians primarily are marine amoebae,

with a few occurring in fresh and brackish water. Most foraminiferans live on the seafloor, whereas radiolarians are usually found in the open sea (**Microbial Diversity & Ecology 27.1**). Foraminiferan tests and radiolarian skeletons have many unique and beautiful shapes (figure 27.4*b,c*). They range in diameter from about 20 mm to several cm.

Finally, there are many symbiotic amoebae, most of which live in other animals. Two common genera are *Endamoeba* and

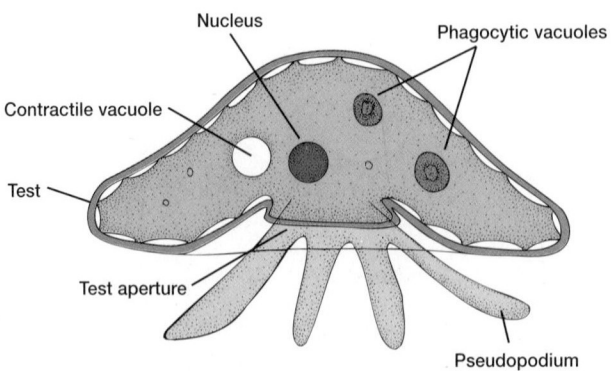

Nucleus

Phagocytic vacuoles

Contractile vacuole

Test

Test aperture

Pseudopodium

(a) *Arcella*

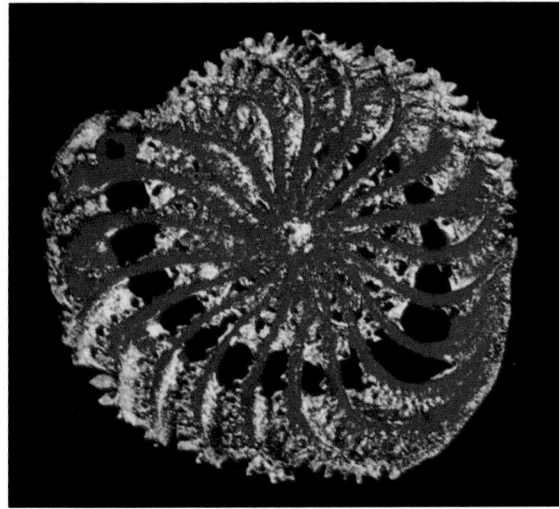

(b) *Elphidium cristum*

(c) Radiolarian shells

Figure 27.4 **Some Free-living Sarcodines.** (**a**) An illustration of *Arcella*, showing the test or shell that is made of chitinlike material secreted by the protist. (**b**) The test of the foraminiferan, *Elphidium cristum* ($\times$100). (**c**) A group of siliceous radiolarian shells, light micrograph ($\times$63).

Microbial Diversity & Ecology

27.1   The Importance of Foraminiferans

Of over 40,000 described species of foraminiferans, about 90% are fossil. During the Tertiary period (about 230 million years ago), the foraminiferans contributed massive shell accumulations to geologic formations. They were so abundant that they formed thick deposits, which became uplifted over time and exposed as great beds of limestone in Europe, Asia, and Africa. The White Cliffs of Dover, the famous landmark of southern England, are made up almost entirely of foraminiferan shells. The Egyptian pyramids of Gizeh, near Cairo, are built of foraminiferan limestone. Currently, foraminiferans are important aids to geologists in identifying and correlating rock layers as they search for oil-bearing strata. The calcareous shells of abundant planktonic foraminiferans are today settling and accumulating over much of the ocean floor as thick deposits called "globigerina ooze"—limestone of the distant future.

Entamoeba. Endamoeba blattae is common in the intestine of cockroaches, and related species are present in termites. *Entamoeba histolytica* (*see figure 40.16*) is an important parasite of humans, in whom it often produces severe amoebic dysentery, which may be fatal. Free-living amoebae from two genera, *Naegleria* and *Acanthamoeba,* can cause disease in humans and other mammals (chapter opening figure; *see also section 40.2*).

1. What characteristics would be exhibited by a protozoan that belongs to the phylum *Sarcomastigophora?* What two subphyla does it contain?
2. How would you characterize a zooflagellate? An amoeba?
3. What two human diseases are caused by zooflagellates?
4. Where can two different symbiotic amoebae be found?

Phylum *Labyrinthomorpha*

The very small phylum *Labyrinthomorpha* consists of protists that have spindle-shaped or spherical nonamoeboid vegetative cells. In some genera, amoeboid cells move within a network of mucous tracks using a typical gliding motion. Most members are marine and either saprozoic or parasitic on algae. Several years ago *Labyrinthula* killed most of the "eel grass" on the Atlantic coast, depriving ducks of their food and starving many of them.

Phylum *Apicomplexa*

The **apicomplexans,** often collectively called the sporozoans, have a spore-forming stage in their life cycle and lack special locomotory organelles (except in the male gametes, and the zygote or ookinete). They are either intra- or intercellular parasites of animals and are distinguished by a unique arrangement of fibrils, microtubules, vacuoles, and other organelles, collectively called the apical complex, which is located at one end of the cell.

The **apical complex** contains several components (figure 27.3*c*). One or two polar rings are at the apical end. The **conoid** consists of a cone of spirally arranged fibers lying next to the polar rings. Subpellicular microtubules radiate from the polar rings and probably serve as support elements. Two or more **rhoptries** extend to the plasma membrane and secrete their contents at the cell surface. These secretions aid in the penetration of the host cell. One or more micropores are thought to function in the intake of nutrients.

Apicomplexans have complex life cycles in which certain stages occur in one host (the mammal) and other stages in a different host (often a mosquito). The life cycle has both asexual and sexual phases and is characterized by an alternation of haploid and diploid generations. At some point an asexual reproduction process called schizogony occurs. **Schizogony** is a rapid series of mitotic events producing many small infective organisms through the formation of uninuclear buds. Sexual reproduction involves the fertilization of a large female macrogamete by a small, flagellated male gamete. The resulting zygote becomes a thick-walled cyst called an **oocyst.** Within the oocyst, meiotic divisions produce infective haploid spores.

The four most important sporozoan parasites are *Plasmodium* (the causative agent of malaria), *Cryptosporidium* (the causative agent of cryptosporidiosis), *Toxoplasma* (the causative agent of toxoplasmosis), and *Eimeria* (the causative agent of coccidiosis).

Malaria (pp. 929–31)

Phylum *Microspora*

The small microsporans (3 to 6 μm) are obligatory intracellular parasites lacking mitochondria. The infective stage is transmitted from host to host as a resistant spore. Traditionally microsporans have been placed with the protozoa, as is done here. However, some recent molecular evidence indicates that they may be specialized fungi that are well adapted for intracellular parasitism. Further research will be required to resolve this issue. Included in this group are several species of some economic importance because they parasitize beneficial insects.

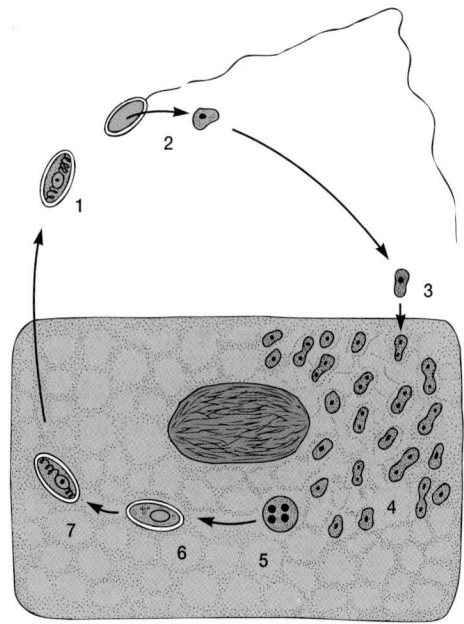

Figure 27.5   The Microsporean *Nosema bombycis,* Which Is Fatal to Silkworms.   (*1*) A typical spore with one coiled filament. (*2*) When ingested, it extrudes the filament. (*3*) The parasite enters an epithelial cell in the intestine of the silkworm and (*4*) divides many times to form small amoebae that eventually fill the cell and kill it. During this phase, some of the amoebae with four nuclei become spores (*5, 6, 7*). Silkworms are infected by eating leaves contaminated by the feces of infected worms.

Nosema bombycis parasitizes silkworms (**figure 27.5**) causing the disease **pebrine,** and *Nosema apis* causes serious dysentery (foul brood) in honeybees. There has been an increased interest in these parasites because of their possible role as biological control agents for certain insects. For example, *Nosema locustae* has been approved and registered by the United States Environmental Protection Agency for use in long-lasting control of rangeland grasshoppers. Recently seven microsporidian genera (*Nosema, Encephalitozoon, Pleistophora, Microsporidium, Vittaforma, Trachipleistophora,* and *Enterocytozoon*) have been implicated in human diseases in immunosuppressed and AIDS patients.

Phylum *Ascetospora*

Ascetospora is a relatively small phylum that consists exclusively of parasitic protists characterized by spores lacking polar caps or polar filaments. Ascetosporans such as *Haplosporidium* are parasitic primarily in the cells, tissues, and body cavities of mollusks.

Phylum *Myxozoa*

The myxozoans are all parasitic, most on freshwater and marine fish. They have a resistant spore with one to six coiled polar filaments. The most economically important myxozoan is *Myxosoma cerebralis,* which infects the nervous system and auditory organ of trout and salmon (salmonids). Infected fish lose their sense of balance and tumble erratically—thus the name whirling or tumbling disease. **Proliferative kidney disease,** caused by an unclassified myxozoan, has become one of the most important diseases of cultured salmon throughout the world.

---

1. Describe a typical apicomplexan (sporozoan) protist, including its apical complex.
2. Summarize the sporozoan life cycle. What is schizogony?
3. What are the four most important sporozoan parasites and the diseases they cause?
4. Give one economically important disease caused by a microsporan. What group of animals do myxosporans usually parasitize?

Phylum *Ciliophora*

The phylum *Ciliophora* is the largest of the seven protozoan phyla. There are about 8,000 species of these unicellular, heterotrophic protists that range from about 10 to 3,000 μm long. As their name implies, ciliates employ many cilia as locomotory organelles. The cilia are generally arranged either in longitudinal rows (figure 27.3d; *see also figure 4.23*) or in spirals around the body of the organism. They beat with an oblique stroke; therefore the protist revolves as it swims. Coordination of ciliary beating is so precise that the protist can go either forward or backward.

There is great variation in ciliate shape, and most do not look like the slipper-shaped *Paramecium* (*see figures 2.8e and 4.1a*). In some species (*Vorticella*) the protozoan attaches itself to the substrate by a long stalk. *Stentor* attaches to a substrate and stretches out in a trumpet shape to feed (*see figure 4.1e*). A few species have tentacles for the capture of prey. Some can discharge toxic threadlike darts called toxicysts, which are used in capturing prey.

A most striking feature of ciliates is their ability to capture many particles in a short time by the action of the cilia around the buccal cavity. Food first enters the cytostome and passes into phagocytic vacuoles that fuse with lysosomes after detachment from the cytostome. A vacuole's contents are digested when the vacuole is acidified and lysosomes release digestive enzymes into it. After the digested material has been absorbed into the cytoplasm, the vacuole fuses with a special region of the pellicle called the **cytoproct** and empties its waste material to the outside. Contractile vacuoles are used for osmoregulation and are present chiefly in freshwater species.

Most ciliates have two types of nuclei: a large macronucleus and a smaller micronucleus. The micronucleus is diploid and contains the normal somatic chromosomes. It divides by mitosis and transmits genetic information through meiosis and sexual reproduction. Macronuclei are derived from micronuclei by a complex series of steps. Within the macronucleus are many chromatin bodies, each containing many copies of only one or two genes. Macronuclei are thus polyploid and divide by elongating and then by constricting. They produce mRNA to direct protein synthesis, maintain routine cellular functions, and control normal cell metabolism.

Some ciliates reproduce asexually by transverse binary fission, forming two equal daughter protozoa. Many ciliates also reproduce by conjugation as previously described.

Although most ciliates are free living, symbiotic forms do exist. Some ciliated protozoa live as harmless commensals. For example, *Entodinium* is found in the rumen of cattle, and *Nyctotherus* occurs in the colon of frogs. Other ciliates are strict parasites. For example, *Balantidium coli* lives in the intestine of mammals, including humans, where it can produce dysentery. *Ichthyophthirius* lives in freshwater where it can attack many species of fish, producing a disease known as "ick."

---

1. Describe the morphology of a typical ciliated protozoan.
2. Describe the food-gathering structures found in the ciliated protozoa.
3. In ciliates, what is the function of the macronucleus? The micronucleus?
4. How do ciliates reproduce?
5. Where can the following ciliates be found: *Entodinium, Nyctotherus, Balantidium, Ichthyophthirius?*

Summary

Protozoa are protists that can be defined as usually motile eucaryotic unicellular microorganisms.

27.1 Distribution
a. Protozoa are found wherever other eucaryotic organisms exist.

27.2 Importance
a. They are important components of food chains and food webs. Many are parasitic in humans and animals (**table 27.1**), and some have become very useful in the study of molecular biology.

27.3 Morphology
a. Because protozoa are eucaryotic cells, in many respects their morphology and physiology resemble those of multicellular animals. However, because all their functions must be performed within the individual protist, many morphological and physiological features are unique to protozoan cells.

27.4 Nutrition
a. Most protozoa are chemoheterotrophic and may be either holozoic or saprozoic.

27.5 Encystment and Excystment
a. Some protozoa can secrete a resistant covering and go into a resting stage (encystation) called a cyst. Cysts protect the organism against adverse environments, function as a site for nuclear reorganization, and serve as a means of transmission in parasitic species.

27.6 Locomotory Organelles

a. Protozoa move by one of three major types of locomotory organelles: pseudopodia, flagella, or cilia. Some have no means of locomotion.

27.7 Reproduction

a. Most protozoa reproduce asexually (**figure 27.1**), some use sexual reproduction (**figure 27.2**), and some employ both methods.

27.8 Classification

a. According to one classical classification scheme, there are seven protozoan phyla (**table 27.2**). The phylum *Sarcomastigophora* is characterized by protists that have a single type of nucleus, possess flagella (subphylum *Mastigophora*), pseudopodia (subphylum *Sarcodina*), or both types of locomotory organelles.

27.9 Representative Types

a. The subphylum *Sarcodina* consists of the amoeboid protists. They are found throughout the world in both fresh and salt water and in the soil. Some species are parasitic.

b. The phylum *Labyrinthomorpha* contains protists that have spindle-shaped or spherical nonamoeboid vegetative cells. Most members are marine and either saprozoic or parasitic on algae.

c. The phylum *Apicomplexa* consists of sporozoan protists that possess an apical complex, a unique arrangement of fibrils, microtubules, vacuoles, and other organelles at one end of the cell. Representative members include the *Plasmodium* parasites, which cause malaria; *Toxoplasma,* which causes toxoplasmosis; and *Eimeria,* the agent of coccidiosis.

d. The phylum *Microspora* consists of very small protists that are intracellular parasites of every major animal group. They are transmitted from one host to the next as a spore, the form from which the group obtains its name (**figure 27.5**).

e. The phylum *Ascetospora* contains protists that produce spores lacking polar capsules. These protists are primarily parasitic in mollusks.

f. The phylum *Myxozoa* consists entirely of parasitic species, usually found in fish. The spore is characterized by one to six polar filaments.

g. The phylum *Ciliophora* comprises a group of protists that have cilia and two types of nuclei. Conjugation in the *Ciliophora* is a form of sexual reproduction that involves exchange of micronuclear material.

Key Terms

amoeboid movement 570	cytostome 567	macronucleus 567	pseudopodia 568
apical complex 573	ectoplasm 567	micronucleus 567	rhoptry 573
apicomplexan 573	encystation 568	oocyst 573	saprozoic nutrition 567
binary fission 568	endoplasm 567	pebrine 573	schizogony 573
conjugant 568	excystation 568	pellicle 567	secretory vacuole 567
conjugation 568	food chain 566	phagocytic vacuole 567	test 571
conoid 573	food web 566	plankton 566	trophozoite 568
contractile vacuole 567	holozoic nutrition 567	proliferative kidney disease 574	trypanosome 570
cyst 568	hydrogenosome 567	protozoa 566	zooflagellate 569
cytoproct 574	kinetoplast 570	protozoology 566	

Questions for Thought and Review

1. What is the economic impact or human relevance of the protozoa?
2. What criteria are now used in the classification of protozoa?
3. Seven protozoan phyla have been discussed. What are their distinguishing characteristics?
4. What are some typical organelles found in the protozoa?

5. What advantage does cyst formation give to protozoa?
6. How do protozoa move? Reproduce?
7. If the diversity within the seven protozoan phyla is considered, which one shows the greatest evolutionary advancement? Defend your answer.
8. The protozoa are said to be grouped together for a negative reason. What does this mean?

9. Describe how DNA is distributed to daughter cells when the ciliate *Paramecium* divides. Include a discussion of both conjugation and binary fission.
10. How does a protozoan cyst differ from a bacterial endospore?

Critical Thinking Questions

1. Why don't we know as much about the basic biology of protozoa as we know about the biology of fungi, viruses, and bacteria?
2. The text suggests that excystation requires the recognition of environmental signals. Following

this line of thinking, suggest some antiprotozoan targets that could be exploited to prevent excystation in patients who have ingested cysts. Once the cysts emerge, are there other targets?

3. Suggest a reason or mechanism why, in some protozoa, the cytoplasmic material (ectoplasm) just under the plasma membrane is so rigid.

For recommended readings on these and related topics, see Appendix V.

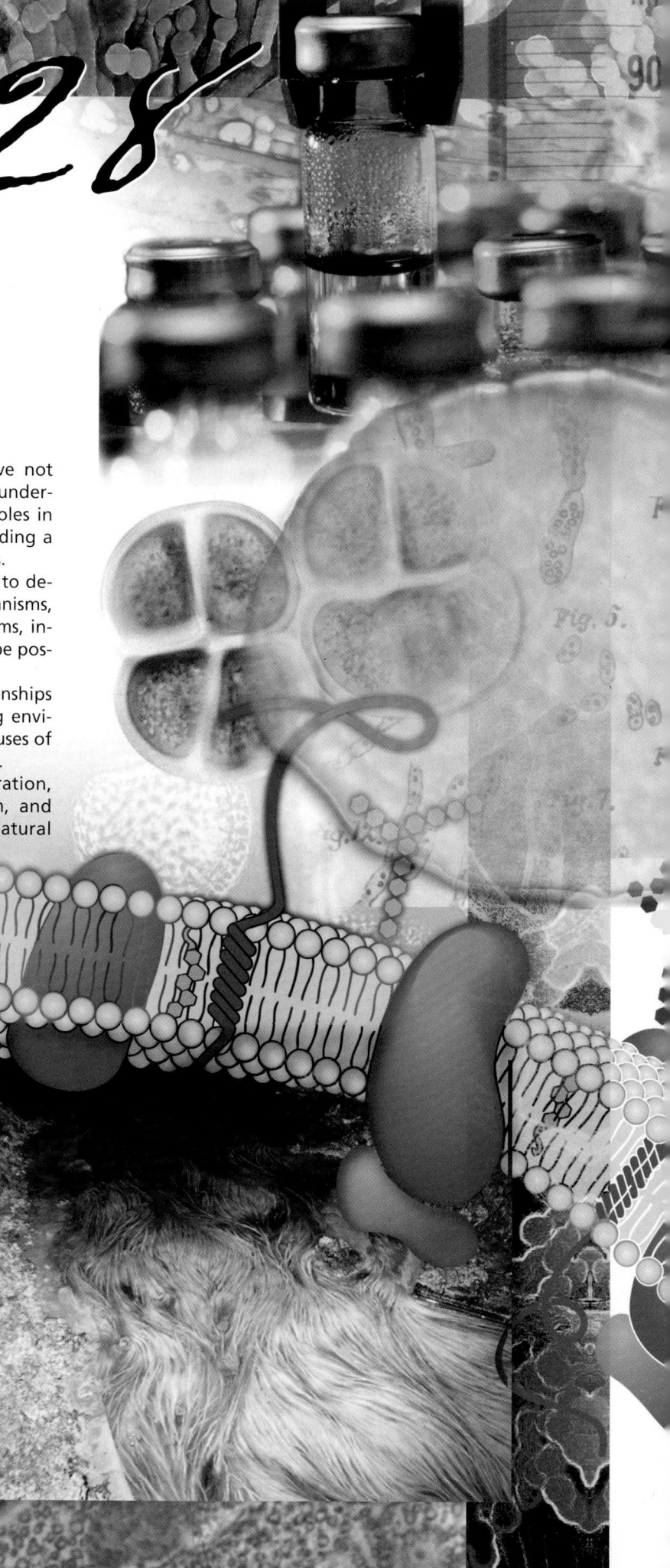

chapter 28

Microorganism Interactions and Microbial Ecology

Concepts

- Most microorganisms in complex communities have not been grown or characterized. This has limited our understanding of microorganism interactions and their roles in nature and disease. Molecular techniques are providing a better understanding of these uncultured organisms.

- The term symbiosis, or "together-life," can be used to describe many of the interactions between microorganisms, and also microbial interactions with higher organisms, including plants and animals. These interactions may be positive or negative.

- Microbial ecology is the study of microbial relationships with other organisms and also with their nonliving environments. These relationships, based on interactive uses of resources, have effects extending to the global scale.

- Symbiotic interactions include mutualism, cooperation, commensalism. parasitism, predation, amensalism, and competition. These interactions are important in natural processes and in the occurrence of disease. The interactions can vary depending on the environment and changes in the interacting organisms.

- Microorganisms, as they interact, can form complex physical assemblages that include particularly biofilms. These form on living and inert surfaces and have major impacts on microbial survival and the occurrence of disease.

- Microorganisms also interact by the use of chemical signal molecules, which allow the microbial population to respond to increased population density. Such responses include quorum sensing, which controls a wide variety of microorganism properties.

Microorganisms living in environments where most known organisms cannot survive are important for understanding microbial diversity. These strands of iron-oxidizing *Ferroplasma* were discovered growing at pH 0 in an abandoned mine near Redding, California. This hardy microorganism has only a plasma membrane to protect itself from the rigors of this harsh environment.

- Energy, electrons, and nutrients must be available in a suitable physical environment for microorganisms to function. Microbes interact with their environment to obtain energy (from light or chemical sources), electrons, and nutrients, which leads to a process called biogeochemical cycling. Microorganisms change the physical state and mobility of many nutrients as they use them in their growth processes.
- Microorganisms are an important part of ecosystems, or self-regulating biological communities and their physical environments. Microorganisms play an important role in succession, or the predictable changes that occur in ecosystems when they are disturbed.
- Extreme environments restrict the range of microbial types able to survive and function. This can be due to physical factors such as temperature, pH, pressure, or salinity. Many microorganisms found in "extreme" environments are especially adapted not only to survive, but to function metabolically under these particular conditions.
- Methods used to study microbial interactions and microbial ecology provide information on environmental characteristics; microbial biomass, numbers, types, activity, and community structure. Microscopic, chemical, enzymatic, and molecular techniques are used in these studies.
- It is now possible to determine the nucleic acid sequences of specific microorganisms or organelles isolated from natural environments and to study the phylogeny of uncultured microorganisms. This should lead to important new advances in the study of microbial ecology.

> *Everything is everywhere, the environment selects.*
> —*M. W. Beijerinck*

*I*n previous chapters microorganisms usually have been considered as isolated entities. The basic characteristics of microorganisms, including the structure and function of microbial cells, metabolism, growth and the control of growth, have been discussed. In addition, metabolism, genetics, and molecular aspects of microorganisms, including genomics, have been described. In this chapter, microbial interactions with both the physical environment and with other organisms will be considered.

28.1 FOUNDATIONS OF MICROBIAL ECOLOGY

Two major themes will be developed in this chapter: (1) the nature of microbial relationships with other living organisms, or the nature of **symbioses,** and (2) the interactions of these organisms with each other and with their nonliving physical environment, or

the area of **microbial ecology** (**Microbial Diversity & Ecology 28.1**). The term symbiosis is used in its original broadest sense, as an association of two or more different species of organisms, as suggested by H. A. deBary in 1879.

Microorganisms function as **populations** or assemblages of similar organisms, and as **communities,** or mixtures of different microbial populations. These microorganisms have evolved while interacting with the inorganic world and with higher organisms, and they largely play beneficial and vital roles; disease-causing organisms are only a minor component of the microbial world. Microorganisms, as they interact with other organisms and their environment, also contribute to the functioning of **ecosystems,** or self-regulating biological communities and their physical environment (pp. 603–4). Knowledge of these interactions is important in understanding both microbial contributions to the natural world and microbial roles in disease processes.

A major problem in understanding microbial interactions is that most microscopically observable microorganisms cannot be grown. The differences between observable and culturable microorganisms, which limit this field even today, have been noted for at least 70 years. This problem was discussed in a 1932 textbook on soil microbiology written by Selman Waksman, the discoverer of streptomycin, and advances continue in this important area (*see section 6.5*). Molecular techniques and sequence data provide valuable information on these uncultured microorganisms. Probes can be used to both identify specific organisms and better understand physical relationships between microbes. Attempting to grow these uncultured microbes remains a central challenge in microbial ecology

28.2 MICROBIAL INTERACTIONS

Microorganisms can be physically associated with other organisms in a variety of ways. One organism can be located on the surface of another, as an **ectosymbiont.** In this case, the ectosymbiont usually is a smaller organism located on the surface of a larger organism. Often, dissimilar organisms of similar size are in physical contact. The term **consortium** can be used to describe this physical relationship. Consortia in aquatic environments are frequently complex, involving multiple layers of similar-looking microorganisms that often have complementary physiological properties. In contrast, one organism can be located within another organism as an **endosymbiont.** There also are many cases in which microorganisms live on both the inside and the outside of another organism, a phenomenon called **ecto/endosymbiosis.** Interesting examples of ecto/endosymbioses include a *Thiothrix* species, a sulfur-using bacterium, which is attached to the surface of a mayfly larva and which itself contains a parasitic bacterium. Fungi associated with plant roots (mycorrhizal fungi) often contain endosymbiotic bacteria, as well as having bacteria living on their surfaces (*see pp. 655–58*).

These physical associations can be intermittent and cyclic or permanent. Examples of intermittent and cyclic associations of microorganisms with plants and marine animals are shown

Microbial Diversity & Ecology

28.1 Microbial Ecology Versus Environmental Microbiology

The term "microbial ecology" is now used in a general way to describe the presence and contributions of microorganisms, through their activities, to the places where they are found. Students of microbiology should be aware that much of the information on microbial presence and contributions to soils, waters, and associations with plants, now described by this term, would have been considered as "environmental microbiology" in the past. Thomas D. Brock, the discoverer of *Thermus aquaticus*, which is known the world over as the source of *Taq* polymerase for the polymerase chain reaction (PCR), has given a definition of microbial ecology that may be useful: "Microbial ecology is the study of the behavior and activities of microorganisms in their natural environments." The important operator in this sentence is *their* environment instead of *the* environment. To emphasize this point, Brock has noted that "microbes are small; their environments also are small." In these small environments or "microenvironments," other kinds of microorganisms (and macroorganisms) often also are present, a critical point that was emphasized by Sergei Winogradsky in 1947.

Environmental microbiology, in comparison, relates primarily to all-over microbial processes that occur in a soil, water, or food, as examples. It is not concerned with the particular "microenvironment" where the microorganisms actually are functioning, but with the broader-scale effects of microbial presence and activities. One can study these microbially mediated processes and their possible global impacts at the scale of "environmental microbiology" without knowing about the specific microenvironment (and the organisms functioning there) where these processes actually take place. However, it is critical to be aware that microbes function in their localized environments and affect ecosystems at greater scales, including causing global-level effects. In the last decades the term "microbial ecology" largely has lost its original meaning, and recently the statement has been made that "microbial ecology has become a 'catch-all' term." As you read various textbooks and scientific papers, possible differences between "microbial ecology" and "environmental microbiology" should be kept in mind.

in **table 28.1.** Important human diseases, including listeriosis, malaria, leptospirosis, legionellosis, and vaginosis also involve such intermittent and cyclic symbioses. These diseases will be discussed in chapters 39 and 40. Interesting permanent relationships also occur between bacteria and animals, as shown in **table 28.2.** Hosts include squid, leeches, aphids, nematodes, and mollusks. In each of these cases, an important characteristic of the host animal is conferred by the permanent bacterial symbiont.

Although it is possible to observe microorganisms in these varied physical associations with other organisms, the fact that there is some type of physical contact provides no information on the types of interactions that might be occurring. These interactions include mutualism, cooperation, commensalism, predation, parasitism, amensalism, and competition (**figure 28.1**). In large part, they were first described and defined by biologists studying

Table 28.1

Intermittent and Cyclical Symbioses of Microorganisms with Plants and Marine Animals

Symbiosis	Host	Cyclical Symbiont
Plant-bacterial	*Gunnera* (tropical angiosperm)	*Nostoc* (cyanobacterium)
	Azolla (rice paddy fern)	*Anabaena* (cyanobacterium)
	Phaseolus (bean)	*Rhizobium* (N_2 fixer)
	Ardisia (angiosperm)	*Protobacterium*
Marine animals	Coral coelenterates	*Symbiodinium* (dinomastigote)
	Luminous fish	*Vibrio, Photobacterium*
	Squid	*Photobacterium fischeri*

Adapted from L. Margulis and M. J. Chapman. 1998. Endosymbioses: Cyclical and permanent in evolution. *Trends in Microbiology* 6(9):342–46, tables 1, 2, and 3.

Table 28.2

Examples of Permanent Bacterial-Animal Symbioses and the Characteristics Contributed by the Bacterium to the Symbiosis

Animal Host	Symbiont	Symbiont Contribution
Sepiolid squid (*Euprymna scolopes*)	Luminous bacterium	Luminescence (*Vibrio fisheri*)
Medicinal leech (*Hirudo medicinalis*)	Enteric bacterium (*Aeromonas veronii*)	Blood digestion
Aphid (*Schizaphis graminum*)	Bacterium (*Buchnera aphidicola*)	Amino acid synthesis
Nematode worm (*Heterorhabditis* spp.)	Luminous bacterium (*Photorhabdus luminescens*)	Predation and antibiotic synthesis
Shipworm mollusk (*Lyrodus pedicellatus*)	Gill cell bacterium	Cellulose digestion and nitrogen fixation

Source: From E. G. Ruby, 1999. Ecology of a benign "infection": Colonization of the squid luminous organ by *Vibrio fischeri*. In *Microbial ecology and infectious disease*, E. Rosenberg, editor, American Society for Microbiology, Washington, D.C., 217–31, table 1.

Interaction type	Interaction example
Mutualism	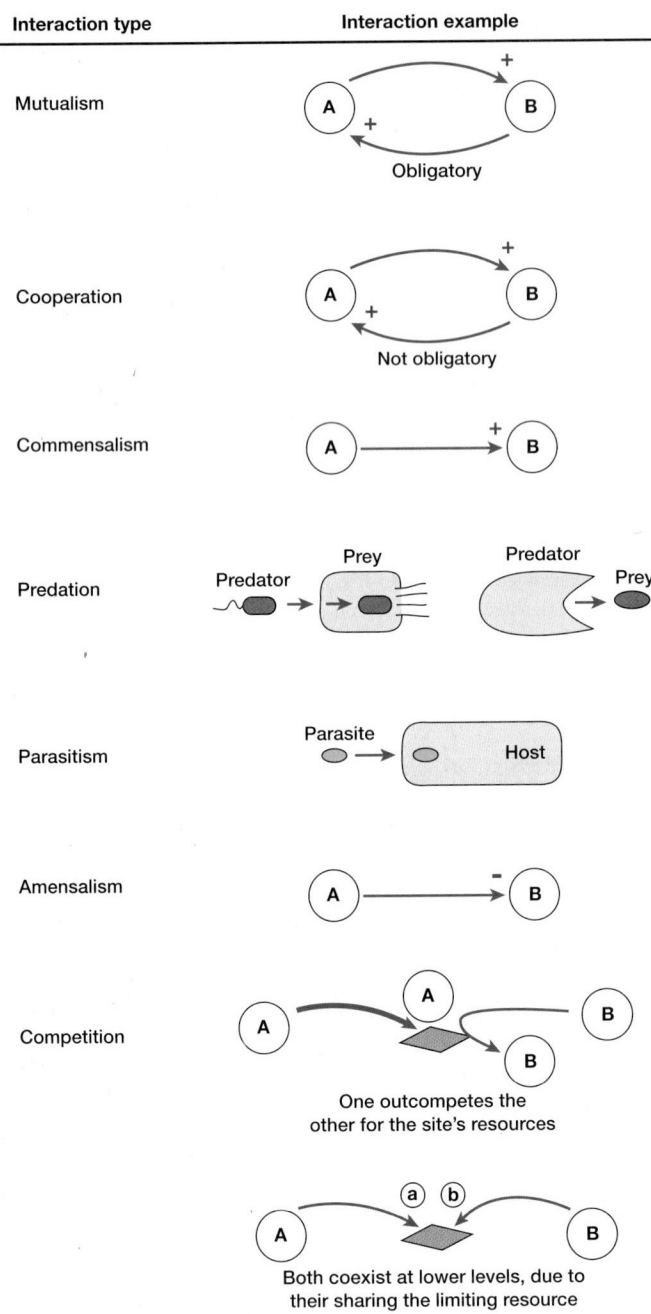

Figure 28.1 **Microbial Interactions.** Basic characteristics of symbiotic interactions that can occur between different organisms.

plants, animals, and easily observable microbial assemblages. H. A. deBary's work with lichens is an excellent example.

1. Define the terms symbiosis and microbial ecology. How are they similar and different?
2. Define the terms population, community, and ecosystem.
3. In what ways can different microorganisms be in physical contact?
4. List several important diseases that involve cyclic and intermittent symbioses.

Mutualism

Mutualism [Latin *mutuus,* borrowed or reciprocal] defines the relationship in which some reciprocal benefit accrues to both partners. This is an obligatory relationship in which the **mutualist** and the host are metabolically dependent on each other. When separated, in many cases, the individual organisms will not survive. Several examples of mutualism are presented next.

Microorganism-Insect Mutualisms

Mutualistic associations are common in the insects. This is related to the foods used by insects, which often include plant sap or animal fluids lacking in essential vitamins and amino acids. The required vitamins and amino acids are provided by bacterial symbionts in exchange for a secure physical habitat and ample nutrients. The aphid is an excellent example of this mutualistic relationship. This insect contains *Buchnera aphidicola* in its cytoplasm, and a mature insect contains literally millions of these bacteria in its body. The *Buchnera* provides its host with amino acids, particularly tryptophan, and if the insect is treated with antibiotics, it dies. *Wolbachia pipientis,* a rickettsia, is a cytoplasmic endosymbiont found in 15 to 20% of insect species and can control the reproduction of its host. This microbial association is thought to be a major factor in the evolution of sex and speciation in the parasitic wasps. *Wolbachia* also can cause cytoplasmic incompatibility in insects, parthenogenesis in butterflies, and the feminization of genetic males in isopods. What could be the advantage to the *Wolbachia?* By limiting sexual variability, the bacterium might benefit by creating a more stable asexual environment for its own longer-term maintenance. Our understanding of microbe-insect mutualisms, including the role of *Wolbachia* in insects, is constantly expanding with the increased use of molecular techniques.

The protozoan-termite relationship is a classic example of mutualism in which the flagellated protozoa live in the gut of termites and wood roaches (**figure 28.2***a*). These flagellates exist on a diet of carbohydrates, acquired as cellulose ingested by their host (figure 28.2*b*). The protozoa engulf wood particles, digest the cellulose, and metabolize it to acetate and other products. Termites oxidize the acetate released by their flagellates. Because the host is almost always incapable of synthesizing cellulases (enzymes that catalyse the hydrolysis of cellulose), it is dependent on the mutualistic protozoa for its existence.

This mutualistic relationship can be readily tested in the laboratory if wood roaches are placed in a bell jar containing wood chips and a high concentration of O_2. Because O_2 is toxic to the flagellates, they die. The wood roaches are unaffected by the high O_2 concentration and continue to ingest wood, but they soon die of starvation due to a lack of cellulases.

Zooxanthellae

Many marine invertebrates (sponges, jellyfish, sea anemones, corals, ciliates) harbor endosymbiotic, spherical algal cells called **zooxanthellae** within their tissue (**figure 28.3***a*). Because the degree of host dependency on the mutualistic alga is somewhat variable, only one well-known example is presented.

(a)

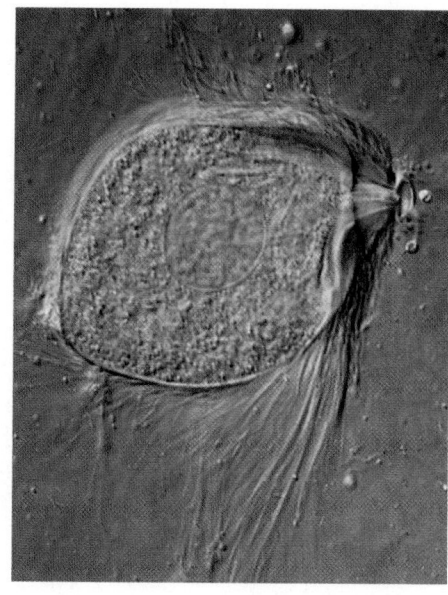

(b)

Figure 28.2 Mutualism. Light micrographs of (**a**) a worker termite of the genus *Reticulitermes* eating wood (×10), and (**b**) *Trichonympha*, a multiflagellated protozoan from the termite's gut (×135). Notice the many flagella that occur over most of its length. The ability of *Trichonympha* to break down cellulose allows termites to use wood as a food source.

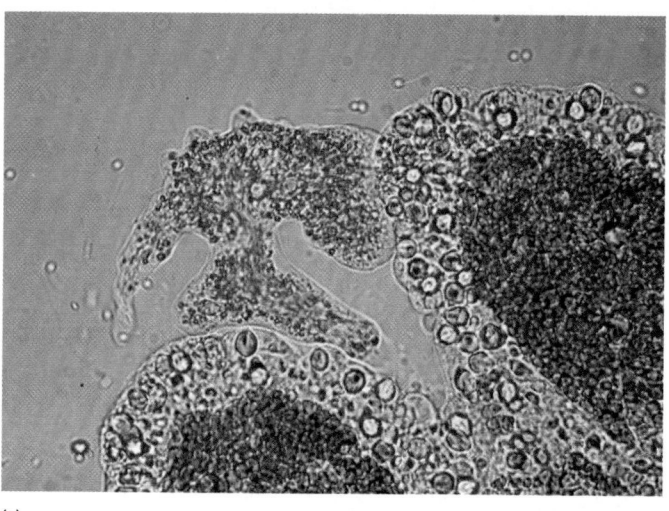

(a)

(b)

Figure 28.3   Zooxanthellae.   (**a**) Zooxanthellae (green) within the tip of a hydra tentacle (×150). (**b**) The green color of this rose coral (*Manilina*) is due to the abundant zooxanthellae within its tissues.

The hermatypic (reef-building) corals (figure 28.3*b*) satisfy most of their energy requirements using their zooxanthellae. Pigments produced by the coral protect the algae from the harmful effects of ultraviolet radiation. Clearly the zooxanthellae also benefit the coral because the calcification rate is at least 10 times greater in the light than in the dark. Hermatypic corals lacking zooxanthellae have a very low rate of calcification. Based on their stable carbon isotopic composition, it has been determined that most of the organic carbon in the tissues of the hermatypic corals has come from the zooxanthellae. Because of this coral-algal mutualistic relationship—capturing, conserving, and cycling nutrients and energy—coral reefs are among the most productive and successful of known ecosystems.

Sulfide-Based Mutualisms

Tube worm-bacterial relationships exist several thousand meters below the surface of the ocean, where the Earth's crustal plates are spreading apart (**figure 28.4**). Vent fluids are anoxic, contain high concentrations of hydrogen sulfide, and can reach a temperature of 350°C. The seawater surrounding these vents has sulfide concentrations around 250 μM and temperatures 10 to 20°C above the normal seawater temperature of 2.1°C.

The giant (>1 m in length), red, gutless tube worms (*Riftia* spp.; **figure 28.5***a*) near these hydrothermal vents provide an example of a unique form of mutualism and animal nutrition in which chemolithotrophic bacterial endosymbionts are maintained within specialized cells of the tube worm host (figure 28.5 *b,c,d*).

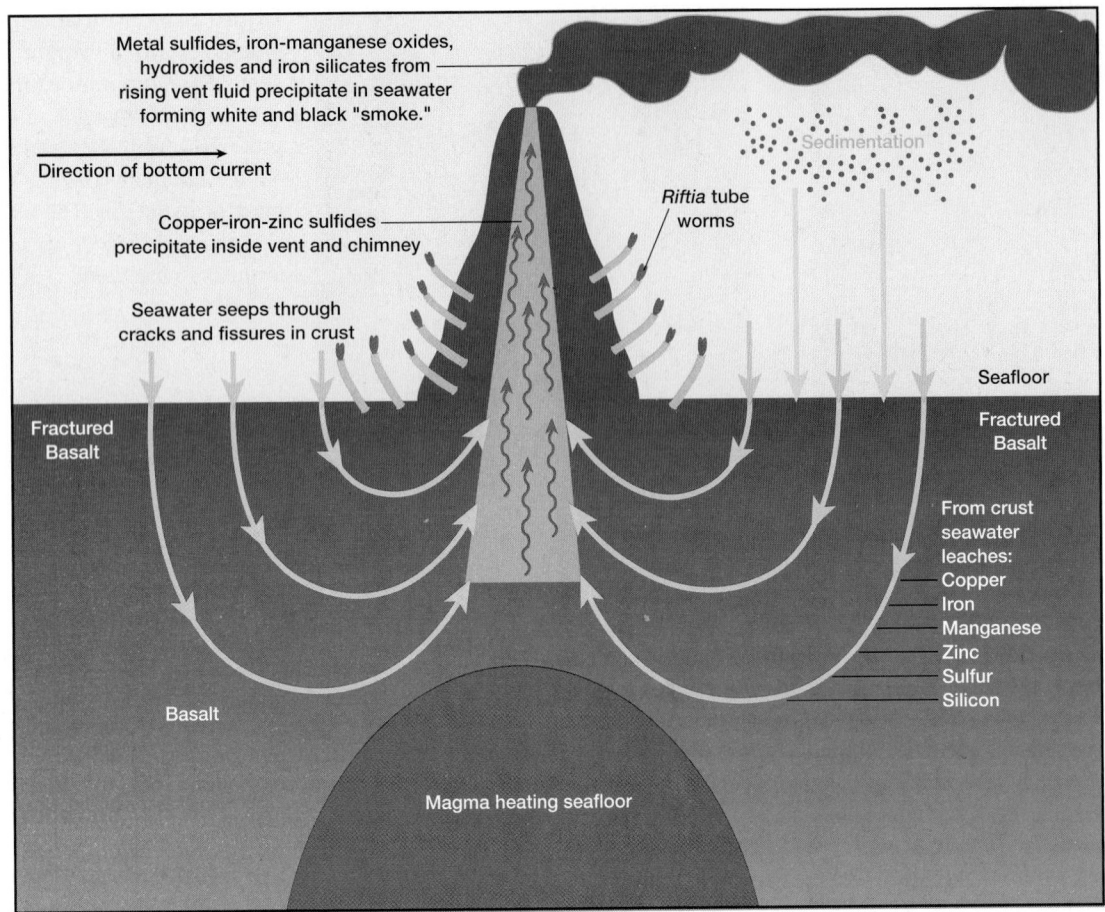

Figure 28.4 Basic Structure of a Hydrothermal Vent with its Mutualistic Microbe-Animal Associations. Reduced chemicals including sulfide are released as seawater penetrates the fractured basaltic ocean floor, is heated, and returns as vent fluid to the ocean, creating environments for growth of the tube worms and their procaryotic mutualists.

To date all attempts to culture these isolated bacteria-like organelles have been unsuccessful.

The tube worm takes up hydrogen sulfide from the seawater and binds it to hemoglobin (the reason the worms are bright red). The hydrogen sulfide is then transported in this form to the bacteria, which use the sulfide-reducing power to fix carbon dioxide in the Calvin cycle (*see figure 10.4*). The CO_2 required for this cycle is transported to the bacteria in three ways: freely dissolved in the blood, bound to hemoglobin, and in the form of organic acids such as malate and succinate. These acids are decarboxylated to release CO_2 in the trophosome, the tissue containing bacterial symbionts. Using these mechanisms, the bacteria synthesize reduced organic material from inorganic substances. The organic material is then supplied to the tube worm through its circulatory system and serves as the main nutritional source for the tissue cells.

A similar mutualistic relationship occurs in *Ridgeia piscesae*, a tube worm found in the northeastern Pacific. This animal maintains its endosymbiotic sulfur-oxidizing bacteria in specialized tissues, similar to the nutritional strategy of *Riftia*.

Methane-Based Mutualisms

Other unique food chains involve methane-fixing microorganisms as the first step in providing organic matter for consumers. Methanotrophs, bacteria capable of using methane, occur as intracellular symbionts of methane-vent mussels. In these mussels the thick fleshy gills are filled with bacteria. In addition, methanotrophic carnivorous sponges have been discovered in a mud volcano at a depth of 4,943 m in the Barbados Trench. Abundant methanotrophic symbionts were confirmed by the presence of enzymes related to methane oxidation in sponge tissues. These sponges are not satisfied to use bacteria to support themselves; they also trap swimming prey to give variety to their diet.

1. What is the critical characteristic of a mutualistic relationship?
2. What are important roles of bacteria, such as *Buchnera* and *Wolbachia*, in insects?
3. How do tube worms obtain energy and organic compounds for their growth?
4. What is the source of the waters released in a deep hydrothermal vent, and how is it heated?

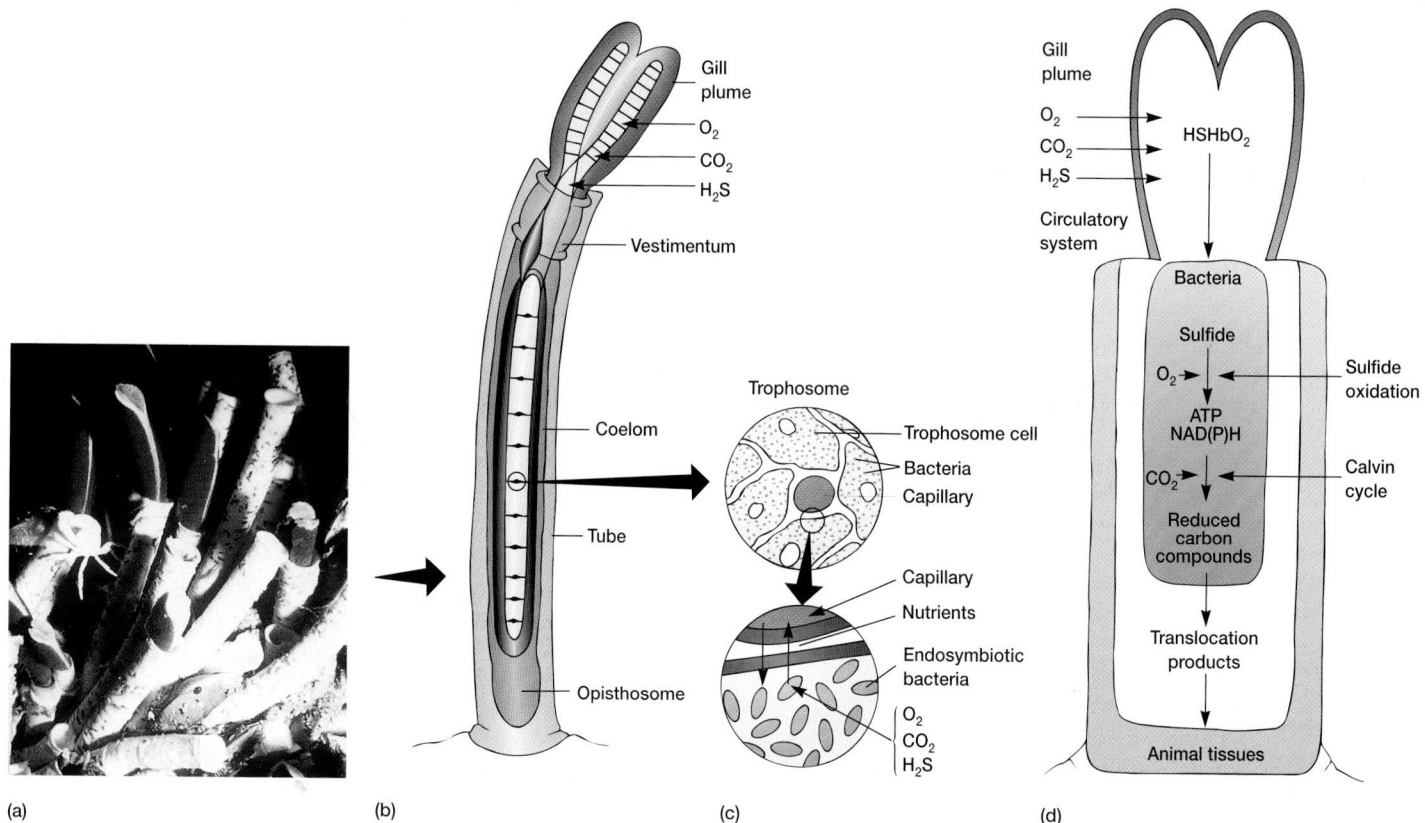

(a)   (b)   (c)   (d)

Figure 28.5 The Tube Worm–Bacterial Relationship. (**a**) A community of tube worms (*Riftia pachyptila*) at the Galápagos Rift hydrothermal vent site (depth 2,550 m). Each worm is more than a meter in length and has a 20 cm gill plume. (**b, c**) Schematic illustration of the anatomical and physiological organization of the tube worm. The animal is anchored inside its protective tube by the vestimentum. At its anterior end is a respiratory gill plume. Inside the trunk of the worm is a trophosome consisting primarily of endosymbiotic bacteria, associated cells, and blood vessels. At the posterior end of the animal is the opisthosome, which anchors the worm in its tube. (**d**) Oxygen, carbon dioxide, and hydrogen sulfide are absorbed through the gill plume and transported to the blood cells of the trophosome. Hydrogen sulfide is bound to the worm's hemoglobin (HSHbO$_2$) and carried to the endosymbiont bacteria. The bacteria oxidize the hydrogen sulfide and use some of the released energy to fix CO$_2$ in the Calvin cycle. Some fraction of the reduced carbon compounds synthesized by the endosymbiont is translocated to the animal's tissues.

The Rumen Ecosystem

Ruminants are a group of herbivorous animals that have a stomach divided into four compartments and chew a cud consisting of regurgitated, partially digested food. Examples include cattle, deer, elk, camels, buffalo, sheep, goats, and giraffes. The plant materials used as their main food source usually are of poor nutritional quality; large volumes of materials must be ingested and processed in the gut, in cooperation with complex microbial communities, to meet the nutritional needs of the animal. As would be expected, the gut in these animals is uniquely adapted for processing plant materials. By using microorganisms to degrade the thick cellulose walls of grass and other vegetation, ruminants digest vast amounts of otherwise unavailable forage. Because ruminants cannot synthesize cellulases, they have established a mutualistic relationship with anaerobic microorganisms that produce these enzymes. Cellulases hydrolyze the β(1→4) linkages between successive D-glucose residues of cellulose and release glucose, which is then fermented to organic acids such as acetate,

butyrate, and propionate (*see figure 9.10*). These organic acids are the true energy source for the ruminant. Organic matter processing can either stop at the acetate level or continue to form methane as a major end product. These two patterns of processing appear to be under genetic and phylogenetic control as discussed in **Microbial Diversity & Ecology 28.2**.

The upper portion of a ruminant's stomach expands to form a large pouch called the **rumen** (**figure 28.6**) and also a smaller honeycomb-like reticulum. The bottom portion of the stomach consists of an antechamber called the omasum, with the "true" stomach (abomasum) behind it.

The insoluble polysaccharides and cellulose eaten by the ruminant are mixed with saliva and enter the rumen. Within the rumen, food is churned in a constant rotary motion and eventually reduced to a pulpy mass, which is partially digested and fermented by microorganisms. Later the food moves into the reticulum. It is then regurgitated as a "cud," which is thoroughly chewed for the first time. The food is mixed with saliva,

Microbial Diversity & Ecology

28.2   Coevolution of Animals and Their Gut Microbial Communities

Organisms with digestive tracts had to make an interesting evolutionary choice: will the microbial community produce methane or not? The use of plant materials as a major food source does not always lead to methane production in the digestive tract. For example, kangaroos do not produce methane, whereas sheep and cattle do. The kangaroo has a distinct advantage in terms of nutrition. When complex plant materials are degraded only to organic acids such as acetate, the animal's digestive system can directly absorb the acids. In sheep and cattle, the microbial community is more complex and converts acetate-level substrates to methane and carbon dioxide, leading to nutrient loss from the original plant material. This loss can be substantial, with 10 to 15% of the organic matter in the feed lost to the atmosphere as methane.

An examination of over 250 reptiles, birds, and mammals showed that their maintenance of methanogenic microorganisms and methane production is under phylogenetic and not dietary control (see **Box figure**). Although low levels of methanogens can be detected in vertebrates that do not produce much of this important greenhouse gas, the lack of methane production seems to result from absence of methanogen receptor sites in the digestive system. As shown in this figure, the ability to maintain methanogens often has been lost. A similar situation occurs with arthropods: methane is produced by only a few organisms, including tropical millipedes, cockroaches, termites, and scarab beetles. If the presence of methanogens results in waste of nutrients, why have so many vertebrates and arthropods maintained this "energetically wasteful" process? It is an interesting question that relates back to the basic nature of vertebrate-gut microbial community evolution.

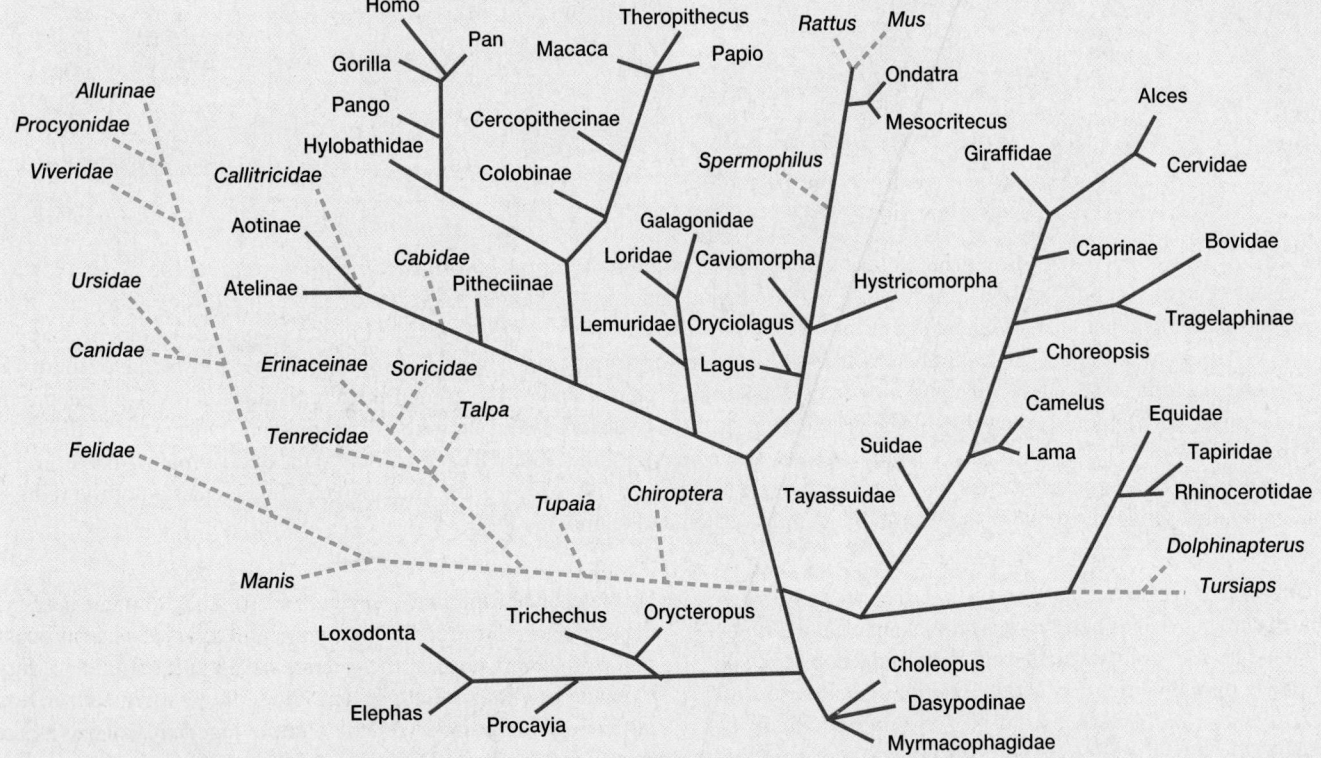

Microbial Diversity and Ecology. Coevolution of animals and their gut microbial communities: the methane choice. Methane-producing vertebrates are noted with solid red lines and roman letters; nonmethane producers are noted with blue dotted lines and italics.

reswallowed, and reenters the rumen while another cud is passed up to the mouth. As this process continues, the partially digested plant material becomes more liquid in nature. The liquid then begins to flow out of the reticulum and into the lower parts of the stomach: first the omasum and then the abomasum. It is in the abomasum that the food encounters the host's normal digestive enzymes and the digestive process continues in the regular mammalian way.

The rumen of methane-producing animals such as cattle contains a large and diverse microbial community (about 10^{12} organisms per milliliter), including procaryotes, anaerobic fungi such as *Neocallimastix*, ciliates, and other protozoans. Food entering the rumen is quickly attacked by the cellulolytic anaerobic procaryotes, fungi, and protozoa. Although the masses of procaryotes and protozoa are approximately equal, the processing of

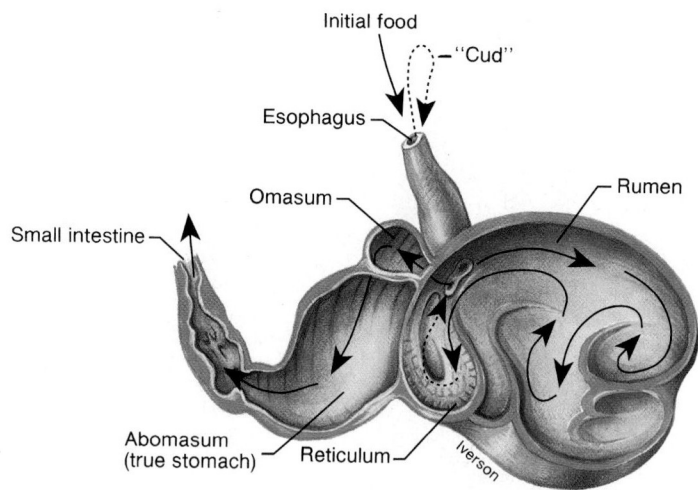

Figure 28.6   Ruminant Stomach. The stomach compartments of a cow. The microorganisms are active mainly in the rumen. Arrows indicate direction of food movement.

rumen contents is carried out mainly by the procaryotes. Microorganisms break down the plant material. Because the reduction potential in the rumen is −30 mV, all indigenous microorganisms engage in anaerobic metabolism. The bacteria ferment carbohydrates to fatty acids, carbon dioxide, and hydrogen. The archaea (methanogens) produce methane (CH_4) from acetate, CO_2, and H_2.

Dietary carbohydrates degraded in the rumen include soluble sugars, starch, pectin, hemicellulose, and cellulose. The largest percentage of each carbohydrate is fermented to volatile fatty acids (acetic, propionic, butyric, formic, and valeric), CO_2, H_2, and methane. Fatty acids produced by the rumen organisms are absorbed into the bloodstream and are oxidized by the animal as its main source of energy. The CO_2 and methane, produced at a rate of 200 to 400 liters per day in a cow, are released by eructation [Latin *eructare*, to belch], a continuous, scarcely audible reflex process similar to belching. ATP produced during fermentation is used to support the growth of rumen microorganisms. These microorganisms in turn produce most of the vitamins needed by the ruminant. In the remaining two stomachs, the microorganisms, having performed their symbiotic task, are digested to yield amino acids, sugars, and other nutrients for ruminant use.

Cooperation

Cooperation and commensalism are two positive but not obligatory types of symbioses found widely in the microbial world. These involve syntrophic relationships. **Syntrophism** [Greek *syn*, together, and *trophe*, nourishment] is an association in which the growth of one organism either depends on or is improved by growth factors, nutrients, or substrates provided by another organism growing nearby. Sometimes both organisms benefit.

As noted in figure 28.1, **cooperation** benefits both microorganisms, but this relationship is not obligatory. Beneficial com-

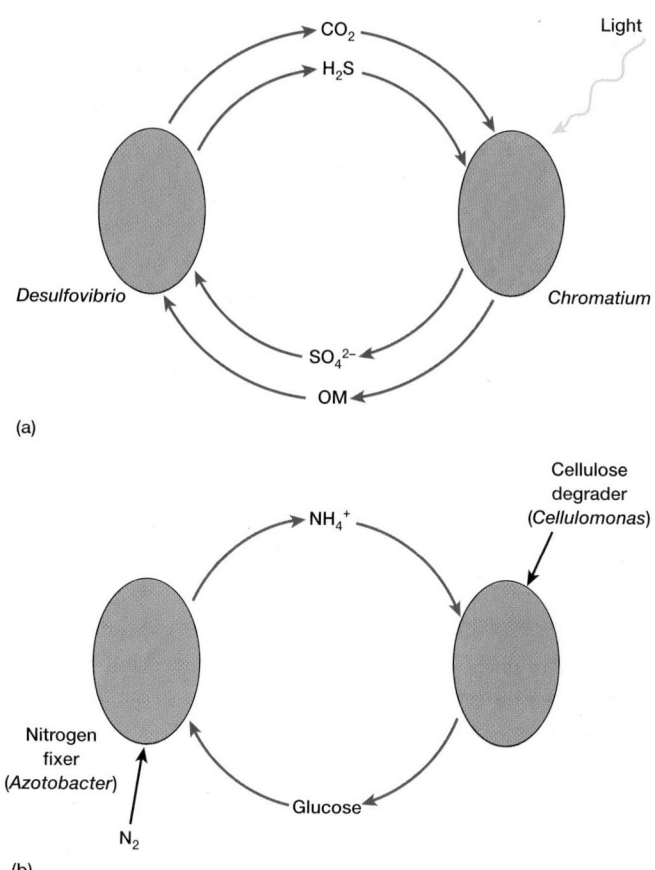

(a)

(b)

Figure 28.7   Examples of Cooperative Symbiotic Processes. (**a**) The organic matter (OM) and sulfate required by *Desulfovibrio* are produced by the *Chromatium* in its photosynthesis-driven reduction of CO_2 to organic matter and oxidation of sulfide to sulfate. (**b**) *Azotobacter* uses glucose provided by a cellulose-degrading microorganism such as *Cellulomonas*, which uses the nitrogen fixed by *Azotobacter*.

plementary resources are provided by each of the paired microorganisms. The organisms involved in this type of relationship can be separated, and if the resources provided by the complementary microorganism are supplied in the growth environment, each microorganism will function independently. Two examples of this type of relationship are the association of *Desulfovibrio* and *Chromatium* (**figure 28.7a**), in which the carbon and sulfur cycles are linked, and the interaction of a nitrogen-fixing microorganism with a cellulolytic organism such as *Cellulomonas* (figure 28.7b). In the second example, the cellulose-degrading microorganism liberates glucose from the cellulose, which can be used by nitrogen-fixing microbes.

An excellent example of a cooperative biodegradative association is shown in **figure 28.8.** In this case 3-chlorobenzoate degradation depends on the functioning of microorganisms with complementary capabilities. If any one of the three microorganisms is not present and active, the degradation of the substrate will not occur.

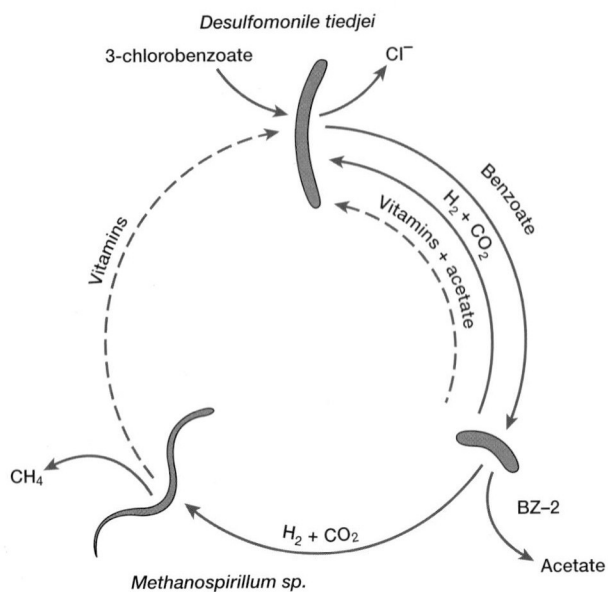

Figure 28.8 Associations in a Defined Three-Membered Cooperative and Commensalistic Community That Can Degrade 3-Chlorobenzoate. If any member is missing, degradation will not take place. The solid arrows demonstrate nutrient flows, and the dashed lines represent hypothesized flows.

Figure 28.9 A Marine Worm-Bacterial Cooperative Relationship. *Alvinella pompejana,* a 10 cm long worm, forms a cooperative relationship with bacteria that grow as long threads on the worm's surface. The bacteria and *Alvinella* are found in tunnels near the black smoker-heated water fonts.

(a)

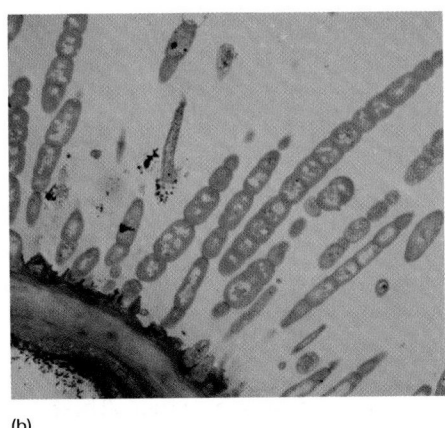

(b)

Figure 28.10 A Marine Crustacean-Bacterial Cooperative Relationship. (a) A picture of the marine shrimp *Rimicaris exoculata* clustering around a hydrothermal vent area, showing the massive development of these crustaceans in the area where chemolithotrophic bacteria grow using sulfide as an electron and energy source. The bacteria, which grow on the vent openings and also on the surface of the crustaceans, fix carbon from CO_2 in their autotrophic metabolism, and serve as the nutrient for the shrimp. (b) An electron micrograph of a thin section across the leg of the marine crustacean *Rimicaris exoculata,* showing the chemolithotrophic bacteria that cover the surface of the shrimp. The filamentous nature of these bacteria, upon which this commensalistic relationship is based, is evident in this thin section.

In other cooperative relations, sulfide-dependent autotrophic filamentous microorganisms fix carbon dioxide and synthesize organic matter that serves as a carbon and energy source for a heterotrophic organism. Some of the most interesting include the polychaete worms *Alvinella pompejana* (**figure 28.9**), the Pompeii worm, and also *Paralvinella palmiformis,* the Palm worm. Both have filamentous bacteria on their dorsal surfaces. These fil-

amentous bacteria can tolerate high levels of metals such as arsenic, cadmium, and copper. When growing on the surface of the animal, they may provide protection from these toxic metals, as well as thermal protection; in addition, they appear to be used as a food source. A deep-sea crustacean has been discovered that uses sulfur-oxidizing autotrophic bacteria as its food source. This shrimp, *Rimicaris exoculata* (**figure 28.10**) has filamentous

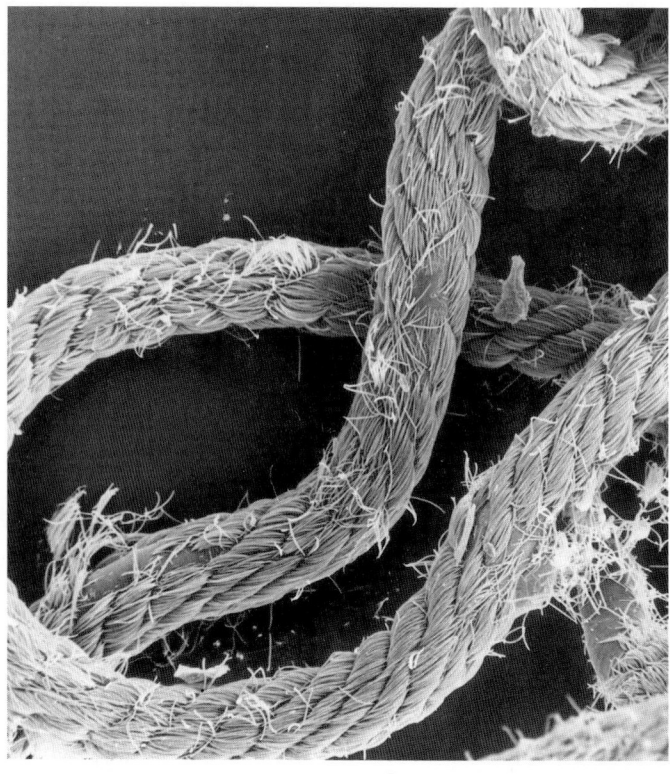

(a)

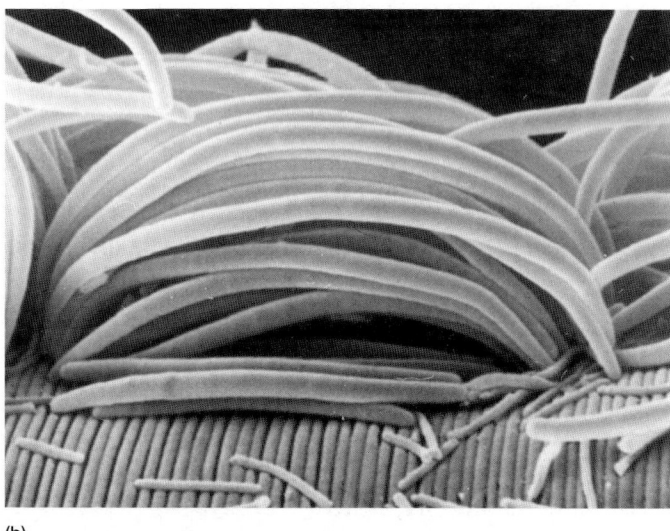

(b)

Figure 28.11   A Marine Nematode-Bacterial Cooperative Relationship.   Marine free-living nematodes, which grow at the oxidized-reduced zone where sulfide and oxygen are present, are covered by sulfide-oxidizing bacteria. The bacteria protect the nematode by decreasing sulfide concentrations near the worm, and the worm uses the bacteria as a food source. (**a**) The marine nematode *Eubostrichus parasitiferus* with bacteria arranged in a characteristic helix pattern. (**b**) The chemolithotrophic bacteria attached to the cuticle of the marine nematode *Eubostrichus parasitiferus*. Cells are fixed to the nematode surface at both ends. *See also figure 28.16.*

sulfur-oxidizing bacteria growing on its surface (figure 28.10*b*). When these are dislodged the shrimp ingests them. This nominally "blind" shrimp can respond to the glow emitted by the black smoker, using a reflective organ on its back. The organ is sensitive to a light wavelength that is not detectable by humans.

Another interesting example of bacterial epigrowth is shown by nematodes, including *Eubostrichus parasitiferus,* that live at the interface between aerobic and anaerobic sulfide-containing marine sediments (**figure 28.11*a,b***). These animals are covered by sulfide-oxidizing bacteria that are present in intricate patterns (figure 28.11*b*). The bacteria not only decrease levels of toxic sulfide, which often surround the nematodes, but they also serve as a food supply.

In 1990, hydrothermal vents were discovered in a freshwater environment, at the bottom of Lake Baikal, the oldest (25 million years old) and deepest lake in the world. This lake is located in the far east of Russia (**figure 28.12*a,b,c***) and has the largest volume of any freshwater lake (not the largest area—which is Lake Superior). The bacterial growths, with long white strands, are in the center of the vent field where the highest temperatures are found (figure 28.12*b*). At the edge of the vent field, where the water temperature is lower, the bacterial mat ends, and sponges, gastropods, and other organisms, which use the sulfur-oxidizing

bacteria as a food source, are present (figure 28.12*c*). Similar although less developed areas have been found in Yellowstone Lake, Wyoming.

A hydrogen sulfide-based ecosystem has been discovered in southern Romania that is closer to the Earth's surface. Caves in the area contain mats of microorganisms that fix carbon dioxide using hydrogen sulfide as the electron donor. Forty-eight species of cave-adapted invertebrates are sustained by this chemoautotrophic base.

A form of cooperation also occurs when a population of similar microorganisms monitors its own density, the process of quorum sensing, which was discussed in section 6.5. The microorganisms produce specific autoinducer compounds, and as the population increases and the concentration of these compounds reaches critical levels, specific genes are expressed. These responses are important for microorganisms that form associations with plants and animals, and particularly for human pathogens.

Mycorrhizal (fungal-plant root interactions) present a special case of a cooperative relationship (see chapter 30). "Normal" chlorophyll-containing plants can grow without the fungus, particularly in the greenhouse, indicating that this is not an obligatory relationship. Some fungi, in contrast, do not survive without being associated with a plant. The achlorophyllous

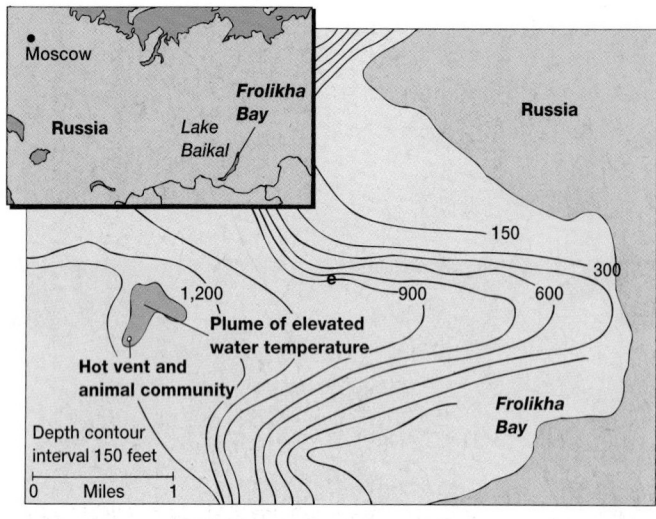

(a)

(b)

(c)

Figure 28.12 Hydrothermal Vent Ecosystems in Freshwater Environments. Lake Baikal (in Russia) has been found to have low temperature hydrothermal vents. (**a**) Location of Lake Baikal, site of the hydrothermal vent field. (**b**) Bacterial mat near the center of the vent field. (**c**) Bacterial filaments and sponges at the edge of the vent field. *(a) Source: Data from the National Geographic Society.*

monotropoid plants (*see p. 656*), however, are different. These plants depend greatly on the mycorrhizal fungal network to provide nutrients "taken" from other chlorophyll-containing plants.

---

1. What structural features of the rumen make it suitable for a herbivorous type of diet? Why does a cow chew its cud?
2. What biochemical roles do the rumen microorganisms play in this type of symbiosis?
3. What is syntrophism? Is physical contact required for this relationship?
4. Why are *Alvinella, Rimicaris,* and *Eubostrichus* good examples of cooperative microorganism-animal interactions?
5. Where is Lake Baikal located, and why is this unique in terms of its microbial communities?
6. Why do chlorophyllous and achlorophyllous plants have different degrees of dependence on mycorrhizal fungi?

Commensalism

Commensalism [Latin *com,* together, and *mensa,* table] is a relationship in which one symbiont, the **commensal,** benefits while the other (sometimes called the host) is neither harmed nor helped, as shown in figure 28.1. This is a unidirectional process. Often both the host and the commensal "eat at the same table." The spatial proximity of the two partners permits the commensal to feed on substances captured or ingested by the host, and the commensal often obtains shelter by living either on or in the host. The commensal is not directly dependent on the host metabolically and causes it no particular harm. When the commensal is separated from its host experimentally, it can survive without being provided some factor or factors of host origin.

Commensalistic relationships between microorganisms include situations in which the waste product of one microorganism is the substrate for another species. One good example is nitrification, the oxidation of ammonium ion to nitrite by microorganisms such as *Nitrosomonas,* and the subsequent oxidation of the nitrite to nitrate by *Nitrobacter* and similar bacteria (*see pp. 188–89*). *Nitrobacter* benefits from its association with *Nitrosomonas* because it uses nitrite to obtain energy for growth. A second example of this type of relationship is found in anaerobic methanogenic ecosystems such as sludge digesters (*see section 29.6*), anaerobic freshwater aquatic sediments, and flooded soils. In these environments, fatty acids can be degraded to produce H_2 and methane by the interaction of two different bacterial groups. Methane production by methanogens depends on **interspecies hydrogen transfer.** A fermentative bacterium generates hydrogen gas, and the methanogen uses it quickly as a substrate for methane gas production.

Various fermentative bacteria produce low molecular weight fatty acids that can be degraded by anaerobic bacteria such as *Syntrophobacter* to produce H_2 as follows:

$$\text{Propionic acid} \rightarrow \text{acetate} + CO_2 + H_2$$

Syntrophobacter uses protons ($H^+ + H^+ \rightarrow H_2$) as terminal electron acceptors in ATP synthesis. The bacterium gains sufficient energy for growth only when the H_2 it generates is consumed. The products H_2 and CO_2 are used by methanogenic archaea (e.g., *Methanospirillum*) as follows:

$$4H_2 + CO_2 \rightarrow CH_4 + 2H_2O$$

By synthesizing methane, *Methanospirillum* maintains a low H_2 concentration in the immediate environment of both bacteria. Continuous removal of H_2 promotes further fatty acid fermentation and H_2 production. Because increased H_2 production and consumption stimulate the growth rates of *Syntrophobacter* and *Methanospirillum,* both participants in the relationship benefit.

Commensalistic associations also occur when one microbial group modifies the environment to make it more suited for another organism. For example, in the intestine the common, nonpathogenic strain of *Escherichia coli* lives in the human colon, but also grows quite well outside the host, and thus is a typical commensal. When oxygen is used up by the facultatively anaerobic *E. coli,* obligate anaerobes such as *Bacteroides* are able to grow in the colon. The anaerobes benefit from their association with the host and *E. coli,* but *E. coli* derives no obvious benefit from the anaerobes. In this case the commensal *E. coli* contributes to the welfare of other symbionts. Commensalism can involve other environmental modifications. The synthesis of acidic waste products during fermentation stimulate the proliferation of more acid-tolerant microorganisms, which are only a minor part of the microbial community at neutral pHs. A good example is the succession of microorganisms during milk spoilage. As another example, when biofilms are formed (section 28.4), the colonization of a newly exposed surface by one type of microorganism (an initial colonizer) makes it possible for other microorganisms to attach to the microbially modified surface.

Commensalism also is important in the colonization of the human body and the surfaces of other animals and plants. The microorganisms associated with an animal skin and body orifices can use volatile, soluble, and particulate organic compounds from the host as nutrients (*see section 31.2*). Under most conditions these microbes do not cause harm, other than possibly contributing to body odor. Sometimes when the host organism is stressed or the skin is punctured, these normally commensal microorganisms may become pathogenic. These interactions will be discussed in chapter 31.

---

1. How does commensalism differ from cooperation?
2. Why is nitrification a good example of a commensalistic process?
3. What is interspecies hydrogen transfer, and why can this be beneficial to both the producer and consumer of hydrogen?
4. Why are commensalistic microorganisms important for humans? Where are these found in relation to the human body?

Predation

Predation is a widespread phenomenon where the predator engulfs or attacks the prey, as shown in figure 28.1. The prey can be

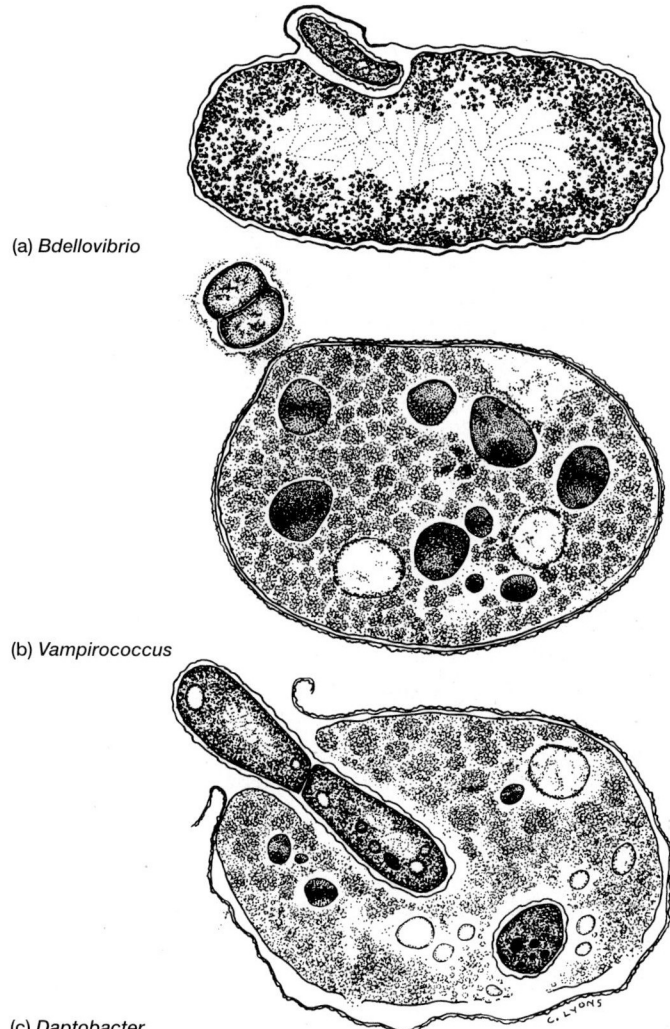

(a) *Bdellovibrio*

(b) *Vampirococcus*

(c) *Daptobacter*

Figure 28.13   Examples of Predatory Bacteria Found in Nature. (**a**) *Bdellovibrio,* a periplasmic predator that penetrates the cell wall and grows outside the plasma membrane, (**b**) *Vampirococcus* with its unique epibiotic mode of attacking a prey bacterium, and (**c**) *Daptobacter* showing its cytoplasmic location as it attacks a susceptible bacterium.

larger or smaller than the predator, and this normally results in the death of the prey.

An interesting array of predatory bacteria are active in nature. Several of the best examples are shown in **figure 28.13,** including *Bdellovibrio, Vampirococcus,* and *Daptobacter.* Each of these has a unique mode of attack against a susceptible bacterium. *Bdellovibrio* penetrates the cell wall and multiplies between the wall and the plasma membrane, a periplasmic mode of attack, followed by lysis of the prey and release of progeny (*see figure 22.33*). This bacterium has an interesting life cycle, which is discussed in section 22.4.

Nonlytic forms also are observed. *Vampirococcus* attaches to the surface of the prey (an epibiotic relationship) and then secretes enzymes to release the cell contents. *Daptobacter*

Table 28.3	
The Many Faces of Predation	

Predation Result	Example
Digestion	The microbial loop. Soluble organic matter from primary producers is normally used by bacteria, which become a particulate food source for higher consumers. Flagellates and ciliates prey on these bacteria and digest them, making the nutrients they contain available again in mineral form for use in primary production, creating the microbial loop. In this way a large portion of the carbon fixed by the photosynthetic microbes is mineralized and recycled (thus the term microbial loop) and does not reach the higher trophic levels of the ecosystem (*see figure 29.4*).
	Predation also can reduce the density-dependent stress factors in prey populations, allowing more rapid growth and turnover of the prey than would occur if the predator were not active.
Retention	Bacteria retained within the predator serve a useful purpose, as in the transformation of toxic hydrogen produced by ciliates in the rumen to harmless methane. Also, trapping of chloroplasts (kleptochloroplasty) by protozoa provides the predator with photosynthate.
Protection and increased fitness	The intracellular survival of *Legionella* ingested by ciliates protects it from stresses such as heating and chlorination. Ingestion also results in increased pathogenicity when the prey is again released to the external environment, and this may be required for infection of humans. The predator serves as a reservoir host.
	Nanoplankton may be ingested by zooplankton and grow in the zooplankton digestive system. They are then released to the environment in a fitter state. Dissemination to new locations also occurs.

penetrates a susceptible host and uses the cytoplasmic contents as a nutrient source.

Ciliates are excellent examples of predators that engulf their prey, and based on work with fluorescently marked prey bacteria, a single ciliate can ingest 60 to 70 prey bacteria per hour! Predation on bacteria is important in the aquatic environment and in sewage treatment where the ciliates remove suspended bacteria that have not settled.

A surprising finding is that predation has many beneficial effects, especially when one considers interactive populations of predators and prey, as summarized in **table 28.3.** Simple ingestion and assimilation of a prey bacterium can lead to increased rates of nutrient cycling, critical for the functioning of the **microbial loop** (*see section 29.1 and figure 29.4*). In this process, organic matter produced through photosynthetic and chemotrophic activity is mineralized before it reaches the higher consumers, allowing the minerals to be made available to the primary producers, in a "loop." This is important in freshwater, marine, and terrestrial environments. Ingestion and short-term retention of bacteria also are critical for ciliate functioning in the rumen, where methanogenic bacteria contribute to the health of the ciliates by decreasing toxic hydrogen levels through using H_2 to produce methane, which then is passed from the rumen.

Predation also can provide a protective, high-nutrient environment for particular prey. Ciliates ingest *Legionella* and protect this important pathogen from chlorine, which often is used in an attempt to control *Legionella* in cooling towers and air-conditioning units. The ciliate serves as a reservoir host. *Legionella pneumophila* also has been found to have a greater potential to invade macrophages and epithelial cells after predation, indicating that ingestion not only provides protection but also may make the bacterium a better pathogen. A similar phenomenon of survival in protozoa has been observed for *Mycobacterium avium,* a pathogen of worldwide concern. These protective aspects of predation have major implications for survival and control of disease-causing microorganisms in the biofilms present in water supplies and air-

conditioning systems. In marine systems the ingestion of nanoplankton by zooplankton provides a nutrient-rich environment that allows nanoplankton reproduction in the digestive tract and promotes dissemination in the environment. A similar process occurs after bacteria are ingested by polychaetes (segmented worms found mostly in marine environments).

Fungi often show interesting predatory skills. Some fungi can trap protozoa by the use of sticky hyphae or knobs, sticky networks of hyphae, or constricting or nonconstricting rings. A classic example is *Arthrobotrys,* which traps nematodes by use of constricting rings. After the nematode is trapped, hyphae grow into the immobilized prey and the cytoplasm is used as a nutrient. Other fungi have conidia that, after ingestion by an unsuspecting prey, grow and attack the susceptible host from inside the intestinal tract. In this situation the fungus penetrates the host cells in a complex interactive process.

Thus predation, which usually has a fatal and final outcome for an individual prey organism, has a wide range of beneficial effects on prey populations, and is critical in the functioning of natural environments.

Parasitism

Parasitism is one of the most complex microbial interactions; the line between parasitism and predation is difficult to define (figure 28.1; *see also section 34.1*). This is a relationship in which one of a pair benefits from the other, and the host is usually harmed. This can involve nutrient acquisition and/or physical maintenance in or on the host. In parasitism there is a degree of coexistence of the parasite in association with the host. Depending on the equilibrium between the two organisms, this may shift and what might have been a stable parasitic relationship may then become a pathogenic one which can be defined as predation.

Some bacterial viruses can establish a lysogenic relationship with their hosts, and the viruses, in their prophage state, can confer positive new attributes on the host bacteria, as occurs with

Figure 28.14   Lichens.   Crustose (encrusting) lichens growing on a granite post.

toxin production by *Corynebacterium diphtheriae* (*see sections 17.5 and 34.3*). Parasitic fungi include *Rhizophydium sphaero-carpum* with the alga *Spyrogyra.* Also, *Rhizoctonia solani* is a parasite of *Mucor* and *Pythium,* which is important in **biocontrol processes,** the use of one microorganism to control another. Human diseases caused by viruses, bacteria, fungi, and protozoa are discussed in chapters 38 through 40.

Lichens are the association between specific ascomycetes (the fungus) and certain genera of either green algae or cyanobacteria. In a lichen, the fungal partner is termed the **mycobiont** and the algal or cyanobacterial partner, the **phycobiont.** Lichens are an excellent example of a controlled parasitism (**figure 28.14**). In the past the lichen symbiosis was considered to be a mutualistic interaction. It recently has been found that a lichen will form only when the two potential partners are nutritionally deprived. Under normal conditions the potential partners aren't interested in each other!

The remarkable aspect of this association is that its morphology and metabolic relationships are so constant that lichens are assigned generic and species names. The characteristic morphology of a given lichen is a property of the association and is not exhibited by either symbiont individually.   Ascomycetes (pp. 544–45); cyanobacteria (pp. 458–64)

Because the phycobiont is a photoautotroph—dependent only on light, carbon dioxide, and certain mineral nutrients—the fungus can get its organic carbon directly from the alga or cyanobacterium. The fungus often obtains nutrients from its partner by haustoria (projections of fungal hyphae) that penetrate the phy-

cobiont cell wall. It also uses the O_2 produced during phycobiont photophosphorylation in carrying out respiration. In turn the fungus protects the phycobiont from high light intensities, provides water and minerals to it, and creates a firm substratum within which the phycobiont can grow protected from environmental stress. The ability of the fungus to parasitize the alga or cyanobacterium under such longer-term conditions also has been described as a form of fungal-directed "algaculture."

An important aspect of parasitism is that over time, the parasite, once it has established a relationship with the host, will tend to discard excess, unused genomic information, a process of massive **genomic reduction.** This has occurred with *Mycobacterium leprae* (*see p. 344*), and with the microsporidium *Encephalitozoon cuniculi.* The latter organism, which parasitizes a wide range of animals, including humans, now can only survive inside the host cell.

1. Define predation and parasitism. How are these similar and different?
2. How can a predator confer positive benefits on its prey? Think of the responses of individual organisms versus populations as you consider this question.
3. What are examples of parasites that are important in microbiology?
4. What is a lichen? Discuss the benefits the phycobiont and mycobiont provide each other.

Amensalism

Amensalism (from the Latin for *not* at the same table) describes the adverse effect that one organism has on another organism as shown in figure 28.1. This is a unidirectional process based on the release of a specific compound by one organism which has a negative effect on another organism. A classic example of amensalism is the production of antibiotics that can inhibit or kill a susceptible microorganism (**figure 28.15a**). The attine ant–fungal mutualistic relationship is promoted by antibiotic-producing bacteria that are maintained in the fungal garden system (figure 28.15b). In this case a streptomycete produces an antibiotic that controls *Escovopsis,* a persistent parasitic fungus that can destroy the ant's fungal garden. This unique amensalistic process appears to have evolved 50 million years ago in South America.

Other important amensalistic relationships involve microbial production of specific organic compounds that disrupt cell wall or plasma membrane integrity. These include the bacteriocins (*see pp. 291, 688*). Bacteriocins are of increasing interest as food additives for controlling growth of undesired pathogens (*see section 41.3*). Antibacterial peptides also can be released by the host in the intestine and other places. These molecules, called cecropins in insects and defensins in mammals, recently have been recognized as effector molecules that play significant roles in innate immunity (*see p. 697*). In animals these molecules are released by phagocytes and intestinal cells, and are as powerful as tetracyclines. Recently it has been found that human sweat is antimicrobial. The sweat gland produces an antimicrobial peptide

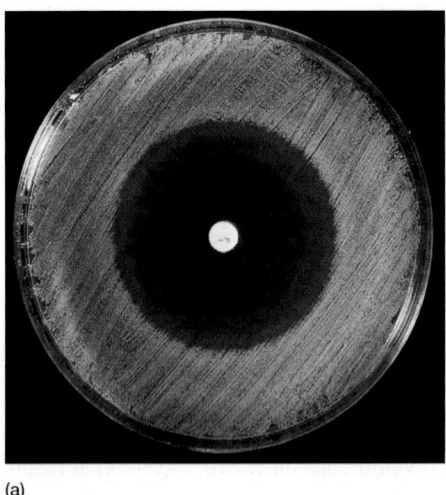

(a)

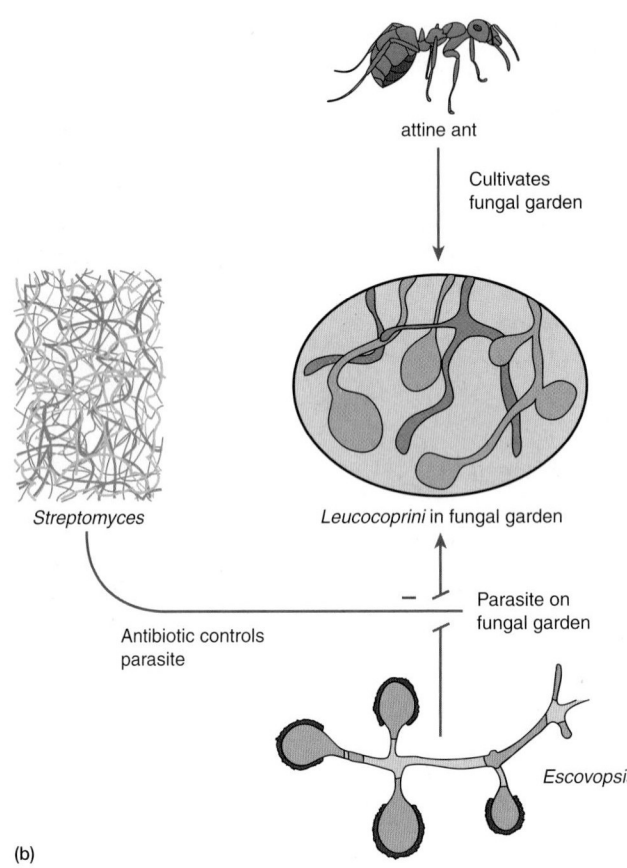

attine ant

Cultivates
fungal garden

Streptomyces

Leucocoprini in fungal garden

Antibiotic controls
parasite

Parasite on
fungal garden

Escovopsis

(b)

Figure 28.15 Amensalism: An Adverse Microbe-Microbe Interaction. (**a**) Antibiotic production and inhibition of growth of a susceptible bacterium on an agar medium. (**b**) A schematic diagram describing the use of antibiotic-producing streptomycetes by ants to control fungal parasites in their fungal garden.

that has been called dermicidin. It has been suggested that the evolution of the sweat gland may be related to the ability to produce this compound and inhibit microbial infection in this sensitive area of the body. The skin also produces similar compounds; a recently discovered antimicrobial peptide has been called cathelicidin. Finally, metabolic products, such as organic acids formed in fermentation, can produce amensalistic effects. These compounds inhibit growth by changing the environmental pH, for example, during natural milk spoilage (*see section 41.2*).

Competition

Competition arises when different microorganisms within a population or community try to acquire the same resource, whether this is a physical location or a particular limiting nutrient (figure 28.1). If one of the two competing organisms can dominate the environment, whether by occupying the physical habitat or by exclusively using a limiting nutrient, it will outgrow the other organism. This phenomenon was studied by E. F. Gause, who in 1934 described it as the **competitive exclusion principle.** He found that if the two competing ciliates overlapped too much in terms of their resource use, one of the two protozoan populations was excluded. In chemostats (*see section 6.3*), we may see competition for a limiting nutrient among microorganisms with transport systems of differing affinity. This can lead to the exclusion of the slower-growing population under a particular set of condi-

tions. If the dilution rate is changed, the previously slower-growing population may become dominant. Often two microbial populations that appear to be similar nevertheless coexist. In this case, they share the limiting resource (space, a limiting nutrient) and coexist while surviving at lower population levels.

1. What is the origin of the term amensalism?
2. What are bacteriocins?
3. What is the competitive exclusion principle?

Symbioses in Complex Systems

It should be emphasized that the symbiotic interactions discussed in this section do not occur independently. Each time a microorganism interacts with other organisms and their environments, a series of feedback responses occurs in the larger biotic community that will impact other parts of ecosystems. As an illustration, the interactions between the nematode *Eubostrichus parasitiferus* and its cooperative sulfide-oxidizing bacterium (p. 587) involves a series of symbiotic interactions, as shown in **figure 28.16.** In this case, the cooperative interaction between the nematodes and the sulfide-oxidizing epibiont is influenced by the population size of the associated bacterium, and whether it is host-associated or free-living. This equilibrium between host-associated and free-living bacteria is controlled by a series of feedback processes that in-

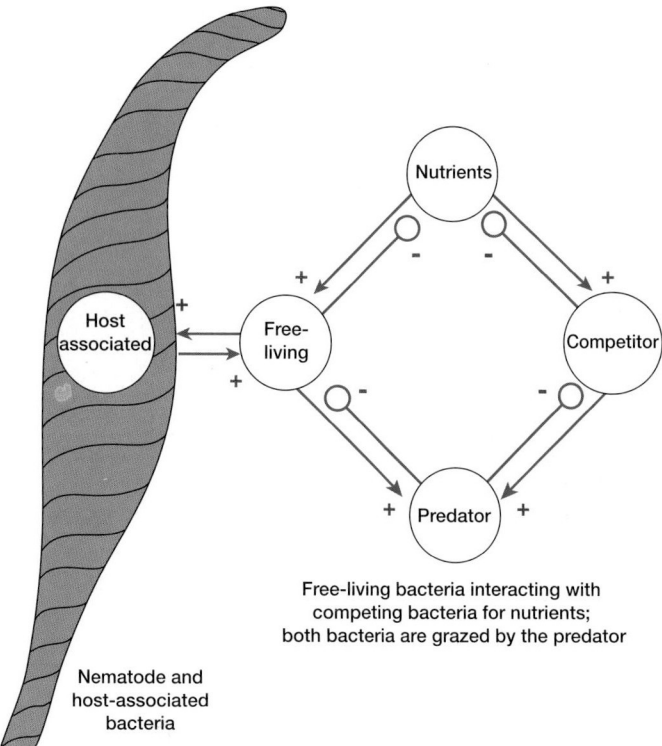

Figure 28.16   Complex Interactions in Microbial Ecology.
Interactions between the marine nematode *Eubostrichus para-sitiferus* and its host-associated and free-living cooperative bac-teria are influenced by levels of nutrients, as well as by other bacteria competing for these nutrients. Predators, in turn, use both the free-living cooperative bacteria and the competing bacteria as food sources, leading to complex feedback responses in this dy-namic ecosystem. Arrows show beneficial or enhancing interac-tions; lines with small circles indicate negative or damping interactions.

volve competition for sulfide with other bacteria, and predation on the epibiont and on the competing bacteria. Thus the equilibrium observed at any time is the result of a series of interactions, in-volving cooperation, predation, and competition.

 28.3   NUTRIENT CYCLING INTERACTIONS

Microorganisms, in the course of their growth and metabolism, in-teract with each other in the cycling of nutrients, including carbon, sulfur, nitrogen, phosphorus, iron, and manganese. This nutrient cycling, called **biogeochemical cycling** when applied to the envi-ronment, involves both biological and chemical processes. Nutri-ents are transformed and cycled, often by oxidation-reduction reactions (*see section 8.5*) that can change the chemical and phys-ical characteristics of the nutrients. All of the biogeochemical cy-cles are linked (**figure 28.17**), and the metabolism-related transformations of these nutrients have global-level impacts.

The major reduced and oxidized forms of the most important elements are noted in **table 28.4,** together with their valence states. Significant gaseous components occur in the carbon and nitrogen cycles and, to a lesser extent, in the sulfur cycles. Thus a soil, aquatic, or marine microorganism often can fix gaseous forms of carbon and nitrogen compounds. In the "sedimentary" cycles, such as that for iron, there is no gaseous component.

Carbon Cycle

Carbon can be present in reduced forms, such as methane (CH_4) and organic matter, and in more oxidized forms, such as carbon monoxide (CO) and carbon dioxide (CO_2). The major pools pres-ent in an integrated basic carbon cycle are shown in **figure 28.18.** Electron donors (e.g., hydrogen, which is a strong reductant) and acceptors (e.g., O_2) influence the course of biological and chem-ical reactions involving carbon. Hydrogen can be produced dur-ing organic matter degradation especially under anaerobic conditions when fermentation occurs. If hydrogen and methane are generated, they can move upward from anaerobic to aerobic areas. This creates an opportunity for aerobic hydrogen and methane oxidizers to function.

Methane levels in the atmosphere have been increasing ap-proximately 1% per year, from 0.7 to 1.6 to 1.7 ppm (volume) in the last 300 years. This methane is derived from a variety of sources. If an aerobic water column is above the anaerobic zone where the methanogens are located, the methane can be oxidized before it reaches the atmosphere. In many situations, such as in rice paddies without an overlying aerobic water zone, the methane will be released directly to the atmosphere, thus con-tributing to global atmospheric methane increases. Rice paddies, ruminants, coal mines, sewage treatment plants, landfills, and marshes are important sources of methane. Anaerobic microor-ganisms such as *Methanobrevibacter* in the guts of termites also can contribute to methane production. Physiology of aerobic hydro-gen and methane utilizers (pp. 188–89, 488)

Carbon fixation occurs through the activities of cyanobacte-ria and green algae, photosynthetic bacteria (e.g., *Chromatium* and *Chlorobium*), and aerobic chemolithoautotrophs.

In the carbon cycle depicted in figure 28.18, no distinction is made between different types of organic matter that are formed and degraded. This is a marked oversimplification because or-ganic matter varies widely in physical characteristics and in the biochemistry of its synthesis and degradation. Organic matter varies in terms of elemental composition, structure of basic re-peating units, linkages between repeating units, and physical and chemical characteristics.

The formation of organic matter is discussed in chapters 10 through 12. The degradation of this organic matter, once formed, is influenced by a series of factors. These include (1) nutrients present in the environment; (2) abiotic conditions (pH, oxidation-reduction potential, O_2, osmotic conditions), and (3) the micro-bial community present.

The major complex organic substrates used by microorgan-isms are summarized in **table 28.5.** Of these, only previously

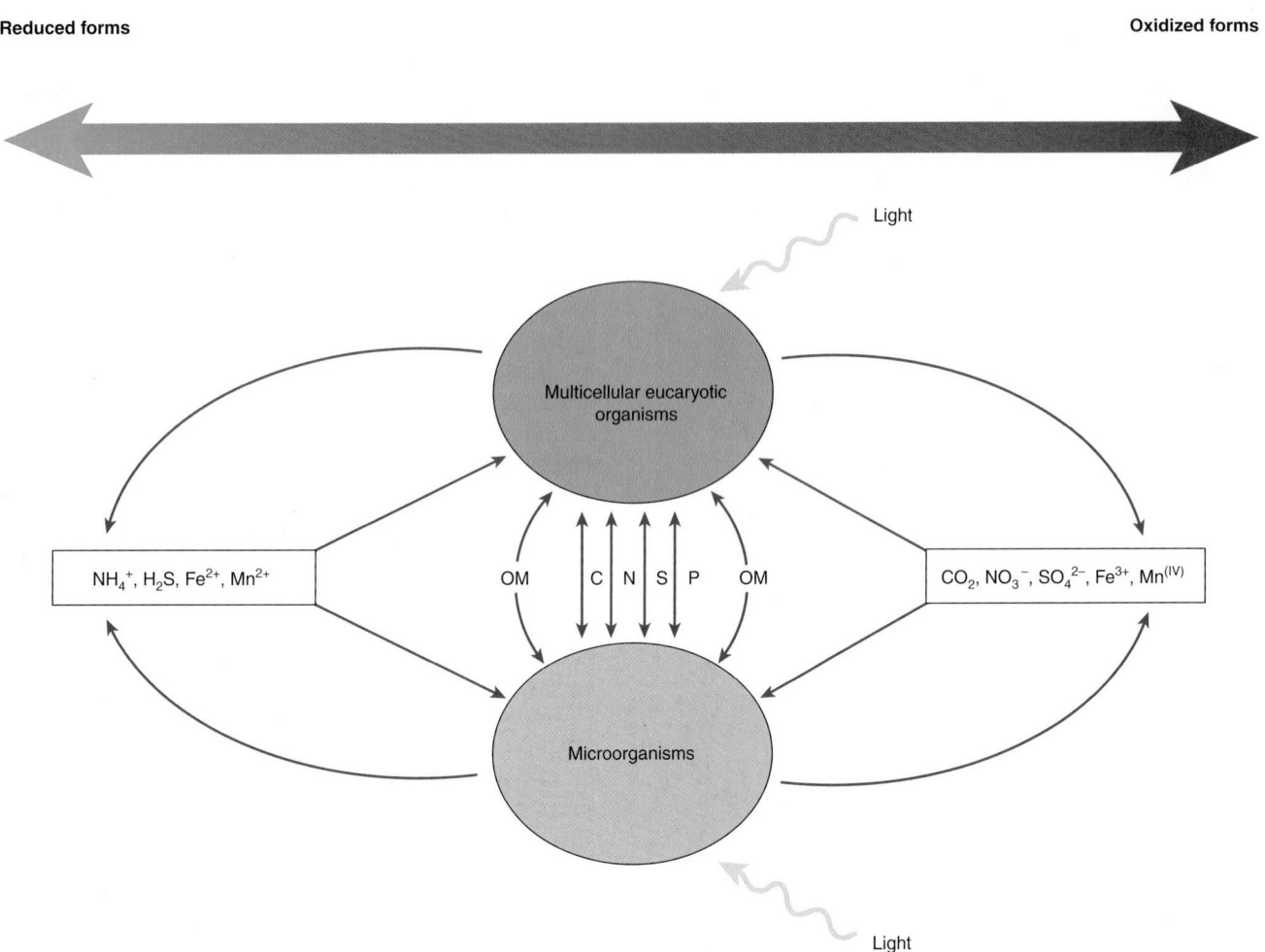

Figure 28.17   **Macrobiogeochemistry: A Cosmic View of Mineral Cycling in Microorganisms, Higher Organisms, and the Abiotic Chemical World.**   All biogeochemical cycles are linked, with energy being obtained from light and pairs of reduced and oxidized compounds. Only major flows are shown. See individual cycles for details of energetic-linked relationships. The forms that move between the microorganisms and multicellular organisms can vary. The broad concept is that all cycles are linked. The biotic components include both living forms and those that have died/senesced and which are being processed. Flows from lithogenic sources are important for phosphorus. Methane transformations are only microbial (see discussion of the carbon cycle). Organic matter (OM).

grown microbial biomass contains all of the nutrients required for microbial growth. Chitin, protein, microbial biomass, and nucleic acids contain nitrogen in large amounts. If these substrates are used for growth, the excess nitrogen and other minerals that are not used in the formation of new microbial biomass will be released to the environment, in the process of **mineralization.** This is the process in which organic matter is decomposed to release simpler, inorganic compounds (e.g., CO_2, NH_4^+, CH_4, H_2).

The other complex substrates in table 28.5 contain only carbon, hydrogen, and oxygen. If microorganisms are to grow by using these substrates, they must acquire the remaining nutrients they need for biomass synthesis from the environment in the process of **immobilization.**

The oxygen relationships for the use of these substrates also are of interest, because most of them can be degraded easily with or without oxygen present. The exceptions are hydrocarbons and

lignin. Hydrocarbons are unique in that microbial degradation, especially of straight-chained and branched forms, involves the initial addition of molecular O_2. Recently, anaerobic degradation of hydrocarbons with sulfate or nitrate as electron acceptors has been observed. With sulfate present, organisms of the genus *Desulfovibrio* are active. This occurs only slowly and with microbial communities that have been exposed to these compounds for extended periods.

Lignin, an important structural component in mature plant materials, is a complex amorphous polymer based on a phenylpropane building block, linked by carbon-carbon and carbon-ether bonds. It makes up approximately 1/3 of the weight of wood. This is a special case in which biodegradability is dependent on O_2 availability. There often is no significant degradation because most filamentous fungi that degrade native lignin in situ can function only under aerobic conditions; oxidases can act by the release of active oxygen

Table 28.4

The Major Forms of Carbon, Nitrogen, Sulfur, and Iron Important in Biogeochemical Cycling

Cycle	Significant Gaseous Component Present?	*Major Forms and Valences*			
		Reduced Forms	Intermediate Oxidation State Forms		Oxidized Forms
C	Yes	CH_4 (-4)	CO $(+2)$		CO_2 $(+4)$
N	Yes	NH_4^+, organic N (-3)	N_2 (0) N_2O $(+1)$ NO_2^- $(+3)$		NO_3^- $(+5)$
S	Yes	H_2S, SH groups in organic matter (-2)	S^0 (0) $S_2O_3^{2-}$ $(+2)$ SO_3^{2-} $(+4)$		SO_4^{2-} $(+6)$
Fe	No	Fe^{2+} $(+2)$			Fe^{3+} $(+3)$

Note: The carbon, nitrogen, and sulfur cycles have significant gaseous components, and these are described as gaseous nutrient cycles. The iron cycle does not have a gaseous component, and this is described as a sedimentary nutrient cycle. Major reduced, intermediate oxidation state, and oxidized forms are noted, together with valences.

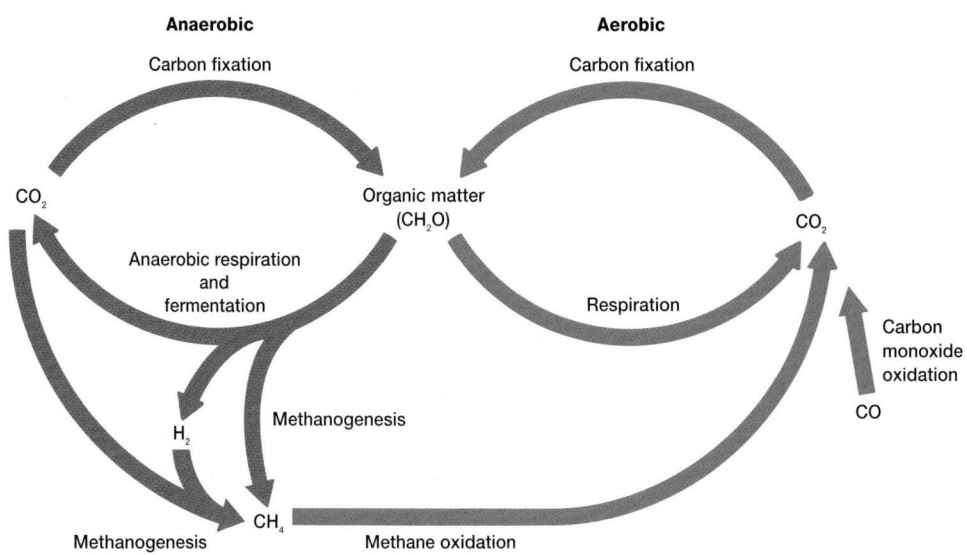

Figure 28.18   The Basic Carbon Cycle in the Environment.   Carbon fixation can occur through the activities of photoautotrophic and chemoautotrophic microorganisms. Methane can be produced from inorganic substrates (CO_2 + H_2) or from organic matter. Carbon monoxide (CO)—produced by sources such as automobiles and industry—is returned to the carbon cycle by CO-oxidizing bacteria. Aerobic processes are noted with blue arrows, and anaerobic processes are shown with red arrows. Reverse methanogenesis will be discussed in chapter 29.

species. Lignin's lack of biodegradability under anaerobic conditions results in accumulation of lignified materials, including the formation of peat bogs and muck soils. This absence of lignin degradation under anaerobic conditions also is important in construction. Large masonry structures often are built on swampy sites by driving in wood pilings below the water table and placing the building footings on the pilings. As long as the foundations remain water-saturated and anaerobic, the structure is stable. If the water table drops, however, the pilings will begin to rot and the structure will be threatened. Similarly, the cleanup of harbors can lead to decomposition of costly docks built with wooden pilings due to increased degradation of wood by aerobic filamentous fungi.

The presence or absence of oxygen also affects the final products that accumulate when organic substrates have been processed by microorganisms and mineralized either under aerobic or anaerobic conditions. Under aerobic conditions, oxidized products such

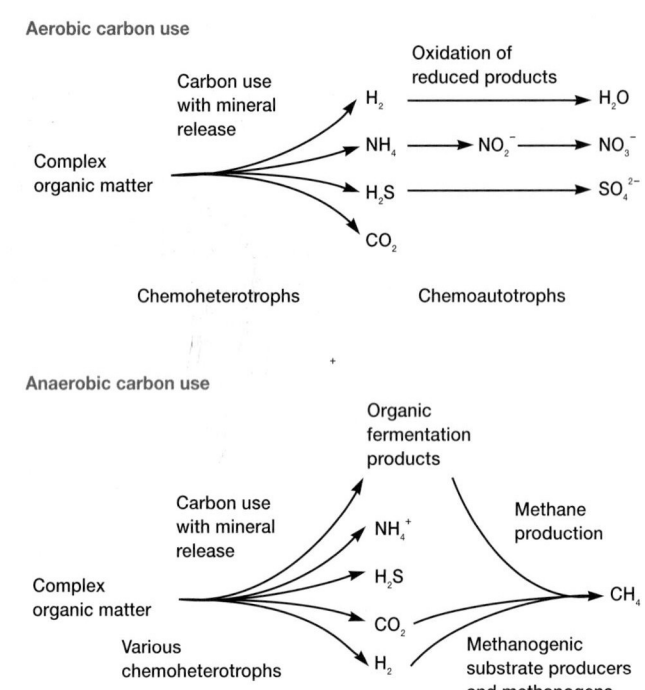

Figure 28.19 The Influence of Oxygen on Organic Matter Decomposition. Microorganisms form different products when breaking down complex organic matter aerobically than they do under anaerobic conditions. Under aerobic conditions oxidized products accumulate, while reduced products accumulate anaerobically. These reactions also illustrate commensalistic transformations of a substrate, where the waste products of one group of microorganisms can be used by a second type of microorganism.

as nitrate, sulfate, and carbon dioxide **(figure 28.19)** will result from microbial degradation of complex organic matter. In comparison, under anaerobic conditions reduced end products tend to accumulate, including ammonium ion, sulfide, and methane.

These oxidized and reduced forms, if they remain in the aerobic or anaerobic environments where they were formed, will usually only serve as nutrients. If mixing occurs, oxidized species might be moved to a more reduced zone or reduced species might be moved to a more oxidized zone. Under such circumstances, additional energetic possibilities (linking of electron acceptors and donors) will be created, leading to succession and further nutrient cycling as these mixed oxidants and reductants are exploited by the microbial community.

1. What is biogeochemical cycling?
2. Which organic polymers discussed in this section do and do not contain nitrogen?
3. Define mineralization and immobilization and give examples.
4. What is unique about lignin and its degradation?
5. What C, N, and S forms will accumulate after anaerobic degradation of organic matter?

Sulfur Cycle

Microorganisms contribute greatly to the sulfur cycle, a simplified version of which is shown in **figure 28.20.** Sulfide can serve as an electron source for both photosynthetic microorganisms and chemolithoautotrophs such as *Thiobacillus* (*see pp. 189–90 and 483–85*); it is converted to elemental sulfur and sulfate. When

Table 28.5										
Complex Organic Substrate Characteristics That Influence Decomposition and Degradability										
			Elements Present in Large Quantity					*Degradation*		
Substrate	**Basic Subunit**	**Linkages (if Critical)**	**C**	**H**	**O**	**N**	**P**	**With O_2**	**Without O_2**	
Starch	Glucose	$\alpha(1 \to 4)$ $\alpha(1 \to 6)$	+	+	+	–	–	+	+	
Cellulose	Glucose	$\beta(1 \to 4)$	+	+	+	–	–	+	+	
Hemicellulose	C6 and C5 monosaccharides	$\beta(1 \to 4), \beta(1 \to 3),$ $\beta(1 \to 6)$	+	+	+	–	–	+	+	
Lignin	Phenylpropane	C–C, C–O bonds	+	+	+	–	–	+	–	
Chitin	*N*-acetylglucosamine	$\beta(1 \to 4)$	+	+	+	+	–	+	+	
Protein	Amino acids	Peptide bonds	+	+	+	+	–	+	+	
Hydrocarbon	Aliphatic, cyclic, aromatic		+	+	–	–	–	+	+/–	
Lipids	Glycerol, fatty acids; some contain phosphate and nitrogen	Esters	+	+	+	+	+	+	+	
Microbial biomass		Varied	+	+	+	+	+	+	+	
Nucleic acids	Purine and pyrimidine bases, sugars, phosphate	Phosphodiester and *N*-glycosidic bonds	+	+	+	+	+	+	+	

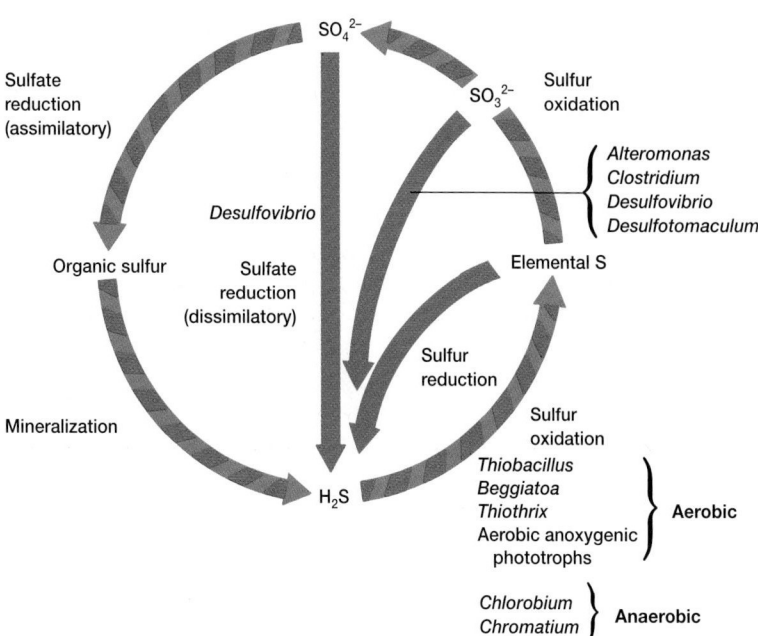

Figure 28.20   The Basic Sulfur Cycle. Photosynthetic and chemosynthetic microorganisms contribute to the environmental sulfur cycle. Sulfate and sulfite reductions carried out by *Desulfovibrio* and related microorganisms, noted with purple arrows, are dissimilatory processes. Sulfate reduction also can occur in assimilatory reactions, resulting in organic sulfur forms. Elemental sulfur reduction to sulfide is carried out by *Desulfuromonas*, thermophilic archaea, or cyanobacteria in hypersaline sediments. Sulfur oxidation can be carried out by a wide range of aerobic chemotrophs and by aerobic and anaerobic phototrophs.

sulfate diffuses into reduced habitats, it provides an opportunity for different groups of microorganisms to carry out **sulfate reduction.** For example, when a usable organic electron donor is present, *Desulfovibrio* uses sulfate as an oxidant (*see pp. 185 and 496*). This use of sulfate as an external electron acceptor to form sulfide, which accumulates in the environment, is an example of a **dissimilatory reduction** process and anaerobic respiration. In comparison, the reduction of sulfate for use in amino acid and protein biosynthesis is described as an **assimilatory reduction** process (*see section 10.4*). Other microorganisms have been found to carry out dissimilatory elemental sulfur reduction. These include *Desulfuromonas* (*see pp. 495–96*), thermophilic archaea (*see chapter 20*), and also cyanobacteria in hypersaline sediments. Sulfite is another critical intermediate that can be reduced to sulfide by a wide variety of microorganisms, including *Alteromonas* and *Clostridium,* as well as *Desulfovibrio* and *Desulfotomaculum. Desulfovibrio* is usually considered as an obligate anaerobe. Recent research, however, has shown that this interesting organism also respires using oxygen, when it is present at a dissolved oxygen level of 0.04%.

In addition to the very important photolithotrophic sulfur oxidizers such as *Chromatium* and *Chlorobium,* which function under strict anaerobic conditions in deep water columns, a large and varied group of bacteria carry out **aerobic anoxygenic photosynthesis.** These aerobic anoxygenic phototrophs use bacteriochlorophyll *a* and carotenoid pigments and are found in marine and freshwater environments; they are often components of microbial mat communities. Important genera include *Erythromonas, Roseococcus, Porphyrobacter,* and *Roseobacter.*

"Minor" compounds in the sulfur cycle play major roles in biology. An excellent example is dimethylsulfoniopropionate (DMSP), which is used by bacterioplankton (floating bacteria) as a sulfur source for protein synthesis, and which is transformed to dimethylsulfide (DMS), a volatile sulfur form that can affect atmospheric processes.

When pH and oxidation-reduction conditions are favorable, several key transformations in the sulfur cycle also occur as the result of chemical reactions in the absence of microorganisms. An important example of such an abiotic process is the oxidation of sulfide to elemental sulfur. This takes place rapidly at a neutral pH, with a half-life of approximately 10 minutes for sulfide at room temperature.

Nitrogen Cycle

Several important aspects of the basic nitrogen cycle will be discussed: the processes of autotrophic nitrification, as well as denitrification and nitrogen fixation (**figure 28.21**). It should be emphasized that this is a "basic" nitrogen cycle. Although not mentioned in the figure, heterotrophs can carry out nitrification, and some of these heterotrophs combine nitrification with anaerobic denitrification, thus oxidizing ammonium ion to N_2O and N_2 with depressed oxygen levels. The occurrence of anoxic ammonium ion oxidation (anammox is the term used for the commercial process) means that nitrification is not only an aerobic process. Thus as we learn more about the biogeochemical cycles, including that of nitrogen, the simple cycles of earlier textbooks are no longer accurate representations of biogeochemical processes.

Nitrification is the aerobic process of ammonium ion (NH_4^+) oxidation to nitrite (NO_2^-) and subsequent nitrite oxidation to nitrate (NO_3^-). Bacteria of the genera *Nitrosomonas* and *Nitrosococcus,* for example, play important roles in the first step, and *Nitrobacter* and related chemolithoautotrophic bacteria carry out the second step. Recently *Nitrosomonas eutropha* has been found to oxidize ammonium ion anaerobically to nitrite and nitric oxide (NO) using nitrogen dioxide

Figure 28.21 The Basic Nitrogen Cycle.
Flows that occur predominantly under aerobic conditions are noted with open arrows. Anaerobic processes are noted with solid bold arrows. Processes occurring under both aerobic and anaerobic conditions are marked with cross-barred arrows. The anammox reaction of NO_2^- and NH_4^+ to yield N_2 is shown. Important genera contributing to the nitrogen cycle are given as examples.

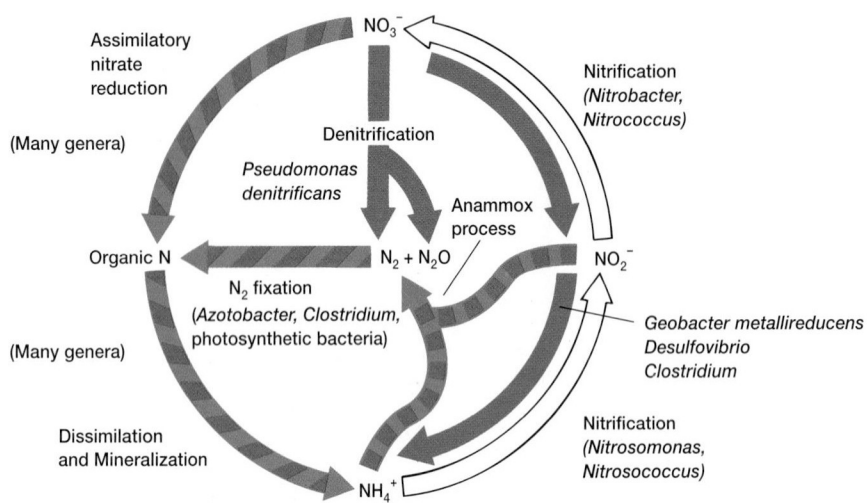

(NO_2) as an acceptor in a denitrification-related reaction. In addition, **heterotrophic nitrification** by bacteria and fungi contributes significantly to these processes in more acidic environments.   Nitrification and nitrifiers (p. 189)

The process of **denitrification** requires a different set of environmental conditions. This dissimilatory process, in which nitrate is used as an electron acceptor in anaerobic respiration, usually involves heterotrophs such as *Pseudomonas denitrificans*. The major products of denitrification include nitrogen gas (N_2) and nitrous oxide (N_2O), although nitrite (NO_2^-) also can accumulate. Nitrite is of environmental concern because it can contribute to the formation of carcinogenic nitrosamines. Finally, nitrate can be transformed to ammonia in dissimilatory reduction by a variety of bacteria, including *Geobacter metallireducens*, *Desulfovibrio* spp., and *Clostridium*.   Denitrification and anaerobic respiration (pp. 185–86)

Nitrogen assimilation occurs when inorganic nitrogen is used as a nutrient and incorporated into new microbial biomass. Ammonium ion, because it is already reduced, can be directly incorporated without major energy costs. However, when nitrate is assimilated, it must be reduced with a significant energy expenditure. In this process nitrite may accumulate as a transient intermediate.   The biochemistry of nitrogen assimilation (pp. 206–9)

Nitrogen fixation can be carried out by aerobic or anaerobic procaryotes and does not occur in eucaryotes. Under aerobic conditions a wide range of free-living microbial genera (*Azotobacter*, *Azospirillum*) contribute to this process. Under anaerobic conditions the most important free-living nitrogen fixers are members of the genus *Clostridium*. Nitrogen fixation by cyanobacteria such as *Anabaena* and *Oscillatoria* can lead to the enrichment of aquatic environments with nitrogen. These nutrient-enrichment processes are discussed in chapter 29. In addition, nitrogen fixation can occur through the activities of bacteria that develop symbiotic associations with plants. These associations include *Rhizobium* and *Bradyrhizobium* with legumes, *Frankia* in association with many woody shrubs, and *Anabaena*, with *Azolla*, a water fern important in rice cultivation.   The establishment of the *Rhizobium*-legume association (pp. 652–55)

The nitrogen-fixation process involves a sequence of reduction steps that require major energy expenditures. Ammonia, the product of nitrogen reduction, is immediately incorporated into organic matter as an amine. Reductive processes are extremely sensitive to O_2 and must occur under anaerobic conditions even in aerobic microorganisms. Protection of the nitrogen-fixing enzymes is achieved by means of a variety of mechanisms, including physical barriers, as occurs with heterocysts in some cyanobacteria (*see section 21.3*), O_2 scavenging molecules, and high rates of metabolic activity.   The biochemistry of nitrogen fixation (pp. 206–9)

As shown in figure 28.21, microorganisms have been isolated that can couple the anaerobic oxidation of NH_4^+ with the reduction of NO_2^-, to produce gaseous nitrogen, in what has been termed the **anammox process** (*anoxic ammonia oxidation*). An example is an anaerobic chemolithoautotroph, found to be a planctomycete (*see section 21.4*), that has been named *Brocadia anammoxidans*. This may provide a means by which nitrogen can be removed from sewage plant effluents to decrease nitrogen flow to sensitive freshwater and marine ecosystems.

---

1. What are the major oxidized and reduced forms of sulfur and nitrogen?
2. Diagram a simple sulfur cycle.
3. What is aerobic anoxygenic photosynthesis?
4. Why is dimethylsulfide (DMS), considered to be a "minor" part of the sulfur cycle, of such environmental importance?
5. What are nitrification, denitrification, nitrogen fixation, and the anammox process?

Iron Cycle

The iron cycle (**figure 28.22**) includes several different genera that carry out iron oxidations, transforming ferrous ion (Fe^{2+}) to ferric ion (Fe^{3+}). *Thiobacillus ferrooxidans* carries out this process under acidic conditions, *Gallionella* is active under neutral pH conditions, and *Sulfolobus* functions under acidic, thermophilic condi-

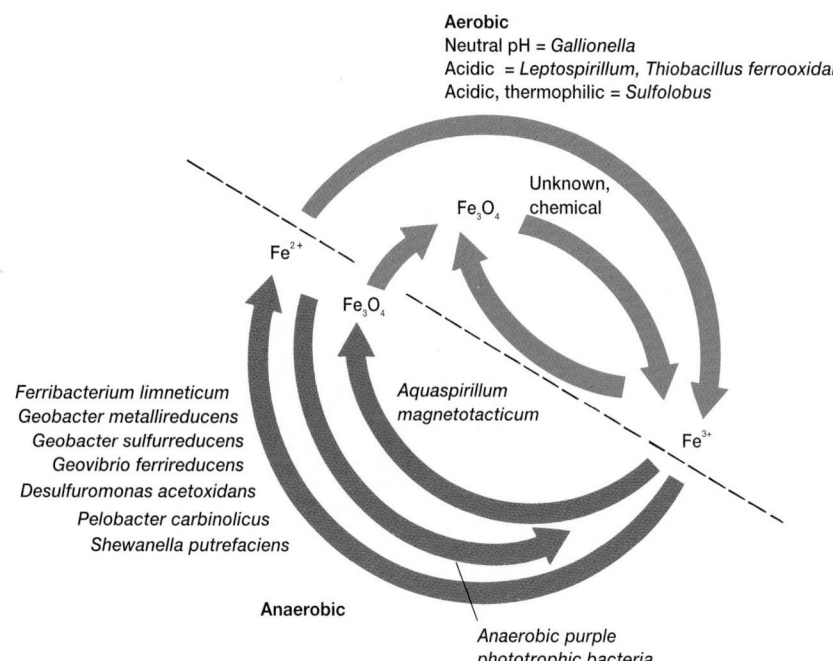

Aerobic
Neutral pH = *Gallionella*
Acidic   = *Leptospirillum, Thiobacillus ferrooxidans*
Acidic, thermophilic = *Sulfolobus*

Unknown, chemical

Fe_3O_4

Fe^{2+}

Fe_3O_4

Ferribacterium limneticum
Geobacter metallireducens
Geobacter sulfurreducens
Geovibrio ferrireducens
Desulfuromonas acetoxidans
Pelobacter carbinolicus
Shewanella putrefaciens

Aquaspirillum magnetotacticum

Fe^{3+}

Anaerobic

Anaerobic purple phototrophic bacteria

Figure 28.22 The Basic Iron Cycle. A simplified iron cycle with examples of microorganisms contributing to these oxidation and reduction processes. In addition to ferrous ion (Fe^{2+}) oxidation and ferric ion (Fe^{3+}) reduction, magnetite (Fe_3O_4), a mixed valence iron compound formed by magnetotactic bacteria is important in the iron cycle. Different microbial groups carry out the oxidation of ferrous ion depending on environmental conditions.

tions. Much of the earlier literature suggested that additional genera could oxidize iron, including *Sphaerotilus* and *Leptothrix*. These two genera are still termed "iron bacteria" by many nonmicrobiologists. Confusion about the role of these genera resulted from the occurrence of the chemical oxidation of ferrous ion to ferric ion (forming insoluble iron precipitates) at neutral pH values, where microorganisms also grow on organic substrates. These microorganisms are now classified as chemoheterotrophs.

Recently microbes have been found that oxidize Fe^{2+} using nitrate as an electron acceptor, including the interesting microorganism *Dechlorosoma suillum*, a mixotroph (*see p. 96*) that can also use chlorate as an electron acceptor. This process occurs in aquatic sediments with depressed levels of oxygen and may be another route by which large zones of oxidized iron have accumulated in environments with lower oxygen levels. Banded iron formation that occurred when atmospheric oxygen levels were beginning to increase at the end of the Precambrian era may be evidence of increased iron bacteria activity.

Iron reduction occurs under anaerobic conditions resulting in the accumulation of ferrous ion. Although many microorganisms can reduce small amounts of iron during their metabolism, most iron reduction is carried out by specialized iron-respiring microorganisms such as *Geobacter metallireducens, Geobacter sulfurreducens, Ferribacterium limneticum,* and *Shewanella putrefaciens,* which can obtain energy for growth on organic matter using ferric iron as an oxidant.

In addition to these relatively simple reductions to ferrous ion, some magnetotactic bacteria such as *Aquaspirillum magnetotacticum* (*see section 3.3*) transform extracellular iron to the mixed valence iron oxide mineral magnetite (Fe_3O_4) and construct intracellular magnetic compasses. Furthermore, dissimilatory iron-reducing bacteria accumulate magnetite as an extracellular product.

Magnetite has been detected in sediments, where it is present in particles similar to those found in bacteria, indicating a longer-term contribution of bacteria to iron cycling processes. Genes for magnetite synthesis have been cloned into other organisms, creating new magnetically sensitive microorganisms. Magnetotactic bacteria are now described as **magneto-aerotactic bacteria,** due to their using magnetic fields to migrate to the position in a bog or swamp where the oxygen level is best suited for their functioning. In the last decade new microorganisms have been discovered that use ferrous ion as an electron donor in anoxygenic photosynthesis. Thus with production of ferric ion in lighted anaerobic zones by iron-oxidizing bacteria, the stage is set for subsequent chemotrophic-based iron reduction, such as by *Geobacter* and *Shewanella,* creating a strictly anaerobic oxidation/reduction cycle for iron.

Manganese Cycle

The importance of microorganisms in manganese cycling is becoming much better appreciated. The manganese cycle (**figure 28.23**) involves the transformation of manganous ion (Mn^{2+}) to MnO_2 (equivalent to manganic ion [Mn^{4+}]), which occurs in hydrothermal vents, bogs, and as an important part of rock varnishes. *Leptothrix, Arthrobacter, Pedomicrobium,* and the incompletely characterized "*Metallogenium*" are important in Mn^{2+} oxidation. *Shewanella, Geobacter,* and other chemoorganotrophs can carry out the complementary manganese reduction process.

Other Cycles and Cycle Links

Microorganisms can use a wide variety of additional metals as electron acceptors. Metals such as europium, tellurium, selenium, and

Figure 28.23 The Manganese Cycle, Illustrated in a Stratified Lake. Microorganisms make many important contributions to the manganese cycle. After diffusing from anaerobic (pink) to aerobic (blue) zones, manganous ion (Mn^{2+}) is oxidized chemically and by many morphologically distinct microorganisms in the aerobic water column to manganic oxide—$MnO_2^{(IV)}$, valence equivalent to $4+$. When the $MnO_2^{(IV)}$ diffuses into the anaerobic zone, bacteria such as *Geobacter* and *Shewanella* carry out the complementary reduction process. Similar processes occur across aerobic/anaerobic transitions in soils, muds, and other environments.

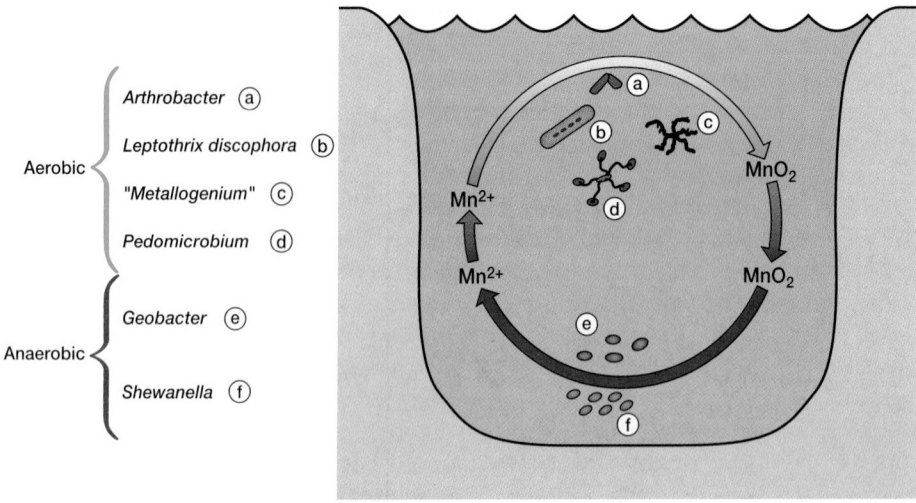

Aerobic
- *Arthrobacter* (a)
- *Leptothrix discophora* (b)
- *"Metallogenium"* (c)
- *Pedomicrobium* (d)

Anaerobic
- *Geobacter* (e)
- *Shewanella* (f)

rhodium can be reduced. Important microorganisms that reduce these metals include the photoorganotrophs *Rhodobacter, Rhodospirillum,* and *Rhodopseudomonas.* For selenium, *Pseudomonas stutzeri, Thauera selenatis,* and *Wolinella succinogenes* are active. Such reductions can decrease the toxicity of a metal.

The microbial transformation of phosphorus involves primarily the transformation of phosphorus ($+5$ valence) from simple orthophosphate to various more complex forms, including polyphosphates found in metachromatic granules (*see* p. 49). A unique (and possibly microbial) product is phosphine (PH_3) with a -3 valence, which is liberated from swamps, soils, and marine regions and which ignites when exposed to air. This can then ignite methane produced in the same environment! The production of methane by anaerobic microorganisms will be discussed in chapter 29.

Having described the sulfur, iron, and manganese cycles as functioning independently, it is important to again emphasize that many microorganisms metabolically link these cycles by using commonly shared oxidants and reductants. For example, some sulfate-reducing microorganisms can reduce Fe^{3+} using H_2 or organic matter as an electron donor and also can oxidize elemental sulfur to sulfate when $Mn^{(IV)}$ is present as an electron acceptor. The $Mn^{(IV)}$-dependent production of sulfate under anaerobic conditions, carried out by *Desulfobulbus propionicus,* provides a way to link sulfur and manganese cycling anaerobically.

1. What major forms of iron, manganese, and phosphorus are important in biogeochemical cycling?
2. Why is *Aquaspirillum* considered to be a magneto-aerotactic bacterium?
3. What are some important microbial genera that contribute to manganese cycling?
4. What is phosphine? Under which conditions will it be produced?
5. How can microorganisms metabolically link the sulfur, iron, and manganese cycles?

Microorganisms and Metal Toxicity

In addition to metals such as iron and manganese, which are largely nontoxic to microorganisms and animals, there are a series of metals that have varied toxic effects on microorganisms and homeothermic animals. Microorganisms play important roles in modifying the toxicity of these metals (**table 28.6**).

The "metals" can be considered in broad categories. The "noble metals" tend not to cross the vertebrate blood-brain barrier but can have distinct effects on microorganisms. Microorganisms also can reduce ionic forms of noble metals to their elemental forms.

The second group includes metals or metalloids that microorganisms can methylate to form more mobile products called organometals. Some organometals can cross the blood-brain barrier and affect the central nervous system of vertebrates. Organometals contain carbon-metal bonds. These bonds are their unique identifying characteristics.

The mercury cycle is of particular interest and illustrates many characteristics of those metals that can be methylated. Mercury compounds were widely used in industrial processes over the centuries. One has only to think of Lewis Carroll's allusion to this problem when he wrote of the Mad Hatter in *Alice in Wonderland.* At that time mercury was used in the shaping of felt hats. Microorganisms methylated some of the mercury, thus rendering it more toxic to the hatmakers.

A devastating situation developed in southwestern Japan when large-scale mercury poisoning occurred in the Minamata Bay region because of industrial mercury released into the marine environment. Inorganic mercury that accumulated in bottom muds of the bay was methylated by anaerobic bacteria of the genus *Desulfovibrio* (**figure 28.24**). Such methylated mercury forms are volatile and lipid soluble, and the mercury concentrations increased in the food chain (by the process of **biomagnification**). The mercury was ultimately ingested by the human population, the "top consumers," through their primary food source—fish—leading to severe neurological disorders.

Table 28.6

Examples of Microorganism-Metal Interactions and Relations to Effects on Microorganisms and Homeothermic Animals

Metal Group	Metal		Interactions and Transformations	
			Microorganisms	Homeothermic Animals
Noble metals	Ag	Silver	Microorganisms can reduce ionic forms to the elemental state. Low levels of ionized metals released to the environment have antimicrobial activity.	Many of these metals can be reduced to elemental forms and do not tend to cross the blood-brain barrier. Silver reduction can lead to inert deposits in the skin.
	Au	Gold		
	Pt	Platinum		
Metals that form stable carbon metal bonds	As	Arsenic	Microorganisms can transform inorganic and organic forms to methylated forms, some of which tend to bioaccumulate in higher trophic levels.	Methylated forms of some metals can cross the blood-brain barrier, resulting in neurological effects or death.
	Hg	Mercury		
	Se	Selenium		
Other metals	Cu	Copper	In the ionized form, at higher concentrations, these metals can directly inhibit microorganisms. They are often required at lower concentrations as trace elements.	At higher levels, clearance from higher organisms occurs by reaction with plasma proteins and other mechanisms. Many of these metals serve as trace elements at lower concentrations.
	Zn	Zinc		
	Co	Cobalt		

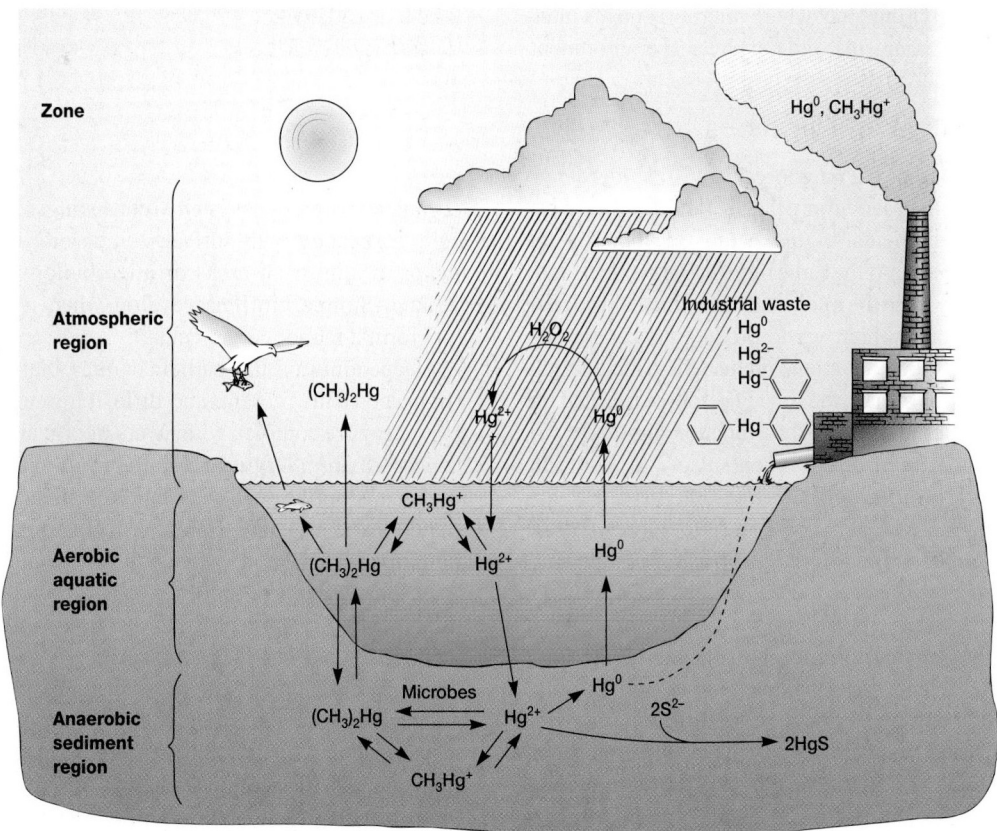

Figure 28.24   The Mercury Cycle.   Interactions between the atmosphere, aerobic water, and anaerobic sediment are critical. Microorganisms in anaerobic sediments, primarily *Desulfovibrio,* can transform mercury to methylated forms that can be transported to water and the atmosphere. These methylated forms also undergo biomagnification. The production of volatile elemental mercury (Hg^0) releases this metal to waters and the atmosphere. Sulfide, if present in the anaerobic sediment, can react with ionic mercury to produce less soluble HgS.

The third group of metals occurs in ionic forms directly toxic to microorganisms. The metals in this group also can affect more complex organisms. However, plasma proteins react with the ionic forms of these metals and aid in their excretion unless excessive long-term contact and ingestion occur. Relatively high doses of these metals are required to cause lethal effects. At lower concentrations many of these metals serve as required trace elements.

1. What are examples of the three groups of metals in terms of their toxicity to microorganisms and homeothermic animals?
2. How can microbial activity render some metals more or less toxic to warm-blooded animals?
3. Why do metals such as mercury have such major effects on higher organisms?

28.4   THE PHYSICAL ENVIRONMENT

Microorganisms, as they interact with each other and with other organisms in biogeochemical cycling, also are influenced by their immediate physical environment, whether this might be soil, water, the deep marine environment, or a plant or animal host. It is important to consider the specific environments where microorganisms interact with each other, other organisms, and the physical environment.

The Microenvironment and Niche

The specific physical location of a microorganism is its **microenvironment.** In this physical microenvironment, the flux of required oxidants, reductants, and nutrients to the actual location of the microorganism can be limited. At the same time, waste products may not be able to diffuse away from the microorganism at rates sufficient to avoid growth inhibition by high waste product concentrations. These fluxes and gradients create a unique **niche,** which includes the microorganism, its physical habitat, the time of resource use, and the resources available for microbial growth and function (**figure 28.25**).

This physically structured environment also can limit the predatory activities of protozoa. If the microenvironment has pores with diameters of 3 to 6 μm, bacteria in the pores will be protected from predation, while allowing diffusion of nutrients and waste products. If the pores are larger, perhaps greater than 6 μm in diameter, protozoa may be able to feed on the bacteria. It is important to emphasize that microorganisms can create their own microenvironments and niches. For example, microorganisms in the interior of a colony have markedly different microenvironments and niches than those of the same microbial populations located on the surface or edge of the colony. Microorganisms also can associate with clays and form inert microhabitats called "clay hutches" for protection (*see section 42.4*).

Biofilms and Microbial Mats

As noted in section 28.3, microorganisms tend to create their own microenvironments and niches, by forming **biofilms.** These are

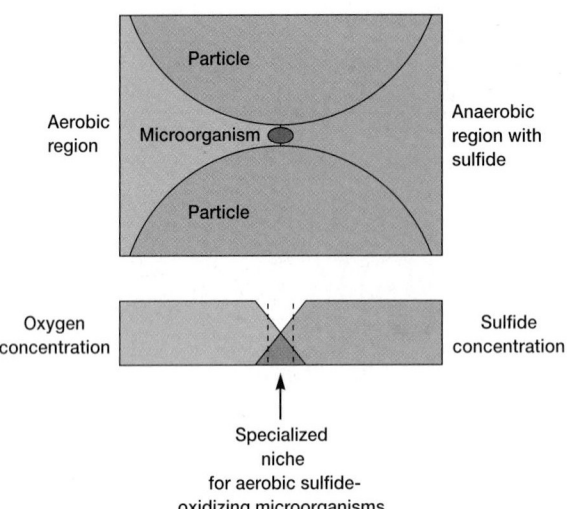

Figure 28.25   The Creation of a Niche from a Microenvironment.   As shown in this illustration, two nearby particles create a physical microenvironment for possible use by microorganisms. Chemical gradients, as with oxygen from the aerobic region, and sulfide from the anaerobic region, create a unique niche. This niche thus is the physical environment and the resources available for use by specialized aerobic sulfide-oxidizing bacteria.

organized microbial systems consisting of layers of microbial cells associated with surfaces. Such biofilms are an important factor in almost all areas of microbiology, as shown in **figure 28.26a.** Simple biofilms develop when microorganisms attach and form a monolayer of cells.

Depending on the particular microbial growth environment (light, nutrients present, and diffusion rates), these biofilms can become more complex with layers of organisms of different types (figure 28.26b). A typical example would involve photosynthetic organisms on the surface, facultative chemoorganotrophs in the middle, and possibly sulfate-reducing microorganisms on the bottom.

More complex biofilms can develop to form a four-dimensional structure (X, Y, Z, and time) with cell aggregates, interstitial pores, and conduit channels (figure 28.26c). This developmental process involves the growth of attached microorganisms, resulting in accumulation of additional cells on the surface, together with the continuous trapping and immobilization of free-floating microorganisms that move over the expanding biofilm. This structure allows nutrients to reach the biomass, and the channels are shaped by protozoa that graze on bacteria.

These more complex biofilms, in which microorganisms create unique environments, can be observed by the use of confocal scanning laser microscopy (CSLM) as discussed in chapter 2. The diversity of nonliving and living surfaces that can be exploited by biofilm-forming microorganisms include surfaces in

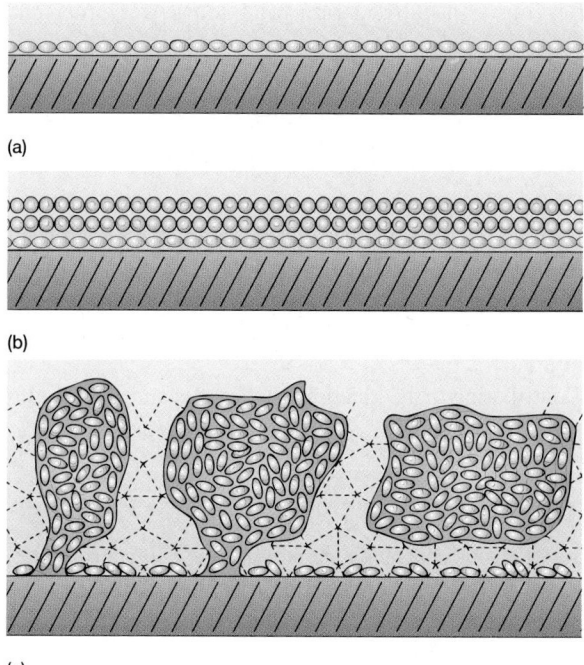

(a)

(b)

(c)

Figure 28.26   The Growth of Biofilms.   Biofilms, or micro-bial growths on surfaces such as in freshwater and marine environments, can develop and become extremely complex, depending on the energy sources that are available. (**a**) Initial colonization by a single type of bacterium. (**b**) Development of a more complex biofilm with layered microorganisms of different types. (**c**) A mature biofilm with cell aggregates, interstitial pores, and conduits.

catheters and dialysis units, which have intimate contact with human body fluids. Control of such microorganisms and their establishment in these sensitive medical devices is an important part of modern hospital care.

Biofilms also can protect pathogens from disinfectants, create a focus for later occurrence of disease, or release microorganisms and microbial products that may affect the immunological system of a susceptible host. Biofilms are critical in ocular diseases because *Chlamydia, Staphylococcus,* and other pathogens survive in ocular devices such as contact lenses and in cleaning solutions.

Depending on environmental conditions, biofilms can become so large that they are visible and have macroscopic dimensions. Bands of microorganisms of different colors are evident as shown in **figure 28.27.** These thick biofilms, called **microbial mats,** are found in many freshwater and marine environments. These mats are complex layered microbial communities that can form at the surface of rocks or sediments in hypersaline and freshwater lakes, lagoons, hot springs, and beach areas. They consist of microbial filaments, including cyanobacteria. A major characteristic of mats is the extreme gradients that are present. Light only penetrates approximately 1 mm into these communities, and below this photosynthetic zone, anaerobic conditions occur and sulfate-reducing bacteria play a major role. The sulfide that these

Figure 28.27   Microbial Mats.   Microorganisms, through their metabolic activities, can create environmental gradients resulting in layered ecosystems. A vertical section of a hot spring (55°C) microbial mat, showing the various layers of microorganisms.

organisms produce diffuses to the anaerobic lighted region, allowing sulfur-dependent photosynthetic microorganisms to grow. Some believe that microbial mats could have allowed the formation of terrestrial ecosystems prior to the development of vascular plants, and fossil microbial mats, called stromatolites, have been dated at over 3.5 billion years old *(see p. 410).* Molecular techniques and stable isotope measurements (see section 28.5) are being used to better understand these unique microbial communities.

---

1. What are the similarities and differences between a microenvironment and a niche?
2. Why might pores in soils, waters, and animals be important for survival of bacteria if protozoa are present?
3. Why might conditions vary for a bacterium on the edge of a colony in comparison with the center of the colony?
4. What are biofilms? What types of surfaces on living organisms can provide a site for biofilm formation?
5. Why are biofilms important in human health?
6. What are microbial mats, and where are they found?

Microorganisms and Ecosystems

Microorganisms, as they interact with each other and other organisms, and influence nutrient cycling in their specific microenvironments and niches, also contribute to the functioning of

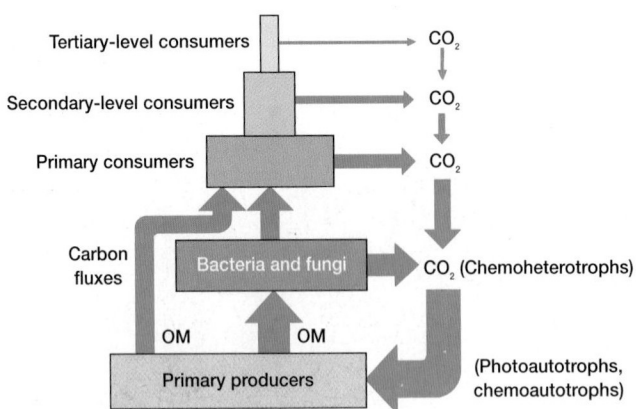

Figure 28.28 **The Vital Role of Microorganisms in Ecosystems.** Microorganisms play vital roles in ecosystems as primary producers, decomposers, and primary consumers. Carbon is fixed by the primary producers, including microorganisms, which use light or chemically bound energy. Chemoheterotrophic bacteria and fungi serve as the main decomposers of organic matter, making minerals again available for use by the primary producers. Ciliates and flagellates, important microbial primary consumers, feed on the bacteria and fungi, recycling nutrients as part of the microbial loop. Organic matter (OM).

ecosystems. Ecosystems have been defined as "communities of organisms and their physical and chemical environments that function as self-regulating units." These self-regulating biological units respond to environmental changes by modifying their structure and function.

Microorganisms in ecosystems can have two complementary roles: (1) the synthesis of new organic matter from CO_2 and other inorganic compounds during **primary production** and (2) decomposition of this accumulated organic matter. The general relationships between the **primary producers** that synthesize organic matter, the heterotrophic **decomposers,** and the consumers are illustrated in **figure 28.28**. Microorganisms of different types contribute to each of these complementary relationships.

In terrestrial environments the primary producers are usually vascular plants. In freshwater and marine environments, the cyanobacteria and algae (*see section 21.3 and chapter 26*) play a similar role. The major energy source driving primary production is light in both habitats although in hydrothermal and hydrocarbon seep areas, chemotrophic ecosystems occur. The higher **consumers,** including humans, are chemoheterotrophs. These consumers depend on the "life support systems" provided by organisms that accumulate and decompose organic matter.

Microorganisms thus carry out many important functions as they interact in ecosystems, including:

1. Contributing to the formation of organic matter through photosynthetic and chemosynthetic processes.
2. Decomposing organic matter, often with the release of inorganic compounds (e.g., CO_2, NH_4^+, CH_4, H_2) in mineralization processes.

3. Serving as a nutrient-rich food source for other chemoheterotrophic microorganisms, including protozoa and animals.
4. Modifying substrates and nutrients used in symbiotic growth processes and interactions, thus contributing to biogeochemical cycling.
5. Changing the amounts of materials in soluble and gaseous forms. This occurs either directly by metabolic processes or indirectly by modifying the environment.
6. Producing inhibitory compounds that decrease microbial activity or limit the survival and functioning of plants and animals.
7. Contributing to the functioning of plants and animals through positive and negative symbiotic interactions.

Microorganism Movement between Ecosystems

Microorganisms constantly are moving and being moved between ecosystems. This often happens naturally in many ways: (1) soil is transported around the Earth by windstorms and falls on land areas and waters far from its origins; (2) rivers transport eroded materials, sewage plant effluents, and urban wastes to the ocean; and (3) insects and animals release urine, feces, and other wastes to environments as they migrate around the Earth. When plants and animals die after moving to a new environment, they decompose and their specially adapted and coevolved microorganisms (and their nucleic acids) are released (*see section 29.2*). The fecal-oral route of disease transmission, often involving foods and waters, and the acquisition of diseases in hospitals (nosocomial infections) are important examples of pathogen movement between ecosystems. Each time a person coughs or sneezes, microorganisms also are being transported to new ecosystems.

Humans also both deliberately and unintentionally move microorganisms between different ecosystems. This occurs when microbes are added to environments to speed up microbially mediated degradation processes (*see bioremediation, section 42.4*) or when a plant-associated inoculum such as *Rhizobium,* is added to a soil to increase the formation of nitrogen-fixing nodules on legumes (*see pp. 652–55*). One of the most important accidental modes of microbial movement is the use of modern transport vehicles such as automobiles, trains, ships, and airplanes. These often rapidly move microorganisms long distances.

The fate of microorganisms placed in environments where they normally do not live, or of microorganisms returned to their original environments, is of theoretical and practical importance. Pathogens that are normally associated with an animal host largely have lost their ability to compete effectively for nutrients with microorganisms indigenous to other environments. Upon moving to a new environment, the population of viable and culturable pathogens gradually decreases. However, more sensitive viability assessment procedures, particularly molecular techniques, indicate that **noncultured microorganisms,** as observed with *Vibrio* (*see section 6.5*), may play critical roles in disease occurrence.

Even microorganisms recovered from a particular environment, after growth in the laboratory on rich media, may lose their ability to survive when placed back in their original environment. The cause may be physical or physiological. From a physical standpoint the microorganisms may find themselves outside their protected physical niche, where they can be consumed by protozoa and other predators. On the other hand, after growth in rich laboratory media, they may have lost the ability to compete physiologically with the native populations.

---

1. Define the following terms: ecosystem, primary production, decomposer.
2. List important functions of higher consumers in natural environments.
3. What are the important functions of microorganisms in ecosystems?
4. How can microorganisms move between different ecosystems?
5. Why might microorganisms isolated from soil or water, after being grown in the laboratory, lose the ability to survive in the environment from which they were taken?

Stress and Ecosystems

Microorganisms function in ecosystems that develop under a wide range of environmental conditions. These have varied pHs, temperatures, pressures, salinity, water availability, and ionizing radiation as summarized in **table 28.7**. Such stress factors have major effects on microbial populations and communities, and can create an **extreme environment,** as shown in **figure 28.29**. In these cases high salt concentrations, extreme temperature, and acidic conditions have affected the microbial communities. The microorganisms that survive in such environments are described as **extremophiles,** and such extreme environments are usually considered to have decreased microbial diversity, as judged by the microorganisms that can be cultured. With the increased use of molecular detection techniques, however, it appears that there is surprising diversity among the microorganisms that cannot be cultured from these extreme environments. Further work to establish relationships between the microorganisms that can be observed and detected by these molecular techniques and culturable microorganisms will be required in the future. The influence of environmental factors on growth (pp. 118–28)

Many microbial genera have specific requirements for survival and functioning in such so-called extreme environments. For example, a high sodium ion concentration is required to maintain membrane integrity in many halophilic bacteria, including members of the genus *Halobacterium*. Halobacteria require a sodium ion concentration of at least 1.5 M, and about 3 to 4 M for optimum growth. Halophilic archaea (pp. 120–21, 447–49)

The bacteria found in deep-sea environments have different pressure requirements, depending on the depth from which they are recovered. These bacteria can be described as **baro-** or **piezotolerant bacteria** (growth from approximately 1 to 500 atm), **moderately barophilic bacteria** (growth optimum 5,000 meters, and still able to grow at 1 atm), and **extreme barophilic bacteria,** which require approximately 400 atm or higher for growth (*see chapter 29*).

Table 28.7

Characteristics of Extreme Environments in which Microorganisms Grow

Stress	Environmental Conditions	Microorganisms Observed
High temperature	121°C	*Geogemma barossii*
	110–113°C, deep marine trenches	*Pyrolobus fumarii*
		Methanopyrus kandleri
		Pyrodictium abyssi
	67–102°C, marine basins	*Pyrococcus abyssi*
	85°C, hot springs	*Thermus*
		Sulfolobus
	75°C, sulfur hot springs	*Thermothrix thiopara*
Low temperature	−12°C, antarctic ice	*Psychromonas ingrahamii*
Osmotic stress	13–15% NaCl	*Chlamydomonas*
	25% NaCl	*Halobacterium*
		Halococcus
Acidic pH	pH 3.0 or lower	*Saccharomyces*
		Thiobacillus
	pH 0.5	*Picrophilus oshimae*
	pH 0.0	*Ferroplasma acidarmanus*
Basic pH	pH 10.0 or above	*Bacillus*
Low water availability	$a_w = 0.6–0.65$	*Torulopsis*
		Candida
Temperature and low pH	85°C, pH 1.0	*Cyanidium*
		Sulfolobus acidocaldarum
Pressure	500–1,035 atm	*Colwellia hadaliensis*
Radiation	1.5 million rads	*Deinococcus radiodurans*

Intriguing changes in basic physiological processes occur in microorganisms functioning under extreme acidic or alkaline conditions. These acidophilic and alkalophilic microorganisms have markedly different problems in maintaining a more neutral internal pH and chemiosmotic processes (*see chapter 9*). Obligately acidophilic microorganisms can grow at a pH of 3.0 or lower, and major pH differences can exist between the interior and exterior of the cell. These acidophiles include members of the genera *Thiobacillus, Sulfolobus,* and *Thermoplasma*. The higher relative internal pH is maintained by a net outward translocation of protons. This may occur as the result of unique membrane lipids, hydrogen ion removal during reduction of oxygen to water, or the pH-dependent characteristics of membrane-bound enzymes.

An archaeal iron-oxidizing acidophile, *Ferroplasma acidarmanus,* capable of growth at pH 0, has been isolated from a sulfide ore body in California. This unique procaryote, capable of massive surface growth in flowing waters in the subsurface (**figure 28.30**), possesses a single peripheral cytoplasmic membrane and no cell wall.

The extreme alkalophilic microorganisms grow at pH values of 10.0 and higher and must maintain a net inward translocation of protons. These obligate alkalophiles cannot grow below a pH of 8.5 and are often members of the genus *Bacillus*; *Micrococcus* and *Exiguobacterium* representatives have also been reported. Some photosynthetic cyanobacteria also have similar characteristics. Increased internal proton concentrations may be maintained by means of coordinated hydrogen and sodium ion fluxes.

(a)

(b)

(c)

Figure 28.29 Microorganisms Growing in Extreme Environments. Many microorganisms are especially suited to survive in extreme environments. (**a**) Salterns turned red by halophilic algae and halobacteria. (**b**) A hot spring colored green and blue by cyanobacterial growth. (**c**) A source of acid drainage from a mine into a stream. The soil and water have turned red due to the presence of precipitated iron oxides caused by the activity of bacteria such as *Thiobacillus*.

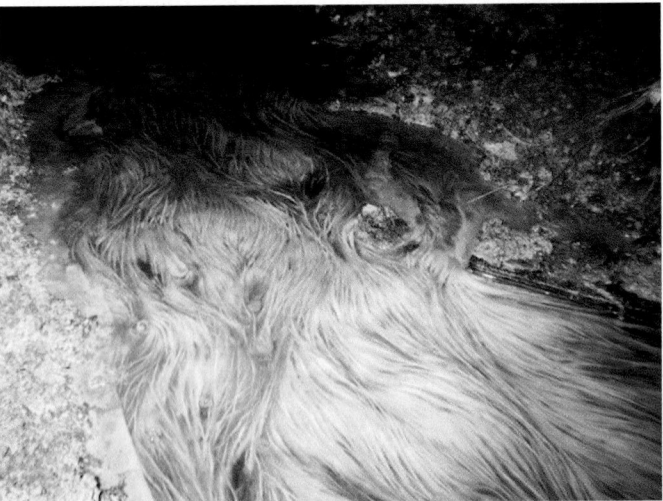

Figure 28.30 Massive Growth of the Extreme Acidophile *Ferroplasma* in a California Mine. Slime streamers of *Ferroplasma acidarmanus,* an archaean, which have developed within pyritic sediments at and near pH 0. This unique procaryote has a single plasma membrane and no cell wall.

Techniques & Applications
28.3   Thermophilic Microorganisms and Modern Biotechnology

There is great interest in the characteristics of procaryotes isolated from the outflow mixing regions above deep hydrothermal vents that release water at 250 to 350°C. This is because these procaryotes can grow at temperatures as high as 121°C. The problems in growing these microorganisms, often archaea, are formidable. For example, to grow some of them, it will be necessary to use special culturing chambers and other specialized equipment to maintain water in the liquid state at these high temperatures.

Such microorganisms, termed hyperthermophiles, with optimum growth temperatures of 80°C or above (*see pp. 124–25*), confront unique challenges in nutrient acquisition, metabolism, nucleic acid replication, and growth. Many of these are anaerobes that depend on

elemental sulfur as an oxidant and reduce it to sulfide. Enzyme stability is critical. Some DNA polymerases are inherently stable at 140°C, whereas many other enzymes are stabilized in vivo with unique thermoprotectants. When these enzymes are separated from their protectant, they lose their unique thermostability.

These enzymes may have important applications in methane production, metal leaching and recovery, and for use in immobilized enzyme systems. In addition, the possibility of selective stereochemical modification of compounds normally not in solution at lower temperatures may provide new routes for directed chemical syntheses. This is an exciting and expanding area of the modern biological sciences to which microbiologists can make significant contributions.

Observations of microbial growth at temperatures approaching 121°C in thermal vent areas, or of **hyperthermophiles (Techniques & Applications 28.3;** *see also Microbial Diversity & Ecology 6.1*), indicate that this area will continue to be a fertile field for investigation. For some successful microorganisms, an extreme environment may not be "extreme" but required and even, perhaps, ideal.   Thermophilic microorganisms (pp. 124, 449–51)

---

1. What stress factors can create extreme environments?
2. Define extremophile and discuss an example of adaptation to stress resulting from water availability, pressure, or pH.
3. Why are molecular techniques possibly changing our view of these environments?
4. What is unique about *Ferroplasma acidarmanus*?

28.5   MICROBIAL ECOLOGY AND ITS METHODS: AN OVERVIEW

As with so many areas of science, an almost bewildering assortment of methods is available to the microbial ecologist. We will consider primarily methods for examining microbial communities as complex assemblages.

Examination of Microbial Communities as Complex Assemblages

The study of microbial communities as complex assemblages has been a fundamental focus of microbial ecology in which microscopic, cultural, physical, chemical, and particularly molecular techniques play important roles. In many studies, bulk nucleic acids are recovered by direct extraction techniques. However, it has been found that the DNA obtained can vary depending on the

method employed. This makes it difficult to state with certainty that the microbial community has specific characteristics based on the use of a single DNA extraction procedure. A second problem, especially with muds and soils, is that one has no knowledge of the source of the bulk-extracted DNA. The DNA recovered by this approach may not even be derived from living organisms. Third, the DNA may have been recovered from nonfunctional propagules such as fungal or bacterial spores, or other resting structures. Thus the genera identified by such cloned DNA libraries may have minimal relevance to the microbial community actually functioning in the particular environment. Major advances are now being made in culturing previously uncultured microorganisms.

It is important to note that all methods have inherent limitations; a critical challenge is to recognize these limitations and to understand what information a particular method will (and will not) provide. Generally, it is best to use more than one method to obtain complementary information on different aspects of a microbial community. Second, some methods are better suited to the study of one type of environment than another. For example, because waters have fewer interfering inert particles than muds or soils, it is easier to view microbes or to extract cellular constituents for use in molecular studies. Finally, one should use approaches suitable for the study of all microbial components of a habitat (bacteria, archaea, and eucaryotes). Otherwise, one may miss critical relationships that arise because of interactions between different groups.

The methods used to characterize complex microbial communities can be directed toward several community features: numbers and types of microbes, microbial community structure and function, microbial viability and stress sensitivity, microbial activity at various time scales and levels of resolution, and mechanisms of communication between microbes. These aspects will now be briefly discussed.

Numbers and Types of Microbes

A variety of standardized growth media is used in direct viable count procedures, which are based on colony formation and enumeration. These methods are inherently biased, as not all microbes will be able to grow under a particular set of conditions. If it is necessary to isolate specific groups of microbes, or to attempt to search for organisms with new capabilities, enrichment techniques (*see pp. 102–5*) also can be used. These techniques are based on expansion of the microenvironment to allow massive growth of an organism formerly restricted to a small ecological niche. This approach still is valuable in studies of microbial ecology and plays a central role in finding new and undescribed microbes. It also can be used in most probable number approaches to estimate populations of specific physiological groups in an environment.

With enrichment techniques, however, microorganisms must be able to grow under the test conditions that are used. Too often microbes can be observed in environments, but enrichment approaches and other cultural techniques do not work—no growth can be observed. There are two alternative explanations: the observed organisms truly are nonviable, or the right conditions for their growth in the lab haven't been created. This has led to the description of such potentially viable microbes as being "nonculturable" (*see p. 129*). With effort, there have been startling successes in this area. It took over 70 years to figure out how to grow *Frankia*, the plant root-associated nitrogen-fixing actinomycete. The basic assumption should be that with more effort, it will be possible to grow these uncooperative microorganisms.

Another critical problem in growing and characterizing microorganisms, particularly protozoa, algae, and cyanobacteria, is that many of these "microorganisms" are actually microbial assemblages. Often cultures of these types of microbes are not axenic; they can have surface-associated commensal partners, and phagotrophs such as protozoa can "trap" other organisms. When such microorganisms are finally studied as axenic cultures, their morphological and physiological characteristics frequently change due to the lack of growth factors and vitamins formerly provided by the commensal organisms.

It is one thing to provide the nutrients and growth media that a recalcitrant microbe requires. Often, the physical environment also must be tailor made—that is, it may be necessary to mimic the actual physical, often protective, niche that a microbe uses in nature.

Microbial Community Structure and Constituents

To assess microbial community structure, much emphasis is given to the use of direct observation of complex microbial communities in nature. This can be carried out in situ using immersed slides or electron microscope grids, placed in a location of interest and then recovered later for observation. Samples taken from an environment also are examined in the laboratory using classical cellular stains, fluorescent stains, or fluorescent molecular probes (*see p. 622*).

By combining direct observational and molecular techniques, microorganisms and their physical relationships can be studied,

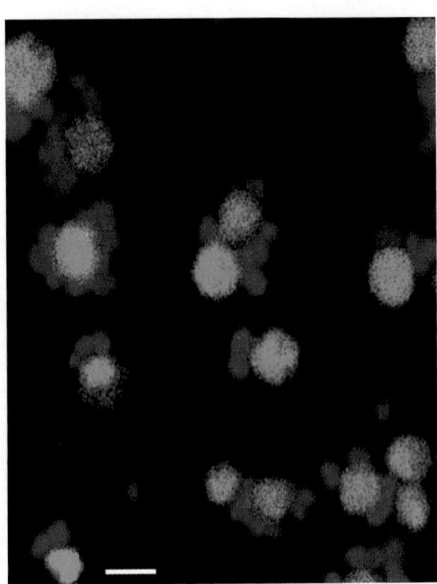

Figure 28.31 The Use of Differential Molecular Probes to Study Microbial Community Characteristics. In this study, specially designed molecular probes were used to study the physical relationships between two archeans: a larger archaeal host, a member of the genus *Ignicoccus* (*green*), and a nanosized (~ 400 nm) hyperthermophilic symbiont named *Nanoarchaeum equitans* (*red*). Bar = 1 μm.

as shown for the *Nanoarchaeum equitans–Ignicoccus* coculture (**figure 28.31**). Using specific molecular probes, a unique archaean, 400 nm in diameter, was found growing on a larger archaean, *Ignicoccus,* in a special relationship. *Nanoarchaeum* has the smallest archaean genome found to date: only 0.5 megabases. Thus through using direct observation and molecular probes it was possible to document the life-style of this unusual archaeal symbiont, apparently dependent on its larger host for survival.

Microbial communities also can be described in terms of their structure and the nutrients contained in the community. Aquatic microorganisms can be recovered directly using filtration; the volume, dry weight, or chemical content of the microorganisms can then be measured. If it is not possible to recover the cells, carbon, nitrogen, phosphorus, or an organic constituent of the cells (such as lipids or ergosterol) can be determined. This will give a single-value estimate of the microbial community. Such single-value chemical measurements either may be used directly or expressed as microbial biomass. Such information is needed to establish Redfield ratios (*see p. 619*) of the microbial biomass, especially in aquatic environments.

Single-value estimates of microbial constituents or biomass, although valuable, provide no information about the physical structure of microbes or of the microbial community. For example, in nature most filamentous fungi consist primarily of empty hyphae. A small amount of cytoplasm moves within the tubular network as the organism penetrates and exploits new substrates. Such microbes, without distinct edges and boundaries, do not have predictable volume-biomass ratios, and are described as indeterminate or **nondiscrete microorganisms.** With such microorganisms, a single-value biomass measurement is of limited value, as the organisms only can be described by direct observation of their physical structure.

Molecular "fingerprinting" techniques are frequently used to characterize microbial communities. Bulk DNA, extracted from an environment, can be analyzed directly or after the use of PCR amplification (*see pp. 316–19*). This provides comparative information; it also is possible to use the isolated DNA in sequencing and probe-type investigations. Again, these approaches are limited by a lack of knowledge of the DNA source, and they yield no information on the physical structure of the microbial community.

Stress and Viability of Microorganisms

Microorganisms are constantly being stressed by physical factors and nutrient limitations. Thus determining whether they are viable or not is an important part of microbial ecology.

Stress can be caused by the absence of needed nutrients, leading to internal starvation stress, or it may be associated with external factors, such as physical environment changes, toxic chemicals, or temperature changes. Microorganisms even may be affected by a combination of internal and external stresses. Sometimes slight temperature changes, or changes in media buffer characteristics, can affect microbes and the outcome of an experiment. Stress can be measured using cultural methods. For example, plates of a growth medium are prepared with varying levels of a stressing agent such as salt. Stressed organisms will not grow well in the presence of the stressing salt. Recovery from stress can be evaluated because the recovering organism gradually grows better in the presence of the stressing agent. Direct observation of individual cells also can be made using special stains (*see p. 26*) as illustrated in **figure 28.32.** Microviability measurements for the analysis of cell viability are briefly described on page 129.

The study of stress and viability are important in many areas of microbial ecology. Even in the absence of visible growth, an organism may show "life signs" when directly observed, or when the microbe is growing in its natural habitat. As an example, *Vibrio* can exist in a nominally viable but nonculturable (VNC) state; it may not grow in conventional laboratory media, but it still can grow in a susceptible host. Thus it may be difficult to monitor a pathogen's survival in waters and in foods such as shellfish, even though it will cause disease if it encounters an appropriate susceptible host.

Microbial Activity and Turnover

Microbial activity measurements can be made over various time intervals, ranging from the essentially instantaneous responses of samples containing active microbes to long-term geological process-related measurements. A few examples of activity measurements are described here.

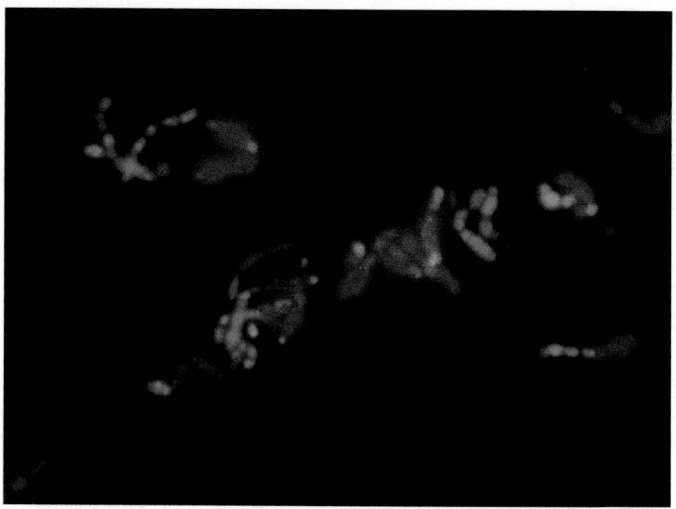

Figure 28.32   Assessment of Microbial Viability by Use of Direct Staining.   By using differential staining methods, it is possible to estimate the portion of cells in a given population that are viable. In the LIVE/DEAD *Bac*Light Bacterial Viability procedure, two stains are used: a membrane-permeable green fluorescent, nucleic acid stain, and propidium iodide (red) that penetrates only cells with damaged membranes. A *Bifidobacterium* culture is shown here. Uninjured cells stain green, while dead cells stain red.

Specific processes, such as nitrification, denitrification, and sulfate reduction are studied by the use of direct chemical measurements. Microarrays (*see pp. 345–55, 986–88*) are now used to measure gene expression and activity in complex microbial communities. Again, the results depend on the source and quality of the nucleic acids that are recovered from a particular microbial community and its physical environment. Stable isotope measurements can indicate whether carbon, nitrogen, or other elements have been processed by an organism (*see p. 668*). Microbes generally prefer the lighter of two stable isotopes, as ^{12}C over ^{13}C. When a microbe uses carbon dioxide or an organic substrate, the cells and metabolic products often have lower concentrations of the heavy isotope than does the original substrate.

Microbial growth rates in complex systems also can be measured directly. Colonization of surfaces can be observed using microscope slides or other materials. Changes in microbial numbers are followed over time, and the frequency of dividing cells (FDC) is also used to estimate production. This approach is especially valuable in studies of aquatic microorganisms. Finally, the incorporation of radiolabelled components such as thymidine (a DNA constituent) into microbial biomass provides information about growth rates and microbial turnover.

Microbial Interactions

Simply seeing microbes in certain physical relationships is only the beginning of the story. The measurement of communication molecules, such as autoinducers involved in quorum sensing (*see*

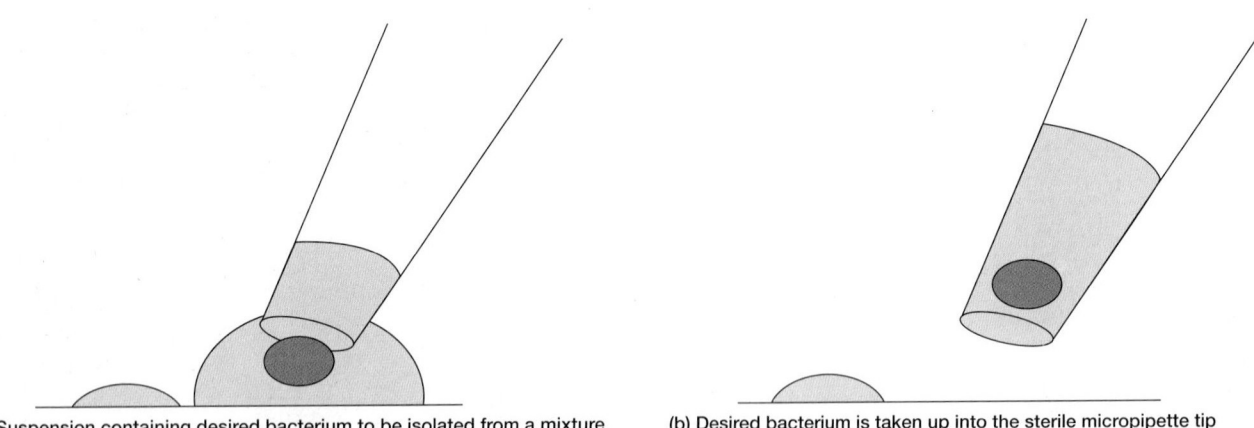

(a) Suspension containing desired bacterium to be isolated from a mixture   (b) Desired bacterium is taken up into the sterile micropipette tip

Figure 28.33   Recovery of Single Cells or Cell Organelles from Complex Natural Mixtures Using Micromanipulation.   By use of an inverted phase-contrast microscope and a micromanipulator, a microorganism can be recovered for direct molecular analysis. (**a**) The bacterium to be isolated is placed below the micromanipulator tip (diameter 5 to 10 μm) and a slight vacuum is drawn. (**b**) The desired single bacterium is drawn up into the micropipette tube and is ready for molecular analysis.

pp. 129–31), as well as metabolic intermediates and toxic microbial products involved in various symbiotic interactions, provides key information for understanding how and why microbial communities respond in specific ways. The field has developed rapidly, particularly in the study of positive and negative microbial interactions with plants and animals. This challenging area of microbial ecology is discussed in chapters 29 and 30.

Recovery or Addition of Individual Microbes

The direct observation of microorganisms in their environments is central to the methodology used in any study of microbial ecology. In recent years, valuable new experimental approaches have been developed to recover and study individual microorganisms from an environment.

Such single-cell isolations can be carried out using **optical tweezers** (a laser beam used to drag a microbe away from its neighbors) and by micromanipulation. With a **micromanipulator,** a desired cell or cellular organelle is drawn up into a micropipette after direct observation (**figure 28.33**). Once the microbe is isolated, PCR amplification of the DNA from the individual cell or cell organelle makes available sequence data for use in phylogenetic analysis (*see chapter 19*). For example, it has been possible to establish the phylogenetic relationship of a mycoplasma recovered from the flagellate *Koruga bonita* by micromanipulation (**figure 28.34**).

Consideration of microbial ecology on the scale of the individual cell has led to important ecological insights. It is now evident that there is surprising heterogeneity in what have been assumed to be homogenous microbial populations. Cells of a genetically uniform population do not have similar phenotypic attributes, the phenomenon of **phenotypic** or **population heterogeneity.** This realization is increasingly important in un-

derstanding responses of microorganisms in complex environments, and particularly in disease processes.

Microbial ecologists now use "reporter" microbes to characterize the physical microenvironment on the scale of an individual bacterium (around 1 to 3 μm). This is done by constructing cells with **reporter genes,** often based on green fluorescent protein (gfp) that change their fluorescence in response to environmental and physiological alterations. Such "reporter" microbes are being used to measure oxygen availability, UV radiation dose, pollutant or toxic chemical effects, and stress. For example, when microbes that contain a moisture stress reporter gene have less available water, there is an increase in gfp-based fluorescence.

In summary, the direct observation of microorganisms in their natural environments, combined with carefully selected classical cultural, chemical, and molecular techniques, is leading to new views of how microorganisms interact with each other, with other organisms such as plants and animals, and with their abiotic environment. Important new advances continue to make microbial ecology one of the most exciting areas of modern science.

---

1. Why are "classic" microscopic and physical methods still used for the study of microorganisms when molecular techniques are available?
2. What are optical tweezers and micromanipulators?
3. What time scales can be used when studying the activity of microorganisms?
4. What important advances have been made in microbial ecology, based on the recovery of individual microbes from complex environmental samples, or by addition of microbes that contain reporter genes?

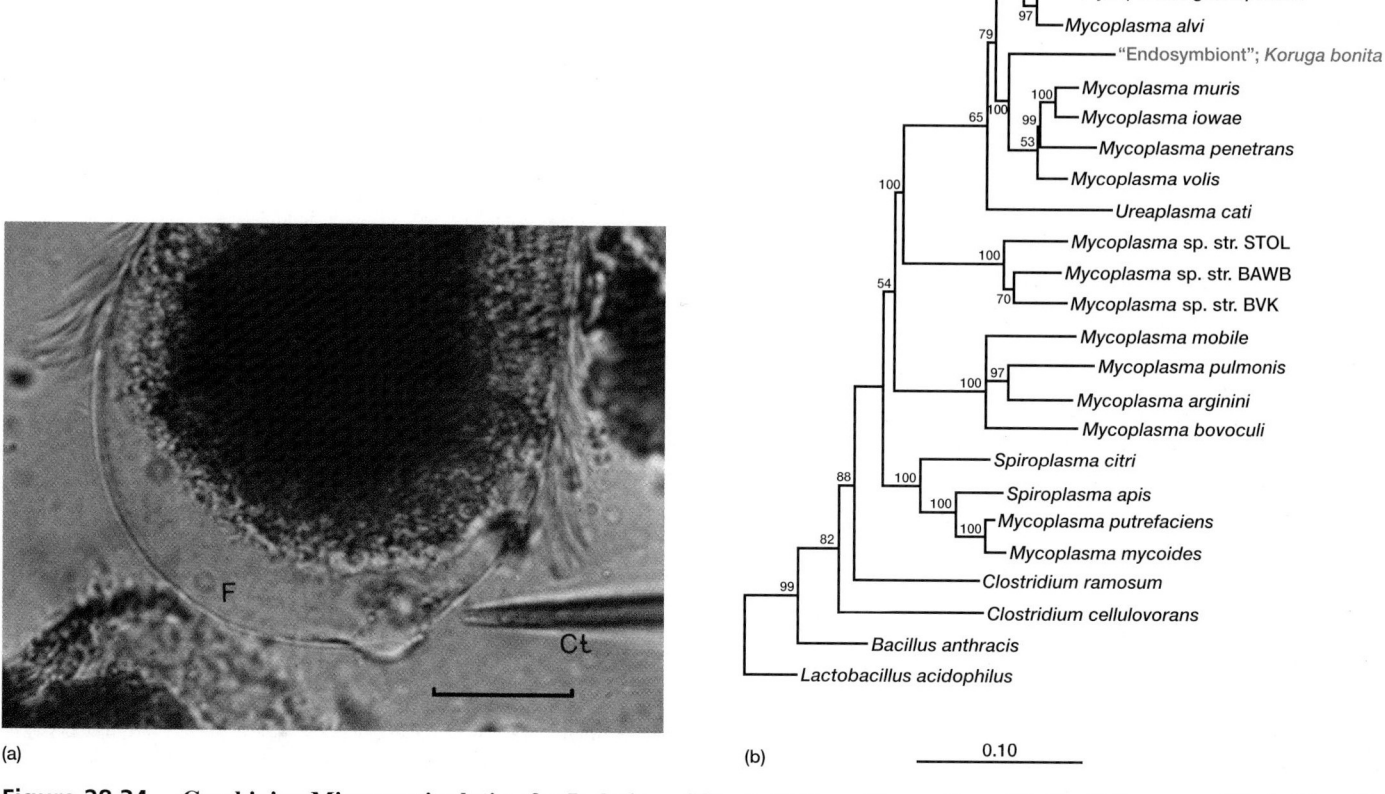

Figure 28.34 Combining Micromanipulation for Isolation of Single Cells or Organelles with the Polymerase Chain Reaction (PCR). (**a**) Recovery of an endosymbiotic mycoplasma from single cell of the flagellate *Koruga bonita* by micromanipulation (bar = 10 μm) and (**b**) phylogenetic analysis of the recovered mycoplasma following PCR amplification and sequencing of the PCR products, with the bar indicating 10% estimated sequence divergence. *Lactobacillus acidophilus* is the outgroup reference. This approach makes it possible to link a specific microorganism or organelle, isolated from a natural environment, to its molecular sequence and phylogenetic information. Flagellate (F), capillary tube (Ct).

Summary

28.1 Foundations of Microbial Ecology

a. Most microorganisms that can be observed in complex natural assemblages under a microscope cannot be grown at the present time. Molecular techniques are making it possible to obtain information on these uncultured microorganisms.

b. Microbial ecology is the study of microorganisms' interactions with their living and nonliving environments. Symbiosis is a narrower term that means "together life," the study of organism-organism interactions.

28.2 Microbial Interactions

a. Symbiotic interactions (**figure 28.1**) include mutualism (mutually beneficial and obligatory), cooperation (mutually beneficial, not obligatory), and commensalistic (product of one organism can be used beneficially by another organism). Predation

involves one organism (the predator) ingesting/killing a larger or smaller prey, parasitism (a longer-term internal maintenance of another organism or acellular infectious agent), and amensalism (a microbial product can inhibit another organism). Competition involves organisms competing for space or a limiting nutrient. This can lead to dominance of one organism, or coexistence of both at lower populations.

b. A consortium is a physical association of organisms that have a mutually beneficial relationship based on positive interactions.

c. Mutual advantage is central to many organism-organism interactions. These interactions can be based on material transfers related to energetics, cell-to-cell communication, or physical protection. With several important mutualistic interactions, chemolithotrophic microorgan-

isms play a critical role in making organic matter available for use by an associated organism (e.g., endosymbionts in *Riftia*).

d. The rumen is an excellent example of a mutualistic interaction between a ruminant and its complex microbial community. In this microbial community, complex plant materials are broken down to simple organic compounds that can be absorbed by the animal, as well as forming waste gases such as methane that are released to the environment (**figure 28.6**).

e. Syntrophism simply means growth together. It does not require physical contact but only a mutually positive transfer of materials, such as interspecies hydrogen transfer.

f. Cooperative interactions are beneficial for both organisms but are not obligatory (**figure 28.7**). Important examples are marine animals, including

Alvinella, Rimicarus, and *Eubostrichus,* that involve interactions with hydrogen sulfide-oxidizing chemotrophs.

g. Predation and parasitism are closely related. Predation has many beneficial effects on populations of predators and prey. These include the microbial loop (returning minerals immobilized in organic matter to mineral forms for reuse by chemotrophic and photosynthetic primary producers), protection of prey from heat and damaging chemicals, and possibly aiding pathogenicity, as with *Legionella.*

h. These varied interactions occur in complex biological systems and lead to feedback responses by members of a biological community.

i. Microbes can be present as individual cells, as populations of similar organisms, or as mixtures of populations or communities. These populations and communities can be parts of self-regulating ecosystems.

28.3 Nutrient Cycling Interactions

a. Microorganisms—functioning with plants, animals, and the environment—play important roles in nutrient cycling, which is also termed biogeochemical cycling. Assimilatory processes involve incorporation of nutrients into the organism's biomass during metabolism; dissimilatory processes, in comparison, involve the release of nutrients to the environment after metabolism.

b. Biogeochemical cycling involves oxidation and reduction processes, and changes in the concentrations of gaseous cycle components, such as carbon, nitrogen, and sulfur can result from microbial activity.

c. Major organic compounds used by microorganisms differ in structure, linkage, elemental composition, and susceptibility to degradation under aerobic and anaerobic conditions. Lignin is degraded only under aerobic conditions, a fact that has important implications in terms of carbon retention in the biosphere.

d. In terms of effects on humans, metals can be considered in three broad groups: (1) the noble metals, which have antimicrobial properties but which do not have negative effects on humans; (2) metals such as mercury and lead, from which toxic organometallic compounds can be formed; and (3) certain other metals, which are antimicrobial in ionic form, such as copper and zinc. The second of these groups is of particular concern.

28.4 The Physical Environment

a. A microorganism functions in a physical location that can be described as its microenvironment. The resources available in a microenvironment and their time of use by a microorganism describe the niche. Pores are important microenvironments that can protect bacteria from predation.

b. Biofilms, or layers of microorganisms, are widespread and are formed on a wide variety of living and nonliving surfaces (**figure 28.26**). These are important in disease occurrence and the survival of pathogens. Biofilms can develop to form complex layered ecosystems.

c. Microorganisms serve as primary producers that accumulate organic matter (**figure 28.28**). Energy sources include hydrogen, sulfide, and methane. In addition, many chemoheterotrophs decompose the organic matter that primary producers accumulate and carry out mineralization, the release of inorganic nutrients from organic matter.

d. Most disease-causing bacteria from the intestinal tract of humans and other higher organisms do not survive in the environment. At lower temperatures, however, increased survival can occur.

e. Decreased species diversity usually occurs in extreme environments, and many microorganisms that can function in such habitats, called extremophiles, have specialized growth requirements. For them, extreme environments can be required environments.

28.5 Microbial Ecology and Its Methods: An Overview

a. Many approaches can be used to study microorganisms in the environment. These include nutrient cycling, biomass, numbers, activity, and community structure analyses. It is now possible to study the genetic characteristics of microorganisms that cannot be grown in the laboratory.

b. Methods presently being used make it possible to study presence, types, and activities of microorganisms in their natural environments (including soils, waters, plants, and animals). Although many microorganisms that can be observed still cannot be grown, molecular techniques are making it possible to obtain information on these noncultured microorganisms.

c. Optical tweezers and micromanipulators can be used to recover individual cells or cell organelles from complex microbial communities. This makes it possible to obtain genomic and phylogenetic information from specific individual microbial cells for use in studies of microbial ecology (**figure 28.34**).

Key Terms

Questions for Thought and Review

1. It is evident that some microorganisms have a wide range of electron donors and acceptors that they can use, and others use only single oxidants and reductants. What are the advantages and disadvantages of each strategy?

2. Why might microorganisms prefer to grow in association with other microorganisms, as in biofilms, when they can have better access to nutrients as single cells?

3. Why is mutualism in association with multicellular organisms a major strategy of microorganisms? Why have these especially developed with sulfide and simple hydrocarbon-based systems?

4. What types of unique information do microscopic and molecular techniques provide and not provide?

5. How might you show that a microorganism found in a particular extreme environment is actually growing there?

6. Why is predation such a very important part of microbial ecology? Can a predator ever completely eliminate all of its prey?

7. How might you attempt to grow a microorganism in the laboratory to increase its chances of being a strong competitor when placed back in a natural habitat?

8. Considering the possibility of microorganisms functioning at temperatures approaching 120°C, what do you think the limiting factor for microbial growth at higher temperatures will be and why?

9. Where in their bodies do most people have noble metals? Why have these been used with more success than materials such as ceramics, which have been tested over many decades?

10. Compare the degradation of lignin and cellulose by microorganisms. What different environmental factors are required for these important polymers to be degraded?

11. Considering the intensive searches for unique microorganisms that have been carried out all over the world, where can we look for new microbes?

Critical Thinking Questions

1. Compare and contrast diversity among microorganisms with diversity among macroorganisms.

2. Describe a naturally occurring niche on this planet that you believe is inhospitable to microbial life. Explain, in light of what is known about extremophiles, why you believe this environment will not support microbial life.

For recommended readings on these and related topics, see Appendix V.

chapter 29

Microorganisms in Aquatic Environments

Concepts

- Oxygen diffuses through water at a slow rate, in comparison with the rate at which it diffuses through air; this is a defining characteristic of water as an environment for microbial growth and functioning.

- Oxygen, once it is dissolved in water, can be used by microorganisms at a faster rate than it can be replenished. This leads to the creation of anaerobic zones. If light penetrates into these anaerobic zones, unique groups of photosynthetic microorganisms can develop. Specialized microbial communities also develop at the interface between anoxic and oxic regions.

- Aquatic environments, both marine and freshwater, include vast volumes of cold water and ice. In addition to fresh water stored in glaciers and the polar regions, sea ice can cover approximately 7% of the Earth's surface. Together with the cold (2 to 3°C), high pressure waters of the Earth's oceans, these cold environments are important sites of microbial survival and functioning.

- Aquatic environments allow development of many unique types of microorganisms. These include microbes at deep hydrothermal vents and hydrocarbon seeps, and other microorganisms that exploit surfaces by gliding motility and attachment. Other microorganisms link spatially separated resources such as nitrate and sulfide. New microorganisms continue to be discovered in aquatic environments.

- The biogeochemical cycles of carbon, nitrogen, sulfur, and phosphorus for lakes and marine environments involve interactions over both local and vast distances.

New procaryotes are discovered at locations where reduced and oxidized nutrients mix. This giant bacterium, *Thiomargarita namibiensis*, about 100 to 300 μm in diameter, accumulates sulfur from sediments in its refractive sulfur granules and nitrate from overlying waters to support its growth. *Thiomargarita*, which resembles a string of pearls, is found off the coast of Namibia in West Africa.

- Phytoplankton (photosynthetic procaryotic and eucaryotic organisms) form the basis of primary production in most marine and freshwater environments. The phytoplankton release dissolved and particulate organic matter, which is used by heterotrophic bacteria, a portion of which are consumed by predators, releasing materials that are recycled for use by the phytoplankton, creating a microbial loop. Iron and nitrogen can limit these activities in different marine environments.

- Disease-causing microorganisms, as well as organic materials, are constantly being added to waters. These can be transported over large distances, especially in rapidly moving rivers, lakes, and marine environments. Atmospheric-borne dusts and other materials, including pollutants, can be carried to the farthest regions of the oceans, freshwater bodies, and ice-covered regions of the Earth.

- Many important human pathogens, such as *Shigella*, *Vibrio*, and *Legionella*, are found in waters. These can occur normally or can survive for various intervals after addition to waters. Often protozoa provide protection for such pathogens, especially when the protozoa are associated with biofilms.

- Aquatic environments can serve as reservoirs and transmission routes for disease-causing microorganisms. A major goal of aquatic system management is to control pathogen survival and transfer.

- Water is an important reservoir for the survival and dissemination of protozoa and viruses. These cannot be reliably controlled by chlorination. Protozoan pathogens include *Cryptosporidium*, *Cyclospora*, and *Giardia*.

- The biological use of organic wastes follows regular and predictable sequences. Once these sequences are understood, more efficient sewage treatment systems can be created.

- Sewage treatment can be carried out using large vessels where mixing and aeration can be controlled (conventional treatment). Constructed wetlands, where aquatic plants and their associated microorganisms are used, now are finding widespread applications in the treatment of liquid wastes.

- Indicator organisms, which usually die off at slower rates than many disease-causing microorganisms, can be used to evaluate the microbiological quality of water.

- Groundwater is an important source of drinking water, especially in suburban and rural areas. In too many cases, this resource is being contaminated by disease-causing microorganisms and nutrients, especially from septic tanks.

Water is a very good servant, but it is a cruel master.
—*John Bullein*

 f all the water found on Earth, 97% is marine. Most of this water is at a temperature of 2 to 3°C and devoid of light; 62% is under high pressure (>100 atm). Microscopic phytoplankton and associated bacteria create a complex food web that can extend over long distances and extreme depths. The marine environment seems so vast that it will not be able to be affected by pollution; however, in coastal areas human activities are increasingly disrupting microbial processes and damaging water quality. People around the world are migrating to coastal areas. This trend will continue in the coming decades, and as much as 70% of the U.S. population will be living near the Great Lakes or the marine coasts.

Fresh waters, although a small part of the waters on Earth, are extremely important as a source of drinking water. In many locations the contamination of surface and subsurface waters by domestic and industrial wastes causes environmental problems.

Marine and freshwater environments create unique niches for many specialized microorganisms needing habitats with a continuous water phase.

29.1 AQUATIC ENVIRONMENTS AND MICROORGANISMS

Aquatic environments have varied surface areas and volumes. They are found in locations as diverse as the human body; drinks and beverages; and the usual places one would expect—rivers, lakes, and the oceans. They also occur in water-saturated zones in materials we usually describe as soils! These environments can range from alkaline to extremely acidic. The temperatures within which microorganisms function in aquatic environments can range from −5 to −15°C at the lower range, to at least 121°C in geothermal areas. Some of the most intriguing microbes have come from the study of high-temperature environments, including the now-classic studies of Thomas D. Brock and his coworkers at Yellowstone National Park which led to the discovery of *Thermus aquaticus*, the source of *Taq* polymerase (*see p. 318*). Hyperthermophilic microorganisms, including *Pyrolobus fumarii*, also have been isolated from hydrothermal vents in deep marine environments.

The mixing and movement of nutrients, O_2, and waste products that occur in freshwater and marine environments are the dominant factors controlling the microbial community. For example, in deep lakes or oceans, organic matter from the surface can sink to great depths, creating nutrient-rich zones where decomposition takes place. Gases and soluble wastes produced by microorganisms in these deep marine zones can move into overlying waters and stimulate the activity of other microbial groups.

A goal of microbiologists is to isolate and culture unique marine microorganisms, especially in the search to find new antibiotic-producing microorganisms (**Disease 29.1**). Techniques for collecting microorganisms without changing temperature or pressure are now available. For example, scientists have constructed equipment to culture microorganisms that grow at 1,000 atm without decompressing them.

Gases and Aquatic Microorganisms

In aquatic environments the distance (on a microbial scale) from an air bubble or the water surface limits oxygen diffusion. Thus aquatic environments are termed **low oxygen diffusion environments.** The flux rate for oxygen through water is approximately

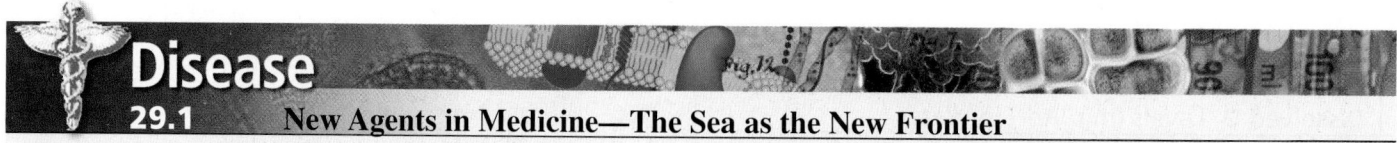

Disease

29.1   New Agents in Medicine—The Sea as the New Frontier

Most currently available antibiotics have been derived from soil microorganisms, primarily from the actinomycetes, but also from a lesser number of nonfilamentous gram-positive bacteria and fungi (*see section 35.2*). Over 100 products are in use as antibiotics, antitumor agents, and agrochemicals.

In recent years, with the need for additional compounds for use in medicine, marine microorganisms are receiving increased attention. Some of the newer chemicals that have been discovered are microalgal metabolites. There also is an interest in the culture of symbiotic marine microorganisms, including *Prochloron,* which

are associated with macroscopic hosts. A variety of interesting compounds of unknown origin have been discovered. Many are assumed to be of microbial origin, but more work will be needed to establish this. Many biologists feel that marine microorganisms may provide unique bioactive compounds, including marine toxins, which do not occur in terrestrial microorganisms. There is a worldwide effort to better characterize the marine microbial community and to harness these often poorly studied microorganisms for modern medicine. Often the first challenge is to culture these microorganisms.

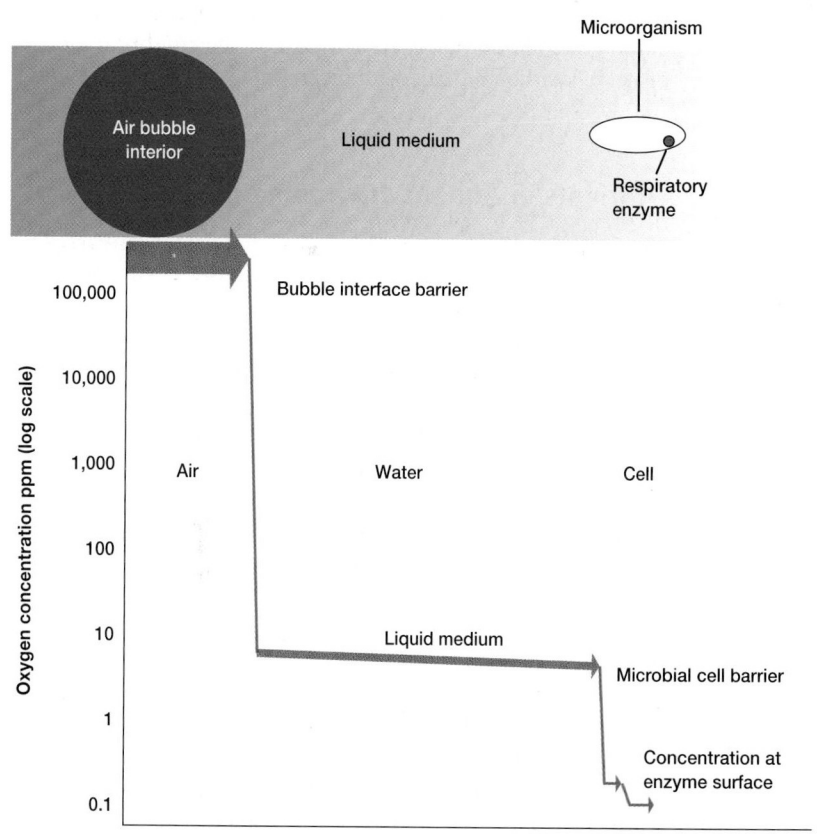

Flux rates and barriers to oxygen transfer

Figure 29.1 Oxygen Transfer in Water Is Limited. Oxygen diffuses at a high rate through air and at approximately 1/10,000 this rate through water. The wide arrow size represents the rapid oxygen flux rate through the air; the decrease in the intensity of the blue color indicates the oxygen concentration. After rapid diffusion through air, oxygen crosses the air/water barrier and then diffuses at a lower rate (the narrow arrow) and lower concentration through the water to the point at which it is used by a microorganism. The oxygen concentration (ppm) is expressed on a logarithmic scale, and the oxygen present at an enzyme can be less than one-millionth of that occurring in the air.

1/10,000 of that which would occur if the microorganisms were in direct association with a gas phase (**figure 29.1**).

Oxygen not only diffuses slowly through water, its solubility is further decreased at higher temperatures and with lower pressure (**table 29.1**). Because of limited solubility and the low oxygen diffusion in water, oxygen can be used by aerobic microorganisms faster than it can be replenished. This frequently leads to the formation of **hypoxic** or **anoxic** zones in aquatic environments. These

zones allow specialized anaerobic microbes, both chemotrophic and phototrophic, to grow in the lower regions of lakes where light can penetrate. In contrast, if the microorganisms are functioning in an extremely thin water film and oxygen-containing air is close to them, they are in **high oxygen diffusion environments.** Examples include soils, which will be discussed in chapter 30.

Low oxygen diffusion environments, which will be discussed in this chapter, can range from an ice cube in a cocktail glass to

Table 29.1				
The Effects of Water Temperature and Elevation on Dissolved O_2 Levels (mg/liter) in Water				
	Elevation above Sea Level (Meters)			
Temperature (°C)	0	1,000	2,000	3,000
0	14.6	12.9	11.4	10.2
5	12.8	11.2	9.9	8.9
10	11.3	9.9	8.8	7.9
15	10.0	8.9	7.9	7.1
25	9.1	8.1	7.1	6.4
30	8.2	7.3	6.4	5.8
35	7.5	6.6	5.9	5.3
40	6.9	6.1	5.4	4.9

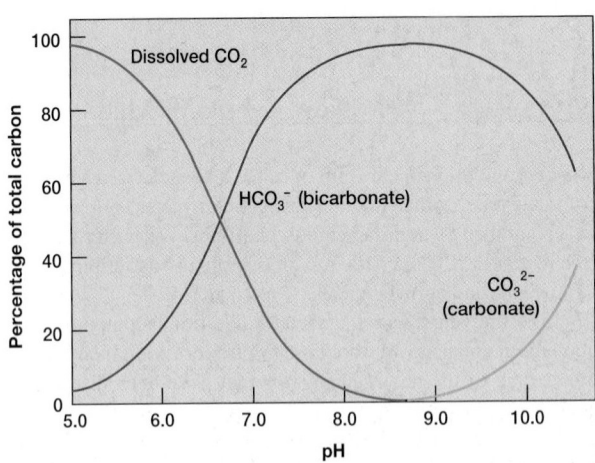

Figure 29.2 The Relationship of pH to Dissolved CO_2. Atmospheric gases can affect the physical characteristics of water. The pH of water is influenced by the amount of dissolved carbon dioxide in the water, and also by the equilibrium of the dissolved carbon dioxide with bicarbonate and carbonate ions.

Arctic and Antarctic ice sheets and the vastness of marine and freshwater systems.

The second major gas in water, CO_2, plays many important roles in chemical and biological processes. The carbon dioxide-bicarbonate-carbonate equilibrium can control the pH in weakly buffered waters, or it can be controlled by the pH of strongly buffered waters. As shown in **figure 29.2,** the pH of distilled water, which is not buffered, is determined by the dissolved CO_2 in equilibrium with the air, and is approximately 5.0 to 5.5. In comparison, water strongly buffered at pH 8.0 contains CO_2 absorbed from the air, which is present primarily as bicarbonate (HCO_3^-). When autotrophic microorganisms such as algae use CO_2, the pH of the water will often be increased.

Other gases also are important in aquatic environments. These include nitrogen gas, used as a nitrogen source by nitrogen fixers; hydrogen, which is both a waste product and a vital substrate; and methane (CH_4). These gases vary in their water solubility, and methane is the least soluble of the three. Methane thus is an example of an ideal microbial waste product: once it is produced under anaerobic conditions, it leaves the microorganism's environment by diffusing up in the water column and being released to the atmosphere. This eliminates the problem of toxic waste accumulation that occurs with many microbial metabolic products, such as organic acids and ammonium ion.

1. What factors exert the most control over aquatic microbial communities?
2. Why is it critical to consider not only concentration but also flux rates of gases in water?
3. How are anoxic zones created in aquatic environments?
4. How much oxygen, in milligrams per liter, is dissolved in water at normal room temperature and pressure? What are the effects of higher temperature and decreased pressure on oxygen solubility?
5. Why is carbon dioxide (CO_2) such a critical gas in aquatic environments? Consider both chemical equilibria and microbial processes.

Nutrients in Aquatic Environments

Nutrient concentrations in aquatic environments can vary from extremely low, in the range of micrograms of organic matter per liter, to levels approaching those in laboratory culture media, creating gradients that microorganisms exploit. High nutrient levels are found in polluted environments and sewage treatment plants, for example. Nutrient turnover rates also vary. In marine environments the turnover time for nutrient processing may range from hundreds to thousands of years. In contrast, marsh and estuarine areas may have rapid rates of nutrient turnover, and a complex, diverse microbial community of rapidly responding microorganisms is present.

As noted in chapter 28, chemotrophically based marine and freshwater ecosystems are important recent discoveries, whether these occur in deep "black smoker" areas (*see pp. 124, 581–82*), in subsurface cave systems, or in shallower methane seep areas.

The **Winogradsky column,** usually constructed using a glass graduated cylinder, illustrates many interactions and gradients that occur in aquatic environments (**figure 29.3**). In this glass cylinder a layer of reduced mud is mixed with sodium sulfate, sodium carbonate, and shredded newspaper—a cellulose source—and additional mud and water are placed in the column, which is then incubated in the light.

In the bottom of the column, cellulose is degraded to fermentation products by the genus *Clostridium* (*see section 23.3*). With these fermentation products available as electron donors and using sulfate as an acceptor, *Desulfovibrio* produces hydrogen sulfide (*see section 22.4*). The hydrogen sulfide diffuses upward toward the oxygenated zone, creating a stable hydrogen sulfide gradient. In this gradient the photoautotrophs *Chlorobium* and *Chromatium* develop as visible olive green and purple zones. These microorganisms use hydrogen sulfide as an electron

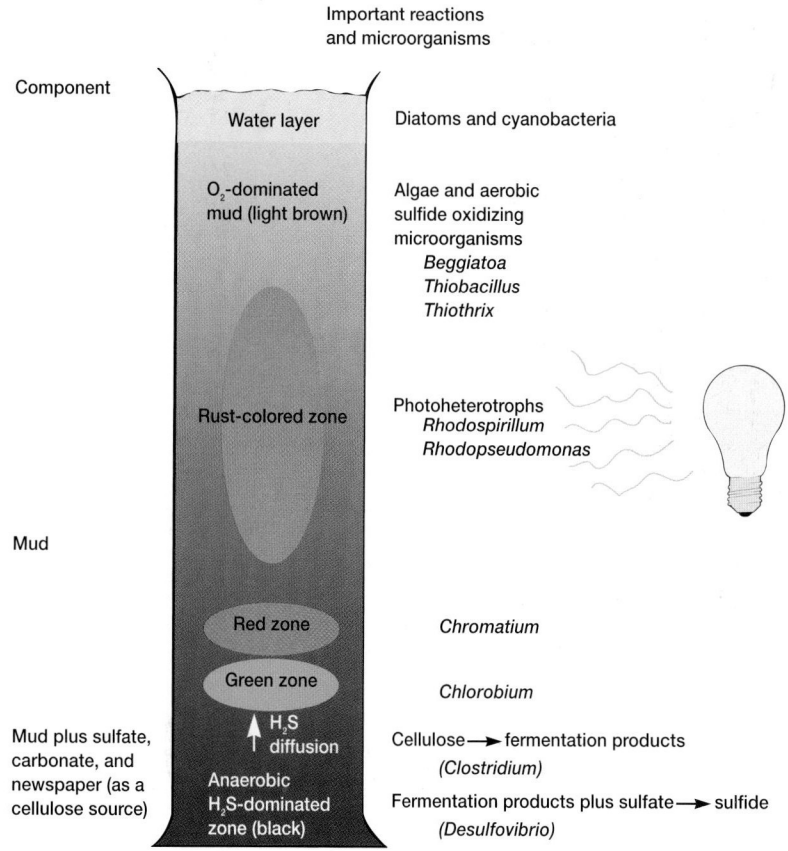

Figure 29.3   The Winogradsky Column.   A microcosm in which microorganisms and nutrients interact over a vertical gradient. Fermentation products and sulfide migrate up from the reduced lower zone, and oxygen penetrates from the surface. This creates conditions similar to those in a lake with nutrient-rich sediments. Light is provided to stimulate the penetration of sunlight into the anaerobic lower region, which allows photosynthetic microorganisms to develop.

source, and CO_2, from sodium carbonate, as a carbon source. Above this region the purple nonsulfur bacteria of the genera *Rhodospirillum* and *Rhodopseudomonas* can grow. These photoheterotrophs use organic matter as an electron donor under anaerobic conditions and function in a zone where the sulfide level is lower. Both O_2 and hydrogen sulfide may be present higher in the column, allowing specially adapted microorganisms to function. These include *Beggiatoa* and *Thiothrix,* which use reduced sulfur compounds as electron donors and O_2 as an acceptor. In the upper portion of the column, diatoms and cyanobacteria may be visible.   Green and purple photosynthetic bacteria (pp. 456–58, 475, 485–86); Cyanobacteria (pp. 458–64)

---

1. What are the sources of energy that support microbially based ecosystems in freshwater and marine environments?
2. What are the reasons for adding cellulose, sodium sulfate, and sodium carbonate to the Winogradsky column? Discuss this in relation to microbial groups that respond to these materials or their products.
3. What major microbial genera occur in the bottom of the Winogradsky column?

Nutrient Cycles in Aquatic Environments

The major source of organic matter in illuminated surface waters is photosynthetic activity, primarily from **phytoplankton** [Greek *phyto,* plant and *planktos,* wandering]. A common planktonic genus is *Synechococcus,* which can reach densities of 10^4 to 10^5 cells per milliliter at the ocean surface. Picocyanobacteria (very small cyanobacteria) may represent 20 to 80% of the total phytoplankton biomass upon which grazers depend.

As they grow and fix carbon dioxide to form organic matter, the phytoplankton acquire needed nitrogen and phosphorus from the surrounding water. The nutrient composition of the water affects the final carbon-nitrogen-phosphorus (C:N:P) ratio of the phytoplankton, which is termed the **Redfield ratio,** named for the aquatic biologist A. C. Redfield. A commonly used value for this ratio is 106 parts C, 16 parts N, and 1 part P. This ratio is important for following nutrient dynamics, especially mineralization and immobilization processes, and for studying the sensitivity of oceanic photosynthesis to atmospheric additions of nitrogen, sulfur, and iron.

Once the phytoplankton have grown, much of the organic matter fixed by these minute photosynthetic organisms then enters the

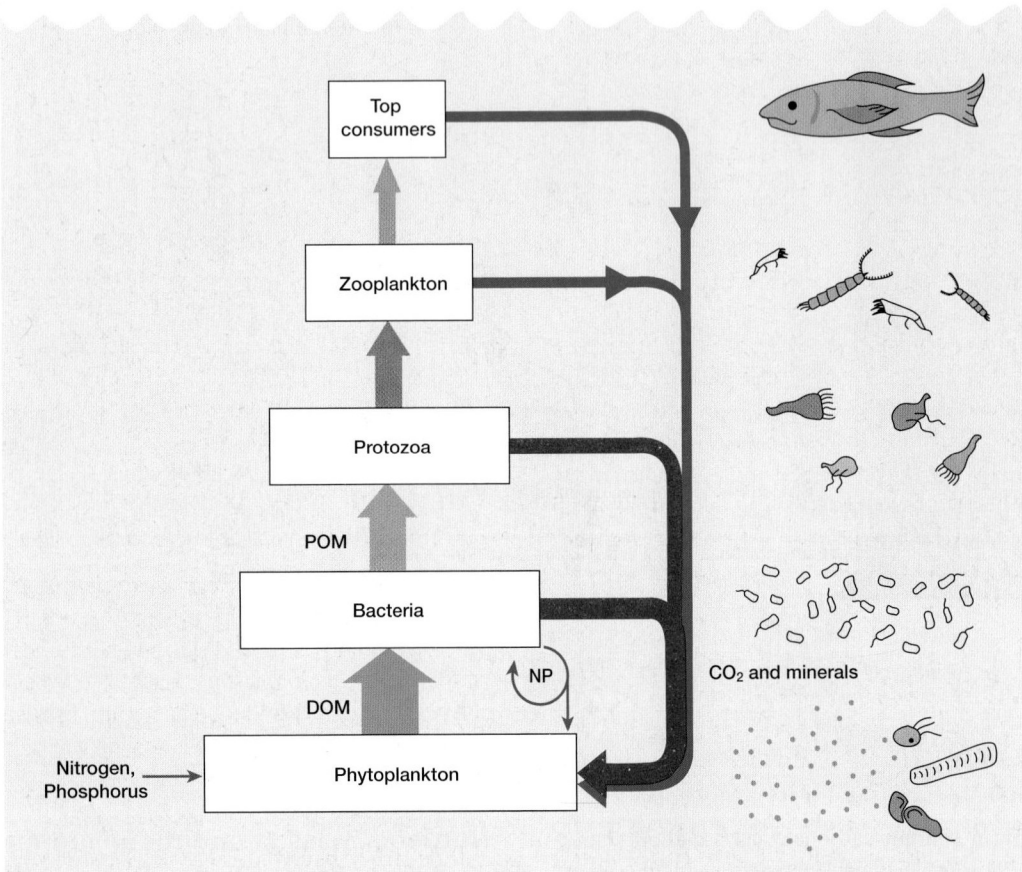

Figure 29.4 The Microbial Loop. A simplified view of the microbial loop, as it occurs in aquatic environments, is noted in red. In this loop a large portion of the organic matter synthesized by the phytoplankton during photosynthesis is released as dissolved organic matter (DOM). This DOM is used by the bacteria, which become part of the particulate organic matter (POM) pool. A portion of these bacteria are then used as a food source by protozoa. After digestion by the protozoa, a portion of the nutrients contained in the bacteria and protozoa is mineralized and "looped back" to the phytoplankton, making it unavailable to the higher trophic levels of the ecosystem. Arrow sizes representing the relative materials flows and organisms types (not to scale) are shown. Nitrogen (N), phosphorus (P).

microbial loop (figure 29.4). In the microbial loop, organic matter is recycled to carbon dioxide and minerals. Dissolved organic matter (DOM) released by the phytoplankton is used by the heterotrophic bacteria. The DOM is transformed to bacteria, which become part of the particulate organic matter (POM) pool.

These bacteria are then consumed and digested by a series of increasingly larger predators, including protozoa and metazoan zooplankters, releasing the carbon as CO_2 and the other nutrients in mineral forms to be cycled through the phytoplankton again.

This results in the rapid cycling of nutrients at a stage between the primary producers and the higher consumers such as fish, leading to a decrease in ecosystem productivity.

1. What does the term phytoplankton mean?
2. Describe the Redfield ratio and its use.
3. What is the microbial loop? What role do protozoans play in this "loop"?

29.2 THE MICROBIAL COMMUNITY

Water provides an environment for a wide variety of microorganisms to survive and function. Microbial diversity depends on available nutrients, their varied concentrations (ranging from extremely low to very high levels), the transitions from aerobic to anaerobic zones, and the mixing of oxidants and reductants in this dynamic environment. In addition, the penetration of light into many anaerobic zones creates environments for certain types of photosynthetic microorganisms. Although important microorganisms found in aquatic environments have been described in the previous chapters on microbial diversity, it is important to discuss adaptations of specific microbes to these particular aquatic environments.   Survey of microbial diversity (chapters 20–27)

One of the most exciting findings in the last few years is the presence of extremely high levels of **ultramicrobacteria** or **nanobacteria** in aquatic environments, particularly the genus *Sphingomonas,* which easily can pass through a 0.2 μm membrane filter. These cells, which have a volume of less than 0.08 μm^3, have been found to be the dominant bacteria in marine systems (and in soils, as will be discussed in chapter 30); they may approach 10^{12} to 10^{13} cells per gram or milliliter of material. They can make up the dominant microbial biovolume in many environments, as has been shown in the northern Pacific Ocean region.

Another unusual marine microbe, found off the coast of Namibia on the west coast of Africa, is *Thiomargarita namibiensis* (**figure 29.5;** *see also Microbial Diversity & Ecology 3.1*), which means the "sulfur pearl of Namibia." This microorganism is considered to be the world's largest bacterium! The individual cells are usually 100 to 300 μm in diameter (750 μm cells occasionally occur with a biovolume of 200,000,000 μm^3) and sulfide and nitrate are used as the reductant and oxidant, respectively. In this case the nitrate, from the overlying seawater, penetrates the anaerobic sulfide-containing muds only during storms. When this short-term mixing occurs, the *Thiomargarita* takes up and stores the nitrate in a huge internal vacuole, which may occupy 98% of the organism's volume. The vacuolar nitrate can approach a concentration of 800 mM. The elemental sulfur granules appear near the cell edge in a thin layer of cytoplasm. Between the storms, the organism lives using the stored nitrate as an electron acceptor. These unique bacteria are important in sulfur and nitrogen cycling in these environments.

A critical adaptation of microorganisms in aquatic systems is the ability to link and use resources that are in separate locations, or that are available at the same location only for short intervals such as during storms. One of the most interesting bacteria linking widely separated resources is *Thioploca* spp., which lives in bundles surrounded by a common sheath (**figure 29.6**). These microbes are found in upwelling areas along the coast of Chile, where oxygen-poor but nitrate-rich waters are in contact with sulfide-rich bottom muds. The individual cells are 15 to 40 μm in diameter and many centimeters long, making them one of the largest bacteria known. They form filamentous sheathed structures, and the individual cells can glide 5 to 15 cm deep into the

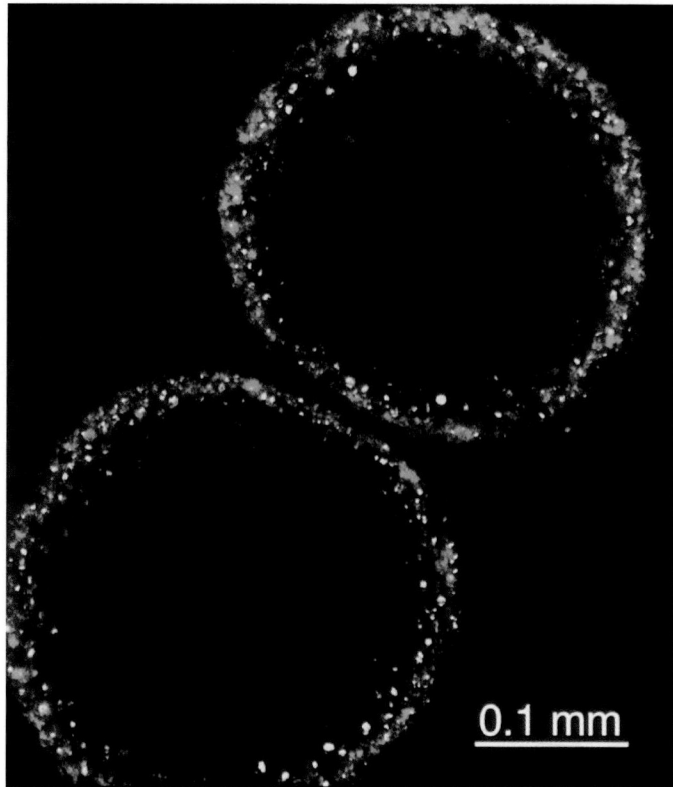

Figure 29.5 *Thiomargarita namibiensis,* **the World's Largest Known Bacterium.** This procaryote, usually 100 to 300 μm in diameter as shown here, occasionally reaches a size of 0.75 mm (larger than a period on this page), 100 times the size of a common bacterium. This unique bacterium uses sulfide from bottom sediments as an energy source and nitrate, which is found in the overlying waters, as an electron acceptor.

sulfide-rich sediments. These unique microorganisms are found in vast fields off the coast of Chile and are "the largest communities of visible bacteria in the world."

Other microorganisms take advantage of surfaces in aquatic systems. These include sessile microorganisms of the genera *Sphaerotilus* and *Leucothrix* and the prosthecate and budding bacteria of the genera *Caulobacter* and *Hyphomicrobium.* There are also a wide range of aerobic gliding bacteria such as the genera *Flexithrix* and *Flexibacter,* which move over surfaces where organic matter is adsorbed. These organisms are characterized by their exploitation of surfaces and nutrient gradients. They are obligate aerobes, although sometimes they can carry out denitrification, as occurs in the genus *Hyphomicrobium.* In addition, bacteria may be primarily colonizers of submerged surfaces, allowing subsequent development of a more complex biofilm. Budding bacterium (pp. 476–79); Sheathed bacteria (pp. 482–83); Gliding bacteria (pp. 470–71)

Microscopic fungi, which usually are thought to be terrestrial organisms living in soils and on fruits and other foods, also grow in freshwater and marine environments. Zoosporic organisms that are adapted to an aquatic existence include both the

(a)

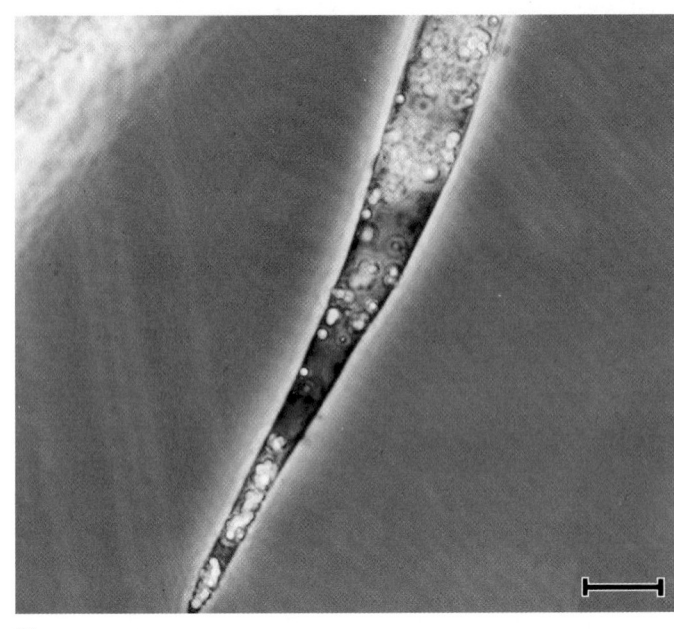

(b)

Figure 29.6 *Thioploca* the "Spaghetti Bacterium." *Thioploca* (sulfur braid) is an unusual microorganism that links separated resources of sulfide from the mud and nitrate from the overlying water. (**a**) Bundles that join the aerobic surface and the lower anaerobic mud. (**b**) An individual *Thioploca,* showing the elemental sulfur globules and tapered ends. Bar = 40 μm.

oomycetes, which have motile asexual reproductive spores with two flagella, and the chytrids, which have motile asexual reproductive spores with a single whiplash flagellum (*see chapter 25*). Most chytrids are important because of their role in decomposing dead organic matter, but they also cause potato wart disease and parasitize invertebrates, particularly nematodes and mosquitoes. In addition, many chytrids attack algae (**figure 29.7**).

Recently chytrids have been reported to live in the skin cells of amphibians, and these have been associated with die-offs of frogs and toads in many parts of the world, including the United States, Australia, and Central America. The chytrid *Batrachochytrium dendrobatidis* is one of the major zoosporic fungi that has been associated with amphibian mortalities.

Another important group includes filamentous fungi that can sporulate under water. These hyphomycetes include the **Ingoldian fungi,** named after C. T. Ingold. In 1942 Ingold discovered fungi that produce unique tetraradiate forms (**figure 29.8a**). The ecology of these aquatic fungi is very interesting (figure 29.8b). The tetraradiate conidium forms on a vegetative mycelium, which grows inside decomposing leaves. Released tetraradiate conidia are transported by the water and often are present in surface foam. When they contact leaves, the conidia attach and establish new centers of growth. These uniquely adapted fungi contribute significantly to the processing of organic matter, especially leaves. Often aquatic insects will feed only on leaves that contain fungi. Without fungal processing, the insects are not interested!

One of the most interesting recent discoveries regarding the marine environment is the presence of large numbers of archaea. By use of rRNA sequence comparisons (*see chapter 19*), it has been determined that approximately 20% of the oceanic picoplankton (cells smaller than 2 μm) are archaea, particularly crenarchaeans, organisms normally associated with hostile environments (hot springs, deep marine "black smoker" areas, saline and thermoacidophilic regions).

Planktonic archaea and bacteria are present in both freshwater and marine environments, and deep in the ocean. Using fluorochrome labeled probes in a fluorescent in situ hybridization (FISH) technique, it is possible to detect these different microbial groups in the same sample by the use of different exciting wavelengths (**figure 29.9**). Bacterial concentrations are higher in the surface zone, but below 100 meters archaea form a greater portion of the total population (**figure 29.10**). More than 90% of the cells react to the stains and can be detected by the use of different probes and wavelengths.

Aquatic environments also harbor large populations of viruses. These are present at 10-fold higher concentrations than the bacteria, and most are bacteriophages. These **virioplankton** are an important part of the aquatic microbial community. They may influence the functioning of the microbial loop, be involved in horizontal gene transfer between procaryotes, and control microbial community diversity. In coastal regions, Norwalk-like viruses (*see p. 868*), family *Caliciviridae,* are of increasing concern in terms of disease occurrence. Some of the caliciviruses appear to have reservoirs in the ocean. Animals that have been documented to be susceptible to these viruses include humans, swine (vesicular exanthema), and felines.

Microorganisms are constantly being added to waters. Aquatic microorganisms are released to the atmosphere and returned to

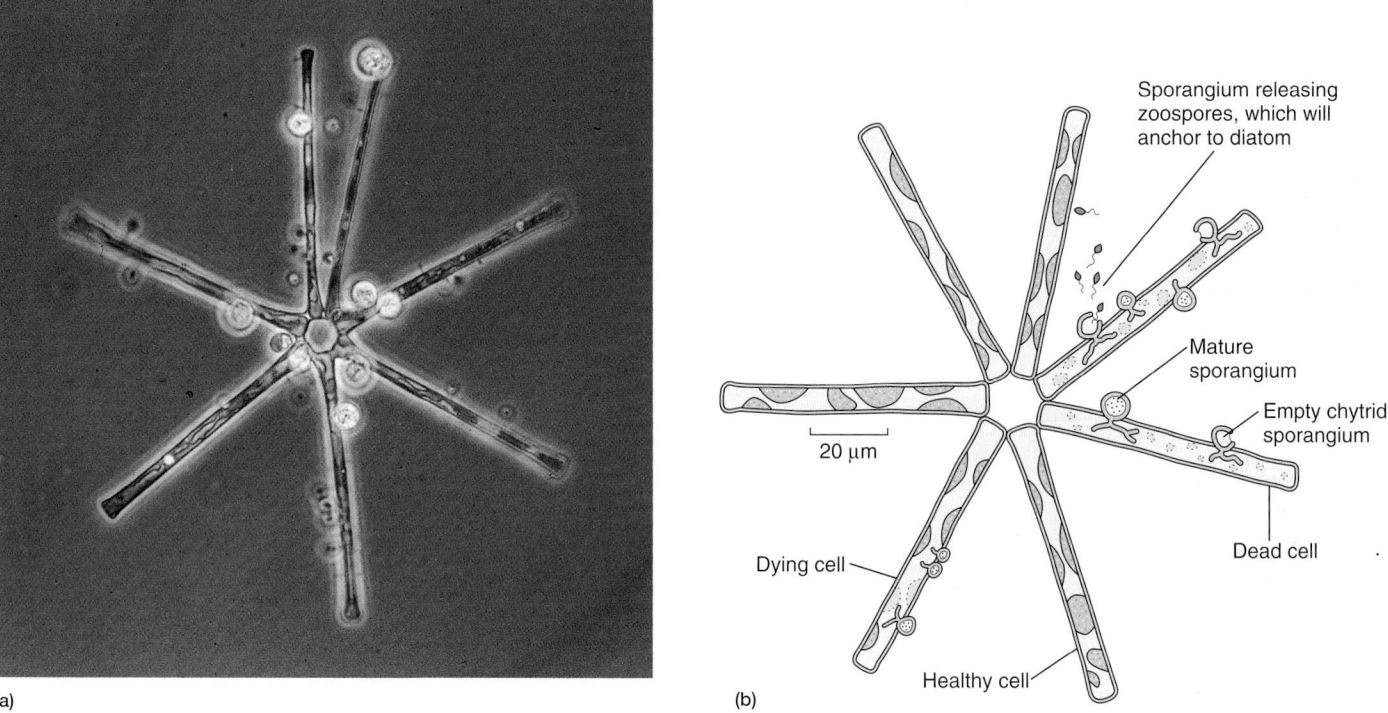

Sporangium releasing zoospores, which will anchor to diatom
Mature sporangium
Empty chytrid sporangium
20 μm
Dead cell
Dying cell
Healthy cell
(a)
(b)

Figure 29.7 Chytrids and Aquatic Environments. Chytrids play important roles in aquatic environments. (**a**) The infection of the diatom by the chytrid *Rhizophydium* is shown in the photograph, and (**b**) in the illustration showing details of the parasitic process.

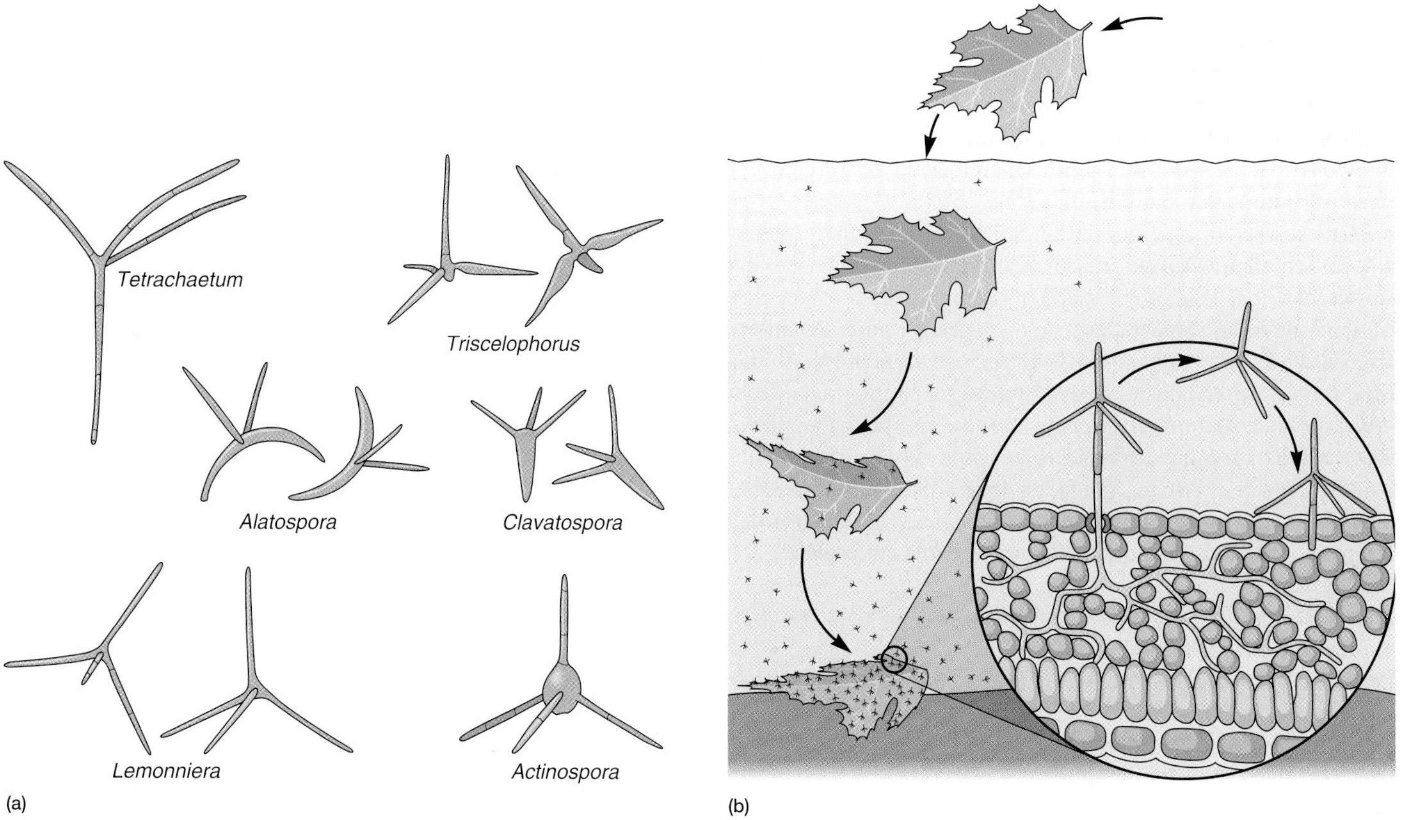

Tetrachaetum
Triscelophorus
Alatospora
Clavatospora
Lemonniera
Actinospora
(a)
(b)

Figure 29.8 Ingoldian Fungi. These aquatic fungi, named after C. T. Ingold, are capable of sporulation under water. They play important roles in the processing of complex organic matter, such as leaves, which fall into streams and lakes. These microbes include types with tetraradiate conidia. (**a**) The tetraradiate conidia are produced as aerial structures (**b**) from the mycelium, which is growing inside the decomposing leaf. The new tetraradiate conidium then can be released and attaches to a leaf surface, repeating the process and accelerating leaf decomposition.

Figure 29.9 Detection of Archaea in the Ocean with Probes. Fluorescent-labeled probes and selective barrier filters can be used to detect archaea (top plate) as a part of a mixed population of procaryotes (bottom plate) in deep marine water samples using microscopy. Bar = 5 μm.

other aquatic locations by air movement, a part of aerobiology. Microorganisms associated with detritus, animals, and fish carcasses also move with their substratum. Occasionally, carcasses of large animals such as whales will drop to the ocean floor over thousands of meters of water depth, creating new opportunities for scavengers and microorganisms.

Dust and sediment from distant terrestrial areas also continually fall on all waters and ice-covered regions, due to wind and storms, another important part of aerobiological processes. Through these atmospheric transport processes, microorganisms are constantly being mixed across all regions of the Earth. In this regard, recent studies indicate that microorganisms cultured from sediments at the bottom of the deep ocean are similar to those from surface soils. Another example is airborne dust from the sub-Saharan region of Africa, which has been linked to the massive die-off of corals in the Caribbean due to fungi and algae contained in soils that have drifted around the world in the atmosphere.

1. What does *Thiomargarita* mean? Where would one go to find this interesting microorganism?
2. Discuss the unusual structure and function of *Thioploca*. What environment is this organism found in?
3. Give the genera of sessile and prosthecate bacteria that are important in aquatic environments.
4. What are chytrids? Why might they be important in aquatic environments?
5. Describe Ingoldian fungi. How do they function in aquatic environments, and what is unique about their shape? What does the term tetraradiate mean?

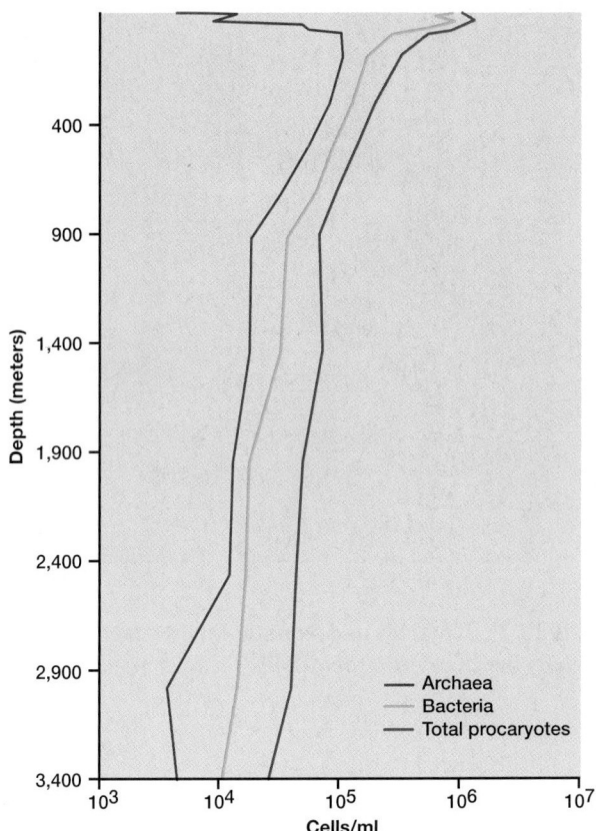

Figure 29.10 Archaea Are Plentiful in Ocean Depths. The distribution of group I archaea and bacteria, together with an estimate of total procaryotes, over a depth of 3,400 meters are shown at a Pacific Ocean location. These results indicate that archaea make up a significant part of the observable picoplankton below the surface zone.

6. What levels of archaea are found in aquatic environments? How have these been detected?
7. How are microorganisms moved within and between various water bodies?

MARINE ENVIRONMENTS

In terms of sheer volume, the marine environment represents a major portion of the biosphere and contains 97% of the Earth's water. Much of this is in the deep sea at a depth greater than 1,000 meters, representing 75% of the ocean's volume. The ocean has been called a "high-pressure refrigerator," with most of the volume below 100 meters at a constant 3°C temperature. The ocean, at its greatest depth, is slightly more than 11,000 meters deep or equivalent to almost 29 Empire State Buildings (each 1,250 feet or 381 meters in height) stacked on top of one another! The pressure in the marine environment increases approximately 1 atm/10

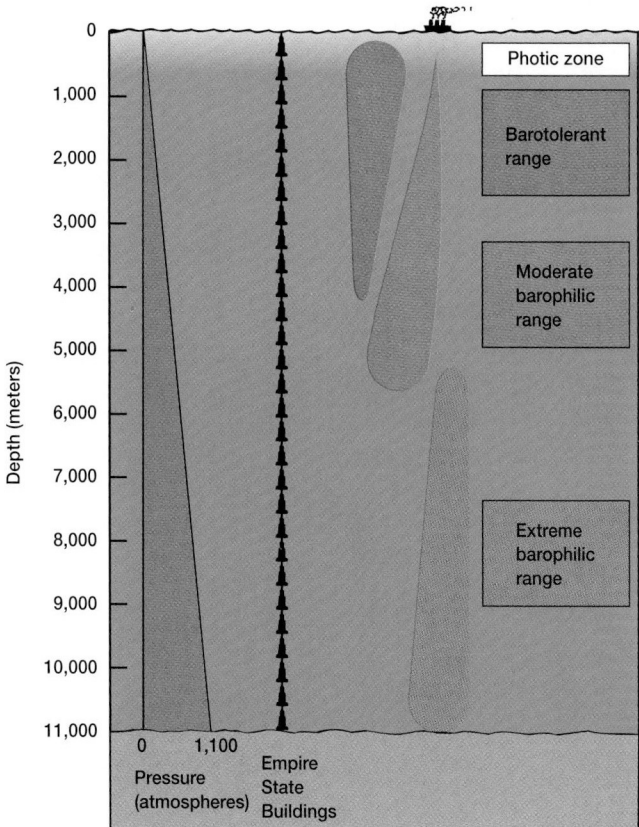

Figure 29.11 The Deep Marine Environment. Microorganisms with specialized pressure relationships are found at various depths. The relative activity and depth relationships of barotolerant, moderate barophilic, and extreme barophilic microorganisms are noted schematically. The pressure can approach 1,100 atmospheres at the deepest ocean depths. The depths are given in meters and "stacked" Empire State Buildings for perspective. Light penetrates only into a relatively shallow surface layer, creating the photic zone.

meters in depth, and pressures are in the vicinity of 1,000 atm at the greatest ocean depths (**figure 29.11**).

A series of pressure relationships is observed among the bacteria growing in this vertically differentiated system. Some bacteria are barotolerant and grow between 0 and 400 atm, but best at normal atmospheric pressure. Many other bacteria are **barophiles** [Greek, *baro,* weight and *philein,* to love], also termed piezophiles, and prefer higher pressures. Moderate barophiles grow optimally at 400 atm, but still grow at 1 atm; extreme barophiles grow only at higher pressures. Pressure differences influence many biological processes including cell division, flagellar assembly, DNA replication, membrane transport, and protein synthesis. Porins, outer membrane proteins that form channels for diffusion of materials into the periplasm (*see figure 3.23*), also function most effectively at specific pressures. Pressure and microbial growth (p. 127)

Most nutrient cycling in oceans occurs in the top 300 meters where light penetrates. Light allows phytoplankton to grow and

fall as a "marine snow" to the seabed. This "trip" can take a month or longer. Most of the organic matter that falls below the 300 meter zone is decomposed, and only 1% of photosynthetically derived material reaches the deep-sea floor unaltered. Because low inputs of organic matter occur in the deep sea, the ability of microorganisms to grow under oligotrophic conditions becomes important (*see section 6.5*).

The carbon cycle within the ocean environment is only poorly understood (**figure 29.12**), but clearly microorganisms can profoundly influence the cycle. It is estimated that the large amount of dissolved organic carbon (DOC) in the ocean has a mean age of greater than 1,000 years. Organic matter in deep ocean waters has a similarly long residence time. Besides DOC, massive deposits of methane hydrate occur in ocean sediments. Under the low-temperature, high-pressure conditions found at the ocean floor below 500 meters, methane accumulates in latticelike cages of crystalline water. There may be up to 10,000 billion metric tons of carbon present as methane hydrate worldwide, 80,000 times the world's current known natural gas reserves.

The nitrogen and sulfur cycles also function at an "ocean scale" and have potentially significant effects on global-level processes, which have not been appreciated until recently. Large volumes of the ocean's water have lower oxygen levels, leading to denitrification and a decrease in the nitrate-phosphate ratio in the water. As a consequence, nitrogen fixation may be favored and increase the nitrogen level in the water. It appears that nitrogen, and not phosphorus, often can limit biological activity in many marine environments. The ocean sulfur cycle also has widespread effects on global processes. Dimethylsulfide (DMS), an algal osmolite, can be released to the atmosphere and comprises 90% of the volatile sulfur compounds in the sulfur cycle. When DMS is oxidized, its end products can influence the acidity of the atmosphere, as well as the Earth's temperature and cloud formation.

Additional cold-temperature microorganisms include those involved in the production and use of methane. As already noted, methane hydrates are found at low temperatures and high pressures in many marine regions of the world. The procaryotes that consume these methane hydrates serve as a food source for ice worms, *Hesiocaeca methanicola.* Archaea, as a part of complex microbial communities, appear to metabolize methane deposits at low hydrogen levels under sulfate reducing conditions. The process is called **reverse methanogenesis.**

Much of the marine environment is covered by sea ice that may comprise up to 7% of the Earth's surface at the winter maxima at the north and south poles! Microorganisms actually grow and reproduce at the interface between the ice and the seawater. **Figure 29.13** shows an ice core taken from the lower zone of an ice sheet at the seawater-ice interface; a band of microorganisms is clearly visible in the ice profile. Pockets of salt brine also form in this region, creating additional environments with varied temperatures, salinities, nutrient concentrations, and light levels for these cold-adapted microorganisms. The microbes that have been recovered from these ice cores have been given intriguing generic names: these include *Polaromonas,*

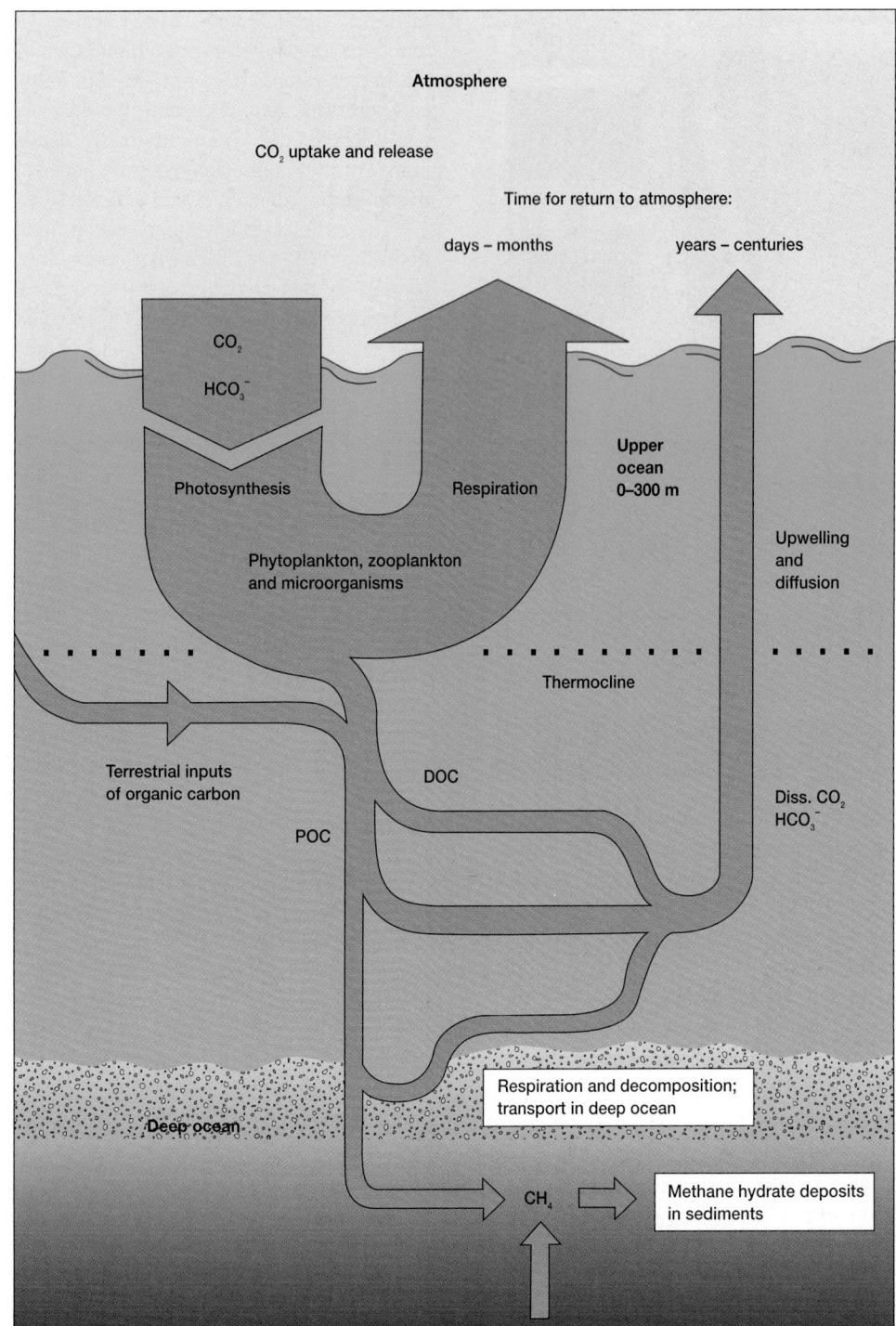

Figure 29.12 Carbon Cycling in the Ocean Environment. Microorganisms in the oceans can influence global carbon cycling and ocean-atmosphere interactions. Most carbon processing occurs in the surface water zone, with particulate organic carbon (POC), dissolved organic carbon (DOC), and methane hydrate (in sediments) being major carbon pools. The ocean also contains bicarbonate and dissolved CO_2 (diss. CO_2) that come from the atmosphere and the degradation of organic carbon. Methane hydrate allows microorganisms and associated animals such as ice worms to develop.

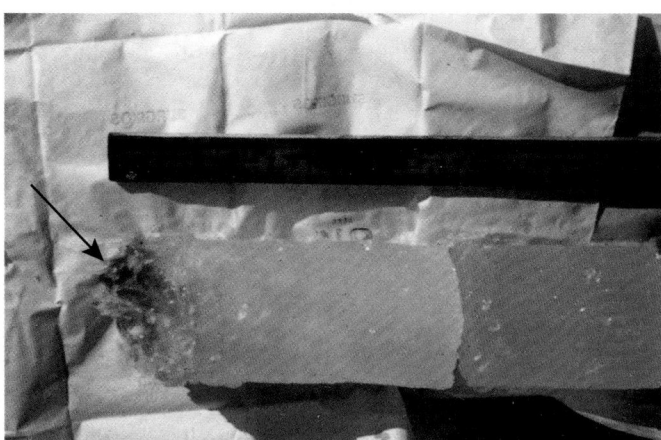

Figure 29.13   Sea Ice as a Habitat for Microorganisms.   Sea ice allows the development of complex microbial communities. The bottom of a sea ice core from the Antarctic, which is in contact with the underlying seawater, is shown. The dark band (see arrow) is the sea ice microbial community.

Marinobacter, Psychroflexus, Iceobacter, Polibacter, and *Psychromonas antarcticus.*

Increased human populations and the urban development occurring in coastal areas around the world is taxing the seemingly inexhaustible ability of oceans to absorb and process pollutants. Coastal areas that have limited mixing with ocean waters (e.g., the Baltic Sea, Long Island Sound, Chesapeake Bay, the Mediterranean) are showing signs of nutrient enrichment and microbial pollution.

In the Gulf of Mexico, at the Mississippi River delta, releases of nutrients, primarily nitrogen, from states that drain into this river have stimulated microbial growth and oxygen depletion. This has created a "dead zone," a hypoxic or anoxic region, that is larger than the state of New Jersey. The lowered oxygen levels occurring in this rich shellfish area have damaged the economy of these aquaculture-dependent regions.

Another problem that relates to ocean waters and water mixing in coastal areas is the occurrence of red tides (*see Disease 26.2*). This has major economic effects when shellfish cannot be harvested or consumed. Recently the occurrence of algal blooms and red tides in the Pacific Ocean off the central California coast has resulted in the die-off of dolphins. The major algal genus responsible for the widespread loss of these aquatic animals is *Pseudo-nitzschia.* The key link is anchovies. This fish, consumed by the dolphins, feeds on algae that contain a high concentration of a neurotoxin, domoic acid; this toxin has been detected in the sick and dead dolphins.

Algae may cause problems other than red tides. An excellent example is *Pfiesteria piscicida,* a dinoflagellate (*see pp. 561–62*) that has produced major fish kills from Maryland southward along the U.S. Atlantic coast. It even can cause dizziness and loss of memory in humans. This primitive protistan alga is a fish predator. Its flagellated vegetative **dinospore** stage appears to detect

Figure 29.14   *Pfiesteria* Lesions.   Lesions on menhaden resulting from parasitism by the dinoflagellate *Pfiesteria piscicida.*

fish compounds in the water and then literally attack the approaching fish, causing massive lesions due to micropredatory feeding (**figure 29.14**). This dinoflagellate apparently once used other algae as its primary food source, and it can use these as a source of photosynthetic organelles. This process of kleptoplastidy makes it possible for *Pfiesteria* to use both light and organic matter to support its growth. It now has become a killer of fish, eels, crabs, and other animals. Increased nutrient levels are suspected as the cause of its increasing prevalence.

Global-level movements of air also affect marine microorganisms. The atmospheric movement of soils and industrial activities influence phytoplankton growth and their Redfield ratio (p. 619). In the northern Pacific, the water is iron-limited, and iron is being added to these waters by desertification and dust storms in central Asia, thus increasing primary production. In contrast, the northern Atlantic Ocean is nitrogen-limited, and a transition from N- to P-limitation is occurring with increased depositions of atmospheric nitrogen from human activities. The N:P ratio in the deep North Atlantic is increasing and altering the phytoplankton Redfield ratio.

1. What portion of the water in the world is marine?
2. Where are methane hydrates found, and what is reverse methanogenesis?
3. What microbially produced volatile sulfur compound can influence the weather?
4. Why is sea ice an important habitat for microorganisms? Where are the microorganisms mostly located?
5. What is hypoxia or anoxia? What are possible causes?
6. Microbiological quality of waters in marine coastal areas is important. What are some major concerns?
7. What link has been found recently between algae, anchovies, and dolphins?
8. What is unique about *Pfiesteria?*
9. How can atmospheric processes influence nutrient limitation in marine waters?

29.4 FRESHWATER ENVIRONMENTS

Most fresh water that is not locked up in ice sheets, glaciers, or groundwaters is found in lakes and rivers. These provide microbial environments that are different from the larger oceanic systems in many important ways. For example, in lakes, mixing and water exchange can be limited. This creates vertical gradients over much shorter distances. Changes in rivers occur over distance and/or time as water flows through river channels.

Lakes

Lakes vary in nutrient status. Some are **oligotrophic** or nutrient-poor (**figure 29.15a**), others are **eutrophic** or nutrient-rich (figure 29.15b). Nutrient-poor lakes remain aerobic throughout the year, and seasonal temperature shifts do not result in distinct oxygen stratification. In contrast, nutrient-rich lakes usually have bottom sediments that contain organic matter. In thermally stratified lakes the **epilimnion** (warm, upper layer) is aerobic, while the **hypolimnion** (deeper, colder, bottom layer) often is anaerobic (particularly if the lake is nutrient-rich). The epilimnion and hypolimnion are separated by a zone of rapid temperature decrease called the **thermocline,** and there is little mixing of water between the two layers. In the spring and fall, the aerobic surface water and the anaerobic subsurface water will turn over as the result of differences in temperature and specific gravity. After such mixing occurs, motile bacteria and algae migrate within the water column to again find their most suitable environment.

When sufficiently large amounts of nitrogen and phosphorus are added to water, **eutrophication** (nutrient enrichment) takes place and stimulates the growth of plants, algae, and bacteria (*see figure 21.11*). Depending on the body of water and the rate of nutrient addition, eutrophication may either require many centuries or occur very rapidly.

If phosphorus is added to oligotrophic fresh water, cyanobacteria play the major role in nutrient accumulation, even in the absence of extra nitrogen. Several genera, notably *Anabaena, Nostoc,* and *Cylindrospermum,* can fix nitrogen under aerobic conditions (*see section 21.3*). The genus *Oscillatoria,* using hydrogen sulfide as an electron donor for photosynthesis, can fix nitrogen under anaerobic conditions. Even with both nitrogen and phosphorus present, cyanobacteria still can compete with algae. Cyanobacteria function more efficiently with higher pH conditions (8.5 to 9.5) and higher temperatures (30 to 35°C). Algae, in comparison, generally prefer a more neutral pH and have lower optimum temperatures. By using CO_2 at rapid rates, cyanobacteria also increase the pH, making the environment less suitable for algae. Algae (chapter 26); Cyanobacteria (pp. 458–64)

Cyanobacteria have additional competitive advantages. Many produce hydroxamates, which bind iron, making this important trace nutrient less available for the algae. Cyanobacteria also often resist predation because of their production of toxins. In addition, some synthesize odor-producing compounds that affect the quality of drinking water.

Both cyanobacteria and algae can contribute to massive blooms in strongly eutrophied lakes (*see figure 21.11*). Lake man-

Figure 29.15 Oligotrophic and Eutrophic Lakes. Lakes can have different levels of nutrients, ranging from low nutrient to extremely high nutrient systems. The comparison of (**a**) an oligotrophic (nutrient-poor) lake, which is oxygen saturated and has a low microbial population, with (**b**) a eutrophic (nutrient-rich) lake. The eutrophic lake has a bottom sediment layer and can have an anaerobic hypolimnion. Sulfur photosynthetic microorganisms can develop in this anaerobic region.

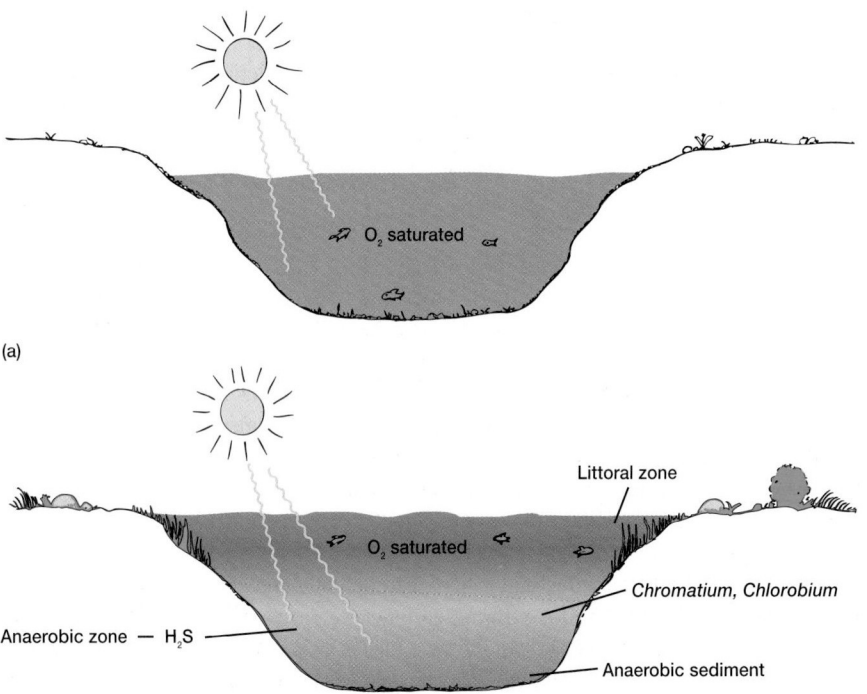

agement can improve the situation by removing or sealing bottom sediments or adding coagulating agents to speed sedimentation.

---

1. What terms can be used to describe the different parts of a lake?
2. What are some important effects of eutrophication on lakes?
3. Why are cyanobacteria so important in waters that have been polluted by phosphorus additions?

Streams and Rivers

Streams and rivers present a different situation from lakes in that there is sufficient horizontal water movement to minimize vertical stratification; in addition, most of the functional microbial biomass is attached to surfaces. Only in the largest rivers will a greater relative portion of the microbial biomass be suspended in the water. Depending on the size of the stream or river, the source of nutrients may vary. The source may be in-stream production based on photosynthetic microorganisms (**figure 29.16a**). Nutrients also may come from outside the stream, including runoff sediment from riparian areas (the edge of a river), or leaves and other organic matter falling directly into the water (figure 29.16b). Chemoorganotrophic microorganisms metabolize the available organic material and provide an energy base for the ecosystem. Under most conditions the amounts of organic matter added to streams and rivers will not exceed the system's oxidative capacity, and productive, aesthetically pleasing streams and rivers will be maintained.

The capacity of streams and rivers to process such added organic matter is, however, limited. If too much organic matter is added, the water may become anaerobic. This is especially the case with urban and agricultural areas located adjacent to streams and rivers. The release of inadequately treated municipal wastes and other materials from a specific location along a river or stream represents **point source pollution.** Such point source additions of organic matter can produce distinct and predictable changes in the microbial community and available oxygen, creating an oxygen sag curve (**figure 29.17**). Runoff from fields and feedlots, which causes algal blooms in eutrophic water bodies, is an example of **nonpoint source pollution.**

When the amount of organic matter added is not excessive, the algae will grow using the minerals released from the organic matter. This leads to the production of O_2 during the daylight hours, and respiration will occur at night farther down the river, resulting in **diurnal oxygen shifts.** Eventually the O_2 level approaches saturation, completing the self-purification process.

Along with the stresses of added nutrients, removal of silicon from rivers by the construction of dams and trapping of sediments causes major ecological disturbances. For example, construction of the dam at the "iron gates" on the Danube (600 miles above the Black Sea) has led to a decrease in silicon to 1/60th of the previous concentration. This decreased silicon availability then inhibits the growth of diatoms (*see pp. 559–60)* because the ratio of silicon to nitrate has been altered (silicon is required for diatom

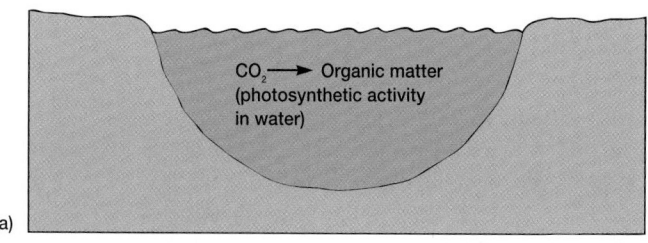

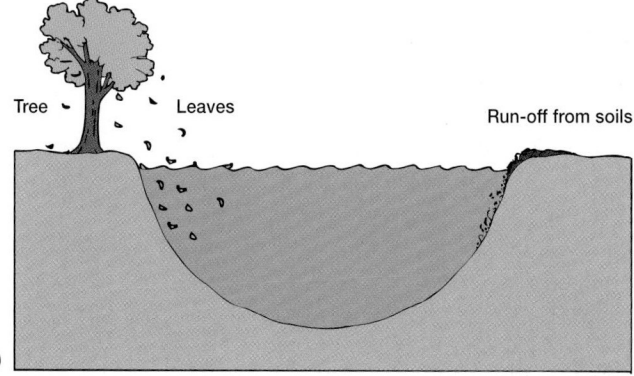

Figure 29.16   Organic Matter Sources for Lakes and Rivers. Organic matter used by microorganisms in lakes and rivers can be synthesized in the water, or can be added to the water from outside. (**a**) Within-stream sources of organic matter, primarily photosynthesis, and (**b**) sources of organic matter which enter streams and rivers from outside the water. Stream cross sections are shown.

frustule formation). With this shift in resources, Black Sea diatoms are not able to grow and immobilize nutrients. The result has been a 600-fold increase in nitrate levels and a massive development of toxic algae. Thus the delicate balance of rivers can be altered in unexpected ways by dams (there are over 36,000 dams in the world, and more being built, including in China), leading to effects on aquatic microbiological processes and whole ecosystems.

With the damaging effects of dams now being recognized, there is a growing trend to attempt to breach these structures to restore normal water and sediment movements, and to allow fish migration into upper regions of rivers where they often have been excluded for decades.

Microorganisms in Freshwater Ice

Fresh waters include glaciers, as well as vast ice sheets in the Arctic and Antarctic polar regions. These regions, although far away from most people, are important as habitats for microorganisms. Of particular interest are deep frozen lakes in the Antarctic region. An excellent example is the McMurdo Dry Valley Lakes, which have 3 to 6 meter thick ice layers with windblown sediments occupying various levels in the ice profile. In the summer,

Figure 29.17 The Dissolved Oxygen Sag Curve. Microorganisms and their activities can create gradients over distance and time when nutrients are added to rivers. An excellent example is the dissolved oxygen sag curve, caused when organic wastes are added to a clean river system. During the later stages of self-purification, the phototrophic community will again become dominant, resulting in diurnal changes in river oxygen levels.

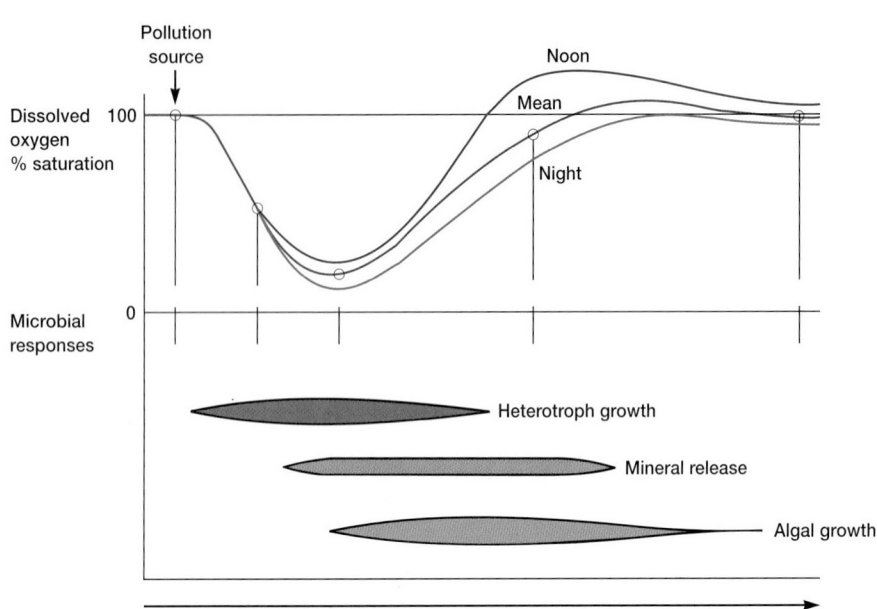

melting of the ice in these sediment-containing zones provides an opportunity for microbial growth. There is interest in studying ice-inhabiting microorganisms from the north and south poles. Are they isolated and different populations? Do some of them show biogeographic distribution?

One of the most interesting deep-frozen habitats lies above Lake Vostok in the Antarctic, where cores have been taken to a depth of 3,600 meters (11,800 feet) in ice that has been frozen for over 420,000 years. The actual Lake Vostok lies below another 120 meters of ice. The lake itself has a size similar to that of Lake Ontario, but it is deeper, being 500 meters in depth. Scientists are actively searching for microorganisms in these unique older freshwater frozen environments. It is hoped that it will be possible to link microbial communities and the time of their deposition.

---

1. What is an oxygen sag curve, and what changes in a river lead to these effects?
2. What are point and nonpoint source pollution? Give examples.
3. Why might dams influence microorganisms and microbial processes in rivers?
4. What is unique about Lake Vostok in the Antarctic?

 29.5 WATERS AND DISEASE TRANSMISSION

Since the beginning of recorded history water has been recognized as a potential carrier of disease. In order to protect his health, Alexander the Great (356–323 B.C.) had his personal drinking water carried in silver urns. Clearly the association between noble heavy metals such as silver and the prevention of waterborne diseases was early established through chance observation. The connection between a freshwater supply and the

health of an urban population was recognized by the time of the Roman Empire (27 B.C.). However, much of the technology for the protection of the water supply was subsequently lost until the middle of the nineteenth century.

With increased use of waters by humans, especially as a receptacle for the disposal of human wastes, the effects of added organic matter and pathogens is of continuing concern for human health.

Waterborne Pathogens and Water Purification

Many important human pathogens are maintained in association with living organisms other than humans, including many wild animals and birds. Some of these bacterial and protozoan pathogens can survive in water and infect humans. As examples, *Vibrio vulnificus, V. parahaemolyticus,* and *Legionella* are of continuing concern. When waters are used for recreation or are a source of seafood that is consumed uncooked, the possibility for disease transmission certainly exists. A waterborne infection may have serious consequences for immunologically compromised individuals.

Some major waterborne diseases are summarized in **table 29.2.** Waterborne bacterial and viral pathogens are extensively discussed in sections 38.4 and 39.4. The protozoan pathogens *Giardia, Cryptosporidium* are described in section 40.2; *Cyclospora* is discussed on pages 632–33.

Another waterborne protozoan disease of increasing concern worldwide is primary amebic meningoencephalitis (PAM), an infection of the central nervous system caused by *Naegleria fowleri.* The disease usually occurs in children or young adults who have been swimming in lakes or pools or have been waterskiing. After nasal infection the protozoan reaches the brain and initiates an inflammatory response. The result usually is fatal. Warm wa-

Table 29.2

Water-Based Microbial Pathogens That Can Be Maintained in the Environment Independent of Humans

Organism	Reservoir	Comments
Bacteria		
Aeromonas hydrophila	Free-living	Sometimes associated with gastroenteritis, cellulitis, and other diseases
Campylobacter	Bird and animal reservoirs	Major cause of diarrhea; common in processed poultry; a microaerophile
Helicobacter pylori	Free-living, not known	Can cause type B gastritis, peptic ulcers, gastric adenocarcinomas
Legionella pneumophila	Free-living and associated with protozoa	Found in cooling towers, evaporators, condensers, showers, and other water sources
Leptospira	Infected animals	Hemorrhagic effects, jaundice
Mycobacterium	Infected animals and free-living	Complex recovery procedure required
Pseudomonas aeruginosa	Free-living	Swimmer's ear and related infections
Salmonella enteriditis	Animal intestinal tracts	Common in many waters
Vibrio cholera	Free-living	Found in many waters and estuaries
Vibrio parahaemolyticus	Free-living in coastal waters	Causes diarrhea in shellfish consumers
Yersinia enterocolitica	Frequent in animals and in the environment	Waterborne gastroenteritis
Protozoa		
Acanthamoeba	Sewage sludge disposal areas	Can cause granulomatous amebic encephalitis (GAE); keratitis, corneal ulcers
Cryptosporidium	Many species of domestic and wild animals	Causes acute enterocolitis; important with immunologically compromised individuals; cysts resistant to chemical disinfection; not antibiotic sensitive
Cyclospora cayetanensis	Waters—does not withstand drying; possibly other reservoirs	Causes long-lasting (43 days average) diarrheal illness; infection self-limiting in immunocompetent hosts; sensitive to prompt treatment with Bactrim
Giardia lamblia	Beavers, sheep, dogs, cats	Major cause of early spring diarrhea; important in cold mountain water
Naegleria fowleri	Warm water (hot tubs), swimming pools, lakes	Inhalation in nasal passages; central nervous system infection; causes primary amebic meningoencephalitis (PAM)

ter and heated industrial effluents promote the growth of this protozoan. Protozoan diseases (pp. 926–34); Food-borne and waterborne viral diseases (pp. 868–70); Food-borne and waterborne bacterial diseases (pp. 905–11)

Water purification is a critical link in controlling disease transmission in waters. As shown in **figure 29.18,** water purification can involve a variety of steps, depending on the type of impurities in the raw water source. Usually municipal water supplies are purified by a process that consists of at least three or four steps. If the raw water contains a great deal of suspended material, it often is first routed to a **sedimentation basin** and held so that sand and other very large particles can settle out. The partially clarified water is then mixed with chemicals such as alum and lime and moved to a **settling basin** where more material precipitates out. This procedure is called **coagulation** or flocculation and removes microorganisms, organic matter, toxic contaminants, and suspended fine particles. After these steps the water is further purified by passing it through **rapid sand filters,** which depend on the physical trapping of fine particles and flocs. This filtration removes up to 99% of the remaining bacteria. After filtration the water is treated with a disinfectant. This step usually involves chlorination, but ozonation is becoming increasingly popular. When chlorination is employed, the chlorine dose must

be large enough to leave residual free chlorine at a concentration of 0.2 to 2.0 mg/liter. A concern is the creation of **disinfection by-products (DBPs)** such as **trihalomethanes (THMs)** that are formed when chlorine reacts with organic matter. Some of these compounds may be carcinogens.

The preceding purification process effectively removes or inactivates disease-causing bacteria and indicator organisms (coliforms). Unfortunately, however, the use of coagulants, rapid filtration, and chemical disinfection often does not consistently and reliably remove *Giardia lamblia* cysts, *Cryptosporidium* oocysts, *Cyclospora,* and viruses. *Giardia,* a cause of human diarrhea, is now recognized as the most common identified waterborne pathogen in the United States. The protozoan, first observed by Leeuwenhoek in 1681, has trophozoite and cyst forms. The disease often is called "backpacker's diarrhea" and is transmitted primarily through untreated stream water or undependable municipal water supplies. More consistent removal of *Giardia* cysts, which are about 7 to 10 by 8 to 12 μm in size, can be achieved with **slow sand filters.** This treatment involves the slow passage of water through a bed of sand in which a microbial layer covers the surface of each sand grain. Waterborne microorganisms are removed by adhesion to the gelatinous surface microbial layer (**Techniques & Applications 29.2**). *Giardia* (pp. 570, 928–29)

Water purification steps

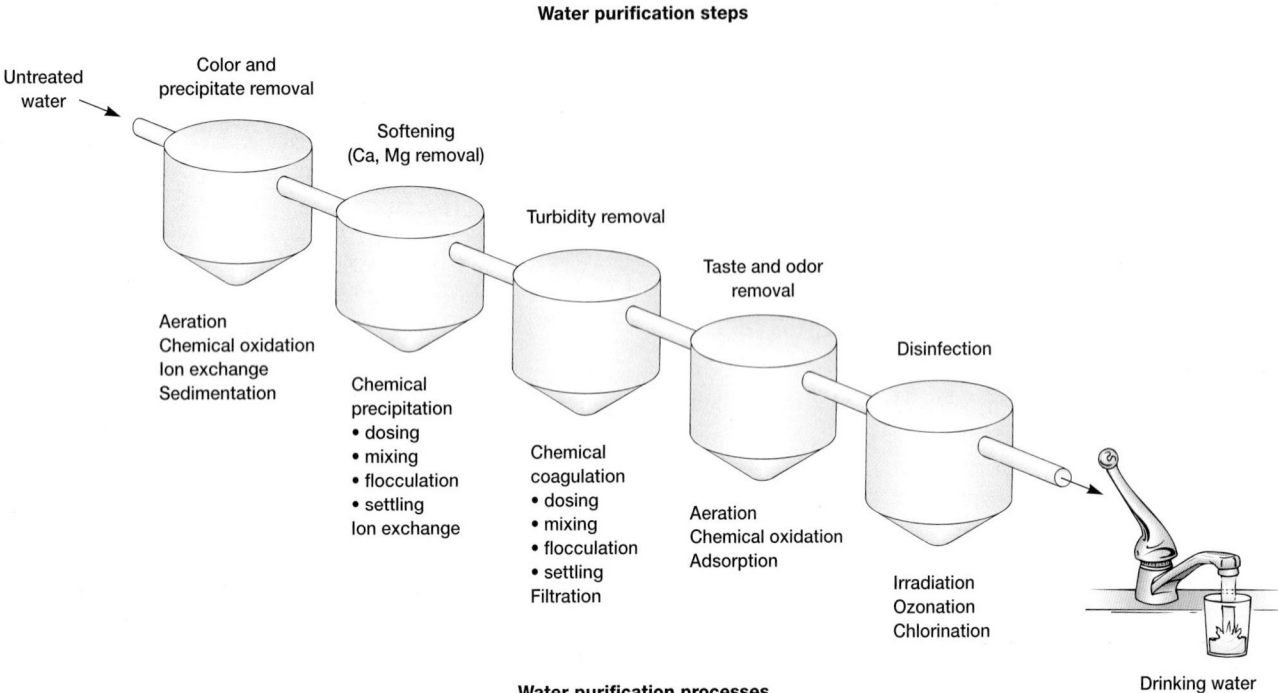

Untreated water →

Color and precipitate removal

Aeration
Chemical oxidation
Ion exchange
Sedimentation

Softening (Ca, Mg removal)

Chemical precipitation
• dosing
• mixing
• flocculation
• settling
Ion exchange

Turbidity removal

Chemical coagulation
• dosing
• mixing
• flocculation
• settling
Filtration

Taste and odor removal

Aeration
Chemical oxidation
Adsorption

Disinfection

Irradiation
Ozonation
Chlorination

Drinking water

Water purification processes

Figure 29.18   Water Purification.   Several alternatives can be used for drinking water treatment depending on the initial water quality. A major concern is disinfection: chlorination can lead to the formation of disinfection byproducts (DBPs), including potentially carcinogenic trihalomethanes (THMs).

Techniques & Applications
29.2   Waterborne Diseases, Water Supplies, and Slow Sand Filtration

Slow sand filtration, in which drinking water is passed through a sand filter that develops a layer of microorganisms on its surface, has had a long and interesting history. After London's severe cholera epidemic of 1849, Parliament, in an act of 1852, required that the entire water supply of London be passed through slow sand filters before use.

The value of this process was shown in 1892, when a major cholera epidemic occurred in Hamburg, Germany, and 10,000 lives were lost. The neighboring town, Altona, which used slow sand filtra-

tion, did not have a cholera epidemic. Slow sand filters were installed in many cities in the early 1900s, but the process fell into disfavor with the advent of rapid sand filters, chlorination, and the use of coagulants such as alum. Slow sand filtration, a time-tested process, is regaining favor because of its filtration effectiveness and lower maintenance costs. Slow sand filtration is particularly effective for the removal of *Giardia* cysts from waters. For this reason slow sand filtration is used in many mountain communities where *Giardia* is a problem.

In the last few years, *Cryptosporidium* has become an even greater problem than *Giardia*. This protozoan parasite is smaller than *Giardia* and is even more difficult to remove from water. *Cryptosporidium* is discussed in section 40.2. A major source of *Giardia* and *Cryptosporidium* contamination of soils, plant materials, and waters is the beautiful Canada goose, protected as a migratory bird. Each goose produces 1.5 lb (0.68 kg) of manure per day, and geese are documented carriers of these important protozoan pathogens. The goose population also is rapidly increasing. For example, in the

Atlantic flyway, there are perhaps a million geese, and these are estimated to be increasing at 17% per year. They gather on golf courses, pastures, croplands, and vulnerable watersheds used for public water supplies, all of which adds to the problem.

Another newly emerging protozoan human pathogen is *Cyclospora*. This coccidian protozoan is larger than *Cryptosporidium* and has two sporocysts, each with two sporozoites (*Cryptosporidium* has four sporozoites in its oocysts). *Cyclospora* causes cyclosporiasis, a self-limiting diarrhea that lasts from 19 to

43 days and can be accompanied by nausea, vomiting, cramps, and fever. *Cyclospora* has recently been implicated in the occurrence of diarrheal diseases due to fecal contamination of imported fruits and vegetables.

Viruses in drinking water also must be inactivated or removed. Coagulation and filtration do reduce virus levels about 90 to 99%. Further inactivation of viruses by chemical oxidants, high pH, and photooxidation may yield a reduction as great as 99.9%. None of these processes, however, is considered to provide sufficient protection. New standards for virus inactivation are being developed. Bacteriophages (*see chapter 17*), which can be easily grown and assayed, now are being used as disinfection surrogates. If sufficient log reductions of bacteriophage infectivity occur with a given disinfection process, it is assumed that viruses capable of infecting humans will be reduced to satisfactory levels.

With the increased concern for drinking water quality in the United States, an Information Collection Rule (ICR) has been initiated. This program has been designed to assess the threat of these pathogens to the waters of cities with populations over 100,000.

---

1. Which important bacterial pathogens can be transmitted by waters?
2. What disease is caused by *Naegleria fowleri*? What is the route of entry to the human body?
3. What steps are usually taken to purify drinking water?
4. Why is chlorination, although beneficial in terms of bacterial pathogen control, of environmental concern?
5. Which important waterborne pathogens are not controlled reliably by chlorination?

Sanitary Analysis of Waters

Monitoring and detection of indicator and disease-causing microorganisms are a major part of sanitary microbiology. Bacteria from the intestinal tract generally do not survive in the aquatic environment, are under physiological stress, and gradually lose their ability to form colonies on differential and selective media. Their die-out rate depends on the water temperature, the effects of sunlight, the populations of other bacteria present, and the chemical composition of the water. Procedures have been developed to attempt to "resuscitate" these stressed coliforms before they are identified using selective and differential media.

A wide range of viral, bacterial, and protozoan diseases result from the contamination of water with human fecal wastes (*see chapters 38, 39, and 40*). Although many of these pathogens can be detected directly, environmental microbiologists have generally used **indicator organisms** as an index of possible water contamination by human pathogens. Researchers are still searching for the "ideal" indicator organism to use in sanitary microbiology. These are among the suggested criteria for such an indicator:

1. The indicator bacterium should be suitable for the analysis of all types of water: tap, river, ground, impounded, recreational, estuary, sea, and waste.

2. The indicator bacterium should be present whenever enteric pathogens are present.
3. The indicator bacterium should survive longer than the hardiest enteric pathogen.
4. The indicator bacterium should not reproduce in the contaminated water and produce an inflated value.
5. The assay procedure for the indicator should have great specificity; in other words, other bacteria should not give positive results. In addition, the procedure should have high sensitivity and detect low levels of the indicator.
6. The testing method should be easy to perform.
7. The indicator should be harmless to humans.
8. The level of the indicator bacterium in contaminated water should have some direct relationship to the degree of fecal pollution.

Coliforms, including *Escherichia coli*, are members of the family *Enterobacteriaceae*. These bacteria make up approximately 10% of the intestinal microorganisms of humans and other animals (*see figure 31.2*) and have found widespread use as indicator organisms. They lose viability in fresh water at slower rates than most of the major intestinal bacterial pathogens. When such "foreign" enteric indicator bacteria are not detectable in a specific volume (100 ml) of water, the water is considered **potable** [Latin *potabilis*, fit to drink], or suitable for human consumption. The Enterobacteriaceae (pp. 492–94)

The coliform group includes *E. coli, Enterobacter aerogenes,* and *Klebsiella pneumoniae.* Coliforms are defined as facultatively anaerobic, gram-negative, nonsporing, rod-shaped bacteria that ferment lactose with gas formation within 48 hours at 35°C. The original test for coliforms that was used to meet this definition involved the presumptive, confirmed, and completed tests, as shown in **figure 29.19.** The presumptive step is carried out by means of tubes inoculated with three different sample volumes to give an estimate of the **most probable number (MPN)** of coliforms in the water. The complete process, including the confirmed and completed tests, requires at least 4 days of incubations and transfers.

Unfortunately the coliforms include a wide range of bacteria whose primary source may not be the intestinal tract. To deal with this difficulty, tests have been developed that allow waters to be tested for the presence of **fecal coliforms.** These are coliforms derived from the intestine of warm-blooded animals, which can grow at the more restrictive temperature of 44.5°C.

To test for coliforms and fecal coliforms, and more effectively recover stressed coliforms, a variety of simpler and more specific tests have been developed. These include the membrane filtration technique, the **presence-absence (P-A) test** for coliforms and the related Colilert **defined substrate test** for detecting both coliforms and *E. coli.*

The **membrane filtration technique** (*see figure 6.6*), has become a common and often preferred method of evaluating the microbiological characteristics of water. The water sample is passed through a membrane filter. The filter with its trapped bacteria is transferred to the surface of a solid medium or to an absorptive pad

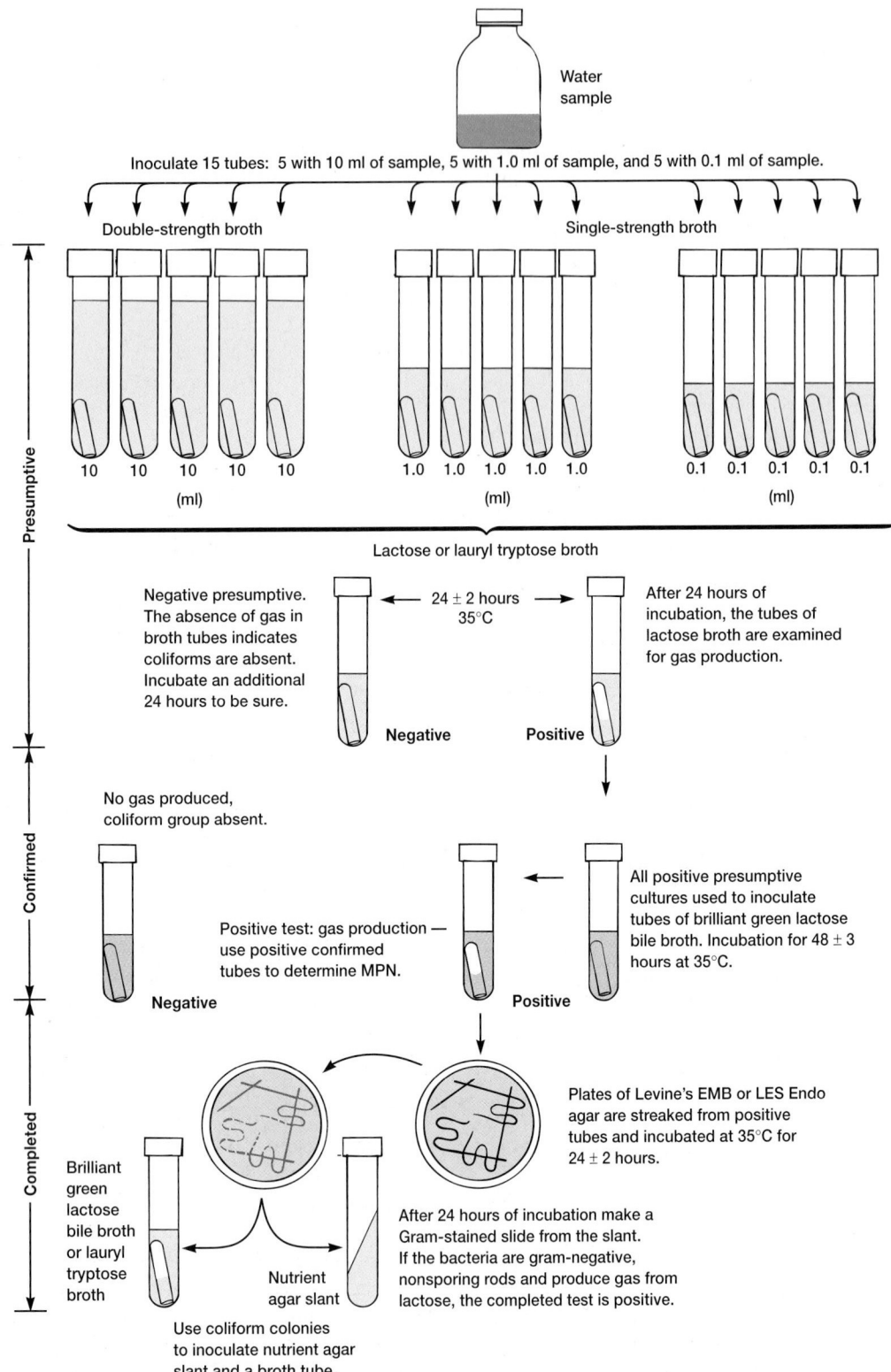

Figure 29.19 The Multiple-Tube Fermentation Test. The multiple-tube fermentation technique has been used for many years for the sanitary analysis of water. Lactose broth tubes are inoculated with different water volumes in the presumptive test. Tubes that are positive for gas production are inoculated into brilliant green lactose bile broth in the confirmed test, and positive tubes are used to calculate the most probable number (MPN) value. The completed test is used to establish that coliform bacteria are present.

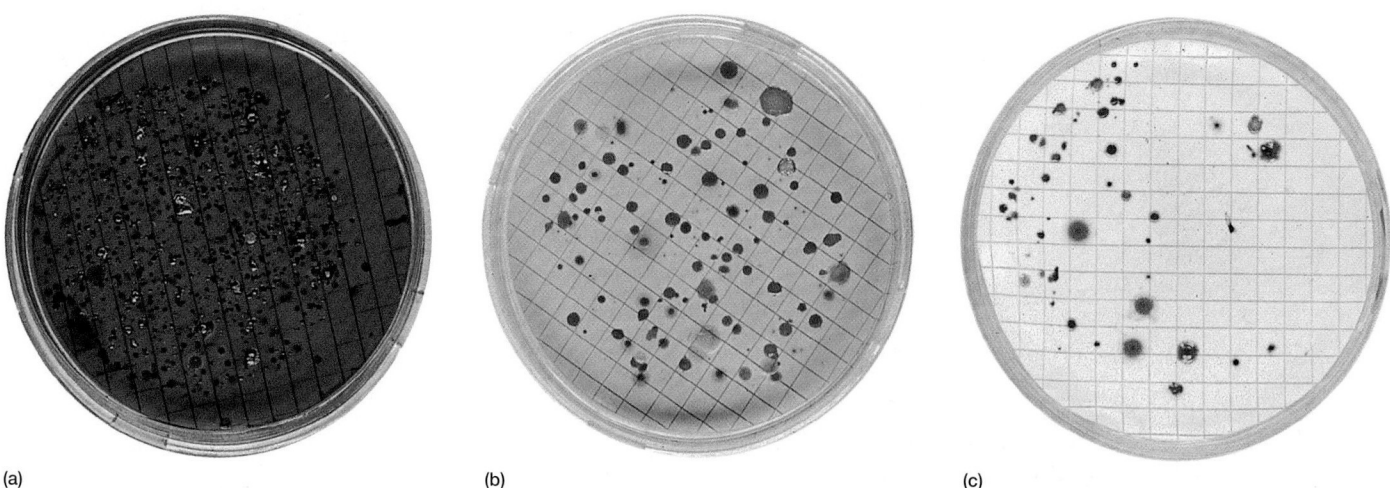

(a)   (b)   (c)

Figure 29.20   Coliform and Enterococcal Colonies.   Membrane filters made it possible to more rapidly test waters for the presence of coliforms, fecal coliforms, and fecal enterococci by the use of differential media. (**a**) Coliform reactions on an Endo medium. (**b**) Fecal coliform growth on a bile salt medium (m-FC agar) containing aniline blue dye. (**c**) Fecal enterococci growing on an azide-containing medium (KF agar) with TTC, an artificial electron acceptor, added to allow better detection of colonies.

containing the desired liquid medium. Use of the proper medium allows the rapid detection of total coliforms, fecal coliforms, or fecal streptococci by the presence of their characteristic colonies (**figure 29.20;** *see also figure 6.7*). Samples can be placed on a less selective resuscitation medium, or incubated at a less stressful temperature, prior to growth under the final set of selective conditions. An example of a resuscitation step is the use of a 2 hour incubation on a pad soaked with lauryl sulfate broth, as is carried out in the LES Endo procedure. A resuscitation step often is needed with chlorinated samples, where the microorganisms are especially stressed. The advantages and disadvantages of the membrane filter technique are summarized in **table 29.3.** Membrane filters have been widely used with water that does not contain high levels of background organisms, sediment, or heavy metals.

More simplified tests for detecting coliforms and fecal coliforms are now available. The presence-absence test (P-A test) can be used for coliforms. This is a modification of the MPN procedure, in which a larger water sample (100 ml) is incubated in a single culture bottle with a triple-strength broth containing lactose broth, lauryl tryptose broth, and bromcresol purple indicator. The P-A test is based on the assumption that no coliforms should be present in 100 ml of drinking water. A positive test results in the production of acid (a yellow color) and constitutes a positive presumptive test requiring confirmation.

To test for both coliforms and *E. coli,* the related Colilert defined substrate test can be used. A water sample of 100 ml is added to a specialized medium containing *o*-nitrophenyl-β-D-galactopyranoside (**ONPG**) and 4-methylumbelliferyl-β-D-glucuronide (**MUG**) as the only nutrients. If coliforms are present, the medium will turn yellow within 24 hours at 35°C due to the hydrolysis of ONPG, which releases *o*-nitrophenol, as shown in **figure 29.21.** To check for *E. coli,* the medium is observed under long-wavelength UV light for fluorescence. When *E. coli* is present, the MUG is modified to yield a flu-

---

Table 29.3

Advantages and Disadvantages of the Membrane Filter Technique for Evaluation of the Microbial Quality of Water

Advantages
  Good reproducibility
  Single-step results often possible
  Filters can be transferred between different media
  Large volumes can be processed to increase assay sensitivity
  Time savings are considerable
  Ability to complete filtrations on site
  Lower total cost in comparison with MPN procedure

Disadvantages
  High-turbidity waters limit volumes sampled
  High populations of background bacteria cause overgrowth
  Metals and phenols can adsorb to filters and inhibit growth

Source: Data from A. E. Greenberg, et al., *Standard Methods for the Examination of Water and Wastewater,* 16th edition, page 886, 1985. American Public Health Association, Washington, D.C.

---

orescent product. If the test is negative for the presence of coliforms, the water is considered acceptable for human consumption. The main change from previous standards is the requirement to have waters free of coliforms and fecal coliforms. If coliforms are present, fecal coliforms or *E. coli* must be tested for.

Molecular techniques are now used routinely to detect coliforms in waters and other environments, including foods. 16 S rRNA gene-targeted primers for coliforms have been developed. Using these primers, it is possible to detect one colony-forming unit (CFU) of *E. coli* per 100 ml of water, if a short enrichment step precedes the use of the PCR amplification. This allows the

Figure 29.21 The Defined Substrate Test. This much simpler test is now being used to detect coliforms and fecal coliforms in single 100 ml water samples. The medium uses ONPG and MUG (see text) as defined substrates. (**a**) Uninoculated control. (**b**) Yellow color due to the presence of coliforms. (**c**) Fluorescent reaction due to the presence of fecal coliforms.

differentiation of nonpathogenic and enterotoxigenic strains, including the shiga-toxin producing *E. coli* O157:H7. The PCR technique (pp. 316–19)

In the United States a set of general guidelines for microbiological quality of drinking waters has been developed, including standards for coliforms, viruses, and *Giardia* (**table 29.4**). If unfiltered surface waters are being used, one coliform test must be run each day when the waters have higher turbidities.

Other indicator microorganisms include **fecal enterococci.** The fecal enterococci are increasingly being used as an indicator of fecal contamination in brackish and marine water. In salt water these bacteria die back at a slower rate than the fecal coliforms, providing a more reliable indicator of possible recent pollution. The genus *Enterococcus* (pp. 516–18)

---

1. What is an indicator organism, and what properties should it have?
2. How is a coliform defined? How does this definition relate to presumptive, confirmed, and completed tests?
3. How does one differentiate between coliforms and fecal coliforms in the laboratory?
4. In what type of environment is it better to use fecal enterococci rather than fecal coliforms as an indicator organism?
5. What are the advantages and disadvantages of membrane filters for microbiological examinations of water?
6. Why has the defined substrate test with ONPG and MUG been accepted as a test of drinking water quality?

Table 29.4

Current Drinking Water Standards in the United States

Agent	Allowable Maximum Contaminant Level Goal (MCLG) or Maximum Contaminant Level (MCL)
Coliforms	MCLG = 0
	MCL = No more than 5% positive total coliform samples/month for water systems that collect > 40 samples/month. For water systems that collect < 40 routine samples/month, no more than 1 can be coliform positive. Every sample that has total coliforms must be analyzed for fecal coliforms. There cannot be any fecal coliforms or *E. coli*.
Cryptosporidium	MCLG = 0
Giardia lamblia	MCLG = 0
Legionella	MCLG = 0
Viruses (enteric)	MCLG = 0

Source: Environmental Protection Agency, USA, July, 2002.

 29.6 WASTEWATER TREATMENT

Waters often contain high levels of organic matter from industrial and agricultural wastes (e.g., from food processing, petrochemical and chemical plants, and plywood plant resin wastes), and from human wastes. It is necessary to remove organic matter by the process of wastewater treatment. Depending on the effort given to this task, it may still produce waters containing nutrients and some microorganisms, which can be released to rivers and streams.

The treatment of human waste-impacted waters, with the many pathogens that may be present, is one of the most important factors in maintaining an advanced society. In fact, the use of modern sewage treatment, together with chlorination, has led to major reductions in the worldwide spread of pathogens. This often unappreciated technology thus is at the front line of maintaining a healthy society. When sewage treatment systems break down, as often occurs with natural disasters or civil unrest, long-forgotten diseases such as cholera reappear.

Measuring Water Quality

To measure carbon removal in these wastewater treatment processes, several approaches can be used. Carbon removal can be measured (1) as **total organic carbon (TOC),** (2) as chemically oxidizable carbon by the **chemical oxygen demand (COD)** test, or (3) as biologically usable carbon by the **biochemical oxygen demand (BOD)** test. The TOC includes all carbon, whether or not it is usable by microorganisms. This is carried out by oxidizing the organic matter in a sample at high temperature in an oxygen stream and measuring the resultant CO_2 by infrared or potentiometric techniques. The COD gives a similar measurement, except that lignin often will not react with the oxidizing chemical, such as permanganate, that is used in this procedure. The

Table 29.5

The Biochemical Oxygen Demand (BOD) Test: A System with Excess and Limiting Components

Components in Excess at the End of the Incubation Period
Nitrogen
Phosphorus
Iron
Trace elements
Microorganisms
Oxygen

Component Limiting at the End of the Incubation Period
Organic matter

(a)

BOD test, in comparison, measures only the portion of the total carbon that can be oxidized by microorganisms in a 5-day period under standard conditions.

The biochemical oxygen demand is an indirect measure of organic matter in aquatic environments. It is the amount of dissolved O_2 needed for microbial oxidation of biodegradable organic matter. When O_2 consumption is measured, the O_2 itself must be present in excess and not limit oxidation of the nutrients (**table 29.5**). To achieve this, the waste sample is diluted to assure that at least 2 mg/liter of O_2 are used while at least 1 mg/liter of O_2 remains in the test bottle. Ammonia released during organic matter oxidation can also exert an O_2 demand in the BOD test, so nitrification or the **nitrogen oxygen demand (NOD)** is often inhibited by 2-chloro-6-(trichloromethyl) pyridine (nitrapyrin). In the normal BOD test, which is run for 5 days at 20°C on untreated samples, nitrification is not a major concern. However, when treated effluents are analyzed, NOD can be a problem.

In terms of speed, the TOC is fastest, but less informative in terms of biological processes. The COD is slower and involves the use of wet chemicals with higher waste chemical disposal costs. The TOC, COD, and BOD provide different but complementary information on the carbon in a water sample. It is critical to note that these measurements, concerned with carbon and carbon removal, do not directly address concerns for removal of minerals such as nitrate, phosphate, and sulfate from waters. These minerals are having worldwide impacts on cyanobacterial and algal growth in lakes, rivers, and the oceans by contributing to the process of eutrophication. The removal of dissolved organic matter and possibly inorganic nutrients, plus inactivation and removal of pathogens, are important parts of wastewater treatment.

1. What are TOC, COD, and BOD and how are these similar and different?
2. What factors can lead to a nitrogen oxygen demand (NOD) in water?
3. What components should limit the reactions in a BOD test, and what components should not limit reaction rates? Why?
4. What minerals can contribute to eutrophication?

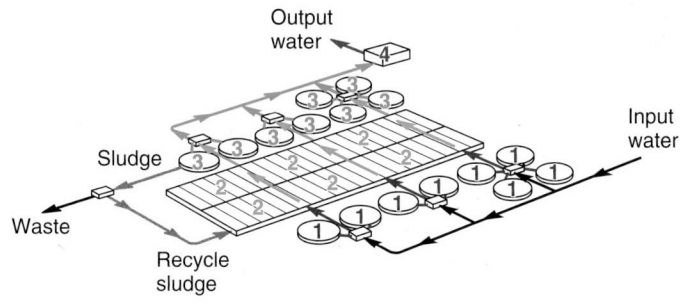

1 = Primary clarifiers
2 = Activated sludge vessels
3 = Final clarifiers
4 = Chlorination vessels

(b)

Figure 29.22 An Aerial View of a Modern Conventional Sewage Treatment Plant. Sewage treatment plants allow natural processes of self-purification that occur in rivers and lakes to be carried out under more intense, managed conditions in large concrete vessels. (**a**) A plant in New Jersey. (**b**) A diagram of flows in the plant.

Water Treatment Processes

The aerobic self-purification sequence that occurs when organic matter is added to lakes and rivers can be carried out under controlled conditions in which natural processes are intensified. This often involves the use of large basins (conventional sewage treatment) where mixing and gas exchange are carefully controlled.

An aerial photograph of a modern sewage treatment plant is shown in **figure 29.22*a***, and a schematic of this plant is shown in figure 29.22*b*. The sequence of physical removal of settleable solids (primary treatment) and secondary treatment (biological transformation of dissolved organic matter to microbial biomass and carbon dioxide) is shown, together with the final clarifiers. The final clarifiers separate the newly formed microbial biomass

Table 29.6

Major Steps in Primary, Secondary, and Tertiary Treatment of Wastes

Treatment Step	Processes
Primary	Removal of insoluble particulate materials by settling, screening, addition of alum and other coagulation agents, and other physical procedures
Secondary	Biological removal of dissolved organic matter
	Trickling filters
	Activated sludge
	Lagoons
	Extended aeration systems
	Anaerobic digesters
Tertiary	Biological removal of inorganic nutrients
	Chemical removal of inorganic nutrients
	Virus removal/inactivation
	Trace chemical removal

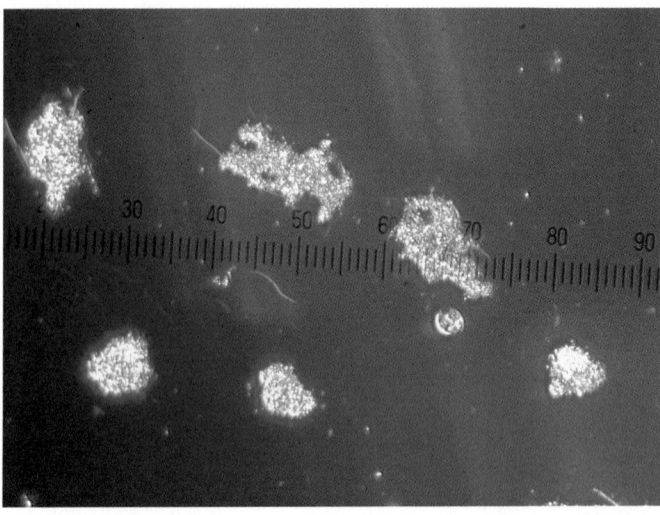

(a)

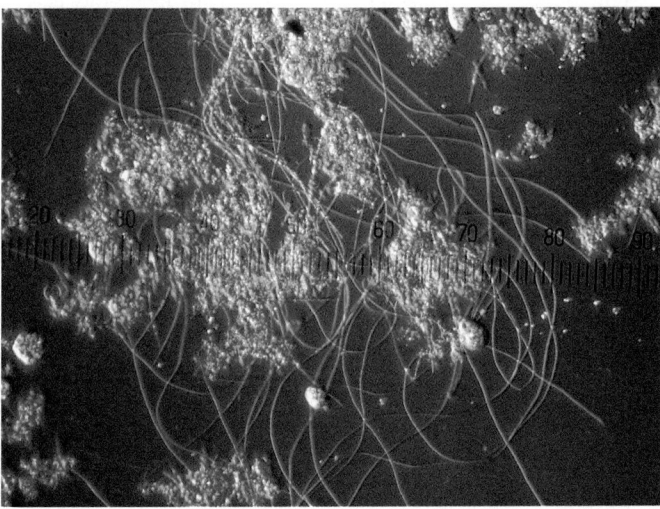

(b)

Figure 29.23 Proper Floc Formation in Activated Sludge. Microorganisms play a critical role in the functioning of activated sludge systems. The operation is dependent on the formation of settleable flocs (**a**). If the plant does not run properly, poorly settling flocs (**b**) can form due to such causes as low aeration, sulfide, and acidic organic substrates. These flocs do not settle properly because of their open or porous structure. As a consequence, the organic material is released with the treated water and lowers the quality of the final effluent.

(sewage sludge) from the processed water stream, which can be returned to the receiving body of water. At the end of the process, the water is usually chlorinated (itself an emerging environmental and human health problem) before it is released.

Conventional **wastewater treatment** normally involves primary, secondary, and tertiary treatment, as summarized in **table 29.6. Primary treatment** can physically remove 20 to 30% of the BOD that is present in particulate form. In this treatment, particulate material is removed by screening, precipitation of small particulates, and settling in basins or tanks. The resulting solid material is usually called **sludge.**

Secondary treatment is used after primary treatment for the biological removal of dissolved organic matter. About 90 to 95% of the BOD and many bacterial pathogens are removed by this process. Several approaches can be used in secondary treatment to biologically remove dissolved organic matter. All of these techniques involve similar microbial activities. Under aerobic conditions, dissolved organic matter will be transformed into additional microbial biomass plus carbon dioxide. When microbial growth is completed, under ideal conditions the microorganisms will aggregate and form a settleable stable floc structure. Minerals in the water also may be tied up in microbial biomass. When microorganisms grow, flocs can form (**figure 29.23**). As shown in figure 29.23*a,* a healthy settleable floc will be compact. In contrast, poorly formed floc can have a network of filamentous microbes that will retard settling (figure 29.23*b*).

When these processes occur with lower O$_2$ levels or with a microbial community that is too young or too old, unsatisfactory floc formation and settling can occur. The result is a **bulking sludge,** caused by the massive development of filamentous bacteria such as *Sphaerotilus* and *Thiothrix,* together with many poorly characterized filamentous organisms. These important filamentous bacteria (*see pp. 483, 488*) form flocs that do not settle well, and thus produce effluent quality problems.

An aerobic **activated sludge** system (**figure 29.24***a*) involves a horizontal flow of materials with a recycle of sludge—the active biomass that is formed when organic matter is oxidized and degraded by microorganisms. Activated sludge systems can be designed with variations in mixing. In addition, the ratio of organic matter added to the active microbial biomass can be varied. A low-rate system (low nutrient input per unit of microbial biomass), with slower growing microorganisms, will produce an effluent with low residual levels of dissolved organic matter. A high-rate

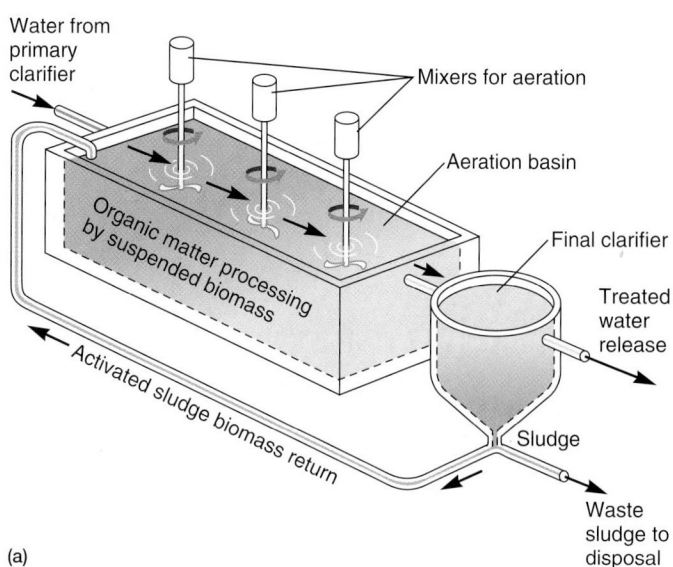

(a)

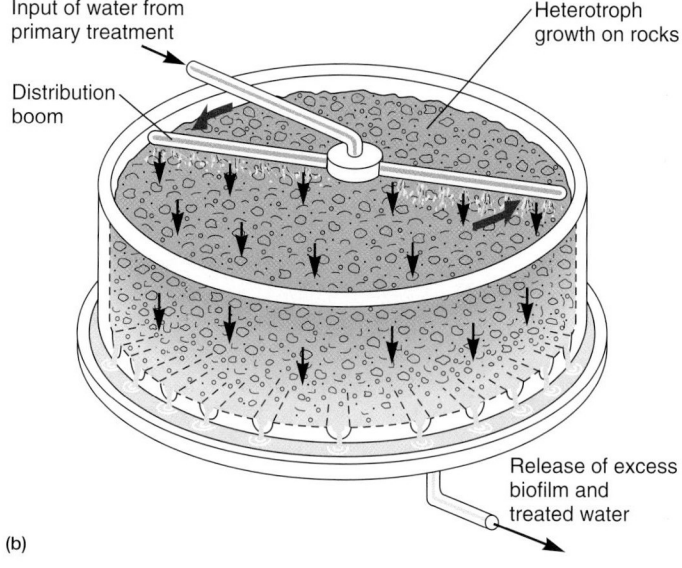

(b)

Figure 29.24 Aerobic Secondary Sewage Treatment.
(**a**) Activated sludge with microbial biomass recycling. The biomass is maintained in a suspended state to maximize oxygen, nutrient, and waste transfer processes. (**b**) Trickling filter, where waste water flows over biofilms attached to rocks or other solid supports, resulting in transformation of dissolved organic matter to new biofilm biomass and carbon dioxide. Excess biomass and treated water flow to a final clarifier. (**c**) An extended aeration process, where aeration is continued beyond the point of microbial growth, allows the microbial biomass to self-consume due to microbial energy of maintenance requirements. (The extended length of the reactor allows this process of biomass self-consumption to occur.) Minerals originally incorporated in the microbial biomass are released to the water as the process occurs.

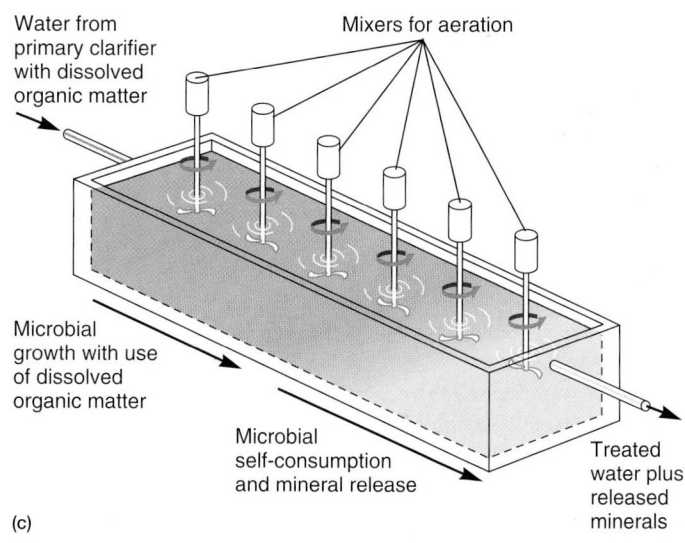

(c)

system (high nutrient input per unit of microbial biomass), with faster growing microorganisms, will remove more dissolved organic carbon per unit time, but produce a poorer quality effluent.

Aerobic secondary treatment also can be carried out with a **trickling filter** (figure 29.24*b*). The waste effluent is passed over rocks or other solid materials upon which microbial films have developed, and the microbial community in these films degrades the organic waste. A sewage treatment plant can be operated to produce less sludge by employing the **extended aeration** process (figure 29.24*c*). Microorganisms grow on the dissolved organic matter, and the newly formed microbial biomass is eventually consumed to meet maintenance energy requirements. This requires extremely large aeration basins and extended aeration times. In addition, with the biological self-utilization of the biomass, minerals originally present in the microorganisms are again released to the water.

All aerobic processes produce excess microbial biomass, or sewage sludge, which contains many recalcitrant organics. Often

the sludges from aerobic sewage treatment, together with the materials settled out in primary treatment, are further treated by **anaerobic digestion** (**figure 29.25**). Anaerobic digesters are large tanks designed to operate with continuous input of untreated sludge and removal of the final, stabilized sludge product. Methane is vented and often burned for heat and electricity production. This digestion process involves three steps: (1) the fermentation of the sludge components to form organic acids, including acetate; (2) production of the methanogenic substrates: acetate, CO_2, and hydrogen; and finally, (3) methanogenesis by the methane producers. These methanogenic processes, summarized in **table 29.7,** involve critical balances between oxidants and reductants. To function most efficiently, the hydrogen concentration must be maintained at a low level. If hydrogen and organic acids accumulate, methane production can be inhibited, resulting in a stuck digester. Methanogenic archaea (pp. 444–47)

Anaerobic digestion has many advantages. Most of the microbial biomass produced in aerobic growth is used for methane

Figure 29.25 High-Efficiency Anaerobic Bioreactors Used for Sludge Digestion and Methane Production. These "egg-shaped" units are so well insulated that they maintain their temperature (39°C) without outside heating, and the methane can be burned to provide electricity. These reactors are located at Kiel, Germany.

Table 29.7			
Sequential Reactions in the Anaerobic Biological Use of Organic Wastes			
Process Step	**Substrates**	**Products**	**Major Microorganisms**
Fermentation	Organic polymers	Butyrate, propionate, lactate, succinate, ethanol, acetate,[a] H_2,[a] CO_2[a]	*Clostridium* *Bacteroides* *Peptostreptococcus* *Peptococcus* *Eubacterium* *Lactobacillus*
Acetogenic reactions	Butyrate, propionate, lactate, succinate, ethanol	Acetate, H_2, CO_2	*Syntrophomonas* *Syntrophobacter* *Acetobacterium*
Methanogenic reactions	Acetate	$CH_4 + CO_2$	*Methanosarcina* *Methanothrix*
	H_2 and HCO_3^-	CH_4	*Methanobrevibacter* *Methanomicrobium* *Methanogenium* *Methanobacterium* *Methanococcus* *Methanospirillum*

[a]Methanogenic substrates produced in the initial fermentation step.

production. The resulting sludge occupies less volume and can be dried easily. However, heavy metals and other environmental contaminants are often concentrated in the sludge. There may be longer-term environmental and public health effects from disposal of this material on land or in water (**Disease 29.3**).

Tertiary treatment purifies wastewaters more than is possible with primary and secondary treatments. It is particularly important to remove nitrogen and phosphorus compounds that can promote eutrophication. Organic pollutants can be removed with activated carbon filters. Phosphate usually is precipitated as calcium or iron phosphate (for example, by the addition of lime). Excess nitrogen may be removed by "stripping," volatilization as NH_3 at high pHs. Ammonia itself can be chlorinated to form dichloramine, which is then converted to molecular nitrogen. In

some cases, biological processes can be used to remove nitrogen and phosphorus. A widely used process for nitrogen removal is denitrification (*see p. 185*). Here nitrate, produced under aerobic conditions, is used as an electron acceptor under conditions of low oxygen with organic matter added as an energy source. Nitrate reduction yields nitrogen gas and nitrous oxide (N_2O) as the major products. Currently there is a great deal of interest in anaerobic nitrogen removal processes. These include the anammox process where ammonium ion (used as the reductant), is reacted with nitrite (the oxidant) produced by partial nitrification (*see p. 189*). By use of this process, as carried out in Holland, the partial nitrification step gave a 53% transformation to nitrite, and no nitrate production. When reacted with the ammonium ion in the anammox process, a total of 80% of the beginning ammonium ion

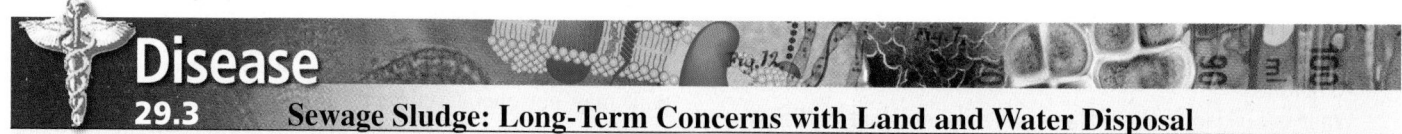

Disease
29.3   Sewage Sludge: Long-Term Concerns with Land and Water Disposal

Sewage treatment plants have allowed large cities to develop near rivers and lakes and still maintain the quality of the water. Large quantities of sludge are produced during sewage treatment and are usually subjected to anaerobic digestion. This process transforms complex organic matter (including microorganisms that grew in the aerobic treatment process) to methane and CO_2. At the same time heavy metals are concentrated in the residual sludge, and viable cysts of free-living protozoa may be present.

The sludge is disposed of on land or, in large urban areas near the ocean, by dumping at designated disposal sites in coastal waters.

When such sludges are dumped offshore, free-living protozoa of the genus *Acanthamoeba* may be released in the water, where they may infect bathers. *Acanthamoeba* infections are more widespread than usually is recognized, and the organisms can be common contaminants in water that tests negative for coliforms and fecal coliforms.

The use of water-based waste disposal systems has, of course, allowed major improvements in urban living and the retirement of chamber pots to museums. A product of this advancement in public health is sewage sludge. NIMBY (not in my backyard) has been the usual means of managing this problem.

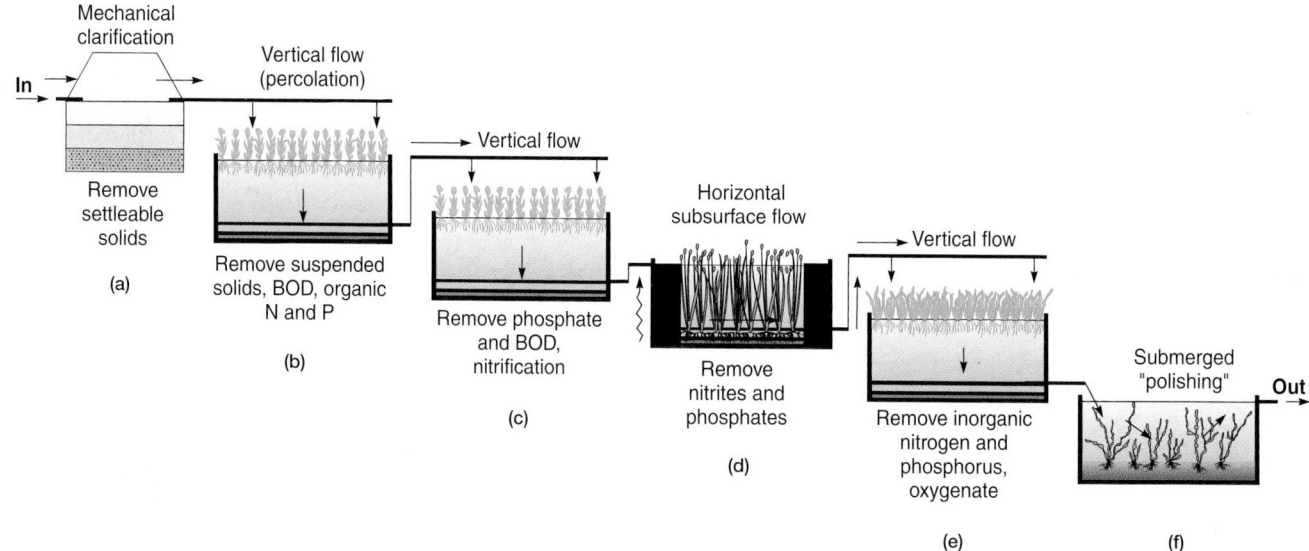

Figure 29.26   Constructed Wetland for Wastewater Treatment. Multistage constructed wetland systems can be used for organic matter and phosphate removal. Free-floating macrophytes (**b,c,e**), such as duckweed and water hyacinth, can be used for a variety of purposes. Emergent macrophytes (**d**) such as bulrush, allow surface flow as well as vertical and horizontal subsurface flow. Submerged vegetation (**f**), such as waterweed, allows final "polishing" of the water. These wetlands also can be designed for nitrification and metal removal from waters.

was converted to N_2 gas, showing the value of this new technology. For phosphorus removal, aerobic and anaerobic conditions are used alternately in a series of treatments, and phosphorus accumulates in specially adapted microbial biomass as polyphosphate. Tertiary treatment is expensive and usually not employed except where necessary to prevent obvious ecological disruption.

Wetlands are a vital natural resource and a critical part of our environment, and increasingly efforts are being made to protect these fragile aquatic communities from pollution. Surprisingly a major means of treatment of wastes is the use of **constructed wetlands** where the basic components of natural wetlands (soils, aquatic plants, waters) are used as a functional waste treatment system. Constructed wetlands now are increasingly employed in the treatment of liquid wastes and for bioremediation (*see pp. 982–84*). This system uses floating, emergent, or submerged plants, as shown in **figure 29.26.** The aquatic plants provide nutrients in the root zone, which can support microbial growth. Especially with the emergent plants, the root zone can be maintained in an anaerobic state in which sulfide, synthesized by *Desulfovibrio* using root zone organic matter as an energy source, can trap metals. Constructed wetlands also are being used to treat acid mine drainage (AMD) in many parts of the world. Higher-strength industrial wastes also can be treated.

1. Explain how primary, secondary, and tertiary treatment are accomplished.
2. What is bulking sludge? Name several important microbial groups that contribute to this problem.
3. What are the steps of organic matter processing that occur in anaerobic digestion? Why is acetogenesis such an important step?
4. After anaerobic digestion is completed, why is sludge disposal still of concern?
5. Why might different aquatic plant types be used in constructed wetlands?

 ## 29.7   GROUNDWATER QUALITY AND HOME TREATMENT SYSTEMS

Groundwater, or water in gravel beds and fractured rocks below the surface soil, is a widely used but often unappreciated water resource. In the United States groundwater supplies at least 100 million people with drinking water, and in rural and suburban areas beyond municipal water distribution systems, 90 to 95% of all drinking water comes from this source.

The great dependence on this resource has not resulted in a corresponding understanding of microorganisms and microbiological processes that occur in the groundwater environment. Increasing attention is now being given to predicting the fate and effects of groundwater contamination on the chemical and microbiological quality of this resource. Pathogenic microorganisms and dissolved organic matter are removed from water during subsurface passage through adsorption and trapping by fine sandy materials, clays, and organic matter. Microorganisms associated with these materials—including predators such as protozoa—can use the trapped pathogens as food. This results in purified water with a lower microbial population.

This combination of adsorption-biological predation is used in home treatment systems (**figure 29.27**). Conventional **septic tank** systems include an anaerobic liquefaction and digestion step that occurs in the septic tank itself (the tank functions as a simple anaerobic digester). This is followed by organic matter ad-

sorption and entrapment of microorganisms in the aerobic leach-field environment where biological oxidation occurs. A septic tank may not operate correctly for several reasons. It will not function properly if the retention time of the waste in the septic tank is too short. As a result, undigested solids move into the leach field, gradually plugging the system. If the leach field floods and becomes anaerobic, biological oxidation does not occur, and effective treatment ceases.

Especially when a suitable soil is not present and the septic tank outflow from a conventional system drains too rapidly to the deeper subsurface, problems can occur. Fractured rocks and coarse gravel materials provide little effective adsorption or filtration. This may result in the contamination of well water with pathogens and the transmission of disease. In addition, nitrogen and phosphorus from the waste can pollute the groundwater. This leads to nutrient enrichment of ponds, lakes, and rivers as the subsurface water enters these environmentally sensitive water bodies.

Domestic and commercial on-site septic systems are now being designed with nitrogen and phosphorus removal steps. Nitrogen is usually removed by nitrification-denitrification processes (p. 640) with organic matter provided by sawdust or a similar material. Currently used systems function with essentially no maintenance for up to 8 years. For phosphorus removal, a reductive iron dissolution process can be used. With the need to control nitrogen and phosphorus releases from septic systems around the world, there will be increased emphasis on use of these and similar technologies in the future.

Subsurface zones (*see section 30.8*) also can become contaminated with pollutants from other sources. Land disposal of sewage sludges, illegal dumping of septic tank pumpage, improper toxic waste disposal, and runoff from agricultural operations all contribute to groundwater contamination with chemicals and microorganisms.

Many pollutants that reach the subsurface will persist and may affect the quality of groundwater for extended periods. Much research is being conducted to find ways to treat groundwater in place—**in situ treatment.** Microorganisms and microbial processes are critical in many of these remediation efforts (*see chapter 42*).

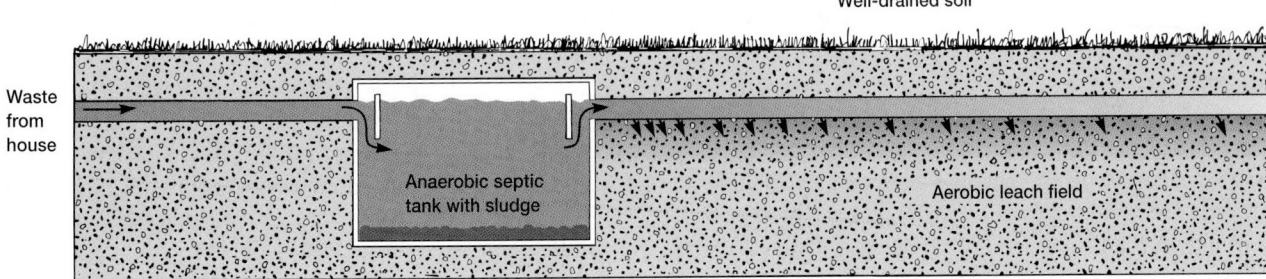

Figure 29.27   The Conventional Septic Tank Home Treatment System.   This system combines an anaerobic waste liquefaction unit (the septic tank) with an aerobic leach field. Biological oxidation of the liquefied waste takes place in the leach field, unless the soil becomes flooded.

1. In rural areas, approximately what percentage of the water used for human consumption is groundwater?
2. What factors can limit microbial activity in subsurface environments? Consider the energetic and nutritional requirements of microorganisms in your answer.
3. How, in principle, are a conventional septic tank system and a leach-field system supposed to work? What alternatives are available to remove nitrogen and phosphorus from effluents? What factors can reduce the effectiveness of this system?

Summary

29.1 Aquatic Environments and Microorganisms

a. The largest portion of waters on the Earth is marine (97%). Most of this is cold (2 to 3°C) and at high pressure. Fresh waters are a minor but important part of the Earth's waters.

b. The movement of materials in aquatic environments extends over long distances and occurs at varied rates. Mixing and diffusion are critical processes in creating unique environments for different microorganisms.

c. Oxygen solubility and diffusion rates in waters are limited; waters are low oxygen diffusion rate environments, in comparison with soils (**figure 29.1**). Carbon dioxide, nitrogen, hydrogen, and methane are also important gases for microbial functioning in waters.

d. If sufficient organic matter is present, heterotrophic microorganisms can use this dissolved O_2 more rapidly than it can be replenished. Respiration or decomposition of algal and cyanobacterial blooms also can cause O_2 depletion.

e. Nutrient concentrations in waters can vary, and diffusion produces unique environments for microorganisms by creating gradients. The Winogradsky column is a model laboratory system that allows gradients and unique microbial communities to develop (**figure 29.3**).

f. The nutrient composition of the ocean influences the C:N:P ratio of the phytoplankton, which is called the Redfield ratio. This ratio is important for predicting nutrient cycling in oceans. Atmospheric additions of minerals, including iron and nitrogen, affect this ratio and global-level oceanic processes.

g. The microbial loop is an important part of the food web in waters (**figure 29.4**). This results from bacterial growth on organic matter released by the phytoplankton, followed by use of the bacteria as a prey by protozoa. The nutrients in a portion of these bacterial prey are then released in mineral forms, making them again available for use by the phytoplankton.

29.2 The Microbial Community

a. The marine microbial community is dominated, in terms of numbers and biomass, by ultramicrobacteria or nanobacteria. Archaea are important components of the microbial community. Viruses are present at high concentrations in many waters, and occur at 10-fold higher levels than the bacteria. In marine systems they may play a major role in controlling cyanobacterial development and nutrient turnover.

b. Many unusual microbial groups are found in waters, especially when oxidants and reductants can be linked. These include *Thioploca* and *Thiomargarita*, both of which are found in coastal areas where nutrient mixing occurs. *Thiomargarita* is the world's largest known bacterium (**figure 29.5**).

c. Aquatic fungi are important parts of the aquatic microbial community. These include the chytrids (**figure 29.7**), with a motile zoosporic stage, and the Ingoldian fungi (**figure 29.8**), which often have tetraradiate structures. Both of these are uniquely adapted to an aquatic existence, and the chytrids may contribute to disease in amphibians.

d. Microorganisms and their metabolic products interact with the atmosphere and are important in the field of aerobiology.

29.3 Marine Environments

a. Barophiles, or piezophiles, are important in deep marine environments. These include barotolerant, barophilic, and extreme barophilic procaryotes.

b. Materials can move in aquatic environments over extended distances and depths. This may include animal and human wastes, carcasses, and plant detritus. Dust is transported over wide distances. All of these processes result in the constant introduction of new organisms.

c. The global cycles of carbon, nitrogen, phosphorus and sulfur are influenced by microbial processes that occur in the vast regions of the world's oceans. This is especially important for the sulfur cycle, where dimethylsulfide (DMS) release from algae can lead to changes in the acidity of the atmosphere and cloud formation.

29.4 Freshwater Environments

a. Eutrophication can be caused by nutrient releases from rural and urban areas. Sources of nitrogen and phosphorus are particularly important in eutrophication. These releases can be from point and nonpoint sources.

b. Lakes can be oligotrophic (low-nutrient) and eutrophic (high-nutrient) environments (**figure 29.15**). The addition of organic matter, especially with the microbial use and release of N and P, can lead to eutrophication.

c. In rivers the addition of organic matter can produce dissolved O_2 sag curves and diurnal O_2 changes in the latter stages of self-purification (**figure 29.17**). This occurs with materials added from point sources.

d. Much of the marine environment is covered by sea ice. Specially adapted microbial communities are found at the ice-water interface. Deep Antarctic lakes (such as Lake Vostok) and dry valley frozen lakes provide unique environments for microbes.

29.5 Waters and Disease Transmission

a. Water purification can involve the use of sedimentation, coagulation, chlorination, and rapid and slow sand filtration. Chlorination may lead to the formation of organic disinfection by-products, including trihalomethane (THM) compounds, which are potential carcinogens.

b. *Cryptosporidium, Cyclospora,* viruses, and *Giardia* are of concern, as conventional water purification and chlorination will not always assure their removal and inactivation to acceptable limits.

c. Indicator organisms are used to indicate the presence of pathogenic microorganisms. Most probable number (MPN) and membrane filtration procedures are employed to estimate the number of indicator organisms present. Presence-absence (P-A) tests for coliforms and defined substrate tests for coliforms and *E. coli* allow 100 ml water volumes to be tested with minimum time and materials.

d. Molecular techniques based on the polymerase chain reaction (PCR) can be used to detect waterborne Shiga-toxin producing *E. coli* O157:H7, when a preenrichment step is used.

29.6 Wastewater Treatment

a. The biochemical oxygen demand (BOD) test is an indirect measure of organic matter that can be oxidized by the aerobic microbial community. In this assay, oxygen should never limit the rate of reaction. The chemical oxygen demand (COD) and total organic carbon (TOC) tests provide information on carbon that is not biodegraded in the 5-day BOD test.

b. Conventional sewage treatment is a controlled intensification of natural self-purification processes, and it can involve primary, secondary, and tertiary treatment (**figure 29.22**).

c. Constructed wetlands involve the use of aquatic plants (floating, emergent, submerged) and their associated microorganisms for the treatment of liquid wastes (**figure 29.26**).

29.7 Groundwater Quality and Home Treatment Systems

a. Home treatment systems operate on general self-purification principles. The conventional septic tank (**figure 29.27**) provides anaerobic liquefaction and digestion whereas the aerobic leach-field allows oxidation of the soluble effluent. These systems are now being designed to provide nitrogen and phosphorus removal, to lessen impacts of on-site sewage treatment systems on vulnerable marine and fresh waters.

b. Groundwater is an important resource that can be affected by pollutants from septic tanks and other sources. This vital water source must be protected and improved.

Key Terms

activated sludge 638
anaerobic digestion 639
anoxic 617
barophile 625
biochemical oxygen demand (BOD) 636
bulking sludge 638
chemical oxygen demand (COD) 636
coagulation 631
coliform 633
constructed wetland 641
defined substrate test 633
dinospore 627
disinfection by-products (DBPs) 631
diurnal oxygen shift 629
epilimnion 628

eutrophic 628
eutrophication 628
extended aeration 639
fecal coliform 633
fecal enterococci 636
groundwater 642
high oxygen diffusion environment 617
hypolimnion 628
hypoxic 617
indicator organism 633
Ingoldian fungi 622
in situ treatment 642
low oxygen diffusion environment 616
membrane filtration technique 633

microbial loop 620
most probable number (MPN) 633
MUG 635
nanobacteria 621
nitrogen oxygen demand (NOD) 637
nonpoint source pollution 629
oligotrophic 628
ONPG 635
phytoplankton 619
point source pollution 629
potable 633
presence-absence (P-A) test 633
primary treatment 638
rapid sand filter 631
Redfield ratio 619
reverse methanogenesis 625

secondary treatment 638
sedimentation basin 631
septic tank 642
settling basin 631
slow sand filter 631
sludge 638
tertiary treatment 640
thermocline 628
total organic carbon (TOC) 636
trickling filter 639
trihalomethanes (THMs) 631
ultramicrobacteria 621
virioplankton 622
wastewater treatment 638
wetlands 641
Winogradsky column 618

Questions for Thought and Review

1. Why might you describe the interaction of many microorganisms with oxygen as a "love-hate" relationship? How do these microorganisms deal with this important gas?
2. How might you measure the activity of nanobacteria or ultramicrobacteria, in spite of the fact that they are so difficult to culture and observe?
3. What are the functional strategies of organisms such as *Thioploca* and *Thiomargarita*? Where would you go to find similar organisms, and how would you search for them?
4. You wish to study the characteristics of microorganisms from a 10,000 meter depth in the ocean. What effects might decompression have on these organisms, and what type of equipment would you design to work with them under their natural conditions?
5. Discuss the advantages and disadvantages for microorganisms that become attached to surfaces.
6. Why are levels of virus particles so high in many marine environments? Consider the requirements for viral replication and destruction.
7. How might one manage waters to control *Pfiesteria piscicida?*
8. How might it be possible to rejuvenate an aging eutrophic lake? Consider physical, chemical, and biological approaches.
9. You wish to develop a constructed wetland for removal of metals from a stream at one site, and at another site, you wish to treat acid mine drainage. How might you approach each of these problems?
10. Why are indicator organisms still being used despite the fact that methods are available for the direct isolation of most pathogens that might occur in water?
11. What alternatives, if any, can one use for protection against microbiological infection when swimming in polluted recreational water? Assume that you are part of a water rescue team.
12. What possible alternatives could be used to eliminate N and P releases from sewage treatment systems? What suggestions could you make that might lead to new technologies?

Critical Thinking Questions

1. If we were able to decrease or increase the activity of the microbial loop in the ocean, which should be done and why? Would either shift be desirable?
2. There have been numerous "fish kills" in the waters of the Chesapeake Bay region. Fishermen blame poultry farmers (poultry waste disposal) and the overuse and subsequent runoff of residential fertilizers by homeowners. What components in poultry waste and fertilizer can create an ecological imbalance that would result in hundreds of dead fish floating in inlet waterways?

For recommended readings on these and related topics, see Appendix V.

chapter 30

Microorganisms in Terrestrial Environments

Concepts

- Terrestrial [Latin *terra,* earth] environments are dominated by inert solid materials. Organic substances, including microorganisms, are usually a minor part of a soil.
- Water, when present, will be predominantly in thin water films on particle surfaces. Microorganisms in these water films have good contact with gases in air, including oxygen.
- Oxygen is much more accessible to microorganisms in soils because its flux into thin water films is not limited by a large continuous water bulk phase. The flux rate of oxygen in air is approximately 10,000 times faster than in water.
- Microorganisms growing in such higher oxygen flux environments have developed complex physical and physiological mechanisms to deal with potentially toxic oxygen species. Fungi even develop structures impermeable to oxygen.
- Soils can have isolated water-saturated zones that become anaerobic. These may be considered as isolated aquatic environments surrounded by high-oxygen flux zones.
- There is a great variety of terrestrial environments: dry, cold, temperate, tropical, and geothermally heated zones.
- Insects, nematodes, and other soil animals interact with microorganisms to influence nutrient cycling and other processes.
- The subsurface supports a varied and diverse microbial community. These microorganisms are maintained by nutrient sources such as materials leaching from the surface, decomposition of buried plant remains, and methane synthesis.
- Fungi play important roles in the functioning of plants. They occur inside and on the surfaces of plants, and can have positive and negative effects. Most plant roots harbor fungi.

Terrestrial plants and filamentous fungi have developed long-term relationships that benefit both partners. The tips of pine tree roots are usually surrounded by dense fungal sheaths that are part of a hyphal network which extends out into the soil. The plant supplies organic matter to maintain the fungus, and the fungus, in turn, provides the plant with nutrients and water.

- Soil microorganisms interact with plants and the atmosphere. Critical microbial products are nitrous oxide and methane, both greenhouse gases. In addition, microorganisms produce chloromethane and cyanide, usually considered as anthropogenic pollutants.

They [the leaves] that waved so loftily, how contentedly, they return to dust again and are laid low, resigned to lie and decay at the foot of the tree and afford nourishment to new generations of their kind, as well as to flutter on high!
—Henry D. Thoreau

oils are dynamic and develop over time. This process may take decades and even centuries; the organic matter in soils can be thousands of years old. Because of changes in plant growth, temperature, rainfall, disturbance and erosion, a soil that has taken hundreds of years to form can be quickly degraded if the microbial community is activated. For example, this can occur when a bog soil is drained, which increases oxygen access to the accumulated organic matter. In most soils the major producers of organic matter are the vascular plants, although algae, cyanobacteria, and photosynthetic bacteria also can contribute to these processes, especially in desert crust environments.

Soil is the habitat for a variety of organisms, including bacteria, fungi, protozoa, insects, nematodes, worms, and many other animals. Viruses also are present in soils. This complex biological community contributes to the formation, maintenance, and in some situations, the degradation and disappearance of soils.

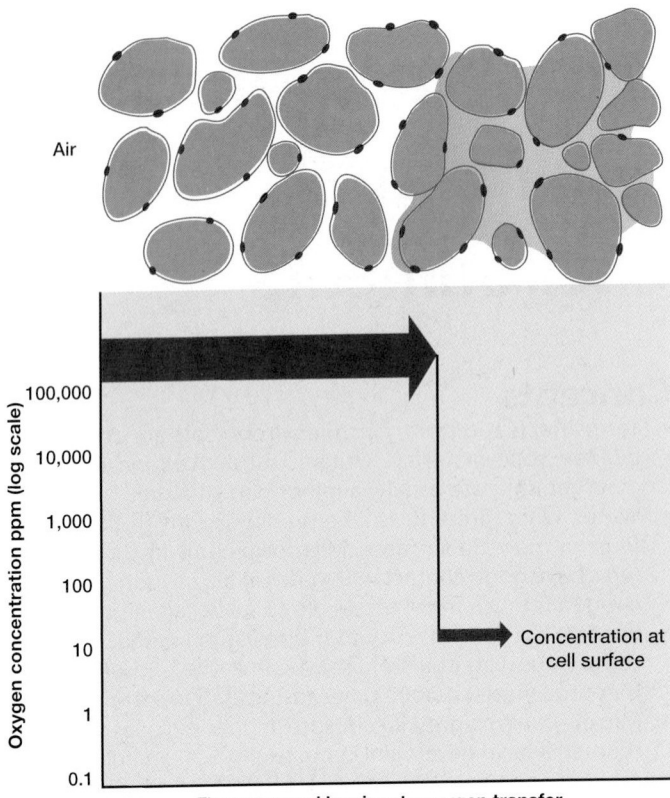

Figure 30.1 Oxygen Concentrations and Fluxes in a Soil. Microorganisms in thin water films on the surface of soil particles have good access to oxygen. In comparison, microbes in isolated water volumes have limited oxygen fluxes, creating miniaquatic environments.

30.1 SOILS AS AN ENVIRONMENT FOR MICROORGANISMS

Soil are formed in a wide variety of environments. These environments range from Arctic tundra regions, where approximately 11% of the Earth's soil carbon pool is stored, to Antarctic dry valleys, where there are no vascular plants. In addition, deeper subsurface zones, where plant roots and their products cannot penetrate, also have microbial communities. Microbial activities in these environments can lead to the formation of minerals such as dolomite; microbial activity also occurs in deep continental oil reservoirs, in stones, and even in rocky outcrops. These microbes are dependent on energy sources from algae, nutrients in rainfall, and dust.

Most soils are dominated by inorganic geological materials, which are modified by the biotic community, including the microorganisms and plants, to form soils. An important defining characteristic of a soil, from a microbial viewpoint, is that the microorganisms are in close physical contact with oxygen; they are located in thin water films on the particle surfaces where oxygen is present at high levels and can be easily replenished from the gaseous phase. When microorganisms use oxygen, it can be replenished rapidly by diffusion, thus maintaining the microbes under aerobic conditions. Oxygen diffusion through air in the soil occurs about 10,000 times faster than it does through water (**figure 30.1**). As shown in this figure, the oxygen concentrations and flux rates in pores and channels is high, whereas within water-filled zones the oxygen flux rate is much lower. As an example, particles as small as about 2.0 mm can be aerobic on the outside and anaerobic on the inside.

Depending on the physical characteristics of the soil, rainfall or irrigation may rapidly change a soil from being ideal, in terms of having reoxygenating thin water films, to an environment with isolated pockets of water, which are "miniaquatic" habitats. If this process of flooding continues, a water-logged soil can be created that is more like a lake sediment.

Shifts in water content and gas fluxes also affect the concentrations of CO_2, CO, and other gases present in the soil atmosphere, as noted in **table 30.1**. These changes will be accentuated

Table 30.1

Concentrations of Oxygen and Carbon Dioxide in the Atmosphere of a Tropical Soil under Wet and Dry Conditions

	Oxygen Content (%)		Carbon Dioxide Content (%)	
Soil Depth (cm)	Wet	Dry	Wet	Dry
10	13.7	20.7	6.5	0.5
25	12.7	19.8	8.5	1.2
45	12.2	18.8	9.7	2.1
90	7.6	17.3	10.0	3.7
120	7.8	16.4	9.6	5.1

From E. W. Russell, *Soil Conditions and Plant Growth*, 10th edition. Copyright © 1973 Longman Group Limited, Essex, United Kingdom. Reprinted by permission.

Note: Normal air contains approximately 21% oxygen and 0.035% carbon dioxide.

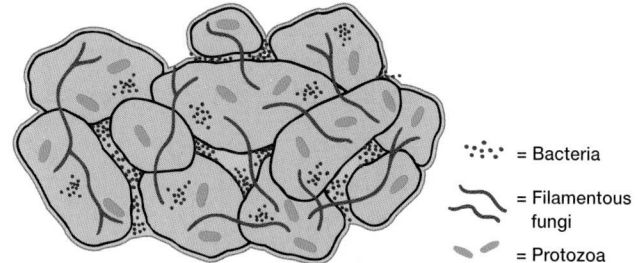

Figure 30.2 The Microenvironment—The World of Microorganisms in Soil. Bacteria tend to be present as isolated microcolonies on surfaces and in pores, which are covered by thin water films. Filamentous fungi are able to grow on and between these aggregated particles, or peds. Protozoa move in water films and graze on bacteria, especially when they are not protected in soil pores.

in the smaller pores where many bacteria are found. At lower depths less oxygen is available, especially in wetter, less permeable soils. Another factor that affects the levels of oxygen and CO_2 in a soil are plant roots. The roots of plants growing in normal aerated soils also consume oxygen and release CO_2, influencing the concentrations of these gases in the root environment.

1. Why are Arctic tundra soils important in terms of carbon storage?
2. Contrast differences in oxygen flux rates and concentrations in a soil with those in a miniaquatic environment.
3. How do the concentrations of oxygen and carbon dioxide differ between the atmosphere and the soil interior?

30.2 MICROORGANISMS IN THE SOIL ENVIRONMENT

If we look at a soil in greater detail (**figure 30.2**), we find that the bacteria, fungi, and protozoa use different functional strategies to take advantage of this complex physical matrix. Most soil bacteria are located on the surfaces of soil particles and require water and nutrients that must be located in their immediate vicinity. Bacteria are found most frequently on surfaces within smaller soil pores (2 to 6 μm in diameter). Here they are probably less liable to be eaten by protozoa, unlike bacteria that are located on the exposed outer surface of a sand grain or organic matter particle.

Terrestrial filamentous fungi, in comparison, bridge open areas between soil particles or aggregates called **peds,** and are exposed to high levels of oxygen (figure 30.2). These fungi will tend to darken and form oxygen-impermeable structures including sclerotia and hyphal cords. This is particularly important for the functioning of basidiomycetes, which form such structures as an oxygen sealing mechanism. Within these structures, the filamentous fungi move nutrients and water over great distances, in-

cluding across air spaces, a unique part of their functional strategy. These oxidatively polymerized, oxygen-impermeable hyphal boundaries do not usually occur in fungi growing in aquatic environments.   Basidiomycetes (pp. 545–46)

The microbial populations in soils can be very high. In a surface soil the bacterial population can approach 10^8 to 10^9 cells per gram dry weight of soil as measured microscopically. Fungi can be present at up to several hundred meters of hyphae per gram of soil. We tend to think that soil fungi are small like the mushrooms sprouting from our lawns. This is natural because most of the fungal thallus lies beneath the soil surface, but such a view is sometimes quite inaccurate. A case in point is the fungus *Armillaria bulbosa,* which lives associated with tree roots in hardwood forests. An individual *Armillaria* clone that covers about 30 acres has been discovered in the Upper Peninsula of Michigan. It is estimated to weigh a minimum of 100 tons (an adult blue whale may weigh 150 tons) and be at least 1,500 years old. Thus some fungal mycelia are among the largest and most ancient living organisms on Earth.

It is important to remember, when discussing soils and their microorganisms, that only a minor portion (approximately 10%) of the microscopically observable organisms making up this biomass have been cultured. We are gradually learning more about these uncultured microorganisms through the use of molecular techniques. For example, novel groups of the *Crenarchaeota (see pp. 442–44)* have been discovered in forest soil and the ocean by extracting microbial DNA and amplifying it with the polymerase chain reaction. Examination of soils from different areas of the world continues to yield surprises. Microorganisms are present and prolific in subsurface environments, including oil reservoirs. Hyperthermophilic archaea have been found in such harsh subsurface environments and are probably indigenous to these poorly understood regions of our world.

The coryneforms, the nocardioforms, and the true filamentous bacteria or actinomycetes (**table 30.2**) are an important and less studied part of the soil microbial community. These bacteria

Table 30.2

Easily Cultured Gram-Positive Irregular Branching and Filamentous Bacteria Common in Soils

Bacterial Group	Representative Genera	Comments and Characteristics
Coryneforms	*Arthrobacter*	Rod-coccus cycle
	Cellulomonas	Important in degradation of cellulose
	Corynebacterium	Club-shaped cells
Mycobacteria	*Mycobacterium*	Acid-fast
Nocardioforms	*Nocardia*	Rudimentary branching
Actinomycetes	*Streptomyces*	Aerobic filamentous bacteria
	Thermoactinomyces	Higher temperature growth

play a major role in the degradation of hydrocarbons, older plant materials, and soil humus. In addition, some members of these groups actively degrade pesticides. The filamentous actino-mycetes, primarily of the genus *Streptomyces,* produce an odor-causing compound called **geosmin,** which gives soils their characteristic earthy odor. Polyprosthecate bacteria such as the genera *Verrucomicrobium, Pedomicrobium,* and *Prosthecobacter* are present in soils at high levels. With their small size, and the difficulties involved in culturing them, they have been largely overlooked in assessments of soil microbial diversity completed to the present time.   Actinobacteria (chapter 24)

As with aquatic environments (*see chapter 29*), microorgan-isms are constantly being added to soils from waters, wind, dust, plants, and animal sources. Most of these added microorganisms will not survive, either being outcompeted by indigenous mi-crobes or being consumed by predators such as the protozoa.

The microbial community in soil makes important contribu-tions to biogeochemical cycling and the carbon, nitrogen, sulfur, iron, and manganese cycles (*see chapter 28*). Because the soil is primarily an oxidized environment, the inorganic forms of these elements will tend to be in the oxidized state. If there are local-ized water-saturated, lower oxygen-flux environments, the bio-geochemical cycles will shift toward reduced species. Biogeochemical cycles, (section 28.3)

Soil insects and other animals such as nematodes and earth-worms also contribute to organic matter transformations in soils. These organisms carry out decomposition, often leading to the re-lease of minerals, and physically "reducing" the size of organic par-ticles such as plant litter. This increases the surface area and makes organic materials more available for use by bacteria and fungi. Earthworms also mix substrates with their internal gut microflora and enzymes; this contributes substantially to decomposition and has major effects on soil structure and the soil microbial community.

Protozoa also can influence nutrient cycling by feeding on "palatable" microorganisms. This process of **microbivory,** or use of microorganisms as a food source, results in higher rates of ni-trogen and phosphorus mineralization, thus increasing the avail-ability of nutrients for plant growth.   Protozoa (chapter 27)

A significant part of the biological activity of soils arises from enzymes released by plants, insects, and other animals, and from lysed microorganisms. These free enzymes contribute to many hydrolytic degradation reactions, such as proteolysis; catalase and peroxidase activities also have been detected.

A microbial loop regenerates nutrients in soils as it does in aquatic environments. In this process carbon and nutrients fixed by plants will be turned over rapidly by bacteria and predatory protozoa and will not be available for transfer to higher trophic levels in the food web. Protozoa, predominantly the naked amoe-bae that glide over surfaces, are the major protozoan predators.

---

1. What are the differences in preferred soil habitats between bacteria and filamentous fungi?
2. What types of archaeans have been detected in soils?
3. How can earthworms, nematodes, and insects influence micro-bial communities?
4. Does the microbial loop function in soils?

 **30.3   MICROORGANISMS AND THE FORMATION OF DIFFERENT SOILS**

Soils form under various environmental conditions. When newly exposed geologic materials begin to weather, as after a volcanic event or a simple soil disturbance, microbial colonization occurs. If only subsurface materials are available, phosphorus may be present, but nitrogen and carbon must be imported by physical or biological processes. Under these circumstances many cyanobac-teria, which can fix atmospheric nitrogen and carbon, are active in pioneer-stage nutrient accumulation.

Once they are formed, most soils are rich sources of nutrients. Nutrients are found in organic matter, microorganisms, soil in-sects, and other animals. Plants grow, senesce, and die—and at each of these phases, they provide nutrients for soil organisms. Different plant parts vary in their nutrient content and biomasses. The components in the plant-soil system with the lowest carbon-nitrogen ratios (most nutrient-rich) are soil organic matter, mi-croorganisms, soil insects, and other soil animals. The soil organic matter contains the greatest portion of the carbon and ni-trogen in a typical soil, but with its slow turnover time (100 to 1,000 years and longer), most of this nutrient resource is not im-mediately available for plant or microbial use. The major types of soils that can be formed are shown in **figure 30.3.**

Tropical and Temperate Region Soils

In moist tropical soils, with their higher mean temperatures, or-ganic matter is decomposed very quickly, and the mobile inor-ganic nutrients can be leached out of the surface soil environment, causing a rapid loss of fertility. To limit nutrient loss, many tropi-cal plant root systems penetrate the rapidly decomposing litter

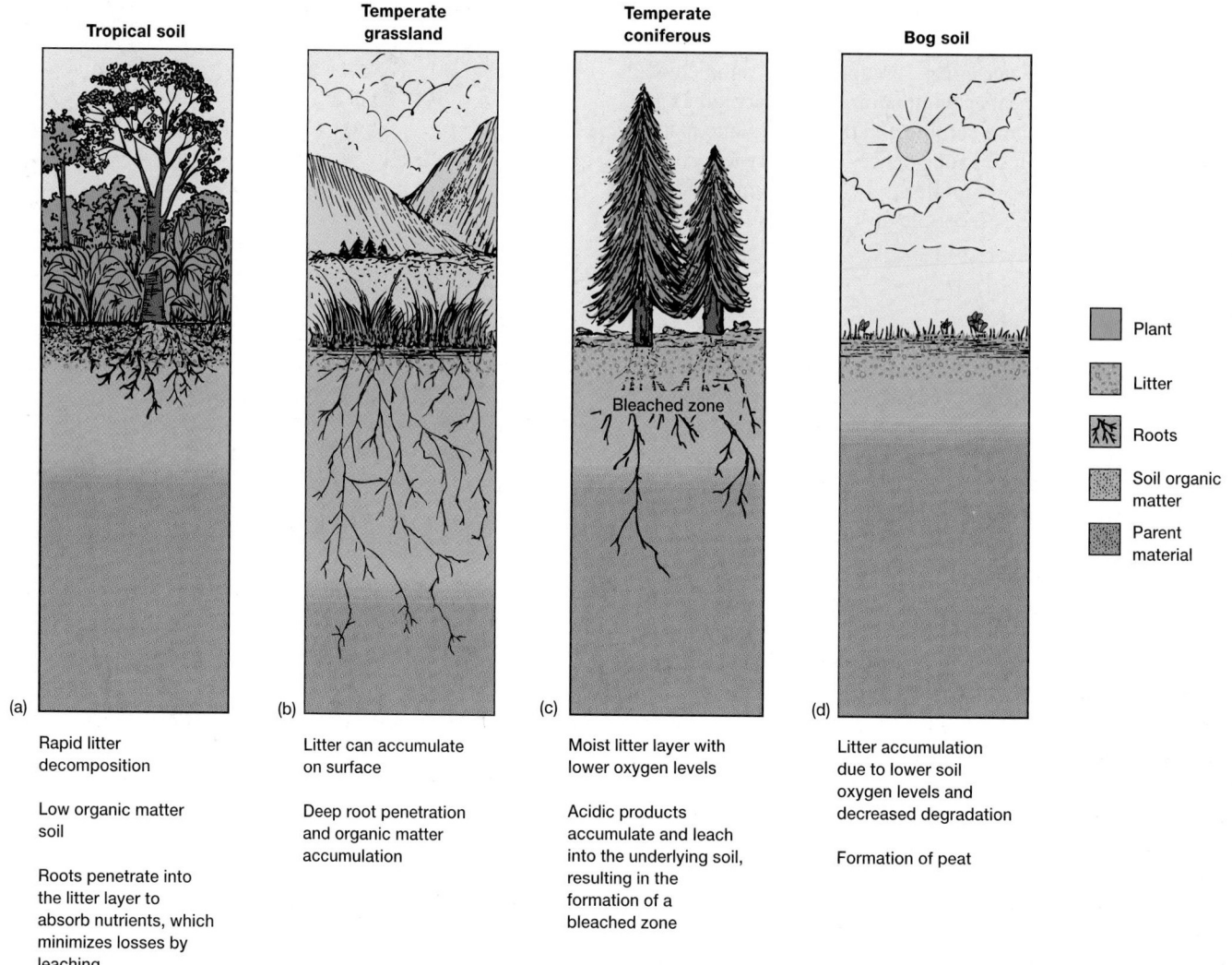

Tropical soil

Temperate grassland

Temperate coniferous

Bleached zone

Bog soil

Plant

Litter

Roots

Soil organic matter

Parent material

(a)

(b)

(c)

(d)

Rapid litter decomposition

Low organic matter soil

Roots penetrate into the litter layer to absorb nutrients, which minimizes losses by leaching

Litter can accumulate on surface

Deep root penetration and organic matter accumulation

Moist litter layer with lower oxygen levels

Acidic products accumulate and leach into the underlying soil, resulting in the formation of a bleached zone

Litter accumulation due to lower soil oxygen levels and decreased degradation

Formation of peat

Figure 30.3   Examples of Tropical and Temperate Region Plant-Soil Systems.   Climate, parent material, plants, topography, and microorganisms interact over time to form different plant-soil systems. In these figures the characteristics of (**a**) tropical, (**b**) temperate grassland, (**c**) temperate coniferous forest, and (**d**) bog soils are illustrated.

layer. As soon as organic material and minerals are released during decomposition, the roots can take them up to avoid losses in leaching. Thus it is possible to "recycle" nutrients before they are lost with water movement through the soil (figure 30.3a). With deforestation, the nutrients are not recycled, leading to their loss from the soil and decreased soil fertility.

Tropical plant-soil communities are often used in **slash-and-burn agriculture.** The vegetation on a site is chopped down and burned to release the trapped nutrients. For a few years, until the minerals are washed from the soils, crops can be grown. When the minerals are lost from these low organic matter soils, the farmer must move to a new area and start over by again cutting and burning the native plant community. This cycle of slash-and-burn agriculture is stable if there is sufficient time for the plant community to regenerate before it is again cut and burned. If the cycle is too short, rapid and almost irreversible degradation of the soil can occur.

In many temperate region soils, in contrast, the decomposition rates are less than that of primary production, leading to litter accumulation. Deep root penetration in temperate grasslands results in the formation of fertile soils, which provide a valuable resource for the growth of crops in intensive agriculture (figure 30.3b).

The soils in many cooler coniferous forest environments suffer from an excessive accumulation of organic matter as plant litter (figure 30.3c). In winter, when moisture is available, the soils are cool, and this limits decomposition. In summer, when the soils are warm, water is not as available for decomposition. Organic acids are produced in the cool, moist litter layer, and they leach into the underlying soil. These acids solubilize soil components such as aluminum and iron, and a bleached zone may form. Litter continues to accumulate, and fire becomes the major means by which nutrient cycling is maintained. Controlled burns are becoming a more important part of environmental management in this type of plant-soil system.

Bog soils provide a unique set of conditions for microbial functioning (figure 30.3d). In these soils the decomposition rate is slowed by the waterlogged, predominantly anoxic conditions, which lead to peat accumulation. When such areas are drained, they become more aerobic and the soil organic matter is degraded, resulting in soil subsidence. Under aerobic conditions the lignin-cellulose complexes of the accumulated organic matter are more susceptible to decomposition by the filamentous fungi.

Cold Moist Area Soils

Soils in cold environments, whether in Arctic, Antarctic, or alpine regions, are of extreme interest because of their wide distribution and impacts on global-level processes. The colder mean soil temperatures at these sites decrease the rates of both decomposition and plant growth. In these cases soil organic matter accumulates, and plant growth can become limited due to the immobilization of nutrients in soil organic matter. Often, below the plant growth zone, these soils are permanently frozen. These permafrost soils hold 11% of the Earth's soil carbon and 95% of its organically bound nutrients. These soils are very sensitive to physical disturbance and pollution, and the widespread exploration of such areas for oil and minerals can have long-term effects on their structure and function.

In water-saturated bog areas, bacteria are more important than fungi in decomposition processes, and there is decreased degradation of lignified materials. As in other soils, the nutrient cycling processes of nitrification, denitrification, nitrogen fixation, and methane synthesis and utilization, although occurring at slower rates, can have major impacts on global gaseous cycles. Lignin degradation (pp. 594–95)

Desert Soils

Soils of hot and cold arid and semiarid deserts are dependent on periodic and infrequent rainfall. When these rainfalls occur, water can puddle in low areas and be retained on the soil surface by microbial communities called **desert crusts.** These consist of cyanobacteria and associated commensalistic microbes, including *Anabaena, Microcoleus, Nostoc,* and *Scytonema.* The depth of the photosynthetic layer is perhaps 1 mm, and the cyanobacterial filaments and slimes link the sand particles (**figure 30.4**), which change the surface soil albedo, water infiltration rate, and susceptibility to erosion. These crusts are quite fragile, and vehicle damage can be evident for decades. After a rainfall, nitrogen fixation will begin within approximately 25 to 30 hours, and when the rain has evaporated or drained into the soil, the crust will dry up and nitrogen will be released for use by other microorganisms and the plant community.

Geologically Heated Hyperthermal Soils

Geologically heated soils are found in such areas as Iceland and the Kamchatka peninsula in eastern Russia, and such

Figure 30.4   A Desert Crust as Observed with the Scanning Electron Microscope.   The crust has been disturbed to show the extracellular sheaths and filaments of the cyanobacterium *Microcoleus vaginatus.* The sand grains are linked by these filamentous growths, creating a unique ecological structure.

heated soils also occur at many mining waste sites. An important microorganism found in heated mining wastes is *Thermoplasma.* These soils are populated by bacterial and archaeal procaryotes, many of which are chemolithoautotrophs. A wide variety of chemoorganotrophic genera also are found in these environments; these include the aerobes *Thermomicrobium, Thermoleophilum,* and also the anaerobes *Thermosipho* and *Thermotoga.* Such geothermal soils have been of great interest as a source of new microbes to use in biotechnology (*see p. 607*), and the search for new, unique microorganisms in these areas is intensifying all over the world. Thermoplasma (p. 449)

1. Characterize each major soil type discussed in this section in terms of the balance between primary production and organic matter decomposition.
2. What is slash-and-burn agriculture? Describe the roles of microorganisms in this process.
3. What is unique about bogs in terms of organic matter degradation?
4. Why can nutrients become limiting for plants and microorganisms in tundra soils?
5. Describe desert crusts. What types of microorganisms function in these unique environments?
6. What unique microbial genera are found in geothermally heated soils?

Table 30.3

Compounds Excreted by Microorganism-Free Wheat Roots

Volatile Compounds	Low-Molecular-Weight Compounds	High-Molecular-Weight Compounds
CO_2	Sugars	Polysaccharides
Ethanol	Amino acids	Enzymes
Isobutanol	Vitamins	
Isoamyl alcohol	Organic acids	
Acetoin	Nucleotides	
Isobutyric acid		
Ethylene		

From J. W. Woldendorp, "The Rhizosphere as Part of the Plant-Soil System" in *Structure and Functioning of Plant Populations* (Amsterdam, Holland: Proceedings, Royal Dutch Academy of Sciences, Natural Sciences Section: 2d Series, 1978) 70:243.

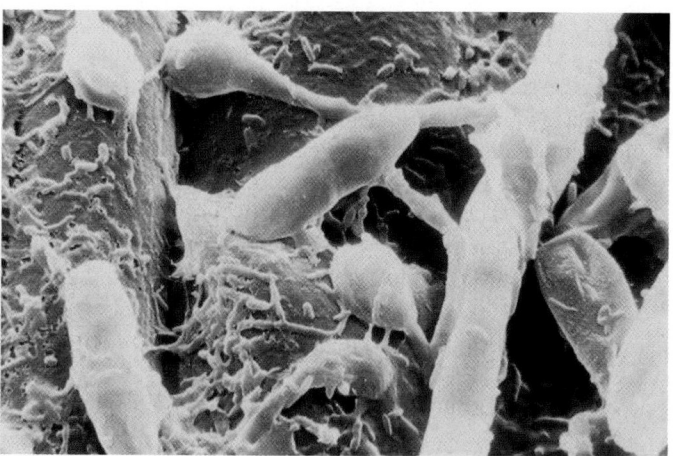

Figure 30.5   Root Surface Microorganisms.   Plant roots release nutrients that allow intensive development of bacteria and fungi on and near the plant root surface, the rhizoplane. A scanning electron micrograph shows bacteria and fungi growing on a root surface.

30.4   SOIL MICROORGANISM ASSOCIATIONS WITH VASCULAR PLANTS

Plants—the major source of organic matter on which most soil microorganisms are dependent—are heavily colonized with microorganisms. Microorganisms of various kinds are associated with the leaves, stems, flowers, seeds, and roots. The microbial community influences plants in many direct and indirect ways. The community includes microorganisms that grow on the plant surface, between cells on the surface of the plant, and within plant cells.

Microorganisms on the Outside of Plants

Plants are covered by microorganisms, on both the above- and belowground surfaces. These microbes usually are simple commensals; in some cases they can become parasitic or pathogenic, especially when the plant surface has been damaged.

Phyllosphere Microorganisms

A wide variety of microorganisms are found on and in the aerial surfaces of plants, called the **phyllosphere.** These include microorganisms that have complex interactions with the plant at various stages of development. The plant leaves and stems release organic compounds, and this can lead to massive development of microbes. The genera present on plant leaves and stems include *Sphingomonas,* which is especially equipped to survive with the high levels of UV irradiation occurring on these plant surfaces. This bacterium, also common in soils and waters, can occur at 10^8 cells per gram of plant tissue. *Sphingomonas* often represents a majority of the culturable species.

Rhizosphere and Rhizoplane Microorganisms

Plant roots release a wide variety of materials to their surrounding soil, including various alcohols, ethylene, sugars, amino and organic acids, vitamins, nucleotides, polysaccharides, and en-

zymes as shown in **table 30.3.** These materials create unique environments for the soil microorganisms. These environments include the **rhizosphere,** first described by Lorenz Hiltner in 1904, the volume of soil around the root influenced by the materials released from the root. The plant root surface, termed the **rhizoplane,** also provides a unique environment for microorganisms, as these gaseous, soluble, and particulate materials move from the plant to the soil. The rhizosphere and rhizoplane microorganisms not only increase their numbers when these newly available substrates become available, but their composition and function also change. Rhizosphere and rhizoplane microorganisms, as they respond to these released materials, also serve as labile sources of nutrients for other organisms, thus creating a soil microbial loop in addition to playing critical roles in organic matter synthesis and degradation.   Microbial loop (chapter 29)

A wide range of bacteria in the rhizosphere can promote plant growth, orchestrated by rhizosphere bacteria that communicate with the plant using complex chemical signals. These chemical signal compounds, including auxins, gibberellins, glycolipids, and cytokinins, are only now beginning to be fully appreciated in terms of their biotechnological potential. Plant growth-promoting rhizobacteria include the genera *Pseudomonas* and *Achromobacter.* These can be added to the plant, even in the seed stage, if the bacteria have the required surface attachment proteins. The genes that control the expression of these attachment proteins are still being identified.

A critical process that occurs on the surface of the plant, and particularly in the root zone, is **associative nitrogen fixation,** in which the nitrogen-fixing microorganism is on the surface of the plant root, the rhizoplane (**figure 30.5**), as well as in the rhizosphere. This process is carried out by representatives of the genera *Azotobacter, Azospirillum* and *Acetobacter.* These bacteria contribute to nitrogen accumulation by tropical grasses. Recent evidence suggests that their major contribution may not be in nitrogen fixation but in production of growth-promoting hormones that increase root hair development and thus greater ability of the plant to take up nutrients. This is an area of research that is particularly important in tropical agricultural areas.

---

1. Define the following terms: rhizosphere, rhizoplane, and associative nitrogen fixation.
2. What unique stresses face a microorganism on a leaf but not in the soil?
3. What microbial genera are considered as plant growth–promoting rhizobacteria?
4. What important genera are involved in associative nitrogen fixation?

Microorganism Growth within Plants

In addition to the microorganisms growing on the surface of a plant, many interesting and important microbes grow within plants. These associations are dependent upon complex chemical signals, which indicates that these relationships are very ancient. It is important to emphasize that the mechanisms by which many of these plant-microbe interactions occur are also found in microorganism-animal interactions (*see chapter 31*).

Rhizobium and Its Relatives

Several microbial genera are able to form nitrogen-fixing nodules with legumes. These include *Allorhizobium, Azorhizobium, Bradyrhizobium, Mesorhizobium, Sinorhizobium,* and *Rhizobium*. The mechanism of this interaction will be illustrated by work with *Rhizobium*.

The genus *Rhizobium* (*see p. 479*) is a prominent member of the rhizosphere community. This bacterium also can establish a symbiotic association with legumes and fix nitrogen for use by the plant. *Rhizobium* infects and nodulates specific legume hosts. The bacterium contains plasmids, including the symbiotic plasmid pSym that codes for nodulation (*nod, nol, noe*) and nitrogen fixation (*nif, fix*) functions, vital for infection and nodulation of the susceptible host plant. Plasmid-encoded genes also influence the range of host plants that *Rhizobium* can nodulate.

The *Rhizobium*-legume symbiosis is based on complex molecular interactions, functioning much like successive locks and doors, that control this process; similar mechanisms are often used in both symbiotic and pathogenic interactions. For example, the *Rhizobium* infection process is controlled by the *bacA* gene that is required to establish the nodule. *Brucella abortus* has a similar *bacA* gene that is required to produce a chronic infection in animal hosts. Another similarity with pathogens is the *Lon* protease, which is important in *Rhizobium* and also influences capsule production in *E. coli, Klebsiella,* and *Erwinia*. Symbiotic development in legumes is controlled by a plant regulator protein that dedifferentiates root cortical cells and restarts cell division to establish nodule primordia within the invaded root. The primordia are invaded by *Rhizobium* using infection threads created by

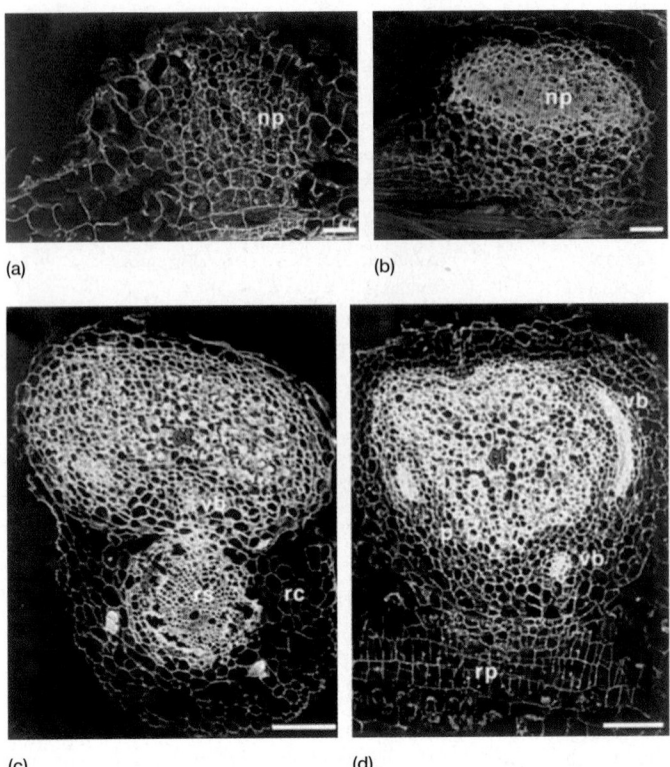

(a)   (b)

(c)   (d)

Figure 30.6 Controlled Initial Development of Legume Root Nodules Incited by *Rhizobium*. Expression of nodule inception (*nin*) activity is a critical step in infection thread formation and initiation of primordia: (**a**) early stage of primordial cell development with low levels of *nin* activity, (**b**) later higher-level expression of *nin* activity, (**c**) transverse nodule section showing extended *nin* expression, (**d**) fully differentiated nitrogen-fixing nodule with *nin* activity in central tissue area. In situ hybridization with probes used to detect activity. Abbreviations used: np, nodule primordium; ct, central tissue; rc, root cortex; rs, root stele; vb, vascular bundle. Bar = 100 μm.

the plant. The establishment of the nodules involves a newly discovered plant gene called *nin*. Expression of this gene during nodule development has been detected by the use of in situ probes in legume root tissue (**figure 30.6**).

The complex infection process (**figure 30.7**), presented here in a simplified form, appears to involve a series of molecules produced by the host plant that lead to the exchange of recognition signals. The first step in the recognition process occurs when a potential root nodule–forming microbe approaches the plant root and is assumed to be an alien invader. The plant responds with an **oxidative burst,** producing a mixture that can contain superoxide radicals, hydrogen peroxide, and N_2O. This redox-based oxidative burst, involving glutathione and homoglutathione, is critical for determining the fate of the infection process and influences whether further steps in the early infection process will occur. The rhizobia, if they are to be effective colonizers, must use antioxidant defenses to survive and continue the infection process.

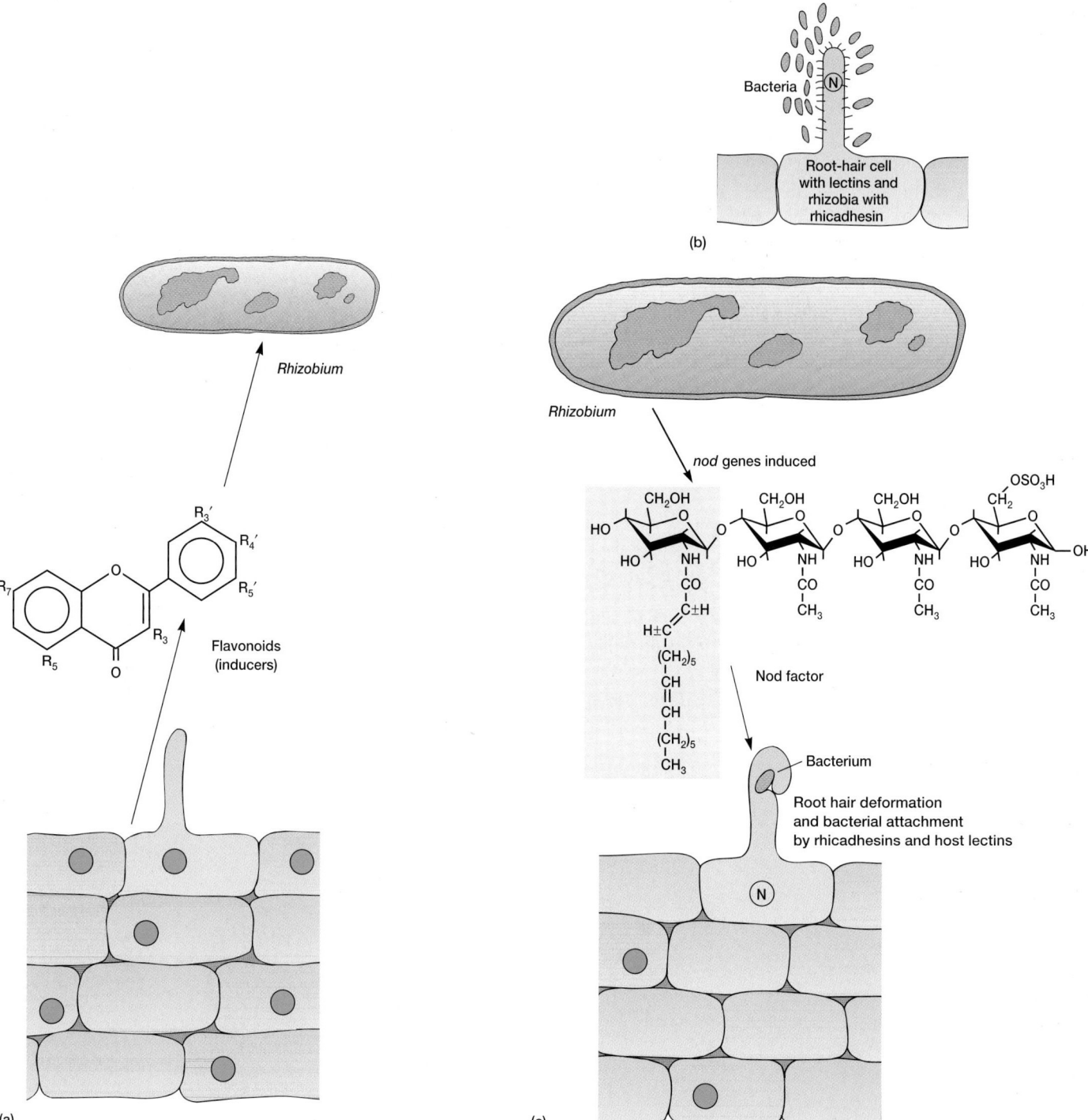

Figure 30.7 **Root Nodule Formation by *Rhizobium*.** Root nodule formation on legumes by *Rhizobium* is a complex process that produces the nitrogen-fixing symbiosis. (**a**) The plant root releases flavonoids that stimulate the production of various Nod metabolites by *Rhizobium*. There are many different Nod factors that control infection specificity. (**b**) Attachment of *Rhizobium* to root hairs involves specific bacterial proteins called rhicadhesins and host plant lectins that affect the pattern of attachment and *nod* gene expression. (**c**) Structure of a typical Nod factor that promotes root hair curling and plant cortical cell division. The bioactive portion (nonreducing *N*-fatty acyl glucosamine) is highlighted. These Nod factors enter root hairs and migrate to their nuclei. (**d**) A plant root hair covered with *Rhizobium* and undergoing curling. (**e**) Initiation of bacterial penetration into the root hair cell and infection thread growth coordinated by the plant nucleus "N." (**f**) A branched infection thread shown in an electron micrograph. (**g**) Cell-to-cell spread of *Rhizobium* through transcellular infection threads followed by release of rhizobia and infection of host cells. (**h**) Formation of bacteroids surrounded by plant-derived peribacteroid membranes and differentiation of bacteroids into nitrogen-fixing symbiosomes. The bacteria change morphologically and enlarge around 7 to 10 times in volume. The symbiosome contains the nitrogen-fixing bacteroid, a peribacteroid space, and the peribacteroid membrane. (**i**) Light micrograph of two nodules that develop by cell division (×5). This section is oriented to show the nodules in longitudinal axis and the root in cross section. (**j**) *Sinorhizobium meliloti* nitrogen-fixing nodules on roots of white sweet clover (*Melilotus alba*).

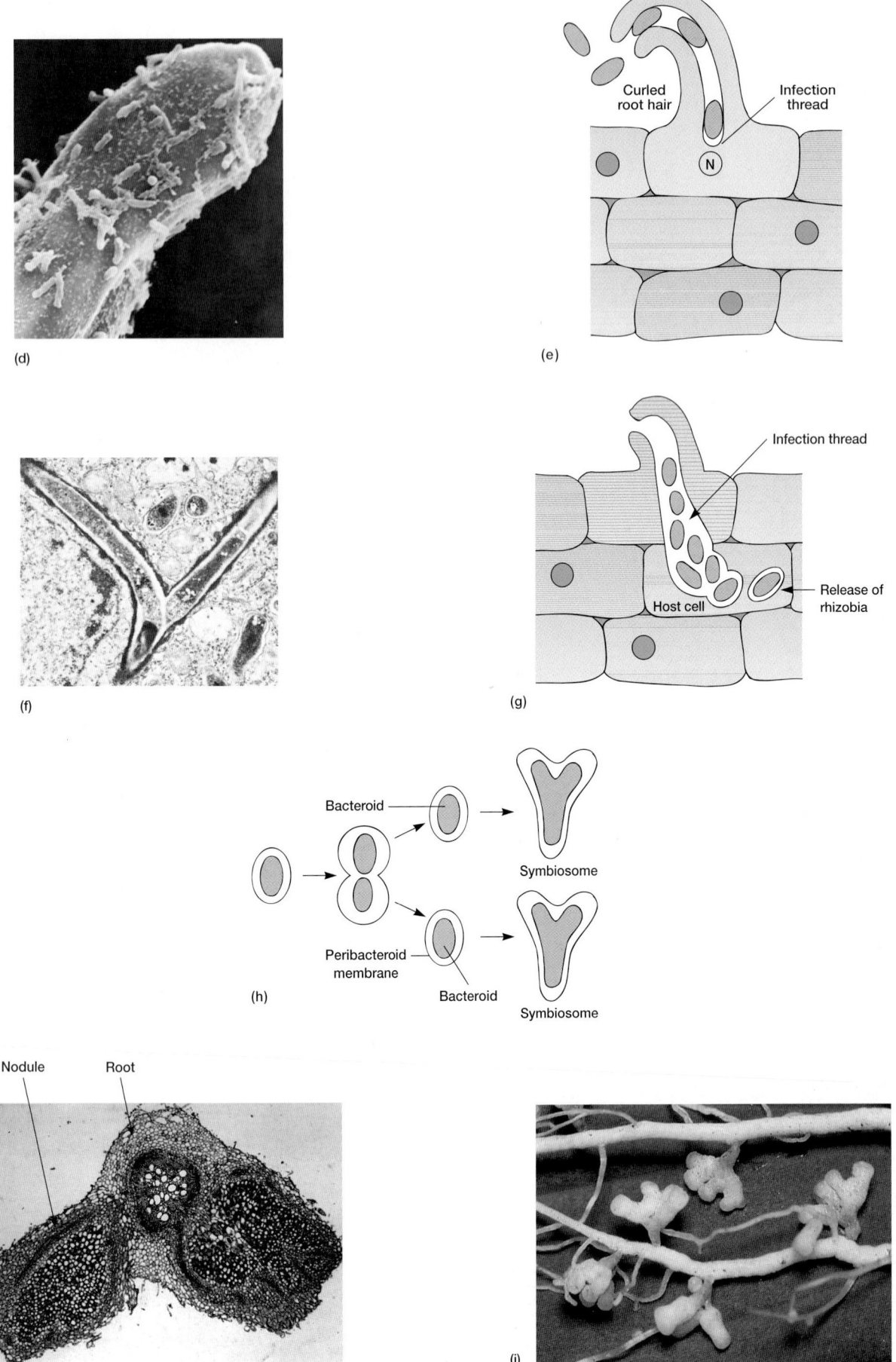

(d)

(e)
Curled root hair · Infection thread · N

(f)

(g)
Infection thread · Host cell · Release of rhizobia

(h)
Bacteroid · Symbiosome · Peribacteroid membrane · Bacteroid · Symbiosome

(i)
Nodule · Root

(j)

Figure 30.7 *(Continued)*

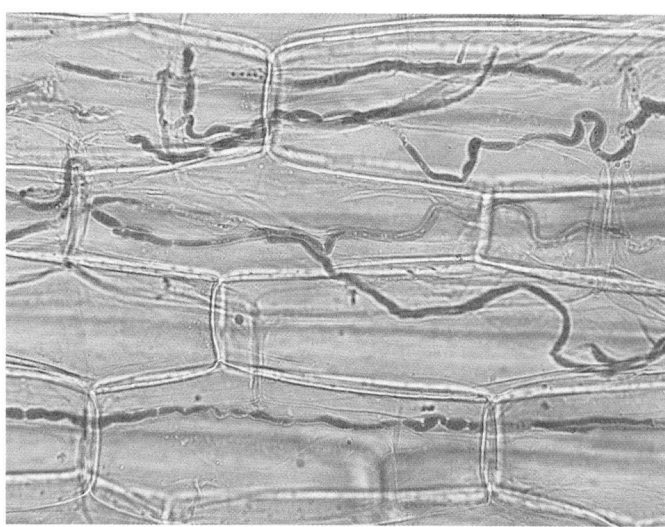

Figure 30.8   Fungal Endophytes.   Fungi can invade the upper parts of some plants. A fungal endophyte growing inside the leaf sheath of a grass, tall fescue, is shown.

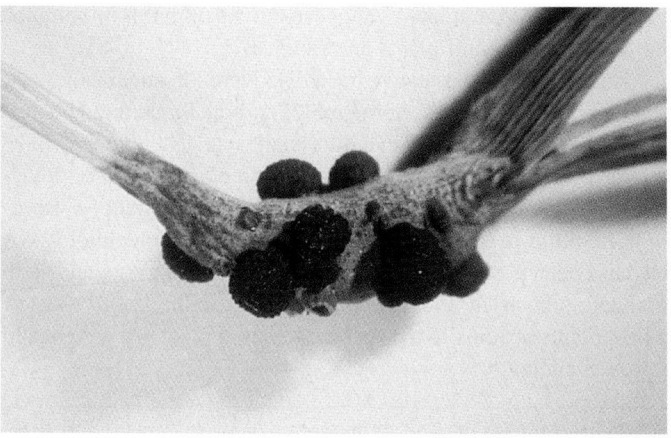

Figure 30.9   Parasitic Castration of Plants by Endophytic Fungi.   Stroma of the fungus *Atkinsonella hypoxylon* infecting *Danthonia compressa* and causing abortion of the terminal spikelets.

Only rhizobia and related genera with sufficient antioxidant abilities are able to proceed to the next steps in the infection process.

Another vital part of this complex interaction process is the formation of communication molecules that are coded on the pSym plasmid. These molecules have not yet been well studied. Flavonoid inducers produced by the plant also play a major role in this process by stimulating the *Rhizobium* to synthesize specific **Nod factors** that activate the host symbiotic processes necessary for root hair infection and nodule development (figure 30.7*a–d*). After bacterial attachment, the root hairs curl and the bacteria induce the plant to form an **infection thread** that grows down the root hair.

The *Rhizobium* then spreads within the infection thread into the underlying root cells as noted in figure 30.7*e*. At no time does the *Rhizobium* actually enter the plant cytoplasm while it is in the infection thread! When the bacteria are released from the infection thread into the host cell (figure 30.7*g*) the *Rhizobium* is enclosed by a plant-derived membrane, called the peribacteroid membrane, to form a **bacteroid** (figure 30.7*h*). Further growth and differentiation lead to the development of a nitrogen-fixing form, a structure called a **symbiosome** (figure 30.7*h*). At this point, specific nodule components such as leghemoglobin, which protect the nitrogen fixation enzymes from oxygen, are produced to complete the nodulation process (figure 30.7*i*). The final **root nodules** are shown in figure 30.7*j*.

The symbiosomes within mature root nodules are the site of nitrogen fixation. Within these nodules, the differentiated bacteroids reduce atmospheric N_2 and form ammonia (the primary product) and alanine; these compounds are released into the host plant cell, assimilated into various other nitrogen-containing organic compounds, and distributed throughout the plant.

After the *Rhizobium* cells have developed into the final differentiated bacteria, they no longer can revert back to bacteria capable of reproduction and have been lost to the "gene pool." The fittest members of the bacterial community, which have achieved

nodulation, thus have sacrificed their own reproduction! This "altruism in the rhizosphere," a concept suggested by I. Olivieri and S. A. Frank, actually may promote increased root exudation and the maintenance of higher numbers of *Rhizobium* in the plant-root environment. The molecular biology of nitrogen fixation by *Rhizobium*, with the complex chemical communications involved, continues to be a subject of intense study around the world.

A major goal of biotechnology is to introduce nitrogen fixation genes into plants that do not normally form such associations. It has been possible to produce modified lateral roots on nonlegumes such as rice, wheat, and oilseed rape; the roots are invaded by nitrogen-fixing bacteria. It appears that the infection begins with bacterial attachment to the root tips. Although these modified root structures have not yet been found to fix useful amounts of nitrogen, they do enhance rice production and intense work is expected to continue in this area.   (Plant biotechnology, pp. 326–27, 330–32)

Fungal and Bacterial Endophytes of Plants

Specialized fungi and bacteria can live within some plants as **endophytes** and may be beneficial. Specialized clavicipitaceous fungi form systemic fungal infections in which the endophyte grows between the plant's cortex cells (**figure 30.8**). Plants infected with these endophytes may be less susceptible to attack by various chewing insects due to the production of alkaloids, a form of "chemical defense." *Rhizobium leguminosarum* bv. *trifolii* can form a natural endophytic association with rice roots. This interaction, first observed in the Nile Delta, is supported by the rotation of clover with the rice. The association promotes rice root and shoot growth, resulting in increased rice grain yield at maturity. The rice/*Rhizobium*/clover association provides approximately one-quarter of the nitrogen needed by the rice crop.

Not all such relationships are mutualistic (beneficial to both partners). Some are parasitic. Parasitic fungal endophytes can actually reduce the genetic variability of the plant by sterilizing their host (**figure 30.9**). This "parasitic castration of plants" by systemic fungi, which promotes increased fungal spread in a less variable plant community, is suggested to be of major importance in the co-evolution of plants and fungi.

Endophytic bacteria have been discovered in sugar cane, cotton, pears, and potatoes. Some are plant pathogens that can survive for extended periods in a quiescent state. The majority have no known positive or deleterious effect on plant growth or development. The use of these bacteria as microbial delivery systems in agriculture is a current topic in agricultural biotechnology. It also has been possible to establish *Azorhizobium*, a root and stem nodulating bacterium for *Sesbana rostrata*, in the lateral roots of wheat plants, leading to a possible increased plant dry weight and nitrogen content.

---

1. Give examples of genes that operate in microbe-plant and also in microbe-animal associations.
2. Describe the *nin* gene and its role in *Rhizobium* infection.
3. Define and relate primordia and infection thread in terms of nodule formation.
4. What are the basic steps in establishing the *Rhizobium*-legume root symbiosis?
5. What are possible effects of endophytic fungi on plants?

Mycorrhizae

Mycorrhizae are fungus-root associations, first discovered by Albert Bernhard Frank in 1885 (**Microbial Diversity & Ecology 30.1**). The term "mycorrhizae" comes from the Greek words meaning fungus and roots. These microorganisms contribute to plant functioning in natural environments, agriculture, and reclamation. The roots of about 95% of all kinds of vascular plants are normally involved in symbiotic associations with mycorrhizae.

Six major types of mycorrhizal associations are recognized (some researchers suggest that there are seven), and are shown in **figure 30.10.** Those that penetrate the plant root cells, or endomycorrhizae, are noted in figure 30.10*a–c*, and those that form sheaths are illustrated in figure 30.10*d–f*. **Endomycorrhizae** are of particular interest, as it has not been possible to grow these fungi, usually members of the zygomycetes, without the plant. In this association the fungal hyphae penetrate the outer cortical cells of the plant root, where they grow intracellularly and form coils, swellings, or minute branches. Endotrophic mycorrhizae

are found in wheat, corn, beans, tomatoes, apples, oranges, and many other commercial crops, as well as most pasture and rangeland grasses.

A characteristic intracellular structure is the **arbuscule** (**figure 30.11**), and thus endomycorrhizae often are called **arbuscular mycorrhizal** or **AM fungi.** However, vesicles are not consistently observed. In addition, the endomycorrhizae include septate types associated with orchids (figure 30.10*b*), as well as those that form endomycorrhizal relationships with ericoid plants such as blueberries (figure 30.10*c*).

The mycorrhizal fungi that form sheaths around the root are shown in figure 30.10*d–f*. The **ectomycorrhizae** (figure 30.10*d*) form a characteristic mantle that covers the root tip area, and an intercellular network of hyphae called the Hartig net forms between the cortical cells. Fungi that form these easily recognized structures (**figure 30.12**) include *Cenococcum, Pisolithus,* and *Amanita*. In addition, the sheath-forming fungi include the arbutoid ectendomycorrhizae (figure 30.10*e*), formed by basidiomycetes (*see pp. 545–46*). These have both sheaths and intracellular coils. Another interesting group is the monotropoid mycorrhizae (figure 30.10*f*) that are associated with achlorophyllous plants (plants that have no chlorophyll), and have characteristic "fungal peg" structures. Some of these achlorophyllous plants use proteases to release amino acids from soil organic matter and obtain nutrients; most, however, are strictly parasitic. They depend on nutrients gained by "tapping into" a mycorrhizal network supported by normal photosynthetic plants. Recently, it has been found that some achlorophyllous plants also can tap into AM fungal networks, increasing their parasitic versatility.

More than 5,000 species of fungi, predominantly basidiomycetes, are involved in ectomycorrhizal symbiotic relationships. Their extensive mycelia extend far out into the soil and greatly aid the transfer of nutrients to the plant. One of the more economically important ectomycorrhizal fungi is *Pisolithus tinctorius,* mentioned earlier. This fungus can be grown in mass culture, with small Styrofoam beads acting as a physical support. The fungal inoculum is then mixed with rooting soil, resulting in improved plant establishment and growth.

Depending on the environment of the plant, mycorrhizae can increase a plant's competitiveness. In wet environments they increase the availability of nutrients, especially phosphorus. In arid

Microbial Diversity & Ecology

30.1 Mycorrhizae and the Evolution of Vascular Plants

Fossil evidence shows that endomycorrhizal symbioses were as frequent in vascular plants during the Devonian period, about 400 million years ago, as they are today. As a result some botanists have suggested that the evolution of this type of association may have been a critical step in allowing colonization of the land by plants. During this period soils were poorly developed, and as a result mycorrhizal

fungi were probably significant in aiding the uptake of phosphorus and other nutrients. Even during current times those plants that start to colonize extremely nutrient-poor soils survive much better if they have endomycorrhizae. Thus it may have been a symbiotic association of plants and fungi that initially colonized the land and led to our modern vascular plants.

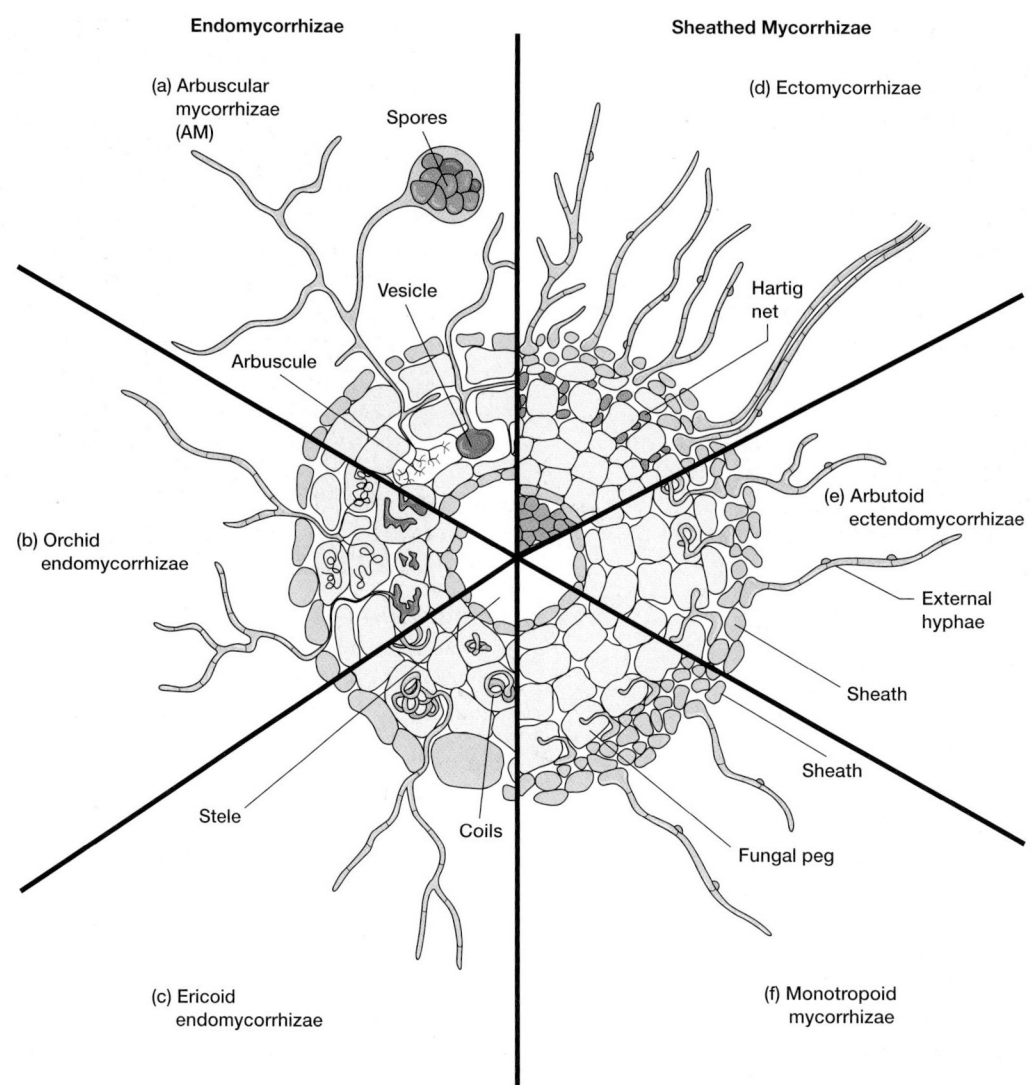

Figure 30.10   Mycorrhizae.   Fungi can establish mutually beneficial relationships with plant roots, called mycorrhizae. Root cross sections illustrate different mycorrhizal relationships. See text for details.

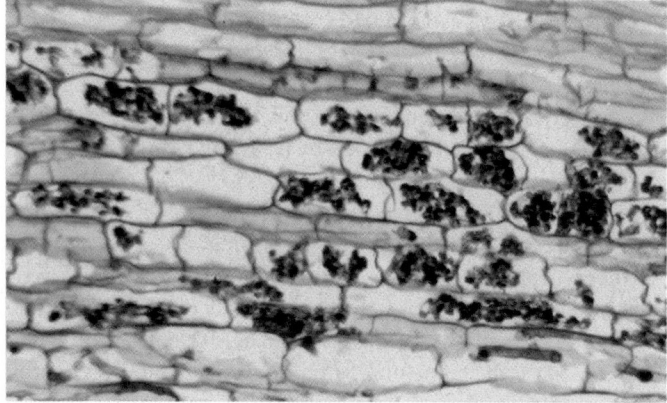

Figure 30.11   Endomycorrhizae.   Endomycorrhizae, or arbuscular mycorrhizae, form characteristic structures within roots. These can be observed with a microscope after the roots are cleared and stained. The arbuscules of *Gigaspora margarita* can be seen inside the root cortex cells of cotton.

Figure 30.12   Ectomycorrhizae as Found on Roots of a Pine Tree.   Typical irregular branching of the white, smooth mycorrhizae is evident.

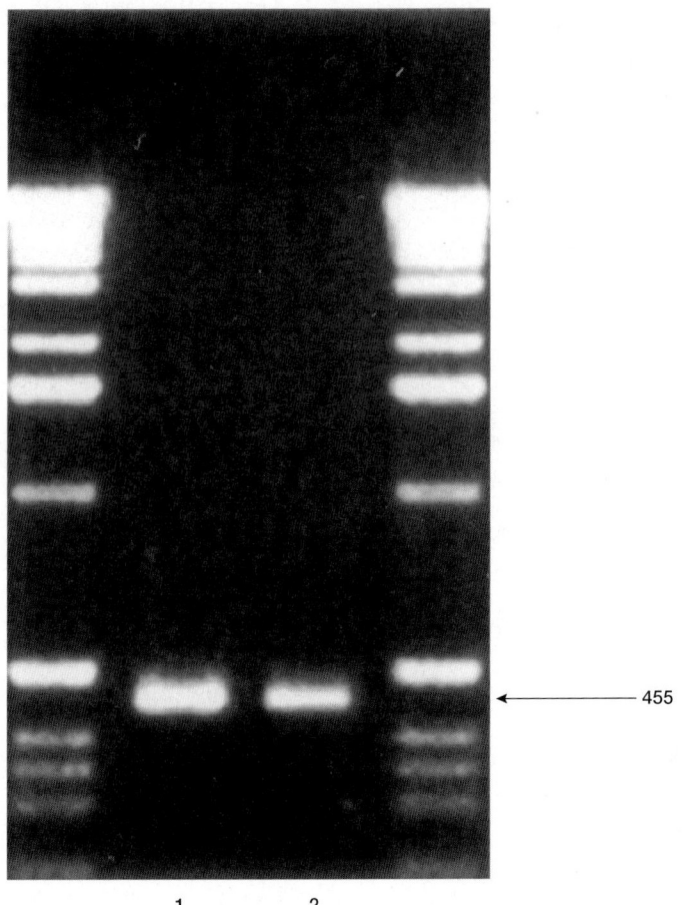

Figure 30.13 Use of the Polymerase Chain Reaction (PCR) to Detect a Mycorrhizal Fungus. Use of PCR to detect *Glomus intraradices* as spores (1) and after growth of the fungus in a root (2). In this example, a nested PCR was used. The second amplification is illustrated with the mycorrhizal fungal PCR product having a 455 base pair size. Molecular size markers are noted in the outer lanes.

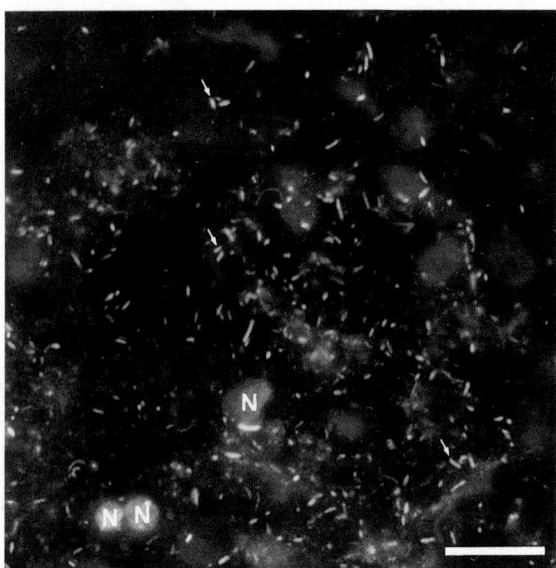

Figure 30.14 Mycorrhization Helper Bacteria. Stained bacterial endosymbionts in unfixed spores of the AM fungus *Gigaspora margarita.* Living bacteria fluoresce bright yellow-green; lipids and fungal nuclei (N) appear as diffuse masses. Bar = 7 μm.

environments, where nutrients do not limit plant functioning to the same degree, the mycorrhizae aid in water uptake, allowing increased transpiration rates in comparison with nonmycorrhizal plants. These benefits have distinct energy costs for the plant in the form of photosynthate required to support the plant's "mycorrhizal habit." Under certain conditions the plant is apparently willing to trade photosynthate—produced with the increased water acquisition—for water.

Mycorrhizal and endophytic fungal communities have been found to influence the development of plant communities. The mycorrhizal fungi also can make it possible to share resources such as carbon minerals and water between different plants.

When studying mycorrhizal and endophytic relationships, molecular techniques are now used to identify spores and also mycorrhizal fungi within plant roots, as shown in **figure 30.13**.

By the use of PCR *(see pp. 316–19)* with specific primers, it is possible to identify the presence of specific mycorrhizal fungi in association with a given plant after extraction of the rDNA from the root!

Bacteria also have relations with the mycorrhizal fungi. As the external hyphal network radiates out into the soil, a **mycorrhizosphere** is formed due to the flow of carbon from the plant into the mycorrhizal hyphal network and then into the surrounding soil. In addition, "mycorrhization helper bacteria" (MHB) can play a role in the development of the mycorrhizal relationships with the ectomycorrhizal fungus. Bacterial symbionts also are found in the cytoplasm of AM fungi, as shown in **figure 30.14**. By using PCR, such bacteria-like organisms (BLOs) appear to be related to *Burkholderia cepacia.* What is the possible function of these symbionts? It has been suggested that these "trapped" bacteria contribute to the nitrogen metabolism of the plant-fungal complex by assisting with the synthesis of essential amino acids.

Actinorhizae

Actinomycete associations with plant roots, called **actinorhizae** or actinorhizal relationships, also occur (**figure 30.15**). These are formed by the association of *Frankia* strains (*see pp. 533–34*) with eight nonleguminous plant families (**table 30.4**). They fix nitrogen and are important, particularly in trees and shrubs. As examples, these associations occur in areas where Douglas fir forests have been clear-cut, and in bog and heath environments where bayberries and alders are dominant. The nodules of some plants (*Alnus, Ceanothus*) have nodules

Figure 30.15　Actinorhizae. *Frankia*-induced actinorhizal nodules in *Ceanothus* (Buckbrush).

as large as baseballs. The nodules of *Casuarina* approach soccer ball size!

Members of the genus *Frankia* are slow-growing, and, until 1978, it was impossible to culture the organisms apart from the plant. Since then, this actinomycete has been grown on specialized media supplemented with metabolic intermediates such as pyruvate. Major advances in understanding the physiology, genetics, and molecular biology of these microorganisms are now taking place.

As in all plant-microbe associations, the actinorhizal relationship costs the plant energy. However, the plant does benefit and is better able to compete in nature. This association provides a unique opportunity for microbial management to improve plant growth processes.

Other associations of nitrogen-fixing microorganisms with plants also occur. A particularly interesting association is caused by **stem-nodulating rhizobia,** found primarily in tropical legumes (**figure 30.16**). These nodules form at the base of adventitious roots branching out of the stem just above the soil surface and, because they contain oxygen-producing photosynthetic tissues, they have unique mechanisms to protect the oxygen-sensitive nitrogen fixation enzymes. Another microorganism that forms such root and stem nodules is *Azorhizobium caulinodans,* which forms nodules on the tropical legume *Sesbania rostrata.*　Nitrogen fixation: biochemical aspects (pp. 206–9); Nitrogen cycle (pp. 597–98)

Table 30.4

Nonleguminous Nodule-Bearing Plants with *Frankia* Symbioses[a]

Family	Genus	*Frankia* Isolated?	Isolated Strains Infective?
Casuarinaceae	Allocasuarina	+	+
	Casuarina	+	+
	Ceuthostoma	–	–
	Gymnostoma	+	+
Coriariaceae	Coriaria	+	+
Datiscaceae	Datisca	+	–
Betulaceae	Alnus	+	+
Myricaceae	Comptonia	+	+
	Myrica	+	+
Elaeagnaceae	Elaeagnus	+	+
	Hippophae	+	+
	Shepherdia	+	+
Rhamnaceae	Adolphia	–	–
	Ceanothus	+	–
	Colletia	+	+
	Discaria	+	+
	Kentrothamnus	–	–
	Retanilla	+	+
	Talguenea	+	+
	Trevoa	+	+
Rosaceae	Cercocarpus	+	–
	Chaemabatia	–	–
	Cowania	+	–
	Dryas	–	–
	Purshia	+	–

Source: Data from Dr. D. Baker, MDS Panlabs and Dr. J. Dawson, University of Illinois. Personal communications.

[a]*Frankia* isolation from nodules and ability of these isolated strains to initiate nodulation are also noted.

Figure 30.16   Stem-Nodulating Rhizobia.   Nitrogen-fixing microorganisms also can form nodules on stems of some tropical legumes. Nodules formed on the stem of a tropical legume by a stem-nodulating *Rhizobium.*

**Figure 30.17   *Agrobacterium.*   *Agrobacterium*-caused tumor on a *Kalanchoe* sp. plant.

Recently, it has been shown that some of these stem-nodulating rhizobia are photosynthetic. Thus they can obtain their energy not only from the plant organic compounds, but also from the light!

1. Describe the major contributions of mycorrhizae and actinorhizae to plant functioning.
2. Discuss the major differences between endomycorrhizae and ectomycorrhizae.
3. What is the major contribution of *Frankia* to plant functioning? Which types of plants are infected?
4. What are stem-nodulating rhizobia?

Agrobacterium

Another exciting plant-soil microorganism interaction is the *Agrobacterium* infection that produces tumorlike growths on plants (**figure 30.17**; *see also Techniques & Applications 14.2*). Gall formation involves *Agrobacterium* strains that contain the **Ti** (tumor-inducing) **plasmid.** Tumorigenesis is a natural form of genetic engineering, in which a portion of the Ti plasmid is excised, transferred from the bacterium to the plant host, and ligated into the host nuclear chromosome. The tumor is produced by an imbalance of phytohormones controlled by enzymes encoded by the introduced bacterial Ti DNA. Biotechnologists have modified this plasmid to allow the transfer of genetic characteristics such as herbicide resistance and bioluminescence to plants. With the ability to modify plant DNA using *Agrobacterium*-borne plasmids, rapid advances are being made in plant molecular biology. *Agrobacterium* (pp. 330–31, 479–80)

Tumor production involves binding of the bacterium to a plant, usually at a wound site, after plant phenolics are detected by the bacterium, which then expresses a tumor-inducing two-component system (**figure 30.18**). This oncogenic interaction is initiated by the transfer of tumor-DNA fragments (T-DNA) from the tumor-inducing (Ti) bacterial plasmid to the plant cell nucleus. One component controls the excision of the T-DNA, leading to the initiation of the tumor. The second component stimulates opine synthesis by the plant. The opine is used as a nutrient by the infecting *Agrobacterium,* but not as efficiently by other competing microorganisms. Thus plant metabolism is switched to produce substances that stimulate growth of the bacterium responsible for initiation of the tumor! This is a delicate interaction that has a genetic basis.

Fungi and Bacteria as Plant Pathogens

It should be noted that fungi can be devastating plant pathogens. Examples are the fungus *Puccinia graminis,* which causes wheat rust, and the fungus *Phytophthora infestans,* which was responsible for the Irish potato famine.

A wide variety of bacteria, including *Agrobacterium,* attack plants. Bacteria cause a bewildering array of spots, blights, wilts, rots, cankers, and galls, as shown in **table 30.5**. In addition, wall-less phytoplasmas are important pathogens of vegetable crops such as sweet potatoes.

Viruses

A wide range of viruses infect plants, as described in section 18.6, and are of worldwide importance in terms of plant disease and economic losses. These include TMV, the first virus to be characterized in 1892. A virus of particular interest in terms of plant-pathogen interactions is a hypovirus that infects the fungus *Cryphonectria parasitica,* the cause of chestnut blight. Based on pioneering studies carried out in Italy and France, workers in Connecticut and West Virginia noted that if they infected the fungus with the hypovirus, the rate and occurrence of infection was

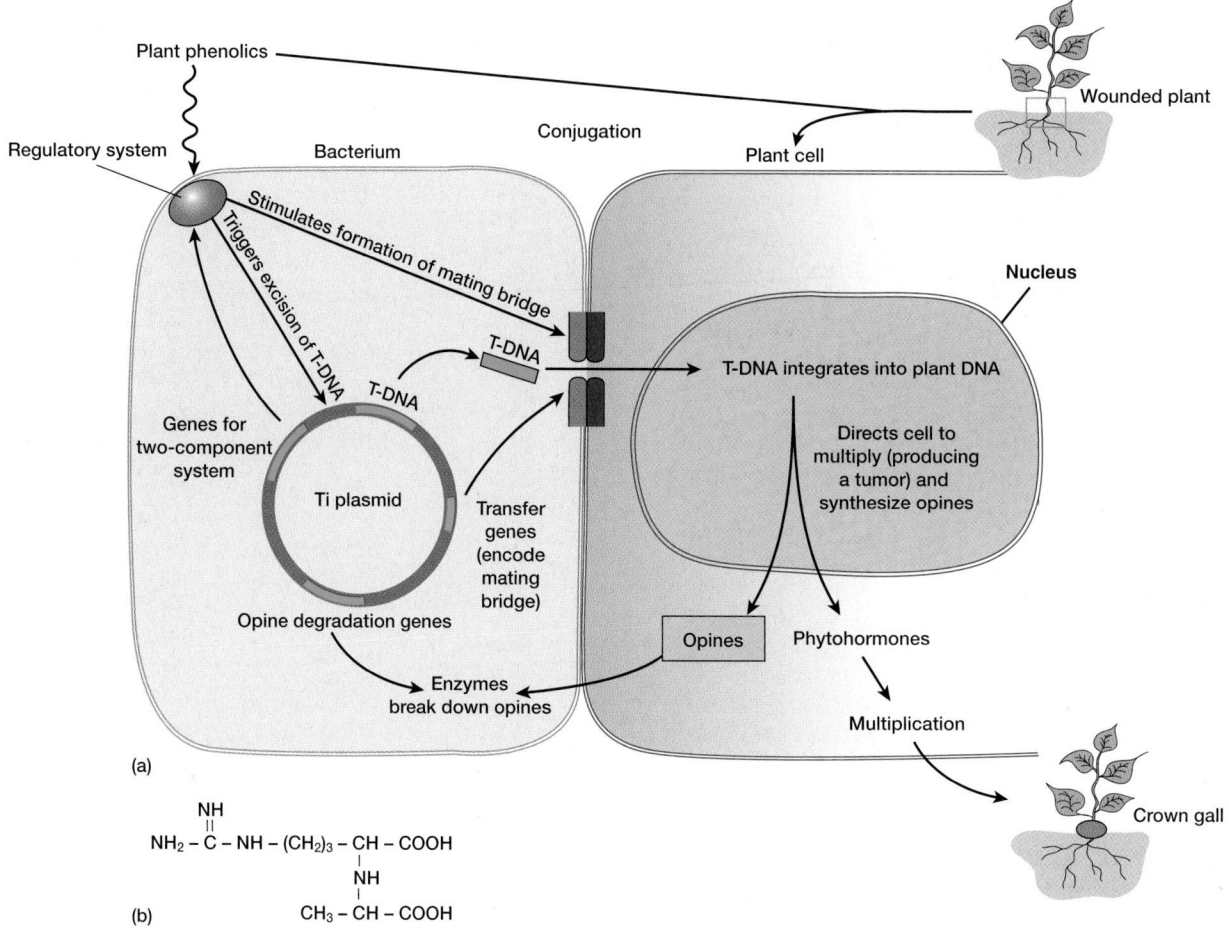

Figure 30.18   Functions of Genes Carried on the *Agrobacterium* Ti Plasmid.   (a) Genes carried on the Ti plasmid of *Agrobacterium* control tumor formation by a two-component regulatory system that stimulates formation of the mating bridge and excision of the T-DNA. The T-DNA is moved by transfer genes, which lead to integration of the T-DNA into the plant nucleus. T-DNA encodes plant hormones that cause the plant cells to divide, producing the tumor. The tumor cells produce opines (shown in **b**) that can serve as a carbon source for the infecting *Agrobacterium*. This results in the formation of a crown gall on the stem of the wounded plant above the soil surface.

decreased. They are hoping to be able to treat trees with the less lethal virus strains and eventually transform the indigenous lethal strains of *Cryphonectria* into more benign fungi.

Tripartite and Tetrapartite Associations

An additional set of interactions occurs when the same plant develops relationships with two or three different types of microorganisms. These more complex interactions are important to a variety of plant types in both temperate and tropical agricultural systems. First described in 1896, these symbiotic associations involve the interaction of the plant-associated microorganisms with each other and the host plant. Several **tripartite associations** are known to occur: the plant plus (1) endomycorrhizae plus rhizobia, including *Rhizobium* and *Bradyrhizobium;* (2) endomycorrhizae and actinorhizae; and (3) ectomycorrhizae and

actinorhizae. Nodulated and mycorrhizal plants are better suited for coping with nutrient-deficient environments. **Tetrapartite associations** also occur. These consist of endomycorrhizae, ectomycorrhizae, *Frankia*, and the host plant. These complex associations, in spite of their additional energy costs, provide important benefits for the plant.

1. Discuss the nature and importance of the Ti plasmid.
2. What are the functions of the members of the two-component system in infection of a plant by *Agrobacterium?* What are the roles of phenolics and opines in this infection process?
3. Give some important fungal and bacterial plant pathogens.
4. How are plant pathologists attempting to control chestnut blight?
5. What are tripartite and tetrapartite associations?

Table 30.5

Major Plant Diseases Caused by Bacteria

Symptoms	Examples	Pathogen
Spots and blights	Wildfire (tobacco)	*Pseudomonas syringae* pv.[a] *tabaci*
	Haloblight (bean)	*P. syringae* pv. *phaseolica*
	Citrus blast	*P. syringae* pv. *syringae*
	Leaf spot (bean)	*P. syringae* pv. *syringae*
	Blight (rice)	*Xanthomonas campestris* pv. *oryzae*
	Blight (cereals)	*X. campestris* pv. *translucens*
	Spot (tomato, pepper)	*X. campestris* pv. *vesicatoria*
	Ring rot (potato)	*Clavibacter michiganensis* pv. *sepedonicum*
Vascular wilts	Wilt (tomato)	*C. michiganensis* pv. *michiganensis*
	Stewart's wilt (corn)	*Erwina stewartii*
	Fire blight (apples)	*E. amylovora*
	Moko disease (banana)	*P. solanacearum*
Soft rots	Black rot (crucifers)	*X. campestris* pv. *campestris*
	Soft rots (numerous)	*E. carotovora* pv. *carotovora*
	Black leg (potato)	*E. carotovora* pv. *atroseptica*
	Pink eye (potato)	*P. marginalis*
	Sour skin (onion)	*P. cepacia*
Canker	Canker (stone fruit)	*P. syringae* pv. *syringae*
	Canker (citrus)	*X. campestris* pv. *citri*
Galls	Crown galls (numerous)	*Agrobacterium tumefaciens*
	Hairy root	*A. rhizogenes*
	Olive knot	*P. syringae* pv. *savastonoi*

Source: From J. W. Lengler, G. Drews, H. G. Schlegel. 1999. *Biology of the prokaryotes*, Blackwell Science, Malden, Mass., table 34.4.

[a]pv., pathover, a variety of microorganisms with phytopathogenic properties.

30.5 SOILS, PLANTS, AND NUTRIENTS

Soil organic matter helps retain nutrients, maintain soil structure, and hold water for plant use. This important resource is subject to gains and losses, depending on changes in environmental conditions and agricultural management practices. Plowing and other similar disturbances expose the soil organic matter to more oxygen, leading to extensive microbiological degradation of organic matter. Irrigation causes periodic wetting and drying, which can also lead to increased degradation of soil organic matter, especially at higher temperatures.

Several approaches have been developed to overcome management practices that stimulate soil microorganisms and lead to increased degradation of soil organic matter. These approaches include "no till" or "minimum till" agriculture, in which surface soil is minimally disturbed and chemicals are used to control weed growth.

Composting is another, much older but still effective approach to maintaining and augmenting the organic matter content of a soil. In this process, plant materials are allowed to decompose under moist, aerobic conditions, in which the readily usable plant fractions are rapidly decomposed. If the compost is too moist or too dry, the desired decomposition process will not occur. The compost pile reaches higher temperatures, allowing thermophilic microorganisms to participate actively in these processes. When the composting is completed, the residual lignin and other more resistant plant materials will have been partially transformed to humus. Such biologically stabilized compost, when added to soil, increases the soil organic matter content and does not stimulate the soil microorganisms. Thermophile characteristics (pp. 124–25)

Throughout the world, soils are increasingly being impacted by mineral nitrogen releases resulting from human activities. This nitrogen is derived from two major sources: (1) agricultural fertilizers containing chemically synthesized nitrogen, and (2) fossil fuel combustion. Fossil fuel-based releases occur especially in eastern North America, Europe, eastern China, and Japan (**figure 30.19**).

Such nitrogen releases have had a wide variety of effects. Soils have a limited ability to tie up added nitrogen compounds in organic matter; they are limited by the amount of carbon that can be incorporated into microbial biomass and can reach the **nitrogen saturation point.** Beyond this point, added nitrogen will not be incorporated into organic matter and remains in a mobile form. The major types of nitrogen fertilizers used in agriculture are liquid ammonia and ammonium nitrate. Ammonium ion usually is added because it will be attracted to the negatively charged clays in a soil and be retained on the negative clay surfaces until used as a nutrient by the plants (**figure 30.20**). However, the nitrifier populations in a soil (*see pp. 597–98*) can oxidize the ammonium ion to nitrite and nitrate, and these anions can be leached from the plant environment and enter surface waters and groundwaters.

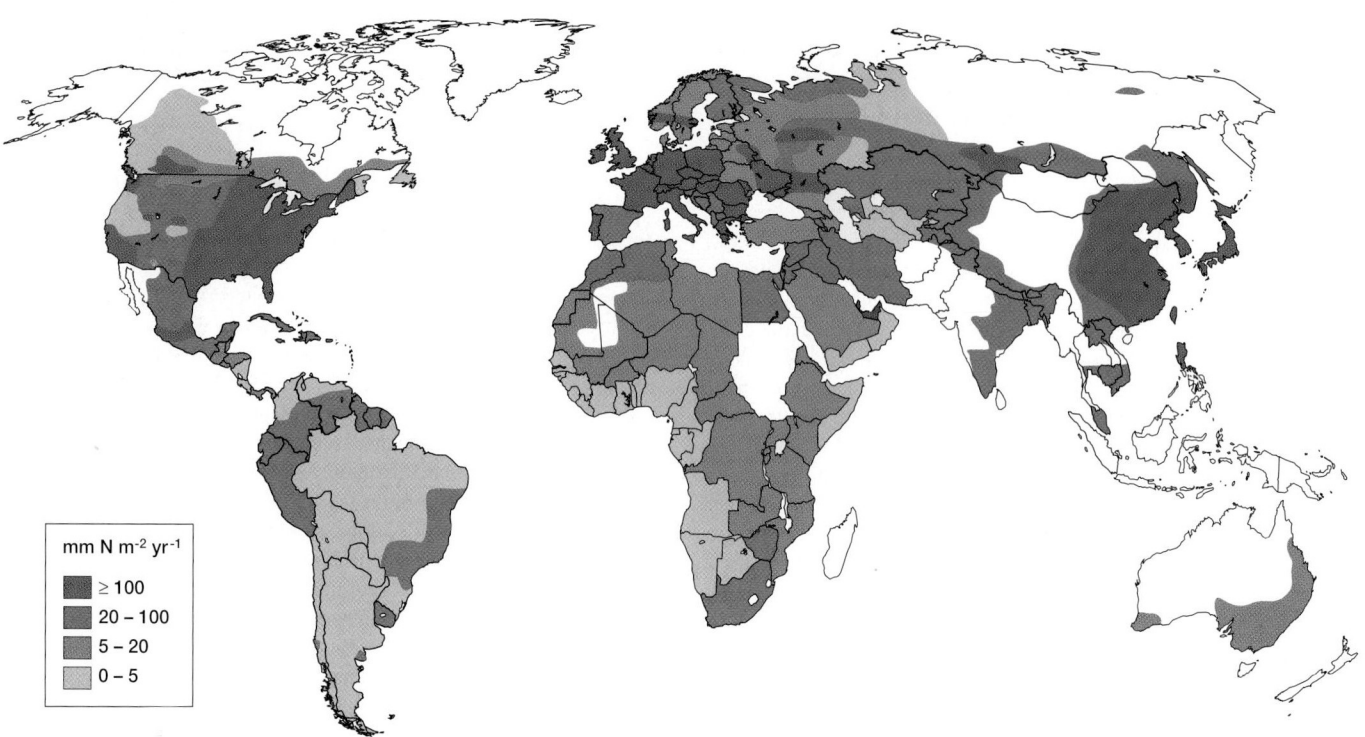

Figure 30.19 Global Releases of Nitrogen to the Environment from Fossil Fuels. Fossil fuel combustion releases, expressed as mmoles N m^{-2}yr^{-1}, are primarily from developed/developing areas of the world with high population densities. *Adapted from: Nixon, S. W. 1995. Coastal marine eutrophication: A definition, social causes, and future concerns. Ophelia 41:199–219.*

(a) Fertilizer-derived ammonium ions enter the soil and are attracted to negatively charged clays in the plant root zone.

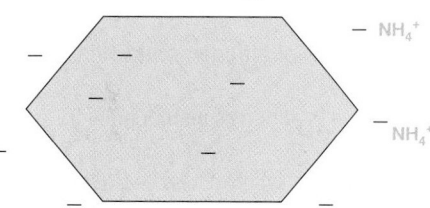

(b) Nitrification occurs in aerobic zones, transforming ammonium ions to the anionic nitrate.

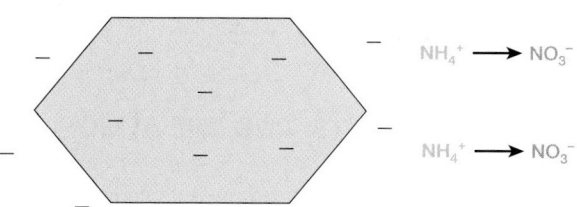

(c) Nitrate, repelled from the negatively charged clays, moves with the soil water out of the plant root zone.

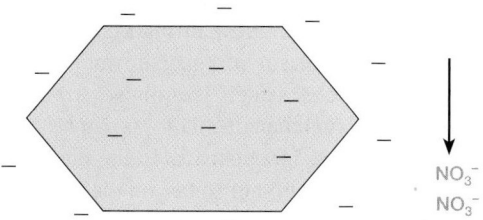

When nitrogen fertilizers are used, an inevitable consequence has been higher nitrate levels in waters, which can contribute to infant respiratory problems, and possibly to the production of nitrites and the formation of nitrosamine carcinogens. In addition, plants grown in high nitrate-concentration soils may accumulate nitrate to a level that is harmful for animals. Cereal grains, many weeds, and grass hay contain high nitrate levels when grown in such soils.

Nitrogen fertilizers also can affect the microbial community structure and function. The use of nitrogen fertilizers has led to decreases in filamentous fungal development in a wide variety of soils. The loss of fungi affects the soil structure because fungi link soil particles to stabilize the soil peds. With a weakened and decreased fungal community, mycorrhizal-dependent plants become more susceptible to stresses such as drought and toxic metals. As a final note, excessive fertilizer use can have many unexpected global-level effects (**Microbial Tidbits 30.2**).

A critical aspect of nitrogen accretion is its possible link to the increased occurrence of invasive plants in many areas of the world. Transient higher levels of mineral nitrogen are a common

Figure 30.20 Fertilizer Mobility and Soil Clays. The role of negatively charged clays in (**a**) retaining ammonium ion-containing fertilizer in a soil, and (**b**) the release of nitrate, formed by nitrification of the ammonium ions, to the soil water and the environment. (**c**) Nitrate, being anionic or negatively charged, is not retained by the negatively charged soil clays.

Microbial Tidbits

30.2 An Unintended Global-Scale Nitrogen Experiment

As discussed in chapter 42 (*see p. 990*), technological advances can have many unexpected consequences. An excellent example is the discovery of the Haber-Bosch Process at the beginning of the twentieth century that made possible the synthesis of ammonium nitrogen from inert dinitrogen gas. This led to the low-cost availability of mineral nitrogen for crop fertilization, which until then was largely dependent on animal-derived nitrogen sources (such as guano, primarily from Chile, and animal manure) and crop rotation schemes, often involving planting nitrogen-fixing legumes. With the availability of ample ammonium ion as fertilizer from the Haber-Bosch Process, there was no further need to continue such "inefficient" agricultural systems; a green cash crop could be generated each year without the tiresome use of manure and crop rotations. This, however, had unexpected consequences; without the addition of manures, organic matter was lost from soils in these more intensive cropping systems, and there were changes in plant and microbial communities, as well as increased water pollution with nitrates.

Nitrogen production by the Haber-Bosch Process, together with burning fossil fuels, adds mineral nitrogen to the atmosphere with other unexpected consequences. Studies of Northern Hemisphere forests indicate that nitrogen leaves soils and drainage waters primarily in the inorganic mineral form of nitrate. It has been assumed that this reflected global-level natural processes as the phenomenon was so widespread. Recently, however, studies of forests in South America, which are far from sources of mineral nitrogen, suggest a very different view. Steven Perakis and Lars Hedin of Cornell University have shown that remote South American forests release only small amounts of nitrate, and instead that most nitrogen is released in organic forms! What appears to have happened? By having large-scale releases of mineral nitrogen in the Northern Hemisphere for almost a century, human activities have dramatically changed the nitrogen cycle of soils and waters in most areas of the populated world. It appears that the entire Earth has been turned into a giant nitrogen-amendment experiment. Agricultural and industrial pollution has strongly impacted the Northern Hemisphere, and is now affecting the Southern Hemisphere as well. Most researchers and policymakers have not been aware of this global-level nitrogen impact! The problem is that we cannot return to the beginning and modify or replicate such a global-scale experiment.

---

occurrence in disturbed soils. First, primarily nonmycorrhizal weeds become dominant, and then are gradually outcompeted by mycorrhizal plants as the more mature plant community again takes over such disturbed sites. At the present time, however, large areas of North America are being invaded by plants such as cheatgrass (*Bromus tectorum*) and knapweeds (*Centaurea biebersteinii, C. diffusa*), that may be responding to increased levels of nitrogen from atmospheric sources (**occult nitrogen accretion**). To control these and similar invasive plants, the technique of **soil impoverishment** is now being used with some success. In this management approach, organic matter lacking nitrogen, such as ground-up waste paper or sawdust, is added to soil. In response, the microbial community immobilizes the excess mineral nitrogen. Such nitrogen immobilization has led to rapid reestablishment of native plant communities in areas that have become overrun with invasive plants. There is an increasing need to understand the contributions of soil biota to soil impoverishment, particularly in relation to the immobilization of excess mineral nitrogen that is impacting many areas of the world.

Phosphorus in fertilizers also is critical. The binding of this anionic fertilizer component to soils is dependent on the cation exchange capacity (CEC) and soil pH. As the soil phosphorus sorption capacity is reached, the excess, together with phosphorus that moves to waters with soil erosion can stimulate the growth of freshwater organisms, particularly cyanobacteria, in the process of eutrophication. The cyanobacteria are then able to fix more nitrogen, because in most freshwater systems, phosphorus is the critical limiting element.

1. What soil management processes can lead to increases in degradation of soil organic matter?
2. Why might composted materials have different effects on soil microorganisms and organic matter than additions of fresh plant materials?
3. What are possible effects of nitrogen-containing fertilizers on microbial communities?
4. Most nitrogen fertilizer is added as ammonium ion. Why is this preferred over nitrate?
5. Why is nitrate of concern when it reaches rivers, lakes, and groundwaters?
6. Why might enrichment of fresh waters with phosphorus be even more critical than nitrogen enrichment?

 30.6 SOILS, PLANTS, AND THE ATMOSPHERE

Soil microorganisms have interesting interactions with the atmosphere, a critical part of aerobiology. Ice-nucleating bacteria from decomposing litter can increase precipitation. Specific strains of *Pseudomonas syringae* synthesize proteins that can serve as nucleation centers for snow formation in cloud chambers. Such ice-nucleating soil microorganisms may significantly influence weather on a global scale. "Ice-minus" strains of *P. syringae*, developed by recombinant technology, can be sprayed on frost-sensitive crops such as strawberries. In the presence of the "ice-minus" bacteria, water will not freeze on the leaves as read-

Techniques & Applications

30.3   Keeping Inside Air Fresh with Soil Microorganisms

A major problem in the development of more energy-efficient homes and office buildings is the potential effect of such closed environments on human health. With many people spending much of their lives in such enclosed environments, the "sick building syndrome" is an increasing concern. While saving energy, these "sick buildings" have higher levels of many volatile compounds, including benzene, trichloroethylene (TCE), formaldehyde, phenolics, and solvents. These are released from rugs, furniture, plastic flooring, paints, and office machines such as photocopiers and printers. An important but still largely unappreciated means of improving the air in such "sick

buildings" is through plants and their associated soil microorganisms. Plants not only produce oxygen, but the soil microorganisms degrade many airborne pollutants. It is recommended that one plant be used per 100 square feet of living area. As noted by B. C. Wolverton, "The ultimate solution to the indoor air pollution problem must involve plants, the plant soils and their associated microorganisms." Soil microorganisms, especially in association with plants, can help keep air in closed environments fresher and more healthful (plants are also nice to look at).

---

ily, making it possible to save frost-sensitive crops from destruction. The release of such genetically modified microorganisms (GEMs) to the environment is of continuing concern (*see pp. 332–33*).

Soil microorganisms also influence the atmosphere by degrading airborne pollutants such as methane, hydrogen, CO, benzene, trichloroethylene (TCE), and formaldehyde. They can substantially improve the air in closed buildings (**Techniques & Applications 30.3**). Although soil microorganisms cannot completely eliminate air-borne pollutants, they can decrease these to equilibrium levels of approximately 1 to 2 ppm.

Soil microorganisms, through their activities in biogeochemical cycling (*see section 28.3*), also can have major effects on global fluxes of a variety of gases. These gases can be considered as those that are "relatively stable" and those that are "reactive gases." Relatively stable gases that are influenced by microbial activities include carbon dioxide, nitrous oxide, nitric oxide, and methane. Microorganisms also contribute to the flow of reactive gases such as ammonia, hydrogen sulfide, and dimethylsulfide, as major examples. These reactive gases tend to be produced in more waterlogged environments.

Addition of nitrogen-containing fertilizers also affects atmospheric gas exchange processes in a soil. Nitrogen additions stimulate the production of the nitrification intermediates NO and N_2O, which are critical as **greenhouse gases.** NO also appears to be required for *Nitrosomonas eutropha* to carry out nitrification. The oxidation of NH_4^+ involves the formation of hydroxylamine (NH_2OH) and NO as intermediates. This NO reacts with oxygen to give NO_2, which then can repeat the process in this reaction:

$$NH_4^+ + NO_2 \rightarrow NH_2OH + NO$$

$$2NO + O_2 \rightarrow NO_2 \text{ (nitrogen dioxide)}$$

In this sequence molecular oxygen does not react with NH_4^+ but with NO. If NO is absent the reaction will not proceed. Nitrification (p. 189)

Atmospheric gases such as carbon dioxide, nitrous oxide, nitric oxide, and methane, that decrease the loss of radiation from the earth and may produce atmospheric warming, are called greenhouse gases. The production and consumption of these greenhouse gases can be influenced by a range of human activity-related factors. These include plant fertilization and automobile use, conversion of soils to agricultural use, and landfills, all of which can cause increases in mineral nitrogen levels.

Methane is a greenhouse gas of increasing concern that can be derived from a variety of sources. These include ruminants, rice paddies, and landfills. Landfills, especially, can release methane to the atmosphere over longer terms (decades, centuries). Another interesting source of atmospheric methane is microorganisms that inhabit the gut of wood-eating termites. As noted in **Microbial Diversity & Ecology 30.4,** with increased deforestation and the accumulation of plant residues, populations of termites (and termite gut-inhabiting methanogenic microorganisms) are increasing.

Methane levels are influenced by microorganisms and their functioning in the environment. Well-drained, aerobic soils are capable of methane oxidation by **methanotrophs** (*see p. 488*), whereas in water-saturated sites of soils and in bogs and wetlands, methane may be produced faster than it can be used by methanotrophs. Based on analyses of gas bubbles in glacier ice cores, the levels of methane in the atmosphere remained essentially constant until about 400 years ago. Since then the methane level has increased 2.5-fold to the present level of 1.7 parts per million (ppm) by volume. Considering these trends, there is a worldwide interest in understanding the factors that control methane synthesis and use by microorganisms.

The processes of methane synthesis and use occur on a variety of scales in upland soils and in wetlands, as shown in **figure 30.21.** In nominally aerobic upland soils, there may be anaerobic hot spots (water-saturated local areas) where methane is produced. If these areas are surrounded by aerobic soils, methanotrophs degrade most of the methane before it can be released to the atmosphere. In more waterlogged areas such as lowland soils, methanogenesis can proceed and there is less opportunity for methanotrophs to function. In spite of this, most of the methane

Microbial Diversity & Ecology
30.4   Soils, Termites, Intestinal Microbes, and Atmospheric Methane

Termites are important components of tropical ecosystems, where their use of cellulose plant materials allows rapid—sometimes too rapid—recycling of plant materials. Termites harbor significant populations of archaea that use products of cellulose digestion, including CO_2 and hydrogen, to produce methane.

Termites occur on 2/3 of the Earth's land surface, and, based on laboratory studies, 0.77% of the carbon ingested by termites can be released as methane. In tropical wet savannas and cultivated areas, termite populations are increasing rapidly. This increase is being ac-

celerated by the destruction of tropical forests, which results in the accumulation of dead plant materials on the soil surface. This provides an ideal environment for the growth of termites. Termites are estimated to be contributing annually at least 1.5×10^{14} grams of methane, together with hydrogen and CO_2, to the atmosphere. This is believed to be contributing to measurable increases in the atmospheric methane level. Thus unseen termites and their associated gut microorganisms may be affecting global warming.

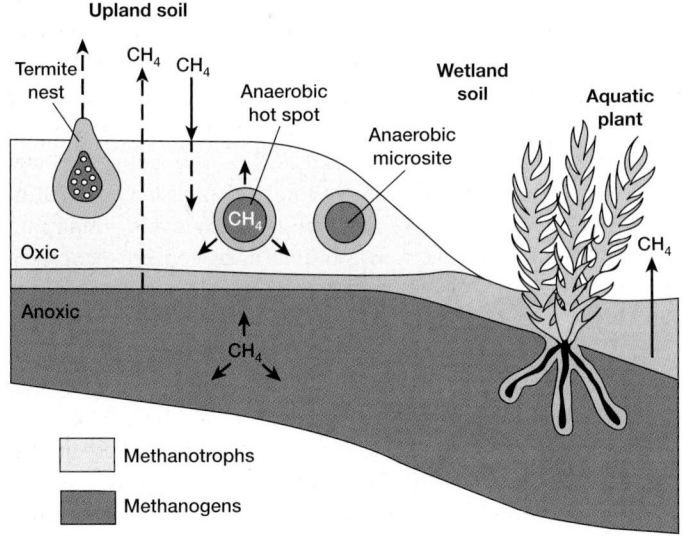

Methanotrophs

Methanogens

Figure 30.21   Methane Production and Use in Soils.
Methane production and degradation can occur in closely located aerobic and anaerobic zones. In upland soils, methane synthesis can take place in localized anaerobic hot spots and termite nests. Much of this methane can be degraded in the surrounding aerobic soils. In wetland areas, methane production dominates in the waterlogged areas, and methane has a greater chance of being released to the atmosphere. If aquatic plants are present and translocate oxygen into anaerobic zones, these create localized aerobic hot spots in the rhizosphere for methane oxidation.

will be degraded. Aquatic plants transport oxygen to their rhizospheres and thus encourage methane oxidation in these localized aerobic hot spots. Methane oxidation also is sensitive to nitrogen-containing fertilizers and increases in atmospheric CO_2 levels. The balance between methane synthesis and degradation is very important. It is estimated that soils of all types provide 60% of the total atmospheric methane; wet areas, water-saturated hot spots, and rice paddies are particularly significant contributors.

1. What is a greenhouse gas? Give examples and discuss their possible importance.
2. Discuss the possible role of termites in the production of greenhouse gases.
3. What microbial processes occur in soils to both produce and degrade methane?
4. Describe factors that might lead to the formation of localized hot spots for the production and consumption of greenhouse gases.

 **30.7   MICROORGANISMS AND PLANT DECOMPOSITION**

Plant growth is important in terrestrial ecosystems; equally important are plant death, decomposition, and recycling. Plants are constantly releasing leaves, branches, and other plant parts which enter the soil. These released soluble materials create a **residuesphere,** an area between decaying plant material and the soil. This has been described as a hotbed of microbial activity, which drives a wide range of microbiological processes. In anaerobic "hot spots" soluble plant materials support denitrification and possibly genetic exchange processes. After soluble materials have leached from the plant, the remaining starch, cellulose, and proteins are degraded both aerobically and anaerobically. Lignin, a random aromatic cyclic polymer that contains no nitrogen, is the one exception: its degradation requires oxygen. The basidiomycetes, which function best aerobically, are the major producers of laccases and the phenoloxidase enzymes required for lignin degradation. Lignin decomposition also is limited by the physical nature of the material: healthy woody plants are saturated with sap, creating an aerobic, low oxygen flux environment similar to that found in waters (*see chapter 29*). In addition, high ethylene and CO_2 levels, as well as the presence of phenolic and terpenoid compounds, retard the growth of lignin-degrading fungi.

Table 30.6	
Global Emissions of Chloromethane	

Source	Inputs to Atmosphere (10⁵ tons/year)
Natural Sources	
Terrestrial processes	0–20
Ocean fluxes	3–20
Biomass burning	4–14
Anthropogenic Sources	0–3

Source: From R. Watling and D. B. Harper. 1998. *Mycol. Res.* 102(7):769–87.
ᵃEstimated atmospheric inputs from natural and anthropogenic sources.

Unusual gases are produced by fungi in this process of woody plant decomposition; these include chloromethanes and cyanide, compounds that we normally associate with industrial pollution. Large amounts of CH_3Cl, an important greenhouse gas, are produced by many fungi, including the basidiomycetes *Phellinus* and *Inonotus,* which are members of the wood-rotting *Hymenochaetaceae.* The global input of CH_3Cl to the atmosphere from plant decomposition is thought to be 160,000 tons, 75% of which is derived from tropical and subtropical soils. It is estimated that 15 to 20% of the chlorine-catalyzed ozone destruction is due to naturally produced chlorohydrocarbons. Basidiomycetes (pp. 545–46)

As noted in **table 30.6,** terrestrial environments, the oceans, and biomass burning are all important sources of atmospheric chloromethane. Large amounts also have been detected in some basidiomycetes. These include concentrations of 74 to 2,400 mg/kg in the basidia of some agarics and bracket fungi. It is of interest that the maximum permissible chlorophenol levels in soils are only 1 to 10 mg/kg! Cyanide is another chemical of widespread concern produced by fungi, especially by basidiomycetes and ascomycetes. This cyanide may be derived from the S-methyl group of L- and D-methionine, or its production may be linked to methyl benzoate synthesis. The best studied cyanide-producing fungi are *Marasmius oreades* (the cause of fairy ring disease) and the snow mold basidiomycetes. Some bacteria also produce cyanide. Cyanide synthesis involves the oxidative decarboxylation of glycine, which is stimulated by methionine or other methyl-group donors in this reaction:

$$NH_2CH_2COOH \rightarrow HCN + CO_2 + 4[H]$$

Cyanide can inhibit respiration. It also can serve as a carbon and nitrogen source for microorganisms, including cyanogenic fungi such as *Marasmius* and *Pholiota,* and some actinomycetes. This illustrates the adaptability of microorganisms in the use of a nominally toxic metabolic product.

---

1. Why are fungi critical for the degradation of woody plants? How does the plant defend itself against these fungi?
2. What is the residuesphere?
3. What critical products of fungal growth on wood are usually considered only to be industrial pollutants?
4. How do microorganisms contribute to the cycling of cyanide?

 30.8   THE SUBSURFACE BIOSPHERE

An important habitat for microbial growth is the **subsurface biosphere,** a source of many questions concerning the origin of life on the Earth. The subsurface biosphere has been studied by examining outcrops, surface excavations, petroleum hydrocarbons, well corings, and materials from deep mine sites. Drillers have reached a depth of 6,100 meters (20,000 feet) at the Inigok well, North Slope, Alaska, and 9,150 meters (30,000 feet) at the Becha Rogers well in the Anadarko Basin, located in northern Texas and Oklahoma. In a South African gold mine, rocks from deeper than 3,350 meters (11,000 feet) below the surface have been recovered by hand and examined in the search for new microorganisms. A long history of interest in the subsurface preceded these efforts, and there have been reports of microorganisms in rocks, particularly from rocks obtained in deep drilling programs. Much of this information has been discounted as being due to drilling contamination or water flows from surface areas. In recent years the use of improved techniques, including molecular approaches, has led to a greater acceptance of the idea of microorganisms occurring far below the Earth's surface, leading to a new field: deep subsurface microbiology.

Microbial processes take place in different subsurface regions, including (1) the shallow subsurface where water flowing from the surface moves below the plant root zone; (2) subsurface regions where organic matter, originating from the Earth's surface in times past, has been transformed by chemical and biological processes to yield coal (from land plants), kerogens (from marine and freshwater microorganisms), and oil and gas; and (3) zones where methane is being synthesized as a result of microbial activity.

In the shallow subsurface, surface waters often move through **aquifers,** porous geological structures below the plant root zone. As shown in **figure 30.22,** in a pristine system with an aerobic surface zone, the oxidants used in catabolism will be distributed from the most oxidized and energetically favorable (oxygen) near the surface to the least favorable (in which CO_2 is used in methanogenesis) in lower zones.

In subsurface regions where organic matter from the Earth's surface has been buried and processed by thermal and possibly biological processes, kerogen and coals break down to yield gas and oil, as noted in **figure 30.23.** After their generation, these mobile products, predominantly hydrocarbons, move upward into the more porous geological structures where microorganisms can be active. Chemical signature molecules from plant and microbial biomass are present in these petroleum hydrocarbons.

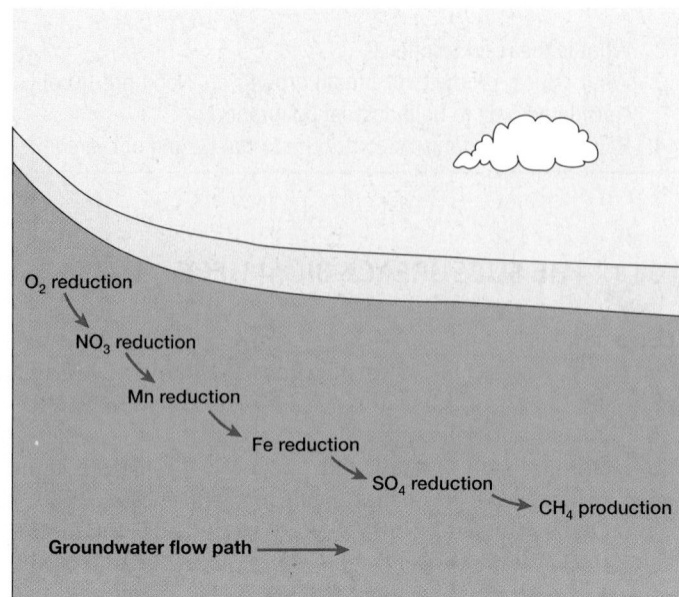

Figure 30.22 The Shallow Subsurface Biosphere. The shallow subsurface, as in a stable sediment, showing the distribution with depth of electron acceptors that can occur in an aerobic pristine aquifer. In aerobic sediments, the oxidants will be distributed with the most energetically favorable (oxygen) near the surface and the least energetically favorable at the lower zones of the geological structure. *Source: Lovley, D. K., 1991. Dissimilatory Fe (III) and Mn (IV) reduction. Microbiol. Rev. 55:259–87.*

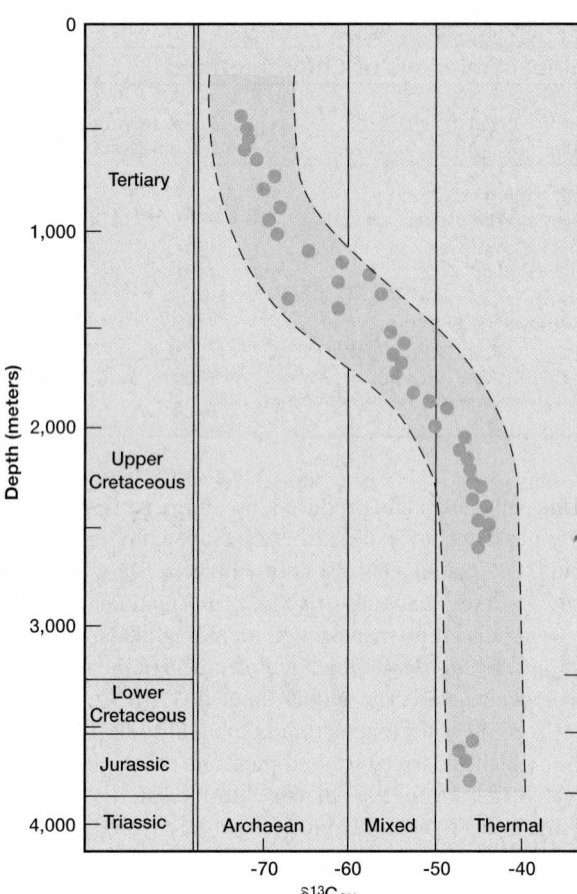

Figure 30.24 Methane Synthesis by Archaeans in the Subsurface. The microbially mediated production of methane has been shown to occur in the subsurface by the use of stable isotope techniques. The decrease in the occurrence of the ^{13}C isotope of carbon indicates that the methane was produced by microorganisms down to a depth of 1,500 meters under the floor of the North Sea. Below a depth of 2,000 meters, the methane does not have a lower frequency of the ^{13}C isotope of carbon, indicating that it was formed by abiotic processes. The δ ^{13}C value gives an indication of the relative proportion of ^{13}C to ^{12}C in the sample. The more negative scale values signify a decreased presence of the heavier isotope in the upper zone at this location.

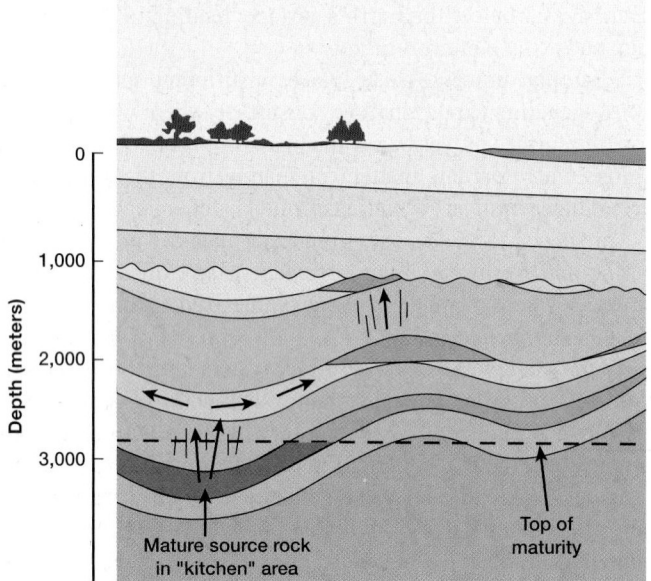

Figure 30.23 The Oil and Gas Region of the Subsurface. Organic matter, originating from the Earth's surface, is transformed to oil, gas and coal by chemical, thermal, and biological processes. Above the higher-temperature "kitchen" area, where chemical changes occur, microorganisms can contribute to the processing of these organic materials. Hydrocarbons migrate through porous strata and fractures, finally accumulating in porous, overlying geological structures. Lines indicate fractures, and arrows indicate hydrocarbon flows.

Below these zones lie vast regions where methane is present in geological structures (**figure 30.24**); this methane is continuously being released to the overlying strata. Based on studies of stable carbon isotopes, in the "biogenic" zone, methane has less ^{13}C isotope, indicating that it was derived from microorganisms using H_2 as an energy source and CO_2 as a carbon source and oxidant. Given a choice, microbes will tend to use the lighter of two isotopes—in this case ^{12}C over the ^{13}C isotope-containing CO_2. In the underlying hotter abiogenic zone, the methane is not depleted of the heavier carbon isotope, indicating that this is of chemical and thermal origin.

It is essential to understand the source of hydrogen for the deep subsurface methanogenesis process. Suggested sources are geological and may include (1) the reaction of water with basaltic rocks, (2) possible radioactive interactions, and (3) so-called magma degassing processes that occur along geological faults. The actual origin of this almost limitless supply of hydrogen, upon which deep surface methanogenesis depends, has not been determined with certainty.

Microorganisms also appear to be growing in intermediate depth oil-bearing structures. Recent studies indicate that active procaryotic assemblages are present in high-temperature (60 to 90°C) oil reservoirs, including such genera as *Thermotoga, Thermoanaerobacter,* and *Thermococcus.* The archaean genera are dominated by methanogens. Thus microbial activities may be occurring above or in the "deep hot biosphere," a term suggested by Thomas Gold to describe this poorly understood region.

1. What types of microbial activities have been observed in the deep subsurface?
2. What happens in terms of microbiological processes when organic matter leaches from the surface into the subsurface?
3. Why are stable isotope analyses so important in studies of microbe-geological interactions?
4. What microbial genera have been observed in oil field materials?

30.9 SOIL MICROORGANISMS AND HUMAN HEALTH

As with waters, humans are in constant contact with soils. This occurs directly as when children or adults play in the "dirt," or even when leafy and root vegetables, covered with soil dust, are eaten. In most cases the contact with soil is harmless. However, soils do contain a wide variety of pathogenic organisms. What is needed is an entry point and favorable conditions within or on the human body. A wide variety of anaerobes, including *Clostridium,* (*see p. 509*) are present in soils. Unless there is a deep puncture wound that will provide the anaerobic environment required for their growth, anaerobes are of little concern. However, puncture wounds, as occur in warfare and accidents, can lead to gangrene. Soils contain other pathogens. Organisms such as *Acanthamoeba,* which can be inhaled from dust, may cause primary amebic meningoencephalitis. When soils are used for surface disposal of human wastes without sewage treatment, the transmission of a wide variety of pathogens, including protozoa such as *Acanthamoeba* and *Cyclospora,* can occur.

Soil and soil-related microorganisms also are of concern when they grow in buildings (**figure 30.25**). This increasingly common problem, often linked to the flooding of houses located in low-lying districts or to moisture accumulation in sink and bathroom areas (even in large and modern homes), has led to major health problems. This is particularly severe when water penetrates into house walls and insulation materials. The problem has reached the point that national newsmagazines have published articles, showing large new expensive homes, entitled "The Mold in Your Home May Be Deadly!" In a recent article it was noted that 50% of homes have mold problems, a major source of

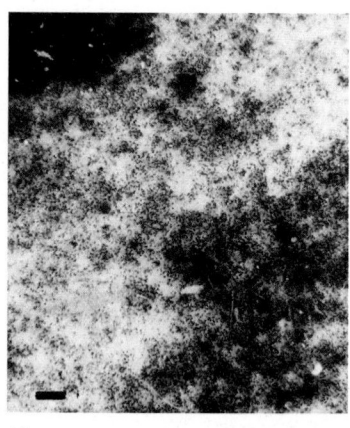

(a)

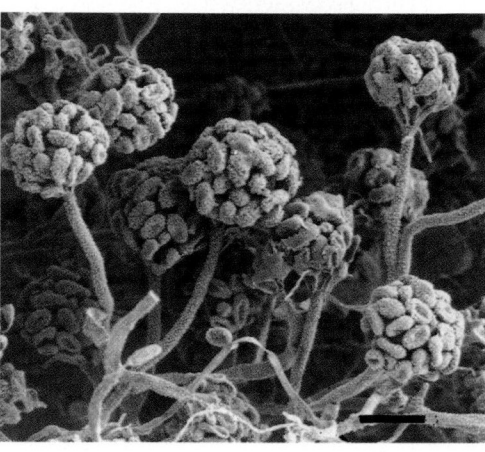

(b)

Figure 30.25   Fungal Growth in a Building.   Fungal growth on sheet rock removed from a water-damaged building. (**a**) Stereo microscopic view showing black discolorations. Bar = 500 μm. (**b**) Scanning electron micrograph of dense mycelia and conidiophores characteristic of *Stachybotrys.* Bar = 10 μm.

chronic sinus infections. These molds also have been related to increases in asthma rates. The major responsible fungi are *Stachybotrys chartarum, Eurotium herbariorum,* and *Aspergillus versicolor.* Fungal growth results in a black slime; when this fungal growth dries, a dry dusty layer remains and the spores can be dispersed into the air. These spores are particularly dangerous for infants, whose lungs are less developed. *Stachybotrys* infection can result in pulmonary hemosiderosis, which causes bleeding of the lungs and sometimes death. Rapid drying of water-damaged buildings is required to control this problem.

It has been reported that wallboards can contain baseline bioburdens of these organisms, which need only high humidity to trigger fungal growth. In addition, *Mycobacterium komossense* and gram-negative endotoxin-producing bacteria have been isolated. There are limited alternatives for controlling and removing such dangerous fungi; the most important are the removal/disinfection of moldy materials and keeping a house dry.

UNDERSTANDING MICROBIAL DIVERSITY IN THE SOIL

As noted in chapter 28, a wide variety of methods are available to attempt to learn more about the diversity of microorganisms that are present in a soil. As with other environments, such as waters, animal bodies, and insects, only 1 to 10% of the microscopically observable microbes can be cultured in the laboratory using conventional techniques. With the availability of molecular techniques, as discussed in chapters 19 and 28, it is now becoming possible to study microbial diversity in a soil sample by extracting and analyzing the nucleic acids that are present. Ultimately, it will be necessary to grow the microorganisms and relate specific nucleic acids to known structures in a soil (*see section 28.5*). Without growth of mi-

croorganisms considered to contain these nucleic acids, it is not possible to do more than note that these specific sequences are present. This makes it doubly important to better understand the physiology and activities of the microorganisms that make up so much of our world.

---

1. Surface soil spreading of untreated human wastes is carried out in many parts of the world. What are some of the possible effects of this practice?
2. How can the growth of fungi affect human health inside of a home?
3. What important fungal genera are involved in mold problems in homes?
4. Why is the growth of microorganisms important for furthering our understanding of microbial diversity?

30.1 Soils as an Environment for Microorganisms

a. Terrestrial environments are dominated by the solid phase, consisting of organic and inorganic components.

b. In an ideal soil, microorganisms function in thin water films that have close contact with air (**figure 30.1**). If isolated water volumes are present within the soil volume, from a microbial viewpoint, these are miniaquatic environments.

30.2 Microorganisms in the Soil Environment

a. Most microorganisms in these environments are associated with surfaces, and these surfaces influence microbial use of nutrients and interactions with plants and other living organisms (**figure 30.2**).

b. Bacteria and fungi in soils have different functional strategies. Fungi tend to develop on the surfaces of aggregates, whereas microcolonies of bacteria are commonly associated with smaller pores.

c. Insects, earthworms, and other soil animals are also important parts of the soil. These decomposer-reducers interact with the microorganisms to influence nutrient cycling and other processes.

30.3 Microorganisms and the Formation of Different Soils

a. Soils form under a bewildering array of conditions (**figure 30.3**). In all cases organic matter accumulation occurs through the direct activities of primary producers or by the import of preformed organic materials. Soils can be formed in regions such as the antarctic where there are no vascular plants.

30.4 Soil Microorganism Associations with Vascular Plants

a. Plants develop associations with many types of microorganisms. These include important asso-

ciations with rhizobia; actinomycetes, forming actinorhizae; and with endophytic fungi and bacteria. *Agrobacterium* forms tumors and is useful in molecular biology (**figure 30.18**).

b. The *Rhizobium*-legume symbiosis is one of the best-studied examples of plant-microorganism interactions. This interaction is mediated by complex chemicals that serve as communication signals (**figure 30.7**).

c. Mycorrhizal relationships (plant-fungal associations) are varied and complex. Six basic types can be observed (**figure 30.10**) including endomycorrhizal and sheathed/ectomycorrhizal types. Specialized monotropoid fungi make it possible for achlorophyllous plants to survive using carbon fixed by green plants. The hyphal network of the mycobiont can lead to the formation of a mycorrhizosphere.

d. The mycorrhizal relationship often is established with the assistance of mycorrhization helper bacteria (MHB). In addition, bacteria may occur inside of the mycorrhizal fungus. These bacteria apparently contribute to the nitrogen cycling of the plant-fungus complex.

e. Some plants form tripartite and tetrapartite associations.

30.5 Soils, Plants, and Nutrients

a. Composting is useful for maintaining and increasing soil fertility.

b. Soil microorganisms can interact with the atmosphere. Some can serve as nucleating agents, whereas others can degrade airborne pollutants.

c. Nitrogen fertilizers can affect soil microbial communities, and the nitrification of ammonium ion-containing fertilizers to nitrate allows the nitrogen to enter ground and surface waters. Nitrogen accretion is having global-level effects, with most impacts occurring in the Northern Hemisphere.

30.6 Soils, Plants, and the Atmosphere

a. Microorganisms can play major roles in the dynamics of greenhouse gases such as carbon dioxide, nitrous oxide, nitric oxide, and methane (**figure 30.21**). Microorganisms can contribute to both the production and consumption of these gases.

30.7 Microorganisms and Plant Decomposition

a. Fungi play important roles in the decomposition of woody plants. The basidiomycetes are the major decomposers of lignin in woody plant materials. The functioning of these fungi requires the presence of oxygen.

b. Fungi, especially, can produce chemicals that are normally considered as anthropogenic pollutants. These include chloromethane and cyanide.

30.8 The Subsurface Biosphere

a. The subsurface includes at least three zones: the shallow subsurface; the zone where gas, oil, and coal have accumulated; and the deep subsurface, where methane synthesis occurs (**figure 30.24**).

30.9 Soil Microorganisms and Human Health

a. Microorganisms, particularly the fungi, can develop in moist areas in houses and cause major health problems for humans, including asthma. An important fungus involved in these problems is *Stachybotrys* (**figure 30.25**).

30.10 Understanding Microbial Diversity in the Soil

a. Only a few soil microorganisms can be cultured, and nucleic acid analysis must be used to study microbial diversity. Thus it is necessary to learn more about microbial physiology and culture techniques.

Key Terms

actinorhizae 658
aquifer 667
arbuscular mycorrhizal (AM)
  fungi 656
arbuscule 656
associative nitrogen fixation 651
bacteroid 655
composting 662
desert crust 650

ectomycorrhiza 656
endomycorrhiza 656
endophyte 655
geosmin 648
greenhouse gas 665
infection thread 655
methanotroph 665
microbivory 648
mycorrhizosphere 658

nitrogen saturation point 662
Nod factors 655
occult nitrogen accretion 664
oxidative burst 652
ped 647
phyllosphere 651
residuesphere 666
rhizoplane 651
rhizosphere 651

root nodule 655
slash-and-burn agriculture 649
soil impoverishment 664
stem-nodulating rhizobia 659
subsurface biosphere 667
symbiosome 655
tetrapartite association 661
Ti plasmid 660
tripartite association 661

Questions for Thought and Review

1. Considering the oxygen gradients, together with nutrient and waste-product gradients present in soils, is there a typical soil microenvironment? Explain.

2. Why might protozoa prefer to use laboratory-grown microorganisms as a nutrient source rather than normal soil microorganisms?

3. Tropical soils throughout the world are under intense pressure in terms of agricultural development. What land use and microbial approaches might be employed to better maintain this valuable resource?

4. Microorganisms interact with plants in many ways. How might it be possible to improve these interactions through molecular biology?

5. Why might vascular plants have developed subtle relationships with so many types of microorganisms? What do these molecular-level interactions, which show so many similarities when microbe-plant and microbe-human interactions are considered, suggest concerning possible common evolutionary relationships?

6. Why might a plant assist in its' parasitism by *Agrobacterium?*

7. What alternatives could be used to minimize the environmental effects of nitrogen-containing fertilizers on the environment?

8. If lignin-containing plant materials didn't decompose, what could happen?

9. How might you maintain organisms from the deep hot subsurface under their in situ conditions? Compare this problem with that of working with microorganisms from deep marine environments.

10. What are some of the potential strong and weak points of using stable isotope-based analyses of microbial processes?

Critical Thinking Questions

1. How might it be possible to minimize the production of pollutants by soil fungi? Is there any possible role for genetic engineering in solving this problem?

2. Soil bacteria such as *Streptomyces* produce the bulk of known antibiotics. Look up the competitors for *Streptomyces,* the types of antibiotics these bacteria produce, and how the compounds are effective against competitors (what are the physiological targets?). Would you expect aquatic/marine bacteria to be major producers of antibiotics? Why or why not?

For recommended readings on these and related topics, see Appendix V.

For recommended readings on these and related topics, see Appendix V.

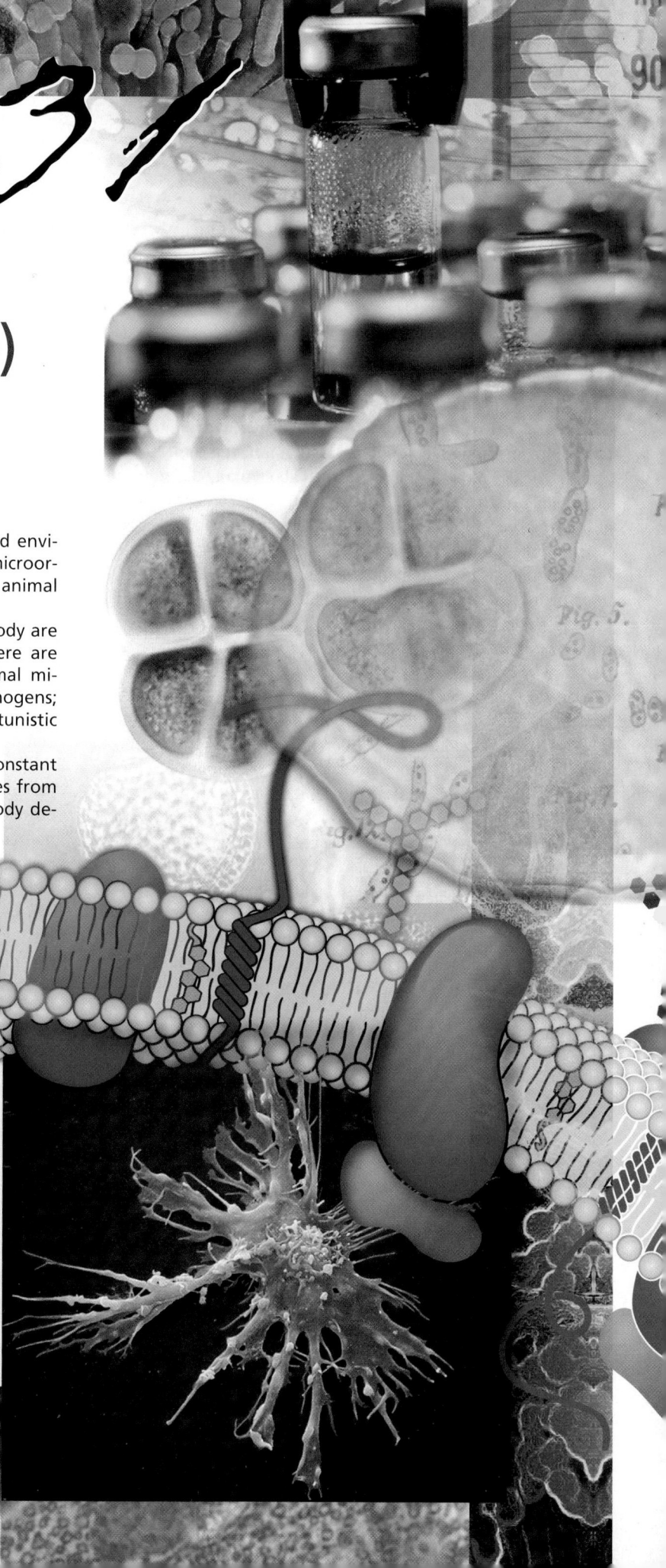

chapter 31

Normal Microbiota and Nonspecific (Innate) Host Resistance

Concepts

- Gnotobiotic refers to a microbiologically monitored environment or animal in which the identities of all microorganisms present are known or to an environment or animal that is germfree.

- Most microorganisms associated with the human body are bacteria; they normally colonize specific sites. There are both positive and negative aspects of these normal microorganisms. Sometimes they compete with pathogens; other times they are capable of producing opportunistic infections.

- The host's ability to resist infection depends on a constant defense against microbial invasion. Resistance arises from both nonspecific (innate) and specific (adaptive) body defense mechanisms.

- Nonspecific host defenses are those innate mechanisms with which a host is genetically endowed. Examples include physical and mechanical barriers: specifically, the skin, mucous membranes, the respiratory system, gastrointestinal tract, genitourinary tract, and the eye. Other nonspecific defenses include chemical barriers such as bacteriocins, beta-lysin, and other polypeptides.

- Inflammation, the alternative complement pathway, phagocytosis, cytokines, fever, and natural killer cells are other examples of nonspecific defenses that help protect the host against microorganisms and cancer.

This dendritic cell (electron micrograph) acquired its name because it is covered with long cell extensions that resemble the dendrites of nerve cells. Dendritic cells are resident in most tissues and are relatively long-lived. Their role is to survey the local environment for pathogens.

Half the secret of resistance is cleanliness, the other half is dirtiness.

—Anonymous

s presented in chapter 28, many microorganisms live much of their lives in a special ecological relationship: an important part of their environment is a member of another species. Previously you learned about many different environments in chapters 28, 29, and 30; this chapter presents the normal microorganisms associated with the human body. Very little is known about the nature of these normal associations, but they are thought to be dynamic interactions rather than associations of mutual indifference. Each member of the symbiosis is thought to derive some benefit from the other, and the associations are for the most part mutualistic or commensalistic (*see section 28.2*). The survival of a host, such as a human, depends upon an elaborate network of defenses that keeps harmful microorganisms and other foreign material from infecting the body. Should they gain access, additional host defenses are summoned to prevent them from establishing another type of relationship, one of parasitism or **pathogenicity.** Pathogenicity is the ability to produce pathologic changes or disease. A **pathogen** [Greek *patho,* and *gennan,* to produce] is any disease-producing microorganism. Nonspecific (innate) host defenses involve general barriers, physical barriers, chemical barriers, and biological barriers such as inflammation and fever. This chapter introduces the normal human microbiota and the main lines of nonspecific host defenses against invasion by harmful microorganisms and cancer cells. We begin, however, with a look at the field of gnotobiotics in which it is possible to have an environment or an animal that is free of microorganisms.

31.1 GNOTOBIOTIC ANIMALS

Environments and animals with known microorganisms increasingly are used in research. For example, to determine the role of the normal microorganisms associated with a host, it is possible to deliver an animal by caesarean section and raise that animal in the absence of microorganisms—that is, germfree. These microorganism-free animals provide suitable experimental models for investigating the interactions of animals and their microorganisms. Comparing animals possessing normal microorganisms with germfree animals permits the elucidation of many of the complex relationships covered in chapters 28, 29,

and 30 between microorganisms, hosts, and specific environments. Germfree experiments also extend the microbiologist's "pure culture concept" to in vivo research.

The term **gnotobiotic** [Greek *gnotos,* known, and *biota,* the flora and fauna of a region] has been defined in two ways. Some think of a gnotobiotic environment or animal as one in which all the microbiota are known; they distinguish it from one that is truly germfree. We shall use the term in a more inclusive sense. Gnotobiotic refers to a microbiologically monitored environment or animal that is germfree (**axenic** [*a,* neg, and Greek *Xenos,* a stranger]) or in which the identities of all microbiota are known.

The human fetus in utero (as is the case in most mammals) is usually free from microorganisms. Within hours after birth, it begins to acquire a normal microbiota, which stabilizes during the first week or two of life. From then on, enormous numbers of microorganisms become associated with the human body.

Louis Pasteur first suggested that animals could not live in the absence of microorganisms. Attempts between 1899 and 1908 to grow germfree chickens had limited success because the birds died within a month. Thus it was believed that intestinal bacteria were essential for the adequate nutrition and health of the chickens. It was not until 1912 that germfree chickens were shown to be as healthy as normal birds when they were fed an adequate diet. Since then, gnotobiotic animals and systems have become commonplace in research laboratories (**figure 31.1**).

Gnotobiotic animals and techniques provide good experimental systems to investigate the interactions of animals and specific microorganisms. The comparison of animals containing normal microbiota with gnotobiotic animals helps one better understand the many complex symbiotic associations that exist between the host and its microorganisms.

Establishing a germfree colony of rats, mice, hamsters, rabbits, guinea pigs, or monkeys begins with caesarean sections on pregnant females. The operation is performed under aseptic conditions in a sterile isolator. The newborn animals are then transferred to other germfree isolators in which all entering air, water, and food is sterile. Once the gnotobiotic animals become established, normal mating among themselves maintains the germfree colony.

Establishment of a germfree colony is much easier with chickens and other birds than with mammals. Fertile eggs from the birds are first sterilized with a germicide and then placed in sterile isolators. When the chick hatches, it is sterile and able to feed by itself. To make sure that a colony is germfree, periodic bacteriologic examination of all exhaust air, waste material, and cages must be done. Microorganisms should not be found.

Germfree animals are usually more susceptible to pathogens. With the normal protective microbiota absent, foreign and pathogenic microorganisms establish themselves very easily. The number of microorganisms necessary to infect a germfree animal and produce a diseased state also is much smaller. Conversely, germfree animals are almost completely resistant to the intestinal protozoan *Entamoeba histolytica* that causes amebic dysentery. This resistance results from the absence of the bacteria that *E. histolytica* uses as a food source. Germfree animals also do not show

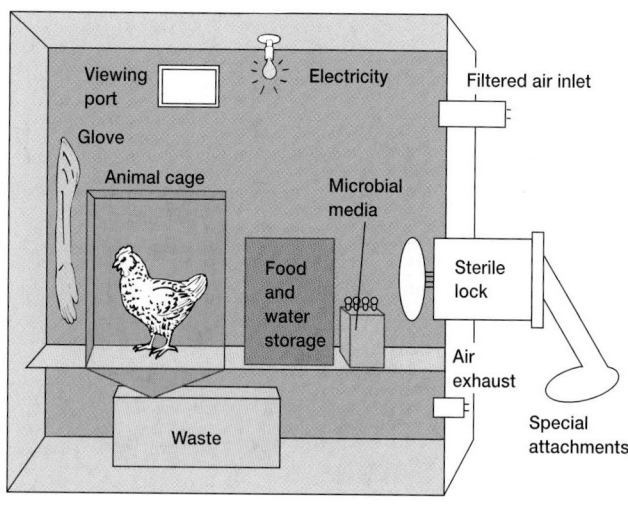

Figure 31.1   Raising Gnotobiotic Animals.   (a) Schematic of a gnotobiotic isolator. The microbiological culture media monitor the sterile environment. If growth occurs on any of the cultures, gnotobiotic conditions do not exist. **(b)** Gnotobiotic isolators for rearing colonies of small mammals.

any dental caries or plaque formation (*see section 39.6*). However, if they are inoculated with cariogenic (caries or cavity-causing) streptococci of the *Streptococcus mutans–Streptococcus gordonii* group and fed a high-sucrose diet, they will develop caries.   *Entamoeba histolytica* (pp. 927–28)

1. Define gnotobiotic.
2. How would you establish a germfree colony of mice? Of chickens?
3. Compare a germfree mouse to a normal one with regard to overall susceptibility to pathogens. What benefits does an animal gain from its microbiota?

31.2   NORMAL MICROBIOTA OF THE HUMAN BODY

In a healthy human the internal tissues (e.g., brain, blood, cerebrospinal fluid, muscles) are normally free of microorganisms. Conversely, the surface tissues (e.g., skin and mucous membranes) are constantly in contact with environmental microorganisms and become readily colonized by certain microbial species. The mixture of microorganisms regularly found at any anatomical site is referred to as the normal microbiota, the indigenous microbial population, the microflora, or the normal flora. For consistency, the term **normal microbiota** is used in this chapter. An overview of the microbiota native to different regions of the body (**figure 31.2**) and an introduction to the microorganisms one can expect to find on culture reports is presented next. Because bacteria make up most of the normal microbiota, they are emphasized over the fungi (mainly yeasts) and protozoa.

There are many reasons to acquire knowledge of the normal human microbiota. Three specific examples include:

1. An understanding of the different microorganisms at specific locations provides greater insight into the possible infections that might result from injury to these body sites.
2. A knowledge of the normal microbiota helps the physician-investigator understand the causes and consequences of colonization and growth by microorganisms normally absent at a specific body site.
3. An increased awareness of the role that these normal microbiota play in stimulating the host immune response can be gained. This awareness is important because the immune system provides protection against potential pathogens.

Distribution of the Normal Microbiota

As noted in chapter 28, three of the most important types of symbiotic relationships are commensalism, mutualism, and parasitism. Within each category the association may be either ectosymbiotic or endosymbiotic. In **ectosymbiosis** one organism remains outside the other. In **endosymbiosis** one organism is present within the other. In the following subsections, examples will be presented of both ecto- and endosymbiotic relationships that exist with the human host. Both commensalistic and mutualistic relationships are considered. Parasitism and pathogenicity will be presented in chapter 34.

Skin

The adult human is covered with approximately 2 square meters of skin. It has been estimated that this surface area supports about 10^{12} bacteria. As noted in chapter 28, a commensal is an organism that participates in commensalism. Commensalism is a symbiotic

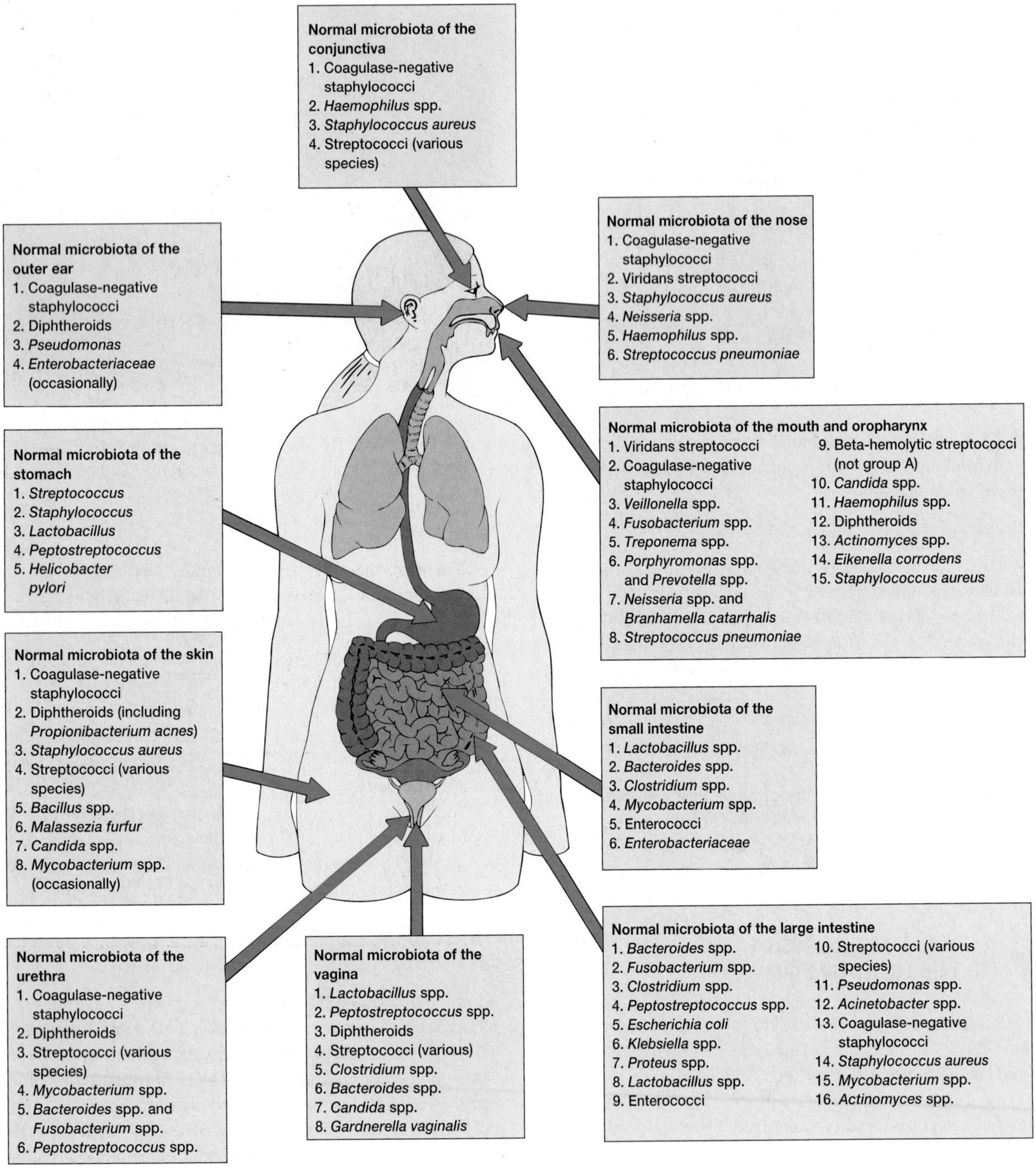

Normal microbiota of the conjunctiva
1. Coagulase-negative staphylococci
2. *Haemophilus* spp.
3. *Staphylococcus aureus*
4. Streptococci (various species)

Normal microbiota of the nose
1. Coagulase-negative staphylococci
2. Viridans streptococci
3. *Staphylococcus aureus*
4. *Neisseria* spp.
5. *Haemophilus* spp.
6. *Streptococcus pneumoniae*

Normal microbiota of the outer ear
1. Coagulase-negative staphylococci
2. Diphtheroids
3. *Pseudomonas*
4. *Enterobacteriaceae* (occasionally)

Normal microbiota of the mouth and oropharynx
1. Viridans streptococci
2. Coagulase-negative staphylococci
3. *Veillonella* spp.
4. *Fusobacterium* spp.
5. *Treponema* spp.
6. *Porphyromonas* spp. and *Prevotella* spp.
7. *Neisseria* spp. and *Branhamella catarrhalis*
8. *Streptococcus pneumoniae*
9. Beta-hemolytic streptococci (not group A)
10. *Candida* spp.
11. *Haemophilus* spp.
12. Diphtheroids
13. *Actinomyces* spp.
14. *Eikenella corrodens*
15. *Staphylococcus aureus*

Normal microbiota of the stomach
1. *Streptococcus*
2. *Staphylococcus*
3. *Lactobacillus*
4. *Peptostreptococcus*
5. *Helicobacter pylori*

Normal microbiota of the skin
1. Coagulase-negative staphylococci
2. Diphtheroids (including *Propionibacterium acnes*)
3. *Staphylococcus aureus*
4. Streptococci (various species)
5. *Bacillus* spp.
6. *Malassezia furfur*
7. *Candida* spp.
8. *Mycobacterium* spp. (occasionally)

Normal microbiota of the small intestine
1. *Lactobacillus* spp.
2. *Bacteroides* spp.
3. *Clostridium* spp.
4. *Mycobacterium* spp.
5. Enterococci
6. *Enterobacteriaceae*

Normal microbiota of the urethra
1. Coagulase-negative staphylococci
2. Diphtheroids
3. Streptococci (various species)
4. *Mycobacterium* spp.
5. *Bacteroides* spp. and *Fusobacterium* spp.
6. *Peptostreptococcus* spp.

Normal microbiota of the vagina
1. *Lactobacillus* spp.
2. *Peptostreptococcus* spp.
3. Diphtheroids
4. Streptococci (various)
5. *Clostridium* spp.
6. *Bacteroides* spp.
7. *Candida* spp.
8. *Gardnerella vaginalis*

Normal microbiota of the large intestine
1. *Bacteroides* spp.
2. *Fusobacterium* spp.
3. *Clostridium* spp.
4. *Peptostreptococcus* spp.
5. *Escherichia coli*
6. *Klebsiella* spp.
7. *Proteus* spp.
8. *Lactobacillus* spp.
9. Enterococci
10. Streptococci (various species)
11. *Pseudomonas* spp.
12. *Acinetobacter* spp.
13. Coagulase-negative staphylococci
14. *Staphylococcus aureus*
15. *Mycobacterium* spp.
16. *Actinomyces* spp.

Figure 31.2 Normal Microbiota of a Human.   A compilation of microorganisms that constitute normal microbiota encountered in various body sites.

relationship in which one species benefits and the other is unharmed. Commensal microorganisms living on or in the skin can be either resident (normal) or transient microbiota. Resident organisms normally grow on or in the skin. Their presence becomes fixed in well-defined distribution patterns. Those microorganisms that are temporarily present are transients. Transients usually do not become firmly entrenched and are unable to multiply.

It should be emphasized at the beginning that the skin is a mechanically strong barrier to microbial invasion. Few microorganisms can penetrate the skin because its outer layer consists of thick, closely packed cells called keratinocytes. These cells produce keratins. Keratins are scleroproteins comprising the main components of hair, nails, and outer skin cells that microorganisms cannot enzymatically attack. In addition to direct resistance to penetration, continuous shedding of the outer epithelial cells removes many of those microorganisms that manage to adhere to the skin surface.

The anatomy and physiology of the skin vary from one part of the body to another, and the normal resident microbiota reflect these variations. The skin surface or epidermis is not a favorable environment for microbial colonization. Several factors are responsible for this hostile microenvironment. First, the skin is subject to periodic drying. Lack of moisture drives many resident microbiota into a dormant state. However, in certain parts of the body (scalp, ears, axillary areas, genitourinary and anal regions, perineum, palms), moisture is sufficiently high to support a resident microbiota. Second, the skin has a slightly acidic pH (around pH 5 to 6) due to the organic acids produced by normal staphylococci and secretions from skin oil and sweat glands, which discourages colonization by many microorganisms. Third, sweat contains a high concentration of sodium chloride. This makes the skin surface hyperosmotic and osmotically stresses most microorganisms. Finally, certain inhibitory substances (bactericidal and/or bacteriostatic) on the skin help control colonization, overgrowth, and infection from microorganisms. For example, the sweat glands release lysozyme (muramidase), an enzyme that lyses *Staphylococcus epidermidis* and other gram-positive bacteria by hydrolyzing the $\beta(1\rightarrow4)$ glycosidic bond connecting *N*-acetylmuramic acid and *N*-acetylglucosamine in the bacterial cell wall peptidoglycan (figure 31.9). Sweat glands also produce antimicrobial peptides called **cathelicidins** (Latin *catharticus,* to purge, and *cida,* to kill) that help protect against infectious agents. One example is dermcidin. Current evidence suggests that this cathelicidin destroys bacterial invaders either by punching holes in their plasma membranes or by solubilizing the membranes through some detergent-like action.

The oil glands secrete complex lipids that may be partially degraded by the enzymes from certain gram-positive bacteria (*Propionibacterium acnes*). These bacteria can change the secreted lipids to unsaturated fatty acids such as oleic acid that have strong antimicrobial activity against gram-negative bacteria and some fungi. Some of these fatty acids are volatile and may be associated with a strong odor. Therefore many body deodorants contain antibacterial substances that act selectively against gram-positive bacteria to reduce the production of volatile unsaturated fatty acids and body odor.

Most skin bacteria are found on the superficial cells, colonizing dead cells, or closely associated with the oil and sweat glands. Secretions from these glands provide the water, amino acids, urea, electrolytes, and specific fatty acids that serve as nutrients primarily for *Staphylococcus epidermidis* and aerobic corynebacteria. Gram-negative bacteria generally are found in the moister regions. The yeasts *Pityrosporum ovale* and *P. orbiculare* normally occur on the scalp. Some dermatophytic fungi may colonize the skin and produce athlete's foot and ringworm. Athlete's foot disease (p. 920)

The most prevalent bacterium in the oil glands is the gram-positive, anaerobic, lipophilic rod *Propionibacterium acnes*. This bacterium usually is harmless; however, it has been associated with the skin disease acne vulgaris. Acne commonly occurs during adolescence when the endocrine system is very active. Hormonal activity stimulates an overproduction of **sebum,** a fluid secreted by the oil glands. A large volume of sebum accumulates within the glands and provides an ideal microenvironment for *P. acnes*. In some individuals this accumulation triggers an inflammatory response that causes redness and swelling of the gland's duct and produces a **comedo** [pl., comedones], a plug of sebum and keratin in the duct. Inflammatory lesions (papules, pustules, nodules) commonly called "blackheads" or "pimples" can result. *P. acnes* produces lipases that hydrolyse the sebum triglycerides into free fatty acids. Free fatty acids are especially irritating because they can enter the dermis and promote inflammation. Because *P. acnes* is extremely sensitive to tetracycline, this antibiotic may aid acne sufferers. Accutane, a synthetic form of vitamin A, is also used.

---

1. What are three reasons why knowledge of the normal human microbiota is important?
2. Why is the skin not usually a favorable microenvironment for colonization by bacteria?
3. How do microorganisms contribute to body odor?
4. What physiological role does *Propionibacterium acnes* play in the establishment of acne vulgaris?

Nose and Nasopharynx

The normal microbiota of the nose is found just inside the nostrils. *Staphylococcus aureus* and *S. epidermidis* are the predominant bacteria present and are found in approximately the same numbers as on the skin of the face.

The nasopharynx, that part of the pharynx lying above the level of the soft palate, may contain small numbers of potentially pathogenic bacteria such as *Streptococcus pneumoniae, Neisseria meningitidis,* and *Haemophilus influenzae*. Diphtheroids, a large group of nonpathogenic gram-positive bacteria that resemble *Corynebacterium* (*see section 24.5*), are commonly found in both the nose and nasopharynx.

Oropharynx

The oropharynx is that division of the pharynx lying between the soft palate and the upper edge of the epiglottis. The most important bacteria found in the oropharynx are the various alpha-hemolytic streptococci (*S. oralis, S. milleri, S. gordonii, S. salivarius*); large numbers of diphtheroids; *Branhamella catarrhalis;* and small gram-negative cocci related to *Neisseria meningitidis.* It should be noted that the palatine and pharyngeal tonsils harbor a similar microbiota, except that within the tonsillar crypts, there is an increase in *Micrococcus* and the anaerobes *Porphyromonas, Prevotella,* and *Fusobacterium.*

Respiratory Tract

The upper and lower respiratory tracts (trachea, bronchi, bronchioles, alveoli) do not have a normal microbiota. This is because microorganisms are removed in at least three ways. First, a continuous stream of mucus is generated by the goblet cells. This mucus entraps microorganisms, and the ciliated epithelial cells continually move the entrapped microorganisms out of the respiratory tract. Second, alveolar macrophages phagocytize and destroy microorganisms. Finally, a bactericidal effect is exerted by the enzyme lysozyme, which is present in the nasal mucus. Mucociliary blanket (p. 687)

Mouth

The normal microbiota of the mouth or oral cavity contains organisms that resist mechanical removal by adhering to surfaces like the gums and teeth. Those that cannot attach are removed by the mechanical flushing of the oral cavity contents to the stomach where they are destroyed by hydrochloric acid. The continuous desquamation (shedding) of epithelial cells also removes microorganisms. Those microorganisms able to colonize the mouth find a very comfortable environment due to the availability of water and nutrients, the suitability of pH and temperature, and the presence of many other growth factors. Biofilms (pp. 602–3)

The oral cavity is colonized by microorganisms from the surrounding environment within hours after a human is born. Initially the microbiota consists mostly of the genera *Streptococcus, Neisseria, Actinomyces, Veillonella,* and *Lactobacillus.* Some yeasts also are present. Most microorganisms that invade the oral cavity initially are aerobes and obligate anaerobes. When the first teeth erupt, the anaerobes (*Porphyromonas, Prevotella,* and *Fusobacterium*) become dominant due to the anaerobic nature of the space between the teeth and gums. As the teeth grow, *Streptococcus parasanguis* and *S. mutans* attach to their enamel surfaces; *S. salivarius* attaches to the buccal and gingival epithelial surfaces and colonizes the saliva. These streptococci produce a glycocalyx and various other adherence factors that enable them to attach to oral surfaces. The presence of these bacteria contributes to the eventual formation of dental plaque, caries, gingivitis, and periodontal disease. Periodontal disease (p. 914)

Eye

At birth and throughout human life, a small number of bacterial commensals are found on the conjunctiva of the eye. The predominant bacterium is *Staphylococcus epidermidis* followed by *S. aureus, Haemophilus* spp., and *Streptococcus pneumoniae.*

External Ear

The norma microbiota of the external ear resemble those of the skin, with coagulase-negative staphylococci and *Corynebacterium* predominating. Mycological studies show the following fungi to be normal microbiota: *Aspergillus, Alternaria, Penicillium, Candida,* and *Saccharomyces.*

Stomach

As noted earlier, many microorganisms are washed from the mouth into the stomach. Owing to the very acidic pH values (2 to 3) of the gastric contents, most microorganisms are killed. As a result the stomach usually contains less than 10 viable bacteria per milliliter of gastric fluid. These are mainly *Streptococcus, Staphylococcus, Lactobacillus, Peptostreptococcus, Helicobacter,* and yeasts such as *Candida* spp. Microorganisms may survive if they pass rapidly through the stomach or if the organisms ingested with food are particularly resistant to gastric pH (mycobacteria).

Small Intestine

The small intestine is divided into three anatomical areas: the duodenum, jejunum, and ileum. The duodenum (the first 25 cm of the small intestine) contains few microorganisms because of the combined influence of the stomach's acidic juices and the inhibitory action of bile and pancreatic secretions. Of the bacteria present, gram-positive cocci and rods comprise most of the microbiota. *Enterococcus faecalis,* lactobacilli, diphtheroids, and the yeast *Candida albicans* are occasionally found in the jejunum. In the distal portion of the small intestine (ileum), the microbiota begin to take on the characteristics of the colon microbiota. It is within the ileum that the pH becomes more alkaline. As a result anaerobic gram-negative bacteria and members of the family *Enterobacteriaceae* become established.

Large Intestine (Colon)

The large intestine or colon has the largest microbial community in the body. Microscopic counts of feces approach 10^{12} organisms per gram wet weight. Over 400 different species have been isolated from human feces. The colon can be viewed as a large fermentation vessel, and the microbiota consist primarily of anaerobic, gram-negative, nonsporing bacteria and gram-positive, spore-forming, and nonsporing rods. Not only are the vast majority of microorganisms anaerobic, but many different species are present in large numbers. Several studies have shown that the ratio of anaerobic to facultative anaerobic bacteria is approximately 300 to 1. Even the most abundant of the latter, *Escherichia coli,* is only about 0.1% of the total population.

Besides the many bacteria in the large intestine, the yeast *Candida albicans* and certain protozoa may occur as harmless commensals. *Trichomonas hominis, Entamoeba hartmanni, Endolimax nana,* and *Iodamoeba butschlii* are common inhabitants. Protozoan diseases (pp. 926–34)

Various physiological processes move the microbiota through the colon so an adult eliminates about 3×10^{13} microorganisms daily. These processes include peristalsis and desquamation of the

Techniques & Applications
31.1   Probiotics for Humans and Animals

The large intestine of humans and animals contains a very complex and balanced microbiota. These microorganisms normally prevent infection and have a positive effect on nutrition. Any abrupt change in diet, stress, or antibiotic therapy can upset this microbial balance, making the host susceptible to disease and decreasing the efficiency of food use. **Probiotics** [Greek *pro*, for, and *bios*, life], the oral administration of either living microorganisms or substances to promote health and growth, has the potential to reestablish the natural balance and return the host to normal health and nutrition.

Probiotic microorganisms are host specific; thus a strain selected as a probiotic in one animal may not be suitable in another species. Furthermore, microorganisms selected for probiotic use should exhibit these characteristics:

1. Adhere to the intestinal mucosa of the host
2. Be easily cultured
3. Be nontoxic and nonpathogenic to the host
4. Exert a beneficial effect on the host
5. Produce useful enzymes or physiological end products that the host can use
6. Remain viable for a long time
7. Withstand HCl in the host's stomach and bile salts in the small intestine

There are several possible explanations of how probiotic microorganisms displace pathogens and enhance the development

and stability of the microbial balance in the large intestine. These include:

1. Competition with pathogens for nutrients and adhesion sites
2. Inactivation of pathogenic bacterial toxins or metabolites
3. Production of substances that inhibit pathogen growth
4. Stimulation of nonspecific immunity

A wide variety of probiotic preparations have been patented for cattle, goats, horses, pigs, poultry, sheep, and domestic animals. Most of these preparations contain lactobacilli and/or streptococci; a few contain bifidobacteria.

Evidence has accumulated that certain probiotic microorganisms also offer considerable health benefits for humans. Potential benefits include:

1. Anticarcinogenic activity
2. Control of intestinal pathogens
3. Improvement of lactose use in individuals who have lactose intolerance
4. Reduction in the serum cholesterol concentration

Although the development of probiotics is still in its early stages, a better understanding of the normal microbiota in the large intestine of both animals and humans will be forthcoming as more microbiologists investigate probiotic activity.   Food microbiology (chapter 41)

---

surface epithelial cells to which microorganisms are attached, and continuous flow of mucus that carries adhering microorganisms with it. To maintain homeostasis of the microbiota, the body must continually replace those lost microorganisms. The bacterial population in the human colon usually doubles once or twice a day. Under normal conditions the resident microbial community is self-regulating. Competition and mutualism between different microorganisms and between the microorganisms and their host serve to maintain a status quo. However, if the intestinal environment is disturbed, the normal microbiota may change greatly. Disruptive factors include stress, altitude changes, starvation, parasitic organisms, diarrhea, and use of antibiotics or probiotics (**Techniques & Applications 31.1**). Finally, it should be emphasized that the actual proportions of the individual bacterial populations within the indigenous microbiota depend largely on a person's diet.

The initial residents of the colon of breast-fed infants are members of the gram-positive *Bifidobacterium* genus, because human milk contains a disaccharide amino sugar that *Bifidobacterium* species require as a growth factor. In formula-fed infants, gram-positive *Lactobacillus* species predominate because formula lacks the required growth factor. With the ingestion of solid food, these initial colonizers of the colon are eventually displaced by a typical gram-negative microbiota. Ultimately the composition of the adult's microbiota is established.

Genitourinary Tract

The upper genitourinary tract (kidneys, ureters, and urinary bladder) is usually free of microorganisms. In both the male and female, a few bacteria (*Staphylococcus epidermidis*, *Enterococcus faecalis*, and *Corynebacterium* spp.) usually are present in the distal portion of the urethra.

In contrast, the adult female genital tract, because of its large surface area and mucous secretions, has a complex microbiota that constantly changes with the female's menstrual cycle. The major microorganisms are the acid-tolerant lactobacilli, primarily *Lactobacillus acidophilus*, often called Döderlein's bacillus. They ferment the glycogen produced by the vaginal epithelium, forming lactic acid. As a result the pH of the vagina and cervix is maintained between 4.4 and 4.6.

---

1. What are the most common microorganisms found in the nose? The oropharynx? The nasopharynx? The tonsillar crypts? The lower respiratory tract? The mouth? The eye? The external ear? The stomach? The small intestine? The colon? The genitourinary tract?
2. Why is the colon considered a large fermentation vessel?
3. What physiological processes move the microbiota through the gastrointestinal tract?

4. How do the initial microbial colonizers of breast-fed infants differ from those of formula-fed infants?
5. Describe the microbiota of the upper and lower female genitourinary tract.

The Relationship between Normal Microbiota and the Host

The interaction between a host and a microorganism is a dynamic process in which each protagonist acts to maximize its survival. In some instances, after a microorganism enters or contacts a host, a positive mutually beneficial relationship occurs that becomes integral to the health of the host. These microorganisms become the normal microbiota. In other instances, the microorganism produces or induces deleterious effects on in the host; the end result may be disease or even death of the host (*see chapter 34*).

Our environment is teeming with microorganisms and we come in contact with many of them every day. Some of these microorganisms are pathogenic—that is, they cause disease. Yet these pathogens are at times prevented from causing disease by the competition provided by the normal microbiota. In general, the normal microbiota use space, resources, nutrients, and may produce chemicals that repel invading pathogens. These normal microbiota prevent colonization of pathogens and possible disease through "bacterial interference." For instance, the lactobacilli in the female genital tract maintain a low pH and inhibit colonization by pathogenic bacteria, and the corynebacteria on the skin produce fatty acids that inhibit colonization by pathogenic bacteria. This is an excellent example of amensalism (*see pp. 591–92*).

Although normal microbiota offer some protection from invading pathogens, they may themselves become pathogenic and produce disease under certain circumstances, and then are termed **opportunistic microorganisms** or **pathogens.** These opportunistic microorganisms are adapted to the noninvasive mode of life defined by the limitations of the environment in which they are living. If removed from these environmental restrictions and introduced into the bloodstream or tissues, disease can result. For example, streptococci of the viridans group are the most common resident bacteria of the mouth and oropharynx. If large numbers of them are introduced into the bloodstream (e.g., following tooth extraction or a tonsillectomy), they may settle on deformed or prosthetic heart valves and cause endocarditis.

Opportunistic microorganisms often cause disease in compromised hosts. A **compromised host** is seriously debilitated and has a lowered resistance to infection. There are many causes of this condition including malnutrition, alcoholism, cancer, diabetes, leukemia, another infectious disease, trauma from surgery or an injury, an altered normal microbiota from the prolonged use of antibiotics, and immunosuppression by various factors (e.g., drugs, viruses [HIV], hormones, and genetic deficiencies). For example, *Bacteroides* species are the most common residents in the large intestine (figure 31.2) and are quite harmless in that lo-

cation. If introduced into the peritoneal cavity or into the pelvic tissue as a result of trauma, they cause suppuration (the formation of pus) and bacteremia (the presence of bacteria in the blood). Many other examples of opportunistic infections will be presented in chapters 38, 39, and 40. The important point here is that the normal microbiota are harmless and may be beneficial in their normal location in the host and in the absence of coincident abnormalities. However, they can produce disease if introduced into foreign locations or compromised hosts.

1. Give one example of the normal microbiota benefitting a host.
2. Provide two examples of how the normal host microbiota prevent the establishment of a pathogen.
3. How would you define an opportunistic microorganism or pathogen? A compromised host?

 ## 31.3 OVERVIEW OF HOST RESISTANCE

To establish an infection, the pathogenic microorganism must first overcome many surface barriers, such as enzymes and mucus, that are either directly antimicrobial or inhibit attachment of the microorganism to the host (section 31.5). Because neither the surface of the skin nor the mucus-lined body cavities are ideal environments for many microorganisms, some pathogens must breach these barriers and get to underlying tissues. Any microorganism that penetrates these barriers encounters the two levels of resistance: other nonspecific resistance mechanisms and the specific immune response.

Vertebrates (including humans) are continuously exposed to microorganisms and their metabolic products that can cause disease. Fortunately these animals are equipped with an immune system that protects them against adverse consequences of this exposure. The **immune system** is composed of widely distributed cells, tissues, and organs that recognize foreign substances and microorganisms and act to neutralize or destroy them. **Immunity** [Latin *immunis,* free of burden] refers to the general ability of a host to resist a particular infection or disease. **Immunology** is the science that is concerned with immune responses to the foreign challenge and how these responses are used to resist infection. It includes the distinction between "self" and "nonself" and all the biological, chemical, and physical aspects of the immune response.

There are two fundamentally different types of immune responses to invading microorganisms and foreign material. The **nonspecific immune response** is also known as **nonspecific resistance** and **innate** or **natural immunity;** it offers resistance to any microorganism or foreign material encountered by the vertebrate host. It includes general mechanisms inherited as part of the innate structure and function of each animal, and acts as a first line of defense. The nonspecific immune response lacks immunological memory—that is, nonspecific responses occur to the same extent each time a microorganism or foreign body is encountered.

In contrast, the **specific immune responses** also known as **acquired, adaptive,** or **specific immunity** resist a particular foreign agent; moreover, specific immune responses improve on repeated exposure to foreign agents such as viruses, bacteria, and toxins. Substances that are recognized as foreign and provoke immune responses are called antigens. The antigens cause specific cells to produce proteins called antibodies. Antibodies bind to and inactivate a specific antigen. Other cells destroy virus-infected cells. The nonspecific and specific responses usually work together to eliminate pathogenic microorganisms and other foreign agents.

Section 31.4 provides an overview of the participants in the immune system, and section 31.5 describes the nonspecific immune response. The specific immune response will be covered in chapter 32.

---

1. Define each of the following terms: immune system, immunity, immunology, antigen, antibody.
2. Briefly describe the nonspecific immune response. What are some of its characteristics?
3. Briefly describe the specific immune response. What are some of its characteristics?

31.4 CELLS, TISSUES, AND ORGANS OF THE IMMUNE SYSTEM

The immune system is an organization of molecules, cells, tissues, and organs with specialized roles in defending against viruses, microorganisms, cancer cells, and nonself proteins (e.g., organ transplants). Immune system cells and tissue will now be considered.

Cells of the Immune System

The cells responsible for both nonspecific and specific immunity are the white blood cells called **leukocytes.** [Greek *leukos*, white, and *kytos*, cell]. All of the leukocytes originate from pluripotent stem cells in the fetal liver and in the bone marrow of the animal host (**figure 31.3**), from which they migrate to other body sites, undergo further development, and perform their various functions. These cells of the immune system are present throughout the host's body. Some become residents within tissues, where they respond to local trauma and sound the alarm; others circulate in body fluids and are recruited to the sites of infection. In defending the host against pathogenic microorganisms, leukocytes cooperate with each other first to recognize the pathogen as an invader and then to destroy it. These different leukocytes are now briefly examined.

Lymphocytes

Lymphocytes [Latin *lympha*, water, and *cyte*, cell] are the major cells of the specific immune system. Lymphocytes can be divided into three populations: T cells, B cells, and null cells (natural killer cells are included in this group). **B cells** or **B lymphocytes** reach maturity within the bone marrow, circulate in the blood, and also settle in various lymphoid organs. **T cells** or **T lymphocytes** mature in the thymus gland; they can remain in the thymus, circulate in the blood, or reside in lymphoid organs such as the lymph nodes and spleen. B cells and T cells will be discussed further in chapter 32 along with their roles in specific immunity. Natural killer cells are important in killing cells infected with either viruses or intracellular pathogens and destroying cancer cells (section 31.10).

Monocytes and Macrophages

Monocytes and macrophages are highly phagocytic and make up the **monocyte-macrophage system (figure 31.4)**. Recall (*see figure 4.10*) that during phagocytosis large particles and even other microorganisms are engulfed and enclosed in a phagocytic vacuole or phagosome.

Monocytes [Greek *monos*, single, and *cyte*, cell] are mononuclear phagocytic leukocytes with an ovoid or kidney-shaped nucleus and granules in the cytoplasm that stain gray-blue (figure 31.3). They are produced in the bone marrow and enter the blood, circulate for about eight hours, enlarge, migrate to the tissues, and mature into macrophages.

As just noted, **macrophages** [Greek *macros*, large, and *phagein*, to eat] are derived from monocytes and are also classified as mononuclear phagocytic leukocytes. However, they may be larger than monocytes, contain more organelles (especially lysosomes and phagolysosomes), and have a plasma membrane covered with ruffles or microvilli (**figure 31.5**). Macrophages have receptors for antibodies and complement; these can coat microorganisms or foreign material and enhance phagocytosis. This enhancement is termed opsonization. Macrophages spread throughout the animal body and take up residence in specific tissues where they are given special names (figure 31.4). Since macrophages are highly phagocytic, their function in nonspecific resistance will be discussed in more detail in the context of phagocytosis.

Granulocytes

Granulocytes have irregular-shaped nuclei with two to five lobes, and the cytoplasmic matrix has granules (figure 31.3) that contain reactive substances that kill microorganisms and enhance inflammation. Because of the irregular-shaped nuclei, granulocytes are also called **polymorphonuclear leukocytes** or **PMNs.** Three types of granulocytes exist: basophils, eosinophils, and neutrophils.

Basophils [Greek *basis*, base, and *philein*, to love] have an irregular-shaped nucleus with two lobes, and the granules stain bluish-black with basic dyes. Basophils are nonphagocytic cells that function by releasing histamine, prostaglandins, serotonin, and leukotrienes from their granules upon appropriate stimulation. Because these physiological mediators influence the tone and diameter of blood vessels, they are termed vasoactive mediators. Basophils (and mast cells) possess high-affinity receptors for an immunoglobulin (IgE; *see figure 32.14*) and thereby

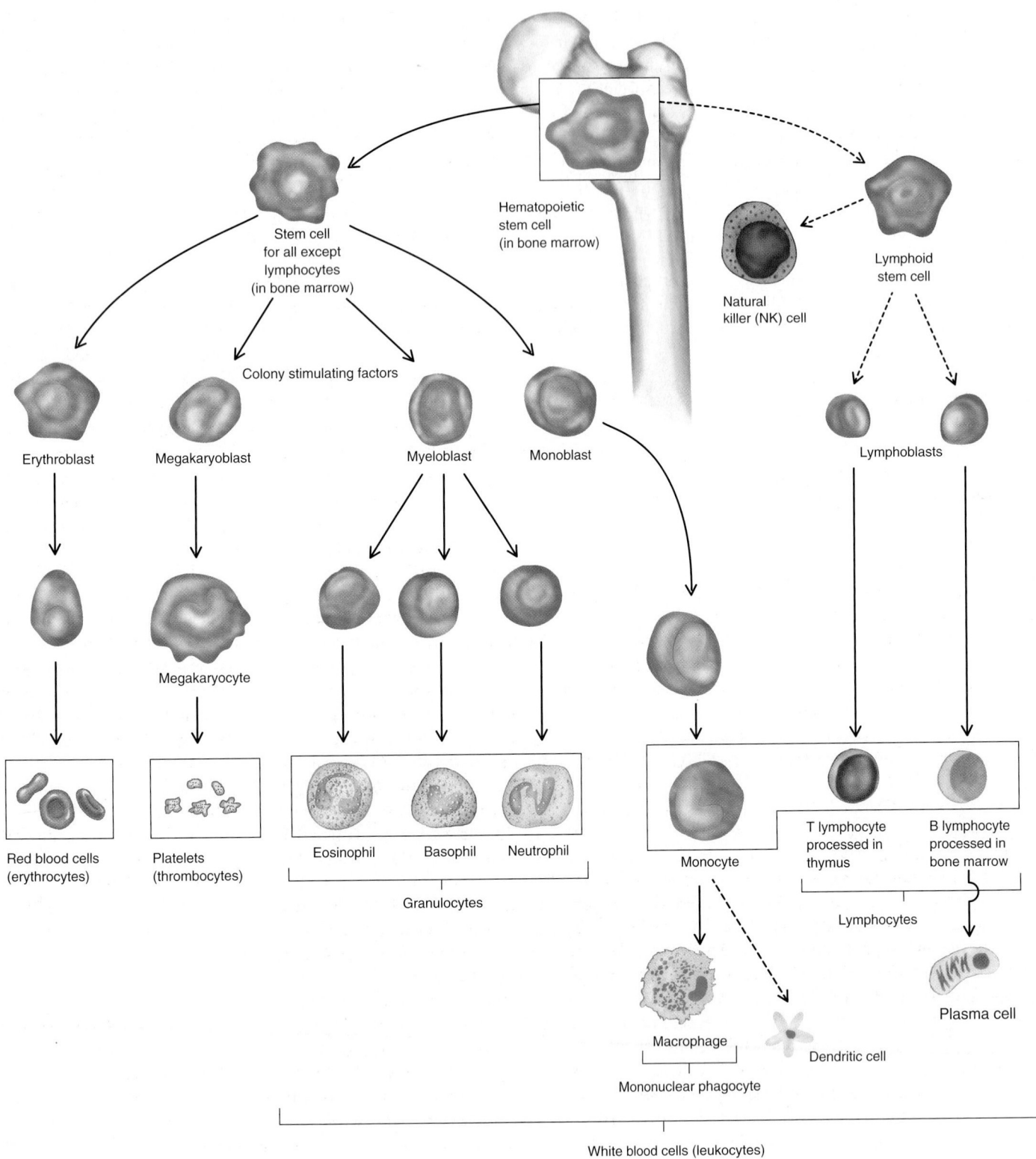

Figure 31.3   The Different Types of Human Blood Cells.   Pluripotent stem cells in the bone marrow divide to form two lineages:
(1) the lymphoid stem cell gives rise to B cells that become antibody-secreting plasma cells, T cells that become activated T cells, and
natural killer cells. (2) The common myeloid progenitor cell gives rise to the granulocytes (neutrophils, eosinophils, basophils), monocytes that give rise to macrophages and dendritic cells, an unknown precursor that gives rise to mast cells, megakaryocytes that produce platelets, and the erythroblast that produces erythrocytes (red blood cells). The dashed lines indicate steps that have not been very
well studied.

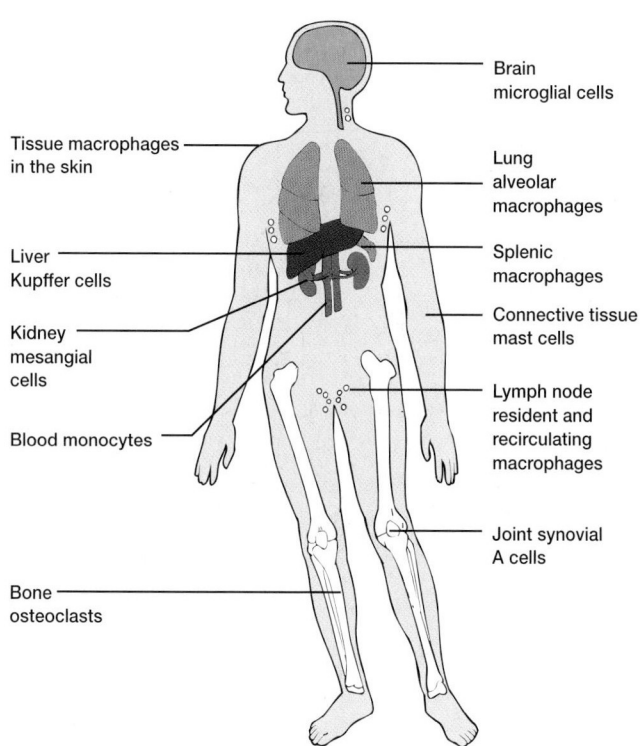

Figure 31.4 **The Monocyte-Macrophage System.** This system consists of tissue (such as found within the liver, spleen, and lymph nodes) containing "fixed" or immobile phagocytes that have specific names depending on their location.

become coated with IgE antibodies. Once coated, antigens can trigger the cell to secrete vasoactive mediators. These inflammatory mediators play a major role in certain allergic responses such as eczema, hay fever, and asthma.

Eosinophils [Greek *eos*, dawn, and *philien*] have a two-lobed nucleus connected by a slender thread of chromatin, and the granules stain red with acidic dyes. Unlike basophils, eosinophils are mobile cells that can migrate from the bloodstream into tissue spaces. Their role is important only in the defense against protozoan and helminth parasites, mainly by releasing cationic proteins and reactive oxygen metabolites into the extracellular fluid to damage the parasite's plasma membranes.

Neutrophils [Latin *neuter*, neither, and *philien*] stain readily at a neutral pH, have a nucleus with three to five lobes connected by slender threads of chromatin, and contain fine primary and secondary inconspicuous granules. Like macrophages, neutrophils have receptors for antibodies and complement proteins and are highly phagocytic cells. However, unlike the macrophage, neutrophils do not reside in healthy tissue but rapidly migrate to the site of tissue damage and infection where they are the principal phagocytic and microbicidal cells. The lytic enzymes and bactericidal substances in neutrophils are contained within large primary and smaller secondary granules. Primary granules contain peroxidase, lysozyme, and various hydrolytic enzymes, whereas secondary granules have collagenase, lactoferrin, and lysozyme.

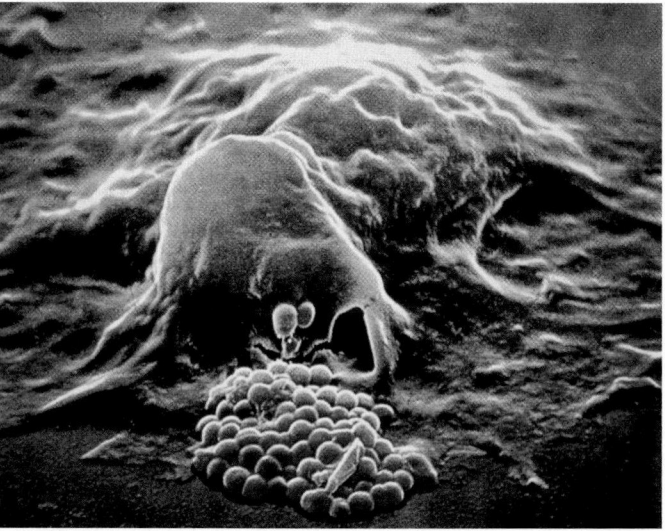

Figure 31.5 **Phagocytosis by a Macrophage.** One type of nonspecific host resistance involves white blood cells called macrophages and the process of phagocytosis. This scanning electron micrograph ($\times 3,000$) shows a macrophage devouring a colony of bacteria. Phagocytosis is but one of many nonspecific defenses humans and other animals have to combat microbial pathogens.

Both of these granules help accomplish intracellular digestion. Neutrophils also use oxygen-dependent and oxygen-independent pathways that generate antimicrobial substances and defensins to kill ingested microorganisms (section 31.8). Neutrophils are described in more detail in the contexts of the inflammatory response and phagocytosis.

Mast Cells

Mast cells are bone marrow–derived cells found in connective tissue. They contain granules with histamine and other pharmacologically active substances that contribute to the inflammatory response. Mast cells, along with basophils, play an important role in the development of allergies and hypersensitivities (*see section 33.2*).

Dendritic Cells

Dendritic cells (chapter opener figure, p. 673) are a relatively scarce class of white blood cells. They constitute only 0.2% of white blood cells in the blood and are present in even smaller numbers in the skin and mucous membranes of the nose, lungs, and intestines. Immature dendritic cells can recognize specific pathogen-associated molecular patterns on microorganisms (see pattern recognition receptors later in this chapter, p. 694) and play an important role in nonspecific resistance. They can distinguish between potentially harmful microorganisms and "self" molecules. After the pathogen is recognized, it binds to the dendritic cell's pattern recognition receptors and then is phagocytosed. Some immature dendritic cells can also kill viruses immediately by secreting interferon-α. After maturing, dendritic cells migrate

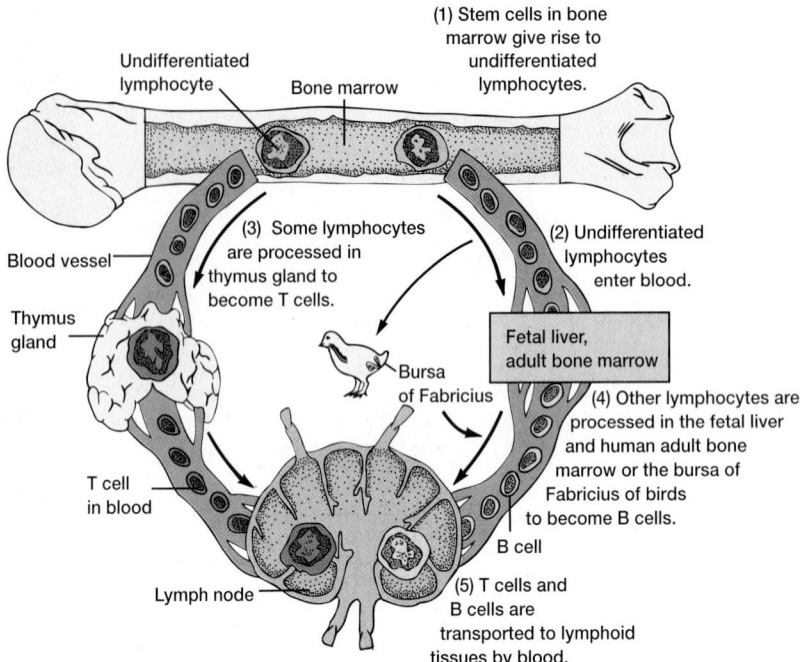

Figure 31.6 Schematic Representation of Lymphocyte Development. Bone marrow releases undifferentiated lymphocytes, which, after processing, become T and B cells.

to the bloodstream, lymph nodes, and spleen where they interact with other cells of the immune system such as B cells, which make antibodies, and natural killer cells, which attack pathogens and infected cells. Since dendritic cells present antigens to T cells, they also play an important accessory role in the specific immune response (*see T-cell receptors, section 32.4*). Interestingly, cancer vaccines composed of dendritic cells bearing tumor antigens are now in clinical trials involving humans.

Organs and Tissues of the Immune System

Based on function, the organs and tissues of the immune system can be divided into primary or secondary lymphoid organs or tissues. The primary organs or tissues are where immature lymphocytes mature and differentiate into antigen-sensitive mature B and T cells. The thymus is the primary lymphoid organ and bone marrow is the primary lymphoid tissue. The secondary organs and tissues serve as areas where lymphocytes may encounter and bind antigen, whereupon they proliferate and differentiate into fully mature, antigen-specific effector cells. The spleen is a secondary lymphoid organ and the lymph nodes and mucosal-associated tissues (GALT, gut-associated lymphoid tissue and SALT, skin-associated lymphoid tissues) are the secondary lymphoid tissues. The thymus, bone marrow, lymph nodes, and spleen will now be discussed in more detail. GALT and SALT are described in section 31.5 under physical and mechanical barriers.

Primary Lymphoid Organs and Tissues

Immature undifferentiated lymphocytes are generated in the bone marrow, and mature and become committed to a particular antigenic specificity within the primary lymphoid organ/tissues.

The two most important of these in mammals are the thymus and bone marrow.

The **thymus** is a lymphoid organ located above the heart. Precursor cells from the bone marrow migrate into the thymus to the outer cortex where they proliferate. As they mature and acquire T-cell surface markers, they move to the inner cortex where approximately 90% die, possibly as part of the acquisition of immune tolerance (*see section 32.8*). The other 10% move into the medulla, become mature T cells, and enter the bloodstream (**figure 31.6**).

In birds, undifferentiated lymphocytes move from the bone marrow to the **bursa of Fabricius** where B cells mature. In mammals, the bone marrow is the site of B-cell maturation. Like thymic selection during T-cell maturation, a selection process within the bone marrow eliminates B cells with self-reactive antibody receptors (the acquisition of tolerance).

Secondary Lymphoid Organ/Tissue

The spleen is the most highly organized secondary lymphoid organ and the lymph nodes the most highly organized tissue.

The **spleen** is the large secondary lymphoid organ located in the abdominal cavity. Whereas lymph nodes are specialized for trapping microorganisms and antigens from local tissues, the spleen specializes in filtering the blood and trapping blood-borne microorganisms and antigens. Once trapped by macrophages and dendritic cells, the pathogen is phagocytosed and antigens are presented to B and T cells, which become activated to carry out their immune functions.

Lymph nodes lie at the junctions of lymphatic vessels where they filter out harmful microorganisms and antigens from the lymph; pathogens and antigens are trapped by phagocytic and dendritic cells. Fixed macrophages then phagocytose the foreign material. It is within the lymph nodes that B cells proliferate into antibody-secreting **plasma cells.** Dendritic and T cells are also found here; dendritic cells serve as antigen-presenting cells and T helper cells promote the B-cell immune response.

---

1. Describe the structure and function of each of the following blood cells: monocytes, macrophages, basophils, eosinophils, neutrophils, mast cells, and dendritic cells.
2. Briefly describe each of the primary lymphoid organs and tissues.
3. What is the function of the spleen? A lymph node? The thymus?

 ## 31.5 PHYSICAL AND CHEMICAL BARRIERS IN NONSPECIFIC (INNATE) RESISTANCE

With few exceptions a potential microbial pathogen invading a human host immediately confronts a vast array of nonspecific (innate) defense mechanisms (**figure 31.7**). Although the effectiveness of some mechanisms is not great, collectively their defense is formidable.

Many direct factors (nutrition, physiology, fever, age, genetics) and equally as many indirect factors (personal hygiene, socioeconomic status, living conditions) influence all host-microbe relationships. At times they favor the establishment of the microorganism; at other times they provide some measure of defense to the host. For example, when the host is either very young or very old, susceptibility to infection increases. Babies are at a particular risk after their maternal immunity has disappeared and before their own immune system has matured. In very old persons there is a decline in the immune system and in the homeostatic functioning of many organs that reduces host defenses. In addition to these direct and indirect factors, a vertebrate host has some very specific physical and mechanical barriers.

Physical and Mechanical Barriers

Physical or mechanical barriers, along with the host's secretions (flushing mechanisms), are the first line of defense against microorganisms. Protection of the most important body surfaces by these mechanisms is discussed next.

Skin

The intact skin contributes greatly to nonspecific host resistance. It forms a very effective mechanical barrier to microbial invasion (p. 677).

Despite the skin's defenses, at times some pathogenic microorganisms gain access to the tissue under the skin. Here they encounter a specialized set of cells called the *skin-associated lymphoid tissue* (**SALT**). The major function of SALT is to confine microbial invaders to the area immediately underlying the skin and to prevent them from gaining access to the bloodstream.

One type of SALT cell is the **Langerhans cell,** a specialized dendritic cell that can phagocytose antigens. Once the Langerhans cell has internalized the antigen, it migrates from the epi-

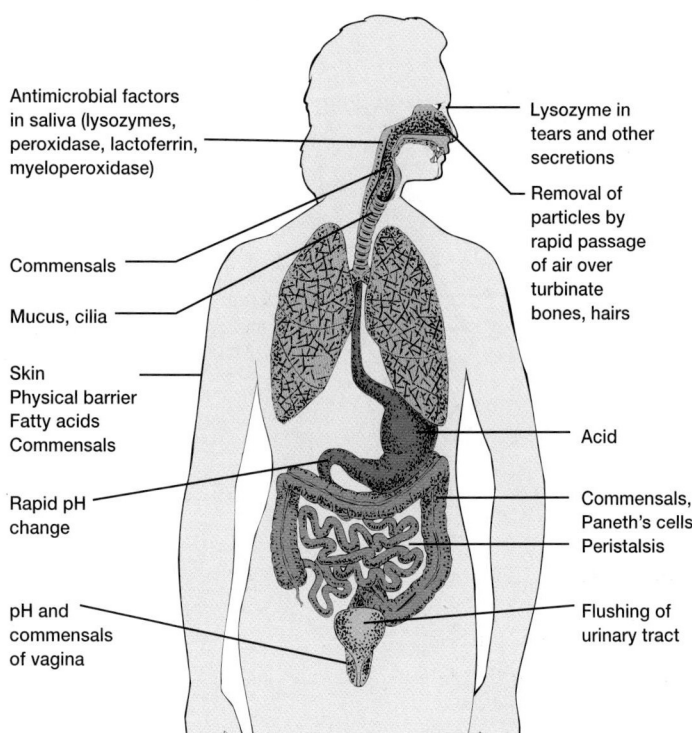

Figure 31.7   Host Defenses.   Some nonspecific (innate) host defense mechanisms that help prevent entry of microorganisms into the host's tissues.

dermis to nearby lymph nodes where the cell differentiates into a specialized cell called the **interdigitating dendritic cell (figure 31.8).** This cell can present the antigen on its surface and activate nearby T cells. The activated T cells can then interact with activated B cells to induce a humoral response that produces specific antibodies.

The epidermis also contains another type of SALT cell called the **intraepidermal lymphocyte** (figure 31.8). These cells are strategically located in the skin so that they can intercept any antigens that breach the first line of defense. These specialized SALT cells function as T cells to destroy the antigen. A large number of tissue macrophages (figure 31.4) are also located in the dermal layer of the skin and phagocytose (figure 31.5) most microorganisms they encounter.

Mucous Membranes

The mucous membranes of the eye (conjunctiva), respiratory, digestive, and urogenital systems withstand microbial invasion because the intact stratified squamous epithelium and mucous secretions form a protective covering that resists penetration and traps many microorganisms. This mechanism contributes to nonspecific immunity. Furthermore, many mucosal surfaces are bathed in specific antimicrobial secretions. For example, cervical mucus, prostatic fluid, and tears are toxic to many bacteria. One antibacterial substance is **lysozyme** (muramidase), an enzyme that lyses bacteria by hydrolyzing the $\beta(1\rightarrow4)$ bond connecting

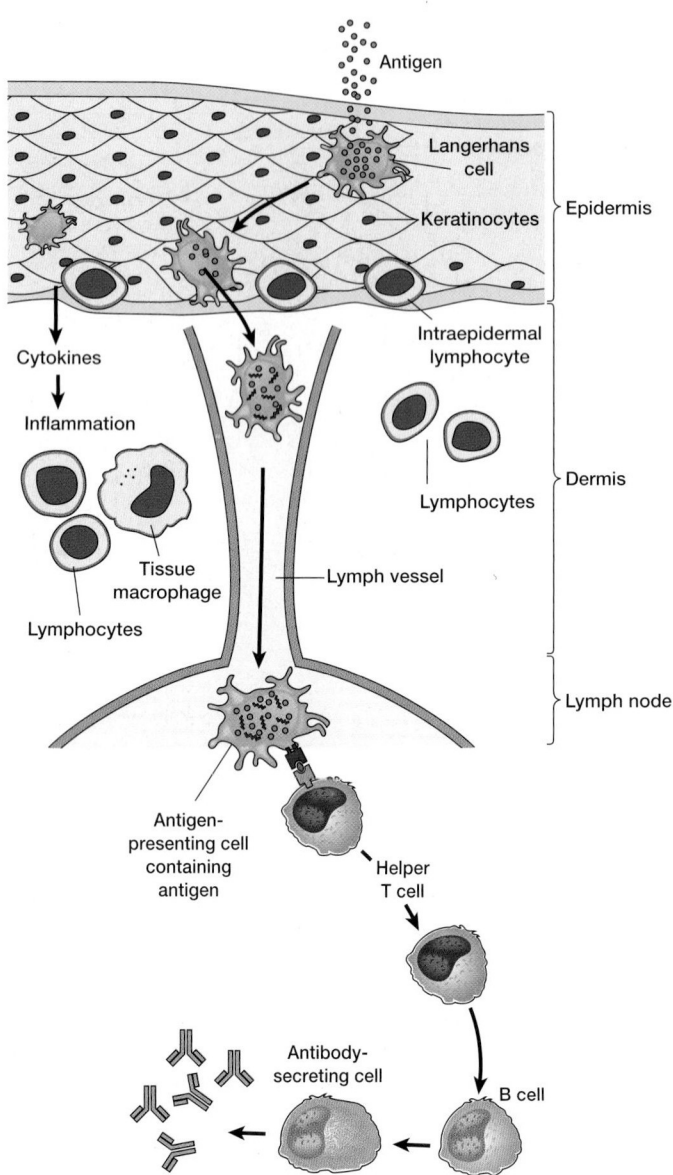

Figure 31.8   Skin-Associated Lymphoid Tissue. Keratinocytes make up 90% of the epidermis. They are capable of secreting cytokines that cause an inflammatory response to invading pathogens. Langerhans cells internalize antigen and move it to a lymph node where they differentiate into interdigitating dendritic cells that present antigen to helper T cells. The intraepidermal lymphocytes may function as T cells that can activate B cells to induce a humoral response.

N-acetylmuramic acid and *N*-acetylglucosamine in the bacterial cell wall peptidoglycan (*see figures 3.16–3.19*)—especially in gram-positive bacteria (**figure 31.9**). These mucous secretions also contain specific immune proteins that help prevent the attachment of microorganisms and significant amounts of iron-binding lactoferrin. **Lactoferrin** is released by activated macrophages and PMNs. It sequesters iron from the plasma. This sequestration reduces the amount of iron available to invading

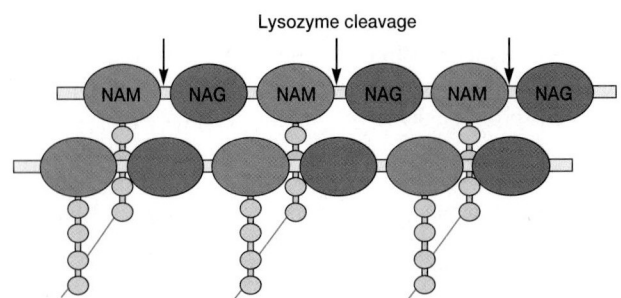

Figure 31.9   Action of Lysozyme on the Cell Wall of Gram-Positive Bacteria. In the structure of the cell wall peptidoglycan backbone, the β(1→4) bonds connect alternating *N*-acetylglucosamine (NAG) and *N*-acetylmuramic acid (NAM) residues. The chains are cross-linked through the tetrapeptide side chains. Lysozyme splits the molecule at the places indicated by the arrows.

microbial pathogens and limits their ability to multiply. Finally, mucous membranes produce lactoperoxidase, an enzyme that produces superoxide radicals (*see p. 125*), a reactive form of oxygen that is toxic to many microorganisms.

Like the skin, mucous membranes also have a specialized immune barrier called **m*ucosal-a*ssociated *l*ymphoid *t*issue (MALT).** There are several types of MALT. The system most studied is the **gut-*a*ssociated *l*ymphoid *t*issue (GALT).** GALT includes the tonsils, adenoids, and Peyer's patches in the intestine. Less well-organized mucosal-associated lymphoid tissue also occurs in the respiratory system and is called **b*ronchial-a*ssociated *l*ymphoid *t*issue (BALT);** the diffuse MALT in the urogenital system does not have a specific name.

MALT can operate by two basic mechanisms. First, when an antigen arrives at the mucosal surface, it contacts a type of cell called the **M cell (figure 31.10*a*).** The M cell does not have the brush border or microvilli found on adjacent columnar epithelial cells but does have a large pocket containing B cells, T cells, and macrophages. When an antigen contacts the M cell, it is phagocytosed and released into the pocket. Macrophages engulf the antigen or pathogen and try to destroy it. An M cell also can phagocytose an antigen and transport it to a cluster of cells called an organized lymphoid follicle (figure 31.10*b*). The B cells within this follicle recognize the antigen and mature into plasma cells. The plasma cells leave the follicle and secrete a class of antibody called secretory IgA. sIgA is then transported into the lumen of the gut where it interacts with the antigen that caused its production.

1. Why is the skin such a good first line of defense against pathogenic microorganisms?
2. How do intact mucous membranes resist microbial invasion of the host?
3. Describe SALT function in the immune response.
4. How do M cells function in MALT?

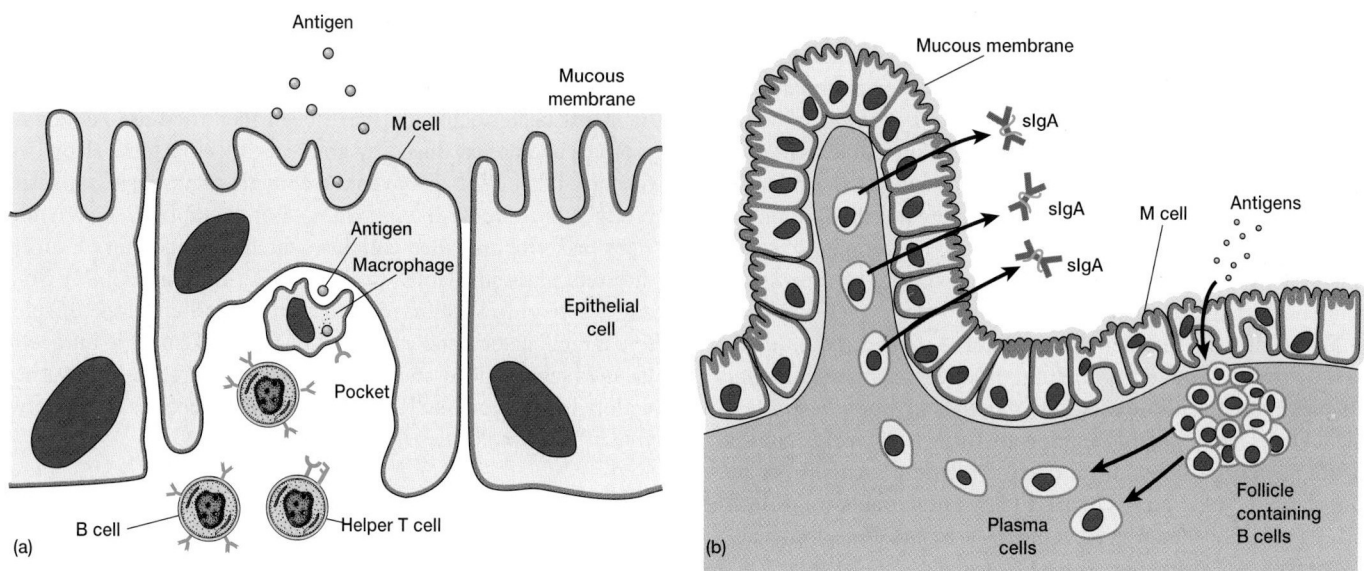

Figure 31.10   Function of M Cells in Mucosal-Associated Immunity.   (a) Structure of an M cell located between two epithelial cells in a mucous membrane. The M cell endocytoses the pathogen and releases it into the pocket containing helper T cells, B cells, and macrophages. It is within the pocket that the pathogen often is destroyed. (b) The antigen is transported by the M cell to the organized lymphoid follicle containing B cells. The activated B cells mature into plasma cells, which produce secretory IgA and release it into the lumen where it reacts with the antigen that caused its production.

Respiratory System

The mammalian respiratory system has formidable defense mechanisms. The average person inhales at least eight microorganisms a minute, or 10,000 each day. Once inhaled, a microorganism must first survive and penetrate the air filtration system of the upper and lower respiratory tracts. Because the airflow in these tracts is very turbulent, microorganisms are deposited on the moist, sticky mucosal surfaces. The cilia in the nasal cavity beat toward the pharynx, so that mucus with its trapped microorganisms is moved toward the mouth and expelled. Humidification of the air within the nasal cavity causes many hygroscopic microorganisms to swell and aids the phagocytic process.

The **mucociliary blanket** of the respiratory epithelium traps microorganisms less than 10 μm in diameter that are deposited on the mucosal surface and transports them by ciliary action away from the lungs (**mucociliary escalator**). Microorganisms larger than 10 μm usually are trapped by hairs and cilia lining the nasal cavity. Coughing and sneezing reflexes clear the respiratory system of microorganisms by expelling air forcefully from the lungs through the mouth and nose, respectively. Salivation also washes microorganisms from the mouth and nasopharyngeal areas into the stomach. Microorganisms that succeed in reaching the alveoli of the lungs encounter a population of fixed phagocytic cells called **alveolar macrophages** (figure 31.4). These cells can ingest and kill most bacteria by phagocytosis.

Gastrointestinal Tract

Once microorganisms reach the stomach, many are killed by its gastric juice (a mixture of hydrochloric acid, proteolytic en-

zymes, and mucus). The very acidic gastric juice (pH 2 to 3) is usually sufficient to destroy most organisms and their toxins, although exceptions exist (protozoan cysts, *Clostridium* and *Staphylococcus* toxins). However, many organisms are protected by food particles and reach the small intestine.

Once in the small intestine, microorganisms often are damaged by various pancreatic enzymes, bile, enzymes in intestinal secretions, and the GALT system. **Peristalsis** [Greek *peri,* around, and *stalsis,* contraction] and the normal loss of columnar epithelial cells act in concert to purge intestinal microorganisms. In addition, the normal microbiota of the large intestine (figure 31.2) is extremely important in preventing the establishment of pathogenic organisms. For example, many normal commensals in the intestinal tract produce metabolic products, such as fatty acids, that prevent "unwanted" microorganisms from becoming established. Other normal microbiota take up attachment sites and compete for nutrients.

The mucous membranes of the intestinal tract contain cells called **Paneth cells.** These cells produce lysozyme (figure 31.9) and a set of peptides called **cryptins.** Cryptins are toxic for some bacteria, although their mode of action is not known.

Genitourinary Tract

Under normal circumstances the kidneys, ureters, and urinary bladder of mammals are sterile. Urine within the urinary bladder also is sterile. However, in both the male and female, a few bacteria are usually present in the distal portion of the urethra (figure 31.2).

The factors responsible for this sterility are complex. For example, urine kills some bacteria due to its low pH and the presence of urea and other metabolic end products (uric acid, hippuric acid, indican, fatty acids, mucin, enzymes). The kidney medulla is so hypertonic that few organisms can survive. The lower urinary tract is flushed with urine and some mucus 4 to 10 times each day, eliminating potential pathogens. In males the anatomical length of the urethra (20 cm) provides a distance barrier that excludes microorganisms from the urinary bladder. Conversely the short urethra (5 cm) in females is more readily traversed by microorganisms; this explains why general urinary tract infections are 14 times more common in females than in males.

The vagina has another unique defense. Under the influence of estrogens, the vaginal epithelium produces increased amounts of glycogen that acid-tolerant *Lactobacillus acidophilus* bacteria called Döderlein's bacilli degrade to form lactic acid. Normal vaginal secretions contain up to 10^8 Döderlein's bacilli per ml. Thus an acidic environment (pH 3 to 5) unfavorable to most organisms is established. Cervical mucus also has some antibacterial activity.

The Eye

The conjunctiva is a specialized mucus-secreting epithelial membrane that lines the interior surface of each eyelid and the exposed surface of the eyeball. It is kept moist by the continuous flushing action of tears (lacrimal fluid) from the lacrimal glands. Tears contain large amounts of lysozyme, lactoferrin, and sIgA (*see p. 714*) and thus provide chemical as well as physical protection.

---

1. Describe the different antimicrobial defense mechanisms that operate within the respiratory system of mammals.
2. What factors operate within the gastrointestinal system that help prevent the establishment of pathogenic microorganisms?
3. Except for the anterior portion of the urethra, why is the genitourinary tract a sterile environment?

Chemical Barriers

Mammalian hosts have a chemical arsenal with which to combat the continuous onslaught of microorganisms. Some of these chemicals (gastric juices, salivary glycoproteins, lysozyme, oleic acid on the skin, urea) have already been discussed with respect to the specific body site(s) they protect. In addition, blood, lymph, and other body fluids contain a potpourri of defensive chemicals such as beta-lysin and other polypeptides.

Bacteriocins

As previously noted, the first line of defense against microorganisms is the host's anatomical barrier, consisting of the skin and mucous membranes. These surfaces are colonized by normal microbiota, which by themselves provide a biological barrier against uncontrolled proliferation of foreign microorganisms. Many of these normal bacteria synthesize and release plasmid-encoded toxic proteins (e.g., colicin, staphylococcin) called **bacteriocins** that are lethal to related species. Bacteriocins may give

their producers an adaptive advantage against other bacteria. Sometimes they may increase bacterial virulence by damaging host cells such as mononuclear phagocytes.

Most bacteriocins that have been identified are peptides or proteins and are produced by gram-negative bacteria. (However, recently it has been discovered that some gram-positive bacteria produce bacteriocin-like peptides). For example, *E. coli* synthesizes bacteriocins called **colicins,** which are coded for by several different plasmids (ColB, ColE1, ColE2, ColI, and ColV). Some colicins bind to specific receptors on the cell envelope of sensitive target bacteria and cause cell lysis, attack specific intracellular sites such as ribosomes, or disrupt energy production. It is now widely recognized that these antimicrobial peptides act as defensive effector molecules in the large intestine.

Beta-Lysin and Other Polypeptides

Beta-lysin is a cationic polypeptide released from blood platelets; it can kill some gram-positive bacteria by disrupting their plasma membranes. Other cationic polypeptides include leukins, plakins, cecropins, and phagocytin. A zinc-containing polypeptide, known as the prostatic antibacterial factor, is an important antimicrobial substance secreted by the prostate gland in males.

---

1. How do bacteriocins function?
2. How does beta-lysin function against gram-positive bacteria?

 31.6 INFLAMMATION

Inflammation [Latin, *inflammatio,* to set on fire] is an important nonspecific defense reaction to tissue injury, such as that caused by a pathogen or wound. Acute inflammation is the immediate response of the body to injury or cell death. The gross features were described over 2,000 years ago and are still known as the cardinal signs of inflammation. These signs include redness (*rubor*), warmth (*calor*), pain (*dolor*), swelling (*tumor*), and altered function (*functio laesa*).

The acute inflammatory response begins when injured tissue cells release chemical signals (inflammatory chemicals) that activate the inner lining (endothelium) of nearby capillaries (**figure 31.11**). Within the capillaries **selectins** (a family of cell adhesion molecules) are displayed on the activated endothelial cells. These adhesion molecules randomly attract and attach neutrophils to the endothelial cells, slow the neutrophils down, and cause them to roll along the endothelium. As the neutrophils (figure 31.3) roll along the endothelium, they encounter the inflammatory chemicals that act as activating signals. These signals activate **integrins** (adhesion receptors) on the neutrophils. The neutrophil integrins then tightly attach to endothelial adhesion molecules. This causes the neutrophils to stick to the endothelium and stop rolling (margination). The neutrophils now undergo dramatic shape changes, squeeze through the endothelial wall (diapedesis) into the interstitial tissue fluid, migrate to the site of injury (extravasation), and attack the pathogen or other cause of the tissue damage. Neu-

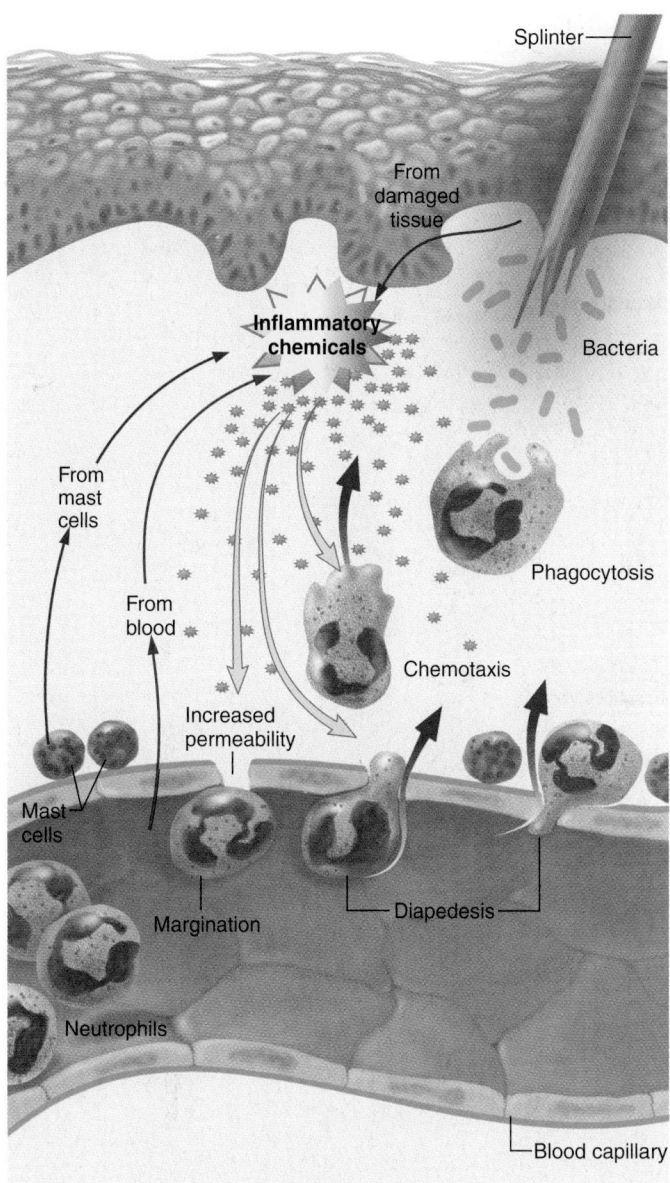

Figure 31.11   Physiological Events of the Acute Inflammatory Response.   At the site of injury (splinter), chemical messengers are released from the damaged tissue, mast cells, and the blood plasma. These inflammatory chemicals stimulate neutrophil migration, diapedesis, chemotaxis, and phagocytosis. See text for details.

trophils and other leukocytes are attracted to the infection site by chemotactic factors or chemotaxins such as substances released by bacteria and mast cells, and tissue breakdown products. Depending on the severity and nature of tissue damage, other types of leukocytes (e.g., lymphocytes, monocytes, and macrophages) may follow the neutrophils.

The inflammatory mediators (chemicals) that are released by the injured tissue cells increase the acidity in the surrounding extracellular fluid. This decrease in pH activates the extracellular

enzyme kallikrein, which splits bradykinin from its long precursor chain. Bradykinin binds to receptors on the capillary wall, opening the junctions between cells and allowing fluid and infection-fighting leukocytes to leave the capillary and enter the infected tissue. Simultaneously bradykinin (**figure 31.12**) binds to mast cells in the connective tissue associated with most small blood vessels. This activates the mast cells by causing an influx of calcium ions, which leads to degranulation and release of preformed mediators such as histamine. If nerves in the infected area are damaged, they release substance P, which also binds to mast cells, boosting preformed-mediator release. Histamine in turn makes the intercellular junctions in the capillary wall wider so that more fluid, leukocytes, kallikrein, and bradykinin precursors move out, causing edema. Bradykinin then binds to nearby capillary cells and stimulates the production of prostaglandins (PGE_2 and $PGF_{2\alpha}$) to promote tissue swelling in the infected area. Prostaglandins also bind to free nerve endings, making them fire and start a pain impulse.

The change in mast cell plasma membrane permeability associated with activation allows phospholipase A2 to release arachidonic acid. Arachidonic acid is then metabolized by the cyclooxygenase or lipoxygenase pathways, depending on mast cell type. The newly synthesized mediators include prostaglandins E_2 and $F_{2\alpha}$, thromboxane A_2, slow-reacting substance (SRS), and leukotrienes (LTC_4 + LTD_4). All play various roles in the inflammatory response.

During acute inflammation, the offending pathogen is neutralized and eliminated by a series of events, the most important of which are these:

1. The increase in blood flow and capillary dilation bring into the area more antimicrobial factors and leukocytes that destroy the pathogen. Dead cells also release antimicrobial factors.
2. The rise in temperature stimulates the inflammatory response and may inhibit microbial growth.
3. A fibrin clot often forms and may limit the spread of the invaders so that they remain localized.
4. Phagocytes collect in the inflamed area and phagocytose the pathogen. In addition, chemicals stimulate the bone marrow to release neutrophils and increase the rate of granulocyte production.

Chronic Inflammation

Chronic inflammation is a slow process characterized by the formation of new connective tissue, and it usually causes permanent tissue damage. Superficially, the difference between acute and chronic inflammation is one of duration. Regardless of the cause, chronic inflammation lasts two weeks or longer. Chronic inflammation can occur as a distinct process without much acute inflammation. The persistence of bacteria by a variety of mechanisms can stimulate chronic inflammation. For example, the mycobacteria have cell walls with a very high lipid and wax content, making them relatively insensitive to phagocytosis. The

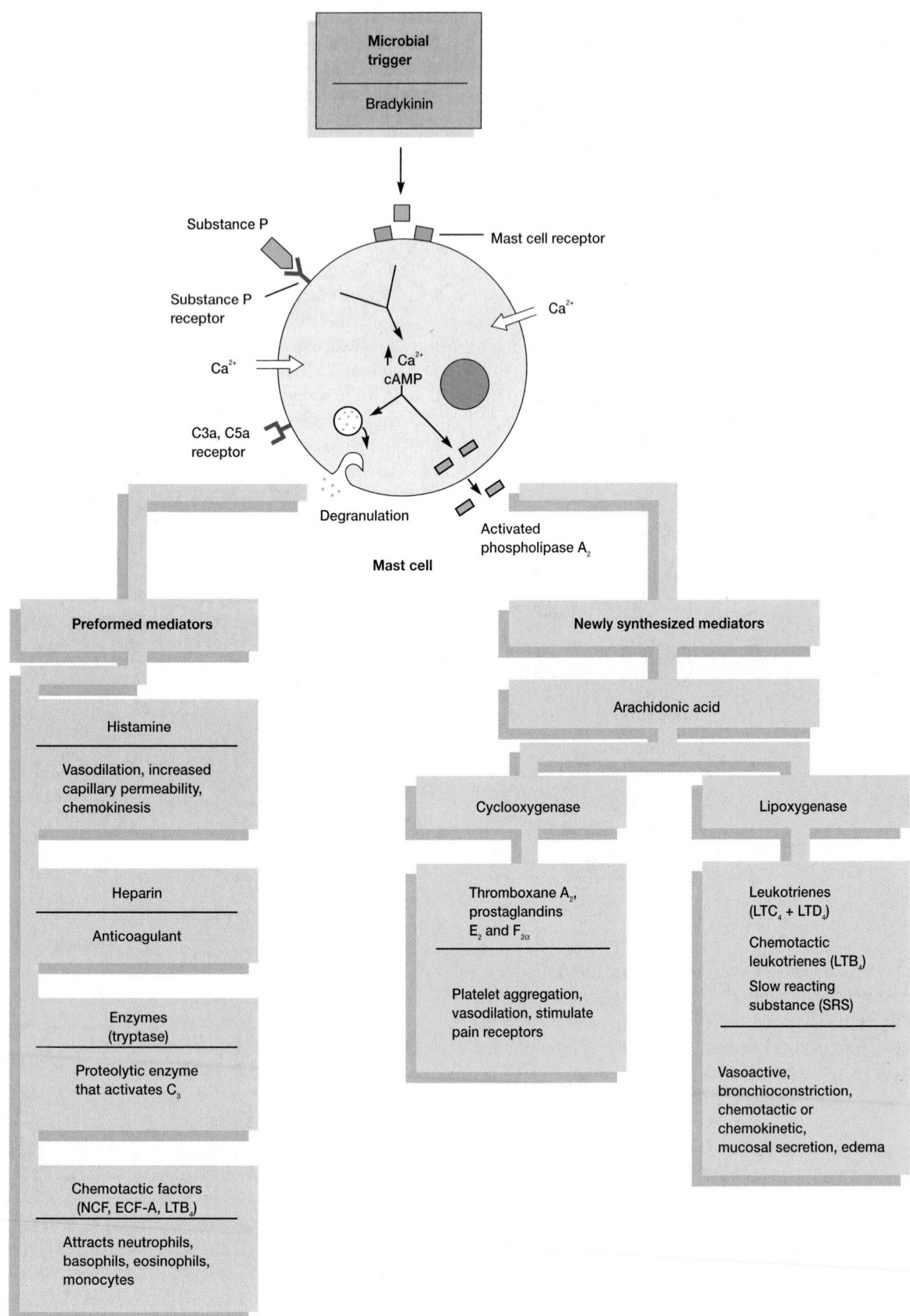

Figure 31.12 Biochemical Events of Inflammation. Mast cell activation and physiological effects of mast cell-derived performed and newly synthesized mediators that lead to the inflammatory response.

bacteria that cause tuberculosis, leprosy, and syphilis often survive within the macrophage. In addition, some bacteria produce toxins that stimulate tissue-damaging reactions even after bacterial death.

Chronic inflammation is characterized by a dense infiltration of lymphocytes and macrophages. If the macrophages are unable to protect the host from tissue damage, the body attempts to wall off and isolate the site by forming a **granuloma** [Latin, *granulum*, a small particle; Greek, *oma*, to form]. Granulomas are formed when neutrophils and macrophages are unable to destroy the microorganism during inflammation. Infections caused by some bacteria (listeriosis, brucellosis), fungi (histoplasmosis, coccidiodomycosis), helminth parasites (leishmaniasis, schistosomiasis), and large antibody-antigen complexes (rheumatoid arthritis) result in granuloma formation and chronic inflammation.

---

1. What major events occur during an inflammatory reaction, and how do they contribute to pathogen destruction?
2. What causes degranulation of mast cells?
3. How does chronic inflammation differ from acute inflammation?

 ## 31.7 THE COMPLEMENT SYSTEM

Complement was discovered many years ago as a heat-labile component of human blood plasma that augments opsonization (section 31.8) of bacteria by antibodies and helps other antibodies kill bacteria. This activity was said to "complement" the antibacterial activity of antibody; hence, the name complement. It is now known that the **complement system** is composed of a large number (over 30) of serum proteins. This system has three major physiologic activities: (1) defending against bacterial infections (opsonization, chemotaxis and activation of leukocytes, lysis of bacterial cell walls); (2) bridging innate and adaptive immunity (augmentation of antibody responses, enhancement of immunologic memory); and (3) disposing of wastes (immune complexes, the products of inflammatory injury, clearance of apoptotic cells).

The complement cascade is made up of complement proteins designated C1 (which has three protein subcomponents) through C9 in addition to Factor B, Factor $\overline{D}$, Factor H, Factor I, C4b binding protein, C1 INH complex, S protein, and properdin (**table 31.1**). The complement system acts in a cascade fashion, the activation of one component resulting in the activation of the next. Collectively the complement proteins make up much of the globulin fraction of serum (*see section 32.3 and figure 32.6*). Within plasma and other body fluids, complement proteins are in an inactive state—they become activated upon cleavage.

There are three pathways of complement activation: the classical, alternative, and lectin pathways. Although they employ similar mechanisms, specific proteins are unique to the first part of each pathway. Since the classical pathway involves an understanding of the specific immune response, it will be discussed in section 32.7.

Table 31.1

Some Important Proteins of the Complement Cascade

Protein	Fragment	Function
Recognition Unit		
C1	q	Binds to the Fc portion of antigen-antibody complexes
	r	Subunit of C1; activates C1s
	s	Cleaves C4 and C2 due to its enzymatic activity
Activation Unit		
C2		Causes viral neutralization
C3	a	Anaphylatoxin, immunoregulatory
	b	Key component of the alternative pathway and major opsonin in serum
	e	Induces leukocytosis
C4	a	Anaphylatoxin
	b	Causes viral neutralization; opsonin
Membrane Attack Unit		
C5	a	Anaphylatoxin; principal chemotactic factor in serum; induces neutrophil attachment to blood vessel walls
	b	Initiates membrane attack
C6 ⎫		Participate with C5b in formation of the membrane attack complex that lyses targeted cells
C7 ⎬		
C8 ⎪		
C9 ⎭		
Alternative Pathway		
Factor B		Causes macrophage spreading on surfaces; precursor of C3 convertase
Factor $\overline{D}$		Cleaves Factor B to form active C3Bb in alternative pathway
Properdin		Stabilizes alternative pathway C3 convertase
Regulatory Proteins		
Factor H		Promotes C3b breakdown and regulates alternative pathway
Factor I		Degrades C3b and regulates alternative pathway
C4b binding protein		Inhibits assembly and accelerates decay of C4bC2a
C1 INH complex		Binds to and dissociates C1r and C1s from C1
S protein		Binds fluid-phase C5b67; prevents membrane attachment

The **alternative complement pathway (figure 31.13)** plays an important role in the innate, nonspecific immune defense against intravascular invasion by bacteria and some fungi. The alternative pathway begins with cleavage of C3 into fragments C3a and C3b by a blood enzyme. These fragments are produced at a slow rate and do not carry out the next step, the cleavage of C5, because free C3b is rapidly cleaved into inactive fragments. However, C3b becomes stable when it binds to the lipopolysaccharide (LPS) of gram-negative bacterial cell walls, to aggregates of IgA or IgE, or to some endotoxins. A protein in blood termed Factor B adsorbs to bound C3b and is cleaved into two fragments by

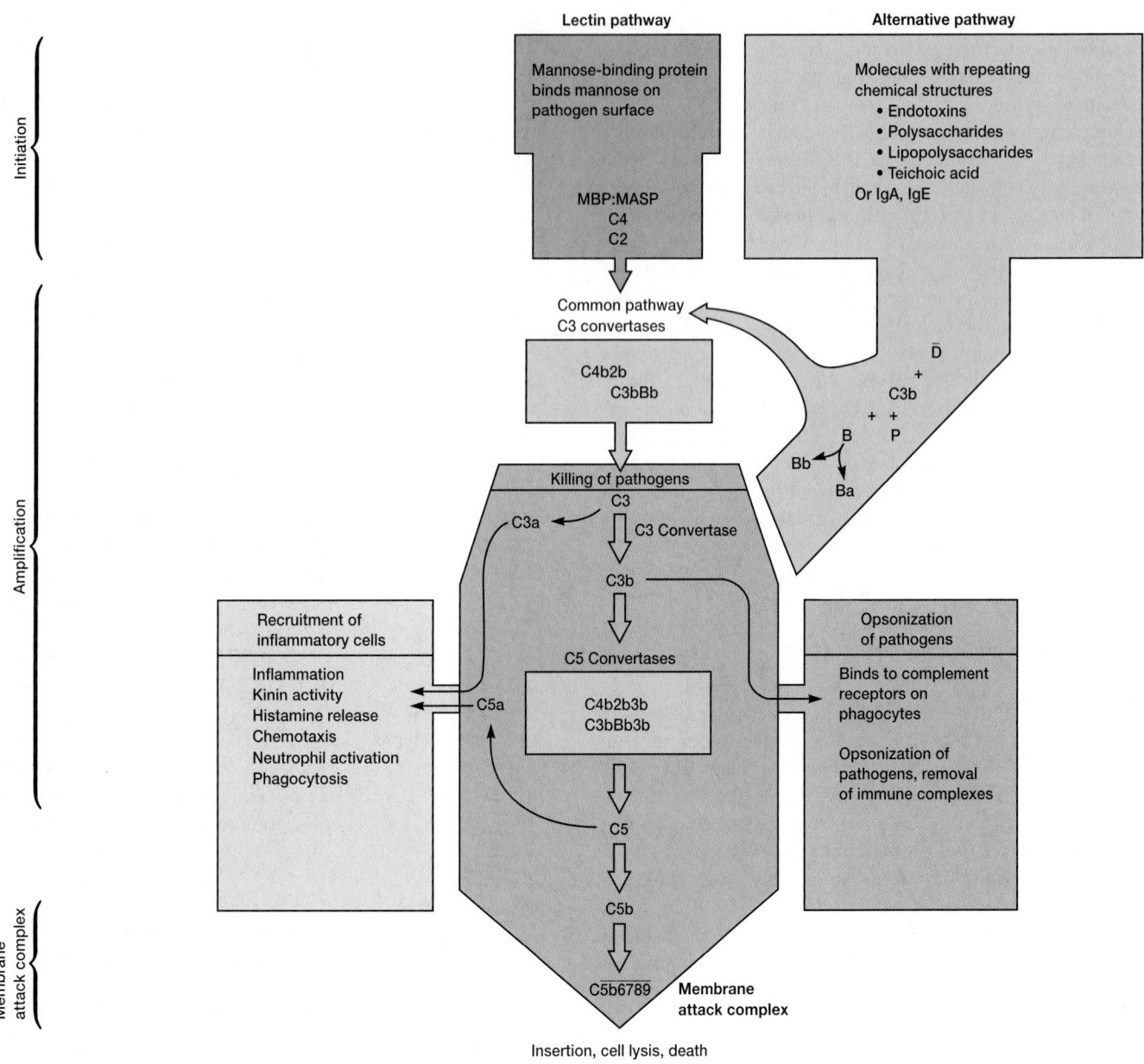

Figure 31.13 Alternative and Lectin Complement Pathways. Within each pathway the components are arranged in order of their activation and aligned opposite their functional and structural analogs in the opposite pathway. Two of the three pathways of complement activation are shown: (1) the lectin-mediated pathway (red), which is triggered by a serum protein that binds to mannose residues of bacteria; and (2) the alternative pathway (blue), which is triggered directly on pathogen surfaces. Both of these generate a crucial enzymatic activity (common pathway of C3 convertases) that in turn generates the effector activity of complement. The three main consequences of complement activation are recruitment of inflammatory cells (left box), direct killing of pathogens by the membrane attack complex (middle box), and opsonization of pathogens (right box).

Factor $\overline{D}$, leading to the formation of active enzyme $\overline{C3bBb}$ (this complex is sometimes called the C3 convertase because it cleaves more C3 to C3a and C3b). $\overline{C3bBb}$ is further stabilized by a second blood protein, properdin, which allows another addition of C3b forming C5 convertase ($\overline{C3bBb3b}$). The convertase then cleaves C5 to C5a and C5b. C6 and C7 rapidly bind to C5b, forming a $\overline{C5b67}$ complex that possesses an unstable membrane-binding site; once bound to a membrane, this complex is stable. C8 and C9 then bind, forming the **membrane attack complex** ($\overline{C5b6789}$) that creates a pore in the plasma membrane of the target cell (**figure 31.14**). It is believed that the actual pore is a doughnut-shaped polymer of C9. If the cell is eucaryotic, Na^+ and H_2O enter through the pore and the cell lyses osmotically. Lysozyme can pass through pores in the outer membrane of gram-

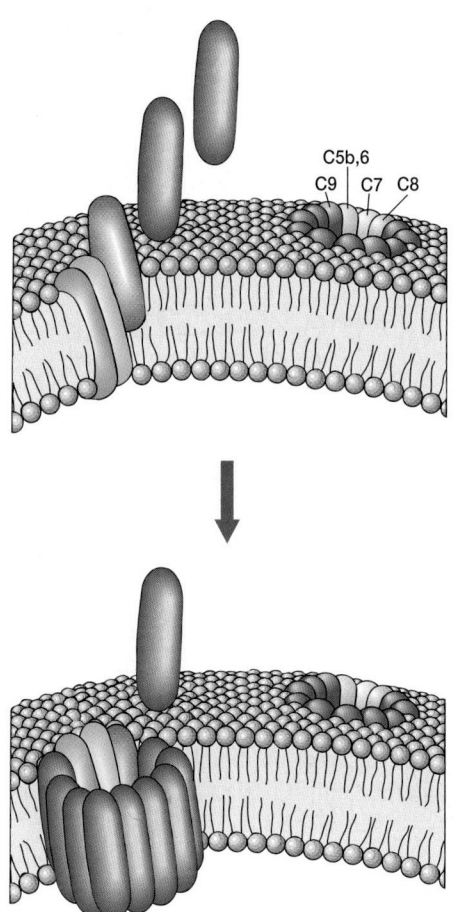

Figure 31.14   The Membrane Attack Complex.   The membrane attack complex (MAC) is a tubular structure that forms a transmembrane pore in the target cell's plasma membrane. This representation is the subunit architecture of the membrane attack complex. The transmembrane channel is formed by a C5b678 complex and 10 to 16 polymerized molecules of C9.

negative cell walls and digest the peptidoglycan cell wall, thus weakening the wall and aiding lysis. In contrast, gram-positive bacteria resist the cytolytic action of the membrane attack complex because they lack an exposed outer membrane and have a thick peptidoglycan that prevents an attack on the plasma membrane. However, gram-positive bacteria are very susceptible to opsonization (figure 31.15) when coated with C3b.

The **lectin complement pathway** (also called the mannan-binding lectin pathway) activates C3 convertase by using a lectin, a special protein that binds to specific carbohydrates. When macrophages ingest viruses, bacteria, and other foreign material, they release chemicals that stimulate liver cells to secrete acute phase proteins such as mannose-binding protein (MBP). Because mannose is a major component of bacterial cell walls and of some virus envelopes and immune complexes, MBP binds to many pathogens and soluble immune complexes. MBP is an opsonin and directly enhances phagocytosis. The lectin also activates complement in two ways (figure 31.13). After binding to the sur-

face, it can trigger the alternative pathway directly. In addition, if MBP is bound to the MBP-associated serine esterase (MASP), it activates the classical complement pathway. Thus although the lectin pathway activates the complement cascade just as the classical and alternative pathways do, it uses a mechanism that is independent of antibody-antigen interactions (the classical pathway) and the interaction of complement with pathogen surfaces (alternative pathway).

This overview of the alternative and lectin complement pathways provides a basis for consideration of the function of complement as an integrated system during an animal's defensive effort. Bacteria arriving at a local tissue site will interact with components of the alternative pathway, resulting in the generation of biologically active fragments, opsonization of the bacteria, and initiation of the lytic sequence. If the bacteria persist or if they invade the animal a second time, antibody responses also will activate the classical pathway (*see section 32.7*).

The generation of complement fragments C3a and C5a leads to several important inflammatory effects. Mast cells release their contents, and the blood supply to the area increases markedly (hyperemia) as blood vessels dilate due to released histamine (figure 31.12). These fragments also cause the release of neutrophils from the bone marrow into the circulation. Neutrophils then make their way to the site of hyperemia where, in the presence of C5a, they attach to the endothelium and leave the blood. C5a induces a directed, chemotactic migration of neutrophils to the site of complement activation. Macrophages in the area can synthesize even more complement components to interact with the bacteria. All these defensive events promote the ingestion and ultimate destruction of the bacteria by the neutrophils and macrophages.

---

1. List the major functions of complement.
2. How is the alternative pathway activated? The lectin pathway?
3. What is the membrane attack complex, and how does its formation lead to cell lysis?
4. What role do complement fragments C3a and C5a play in an animal's defense against gram-negative bacteria?

 31.8   PHAGOCYTOSIS

During their lifetimes, humans and other vertebrates encounter many microbial species, but only a few of these species can grow and cause serious disease in otherwise healthy hosts. Phagocytic cells (monocytes, tissue macrophages, dendritic cells, and neutrophils) are an important early defense against invading microorganisms. These phagocytic cells recognize, ingest, and kill many extracellular microbial species by the process called **phagocytosis** [Greek *phagein*, to eat, and *cyte*, cell, and *osis*, a process]. The concept of phagocytosis was briefly introduced in section 4.5 within the discussion of lysosomes and how certain cells obtain nutrients by endocytosis. Phagocytosis will now be considered in more detail in the context of nonspecific host resistance.

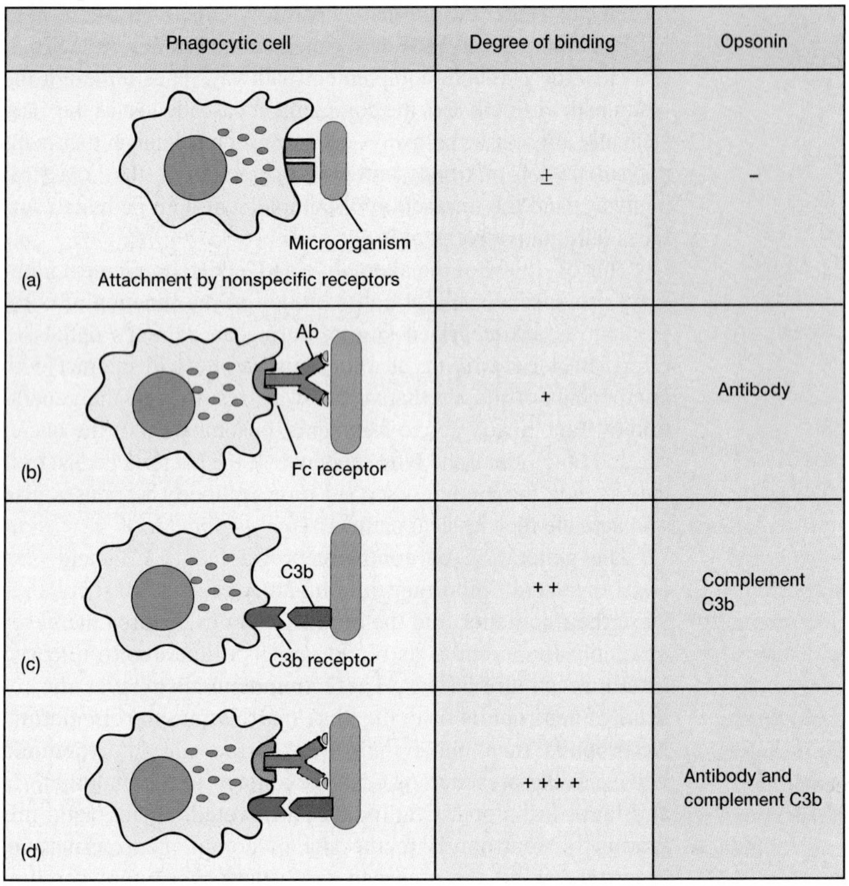

Phagocytic cell	Degree of binding	Opsonin
(a) Attachment by nonspecific receptors — Microorganism	±	–
(b) Ab — Fc receptor	+	Antibody
(c) C3b — C3b receptor	+ +	Complement C3b
(d)	+ + + +	Antibody and complement C3b

Figure 31.15 Opsonization. (a) A phagocytic cell has some intrinsic ability to bind directly to a microorganism through nonspecific receptors. (b) This binding ability is enhanced if the microorganism elicits the formation of antibodies (Ab) that act as a bridge to attach the microorganism to the Fc receptor on the phagocytic cell. (c) If the microorganism has activated complement (C3b), the degree of binding is further enhanced by the C3b receptor. (d) If both antibody and C3b opsonize, binding is greatly enhanced.

Phagocytic cells use two basic molecular mechanisms for the recognition of microorganisms: (1) opsonin-dependent (opsonic) and (2) opsonin-independent (nonopsonic) recognition. The phagocytic process can be greatly enhanced by opsonization. **Opsonization** [Greek *opson,* to prepare victims for] is a process in which microorganisms or other particles are coated by serum components (antibodies, mannose binding proteins, and/or complement C3b) thereby preparing them for recognition and ingestion by phagocytic cells. In the opsonin-dependent recognition mechanism, it is the serum components that function as a bridge between the microorganisms and the phagocyte. They act by binding to the surface of the microorganism at one end and to specific receptors on the phagocyte surface at the other (**figure 31.15b–d**).

The opsonin-independent mechanism does not require opsonins and uses other nonspecific (figure 31.15a) and specific receptors (**figure 31.16a**) on phagocytic cells that recognize structures expressed on the surface of different microorganisms. Three main forms of recognition in opsonin-independent phagocytosis and one form of signaling have been identified in phagocytic cells (**table 31.2**). One mode, termed lectin phagocytosis, is based on the recognition between surface lectins on one cell and surface carbohydrates on the opposing cell. The second mode is the result of protein-protein interactions between the Arg-Gly-Asp peptide sequence of microorganisms and phagocytic cell receptors. Third, hydrophobic interactions between bacteria and phagocytic cells also promote phagocytosis. It should be noted that a particular microbial species can express multiple adhesins, each recognized by a distinct receptor present on phagocytic cells.

The fourth type of interaction also involves the recognition of microbial nonself and plays a crucial role in nonspecific host resistance. This recognition strategy is based on the detection of conserved molecular structures that occur in patterns and are the essential products of normal microbial physiology. These invariant structures are called **pathogen-associated molecular patterns (PAMPs).** PAMPs are unique to microorganisms, invariant among microorganisms of a given class, and not produced by the host. The most well-known examples of PAMPs are the lipopolysaccharide (LPS) of gram-negative bacteria and the peptidoglycan of gram-positive bacteria. These and other PAMPs are recognized by receptors on phagocytic cells called **pattern recognition receptors (PRRs).** Since, as just noted, PAMPs are produced only by microorganisms, they are perceived by the phagocytic cells of the innate immune system as molecular signatures of infection. This allows the innate immune system to distinguish self from microbial nonself.

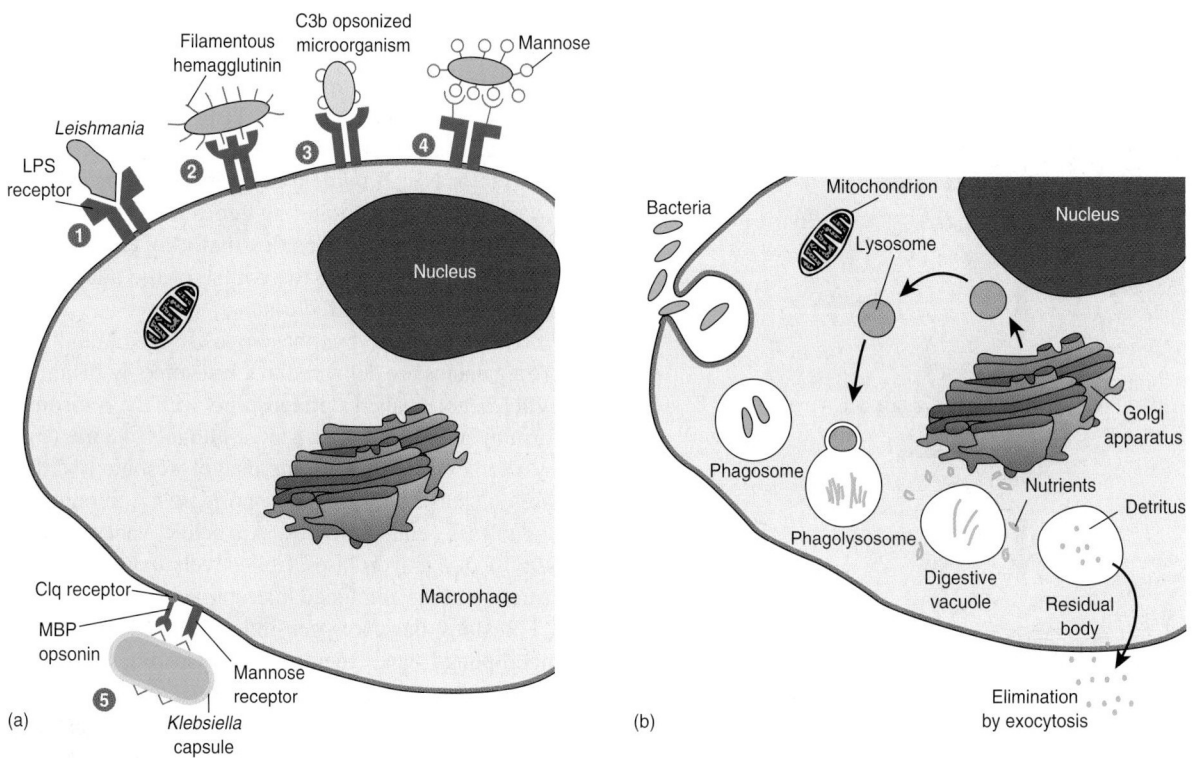

Figure 31.16   Phagocytosis.   (a) Drawing shows receptors on a phagocytic cell, such as a macrophage, and the corresponding adhesins on microbial surfaces participating in phagocytosis. (1) The LPS binding site for lipophosphoglycan of *Leishmania* spp; (2) filamentous hemagglutinin of *Bordetella pertussis;* (3) binding to C3b on an opsonized cell; (4) the mannose-containing oligosaccharide side chain for lectin phagocytosis mediated by type 1 fimbriated bacteria; and (5) the participation of capsular polysaccharide of *Klebsiella pneumoniae* in nonopsonic binding mediated by the mannose receptor. **(b)** Drawing of phagocytosis, showing ingestion, intracellular digestion, and exocytosis.

Table 31.2

Nonopsonic Modes of Recognition and Signaling by Phagocytes

Type of Interaction	Bacterial Ligand (and Example)	Phagocytic Receptor (and Example)
Lectin-carbohydrate	Lectin (Type I fimbriae; *see figure 3.31*)	Glycoprotein (integrins)
	Polysaccharide (capsule; *see figure 3.28*)	Lectin (Man/GlcNAc receptors)
Protein-protein	Arginine-glycine-aspartic acid;	RGC receptor (integrins)
	RGD-containing proteins (filamentous	
	hemagglutinin; *see table 34.3*)	
Hydrophobic protein	Glycolipid (lipoteichoic acid: *see figure 3.21*)	Lipid receptors (integrins)
Signaling	Proteolycans (gram-positive bacteria)	TLR-2
	LPS (gram-negative bacteria, heat-shock proteins)	TLR-4
	Flagellin	TLR-5
	Bacterial CpG DNA	TLR-9

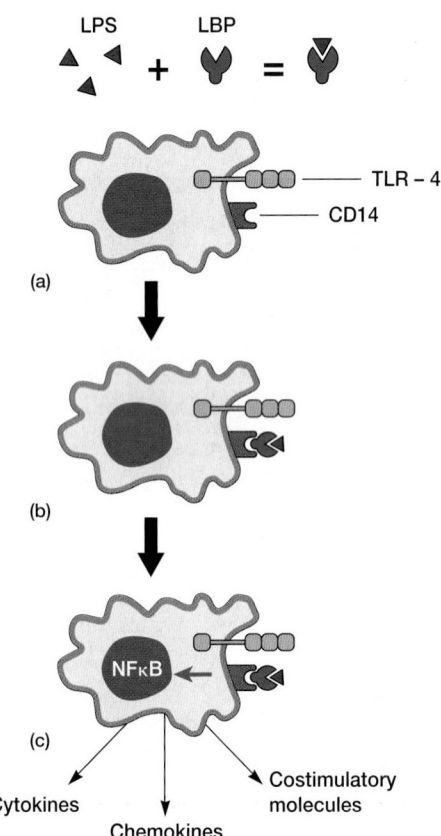

Figure 31.17 Toll-like Receptor 4 on a Macrophage Signals the Presence of a Pathogen. (a) Bacterial lipopolysaccharide (LPS) from a gram-negative bacterial pathogen binds to LPS-binding protein (LPB) found in blood plasma and other body fluids. (b) The LPS-LBP complex binds to the receptor protein CD14 on the surface of the macrophage. (c) The bound LPS-LBP interacts with the Toll-like receptor 4 resulting in a signal (blue arrow in the cytoplasm) to the nucleus. This activates the transcription factor NFκB, which in turn activates genes that produce cytokines, chemokines, and costimulatory molecules that play a role in defense against the gram-negative bacterial pathogen.

Several structurally and functionally distinct classes of PRRs evolved in phagocytic cells to recognize PAMPs and to induce various host defensive pathways. For example, secreted PRRs bind to microbial cells and mark them for destruction by either the complement system or phagocytosis. Another class of PRRs function exclusively as signaling receptors (**figure 31.17**). These receptors are known as **Toll-like receptors (TLRs).** TLRs appear not to recognize and bind pathogens directly, but are involved in signaling the appropriate response to different classes of pathogens (viruses, bacteria, or fungi). In mammals, TLR-4 signals the presence of bacterial lipopolysaccharide (LPS) and heat-shock proteins. TLR-9 signals the dinucleotide CpG motif present on DNA re-

leased by dying bacteria. TLR-2 signals the presence of peptidoglycans of gram-positive bacteria. (There are at least nine distinct proteins in this newly discovered family of receptors in mammals.) The binding of TLR-4 or TLR-2 triggers an evolutionarily ancient signaling pathway that activates transcription factor NFκB by degrading its inhibitor IκB. This induces a variety of genes, including genes for cytokines, chemokines, and costimulatory molecules that play essential roles in calling forth and directing the course of the adaptive immune response later in the infection.

Once ingested by phagocytosis, membrane-enveloped microorganisms are delivered to a lysosome by fusion of the phagocytic vacuole, called a **phagosome,** with the lysosome membrane, forming a new vacuole called a **phagolysosome** (figure 31.16b). Lysosomes contribute to the phagolysosome a variety of hydrolases such as lysozyme, phospholipase A_2, ribonuclease, deoxyribonuclease, and proteases. An acidic vacuolar pH favors the activity of the hydrolases. Collectively these participate in the destruction of the entrapped microorganisms.

Besides these oxygen-independent lysosomal hydrolases, macrophage and neutrophil lysosomes contain oxygen-dependent enzymes that can produce toxic **reactive oxygen intermediates (ROIs)** such as the superoxide radical ($O_2^- \cdot$), hydrogen peroxide (H_2O_2), singlet oxygen ($^1O_2$), and hydroxyl radical (OH·). The NADPH required for this process is supplied by a large increase in pentose phosphate pathway activity (*see figure 9.6*). Neutrophils also contain myeloperoxidase and produce hypochlorous acid. Some reactions forming these toxic products are shown here.

Superoxide formation

$$\text{NADPH} + 2O_2 \xrightarrow{\text{NADPH oxidase}} 2O_2^- \cdot + H^+ + \text{NADP}^+$$

Hydrogen peroxide formation

$$2O_2^- \cdot + 2H^+ \xrightarrow{\text{superoxide dismutase}} H_2O_2 + O_2$$

Hypochlorous acid formation

$$H_2O_2 + Cl^- \xrightarrow{\text{myeloperoxidase}} \text{HOCl} + OH^-$$

Singlet oxygen formation

$$ClO^- + H_2O_2 \rightarrow {}^1O_2 + Cl^- + H_2O$$

Hydroxyl radical formation

$$O_2^- \cdot + H_2O_2 \rightarrow \text{OH} \cdot + OH^- + O_2$$

These reactions result from the **respiratory burst** that accompanies the increased oxygen consumption and ATP generation needed for phagocytosis. Because these reactions occur as soon as the phagosome is formed, lysosome fusion is not necessary for the respiratory burst. The toxic oxygen products produced are ef-

Table 31.3

The Four Cytokine Families

Family	Examples	Functions
Chemokines	IL-8, RANTES, MIP (macrophage inflammatory protein)	Cytokines that are chemotactic and chemokinetic for leukocytes. They stimulate cell migration and attract phagocytic cells and lymphocytes. Chemokines play a central role in the inflammatory response.
Hematopoietins	Epo (erythropoietin), various colony-stimulating factors	Cytokines that stimulate and regulate the growth and differentiation processes involved in blood cell formation (hematopoiesis).
Interleukins	IL-1 to IL-18	Cytokines produced by lymphocytes and monocytes that regulate the growth and differentiation of other cells, primarily lymphocytes and hematopoietic stem cells. They often also have other biological effects.
Tumor necrosis factor (TNF) family	TNF-α, TNF-β, Fas ligand	Cytokines that are cytotoxic for tumor cells and have many other effects such as promoting inflammation, fever, and shock; some can induce apoptosis.

fective in killing invading microorganisms.   Oxygen and microbial growth (pp. 125–27)

Recently macrophages, neutrophils, and mast cells have been shown to form **reactive nitrogen intermediates (RNIs).** These molecules include nitric oxide (NO) and its oxidized forms, nitrite (NO_2^-) and nitrate (NO_3^-). The RNIs are very potent cytotoxic agents, and may be either released from cells or generated within cell vacuoles. Nitric oxide is probably the most effective RNI. Macrophages produce it from the amino acid arginine. Nitric oxide can block cellular respiration by complexing with the iron in electron transport proteins. Macrophage destruction of the herpes simplex virus, the protozoa *Toxoplasma gondii* and *Leishmania major,* the opportunistic fungus *Cryptococcus neoformans,* the metazoan pathogen *Schistosoma mansoni,* and tumor cells involves RNIs.

Neutrophil granules also contain a variety of other microbicidal substances such as several cationic proteins, the bactericidal permeability increasing protein (BPI), and the family of broad-spectrum antimicrobial peptides called **defensins.** There are four human defensins, called human neutrophil proteins (HNPs): HNP-1, 2, 3, and 4. These defensins are synthesized by myeloid precursor cells during their sojourn in the bone marrow, and are then stored in the cytoplasmic granules of mature cells. This compartmentation strategically locates defensins for extracellular secretion or delivery to phagocytic vacuoles. Susceptible microbial targets include a variety of gram-positive and gram-negative bacteria, yeasts and molds, and some viruses. Defensins act against bacteria and fungi by permeabilizing cell membranes. They form voltage-dependent membrane channels that allow ionic efflux. Antiviral activity involves direct neutralization of enveloped viruses; nonenveloped viruses are not affected by defensins.

1. What is the difference between opsonic and nonopsonic phagocytosis?

2. Once a phagolysosome forms, how is the entrapped microorganism destroyed?
3. What is the purpose of the respiratory burst that occurs within macrophages? Describe the nature and function of reactive oxygen and nitrogen intermediates.
4. What are defensins? How do they function?

 ### 31.9 CYTOKINES

Defense against viruses, microorganisms and their products, parasites, and cancer cells is mediated by both nonspecific and specific immunity. Cytokines are required for immunoregulation of both of these immune responses.

The term **cytokine** [Greek *cyto,* cell, and *kinesis,* movement] is a generic term for the soluble protein or glycoprotein released by one cell population that acts as an intercellular (between cells) mediator or signaling molecule. When released from mononuclear phagocytes, these proteins are called **monokines;** when released from T lymphocytes they are called **lymphokines;** when produced by a leukocyte and the action is on another leukocyte, they are **interleukins;** and if their effect is to stimulate the growth and differentiation of immature leukocytes in the bone marrow, they are called **colony-stimulating factors (CSFs).** Recently cytokines have been grouped into the following categories or families: chemokines, hematopoietins, interleukins, and members of the tumor necrosis factor (TNF) family. Some examples of these cytokine families are provided in **table 31.3.**

Cytokines can affect the same cell responsible for their production (an autocrine function), nearby cells (a paracrine function), or can be distributed by the circulatory system to their target cells (an endocrine function). Their production is induced by nonspecific stimuli such as a viral, bacterial, or parasitic infection; cancer; inflammation; or the interaction between a T cell and antigen. Some cytokines also can induce the production of other cytokines.

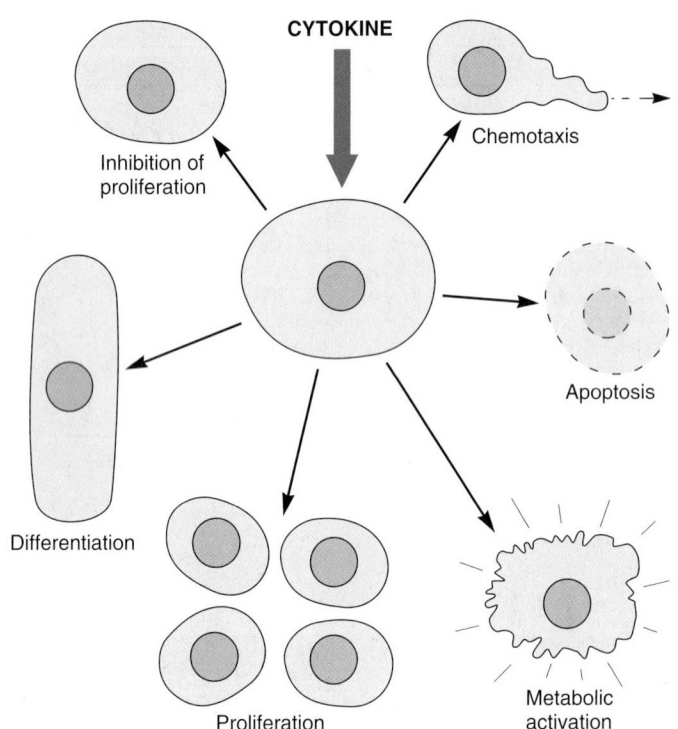

Figure 31.18 Range of Biological Actions That Cytokines Have on Eucaryotic Cells. Chemokines are one family of cytokines that induce leukocyte chemotaxis and migration. Other cytokines activate cell metabolism and synthesis. This can lead to the synthesis of a wide range of proteins including cyclooxygenase II, proteolytic enzymes, NO synthase, and various adhesion receptors. In addition, other cytokines can cause proliferation, inhibition of cell proliferation, or apoptosis.

Knowledge of the biological actions of cytokines has grown enormously over the past two decades. This is due in part to the incredible potency and range of effects on eucaryotic cells exhibited by cytokines, and to the fact that they are involved in all aspects of disease. Cytokines produce biological actions only when they act as ligands and bind to specific, high-affinity receptors called CDs (cell-associated differentiation antigens [*see section 32.2*]) on the surface of target cells. The affinity of cytokine receptors for their ligands is very high, and consequently cytokines are effective at concentrations of 10^{-10} to 10^{-15}M. Most cells have relatively few receptors, hundreds to a few thousand, and a maximal cellular response results when only a small number of these are occupied by the cytokine. This binding activates specific intracellular signaling pathways and switches on genes that encode proteins essential to the appropriate target cell functions. Examples of these proteins include other cytokines, cell-to-cell adhesion receptors, proteases, lipid-synthesizing enzymes, and nitric oxide synthase. Cytokines can activate cell proliferation and/or cell differentiation (**figure 31.18**). They also

can inhibit cell division and cause apoptosis (programmed cell death). Chemokines, one type of cytokine, stimulate chemotaxis and chemokinesis (i.e., they direct cell movement) and thus play an important role in the acute inflammatory response (figure 31.11). Some examples of important cytokines and their functions are given in **table 31.4**.

Interferons

Interferons (IFNs) are a group of related low molecular weight, regulatory cytokines produced by certain eucaryotic cells in response to a viral infection. Besides defending against viruses, they also help regulate the immune response. Interferons usually are species specific but virus nonspecific. Several classes of interferons are recognized: IFN-α is a family of 20 different molecules that can be synthesized by virus-infected leukocytes (table 31.4). IFN-β is derived from virus-infected fibroblasts; and IFN-γ is produced by antigen-stimulated T cells and natural killer cells. Some of the ways in which interferon renders cells resistant to virus infections are described in **figure 31.19**.

Fever

From a physiological point of view, fever results from disturbances in hypothalamic thermoregulatory activity, leading to an increase of the thermal "set point." In adult humans **fever** is defined as an oral temperature above 98.6°F (37°C) or a rectal temperature above 99.5°F (37.5°C). The most common cause of a fever is a viral or bacterial infection (or bacterial toxins). In almost every instance there is a specific constituent, the endogenous pyrogen, that directly triggers fever production. Examples of these pyrogens includes interleukin-1, IL-6, and tissue necrosis factor that are produced by host macrophages in response to pathogenic microorganisms. After their release, these pyrogens circulate to the hypothalamus and induce neurons to secrete prostaglandins. Prostaglandins reset the hypothalamic thermostat at a higher temperature, and temperature-regulating reflex mechanisms then act to bring the core body temperature up to this new setting.

The fever induced by a microorganism augments the host's defenses by three complementary pathways (**figure 31.20**): (1) it stimulates leukocytes so that they can destroy the microorganism, (2) it enhances the specific activity of the immune system, and (3) it enhances microbiostasis (growth inhibition) by decreasing available iron to the microorganism. Evidence suggests that some hosts are able to redistribute the iron during a fever in an attempt to withhold it (**hypoferremia**) from the microorganism. Conversely, the virulence of many microorganisms is enhanced with increased iron availability (**hyperferremia**). Gonococci, for example, spread most often during menstruation, a time in which there is an increased concentration of free iron available to these bacteria.

Table 31.4

Some of the Cytokines That Mediate Immune Responses

Cytokine	Cell Source	Functions
IL-1 (interleukin-1)	Monocytes/macrophages, endothelial cells, fibroblasts, neuronal cells, glial cells, keratinocytes, epithelial cells	Produces a wide variety of effects on the differentiation and function of cells involved in inflammatory and immune responses; also affects central nervous and endocrine systems; it is an endogenous pyrogen
IL-2 (interleukin-2, T-cell growth factor)	T cells (T_H1)	Stimulates T-cell proliferation and differentiation, enhances cytolytic activity of NK cells, promotes proliferation and immunoglobin secretion of activated B cells
IL-3 (interleukin-3)	T cells, keratinocytes, neuronal cells, mast cells	Stimulates the production and differentiation of macrophages, neutrophils, eosinophils, basophils, mast cells
IL-4 (interleukin-4, B-cell growth factor-1 [BCGF-1], B-cell stimulatory factor-1 [BCSF-1])	T cells (T_H2), macrophages, mast cells, basophils, B cells	Induces the differentiation of naive CD4$^+$ T cells into T-helper-like cells; induces the proliferation and differentiation of B cells; exhibits diverse effects on T cells, monocytes, granulocytes, fibroblasts, endothelial cells
IL-5 (interleukin-5)	T cells (T_H2)	Growth and activation of B cells and eosinophils; activation of eosinophil function; chemotactic for eosinophils
IL-6 (interleukin-6, cytotoxic T-cell differentiation factor, B-cell differentiation factor)	T_H2 cells, monocytes/macrophages, fibroblasts, hepatocytes, endothelial cells, neuronal cells	Activates hematopoietic cells; induces growth of T cells, B cells, hepatocytes, keratinocytes, and nerve cells; stimulates the production of acute-phase proteins
IL-8 (interleukin-8)	Monocytes, endothelial cells, fibroblasts, alveolar epithelium, T cells, keratinocytes, neutrophils, hepatocytes	Chemoattractant for PMNs; causes degranulation and expression of receptors; inhibits adhesion of PMNs to endothelium; promotes migration of PMNs through endothelium; induces adhesion of neutrophils to endothelial cells
IL-10 (interleukin-10)	T cells (T_H2), B cells, macrophages, keratinocytes	Reduces the production of IFN-γ, IL-1, TNF-α, and IL-6 by macrophages; in combination with IL-3 and IL-4, causes mast cell growth; in combination with IL-2, causes growth of cytotoxic T cells and differentiation of CD8$^+$ cells
IFNs α/β (interferons α/β)	T cells, B cells, monocytes/macrophages, fibroblasts	Antiviral activity, stimulates macrophage activity, modulates MHC class I and II protein expression on various cells, regulates the development of the specific immune response
IFN-γ (interferon-γ)	T cells (T_H1, T_C), NK cells	Activation of T cells, macrophages, neutrophils, and NK cells; increases class I and II MHC molecules
TNF-α (tumor necrosis factor-α [cachectin])	T cells, macrophages and NK cells	A wide variety of effects due to its ability to mediate expression of genes for growth factors and cytokines, transcription factors, receptors, inflammatory mediators, and acute-phase proteins; plays a role in host resistance to infection by serving as an immunostimulant and mediator of the inflammatory response; cytotoxic for tumor cells
TNF-β (tumor necrosis factor-β [lymphotoxin])	T cells, B cells	Same as TNF-α
G-CSF (granulocyte colony-stimulating factor)	T cells, macrophages, neutrophils	Enhances the differentiation and activation of neutrophils
M-CSF (macrophage colony-stimulating factor)	T cells, neutrophils, macrophages, fibroblasts, endothelial cells	Stimulates various functions of monocytes and macrophages, promotes the growth and development of macrophage colonies from undifferentiated precursors

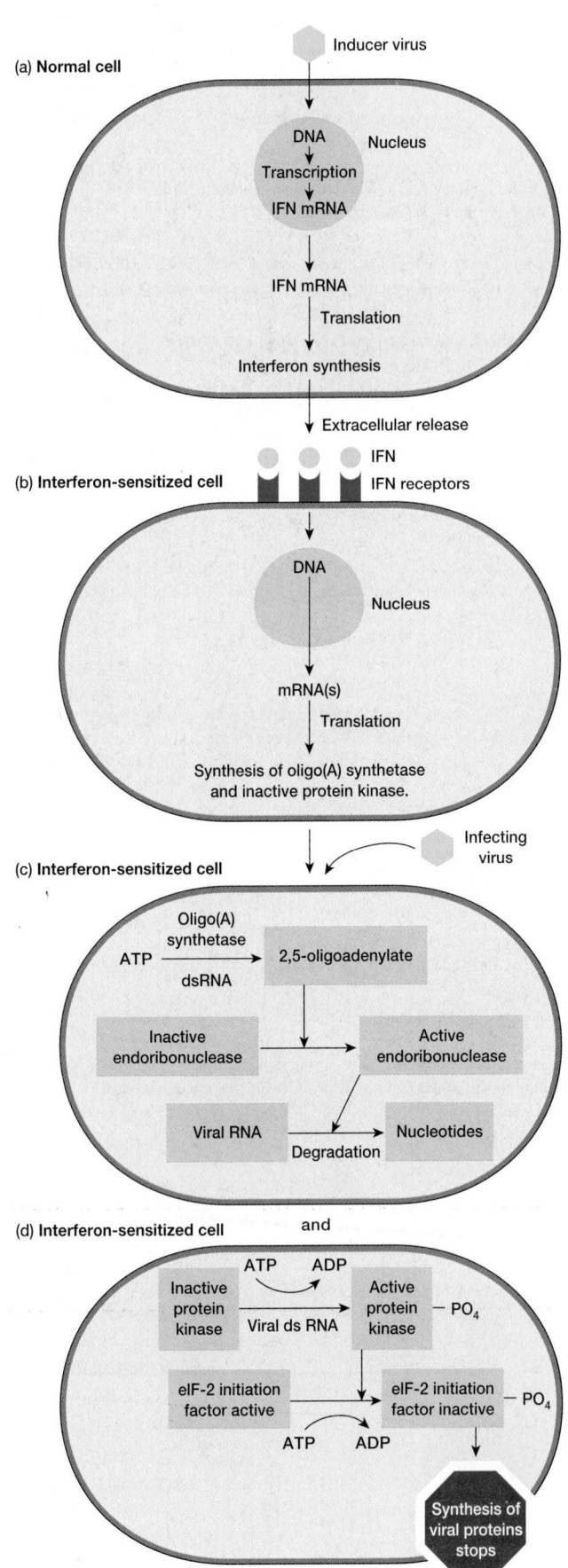

1. What is a cytokine?
2. Define monokine, lymphokine, interleukin, and colony-stimulating factor.
3. Into what four families are cytokines sometimes placed (table 31.3)?
4. How do cytokines play a role as mediators of natural immunity?
5. How do interferons render cells resistant to viruses?
6. How can a fever be beneficial to a host?

31.10 NATURAL KILLER CELLS

Natural killer (NK) cells are a small population of large non-phagocytic granular lymphocytes (figure 31.3). Their major function is to destroy malignant cells and cells infected with microorganisms. They recognize their targets in one of two ways. Like many cells, they possess Fc receptors that bind IgG antibody. These receptors link the NK cells to IgG-coated target cells, which they kill by a process called **antibody-dependent cell-mediated cytotoxicity (ADCC) (figure 31.21).** The second way NK cells recognize infected cells and cancer cells relies on the killer-activating receptors and killer-inhibitory receptors of the NK cells (**figure 31.22**). The killer-activating receptors recognize a number of different molecules present on the surface of all nucleated cells. The killer inhibitory receptors recognize a surface marker known as the major histocompatibility complex (MHC) class I molecule (*see section 32.4*) that is also present on all nucleated cells and is an indicator of "self." If the killer-activating receptors bind to a ubiquitous surface molecule, a "kill" instruction is issued to the NK cell. Conversely, when this "kill" signal is overridden by an inhibitory signal sent by the killer inhibitory receptor binding to the MHC class I molecule (which indicates "self"), no killing is done.

Although all nucleated cells normally have the MHC class I molecule on their plasma membrane surface, they can lose this

Figure 31.19 The Antiviral Action of Interferon. The arrows indicate the sequence of events. (**a**) Interferon (IFN) synthesis and release is often induced by a virus infection or double-stranded RNA (dsRNA). (**b**) Interferon binds to a ganglioside receptor on the plasma membrane of a second cell and triggers the production of enzymes that render the cell resistant to virus infection. The two most important such enzymes are oligo(A) synthetase and a special protein kinase. (**c**) When an interferon-stimulated cell is infected, viral protein synthesis is inhibited by an active endoribonuclease that degrades viral RNA. (**d**) An active protein kinase phosphorylates and inactivates the initiation factor eIF-2 required for viral protein synthesis.

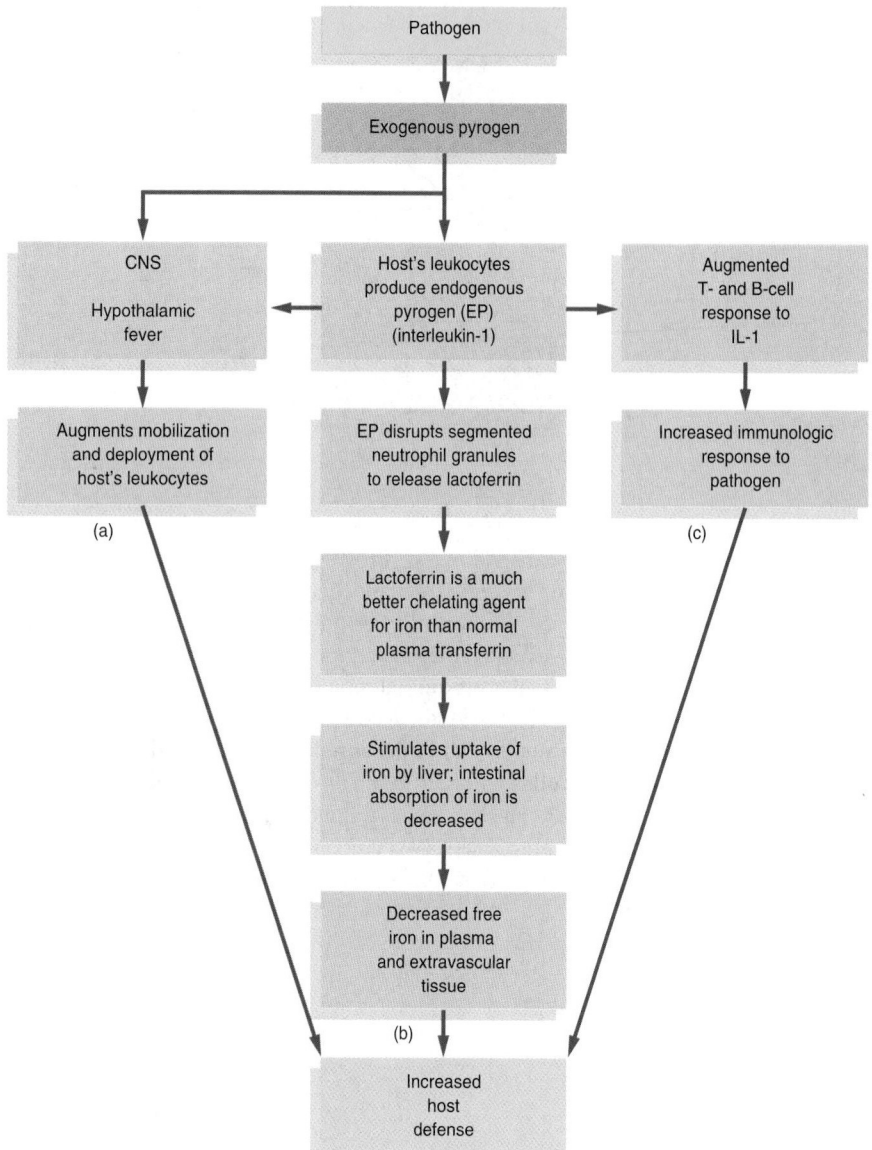

Figure 31.20   Effects of Fever in a Mammalian Host.   (a) One of the first effects of a fever is to accelerate and augment the mobilization and deployment of the host's leukocytes, which in turn isolate and help destroy the pathogen. **(b)** A fever leads to a reduction of free plasma iron, which limits the growth of those organisms that require a narrow, crucial concentration of iron for their replication and synthesis of toxins. **(c)** One of the immunologic consequences of fever is the production of endogenous pyrogen (IL-1). IL-1 causes the proliferation, maturation, and activation of lymphocytes (T and B cells), which in turn augment the immunologic response of the host to the pathogen.

molecule. This loss can occur as a result of a microbial pathogen interfering with the expression mechanism (e.g., after a herpesvirus infection) or from a malignant transformation. Cells that lack this MHC class I molecule are perceived as being abnormal and NK cells do not receive an inhibitory signal. In the absence of an inhibitory signal, NK cells kill these abnormal cells by inserting the pore-forming protein, perforin, into the plasma membrane of the target cell and then injecting it with cytotoxic enzymes called granzymes.

1. Discuss the role of NK cells in protecting the host.
2. What is the purpose of antibody-dependent cell-mediated cytotoxicity?

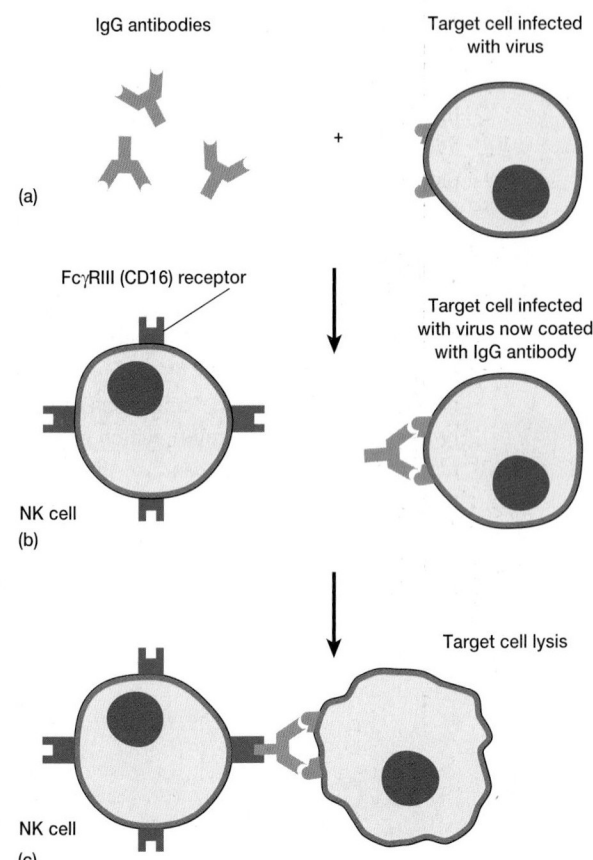

Figure 31.21 Antibody-Dependent Cell-Mediated Cytotoxicity. (**a**) In this mechanism, IgG antibodies bind to a target cell infected with a virus. (**b**) NK cells have FcγRIII (CD16) receptors on their surface. (**c**) When the NK cells encounter infected virus cells coated with IgG antibody, they kill the target cell by releasing cytotoxic mediators and/or membrane damage.

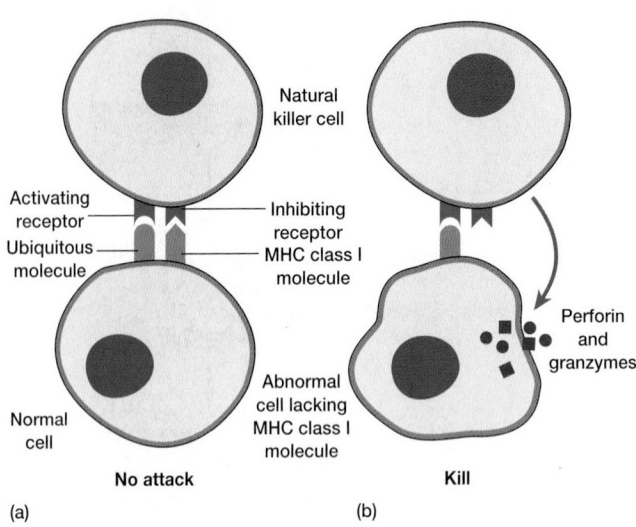

Figure 31.22 The System Used by Natural Killer Cells to Recognize Normal Cells and Abnormal Cells That Lack the Major Histocompatability Complex Class I Surface Molecule. (**a**) The killer-activating receptor recognizes a normal ubiquitous molecule on the plasma membrane of a normal cell. Since the killer-inhibitory receptor recognizes the MHC class I molecule, there is no attack. (**b**) In the absence of the inhibitory signal, the receptor issues an order to the NK cell to attack and kill the abnormal cell. The cytotoxic granules of the NK cell contain perforin and granzymes. With no inhibitory signal, the granules release their contents, killing the abnormal cell.

Summary

31.1 Gnotobiotic Animals

a. Animals and environments that are germfree or have one or more known microorganisms are termed gnotobiotic. Methods are available for rearing gnotobiotic organisms (**figure 31.1**). Gnotobiotic animals and techniques provide good experimental systems with which to investigate the interactions of animals and specific species or microorganisms.

31.2 Normal Microbiota of the Human Body

a. Commensal microorganisms living on or in the skin can be characterized as either transients or residents (**figure 31.2**).

b. The normal microbiota of the oral cavity is composed of those microorganisms able to resist mechanical removal.

c. The stomach contains very few microorganisms due to its acidic pH.

d. The distal portion of the small intestine and the entire large intestine have the largest microbial community in the body. Over 400 species have been identified, the vast majority of them anaerobic.

e. The upper genitourinary tract is usually free of microorganisms. In contrast, the adult female genital tract has a complex microbiota.

f. In some instances, after a microorganism contacts or enters a host, a positive mutually beneficial relationship occurs and becomes integral to the health of the host. In other instances, the microorganism may produce disease or even death of the host.

g. Many of the normal host microbiota compete with pathogenic microorganisms.

h. An opportunistic microorganism is generally harmless in its normal environment but becomes pathogenic in a compromised host.

31.3 Overview of Host Resistance

a. There are two fundamentally different types of immune response to invading microorganisms and foreign material. The nonspecific (innate) response offers resistance to any microorganism or foreign material. It includes general mechanisms that are a part of the animal's innate structure and function. The nonspecific system has no immunological memory—that is, nonspecific responses occur to the same extent each time. In contrast, the specific (adaptive) response resists a particular foreign agent; moreover, specific immune responses improve on repeated exposure to the agent.

31.4 Cells, Tissues, and Organs of the Immune System

a. The cells responsible for both nonspecific and specific immunity are the white blood cells called leukocytes (**figure 31.3**). Examples include lymphocytes, monocytes and macrophages (**figures 31.4** and **31.5**), granulocytes, mast cells, and dendritic cells.

b. Immature undifferentiated lymphocytes generated in the bone marrow mature and become committed to a particular antigenic specificity within the primary lymphoid organs and tissues (**figure 31.6**). In mammals, T cells mature in the thymus and B cells in the bone marrow. The thymus is the primary lymphoid organ and the bone marrow the primary lymphoid tissue.

c. The secondary lymphoid organs and tissues serve as areas where lymphocytes may encounter and bind antigens, then they proliferate and differentiate into fully mature, antigen-specific effector cells. The spleen is a secondary lymphoid organ and the lymph nodes and mucosal-associated tissues (GALT and SALT) are the secondary lymphoid tissues.

31.5 Physical and Chemical Barriers in Nonspecific (Innate) Resistance

a. Many direct factors (age, nutrition) or general barriers contribute in some degree to all host-microbe relationships. At times they favor the establishment of the microorganism; at other times they provide some measure of general defense to the host.

b. Physical and mechanical barriers along with host secretions are the host's first line of defense against pathogens (**figure 31.7**). Examples include the skin (**figure 31.8**) and mucous membranes; the epithelia of the respiratory, gastrointestinal (**figure 31.10**), and genitourinary systems.

c. Mammalian hosts have specific chemical barriers that help combat the continuous onslaught of pathogens. Examples include general chemicals, bacteriocins, and beta-lysins.

31.6 Inflammation

a. Inflammation is one of the host's nonspecific defense mechanisms to tissue injury that may be caused by a pathogen (**figures 31.11** and **31.12**). Inflammation can either be acute or chronic.

31.7 The Complement System

a. The complement system is composed of a large number of serum proteins (**table 31.1**) that play a major role in the animal's defensive immune response. There are three pathways of complement activation: the classical, alternative, and lectin pathways (**figure 31.13**).

31.8 Phagocytosis

a. Phagocytosis involves the recognition, ingestion, and destruction of pathogens by lysosomal enzymes, superoxide radicals, hydrogen peroxide, defensins, RNIs, and metallic ions (**figure 31.16**). Phagocytic cells use two basic mechanisms for the recognition of microorganisms: opsonin-dependent and opsonin-independent.

31.9 Cytokines

a. Cytokines (**tables 31.3** and **31.4**) are required for immunoregulation of both the nonspecific and specific immune responses. Cytokines have a broad range of actions on eucaryotic cells (**figure 31.18**).

b. Interferons are a group of cytokines that respond in a defensive way to viral infections, double-stranded RNA, endotoxins, antigenic stimuli, mitogenic agents, and many pathogens capable of intracellular growth (**figure 31.19**).

c. Fever induced by a microorganism augments the host's defenses in three ways (**figure 31.20**): it stimulates leukocytes so they can destroy the invading microorganism; it enhances microbiostasis by decreasing iron available to the microorganism; and it enhances the specific activity of the immune system.

31.10 Natural Killer Cells

a. Natural killer cells are a small population of large, nonphagocytic lymphocytes (**figures 31.21** and **31.22**) that destroy cancer cells and cells infected with microorganisms.

Key Terms

alternative complement pathway 691
alveolar macrophage 687
antibody-dependent cell-mediated cytotoxicity (ADCC) 700
axenic 674
bacteriocin 688
basophil 681
B cell 681
B lymphocyte 681
bronchial-associated lymphoid tissue (BALT) 686
bursa of Fabricius 684
cathelicidins 677
colicin 688
colony-stimulating factor (CSF) 697
comedo 677
complement system 691
compromised host 680
cryptin 687
cytokine 697
defensin 697
dendritic cell 683
ectosymbiosis 675
endosymbiosis 675
eosinophil 683

fever 698
gnotobiotic 674
granuloma 691
gut-associated lymphoid tissue (GALT) 686
hyperferremia 698
hypoferremia 698
immune system 680
immunity 680
immunology 680
inflammation 688
innate or natural immunity 680
integrin 688
interdigitating dendritic cell 685
interferon (IFN) 698
interleukin 697
intraepidermal lymphocyte 685
lactoferrin 686
Langerhans cell 685
lectin complement pathway 693
leukocyte 681
lymphocyte 681
lymphokine 697
lymph node 684
lysozyme 685

M cell 686
macrophage 681
mast cell 683
membrane attack complex 692
monocyte 681
monocyte-macrophage system 681
monokine 697
mucociliary blanket 687
mucociliary escalator 687
mucosal-associated lymphoid tissue (MALT) 686
natural killer (NK) cell 700
neutrophil 683
nonspecific immune response 680
nonspecific resistance 680
normal microbiota 675
opportunistic microorganism or pathogen 680
opsonization 694
Paneth cell 687
pathogen 674
pathogen-associated molecular pattern (PAMP) 694
pathogenicity 674
pattern recognition receptor (PRR) 694

peristalsis 687
phagocytosis 693
phagolysosome 696
phagosome 696
plasma cell 685
polymorphonuclear leukocyte (PMN) 681
probiotic 679
reactive nitrogen intermediate (RNI) 697
reactive oxygen intermediate (ROI) 696
respiratory burst 696
sebum 677
selectin 688
skin-associated lymphoid tissue (SALT) 685
specific immune response (acquired, adaptive, or specific immunity) 681
spleen 684
T cell 681
thymus 684
T lymphocyte 681
Toll-like receptor (TLR) 696

Questions for Thought and Review

1. What are some uses of gnotobiotic animals in microbiology?
2. What has the branch of microbiology known as gnotobiotics taught us about the relationship between the human body and the bacteria found on/in it?
3. Describe why the microenvironment of the skin is both favorable to some microorganisms and unfavorable to others. Explain your answer.
4. What is the relationship between a compromised host and an opportunistic microorganism?
5. Compare and contrast the nonspecific (innate) and specific (adaptive) immune responses.
6. Describe the different types of human blood cells and give a function for each one.
7. Describe how each of the following systems contributes to the defense of the host: respiratory, gastrointestinal, and genitourinary.
8. How do the SALT and MALT systems operate?
9. How can inflammation be both beneficial and detrimental to a microorganism?
10. What is the relationship between the respiratory burst and phagocytosis?
11. Why is it difficult to group cytokines into specific categories?
12. Why are most fevers caused by either viruses or bacteria?

Critical Thinking Questions

1. Some patients who take antibiotics for acne develop yeast infections. Explain.
2. Some microorganisms are classified as intracellular pathogens. Explain why the nonspecific immune response to these pathogens differs from the response to pathogens that occur on or in the skin.

**For recommended readings on these
and related topics, see Appendix V.**

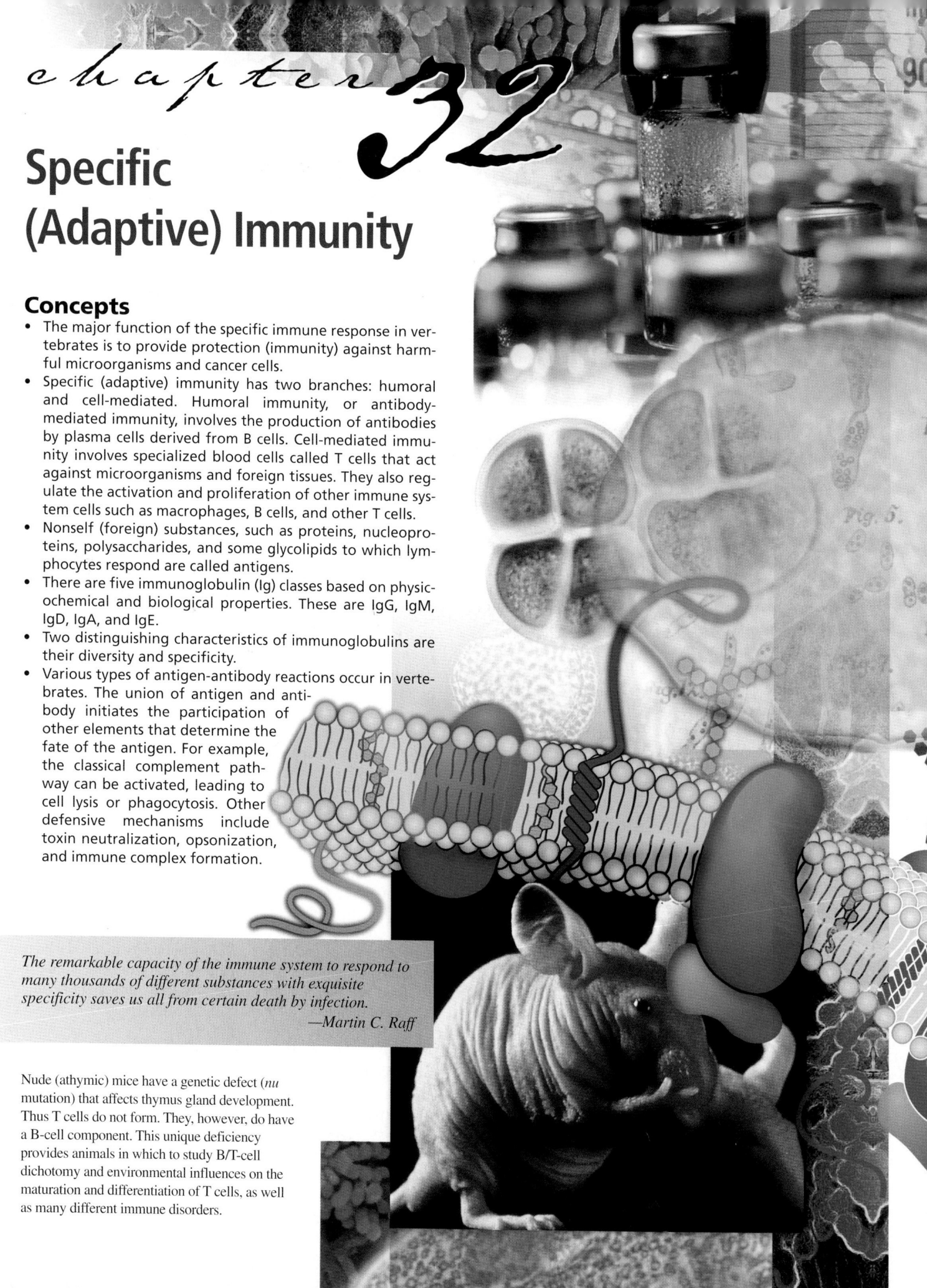

chapter 32

Specific (Adaptive) Immunity

Concepts

- The major function of the specific immune response in vertebrates is to provide protection (immunity) against harmful microorganisms and cancer cells.

- Specific (adaptive) immunity has two branches: humoral and cell-mediated. Humoral immunity, or antibody-mediated immunity, involves the production of antibodies by plasma cells derived from B cells. Cell-mediated immunity involves specialized blood cells called T cells that act against microorganisms and foreign tissues. They also regulate the activation and proliferation of other immune system cells such as macrophages, B cells, and other T cells.

- Nonself (foreign) substances, such as proteins, nucleoproteins, polysaccharides, and some glycolipids to which lymphocytes respond are called antigens.

- There are five immunoglobulin (Ig) classes based on physicochemical and biological properties. These are IgG, IgM, IgD, IgA, and IgE.

- Two distinguishing characteristics of immunoglobulins are their diversity and specificity.

- Various types of antigen-antibody reactions occur in vertebrates. The union of antigen and antibody initiates the participation of other elements that determine the fate of the antigen. For example, the classical complement pathway can be activated, leading to cell lysis or phagocytosis. Other defensive mechanisms include toxin neutralization, opsonization, and immune complex formation.

The remarkable capacity of the immune system to respond to many thousands of different substances with exquisite specificity saves us all from certain death by infection.
— *Martin C. Raff*

Nude (athymic) mice have a genetic defect (*nu* mutation) that affects thymus gland development. Thus T cells do not form. They, however, do have a B-cell component. This unique deficiency provides animals in which to study B/T-cell dichotomy and environmental influences on the maturation and differentiation of T cells, as well as many different immune disorders.

*T*he immune system is responsible for the ability of a host to resist foreign invaders. It includes an array of cells and molecules with specialized roles in defending the host against the continuous onslaught of microbial invaders, and cancer cells. Chapter 31 presents a basic discussion of nonspecific host resistance or innate immunity. Chapter 32 continues this discussion of the immune response by describing the system of specific (adaptive) immunity used to protect the host. Specific immunity can be acquired either naturally or artificially and involves antigens, antibodies, T and B cells, and the classical complement pathway. These aspects of immune system function are presented in this chapter in the continuing discussion of the immune system.

32.1 OVERVIEW OF SPECIFIC (ADAPTIVE) IMMUNITY

The specific (adaptive) immune system of vertebrates has three major functions: to recognize anything that is foreign to the body ("nonself"), to respond to this foreign material, and to remember the foreign invader. The recognition response is highly specific. The immune system is able to distinguish one pathogen from another, to identify cancer cells, and to recognize the body's own "self" proteins and cells as different from "nonself" proteins, cells, tissues, and organs. After recognition of an invader has occurred, the specific immune system responds by recruiting its defensive molecules and cells to attack the invader. This is called an **effector response.** A successful effector response either eliminates the foreign material or renders it harmless to the host, thus preventing disease. If the invader is encountered at a later time, the immune system remembers, and mounts a more intense and rapid memory or anamnestic response that eliminates the invader once again and protects the host from disease (p. 719).

Four characteristics distinguish specific (adaptive) immunity from nonspecific (innate) resistance:

1. *Specificity.* Immunity is directed against a particular pathogen or foreign substance, and the immunity to this pathogen or substance usually does not confer immunity to others.
2. *Memory.* When reexposed to the same pathogen or substance, the body reacts so quickly that there is no noticeable pathogenesis. By contrast, the reaction time for inflammation and other nonspecific (innate) defenses is just as long for a later exposure as it was for the initial one.
3. *Diversity.* The system is able to generate an enormous diversity of molecules such as antibodies that recognize billions of different antigens.
4. *Discrimination between self and nonself.* The (adaptive) specific immune system almost always responds only to nonself antigens and thus does not destroy the organism it is defending.

Two branches or arms of specific (adaptive) immunity are recognized (**figure 32.1**): humoral (antibody-mediated) immunity and cellular (cell-mediated) immunity. **Humoral (antibody-mediated) immunity,** named for the fluids or "humors" of the body, is based on the action of soluble proteins called antibodies that occur in the body fluids and on the plasma membranes of B lymphocytes. Circulating antibodies bind to bacteria, toxins, and extracellular viruses, neutralizing them or "tagging or marking" them for destruction by mechanisms described in section 32.6.

Cellular (cell-mediated) immunity is based on the action of specific kinds of T lymphocytes (*see section 31.4 and figure 31.3*) that directly attack cells infected with viruses or parasites, transplanted cells or organs, and cancer cells. T cells can lyse these cells or release chemicals (cytokines) that enhance specific immunity and nonspecific (innate) defenses such as phagocytosis and inflammation.

Types of Acquired Immunity

Acquired immunity refers to the type of specific (adaptive) immunity a host develops after exposure to a suitable antigen, or after transfer of antibodies or lymphocytes from an immune donor. Acquired immunity can be obtained by natural or artificial means and actively or passively (**figure 32.2**).

Naturally Acquired Immunity

Naturally acquired active immunity occurs when an individual's immune system contacts an antigenic stimulus such as a pathogen that causes an infection. The immune system responds by producing antibodies and activated lymphocytes that inactivate or destroy the antigen (pathogen). The immunity produced can be either lifelong, as with measles or chickenpox, or last for only a few years, as with tetanus.

Naturally acquired passive immunity involves the transfer of antibodies from one host to another. For example, some of a pregnant woman's antibodies pass across the placenta to her fetus. If the female is immune to diseases such as polio or diphtheria, this placental transfer also gives the fetus and newborn immunity to these diseases. Certain other antibodies can pass from the female to her offspring in the first secretions (called colostrum) from the mammary glands. These maternal antibodies are essential for providing immunity to the newborn until its own immune system matures. Unfortunately naturally acquired passive immunity generally lasts only a short time (weeks or months at the most).

Artificially Acquired Immunity

Artificially acquired active immunity results when an animal is given an antigen preparation to induce the formation of antibodies and activated lymphocytes. This preparation is called a vaccine and the procedure is vaccination (immunization). A vaccine consists of a preparation of killed microorganisms; living, weakened (attenuated) microorganisms; or inactivated bacterial toxins (toxoids) that are administered to an animal to induce immunity artificially. Vaccines and immunizations are discussed in detail in section 33.1.

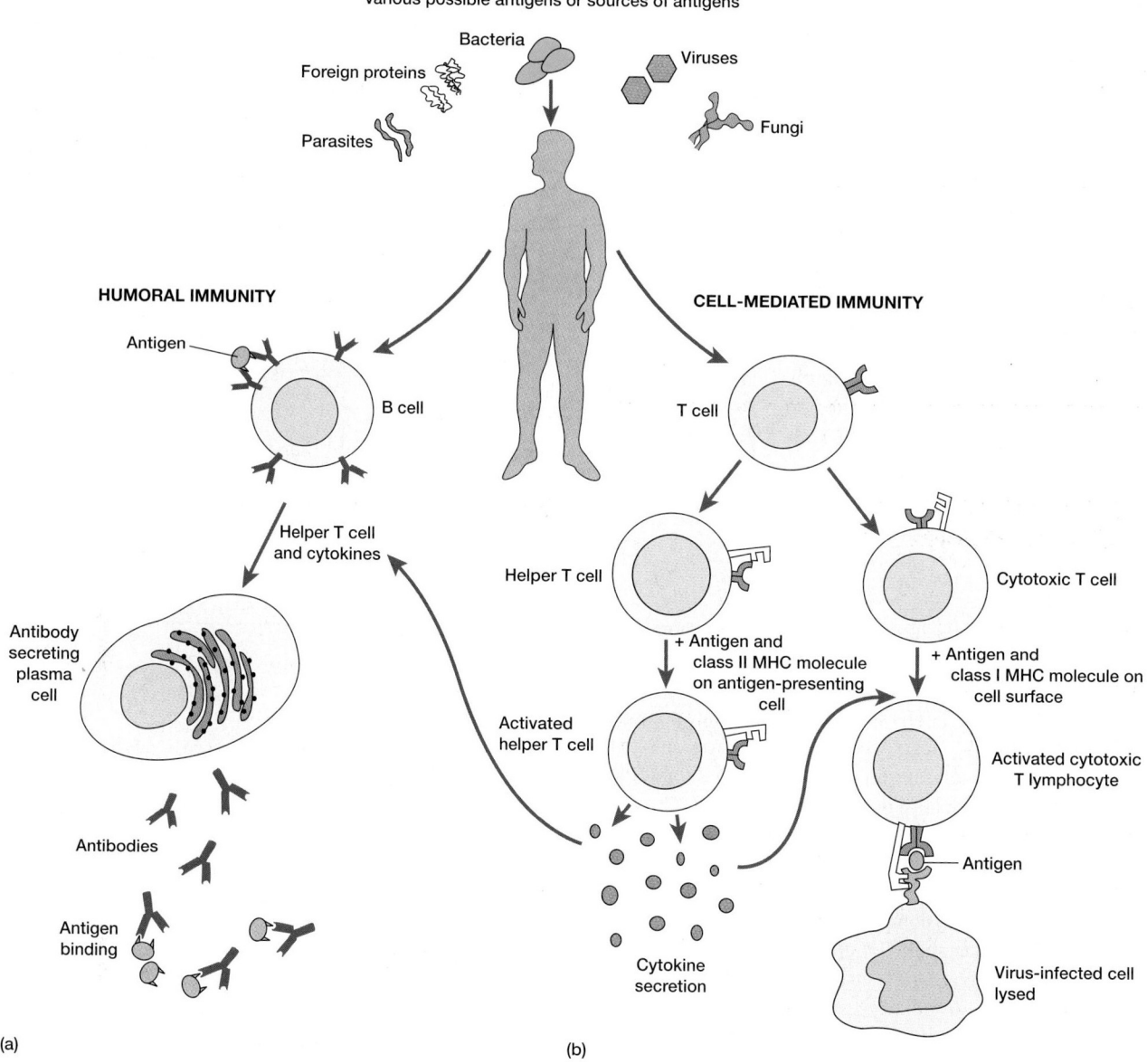

Various possible antigens or sources of antigens

Figure 32.1   The Humoral and Cell-Mediated Branches of Immunity.   (a) In the humoral branch, B cells interact with antigen and differentiate into antibody-secreting plasma cells. The antibody binds to the antigen and tags it for destruction by other mechanisms. (b) In the cell-mediated response, subpopulations of T cells are activated by antigen presented in the presence of a major histocompatibility complex (MHC). Activated helper T cells respond by producing a group of specific chemicals called cytokines that facilitate both the humoral and cell-mediated responses. Cytotoxic T cells respond to the antigen by developing into cytotoxic T lymphocytes that kill virus-infected cells and other altered cells. B cells and T cells also differentiate into memory cells (not shown) that are involved in subsequent responses. B cells and T cells also differentiate into memory cells involved in subsequent secondary responses (not shown).

Artificially acquired passive immunity results when antibodies that have been produced either in an animal or by specific methods in vitro are introduced into a host. Although this type of immunity is immediate, it is short lived and lasts only a few weeks to a few months. An example would be botulinum antitoxin produced in a horse and given to a human suffering from botulism food poisoning.

1. The specific (adaptive) immune response can be divided into three related activities. What are these activities?
2. What distinguishes specific immunity from nonspecific (innate) resistance?
3. What are the two arms of specific (adaptive) immunity?
4. How does naturally acquired immunity occur? Contrast active and passive immunity.

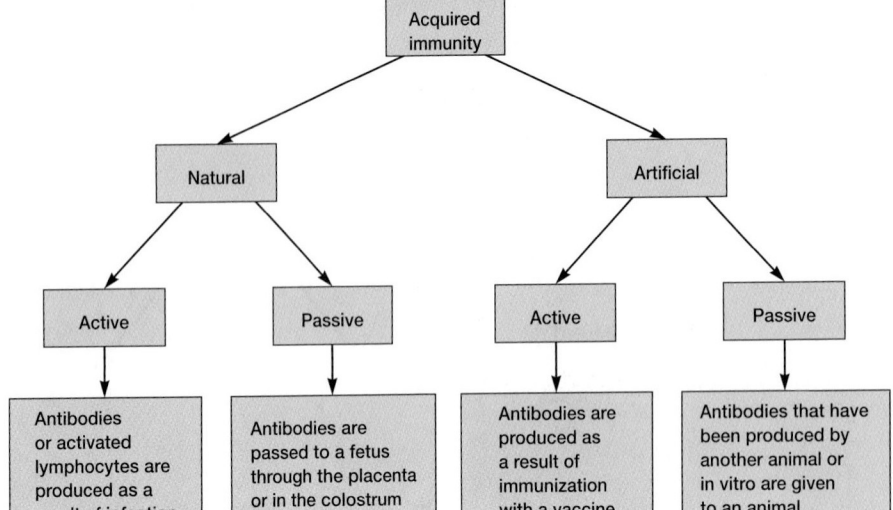

Figure 32.2 Types of Acquired Immunity. This illustration can be used as a guide to the text description of acquired immunity. The details of the immune system are astonishingly complex; many events have been omitted so that the overall organization of the immune response can be clearly seen.

32.2 ANTIGENS

The immune system distinguishes between "self" and "nonself" through an elaborate recognition process. Prior to birth the body inventories the proteins and various other large molecules present (self) and removes most T cells specific for self-determinants. Subsequently self-substances can be distinguished from nonself substances, and lymphocytes can produce specific immunologic reactions against the latter, leading to their removal.

Substances, such as proteins, nucleoproteins, polysaccharides, and some glycolipids, that elicit an immune response and react with the products of that response are called **antigens** (**anti**body **gen**erator). Most antigens are large, complex molecules with a molecular weight generally greater than about 10,000. The ability of a molecule to function as an antigen depends on its size, structural complexity, chemical nature, and degree of foreignness to the host.

Each antigen can have several **antigenic determinant sites** or **epitopes** (**figure 32.3**). Epitopes are the regions or sites in the antigen that bind to the antigen-binding site of a specific antibody or with a T cell receptor. Antibodies are formed most readily in response to determinants that project from the foreign molecule or to terminal residues of a specific polymer chain. Chemically, determinants include sugars, organic acids and bases, amino acid side chains, hydrocarbons, and aromatic groups.

The number of antigenic determinant sites on the surface of an antigen is its **valence.** The valence determines the number of antibody molecules that can combine with the antigen at one time. If one determinant site is present, the antigen is monovalent. Most antigens, however, have more than one determinant site or more than one copy of the same epitope, and are termed multivalent. Multivalent antigens generally elicit a stronger immune response than do monovalent antigens.

Haptens

Many small organic molecules are not antigenic by themselves but can become antigenic if they bond to a larger carrier molecule such

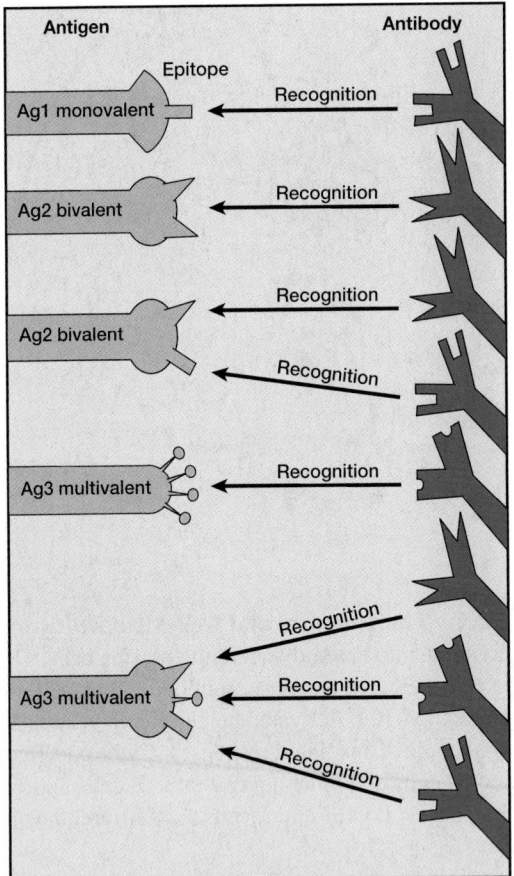

Figure 32.3 The Number of Antigenic Determinant Sites or Epitopes on an Antigen Is Its Valence. Antigen (Ag) molecules each have a set of antigenic determinants or epitopes. There is a single epitope on a monovalent antigen (Ag1). A bivalent antigen (Ag2) may have two of the same epitopes or two different ones, and a multivalent epitope (Ag3) may have many of the same epitopes or three or more different epitopes. In each case, however, a single antibody recognizes only the epitope rather than the whole bivalent or multivalent antigen.

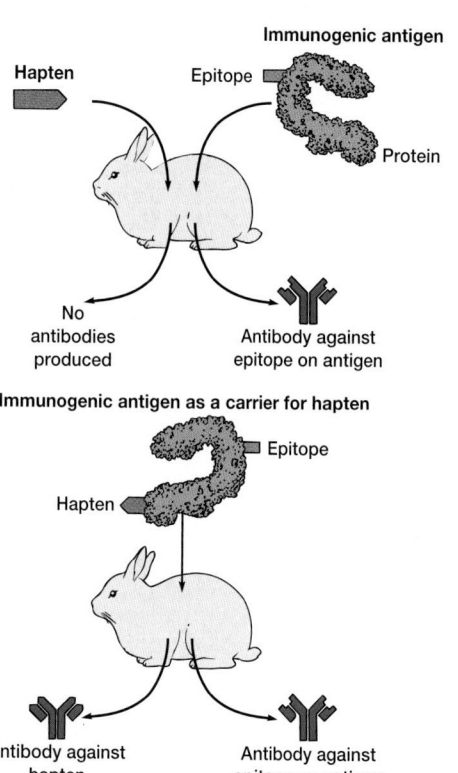

Figure 32.4 Effect of Carrier On Immunogenicity of Hapten.
Haptens, as small molecules, are incapable of inducing an immune response when presented alone. This is markedly different from what occurs when a more complex, immunogenic antigen is presented. In this instance antibody directed against the specific epitope(s) of that antigen can be produced. However, if hapten is linked to a large molecule (which is used as a carrier), the hapten may then be immunogenic. Under these conditions antibody is produced, not only against the epitope of the immunogenic carrier molecule but against the epitope of the hapten as well.

as a protein (**figure 32.4**). They cannot stimulate antibody formation or T-cell responses by themselves, but can react with antibodies once formed. Such small molecules are termed **haptens** [Latin *haptein,* to grasp]. When lymphocytes are stimulated by the combined molecule, they can react to either the hapten or the larger carrier molecule. Conjugation of a hapten to a carrier protein makes the hapten immunogenic because the carrier protein can be processed and presented to specific T cells. As a result, both hapten-specific and carrier-specific antibodies can be made. One example of a hapten is penicillin. By itself penicillin is not antigenic. However, when it combines with certain serum proteins of sensitive individuals, the resulting molecule does initiate a severe and sometimes fatal allergic immune reaction. In these instances the hapten is acting as an antigenic determinant on the carrier molecule.

Superantigens

Certain antigens provoke such a drastic immune response that they are termed **superantigens.** Superantigens are bacterial

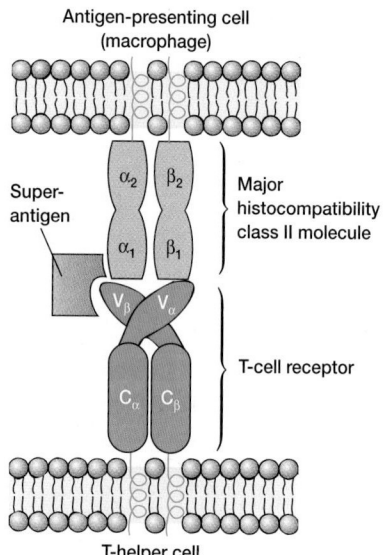

Figure 32.5 Superantigens. The most prevalent antigen receptor of T cells is the T-cell receptor. Superantigens stimulate T cells bearing a particular V_β element of the T-cell receptor, regardless of the nature of the V_α receptor element. They bind directly to the outer surface of the MHC class II molecule without being processed by the antigen-presenting cell. Thus a superantigen can activate many more T cells than a normal antigen does.

proteins. These superantigens nonspecifically stimulate T cells to proliferate by interacting with both class II major histocompatibility complex (section 32.4) products on antigen-presenting cells and the T-cell receptor (TCR) outside of the antigen-binding site. This binding allows many different T cells with different TCRs to become activated (**figure 32.5**). Good examples of superantigens are the staphylococcal enterotoxins that cause food poisoning and the toxin that causes toxic shock syndrome. Superantigens cause symptoms by stimulating the release of massive quantities of cytokines from CD4+ T cells. These cytokines have two effects on the host: systemic toxicity and suppression of the specific immune response. Both of these effects contribute to microbial pathogenicity. Superantigens should be considered possible chronic associates in such diseases as rheumatic fever, arthritis, Kawasaki syndrome, atopic dermatitis, and one type of psoriasis.

Cluster of Differentiation Molecules (CDs)

The involvement of lymphocytes with the immune response and various diseases can be assessed by enumeration of specific cells bearing particular membrane proteins called **cluster of *d*ifferentiation molecules (CDs).** CDs are functional cell surface proteins or receptors that can be measured in situ and from peripheral blood, biopsy samples, or other body fluids. They often are used in a classification system to differentiate between leukocyte subpopulations. To date, over 250 CDs have been characterized. **Table 32.1** summarizes some of the functions of several CDs.

Table 32.1	
Some Functional Examples of Cluster of Differentiation Molecules (CDs)	

CD Molecule	Function
CD2	The CD2 molecule is a glycoprotein present on T cells, thymocytes, and NK cells. It functions as an intercellular adhesion molecule and binds to the human leukocyte function-associated antigen-3 or LFA-3 (CD58). CD2 may be involved in the attachment of T_H cells to antigen-presenting cells and cytotoxic T cells to target cells, and in other cellular interactions. It also is a signal-transducing molecule and may help stimulate T cells to proliferate and secrete cytokines.
CD4	The CD4 molecule is a glycoprotein found primarily on T cells associated with helper/inducer regulatory functions and to a lesser degree on monocytes and macrophages. It is a cell adhesion molecule with great affinity for class II MHC.
CD8	The CD8 molecule is a polypeptide found on the surface of cytotoxic T cells. It is a molecule that participates in the interaction of class I MHC found on target cells.
CD23	CD23 is a protein found on the surface of IgM-bearing B cells, eosinophils, macrophages, and platelets. Soluble CD23 is a B-cell growth factor and upregulates IgE synthesis in conjunction with interleukin-4 released from T cells.
CD35	CD35 is a glycoprotein present on erythrocytes, PMNs, monocytes, all B and some T cells, mast cells, and in soluble form in the plasma. It is the receptor for complement C3b/C4b. The functions of CD35 vary according to the cell type possessing the receptor. For example, CD35 on erythrocytes binds immune complexes bearing C3b or C4b and promotes their clearance from the circulation. On B cells, CD35 augments the maturation of these cells into antibody-secreting cells. It also aids leukocytes in binding and phagocytosing C3b- or C4b-coated objects.
CD54	CD54, also known as intercellular adhesion molecule-1 (ICAM-1), is a protein that is an important early marker of immune activation and response. It is a ligand for the lymphocyte function-associated antigen-1 (LFA-1) found on lymphocytes, monocytes, and neutrophils. The CD54/LFA-1 interaction is important in T-cell–antigen-specific responses and lymphocyte emigration into inflammatory sites. CD54 also is found on mucosal epithelial cells and erythrocytes. Recent evidence suggests CD54 is the cellular receptor for the human rhinovirus and *Plasmodium falciparum*–infected erythrocytes.

CDs have both biological and diagnostic significance. The presence of various CDs on the cell's surface can be used to determine the cell's identity. In normal individuals the concentration of these molecules in serum is very low. When one's immune system is activated in response to disease, the concentration of these molecules usually rises and fluctuates. Monitoring CD levels may help in the management of the disease. For example, circulating levels of CD54 are directly related to the progress and prognosis of the skin cancer metastatic melanoma. CD23 is a B-cell growth factor, and elevated levels are associated with chronic lymphocytic leukemia. CD35 is a potent agent for the suppression of complement-dependent tissue injury in autoimmune inflammatory disease. Elevated levels of CD8 have been found in childhood lymphoid malignancies and in HIV-infected individuals. It has been established that the CD4 molecule is a cell surface receptor for HIV-1 (*see figure 38.9*), and considerable research is currently being done on the role of both cell-bound and soluble CD4 in AIDS.

1. Distinguish between "self" and "nonself" substances.
2. Define and give several examples of an antigen. What is an antigenic-determinant or epitope?
3. Describe a hapten, a superantigen.
4. Give some examples of the biological significance of cluster of differentiation molecules (CDs).

32.3 ANTIBODIES

An **antibody** or **immunoglobulin (Ig)** is a glycoprotein that is made by plasma cells in response to an antigen, and can recognize and bind to the antigen that caused its production. Antibod-

ies are present in the blood serum, tissue fluids, and mucosal surfaces of vertebrate animals. Serum glycoproteins can be separated according to their charge by movement through a gel in an electric field and classified as albumin, alpha-1 globulin, alpha-2 globulin, beta globulin, and gamma globulin (**figure 32.6**). The gamma globulin band is relatively wide because it contains a heterogeneous mixture of immunoglobulins—namely, IgG, IgA, IgM, and IgD. These differ from each other in molecular size,

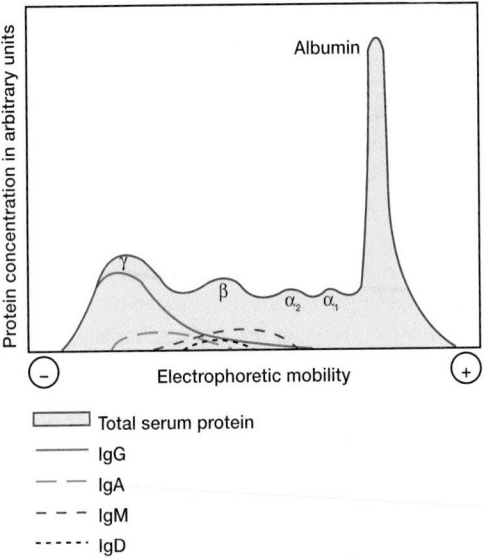

Figure 32.6 Electrophoresis of Human Serum. Schematic representation of electrophoretic results illustrating the distribution of serum proteins and four major classes of immunoglobulins.

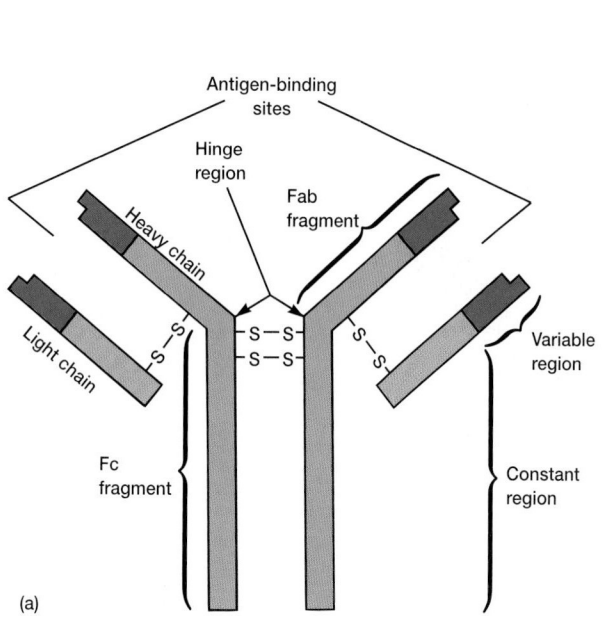

(a)

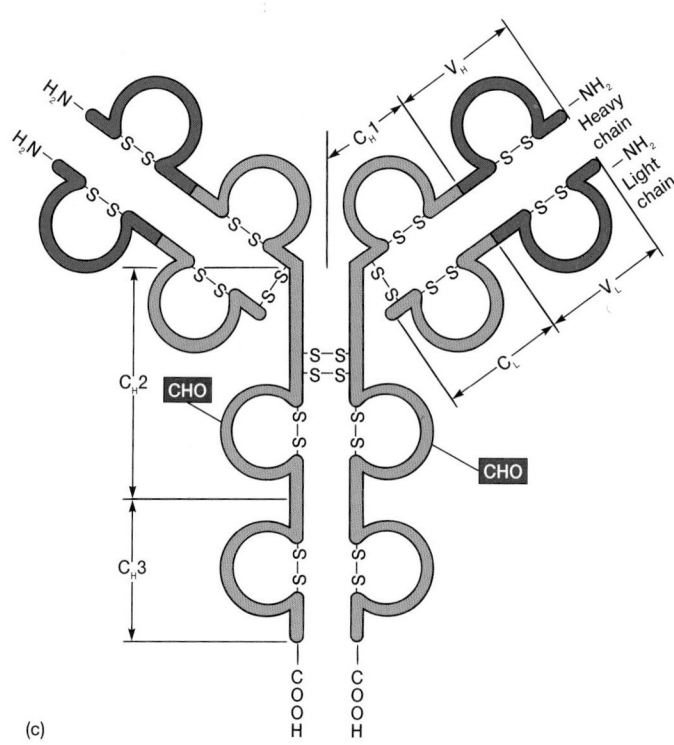

(c)

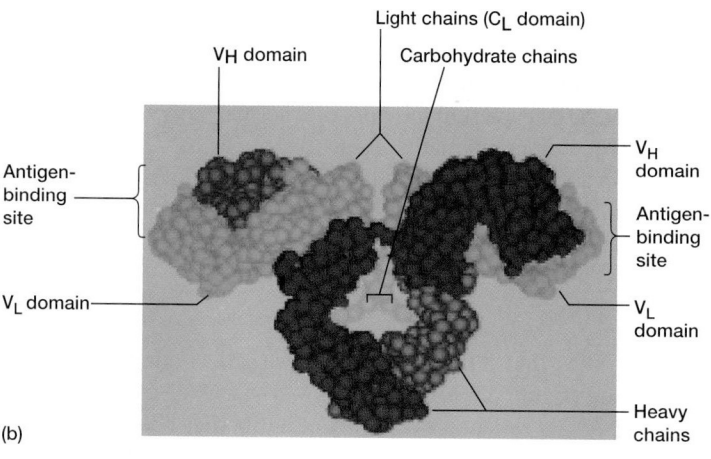

(b)

Figure 32.7   Immunoglobulin (Antibody) Structure.
(**a**) An immunoglobulin molecule. The molecule consists of two identical light chains and two identical heavy chains held together by disulfide bonds. (**b**) A computer-generated model of antibody structure showing the arrangement of the four polypeptide chains. (**c**) Within the immunoglobulin unit structure, intrachain disulfide bonds create loops that form domains. All light chains contain a single variable domain (V_L) and a single constant domain (C_L). Heavy chains contain a variable domain (V_H) and either three or four constant domains (C_H1, C_H2, C_H3, and C_H4). The variable regions (V_H, V_L), when folded together in three-dimensions, form the antigen-binding sites.

structure, charge, amino acid composition, and carbohydrate content. Most antibodies in serum are in the IgG class. A fifth immunoglobulin class, the IgE class, has a similar mobility to IgD but cannot be represented quantitatively because of its low concentration in serum.

Immunoglobulin Structure

Antibodies have more than one antigen-combining site. For example, most human antibodies have two combining sites or are bivalent (figure 32.3). Some bivalent antibody molecules can combine to form multimeric antibodies that have up to 10 combining sites.

All immunoglobulin molecules have a basic structure composed of four polypeptide chains (**figure 32.7a,b**) connected to

each other by disulfide bonds. Each light chain usually consists of about 220 amino acids and has a mass of approximately 25 kDa. Each heavy chain consists of about 440 amino acids and has a mass of about 50 to 70 kDa. The heavy chains are structurally distinct for each immunoglobulin class or subclass. Both light and heavy chains contain two different regions. **Constant regions** (C_L and C_H) have amino acid sequences that do not vary significantly between antibodies of the same class. The **variable regions** (V_L and V_H) from different antibodies do have different sequences (figure 32.7a). It is the variable regions (V_L, V_H) that when folded together form the antigen binding sites.

The four chains are arranged in the form of a flexible Y with a hinge region. This hinge region allows the antibody molecule to assume a T shape. The stalk of the Y is termed the **crystallizable**

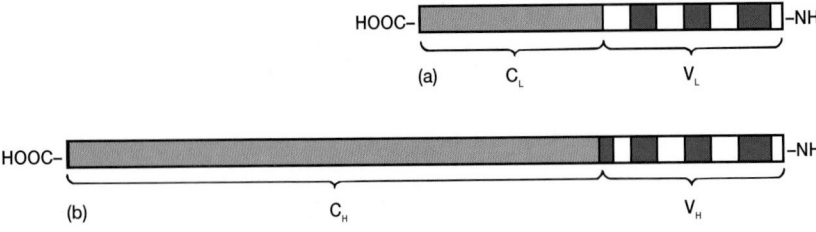

Figure 32.8 Constant and Variable Domains.
Location of constant (C) and variable (V) domains within (**a**) light chains and (**b**) heavy chains. The dark blue bands represent hypervariable regions or complementarity-determining regions within the variable domains.

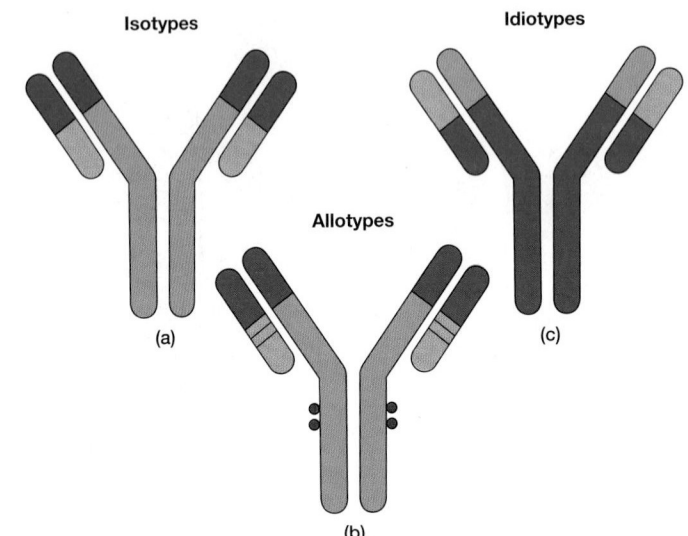

Figure 32.9 Variants of Immunoglobulins. (**a**) Isotypes represent variants present in serum of a normal individual. (**b**) Allotypes represent alternative forms coded for by different alleles and so are not present in all individuals of a population. (**c**) Idiotypes are individually specific to each immunoglobulin molecule. Carbohydrate side chains are in red.

fragment (**Fc**) and contains the site at which the antibody molecule can bind to a cell. The top of the Y consists of two **antigen-binding fragments (Fab)** that bind with compatible epitopes (or antigenic determinant sites). The Fc fragments are composed only of constant regions, whereas the Fab fragments have both constant and variable regions. Both the heavy and light chains contain several homologous units of about 100 to 110 amino acids. Within each unit, called a **domain,** disulfide bonds form a loop of approximately 60 amino acids (figure 32.7c). Interchain disulfide bonds also link heavy and light chains together.

More specifically the light chain may be either of two distinct forms called kappa (κ) and lambda (λ). These can be distinguished by the amino acid sequence of the carboxyl portion of the chain (**figure 32.8a**). In human immunoglobulins, the carboxyl-terminal portion of all κ chains is identical; thus this region is termed the constant (C_L) domain. With respect to the lambda chains, there are four very similar sequences that define the subtypes λ_1, λ_2, λ_3, and λ_4 with their corresponding constant regions $C_\lambda 1$, $C_\lambda 2$, $C_\lambda 3$, and $C_\lambda 4$. Within the light chain variable domain are hypervariable regions or complementarity-determining regions (CDRs) that differ in amino acid sequence more frequently than the rest of the variable domain.

In the heavy chain the NH2-terminal domain has a pattern of variability similar to that of the V_κ and V_λ domains and is termed the V_H domain. However, notice in figure 32.8b that this domain contains four hypervariable regions. The other domains of the heavy chains are termed constant domains and are numbered $C_H 1$, $C_H 2$, $C_H 3$, and sometimes $C_H 4$, starting with the domain next to the variable domain (figure 32.7c). The constant domains of the heavy chain form the constant (C_H) region. The amino acid sequence of this region determines the classes of heavy chains. In humans there are five classes of heavy chains designated by lowercase Greek letters: gamma (γ), alpha (α), mu (μ), delta (δ), and epsilon (ε). The properties of these heavy chains determine, respectively, the five immunoglobulin classes—IgG, IgA, IgM, IgD, and IgE. Each immunoglobulin class differs in its general properties, half-life, distribution in the body, and interaction with other components of the host's defensive systems.

Within two of the major human immunoglobulin classes, there are variants. These variations can be classified as (1) **isotypes,** referring to the variations in the heavy chain constant regions associated with the different classes that are normally present in all individuals (**figure 32.9a**); (2) **allotypes,** the genetically controlled allelic forms of immunoglobulin molecules that are not present in all individuals (figure 32.9b); and (3) **idiotypes,** referring to individual specific immunoglobulin

molecules that differ in the hypervariable region of the Fab portion (figure 32.9c). The many variations of immunoglobulin structure reflect the diversity of antibodies generated by the immune response.

1. What is the variable region of an antibody? The hypervariable or complementarity-determining region? The constant region?
2. What is the function of the Fc region of an antibody? The Fab region?
3. Name the two types of antibody light chains.
4. What determines the class of heavy chain within an antibody?
5. Name the five immunoglobulin classes.
6. Distinguish between isotype, allotype, and idiotype.

Immunoglobulin Function

Each end of the immunoglobulin molecule has a different role. The Fab region is concerned with binding to antigen, whereas the Fc region mediates binding to host tissue, various cells of the im-

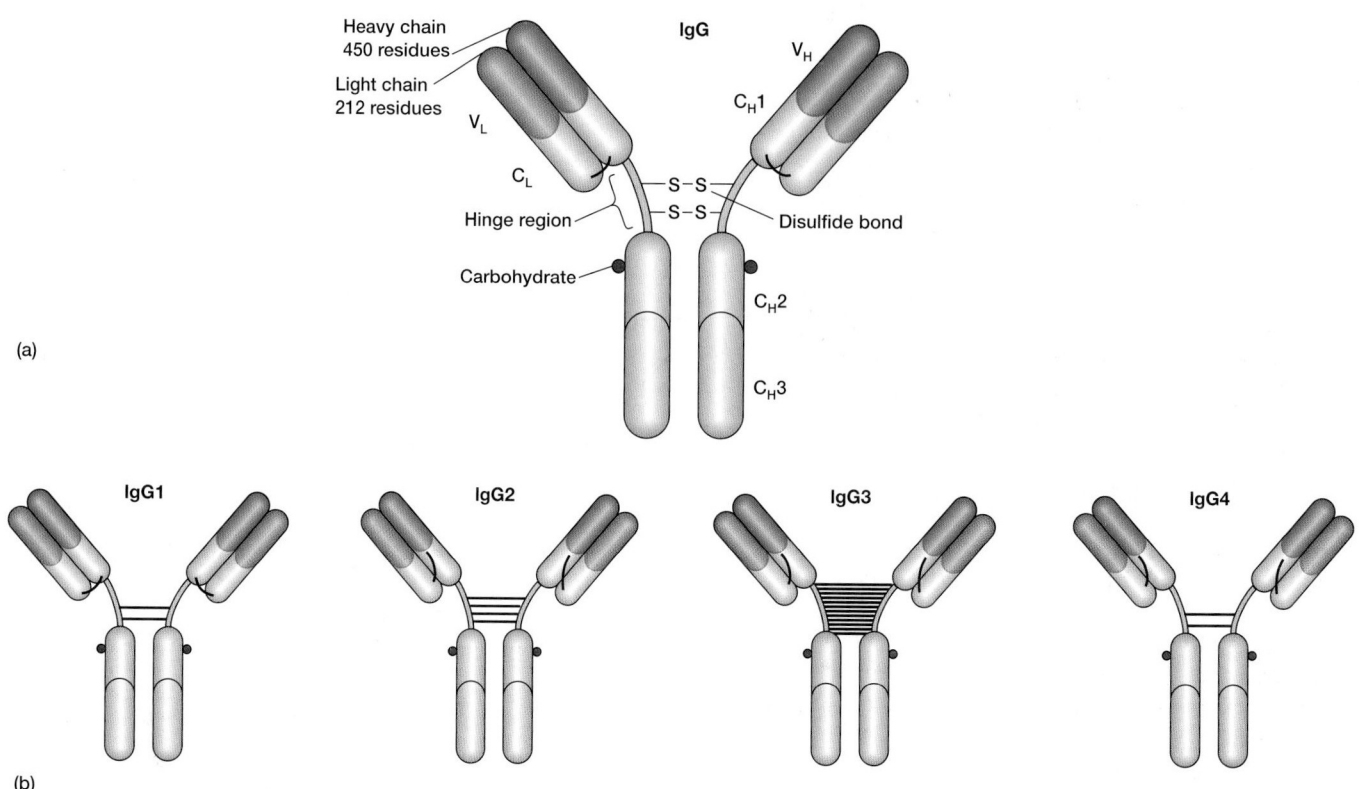

Figure 32.10 Immunoglobulin G. (**a**) The basic structure of human IgG. (**b**) The structure of the four human IgG subclasses. Note the arrangement and numbers of disulfide bonds (shown as thin black lines). Carbohydrate side chains are shown in red.

mune system, some phagocytic cells, or the first component of the complement system. The binding of an antibody with an antigen usually does not cause destruction of the antigen or of the microorganism, cell, or agent to which it is attached. Rather the antibody serves to mark and identify the target for immunologic attack and to activate nonspecific immune responses that can destroy the target. For example, bacteria that are covered with antibodies are better targets for phagocytosis by neutrophils and macrophages. The alteration of the surface of bacteria, viruses, and other particles so that they can be more readily phagocytized is termed opsonization (*see section 31.8*). Immune destruction also is promoted by antibody-induced activation of the classical complement system as discussed in section 32.7.

Immunoglobulin Classes

IgG is the major immunoglobulin in human serum, accounting for 80% of the immunoglobulin pool (**figure 32.10***a*). IgG is present in blood plasma and tissue fluids. The IgG class acts against bacteria and viruses by opsonizing the invaders and neutralizing toxins. It is also one of the two immunoglobulin classes that activate complement by the classical pathway (section 32.7; *see also section 31.7*). IgG is the only immunoglobulin molecule able to cross the placenta and provide natural immunity in utero and to the neonate at birth.

There are four human IgG subclasses (IgG1, IgG2, IgG3, and IgG4) that vary chemically in their heavy chain composition and the number and arrangement of interchain disulfide bonds (figure 32.10*b*). About 65% of the total serum IgG is IgG1, and 23% is IgG2. Differences in biological function have been noted in these subclasses. For example, IgG2 antibodies are opsonic and develop in response to antitoxins. Anti-Rh antibodies are of the IgG1 or IgG3 subclass. IgG1 and IgG3, upon recognition of their specific antigens, bind to receptors expressed on monocytes and macrophages and make them better phagocytes. These receptors are termed Fc receptors. The IgG4 antibodies function as skin-sensitizing immunoglobulins (*see section 33.2*).

IgM accounts for about 10% of the immunoglobulin pool. It is usually a polymer of five monomeric units (pentamer), each composed of two heavy chains and two light chains (**figure 32.11**). The monomers are arranged in a pinwheel array with the Fc ends in the center, held together by disulfide bonds and a special **J** (joining) **chain.** IgM is the first immunoglobulin made during B-cell maturation and is expressed as membrane-bound antibody on B cells. IgM is secreted into serum during a primary antibody response (figure 32.19*b*). Since IgM is so large, it does not leave the bloodstream or cross the placenta. IgM agglutinates bacteria, activates complement by the classical pathway, and enhances the ingestion of pathogens by phagocytic cells.

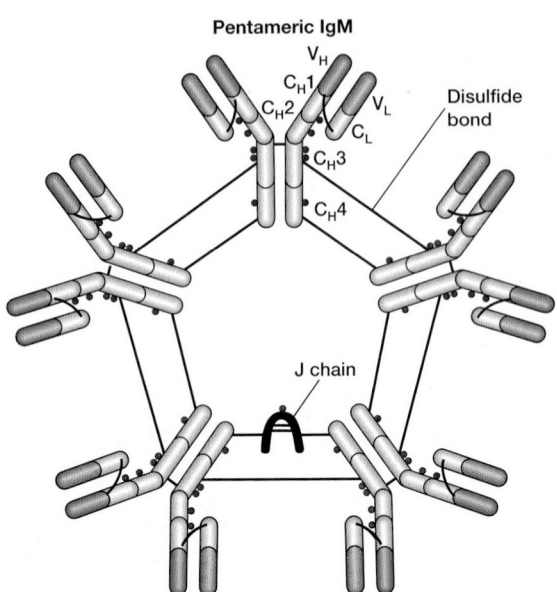

Figure 32.11 Immunoglobulin M. The pentameric structure of human IgM. The disulfide bonds linking peptide chains are shown in black; carbohydrate side chains are in red. Note that 10 antigen-binding sites are present.

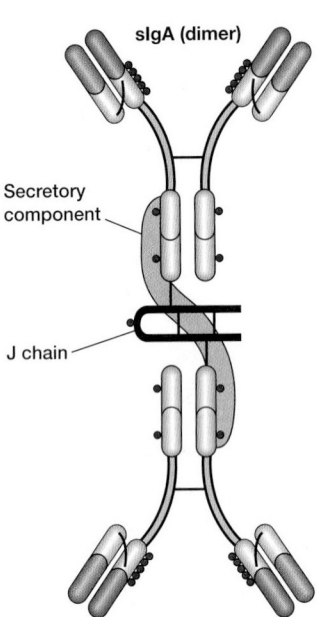

Figure 32.12 Immunoglobulin A. The dimeric structure of human secretory IgA. Notice the secretory component (tan) wound around the IgA dimer and attached to the constant domain of each IgA monomer. Carbohydrate side chains are shown in red.

Although most IgM appears to be pentameric, around 5% or less of human serum IgM exists in a hexameric form. This molecule contains six monomeric units but seems to lack a J chain. Hexameric IgM activates complement up to twentyfold more effectively than does the normal pentameric form. It has been suggested that bacterial cell wall antigens such as gram-negative lipopolysaccharides may directly stimulate B cells to form hexameric IgM without a J chain. If this is the case, the immunoglobulins formed during primary immune responses (figure 32.19a) are less homogeneous than previously thought.

IgA accounts for about 15% of the immunoglobulin pool. Some IgA is present in the serum as a monomer of two heavy and two light chains. Most IgA, however, occurs in mucous secretions as a polymerized dimer held together by a J chain (**figure 32.12**). IgA has special features that are associated with secretory mucosal surfaces. IgA, when transported from the mucosal-associated lymphoid tissue to mucosal surfaces, acquires a protein termed the secretory component. **Secretory IgA (sIgA),** as the modified molecule is now called, is the primary immunoglobulin of mucosal-associated lymphoid tissue (*see figure 31.10*). Secretory IgA is also found in saliva, tears, and breast milk. In these fluids and related body areas, sIgA plays a major role in protecting surface tissues against infectious microorganisms by the formation of an immune barrier. For example, in breast milk sIgA helps protect nursing newborns. In the intestine, sIgA attaches to viruses, bacteria, and protozoan parasites such as *Entamoeba histolytica*. This prevents pathogen adherence to mucosal surfaces and invasion of

host tissues, a phenomenon known as immune exclusion. In addition, sIgA binds to antigens within the mucosal layer of the small intestine; subsequently the antigen-sIgA complexes are excreted through the adjacent epithelium into the gut lumen. This rids the body of locally formed immune complexes and decreases their access to the circulatory system. Secretory IgA also plays a role in the alternative complement pathway (*see section 31.7*).

IgD is an immunoglobulin found in trace amounts in the blood serum. It has a monomeric structure (**figure 32.13**) similar to IgG. IgD antibodies do not fix complement and cannot cross the placenta, but they are abundant in combination with IgM on the surface of B cells and bind antigens, thus signaling the B cell to start antibody production.

IgE (figure 32.14) makes up only a small percent of the total immunoglobulin pool. The classic skin-sensitizing and anaphylactic antibodies belong to this class. IgE molecules have four constant region domains ($C_\epsilon 1$, $C_\epsilon 2$, $C_\epsilon 3$, and $C_\epsilon 4$) on their heavy chains. The Fc portion of the $C_\epsilon 4$ chain can bind to special Fc_ϵ receptors on mast cells and basophils. When two IgE molecules on the surface of these cells are cross-linked by binding to the same antigen, the cells degranulate. This degranulation releases histamine and other pharmacological mediators of anaphylaxis. It also stimulates eosinophilia and gut hypermotility (increased rate of movement of the intestinal contents) that aid in the elimination of helminthic parasites. Thus though IgE is present in small amounts, this class of antibodies has very potent biological capabilities. Anaphylaxis (pp. 744–46)

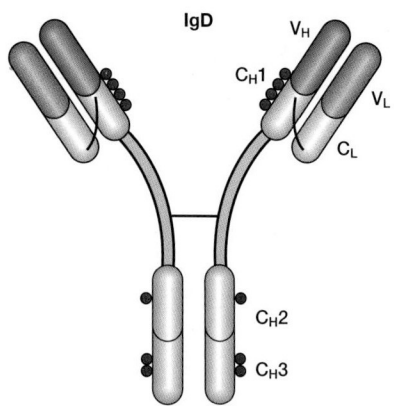

Figure 32.13   Immunoglobulin D. The structure of human IgD. The disulfide bonds linking protein chains are shown in black; carbohydrate side chains are in red.

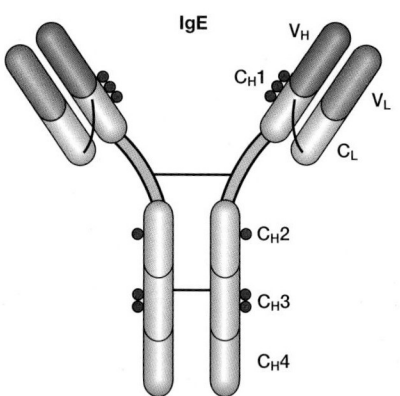

Figure 32.14   Immunoglobulin E. The structure of human IgE.

Table 32.2

Physicochemical Properties of Human Immunoglobulin Classes

	Immunoglobulin Classes				
Property	**IgG[a]**	**IgM**	**IgA[b]**	**IgD**	**IgE**
Heavy chain	γ_1	μ	α_1	δ	ε
Mean serum concentration (mg/ml)	9	1.5	3.0	0.03	0.00005
Percent of total antibody	80–85	5–10	5–15	<1	0.002–0.05
Valency	2	5(10)	(2–4)	2	2
Mass of heavy chain (kDa)	51	65	56	70	72
Mass of entire molecule (kDa)	146	970	160[c]	184	188
Placental transfer	+	–	–	–	–
Half-life in serum (days)[d]	23	5	6	3	2
Complement activation					
Classical pathway	++	+++	–	–	–
Alternative pathway	–	–	+	–	–
Induces mast cell degranulation	–	–	–	–	+
Major characteristics	Most abundant Ig in body fluids; neutralizes toxins, opsonizes bacteria, activates complement, maternal antibody	First to appear after antigen stimulation; very effective agglutinator; expressed as membrane-bound antibody on B cells	Secretory antibody; protects external surfaces	Present on B-cell surface; B-cell recognition of antigen	Anaphylactic-mediating antibody: resistance to helminths
% carbohydrate	3	7–10	7	12	11

[a]Properties of IgG subclass 1.

[b]Properties of IgA subclass 1.

[c]sIgA = 360 – 400 kDa

[d]Time required for half of the antibodies to disappear.

Table 32.2 summarizes some of the more important physicochemical properties of the human immunoglobulin classes.

1. What does antigen binding accomplish in a host?
2. Give the major functions of each immunoglobulin class.
3. Why is the structure of IgG considered the model for all five immunoglobulin classes?
4. Which immunoglobulin can cross the placenta?
5. Which immunoglobulin is most prevalent in the immunoglobulin pool? The least prevalent?

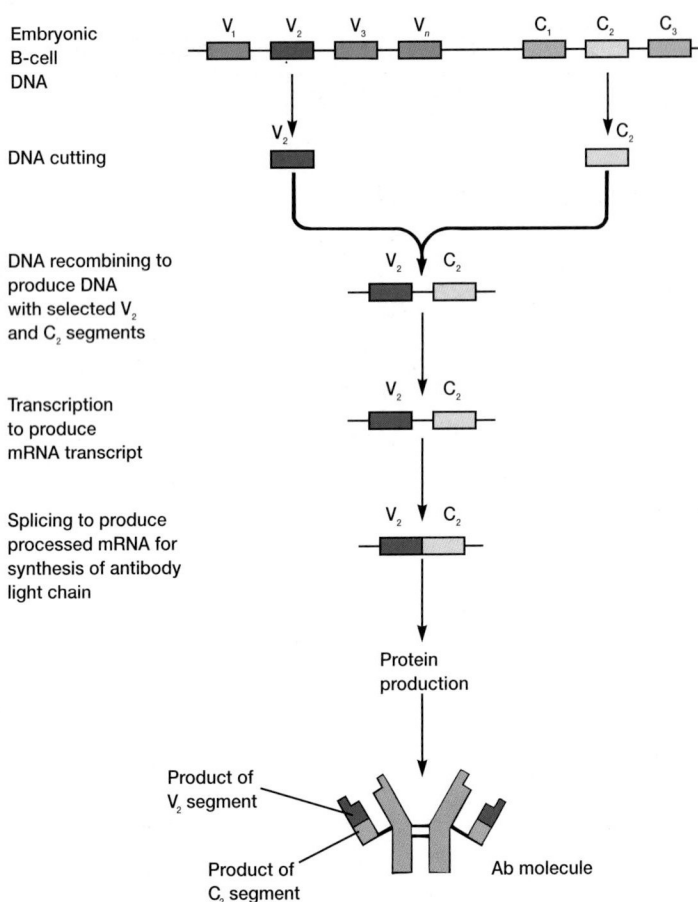

Figure 32.15   Gene Shuffling and Antibody Diversity. Antibody diversity is partly the result of the shuffling of gene sequences that code for both heavy and light chains. This drawing shows the shuffling, cutting, and splicing process used to produce an assembled light chain of the antibody molecule.

Diversity of Antibodies

One unique property of antibodies is their remarkable diversity. According to current estimates each human can synthesize more than 10^{11} (100 billion) different kinds of antibodies. How is this diversity generated? The answer is threefold: rearrangement of antibody gene segments, somatic mutations, and generation of different codons during antibody gene splicing.

Immunoglobulin genes are split or interrupted genes with many gene segments (*see section 12.1*). Embryonic B cells contain a small number of gene segments, close together on the same chromosome, that determine the constant (C) region of the light chains (**figure 32.15**). Separated from them, but on the same chromosome, is a larger cluster of segments that determines the variable (V) region of the light chains. During B-cell differentiation, one segment for the constant region is joined by a process of recombination mediated by specific proteins to one segment for the variable region. This splicing produces a complete light-chain antibody gene. A similar splicing mechanism also occurs to join the constant and variable segments of the heavy chains.

Because the light-chain genes actually consist of three parts, and the heavy-chain genes consist of four, the formation of a finished antibody molecule is slightly more complicated than previously outlined (**figure 32.16**). The germ line DNA for the light-chain gene contains multiple coding sequences called V and J (joining) regions. During the differentiation of a B cell, a deletion (which is variable in length) occurs that joins one V gene segment with one J segment. This DNA joining process is termed combinatorial joining because it can create many combinations of the V and J regions. When the light-chain gene is transcribed, transcription continues through the DNA region that encodes for the constant portion of the gene. RNA splicing subsequently joins the VJ and C regions, creating mRNA.

Combinatorial joining in the formation of a heavy-chain gene occurs by means of DNA splicing of the heavy-chain counterparts of V and J along with a third set of D (diversity) sequences (**figure 32.17a**). Initially, all heavy chains have the μ

Figure 32.16   Light Chain Production in a Mouse. One V segment is randomly joined with one J-C region by deletion of the intervening DNA. The remaining J segments are eliminated from the RNA transcript during RNA processing. An intron is a segment of DNA occurring between expressed regions of genes.

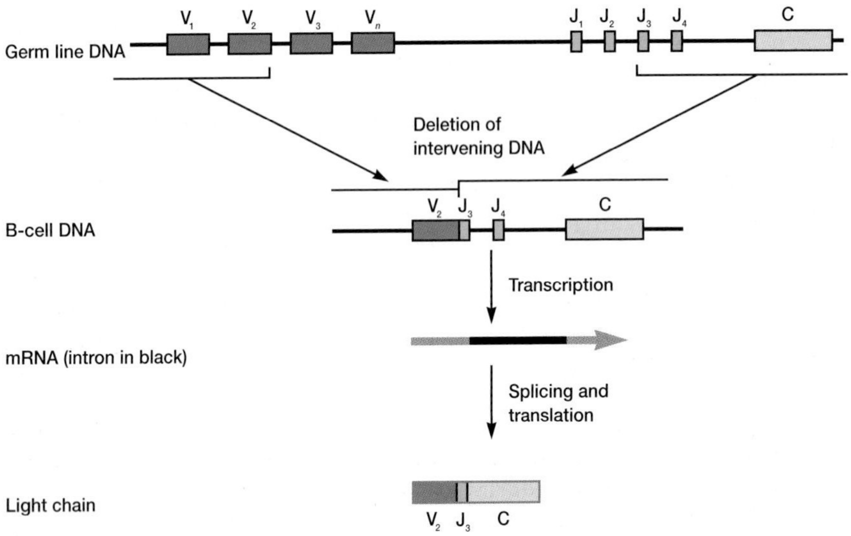

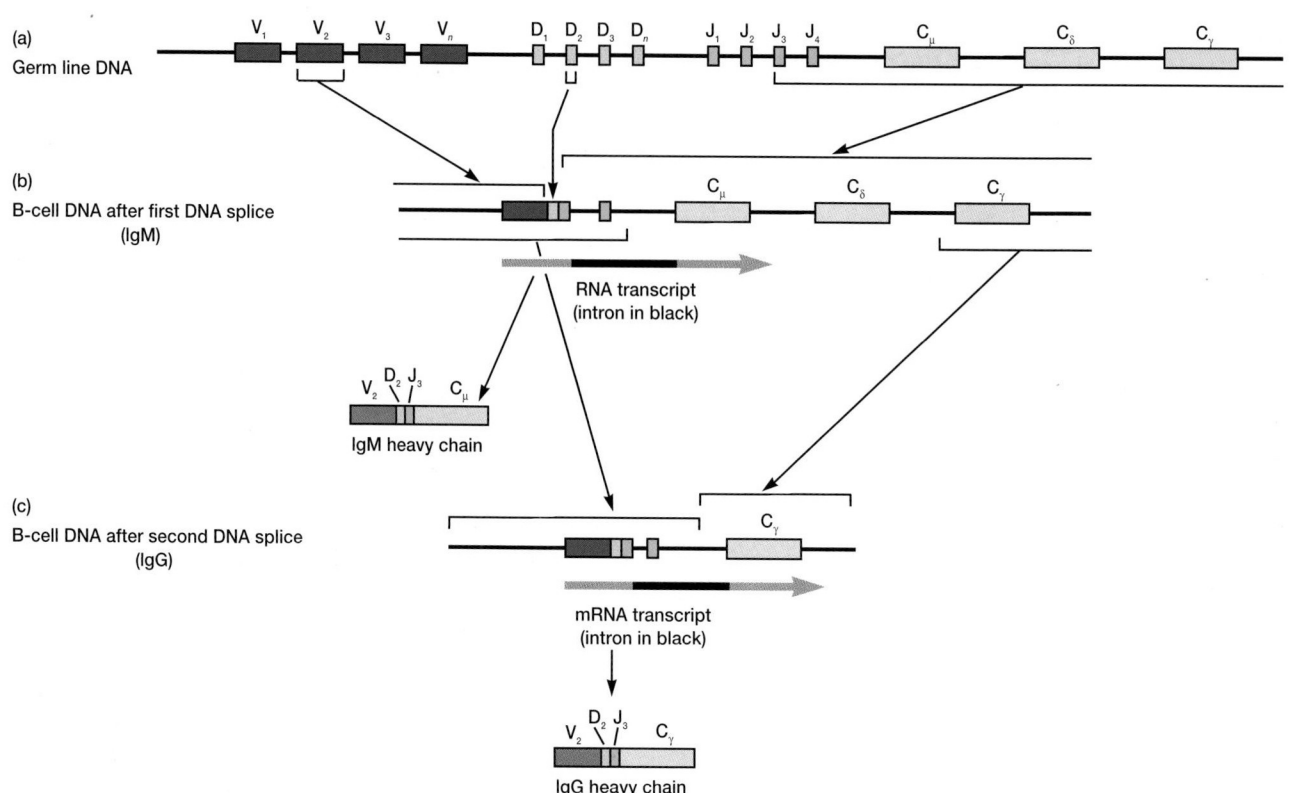

Figure 32.17 **The Formation of a Gene for the Heavy Chain of an Antibody Molecule.** See text for further details.

type of constant region. This corresponds to antibody class IgM (figure 32.17*b*). Another DNA splice joins the VDJ region with a different constant region that can subsequently change the class of antibody produced by the B cell (figure 32.17*c*).

The amount of antibody diversity in the mouse that can be generated by combinatorial joining is shown in **table 32.3**. In this animal the κ light chains can be formed from any combination of about 250 to 350 V_κ and 4 J_κ regions giving a maximum of approximately 1,400 different κ chains. The λ chains have their own V_λ and J_λ regions but smaller in number than their κ counterparts (6 different λ chains). The heavy chains have approximately 250 to 1,000 V_H, 10 to 30 D, and 4 J_H regions, giving a maximum of 120,000 different combinations. Because any light chain can combine with any heavy chain, there will be a maximum of 2×10^8 possible κ chain antibody types.

The value of 2×10^8 different antibodies is actually an underestimate because antibody diversity is further augmented by two processes:

1. The V regions of germ line DNA are susceptible to a high rate of somatic mutation during B-cell development in response to an antigen challenge. These mutations allow B-cell clones to produce antibodies with somewhat different polypeptide sequences.
2. The junction for either VJ or VDJ splicing in combinatorial joining can occur between different nucleotides and thus generate different codons in the spliced gene. For example, one

Table 32.3	
Number of Antibodies Possible through the Combinatorial Joining of Mouse Germ Line Genes[a]	
λ light chains	V regions = 2
	J regions = 3
	Combinations = $2 \times 3 = 6$
κ light chains	V_κ regions = 250–350
	J_κ regions = 4
	Combinations = $250 \times 4 = 1,000$
	$= 350 \times 4 = 1,400$
Heavy chains	$V_H = 250$–1,000
	D = 10–30
	$J_H = 4$
	Combinations = $250 \times 10 \times 4 = 10,000$
	$= 1,000 \times 30 \times 4 = 120,000$
Diversity of antibodies	κ-containing: $1,000 \times 10,000 = 10^7$
	$1,400 \times 120,000 = 2 \times 10^8$
	λ-containing: $6 \times 10,000 = 6 \times 10^4$
	$6 \times 120,000 = 7 \times 10^5$

[a]Approximate values.

VJ splicing event can join the V sequence CCTCCC with the J sequence TGGTGG in two ways:

$$CCTCCC + TGGTGG = CCGTGG,$$

that codes for the amino acids proline and tryptophan; and

$$CCTCCC + TGGTGG = CCTCGG,$$

that codes for proline and arginine. Thus the same VJ joining could produce polypeptides differing in a single amino acid.

---

1. How many chromosomes encode for antibody production in humans?
2. What is the name of each gene segment set that encodes for the different regions of antibody chains?
3. Describe what is meant by combinatorial joining of VJ and C regions of a chromosome.
4. In addition to combinatorial joining, what other two processes play a role in antibody diversity?

Specificity of Antibodies

As noted previously, combinatorial joinings, somatic mutations, and variations in the splicing process generate the great variety of antibodies produced by mature B cells. From a large, diverse B-cell pool, specific cells are stimulated by antigens to reproduce and form a B-cell clone whose cells contain the same genetic information. This is known as the **clonal selection theory,** a hypothesis to explain immunologic specificity and memory.

The existence of a B-cell **clone** (a population of cells derived asexually from a single parent) that can respond to an antigen by producing the correct antibody is the first tenet of this theory. The lymphoid system is thus considered to contain many B-cell clones, each clone able to recognize a specific antigen. The antigen selects the appropriate B cell to form a clone (hence the phrase "clonal selection"), and the cells with other antigen specificities are unaffected.

According to the second tenet, each B cell is genetically programmed to respond to its own distinctive antigen before the antigen is introduced. The particular antibody an individual B cell can produce is integrated into the plasma membrane of that B cell and acts as a specific surface receptor for the corresponding antigen molecule. The reaction of the antibody and antigen initiates the differentiation and multiplication of the B cell to form two different cell populations: plasma cells and **memory B cells (figure 32.18).**

Plasma cells are literally protein factories that produce about 2,000 antibodies per second in their brief five- to seven-day life span. Memory B cells can initiate the antibody-mediated immune response upon detecting the particular antigen specific for their antibody molecules. These memory cells circulate more actively from blood to lymph and live much longer (years or even decades) than plasma cells. Memory cells are responsible for the immune system's rapid secondary antibody response (figure 32.19) to the same antigen. Finally, memory B cells and plasma cells are usually not produced unless the B cell has interacted with, and received cytokine signals from, activated helper T cells (figure 32.1).

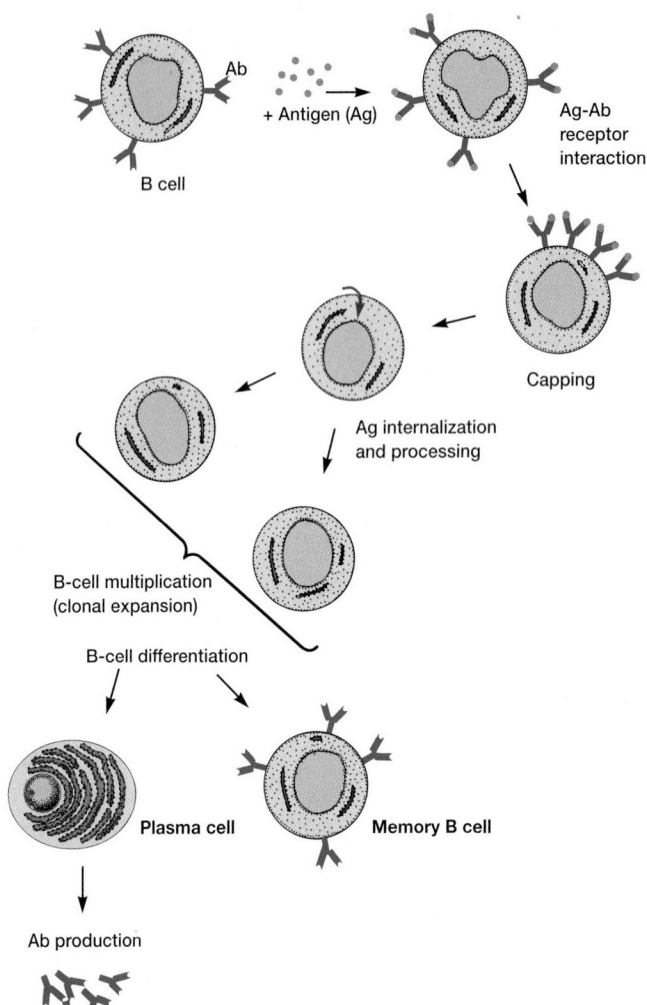

Figure 32.18   Clonal Selection.   The immune system can respond specifically to a myriad of possible antigens, whether they are individual molecules or are attached to pathogens and abnormal cells such as cancer cells. B cells or B lymphocytes constantly roam the body, particularly the blood and lymphoid tissues. Each B cell synthesizes only one of the millions of possible antibodies and displays this antibody on its surface. When the antigen meets a B cell having a surface antibody of the proper specificity (top left), it complexes with the antibody and capping occurs. (Capping is the regional aggregation of antibodies on the surface of the cell following Ag-Ab interaction.) The antigen is then internalized; the B cell swells and begins to divide rapidly, producing a B-cell clone. The activated B-cell clone differentiates into plasma cells and memory cells. Plasma cells form the specific antibody that immediately attacks the antigen that provoked its formation. Memory B cells persist in the body and boost the immune system's readiness to eliminate the same antigen if it presents itself in the future. Most antibody responses also require signals from T-helper cells (not shown).

Sources of Antibodies

The need for pure homogeneous antibodies has increased dramatically in recent years. Currently antibodies are produced either naturally by immunization or artificially through hybridoma formation.

Immunization

Specific antibodies can be produced naturally by the immunization of domestic animals or human volunteers. Whatever the source, purified antigen is injected into the host. The host's immune system recognizes and responds to the antigen, and its B cells proliferate and differentiate to produce specific antibodies. To promote the efficiency of antigen stimulation of antibody production, the antigen may be mixed with an **adjuvant** [Latin *adjuvans,* aiding], which enhances the rate and quantity of antibody produced. Following repeated antigen injections at regular intervals, blood is withdrawn from the host and allowed to clot. The fluid that remains after the blood clots is the **serum** (Latin "whey"). Because this serum has been obtained from an immunized host and contains the desired antibodies, it is called **antiserum.**

Although antiserum is a major and convenient source of antibodies, its usefulness is limited in three ways:

1. Antibodies obtained by this method are polyclonal; they are produced by several B-cell clones and have different specificities. This decreases their sensitivity to particular antigens and results in some degree of cross-reaction with closely related antigen molecules.
2. Second or repeated injections of antiserum from one species into another can cause serious allergic or hypersensitivity reactions.
3. Antiserum contains a mixture of antibodies (figure 32.6), not all of which are of interest in a given immunization.

The Primary Antibody Response

During immunization procedures (and also in naturally acquired immunity), there is an initial lag phase of several days following a primary challenge with an antigen. During the lag phase no antibody can be detected in the blood (**figure 32.19***a*). The antibody **titer,** which is the reciprocal of the highest dilution of an antiserum that gives a positive reaction in the test being used, rises logarithmically to a plateau during the second or log phase. In the plateau phase the antibody titer stabilizes. This is followed by a decline phase, during which antibodies are naturally metabolized or bound to the antigen and cleared from the circulation. During the primary antibody response, IgM appears first, then IgG (figure 32.19*b*). The affinity of the antibodies for the antigen's determinants is low to moderate during this primary antibody response.

The Secondary Antibody Response

The primary antibody response primes the immune system so that it possesses specific immunologic memory through its clones of memory B cells. Upon secondary antigen challenge (figure 32.19*b*), the B cells mount a heightened secondary or **anamnes-**

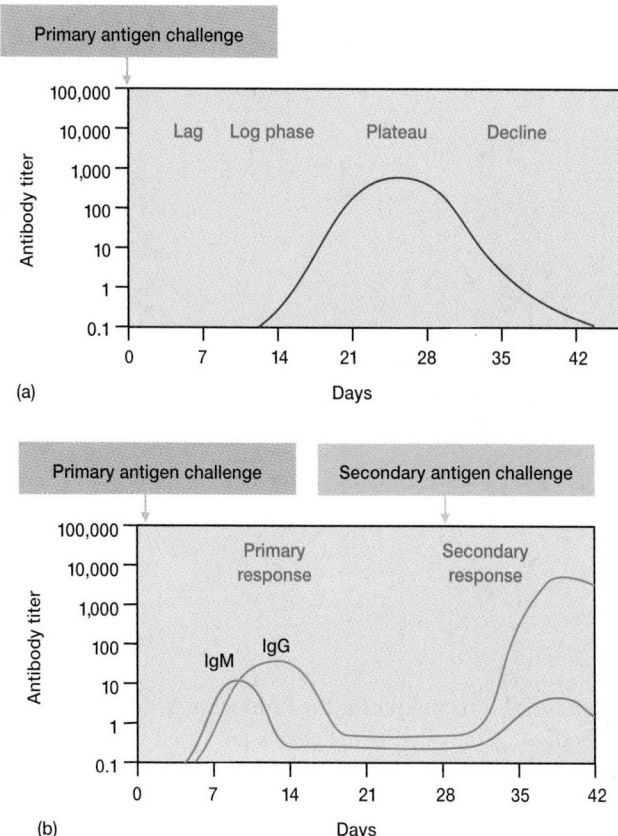

Figure 32.19 Primary and Secondary Antibody Responses. (**a**) The four phases of a primary antibody response. (**b**) Comparison of primary and secondary antibody responses. Note that the secondary response to the antigen is both faster and greater than the primary response. T-cell–mediated responses exhibit similar immunologic memory.

tic [Greek *anamnesis,* remembrance] **response** to the same antigen. Compared to the primary antibody response, the secondary antibody response has a shorter lag phase, a more rapid log phase, persists for a longer plateau period, attains a higher IgG titer, and produces antibodies with a higher affinity for the antigen (affinity maturation through somatic hypermutation).

Hybridomas

The limitations of antiserum as a source of antibodies have been overcome with the development of hybridoma techniques to manipulate and culture various mammalian cells that synthesize antibodies in vitro. Each cell and its progeny normally produce a **monoclonal antibody (MAb)** of a single specificity.

The methodology of one of these techniques is illustrated in **figure 32.20.** Animals (usually mice or rats) are immunized with antigens as discussed previously. Once the animals are producing a large quantity of antibodies, their spleens are removed and antibody-producing B and plasma cells (lymphocytes) isolated. These lymphocytes are then fused with myeloma cells by the

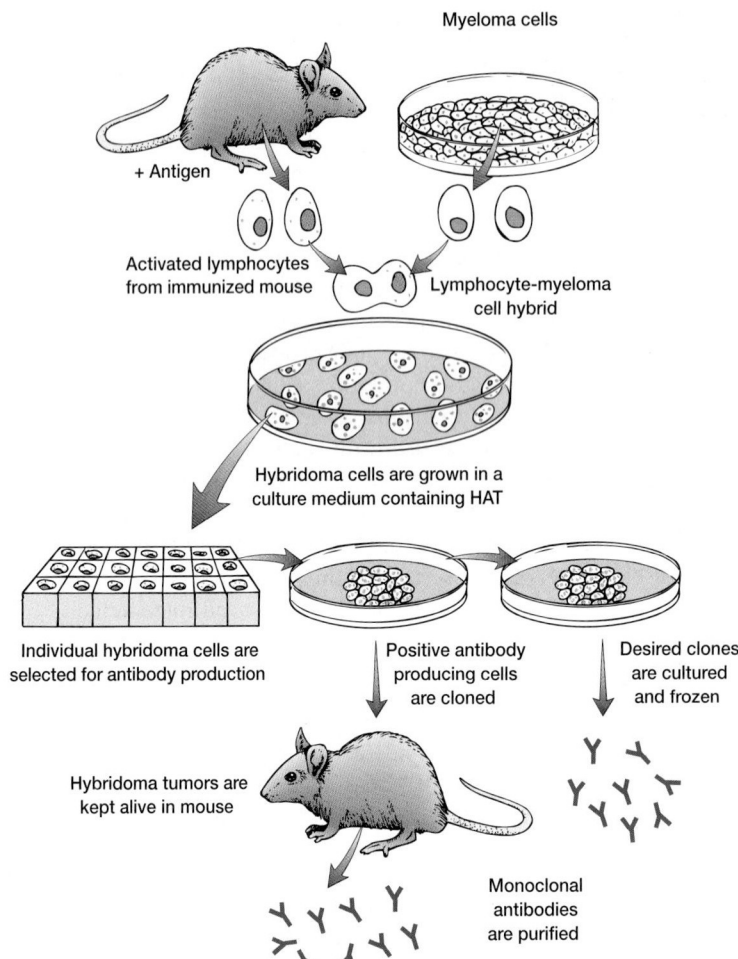

Myeloma cells

+ Antigen

Activated lymphocytes from immunized mouse

Lymphocyte-myeloma cell hybrid

Hybridoma cells are grown in a culture medium containing HAT

Individual hybridoma cells are selected for antibody production

Positive antibody producing cells are cloned

Desired clones are cultured and frozen

Hybridoma tumors are kept alive in mouse

Monoclonal antibodies are purified

Figure 32.20 Technique for the Production of Monoclonal Antibodies. Activated lymphocytes are fused with special mutant myeloma cells, yielding hybridomas. Each of them secretes a single, "monoclonal" antibody. Once the hybridoma secreting the desired antigen is identified, it is cloned to generate many antibody-secreting cells that yield the huge quantity of a single antibody needed in medicine or science. Some hybridoma cells may be stored frozen and later cloned for antibody production or kept alive in laboratory animals.

addition of polyethylene glycol, which promotes membrane fusion. **Myeloma cells** are cancerous plasma cells that can readily be cultivated; mutant myeloma cells incapable of producing immunoglobulins are used. These fused cells, derived from lymphocytes and myeloma cells, are called **hybridomas** (they are hybrids of the two cells).

The fusion mixture is then transferred to a culture medium containing a combination of **h**ypoxanthine, **a**minopterin, and **t**hymidine (HAT). Aminopterin is a poison that blocks a specific metabolic pathway in cells. Myeloma cells lack an enzyme that allows their growth in the presence of aminopterin. However, the pathway is bypassed in lymphoid cells provided with the intermediate metabolites hypoxanthine and thymidine. As a result the hybridomas grow in the HAT medium but the myeloma cells die because they have a metabolic defect and cannot employ the bypass or salvage pathway.

When the culture is initially established using the HAT medium, it contains lymphocytes, myeloma cells, and hybridomas. The unfused lymphoid cells die naturally in culture within a week or two, and the myeloma cells die in the HAT as just described. In contrast, the fused cells survive because they have the immortality of the myeloma and the metabolic bypass of the lymphoid cells.

Hybridomas that have the antibody-producing capacity of the original lymphoid cells are randomly placed in culture wells. The wells are individually tested for production of the desired antibody, and, if positive, the cells within the well provide clones of immortal cells, all producing the same monoclonal antibody.

Monoclonal antibodies currently have many applications. For example, they are routinely used in the typing of tissue, in the identification and epidemiological study of infectious microorganisms, in the identification of tumor and other surface antigens, in the classification of leukemias, and in the identification of functional populations of different types of T cells. Anticipated future uses include (1) passive immunizations against infectious agents and toxic drugs, (2) tissue and organ graft protection, (3) stimulation of tumor rejection and elimination, (4) manipulation of the immune response, (5) preparation of more specific and sensitive diagnostic procedures, and (6) delivery of antitumor agents (immunotoxins) to tumor cells (**Techniques & Applications 32.1**).

---

1. Describe the major two tenets of the clonal selection theory.
2. What two cell populations are produced by a B cell after it initially reacts with an antigen?

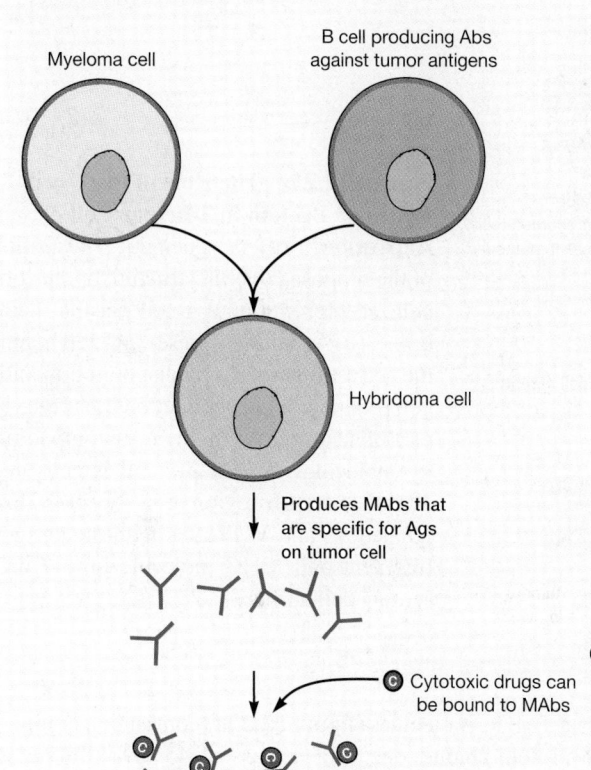

Myeloma cell

B cell producing Abs
against tumor antigens

Hybridoma cell

Produces MAbs that
are specific for Ags
on tumor cell

Ⓒ Cytotoxic drugs can
be bound to MAbs

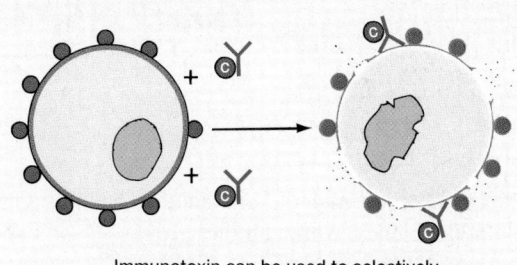

Immunotoxin can be used to selectively
bring about destruction of tumor cells

One result of hybridoma research is the production of **immunotoxins** (see **Box figure**). Immunotoxins are monoclonal antibodies that have been attached to a specific toxin or toxic agent (antibody + toxin = immunotoxin). Immunotoxins kill target cells and no others, because the antibody binds specifically to plasma membrane surface antigens found only on the target cells. This approach is being used to treat certain types of cancer.

In this procedure cancer cells from a person are injected into mice or rats to stimulate the production of specific antibodies against their plasma membrane antigens. Monoclonal antibodies are produced using hybridomas, purified, and attached to an agent toxic to the cancer cells.

When the immunotoxin is given to a cancer patient, it circulates through the body and binds only to the cancer cells that have the appropriate surface antigens. After binding to the surface, the immunotoxin is taken into cancer cells by receptor-mediated endocytosis, and released inside. The immunotoxin then interferes with the metabolism of the target cells and kills them. Although this procedure is still experimental, it holds great promise in the treatment of certain types of cancer.

---

3. What is the function of an adjuvant?
4. Describe the difference between serum and antiserum.
5. What is a hybridoma? How is it made?

 32.4   **T-CELL BIOLOGY**

T cells are major players in the cell-mediated immune response (figure 32.1), and also have a major role in B-cell activation. They are immunologically specific, can carry a vast repertoire of immunologic memory, and can function in a variety of regulatory and effector ways.

T-Cell Receptors

T cells have specific **T-cell receptors (TCRs)** for antigens on their plasma membrane surface. The receptor site is composed of two parts, an alpha polypeptide chain and a beta polypeptide chain (**figure 32.21***a*). Each chain is stabilized by disulfide bonds. The re-

ceptor is anchored in the plasma membrane and parts of the α and β chains extend into the cytoplasm. The recognition sites of the T-cell receptor extend out from the membrane and have a terminal variable section complementary to antigen fragments. T cells respond to antigen fragments exposed on the surfaces of **antigen-presenting cells (APCs),** most of which are macrophages, dendritic cells, and B cells. These cells scavenge foreign materials (antigens or pathogens) from the blood and tissues and digest them. Inside the cells small pieces of the antigens bind to a special class of proteins called major histocompatibility complex (MHC) class II molecules (section 32.5). The MHC-antigen complex is then transported to the surface of the presenting cell and displayed to passing lymphocytes known as T-helper cells (figure 32.21*b*). Some knowledge of these histocompatibility molecules is thus required before T-cell–macrophage interactions and T-cell biology can be understood.

Major Histocompatibility Complex (MHC)

A woman dies because her body has rejected a kidney transplant; a man is crippled with rheumatoid arthritis; an African child goes

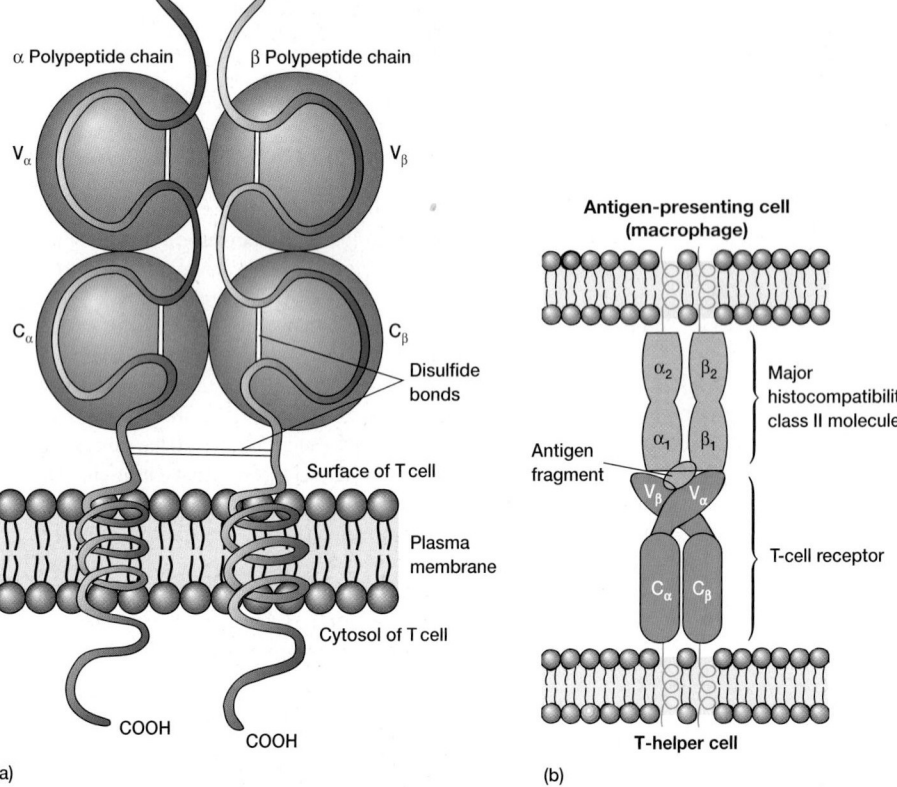

α Polypeptide chain

β Polypeptide chain

V_α

V_β

C_α

C_β

Disulfide bonds

Surface of T cell

Plasma membrane

Cytosol of T cell

COOH

COOH

(a)

Antigen-presenting cell (macrophage)

α_2 β_2

Major histocompatibility class II molecule

Antigen fragment

α_1 β_1

V_β V_α

T-cell receptor

C_α C_β

T-helper cell

(b)

Figure 32.21 The Role of the T-cell Receptor Protein in T-helper Cell Activation. (**a**) A schematic illustration of the proposed overall structure of the antigen receptor site on a T-cell plasma membrane. (**b**) An antigen-presenting cell begins the activation process by displaying an antigen fragment (e.g., peptide) on its surface as part of a complex with the histocompatibility molecules. A T-helper cell is activated after the variable region of its receptor (designated V_α and V_β) reacts with the antigen fragment and MHC molecule on the presenting cell surface.

into a coma that is brought on by cerebral malaria; and another child dies because of an immunodeficiency. Although these four clinical situations are diverse, they all have one thing in common: the cause of all of them involves the human version of the major histocompatibility complex. Malfunction of this system, which is at the basis of these situations, has such wide-ranging effects not only because of the system's role in the specific immune response, but also because of its genetic complexity.

The **major histocompatibility complex (MHC)** is a collection of genes on chromosome 6 in humans and chromosome 17 in mice. This term is derived from the Greek word for tissue [*histo*] and the ability to get along [compatibility]. The MHC is called the **human leukocyte antigen (HLA) complex** in humans and **H-2 complex** in mice. Almost all human tissue cells contain MHC molecules on their plasma membranes. MHC molecules can be divided into three classes: class I molecules are found on almost all types of nucleated body cells; class II molecules appear only on those leukocytes involved in T helper cell-related immune responses (macrophages, dendritic cells, and B cells); and class III molecules include various secreted proteins that have immune functions (complement components C2, C4a, and factor B), two steroid 21-hydroxylase enzymes (21-OHA and 21-OHB), the inflammatory cytokines, tumor necrosis factors α and β, and two heat-shock proteins. Unlike class I and II MHC molecules, the class III molecules are not membrane proteins, are not related to class I or II molecules, and have no role in antigen presentation. Thus they will not be discussed further.

Class I MHC molecules (**figure 32.22a**) consist of a complex of two protein chains, one with a mass of 45 kDa (the heavy chain) and the other with a mass of 12 kDa (β_2-microglobulin). The two chains contain four regions. The outer segment of the heavy chain can be divided into three functional domains, designated α_1, α_2, and α_3. The β_2-microglobulin (β_2m) protein and α_3 segment of the heavy chain are noncovalently associated with one another and are close to the plasma membrane. A small segment of the heavy chain is attached to the membrane by a short amino acid sequence that extends into the cell interior, but the rest of the protein protrudes to the outside. The α_1 and α_2 domains lie to the outside and form the antigen-binding pocket.

Class II MHC molecules are also transmembrane proteins consisting of α and β chains of mass 34 kDa and 28 kDa, respectively. Both chains are folded to give two domains.

Although MHC class I and class II molecules are structurally distinct, both fold into very similar shapes (figure 32.22b,c). Each MHC molecule has a deep groove into which a short peptide can bind. Because this peptide is not part of the MHC molecule, it can vary from one MHC molecule to the next. On healthy cells all of these peptides come from self-proteins. The presence of foreign peptides (antigen fragments) in the MHC groove alerts the immune system and activates T cells, which in turn activate macrophages.

Class I and class II molecules present peptides that arise in different places within cells as the result of a process known as antigen processing. Class I molecules bind to peptides that originate in the cytoplasm (e.g., antigens from replicating viruses).

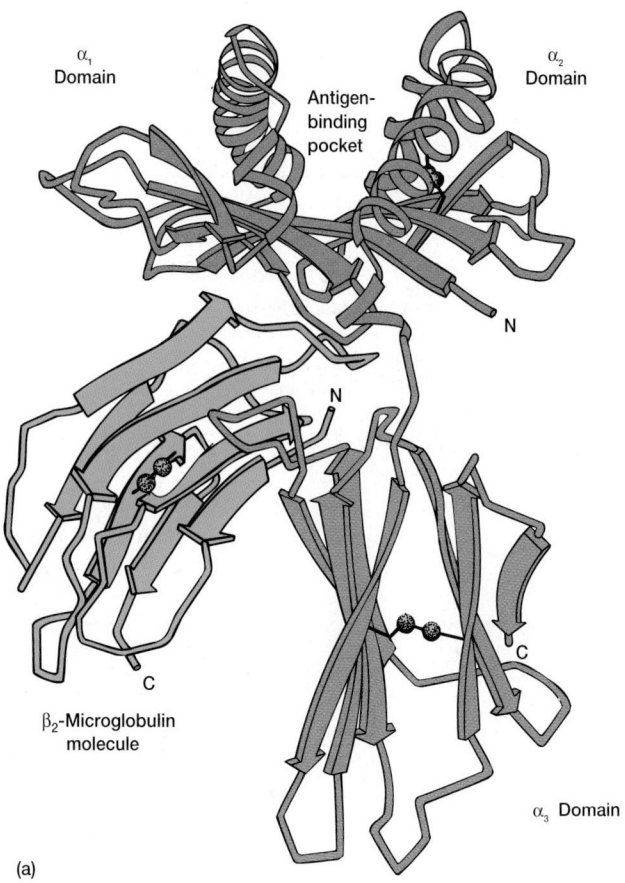

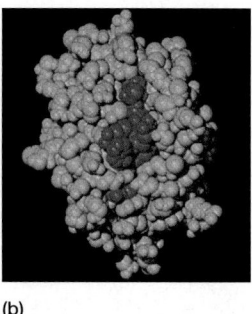

(b)

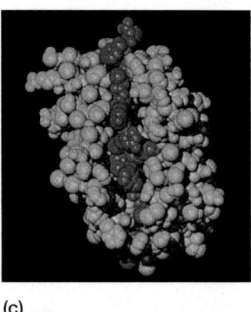

(c)

Figure 32.22   Class I Major Histocompatibility Molecule Structure. **(a)** This side view of the molecule shows the arrangement of its three domains (α_1, α_2, α_3) and the β_2-microglobulin molecule. **(b)** Class I MHC molecules (space-filling model) can hold only short peptides (shown in blue) because the binding site is closed off. **(c)** Since the binding site is open on both ends, class II MHC molecules (space-filling model) can bind to peptides (shown in blue) of different lengths.

Endogenous antigenic proteins are digested inside the cell as part of the natural process by which a cell continually renews its protein contents. The short peptide fragments that result from this process are pumped by a specific transporter protein from the cytoplasm into the endoplasmic reticulum. Within the endoplasmic reticulum the class I MHC alpha chain is synthesized and associates with β_2-microglobulin. This dimer appears to bind peptide as soon as it enters the endoplasmic reticulum. The MHC molecule and peptide are then carried to, and anchored in the plasma membrane. If the peptide is foreign (e.g., short pieces of viral protein), a passing CD8$^+$ T cell (cytotoxic T lymphocyte) whose T-cell receptor is specific for the antigenic peptide will bind to the peptide-MHC complex and ultimately kill the infected cell.

Class II MHC molecules bind to fragments that arise from exogenous antigens. This pathway would function with bacteria and viruses that have been taken up endocytotically. The antigen-presenting cell takes in the antigen or pathogen by receptor-mediated endocytosis or phagocytosis and produces antigen fragments by digestion in the phagolysosome. Fragments then combine with preformed class II MHC molecules and are delivered to the cell surface. It is here that the peptide is recognized by

CD4$^+$ T-helper cells. Unlike CD8$^+$ T cells, CD4$^+$ T cells do not directly kill target cells. Instead they respond in two distinct ways. One is to enlarge and divide, thereby increasing the number of CD4$^+$ cells that can react to the antigen. The second response is to secrete cytokines (e.g., interleukin-2) that either directly inhibit the pathogen that produced the antigen or recruit and stimulate other cells to join in the immune response.

As previously noted, MHC molecules are coded by a group of genes termed the major histocompatibility complex (**figure 32.23**). The MHC of humans is located on chromosome 6 and contains three major classes of genes. Many forms of MHC genes exist because of the presence of multiple alleles that have arisen by high gene mutation rates, gene recombination, and other mechanisms. Each individual has two sets of these genes, one from each parent, and both are expressed (i.e., they are codominant). Thus a person expresses many different MHC products. The MHC proteins will differ between individuals; the closer two people are related, the more similar are their MHC molecules.

Class I molecules are made by all cells of the body except red blood cells and comprise MHC types A, B, and C. Class I molecules serve to identify almost all cells of the body as "self." They also

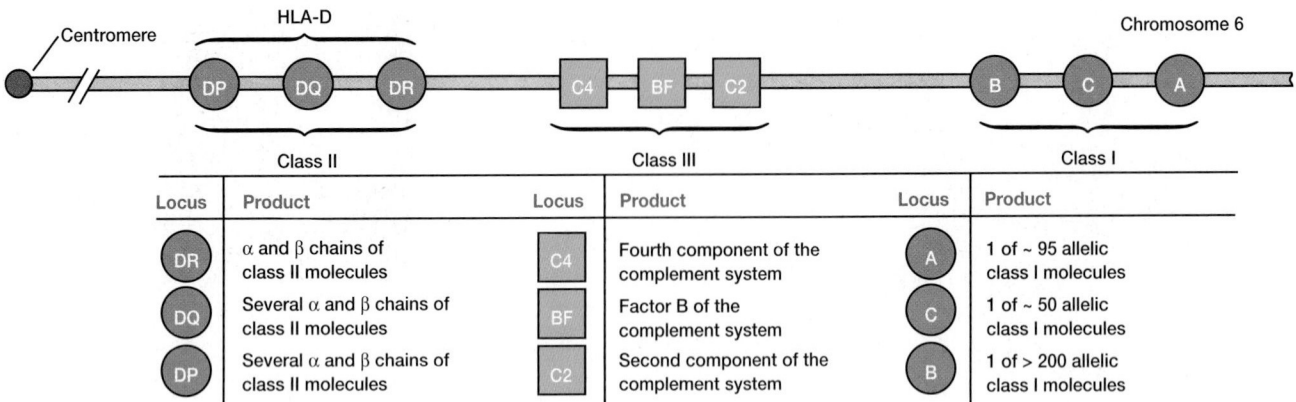

Figure 32.23 Major Histocompatibility Complex. The MHC region of human chromosome 6 and the gene products associated with each locus.

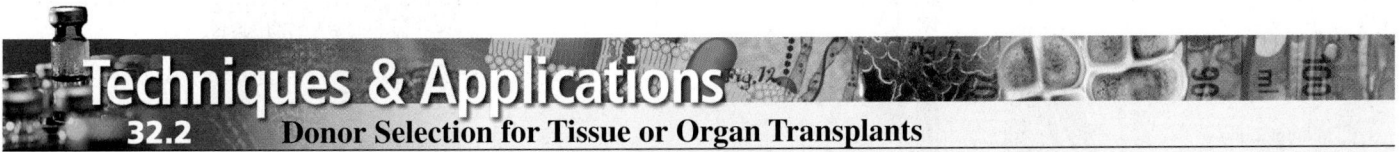

32.2 Donor Selection for Tissue or Organ Transplants

The likelihood of tissue or organ transplant acceptance can be increased by using persons who have a great degree of genetic similarity. The greater the similarity, the more likely the persons will have similar MHC complexes. The reason for this is that 77 histocompatibility alleles are determined by just four human histocompatibility genes—A, B, C, and D (which includes DQ, DR, and DP). These genes are transmitted from parents to offspring according to Mendelian genetics. For example, two siblings (brother and sister) may have approximately a 25% chance of being B identical, a 50% chance of sharing one B gene, and only a 25% chance of having completely different B genes. Therefore, when tissue or organ transplants are performed, a deliberate attempt is made to use tissues from siblings or other genetically related, histocompatible people.

Once a tissue donor has been selected, various in vitro tests are performed to find how closely the MHCs match. The greater the degree of similarity, the greater the chances are that the transplant will be successful. One such test is the lymphocytotoxic cross-match test. In this test the recipient's serum is mixed with the donor's lymphocytes to determine whether antibodies are present that will cause tissue rejection. The cell-mediated lymphocytosis test, in which recipient leukocytes are mixed with donor serum, is used to test for acute rejection. After a prolonged incubation period, the presence of activated killer cells is an indication that rejection will occur.

stimulate antibody production when introduced into a host with different class I molecules. This is the basis for MHC typing when a patient is being prepared for an organ transplant (**Techniques & Applications 32.2**).

Class II molecules comprise the D group of the MHC and are produced only by activated macrophages, dendritic cells, mature B cells, some T cells, and certain cells of other tissues. Class II molecules are required for T-cell communication with macrophages and B cells. As discussed later in more detail, part of the T-cell receptor must recognize a peptide and a class II molecule on the antigen-presenting cell before the T cell can secrete cytokines necessary for the immune response.

The class III genes encode the second component of complement (C2), factor B, two cleavage products (C4a and C4b) of complement C4, tumor necrosis factor, heat-shock proteins, and other proteins. C2, C4a, and C4b participate in the classical pathway and factor B in the alternative pathway (*see figure 31.13*).

Classes I and II of the MHC are involved in not only the immunologic recognition of microorganisms but also an individual's susceptibility to particular noninfectious diseases. For example, there is evidence of MHC-linked determinants in multiple sclerosis and acute glomerulonephritis. The effect of MHC molecule expression on the frequency of these diseases is just beginning to be understood.

Types of T Cells

There are several classes or subpopulations of T cells (**table 32.4**). One class (effector) causes cytolysis and cell death of target cells. The other classes regulate specific cells. These effector and regulatory T cells are now discussed in more detail.

Effector Cells

Cytotoxic T (T_C) cells (CD8$^+$ cells) are activated in a complex process that involves at least three signals. They attach by T-cell

Table 32.4
Classes of T Cells

Classes	Function
Class That Kills	
T_C (cytotoxic) cells; also called CD8$^+$ cells	Cause cytolysis and cell death of target cells—virus-infected cells and tumor cells; important in host defense against cytosolic pathogens
Classes That Regulate Other Cells	
T_H (helper) cells; also called CD4$^+$ cells	Help B cells make antibody in response to antigenic challenge; stimulate cell-mediated immunity
T_S (suppressor) cells	When mixed with naive or effector T cells, suppress their activity

receptor proteins to viral antigen–class I MHC complexes on the surface of virus-infected cells (**figure 32.24**). A CD28-B7 costimulatory signal also is involved. Finally the T_C cells require exposure to interleukin-2 from T_H1 cells. The T_C cells then proliferate and differentiate to form active **cytotoxic T lymphocytes (CTLs),** which can attack virus-infected cells. CTLs and T_H1 cells release cytokines such as gamma-interferon (IFN-γ), and both macrophages and T cells release tumor necrosis factor-alpha (TNF-α). These cytokines limit virus reproduction, while also activating macrophages and other phagocytes that can destroy the infected cell.

T_H2 cells secrete cytokines such as IL-4, IL-5, and IL-6, which stimulate B-cell proliferation and differentiation into antibody-forming plasma cells (figure 32.24). Thus activation of T-helper cells by macrophages is required for the production of both CTLs and antibodies.

Cytotoxic T lymphocytes can destroy target cells in at least two ways. The CD95 pathway triggers apoptosis or programmed cell death (*see pp. 765, 858–59*). The perforin pathway is probably more often employed by CTLs and uses perforin and granzymes to induce both osmotic cytolysis and apoptosis.

The **CD95 pathway** depends on the CD95 transmembrane Fas protein receptor, which is found on many eucaryotic cells. CD95 or Fas is coded for by the *fas* gene, a member of the tumor necrosis factor (TNF) family of genes. When the Fas ligand (CD95L) binds to the target cell Fas protein, the CD95-CD95L complex activates several cytoplasmic proteins that initiate a cellular suicide cascade, leading to apoptosis. CTLs have high concentrations of the Fas ligand or CD95L on their surface. The target cell undergoes apoptosis upon binding of the ligand to Fas.

The **perforin pathway** begins when the CTL binds by its T-cell receptor to a viral peptide–class I MHC complex on the target cell surface (**figure 32.25a**). This complex series of events now occur:

1. A Ca^{2+}-dependent sequence first takes place: (a) microtubule assembly; (b) movement of cytoplasmic granules toward the part of the plasma membrane that is in contact with the target cell; (c) reorientation of the Golgi apparatus to the target cell; and (d) movement of the microtubule-organizing center and other cytoskeletal components into the region of the cytoplasm adjacent to the bound target cell.

2. Next the CTL secretes a pore-forming protein, perforin, into the target cell's plasma membrane. Perforin polymerizes in the target cell's membrane to form circular transmembrane pores (figure 32.25c). This can lead to osmotic lysis.

3. Finally the CTL secretes granzymes, proteolytic enzymes that enter the target cell and induce apoptosis.

This sequence of events will cause either cytolysis or apoptosis of the target cell. The CTL then disengages from the dying target cell and recycles its cytoplasmic components in preparation for another attack.

Regulator T Cells

Regulator T cells control the development of effector T cells. Two subpopulations exist: T-helper cells (CD4$^+$ cells) and T-suppressor cells. There are three subsets of **T-helper (T_H) cells:** T_H1, T_H2, and T_H0. The T_H1 and T_H2 subsets produce and secrete a specific mixture of cytokines. **T_H1 cells** produce IL-2, IFN-γ, and TNF-β that are involved in cellular immunity. These cytokines are responsible for delayed-type hypersensitivity reactions and macrophage activation. **T_H2 cells** also produce various cytokines (e.g., IL-4, IL-5, IL-10, IL-13) and are involved in humoral immunity. T_H2 cells are helpers for B-cell antibody responses and defense against helminths. **T_H0 cells** are simply undifferentiated precursors of T_H1 and T_H2 cells.

T-helper cells (T_H1) need two signals (costimulation) to be stimulated by an antigen. The first signal is when antigen fragments are effectively presented to a T cell by a macrophage, dendritic cell, or an activated B cell. For example, when a virus infects a host, some particles may be phagocytosed by a macrophage and partially digested (figure 32.24). Some of the antigenic viral protein fragments are then moved to the surface of the macrophage and presented in close association with class II MHC molecules. This combination of viral antigen fragment and class II MHC molecule is needed for recognition by the T-cell receptor protein on the surface of the T_H1 cell and represents activation signal 1. (The abbreviation CD is from cell-associated differentiation antigen; CDs are functional molecules or receptors that serve as surface markers; section 32.2.)

The second signal for activation involves the CD28 protein receptor on T_H1 cells (**figure 32.26**). When CD28 binds to a protein called B7 (CD80) on the surface of a macrophage (an APC),

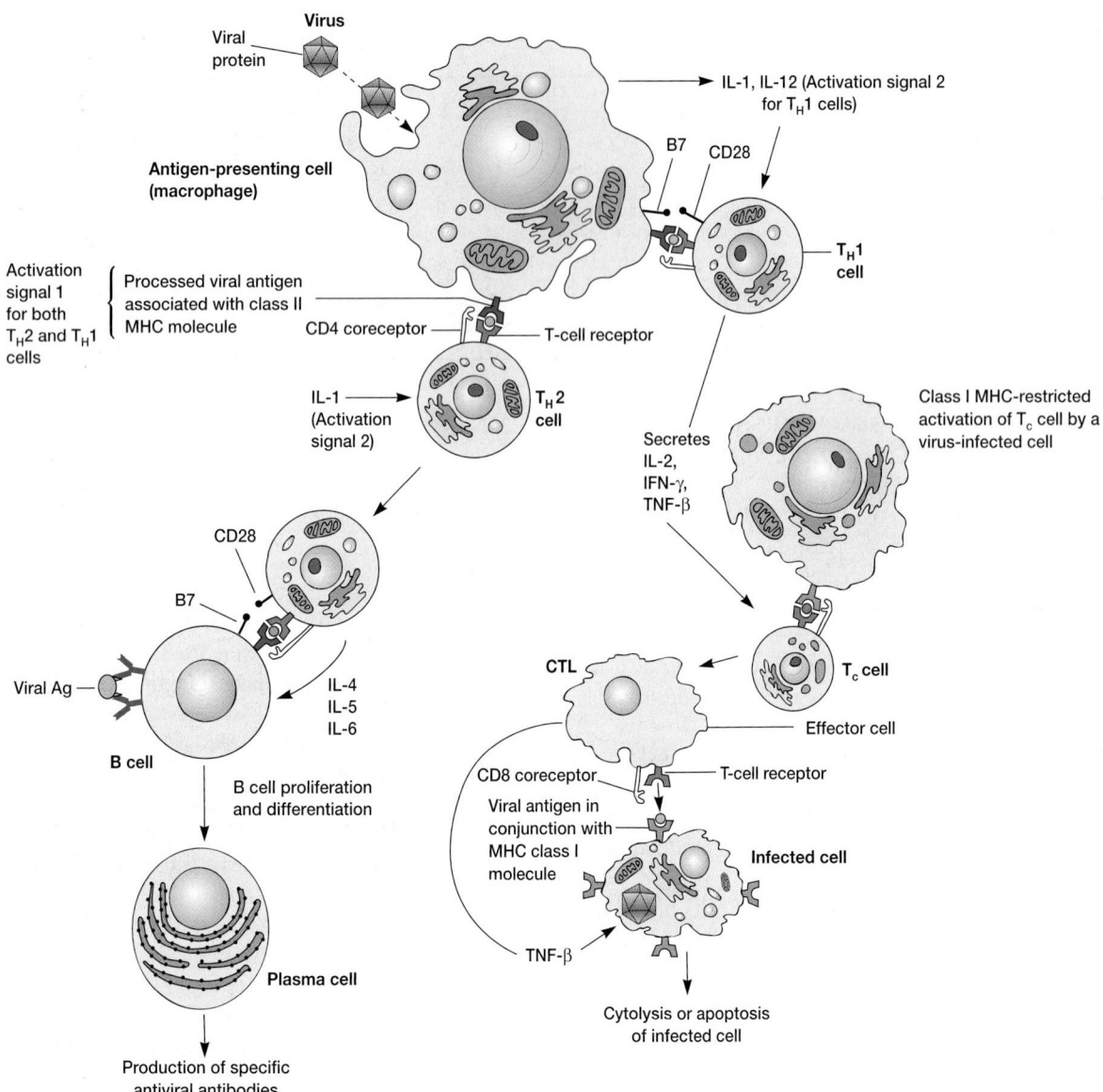

Figure 32.24   Helper T-Cell Responses In Cell-Mediated Immunity. A virus is phagocytosed by a macrophage and a small antigen fragment (peptide) presented to T-helper cells (T_H1 and T_H2) in association with class II MHC molecules. Once activated (activation signals 1 and 2) the T_H2 cell secretes the cytokines IL-4, IL-5, IL-10, and IL-13 that cause B-cell proliferation and secretion of specific antiviral antibodies. Activation signals 1 and 2 also cause the T_H1 cell to secrete IL-2, IFN-γ, and TNF-β. IL-2 regulates the proliferation of cytotoxic T cells. Once the effector T_C cell proliferates, it attacks and causes lysis or programmed cell death (apoptosis) of the virus-infected cell by either the Fas or perforin pathways.

it delivers a second signal to the T_H1 cell. Both of these signals go into the cytoplasm of the T_H1 cell.

Signal 1 activates a tyrosine kinase located in the cytoplasm. Tyrosine kinase adds phosphate groups to the amino acid tyrosine in proteins. When this phosphorylation occurs, the enzyme phospholipase Cγ1 cleaves the phosphatidylinositol bisphosphate located in the T-helper cell plasma membrane. Two cleavage products are formed and two specific pathways now operate.

One of the cleavage products, diacylglycerol, activates protein kinase C. Protein kinase C moves into the nucleus where it catalyzes the formation of a protein complex called AP-1 from separate components. The other cleavage product, inositol trisphosphate, causes a calcium channel to open; calcium ions to rush into the cytoplasm, leading to further enzymatic activity and the activation of calmodulin, calcineurin, and nuclear factor of activated T_H1 cells (NF-AT). NF-AT then migrates into the nu-

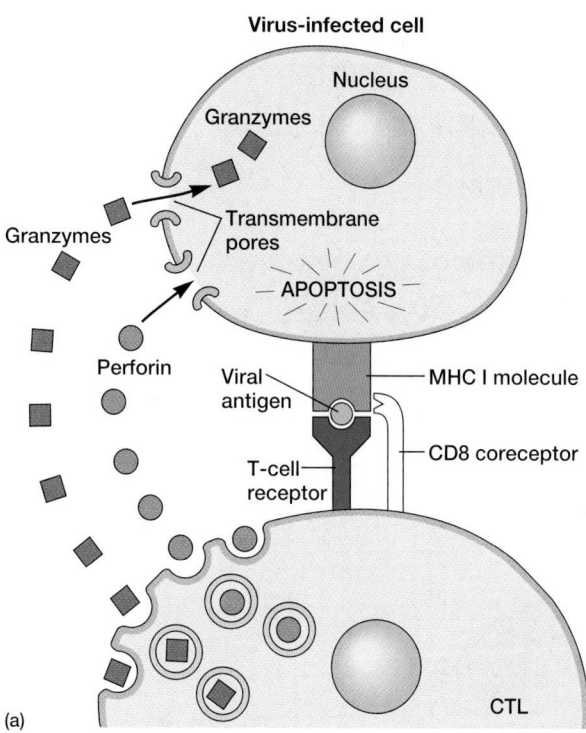

Virus-infected cell

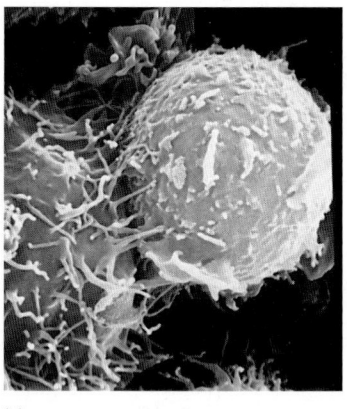

(b)

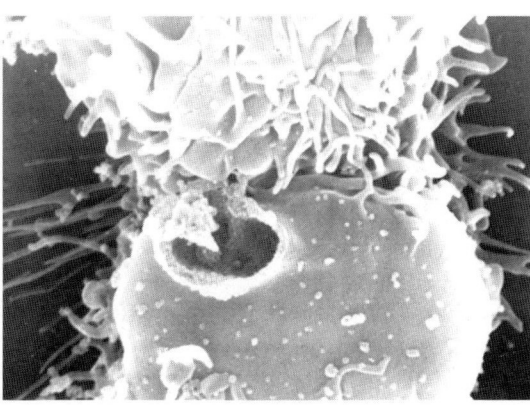

(c)

(a)

Figure 32.25   How an Effector Cytotoxic T Cell Destroys a Virus-Infected Target Cell. (a) T-cell cytotoxicity often involves the perforin pathway and leads to apoptosis and cytolysis. (b) A cytotoxic T cell (left) contacts a target cell (right) ($\times$5,700). (c) The T cell secretes perforin that forms transmembrane pores in the target cell's plasma membrane. These pores allow the contents of the target cell to leak out (cytolysis).

cleoplasm where it binds to the newly formed AP-1, forming a NF-AT/AP-1 complex (a transcription factor). This transcription factor binds tightly to a specific gene sequence in the DNA molecule, causing interleukin-2 mRNA to be transcribed. Interleukin-2 mRNA moves out of the nucleus to the ribosomes where the protein interleukin-2 is produced.

Signal 2 is mediated by the CD28 receptor and B7 molecule. This signal activates another tyrosine kinase, causes the formation of the transcription factor CD28RC, and also stabilizes the mRNA transcribed from the interleukin-2 gene. This increases the concentration of interleukin-2.

T_H1 cells that are activated by these two signals secrete large amounts of interleukin-2, which activates cytotoxic T cells (figure 32.24). T_H1 cells also secrete gamma-interferon (IFN-γ), which is capable of activating macrophages and increasing microbicidal activity.

The activity of T_H2 cells is depicted in figure 32.24. They are costimulated by antigen presentation and interleukin-1. The T_H2 cells then release several cytokines that stimulate B-cell proliferation and differentiation into antibody-forming plasma cells.

Although the evidence is not conclusive, many immunologists believe in the existence of **T-suppressor (T_S) cells** that can suppress B-cell and T-cell responses. Subpopulations of T_S cells specific to a given antigen can be stimulated to proliferate by the interleukin-2 from activated T-helper cells. This proliferation occurs at a slow rate to provide negative feedback control for that part of the immune response known as acquired immune tolerance.

1. What is the function of an antigen-presenting cell? What is a T-cell receptor and how is it involved in T-cell activation?
2. What is a histocompatibility molecule? What are MHCs and HLAs? Describe the roles of the three MHC classes.
3. Describe antigen processing. How does this process differ for endogenous and exogenous antigens?
4. Briefly describe the cytotoxic T cell, its general role, how it is activated, and the two ways in which it destroys target cells.
5. Outline the functions of a T-helper cell. How do T_H1 and T_H2 cells differ in function? Briefly describe how T_H cells are activated by costimulation.

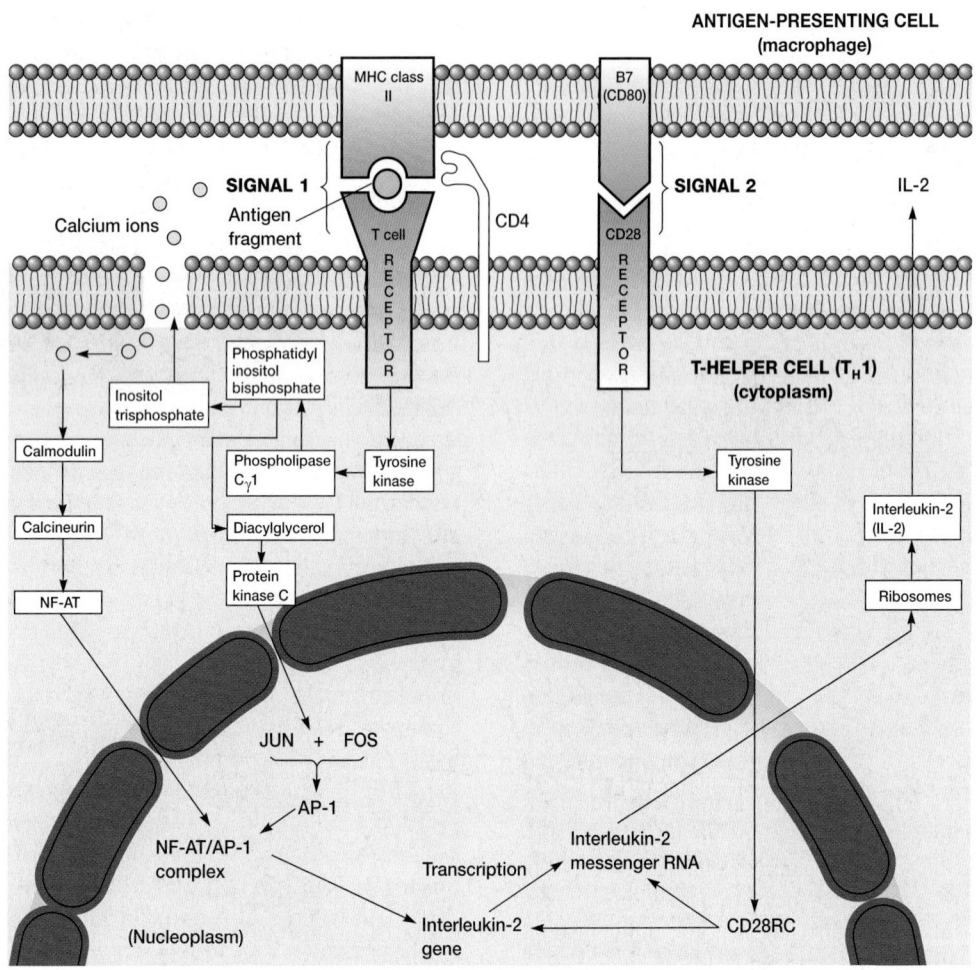

Figure 32.26   Two Signals (Costimulation) Are Essential for T-helper (T$_H$1) Cell Activation.   The first signal is the presentation of the antigen fragment by a macrophage or other antigen-presenting cell along with the MHC class II molecule to the T-helper cell receptor and CD4 protein. The second signal occurs when the macrophage presents the B7 (CD80) protein to the T-helper cell with its CD28 protein receptor. Both signals send information into the cytoplasm of the T-helper cell. The first signal causes interleukin-2 mRNA to be produced. The second signal boosts the production of interleukin-2 to effective concentrations. It now is known that the gene for interleukin-2 is under very tight regulation. It cannot be transcribed unless the NF-AT/AP-1 complex, AP-1, and other transcription factors (e.g., CD28RC) are present. All these factors must be produced anew or activated when the T-helper cell is stimulated through its antigen-specific receptor.

 **B-CELL BIOLOGY**

Stem cells in the bone marrow produce B cell precursors (*see figure 31.3*). Mature B cells contain cell surface, transmembrane immunoglobulin molecules associated with Ig-α/Ig-β heterodimer proteins; these complexes are called **B-cell antigen receptors (BCRs).** BCRs can contain IgM and IgD. It is the IgM, however, that is the major receptor for its appropriate antigen. B cells also have receptors for the Fc part of some immunoglobulin classes and may have receptors for complement (C3b). Each B cell may have as many as 50,000 of these membrane-bound immunoglobulins. The total mature B-cell population of each individual thus carries BCRs specific for many antigens; however, each mature B cell possesses BCRs specific only for a particular antigenic determinant.

Antibodies are secreted by plasma cells and bind antigens in solution. The various classes of antibodies (IgG, IgM, IgA, IgD, IgE; figures 32.10–32.14) are related to the membrane-bound BCRs. In addition to antigen binding, BCRs also trigger the endocytosis of an antigen leading to it being processed inside the B cell. A small antigen fragment is presented on the surface of the cell in association with major histocompatibility class II molecules. Thus B cells have two immunological roles: (1) they proliferate and differentiate into plasma cells and respond to antigens by making antibodies, and at the same time (2) can act as antigen-presenting cells.

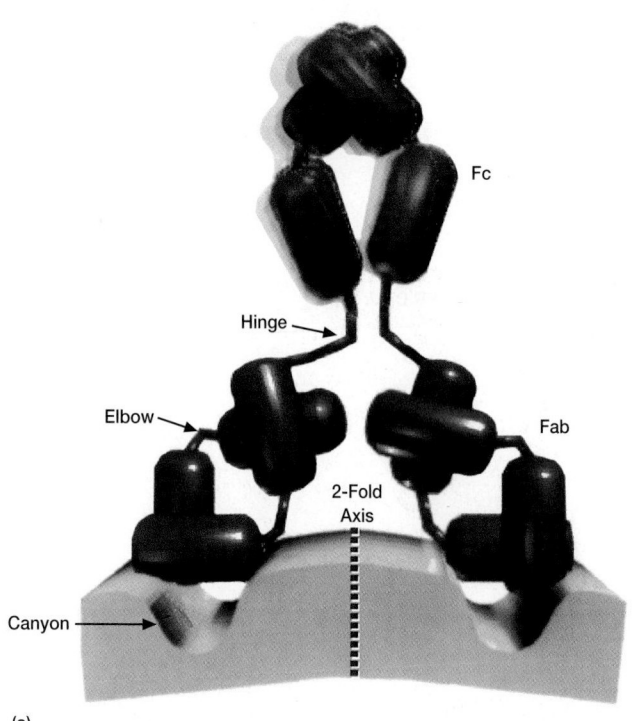

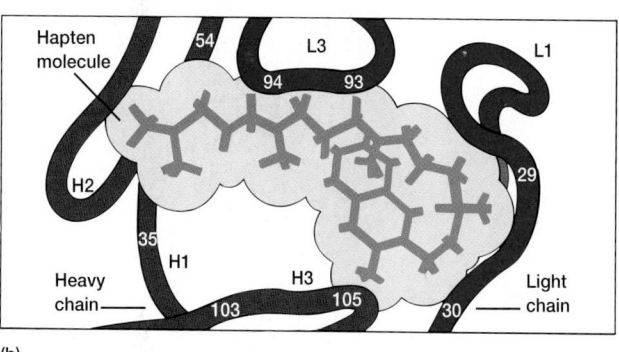

Figure 32.27 Antigen-Antibody Binding. (a) An example of antigen-antibody binding is represented in this model of the monoclonal antibody MAb17-IA bound to the surface of the human rhinovirus. The heavy chains are red, light chains are blue, and the antigen capsid protein is yellow. The RNA interior of the virus would be toward the bottom of the diagram. The antibody is bound bivalently across icosahedral twofold axes of the virus. **(b)** Based on X-ray crystallography, the hapten molecule nestles in a pocket formed by the antibody combining site. In the illustration the hapten makes contact with only 10–12 amino acids in the hypervariable regions of the light and heavy chains. The numbers represent contact amino acids.

Antigen-Antibody Binding

An antigen binds to an antibody at the antigen-binding site within the Fab region of the antibody. More specifically, a pocket is formed by the folding of the V_H and V_L regions (figure 32.7). At this site, specific amino acids contact the antigen's epitope or haptenic groups and form multiple noncovalent bonds between the antigen and amino acids of the binding site (**figure 32.27**).

Because binding is due to weak, noncovalent bonds such as hydrogen bonds and electrostatic attractions, the antigen's shape must exactly match that of the antigen-binding site. If the shape of the epitope (figure 32.3) and binding site are not truly complementary, the antibody will not effectively bind the antigen. Although a lock-and-key mechanism normally may operate, in at least one case the antigen-binding site does change shape when it complexes with the antigen (an induced fit mechanism). Regardless of the precise mechanism, antibody specificity results from the nature of antibody-antigen binding.

B-Cell Activation

Upon activation by appropriate signals, the mature B cell proliferates and differentiates into a plasma cell that secretes antibody. Activation is typically antigen specific, activating a particular clone of cells, but can be nonspecific and polyclonal through treatment with B-cell mitogens.

T-Dependent Antigen Triggering

Most antigens have more than one type of antigenic determinant site (epitope) on each molecule (figure 32.3). B cells specific for a given epitope on the antigen (e.g., epitope X) often cannot develop into plasma cells secreting antibody (anti-X) without the collaboration of helper T cells. In other words, binding of the epitope X to the B cell may be necessary, but it is not sufficient for B-cell activation. Antigens that elicit a response with the aid of T-helper cells are called **T-dependent antigens.** Examples include bacteria, foreign red blood cells, certain proteins, and hapten-carrier combinations (figure 32.4).

The basic mechanism for T-dependent antigen triggering of a B cell is illustrated in **figure 32.28** and involves three cells: (1) a macrophage or other antigen-presenting cell to process and present the antigen; (2) a T-helper cell able to recognize the antigen and respond to it; and (3) a B cell specific for the antigen. When all cells and the antigen are present, the following sequence takes place. The macrophage presents part of the antigen and its own class II MHC to the T-helper cell (signal #1 to the T cell). Costimulation is provided by the B7-CD28 interaction (signal #2). Interleukins 1 and 6 stimulate T cells to divide and make interleukins 2 and 4–6. Interleukin-1 also stimulates the hypothalamus to raise the body temperature (causing a fever), which enhances the activity of T cells. Interleukins 2 and 4–6 stimulate T-helper cells to multiply. These helper cells directly associate with B cells that display the correct antigen–class II MHC complex and secrete cytokines called B-cell growth factors (BCGFs), which cause B cells to multiply. As the number of B cells increase, T-helper cells produce other cytokines, the B-cell differentiation factors (BCDFs). These instruct some B cells to stop replicating, to differentiate into plasma cells, and to start producing antibodies. B-cell activation involves more than T-helper cell activity. The B cell also recognizes the antigen through its surface IgM receptors for the antigen (signal #1 for

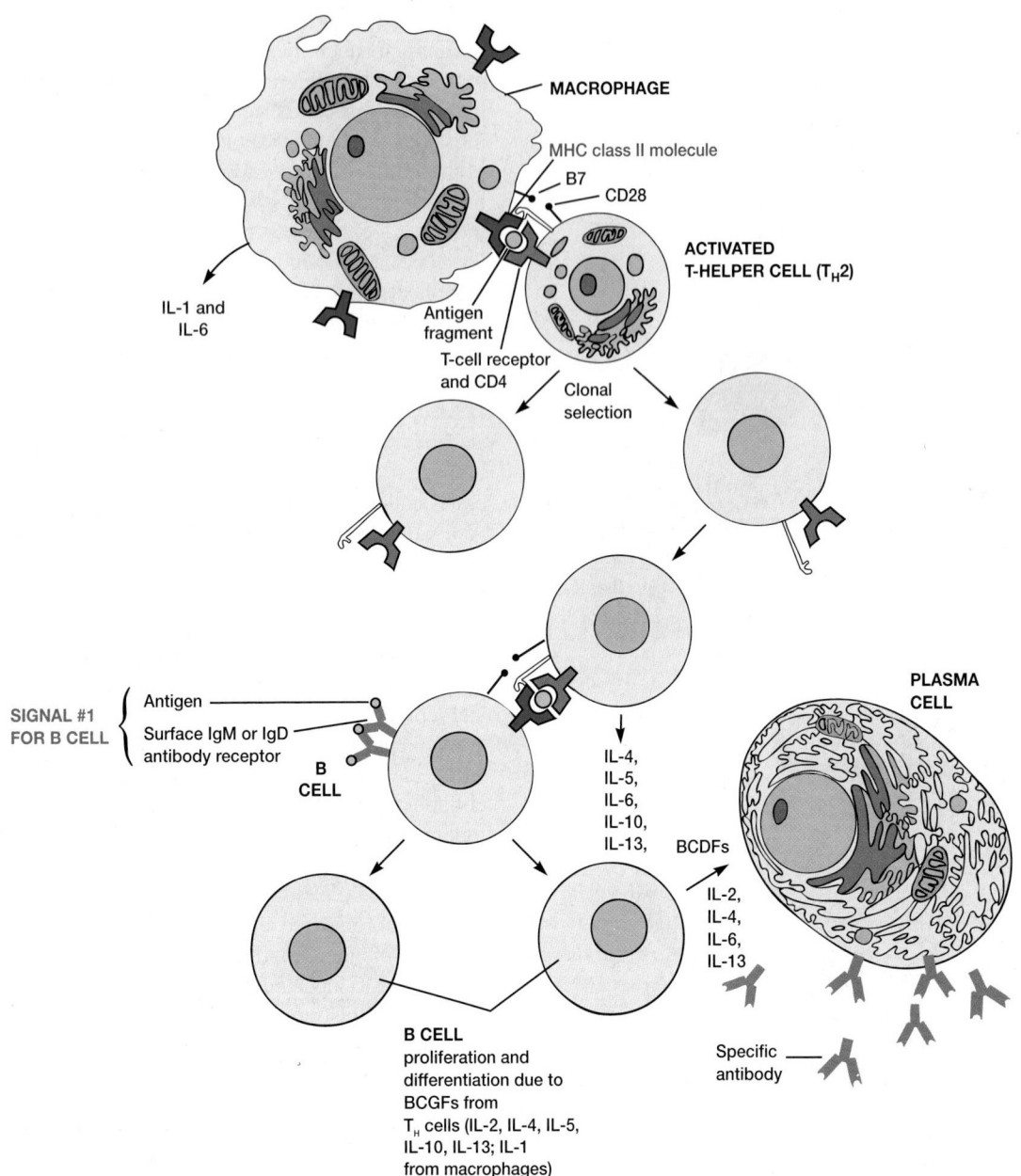

Figure 32.28   T-Dependent Antigen Triggering of a B Cell.   Schematic diagram of the events occurring in the interactions of macrophages, T-helper cells, and B cells that produce humoral immunity. Many cytokines (e.g., IL-1, IL-4, IL-5, IL-10, IL-13) stimulate B-cell proliferation and may be called B-cell growth factors (BCGFs). Cytokines such as IL-2, IL-4, IL-6, and IL-13 stimulate B-cell differentiation and are called B-cell differentiation factors (BCDFs).

the B cell) and is subsequently triggered to proliferate and differentiate into plasma cells secreting antibody. (One plasma cell can synthesize more than 10 million antibody molecules per hour!) It should be noted here that B cells are more effective antigen presenters than macrophages at low antigen concentrations if the antigen can bind to surface immunoglobulins. The B cells bind antigen, take it up by receptor-mediated endocytosis, and present it to T-helper cells to activate them. In this situation B cells and T_H cells activate each other. All the immune responses

that produce IgG, IgA, and IgE involve this T-dependent antigen triggering. Fever (p. 698)

T-Independent Antigen Triggering

Not all antibody responses require T-cell help. There are antigens that trigger B cells into immunoglobulin production without T-cell cooperation. These are called **T-independent antigens.** Examples include certain tumor-promoting agents, anti-immunoglobulin (anti-Ig), antibodies to certain B-cell dif-

Table 32.5

Comparison of Lymphocytes Involved in the Immune Response

Property	T Cells	B Cells
Origin	Bone marrow in adults	Bone marrow in adults
Maturation and differentiation	Thymus	Lymphoid tissue; bursa of Fabricius in birds
Mobility	Great	Very little (some stages circulate)
Complement receptors	Absent	Present
Surface immunoglobulins	Absent	Present
Proliferation	Upon antigenic stimulation, differentiate into effector and memory cells	Upon antigenic stimulation, proliferate and differentiate into plasma and memory cells
Immunity type	Cell mediated and humoral (B cell activation by T_H cells)	Humoral
Distribution	High in blood, lymph, and lymphoid tissue	High in spleen, lymph nodes, bone marrow, and other lymphoid tissue; low in blood
Secretory product	Cytokines	Antibodies
Subsets and functions	T-helper (T_H) cell: necessary for B-cell activation by T-dependent antigens and T-effector cells. There are three types of T-helper cells: T_H1, T_H2, and T_H0.	Plasma cell: a cell arising from a B cell that manufactures specific antibodies
	T-suppressor (T_S) cell: blocks induction and/or activation of T_H cells and B cells; helps maintain tolerance	Memory cell: a long-lived cell responsible for the anamnestic response
	Cytotoxic T (T_C) cell: differentiates into a CTL that lyses cells recognized as nonself and virus or parasite-infected cells	
	Memory cells	

Table 32.6

Antigen Recognition by T and B Cells

Characteristic	T Cells	B Cells
Binds soluble antigen	No	Yes
Biochemistry of the antigens	Mostly proteins, but some glycolipids presented on MHC molecules	Proteins, glycolipids, polysaccharides
Antigen recognition	Antigens processed internally and presented as linear peptides bound to MHC molecules	Accessible areas of protein structure containing sequential amino acids and nonsequential amino acids
	Involves three partners: T-cell receptor, processed antigen, and MHC molecule	Immunoglobulin receptor binds antigen in native conformation
		Involves two partners: antigen and membrane immunoglobulin

ferentiation antigens, and bacterial lipopolysaccharides. The T-independent antigens are polymeric—that is, composed of repeating polysaccharide or protein subunits. They elicit almost exclusively IgM antibody formation, and there is little switching to IgG; the resulting antibody generally has a low affinity for antigen.

The mechanism for activation by T-independent antigens probably depends on their polymeric structure. These large molecules present a large array of identical epitopes to a B cell specific for that determinant. The multivalent nature of the cell receptor (surface IgM) is such that cell activation occurs and IgM is secreted. Because there is no T-cell help, the B cell cannot switch to high-affinity IgG production, and no memory B cells are formed.

Section 32.4 presented the biology of T-cells and this current section describes B-cell biology. **Tables 32.5** and **32.6** summarize and compare many of the important properties of lymphocytes that were discussed in these two sections.

1. What are B-cell antigen receptors, and how are they involved in B-cell activation?
2. Briefly compare and contrast B cells and T cells with respect to their formation, structure, and roles in the immune response.
3. How does antigen-antibody binding occur? What is the basis for antibody specificity?
4. How does T-independent antigen triggering of B cells differ from T-dependent triggering?

 ## 32.6   ACTION OF ANTIBODIES

The antigen-antibody interaction is a bimolecular association that exhibits exquisite specificity. The in vivo interactions that occur in vertebrate animals are absolutely essential in protecting the animal against the continuous onslaught of viruses, microorganisms and their products, certain macromolecules, and cancer cells. Outside the animal body (in vitro), this same specificity has led to the development of a variety of immunological assays that can detect the presence of either antibody or antigen. They are important in the diagnosis of diseases; in the identification of specific viruses, bacteria, and parasites; in monitoring the level of the humoral response and immunologic problems; and in identifying molecules of medical and biological interest. Immunological assays differ in their speed and sensitivity; some are qualitative whereas others are quantitative. The rest of this section examines those antigen-antibody interactions that occur in the animal body (in vivo). Antigen-antibody reactions that are routinely dealt with in the laboratory (in vitro) are covered in chapter 33, which is concerned with medical immunology.

Toxin Neutralization

Immunity to a disease like diphtheria depends on the production of specific antibodies that inactivate the toxins produced by the bacteria. This process is termed **toxin neutralization (figure 32.29).** Once neutralized, the toxin-antibody complex is either unable to attach to receptor sites on host target cells, unable to enter the cell, or it is ingested by macrophages. For example, diphtheria toxin inhibits protein synthesis after binding to the cell surface by the B fragment and subsequent passage of the active A fragment into the cytoplasm of the target cell (*see figure 34.5*). Thus the antibody blocks the toxic effect by inhibiting the entry of the A fragment or the binding of the B fragment. An antibody capable of neutralizing a toxin or antiserum containing neutralizing antibody against a toxin is called **antitoxin** (figure 32.29*b*).   Exotoxins (pp. 769–73)

Viral Neutralization

IgG, IgM, and IgA antibodies can bind to some viruses during their extracellular phase and inactivate them. This antibody-mediated viral inactivation is called **viral neutralization.** Fixation of the classical pathway (section 32.7) complement component C3b to the virus aids the neutralization process. Viral neutralization prevents a viral infection due to the inability of the virus to bind to and enter its target cell (figure 32.29*c*).

Adherence Inhibition

The capacity of bacteria to colonize the mucosal surfaces of mammalian hosts is dependent in part on their ability to adhere to mucosal epithelial cells. Secretory IgA (sIgA; p. 714) antibodies inhibit certain bacterial adherence promoting factors. Thus sIgA has a unique role in protecting the host against infection by some pathogenic bacteria, and perhaps by other microorganisms on mucosal surfaces.

IgE and Parasitic Infections

Immune reactions against protozoan and helminthic parasites are only partially understood. Parasites that have a tissue invasive phase in their life cycle often are associated with both eosinophilia (an excessive number of eosinophils in the blood) and elevated IgE levels. Recent evidence shows that, in the presence of elevated IgE, eosinophils can bind to the parasites and discharge their granules. Degranulation releases lytic and inflammatory mediators that destroy the parasites (*see figure 33.2*).

Opsonization

Phagocytes have an intrinsic ability to bind directly to microorganisms by nonspecific cell surface receptors, engulf the microorganisms, form phagosomes, and digest the microorganisms (*see figure 31.16*). This phagocytic process can be greatly enhanced by opsonization. As noted in section 31.8, opsonization is the process in which microorganisms or other particles are coated by antibody and/or complement, and thus prepared for "recognition" and ingestion by phagocytic cells. Opsonizing antibodies, especially IgG1 and IgG3, bind to Fc receptors on the surface of macrophages and neutrophils. This binding forms a bridge between the phagocyte and the antigen.   Phagocytosis (pp. 693–97)

Immune Complex Formation

Because antibodies have at least two antigen-binding sites and most antigens have at least two antigenic determinants, crosslinking can occur, producing large aggregates termed **immune complexes (figure 32.30).** If the antigens are soluble molecules and the complex becomes large enough to settle out of solution, a **precipitation** [Latin *praecipitare,* to cast down] or **precipitin reaction** has occurred, and is caused by a **precipitin** antibody. When the immune complex involves the cross-linking of cells or particles, an **agglutination reaction** occurs and the responsible antibody is an **agglutinin.** Agglutination specifically involving red blood cells is a **hemagglutination** reaction and is caused by a **hemagglutinin.** These immune complexes are more rapidly phagocytosed in vivo than are free antigens.

The extent of immune complex formation, whether within an animal or in vitro, depends on the relative concentrations of the precipitin antibody and antigen. If there is a large excess of antibody, separate antibody molecules usually bind to each antigenic determinant and a less insoluble network or lattice forms (*see figure 33.17a*). When antigen is present in excess, two separate antigen molecules tend to bind to each antibody and network development or crosslinking also is inhibited. In the equivalence zone the ratio of antibody and antigen concentrations is optimal for the formation of a large network of interconnected antibody and antigen molecules. All antibody and antigen molecules precipitate or agglutinate as an insoluble complex. Precipitin reactions can occur in both solutions and agar gel media. In either case, antibody-antigen equivalence is required for optimal results.

---

1. How does toxin neutralization occur? Viral neutralization?
2. Describe the role of IgE in resistance to parasitic infections.
3. How does adherence inhibition occur?
4. Describe an immune complex. What two types are formed?

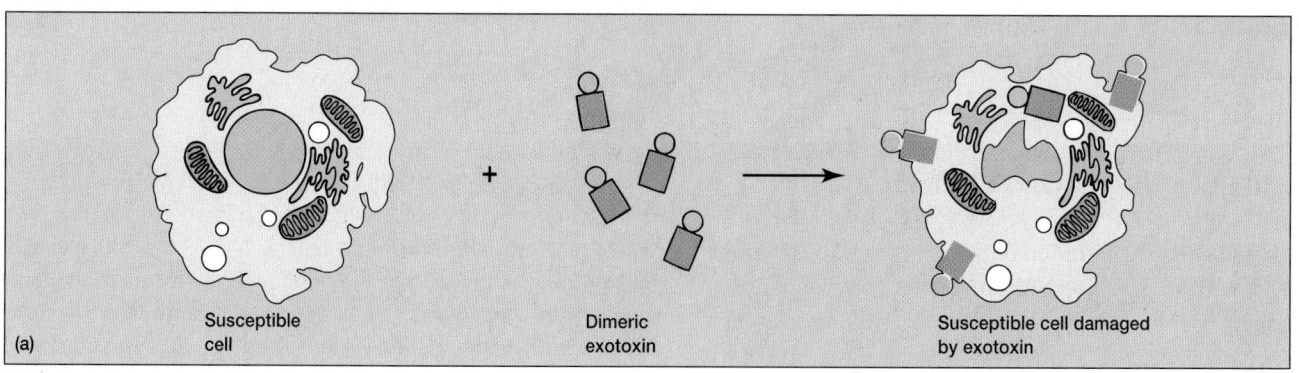

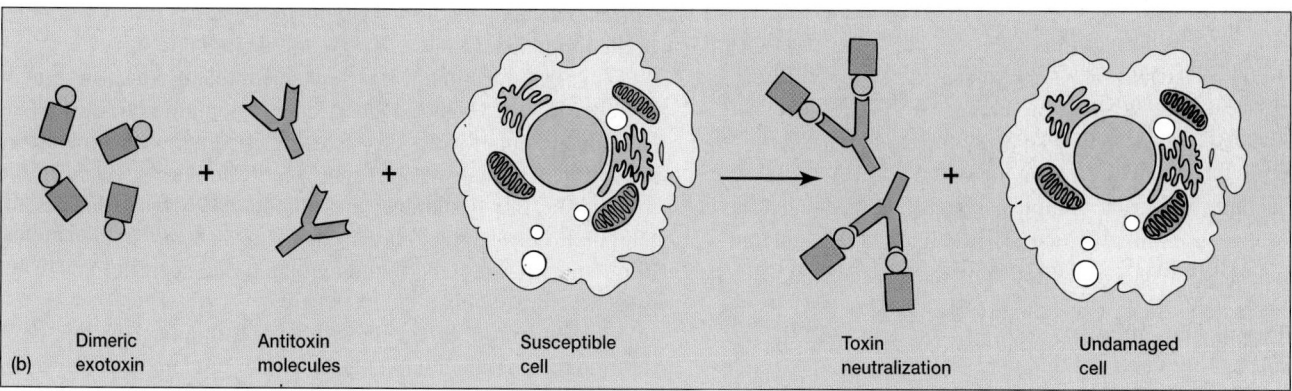

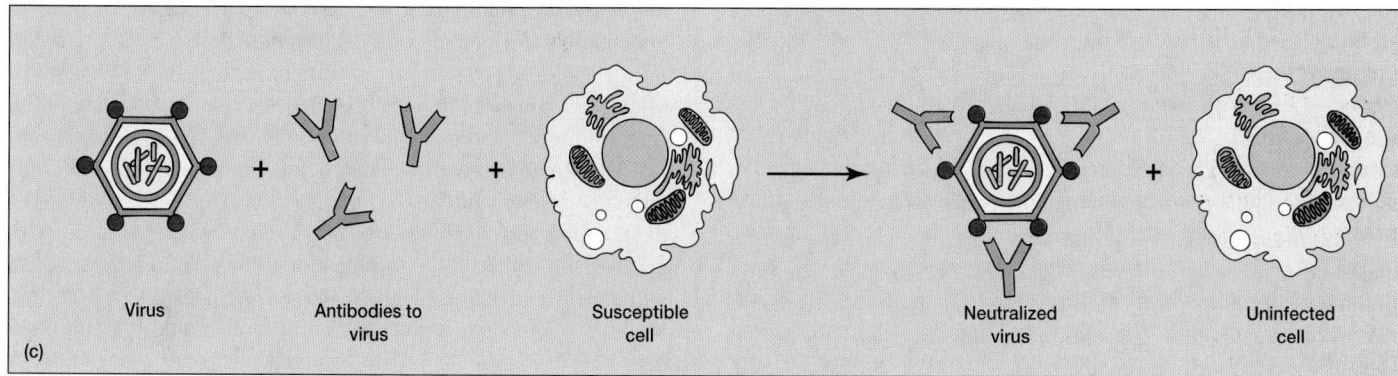

Figure 32.29 Neutralization Reactions. (**a**) Effects of a dimeric exotoxin on a susceptible cell. (**b**) Neutralization of the toxin by antitoxin. (**c**) In viral neutralization the specific antibodies neutralize the virus and prevent it from attaching to the susceptible cells.

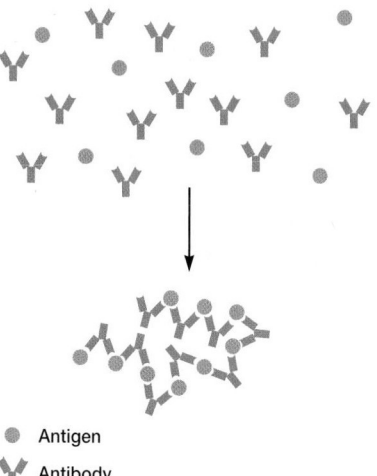

Figure 32.30 Immune Complex Formation. Antibodies cross-link antigens forming an aggregate of antibody and antigen termed an immune complex.

733

32.7 THE CLASSICAL COMPLEMENT PATHWAY

Activation of the **classical complement pathway** requires initiation by the interaction of antibodies with an antigen that is usually cell bound (**figure 32.31**). The order of effectiveness in activating complement is as follows: IgM > IgG3 > IgG1 > IgG2. However, some microbial products (lipid A of endotoxin and staphylococcal protein A) or plasmin (a proteolytic enzyme that dissolves the fibrin of blood clots) may activate C1 directly without antibody participation. Following binding of antigen to antibody, the C1 complement component, which is composed of three proteins (q, r, and s), attaches to the Fc portion of the antibody molecule through its C1q subcomponent. In the presence of calcium ions, a trimolecular complex (C1qrs · Ag · Ab) that has esterase activity is rapidly formed. The activated C1's subcomponent attacks and cleaves its natural substrates in serum (C2 and C4). This leads to binding of a portion of each molecule (C2b and C4b) to the antigen-antibody-complement complex with the release of small C4a and C2a fragments. (The released C2 fragment traditionally has been called C2a. For consistency, this text will label all larger complement fragments that are bound to the target cell as b fragments.) With the binding of C2b to C4b, an enzyme with trypsinlike proteolytic activity is generated. The natural substrate for this enzyme is C3; thus it is termed a C3 convertase. Through the activity of $\overline{C4b2b}$ (the bar indicates an active enzyme complex), C3 is cleaved into a bound subcomponent C3b and a C3a soluble component. C3b then absorbs to bound C4b2b, forming the complex $\overline{C4b2b3b}$, which acts as a C5 convertase and cleaves C5 into fragments C5a and C5b. C6 and C7 rapidly bind to C5b, forming a $\overline{C5b67}$ complex that possesses an unstable membrane-binding site; once bound to a membrane, this complex is stable. C8 and C9 then bind, forming the **membrane attack complex** ($\overline{C5b6789}$) that creates a pore in the plasma membrane of the target cell (*see figure 31.14*). It is believed that the actual pore is a doughnut-shaped polymer of C9. (The perforin pores generated by cytotoxic T lymphocytes [figure 32.25] are somewhat similar to complement pores.) If the cell is eucaryotic, Na^+ and H_2O enter through the pore leading to osmotic lysis of the cell. If the cell is a gram-negative bacterium, lysozyme from the blood enters through the pore and digests the peptidoglycan cell wall causing the bacterium to lyse osmotically. In contrast, gram-positive bacteria are resistant to the cytolytic action of the membrane attack complex because they lack an exposed outer membrane and the thick peptidoglycan prevents an attack on the plasma membrane. The complement system and its functions (pp. 691–93)

1. How is the classical complement pathway activated?
2. What effect does the membrane attack complex have on eucaryotic cells? Procaryotic cells?

 32.8 ACQUIRED IMMUNE TOLERANCE

Acquired immune tolerance refers to the body's ability to produce antibodies against nonself antigens such as microbial antigens, while "tolerating" (not producing antibodies against) self-antigens. Some of this tolerance arises early in embryonic life when immunologic competence is being established. Three general tolerance mechanisms have been proposed: negative selection by clonal deletion, the induction of anergy, and inhibition of the immune response by T-suppressor cells (p. 727).

Negative selection is one mechanism that produces immunologic tolerance. Negative selection by clonal deletion removes from the immune system T cells within the thymus that recognize any of the body's own antigens. T-cell tolerance induced in the thymus and B-cell tolerance in the bone marrow is called central tolerance. However, another mechanism is needed to prevent autoimmunity because many antigens are tissue-specific and are not present in the thymus or bone marrow.

Mechanisms occurring elsewhere in the body are collectively referred to as peripheral tolerance, which supplements central tolerance. Peripheral tolerance is thought to be based largely on incomplete activation signals given to the lymphocyte when it encounters self-antigens in the periphery of the body. This mechanism leads to a state of unresponsiveness called **anergy** [immunologists describe an inactive lymphocyte as "anergic" from the Greek *an,* negative, and *ergon,* without] that is associated with impaired intracellular signaling or apoptosis.

Many autoreactive B cells undergo clonal deletion or become anergic as they mature in the bone marrow. Negative selection occurs in the bone marrow if the B cells encounter large amounts of self-antigen, either in the soluble phase or as cell membrane constituents. The deletion of self-reactive B cells also takes place in secondary lymphoid tissue such as the spleen and lymph nodes. Since B cells recognize native antigen, there is no need for the MHC molecules in any of these processes. For those self-antigens that are present at relatively low concentrations, immunologic tolerance is often maintained only within the T-cell population. Nevertheless, this is sufficient to maintain tolerance because it denies the help essential for antibody production by self-reactive B cells.

1. Describe the three ways acquired immune tolerance develops in the vertebrate host.
2. How would you define the word anergy?

 32.9 SUMMARY: THE ROLE OF ANTIBODIES AND LYMPHOCYTES IN RESISTANCE

The last part of chapter 31 presented nonspecific (innate) host immunity and this chapter described specific (adaptive) immunity. Although both the humoral and cellular arms of the specific immune response have been considered separately, it is important to

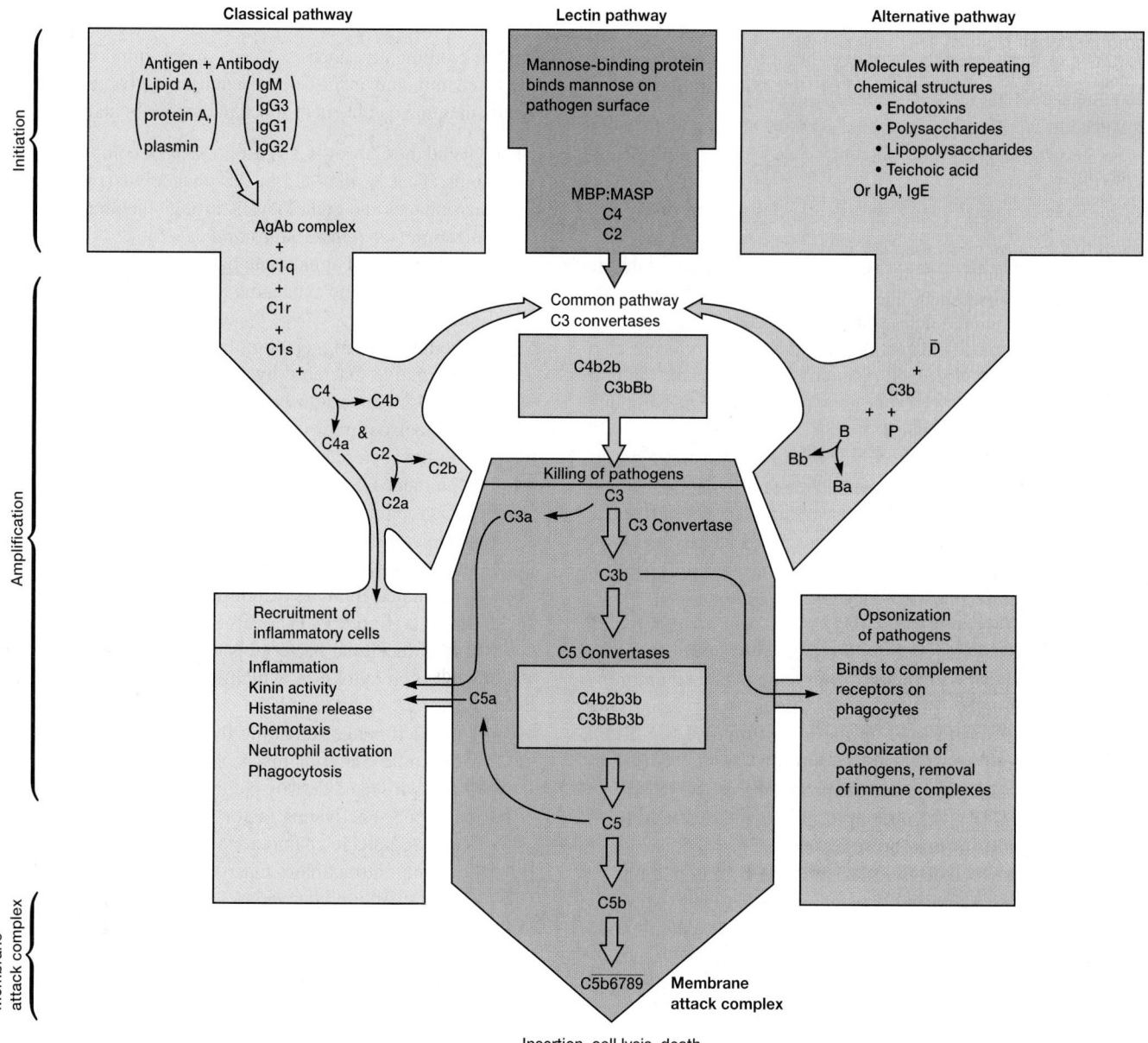

Figure 32.31   Complement Activation.   Within each pathway the components are arranged in order of their activation and aligned opposite their functional and structural analogs in the opposite pathway. There are three pathways of complement activation: (1) the classical pathway (yellow), which is mediated by antibody; (2) the lectin-mediated pathway (red), which is triggered by a serum protein that binds to mannose residues of bacteria; and (3) the alternative pathway (blue), which is triggered directly on pathogen surfaces. All three generate a crucial enzymatic activity (common pathway of C3 convertases) that in turn generates the effector activity of complement. The three main consequences of complement activation are recruitment of inflammatory cells (left box), direct killing of pathogens by the membrane attack complex (middle box), and opsonization of pathogens (right box).

understand that the response of a host to any particular pathogen may involve a complex interaction between both the host and pathogen, as well as the components of both nonspecific and specific immunity to ensure a maximal survival advantage for the host. This last section summarizes the defense mechanisms ver- tebrate hosts have against viral and bacterial pathogens. At times, however, these defenses are not enough to protect the host. Chap- ter 34 continues this discussion by presenting those mechanisms that pathogens have evolved to circumvent many of the follow- ing defenses.

Immunity to Viral Infections

Resistance to virus infections involves humoral immunity, interferon sensitization of host cells, and cell-mediated immunity.

1. Antibodies can neutralize viruses by combining with them and interfering with their adsorption and entrance into cells (figure 32.29).
2. Antibodies enhance phagocytosis and destruction of viruses in much the same way as for bacteria (*see figure 31.15*).
3. Interferons are important in resistance when the target cell is reached immediately as in the case of colds and influenza. Interferon-stimulated cells shut down protein synthesis (*see figure 31.19*) and destroy viral mRNA. Some interferons also stimulate the activity of T cells (figure 32.24) and natural killer cells (*see figure 31.22*), thus accelerating the immune response to a viral infection.
4. Cell-mediated immunity to viruses is a major resistance mechanism when enveloped viruses modify host cell membranes and bud off from the surface (e.g., herpesvirus, poxvirus, influenza, mumps, measles, rabies, and rubella viruses). Activated lymphocytes can recognize and destroy virus-infected cells because the target cell's plasma membrane has been altered.
   - Cytotoxic T lymphocytes (CTLs) destroy virus-infected cells through FasL, and the production of granzymes and perforin that form channels through the plasma membrane of infected cells causing apoptosis and cytolysis (figure 32.25). The class I MHC proteins are involved in T-cell recognition of infected cells (figure 32.24). Cells displaying both viral antigens and the proper class I MHC will be destroyed. CTLs are also involved in the destruction of cancer cells (**immune surveillance**).
   - Natural killer cells (NK cells) are non-B, non-T lymphocytes that are active without any prior antigen exposure. Interferon and antibodies, however, will stimulate them to greater activity. They also are capable of destroying virus-infected and cancer cells (*see figure 31.22*). They are components of innate immunity and possess no antigen receptors.

Immunity to Bacterial Infections

Humoral immunity appears to be more important than cell-mediated immunity in the defense against most bacterial pathogens. Antibodies and complement attack pathogens in many ways.

1. IgG, and the C3b and C4b components of complement are opsonins. That is, they aid in the phagocytosis of bacteria by macrophages and granulocytes through the process called opsonization (*see figures 31.15 and 31.16*).
2. IgM and IgG will agglutinate bacterial pathogens, thus limiting their spread and enhancing the efficiency of phagocytosis (figure 32.30).
3. Antibodies can trigger complement attack on the wall of gram-negative bacteria by the classical pathway (figure 32.31). Once this pathway is activated, it forms a C5b-9 membrane attack complex (*see figure 31.14*) that forms channel or pore in the bacterial cell, leading to lyses.
4. Complement C3a, C5a, and C5b67 also attract neutrophils and macrophages to a site of infection (*see figure 31.11*).
5. Antibodies called antitoxins (figure 32.29) bind to bacterial exotoxins and block their action (neutralization).
6. Cell-mediated immune responses by activated macrophages and T-cells (figure 32.1) are also important, particularly in resisting intracellular bacterial pathogens. Activated T cells secrete several cytokines, which have a variety of effects (*see figure 31.18 and table 31.4*).
   - The macrophage activating factor stimulates macrophages to become "angry" macrophages and more effectively phagocytose and destroy pathogens. Interferon-γ is a major macrophage activating factor.
   - The macrophage chemotactic factor and migration inhibition factor attract more macrophages and keep them in the area of infection after arrival.
   - Interleukin-2 (IL-2) stimulates the proliferation of activated T cells to increase the population of cells involved in the cell-mediated immune response. It also increases the effectiveness of cytotoxic T cells and NK cells by promoting the synthesis of immune interferon (γ-interferon) by T cells.

32.1 Overview of Specific (Adaptive) Immunity

a. The specific (adaptive) immune response system consists of lymphocytes that can recognize foreign molecules (antigens) and respond to them. Two branches or arms of immunity are recognized: humoral (antibody-mediated) immunity and cellular (cell-mediated) immunity (**figure 32.1**).
b. Acquired immunity refers to the type of specific (adaptive) immunity that a host develops after exposure to a suitable antigen. It can be obtained actively or passively by natural or artificial means (**figure 32.2**).

32.2 Antigens

a. An antigen is a substance that stimulates an immune response and reacts with the products of that response. Each antigen can have several antigenic determinant sites or epitopes that stimulate production of and combine with specific antibodies (**figure 32.3**). Haptens are small organic molecules that are not antigenic by themselves but can become antigenic if bonded to a larger carrier molecule (**figure 32.4**).

32.3 Antibodies

a. Antibodies or immunoglobulins are a group of glycoproteins present in the blood, tissue fluids, and mucous membranes of vertebrates. All immunoglobulins have a basic structure composed of four polypeptide chains (two light and two heavy) connected to each other by disulfide bonds (**figure 32.7**). In humans five immunoglobulin classes exist: IgG, IgA,

IgM, IgD, and IgE (**figures 32.10–32.14** and **table 32.2**).

b. Antibody diversity results from the rearrangement and splicing of the individual gene segments on the antibody coding chromosomes, somatic mutations, the generation of different codons during splicing, and the independent assortment of light- and heavy-chain genes (**figure 32.15** and **table 32.3**).

c. Immunologic specificity and memory is partly explained by the clonal selection theory (**figure 32.18**).

d. Specific antibodies can be produced naturally by immunized animals. The primary antibody response in a host occurs following initial exposure to the antigen. This response has lag, log, plateau, and decline phases. Upon secondary antigen challenge, the B cells mount a heightened and accelerated anamnestic response (**figure 32.19**).

e. Hybridomas result from the fusion of lymphocytes with myeloma cells. These cells produce a single monoclonal antibody (**figure 32.20**). Monoclonal antibodies have many uses.

32.4 T-Cell Biology

a. T cells are pivotal elements of the immune response. T cells have antigen-specific receptor proteins (**figure 32.21**).

b. Antigen-presenting cells—most of which are macrophages, dendritic cells, and B cells—take in foreign antigens or pathogens, process them, and present antigenic fragments complexed with MHC molecules (**figure 32.22**) to T-helper cells. These MHC molecules are proteins coded by a group of genes termed the major histocompatibility complex.

c. The cytotoxic T lymphocyte recognizes target cells such as virus-infected cells that have foreign antigens and class I MHC molecules on their surface. The CTLs then attack and destroy the target cells using the CD95 pathway and/or the perforin pathway (**figures 32.24** and **32.25**).

d. Regulator T cells control the development of effector T cells. Two types exist: T-helper cells (CD4$^+$) and T-suppressor cells. There are three subsets of T-helper cells: T_H1, T_H2, and T_H0. T_H1 cells produce various cytokines and are involved in cellular immunity (**figure 32.24**). The T_H2 cells also produce various cytokines but are involved in humoral immunity. T_H0 cells are simply undifferentiated precursors of T_H1 and T_H2 cells.

32.5 B-Cell Biology

a. B cells defend against antigens by differentiating into plasma cells that secrete antibodies into the blood and lymph, providing humoral or antibody-mediated immunity.

b. B cells can be stimulated to divide and/or differentiate to secrete antibody when triggered by the appropriate signals (**figures 32.24** and **32.27**).

c. B cells have receptor immunoglobulins on their plasma membrane surface that are specific for given antigenic determinants (**figure 32.28**). Contact with the antigenic determinant is required for the B cell to divide and differentiate into plasma cells and memory cells.

32.6 Action of Antibodies

a. Various types of antigen-antibody reactions occur in vertebrates and initiate the participation of

other body processes that determine the ultimate fate of the antigen. For example, the complement system can be activated, leading to cell lysis, phagocytosis, chemotaxis, or stimulation of the inflammatory response. Other defensive antigen-antibody interactions include toxin neutralization, viral neutralization, adherence inhibition, opsonization, and immune complex formation (**figures 32.29** and **32.30**).

32.7 The Classical Complement Pathway

a. Once activated, the classical complement pathway (**figure 32.31**) forms the membrane attack complex that can lyse antibody-coated cells.

32.8 Acquired Immune Tolerance

a. Acquired immune tolerance is the ability of a host to react against nonself antigens while tolerating self-antigens. It can be induced in several ways.

32.9 Summary: The Role of Antibodies and Lymphocytes in Resistance

a. Although both the humoral and cellular arms of the specific (adaptive) immune response have been considered separately, it is important to understand that the response of a vertebrate host to any particular pathogen may involve a complex set of responses. Both humoral and cellular immune responses can join nonspecific (innate) defenses to ensure a maximal survival advantage against viral and bacterial pathogens.

Key Terms

acquired immune tolerance   734
acquired immunity   706
adjuvant   719
agglutination reaction   732
agglutinin   732
allotypes   712
anamnestic response   719
anergy   734
antibody or immunoglobulin (Ig)   710
antigen   708
antigen-binding fragment (Fab)   712
antigen-presenting cell (APC)   721
antigenic determinant site   708
antiserum   719
antitoxin   732
artificially acquired active immunity   706
artificially acquired passive immunity   707
B-cell antigen receptor (BCR)   728
CD95 pathway   725
cellular (cell-mediated) immunity   706

classical complement pathway   734
clonal selection theory   718
clone   718
cluster of differentiation molecule (CD)   709
constant regions (C_L and C_H)   711
crystallizable fragment (Fc)   711
cytotoxic T (T_C) cell   724
cytotoxic T lymphocyte (CTL)   725
domain   712
effector response   706
epitope   708
fas gene   725
H-2 complex   722
hapten   709
hemagglutination   732
hemagglutinin   732
human leukocyte antigen complex (HLA)   722
humoral (antibody-mediated) immunity   706
hybridoma   720

idiotype   712
IgA   714
IgD   714
IgE   714
IgG   713
IgM   713
immune complex   732
immune surveillance   736
immunotoxin   721
isotype   712
J chain   713
major histocompatibility complex (MHC)   722
membrane attack complex   734
memory B cell   718
monoclonal antibody (MAb)   719
myeloma cell   720
naturally acquired active immunity   706
naturally acquired passive immunity   706

perforin pathway   725
precipitation   732
precipitin   732
precipitin reaction   732
regulator T cell   725
secretory IgA (sIgA)   714
serum   719
superantigen   709
T-cell antigen receptor (TCR)   721
T-dependent antigen   729
T-helper (T_H) cell   725
T_H1 cell   725
T_H2 cell   725
T_H0 cell   725
T-independent antigen   730
titer   719
toxin neutralization   732
T-suppressor (T_S) cell   727
valence   708
variable regions (V_L and V_H)   711
viral neutralization   732

Questions for Thought and Review

1. What is the difference between a nonspecific (innate) host defense mechanism and a specific (adaptive) host defense mechanism? Give examples.
2. What are the major differences between natural and artificial immunity and between active and passive immunity?
3. Compare humoral and cell-mediated immunity.
4. A person with AIDS has a low T-helper/T-cytotoxic cell ratio. What problem(s) does this create?

5. How do B and T cells cooperate in the immune response? What is the role of macrophages?
6. All antibodies are proteins. What protein property has favored their use as antibodies in preference to some other type of biomolecule?
7. How did the clonal selection theory inspire the development of monoclonal antibody techniques?
8. What is the difference in the kinetics of antibody formation in response to a first and second exposure to the same antigen?

9. What is the basis of immunologic memory?
10. Describe the derivation of B and T cells.
11. Discuss the role of the classical complement pathway in host defense.
12. What is the purpose of acquired immune tolerance?
13. Why are T-cell receptors an important link in T-cell functioning?

Critical Thinking Questions

1. The placenta has many functions and one of them is as an immunosuppressive organ. Why is this necessary?
2. Why, evolutionarily, would sperm cells not express MHC on their surfaces?

3. What functionally important proteins are on the surface of B cells? T cells?
4. One of the main differences between an immature B cell and a plasma cell is the amount of en-

doplasmic reticulum. Explain which has more and why.
5. How would you distinguish a stem cell from a B cell at the DNA level?

For recommended readings on these and related topics, see Appendix V.

chapter 33

Medical Immunology

Concepts

- Vaccines and immunizations are among the most cost-effective weapons for microbial disease prevention. Vaccines constitute one of the greatest achievements of modern medicine.
- Some individuals experience harmful overreactions of the immune system known as hypersensitivities. There are four types: type I is characterized by the release of physiological mediators from IgE-bound mast cells and basophils; type II results from complement-dependent lysis of cells; type III involves the formation of immune complexes that are deposited on basement membranes; and type IV arises from the reaction of T_H1 cells, cytokines, and macrophages.
- The union of antigen and antibody in vitro produces either a visible reaction or one that can be made visible in a variety of ways. These techniques can be used to identify viruses, microorganisms, and their products; to quantitate and identify antigens and antibodies; to follow the course of a disease; to determine the serotype of a microorganism; and to determine the amount of protection from disease an animal possesses.

The three greatest discoveries of medicine in the last 200 years are antisepsis, antibiotics, and vaccines.

—Stanley A. Plotkin

This illustration is a wood engraving that first appeared in *Harper's Weekly* in 1885. During this year, Louis Pasteur (left) oversaw the administration of a rabies vaccine by a colleague (right) to Joseph Meister (center), a young boy who had been bitten repeatedly by a dog with rabies. As a result of the vaccination, Joseph Meister survived the bite and became a custodian in Louis Pasteur's laboratory—now called the Pasteur Institute.

Historical Highlights

33.1 The First Immunizations

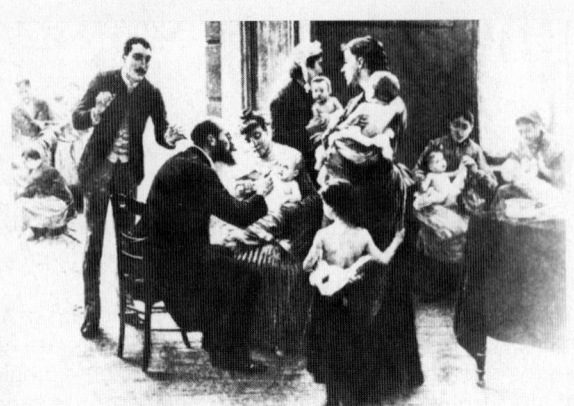

Since the time of the ancient Greeks, it has been recognized that people who have recovered from plague, smallpox, yellow fever, and various other infectious diseases rarely contract the diseases again. The first scientific attempts at artificial immunizations were made in the late eighteenth century by Edward Jenner (1749–1823), who was a country doctor from Berkley, Gloucestershire, England. Jenner investigated the basis for the widespread belief of the English peasants that anyone who had vaccinia (cowpox) never contracted smallpox. Smallpox was often fatal—10 to 40% of the victims died—and those who recovered had disfiguring pockmarks. Yet most English milkmaids, who were readily infected with cowpox, had clear skin because cowpox was a relatively mild infection that left no scars.

It was on May 14, 1796, that Jenner extracted the contents of a pustule from the arm of a cowpox-infected milkmaid, Sarah Nelmes, and injected it into the arm of eight-year-old James Phipps. As Jenner expected, immunization with the cowpox virus caused only mild symptoms in the boy. When he subsequently inoculated the boy with smallpox virus, the boy showed no symptoms of the disease.

Jenner then inoculated large numbers of his patients with cowpox pus, as did other physicians in England and on the European continent (see **Box figure**). By 1800 the practice known as vaccination had begun in America, and by 1805 Napoleon Bonaparte had ordered all French soldiers to be vaccinated.

Further work on immunization was carried out by Louis Pasteur (1822–1895). Pasteur discovered that if cultures of chicken cholera bacteria were allowed to age for two or three months the bacteria pro-

Nineteenth-Century Physicians Performing Vaccinations on Children.

duced only a mild attack of cholera when inoculated into chickens. Somehow the old cultures had become less pathogenic (attenuated) for the chickens. He then found that fresh cultures of the bacteria failed to produce cholera in chickens that had been previously inoculated with old, attenuated cultures. To honor Jenner's work with cowpox, Pasteur gave the name vaccine to any preparation of a weakened pathogen that was used (as was Jenner's "vaccine virus") to immunize against infectious disease.

---

 hapters 31 and 32 describe the various ways in which the immune system responds in vivo to protect the host by nonspecific (innate) resistance and specific (adaptive) immunity. Given the spectrum of diseases that result from inadequate or inappropriate immunity, clinical microbiologists are continually searching for ways to manipulate the immune response to benefit humankind. This chapter examines three areas of medical immunology in which this quest is being pursued: vaccines and immunizations, immune disorders, and antigen-antibody interactions in vitro.

33.1 VACCINES AND IMMUNIZATIONS

Active immunization is the protection of susceptible humans and domestic animals from communicable diseases by the administration of vaccines (vaccination). A **vaccine** [Latin *vacca,* cow] is a preparation from an infectious agent that is administered to humans and other animals to induce protective immunity. It may consist of a preparation of killed microorganisms; living, weakened (attenuated) microorganisms; inactivated bacterial toxins (toxoids); puri-

fied macromolecules; recombinant vectors (e.g., modified polio vaccine); or DNA vaccines (none approved yet) that are administered to an animal to induce immunity artificially.

The modern era of vaccines and vaccination began in 1798 with Edward Jenner's use of cowpox as a vaccine against smallpox (**Historical Highlights 33.1**) and in 1881 with Louis Pasteur's anthrax vaccine. Vaccines for other diseases did not emerge until the latter part of the nineteenth century, when largely through a process of trial and error, methods for inactivating and attenuating microorganisms were developed and vaccines were produced. Vaccines were eventually developed against most of the epidemic diseases that had plagued western Europe and North America (diphtheria, measles, mumps, whooping cough, German measles, polio). Indeed, toward the end of the twentieth century it began to seem that the combination of vaccines and antibiotics would temper the problem of microbial infections. Such optimism was cut short by the emergence of new or previously unrecognized diseases (*see section 37.8*) and antibiotic resistance to old ones (*see section 35.7*). Nevertheless, vaccination is still one of the most cost-effective weapons for microbial disease prevention, and vaccines arguably constitute one of the greatest achievements of modern medicine. **Table 33.1** summarizes the principal vaccines used to prevent viral and bacterial diseases in humans.

Table 33.1

Examples of Vaccines Used in the United States to Prevent Viral and Bacterial Diseases in Humans

Disease	Vaccine	Booster	Recommendation
Viral Diseases			
Brain infection	Inactivated Japanese encephalitus virus	None	Residents of and travelers to areas of endemic disease
Chickenpox	Attenuated Oka strain (Varivax)	None	Children 12–18 months; older children who have not had chickenpox
Hepatitis A	Inactivated virus (Havrix)	6–12 months	International travelers
Hepatitis B	HB viral antigen (Engerix-B, Recombivax HB)	None	High-risk medical personnel; children, birth to 18 months and 11–12 years of age
Influenza A	Inactivated virus or viral components	Yearly	All persons
Measles, Mumps, Rubella	Attenuated viruses (combination MMR vaccine)	None	Children 15–19 months old
Poliomyelitis	Attenuated (oral poliomyelitis vaccine, OPV) or inactivated vaccine	Adults as needed	Children 2–3 years old
Rabies	Inactivated virus	None	For individuals in contact with wildlife, animal control personnel, veterinarians
Respiratory disease	Live attenuated adenovirus	None	Military personnel
Smallpox	Live attenuated vaccinia virus	None	Laboratory and health-care personnel
Yellow fever	Attenuated virus	10 years	Military personnel and individuals traveling to endemic areas
Bacterial Diseases			
Anthrax	Extracellular components of unencapsulated *B. anthracis*	None	Agricultural workers, veterinary personnel; to limit biological warfare
Cholera	Fraction of *Vibrio cholerae*	6 months	Individuals in endemic areas, travelers
Diphtheria, Pertussis, Tetanus	Diphtheria toxoid, acellular or killed *Bordetella pertussis*, tetanus toxoid (DPT vaccine)	10 years	Children from 2–3 months old to 12 years, and adults
Haemophilus influenzae type b	Polysaccharide-protein conjugate (HbCV) or bacterial polysaccharide (HbPV)	None	Children under 5 years of age
Meningococcal infections	Bacterial polysaccharides of serotypes A/C/Y/W-135	None	Military; high-risk individuals
Plague	Fraction of *Yersinia pestis*	Yearly	Individuals in contact with rodents in endemic areas
Pneumococcal pneumonia	Purified *S. pneumoniae* polysaccharide of 23 pneumococcal types	None	Adults over 50 with chronic disease
Q fever	Inactivated *Coxiella burnetii*	None	Workers in slaughter houses and meat-processing plants
Tuberculosis	Attenuated *Mycobacterium bovis* (BCG vaccine)	3–4 years	Individuals exposed to TB for prolonged periods of time
Typhoid fever	Ty21a (live attenuated, polysaccharide)	None	Residents of and travelers to areas of endemic disease
Typhus fever	Killed *Rickettsia prowazekii*	Yearly	Scientists and medical personnel in areas where typhus is endemic

Vaccination of most children should begin at about 2 months of age. Before that age, they are protected by passive natural immunity. The recommended program of childhood immunizations for the United States is outlined in **figure 33.1.** Vaccination of adults depends on their risk group. For individuals living in close quarters (military personnel), the elderly, and those individuals with reduced immunity (transplant patients, people with sickle-cell anemia), vaccines for influenza, meningitis, and pneumonia are given. Depending on the country of travel, international travelers are routinely immunized against cholera, hepatitis, plague, polio, typhoid, typhus, and yellow fever. Health-care personnel and people exposed through life-style patterns are immunized against hepatitis B. Routine immunization against anthrax is re-

served for individuals (veterinarians or agricultural workers) who might come into contact with infected animals or products from the animals. Veterinarians, forest rangers, and others whose jobs involve contact with animals are vaccinated against rabies and plague. The recent use of anthrax spores by terrorists (*see section 37.10*) has prompted the vaccination of military personnel against anthrax.

Passive immunization, artificially acquired passive immunity (*see figure 32.2*), can be produced by injecting an animal or human with preformed antibodies that have been produced in another animal, in another human, or in vitro (*see section 32.3*). This type of immunization is called passive because protection does not require participation of the recipient's immune system.

Figure 33.1 Recommended Regimen and Indications for Vaccinations of Children in the United States. *(Data from The American Academy of Pediatrics.)*

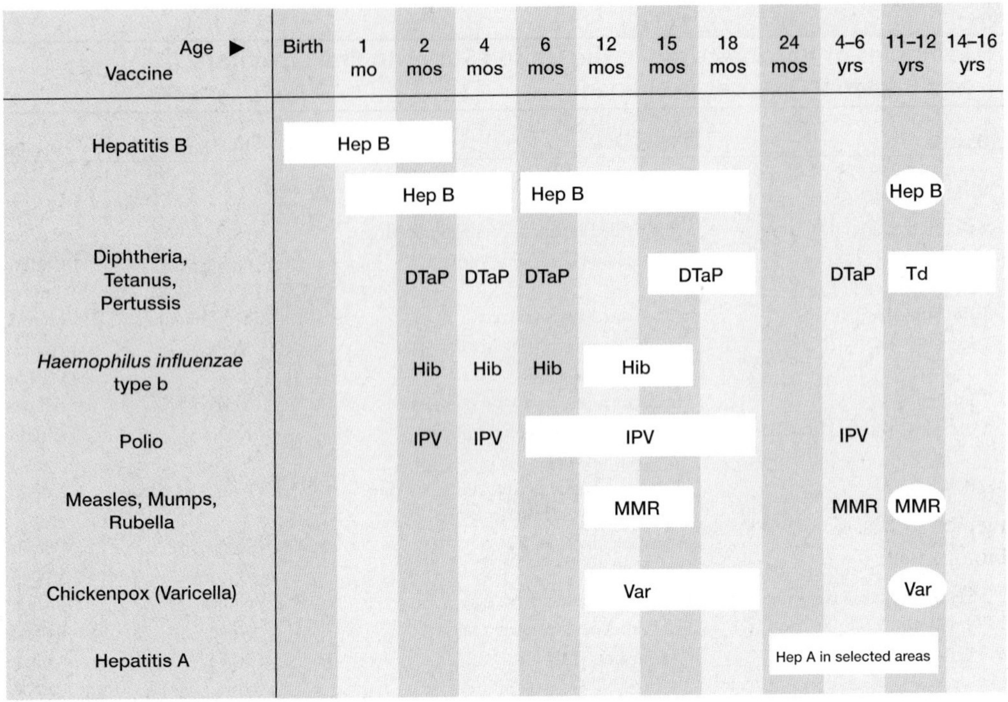

Bars indicate range of recommended ages for immunization. Ovals indicate vaccine to be given if previous recommended doses were missed or given earlier than the recommended age. See text for details.

Table 33.2

Immune Globulins and Antitoxins Used for Passive Immunization against Specific Organisms and Diseases in the United States

Organism/Disease	Agent[a]
Black widow spider bite	Horse antivenom
Snakebite	Horse antivenom
Acute respiratory failure (respiratory syncytial virus)	Monoclonal antibodies
Botulism	Botulinum antitoxin (equine)
Diphtheria	Diphtheria antitoxin (equine)
Hepatitis A and B	Pooled human immune gamma globulin
Measles	Pooled human immune gamma globulin
Rabies	Rabies immune globulin administered around the wound in addition to being injected intramuscularly
Tetanus	Tetanus immune globulin
Eczema	Vaccinia immune globulin
Immunocompromised individuals	Varicella-zoster immune globulin

[a]Immune globulins and antitoxins are administered intramuscularly unless otherwise noted.

Passive immunization is routinely administered to individuals exposed to certain microbial pathogens that cause diseases such as botulism, diphtheria, hepatitis, measles, rabies, and tetanus as well as to protect them against snake and spider bites (**table 33.2**). However, this form of immunization should be used only when absolutely necessary because of the risks involved such as devel-

oping anaphylaxis, serum sickness, or a type III hypersensitivity reaction. Furthermore, the protection lasts only as long as the antibody molecules survive in the recipient—months with antibodies from another human, but only weeks with antibodies from animals or in vitro methods.

Types of Vaccines and Their Characteristics

As previously noted, the diseases for which we already have vaccines are ones in which the infection is acute and either resolves in several weeks or causes death of the infected individual. In the absence of vaccination, many individuals would survive as a result of the immune system defeating the invader. For these diseases, the vaccine mimics the pathogen and causes an immune response similar to that raised by the pathogen. These vaccines have eradicated smallpox, pushed polio to the brink of extinction, and spared countless individuals from hepatitis A and B, measles, rotavirus disease, tetanus, typhus, and other dangerous diseases.

In contrast, successful vaccines have yet to be developed for too many deadly or debilitating diseases (AIDS, herpes, hepatitis C, malaria, tuberculosis) because they are due to chronic infections. In a chronic infection, the pathogen is able to evade the immune system or subvert it into making ineffective responses. Successful vaccines against these chronic diseases must be able to stimulate immune responses that are similar to those resulting from most natural exposures to the pathogen. To date, this has been an unsolvable challenge in medical immunology and vaccine development.

In the remainder of this section, various approaches to the design of vaccines, both currently used vaccines and experimental

Table 33.3

A Comparison of Inactivated (Killed) and Attenuated (Live) Vaccines

Major Characteristic	Inactivated Vaccine	Attenuated Vaccine
Booster shots	Multiple boosters required	Only a single booster
Production	Virulent microorganism inactivated by chemicals or irradiation	Virulent microorganism grown under adverse conditions or passed through different hosts until avirulent
Reversion tendency	None	May revert to a virulent form
Stability	Very stable, even where refrigeration is unavailable	Less stable
Type of immunity induced	Humoral	Humoral and cell-mediated

Source: Adapted from Goldsby, T. J. Kindt, and B. A. Osborne, *Kuby Immunology.* 2000, New York: W. H. Freeman.

ones, are described and examined in terms of their ability to induce humoral and cell-mediated immunity and the production of memory cells. **Vaccinomics,** the application of genomics and bioinformatics to vaccine development, is bringing a fresh approach to the Herculean problem of making vaccines against various microorganisms and parasites.

Whole-Organism Vaccines

Many of the current vaccines in use for humans that are effective against viral and bacterial diseases consist of whole microorganisms that are either inactivated (killed) or **attenuated** (live but avirulent) (table 33.1). These are termed **whole-organism vaccines.** The major characteristics of these vaccines are compared and contrasted in **table 33.3.** Inactivated vaccines are effective, but they often require several boosters and normally do not adequately stimulate cell-mediated immunity or secretory IgA production. In contrast, attenuated vaccines usually are given in a single dose and stimulate both humoral and cell-mediated immunity.

Even though whole-organism vaccines are considered the "gold standard" of existing vaccines, they can be problematic in their own way. For example, whole-organism vaccines fail to shield against some diseases. Attenuated vaccines that do work can also cause full-blown illness in individuals whose immune system is compromised (AIDS patients, cancer patients undergoing chemotherapy, the elderly). These same individuals may also contract the disease from healthy people who have been vaccinated recently. Moreover, attenuated viruses can at times mutate in ways that restore virulence, as has happened in some monkeys given an attenuated simian form of the AIDS virus. In the case of very lethal diseases, the risk of reversion to virulence is intolerable. Whole-organism vaccines, whether live or dead, have another big drawback. Since they are composed of complete pathogens, they retain molecules that are not involved in evoking immunity. These molecules, as well as contaminants that are unavoidable by-products of the manufacturing process, can trigger allergic or other disruptive reactions.

Purified Macromolecules as Vaccines

A few of the common risks associated with whole-organism vaccines can be avoided by using only specific, purified macromolecules derived from pathogenic microorganisms. Currently, there are three general forms of **macromolecule vaccines:** (1) capsu-

Table 33.4

Purified Macromolecule Vaccines Currently Available for Human Use

Type of Purified Macromolecule (Disease or Microorganism)	Form of Vaccine
Capsular polysaccharide	
Haemophilus influenzae type b	Polysaccharide-protein conjugate (HbCV) or bacterial polysaccharide (HbPV)
Neisseria meningitidis	Polysaccharides of serotypes A/C/Y/W-135
Streptococcus pneumoniae	23 distinct capsular polysaccharides
Surface antigen	
Hepatitis B	Recombinant surface antigen (HbsAg)
Toxoids	
Diphtheria	Inactivated exotoxin
Tetanus	Inactivated exotoxin

lar polysaccharides, (2) recombinant surface antigens, and (3) inactivated exotoxins called **toxoids** (*see p. 769*). The purified macromolecule vaccines that are currently available for humans are listed in **table 33.4.**

Recombinant-Vector Vaccines

Genetic vaccines are quite different in structure from whole-organism vaccines. It is now possible to isolate genes that encode major antigens from a pathogen and insert them into nonvirulent viruses or bacteria. The vaccines are usually delivered by needle injection or by a device called a gene gun. The attenuated microorganism serves as a vector, replicating within the host and expressing the gene product of the pathogen-encoded antigenic proteins. The antigens can elicit humoral immunity when they escape from the vector, and they can also elicit cellular immunity when they are broken down and properly displayed on the cell surface (just as occurs when cells harbor an active pathogen).

Recently several microorganisms have been used in the production of these **recombinant-vector vaccines.** Examples include adenovirus, vaccinia virus, canarypox virus, attenuated poliovirus, and attenuated strains of *Salmonella* and *Mycobacterium.*

DNA Vaccines

A more complicated genetic vaccine to emerge in recent years is the DNA vaccine. A **DNA vaccine** elicits protective immunity against a microbial pathogen by activating both branches of the immune system: humoral and cellular. Long-lasting memory cells also are generated.

The immunization procedure begins with the injection into muscle of a plasmid preparation that contains genes for pathogen antigens. The plasmids are taken up by muscle cells, enter the cell nuclei, and express their antigen genes. The muscle cells commence protein synthesis and produce the pathogen's antigenic proteins.

In the humoral response, the antigenic proteins are released from the muscle cell and bind to B-cell antigen receptors. At the same time, antigen-presenting cells ingest the antigenic proteins, break them down, and display the fragments on a class II MHC molecule. Helper T cells then recognize the antigen and secrete T_H2 cytokines that activate B cells containing bound antigen. The activated B cells multiply and differentiate into plasma cells that release antibodies which bind to the pathogen and neutralize it or mark it for destruction. Other B cells become memory cells that will protect the host against future infections with the same pathogen.

The cellular immune response begins when the muscle cells display the same antigenic proteins or protein fragments (peptides) on MHC class I molecules. The plasmid vaccine vector also becomes incorporated into antigen-presenting cells, which synthesize the encoded antigens and express fragments of the antigens on MHC I molecules along with co-stimulatory molecules. Cytotoxic T cells recognize these signals and are additionally stimulated by cytokines (this time, T_H1 cytokines) from helper T cells. The activated cytotoxic T cells multiply and attack cells infected by the pathogen. Some activated T cells also develop into memory T cells and protect against future infections.

At present, there are human trials under way with several different DNA vaccines against malaria, AIDS, influenza, hepatitis B, and herpesvirus. Vaccines against a number of cancers (lymphomas, prostate, colon) are also being tested.

---

1. What is a vaccine? Briefly describe and contrast active and passive immunization.
2. Describe the major differences between inactivated and attenuated vaccines. Discuss the advantages and disadvantages of using attenuated microorganisms as vaccines.
3. Give the three types of purified macromolecules that are currently used as vaccines.
4. What are some advantages of genetically engineered vaccines? Some disadvantages?

33.2   IMMUNE DISORDERS

Chapters 31 and 32 presented the awesome power of the immune system in protecting the host against the continual onslaught of microbial invaders and cancer cells. Like any system in a vertebrate host, disorders (malfunctions) also occur in the immune system.

Immune disorders can be categorized as hypersensitivities, autoimmune diseases, transplantation (tissue) rejection, and immunodeficiencies. Each of these immune disorders is now discussed.

Hypersensitivities

Hypersensitivity is an exaggerated immune response that results in tissue damage and is manifested in the individual on second or subsequent contact with an antigen. Hypersensitivity reactions can be classified as either immediate or delayed. Obviously immediate reactions appear faster than delayed ones, but the main difference between them is in the nature of the immune response to the antigen. Realizing this fact, Peter Gell and Robert Coombs developed a classification system for reactions responsible for hypersensitivities in 1963. Their system correlates clinical symptoms with information about immunologic events that occur during hypersensitivity reactions. The Gell-Coombs classification system divides hypersensitivity into four types: I, II, III, and IV.

Type I Hypersensitivity

Allergic reactions occur when an individual who has produced IgE antibody in response to an inocuous antigen (**allergen**) subsequently encounters the same allergen. An **allergy** [Greek *allos*, other and *ergon*, work] is one kind of type I hypersensitivity reaction.

Type I hypersensitivity is IgE-mediated. It is characterized by an allergic reaction occurring immediately following an individual's second contact with the responsible antigen (the allergen). Upon initial exposure to a soluble allergen, B cells are stimulated to differentiate into plasma cells and produce specific IgE with the help of T cells (**figure 33.2**). This IgE is sometimes called a **reagin**, and the individual has a hereditary predisposition for its production. Once synthesized, IgE binds to the Fc receptors of mast cells (basophils and eosinophils can also be activated) and sensitizes these cells, making the individual allergic to the allergen. When a second exposure to the allergen occurs, the allergen attaches to the surface-bound IgE on the sensitized mast cells causing degranulation. Degranulation releases physiological mediators such as histamine, leukotrienes, heparin, prostaglandins, PAF (platelet-activation factor), ECF-A (eosinophil chemotactic factor of anaphylaxis), and proteolytic enzymes. These mediators trigger smooth muscle contractions, vasodilation, increased vascular permeability, and mucous secretion. The inclusive term for these responses is **anaphylaxis** [Greek *ana*, up, back, again; and *phylaxis*, protection]. Anaphylaxis can be divided into systemic and localized reactions.

Systemic anaphylaxis is a generalized response that occurs when an individual sensitized to an allergen receives a subsequent exposure to it. The reaction is immediate due to the large amount of mast cell mediators released over a short period. Usually there is respiratory impairment caused by smooth muscle constriction in the bronchioles. The arterioles dilate, which greatly reduces arterial blood pressure and increases capillary permeability with rapid loss of fluid into the tissue spaces. Be-

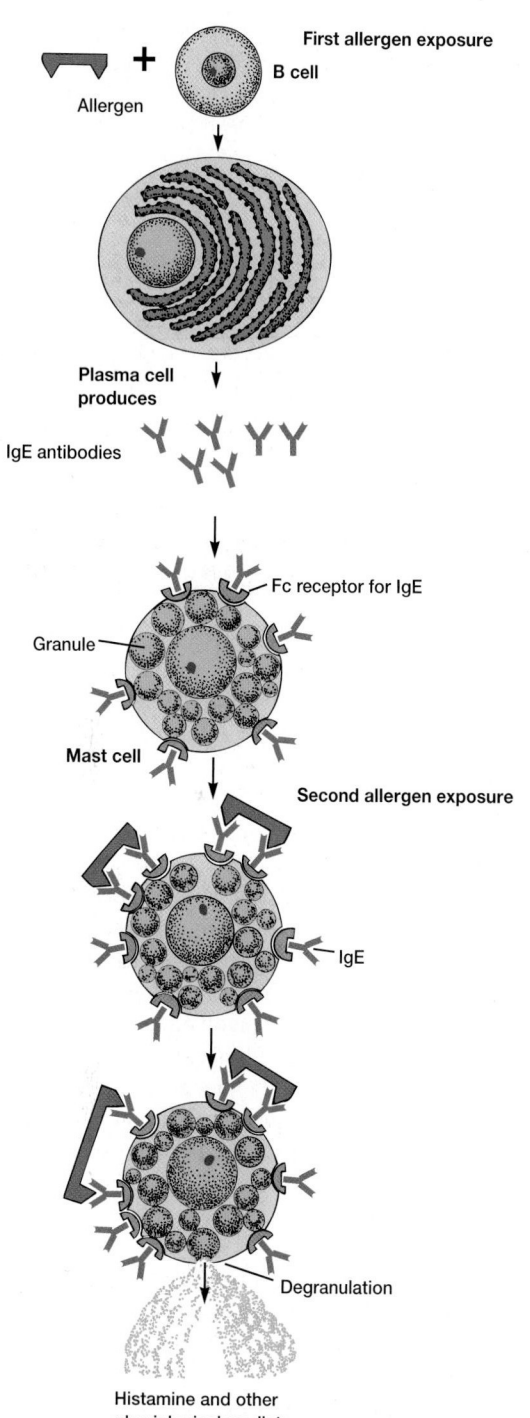

Figure 33.2 Type I Hypersensitivity. This type of hypersensitivity occurs when IgE antibodies attach to mast cells. The combination of these antibodies with allergens stimulates the mast cell (or basophil) to degranulate and produce the physiological mediators that cause the anaphylactic reaction, asthma, or hay fever.

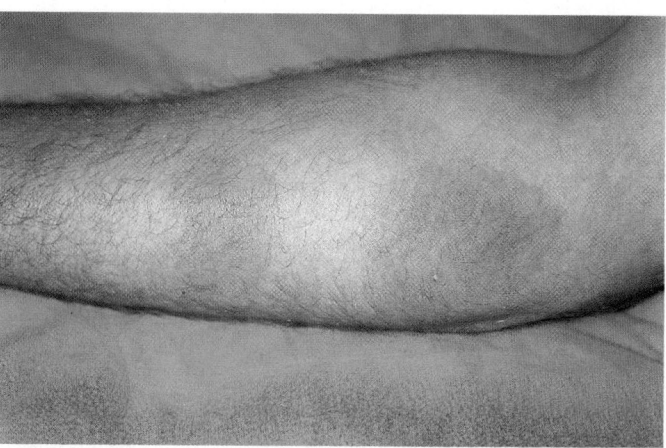

Figure 33.3   The Anaphylactic Response to Bee Venom. This person has been stung on the arm by a bee, leading to type I hypersensitivity in a generalized area.

cause of these reactions the individual can die within a few minutes from reduced venous return, asphyxiation, reduced blood pressure, and circulatory shock. Common examples of allergens that can produce systemic anaphylaxis include drugs (penicillin),

passively administered antisera, peanuts, and insect venom from the stings or bites of wasps, hornets, or bees (**figure 33.3**).

Localized anaphylaxis is called an atopic ("out of place") allergy. The symptoms that develop depend primarily on the route by which the allergen enters the body. **Hay fever** (allergic rhinitis) is a good example of an atopic allergy involving the upper respiratory tract. Initial exposure involves airborne allergens—such as plant pollen, fungal spores, animal hair and dander, and house dust mites—that sensitize mast cells located within the mucous membranes. Reexposure to the allergen causes the typical localized anaphylactic response: itchy and tearing eyes, congested nasal passages, coughing, and sneezing. Antihistamine drugs are used to help alleviate these symptoms.

Bronchial asthma (asthma means panting) is an example of an atopic allergy involving the lower respiratory tract. Common allergens are the same as for hay fever. In bronchial asthma, however, the air sacs (alveoli) become overdistended and fill with fluid and mucus; the smooth muscle contracts and narrows the walls of the bronchi. Bronchial constriction produces a wheezing or whistling sound during exhalation. Symptomatic relief is obtained from bronchodilators that help relax the bronchial muscles, and from expectorants and liquefacients that dissolve and expel mucous plugs that accumulate.

Allergens that enter the body through the digestive system may cause food allergies. **Hives** (eruptions of the skin) are a good diagnostic sign of a true food allergy. Once established, type I food allergies are usually permanent but can be partially controlled with antihistamines or by avoidance of the allergen.

Skin testing can be used to identify the allergen responsible for allergies. These tests involve inoculating small amounts of suspect allergen(s) into the skin. Sensitivity to the antigen is shown by a rapid inflammatory reaction characterized by redness, swelling, and itching at the site of inoculation (**figure 33.4**). The affected area in which the allergen-mast cell reaction takes place is called a wheal and flare reaction site.

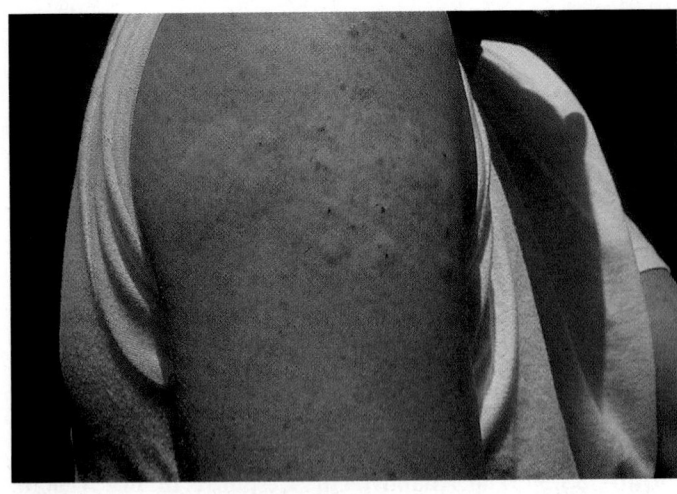

(a)

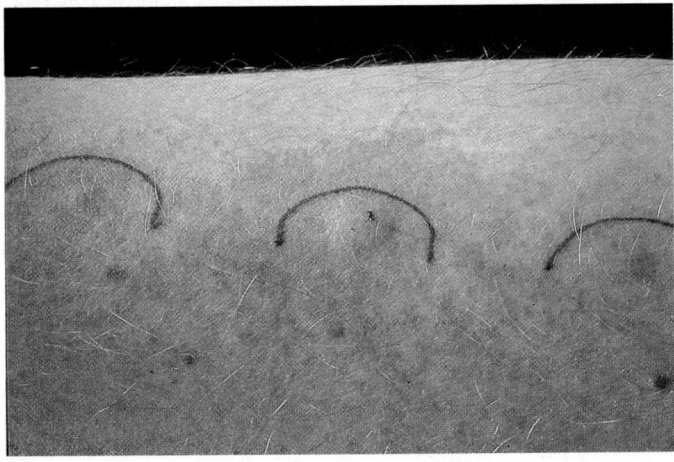

(b)

Figure 33.4 In Vivo Skin Testing. (a) Skin prick tests with grass pollen in a person with summer hay fever. Notice the various reactions with increasing dosages (from top to bottom). (b) Skin patch test. The surface of the skin (left) is abraded and the suspect allergic extract placed on the skin. After 48 hours (center) it is eczematous and positive for the suspect antigen.

Once the responsible allergen has been identified, the individual should avoid contact with it. At times this is not possible, and **desensitization** is warranted. This procedure consists of a series of allergen doses injected beneath the skin to stimulate the production of IgG antibodies rather than IgE antibodies. The circulating IgG antibodies can then act as blocking antibodies to intercept and neutralize allergens before they have time to react with mast cell-bound IgE. Recent evidence suggests that suppressor T-cell activity also may cause a decrease in IgE synthesis. Desensitizations are about 65 to 75% effective in individuals whose allergies are caused by inhaled allergens.

Type II Hypersensitivity

Type II hypersensitivity is generally called a cytolytic or cytotoxic reaction because it results in the destruction of host cells, ei-

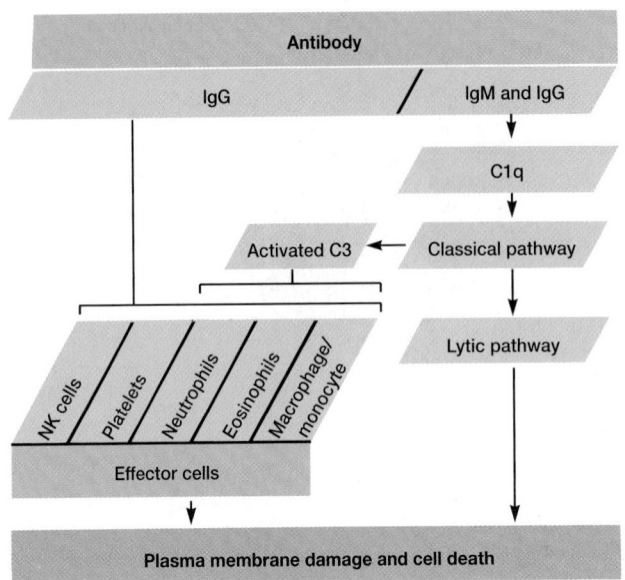

Figure 33.5 Type II Hypersensitivity. The action of antibody occurs through effector cells or the membrane attack complex, which damages target cell plasma membranes, causing cell destruction.

ther by lysis or toxic mediators. In type II hypersensitivity, IgG or IgM antibodies are directed against cell surface or tissue-associated antigens. They usually stimulate the complement pathway and a variety of effector cells (**figure 33.5**). The antibodies interact with complement (Clq) and the effector cells through their Fc regions. The damage mechanisms are a reflection of the normal physiological processes involved in interaction of the immune system with pathogens. Classical examples of type II hypersensitivity reactions are (1) the response when a person receives a transfusion with blood from a donor with a different blood group; (2) erythroblastosis fetalis; and (3) autoimmune hemolytic anemia.

Type III Hypersensitivity

Type III hypersensitivity involves the formation of immune complexes (**figure 33.6a**). Normally these complexes are removed effectively by the fixed monocytes and macrophages of the monocyte-macrophage system. In the presence of excess amounts of some soluble antigens, the antigen-antibody complexes may not be efficiently removed. Their accumulation can lead to a hypersensitivity reaction from complement that triggers a variety of inflammatory processes. This inflammation causes damage, especially of blood vessels (vasculitis; figure 33.6b), kidney glomerular basement membranes (glomerulonephritis), joints (arthritis), and skin (systemic lupus erythematosus).

Diseases resulting from type III reactions can be placed into three groups. First, a persistent viral, bacterial, or protozoan infection, together with a weak antibody response, leads to chronic immune complex formation and eventual deposition of the complex in host tissues. Second, the continued production of autoantibody to self-antigen during an autoimmune disease can lead to

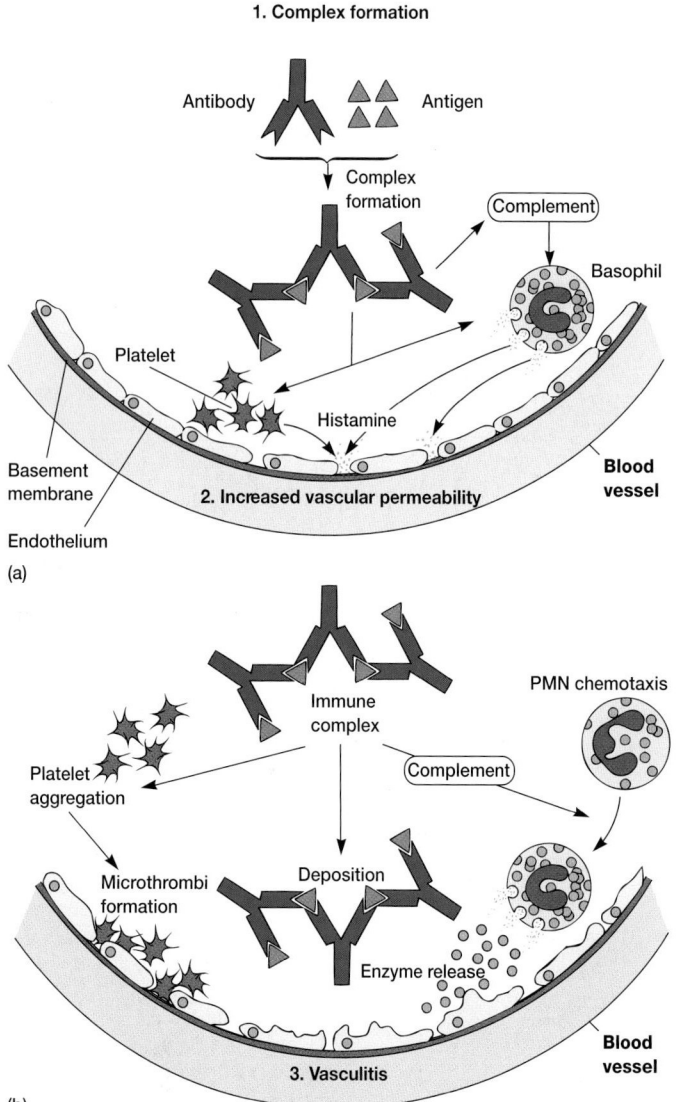

1. Complex formation

Antibody   Antigen

Complex
formation

Complement

Basophil

Platelet

Histamine

Basement
membrane

Blood
vessel

Endothelium

2. Increased vascular permeability

(a)

PMN chemotaxis

Immune
complex

Complement

Platelet
aggregation

Microthrombi
formation

Deposition

Enzyme release

Blood
vessel

3. Vasculitis

(b)

Figure 33.6   Type III Hypersensitivity.   Deposition of immune complexes in blood vessel walls. (**a**) Antibody and antigen combine to form immune complexes. These activate complement, which causes basophils and platelets to degranulate and release histamine and other mediators. These mediators increase vascular permeability. (**b**) The increased permeability allows the immune complexes to be deposited in the blood vessel wall. This induces platelet aggregation to form microthrombi (blood clots) on the vessel wall. PMNs, stimulated by complement, degranulate causing enzymatic damage to the blood vessel wall.

prolonged immune complex formation. This overloads the monocyte-macrophage system, and tissue deposition of the complexes occurs, as in the disease **systemic lupus erythematosus.** Third, immune complexes can form at body surfaces (such as the lungs), following repeated inhalation of allergens from molds, plants, or animals. For example, in farmer's lung disease, an individual has circulating antibodies to fungi after being exposed

repeatedly to moldy hay. These antibodies are primarily IgG. When the allergens (fungal spores) enter the alveoli of the lungs, local immune complexes form, leading to inflammation.

Some group A streptococcal infections can produce an immunologically mediated acute glomerulonephritis. Although the mechanism is not completely understood, it is believed that complexes of antibody and streptococcal antigen are deposited within the kidney glomeruli and generate a type III hypersensitivity reaction.

Type IV Hypersensitivity

Type IV hypersensitivity involves delayed T-cell-mediated immune reactions. A major factor in the type IV reaction is the time required for a special subset of T_H1 cells (often called delayed-type hypersensitivity [T_{DTH}] cells) to migrate to and accumulate near the antigens. This usually takes a day or more.

Type IV reactions occur when antigens, especially those binding to tissue cells, are phagocytosed by macrophages and then presented to receptors on the T_H1 cell surface in the context of class I MHC. Contact between the antigen and T_H1 cell causes the cell to proliferate and release cytokines. Cytokines attract lymphocytes, macrophages, and basophils to the affected tissue. Extensive tissue damage may result. Examples of type IV hypersensitivities include tuberculin hypersensitivity (the **TB skin test**), allergic contact dermatitis, some autoimmune diseases, transplantation rejection, and killing of cancer cells.

In tuberculin hypersensitivity a partially purified protein called tuberculin, which is obtained from the bacillus that causes tuberculosis (*Mycobacterium tuberculosis; see section 39.1*), is injected into the skin of the forearm (**figure 33.7a**). The response in a tuberculin-positive individual begins in about 8 hours, and a reddened area surrounding the injection site becomes indurated (firm and hard) within 12 to 24 hours. The T_H1 cells that migrated into the injection site are responsible for the induration. The reaction reaches its peak in 48 hours and then subsides. The size of the induration is directly related to the amount of antigen that was introduced and to the degree of hypersensitivity of the tested individual. Other microbial products used in type IV skin testing are histoplasmin for histoplasmosis, coccidioidin for coccidioidomycosis, lepromin for leprosy, and brucellergen for brucellosis.

Allergic contact dermatitis is caused by haptens (*see figure 32.4*) that combine with proteins in the skin to form the allergen that elicits the immune response. The haptens are the antigenic determinants, and the skin proteins are the carrier molecules for the haptens. Examples of these haptens include cosmetics, plant materials (catechol molecules from poison ivy and poison oak; figure 33.7b), topical chemotherapeutic agents, metals, and jewelry (especially jewelry containing nickel).

Several important chronic diseases involve cell and tissue destruction by type IV hypersensitivity reactions. These diseases are caused by viruses, mycobacteria, protozoa, and fungi that produce chronic infections in which the macrophages and T cells are continually stimulated. Examples are leprosy, tuberculosis, leishmaniasis, candidiasis, and herpes simplex lesions.

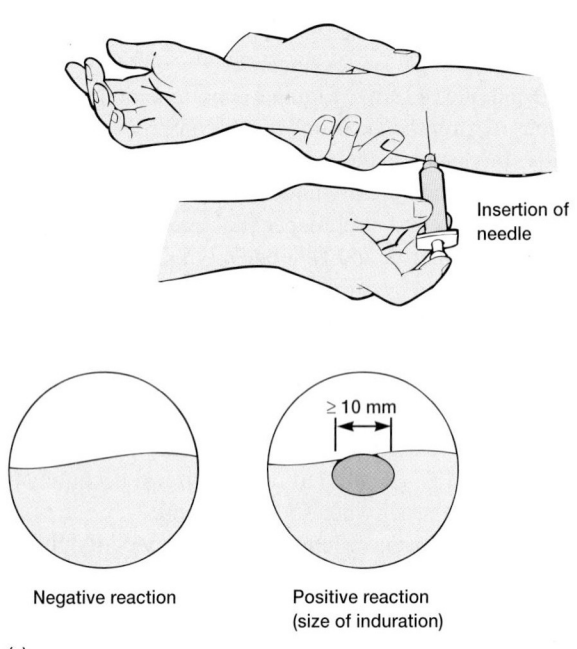

(a)

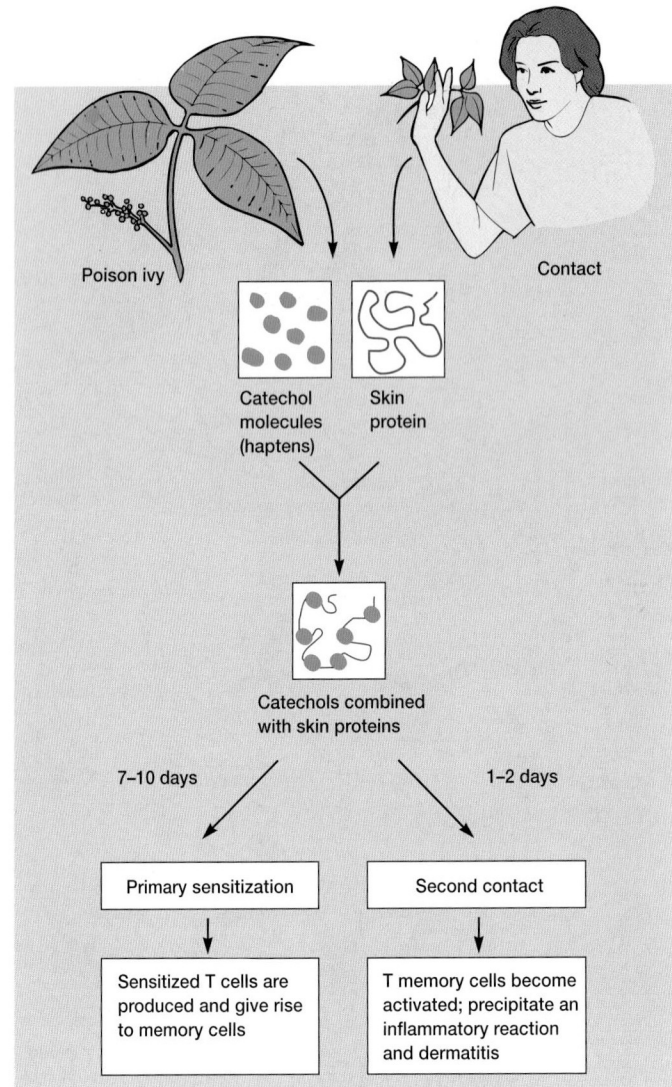

(b)

Figure 33.7 Mechanism of Type IV Hypersensitivity.
(**a**) The tuberculin skin test. A positive reaction is one in which the diameter of induration is 10 mm or more. (**b**) In this example of contact dermatitis to poison ivy, a person initially becomes exposed to the allergen, predominantly 3-n-pentadecyl-catechol found in the resinous sap material uroshiol, which is produced by the leaves, fruit, stems, and bark of the poison ivy plant. The catechol molecules combine with high molecular weight skin proteins and act as haptens. After 7–10 days sensitized T cells are produced and give rise to memory T cells. Upon second contact, the catechols bind to the same skin proteins, and the T memory cells become activated in only 1–2 days, leading to the inflammatory reaction (contact dermatitis).

---

1. Discuss the mechanism of type I hypersensitivity reactions and how these can lead to systemic and localized anaphylaxis.
2. What causes a wheal and flare reaction site?
3. Why are type II hypersensitivity reactions called cytolytic or cytotoxic?
4. What characterizes a type III hypersensitivity reaction? Give an example.
5. Characterize a type IV hypersensitivity reaction.
6. What is the TB skin test used for?

Autoimmune Diseases

As discussed earlier, the body is normally able to distinguish its own self-antigens from foreign nonself antigens and does not mount an immunologic attack against the former. This phenomenon is called immune tolerance. At times the body loses tolerance and mounts an abnormal immune attack, either with antibodies or T cells, against a person's own self-tissue antigens.

It is important to distinguish between autoimmunity and autoimmune disease. Autoimmunity often is benign, whereas autoimmune disease often is fatal. **Autoimmunity** is characterized only by the presence of serum autoantibodies. It is a normal consequence of aging; is readily inducible by infectious agents, organisms, or drugs; and is potentially reversible (it disappears when the offending "agent" is removed or eradicated). **Autoimmune disease** results from the activation of self-reactive T and B cells that, following stimulation by genetic or environmental triggers, cause actual tissue damage (**table 33.5**).

Four factors influence the development of autoimmune disease. Two major factors are genetic and viral. The third factor is endocrine; the hormone estrogen promotes autoimmune disease, and androgens act as natural immunosuppressive agents. These physiological modulatory effects of sex hormones acting on nor-

Table 33.5
Some Autoimmune Diseases in Humans

Disease	Autoantigen	Pathophysiology
Type II Hypersensitivity: Antibody to Cell-Surface or Cell Matrix Antigens (Cytotoxic)		
Acute rheumatic fever	Streptococcal cell wall antigens; antibodies cross-react with cardiomyocytes	Arthritis, scarring of heart valves, myocarditis
Autoimmune hemolytic anemia	Rh blood group, I antigen	Red blood cells are destroyed by complement and phagocytosis, anemia
Autoimmune thrombocytopenia purpura	Platelet integrin	Perfuse bleeding
Goodpasture's syndrome	Basement membrane collagen	Glomerulonephritis, pulmonary hemorrhage
Graves' disease	Thyroid-stimulating hormone receptor	Hyperthyroidism
Myasthenia gravis	Acetycholine receptor	Progressive muscular weakness
Pemphigus vulgaris	Cadherin in epidermis	Skin blisters
Type III Hypersensitivity: Immune Complex Disease		
Systemic lupus erythematosus	DNA, histones, ribosomes	Arthritis, glomerulonephritis, vasculitis, rash
Type IV Hypersensitivity: T-Cell–Mediated Disease		
Type 1 diabetes mellitus	Pancreatic beta cell antigen	Beta cell destruction
Multiple sclerosis	Myelin basic protein	Demyelination of axons
Rheumatoid arthritis[a]	Unknown synovial joint antigen	Joint inflammation and destruction

[a] Type III hypersensitivity processes also may be involved.

mal immune responses partially explain the marked female predominance of autoimmune diseases. The fourth factor is psychoneuroimmunological—that is, the influence of stress and neurochemicals on the immune response. Overall, all four of these factors can affect gene expression, which directly or indirectly interferes with important immunoregulatory actions.

Although their causal mechanism is not well known, these diseases are more common in older people and may involve viral or bacterial infections. Some investigators believe that the release of abnormally large quantities of antigens may occur when the infectious agent causes tissue damage. The same agents also may cause body proteins to change into forms that stimulate antibody production or T-cell activation. Simultaneously the activity of T-suppressor cells, which normally limits this type of reaction, seems to be repressed. Many autoimmune diseases have a genetic component. For example, there is a well-known association between an individual's susceptibility to Graves' disease and multiple sclerosis and specific determinants on the major histocompatibility complex.

Transplantation (Tissue) Rejection

Tissue transplant rejection is the third area (after hypersensitivity and autoimmunity) in which the immune system can act detrimentally. It is occasionally desirable to replace a nonfunctional or damaged body part by transplanting a tissue or organ from one person to another. Transplants between genetically different individuals within a species are termed **allografts** [Greek *allos,* other]. Some transplants do not stimulate an immune response. For example, a transplanted cornea is rarely rejected since lymphocytes do not circulate into the anterior chamber of the eye. This site is considered an immunologically privileged site. Another example of a privileged tissue transplant is the heart valve transplanted from a pig to a human. Such a graft between different species is termed a **xenograft** [Greek *xenos,* strayed].

With allografts, there is the possibility that the recipient's cells will recognize the donor's tissues as foreign. This triggers the recipient's immune mechanisms, which may destroy the donor tissue. Such a response is called a tissue rejection reaction. A tissue rejection reaction can occur by two different mechanisms. First, foreign major histocompatibility complex (MHC) class II molecules on the graft stimulate host T-helper cells to aid cytotoxic T cells in graft destruction (**figure 33.8a**). Cytotoxic T cells recognize the graft through the foreign MHC class I molecules. A second mechanism involves the T-helper cells reacting to the graft and releasing cytokines (figure 33.8b). The cytokines stimulate macrophages to enter the graft and destroy it.

As presented in figure 33.8, the major histocompatibility complex molecules play a dominant role in tissue rejection reactions because of their unique association with the recognition system of T cells. Unlike antibodies, T cells cannot recognize or react directly with non-MHC molecules (viruses, allergens).

Figure 33.8   Graft Cell Destruction. (a) Foreign MHC class II molecules on the graft cell stimulate host T-helper cells to help cytotoxic T cells destroy the target graft cell. Cytotoxic T cells recognize the graft cell by its foreign MHC class I molecules. (b) T-helper cells reacting to the graft cell release cytokines that stimulate macrophages to enter the graft and destroy it by cytotoxic action.

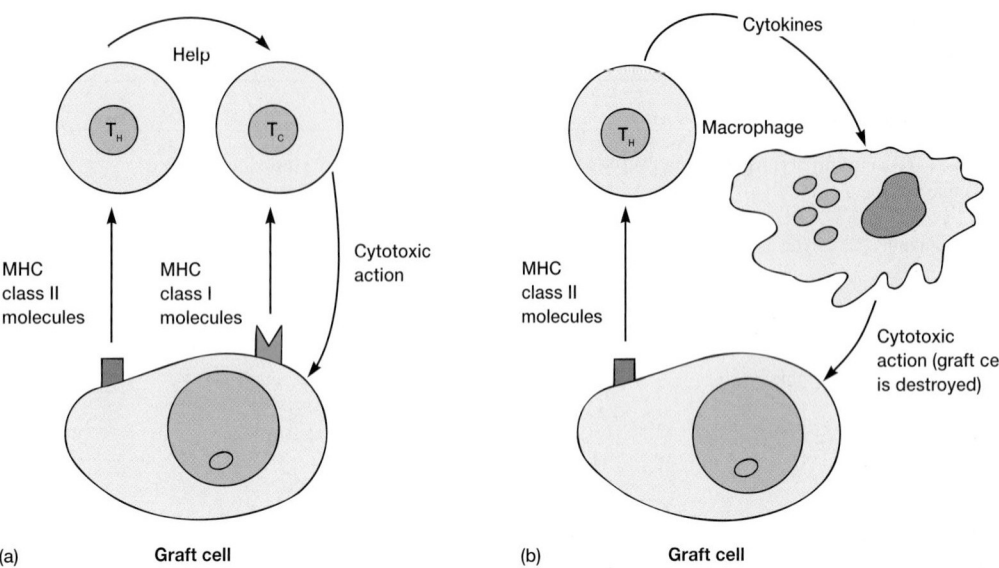

Table 33.6		
Some Congenital Immune Deficiencies in Humans		
Condition	**Symptoms**	**Cause**
Chronic granulomatous disease	Defective monocytes and neutrophils leading to recurrent bacterial and fungal infections	Failure to produce reactive oxygen intermediates due to defective NADPH oxidase
X-linked agammaglobulinemia	Plasma cell or B-cell deficiency and inability to produce adequate specific humoral antibodies	Defective B-cell differentiation due to loss of tyrosine kinase
DiGeorge syndrome	T-cell deficiency and very poor cell-mediated immunity	Lack of thymus or a poorly developed thymus
Severe combined immunodeficiency disease (SCID)	Both antibody production and cell-mediated immunity impaired due to a great reduction of B- and T-cell levels	Various mechanisms (e.g., defective B- and T-cell maturation because of X-linked gene mutation; absence of adenosine deaminase in lymphocytes)

They recognize these molecules only in association with, or complexed to, an MHC molecule. There are two classes of MHC molecules. Class I MHC molecules are present on every cell in the body and consequently are important targets of the rejection reaction. Class II MHC molecules are involved in T-helper cell activation as previously discussed. The greater the antigenic difference between class I molecules of the recipient and donor tissues, the more rapid and severe the rejection reaction is likely to be. However, the reaction can sometimes be minimized if recipient and donor tissues are matched as closely as possible.

Immunosuppression in organ transplant cases also can lead to **graft-versus-host disease.** This occurs when the transplanted tissue contains immunocompetent cells that recognize host antigens and attack the host. The immunosuppressed recipient cannot control the response of the grafted tissue. Graft-versus-host disease is a common problem in allogenic bone marrow transplants. The transplanted bone marrow contains many mature post-thymic T cells. These cells recognize the host MHC antigens and attack the immunosuppressed recipient's normal tissue cells. Currently one way to prevent graft-versus-host disease is to deplete the bone

marrow of mature T cells by using immunosuppressive techniques. Examples include drugs that attack T cells (azathioprine, methotrexate, and cyclophosphamide), immunosuppressive drugs (cyclosporin, tacrolimus, and rapamycin), anti-inflammatory drugs (corticosteroids), irradiation of the lymphoid tissue, and antibodies directed against T-cell antigens.

Immunodeficiencies

Defects in one or more components of the immune system can result in its failing to recognize and respond properly to antigens. Such **immunodeficiencies** can make a person more prone to infection than those people capable of a complete and active immune response. Despite the increase in knowledge of functional derangements and cellular abnormalities in the various immunodeficiency disorders, the fundamental biological errors responsible for them remain largely unknown. To date, most genetic errors associated with these immunodeficiencies are located on the X chromosome and produce primary or congenital immunodeficiencies (**table 33.6**). Other immunodeficiencies can be acquired because of infections by immunosuppressive microorganisms (chronic mucocutaneous candidiasis) or by some viruses (HIV).

1. What is an autoimmune disease and how might it develop?
2. What is an immunologically privileged site and how is it related to transplantation success?
3. How does a tissue rejection reaction occur?
4. Describe an immunodeficiency. How might immunodeficiencies arise?

 ### 33.3 ANTIGEN-ANTIBODY INTERACTIONS IN VITRO

Chapter 32 (*see pp. 731–33*) deals with those antigen-antibody interactions that occur within an animal (in vivo). Many of these same reactions can take place outside the animal (in vitro) under controlled laboratory conditions and are extensively used in diagnostic testing. The branch of medical immunology concerned with antigen-antibody reactions in vitro is **serology** [serum and -ology].

Due to current advances in medical immunology, there has been a marked increase in the number, sensitivity, and specificity of serological tests. The increase results from a better understanding of the cell surface of various lymphocytes, the production of monoclonal antibodies, the development of radioactive and enzyme-linked assays, and the use of fluorescence technology. In this section some more common serological tests employed in the diagnosis of microbial and immunologic diseases are presented.

Agglutination

As noted in figure 32.30, when an immune complex is formed by cross-linking cells or particles with specific antibodies, it is called an agglutination reaction. Agglutination reactions usually form visible aggregates or clumps (**agglutinates**) that can be seen with the unaided eye. Direct agglutination reactions are very useful in the diagnosis of certain diseases. For example, the **Widal test** is a reaction involving the agglutination of typhoid bacilli when they are mixed with serum containing typhoid antibodies from an individual who has typhoid fever. Immune complex formation (pp. 732–33)

Techniques have also been developed that employ microscopic synthetic latex spheres coated with antigens. These coated microspheres are extremely useful in diagnostic agglutination reactions. For example, the modern pregnancy test detects the elevated level of human chorionic gonadotropin (HCG) hormone that occurs in female urine and blood early in pregnancy. Latex agglutination tests are also used to detect antibodies that develop during certain mycotic, helminthic, and bacterial infections, and in drug testing.

Hemagglutination usually results from antibodies cross-linking red blood cells through attachment to surface antigens and is routinely used in blood typing. In addition, some viruses can

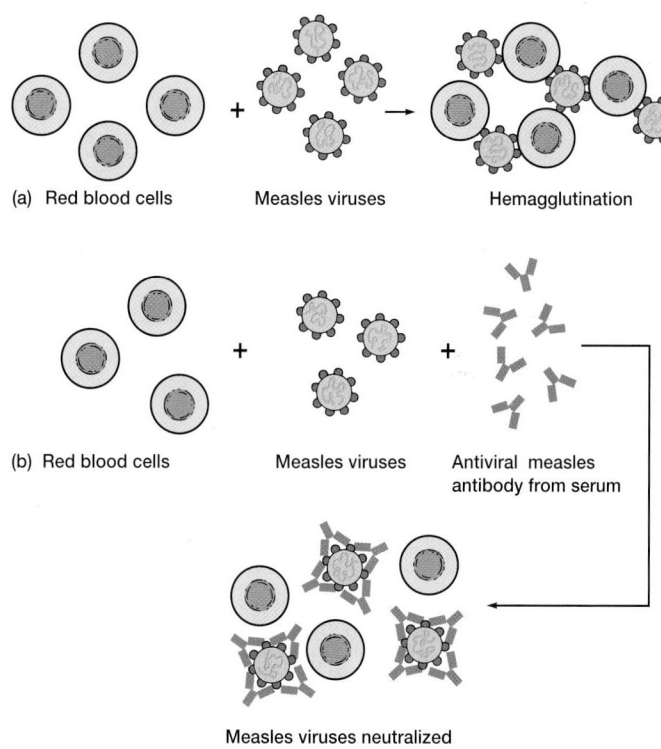

(a) Red blood cells   Measles viruses   Hemagglutination

(b) Red blood cells   Measles viruses   Antiviral measles antibody from serum

Measles viruses neutralized and hemagglutination inhibited

Figure 33.9 Viral Hemagglutination. (**a**) Certain viruses can bind to red blood cells causing hemagglutination. (**b**) If serum containing specific antibodies to the virus is mixed with the red blood cells, the antibodies will neutralize the virus and inhibit hemagglutination (a positive test).

accomplish **viral hemagglutination.** For example, if a person has a certain viral disease, such as measles, antibodies will be present in the serum to react with the measles viruses and neutralize them. Normally, hemagglutination occurs when measles viruses and red blood cells are mixed but is not seen when the person's serum is added to the mixture; this shows that the serum antibodies have neutralized the measles viruses and is considered a positive test (**figure 33.9**). This hemagglutination inhibition test is widely used to diagnose influenza, measles, mumps, mononucleosis, and other viral infections (*see chapter 38*).

Agglutination tests are also used to measure antibody titer (*see figure 32.19*). In the tube or well agglutination test, a specific amount of antigen is added to a series of tubes (**figure 33.10a**) or shallow wells in a microtiter plate (figure 33.10b). Serial dilutions of serum (1/20, 1/40, 1/80, 1/160, etc.) containing the antibody are then added to each tube or well. The greatest dilution of serum showing an agglutination reaction is determined, and the reciprocal of this dilution is the serum antibody titer.

1. What is serology?
2. When would you use the Widal test?
3. Why does hemagglutination occur and how can it be used in the clinical laboratory?

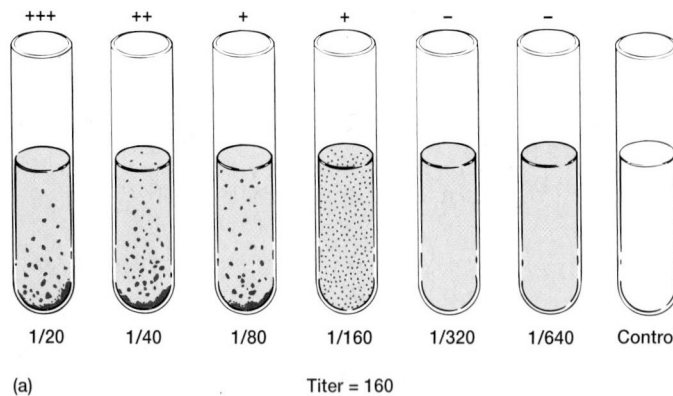

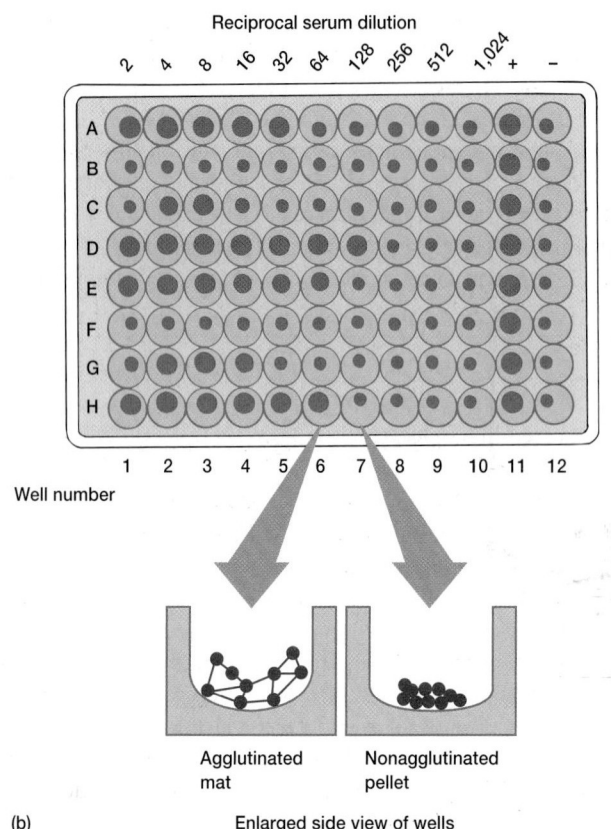

Figure 33.10   Agglutination Tests.   (a) Tube agglutination test for determining antibody titer. The titer in this example is 160 since there is no agglutination in the next tube in the dilution series (1/320). The blue in the dilution tubes indicates the presence of the patient's serum. **(b)** A microtiter plate illustrating hemagglutination. The antibody is placed in the wells (1–10). Positive controls (row 11) and negative controls (row 12) are included. Red blood cells are added to each well. If sufficient antibody is present to agglutinate the cells, they sink as a mat to the bottom of the well. If insufficient antibody is present, they form a pellet at the bottom. Can you read the different titers in rows A–H?

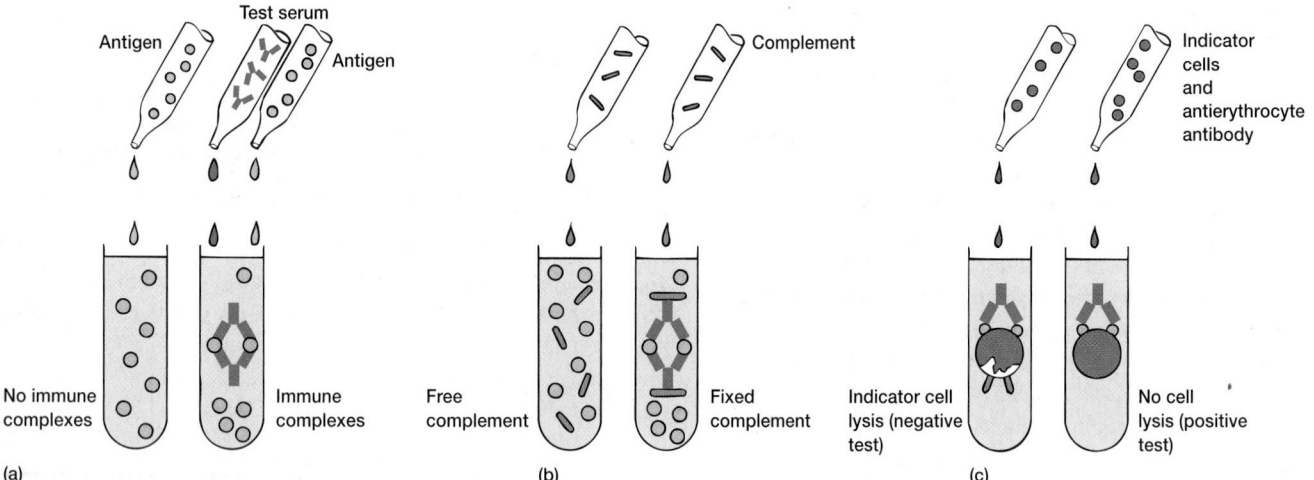

Figure 33.11   Complement Fixation.   (a) Test serum is added to one test tube. A fixed amount of antigen is then added to both tubes. If antibody is present in the test serum, immune complexes form. **(b)** When complement is added, if complexes are present, they fix complement and consume it. **(c)** Indicator cells and a small amount of antierythrocyte antibody are added to the two tubes. If there is complement present, the indicator cells will lyse (a negative test); if the complement is consumed, no lysis will occur (a positive test).

Complement Fixation

When complement binds to an antigen-antibody complex, it becomes "fixed" and "used up." Complement fixation tests are very sensitive and can be used to detect extremely small amounts of an antibody for a suspect microorganism in an individual's serum. A known antigen is mixed with test serum lacking complement (**figure 33.11a**). When immune complexes have had time to form, complement is added (figure 33.11b) to the mixture. If immune complexes are present, they will fix and consume complement. Afterward, sensitized indicator cells, usually sheep red blood cells previously coated with complement-fixing antibodies, are added to the mixture. Lysis of the indicator cells (figure 33.11*c*) results if immune complexes do not form in part *a* of the test because the antibodies are not present in the test serum. In the absence of antibodies, complement remains and lyses the indicator cells. On the other hand, if the specific antibodies are present in the test serum and complement is consumed by the immune complexes, insufficient amounts of complement will be available to lyse the indicator cells. Absence of lysis shows that specific antibodies are present in the test serum.

Complement fixation was once used in the diagnosis of syphilis (the Wassermann test) and is currently used in the diagnosis of certain viral, fungal, rickettsial, chlamydial, and protozoan diseases.

Enzyme-Linked Immunosorbent Assay

The ***enzyme-*l*inked *i*mmunosorbent *a*ssay* (**ELISA**) has become one of the most widely used serological tests for antibody or antigen detection. This test involves the linking of various "label" enzymes to either antigens or antibodies. Two basic methods are used: the double antibody sandwich assay and the indirect immunosorbent assay.

The double antibody sandwich assay is used for the detection of antigens (**figure 33.12***a*). In this assay, specific antibody is placed in wells of a microtiter plate (or it may be attached to a membrane). The antibody is absorbed onto the walls, coating and sensitizing the plate. A test antigen is then added to each well. If the antigen reacts with the antibody, the antigen is retained when the well is washed to remove unbound antigen. An antibody-enzyme conjugate specific for the antigen is then added to each well. The final complex is formed of an outer antibody-enzyme, middle antigen, and inner antibody—that is, it is a layered (Ab-Ag-Ab) sandwich. A substrate that the enzyme will convert to a colored product is then added, and any resulting product is quantitatively measured by optical density scanning of the plate. If the antigen has reacted with the absorbed antibodies in the first step, the ELISA test is positive. If the antigen is not recognized by the absorbed antibody, the ELISA test is negative because the unattached antigen has been washed away, and no antibody-enzyme is bound. This assay is currently being used for the detection of *Helicobacter pylori* infections, brucellosis, salmonellosis, and cholera. Many other antigens also can be detected by the sandwich method. For example, there are ELISA kits on the market that can test for many different food allergens.

The indirect immunosorbent assay detects antibodies rather than antigens. In this assay, antigen in appropriate buffer is incubated in the wells of a microtiter plate (figure 33.12*b*) and is absorbed onto the walls of the wells. Free antigen is washed away. Test antiserum is added, and if specific antibody is present, it binds to the antigen. Unbound antibody is washed away. Alternatively the test sample can be incubated with a suspension of latex beads that have the desired antigen attached to their surface. After allowing time for antibody-antigen complex formation, the beads are trapped on a filter and unbound antibody is washed away. An anti-antibody that has been covalently coupled to an enzyme, such as horseradish peroxidase, is added next. The antibody-enzyme complex (the conjugate) binds to the test antibody, and after unbound conjugate is washed away, the attached

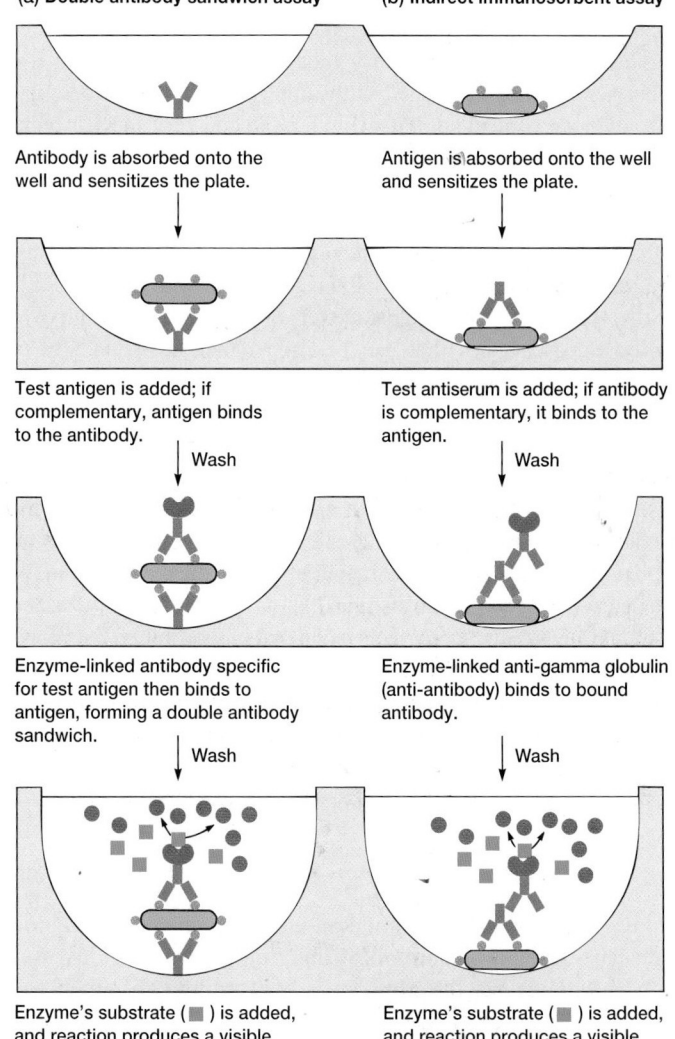

(a) Double antibody sandwich assay

Antibody is absorbed onto the well and sensitizes the plate.

Test antigen is added; if complementary, antigen binds to the antibody.

Wash

Enzyme-linked antibody specific for test antigen then binds to antigen, forming a double antibody sandwich.

Wash

Enzyme's substrate (■) is added, and reaction produces a visible color change (●) that is measured spectrophotometrically.

(b) Indirect immunosorbent assay

Antigen is absorbed onto the well and sensitizes the plate.

Test antiserum is added; if antibody is complementary, it binds to the antigen.

Wash

Enzyme-linked anti-gamma globulin (anti-antibody) binds to bound antibody.

Wash

Enzyme's substrate (■) is added, and reaction produces a visible color change (●) that is measured spectrophotometrically.

Figure 33.12   The ELISA or EIA Test. (**a**) The double antibody sandwich method for the detection of antigens. (**b**) The indirect immunosorbent assay for detecting antibodies. See text for details.

ligand is visualized by the addition of a chromogen. A **chromogen** is a colorless substrate acted on by the enzyme portion of the ligand to produce a colored product. The amount of test antibody is quantitated in the same way as an antigen is in the double antibody sandwich method. The indirect immunosorbent assay currently is being used to test for antibodies to human immunodeficiency virus (the causative agent of AIDS) and rubella virus (German measles), and to detect certain drugs in serum. For example, antigen-coated latex beads are used in the SUDS HIV-1 test to detect HIV serum antibodies in about 10 minutes.

Flow Cytometry and Fluorescence

Classical microbiological techniques (*see section 36.2*) are relatively slow in comparison to analytical techniques, in many cases due to the need to culture microorganisms. Furthermore, classical techniques are not useful in identifying those microorganisms that can't be cultured. The recent emergence of molecular biological techniques, particularly those using antibodies and nucleic acid probes combined with amplification techniques, has provided the clinical laboratory with rapid and very specific techniques for microbiological diagnosis. One such new technique is flow cytometry and fluorescence.

Flow cytometry allows single- or multiple-microorganism detection in clinical samples (*see section 36.1*) in an easy, reliable, fast way. In flow cytometry, microorganisms are identified on the basis of their peculiar cytometric parameters or by means of certain dyes called fluorochromes that can be used either independently or bound to specific antibodies or oligonucleotides. The flow cytometer forces a suspension of cells through a laser beam and measures the light they scatter or the fluorescence the cells emit as they pass through the beam. For example, cells can be tagged with a fluorescent antibody directed against a specific surface antigen. As the stream of cells flows past the laser beam, the fluorescent cells can be detected, counted, and even separated from the other cells in the suspension. The cytometer also can measure a cell's shape, size, and content of DNA and RNA. This technique has permitted the development of quantitative methods to access antimicrobial susceptibility and drug cytotoxicity in a rapid, accurate, and highly reproducible way. The most outstanding contribution of this technique is the ability to detect the presence of heterogenous microbial populations with different responses to antimicrobial treatments.

Immunoblotting (Western Blot)

Another immunologic technique used in the clinical microbiology laboratory is **immunoblotting.** Immunoblotting involves polyacrylamide gel electrophoresis of a protein specimen followed by transfer of the separated proteins to nitrocellulose sheets (*see figure 14.5*). Protein bands are then visualized by treating the nitrocellulose sheets with solutions of enzyme-tagged antibodies. This procedure demonstrates the presence of common and specific proteins among different strains of microorganisms (**figure 33.13**). Immunoblotting also can be used to show strain-specific immune responses to microorganisms, to serve as an important diagnostic indicator of a recent infection with a particular strain of microorganism, and to allow for prognostic implications with severe infectious diseases.

Immunodiffusion

Immunodiffusion refers to a precipitation reaction that occurs between an antibody and antigen in an agar gel medium. Two techniques are routinely used: single radial immunodiffusion and double diffusion in agar.

The **single radial immunodiffusion (RID) assay** or Mancini technique quantitates antigens. Monospecific antibody is added to agar, then the mixture is poured onto slides and allowed to set. Wells are cut in the agar and known amounts of standard antigen added. The unknown test antigen is added to a separate well (**figure 33.14a**). The slide is left for 24 hours or until equilibrium has been reached, during which time the antigen diffuses out of the

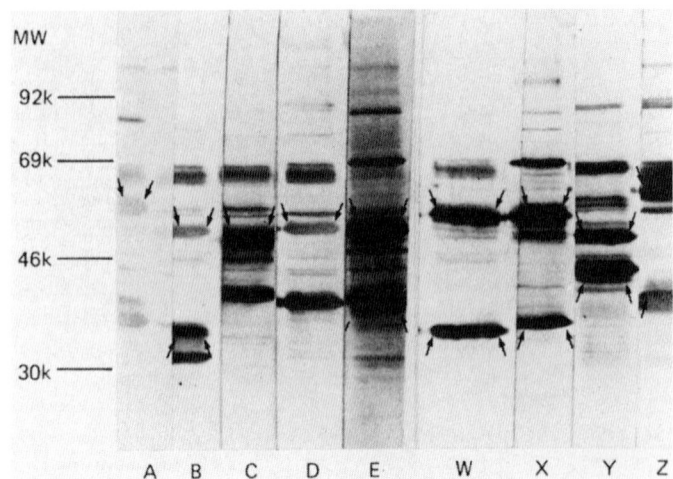

Figure 33.13   Immunoblotting (Western Blot).   Immunoblot of the standard strains of *Clostridium difficile.* Arrows indicate strain-specific bands (specific molecular weights) of the various proteins in different lanes (A–E). The molecular weight of the protein is indicated on the left.

wells to form insoluble complexes. The size of the resulting precipitation ring surrounding various dilutions of antigen selected is proportional to the amount of antigen in the well (the wider the ring, the greater the antigen concentration). This is because the antigen's concentration drops as it diffuses farther out into the agar. The antigen forms a precipitin ring in the agar when its level has decreased sufficiently to reach equivalence and combine with the antibody to produce a large, insoluble network. This method is commonly used to quantitate serum immunoglobulins, complement proteins, and other substances.

The **double diffusion agar assay (Ouchterlony technique)** is based on the principle that diffusion of both antibody and antigen (hence, double diffusion) through agar can form stable and easily observable immune complexes. Test solutions of antigen and antibody are added to the separate wells punched in agar. The solutions diffuse outward, and when antigen and the appropriate antibody meet, they combine and precipitate at the equivalence zone, producing an indicator line (or lines) (figure 33.14b). The visible line of precipitation permits a comparison of antigens for identity (same antigenic determinants), partial identity (cross-reactivity), or nonidentity against a given selected antibody. For example, if a V-shaped line of precipitation forms, this demonstrates that the antibodies bind to the same antigenic determinants in each antigen sample and are identical. If one well is filled with a different antigen that shares some but not all determinants with the first antigen, a Y-shaped line of precipitation forms, demonstrating partial identity. In this reaction the stem of the Y, called a spur, is formed if those antigen or antigenic determinants absent in the first well but present in the second one (antigen a in figure 33.14b) react with the diffusing antibodies. If two completely unrelated antigens are added to the wells, either a single straight line of precipitation forms between the two wells, or two separate lines of precipitation form, creating an X-shaped pattern, a reaction of nonidentity.

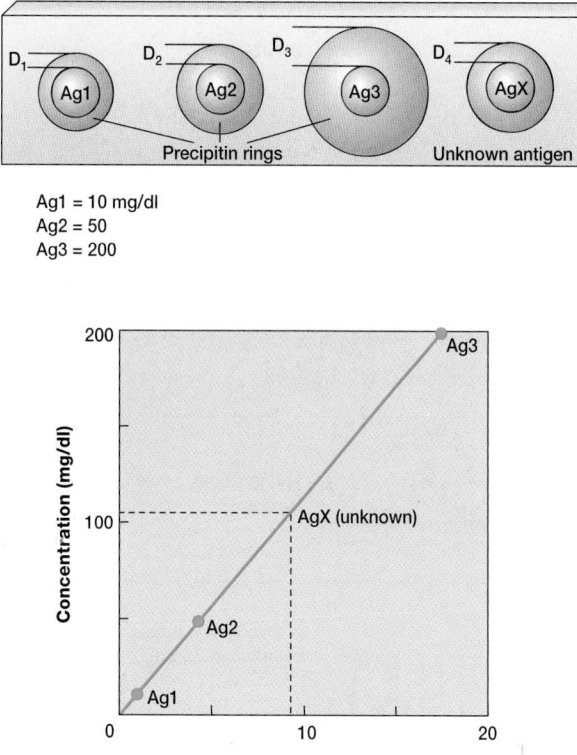

Ag1 = 10 mg/dl
Ag2 = 50
Ag3 = 200

(a)

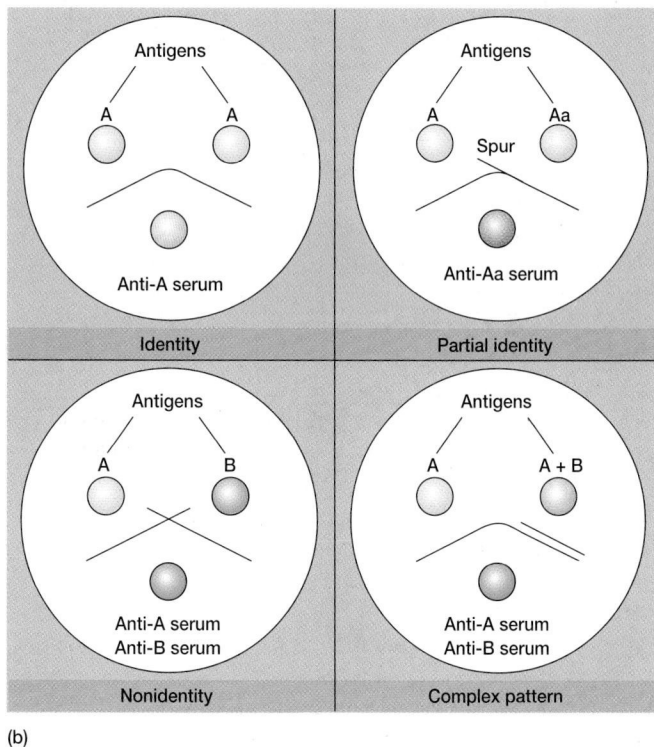

(b)

Figure 33.14　Immunodiffusion. (**a**) Single radial immunodiffusion assay. Three standard solutions of different antigen concentrations (Ag1, Ag2, Ag3) and an unknown (AgX) are placed on agar. After equilibration the ring diameters are measured. Usually the square of the diameter of the standard rings is plotted on the x-axis and the antigen concentration on the y-axis. From this standard curve the concentration of an unknown can be determined. (**b**) Double diffusion agar assay showing characteristics of identity (top left), reaction of nonidentity (bottom left), partial identity (top right), and a complex pattern (bottom right).

Immunoelectrophoresis

Some antigen mixtures are too complex to be resolved by simple diffusion and precipitation. Greater resolution is obtained by the technique of classical **immunoelectrophoresis** in which antigens are first separated based on their electrical charge, then visualized by the precipitation reaction. In this procedure antigens are separated by electrophoresis in an agar gel. Positively charged proteins move to the negative electrode, and negatively charged proteins move to the positive electrode (**figure 33.15a**). A trough is then cut next to the wells (figure 33.15b) and filled with antibody. If the plate is incubated, the antibodies and antigens will diffuse and eventually form precipitation bands or arcs (figure 33.15c) that can be better visualized by staining (figure 33.15d). This assay is used to separate the major blood proteins in serum for certain diagnostic tests.　Electrophoresis (pp. 319–20)

Immunofluorescence

Immunofluorescence is a process in which fluorochromes are exposed to UV, violet, or blue light to make them fluoresce or emit visible light. Dyes such as rhodamine B or fluorescein isothiocyanate (FITC) can be coupled to antibody molecules without

changing the antibody's capacity to bind to a specific antigen. Fluorochrome dyes also can be attached to antigens. There are two main kinds of fluorescent antibody assays: direct and indirect.

Direct immunofluorescence involves fixing the specimen (cell or microorganism) containing the antigen of interest onto a slide (**figure 33.16a**). Fluorescein-labeled antibodies are then added to the slide and incubated. The slide is washed to remove any unbound antibody and examined with the fluorescence microscope (*see figure 2.12*) for a yellow-green fluorescence. The pattern of fluorescence reveals the antigen's location. Direct immunofluorescence is used to identify antigens such as those found on the surface of group A streptococci and to diagnose enteropathogenic *Escherichia coli, Neisseria meningitidis, Salmonella typhi, Shigella sonnei, Listeria monocytogenes, Haemophilus influenzae* type b, and the rabies virus.　The fluorescence microscope (pp. 25–26)

Indirect immunofluorescence (figure 33.16b) is used to detect the presence of antibodies in serum following an individual's exposure to microorganisms. In this technique a known antigen is fixed onto a slide. The test antiserum is then added, and if the specific antibody is present, it reacts with antigen to form a complex. When fluorescein-labeled anti-immunoglobulin is added, it reacts with the fixed antibody. After incubation and washing, the slide is examined with the fluorescence microscope. The occurrence of

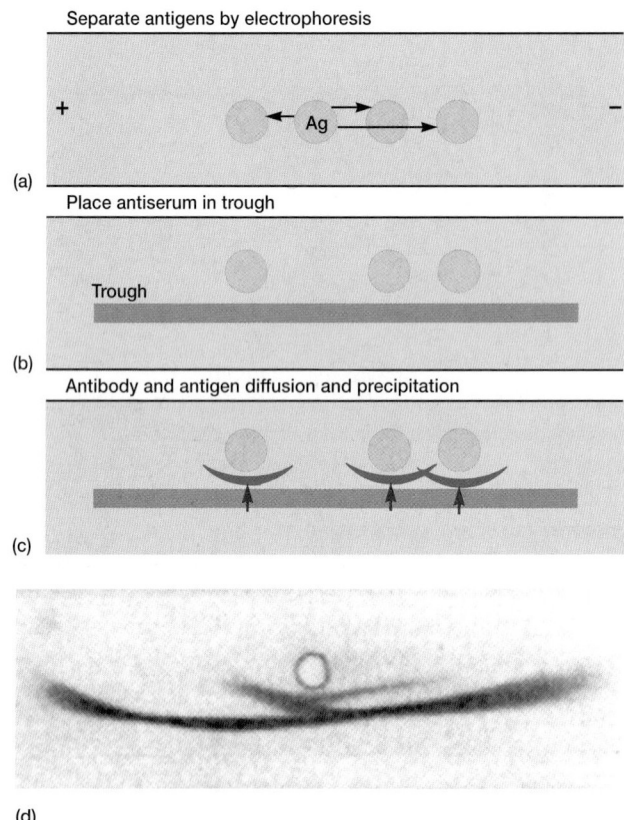

Separate antigens by electrophoresis

(a)

Place antiserum in trough

Trough

(b)

Antibody and antigen diffusion and precipitation

(c)

(d)

Figure 33.15 Classical Immunoelectrophoresis. (a) Antigens are separated in an agar gel by an electrical charge. (b) Antibody (antiserum) is then placed in a trough cut parallel to the direction of the antigen migration. (c) The antigens and antibodies diffuse through the agar and form precipitin arcs. (d) After staining, better visualization is possible.

fluorescence shows that antibody specific to the test antigen is present in the serum. Indirect immunofluorescence is used to identify the presence of *Treponema pallidum* antibodies in the diagnosis of syphilis (treponemal antibody absorption, FTA-ABS; *see figure 39.21*) as well as antibodies produced in response to other microorganisms.

Immunoprecipitation

The **immunoprecipitation** technique detects soluble antigens that react with antibodies called precipitins. The precipitin reaction occurs when bivalent or multivalent antibodies and antigens are mixed in the proper proportions. The antibodies link the antigen to form a large antibody-antigen network or lattice that settles out of solution when it becomes sufficiently large (**figure 33.17a**). Immunoprecipitation reactions occur only at the equivalence zone when there is an optimal ratio of antigen to antibody so that a lattice forms. If the precipitin reaction takes place in a test tube (figure 33.17b), a precipitation ring forms in the area in which the optimal ratio or equivalence zone develops. Immune complex formation (pp. 732–33)

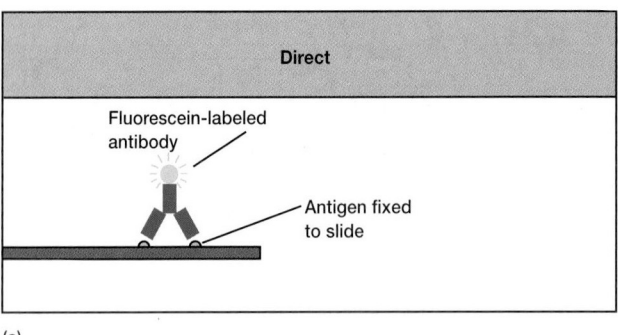

(a)

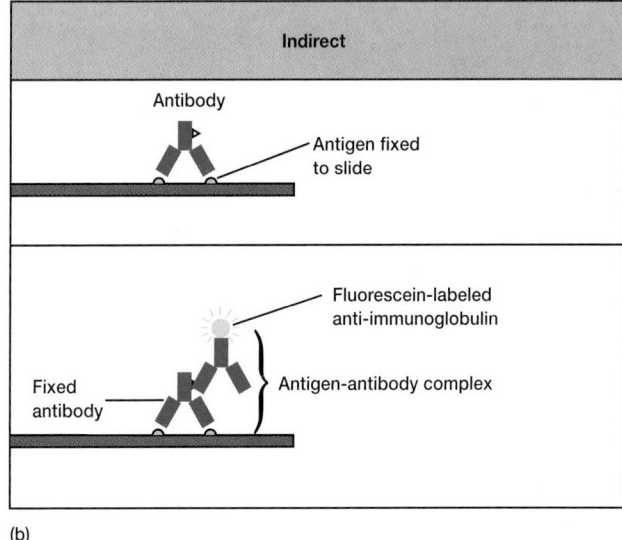

(b)

Figure 33.16 Direct and Indirect Immunofluorescence.
(a) In the direct fluorescent-antibody (FA) technique, the specimen containing antigen is fixed to a slide. Fluorescenated antibodies that recognize the antigen are then added, and the specimen is examined under a UV microscope for yellow-green fluorescence. (b) Indirect fluorescent-antibody technique (IFA). The antigen on a slide reacts with an antibody directed against it. The antigen-antibody complex is located with a fluorescent antibody that recognizes immunoglobulins.

1. What does a negative complement fixation test show? A positive test?
2. What are the two types of ELISA methods and how do they work? What is a chromogen?
3. What are some advantages of using flow cytometry in comparison to classical microbiological techniques to detect microorganisms?
4. Name two types of immunodiffusion tests and describe how they operate.
5. Describe the classical immunoelectrophoresis technique.
6. What is a fluorochrome?
7. Name and describe the two kinds of fluorescent antibody assays.
8. Specifically, when do immunoprecipitation reactions occur?

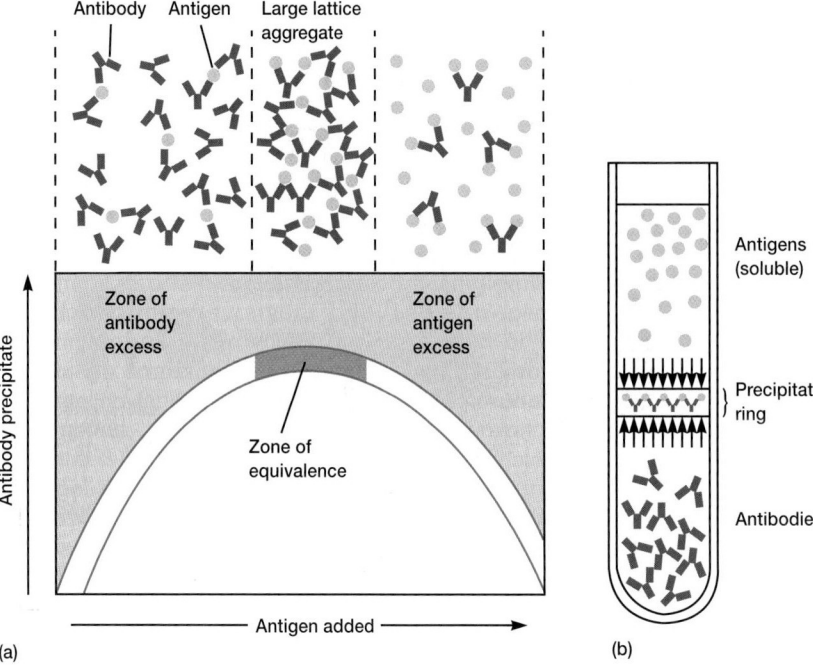

Figure 33.17 Immunoprecipitation. (a) Graph showing that a precipitation curve is based on the ratio of antigen to antibody. The zone of equivalence represents the optimal ratio for precipitation. **(b)** A precipitation ring test. Antibodies and antigens diffuse toward each other in a test tube. A precipitation ring is formed at the zone of equivalence.

Techniques & Applications

33.2 History and Importance of Serotyping

In the early 1930s Rebecca Lancefield (1895–1981) recognized the importance of serological tests. She developed a classification system for the streptococci based on the antigenic nature of cell wall carbohydrates. Her system is now known as the **Lancefield system** in which each different serotype is a Lancefield group and identified by a letter (A through T). This scheme is based on specific antibody agglutination reactions with cell wall carbohydrate antigens (C polysaccharides) extracted from the streptococci. Lancefield also showed that further subdividing of the group A streptococci into specific serological types was possible, based on the presence of type-specific M (protein) antigens.

More recently *Escherichia coli, Salmonella,* and other bacteria have been serotyped with specific antigen-antibody reactions involving flagella (H) antigens, capsular (K) antigens, and somatic (cell wall or O) antigens. Within *E. coli* there are over 167 different O antigens.

The current value of serotyping may be seen in the fact that *E. coli* O55, O111, and O127 serotypes are the ones most frequently associated with infantile diarrhea. Thus the serotype of *E. coli* from stool samples of infants with infantile diarrhea is of diagnostic value and aids in identifying the source of the infection.

Radioimmunoassay

The *radioimmuno*assay **(RIA)** technique has become an extremely important tool in biomedical research and clinical practice (e.g., in cardiology, blood banking, diagnosis of allergies, and endocrinology). Indeed, Rosalyn Yalow won the 1977 Nobel Prize in Physiology or Medicine for its development. RIA uses a purified antigen that is radioisotope-labeled and competes for antibody with unlabeled standard or antigen in experimental samples. The radioactivity associated with the antibody is then detected by means of radioisotope analyzers and autoradiography (photographic emulsions that show areas of radioactivity). If

there is much antigen in an experimental sample, it will compete with the radioisotope-labeled antigen for antigen-binding sites on the antibody, and little radioactivity will be bound. A large amount of bound radioactivity indicates that there is little antigen present in the experimental sample.

Serotyping

Serotyping refers to serological procedures used to differentiate strains (serovars or serotypes) of microorganisms that differ in the antigenic composition of a structure or product (**Techniques & Applications 33.2**). The serological identification of a pathogen strain has diagnostic value. Often the symptoms of infections depend on the nature of cell products released by a pathogen. Genes for virulence often occur in the same clone with genes for antigenic cell wall material. Therefore it is possible to identify a pathogen serologically by testing for cell wall antigens. For example, there are 84 strains of *Streptococcus pneumoniae,* each differing in the nature of its capsular material. These differences can be detected by capsular swelling (termed the **Quellung reaction**) if antisera specific for the capsular types are used (**figure 33.18**).

1. Describe the RIA technique.
2. What is serotyping?

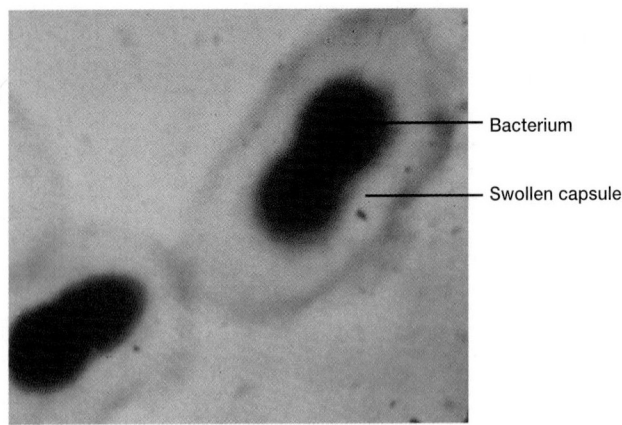

Figure 33.18 Serotyping. *Streptococcus pneumoniae* has reacted with a specific pneumococcal antiserum leading to capsular swelling (the Quelling reaction). The capsules seen around the bacteria indicate potential virulence.

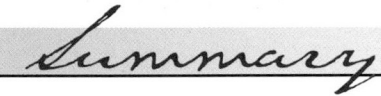

33.1 Vaccines and Immunizations

a. Vaccination is one of the most cost-effective weapons for microbial disease prevention, and vaccines constitute one of the greatest achievements of modern medicine.

b. Artificially acquired immunity to pathogens can be accomplished by either active or passive immunization (**tables 33.1, 33.2,** and **figure 33.1**).

c. Many of the current vaccines in use for humans (**table 33.3**) consists of whole organisms that are either inactivated (killed) or attenuated (live but avirulent).

d. Some of the risks associated with whole-organism vaccines can be avoided by using only specific purified macromolecules derived from pathogenic microorganisms (**table 33.4**). Currently, there are three general forms of macromolecule vaccines: capsular polysaccharides, recombinant surface antigens, and inactivated exotoxins (toxoids).

e. Recently a number of microorganisms have been used for recombinant-vector vaccines. The attenuated microorganism serves as a vector, replicating within the host, and expressing the gene product of the pathogen-encoded antigenic proteins. The proteins can elicit humoral immunity when the proteins escape from the cells and cellular immunity when they are broken down and properly displayed on the cell surface.

f. More complicated genetic vaccines are termed DNA vaccines. DNA vaccines elicit protective immunity against a pathogen by activating both branches of the immune system: humoral and cellular.

33.2 Immune Disorders

a. When the immune response occurs in an exaggerated form and results in tissue damage to the individual, the term hypersensitivity is applied. There are four types of hypersensitivity reactions, designated as types I through IV (**figures 33.2–33.7**).

b. Autoimmune diseases result when self-reactive T and B cells attack the body and cause tissue damage (**table 33.5**). A variety of factors can influence the development of autoimmune disease.

c. The immune system can act detrimentally and reject tissue transplants. There are different types of transplants. Xenografts involve transplants of privileged tissue between different species, and allografts are transplants between genetically different individuals of the same species.

d. Immunodeficiency diseases are a diverse group of conditions in which an individual's susceptibility to various infections is increased; several severe diseases can arise because of one or more defects in the specific (adaptive) or nonspecific (innate) immune response.

33.3 Antigen-Antibody Interactions In Vitro

a. Agglutination reactions in vitro usually form aggregates or clumps (agglutinates) that are visible with the naked eye. Tests have been developed, such as the Widal test, latex microsphere agglutination reaction, hemagglutination, and viral hemagglutination (**figure 33.9**), to detect antigen as well as to determine antibody titer.

b. The complement fixation test can be used to detect a specific antibody for a suspect microorganism in an individual's serum (**figure 33.11**).

c. The enzyme-linked immunosorbent assay (ELISA) involves the linking of various enzymes to either antigens or antibodies. Two basic methods are involved: the double antibody sandwich method and the indirect immunosor-

bent assay (**figure 33.12**). The first method detects antigens and the latter, antibodies.

d. Flow cytometry and fluorescence allow single- or multiple-microorganism detection based on their peculiar cytometric parameters or by means of certain dyes called fluorochromes.

e. Immunoblotting involves polyacrylamide gel electrophoresis of a protein specimen followed by transfer of the separated proteins to nitrocellulose sheets and identification of specific bands (**figure 33.13**).

f. Immunodiffusion refers to a precipitation reaction that occurs between antibody and antigen in an agar gel medium. Two techniques are routinely used: double diffusion in agar and single radial diffusion (**figure 33.14**).

g. In classical immunoelectrophoresis antigens are separated based on their electrical charge, then visualized by precipitation and staining (**figure 33.15**).

h. Immunofluorescence is a process in which fluorochromes are irradiated with UV, violet, or blue light to make them fluoresce. These dyes can be coupled to an antibody. There are two main kinds of fluorescent antibody assays: direct and indirect (**figure 33.16**).

i. Immunoprecipitation reactions occur only when there is an optimal ratio of antigen and antibody to produce a lattice at the zone of equivalence, which is evidenced by a visible precipitate (**figure 33.17**).

j. Radioimmunoassays use a purified antigen that is radioisotope-labeled and used to compete with unlabeled standard or antigen in experimental samples for a specific antibody.

k. Serotyping refers to serological procedures used to differentiate strains (serovars or serotypes) of microorganisms that have differences in the antigenic composition of a structure or product (**figure 33.18**).

Key Terms

active immunization 740
agglutinates 751
allergen 744
allergic contact dermatitis 747
allergy 744
allograft 749
anaphylaxis 744
attenuated 743
autoimmune disease 748
autoimmunity 748
bronchial asthma 745
chromogen 753
desensitization 746
DNA vaccine 744

double diffusion agar assay
  (Ouchterlony technique) 754
enzyme-linked immunosorbent assay
  (ELISA) 753
flow cytometry 754
graft-versus-host disease 750
hay fever 745
hives 745
hypersensitivity 744
immunoblotting 754
immunodeficiencies 750
immunodiffusion 754
immunoelectrophoresis 755

immunofluorescence 755
immunoprecipitation 756
Lancefield system 757
macromolecule vaccine 743
passive immunization 741
Quellung reaction 758
radioimmunoassay (RIA) 757
reagin 744
recombinant-vector vaccine 743
serology 751
serotyping 757
single radial immunodiffusion (RID)
  assay 754

systemic lupus erythematosus 747
TB skin test 747
toxoid 743
type I hypersensitivity 744
type II hypersensitivity 746
type III hypersensitivity 746
type IV hypersensitivity 747
vaccine 740
vaccinomics 743
viral hemagglutination 751
whole-organism vaccine 743
Widal test 751
xenograft 749

Questions for Thought and Review

1. Describe how IgE can have both beneficial and detrimental effects in the same host. Why have its detrimental effects persisted? Explain your reasoning.
2. In desensitization procedures, the allergist injects more of the same allergen to which the person is allergic. How can this be beneficial?
3. If excess complement is added in a complement fixation test, how will this affect the results?
4. Specific chemical markers are conjugated to antibodies or antigens in some immunologic tests. What are the advantages of this technique over such tests as agglutination or complement fixation?
5. What is the importance of serotyping?
6. What are some features of an in vitro antigen-antibody reaction that makes the reaction so useful for identification and monitoring tests?
7. How are chromogens used as biological markers?
8. Even though *Streptococcus pneumoniae* has over 80 different serotypes, serological reactions are sensitive enough to differentiate one serotype from another. How is this possible?
9. How can electrophoresis be used in immunologic testing procedures?
10. Describe each of the four types of hypersensitivities.
11. Are attenuated vaccines more likely than inactivated vaccines to induce cell-mediated immunity?
12. Do DNA vaccines generate immunological memory? Explain your answer.
13. List the three types of purified molecules that are used as vaccines.
14. Define vaccine immunization.

Critical Thinking Questions

1. How does an inactivated vaccine induce only a humoral response, whereas an attenuated vaccine induces both humoral and cell-mediated responses?
2. Why is a DNA vaccine delivered intramuscularly and not by intravenous or oral routes?
3. ELISA tests usually use a primary and secondary antibody. Why? What are the necessary controls one would need to perform to ensure that the antibody specificity is valid (no false positives or negatives)?

For recommended readings on these and related topics, see Appendix V.

chapter 34

Pathogenicity of Microorganisms

Concepts

- If a microorganism (symbiont) either harms or lives at the expense of another organism, it is called a parasitic organism and the relationship is termed parasitism. In this relationship the body of an animal is referred to as the host.
- Those organisms capable of causing disease are called pathogens. Disease is any change in the host from a healthy to an unhealthy, abnormal state in which part or all of the host's body is not capable of carrying on its normal functions.
- The steps for the infectious process involving viral diseases usually include the following: a virus enters a host, comes into contact with susceptible cells, replicates within the cells, spreads to adjacent cells, causes cellular injury, engenders a host immune response, is cleared from the body of the host or establishes a persistent infection, and is shed back into the environment.
- The steps for the infectious process involving bacterial diseases usually include the following: the bacterium is transmitted to a suitable host, attaches to and/or colonizes the host, grows and multiplies within or on the host, and interferes with or impairs the normal physiological activities of the host.
- One important result of the conservation of chromosomal genes is that bacteria are clonal. Only one or a few clonal types of some bacterial pathogens exist in the environment.
- During coevolution with human hosts, some pathogenic bacteria have evolved complex signal transduction pathways to regulate the genes necessary for virulence.
- Many bacteria are pathogenic because they have large segments of DNA, called pathogenicity islands, that carry genes responsible for virulence.
- Two distinct categories of disease can be recognized based on the role bacteria play in the disease causing process: infections (invasions) and intoxications.

Three *Streptococcus pneumoniae*, each surrounded by a slippery mucoid capsule (shown as a layer of white spheres around the diplococcus bacteria). The polysaccharide capsule is vital to the pathogenicity of this bacterium since it prevents phagocytic cells from accomplishing phagocytosis.

- Toxins produced by pathogenic bacteria are either exotoxins or endotoxins.
- Viruses and bacteria are continuously evolving and producing unique mechanisms that enable them to escape the host's arsenal of defenses.

> *Pathogenicity is not the rule. Indeed, it occurs so infrequently and involves such a relatively small number of species, considering the huge population of bacteria on earth, that it has a freakish aspect. Disease usually results from inconclusive negotiations for symbiosis, an overstepping of the line by one side or the other, a biological misinterpretation of borders.*
>
> —Lewis Thomas

hapter 28 introduces the concept of symbiosis and deals with several of its subordinate categories, including commensalism and mutualism. In this chapter the third category, parasitism, is presented along with one of its possible consequences—pathogenicity. The parasitic way of life is so successful that it has evolved independently in nearly all groups of organisms. In recent years concerted efforts to understand organisms and their relationships with their hosts have developed within the disciplines of virology, rickettsiology, chlamydiology, bacteriology, mycology, parasitology (protozoology and helminthology), entomology, and zoology. This chapter examines the parasitic way of life in terms of health and disease in the animal body and emphasizes viral and bacterial disease mechanisms. The chapter concludes with some viral and bacterial mechanisms used to escape host defenses.

 ## 34.1 HOST-PARASITE RELATIONSHIPS

If a symbiont either harms or lives at the expense of another organism (the **host**), it is a **parasitic organism,** and the relationship is called **parasitism.** In this relationship the body of the host can be viewed as a microenvironment that shelters and supports the growth and multiplication of the parasitic organism. The parasitic organism is usually the smaller of the two partners and is metabolically dependent on the host. There are many parasitic agents or organisms among the viruses, bacteria, fungi, plants, and animals (**table 34.1**). By convention, when the word **parasite** is used without qualification, it refers specifically to a protozoan or helminthic (nematode, trematode, cestode) organism.

Several types of parasitism are recognized. If an organism lives on the surface of its host, it is an **ectoparasite;** if it lives internally, it is an **endoparasite.** The host on or in which the parasitic organism either attains sexual maturity or reproduces is the **final host.** A host that serves as a temporary but essential environment for some stages of development is an **intermediate host.** In contrast, a **transfer host** is not necessary for the completion of the organism's life cycle but is used as a vehicle for reaching a fi-

Table 34.1

Categorization of Parasitic Organisms and Agents by Size

Discipline	Parasitic Group		Approximate Size
Virology	Prions		350 kDa
	Viroids	Agents	130 kDa
	Viruses	25–400 nm	20–400 nm
Bacteriology	Chlamydiae		0.2–1.5 μm
	Mycoplasmas		0.3–0.8 μm
	Rickettsias		0.5–2 μm
	Other bacteria		1–10 μm
		Microorganisms (microbiota)	
Mycology	Fungi		3–15 μm diameter (hyphae)
Protozoology	Protozoa		1–150 μm
Helminthology	Nematodes		3 mm–30 cm
	Platyhelminthes (cestodes, trematodes)	Parasites	1 mm–10 m
Entomology	Ticks and mites		0.1–15 mm
Zoology	Horsehair worms	Ectoparasites	10–20 cm
	Mesozoa		Up to 100 cm
	Leeches		1–5 cm

(Parasitology brackets Protozoology and Helminthology)

Table 34.2

Various Types of Infections Associated with Parasitic Organisms

Type	Definition
Abscess	A localized infection with a collection of pus surrounded by an inflamed area
Acute	Short but severe course
Bacteremia	Presence of viable bacteria in the blood
Chronic	Persists over a long time
Covert	Subclinical, no symptoms
Cross	Transmitted between hosts infected with different organisms
Focal	Exists in circumscribed areas
Fulminating	Infectious agent multiplies with great intensity
Iatrogenic	Caused as a result of health care
Latent	Persists in tissues for long periods, during most of which there are no symptoms
Localized	Restricted to a limited region or to one or more anatomical areas
Mixed	More than one organism present simultaneously
Nosocomial	Develops during a stay at a hospital or other clinical care facility
Opportunistic	Due to an agent that does not harm a healthy host but takes advantage of an unhealthy one
Overt	Symptomatic
Phytogenic	Caused by plant pathogens
Primary	First infection that often allows other organisms to appear on the scene
Pyogenic	Results in pus formation
Secondary	Caused by an organism following an initial or primary infection
Sepsis	(1) The condition resulting from the presence of bacteria or their toxins in blood or tissues; the presence of pathogens or their toxins in the blood or other tissues
	(2) Systemic response to infection; this systemic response is manifested by two or more of the following conditions as a result of infection: temperature, >38 or <36°C; heart rate, >90 beats per min; respiratory rate, >20 breaths per min, or pCO_2, <32 mm Hg; leukocyte count, >12,000 cells per ml^3, or >10% immature (band) forms
Septicemia	Blood poisoning associated with persistence of pathogenic organisms or their toxins in the blood
Septic shock	Sepsis with hypotension despite adequate fluid resuscitation, along with the presence of perfusion abnormalities that may include, but are not limited to, lactic acidosis, oliguria, or an acute alteration in mental status
Severe sepsis	Sepsis associated with organ dysfunction, hypoperfusion, or hypotension; hypoperfusion and perfusion abnormalities may include, but are not limited to, lactic acidosis, oliguria, or an acute alteration in mental status
Sporadic	Occurs only occasionally
Subclinical (inapparent or covert)	No detectable symptoms or manifestations
Systemic	Spread throughout the body
Toxemia	Condition arising from toxins in the blood
Zoonosis	Caused by a parasitic organism that is normally found in animals other than humans

Figure 34.1 Symbiosis. All symbiotic relationships are dynamic, and shifts among them can occur as indicated by the arrows. The most beneficial relationship is mutualism; the most destructive is parasitism. Host susceptibility, virulence of the parasitic organism, and number of parasites are factors that influence these relationships. Disease can result from a shift from either mutualism or commensalism to parasitism. Health may be regained by the reestablishment of mutualism or commensalism.

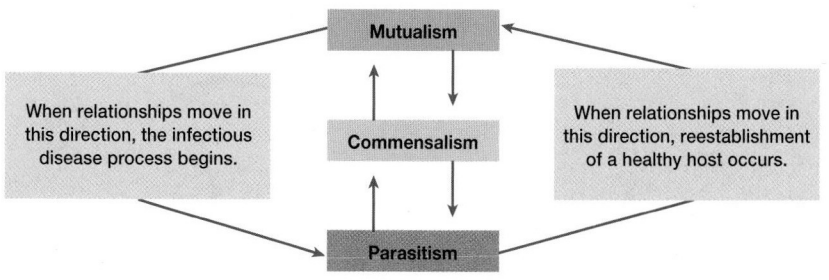

nal host. A host infected with a parasitic organism that also can infect humans is called a **reservoir host.**

Because, by definition, parasitic organisms are dependent on their hosts, the symbiotic relationship between the host and parasite is a dynamic one (**figure 34.1**). When a parasite is growing and multiplying within or on a host, the host is said to have an **infection.** The nature of an infection can vary widely with respect to severity, location, and number of organisms involved (**table 34.2**). An infection may or may not result in overt disease. An **infectious disease** is any change from a state of health in which part or all of the host body is not capable of carrying on its normal functions due to the presence of an organism or its products. Any organism or agent that produces such a disease is a **pathogen** [Greek *patho,* disease, and *gennan,* to produce]. Its ability to cause disease is called **pathogenicity**. A **primary (frank) pathogen** is any organism that causes disease in a healthy host by a direct interaction. Conversely, an **opportunistic pathogen** is an organism that is either usually free-living, or a part of the host's normal microbiota, but which may adopt a pathogenic role under certain circumstances, such as when the immune system is compromised.

At times an infectious organism can enter a latent state in which there is no shedding of the organism and no symptoms present within the host. This latency can be either intermittent or quiescent. Intermittent latency is exemplified by the herpesvirus that causes cold sores (fever blisters). After an initial infection, the symptoms subside. However, the virus remains in nerve tissue and can be activated weeks or years later by factors such as stress or sunlight. In a quiescent latency the organism persists but remains inactive for long periods of time, usually for years. For example, the varicella-zoster virus causes chickenpox in children and remains after the disease has subsided. In adulthood, under certain conditions, the same virus may erupt into a disease called shingles. Cold sores (pp. 860–61); Chickenpox (varicella) and shingles (herpes zoster) (pp. 846–47)

The outcome of most host-parasite relationships is dependent on three main factors: (1) the number of organisms present in or on the host, (2) the virulence of the organism, and (3) the host's defenses or degree of resistance. Usually the greater the number of organisms within a given host, the greater the likelihood of disease. However, a few organisms can cause disease if they are extremely virulent or if the host's resistance is low. A host's resistance can drop so much that its own microbiota may cause disease. Such a disease is sometimes called an endogenous disease because the agent originally comes from within the host's own body. Endogenous diseases can be a serious problem among hospitalized patients with very low resistance.

The term **virulence** [Latin *virulentia*, from *virus*, poison] refers to the degree or intensity of pathogenicity. It is determined by three characteristics of the pathogen: invasiveness, infectivity, and pathogenic potential. **Invasiveness** is the ability of the organism to spread to adjacent or other tissues. **Infectivity** is the ability of the organism to establish a focal point of infection. **Pathogenic potential** refers to the degree that the pathogen causes damage. A major aspect of pathogenic potential is toxigenicity. **Toxigenicity** is the pathogen's ability to produce toxins, chemical substances that will damage the host and produce disease. Virulence is often measured experimentally by determining the **lethal dose 50 (LD_{50})** or the **infectious dose 50 (ID_{50})**. These values refer to the dose or number of pathogens that will either kill or infect, respectively, 50% of an experimental group of hosts within a specified period (**figure 34.2**).

It should be noted that disease can result from causes other than toxin production. Sometimes a host will trigger exaggerated immunological responses (**immunopathology**) upon a second exposure or chronic exposure to a microbial antigen. These hypersensitivity reactions damage the host even though the pathogen doesn't produce a toxin. Tuberculosis is a good example of the involvement of hypersensitivity reactions in disease (*see pp. 882–84*). Some diseases also might be due to autoimmune re-

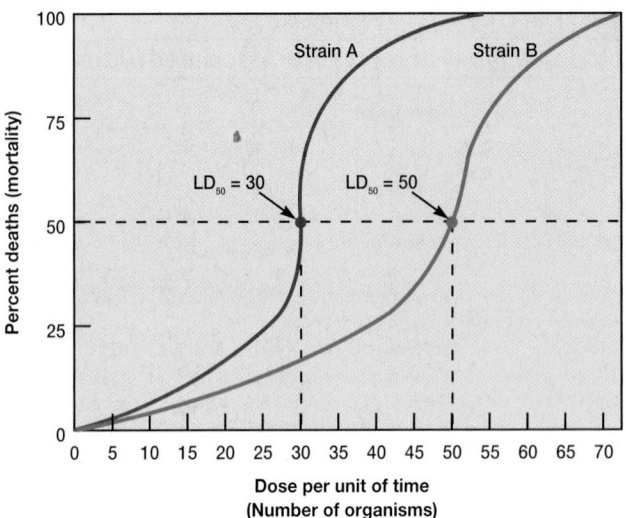

Figure 34.2 Determination of the LD_{50} of a Pathogenic Microorganism. Various doses of a specific pathogen are injected into experimental host animals. Deaths are recorded and a graph constructed. In this example, the graph represents the susceptibility of host animals to two different strains of a pathogen—strain A and strain B. For strain A the LD_{50} is 30, and for strain B it is 50. Hence strain A is more virulent than strain B.

sponses. For instance, a viral or bacterial pathogen may stimulate the immune system to attack host tissues because it carries antigens that resembled those of the host, a phenomenon known as molecular mimicry. Streptococcal infections may cause rheumatic fever in this way (*see p. 881*). Hypersensitivity reactions (pp. 744–48)

---

1. Define parasitic organism, parasitism, infection, infectious disease, pathogenicity, virulence, invasiveness, infectivity, pathogenic potential, and toxigenicity.
2. What factors determine the outcome of most host-parasite relationships?

 34.2 PATHOGENESIS OF VIRAL DISEASES

The fundamental process of viral infection is the expression of the viral replicative cycle (*see figures 17.6, 18.4–18.7*) in a host cell. The steps for the infectious process involving viruses usually include the following:

1. Enter a host
2. Contact and enter susceptible cells
3. Replicate within the cells
4. Spread to adjacent cells
5. Cause cellular injury
6. Engender a host immune response
7. Be either cleared from the body of the host, establish a persistent infection, or kill the host
8. Be shed back into the environment

The determinants of pathogenicity are now discussed in more detail.

Entry, Contact, and Primary Replication

The first step in the infectious process is the attachment and entrance of the virus into a susceptible host and the host's cells.

Entrance may be accomplished through one of the body surfaces (skin, respiratory system, gastrointestinal system, urogenital system, or the conjunctiva of the eye). Other viruses enter the host by needle sticks, blood transfusions and organ transplants, or by insect **vectors** (organisms that transmit the pathogen from one host to another). Some viruses replicate at the site of entry, cause disease at the same site (e.g., respiratory and gastrointestinal infections), and do not spread throughout the body. Others spread to sites distant from the point of entry and replicate at these sites. For example, the enteroviruses enter through the gastrointestinal tract but produce disease in the central nervous system. Mucous membranes (pp. 685–86)

Viral Spread and Cell Tropism

Mechanisms of viral spread vary, but the most common routes are the bloodstream and lymphatic system. The presence of viruses in the blood is called **viremia.** In some instances, spread is by way of nerves (e.g., rabies, herpes simplex, and varicella-zoster viruses; *see figure 38.2*).

Viruses exhibit cell, tissue, and organ specificities. These specificities are called **tropisms** (Greek *trope,* turning). A tropism by a specific virus usually reflects the presence of specific cell surface receptors on the eucaryotic host cell for that virus (*see figures 18.4 and 38.14*).

Cell Injury and Clinical Illness

Destruction of the virus-infected cells in the target tissues and alterations in host physiology are responsible for the development of viral disease and clinical illness. Some tissue (e.g., intestinal epithelium) can rapidly regenerate after a viral attack and withstand extensive damage. Other tissues, such as nervous system tissues, are not able to regenerate and may never resume normal functioning after damage has occurred.

The potential effects of viruses on individual host cells are the result of a complex series of events. There are four generally accepted patterns of a viral infection. (1) In lytic infections the virus multiplies and kills the host cell immediately and new virions are released. (2) In persistent viral infections the virus lives in the host cell and releases small numbers of virions over a long period of time. This causes little damage to the host cell. (3) In latent infections, the virus resides in the cell but produces no virions. At some later time the virus can be activated and a lytic infection occurs (*see section 18.4*). (4) Some viruses can transform the host cell into a cancer cell that becomes the focal point for a tumor (*see section 18.5*).

Viruses such as the influenza virus, HIV, adenovirus, and the measles virus can destroy host cells by triggering **apoptosis**

[Greek "a falling off" from *apo,* off, and *ptosis,* fall] or programmed cell death (*see pp. 725, 858*). Apoptosis often is part of normal tissue development during which it removes cells that are no longer needed. The process involves nuclear degeneration, the partial digestion of many cell proteins by proteolytic enzymes called caspases, and the formation of apoptotic bodies which are phagocytosed by macrophages (*see figure 38.12*). Unlike the case with normal necrosis, the cell does not lyse and release its contents. Although apoptosis kills host cells, it also defends the host by preventing virus reproduction and spread. Some viruses take advantage of this defense mechanism. They deliberately stimulate the host cell and trigger apoptosis in order to use the cell's activated machinery to reproduce themselves. After apoptosis has commenced, special viral proteins block its progress long enough to complete viral replication and construct new virions. In summary, apoptosis appears to be involved in both the pathogenesis of viral infections and in host defense.

Host Immune Response

Both humoral and cellular components of the immune response are involved in the control of viral infections and were discussed in detail in chapters 31 and 32.

Recovery from Infection

The host will either succumb or recover from a viral infection. Recovery mechanisms involve nonspecific defense mechanisms and specific humoral and cellular immunity. The relative importance of each of these factors varies with the virus and the disease, and will be covered in chapter 38.

Virus Shedding

The last step in the infectious process is shedding of the infectious virus back into the environment. This is necessary to maintain a source of viruses in a population of hosts. Shedding often occurs from the same body surface used for entry. During this period, an infected host is infectious and can spread the virus. In some viral infections, such as a rabies infection, humans are dead-end hosts because virus shedding does not occur.

---

1. For a virus to cause disease, certain steps are usually accomplished. Briefly describe each of these steps.
2. What are the four most common patterns of viral infections? Describe apoptosis and its role in viral infections.

 34.3 PATHOGENESIS OF BACTERIAL DISEASES

The steps for infections by pathogenic bacteria usually include the following:

1. Maintain a reservoir. A **reservoir** is a place to live before and after causing an infection (*see p. 828*).
2. Initially be transported to the host.

Table 34.3

Bacterial Adherence Factors (Adhesins) That Play a Role in Infectious Diseases

Adherence Factor	Description
Fimbriae	Filamentous structures that help attach bacteria to other bacteria or to solid surfaces
Glycocalyx or capsule	A layer of exopolysaccharide fibers with a distinct outer margin that surrounds many cells; it inhibits phagocytosis and aids in adherence; when the layer is well organized and not easily washed off it is called a capsule
Pili	Filamentous structures that bind procaryotes together for the transfer of genetic material
S layer	The outermost regularly structured layer of cell envelopes of some bacteria that may promote adherence to surfaces
Slime layer	A bacterial film that is less compact than a capsule and is removed easily
Teichoic and lipoteichoic acids	Cell wall components in gram-positive bacteria that aid in adhesion

3. Adhere to, colonize, and/or invade the host.

4. Initially evade host defense mechanisms.

5. Multiply (grow) or complete its life cycle on or in the host or the host's cells.

6. Possess the ability to damage the host.

7. Leave the host and return to the reservoir or enter a new host.

The first five factors influence the degree of infectivity and invasiveness. Toxigenicity plays a major role in the sixth. The determinants are now discussed in more detail.

Maintaining a Reservoir of the Bacterial Pathogen

All bacterial pathogens must have at least one reservoir. The most common reservoirs for human pathogens are other humans, animals, and the environment. Since the source and/or reservoir of the pathogen is part of the infectious disease cycle, this aspect of pathogenicity is discussed in detail in chapter 37, which covers the epidemiology of infectious diseases. The infectious disease cycle (pp. 827–31)

Transport of the Bacterial Pathogen to the Host

An essential feature in the development of an infectious disease is the initial transport of the bacterial pathogen to the host. The most obvious means is direct contact—from host to host (coughing, sneezing, body contact). Bacteria also are transmitted indirectly in a variety of ways. Infected hosts shed bacteria into their surroundings. Once in the environment bacteria can be deposited on various surfaces, from which they can be either resuspended into the air or indirectly transmitted to a host later. Soil, water, and food are indirect vehicles that harbor and transmit bacteria to hosts. Vectors and **fomites** (inanimate objects that harbor and transmit pathogens) also are involved in the spread of many bacteria.

Attachment and Colonization by the Bacterial Pathogen

After being transmitted to an appropriate host, the bacterial pathogen must be able to adhere to and colonize host cells and tissues. In this context **colonization** means the establishment of a site of microbial reproduction on or within a host. It does not necessarily result in tissue invasion or damage. Colonization depends on the ability of the bacteria to compete successfully with the host's normal microbiota for essential nutrients. Specialized structures that allow bacteria to compete for surface attachment sites also are necessary for colonization.

Bacterial pathogens and many nonpathogens adhere with a high degree of specificity to particular tissues. Adherence factors called **adhesins (table 34.3)** are one reason for this specificity. Adhesins are specialized molecules or structures on the bacteria's cell surface that bind to complementary receptor sites on the host cell surface (**figure 34.3**). They are one type of virulence factor. **Virulence factors** are bacterial products or structural components (e.g., capsules and adhesins) that contribute to virulence or pathogenicity. Bacterial capsules and fimbriae (pp. 61–63)

Invasion of the Bacterial Pathogen

Entry into host cells and tissues is a specialized strategy used by many bacterial pathogens for survival and multiplication. Pathogens often actively penetrate the host's mucous membranes and epithelium after attachment to the epithelial surface. This may be accomplished through production of lytic substances that alter the host tissue by (1) attacking the ground substance and basement membranes of integuments and intestinal linings, (2) degrading carbohydrate-protein complexes between cells or on the cell surface (the glycocalyx), or (3) disrupting the cell surface.

At times a bacterial pathogen can penetrate the epithelial surface by passive mechanisms not related to the pathogen itself. Examples include (1) small breaks, lesions, or ulcers in a mucous membrane that permit initial entry; (2) wounds, abrasions, or burns on the skin's surface; (3) arthropod vectors that create small wounds while feeding; (4) tissue damage caused by other organisms; and (5) existing eucaryotic internalization pathways (e.g., endocytosis and phagocytoses; *see figure 31.16*).

Once under the mucous membrane, the bacterial pathogen may penetrate to deeper tissues and continue disseminating throughout the body of the host. One way the pathogen accomplishes this is by producing specific products and/or enzymes that promote spreading (**table 34.4**). These products are virulence factors. Bacteria may also enter the small terminal lymphatic capillaries that surround epithelial cells. These capillaries merge into large lymphatic vessels that eventually drain into the circulatory system. Once the circulatory system is reached, the bacteria have access to all organs and systems of the host.

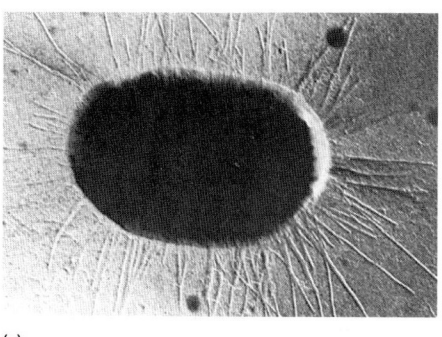
(a)

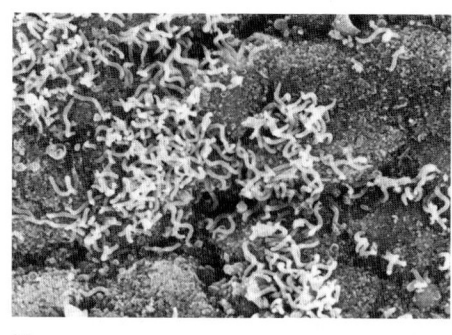
(b)

(c)

Figure 34.3   Microbial Adherence. (**a**) Transmission electron micrograph of fimbriated *Escherichia coli* (×16,625). (**b**) Scanning electron micrograph of epithelial cells with adhering vibrios (×1,200). (**c**) *Candida albicans* fimbriae (arrow) are used to attach the fungus to vaginal epithelial cells.

Table 34.4

Microbial Products (Virulence Factors) Involved in Bacterial Pathogen Dissemination Throughout a Mammalian Host

Product	Organism Involved	Mechanism of Action
Coagulase	*Staphylococcus aureus*	Coagulates (clots) the fibrinogen in plasma. The clot protects the pathogen from phagocytosis and isolates it from other host defenses.
Collagenase	*Clostridium* spp.	Breaks down collagen that forms the framework of connective tissues; allows the pathogen to spread.
Deoxyribonuclease (along with calcium and magnesium)	Group A streptococci, staphylococci, *Clostridium perfringens*	Lowers viscosity of exudates, giving the pathogen more mobility.
Elastase and alkaline protease	*Pseudomonas aeruginosa*	Cleaves laminin associated with basement membranes.
Hemolysins	Staphylococci, streptococci, *Escherichia coli*, *Clostridium perfringens*	Lyse erythrocytes, causing anemia and weakened host defenses; make iron available for microbial growth.
Hyaluronidase	Groups A, B, C, and G streptococci, staphylococci, clostridia	Hydrolyzes hyaluronic acid, a constituent of the intercellular ground substance that cements cells together and renders the intercellular spaces amenable to passage by the pathogen.
Hydrogen peroxide (H_2O_2) and ammonia (NH_3)	*Mycoplasma* spp., *Ureaplasma* spp.	Are produced as metabolic wastes. These are toxic and damage epithelia in respiratory and urogenital systems.
Immunoglobulin A protease	*Streptococcus pneumoniae*	Cleaves immunoglobulin A into Fab and Fc fragments.
Lecithinase or phospholipase	*Clostridium* spp.	Destroys the lecithin (phosphatidycholine) component of plasma membranes, allowing pathogen to spread.
Leukocidins	Staphylococci, pneumococci, streptococci	Pore-forming exotoxins that kill leukocytes; cause degranulation of lysosomes within leukocytes, which decreases host resistance.
Porins	*Salmonella typhimurium*	Inhibit leukocyte phagocytosis by activating the adenylate cyclase system.
Protein A	*Staphylococcus aureus*	Located on cell wall. Immunoglobulin G (IgG) binds to protein A by its Fc end, thereby preventing complement from interacting with bound IgG.
Pyrogenic exotoxin B (cysteine protease)	Group A streptococci, *Streptococcus pyogenes*	Degrades proteins.
Streptokinase (fibrinolysin, staphylokinase)	Group A, C, and G streptococci, staphylococci	A protein that binds to plasminogen and activates the production of plasmin, thus digesting fibrin clots; this allows the pathogen to move from the clotted area.

Growth and Multiplication of the Bacterial Pathogen

For a bacterial pathogen to be successful in growth and reproduction (colonization), it must find an appropriate environment (nutrients, pH, temperature, redox potential) within the host. Those areas of the host's body that provide the most favorable conditions will harbor the pathogen and allow it to grow and multiply to produce an infec-

tion. Some bacteria invade specific cells in which they grow and multiply. Many of these intracellular pathogens have evolved such elaborate nutrient-gathering mechanisms that they have become totally dependent on the host's cells. Finally, some bacteria can actively grow and multiply in the blood plasma. The presence of viable bacteria in the bloodstream is called **bacteremia.** The presence of bacteria or their toxins in the blood often is termed **septicemia** [Greek *septikos,* produced by putrefaction, and *haima,* blood].

Leaving the Host

The last determinant of a successful bacterial pathogen is its ability to leave the host and enter either a new host or a reservoir. Unless a successful escape occurs, the disease cycle will be interrupted and the microorganism will not be perpetuated. Most bacteria employ passive escape mechanisms. Passive escape occurs when a pathogen or its progeny leave the host in feces, urine, droplets, saliva, or desquamated cells.

The Clonal Nature of Bacterial Pathogens

The main mechanism bacteria use to exchange genetic information is the transfer of extrachromosomal genetic elements, plasmids, and phages (*see chapter 13*). Many genes that code for bacterial virulence factors are found on plasmids or phages. These mobile genetic elements can transfer virulence factors between members of the same species or different species by horizontal gene transfer (*see section 13.1*). At times the genetic elements are part of highly mobile DNA (transposons, *see section 13.3*), and there is a recombination between the extrachromosomal DNA and the chromosome. When this recombination occurs, the genes coding for virulence may become chromosomal.

One important result of the conservation of these chromosomal genes is that the bacteria are clonal. Some bacterial pathogens have only one or a few clonal types existing in the environment. For example, the bacterium that causes typhoid fever (*Salmonella typhi*) has two clonal types, whereas there are over 100 clonal types of *Haemophilus influenzae*, but only a small number are associated with bacterial pneumonia.

Regulation of Bacterial Virulence Factors

As noted in many chapters, some pathogenic bacteria have adapted to both the free-living state and to an environment within a human host. In the adaptive process, these pathogens have evolved complex signal transduction pathways to regulate the genes necessary for virulence. A virulence factor may be present simply because the bacterium has been infected by a phage. Often environmental factors control the expression of the virulence genes. Common signals include temperature, osmolality, available iron, pH, specific ions, and other nutrient factors. Several examples are now presented.

The gene for diphtheria toxin (figure 34.5b) from *Corynebacterium diphtheriae* (the pathogen that causes diphtheria) is carried on the temperate bacteriophage β, and its expression is regulated by iron. The toxin is produced only by strains lysogenized by the phages. Expression of the virulence gene of *Bordetella pertussis* (the pathogen that causes whooping cough) is enhanced when the bacteria grow at body temperature (37°C) and suppressed when grown at a lower temperature. Finally, the virulence factors of *Vibrio cholerae* (the pathogen that causes cholera) are regulated at various levels by many environmental factors. Expression of the cholera toxin is higher at pH 6 than at pH 8 and higher at 30 than at 37°C. Osmolality and available amino acids are also important.

Pathogenicity Islands

Many bacteria (*Yersinia* spp., *Pseudomonas aeruginosa*, *Shigella flexneri*, *Salmonella typhimurium*, enteropathogenic *E. coli*) are pathogenic because they have large segments of DNA, called **pathogenicity islands,** that carry genes responsible for virulence. These pathogenicity islands have been acquired during evolution by horizontal gene transfer. A pathogen may have more than one pathogenicity island. An excellent example of virulence genes carried in a pathogenicity island are those involved in protein secretion. So far, five pathways of protein secretion (types I to V) have been described in gram-negative bacteria. A set of approximately 20 genes encode a pathogenicity mechanism termed the **type III secretion pathway** that enables gram-negative bacteria to secrete and inject virulence proteins into the cytoplasm of eucaryotic host cells (**figure 34.4;** *see also figure 3.27*). Unlike other bacterial secretory systems, the type III system is triggered specifically by contact with host cells, which helps avoid inappropriate activation of host defenses. Secretion of these virulence proteins into a host cell initiates sophisticated "biochemical cross-talk" between the pathogen and the host. The injected proteins resemble eucaryotic factors that signal transduction functions and they are capable of interfering with eucaryotic signaling pathways. Redirection of cellular signaling transduction may disarm host immune responses or reorganize the cytoskeleton, thus establishing subcellular niches for bacterial colonization and facilitating "stealth and interdiction" of host defense communication lines.   Protein secretion in procaryotes (pp. 59–61)

Pathogenicity islands generally increase microbial virulence and are not present in nonpathogenic members of the same genus or species. One specific example is found in *E. coli*. The enteropathogenic *E. coli* possesses large DNA fragments, 35 to 170 kilobases in size, that contain several virulence genes absent from commensal *E. coli*. Some of these genes code for proteins that alter actin microfilaments within a host intestinal cell. As a consequence, the host cell surface bulges and develops into a cuplike pedestal to which the bacterium tightly binds.

---

1. What seven steps are involved in the infection process and pathogenesis of bacterial diseases?
2. What are some ways in which bacterial pathogens are transmitted to their hosts? Define vector and fomite.
3. Describe several specific adhesins by which bacterial pathogens attach to host cells.
4. Once under the mucous and epithelial surfaces, what are some mechanisms that bacterial pathogens possess to promote their dissemination throughout the body of a host?
5. What is the significance of the clonal nature of bacterial pathogens? The regulation of virulence factors?
6. What are virulence factors? Pathogenicity islands?

Toxigenicity

Two distinct categories of disease can be recognized based on the role of the bacteria in the disease-causing process: infections and intoxications. An infectious disease results partly from the

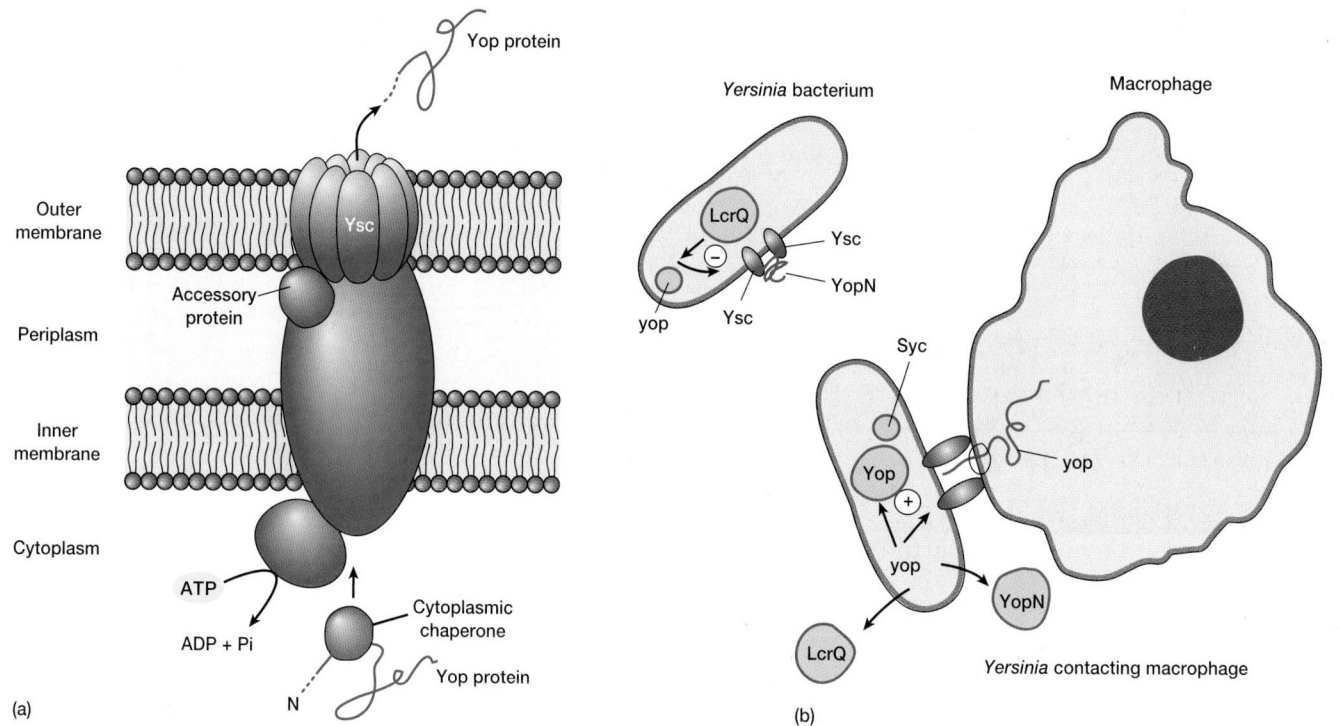

Figure 34.4   Type III Secretion System and Mode of Action of Yop Virulence Protein.   (a) Schematic drawing of the type III secretion system and Yop (a specific virulence protein) secretion by *Yersinia* spp. (b) In this schematic, prior to cell contact, the type III secretion channels (Ysc) are kept shut by YopN, and cytoplasmic accumulation of the negative regulatory factor LcrQ leads to transcription repression of *yop* genes (negative sign in circle). After *Yersinia* contacts the macrophage, YopN is released, allowing rapid secretion of LcrQ, which in turn relieves the block of *yop* gene expression (positive sign in circle). Anti-host Yop proteins, protected by cognate chaperones (Syc), are translocated into the macrophage by the type III secretion machinery (as shown in part *a*). After contact of the *Yersinia* bacterium with a macrophage, Yop is injected into the cytoplasm of the target cell where it catalyzes a rapid and specific dephosphorylation of several macrophage proteins that are required for normal phagocytosis.

pathogen's growth and reproduction (or invasiveness) that often produce tissue alterations.

    Intoxications are diseases that result from the entrance of a specific preformed toxin (e.g., botulinum toxin) into the body of a host. Toxins can even induce disease in the absence of the organism that produced them. A **toxin** [Latin *toxicum,* poison] is a substance, such as a metabolic product of the organism, that alters the normal metabolism of host cells with deleterious effects on the host. The term **toxemia** refers to the condition caused by toxins that have entered the blood of the host. Toxins produced by bacteria can be divided into two main categories: exotoxins and endotoxins. The primary characteristics of the two groups are compared in **table 34.5.**

Exotoxins

Exotoxins are soluble, heat-labile, proteins (a few are enzymes) that usually are released into the surroundings as the bacterial pathogen grows. Often exotoxins may travel from the site of infection to other body tissues or target cells in which they exert their effects. Exotoxins usually are

1. Synthesized by specific bacteria that often have plasmids or prophages bearing the exotoxin genes
2. Heat-labile proteins inactivated at 60 to 80°C

3. Among the most lethal substances known (toxic in very small doses [microgram per kilogram amounts]; e.g., the botulinum toxin)
4. Associated with specific diseases and have specific mechanisms of action
5. Highly immunogenic and stimulate the production of neutralizing antibodies called **antitoxins**
6. Easily inactivated by formaldehyde, iodine, and other chemicals to form immunogenic **toxoids**
7. Unable to produce a fever in the host directly
8. Often given the name of the disease they produce (e.g., the diphtheria toxin)

    Exotoxins can be divided into four types based on their structure and physiological activities. (1) One type is the AB toxin, which gets its name from the fact that the portion of the toxin (B) that binds to a host cell receptor is separate from the portion (A) that has the enzyme activity that causes the toxicity. (2) A second type, which also may be an AB toxin, consists of those toxins that affect a specific host site (nervous tissue [neurotoxins], the intestines [enterotoxins], general tissues [cytotoxins]) by acting extracellularly or intracellularly on the host cells. (3) A third type does not have separable A and B portions and acts by disorganizing host cell membranes. Examples

Table 34.5

Characteristics of Exotoxins and Endotoxins

Characteristic	Exotoxins	Endotoxins
Chemical composition	Protein, often with two components (A and B)	Lipopolysaccharide complex on outer membrane; lipid A portion is toxic
Disease examples	Botulism, diphtheria, tetanus	Gram-negative infections, meningococcemia
Effect on host	Highly variable between different toxins	Similar for all endotoxins
Fever	Usually do not produce fever	Produce fever by release of interleukin-1
Genetics	Frequently carried by extrachromosomal genes such as plasmids	Synthesized directly by chromosomal genes
Heat stability	Most are heat sensitive and inactivated at 60–80°C	Heat stable
Immune response	Antitoxins provide host immunity; highly antigenic	Limited antibodies produced; weakly immunogenic
Location	Usually excreted outside the living cell	Part of outer membrane of gram-negative bacteria
Production	Produced by both gram-positive and gram-negative bacteria	Found only in gram-negative bacteria; Released on bacterial death and some liberated during growth
Toxicity	Highly toxic and fatal in microgram quantities	Moderate toxicity
Toxoid production	Converted to antigenic, nontoxic toxoids; toxoids are used to immunize (e.g., tetanus toxoid)	Toxoids cannot be made

Table 34.6

Properties of Some AB Model Bacterial Exotoxins

Toxin	Organism	Genetic Control	Subunit Structure	Target Cell Receptor	Enzymatic Activity	Biologic Effects
Anthrax toxins	B. anthracis	Plasmid	Three separate proteins (EF, LF, PA)[a]	Unknown, probably glycoprotein	EF is a calmodulin-dependent adenylate cyclase; LF enzyme activity is unknown	EF + PA: increase in target cell cAMP level, localized edema; LF + PA: death of target cells and experimental animals
Bordetella adenylate cyclase toxin	Bordetella spp.	Chromosomal	A-B[b]	Unknown, probably glycolipid	Calmodulin-activated cyclase	Increase in target cell cAMP level; modified cell function or cell death
Botulinum toxin	C. botulinum	Phage	A-B[c]	Possibly ganglioside (GD_{1b})	Zinc-dependent endopeptidase cleavage of synaptobrevin	Decrease in peripheral, presynaptic acetylcholine release; flaccid paralysis
Cholera toxin	V. cholera	Chromosomal	A-5B[d]	Ganglioside (GM_1)	ADP ribosylation of adenylate cyclase regulatory protein, G_S	Activation of adenylate cyclase, increase in cAMP level; secretory diarrhea
Diphtheria toxin	C. diphtheriae	Phage	A-B[e]	Heparin-binding, EGF-like growth factor precursor	ADP ribosylation of elongation factor 2	Inhibition of protein synthesis; cell death
Heat-labile enterotoxins[f]	E. coli	Plasmid	——————— Similar or Identical to Cholera Toxin ———————			
Pertussis toxin	B. pertussis	Chromosomal	A-5B[g]	Unknown, probably glycoprotein	ADP ribosylation of signal-transducing G proteins	Block of signal transduction mediated by target G proteins
Pseudomonas exotoxin A	P. aeruginosa	Chromosomal	A-B	α_2-Macroglobulin/LDL receptor	——— Similar or Identical to Diphtheria Toxin ———	
Shiga toxin	S. dysenteriae	Chromosomal	A-5B[h]	Globotriaosylceramide (Gb_3)	RNA N-glycosidase	Inhibition of protein synthesis, cell death
Shiga-like toxin 1	Shigella spp., E. coli	Phage	——————— Similar or Identical to Shiga Toxin ———————			
Tetanus toxin	C. tetani	Plasmid	A-B[c]	Ganglioside (GT_1 and/or GD_{1b})	Zinc-dependent endopeptidase cleavage of synaptobrevin	Decrease in neurotransmitter release from inhibitory neurons; spastic paralysis

Adapted from G. L. Mandell, et al., *Principles and Practice of Infectious Diseases*, 3d edition Copyright © 1990 Churchill-Livingstone, Inc., Medical Publishers, New York, NY. Reprinted by permission.

[a]The binding component (known as protective antigen [PA]) catalyzes/facilitates the entry of either edema factor (EF) or lethal factor (LF).

[b]Apparently synthesized as a single polypeptide with binding and catalytic (adenylate cyclase) domains.

[c]Holotoxin is apparently synthesized as a single polypeptide and cleaved proteolytically as diphtheria toxin; subunits are referred to as L: light chain, A equivalent; H: heavy chain, B equivalent.

[d]The A subunit is proteolytically cleaved into A_1 and A_2, with A_1 possessing the ADP-ribosyl transferase activity; the binding component is made up of five identical B units.

[e]Holotoxin is synthesized as a single polypeptide and cleaved proteolytically into A and B components held together by disulfide bonds.

[f]The heat-labile enterotoxins of *E. coli* are now recognized to be a family of related molecules with identical mechanisms of action.

[g]The binding portion is made up of two dissimilar heterodimers labeled S2-S3 and S2-S4 that are held together by a bridging peptide, SS.

[h]Subunit composition and structure similar to cholera toxin.

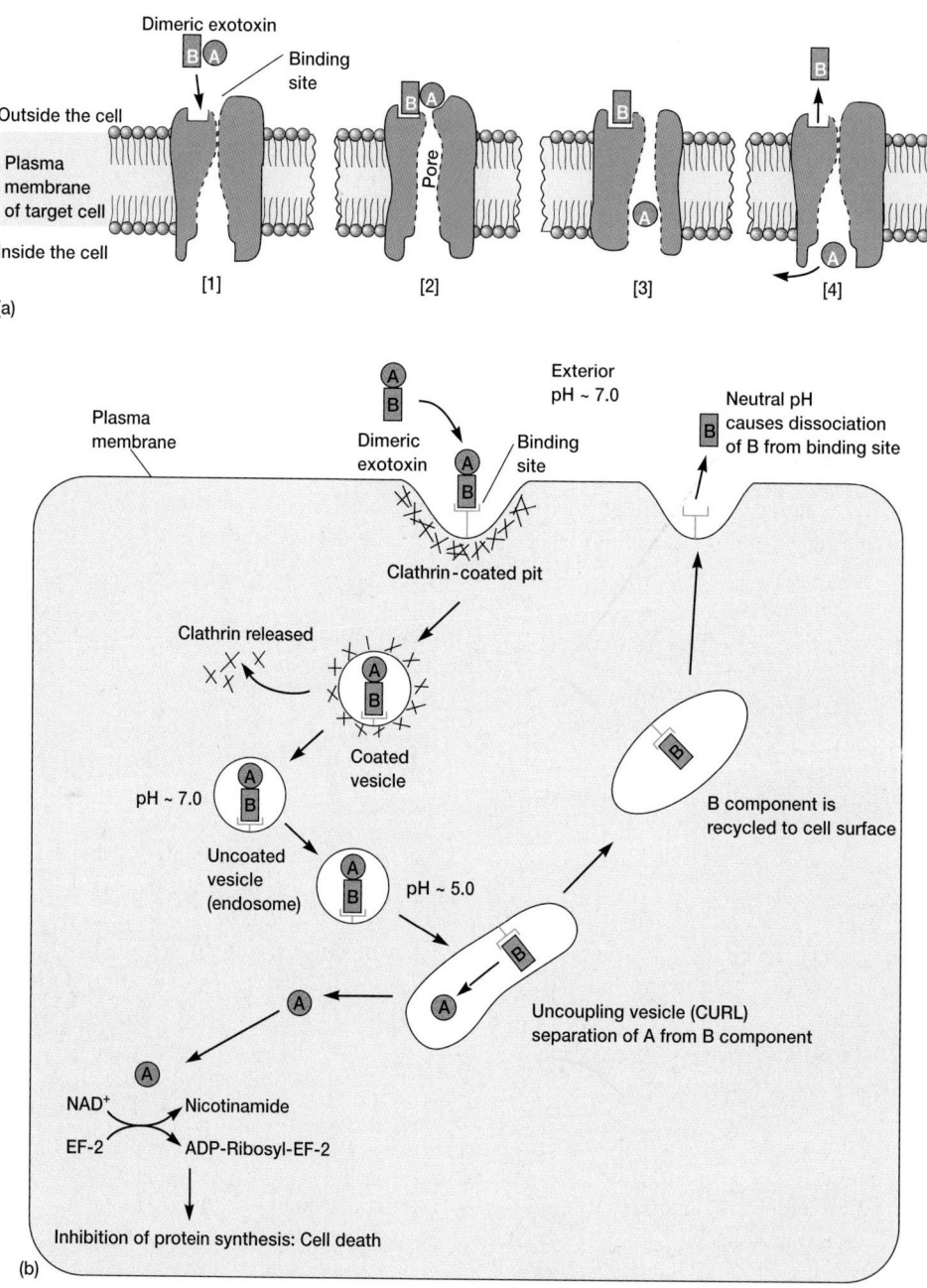

Figure 34.5   Diagrammatic Representation of Two AB Exotoxin Transport Mechanisms.   (a) Domain B of the dimeric exotoxin (AB) binds to a specific membrane receptor of a target cell [*1*]. A conformational change [*2*] generates a pore [*3*] through which the A domain crosses the membrane and enters the cytosol, followed by recreation [*4*] of the binding site. **(b)** Receptor-mediated endocytosis of the diphtheria toxin involves the dimeric exotoxin binding to a receptor-ligand complex that is internalized in a clathrin-coated pit that pinches off to become a coated vesicle. The clathrin coat depolymerizes resulting in an uncoated endosome vesicle. The pH in the endosome decreases due to the H^+-ATPase activity. The low pH causes A and B components to separate. An endosome in which this separation occurs is sometimes called a CURL (compartment of *u*ncoupling of *r*eceptor and *l*igand). The B domain is then recycled to the cell surface. The A domain moves through the cytosol, catalyzes the ADP-ribosylation of EF-2 (elongation factor 2) and inhibits protein synthesis, leading to cell death.

include the leukocidins, hemolysins, and phospholipases. (4) A fourth type is the superantigen that acts by stimulating T cells to release cytokines. Examples of the first three types are now discussed. Superantigens are discussed in detail in section 32.2. The general properties of some AB exotoxins are presented in **table 34.6**.

AB Exotoxins.   AB toxins are composed of an enzymatic subunit or fragment (A) that is responsible for the toxic effect once inside the host cell and a binding subunit or fragment (B) (**figure 34.5**). Isolated A subunits are enzymatically active but lack binding and cell entry capability, whereas isolated B subunits bind to target cells but are nontoxic and biologically inactive. The B subunit interacts with specific receptors on the target cell or tissue such as the gangliosides GM_1 for cholera toxin, GT_1 and/or GD_1 for tetanus toxin, and GD_1 for botulinum toxin.

Several mechanisms for the entry of A subunits or fragments into target cells have been proposed. In one mechanism the B subunit inserts into the plasma membrane and creates a pore through which the A subunit enters (figure 34.5*a*). In another mechanism entry is by receptor-mediated endocytosis (figure 34.5*b*).

The mechanism of action of an AB toxin can be quite complex, as shown by the example of diphtheria toxin (figure 34.5*b*). The diphtheria toxin is a protein of about 62,000 mol wt. It binds to cell surface receptors by the B fragment portion and is taken into the cell through the formation of a clathrin-coated vesicle (*see p. 80*). The toxin then enters the vesicle membrane and is cleaved into two parts, one of which, the A fragment, escapes into the cytosol. The A fragment is an enzyme that catalyzes the addition of an ADP-ribose group to the eucaryotic elongation factor EF2 that aids in translocation during protein synthesis (*see pp. 265–66*). The substrate for this reaction is the coenzyme NAD^+.

$$NAD^+ + EF2 \rightarrow ADP\text{-ribosyl-}EF2 + \text{nicotinamide}$$

The modified EF2 protein cannot participate in the elongation cycle of protein synthesis, and the cell dies because it can no longer synthesize proteins.

AB exotoxins vary widely in their relative contribution to the disease process with which they are associated.

Specific Host Site Exotoxins. The second type of exotoxin is categorized on the basis of the site affected: **neurotoxins** (nerve tissue), **enterotoxins** (intestinal mucosa), and **cytotoxins** (general tissues). Some of the bacterial pathogens that produce these exotoxins are presented in table 34.6: neurotoxins (botulinum toxin and tetanus toxin), enterotoxins (cholera toxin, *E. coli* heat-labile toxins), and cytoxins (diphtheria toxin, Shiga toxin).

Neurotoxins usually are ingested as preformed toxins that affect the nervous system and indirectly cause enteric (pertaining to the small intestine) symptoms. Examples include staphylococcal enterotoxin B, *Bacillus cereus* emetic toxin [Greek *emetos*, vomiting], and botulinum toxin.

True enterotoxins [Greek *enter*, intestine] have a direct effect on the intestinal mucosa and elicit profuse fluid secretion. The classic enterotoxin, cholera toxin (choleragen), has been studied extensively. It is an AB toxin. The B subunit is made of five parts arranged as a donut-shaped ring. The B subunit ring anchors itself to the epithelial cell's plasma membrane and then inserts the smaller A subunit into the cell. The A subunit activates tissue adenylate cyclase to increase intestinal cyclic AMP (cAMP) concentrations. High concentrations of cAMP provoke the movement of massive quantities of water and electrolytes across the intestinal cells into the lumen of the gut. The genes for this enterotoxigenicity reside on the *Vibrio cholera* chromosome.

Toxins in food (pp. 905–11, 942–43, 948–49); Toxins as superantigens (p. 709)

Cytotoxins have a specific toxic action upon cells/tissues of special organs and are named according to the type of cell/tissue

or organ for which they are specific. Examples include nephrotoxin (kidney), hepatotoxin (liver), and cardiotoxin (heart).

Membrane-Disrupting Exotoxins. The third type of exotoxin lyses host cells by disrupting the integrity of the plasma membrane. There are two subtypes of **membrane-disrupting exotoxins.** The first (**figure 34.6***a*), is a protein that binds to the cholesterol portion of the host cell plasma membrane, inserts itself into the membrane, and forms a channel (pore). This causes the cytoplasmic contents to leak out. Also, because the osmolality of the cytoplasm is higher than the extracellular fluid, this causes a sudden influx of water into the cell, causing it to swell and rupture. Two specific examples of this type of membrane-disrupting exotoxin are now presented.

Some pathogens produce membrane-disrupting toxins that kill phagocytic leukocytes. These are termed **leukocidins** [*leuko*cyte and Latin *caedere*, to kill]. Most leukocidins are produced by pneumonococci, streptococci, and staphylococci. Since the pore-forming exotoxin produced by these bacteria destroys leukocytes, this in turn decreases host resistance. Other toxins, called **hemolysins** [*haima*, blood, and Greek *lysis*, dissolution], also can be secreted by pathogenic bacteria. Many hemolysins probably form pores in the plasma membrane of erythrocytes through which hemoglobin and/or ions are released (the erythrocytes lyse or, more specifically, hemolyze). **Streptolysin-O (SLO)** is a hemolysin, produced by *Streptococcus pyogenes*, that is inactivated by O_2 (hence the "O" in its name). SLO causes beta hemolysis of erythrocytes on agar plates incubated anaerobically (*see figure 23.17*). A complete zone of clearing around the bacterial colony growing on blood agar is called **beta hemolysis,** and a partial clearing of the blood is called **alpha hemolysis. Streptolysin-S (SLS)** is also produced by *S. pyogenes* but is insoluble and bound to the bacterial cell. It is O_2 stable (hence the "S" in its name) and causes beta hemolysis on aerobically incubated blood-agar plates. SLO and SLS act as leukocidins and kill leukocytes. It should also be noted that hemolysins attack the plasma membranes of many cells, not just erythrocytes and leukocytes.

The second subtype of membrane-disrupting toxins are the **phospholipase** enzymes. Phospholipases remove the charged head group (figure 34.6*b*) from the lipid portion of the phospholipids in the host-cell plasma membrane. This destabilizes the membrane and the cell lyses and dies. One example of the pathogenesis caused by phospholipases is the disease gas gangrene. In this disease, the *Clostridium perfringens* α-toxin almost totally destroys the local population of white blood cells that are drawn in by inflammation to fight the infection.

Roles of Exotoxins in Disease. Humans are exposed to bacterial exotoxins in three main ways: (1) ingestion of preformed exotoxin, (2) colonization of a mucosal surface followed by exotoxin production, and (3) colonization of a wound or abscess followed by local exotoxin production. Each of these is now briefly discussed.

In the first example (**figure 34.7***a*), the exotoxin is produced by bacteria growing in food. When food is consumed, the preformed exotoxin is also consumed. The classical example is

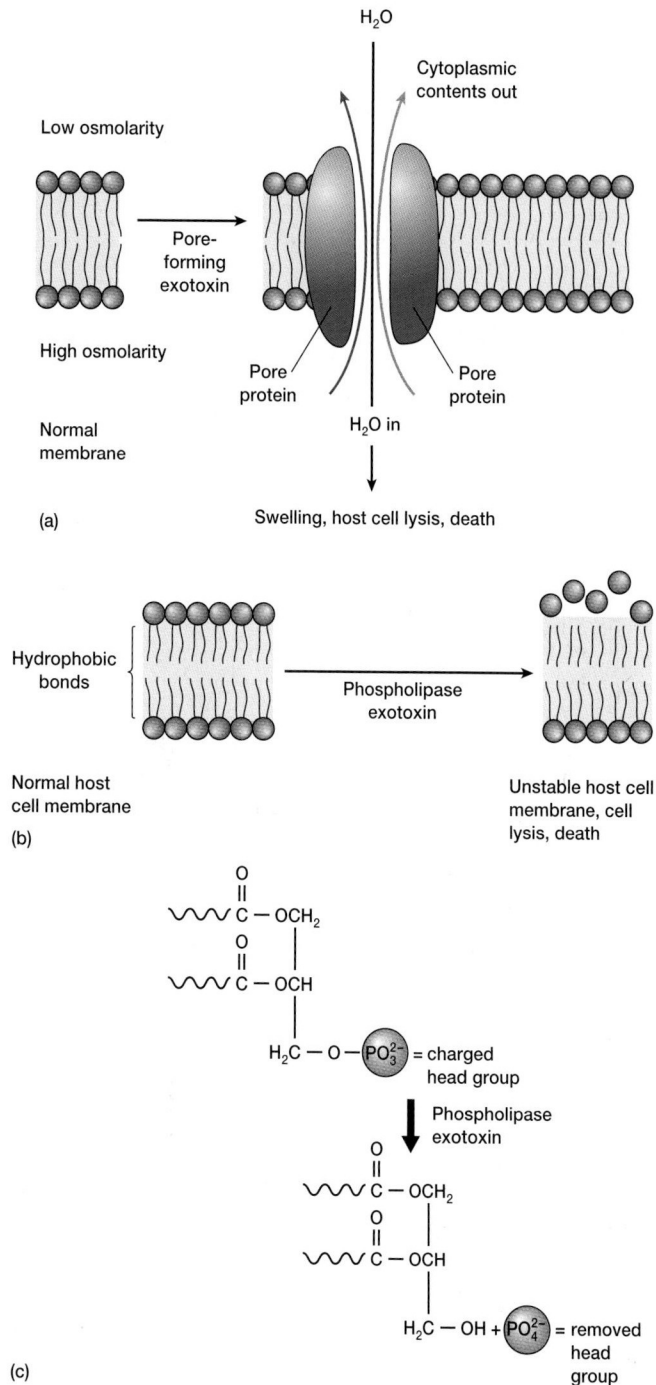

Figure 34.6   Two Subtypes of Membrane-Disrupting Exotoxins.   (**a**) A channel-forming (pore-forming) type of exotoxin that inserts itself into the normal host cell membrane and makes an open channel (pore). Formation of multiple pores causes cytoplasmic contents to leave the cell and water to move in, leading to cellular lysis and death of the host cell. (**b**) A phospholipid-hydrolyzing phospholipase exotoxin destroys membrane integrity. (**c**) The exotoxin removes the charged polar head groups from the phospholipid part of the host cell membrane. This destabilizes the membrane and causes the host cell to lyse.

staphylococcal food poisoning (*see section 39.4*) caused solely by the ingestion of preformed enterotoxin. Since the bacteria (*Staphylococcus aureus*) cannot colonize the gut, they pass through the body without producing any more exotoxin; thus, this type of bacterial disease is self-limiting.

In the second example (figure 34.7*b*), bacteria colonize a mucosal surface but do not invade underlying tissue or enter the bloodstream. The toxin either causes disease locally or enters the bloodstream and is distributed systemically where it can cause disease at distant sites. The classical example here is the disease cholera caused by *Vibrio cholerae* (*see section 39.4*). Once the bacteria enter the body, they adhere to the intestinal mucosa. They are not invasive but secrete the cholera toxin, which is an AB exotoxin that catalyzes an ADP-ribosylation similar to that of diphtheria exotoxin (figure 34.5*b*). As a result, cholera toxin stimulates hypersecretion of water and chloride ions and the patient loses massive quantities of water.

The third example of exotoxins in disease pathogenesis occurs when bacteria grow in a wound or abscess (figure 34.7*c*). The exotoxin causes local tissue damage or kills phagocytes that enter the infected area. A disease of this type is gas gangrene (*see section 39.3*) in which the exotoxin (α-toxin) of *Clostridium perfringens* causes the tissue destruction in the wound.

---

1. What is the difference between an infectious disease and an intoxication? Define toxemia.
2. Describe some general characteristics of exotoxins.
3. How do exotoxins get into host cells?
4. Describe the biological effects of several bacterial exotoxins.
5. Discuss the mechanisms by which exotoxins can damage cells.
6. What are the four types of exotoxins?
7. What is the mode of action of a leukocidin? Of a hemolysin?
8. Name two specific hemolysins.
9. What are the three main roles exotoxins have in human disease pathogenesis?

Endotoxins

Gram-negative bacteria have lipopolysaccharide (LPS) in the outer membrane of their cell wall that, under certain circumstances, is toxic to specific hosts. This LPS (*see figures 3.23–3.25*) is called an **endotoxin** because it is bound to the bacterium and is released when the microorganism lyses (**Techniques & Applications 34.1**). Some is also released during bacterial multiplication. The toxic component of the LPS is the lipid portion, called lipid A. Lipid A is not a single macromolecular structure but appears to be a complex array of lipid residues. The lipid A component exhibits all the properties (see characteristic 5 on p. 775) associated with endotoxicity and gram-negative bacteremia. Gram-negative cell wall (pp. 56–58)

Besides the preceding characteristics, bacterial endotoxins are

1. Heat stable
2. Toxic only at high doses (milligram per kilogram amounts)
3. Weakly immunogenic
4. Generally similar, despite source

Bacterial endotoxins have plagued the pharmaceutical industry and medical device producers for years. For example, administration of drugs contaminated with endotoxins can result in complications—even death—to patients. Recently endotoxins have become a problem for individuals and firms working with cell cultures and genetic engineering. The result has been the development of sensitive tests and methods to identify and remove these endotoxins. The procedures must be very sensitive to trace amounts of endotoxins. Most firms have set a limit of 0.25 **endotoxin units (E.U.),** 0.025 ng/ml, or less as a release standard for their drugs, media, or products.

One of the most accurate tests for endotoxins is the in vitro *Limulus* amoebocyte lysate (LAL) assay. The assay is based on the observation that when an endotoxin contacts the clot protein from circulating amoebocytes of *Limulus,* a gel-clot forms. The assay kits available today contain calcium, proclotting enzyme, and procoagulogen. The proclotting enzyme is activated by bacterial endotoxin lipopolysaccharide and calcium to form active clotting enzyme (see **Box figure**). Active clotting enzyme then catalyzes the cleavage of procoagulogen into polypeptide subunits (coagulogen). The subunits join by disulfide bonds to form a gel-clot. Spectrophotometry is then used to measure the protein precipitated by the lysate. The LAL test is sensitive at the nanogram level but must be standardized against Food and Drug Administration Bureau of Biologics endotoxin reference standards. Results are reported in endotoxin units per milliliter and reference made to the particular reference standards used.

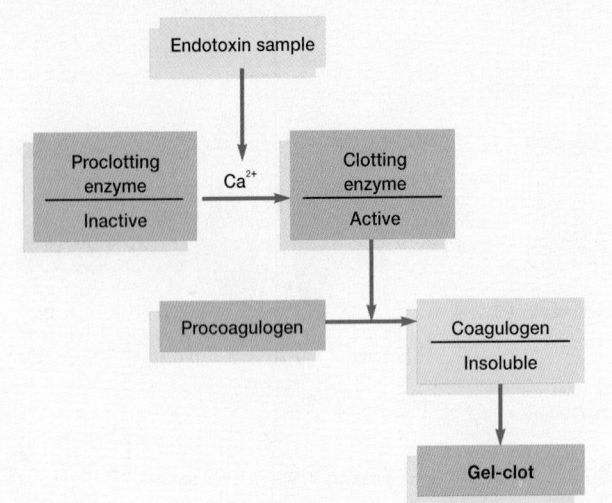

Removal of endotoxins presents more of a problem than their detection. Those present on glassware or medical devices can be inactivated if the equipment is heated at 250°C for 30 minutes. Soluble endotoxins range in size from 20 kDa to large aggregates with diameters up to 0.1 μm. Thus they cannot be removed by conventional filtration systems. Manufacturers are currently developing special filtration systems and filtration cartridges that retain these endotoxins and help alleviate contamination problems.

Figure 34.7   Roles of Exotoxins in Disease Pathogenesis.   Three ways (**a, b, c**) in which bacterial exotoxins can contribute to the progression of disease in a human.

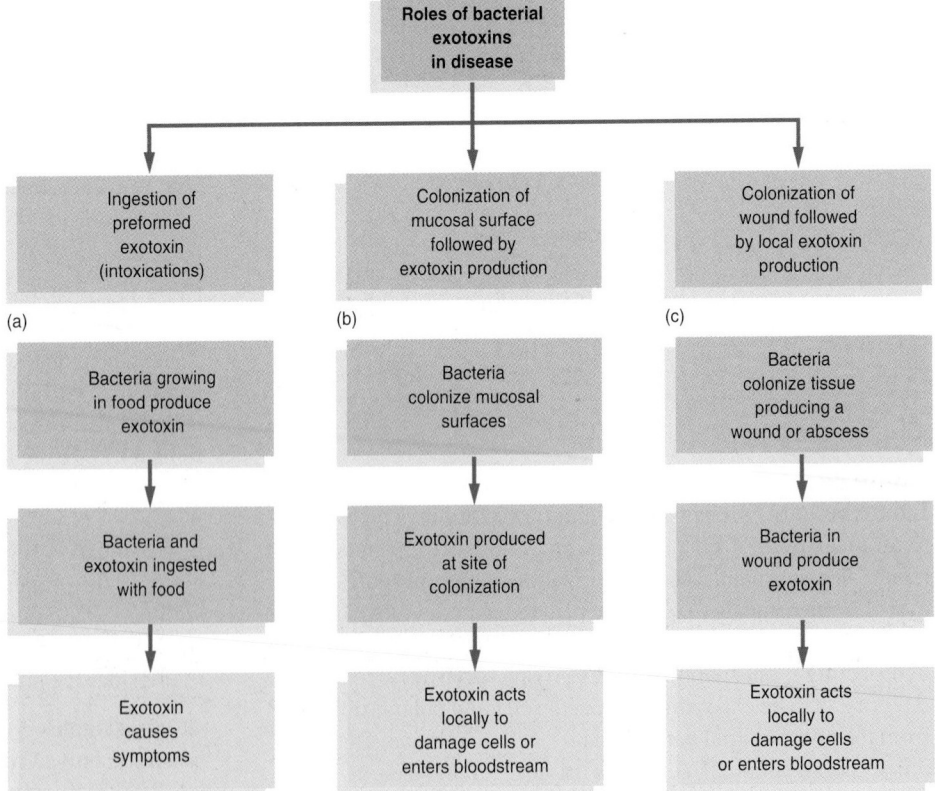

5. Usually capable of producing general systematic effects: fever (are pyrogenic), shock, blood coagulation, weakness, diarrhea, inflammation, intestinal hemorrhage, and fibrinolysis (enzymatic breakdown of fibrin, the major protein component of blood clots)

The characteristics of endotoxins and exotoxins are contrasted in table 34.5.

The main biological effect of LPS is an indirect one, being mediated by host molecules and systems rather than by LPS directly. For example, endotoxins can initially activate Hageman Factor (blood clotting factor XII), which in turn activates up to four humoral systems: coagulation, complement, fibrinolytic, and kininogen systems.

Gram-negative endotoxins also indirectly induce a fever in the host by causing macrophages to release **endogenous pyrogens** that reset the hypothalamic thermostat. One important endogenous pyrogen is the cytokine interleukin-1 (*see figure 31.20*). Other cytokines released by macrophages, such as the tumor necrosis factor, also produce fever.

Recent evidence indicates that LPS affects macrophages and monocytes by binding to special plasma proteins called **LPS-binding proteins.** The LPS-LPS-binding protein complex then attaches to receptors on monocytes, macrophages, and other cells. This triggers several events, including the production of cytokines IL-1, IL-6, and tumor necrosis factor. As mentioned previously, IL-1 and tumor necrosis factor induce fever. These cytokines also promote other endotoxin effects: complement activation, coagulation, prostaglandin formation, and so forth.

---

1. Describe the chemical structure of the LPS endotoxin.
2. List some general characteristics of endotoxins.
3. How do gram-negative endotoxins induce fever in a mammalian host?

34.4 MICROBIAL MECHANISMS FOR ESCAPING HOST DEFENSES

So far, we have discussed some of the ways viral and bacterial pathogens cause disease in a host. During the course of microbe and human evolution, these same pathogens have evolved ways for escaping host defenses. Many of these mechanisms are found throughout the microbial world and several are now discussed.

Evasion of Host Defenses by Viruses

As noted earlier in this chapter, the pathology arising from a viral infection is due to either (1) the host's immune response, which attacks virus-infected cells or produces hypersensitivity reactions (*see section 33.2*), or (2) the direct consequence of viral multiplication within host cells. Viruses have evolved a variety of ways to suppress or evade the host's immune response. These mechanisms are just now becoming recognized through genomics and the functional analysis of specific gene products.

Some viruses may mutate and change antigenic sites (antigenic drift, *see section 37.5*) on the virion proteins (the influenza virus) or may down-regulate the level of expression of viral cell surface proteins (the herpesvirus). Other viruses (HIV) may infect cells (T cells) of the immune system and diminish their function. HIV as well as the measles virus and cytomegalovirus cause the fusion of host cells. This allows these viruses to move from an infected cell to an uninfected cell without exposure to the antibody-containing fluids of the host. The herpesvirus may infect neurons that express little or no major histocompatibility complex molecules (*see section 32.4*). The adenovirus produces proteins that inhibit major histocompatibility complex function. Finally, hepatitis B virus-infected cells produce large amounts of antigens not associated with the complete virus. These antigens bind the available neutralizing antibody (*see section 32.6*) so that there is insufficient free antibody to bind with the complete viral particle.

Evasion of Host Defenses by Bacteria

Bacteria also have evolved many mechanisms to evade host defenses. Because bacteria would not be well served either by the death of their host or their own death, their survival strategy is protection against host defenses rather than host destruction. Several of these evasive mechanisms are now discussed.

Evading the Complement System

To evade the activity of complement, some bacteria have capsules (see chapter opening figure) that prevent complement activation. Some gram-negative bacteria can lengthen the O chains in their lipopolysaccharide to prevent complement activation. Others such as *Neisseria gonorrhoeae* generate **serum resistance.** These bacteria have modified lipooligosaccharides on their surface that interfere with proper formation of the membrane attack complex (*see figure 31.14*) during the complement cascade. The virulent forms of *N. gonorrhoeae* that possess serum resistance are able to spread throughout the body of the host and cause systemic disease, whereas those *N. gonorrhoeae* that lack serum resistance remain localized in the genital tract.   Complement activation (pp. 691–93, 734–35)

Resisting Phagocytosis

As noted several times previously, before a phagocytic cell can phagocytose a bacterium, it must first directly contact the bacterium's surface. Some bacteria such as *Streptococcus pneumoniae*, *Neisseria meningitidis*, and *Haemophilus influenzae* can produce a slippery mucoid capsule that prevents the phagocyte from effectively contacting the bacterium. Other bacteria evade phagocytosis by producing specialized surface proteins such as the M protein on *S. pyogenes*. Like capsules, these proteins interfere with adherence between a phagocytic cell and the bacterium.

Bacterial pathogens can resist phagocytosis in quite different ways. For example, *Staphylococcus* produces leukocidins that destroy phagocytes before phagocytosis can occur. *Streptococcus pyogenes* releases a protease that cleaves the C5a complement factor and thus inhibits complement's ability to attract phagocytes to the infected area.

Survival Inside Phagocytic Cells

Some bacteria have evolved the ability to survive inside neu-trophils, monocytes, and macrophages. They are very pathogenic because they are impervious to a most important host protective mechanism. One method of evasion is to escape from the phago-some before it merges with the lysosome, as seen with *Listeria monocytogenes, Shigella,* and *Rickettsia.* These bacteria use actin-based motility to move within mammalian host cells and spread between them (*see figure 4.4*). Upon lysing the phagosome, they gain access to the cytoplasm. Each bacterium then recruits to its surface host cell actin and other cytoskeletal proteins and activates the assembly of an actin tail. The actin tails propel the bacteria through the cytoplasm of the infected cell to its surface where they push out against the plasma membrane and form protrusions. The protrusions are engulfed by adjacent cells, and the bacteria once again enter phagosomes and escape into the cytoplasm. In this way the infection spreads to adjacent cells. The lysosomes never do have a chance to merge with the phagosomes (*see figure 31.16*). Another approach is to resist the toxic products released into the phagolysosome after fusion occurs. A good example of a bac-terium that is resistant to the lysosomal enzymes is *Mycobac-terium tuberculosis,* probably at least partly because of its waxy external layer. Still other bacteria prevent fusion of phagosomes with lysosomes (*Chlamydia*). Phagocytosis (pp. 693–97)

Evading the Specific Immune Response

To evade the specific immune response, some bacteria (*Strepto-coccus pyogenes*) produce capsules that are not antigenic since they resemble host tissue components. *N. gonorrhoeae* can also evade the specific immune response by two mechanisms: (1) it makes ge-netic variations in its pili (phase variation) so that specific antibod-ies are useless against the new pili and adherence to host tissue occurs, and (2) it produces IgA proteases that destroy secretory IgA and allow adherence. Finally, some bacteria produce proteins (such as staphylococcal protein A and protein G of *Streptococcus pyo-genes*) that interfere with antibody-mediated opsonization (*see fig-ure 31.15*) by binding to the Fc portion of immunoglobulins.

---

1. What are some mechanisms viruses use to evade host defenses?
2. How do bacteria evade each of the following host defenses: the complement system, phagocytosis, and the specific im-mune response?

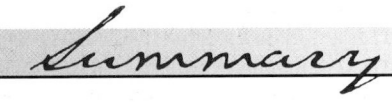

Summary

34.1 Host-Parasite Relationships

a. Parasitism is a type of symbiosis between two species in which the smaller organism is physi-ologically dependent on the larger one, termed the host. The parasitic organism usually harms its host in some way.

b. An infection is the colonization of the host by a parasitic organism. An infectious disease is the result of the interaction (**figure 34.1**) between the parasitic organism and its host, causing the host to change from a state of health to a dis-eased state. Any organism that produces such a disease is a pathogen.

c. Pathogenicity refers to the quality or ability of an organism to produce pathological changes or disease. Virulence refers to the degree or in-tensity of pathogenicity of an organism and is measured experimentally by the LD_{50} or ID_{50} (**figure 34.2**).

34.2 Pathogenesis of Viral Diseases

a. The fundamental process of viral infection is the expression of the viral replicative cycle in a host cell. To produce disease a virus enters a host; comes into contact with susceptible cells; repro-duces; spreads to adjacent cells; causes cellular injury; engenders a host immune response; is ei-ther cleared from the body of the host, estab-lishes a persistent infection, or kill the host; and is shed back into the environment.

34.3 Pathogenesis of Bacterial Diseases

a. Pathogens or their products can be transmitted to a host by either direct or indirect means. Trans-missibility is the initial requisite in the estab-lishment of an infectious disease.

b. Special adherence factors (**table 34.3**) allow pathogens to bind to specific receptor sites on host cells and colonize the host (**figure 34.3**).

c. Pathogens can enter host cells by both active and passive mechanisms. Once inside, they can produce specific products and/or enzymes that promote dis-semination throughout the body of the host. These are termed virulence factors (**table 34.4**).

d. The pathogen generally is found in the area of the host's body that provides the most favorable conditions for its growth and multiplication.

e. In bacteria, one important result of the conserva-tion of chromosomal genes is that bacteria are clonal. For most pathogenic bacteria, there are only a few clonal types that exist in the environment.

f. During coevolution with human hosts, some pathogenic bacteria have evolved complex sig-nal transduction pathways to regulate the genes necessary for virulence.

g. Many bacteria are pathogenic because they have large segments of DNA called pathogenicity is-lands that carry genes responsible for virulence.

h. Intoxications are diseases that result from the entrance of a specific toxin into a host. The toxin can induce the disease in the absence of the toxin-producing organism. Toxins produced by pathogens can be divided into two main cate-gories: exotoxins and endotoxins (**table 34.5**).

i. Exotoxins are soluble, heat-labile, potent, toxic proteins produced by the pathogen as a result of its normal metabolism. They have very specific effects and can be categorized as neurotoxins, cytotoxins, or enterotoxins. Most exotoxins con-form to the AB model in which the A subunit or fragment is enzymatic and the B subunit or frag-ment, the binding portion (**table 34.6**). Several mechanisms exist by which the A component en-ters target cells (**figure 34.5**).

j. Exotoxins can be divided into four types: (a) the AB toxins, (b) specific host site toxins (neuro-toxins, enterotoxins, cytotoxins), (c) toxins that disrupt plasma membranes of host cells (leuko-cidins, hemolysins, and phospholipases), and (d) superantigens.

k. Bacterial exotoxins cause disease in a human host in three main ways: (1) ingestion of pre-formed exotoxin, (2) colonization of a mucosal surface followed by exotoxin production, and (3) colonization of a wound followed by local exotoxin production.

l. Endotoxins are heat-stable, toxic substances that are part of the cell wall lipopolysaccharide of some gram-negative bacteria. Most endotoxins function by initially activating Hageman Factor, which in turn activates one to four humoral systems. These include the intrinsic blood clotting cascade, complement activation, fibrinolytic system, and kininogen system. Endotoxins also stimulate macrophages to release cytokines such as IL-1, IL-6, and TNF-α.

34.4 Microbial Mechanisms for Escaping Host Defenses

a. During the course of microbe and human evolution, some pathogens have evolved ways for escaping host defenses. Viruses have mechanisms that either suppress or evade the host's immune response. Bacteria have evolved mechanisms to evade the complement system, phagocytosis, and the specific immune response.

Key Terms

AB toxins 771	exotoxin 769	LPS-binding protein 775	septicemia 767
adhesin 766	final host 762	membrane-disrupting exotoxin 772	serum resistance 775
alpha hemolysis 772	fomite 766	neurotoxin 772	streptolysin-O (SLO) 772
antitoxin 769	hemolysin 772	opportunistic pathogen 764	streptolysin-S (SLS) 772
apoptosis 765	host 762	parasite 762	toxemia 769
bacteremia 767	immunopathology 764	parasitic organism 762	toxigenicity 764
beta hemolysis 772	infection 763	parasitism 762	toxin 769
colonization 766	infectious disease 763	pathogen 763	toxoid 769
cytotoxin 772	infectious dose 50 (ID$_{50}$) 764	pathogenicity 764	transfer host 762
ectoparasite 762	infectivity 764	pathogenicity island 768	tropism 765
endogenous pyrogen 775	intermediate host 762	pathogenic potential 764	type III secretion pathway 768
endoparasite 762	intoxication 769	phospholipase 772	vector 765
endotoxin 773	invasiveness 764	primary (frank) pathogen 764	viremia 765
endotoxin unit (E.U.) 774	lethal dose 50 (LD$_{50}$) 764	reservoir 765	virulence 764
enterotoxin 772	leukocidin 772	reservoir host 763	virulence factor 766

Questions for Thought and Review

1. Why does a parasitic organism not have to be a parasite?
2. In general, infectious diseases that are commonly fatal are newly evolved relationships between the parasitic organism and the host. Why is this so?
3. What does an organism require to be parasitic?
4. What are some bacterial determinants that provide the organism with the ability to colonize and invade the host?
5. What is the significance of the clonal nature of bacterial pathogens?
6. How do some bacteria regulate their virulence factors?
7. Describe a pathogenicity island.
8. What are four types of exotoxins based on their structural and physiological activities?
9. What is the difference between the general properties of endotoxins and exotoxins?
10. How do some viruses evade the defenses of a human host? How do some bacteria evade the defenses of a human host?

Critical Thinking Questions

1. Explain the observation that different pathogens infect different parts of the host.
2. Intracellular bacterial infections present a particular difficulty for the host. Why is it harder to defend against these infections than against viral infections and extracellular bacterial infections?

For recommended readings on these and related topics, see Appendix V.

chapter 35

Antimicrobial Chemotherapy

Concepts

- Many diseases are treated with chemotherapeutic agents, such as antibiotics, that inhibit or kill the pathogen while harming the host as little as possible.
- Ideally, antimicrobial agents disrupt microbial processes or structures that differ from those of the host. They may damage pathogens by hampering cell wall synthesis, inhibiting microbial protein and nucleic acid synthesis, disrupting microbial membrane structure and function, or blocking metabolic pathways through inhibition of key enzymes.
- The effectiveness of chemotherapeutic agents depends on many factors: the route of administration and location of the infection, the presence of interfering substances, the concentration of the drug in the body, the nature of the pathogen, the presence of drug allergies, and the resistance of microorganisms to the drug.
- The increasing number and variety of drug-resistant pathogens is a serious public health problem.
- Although antibacterial chemotherapy is more advanced, drugs for the treatment of fungal and viral infections are also becoming increasingly available.

It was the knowledge of the great abundance and wide distribution of actinomycetes, which dated back nearly three decades, and the recognition of the marked activity of this group of organisms against other organisms that led me in 1939 to undertake a systematic study of their ability to produce antibiotics.

—Selman A. Waksman

Many antimicrobial medications are available to combat infections.

*T*he control of microorganisms is critical for the prevention and treatment of disease. Chapter 7 is concerned principally with the chemical and physical agents used to treat inanimate objects in order to destroy microorganisms or inhibit their growth. In addition, the use of antiseptics is briefly reviewed. Microorganisms also grow on and within other organisms, and microbial colonization can lead to disease, disability, and death. Thus the control or destruction of microorganisms residing within the bodies of humans and other animals is of great importance.

When disinfecting or sterilizing an inanimate object, one naturally must use procedures that do not damage the object itself. The same is true for the treatment of living hosts. The most successful drugs interfere with processes that differ between the pathogen and host, and seriously damage the target microorganism while harming its host as little as possible. This chapter introduces the principles of chemotherapy and briefly reviews the characteristics of selected antibacterial, antifungal, and antiviral drugs.

Modern medicine is dependent on **chemotherapeutic agents,** chemical agents that are used to treat disease. Chemotherapeutic agents destroy pathogenic microorganisms or inhibit their growth at concentrations low enough to avoid undesirable damage to the host. Most of these agents are **antibiotics** [Greek *anti,* against, and *bios,* life], microbial products or their derivatives that can kill susceptible microorganisms or inhibit their growth. Drugs such as the sulfonamides are sometimes called antibiotics although they are synthetic chemotherapeutic agents, not microbially synthesized.

35.1 THE DEVELOPMENT OF CHEMOTHERAPY

The modern era of chemotherapy began with the work of the German physician Paul Ehrlich (1854–1915). Ehrlich reasoned that a chemical with selective toxicity that would kill pathogens and not human cells might be effective in treating disease. He hoped to find a toxic dye molecule, a "magic bullet," that would specifically bind to pathogens and destroy them; therefore he began experimenting with dyes. By 1904 Ehrlich found that the dye trypan red was active against the trypanosome that causes African sleeping sickness (*see figure 27.3*) and could be used therapeutically. Subsequently Ehrlich and a young Japanese scientist named Sahachiro Hata tested a variety of arsenicals on syphilis-infected rabbits and found that compound number 606, arsphenamine, was active against the syphilis spirochete (*see figure 21.15b*). Arsphenamine was made available in 1910 under the trade name Salvarsan. Ehrlich's successes in the chemotherapy of sleeping sickness and syphilis established his concept of selective toxicity and led to the testing of hundreds of compounds for their therapeutic potential.

In 1927 the German chemical industry giant, I. G. Farbenindustrie, began a long-term search for chemotherapeutic agents under the direction of Gerhard Domagk. The company provided vast numbers of dyes and other chemicals that Domagk tested for activity against pathogenic bacteria and for toxicity in animals. During this screening program Domagk discovered that Prontosil Red, a new dye for staining leather, was nontoxic for animals and completely protected mice against pathogenic streptococci and staphylococci. These results were published in 1935, and in the same year the French scientists Jacques and Therese Trefouel showed that Prontosil Red was converted in the body to sulfanilamide, the true active factor. Domagk had actually discovered sulfonamides or sulfa drugs and for this discovery he received the Nobel Prize in 1939.

The story of the discovery and development of penicillin, the first antibiotic to be used therapeutically, is complex and fascinating. Although penicillin was actually discovered in 1896 by a 21-year-old French medical student named Ernest Duchesne, his work was forgotten, and penicillin was rediscovered and brought to the attention of scientists by the Scottish physician Alexander Fleming. Fleming had been interested in finding something that would kill pathogens ever since working on wound infections during the First World War. One day in September 1928, a *Penicillium notatum* spore accidentally landed on the surface of an exposed petri dish before it had been inoculated with staphylococci, and a new medical era was born. Although the precise events are still unclear, Ronald Hare has suggested that Fleming left the contaminated plate on a laboratory bench while he was on vacation. Because the first few days of the vacation were cool, the fungus grew more rapidly than the bacteria and produced penicillin. When the weather then turned warm, the bacteria began to grow and were lysed. On his return, Fleming noticed that a *Penicillium* colony was growing at one edge and that the staphylococci surrounding it had been destroyed. Rather than discarding the contaminated plate, he correctly deduced that his mold contaminant was producing a diffusible substance lethal to staphylococci. He began efforts to characterize what he called penicillin. He found that broth from a *Penicillium* culture contained penicillin and that the antibiotic could destroy several pathogenic bacteria. Unfortunately Fleming's next experiments convinced him that penicillin would not remain active in the body long enough after injection to destroy pathogens. After giving a talk on penicillin and publishing several papers on the subject between 1929 and 1931, he dropped the research.

In 1939 Howard Florey, a professor of pathology at Oxford University, was in the midst of testing the bactericidal activity of many substances, including lysozyme and the sulfonamides. After reading Fleming's paper on penicillin, one of Florey's coworkers, Ernst Chain, obtained the *Penicillium* culture from Fleming and set about culturing it and purifying penicillin. Florey and Chain were greatly aided in this by the biochemist Norman Heatley. Heatley devised the original assay, culture, and purification techniques needed to produce crude penicillin for further experimentation. When the purified penicillin was injected into mice infected with streptococci or staphylococci, practically all the mice survived. Florey and Chain's success was reported in 1940, and subsequent human trials were equally successful. Fleming, Florey, and Chain received the Nobel Prize in 1945 for the discovery and production of penicillin.

Techniques & Applications
35.1　　The Use of Antibiotics in Microbiological Research

Although the use of antibiotics in the treatment of disease is emphasized in this chapter, it should be noted that antibiotics are extremely important research tools. For example, they aid the cultivation of viruses by preventing bacterial contamination. When eggs are inoculated with a virus sample, antibiotics often are included in the inoculum to maintain sterility. Usually a mixture of antibiotics (e.g., penicillin, amphotericin, and streptomycin) also is added to tissue cultures used for virus cultivation and other purposes.

Researchers often use antibiotics as instruments to dissect metabolic processes by inhibiting or blocking specific steps and observing the consequences. Although selective toxicity is critical when antibiotics are employed therapeutically, specific toxicity is more important in this context: the antibiotic must act by a specific and precisely understood mechanism. A clinically useful antimicrobial agent such as penicillin sometimes may be employed in research, but often an agent with specific toxicity and excellent research potential is too toxic for therapeutic use. The actinomycins, discovered in 1940 by Selman Waksman, are a case in point. They are so toxic to higher organisms that it was suggested they be used as rat poison. Today actinomycin D is a standard research tool specifically used to block RNA

synthesis. Other examples of antibiotics useful in research, with the process inhibited, are the following: chloramphenicol (bacterial protein synthesis), cycloserine (peptidoglycan synthesis), nalidixic acid and novobiocin (bacterial DNA synthesis), rifampin (bacterial RNA synthesis), cycloheximide (eucaryotic protein synthesis), daunomycin (fungal RNA synthesis), mitomycin C (DNA synthesis), polyoxin D (fungal cell wall chitin synthesis), and cerulenin (fatty acid synthesis).

In practice, the antibiotic is administered and changes in cell function are monitored. If one desired to study the dependence of bacterial flagella synthesis on RNA transcription, the flagella could be removed by high-speed mixing in a blender, followed by actinomycin D addition to the incubation mixture. The bacterial culture would then be observed for flagella regeneration in the absence of RNA synthesis. The results of such experiments must be interpreted with caution. Flagella synthesis may have been blocked because actinomycin D inhibited some other process, thus affecting flagella regeneration indirectly rather than simply inhibiting transcription of a gene required for flagella synthesis. Furthermore, not all microorganisms respond in the same way to a particular drug.

The discovery of penicillin stimulated the search for other antibiotics. Selman Waksman announced in 1944 that he and his associates had found a new antibiotic, streptomycin, produced by the actinomycete *Streptomyces griseus*. This discovery arose from the patient screening of about 10,000 strains of soil bacteria and fungi. Waksman received the Nobel Prize in 1952, and his success led to a worldwide search for other antibiotic-producing soil microorganisms. Microorganisms producing chloramphenicol, neomycin, terramycin, and tetracycline were isolated by 1953.

The discovery of chemotherapeutic agents and the development of newer, more powerful drugs has transformed modern medicine and greatly alleviated human suffering. Furthermore, antibiotics have proven exceptionally useful in microbiological research (**Techniques & Applications 35.1**).

---

1. What are chemotherapeutic agents? Antibiotics?
2. What contributions to chemotherapy were made by Ehrlich, Domagk, Fleming, Florey and Chain, and Waksman?

GENERAL CHARACTERISTICS OF ANTIMICROBIAL DRUGS

As Ehrlich so clearly saw, to be successful a chemotherapeutic agent must have **selective toxicity:** it must kill or inhibit the microbial pathogen while damaging the host as little as possible. The degree of selective toxicity may be expressed in terms of (1) the therapeutic dose, the drug level required for clinical treat-

ment of a particular infection, and (2) the toxic dose, the drug level at which the agent becomes too toxic for the host. The **therapeutic index** is the ratio of the toxic dose to the therapeutic dose. The larger the therapeutic index, the better the chemotherapeutic agent (all other things being equal).

A drug that disrupts a microbial function not found in eucaryotic animal cells often has a greater selective toxicity and a higher therapeutic index. For example, penicillin inhibits bacterial cell wall peptidoglycan synthesis but has little effect on host cells because they lack cell walls; therefore penicillin's therapeutic index is high. A drug may have a low therapeutic index because it inhibits the same process in host cells or damages the host in other ways. These undesirable effects on the host, called **side effects,** are of many kinds and may involve almost any organ system (**table 35.1**). Because side effects can be severe, chemotherapeutic agents should be administered with great care.

Drugs vary considerably in their range of effectiveness. Many are **narrow-spectrum drugs**—that is, they are effective only against a limited variety of pathogens (table 35.1). Others are **broad-spectrum drugs** that attack many different kinds of pathogens. Drugs may also be classified based on the general microbial group they act against: antibacterial, antifungal, antiprotozoan, and antiviral. Some agents can be used against more than one group; for example, sulfonamides are active against bacteria and some protozoa.

Chemotherapeutic agents can be synthesized by microorganisms or manufactured by chemical procedures independent of microbial activity. A number of the most commonly employed antibiotics are natural—that is, totally synthesized by one of a

Table 35.1

Table 35.1

Properties of Some Common Antibacterial Drugs

Drug	Primary Effect	Spectrum	Side Effects[a]
Ampicillin	Cidal	Broad (gram +, some −)	Allergic responses (diarrhea, anemia)
Bacitracin	Cidal	Narrow (gram +)	Renal injury if injected
Carbenicillin	Cidal	Broad (gram +, many −)	Allergic responses (nausea, anemia)
Cephalosporins	Cidal	Broad (gram +, some −)	(Allergic responses, thrombophlebitis, renal injury)
Chloramphenicol	Static	Broad (gram +, −; rickettsia and chlamydia)	Depressed bone marrow function, allergic reactions
Ciprofloxacin	Cidal	Broad (gram +, −)	Gastrointestinal upset, allergic responses
Clindamycin	Static	Narrow (gram +, anaerobes)	Diarrhea
Dapsone	Static	Narrow (mycobacteria)	(Anemia, allergic responses)
Erythromycin	Static	Broad (gram +, some −)	(Gastrointestinal upset, hepatic injury)
Gentamicin	Cidal	Narrow (gram −)	(Allergic responses, nausea, loss of hearing, renal damage)
Isoniazid	Static or cidal	Narrow (mycobacteria)	(Allergic reactions, gastrointestinal upset, hepatic injury)
Methicillin	Cidal	Narrow (gram +)	Allergic responses (renal toxicity, anemia)
Penicillin	Cidal	Narrow (gram +)	Allergic responses (nausea, anemia)
Polymyxin B	Cidal	Narrow (gram −)	(Renal damage, neurotoxic reactions)
Rifampin	Static	Broad (gram +, mycobacteria)	(Hepatic injury, nausea, allergic responses)
Streptomycin	Cidal	Broad (gram +, −; mycobacteria)	(Allergic responses, nausea, loss of hearing, renal damage)
Sulfonamides	Static	Broad (gram +, −)	Allergic responses (renal and hepatic injury, anemia)
Tetracyclines	Static	Broad (gram +, −; rickettsia and chlamydia)	Gastrointestinal upset, teeth discoloration (renal and hepatic injury)
Trimethoprim	Cidal	Broad (gram +, −)	(Allergic responses, rash, nausea, leukopenia)
Vancomycin	Cidal	Narrow (gram +)	Hypotension, neutropenia, kidney damage, allergic reactions

[a]Occasional side effects are in parentheses. Other side effects not listed may also arise.

few bacteria or fungi (**table 35.2**). In contrast, several important chemotherapeutic agents are completely synthetic. The synthetic antibacterial drugs in table 35.1 are the sulfonamides, trimethoprim, chloramphenicol, ciprofloxacin, isoniazid, and dapsone. Many antiviral and antiprotozoan drugs are synthetic. An increasing number of antibiotics are semisynthetic. Semisynthetic antibiotics are natural antibiotics that have been chemically modified by the addition of extra chemical groups to make them less susceptible to inactivation by pathogens. Ampicillin, carbenicillin, and methicillin (table 35.1) are good examples.

Chemotherapeutic agents, like disinfectants, can be either **cidal** or **static.** As mentioned earlier (*see section 7.1*), static agents reversibly inhibit growth; if the agent is removed, the microorganisms will recover and grow again.

Although a cidal agent kills the target pathogen, its activity is concentration dependent and the agent may be only static at low levels. The effect of an agent also varies with the target species: an agent may be cidal for one species and static for another. Because static agents do not directly destroy the pathogen, elimination of the infection depends on the host's own resistance mechanisms. A static agent may not be effective if the host's resistance is too low. Host resistance and the immune response (chapters 31–33)

Some idea of the effectiveness of a chemotherapeutic agent against a pathogen can be obtained from the **minimal inhibitory concentration (MIC).** The MIC is the lowest concentration of a drug that prevents growth of a particular pathogen. The **minimal**

Table 35.2

Microbial Sources of Some Antibiotics

Microorganism	Antibiotic
Bacteria	
Streptomyces spp.	Amphotericin B
	Chloramphenicol (also synthetic)
	Erythromycin
	Kanamycin
	Neomycin
	Nystatin
	Rifampin
	Streptomycin
	Tetracyclines
	Vancomycin
Micromonospora spp.	Gentamicin
Bacillus spp.	Bacitracin
	Polymyxins
Fungi	
Penicillium spp.	Griseofulvin
	Penicillin
Cephalosporium spp.	Cephalosporins

lethal concentration (MLC) is the lowest drug concentration that kills the pathogen. A cidal drug kills pathogens at levels only two to four times the MIC, whereas a static agent kills at much higher concentrations (if at all).

1. Define the following terms: selective toxicity, therapeutic index, side effect, narrow-spectrum drug, broad-spectrum drug, synthetic and semisynthetic antibiotics, cidal and static agents, minimal inhibitory concentration (MIC), and minimal lethal concentration (MLC).

35.3 DETERMINING THE LEVEL OF ANTIMICROBIAL ACTIVITY

Determination of antimicrobial effectiveness against specific pathogens is essential to proper therapy. Testing can show which agents are most effective against a pathogen and give an estimate of the proper therapeutic dose.

Dilution Susceptibility Tests

Dilution susceptibility tests can be used to determine MIC and MLC values. In the broth dilution test, a series of broth tubes (usually Mueller-Hinton broth) containing antibiotic concentrations in the range of 0.1 to 128 μg/ml is prepared and inoculated with standard numbers of the test organism. The lowest concentration of the antibiotic resulting in no growth after 16 to 20 hours of incubation is the MIC. The MLC can be ascertained if the tubes showing no growth are subcultured into fresh medium lacking antibiotic. The lowest antibiotic concentration from which the microorganisms do not recover and grow when transferred to fresh medium is the MLC. The agar dilution test is very similar to the broth dilution test. Plates containing Mueller-Hinton agar and various amounts of antibiotic are inoculated and examined for growth. Several automated systems for susceptibility testing and MIC determination with broth or agar cultures have been developed.

Disk Diffusion Tests

If a rapidly growing aerobic or facultative pathogen like *Staphylococcus* or *Pseudomonas* is being tested, a disk diffusion technique may be used to save time and media. The principle behind the assay technique is fairly simple. When an antibiotic-impregnated disk is placed on agar previously inoculated with the test bacterium, the disk picks up moisture and the antibiotic diffuses radially outward through the agar, producing an antibiotic concentration gradient. The antibiotic is present at high concentrations near the disk and affects even minimally susceptible microorganisms (resistant organisms will grow up to the disk). As the distance from the disk increases, the antibiotic concentration drops and only more susceptible pathogens are harmed. A clear zone or ring is present around an antibiotic disk after incubation if the agent inhibits bacterial growth. The wider the zone surrounding a disk, the more susceptible the pathogen is. Zone width also is a function of the antibiotic's initial concentration, its solubility, and its diffusion rate through agar. Thus

zone width cannot be used to compare directly the effectiveness of two different antibiotics.

Currently the disk diffusion test most often used is the **Kirby-Bauer method,** which was developed in the early 1960s at the University of Washington Medical School by William Kirby, A. W. Bauer, and their colleagues. An inoculating loop or needle is touched to four or five isolated colonies of the pathogen growing on agar and then used to inoculate a tube of culture broth. The culture is incubated for a few hours at 35°C until it becomes slightly turbid and is diluted to match a turbidity standard. A sterile cotton swab is dipped into the standardized bacterial test suspension and used to evenly inoculate the entire surface of a Mueller-Hinton agar plate. After the agar surface has dried for about 5 minutes, the appropriate antibiotic test disks are placed on it, either with sterilized forceps or with a multiple applicator device (**figure 35.1**). The plate is immediately placed in a 35°C incubator. After 16 to 18 hours of incubation, the diameters of the zones of inhibition are measured to the nearest mm.

Kirby-Bauer test results are interpreted using a table that relates zone diameter to the degree of microbial resistance (table 35.3). The values in **table 35.3** were derived by finding the MIC values and zone diameters for many different microbial strains. A plot of MIC (on a logarithmic scale) versus zone inhibition diameter (arithmetic scale) is prepared for each antibiotic (**figure 35.2**). These plots are then used to find the zone diameters corresponding to the drug concentrations actually reached in the body. If the zone diameter for the lowest level reached in the body is smaller than that seen with the test pathogen, the pathogen should have an MIC value low enough to be destroyed by the drug. A pathogen with too high an MIC value (too small a zone diameter) is resistant to the agent at normal body concentrations.

The Etest

The Etest from AB Biodisk, Solna, Sweden, has been used in sensitivity testing under some conditions. It is particularly convenient for use with anaerobic pathogens. A petri dish of the proper agar is streaked in three different directions with the test organism and special plastic Etest® strips are placed on the surface so that they extend out radially from the center (**figure 35.3**). Each strip contains a gradient of an antibiotic and is labeled with a scale of minimal inhibitory concentration values. The lowest concentration in the strip lies at the center of the plate. After 24 to 48 hours of incubation, an elliptical zone of inhibition appears. As shown in the figure, MICs are determined from the point of intersection between the inhibition zone and the strip's scale of MIC values.

Measurement of Drug Concentrations in the Blood

A drug must reach a concentration at the site of infection above the pathogen's MIC to be effective. In cases of severe, life-threatening disease, it often is necessary to monitor the concentration of drugs in the blood and other body fluids. This may

(a)

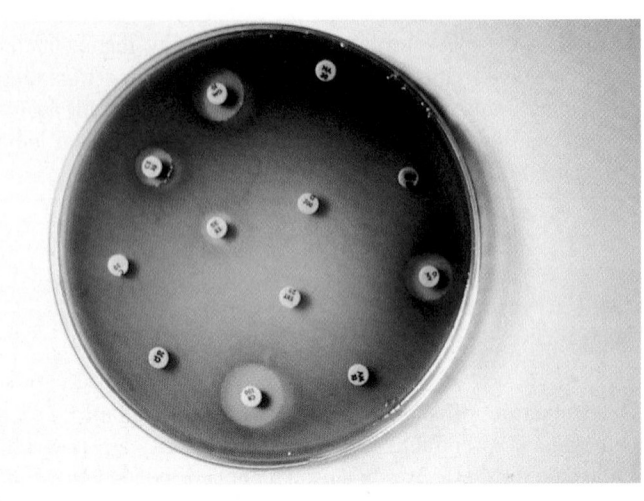

(b)

Figure 35.1   The Kirby-Bauer Method.   (a) A multiple antibiotic disk dispenser and **(b)** disk diffusion test results.

Table 35.3					
Inhibition Zone Diameter of Selected Chemotherapeutic Drugs					
			Zone Diameter (Nearest mm)		
Chemotherapeutic Drug	**Disk Content**		**Resistant**	**Intermediate**	**Susceptible**
Carbenicillin (with *Proteus* spp. and *E. coli*)	100 µg		≤17	18–22	≥23
Carbenicillin (with *Pseudomonas aeruginosa*)	100 µg		≤13	14–16	≥17
Ceftriaxone	30 µg		≤13	14–20	≥21
Chloramphenicol	30 µg		≤12	13–17	≥18
Erythromycin	15 µg		≤13	14–17	≥18
Penicillin G (with staphylococci)	10 Uᵃ		≤20	21–28	≥29
Penicillin G (with other microorganisms)	10 U		≤11	12–21	≥22
Streptomycin	10 µg		≤11	12–14	≥15
Sulfonamides	250 or 300 µg		≤12	13–16	≥17
Tetracycline	30 µg		≤14	15–18	≥19

ᵃOne milligram of penicillin G sodium = 1,600 units.

be achieved by microbiological, chemical, immunologic, enzymatic, or chromatographic assays.

1. How can dilution susceptibility tests and disk diffusion tests be used to determine microbial drug sensitivity?
2. Briefly describe the Kirby-Bauer test and its purpose.
3. How is the Etest carried out?

35.4   MECHANISMS OF ACTION OF ANTIMICROBIAL AGENTS

The mechanisms of action of specific chemotherapeutic agents are taken up in more detail when individual drugs and groups of drugs are discussed later in this chapter. A few general observations are offered at this point. It is important to know something

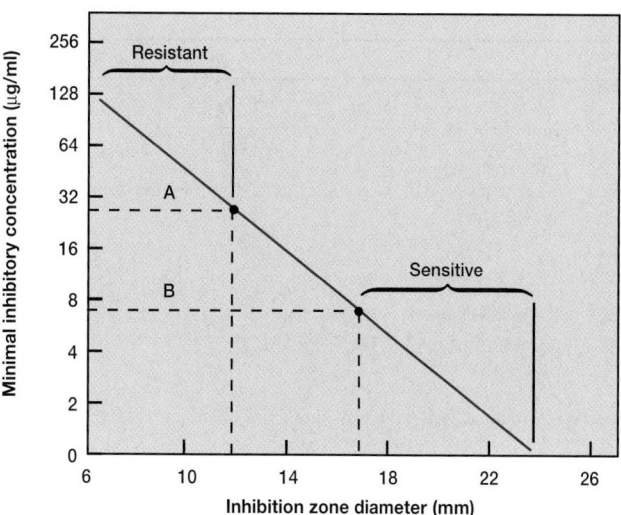

Figure 35.2   Interpretation of Kirby-Bauer Test Results.
The relationship between the minimal inhibitory concentrations with a hypothetical drug and the size of the zone around a disk in which microbial growth is inhibited. As the sensitivity of microorganisms to the drug increases, the MIC value decreases and the inhibition zone grows larger. Suppose that this drug varies from 7–28 μg/ml in the body during treatment. Dashed line A shows that any pathogen with a zone of inhibition less than 12 mm in diameter will have an MIC value greater than 28 μg/ml and will be resistant to drug treatment. A pathogen with a zone diameter greater than 17 mm will have an MIC less than 7 μg/ml and will be sensitive to the drug (see line B). Zone diameters between 12 and 17 mm indicate intermediate sensitivity and usually signify resistance.

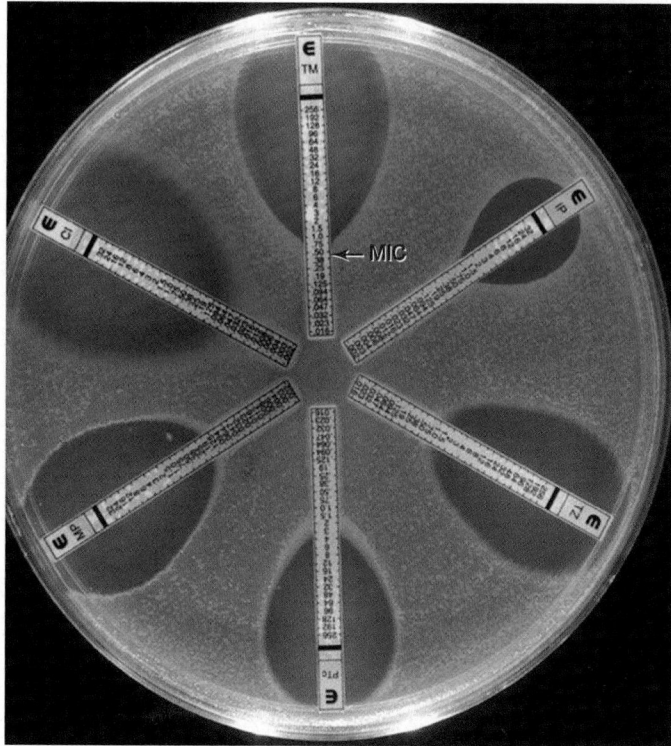

Figure 35.3   The Etest®.   An example of a bacterial culture plate with Etest strips arranged radially on it. The strips are arranged so that the lowest antibiotic concentration in each is at the center. The MIC concentration is read from the scale at the point it intersects the zone of inhibition as shown by the arrow in this example. See text for details.

about the mechanisms of drug action because such knowledge helps explain the nature and degree of selective toxicity of individual drugs and sometimes aids in the design of new chemotherapeutic agents.

Antimicrobial drugs can damage pathogens in several ways, as can be seen in **table 35.4,** which summarizes the mechanisms of the antibacterial drugs listed in table 35.1. The most selective antibiotics are those that interfere with the synthesis of bacterial cell walls (e.g., penicillins, cephalosporins, vancomycin, and bacitracin). These drugs have a high therapeutic index because bacterial cell walls have a unique structure (*see section 3.5*) not found in eucaryotic cells.

Streptomycin, gentamicin, spectinomycin, clindamycin, chloramphenicol, tetracyclines, erythromycin, and many other antibiotics inhibit protein synthesis by binding with the procaryotic ribosome. Because these drugs discriminate between procaryotic and eucaryotic ribosomes, their therapeutic index is fairly high, but not as favorable as that of cell wall synthesis inhibitors. Some drugs bind to the 30S (small) subunit, while others attach to the 50S (large) ribosomal subunit. Several different steps in the protein synthesis mechanism can be affected: aminoacyl-tRNA binding, peptide bond formation, mRNA

reading, and translocation. For example, fusidic acid binds to EF-G and blocks translocation, whereas mucopirocin inhibits isoleucyl-tRNA synthetase (*see section 12.2*).

The antibacterial drugs that inhibit nucleic acid synthesis or damage cell membranes often are not as selectively toxic as other antibiotics. This is because procaryotes and eucaryotes do not differ as greatly with respect to nucleic acid synthetic mechanisms or cell membrane structure. Good examples of drugs that affect nucleic acid synthesis or membrane structure are quinolones and polymyxins. Quinolones inhibit the DNA gyrase and thus interfere with DNA replication, repair, and transcription. Polymyxins act as detergents or surfactants and disrupt the bacterial plasma membrane.

Several valuable drugs act as **antimetabolites:** they block the functioning of metabolic pathways by competitively inhibiting the use of metabolites by key enzymes. Sulfonamides and several other drugs inhibit folic acid metabolism. Sulfonamides (e.g., sulfanilamide, sulfamethoxazole, and sulfacetamide) have a high therapeutic index because humans cannot synthesize folic acid and must obtain it in their diet. Most bacterial pathogens synthesize their own folic acid and are therefore susceptible to inhibitors of folate metabolism. Antimetabolite drugs also can inhibit other

Table 35.4

Mechanisms of Antibacterial Drug Action

Drug	Mechanism of Action
Cell Wall Synthesis Inhibition	
Penicillin	Inhibit transpeptidation enzymes involved in the cross-linking of the polysaccharide chains of the bacterial
Ampicillin	cell wall peptidoglycan. Activate cell wall lytic enzymes.
Carbenicillin	
Methicillin	
Cephalosporins	
Vancomycin	Binds directly to the D-Ala-D-Ala terminus and inhibits the transpeptidation step.
Bacitracin	Inhibits cell wall synthesis by interfering with action of the lipid carrier that transports wall precursors across the plasma membrane.
Protein Synthesis Inhibition	
Streptomycin	Binds with the 30S subunit of the bacterial ribosome to inhibit protein synthesis and causes misreading
Gentamicin	of mRNA.
Chloramphenicol	Binds to the 50S ribosomal subunit and blocks peptide bond formation through inhibition of peptidyl transferase.
Tetracyclines	Bind to the 30S ribosomal subunit and interfere with aminoacyl-tRNA binding.
Erythromycin and clindamycin	Bind to the 50S ribosomal subunit and inhibit peptide chain elongation.
Fusidic acid	Binds to EF-G and blocks translocation.
Nucleic Acid Synthesis Inhibition	
Ciprofloxacin and other quinolones	Inhibit bacterial DNA gyrase and thus interfere with DNA replication, transcription, and other activities involving DNA.
Rifampin	Blocks RNA synthesis by binding to and inhibiting the DNA-dependent RNA polymerase.
Cell Membrane Disruption	
Polymyxin B	Binds to the plasma membrane and disrupts its structure and permeability properties.
Metabolic Antagonism	
Sulfonamides	Inhibit folic acid synthesis by competition with *p*-aminobenzoic acid.
Trimethoprim	Blocks tetrahydrofolate synthesis through inhibition of the enzyme dihydrofolate reductase.
Dapsone	Interferes with folic acid synthesis.
Isoniazid	May disrupt pyridoxal or NAD metabolism and functioning. Inhibits the synthesis of the mycolic acid "cord factor."

pathways. For example, isoniazid interferes with either pyridoxal or NAD metabolism.

---

1. Give five ways in which chemotherapeutic agents kill or damage bacterial pathogens.
2. What are antimetabolites?

35.5 FACTORS INFLUENCING THE EFFECTIVENESS OF ANTIMICROBIAL DRUGS

It is crucial to recognize that drug therapy is not a simple matter. Drugs may be administered in several different ways, and they do not always spread rapidly throughout the body or immediately kill all invading pathogens. A complex array of factors influence the effectiveness of drugs.

First, the drug must actually be able to reach the site of infection. The mode of administration plays an important role. A drug such as penicillin G is not suitable for oral administration because it is relatively unstable in stomach acid. Some antibiotics—for example, gentamicin and other aminoglycosides—are not well absorbed from the intestinal tract and must be injected intramuscularly or given intravenously. Other antibiotics (neomycin, bacitracin) are applied topically to skin lesions. Nonoral routes of administration often are called **parenteral routes.** Even when an agent is administered properly, it may be excluded from the site of infection. For example, blood clots or necrotic tissue can protect bacteria from a drug, either because body fluids containing the agent may not easily reach the pathogens or because the agent is absorbed by materials surrounding it.

Second, the pathogen must be susceptible to the drug. Bacteria in abscesses may be dormant and therefore resistant to chemotherapy, because penicillins and many other agents affect pathogens only if they are actively growing and dividing. A pathogen, even though growing, may simply not be susceptible to a particular agent. For example, penicillins and cephalosporins, which inhibit cell wall synthesis (table 35.4), do not harm mycoplasmas, which lack cell walls.

Third, the chemotherapeutic agent must exceed the pathogen's MIC value if it is going to be effective. The concentration reached will depend on the amount of drug administered, the route of administration and speed of uptake, and the rate at which the drug is cleared or eliminated from the body. It makes sense that a drug will remain at high concentrations longer if it is absorbed over an extended period and excreted slowly.

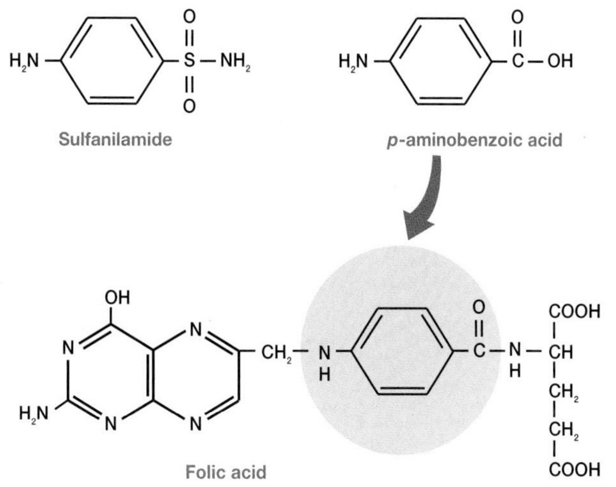

Figure 35.4   Sulfanilamide.   Sulfanilamide and its relationship to the structure of folic acid.

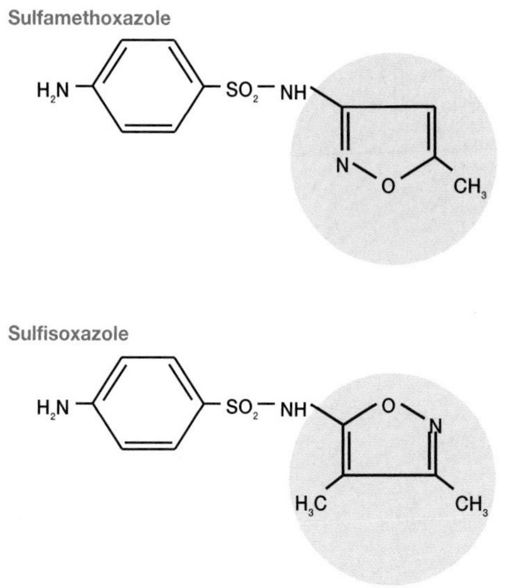

Figure 35.5   Two Sulfonamide Drugs.   The blue shaded areas are side chains substituted for a hydrogen in sulfanilamide (figure 35.4).

Finally, chemotherapy has been rendered less effective and much more complex by the spread of drug resistance plasmids. The nature of drug resistance is dealt with later in this chapter.

---

1. Briefly discuss the factors that influence the effectiveness of antimicrobial drugs.
2. What is parenteral administration of a drug?

35.6   ANTIBACTERIAL DRUGS

With the background provided by the preceding introduction to chemotherapy, a few of the more important chemotherapeutic agents can be briefly described. Tables 35.1 and 35.4 summarize the properties and mechanism of action of a variety of drugs.

Sulfonamides or Sulfa Drugs

A good way to inhibit or kill pathogens is by use of compounds that are **structural analogues,** molecules structurally similar to metabolic intermediates. These analogues compete with metabolites in metabolic processes because of their similarity, but are just different enough so that they cannot function normally in cellular metabolism. The first antimetabolites to be used successfully as chemotherapeutic agents were the sulfonamides, discovered by G. Domagk. **Sulfonamides** or sulfa drugs are structurally related to sulfanilamide, an analogue of *p*-aminobenzoic acid (**figures 35.4** and **35.5**). The latter substance is used in the synthesis of the cofactor folic acid.

When sulfanilamide or another sulfonamide enters a bacterial cell, it competes with *p*-aminobenzoic acid for the active site of an enzyme involved in folic acid synthesis, and the folate concentration decreases. The decline in folic acid is detrimental to the bacterium because folic acid is essential to the synthesis of

purines and pyrimidines, the bases used in the construction of DNA, RNA, and other important cell constituents. The resulting inhibition of purine and pyrimidine synthesis leads to cessation of bacterial growth or death of the pathogen. Sulfonamides are selectively toxic for many pathogens because these bacteria manufacture their own folate and cannot effectively take up the cofactor. In contrast, humans cannot synthesize folate and must obtain it in the diet; therefore sulfonamides will not affect the host. Sulfonamides and the competitive inhibition of enzymes (pp. 159–60); Purine and pyrimidine structure and synthesis (pp. 211–14)

The effectiveness of sulfonamides is limited by the increasing sulfonamide resistance of many bacteria. Furthermore, as many as 5% of the patients receiving sulfa drugs experience adverse side effects, chiefly in the form of allergic responses (fever, hives, and rashes).

Quinolones

A second group of synthetic antimicrobial agents are increasingly used to treat a wide variety of infections. The **quinolones** are synthetic drugs that contain the 4-quinolone ring. The first quinolone, nalidixic acid (**figure 35.6**), was synthesized in 1962. More recently a family of fluoroquinolones has been produced. Three of these—ciprofloxacin, norfloxacin, and ofloxacin—are currently used in the United States, and more fluoroquinolones are being synthesized and tested. Quinolones are effective when administered orally. They sometimes cause adverse side effects, particularly gastrointestinal upset.

Quinolones act by inhibiting the bacterial DNA gyrase or topoisomerase II, probably by binding to the DNA gyrase complex. This enzyme introduces negative twists in DNA and helps separate its strands (**figure 35.7**). DNA gyrase inhibition disrupts

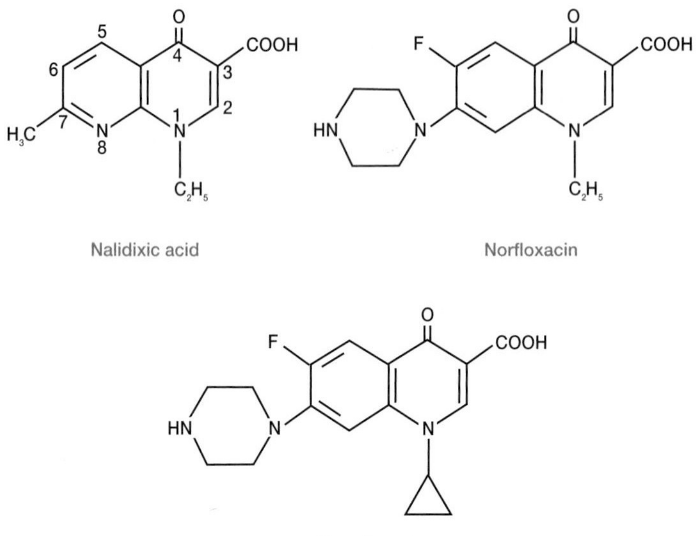

Nalidixic acid

Norfloxacin

Ciprofloxacin

Figure 35.6 Quinolone Antimicrobial Agents. Ciprofloxacin and norfloxacin are newer generation fluoroquinolones. The 4-quinolone ring in nalidixic acid has been numbered.

DNA replication and repair, transcription, bacterial chromosome separation during division, and other cell processes involving DNA. Fluoroquinolones also inhibit topoisomerase IV, another enzyme that untangles DNA during replication. It is not surprising that quinolones are bactericidal. The mechanism of DNA replication (pp. 229–33)

The quinolones are broad-spectrum drugs. They are highly effective against enteric bacteria such as *E. coli* and *Klebsiella pneumoniae.* They can be used with *Haemophilus, Neisseria, Pseudomonas aeruginosa,* and other gram-negative pathogens. The quinolones also are active against gram-positive bacteria such as *Staphylococcus aureus, Streptococcus pyogenes,* and *Mycobacterium tuberculosis.* Currently they are used in treating urinary tract infections, sexually transmitted diseases caused by *Neisseria* and *Chlamydia,* gastrointestinal infections, respiratory tract infections, skin infections, and osteomyelitis.

Penicillins

Penicillin G or benzylpenicillin, the first antibiotic to be widely used in medicine, has the structural properties characteristic of the penicillin family (**figure 35.8**). Most **penicillins** are derivatives of 6-aminopenicillanic acid and differ from one another only with respect to the side chain attached to its amino group. The most crucial feature of the molecule is the β-lactam ring, which appears to be essential for activity. **Penicillinase,** the enzyme synthesized by many penicillin-resistant bacteria, destroys penicillin activity by hydrolyzing a bond in this ring (figure 35.8).

The mechanism of action of penicillins is still not completely known. Their structures do resemble that of the terminal D-alanyl-D-alanine found on the peptide side chain of the peptidoglycan subunit. It has been proposed that penicillins inhibit the enzyme catalyzing the transpeptidation reaction because of their structural

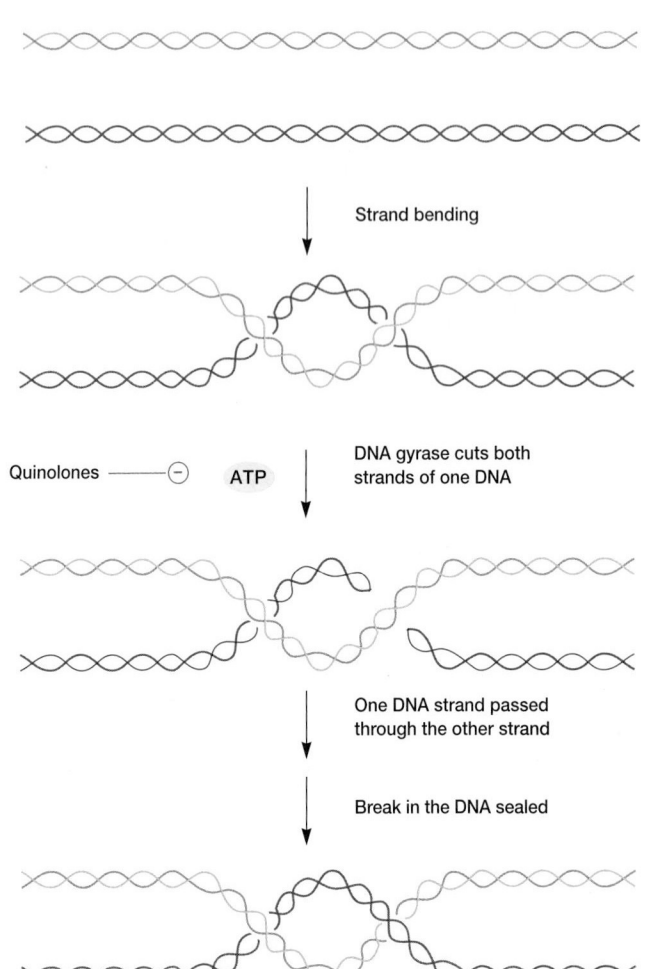

Strand bending

Quinolones ——⊖ ATP

DNA gyrase cuts both strands of one DNA

One DNA strand passed through the other strand

Break in the DNA sealed

Figure 35.7 DNA Gyrase Action and Quinolone Inhibition.

similarity, which would block the synthesis of a complete, fully cross-linked peptidoglycan and lead to osmotic lysis. The mechanism is consistent with the observation that penicillins act only on growing bacteria that are synthesizing new peptidoglycan. However, more recently it has been discovered that penicillins bind to several penicillin-binding proteins and may destroy bacteria by activating their own autolytic enzymes. There now is some evidence that penicillin kills bacteria even in the absence of autolysins or murein hydrolases. Lysis could occur after bacterial viability has already been lost. Penicillin may stimulate special proteins called bacterial holins to form holes or lesions in the plasma membrane. This would lead directly to membrane leakage and death; murein hydrolases also could move through the holes, disrupt the peptidoglycan, and lyse the cell. Apparently the mechanism of penicillin action is more complex than previously imagined. Bacterial peptidoglycan structure (pp. 54–55): Peptidoglycan synthesis (pp. 216–17)

Penicillins differ from each other in several ways. Penicillin G is effective against gonococci, meningococci, and several gram-positive pathogens such as streptococci and staphylococci, but it must be administered parenterally because it is destroyed by stomach acid. Penicillin V is similar to penicillin G, but it is more acid resistant and can be given orally (figure 35.8). Ampicillin

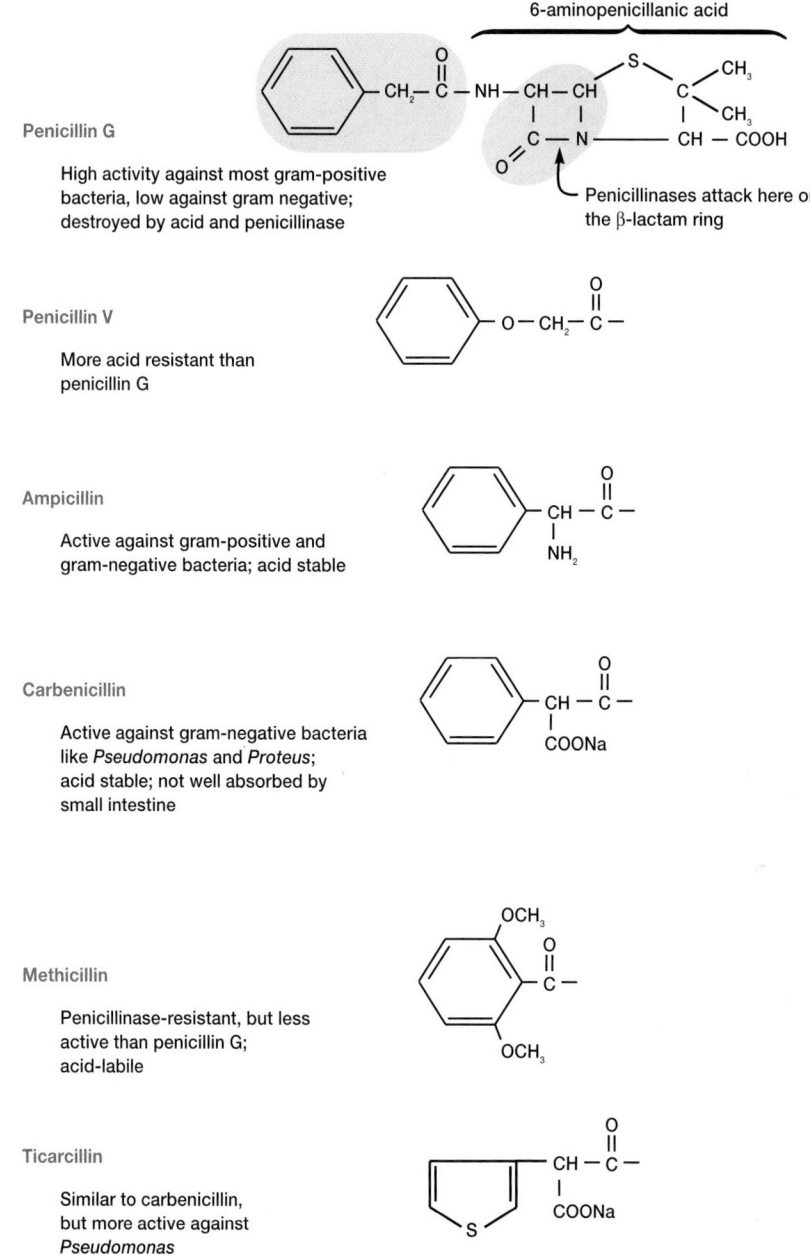

Figure 35.8   Penicillins.   The structures and characteristics of representative penicillins. All are derivatives of 6-aminopenicillanic acid; in each case the purple shaded portion of penicillin G is replaced by the side chain indicated. The β-lactam ring is also shaded (blue), and an arrow points to the bond that is hydrolyzed by penicillinase.

can be administered orally and has a broader spectrum of activity as it is effective against gram-negative bacteria such as *Haemophilus, Salmonella,* and *Shigella.* Carbenicillin and ticarcillin also are broad spectrum and particularly potent against *Pseudomonas* and *Proteus.*

An increasing number of bacteria are penicillin resistant. Penicillinase-resistant penicillins such as methicillin (figure 35.8), nafcillin, and oxacillin are frequently employed against these bacterial pathogens.

Although penicillins are the least toxic of the antibiotics, about 1 to 5% of the adults in the United States are allergic to

them. Occasionally a person will die of a violent allergic response; therefore patients should be questioned about penicillin allergies before treatment is begun.

Cephalosporins

Cephalosporins are a family of antibiotics originally isolated in 1948 from the fungus *Cephalosporium,* and their β-lactam structure is very similar to that of the penicillins (**figure 35.9**). As might be expected from their structural similarities, cephalosporins resemble penicillins in inhibiting the

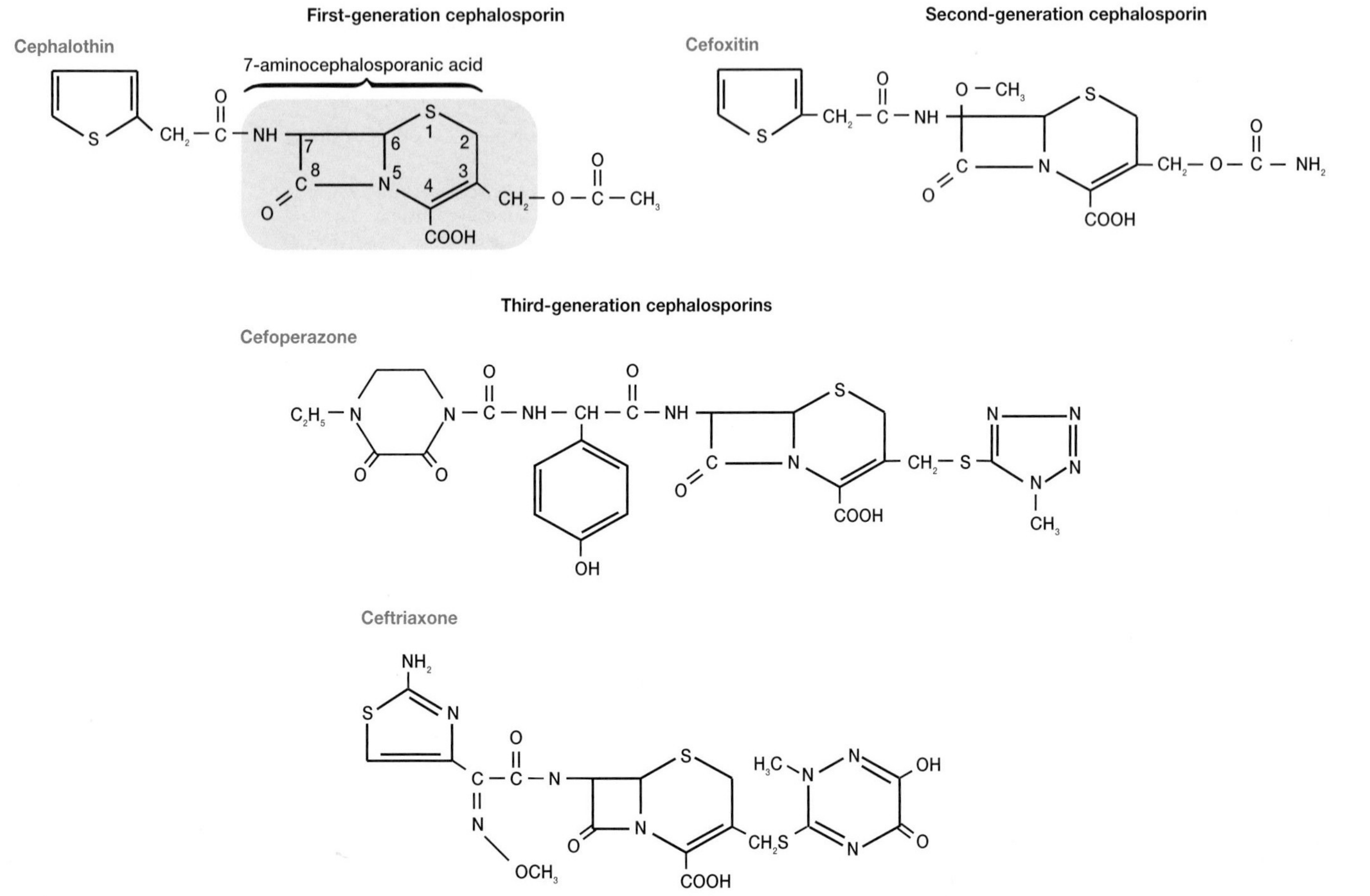

Figure 35.9 **Cephalosporin Antibiotics.** These drugs are derivatives of 7-aminocephalosporanic acid and contain a β-lactam ring.

transpeptidation reaction during peptidoglycan synthesis. They are broad-spectrum drugs frequently given to patients with penicillin allergies.

Many cephalosporins are in use. There are three groups or generations of these drugs that differ in their spectrum of activity (figure 35.9). First-generation cephalosporins are more effective against gram-positive than gram-negative pathogens. Second-generation drugs act against many gram-negative as well as gram-positive pathogens. Third-generation drugs are particularly effective against gram-negative pathogens, and often also reach the central nervous system.

Most cephalosporins (including cephalothin, cefoxitin, ceftriaxone, and cefoperazone) are administered parenterally. Cefoperazone is resistant to destruction by β-lactamases and effective against many gram-negative bacteria, including *Pseudomonas aeruginosa.* Cephalexine and cefixime are given orally rather than by injection.

The Tetracyclines

The **tetracyclines** are a family of antibiotics with a common four-ring structure to which a variety of side chains are attached (**figure 35.10**). Oxytetracycline and chlortetracycline are natu-

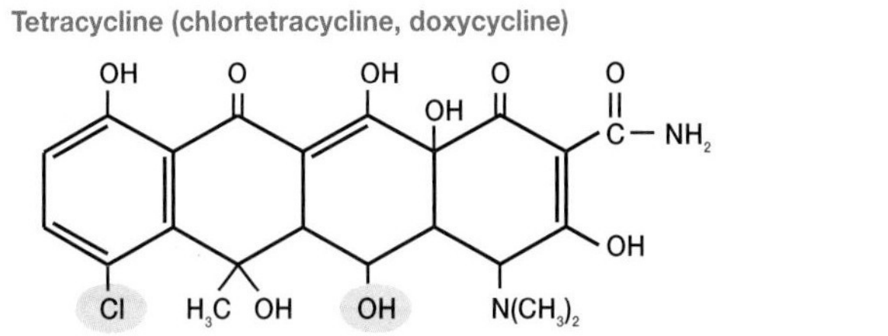

Figure 35.10 **Tetracyclines.** Three members of the tetracycline family. Tetracycline lacks both of the groups that are shaded. Chlortetracycline (aureomycin) differs from tetracycline in having a chlorine atom (blue): doxycycline consists of tetracycline with an extra hydroxyl (purple).

rally produced by some species of the actinomycete genus *Streptomyces;* others are semisynthetic drugs. These antibiotics inhibit protein synthesis by combining with the small (30S) subunit of the ribosome and inhibiting the binding of aminoacyl-tRNA molecules to the ribosomal A site. Because their action is only bacteriostatic, the effectiveness of treatment depends on active host resistance to the pathogen. The mechanism of protein synthesis (pp. 260–70)

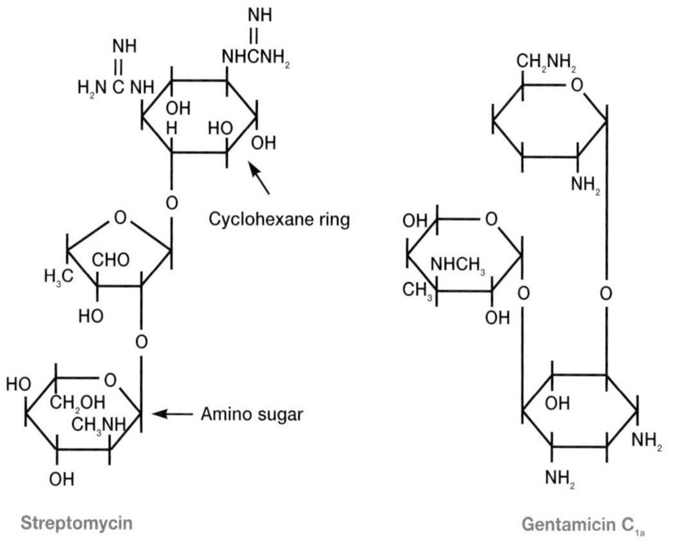

Figure 35.11   Representative Aminoglycoside Antibiotics.

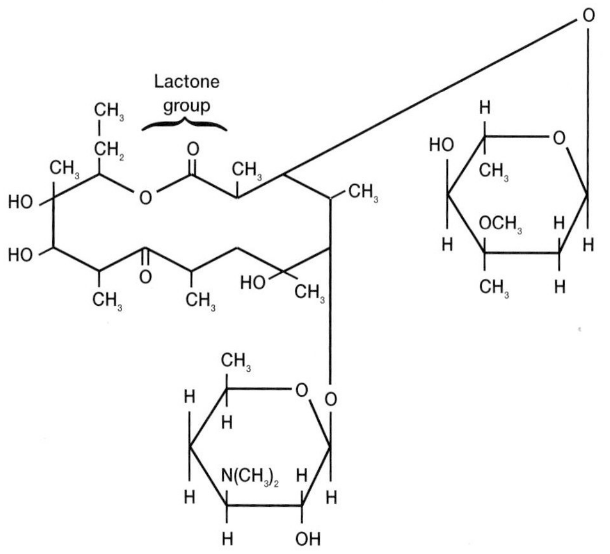

Figure 35.12   Erythromycin, a Macrolide Antibiotic. The 14-member lactone ring is connected to two sugars.

Tetracyclines are broad-spectrum antibiotics active against gram-negative bacteria, gram-positive bacteria, rickettsias, chlamydiae, and mycoplasmas. High doses may result in nausea, diarrhea, yellowing of teeth in children, and damage to the liver and kidneys.

Aminoglycoside Antibiotics

There are several important **aminoglycoside antibiotics. Streptomycin,** kanamycin, neomycin, and tobramycin are synthesized by *Streptomyces,* whereas gentamicin comes from a related bacterium, *Micromonospora purpurea.* Although there is considerable variation in structure among the different aminoglycosides, all contain a cyclohexane ring and amino sugars (**figure 35.11**). Aminoglycosides bind to the small ribosomal subunit and interfere with protein synthesis in at least two ways. They directly inhibit protein synthesis and also cause misreading of the genetic message carried by mRNA.

The aminoglycosides are bactericidal and tend to be most active against gram-negative pathogens. Streptomycin's usefulness has decreased greatly due to widespread drug resistance, but it is still effective against tuberculosis and plague. Gentamicin is used to treat *Proteus, Escherichia, Klebsiella,* and *Serratia* infections. Aminoglycosides are quite toxic and can cause deafness, renal damage, loss of balance, nausea, and allergic responses.

Erythromycin and Other Macrolides

Erythromycin, the most frequently used **macrolide antibiotic,** is synthesized by *Streptomyces erythraeus.* The macrolides contain a 12- to 22-carbon lactone ring linked to one or more sugars (**figure 35.12**). Erythromycin is usually bacteriostatic and binds with the 23S rRNA of the 50S ribosomal subunit to inhibit peptide chain elongation during protein synthesis.

Erythromycin is a relatively broad-spectrum antibiotic effective against gram-positive bacteria, mycoplasmas, and a few

gram-negative bacteria. It is used with patients allergic to penicillins and in the treatment of whooping cough, diphtheria, diarrhea caused by *Campylobacter,* and pneumonia from *Legionella* or *Mycoplasma* infections.

Newer macrolides are now in use. Clindamycin is effective against a variety of bacteria including staphylococci and anaerobes such as *Bacteroides.* Azithromycin is particularly effective against *Chlamydia trachomatis.*

Vancomycin and Teicoplanin

Vancomycin is a glycopeptide antibiotic produced by *Streptomyces orientalis.* It is a cup-shaped molecule composed of a peptide linked to a disaccharide. The antibiotic blocks peptidoglycan synthesis by inhibiting the transpeptidation step that cross-links adjacent peptidoglycan strands. The resulting peptidoglycan is mechanically weak and the cells osmotically lyse. Vancomycin's peptide portion binds specifically to the D-alanine-D-alanine terminal sequence on the pentapeptide portion of peptidoglycan. This complex blocks transpeptidase action.

The antibiotic is bactericidal for *Staphylococcus* and some members of the genera *Clostridium, Bacillus, Streptococcus,* and *Enterococcus.* It is given both orally and intravenously, and has been particularly important in the treatment of antibiotic-resistant staphylococcal and enterococcal infections. Vancomycin-resistant strains of *Enterococcus* have become widespread and recently a few cases of resistant *Staphylococcus aureus* have appeared.

Teicoplanin is a glycopeptide antibiotic from *Actinoplanes teichomyceticus* that is similar in structure and mechanism of action to vancomycin. It is active against staphylococci, enterococci, streptococci, clostridia, *Listeria,* and many other gram-positive pathogens. This antibiotic presently is used in Europe and elsewhere, but not in the United States.

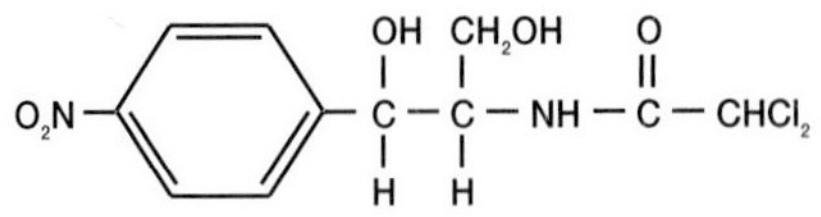

Figure 35.13 Chloramphenicol.

Chloramphenicol

Although **chloramphenicol (figure 35.13)** was first produced from cultures of *Streptomyces venezuelae*, it is now made through chemical synthesis. Like erythromycin, chloramphenicol binds to 23S rRNA on the 50S ribosomal subunit. It inhibits the peptidyl transferase and is bacteriostatic.

This antibiotic has a very broad spectrum of activity but unfortunately is quite toxic. One may see allergic responses or neurotoxic reactions. The most common side effect is a temporary or permanent depression of bone marrow function, leading to aplastic anemia and a decreased number of blood leukocytes. Chloramphenicol is used only in life-threatening situations when no other drug is adequate.

---

1. For each class of antibacterial drugs presented, give the following information: general chemical composition or structure, mechanism of action, behavioral properties (cidal or static, broad spectrum or narrow spectrum), route of administration and therapeutic uses, significant problems with side effects, and drug resistance.
2. Define structural analogue, 6-aminopenicillanic acid, penicillinase, aminoglycoside, and macrolide.

35.7 DRUG RESISTANCE

The spread of drug-resistant pathogens is one of the most serious threats to the successful treatment of microbial disease (**Disease 35.2**). This section describes the ways in which bacteria acquire drug resistance and how resistance spreads within a bacterial population.

Mechanisms of Drug Resistance

Bacteria become drug resistant in several different ways. It should be noted at the beginning that a particular type of resistance mechanism is not confined to a single class of drugs. Two bacteria may use different resistance mechanisms to withstand the same chemotherapeutic agent. Furthermore, resistant mutants arise spontaneously and are then selected. Mutants are not created directly by exposure to a drug.

Pathogens often become resistant simply by preventing entrance of the drug. Many gram-negative bacteria are unaffected by penicillin G because it cannot penetrate the envelope's outer membrane. Changes in penicillin binding proteins also render a cell resistant. A decrease in permeability can lead to sulfonamide resistance. Mycobacteria resist many drugs because of the high content of mycolic acids (*see p. 528*) in a complex lipid layer outside their peptidoglycan. This layer is impermeable to most drugs.

A second resistance strategy is to pump the drug out of the cell after it has entered. Some pathogens have plasma membrane translocases, often called efflux pumps, that expel drugs. Because they are relatively nonspecific and can pump many different drugs, these transport proteins often are called multidrug-resistance pumps. Many are drug/proton antiporters—that is, protons enter the cell as the drug leaves. Such systems are present in *E. coli, Pseudomonas aeruginosa, Mycobacterium smegmatis,* and *Staphylococcus aureus.*

Many bacterial pathogens resist attack by inactivating drugs through chemical modification. The best-known example is the hydrolysis of the β-lactam ring of many penicillins by the enzyme penicillinase. Drugs also are inactivated by the addition of groups. For example, chloramphenicol contains two hydroxyl groups (figure 35.13) that can be acetylated in a reaction catalyzed by the enzyme chloramphenicol acyltransferase with acetyl CoA as the donor. Aminoglycosides (figure 35.11) can be modified and inactivated in several ways. Acetyltransferases catalyze the acetylation of amino groups. Some aminoglycoside-modifying enzymes catalyze the addition to hydroxyl groups of either phosphates (phosphotransferases) or adenyl groups (adenyltransferases).

Because each chemotherapeutic agent acts on a specific target, resistance arises when the target enzyme or organelle is modified so that it is no longer susceptible to the drug. For example, the affinity of ribosomes for erythromycin and chloramphenicol can be decreased by a change in the 23S rRNA to which they bind. Enterococci become resistant to vancomycin by changing the terminal D-alanine-D-alanine in their peptidoglycan to a D-alanine-D-lactate. This drastically reduces antibiotic binding. Antimetabolite action may be resisted through alteration of susceptible enzymes. In sulfonamide-resistant bacteria the enzyme that uses *p*-aminobenzoic acid during folic acid synthesis (the tetrahydropteroic acid synthetase) often has a much lower affinity for sulfonamides. *Mycobacterium tuberculosis* becomes resistant to the drug rifampin due to mutations that alter the β subunit of its RNA polymerase. Rifampin cannot bind to the mutant RNA polymerase and block the initiation of transcription.

Resistant bacteria may either use an alternate pathway to bypass the sequence inhibited by the agent or increase the production of the target metabolite. For example, some bacteria are resistant to sulfonamides simply because they use preformed folic acid from their surroundings rather than synthesize it themselves. Other strains increase their rate of folic acid production and thus counteract sulfonamide inhibition.

The Origin and Transmission of Drug Resistance

The genes for drug resistance are present on both the bacterial chromosome and **plasmids,** small DNA molecules that can exist separate from the chromosome or be integrated in it. Plasmids (pp. 288–91)

Spontaneous mutations in the bacterial chromosome, although they do not occur very often, will make bacteria drug resistant. Usually such mutations result in a change in the drug

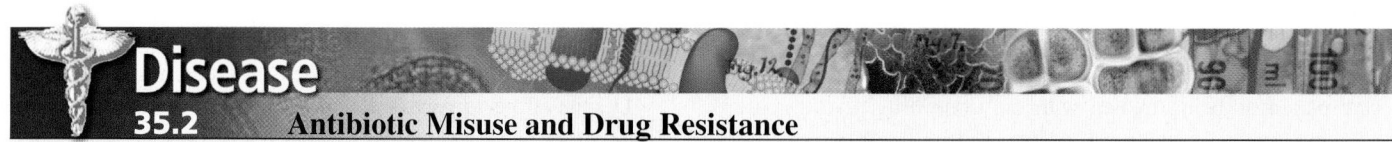

Disease

35.2   Antibiotic Misuse and Drug Resistance

The sale of antimicrobial drugs is big business. In the United States millions of pounds of antibiotics valued at billions of dollars are produced annually. As much as 70% of these antibiotics are added to livestock feed.

Because of the massive quantities of antibiotics being prepared and used, an increasing number of diseases are resisting treatment due to the spread of drug resistance. A good example is *Neisseria gonorrhoeae,* the causative agent of gonorrhea. Gonorrhea was first treated successfully with sulfonamides in 1936, but by 1942 most strains were resistant and physicians turned to penicillin. Within 16 years a penicillin-resistant strain had emerged in the Far East. A penicillinase-producing gonococcus reached the United States in 1976 and is still spreading in this country.

In late 1968 an epidemic of dysentery caused by *Shigella* broke out in Guatemala and affected at least 112,000 persons; 12,500 deaths resulted. The strains responsible for this devastation carried an R plasmid giving them resistance to chloramphenicol, tetracycline, streptomycin, and sulfonamide. In 1972 a typhoid epidemic swept through Mexico producing 100,000 infections and 14,000 deaths. It was due to a *Salmonella typhi* strain with the same multiple-drug-resistance pattern seen in the previous *Shigella* outbreak.

Haemophilus influenzae type b is responsible for many cases of childhood pneumonia and middle ear infections, as well as respiratory infections and meningitis. It is now becoming increasingly resistant to tetracyclines, ampicillin, and chloramphenicol. The same situation is occurring with *Streptococcus pneumoniae.* It has been estimated that by sometime in 2004, as much as 40% of *S. pneumoniae* may be resistant to both penicillin and erythromycin.

In 1946 almost all strains of *Staphylococcus* were penicillin sensitive. Today most hospital strains are resistant to penicillin G, and some are now also resistant to methicillin and/or gentamicin and only can be treated with vancomycin. Some strains of *Enterococcus* have become resistant to most antibiotics, including vancomycin. Recently a few cases of vancomycin-resistant *S. aureus* have been reported in the United States and Japan. At present these strains are only intermediately resistant to vancomycin. If full vancomycin resistance develops and spreads, *S. aureus* may become untreatable.

It is clear from these and other examples (e.g., multiresistant *Mycobacterium tuberculosis*) that drug resistance is an extremely serious public health problem. Much of the difficulty arises from drug misuse. Drugs frequently have been overused in the past. It has been estimated that over 50% of the antibiotic prescriptions in hospitals are given without clear evidence of infection or adequate medical indication. Many physicians have administered antibacterial drugs to patients with colds, influenza, viral pneumonia, and other viral diseases. A recent study showed that over 50% of the patients diagnosed with colds and upper respiratory infections and 66% of those with chest colds (bronchitis) are given antibiotics, even though over 90% of these cases are caused by viruses. Frequently antibiotics are prescribed without culturing and identifying the pathogen or without determining bacterial sensitivity to the drug. Toxic, broad-spectrum antibiotics are sometimes given in place of narrow-spectrum drugs as a substitute for culture and sensitivity testing, with the consequent risk of dangerous side effects, superinfections, and the selection of drug-resistant mutants. The situation is made worse by patients not completing their course of medication. When antibiotic treatment is ended too early, drug-resistant mutants may survive. Drugs are available to the public in many countries; people may practice self-administration of antibiotics and further increase the prevalence of drug-resistant strains.

The use of antibiotics in animal feeds is undoubtedly another contributing factor to increasing drug resistance. The addition of low levels of antibiotics to livestock feeds does raise the efficiency and rate of weight gain in cattle, pigs, and chickens (partially because of infection control in overcrowded animal populations). However, this also increases the number of drug-resistant bacteria in animal intestinal tracts. There is evidence for the spread of bacteria such as *Salmonella* from animals to human populations. In 1983, 18 people in four midwestern states were infected with a multiple-drug-resistant strain of *Salmonella newport.* Eleven were hospitalized for salmonellosis and one died. All 18 patients had recently been infected by eating hamburger from beef cattle fed subtherapeutic doses of chlortetracycline for growth promotion. Resistance to some antibiotics has been traced to the use of specific farmyard antibiotics. Avoparcin resembles vancomycin in structure, and virginiamycin resembles Synercid. There is good circumstantial evidence that extensive use of these two antibiotics in animal feed has led to an increase in vancomycin and Synercid resistance among enterococci. The use of the quinolone antibiotic enrofloxacin in swine herds appears to have promoted ciprofloxacin resistance in pathogenic strains of *Salmonella.* Elimination of antibiotic food supplements might well slow the spread of drug resistance.

The spread of antibiotic resistance can be due to quite subtle factors. For example, products such as soap and deodorants often now contain triclosan and other germicides. There is increasing evidence that the widespread use of triclosan actually favors an increase in antibiotic resistance (*see section 7.5*).

receptor; therefore the antibiotic cannot bind and inhibit (e.g., the streptomycin receptor protein on bacterial ribosomes). Many mutants are probably destroyed by natural host resistance mechanisms. However, when a patient is being treated extensively with antibiotics, some resistant mutants may survive and flourish because of their competitive advantage over nonresistant strains.

Frequently a bacterial pathogen is drug resistant because it has a plasmid bearing one or more resistance genes; such plasmids are called **R plasmids** (*resistance plasmids*). Plasmid resistance genes often code for enzymes that destroy or modify drugs; for example, the hydrolysis of penicillin or the acetylation of chloramphenicol and aminoglycoside drugs. Plasmid-associated genes have been implicated in resistance to the aminoglycosides, choramphenicol, penicillins and cephalosporins, erythromycin, tetracyclines, sulfonamides, and others. Once a bacterial cell possesses an R plasmid, the plasmid may be transferred to other cells quite rapidly through normal gene exchange processes such as conjugation, transduction, and transformation. Because a single plasmid may carry genes for resistance to several drugs, a pathogen population can become resistant to several antibiotics simultaneously, even though the infected patient is being treated with only one drug. Conjugation (pp. 296–99); Transduction (p. 301–5); Transformation (pp. 299–301)

Antibiotic resistance genes are located on genetic elements other than plasmids. Many composite transposons contain genes for antibiotic resistance, and some bear more than one resistance gene. They are found in both gram-negative and gram-positive bacteria. Some examples and their resistance markers are Tn5 (kanamycin, bleomycin, streptomycin), Tn9 (chloramphenicol), Tn10 (tetracycline), Tn21 (streptomycin, spectinomycin, sulfonamide), Tn551 (erythromycin), and Tn4001 (gentamicin, tobramycin, kanamycin). Resistance genes on composite transposons can move rapidly between plasmids and through a bacterial population. Often several resistance genes are carried together as gene cassettes in association with a genetic element known as an integron. An **integron** has an attachment site for site-specific recombination into which genes can be inserted (see p. 286) and an integrase gene. Thus integrons can capture genes and gene cassettes. **Gene cassettes** are genetic elements that may exist as circular nonreplicating DNA when moving from one site to another, but which normally are a linear part of a transposon, plasmid, or bacterial chromosome. Cassettes usually carry one or two genes and a recombination site. Several cassettes can be integrated sequentially in an integron. Thus integrons also are important in spreading resistance genes. Finally, conjugative transposons (see p. 295), like composite transposons, can carry resistance genes. Since they are capable of moving between bacteria by conjugation, they are also effective in spreading resistance. Transposable elements (pp. 291–95)

Extensive drug treatment favors the development and spread of antibiotic-resistant strains because the antibiotic destroys normal, susceptible bacteria that would usually compete with drug-resistant strains. The result may be the emergence of drug-resistant pathogens leading to a **superinfection.** Superinfections are a significant problem because of the existence of multiple-drug-resistant bacteria that often produce drug-resistant respiratory and urinary tract infections. A classic example of a superinfection resulting from antibiotic administration is the disease pseudomembranous enterocolitis caused by *Clostridium difficile.* When a patient is treated with clindamycin, ampicillin, or cephalosporin, many intestinal bacteria are killed, but *C. difficile* is not. This intestinal inhabitant, which is normally a minor constituent of the population, flourishes in the absence of competition and produces a toxin that stimulates the secretion of a pseudomembrane by intestinal cells. If the superinfection is not treated early with vancomycin, the pseudomembrane must be surgically removed or the patient will die. Fungi, such as the yeast *Candida albicans,* also produce superinfections when bacterial competition is eliminated by antibiotics.

Several strategies can be employed to discourage the emergence of drug resistance. The drug can be given in a high enough concentration to destroy susceptible bacteria and most spontaneous mutants that might arise during treatment. Sometimes two different drugs can be administered simultaneously with the hope that each drug will prevent the emergence of resistance to the other. Finally, chemotherapeutic drugs, particularly broad-spectrum drugs, should be used only when definitely necessary.

If possible, the pathogen should be identified, drug sensitivity tests run, and the proper narrow-spectrum drug employed.

Despite efforts to control the emergence and spread of drug resistance, the situation continues to worsen. Of course, as mentioned previously, antibiotics should be used in ways that reduce the development of resistance. Another approach is to search for new antibiotics that microorganisms have never encountered (*see Disease 29.1*). Pharmaceutical companies collect and analyze samples from around the world in a search for completely new antimicrobial agents. Structure-based or rational drug design is a third option. If the three-dimensional structure of a susceptible target molecule such as an enzyme essential to microbial function is known, computer programs can be used to design drugs that precisely fit the target molecule. These drugs might be able to bind to the target and disrupt its function sufficiently to destroy the pathogen. Pharmaceutical companies are using this approach to attempt to develop drugs for the treatment of AIDS, cancer, and the common cold. At least one company is developing "enhancers." These are cationic peptides that disrupt bacterial membranes by displacing their magnesium ions. Antibiotics then penetrate and rapidly exert their effects. Other pharmaceutical companies are developing efflux-pump inhibitors to administer with antibiotics and prevent their expulsion by the resistant pathogen.

There has been some recent progress in developing new antibiotics that are effective against drug-resistant pathogens. Two new drugs are fairly effective against vancomycin-resistant enterococci. Synercid is a mixture of the streptogramin antibiotics quinupristin and dalfopristin that inhibits protein synthesis. A second drug, linezolid (Zyvox), is the first drug in a new family of antibiotics, the oxazolidinones. It inhibits protein synthesis and is active against both vancomycin-resistant enterococci and methicillin-resistant *Staphylococcus aureus.*

Information that is coming from the sequencing and analysis of pathogen genomes almost certainly will be useful in identifying new targets for antimicrobial drugs. For example, genomics studies are providing data for research on inhibitors of both aminoacyl-tRNA synthetases (*see p. 260*) and the enzyme that removes the formyl group from the N-terminal methionine during bacterial protein synthesis (*see p. 262*). Bacteria must synthesize the fatty acids they require for growth rather than acquiring the acids from their environment. The drug susceptibility of enzymes in the fatty acid synthesis system is being analyzed by screening pathogens for potential targets.

A most interesting response to the current crisis is the renewed interest in an idea first proposed early in the twentieth century by Felix d'Herelle, one of the discoverers of bacterial viruses or bacteriophages (*see pp. 352–53*). d'Herelle proposed that bacteriophages could be used to treat bacterial diseases. Although most microbiologists did not pursue his proposal actively due to technical difficulties and the advent of antibiotics, Russian scientists developed the medical use of bacteriophages. Currently Russian physicians use bacteriophages to treat many bacterial infections. Bandages are saturated with phage solutions, phage mixtures are administered orally, and phage preparations are given intravenously to treat *Staphylococcus* infections. Three American companies are actively conducting research on phage therapy and preparing to carry out clinical trials. Bacteriophages (chapter 17)

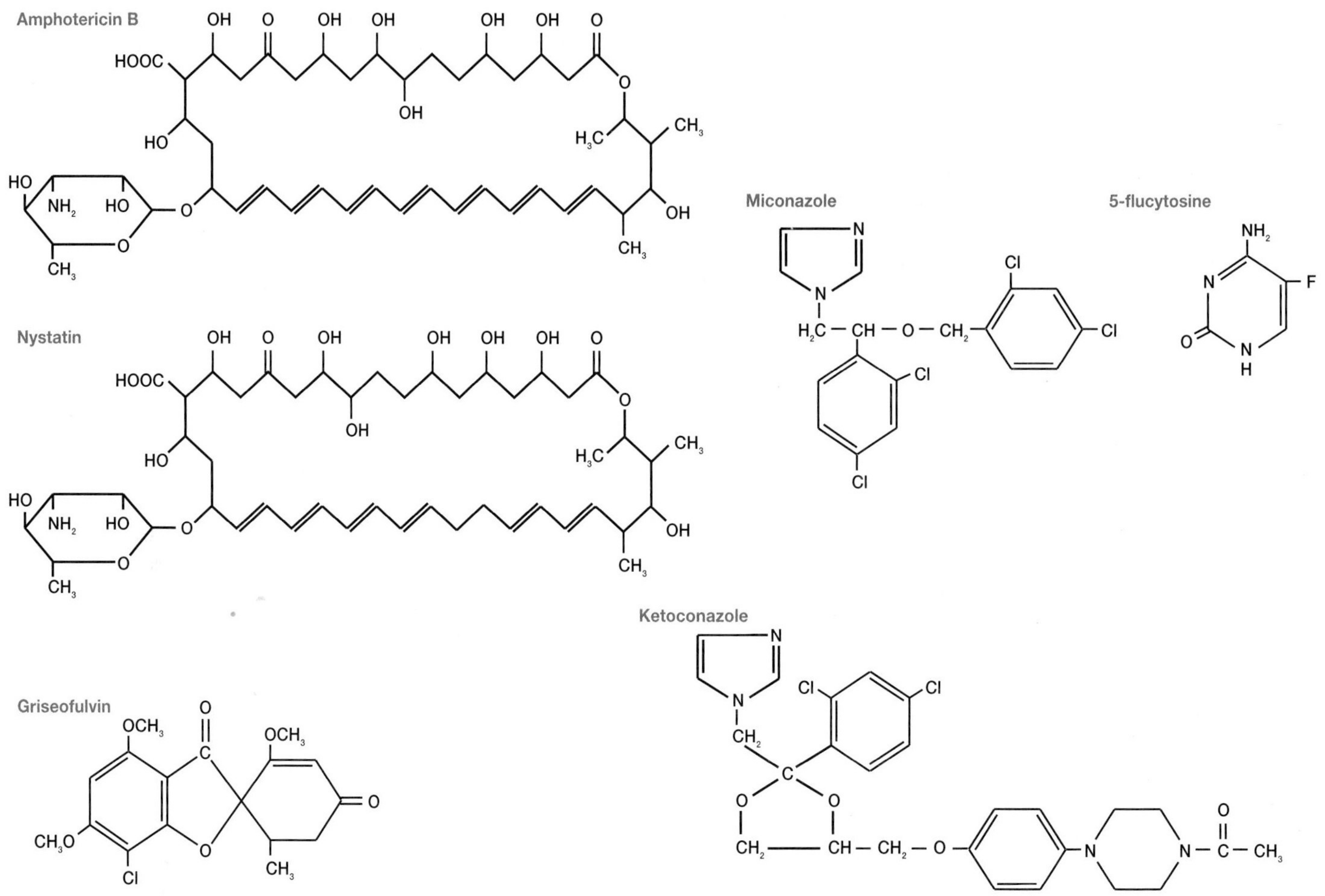

Amphotericin B

Nystatin

Griseofulvin

Miconazole

5-flucytosine

Ketoconazole

Figure 35.14 Antifungal Drugs. Six commonly used drugs are shown here.

1. Briefly describe the five major ways in which bacteria become resistant to drugs and give an example of each.
2. Define plasmid, R plasmid, integron, gene cassette, and superinfection. How are R plasmids involved in the spread of drug resistance?

 35.8 ANTIFUNGAL DRUGS

Treatment of fungal infections generally has been less successful than that of bacterial infections largely because eucaryotic fungal cells are much more similar to human cells than are bacteria. Many drugs that inhibit or kill fungi are therefore quite toxic for humans. In addition, most fungi have a detoxification system that modifies many antibiotics, probably by hydroxylation. As a result the added antibiotics are fungistatic only as long as repeated application maintains high levels of unmodified antibiotic. Despite their relatively low therapeutic index, a few drugs are useful in treating many major fungal diseases. Effective antifungal agents frequently either extract membrane sterols or prevent their synthesis. Similarly, because animal cells do not have cell walls, the enzyme chitin synthase is the target for fungal-active antibiotics such as polyoxin D and nikkomycin.

Fungal infections are often subdivided into infections of superficial tissues or superficial mycoses and systemic mycoses. Treatment for these two types of disease is very different. Several drugs are used to treat superficial mycoses. Three drugs containing imidazole—miconazole, ketoconazole (**figure 35.14**), and clotrimazole—are broad-spectrum agents available as creams and solutions for the treatment of dermatophyte infections such as athlete's foot, and oral and vaginal candidiasis (*see chapter 40*). They are thought to disrupt fungal membrane permeability and inhibit sterol synthesis. Tolnaftate is used topically for the treatment of cutaneous infections, but is not as effective against infections of the skin and hair. **Nystatin** (figure 35.14), a polyene antibiotic from *Streptomyces,* is used to control *Candida* infections of the skin, vagina, or alimentary tract. It binds to sterols and damages the membrane, leading to membrane leakage. **Griseofulvin** (figure 35.14), an antibiotic formed by *Penicillium,* is given orally to treat chronic dermatophyte infections. It is thought to disrupt the mitotic spindle and inhibit cell division; it also may inhibit protein and nucleic acid synthesis. Side effects of griseofulvin include headaches, gastrointestinal upset, and allergic reactions. Superficial and systemic mycoses (pp. 918–19, 921–24)

Systemic infections are very difficult to control and can be fatal. Three drugs commonly used against systemic mycoses are **amphotericin B,** 5-flucytosine, and fluconazole (figure 35.14). Amphotericin B from *Streptomyces* spp. binds to the sterols in fungal membranes, disrupting membrane permeability and causing leakage of cell constituents. It is quite toxic and used only for serious, life-threatening infections. The synthetic oral antimycotic agent 5-flucytosine (5-fluorocytosine) is effective against most systemic fungi, although drug resistance often develops rapidly. The drug is converted to 5-fluorouracil by the fungi, incorporated into RNA in place of uracil, and disrupts RNA function. Its side effects include skin rashes, diarrhea, nausea, aplastic anemia, and liver damage. Fluconazole is used in the treatment of candidiasis, cryptococcal meningitis, and coccidioidal meningitis. Adverse effects are relatively uncommon.

Just as with antibacterial drugs, overuse of antifungal agents leads to an increase in drug resistance. For example, *Candida* infections are becoming more frequent and drug resistant.

---

1. Summarize the mechanism of action and the therapeutic use of the following antifungal drugs: miconazole, nystatin, griseofulvin, amphotericin B, and 5-flucytosine.

35.9 ANTIVIRAL DRUGS

For many years the possibility of treating viral infections with drugs appeared remote because viruses enter host cells and make use of host cell enzymes and constituents to a large extent. A drug that would block virus reproduction also was thought to be toxic for the host. Inhibitors of virus-specific enzymes and life cycle processes have now been discovered, and several drugs are used therapeutically. Some important examples are shown in **figure 35.15.**   Animal virus life cycles (pp. 388–98)

Most antiviral drugs disrupt either critical stages in the virus life cycle or the synthesis of virus-specific nucleic acids. **Amantadine** and rimantadine can be used to prevent influenza A infections. When given in time, it will reduce the incidence of influenza by 50 to 70% in an exposed population. Amantadine blocks the penetration and uncoating of influenza virus particles (*see section 18.2*). **Adenine arabinoside** or **vidarabine** disrupts the activity of DNA polymerase and several other enzymes involved in DNA and RNA synthesis and function. It is given intravenously or applied as an ointment to treat herpes infections. A third drug, **acyclovir,** is also used in the treatment of herpes in-

fections. Upon phosphorylation, acyclovir resembles deoxy-GTP and inhibits the virus DNA polymerase. Unfortunately acyclovir-resistant strains of herpes are already developing. Effective acyclovir derivatives and relatives are now available. Valacyclovir is an orally administered prodrug form of acyclovir. Ganciclovir, penciclovir, and its oral form famciclovir are effective in treatment of herpesviruses. Another kind of drug, foscarnet, inhibits the virus DNA polymerase in a different way. Foscarnet is an organic analogue of pyrophosphate (figure 35.15) that binds to the polymerase active site and blocks the cleavage of pyrophosphate from nucleoside triphosphate substrates. It is used in treating herpes and cytomegalovirus infections.

Several broad-spectrum anti-DNA virus drugs have been developed. A good example is the drug HPMPC or cidofovir (figure 35.15). It is effective against papovaviruses, adenoviruses, herpesviruses, iridoviruses, and poxviruses. The drug acts on the viral DNA polymerase as a competitive inhibitor and alternative substrate of dCTP. It has been used primarily against cytomegalovirus but also against herpes simplex and human papillomavirus infections.

Research on anti-HIV drugs has been particularly active. Many of the first drugs to be developed were reverse transcriptase inhibitors such as **azidothymidine (AZT)** or **zidovudine,** lamivudine (3TC), didanosine (ddI), zalcitabine (ddC), and stavudine (d4T) (figure 35.15). These interfere with reverse transcriptase activity and therefore block HIV reproduction. More recently **HIV protease inhibitors** have been developed. Three of the most used are saquinvir, indinavir, and ritonavir (figure 35.15). These mimic the peptide bond that is normally attacked by the protease. The most successful treatment regimen involves a cocktail of agents given at high dosages to prevent the development of drug resistance. For example, the combination of AZT, 3TC, and ritonavir is very effective in reducing HIV plasma concentrations almost to zero. However, the treatment does not seem able to eliminate latent proviral HIV DNA that still resides in memory T cells, and possibly elsewhere.   The biology and replication of HIV (pp. 395–97, 855–60)

Probably the most publicized antiviral agents are **interferons.** These small proteins, produced by the host, inhibit virus replication and may be clinically useful in the treatment of influenza, hepatitis, herpes, and colds. The mechanism of interferon action is discussed in more detail in *section 31.9.*

---

1. Describe two different ways in which antiviral drugs disrupt virus reproduction and give an example of each.
2. Briefly summarize the way in which HIV is currently treated and how the two general classes of anti-HIV drugs act.

$\mathscr{Summary}$

Chemotherapeutic agents are compounds that destroy pathogenic microorganisms or inhibit their growth and are used in the treatment of disease. Most are antibiotics: microbial products or their derivatives that can kill susceptible microorganisms or inhibit their growth.

35.1 The Development of Chemotherapy

a. The modern era of chemotherapy began with Paul Ehrlich's work on drugs against African sleeping sickness and syphilis. Other early pioneers were Gerhard Domagk, Alexander Fleming, Howard Florey, Ernst Chain, Norman Heatley, and Selman Waksman.

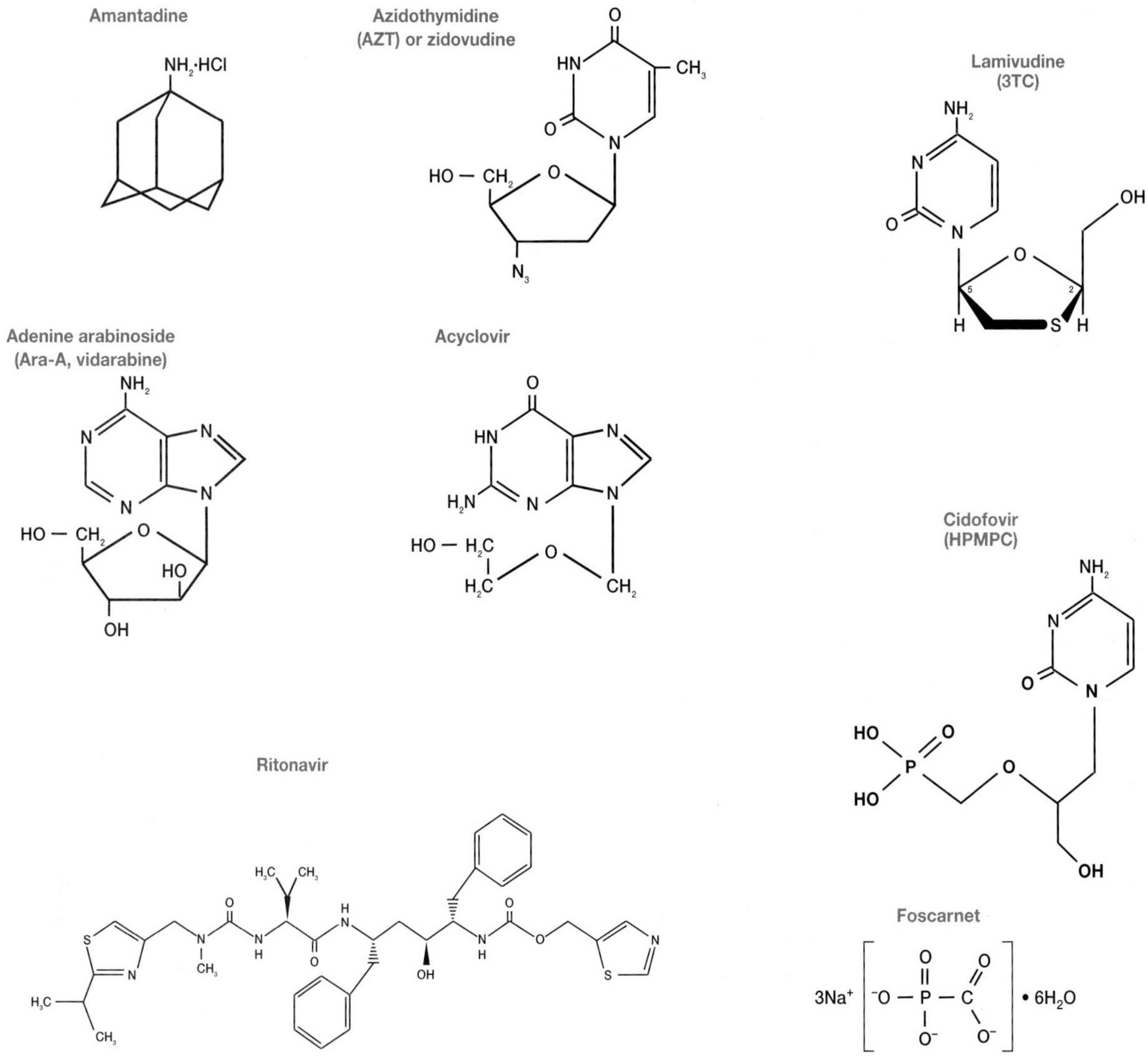

Figure 35.15   Representative Antiviral Drugs.

35.2 General Characteristics of Antimicrobial Drugs

a. An effective chemotherapeutic agent must have selective toxicity. A drug with great selective toxicity has a high therapeutic index and usually disrupts a structure or process unique to the pathogen. It has fewer side effects.

b. Antibiotics can be classified in terms of the range of target microorganisms (narrow spectrum versus broad spectrum); their source (natural, semisynthetic, or synthetic); and their general effect (static versus cidal) (**table 35.1**).

35.3 Determining the Level of Antimicrobial Activity

a. Antibiotic effectiveness can be estimated through the determination of the minimal inhibitory concentration and the minimal lethal concentration with dilution susceptibility tests. Tests like the Kirby-Bauer test (a disk diffusion test) and the Etest are often used to estimate a pathogen's susceptibility to drugs quickly (**figures 35.2** and **35.3**).

35.4 Mechanisms of Action of Antimicrobial Agents

a. Chemotherapeutic agents can damage bacterial pathogens in several ways: inhibition of cell wall synthesis, protein synthesis, or nucleic acid synthesis; disruption of membrane structure; and inhibition of key enzymes (**table 35.4**).

35.5 Factors Influencing the Effectiveness of Antimicrobial Drugs

a. A variety of factors can greatly influence the effectiveness of antimicrobial drugs during actual use.

35.6 Antibacterial Drugs

a. Sulfonamides or sulfa drugs resemble *p*-aminobenzoic acid and competitively inhibit folic acid synthesis (**figure 35.4**).

b. Quinolones are a family of bactericidal synthetic drugs that inhibit DNA gyrase and thus inhibit such processes as DNA replication (**figure 35.7**).

c. Members of the penicillin family contain a β-lactam ring and disrupt bacterial cell wall synthesis, resulting in cell lysis (**figure 35.8**). Some, like penicillin G, are usually administered by injection and are most effective against gram-positive bacteria. Others can be given orally (penicillin V), are broad spectrum (ampicillin, carbenicillin), or are penicillinase resistant (methicillin).

d. Cephalosporins are similar to penicillins, and are given to patients with penicillin allergies (**figure 35.9**).

e. Tetracyclines are broad-spectrum antibiotics having a four-ring nucleus with attached groups (**figure 35.10**). They bind to the small ribosomal subunit and inhibit protein synthesis.

f. Aminoglycoside antibiotics like streptomycin and gentamicin bind to the small ribosomal subunit, inhibit protein synthesis, and are bactericidal (**figure 35.11**).

g. Erythromycin is a bacteriostatic macrolide antibiotic that binds to the large ribosomal subunit and inhibits protein synthesis (**figure 35.12**).

h. Vancomycin is a glycopeptide antibiotic that inhibits the transpeptidation reaction during peptidoglycan synthesis. It is used against drug-resistant staphylococci, enterococci, and clostridia.

i. Chloramphenicol is a broad-spectrum, bacteriostatic antibiotic that inhibits protein synthesis (**figure 35.13**). It is quite toxic and used only for very serious infections.

35.7 Drug Resistance

a. Bacteria can become resistant to a drug by excluding it from the cell, pumping the drug out of the cell, enzymatically altering it, modifying the target enzyme or organelle to make it less drug sensitive, as examples. The genes for drug resistance may be found on the bacterial chromosome, a plasmid called an R plasmid, or other genetic elements such as transposons.

b. Chemotherapeutic agent misuse fosters the increase and spread of drug resistance, and may lead to superinfections.

35.8 Antifungal Drugs

a. Because fungi are more similar to human cells than bacteria, antifungal drugs generally have lower therapeutic indexes than antibacterial agents and produce more side effects.

b. Superficial mycoses can be treated with miconazole, ketoconazole, clotrimazole, tolnaftate, nystatin, and griseofulvin (**figure 35.14**). Amphotericin B, 5-flucytosine, and fluconazole are used for systemic mycoses.

35.9 Antiviral Drugs

a. Antiviral drugs interfere with critical stages in the virus life cycle (amantadine, rimantadine, and ritonavir) or inhibit the synthesis of virus-specific nucleic acids (zidovudine, adenine arabinoside, acyclovir) (**figure 35.15**). Interferon proteins inhibit virus replication and may be therapeutically useful in the future.

Key Terms

acyclovir   796
adenine arabinoside or vidarabine   796
amantadine   796
aminoglycoside antibiotic   791
amphotericin B   796
antibiotic   780
antimetabolites   785
azidothymidine (AZT)   796
broad-spectrum drugs   781
cephalosporin   789
chemotherapeutic agent   780
chloramphenicol   792

cidal   782
dilution susceptibility tests   783
erythromycin   791
gene cassette   794
griseofulvin   795
HIV protease inhibitors   796
integron   794
interferon   796
Kirby-Bauer method   783
macrolide antibiotic   791
minimal inhibitory
   concentration (MIC)   782

minimal lethal concentration
   (MLC)   782
narrow-spectrum drugs   781
nystatin   795
parenteral route   786
penicillinase   788
penicillins   788
plasmids   792
quinolones   787
R plasmids   793
selective toxicity   781

side effects   781
static   782
streptomycin   791
structural analogues   787
sulfonamide   787
superinfection   794
tetracycline   790
therapeutic index   781
vancomycin   791
zidovudine   796

Questions for Thought and Review

1. Why do penicillins and cephalosporins have a much higher therapeutic index than most other antibiotics? What are antimetabolites?
2. Would there be any advantage to administering a bacteriostatic agent along with penicillin? Any disadvantage?

3. Why do antifungal drugs have a much lower therapeutic index than do most antibacterial drugs? Why does one often have to apply antifungal medications frequently?
4. What advantages would there be from administering two chemotherapeutic agents simultaneously?

5. Give several ways in which the development of antibiotic-resistant pathogens can be slowed or prevented.

Critical Thinking Questions

1. What advantage might soil bacteria and fungi gain from the synthesis of antibiotics?

2. Why might it be desirable to prepare a variety of semisynthetic antibiotics?

3. Why is it so difficult to find or synthesize effective antiviral drugs?

**For recommended readings on these
and related topics, see Appendix V.**

chapter 36

Clinical Microbiology

Concepts

- Clinical microbiologists and clinical microbiology laboratories perform many services, all related to the identification and control of pathogens.
- Success in clinical microbiology depends on (1) using the proper aseptic technique; (2) correctly obtaining the clinical specimen from the infected patient by swabs, needle aspiration, intubation, or catheters; (3) correctly handling the specimen; and (4) quickly transporting the specimen to the laboratory.
- Once the clinical specimen reaches the laboratory, it is cultured and identified. Identification measures include microscopy; growth on enrichment, selective, differential, or characteristic media; specific biochemical tests; rapid test methods; immunologic techniques; bacteriophage typing; and molecular methods such as nucleic acid-based detection techniques, gas-liquid chromatography, and plasmid fingerprinting.
- After the microorganism has been isolated, cultured, and/or identified, samples are used in susceptibility tests to find which method of control will be most effective. The results are provided to the physician as quickly as possible.
- Computer systems in clinical microbiology are designed to speed identification of the pathogen and communication of results back to the physician.

The specimen is the beginning. All diagnostic information from the laboratory depends upon the knowledge by which specimens are chosen and the care with which they are collected and transported.

—Cynthia A. Needham

The major objective of the clinical microbiologist is to isolate and identify pathogens from clinical specimens rapidly. In this illustration, a clinical microbiologist is picking up suspect bacterial colonies for biochemical, immunologic, or molecular testing.

athogens, particularly bacteria and yeasts, coexist with harmless microorganisms on or in the host. These pathogens must be properly identified as the actual cause of infectious diseases. This is the purpose of clinical microbiology. The clinical microbiologist identifies agents and organisms (hereafter referred to as microorganisms) based on morphological, biochemical, immunologic, and molecular procedures. Time is a significant factor in the identification process, especially in life-threatening situations. Computers and advances in technology for rapid identification, some commercially available, have greatly aided the clinical microbiologist. Molecular methods allow identification of microorganisms based on highly specific genomic and biochemical properties. Once isolated and identified, the microorganism can then be subjected to antimicrobial sensitivity tests. In the final analysis the patient's well-being and health can benefit significantly from information provided by the clinical microbiology laboratory—the subject of this chapter.

36.1 SPECIMENS

The major focus of the **clinical microbiologist** is to isolate and identify microorganisms from clinical specimens rapidly. The purpose of the clinical microbiology laboratory is to provide the physician with information concerning the presence or absence of microorganisms that may be involved in the infectious disease process (**figure 36.1**). These individuals and facilities also determine the susceptibility of microorganisms to antimicrobial agents. Clinical microbiology makes use of information obtained from research on such diverse topics as microbial biochemistry and physiology, immunology, molecular biology, genomics, and the host-parasite relationships involved in the infectious disease process.

In clinical microbiology a clinical specimen (hereafter, specimen) represents a portion or quantity of human material that is tested, examined, or studied to determine the presence or absence of particular microorganisms. Safety for the patients, hospital, and laboratory staff is very important. The guidelines presented in **Techniques & Applications 36.1** (Universal Precautions for Health-Care Professionals) were established by the Centers for Disease Control and Prevention (CDC) to address areas of specimen handling. Other important concerns regarding specimens need emphasis:

1. The specimen selected should adequately represent the diseased area and also may include additional sites (e.g., liver and blood specimens) in order to isolate and identify potential agents of the particular disease process.
2. A quantity of specimen adequate in amount to allow a variety of diagnostic testing should be obtained.
3. Attention must be given to specimen collection in order to avoid contamination from the many varieties of microorganisms indigenous to the skin and mucous membranes (*see figure 31.2*).
4. The specimen should be forwarded promptly to the clinical laboratory.
5. If possible, the specimen should be obtained before antimicrobial agents have been administered to the patient.

Collection

Overall, the results obtained in the clinical laboratory are only as good as the quality of the specimen collected for analysis. Specimens may be collected by several methods using aseptic technique. Aseptic technique refers to specific procedures used to prevent unwanted microorganisms from contaminating the clinical specimen. Each method is designed to ensure that only the proper material will be sent to the clinical laboratory.

The most common method used to collect specimens from the anterior nares or throat is the sterile **swab.** A sterile swab is a rayon-, calcium alginate, or dacron-tipped polystyrene applicator. Manufacturers of swabs have their own unique container design and instructions for proper use. For example, many commercially manufactured swabs contain a transport medium designed to preserve a variety of microorganisms and to prevent multiplication of rapidly growing members of the population (**figure 36.2a**). However, with the exception of the nares or throat, the use of swabs for the collection of specimens is of little value and should be discouraged for two major reasons: swabs are associated with a greater risk of contamination with surface and subsurface microorganisms, and they have a limited volume capacity (<0.1 ml).

Needle aspiration is used to collect specimens aseptically (e.g., anaerobic bacteria) from cerebrospinal fluid, pus, and blood. For both samples stringent antiseptic techniques are used to avoid skin contamination. To prevent blood from clotting and entrapping microorganisms, various anticoagulants (e.g., heparin, sodium citrate) are included within the specimen bottle or tube (figure 36.2b).

Intubation [Latin *in*, into, and *tuba*, tube] is the inserting of a tube into a body canal or hollow organ. For example, intubation can be used to collect specimens from the stomach. In this procedure a long sterile tube is attached to a syringe, and the tube is either swallowed by the patient or passed through a nostril (figure 36.2c) into the patient's stomach. Specimens are then withdrawn periodically into the sterile syringe. The most common intubation tube is the Levin tube.

A **catheter** is a tubular instrument used for withdrawing or introducing fluids from or into a body cavity. For example, urine specimens may be collected with catheters to detect urinary tract infections caused by bacteria and from newborns and neonates who cannot give a voluntary urinary specimen. Three types are commonly used for urine. The hard catheter is used when the urethra is very narrow or has strictures. The French catheter is a soft tube used to obtain a single specimen sample. If multiple samples are required over a prolonged period, a Foley catheter is used (figure 36.2d).

The most common method used for the collection of urine is the clean-catch method. After the patient has cleansed the

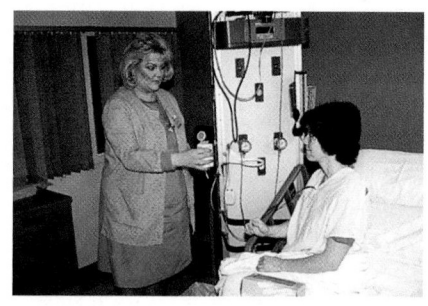

(**a**) The identification of the microorganism begins at the patient's bedside. The nurse is giving instructions to the patient on how to obtain a sputum specimen.

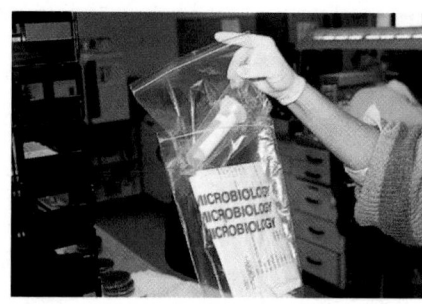

(**b**) The specimen is sent to the laboratory to be processed. Notice that the specimen and worksheet are in different Ziploc bags.

(**c**) Specimens such as sputum are plated on various types of media under a laminar airflow hood. This is to prevent specimen aerosols from coming in contact with the microbiologist.

(**d**) Sputum and other specimens are usually Gram stained to determine whether or not bacteria are present and to obtain preliminary results on the nature of any bacteria found.

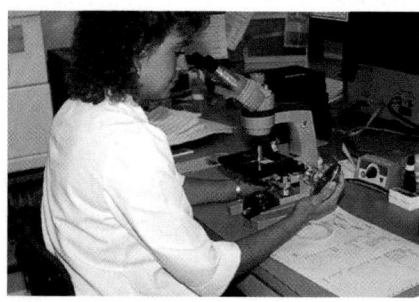

(**e**) After incubation, the plates are examined for significant isolates. The Gram stain may be reexamined for correlation.

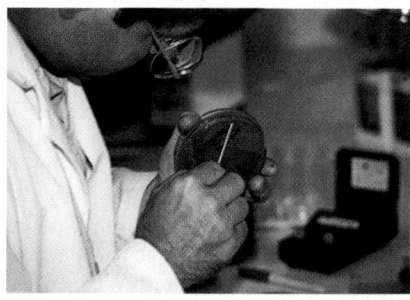

(**f**) Suspect colonies are picked for biochemical, immunologic, or molecular testing.

(**g**) Colonies are prepared for identification by rapid test systems.

(**h**) In a short period of time, sometimes 4 hours, computer-generated information is obtained that will consist of biochemical identification and antibiotic susceptibility results.

(**i**) All information about the specimen is now entered into a computer and the data are transmitted directly to the hospital ward.

Figure 36.1   Isolation and Identification of Microorganisms in a Clinical Laboratory.

Techniques & Applications
36.1   Universal Precautions for Health-Care Professionals

Since medical history and examination cannot reliably identify all patients infected with HIV or other blood-borne pathogens, blood and body-fluid precautions should be consistently used for all patients. Workers in microbiological research laboratories also are exposed to health hazards and need to employ universal precautions (*see Techniques & Applications 7.2, p. 142*).

1. All health-care workers should routinely use appropriate barrier precautions to prevent skin and mucous-membrane exposure when contact with blood or other body fluids of any patient is anticipated. Gloves should be worn for touching blood and body fluids, mucous membranes, or non-intact skin of all patients, for handling items or surfaces soiled with blood or body fluids, and for performing venipuncture and other vascular access procedures. Gloves should be changed after contact with each patient. Masks and protective eyewear or face shields should be worn during procedures that are likely to generate droplets of blood or other body fluids to prevent exposure of mucous membranes of the mouth, nose, and eyes. Gowns, aprons, or lab coats should be worn during procedures that are likely to generate splashes of blood or other body fluids.
2. Hands and other skin surfaces should be washed immediately and thoroughly if contaminated with blood or other body fluids. Hands should be washed immediately after gloves are removed.
3. All health-care workers should take precautions to prevent injuries caused by needles, scalpels, and other sharp instruments or devices during procedures; when cleaning used instruments; during disposal of used needles; and when handling sharp instruments after procedures. To prevent needlestick injuries, needles

should not be recapped, purposely bent or broken by hand, removed from disposable syringes, or otherwise manipulated by hand. After they are used, disposable syringes and needles, scalpel blades, and other sharp items should be placed in puncture-resistant containers for disposal.

4. Although saliva has not been implicated in HIV transmission, to minimize the need for emergency mouth-to-mouth resuscitation, mouthpieces, resuscitation bags, or other ventilation devices should be available for use in areas in which the need for resuscitation is predictable.
5. Health-care workers who have exudative lesions or weeping dermatitis should refrain from all direct patient care and from handling patient-care equipment.
6. The following procedure should be used to clean up spills of blood or blood-containing fluids. (a) Put on gloves and any other necessary barriers. (b) Wipe up excess material with disposable towels and place the towels in a container for sterilization. (c) Disinfect the area with either a commercial EPA-approved germicide or household bleach (sodium hypochlorite). The latter should be diluted from 1/100 (smooth surfaces) to 1/10 (porous or dirty surfaces); the dilution should be no more than 24 hours old. When dealing with large spills or those containing sharp objects such as broken glass, first cover the spill with disposable toweling. Then saturate the toweling with commercial germicide or a 1/10 bleach solution and allow it to stand for at least 10 minutes. Finally clean as just described.

Source: Adapted from *Morbidity and Mortality Weekly Report,* 36 (Suppl. 2S) 5S–10S. Centers for Disease Control and Prevention Guidelines, Atlanta, GA.

---

urethral meatus (opening), a small container is used to collect the urine. The optimal time to use the clean-catch method is early morning because the urine contains more microorganisms as a result of being in the bladder overnight. In the clean-catch midstream method, the first urine voided is not collected because it will be contaminated with those transient microorganisms normally occurring in the lower portion of the urethra. Only the midstream portion is collected since it most likely will contain those microorganisms found in the urinary bladder. If warranted for some patients, needle aspirations also are done directly into the urinary bladder.

Sputum is the most common specimen collected in suspected cases of lower respiratory tract infections. Specifically, **sputum** is the mucous secretion expectorated from the lungs, bronchi, and trachea through the mouth, in contrast to saliva, which is the secretion of the salivary glands. Sputum is collected in specially designed sputum cups (figure 36.2*e*).

Handling

Immediately after collection the specimen must be properly labeled and handled. The person collecting the specimen is respon-

sible for ensuring that the name, hospital, registration number, location in the hospital, diagnosis, current antimicrobial therapy, name of attending physician, admission date, and type of specimen are correctly and legibly written or imprinted on the culture request form. This information must correspond to that written or imprinted on a label affixed to the specimen container. The type or source of the sample and the choice of tests to be performed also must be specified on the request form.

Transport

Speed in transporting the specimen to the clinical laboratory after it has been obtained from the patient is of prime importance. Some laboratories refuse to accept specimens if they have been in transit too long.

Microbiological specimens may be transported to the laboratory by various means (figure 36.1*b*). For example, certain specimens should be transported in a medium that preserves the microorganisms and helps maintain the ratio of one organism to another. This is especially important for specimens in which normal microorganisms may be mixed with microorganisms foreign to the body location.

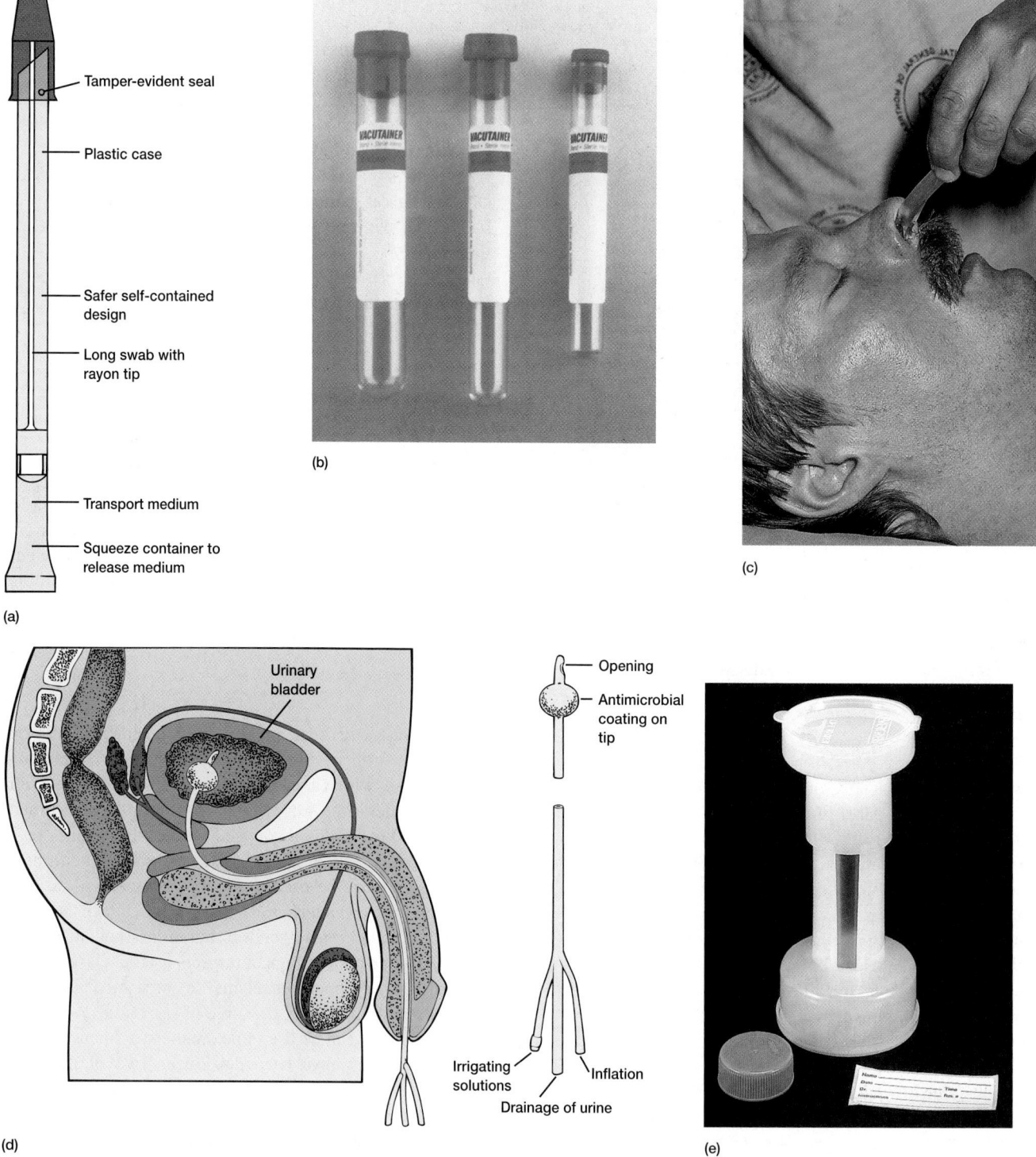

(a)

Tamper-evident seal

Plastic case

Safer self-contained design

Long swab with rayon tip

Transport medium

Squeeze container to release medium

(b)

(c)

Urinary bladder

(d)

Opening

Antimicrobial coating on tip

Irrigating solutions

Drainage of urine

Inflation

(e)

Figure 36.2   Collection of Clinical Specimens.   (a) A drawing of a sterile swab with a specific transport medium. (b) Sterile Vacutainer tubes for the collection of blood. (c) Nasotracheal intubation. (d) A drawing of a Foley catheter. Notice that three separate lumens are incorporated within the round shaft of the catheter for drainage of urine, inflation, and introducing irrigating solutions into the urinary bladder. After the Foley catheter has been introduced into the urinary bladder, the tip is inflated to prevent it from being expelled. (e) This specially designed sputum cup allows the patient to expectorate a clinical specimen directly into the cup. In the laboratory, the cup can be opened from the bottom to reduce the chance of contamination from extraneous pathogens.

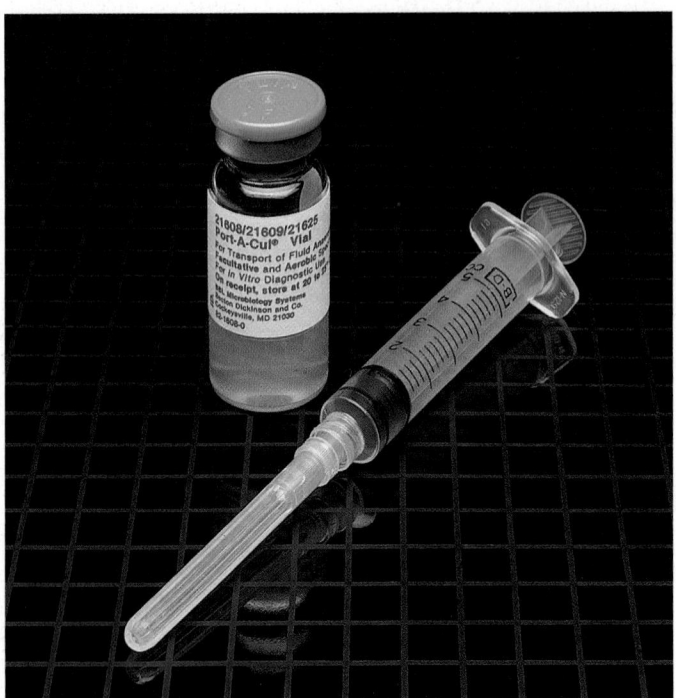

Figure 36.3   Some Anaerobic Transport Systems.   A vial and syringe. These systems contain a nonnutritive transport medium that retards diffusion of oxygen after specimen addition and helps maintain microorganism viability up to 72 hours. A built-in color indicator system is clear and turns lavender in the presence of oxygen.

Special treatment is required for specimens when the microorganism is thought to be anaerobic. The material is aspirated with a needle and syringe. Most of the time it is practical to remove the needle, cap the syringe with its original seal, and bring the specimen directly to the clinical laboratory. Transport of these specimens should take no more than 10 minutes; otherwise, the specimen must be injected immediately into an anaerobic transport vial (**figure 36.3**). Vials should contain a transport medium with an indicator, such as resazurin, to show that the interior of the vial is anaerobic at the time the specimen is introduced. Swabs for anaerobic culture usually are less satisfactory than aspirates or tissues, even if they are transported in an anaerobic vial.

Many clinical laboratories insist that stool specimens (the fecal discharge from the bowels) for culture be transported in special buffered preservatives. Preparation of these transport media is described in various manuals (*see appendix V*).

Transport of urine specimens to the clinical laboratory must be done as soon as possible. No more than 1 hour should elapse between the time the specimen is obtained and the time it is examined. If this time schedule cannot be followed, the urine sample must be refrigerated immediately.

Cerebrospinal fluid (CSF) from patients suspected of having meningitis should be examined immediately by skilled personnel in the clinical microbiology laboratory. CSF is obtained by lumbar puncture under conditions of strict asepsis, and the sample is transported to the laboratory within 15 minutes. Specimens for the isolation of viruses are iced before transport, and can be kept at 4°C for up to 72 hours; if the sample will be stored longer than 72 hours, it should be frozen at −72°C.

1. What is the function of the clinical microbiologist? The clinical microbiology laboratory?
2. What general guidelines should be followed in collecting and handling clinical specimens?
3. Define the following terms: specimen, swab, catheter, and sputum.
4. What are some transport problems associated with stool specimens? Anaerobic cultures? Urine specimens?

36.2   IDENTIFICATION OF MICROORGANISMS FROM SPECIMENS

The clinical microbiology laboratory can provide preliminary or definitive identification of microorganisms based on (1) microscopic examination of specimens, (2) study of the growth and biochemical characteristics of isolated microorganisms (pure cultures), (3) immunologic tests that detect antibodies or microbial antigens, (4) bacteriophage typing (restricted to research settings and the CDC), and (5) molecular methods.

Microscopy

Wet-mount, heat-fixed, or chemically fixed specimens can be examined with an ordinary bright-field microscope. Examination can be enhanced with either phase-contrast or dark-field microscopy. The latter is the procedure of choice for the detection of spirochetes in skin lesions associated with early syphilis or in blood specimens of people with early leptospirosis. The fluorescence microscope can be used to identify certain acid-fast microorganisms (*Mycobacterium tuberculosis*) after they are stained with fluorochromes such as auramine-rhodamine (*see section 2.2*). (Some morphological features used in classification and identification of microorganisms are presented in section 19.5 and in table 19.3.)   The light microscope (pp. 18–26)

Many stains that can be used to examine specimens for specific microorganisms have been described. Two of the more widely used are the Gram stain and acid-fast stain. Because these stains are based on the chemical composition of cell walls, they are not useful in identifying bacteria without walls. Refer to standard references, such as the *Manual of Clinical Microbiology* published by the American Society for Microbiology, for details about other reagents and staining procedures.

Growth and Biochemical Characteristics

Typically microorganisms have been identified by their particular growth patterns and biochemical characteristics. These characteristics vary depending on whether the clinical microbiologist is dealing with viruses, fungi (yeasts, molds), parasites (protozoa, helminths), common gram-positive or gram-negative bacteria, rickettsias, chlamydiae, or mycoplasmas.

Viruses

Viruses are identified by isolation in conventional cell (tissue) culture, by immunodiagnosis (fluorescent antibody, enzyme immunoassay, radioimmunoassay, latex agglutination, and immunoperoxidase) tests, and by molecular detection methods such as nucleic acid probes and amplification assays. Several types of systems are available for virus cultivation: cell cultures, embryonated hen's eggs, and experimental animals. Immunological tests (section 33.3)

Cell cultures are divided into three general classes:

1. Primary cultures consist of cells derived directly from tissues such as monkey kidney and mink lung cells that have undergone one or two passages since harvesting.
2. Semicontinuous cell cultures or low-passage cell lines are obtained from subcultures of a primary culture and usually consist of diploid fibroblasts that undergo a finite number of divisions.
3. Continuous cell cultures, such as HEp-2 cells, are derived from transformed cells that are generally epithelial in origin. These cultures grow rapidly, are heteroploid (having a chromosome number that is not a simple multiple of the haploid number), and can be subcultured indefinitely.

Each type of cell culture favors the growth of a different array of viruses, just as bacterial culture media have differing selective and restrictive properties for growth of bacteria.

Viral replication in cell cultures is detected in two ways: (1) by observing the presence or absence of cytopathic effects (CPEs), and (2) by hemadsorption.

A **cytopathic effect** is an observable morphological change that occurs in cells because of viral replication (*see section 16.3*). Examples include ballooning, binding together, clustering, or even death of the culture cells (*see figure 16.3*). During the incubation period of a cell culture, red blood cells can be added. Several viruses alter the plasma membrane of infected culture cells so that red blood cells adhere firmly to them. This phenomenon is called **hemadsorption** (*see figure 33.9*).

Embryonated hen's eggs can be used for virus isolation. There are three main routes of egg inoculation for virus isolation: (1) the allantoic cavity, (2) the amniotic cavity, and (3) the chorioallantoic membrane (*see figure 16.1*). Virus replication is recognized by the development of pocks on the chorioallantoic membrane, by the development of hemagglutinins (*see figure 33.9*) in the allantoic and amniotic fluid, and by death of the embryo.

Laboratory animals, especially suckling mice, are used for virus isolation. Inoculated animals are observed for specific signs of disease or death.

Several new serological tests for viral identification make use of monoclonal antibody-based immunofluorescence. These tests (**figure 36.4**) detect viruses such as the cytomegalovirus and herpes simplex virus in tissue-vial cultures.

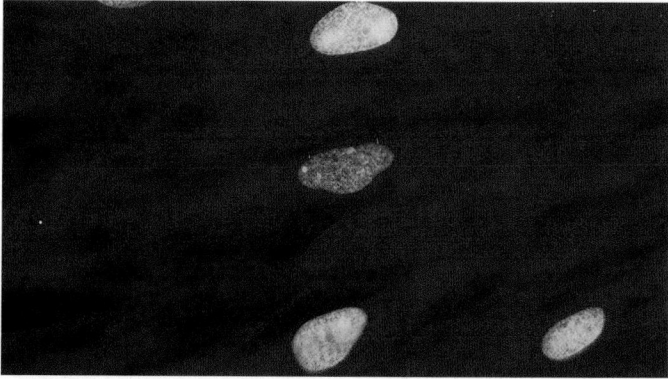

(a)

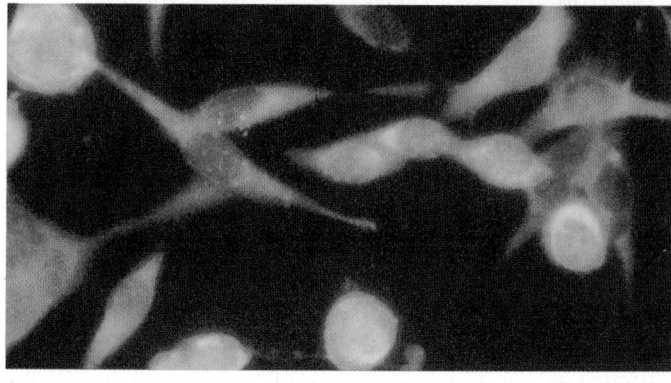

(b)

Figure 36.4 Viral Identification Test Using Immunofluorescence in Tissue Culture. (a) Three infected nuclei in a cytomegalovirus (CMV) positive tissue culture. (b) Several infected cells in a herpes simplex virus positive tissue culture.

1. Name two specimens for which microscopy would be used in the initial diagnosis of an infectious disease.
2. Name three general classes of cell cultures.
3. Give two ways by which the presence of viral replication is detected in cell culture.
4. What are the three main routes of egg inoculation for virus isolation?

Fungi

Fungal infections (i.e., mold and yeast infections) often are diagnosed by direct microscopic (fluorescence) examination of specimens. For example, the identification of molds often can be made if a portion of the specimen is mixed with a drop of 10% Calcofluor White stain on a glass slide. Fungal cultures remain as the standard for the recovery of fungi from patient specimens; however, the time needed to culture fungi varies anywhere from a few days to several weeks depending on the organism. Fungal serology (e.g., complement fixation and immunodiffusion) is

Table 36.1
Isolation of Pure Bacterial Cultures from Specimens

Selective Media

A selective medium is prepared by the addition of specific substances to a culture medium that will permit growth of one group of bacteria while inhibiting growth of some other groups. These are examples:

Salmonella-Shigella agar (SS) is used to isolate *Salmonella* and *Shigella* species. Its bile salt mixture inhibits many groups of coliforms. Both *Salmonella* and *Shigella* species produce colorless colonies because they are unable to ferment lactose. Lactose-fermenting bacteria will produce pink colonies.

Mannitol salt agar (MS) is used for the isolation of staphylococci. The selectivity is obtained by the high (7.5%) salt concentration that inhibits growth of many groups of bacteria. The mannitol in this medium helps in differentiating the pathogenic from the nonpathogenic staphylococci, as the former ferment mannitol to form acid while the latter do not.

Bismuth sulfite agar (BS) is used for the isolation of *Salmonella typhi,* especially from stool and food specimens. *S. typhi* reduces the sulfite to sulfide, resulting in black colonies with a metallic sheen.

Differential Media

The incorporation of certain chemicals into a medium may result in diagnostically useful growth or visible change in the medium after incubation. These are examples:

Eosin methylene blue agar (EMB) differentiates between lactose fermenters and nonlactose fermenters. EMB contains lactose, salts, and two dyes—eosin and methylene blue. *E. coli,* which is a lactose fermenter, will produce a dark colony or one that has a metallic sheen. *S. typhi,* a nonlactose fermenter, will appear colorless.

MacConkey agar is used for the selection and recovery of *Enterobacteriaceae* and related gram-negative rods. The bile salts and crystal violet in this medium inhibit the growth of gram-positive bacteria and some fastidious gram-negative bacteria. Because lactose is the sole carbohydrate, lactose-fermenting bacteria produce colonies that are various shades of red, whereas nonlactose fermenters produce colorless colonies.

Hektoen enteric agar is used to increase the yield of *Salmonella* and *Shigella* species relative to other microbiota. The high bile salt concentration inhibits the growth of gram-positive bacteria and retards the growth of many coliform strains.

Enrichment Media

The addition of blood, serum, or extracts to tryptic soy agar or broth will support the growth of many fastidious bacteria. These media are used primarily to isolate bacteria from cerebrospinal fluid, pleural fluid, sputum, and wound abscesses. These are examples:

Blood agar (can also be a differential medium): addition of citrated blood to tryptic soy agar makes possible variable hemolysis, which permits differentiation of some species of bacteria. Three hemolytic patterns can be observed on blood agar.
1. α-hemolysis—greenish to brownish halo around the colony (e.g., *Streptococcus gordonii, Streptococcus pneumoniae*).
2. β-hemolysis—complete lysis of blood cells resulting in a clearing effect around growth of the colony (e.g., *Staphylococcus aureus* and *Streptococcus pyogenes*).
3. Nonhemolytic—no change in medium (e.g., *Staphylococcus epidermidis* and *Staphylococcus saprophyticus*).

Chocolate agar is made from heated blood, which provides necessary growth factors to support bacteria such as *Haemophilus influenzae* and *Neisseria gonorrhoeae.*

Characteristic Media

Characteristic media are used to test bacteria for particular metabolic activities, products, or requirements. These are examples:

Urea broth is used to detect the enzyme urease. Some enteric bacteria are able to break down urea, using urease, into ammonia and CO_2.

Triple sugar iron (TSI) agar contains lactose, sucrose, and glucose plus ferrous ammonium sulfate and sodium thiosulfate. TSI is used for the identification of enteric organisms based on their ability to attack glucose, lactose, or sucrose and to liberate sulfides from ammonium sulfate or sodium thiosulfate.

Citrate agar contains sodium citrate, which serves as the sole source of carbon, and ammonium phosphate, the sole source of nitrogen. Citrate agar is used to differentiate enteric bacteria on the basis of citrate utilization.

Lysine iron agar (LIA) is used to differentiate bacteria that can either deaminate or decarboxylate the amino acid lysine. LIA contains lysine, which permits enzyme detection, and ferric ammonium citrate for the detection of H_2S production.

Sulfide, indole, motility (SIM) medium is used for three different tests. One can observe the production of sulfides, formation of indole (a metabolic product from tryptophan utilization), and motility. This medium is generally used for the differentiation of enteric organisms.

---

designed to detect serum antibody but is limited to a few fungi (*Blastomyces dermatitidis, Coccidioides immitis, Histoplasma capsulatum*). The cryptococcal latex antigen test is routinely used for the direct detection of *Cryptococcus neoformans* in serum and cerebrospinal fluid. In the clinical laboratory, nonautomated (conventional kits) and automated methods for rapid identification (4 to 24 hours) are used to detect most yeasts. Any biochemical methods used to detect fungi should always be accompanied by morphological studies examining for pseudohyphae, yeast cell structure, chlamydospores, and so on.

Parasites

Concentrated wet mounts of blood, stool, or urine specimens can be examined microscopically for the presence of eggs, cysts, larvae, or vegetative cells of parasites. Blood smears for sporozoan (malaria) and flagellate (trypanosome) parasites are stained with Giemsa. Some serological tests also are available.

Bacteria

Isolation and growth of bacteria are required before many diagnostic tests can be used to confirm the identification of the pathogen.

Table 36.2

Some Biochemical Tests Used by Clinical Microbiologists in the Diagnosis of Bacteria from the Patient's Specimen

Biochemical Test	Description	Laboratory Application
Carbohydrate fermentation (figure 36.5m)	Acid and/or gas are produced during fermentative growth with sugars or sugar alcohols.	Fermentation of specific sugars used to differentiate enteric bacteria as well as other genera or species.
Casein hydrolysis	Detects the presence of caseinase, an enzyme able to hydrolyze milk protein casein. Bacteria that use casein appear as colonies surrounded by a clear zone.	Used to cultivate and differentiate aerobic actinomycetes based on casein utilization. For example, *Streptomyces* uses casein and *Nocardia* does not.
Catalase (figure 36.5d,e)	Detects the presence of catalase, which converts hydrogen peroxide to water and O_2.	Used to differentiate *Streptococcus* ($-$) from *Staphylococcus* ($+$) and *Bacillus* ($+$) from *Clostridium* ($-$).
Citrate utilization	When citrate is used as the sole carbon source, this results in alkalinization of the medium.	Used in the classification of enteric bacteria. *Klebsiella* ($+$), *Enterobacter* ($+$), *Salmonella* (often $+$); *Escherichia* ($-$), *Edwardsiella* ($-$).
Coagulase	Detects the presence of coagulase. Coagulase causes plasma to clot.	This is an important test to differentiate *Staphylococcus aureus* ($+$) from *S. epidermidis* ($-$).
Decarboxylases (arginine, lysine, ornithine) (figure 36.5g)	The decarboxylation of amino acids releases CO_2 and amine.	Used in the classification of enteric bacteria.
Esculin hydrolysis	Tests for the cleavage of a glycoside.	Used in the differentiation of *Staphylococcus aureus*, *Streptococcus mitis*, and others ($-$) from *S. bovis*, *S. mutans*, and enterococci ($+$).
β-galactosidase (ONPG) test	Demonstrates the presence of an enzyme that cleaves lactose to glucose + galactose.	Used to separate enterics (*Citrobacter*+, *Salmonella*−) and to identify pseudomonads.
Gelatin liquefaction (figure 36.5i)	Detects whether or not a bacterium can produce proteases that hydrolyze gelatin and liquify solid gelatin medium.	Used in the identification of *Clostridium*, *Serratia*, *Pseudomonas*, and *Flavobacterium*.
Hydrogen sulfide (H_2S)	Detects the formation of hydrogen sulfide from the amino acid cysteine due to cysteine desulfurase.	Important in the identification of *Edwardsiella*, *Proteus*, and *Salmonella*.
IMViC (indole; methyl red; Voges-Proskauer; citrate) (figure 36.5a,b,h)	The indole test detects the production of indole from the amino acid tryptophan. Methyl red is a pH indicator to determine whether the bacterium has produced acid. VP (Voges-Proskauer) detects the production of acetoin. The citrate test determines whether or not the bacterium can use sodium citrate as a sole source of carbon.	Used to separate *Escherichia* (MR+, VP−, indole+) from *Enterobacter* (MR−, VP+, indole−) and *Klebsiella pneumoniae* (MR−, VP+, indole−); also used to characterize members of the genus *Bacillus*.
Lipid hydrolysis	Detects the presence of lipase, which breaks down lipids into simple fatty acids and glycerol.	Used in the separation of clostridia.
Nitrate reduction (figure 36.5f)	Detects whether a bacterium can use nitrate as an electron acceptor.	Used in the identification of enteric bacteria which are usually $+$.
Oxidase (figure 36.5n,p)	Detects the presence of cytochrome *c* oxidase that is able to reduce O_2 and artificial electron acceptors.	Important in distinguishing *Neisseria* and *Moraxella* spp. ($+$) from *Acinetobacter* ($-$), and enterics (all $-$) from pseudomonads ($+$).
Phenylalanine deaminase (figure 36.5j)	Deamination of phenylalanine produces phenylpyruvic acid, which can be detected colorimetrically.	Used in the characterization of the genera *Proteus* and *Providencia*.
Starch hydrolysis (figure 36.5c)	Detects the presence of the enzyme amylase, which hydrolyzes starch.	Used to identify typical starch hydrolyzers such as *Bacillus* spp.
Urease (figure 36.5r)	Detects the enzyme that splits urea to NH_3 and CO_2.	Used to distinguish *Proteus*, *Providencia rettgeri*, and *Klebsiella pneumoniae* ($+$) from *Salmonella*, *Shigella* and *Escherichia* ($-$).

The presence of bacterial growth usually can be recognized by the development of colonies on solid media or turbidity in liquid media. The time for visible growth to occur is an important variable in the clinical laboratory. For example, most pathogenic bacteria require only a few hours to produce visible growth, whereas it may take weeks for colonies of mycobacteria or mycoplasmas to become evident. The clinical microbiologist as well as the clinician should be aware of reasonable reporting times for various cultures.

The initial identity of a bacterial organism may be suggested by (1) the source of the culture specimen; (2) its microscopic appearance and Gram stain; (3) its pattern of growth on selective, differential, enrichment, or characteristic media (**table 36.1;** *see also figure 5.10*); and (4) its hemolytic, metabolic, and fermentative properties on the various media (table 36.1; *see also table 19.4*). After the microscopic and growth characteristics of a pure culture of bacteria are examined, specific biochemical tests can be performed. Some of the most common biochemical tests used to identify bacterial isolates are given in **table 36.2** and **figure 36.5.** Microbial nutrition and types of media (chapter 5)

(a) Methyl Red Test. This is a qualitative test of the acidity produced by bacteria grown in MR-VP broth. After the addition of several drops of methyl red solution, a bright red color (left tube) is a positive test; a yellow or orange color (right tube) is a negative test. Used to separate enterics, for example *E. coli* is MR positive and *Enterobacter aerogenes* is MR negative.

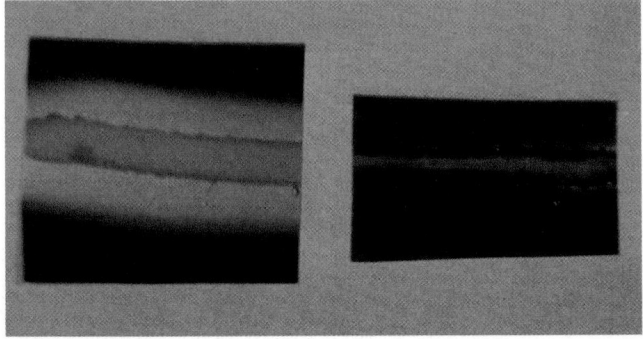

(c) Starch Hydrolysis. After incubation on starch agar, plates are flooded with iodine solution. A positive test is indicated by the colorless area around the growth (left); a negative test is shown on the right. Used to detect the production of α-amylase by certain bacteria.

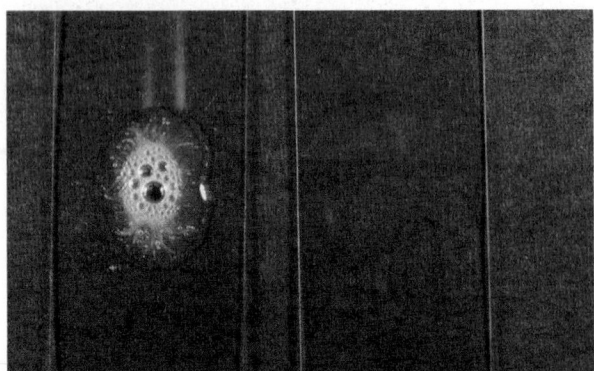

(e) Slide Catalase Test. A wooden applicator stick (or nichrome wire loop) is used to pick up a colony from a culture plate and place it in a drop of hydrogen peroxide on a glass slide. A positive catalase reaction (left slide) shows gas bubbles; a negative catalase reaction reveals an absence of gas bubbles (right slide). Used to differentiate *Streptococcus* (catalase negative) from *Staphylococcus* (catalase positive).

(b) Voges-Proskauer Reaction. VP-positive bacteria produce acetylmethylcarbinol or acetoin, which reacts with the reagents to produce a red color (left tube); a VP-negative tube is shown on the right. Used to differentiate *Bacillus* species and enterics. For example, *E. coli* is VP negative and *Enterobacter aerogenes* is VP positive.

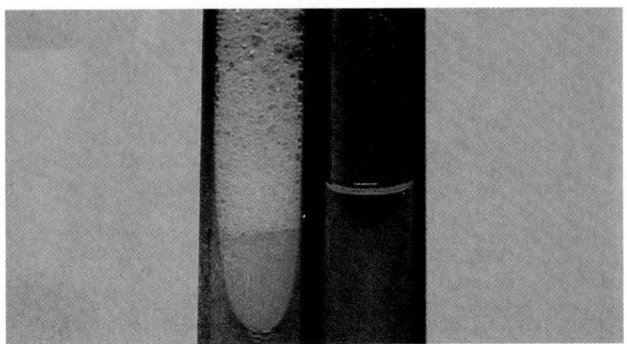

(d) Tube Catalase Test. After incubation of slant cultures, 1 ml of 3% hydrogen peroxide is trickled down the slants. Catalase converts hydrogen peroxide to water and oxygen bubbles (left tube); a negative catalase test is shown in the right tube. Used to differentiate *Streptococcus* (catalase negative) from *Staphylococcus* (catalase positive).

(f) Nitrate Reduction. After 24–48 hours of incubation, nitrate reagents are added to culture tubes. The tube on the left illustrates gas formation (a positive reaction for nitrate reduction); the tube in the middle is a positive reaction for nitrate reduction to nitrite as indicated by the red color; the tube on the right is a negative broth control. Used to test for miscellaneous gram-negative bacteria.

Figure 36.5   Some Common Diagnostic Tests Used in Microbiology.

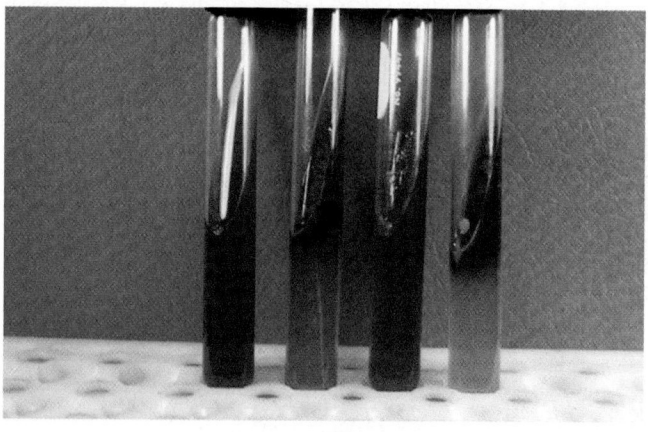

(g) Test for Amino Acid Decarboxylase. The tube on the left is an uninoculated control; the second tube from the left is lysine decarboxylase negative; the third tube is lysine decarboxylase positive; and the tube on the right is lysine deaminase positive. Used to classify enteric bacteria. Recommended for detection of arginine dihydrolase and lysine and ornithine decarboxylase activities.

(h) Test for Indole. Tryptophan can be broken down to indole by some bacteria. The presence of indole is detected by adding Kovacs' reagent. A red color on the surface is a positive test for indole (left tube) and an orange-yellow color is a negative test for indole (right tube). Used to separate enterics such as *E. coli* (indole positive) and *Enterobacter* (indole negative).

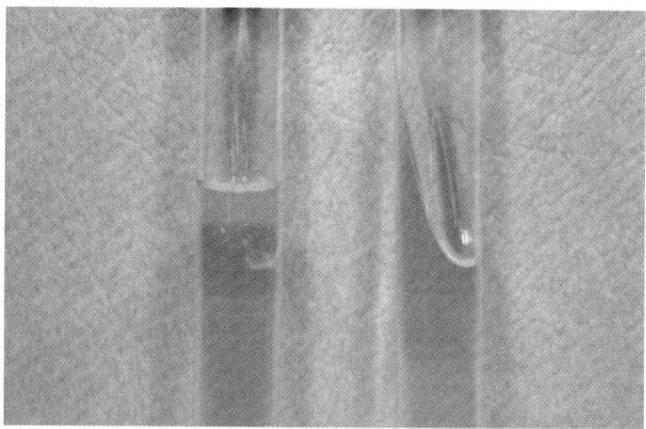

(i) Gelatin Hydrolysis or Liquefaction. If gelatin is hydrolyzed by the enzyme gelatinase, it does not gel when cooled but remains a liquid. Thus it flows when the culture is tilted backward (right tube). A negative control is on the left. Note that the solid gelatin does not flow when the tube is tilted. Used to differentiate a variety of heterotrophic bacteria, such as clostridia.

Figure 36.5 *continued*

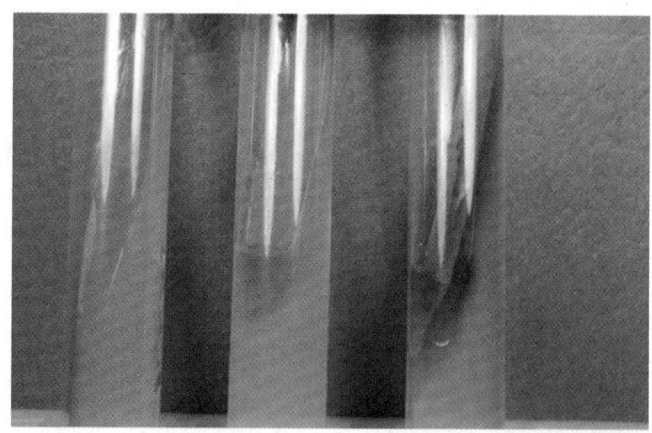

(j) Phenylalanine Deamination Test. When 10% ferric chloride is added to a phenylalanine deaminase agar slant culture, a dark green color (tube on the right) is a positive test for the enzyme. The tube on the left is an uninoculated control, and the tube in the middle is a negative test. Used to differentiate members of the *Proteus* group (*Proteus vulgaris*, positive, from *E. coli*, negative).

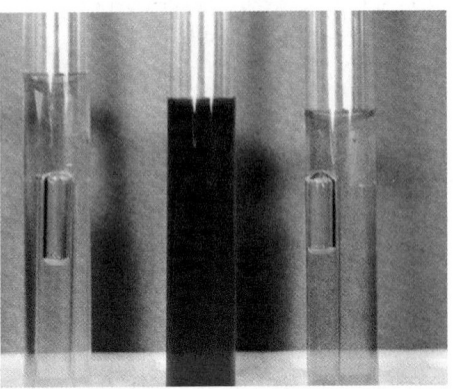

(k) Bile Solubility for Pneumococcus. The tube on the left and the one in the middle contain cultures of pneumococcus (*Streptococcus pneumoniae*). The tube on the right contains α-streptococcus. One-tenth ml of deoxycholate was added to the middle and right tubes, and 0.1 ml of distilled water was added to the left tube. The pneumococci in the center are bile soluble, as indicated by the clear suspension. The bacteria in the right tube are not bile soluble, as indicated by the turbidity.

(l) Stormy Fermentation of Litmus Milk. The tube on the left shows fermentation; the tube on the right is negative for stormy fermentation. Used for the identification of *Clostridium* species.

(m) Fermentation Reactions. The tube on the left shows acid (yellow color) and gas in the Durham tube. The center tube shows no carbohydrate utilization to produce acid or gas. The tube on the right has less acid formation than that on the left. Used to separate enterics.

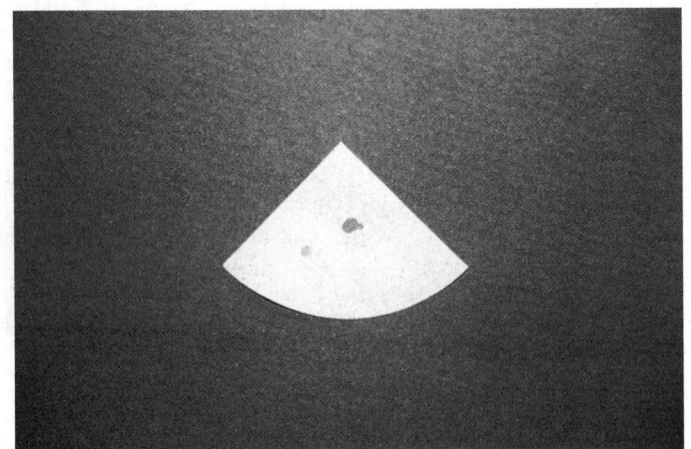

(n) Oxidase Test. Filter paper is moistened with a few drops of 1% tetramethyl-*p*-phenylenediamine dihydrochloride. With a wooden applicator, growth from an agar medium is smeared on the paper. A positive test is the development of a purple color within 10 seconds. Used to separate enterics from pseudomonads. For example, *Pseudomonas aeruginosa* is positive and *E. coli* is negative.

(o) Bacitracin Sensitivity Test. The bacteria on the left are presumptively identified as group A streptococci because of inhibition by the antibiotic bacitracin. The bacteria on the right (*Enterococcus faecalis*) are bacitracin resistant.

Figure 36.5 *continued*

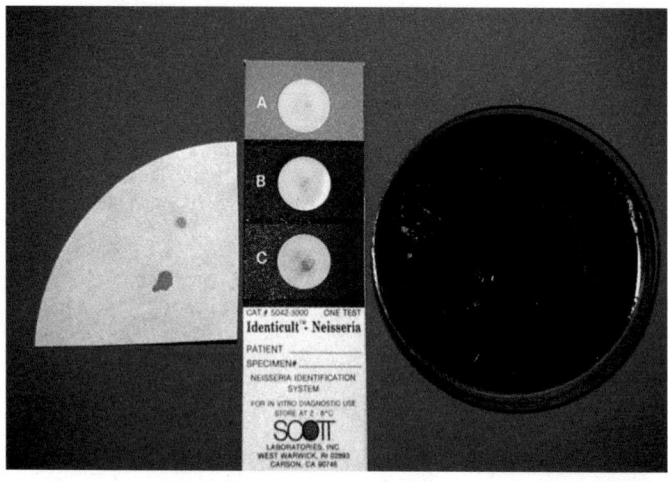

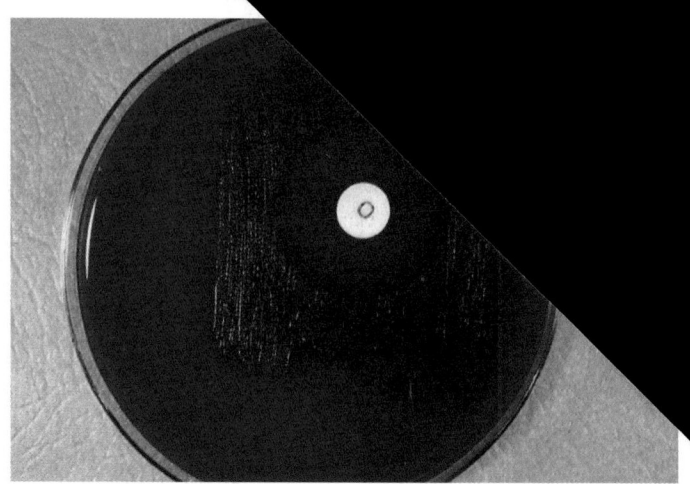

(p) Identification of *Neisseria gonorrhoeae.* Oxidase-positive test on filter paper is shown on the left. Identicult reaction for *N. gonorrhoeae* (center). The plate on the right shows characteristic growth on modified Thayer-Martin medium.

(q) Optochin Sensitivity. An optochin disk on blood agar specifically inhibits pneumococci. The optochin test for *Streptococcus pneumoniae* is based on the zone of inhibition.

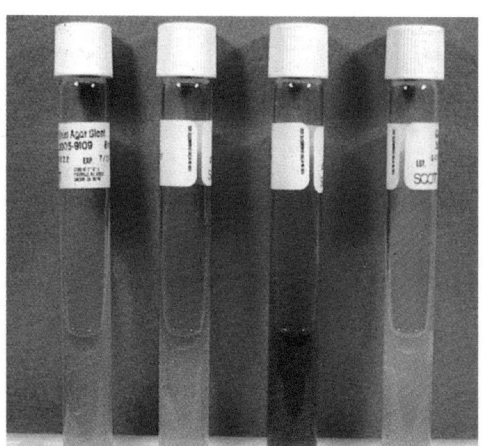

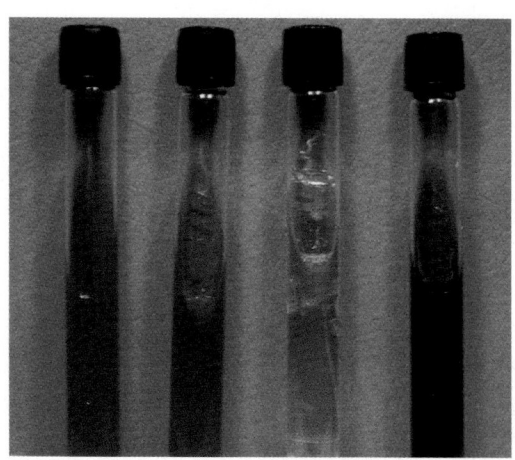

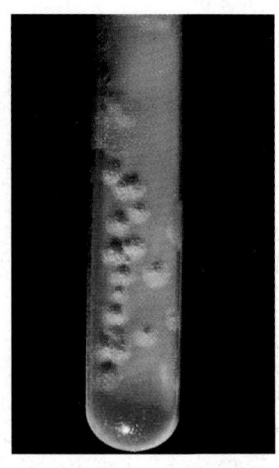

(r) Urease Production. Christensen urea agar slants. Urease-positive bacteria hydrolyze urea to ammonia, which turns the phenol red indicator red-violet. From left to right: uninoculated control, delayed positive (>24 hrs), rapidly positive (<4 hrs), negative reaction. Used to separate enterics.

(s) Triple Sugar Iron Agar Reactions. Left to right: left tube is an uninoculated control, second tube is K/K (nonfermenter), third tube is A/A with gas indicating lactose or sucrose fermentation, and the right tube is K/A plus H₂S production. A stands for acid production, and K indicates that the medium becomes alkaline. Used for the differentiation of members of the *Enterobacteriaceae* based on the fermentation of lactose, sucrose, glucose, and the production of H₂S.

(t) Lowenstein-Jensen Medium. Growth of *Mycobacterium tuberculosis* showing nodular and non-pigmented growth.

Figure 36.5 *continued*

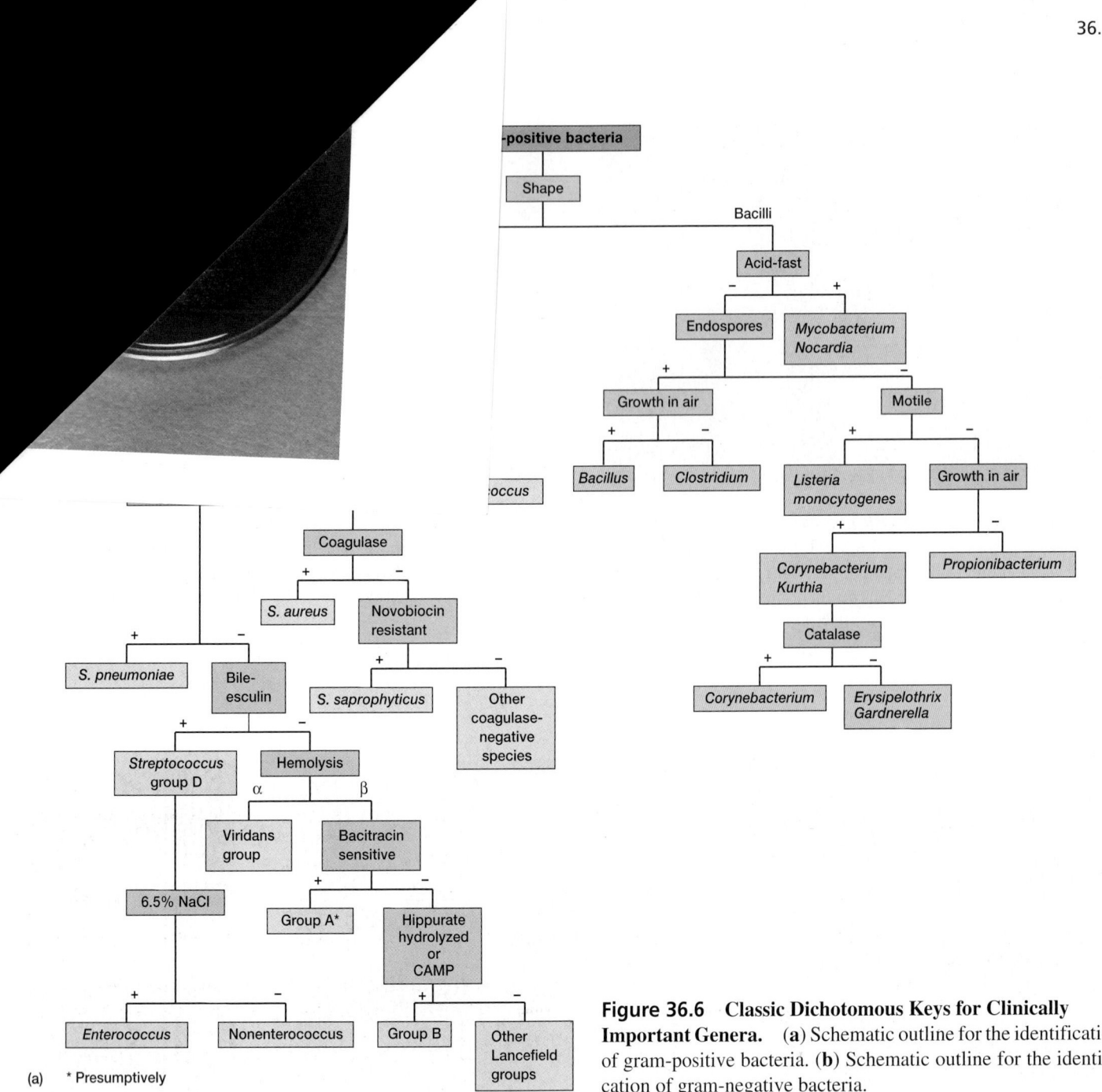

Figure 36.6 Classic Dichotomous Keys for Clinically Important Genera. (**a**) Schematic outline for the identification of gram-positive bacteria. (**b**) Schematic outline for the identification of gram-negative bacteria.

Classic dichotomous keys are coupled with the biochemical tests for the identification of bacteria from specimens. Generally, fewer than 20 tests are required to identify clinical bacterial isolates to the species level (**figure 36.6**).

Rickettsias

Although rickettsias, chlamydiae, and mycoplasmas are bacteria, they differ from other bacterial pathogens in a variety of ways. Therefore the identification of these three groups is discussed separately. Rickettsias can be diagnosed by immunoassays or by isolation of the microorganism. Because isolation is both hazardous to the clinical microbiologist and expensive, immunological methods are preferred. Isolation of rickettsias and diagnosis of rickettsial diseases is generally confined to reference and specialized research laboratories.

Chlamydiae

Chlamydiae can be demonstrated in tissues and cell scrapings with Giemsa staining, which detects the characteristic intracellular inclusion bodies (*see figure 21.13*). Immunofluorescent staining of tissues and cells with monoclonal antibody reagents is a more sensitive and specific means of diagnosis (**Techniques & Applications 36.2**; *see also figure 32.20*). The most sensitive methods for demonstrating chlamydiae in clinical specimens are DNA probes (*see p. 314*) and PCR methods (*see section 14.3*).

Mycoplasmas

The most routinely used techniques for identification of the mycoplasmas are immunological (hemagglutinin) or complement-fixing antigen-antibody reactions using the patient's sera. These microorganisms are slow growing; therefore positive results from

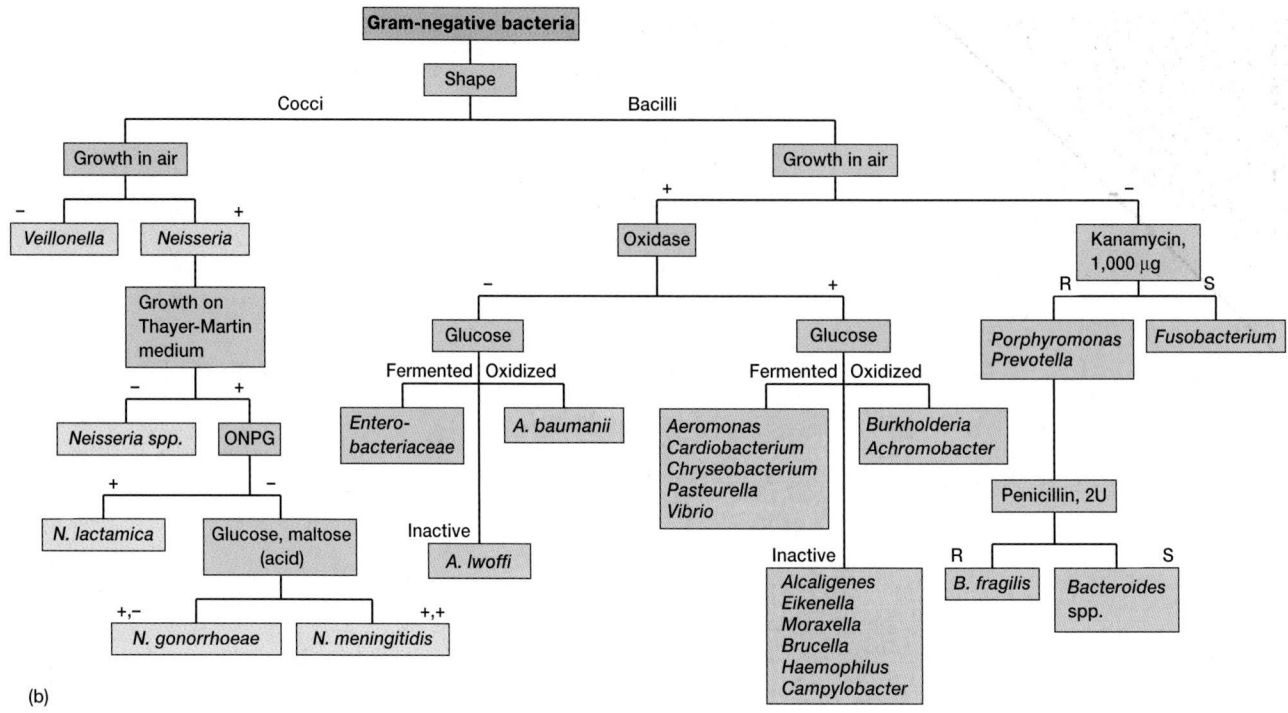

Figure 36.6   *continued*

Techniques & Applications

36.2   Monoclonal Antibodies in Clinical Microbiology

An important application of monoclonal antibody technology (*see hybridomas, section 32.3*) is the identification of microorganisms. Monoclonal antibodies have been prepared for a wide range of viruses, bacteria, fungi, and parasites; however, many remain as research tools and are not commercially available. If specific monoclonal antibodies are selected, immunologic assays can be created for different types of analyses. For example, monoclonal antibodies of cross-species or cross-genus reactivity have applications in the taxonomy of microorganisms. Those monoclonal antibodies that define species-specific antigens are extremely valuable in diagnostic reagents. Monoclonal antibodies that exhibit more restrictive specificity can be used to identify strains or biotypes within a species, to aid in studies of antigenic drift, and in epidemiological studies involving the matching of microbial

strains. In addition, individual antigenic determinants on protein molecules can be mapped.

In the clinical microbiology laboratory, monoclonal antibodies to viral or bacterial antigens are replacing polyclonal antibodies for use in culture confirmation when accurate, rapid identification is required. With the use of sensitive techniques such as fluorescent antibody assays it is possible to perform culture identifications with improved accuracy, speed, and fewer microorganisms. The formulation of direct assays with monoclonal antibody reagents, which contain no contaminating antibodies and produce a minimum of artifacts, is now reality. The highly defined and reproducible properties of monoclonal antibodies invite their incorporation into immunoassays being developed for the next generation of instruments that will detect microbial antigens and serum antibodies for the clinical microbiologist.

---

isolation procedures are rarely available before 30 days—a long delay with an approach that offers little advantage over standard techniques. Recently DNA probes have been applied to the detection of *Mycoplasma pneumoniae* in clinical specimens.

1. How can fungi and parasites be detected in a clinical specimen? Rickettsias? Chlamydiae? Mycoplasmas?

2. Why must the clinical microbiologist know what are reasonable reporting times for various microbial specimens?

3. How can a clinical microbiologist determine the initial identity of a bacterium?

4. Describe a dichotomous key that could be used to identify a bacterium.

Figure 36.7 A "Kit Approach" to Bacterial Identification. The API 20E manual biochemical system for microbial identification. (**a**) Positive and (**b**) negative results.

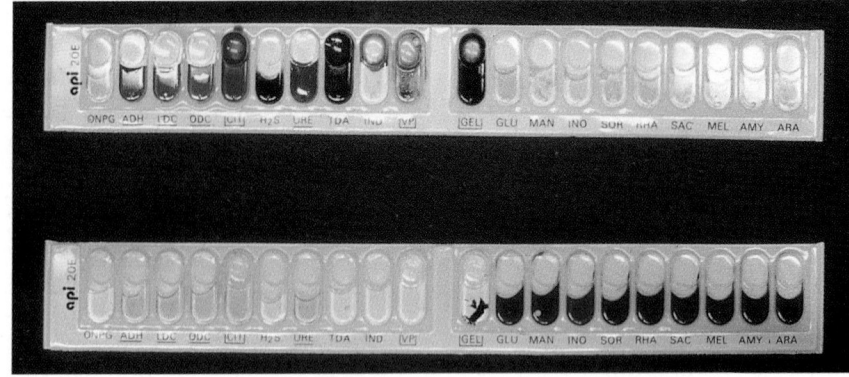

(a)

(b)

Rapid Methods of Identification

Clinical microbiology has benefited greatly from technological advances in equipment, computer programs and data bases, molecular biology, and immunochemistry. With respect to the detection of microorganisms in specimens, there has been a shift from the multistep methods previously discussed to unitary procedures and systems that incorporate standardization, speed, reproducibility, miniaturization, mechanization, and automation. These rapid identification methods can be divided into three categories: (1) manual biochemical systems, (2) mechanized/automated systems, and (3) immunologic systems.

One example of a "kit approach" biochemical system for the identification of members of the family *Enterobacteriaceae* and other gram-negative bacteria is the API 20E system. It consists of a plastic strip (**figure 36.7**) with 20 microtubes containing dehydrated biochemical substrates that can detect certain biochemical characteristics. The biochemical substrates in the 20 microtubes are inoculated with a pure culture of bacteria evenly suspended in sterile physiological saline. After 5 hours or overnight incubation, the 20 test results are converted to a seven- or nine-digit profile (**figure 36.8**). This profile number can be used with a computer or a book called the *API Profile Index* to find the name of the bacterium.

Immunologic Techniques

The culturing of certain viruses, bacteria, fungi, and parasites from clinical specimens may not be possible because the methodology remains undeveloped (*Treponema pallidum;* hepatitis A, B, C; and Epstein-Barr virus), is unsafe (rickettsias), or is impractical for all but a few clinical microbiology laboratories (mycobacteria). Cultures also may be negative because of prior antimicrobial therapy. Under these circumstances, detection of antibodies or antigens may be quite valuable diagnostically.

Immunologic systems for the detection and identification of pathogens from clinical specimens are easy to use, give relatively rapid reaction endpoints, and are sensitive and specific (they give a low percentage of false positives and negatives). Some of the more popular immunologic rapid test kits for viruses and bacteria are presented in **table 36.3**.

Table 36.3

Some Common Rapid Immunologic Test Kits for the Detection of Bacteria and Viruses in Clinical Specimens

Bactigen (Wampole Laboratories, Cranburg, N.J.)
The Bactigen kit is used for the detection of *Streptococcus pneumoniae, Haemophilus influenzae* type b, and *Neisseria meningitidis* groups A, B, C, and Y from cerebrospinal fluid, serum, and urine.

Culturette Group A Strep ID Kit (Marion Scientific, Kansas City, Mo.)
The Culturette kit is used for the detection of group A streptococci from throat swabs.

Directigen (Hynson, Wescott, and Dunning, Baltimore, Md.)
The Directigen Meningitis Test kit is used to detect *H. influenzae* type b, *S. pneumoniae,* and *N. meningitidis* groups A and C.
The Directigen Group A Strep kit is used for the direct detection of group A streptococci from throat swabs.

Gono Gen (Micro-Media Systems, San Jose, Calif.)
The Gono Gen kit detects *Neisseria gonorrhoeae.*

QuickVue *H. pylori* Test (Quidel, San Diego, Calif.)
A 7-minute test for detection of IgG antibodies against *Helicobacter pylori* in human serum or plasma.

Staphaurex (Wellcome Diagnostics, Research Triangle Park, N.C.)
Staphaurex screens and confirms *Staphylococcus aureus* in 30 seconds.

Directigen RSV (Becton Dickinson Microbiology Systems, Cockeysville, Md.)
By using a nasopharyngeal swab, the respiratory syncytial virus can be detected in 15 minutes.

SureCell Herpes (HSV) Test (Kodak, Rochester, N.Y.)
Detects the herpes (HSV) 1 and 2 viruses in minutes.

SUDS HIV-1 Test (Murex Corporation, Norcross, Ga.)
Detects antibodies to HIV-1 antigens in about 10 minutes.

Each individual's immunologic response to a microorganism is quite variable. As a result the interpretation of immunologic tests is sometimes difficult. For example, a single elevated antibody IgM titer usually does not distinguish between active and past infections. Furthermore, the lack of a measurable antibody titer may reflect either a microorganism's lack of immunogenicity or an insufficient time for an antibody response to develop following the onset of the infectious disease. Some patients also are immunosuppressed due to other disease processes and/or treatment procedures (e.g., cancer and AIDS patients) and therefore do not respond. For these reasons, test selection and timing of specimen collection are essential to the proper interpretation of immunologic tests. Antibody titer (pp. 719, 751)

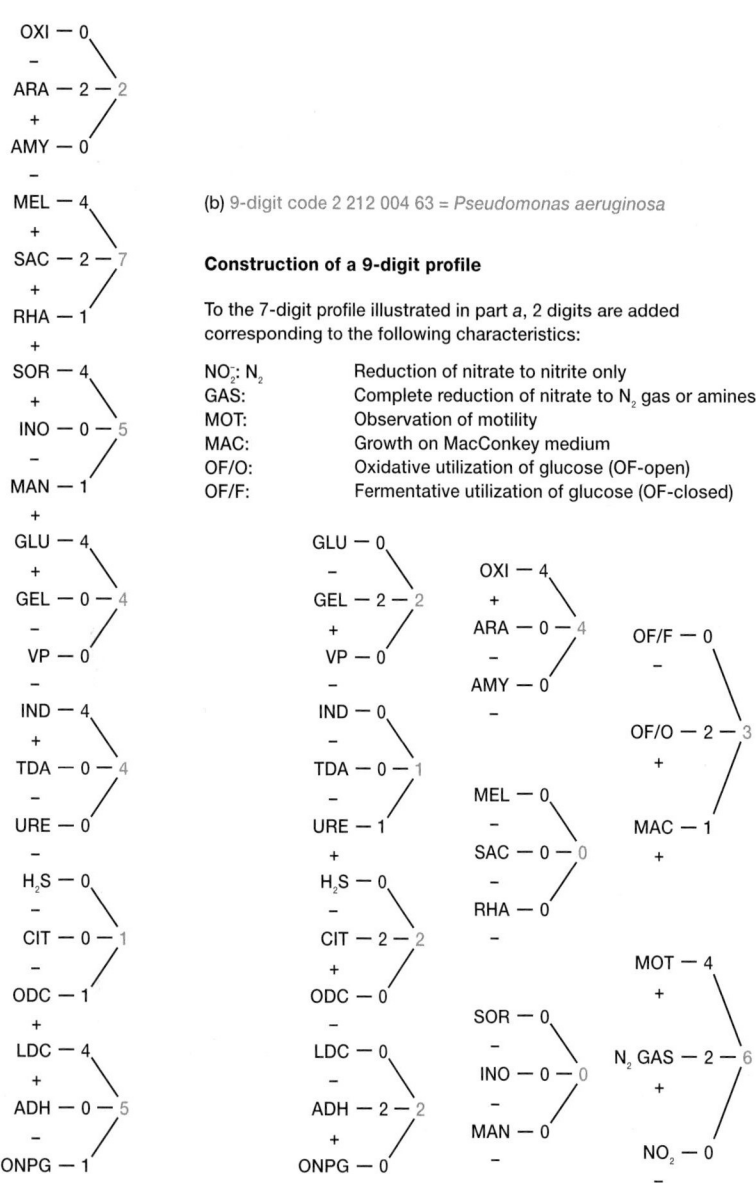

(a) Normal 7-digit code 5 144 572 = *E. coli.*

(b) 9-digit code 2 212 004 63 = *Pseudomonas aeruginosa*

Construction of a 9-digit profile

To the 7-digit profile illustrated in part *a*, 2 digits are added corresponding to the following characteristics:

NO$_2$; N$_2$	Reduction of nitrate to nitrite only
GAS:	Complete reduction of nitrate to N$_2$ gas or amines
MOT:	Observation of motility
MAC:	Growth on MacConkey medium
OF/O:	Oxidative utilization of glucose (OF-open)
OF/F:	Fermentative utilization of glucose (OF-closed)

Figure 36.8 The API 20E Profile Number. The conversion of API 20E test results to the codes used in identification of unknown bacteria. The test results read top to bottom (and right to left in part *b*) correspond to the 7- and 9-digit codes when read in the right-to-left order. The tests required for obtaining a 7-digit code take an 18–24 hour incubation and will identify most members of the *Enterobacteriaceae*. The longer procedure that yields a 9-digit code is required to identify many gram-negative nonfermenting bacteria. The following tests are common to both procedures: ONPG (β-galactosidase); ADH (arginine dihydrolase); LDC (lysine decarboxylase); ODC (ornithine decarboxylase); CIT (citrate utilization); H$_2$S (hydrogen sulfide production); URE (urease); TDA (tryptophane deaminase); IND (indole production); VP (Voges-Proskauer test for acetoin); GEL (gelatin liquefaction); the fermentation of glucose (GLU), mannitol (MAN), inositol (INO), sorbitol (SOR), rhamnose (RHA), sucrose (SAC), melibiose (MEL), amygdalin (AMY), and arabinose (ARA); and OXI (oxidase test).

The most widely used immunologic techniques available to detect microorganisms in clinical specimens are covered in detail in section 33.3. No single technique is universally applicable for measuring an individual's immunologic response to all microorganisms. Techniques are therefore chosen based on their selectivity, specificity, ease, speed of performance, and cost-effectiveness.

1. Describe in general how biochemical tests are used in the API 20E system to identify bacteria.
2. Name the two basic immunologic procedures used in kits to identify microorganisms.
3. Why might cultures for some microorganisms be unavailable?
4. Why are test selection and timing of specimen collection essential to the proper interpretation of immunologic tests?

Bacteriophage Typing

Bacteriophages (phages) are viruses that attack members of a particular bacterial species, or strains within a species (*see chapter 17*). **Bacteriophage (phage) typing** is based on the specificity of phage surface receptors for cell surface receptors. Only those bacteriophages that can attach to these surface receptors can infect bacteria and cause lysis. On a petri dish culture, lytic bacteriophages cause plaques on lawns of sensitive bacteria. These plaques represent infection by the virus (*see figure 16.4*).

In bacteriophage typing the clinical microbiologist inoculates the bacterium to be tested onto a petri plate. The plate is heavily and uniformly inoculated with a cotton swab so that the bacteria will grow to form a solid sheet or lawn of cells. No uninoculated areas should be left. The plate is then marked off into squares

(15 to 20 mm per side), and each square is inoculated with a drop of suspension from the different phages available for typing. After the plate is incubated for 24 hours, it is observed for plaques. The phage type is reported as a specific genus and species followed by the types that can infect the bacterium. For example, the series 10/16/24 indicates that this bacterium is sensitive to phages 10, 16, and 24, and belongs to a collection of strains, called a **phagovar,** that have this particular phage sensitivity. Bacteriophage typing remains a tool of the research and reference laboratory.

Molecular Methods and Analysis of Metabolic Products

With the application of new molecular technology, it is now possible to analyze the molecular characteristics of microorganisms in the clinical laboratory. Some of the most accurate approaches to microbial identification are through the analysis of proteins and nucleic acids. Examples previously discussed (*see chapter 19*) include comparison of proteins; physical, kinetic, and regulatory properties of microbial enzymes; nucleic acid–base composition (*see table 19.5*); nucleic acid hybridization; and nucleic acid sequencing. Three other molecular methods being widely used are nucleic acid probes, gas-liquid chromatography, and plasmid fingerprinting.

Nucleic Acid–Based Detection Methods

A recent development in clinical microbiology is the use of nucleic acid–based diagnostic methods for the detection and identification of microorganisms. For example, DNA probe technology (*see section 14.4*) identifies a microorganism by probing its genetic composition. The use of cloned DNA as a probe is based upon the capacity of single-stranded DNA to bind (hybridize) with a complementary nucleic acid sequence present in test specimens to form a double-stranded DNA hybrid (**figure 36.9**). Thus a single-stranded sequence derived from one microorganism (the probe) is used to search for others containing the same sequence. This hybridization reaction may be applied to purified DNA preparations, to bacterial colonies, or to clinical specimens such as tissue, serum, sputum, and pus. Recently DNA probes have been developed that bind to complementary strands of ribosomal RNA. These DNA:rRNA hybrids are more sensitive than conventional DNA probes, give results in 2 hours or less, and require the presence of fewer microorganisms. DNA probe sensitivity can be increased by over one million-fold if the target DNA is first amplified using the polymerase chain reaction (*see section 14.3*). DNA:rRNA probes are available or are currently being developed for most clinically important microorganisms.

Ribosomal RNA from *E. coli* can be used to type bacterial strains by probing chromosomal DNA in Southern blots (*see figure 14.5*). This method of strain typing, called **ribotyping,** makes use of the fact that rRNA genes are scattered throughout the chromosome of most bacteria. When the chromosomes of several strains are cleaved using restriction endonucleases and the digests

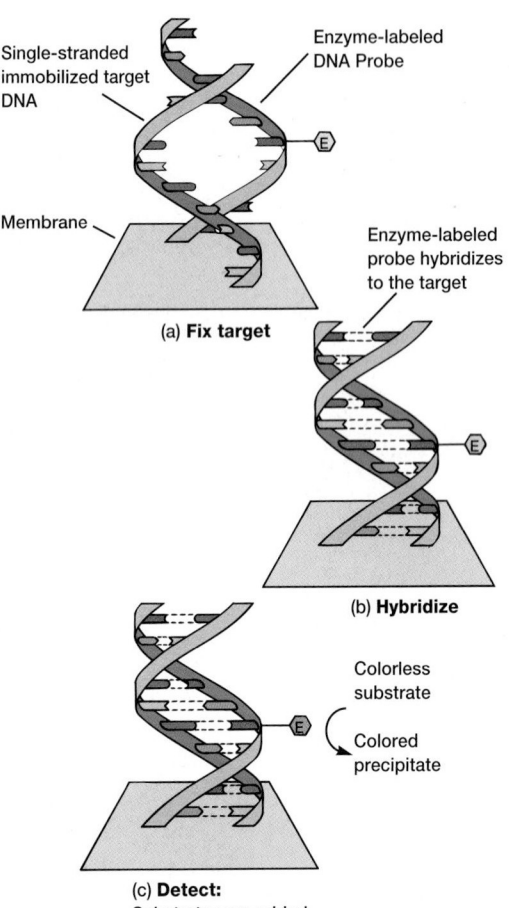

(a) **Fix target**

(b) **Hybridize**

Colorless substrate

Colored precipitate

(c) **Detect:** Substrates are added

Figure 36.9 Basic Steps in a DNA Probe Hybridization Assay. (**a**) Single-stranded target nucleic acid is bound to a membrane. A DNA probe with attached enzyme (E) also is employed. (**b**) The probe is added to the membrane. If the probe hybridizes to the target DNA, a double-stranded DNA hybrid is formed. (**c**) A colorless substrate is added. The enzyme attached to the probe converts the substrate to a colored precipitate. This detection system is semiquantitative, in that color intensity is proportional to the quantity of hybridized target nucleic acid present.

are analyzed with the Southern blotting procedure, rRNA probes will produce different patterns with different strains. Cloned rRNA genes also can be used as probes instead of *E. coli* rRNA, and similar banding patterns result.

Gas-Liquid Chromatography

During chromatography a chemical mixture carried by a liquid or gas is separated into its individual components because of processes such as adsorption, ion-exchange, and partitioning between different solvent phases. In gas-liquid chromatography (GLC), specific microbial metabolites, cellular fatty acids, and products from the pyrolysis (a chemical change caused by heat) of whole bacterial cells are analyzed and identified. These com-

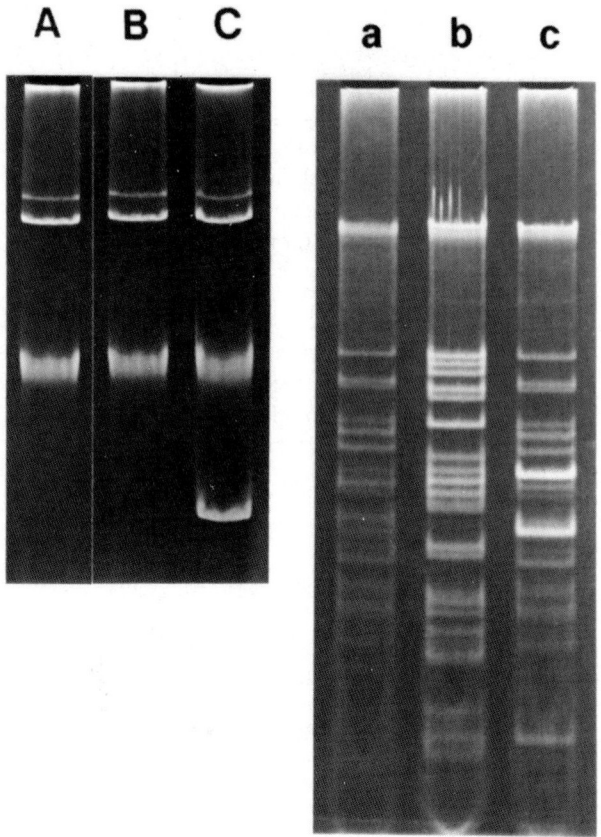

Figure 36.10   Plasmid Fingerprinting.   Agarose gel electrophoresis of plasmid DNA.

pounds are easily removed from growth media by extraction with an organic solvent such as ether. The ether extract is then injected into the GLC system. Both volatile and nonvolatile acids can be identified. Based on the pattern of fatty acid production, common bacteria isolated from clinical specimens can be identified.

The reliability, precision, and accuracy of GLC have been improved significantly with continued advances in instrumentation; the introduction of instruments for high-performance liquid chromatography; and the use of mass spectrometry, nuclear magnetic resonance spectroscopy, and associated analytical techniques for the identification of components separated by the chromatographic process. These combined techniques have recently been used to discover specific chemical markers of various infectious disease agents by direct analysis of body fluids.

Plasmid Fingerprinting

As presented in section 13.2, a plasmid is an autonomously replicating extrachromosomal molecule of DNA in bacteria. **Plasmid fingerprinting** identifies microbial isolates of the same or similar strains; related strains often contain the same number of plasmids with the same molecular weights and similar phenotypes. In contrast, microbial isolates that are phenotypically distinct have

different plasmid fingerprints. Plasmid fingerprinting of many *E. coli, Salmonella, Campylobacter,* and *Pseudomonas* strains and species has demonstrated that this method often is more accurate than other phenotyping methods such as biotyping, antibiotic resistance patterns, phage typing, and serotyping.

The technique of plasmid fingerprinting involves five steps:

1. The bacterial strains are grown in broth or on agar plates.
2. The cells are harvested and lysed with a detergent.
3. The plasmid DNA is separated from the chromosomal DNA.
4. The plasmid DNA is applied to agarose gels and electrophoretically separated.
5. The gel is stained with ethidium bromide, which binds to DNA, causing it to fluoresce under UV light. The plasmid DNA bands are then located.

Because the migration rate of plasmid DNA in agarose is inversely proportional to the molecular weight, plasmids of a different size appear as distinct bands in the stained gel. The molecular weight of each plasmid species can then be determined from a plot of the distance that each species has migrated versus the log of the molecular weights of plasmid markers of known size that have been electrophoresed simultaneously in the same gel (**figure 36.10**).

 SUSCEPTIBILITY TESTING

Many clinical microbiologists believe that determining the susceptibility of a microorganism to specific antibiotics is one of the most important tests performed in the clinical microbiology laboratory. Results (figure 36.5*o*) can show the antibiotics to which a microorganism is most susceptible and the proper therapeutic dose needed to treat the infectious disease. (Dilution susceptibility tests, disk diffusion tests [Kirby-Bauer method], the Etest, and drug concentration measurements in the blood are discussed in detail in section 35.3.)

---

1. What is the basis for bacteriophage typing?
2. How can nucleic acid–based detection methods be used by the clinical microbiologist? Gas-liquid chromatography?
3. How can a suspect bacterium be plasmid fingerprinted? Why is susceptibility testing so important in clinical microbiology?

 COMPUTERS IN CLINICAL MICROBIOLOGY

Computer systems in the clinical microbiology laboratory are designed primarily to replace the handwritten mode of information acquisition and transmission. Computers improve the efficiency of the laboratory operation and increase the speed and clarity with which results can be reported to physicians. From a work-flow standpoint, the major functions involving the computer are test ordering, result entry, analysis of results, and report preparation (figure 36.1*h,i*).

Test orders may be entered into the computer from the hospital unit or laboratory. Standard order practices should include specific requests (e.g., rule out *Nocardia* and diphtheria), all pertinent patient data, and an accession number. Once the test order has been placed, the system should allow the usual work flow to proceed with the labeled clinical specimen, date, test order, and computer accession number.

After clinical results are obtained in the laboratory, they are entered into a written log and then into the computer. To meet the many needs of microbiological entry, the computer system must be rapid and flexible in its entry modes.

Printed reports of the patient's laboratory findings are the product of the computer system. Print programs also should permit flexible formatting of reports so that additional data can be generated. For example, the computer system should be able to generate cumulative reports that summarize days to weeks of an inpatient stay.

Besides reporting laboratory tests, computers manage specimen logs, reports of overdue tests, quality control statistics, antimicrobial susceptibility probabilities, hospital epidemiological data, and many other items. The computer can be interfaced with various automated instruments for rapid and accurate calculation and transfer of clinical data.

---

1. What are some different ways in which computers can be used in the clinical microbiology laboratory? What are their major functions from the standpoint of work flow?

Summary

36.1 Specimens

a. The major focus of the clinical microbiologist is to isolate and identify microorganisms from clinical specimens accurately and rapidly (**figure 36.1**). A clinical specimen represents a portion or quantity of biological material that is tested, examined, or studied to determine the presence or absence of specific microorganisms.

b. Specimens may be collected by various methods (**figure 36.2**) that include swabs, needle aspiration, intubation, catheters, and clean-catch techniques. Each method is designed to ensure that only the proper material will be sent to the clinical laboratory.

c. Immediately after collection the specimen must be properly handled and labeled. Speed in transporting the specimen to the clinical laboratory after it has been collected is of prime importance.

36.2 Identification of Microorganisms from Specimens

a. The clinical microbiology laboratory can provide preliminary or definitive identification of microorganisms based on (1) microscopic examination of specimens; (2) growth and biochemical characteristics of microorganisms isolated from cultures (**figure 36.5**); and (3) immunologic techniques that detect antibodies or microbial antigens.

b. Viruses are identified by isolation in living cells or immunologic tests. Several types of living cells are available: cell culture, embryonated hen's eggs, and experimental animals. Rickettsial disease can be diagnosed immunologically or by isolation of the organism. Chlamydiae can be demonstrated in tissue and cell scrapings with Giemsa stain, which detects the characteristic intracellular inclusion bodies. The most routinely used techniques for identification of the mycoplasmas are immunologic. Identification of fungi often can be made if a portion of the specimen is mixed with a drop of 10% Calcofluor White stain. Wet mounts of stool specimens or urine can be examined microscopically for the presence of parasites.

c. The initial identity of a bacterial organism may be suggested by (1) the source of the culture specimen; (2) its microscopic appearance; (3) its pattern of growth on selective, differential, enrichment, or characteristic media; and (4) its hemolytic, metabolic, and fermentative properties.

d. Rapid methods for microbial identification can be divided into three categories: (1) manual biochemical systems (**figure 36.7**), (2) mechanized/automated systems, and (3) immunologic systems.

e. Bacteriophage typing for bacterial identification is based on the fact that phage surface receptors bind to specific cell surface receptors. On a petri plate culture, bacteriophages cause plaques on lawns of bacteria with the proper receptors.

f. Various molecular methods and analyses of metabolic products also can be used to identify microorganisms. Examples include nucleic acid-based detection, gas-liquid chromatography, and plasmid fingerprinting.

36.3 Susceptibility Testing

a. After the microorganism has been isolated, cultured, and/or identified, samples are used in susceptibility tests to find which method of control will be most effective. The results are provided to the physician as quickly as possible.

36.4 Computers in Clinical Microbiology

a. Computer systems in clinical microbiology are designed to replace handwritten information exchange and to speed data evaluation and report preparation.

Key Terms

Questions for Thought and Review

1. How can clinical specimens be taken from a patient with various infectious diseases? Give specific examples of procedures used.
2. What precaution must be observed when a culture is obtained from the respiratory system?
3. Why is gas-liquid chromatography a useful approach to the identification of anaerobes?
4. How does a clinical microbiologist convert an API 20E test result to a numerical code for bacterial identification?
5. How is a dichotomous key used in bacterial identification?
6. What are some different ways in which biochemical reactions can be used to identify microorganisms?
7. What are some advantages of automation in the clinical microbiology laboratory?
8. When should laboratory animals be used in the identification of microorganisms?
9. Why is plasmid fingerprinting such an accurate method for the identification of microorganisms?

Critical Thinking Questions

1. As more new ways of identifying the characteristics of microorganisms emerge, the number of distinguishable microbial strains also seems to increase. Why do you think this occurs?
2. Why are miniaturized identification systems used in clinical microbiology? Describe one such system and its advantage over classic dichotomous keys.

For recommended readings on these and related topics, see Appendix V.

The Epidemiology of Infectious Disease

Concepts

- The science of epidemiology deals with the occurrence and distribution of disease within a given population. Infectious disease epidemiology is concerned with organisms or agents responsible for the spread of infectious diseases in human and other animal populations.

- Because numbers and time are major epidemiological parameters, statistics is an important working tool in this discipline. Statistics are used to determine morbidity, frequency, and mortality rates.

- To trace the origin and manner of spread of an infectious disease outbreak, it is necessary to learn what pathogen is responsible.

- Epidemiologists investigate five links in the infectious disease cycle: (1) characteristics of the pathogen, (2) source and/or reservoir of the pathogen, (3) mode of transmission, (4) susceptibility of the host, and (5) exit mechanisms.

- Emerging and reemerging diseases and pathogens are a major global concern, as is the threat of bioterrorism.

- Global travel requires global health considerations.

- The control of nosocomial (hospital acquired) infections has received increasing attention in recent years because of the number of individuals involved, increasing costs, and the length of hospital stays.

Epidemics of infectious disease are often compared with forest fires. Once fire has spread through an area, it does not return until new trees have grown up. Epidemics in humans develop when a large population of susceptible individuals is present. If most individuals are immune, then an epidemic will not occur.

—Andrew Cliff and Peter Haggett

This laboratory worker at the Centers for Disease Control and Prevention (CDC) is in the highest level of isolation (Level 4) to avoid contact with microorganisms and to prevent their escape into the environment.

his chapter describes the epidemiological parameters that are studied in the infectious disease cycle. The practical goal of epidemiology is to establish effective recognition, control, prevention, and eradication measures within a given population. Because emerging and reemerging diseases and pathogens as well as the threat of bioterrorism are worldwide concerns, these topics are covered here. Global travel requires global health considerations that are continually monitored by epidemiologists. Nosocomial (hospital) acquired infections have increased in recent years, and a brief synopsis of their epidemiology also is presented.

The science of epidemiology originated and evolved in response to the great epidemic diseases such as cholera, typhoid fever, smallpox, and yellow fever (**Historical Highlights 37.1**). Today its scope encompasses all diseases: infectious diseases and those resulting from anatomical deformities, genetic abnormalities, metabolic dysfunction, malnutrition, neoplasms, psychiatric disorders, and aging. This chapter emphasizes only infectious disease epidemiology.

By definition, **epidemiology** [Greek *epi,* upon, and *demos,* people or population, and *logy,* study] is the science that evaluates the occurrence, determinants, distribution, and control of health and disease in a defined human population. **Health** is the condition in which the organism (and all of its parts) performs its vital functions normally or properly. It is a state of physical and mental well-being and not merely the absence of disease. A **disease** [French *des,* from, and *aise,* ease] is an impairment of the normal state of an organism or any of its components that hinders the performance of vital functions. It is a response to environmental factors (e.g., malnutrition, industrial hazards, climate), specific infective agents (e.g., viruses, bacteria, fungi, protozoa, helminths), inherent defects of the body (e.g., various genetic or immunologic anomalies), or combinations of these.

Any individual who practices epidemiology is an **epidemiologist.** Epidemiologists are, in effect, disease detectives. Their major concerns are the discovery of the factors essential to disease occurrence and the development of methods for disease prevention.

37.1 EPIDEMIOLOGICAL TERMINOLOGY

When a disease occurs occasionally, and at irregular intervals in a human population, it is a **sporadic disease** (e.g., typhoid fever). When it maintains a steady, low-level frequency at a moderately regular interval, it is an **endemic** [Greek *endemos,* dwelling in the same people] **disease** (e.g., the common cold). **Hyperendemic diseases** gradually increase in occurrence frequency beyond the endemic level but not to the epidemic level (e.g., the common cold during winter months). An **epidemic** [Greek *epidemios,* upon the people] is a sudden increase in the occurrence of a disease above the expected level (**figure 37.1**). Influenza is an

Historical Highlights

37.1   John Snow—The First Epidemiologist

Much of what we know today about the epidemiology of cholera is based on the classic studies conducted by the British physician John Snow between 1849 and 1854. During this period a series of cholera outbreaks occurred in London, England, and Snow set out to find the source of the disease. Some years earlier when he was still a medical apprentice, Snow had been sent to help during an outbreak of cholera among coal miners. His observations convinced him that the disease was usually spread by unwashed hands and shared food, not by "bad" air or casual direct contact.

Thus when the outbreak of 1849 occurred, Snow believed that cholera was spread among the poor in the same way as among the coal miners. He suspected that water, and not unwashed hands and shared food, was the source of the cholera infection among the wealthier residents. Snow examined official death records and discovered that most of the victims in the Broad Street area had lived close to the Broad Street pump or had been in the habit of drinking from it. He concluded that cholera was spread by drinking water from the Broad Street pump, which was contaminated with raw sewage containing the disease agent. When the pump handle was removed, the number of cholera cases dropped dramatically.

In 1854 another cholera outbreak struck London. Part of the city's water supply came from two different suppliers: the Southwark and Vauxhall Company and the Lambeth Company. Snow interviewed cholera patients and found that most of them purchased their drinking water from the Southwark and Vauxhall Company. He also discovered that this company obtained its water from the Thames River below locations where Londoners had discharged their sewage. In contrast, the Lambeth Company took its water from the Thames before the river reached the city. The death rate from cholera was over eightfold lower in households supplied with Lambeth Company water. Water contaminated by sewage was transmitting the disease. Finally, Snow concluded that the cause of the disease must be able to multiply in water. Thus he nearly recognized that cholera was caused by a microorganism, though Robert Koch didn't discover the causative bacterium (*Vibrio cholerae*) until 1883.

To commemorate these achievements, the John Snow Pub now stands at the site of the old Broad Street pump. Those who complete the Epidemiologic Intelligence Program at the Centers for Disease Control and Prevention receive an emblem bearing a replica of a barrel of Whatney's Ale—the brew dispensed at the John Snow Pub.

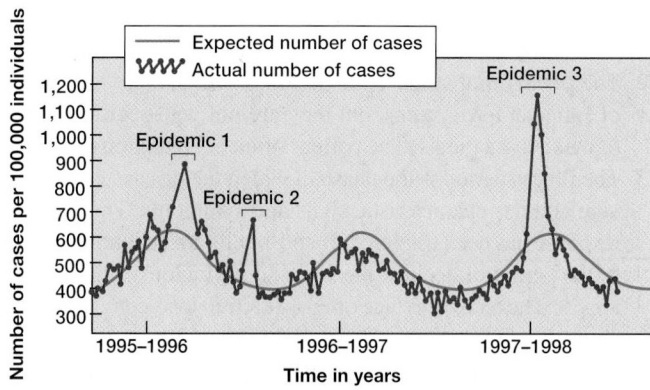

Figure 37.1 A Graph Illustrating Three Epidemics. The solid blue line indicates the expected number of endemic cases. The connected red dots indicate the actual number of cases. Epidemics (marked by brackets) are sharp increases in the number of cases of a disease above that which is normally expected (solid line).

example of a disease that often achieves epidemic status. The first case in an epidemic is called the **index case.** An **outbreak,** on the other hand, is the sudden, unexpected occurrence of a disease, usually focally or in a limited segment of a population (e.g., Legionnaires' disease). Although the epidemiology of an outbreak may be no different from that of an epidemic, the community regards the outbreak as less serious. A **pandemic** [Greek *pan,* all] is an increase in disease occurrence within a large population over a very wide region (usually the world). Usually, pandemic diseases spread among continents. The influenza outbreak of the 1960s and AIDS in the 1980s are good examples.

The discipline that deals with the factors that influence the frequency of a disease in an animal population is known as **epizootiology.** Moderate prevalence of a disease in animals is termed **enzootic,** a sudden outbreak is **epizootic,** and wide dissemination is **panzootic.** Animal diseases that can be transmitted to humans are termed **zoonoses** [Greek *zoon,* animal, and *nosos,* disease].

 37.2 MEASURING FREQUENCY: THE EPIDEMIOLOGIST'S TOOLS

An important tool used in the study of modern epidemiology is statistics. **Statistics** is the branch of mathematics dealing with the collection, organization, and interpretation of numerical data. As a science particularly concerned with rates and the comparison of rates, epidemiology was the first medical field in which statistical methods were extensively used.

Measures of frequency usually are expressed as fractions. The numerator is the number of individuals experiencing the event—infection or other problem—and the denominator is the number of individuals in whom the event could have occurred, that is, the population at risk. The fraction is a proportion or ratio but is commonly called a rate because a time period is always specified. (A rate also can be expressed as a percentage.) In population statistics, rates

usually are stated per 1,000 individuals, although other powers of 10 may be used for particular diseases (e.g., per 100 for very common diseases and per 10,000 or 100,000 for uncommon diseases).

A **morbidity rate** measures the number of individuals that become ill due to a specific disease within a susceptible population during a specific time interval. It is an incidence rate and reflects the number of new cases in a period. The rate is commonly determined when the number of new cases of illness in the general population is known from clinical reports. It is calculated as follows:

$$\text{Morbidity rate} = \frac{\substack{\text{number of new cases of a disease} \\ \text{during a specific period}}}{\text{number of individuals in the population}}$$

For example, if there were 700 new cases of influenza per 100,000 individuals, then the morbidity rate would be expressed as 700 per 100,000 or 0.7%.

The **prevalence rate** refers to the total number of individuals infected in a population at any one time no matter when the disease began. The prevalence rate depends on both the incidence rate and the duration of the illness.

The **mortality rate** is the relationship of the number of deaths from a given disease to the total number of cases of the disease. The mortality rate is a simple statement of the proportion of all deaths that are assigned to a single cause. It is calculated as follows:

$$\text{Mortality rate} = \frac{\substack{\text{number of deaths due} \\ \text{to a given disease}}}{\substack{\text{size of the total population} \\ \text{with the same disease}}}$$

For example, if there were 15,000 deaths due to AIDS in a year, and the total number of people infected was 30,000, the mortality rate would be 15,000 per 30,000 or 1 per 2 or 50%.

The determination of morbidity, prevalence, and mortality rates aids public health personnel in directing health-care efforts to control the spread of infectious diseases. For example, a sudden increase in the morbidity rate of a particular disease may indicate a need for the implementation of preventive measures designed to reduce mortality.

---

1. What is epidemiology?
2. How would you define a disease? Health?
3. What terms are used to describe the occurrence of a disease in a human population? In an animal population?
4. Define morbidity rate, prevalence rate, and mortality rate.

 37.3 INFECTIOUS DISEASE EPIDEMIOLOGY

An infectious disease is a disease resulting from an infection by microbial agents such as viruses, bacteria, fungi, protozoa, and helminths that can be transmitted from one host to another. The manifestations of an infectious disease can range from mild to severe to deadly depending on the agent and host. An epidemiologist studying an infectious disease is concerned with the causative agent, the source and/or reservoir of the disease agent (p. 828), how it was transmitted, what host and environmental factors could have aided development of the disease within a defined population, and how best to control or eliminate the disease. These factors describe the natural history or cycle of an infectious disease.

37.4 RECOGNITION OF AN INFECTIOUS DISEASE IN A POPULATION

Epidemiologists can recognize an infectious disease in a population by using various surveillance methods. Surveillance is a dynamic activity that includes gathering information on the development and occurrence of a disease, collating and analyzing the data, summarizing the findings, and using the information to select control methods. Some combination of these surveillance methods is used most often:

1. Generation of morbidity data from case reports
2. Collection of mortality data from death certificates
3. Investigation of actual cases
4. Collection of data from reported epidemics
5. Field investigation of epidemics
6. Review of laboratory results: surveys of a population for antibodies against the agent and specific microbial serotypes, skin tests, cultures, stool analyses, etc.
7. Population surveys using valid statistical sampling to determine who has the disease
8. Use of animal and vector disease data
9. Collection of information on the use of specific biologics—antibiotics, antitoxins, vaccines, and other prophylactic measures
10. Use of demographic data on population characteristics such as human movements during a specific time of the year
11. Use of remote sensing and geographic information systems

As noted, surveillance may not always require the direct examination of cases. However, to accurately interpret surveillance data and study the course of a disease in individuals, epidemiologists and other medical professionals must be aware of the pattern of infectious diseases. Often infectious diseases have characteristic signs and symptoms. **Signs** are objective changes in the body, such as a fever or rash, that can be directly observed. **Symptoms** are subjective changes, such as pain and loss of appetite, that are personally experienced by the patient. The term symptom is often used in a broader scope to include the clinical signs. A **disease syndrome** is a set of signs and symptoms that are characteristic of the disease. Frequently additional laboratory tests are required for an accurate diagnosis because symptoms and readily observable signs are not sufficient for diagnosis.

The course of an infectious disease usually has a characteristic pattern and can be divided into several phases. Knowledge of the pattern is essential in accurately diagnosing the disease.

1. The **incubation period** is the period between pathogen entry and the expression of signs and symptoms. The pathogen is spreading but has not reached a sufficient level to cause clinical manifestations. This period's length varies with disease.
2. The **prodromal stage** is the period in which there is an onset of signs and symptoms, but they are not yet specific enough to make a diagnosis. The patient often is contagious.
3. The illness period is the phase in which the disease is most severe and has characteristic signs and symptoms. The immune response has been triggered; B and T cells are becoming active.
4. In the period of decline, the signs and symptoms begin to disappear. The recovery stage often is referred to as convalescence.

Remote Sensing and Geographic Information Systems: Charting Infectious Diseases

Remote sensing and geographic information systems are map-based tools that can be used to study the distribution, dynamics, and environmental correlates of microbial diseases. **Remote sensing (RS)** is the gathering of digital images of the Earth's surface from satellites and transforming the data into maps. A **geographic information system (GIS)** is a data management system that organizes and displays digital map data from RS and facilitates the analysis of relationships between mapped features. Statistical relationships often exist between mapped features and diseases in natural host or human populations. Examples include the location of the habitats of the malaria parasite and mosquito vectors in Mexico and Asia, Rift Valley fever in Kenya, Lyme disease in the United States, and African trypanosomiasis and schistosomiasis in both humans and livestock in the southeastern United States. RS and GIS may also permit the assessment of human risk from pathogens such as Sin Nombre virus (the virus that causes hantavirus pulmonary syndrome in North America). RS and GIS are most useful if disease dynamics and distributions are clearly related to mapped environmental variables. For example, if a microbial disease is associated with certain vegetation types or physical characteristics (elevation, precipitation), RS and GIS can identify regions where risk is relatively high.

Correlation with a Single Causative Agent

After an infectious disease has been recognized in a population, epidemiologists correlate the disease outbreak with a specific organism—its exact cause must be discovered (**Historical Highlights 37.2**). At this point the clinical or diagnostic microbiology laboratory enters the investigation. Its purpose is to isolate and identify the organism responsible for the disease.

37.5 RECOGNITION OF AN EPIDEMIC

As previously noted, an infectious disease epidemic is usually a short-term increase in the occurrence of the disease in a particular population (figure 37.1). Two major types of epidemic are recognized: common source and propagated.

A **common-source epidemic** is characterized as having reached a peak level within a short period of time (1 to 2 weeks)

Historical Highlights
37.2   "Typhoid Mary"

In the early 1900s there were thousands of typhoid fever cases, and many died of the disease. Most of these cases arose when people drank water contaminated with sewage or ate food handled by or prepared by individuals who were shedding the typhoid fever bacterium *(Salmonella typhi)*. The most famous carrier of the typhoid bacterium was Mary Mallon.

Between 1896 and 1906 Mary Mallon worked as a cook in seven homes in New York City. Twenty-eight cases of typhoid fever occurred in these homes while she worked in them. As a result the New York City Health Department had Mary arrested and admitted to an isolation hospital on North Brother Island in New York's East River. Examination of

Mary's stools showed that she was shedding large numbers of typhoid bacteria though she exhibited no external symptoms of the disease. An article published in 1908 in the *Journal of the American Medical Association* referred to her as "Typhoid Mary," an epithet by which she is still known today. After being released when she pledged not to cook for others or serve food to them, Mary changed her name and began to work as a cook again. For five years she managed to avoid capture while continuing to spread typhoid fever. Eventually the authorities tracked her down. She was held in custody for 23 years until she died in 1938. As a lifetime carrier, Mary Mallon was positively linked with 10 outbreaks of typhoid fever, 53 cases, and 3 deaths.

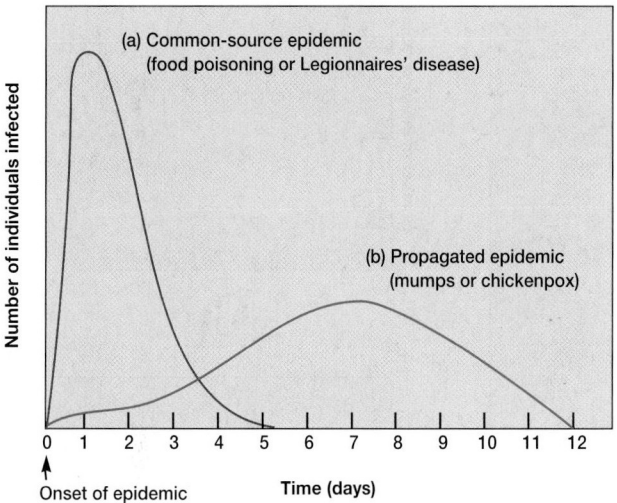

Figure 37.2   Epidemic Curves.   (**a**) In a common-source epidemic, there is a rapid increase up to a peak in the number of individuals infected and then a rapid but more gradual decline. Cases usually are reported for a period that equals approximately one incubation period of the disease. (**b**) In a propagated epidemic the curve has a gradual rise and then a gradual decline. Cases usually are reported over a time interval equivalent to several incubation periods of the disease.

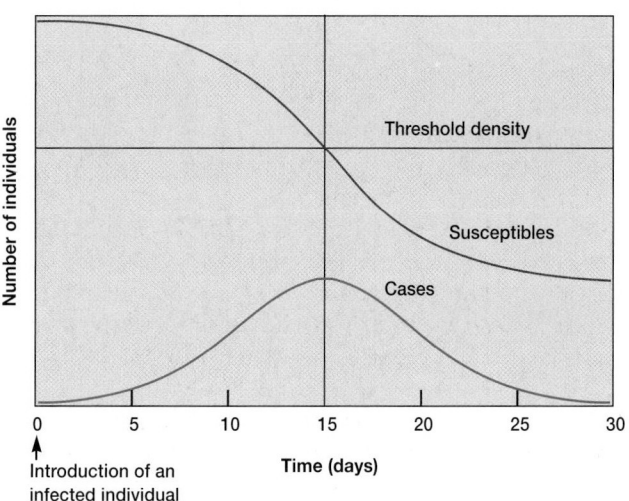

Figure 37.3   Diagrammatic Representation of the Spread of an Imaginary Propagated Epidemic.   The lower curve represents the number of cases and the upper curve the number of susceptible individuals. Notice the coincidence of the peak of the epidemic wave with the threshold density of susceptibles.

followed by a moderately rapid decline in the number of infected patients (**figure 37.2***a*). This type of epidemic usually results from a single common contaminated source such as food (food poisoning) or water (Legionnaires' disease).

A **propagated epidemic** is characterized by a relatively slow and prolonged rise and then a gradual decline in the number of individuals infected (figure 37.2*b*). This type of epidemic usually results from the introduction of a single infected individual into a susceptible population. The initial infection is then propagated to others in a gradual fashion until many individuals within the population are infected. An example is the increase in mumps or

chickenpox cases that coincides with new populations of sensitive children who arrive in classrooms each fall. Only one infected child is necessary to initiate the epidemic.

To understand how epidemics are propagated, consider **figure 37.3.** At time 0, all individuals in this population are susceptible to a hypothetical pathogen. The introduction of an infected individual initiates the epidemic outbreak (lower curve), which spreads and reaches a peak (day 15). As individuals recover from the disease, they become immune and no longer transmit the pathogen (upper curve). The number of susceptible individuals therefore decreases. The decline in the number of susceptibles to the threshold density (the minimum number of individuals necessary to continue propagating the disease) coincides with the

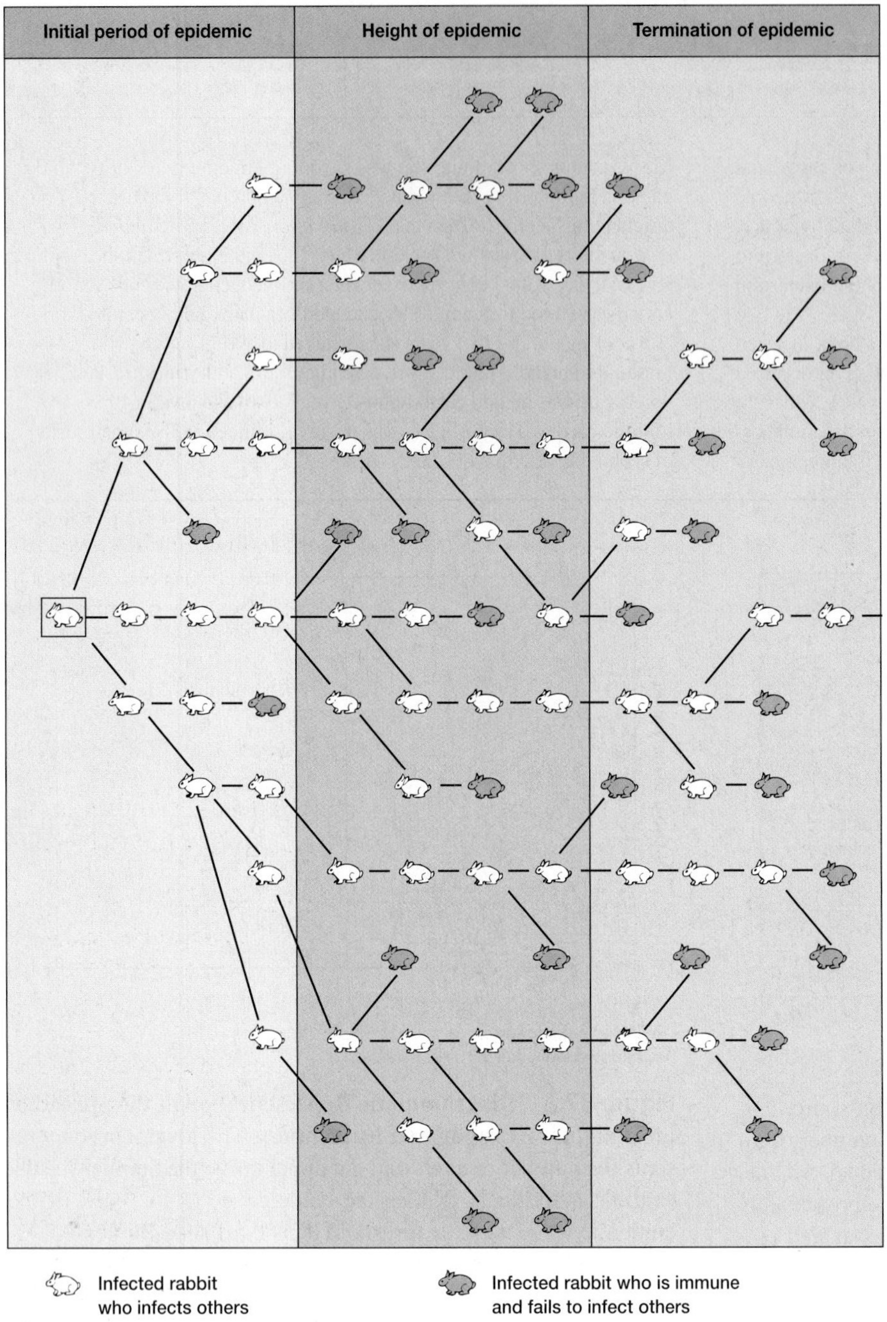

Initial period of epidemic	Height of epidemic	Termination of epidemic

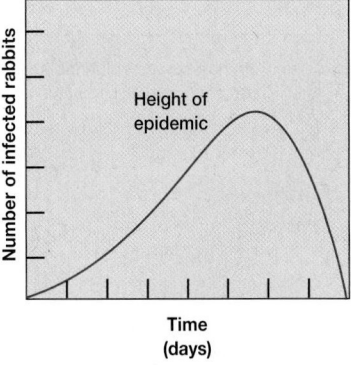

Figure 37.4 Herd Immunity. The kinetics of the spread of an infectious disease and the effect of increasing the number of immune individuals in the population in limiting the disease. On day 1, a single infected individual enters the population. The incubation period is 1 day, and recovery occurs in 2 days. The number of susceptibles is the total population on day 1. The number of infected and recovered are illustrated in the two graphs.

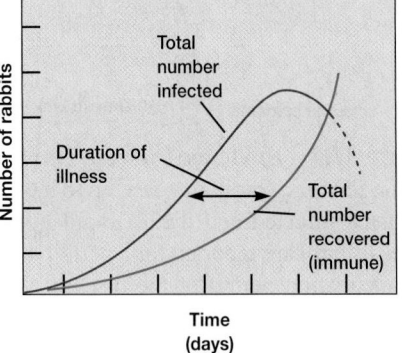

Infected rabbit who infects others

Infected rabbit who is immune and fails to infect others

peak of the epidemic wave, and the incidence of new cases declines because the pathogen cannot propagate itself.

Herd immunity is the resistance of a population to infection and pathogen spread because of the immunity of a large percentage of the population (**figure 37.4**). The larger the proportion of those immune, the smaller the probability of effective contact between infective and susceptible individuals—that is, many contacts will be with immunes, and thus the population will exhibit a group resistance. A susceptible member of such an immune population enjoys an immunity that is not of his or her own making (not self-made) but instead arises because of membership in the group.

At times public health officials immunize large portions of the susceptible population in an attempt to maintain a high level of herd immunity. Any increase in the number of susceptibles may result in an endemic disease becoming epidemic. The proportion of immune to susceptible individuals must be constantly monitored because new susceptible individuals continually enter a population through migration and birth. In addition, pathogens can change so much through processes such as antigenic shift (see next paragraph) that immune individuals become susceptible again.

Many pathogens do not ordinarily change in nature. They cause endemic diseases because infected humans continually

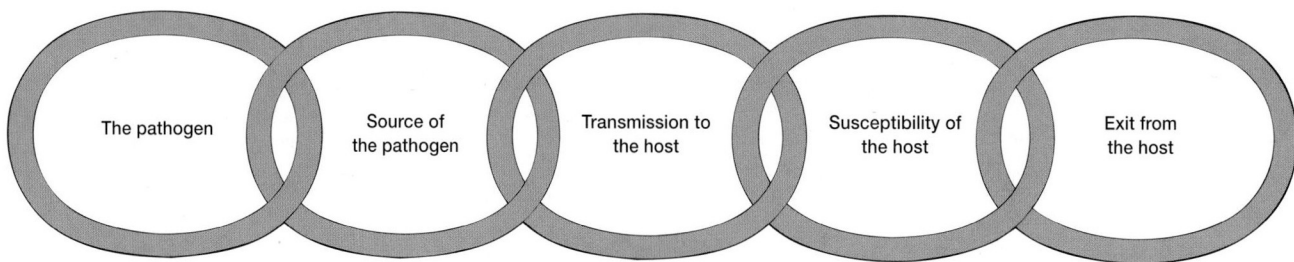

Figure 37.5   Infectious Disease Cycle or Chain of Infection.   See text for further details.

transfer them to others (e.g., sexually transmitted diseases) or because they continually reenter the human population from animal reservoirs (e.g., rabies). Other pathogens continue to evolve and may produce epidemics (e.g., AIDS, influenza virus [A strain], and *Legionella* bacteria). One way in which a pathogen changes is by **antigenic shift,** a major genetically determined change in the antigenic character of a pathogen. An antigenic shift can be so extensive that the pathogen is no longer recognized by the host's immune system. For example, influenza viruses frequently change by recombination from one antigenic type to another. Antigenic shift also occurs through the hybridization of different influenza virus serovars; two serovars of a virus intermingle to form a new antigenic type. Hybridization may occur between an animal strain and a human strain of the virus. Even though resistance in the human population becomes so high that the virus can no longer spread (herd immunity), it can be transmitted to animals, where the hybridization takes place. Smaller antigenic changes also can take place by a mutation in pathogen strains and help the pathogen avoid host immune responses. These smaller changes are called **antigenic drift.**

Whenever antigenic shift or drift occurs, the population of susceptibles increases because the immune system has not been exposed to the new mutant strain. If the percentage of susceptibles is above the threshold density (figure 37.3), the level of protection provided by herd immunity will decrease and the morbidity rate will increase. For example, the morbidity rates of diphtheria and measles among school children may reach epidemic levels if the number of susceptibles rises above 30% for the whole population. As a result the goal of public health agencies is to make sure that at least 70% of the population is immunized against these diseases to provide the herd immunity necessary for protection of those who are not immunized.

---

1. How can epidemiologists recognize an infectious disease in a population? Define sign, symptom, and disease syndrome. What are the four phases seen during the course of an infection?
2. How can remote sensing and geographic information systems chart infectious diseases?
3. Differentiate between common-source and propagated epidemics.
4. Explain herd immunity.
5. What is the significance of antigenic shift and drift in epidemiology?

37.6 THE INFECTIOUS DISEASE CYCLE: STORY OF A DISEASE

To continue to exist, a pathogen must reproduce and be disseminated among its hosts. Thus an important aspect of infectious disease epidemiology is a consideration of how reproduction and dissemination occur. The **infectious disease cycle** or **chain of infection** represents these events in the form of an intriguing epidemiological mystery story (**figure 37.5**).

What Pathogen Caused the Disease?

The first link in the infectious disease cycle is the pathogen. After an infectious disease has been recognized in a population, epidemiologists must correlate the disease outbreak with a specific pathogen. The disease's exact cause must be discovered. This is where Koch's postulates *(see section 1.3),* and modifications of them, are used to determine the etiology or cause of an infectious disease. At this point the clinical or diagnostic microbiology laboratory enters the investigation *(see section 36.1).* Its purpose is to isolate and identify the pathogen that caused the disease and to determine the pathogen's susceptibility to antimicrobial agents or methods that may assist in its eradication.

Many pathogens can cause infectious diseases in humans and will be discussed in detail in chapters 38 to 40. Often these pathogens are transmissible from one individual to another. A **communicable disease** is an illness caused by a pathogen or its products that has been transmitted from an infected person or a reservoir, either directly or indirectly. Pathogens have the potential to produce disease (pathogenicity); this potential is a function of such factors as the number of pathogens, their virulence, and the nature and magnitude of host defenses.

What Was the Source and/or Reservoir of the Pathogen?

The source and/or reservoir of a pathogen is the second link in the infectious disease cycle. Identifying the source and/or reservoir is an important aspect of epidemiology. If the source or reservoir of the infection can be eliminated or controlled, the infectious disease cycle itself will be interrupted and transmission of the pathogen will be prevented (Historical Highlights 37.1 and 37.2).

Infectious Organisms in Nonhuman Reservoirs That May Be Transmitted to Humans

Disease	Etiologic Agent	Usual or Suspected Nonhuman Host	Usual Method of Human Infection
Anthrax	*Bacillus anthracis*	Cattle, horses, sheep, swine, goats, dogs, cats, wild animals, birds	Inhalation or ingestion of spores; direct contact
Babesiosis	*Babesia bovis, B. divergens, B. microti, B. equi*	*Ixodes* ticks of various species	Bite of infected tick
Brucellosis (undulant fever)	*Brucella melitensis, B. abortus, B. suis*	Cattle, goats, swine, sheep, horses, mules, dogs, cats, fowl, deer, rabbits	Milk; direct or indirect contact
Campylobacteriosis	*Campylobacter fetus, C. jejuni*	Cattle, sheep, poultry, swine, pets, other animals	Contaminated water and food
Cat-scratch disease	*Bartonella henselae*	Cats, dogs	Cat or dog scratch
Colorado tick fever	*Coltivirus*	Squirrels, chipmunks, mice, deer	Tick bite
Cowpox	Cowpox virus	Cattle, horses	Skin abrasions
Cryptosporidiosis	*Cryptosporidium* spp.	Calves	Contact with infected calves
Encephalitis (California)	Arbovirus	Rats, squirrels, horses, deer, hares, cows	Mosquito
Encephalitis (St. Louis)	Arbovirus	Birds	Mosquito
Encephalomyelitis (Eastern equine)	Arbovirus	Birds, ducks, fowl, horses	Mosquito
Encephalomyelitis (Venezuelan equine)	Arbovirus	Rodents, horses	Mosquito
Encephalomyelitis (Western equine)	Arbovirus	Birds, snakes, squirrels, horses	Mosquito
Giardiasis	*Giardia lamblia*	Rodents, deer, cattle, dogs, cats	Contaminated water
Glanders	*Pseudomonas mallei*	Horses	Skin contact; inhalation
Hantavirus pulmonary syndrome	Pulmonary syndrome hantavirus	Deer mice	Contact with the saliva, urine, or feces of deer mice; aerosolized viruses
Herpes B viral encephalitis	*Herpesvirus simiae*	Monkeys	Monkey bite; contact with material from monkeys
Leptospirosis	*Leptospira interrogans*	Dogs, rodents, wild animals	Direct contact with urine, infected tissue, and contaminated water
Listeriosis	*Listeria monocytogenes*	Sheep, cattle, goats, guinea pigs, chickens, horses, rodents, birds, crustaceans	Food-borne
Lyme disease	*Borrelia burgdorferi*	Ticks (*Ixodes scapularis*) or related ticks	Bite of infected tick
Lymphocytic choriomeningitis	Arenavirus	Mice, rats, dogs, monkeys, guinea pigs	Inhalation of contaminated dust; ingestion of contaminated food

A **source** is the location from which the pathogen is immediately transmitted to the host, either directly through the environment or indirectly through an intermediate agent. The source can be either animate (e.g., humans or animals) or inanimate (e.g., water, soil, or food). The **period of infectivity** is the time during which the source is infectious or is disseminating the pathogen.

The **reservoir** is the site or natural environmental location in which the pathogen is normally found living and from which infection of the host can occur. Thus a reservoir sometimes functions as a source. Reservoirs also can be animate or inanimate.

Much of the time, human hosts are the most important animate sources of the pathogen and are called carriers. A **carrier** is an infected individual who is a potential source of infection for others. Carriers play an important role in the epidemiology of disease. Four types of carriers are recognized:

1. An **active carrier** is an individual who has an overt clinical case of the disease.

2. A **convalescent carrier** is an individual who has recovered from the infectious disease but continues to harbor large numbers of the pathogen.

3. A **healthy carrier** is an individual who harbors the pathogen but is not ill.

4. An **incubatory carrier** is an individual who is incubating the pathogen in large numbers but is not yet ill.

Convalescent, healthy, and incubatory carriers may harbor the pathogen for only a brief period (hours, days, or weeks) and then are called **casual, acute,** or **transient carriers.** If they harbor the pathogen for long periods (months, years, or life), they are called **chronic carriers.**

As noted earlier, infectious diseases called zoonoses occur in animals and are occasionally transmitted to humans; thus these animals also can serve as reservoirs. Humans contract the pathogen by several mechanisms: coming into direct contact with diseased animal flesh (tularemia); drinking contaminated cow's

Table 37.1   Continued

Disease	Etiologic Agent	Usual or Suspected Nonhuman Host	Usual Method of Human Infection
Mediterranean fever (boutonneuse fever, African tick typhus)	*Rickettsia conorii*	Dogs	Tick bite
Melioidosis	*Pseudomonas pseudomallei*	Rats, mice, rabbits, dogs, cats	Arthropod vectors, water, food
Orf (contagious ecthyma)	Virus	Sheep, goats	Through skin abrasions
Pasteurellosis	*Pasteurella multocida*	Fowl, cattle, sheep, swine, goats, mice, rats, rabbits	Animal bite
Plague (bubonic)	*Yersinia pestis*	Domestic rats, many wild rodents	Flea bite
Psittacosis	*Chlamydia psittaci*	Birds	Direct contact, respiratory aerosols
Q fever	*Coxiella burnetii*	Cattle, sheep, goats	Inhalation of infected soil and dust
Rabies	Rabies virus (rhabdovirus group)	Dogs, bats, opposums, skunks, foxes, cats, cattle	Bite of rabid animal
Rat bite fever	*Spirillum minus*	Rats, mice, cats	Rat bite
	Streptobacillus moniliformis	Rats, mice, squirrels, weasels, turkeys, contaminated food	Rat bite
Relapsing fever (borreliosis)	*Borrelia* spp.	Rodents, porcupines, opposums, armadillos, ticks, lice	Tick or louse bite
Rickettsialpox	*Rickettsia akari*	Mice	Mite bite
Rocky Mountain spotted fever	*Rickettsia rickettsii*	Rabbits, squirrels, rats, mice, groundhogs	Tick bite
Salmonellosis	*Salmonella* spp. (except *S. typhosa*)	Fowl, swine, sheep, cattle, horses, dogs, cats, rodents, reptiles, birds, turtles	Direct contact; food
Scrub typhus	*Rickettsia tsutsugamushi*	Wild rodents, rats	Mite bite
Tuberculosis	*Mycobacterium bovis*	Cattle, horses, cats, dogs	Milk; direct contact
Tularemia	*Francisella tularensis*	Wild rabbits, most other wild and domestic animals	Direct contact with infected carcass, usually rabbit; tick bite, biting flies
Typhus fever (endemic)	*Rickettsia mooseri*	Rats	Flea bite
Vesicular stomatitis	Virus (rhabdovirus group)	Cattle, swine, horses	Direct contact
Weil's disease (leptospirosis)	*Leptospira interrogans*	Rats, mice, skunks, opposums, wildcats, foxes, raccoons, shrews, bandicoots, dogs, cattle, swine	Through skin, drinking water, eating food
Yellow fever (jungle)	Yellow fever virus	Monkeys, marmosets, lemurs, mosquitoes	Mosquito

Modified from Guy Youmans, et al., *The Biologic and Clinical Basis of Infectious Diseases.* Copyright © 1985 W.B. Saunders, Philadelphia, PA. Reprinted by permission.

milk (tuberculosis and brucellosis); inhaling dust particles contaminated by animal excreta or products (Q fever, anthrax); or eating insufficiently cooked infected flesh (anthrax, trichinosis). In addition, being bitten by arthropod **vectors** (organisms that spread disease from one host to another) such as mosquitoes, ticks, fleas, mites, or biting flies (equine encephalomyelitis and malaria, Lyme disease, Rocky Mountain spotted fever, plague, scrub typhus, and tularemia); or being bitten by a diseased animal (rabies) can lead to infection.

Table 37.1 lists some more common zoonoses found in the Western Hemisphere. This table is noninclusive in scope; it merely abbreviates the enormous spectrum of zoonotic diseases that are relevant to human epidemiology. Domestic animals are the most common source of zoonoses because they live in greater proximity to humans than do wild animals. Diseases of wild animals that are transmitted to humans tend to occur sporadically because close contact is infrequent. Other major reservoirs of pathogens are water, soil, and food. These reservoirs are discussed in detail in chapters 29, 30, and 41.

How Was the Pathogen Transmitted?

To maintain an active infectious disease in a human population, the pathogen must be transmitted from one host or source to another. Transmission is the third link in the infectious disease cycle and occurs by four main routes: airborne, contact, vehicle, and vector-borne.

Airborne Transmission

Because air is not a suitable medium for the growth of a pathogen, any pathogen that is airborne must have originated from a source such as humans, other animals, plants, soil, food, or water. In **airborne transmission** the pathogen is truly suspended in the air and travels over a meter or more from the source to the host. The pathogen can be contained within droplet nuclei or dust. **Droplet nuclei** are small particles, 1 to 4 μm in diameter, that result from the evaporation of larger particles (10 μm or more in diameter) called droplets. Droplet nuclei can remain airborne for hours or days and travel long distances.

Figure 37.6   A Sneeze.   High-speed photograph of an aerosol generated by an unstifled sneeze. The particles seen are comprised of saliva and mucus laden with microorganisms. These airborne particles may be infectious when inhaled by a susceptible host. Even a surgical mask will not prevent the spread of all particles.

Table 37.2
Some Airborne Pathogens and the Diseases They Cause in Humans

Microorganism	Disease	Microorganism	Disease
Viruses		**Bacteria**	
Varicella	Chickenpox	*Actinomyces* spp.	Lung infections
Influenza	Flu	*Bordetella pertussis*	Whooping cough
Rubeola	Measles	*Chlamydia psittaci*	Psittacosis
Rubella	German measles	*Corynebacterium diphtheriae*	Diphtheria
Mumps	Mumps	*Mycoplasma pneumoniae*	Pneumonia
Poliomyelitis	Polio	*Mycobacterium tuberculosis*	Tuberculosis
Acute respiratory viruses	Viral pneumonia	*Neisseria meningitidis*	Meningitis
Pulmonary syndrome hantavirus	Hantavirus pulmonary syndrome	*Streptococcus* spp.	Pneumonia, sore throat
		Fungi	
		Blastomyces spp.	Lung infections
		Coccidioides spp.	Coccidioidomycosis
		Histoplasma capsulatum	Histoplasmosis

When animals or humans are the source of the airborne pathogen, it usually is propelled from the respiratory tract into the air by an individual's coughing, sneezing, or vocalization. For example, enormous numbers of moisture droplets are aerosoled during a typical sneeze (**figure 37.6**). Each droplet is about 10 μm in diameter and initially moves about 100 m/second or more than 200 mi/hour!

Dust also is an important route of airborne transmission. At times a pathogen adheres to dust particles and contributes to the number of airborne pathogens when the dust is resuspended by some disturbance. A pathogen that can survive for relatively long periods in or on dust creates an epidemiological problem, particularly in hospitals, where dust can be the source of hospital-acquired infections. **Table 37.2** summarizes some human airborne pathogens and the diseases they cause.

Contact Transmission

Contact transmission implies the coming together or touching of the source or reservoir of the pathogen and the host (**Historical Highlights 37.3**). Contact can be direct, indirect, or by droplet spread. Direct contact implies an actual physical interaction with the infectious source. This route is frequently called person-to-person contact. Person-to-person transmission occurs primarily by touching, kissing, or sexual contact (sexually transmitted diseases); by contact with oral secretions or body lesions (herpes and boils); by nursing mothers (staphylococcal infections); and through the placenta (AIDS, syphilis). Some infectious pathogens also can be transmitted by direct contact with animals or animal products (*Salmonella* and *Campylobacter*).

Indirect contact refers to the transmission of the pathogen from the source to the host through an intermediary—most often an inanimate object. The intermediary is usually contaminated by an animate source. Common examples of intermediary inanimate objects include thermometers, eating utensils, drinking cups, and bedding. *Pseudomonas* bacteria are easily transmitted by this route. This mode of transmission is often also considered a form of vehicle transmission (see next section).

In droplet spread the pathogen is carried on particles larger than 5 μm. The route is through the air but only for a very short distance—usually less than a meter. Because these particles are

Historical Highlights
37.3 The First Indications of Person-to-Person Spread of an Infectious Disease

In 1773 Charles White, an English surgeon and obstetrician, published his "Treatise on the Management of Pregnant and Lying-In Women." In it, he appealed for surgical cleanliness to combat childbed or puerperal fever. (**Puerperal fever** is an acute febrile condition that can follow childbirth and is caused by streptococcal infection of the uterus and/or adjacent regions.) In 1795 Alexander Gordon, a Scottish obstetrician, published his "Treatise on the Epidemic Puerperal Fever of Aberdeen," which demonstrated for the first time the contagiousness of the disease. In 1843 Oliver Wendell Holmes, a noted physician and anatomist in the United States, published a paper entitled "On the Contagiousness of Puerperal Fever" and also appealed for surgical cleanliness to combat this disease.

However, the first person to realize that a pathogen could be transmitted from one person to another was the Hungarian physician Ignaz Phillip Semmelweis. Between 1847 and 1849 Semmelweis observed that women who had their babies at the hospital with the help of medical students and physicians were four times as likely to contract puerperal fever as those who gave birth with the help of midwives. He concluded that the physicians and students were infecting women with material remaining on their hands after autopsies and other activities. Semmelweis thus began washing his hands with a calcium chloride solution before examining patients or delivering babies. This simple procedure led to a dramatic decrease in the number of cases of puerperal fever and saved the lives of many women. As a result Semmelweis is credited with being the pioneer of antisepsis in obstetrics. Unfortunately, in his own time, most of the medical establishment refused to acknowledge his contribution and adopt his procedures. After years of rejection Semmelweis had a nervous breakdown in 1865. He died a short time later of a wound infection. It is very probable that it was a streptococcal infection, arising from the same pathogen he had struggled against his whole professional life.

---

large, they quickly settle out of the air. As a result droplet transmission of a pathogen depends on the proximity of the source and the host. Measles is an example of a droplet-spread disease.

Vehicle Transmission

Inanimate materials or objects involved in pathogen transmission are called **vehicles**. In **common vehicle transmission** a single inanimate vehicle or source serves to spread the pathogen to multiple hosts but does not support its reproduction. Examples include surgical instruments, bedding, and eating utensils. In epidemiology these common vehicles are called **fomites** [s., fomes or fomite]. A single source containing pathogens (blood, drugs, IV fluids) can contaminate a common vehicle that causes multiple infections. Food and water are important common vehicles for many human diseases (*see tables 39.5 and 41.6*).

Vector-Borne Transmission

As noted earlier, living transmitters of a pathogen are called **vectors**. Most vectors are arthropods (insects, ticks, mites, fleas) or vertebrates (dogs, cats, skunks, bats). **Vector-borne transmission** can be either external or internal. In external (mechanical) transmission the pathogen is carried on the body surface of a vector. Carriage is passive, with no growth of the pathogen during transmission. An example would be flies carrying *Shigella* organisms on their feet from a fecal source to a plate of food that a person is eating.

In internal transmission the pathogen is carried within the vector. Here it can go into either a harborage or biologic transmission phase. In **harborage transmission** the pathogen does not undergo morphological or physiological changes within the vector. An example would be the transmission of *Yersinia pestis* (the etiologic agent of plague) by the rat flea from rat to human. **Biologic transmission** implies that the pathogen does go through a morphological or physiological change within the vector. An example would be the developmental sequence of the malarial parasite inside its mosquito vector. Malaria (pp. 929–31)

Why Was the Host Susceptible to the Pathogen?

The fourth link in the infectious disease cycle is the host. The susceptibility of the host to a pathogen depends on both the pathogenicity of the organism and the nonspecific and specific defense mechanisms of the host. These susceptibility factors are the basis for chapters 31 and 32 and are outside the realm of epidemiology.

How Did the Pathogen Leave the Host?

The fifth and last link in the infectious disease cycle is release or exit of the pathogen from the host. It is equally important that the pathogen escapes from its host as it is that the pathogen originally contacts and enters the host. Unless a successful escape occurs, the disease cycle will be interrupted and the pathogenic species will not be perpetuated. Escape can be active or passive, although often a combination of the two occurs. Active escape takes place when a pathogen actively moves to a portal of exit and leaves the host. Examples include the many parasitic helminths that migrate through the body of their host, eventually reaching the surface and exiting. Passive escape occurs when a pathogen or its progeny leaves the host in feces, urine, droplets, saliva, or desquamated cells. Microorganisms usually employ passive escape mechanisms.

---

1. What are some epidemiologically important characteristics of a pathogen? What is a communicable disease?
2. Define source, reservoir, period of infectivity, and carrier.

3. What types of infectious disease carriers does epidemiology recognize?
4. Describe the four main types of infectious disease transmission and give examples of each.
5. Define the terms droplet nuclei, vehicle, fomite, and vector.

37.7 VIRULENCE AND THE MODE OF TRANSMISSION

There is evidence that a pathogen's virulence may be strongly influenced by its mode of transmission and ability to live outside its host. When the pathogen uses a mode of transmission such as direct contact, it cannot afford to make the host so ill that it will not be transmitted effectively. This is the case with the common cold, which is caused by rhinoviruses and several other respiratory viruses. If the virus reproduced too rapidly and damaged its host extensively, the person would be bedridden and not contact others. The efficiency of transmission would drop because rhinoviruses shed from the cold sufferer could not contact new hosts and would be inactivated by exposure. Cold sufferers must be able to move about and directly contact others. Thus virulence is low and people are not incapacitated by the common cold.

On the other hand, if a pathogen uses a mode of transmission not dependent on host health and mobility, then the person's health will not be a critical matter. The pathogen might be quite successful—that is, transmitted to many new hosts even though it kills its host relatively quickly. Host death will mean the end of any resident pathogens, but the species as a whole can spread and flourish as long as the increased transmission rate outbalances the loss due to host death. This situation may arise in several ways.

When a pathogen is transmitted by a vector, it will be benefitted by extensive reproduction and spread within the host. If pathogen levels are very high in the host, a vector such as a biting insect has a better chance of picking up the pathogen and transferring it to a new host. Indeed, pathogens transmitted by biting arthropods such as mosquitoes often are very virulent (e.g., malaria, typhus, sleeping sickness). It is important that such pathogens take good care of their vectors, and the vector generally remains healthy, at least long enough for pathogen transmission.

Virulence also is often directly correlated with a pathogen's ability to survive in the external environment. If a pathogen cannot survive well outside its host and does not use a vector, it depends on host survival and will tend to be less virulent. When a pathogen can survive for long periods outside its host, it can afford to leave the host and simply wait for a new one to come along. This seems to promote increased virulence. Host health is not critical, but extensive multiplication within the host will increase the efficiency of transmission. Good examples are tuberculosis and diphtheria. *Mycobacterium* and *Corynebacterium* survive for a long time, at least weeks to months, outside human hosts.

Human cultural patterns and behavior almost certainly also affect pathogen virulence. Waterborne pathogens such as *Vibrio*

cholerae (which causes diarrhea) are transmitted through drinking water systems. They can be virulent because immobile hosts still shed pathogens, which frequently reach the water. By this argument, establishment of uncontaminated drinking water supplies should reduce cholera virulence, and this seems to be the case. The same appears to be true of *Shigella* and bacillary dysentery. Often one of the best ways to reduce virulence may be to reduce the frequency of transmission.

37.8 EMERGING AND REEMERGING INFECTIOUS DISEASES AND PATHOGENS

Only a few decades ago, a grateful public trusted that science had triumphed over infectious diseases by building a fortress of health protection. Antibiotics, vaccines, and aggressive public health campaigns had yielded a string of victories over old enemies like whooping cough, pneumonia, polio, and smallpox. In developed countries, people were lulled into believing that microbial threats were a thing of the past. Trends in the number of deaths caused by infectious diseases in the United States from 1900 through 1982 supported this conclusion (**figure 37.7**). However, this downward trend ended in 1982 and the death rate has risen in the past 20 years. In these 20 years, the world has seen the global spread of AIDS, the resurgence of tuberculosis, and the appearance of new enemies like hantavirus pulmonary syndrome, hepatitis C and E, Ebola virus, Lyme disease, cryptosporidiosis, and the deadly *E. coli* O157:H7. In addition, during this same time period:

- A "bird flu" virus that had never before attacked humans began to kill people in Hong Kong.
- A new variant of a fatal brain disease, Creutzfeldt-Jakob disease *(see section 38.5 and table 38.6)*, was identified in the United Kingdom, apparently transmitted by beef from animals with "mad cow disease."

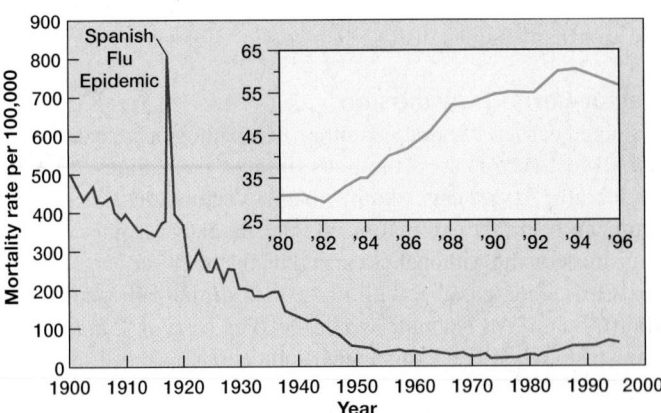

Figure 37.7 Infectious Disease Mortality in the United States Decreased Greatly during Most of the Twentieth Century. The insert is an enlargement of the right-hand portion of the graph and shows that the death rate from infectious diseases increased between 1980 and 1994.

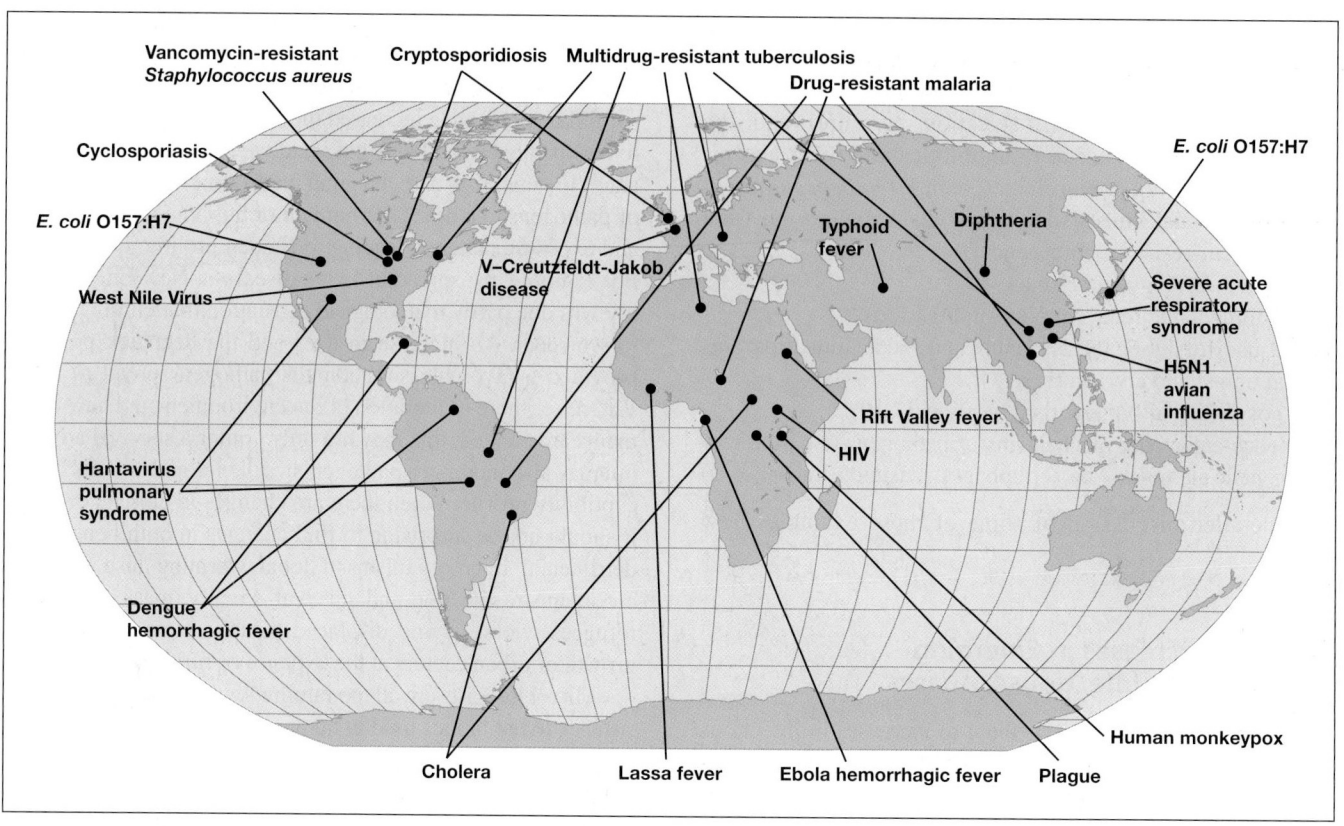

Figure 37.8   Some Examples of Emerging and Reemerging Infectious Diseases.   Although diseases such as HIV are indicated in only one or two significant locations, they are very widespread and a threat in many regions. See text for further details.

- *Staphylococcus* bacteria with increased resistance to vancomycin, long the antibiotic of first choice, were seen for the first time.
- Several major multistate foodborne outbreaks occurred in the United States, including those caused by parasites on raspberries, viruses on strawberries, and bacteria in produce, ground beef, cold cuts, and breakfast cereal.
- A new strain of tuberculosis that is resistant to many drugs, and occurs most often in people infected with HIV, arose in the city of New York and other large cities.

By the 1990s, the idea that infectious diseases no longer posed a serious threat to human health was obsolete. It is now clear that globally, humans will continually be faced with both new infectious diseases and the reemergence of older diseases once thought to be conquered (e.g., tuberculosis, dengue hemorrhagic fever, yellow fever). William H. McNeill, in *Plagues and Peoples* (1976), addresses this problem as follows: "Ingenuity, knowledge, and organization alter but cannot cancel humanity's vulnerability to invasion by parasitic forms of life. Infectious disease which antedated the emergence of humankind will last as long as humanity itself, and will surely remain, as it has been hitherto, one of the fundamental parameters and determinants of human history."

The Centers for Disease Control and Prevention (CDC) has defined these diseases as "new, reemerging, or drug-resistant infections whose incidence in humans has increased within the past two decades or whose incidence threatens to increase in the near future." Some of the most recent examples of these diseases are shown in **figure 37.8.** The increased importance of emerging and reemerging infectious diseases has stimulated the establishment of a field called **systematic epidemiology,** which focuses on the ecological and social factors that influence the development of these diseases.

After a century marked by dramatic advances in medical research and drug discovery, technology development, and sanitation, why are viruses, bacteria, fungi, and parasites posing such a problem and challenge? Many factors characteristic of the modern world undoubtedly favor the development and spread of these microorganisms and their diseases. Examples include:

1. Unprecedented worldwide population growth, population shifts (demographics), and urbanization
2. Increased international travel
3. Increased worldwide transport (commerce), migration, and relocation of animals and food products
4. Changes in food processing, handling, and agricultural practices

5. Changes in human behavior, technology, and industry
6. Human encroachment on wilderness habitats that are reservoirs for insects and animals that harbor infectious agents
7. Microbial evolution (e.g., selection pressure) and the development of resistance to antibiotics and other antimicrobial drugs (e.g., penicillin-resistant *Streptococcus pneumoniae,* methicillin-resistant *Staphylococcus aureus,* and vancomycin-resistant enterococci)
8. Changes in ecology and climate
9. Modern medicine (e.g., immunosuppression)
10. Inadequacy of public infrastructure and vaccination programs
11. Social unrest and civil wars
12. The possibility of bioterrorism (section 37.10)
13. Virulence-enhancing mechanisms of pathogens (e.g., the mobile genetic elements—bacteriophages, plasmids, transposons)

A more detailed discussion of some of these examples now follows.

Reasons for Increases in Emerging and Reemerging Infectious Diseases

As just noted, many factors contribute to increases in infectious disease. Economic or military forces that cause population shifts create fertile ground for diseases to emerge and spread. Changes in human behavior, whether in sexual practices, the use of IV drugs, or food preferences, affect microbial spread.

Food practices that include handling, cutting, refrigeration, and other treatments to maintain quality also disseminate pathogenic microorganisms and select for those that grow under these conditions. The globalization of food processing and supply centers has created the potential for widespread outbreaks of foodborne microbial diseases originating from a single batch of tainted processed foods or contaminated raw foods, mishandling of food shipments, or distribution of microbe-laden fresh produce. The CDC estimates that food-borne diseases cause approximately 76 million illnesses, 350,000 hospitalizations, and 5,000 deaths each year in the United States alone. Known pathogens are responsible for only 18% of these illnesses, and *Listeria, Salmonella,* and *Toxoplasma* are responsible for 75% of food-related deaths globally. Among the incidences in recent years are the following: (1) the 1994 outbreak of salmonellosis due to contaminated ice cream; (2) the 1996 and 1997 outbreaks of *Cyclospora* infections linked to Guatemalan raspberries; and (3) the periodic outbreaks in recent years of *E. coli* O157:H7 related to consumption of raw or undercooked beef and unpasteurized apple juice (it should be noted that this pathogen causes hemolytic uremic syndrome in children; *see section 39.4*).

As population density increases in cities, the dynamics of microbial exposure and evolution increase in humans themselves. Urbanization often crowds humans and increases exposure to microorganisms (**Disease 37.4**). Crowding leads to unsanitary conditions and hinders the effective implementation of adequate medical care, enabling more widespread transmission and propagation of pathogens. In modern societies, crowded workplaces,

community-living settings, day-care centers, large hospitals, and public transportation all facilitate microbial transmission. Furthermore, land development and the exploration and destruction of natural habitats have increased the likelihood of human exposure to new pathogens and may put selective pressures on pathogens to adapt to new hosts and changing environments. The introduction of pathogens to a new environment or host can alter transmission and exposure patterns, leading to sudden proliferation of disease. For example, the spread of Lyme disease in New England probably was due partly to ecological disruption that eliminated predators of deer. An increase in deer and the deer tick populations provided a favorable situation for pathogen spread to humans. Whenever there is alteration of the environment and new environments are created, this may not only confer a survival advantage, but may also increase a pathogen's virulence and alter its drug susceptibility profile. When there are changes in climate or ecology, it should not be surprising to find changes in both beneficial and detrimental microorganisms. Global warming also affects microorganism selection and survival. Finally, mass migrations of refugees, workers, and displaced persons have led to a steady growth of urban centers at the expense of rural areas.

Microbiologists are all too familiar with the development of resistance to antibiotics used in human medicine. The distribution of nosocomial pathogens (**figure 37.9**) has changed throughout the antibiotic era. Hospital acquired infections were dominated early on by staphylococci, which initially responded to penicillin. During subsequent years, the emergence of methicillin-resistant *Staphylococcus aureus* (MRSA) increased from 2% in 1975, to 14% in 1987, and to over 40% in 1999; similar patterns are emerging for penicillin-resistant *Streptococcus pneumoniae. S. aureus* with intermediate resistance to vancomycin began to appear in 1997 and the CDC believes that glycopeptide-resistant *S. aureus* infections are inevitable. Glycopeptide-resistant *Enterococcus faecium* was first reported in the late 1980s; in 1993 nearly 14% of enterococci-associated nosocomial infections were caused by vancomycin-resistant enterococci and that number rose to 25% in 1998. In recent years, gram-positive cocci have reemerged as significant nosocomial pathogens. Newly recognized nosocomial, gram-positive species include *Corynebacterium jeikeium* and *Rhodococcus equi.* The incidences of infections by the gram-negative pathogens *P. aeruginosa* and *Acinetobacter* have increased. With respect to the extended-spectrum β-lactamase-resistant gram-negative bacilli, bacteria such as *Klebsiella pneumoniae, Escherichia coli,* other *Klebsiella* spp., *Proteus* spp., *Morganella* spp., *Citrobacter* spp., *Salmonella* spp., and *Serratia marcescens* are resistant to penicillins; first-generation cephalosporins; and some third-generation cephalosporins such as cefotaxime (Claforan), ceftriaxone (Rocephin), ceftazidime, and aztreonam (Azactam). Newly renamed gram-negative bacterial pathogens include *Burkholderia cepacia* and *Stenotrophomonas maltophilia.*

Without a doubt, the key factors responsible for the rise in drug-resistant pathogens have been the excessive or inappropriate use of antimicrobial therapy and the sometimes indiscriminant use of broad-spectrum antibiotics. Although the CDC acknowledges that it is too late to solve the resistance problem

Disease

37.4 The SARS Epidemic of 2003

A worldwide outbreak of SARS (*s*evere *a*cute *r*espiratory *s*yndrome) began with a single ill health-care worker from the Guangdong Province of China in November 2002. Since the initial index case, the SARS virus infected more than 8,500 people and killed some 800 patients in 27 countries, including up to 115 in the United States by mid-2003. The global spread proceeded with unprecedented speed, overwhelming many hospitals and some public health systems in a matter of weeks. For the first time in its 55-year history, the World Health Organization (WHO) declared a global alert and advised against travel to mainland China and Hong Kong, Singapore, Hanoi, and Toronto. As a result, SARS rocked Asian markets, ruined the tourist trade of an entire region, nearly bankrupted vulnerable airlines, and spread panic through some of the world's largest countries. SARS is an excellent example of the threat that infective agents pose and the rapidity with which they can move around the world. Fortunately, the concerted effort of the WHO and the CDC controlled the SARS epidemic by the summer of 2003, only a few months after the alert had been sounded.

SARS is caused by a coronavirus (see **Box figure**). These are large (about 120 to 150 nm), more-or-less spherical viruses with a membranous envelope, a helical nucleocapsid, and a characteristic external layer or halo (much like a solar corona) of large spikes or peplomers. The peplomers aid in attachment and entry into host cells. SARS genetic material is single-stranded RNA (+ssRNA); it encodes five major proteins, and perhaps more. Coronaviruses cause respiratory and enteric infections in birds and many mammals; the liver and nervous system also may be involved. They are familiar to veterinary medicine because they routinely infect livestock, ducks, and other domestic animals. Several human coronaviruses cause mild upper respiratory tract illness, and around 10 to 30% of common colds are coronavirus related. Usually coronaviruses are quite host specific, but sometimes they can infect other species.

The Guangdong Province has a high population density in close association with a tremendous variety of animals, many of them used in food preparation. Indeed, two of China's largest cities—Canton and Hong Kong—are within or near the province. SARS is probably a true zoonosis. It is thought that the SARS virus developed when a coronavirus from an animal host mutated and became able to infect humans. Some evidence indicates that the original animal host was a civet cat (a member of the mongoose family), which is considered a food delicacy in southern China. The virus could then have moved to workers in the food industry, and from there to health-care professionals. Transmission through droplet nuclei seems to be the major way the SARS virus is spread, although it might be transmitted through water, sewage, or contaminated objects like doorknobs. China is a particularly good source of all sorts of new pathogen strains because of its concentrated mixture of humans and other animals.

The illness is a potentially severe form of pneumonia that begins in the same way as the flu: a fever of 100.4°F or higher (sometimes with

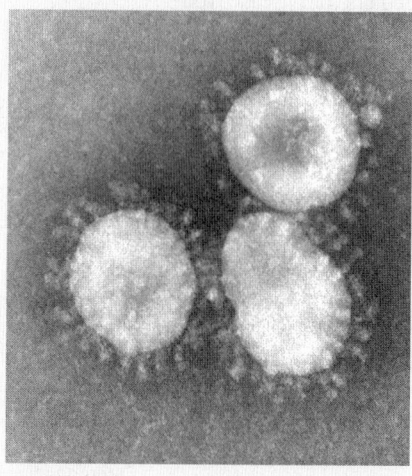

The SARS Coronavirus. Note the faint corona or halo of peplomers at the surface of these negatively stained virus particles. The CDC has proposed that this new virus be named the Urbani SARS-associated virus, after the WHO scientist Carlo Urbani, who died from SARS while investigating the epidemic.

chills), headache, body aches, and malaise. Within a week patients develop a dry cough and difficulty breathing; about 10 to 20% require a ventilator. Some also get severe diarrhea. Most patients start to recover after five or six days. An effective treatment is not known, but usually the patient is given the same antibiotics, antivirals, and steroids used to treat atypical pneumonia. Patients are isolated in specially ventilated rooms; medical workers wear masks, gowns, gloves, goggles, and vigorously wash their hands with either soap and water or alcohol-based rubs. Children and adults of all ages and every health condition have been infected. As with most other respiratory infections, the very old and patients with preexisting conditions tend to suffer the most.

The successful defense against the SARS outbreak illustrates the need for appropriate laboratory and epidemiologic responses to future infectious disease epidemics. Specifically, the rapid isolation and characterization of the SARS virus enabled the development of diagnostic tests that aided our understanding of the epidemiology of SARS and how to prevent its transmission. Effective quarantine measures were established quickly and slowed the pathogen's spread. Early recognition also made it possible to investigate antiviral compounds promptly and to begin developing vaccines. The speed with which this novel virus was detected characterized, and linked to SARS was a tribute to the power of prompt exchange of information among the WHO, CDC, and many government and private laboratories. This international collaborative approach improves understanding and control of emerging public health threats.

Figure 37.9   Nosocomial Infections.
Relative frequency by body site. These
data are from the National Nosocomial
Infections Surveillance, which is conducted by the Centers for Disease Control and Prevention (CDC).

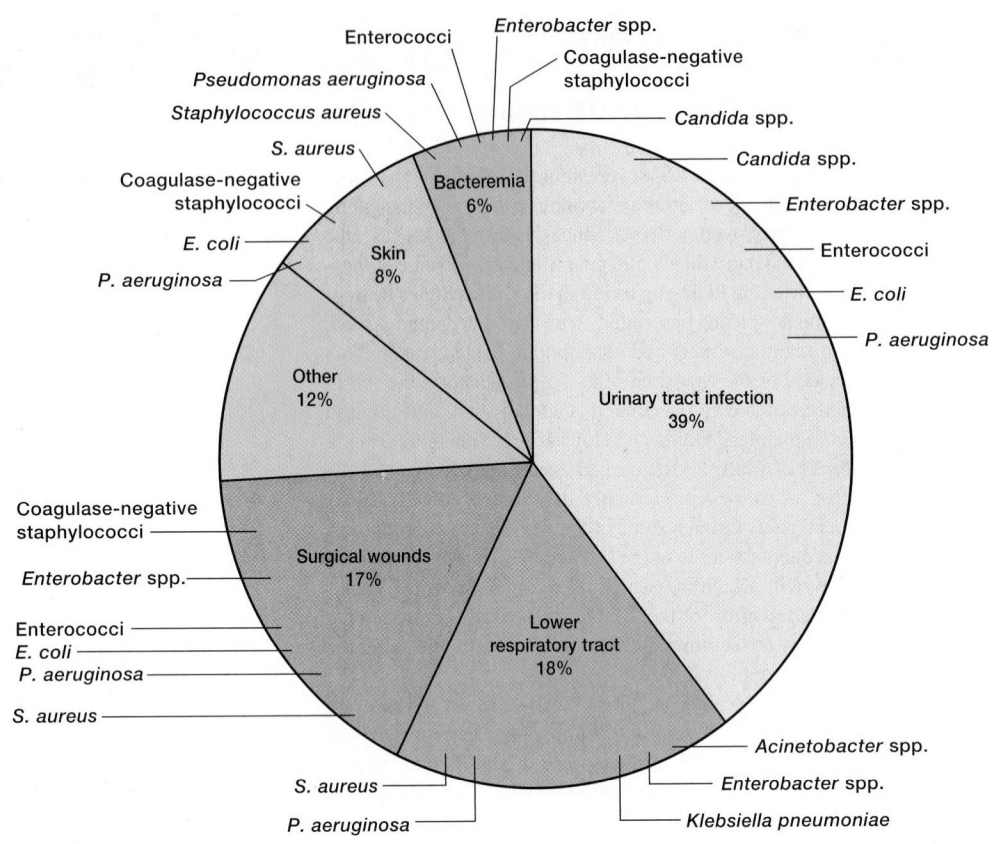

simply by using antimicrobial agents more prudently, it is no less true that the problem of drug resistance will continue to worsen if this does not occur. Also needed (especially in underdeveloped countries) is a renewed emphasis on alternate prevention and control strategies that prevailed in the years before antimicrobial chemotherapy. These include improved sanitation and hygiene, isolation of infected persons, antisepsis, and vaccination.   Drug resistance (pp. 792–94)

One of the paradoxes of modern medicine is that while we are managing and even eradicating some infectious diseases, others are emerging. Immunosuppression, either by another disease agent such as the AIDS virus or by drugs taken upon organ transplantation, increases the number of individuals susceptible to new pathogens. Immunocompromised patients represent an expanding intermediate habitat, providing the opportunity (not available in immunocompetent patients) for a pathogen to adapt to the human host and acquire the capability to circumvent the immune system of a healthy person. These immunocompromised populations can then serve as a reservoir for emerging and reemerging infections, which may eventually spread beyond the confines of institutional settings to the community at large.

In this new millennium, the speed and volume of international travel are major factors contributing to the global emergence of infectious diseases. The spread of a new disease often used to be limited by the travel time needed to reach a new host population.

If the travel time was sufficiently long, as when a ship crossed the ocean, the infected travelers would either recover or die before reaching a new population. Because travel by air has obliterated time between exposure and disease outbreak, a traveler can spread virtually any human disease in a matter of hours. Vehicles of human transport, such as aircraft and ships, also transport the infectious agents and their vectors.

It is probably best to view emerging and reemerging pathogens and their diseases as the outcome of many different factors. Because the world is now so interrelated, we cannot isolate ourselves from other countries and continents. Changes in the disease status of one part of the world may well affect the health of the remainder. As Nobel laureate Joshua Lederberg has so eloquently stated, "The microbe that felled one child in a distant continent yesterday can reach your child today and seed a global pandemic tomorrow."

1. Describe how virulence and the mode of transmission may be related. What might cause the development of new human diseases?
2. How would you define emerging or reemerging infectious diseases?
3. What are some of the factors responsible for the emergence or reemergence of pathogens?
4. What are some key factors responsible for the rise in drug-resistant bacteria?

37.9 CONTROL OF EPIDEMICS

The development of an infectious disease is a complex process involving many factors, as is the design of specific epidemiological control measures. Epidemiologists must consider available resources and time constraints, adverse effects of potential control measures, and human activities that might influence the spread of the infection. Many times control activities reflect compromises among alternatives. To proceed intelligently, one must identify components of the infectious disease cycle that are primarily responsible for a particular epidemic. Control measures should be directed toward that part of the cycle that is most susceptible to control—the weakest link in the chain.

There are three types of control measures. The first type is directed toward reducing or eliminating the source or reservoir of infection:

1. Quarantine and isolation of cases and/or carriers
2. Destruction of an animal reservoir of infection
3. Treatment of sewage to reduce water contamination
4. Therapy that reduces or eliminates infectivity of the individual

The second type of control measure is designed to break the connection between the source of the infection and susceptible individuals. Examples include general sanitation measures:

1. Chlorination of water supplies
2. Pasteurization of milk
3. Supervision and inspection of food and food handlers
4. Destruction of vectors by spraying with insecticides

The third type of control measure reduces the number of susceptible individuals and raises the general level of herd immunity by immunization. Examples include:

1. Passive immunization to give a temporary immunity following exposure to a pathogen or when a disease threatens to take an epidemic form
2. Active immunization to protect the individual from the pathogen and the host population from the epidemic

The Role of the Public Health System: Epidemiological Guardian

The control of an infectious disease relies heavily on a well-defined network of clinical microbiologists, nurses, physicians, epidemiologists, and infection control personnel who supply epidemiological information to a network of local, state, national, and international organizations. These individuals and organizations comprise the public health system. For example, each state has a public health laboratory that is involved in infection surveillance and control. The communicable disease section of a state laboratory includes specialized laboratory services for the examination of specimens or cultures submitted by physicians, the local health department, hospital laboratories, sanitarians, epidemiologists, and others. These groups share their findings with other health agencies in the state, the Centers for Disease Control and Prevention, and the World Health Organization.

 ## 37.10 THE EMERGING THREAT OF BIOTERRORISM

Today, **bioterrorism** (Greek *bios,* life, and terrorism, the systematic use of terror to demoralize, intimidate, and subjugate) is a reality. The Centers for Disease Control and Prevention recently defined bioterrorism as "the intentional or threatened use of viruses, bacteria, fungi, or toxins from living organisms to produce death or disease in humans, animals, and plants." At the end of this past century, Iraq used chemical weapons on Iran and on its own citizens, and concealed a biological weapons program. During this same time period, the hitherto-unknown Japanese cult Aum Shinrikyo used sarin nerve gas in a Tokyo subway. In October 2001, letters containing a powder with anthrax spores were sent through the U.S. mail to unsuspecting recipients. Several cases of inhalational and cutaneous anthrax were caused and some patients died *(see pp. 889–91).* Currently terrorist incidents and hoaxes involving toxic or infectious agents have been on the rise (**Historical Highlights 37.5**). Somewhere, sometime in the future, terrorists may well threaten to use, or attempt to use, a biological weapon against the United States or another country. Although the arsenal of the terrorist is potentially large, this discussion of bioterrorism focuses only on the biological threats.

Among weapons of mass destruction, biological weapons are more destructive than chemical weapons, including nerve gas. In certain circumstances, biological weapons can be as devastating as a nuclear explosion—a few kilograms of anthrax can kill as many people as a Hiroshima-size nuclear bomb. In 1998, the U.S. government launched the first national effort to create a biological weapons defense. The initiatives include (1) the first-ever procurement of specialized vaccines and medicines for a national civilian protection stockpile; (2) invigoration of research and development in the science of biodefense; (3) investment of more time and money in genome sequencing, new vaccine research, and new therapeutic research; (4) development of improved detection and diagnostic systems; and (5) preparation of clinical microbiologists and the clinical microbiology laboratory as members of the "first responder" team, which is to respond in a timely manner to acts of bioterrorism. In 2003, the U.S. government established a Department of Homeland Security to coordinate the defense of the United States against terrorist attacks.

The list of biological agents that could pose the greatest public health risk in the event of a bioterrorist attack is short and includes viruses, bacteria, parasites, and toxins (**table 37.3**). Although short, the list includes agents that, if acquired and properly disseminated, could become a difficult public health challenge in terms of limiting the numbers of casualties and controlling damage.

As noted in section 33.1, vaccination has been the single most cost-effective public health intervention that has shielded populations from a dozen serious and sometimes fatal naturally transmitted illnesses. In evaluating the role of vaccines for protecting

Historical Highlights

37.5   1346—The First Recorded Biological Warfare Attack

The Black Death *(see pp. 887–88),* which swept through Europe, the Near East, and North Africa in the mid-fourteenth century, was probably the greatest public health disaster in recorded history. Europe, for example, lost an estimated quarter to third of its population. This is not only of great historical interest but also relevant to current efforts to evaluate the threat of military or terrorist use of biological weapons.

Mark Wheelis *(see appendix V)* believes that evidence for the origin of the Black Death in Europe is found in a memoir by the Genoese Gabriele de' Mussi. According to this fourteenth-century memoir, the Black Death reached Europe from the Crimea (a region of the Ukraine) in 1346 as a result of a biological warfare attack. The Mongol army hurled plague-infected cadavers into the besieged Crimean city of Caffa (now Feodosija, Ukraine), thereby transmitting the disease to the inhabitants; fleeing survivors then spread the plague from

Caffa to the Mediterranean Basin. Such transmission was especially likely at Caffa where cadavers would have been badly mangled by being hurled, and the defenders probably often had cut or abraded hands from coping with the bombardment. Because many cadavers were involved, the opportunity for disease transmission was greatly increased. Disposal of victims' bodies in a major disease outbreak is always a problem, and the Mongol army used their hurling machines as a solution to limited mortuary facilities. It is possible that thousands of cadavers were disposed of this way; de' Mussi's description of "mountains of dead" might have been quite literally true. Indeed, Caffa could be the site of the most spectacular incident of biological warfare ever, with the Black Death as its disastrous consequence. It is a powerful reminder of the horrific consequences that can result when disease is successfully used as a weapon.

Table 37.3

Biological Agents Associated with Bioterrorism or Biocrimes

	Traditional Biological Warfare Agents	Agents of Biocrimes and Bioterrorism
Pathogens	Smallpox virus (variola)	*Ascaris suum*
	Viral encephalitides	*Bacillus anthracis*
	Viral hemorrhagic fevers	*Coxiella burnetii*
	Bacillus anthracis	*Giardia lamblia*
	Brucella suis	HIV
	Coxiella burnetii	*Rickettsia prowazekii* (typhus)
	Francisella tularensis	*Salmonella typhi*
	Yersinia pestis	*Schistosoma* spp.
		Vibrio cholerae
		Viral hemorrhagic fevers (Ebola)
		Yellow fever virus
		Yersinia enterocolitica
		Yersinia pestis (plague)
Toxins	Botulinum	Botulinum
	Ricin	Cholera endotoxin
	Staphylococcal enterotoxin B	Diphtheria toxin
		Ricin
		Snake toxin
		Tetrodotoxin
Anticrop agents	Rice blast	
	Rice stem rust	
	Wheat stem rust	

Source: Adapted from the 2000 *NATO Handbook on the Medical Aspects of NBC Defensive Operations.*

Table 37.4

Activities of the U.S. Food and Drug Administration to Counter Bioterrorism

1. Enhancing the development and licensure of new vaccines and biological therapeutics through research and review activities—anthrax vaccine and antisera to botulinum toxin, for example
2. Enhancing the timeliness of application reviews of new drugs and biological products and new and existing products
3. Participation in the planning and coordination of public health and medical responses to a terrorist attack involving biological or chemical agent(s)
4. Participation in the development of rapid detection and decontamination of bioterrorism agents such as *Clostridium botulinum* toxins, *Yersinia pestis,* and *Bacillus anthracis*
5. Ensuring the safety of regulated foods, drugs, medical devices, and biological products; arranging for seizure and disposal of affected products
6. Developing techniques for detection of genetic modifications of microorganisms to make them more toxic or more antibiotic- or vaccine-resistant
7. Rapidly determining a microorganism's sensitivity to drug therapies
8. Determining the mechanism of replication and pathogenicity or virulence of identified microorganisms, including elements that can be transferred to other organisms to circumvent detection, prevention, or treatment
9. Enhancing bioterrorism agent reporting and surveillance capabilities

potential agents, make it very difficult to use vaccines to protect a population against bioterrorism.

In the United States, the Department of Homeland Security and the U.S. Food and Drug Administration have been charged with fostering the development of vaccines, drugs and diagnostic products, food supply safeguards, and other measures needed to respond to bioterrorist threats (**table 37.4**). The expected outcomes of these activities include safe and effective products to treat or prevent toxicity of biological and chemical agents; methods to rapidly detect, identify, and decontaminate hazardous microorganisms; a greater ability to ensure the safety of the food supply; and greater capacity to provide appropriate medical care and a public health response.

civilian populations from bioterrorism, new problems arise. Despite the protective efficacy of vaccines against individual microorganisms, the very high costs and the problems involved in vaccinating large populations, along with the broad spectrum of

1. In what three general ways can epidemics be controlled? Give one or two specific examples of each type of control measure.
2. Name some of the microorganisms that can be used to commit biocrimes. From this list, which two pose the greatest risk for causing large numbers of casualties?
3. Why are biological weapons more destructive than chemical weapons?
4. What can be done to defend a population against bioterrorism?

37.11 GLOBAL TRAVEL AND HEALTH CONSIDERATIONS

From a global health perspective, developed countries such as Australia, the European countries, Israel, New Zealand, and the United States have highly effective public health systems. Approximately 25% of the over 6 billion people on our planet Earth live in these countries. As a result, of the approximately 12 million deaths in these countries per year, only about 500,000 are due to infectious diseases. Less developed areas such as Africa, Central and South America, India, the New Independent States of the former Soviet Union, and Asia have less developed public health systems and represent the other 75% of the human population. It is in these regions that infectious diseases are the major cause of death; for example, of approximately 38.5 million deaths per year, about 18 million are attributed to infectious microbial diseases.

The high incidence of infectious microbial diseases in less developed countries must be of great concern for people traveling to these destinations. Each year 1 billion passengers travel by air, and over 50 million people from developed countries visit less developed countries. Furthermore, the time required to circumnavigate the globe has decreased from 365 days to fewer than 3 days.

Several kinds of precautions can be taken by individuals to prevent travel-related infectious diseases. Examples include:

1. If one is traveling to an area where malaria is endemic, weekly prophylaxis before entering the area and after leaving the area is recommended.
2. When traveling, individuals often have a sense of anonymity; they may feel less inhibited sexually and place themselves at greater risk of acquiring sexually transmitted diseases. Travelers should recall the benefits of abstinence or protective sexual practices, especially the use of condoms. Hepatitis B vaccine should be administered if it is indicated.
3. Travelers should avoid uncooked food, nonbottled water and beverages, and unpasteurized dairy products. Use bottled water for drinking, making ice cubes, and brushing teeth.
4. Wash hands with soap and water frequently, especially before each meal.
5. To prevent respiratory infections, avoid excessive outdoor activities in areas of heavy air pollution during hot or humid parts of the day. Consider tuberculin skin testing before and after travel.
6. Minimize skin exposure and use repellents to prevent arthropod-borne illnesses (e.g., malaria, dengue, yellow fever, Japanese encephalitis).
7. Avoid skin-perforating procedures (e.g., acupuncture, body piercing, tattooing, venipuncture, sharing of razors).
8. Do not pet or feed animals, especially dogs and monkeys.
9. Avoid swimming or wading in nonchlorinated fresh water.

Vaccinations are one of the most important strategies of prophylaxis in travel medicine. A medical consultation before travel is an excellent opportunity to update routine immunizations. Selection of immunizations should be based on requirements and risk of infection at the travel destination. According to International Health Regulations, many countries require proof of yellow fever vaccination on the International Certificate of Vaccination. Additionally, a few countries still require proof of vaccination against cholera, diphtheria, and meningococcal disease. The basic CDC immunization recommendations for those traveling abroad are presented in **table 37.5.** In the not-too-distant future, travelers will be offered a variety of oral vaccines against the microorganisms causing dengue fever and travelers' diarrhea (e.g., enterotoxigenic *Escherichia coli, Campylobacter, Shigella*); vaccines against malaria and AIDS, the infections causing most deaths in travelers, are much farther in the future.

In addition, a traveler should:

1. Read carefully CDC information about the destination and follow recommendations
2. Begin the vaccination process early
3. Find a travel clinic for information and specialized immunizations
4. Plan ahead when traveling with children or if there are special needs
5. Learn about safe food and water (contaminated food and water are the major sources of stomach or intestinal illness while traveling), protection against insects, and other precautions
6. Prepare for medical emergencies and for nonmedical emergencies such as crime and natural disasters

Space Travel

Space travel entails unique circumstances that influence the health of astronauts, most notably the limitations imposed by zero-gravity environments. The impact of weightlessness on human physiology is substantial and affects various body systems: neurologic, psychologic, vestibular, cardiovascular, musculoskeletal, endocrine, hematologic, and immunologic. For example, recent studies on the immune system have documented a decrease in cell-mediated immunity. The space environment also challenges the growth of microorganisms. Increased antimicrobial resistance has been noted among *E. coli* and *Staphylococcus aureus* bacteria in space, and fungal overgrowths can be a problem because of changes in humidity and pockets of increased condensation aboard spacecraft and space stations.

The primary strategies used by NASA to prevent health-related mishaps associated with space travel include selecting exceptionally fit, healthy crew members and imposing preflight protective quarantines. Access to astronauts is restricted to family members beginning 10 days before departure and a strict 7-day pre-launch quarantine is imposed.

Table 37.5
Vaccine Recommendations for Travelers Aged 2 Years or Older

For Travelers Aged 2 Years or Older, the Following Immunizations Normally Given During Childhood (*see figure 33.1*) **Should Be Up-to-Date:**

1. Measles, mumps, and rubella (MMR) vaccine—at least 1 dose given on or after 12 months of age
2. Diphtheria, tetanus, and acellular pertussis (DtaP)—4 or 5 doses until age 7; after age 7, 1 dose of adult tetanus and diphtheria (Td) vaccine every 10 years
3. Polio vaccine—at least 3 doses
4. *Haemophilus influenzae* type b (Hib) vaccine—not recommended after 5 years of age
5. Hepatitis B vaccine—3 doses
6. Varicella vaccine (for persons who have never had chickenpox)

In Addition, Depending on the Country, Travelers Should Consider:

1. Influenza (flu) vaccine—recommended for adults 65 years or older and for other high-risk individuals
2. Pneumococcal vaccine—recommended for adults 65 years or older and for other high-risk individuals

Booster or Additional Doses:

1. A booster dose of adult tetanus-diphtheria (Td) recommended every 10 years
2. For persons who have received a complete series of polio vaccine, an additional single dose of inactivated polio vaccine should be given to persons over 18 traveling to the following areas: Africa, Asia, the Middle East, India, and the New Independent States of the former Soviet Union
3. Persons born after 1957 should have a second dose of measles vaccine before traveling abroad

The Following Immunizations Are Recommended Based on Destination:

Immunization	*Destination*
Yellow fever	Africa and South America
Hepatitis A or immune globulin (IG)	To all areas except Japan, Australia, New Zealand, northern and western Europe, and North America (excluding Mexico)
Hepatitis B (if staying 6 or more months)	Southeast Asia, Africa, the Middle East, the islands of South and Western Pacific, and the Amazon region of South America
Typhoid	In all developing countries and for those travelers spending time in areas where food and water precautions are recommended
Meningococcus	Sub-Saharan Africa
Japanese encephalitis or tick-borne encephalitis	Areas of risk
Cholera	The risk of cholera to U.S. travelers is so low that it is questionable whether cholera vaccine is of benefit

Spacing of Immunobiologics

All vaccines (except cholera and yellow fever) may be safely administered simultaneously without any decrease in effectiveness. Immune globulin (IG) may be simultaneously administered at different body locations with an inactivated vaccine such as DtaP, IPV, Hib, and hepatitis A and B vaccines. However, IG diminishes the effectiveness of live virus MMR and varicella vaccines if IG is given simultaneously. IG does not interfere with either oral polio vaccine (OPV) or yellow fever vaccine when given simultaneously.

Women who are pregnant or who are likely to become pregnant within 3 months should not receive MMR or varicella vaccines. Yellow fever vaccine or OPV should be given to pregnant women only if there is a substantial risk of exposure.

Source: *National Center for Infectious Diseases, Travelers' Health,* CDC Vaccine Recommendations (Year 2000).

---

1. Give some examples where population movements affect microbial disease transmission.
2. In addition to vaccinations, what are some additional precautions global travelers should take?
3. What are some of the health problems associated with space travel?

37.12 NOSOCOMIAL INFECTIONS

Nosocomial infections [Greek *nosos,* disease, and *komeion,* to take care of] result from pathogens that develop within a hospital or other type of clinical care facility and are acquired by patients while they are in the facility. Besides harming patients, nosocomial infections can affect nurses, physicians, aides, visitors,

salespeople, delivery personnel, custodians, and anyone who has contact with the hospital. Most nosocomial infections become clinically apparent while patients are still hospitalized; however, disease onset can occur after patients have been discharged. Infections that are incubating when patients are admitted to a hospital are not nosocomial; they are community acquired. However, because such infections can serve as a ready source or reservoir of pathogens for other patients or personnel, they are also considered in the total epidemiology of nosocomial infections.

The Centers for Disease Control and Prevention estimate that 5 to 10% of all hospital patients acquire some type of nosocomial infection. Because approximately 40 million people are admitted to hospitals annually, about 2 to 4 million people may develop an infection they did not have upon entering the hospital. Thus nosocomial infections represent a significant proportion of all infectious diseases acquired by humans.

Nosocomial diseases are usually caused by bacteria, most of which are noninvasive and part of the normal microbiota; viruses, protozoa, and fungi are rarely involved. Figure 37.9 summarizes the most common types of nosocomial infections and the most common nosocomial pathogens.

Source

The nosocomial pathogens that cause diseases come from either endogenous or exogenous sources. Endogenous sources are the patient's own microbiota; exogenous sources are microbiota other than the patient's. Endogenous pathogens are either brought into the hospital by the patient or are acquired when the patient becomes colonized after admission. In either case the pathogen colonizing the patient may subsequently cause a nosocomial disease (e.g., when the pathogen is transported to another part of the body or when the host's resistance drops). If it cannot be determined that the specific pathogen responsible for a nosocomial disease is exogenous or endogenous, then the term autogenous is used. An **autogenous infection** is one that is caused by an agent derived from the microbiota of the patient, despite whether it became part of the patient's microbiota following his or her admission to the hospital.

There are many potential exogenous sources in a hospital. Animate sources are the hospital staff, other patients, and visitors. Some examples of inanimate exogenous sources are food, computer keyboards, urinary catheters, intravenous and respiratory therapy equipment, and water systems (softeners, dialysis units, and hydrotherapy equipment).

Control, Prevention, and Surveillance

In the United States nosocomial infections prolong hospital stays by 4 to 13 days, result in over 4.5 billion dollars a year in direct hospital charges, and lead to over 20,000 direct and 60,000 indirect deaths annually. The enormity of this problem has led most hospitals to allocate substantial resources to the development of methods and programs for the surveillance, prevention, and control of nosocomial infections.

All personnel involved in the care of patients should be familiar with basic infection control measures such as isolation policies of the hospital; aseptic techniques; proper handling of equipment, supplies, food, and excreta; and surgical wound care and dressings. To adequately protect their patients, hospital personnel must practice proper aseptic technique and handwashing procedures, and must wear gloves when contacting mucous membranes, secretions, and "moist body substances." Patients should be monitored with respect to the frequency, distribution, symptomatology, and other characteristics common to nosocomial infections. A dynamic control and surveillance program can be invaluable in preventing many nosocomial infections, patient discomfort, extended stays, and further expense.

The Hospital Epidemiologist

Because of nosocomial infections, all hospitals desiring accreditation by the Joint Commission on Accreditation of Healthcare Organizations (JCAHO) must have a designated individual directly responsible for developing and implementing policies governing control of infections and communicable diseases. This individual may be a registered nurse and is known as a hospital epidemiologist, nurse epidemiologist, infection control nurse, infection control practitioner, or a clinical microbiologist/technologist. In larger hospitals, a physician is the hospital epidemiologist and should be trained in infectious diseases. He or she oversees a staff that includes nurse epidemiologists, quality assurance specialists, fellows in infectious disease/hospital epidemiology, and epidemiology technicians. The hospital epidemiologist must meet with an infection control committee composed of various professionals who have expertise in the different aspects of infection control and hospital operation. The infection control committee periodically evaluates laboratory reports, patients' charts, and surveys done by the hospital epidemiologist to determine whether there has been any increase in the frequency of particular infectious diseases or potential pathogens.

Overall, the services provided by the hospital epidemiologist should include at least the following:

1. Research in infection control
2. Evaluation of disinfectants, rapid test systems, and other products
3. Efforts to encourage appropriate legislation related to infection control, particularly at the state level
4. Efforts to contain hospital operating costs, especially those related to fixed expenses such as the DRGs (diagnosis-related groups)
5. Surveillance and comparison of endemic and epidemic infection frequencies
6. Direct participation in a variety of hospital activities relating to infection control and maintenance of employee health
7. Education of hospital personnel in communicable disease control and disinfection and sterilization procedures
8. Establishment and maintenance of a system for identifying, reporting, investigating, and controlling infections and communicable diseases of patients and hospital personnel
9. Maintenance of a log of incidents related to infections and communicable diseases
10. Monitoring trends in the antimicrobial drug resistance of infectious agents

Recently computer software packages have been developed to aid the infection control practitioner. Such packages generate standard reports, cause-and-effect tabulations, and graphics for the daily epidemiological monitoring that must be done.

---

1. Describe a nosocomial infection.
2. What two general sources are responsible for nosocomial infections? Give some specific examples of each general source.
3. Why are nosocomial infections important?
4. What does a hospital epidemiologist do to control nosocomial infections?

Summary

37.1 Epidemiological Terminology

a. Epidemiology is the science that evaluates the determinants, occurrence, distribution, and control of health and disease in a defined population.

b. Specific epidemiological terminology is used to communicate disease incidence in a given population. Frequently used terms include sporadic disease, endemic disease, hyperendemic disease, epidemic, index case, outbreak, and pandemic.

37.2 Measuring Frequency: The Epidemiologist's Tools

a. An important tool used in the study of modern epidemiology is statistics.

b. Epidemiological data can be obtained from such factors as morbidity, prevalence, and mortality rates.

37.3 Infectious Disease Epidemiology

a. An infectious disease is caused by microbial agents such as viruses, bacteria, fungi, protozoa, and helminths that can be transmitted from one host to another.

b. The manifestations of an infectious disease can range from mild to severe to deadly, depending on the agent and host.

37.4 Recognition of an Infectious Disease in a Population

a. Surveillance is necessary for recognizing a specific infectious disease within a given population. This consists of gathering data on the occurrence of the disease, collating and analyzing the data, summarizing the findings, and applying the information to control measures.

b. It is important to recognize the signs and symptoms that compose a disease syndrome. This includes the characteristic course of the disease, such aspects as the incubation period and prodromal stage.

c. Remote sensing and geographic information systems can be used to gather epidemiological data on the environment.

37.5 Recognition of an Epidemic

a. A common-source epidemic is characterized by a sharp rise to a peak and then a rapid, but not as pronounced, decline in the number of individuals infected (**figure 37.2**). A propagated epidemic is characterized by a relatively slow and prolonged rise and then a gradual decline in the number of individuals infected.

b. Herd immunity is the resistance of a population to infection and pathogen spread because of the immunity of a large percentage of the individuals within the population (**figure 37.4**).

37.6 The Infectious Disease Cycle: Story of a Disease

a. The infectious disease cycle or chain involves the characteristics of the pathogen, the source and/or reservoir of the pathogen, the transmission of the pathogen, the susceptibility of the host, the exit mechanism of the pathogen from the body of the host, and its spread to a new reservoir or host (**figure 37.5**).

b. There are four major modes of transmission: airborne, contact, vehicle, and vector-borne.

37.7 Virulence and the Mode of Transmission

a. The degree of virulence may be influenced by the pathogen's preferred mode of transmission. New human diseases may arise and spread because of ecosystem disruption, rapid transportation, human behavior, and other factors.

37.8 Emerging and Reemerging Infectious Diseases and Pathogens

a. It is now clear that globally, humans will continually be faced with both new infectious diseases and the reemergence of older diseases once thought to be conquered.

b. CDC has defined these diseases as "new, reemerging, or drug-resistant infections whose incidence in humans has increased within the past two decades or whose incidence threatens to increase in the near future" (**figure 37.8**).

c. Many factors characteristic of the modern world undoubtedly favor the development and spread of these microorganisms and their diseases.

37.9 Control of Epidemics

a. The public health system consists of individuals and organizations that function in the control of infectious diseases and epidemics.

b. Epidemiological control measures can be directed toward reducing or eliminating infection sources, breaking the connection between sources and susceptible individuals, or isolating the susceptible individuals and raising the general level of herd immunity by immunization.

37.10 The Emerging Threat of Bioterrorism

a. Today bioterrorism is a reality. Terrorist incidents and hoaxes involving toxic or infectious agents have been on the rise.

b. Among weapons of mass destruction, biological weapons are more destructive than chemical weapons. The list of biological agents that could pose the greatest public health risk in the event of a bioterrorist attack is short and includes viruses, bacteria, parasites, and toxins (**table 37.3**).

37.11 Global Travel and Health Considerations

a. Certain precautions and health considerations should be taken into consideration when traveling globally.

b. Vaccinations are one of the most important strategies of prophylaxis in travel medicine (**table 37.5**).

c. Space travel entails unique considerations that influence the health of astronauts, most notably limitations imposed by zero-gravity environments. The space environment also challenges the growth of microorganisms.

37.12 Nosocomial Infections

a. Nosocomial infections are infections acquired during hospitalization and are produced by a pathogen acquired during a patient's stay. These infections come from either endogenous or exogenous sources (**figure 37.9**).

b. Hospitals must designate an individual to be responsible for identifying and controlling nosocomial infections. This person is known as a hospital epidemiologist, nurse epidemiologist, infection control nurse, or infection control practitioner.

Key Terms

active carrier 828
acute carrier 828
airborne transmission 829
antigenic drift 827
antigenic shift 827
autogenous infection 841

biologic transmission 831
bioterrorism 837
carrier 828
casual carrier 828
chronic carrier 828
common-source epidemic 824

common vehicle transmission 831
communicable disease 827
contact transmission 830
convalescent carrier 828
disease 822
disease syndrome 824

droplet nuclei 829
endemic disease 822
enzootic 823
epidemic 822
epidemiologist 822
epidemiology 822

Questions for Thought and Review

1. Why is international cooperation a necessity in the field of epidemiology?
2. Are any risks involved in attempting to eliminate a pathogen from the world?
3. Why is a knowledge of statistics important to an epidemiologist?
4. What is the role of the Centers for Disease Control and Prevention in epidemiology?
5. What common sources of infectious disease are found in your community? How can the etiologic agents of such infectious diseases spread from their source or reservoir to members of your community?
6. How could you experimentally prove the cause of an infectious disease?
7. How do epidemiologists recognize an infectious disease within a given population?
8. How could you prove that an epidemic of a given infectious disease was occurring?
9. Why is the infectious disease cycle also referred to as a chain?
10. How can epidemics be controlled?
11. Why are nosocomial infections so important to humans?
12. Why do some epidemiologists believe it is impossible to prevent all nosocomial infections?
13. What is the difference between a fomite and a vector? A reservoir and a source?
14. How can changes in herd immunity contribute to an outbreak of a disease on an island?
15. What is an index case?
16. Where in the world would you expect new human diseases to arise? Why?
17. Contrast mortality due to infectious diseases in developing and developed countries.
18. What are some of the factors that are important in the reemergence of a potential pathogen?
19. Why should we be worried about bioterrorism?
20. What precautions should you take when traveling abroad?

Critical Thinking Questions

1. College dormitories are notorious for outbreaks of flu and other infectious diseases. This is particularly prevalent during final exam weeks. Using your knowledge of the immune response and epidemiology, suggest practices that could be adopted to minimize the risks at such a critical time of the term.
2. Space exploration presents an area of concern. What sort of precautions should have been taken to prevent contamination of the Martian terrain with Earth microorganisms when the Mars Lander touched down on the planet?

**For recommended readings on these
and related topics, see Appendix V.**

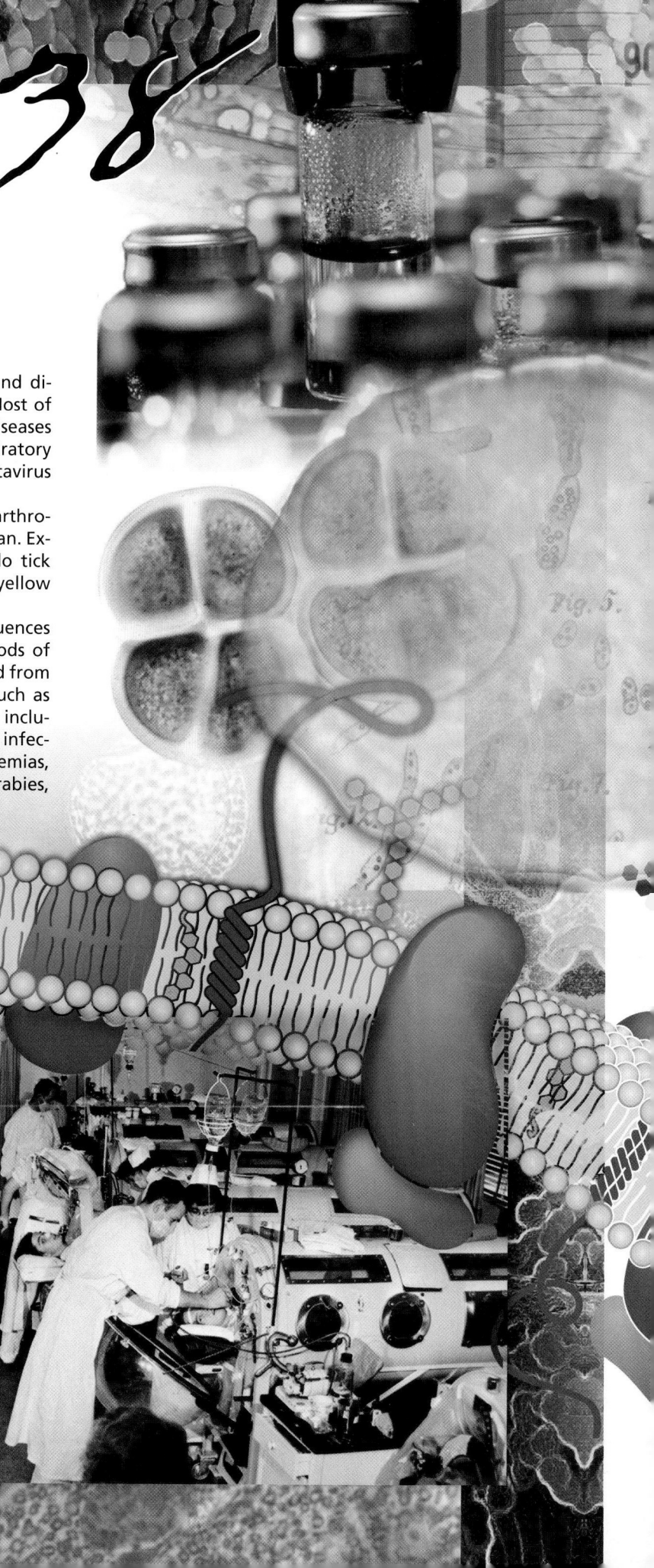

chapter 38

Human Diseases Caused by Viruses

Concepts

- Some viruses can be transmitted through the air and directly or indirectly involve the respiratory system. Most of these viruses are highly communicable and cause diseases such as chickenpox, influenza, measles, mumps, respiratory syndromes and viral pneumonia, rubella, and hantavirus pulmonary syndrome.

- The arthropod-borne diseases are transmitted by arthropod vectors from human to human or animal to human. Examples include the various encephalitides, Colorado tick fever, West Nile fever, and historically important yellow fever.

- Some viruses are so sensitive to environmental influences that they are unable to survive for significant periods of time outside their hosts. These viruses are transmitted from host to host by direct contact and cause diseases such as AIDS, cold sores, the common cold, cytomegalovirus inclusion disease, genital herpes, human herpesvirus 6 infections, human parvovirus B19 infections, certain leukemias, infectious mononucleosis, human papillomavirus, rabies, and viral hepatitides.

- Viruses that can be transmitted by food and water and usually either grow in or pass through the intestinal system leave the body in the feces and are acquired through the oral route. Examples of such diseases include viral gastroenteritis, hepatitis A and E, and poliomyelitis.

- The slow virus diseases represent progressive pathological processes caused by viruses or prions that remain clinically silent during a prolonged period of months or years, after which progressive clinical disease becomes apparent, usually ending months later in profound disability or

These mechanical respirators, also called iron lungs, were used to keep patients alive during the polio epidemics of the 1940s through the mid 1950s. In 1955 the inactivated poliovirus vaccine (IPV) developed by Jonas Salk was licensed for use in the United States, and in 1962 the oral poliovirus vaccine (OPV) developed by Albert Sabin was also licensed. As a result, polio has been eradicated in the United States and iron lungs are part of our history.

death. Examples include new variant Creutzfeldt-Jakob disease, kuru, progressive multifocal leukoencephalopathy, Gerstmann-Sträussler-Scheinker disease, and subacute sclerosing panencephalitis.
- One other disease associated with viruses but that does not fit into any of the foregoing categories is warts.

Only once in human history have we witnessed the total eradication of a dreaded disease, and that was smallpox more than two decades ago. Now humanity stands on the brink of a second: the global eradication of polio.
—*From UNICEF's Polio Website*

 hapters 16, 17, and 18 provide a review of the general biology of viruses and an introduction to basic virology. Chapter 38 continues this coverage by discussing viruses that are pathogenic to humans. Viruses are grouped according to their mode of acquisition and transmission, and viral diseases that occur in the United States are emphasized.

More than 400 different viruses can infect humans. Human diseases caused by viruses are unusually interesting, considering the small amount of genetic information introduced into a host cell. This apparent simplicity belies the severe pathological features, clinical consequences, and death that result from many viral diseases. With few exceptions, only prophylactic or supportive treatment is available. Collectively these diseases are some of the most common and yet most puzzling of all infectious diseases. The resulting frustration is compounded when year after year familiar diseases of unknown etiology or new diseases become linked to virus infections (**table 38.1**).

38.1 AIRBORNE DISEASES

Because air does not support virus growth, any virus that is airborne must have originated from a source such as another human. When humans are the source of the airborne virus, it usually is propelled from the respiratory tract by an individual's coughing, sneezing, or vocalizing.

Chickenpox (Varicella) and Shingles (Herpes Zoster)

Chickenpox (varicella) is a highly contagious skin disease primarily of children 2 to 7 years of age. It is estimated that about 4 million cases occur annually in the United States. The causative agent is the varicella-zoster virus, a member of the family *Herpesviridae,* which is acquired by droplet inhalation into the respi-

Table 38.1

Some Examples of Human Viral Diseases Recognized Since 1967

Year	Virus	Disease
1967	Marburg virus	Hemorrhagic fever
1973	Rotavirus	Major cause of infantile diarrhea worldwide
1974	Parvovirus B19	Aplastic crisis in chronic hemolytic anemia
1977	Ebola virus	Ebola hemorrhagic fever
1977	Hanta virus	Hemorrhagic fever with renal syndrome
1980	Human T-cell lymphotrophic virus 1 (HTLV-1)	Adult T-cell leukemia
1982	Human T-cell lymphotrophic virus 2 (HTLV-2)	Hairy-cell leukemia
1983	Human immunodeficiency virus (HIV)	Acquired immunodeficiency syndrome (AIDS)
1988	Human herpesvirus 6 (HHV-6)	Sixth disease (roseola subitum); may be associated with multiple sclerosis
1988	Hepatitis E	Enterically transmitted non-A, non-B hepatitis
1989	Hepatitis C	Parenterally transmitted non-A, non-B liver infection
1991	Guanarito virus	Venezuelan hemorrhagic fever
1992	Lymphocytic choriomeningitis virus	Central nervous system infection often leading to meningitis, encephalomyelitis, or other diseases
1993	Sin Nombre virus	Adult respiratory distress syndrome
1994	Sabia virus	Brazilian hemorrhagic fever
1994	Ross River virus	Ross River viral disease (Australia)
1994	Human herpesvirus 8 (HHV-8)	Associated with Kaposi's sarcoma in AIDS patients
1996	O'nyoung-nyong virus	Epidemic O'nyong fever
1997	Deer tick virus	Enzootic tick-borne encephalitis
1997	Chicken flu (H5N1) virus	Influenzalike illness
1997	Transfusion-transmitted virus (TTV)	Hepatitis
1999	Australian bat lyssavirus (ABL)	ABL infection
2000	Hepatitis G	Not clear
2002	Metapneumovirus	Respiratory tract infections
2003	SARS-associated virus	Severe acute respiratory syndrome (SARS)

ratory system. The virus produces at least six glycoproteins that play a role in viral attachment to specific receptors on respiratory epithelial cells, and their recognition by the human immune system results in humoral and cellular immunity. Following an incubation period of from 10 to 23 days, small vesicles erupt on the face or upper trunk, fill with pus, rupture, and become covered by scabs (**figure 38.1**). Healing of the vesicles occurs in about 10 days. During this time intense itching often occurs. Chickenpox can be prevented or the infection shortened with an attenuated varicella vaccine (Varivax; *see table 33.1*) or the drug acyclovir (Zovirax or Valtrex). It should be noted that Valtrex (valacy-

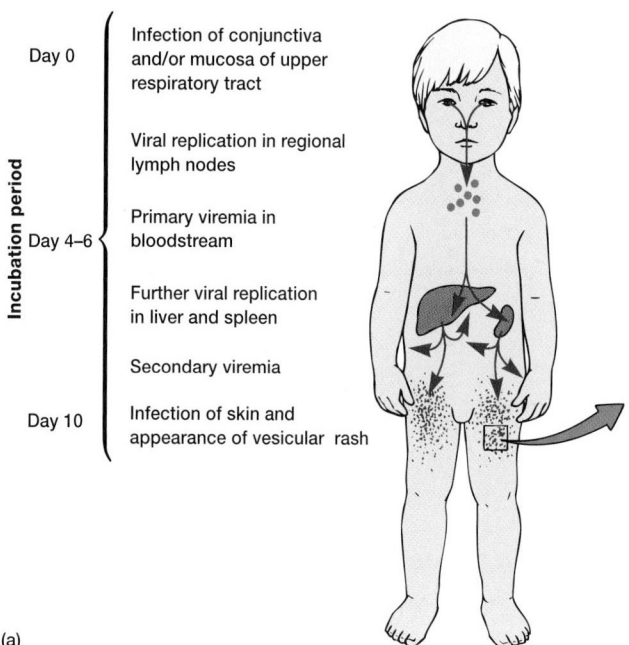

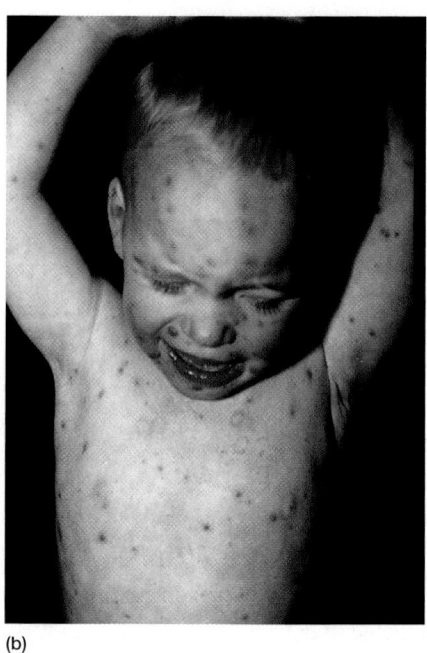

Day 0 — Infection of conjunctiva and/or mucosa of upper respiratory tract

Viral replication in regional lymph nodes

Day 4–6 — Primary viremia in bloodstream

Further viral replication in liver and spleen

Secondary viremia

Day 10 — Infection of skin and appearance of vesicular rash

Incubation period

(a)

(b)

Figure 38.1　Chickenpox (Varicella).　(**a**) Pathogenesis. (**b**) Typical vesicular skin rash. This rash occurs all over the body, but is heaviest on the trunk and diminishes in intensity toward the periphery.

clovir) is an orally administered prodrug of Zovirax or acyclovir (*see figure 35.15*). Valtrex is the valyl ester of acyclovir and is rapidly hydrolyzed to acyclovir in the body.

Individuals who recover from chickenpox are subsequently immune to this disease; however, they are not free of the virus, as viral DNA resides in a dormant (latent) state within the nuclei of cranial nerves and sensory neurons in the dorsal root ganglia. This viral DNA is maintained in infected cells but virions cannot be detected (**figure 38.2***a*). When the infected person becomes immunocompromised by such factors as age, neoplastic diseases, organ transplants, AIDS, or psychological or physiological stress, the viruses may become activated (figure 38.2*b*). They migrate down sensory nerves, initiate viral replication, and produce painful vesicles (figure 38.2*c*) because of sensory nerve damage. This syndrome is called **postherpetic neuralgia.** To manage the intense pain, the drug gabapentin (Neurontin) can be prescribed. The reactivated form of chickenpox is called **shingles (herpes zoster).** Most cases occur in people over 50 years of age. Except for the pain of postherpetic neuralgia, shingles does not require specific therapy; however, in immunocompromised individuals, acyclovir, valacyclovir, vidarabine (Vira-A), or famciclovir (Famvir) are recommended. More than 500,000 cases of herpes zoster occur annually in the United States.

Influenza (Flu)

Influenza [Italian, to be influenced by the stars—*un influenza di freddo*], or the **flu,** is a respiratory system disease caused by orthomyxoviruses. Influenza viruses (*see figures 16.10i, 16.17a,b*) are classified into A, B, and C groups based on the antigens (HA

for hemagglutinin and NA for neuraminidase) of the envelope. HA and NA are used to subtype the viruses and function in their attachment and virulence. One of the unique features of the influenza viruses is the frequency with which changes in antigenicity occur. If the variation is small, it is called **antigenic drift.** Antigenic drift results from the accumulation of mutations of HA and NA in a single strain of flu virus within a geographic region. This usually occurs every 2 to 3 years, causing local increases in the number of flu infections. **Antigenic shift** is a large antigenic change resulting from the reassortment of genomes when two different strains of flu viruses (from both animals and humans) infect the same cell and are incorporated into a single new capsid. Because there is a greater change with antigenic shift than antigenic drift, antigenic shifts can yield major epidemics or pandemics. Antigenic variation occurs almost yearly with influenza A virus, less frequently with the B virus, and has not been demonstrated in the C virus.

The standard nomenclature system for influenza virus subtypes include the following information: group (A, B, or C), host of origin, geographic location, strain number, and year of original isolation. Antigenic descriptions of the HA and NA are given in parentheses for type A. The host of origin is not indicated for human isolates—for example, A/Hong Kong/03/68 (H3N2). However, the host origin is given for others—for example, A/swine/Iowa/15/30 (H1N1).

Animal reservoirs are critical to the epidemiology of human influenza. For example, rural China is one region of the world where chickens, pigs, and humans live in close, crowded conditions. Influenza is widespread in chickens; although chickens can't transmit the virus to humans, they can transfer it to pigs.

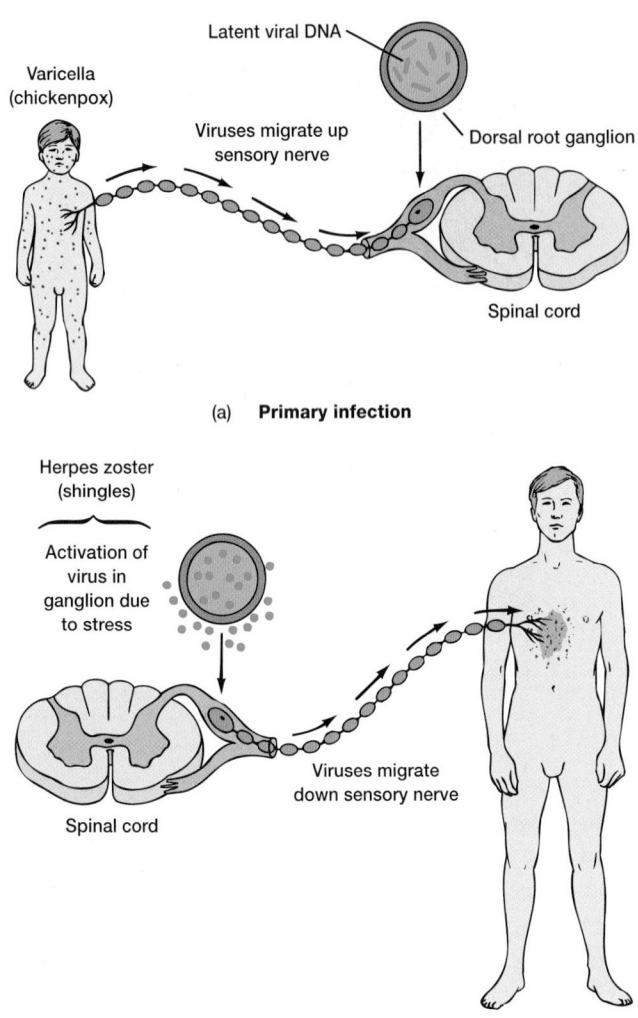

(a) **Primary infection**

(b) **Recurrence**

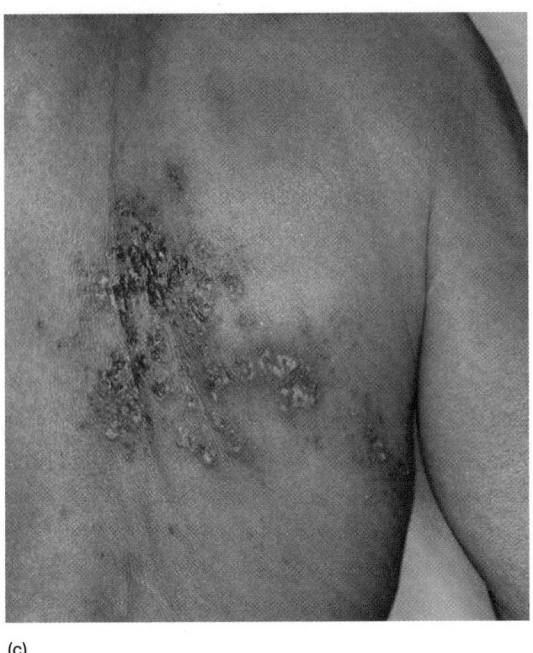

(c)

Figure 38.2 Pathogenesis of the Varicella-Zoster Virus.
(**a**) After an initial infection with varicella (chickenpox), the viruses migrate up sensory peripheral nerves to their dorsal root ganglia, producing a latent infection. (**b**) When a person becomes immunocompromised or is under psychological or physiological stress, the viruses may be activated. (**c**) They migrate down sensory nerve axons, initiate viral replication, and produce painful vesicles. Since these vesicles usually appear around the trunk of the body, the name *zooster* (Greek for girdle) was originally used.

Pigs can transfer it to humans and humans, back to pigs. Recombination between human and avian strains thus occurs in pigs, leading to major antigenic shifts. This explains why influenza continues to be a major epidemic disease and frequently produces worldwide pandemics. Hippocrates described influenza in 412 B.C. The first well-documented global epidemic of influenza-like disease occurred in 1580. Since then, 31 possible influenza pandemics have been documented, with four occurring in the twentieth century. The worst pandemic on record occurred in 1918 and killed more than 20 million people. This disaster, traced to the Spanish influenza virus (*see figure 37.7*), was followed by pandemics of Asian flu (1957), Hong Kong flu (1968), and Russian flu (1977). (The names reflect popular impressions of where the episodes began, although all are now thought to have originated in China.)

The virus (*see figures 16.17a, 18.7, 18.9*) is acquired by inhalation or ingestion of virus-contaminated respiratory secretions. During an incubation period of 1 to 2 days, the virus adheres to the epithelium of the respiratory system (the neuraminidase present in envelope spikes may hydrolyze the mucus that covers the epithelium). The virus attaches to the epithelial cell by its hemagglutinin spike protein, causing part of the cell's plasma membrane to bulge inward, seal off, and form a vesicle (receptor-mediated endocytosis). This encloses the virus in an endosome. The hemagglutinin molecule in the virus envelope undergoes a dramatic conformational change when the endosomal pH decreases. The hydrophobic ends of the hemagglutinin spring outward and extend toward the endosomal membrane. After they contact the membrane, fusion occurs and the RNA nucleocapsid is released into the cytoplasmic matrix.

Influenza is characterized by chills, fever, headache, malaise, and general muscular aches and pains. These symptoms arise from the death of respiratory epithelial cells, probably due to attacks by activated T cells. Recovery usually occurs in 3 to 7 days, during which coldlike symptoms appear as the fever subsides. Influenza alone usually is not fatal. However, death may result from pneumonia caused by secondary bacterial invaders such as *Staphylococcus aureus*, *Streptococcus pneumoniae*, and *Haemophilus influenzae*. A commercially available identification technique is Directigen FLU-A (an enzyme immunoassay [EIA] rapid test). This test can detect influenza A virus in clinical specimens in less than 15 minutes.

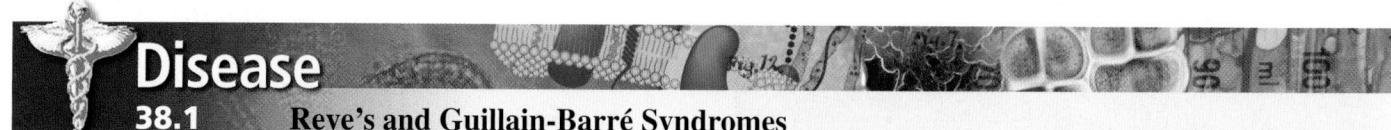

An occasional complication of influenza, chickenpox, and a few other viral diseases in children under 14 years of age is **Reye's syndrome.** After the initial infection has disappeared, the child suddenly begins to vomit persistently and experiences convulsions followed by delirium and a coma. Pathologically, the brain swells with injury to the neuronal mitochondria, fatty infiltration into the liver occurs, blood ammonia is elevated, and both serum glutamic oxaloacetic transaminase (SGOT) and serum glutamic pyruvic transaminase (SGPT) are elevated in the blood. Diagnosis is made by the measurement of the levels of these enzymes and ammonia.

The relationship between the initial viral infection and the brain and liver damage is unknown. Treatment is nonspecific and directed toward reducing intracranial pressure and correcting metabolic and electrolyte abnormalities. Some children who recover have residual neurological deficits—impaired mental capacity, seizures, and hemi-

plegia (paralysis on one side of the body). Mortality ranges from 10 to 40%. It is suspected that the use of aspirin or salicylate-containing products to lower the initial viral fever increases one's chances of acquiring Reye's syndrome.

Another condition that involves the central nervous system and is associated with influenza infections is **Guillain-Barré syndrome** (sometimes called **French polio**). In this disorder the individual suffers a delayed reaction (usually within 8 weeks) either to the actual virus infection or to vaccines against influenza. The virus or viral antigen in the vaccine damages the Schwann cells that myelinate the peripheral nerves and thus causes demyelination. As a result a prominent feature of this syndrome is a symmetric weakness of the extremities and sensory loss. Fortunately recovery usually is complete because the remaining undamaged Schwann cells eventually proliferate and wrap around the demyelinated nerves.

As with many other viral diseases, only the symptoms of influenza usually are treated. However, the antiviral drugs amantadine (Symmetrel) (*see figure 35.15*), rimantadine (Flumadine), zanamivir (Relenza), and oseltamivir (Tamiflu) have been shown to reduce the duration and symptoms of type A influenza if administered during the first two days of illness. All four of these drugs attack the virus directly by plugging the catalytic site of the enzyme neuramidase. With the enzyme inactivated, viral particles can't travel from cell to cell. Aspirin (salicylic acid) should be avoided in children younger than 14 years to reduce the risk of Reye's syndrome (**Disease 38.1**). The mainstay for prevention of influenza since the late 1940s has been inactivated virus vaccines, (*see table 33.1*), especially for the chronically ill, individuals over age 65, residents of nursing homes, and health-care workers in close contact with people at risk. Antiviral drugs (pp. 796–97)

Currently three influenza A virus subtypes are epidemic in humans: those with H1N1, H2N1, and H3N2 hemagglutinin and neuraminidase surface glycoproteins. Influenza A infections usually peak in the winter and involve 10% or more of the population, with rates of 50 to 75% in school-age children. Influenza B usually accounts for only 3% of all flu cases in the United States.

Measles (Rubeola)

Measles [rubeola: Latin *rubeus,* red] is a highly contagious skin disease that is endemic throughout the world. The measles virus is a member of the genus *Morbillivirus* and the family *Paramyxoviridae.* The measles virus is monotypic, but small variations at the epitope level have been described. The variations are based on genetic

variability in the virus genes. Such variations, however, have no effect on protective function since a measles infection still provides a lifelong immunity against reinfection. The virus enters the body through the respiratory tract or the conjunctiva of the eyes. The receptor for the measles virus has recently been shown to be the complement regulator CD46, also known as membrane cofactor protein.

The incubation period is usually 10 to 21 days, and the first symptoms begin about the tenth day with a nasal discharge, cough, fever, headache, and conjunctivitis. Within 3 to 5 days skin eruptions occur as faintly pink maculopapular lesions that are at first discrete, but gradually become confluent (**figure 38.3**). The rash normally lasts about 5 to 10 days. Lesions of the oral cavity include the diagnostically useful bright-red **Koplik's spots** with a bluish-white speck in the center of each. Koplik's spots represent a viral exanthem (a skin eruption) occurring in the form of macules or papules as a result of the viral infection. Very infrequently a progressive degeneration of the central nervous system called **subacute sclerosing panencephalitis** occurs (table 38.6). No specific treatment is available for measles. The use of attenuated measles vaccine (Attenuvax) or in combination (MMR vaccine; **m**easles, **m**umps, **r**ubella) is recommended for all children (*see table 33.1*). Since public health immunization programs began in 1963, there has been over a 99% decrease in measles cases. Currently there are about 2,000 to 3,000 cases a year in the United States, and 90% of these are among unvaccinated individuals. In less well-developed countries, however, the morbidity and mortality in young children from measles infection remain high. It has been estimated that measles infects 50 million people and kills about 4 million a year worldwide. Serious outbreaks of measles are still reported in North America and Europe, especially among college students.

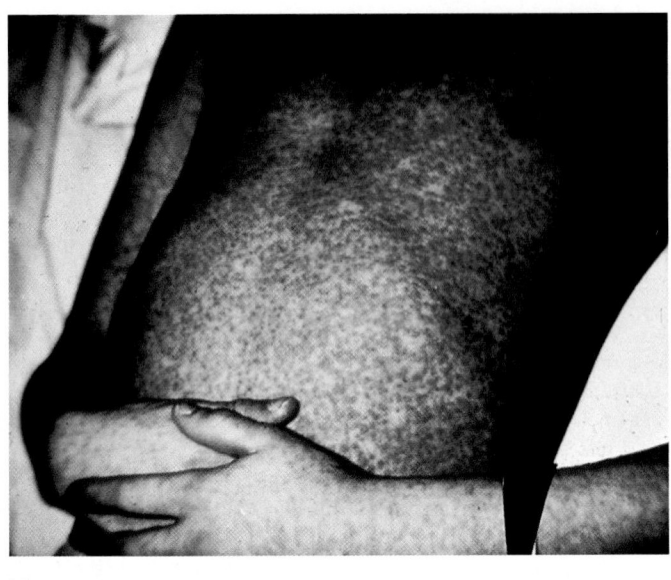

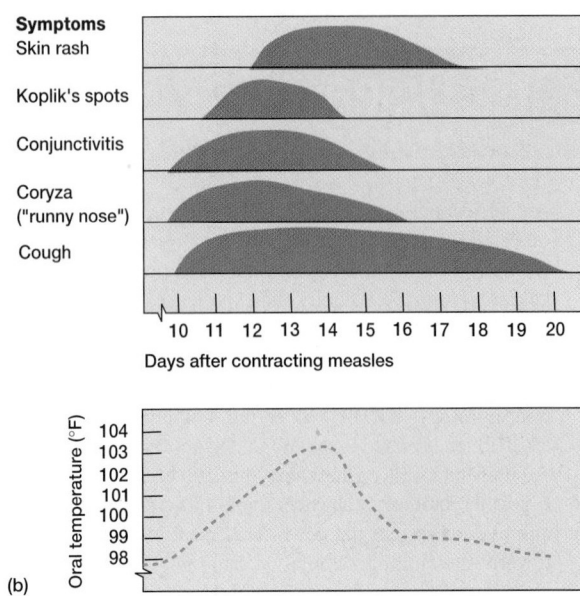

Figure 38.3   Measles (Rubeola).   (**a**) The rash of small, raised spots is typical of measles. The rash usually begins on the face and moves downward to the trunk. (**b**) Signs and symptoms of a measles infection.

Mumps

Mumps is an acute generalized disease that occurs primarily in school-age children. The mumps virus is a member of the genus *Rubulavirus* in the family *Paramyxoviridae*. This virus (*see figure 16.10b*) is a pleomorphic, enveloped virus that contains a helical nucleocapsid composed of negative-sense, single-stranded RNA and three nucleocapsid-associated proteins (L, NP, and P). The virus is transmitted in saliva and respiratory droplets to nonimmune contacts. The portal of entry is the respiratory tract. The most prominent manifestations of mumps are swelling and tenderness of the salivary (parotid) glands 16 to 18 days after infection of the host by the virus (**figure 38.4**). The swelling usually lasts for 1 to 2 weeks and is accompanied by a low grade fever. Meningitis and inflammation of the epididymis and testes (**orchitis**) can be important complications associated with this disease—especially in the postpubescent male. Therapy of mumps is limited to symptomatic and supportive measures. A live, attenuated mumps virus vaccine is available. It usually is given as part of the triple MMR vaccine (*see table 33.1*). There are currently about 3,000 cases of mumps in the United States each year. Recently a few outbreaks have occurred among unvaccinated children in states without a comprehensive school immunization law. Prevention and control involves keeping children with mumps from school or other associated activities for about 2 weeks.

Respiratory Syndromes and Viral Pneumonia

Acute viral infections of the respiratory system are among the most common causes of human disease. The infectious agents are called the acute respiratory viruses and collectively produce a variety of clinical manifestations, including rhinitis (inflammation

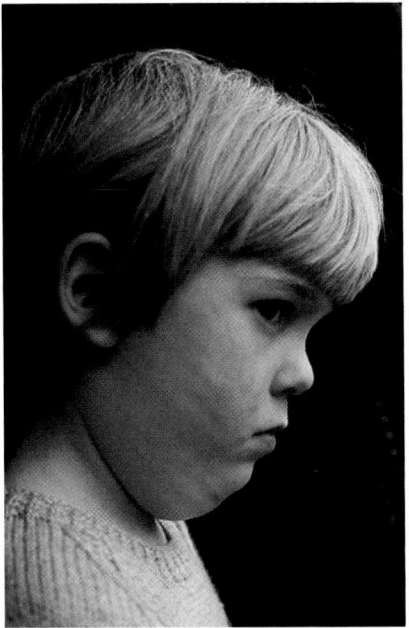

Figure 38.4   Mumps.   Child with diffuse swelling of the salivary (parotid) glands due to the mumps virus.

of the mucous membrane of the nose), tonsillitis, laryngitis, and bronchitis. The adenoviruses, coxsackievirus A, coxsackievirus B, echovirus, influenza viruses, parainfluenza viruses, poliovirus, respiratory syncytial virus, and reovirus are thought to be responsible. It should be emphasized that for most of these viruses there is a lack of specific correlation between the agent and the clinical manifestation—hence the term syndrome. Immunity is

not complete, and reinfection is common. The best treatment is
rest.   Disease syndromes, signs, and symptoms (p. 824); Classification of animal viruses (pp. 388–91, appendix IV)

In those cases of pneumonia for which no cause can be identified, viral pneumonia may be assumed if mycoplasmal pneumonia has been ruled out. The clinical picture is nonspecific.
Symptoms may be mild, or there may be severe illness and death.

Respiratory syncytial virus (RSV) often is described as the
most dangerous cause of lower respiratory infections in young
children. In the United States over 90,000 infants are hospitalized each year and over 4,000 die. The RSV is a member of the
RNA virus family, *Paramyxoviridae.* It is a negative single-
stranded RNA virus that is enveloped with two virally specific
glycoproteins as part of the structure. One of them, the large glycoprotein or G, is responsible for the binding of the virus to the
host cell; the other, the fusion protein or F, permits fusion of the
viral envelope with the host cell plasma membrane, leading to
entry of the virus. The F protein also induces the fusion of the
plasma membranes of infected cells. RSV thus gets its name
from the resulting formation of a syncytium or multinucleated
mass of fused cells. The multinucleated syncytia are responsible
for inflammation, alveolar thickening, and the filling of alveolar
spaces with fluid. The source of the RSV is hand contact and respiratory secretions of humans. Clinical manifestations consist
of an acute onset of fever, cough, rhinitis, and nasal congestion.
In infants and young children, this often progresses to severe
bronchitis and viral pneumonia. Diagnosis is by either Directigen RSV or Test-Pack RSV rapid test kits. The virus is found
worldwide and causes seasonal (November to March) outbreaks
lasting several months. Treatment is with inhaled ribavirin (Vi-
razole). A recently developed series of antibody (RSV-immune
globulin) injections has been shown to reduce the severity of this
disease in infants by 75%. Prevention and control consists of isolation for RSV-infected individuals, use of gowns when contact
with secretions is likely, and strict attention to good handwashing practices.

Rubella (German Measles)

Rubella [Latin *rubellus,* reddish] was first described in Germany
in the 1800s and was subsequently called **German measles.** It is
a moderately contagious disease that occurs primarily in children
5 to 9 years of age. It is caused by the rubella virus, a single-
stranded RNA virus that is a member of the family *Togaviridae.*
Rubella is worldwide in distribution and occurs more frequently
during the winter and spring months. This virus is spread in
droplets that are shed from the respiratory secretions of infected
individuals. Once the virus is inside the body, the incubation period ranges from 12 to 23 days. A rash of small red spots (**figure
38.5**), usually lasting no more than 3 days, and a light fever are
the normal symptoms. (This is why rubella is sometimes referred
to as the "three-day measles.") The rash appears as immunity develops and the virus disappears from the blood, suggesting that
the rash is immunologically mediated and not caused by the virus
infecting skin cells.

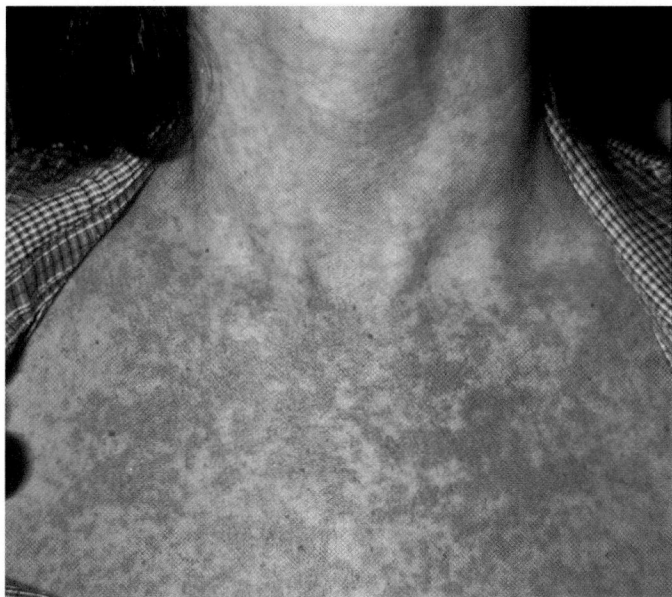

Figure 38.5   German Measles (Rubella).   This disease is
characterized by a rash of red spots. Notice that the spots are not
raised above the surrounding skin as in measles (rubeola; see
figure 38.3).

Rubella can be a disastrous disease (**congenital rubella syndrome**) in the first trimester of pregnancy and can lead to fetal
death, premature delivery, or a wide array of congenital defects
that affect the heart, eyes, and ears.

Because rubella is usually such a mild infection, no treatment
is indicated. All children and women of childbearing age who
have not been previously exposed to rubella should be vaccinated. The live attenuated rubella vaccine (part of MMR, *see
table 33.1*) is recommended. Because routine vaccination began
in the United States in 1969, fewer than 1,000 cases of rubella and
10 cases of congenital rubella occur annually.

Smallpox (Variola)

Smallpox (variola) was once one of the most prevalent of all diseases. It was a universally dreaded scourge for more than 3 millennia, with case fatality rates of 20 to 50%. First subjected to
some control by variolation in tenth-century India and China, it
was gradually suppressed in the industrialized world after Edward Jenner's 1796 landmark discovery that infection with the
harmless cowpox (vaccinia) virus renders humans immune to the
smallpox virus (*see Historical Highlights 33.1*).

The variola virus belongs to the family *Poxviridae,* which
includes vaccinia (smallpox vaccine), monkeypox virus, and
molluscum contagiosum virus. The virion is large, brick-
shaped, and contains a dumbell-shaped core (*see figure 16.18*).
Interestingly, the size of the smallpox virus (300 by 250 to 200
nm) is slightly larger than that of the smallest bacterium—for
example, *Chlamydia.* The genome inside the core consists of a

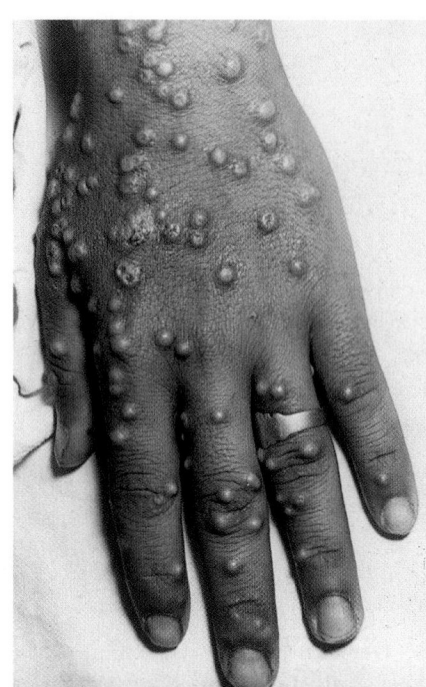

Figure 38.6 Smallpox. Back of hand showing single crop of smallpox vesicles.

single linear molecule of double-stranded DNA and replicates in the cell's cytoplasm.

Variola is transmitted between humans by aerosol or contact. The virus enters the respiratory tract, seeding the mucous membranes and passing rapidly into regional lymph nodes. After a brief period of viremia, there is a latent period of 4 to 14 days, during which the virus multiplies in the monocyte-macrophage system. Another brief period of viremia precedes the prodromal phase (see p. 824). The prodromal phase lasts for 2 to 3 days and is characterized by severe headaches, backache, and fever (over 40°C), all beginning abruptly. During the prodromal phase, the mucous membranes in the mouth and pharynx become infected. The virus then invades the capillary epithelium of the skin, leading to the development of the following sequence of lesions in/on the skin: eruptions, papules, vesicles (**figure 38.6**), pustules, crusts, and desquamation. Oropharyngeal and skin lesions contain abundant virus particles, particularly early in the disease process. Death from smallpox is due to toxemia, associated with immune complexes (see figure 33.6) and hypotension.

A suspect case of smallpox should be managed in a negative-pressure room, if possible, and the patient should be vaccinated, particularly if the disease is in the early stage. When given very early in the incubation period, vaccination can markedly attenuate or even prevent the clinical manifestations of smallpox. Strict respiratory and contact isolation is important. There is no treatment approved by the Food and Drug Administration for smallpox.

Since the advent of immunization with the vaccinia virus, and because of concerted efforts by the World Health Organization, the disease smallpox has been eradicated throughout the world—

the greatest public health achievement ever. (The last case from a natural infection occurred in Somalia in 1977.) Eradication was possible because smallpox has obvious clinical features, virtually no asymptomatic carriers, only human hosts and reservoirs, and a short period of infectivity (3 to 4 weeks). The disease was successfully eliminated by simply preventing the spread of smallpox until no new cases developed. The virus, however, is still kept in two locations: the CDC in Atlanta, Georgia, and the Russian State Center for Research on Virology and Biotechnology, Koltsovo, Novosibrisk region, Russia.

There is the concern and possibility that the smallpox virus could reside outside of these laboratories and could be used as a weapon by terrorists. Potential sources of the virus are countries that falsely claim they have destroyed their stocks and virus stocks acquired from laboratories in the former Soviet Union. An accidental or deliberate release of smallpox virus would be catastrophic in an unimmunized population and could cause a major pandemic. Because smallpox vaccination has not been performed routinely in the United States since about 1972, there is now a large population of susceptible persons. Currently, less than half the world's population has been exposed to either smallpox (variola virus) or to the vaccine. Thus, if an outbreak occurred, prompt recognition and institution of control measures would be important.   Disease (smallpox) and the early colonization of America (p. 353)

1. Why are chickenpox and shingles discussed together? What is their relationship?
2. Briefly describe the course of an influenza infection and how the virus causes the symptoms associated with the flu. Why has it been difficult to develop a single flu vaccine?
3. What are some common symptoms of measles?
4. What are Koplik's spots?
5. What is one side effect that mumps can cause in a young post-pubescent male?
6. Describe some clinical manifestations caused by the acute respiratory viruses.
7. Is viral pneumonia a specific disease? Explain.
8. When is a German measles infection most dangerous and why?

 ARTHROPOD-BORNE DISEASES

The **ar**thropod-**bo**rne viruses (arboviruses) are transmitted by bloodsucking arthropods from one vertebrate host to another. They multiply in the tissues of the arthropod without producing disease, and the vector acquires a lifelong infection. Approximately 150 of the recognized arboviruses cause illness in humans. Diseases produced by the arboviruses can be divided into three clinical syndromes: (1) fevers of an undifferentiated type with or without a rash; (2) encephalitis (inflammation of the brain), often with a high fatality rate; and (3) hemorrhagic fevers, also frequently severe and fatal (**Disease 38.2**). **Table 38.2** sum-

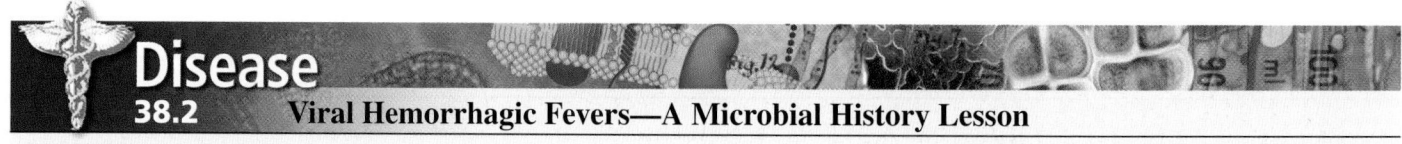

Disease

38.2   Viral Hemorrhagic Fevers—A Microbial History Lesson

Scientists know of several viruses lurking in the tropics that—with a little help from nature—could wreak far more loss of life than will likely result from the AIDS pandemic. Collectively these viruses produce what is known as the **hemorrhagic fevers.** The viruses are passed among wild vertebrates, which serve as reservoir hosts. Arthropods transmit the viruses among vertebrates, and humans are infected when they invade the environment of the natural host. These diseases, distributed throughout the world, are known by over 70 names, usually denoting the geographic area where they were first described.

Viral hemorrhagic fevers can be fatal. Patients suffer headache, muscle pain, flushing of the skin, massive hemorrhaging either locally or throughout the body, circulatory shock, and death.

Despite the lack of public awareness, recent outbreaks may foreshadow much broader outbreaks in the future. For example, in the late 1960s dozens of scientists in West Germany fell seriously ill, and several died, from a mysterious new disease. Victims suffered from a breakdown of liver function and a bizarre combination of bleeding and blood clots. The World Health Organization traced the outbreak to a batch of fresh monkey cells the scientists had used to grow polioviruses. The cells from the imported Ugandan monkeys were infected with the lethal tropical Marburg virus (see **Box figure**) and the scientists suffered from **Marburg viral hemorrhagic fever.**

In 1977 the *Plebovirus* causing Rift Valley Fever in sheep and cattle moved from these animals into the South African population. The virus, which causes severe weakness, incapacitating headaches, damage to the retina, and hemorrhaging, then made its way to Egypt, where millions of humans became infected and thousands died.

Among the most frightening hemorrhagic outbreak was that of the **Ebola virus hemorrhagic fever** in Zaire and Sudan in 1976. This disease infected more than 1,000 people and left over 500 dead. It became concentrated in hospitals, where it killed many of the Belgian physicians and nurses treating infected patients. A similar outbreak occurred in this same area in 1995 but was quickly contained.

In the United States in 1989, epidemiologists provided new evidence that rats infected with a potentially deadly hemorrhagic virus are prevalent in Baltimore slums. The virus appears to be taking a previously unrecognized toll on the urban poor by causing **Korean hemorrhagic fever.**

In the summer of 1993, reports appeared in the news media about a mysterious illness that had caused over 30 deaths among the Navajo Nation in the Four-Corners area of the southwestern United States. The CDC finally determined the causative agent to be a hantavirus, a negative single-stranded RNA virus that is a member of the family *Bunyaviridae.* Hantaviruses are endemic in rodents, such as deer

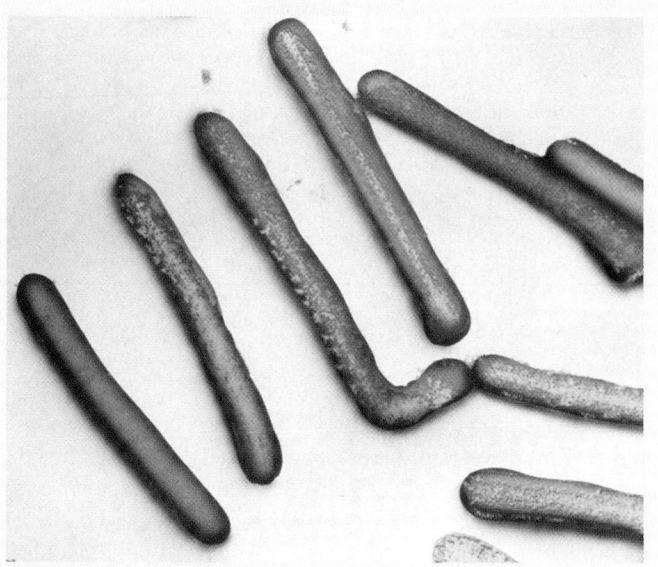

Sinister Foe.   The deadly Marburg virus was first isolated in 1967 at the Institute for Hygiene and Microbiology in Marburg, Germany.

mice, in many areas of the world. Deer mice shed the virus in their saliva, feces, and urine. Humans contract the disease when they inhale aerosolized particles containing the excreted virus. Throughout Asia and central Europe, hantaviruses cause hemorrhagic fever with renal syndrome in humans. But the type of virus found in the Southwest had not been previously recognized, and no hantavirus anywhere in the world has been associated with the clinical syndrome initially seen among the Navajo; namely, the **hantavirus pulmonary syndrome** in which the virus destroys the lungs. In 1993 the CDC named this virus **pulmonary syndrome hantavirus** (sometimes called the Sin Nombre or no-name virus) and to date, isolated cases have been reported from almost every state. Prevention involves wearing gloves when handling mice and spraying the feces and urine of all mice with a disinfectant.

Although to date, these epidemics have not become global, they do provide a humbling vision of humankind's viral vulnerability. History shows that the life-threatening viral hemorrhagic outbreaks often have arisen when humans moved into unexplored terrain or when living conditions deteriorated in ways that generated new viral hosts. In each case medical and scientific resources have been reactive, not proactive.

---

marizes the seven major human arbovirus diseases that occur in the United States. For all these diseases, immunity is believed to be permanent after a single infection. No vaccines are available for the human arthropod-borne diseases listed in table 38.2, although supportive treatment is beneficial.

Colorado Tick Fever

Colorado tick fever is caused by an RNA virus of the genus *Coltivirus.* The tick, *Dermacentor andersoni,* is the main vector that transmits the disease to humans. Ground squirrels, rabbits,

Table 38.2

Summary of the Seven Major Human Arbovirus Diseases That Occur in the United States

Disease	Distribution	Vectors	Natural Host	Mortality Rate
California encephalitis (La Crosse)	North Central, Atlantic, South	Mosquitoes (*Aedes* spp.)	Birds	Fatalities rare
Colorado tick fever	Pacific Coast (mountains)	Ticks (*Dermacentor andersoni*)	Small mammals, deer	Fatalities rare
Eastern equine encephalitis (EEE)	Atlantic, Southern Coast	Mosquitoes (*Aedes* spp.)	Birds	50–70%
St. Louis encephalitis (SLE)	Widespread	Mosquitoes (*Culex* spp.)	Birds	10–30%
Venezuelan equine encephalitis (VEE)	Southern United States	Mosquitoes (*Aedes* spp. and *Culex* spp.)	Rodents, horses	20–30% (children) <10% (adults)
Western equine encephalitis (WEE)	Mountains of the West	Mosquitoes (*Culex* spp.)	Birds	3–7%
West Nile fever	Most of United States	Mosquitoes (various spp.)	Birds	?

and deer are the principal reservoirs. The disease occurs in mountainous regions in western states. Approximately 300 cases are reported annually in the United States, but the actual incidence is probably much higher. After a 3 to 6 day incubation period, there is an abrupt onset of fever, chills, severe headache, photophobia, rash, and muscle pain. These symptoms persist for 5 to 8 days and then disappear; complications are rare. Serology is used to confirm the diagnosis. No specific treatments exist for Colorado tick fever; therapy is limited to supportive care. Prevention involves common tick precautions (*see p. 887*).

West Nile Fever (Encephalitis)

West Nile fever (encephalitis) is caused by a flavivirus that occurs primarily in the Middle East, Africa, and Southwest Asia. The disease was first discovered in 1937 in the West Nile district of Uganda. In 1999, the virus appeared unexpectedly in the United States (New York), causing seven deaths among 60 confirmed human cases and extensive mortality in a range of domestic and exotic birds. It probably crossed the Atlantic in an infected bird, mosquito, or human traveler. The disease became a significant problem in the United States by 2002 with almost every state reporting cases and deaths.

Mosquitoes (*Culex* spp.) become infected by feeding on birds such as sparrows and crows that are infected with the virus. It is now known that in addition to mosquitoes, the virus can be transmitted by blood transfusions and organ transplants. The virus produces a viremia and an acute, mild febrile disease with lymphadenopathy and rash 3 to 14 days after exposure. Transitory involvement of the brain membranes may occur during the acute stage. The virus can produce fatal encephalitis in older people and in younger people who have diabetes, emphysema, AIDS, and chronic diseases of the liver, kidneys, or heart. However, only one in five people who contract the virus has any symptoms and less than 1 in 100 develops the serious complications.

The virus can be recovered from *Culex* mosquitoes, birds, and blood taken in the acute stage of a human infection. Diagnosis is by a rise in neutralizing antibody in a patient's serum. Only one

antigenic type exists and immunity is presumed permanent. There is no treatment other than hospitalization and intravenous fluids.

Mosquito abatement and the use of repellents such as DEET appear to be the only control measures. The risk of infection is highest during the mosquito season.

Yellow Fever

Yellow fever is not now endemic to the United States, but because of its historical importance, it is briefly discussed here. Yellow fever was the first human disease found to be caused by a virus. (Walter Reed discovered this in 1901.) It also provided the first confirmation (by Carlos Juan Finley) that an insect could transmit a virus. Yellow fever is caused by a flavivirus that is endemic in many tropical areas, such as Mexico, South America, and Africa.

The disease received its first name, yellow jack, because jaundice is a prominent sign in severe cases. The jaundice is due to the deposition of bile pigments in the skin and mucous membranes because of damage to the liver. The disease is spread through a population in two epidemiological patterns. In the urban cycle human-to-human transmission is by *Aedes aegypti* mosquitoes. In the sylvatic cycle the mosquitoes transmit the virus between monkeys and from monkeys to humans (sylvatic means in the woods or affecting wild animals).

Once inside a person the virus spreads to local lymph nodes and multiplies; from this site it moves to the liver, spleen, kidneys, and heart, where it can persist for days. In the early stages of the disease, the infected person experiences fever, chills, headache, and backache, followed by nausea and vomiting. In severe cases the virus produces lesions in the infected organs and hemorrhaging occurs.

There is no specific treatment for yellow fever. Diagnosis is by serology. An active immunity to yellow fever results from an initial infection or from vaccines containing the attenuated yellow fever 17D strain or the Dakar strain virus. Prevention and control of this disease involves vaccination (*see table 33.1*) and control of the insect vector.

38.3   DIRECT CONTACT DISEASES

Acquired Immune Deficiency Syndrome (AIDS)

It is now recognized that **AIDS (acquired immune deficiency syndrome)** was the great pandemic of the second half of the twentieth century. First described in 1981 AIDS is the result of an infection by the **human immunodeficiency virus (HIV),** a lentivirus within the family *Retroviridae.* The disease appears to have begun in central Africa as early as the 1950s; HIV may have developed in the human population in the 1930s, or even earlier. Simian immunodeficiency viruses (SIVs) related to HIV-1 and HIV-2, the strains primarily responsible for AIDS, have been isolated from African primates. The SIV from chimpanzees seems to have infected humans and developed into HIV-1; HIV-2 may have arisen from the SIV that infects sooty mangabeys. Once established, HIV-1 spread to the Caribbean and then to the United States and Europe.

Epidemiologically AIDS occurs worldwide (**figure 38.7**). The groups most at risk in acquiring AIDS are (in descending order of risk) homosexual/bisexual men; intravenous (IV) drug users; heterosexuals who have intercourse with drug users, prostitutes (sex trade workers), and bisexuals; transfusion patients or hemophiliacs who must receive clotting factor preparations made from donated blood; and children born of infected mothers. The mortality rate from AIDS is extremely high.

In the United States, AIDS is caused primarily by the HIV-1 virus (some cases result from an HIV-2 infection). This virus is a retrovirus and closely related to HTLV-1, the cause of adult T-cell leukemia, and HTLV-2, which has been isolated from individuals with hairy-cell leukemia (*see leukemia, pp. 400–1).* HIV-1 is an enveloped lentivirus and a member of the family *Retroviridae (see section 18.1)* with a cylindrical core inside its capsid (**figure 38.8**). The core contains two copies of its plus single-stranded RNA genome and several enzymes. Thus far 10 virus-specific proteins have been discovered. One of them, the gp120 envelope protein, participates in HIV-1 attachment to $CD4^+$ cells (T-helper cells; see figure 38.9).

The AIDS virus is acquired by direct exposure of a person's bloodstream to body fluids (blood, semen, vaginal secretions)

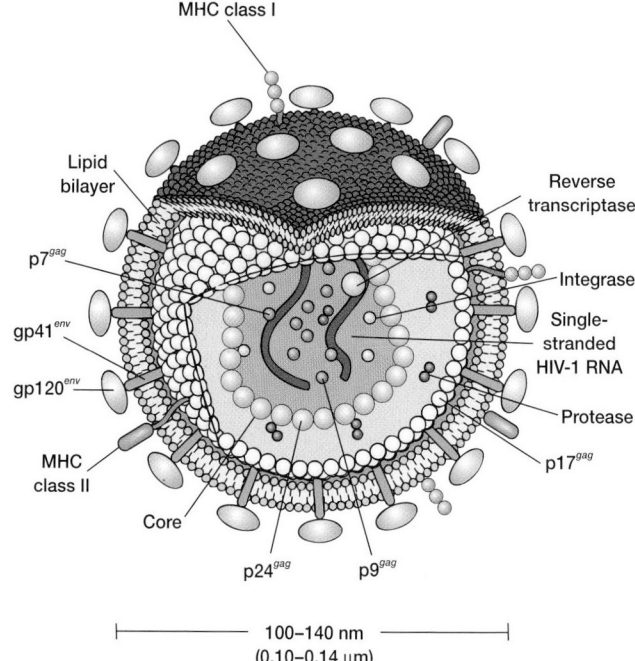

Figure 38.8   **Schematic Diagram of the HIV-1 Virion.** The HIV-1 virion is an enveloped structure containing 72 external spikes. These spikes are formed by the two major viral-envelope proteins, gp120 and gp41. (Gp stands for glycoprotein—the proteins are linked to sugars—and the number refers to the mass of the protein, in thousands of daltons.) The HIV-1 lipid bilayer is also studded with various host proteins, including class I and class II major histocompatibility complex molecules, acquired during virion budding. The cone-shaped core of HIV-1 contains four nucleocapsid proteins (p24, p9, p7) each of which is proteolytically cleaved from a 53 kDa *gag* precursor by the HIV-1 protease. The phosphorylated p24 polypeptide forms the chief component of the inner shelf of the nucleocapsid, whereas the p17 protein is associated with the inner surface of the lipid bilayer and stabilizes the exterior and interior components of the virion. The p7 protein binds directly to the genomic RNA through a zinc-finger structural motif and together with p9 forms the nucleoid core. The retroviral core contains two copies of the single-stranded HIV-1 genomic RNA that is associated with the various preformed viral enzymes, including the reverse transcriptase, integrase, ribonuclease, and protease.

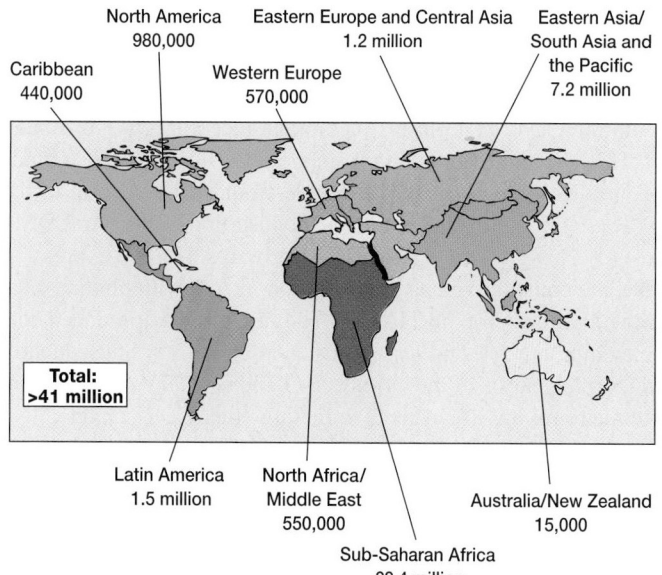

Figure 38.7   **Distribution of HIV/AIDS in Adults by Continent or Region.** The figure shows data from a 2002 United Nations report. According to recent estimates by the UN, the number of HIV/AIDS cases may be over 41 million. *Source of data: UNAIDS.*

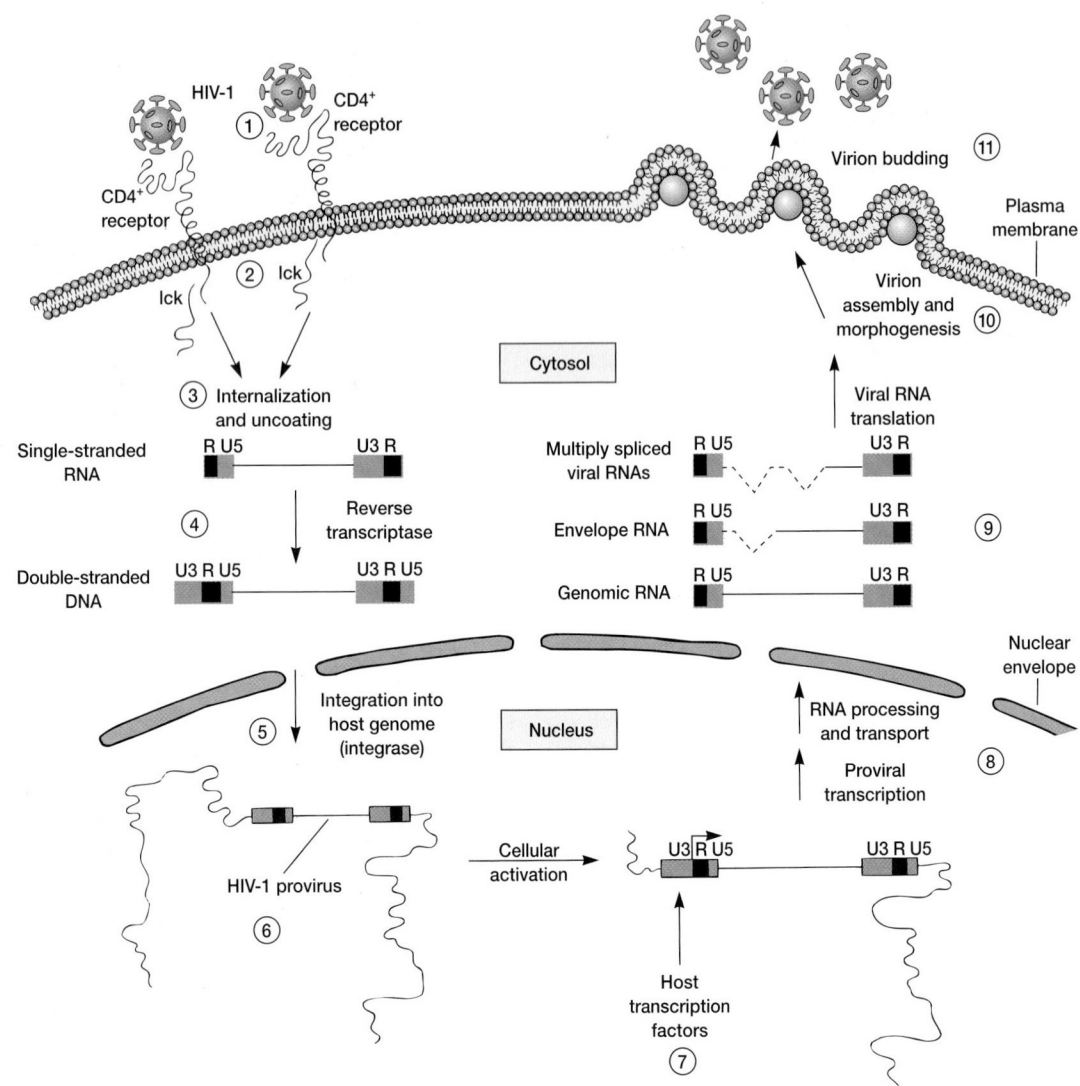

Figure 38.9   Life Cycle of HIV-1.   (*1*) After interaction of gp120 with the CD4$^+$ cell plasma membrane receptor, gp41-mediated membrane fusion occurs. (*2*) This leads to the entry of HIV-1 into the cell. The lck denotes a lymphoid-specific tyrosine kinase that binds to CD4. (*3*) After internalization and uncoating, reverse transcription of viral RNA begins. (*4*) The double-stranded DNA form of the virus genome is produced in the presence of appropriate host factors. (*5*) The HIV-1 integrase promotes the insertion of this viral DNA duplex into the CD4 cell's genome after the DNA has entered the nucleus. (*6*) This gives rise to the HIV-1 provirus. (*7*) The expression of the HIV-1 gene is stimulated initially by the action of specific inducible and constitutive host transcription factors with binding sites in the long terminal repeat. Their binding leads to the sequential production of various viral mRNAs. (*8*) The first mRNAs produced correspond to the multiply spliced species of approximately 2.0 kilobases encoding *tat, rev,* and *nef* regulatory proteins. (*9*) Subsequently the viral structural proteins are produced, allowing the (*10*) assembly and morphogenesis of the virions. (*11*) The new HIV-1 virions that are produced by viral budding from the host CD4$^+$ cell can then reinitiate the retroviral life cycle by infecting other CD4$^+$ target cells.

containing the virus, through sexual contact, or perinatally from an infected mother to her fetus. It also is possible that a newborn can be infected through breast-feeding. Once inside the body, the virus gp120 envelope protein (figure 38.8) binds to the CD4 glycoprotein plasma membrane receptor on CD4$^+$ T cells, macrophages, dendritic cells, and monocytes (**figure 38.9**). (Dendritic cells are present throughout the body's mucosal surfaces and bear the CD4 protein. Thus it is possible that these are the

first cells infected by HIV in sexual transmission.) Recent evidence shows that the virus requires a coreceptor in addition to the CD4 receptor. Macrophage-tropic strains, which seem to predominate early in the disease and infect both macrophages and T cells, require the CCR5 (CC-CKR-5) chemokine receptor protein as well as CD4. A second chemokine coreceptor, called CXCR-4 or fusin, is T cell–tropic and used by an HIV strain that is active at later stages of the infection. This strain induces the formation

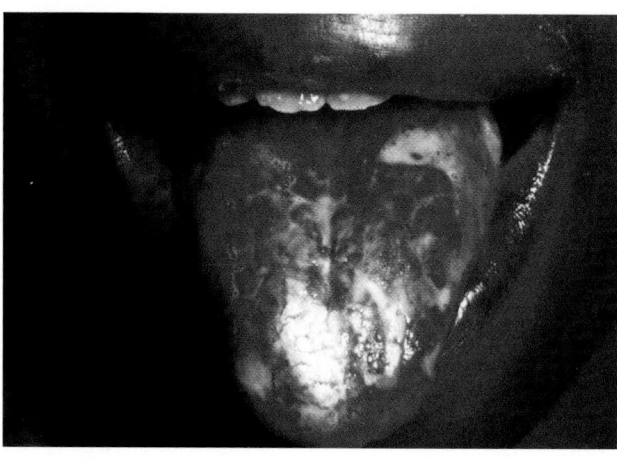

(a)

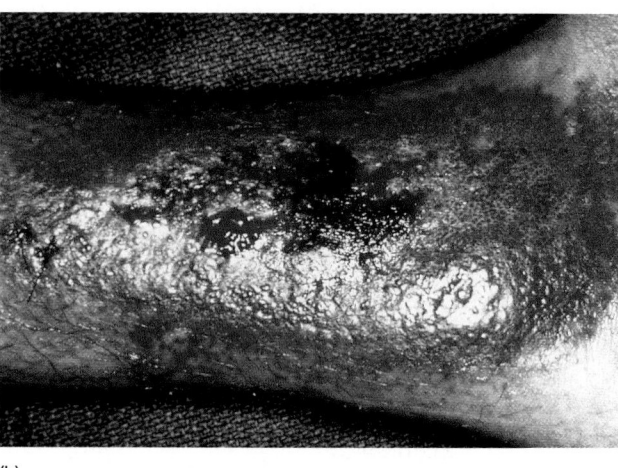

(b)

Figure 38.10   Some Diseases Associated with AIDS.   (**a**) Candidiasis of the oral cavity and tongue (thrush) caused by *Candida albicans.* (**b**) Kaposi's sarcoma on the arm of an AIDS patient. The flat purple tumors can occur in almost any tissue and are frequently multiple.

of syncytia, as described later. Individuals with two defective copies of the *CCR5* gene do not seem to get AIDS; apparently the virus cannot infect their T cells. People with one good copy of the *CCR5* gene do get AIDS but survive several years longer than those with no mutation.   Host cell receptors and virion adsorption (pp. 388–92); Replication and transcription in retroviruses (pp. 395–97)

Entry into the host cell begins when the envelope fuses with the plasma membrane, and the virus releases its core and two RNA strands into the cytoplasm. Inside the infected cell, the core protein remains associated with the RNA as it is copied into a single strand of DNA by the RNA/DNA-dependent DNA polymerase activity of the reverse transcriptase enzyme. The RNA is next degraded by another reverse transcriptase component, ribonuclease H, and the DNA strand is duplicated to form a double-stranded DNA copy of the original RNA genome. A complex of the double-stranded DNA (the provirus) and the integrase enzyme moves into the nucleus. Then the proviral DNA is integrated into the cell's DNA through a complex sequence of reactions catalyzed by the integrase (figure 38.9). The integrated provirus can remain latent, giving no sign of its presence. Alternatively the provirus can force the cell to synthesize viral mRNA. Some of the RNA is translated to produce viral proteins by the cell's own ribosomes. Viral proteins and the complete HIV-1 RNA genome are then assembled into new virions that bud from the infected host cell (*see figure 18.10*). Eventually the host cell lyses.

The precise mechanism of AIDS pathogenesis still is not known, and many hypotheses exist. Many believe that AIDS is caused primarily by depletion of T cells and disruption of their function, although the exact mechanisms are unclear. The cytopathic effect may be due to the disruption of plasma membrane permeability and function by excessive virus budding. Free gp120 proteins may bind to CD4 proteins on uninfected cells, making them targets for attack by immune system cells. Infected cells do fuse with other cells to form large, multinucleate syncy-

tia that eventually die, and this may contribute greatly to cell destruction. Possibly, insertion of the provirus DNA into the cell's genome and the transposition of the integrated provirus disrupt cell function and destroy the host T cell. Recent research shows that the virus replicates at a furious pace, about a billion virus particles a day. The immune system often manages to hold it off for years but eventually succumbs in almost all cases.

Once a human's CD4$^+$ cells are infected with HIV-1, four types of pathological changes may ensue. First, a mild form of AIDS may develop with symptoms that include fever, malaise, headache, macular rash, weight loss, lymph node enlargement (lymphadenopathy), oral candidiasis (**figure 38.10***a*), and the presence of antibodies to HIV-1 (**figure 38.11**). These symptoms may occur in the first few months after infection, last for 1 to 3 weeks, and recur. This is known as **AIDS-related complex (ARC).** ARC develops into a full-blown case of AIDS in an undetermined proportion of cases.

Second, a true case of AIDS can develop directly upon infection. The mean interval between HIV infection and the onset of AIDS appears to be about 8 to 10 years, although it varies considerably with each individual. At first, a person's immune system responds to the HIV-1 infection by manufacturing HIV-1 antibodies, but not in sufficient quantities to stop the viral attack. The virus becomes established within primarily CD4$^+$ T-helper cells, and HIV accumulates in lymphoid organs in large quantities even before symptoms appear. Initially, CD4$^+$ T-helper cells proliferate abnormally in the lymph nodes. Thereafter the lymph nodes' internal structure collapses due to viral replication. This leads to a decline in the number of lymphocytes within the lymph nodes and results in a selective depletion of the CD4$^+$ T-cell subset that is critical to the propagation of the entire T-cell pool. When this CD4$^+$ population declines, interleukin-2 (IL-2) production also decreases. Because IL-2 stimulates the production of T cells in general, the whole T-cell population may decline. This

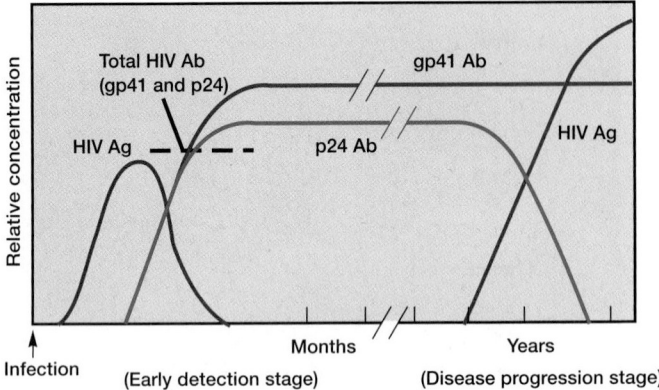

Figure 38.11 The Typical Serological Pattern in an HIV-1 Infection. HIV-1 antigen (HIV Ag) is detectable as early as 2 weeks after infection and typically declines at seroconversion. Seroconversion occurs when HIV-1 antibodies have risen to detectable levels. This usually takes place several weeks to months after the HIV-1 infection. The period between HIV-1 infection and seroconversion often is associated with an acute illness. Whether or not the individual has flulike symptoms, the appearance of circulating HIV-1 antigens typically occurs before IgG antibodies against gp41 and p24 develop. HIV-1 antigen then usually disappears following seroconversion but reappears in the latter stages of the disease. The reappearance of antigen usually indicates impending clinical deterioration. An asymptomatic HIV-1 antigen-positive individual is six times more likely to develop AIDS within 3 years than a similar individual who is HIV-1 antigen negative. Thus testing for the presence of the HIV-1 antigen assists clinicians in monitoring the progression of the disease.

Table 38.3

Disease Processes Associated with AIDS

Candidiasis of bronchi, trachea, or lungs
Candidiasis, esophageal
Cervical cancer, invasive
Coccidioidomycosis, disseminated or extrapulmonary
Cryptosporidiosis, chronic intestinal (>1 month's duration)
Cyclospora, diarrheal disease
Cytomegalovirus disease (other than liver, spleen, or lymph nodes)
Cytomegalovirus retinitis (with loss of vision)
Encephalopathy, HIV-related
Herpes simplex: chronic ulcer(s) (>1 month's duration); or bronchitis, pneumonitis, or esophagitis
Histoplasmosis, disseminated or extrapulmonary
Isosporiasis, chronic intestinal (>1 month's duration)
Kaposi's sarcoma
Lymphoma, Burkitt's (or equivalent term)
Lymphoma, immunoblastic (or equivalent term)
Lymphoma, primary, of brain
Mycobacterium avium complex or *M. kansasii*
Mycobacterium tuberculosis, any site
Mycobacterium, other species or unidentified species
Pneumocystis pneumonia
Pneumonia, recurrent
Progressive multifocal leukoencephalopathy
Salmonella septicemia, recurrent
Toxoplasmosis of brain
Wasting syndrome due to AIDS

Source: Data from *MMWR* 41 (No. RR17). 1993 Revised Classification System for HIV Infection and Expanded Surveillance Case Definition for AIDS Among Adolescents and Adults.

leaves the infected person open to opportunistic infections: invasion by pathogens that proliferate widely only because the immune system is defective.

It should be noted that factors other than direct T-cell destruction also may be involved in AIDS pathogenesis. HIV may reduce the immune response by destroying or disabling dendritic cells, which present foreign antigens to T cells. HIV also mutates exceptionally rapidly and thus could evade and eventually overwhelm the immune system. HIV may disrupt the balance between different types of T-helper cells and consequently decrease the killer T-cell population. It is possible that several different mechanisms contribute to T-cell destruction.

New findings suggest still another potential mechanism for the depletion of CD4$^+$ cells (**figure 38.12**). In HIV-infected individuals the loss of CD4$^+$ cells is associated with lymphocyte activation; however, this activation does not result in cell proliferation, as it does normally, but rather in cell death by a mechanism known as programmed cell death or apoptosis (*see pp. 725, 765*).

In 1993 the CDC revised its definition of AIDS to include all HIV-infected persons who have fewer than 200 CD4$^+$ cells per μl or a CD4$^+$ cell percentage of total lymphocytes of less than 14. The reason for this definition is that the development of a partic-

ular opportunistic infection is related to the concentration of CD4$^+$ cells in the blood. Healthy individuals have about 1,000 such cells in every μl of blood. In HIV-1 infected individuals, the number declines by an average of about 40 to 80 cells per μl each year. When the CD4$^+$ cell count falls to between 400 and 200 per μl, the first opportunistic infections and disease processes usually appear (**table 38.3**). Examples of such opportunistic infections and diseases include *Pneumocystis* pneumonia; *Mycobacterium avium-intracellulare* pneumonia (*see pp. 878–79*); toxoplasmosis; herpes zoster infection (figure 38.2b); chronic diarrhea caused by *Cyclospora;* cryptococcal meningitis; *Histoplasma capsulatum* infection; and tuberculosis (*see figure 39.7*).

The third main type of disease caused by HIV-1 involves the central nervous system since virus-infected macrophages can cross the blood-brain barrier. The classical symptoms of central nervous system disease in AIDS patients are headaches, fevers, subtle cognitive changes, abnormal reflexes, and ataxia (irregularity of muscular action). Dementia and severe sensory and motor changes characterize more advanced stages of the disease. Autoimmune neuropathies, cerebrovascular disease, and brain tumors also are common. Histological changes include inflammation of neurons, nodule formation, and demyelination. All evidence indicates that these neurological changes are correlated with higher levels of HIV-1 antigen (figure 38.11) and/or the

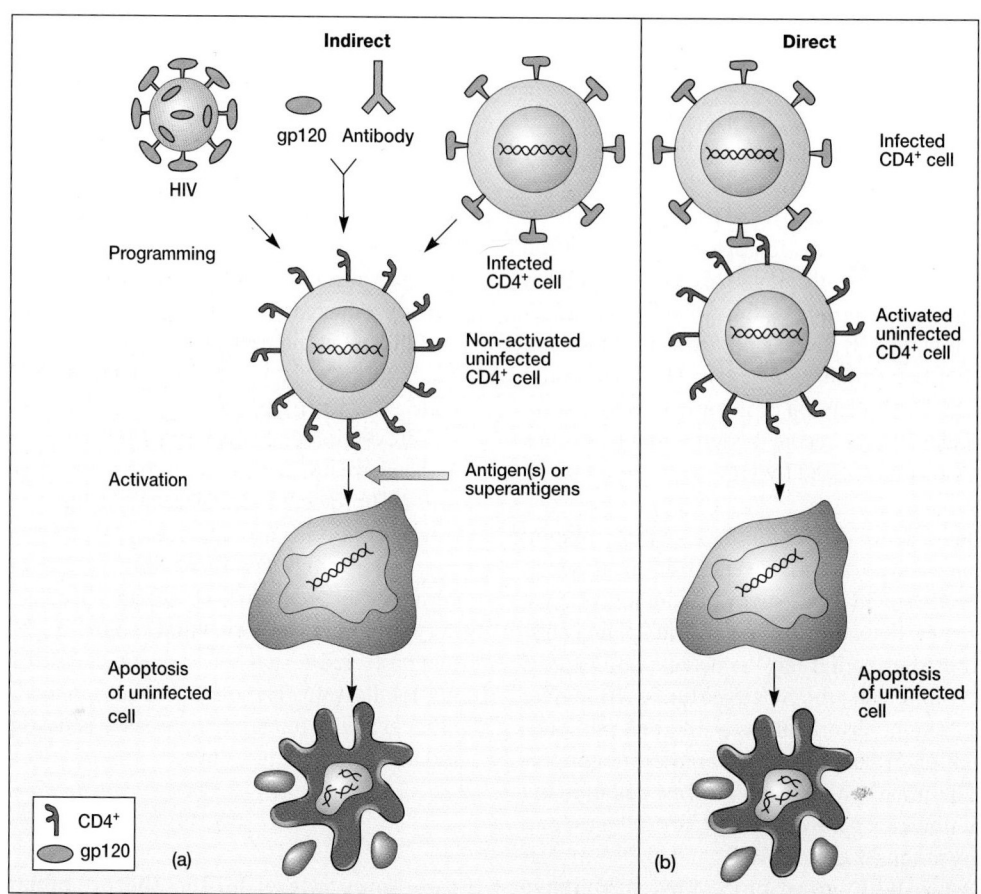

Figure 38.12   Apoptosis and AIDS.   Apoptosis is a homeostatic physiological suicide mechanism in which cell death occurs naturally during normal tissue turnover. Usually apoptosis occurs after activation of a calcium-dependent endogenous endonuclease. Cells undergoing apoptosis display profound structural changes such as a decrease in cell volume, blebbing of the plasma membrane, and nuclear fragmentation. The nuclear DNA is cleaved into short oligonucleosomal length DNA fragments. The dying cell sheds small membrane-bound apoptotic bodies, which are phagocytosed and digested. (**a**) There may be several ways in which an HIV infection can indirectly trigger apoptosis. In all cases an initial event would program or prime the target cell so that apoptosis would be triggered by the binding of antigens or superantigens to the cell's T-cell receptors. Possibly the external gp120 envelope glycoprotein of the HIV virion binds to the CD4 protein on lymphocytes and programs the lymphocyte. A combination of free gp120 and antibodies to gp120 also could stimulate programmed cell death. First, the gp120 would bind to CD4 receptors. Then antibodies would attach to the gp120 and cause clustering of the receptors, thus priming the uninfected CD4$^+$ cell. It also is possible that binding of the infected cell's surface gp120 proteins to the CD4 receptors on an uninfected cell will program the uninfected cell for apoptosis in response to antigens. (**b**) Apoptosis may be directly triggered in an uninfected cell. The gp120 envelope proteins on the surface of an infected cell may combine with the CD4 proteins of an uninfected cell and directly stimulate programmed cell death without activation by antigens.

HIV-1 genome in central nervous system tissue. In AIDS dementia, macrophages and glial cells (supporting cells of the nervous system) are primarily infected and bud new viruses. However, it is unlikely that direct infection of these cells by HIV-1 is responsible for the symptoms. More probably the symptoms arise through either the secretion of viral proteins or viral induction of cytokines that bind to glial cells and neurons. HIV-1 induction of interleukin-1 and tumor necrosis factor-α (TNF-α) may stimulate further viral reproduction and the induction of other cytokines (e.g., interleukin-6, granulocyte-macrophage colony-stimulating factor [GMCSF]). IL-1 and TNF-α in combination with IL-6 and

GMCSF could account for the many clinical and histopathological findings in the central nervous system of AIDS individuals.

The fourth result of an HIV infection is cancer. Individuals infected with HIV-1 have an increased risk of three types of tumors: (1) Kaposi's sarcoma (figure 38.10*b*), (2) carcinomas of the mouth and rectum, and (3) B-cell lymphomas or lymphoproliferative disorders. It seems likely that the depression of the initial immune response enables secondary tumor-causing agents to initiate the cancers.

In 1994 it was discovered that Kaposi's sarcoma–associated herpesvirus (KSHV, also known as human herpesvirus 8, or

HHV-8) is a virus that is consistently present in Kaposi's sarcoma and in primary effusion (body cavity–based) lymphomas. These cancers occur most frequently in AIDS patients. KSHV is a gamma herpesvirus with homology to herpesvirus saimiri and Epstein-Barr virus, both of which can transform lymphocytes. It is now known that this virus is spread in saliva through kissing.

A harmless virus (GB virus C) discovered in 1995 and carried by tens of millions of people worldwide appears to prolong the lives of those who are infected with HIV. This virus decreases mortality, slows damage to the immune system, and boosts the effects of AIDS drugs. How the GB virus C performs its beneficial effects is not known but two possibilities exist. The virus could be directly suppressing HIV replication or it could be stimulating the immune system. There is some recent evidence that it attaches to the same CD4 T-cell receptors used by HIV. This would block virion entrance.

The laboratory diagnosis of AIDS can be by viral isolation and culture or by using assays for viral reverse transcriptase activity or viral antigens (figure 38.11). However, diagnosis is best accomplished through the detection of specific anti-HIV antibodies in the blood. These antibodies may be detected by GENIE or HIVAG-1 (Abbott) rapid tests, the SUDS (Murex) 10 minute assay, enzyme-linked immunoabsorbent assay, indirect immunofluorescence, immunoblot (Western Blot), and radioimmunoprecipitation methods. The most sensitive HIV assay employs the polymerase chain reaction (*see pp. 316–19*). PCR can be used to amplify and detect tiny amounts of viral RNA and cDNA in virions or infected host cells. With proper procedures quantitative PCR assays will provide an estimate of a patient's viral load. This is particularly significant because the level of virions in the blood, as well as the concentration of CD4$^+$ cells, is very predictive of the clinical course of the infection. The probable time to development of AIDS can be estimated from the patient's blood virion level and CD4$^+$ cell count.

At present there is no complete cure for AIDS. Primary treatment is directed at reducing the viral load and disease symptoms, and at treating opportunistic infections and malignancies. The antivirals currently approved for use in HIV disease are of three types. (1) Most reverse transcriptase inhibitors are nucleoside analogues that inhibit the enzyme reverse transcriptase as it synthesizes DNA. Examples include AZT or zidovudine (Retrovir), didanosine (Videx), ddC or zalcitabine (HIVID), stavudine (Zerit), and lamivudine or 3TC (Epivir). (2) The nonnucleoside inhibitors of reverse transcriptase include delavirdine (Rescriptor) and nevirapine (Viramune). (3) The protease inhibitors work by blocking the activity of the HIV protease and thus interfere with virion assembly. Examples include indinavir (Crixivan), ritonavir (Norvir), nelfinavir (Viracept), and saquinavir (Invirase). The most successful treatment approach is to use drug combinations. An effective combination is a cocktail of AZT, lamivudine, and a protease inhibitor such as ritonavir. In many patients the virus disappears from the patient's blood with proper treatment and drug-resistant strains do not seem to arise. Recently it has been discovered that HIV is dormant in the memory T cells, survives the drug cocktail, and is capable of reactivation. Thus patients cannot be completely cured with this drug treatment protocol. It should be noted that side effects can be very severe, and not everyone can tolerate the drugs. If the treatment is not started soon enough, the immune system does not fully recover after HIV levels have fallen.

Another avenue of current research is the development of a vaccine that can (1) stimulate the production of neutralizing antibodies which can bind to the virus envelope and prevent it from entering host cells, and (2) promote the destruction of those cells already infected with the virus. The production of an effective vaccine, if possible, is not yet in sight. One difficulty is that the envelope proteins of the virus (figure 38.8) continually change their antigenic properties.

Prevention and control of AIDS involves screening of blood and heat treatment of blood products to destroy the virus. Education and protected sexual behavior and practices, including the use of condoms, are keys to comprehensive community-based prevention programs. Education of intravenous drug users concerning the need to avoid sharing needles and syringes is also very important in prevention programs. It must be emphasized that protected sexual behavior and the use of condoms is safer but not totally safe. The only absolute protection against AIDS and other sexually transmitted diseases is abstinence and completely monogamous relationships.

Cold Sores

Cold sores or **fever blisters (herpes labialis)** are caused by the herpes simplex virus type 1 (HSV-1). Like all herpesviruses, it is a double-stranded DNA virus with an enveloped, icosahedral capsid. The term herpes is derived from the Greek word meaning "to creep," and clinical descriptions of herpes labialis (lips) go back to the time of Hippocrates (circa 400 B.C.). Transmission is through direct contact of epithelial tissue surfaces with the virus (*see figure 18.5*). A blister(s) develops at the inoculation site (**figure 38.13**) because of host- and viral-mediated tissue de-

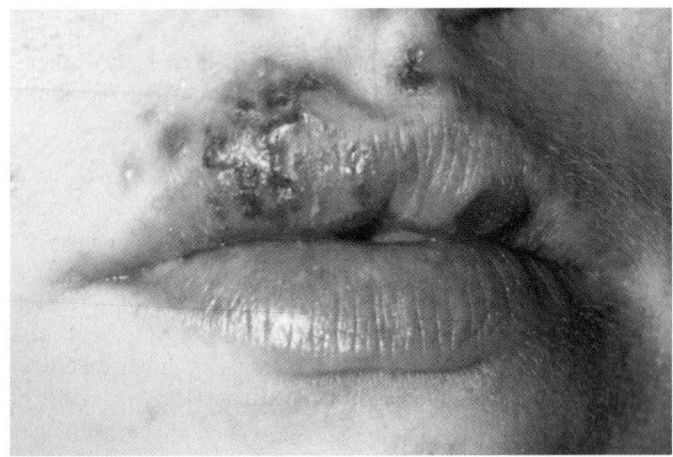

Figure 38.13 Cold Sores. Herpes simplex fever blisters on the lip, caused by herpes simplex type 1 virus.

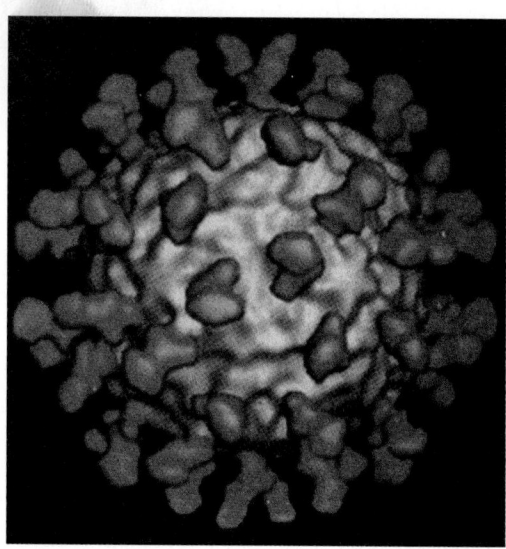

(a)

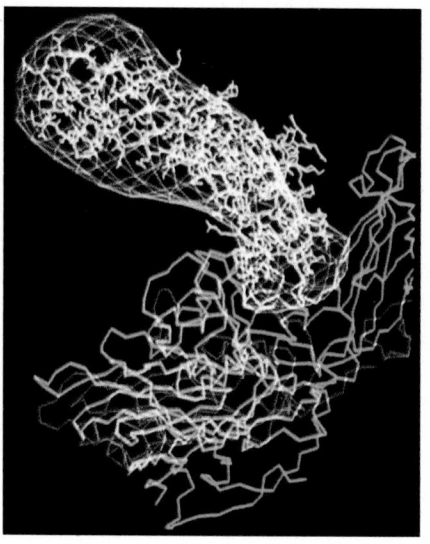

(b)

Figure 38.14 Role of an Integrin (Intercellular Adhesion Molecule-1) in Rhinovirus Uptake Into Cells of the Upper Respiratory Tract. Rhinoviruses are a large family of viruses that are the single major cause of acute respiratory infections in humans. ICAM-1 (intercellular adhesion molecule-1) has been identified as the cell surface receptor for the majority of these rhinoviruses. (a) In this computer-generated model, the virus-binding portion of ICAM-1 is shown in orange-red attaching to a rhinovirus (gray icosahedral protein capsid in the center). (b) This diagram indicates how the rhinovirus surface proteins (shown in blue-green) bind to the ICAM-1 molecule (red-yellow).

struction. Most blisters involve the epidermis and surface mucous membranes of the lips, mouth, and gums (**gingivostomatitis**). The blisters generally heal within a week. However, after a primary infection, the virus travels to the trigeminal nerve ganglion, where it remains in a latent state for the lifetime of the infected person. Stressful stimuli such as excessive sunlight, fever, trauma, chilling, emotional stress, and hormonal changes can reactivate the virus. Once reactivated, the virus moves from the trigeminal ganglion down a peripheral nerve to the border of the lip or other parts of the face to produce another fever blister. Primary and recurring infections also may occur in the eyes, causing **herpetic keratitis** (inflammation of the cornea)—currently a major cause of blindness in the United States. The drugs vidarabine (Vira-A) and acyclovir (Zovirax) are effective against cold sores. By adulthood, 70 to 90% of all people in the United States have been infected and have type 1 herpes antibodies. The "gold standard" for the detection of herpes is cell culture. However, cell culture requires up to 14 days for results. The rapid detection of herpes simplex virus (HSV) antigen is by several enzyme immunoassay kits (HERPCHEK, VIDAS, IDEIA, Sure-Cell). This rapid test is especially useful in cases of pregnant women with genital infections and individuals who are particularly susceptible to severe infections.

Common Cold

The **common cold** [**coryza:** Greek *koryza,* discharge from the nostrils] is one of the most frequent infections experienced by humans of all ages. About 50% of the cases are caused by rhinoviruses [Greek *rhinos,* nose], single-stranded RNA viruses in the family *Picornaviridae* (*see figure 16.10*k). There are over 115 distinct serotypes, and each of these antigenic types has a varying capacity to infect the nasal mucosa and to cause a cold. In addition, immunity to many of them is transitory. Several other respiratory viruses are also associated with colds (e.g., coronaviruses

and parainfluenza viruses). Thus colds are common because of the diversity of rhinoviruses, the involvement of other respiratory viruses, and the lack of a durable immunity.

Rhinoviruses provide an excellent example of the medical relevance of research on virus morphology. The complete rhinovirus capsid structure has recently been elucidated with the use of X-ray diffraction techniques. The results help explain rhinovirus resistance to human immune defenses. Human antibodies against the virus combine with four types of surface proteins. Unfortunately there are at least 89 variants of these four antigens, making the production of a cold vaccine from them unlikely. The structural studies have shown that the site that recognizes and binds to cell surface molecules during infection lies at the bottom of a surface cleft about 12 Å deep and 15 Å wide. Thus the binding site is well protected from the immune system while it carries out its functions, and the preparation of a vaccine directed against the binding site is not feasible. Possibly drugs that could fit in the cleft and interfere with virus attachment can be designed.

Viral invasion of the upper respiratory tract is the basic mechanism in the pathogenesis of a cold. The virus enters the body's cells by binding to the adhesion molecule ICAM-1 (**figure 38.14**). The clinical manifestations include the familiar nasal stuffiness and/or partial obstruction, sneezing, scratchy throat, and a watery discharge from the nose. The discharge becomes thicker and assumes a yellowish appearance over several days. General malaise is commonly present. The disease usually runs its course in about a week. Diagnosis of the common cold is made from observations of clinical symptoms. There are no procedures for direct examination of clinical specimens or for serological diagnosis.

The source of the cold viruses may be infected individuals excreting viruses in nasal secretions, airborne transmission over short distances by way of moisture droplets, or transmission on contaminated hands or fomites. Epidemiological studies of rhinovirus colds have shown that the familiar explosive, noncontained sneeze (*see figure 37.6*) may not play an important role in

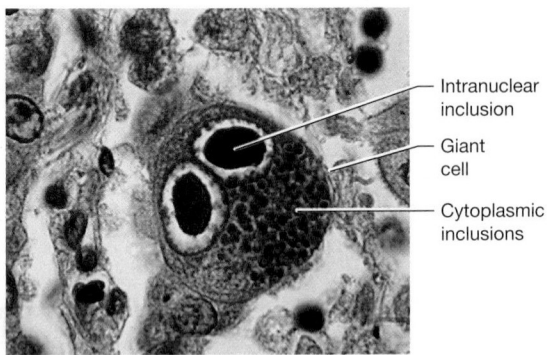

- Intranuclear inclusion
- Giant cell
- Cytoplasmic inclusions

Figure 38.15 Cytomegalovirus Inclusion Disease. Light micrograph of a giant cell in a lung section infected with the cytomegalovirus (×480). The intranuclear inclusion body has a typical "owl-eyed" appearance because of its surrounding clear halo. Sometimes virus inclusions also are visible in the cytoplasm.

virus spread. Rather, hand-to-hand contact between a rhinovirus "donor" and a susceptible "recipient" is more likely. The common cold occurs worldwide with two main seasonal peaks, spring and early autumn. Infection is most common early in life and generally decreases with an increase in age. Nothing is available for treating the common cold except additional rest, extra fluids, and the use of aspirin for alleviating local and systemic discomfort.

Cytomegalovirus Inclusion Disease

Cytomegalovirus inclusion disease is caused by the human cytomegalovirus (HCMV), a member of the family *Herpesviridae*. The virus contains a core with double-stranded DNA, an icosahedral capsid, and a phospholipid-rich envelope. Most people become infected with this virus at some time during their life; in the United States, as many as 80% of individuals older than 35 years have been exposed to this virus and carry a lifelong infection. Although most HCMV infections are asymptomatic, certain patient groups are at risk to develop serious illness and long-term effects from an HCMV infection. For example, this virus remains the leading cause of congenital virus infection in the United States, a significant cause of transfusion-acquired infections, and a frequent contributor to morbidity and mortality among organ transplant recipients and immunocompromised individuals (especially AIDS patients). Because the virus persists in the body, it is shed for several years in saliva, urine, semen, and cervical secretions.

The HCMV can infect any cell of the body, where it multiplies slowly and causes the host cell to swell in size—hence the prefix cytomegalo, which means "an enlarged cell." Infected cells contain the unique **intranuclear inclusion bodies** and cytoplasmic inclusions (**figure 38.15**). In fatal cases, cell damage is seen in the gastrointestinal tract, lungs, liver, spleen, and kidneys. Normally cytomegalovirus inclusion disease symptoms resemble those of infectious mononucleosis.

Laboratory diagnosis is by viral isolation from urine, blood, lung, semen, or tissue. Serological tests (e.g., immunofluores-

cence, complement fixation, ELISA, and immunohistologic staining) and a rapid test kit (CMV-vue kit) also are available.

Epidemiologically the virus has a worldwide distribution, especially in developing countries where infection is universal by childhood. The prevalence of this disease increases with a lowering of socioeconomic status and hygienic practices. The only drugs available, ganciclovir (Cytovene-IV) and cidofovir (Vistide), are used only for high-risk patients. Infection can be prevented by avoiding close personal contact (including sexual) with an actively infected individual. Transmission by blood transfusion or organ transplantation can be avoided by using blood or organs from seronegative donors.

Genital Herpes

Genital herpes is caused by the herpes simplex virus type 2 (HSV-2). The HSV-2 virus (**figure 38.16a**) is classified in the alphaherpes subfamily of the family *Herpesviridae*. All members of this subfamily have a very short replication cycle. The core DNA is linear and double stranded. The HSV-2 envelope contains at least eight glycoproteins. HSV-2 is most frequently transmitted by sexual contact (*see table 39.4, p. 904 for a summary of the major sexually transmitted diseases*). Infection begins when the virus is introduced into a break in the skin or mucous membranes. The virus infects the epithelial cells of the external genitalia, the urethra, and the cervix. Rectal and pharyngeal herpes are also caused by sexual contact.

The HSV-2 genome must first enter an epithelial cell for the initiation of infection. The initial association is between the proteoglycans of the epithelial cell surface and viral glycoprotein C. This is followed by a specific interaction with one of several cellular receptors collectively termed HVEMs for "herpesvirus entry mediators." The association also requires the specific interaction with glycoprotein D. Fusion with the epithelial cell plasma membrane follows. This requires the action of a number of other viral glycoproteins. The viral capsid along with some viral tegument proteins then migrates to nuclear envelope pores along the cellular microtubule transport machinery. This "docking" is thought to result in the viral DNA being injected through the nuclear envelope pores while the capsid remains in the cytoplasm. Some tegument proteins also enter the nucleus with the viral genome.

In the infected person, there is an active and latent phase. After an incubation period of about a week, the active phase begins. During the active phase, the virus multiplies explosively—between 50,000 and 200,000 new virions are produced from each infected cell. During this replication cycle, HSV-2 inhibits its host cell's metabolism and degrades its DNA. As a result, the cell dies. Such an active infection may be symptom-free, or painful blisters in the infected tissue (figure 38.16b) may occur. The blisters are the result of cell lysis and the development of a local inflammatory response; they contain fluid and infectious viruses. A fever, headache, muscle aches and pains, a burning sensation, and genital soreness are frequently present during the active phase. Although blisters generally heal spontaneously in a few weeks,

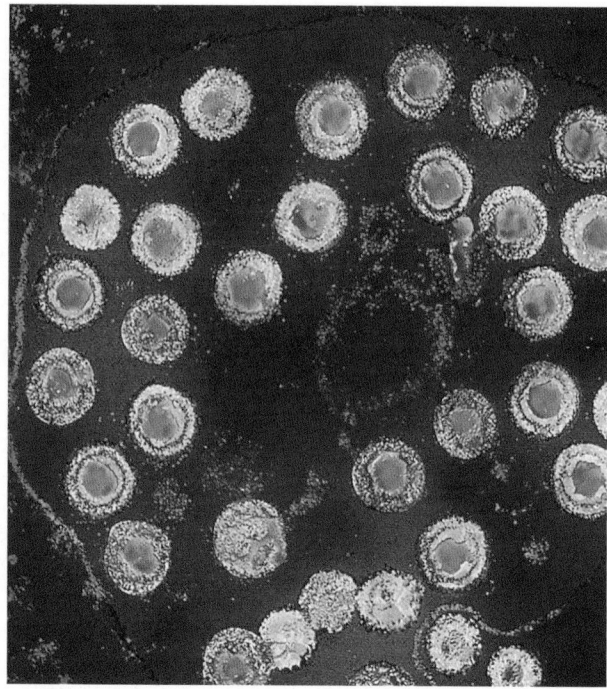

(a)

Vesicles

(b)

Figure 38.16 Genital Herpes. (a) Herpes simplex virus type 2 (yellow and green) inside an infected cell (*see also figure 16.17e*). (b) Herpes vesicles on the penis. The vesicles contain fluid that is infectious.

the viruses retreat to nerve cells in the sacral plexus, where they remain in a latent form. During the latent phase, the viral genome becomes incorporated into the nucleus of the host cell. During this phase, the host cell does not die. Because the viral genes are not expressed, the infected person is symptom-free. However, periodically the viruses multiply and migrate down nerve fibers to the skin or mucous membranes that the nerve supplies, where

they produce new blisters. Activation may be due to sunlight, sexual activity, illness accompanied by fever, hormones, or stress. It should be noted that both the primary infection and reactivation can occur without any symptoms and apparently healthy people can transmit HSV-2 to their sexual partners or their newborns.

Besides being transmitted by sexual contact, herpes can be spread to an infant during vaginal delivery, leading to **congenital (neonatal) herpes.** Congenital herpes is one of the most life-threatening of all infections in newborns, affecting approximately 1,500 to 2,200 babies per year in the United States. It can result in neurological involvement as well as blindness. As a result any female who has had genital herpes should have a caesarean section instead of delivering vaginally. For unknown reasons, the HSV-2 virus is also associated with a higher-than-normal rate of cervical cancer and miscarriages.

Although there is no cure for genital herpes, oral use of the antiviral drugs acyclovir (Zovirax or Valtrex) and famciclovir (Famvir) has proven to be effective in ameliorating the recurring blister outbreaks. Topical acyclovir is also effective in reducing virus shedding, the time until the crusting of blisters occurs, and new lesion formation. Idoxuridine and trifluridine are used to treat herpes infections of the eye.

In the United States the incidence of genital herpes has increased so much during the past decade that it is now a very common sexually transmitted disease. It is estimated that over 25 million Americans (20% of adults) are infected with the herpes simplex virus type 2.

Human Herpesvirus 6 Infections

Human herpesvirus 6 (HHV-6) is the etiologic agent of **exanthem subitum** [Greek *exanthema,* rash] in infants. HHV-6 is a unique member of the family *Herpesviridae* that is distinct serologically and genetically from the other herpesviruses. The virus envelope encloses an icosahedral capsid and a core containing double-stranded DNA. The disease caused by HHV-6 was originally termed **roseola infantum** and then given the ordinal designation **sixth disease** to differentiate it from other exanthems and roseolas. Exanthem subitum is a short-lived disease characterized by a high fever of 3 to 4 days' duration, after which the temperature suddenly drops to normal and a macular rash appears on the trunk and then spreads to other areas of the body. HHV-6 infects over 95% of the United States infant population, and most children are seropositive for HHV-6 by 3 years of age. CD4$^+$ T cells are the main site of viral replication, whereas monocytes are in an infected, latent state. The tropism of HHV-6 appears to be wide, including CD8$^+$ T cells, natural killer cells, and probably epithelial cells. In adults HHV-6 is commonly found in peripheral-blood mononuclear cells and saliva, suggesting that the infection is lifelong. Since the salivary glands are the major site of latent infection, transmission is probably by way of saliva.

HHV-6 also produces latent and chronic infections and is occasionally reactivated in immunocompromised hosts leading to pneumonitis. Furthermore, HHV-6 has been implicated in several other diseases (lymphadenitis, multiple sclerosis, and infectious mononucleosis-like syndrome or chronic fatigue syndrome) in immunocompetent adults. Diagnosis is either by immunofluorescence or enzyme immunoassay. To date, there is no antiviral therapy or prevention.

Human Parvovirus B19 Infections

Since its discovery in 1974, **human parvovirus B19** (family *Parvoviridae,* genus *Parvovirus*) has emerged as a significant human pathogen. B19 virions are uniform, icosahedral, unenveloped particles approximately 23 nm in diameter. The genome of B19 is a single-stranded DNA. There is a spectrum of disease caused by parvovirus B19 infection, ranging from mild symptoms (fever, headache, chills, malaise) in normal persons, **erythema infectiosum** in children (**fifth disease**), and a joint disease syndrome in adults; to more serious diseases such as aplastic crisis in persons with sickle cell disease and autoimmune hemolytic anemia, and pure red cell aplasia due to persistent B19 virus infection in immunocompromised individuals. The B19 parvovirus can also infect the fetus, resulting in anemia and fetal hydrops (the accumulation of fluid in tissues). It is assumed that the natural mode of infection is by the respiratory route. A variety of techniques are available for the detection of the B19 virus. Antiviral antibodies appear to represent the principal means of defense against B19 parvovirus infection and disease. The treatment of individuals suffering from acute and persistent B19 infections with commercial immunoglobins containing anti-B19 and human monoclonal antibodies to B19 have shown to be effective therapy. Most of the time infection is followed by lifelong immunity. Vaccine phase I trials show promising results.

Leukemia

Certain **leukemias** in humans are caused by two retroviruses: human T-cell lymphotropic virus I (HTLV-I) and HTLV-II. HTLV-I and HTLV-II are members of the family *Retroviridae.* They have a nuclear core containing two positive-sense single-stranded RNA genomes. Once a cell is infected, the RNA genome is converted by reverse transcriptase to DNA and integrates into the host's genome. The viruses are transmitted by transfusions of contaminated blood, among drug addicts sharing needles, by sexual contact, across the placenta, from the mother's milk, or by mosquitoes. *Retroviruses and their replication (pp. 395–97); Viruses and cancer (pp. 400–1)*

HTLV-1 causes **adult T-cell leukemia.** Once within the body the HTLV-I virus enters white blood cells and integrates into the cellular genome, where it activates the growth-promoting genes. The transformed cell proliferates extensively, and death generally results from the explosive proliferation of the leukemia cells or from opportunistic infections. To date, no effective treatment exists.

In 1982 the second human retrovirus (HTLV-II) was shown to be the agent responsible for hairy-cell leukemia. This virus shares the same disease-causing mechanism as HTLV-I. Hairy-cell leukemia gets its name from the many membrane-derived protrusions that give white blood cells the appearance of being "hairy" (**figure 38.17**). This leukemia is a chronic, progressive lympho-

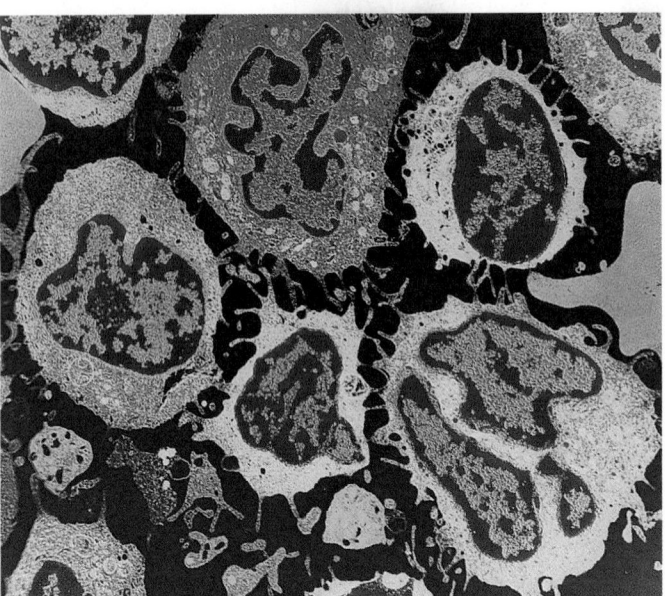

Figure 38.17   Hairy-Cell Leukemia.   False-color transmission electron micrograph ($\times 3{,}100$) of abnormal B lymphocytes. Notice that the lymphocytes are covered with characteristic hairlike membrane-derived protrusions.

proliferative disease. The malignancy is believed to originate in a stage of B-cell development. The bone marrow, spleen, and liver become infiltrated with malignant cells. This lowers the person's immunity. The primary cause of mortality is bacterial and other opportunistic infections. IFN-α n3 (Alferon N) has shown some promise for treatment in certain cases.

Mononucleosis (Infectious)

The Epstein-Barr virus (EBV) is a member of the family *Herpesviridae.* As a member of the family *Herpesviridae,* EBV has the characteristic 120 nm enveloped morphology, with 162 capsomeres in icosahedral arrangement. Its double-stranded DNA exists both as a linear form in the mature viron and a circular episomal form in latently infected cells. EBV is the etiologic agent of **infectious mononucleosis (mono),** a disease whose symptoms closely resemble those of cytomegalovirus-induced mononucleosis. Because the Epstein-Barr virus occurs in oropharyngeal secretions, it can be spread by mouth-to-mouth contact (hence the terminology infectious and kissing disease) or shared drinking bottles and glasses. A person gets infected when the virus from someone else's saliva makes its way into epithelial cells lining the throat. After a brief bout of replication in the epithelial cells, the new viruses are shed and infect memory B cells. Infected B cells rapidly proliferate and take on an atypical appearance (Downey cells) that is useful in diagnosis. The disease is manifested by enlargement of the lymph nodes and spleen, sore throat, headache, nausea, general weakness and tiredness, and a mild fever that usually peaks in the early evening. The disease lasts for 1 to 6

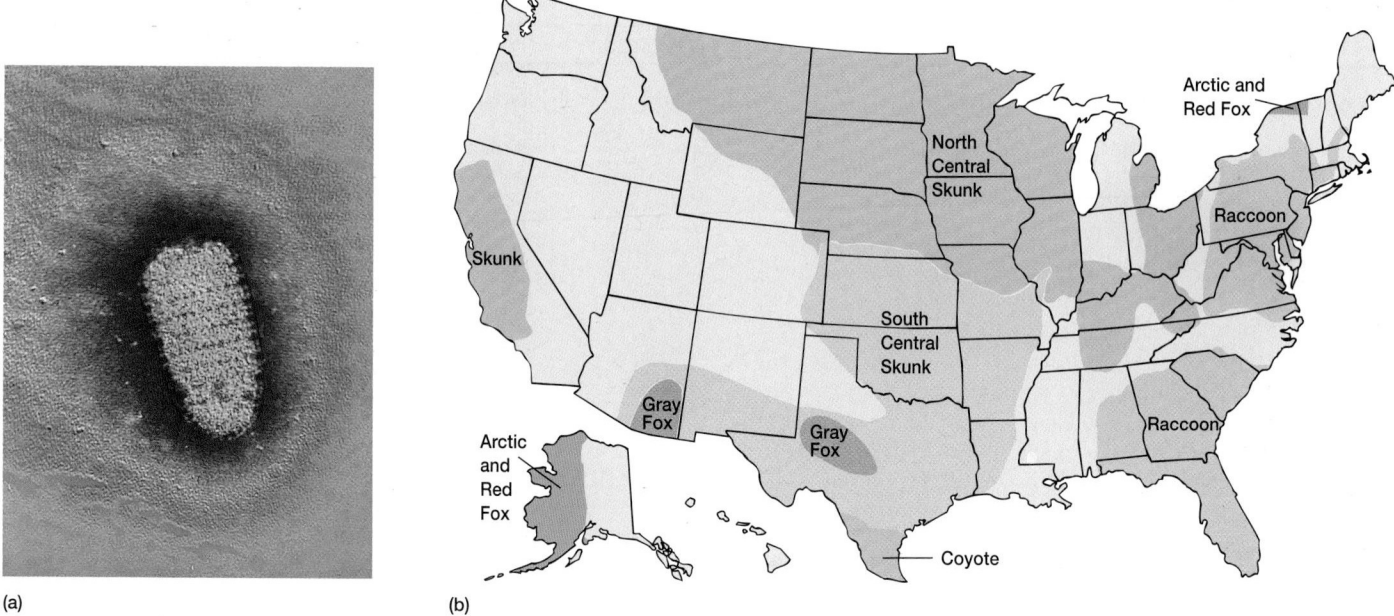

(a)   (b)

Figure 38.18   Rabies.   (a) Electron micrograph of the rabies virus (yellow) (×36,700). Note the bullet shape. The external surface of the virus contains spikelike glycoprotein projections that bind specifically to cellular receptors. **(b)** In the United States rabies is found in terrestrial animals in 10 distinct geographic areas. In each area a particular species is the reservoir and one of five antigenic variants of the virus predominates as illustrated by the five different colors. Although not shown, another eight viral variants are found in insectivorous bats and cause sporadic cases of rabies in terrestrial animals throughout the country. Absence of a strain does not imply absence of rabies.

weeks and is self-limited. Like other herpesviruses, EBV does become latent in the memory B-cell population.

Treatment of mononucleosis is largely supportive and includes plenty of rest. Diagnosis of mononucleosis is made with a serological test for nonspecific (heterophile) antibodies. Several rapid tests are on the market.

The peak incidence of mononucleosis occurs in people 15 to 25 years of age. Collegiate populations, particularly those in the upper-socioeconomic class, have a high incidence of the disease. About 50% of college students have no immunity, and approximately 15% of these can be expected to contract mononucleosis. People in lower-socioeconomic classes tend to acquire immunity to the disease because of early childhood infection. The Epstein-Barr virus may well be the most common virus in humans as it infects 80 to 90% of all adults worldwide. EBV infections are associated with the cancers Burkitt's lymphoma in tropical Africa and nasopharyngeal carcinoma in Southeast Asia, East and North Africa, and in Eskimos.

Rabies

Rabies [Latin *rabere,* rage or madness] has been the object of human fascination, torment, and fear since the disease was first recognized. Prior to Pasteur's development of an antirabies vaccine, few words were more terrifying than the cry of "mad dog!" Improvements in prevention during the past 50 years have led to the almost complete elimination of indigenously acquired rabies in

the United States. Worldwide, virtually all cases of human rabies are attributable to dog bites in developing countries where canine rabies is still endemic, accounting for up to 40,000 deaths per year.

Rabies is caused by a number of different strains of highly neurotropic viruses. Most belong to a single serotype in the genus *Lyssavirus* [Greek *lyssa,* rage or rabies], family *Rhabdoviridae.* The bullet-shaped virion (**figure 38.18***a; see also figure 16.17*c) contains a single-stranded, negative-sense RNA genome.

As previously noted, in the United States rabies is primarily a disease of animals. Most wild animals can become infected with rabies, but susceptibility varies according to species. Foxes, coyotes, and wolves are the most susceptible; intermediate are skunks, raccoons, insectivorous bats, and bobcats; and opossums are quite resistant.

The virus multiplies in the salivary glands of an infected host. It is transmitted to humans or other animals by the bite of an infected animal whose saliva contains the virus; by aerosols of the virus that can be spread in caves where bats roost; or by contamination of scratches, abrasions, open wounds, and mucous membranes with saliva from an infected animal.

After inoculation, a region of the virions' glycoprotein envelope spike attaches to the plasma membrane of nearby skeletal muscle cells, the virus enters the cells, and multiplication of the virus occurs. When the concentration of the muscle virus is sufficient, the virus enters the nervous system through unmyelinated sensory and motor terminals; the reported binding site is the nicotinic acetylcholine receptor. Since the virus is sequestered from the

Table 38.4

Characteristics of Hepatitides Caused by Hepatotropic Viruses[a]

Disease (Virus)	Genome	Classification	Transmission	Outcome	Prevention
Hepatitis A (hepatitis A)	RNA	*Picornaviridae, Hepatovirus*	Fecal-oral	Subclinical, acute infection	Killed HAV (Havrix vaccine)
Hepatitis B (hepatitis B)	DNA	*Hepadnaviridae, Orthohepadnavirus*	Blood, needles, body secretions, placenta, sexually	Subclinical, acute chronic infection; cirrhosis; primary hepatocarcinoma	Recombinant HBV vaccines
Hepatitis C (hepatitis C)	RNA	*Flaviviridae, Pestivirus, or Flavivirus* (?)	Blood, sexually	Subclinical, acute chronic infection; primary hepatocarcinoma	Routine screening of blood
Hepatitis D (hepatitis D)	RNA	Unclassified	Blood, sexually	Superinfection or coinfection with HBV	HBV vaccine
Hepatitis E (hepatitis E)	RNA	*Caliciviridae* (?)	Fecal-oral	Subclinical, acute infection (but high mortality in pregnant women)	Improve sanitary conditions
Hepatitis G	RNA	*Flaviviridae*	Sexually, parenterally	Unknown	Unknown

[a]Hepatitis TTV has been discovered but not well characterized. Thus it is not included in this table.

immune system, infection can no longer be halted by immunization. The virus spreads by retrograde axonal flow at 8 to 20 mm per day until it reaches the spinal cord, when the first specific symptoms of the disease—pain or paresthesia at the wound site—may occur. A rapidly progressive encephalitis develops as the virus quickly disseminates through the central nervous system. The virus then spreads throughout the body along the peripheral nerves, including those in the salivary glands, where it is shed in the saliva.

Within brain neurons the virus produces characteristic **Negri bodies,** masses of viruses or unassembled viral subunits that are visible in the light microscope. In the past the diagnosis of rabies consisted solely of examining nervous tissue for the presence of these bodies. Today diagnosis is based on direct immunofluorescent-antibody (dIFA) of brain tissue, virus isolation, detection of Negri bodies, and a rapid rabies enzyme-mediated immunodiagnosis test.

Symptoms of rabies usually begin 2 to 16 weeks after viral exposure and include anxiety, irritability, depression, fatigue, loss of appetite, fever, and a sensitivity to light and sound. The disease quickly progresses to a stage of paralysis. In about 50% of all cases, intense and painful spasms of the throat and chest muscles occur when the victim swallows liquids. The mere sight, thought, or smell of water can set off spasms. Consequently rabies has been called hydrophobia (fear of water). Death results from destruction of the regions of the brain that regulate breathing.

Safe and effective vaccines (human diploid-cell rabies vaccine HDCV [Imovax Rabies] or rabies vaccine adsorbed [RVA]) against rabies are available; however, to be effective they must be given soon after the person has been infected. Veterinarians and laboratory personnel, who have a high risk of exposure to rabies, usually are immunized every 2 years and tested for the presence of suitable antibody titer. About 30,000 people annually receive this treatment. In the United States fewer than 10 cases of rabies occur yearly in humans; about 8,000 cases of animal rabies are reported each year from various sources (figure 38.18b). Prevention and control involves preexposure vaccination of dogs and cats, postexposure vaccination of humans, and preexposure vaccination of humans at special risk (persons spending a month or more in countries where rabies is common in dogs).

Viral Hepatitides

Any infection that results in inflammation of the liver is called **hepatitis** [pl., hepatitides] [Greek *hepaticus,* liver]. Currently 11 viruses are recognized as causing hepatitis. Two are herpesviruses (cytomegalovirus [CMV] and Epstein-Barr virus [EBV]) and 9 are hepatotropic viruses.

EBV and CMV cause mild, self-resolving forms of hepatitis with no permanent hepatic damage. Both viruses cause the typical infectious mononucleosis syndrome of fatigue, nausea, and malaise.

Of the nine human hepatotropic viruses, only five are well characterized; hepatitis G (**table 38.4**) and TTV (transfusion-transmitted virus) are newly discovered viruses. Hepatitis A (sometimes called infectious hepatitis), and hepatitis E (formerly enteric-transmitted NANB hepatitis), are transmitted by fecal-oral contamination and discussed in the section on food- and water-borne diseases (section 38.4). The other major types include hepatitis B (sometimes called serum hepatitis), hepatitis C (formerly non-A, non-B hepatitis), and hepatitis D (formerly delta hepatitis).

Hepatitis B (serum hepatitis) is caused by the hepatitis B virus (HBV), a double-stranded circular DNA virus of complex structure. HBV is classified as an *Orthohepadnavirus* within the family *Hepadnaviridae.* Serum from individuals infected with hepatitis B contains three distinct antigenic particles: a spherical 22 nm particle, a 42 nm spherical particle (containing DNA and DNA polymerase) called the **Dane particle,** and tubular or filamentous particles that vary in length (**figure 38.19**). The small spherical and tubular particles are the unassembled components of the Dane particle—the infective form of the virus. The unassembled particles contain **hepatitis B surface antigen**

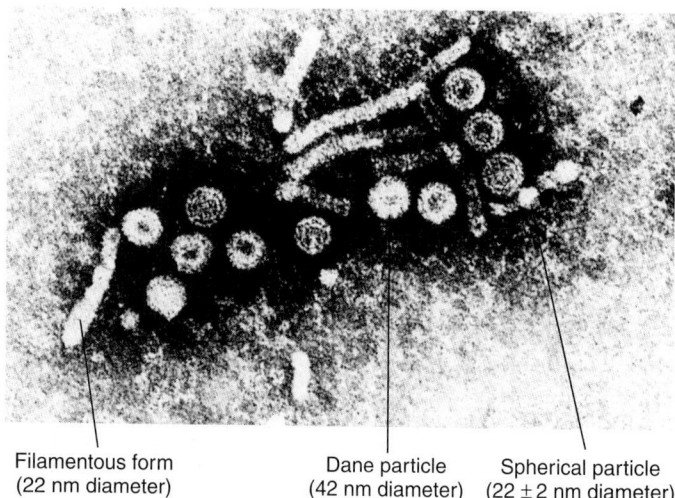

Filamentous form (22 nm diameter)   Dane particle (42 nm diameter)   Spherical particle (22 ± 2 nm diameter)

Figure 38.19   Hepatitis B Virus in Serum.   Electron micrograph (×210,000) showing the three distinct types of hepatitis B antigenic particles. The spherical particles and filamentous forms are small spheres or long filaments without an internal structure, and only two of the three characteristic viral envelope proteins appear on their surface. Dane particles are the complete, infectious virion.

(HBsAg) whose presence in the blood is (1) an indicator of hepatitis B infection, (2) the basis for the large-scale screening of blood for the hepatitis B virus, and (3) the basis for the first vaccine for human use developed by recombinant DNA technology.

The hepatitis B virus is normally transmitted through blood transfusions, contaminated equipment, drug users' unsterile needles, or any body secretion (saliva, sweat, semen [*see table 39.4, p. 904*], breast milk, urine, feces). The virus also can pass from the blood of an infected mother through the placenta to infect the fetus. Each year an estimated 300,000 people in the United States are infected with HBV. About 5,000 persons die yearly from hepatitis-related cirrhosis and about 1,000 die from HBV-related liver cancer. (HBV is second only to tobacco as a known cause of human cancer.) Worldwide, HBV infects over 200 million people.

The clinical signs of hepatitis B vary widely. Most cases are asymptomatic. However, sometimes fever, loss of appetite, abdominal discomfort, nausea, fatigue, and other symptoms gradually appear following an incubation period of 1 to 3 months. The virus infects liver hepatic cells and causes liver tissue degeneration and the release of liver-associated enzymes (transaminases) into the bloodstream. This is followed by jaundice, the accumulation of bilirubin (a breakdown product of hemoglobin) in the skin and other tissues with a resulting yellow appearance. Chronic hepatitis B infection also causes the development of primary liver cancer, hepatocellular carcinoma.

General measures for prevention and control involve (1) excluding contact with HBV-infected blood and secretions, and

minimizing needle-sticks by scrupulous technique; (2) passive prophylaxis with intramuscular injection of hepatitis B immune globulin within 7 days of exposure; and (3) active prophylaxis with two recombinant vaccines: Engerix-B, and Recombivax HB. These vaccines are widely used by health professionals who are at increased risk of contacting the hepatitis B virus. Other individuals who should be vaccinated include:

1. Contacts of HBV carriers (e.g., household members, sex partners, institutional inmates)
2. International travelers (*see table 37.5*)
3. Sexually active homosexual males
4. Hemodialysis patients
5. Recipients of blood and related products that are possibly HBV contaminated

Another form of hepatitis is called **hepatitis C.** The virion has an 80 nm diameter, has a lipid coat, contains a single strand of linear ssRNA, and has been designated the hepatitis C virus (HCV) within the family *Flaviviridae.* HCV has recently been classified into multiple genotypes. This virus is transmitted by contact with virus-contaminated blood, by the fecal-oral route, by in utero transmission from mother to fetus, sexually, or through organ transplantation. Diagnosis is made by a first-generation enzyme-linked immunosorbent assay (ELISA), which detects serum antibody to a recombinant antigen of HCV. Hepatitis C is found worldwide. In recent years hepatitis C has accounted for more than 90% of hepatitis cases developing after a blood transfusion. Worldwide, hepatitis C has reached epidemic proportions, with more than 1 million new cases reported annually. In the United States alone, nearly 4 million persons are infected and 30,000 new cases occur annually. Currently, HCV is responsible for 8,000 to 10,000 deaths annually in the United States, and without effective intervention, that number is predicted to triple in the next 10 years. Furthermore, HCV is the leading reason for liver transplantation in the United States. Treatment is with recombinant IFN-α 2b or IFN-α 2a (Intron A, Roferon-A) three times weekly for 6 months.

In 1977 a cytopathic hepatitis agent termed the **Delta agent** was discovered. Later it was called the hepatitis D virus (HDV) and the disease **hepatitis D** designated. HDV is a unique agent in that it is dependent on the hepatitis B virus to provide the envelope protein (HBsAg) for its RNA genome. Thus HDV only replicates in liver cells in which the HBV also is actively replicating. Furthermore, the RNA of the HDV is smaller than the RNA of the smallest picornaviruses and its circular conformation differs from the linear structure typical of animal RNA viruses. (The nearest relatives of the HDV are in the plant kingdom—the viroids, virusoids, and satellite RNAs [*see chapter 18*].) HDV is spread only to persons who are already infected with HBV (superinfection) or to individuals who get both viruses at once (coinfection). The primary laboratory tools for the diagnosis of an HDV infection are serological tests for anti-delta antibodies. Treatment of patients with chronic HDV remains difficult. Early studies indicate that some positive results can be obtained with alpha interferon for 3 months to 1 year. Worldwide, there are approximately 300 million HBV carriers, and available data indicate that no fewer than

5% of these are infected with HDV. Thus because of the propensity of HDV to cause fulminant as well as chronic liver disease, continued incursion of HDV into areas of the world where persistent hepatitis B infection is endemic will have serious implications. Prevention and control involves the widespread use of the hepatitis B vaccine.

Recently the presence of two other forms of hepatitis have been identified: Hepatitis F (causing fulminant, posttransfusion hepatitis) and a syncytial giant-cell hepatitis (hepatitis G) with viruslike particles resembling the measles virus. **Hepatitis G** (HGV) is a member of the *Flaviviridae* family. It has been cloned and is widely distributed in humans. HGV can be transmitted parenterally or sexually. The significance of HGV infections in liver disease is not clear. Further virologic, epidemiological, and molecular efforts to characterize these new agents and their diseases are currently being undertaken.

1. Describe the AIDS virus and how it cripples the immune system. How is the virus transmitted? What types of pathological changes can result?
2. Why do people periodically get cold sores? Describe the causative agent.
3. Why do people get the common cold so frequently? How are cold viruses spread?
4. Give two major ways in which herpes simplex virus type 2 is spread. Why do herpes infections become active periodically?
5. What two types of leukemias are caused by viruses?
6. Describe the causative agent and some symptoms of mononucleosis and exanthem subitum.
7. How does the rabies virus cause death in humans?
8. What are the different causative viruses of hepatitis and how do they differ from one another? How can one avoid hepatitis? Do you know anyone who is a good candidate for infection with these viruses?

38.4 FOOD-BORNE AND WATERBORNE DISEASES

Food and water have been recognized as potential carriers of disease since the beginning of recorded history. Collectively more infectious diseases occur by these two routes than any other. A few human viral diseases that are food- and waterborne are now discussed.  Water-based diseases (section 29.5); Diseases and food (section 41.4)

Gastroenteritis (Viral)

Acute viral gastroenteritis (inflammation of the stomach or intestines) is caused by four major categories of viruses: rotaviruses (**figure 38.20**), Norwalk and Norwalk-like viruses, other caliciviruses, and astroviruses. The medical importance of these viruses is summarized in **table 38.5**.

The viruses responsible for gastroenteritis are probably transmitted by the fecal-oral route. Infection is most common during

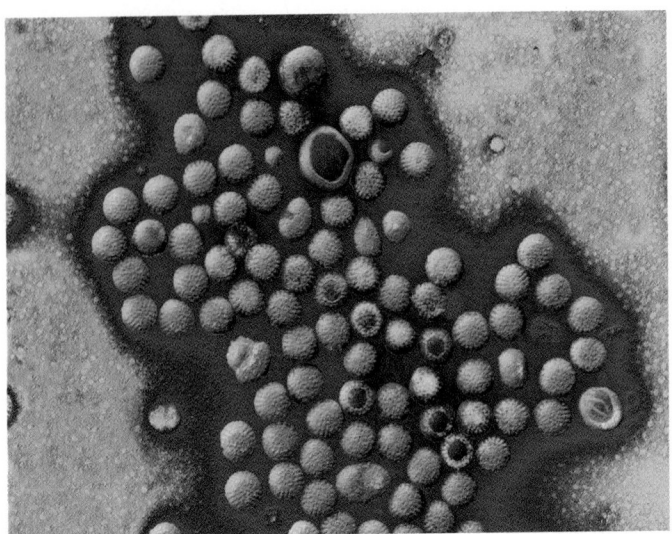

Figure 38.20 Viral Gastroenteritis. Electron micrograph of rotaviruses (reoviruses) in a human gastroenteritis stool filtrate (×90,000). Note the spokelike appearance of the icosahedral capsids that surround double-stranded RNA within each virion.

the cooler months in contrast to bacteria-caused diarrheal diseases, which usually occur in the warmer months of the year. Diarrheal diseases are the leading cause of childhood deaths (5 to 10 million deaths per year) in developing countries where malnutrition is common. Current estimates are that viral gastroenteritis produces 30 to 40% of the cases of infectious diarrhea in the United States, far outnumbering documented cases of bacterial and parasitic diarrhea (the cause of approximately 40% of presumed cases of diarrhea remains unknown). In the United States rotaviruses account for about 3.5 million cases of illness, resulting in 35% of the hospitalizations for gastroenteritis, and 75 to 150 deaths each year.

Viral gastroenteritis is seen most frequently in infants 1 to 11 months of age, where the virus attacks the upper intestinal epithelial cells of the villus, causing malabsorption, impairment of sodium transport, and diarrhea. The clinical manifestations range from asymptomatic to a relatively mild diarrhea with headache and fever to a severe and occasionally fatal dehydrating disease. Vomiting is almost always present.

Viral gastroenteritis is usually self-limited. Treatment is designed to provide relief through the use of oral fluid replacement with isotonic liquids, analgesics, and antiperistaltic agents.

Hepatitis A

Hepatitis A (infectious hepatitis) usually is transmitted by fecal-oral contamination of food, drink, or shellfish that live in contaminated water and contain the virus in their digestive system. The disease is caused by the hepatitis A virus (HAV). HAV has recently been reclassified as the type species of a new genus (*Hepatovirus*) in the family *Picornaviridae*. The hepatitis A virus is

Table 38.5		
Medically Important Gastroenteritis Viruses		
Virus	**Epidemiological Characteristics**	**Clinical Characteristics**
Rotaviruses		
Group A	Endemic diarrhea in infants worldwide	Dehydrating diarrhea for 5–7 days; fever, abdominal cramps, nausea, and vomiting common
Group B	Large outbreaks in adults and children in China	Severe watery diarrhea for 3–5 days
Group C	Sporadic cases in children in Japan	Similar to group A
Norwalk and Norwalk-like viruses	Epidemics of vomiting and diarrhea in older children and adults; occurs in families, communities, and nursing homes; often associated with shellfish, other food, or water and infected food handlers	Acute vomiting, fever, myalgia, and headache lasting 1–2 days, diarrhea
Caliciviruses other than the Norwalk group	Pediatric diarrhea; associated with shellfish and other foods in adults	Rotavirus-like illness in children; Norwalk-like in adults
Astroviruses	Pediatric diarrhea; reported in nursing homes	Watery diarrhea for 1–3 days

an icosahedral, single-stranded linear, positive-sense RNA virus that lacks an envelope. Once in the digestive system, the viruses multiply within the intestinal epithelium. Usually only mild intestinal symptoms result. Occasionally viremia (the presence of viruses in the blood) occurs and the viruses may spread to the liver. The viruses reproduce in the liver, enter the bile, and are released into the small intestine. This explains why feces are so infectious. Symptoms last from 2 to 20 days and include anorexia, general malaise, nausea, diarrhea, fever, and chills. If the liver becomes infected, jaundice ensues. Laboratory diagnosis is by detection of the hepatitis A antibody. About 30,000 cases are reported annually in the United States. Fortunately the mortality rate is low (less than 1%), and infections in children are usually asymptomatic. Most cases resolve in 4 to 6 weeks and yield a strong immunity. Approximately 40 to 80% of the United States population have serum antibodies though few have been aware of the disease. Control of infection is by simple hygienic measures, the sanitary disposal of excreta, and the killed HAV vaccine (Havrix). This vaccine is recommended for travelers (*see table 37.5*) going to regions with high evidence rates of hepatitis A.

Hepatitis E

Hepatitis E is implicated in many epidemics in certain developing countries in Asia, Africa, and Central and South America. The virus has been cloned and sequenced. The genomic organization suggests it is the prototype human pathogen for a new class of single-stranded linear RNA viruses or perhaps a separate genus within the family *Caliciviridae*. Infection usually is associated with feces-contaminated drinking water. Presumably HEV enters the blood from the gastrointestinal tract, replicates in the liver, is released from hepatocytes into the bile, and is subsequently excreted in the feces. Like hepatitis A, an HEV infection usually runs a benign course and is self-limiting. One exception is that at least 10% of women infected in their last 3 months of pregnancy die of fulminant hepatic failure. Antibody assay systems are un-

der development using HEV cDNA products. There are no specific measures for preventing HEV infections other than those aimed at improving the level of health and sanitation in affected areas. There are only a few documented cases of hepatitis E in the United States and all of them imported.

Poliomyelitis

Poliomyelitis [Greek *polios,* gray, and *myelos,* marrow or spinal cord], **polio,** or **infantile paralysis** is caused by the poliovirus, a member of the family *Picornaviridae* (**Historical Highlights 38.3**). The poliovirus is a plus-stranded RNA virus with three different serotypes—1, 2, and 3. The virus is very stable and can remain infectious for relatively long periods in food and water—its main routes of transmission. Once ingested, the virus multiplies in the mucosa of the throat and/or small intestine. From these sites the virus invades the tonsils and lymph nodes of the neck and terminal portion of the small intestine. Generally, there are either no symptoms or a brief illness characterized by fever, headache, sore throat, vomiting, and loss of appetite. The virus sometimes enters the bloodstream and causes a viremia. In most cases (more than 99%), the viremia is transient and clinical disease does not result. In the minority of cases (less than 1%), the viremia persists and the virus enters the central nervous system and causes paralytic polio. The virus has a high affinity for anterior horn motor nerve cells of the spinal cord. Once inside these cells, it multiplies and destroys the cells; this results in motor and muscle paralysis. Since the licensing of the formalin-inactivated Salk vaccine (1955) and the attenuated virus Sabin vaccine (1962), the incidence of polio has decreased markedly. No endogenous reservoir of polioviruses exists in the United States. However, there is a continuing need for vaccination programs in all population groups to limit the spread of poliovirus when it is introduced from other countries. Prevention and control is by vaccination; global eradication of polio is possible in the next few years (see the quote at the beginning of this chapter). At the

Historical Highlights

38.3 A Brief History of Polio

Like many other infectious diseases, polio is probably of ancient origin. Various Egyptian hieroglyphics dated approximately 2000 B.C. depict individuals with wasting, withered legs and arms (see **Box figure**). In 1840 the German orthopedist Jacob von Heine described the clinical features of poliomyelitis and identified the spinal cord as the problem area. Little further progress was made until 1890, when Oskar Medin, a Swedish pediatrician, portrayed the natural history of the disease as epidemic in form. He also recognized that a systemic phase, characterized by minor symptoms and fever, occurred early and was complicated by paralysis only occasionally. Major progress occurred in 1908, when Karl Landsteiner and William Popper successfully transmitted the disease to monkeys. In the 1930s much public interest in polio occurred because of the polio experienced by Franklin D. Roosevelt. This led to the founding of the March of Dimes campaign in 1938; the sole purpose of the March of Dimes was to collect money for research on polio. In 1949 John Enders, Thomas Weller, and Frederick Robbins discovered that the poliovirus could be propagated in vitro in cultures of human embryonic tissues of nonneural origin. This was the keystone that later led to the development of vaccines.

In 1952 David Bodian recognized that there were three distinct serotypes of the poliovirus. Jonas Salk successfully immunized humans with formalin-inactivated poliovirus in 1952, and this vaccine (IPV) was licensed in 1955. The live attenuated poliovirus vaccine (oral polio vaccine, OPV) developed by Albert Sabin and others had been employed in Europe since 1960 and was licensed for U.S. use in 1962. Both the Salk and Sabin vaccines led to a dramatic decline of paralytic poliomyelitis in most developed countries and, as such, have been rightfully hailed as two of the great accomplishments of medical science.

Ancient Egyptian with Polio. Note the withered leg.

---

close of 2001, wild polioviruses remained endemic in only 10 developing countries, and fewer than 1,000 cases were reported during that year.

1. What two virus groups are associated with acute viral gastroenteritis? How do they cause the disease's symptoms?
2. Describe some symptoms of hepatitis A.
3. Why was hepatitis A called infectious hepatitis?
4. At what specific sites within the body can the poliomyelitis virus multiply? What is the usual outcome of an infection?

 ## 38.5 SLOW VIRUS AND PRION DISEASES

An introduction to slow diseases caused by viruses and prions appears on pp. 399–400 and 405–7. A **slow virus disease** can be defined as a progressive pathological process caused by a transmissible agent—virus or prion—that remains clinically silent during a prolonged incubation period of months to years, after which progressive clinical disease becomes apparent. This usually ends months later in disability or death. It is probably inappropriate to say that a slow virus disease or slow disease is caused by a slow virus for two reasons. First, the causative agent of some of these diseases does not fit the conventional definition of a virus. Second, even in those diseases caused by viruses, it is the disease process and not the virus that is slow. Six of these diseases are summarized in **table 38.6**. From this list, the first four diseases are caused by prions and they will now be briefly discussed.

Prion diseases, also called **spongiform encephalopathies,** are fatal neurodegenerative disorders that have attracted enormous attention not only for their unique biological features but also for their impact on public health. This group of diseases includes kuru, Creutzfeldt-Jakob disease (CJD), Gerstmann-Sträussler disease (GSD), and fatal familial insomnia (FFI). The primary symptom of the human disorders is dementia, usually accompanied by manifestations of motor dysfunction such as cerebral ataxia (inability to coordinate muscle activity) and

Table 38.6
Slow Virus and Prion Diseases of Humans

Disease	Agent	Incubation Period	Nature of Disease
Creutzfeldt-Jakob disease (CJD) (sporadic, iatrogenic, familial, new-variant)	Prion	Months to years	Spongiform encephalopathy (degenerative changes in the central nervous system)
Kuru	Prion	Months to years	Spongiform encephalopathy
Gerstmann-Sträussler-Scheinker disease (GSD)	Prion	Months to years	Genetic neurodegenerative disease
Fatal familial insomnia (FFI)	Prion	Months to years	Genetic neurodegenerative disease with progressive, untreatable insomnia
Progressive multifocal leukoencephalopathy	Papovavirus	Years	Central nervous system demyelination
Subacute sclerosing panencephalitis (SSPE)	Measle virus variant	2–20 years	Chronic sclerosing (hard tissue) panencephalitis (involving both white and gray matter of the brain)

myoclonus (shocklike contractions of muscle groups). FFI is also characterized by dysautonomia (abnormal functioning of the autonomic nervous system) and sleep disturbances. These symptoms appear insidiously in middle to late adult life and last from months (CJD, FFI, and kuru) to years (GSD) prior to death. Neuropathologically, these disorders produce a characteristic spongiform degeneration of the brain, as well as deposition of amyloid plaques. Prion diseases thus share important clinical, neuropathological, and cell biological features with another, more common cerebral amyloidosis, Alzheimer's disease.

Recently, two new forms of CJD have arisen. New-variant CJD (vCJD) is transmitted from cattle that have spongiform encephalopthy (mad cow disease) as described in section 41.4. Iatrogenic CJD is induced by a physician or surgeon, a medical treatment, or diagnostic procedures. It commonly is transmitted by prion-contaminated human growth hormone and grafts of dura mater (tissue surrounding the brain).

 38.6 OTHER DISEASES

Several other human diseases (e.g., diabetes mellitus, viral arthritis) have been associated with viruses but do not fit into any of the previous categories. Another example is warts.

Warts

Warts or verrucae [Latin *verruca,* wart] are horny projections on the skin caused by the human papillomaviruses (*see figure 16.12*d). The papillomaviruses are placed in the family *Papillomaviridae* (formerly they were in the *Papovaviridae*). These viruses have naked icosahedral capsids with a double-stranded, supercoiled, circular DNA genome. At least eight distinct genotypes produce benign epithelial tumors that vary in respect to their location, clinical appearance, and histopathologic features. Warts occur principally in children and young adults and are lim-

ited to the skin and mucous membranes. The viruses are spread between people by direct contact; autoinoculation occurs through scratching. Four major kinds of warts are **plantar warts, verrucae vulgaris, flat** or **plane warts,** and **anogenital condylomata (venereal warts) (figure 38.21).** Treatment includes physical destruction of the wart(s) by electrosurgery, cryosurgery with liquid nitrogen or solid CO_2, laser fulguration (drying), direct application of the drug podophyllum to the wart(s), or injection of IFN-α (Intron A, Alferon N).

Anogenital condylomata (venereal warts) are sexually transmitted and caused by types 6, 11, and 42 **h**uman **p**apillomavirus (HPV). Once the virus enters the body, the incubation period is 1 to 6 months. The warts (figure 38.21*d*) are soft, pink, cauliflowerlike growths that occur on the external genitalia, in the vagina, on the cervix, or in the rectum. They often are multiple and vary in size. In addition to being a common sexually transmitted disease (*see table 39.4, p. 904*), genital infection with HPV is of considerable importance because specific types of genital HPV play a major role in the pathogenesis of epithelial cancers of the male and female genital tracts. Over the last decade many studies have convincingly demonstrated that specific types of HPV are the causal agents of at least 90% of cervical cancers. The most common types conferring a high risk for cervical cancer include HPV types 16, 18, 31, 33, 35, 45, 51, 52, and 56. There is also a possible link between papillomaviruses and nonmelanoma squamous and basal cell cancers. Thirty percent of humans with a rare syndrome of persistent warts (not common warts, but a particular type of warty growth) eventually develop skin cancer, and HPV viral DNA is found in the malignant cells. However, the epidemiology, molecular biology, and role of HPV in the development of such cancers are largely unknown.

1. Name six slow virus diseases of humans.
2. What kind of viruses cause the formation of warts? Describe venereal warts or anogenital condylomata.

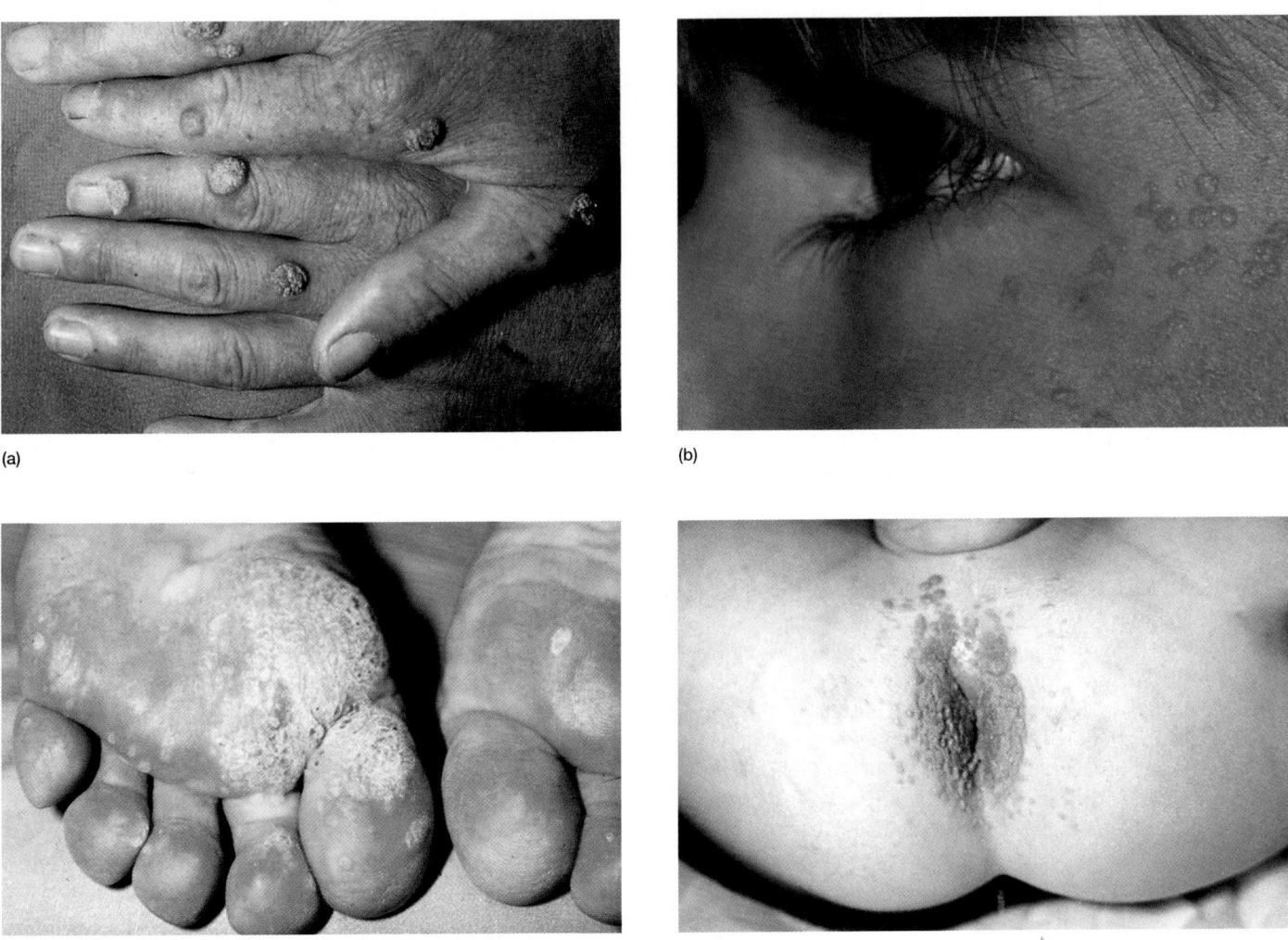

Figure 38.21   Warts.   (**a**) Common warts on fingers. (**b**) Flat warts on the face. (**c**) Plantar warts on the feet. (**d**) Perianal condyloma acuminata.

Summary

38.1 Airborne Diseases

a. More than 400 different viruses can infect humans. These viruses can be grouped and discussed according to their mode of transmission and acquisition.

b. Most airborne viral diseases involve either directly or indirectly the respiratory system. Examples include chickenpox (varicella, **figure 38.1**), shingles (herpes zoster, **figure 38.2**), rubella (German measles, **figure 38.5**), influenza (flu), measles (rubeola, **figure 38.3**), mumps (**figure 38.4**), the acute respiratory viruses such as the respiratory syncytial virus, the eradicated smallpox (variola, **figure 38.6**), and viral pneumonia.

38.2 Arthropod-Borne Diseases

a. The arthropod-borne viral diseases are transmitted by arthropod vectors from human to human

or animal to human (**table 38.2**). Examples include California encephalitis; Colorado tick fever; St. Louis encephalitis; eastern, western, and Venezuelan equine encephalitis; West Nile fever; and yellow fever. All these diseases are characterized by fever, headache, nausea, vomiting, and the characteristic encephalitis.

38.3 Direct Contact Diseases

a. Person-to-person contact is another way of acquiring or transmitting a viral disease. Examples of such diseases include AIDS, cold sores (**figure 38.13**), the common cold, cytomegalovirus inclusion disease, genital herpes (**figure 38.1b**), human herpesvirus 6 infections, human parvovirus B19 infection, certain leukemias, infectious mononucleosis, rabies, and the five well-characterized types of hepatitis (**table 38.4**): hepatitis A (infectious hepatitis); hepatitis

B (serum hepatitis); hepatitis C; hepatitis D (delta hepatitis); hepatitis E (enteric-transmitted hepatitis); and hepatitis G.

38.4 Food-Borne and Waterborne Diseases

a. The viruses that are transmitted in food and water usually grow in the intestinal system and leave the body in the feces (**table 38.5**). Acquisition is generally by the oral route. Examples of diseases caused by these viruses include acute viral gastroenteritis (rotavirus and others), infectious hepatitis A, and poliomyelitis.

38.5 Slow Virus and Prion Diseases

a. A slow virus or prion disease is a pathological process caused by a transmissible agent (a prion or virus) that remains clinically silent for a prolonged period, after which the clinical disease becomes apparent. Examples include (**table 38.6**) Creutzfeldt-Jakob disease, kuru, subacute sclerosing panencephalitis, Gerstmann-Sträussler-Scheinker disease, fatal familial insomnia, and progressive multifocal leukoencephalopathy. These diseases are chronic infections of the central nervous system that result in progressive degenerative changes and eventual death.

38.6 Other Diseases

a. Common skin warts (**figure 38.21**) are caused by viruses and can be spread by autoinoculation through scratching or by direct or indirect contact. Anogenital condylomata (venereal warts) are sexually transmitted.

Key Terms

acute viral gastroenteritis 868
adult T-cell leukemia 864
AIDS (acquired immune deficiency syndrome) 855
AIDS-related complex (ARC) 857
anogenital condylomata (venereal warts) 871
antigenic drift 847
antigenic shift 847
chickenpox (varicella) 846
cold sore 860
Colorado tick fever 853
common cold 861
congenital (neonatal) herpes 863
congenital rubella syndrome 851
coryza 861
cytomegalovirus inclusion disease 862
Dane particle 866
Delta agent 867

Ebola virus hemorrhagic fever 853
erythema infectiosum 864
exanthem subitum 863
fever blister 860
fifth disease 864
flat or plane warts 871
genital herpes 862
gingivostomatitis 861
Guillain-Barré syndrome (French polio) 849
hantavirus pulmonary syndrome 853
hemorrhagic fevers 853
hepatitis 866
hepatitis A 868
hepatitis B 866
hepatitis C 867
hepatitis D 867
hepatitis E 869
hepatitis G 868
herpes labialis 860

herpetic keratitis 861
human herpesvirus 6 863
human immunodeficiency virus (HIV) 855
human parvovirus B19 864
infantile paralysis 869
infectious mononucleosis (mono) 864
influenza or flu 847
intranuclear inclusion body 862
Koplik's spots 849
Korean hemorrhagic fever 853
leukemia 864
Marburg viral hemorrhagic fever 853
measles (rubeola) 849
mumps 850
Negri bodies 866
orchitis 850
plantar warts 871
polio 869

poliomyelitis 869
postherpetic neuralgia 847
pulmonary syndrome hantavirus 853
rabies 865
respiratory syncytial virus (RSV) 851
Reye's syndrome 849
roseola infantum 863
rubella (German measles) 851
shingles (herpes zoster) 847
sixth disease 863
slow virus disease 870
smallpox (variola) 851
spongiform encephalopathies 870
subacute sclerosing panencephalitis 849
verrucae vulgaris 871
wart 871
West Nile fever (encephalitis) 854
yellow fever 854

Questions for Thought and Review

1. Briefly describe each of the major or most common viral diseases in terms of its causative agent, signs and symptoms, the course of infection, mechanism of pathogenesis, epidemiology, and prevention and/or treatment.
2. Which virus is responsible for each of the following: shingles, German measles, influenza, measles, mumps, and smallpox?
3. What are respiratory syndromes?

4. From an epidemiological perspective, why are most arthropod-borne viral diseases hard to control?
5. In terms of molecular genetics, why is the common cold such a prevalent viral infection in humans?
6. What are the differences between the five well characterized types of hepatitis?
7. Which viral diseases can be transmitted by sexual contact?

8. Why are the herpesviruses called persistent viruses?
9. What viruses specifically attack the nervous system?
10. Why is rabies such a feared disease?
11. Which viruses can cause encephalitis? Hepatitis?
12. What are two types of mononucleosis?
13. Will it be possible to eradicate many viral diseases in the same way as smallpox? Why or why not?

Critical Thinking Questions

1. Explain why antibiotics are not effective against viral infections. Advise a person about what can be done to relieve symptoms of a viral infection and recover most quickly. Address your advice to (a) someone who has had only a basic course in high school biology, and (b) a third-grade student.

2. Several characteristics of AIDS render it particularly difficult to detect, prevent, and treat effectively. Discuss two of them. Contrast the disease with polio and smallpox.

For recommended readings on these and related topics, see Appendix V.

chapter 39

Human Diseases Caused by Bacteria

Concepts

- Bacterial diseases of humans can be discussed according to their mode of acquisition/transmission.
- Most of the airborne diseases caused by bacteria involve the respiratory system. Examples include diphtheria, Legionnaires' disease and Pontiac fever, *Mycobacterium avium–M. intracellulare* pulmonary infections, pertussis, streptococcal diseases, and tuberculosis. Other airborne bacteria can cause skin diseases, including cellulitis, erysipelas, and scarlet fever, or systemic diseases such as meningitis, glomerulonephritis, and rheumatic fever.
- Although arthropod-borne bacterial diseases are generally rare, they are of interest either historically (plague) or because they have been newly introduced into humans (Lyme disease). Most of the rickettsial diseases are arthropod-borne. The rickettsias found in the United States can be divided into the typhus group (epidemic typhus caused by *R. prowazekii* and murine typhus caused by *R. typhi*) and the spotted fever group (Rocky Mountain spotted fever caused by *R. rickettsii* and ehrlichiosis caused by *Ehrlichia chaffeensis*), with Q fever (caused by *Coxiella burnetti*) being an exception because it forms endospore-like structures and does not have to use an insect vector as with other rickettsias.
- Most of the direct contact bacterial diseases involve the skin, mucous membranes, or underlying tissues. Examples include anthrax, bacterial vaginosis, cat-scratch disease, chancroid, gas gangrene, leprosy, peptic ulcer disease and gastritis, staphylococcal diseases, and syphilis. Others can become disseminated throughout specific regions of the body—for example, gonorrhea, staphylococcal diseases, syphilis, tetanus, and tularemia. Three chlamydial species cause direct contact disease: *Chlamydia pneumoniae* causes chlamydial pneumonia; *Chlamydia trachomatis* causes inclusion conjunctivitis, lymphogranuloma venereum, nongonococcal urethritis, and trachoma; and *C. psittaci*

The toll of tetanus. The bacterial genus *Clostridium* contains many pathogenic species, including the species responsible for tetanus (*C. tetani*). Sir Charles Bell's portrait (c. 1821) of a soldier wounded in the Peninsular War in Spain shows the suffering from generalized tetanus.

causes psittacosis. Three species of mycoplasmas are human pathogens: *Mycoplasma hominis* and *Ureaplasma urealyticum* cause genitourinary tract disease, whereas *M. pneumoniae* is a major cause of acute respiratory disease and pneumonia.

- The food-borne and waterborne bacterial diseases are contracted when contaminated food or water is ingested. These diseases are essentially of two types: infections and intoxications. An infection occurs when a pathogen enters the gastrointestinal tract and multiplies. Examples include *Campylobacter* gastroenteritis, salmonellosis, listerosis, shigellosis, traveler's diarrhea, *Escherichia coli* infections, and typhoid fever. An intoxication occurs because of the ingestion of a toxin produced outside the body. Examples include botulism, cholera, and staphylococcal food poisoning.
- Some microbial diseases and their effects cannot be related to a specific mode of transmission. Two important examples are sepsis and septic shock. Gram-positive bacteria, fungi, and endotoxin-containing gram-negative bacteria can initiate the pathogenic cascade of sepsis leading to septic shock.
- Several bacterial odontopathogens are responsible for the most common bacterial diseases in humans—tooth decay and periodontal disease. Both are the result of plaque formation and the production of lactic and acetic acids by the odontopathogens.

Soldiers have rarely won wars. They more often mop up after the barrage of epidemics. And typhus, with its brothers and sisters—plague, cholera, typhoid, dysentery—has decided more campaigns than Caesar, Hannibal, Napoleon, and all the . . . generals of history. The epidemics get the blame for the defeat, the generals the credit for victory. It ought to be the other way around. . . .

—*Hans Zinsser*

he first four parts of this textbook cover the general biology of bacteria. Chapters 19 through 24 specifically review bacterial morphology and taxonomy. Chapter 39 continues the coverage of bacteria by discussing some of the more important gram-positive and gram-negative bacteria, chlamydiae, mycoplasmas, and rickettsias that are pathogenic to humans. The microorganisms involved in dental infections are also described.

Of all the known bacterial species, only a few are pathogenic to humans. Some human diseases have been only recently recognized (**table 39.1**); others have been known since antiquity. In the following sections the more important disease-causing bacteria are discussed according to their mode of acquisition/transmission.

Table 39.1
Some Examples of Human Bacterial Diseases Recognized Since 1973

Year	Bacterium	Disease
1977	*Legionella pneumophila*	Legionnaires' disease
1977	*Campylobacter jejuni*	Enteric disease (gastroenteritis)
1981	*Staphylococcus aureus*	Toxic shock syndrome
1982	*Escherichia coli* O157:H7	Hemorrhagic colitis; hemolytic uremic syndrome
1982	*Borrelia burgdorferi*	Lyme disease
1982	*Helicobacter pylori*	Peptic ulcer disease
1986	*Ehrlichia chaffeensis*	Human ehrlichiosis
1992	*Vibrio cholerae* O139	New strain associated with epidemic cholera in Asia
1992	*Bartonella henselae*	Cat-scratch disease; bacillary angiomatosis
1993	*Enterococcus faecium;* vancomycin-resistant enterococci	Colitis and enteritis
1994	*Ehrlichia* spp.	Human granulocytic ehrlichiosis
1995	*Neisseria meningitidis*	Meningococcal supraglottitis
1997	*Kingella kingae*	Pediatric infections
2000	*Tropheryma whipplei*	Whipple's disease

39.1   AIRBORNE DISEASES

Most airborne diseases caused by bacteria involve the respiratory system. Other airborne bacteria can cause skin diseases. Some of the better known of these diseases are now discussed.

Diphtheria

Diphtheria [Greek *diphthera*, membrane, and *ia*, condition] is an acute contagious disease caused by the gram-positive *Corynebacterium diphtheriae* (*see figure 24.8*). *C. diphtheriae* is well adapted to airborne transmission by way of nasopharyngeal secretions and is very resistant to drying. Diphtheria mainly affects poor people living in crowded conditions. Once within the respiratory system, bacteria that carry the prophage β containing the *tox* gene (*see section 17.5*) produce diphtheria toxin, an exotoxin that causes an inflammatory response and the formation of a grayish pseudomembrane on the respiratory mucosa (**figure 39.1**). The pseudomembrane consists of dead host cells and cells of *C. diphtheriae*. The exotoxin is also absorbed into the circulatory system and distributed throughout the body, where it may cause destruction of cardiac, kidney, and nervous tissues by inhibiting protein synthesis (*see table 34.6 and figure 34.5*).

Typical symptoms of diphtheria include a thick mucopurulent (containing both mucus and pus) nasal discharge, fever, and cough. Diagnosis is made by observation of the pseudomembrane in the throat and by bacterial culture. Diphtheria antitoxin is given to neutralize any unabsorbed exotoxin in the patient's tissues; penicillin and erythromycin are used to treat the infection. Pre-

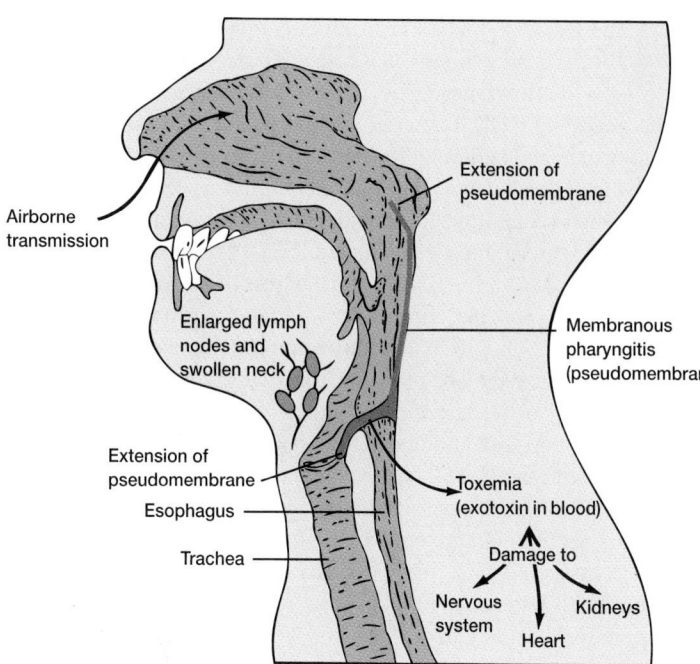

Figure 39.1 Diphtheria Pathogenesis. Diphtheria is a well-known exotoxin-mediated infectious disease caused by *Corynebacterium diphtheriae.* The disease is an acute, contagious, febrile illness characterized by local oropharyngeal inflammation and pseudomembrane formation. If the exotoxin gets into the blood, it is disseminated and can damage the peripheral nerves, heart, and kidneys.

vention is by active immunization with the **DPT** (**d**iphtheria-**p**ertussis-**t**etanus) **vaccine** (*see table 33.1*).

C. diphtheriae can also infect the skin, usually at a wound or skin lesion, causing a slow-healing ulceration termed **cutaneous diphtheria.** Most cases involve people over 30 years of age who have a weakened immunity to the diphtheria toxin and live in tropical areas.

Fewer than 100 diphtheria cases are reported annually in the United States, and most occur in nonimmunized individuals.

Legionnaires' Disease and Pontiac Fever

In 1976 the term **Legionnaires' disease,** or **legionellosis,** was coined to describe an outbreak of pneumonia that occurred at the Pennsylvania State American Legion Convention in Philadelphia. The bacterium responsible for the outbreak was described as *Legionella pneumophila,* a nutritionally fastidious aerobic gram-negative rod (**figure 39.2**). It is now known that this bacterium is part of the natural microbial community of soil and aquatic ecosystems, and it has been found in large numbers in air-conditioning systems and shower stalls.

An increasing body of evidence suggests that environmental protozoa are the most important factor for the survival and growth of *Legionella* in nature (*see p. 590*). A variety of free-living amoebae and ciliated protozoa that contain *Legionella*

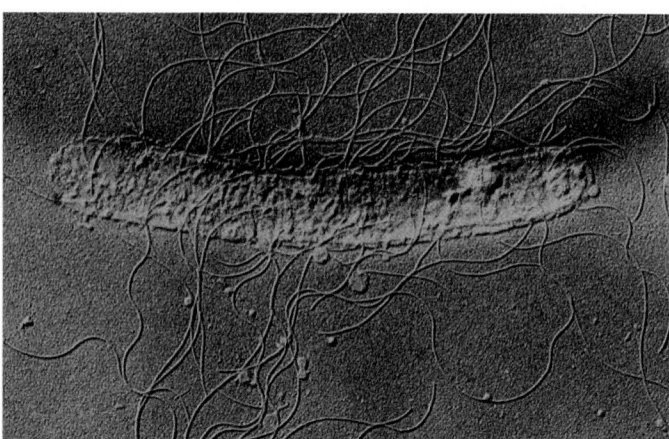

Figure 39.2 Legionnaires' Disease. *Legionella pneumophila,* the causative agent of Legionnaires' disease, with many lateral flagella; SEM (×10,000).

spp. have been isolated from water sites suspected as sources of *Legionella* infections. *Legionella* spp. multiply intracellularly within the amoebae, just as they do within human monocytes and macrophages. This might explain why there is no human-to-human spread of legionellosis.

Infection with *L. pneumophila* results from the airborne spread of bacteria from an environmental reservoir to the human respiratory system. Males over 50 years of age most commonly contract the disease, especially if their immune system is compromised by heavy smoking, alcoholism, or chronic illness. The bacteria reside within the phagosomes of alveolar macrophages, where they multiply and produce localized tissue destruction through export of a cytotoxic exoprotease. Symptoms include a high fever, nonproductive cough (respiratory secretions are not brought up during coughing), headache, neurological manifestations, and severe bronchopneumonia. Diagnosis depends on isolation of the bacterium, documentation of a rise in antibody titer over time, or a rapid test kit using urine to detect antigens. Treatment begins with supportive measures and the administration of erythromycin or rifampin.

Prevention of Legionnaires' disease depends on the identification and elimination of the environmental source of *L. pneumophila* contamination. Chlorination, the heating of water, and the cleaning of water-containing devices can help control the multiplication and spread of *Legionella.* These control measures are effective because the pathogen does not appear to be spread from person to person.

Since the initial outbreak of this disease in 1976, many outbreaks during summer months have been recognized in all parts of the United States. About 1,000 to 1,400 cases are diagnosed each year, and about 30,000 or more additional mild or subclinical cases are thought to occur. It is estimated that 3 to 6% of all nosocomial pneumonias are due to *L. pneumophila,* especially among immunocompromised patients.

L. pneumophila also causes an illness called **Pontiac fever.** This disease, which resembles an allergic disease more than an

Table 39.2
Causative Agents of Meningitis by Diagnostic Category

Type of Meningitis	Causative Agent
Bacterial (Septic) Meningitis	
	Streptococcus pneumoniae
	Neisseria meningitidis
	Haemophilus influenzae type b
	Group B streptococci
	Listeria monocytogenes
	Mycobacterium tuberculosis
	Nocardia asteroides
	Staphylococcus aureus
	Staphylococcus epidermidis
Aseptic Meningitis Syndrome	
Agents Requiring Antimicrobials	Fungi
	Amoebae
	Syphilis
	Mycoplasmas
	Leptospires
Agents Requiring Other Treatments	Viruses
	Cancers
	Parasitic cysts
	Chemicals

infection, is characterized by an abrupt onset of fever, headache, dizziness, and muscle pains. It is indistinguishable clinically from the various respiratory syndromes caused by viruses. Pneumonia does not occur. The disease resolves spontaneously within 2 to 5 days. No deaths from Pontiac fever have been reported.

Pontiac fever was first described from an outbreak in a county health department in Pontiac, Michigan. Ninety-five percent of the employees became ill and eventually showed elevated serum titers against *L. pneumophila*. These bacteria were later isolated from the lungs of guinea pigs exposed to the air of the building. The likely source was water from a defective air conditioner.

Meningitis

Meningitis [Greek *meninx,* membrane, and *-itis,* inflammation] is an inflammation of the brain or spinal cord meninges (membranes). Based on the specific cause, it can be divided into **bacterial (septic) meningitis** and the **aseptic meningitis syndrome** (**table 39.2**). As shown by the table, there are many causes of the aseptic meningitis syndrome, only some of which can be treated with antimicrobials. Thus accurate identification of the causative agent is essential to proper treatment of the disease. The immediate sources of the bacteria responsible for meningitis are respiratory secretions from carriers or active cases. The bacteria initially colonize the nasopharynx after which they cross the mucosal barrier and enter the bloodstream and cerebrospinal fluid, where they produce inflammation of the meninges.

The usual symptoms of meningitis include an initial respiratory illness or sore throat interrupted by one of the meningeal syndromes: vomiting, headache, lethargy, confusion, and stiffness in the neck and back. Bacterial meningitis can be diagnosed by a Gram stain and culture of the bacteria from cerebrospinal fluid or rapid tests (*see table 36.3*). Once meningitis is suspected, specific antibiotics (penicillin, chloramphenicol, cefotaxime, ceftriaxone, ofloxacin) are administered immediately. A sharp reduction in the incidence of *H. influenzae* serotype b infections began in the mid-1980s due to vaccine administration, rifampin prophylaxis of disease contacts, and the availability of more efficacious therapeutic agents. From 1987 through 1999, the incidence of invasive infection among U.S. children less than 5 years of age declined by 95%. Currently, all children at the age of 2 months should be vaccinated with the *H. influenzae* type b conjugate vaccine.

A person may have meningitis symptoms, but show no microbial agent in Gram-stained specimens, and have negative cultures. In such a case the diagnosis often is aseptic meningitis syndrome. Aseptic meningitis is more difficult to treat, and the prognosis is usually poor.

Mycobacterium avium–M. intracellulare Pulmonary Infections

During the past decade it has been discovered that there is an extremely large group of mycobacteria that are normal inhabitants of soil, water, and house dust. Two of these have become noteworthy pathogens in the United States. The two, *Mycobacterium avium* and *Mycobacterium intracellulare,* are so closely related that they are referred to as the *M. avium* complex (MAC).

These mycobacteria are found worldwide and infect a variety of insects, birds, and animals. Both the respiratory and the gastrointestinal tracts have been proposed as entry portals for the *M. avium* complex. The gastrointestinal tract is thought to be the most common site of colonization and dissemination in AIDS patients. MAC causes a pulmonary infection in humans similar to *M. tuberculosis.* Pulmonary MAC is more common in non-AIDS patients, particularly in elderly persons with preexisting pulmonary disease.

Shortly after the recognition of AIDS and the associated opportunistic infections (*see table 38.3*), it became apparent that one of the more common infections was caused by MAC. Disseminated infection with MAC occurs in 15 to 40% of persons with AIDS in the United States with CD4$^+$ cell counts of less than 100 per cubic millimeter. Disseminated infection with MAC produces disabling symptoms, including fever, malaise, weight loss, night sweats, and diarrhea. Carefully controlled epidemiological studies have shown that MAC shortens survival by 5 to 7 months among persons with AIDS. With more effective antiviral therapy for AIDS and with prolonged survival, the number of cases of disseminated MAC is likely to increase substantially, and its contribution to AIDS mortality will increase.

MAC can be isolated from sputum, blood, and aspirates of bone marrow. Acid-fast stains are of value in making a diagnosis. The most sensitive method for detection is the commercially available lysis-centrifugation blood culture system (Wampole Laboratories). Although no drugs are currently approved by the FDA for the therapy of MAC, every regimen should contain either azithromycin or clarithromycin or ethambutol as a second drug, and one or more of the following: closfazimine, rifabutin, rifampin, ciprofloxacin, and amikacin.

Pertussis

Pertussis [Latin *per,* intensive, and *tussis,* cough], sometimes called "whooping cough," is caused by the gram-negative bacterium *Bordetella pertussis.* (*B. parapertussis* is a closely related species that causes a milder form of the disease.) Pertussis is a highly contagious disease that primarily affects children. It has been estimated that over 95% of the world's population has experienced either mild or severe symptoms of the disease. Around 500,000 die from the disease each year. However, there are less than around 5,000 cases and less than 10 deaths annually in the United States.

Transmission occurs by inhalation of the bacterium in droplets released from an infectious person. The incubation period is 7 to 14 days. Once inside the upper respiratory tract, the bacteria attach to the ciliated epithelial cells by producing adhesins such as the factor called filamentous hemagglutinin, which recognizes a complementary molecule on the cells. After attachment, the bacteria synthesize several toxins (*see table 34.6*) that are responsible for the symptoms. The most important toxin is pertussis toxin, which causes increased tissue susceptibility to histamine and serotonin, and an increased lymphocyte response. *B. pertussis* also produces an extracytoplasmic invasive adenylate cyclase, and tracheal cytotoxin and dermonecrotic toxin, which destroy epithelial tissue. Working together, the tracheal cytotoxin and pertussis toxin also provoke the secretory cells in the respiratory tract to produce nitric oxide, which kills the nearby ciliated cells. In addition, the secretion of a thick mucus impedes ciliary action, and often, ciliated epithelial cells die.

Pertussis is divided into three stages. (1) The catarrhal stage, so named because of the mucous membrane inflammation, which is insidious and resembles the common cold. (2) Prolonged coughing sieges characterize the paroxysmal stage. During this stage the infected person tries to cough up the mucous secretions by making 5 to 15 rapidly consecutive coughs followed by the characteristic whoop—a hurried deep inspiration. The catarrhal and paroxysmal stages last about 6 weeks. (3) Final recovery may take several months (the convalescent stage).

Laboratory diagnosis of pertussis is by culture of the bacterium, fluorescent antibody staining of smears from nasopharyngeal swabs, and serological tests. The development of a strong, lasting immunity takes place after an initial infection. Treatment is with erythromycin, tetracycline, or chloramphenicol. Treatment ameliorates clinical illness when begun during the catarrhal phase and may also reduce the severity of the disease when begun within 2 weeks of the onset of the paroxysmal cough. Prevention is with the DPT vaccine (*see p. 741*); vaccination of children is recommended when they are 2 to 3 months old (*see table 33.1 and figure 33.1*).

Streptococcal Diseases

Streptococci, commonly called strep, are a heterogeneous group of gram-positive bacteria. In this group *Streptococcus pyogenes* (group A β-hemolytic streptococci; *see pp. 516–18*) is one of the most important bacterial pathogens. The different serotypes produce (1) extracellular enzymes that break down host molecules; (2) streptokinases, enzymes that activate a host-blood factor that dissolves blood clots; (3) the cytolysins streptolysin O and streptolysin S, which kill host leukocytes; and (4) capsules and M protein that help retard phagocytosis.

S. pyogenes is widely distributed among humans, but usually people are asymptomatic carriers. Individuals with acute infections may spread the pathogen, and transmission can occur through respiratory droplets, direct, or indirect contact. When highly virulent strains appear in schools, they can cause sharp outbreaks of sore throats and scarlet fever. Due to the cumulative buildup of antibodies to many different *S. pyogenes* serotypes over the years, outbreaks among adults are less frequent.

Diagnosis of a streptococcal infection is based on both clinical and laboratory findings. Several rapid tests are available (*see table 36.3*). Treatment is with penicillin or erythromycin. Vaccines are not available for streptococcal diseases other than streptococcal pneumonia because of the large number of serotypes.

The best control measure is prevention of bacterial transmission. Individuals with a known infection should be isolated and treated. Personnel working with infected patients should follow standard aseptic procedures.

In the following sections some of the more important human streptococcal diseases are discussed (**figure 39.3**).

Cellulitis and Erysipelas

Cellulitis is a diffuse, spreading infection of subcutaneous skin tissue. The resulting inflammation is characterized by a defined area of redness (erythema) and the accumulation of fluid (edema).

The most frequently diagnosed skin infection caused by *S. pyogenes* is **impetigo** (impetigo also can be caused by *S. aureus* [figure 39.19*e*]). Impetigo is a superficial cutaneous infection, most commonly seen in children, usually located on the face, and characterized by crusty lesions and vesicles surrounded by a red border. Impetigo is most common in late summer and early fall. The drugs of choice for impetigo are penicillin or erythromycin in those individuals who are allergic to penicillin.

Erysipelas [Greek *erythros,* red, and *pella,* skin] is an acute infection and inflammation of the dermal layer of the skin. It occurs primarily in infants and people over 30 years of age with a history of streptococcal sore throat. The skin often develops painful reddish patches that enlarge and thicken with a sharply defined edge (**figure 39.4**). Recovery usually takes a week or longer if no treatment is given. The drugs of choice for the treatment of erysipelas are erythromycin and penicillin. Erysipelas may recur periodically at the same body site for years.

Invasive Streptococcus A Infections

In the nineteenth century invasive *Streptococcus pyogenes* infections were a major cause of morbidity and mortality. However,

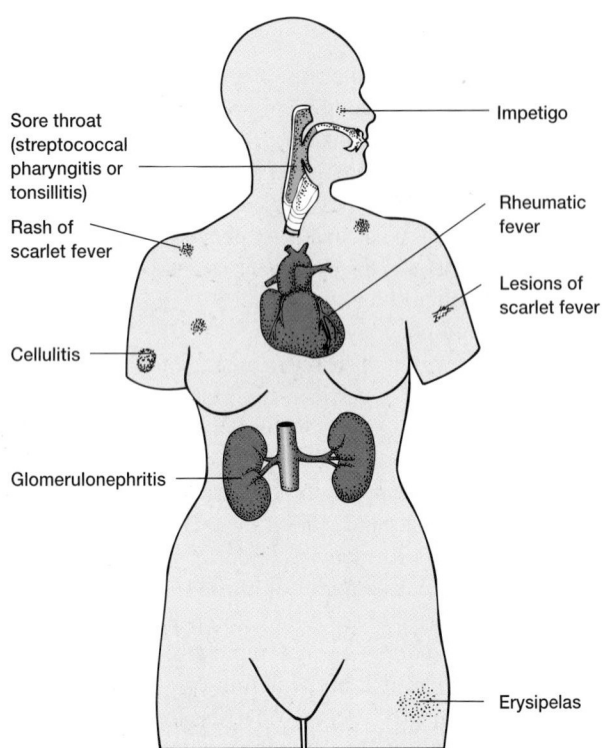

Figure 39.3 Streptococcal Diseases. Some of the more prominent diseases associated with group A streptococcal infections, and the body sites affected.

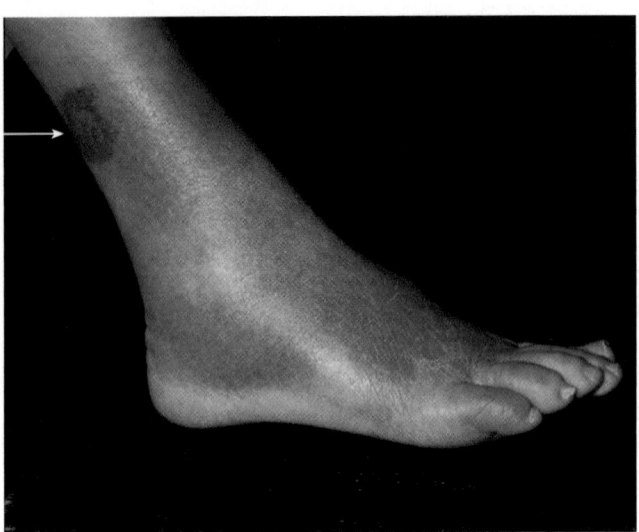

Figure 39.4 Erysipelas. Notice the bright, raised, rubbery lesion at the site of initial entry (white arrow) and the spread of the inflammation to the foot. The reddening is caused by toxins produced by the streptococci as they invade new tissue.

during the twentieth century the incidence of severe group A streptococcal infections declined, especially with the arrival of antibiotic therapy. In the mid-1980s there was a worldwide increase in group A streptococcal sepsis; clusters of rheumatic fever were reported from locations within the United States, and a streptococcal toxic shock–like syndrome emerged. (A virulent strep A infection killed *Sesame Street* Muppeteer Jim Henson in 1990, and in 1994 the press made headlines with articles on "the flesh-eating invasive disease.")

The development of invasive strep A disease appears to depend on the presence of specific virulent strains (M-1 and M-3 serotypes) and predisposing host factors (surgical or nonsurgical wounds, diabetes, and other underlying medical problems). A life-threatening infection begins when invasive strep A strains penetrate a mucous membrane or take up residence in a skin lesion such as a bruise. This infection can quickly lead either to **necrotizing fasciitis** [Greek *nekrosis*, deadness, Latin *fascis*, band or bandage, and *itis*, inflammation] that destroys the sheath covering skeletal muscles or to **myositis** [Greek *myos*, muscle, and *itis*], the inflammation and destruction of skeletal muscle and fat tissue. Because necrotizing fasciitis and myositis arise and spread so quickly, they have been colloquially called "galloping gangrene."

Rapid treatment is necessary to reduce the risk of death, and penicillin G remains the treatment of choice. In addition, surgical removal of dead and dying tissue usually is needed in more ad-

vanced cases of necrotizing fasciitis. It is estimated that 10,000 to 15,000 cases of invasive strep A infections occur annually in the United States, and between 5 and 10% of them are associated with necrotizing conditions.

One reason invasive strep A strains are so deadly is that about 85% of them carry the genes for the production of **s**treptococcal **p**yrogenic **e**xotoxins A and B (Spe exotoxins). Exotoxin A acts as a superantigen. This superantigen quickly stimulates T cells to begin producing abnormally large quantities of cytokines. These cytokines damage the endothelial cells that line blood vessels, causing fluid loss and rapid tissue death from a lack of oxygen. Another pathogenic mechanism involves the secretion of exotoxin B (cysteine protease). Cysteine protease is an enzyme that rapidly destroys tissue by breaking down proteins.

Since 1986 it has been recognized that invasive strep A infections can also trigger a *toxic shock–like syndrome* (**TSLS),** characterized by a precipitous drop in blood pressure, failure of multiple organs, and a very high fever. TSLS is caused by an invasive strep A that produces one or more of the streptococcal pyrogenic exotoxins. TSLS has a mortality rate of over 30%.

Because group A streptococci are less contagious than cold or flu viruses, infected individuals do not pose a major threat to people around them. The best preventive measures are simple ones such as covering food, washing hands, and cleansing and medicating wounds.

Poststreptococcal Diseases

The poststreptococcal diseases are glomerulonephritis and rheumatic fever. They occur 1 to 4 weeks after an acute streptococcal infection (hence the term post). Today these two diseases are the most serious problems associated with streptococcal infections in the United States.

Glomerulonephritis or **Bright's disease** is an inflammatory disease of the renal glomeruli, membranous structures within the kidney where blood is filtered. Damage probably results from the deposition of antigen-antibody complexes, possibly involving the streptococcal M protein, in the glomeruli. Thus the disease arises from a type III hypersensitivity reaction (*see figure 33.6*). The complexes cause destruction of the glomerular membrane, allowing proteins and blood to leak into the urine. Clinically the affected person exhibits edema, fever, hypertension, and hematuria (blood in the urine). The disease occurs primarily among school-age children. Diagnosis is based on the clinical history, physical findings, and confirmatory evidence of prior streptococcal infection. The incidence of glomerulonephritis in the United States is less than 0.5% of streptococcal infections. Penicillin G or erythromycin can be given for any residual streptococci. However, there is no specific therapy once kidney damage has occurred. About 80 to 90% of all cases undergo slow spontaneous healing of the damaged glomeruli, whereas the others develop a chronic form of the disease. The latter may require a kidney transplant or lifelong renal dialysis.

Rheumatic fever is an autoimmune disease characterized by inflammatory lesions involving the heart valves, joints, subcutaneous tissues, and central nervous system. It usually results from a prior streptococcal sore throat infection. The exact mechanism of rheumatic fever development remains unknown. The disease occurs most frequently among children 6 to 15 years of age and manifests itself through a variety of signs and symptoms, making diagnosis difficult. In the United States rheumatic fever has become very rare (less than 0.05% of streptococcal infections), but it occurs 100 times more frequently in tropical countries. Therapy is directed at decreasing the inflammation and fever, and controlling cardiac failure. Salicylates and corticosteroids are the mainstays of treatment. Though rheumatic fever is rare, it is still the most common cause of permanent heart valve damage in children.

Scarlet Fever

Scarlet fever (scarlatina) results from a throat infection with a strain of *S. pyogenes* that carries a lysogenic bacteriophage. This codes for the production of an erythrogenic or rash-inducing toxin that causes shedding of the skin. Scarlet fever is a communicable disease spread by inhalation of infective respiratory droplets. After a 2-day incubation period, a scarlatinal rash appears on the upper chest and then spreads to the remainder of the body. This rash represents the skin's generalized reaction to the circulating toxin. Along with the rash, the infected individual experiences a sore throat, chills, fever, headache, and a strawberry-colored tongue (**figure 39.5**). Treatment is with penicillin.

Streptococcal Sore Throat

Streptococcal sore throat is one of the most common bacterial infections of humans and is commonly called strep throat. The β-hemolytic group A streptococci are spread by droplets of saliva or nasal secretions. The incubation period in humans is 2 to 4 days. The incidence of sore throat is greater during the winter and spring months.

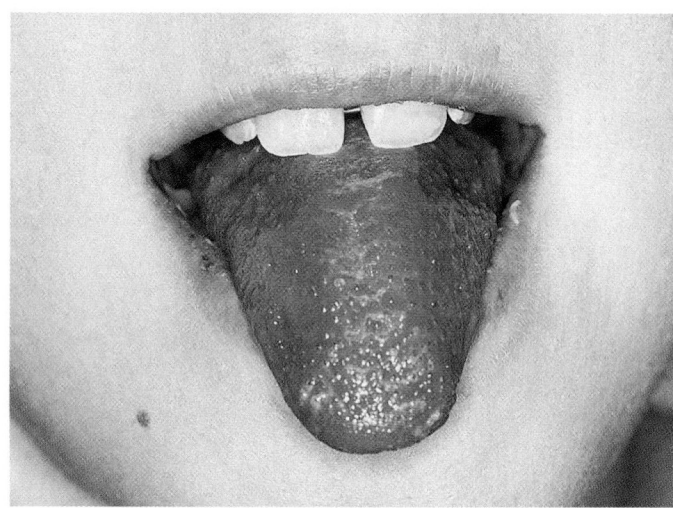

Figure 39.5 Scarlet Fever.   The strawberry-colored tongue characteristic of this streptococcal disease.

The action of the strep bacteria in the throat (**pharyngitis**) or on the tonsils (**tonsillitis**) stimulates an inflammatory response and the lysis of leukocytes and erythrocytes. An inflammatory exudate consisting of cells and fluid is released from the blood vessels and deposited in the surrounding tissue. This is accompanied by a general feeling of discomfort or malaise, fever (usually above 101°F), and headache. Prominent physical manifestations include redness, edema, and lymph node enlargement in the throat. Several common rapid test kits are available for diagnosing strep throat. In the absence of complications, the disease is self-limited and disappears within a week. However, treatment with penicillin G benzathine (or erythromycin for penicillin-allergic people) can shorten the infection and clinical syndromes, and is especially important in children for the prevention of complications such as rheumatic fever and glomerulonephritis. Infections in older children and adults tend to be milder and less frequent due in part to the immunity they have developed against the many serotypes encountered in early childhood. Prevention and control measures include proper disposal or cleansing of objects (e.g., facial tissue, handkerchiefs) contaminated by discharges from the infected individual.

Streptococcal Pneumonia

Streptococcal pneumonia is now considered an **endogenous infection**—that is, it is contracted from one's own normal microbiota (*see figure 31.2*). It is caused by the gram-positive *Streptococcus pneumoniae,* found in the upper respiratory tract (**figure 39.6**). However, disease usually occurs only in those individuals with predisposing factors such as viral infections of the respiratory tract, physical injury to the tract, alcoholism, or diabetes. About 60 to 80% of all respiratory diseases known as pneumonia are caused by *S. pneumoniae.* An estimated 150,000 to 300,000 people in the United States contract this form of pneumonia annually, and between 13,000 to 66,000 deaths result.

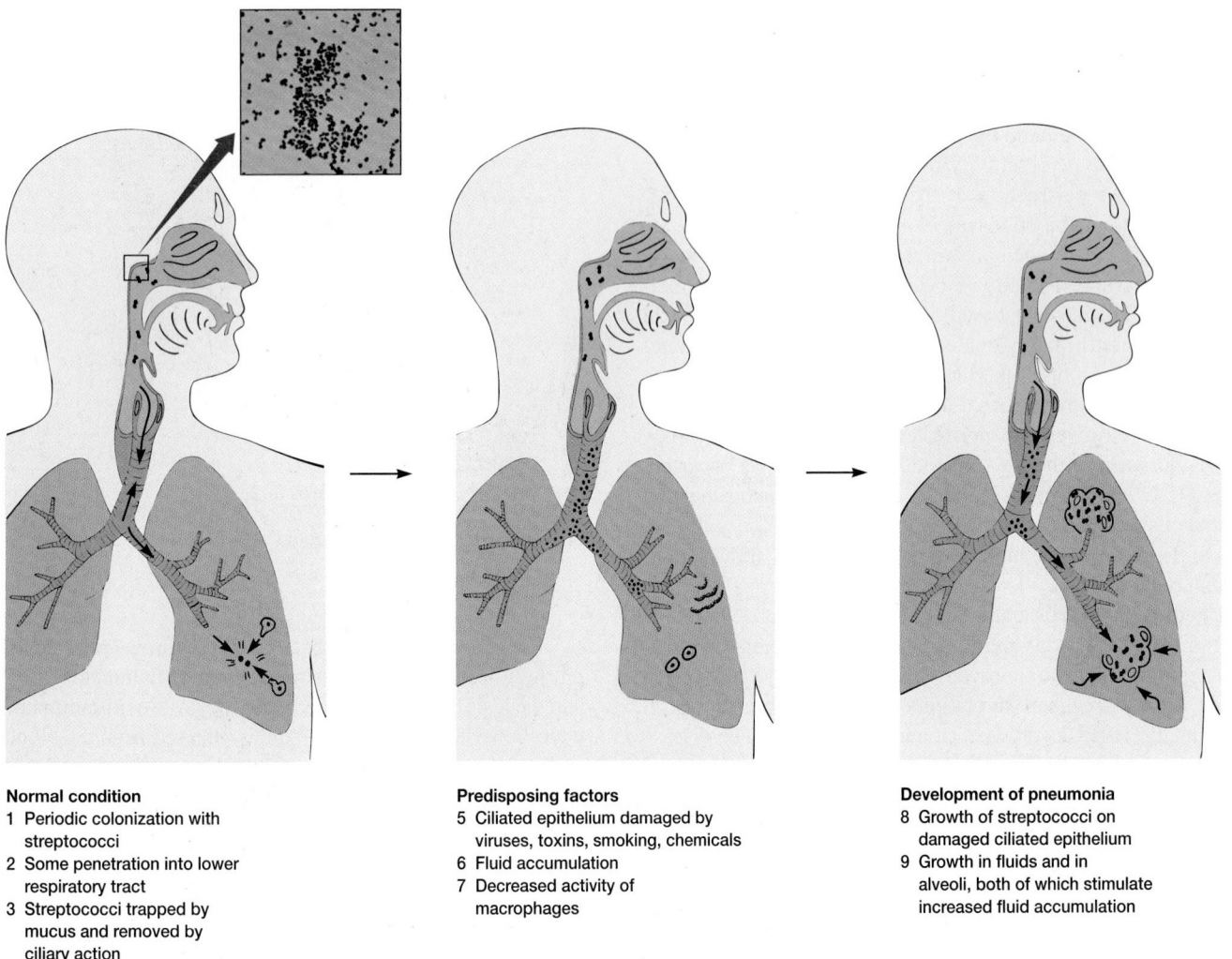

Normal condition
1 Periodic colonization with streptococci
2 Some penetration into lower respiratory tract
3 Streptococci trapped by mucus and removed by ciliary action
4 Phagocytosed by macrophages

Predisposing factors
5 Ciliated epithelium damaged by viruses, toxins, smoking, chemicals
6 Fluid accumulation
7 Decreased activity of macrophages

Development of pneumonia
8 Growth of streptococci on damaged ciliated epithelium
9 Growth in fluids and in alveoli, both of which stimulate increased fluid accumulation

Figure 39.6 Predisposition to and the Development of Streptococcal Pneumonia. The insert shows the morphology of *Streptococcus pneumoniae* (× 1,000). Streptococci are gram-positive bacteria of spherical to ovoid shape that characteristically form chains when grown in liquid media or as diplococci on solid media.

The primary virulence factor of *S. pneumoniae* is its capsular polysaccharide. The capsular polysaccharide is composed of hyaluronic acid. It is the production of large amounts of hyaluronic capsular polysaccharide that plays an important role in protecting the organism from ingestion and killing by phagocytes (*see photo, p. 761*). The pathogenesis is due to the rapid multiplication of the bacteria in alveolar spaces. The bacteria also produce the toxin pneumolysin that destroys host cells. The alveoli fill with blood cells and fluid and become inflamed. The sputum is often rust colored because of blood coughed up from the lungs. The onset of clinical symptoms is usually abrupt, with chills, hard labored breathing, and chest pain. Diagnosis is by chest X ray, biochemical tests, and culture. Penicillin G, cefotaxime, ofloxacin, and ceftriaxone have contributed to a greatly reduced mortality rate. For individuals who are sensitive to penicillin, erythromycin, or tetracycline can be used. Recently a peni-

cillin- and tetracyline-resistant strain of *S. pneumoniae* has appeared in the United States. Pneumococcal vaccines (Pneumovax 23, Pnu-Imune 23) are available for people who are debilitated (e.g., people in chronic-care facilities). The efficacy of the Pneumovax vaccines (pooled collections of 23 different *S. pneumoniae* capsular polysaccharides) is that they generate antibodies to the capsule. When these antibodies are deposited on the surface of the capsule, they become opsonic and enhance phagocytosis. Preventive and control measures include immunization and adequate treatment of infected persons.

Tuberculosis

Over a century ago Robert Koch (*see figure 1.4*) identified *Mycobacterium tuberculosis* as the causative agent of **tuberculosis (TB).** At the time, TB was rampant, causing 1/7 of all deaths in

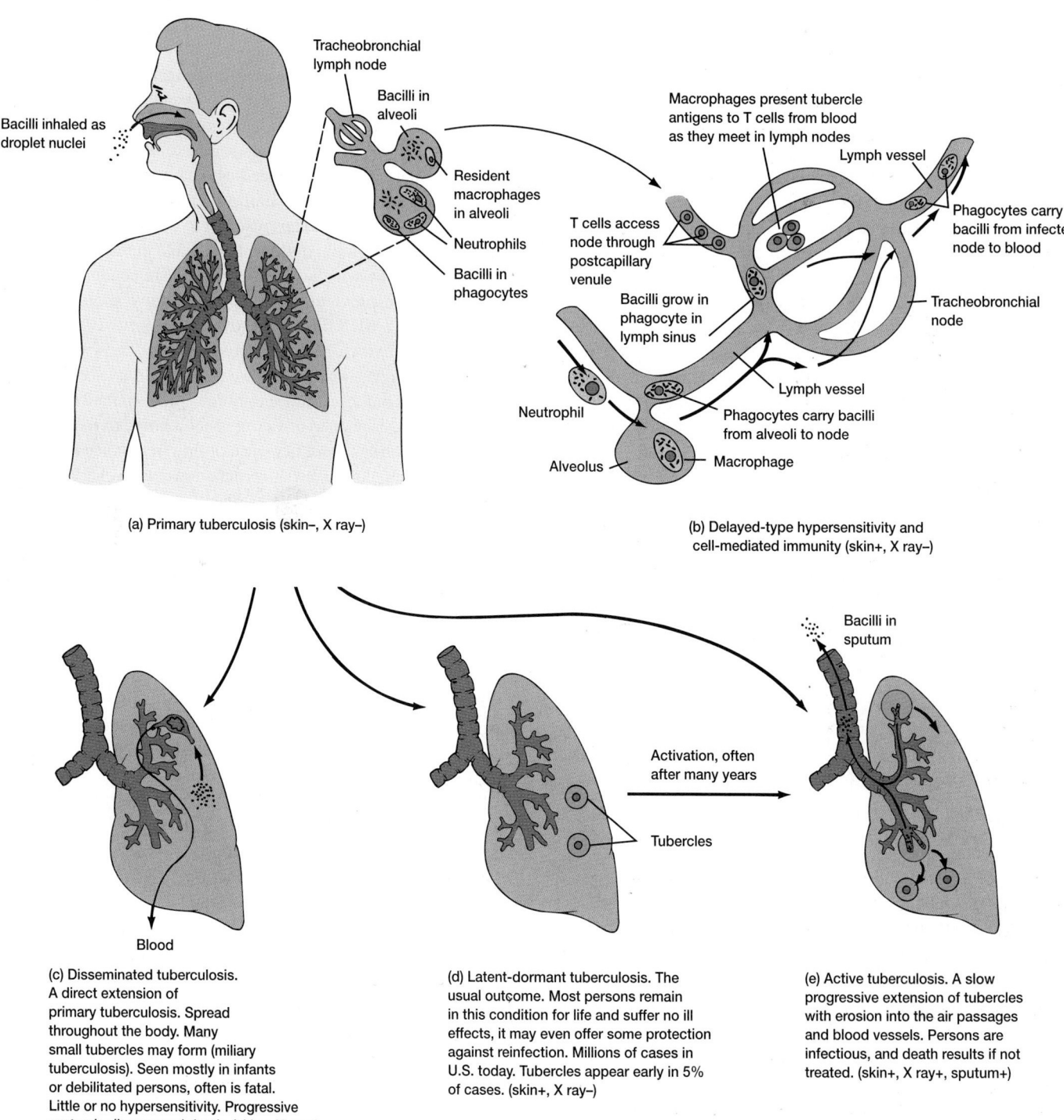

Figure 39.7   Tuberculosis.   The history of untreated tuberculosis in the human body.

Europe and 1/3 of deaths among productive young adults. Today TB remains a global health problem of enormous dimension. It is estimated that there are 1 billion (20% of the world's human population) infected worldwide, with 10 million new cases and over 3 million deaths per year.

In the United States this disease occurs most commonly among the homeless, elderly, malnourished, or alcoholic poor males, minorities, immigrants, prison populations, morticians, and Native Americans. More than 26,000 new cases of tuberculosis and over 12,000 deaths are reported annually. Most cases in the United States are caused by the acid-fast *Mycobacterium tuberculosis,* acquired from other humans through droplet nuclei and the respiratory route (**figure 39.7**). It appears that about 1/4 to 1/3 of active TB cases in the United States may be due to recent transmission. The majority of active cases result from the reactivation of old dormant infections. Worldwide, *M. bovis* and *M. africanum* also cause TB. Transmission to humans from susceptible animal species and their products (e.g., milk) is also possible. Recently there has been a steady yearly increase in the number of TB cases as a result of the AIDS epidemic. Available statistics indicate that a close association exists between AIDS and TB. Therefore further spread of HIV infection among the population with a high prevalence of TB infection is resulting in dramatic increases in TB.

Once in the lungs the bacteria are phagocytosed by macrophages and a hypersensitivity response forms small, hard nodules called **tubercles,** which are characteristic of tuberculosis and give the disease its name. The disease process usually stops at this stage, but the bacteria often remain alive within macrophage phagosomes. Resistance to oxidative killing, inhibition of phagosome-lysosome fusion, and inhibition of diffusion of lysosomal enzymes are some of the mechanisms that may explain the survival of *M. tuberculosis* inside macrophages. In time the tubercle may change to a cheeselike consistency and is then called a **caseous lesion.** If such lesions calcify, they are termed **Ghon complexes,** which show up prominently in a chest X ray. (Often the primary lesion is called the Ghon's tubercle or Ghon's focus.) Sometimes the tubercle lesions liquefy and form air-filled **tuberculous cavities.** From these cavities the bacteria can spread to new foci of infection throughout the body. This spreading is often called **miliary tuberculosis** due to the many tubercles the size of millet seeds that are formed in the infected tissue. It also may be called **reactivation tuberculosis** because the bacteria have been reactivated in the initial site of infection.

Persons infected with *M. tuberculosis* develop a cell-mediated immunity due to the bacteria being phagocytosed by macrophages. This immunity involves sensitized T cells (figure 39.7b) and is the basis for the tuberculin skin test (*see figure 33.7*a). In this test a purified protein derivative (PPD) of *M. tuberculosis* is injected intracutaneously (the Mantoux test). If the person has had tuberculosis, sensitized T cells react with these proteins, and a delayed hypersensitivity reaction occurs within 48 hours. This positive skin reaction appears as an induration (hardening) and reddening of the area around the injection site. Multiple puncture tests such as the Tine test are more convenient but not as accurate.

In a young person a positive skin test possibly indicates active tuberculosis. In older persons it may result from previous disease, vaccination, or a false-positive test. In both cases X rays and bacterial isolation should be completed.

The incubation period is about 4 to 12 weeks, and the disease develops slowly. The symptoms of tuberculosis are fever, fatigue, and weight loss. A cough, which is characteristic of pulmonary involvement, may result in expectoration of bloody sputum.

Laboratory diagnosis of tuberculosis is by isolation of the acid-fast bacterium, chest X ray, commercially available DNA probes, the BACTEC NAP test, and the Mantoux or tuberculin skin test. Both chemotherapy and chemoprophylaxis are carried out by administering isoniazid (INH), plus rifampin, ethambutol, and pyrazinamide. These drugs are administered simultaneously for 12 to 24 months as a way of decreasing the possibility that the patient develops drug resistance.

Recently, new **multi-drug-resistant strains of *tuberculosis* (MDR-TB)** have developed and are spreading. A multi-drug-resistant strain is defined as *M. tuberculosis* resistant to isoniazid and rifampin, with or without resistance to other drugs. Within the United States drug resistance has increased from 2 to 9% in the past three decades, and similar increases have occurred in many other countries. This has resulted in many cases of marginally treatable, often fatal, disease. Inadequate therapy is the most common means by which resistant bacteria are acquired, and patients who have previously undergone therapy should be presumed to harbor MDR-TB until proved otherwise.

The way in which MDR-TB arises is now known. Tubercle bacilli have spontaneous, predictable rates of chromosomally born mutations that confer resistance to drugs. These mutations are unlinked; hence resistance to one drug is not associated to resistance to an unrelated drug. The emergence of drug resistance represents the survival of random preexisting mutations, not a change caused by exposure to the drug—that the mutations are not linked is the cardinal principle underlying TB chemotherapy. For example, mutations causing resistance to isoniazid or rifampin occur roughly 1 in 10^8 replications of *M. tuberculosis*. The likelihood of spontaneous mutations causing resistance to both isoniazid and rifampin is the sum of these probabilities, or 1 in 10^{16}. However, these biological mechanisms of resistance break down when chemotherapy is inadequate. In the circumstances of monotherapy, erratic drug ingestion, omission of one or more drugs, suboptimal dosage, poor drug absorption, or an insufficient number of active drugs in a regimen, a susceptible strain on *M. tuberculosis* may become resistant to multiple drugs within a matter of months.

Prevention and control of tuberculosis requires rapid specific therapy to interrupt infectious spread. Retreatment of patients who have multi-drug-resistant tuberculosis should be carried out in programs with comprehensive microbiological, pharmacokinetic, psychosocial, and nutritional support systems. In many countries individuals, especially infants and children, are vaccinated with **b**acille **C**almette-**G**uérin (BCG) vaccine to prevent complications such as meningitis. The BCG vaccine appears to protect about half of those inoculated. Tuberculosis rates also can be lowered by better public health measures and social conditions, for example, a reduction in homelessness and drug abuse.

---

1. What causes the typical symptoms of diphtheria and how are individuals protected against this disease?
2. What is the environmental source of the bacterium that causes Legionnaires' disease? Pontiac fever?
3. What are the two major types of meningitis? Why is it so important to determine which type a person has?
4. Name the three stages of pertussis.
5. Name the most important human diseases caused by *Streptococcus pyogenes*. How do they differ from one another?
6. How is tuberculosis diagnosed? Describe the various types of lesions and how they are formed. How do multi-drug-resistant strains of tuberculosis develop?

The investigation of human pathogens often is a very dangerous matter, and several microbiologists have been killed by the microorganisms they were studying. The study of typhus fever provides a classic example. In 1906 Howard T. Ricketts (1871–1910), an associate professor of pathology at the University of Chicago, became interested in Rocky Mountain spotted fever, a disease that had decimated the Nez Percé and Flathead Indians of Montana. By infecting guinea pigs, he established that a small bacterium was the disease agent and was transmitted by ticks. In late 1909 Ricketts traveled to Mexico to study Mexican typhus.

He discovered that a microorganism similar to the Rocky Mountain spotted fever bacillus could cause the disease in monkeys and be transmitted by lice. Despite his careful technique, he was bitten while transferring lice in his laboratory and died of typhus fever on May 3, 1910. The causative agent of typhus fever was fully described in 1916 by the Brazilian scientist H. da Roche-Lima and named *Rickettsia prowazekii* in honor of Ricketts and Stanislaus von Prowazek, a Czechoslovakian microbiologist who died in 1915 while studying typhus.

Today modern equipment to control microorganisms, such as laminar airflow hoods, have greatly reduced the risks of research on microbial pathogens.

 39.2   ARTHROPOD-BORNE DISEASES

Although arthropod-borne bacterial diseases are generally rare, they are of interest either historically (plague, typhus [see chapter opener quote]) or because they have been newly introduced into humans (ehrlichiosis, Q fever, Lyme disease). In the next sections, only diseases that occur in the United States are discussed.

Ehrlichiosis

In 1986 the first case of **ehrlichiosis** was diagnosed in the United States and shown to be caused by a new bacterial species (table 39.1), *Ehrlichia chaffeensis*. Members of the genus *Ehrlichia* are related to the genus *Rickettsia* and placed in the order *Rickettsiales* of the α-proteobacteria. Since the initial discovery, more than 400 cases have been reported in the United States. *E. chaffeensis* is transmitted from unknown animal vectors to humans by the Lone Star tick (*Amblyomma americanum*). Once inside the human body, *E. chaffeensis* infects circulating monocytes causing a nonspecific febrile illness (**h**uman **m**onocytic **e**hrlichiosis, HME) that resembles Rocky Mountain spotted fever. Diagnosis involves serological tests and tetracycline is the drug of choice.

In 1994 a new form of ehrlichiosis was discovered. **H**uman **g**ranulocytic **e**hrlichiosis (HGE) is transmitted by deer ticks (*Ixodes scapularis*) and possibly dog ticks (*Dermacentor variabilis*), and has been found in 30 states, particularly in the southeastern and south central United States. The causative agent is an *Ehrlichia* species different from *E. chaffeensis*. The disease is characterized by the rapid onset of fever, chills, headaches, and muscle aches. Treatment is with doxycycline.

Epidemic (Louse-Borne) Typhus

Epidemic (louse-borne) typhus is caused by the rickettsia *Rickettsia prowazekii,* which is transmitted from person to person by

the body louse (**Historical Highlights 39.1**). In the United States a reservoir of *R. prowazekii* also exists in the southern flying squirrel. When a louse feeds on an infected rickettsemic person, the rickettsias infect the insect's gut and multiply, and large numbers of organisms appear in the feces in about a week. When a louse takes a blood meal, it defecates. The irritation causes the affected individual to scratch the site and contaminate the bite wound with rickettsias. The rickettsias then spread by way of the bloodstream and infect the endothelial cells of the blood vessels, causing a **vasculitis** (inflammation of the blood vessels). This produces an abrupt headache, fever, and muscle aches. A rash begins on the upper trunk, and spreads. Without treatment, recovery takes about 2 weeks, though mortality rates are very high (around 50%), especially in the elderly. Recovery from the disease gives a solid immunity and also protects the person from murine typhus.

Diagnosis is by the characteristic rash, symptoms, and immunofluorescence testing (*see pp. 755–56*). Chloramphenicol and tetracycline are effective against typhus. Control of the human body louse (*Pediculus humanus corporis*) and the conditions that foster its proliferation are mainstays in the prevention of epidemic typhus, although a typhus vaccine is available for high-risk individuals. The importance of louse control and good public hygiene is shown by the prevalence of typhus epidemics during times of war and famine when there is crowding and little attention to the maintenance of proper sanitation. For example, around 30 million cases of typhus fever and 3 million deaths occurred in the Soviet Union and Eastern Europe between 1918 and 1922. The bacteriologist Hans Zinsser believes that Napoleon's retreat from Russia in 1812 may have been partially provoked by typhus and dysentery epidemics that ravaged the French army. Fewer than 25 cases of epidemic typhus are reported in the United States each year.

Endemic (Murine) Typhus

The etiologic agent of **endemic (murine) typhus** is the rickettsia *Rickettsia typhi*. It occurs in isolated areas around the world, including

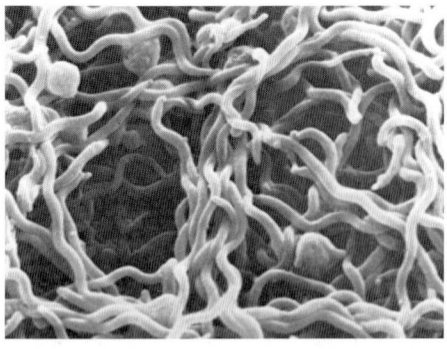

(a)

Figure 39.8 Lyme Disease. (a) The etiological agent is the spirochete *Borrelia burgdorferi;* SEM. (b) The vector in the northeastern U.S. is the tick *Ixodes scapularis.* The youngest (nymphal stage, top) is about the size of a poppy seed. An unengorged adult (bottom), and an engorged adult (center) can reach the size of a jelly bean. (c) The typical rash (erythema migrans) showing concentric rings around the initial site of the tick bite.

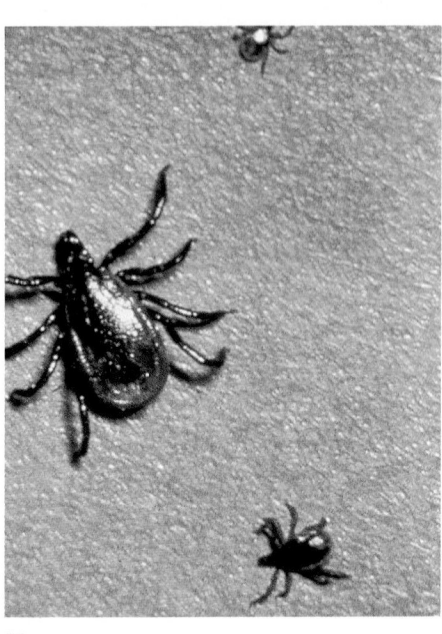

(b)

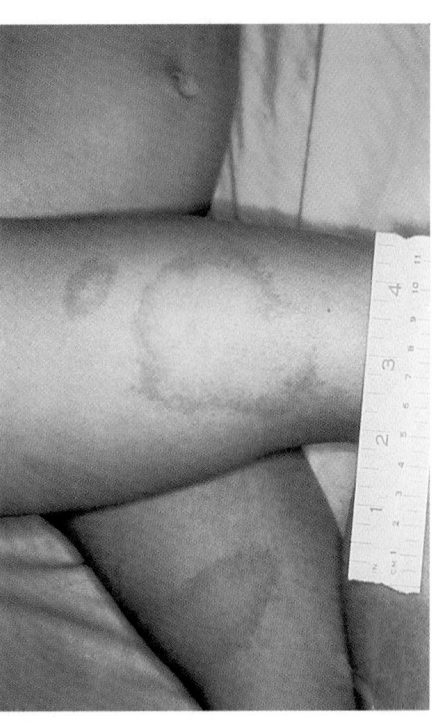

(c)

southeastern and Gulf Coast states, especially Texas. The disease occurs sporadically in individuals who come into contact with rats and their fleas (*Xenopsylla cheopi*). The disease is nonfatal in the rat and is transmitted from rat to rat by fleas. When an infected flea takes a human blood meal, it defecates. Its feces are heavily laden with rickettsias, which infect humans by contaminating the bite wound.

The clinical manifestations of murine [Latin *mus, muris,* mouse or rat] typhus are similar to those of epidemic typhus except that they are milder in degree and the mortality rate is much lower: less than 5%. Diagnosis and treatment also are the same. Rat control and avoidance of rats are preventive measures for the disease. Fewer than 100 cases of endemic typhus are reported in the United States each year.

Lyme Disease

Lyme disease (LD, Lyme borreliosis) was first observed and described in 1975 among people of Old Lyme, Connecticut. It has become the most common tick-borne zoonosis in the United States, with more than 20,000 cases being reported annually. In fact, Lyme disease has reached epidemic proportions, and if it were not for AIDS, Lyme disease would be the most important "new" infectious disease (*table 39.1*) of humans in the United States. The disease is also present in Europe and Asia.

The Lyme spirochetes responsible for this disease comprise at least three species, currently designated *Borrelia burgdorferi* (**figure 39.8a**), *B. garinii,* and *B. afzelii.* Deer and field mice are the natural hosts. In the northeastern United States, *B. burgdorferi* is transmitted to humans by the bite of the infected deer tick (*Ixodes scapularis;* figure 39.8b). On the Pacific Coast, espe-

cially in California, the reservoir is a dusky-footed woodrat, and the tick, *I. pacificus.*

Clinically Lyme disease is a complex illness with three major stages. The initial, localized stage occurs a week to 10 days after an infectious tick bite. The illness usually begins with an expanding, ring-shaped, skin lesion with a red outer border and partial central clearing (figure 39.8c). This often is accompanied by flu-like symptoms (malaise and fatigue, headache, fever, and chills). Often the tick bite is unnoticed, and the skin lesion may be missed due to skin coloration or its obscure location such as on the scalp. Thus treatment, which is usually effective at this stage, may not be given because the illness is passed off as "just a touch of the flu."

The second, disseminated stage may appear weeks or months after the initial infection. It consists of several symptoms such as neurological abnormalities, heart inflammation, and bouts of arthritis (usually in the major joints such as the elbows or knees). Current research indicates that Lyme arthritis might be an autoimmune response to joint cell HLA molecules that are similar to the bacterial antigens. The inflammation that produces organ damage is initiated and possibly perpetuated by the immune response to one or more spirochetal proteins.

Finally, like syphilis, years later the late stage may appear. Infected individuals may develop neuron demyelination with symptoms resembling Alzheimer's disease and multiple sclerosis. Behavioral changes also can occur.

Laboratory diagnosis of Lyme's disease is based on (1) the recovery of the spirochete from the patient, (2) use of the polymerase chain reaction (*see section 14.3*) for detection of *B. burgdorferi* DNA in urine, or (3) serological testing (Lyme ELISA or Western Blot) for IgM and IgG antibodies to the pathogen. Treatment with amoxicillin or tetracycline early in the illness results in prompt recovery and prevents arthritis and other complications. If nervous system involvement is suspected, ceftriaxone is used since it can cross the blood-brain barrier.

Prevention and control of Lyme disease involves environmental modification (clearing and burning tick habitat) and the application of acaricidal compounds (agents that destroy mites and ticks). An individual's risk of acquiring Lyme disease may be greatly reduced by education and personal protection. The following points should be kept in mind whenever a person is active in an area where Lyme disease or other tick-borne zoonoses occur:

1. It takes a minimum of 24 hours of attachment and feeding for transmission to occur; thus prompt removal of attached ticks will greatly reduce the risk of infection. To remove an embedded tick, one should use tweezers to grasp the tick as close as possible to the skin and then pull with slow, steady pressure in a direction perpendicular to the skin.
2. Because each deer tick life cycle stage is most abundant at a certain time, there are periods when an individual should be most aware of the risk of infection. The most dangerous times are May through July, when the majority of nymphal deer ticks are present and the risk of transmission is greatest.
3. If you must be in the woods, dress accordingly. Wear light-colored pants and good shoes. Tuck the cuffs of your pants into long socks to deny ticks easy entry under your clothes. After coming out of the woods, check all clothes for ticks.
4. Repellents containing high concentrations of DEET (diethyltoluamide) or permanone are available over the counter and are very noxious to ticks. Premethrin kills ticks on contact but is approved only for use on clothing.
5. Immediately after being in a high-risk area, examine your body for bites or itches. Taking a shower and using lots of soap aids in this examination. Areas such as the scalp, armpits, and groin are difficult to examine effectively but are preferred sites for tick attachment. Special attention should be given to these parts of the body.

Plague

In the southwestern part of the United States, **plague** [Latin *plaga,* pest] occurs primarily in wild rodents (ground squirrels and prairie dogs). However, massive human epidemics occurred during the Middle Ages, and the disease was known as the Black Death because one of its characteristics is blackish areas on the skin caused by subcutaneous hemorrhages. Infections now occur in humans only sporadically or in limited outbreaks. In the United States approximately 25 cases are reported annually, and the mortality rate is about 15%.

The disease is caused by the gram-negative bacterium *Yersinia pestis.* It is transmitted from rodent to human by the bite of an infected flea, direct contact with infected animals or their products, or inhalation of contaminated airborne droplets (**figure 39.9**). Once in the human body, the bacteria multiply in the blood and lymph. An important factor in the virulence of *Y. pestis* is its

ability to survive and proliferate inside phagocytic cells rather than being killed by them. One of the ways this is accomplished is by the YOPS (**y**ersinal plasmid-encoded **o**uter membrane **p**roteins) that are secreted by the bacterium and act as antiphagocytic proteins to counteract natural defense mechanisms and help the bacteria multiply and disseminate in the host (*see figure 34.4*).

Symptoms—besides the subcutaneous hemorrhages—include fever and the appearance of enlarged lymph nodes called **buboes** (hence the old name, **bubonic plague**). In 50 to 70% of the untreated cases, death follows in 3 to 5 days from toxic conditions caused by the large number of bacilli in the blood.

Laboratory diagnosis of plague is by direct microscopic examination, culture of the bacterium, serological tests, the PCR for detection of bacteria in infected fleas, and phage testing. Treatment is with streptomycin, chloramphenicol, or tetracycline, and recovery from the disease gives a good immunity.

Pneumonic plague arises (1) from primary exposure to infectious respiratory droplets from a person or cat with respiratory plague or (2) secondary to hematogenous spread in a patient with bubonic or septicemic plague. Pneumonic plague can also arise from accidental inhalation of *Y. pestii* in the laboratory. The mortality rate for this kind of plague is almost 100% if it is not recognized within 12 to 24 hours. Obviously great care must be taken to prevent the spread of airborne infections to personnel taking care of pneumonic plague patients.

Prevention and control involves ectoparasite and rodent control, isolation of human patients, prophylaxis or abortive therapy of exposed persons, and vaccination (USP Plague vaccine) of persons at high risk.

Q Fever

Q fever (Q for query because the cause of the fever was not known for some time) is an acute zoonotic disease caused by the γ-proteobacterium *Coxiella burnetii,* a strictly intracellular, gram-negative bacterium. *C. burnetii* is different from *Rickettsia* in its ability to survive outside host cells by forming a resistant endosporelike body. This bacterium infects both wild animals and livestock. In animals, ticks (many species) transmit *C. burnetii,* whereas in humans transmission is primarily by inhalation of dust contaminated with bacteria from dried animal feces, urine, or milk. The disease is apt to occur in epidemic form among slaughterhouse workers and sporadically among farmers and veterinarians. Each year, fewer than 100 cases of Q fever are reported in the United States.

In humans, after inhalation of the bacteria, local proliferation occurs in the lungs. This may result in mild respiratory symptoms similar to those of atypical pneumonia or influenza. Q fever itself is an acute illness characterized by the sudden onset of severe headache, myalgia (muscle pain), and fever, which may remain very high for more than a month if not treated. Unlike rickettsial diseases, Q fever is not accompanied by a rash. It is rarely fatal, but endocarditis—inflammation of the heart muscle—occurs in about 10% of the cases. Five to ten years may elapse between the initial infection and the appearance of the endocarditis. During

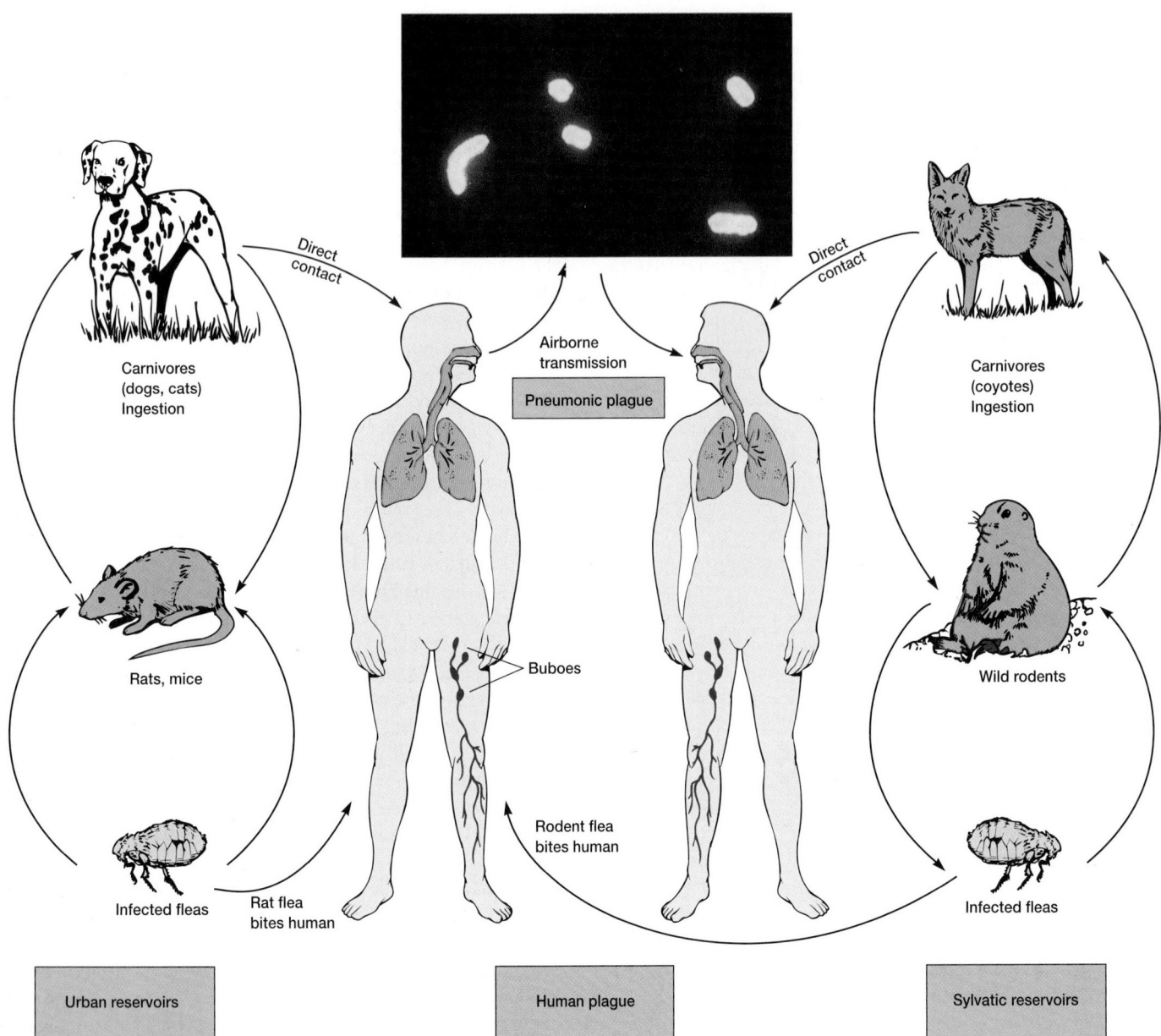

Figure 39.9   Plague.   Plague is spread to humans through (1) the urban cycle and rat fleas, (2) the sylvatic cycle centered on wild rodents and their fleas, or (3) by airborne transmission from an infected person leading to pneumonic plague. Dogs, cats, and coyotes can also acquire the bacterium by ingestion of infected animals. The insert shows *Yersinia pestis* stained with fluorescent antibodies.

this interval the bacteria reside in the liver and often cause hepatitis. Diagnosis is most commonly made serologically. Treatment is with chloramphenicol and tetracycline. Prevention and control consists of vaccinating researchers and others at high occupational risk and in areas of endemic Q fever; cow and sheep milk should be pasteurized before consumption.

Rocky Mountain Spotted Fever

Rocky Mountain spotted fever is caused by the rickettsia *Rickettsia rickettsii.* Although originally detected in the Rocky

Mountain area, most cases of this disease now occur east of the Mississippi River. The disease is transmitted by ticks and usually occurs in people who are or have been in tick-infested areas. There are two principal vectors: *Dermacentor andersoni,* the wood tick, is distributed in the Rocky Mountain states and is active during the spring and early summer. *D. variabilis,* the dog tick, has assumed greater importance and is almost exclusively confined to the eastern half of the United States. Unlike the other rickettsias discussed, *R. rickettsii* can pass from generation to generation of ticks through their eggs in a process known as **transovarian passage.** No humans or mammals are

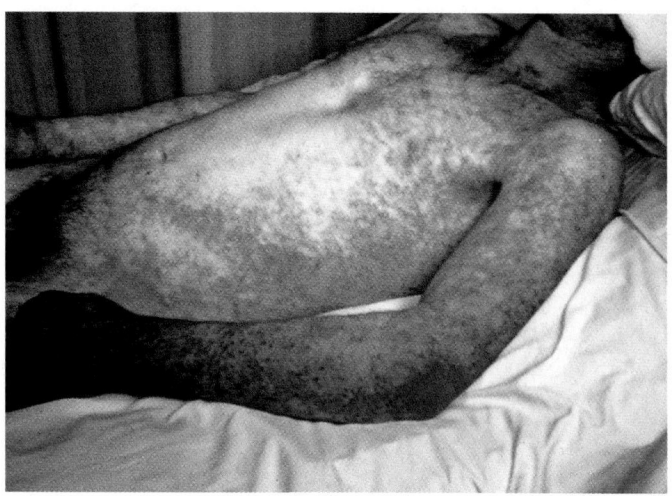

Figure 39.10   Rocky Mountain Spotted Fever.   Typical rash occurring on the arms and chest consists of generally distributed, sharply defined macules.

needed as reservoirs for the continued propagation of this rickettsia in the environment.

When humans contact infected ticks, the rickettsias are either deposited on the skin (if the tick defecates after feeding) and then subsequently rubbed or scratched into the skin, or the rickettsias are deposited into the skin as the tick feeds. Once inside the skin, the rickettsias enter the endothelial cells of small blood vessels, where they multiply and produce a characteristic vasculitis (inflammation of blood vessels).

The disease is characterized by the sudden onset of a headache, high fever, chills, and a skin rash (**figure 39.10**) that initially appears on the ankles and wrists and then spreads to the trunk of the body. If the disease is not treated, the rickettsias can destroy the blood vessels in the heart, lungs, or kidneys and cause death. Usually, however, severe pathological changes are avoided by antibiotic therapy (chloramphenicol, chlortetracycline), the development of immune resistance, and supportive therapy. Diagnosis is made through observation of symptoms and signs such as the characteristic rash and by serological tests. The best means of prevention remains the avoidance of tick-infested habitats and animals (see preventive methods for Lyme disease, p. 887). There are approximately 1,000 reported cases of Rocky Mountain spotted fever annually in the United States.

1. How is epidemic typhus spread? Ehrlichiosis? Murine typhus? What are their symptoms?
2. What two antibiotics are used against most rickettsial infections?
3. What is the causative agent of Lyme disease and how is it transmitted to humans? How does the illness begin? Describe the three stages of Lyme disease.
4. Why is plague sometimes called the Black Death? How is it transmitted? Distinguish between bubonic and pneumonic plague.

5. What is unique about *Coxiella burnetii* compared to the other rickettsias?
6. Describe the symptoms of Rocky Mountain spotted fever.
7. How does transovarian passage occur?

 39.3   DIRECT CONTACT DISEASES

Most of the direct contact bacterial diseases involve the skin or underlying tissues. Others can become disseminated through specific regions of the body. Some of the better-known of these diseases are now discussed.

Anthrax

Anthrax (Greek *anthrax,* coal) is a highly infectious animal disease that can be transmitted to humans by direct contact with infected animals (cattle, goats, sheep) or their products, especially hides. The causative bacterium is the large, gram-positive, aerobic spore-forming *Bacillus anthracis,* which has a nearly worldwide distribution. Its endospores can remain viable in soil and animal products for decades (*see figure 3.42*). Although *B. anthracis* is one of the most molecularly monomorphic (of one shape) bacteria, it is now possible to separate all known strains into five categories (providing some clues to their geographic sites of origin) based on the variable numbers of tandem repeats in the variable region of the *vvvA* gene.

Human infection is usually through a cut or abrasion of the skin, resulting in **cutaneous anthrax;** however, inhaling endospores may result in **pulmonary anthrax,** also known as woolsorter's disease. If endospores reach the gastrointestinal tract, **gastrointestinal anthrax** may result. The principal virulence factors of *B. anthracis* are encoded on two plasmids—one involved in the synthesis of a polyglutamyl capsule that inhibits phagocytosis of the vegetative form and the other bearing the genes for the synthesis of its exotoxins (a complex exotoxin system composed of three proteins: protective antigen [PA], edema factor [EF], and lethal factor [LF]).

In humans, more than 95% of naturally occurring anthrax is the cutaneous form. The incubation period for cutaneous anthrax is 1 to 15 days. Infection is initiated with the introduction of the spores through a break in the skin. After ingestion by macrophages at the site of entry, the spores germinate and give rise to the vegetative form, which multiplies extracellularly and forms a capsule and exotoxins. In the skin the infection produces a skin papule that ulcerates and is called an **eschar (figure 39.11a).** The eschar dries and falls off in 1 to 2 weeks with little ultimate scarring. Without antibiotic treatment, mortality can be as high as 20 percent. Therapy is with either ciprofloxacin or doxycycline plus one or two additional antimicrobial agents with in vitro activity against *B. anthracis.* Treatment should continue for 60 days in the context of bioterrorism, or for 7 to 10 days with the naturally acquired disease. With proper antibiotic treatment, mortality for cutaneous anthrax is very rare.

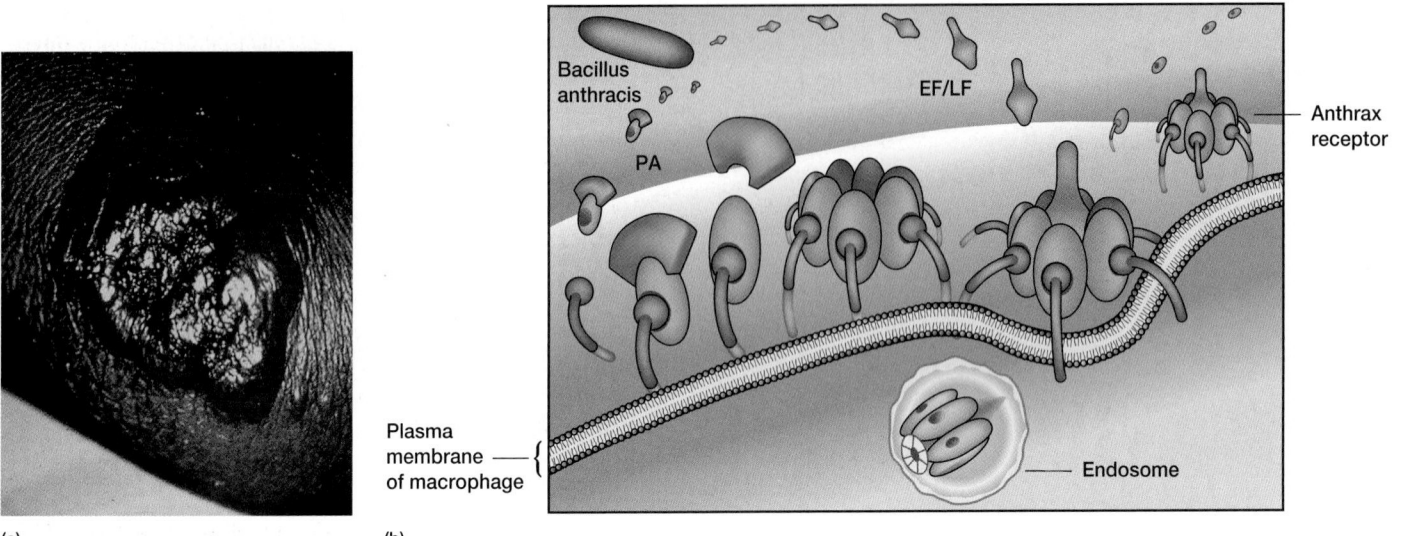

Figure 39.11   Anthrax.   (**a**) The malignant (destructive) eschar pustules of anthrax on the arm of an infected person. (**b**) A protein called protective antigen (PA) delivers anthrax's deadly cargo of edema factor (EF) and lethal factor (LF) into the cell's endosome causing death of the macrophage and also possibly the host.

In inhalation anthrax, the spores (1 to 2 µm in diameter) are inhaled and lodge in the alveolar spaces where they are engulfed by alveolar macrophages. The spores survive phagocytosis and germinate within the endosome; the bacteria then spread to regional lymph nodes and eventually the bloodstream. Once the bacteria enter the bloodstream, they begin producing exotoxin. The medial lethal inhalation dose for humans has been estimated to be 2,500 to 55,000 spores.

For a successful infection, *B. anthracis* must evade the host's innate immune system by killing macrophages. Macrophages have many anthrax toxin receptors (ATRs) on their plasma membranes that the protective antigen (PA) portion of the exotoxin system attaches to. Attachment continues until seven PA-ATR complexes gather in a doughnut-shaped ring (figure 39.11*b*). The ring of seven complexes acts like a syringe, boring through the plasma membrane of the macrophage. The ring then binds edema factor (EF) and lethal factor (LF) after which the entire complex is engulfed by the macrophages' plasma membrane and shuttled to an endosome (*see figure 4.11*) inside the cell. Once there, the protective antigen molecules form a special pore that pierces the endosome's membrane and lets edema factor and lethal factor out into the cytoplasm to do their grisly work. Edema factor interferes with the cell's communications system, preventing it from signaling to the immune system for help; thus other macrophages do not phagocytose the bacteria. The lethal factor inactivates the p38 mitogen-activated kinase signaling cascade. This inhibition prevents the expression of target genes of a transcription factor, nuclear factor B, that promote macrophage survival. As thousands of macrophages die, they release toxic chemicals, leading to fever, internal bleeding, septic shock, and rapid death.

The classic clinical description of inhalation anthrax is that of a two-phase illness. In the initial phase that follows an incu-

bation period of 1 to 6 days, the disease appears as a nonspecific illness characterized by mild fever, malaise, nonproductive cough, and some chest pain. The second phase begins abruptly and involves a higher fever, acute dyspnea (shortness of breath), and cyanosis (oxygen deficiency). This stage progresses rapidly, with septic shock, associated hypothermia, and death occurring within 24 to 36 hours from respiratory failure. Treatment is successful only if begun soon enough (the same antibiotics as for cutaneous anthrax). One reason this form of anthrax is so difficult to treat is that the symptoms appear after *B. anthracis* has already multiplied and started to produce large amounts of the tripartite exotoxin. Thus, although antibiotics may kill the bacterium or suppress its growth, the exotoxin can still eventually kill the human patient. Sixteen of the 18 cases of inhalation anthrax reported in the United States between 1900 and 1978 were fatal.

The symptoms of gastrointestinal anthrax appear 2 to 5 days after the ingestion of undercooked meat containing endospores and include nausea, vomiting, fever, and abdominal pain. The manifestations progress rapidly to severe, bloody diarrhea. The primary lesions are ulcerative. Mortality is greater than 50 percent.

B. anthracis bacteremia can develop in any form of anthrax and occurs throughout the body.

In the past, the diagnosis of anthrax was made on the basis of clinical findings and the history of exposure to animal products. The significance of a history of exposure has changed with the 2001 delivery of anthrax endospores through the mail and concern about biological warfare involving the airborne delivery of weaponized anthrax spores capable of widespread dissemination. Presumptive identification in a hospital laboratory is based on the direct Gram-stained smear of a skin lesion, cerebrospinal fluid, or blood that shows encapsulated, broad, gram-positive bacilli.

Diagnosis is also made on the basis of growth characteristics apparent on sheep's blood agar cultures: nonhemolytic colonies, nonmotile, nonencapsulated, gram-positive, spore-forming rods. Confirmatory diagnosis is performed at a level B laboratory of the Laboratory Response Network for Bioterrorism (LRN).

Between 20,000 and 100,000 cases of anthrax have been estimated to occur worldwide annually, but in the United States the annual incidence was 127 cases in the early part of the twentieth century; it subsequently declined to less than 1 case per year—a rate maintained for the past 20 years. Until 2001, there had not been a case of inhalation anthrax in the United States for more than 20 years. Thus, the recent occurrence of 12 cases of anthrax after the 9/11 attack (when the twin towers of the World Trade Center in New York City were destroyed), six cases involving inhalation and none with the conventional exposure to infected animals or animal products, has spotlighted the current concern about anthrax as a weapon of bioterrorism (*see section 37.10*).

Vaccination of animals, primarily cattle, is an important control measure. However, people with a high occupational risk, such as those who handle infected animals or their products, including hides and wool, should be immunized with the cell-free vaccine obtainable from the CDC.

Bacterial Vaginosis

Bacterial vaginosis is considered a sexually transmitted disease (table 39.4). It has a polymicrobic etiology that includes *Gardnerella vaginalis* (a gram-negative to gram-variable, pleomorphic nonmotile rod), *Mobiluncus* spp., *Mycoplasma hominis,* and various anaerobic bacteria. The recent finding that these microorganisms inhabit the rectum of 20 to 40% of healthy women indicates a potential source of autoinfection in addition to sexual transmission. Although a mild disease it is a risk factor for obstetric infections, various adverse outcomes of pregnancy, and pelvic inflammatory disease. The vaginosis is characterized by a copious, frothy, fishy-smelling discharge without pain or itching. Diagnosis is based on this fishy odor and the microscopic observation of clue cells in the discharge. **Clue cells** are sloughed-off vaginal epithelial cells covered with bacteria, mostly *G. vaginalis.* Treatment for bacterial vaginosis is with metronidazole (Flagyl, MetroGel-Vaginal), a drug that kills the anaerobes that are needed for the continuation of the disease.

Cat-Scratch Disease

Cat-scratch disease (CSD) is a loosely defined syndrome, the cause of which eluded microbiologists for decades. Currently the etiology centers on a recently described gram-negative bacillus, *Bartonella henselae* (table 39.1).

The diagnosis of CSD is based on the combination of the clinical history of either a cat scratch or bite at the skin surface (primary lesion) and subsequent swelling of the lymph node(s) that drain the inoculation site, and on detection by PCR techniques. This is accompanied by malaise and fever. The typical case of CSD is self-limiting, with abatement of symptoms over a period of days to weeks, and the resolution of the lymphadenopathy over several months.

Chancroid

Chancroid [French *chancre,* a destructive sore, and Greek *eidos,* to form], also known as **genital ulcer disease,** is a sexually transmitted disease caused by the gram-negative bacillus *Haemophilus ducreyi* (table 39.4). The bacterium enters the skin through a break in the epithelium. After an incubation period of 4 to 7 days, a papular lesion develops within the epithelium, causing swelling and white blood cell infiltration. Within several days a pustule forms and ruptures, producing a painful circumscribed ulcer with a ragged edge; hence the term genital ulcer disease. Most of the ulcers in males are on the penis and in females at the entrance of the vagina. Genital ulcer disease occurs commonly in the tropics; however, in the past decade there have been major outbreaks in the United States. Worldwide, genital ulcer disease is an important cofactor in the transmission of the AIDS virus; thus it could be an important cofactor in the United States as well. Diagnosis is by isolating *H. ducreyi* from the ulcers; treatment is with erythromycin or ceftriaxone. Control is by the use of condoms or abstinence.

Chlamydial Pneumonia

Chlamydial pneumonia is caused by *Chlamydia pneumoniae* (TWAR). Clinically, infections are generally mild; pharyngitis, bronchitis, and sinusitis commonly accompany some lower respiratory tract involvement. Symptoms include fever, a productive cough, sore throat, hoarseness, and pain on swallowing. Infections with *C. pneumoniae* are common but sporadic; about 50% of adults have antibody to the chlamydiae. Evidence suggests that *C. pneumoniae* is primarily a human pathogen directly transmitted from human to human by respiratory secretions. Identification of chlamydial pneumonia is based on symptoms and a microimmunofluorescence test. Tetracycline and erythromycin are used for treatment.

In seroepidemiological studies, *C. pneumoniae* infections have been linked with coronary artery disease as well as vascular disease at other sites. Following a demonstration of *C. pneumoniae*-like particles in artheromatous tissue by electron microscopy, *C. pneumoniae* genes and antigens have been detected in artheromas. Rarely, the microorganism has been recovered in cultures of atheromatous tissue. As a result of these findings, the possible etiologic role of *C. pneumoniae* in coronary artery disease and systemic atherosclerosis is currently under intense scrutiny.

Gas Gangrene or Clostridial Myonecrosis

Clostridium perfringens, C. novyi, and *C. septicum* are gram-positive spore-forming rods termed the histotoxic clostridia. They can produce a necrotizing infection of skeletal muscle called **gas gangrene** [Greek *gangraina,* an eating sore] or **clostridial myonecrosis** [*myo,* muscles, and *necrosis,* death].

Histotoxic clostridia occur in the soil worldwide and also are part of the normal endogenous microflora of the human large intestine. Contamination of injured tissue with spores from soil containing histotoxic clostridia or bowel flora is the usual means of transmission. Infections are commonly associated with wounds resulting from abortions, automobile accidents, military combat, or frostbite.

If the spores germinate in anaerobic tissue, the bacteria grow and secrete α-toxin, which breaks down muscle tissue. Growth often results in the accumulation of gas (mainly hydrogen as a result of carbohydrate fermentation), and of the toxic breakdown products of skeletal muscle tissue.

Clinical manifestations include severe pain, edema, drainage, and muscle necrosis. The pathology arises from progressive skeletal muscle necrosis due to the effects of α-toxin. Other enzymes produced by the bacteria degrade collagen and tissue, facilitating spread of the disease.

Gas gangrene is a medical emergency. Laboratory diagnosis is through recovery of the appropriate species of clostridia accompanied by the characteristic disease symptoms. Treatment is extensive surgical debridement (removal of all dead tissue), the administration of polyvalent antitoxin, and antimicrobial therapy with penicillin and tetracycline. Hyperbaric oxygen therapy (the use of high concentrations of oxygen at elevated pressures) also is considered effective. The oxygen saturates the infected tissue and thereby prevents the growth of the obligately anaerobic clostridia.

Prevention and control includes debridement of contaminated traumatic wounds plus antimicrobial prophylaxis and prompt treatment of all wound infections. Amputation of limbs often is necessary to prevent further spread of the disease.

Genitourinary Mycoplasmal Diseases

The mycoplasmas *Ureaplasma urealyticum* and *M. hominis* are common parasitic microorganisms of the genital tract and their transmission is related to sexual activity (table 39.4). Both mycoplasmas can opportunistically cause inflammation of the reproductive organs of males and females. Because mycoplasmas are not usually cultured by clinicians, management and treatment of these infections depend on a recognition of clinical syndromes and provision for adequate therapy. Tetracyclines are active against most strains; resistant organisms can be treated with erythromycin.

Gonorrhea

Gonorrhea [Greek *gono,* seed, and *rhein,* to flow] is an acute, infectious, sexually transmitted disease of the mucous membranes of the genitourinary tract, eye, rectum, and throat (table 39.4). It is caused by the gram-negative, oxidase-positive, diplococcus, *Neisseria gonorrhoeae.* These bacteria are also referred to as **gonococci** [pl. of gonococcus; Greek *gono,* seed, and *coccus,* berry] and have a worldwide distribution.

Once inside the body the gonococci attach to the microvilli of mucosal cells by means of pili and protein II, which function as ad-

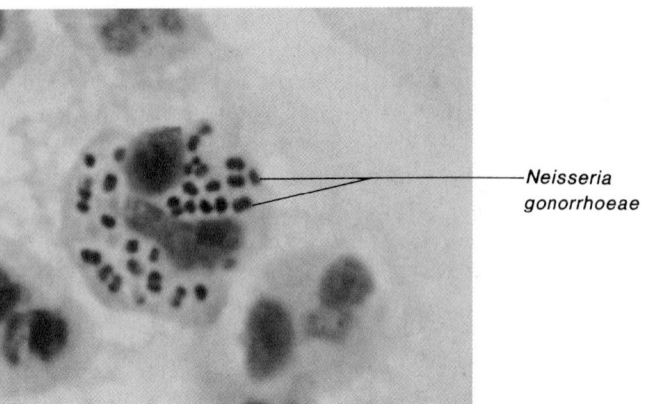

Figure 39.12 Gonorrhea. Gram stain of male urethral exudate showing *Neisseria gonorrhoeae* (diplococci) inside a PMN; light micrograph (×500). Although the presence of gramnegative diplococci in exudates is a probable indication of gonorrhea, the bacterium should be isolated and identified.

hesins. This attachment prevents the bacteria from being washed away by normal vaginal discharges or by the strong flow of urine. They are then phagocytosed by the mucosal cells and may even be transported through the cells to the intercellular spaces and subepithelial tissue. Phagocytes, such as neutrophils, also may contain gonococci (**figure 39.12**) inside vesicles. Because the gonococci are intracellular at this time, the host's defenses have little effect on the bacteria. Following penetration of the bacteria, the host tissue responds locally by the infiltration of mast cells, more PMNs, and plasma cells. These cells are later replaced by fibrous tissue that may lead to urethral closing, or stricture, in males.

In males the incubation period is 2 to 8 days. The onset consists of a urethral discharge of yellow, creamy pus and frequent, painful urination that is accompanied by a burning sensation. In females the disease is more insidious in that few individuals are aware of any symptoms. However, some symptoms may begin 7 to 21 days after infection. These are generally mild; some vaginal discharge may occur. The gonococci also can infect the uterine tubes and surrounding tissues, leading to **pelvic inflammatory disease (PID).** This occurs in 10 to 20% of infected females. Gonococcal PID is a major cause of sterility and ectopic pregnancies because of scar formation in the uterine tubes. Gonococci disseminate most often during menstruation, a time in which there is an increased concentration of free iron available to the bacteria. In both sexes disseminated gonococcal infection with bacteremia may occur. This can lead to involvement of the joints (gonorrheal arthritis), heart (gonorrheal endocarditis), or pharynx (gonorrheal pharyngitis). Gonorrheal eye infections occur most often in newborns as they pass through an infected birth canal. The resulting disease is called **ophthalmia neonatorum** or **conjunctivitis of the newborn,** which was once a leading cause of blindness in many parts of the world. To prevent this, tetracycline, erythromycin, povidone-iodine, or silver nitrate in dilute solution is placed in the eyes of newborns. This type of treatment is required by law in the United States and many other nations.

Laboratory diagnosis of gonorrhea relies on the successful growth of *N. gonorrhoeae* in culture to determine oxidase reaction, Gram stain reaction, and colony and cell morphology. The performance of confirmation tests also is necessary. Because the gonococci are very sensitive to adverse environmental conditions and survive poorly outside the body, special transport media (e.g., PACE) are necessary. A DNA probe (Gen-Probe Pace) for *N. gonorrhoeae* has been developed and is used to supplement other diagnostic techniques.

The Centers for Disease Control and Prevention consider four treatment regimens to be coequal after sensitivity testing has been done: (1) penicillin G plus probenecid, (2) ampicillin plus probenecid, (3) ceftriaxone or ofloxacin plus doxycycline for 7 days, or (4) spectinomycin.

Penicillin-resistant strains of gonococci have now developed and occur worldwide. Most of these strains carry a plasmid that directs the formation of penicillinase, a β-lactamase enzyme able to inactivate penicillin G and ampicillin. Since 1980 strains of *N. gonorrhoeae* with chromosomally mediated penicillin resistance have developed. Instead of producing a penicillinase, these strains have altered penicillin-binding proteins. Since 1986 tetracycline-resistant *N. gonorrhoeae* also have developed.

The most effective method for control of this sexually transmitted disease is public education, diagnosing and treating the asymptomatic patient, condom use, and treating infected individuals quickly to prevent further spread of the disease. Slightly under a million cases of gonorrhea are reported in the United States each year, but the actual number of cases is thought to be three to four times greater than that. More than 60% of all cases occur in the 15- to 24-year-old age group. Repeated gonococcal infections are common. Protective immunity to reinfection does not arise because of antigenic variation in which a single strain changes its pilin gene by a recombinational event and alters the expression of the various protein II genes by slipped strand mispairing. This can thus be viewed as a programmed evasion technique employed by the bacterium rather than a mere reflection of strain variation.

Inclusion Conjunctivitis

Inclusion conjunctivitis is an acute infectious disease caused by *C. trachomatis* serotypes D–K, and it occurs throughout the world. It is characterized by a copious mucous discharge from the eye, an inflamed and swollen conjunctiva, and the presence of large inclusion bodies in the host cell cytoplasm. In inclusion conjunctivitis of the newborn, the chlamydiae are acquired during passage through an infected birth canal. The disease appears 7 to 12 days after birth. If the chlamydiae colonize an infant's nasopharynx and tracheobronchial tree, pneumonia may result. Adult inclusion conjunctivitis is acquired by contact with infective genital tract discharges.

Without treatment, recovery usually occurs spontaneously over several weeks or months. Therapy involves treatment with tetracycline, erythromycin, or a sulfonamide. The specific diagnosis of *C. trachomatis* can be made by direct immunofluorescence, Giemsa stain, nucleic acid probes, and culture. Genital chlamydial infections and inclusion conjunctivitis are sexually

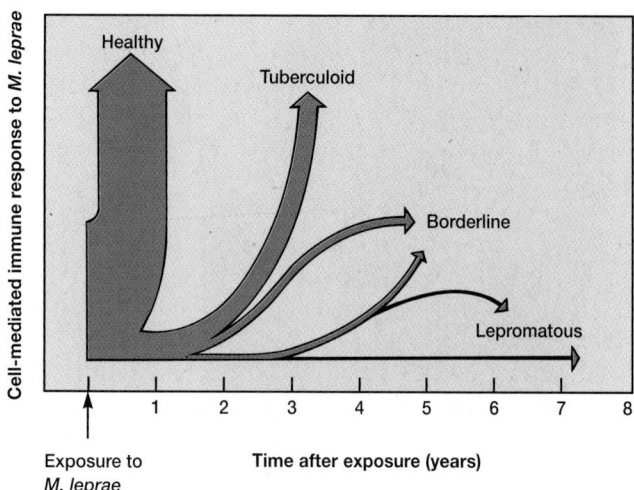

Figure 39.13   Development of Leprosy.   A schematic representation of the hypothesis of how the development of subclinical infection and various types of leprosy is related to the time of onset of cell-mediated immune response to *M. leprae* antigens after the initial exposure. The thickness of the lines indicates the proportion of individuals from the exposed population that is likely to fall into each category.

transmitted diseases that are spread by indiscriminate contact with multiple sex partners. Prevention depends upon diagnosis and treatment of all infected individuals.

Leprosy

Leprosy [Greek *Lepros,* scaly, scabby, rough] or **Hansen's disease** is a severely disfiguring skin disease caused by *Mycobacterium leprae* (*see figure 24.9*). The only reservoirs of proved significance are humans. The disease most often occurs in tropical countries, where there are more than 14 million cases. An estimated 4,000 cases exist in the United States, with approximately 200 to 300 new cases reported annually.

Transmission of leprosy is most likely to occur when individuals are exposed for prolonged periods to infected individuals who shed large numbers of *M. leprae*. Nasal secretions probably are the infectious material for family contacts.

The incubation period is about 3 to 5 years but may be much longer, and the disease progresses slowly. The bacterium invades peripheral nerve and skin cells and becomes an obligately intracellular parasite. It is most frequently found in the Schwann cells that surround peripheral nerve axons and in mononuclear phagocytes. The earliest symptom of leprosy is usually a slightly pigmented skin eruption several centimeters in diameter. Approximately 75% of all individuals with this early solitary lesion heal spontaneously because of the cell-mediated immune response to *M. leprae*. However, in some individuals this immune response may be so weak that one of two distinct forms of the disease occurs: tuberculoid or lepromatous leprosy (**figure 39.13**).

Tuberculoid (neural) leprosy is a mild, nonprogressive form of leprosy associated with a delayed-type hypersensitivity

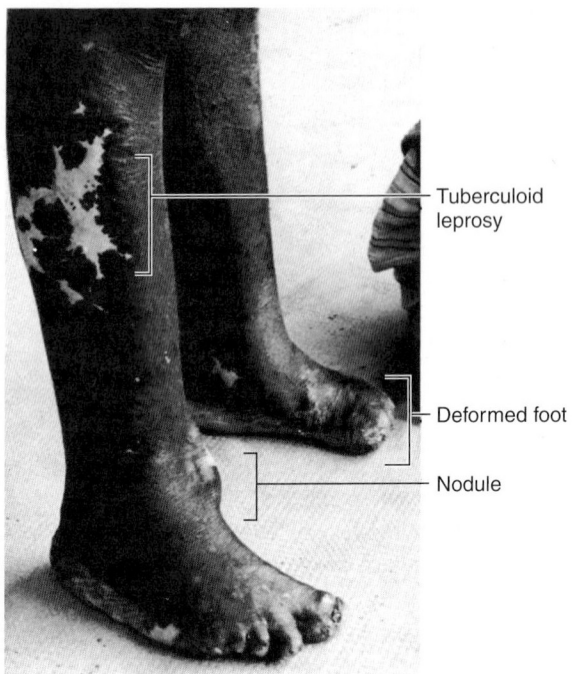

Figure 39.14 captions:
- Tuberculoid leprosy
- Deformed foot
- Nodule

Figure 39.14 Leprosy. In tuberculoid leprosy the skin within the nodule is completely without sensation. The deformed foot is associated with lepromatous leprosy. Note the disfiguring nodule on the ankle.

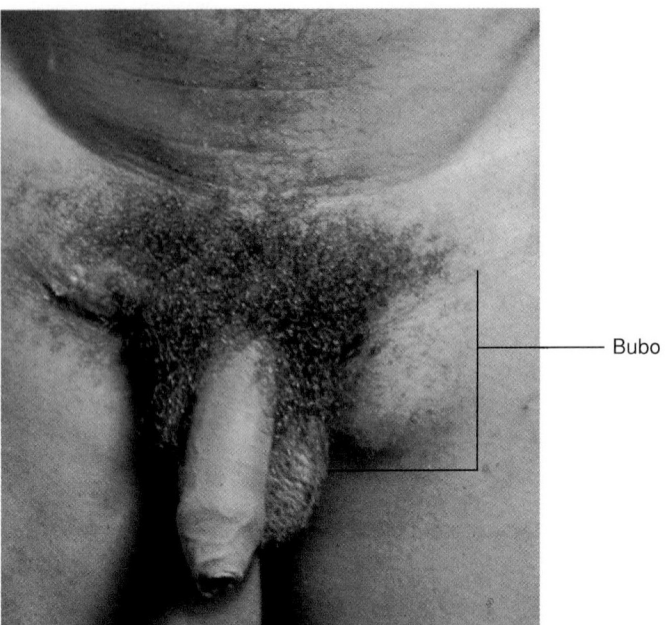

- Bubo

Figure 39.15 Lymphogranuloma Venereum. The bubo in the left inguinal area is draining.

reaction (*see section 33.2*) to antigens on the surface of *M. leprae.* It is characterized by damaged nerves and regions of the skin that have lost sensation and are surrounded by a border of nodules (**figure 39.14**). Afflicted individuals who do not develop hypersensitivity have a relentlessly progressive form of the disease, called **lepromatous (progressive) leprosy,** in which large numbers of *M. leprae* develop in the skin cells. The bacteria kill skin tissue, leading to a progressive loss of facial features, fingers, toes, and other structures. Moreover, disfiguring nodules form all over the body. Nerves are also infected, but usually are less damaged than in tuberculoid leprosy.

Because the leprosy bacillus cannot be cultured in vitro, laboratory diagnosis is supported by the demonstration of the bacterium in biopsy specimens and by acid-fast staining. Serodiagnostic methods, such as the fluorescent leprosy antibody absorption test, DNA amplification, and ELISA have recently been developed.

Treatment is long-term with the sulfone drug diacetyl dapsone and rifampin, with or without clofazimine. Alternative drugs are ethionamide or protionamide. Use of *Mycobacterium* vaccine in conjunction with the drugs shortens the duration of drug therapy and speeds recovery from the disease.

There is good evidence that the nine-banded armadillo is an animal reservoir for the leprosy bacillus in the United States but plays no role in transmission of leprosy to humans. Identification and treatment of patients with leprosy is the key to control. Children of presumably contagious parents should be given chemo-

prophylactic drugs until treatment of the parents has made them noninfectious.

---

1. How can humans acquire anthrax? Gas gangrene? Gonorrhea? Chancroid? Leprosy? Describe the major symptoms of each.
2. Define the following terms: cutaneous anthrax, pulmonary anthrax, bacterial vaginosis, pelvic inflammatory disease, ophthalmia neonatorum, tuberculoid and lepromatous leprosy.
3. How do humans contract chlamydial pneumonia?
4. How does an infant acquire inclusion conjunctivitis?

Lymphogranuloma Venereum

Lymphogranuloma venereum (LGV) is a sexually transmitted disease (table 39.4) caused by *Chlamydia trachomatis* serotypes L_1–L_3. It has a worldwide distribution but is more common in tropical climates.

LGV proceeds through three phases. (1) In the primary phase a small ulcer appears several days to several weeks after a person is exposed to the chlamydiae. The ulcer may appear on the penis in males or on the labia or vagina in females. The ulcer heals quickly and leaves no scar. (2) The secondary phase begins 2 to 6 weeks after exposure, when the chlamydiae infect lymphoid cells, causing the regional lymph nodes to become enlarged and tender; such nodes are called buboes (**figure 39.15**). Systemic symptoms such as fever, chills, and anorexia are common. (3) If the disease is not treated, a late phase ensues. This results from fibrotic changes and abnormal lymphatic drainage that produces fistulas (abnormal passages leading from an abscess or a hollow organ to the body surface or from one hollow organ to another) and urethral or rectal strictures (a decrease in size). An untreatable fluid accumulation in the penis, scrotum, or vaginal area may result.

The disease is detected by staining infected cells with iodine to observe inclusions (chlamydia-filled vacuoles), culture of the chlamydiae from a bubo, nucleic acid probes, or by the detection of a high antibody titer to LGV. Treatment in the early phases consists of aspiration of the buboes and administration of drugs: tetracycline, doxycycline, erythromycin, or cefotoxin. The late phase may require surgery. The methods used for the control of LGV are the same as for other sexually transmitted diseases: reduction in promiscuity, use of condoms, and early diagnosis and treatment of infected individuals. About 300 cases of LGV occur annually in the United States.

Mycoplasmal Pneumonia

Typical pneumonia has a bacterial origin. If a bacterial pathogen cannot be isolated, the pneumonia is termed atypical and a virus is usually suspected. If viruses can't be detected, then **mycoplasmal pneumonia** can be considered. This pneumonia is caused by *Mycoplasma pneumoniae* (*see figure 23.2*), a mycoplasma with worldwide distribution. Spread involves close contact and sometimes airborne droplets. The disease is fairly common and mild in infants and small children; serious disease is seen principally in older children and young adults.

M. pneumoniae usually infects the upper respiratory tract, and subsequently moves to the lower respiratory tract, where it attaches to respiratory mucosal cells. It then produces peroxide, which may be a toxic factor, but the exact mechanism of pathogenesis is unknown. A change in mucosal cell nucleic acid synthesis has been observed. The manifestations of this disease vary in severity from asymptomatic to a serious pneumonia. The latter is accompanied by death of the surface mucosal cells, lung infiltration, and congestion. Initial symptoms include headache, weakness, a low fever, and a predominant cough. The disease and its symptoms usually persist for weeks. The mortality rate is less than 1%.

Several rapid tests using latex agglutination for *M. pneumoniae* antibodies are available for diagnosis of mycoplasmal pneumonia. When isolated from respiratory secretions, the mycoplasmas form distinct colonies with a "fried-egg" appearance (*see figure 23.3*). During the acute stage of the disease, diagnosis must be made by clinical observations. Tetracyclines or erythromycin are effective in treatment. There are no preventive measures.

Nongonococcal Urethritis

Nongonococcal urethritis (NGU) is any inflammation of the urethra not due to the bacterium *Neisseria gonorrhoeae*. This condition is caused both by nonmicrobial factors such as catheters and drugs and by infectious microorganisms. The most important causative agents are *C. trachomatis, Ureaplasma urealyticum, Mycoplasma hominis, Trichomonas vaginalis, Candida albicans,* and herpes simplex viruses. Most infections are acquired sexually (table 39.4), and of these, approximately 50% are

Chlamydia infections. NGU caused by chlamydia is probably the most common sexually transmitted disease in the United States, with over 10 million Americans infected. It is endemic throughout the world.

Symptoms of NGU vary widely. Males may have few or no manifestations of disease; however, complications can exist. These include a urethral discharge, itching, and inflammation of the male reproductive structures. Females may be asymptomatic or have a severe infection called pelvic inflammatory disease (PID) that often leads to sterility. *Chlamydia* may account for as many as 200,000 to 400,000 cases of PID annually in the United States. In the pregnant female, a chlamydial infection is especially serious because it is directly related to miscarriage, stillbirth, inclusion conjunctivitis, and infant pneumonia.

Diagnosis of NGU requires the demonstration of a leukocyte exudate and exclusion of urethral gonorrhea by Gram stain and culture. Several rapid tests for detecting *Chlamydia* in urine specimens are also available. Treatment is with tetracycline, doxycycline, erythromycin, or sulfisoxazole.

Peptic Ulcer Disease and Gastritis

A gram-negative, microaerophilic spiral bacillus found in gastric biopsy specimens from patients with histologic **gastritis** [Greek *gaster,* stomach, and *itis,* inflammation] was successfully cultured in Perth, Australia, in 1982 and named *Campylobacter pylori.* In 1993 its name was changed to *Helicobacter pylori.* It now appears that this bacterium is responsible for most cases of chronic gastritis not associated with another known primary cause (e.g., autoimmune gastritis or eosinophilic gastritis), and it is the leading factor in the pathogenesis of **peptic ulcer disease.** In addition, there are strong positive correlations between gastric cancer rates and *H. pylori* infection rates in certain populations, and it has been classified as a Class I carcinogen by the World Health Organization.

The evidence for *H. pylori* as a gastrointestinal pathogen is now very strong, if not overwhelming. For example, *H. pylori* has been isolated from the gastric mucosa (**figure 39.16**) of 95% of patients with gastric ulcer disease and virtually 100% of those patients with chronic gastritis, but not from healthy tissue.

H. pylori colonizes only gastric mucus-secreting cells, beneath the gastric mucous layers, and surface fimbriae are believed to be one of the adhesins associated with this process. *H. pylori* binds to Lewis B antigens (which are part of the blood group antigens that determine blood group O) and to the monosaccharide sialic acid, also found in the glycoproteins on the surface of gastric epithelial cells. After attachment, the bacterium moves into the mucous layer.

H. pylori is also a strong producer of urease. Urease activity may create an alkaline environment by urea hydrolysis to produce ammonia that protects the bacterium from gastric acid until it is able to grow under the layer of mucus in the stomach. The potential virulence factors responsible for epithelial cell damage and inflammation probably include proteases, phospholipases, cytokines, and cytotoxins.

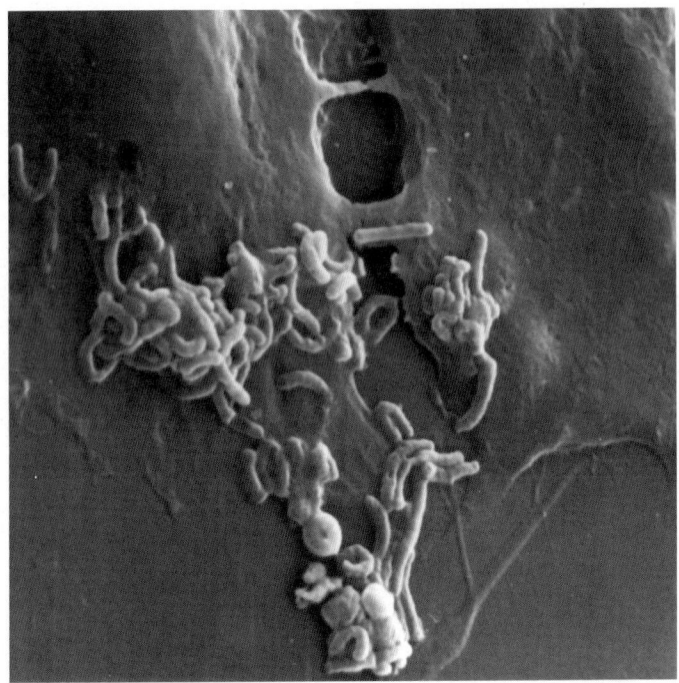

Figure 39.16 Peptic Ulcer Disease. Scanning electron micrograph (×3,441) of *Helicobacter pylori* adhering to gastric cells.

Approximately 50% of the world's population is estimated to be infected with *H. pylori*. *H. pylori* is most likely transmitted from person to person (usually acquired in childhood), although infection from a common exogenous source cannot be completely ruled out, and some think that it is spread by food or water. Support for the person-to-person transmission comes from evidence of clustering within families and from reports of higher than expected prevalences in residents of custodial institutions and nursing homes.

Laboratory identification of *H. pylori* is by culture of gastric biopsy specimens, examination of stained biopsies for the presence of bacteria, detection of serum IgG (Pyloriset EIA-G, Malakit *Helicobacter pylori*), the urea breath test, urinary excretion of [^{15}N] ammonia, stool antigen assays, or detection of urease activity in the biopsies. The goal of *H. pylori* treatment is the complete elimination of the organism. Treatment is with bismuth subsalicylate (Pepto-Bismol) combined with metronidazole and either tetracycline or amoxicillin; clarithromycin (Biaxin), ranitidine, and bismuth citrate; or clarithromycin, amoxicillin, and lansoprazole (Prevacid).

Psittacosis (Ornithosis)

Psittacosis (ornithosis) is a worldwide infectious disease of birds that is transmissible to humans. It was first described in association with parrots and parakeets, both of which are psittacine birds. The disease is now recognized in many other birds—among them, pigeons, chickens, ducks, and turkeys—and the general term ornithosis [Latin *ornis,* bird] is used.

Ornithosis is caused by *Chlamydia psittaci*. Humans contract this disease either by handling infected birds or by inhaling dried bird excreta that contains viable *C. psittaci*. Ornithosis is recognized as an occupational hazard within the poultry industry, particularly to workers in turkey processing plants.

After entering the respiratory tract, the chlamydiae are transported to the cells of the liver and spleen. They multiply within these cells and then invade the lungs, where they cause inflammation, hemorrhaging, and pneumonia.

Laboratory diagnosis is either by isolation of *C. psittaci* from blood or sputum, or by serological studies. Treatment is with tetracycline. Because of antibiotic therapy, the mortality rate has dropped from 20 to 2%. Between 100 and 200 cases of ornithosis are reported annually in the United States. Prevention and control is by chemoprophylaxis (tetracycline) for pet birds and poultry, although this can lead to the development of antibiotic resistance and should be discouraged.

Staphylococcal Diseases

The genus *Staphylococcus* consists of gram-positive cocci, 0.5 to 1.5 mm in diameter, occurring singly, in pairs, and in tetrads, and characteristically dividing in more than one plane to form irregular clusters. The cell wall contains peptidoglycan and teichoic acid. Staphylococci are facultative anaerobes and usually catalase positive.

Staphylococci are among the most important bacteria that cause disease in humans. They are normal inhabitants of the upper respiratory tract, skin, intestine, and vagina (*see figure 31.2*). Staphylococci, with pneumococci and streptococci, are members of a group of invasive gram-positive bacteria known as the pyogenic (or pus-producing) cocci. These bacteria cause various suppurative, or pus-forming diseases (e.g., boils, carbuncles, folliculitis, impetigo contagiosa, scalded-skin syndrome) in humans.

Staphylococci can be divided into pathogenic and relatively nonpathogenic strains based on the synthesis of the enzyme coagulase. Coagulase-positive strains, such as *S. aureus* (*see figure 23.12*), often produce a yellow carotenoid pigment—which has led to their being commonly called golden staph (*see photo, p. 93*)—and cause severe chronic infections. Coagulase-negative staphylococci (CoNS) such as *S. epidermidis* do not produce coagulase, are nonpigmented, and are generally less invasive but have increasingly been associated (as opportunistic pathogens) with serious nosocomial infections.

Staphylococci are further classified into **s**lime **p**roducers (SP) and **n**on-**s**lime **p**roducers (NSP). The ability to produce slime has been proposed as a marker for pathogenic strains of staphylococci (**figure 39.17***a*).

Slime is a viscous extracellular glycoconjugate that allows these bacteria to adhere to smooth surfaces such as prosthetic medical devices and catheters. Scanning electron microscopy has clearly demonstrated that biofilms (figure 39.17*b*) consisting of staphylococci encased in a slimy matrix are formed in association with biomaterial-associated infections (**Disease 39.2**). Slime also appears to inhibit neutrophil chemotaxis, phagocytosis, and the antimicrobial agents vancomycin and teicoplanin.

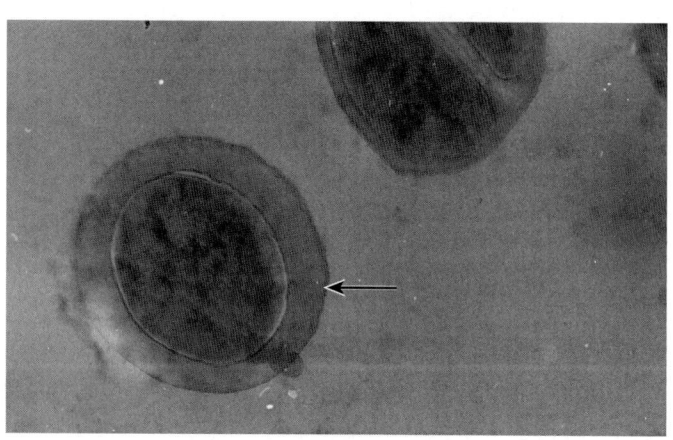

(a)

(b)

Figure 39.17 Slime and Biofilms. *S. aureus* and certain coagulase-negative staphylococci produce a viscous extracellular glyco-conjugate called slime. (**a**) Cells of *S. aureus,* one of which produces a slime layer (arrowhead; transmission electron microscopy, ×10,000). (**b**) A biofilm on a venous catheter consisting of *S. epidermidis* and slime. The slime encases and binds the bacterial colonies to the catheter (scanning electron micrograph, ×6,000).

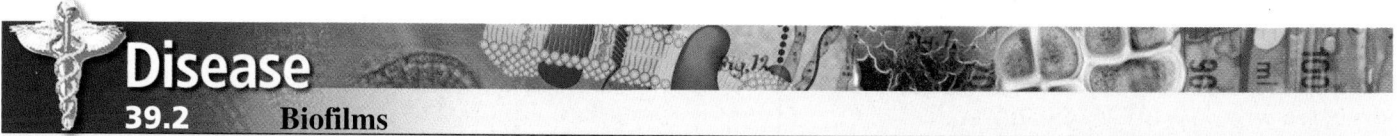

Disease
39.2 Biofilms

Biofilms consist of microorganisms immobilized at a substratum surface and typically embedded in an organic polymer matrix of microbial origin (*see section 28.4*). They develop on virtually all surfaces immersed in natural aqueous environments, including both biological (aquatic plants and animals) and abiological (concrete, metal, plastics, stones). Biofilms form particularly rapidly in flowing aqueous systems where a regular nutrient supply is provided to the microorganisms. Extensive microbial growth, accompanied by excretion of copious amounts of extracellular organic polymers, thus leads to the formation of visible slimy layers (biofilms) on solid surfaces.

Most of the human gastrointestinal tract is colonized by specific microbial groups (the normal indigenous microbiota; *see section 31.2*) that give rise to natural biofilms. At times, these natural biofilms provide protection for pathogenic species, allowing them to colonize the host.

Insertion of a prosthetic device into the human body often leads to the formation of biofilms on the surface of the device. The microorganisms primarily involved are *Staphylococcus epidermidis* (figure 39.17*b*), other coagulase-negative staphylococci, and gram-negative bacteria. These normal skin inhabitants possess the ability to tenaciously adhere to the surfaces of inanimate prosthetic devices. Within the biofilms they are protected from the body's normal defense mechanisms and also from antibiotics; thus the biofilm also provides a source of infection for other parts of the body as bacteria detach during biofilm sloughing.

Some examples of biofilms of medical importance include:

1. The deaths following massive infections of patients receiving Jarvik 7 artificial hearts
2. Cystic fibrosis patients harboring great numbers of *Pseudomonas aeruginosa* that produce large amounts of alginate polymers, which inhibit the diffusion of antibiotics
3. Teeth, where biofilm forms plaque that leads to tooth decay (figure 39.25)
4. Contact lenses, where bacteria may produce severe eye irritation, inflammation, and infection
5. Air-conditioning and other water retention systems where potentially pathogenic bacteria, such as *Legionella* species, may be protected from the effects of chlorination by biofilms

Staphylococci, harbored by either an asymptomatic carrier or a person with the disease, can be spread by the hands, expelled from the respiratory tract, or transported in or on animate and inanimate objects. Staphylococci can produce disease in almost every organ and tissue of the body (**figure 39.18**). However, it should be emphasized that staphylococcal disease, for the most part, occurs in people whose defensive mechanisms have been compromised, such as those in hospitals.

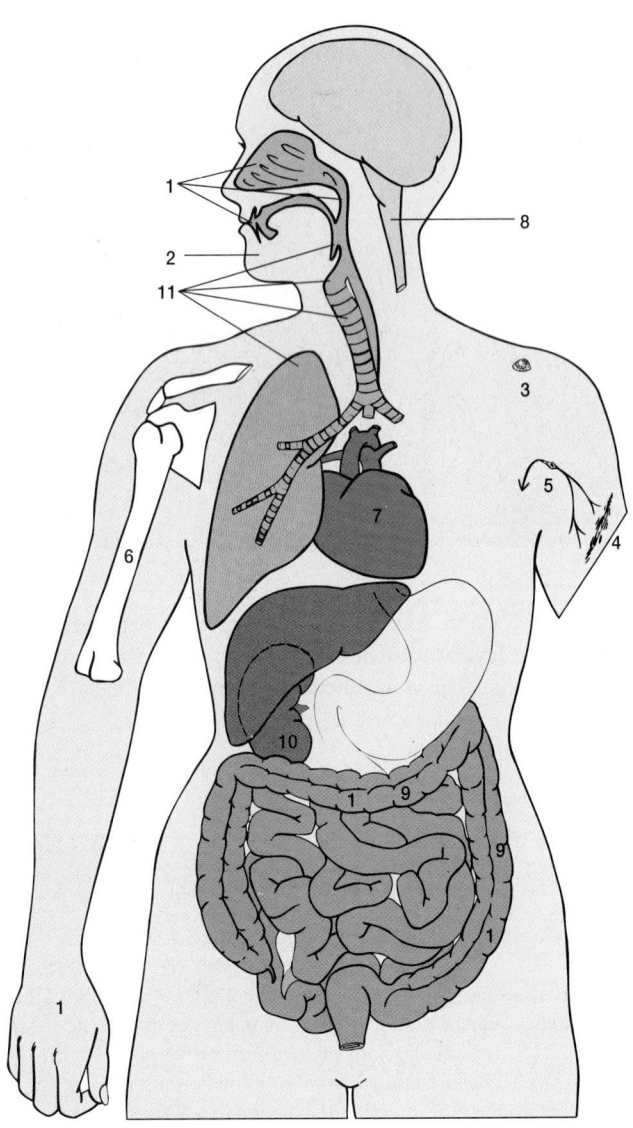

1 Tissue where *S. aureus*
   is often found but does not
   normally cause disease

Diseases that may be
caused by *S. aureus* are:

2 Pimples and impetigo

3 Boils and carbuncles
   on any surface area

4 Wound infections and
   abscesses

5 Spread to lymph nodes
   and to blood (septicemia),
   resulting in widespread
   seeding

6 Osteomyelitis

7 Endocarditis

8 Meningitis

9 Enteritis and enterotoxin
   poisoning (food poisoning)

10 Nephritis

11 Respiratory infections:
   Pharyngitis
   Laryngitis
   Bronchitis
   Pneumonia

Figure 39.18   Staphylococcal Diseases.   The sites of the major staphylococcal infections of humans are indicated by the above numbers.

Staphylococci produce disease through their ability to multiply and spread widely in tissues and through their production of many extracellular substances (**table 39.3**). Some of these substances are exotoxins, and others are enzymes thought to be involved in staphylococcal invasiveness. Many toxin genes are carried on plasmids; in some cases genes responsible for pathogenicity reside on both a plasmid and the host chromosome.

The pathogenic capacity of a particular *S. aureus* strain is due to the combined effect of extracellular factors and toxins, together with the invasive properties of the strain. At one end of the disease spectrum is staphylococcal food poisoning, caused solely by the ingestion of preformed enterotoxin (table 39.3). At the other end of the spectrum are staphylococcal bacteremia and disseminated abscesses in most organs of the body.

The classic example of the staphylococcal lesion is the localized abscess (**figure 39.19***a–d*). When *S. aureus* becomes established in a hair follicle, tissue necrosis results. Coagulase is produced and forms a fibrin wall around the lesion that limits the spread. Within the center of the lesion, liquefaction of necrotic tissue occurs, and the abscess spreads in the direction of least resistance. The abscess may be either a furuncle (boil) or a carbuncle. The central necrotic tissue drains, and healing eventually occurs. However, the bacteria may spread from any focus by the lymphatics and bloodstream to other parts of the body.

Newborn infants and children can develop a superficial skin infection characterized by the presence of encrusted pustules (figure 39.19*e*). This disease, called impetigo contagiosa, is caused by *S. aureus* and group A streptococci. It is contagious and can spread rapidly through a nursery or school. It usually occurs in areas where sanitation and personal hygiene are poor.

Toxic shock syndrome (TSS) is a staphylococcal disease with potentially serious consequences. Most cases of this syndrome have occurred in females who use superabsorbent tampons during menstruation. However, the toxin associated with this syndrome is

Table 39.3
Various Enzymes and Toxins Produced by Staphylococci

Product	Physiological Action
β-lactamase	Breaks down penicillin
Catalase	Converts hydrogen peroxide into water and oxygen and reduces killing by phagocytosis
Coagulase	Reacts with prothrombin to form a complex that can cleave fibrinogen and cause the formation of a fibrin clot; fibrin may also be deposited on the surface of staphylococci, which may protect them from destruction by phagocytic cells; coagulase production is synonymous with invasive pathogenic potential
DNase	Destroys DNA
Enterotoxins	Are divided into heat-stable toxins of six known types (A, B, C1, C2, D, E); responsible for the gastrointestinal upset typical of food poisoning
Exfoliative toxins A and B (superantigens)	Causes loss of the surface layers of the skin in scalded-skin syndrome
Hemolysins	Alpha hemolysin destroys erythrocytes and causes skin destruction
	Beta hemolysin destroys erythrocytes and sphingomyelin around nerves
	Gamma hemolysin destroys erythrocytes
	Delta hemolysin destroys erythrocytes
Hyaluronidase	Also known as spreading factor; breaks down hyaluronic acid located between cells, allowing for penetration and spread of bacteria
Panton-Valentine leukocidin	Inhibits phagocytosis by granulocytes and can destroy these cells by forming pores in their phagosomal membranes
Lipases	Break down lipids
Nuclease	Breaks down nucleic acids
Protein A	Is antiphagocytic by competing with neutrophils for the Fc portion of specific opsonins
Proteases	Break down proteins
Toxic shock syndrome toxin-1 (a superantigen)	Is associated with the fever, shock, and multisystem involvement of toxic shock syndrome

also produced in men and in nonmenstruating women by *S. aureus* present at sites other than the genital area (e.g., in surgical wound infections). Toxic shock syndrome is characterized by low blood pressure, fever, diarrhea, an extensive skin rash, and shedding of the skin. These symptoms are caused by the **toxic shock syndrome toxin-1** (TSST-1 is a superantigen; *see section 32.2*) released by the *S. aureus* (table 39.3), but several other enterotoxins (SEB and SEC$_1$) also may be involved. Several hundred cases of toxic shock syndrome are reported annually in the United States.

Staphylococcal scalded skin syndrome (SSSS) is a third example of a common staphylococcal disease (figure 39.19*f*). SSSS is caused by strains of *S. aureus* that carry a plasmid-borne gene for the **exfoliative toxin** or **exfoliatin** (sometimes the toxin gene is on the bacterial chromosome instead). Like TSST-1, exfoliatin is a superantigen. In this disease the epidermis peels off to reveal a red area underneath—thus the name of the disease. SSSS is seen most commonly in infants and children, and neonatal nurseries occasionally suffer large outbreaks of the disease.

The definitive diagnosis of staphylococcal disease can be made only by isolation and identification of the staphylococcus involved. This requires culture, catalase, and coagulase tests; serology; DNA fingerprinting; and phage typing. Commercial rapid test kits also are available. There is no specific prevention for staphylococcal disease. The mainstay of treatment is the administration of specific antibiotics: penicillin, cloxacillin, methicillin, vancomycin, oxacillin, cefotaxime, ceftriaxone, a cephalosporin, or rifampin and others. Because of the prevalence of drug-resistant strains (e.g., methicillin-resistant staph), all staphylococcal isolates should be tested for antimicrobial susceptibility (**Disease 39.3**). Cleanliness, hygiene, and aseptic management of lesions are the best means of control.

Syphilis

Venereal syphilis [Greek *syn,* together, and *philein,* to love] is a contagious sexually transmitted disease (table 39.4) caused by the spirochete *Treponema pallidum* subsp. *pallidum* (*T. pallidum, see figure 21.15*b). **Congenital syphilis** is the disease acquired in utero from the mother.

T. pallidum enters the body through mucous membranes or minor breaks or abrasions of the skin. It migrates to the regional lymph nodes and rapidly spreads throughout the body. The disease is not highly contagious, and there is only about a 1 in 10 chance of acquiring it from a single exposure to an infected sex partner.

Three recognizable stages of syphilis occur in untreated adults. In the primary stage, after an incubation period of about 10 days to 3 weeks or more, the initial symptom is a small, painless, reddened ulcer, or **chancre** [French *canker,* a destructive sore] with a hard ridge that appears at the infection site (**figure 39.20***a*) and contains spirochetes. Contact with the chancre during sexual intercourse may result in disease transmission. In about 1/3 of the cases, the disease does not progress further and the chancre disappears. Serological tests are positive in about 80% of the individuals during this stage (**figure 39.21**). In the remaining cases the spirochetes enter the bloodstream and are distributed throughout the body.

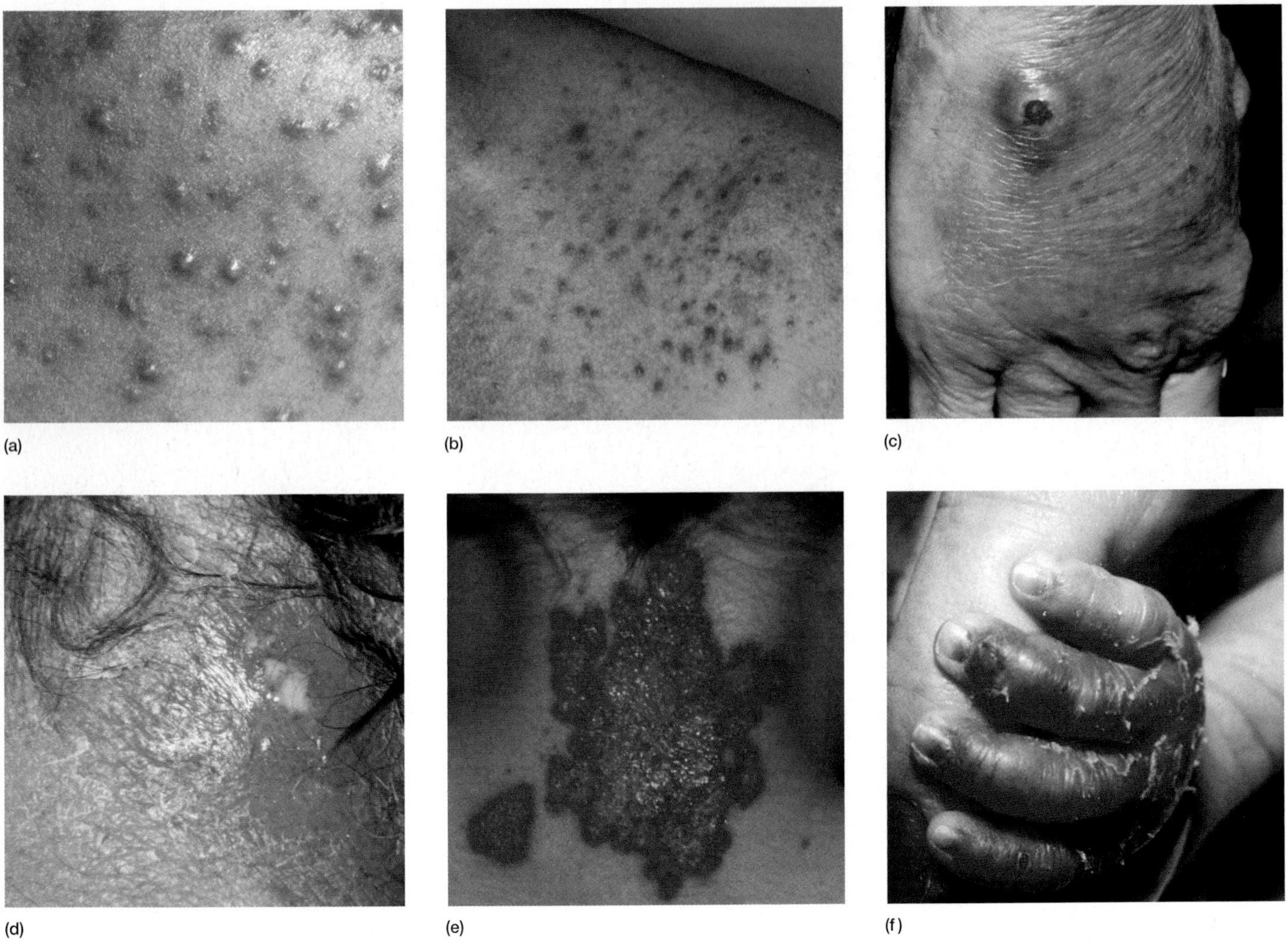

Figure 39.19 Staphylococcal Skin Infections. (a) Superficial folliculitis in which raised, domed pustules form around hair follicles. (b) In deep folliculitis the microorganism invades the deep portion of the follicle and dermis. (c) A furuncle arises when a large abscess forms around a hair follicle. (d) A carbuncle consists of a multilocular abscess around several hair follicles. (e) Impetigo on the neck of 2-year-old male. (f) Scalded skin syndrome in a 1-week-old premature male infant. Reddened areas of skin peel off, leaving "scalded"-looking moist areas.

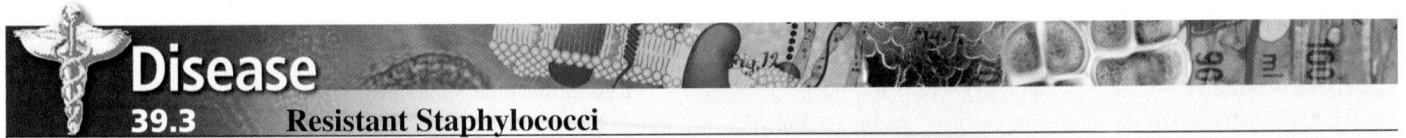

Disease
39.3   Resistant Staphylococci

During the late 1950s and early 1960s, *Staphylococcus aureus* caused considerable morbidity and mortality as a nosocomial, or hospital-acquired, pathogen. Since then, penicillinase-resistant, semisynthetic penicillins have proved to be successful antimicrobial agents in the treatment of staphylococcal infections. Unfortunately methicillin-resistant *S. aureus* (MRSA) strains have recently emerged as a major nosocomial problem. One way in which staphylococci become resistant is through acquisition of a chromosomal gene (*mecA*) that encodes an alternate target protein which is not inactivated by methicillin. The majority of the strains are resistant to several of the most commonly used antimicrobial agents, including macrolides, aminoglycosides, and the beta-lactam antibiotics, including the latest generation of cephalosporins. Serious infections by

methicillin-resistant strains have been most often successfully treated with an older, potentially toxic antibiotic, vancomycin. However, strains of *Enterococcus* and *Staphylococcus* recently have become resistant to vancomycin.

Recently methicillin-resistant *S. epidermidis* strains also have emerged as a nosocomial problem, especially in individuals with prosthetic heart valves or in people who have undergone other forms of cardiac surgery. Resistance to methicillin also may extend to the cephalosporin antibiotics. Difficulties in performing in vitro tests that adequately recognize cephalosporin resistance of these strains continue to exist. Serious infections due to methicillin-resistant *S. epidermidis* have been successfully treated with combination therapy, including vancomycin plus rifampin or an aminoglycoside.

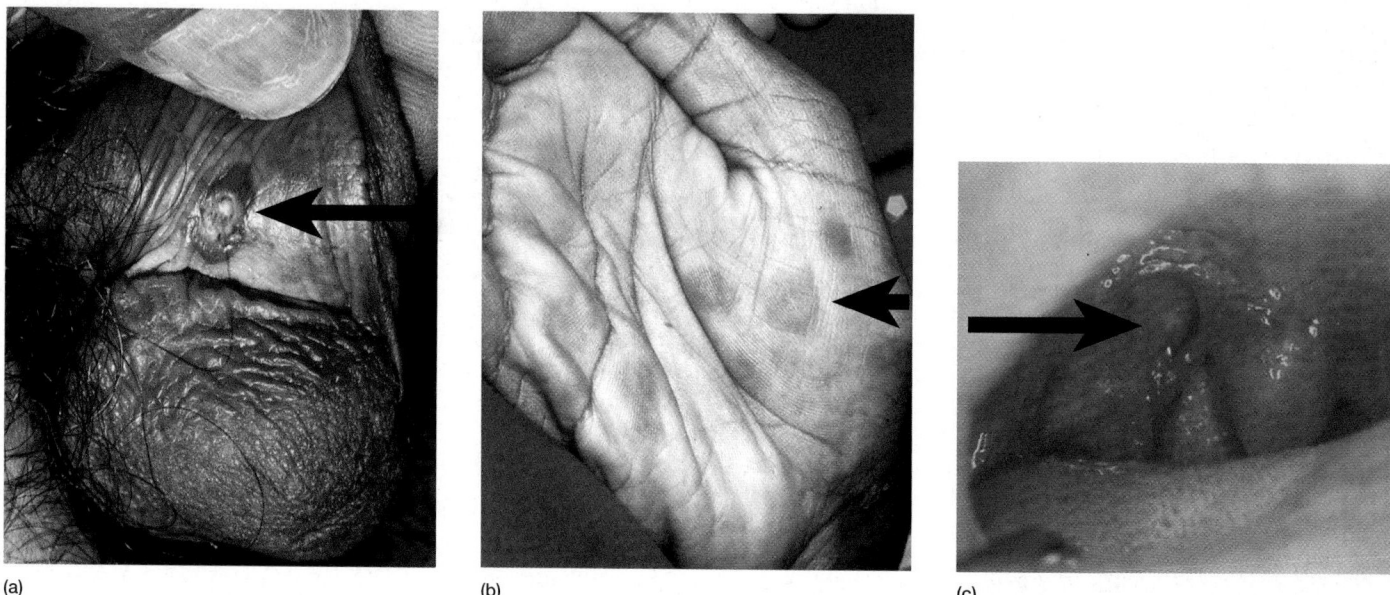

(a)   (b)   (c)

Figure 39.20   Syphilis.   (a) Primary syphilitic chancre of the penis. **(b)** Palmar lesions of secondary syphilis. **(c)** Ruptured gumma and ulcer of upper hard palate of the mouth.

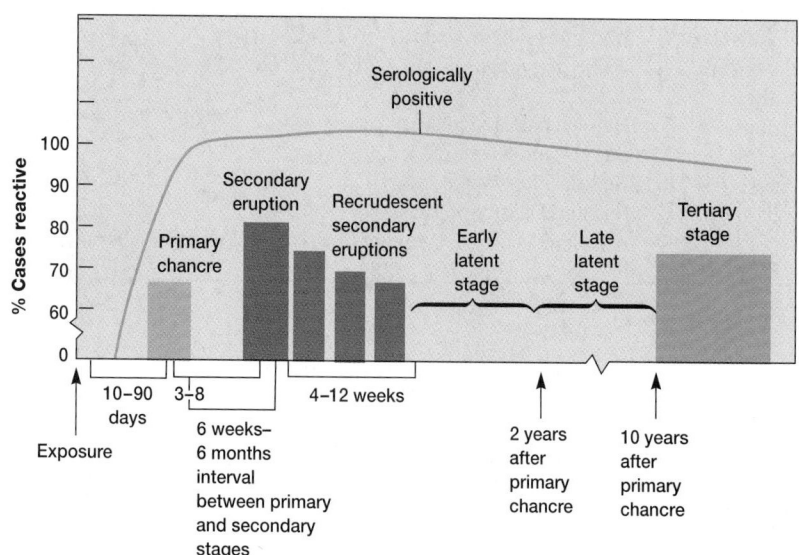

Figure 39.21   The Course of Untreated Syphilis. See text for further details.

Within 2 to 10 weeks after the primary lesion, the disease may enter the secondary stage, which is characterized by a skin rash (figure 39.20*b*). By this time 100% of the individuals are serologically positive. Other symptoms during this stage include the loss of hair patches, malaise, and fever. Both the chancre and the rash lesions are infectious.

After several weeks the disease becomes latent. During the latent period the disease is not normally infectious, except for possible transmission from mother to fetus (congenital syphilis). After many years a tertiary stage develops in about 40% of untreated individuals with secondary syphilis. During this stage degenerative lesions called **gummas** (figure 39.20*c*) form in the

skin, bone, and nervous system as the result of hypersensitivity reactions. This stage also is characterized by a great reduction in the number of spirochetes in the body. Involvement of the central nervous system may result in tissue loss that can lead to mental retardation, blindness, a "shuffle" walk (tabes), or insanity. Many of these symptoms have been associated with such well-known people as Al Capone, Francisco Goya, Henry VIII, Adolf Hitler, Scott Joplin, Friedrich Nietzsche, Franz Schubert, Oscar Wilde, and Kaiser Wilhelm (**Disease 39.4**).

Diagnosis of syphilis is through a clinical history, a thorough physical examination, and dark-field and immunofluorescence examination of lesion fluids (except oral lesions) for typical

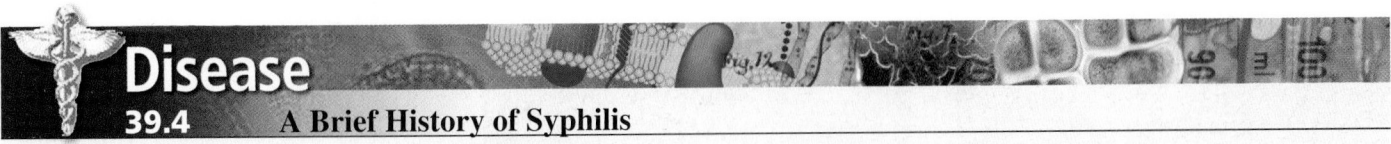

Disease
39.4   A Brief History of Syphilis

Syphilis was first recognized in Europe near the end of the fifteenth century. During this time the disease reached epidemic proportions in the Mediterranean areas. According to one hypothesis, syphilis is of New World origin and Christopher Columbus (1451–1506) and his crew acquired it in the West Indies and introduced it into Spain after returning from their historic voyage. Another hypothesis is that syphilis had been endemic for centuries in Africa and may have been transported to Europe at the same time that vast migrations of the civilian population were occurring (1500). Others believe that the Vikings, who reached the New World well before Columbus, were the original carriers.

Syphilis was initially called the Italian disease, the French disease, and the great pox as distinguished from smallpox. In 1530 the Italian physician and poet Girolamo Fracastoro wrote *Syphilis sive Morbus Gallicus* (Syphilis or the French Disease). In this poem a Spanish shepherd named Syphilis is punished for being disrespectful to the gods by being cursed with the disease. Several years later Fracastoro published a series of papers in which he described the possible mode of transmission of the "seeds" of syphilis through sexual contact.

Its venereal transmission was not definitely shown until the eighteenth century. The term venereal is derived from the name Venus, the Roman goddess of love. Recognition of the different stages of syphilis was demonstrated in 1838 by Philippe Ricord, who reported his observations on more than 2,500 human inoculations. In 1905 Fritz Schaudinn and Erich Hoffmann discovered the causative bacterium, and in 1906 August von Wassermann introduced the diagnostic test that bears his name. In 1909 Paul Ehrlich introduced an arsenic derivative, arsphenamine or salvarsan, as therapy. During this period, an anonymous limerick aptly described the course of this disease:

There was a young man from Black Bay
Who thought syphilis just went away
He believed that a chancre
Was only a canker
That healed in a week and a day.

But now he has "acne vulgaris"—
(Or whatever they call it in Paris);
On his skin it has spread
From his feet to his head,
And his friends want to know where his hair is.

There's more to his terrible plight:
His pupils won't close in the light
His heart is cavorting,
His wife is aborting,
And he squints through his gun-barrel sight.

Arthralgia cuts into his slumber;
His aorta is in need of a plumber;
But now he has tabes,
And saber-shinned babies,
While of gummas he has quite a number.

He's been treated in every known way,
But his spirochetes grow day by day;
He's developed paresis,
Has long talks with Jesus,
And thinks he's the Queen of the May.

---

motile or fluorescent spirochetes. Because humans respond to *T. pallidum* with the formation of antitreponemal antibody and a complement-fixing reagin, serological tests are very informative. Examples include tests for nontreponemal antigens (VDRL, **V**enereal **D**isease **R**esearch **L**aboratories test; RPR, **R**apid **P**lasma **R**eagin test; complement fixation or the Wassermann test) and treponemal antibodies (FTA-ABS, **f**luorescent **t**reponemal **a**ntibody-**abs**orption test; TPI, *T. p*allidum **i**mmobilization; *T. pallidum* complement fixation; TPHA, *T. p*allidum **h**em**a**gglutination).

Treatment in the early stages of the disease is easily accomplished with long-acting benzathine penicillin G or aqueous procaine penicillin. Later stages of syphilis are more difficult to treat with drugs and require much larger doses over a longer period. For example, in neurosyphilis cases, treponemes occasionally survive such drug treatment. Immunity to syphilis is not complete, and subsequent infections can occur once the first infection has spontaneously disappeared or has been eliminated with antibiotics.

Prevention and control of syphilis depends on (1) public education (2) prompt and adequate treatment of all new cases, (3) follow-up on sources of infection and contact so they can be treated, (4) sexual hygiene, and (5) prophylaxis (condoms) to prevent exposure. At present, the incidence of syphilis, as well as other sexually transmitted diseases, is rising in most parts of the world. In the United States around 50,000 cases of primary and secondary syphilis in the civilian population and about 1,000 cases of congenital syphilis are reported annually. The highest incidence is among those 20 to 39 years of age.

Tetanus

Tetanus [Greek *tetanos,* to stretch] is caused by *Clostridium tetani,* an anaerobic gram-positive spore former (*see figure 23.6b*). The endospores of *C. tetani* are commonly found in hospital environments, in soil and dust, and in the feces of many farm animals and humans.

Transmission to humans is associated with skin wounds. Any break in the skin can allow *C. tetani* endospores to enter, and if the oxygen tension is low enough, the endospores germinate. When the bacteria die and lyse, the neurotoxin tetanospasmin is released. **Tetanospasmin** is an endopeptidase that selectively cleaves the synaptic vesicle membrane protein synaptobrevin.

This prevents exocytosis and release of inhibitory neurotransmitters (gamma-aminobutyric acid and glycine) at synapses within the spinal cord motor nerves. The result is uncontrolled stimulation of skeletal muscles.

Early in the course of the disease, tetanospasmin causes tension or cramping and twisting in skeletal muscles surrounding the wound and tightness of the jaw muscles. With more advanced disease, there is trismus ("lockjaw"), an inability to open the mouth because of the spasm of the masseter muscles. Facial muscles may go into spasms producing the characteristic expression known as risus sardonicus. Spasms or contractions of the trunk and extremity muscles may be so severe that there is boardlike rigidity, painful tonic convulsions, and opisthotonos (backward bowing of the back so that the heels and back approach each other [see chapter opening figure]). Death usually results from spasms of the diaphragm and intercostal respiratory muscles. A second toxin, **tetanolysin,** is a hemolysin that aids in tissue destruction.

Testing for tetanus is suggested whenever an individual has a history of wound infection and muscle stiffness. Prevention of tetanus involves the use of the tetanus toxoid. The toxoid, which incorporates an adjuvant (aluminum salts) to increase its immunizing potency, is given routinely with diphtheria toxoid and pertussis vaccine. An initial dose is normally administered a few months after birth, a second dose 4 to 6 months later, and finally a reinforcing dose 6 to 12 months after the second injection. A final booster is given between the ages of 4 to 6 years. For many years booster doses of tetanus toxoid were administered every 3 to 5 years. However, that practice has been discontinued since it has been shown that a single booster dose can provide protection for 10 to 20 years. Serious hypersensitivity reactions have occurred when too many doses of toxoid were administered over a period of years. Booster doses today are generally given only when an individual has sustained a wound infection.

Control measures for tetanus are not possible because of the wide dissemination of the bacterium in the soil and the long survival of its endospores. The case fatality rate in generalized tetanus ranges from 30 to 90% because tetanus treatment is not very effective. Therefore prevention is all important and depends on (1) active immunization with toxoid, (2) proper care of wounds contaminated with soil, (3) prophylactic use of antitoxin, and (4) administration of penicillin. Around 100 cases of tetanus are reported annually in the United States—the majority of which are intravenous drug users.

Trachoma

Trachoma [Greek *trachoma,* roughness] is a contagious disease created by *Chlamydia trachomatis* serotypes A–C. It is one of the oldest known infectious diseases of humans and is the greatest single cause of blindness throughout the world. Probably over 500 million people are infected and 20 million blinded each year by this chlamydia. In endemic areas most children are chronically infected within a few years of birth. Active disease in adults over age 20 is three times as frequent in females as in males because of mother-child contact. Although uncommon in the United

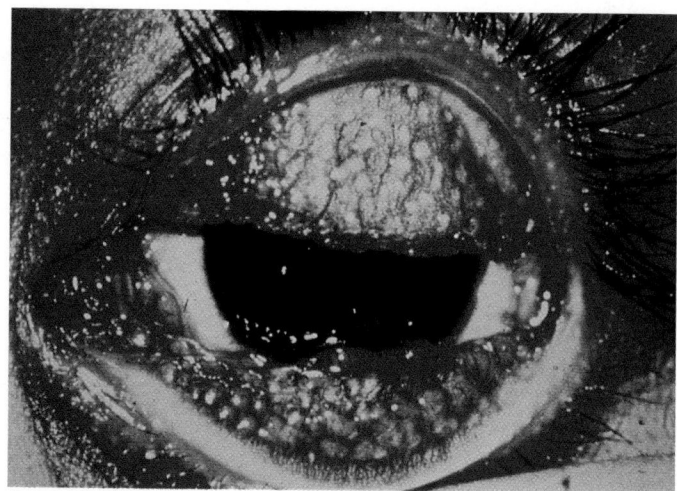

Figure 39.22 Trachoma. An active infection showing marked follicular hypertrophy of both eyelids. The inflammatory nodules cover the thickened conjunctiva of the eye.

States, except among American Indians in the Southwest, trachoma is widespread in Asia, Africa, and South America.

Trachoma is transmitted by contact with inanimate objects such as soap and towels, by hand-to-hand contact that carries *C. trachomatis* from an infected eye to an uninfected eye, or by flies. The disease begins abruptly with an inflamed conjunctiva. This leads to an inflammatory cell exudate and necrotic eyelash follicles (**figure 39.22**). The disease usually heals spontaneously. However, with reinfection, vascularization of the cornea, or **pannus** formation, occurs, leading to scarring of the conjunctiva. If scar tissue accumulates over the cornea, blindness results.

Diagnosis and treatment of trachoma are the same as for inclusion conjunctivitis (previously discussed). However, prevention and control of trachoma lies more in health education and personal hygiene—such as access to clean water for washing—than in treatment.

Tularemia

The gram-negative bacterium *Francisella tularensis* is widely found in animal reservoirs in the United States and causes the disease **tularemia** (from Tulare, a county in California where the disease was first described). It may be transmitted to humans by biting arthropods (ticks, deer flies, or mosquitoes), direct contact with infected tissue (rabbits), inhalation of aerosolized bacteria, or ingestion of contaminated food or water. After an incubation period of 2 to 10 days, a primary ulcerative lesion appears at the infection site, lymph nodes enlarge, and a high fever develops. Diagnosis is by PCR or culture of the bacterium and fluorescent antibody and agglutination tests; treatment is with streptomycin, tetracycline, or aminoglycoside antibiotics. Prevention and control involves public education, protective clothing, and vector control. An attenuated live vaccine is available from the United States Army for high-risk laboratory workers. Fewer than 300 cases of tularemia are reported annually in the United States.

Table 39.4

Summary of the Major Sexually Transmitted Diseases (STDs)

Microorganism	Disease	Comments	Treatment
Viruses			
Human immunodeficiency virus (HIV)	Acquired immune deficiency syndrome (AIDS)	Pandemic in many parts of the world	A cocktail of reverse transcriptase and protease inhibitors
Herpes simplex virus (HSV-2)	Genital herpes	Painful blisters; enters latent stage, with reactivation due to stress; also oral, pharyngeal, and rectal herpes; no cure; very prevalent in the U.S.	Acyclovir and similar drugs alleviate the symptoms
Human papillomavirus type 6 (HPV-6)	Condyloma acuminata (genital warts)	Predisposes to cervical cancer; no cure; very common in the U.S.	Removal by various mechanical and chemical means; interferon injection
Hepatitis B virus (HBV)	Hepatitis B (serum hepatitis)	Transmitted in semen; cirrhosis, primary hepatocarcinoma	No treatment; recombinant HBV vaccine for prevention
Cytomegalovirus (CMV)	Congenital cytomegalic inclusion disease	Avoid sexual contact with an infected person	Ganciclovir and cidofovir for high-risk patients
Molluscum contagiosum	Genital molluscum contagiosum	Localized wartlike skin lesions	None
Bacteria			
Calymmatobacterium granulomatis	Granuloma inguinale (donovanosis)	Rare in the U.S.; draining ulcers can persist for years	Tetracycline, erythromycin, newer quinolones
Campylobacter (Heliobacter) cinaedi, C. fennelliae	Diarrhea and rectal inflammation in homosexual men	Common in immunocompromised individuals	Metronidazole, macrolides
Chlamydia trachomatis	Nongonococcal urethritis (NGU); cervicitis, pelvic inflammatory disease (PID), lymphogranuloma venereum	Serovars D-K cause most of the STDs in the U.S.; lymphogranuloma venereum rare in the U.S.	Tetracyclines, erythromycin, doxycycline, ceftriaxone
Gardnerella vaginalis	Bacterial vaginosis	Clue cells present	Metronidazole
Haemophilus ducreyi	Chancroid ("soft chancre")	Open sores on the genitals can lead to scarring without treatment; on the rise in the U.S.	Erythromycin or ceftriaxone
Mycoplasma genitalium	Implicated in some cases of NGU	Only recently described as an STD	Tetracyclines or erythromycin
Mycoplasma hominis	Implicated in some cases of PID	Widespread, often asymptomatic but can cause PID in women	Tetracyclines or erythromycin
Neisseria gonorrhoeae	Gonorrhea, PID	Most commonly reported STD in the U.S.; usually symptomatic in men and asymptomatic in women; new antibiotic-resistant strains	Third-generation cephalosporins
Treponema pallidum subsp. *pallidum*	Syphilis, congenital syphilis	Manifests many clinical syndromes	Benzathine penicillin G
Ureaplasma urealyticum	Urethritis	Widespread, often asymptomatic but can cause PID in women and NGU in men	Tetracyclines or erythromycin
Yeasts			
Candida albicans	Candidiasis (moniliasis)	Produces a thick white vaginal discharge and severe itching	Nystatin, terconazole
Protozoa			
Trichomonas vaginalis	Trichomoniasis	Produces a frothy vaginal discharge; very common in the U.S.	Oral metronidazole

F. tularensis is also a microorganism of concern as a biological threat agent. Since recent effort in the United States has shifted toward defending against biological terrorism, public health and medical management protocols following a potential release of tularemia are now in place.

Sexually Transmitted Diseases

Sexually transmitted diseases (STDs) represent a worldwide public health problem. The various viruses that cause STDs are presented in chapter 38, the responsible bacteria in chapter 39, and the yeasts and protozoa in chapter 40. **Table 39.4** lists the various microorganisms that can be sexually transmitted and the diseases they cause.

The spread of most sexually transmitted diseases is currently out of control. In the United States alone, at least 8 to 10 million persons contract some type of STD each year. Some of the STDs that are prevalent in the United States are gonorrhea, syphilis, genital herpes, chlamydial infections, trichomoniasis, hepatitis B, and the most serious of all because of its high mortality, AIDS.

STDs were formerly called venereal diseases (from Venus, the Roman goddess of love). They occur most frequently in the most sexually active age group—15 to 30 years of age—but anyone who has sexual contact with an infected individual is at increased risk. In general, the more sexual partners a person has, the more likely that the person is to acquire an STD.

As noted in previous chapters, some of the microorganisms that cause STDs can also be transmitted by nonsexual means. Examples include transmission by contaminated hypodermic needles and syringes shared among intravenous drug users, contaminated blood transfusions, and infected mothers to their infants.

Some STDs can be cured quite easily, but others, especially those caused by viruses, are presently difficult or impossible to cure. Because treatments are often inadequate, prevention is essential. Preventive measures are based mainly on better education of the total population and when possible, control of the sources of infection and treatment of infected individuals with chemotherapeutic agents.

---

1. Describe the three phases of lymphogranuloma venereum.
2. Briefly describe the following diseases: mycoplasmal pneumonia, peptic ulcer disease and gastritis, and psittacosis.
3. What is nongonococcal urethritis and what agents can cause it? Describe complications that may develop in the absence of treatment.
4. Describe several diseases caused by the staphylococci.
5. Name and describe the three stages of syphilis.
6. How is the disease tetanus acquired? What are its symptoms and how do they arise?
7. How does *C. trachomatis*, serotypes A–C, cause trachoma? Describe how it is transmitted, and the way in which blindness may result. What happens if it is left untreated?
8. From what animal can tularemia be contracted?
9. Name four ways in which a person may contract an STD.

39.4 FOOD-BORNE AND WATERBORNE DISEASES

Many microorganisms contaminating food and water can cause acute gastroenteritis or inflammation of the stomach and intestinal lining. When food is the source of the pathogen, the condition is often called **food poisoning.** Gastroenteritis can arise in two ways. The microorganisms may actually produce a **food-borne infection.** That is, they may first colonize the gastrointestinal tract and grow within it, then either invade host tissues or secrete exotoxins. Alternatively the pathogen may secrete an exotoxin that contaminates the food and is then ingested by the host. This is sometimes referred to as a **food intoxication** because the toxin is ingested and the presence of living microorganisms is not required. Because these toxins disrupt the functioning of the intestinal mucosa they are called **enterotox-**

ins. Common symptoms of enterotoxin poisoning are nausea, vomiting, and diarrhea.

Worldwide, diarrheal diseases are second only to respiratory diseases as a cause of adult death; they are the leading cause of childhood death, and in some parts of the world they are responsible for more years of potential life lost than all other causes combined. For example, each year around 5 million children (more than 13,600 a day) die from diarrheal diseases in Asia, Africa, and South America. In the United States estimates exceed 10,000 deaths per year from diarrhea, and an average of 500 childhood deaths are reported.

This section describes several of the more common bacteria associated with gastrointestinal infections, food intoxications, and waterborne diseases. **Table 39.5** summarizes many of the bacterial pathogens responsible for food poisoning. The protozoa responsible for food- and waterborne diseases are covered in chapter 40.　Diseases transmitted by foods (pp. 946–49); Food spoilage (pp. 940–43); Waterborne pathogens (pp. 630–36)

Botulism

Food-borne **botulism** [Latin *botulus,* sausage] is a form of food poisoning caused by *Clostridium botulinum,* which is an obligately anaerobic endospore-forming, gram-positive rod that is found in soil and aquatic sediments. The most common source of infection is home-canned food that has not been heated sufficiently to kill contaminating *C. botulinum* endospores. The endospores can germinate, and a toxin is produced during vegetative growth. If the food is then eaten without adequate cooking, the toxin remains active and the disease results.

The botulinum toxin is a neurotoxin that binds to the synapses of motor neurons. It selectively cleaves the synaptic vesicle membrane protein synaptobrevin, thus preventing exocytosis and release of the neurotransmitter acetylcholine. As a consequence, muscles do not contract in response to motor neuron activity, and flaccid paralysis results (**Techniques & Applications 39.5**). Symptoms of botulism occur within 18 to 24 hours of toxin ingestion and include blurred vision, difficulty in swallowing and speaking, muscle weakness, nausea, and vomiting. Without adequate treatment, 1/3 of the patients may die within a few days of either respiratory or cardiac failure.

Laboratory diagnosis is by a hemagglutination test or inoculation of mice with the patient's serum, stools, or vomitus to prove toxigenicity. Treatment relies on supportive care and polyvalent antitoxin. Fewer than 100 cases of botulism occur in the United States annually.

Infant botulism is the most common form of botulism in the United States and is confined to infants under a year of age. Approximately 100 cases are reported each year. It appears that ingested endospores, which may be naturally present in honey or house dust, germinate in the infant's intestine. *C. botulinum* then multiplies and produces the toxin. The infant becomes constipated, listless, generally weak, and eats poorly. Death may result from respiratory failure.

Table 39.5

Bacteria That Cause Acute Bacterial Diarrheas and Food Poisonings

Organism	Incubation Period (Hours)	Vomiting	Diarrhea	Fever	Epidemiology	Pathogenesis
Staphylococcus aureus	1–8 (rarely, up to 18)	+++	+	–	Staphylococci grow in meats, dairy and bakery products and produce enterotoxins.	Enterotoxins act on gut receptors that transmit impulses to medullary centers; may also act as superantigens.
Bacillus cereus	2–16	+++	++	–	Reheated fried rice causes vomiting or diarrhea.	Enterotoxins formed in food or in gut from growth of B. cereus.
Clostridium perfringens	8–16	±	+++	–	Clostridia grow in rewarmed meat dishes. Huge numbers ingested.	Enterotoxin produced during sporulation in gut, causes hypersecretion.
Clostridium botulinum	18–24	±	Rare	–	Clostridia grow in anaerobic foods and produce toxin.	Toxin absorbed from gut and blocks acetylcholine release at neuromuscular junction.
Escherichia coli (enterotoxigenic strain)	24–72	±	++	–	Organisms grow in gut and are a major cause of traveler's diarrhea.	Heat-labile (LT) and heat-stable (ST) enterotoxins cause hypersecretion in small intestine.
Vibrio parahaemolyticus	6–96	+	++	±	Organisms grow in seafood and in gut and produce toxin, or invade.	Toxin causes hypersecretion; vibrios invade epithelium; stools may be bloody.
Vibrio cholerae	24–72	+	+++	–	Organisms grow in gut and produce toxin.	Toxin causes hypersecretion in small intestine. Infective dose $>10^5$ vibrios.
Shigella spp. (mild cases)	24–72	±	++	+	Organisms grow in superficial gut epithelium. S. dysenteriae produces toxin.	Organisms invade epithelial cells; blood, mucus, and neutrophils in stools. Infective dose $<10^3$ organisms.
Salmonella spp. (gastroenteritis)	8–48	±	++	+	Organisms grow in gut.	Superficial infection of gut, little invasion. Infective dose $>10^5$ organisms.
Salmonella typhi (typhoid fever)	10–14 days	±	±	++	Bacteria invade the gut epithelium and reach the lymph nodes, liver, spleen, and gallbladder.	Symptoms probably due to endotoxins and tissue inflammation. Infective dose $\geq 10^7$ organisms.
Clostridium difficile	Days to weeks after antibiotic therapy	–	+++	+	Antibiotic-associated colitis.	Toxin causes epithelial necrosis in colon; pseudomembranous colitis.
Campylobacter jejuni	2–10 days	–	+++	++	Infection by oral route from foods, pets. Organism grows in small intestine.	Invasion of mucous membrane. Toxin production uncertain.
Yersinia enterocolitica	4–7 days	±	++	+	Fecal-oral transmission. Food-borne. Animals infected.	Gastroenteritis or mesenteric adenitis. Occasional bacteremia. Toxin produced occasionally.

Adapted from Geo. F. Brooks, et al., *Medical Microbiology*, 21st edition. Copyright 1998 Appleton & Lange, Norwalk, CT. Reprinted by permission.

Clinical Features
Abrupt onset, intense vomiting for up to 24 hours, recovery in 24–48 hours. Occurs in persons eating the same food. No treatment usually necessary except to restore fluids and electrolytes.
With incubation period of 2–8 hours, mainly vomiting. With incubation period of 8–16 hours, mainly diarrhea.
Abrupt onset of profuse diarrhea; vomiting occasionally. Recovery usual without treatment in 1–4 days. Many clostridia in cultures of food and feces of patients.
Diplopia, dysphagia, dysphonia, difficulty breathing. Treatment requires clearing the airway, ventilation, and intravenous polyvalent antitoxin. Exotoxin present in food and serum. Mortality rate high.
Usually abrupt onset of diarrhea; vomiting rare. A serious infection in newborns. In adults, "traveler's diarrhea" is usually self-limited in 1–3 days.
Abrupt onset of diarrhea in groups consuming the same food, especially crabs and other seafood. Recovery is usually complete in 1–3 days. Food and stool cultures are positive.
Abrupt onset of liquid diarrhea in endemic area. Needs prompt replacement of fluids and electrolytes IV or orally. Tetracyclines shorten excretion of vibrios. Stool cultures positive.
Abrupt onset of diarrhea, often with blood and pus in stools, cramps, tenesmus, and lethargy. Stool cultures are positive. Give trimethoprim sulfamethoxazole, ampicillin, or chloramphenicol in severe cases. Do not give opiates. Often mild and self-limited. Restore fluids.
Gradual or abrupt onset of diarrhea and low-grade fever. Nausea, headache, and muscle aches common. Administer no antimicrobials unless systemic dissemination is suspected. Stool cultures are positive. Prolonged carriage is frequent.
Initially fever, headache, malaise, anorexia, and muscle pains. Fever may reach 40°C by the end of the first week of illness and lasts for 2 or more weeks. Diarrhea often occurs, and abdominal pain, cough, and sore throat may be prominent. Antibiotic therapy shortens duration of the illness.
Especially after abdominal surgery, abrupt bloody diarrhea and fever. Toxins in stool. Oral vancomycin useful in therapy.
Fever, diarrhea; PMNs and fresh blood in stool, especially in children. Usually self-limited. Special media needed for culture at 43°C. Give erythromycin in severe cases with invasion. Usual recovery in 5–8 days.
Severe abdominal pain, diarrhea, fever; PMNs and blood in stool; polyarthritis, erythema nodosum, especially in children. If severe, treat with gentamicin. Keep stool specimen at 4°C before culture.

Prevention and control of botulism involves (1) strict adherence to safe food-processing practices by the food industry, (2) educating the public on safe home-preserving (canning) methods for foods, and (3) not feeding honey to infants younger than 1 year of age.

Campylobacter jejuni Gastroenteritis

Campylobacter jejuni is a gram-negative curved rod found in the intestinal tract of animals. Studies with chickens, turkeys, and cattle have shown that as much as 50 to 100% of a flock or herd of these birds or animals secrete *C. jejuni.* These bacteria also can be isolated in high numbers from surface waters. They are transmitted to humans by contaminated food and water, contact with infected animals, or anal-oral sexual activity. *C. jejuni* causes an estimated 2 million cases of *Campylobacter* **gastroenteritis**—inflammation of the intestine—or **campylobacteriosis** and subsequent diarrhea in the United States each year.

The incubation period is 2 to 10 days. *C. jejuni* invades the epithelium of the small intestine, causing inflammation, and also secretes an exotoxin that is antigenically similar to the cholera toxin. Symptoms include diarrhea, high fever, severe inflammation of the intestine along with ulceration, and bloody stools.

Laboratory diagnosis is by culture in an atmosphere with reduced O_2 and added CO_2. The disease is self-limited, and treatment is supportive; fluids, electrolyte replacement, and erythromycin is used in severe cases. Recovery usually takes from 5 to 8 days. Prevention and control involves good personal hygiene and food handling precautions, including pasteurization of milk and thorough cooking of poultry.

Cholera

Throughout recorded history **cholera** [Greek *chole,* bile] has caused seven pandemics in various areas of the world, especially in Asia, the Middle East, and Africa. The disease has been rare in the United States since the 1800s, but an endemic focus is believed to exist on the gulf coast of Louisiana and Texas.

Cholera is caused by the gram-negative *Vibrio cholerae* bacterium of the family *Vibrionaceae* (**figure 39.23**). Although there are many serogroups, only O1 and O139 have exhibited the ability to cause epidemics. *V. cholerae* O1 is divided into two serotypes, Inaba and Ogawa, and two biotypes, classic and El Tor.

Cholera is acquired by ingesting food or water contaminated by fecal material from patients or carriers. (Shellfish and copepods are natural reservoirs.) In 1961 the El Tor biotype emerged as an important cause of cholera pandemics, and in 1992 the newly identified strain *V. cholerae* O139 emerged in Asia. This novel toxigenic strain does not agglutinate with O1 antiserum but possesses epidemic and pandemic potential. In Calcutta, India, serogroup O139 of *Vibrio cholerae* has displaced El Tor *V. cholerae* serogroup O1, an event that has never happened in the recorded history of cholera.

Once the bacteria enter the body, the incubation period is from 24 to 72 hours. The bacteria adhere to the intestinal mucosa

Some toxins are currently being used for the treatment of human disease. Specifically, botulinum toxin, the most poisonous biological substance known, is being used for the treatment of specific neuromuscular disorders characterized by involuntary muscle contractions. Since approval of type-A botulinum toxin (Botox) by the FDA in 1989 for three disorders (strabismus [crossing of the eyes], blepharospasm [spasmotic contractions of the eye muscles], and hemifacial spasm [contractions of one side of the face]), the number of neuromuscular problems being treated has increased to include other tremors, migraine and tension headaches, and other maladies. In 2000, dermatologists and plastic surgeons began using Botox to eradicate wrinkles caused by repeated muscle con-

tractions as we laugh, smile, or frown. The remarkable therapeutic utility of botulinum toxin lies in its ability to specifically and potently inhibit involuntary muscle activity for an extended duration. Overall, the clostridia (currently one of the largest and most diverse genera of bacteria containing about 130 "official" species) produce more protein toxins than any other bacterial genus and are a rich reservoir of toxins for research and medicinal uses. For example, research is underway to use clostridial toxins or toxin domains for drug delivery, prevention of food poisoning, and the treatment of cancer and other diseases. The remarkable success of botulinum toxin as a therapeutic agent has thus created a new field of investigation in microbiology.

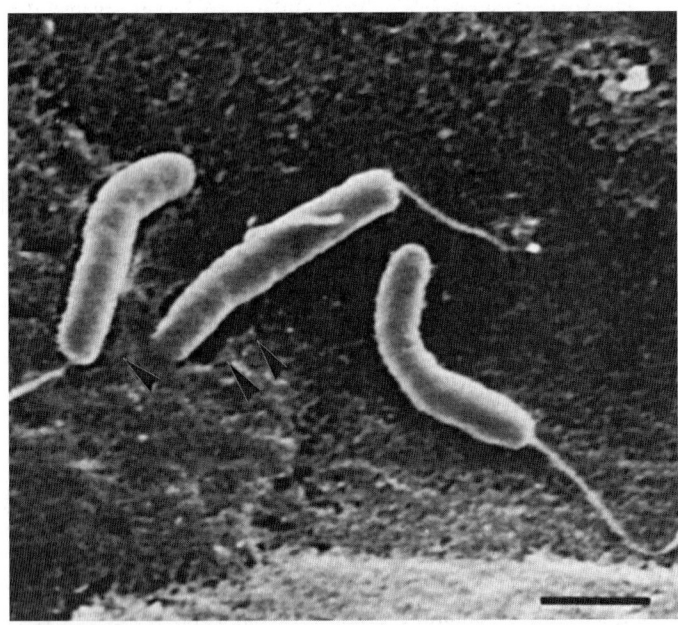

Figure 39.23 Cholera. *Vibrio cholerae* adhering to intestinal epithelium; scanning electron micrograph (×12,000). Notice that the bacteria is slightly curved with a single polar flagellum.

of the small intestine, where they are not invasive but secrete **choleragen,** a cholera toxin. Choleragen is a protein composed of two functional units, an enzymatic A subunit and an intestinal receptor-binding B subunit. The A subunit enters the intestinal epithelial cells and activates the enzyme adenylate cyclase by the addition of an ADP-ribosyl group in a way similar to that employed by diphtheria toxin (*see figure 34.5b*). As a result choleragen stimulates hypersecretion of water and chloride ions while inhibiting absorption of sodium ions. The patient loses massive quantities of fluid and electrolytes, which is associated with abdominal muscle cramps, vomiting, fever, and watery diarrhea. The diarrhea can be so profuse that a person can lose 10 to 15

liters of fluid during the infection. Death may result from the elevated concentrations of blood proteins, caused by reduced fluid levels, which leads to circulatory shock and collapse. There is now evidence that the cholera toxin gene is carried by the CTX filamentous bacteriophage. The phage binds to the pilus used to colonize the host's gut, enters the bacterium, and incorporates its genes into the bacterial chromosome.

Recent evidence indicates that passage through the human host enhances infectivity. Before *V. cholerae* exits the body in watery stools, something about the intestinal environment stimulates the activity of certain genes. These genes, in turn, seem to prepare the bacteria for ever more effective colonization of their next victims, possibly fueling epidemics.

Laboratory diagnosis is by culture of the bacterium from feces and subsequent identification by agglutination reactions with specific antisera. Treatment is by oral rehydration therapy with NaCl plus glucose to stimulate water uptake by the intestine; the antibiotics of choice are a tetracycline, trimethoprim-sulfamethoxazole, or ciprofloxacin. The most reliable control methods are based on proper sanitation, especially of water supplies. The mortality rate without treatment is often over 50%; with treatment and supportive care, it is less than 1%. Fewer than 20 cases of cholera are reported each year in the United States.

Listeriosis

Listeria monocytogenes is a gram-positive rod that can be isolated from soil, vegetation, and many animal reservoirs. Human disease due to *L. monocytogenes* generally occurs in the setting of pregnancy or immunosuppression caused by illness or medication. Recent evidence suggests that a substantial number of cases of human **listeriosis** are attributable to the food-borne transmission of *L. monocytogenes*. *Listeria* outbreaks have been traced to sources such as contaminated milk, soft cheeses, vegetables, and meat. Unlike many of the food-borne pathogens, which cause primarily gastrointestinal illness, *L. monocytogenes* causes invasive syndromes such as meningitis, sepsis, and stillbirth.

L. monocytogenes is an intracellular pathogen, a characteristic consistent with its predilection for causing illness in persons with deficient cell-mediated immunity. This bacterium can be found as part of the normal gastrointestinal microbiota in healthy individuals. In immunosuppressed individuals, invasion, intracellular multiplication, and cell-to-cell spread of the bacterium appears to be mediated through proteins such as internalin, the hemolysin listeriolysin O, and phospholipase C. *Listeria* also uses host cell actin filaments to move within and between cells (*see figure 4.4*). The increased risk of infection in pregnant women may be due to both systemic and local immunological changes associated with pregnancy. For example, local immunosuppression at the maternal-fetal interface of the placenta may facilitate intrauterine infection following transient maternal bacteremia.

Diagnosis of listeriosis is by culture of the bacterium. Treatment is intravenous administration of either ampicillin or penicillin. Since *L. monocytogenes* is frequently isolated from food, the USDA (the U.S. Department of Agriculture) and manufacturers are pursuing measures to reduce the contamination of food products by this bacterium.

Salmonellosis

Salmonellosis (*Salmonella* gastroenteritis) is caused by over 2,000 *Salmonella* serovars (strains; a subspecies category). The most frequently reported one from humans is *S.* serovar *typhimurium.* This bacterium is a gram-negative, motile, non-spore-forming rod.

The initial source of the bacterium is the intestinal tracts of birds and other animals. Humans acquire the bacteria from contaminated foods such as beef products, poultry, eggs, egg products, or water. Around 45,000 cases a year are reported in the United States, but there actually may be as many as 2 to 3 million cases annually.

Once the bacteria are in the body, the incubation time is only about 8 to 48 hours. The disease results from a true food-borne infection because the bacteria multiply and invade the intestinal mucosa where they produce an enterotoxin and cytotoxin that destroy the epithelial cells. Abdominal pain, cramps, diarrhea, nausea, vomiting, and fever are the most prominent symptoms, which usually persist for 2 to 5 days but can last for several weeks. During the acute phase of the disease, as many as 1 billion salmonella can be found per gram of feces. Most adult patients recover, but the loss of fluids can cause problems for children and elderly people.

Laboratory diagnosis is by isolation of the bacterium from food or patients' stools. Treatment is with fluid and electrolyte replacement. Prevention depends on good food-processing practices, proper refrigeration, and adequate cooking.

Shigellosis

Shigellosis or bacillary dysentery is a diarrheal illness resulting from an acute inflammatory reaction of the intestinal tract caused by the four species of the genus *Shigella* (gram-negative, non-motile, facultative rods). About 25,000 to 30,000 cases a year are reported in the United States, and around 600,000 deaths a year worldwide are due to bacillary dysentery.

Shigella is restricted to human hosts. *S. sonnei* is the usual pathogen in the United States and Britain, but *S. flexneri* is also fairly common. The organism is transmitted by the fecal-oral route—primarily by food, fingers, feces, and flies (the four "F's")—and is most prevalent among children, especially 1- to 4-year-olds. The infectious dose is only around 10 to 100 bacteria. In the United States shigellosis is a particular problem in day care centers and custodial institutions where there is crowding.

The shigellae are facultative intracellular parasites that multiply within the cells of the colon epithelium. The bacteria induce the mucosal cell to phagocytose them and then disrupt the phagosome membrane. After reproducing in the cytoplasmic matrix, the shigellae invade adjacent mucosal cells. They may produce both endotoxins and exotoxins but do not usually spread beyond the colon epithelium. The watery stools often contain blood, mucus, and pus. In severe cases the colon can become ulcerated.

The incubation period usually ranges from 1 to 3 days and the organisms are shed over a period of 1 to 2 weeks. Identification of isolates is based on biochemical characteristics and serology. The disease normally is self-limiting in adults and lasts an average of 4 to 7 days; in infants and young children it may be fatal. Usually fluid and electrolyte replacement are sufficient, and antibiotics may not be required in mild cases although they can shorten the duration of symptoms. Sometimes, particularly in malnourished infants and children, neurological complications and kidney failure result. When necessary, treatment is with trimethoprim-sulfamethoxazole or fluoroquinolones. Antibiotic-resistant strains are becoming a problem. Prevention is a matter of good personal hygiene and the maintenance of a clean water supply.

Staphylococcal Food Poisoning

Staphylococcal food poisoning is the major type of food intoxication in the United States. It is caused by ingestion of improperly stored or cooked food (particularly foods such as ham, processed meats, chicken salad, pastries, ice cream, and hollandaise sauce) in which *Staphylococcus aureus* has grown.

S. aureus (a gram-positive coccus) is very resistant to heat, drying, and radiation; it is found in the nasal passages and on the skin of humans and other mammals worldwide. From these sources it can readily enter food. If the bacteria are allowed to incubate in certain foods, they produce heat-stable enterotoxins that render the food dangerous even though it appears normal. Once the bacteria have produced the toxin, the food can be extensively and properly cooked, killing the bacteria but without destroying the toxin. Intoxication can therefore result from food that has been thoroughly cooked. Six different enterotoxins have been identified and are designated A, B, C1, C2, D, and E. These toxins appear to act as neurotoxins that stimulate vomiting through the vagus nerve. At least some are superantigens and trigger the release of IL-2 and other lymphokines.

Typical symptoms include severe abdominal pain, cramps, diarrhea, vomiting, and nausea. The onset of symptoms is rapid (usually 1 to 8 hours) and of short duration (usually less than 24 hours). The mortality rate of staphylococcal food poisoning is negligible among healthy individuals.

Diagnosis is based on the symptoms or laboratory diagnosis of the bacteria from foods. Enterotoxins may be detected in foods by animal toxicity tests. Treatment is with fluid and electrolyte replacement. Prevention and control involve avoidance of food contamination, and control of personnel responsible for food preparation and distribution.

Traveler's Diarrhea and *Escherichia coli* Infections

Millions of people travel yearly from country to country (*see section 37.11*). Unfortunately a large percentage of these travelers acquire a rapidly acting, dehydrating condition called **traveler's diarrhea.** This diarrhea results from an encounter with certain viruses, bacteria, or protozoa usually absent from the traveler's normal environment. One of the major causative agents is *E. coli.* This bacterium circulates in the resident population, typically without causing symptoms due to the immunity afforded by previous exposure. Because many of these bacteria are needed to initiate infection, contaminated food and water are the major means by which the bacteria are spread. This is the basis for the popular warnings to international travelers: "Don't drink the local water" and "Boil it, peel it, cook it, or forget it."

E. coli may cause diarrheal disease by several mechanisms, and six categories or strains of diarrheagenic *E. coli* are now recognized: enterotoxigenic *E. coli* (ETEC), enteroinvasive *E. coli* (EIEC), enteroahemorrhagic *E. coli* (EHEC), enteropathogenic *E. coli* (EPEC), enteroaggregative *E. coli* (EAggEC), and diffusely adhering *E. coli* (DAEC).

The **enterotoxigenic *E. coli* (ETEC)** strains produce one or both of two distinct enterotoxins (*see Disease 13.1*), which are responsible for the diarrhea and distinguished by their heat stability: heat-stable enterotoxin (ST) and heat-labile enterotoxin (LT). The genes for ST and LT production and for colonization factors are usually plasmid-borne and acquired by horizontal gene transfer. ST binds to a glycoprotein receptor that is coupled to guanylate cyclase on the surface of intestinal epithelial cells. Activation of guanylate cyclase stimulates the production of cyclic guanosine monophosphate (cGMP), which leads to the secretion of electrolytes and water into the lumen of the small intestine, manifested as the watery diarrhea characteristic of an ETEC infection. LT binds to specific gangliosides on the epithelial cells and activates membrane-bound adenylate cyclase, which leads to increased production of cyclic adenosine monophosphate (cAMP) through the same mechanism employed by cholera toxin. Again, the result is hypersecretion of electrolytes and water into the intestinal lumen.

The **enteroinvasive *E. coli* (EIEC)** strains cause diarrhea by penetrating and multiplying within the intestinal epithelial cells. The ability to invade the epithelial cells is associated with the presence of a large plasmid; EIEC may also produce a cytotoxin and an enterotoxin.

The **enteropathogenic *E. coli* (EPEC)** strains attach to the brush border of intestinal epithelial cells and cause a specific type of cell damage called effacing lesions. **Effacing lesions** or attaching-effacing (AE) lesions represent destruction of brush border microvilli adjacent to adhering bacteria. This cell destruction leads to the subsequent diarrhea. As a result of this pathology, the term AE *E. coli* is used to describe true EPEC strains. It is now known that AE *E. coli* is an important cause of diarrhea in children residing in developing countries.

The **enterohemorrhagic *E. coli* (EHEC)** strains carry the genetic determinants for attaching-effacing lesions and Shiga-like toxin production. The attaching-effacing lesion causes hemorrhagic colitis with severe abdominal pain and cramps followed by bloody diarrhea. The Shiga-like toxins I and II (also called verotoxins 1 and 2) have also been implicated in two extraintestinal diseases; **hemolytic uremic syndrome** and thrombotic thrombocytopenic purpura. It is believed these toxins kill vascular endothelial cells. A major form of EHEC is the *E. coli* O157:H7 that has caused many outbreaks of hemorrhagic colitis in the United States since it was first recognized in 1982. Currently there are a minimum of 20,000 *E. coli* O157:H7 cases and 250 deaths in the United States each year.

The **enteroaggregative *E. coli* (EAggEC)** strains adhere to epithelial cells in localized regions, forming clumps of bacteria with a "stacked brick" appearance. Conventional extracellular toxins have not been detected in EAggEC, but unique lesions are seen in epithelial cells, suggesting the involvement of toxins.

The **diffusely adhering *E. coli* (DAEC)** strains adhere over the entire surface of epithelial cells and usually cause disease in immunologically naive or malnourished children. It has been suggested that DAEC may have an as yet undefined virulence factor.

Diagnosis of traveler's diarrhea caused by *E. coli* is based on past travel history and symptoms. Laboratory diagnosis is by isolation of the specific type of *E. coli* from feces and identification using DNA probes, the determination of virulence factors, and the polymerase chain reaction. Treatment is with fluid and electrolytes plus doxycycline and trimethroprim-sulfamethoxazole. Recovery is usually without complications. Prevention and control involve avoiding contaminated food and water.

Typhoid Fever

Typhoid [Greek *typhodes*, smoke] **fever** is caused by several virulent serovars of *Salmonella typhi* and is acquired by ingestion of food or water contaminated by feces of infected humans or animals. In earlier centuries the disease occurred in great epidemics.

Once in the small intestine the incubation period is about 10 to 14 days. The bacteria colonize the small intestine, penetrate the epithelium, and spread to the lymphoid tissue, blood, liver, and gallbladder. Symptoms include fever, headache, abdominal pain, anorexia, and malaise, which last several weeks. After approximately 3 months, most individuals stop shedding bacteria in their feces. However, a few individuals continue to shed *S. typhi* for extended periods but show no symptoms. In these carriers, the bacteria continue to grow in the gallbladder and reach the intestine through the bile duct (*see Historical Highlights 37.2 on "Typhoid Mary," one of the most famous typhoid carriers*).

Laboratory diagnosis of typhoid fever is by demonstration of typhoid bacilli in the blood, urine, or stools and serology (the Widal test). Treatment with ceftriaxone, trimethoprim-sulfamethoxazole, or ampicillin has reduced the mortality rate to less than 1%. Recovery from typhoid confers a permanent immunity. Purification of drinking water, milk pasteurization, prevention of food handling by carriers, and complete isolation of patients are the most successful prophylactic measures. There is a vaccine for high-risk individuals (*see table 33.1*). About 400 to 500 cases of typhoid fever occur annually in the United States.

---

1. Distinguish between food intoxication, food poisoning, and food-borne infection. What is an enterotoxin?
2. How does one acquire botulism? Describe how botulinum toxin causes flaccid paralysis.
3. What is the most common form of gastroenteritis in the United States and how are the symptoms caused?
4. Why is cholera the most severe form of gastroenteritis?
5. What is a common source of *Listeria* infections? How is the intracellular growth of *Listeria* related to the symptoms it produces and the observation that immunocompromised individuals are most at risk?
6. What is the usual source of the bacterium responsible for salmonellosis? Shigellosis? Where and how does *Shigella* infect people?
7. Describe the most common type of food intoxication in the United States and how it arises.
8. What are some specific causes of traveler's diarrhea? Briefly describe the six major types of pathogenic *E. coli*.
9. Describe a typhoid carrier. How does one become a carrier?

39.5   SEPSIS AND SEPTIC SHOCK

Some microbial diseases and their effects cannot be categorized under a specific mode of transmission. Two important examples are sepsis and septic shock. Septic shock is the most common cause of death in intensive care units and the thirteenth most common cause of death in the United States. Unfortunately the incidence of these two disorders continues to rise: 400,000 cases of sepsis and 200,000 episodes of septic shock are estimated to occur annually in the United States, resulting in more than 100,000 deaths.

Sepsis recently has been redefined by physicians as the systemic response to a microbial infection (*see also table 34.2*). This response is manifested by two or more of the following conditions: temperature above 38 or below 36°C; heart rate above 90 beats per minute; respiratory rate above 20 breaths per minute or a pCO_2 below 32 mmHg; leukocyte count above 12,000 cells per ml^3 or below 4,000 cells per ml^3. **Septic shock** is sepsis associated with severe hypotension (low blood pressure due to shock) despite adequate fluid replacement. Gram-positive bacteria, fungi, and endotoxin-containing gram-negative bacteria can initiate the pathogenic cascade of sepsis leading to septic shock. Gram-negative sepsis is most commonly caused by *Escherichia*

coli, followed by *Klebsiella* spp., *Enterobacter* spp., and *Pseudomonas aeruginosa.* Endotoxin, or lipopolysaccharide (LPS; *see figure 3.25*), an integral component of the outer membrane of gram-negative bacteria, has been implicated as a primary initiator of the pathogenesis of gram-negative septic shock.

The pathogenesis of sepsis and septic shock begins with the proliferation of the microorganism at the infection site (**figure 39.24**). The microorganism may invade the bloodstream directly or may proliferate locally and release various products into the bloodstream. These products include both structural components of the microorganisms (endotoxin, teichoic acid antigen) and exotoxins synthesized by the microorganism. All of these products can stimulate the release of the endogenous mediators of sepsis from endothelial cells, plasma cells (monocytes, macrophages, neutrophils), and plasma cell precursors.

The endogenous mediators have profound physiological effects on the heart, vasculature, and other body organs. The consequences are either recovery or septic shock leading to death. Death usually ensues if one or more organ systems fail completely.

DENTAL INFECTIONS

Some microorganisms found in the oral cavity are discussed in section 31.2 and presented in figure 31.2. Of this large number, only a few bacteria can be considered true dental pathogens or **odontopathogens.** These few odontopathogens are responsible for the most common bacterial diseases in humans: tooth decay and periodontal disease.

Dental Plaque

The human tooth has a natural defense mechanism against bacterial colonization that complements the protective role of saliva. The hard enamel surface selectively absorbs acidic glycoproteins (mucins) from saliva, forming a membranous layer called the **acquired enamel pellicle.** This pellicle, or organic covering, contains many sulfate (SO_4^{2-}) and carboxylate (—COO^-) groups that confer a net negative charge to the tooth surface. Because most bacteria also have a net negative charge, there is a natural repulsion between the tooth surface and bacteria in the oral cavity. Unfortunately this natural defense mechanism breaks down when dental plaque formation occurs.

Dental plaque formation begins with the initial colonization of the pellicle by *Streptococcus gordonii, S. oralis,* and *S. mitis.* These bacteria selectively adhere to the pellicle by specific ionic, hydrophobic, and lectinlike interactions. Once the tooth surface is colonized, subsequent attachment of other bacteria results from a variety of specific coaggregation reactions (**figure 39.25**). **Coaggregation** is the result of cell-to-cell recognition between genetically distinct bacteria. Many of these interactions are mediated by a lectin on one bacterium that interacts with a complementary carbohydrate receptor on the other bacterium. The most important species at this stage are *Actinomyces viscosus, A. naeslundii,* and

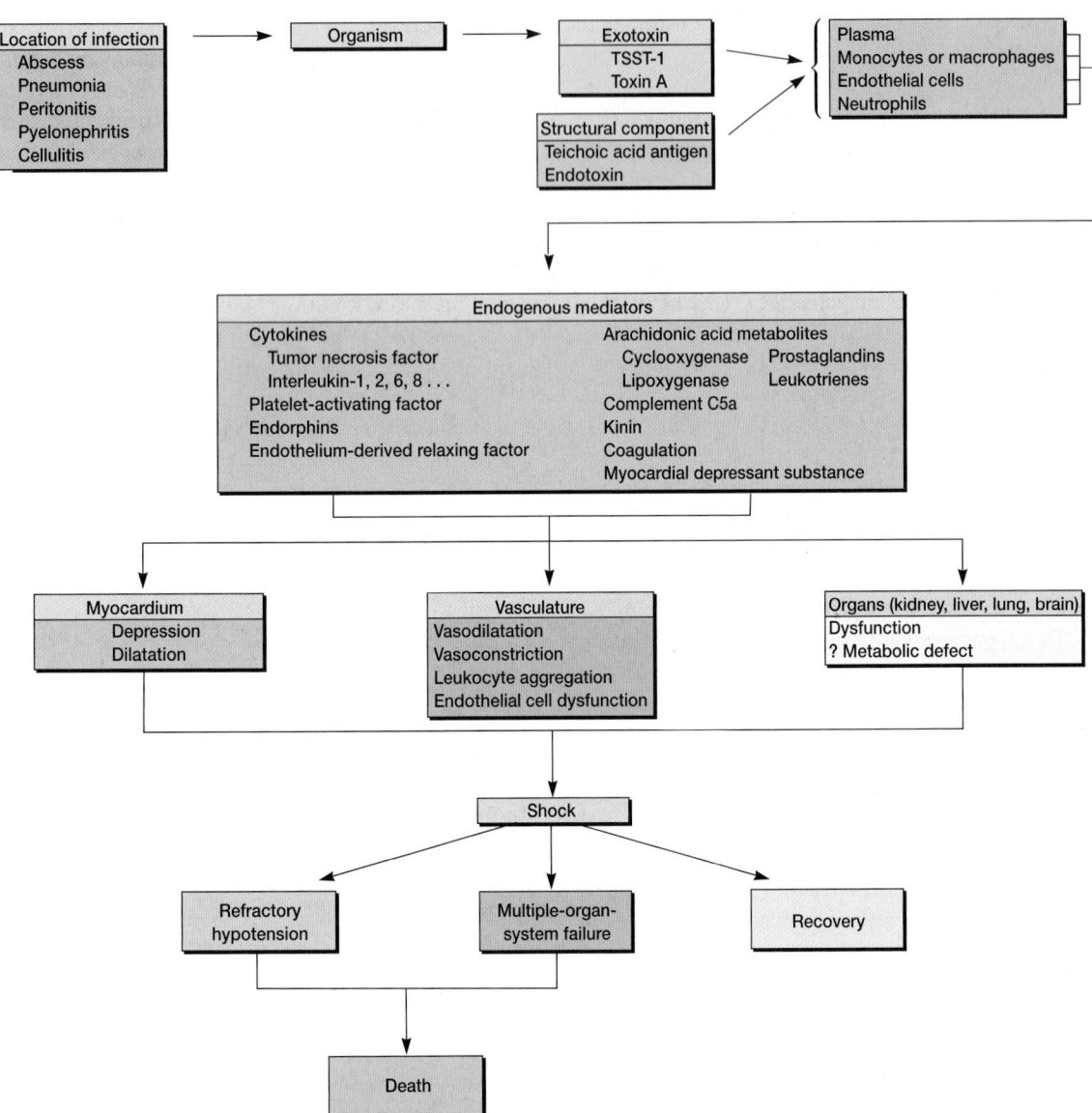

Figure 39.24   The Septic Shock Cascade.   Microbial exotoxins (toxic shock syndrome toxin-1 [TSST-1], *Pseudomonas aeruginosa* toxin A [toxin A]) and structural components of the microorganism (teichoic acid antigen, endotoxin) trigger the biochemical events that lead to such serious complications as shock, adult respiratory distress syndrome, and disseminated intravascular coagulation.

S. gordonii. After these species colonize the pellicle, a microenvironment is created that allows *Streptococcus mutans* and *S. sobrinus* to become established on the tooth surface by attaching to these initial colonizers.   Biofilms (pp. 602–3)

These streptococci produce extracellular enzymes (glucosyltransferases) that polymerize the glucose moiety of sucrose into a heterogeneous group of extracellular water-soluble and water-insoluble glucan polymers and other polysaccharides. The fructose by-product can be used in fermentation. **Glucans** are branched-chain polysaccharides composed of glucose units; many glucans synthesized by oral streptococci have glucoses held together by $\alpha(1\rightarrow6)$ or $\alpha(1\rightarrow3)$ linkages (figure 39.25c). They act like a cement to bind bacterial cells together, forming a

plaque ecosystem. (Dental plaque is one of the most dense collections of bacteria in the body—and perhaps the source of the first human microorganisms to be seen under a microscope by Anton van Leeuwenhoek in the seventeenth century.) Once plaque becomes established, a low oxidation reduction potential is created on the surface of the tooth. This leads to the growth of strict anaerobic bacteria (*Bacteroides melaninogenicus, B. oralis,* and *Veillonella alcalescens*), especially between opposing teeth and the dental-gingival crevices.

After the microbial plaque ecosystem develops, bacteria produce lactic and possibly acetic and formic acids from sucrose and other sugars. Because plaque is not permeable to saliva, the acids are not diluted or neutralized, and they demineralize the enamel

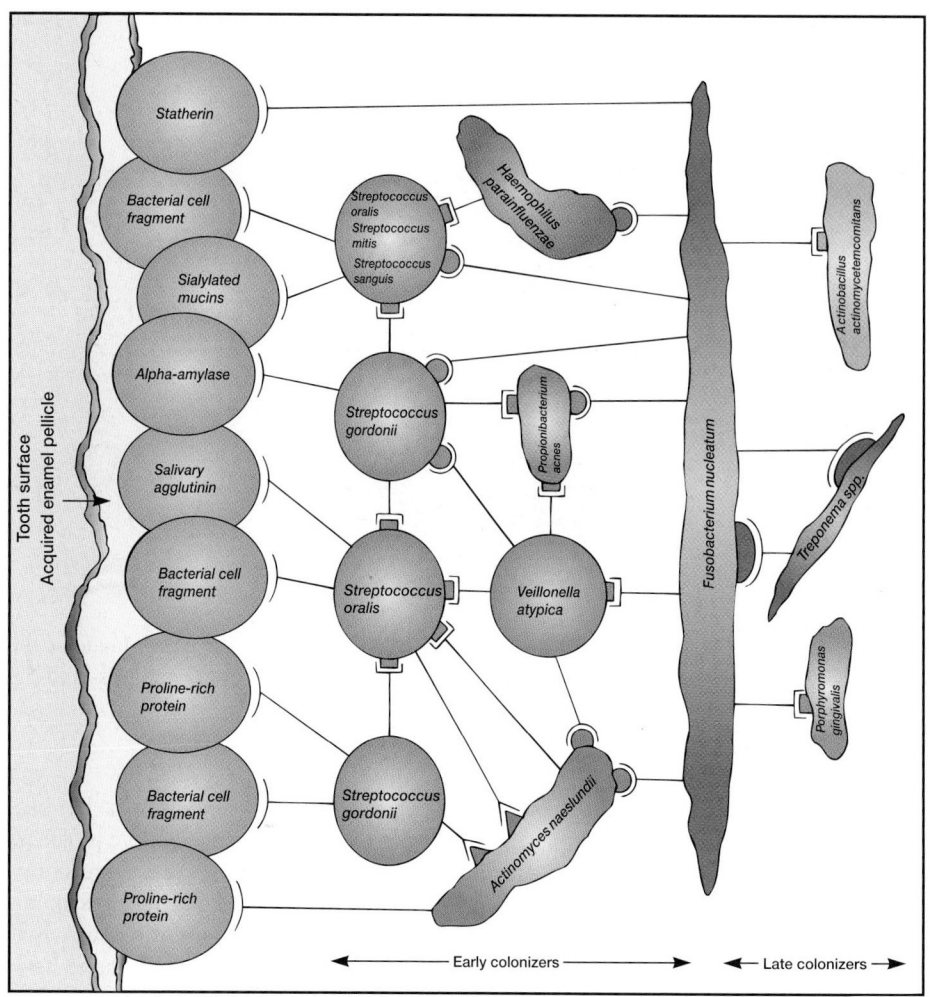

(a)

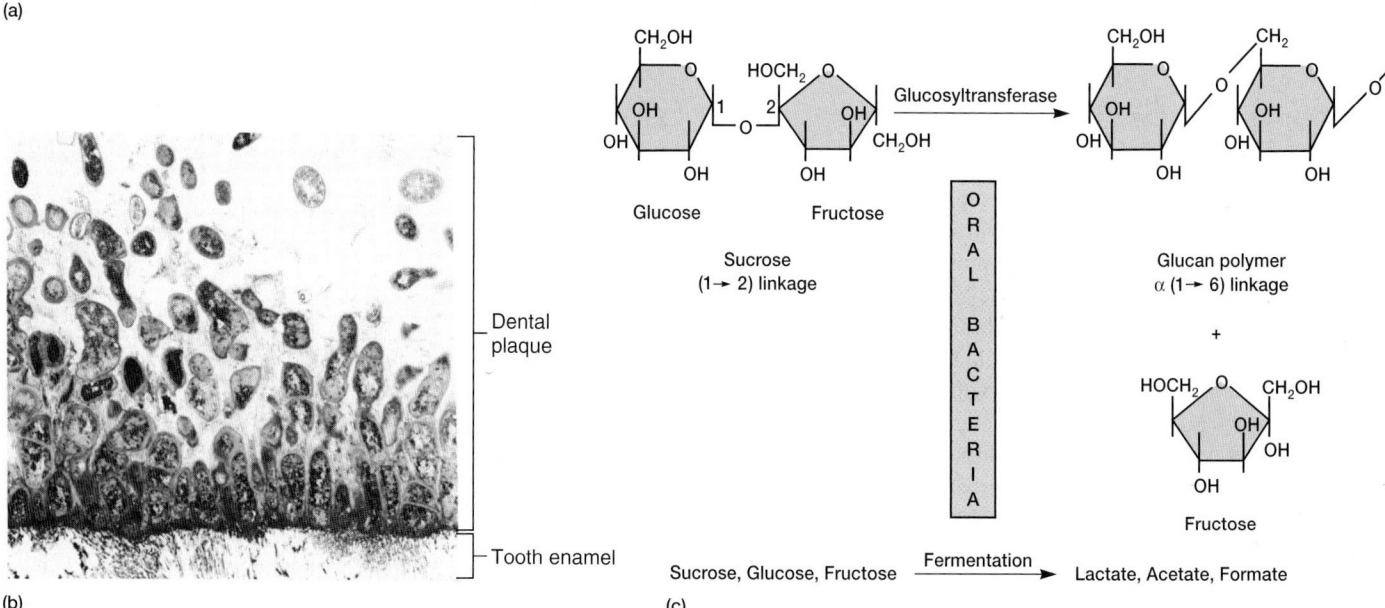

(b)

(c)

Figure 39.25 The Formation of Dental Plaque on a Freshly Cleaned Tooth Surface. (a) Diagrammatic representation of the proposed temporal relationship of bacterial accumulation and multigeneric coaggregation during the formation of dental plaque on the acquired enamel pellicle. Early tooth surface colonizers coaggregate with each other, and late colonizers coaggregate with each other. With a few exceptions, early colonizers do not recognize late colonizers. After the tooth surface is covered with the earliest colonizers, each newly added bacterium becomes a new surface for recognition by unattached bacteria. (b) Dental plaque consisting of bacteria plus polysaccharides such as glucan is shown attached to the enamel surface of a tooth; transmission electron micrograph ($\times 13,600$). (c) The enzyme glucosyltransferase or dextransucrase (produced by the oral bacteria) cause the assembly of glucose units from sucrose into glucans, and fructose is released. Sucrose, glucose, and fructose molecules also are metabolized by the oral bacteria to produce lactate and other acids. Lactate is responsible for dental caries.

913

to produce a lesion on the tooth. It is this chemical lesion that initiates dental decay.

Dental Decay (Caries)

As described previously, a histologically undetectable chemical lesion caused by the diffusion into the tooth's enamel of undissociated fermentation acids initiates the decay process. Once these acids move below the enamel surface, they dissociate and react with the hydroxyapatite of the enamel to form soluble calcium and phosphate ions. As the ions diffuse outward, some reprecipitate as calcium phosphate salts in the tooth's surface layer to create a histologically sound outer layer overlying a porous subsurface area. Between meals and snacks, the pH returns to neutrality and some calcium phosphate reenters the lesion and crystallizes. The result is a demineralization-remineralization cycle.

When fermentable foods high in sucrose are eaten for prolonged periods, acid production overwhelms the repair process and demineralization is greater than remineralization. This leads to dental decay or **caries** [Latin, rottenness]. Once the hard enamel has been breached, bacteria can invade the dentin and pulp of the tooth and cause its death.

No drugs are available to prevent dental caries. The main strategies for prevention include minimal ingestion of sucrose; daily brushing, flossing, and mouthwashes; and professional cleaning at least twice a year to remove plaque. The use of fluorides in toothpaste, drinking water, mouthwashes, or professionally applied to the teeth protects against lactic and acetic acids and reduces tooth decay.

Periodontal Disease

Periodontal disease refers to a diverse group of diseases that affect the periodontium, and is the most common chronic infection in adults. The **periodontium** is the supporting structure of a tooth and includes the cementum, the periodontal membrane, the bones of the jaw, and the gingivae (gums). The gingiva is dense fibrous tissue and its overlying mucous membrane, that surrounds the necks of the teeth. The gingiva helps hold the teeth in place. Disease is initiated by the formation of **subgingival plaque,** the plaque that forms at the dentogingival margin and extends down into the gingival tissue. Colonization of the subgingival region is aided by the ability of *Porphyromonas gingivalis* to adhere to substrates such as adsorbed salivary molecules, matrix proteins, epithelial cells, and bacteria in biofilms on teeth and epithelial surfaces. Binding to these substrates is mediated by *P. gingivalis* fimbrillin, the structural subunit of the major fimbriae. *P. gingivalis* does not use sugars as an energy source, but requires hemin

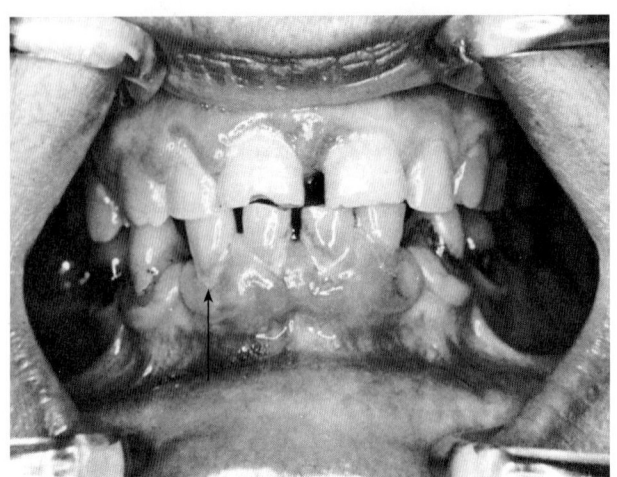

Figure 39.26 Periodontal Disease.   Notice the plaque on the teeth (arrow), especially at the gingival (gum) margins, and the inflamed gingiva.

as a source of iron and peptides for energy and growth. The bacterium produces at least three hemagglutinins and five proteases to satisfy these requirements. It is the proteases that are responsible for the breakdown of the gingival tissue. The result is an initial inflammatory reaction known as **periodontitis,** which is caused by the host's immune response to both the plaque bacteria and the tissue destruction. This leads to swelling of the tissue and the formation of periodontal pockets. Bacteria colonize these pockets and cause more inflammation, which leads to the formation of a periodontal abscess, bone destruction or **periodontosis,** inflammation of the gingiva or **gingivitis,** and general tissue necrosis (**figure 39.26**). If the condition is not treated, the tooth may fall out of its socket.

Periodontal disease can be controlled by frequent plaque removal; by brushing, flossing, and mouthwashes; and at times, by oral surgery of the gums and antibiotics.

1. Define sepsis and septic shock. How are microorganisms thought to cause shock?
2. Name some common odontopathogens that are responsible for dental caries, dental plaque, and periodontal disease. Be specific.
3. What is the function of the acquired enamel pellicle?
4. How does plaque formation occur? Dental decay?
5. Describe some pathological manifestations of periodontal disease.
6. How can caries and periodontal diseases be prevented?

Summary

Although only a small percentage of all bacteria are responsible for human illness, the suffering and death they cause are significant. Each year, millions of people are infected by pathogenic bacteria using the four major modes of transmission: airborne, arthropod-borne, direct contact, and food-borne and waterborne.

As the fields of microbiology, immunology, pathology, pharmacology, and epidemiology have expanded current understanding of the disease process, the incidence of many human illnesses has decreased. Many bacterial infections, once leading causes of death, have been successfully brought under control in most developed countries. Alternatively, several are increasing in incidence throughout the world.

The bacteria emphasized in this chapter and the diseases they cause are as follows:

39.1 Airborne Diseases

diphtheria (*Corynebacterium diphtheriae*) (**figure 39.1**)
Legionnaires' disease and Pontiac fever (*Legionella pneumophila*)
meningitis (*Haemophilus influenzae* type b, *Neisseria meningitidis,* and *Streptococcus pneumoniae*)
M. avium–M. intracellulare pneumonia
pertussis (*Bordetella pertussis*)
Streptococcal diseases (*Streptococcus pyogenes*) (**figure 39.3**)
tuberculosis (*Mycobacterium tuberculosis*) (**figure 39.7**)

39.2 Arthropod-Borne Diseases

ehrlichiosis (*Ehrlichia chaffeensis*)
epidemic (louse-borne) typhus (*Rickettsia prowazekii*)
endemic (murine) typhus (*Rickettsia typhi*)
Lyme disease (*Borrelia burgdorferi*)

plague (*Yersinia pestis*) (**figure 39.9**)
Q fever (*Coxiella burnetii*)
Rocky Mountain spotted fever (*Rickettsia rickettsii*)

39.3 Direct Contact Diseases

anthrax (*Bacillus anthracis*)
bacterial vaginosis (*Gardnerella vaginalis*)
cat-scratch disease (*Bartonella henselae*)
chancroid (*Haemophilus ducreyi*)
chlamydial pneumonia (*Chlamydia pneumoniae*)
gas gangrene or clostridial myonecrosis (*Clostridium perfringens*)
genitourinary mycoplasmal diseases (*Ureaplasma urealyticum, Mycoplasma hominis*)
gonorrhea (*Neisseria gonorrhoeae*)
inclusion conjunctivitis (*Chlamydia trachomatis*)
leprosy (*Mycobacterium leprae*)
lymphogranuloma venereum (*Chlamydia trachomatis*)
mycoplasmal pneumonia (*Mycoplasma pneumoniae*)
nongonococcal urethritis (various microorganisms)
peptic ulcer disease (*Helicobacter pylori*)
psittacosis (ornithosis) (*Chlamydia psittaci*)
staphylococcal diseases (*Staphylococcus aureus*) (**figure 39.18**)
syphilis (*Treponema pallidum*)
tetanus (*Clostridium tetani*)
trachoma (*Chlamydia trachomatis*)
tularemia (*Francisella tularensis*)

39.4 Food-Borne and Waterborne Diseases

botulism (*Clostridium botulinum*)

gastroenteritis (*Campylobacter jejuni* and other bacteria)
cholera (*Vibrio cholerae*)
salmonellosis (*Salmonella* serovar *typhimurium*)
shigellosis (*Shigella* spp.)
staphylococcal food poisoning (*Staphylococcus aureus*)
traveler's diarrhea (*Escherichia coli*)
typhoid fever (*Salmonella typhi*)

39.5 Sepsis and Septic Shock

a. Gram-positive bacteria, fungi, and endotoxin-containing gram-negative bacteria can initiate the pathogenic cascade of sepsis leading to septic shock (**figure 39.24**). Gram-negative sepsis is most commonly caused by *E. coli,* followed by *Klebsiella* spp., *Enterobacter* spp., and *Pseudomonas aeruginosa.*

39.6 Dental Infections

a. Dental plaque formation begins on a tooth with the initial colonization of the acquired enamel pellicle by *Streptococcus gordonii, S. oralis,* and *S. mitis.* Other bacteria then become attached and form a plaque ecosystem (**figure 39.25**). The bacteria produce acids that cause a chemical lesion on the tooth and initiate dental decay or caries. Periodontal disease is a group of diverse clinical entities that affect the periodontium. Disease is initiated by the formation of subgingival plaque, which leads to tissue inflammation known as periodontitis and to periodontal pockets. Bacteria that colonize these pockets can cause an abscess, periodontosis, gingivitis, and general tissue necrosis.

Key Terms

impetigo 879
inclusion conjunctivitis 893
Legionnaires' disease
(legionellosis) 877
lepromatous (progressive)
leprosy 894
leprosy 893
listeriosis 908
Lyme disease (LD, Lyme
borreliosis) 886
lymphogranuloma venereum
(LGV) 894
meningitis 878
miliary tuberculosis 884
multi-drug-resistant strains of
tuberculosis (MDR-TB) 884
mycoplasmal pneumonia 895
myositis 880
necrotizing fasciitis 880

nongonococcal urethritis (NGU) 895
odontopathogens 911
ophthalmia neonatorum (conjunctivitis
of the newborn) 892
ornithosis 896
pannus 903
pelvic inflammatory disease
(PID) 892
peptic ulcer disease 895
periodontal disease 914
periodontitis 914
periodontium 914
periodontosis 914
pertussis 879
pharyngitis 881
plague 887
pneumonic plague 887
Pontiac fever 877

psittacosis 896
pulmonary anthrax 889
Q fever 887
reactivation tuberculosis 884
rheumatic fever 881
Rocky Mountain spotted fever 888
salmonellosis 909
scarlet fever (scarlatina) 881
sepsis 911
septic shock 911
shigellosis 909
slime 896
staphylococcal food poisoning 909
staphylococcal scalded skin syndrome
(SSSS) 899
streptococcal pneumonia 881
streptococcal sore throat 881
subgingival plaque 914

tetanolysin 903
tetanospasmin 902
tetanus 902
tonsillitis 881
toxic shock–like syndrome
(TSLS) 880
toxic shock syndrome (TSS) 898
trachoma 903
transovarian passage 888
traveler's diarrhea 910
tubercles 884
tuberculoid (neural) leprosy 893
tuberculosis (TB) 882
tuberculous cavities 884
tularemia 903
typhoid fever 910
vasculitis 885
venereal syphilis 899

Questions for Thought and Review

1. Briefly describe each of the major or most common bacterial diseases in terms of its causative agent, signs and symptoms, the course of infection, mechanism of pathogenesis, epidemiology, and prevention and/or treatment.
2. Because there are many etiologies of gastroenteritis, how is a definitive diagnosis usually made?
3. Differentiate between the following factors of bacterial intoxication and infection: etiologic agents, onset, duration, symptoms, and treatment.
4. How would you differentiate between salmonellosis and botulism?
5. Why do urinary tract infections frequently occur after catheterization procedures?
6. Why is botulism the most serious form of bacterial food poisoning?

7. Why is tuberculosis still a problem in underdeveloped countries? In the United States?
8. What is the standard treatment for most bacterial diarrheas?
9. Why are most cases of gastroenteritis not treated with antibiotics?
10. How are tetanus, gas gangrene, and botulism related?
11. Give several reasons why treatment of tuberculosis is so difficult.
12. Describe the various bacteria that cause food poisoning and give each one's most probable source.
13. Explain why Lyme disease is receiving increased attention in the United States.
14. Why is septic shock caused by gram-negative sepsis increasing in prevalence in the United States?

15. What is the major difference between trachoma and inclusion conjunctivitis?
16. Why is Rocky Mountain spotted fever the most important rickettsial disease in the United States?
17. As a group of diseases, what are some of the common characteristic symptoms and pathological results of rickettsial disease?
18. Why is it difficult to definitely isolate the microorganism responsible for nongonococcal urethritis?
19. Why are dental diseases relatively new infectious diseases of humans?
20. What makes Q fever a unique rickettsial disease?
21. What specific chlamydial and mycoplasmal diseases are transmitted sexually?

Critical Thinking Questions

1. Why is tetanus a concern only when one has a deep puncture-type wound and not a surface cut or abrasion?
2. Think about our modern, Western lifestyles. Can you name and describe bacterial diseases

that result from this life of relative luxury? Refer to *Infections of Leisure*, second edition, edited by David Schlossberg (1999), published by the American Society for Microbiology Press.

3. You have been assigned the task of eradicating gonorrhea in your community. Explain how you would accomplish this.
4. You are a park employee. How would you prevent people from acquiring arthropod-borne diseases?

For recommended readings on these and related topics, see Appendix V.

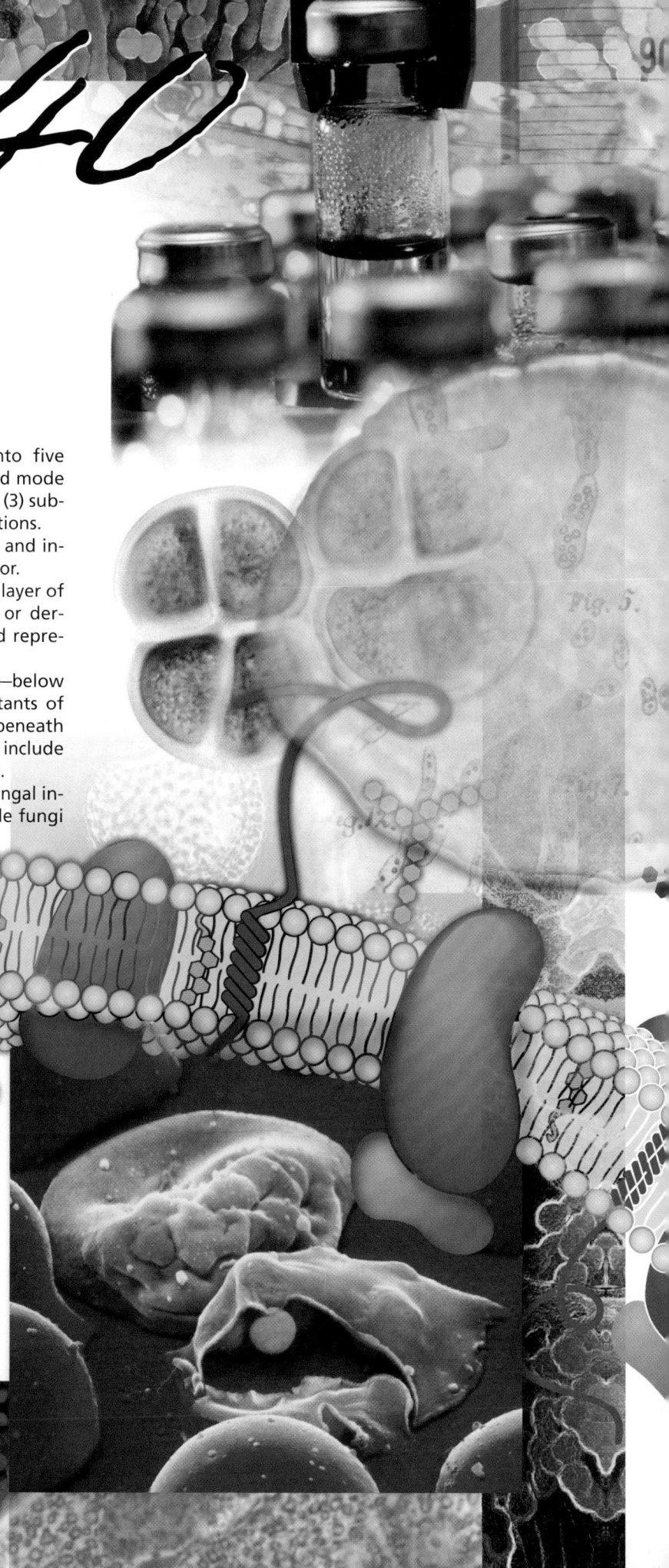

chapter 40

Human Diseases Caused by Fungi and Protozoa

Concepts

- Fungal diseases (mycoses) are usually divided into five groups according to the level of infected tissue and mode of entry into the host: (1) superficial, (2) cutaneous, (3) subcutaneous, (4) systemic, and (5) opportunistic infections.

- The superficial mycoses occur mainly in the tropics and include black piedra, white piedra, and tinea versicolor.

- The cutaneous mycoses—that is, those of the outer layer of the skin—are generally called ringworms, tineas, or dermatomycoses. These diseases occur worldwide and represent the most common fungal diseases in humans.

- The dermatophytes that cause the subcutaneous—below the skin—mycoses are normal saprophytic inhabitants of the soil. They must be introduced into the body beneath the cutaneous layer. Examples of these diseases include chromomycosis, maduromycosis, and sporotrichosis.

- The systemic mycoses are the most serious of the fungal infections in the normal host because the responsible fungi can disseminate throughout the body. Examples include blastomycosis, coccidioidomycosis, cryptococcosis, and histoplasmosis.

- The opportunistic mycoses can create life-threatening situations in the compromised host. Examples of these diseases include aspergillosis, candidiasis, and *Pneumocystis* pneumonia.

- About 20 different protozoa cause human diseases that afflict hundreds of millions of people throughout the world. Examples include amebiasis, cryptosporidiosis, giardiasis, malaria, the hemoflagellate diseases, toxoplasmosis, and trichomoniasis.

This photograph shows malaria parasites (yellow) bursting out of red blood cells. Malaria is one of the worst scourges of humanity. Indeed, malaria has played an important part in the rise and fall of nations (see the chapter opening quote), and has killed untold millions the world over. Despite the combined efforts of 102 countries to eradicate malaria, it remains the most important disease in the world today in terms of lives lost and economic burden.

• Certain fungal and protozoan diseases are increasing in incidence because of organ transplants, immunosuppressive drugs, and AIDS.

Historians believe that malaria has probably had a greater impact on world history than any other infectious disease, influencing the outcome of wars, various population movements, and the development and decline of various civilizations.

—Lynne S. Garcia

he purpose of this chapter is to describe some of the fungal and protozoan microorganisms that are pathogenic to humans and to discuss the clinical manifestations, diagnosis, epidemiology, pathogenesis, and treatment of the diseases caused by them.

Besides the viruses and bacteria, two other major groups of microorganisms cause infectious diseases in humans: fungi and protozoa. The biology of these organisms is covered in chapters 25 and 27, respectively. This chapter emphasizes the diseases they cause.

 40.1   FUNGAL DISEASES

Although hundreds of thousands of fungal species are found in the environment, only about 50 can produce disease in humans. **Medical mycology** is the discipline that deals with the fungi that cause human disease. These fungal diseases, known as **mycoses** [s., mycosis; Greek *mykes,* fungus], are divided into five groups according to the type of infected tissue in the host: superficial, cutaneous, subcutaneous, systemic, and opportunistic mycoses (**tables 40.1** and **40.2**). It should be noted that the importance of opportunistic fungal pathogens is increasing because of the expansion of the immunocompromised patient population.

Superficial Mycoses

The superficial mycoses are extremely rare in the United States, and most occur in the tropics. The fungi responsible are limited

---

Table 40.1

Examples of Some Medically Important Fungi

Group	Pathogen	Location	Disease
Superficial mycoses	*Piedraia hortae*	Scalp	Black piedra
	Trichosporon beigelii	Beard, mustache	White piedra
	Malassezia furfur	Trunk, neck, face, arms	Tinea versicolor
Cutaneous mycoses	*Trichophyton mentagrophytes, T. verrucosum, T. rubrum*	Beard hair	Tinea barbae
	Trichophyton, Microsporum canis	Scalp hair	Tinea capitis
	Trichophyton rubrum, T. mentagrophytes, Microsporum canis	Smooth or bare parts of the skin	Tinea corporis
	Epidermophyton floccosum, T. mentagrophytes, T. rubrum	Groin, buttocks	Tinea cruris (jock itch)
	T. rubrum, T. mentagrophytes, E. floccosum	Feet	Tinea pedis (athlete's foot)
	T. rubrum, T. mentagrophytes, E. floccosum	Nails	Tinea unguium (onychomycosis)
Subcutaneous mycoses	*Phialophora verrucosa, Fonsecaea pedrosoi*	Legs, feet	Chromoblastomycosis
	Madurella mycetomatis	Feet, other areas of body	Maduromycosis
	Sporothrix schenckii	Puncture wounds	Sporotrichosis
Systemic mycoses	*Blastomyces dermatitidis*	Lungs, skin	Blastomycosis
	Coccidioides immitis	Lungs, other parts of body	Coccidioidomycosis
	Cryptococcus neoformans	Lungs, skin, bones, viscera, central nervous system	Cryptococcosis
	Histoplasma capsulatum	Within phagocytes	Histoplasmosis
Opportunistic mycoses	*Aspergillus fumigatus, A. flavus*	Respiratory system	Aspergillosis
	Candida albicans	Skin or mucous membranes	Candidiasis
	Pneumocystis jiroveci	Lungs, sometimes brain	*Pneumocystis* pneumonia

Table 40.2

Examples of Some Human Fungal Diseases Recognized Since 1973

Year	Fungus	Disease
	Molds	
1974	*Phialophora parasitica*	Phaeohyphomycosis
1992	*Penicillium marneffei*	Disseminated infection
	Yeasts	
1989	*Candida lusitaniae*	Fungemia
1989	*Malassezia furfur*	Fungemia
1990	*Rhodotorula rubra*	Fungemia
1991	*Candida ciferrii*	Fungemia
1993	*Hansenula anomala*	Fungemia
1993	*Trichosporon beigelii*	Fungemia

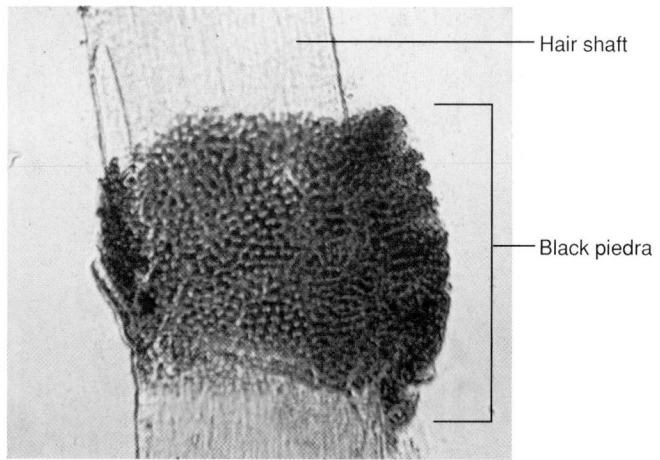

Figure 40.1 Superficial Mycosis: Black Piedra. Hair shaft infected with *Piedraia hortae;* light micrograph (×200).

to the outer surface of hair and skin and hence are called superficial. Infections of the hair shaft are collectively called **piedras** (Spanish for stone because they are associated with the hard nodules formed by mycelia [*see figure 25.3*] on the hair shaft). For example, **black piedra** is caused by *Piedraia hortae* and forms hard black nodules on the hairs of the scalp (**figure 40.1**). **White piedra** is caused by the yeast *Trichosporon beigelii* and forms light-colored nodules on the beard and mustache. Some superficial mycoses are called **tineas** [Latin for grub, larva, worm], the specific type being designated by a modifying term. Tineas are superficial fungal infections involving the outer layers of the skin, nails, and hair. **Tinea versicolor** is caused by the yeast *Malassezia furfur* and forms brownish-red scales on the skin of the trunk, neck, face, and arms. Treatment involves removal of the skin scales with a cleansing agent and removal of the infected hairs. Good personal hygiene prevents these infections.

Cutaneous Mycoses

Cutaneous mycoses—also called **dermatomycoses, ringworms,** or tineas—occur worldwide and represent the most common fungal diseases in humans. Three genera of cutaneous fungi, or **dermatophytes,** are involved in these mycoses: *Epidermophyton, Microsporum,* and *Trichophyton.* Diagnosis is by microscopic examination of biopsied areas of the skin cleared with 10% potassium hydroxide and by culture on Sabouraud dextrose agar. Treatment is with topical ointments such as miconazole (Monistat-Derm), tolnaftate (Tinactin), or clotrimazole (Lotrimin) for 2 to 4 weeks. Griseofulvin (Grifulvin V) and itraconazole (Sporanox) are the only oral antifungal agents currently approved by the FDA for treating dermatophytoses. The mode of action of ketoconazole, miconazole, tolnaftate, and griseofulvin (p. 795)

Tinea barbae [Latin *barba,* the beard] is an infection of the beard hair caused by *Trichophyton mentagrophytes* or *T. verrucosum.* It is predominantly a disease of men who live in rural areas and acquire the fungus from infected animals.

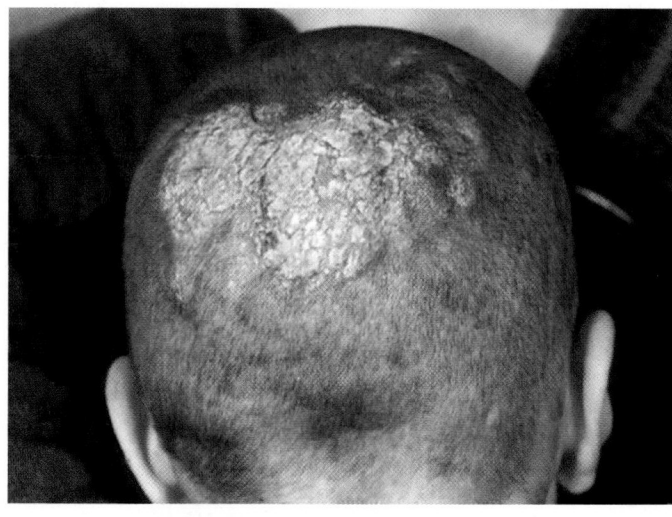

(a)

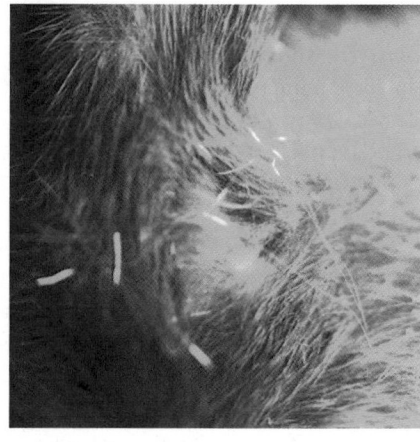

(b)

Figure 40.2 Cutaneous Mycosis: Tinea Capitis. (**a**) Ringworm of the head caused by *Microsporum audouinii.* (**b**) Close-up using a Wood's light (a UV lamp).

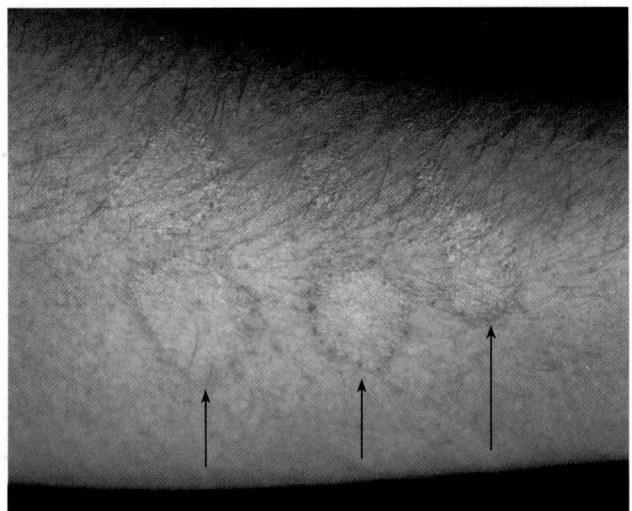

Figure 40.3 Cutaneous Mycosis: Tinea Corporis. Ringworm of the body—in this case, the forearm—caused by *Trichophyton mentagrophytes.* Notice the circular patches (arrows).

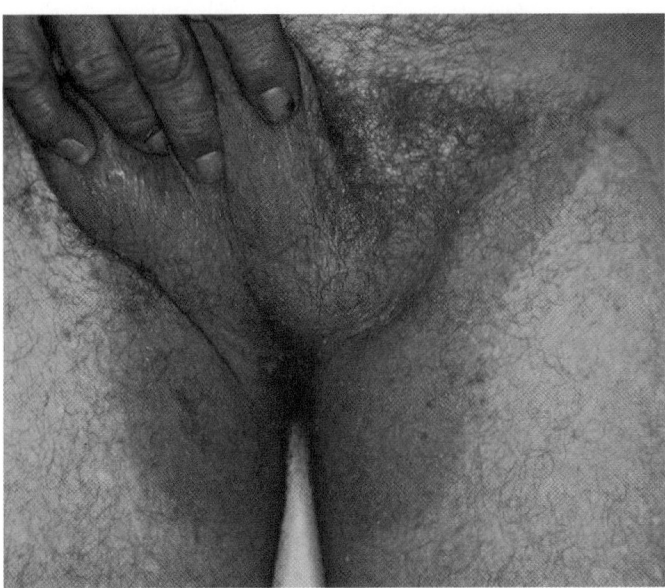

Figure 40.4 Cutaneous Mycosis: Tinea Cruris. Ringworm of the groin caused by *Epidermophyton floccosum.*

Tinea capitis [Latin *capita,* the head] is an infection of the scalp hair (**figure 40.2a**). It is characterized by loss of hair, inflammation, and scaling. Tinea capitis is primarily a childhood disease caused by *Trichophyton* or *Microsporum* species. Person-to-person transmission of the fungus occurs frequently when poor hygiene and overcrowded conditions exist. The fungus also occurs in domestic animals, from whom it can be transmitted to humans. A Wood's light (a UV light) can help with the diagnosis of tinea capitis because fungus-infected hair fluoresces when illuminated by UV radiation (figure 40.2b).

Tinea corporis [Latin *corpus,* the body] is a dermatophytic infection that may occur on any part of the skin (**figure 40.3**). The disease is characterized by circular, red, well-demarcated, scaly, vesiculopustular lesions accompanied by itching. Tinea corporis is caused by *Trichophyton rubrum, T. mentagrophytes,* or *Microsporum canis.* Transmission of the disease agent is by direct contact with infected animals and humans or by indirect contact through fomites.

Tinea cruris [Latin *crura,* the leg] is a dermatophytic infection of the groin (**figure 40.4**). The pathogenesis and clinical manifestations are similar to those of tinea corporis. The responsible fungi are *Epidermophyton floccosum, T. mentagrophytes,* or *T. rubrum.* Factors predisposing one to recurrent disease are moisture, occlusion, and skin trauma. Wet bathing suits, athletic supporters (**jock itch**), tight-fitting slacks, panty hose, and obesity are frequently contributing factors.

Tinea pedis [Latin *pes,* the foot], also known as **athlete's foot,** and **tinea manuum** [Latin *mannus,* the hand] are dermatophytic infections of the feet (**figure 40.5**) and hands, respectively. Clinical symptoms vary from a fine scale to a vesiculopustular eruption. Itching is frequently present. Warmth, humidity, trauma, and occlusion increase susceptibility to infec-

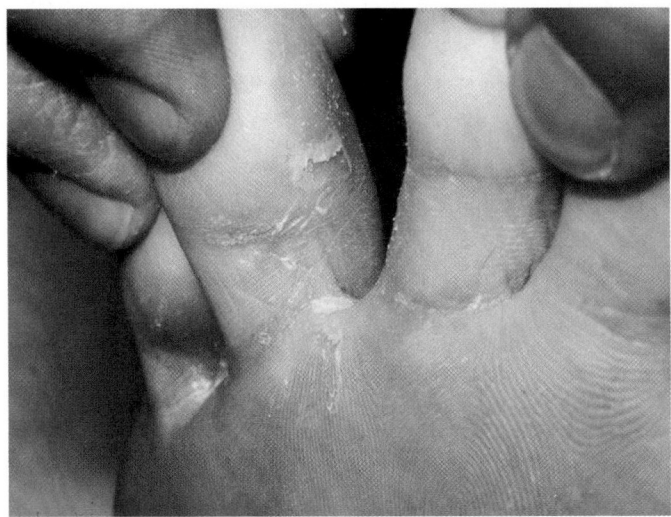

Figure 40.5 Cutaneous Mycosis: Tinea Pedis. Ringworm of the foot caused by *Trichophyton rubrum, T. mentagrophytes,* or *Epidermophyton floccosum.*

tion. Most infections are caused by *T. rubrum, T. mentagrophytes,* or *E. floccosum.* Tinea pedis and tinea manuum occur throughout the world, are most commonly found in adults, and increase in frequency with age.

Tinea unguium [Latin *unguis,* nail] is a dermatophytic infection of the nail bed (**figure 40.6**). In this disease the nail becomes discolored and then thickens. The nail plate rises and separates from the nail bed. *Trichophyton rubrum* or *T. mentagrophytes* are the causative fungi.

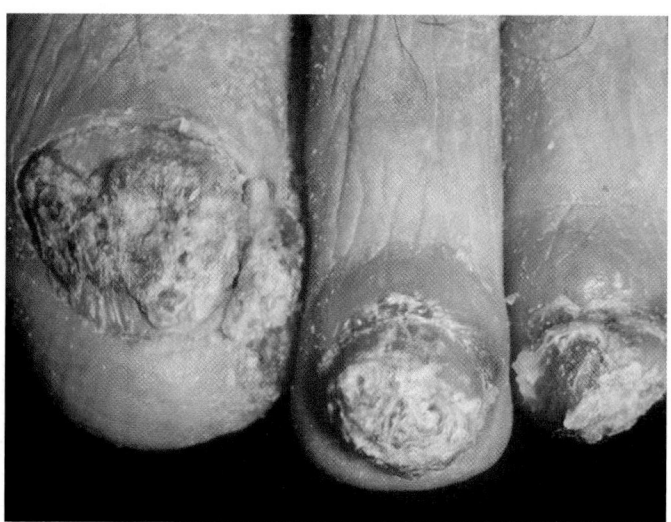

Figure 40.6   Cutaneous Mycosis: Tinea Unguium.   Ringworm of the nails caused by *Trichophyton rubrum.*

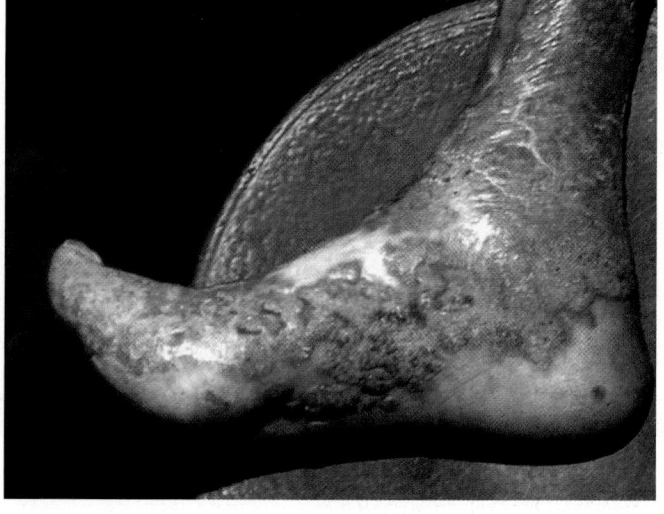

Figure 40.7   Subcutaneous Mycosis.   Chromoblastomycosis of the foot caused by *Fonsecaea pedrosoi.*

Subcutaneous Mycoses

The dermatophytes that cause subcutaneous mycoses are normal saprophytic inhabitants of soil and decaying vegetation. Because they are unable to penetrate the skin, they must be introduced into the subcutaneous tissue by a puncture wound that has been contaminated with soil containing the fungi. Most infections involve barefooted agricultural workers. Once in the subcutaneous tissue, the disease develops slowly—often over a period of years. During this time the fungi produce a nodule that eventually ulcerates and the organisms spread along lymphatic channels producing more subcutaneous nodules. At times such nodules drain to the skin surface. The administration of oral 5-fluorocytosine, iodides, amphotericin B, and surgical excision are the usual treatments. Diagnosis is accomplished by culture of the infected tissue.   The mode of action of amphotericin B (p. 796)

One type of subcutaneous mycosis is **chromoblastomycosis.** The nodules are pigmented a dark brown. This disease is caused by the black molds *Phialophora verrucosa* or *Fonsecaea pedrosoi.* These fungi exist worldwide, especially in tropical and subtropical regions. Most infections involve the legs and feet (**figure 40.7**).

Another subcutaneous mycosis is **maduromycosis,** caused by *Madurella mycetomatis,* which is distributed worldwide and is especially prevalent in the tropics. Because the fungus destroys subcutaneous tissue and produces serious deformities, the resulting infection is often called a **eumycotic mycetoma** or fungal tumor (**figure 40.8**). One form of mycetoma, known as Madura foot, occurs through skin abrasions acquired while walking barefoot on contaminated soil.

Sporotrichosis is the subcutaneous mycosis caused by the dimorphic (*see table 25.2*) fungus *Sporothrix schenckii.* The disease occurs throughout the world and is the most common subcutaneous mycotic disease in the United States. The fungus can be

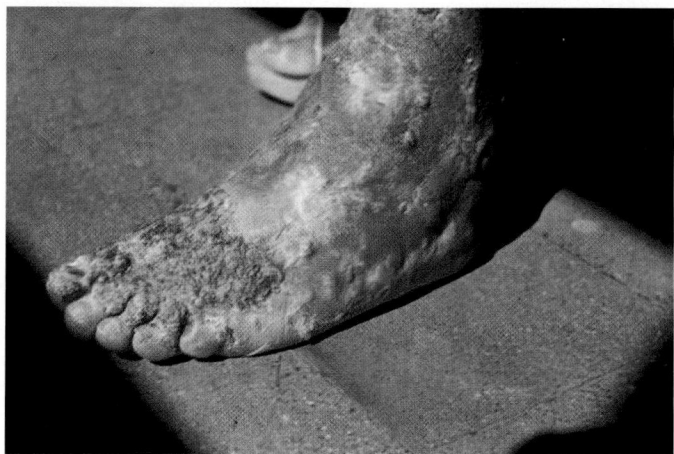

Figure 40.8   Subcutaneous Mycosis.   Eumycotic mycetoma of the foot caused by *Madurella mycetomatis.*

found in the soil, on living plants, such as barberry shrubs and roses, or in plant debris, such as sphagnum moss and pine-bark mulch. Infection occurs by a puncture wound from a thorn or splinter contaminated with the fungus. The disease is an occupational hazard for florists, gardeners, and forestry workers. After an incubation period of 1 to 12 weeks, a small red papule arises and begins to ulcerate (**figure 40.9**). New lesions appear along lymph channels and can remain localized or spread throughout the body, producing **extracutaneous sporotrichosis.**

Systemic Mycoses

Except for *Cryptococcus neoformans,* which has only a yeast form, the fungi that cause the systemic or deep mycoses are dimorphic—that is, they exhibit a parasitic yeastlike phase (Y) and

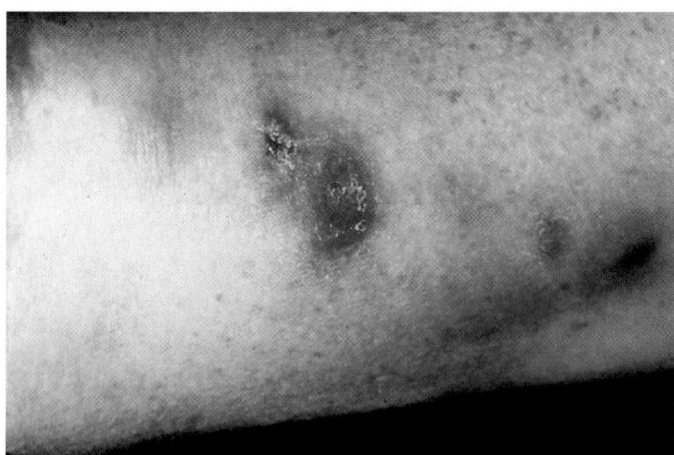

Figure 40.9 Subcutaneous Mycosis. Sporotrichosis of the arm caused by *Sporothrix schenckii*.

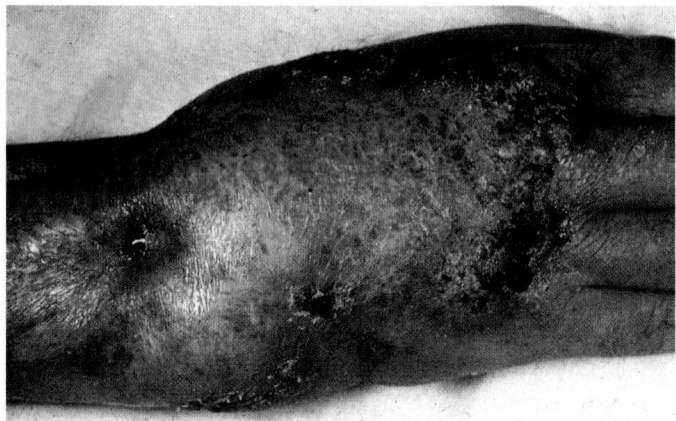

Figure 40.10 Systemic Mycosis. Blastomycosis of the forearm caused by *Blastomyces dermatitidis*.

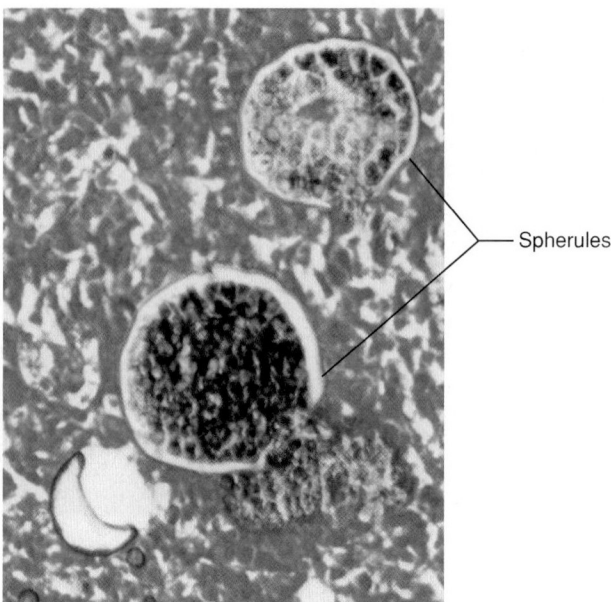

Figure 40.11 Systemic Mycosis: Coccidioidomycosis.
Coccidioides immitis mature spherules filled with endospores within a tissue section; light micrograph (×400).

a saprophytic mold or mycelial phase (M). (*See the YM shift, section 25.3.*) Most systemic mycoses are acquired by the inhalation of spores from soil in which the mold-phase of the fungus resides. If a susceptible person inhales enough spores, an infection begins as a lung lesion, becomes chronic, and spreads through the bloodstream to other organs (the target organ varies with the species).

Blastomycosis is the systemic mycosis caused by *Blastomyces dermatitidis,* a fungus that grows as a budding yeast in humans but as a mold on culture media and in the environment. It is found predominately in the soil of the Mississippi and Ohio River basins. The disease occurs in three clinical forms: cutaneous, pulmonary, and disseminated. The initial infection begins when blastospores (*see figure 25.6f*) are inhaled into the lungs. The fungus can then spread rapidly, especially to the skin, where cutaneous ulcers and abscess formation occur (**figure 40.10**). *B. dermatitidis* can be isolated from pus and biopsy sections. Diagnosis requires the demonstration of thick-walled, yeastlike cells 8

to 15 μm in diameter. Complement-fixation, immunodiffusion, and skin (blastomycin) tests are also useful. Amphotericin B (Fungizone), itraconazole (Sporanox), or ketoconazole (Nizoral) are the drugs of choice for treatment. Surgery may be necessary for the drainage of large abscesses. Approximately 30 to 60 deaths are reported each year in the United States. There are no preventive or control measures.

Coccidioidomycosis, also known as valley fever, San Joaquin fever, or desert rheumatism because of the geographical distribution of the fungus, is caused by *Coccidioides immitis. C. immitis* exists in the dry, highly alkaline soils of North, Central, and South America. It has been estimated that in the United States about 100,000 people are infected annually with 50 to 100 deaths. Endemic areas have been defined by massive skin testing with the antigen coccidioidin. In the soil and on culture media, this fungus grows as a mold that forms arthroconidia (*see figure 25.6b*) at the tips of hyphae. Arthroconidia are so abundant in these endemic areas that by simply moving through such an area, one can acquire the disease by inhaling them. Wind turbulence and even construction of outdoor structures have been associated with increased exposure and infection. In humans the fungus grows as a yeast-forming, thick-walled spherule filled with endospores (**figure 40.11**). Most cases of coccidioidomycosis are asymptomatic or indistinguishable from ordinary upper respiratory infections. Almost all cases resolve themselves in a few weeks, and a lasting immunity results. A few infections result in a progressive chronic pulmonary disease. The fungus also can spread throughout the body, involving almost any organ or site. Diagnosis is accomplished by aspiration and identification of the large spherules (approximately 80 μm in diameter) in pus, sputum, and aspirates.

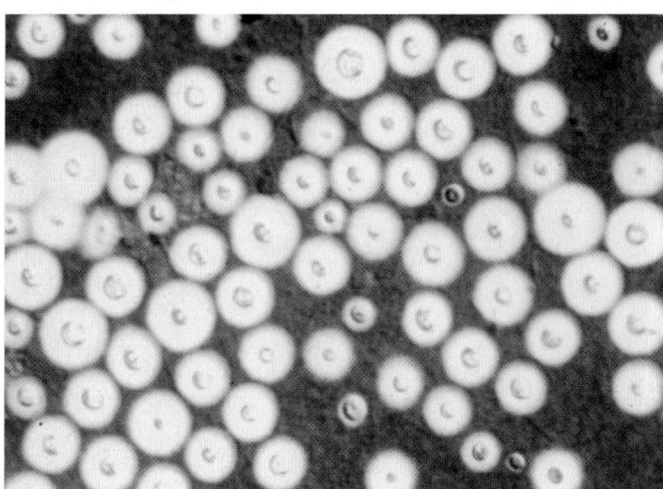

Figure 40.12   Systemic Mycosis: Cryptococcosis.   India ink preparation showing *Cryptococcus neoformans.* Although these microorganisms are not budding, they can be differentiated from artifacts by their doubly refractile cell walls, distinctly outlined capsules surrounding all cells, and refractile inclusions in the cytoplasm; light micrograph ($\times 150$).

Culturing clinical samples in the presence of penicillin and streptomycin on Sabouraud agar also is diagnostic. Newer methods of rapid confirmation include the testing of supernatants of liquid media cultures for antigens, serology, and skin testing. Miconazole (Lotrimin), itraconazole, ketoconazole, and amphotericin B are the drugs of choice for treatment. Prevention involves reducing exposure to dust (soil) in endemic areas.

Cryptococcosis is a systemic mycosis caused by *Cryptococcus neoformans.* This fungus always grows as a large budding yeast. In the environment *C. neoformans* is a saprophyte with a worldwide distribution. Aged, dried pigeon droppings are an apparent source of infection. Cryptococcosis is found in approximately 15% of AIDS patients. The fungus enters the body by the respiratory tract, causing a minor pulmonary infection that is usually transitory. Some pulmonary infections spread to the skin, bones, viscera, and central nervous system. Once the nervous system is involved, cryptococcal meningitis usually results. Diagnosis is accomplished by detection of the thick-walled spherical yeast cells (**figure 40.12**) in pus, sputum, or exudate smears using India ink to define the organism. The fungus can be easily cultured on Sabouraud dextrose agar. Identification of the fungus in body fluids is made by immunologic procedures. Treatment includes amphotericin B or itraconazole. There are no preventive or control measures.

Histoplasmosis is caused by *Histoplasma capsulatum* var. *capsulatum,* a facultative parasitic fungus that grows intracellularly. It appears as a small budding yeast in humans and on culture media at 37°C. At 25°C it grows as a mold, producing small microconidia (1 to 5 μm in diameter) that are borne singly at the tips of short conidiophores. Large macroconidia (8 to 16 μm in diameter) are also formed on conidiophores (**figure 40.13***a*). In humans the yeastlike

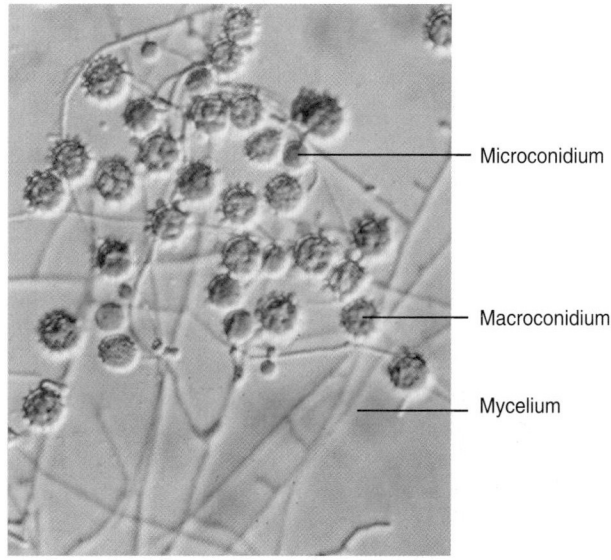

(a)

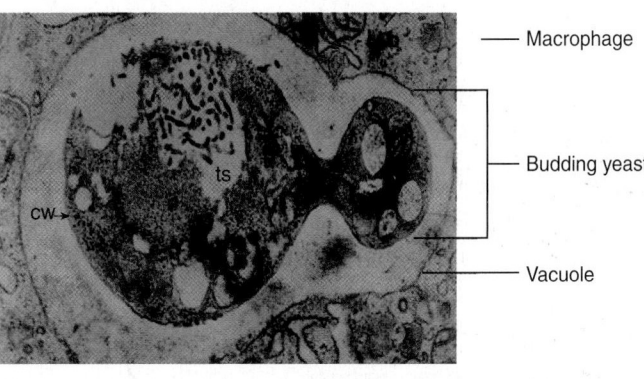

(b)

Figure 40.13   Morphology of *Histoplasma capsulatum* var. *capsulatum.*   (a) Mycelia, microconidia, and chlamydospores as found in the soil. These are the infectious particles; light micrograph ($\times 125$). **(b)** Yeastlike cells in a macrophage. Budding *H. capsulatum* within a vacuole. Tubular structures, ts, are observed beneath the cell wall, cw; electron micrograph ($\times 23{,}000$).

form grows within phagocytic cells (figure 40.13*b*). *H. capsulatum* var. *capsulatum* is found as the mycelial form in soils throughout the world and is localized in areas that have been contaminated with bird or bat excrement. The microconidia are easily spread by air currents. Within the United States, histoplasmosis is endemic within the Mississippi, Kentucky, Tennessee, Ohio, and Rio Grande River basins. More than 75% of the people who reside in parts of these areas have antibodies against the fungus. It has been estimated that in endemic areas of the United States, about 500,000 individuals are infected annually: 50,000 to 200,000 become ill; 3,000 require hospitalization; and about 50 die. The total number of infected individuals may be over 40 million in the United States alone. Histoplasmosis is a common disease among spelunkers (people who explore caves) and bat guano miners.

Humans acquire histoplasmosis from airborne microconidia that are produced under favorable environmental conditions. When inhaled into the alveolar spaces, the microconidia germinate and form small budding yeasts. The yeasts are found within cells of the monocyte macrophage system (figure 40.13*b*). Microconidia are most prevalent where bird droppings, or guanos—especially from starlings, crows, blackbirds, cowbirds, sea gulls, turkeys, and chickens—have accumulated. It is noteworthy that the birds themselves are not infected because of their high body temperature; their droppings simply provide the nutrients for this fungus. Only bats and humans demonstrate the disease and harbor the fungus.

Because histoplasmosis is a disease of the monocyte-macrophage system, many organs of the body can be infected (*see figure 31.4*). More than 95% of "histo" cases have either no symptoms or mild symptoms such as coughing, fever, and joint pain. Lesions may appear in the lungs and show calcification; most infections resolve on their own. Only rarely does the disease disseminate.

Laboratory diagnosis is accomplished by complement-fixation tests and isolation of the fungus from tissue specimens. Most individuals with this disease exhibit a hypersensitive state that can be demonstrated by the histoplasmin skin test. Currently the most effective treatment is with amphotericin B, ketoconazole, or itraconazole. Prevention and control involve wearing protective clothing and masks before entering or working in infested habitats. Soil decontamination with 3 to 5% formalin is effective where economically and physically feasible.

Opportunistic Mycoses

An **opportunistic microorganism** is generally harmless in its normal environment but becomes pathogenic in a compromised host. A **compromised host** is seriously debilitated and has a lowered resistance to infection. There are many causes of this condition, among them the following: malnutrition; alcoholism; cancer, diabetes, leukemia, or another infectious disease; trauma from surgery or injury; an altered microbiota from the prolonged use of antibiotics (e.g., in vaginal candidiasis); and immunosuppression by drugs, viruses (HIV), hormones, genetic deficiencies, chemotherapy of patients, and old age. The most important opportunistic mycoses include systemic aspergillosis, candidiasis, and *Pneumocystis* pneumonia.

Of all the fungi that cause disease in compromised hosts, none are as widely distributed as the *Aspergillus* species. *Aspergillus* is omnipresent in nature, being found wherever organic debris occurs. *Aspergillus fumigatus* is the usual cause of **aspergillosis**. *A. flavus* is the second most important species, particularly in invasive disease of immunosuppressed patients.

The major portal of entry for *Aspergillus* is the respiratory tract. Inhalation of conidiospores (*see figure 25.6*e) can lead to several types of pulmonary aspergillosis. One type is allergic aspergillosis. Infected individuals may develop an immediate allergic response and suffer typical asthmatic attacks when exposed to fungal antigens on the conidiospores. In bronchopulmonary as-

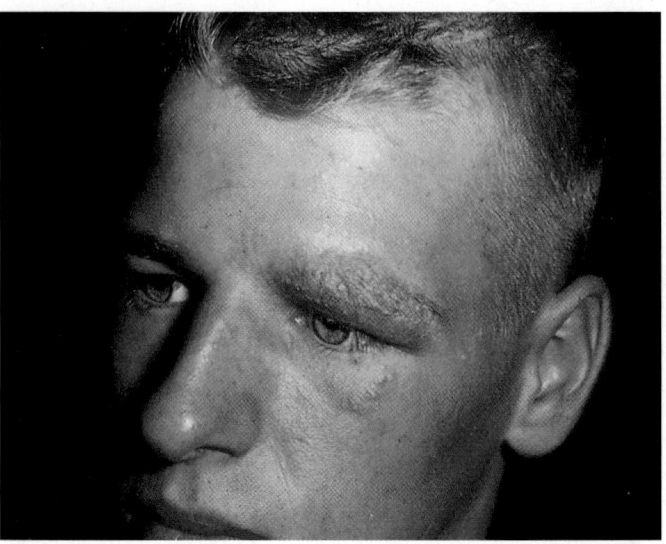

Figure 40.14   An Opportunistic Mycosis.   Aspergillosis of the eye caused by *Aspergillus fumigatus.*

pergillosis the major clinical manifestation of the allergic response is a bronchitis resulting from both type I and type III hypersensitivities (*see figures 33.2* and *33.6*). Although tissue invasion seldom occurs in bronchopulmonary aspergillosis, *Aspergillus* often can be cultured from the sputum. A most common manifestation of pulmonary involvement is the occurrence of colonizing aspergillosis, in which *Aspergillus* forms colonies within the lungs that develop into "fungus balls" called aspergillomas. These consist of a tangled mass of mycelia growing in a circumscribed area. From the pulmonary focus, the fungus may spread, producing disseminated aspergillosis in a variety of tissues and organs (**figure 40.14**). In patients whose resistance has been severely compromised, invasive aspergillosis may occur and fill the lung with fungal mycelia.

Laboratory diagnosis of aspergillosis depends on identification, either by direct examination of pathological specimens or by isolation and characterization of the fungus. Successful therapy depends on treatment of the underlying disease so that host resistance increases. Treatment is with itraconazole.

Candidiasis is the mycosis caused by the dimorphic fungus *Candida albicans* (**figure 40.15***a*). In contrast to the other pathogenic fungi, *C. albicans* is a member of the normal microbiota within the gastrointestinal tract, respiratory tract, vaginal area, and mouth (*see figure 31.2*). In healthy individuals *C. albicans* does not produce disease. Growth is suppressed by other microbiota. However, if anything upsets the normal microbiota, *Candida* may multiply rapidly and produce candidiasis. Recently *Candida* species have become important nosocomial pathogens. In some hospitals they may represent almost 10% of nosocomial bloodstream infections. Because *Candida* can be transmitted sexually, it is also listed by the CDC as a sexually transmitted disease (*see table 39.4*).

No other mycotic pathogen produces as diverse a spectrum of disease in humans as does *C. albicans* (**Disease 40.1**). Most in-

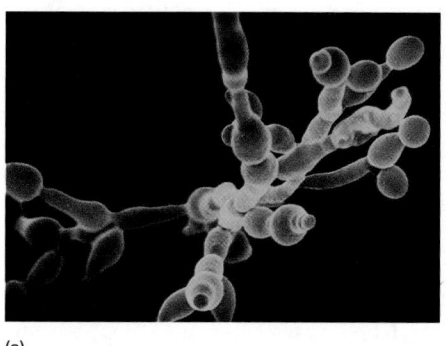
(a)

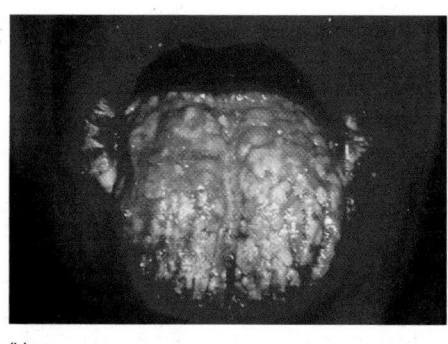
(b)

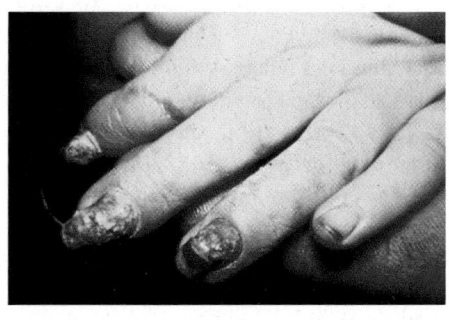
(c)

Figure 40.15   Opportunistic Mycoses Caused by *Candida albicans*.   (**a**) Scanning electron micrograph of the yeast form (×10,000). Notice that some of the cells are reproducing by budding. (**b**) Thrush, or oral candidiasis, is characterized by the formation of white patches on the mucous membranes of the tongue and elsewhere in the oropharyngeal area. These patches form a pseudomembrane composed of spherical yeast cells, leukocytes, and cellular debris. (**c**) Paronychia and onychomycosis of the hands.

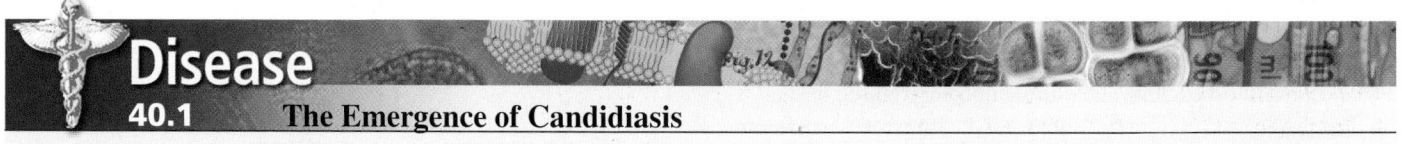

Disease
40.1   The Emergence of Candidiasis

Written descriptions of oral lesions that were probably thrush date back to the early 1800s. In 1839 Bernard Langenbeck in Germany described the organism he found in oral lesions of a patient as "Typhus-Leichen" (typhus bodies). By 1841 Emil Berg established the fungal etiology of thrush by infecting healthy babies with what he called "aphthous membrane material." In 1843 Charles Robin gave the organism its first name: *Oidium albicans.* Since then, more than 100 synonyms have been used for this fungus; of them all, *Candida albicans,* the name Roth Berkhout proposed in 1923, has persisted.

In 1861 Albert Zenker described the first well-documented case of systemic candidiasis. Historically, the most interesting period for candidiasis research coincided with the introduction of antibiotics. Since then there have been documented cases of this fungus involving all tissues and organs of the body, as well as an increase in the overall incidence of candidiasis. Some *Candida*-associated infections and diseases include arthritis, endophthalmitis, meningitis, myocarditis, myositis, and peritonitis. Besides the widespread use of antibiotics, other therapeutic and surgical procedures such as organ transplants and prostheses have been important in the expanding worldwide incidence of candidiasis.

fections involve the skin or mucous membranes. This occurs because *C. albicans* is a strict aerobe and finds such surfaces very suitable for growth. Cutaneous involvement usually occurs when the skin becomes overtly moist or damaged.

Oral candidiasis, or **thrush** (figure 40.15*b*), is a fairly common disease in newborns. It is seen as many small white flecks that cover the tongue and mouth. At birth, newborns do not have a normal microbiota in the oropharyngeal area. If the mother's vaginal area is heavily infected with *C. albicans,* the upper respiratory tract of the newborn becomes colonized during passage through the birth canal. Thrush occurs because growth of *C. albicans* cannot be inhibited by the other microbiota. Once the newborn has developed its own normal oropharyngeal microbiota, thrush becomes uncommon. **Paronychia** and **onychomycosis** are associated with *Candida* infections of the subcutaneous tissues of the digits and nails, respectively (figure 40.15*c*). These infections usually result from continued immersion of the appendages in water.

Intertriginous candidiasis involves those areas of the body, usually opposed skin surfaces, that are warm and moist: axillae,

groin, skin folds. **Napkin (diaper) candidiasis** is typically found in infants whose diapers are not changed frequently and therefore are not kept dry. **Candidal vaginitis** can result from diabetes, antibiotic therapy, oral contraceptives, pregnancy, or any other factor that compromises the female host. Normally the omnipresent lactobacilli (Döderlein's bacilli) can control *Candida* in this area by the low pH they create. However, if their numbers are decreased by any of the aforementioned factors, *Candida* may proliferate, causing a curdlike, yellow-white discharge from the vaginal area. *Candida* can be transmitted to males during intercourse and lead to **balanitis;** thus it also can be considered a sexually transmitted disease. Balanitis is a *Candida* infection of the male glans penis and occurs primarily in uncircumcised males. The disease begins as vesicles on the penis that develop into patches and are accompanied by severe itching and burning.

Diagnosis of candidiasis is difficult because (1) this fungus is a frequent secondary invader in diseased hosts, (2) a mixed microbiota is most often found in the diseased tissue, and (3) no completely specific immunologic procedures for the identification of

Candida currently exist. There is no satisfactory treatment for candidiasis. Cutaneous lesions can be treated with topical agents such as sodium caprylate, sodium propionate, gentian violet, nystatin, miconazole, and trichomycin. Ketoconazole, amphotericin B, fluconazole, itraconazole, and flucytosine also can be used for systemic candidiasis.

Pneumocystis is a protist found in the lungs of a wide variety of mammals. The natural history of *Pneumocystis* is poorly understood, in large part because of a lack of a continuous in vitro culture system for its propagation. Although it was once considered a protozoan parasite, recent comparisons of rRNA and DNA sequences from several genes have shown that *Pneumocystis* is more closely related to fungi than protozoa.

The disease that this protist causes has been called *Pneumocystis* pneumonia or *Pneumocystis carinii* pneumonia (PCP). Recently its name was changed because in 1999 the microorganism that causes human PCP was renamed *Pneumocystis jiroveci* in honor of the Czech parasitologist, Otto Jirovec, who first described this pathogen in humans. In recent medical literature, the acronym PCP for the disease has been retained despite the loss of the old species name (*carinii*). PCP now stands for **Pneumocystis pneumonia.** PCP occurs almost exclusively in immunocompromised hosts.

Extensive use of immunosuppressive drugs and irradiation for the treatment of cancers and following organ transplants accounts for the formidable prevalence rates noted recently. This pneumonia also occurs in more than 80% of AIDS patients. Both the organism and the disease remain localized in the lungs—even in fatal cases. Within the lungs *Pneumocystis* causes the alveoli to fill with a frothy exudate.

Laboratory diagnosis of *Pneumocystis* pneumonia can be made definitively only by microscopically demonstrating the presence of the microorganisms in infected lung material or by a PCR analysis. Treatment is by means of oxygen therapy and either a combination of trimethoprim (Bactrim) and sulfamethoxazole (Septra), atovaquone (Mepron), or trimetrexate (Neutrexin). Prevention and control is through prophylaxis with drugs in susceptible persons.

1. How are human fungal diseases categorized?
2. What are two piedras that infect humans?
3. Briefly describe the major tineas that occur in humans.
4. Describe the three types of subcutaneous mycoses that affect humans.
5. Why is *Histoplasma capsulatum* found in bird feces but not within the birds themselves?
6. Why are some mycotic diseases of humans called opportunistic mycoses?
7. What parts of the human body can be affected by *Candida* infections?
8. When is *Pneumocystis* pneumonia likely to occur in humans?

 40.2   PROTOZOAN DISEASES

Protozoa have become adapted to practically every type of habitat on the face of the Earth, including the human body. Though fewer than 20 genera of protozoa cause disease in humans (**tables 40.3** and **40.4**), their impact is formidable. For example, there are over

Table 40.3
Examples of Some Medically Important Protozoa

Phylum	Group	Pathogen	Disease
Sarcomastigophora	Amoebae	*Entamoeba histolytica*	Amebiasis, amebic dysentery
		Acanthamoeba spp., *Naegleria fowleri*	Amebic meningoencephalitis
Apicomplexa	Coccidia	*Cryptosporidium parvum*	Cryptosporidiosis
		Cyclospora cayetanensis	Cyclosporidiosis
		Isospora belli	Isosporiasis
		Toxoplasma gondii	Toxoplasmosis
Ciliophora	Ciliates	*Balantidium coli*	Balantidiasis
Sarcomastigophora	Blood and tissue flagellates	*Leishmania tropica*	Cutaneous leishmaniasis
		L. braziliensis	Mucocutaneous leishmaniasis
		L. donovani	Kala-azar (visceral leishmaniasis)
		Trypanosoma cruzi	American trypanosomiasis
		T. brucei gambiense, T. brucei rhodesiense	African sleeping sickness
Sarcomastigophora	Digestive and genital organ flagellates	*Giardia lamblia*	Giardiasis
		Trichomonas vaginalis	Trichomoniasis
Apicomplexa	Sporozoa	*Plasmodium falciparum, P. malariae, P. ovale, P. vivax*	Malaria
Microspora	Microsporidia	*Encephalitozoon, Nosema, Vitta forma, Pleistophora, Enterocytozoon, Trachipleistophora, Microsporidium*	Microsporidiosis

Table 40.4

Examples of Some Recently Recognized Human Protozoan Diseases

Year	Protozoan	Disease
1976	*Cryptosporidium parvum*	Acute and chronic diarrhea, cryptosporidiosis
1985	*Enterocytozoon bieneusi*	Diarrhea, microsporidiosis
1986	*Cyclospora cayatanensis*	Persistent diarrhea
1991	*Encephalitozoon hellem*	Conjunctivitis, disseminated microsporidiosis
1991	*Babesia* spp.	Atypical babesiosis
1993	*Encephalitozoon cuniculi*	Disseminated microsporidiosis
1996	*Trachipleistophora hominis*	Disseminated microsporidiosis
1998	*Brachiola vesicularum*	Myositis

150 million cases of malaria in the world each year. In tropical Africa alone, malaria is responsible each year for the deaths of more than a million children under the age of 14. It is estimated that there are at least 8 million cases of trypanosomiasis, 12 million cases of leishmaniasis, and over 500 million cases of amebiasis yearly. There is an increasing problem with *Cryptosporidium* and *Cyclospora* contamination of food and water supplies. More of our population is elderly, and a growing number of persons are immunosuppressed due to the human immunodeficiency virus, organ transplantation, or cancer chemotherapy. These populations are at increased risk for protozoan infections. The remainder of this chapter discusses some of these important protozoan diseases of humans.

Amebiasis

It is now accepted that two species of *Entamoeba* infect humans: the nonpathogenic *E. dispar* and the pathogenic *E. histolytica*. *E. histolytica* is responsible for **amebiasis (amebic dysentery).** This very common parasite is endemic in warm climates where adequate sanitation and effective personal hygiene is lacking. Within the United States about 3,000 to 5,000 cases are reported annually. However, it is a major cause of parasitic death worldwide; about 500 million people are infected and as many as 100,000 die of amebiasis each year.

Infection occurs by ingestion of mature cysts. After excystation in the lower region of the small intestine, the metacyst divides rapidly to produce eight small trophozoites (**figure 40.16**). These trophozoites move to the large intestine where they can invade the host tissue, live as commensals in the lumen of the intestine, or undergo encystation.

If the infective trophozoites invade the intestinal tissues, they multiply rapidly and spread laterally, while feeding on erythrocytes, bacteria, and yeasts. The invading trophozoites destroy the epithelial lining of the large intestine by producing cysteine proteinases. Cysteine proteinases possibly are a virulence factor of *E. histolytica* and may play a role in intestinal invasion by degrading extracellular matrix and circumventing the host immune response through cleavage of secretory immunoglobulin A (sIgA), IgG, and

complement factors. Cysteine proteinases are encoded by at least seven genes, several of which are found in *E. histolytica* but not *E. dispar*. Lesions (ulcers) are characterized by minute points of entry into the mucosa and extensive enlargement of the lesion after penetration into the submucosa. *E. histolytica* also may invade and produce lesions in extraintestinal foci, especially the liver, to cause hepatic amebiasis. However, all extraintestinal amebic lesions are secondary to the ones established in the large intestine.

The symptoms of amebiasis are highly variable, ranging from an asymptomatic infection to fulminating dysentery, exhaustive diarrhea accompanied by blood and mucus, appendicitis, and abscesses in the liver, lungs, or brain.

Laboratory diagnosis of amebiasis is based on finding trophozoites in fresh warm stools and cysts in ordinary stools. Serological testing also should be done. The therapy for amebiasis is complex and depends on the location of the infection within the host and the host's condition. Asymptomatic carriers that are passing cysts should always be treated with iodoquinol or paromomycin because they represent the most important reservoir of the parasite in the population. In symptomatic intestinal amebiasis, metronidazole (Flagyl) or iodoquinol (Yodoxin) are the drugs of choice. Prevention and control of amebiasis is achieved by avoiding water or food that might be contaminated with human feces in endemic areas. Viable cysts in water can be destroyed by hyperchlorination or iodination.

Cryptosporidiosis

The first case of human **cryptosporidiosis** was reported in 1976 (table 40.4). The protozoan responsible was identified as *Cryptosporidium parvum*, a parasite classified as an emerging pathogen by the CDC. In 1993 *C. parvum* contaminated the Milwaukee, Wisconsin, water supply and caused severe diarrheal disease in about 400,000 individuals, the largest recognized outbreak of waterborne illness in U.S. history. *Cryptosporidium* ("hidden spore cysts") is found in about 90% of sewage samples, in 75% of river waters, and in 28% of drinking waters.

C. parvum is a common coccidial apicomplexan parasite found in the intestine of many birds and mammals. When these animals defecate, oocysts are shed into the environment. If a human ingests food or water that is contaminated with the oocysts, excystation occurs within the small intestine and sporozoites enter epithelial cells and develop into merozoites. Some of the merozoites subsequently undergo sexual reproduction to produce zygotes, and the zygotes differentiate into thick-walled oocysts. Oocyst release into the environment begins the life cycle again. A major problem for public health arises from the fact that the oocysts are only 4 to 6 μm in diameter, much too small to be easily removed by the sand filters used to produce drinking water. *Cryptosporidium* also is extremely resistant to disinfectants such as chlorine. The problem is made worse by the low infectious dose, around 10 to 100 oocysts, and the fact that the oocysts may remain viable for 2 to 6 months in a moist environment. Water purification (pp. 630–33)

The incubation period for cryptosporidiosis ranges from 5 to 28 days. Diarrhea, which characteristically may be cholera-like, is the most common symptom. Other symptoms include abdominal

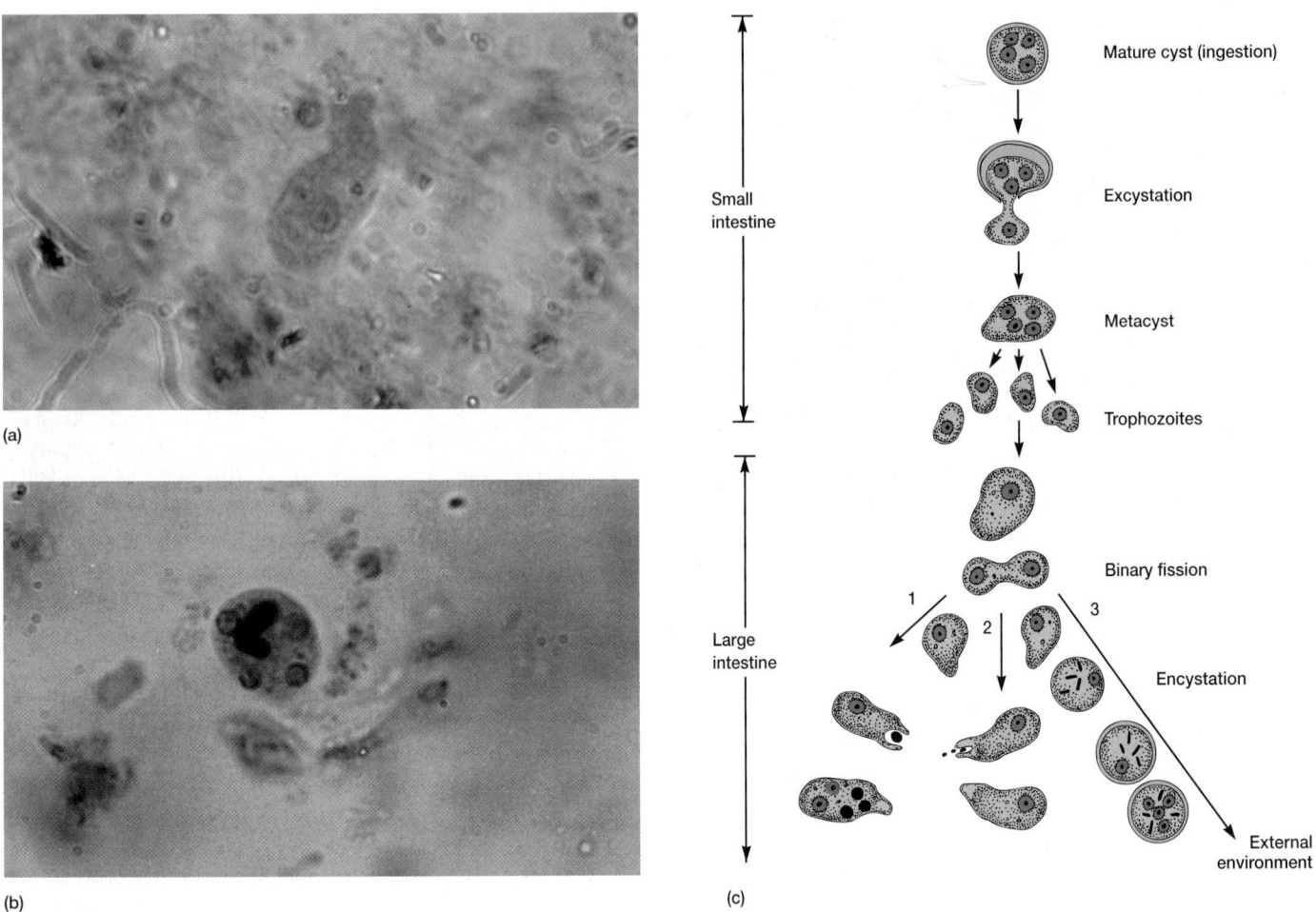

(a)

(b)

(c)

Small intestine

Large intestine

Mature cyst (ingestion)

Excystation

Metacyst

Trophozoites

Binary fission

Encystation

External environment

Figure 40.16 Amebiasis Caused by *Entamoeba histolytica.* (a) Light micrographs of a trophozoite (×1,000) and (b) a cyst (×1,000). (c) Life cycle. Infection occurs by the ingestion of a mature cyst of the parasite. Excystation occurs in the lower region of the small intestine and the metacyst rapidly divides to give rise to eight small trophozoites (only four are shown). These enter the large intestine, undergo binary fission, and may (1) invade the host tissues, (2) live in the lumen of the large intestine without invasion, or (3) undergo encystation and pass out of the host in the feces.

pain, nausea, fever, and fatigue. The pathogen is routinely diagnosed by fecal concentration and acid-fast stained smears. No chemotherapy is available and patients are simply rehydrated. Although the disease usually is self-limiting in healthy individuals, patients that have late-stage AIDS or are immunocompromised in other ways may develop prolonged, severe, and life-threatening diarrhea.

Freshwater Amoeba Diseases

Free-living amoebae of the genera *Naegleria* and *Acanthamoeba* are facultative parasites responsible for causing **primary amebic meningoencephalitis** (*see table 39.2*) in humans. They are among the most common protozoa found in freshwater and moist soil. In addition, several *Acanthamoeba* spp. are known to infect the eye, causing a chronically progressive ulcerative *Acanthamoeba* **keratitis,** inflammation of the cornea, which

may result in blindness. Wearers of soft contact lenses may be predisposed to this infection and should take care to prevent contamination of their lens cleaning and soaking solutions. Diagnosis of these infections is by demonstration of the amoebae in clinical specimens. Most freshwater amoebae are resistant to commonly used antibiotic agents. These amoebae are reported in fewer than 100 human disease cases annually in the United States although the incidence (especially of *Acanthamoeba* keratitis) is likely higher.

Giardiasis

Giardia lamblia [syn., *G. duodenalis, G. intestinalis*] is a flagellated protozoan (**figure 40.17a**) that causes the very common intestinal disease **giardiasis.** (It was discovered by van Leeuwenhoek in 1681 when he examined his own stools.) *G. lamblia* is worldwide in distribution, and it affects children more seriously than it does adults.

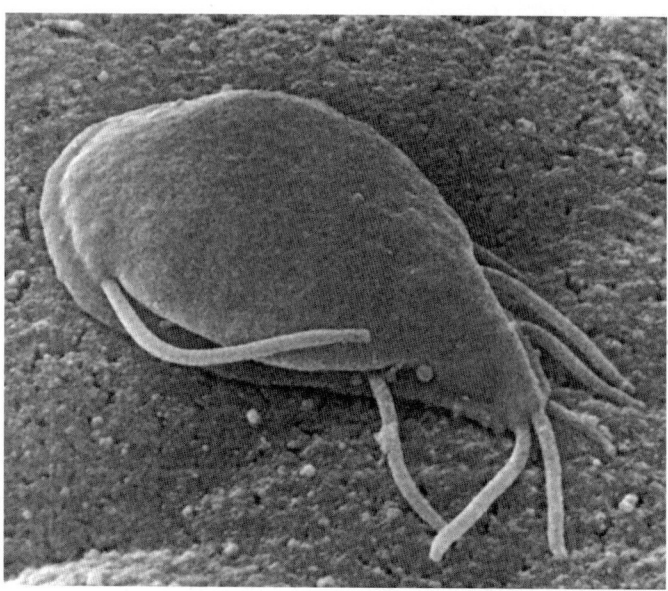

(a)

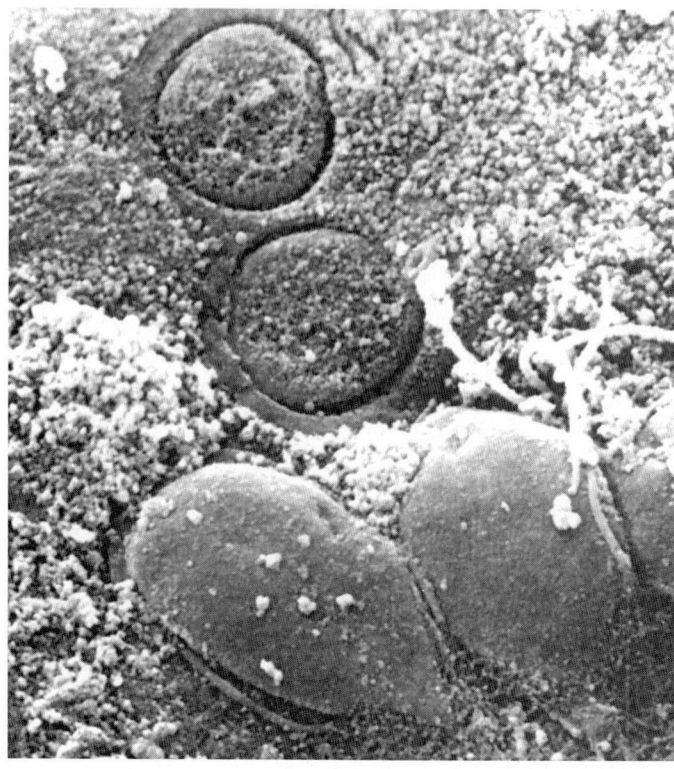

(b)

Figure 40.17 Giardiasis. (**a**) *Giardia lamblia* adhering to the epithelium by its sucking disk; scanning electron micrograph. (**b**) Upon detachment from the epithelium, the protozoa often leave clear impressions on the microvillus surface (upper circles); scanning electron micrograph.

In the United States this protozoan is the most common cause of epidemic waterborne diarrheal disease (about 30,000 cases yearly). Approximately 7% of the population are healthy carriers and shed cysts in their feces. *G. lamblia* is endemic in child day-care centers in the United States, with estimates of 5 to 15% of diapered children being infected.

Transmission is most frequent by cyst-contaminated water supplies. Epidemic outbreaks have been recorded in wilderness areas, suggesting that humans may be infected from "clean water" with *Giardia* harbored by rodents, deer, cattle, or household pets. This implies that human infections also can be a zoonosis (*see table 37.1*). As many as 200 million humans may be infected worldwide.

Following ingestion, the cysts undergo excystation in the duodenum, forming trophozoites. The trophozoites inhabit the upper portions of the small intestine, where they attach to the intestinal mucosa by means of their sucking disks (figure 40.17). The ability of the trophozoites to adhere to the intestinal epithelium accounts for the fact that they are rarely found in stools. It is thought that the trophozoites feed on mucous secretions and reproduce to form such a large population that they interfere with nutrient absorption by the intestinal epithelium.

Giardiasis varies in severity, and asymptomatic carriers are common. The disease can be acute or chronic. Acute giardiasis is characterized by severe diarrhea, epigastric pain, cramps, voluminous flatulence ("passing gas"), and anorexia. Chronic giardiasis is characterized by intermittent diarrhea, with periodic appearance and remission of symptoms.

Laboratory diagnosis is based on the identification of trophozoites—only in the severest of diarrhea—or cysts in stools. A commercial ELISA test is also available for the detection of *G. lamblia* antigen in stool specimens. Quinacrine hydrochloride (Atabrine) and metronidazole (Flagyl) are the drugs of choice for adults, and furazolidone is used for children because it is available in a pleasant-tasting liquid suspension. Prevention and control involves proper treatment of community water supplies, especially the use of slow sand filtration (*see section 29.5*) because the cysts are highly resistant to chlorine treatment.

Malaria

The most important human parasite among the sporozoa is *Plasmodium*, the causative agent of **malaria (Disease 40.2).** It has been estimated that more than 300 million people are infected each year, and over 1 million die annually of malaria in Africa alone. About 1,000 cases are reported each year in the United States, divided between returning U.S. travelers and non-U.S. citizens.

Human malaria is caused by four species of *Plasmodium: P. falciparum, P. malariae, P. vivax,* and *P. ovale.* The life cycle of *P. vivax* is shown in **figure 40.18.** The parasite first enters the bloodstream through the bite of an infected female *Anopheles* mosquito. As she feeds, the mosquito injects a small amount of saliva containing an anticoagulant along with small haploid sporozoites (*see figure 27.3c*). The sporozoites in the bloodstream immediately enter hepatic cells of the liver. In the liver they undergo multiple asexual fission (schizogony) and produce merozoites. After being

No other single infectious disease has had the impact on humans that malaria has. The first references to its periodic fever and chills can be found in early Chaldean, Chinese, and Hindu writings. In the late fifth century B.C., Hippocrates described certain aspects of malaria. In the fourth century B.C., the Greeks noted an association between individuals exposed to swamp environments and the subsequent development of periodic fever and enlargement of the spleen (splenomegaly). In the seventeenth century the Italians named the disease *mal' aria* (bad air) because of its association with the ill-smelling vapors from the swamps near Rome. At about the same time, the bark of the quinaquina (cinchona) tree of South America was used to treat the intermittent fevers, although it was not until the mid-nineteenth century that quinine was identified as the active alkaloid. The major epidemiological breakthrough came in 1880, when French army surgeon Charles Louis Alphonse Laveran observed exflagellated gametocytes in fresh blood. Five years later the Italian histologist Camillo Golgi observed the multiplication of the asexual blood forms. In the late 1890s Patrick Manson postulated that malaria was transmitted by mosquitoes. Sir Ronald Ross, a British army surgeon in the Indian Medical Service, subsequently observed developing plasmodia in the intestine of mosquitoes, supporting Manson's theory. Using birds as experimental models, Ross definitively established the major features of the life cycle of *Plasmodium* and received the Nobel Prize in 1902.

Human malaria is known to have contributed to the fall of the ancient Greek and Roman empires. Troops in both the United States Civil War and the Spanish-American War were severely incapacitated by the disease. More than 25% of all hospital admissions during these wars were malaria patients. During World War II malaria epidemics severely threatened both the Japanese and Allied forces in the Pacific. The same can be said for the military conflicts in Korea and Vietnam.

In the twentieth century efforts have been directed toward understanding the biochemistry and physiology of malaria, controlling the mosquito vector, and developing antimalarial drugs. In the 1960s it was demonstrated that resistance to *P. falciparum* among West Africans was associated with the presence of hemoglobin-S (Hb-S) in their erythrocytes. Hb-S differs from normal hemoglobin-A by a single amino acid, valine, in each half of the Hb molecule. Consequently these erythrocytes—responsible for sickle cell disease—have a low binding capacity for oxygen. Because the malarial parasite has a very active aerobic metabolism, it cannot grow and reproduce within these erythrocytes.

In 1955 the World Health Organization began a worldwide malarial eradication program that finally collapsed by 1976. Among the major reasons for failure were the development of resistance to DDT by the mosquito vectors and the development of resistance to chloroquine by strains of *Plasmodium*. Currently scientists are exploring new approaches, such as the development of vaccines and more potent drugs. For example, in 1984 the gene encoding the sporozoite antigen was cloned, permitting the antigen to be mass-produced by genetic engineering techniques. In 2002, the complete DNA sequences of *P. falciparum* and *Anopheles gambiae* (the mosquito that most efficiently transmits this parasite to humans in Africa) were determined. Together with the human genome sequence, researchers now have in hand the genetic blueprints for the parasite, its vector, and its victim. This will make possible a holistic approach to understanding how the parasite interacts with the human host, leading to new antimalarial strategies. Overall, no greater achievement for molecular biology could be imagined than the control of malaria—a disease that has caused untold misery throughout the world since antiquity and remains one of the world's most serious infectious diseases.

---

released from the liver cells, the merozoites attach to erythrocytes and penetrate these cells.

Once inside the erythrocyte, the *Plasmodium* begins to enlarge as a uninucleate cell termed a trophozoite. The trophozoite's nucleus then divides asexually to produce a schizont that has 6 to 24 nuclei. The schizont divides and produces mononucleated merozoites. Eventually the erythrocyte lyses, releasing the merozoites into the bloodstream to infect other erythrocytes. This erythrocytic stage is cyclic and repeats itself approximately every 48 to 72 hours or longer, depending on the species of *Plasmodium* involved. This sudden release of merozoites, toxins, and erythrocyte debris triggers an attack of the chills and fever, so characteristic of malaria. Occasionally, merozoites differentiate into macrogametocytes and microgametocytes, which do not rupture the erythrocyte. When these are ingested by a mosquito, they develop into female and male gametes, respectively. In the mos-

quito's gut the infected erythrocytes lyse and the gametes fuse to form a diploid zygote called the ookinete. The ookinete migrates to the mosquito's gut wall, penetrates, and forms an oocyst on its outer surface. In a process called sporogony, the oocyst undergoes meiosis and forms sporozoites that migrate to the salivary glands of the mosquito. The cycle is now complete, but when the mosquito bites another human host, the cycle begins anew.

The pathological changes caused by malaria involve not only the erythrocytes but also the spleen and other visceral organs. Classic symptoms first develop with the synchronized release of merozoites and erythrocyte debris into the bloodstream, resulting in the malarial paroxysms—shaking chills, then burning fever followed by sweating. It may be that the fever and chills are caused partly by a malarial toxin that induces macrophages to release TNF-α and interleukin-1. Several of these paroxysms constitute an attack. After one attack there is a remission that lasts

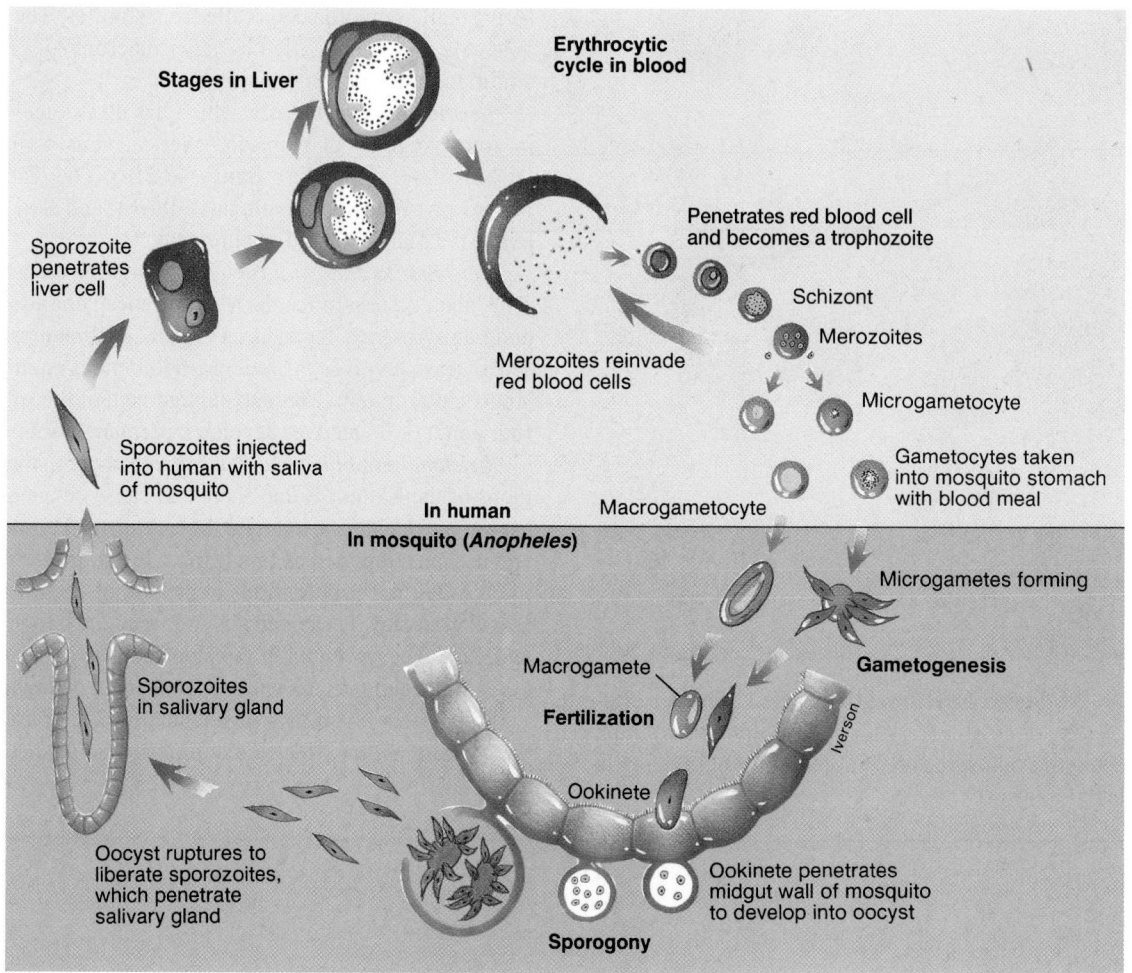

Figure 40.18   Malaria.   Life cycle of *Plasmodium vivax*. See text for details.

from a few weeks to several months. Then there is a relapse. Between paroxysms, the patient feels normal. Anemia can result from the loss of erythrocytes, and the spleen and liver often hypertrophy. Children and nonimmune individuals can die of cerebral malaria.

Diagnosis of malaria is made by demonstrating the presence of parasites within Wright- or Giemsa-stained erythrocytes (**figure 40.19**). When blood smears are negative, serological testing can establish a diagnosis of malaria in individuals. Specific recommendations for treatment are region dependent. Treatment includes administration of chloroquine, amodiaquine, or mefloquine. These suppressant drugs are effective in eradicating erythrocytic asexual stages. Primaquine has proved satisfactory in eradicating the exoerythrocytic stages. However, because resistance to these drugs is occurring rapidly, more expensive drug combinations are being used. One example is Fansidar, a combination of pyrimethamine and sulfadoxine. It is worth noting that individuals who are traveling to areas where malaria is endemic (**figure 40.20**) should receive chemoprophylactic treatment with chloroquine. Efforts to develop a vaccine are under way.

Hemoflagellate Diseases

The flagellated protozoa that are transmitted by the bites of infected arthropods and infect the blood and tissues of humans are called **hemoflagellates.** There are two major groups of pathogens: the leishmanias and the trypanosomes.

Leishmaniasis

Leishmanias are flagellated protozoa that cause a group of human diseases collectively called **leishmaniasis.** Worldwide, there are 2 million new cases each year, about 60,000 deaths, and one-tenth of the world's population is at risk of infection. The primary reservoirs of these parasites are canines and rodents. All species of *Leishmania* use sand flies such as those of the genera *Lutzomyia* and *Phlebotomus* as intermediate hosts. The leishmanias are transmitted from animals to humans or between humans by these sand flies. When an infected sand fly takes a human blood meal, it introduces flagellated promastigotes into the skin of the definitive host. Within the skin the promastigotes are engulfed by macrophages, multiply by binary fission and

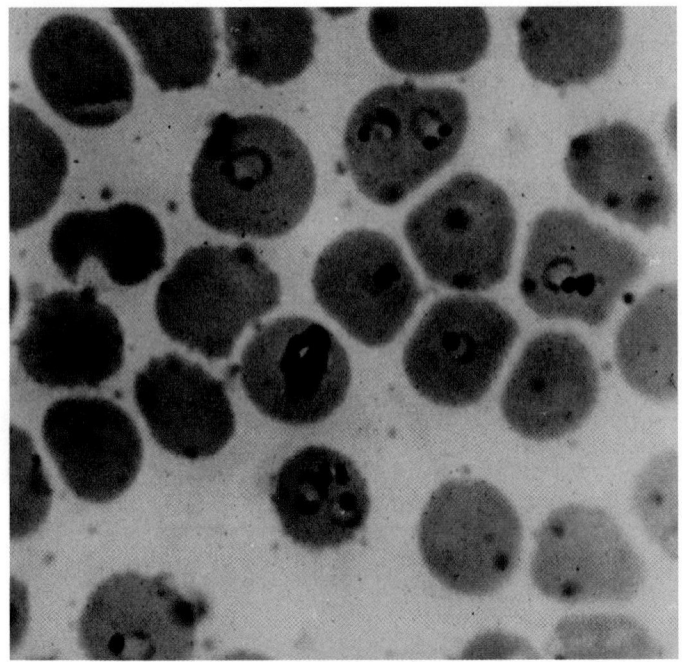

Figure 40.19 Malaria: Erythrocytic Cycle. Trophozoites of *P. falciparum* in circulating erythrocytes; light micrograph (×1,100). The young trophozoites resemble small rings resting in the erythrocyte cytoplasm.

form small nonmotile cells called amastigotes. These destroy the host cell, and are engulfed by other macrophages in which they continue to develop and multiply.

Leishmania braziliensis, which has an extensive distribution in forest regions of tropical America, causes mucocutaneous leishmaniasis (espundia) (**figure 40.21a**). The disease produces lesions involving the mouth, nose, throat, and skin and results in extensive scarring and disfigurement.

Leishmania donovani is endemic in large areas within northern China, eastern India, the Mediterranean countries, the Sudan, and Latin America. It produces visceral leishmaniasis (kala-azar). The disease involves the reticuloendothelial system and often results in intermittent fever and enlargement of the spleen and liver. Individuals who recover develop a permanent immunity.

Leishmania tropica and *L. mexicana* occur in the more arid regions of the Eastern Hemisphere and cause cutaneous leishmaniasis. *L. mexicana* is also found in the Yucatan Peninsula (Mexico) and has been reported as far north as Texas. In this disease a relatively small red papule forms at the site of each insect bite—the inoculation site. These papules are frequently found on the face and ears. They eventually develop into crustated ulcers (figure 40.21b). Healing occurs with scarring and a permanent immunity.

Laboratory diagnosis of leishmaniasis is based on finding the parasite within infected macrophages in stained smears from

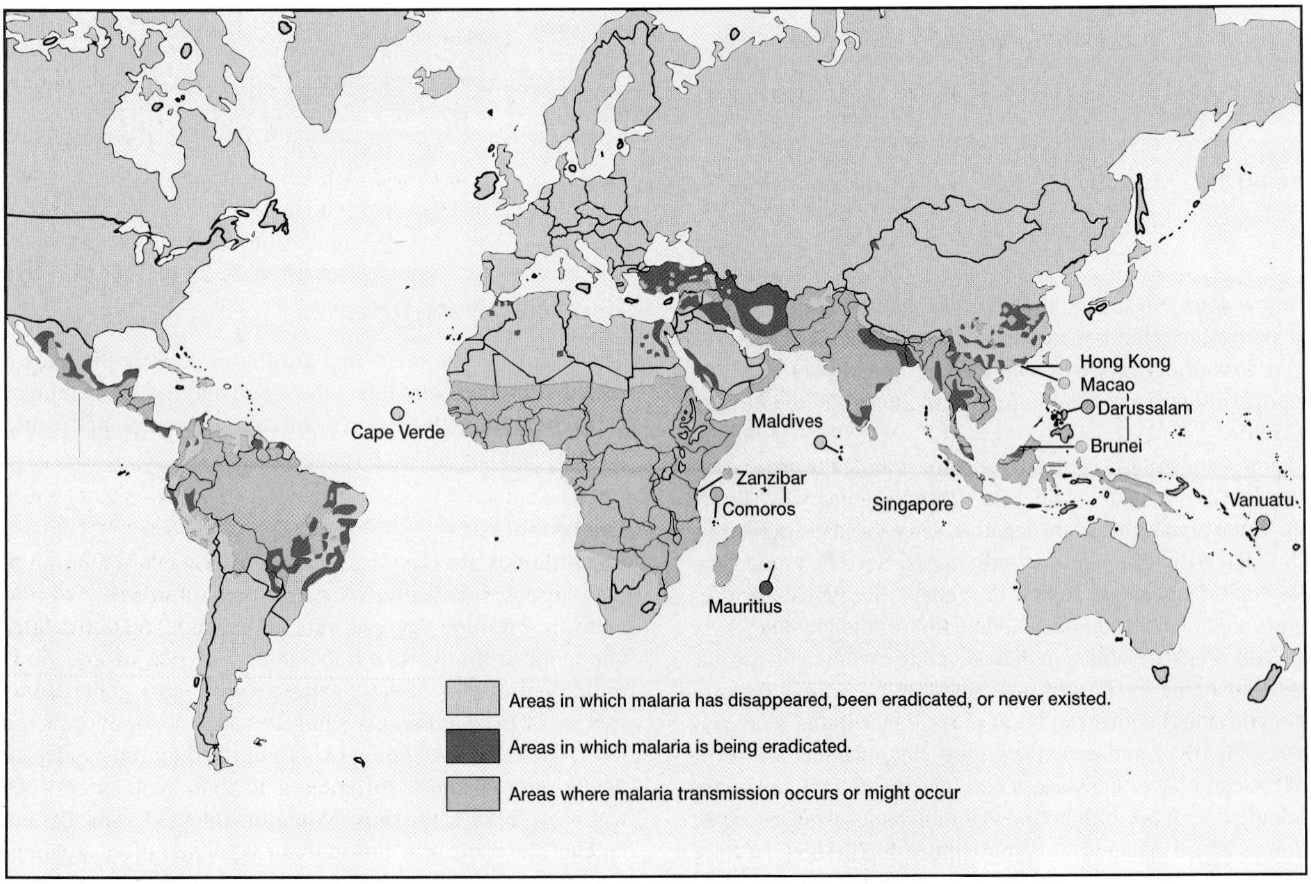

Areas in which malaria has disappeared, been eradicated, or never existed.

Areas in which malaria is being eradicated.

Areas where malaria transmission occurs or might occur

Figure 40.20 Geographic Distribution of Malaria. Notice that malaria is endemic around the equator. Source: Data from the *World Health Statistics Quarterly,* 41:69, 1988, World Health Organization, Switzerland.

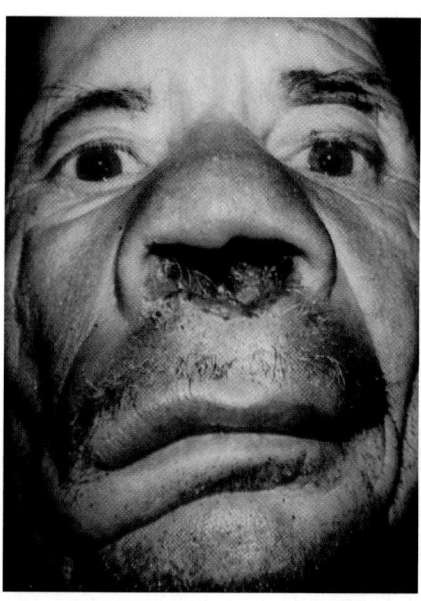

(a)

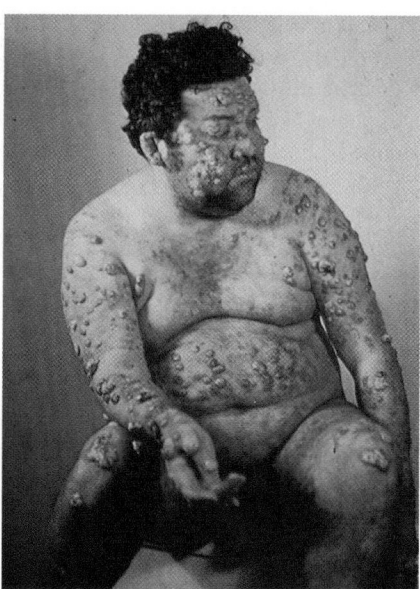

(b)

Figure 40.21 Leishmaniasis. (**a**) A person with mucocutaneous leishmaniasis, which has destroyed the nasal septum and deformed the nose and lips. (**b**) A person with diffuse cutaneous leishmaniasis.

lesions or infected organs. Treatment includes pentavalent antimonial compounds (Pentostam, Glucantime). Vector and reservoir control and aggressive epidemiological surveillance are the best options for prevention and containment of this disease. To date, there are no vaccines against leishmaniasis.

Trypanosomiasis

Another group of flagellated protozoa called **trypanosomes** (*see figure 27.3*a) cause the aggregate of diseases termed **trypanosomiasis.** *Trypanosoma brucei gambiense,* found in the rain forests of west and central Africa, and *T. brucei rhodesiense,* found in the upland savannas of east Africa, cause African trypanosomiasis. Reservoirs for these trypanosomes are domestic cattle and wild animals, within which the parasites cause severe malnutrition. Both species use tsetse flies as intermediate hosts. The parasites are transmitted through the bite of the fly to humans. Once the protozoa enter the bloodstream, they begin to multiply. They cause interstitial inflammation and necrosis within the lymph nodes and small blood vessels of the brain and heart. In *T. brucei rhodesiense* infection the disease develops so rapidly that infected individuals often die within a year. In *T. brucei gambiense* infection the parasites invade the central nervous system, where necrotic damage causes a variety of nervous disorders, including the characteristic sleeping sickness. The name sleeping sickness is derived from the lethargy of the host—characteristically lying prostrate, drooling from the mouth, and insensitive to pain. Usually the victim dies in 2 to 3 years. Trypanosomiasis is such a problem in parts of Africa that millions of square miles are not fit for human habitation. There are over 20,000 new cases each year.

T. cruzi causes **American trypanosomiasis (Chagas' disease),** which occurs in the tropics and subtropics of continental America. The parasite uses the triatomid (kissing) bug as a vector. As the triatomid bug takes a blood meal, the parasites are discharged in the insect's feces. Some trypanosomes enter the bloodstream through the wound and invade the liver, spleen, lymph nodes, and central nervous system, where they multiply and destroy the parasitized cells. In some parts of Latin America, a high percentage of heart disease is due to parasitized cardiac cells. There are 16 to 18 million new cases each year and over 50,000 deaths.

Trypanosomiasis is diagnosed by finding motile parasites in fresh blood and by serological testing. Treatment for African trypanosomiasis uses suramin and pentamidine for non-nervous-system involvement and melarsoprol when the nervous system is involved. Currently there is no drug suitable for Chagas' disease, although nifurtimox (Lampit) and benznidazole have shown some value. Vaccines are not useful because the parasite is able to change its protein coat and evade the immunologic response.

Toxoplasmosis

Toxoplasmosis is a disease caused by the protozoan *Toxoplasma gondii.* This apicomplexan protozoan has been found in nearly all animals and most birds; cats are the definitive host and required for completion of the sexual cycle. Animals shed oocysts in the feces; the oocysts enter another host by way of the nose or mouth; and the parasites colonize the intestine. Toxoplasmosis also can be transmitted by the ingestion of raw or undercooked meat, congenital transfer, blood transfusion, or a tissue transplant. Originally toxoplasmosis gained public notice when it was discovered that in pregnant women the protozoan might also infect the fetus, causing serious congenital defects or death.

Most cases of toxoplasmosis are asymptomatic. Adults usually complain of an "infectious mononucleosis-like" syndrome. In immunoincompetent or immunosuppressed individuals, it frequently results in fatal disseminated disease with a heavy cerebral involvement.

Acute toxoplasmosis is usually accompanied by lymph node swelling (lymphadenopathy) with reticular cell enlargement or hyperplasia. Pulmonary necrosis, myocarditis, and hepatitis caused by tissue necrosis are common. Retinitis (inflammation of the retina of the eye) is associated with the necrosis due to the proliferation of the parasite within retinal cells. Currently toxoplasmosis has become a major cause of death in AIDS patients from a unique encephalitis with necrotizing lesions accompanied by inflammatory infiltrates. It continues to cause more than 3,000 congenital infections per year in the United States.

Laboratory diagnosis of toxoplasmosis is by serological tests. Epidemiologically, toxoplasmosis is ubiquitous in all higher animals. Treatment of toxoplasmosis is with a combination of pyrimethamine (Daraprim) and sulfadiazine. Prevention and control requires minimizing exposure by the following: avoiding eating raw meat and eggs, washing hands after working in the soil, cleaning cat litterboxes daily, keeping personal cats indoors if possible, and feeding them commercial food.

Trichomoniasis

Trichomoniasis is a sexually transmitted disease (*see table 39.4*) caused by the protozoan flagellate *Trichomonas vaginalis* (**figure 40.22**). It is one of the most common sexually transmitted diseases, with an estimated 7 million cases annually in the United States and 180 million cases annually worldwide. In response to the parasite, the body accumulates leukocytes at the site of the infection. In females this usually results in a profuse purulent vaginal discharge that is yellowish to light cream in color and characterized by a disagreeable odor. The discharge is accompanied by itching. Males are generally asymptomatic because of the trichomonacidal action of prostatic secretions; however, at times a burning sensation occurs during urination. Diagnosis is made in females by microscopic examination of the discharge and identification of the parasite. Infected males will demonstrate the parasite in semen or urine. Treatment is by administration of metronidazole (Flagyl).

1. How do infections caused by *Entamoeba histolytica* occur?
2. What is the most common cause of epidemic waterborne diarrheal disease?

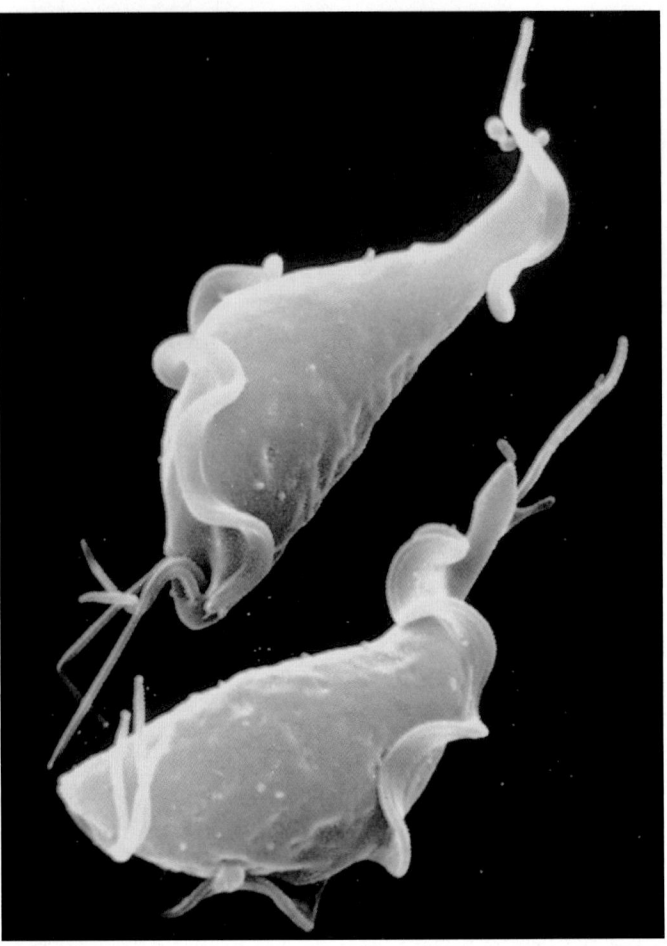

Figure 40.22 Trichomoniasis. *Trichomonas vaginalis,* showing the characteristic undulating membranes and flagella; scanning electron micrograph ($\times$12,000).

3. Describe in detail the life cycle of the malarial parasite.
4. What protozoa are represented within the hemoflagellates? What diseases do they cause?
5. In what two ways does *Toxoplasma* affect human health?
6. How would you diagnose trichomoniasis in a female? In a male?

Summary

40.1 Fungal Diseases

a. Human fungal diseases, or mycoses, can be divided into five groups according to the level and mode of entry into the host. These are the superficial, cutaneous, subcutaneous, systemic, and opportunistic mycoses (**table 40.1**).

b. Superficial mycoses of the hair shaft are collectively called piedras. Two major types are black

piedra (**figure 40.1**) and white piedra. Tinea versicolor is a third common superficial mycosis.

c. The cutaneous fungi that parasitize the hair, nails, and outer layer of the skin are called dermatophytes, and their infections are termed dermatophytoses, ringworms, or tineas. At least seven types can occur in humans: tinea barbae (ringworm of the beard), tinea capitis (ringworm

of the scalp; **figure 40.2**), tinea corporis (ringworm of the body; **figure 40.3**), tinea cruris (ringworm of the groin; **figure 40.4**), tinea pedis (ringworm of the feet; **figure 40.5**), tinea manuum (ringworm of the hands), and tinea unguium (ringworm of the nails; **figure 40.6**).

d. The dermatophytes that cause the subcutaneous mycoses are normal saprophytic inhabitants of soil and decaying vegetation. Three types of subcutaneous mycoses can occur in humans: chromoblastomycosis (**figure 40.7**), maduromycosis (**figure 40.8**), and sporotrichosis (**figure 40.9**).

e. Most systemic mycoses that occur in humans are acquired by inhaling the spores from the soil where the free-living fungi are found. Four types can occur in humans: blastomycosis (**figure 40.10**), coccidioidomycosis, cryptococcosis, and histoplasmosis.

f. An opportunistic organism is one that is generally harmless in its normal environment but that can become pathogenic in a compromised host. The most important opportunistic mycoses affecting humans include systemic aspergillosis (**figure 40.14**), candidiasis (**figure 40.15*b***), and *Pneumocystis* pneumonia.

40.2 Protozoan Diseases

a. Protozoa are responsible for some of the most serious human diseases that affect hundreds of millions of people worldwide (**table 40.3**).

b. *Entamoeba histolytica* is the amoeboid protozoan responsible for amebiasis. This is a very common disease in warm climates throughout the world. It is acquired when cysts are ingested with contaminated food or water (**figure 40.16**).

c. *Cryptosporidium parvum* is a common coccidial apicomplexan parasite that causes severe diarrheal disease. It is acquired from contaminated food or water.

d. *Giardia lamblia* is a flagellated protozoan that causes the common intestinal disease giardiasis (**figure 40.17**). This disease is distributed throughout the world, and in the United States it is the most common cause of waterborne diarrheal disease.

e. The most important human parasite among the sporozoa is *Plasmodium*, the causative agent of malaria (**figure 40.18**). Human malaria is caused by four species of *Plasmodium: P. falciparum, P. vivax, P. malariae,* and *P. ovale*.

f. The flagellated protozoa that are transmitted by arthropods and infect the blood and tissues of humans are called hemoflagellates. Two major groups occur: the leishmanias, which cause the diseases collectively termed leishmaniasis (**figure 40.21**), and the trypanosomes, which cause trypanosomiasis.

g. Toxoplasmosis is a disease caused by the protozoan *Toxoplasma gondii*. It is one of the major causes of death in AIDS patients.

h. Trichomoniasis is a sexually transmitted disease caused by the protozoan flagellate *Trichomonas vaginalis* (**figure 40.22**).

Key Terms

amebiasis (amebic dysentery) 927
American trypanosomiasis 933
aspergillosis 924
athlete's foot 920
balanitis 925
black piedra 919
blastomycosis 922
candidal vaginitis 925
candidiasis 924
Chagas' disease 933
chromoblastomycosis 921
coccidioidomycosis 922
compromised host 924
cryptococcosis 923
cryptosporidiosis 927

dermatomycosis 919
dermatophyte 919
eumycotic mycetoma 921
extracutaneous sporotrichosis 921
giardiasis 928
hemoflagellate 931
histoplasmosis 923
intertriginous candidiasis 925
jock itch 920
keratitis 928
leishmania 931
leishmaniasis 931
maduromycosis 921
malaria 929

medical mycology 918
mycosis 918
napkin (diaper) candidiasis 925
onychomycosis 925
opportunistic microorganism 924
oral candidiasis 925
paronychia 925
piedra 919
Pneumocystis pneumonia (PCP) 926
primary amebic
   meningoencephalitis 928
ringworm 919
sporotrichosis 921
thrush 925

tinea 919
tinea barbae 919
tinea capitis 920
tinea corporis 920
tinea cruris 920
tinea manuum 920
tinea pedis 920
tinea unguium 920
tinea versicolor 919
toxoplasmosis 933
trichomoniasis 934
trypanosome 933
trypanosomiasis 933
white piedra 919

Questions for Thought and Review

1. Briefly describe each of the major or most common fungal and protozoan diseases in terms of its causative agent, signs and symptoms, the course of infection, mechanism of pathogenesis, epidemiology, and prevention and/or treatment.

2. There are no antibiotics—nothing equivalent to penicillin or streptomycin—to control fungi in the human body. Recalling the mechanisms by which antibiotics affect bacteria, why do you think there has been no similar success in the control of fungal pathogens?

3. Give several reasonable explanations why most fungal diseases in humans are not contagious.

4. What factors determine one's susceptibility to fungal diseases?

5. What is the relationship between AIDS and mycotic and protozoan diseases?

6. Why do most protozoan diseases occur in the tropics?

7. Give several reasons why malaria is still one of the most serious of all infectious diseases that affect humans.

8. What morphological property is most important in the identification of fungi?

9. Compared to fungal parasites, how are protozoan parasites transmitted?

10. What are dermatomycoses?

Critical Thinking Questions

1. Compare and contrast treatment of diseases caused by fungi with those caused by viruses or bacteria.

2. What is one distinct feature of fungi that could be exploited for antibiotic therapy?

3. Trypanosomes are notorious for their ability to change their surface antigens frequently. Given the kinetics of a primary immune response (primary antibody production), how often would the surface antigen need to be changed in order to stay "ahead" of the antibody specificity? Why shouldn't it change the expression every time transcription occurs?

For recommended readings on these and related topics, see Appendix V.

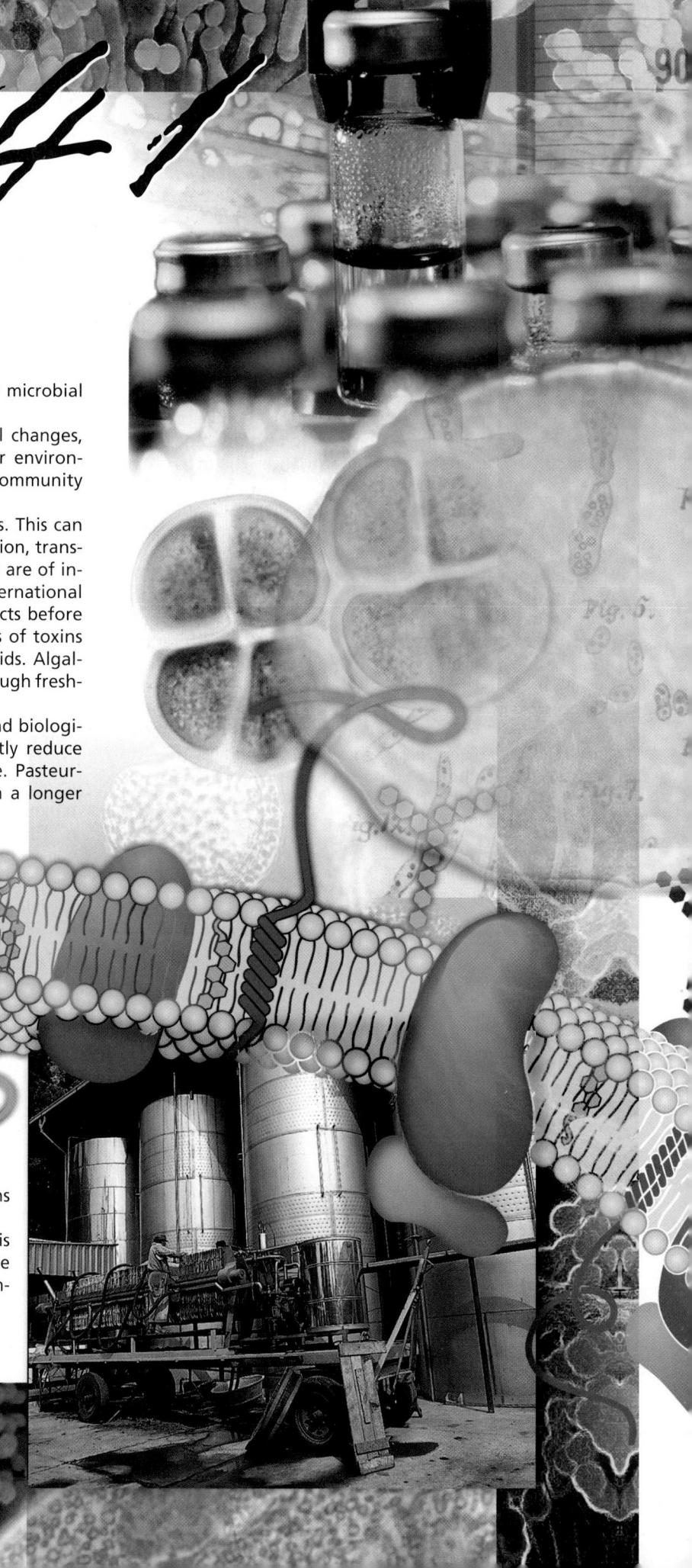

chapter 41

Microbiology of Food

Concepts

- Foods often provide an ideal environment for microbial survival and growth.
- Microbial growth in foods involves successional changes, with intrinsic, or food-related, and extrinsic, or environmental, factors interacting with the microbial community over time.
- Food spoilage is a major problem in all societies. This can occur at any point in the course of food production, transport, storage, or preparation. Food-borne toxins are of increasing concern, especially with increases in international shipments and extended storage of food products before use. Growth of fungi can result in the synthesis of toxins such as aflatoxins, fumonisins, and ergot alkaloids. Algal-derived toxins can be transmitted to humans through freshwater and marine-derived food products.
- Foods can be preserved by physical, chemical, and biological processes. Refrigeration does not significantly reduce microbial populations but only retards spoilage. Pasteurization results in a pathogen-free product with a longer shelf life.
- Chemicals can be added to foods to control microbial growth. Such chemicals include sugar, salt (decreasing water availability), and many organic chemicals that affect specific groups of microorganisms. Microbial products, such as bacteriocins, can be added to foods to control spoilage organisms.
- Foods can transmit a wide range of diseases to humans. In a food infection, the food serves as a vehicle for the transfer of the pathogen to the consumer, in whom the pathogen grows and causes disease. With a food intoxication, the microorganisms grow in the food and produce toxins that can then affect the consumer.
- New pathogens continue to be discovered. This may be due to an increased ability to detect these pathogens as well as to better reporting and monitoring of disease occurrence.

Large tanks used for wine production. Fermentations can be carried out in such open-air units in temperate regions. After completion of the fermentation, the fresh wine will be transferred to barrels for storage and aging.

- Foods that are consumed raw, such as sprouts, shellfish/finfish, and fruits, are increasingly popular. There is a risk of disease transmission if care is not taken in production, storage, and transport of these perishable foods. Without adequate care, major disease outbreaks can occur because foods travel around the globe in extremely short times.
- Detection of food-borne pathogens is carried out using classic culture techniques, as well as immunological and molecular procedures. Modern techniques make it possible to quickly link food-borne disease occurrences to common pathogen sources. With kit-type rapid diagnostic systems, pathogens often can be detected in 1 to 2 days.
- Dairy products, grains, meats, fruits, and vegetables can be fermented. It has been suggested that some fermented dairy products have antimicrobial and anticancer characteristics, especially when particular lactic acid bacteria are used.
- Wines are produced by the direct fermentation of fruit juices or musts. For fermentation of cereals and grains, starches and proteins contained in these substrates must first be hydrolyzed to provide substrates for the alcoholic fermentation.
- The making of bread, sauerkraut, sufu, pickles, and many other foods also involves the use of complex fermentation processes. When chopped plant materials are fermented, silages are created, which can be stored and used by animals.
- Microbial cells can be used as food sources and food amendments. These include mushrooms, cyanobacteria such as *Spirulina,* and yeasts. There is an increasing interest in probiotics, or the use of microorganisms to change the microbial community in the intestine. Microbial colonization of surfaces in the intestine plays a critical role in these processes.

> *Tell me what you eat, and I will tell you what you are.*
> —*Brillat-Savarin*

oods, microorganisms, and humans have had a long and interesting association that developed long before the beginning of recorded history. Foods are not only of nutritional value to those who consume them but often are ideal culture media for microbial growth. Fermentation by some microorganisms can lead to food preservation instead of food spoilage.

Microorganisms can be used to transform raw foods into gastronomic delights, including cheeses, pickles, sausages, and soy sauce. Wines, beers, and other alcoholic products also are produced through microbial activity. On the other hand, foods also can serve as vehicles for disease transmission, and the detection and control of pathogens and food spoilage microorganisms are important parts of food microbiology. During the entire sequence of food handling, from the producer to the final consumer, microorganisms can affect food quality and human health.

Wholesome, nutrient-rich foods are important for all people. Microbial growth in foods can result in either preservation or spoilage, depending on the microorganisms involved and the food storage conditions. Contamination by disease-causing microorganisms can occur at any point in the food-handling sequence.

 MICROORGANISM GROWTH IN FOODS

Foods, because they provide nutrients for us, also are excellent environments for the growth of microorganisms. Microbial growth is controlled by factors related to the food itself, or **intrinsic factors,** and also to the environment where the food is being stored, or what are described as **extrinsic factors,** as shown in **figure 41.1.**

The intrinsic or food-related factors include pH, moisture content, water activity or availability, oxidation-reduction potential, physical structure of the food, available nutrients, and the possible presence of natural antimicrobial agents. Extrinsic or environmental factors include temperature, relative humidity, gases (CO_2, O_2) present, and the types and numbers of microorganisms present in the food.

Intrinsic Factors

Food composition is a critical intrinsic factor that influences microbial growth. If a food consists primarily of carbohydrates, spoilage does not result in major odors. Thus foods such as breads, jams, and some fruits first show spoilage by fungal

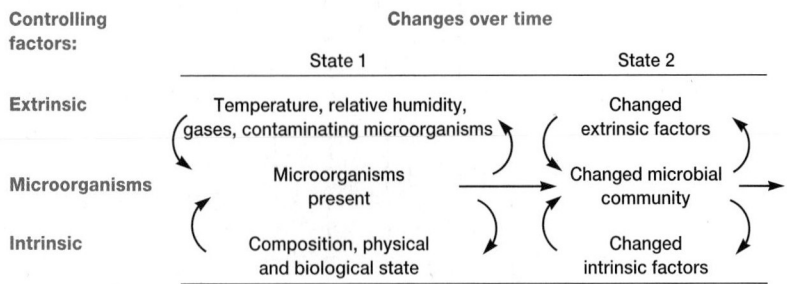

Figure 41.1   Intrinsic and Extrinsic Factors.   A variety of intrinsic and extrinsic factors can influence microbial growth in foods. Time-related successional changes occur in the microbial community and the food.

Table 41.1

Differences in Spoilage Processes in Relation to Food Characteristics

Substrate	Food Example	Chemical Reactions or Processes[a]	Typical Products and Effects
Pectin	Fruits	Pectinolysis	Methanol, uronic acids (loss of fruit structure, soft rots)
Proteins	Meat	Proteolysis, deamination	Amino acids, peptides, amines, H_2S, ammonia, indole (bitterness, souring, bad odor, sliminess)
Carbohydrates	Starchy foods	Hydrolysis, fermentations	Organic acids, CO_2, mixed alcohols (souring, acidification)
Lipids	Butter	Hydrolysis, fatty acid degradation	Glycerol and mixed fatty acids (rancidity, bitterness)

[a]Other reactions also occur during the spoilage of these substrates.

Table 41.2

Approximate Minimum Water Activity Relationships of Microbial Groups and Specific Organisms of Important in Food Spoilage

Organisms	a_w	Organisms	a_w
Groups		**Groups**	
Most spoilage bacteria	0.9	Halophilic bacteria	0.75
Most spoilage yeasts	0.88	Xerophilic molds	0.61
Most spoilage molds	0.80	Osmophilic yeasts	0.61
Specific Microorganisms		**Specific Microorganisms**	
Clostridium botulinum, type E	0.97	*Candida scottii*	0.92
Pseudomonas spp.	0.97	*Trichosporon pullulans*	0.91
Acinetobacter spp.	0.96	*Candida zeylanoides*	0.90
Escherichia coli	0.96	*Geotrichum candidum*	~ 0.90
Enterobacter aerogenes	0.95	*Trichothecium* spp.	~ 0.90
Bacillus subtilis	0.95	*Byssochlamys nivea*	~ 0.87
Clostridium botulinum, types A and B	0.94	*Staphylococcus aureus*	0.86
Candida utilis	0.94	*Alternaria citri*	0.84
Vibrio parahaemolyticus	0.94	*Pencillium patulum*	0.81
Botrytis cinerea	0.93	*Eurotium repens*	0.72
Rhizopus stolonifer	0.93	*Aspergillus conicus*	0.70
Mucor spinosus	0.93	*Aspergillus echinulatus*	0.64
		Zygosaccharomyces rouxii	0.62
		Xeromyces bisporus	0.51

Adapted from James M. Jay. 2000. *Modern Food Microbiology*, 6th edition. Reprinted by permission of Aspen Publishers, Inc. Gaithersburg, MD. Tables 3–5, p. 42.

growth. In contrast, when foods contain large amounts of proteins and/or fats (for example, meat and butter), spoilage can produce a variety of foul odors. One only need think of rotting eggs. This proteolysis and anaerobic breakdown of proteins that yields foul-smelling amine compounds is called **putrefaction.** One major source of odor is the organic amine cadaverine (imagine the origin of that name). Degradation of fats ruins food as well. The production of short-chained fatty acids from fats renders butter rancid and unpleasant.

The pH of a food also is critical because a low pH favors the growth of yeasts and molds (*see section 6.4*). In neutral or alkaline pH foods, such as meats, bacteria are more dominant in spoilage and putrefaction. Depending on the major substrate present in a food, different types of spoilage may occur (**table 41.1**).

The presence and availability of water also affect the ability of microorganisms to colonize foods. Simply by drying a food, one can control or eliminate spoilage processes. Water, even if present, can be made less available by adding solutes such as sugar and salt. Water availability is measured in terms of water activity (a_w). This represents the ratio of relative humidity of the air over a test solution compared with that of distilled water. When large quantities of salt or sugar are added to food, most microorganisms are dehydrated by the hypertonic conditions and cannot grow (**table 41.2;** *see also table 6.4*). Even under these adverse conditions, osmophilic and xerophilic microorganisms may spoil food. **Osmophilic** [Greek *osmus,* impulse, and *philein,* to love] **microorganisms** grow best in or on media with a high osmotic concentration, whereas **xerophilic** [Greek *xerosis,* dry, and *philein,* to love] **microorganisms** prefer a low a_w environment and may not grow under high a_w conditions. Water activity and microbial growth (pp. 118–21)

The oxidation-reduction potential of a food also influences spoilage. When meat products, especially broths, are cooked, they often have lower oxidation-reduction potentials. These products with their readily available amino acids, peptides, and growth factors are ideal media for the growth of anaerobes, including *Clostridium* (*see table 39.5*).

The physical structure of a food also can affect the course and extent of spoilage. The grinding and mixing of foods such as sausage and hamburger not only increase the food surface area and alter cellular structure, but also distribute contaminating microorganisms throughout the food. This can result in rapid spoilage if such foods are stored improperly. Vegetables and fruits have outer skins (peels and rinds) that protect them from spoilage. Often spoilage microorganisms have specialized enzymes that help them weaken and penetrate protective peels and rinds, especially after the fruits and vegetables have been bruised.

Many foods contain natural antimicrobial substances, including complex chemical inhibitors and enzymes. Coumarins found in fruits and vegetables exhibit antimicrobial activity. Cow's milk and eggs also contain antimicrobial substances. Eggs are rich in the enzyme lysozyme that can lyse the cell walls of contaminating gram-positive bacteria (*see figure 31.9*). Other interesting foods with antimicrobial activities include the hot sauces used on raw oysters and other seafoods. Tabasco and other hot red pepper sauces apparently have particularly desirable antimicrobial characteristics.

Herbs and spices often possess significant antimicrobial substances; generally fungi are more sensitive than most bacteria. Sage and rosemary are two of the most antimicrobial spices. Aldehydic and phenolic compounds that inhibit microbial growth are found in cinnamon, mustard, and oregano. These compounds inhibit microbial growth. Other important inhibitors are garlic, which contains allicin, cloves, which have eugenol, and basil, which contains rosmarinic acid. However, spices also can contain pathogenic and spoilage organisms. Coliforms, *B. cereus, C. perfringens,* and *Salmonella* species have been detected in most spices. Microorganisms can be eliminated or reduced by ethylene oxide sterilization. This treatment can result in *Salmonella*-free spices and herbs and a 90% reduction in the levels of general spoilage organisms.

Unfermented green and black teas also have well-documented antimicrobial properties because of their polyphenol contents, which apparently are diminished when the teas are fermented. Such unfermented teas are active against bacteria, viruses, and fungi and may have anticancer properties.

Extrinsic Factors

Temperature and relative humidity are important extrinsic factors in determining whether a food will spoil. At higher relative humidities microbial growth is initiated more rapidly, even at lower temperatures (especially when refrigerators are not maintained in a defrosted state). When drier foods are placed in moist environments, moisture absorption can occur on the food surface, eventually allowing microbial growth.

The atmosphere in which the food is stored also is important. This is especially true with shrink-packed foods because many plastic films allow oxygen diffusion, which results in increased growth of surface-associated microorganisms. Excess CO_2 can decrease the solution pH, inhibiting microbial growth. Storing meat in a high CO_2 atmosphere inhibits gram-negative bacteria, resulting in a population dominated by the lactobacilli.

The observation that food storage atmosphere is important has led to the development **modified atmosphere packaging (MAP).** By the use of modern shrink-wrap materials and vacuum technology, it is possible to package foods with controlled atmospheres. With a carbon dioxide content of 60% or greater in the atmosphere surrounding a food, spoilage fungi will not grow, even if low levels of oxygen are present. Recently, it has been found that high-oxygen MAP also may be effective. This is due to the formation of the superoxide (O_2^-) anion inside cells under these conditions. The superoxide anion is then transformed to the highly toxic peroxide and oxygen, resulting in antimicrobial effects.

---

1. What are some intrinsic factors that influence food spoilage and how do they exert their effects?
2. What are the effects of food composition on spoilage processes?
3. Why might sausage and other ground meat products provide a better environment for the growth of food spoilage organisms than solid cuts of meats?
4. List some antimicrobial substances found in foods. What is the mechanism of action of lysozyme?
5. What primary extrinsic factors can determine whether food spoilage will occur?
6. What are the major gases involved in MAP? How are their concentrations varied to inhibit microbial growth?

41.2 MICROBIAL GROWTH AND FOOD SPOILAGE

Because foods are such excellent sources of nutrients, if the intrinsic and extrinsic conditions are appropriate, microorganisms grow rapidly and make what once was an attractive and appealing food into a sour, foul-smelling or fungus-covered mass suitable only for the garbage can. Microbial growth in foods can lead to visible changes, including a variety of colors caused by spoilage organisms, which often have been associated with "miracles" and "witchcraft." One of the most famous is the report of "blood" on communion wafers and other bread, called the "Miracle of Bolsena," which occurred in 1263. The riddle was eventually solved by Bartolomeo Bizio in 1879, when he described the bacterium responsible for this phenomenon. He also named the bacterium *Serratia marcescens.*

Meat and dairy products, with their high nutritional value and the presence of easily usable carbohydrates, fats, and proteins, provide ideal environments for microbial spoilage. Proteolysis and putrefaction are typical results of microbial spoilage of such high-protein materials. Unpasteurized milk undergoes a

(a)

(b)

Figure 41.2 Spoilage of a Dairy Product.   Fresh (left) and curdled (right) milk are shown. The curdled milk has undergone a natural four-step sequence of spoilage organism activity. The commensalistic spoilage process has produced acidic conditions that have denatured and precipitated the milk casein to yield typical, separated curds and whey.

Figure 41.3 Food Spoilage.   When foods are not stored properly, microorganisms can cause spoilage. Typical examples are fungal spoilage of (**a**) bread and (**b**) corn. Such spoilage of corn is called ear rot. This can result in major economic losses.

predictable four-step succession during spoilage; acid production by *Lactococcus lactis* subsp. *lactis* is followed by additional acid production associated with the growth of more acid tolerant organisms such as *Lactobacillus*. At this point yeasts and molds become dominant and degrade the accumulated lactic acid, and the acidity gradually decreases. Eventually protein-digesting bacteria become active, resulting in a putrid odor and bitter flavor. The milk, originally opaque, can eventually become clear (**figure 41.2**).

In comparison with meat and dairy products, most fruits and vegetables have a much lower protein and fat content and undergo a different kind of spoilage. Readily degradable carbohydrates favor vegetable spoilage by bacteria, especially bacteria that cause soft rots, such as *Erwinia carotovora*, which produces hydrolytic enzymes. The high oxidation-reduction potential and lack of reduced conditions permits aerobes and facultative anaerobes to contribute to the decomposition processes. Bacteria do not seem important in the initial spoilage of whole fruits; instead such spoilage often is initiated by molds. These organisms have enzymes that contribute to the weakening and penetration of the protective outer skin.

Food spoilage problems occur with minimally processed, concentrated frozen citrus products. These are prepared with little or no heat treatment, and major spoilage can be caused by *Lactobacillus* and *Leuconostoc* spp., which produce diacetyl-butter flavors. *Saccharomyces* and *Candida* can also spoil juices. Concentrated juice has a decreased water activity ($a_w = 0.8$ to 0.83), and when kept frozen at about $-9°C$, the juices can be stored for long periods. However, when concentrated juices are diluted with

water that contains spoilage organisms, or if the juice is stored in improperly washed containers, problems can occur. Also, microorganisms in the frozen concentrated juices can begin the spoilage process after addition of water. Ready-to-serve (RTS) juices present other problems as the a_w values are sufficiently high to allow microbial growth. This is especially true with extended storage at refrigeration temperatures. Although pasteurization can be used, most consumers are sensitive to the loss of flavor that this process entails.

Molds are a special problem for tomatoes. Even the slightest bruising of the tomato skin, exposing the interior, will result in rapid fungal growth. Frequently observed genera include *Alternaria, Cladosporium, Fusarium,* and *Stemphylium*. This growth affects the quality of tomato products, including tomato juices and ketchups.

Molds can rapidly grow on grains and corn when these products are held under moist conditions. The moldy bread pictured in **figure 41.3a** shows extensive fungal hyphal development and sporulation. The green growth most likely is *Penicillium;* the black

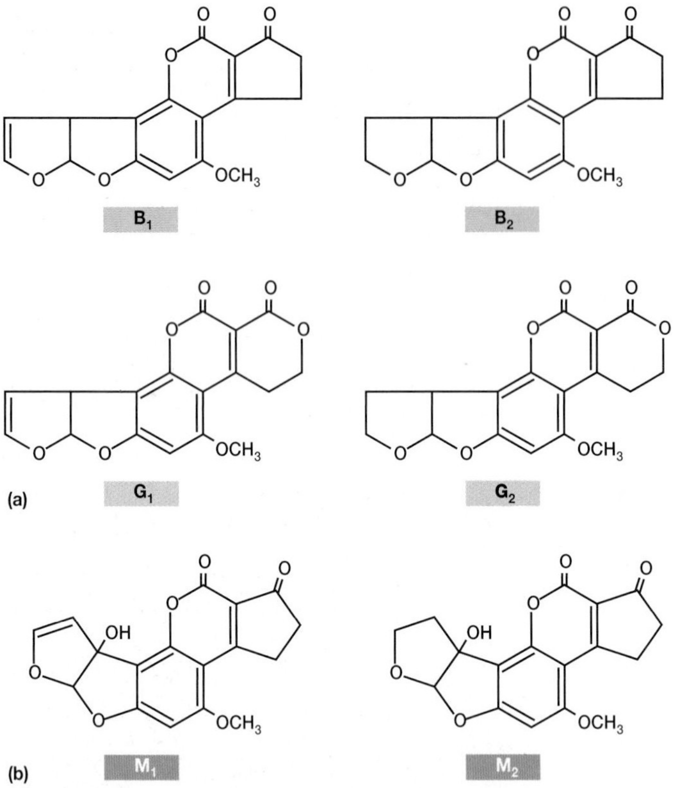

Figure 41.4 Aflatoxins. When *Aspergillus flavus* and related fungi grow on foods, carcinogenic aflatoxins can be formed. These have four basic structures. (**a**) The letter designations refer to the color of the compounds under ultraviolet light after extraction from the grain and separation by chromatography. The B_1 and B_2 compounds fluoresce with a blue color, and the G_1 and G_2 appear green. (**b**) The two type M aflatoxins are found in the milk of lactating animals that have ingested type B aflatoxins.

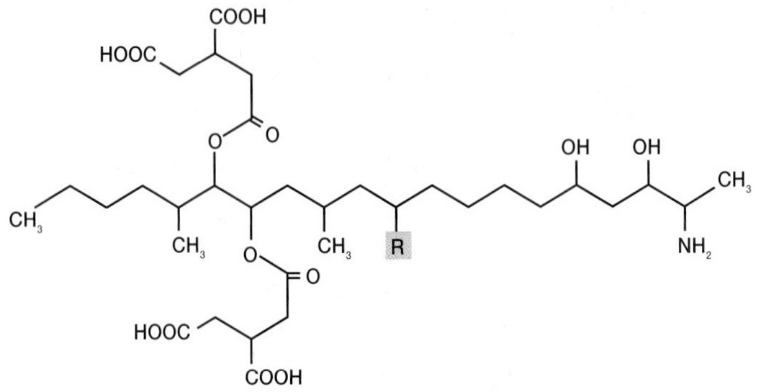

Figure 41.5 Fumonisin Structure. The basic structure of fumonisins FB1 and FB2 produced by *Fusarium moniliforme,* a fungal contaminant that can grow in improperly stored corn. A total of at least ten different fumonisins have been isolated. These are strongly polar compounds that cause diseases in domestic animals and also in humans. FB1, R = −OH; FB2, R = H.

growth is characteristic of *Rhizopus stolonifer.* Infection of grains by the ascomycete *Claviceps purpura* causes **ergotism,** a toxic condition. Hallucinogenic alkaloids produced by this fungus can lead to altered behavior, abortion, and death if infected grains are eaten. Ergotism is further discussed in chapter 25 (*see p. 544*).

Fungus-derived carcinogens include the aflatoxins and fumonisins. **Aflatoxins** are produced most commonly in moist grains and nut products. Aflatoxins were discovered in 1960, when 100,000 turkey poults died from eating fungus-infested peanut meal. *Aspergillus flavus* was found in the infected peanut meal, together with alcohol-extractable toxins termed aflatoxins. These flat-ringed planar compounds intercalate with the cells' nucleic acids and act as frameshift mutagens and carcinogens. This occurs primarily in the liver, where they are converted to unstable derivatives. At the present time, a total of 18 aflatoxins are known. The most important are shown in **figure 41.4.** Of these, aflatoxin B_1 is the most common and the most potent carcinogen.

Aflatoxins B_1 and B_2, after ingestion by lactating animals, will be modified in the animal body to yield the aflatoxins M_1 and M_2. If cattle consume aflatoxin-contaminated feeds, these also

can appear in milk and dairy products. The aflatoxins are potent hepatocarcinogens, which have been linked to effects on immunocompetence, growth, and disease resistance in livestock and laboratory animals. The major aflatoxin types and their derivatives can be separated by chromatographic procedures and can be recognized under UV light by their characteristic fluorescence. Besides their importance in grains, they have also been observed in beer, cocoa, raisins, and soybean meal.

Ultimately, the critical concern is the amounts of aflatoxins that are ingested. Diet preferences appear to be related to aflatoxin exposure: the average aflatoxin intake in the typical European-style diet is 19 ng/day, whereas for Far Eastern diets it is estimated to be 103 ng/day. Aflatoxin sensitivity also can be influenced by prior disease exposure. It has been found that individuals who have had hepatitis B have a 30-fold higher risk of liver cancer upon exposure to aflatoxins than individuals who have not had this disease. Based on this linkage, it has been observed that prevention of hepatitis B infections by vaccination and reduction of carrier populations will help to significantly control the potential effects of aflatoxins in foodstuffs.

More recently discovered fungal contaminants of corn are the **fumonisins,** first isolated in 1988. These are produced by *Fusarium moniliforme* and cause leukoencephalomalacia in horses, pulmonary edema in pigs, and esophageal cancer in humans. The fumonisins function by disrupting the synthesis and metabolism of sphingolipids, important biochemically active compounds, which influence a wide variety of cell functions.

There are at least ten different fumonisins; the basic structures of fumonisins FB1 and FB2 are shown in **figure 41.5.** Corn and corn-based feeds and foods, including cornmeal and corn grits, often are contaminated. The fumonisins inhibit ceramide synthase, a key enzyme for the proper use of fatty subtances in the cell. Thus it is extremely important to store corn and corn products under dry conditions where these fungi cannot develop.

Eucaryotic microorganisms can synthesize potent toxins other than aflatoxins and fumonisins. **Algal toxins** contaminate

Table 41.3
Toxic Syndromes Associated with Marine Algal Toxins

Syndrome	Causative Organism(s)	Primary Vector	Toxin Type
Parasitic shellfish poisoning	*Alexandrium* spp.	Shellfish	Saxitoxins
	Gymnodinium spp.		
	Pyrodinium spp.		
Neurotoxic shellfish poisoning	*Gymnodinium breve*	Shellfish	Brevitoxins
Ciguatera fish poisoning	*Gambierdiscus toxicus*	Reef fish	Ciguatoxins
Amnesic shellfish poisoning	*Pseudo-nitzchia* spp.	Shellfish	Domoic acid
Diarrhetic shellfish poisoning	*Dinophysis* spp.	Shellfish	Dinophysistoxins
	Prorocentrum spp.		Okadaic acid
Estuary syndrome	*Pfiesteria piscicida*	Water	Unknown

Source: F. M. van Dolah, 2000. Marine algal toxins: Origins, health effects, and their increased occurrence. *Environ. Health Perspect.* 108(Suppl. 1):133–141. Table 1, p. 134.

fish and thus affect the health of marine animals higher in the food chain; they also can contaminate shellfish and fin fish, which are later consumed by humans. Most toxins are produced by dinoflagellates, but some diatoms also are toxic. Major human diseases that result from algal toxins in marine products include amnesic, diarrhetic, and neurotoxic shellfish poisoning (**table 41.3**). These complex algal toxins, most of which are temperature stable, are known to cause peripheral neurological system effects, often in less than one hour after ingestion. Toxic algal blooms (Disease 26.2)

1. Describe in general how food spoilage occurs. What factors influence the nature of the spoilage organisms responsible?
2. Why do concentrated citrus juices present such interesting spoilage problems?
3. What fungal genus produces ergot alkaloids? What conditions are required for the synthesis of these substances?
4. Aflatoxins are produced by which fungal genus? How do they damage animals that eat the contaminated food?
5. What microbial genus produces fumonisins and why are these compounds of concern? If improperly stored, what are the major foods and feeds in which these chemicals might be found?
6. What is the usual route by which humans consume algal toxins? What are the major groups of algae that produce these complex substances?

41.3   CONTROLLING FOOD SPOILAGE

With the beginning of agriculture and a decreasing dependence on hunting and gathering, the need to preserve surplus foods became essential to survival. The use of salt as a meat preservative and the production of cheeses and curdled milks was introduced in Near Eastern civilization as early as 3000 B.C. The production of wines and the preservation of fish and meat by smoking also were common by this time. Despite a long tradition of efforts to preserve food from spoilage, it was not until the nineteenth cen-

Table 41.4
Basic Approaches to Food Preservation

Approach	Examples of Process
Removal of microorganisms	Avoidance of microbial contamination; physical filtration, centrifugation
Low temperature	Refrigeration, freezing
High temperature	Partial or complete heat inactivation of microorganisms (pasteurization and canning)
Reduced water availability	Water removal, as with lyophilization or freeze drying; use of spray dryers or heating drums; decreasing water availability by addition of solutes such as salt or sugar
Chemical-based preservation	Addition of specific inhibitory compounds (e.g., organic acids, nitrates, sulfur dioxide)
Radiation	Use of ionizing (gamma rays) and nonionizing (UV) radiation
Microbial product–based inhibition	The addition of substances such as bacteriocins to foods to control food-borne pathogens

tury that the microbial spoilage of food was studied rigorously. Louis Pasteur established the modern era of food microbiology in 1857, when he showed that microorganisms cause milk spoilage. Pasteur's work in the 1860s proved that heat could be used to control spoilage organisms in wines and beers (*see section 1.4*). Control of microorganism growth (chapter 7)

Foods can be preserved by a variety of methods (**table 41.4**). It is vital to eliminate or reduce the populations of spoilage and disease-causing microorganisms and to maintain the microbiological quality of a food with proper storage and packaging. Contamination often occurs after a package or can is opened and just before the food is served. This can provide an ideal opportunity for growth and transmission of pathogens, if care is not taken.

Removal of Microorganisms

Microorganisms can be removed from water, wine, beer, juices, soft drinks, and other liquids by filtration. This can keep bacterial populations low or eliminate them entirely. Prefilters and

The movement and maintenance of large numbers of military personnel have always been limited by food supplies. The need to maintain large numbers of troops under hostile and inclement conditions led the French government in 1795 to offer a prize of 12,000 francs to the individual who could preserve foods for use under field conditions. Eventually the prize was awarded to Nicholas Appert, a candy maker, for his development of a heating process in which meats and other products could be preserved under sealed conditions.

Appert's work was based on the assumptions that heating and boiling control "ferments" and that sealing the food in bottles before heating it avoids the effects of air on spoilage. Despite Leeuwenhoek's earlier work, Appert did not have the concept of microorganisms to assist him in explaining the effectiveness of his process. His containers were large glass bottles, sealed with laminated corks and fish glue. With extreme care and attention to detail, he was able to heat these bottles in boiling water to provide food that could be stored for several years. Appert's work was an important foundation for the later studies of Louis Pasteur.

centrifugation often are used to maximize filter life and effectiveness. Several major brands of beer are filtered rather than pasteurized to better preserve the flavor and aroma of the original product.

Low Temperature

Refrigeration at 5°C retards microbial growth, although with extended storage, microorganisms will eventually grow and produce spoilage. Slow microbial growth at temperatures below −10°C has been described, particularly with fruit juice concentrates, ice cream, and some fruits. Some microorganisms are very sensitive to cold and their numbers will be reduced, but cold does not lead to significant decreases in overall microbial populations.

Temperature effects on microbial growth (pp. 122–25)

High Temperature

Controlling microbial populations in foods by means of high temperatures can significantly limit disease transmission and spoilage.

Heating processes, first used by Nicholas Appert in 1809 (**Historical Highlights 41.1**), provide a safe means of preserving foods, particularly when carried out in commercial **canning** operations (**figure 41.6**). Canned food is heated in special containers called retorts at about 115°C for intervals ranging from 25 to over 100 minutes. The precise time and temperature depend on the nature of the food. Sometimes canning does not kill all the microorganisms, but only those that will spoil the food (remaining bacteria are unable to grow due to acidity of the food, as an example). After heat treatment the cans are cooled as rapidly as possible, usually with cold water. Quality control and processing effectiveness are sometimes compromised, however, in home processing of foods, especially with less acidic (pH values greater than 4.6) products such as green beans or meats. Destruction of microorganisms by heat (pp. 136–39)

Pasteurization involves heating food to a temperature that kills disease-causing microorganisms and substantially reduces the levels of spoilage organisms. In the processing of milk, beers,

Figure 41.6 Food Preparation for Canning. Microbial control is important in the processing and preservation of many foods. Worker pouring peas into a large, clean vat during the preparation of vegetable soup. After preparation the soup is transferred to cans. Each can is heated for a short period, sealed, processed at temperatures around 110–121°C in a canning retort to destroy spoilage microorganisms, and finally cooled.

and fruit juices by conventional low-temperature holding (LTH) pasteurization, the liquid is maintained at 62.8°C for 30 minutes. Products can also be held at 71°C for 15 seconds, a high-temperature, short-time (HTST) process; milk can be treated at 141°C for 2 seconds for ultra-high-temperature (UHT) processing. Shorter-term processing results in improved flavor and extended product shelf life.

Such heat treatment is based on a statistical probability that the number of remaining viable microorganisms will be below a certain level after a particular heating time at a specific temperature. This process is discussed in detail on pp. 135–37.

Despite efforts to eliminate spoilage microorganisms during canning, sometimes canned foods become spoiled. This may be due to spoilage before canning, underprocessing during canning, and leakage of contaminated water through can seams during

Table 41.5

Major Groups of Chemicals Used in Food Preservation

Preservatives	Approximate Maximum Use Range	Organisms Affected	Foods
Propionic acid/propionates	0.32%	Molds	Bread, cakes, some cheeses, inhibitor of ropy bread dough
Sorbic acid/sorbates	0.2%	Molds	Hard cheeses, figs, syrups, salad dressings, jellies, cakes
Benzoic acid/benzoates	0.1%	Yeasts and molds	Margarine, pickle relishes, apple cider, soft drinks, tomato ketchup, salad dressings
Parabens[a]	0.1%	Yeasts and molds	Bakery products, soft drinks, pickles, salad dressings
SO_2/sulfites	200–300 ppm	Insects and microorganisms	Molasses, dried fruits, wine, lemon juice (not to be used in meats or other foods recognized as sources of thiamine)
Ethylene/propylene oxides	700 ppm	Yeasts, molds, vermin	Fumigant for spices, nuts
Sodium diacetate	0.32%	Molds	Bread
Dehydroacetic acid	65 ppm	Insects	Pesticide on strawberries, squash
Sodium nitrite	120 ppm	Clostridia	Meat-curing preparations
Caprylic acid	—	Molds	Cheese wraps
Ethyl formate	15–200 ppm	Yeasts and molds	Dried fruits, nuts

From James M. Jay. 2000. *Modern Food Microbiology*, 6th edition. Reprinted by permission of Aspen Publishing, Frederick, Md.

[a]Methyl-, propyl-, and heptyl-esters of *p*-hydroxybenzoic acid.

cooling. Spoiled food can be altered in such characteristics as color, texture, odor, and taste. Organic acids, sulfides, and gases (particularly CO_2 and H_2S) may be produced. If spoilage microorganisms produce gas, both ends of the can will bulge outward to give a swell. Sometimes the swollen ends can be moved by thumb pressure (soft swells); in other cases the gas pressure is so great that the ends cannot be dented by hand (hard swells). It should be noted that swelling is not always due to microbial spoilage. Acid in high-acid foods may react with the iron of the can to release hydrogen and generate a hydrogen swell. Hydrogen sulfide production by *Desulfotomaculum* can cause "sulfur stinkers."

Water Availability

Dehydration, such as lyophilization to produce freeze-dried foods, is now a common means of eliminating microbial growth. The modern process is simply an update of older procedures in which grains, meats, fish, and fruits were dried. The combination of free-water loss with an increase in solute concentration in the remaining water makes this type of preservation possible.

Chemical-Based Preservation

Various chemical agents can be used to preserve foods, and these substances are closely regulated by the U.S. Food and Drug Administration and are listed as being "generally recognized as safe" or **GRAS (table 41.5)**. They include simple organic acids, sulfite, ethylene oxide as a gas sterilant, sodium nitrite, and ethyl formate. These chemical agents may damage the microbial plasma membrane or denature various cell proteins. Other compounds interfere with the functioning of nucleic acids, thus inhibiting cell reproduction.

The effectiveness of many of these chemical preservatives depends on the food pH. As an example, sodium propionate is most effective at lower pH values, where it is primarily undissociated and able to be taken up by lipids of microorganisms. Breads, with their low pH values, often contain sodium propionate as a preservative. Chemical preservatives are used with grain, dairy, vegetable, and fruit products. Sodium nitrite is an important chemical used to help preserve ham, sausage, bacon, and other cured meats by inhibiting the growth of *Clostridium botulinum* and the germination of its spores. This protects against botulism and reduces the rate of spoilage. Besides increasing meat safety, nitrite decomposes to nitric acid, which reacts with heme pigments to keep the meat red in color. Current concern about nitrite arises from the observation that it can react with amines to form carcinogenic nitrosamines.

Radiation

Radiation, both ionizing and nonionizing, has an interesting history in relation to food preservation. Ultraviolet radiation is used to control populations of microorganisms on the surfaces of laboratory and food-handling equipment, but it does not penetrate food. The major method used for radiation sterilization of food is gamma irradiation from a cobalt-60 source (*see p. 141*). Such electromagnetic radiation has excellent penetrating power and must be used with moist foods because the radiation produces peroxides from water in the microbial cells, resulting in oxidation of sensitive cellular constituents. This process of **radappertization,** named after Nicholas Appert, can extend the shelf life of seafoods, fruits, and vegetables. To sterilize meat products, commonly 4.5 to 5.6 megarads are used.

Among the more interesting radiation-resistant bacteria that have been studied is *Deinococcus radiodurans* (*see section 21.2*). This bacterium has a complex cell wall structure and tetrad-forming growth patterns (*see figure 21.2*). It also has an extraordinary capacity to withstand high doses of radiation, at least partly because of its ability to mend hundreds of double-stranded DNA breaks by template-independent repair mechanisms.

Microbial Product-Based Inhibition

There is increasing interest in the use of **bacteriocins** (*see p. 688*) for the preservation of foods. Bacteriocins are bactericidal proteins active against closely related bacteria, which bind to specific sites on the cell, and often affect cell membrane integrity and function. The only currently approved product is nisin. Nisin, produced by some strains of *Streptococcus lactis,* is a small hydrophobic protein. It is nontoxic to humans and affects mainly gram-positive bacteria, especially *Enterococcus faecalis.* Nisin can be used particularly in low-acid foods to improve inactivation of *Clostridium botulinum* during the canning process or to inhibit germination of any surviving spores.

Bacteriocins have a wide variety of names, depending on the organisms that produce them. They can function by several mechanisms. They may dissipate the proton motive force (PMF) of a susceptible bacterium. Some form hydrophobic pores in plasma membranes and promote the release of low-molecular-weight molecules. Bacteriocins may also inhibit protein or RNA synthesis. Bacteriocin addition to foods such as cheddar cheese can lead to a two- to threefold reduction in *Listeria monocytogenes* in 180-day-old cheeses. Similar compounds also occur in the eucaryotes.

---

1. Describe the major approaches used in food preservation.
2. What types of chemicals can be used to preserve foods?
3. Nitrite is often used to improve the storage characteristics of prepared meats. What toxicological problems may result from the use of this chemical?
4. Under what conditions can ultraviolet light and gamma radiation be used to control microbial populations in foods and in food preparation? What is radappertization?
5. In principle, how do bacteriocins such as nisin function? What bacterial genus produces this important polypeptide?

41.4   FOOD-BORNE DISEASES

Food-borne illnesses impact the entire world. In the United States, based on recent information from the Centers for Disease Control and Prevention, annual incidences of food-related diseases involve 76 million cases, of which only 14 million can be attributed to known pathogens. Food-borne diseases result in

325,000 hospitalizations and at least 5,000 deaths per year. Since 1942, the number of recognized food-borne pathogens has increased over fivefold! Are these new microorganisms? In most cases, these pathogens are simply agents that we now can describe, based on an improved understanding of microbial diversity. Recent estimates indicate that Norwalk-like viruses, *Campylobacter jejuni* and *Salmonella* are the major causes of food-borne diseases. In addition, *Escherichia coli* O157:H7 and *Listeria* are important food-related pathogens.

Many diseases transmitted by foods, or food poisonings, are discussed in chapters 38 and 39, and only a few of the more important food-borne bacterial pathogens are mentioned here. There are two primary types of food-related diseases: food-borne infections and food intoxications. Food-borne and waterborne diseases (pp. 868–70, 905–11)

All these food-borne diseases are associated with poor hygienic practices. Whether by water or food transmission, the fecal-oral route is maintained, with the food providing the vital link between hosts. Fomites, such as sink faucets, drinking cups, and cutting boards, also play a role in the maintenance of the fecal-oral route of contamination.

Food-Borne Infection

A **food-borne infection** involves the ingestion of the pathogen, followed by growth in the host, including tissue invasion and/or the release of toxins. The major diseases of this type are summarized in **table 41.6** (*see also table 39.5*).

Salmonellosis results from ingestion of a variety of *Salmonella* serovars, particularly *typhimurium* and *enteritidis* (*see also section 39.4*). Gastroenteritis is the disease of most concern in relation to foods such as meats, poultry, and eggs, and the onset of symptoms occurs after an incubation time as short as 8 hours. *Salmonella* infection can arise from contamination by workers in food-processing plants and restaurants, as well in canning processes (**Historical Highlights 41.2**).

Campylobacter jejuni is considered a leading cause of acute bacterial gastroenteritis in humans and can affect persons of all ages. This important pathogen is often transmitted by uncooked or poorly cooked poultry products. For example, transmission often occurs when kitchen utensils and containers are used for chicken preparation and then for salads. Contamination with as few as 10 viable *Campylobacter jejuni* cells can lead to the onset of diarrhea. *Campylobacter jejuni* also is transmitted by raw milk, and the organism has been found on various red meats. Thorough cooking of food prevents this disease transmission problem.

Listeriosis, caused by *Listeria monocytogenes* (*see section 39.4*), is of continuing interest, as shown by the outbreak that occurred in Southern California in 1985. This outbreak was caused by improper pasteurization of milk used in the commercial production of Mexican-style cheeses. At least 86 cases of infection occurred, including 58 cases involving mother-infant pairs. Forty-seven people died. The outbreak was traced to pinhole leaks in the heat exchangers of a pasteurizing unit. The leaks al-

Table 41.6

Major Food-Borne Infectious Diseases

Disease	Organism	Incubation Period and Characteristics	Major Foods Involved
Salmonellosis	*S. typhimurium, S. enteritidis*	8–48 hr Enterotoxin and cytotoxins	Meats, poultry, fish, eggs, dairy products
Arcobacter diarrhea	*Arcobacter butzleri*	Severe diarrhea, recurrent cramps	Meat products, especially poultry
Campylobacteriosis	*Campylobacter jejuni*	Usually 2–10 days Most toxins are heat-labile	Milk, pork, poultry products, water
Listeriosis	*L. monocytogenes*	Varying periods Related to meningitis and abortion; newborns and the elderly especially susceptible	Meat products, especially pork and milk
Escherichia coli diarrhea and colitis	*E. coli,* including serotype O157:H7	24–72 hr Enterotoxigenic positive and negative strains; hemorrhagic colitis	Undercooked ground beef, raw milk
Shigellosis	*Shigella sonnei, S. flexneri*	24–72 hr	Egg products, puddings
Yersiniosis	*Yersinia enterocolitica*	16–48 hr Some heat-stable toxins	Milk, meat products, tofu
Plesiomonas diarrhea	*Plesiomonas shigelloides*	1–2 hr	Uncooked mollusks and foreign travel
Vibrio parahaemolyticus gastroenteritis	*V. parahaemolyticus*	16–48 hr	Seafood, shellfish

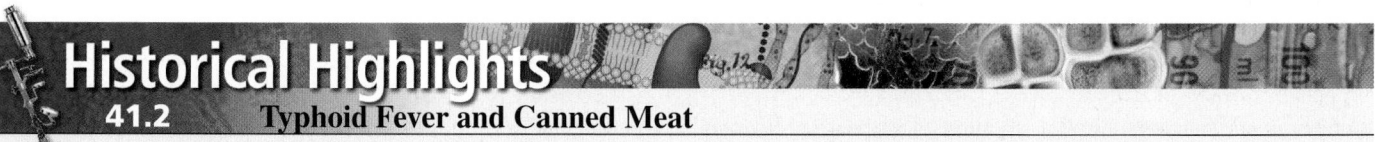

Historical Highlights

41.2   Typhoid Fever and Canned Meat

Minor errors in canning have led to major typhoid outbreaks. In 1964 canned corned beef produced in South America was cooled, after sterilization, with nonchlorinated water; the vacuum created when the cans were cooled drew *S. typhi* into some of the cans, which were not completely sealed. This contaminated product was later sliced in an Aberdeen, Scotland, food store, and the meat slicer became a continuing contamination source; the result was a major epidemic that in-

volved 400 people. The *S. typhi* was a South American strain, and eventually the contamination was traced to the contaminated water used to cool the cans.

This case emphasizes the importance of careful food processing and handling to control the spread of disease during food production and preparation.

lowed incoming raw milk to contaminate the pasteurized milk before production of the cheese. *Listeria* is difficult to work with because an extended incubation of samples is required for growth and detection.

Escherichia coli is now recognized as an important food-borne disease organism. Enteropathogenic, enteroinvasive, and enterotoxigenic types can cause diarrhea. *E. coli* O157:H7 (*see p. 910*) with its specific somatic (O) and flagellar (H) antigens (*see Techniques & Applications 33.2*), is thought to have acquired enterohemorrhagic genes from *Shigella,* including the genes for shigalike verocytotoxins. This produced a new pathogenic strain, first discovered in 1982 and now known around the world. The pathogen is spread by the fecal-oral route, and an infectious dose appears to be only 500 bacteria. Enterohemorrhagic *E. coli* has been found in meat products such as hamburger and salami, in unpasteurized fruit drinks, on fruits and vegetables, and in untreated well water. The newspapers are filled with report after re-

port of million-pound lots of beef being recalled, due to *E. coli* contamination. Even if the contaminated beef does not reach consumers, this economic loss, due to poor hygiene, has many negative effects on cattle producers and meat processors.

Prevention of food contamination by *E. coli* O157:H7 is essential from the time of production until consumption. Hygiene must be monitored carefully in larger-volume slaughterhouses where contact of meat with fecal material can occur. Even fruits and vegetables should be handled with care because disease outbreaks have been caused by imported produce. Caution also is essential at the point of use. For example, avoidance of food contamination by hands and utensils is critical. Utensils used with raw foods should not contact cooked food; proper cleaning of cutting boards and utensils minimizes contamination.

Virus contamination is always a potential problem. This is based on transmission by waters (*see chapter 29*) or by lack of hygiene in food preparation and direct contamination by food

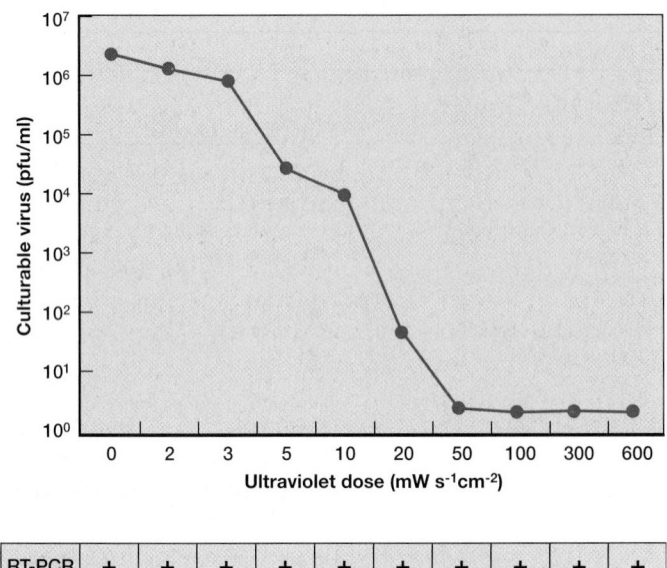

Figure 41.7 **Cultural versus Molecular-Based Virus Detection.** Comparison of plaque-forming ability and recovery of RNA from poliovirus type 2 by RT-PCR with varying UV doses. Even when plaque-forming ability is lost, nucleic acids are still detectable.

processors and handlers. Similar situations occur with protozoan pathogens, discussed in chapter 40. Virus contamination has become a severe problem on many cruise line ships, where Norwalk-like viruses recently have been involved in major sickness bouts, with person-to-person contact and possibly foods implicated in these ultimately avoidable occurrences.

An infectious agent of increasing worldwide concern with respect to food safety is a prion that causes the poorly understood new variant Creutzfeldt-Jakob disease (vCJD). This is one of a group of progressively degenerative neuronal diseases termed transmissible spongiform encephalopathies (TSEs), and is associated with beef cattle. It is often called the "mad cow disease." Major problems in controlling new variant CJD include the long incubation period, which can last many years before onset of the invariably fatal disease, and the lack of reliable detection methods. The major means of vCJD transmission between animals is the use of mammalian tissue in ruminant animal feeds; at the present time, there are significant problems in detecting such prohibited animal products in ruminant feeds. Creutzfeldt-Jakob Disease (CJD), section 38.5; Prions, section 18.9

Foods that are transported and consumed in an uncooked state are an increasingly important source of food-borne infection. The problem becomes more serious because of rapid movement of people and products around the world. International trade in uncooked foods, aided by rapid air transport, provides many opportunities for disease transmission. Fresh foods such as sprouts, seafood, and raspberries pose significant hazards, which will be discussed here.

Sprouts are an increasingly popular part of the new "healthy" life-style. They are fresh, delicate and form an exciting garnish to complement a variety of foods. Unfortunately, if these are not germinated in pathogen-free waters and grown under sanitary conditions, major growth of pathogens can occur. Sprouts produced in areas of the world where there is poor control of water quality and sanitation should be used with caution. Contaminated alfalfa, beans, watercress, mungbean, mustard, and soybean sprouts can be major sources of typhoid and cholera.

Shellfish and finfish also present major concerns. Raw sewage can contaminate shellfish-growing areas; in addition, waterborne pathogens such as *Vibrio* are more prevalent in the water column during the warm months (e.g., in Chesapeake Bay on the mid-Atlantic coast of the United States). Viruses also can be a problem. Oysters are filter feeders that process several liters of water per day, leading to the potential concentration of at least 100 types of enteric viruses. Reverse transcriptase PCR (RT-PCR) can be used to detect RNA viruses in oysters based on the presence of their nucleic acids. However, the inability of molecular techniques to differentiate between infectious and noninfectious particles is a major problem. For example, UV treatment can inactivate many RNA viruses without eliminating the RT-PCR signal, although the virions no longer replicate in a suitable tissue culture environment (**figure 41.7**). Heavy rainfall in shellfish areas can cause runoff of pathogens from adjacent summer cottage septic systems and contaminate coastal waters. Often it is necessary to ban shellfish harvesting until the animals void pathogens from their digestive systems. Alternatively, shellfish from contaminated areas can be moved to clean waters to allow them to clean their digestive systems. The polymerase chain reaction (pp. 316–19)

Raspberries provide an important example of another major problem: the rapid air transport of raw agricultural products around the world. A major recent outbreak of *Cyclospora cayetanensis* poisoning was traced to raspberries imported from Central America into the United States and Canada. In the growth and harvesting process, the raspberries apparently became contaminated, resulting in serious diarrhea in affected individuals. This organism has a complex life cycle, which is not fully understood at the present time. In comparison with *Giardia* and *Cryptosporidium*, which are infective immediately after being shed in feces, *Cyclospora* is not immediately infectious; sporulation or maturing requires 12 hours after release from the body. The mature infective cyst has two sporocysts (**figure 41.8**), an important criterion for confirming the presence of this parasite on foods or in the environment. *Cyclospora* (pp. 632–33)

Food-Borne Intoxications

Microbial growth in food products also can result in a **food intoxication,** as summarized in table 39.5 (*see pp. 906–7*). Intoxication produces symptoms shortly after the food is consumed because growth of the disease-causing microorganism is not required. Toxins produced in the food can be associated with microbial cells or can be released from the cells.

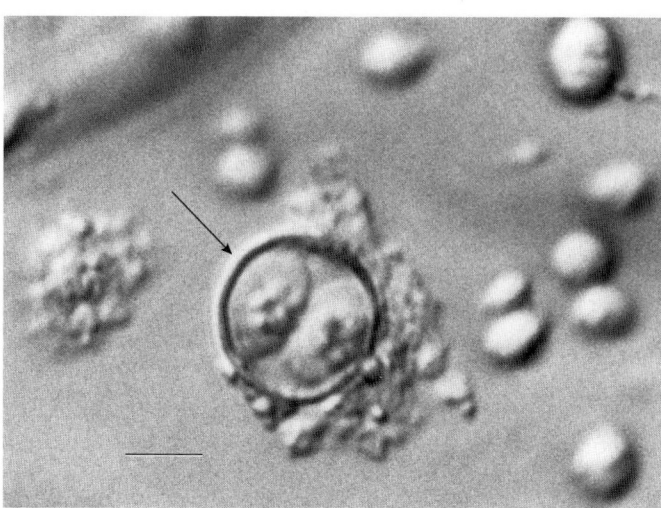

Figure 41.8 *Cyclospora cayetanensis,* **an Important Contaminant of Raw Foods.** *Cyclospora cayetanensis* can be recognized in waste waters and after recovery from contaminated foods due to the occurrence of an oocyst with two sporocysts. Bar = 5 μm.

Most *Staphylococcus aureus* strains cause a staphylococcal enteritis related to the synthesis of extracellular toxins (*see section 39.4*). These are heat-resistant proteins, and heating will not usually render the food safe. The effects of the toxins are quickly felt, with disease symptoms occurring within 2 to 6 hours. The main reservoir of *S. aureus* is the human nasal cavity. Frequently *S. aureus* is transmitted to a person's hands and then is introduced into food during preparation. Growth and enterotoxin production usually occur when contaminated foods are held at room temperature for several hours.

Three gram-positive rods are known to cause food intoxications: *Clostridium botulinum, C. perfringens,* and *Bacillus cereus* (*see table 39.5*). *C. botulinum* poisoning is discussed in chapter 39, and *C. perfringens* intoxication is described here.

Clostridium perfringens food poisoning is one of the more widespread food intoxications. These microorganisms, which produce exotoxins, must grow to levels of approximately 10^6 bacteria per gram or higher in a food to cause disease. At least 10^8 bacteria must be ingested. They are common inhabitants of soil, water, food, spices, and the intestinal tract. Upon ingestion the cells sporulate in the intestine. The enterotoxin is a spore-specific protein and is produced during the sporulation process. Enterotoxin can be detected in the feces of affected individuals. *Clostridium perfringens* food poisoning is common and occurs after meat products are heated, which results in O_2 depletion. If the foods are cooled slowly, growth of the microorganism can occur. At 45°C, enterotoxin can be detected 3 hours after growth is initiated. Onset of the symptoms—watery diarrhea, nausea, and abdominal cramps—usually occurs in about 8 to 16 hours.

Baked potatoes served in aluminum foil can provide a unique environment for pathogenic microorganisms. Potatoes, even after washing, are covered by *Clostridium botulinum,* which naturally occurs in the soil. If the aluminum foil-covered potatoes are not heated sufficiently in the baking process, surviving clostridia can proliferate after removal of the potatoes from the oven and rapidly produce toxins.

Bacillus cereus also is of concern in starchy foods. It can cause two distinct types of illnesses depending on the type of toxin produced: an emetic illness characterized by nausea and vomiting with an incubation time of 1 to 6 hours, and a diarrheal type, with an incubation of 4 to 16 hours. The emetic type is often associated with boiled or fried rice, while the diarrheal type is associated with a wider range of foods.

1. What are the major food-borne diseases in the United States?
2. Discuss the major characteristic of a food-borne infection in terms of the time required between ingestion of the pathogen and the onset of the disease. Why does this occur?
3. What common food is often related to *Campylobacter*-caused gastroenteritis? What means can be used to control the occurrence of this disease from this source?
4. Give some of the sources of *E. coli* O157:H7 that have been of concern in terms of disease transmission.
5. Why is new variant Creutzfeldt-Jakob disease (vCJD) of such concern?
6. What are some uncooked foods that have been implicated in food-borne disease transmission?
7. What are some of the major genera involved in food-borne intoxications?
8. How does a food-borne intoxication differ from a food-borne infection?

 **41.5   DETECTION OF FOOD-BORNE PATHOGENS**

A major problem in maintaining food safety is the need to rapidly detect microorganisms in order to curb outbreaks that can affect large populations. This is especially important because of widescale distribution of perishable foods. Standard culture techniques (*see chapter 5*) may require days to weeks for positive identification of pathogens. Identification is often complicated by the low numbers of pathogens compared with the background microflora. Furthermore, the varied chemical and physical composition of foods can make isolation difficult. Fluorescent antibody, enzyme-linked immunoassays (ELISAs), and radioimmunoassay techniques have proven of value (*see section 33.3*). These can be used to detect small amounts of pathogen-specific antigens.

Molecular techniques also are increasingly used in identification. The basic methods of analysis and manipulation of DNA and RNA are discussed in chapter 14. These methods are valuable for three purposes: (1) to detect the presence of a single, specific pathogen; (2) to detect viruses that cannot be grown conveniently; and (3) to identify slow-growing or noncultured pathogens.

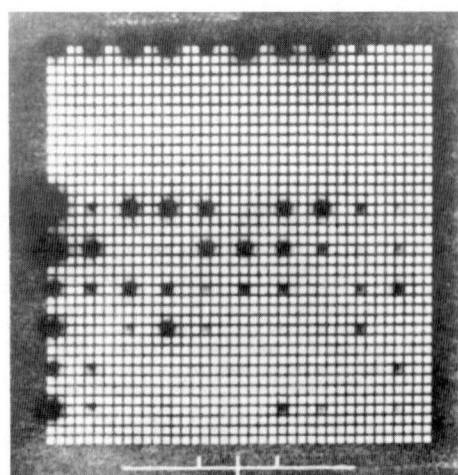

Figure 41.9 Molecular Probes and Food Microbiology. Molecular techniques are finding increasing use in microbial analysis of foods. Autoradiogram of a radioactive-labeled *Listeria monocytogenes* probe against 100 *Listeria* cultures. Only the *Listeria monocytogenes* cultures show sequence homology and binding with the DNA probe, darkening the autoradiogram film. The other *Listeria* spp. do not react with the probe.

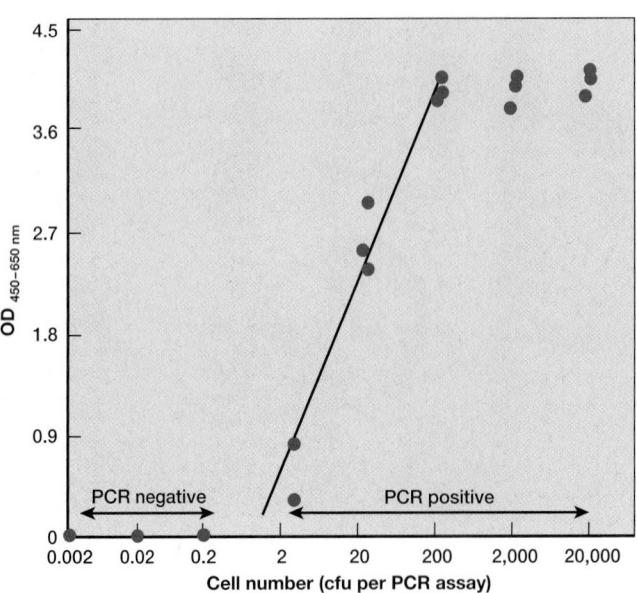

Figure 41.10 Polymerase Chain Reaction (PCR)-Based Pathogen Detection. Comparison of PCR sensitivity and growth for *Salmonella enterica* subsp. *enterica* serovar *agona* detection. The Probalia PCR system can detect as few as two colony forming units (CFU) of the pathogen. OD = optical density.

Pathogens now can be identified by detecting specific DNA or RNA base sequences with **probes.** These usually are 14 to 40 bases in length. They may be created by generating fragments with restriction endonucleases or through direct chemical synthesis. Probes are labeled by linking them to a variety of enzymatic, isotopic, chromogenic, or luminescent/fluorescent markers. A major advantage of their use is the speed with which specific microorganisms can be detected in a set of cultures, as shown in **figure 41.9.** In this example, a hydrophobic grid-membrane system has been used. The *Listeria monocytogenes* cultures are radioactive, indicating that they have bound the probe, while other *Listeria* species do not show probe binding.

Another example of the use of molecular techniques is provided by pathogenic *E. coli*. Currently *E. coli* O157:H7 is isolated and identified using selective culture media, rapid identification kits, rapid probe-based identification procedures, serotype-specific probes, and PCR techniques. These techniques also have been extended to allow the detection of a few target cells in large populations of background microorganisms. For example, by using the polymerase chain reaction, as few as 10 toxin-producing *E. coli* cells can be detected in a population of 100,000 cells isolated from soft-cheese samples.

As few as two colony forming units of *Salmonella* can be detected with a special PCR system (**figure 41.10**). This procedure makes it possible to confirm *Salmonella* presence within 24 to 48 hours, whereas 3 to 4 days is needed for presumptive identification with standard culture procedures. Confirmation of *Salmonella* presence would then require additional time. Frequently, to improve the sensitivity and increase the speed of this

method, a preenrichment step will be used before PCR. In this type of analysis, especially for well-characterized pathogens, all culture and PCR reagents are provided in "pill" form to speed the entire process. PCR is now being used for rapid detection of food-borne pathogens. In recent studies, the construction of specific 159 and 1,223 base pair PCR products, which can be separated electrophoretically, has made it possible to detect *Campylobacter jejuni* and *Arcobacter butzleri* in the same sample within 8 hours.

A major advance in the detection of food-borne pathogens is the use of standardized pathogen DNA patterns, or "food-borne pathogen fingerprinting." The Centers for Disease Control and Prevention in the United States has established a program, called **PulseNet,** in which pulsed-field gel electrophoresis (PFGE) is used under carefully controlled and duplicated conditions to determine the distinctive DNA pattern of each bacterial pathogen. With this uniform procedure, it is possible to link pathogens associated with disease outbreaks in different parts of the world to a specific food source. Data from around the world are being used in **FoodNet,** an active surveillance network, to follow nine major food-borne diseases. Using the FoodNet approach, it is possible to trace the course and cause of infection in days and not weeks. As an example, a *Shigella* outbreak in three different areas of North America was traced to Mexican parsley that had been tainted with polluted irrigation water. This program has resulted in more rapid establishment of epidemiological linkages and a decreased occurrence of many of these important food-borne diseases.

Techniques & Applications
41.3   Starter Cultures, Bacteriophage Infections, and Plasmids

Cultures of lactic acid bacteria, called **starter cultures,** are added to milk during the preparation of many dairy products. For example, *Streptococcus lactis* and *S. cremoris* are used in the production of cheese. One of the greatest problems for the dairy industry is the presence of bacteriophages that destroy these starter cultures. Lactic acid production by a heavily phage-infected starter culture can come to a halt within 30 minutes. The industry has tried to overcome this problem by practicing aseptic techniques in order to reduce phage contamination, and by selecting for phage-resistant bacterial cultures.

Most efforts at control have not been successful in the longer term. It has been found that the very aseptic techniques and phage-resistant pure cultures used to attempt to solve this problem actually were parts of the problem. The most stable and dependable cultures,

called P starter cultures, contain the bacteriophages in a pseudolysogenic state. When cultures are grown without phages or under aseptic conditions (L starters), they lose their phage resistance. The key to this riddle appears to be plasmids, which can block phage adsorption. The loss of the plasmids in a subpopulation of the bacteria allows the phage carrier state to be established. New phages can develop by acquiring restriction enzymes (*see section 14.1*) from plasmids. These modified phages again become lytic, establishing a new equilibrium in the population.

Other control approaches are being tested. Antisense RNA is now being used in an attempt to provide an agent against bacteriophage genes to help in the constant struggle between lactic acid bacteria and their phages.

---

1. What is a "probe" and how is it used for detection of food-borne pathogens?
2. How is the polymerase chain reaction used in pathogen detection?
3. How are PulseNet and FoodNet used in the surveillance of food-borne diseases?

41.6   MICROBIOLOGY OF FERMENTED FOODS

Over the last several thousand years, fermentation has been a major way of preserving food. Microbial growth, either of natural or inoculated populations, causes chemical and/or textural changes to form a product that can be stored for extended periods. The fermentation process also is used to create new, pleasing food flavors and odors.

The major fermentations used in food microbiology are the lactic, propionic, and ethanolic fermentations, as discussed in section 9.3. These fermentations are carried out with a wide range of cultures, many of which have not been characterized.

Fermented Milks

Throughout the world, at least 400 different fermented milks are produced. These fermentations are carried out by mesophilic, thermophilic, and therapeutic lactic acid bacteria, as well as by yeasts and molds as noted in **table 41.7.** Only major examples of these fermentation types will be discussed in this section.

Mesophilic

Mesophilic milk fermentations result from similar manufacturing techniques, in which acid produced through microbial activity causes protein denaturation. To carry out the process, one usually inoculates milk with the desired starter culture (**Techniques &**

Applications 41.3); incubates it at optimum temperature (approximately 20 to 30°C), and then stops microbial growth by cooling. *Lactobacillus* spp. and *Lactococcus lactis* cultures are used for aroma and acid production. The organism *Lactococcus lactis* subsp. *diacetilactis* converts milk citrate to diacetyl, which gives a special buttery flavor to the finished product. The use of these microorganisms with skim milk produces cultured buttermilk, and when cream is used, sour cream is the result.   The genus *Lactobacillus* (p. 515); The genus *Lactococcus* (pp. 516–18)

Thermophilic

In addition to mesophilic milk fermentations, thermophilic fermentations can be carried out at temperatures around 45°C. An important example is yogurt production. **Yogurt** is made using a

---

Table 41.7
Major Categories and Examples of Fermented Milk Products

Category	Typical Examples
I. Lactic fermentations	
Mesophilic	Buttermilk
	Cultured buttermilk
	Långofil
	Tëtmjolk
	Ymer
Thermophilic	Yogurt, laban, zabadi, labneh, skyr
	Bulgarian buttermilk
Therapeutic	Biogarde, Bifighurt
	Acidophilus milk, yakult
	Cultura-AB
II. Yeast-lactic fermentations	Kefir, koumiss, acidophilus-yeast milk
III. Mold-lactic fermentations	Viili

Source: Table 3.1, p. 58. In B. A. Law, editor. 1997. *Microbiology and Biochemistry of Cheese and Fermented Milk*, 2nd ed. New York: Chapman and Hall.

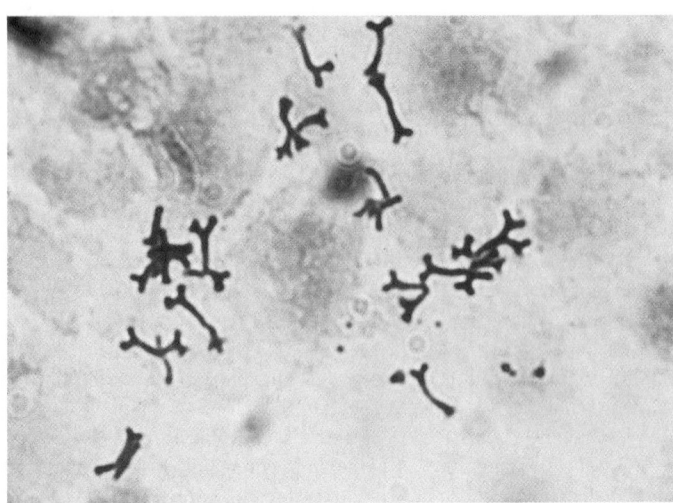

Figure 41.11 Bifidobacteria. Cultured milks are increasing in popularity. A light micrograph of *Bifidobacterium,* a microorganism suggested to provide many health benefits.

Figure 41.12 Examples of Bifid-Amended Dairy Products. These are produced in many countries.

special starter culture in which two major bacteria are present in a 1:1 ratio: *S. thermophilus* and *L. bulgaricus*. With these organisms growing in concert, acid is produced by *Streptococcus,* and aroma components are formed by the *Lactobacillus*. Freshly prepared yogurt contains 10^9 bacteria per gram.

Therapeutic

Fermented milks may have beneficial therapeutic effects. Acidophilus milk is produced by using *Lactobacillus acidophilus. L. acidophilus* may modify the microbial flora in the lower intestine, thus improving general health, and it often is used as a dietary adjunct. Many microorganisms in fermented dairy products stabilize the bowel microflora, and some appear to have antimicrobial properties. The exact nature and extent of health benefits are still unclear but may involve minimizing lactose intolerance, lowering serum cholesterol, and possibly exhibiting anticancer activity. Several lactobacilli have antitumor compounds in their cell walls. Such findings suggest that diets including lactic acid bacteria, especially *L. acidophilus,* may contribute to the control of colon cancer.

Another interesting group used in milk fermentations are the bifidobacteria. The genus *Bifidobacterium* contains irregular, nonsporing, gram-positive rods that may be club-shaped or forked at the end (**figure 41.11**). Bifidobacteria are nonmotile, anaerobic, and ferment lactose and other sugars to acetic and lactic acids. They are typical residents of the human intestinal tract and were discovered in 1906. Many beneficial properties are attributed to them. Bifidobacteria are thought to help maintain the normal intestinal balance, while improving lactose tolerance; to possess antitumorigenic activity; and to reduce serum cholesterol levels. In addition, some believe that they promote calcium absorption and the synthesis of B-complex vitamins. It has also been suggested that bifidobacteria reduce or prevent the excretion of rotaviruses, a cause of diarrhea among children. *Bifidobacterium*-amended

fermented milk products, including yogurt, are now available in various parts of the world (**figure 41.12**).

Yeast Lactic

Yeast-lactic fermentations include **kefir,** a product with an ethanol concentration of up to 2%. This unique fermented milk originated in the Caucasus Mountains and it is produced east into Mongolia. Kefir products tend to be foamy and frothy, due to active carbon dioxide production. This process is based on the use of kefir "grains" as an inoculum. These are coagulated lumps of casein that contain yeasts, lactic acid bacteria, and acetic acid bacteria. In this fermentation, the grains are used to inoculate the fresh milk and then recovered at the end of the fermentation. Originally, kefir was produced in leather sacks hung by the front door during the day, and passersby were expected to push and knead the sack to mix and stimulate the fermentation. Fresh milk could be added occasionally to maintain activity.

Mold Lactic

Mold-lactic fermentation results in a unique Finnish fermented milk called viili. The milk is placed in a cup and inoculated with a mixture of the fungus *Geotrichium candidum* and lactic acid bacteria. The cream rises to the surface, and after incubation at 18 to 20°C for 24 hours, lactic acid reaches a concentration of 0.9%. The fungus forms a velvety layer across the top of the final product, which also can be made with a bottom fruit layer to add extra flavor.

Cheese Production

Cheese is one of the oldest human foods and is thought to have been developed approximately 8,000 years ago. About 2,000 distinct varieties of cheese are produced throughout the world, representing approximately 20 general types (**table 41.8** and **figure 41.13**). Often cheeses are classified based on texture or hardness

Major Types of Cheese and Microorganisms Used in Their Production

Cheese (Country of Origin)	Contributing Microorganisms[a]	
	Earlier Stages of Production	**Later Stages of Production**
Soft, unripened		
Cottage	*Lactococcus lactis*	*Leuconostoc cremoris*
Cream	*L. cremoris, L. diacetylactis, S. thermophilus, L. bulgaricus*	
Mozzarella (Italy)	*S. thermophilus, L. bulgaricus*	
Soft, ripened		
Brie (France)	*Lactococcus lactis, L. cremoris*	*Penicillium camemberti, P. candidum, Brevibacterium linens*
Camembert (France)	*L. lactis, L. cremoris*	*Penicillium camemberti, Brevibacterium linens*
Semisoft		
Blue, Roquefort (France)	*Lactococcus lactis, L. cremoris*	*Penicillium roqueforti*
Brick, Muenster (United States)	*L. lactis, L. cremoris*	*Brevibacterium linens*
Limburger (Belgium)	*L. lactis, L. cremoris*	*Brevibacterium linens*
Hard, ripened		
Cheddar, Colby (Britain)	*Lactococcus lactis, L. cremoris*	*Lactobacillus casei, L. plantarum*
Swiss (Switzerland)	*L. lactis, L. helveticus, S. thermophilus*	*Propionibacterium shermanii, P. freudenreichii*
Very hard, ripened		
Parmesan (Italy)	*Lactococcus lactis, L. cremoris, S. thermophilus*	*Lactobacillus bulgaricus*

[a]*Lactococcus lactis* stands for *L. lactis* subsp. *lactis. Lactococcus cremoris* is *L. lactis* subsp. *cremoris,* and *Lactococcus diacetilactis* is *L. lactis* subsp. *diacetilactis.*

as soft cheeses (cottage, cream, Brie), semisoft cheeses (Muenster, Limburger, blue), hard cheeses (cheddar, Colby, Swiss), or very hard cheeses (Parmesan). All cheese results from a lactic acid fermentation of milk, which results in coagulation of milk proteins and formation of a curd. Rennin, an enzyme from calf stomachs, but now produced by genetically engineered microorganisms, can also be used to promote curd formation. After the curd is formed, it is heated and pressed to remove the watery part of the milk or whey, salted, and then usually ripened (**figure 41.14**). The cheese curd can be packaged for ripening with or without additional microorganisms. Cheese curd inoculation is used in the manufacture of Roquefort and blue cheese. In this case *Penicillium roqueforti* spores are added to the curds just before the final cheese processing. Sometimes the surface of an already formed cheese is inoculated at the start of ripening; for example, Camembert cheese is inoculated with spores of *Penicillium camemberti.* The final hardness of the cheese is partially a function of the length of ripening. Soft cheeses are ripened for only about 1 to 5 months, whereas hard cheeses need 3 to 12

months, and very hard cheeses like Parmesan require 12 to 16 months ripening.

The ripening process also is critical for Swiss cheese. Gas production by *Propionibacterium* contributes to final flavor development and hole or eye formation in this cheese. Some cheeses are soaked in brine to stimulate the development of specific fungi and bacteria; Limburger is one such cheese.

Meat and Fish

Besides the fermentation of dairy products, a variety of meat products, especially sausage, can be fermented: country-cured hams, summer sausage, salami, cervelat, Lebanon bologna, fish sauces (processed by halophilic *Bacillus* species), izushi, and katsuobushi. *Pediococcus cerevisiae* and *Lactobacillus plantarum* are most often involved in sausage fermentations. Izushi is based on the fermentation of fresh fish, rice, and vegetables by *Lactobacillus* spp.; katsuobushi results from the fermentation of tuna by *Aspergillus glaucus.* Both meat fermentations originated in Japan.

(a)

(b)

(c)

(d)

(e)

Figure 41.13   Cheese.   A bewildering array of cheeses are produced around the world using microorganisms. (**a**) Gouda (top left) and cheddar cheese (lower right). Note the typical indentations on the surface of Gouda caused by the cheesecloth and red wax covering. (**b**) Roquefort cheese crumbled for use in salad dressing. The dark areas are the result of extensive *Penicillium* growth. (**c**) Swiss cheese. This hard, ripened cheese contains holes formed by carbon dioxide from a *Propionibacterium* fermentation. (**d**) Brie (left) and Limburger (right) cheeses are soft, ripened cheeses. Ripening results from the surface growth of microorganisms like *Penicillium camemberti* (Brie) and *Brevibacterium linens* (Limburger). (**e**) Cottage cheese and cream cheese (spread on crackers) are soft, unripened cheeses. They are sold immediately after production, and the curd is consumed without further modification by microorganisms.

Figure 41.14   Cheddar Cheese Production.   Cheddar, a village in England, has given its name to a cheese made in many parts of the world. "Cheddaring" is the process of turning and piling the curd to express whey and develop desired cheese texture.

---

1. What are the major types of milk fermentations?
2. Briefly describe how buttermilk, sour cream, and yogurt are made.
3. What is unique about the morphology of *Bifidobacterium?* Why is it used in milk?
4. How and where are kefir and viili made?
5. What major steps are used to produce cheese? How is the cheese curd formed in this process? What is whey? How does Swiss cheese get its holes?
6. Which fungal genus is often used in cheese making? What cheeses are produced using this genus?
7. Give a microbial genus used in meat fermentations.

Production of Alcoholic Beverages

A variety of plants that contain adequate carbohydrates can be used to produce alcoholic beverages. When carbohydrates are available in readily fermentable form, the fermentation can be started immediately. For example, grapes are crushed to release the juice or **must,** which can be allowed to ferment without further delay. The must also can be sterilized by pasteurization or the use of sulfur dioxide, and then the desired microbial culture added.

In contrast, before cereals and other starchy materials can be used as substrates for the production of alcohol, their complex carbohydrates must be hydrolyzed. They are mixed with water and incubated in a process called **mashing.** The insoluble material is then removed to yield the **wort,** a clear liquid containing

fermentable sugars and other simple molecules. Much of the art of beer and ale production involves the controlled hydrolysis of protein and carbohydrates to provide the desired body and flavor of the final product.

Wines and Champagnes

Wine production, or the science of **enology** [Greek *oinos,* wine, and *ology,* the science of], starts with the collection of grapes, continues with their crushing and the separation of the liquid (must) before fermentation, and concludes with a variety of storage and aging steps (**figure 41.15**). All grapes have white juices. To make a red wine from a red grape, the grape skins are allowed to remain in contact with the must before fermentation to release their skin-coloring components. Wines can be created by using the natural grape skin microorganisms. This natural mixture of bacteria and yeasts gives unpredictable fermentation results. To avoid such problems, one can treat the fresh must with a sulfur dioxide fumigant and add a desired strain of *Saccharomyces cerevisiae* or *S. ellipsoideus.* After inoculation the juice is fermented for 3 to 5 days at temperatures varying between 20 and 28°C. Depending on the alcohol tolerance of the yeast strain, the final product may contain 10 to 18% alcohol. Clearing and development of flavor occur during the aging process. The malolactic fermentation is an important part of wine production. Grape juice contains high levels of organic acids, including malic and tartaric acids. If the levels of these acids are not decreased during the fermentation process, the wine will be too acidic, and have poor stability and "mouth feel." This essential fermentation is carried out by *Leuconostoc oenos, L. plantarum, L. hilgardii, L. brevis,* and *L. casei.* The activities of these microbes transform malic acid (a four-carbon tricarboxylic acid) to lactic acid (a three-carbon monocarboxylic acid) and carbon dioxide. This results in deacidification (pH increase), improvement of flavor stability, and in some cases the possible accumulation of bacteriocins in the wines.   Yeasts (pp. 539, 544–45)

A critical part of wine making involves the choice of whether to produce a dry (no remaining free sugar) or a sweeter (varying amounts of free sugar) wine. This can be controlled by regulating the initial must sugar concentration. With higher levels of sugar, alcohol will accumulate and inhibit the fermentation before the sugar can be completely used, thus producing a sweeter wine. During final fermentation in the aging process, flavoring compounds accumulate and influence the bouquet of the wine.

Microbial growth during the fermentation process produces sediments, which are removed during **racking.** Racking can be carried out at the time the fermented wine is transferred to bottles or casks for aging or even after the wine is placed in bottles.

Many processing variations can be used during wine production. The wine can be distilled to make a "burned wine" or brandy. *Acetobacter* and *Gluconobacter* can be allowed to oxidize the ethanol to acetic acid and form a **wine vinegar.** In the past an acetic acid generator was used to recirculate the wine over a bed of wood chips, where the desired microorganisms developed as a surface growth. Today the process is carried out in large aerobic submerged cultures under much more controlled conditions.

Processing step

Biological change

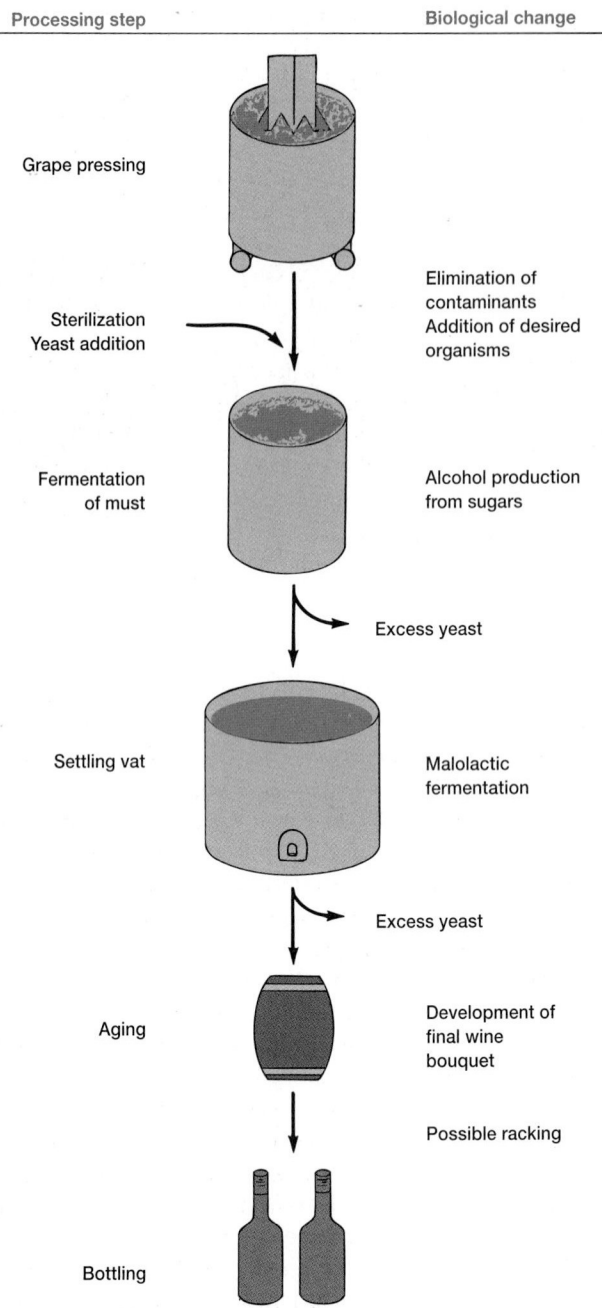

Grape pressing

Sterilization
Yeast addition

Elimination of
contaminants
Addition of desired
organisms

Fermentation
of must

Alcohol production
from sugars

Excess yeast

Settling vat

Malolactic
fermentation

Excess yeast

Aging

Development of
final wine
bouquet

Possible racking

Bottling

Figure 41.15   Wine Making.   Once grapes are pressed, the sugars in the juice (the must) can be immediately fermented to produce wine. Must preparation, fermentation, and aging are critical steps.

Natural champagnes are produced by continuing the fermentation in bottles to produce a naturally sparkling wine. Sediments that remain are collected in the necks of inverted champagne bottles after the bottles have been carefully turned. The necks of the bottles are then frozen and the corks removed to disgorge the accumulated sediments. The bottles are refilled with clear champagne from another disgorged bottle, and the product is ready for final packaging and labeling.

Beers and Ales

Beer and ale production uses cereal grains such as barley, wheat, and rice. The complex starches and proteins in these grains must be changed to a more readily usable mixture of simpler carbohydrates and amino acids. This process, shown in **figure 41.16,** involves germination of the barley grains and activation of their enzymes to produce a **malt.** The malt is then mixed with water and the desired grains, and the mixture is transferred to the mash tun or cask in order to hydrolyze the starch to usable carbohydrates. Once this process is completed, the **mash** is heated with hops (dried flowers of the female vine *Humulus lupulis*), which were originally added to the mash to inhibit spoilage microorganisms (**figure 41.17**). The hops also provide flavor and assist in clarification of the wort. In this heating step the hydrolytic enzymes are inactivated and the wort can be **pitched**—inoculated—with the desired yeast.

Most beers are fermented with **bottom yeasts,** related to *Saccharomyces pastorianus,* which settle at the bottom of the fermentation vat. The beer flavor also is influenced by the production of small amounts of glycerol and acetic acid. Bottom yeasts produce beer with a pH of 4.1 to 4.2 and requiring 7 to 12 days of fermentation. With a top yeast, such as *Saccharomyces cerevisiae,* the pH is lowered to 3.8 to produce ales. Freshly fermented (green) beers are aged or **lagered,** and when they are bottled, CO_2 is usually added. Beer can be pasteurized at 140°F or higher or sterilized by passage through membrane filters to minimize flavor changes.

In many places there is increased interest in specialty beers. Local "braumeisters" develop unique products with special brewing techniques and ingredients. Often unusual names are given to these creations. This has led to local tasting festivals and increased interest in the art and craft of brewing.

Distilled Spirits

Distilled spirits are produced by an extension of beer production processes. The fermented liquid is boiled, and the volatile components are condensed to yield a product with a higher alcohol content than beer. Rye and bourbon are examples of whiskeys. Rye whiskey must contain at least 51% rye grain, and bourbon must contain at least 51% corn. Scotch whiskey is made primarily of barley. Usually a **sour mash** is used; the mash is inoculated with a homolactic (lactic acid is the major fermentation product) bacterium such as *Lactobacillus delbrueckii,* which can lower the mash pH to around 3.8 in 6 to 10 hours. This limits the development of undesirable organisms. Vodka and grain alcohols are also produced by distillation. Gin is vodka to which resinous flavoring agents—often juniper berries—have been added to provide a unique aroma and flavor.

Production of Breads

Bread is one of the most ancient of human foods, and is produced with the help of microorganisms. The use of yeasts to leaven bread is carefully depicted in paintings from ancient Egypt. A bakery at the Giza Pyramid area, from the year 2575 B.C., has been excavated. It is

Processing step   Biological change

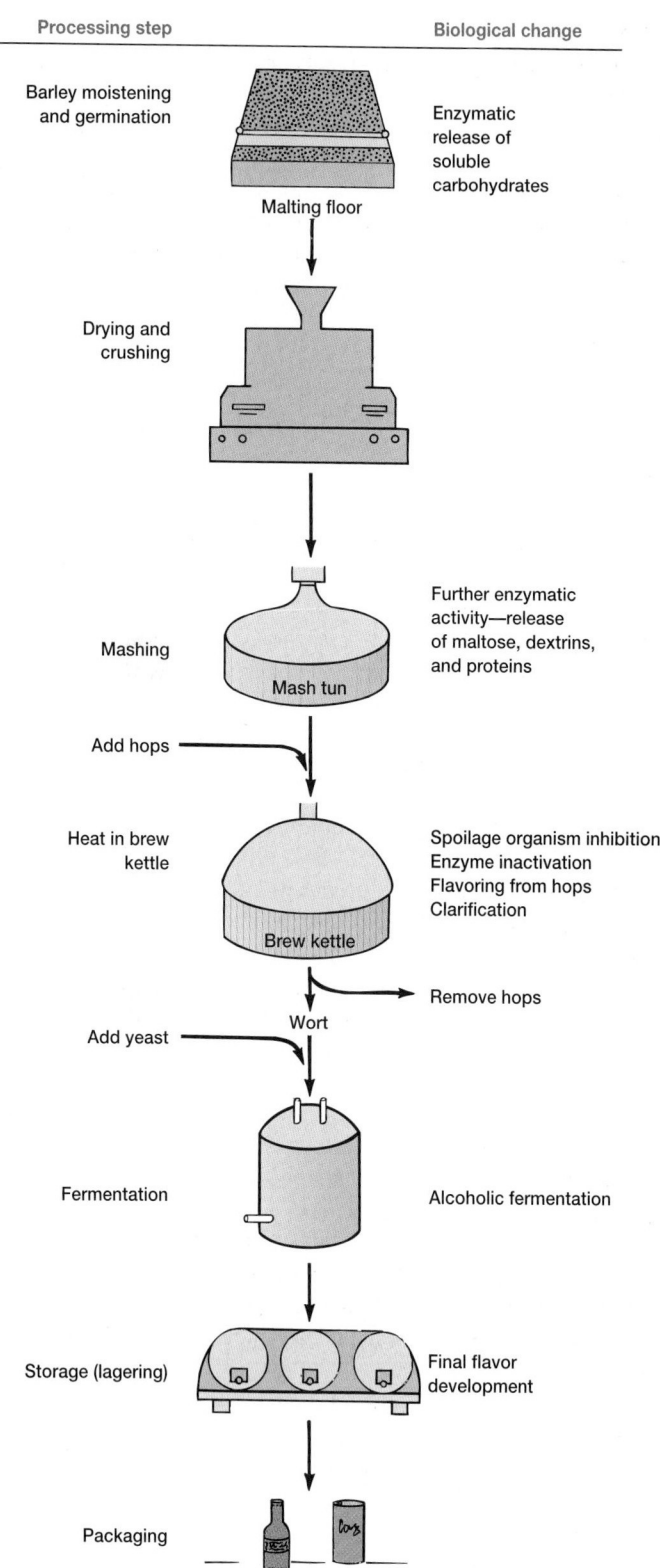

Barley moistening
and germination

Enzymatic
release of
soluble
carbohydrates

Malting floor

Drying and
crushing

Further enzymatic
activity—release
of maltose, dextrins,
and proteins

Mashing

Mash tun

Add hops

Heat in brew
kettle

Spoilage organism inhibition
Enzyme inactivation
Flavoring from hops
Clarification

Brew kettle

Remove hops

Wort

Add yeast

Fermentation

Alcoholic fermentation

Storage (lagering)

Final flavor
development

Packaging

Figure 41.16 Producing Beer. To make beer, the complex carbohydrates in the grain must first be transformed into a fermentable substrate. Beer production thus requires the important steps of malting, and the use of hops and boiling for clarification, flavor development, and inactivation of malting enzymes, to produce the wort. Only after completion of these steps can the actual fermentation be carried out.

Figure 41.17 Brew Kettles Used for Preparation of Wort. In large-scale processes, copper brew kettles can be used for wort preparation, as shown here at the Carlsberg Brewery, Copenhagen, Denmark.

estimated that 30,000 people a day were provided with bread from this bakery. Samples of bread from 2100 B.C. are on display in the British Museum. In breadmaking, yeast growth is carried out under aerobic conditions. This results in increased CO_2 production and minimum alcohol accumulation. The fermentation of bread involves several steps: alpha- and beta-amylases present in the moistened dough release maltose and sucrose from starch. Then a baker's strain of the yeast *Saccharomyces cerevisiae,* which contains maltase, invertase, and zymase enzymes, is added. The CO_2 produced by the yeast results in the light texture of many breads, and traces of fermentation products contribute to the final flavor. Usually bakers add sufficient yeast to allow the bread to rise within 2 hours—the longer the rising time, the more additional growth by contaminating bacteria and fungi can occur, making the product less desirable.

By using more complex assemblages of microorganisms, bakers can produce special breads such as sour doughs. The yeast *Saccharomyces exiguus,* together with a *Lactobacillus* species, produces the characteristic acidic flavor and aroma of such breads.

Bread products can be spoiled by *Bacillus* species that produce ropiness. If the dough is baked after these organisms have grown, stringy and ropy bread will result, leading to decreased consumer acceptance.

Other Fermented Foods

Many other plant products can be fermented, as summarized in **table 41.9.** These include sufu, which is produced by the fermentation of tofu, a chemically coagulated soybean milk product. To carry out the fermentation, the tofu curd is cut into small chunks and dipped into a solution of salt and citric acid. After the cubes are heated to pasteurize their surfaces, *Actinimucor elegans* and other *Mucor* species are added. When a white mycelium develops, the cubes, now called pehtze, are aged in salted rice wine. This product has achieved the status of a delicacy in many parts

Table 41.9

Fermented Foods Produced from Fruits, Vegetables, Beans, and Related Substrates

Foods	Raw Ingredients	Fermenting Microorganisms	Area
Coffee	Coffee beans	*Erwinia dissolvens, Saccharomyces* spp.	Brazil, Congo, Hawaii, India
Gari	Cassava	*Corynebacterium manihot, Geotrichum* spp.	West Africa
Kenkey	Corn	*Aspergillus* spp., *Penicillium* spp., lactobacilli, yeasts	Ghana, Nigeria
Kimchi	Cabbage and other vegetables	Lactic acid bacteria	Korea
Miso	Soybeans	*Aspergillus oryzae, Zygosaccharomyces rouxii*	Japan
Ogi	Corn	*Lactobacillus plantarum, Lactococcus lactis, Zygosaccharomyces rouxii*	Nigeria
Olives	Green olives	*Leuconostoc mesenteroides, Lactobacillus plantarum*	Worldwide
Ontjom	Peanut presscake	*Neurospora sitophila*	Indonesia
Peujeum	Cassava	Molds	Indonesia
Pickles	Cucumbers	*Pediococcus cerevisiae, L. plantarum*	Worldwide
Poi	Taro roots	Lactic acid bacteria	Hawaii
Sauerkraut	Cabbage	*L. mesenteroides, L. plantarum*	Worldwide
Soy sauce	Soybeans	*Aspergillus oryzae* or *A. soyae, Z. rouxii, Lactobacillus delbrueckii*	Japan
Sufu	Soybeans	*Mucor* spp.	China
Tao-si	Soybeans	*A. oryzae*	Philippines
Tempeh	Soybeans	*Rhizopus oligosporus, R. oryzae*	Indonesia, New Guinea, Surinam

Adapted from James M. Jay. 2000. *Modern Food Microbiology,* 6th edition. Reprinted by permission of Aspen Publishing, Frederick, Md.

of the Western world. Another popular product is tempeh, a soybean mash fermented by *Rhizopus.*

Sauerkraut or sour cabbage is produced from wilted, shredded cabbage, as shown in **figure 41.18.** Usually the mixed microbial community of the cabbage is used. A concentration of 2.2 to 2.8% sodium chloride restricts the growth of gram-negative bacteria while favoring the development of the lactic acid bacteria. The primary microorganisms contributing to this product are *Leuconostoc mesenteroides* and *Lactobacillus plantarum.* A predictable microbial succession occurs in sauerkraut's development. The activities of the lactic acid-producing cocci usually cease when the acid content reaches 0.7 to 1.0%. At this point *Lactobacillus plantarum* and *Lactobacillus brevis* continue to function. The final acidity is generally 1.6 to 1.8, with lactic acid comprising 1.0 to 1.3% of the total acid in a satisfactory product.

Pickles are produced by placing cucumbers and such components as dill seeds in casks filled with a brine. The sodium chloride concentration begins at 5% and rises to about 16% in 6 to 9 weeks. The salt not only inhibits the growth of undesirable bacteria but also extracts water and water-soluble constituents from the cucumbers. These soluble carbohydrates are converted to lactic acid. The fermentation, which can require 10 to 12 days, involves the development of *Leuconostoc mesenteroides, Enterococcus faecalis, Pediococcus cerevisiae, Lactobacillus brevis,* and *L. plantarum. L. plantarum* plays the dominant role in this fermentation process. Sometimes, to achieve more uniform pickle quality, natural microorganisms are first destroyed and the cucumbers are fermented using pure cultures of *P. cerevisiae* and *L. plantarum.*

Grass, chopped corn, and other fresh animal feeds, if stored under moist anaerobic conditions, will undergo a lactic-type mixed fermentation that produces pleasant-smelling **silage.** Trenches or more traditional vertical steel or concrete silos are used to store the silage. The accumulation of organic acids in silage can cause rapid deterioration of these silos. Older wooden stave silos, if not properly maintained, allow the outer portions of the silage to become aerobic, resulting in spoilage of a large portion of the plant material.

1. Describe and contrast the processes of wine and beer production. How are red wines produced when the juice of all grapes is white?
2. How do champagnes differ from wines?
3. Describe how distilled spirits like whiskey are produced.
4. How are bread, sauerkraut, and pickles produced? What microorganisms are most important in these fermentations?

41.7 MICROORGANISMS AS FOODS AND FOOD AMENDMENTS

Besides microorganisms' actions in fermentation as agents of physical and biological change, they themselves can be used as a food source. A variety of bacteria, yeasts, and other fungi have been used as animal and human food sources. Mushrooms (*Agaricus bisporus*) are one of the most important fungi used directly as a food source. Large caves provide optimal conditions for the production of this delicacy (**figure 41.19**). Microorganisms can

Processing step   Biological change

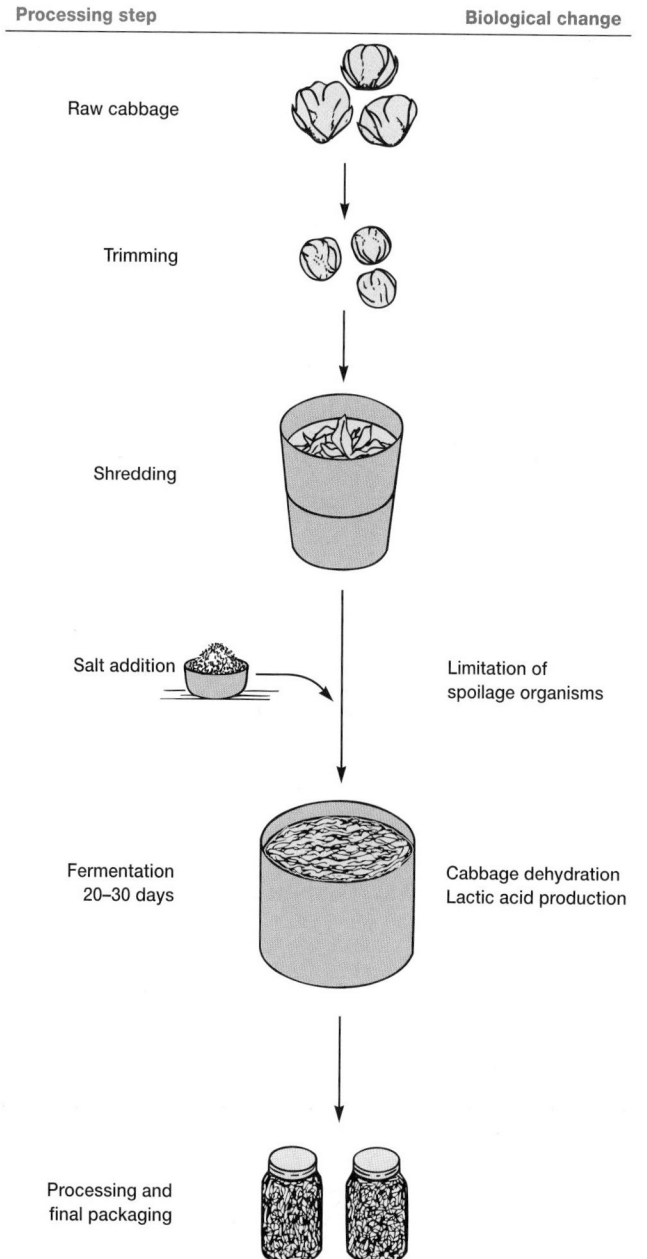

Raw cabbage

Trimming

Shredding

Salt addition   Limitation of
                                              spoilage organisms

Fermentation   Cabbage dehydration
20–30 days   Lactic acid production

Processing and
final packaging

Figure 41.18 Sauerkraut.   Sauerkraut production employs a lactic acid fermentation. The basic process involves fermentation of shredded cabbage in the presence of 2.25–2.5% by weight of salt to inhibit spoilage organisms.

Figure 41.19 Mushroom Farming.   Growing mushrooms requires careful preparation of the growth medium and control of environmental conditions. The mushroom bed is a carefully developed compost, which can be steam sterilized to improve mushroom growth.

benefits beyond basic nutritive value. The possible health benefits of the use of such microbial dietary adjuvants include immunodilation, control of diarrhea, anticancer effects, and possible modulation of the severity of Crohn's disease (inflammatory bowel disease). These bacteria may also influence antigen presentation, uptake, and possible degradation.

An interesting application is the use of probiotic microbes (primarily *Lactobacillus acidophilus*) to decrease *E. coli* occurrence in beef cattle. The desired bacteria are sprayed on feed, and the cattle have been shown to have markedly lower (60% in some experiments) carriage of *E. coli* O157:H7. This can make it easier to produce beef that will meet current standards for microbiological quality at the time of slaughter. Probiotics (Techniques & Applications 31.1)

In addition, there is a greater appreciation of the role of oligosaccharide polymers or **prebiotics,** which are not processed until they enter the large intestine. The combination of prebiotics and probiotic microorganisms is described as a **synbiotic** system. This synbiotic combination can result in an increase in the levels of butyric and propionic acids, as well as an increase in *Bifidobacterium* in the human intestine. The butryrate, especially, may play a role in the possible beneficial effects of probiotics on intestinal processes.

Probiotics are being used successfully with poultry. Recently, the USDA has designated a probiotic *Bacillus* strain for use with chickens as GRAS (p. 945). Feeding chickens a strain of *Bacillus subtilis* (Calsporin) led to increased body weight and feed conversion. There was also a reduction in coliforms and *Campylobacter* in the processed carcasses. It has been suggested that this probiotic decreases the need for antibiotics in poultry production and pathogen levels on farms. *Salmonella*

be used directly as a food source or as a supplement to other foods and are then called single-cell protein. One of the more popular microbial food supplements is the cyanobacterium *Spirulina.* It is used as a food source in Africa and is now being sold in United States health food stores as a dried cake or powdered product.

Microorganisms such as *Lactobacillus* and *Bifidobacterium* are being used in the rapidly developing area of **probiotics,** the addition of microorganisms to the diet in order to provide health

can be controlled by spraying a patented blend of 29 bacteria, isolated from the chicken cecum, on day-old chickens. As they preen themselves, the chicks ingest the bacterial mixture, establishing a functional microbial community in the cecum and limiting *Salmonella* colonization of the gut in a process called **competitive exclusion.** In 1998, this product, called PRE-EMPT, was approved for use in the United States by the Food and Drug Administration.

1. What conditions are needed to have most efficient production of edible mushrooms? What is the scientific name of the most important fungus used for this purpose?
2. A cyanobacterium is widely used as a food supplement. What is this genus, and in what part of the world was it first used as a significant food source?
3. What are prebiotics, probiotics, and synbiotics?
4. What probiotic has recently been recognized as GRAS?

Summary

41.1 Microorganism Growth in Foods

a. Most foods, especially when raw, provide an excellent environment for microbial growth. This growth can lead to spoilage or preservation, depending on the microorganisms present and environmental conditions.

b. The course of microbial development in a food is influenced by the intrinsic characteristics of the food itself—pH, salt content, substrates present, water presence and availability—and extrinsic factors, including temperature, relative humidity, and atmospheric composition (**figure 41.1**).

c. Microorganisms can spoil meat, dairy products, fruits, vegetables, and canned goods in several ways. Spices, with their antimicrobial compounds, sometimes protect foods.

d. Modified atmosphere packaging (MAP) is being used to control microbial growth in foods and to extend product shelf life. This process involves decreasing oxygen and increased carbon dioxide levels in the space between the food surface and the wrapping material.

41.2 Microbial Growth and Food Spoilage

a. Food spoilage is a major concern throughout the world. This can occur at any point in the food production process: growth, harvesting, transport, storage, or final preparation. Spoilage also can occur if foods are not stored properly.

b. Fungi, if they can grow in foods, especially cereals and grains, can produce important disease-causing chemicals, including aflatoxins and fumonisins (carcinogens), and ergot alkaloids (mind-altering drugs).

c. Prior illness with hepatitis B can increase susceptibility to liver cancer from aflatoxins. Control of hepatitis B is suggested to be more critical than control of aflatoxins.

d. Algal toxins can be transmitted to humans by the food chain through marine products. These can have severe amnesic, diarrhetic and neurotoxic effects.

41.3 Controlling Food Spoilage

a. Foods can be preserved in a variety of physical and chemical ways, including filtration, alteration of temperature (cooling, pasteurization, sterilization), drying, the addition of chemicals, radiation, and fermentation (**table 41.4**).

b. There is an increasing interest in using bacteriocins for food preservation. Nisin, a product of *Streptococcus lactis,* is the major substance approved for use in foods. Bacteriocins are especially important for control of *Listeria monocytogenes.*

41.4 Food-Borne Diseases

a. Foods can be contaminated by pathogens at any point in the food production, storage, or preparation processes. Pathogens such as *Salmonella, Campylobacter, Listeria,* and *E. coli* can be transmitted by the food to the susceptible consumer, where they grow and cause disease, or a food-borne infection (**table 41.6**). If the pathogen grows in the food before consumption and forms toxins that affect the food consumer without further microbial growth, the disease is a food-borne intoxication. Examples are intoxications caused by *Staphylococcus, Clostridium,* and *Bacillus.*

b. Norwalk-type viruses, *Campylobacter,* and *Salmonella* are thought to be the most important causes of food-borne illness. The cause of most food-borne illnesses is not known.

c. *E. coli* O157:H7 is an enterohemorrhagic bacterium that apparently can produce shigalike verocytotoxins, which especially affect the young. Proper food handling and thorough cooking are critical in control.

d. New variant Creutzfeldt-Jakob disease (vCJD) is of increasing worldwide concern as a possible food-borne infectious agent, which is related to the occurrence of "mad cow disease." The major means of vCJD transmission between animals is the use of mammalian tissue in ruminant animal feeds. There are difficulties in detecting such prohibited animal products in ruminant feeds.

e. The increasing popularity of raw foods such as sprouts, seafood, and raspberries has provided new routes for disease transmission. Increases in the international shipment of fresh foods contribute to this problem. *Cyclospora* is a protozoan of concern, and the major source is contaminated waters.

41.5 Detection of Food-Borne Pathogens

a. Detection of food-borne pathogens is a major part of food microbiology. The use of immunological and molecular techniques such as probes, PCR, and pulsed field electrophoresis is making it possible to link disease occurrences to a common infection source (**figure 41.10**). PulseNet and InfoNet programs are being used to coordinate these control efforts.

41.6 Microbiology of Fermented Foods

a. Dairy products can be fermented to yield a wide variety of cultured milk products (**table 41.7**). These include mesophilic, therapeutic, thermophilic, lacto-ethanolic, and mold-lactic products.

b. Growth of lactic acid-forming bacteria, often with the additional use of rennin, can coagulate milk solids. These solids can be processed to yield a wide variety of cheeses, including soft unripened, soft ripened, semisoft, hard, and very hard types (**table 41.8** and **figure 41.13**). Both bacteria and fungi are used in these cheese production processes.

c. Wines are produced from pressed grapes and can be dry or sweet, depending on the level of free sugar that remains at the end of the alcoholic fermentation (**figure 41.15**). Champagne is produced when the fermentation, resulting in CO_2 formation, is continued in the bottle.

d. Beer and ale are produced from cereals and grains. The starches in these substrates are hydrolyzed, in the processes of malting and mashing, to produce a fermentable wort. *Saccharomyces cerevisiae* is a major yeast used in the production of beer and ale (**figure 41.16**).

e. Many plant products can be fermented with bacteria, yeasts, and fungi. Important products are breads, soy sauce, sufu, and tempeh (**table 41.9**). Sauerkraut and pickles are produced in a fermentation process in which natural populations of lactobacilli play a major role (**figure 41.18**).

41.7 Microorganisms as Foods and Food Amendments

a. Microorganisms themselves can serve as an important food source. Mushrooms (*Agaricus bisporus*) are one of the most important fungi used as a food source. *Spirulina,* a cyanobacterium, also is a popular food source sold in specialty stores.

b. Many microorganisms, including some of those used to ferment milks, can be used as food amendments or microbial dietary adjuvants. Microorganisms such as *Lactobacillus* and *Bifidobacterium,* termed probiotics, can be used with oligopolysaccharides, termed prebiotics, to yield synbiotics. Several types of probiotic microorganisms are being used successfully in poultry production.

Key Terms

aflatoxin 942
algal toxin 942
bacteriocin 946
bottom yeast 956
canning 944
competitive exclusion 960
enology 955
ergotism 942
extrinsic factor 938
food-borne infection 946
food intoxication 948

FoodNet 950
fumonisin 942
GRAS 945
intrinsic factor 938
kefir 952
lager 956
malt 956
mash 956
mashing 955
modified atmosphere packaging (MAP) 940

must 955
osmophilic microorganism 939
pasteurization 944
pitching 956
prebiotic 959
probe 950
probiotic 959
PulseNet 950
putrefaction 939
racking 955

radappertization 946
silage 958
sour mash 956
starter culture 951
synbiotic 959
wine vinegar 955
wort 955
xerophilic microorganism 939
yogurt 951

Questions for Thought and Review

1. You are going through a salad line in a cafeteria at the end of the day. Which types of foods would you tend to avoid, and why?
2. Why does the law only allow foods to be pasteurized a single time? What could be the consequences of repeated pasteurization?
3. Why is it recommended that frozen foods, once thawed, not be frozen again? Suggest some of the reasons for this important principle of food handling.
4. Hamburger is notorious for having relatively high populations of microorganisms. How can these populations be controlled in preparation and storage?
5. Coffee cream is now packaged in small-portion servings that can be held at room temperature for extended periods. How do you think this product is produced and packaged?
6. What factors might limit the application of DNA- and RNA-based molecular techniques to the detection of pathogens in foods? How might these problems be overcome?
7. Why were aflatoxins not discovered before the 1960s? Do you think this was the first time they had grown in a food product to cause disease?
8. What advantage might the shiga-like toxin give *E. coli* O157:H7? Can we expect to see other "new" pathogens appearing, and what should we do, if anything?
9. Because of the possible beneficial effects of probiotic microorganisms on the intestinal microflora, scientists are becoming increasingly interested in their use as dietary supplements. How might modern biotechnology be used to improve the effectiveness of these microorganisms?
10. "New" food-borne pathogens continue to be discovered. How much of this phenomenon do you think is due to new techniques for recognizing existing organisms rather than the actual appearance of new organisms? How would you sort out the answer to this question?
11. How might microbial ecology be used to better protect humans from food-borne diseases?

Critical Thinking Questions

1. Fresh lemon slices are often served with raw or steamed seafood (oysters, crab, shrimp). From a food microbiology perspective, provide an explanation for their being served. Are there other examples in either the cooking or the serving of foods that not only enhance flavor, but might have an antimicrobial strategy? Consider the example of marinades.
2. Keep a record of what you eat for a day or two. Determine if the food, beverages, and snacks you ate could have been produced (at any level) without the aid of microorganisms. Indicate at what level(s) microorganisms were deliberately used. Be sure to consider ingredients such as citric acid, which is produced at the industrial level by several species of fungi.
3. Colonization of a susceptible human is critical to food-borne disease microorganisms. How might it be possible to modify foods to decrease these attachment processes?

For recommended readings on these and related topics, see Appendix V.

chapter 42

Industrial Microbiology and Biotechnology

Concepts

- Microorganisms are used in industrial microbiology and biotechnology to create a wide variety of products and to assist in maintaining and improving the environment.
- Most work in industrial microbiology has been carried out using microorganisms isolated from nature or modified through mutations. In biotechnology, microorganisms with specific genetic characteristics can be constructed to meet desired objectives.
- Most microorganisms have not been grown or described. A major challenge in biotechnology is to be able to grow and characterize these observed but uncultured microorganisms in what is called "bioprospecting."
- Forced evolution and adaptive mutations now are part of biotechnology. These can be carried out in processes termed "natural genetic engineering."
- The development of media and specific conditions for the growth of microorganisms is a large part of industrial microbiology and biotechnology. Microorganisms can be grown in controlled environments with specific limitations to maximize the synthesis of desired products.
- Microbial growth in soils, waters, and other environments, where complex microbial communities already are present, cannot be completely controlled, and it is not possible to precisely define limiting factors or physical conditions.
- Microbial growth in controlled environments is expensive; it is used to synthesize high-value microbial metabolites and other products for use in animal and

Biodegradation often can be facilitated by changing environmental conditions. Polychlorinated biphenyls (PCBs) are widespread industrial contaminants that accumulate in anaerobic river muds. Although reductive dechlorination occurs under these anaerobic conditions, oxygen is required to complete the degradation process. In this experiment, muds are being aerated to allow the final biodegradation steps to occur.

human health. In comparison, microbial growth in natural environments usually does not involve the creation of specific microbial products; microorganisms are used to carry out lower-value environmental and agriculture-related processes.

- In controlled growth systems, different products are synthesized during growth and after growth is completed. Most antibiotics are produced after the completion of active growth.
- Antibiotics and other microbial products continue to contribute to animal and human welfare. Newer products include anticancer drugs. Combinatorial biology is making it possible to produce hybrid antibiotics with unique properties.
- The products of industrial microbiology impact all aspects of our lives. These often are bulk chemicals that are used as food supplements and acidifying agents. Other products are used as biosurfactants and emulsifiers in a wide variety of applications.
- Degradation is critical for understanding microbial contributions to natural environments. The chemical structure of substrates and microbial community characteristics play important roles in determining the fate of chemicals. Anaerobic degradation processes are important for the initial modification of many compounds, especially those with chlorine and other halogenated functions. Degradation can produce simpler or modified compounds that may not be less toxic than the original compound.
- Biosensors are undergoing rapid development, which is limited only by the advances that are occurring in molecular biology and other areas of science. It is now possible, especially with streptavidin-biotin-linked systems, to have essentially real-time detection of important pathogens.
- Gene arrays, based on recombinant DNA technology, allow gene expression to be monitored. These systems are being used in the analysis of complex microbial systems.
- Bacteria, fungi, and viruses are increasingly employed as biopesticides, thus reducing dependence on chemical pesticides.
- Application of microorganisms and their technology has both positive and negative aspects. Possible broader impacts of applications of industrial microbiology and biotechnology should be considered in the application of this rapidly expanding area.

The microbe will have the last word.

—*Louis Pasteur*

Industrial microbiology and biotechnology both involve the use of microorganisms to achieve specific goals, whether creating new products with monetary value or improving the environment. Industrial microbiology, as it has traditionally developed, focuses on products such as pharmaceutical and medical compounds (antibiotics, hormones, transformed steroids), solvents, organic acids, chemical feedstocks,

amino acids, and enzymes that have direct economic value. The microorganisms employed by industry have been isolated from nature, and in many cases, were modified using classic mutation-selection procedures.

Biotechnology has a long history, involving the use of living organisms for commercial benefit. In the last several decades, the term has become associated with the modification of microorganisms through the use of molecular biology, including the use of recombinant DNA technology (*see chapter 14*). It is now possible to manipulate genetic information and design products such as proteins, or to modify microbial gene expression. In addition, genetic information can be transferred between markedly different groups of organisms, such as between bacteria and plants.

The use of microorganisms in industrial microbiology and biotechnology follows a logical sequence. It is necessary first to identify or create a microorganism that carries out the desired process in the most efficient manner. This microorganism then is used, either in a controlled environment such as a fermenter or in complex natural systems, such as in soils or waters, to achieve specific goals.

 **42.1 CHOOSING MICROORGANISMS FOR INDUSTRIAL MICROBIOLOGY AND BIOTECHNOLOGY**

The first task for an industrial microbiologist is to find a suitable microorganism, one that is genetically stable, easy to maintain and grow, and well suited for extraction or separation of desired products. A wide variety of alternative approaches are available, ranging from isolating microorganisms from the environment to using sophisticated molecular techniques to modify an existing microorganism.

Finding Microorganisms in Nature

Until relatively recently, the major sources of microbial cultures for use in industrial microbiology were natural materials such as soil samples, waters, and spoiled bread and fruit. Cultures from all areas of the world continue to be examined to identify strains with desirable characteristics. Interest in hunting for new microorganisims, or **bioprospecting,** continues today.

Because only a minor portion of the microbial species estimated to exist in most environments has been isolated or cultured (**table 42.1**), there is a continuing effort throughout the world to grow these "uncultured" microorganisms. In terms of the major environments, most microbes that can be observed have not been cultured or identified (**table 42.2**). With increased interest in microbial diversity, microbial ecology, and especially in microorganisms from extreme environments (**Techniques & Applications 42.1**), microbiologists are learning how to grow

Techniques & Applications
42.1   The Potential of Thermophilic Archaea in Biotechnology

There is great interest in the characteristics of archaea isolated from the outflow mixing regions above deep hydrothermal vents that release water at 250 to 350°C. This is because these hardy organisms can grow at temperatures as high as 121°C. The problems in growing these microorganisms in a laboratory are formidable. For example, to grow some of them, it is necessary to use special culturing chambers and other specialized equipment to maintain water in the liquid state at these high temperatures.

Such microorganisms, termed hyperthermophiles, with optimum growth temperatures of 80°C or above (*see pp. 124–125*), confront unique challenges in nutrient acquisition, metabolism, nucleic acid replication, and growth. Many of these are anaerobes that depend on

elemental sulfur or ferric ion as oxidants. Enzyme stability is critical. Some DNA polymerases are inherently stable at 140°C, whereas many other enzymes are stabilized in vivo with unique thermoprotectants. When these enzymes are separated from their protectant, they lose their unique thermostability.

These enzymes may have important applications in methane production, metal leaching and recovery, and in immobilized enzyme systems. In addition, the possibility of selective stereochemical modification of compounds normally not in solution at lower temperatures may provide new routes for directed chemical syntheses. This is an exciting and expanding area of the modern biological sciences to which environmental microbiologists can make significant contributions.

Table 42.1
Estimated Total and Known Species from Different Microbial Groups

Group	Estimated Total Species	Known Species[a]	Percent Known
Viruses	130,000[b]	5,000	[4][c]
Archaea	?[d]	<500	?
Bacteria	40,000[b]	4,800	[12]
Fungi	1,500,000	69,000	5
Algae	60,000	40,000	67

[a]Mid-1990 values and should be increased 10–50%.

[b]These values are substantially underestimated, perhaps by 1–2 orders of magnitude.

[c][ ] indicates that these values are probably gross overestimates.

[d]Small subunit (SSU) rRNA data indicate much higher in situ diversity than given by culture-based studies.

Adapted from: D. A. Cowan. 2000. Microbial genomes—the untapped resource. *Tibtech* 18:14–16. Table 1, p. 15.

Table 42.2
Estimates of the Percent "Cultured" Microorganisms in Various Environments

Environment	Estimated Percent Cultured
Seawater	0.001–0.100
Fresh water	0.25
Mesotrophic lake	0.1–1.0
Unpolluted estuarine waters	0.1–3.0
Activated sludge	1–15
Sediments	0.25
Soil	0.3

Source: D. A. Cowan. 2000. Microbial genomes—the untapped resource. *Tibtech* 18:14–16. Table 2, p. 15.

these previously "uncultured" microbes, including the use of single-cell gel microencapsulation approaches that allow nutrient diffusion and microbial communication, while minimizing overgrowth by more rapid-growing competitors. Uncultured microorganisms and microbial diversity (section 6.5)

Genetic Manipulation of Microorganisms

Genetic manipulations are used to produce microorganisms with new and desirable characteristics. The classical methods of microbial genetics (*see chapter 13*) play a vital role in the development of cultures for industrial microbiology.

Mutation

Once a promising culture is found, a variety of techniques can be used for culture improvement, including chemical mutagens and

ultraviolet light (*see chapter 11*). As an example, the first cultures of *Penicillium notatum*, which could be grown only under static conditions, yielded low concentrations of penicillin. In 1943 a strain of *Penicillium chrysogenum* was isolated—strain NRRL 1951—which was further improved through mutation (**figure 42.1**). Today most penicillin is produced with *Penicillium chrysogenum*, grown in aerobic stirred fermenters, which gives 55-fold higher penicillin yields than the original static cultures.

Protoplast Fusion
Protoplast fusion is now widely used with yeasts and molds. Most of these microorganisms are asexual or of a single mating type, which decreases the chance of random mutations that could lead to strain degeneration. To carry out genetic studies with these microorganisms, protoplasts are prepared by growing the cells in an isotonic solution while treating them with enzymes, including

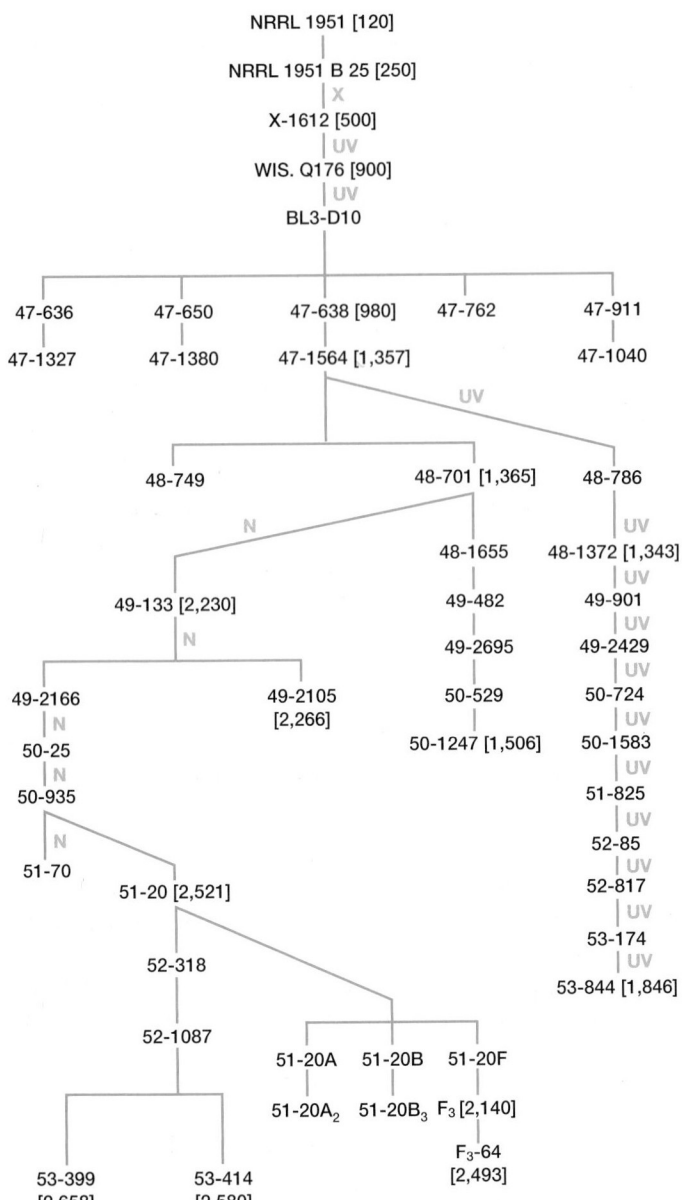

Figure 42.1 Mutation Makes It Possible to Increase Fermentation Yields. A "genealogy" of the mutation processes used to increase penicillin yields with *Penicillium chrysogenum* using X-ray treatment (X), UV treatment (UV), and mustard gas (N). By use of these mutational processes, the yield was increased from 120 International Units (IU) to 2,580 IU, a 20-fold increase. Unmarked transfers were used for mutant growth and isolation. Yields in international units/ml in brackets.

cellulase and beta-galacturonidase. The protoplasts are then regenerated using osmotic stabilizers such as sucrose. After regeneration of the cell wall, the new protoplasm fusion product can be used in further studies.

A major advantage of the protoplast fusion technique is that protoplasts of different microbial species can be fused, even if they are not closely linked taxonomically. For example, protoplasts of

Penicillium roquefortii have been fused with those of *P. chrysogenum*. Even yeast protoplasts and erythrocytes can be fused.

Insertion of Short DNA Sequences

Short lengths of chemically synthesized DNA sequences can be inserted into recipient microorganisms by the process of **site-directed mutagenesis.** This can create small genetic alterations leading to a change of one or several amino acids in the target protein. Such minor amino acid changes have been found to lead, in many cases, to unexpected changes in protein characteristics, and have resulted in new products such as more environmentally resistant enzymes and enzymes that can catalyze desired reactions. These approaches are part of the field of **protein engineering.** Site-directed mutagenesis (pp. 315–16)

---

1. How are industrial microbiology and biotechnology similar and different?
2. What approach is being used to increase the recovery of previously "uncultured" microorganisms from the environment?
3. What is protoplast fusion and what types of microorganisms are used in this process?
4. Describe site-directed mutagenesis and how it is used in biotechnology.
5. What is protein engineering?

Transfer of Genetic Information between Different Organisms

New alternatives have arisen through the transfer of nucleic acids between different organisms, which is part of the rapidly developing field of **combinatorial biology** (**table 42.3**). This involves the transfer of genes from one organism into another, giving the recipient varied capabilities such as an increased capacity to carry out hydrocarbon degradation. An important early example of this approach was the creation of the "superbug," patented by A. M. Chakarabarty in 1974, which had an increased capability of hydrocarbon degradation. The genes for antibiotic production can be transferred to a microorganism that produces another antibiotic, or even to a non-antibiotic-producing microorganism. Other examples are the expression, in *E. coli,* of the enzyme creatininase from *Pseudomonas putida* and the production of pediocin, a bacteriocin, in a yeast used in wine fermentation for the purpose of controlling bacterial contaminants. Bacteriocins (pp. 291, 688)

DNA expression in different organisms can improve production efficiency and minimize the purification steps required before the product is ready for use. For example, recombinant baculoviruses (*see pp. 404–5*) can be replicated in insect larvae to achieve rapid large-scale production of a desired virus or protein. Transgenic plants (*discussed on pp. 326–32*) may be used to manufacture large quantities of a variety of metabolic products. A most imaginative way of incorporating new DNA into a plant is to simply shoot it in using DNA-coated microprojectiles and a gene gun (*see section 14.6*).

A wide range of genetic information also can be inserted into microorganisms using vectors and recombinant DNA techniques.

Table 42.3		
Combinatorial Biology in Biotechnology: The Expression of Genes in Other Organisms to Improve Processes and Products		
Property or Product Transferred	**Microorganism Used**	**Combinatorial Process**
Ethanol production	*Escherichia coli*	Integration of pyruvate decarboxylase and alcohol dehydrogenase II from *Zymomonas mobilis.*
1,3-Propanediol production	*E. coli*	Introduction of genes from the *Klebsiella pneumoniae* dha region into *E. coli* made possible anaerobic 1,3-propanediol production.
Cephalosporin precursor synthesis	*Penicillium chrysogenum*	Production 7-ADC and 7-ADCA[a] precursors by incorporation of the expandase gene of *Cephalosoporin acremonium* into *Penicillium* by transformation.
Lactic acid production	*Saccharomyces cerevisiae*	A muscle bovine lactate dehydrogenase gene (LDH-A) expressed in *S. cerevisiae.*
Xylitol production	*S. cerevisiae*	95% xylitol conversion from xylose was obtained by transforming the XYLI gene of *Pichia stipitis* encoding a xylose reductase into *S. cerevisiae,* making this organism an efficient organism for the production of xylitol, which serves as a sweetener in the food industry.
Creatininase[b]	*E. coli*	Expression of the creatininase gene from *Pseudomonas putida* R565. Gene inserted with a pUC18 vector.
Pediocin[c]	*S. cerevisiae*	Expression of bacteriocin from *Pediococcus acidilactici* in *S. cerevisiae* to inhibit wine contaminants.
Acetone and butanol production	*Clostridium acetobutylicum*	Introduction of a shuttle vector into *C. acetobutylicum* by an improved electrotransformation protocol, which resulted in acetone and butanol formation.

[a]7-ACA = 7-aminocephalosporanic acid; 7-ADCA = 7-aminodecacetoxycephalosporonic acid.

[b]T.-Y. Tang; C.-J. Wen; and W.-H. Liu. 2000. Expression of the creatinase gene from *Pseudomonas putida* RS65 in *Escherichia coli. J. Ind. Microbiol. Biotechnol.* 24:2–6.

[c]H. Schoeman; M. A. Vivier; M. DuToit; L. M. Y. Dicks; and I. S. Pretorius. 1999. The development of bactericidal yeast strains by expressing the *Pediococcus acidilactici* pediocin gene (pedA) in *Saccharomyces cerevisiae. Yeast* 15:647–656.

Adapted from S. Ostergaard; L. Olsson; and J. Nielson. 2000. Metabolic engineering of *Saccharomyces cerevisiae.* Microbiol. Mol. Biol. Rev. 64(1):34–50.

Vectors (*see section 14.5*) include artificial chromosomes such as those for yeasts (YACs), bacteria (BACs), P1 bacteriophage-derived chromosomes (PACs), and mammalian artificial chromosomes (MACs). YACs are especially valuable because large DNA sequences (over 100 kb) can be maintained in the YAC as a separate chromosome in yeast cells. A good example of vector use is provided by the virus that causes foot-and-mouth disease of cattle and other livestock. Genetic information for a foot-and-mouth disease virus antigen can be incorporated into *E. coli,* followed by the expression of this genetic information and synthesis of the gene product for use in vaccine production (**figure 42.2**).

Genetic information transfer allows the production of specific proteins and peptides without contamination by similar products that might be synthesized in the original organism. This approach can decrease the time and cost of recovering and purifying a product. Another major advantage of peptide production with biotechnology is that only biologically active stereoisomers are produced. This specificity is required to avoid the possible harmful side effects of inactive stereoisomers, as occurred in the thalidomide disaster.

---

1. What is combinatorial biology and what is the basic approach used in this technique? What types of major products have been created using combinatorial biology?
2. Why might one want to insert a gene in a foreign cell and how is this done?
3. Why is it important to produce specific isomers of products for use in animal and human health?

Modification of Gene Expression

In addition to inserting new genes in organisms, it also is possible to modify gene regulation by changing gene transcription, fusing proteins, creating hybrid promoters, and removing feedback regulation controls. These approaches make it possible to overproduce a wide variety of products, as shown in **table 42.4.**

The modification of gene expression also can be used to intentionally alter metabolic pathways by inactivation or deregulation of specific genes, which is the field of **pathway architecture.** Understanding pathway architecture makes it possible to design a pathway that will be most efficient by avoiding slower or energetically more costly routes. This approach has been used to improve penicillin production by *metabolic pathway engineering* (MPE).

An interesting development in modifying gene expression, which illustrates **metabolic control engineering,** is that of altering controls for the synthesis of lycopene, an important antioxidant normally present at high levels in tomatoes and tomato products. In this case, an engineered regulatory circuit was designed to control lycopene synthesis in response to the internal metabolic state of *E. coli.* Another recent development is the use of modified gene expression to produce variants of the antibiotic erythromycin. Blocking specific biochemical steps (**figure 42.3**) in pathways for the synthesis of an antibiotic precursor resulted in modified final products. These altered products, which have slightly different structures, can be tested for their possible antimicrobial effects. In addition, by the use of this approach, it is possible to better establish the structure-function relationships of antibiotics.

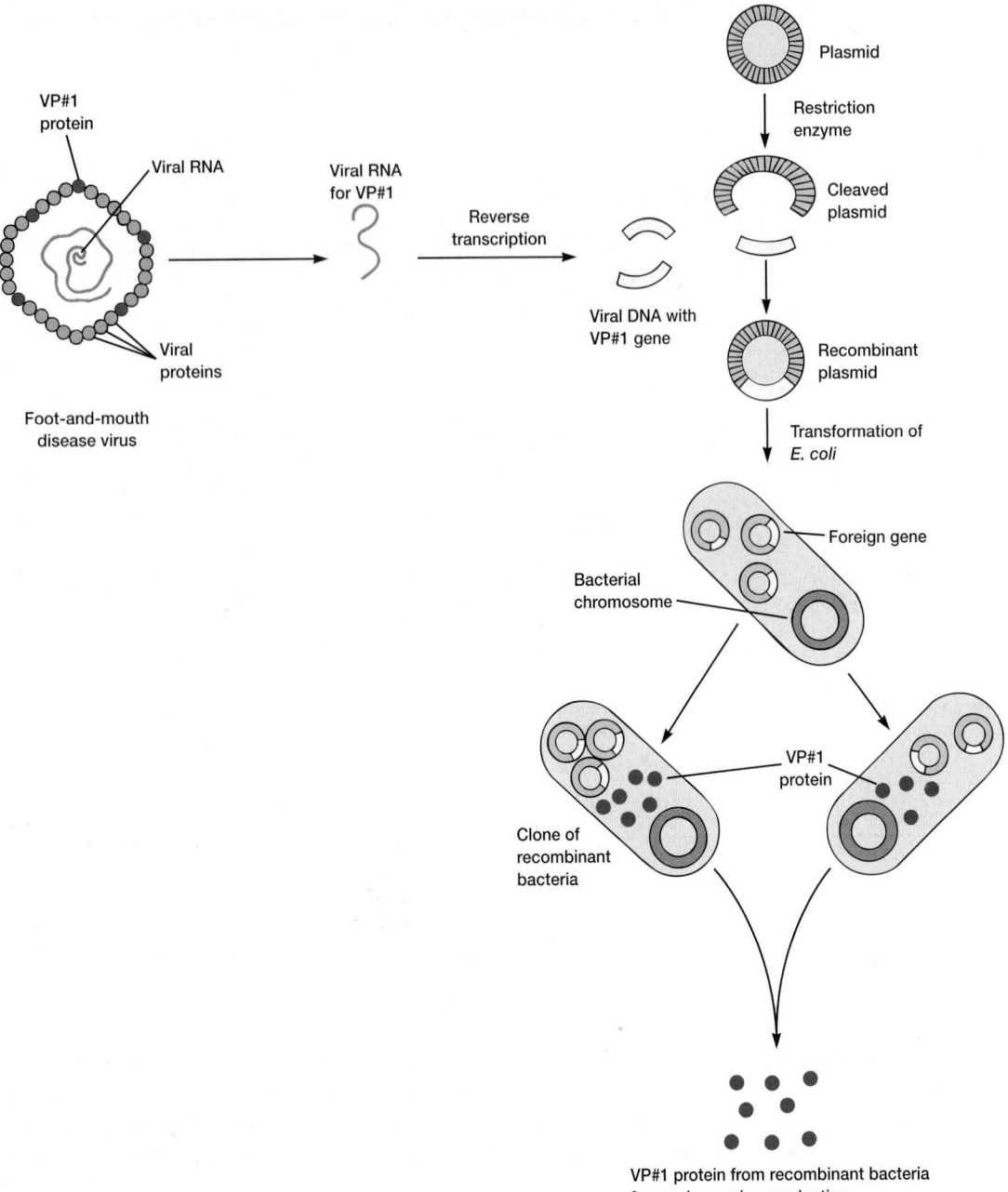

Figure 42.2 Recombinant Vaccine Production. Genes coding for desired products can be expressed in different organisms. By the use of recombinant DNA techniques, a foot-and-mouth disease vaccine is produced through cloning the vaccine genes into *Escherichia coli*.

Table 42.4

Examples of Recombinant DNA Systems Used to Modify Gene Expression

Product	Microorganism	Change
Actinorhodin	*Streptomyces coelicolor*	Modification of gene transcription
Cellulase	*Clostridium* genes in *Bacillus*	Amplification of secretion through chromosomal DNA amplification
Recombinant protein albumin	*Saccharomyces cerevisiae*	Fusion to a high-production protein
Heterologous protein	*Saccharomyces cerevisiae*	Use of the inducible strong hybrid promoter UAS_{gal}/CYCl
Enhanced growth rate[a]	*Aspergillus nidulans*	Overproduction of glyceraldehyde-3-phosphate dehydrogenase
Amino acids[b]	*Corynebacterium*	Isolation of biosynthetic genes that lead to enhanced enzyme activities or removal of feedback regulation

[a,b]S. Ostergaard; L. Olsson; and J. Nielson. 2000. Metabolic engineering of *Saccharomyces cerevisiae*. *Microbiol. Mol. Biol. Rev.* 64(1):34–50. Table 1, p. 35.

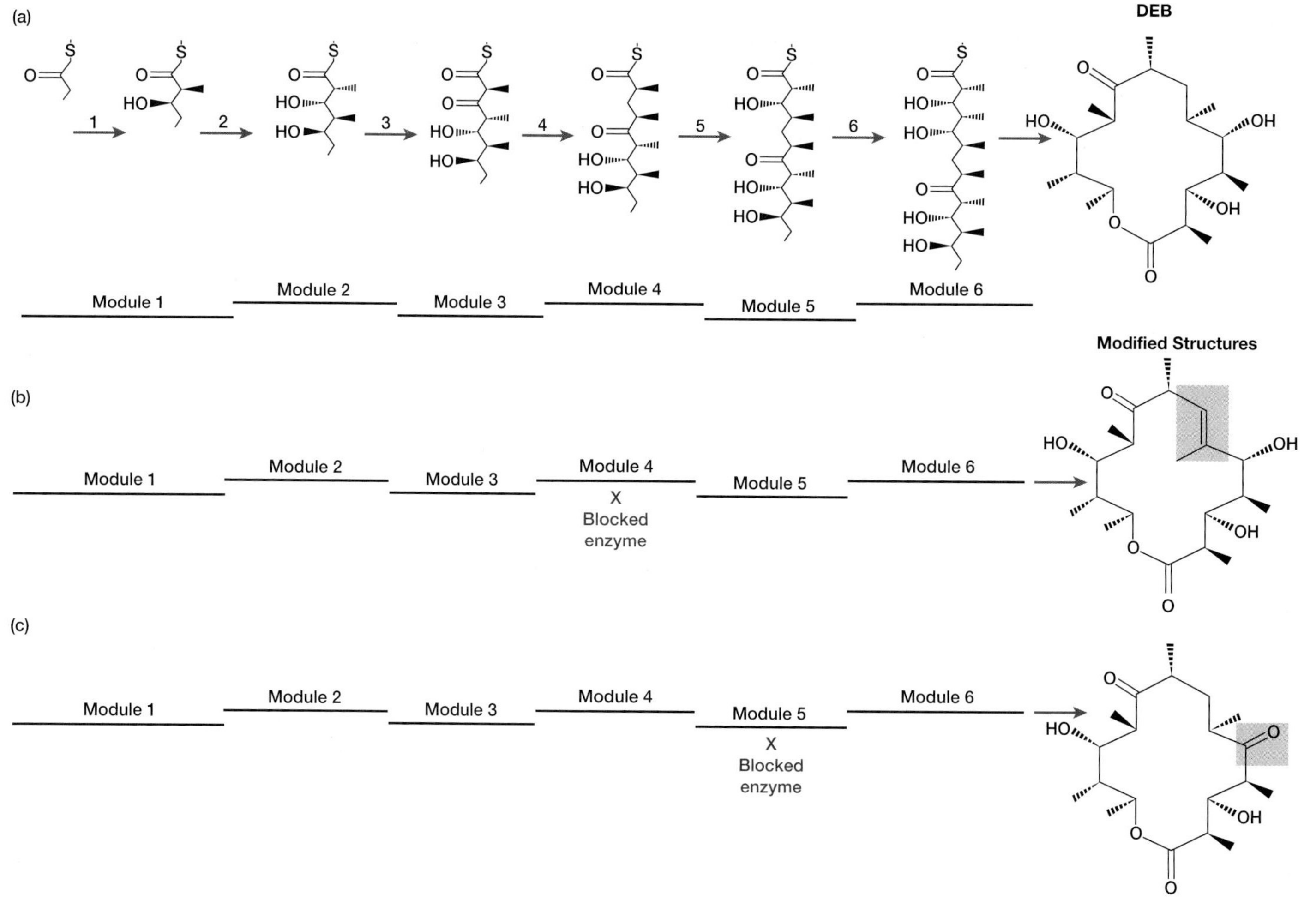

Figure 42.3   Metabolic Engineering to Create Modified Antibiotics.   (**a**) Model for six elongation cycles (modules) in the normal synthesis of 6-deoxyerythonilide B (DEB), a precursor to the important antibiotic erythromycin. (**b**) Changes in structure that occur when the enoyl reductase enzyme of module 4 is blocked. (**c**) Changes in structure that occur when the keto reductase enzyme of module 5 is blocked. These changed structures (the highlighted areas) may lead to the synthesis of modified antibiotics with improved properties.

Other examples include the increased synthesis of antibiotics and cellulases, modification of gene expression, DNA amplification, greater protein synthesis, and interactive enzyme overproduction or removal of feedback inhibition. Recombinant plasminogen, for example, may comprise 20 to 40% of the soluble protein in a modified strain, a tenfold increase in concentration over that in the original strain.

Natural Genetic Engineering

The newest approach for creating new metabolic capabilities in a given microorganism is the area of **natural genetic engineering,** which employs **forced evolution** and **adaptive mutations** (*see pp. 239–40*). This is the process of using specific environmental stresses to "force" microorganisms to mutate and adapt, thus creating microorganisms with new biological capabilities. The mechanisms of these adaptive mutational processes include DNA rearrangements in which transposable elements and various types of recombination play critical roles, as shown in **table 42.5.**

Although there is much controversy concerning this area, the responses of microorganisms to stress provide the potential of generating microorganisms with new microbial capabilities for use in industrial microbiology and biotechnology.

Preservation of Microorganisms

Once a microorganism or virus has been selected or created to serve a specific purpose, it must be preserved in its original form for further use and study. Periodic transfers of cultures have been used in the past, although this can lead to mutations and phenotypic changes in microorganisms. To avoid these problems, a variety of culture preservation techniques may be used to maintain desired culture characteristics (**table 42.6**). **Lyophilization,** or freeze-drying, and storage in liquid nitrogen are frequently employed with microorganisms. Although lyophilization and liquid nitrogen storage are complicated and require expensive equipment, they do allow microbial cultures to be stored for years without loss of viability or an accumulation of mutations.

Table 42.5

Natural Genetic Engineering Systems in Bacteria

Genetic Engineering Mechanisms	DNA Changes Mediated
Localized SOS mutagenesis	Base substitutions, frameshifts
Adapted frameshifting	-1 frameshifting
Tn5, Tn9, Tn10 precise excision	Reciprocal recombination of flanking 8/9 bp repeats; restores original sequence
In vivo deletion, inversion, fusion, and duplication formation	Generally reciprocal recombination of short sequence repeats; occasionally nonhomologous
Type II topoisomerase recombination	Deletions and fusions by nonhomologous recombination, sometimes at short repeats
Site-specific recombination (type I topoisomerases)	Insertions, excisions/deletions, inversions by concerted or successive cleavage-ligation reactions at short sequence repeats; tolerates mismatches
Transposable elements (many species)	Insertions, transpositions, replicon fusions, adjacent deletions/excisions, adjacent inversions by ligation of 3′ OH transposon ends of 5′ PO_4 groups from staggered cuts at nonhomologous target sites
DNA uptake (transformation competence)	Uptake of single strand independent of sequence, or of double-stranded DNA carrying species identifier sequence

Adapted from J. A. Shapiro. 1999. Natural genetic engineering, adaptive mutation, and bacterial evolution. In *Microbial Ecology of Infectious Disease*, E. Rosenberg, editor, 259–75. Washington, D.C.: American Society for Microbiology. Derived from Table 2, pp. 263–64.

Table 42.6

Methods Used to Preserve Cultures of Interest for Industrial Microbiology and Biotechnology

Method	Comments
Periodic transfer	Variables of periodic transfer to new media include transfer frequency, medium used, and holding temperature; this can lead to increased mutation rates and production of variants
Mineral oil slant	A stock culture is grown on a slant and covered with sterilized mineral oil; the slant can be stored at refrigerator temperature
Minimal medium, distilled water, or water agar	Washed cultures are stored under refrigeration; these cultures can be viable for 3 to 5 months or longer
Freezing in growth media	Not reliable; can result in damage to microbial structures; with some microorganisms, however, this can be a useful means of culture maintenance
Drying	Cultures are dried on sterile soil (soil stocks), on sterile filter paper disks, or in gelatin drops; these can be stored in a desiccator at refrigeration temperature, or frozen to improve viability
Freeze-drying (lyophilization)	Water is removed by sublimation, in the presence of a cryoprotective agent; sealing in an ampule can lead to long-term viability, with 30 years having been reported
Ultrafreezing	Liquid nitrogen at $-196°C$ is used, and cultures of fastidious microorganisms have been preserved for more than 15 years

1. What types of recombinant DNA techniques are being used to modify gene expression in microorganisms?
2. Define metabolic control engineering, metabolic pathway engineering, forced evolution, and adaptive mutations.
3. Why might natural genetic engineering be useful in modern microbial biotechnology?
4. What approaches can be used for the preservation of microorganisms?

42.2 MICROORGANISM GROWTH IN CONTROLLED ENVIRONMENTS

For many industrial processes, microorganisms must be grown using specifically designed media under carefully controlled conditions, including temperature, aeration, and nutrient feeding. The development of appropriate culture media and the growth of microorganisms under industrial conditions are the subjects of this section.

Before proceeding, it is necessary to clarify terminology. The term **fermentation,** used in a physiological sense in earlier sections of the book, is employed in a much more general way in relation to industrial microbiology and biotechnology. As noted in **table 42.7,** the term can have several meanings, including the mass culture of microorganisms (or even plant and animal cells). The development of industrial fermentations requires appropriate culture media and the large-scale screening of microorganisms. Often years are needed to achieve optimum product yields. Many isolates are tested for their ability to synthesize a new product in the desired quantity. Few are successful.   Fermentation as a physiological process (pp. 174–77)

Medium Development

The medium used to grow a microorganism is critical because it can influence the economic competitiveness of a particular process. Frequently, lower-cost crude materials are used as sources of carbon,

Table 42.7
Fermentation: A Word with Many Meanings for the Microbiologist

1. Any process involving the mass culture of microorganisms, either aerobic or anaerobic
2. Any biological process that occurs in the absence of O_2
3. Food spoilage
4. The production of alcoholic beverages
5. Use of an organic substrate as the electron donor and acceptor
6. Use of an organic substrate as a reductant, and of the same partially degraded organic substrate as an oxidant
7. Growth dependent on substrate-level phosphorylation

Table 42.8	
Major Components of Growth Media Used in Industrial Processes	
Source	**Raw Material**
Carbon and energy	Molasses
	Whey
	Grains
	Agricultural wastes (corncobs)
Nitrogen	Corn-steep liquor
	Soybean meal
	Stick liquor (slaughterhouse products)
	Ammonia and ammonium salts
	Nitrates
	Distiller's solubles
Vitamins	Crude preparations of plant and animal products
Iron, trace salts	Crude inorganic chemicals
Buffers	Chalk or crude carbonates
	Fertilizer-grade phosphates
Antifoam agents	Higher alcohols
	Silicones
	Natural esters
	Lard and vegetable oils

nitrogen, and phosphorus (**table 42.8**). Crude plant hydrolysates often are used as complex sources of carbon, nitrogen, and growth factors. By-products from the brewing industry frequently are employed because of their lower cost and greater availability. Other useful carbon sources include molasses and whey from cheese manufacture.   Microbial growth media (pp. 102–4)

The levels and balance of minerals (especially iron) and growth factors can be critical in medium formulation. For example, biotin and thiamine, by influencing biosynthetic reactions, control product accumulation in many fermentations. The medium also may be designed so that carbon, nitrogen, phosphorus, iron, or a specific growth factor will become limiting after a given time during the fermentation. In such cases the limitation often causes a shift from growth to production of desired metabolites.

Growth of Microorganisms in an Industrial Setting

Once a medium is developed, the physical environment for microbial functioning in the mass culture system must be defined. This often involves precise control of agitation, temperature, pH changes, and oxygenation. Phosphate buffers can be used to control pH while also functioning as a source of phosphorus. Oxygen limitations, especially, can be critical in aerobic growth processes.

The O_2 concentration and flux rate must be sufficiently high to have O_2 in excess within the cells so that it is not limiting. This is especially true when a dense microbial culture is growing. When filamentous fungi and actinomycetes are cultured, aeration can be even further limited by filamentous growth (**figure 42.4**). Such filamentous growth results in a viscous, plastic medium, known as a **non-Newtonian broth,** which offers even more resistance to stirring and aeration.

It is essential to assure that these physical factors are not limiting microbial growth. This is most critical during **scaleup,** where a successful procedure developed in a small shake flask is modified for use in a large fermenter. One must understand the microenvironment of the culture and maintain similar conditions near the individual cell despite increases in the culture volume. If a successful transition can be made from a process originally

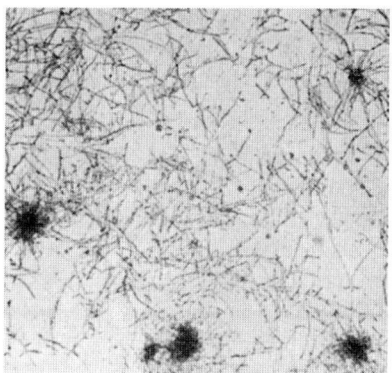

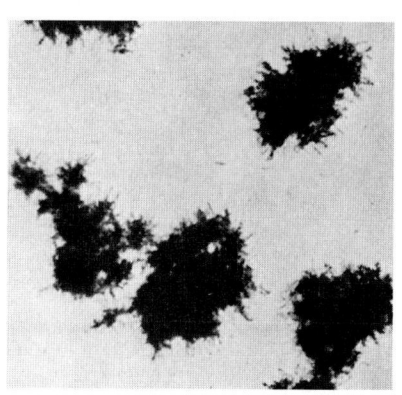

(a)   (b)

Figure 42.4   Filamentous Growth During Fermentation.   Filamentous fungi and actinomycetes can change their growth form during the course of a fermentation. The development of pelleted growth by fungi has major effects on oxygen transfer and energy required to agitate the culture. (**a**) Initial culture. (**b**) After 18 hours growth.

(a)

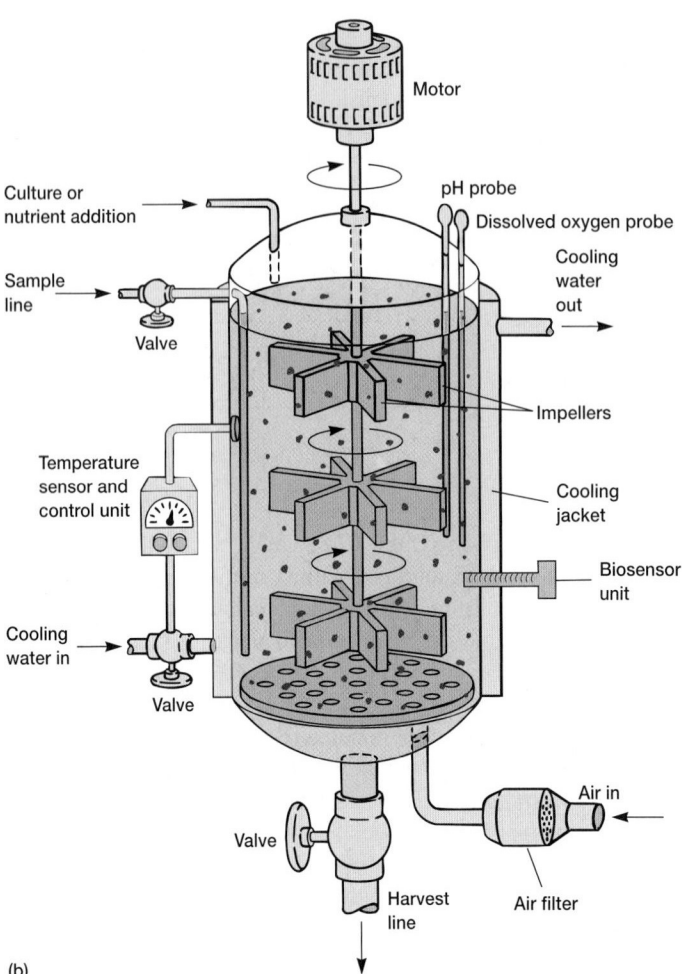

(b)

Figure 42.5 Industrial Stirred Fermenters. (**a**) Large fermenters used by a pharmaceutical company for the microbial production of antibiotics. (**b**) Details of a fermenter unit. This unit can be run under aerobic or anaerobic conditions, and nutrient additions, sampling, and fermentation monitoring can be carried out under aseptic conditions. Biosensors and infrared monitoring can provide real-time information on the course of the fermentation. Specific substrates, metabolic intermediates, and final products can be detected.

developed in a 250 ml Erlenmeyer flask to a 100,000 liter reactor, then the process of scaleup has been carried out properly.

Microorganisms can be grown in culture tubes, shake flasks, and stirred fermenters or other mass culture systems. Stirred fermenters can range in size from 3 or 4 liters to 100,000 liters or larger, depending on production requirements (**figure 42.5**). A typical industrial stirred fermentation unit is illustrated in figure 42.5b. Not only must the medium be sterilized but aeration, pH adjustment, sampling, and process monitoring must be carried out under rigorously controlled conditions. When required, foam control agents must be added, especially with high-protein media. Computers are commonly used to monitor outputs from probes that determine microbial biomass, levels of critical metabolic products, pH, input and exhaust gas composition, and other parameters. Environmental conditions can be changed or held constant over time, depending on the goals for the particular process.

Frequently a critical component in the medium, often the carbon source, is added continuously—**continuous feed**—so that the microorganism will not have excess substrate available at any given time. This is particularly important with glucose and other carbohydrates. If excess glucose is present at the beginning of a fermentation, it can be catabolized to yield ethanol, which is lost

as a volatile product and reduces the final yield. This can occur even under aerobic conditions.

Besides the traditional stirred aerobic or anaerobic fermenter, other approaches can be used to grow microorganisms. These alternatives, illustrated in **figure 42.6**, include lift-tube fermenters (figure 42.6a), which eliminate the need for stirrers that can be fouled by filamentous fungi. Also available is solid-state fermentation (figure 42.6b), in which a particulate substrate is kept moist to maintain a thin surface water film where microbes can grow and oxygen is available. In various types of fixed- (figure 42.6c) and fluidized-bed reactors (figure 42.6d), the microorganisms are associated with inert surfaces as biofilms (see pp. 602–3), and medium flows past the fixed or suspended particles.

Dialysis culture units also can be used (figure 42.6e). These units allow toxic waste metabolites or end products to diffuse away from the microbial culture and permit new substrates to diffuse through the membrane toward the culture. Continuous culture techniques using chemostats (figure 42.6f) can markedly improve cell outputs and rates of substrate use because microorganisms can be maintained in a continuous logarithmic phase. However, continuous maintenance of an organism in an active growth phase is undesirable in many industrial processes.

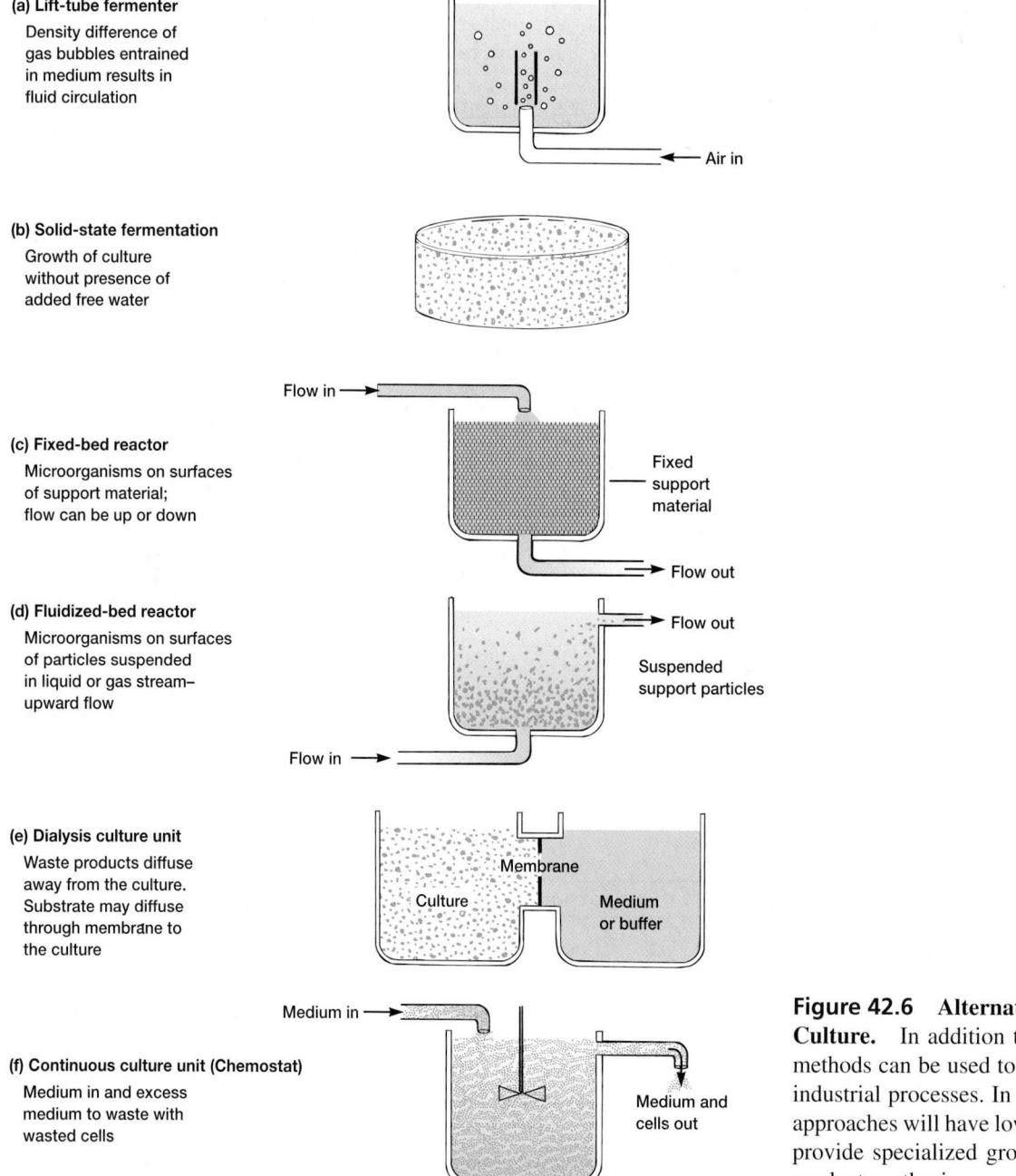

(a) Lift-tube fermenter
Density difference of
gas bubbles entrained
in medium results in
fluid circulation

Air in

(b) Solid-state fermentation
Growth of culture
without presence of
added free water

Flow in

(c) Fixed-bed reactor
Microorganisms on surfaces
of support material;
flow can be up or down

Fixed
support
material

Flow out

(d) Fluidized-bed reactor
Microorganisms on surfaces
of particles suspended
in liquid or gas stream–
upward flow

Flow out

Suspended
support particles

Flow in

(e) Dialysis culture unit
Waste products diffuse
away from the culture.
Substrate may diffuse
through membrane to
the culture

Membrane

Culture

Medium
or buffer

Medium in

(f) Continuous culture unit (Chemostat)
Medium in and excess
medium to waste with
wasted cells

Medium and
cells out

Figure 42.6 Alternate Methods for Mass Culture. In addition to stirred fermenters, other methods can be used to culture microorganisms in industrial processes. In many cases these alternate approaches will have lower operating costs and can provide specialized growth conditions needed for product synthesis.

Microbial products often are classified as primary and secondary metabolites. As shown in **figure 42.7, primary metabolites** consist of compounds related to the synthesis of microbial cells in the growth phase. They include amino acids, nucleotides, and fermentation end products such as ethanol and organic acids. In addition, industrially useful enzymes, either associated with the microbial cells or exoenzymes, often are synthesized by microorganisms during growth.

Secondary metabolites usually accumulate during the period of nutrient limitation or waste product accumulation that follows the active growth phase. These compounds have no direct relationship to the synthesis of cell materials and normal growth. Most antibiotics and the mycotoxins fall into this category.

---

1. How is the cost of media reduced during industrial operations? Discuss the effect of changing balances in nutrients such as minerals, growth factors, and the sources of carbon, nitrogen, and phosphorus.
2. What are non-Newtonian broths and why are these important in fermentations?
3. Discuss scaleup and the objective of the scaleup process.

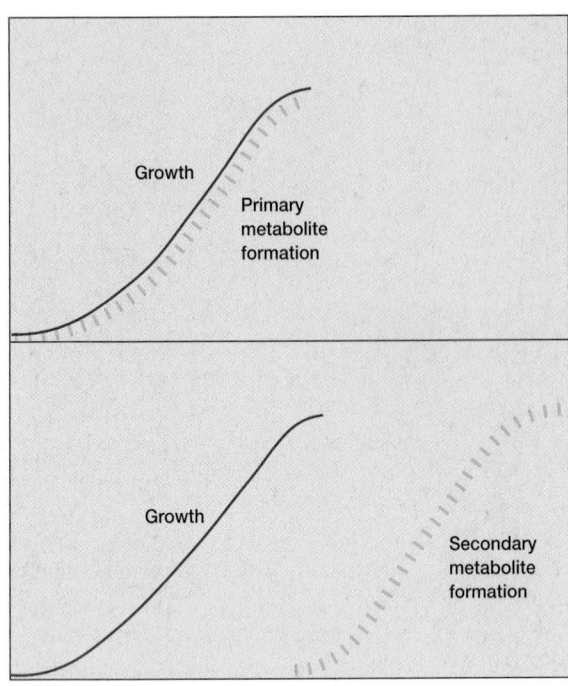

Figure 42.7 Primary and Secondary Metabolites. Depending on the particular organism, the desired product may be formed during or after growth. Primary metabolites are formed during the active growth phase, whereas secondary metabolites are formed after growth is completed.

4. What parameters can be monitored in a modern, large-scale industrial fermentation?

5. Besides the aerated, stirred fermenter, what other alternatives are available for the mass culture of microorganisms in industrial processes? What is the principle by which a dialysis culture system functions?

 **MAJOR PRODUCTS OF INDUSTRIAL MICROBIOLOGY**

Industrial microbiology has provided products that have in profoundly changed our lives and life spans. They include industrial and agricultural products, food additives, medical products for human and animal health, and biofuels (**table 42.9**). Particularly, in the last few years, nonantibiotic compounds used in medicine and health have made major contributions to the improved well-being of animal and human populations. Only major products in each category will be discussed here.

Antibiotics

Many antibiotics are produced by microorganisms, predominantly by actinomycetes in the genus *Streptomyces* and by filamentous fungi (*see table 35.2*). In this chapter, the synthesis of several of the most important antibiotics will be discussed to il-

Table 42.9	
Major Microbial Products and Processes of Interest in Industrial Microbiology and Biotechnology	
Substances	**Microorganisms**
Industrial Products	
Ethanol (from glucose)	*Saccharomyces cerevisiae*
Ethanol (from lactose)	*Kluyveromyces fragilis*
Acetone and butanol	*Clostridium acetobutylicum*
2,3-butanediol	*Enterobacter, Serratia*
Enzymes	*Aspergillus, Bacillus, Mucor, Trichoderma*
Agricultural Products	
Gibberellins	*Gibberella fujikuroi*
Food Additives	
Amino acids (e.g., lysine)	*Corynebacterium glutamicum*
Organic acids (citric acid)	*Aspergillus niger*
Nucleotides	*Corynebacterium glutamicum*
Vitamins	*Ashbya, Eremothecium, Blakeslea*
Polysaccharides	*Xanthomonas*
Medical Products	
Antibiotics	*Penicillium, Streptomyces, Bacillus*
Alkaloids	*Claviceps purpurea*
Steroid transformations	*Rhizopus, Arthrobacter*
Insulin, human growth hormone, somatostatin, interferons	*Escherichia coli, Saccharomyces cerevisiae,* and others (recombinant DNA technology)
Biofuels	
Hydrogen	Photosynthetic microorganisms
Methane	*Methanobacterium*
Ethanol	*Zymomonas, Thermoanaerobacter*

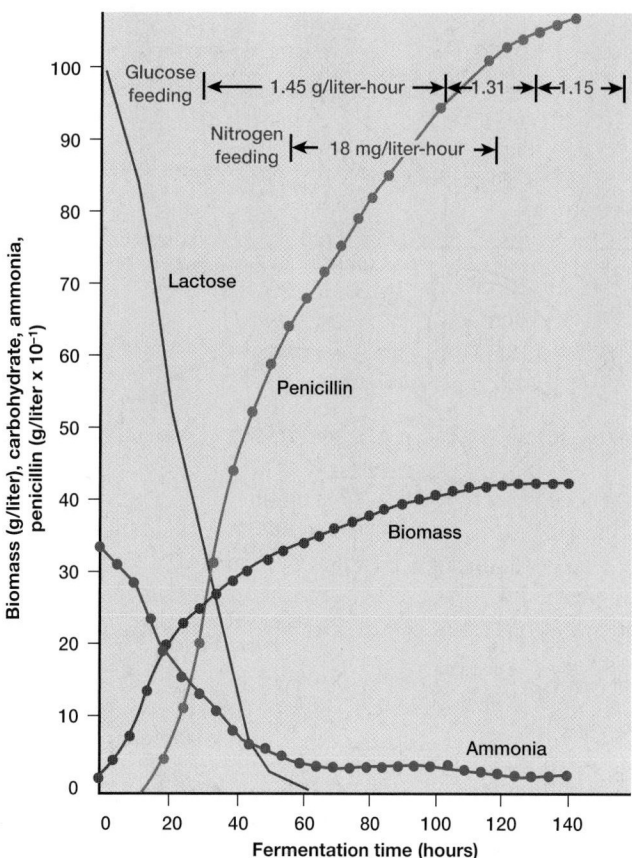

Figure 42.8 Penicillin Fermentation Involves Precise Control of Nutrients.   The synthesis of penicillin begins when nitrogen from ammonia becomes limiting. After most of the lactose (a slowly catabolized disaccharide) has been degraded, glucose (a rapidly used monosaccharide) is added along with a low level of nitrogen. This stimulates maximum transformation of the carbon sources to penicillin. The scale factor is presented using the convention recommended by the ASM. That is, a number on the axis should be multiplied by 0.10 to obtain the true value.

lustrate the critical role of medium formulation and environmental control in the production of these important compounds.   Antibiotics in medicine (chapter 35)

Penicillin

Penicillin, produced by *Penicillium chrysogenum,* is an excellent example of a fermentation for which careful adjustment of the medium composition is used to achieve maximum yields. Provision of the slowly hydrolyzed disaccharide lactose, in combination with limited nitrogen availability, stimulates a greater accumulation of penicillin after growth has stopped (**figure 42.8**). The same result can be achieved by using a slow continuous feed of glucose. If a particular penicillin is needed, the specific precursor is added to the medium. For example, phenylacetic acid is added to maximize production of penicillin G, which has a benzyl side chain (*see figure 35.7*). The fermentation pH is maintained around neutrality by the addition of sterile alkali, which assures maximum stability of the newly synthesized penicillin. Once the fermentation is completed, normally in 6 to 7 days, the broth is separated from the fungal mycelium and processed by absorption, precipitation, and crystallization to yield the final product. This basic product can then be modified by chemical procedures to yield a variety of **semisynthetic penicillins.**

Streptomycin

Streptomycin is a secondary metabolite produced by *Streptomyces griseus.* In this fermentation a soybean-based medium is used with glucose as a carbon source. The nitrogen source is thus in a combined form (soybean meal), which limits growth. After growth the antibiotic levels in the culture begin to increase (**figure 42.9**) under conditions of controlled nitrogen limitation.

Amino Acids

Amino acids such as lysine and glutamic acid are used in the food industry as nutritional supplements in bread products and as flavor-enhancing compounds such as monosodium glutamate (MSG).

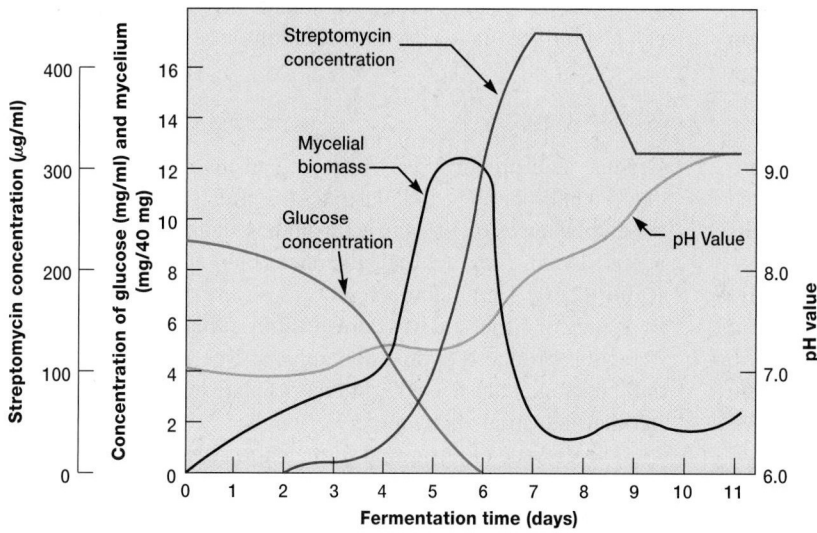

Figure 42.9   Streptomycin Production by *Streptomyces griseus.*   Depletion of glucose leads to maximum antibiotic yields.

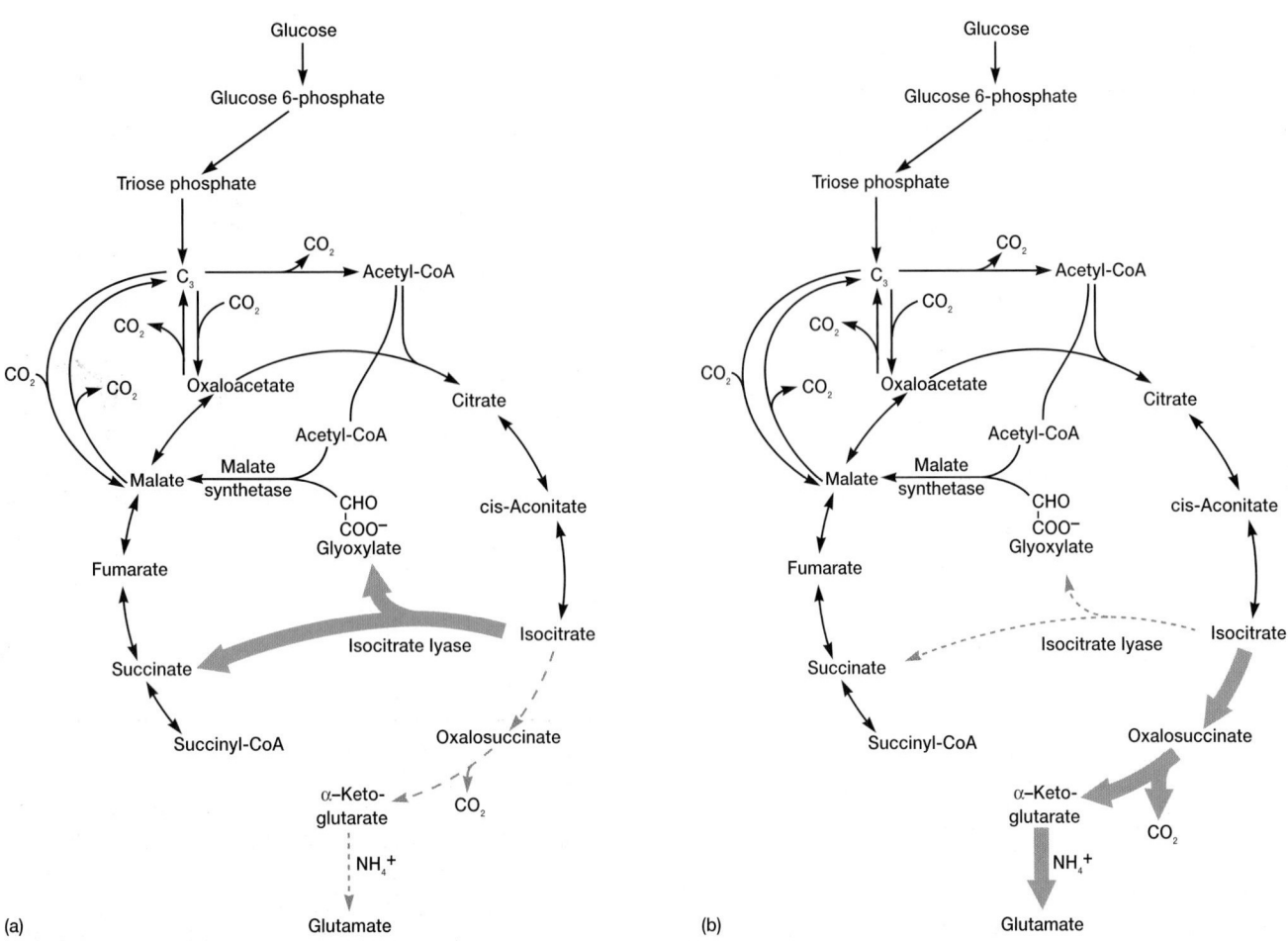

Figure 42.10 Glutamic Acid Production. The sequence of biosynthetic reactions leading from glucose to the accumulation of glutamate by *Corynebacterium glutamicum.* Major carbon flows are noted by bold arrows. (**a**) Growth with use of the glyoxylate bypass to provide critical intermediates in the TCA cycle. (**b**) After growth is completed, most of the substrate carbon is processed to glutamate (note shifted bold arrows). The dashed lines indicate reactions that are being used to a lesser extent.

Amino acid production is typically carried out by means of **regulatory mutants,** which have a reduced ability to limit synthesis of an end product. The normal microorganism avoids overproduction of biochemical intermediates by the careful regulation of cellular metabolism. Production of glutamic acid and several other amino acids is now carried out using mutants of *Corynebacterium glutamicum* that lack, or have only a limited ability to process, the TCA cycle intermediate α-ketoglutarate (*see appendix II*) to succinyl-CoA as shown in **figure 42.10.** A controlled low biotin level and the addition of fatty acid derivatives results in increased membrane permeability and excretion of high concentrations of glutamic acid. The impaired bacteria use the glyoxylate pathway (*see section 10.6*) to meet their needs for essential biochemical intermediates, especially during the growth phase. After growth becomes limited because of changed nutrient availability, an almost complete molar conversion (or 81.7% weight conversion) of isocitrate to glutamate occurs.

Lysine, an essential amino acid used to supplement cereals and breads, was originally produced in a two-step microbial process. This has been replaced by a single-step fermentation in which the bacterium *Corynebacterium glutamicum,* blocked in the synthesis of homoserine, accumulates lysine. Over 44 g/liter can be produced in a 3 day fermentation.

Organic Acids

Organic acid production by microorganisms is important in industrial microbiology and illustrates the effects of trace metal levels and balances on organic acid synthesis and excretion. Citric, acetic, lactic, fumaric, and gluconic acids are major products (**table 42.10**). Until microbial processes were developed, the major source of citric acid was citrus fruit from Italy. Today most citric acid is produced by microorganisms; 70% is used in the food and beverage industry, 20% in pharmaceuticals, and the balance in other industrial applications.

The essence of citric acid fermentation involves limiting the amounts of trace metals such as manganese and iron to stop *Aspergillus niger* growth at a specific point in the fermentation. The

Table 42.10

Major Organic Acids Produced by Microbial Processes

Product	Microorganism Used	Representative Uses	Fermentation Conditions
Acetic acid	*Acetobacter* with ethanol solutions	Wide variety of food uses	Single-step oxidation, with 15% solutions produced; 95–99% yields
Citric acid	*Aspergillus niger* in molasses-based medium	Pharmaceuticals, as a food additive	High carbohydrate concentrations and controlled limitation of trace metals; 60–80% yields
Fumaric acid	*Rhizopus nigricans* in sugar-based medium	Resin manufacture, tanning, and sizing	Strongly aerobic fermentation; carbon-nitrogen ratio is critical; zinc should be limited; 60% yields
Gluconic acid	*Aspergillus niger* in glucose-mineral salts medium	A carrier for calcium and sodium	Uses agitation or stirred fermenters; 95% yields
Itaconic acid	*Aspergillus terreus* in molasses-salts medium	Esters can be polymerized to make plastics	Highly aerobic medium, below pH 2.2; 85% yields
Kojic acid	*Aspergillus flavus-oryzae* in carbohydrate-inorganic N medium	The manufacture of fungicides and insecticides when complexed with metals	Iron must be carefully controlled to avoid reaction with kojic acid after fermentation
Lactic acid	Homofermentative *Lactobacillus delbrueckii*	As a carrier for calcium and as an acidifier	Purified medium used to facilitate extraction

medium often is treated with ion exchange resins to ensure low and controlled concentrations of available metals, and is carried out in aerobic stirred fermenters. Generally, high sugar concentrations (15 to 18%) are used, and copper has been found to counteract the inhibition of citric acid production by iron above 0.2 ppm. The success of this fermentation depends on the regulation and functioning of the glycolytic pathway and the tricarboxylic acid cycle (*see section 9.4*). After the active growth phase, when the substrate level is high, citrate synthase activity increases and the activities of aconitase and isocitrate dehydrogenase decrease. This results in citric acid accumulation and excretion by the stressed microorganism.

In comparison, the production of gluconic acid involves a single microbial enzyme, glucose oxidase, found in *Aspergillus niger*. *A. niger* is grown under optimum conditions in a corn-steep liquor medium. Growth becomes limited by nitrogen, and the resting cells transform the remaining glucose to gluconic acid in a single-step reaction. Gluconic acid is used as a carrier for calcium and iron and as a component of detergents.

Specialty Compounds for Use in Medicine and Health

In addition to the bulk products that have been produced over the last 30 to 40 years, such as antibiotics, amino acids, and organic acids, microorganisms are used for the production of nonantibiotic specialty compounds. These include sex hormones, antitumor agents, ionophores, and special compounds that influence bacteria, fungi, amoebae, insects, and plants (**table 42.11**). In all cases, it is necessary to produce and recover the products under carefully controlled conditions to assure that these medically important compounds reach the consumer in a stable, effective condition.

1. What critical limiting factors are used in the penicillin and streptomycin fermentations?
2. What are regulatory mutants and how were they used to increase the production of glutamic acid by *Corynebacterium?*
3. What is the principal limitation created to stimulate citric acid accumulation by *Aspergillus niger?*
4. Give some important specialty compounds that are produced by the use of microorganisms.

Biopolymers

Biopolymers are microbially produced polymers, primarily polysaccharides, used to modify the flow characteristics of liquids and to serve as gelling agents. These are employed in many areas of the pharmaceutical and food industries. Bacterial exopolysaccharides (pp. 61–62)

At least 75% of all polysaccharides are used as stabilizers, for the dispersion of particulates, as film-forming agents, or to promote water retention in various products. Polysaccharides help maintain the texture of many frozen foods, such as ice cream, that are subject to drastic temperature changes. These polysaccharides must maintain their properties under the pH conditions in the particular food and be compatible with other polysaccharides. They should not lose their physical characteristics if heated.

Biopolymers also include (1) dextrans, which are used as blood expanders and absorbents; (2) *Erwinia* polysaccharides that are used in paints; and (3) polyesters, derived from *Pseudomonas oleovorans,* which are a feedstock for specialty plastics. Cellulose microfibrils, produced by an *Acetobacter* strain, are used as a food thickener. Polysaccharides such as scleroglucan are used by the oil industry as drilling mud additives.

Table 42.11

Nonantibiotic Specialty Compounds Produced by Microorganisms

Compound Type	Source	Specific Product	Process/Organism Affected
Polyethers	*Streptomyces cinnamonensis*	Monensin	Coccidiostat, rumenal growth promoter
	S. lasaliensis	Lasalocid	Coccidiostat, ruminal growth promoter
	S. albus	Salinomycin	Coccidiostat, ruminal growth promoter
Avermectins	*S. avermitilis*		Helminths and arthropods
Statins	*Aspergillus terreus*	Lovastatin	Cholesterol-lowering agent
	Penicillium citrinum + actinomycete[a]	Pravastatin	Cholesterol-lowering agent
Enzyme inhibitors	*S. clavaligerus*	Clavulanic acid	Penicillinase inhibitor
	Actinoplanes sp.	Acarbose	Intestinal glucosidase inhibitor (decreases hyperglycemia and triglyceride synthesis)
Bioherbicides	*S. hygroscopicus*	Bialaphos	
Immunosuppressants	*Tolypocladium inflatum*	Cyclosporin A	Organ transplants
	S. tsukabaensis	FK-506	Organ transplants
	S. hygroscopicus	Rapamycin	Organ transplants
Anabolic agents	*Gibberella zeae*	Zearalenone	Farm animal medication
Uterocontractants	*Claviceps purpurea*	Ergot alkaloids	Induction of labor
Antitumor agents	*S. peuceticus* subsp. *caesius*	Doxorubicin	Cancer treatment
	S. peuceticus	Daunorubicin	Cancer treatment
	S. caespitosus	Mitomycin	Cancer treatment
	S. verticillus	Bleomycin	Cancer treatment

[a]Compactin, produced by *Penicillium citrinum,* is changed to pravastatin by an actinomycete bioconversion.

Based on: A. L. Demain. 2000. Microbial biotechnology. *Tibtech* 18:26–31; A. L. Demain. 2000. Pharmaceutically active secondary metabolites of microorganisms. *App. Microbiol. Biotechnol.* 52:455–463; G. Lancini; A. L. Demain. 1999. Secondary metabolism in bacteria: Antibiotic pathways regulation and function. In *Biology of the Prokaryotes,* J. W. Lengeler, G. Drews, and H. G. Schlegel, editors, 627–51. New York: Thieme.

Xanthan polymers enhance oil recovery by improving water flooding and the displacement of oil. This use of xanthan gum, produced by *Xanthomonas campestris,* represents a large potential market for this microbial product.

The cyclodextrins have a unique structure, as shown in **figure 42.11** for α-cyclodextrin. They are cyclic oligosaccharides whose sugars are joined by α-1,4 linkages. Cyclodextrins can be used for a wide variety of purposes because these cyclical molecules bind with substances and modify their physical properties. For example, cyclodextrins will increase the solubility of pharmaceuticals, reduce their bitterness, and mask chemical odors. Cyclodextrins also can be used as selective adsorbents to remove cholesterol from eggs and butter, to protect spices from oxidation, or as stationary phases in gas chromatography.

Biosurfactants

Biosurfactants are especially important for environmental applications where biodegradability is a major requirement. They are used for emulsification, increasing detergency, wetting and phase dispersion, as well as for solubilization. These properties are especially important in bioremediation, oil spill dispersion, and enhanced oil recovery (EOR). The most widely used microbially produced biosurfactants are glycolipids. These compounds have distinct hydrophilic and hydrophobic regions, and are excellent dispersing agents. They have been used with the *Exxon Valdez* and other oil spills.

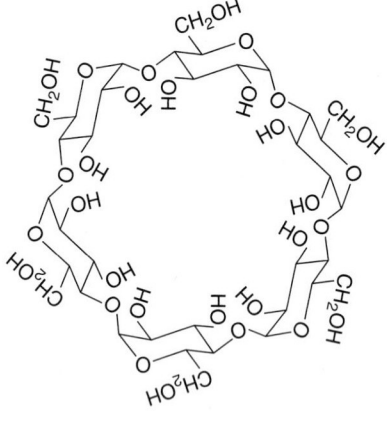

α-Cyclodextrin

Figure 42.11 A Cyclodextrin. The structure of α-cyclodextrin, produced by *Thermoanaerobacter.* There also are β and γ forms, of similar structure but larger; these compounds have many applications in medicine and industry.

Bioconversion Processes

Bioconversions, also known as **microbial transformations** or **biotransformations,** are minor changes in molecules, such as the insertion of a hydroxyl or keto function or the saturation/desaturation of a complex cyclic structure, that are carried out by nongrowing

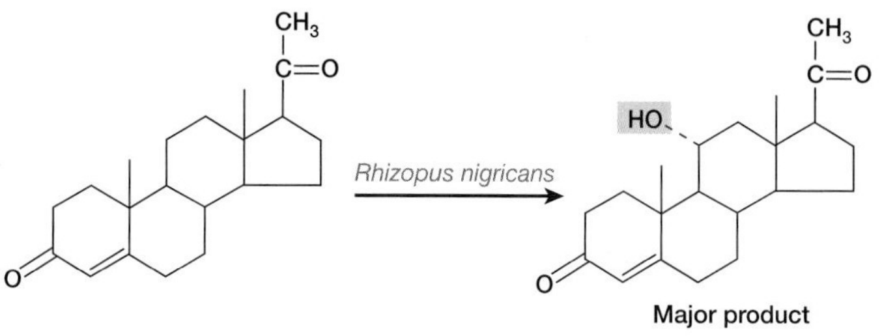

Figure 42.12   Biotransformation to Modify a Steroid.   Hydroxylation of progesterone in the 11α position by *Rhizopus nigricans*. The steroid is dissolved in acetone before addition to the pregrown fungal culture.

microorganisms. The microorganisms thus act as **biocatalysts.** Bioconversions have many advantages over chemical procedures. A major advantage is stereochemical; the biologically active form of a product is made. Enzymes also carry out very specific reactions under mild conditions, and larger water-insoluble molecules can be transformed. Unicellular bacteria, actinomycetes, yeasts, and molds have been used in various bioconversions. The enzymes responsible for these conversions can be intracellular or extracellular. Cells can be produced in batch or continuous culture and then dried for direct use, or they can be prepared in more specific ways to carry out desired bioconversions.

A typical bioconversion is the hydroxylation of a steroid (**figure 42.12**). In this example, the water-insoluble steroid is dissolved in acetone and then added to the reaction system that contains the pregrown microbial cells. The course of the modification is monitored, and the final product is extracted from the medium and purified.

Biotransformations carried out by free enzymes or intact nongrowing cells do have limitations. Reactions that occur in the absence of active metabolism—without reducing power or ATP being available continually—are primarily exergonic reactions (*see section 8.3*). If ATP or reductants are required, an energy source such as glucose must be supplied under carefully controlled nongrowth conditions.

---

1. Discuss the major uses for biopolymers and biosurfactants.
2. What are cyclodextrins and why are they important additives?
3. What are bioconversions or biotransformations? Describe the changes in molecules that result from these processes.

42.4   MICROBIAL GROWTH IN COMPLEX NATURAL ENVIRONMENTS

Microbial management also can be carried out in complex natural environments such as waters, soils, or high organic matter-containing composts where the physical and nutritional conditions for microbial growth cannot be completely controlled, and a largely unknown microbial community is present. Exam-

ples are (1) the use of microbial communities to carry out biodegradation, bioremediation, and environmental maintenance processes; and (2) the addition of microorganisms to soils or plants for the improvement of crop production. Both of these applications will be discussed in this section.

Biodegradation Using Natural Microbial Communities

Before discussing **biodegradation** processes carried out by natural microbial communities, it is important to consider definitions. Biodegradation has at least three definitions (**figure 42.13**): (1) a minor change in an organic molecule leaving the main structure still intact, (2) fragmentation of a complex organic molecule in such a way that the fragments could be reassembled to yield the original structure, and (3) complete mineralization. As mentioned previously (*see p. 594*), mineralization is the transformation of organic molecules to mineral forms.

Originally it was assumed, given time and the almost infinite variety of microorganisms, that all organic compounds, including those synthesized in the laboratory, would eventually degrade. Observations of natural and synthetic organic compound accumulation in natural environments, however, began to raise questions about the ability of microorganisms to degrade these varied substances and the role of the environment (clays, anaerobic conditions) in protecting some chemicals. With the development of synthetic pesticides, it became distressingly evident that not all organic compounds are immediately biodegradable. This chemical **recalcitrance** (resisting authority or control) resulted from the apparent fallibility of microorganisms, or their inability to degrade some industrially synthesized chemical compounds.

Degradation of a complex compound takes place in several stages. In the case of halogenated compounds, dehalogenation often occurs early in the overall process. Dehalogenation of many compounds containing chlorine, bromine, or fluorine occurs faster under anaerobic than under aerobic conditions. The study of **reductive dehalogenation,** especially its commercial applications, is expanding rapidly. Research on the dehalogenation of PCBs shows that this coreductive process can use electrons derived from water; other studies indicate that hydrogen can be the source of reductant for the dehalogenation of different chlorinated compounds. Major genera that carry out this process include *Desulfitobacterium, Dehalospirillum,* and *Desulfomonile.*

Humic acids, brownish polymeric residues of lignin decomposition that accumulate in soils and waters, have been found to play a role in anaerobic biodegradation processes. They can serve as electron acceptors under what are called "humic-acid-reducing conditions." The use of humic acids as electron acceptors has been observed with the anaerobic dechlorination of vinyl chloride and dichloroethylene.

Once the anaerobic dehalogenation steps are completed, degradation of the main structure of many pesticides and other xenobiotics often proceeds more rapidly in the presence of O_2.

Structure and stereochemistry are critical in predicting the fate of a specific chemical in nature. When a constituent is in the

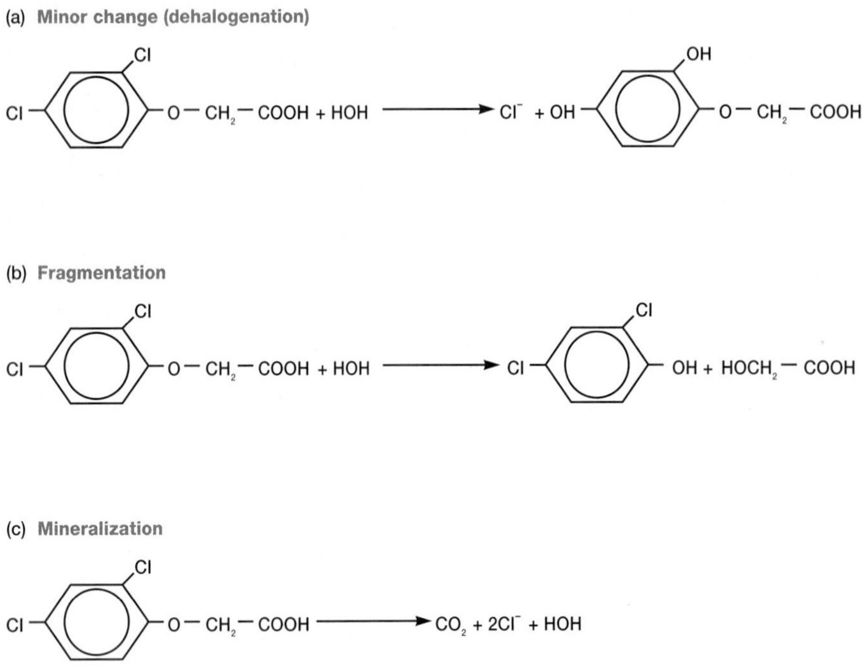

Figure 42.13 Biodegradation Has Several Meanings. Biodegradation is a term that can be used to describe three major types of changes in a molecule. (**a**) A minor change in the functional groups attached to an organic compound, as the substitution of a hydroxyl group for a chlorine group. (**b**) An actual breaking of the organic compound into organic fragments in such a way that the original molecule could be reconstructed. (**c**) The complete degradation of an organic compound to minerals.

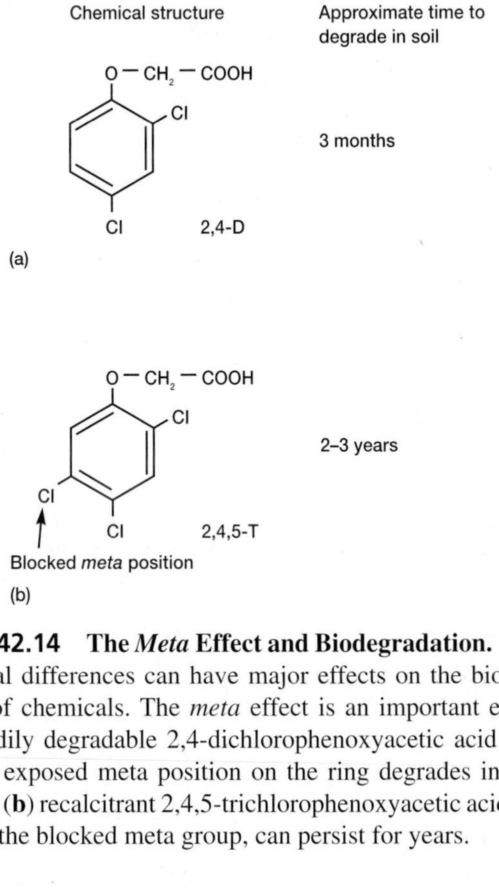

(a) **Minor change (dehalogenation)**

$$Cl-\!\!\bigcirc\!\!-O-CH_2-COOH + HOH \longrightarrow Cl^- + OH-\!\!\bigcirc\!\!-O-CH_2-COOH$$

(b) **Fragmentation**

$$Cl-\!\!\bigcirc\!\!-O-CH_2-COOH + HOH \longrightarrow Cl-\!\!\bigcirc\!\!-OH + HOCH_2-COOH$$

(c) **Mineralization**

$$Cl-\!\!\bigcirc\!\!-O-CH_2-COOH \longrightarrow CO_2 + 2Cl^- + HOH$$

Chemical structure	Approximate time to degrade in soil
2,4-D	3 months
2,4,5-T	2–3 years

(a)

Blocked *meta* position

(b)

Figure 42.14 The *Meta* Effect and Biodegradation. Minor structural differences can have major effects on the biodegradability of chemicals. The *meta* effect is an important example. (**a**) Readily degradable 2,4-dichlorophenoxyacetic acid (2,4-D) with an exposed meta position on the ring degrades in several months; (**b**) recalcitrant 2,4,5-trichlorophenoxyacetic acid (2,4,5-T) with the blocked meta group, can persist for years.

meta as opposed to the ortho position, the compound will be degraded at a much slower rate. The *meta* **effect** is shown in **figure 42.14.** This stereochemical difference is the reason that the common lawn herbicide 2,4-dichlorophenoxyacetic acid (2,4-D), with a chlorine in the ortho position, will be largely degraded in a single summer. In contrast, 2,4,5-trichlorophenoxyacetic acid, with a constituent in the meta position, will persist in the soils for several years, and thus is used for long-term brush control.

An important aspect of managing biodegradation is the recognition that many of the compounds that are added to environments are **chiral,** or possess asymmetry and handedness. Microorganisms often can degrade only one isomer of a substance; the other isomer will remain in the environment. At least 25% of herbicides are chiral. Thus it is critical to add the herbicide isomer that is effective and also degradable. Recent studies have shown that microbial communities in different environments will degrade different enantiomers. Changes in environmental conditions and nutrient supplies can alter the patterns of chiral form degradation.

Microbial communities change their characteristics in response to the addition of inorganic or organic substrates. If a particular compound, such as a herbicide, is added repeatedly to a microbial community, the community adapts and faster rates of degradation can occur—a process of acclimation (**figure 42.15**). The adaptive process often is so effective that this enrichment culture-based approach, established on the principles elucidated by Beijerinck (*see p. 11*) can be used to isolate organisms with a desired set of capabilities. For example, a microbial community can become so efficient at rapid herbicide degradation that herbicide effectiveness is diminished. To counteract this process, herbicides can be changed to throw the microbial community off balance, thus preserving the effectiveness of the chemicals.

Degradation processes that occur in soils also can be used in large-scale degradation of hydrocarbons or wastes from agricultural operations, in a technique called **land farming.** The waste

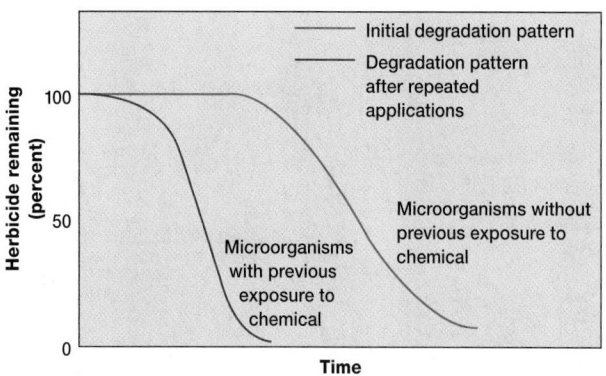

Figure 42.15   Repeated Exposure and Degradation Rate. Addition of an herbicide to a soil can result in changes in the degradative ability of the microbial community. Relative degradation rates for an herbicide after initial addition to a soil, and after repeated exposure to the same chemical.

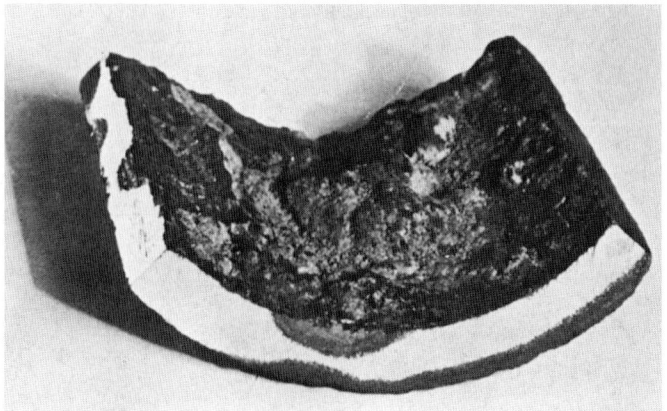

Figure 42.16   Microbial-Mediated Metal Corrosion. The microbiological corrosion of iron is a major problem. The graphitization of iron under a rust bleb on the pipe surface allows microorganisms, including *Desulfovibrio,* to corrode the inner surface.

Microbial Diversity & Ecology

42.2   Methanogens—A New Role for a Unique Microbial Group

The methanogens, an important group of the archaea that can produce methane, have had widespread impact on the Earth. Despite intensive research, new discoveries are still being made concerning these microorganisms. Methanogens have now been found to contribute to the anaerobic corrosion of soft iron. Previously the microbial group usually considered the major culprit in the anaerobic corrosion process was the genus *Desulfovibrio*, which can use sulfate as an oxidant and hydrogen produced in the corrosion process as a reductant. Methanogens can use elemental iron as an electron source in their

metabolism. It appears that corrosion may occur even without the presence of sulfate, which is required for functioning of *Desulfovibrio*. Rates of iron removal by the methanogens are around 79 mg/1,000 cm^2 of surface area in a 24-hour period. This may not seem a high rate, but in relation to the planned service life of metal structures in muds and subsurface soils—possibly years and decades—such corrosion can become a major problem. Continuous efforts to improve protection of iron structures will be required in view of the diversity of iron-corroding microorganisms.

material is incorporated into the soil or allowed to flow across the soil surface, where degradation occurs.

It is important to emphasize that such degradation processes do not always reduce environmental problems. In fact, the partial degradation or modification of an organic compound may not lead to decreased toxicity. An example of this process is the microbial metabolism of 1,1,1-trichloro-2,2-*bis*-(p-chlorophenyl) ethane (DDT), a xenobiotic or foreign (chemically synthesized) organic compound. Degradation removes a chlorine function to give 1,1-dichloro-2,2-*bis*(p-chlorophenyl)ethylene (DDE), which is still of environmental concern. Another important example is the degradation of trichloroethylene (TCE), a widely used solvent. If this is degraded under anaerobic conditions, the dangerous carcinogen vinyl chloride can be synthesized.

$$Cl_2 = CHCl \rightarrow ClHC = CH_2$$

Biodegradation also can lead to widespread damages and financial losses. Metal corrosion is a particularly important example. The microbially mediated corrosion of metals is particularly critical where iron pipes are used in waterlogged anaerobic environments or in secondary petroleum recovery processes carried out at older oil fields. In these older fields water is pumped down a series of wells to force residual petroleum to a central collection point. If the water contains low levels of organic matter and sulfate, anaerobic microbial communities can develop in rust blebs or tubercles (**figure 42.16**), resulting in punctured iron pipe and loss of critical pumping pressure. Microorganisms that use elemental iron as an electron donor during the reduction of CO_2 in methanogenesis have recently been discovered (**Microbial Diversity Ecology 42.2**). Because of the wide range of interactions that occur between microorganisms and metals, the need to develop strategies to deal with microbial corrosion problems is critical.

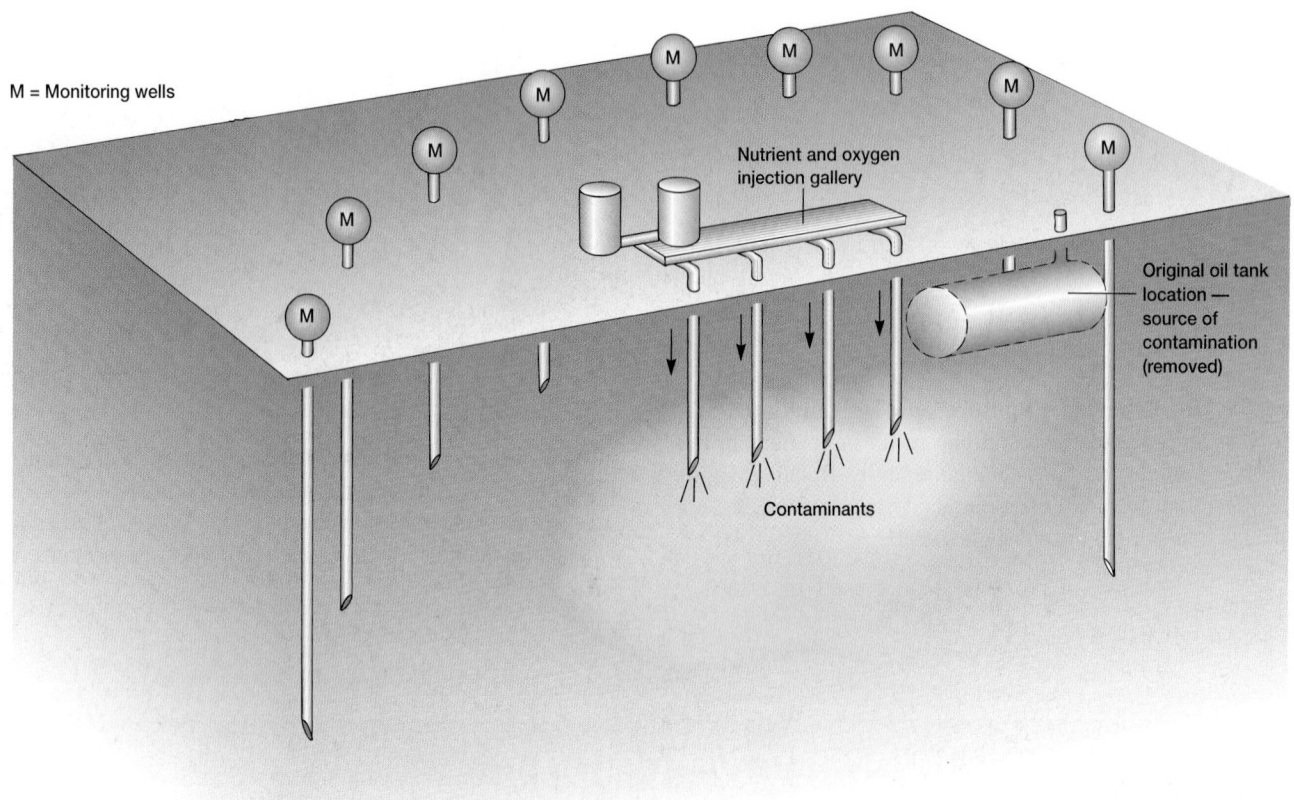

M = Monitoring wells

Nutrient and oxygen
injection gallery

Original oil tank
location —
source of
contamination
(removed)

Contaminants

Figure 42.17   A Subsurface Engineered Bioremediation System.   Monitoring and recovery wells are used to monitor the plume and its possible movement. Nutrients and oxygen (as air, peroxide, or other oxygen-releasing compounds) are added to the soil and groundwater to promote more efficient degradation of contaminants.

---

1. Give alternative definitions for the term biodegradation.
2. What is reductive dehalogenation? Describe humic acids and the role they can play in anaerobic degradation processes.
3. Why is the "*meta* effect" important for understanding biodegradation?
4. Discuss chirality and its importance for understanding degradation effects in the environment.
5. What are some of the important microbial groups involved in iron corrosion?

Changing Environmental Conditions to Stimulate Biodegradation

Often natural microbial communities will not be able to carry out biodegradation processes at a desired rate due to limiting physical or nutritional factors. For example, biodegradation often will be limited by low oxygen levels. Hydrocarbons, nitrogen, phosphorus, and other needed nutrients also may be absent or available only at slow flux rates, thus limiting rates of degradation. In these

cases, it is necessary to determine the limiting factors, based on Liebig's and Shelford's laws, and then to supply needed materials or modify the environment.   Liebig's and Shelford's laws (p. 128)

Most of the early efforts to stimulate the degradative activities of microorganisms involved the modification of waters and soils by the addition of oxygen or nutrients, called **bioremediation** and shown in **figure 42.17.** Contact between the microbes and the substrate; the proper physical environment, nutrients, oxygen (in most cases); and the absence of toxic compounds are critical in this managed process.

Often it is found that the addition of easily metabolized organic matter such as glucose increases biodegradation of recalcitrant compounds that are usually not used as carbon and energy sources by microorganisms. This process, termed **cometabolism,** is finding widespread applications in biodegradation management. Cometabolism can be carried out by simply adding easily catabolized organic matter such as glucose or cellulose and the compound to be degraded to a complex microbial community. Plants also may be used to provide the organic matter. Cometabolism is important in many different biodegradation systems.

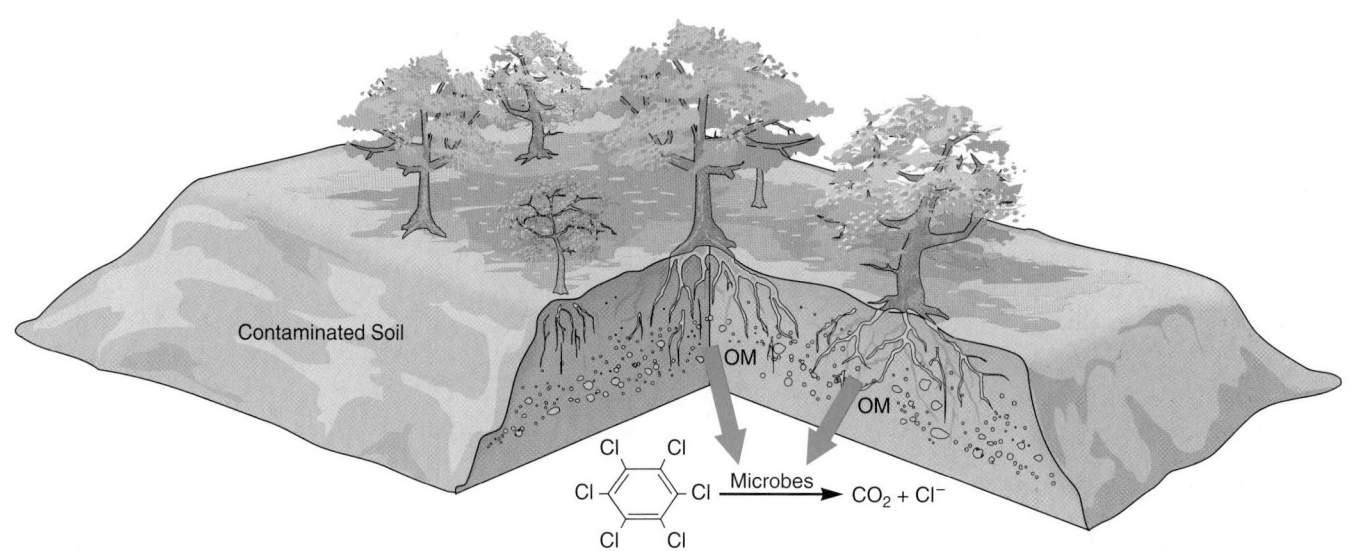

Figure 42.18   Phytoremediation.   A conceptual view of a phytoremediation system, with a cut-away section of the root-soil zone. When organic matter (OM) is released from the plant roots, cometabolic processes can be carried out more efficiently by microbes, leading to enhanced degradation of contaminants. The degradation of hexachlorobenzene is shown as an example.

Stimulating Hydrocarbon Degradation in Waters and Soils

Experiences with oil spills in marine environments illustrate these principles. When working with dispersed hydrocarbons in the ocean, contact between the microorganism, the hydrocarbon substrate, and other essential nutrients must be maintained. To achieve this, pellets containing nutrients and an oleophilic (hydrocarbon soluble) preparation have been used. This technique has accelerated the degradation of different crude oil slicks by 30 to 40%, in comparison with control oil slicks where the additional nutrients were not available.

A major challenge for this technology was the *Exxon Valdez* oil spill, which occurred in Alaska in March 1989. Several different approaches were used to increase biodegradation. These included nutrient additions, chemical dispersants, biosurfactant additions, and the use of high-pressure steam. The use of a microbially produced glycolipid emulsifier has proven helpful.

These bioremediation approaches are also used in soils and sediments. For example, a unique two-stage process can be used to degrade PCBs in river sediments. First, partial dehalogenation of the PCBs occurs naturally under anaerobic conditions. Then the muds are aerated to promote the complete degradation of the less chlorinated residues (chapter opening figure).

Stimulating Degradation with Plants

Phytoremediation, or the use of plants to stimulate the extraction, degradation, adsorption, stabilization or volatilization of contaminants is becoming an important part of biodegradation technology. A plant provides nutrients that allow cometabolism to occur in the plant root zone or rhizosphere (**figure 42.18**). Phytoremediation also includes plant contributions to degradation, immobi-

Table 42.12
Types of Phytoremediation

Process	Function
Phytoextraction	Use of pollutant-accumulating plants to remove metals or organics from soil by concentrating them in the harvestable plant parts
Phytodegradation	Use of plants and associated microorganisms to degrade organic pollutants
Rhizofiltration	Use of plant roots to absorb and adsorb pollutants, mainly metals, from water and aqueous waste streams
Phytostabilization	Use of plants to reduce the bioavailability of pollutants in the environment
Phytovolatilization	Use of plants to volatilize pollutants

Based on T. Macek; M. Mackova; and J. Kás. 2000. Exploitation of plants for the removal of organics in environmental remediation. *Biotechnol. Adv.* 18:23–34. P. 25.

lization, and volatilization processes, as noted in **table 42.12.** Transgenic plants may be employed in phytoremediation. Using cloning techniques with *Agrobacterium* (*see pp. 331, 479–80, 660*), the *mer*A and *mer*B genes have been integrated into a plant (*Arabidopsis thaliana*), thus making it possible to transform extremely toxic organic mercury forms to elemental mercury, which is less of an environmental hazard. Recently transgenic tobacco plants have been constructed that express tetranitrate reductase, an enzyme from an explosive-degrading bacterium, thereby enabling the transgenic plants to degrade nitrate ester and nitro aromatic explosives. The genetically modified plants grow in solutions of explosives that control plants cannot tolerate. Other plants have been

Figure 42.19   Copper Leaching from Low-Grade Ores.   The chemistry and microbiology of copper ore leaching involve interesting complementary reactions. The microbial contribution is the oxidation of ferrous ion (Fe^{2+}) to ferric ion (Fe^{3+}). *Leptospirillum ferrooxidans* and related microorganisms are very active in this oxidation. The ferric ion then reacts chemically to solubilize the copper. The soluble copper is recovered by a chemical reaction with elemental iron, which results in an elemental copper precipitate.

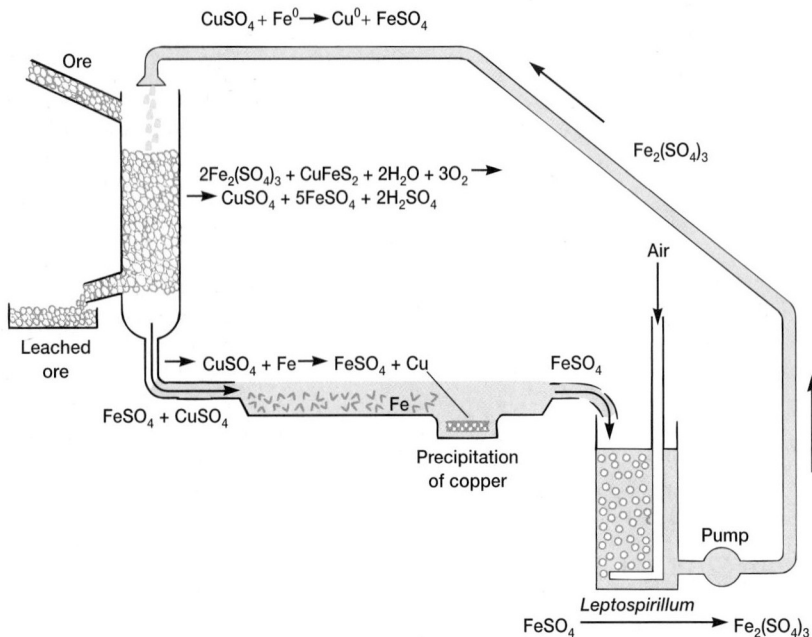

Stimulation of Metal Bioleaching from Minerals

Bioleaching is the use of microorganisms, which produce acids from reduced sulfur compounds, to create acidic environments that solubilize desired metals for recovery. This approach is used to recover metals from ores and mining tailings with metal levels too low for smelting. Bioleaching carried out by natural populations of *Leptospirillum*-like species, *Thiobacillus thiooxidans,* and related thiobacilli, for example, allows recovery of up to 70% of the copper in low-grade ores. As shown in **figure 42.19,** this involves the biological oxidation of copper present in these ores to produce soluble copper sulfate.

It is apparent that nature will assist in bioremediation if given a chance. The role of natural microorganisms in biodegradation is now better appreciated. An excellent example is the recent work with the very versatile fungus *Phanerochaete chrysosporium* (**Microbial Diversity & Ecology 42.3**).

Often biodegradation and biodeterioration have major negative effects, and it becomes important to control and limit these processes by environmental management. Problems include unwanted degradation of paper, jet fuels, textiles, and leather goods. A global concern is microbial-based metal corrosion.

---

1. What is cometabolism and why is this important for degradation processes?
2. What factors must one consider when attempting to stimulate the microbial degradation of a massive oil spill in a marine environment?
3. Describe the major types of phytoremediation. What is the role of microorganisms in each of these processes?

engineered in the same way to degrade trichloroethylene, an environmental contaminant of worldwide concern.

4. How is bioleaching carried out and what microbial genera are involved?
5. What is unique about *Phanerochaete chrysosporium?* What does its name mean?

Addition of Microorganisms to Complex Microbial Communities

Both in laboratory and field studies, attempts have been made to speed up existing microbiological processes by adding known active microorganisms to soils, waters, or other complex systems as a part of **bioaugmentation** studies. The microbes used in these experiments have been isolated from contaminated sites, taken from culture collections, or derived from uncharacterized enrichment cultures. For example, commercial culture preparations are available to facilitate silage formation and to improve septic tank performance.

Addition of Microorganisms without Considering Protective Microhabitats

With the development of the "superbug" by A. M. Chakrabarty in 1974, there was initial excitement due to the hope that such an improved microorganism might be able to degrade hydrocarbon pollutants very effectively. A critical point, which was not considered, was the actual location, or microhabitat, where the microbe had to survive and function. Engineered microorganisms were added to soils and waters with the expectation that rates of degradation would be stimulated as these microorganisms established themselves. Generally such additions led to short-term increases in rates of the desired activity, but typically after a few days the microbial community responses were similar in treated

Microbial Diversity & Ecology

42.3   *A Fungus with a Voracious Appetite*

The basidiomycete *Phanerochaete chrysosporium* (the scientific name means "visible hair, golden spore") is a fungus with unusual degradative capabilities. This organism is termed a "white rot fungus" because of its ability to degrade lignin, a randomly linked phenylpropane-based polymeric component of wood (*see section 28.3*). The cellulosic portion of wood is attacked to a lesser extent, resulting in the characteristic white color of the degraded wood. This organism also degrades a truly amazing range of xenobiotic compounds (nonbiological foreign chemicals) using both intracellular and extracellular enzymes.

As examples, the fungus degrades benzene, toluene, ethylbenzene, and xylenes (the so-called BTEX compounds), chlorinated compounds such as 2,4,5-trichloroethylene (TCE), and trichlorophenols. The latter are present as contaminants in wood preservatives and also are used as pesticides. In addition, other chlorinated benzenes can be degraded with or without toluenes being present. Even the insecticide Hydramethylnon is degraded!

How does this microorganism carry out such feats? Apparently most degradation of these xenobiotic compounds occurs after active growth, during the secondary metabolic lignin degradation phase. Degradation of some compounds involves important extracellular enzymes including lignin peroxidase, manganese-dependent peroxidase, and glyoxal oxidase. A critical enzyme is pyranose oxidase, which releases H_2O_2 for use by the manganese-dependent peroxidase enzyme. The H_2O_2 also is a precursor of the highly reactive hydroxyl radical, which participates in wood degradation. Apparently the pyranose oxidase enzyme is located in the interperiplasmic space of the fungal cell wall, where it can function either as a part of the fungus or be released and penetrate into the wood substrate. It appears that the nonspecific enzymatic system that releases these oxidizing products degrades many cyclic, aromatic, and chlorinated compounds related to lignins.

We can expect to continue hearing of many new advances in work with this organism. Potentially valuable applications being studied include growth in bioreactors where intracellular and extracellular enzymes can be maintained in the bioreactor while liquid wastes flow past the immobilized fungi.

and control systems. After many unsuccessful attempts, it was found that the lack of effectiveness of such added cultures was due to at least three factors: (1) the attractiveness of laboratory-grown microorganisms as a food source for predators such as soil protozoa, (2) the inability of these added microorganisms to contact the compounds to be degraded, and (3) the failure of the added microorganisms to survive and compete with indigenous microorganisms. Such a modified microorganism may be less fit to compete and survive because of the additional energetic burden required to maintain the extra DNA.

Attempts have been made to make such laboratory-grown cultures more capable of survival in a natural environment by growing them in low-nutrient media or starving the microorganisms before adding them to an environment. These "toughening" approaches have improved microbial survival and function somewhat, but have not solved the problem. In recent years, there has been less interest in simply adding microorganisms to environments without considering the specific niche or microenvironment in which they are to survive and function. This has led to the field of **natural attenuation,** which emphasizes the use of natural microbial communities in the environmental management of pollutants.

Addition of Microorganisms Considering Protective Microhabitats

Microorganism additions to natural environments can be more successful if the microorganism is added together with a microhabitat that gives the organism physical protection, as well as possibly supplying nutrients. This makes it possible for the microorganism to survive in spite of the intense competitive pressures that exist in the natural environment, including pressure from protozoan predators such as ciliates, flagellates, and amoebae. Microhabitats may be either living or inert. Predation (pp. 589–90)

Living Microhabitats. Specialized living microhabitats include the surface of a seed, a root, or a leaf, which, with their higher nutrient flux rate and the chance for initial colonization by the added microorganisms, can protect the added microbe from the fierce competitive conditions in the natural environment. Examples include the use of *Rhizobium* and *Bacillus thuringiensis.* In order to ensure that *Rhizobium* is in close association with the legume, seeds are coated with the microbe using an oil-organism mixture, or *Rhizobium* is placed in a band under the seed where the newly developing primary root will penetrate. In contrast, *Bacillus thuringiensis* (BT) is placed on the surface of the plant leaf, or the plant is engineered to contain the BT genes that allow the production of the toxic protein in situ, once it is ingested. After ingestion by the target organism, the toxic protein will be within the digestive tract where it is most effective. *Bacillus thuringiensis* (pp. 512, 988–89); *Rhizobium* (sections 22.1 and 30.4)

Inert Microhabitats. Recently it has been found that microorganisms can be added to natural communities together with protective inert microhabitats! As an example, if microbes are added to a soil with microporous glass, the survival of added microorganisms can be markedly enhanced. Other microbes have been observed to create their own microhabitats! Microorganisms in the water column overlying PCB-contaminated sand-clay soils have been observed to create their own "clay hutches" by binding clays to their outer surfaces with exopolysaccharides. These illustrations show that with the application of principles of microbial ecology it may be possible to more successfully manage microbial communities in nature.

---

1. What factors might limit the ability of microorganisms, after addition to a soil or water, to be able to persist and carry out desired functions?
2. What types of microhabitats can be used with microorganisms when they are added to a complex natural environment?
3. Why are plants inoculated with *Bacillus thuringiensis*?

 42.5 BIOTECHNOLOGICAL APPLICATIONS

Microorganisms and parts of microorganisms, especially enzymes, are used in a wide variety of biotechnological applications to monitor the levels of critical compounds in the environment and in animals and humans. These techniques have wide applications in environmental science, animal and human health, and in basic science.

Biosensors

A rapidly developing area of biotechnology, arousing intense international scientific interest, is that of **biosensor** production. In this field of bioelectronics, living microorganisms (or their enzymes or organelles) are linked with electrodes, and biological reactions are converted into electrical currents by these biosensors (**figure 42.20**). Biosensors are being developed to measure specific components in beer, to monitor pollutants, to detect flavor compounds in food, and to study environmental

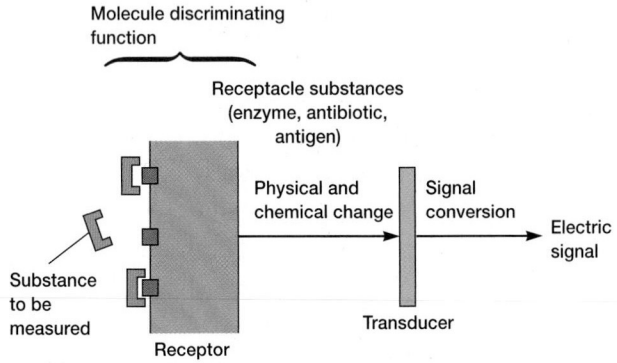

Figure 42.20 Biosensor Design. Biosensors are finding increasing applications in medicine, industrial microbiology, and environmental monitoring. In a biosensor a biomolecule or whole microorganism carries out a biological reaction, and the reaction products are used to produce an electrical signal.

processes such as changes in biofilm concentration gradients. It is possible to measure the concentration of substances from many different environments (**table 42.13**). Applications include the detection of glucose, acetic acid, glutamic acid, ethanol, and biochemical oxygen demand. In addition, the application of biosensors to measure cephalosporin, nicotinic acid, and several B vitamins has been described. Recently biosensors have been developed using immunochemical-based detection systems. These new biosensors will detect pathogens, herbicides, toxins, proteins, and DNA. Many of these biosensors are based on the use of a streptavidin-biotin recognition system (**Techniques & Applications 42.4**).

Rapid advances are being made in all areas of biosensor technology. These include major improvements in the stability and durability of these units, which are being made more portable and sensitive. Microorganisms and metabolites such as glucose can be measured, thus meeting critical needs in modern medicine.

Microarrays

A large part of the new and developing microbial biotechnology involves the use of DNA sequences in **gene arrays** to monitor gene expression in complex biological systems (*see section 15.6*). The rapid advances that have occurred in this area are the result of progress in genomics, recombinant DNA technology, optics, fluid flow systems, and high-speed data acquisition and processing. This **microarray technology** has been suggested to provide the equivalent of the chemist's periodic table. It offers the potential of assaying all genes used to assemble an organism and can monitor expression of tens of thousands of genes based on the principles shown in **figure 42.21**. In this technique, 100 to 200 μl volumes, containing desired sequences, are spotted onto glass slides or other inert materials and dried. These arrays are then mixing with cDNAs from gene expression (*see p. 313*). Binding of the cDNA for various genes is measured using rapid photometric monitoring techniques. Genomics (chapter 15); Nucleic acid hybridization (pp. 419–20)

Commercial microarray products are now available that contain 6,144 open frames for screening gene expression in *Saccharomyces cerevisiae*. For *E. coli*, 4,290 open reading frames can be scanned in a microarray format, representing the entire *E coli*

Table 42.13
Biosensors: Potential Biomedical, Industrial, and Environmental Applications

Clinical diagnosis and biomedical monitoring
Agricultural, horticultural, veterinary analysis
Detection of pollution, and microbial contamination of water
Fermentation analysis and control
Monitoring of industrial gases and liquids
Measurement of toxic gas in mining industries
Direct biological measurement of flavors, essences, and pheromones

Egg white contains many proteins and glycoproteins with unique properties. One of the most interesting, which binds tenaciously to biotin, was isolated in 1963. This glycoprotein, called avidin due to its "avid" binding of biotin, was suggested to play an important role: making egg white antimicrobial by "tying up" the biotin needed by many microorganisms. Avidin, which functions best under alkaline conditions, has the highest known binding affinity between a protein and a ligand. Several years later, scientists at Merck & Co., Inc. discovered a similar protein produced by an actinomycete, *Streptomyces avidini,* which binds biotin at a neutral pH and which does not contain carbohydrates. These characteristics make streptavidin an ideal binding agent for biotin, and it has been used in an almost unlimited range of applications, as shown in the **Box figure.** The streptavidin protein is joined to a probe. When a sample is incubated with the biotinylated binder, the binder attaches to any available target molecules. The presence and location of target molecules can be determined by treating the sample with a streptavidin probe because the streptavidin binds to the biotin on the biotinylated binder, and the probe is then visualized. This detection system is being employed in a wide variety of biotechnological applications, including use as a nonradioactive probe in hybridization studies and as a critical component in biosensors for a wide range of environmental monitoring and clinical applications. Not bad for a protein from a "simple" filamentous bacterium!

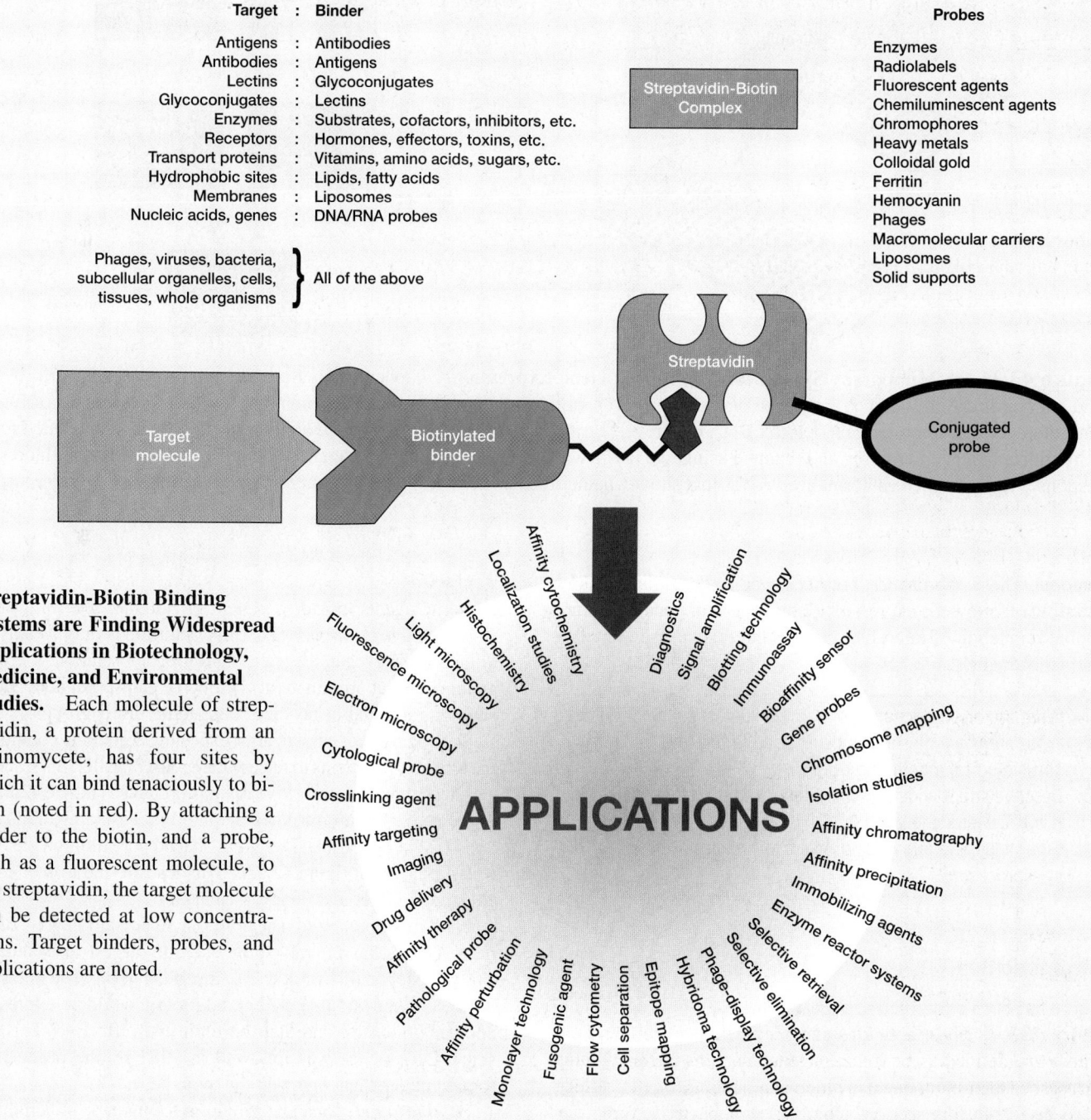

Target	:	Binder
Antigens	:	Antibodies
Antibodies	:	Antigens
Lectins	:	Glycoconjugates
Glycoconjugates	:	Lectins
Enzymes	:	Substrates, cofactors, inhibitors, etc.
Receptors	:	Hormones, effectors, toxins, etc.
Transport proteins	:	Vitamins, amino acids, sugars, etc.
Hydrophobic sites	:	Lipids, fatty acids
Membranes	:	Liposomes
Nucleic acids, genes	:	DNA/RNA probes

Phages, viruses, bacteria, subcellular organelles, cells, tissues, whole organisms } All of the above

Streptavidin-Biotin Complex

Probes

Enzymes
Radiolabels
Fluorescent agents
Chemiluminescent agents
Chromophores
Heavy metals
Colloidal gold
Ferritin
Hemocyanin
Phages
Macromolecular carriers
Liposomes
Solid supports

Streptavidin

Target molecule

Biotinylated binder

Conjugated probe

Streptavidin-Biotin Binding Systems are Finding Widespread Applications in Biotechnology, Medicine, and Environmental Studies. Each molecule of streptavidin, a protein derived from an actinomycete, has four sites by which it can bind tenaciously to biotin (noted in red). By attaching a binder to the biotin, and a probe, such as a fluorescent molecule, to the streptavidin, the target molecule can be detected at low concentrations. Target binders, probes, and applications are noted.

APPLICATIONS

Affinity cytochemistry
Localization studies
Histochemistry
Light microscopy
Fluorescence microscopy
Electron microscopy
Cytological probe
Crosslinking agent
Affinity targeting
Imaging
Drug delivery
Affinity therapy
Pathological probe
Affinity perturbation
Monolayer technology
Fusogenic agent
Flow cytometry
Cell separation
Epitope mapping
Hybridoma technology
Phage-display technology
Selective elimination
Selective retrieval
Enzyme reactor systems
Immobilizing agents
Affinity precipitation
Affinity chromatography
Isolation studies
Chromosome mapping
Gene probes
Bioaffinity sensor
Immunoassay
Blotting technology
Signal amplification
Diagnostics

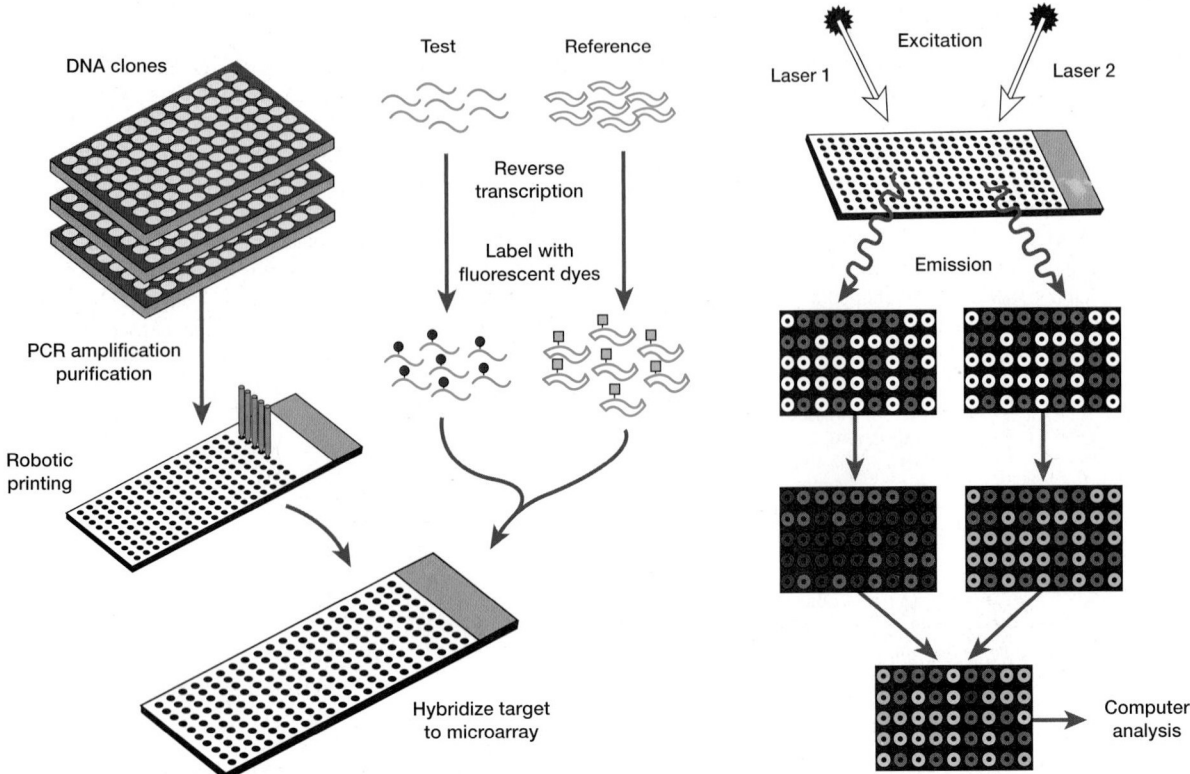

Figure 42.21 A Microarray System for Monitoring Gene Expression. Cloned genes from an organism are amplified by PCR, and after purification, samples are placed on a support in a pattern using a robotic printer. To monitor enzyme expression, RNA from test and reference cultures are converted to cDNA by a reverse transcriptase and labeled with two different fluor dyes. The labeled mixture is hybridized to the microarray and scanned using two lasers with different exciting wavelengths. After pseudocoloring, the fluorescence responses are measured as normalized ratios that show whether the test gene response is higher or lower than that of the reference.

genome. These approaches, both now and in the future, make it possible to follow the expression of thousands of genes and study global regulation of microbial growth and responses to environmental changes.

---

1. What are biosensors and how do they detect substances?
2. What areas are biosensors being used in to assist in chemical and biological monitoring efforts?
3. Describe streptavidin-biotin systems and how they work. Why is this technique important?
4. What is a gene array? What basic techniques are used in this new procedure?

Biopesticides

There has been a long-term interest in the use of bacteria, fungi, and viruses as **bioinsecticides** and **biopesticides** (**table 42.14**). These are defined as biological agents, such as bacteria, fungi, viruses, or their components, which can be used to kill a susceptible insect. In this section, major uses of bacteria, fungi, and viruses to control populations of insects will be discussed.

Bacteria

Bacterial agents include a variety of *Bacillus* species, primarily *B. thuringiensis* (*see p. 512*). This bacterium is only weakly toxic to insects as a vegetative cell, but during sporulation, it produces an intracellular protein toxin crystal, the parasporal body, that can act as a microbial insecticide for specific insect groups.

The parasporal crystal, after exposure to alkaline conditions in the hindgut, fragments to release the protoxin. After this reacts with a protease enzyme, the active toxin is produced (**figure 42.22**). Six of the active toxin units integrate into the plasma membrane (figure 42.22b,c) to form a hexagonal-shaped pore through the midgut cell, as shown in figure 42.22d. This leads to the loss of osmotic balance and ATP, and finally to cell lysis.

The most recent advances in our understanding of *Bacillus thuringiensis* have involved the creation of pest-resistant plants. The first step in this work was to insert the toxin gene into *E. coli*. This work showed that the crystal protein could be expressed in another organism, and that the toxin was effective. This major scientific advance was followed in 1987 by the production of tomato plants that contained the toxin gene.

B. thuringiensis can be grown in fermenters. When the cells lyse, the spores and crystals are released into the medium. The

Microbial Group	Major Organisms and Applications
Bacteria	*Bacillus thuringiensis* and *Bacillus popilliae* are the two major bacteria of interest. *Bacillus thuringiensis* is used on a wide variety of vegetable and field crops, fruits, shade trees, and ornamentals. *B. popilliae* is used primarily against Japanese beetle larvae. Both bacteria are considered harmless to humans. *Pseudomonas fluorescens,* which contains the toxin-producing gene from *B. thuringiensis,* is used on maize to suppress black cutworms.
Viruses	Three major virus groups that do not appear to replicate in warm-blooded animals are used: nuclear polyhedrosis virus (NPV), granulosis virus (GV), and cytoplasmic polyhedrosis virus (CPV). These occluded viruses are more protected in the environment.
Fungi	Over 500 different fungi are associated with insects. Infection and disease occur primarily through the insect cuticle. Four major genera have been used. *Beauveria bassiana* and *Metarhizium anisopliae* are used for control of the Colorado potato beetle and the froghopper in sugarcane plantations, respectively. *Verticillium lecanii* and *Entomophthora* spp., have been associated with control of aphids in greenhouse and field environments. *Coelomyces*, a chytrid, also is being used for control of mosquitoes.

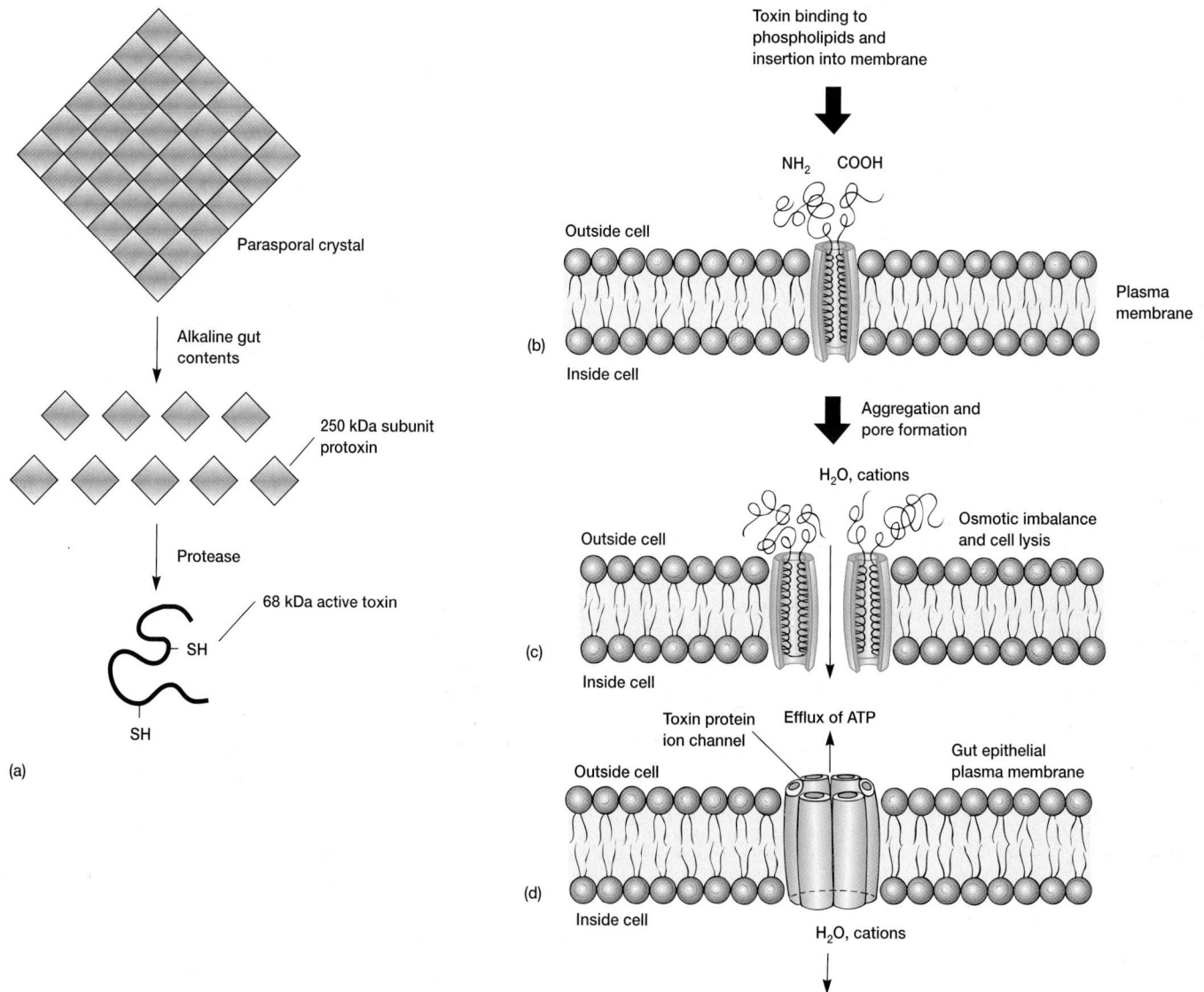

Figure 42.22 The Mode of Action of the *Bacillus thuringiensis* Toxin. (a) Release of the protoxin from the parasporal body and modification by proteases in the hindgut. (b) Insertion of the 68 kDa active toxin molecules into the membrane. (c) Aggregation and pore formation, showing a cross section of the pore. (d) Final creation of the hexagonal pore which causes an influx of water and cations as well as a loss of ATP, resulting in cell imbalance and lysis.

medium is then centrifuged and made up as a dust or wettable powder for application to plants.

A related bacterium, *Bacillus popilliae,* is used to combat the Japanese beetle. This bacterium, however, cannot be grown in fermenters, and inocula must be grown in the living host. The microorganism controls development of larvae, but destruction of the adult beetle requires chemical insecticides.

Viruses

Viruses that are pathogenic for specific insects include nuclear polyhedrosis viruses (NPVs), granulosis viruses (GVs), and cytoplasmic polyhedrosis viruses (CPVs). Currently over 125 types of NPVs are known, of which approximately 90% affect the *Lepidoptera*—butterflies and moths. Approximately 50 GVs are known, and they, too, primarily affect butterflies and moths. CPVs are the least host-specific viruses, affecting about 200 different types of insects. An important commercial viral pesticide is marketed under the trade name Elcar for control of the cotton bollworm *Heliothis zea.*

One of the most exciting advances involves the use of baculoviruses that have been genetically modified to produce a potent scorpion toxin active against insect larvae. After ingestion by the larvae, viruses are dissolved in the midgut and are released. Because the recombinant baculovirus produces this insect-selective neurotoxin, it acts more rapidly than the parent virus, and leaf damage by insects is markedly decreased.   Characteristics of insect viruses (pp. 404–5)

Fungi

Fungi also can be used to control insect pests. Fungal bioinsecticides, as listed in table 42.14, are finding increasing use in agriculture. The development of biopesticides is progressing rapidly.

Available bioinsecticides which are derived from fungi include kasugamycin and the polyoxins; in addition, special microbiological metabolites such as nikkomycin and the spinosyns are active against insects.

---

1. What two important bacteria have been used as bioinsecticides?
2. Briefly describe how the *Bacillus thuringiensis* toxin kills insects.
3. What types of viruses are being used to attempt to control insects? What is a trade name for one of these products?
4. Which fungi presently are being used as biopesticides?

42.6 IMPACTS OF MICROBIAL BIOTECHNOLOGY

The use of microorganisms in industrial microbiology and biotechnology, as discussed in this chapter, does not take place in an ethical and ecological vacuum. Decisions to make a particular product, and also the methods used, can have long-term and often unexpected effects, as with the appearance of antibiotic-resistant pathogens around the world.

Microbiology is a critical part of the area of **industrial ecology,** concerned with tracking the flow of elements and compounds though the natural and social worlds, or the **biosphere** and the **anthrosphere.** Microbiology, especially as an applied discipline, should be considered within its supporting social world.

Microorganisms have been of immense benefit to humanity through their role in food production and processing, the use of their products to improve human and animal health, in agriculture, and for the maintenance and improvement of environmental quality. Other microorganisms, however, are important pathogens and agents of spoilage, and microbiologists have helped control or limit the activities of these harmful microorganisms. The discovery and use of beneficial microbial products, such as antibiotics, have contributed to a doubling of the human life span in the last century.

A microbiologist who works in any of these areas of biotechnology should consider the longer-term impacts of possible technical decisions. An excellent introduction to the relationship between technology and possible societal impacts is given by Samuel Florman (*see appendix V*). Our first challenge, as microbiologists, is to understand, as much as is possible, the potential impacts of new products and processes on the broader society as well as on microbiology. An essential part of this responsibility is to be able to communicate effectively with the various "societal stakeholders" about the immediate and longer-term potential impacts of microbial-based (and other) technologies.

---

1. Discuss possible ethical and ecological impacts of a particular product or process discussed in this chapter. Think in terms of the broadest possible impacts in your discussion of this problem.
2. Define industrial ecology.
3. What are the biosphere and anthrosphere? Why might you think the term anthrosphere was coined?

Summary

42.1 Choosing Microorganisms for Industrial Microbiology and Biotechnology

a. Industrial microbiology has been used to manufacture such products as antibiotics, amino acids, and organic acids and has had many important positive effects on animal and human health. Most work in this area has been carried out using microorganisms isolated from nature or modified by the use of classic mutation techniques. Biotechnology involves the use of molecular techniques to modify and improve microorganisms.

b. Finding new microorganisms in nature for use in biotechnology is a continuing challenge. For most environments, only a very small part of the observable microbial community has been grown (**tables 42.1** and **42.2**), but major advances in growing "uncultured" microbes are being made.

c. Selection and mutation continue to be important approaches for identifying new microorganisms. These well-established procedures are now being complemented by molecular techniques, including metabolic engineering and combinatorial biology. With combinatorial biology (**table 42.3**), it is possible to transfer genes from one organism to another organism, and to form new products.

d. Site-directed mutagenesis and protein engineering are used to modify gene expression. These approaches are leading to new and often different products with new properties (**figure 42.3**).

e. Natural genetic engineering is of increasing interest. This involves exploiting microbial responses to stress in adaptive mutation and forced evolution, with the hope of identifying microorganisms with new properties.

42.2 Microorganism Growth in Controlled Environments

a. Microorganisms can be grown in controlled environments of various types using fermenters and other culture systems. If defined constituents are used, growth parameters can be chosen and varied in the course of growing a microorganism. This approach is used particularly for the production of amino acids, organic acids, and antibiotics.

42.3 Major Products of Industrial Microbiology

a. A wide variety of compounds are produced in industrial microbiology that impact our lives in many ways (**table 42.9**). These include antibiotics, amino acids, organic acids, biopolymers such as the cyclodextrins, and biosurfactants (**figures 42.8, 42.9, and 42.11**). Microorganisms also can be used as biocatalysts to carry out specific chemical reactions (**figure 42.12**).

b. Specialty nonantibiotic compounds are an important part of industrial microbiology and biotechnology. These include widely used antitumor agents (**table 42.11**).

42.4 Microbial Growth in Complex Natural Environments

a. Microorganism growth in complex natural environments such as soils and waters is not used to create microbial products but to carry out environmental management processes, including bioremediation, plant inoculation, and other related activities. In these cases, the microbes themselves are not final products.

b. Biodegradation is a critical part of natural systems mediated largely by microorganisms. This can involve minor changes in a molecule, fragmentation, or mineralization (**figure 42.13**).

c. Biodegradation can be influenced by many factors, including oxygen presence or absence, humic acids, and the presence of readily usable organic matter. Reductive dehalogenation proceeds best under anaerobic conditions, and the presence of organic matter can facilitate modification of recalcitrant compounds in the process of cometabolism.

d. The structure of organic compounds influences degradation. If constituents are in specific locations on a molecule, as in the *meta* position (**figure 42.14**), or if varied structural isomers are present degradation can be affected.

e. Degradation management can be carried out in place, whether this be large marine oil spills, soils, or the subsurface (**figure 42.17**). Such large-scale efforts usually involve the use of natural microbial communities.

f. Degradation can lead to increased toxicity in many cases. If not managed carefully, widespread pollution can occur. Iron corrosion is a particular concern with methanogens and *Desulfovibrio* playing important roles in this process.

g. Plants can be used to stimulate biodegradation processes during phytoremediation. This can involve extraction, filtering, stabilization, and volatilization of pollutants (**figure 42.18** and **table 42.12**).

h. Microorganisms can be added to environments that contain complex microbial communities with greater success if living or inert microhabitats are used. These can include living plant surfaces (seeds, roots, leaves) or inert materials such as microporous glass. *Rhizobium* is an important example of a microorganism added to a complex environment using a living microhabitat (the plant root).

42.5 Biotechnological Applications

a. Microorganisms are being used in a wide range of biotechnological applications such as biosensors (**figure 42.20**). Microarrays are used to monitor gene expression in complex systems (**figure 42.21**).

b. Bacteria, viruses, and fungi can be used as bioinsecticides and biopesticides (**table 42.14**). *Bacillus thuringiensis* is an important biopesticide, and the BT gene has been incorporated into corn.

42.6 Impacts of Microbial Biotechnology

a. Industrial microbiology and biotechnology can have long-term and possibly unexpected positive and negative effects on the environment, and on animals and humans impacted by these technologies. Advances in biotechnology should be considered in a broad ecological and societal context, which is the focus of industrial ecology.

Key Terms

adaptive mutation 969	biosphere 990	lyophilization 969	phytoremediation 983
anthrosphere 990	biotransformation 978	*meta* effect 980	primary metabolite 973
bioaugmentation 984	chiral 980	metabolic control engineering 967	protein engineering 966
biocatalyst 979	combinatorial biology 966	metabolic pathway engineering	protoplast fusion 965
biodegradation 979	cometabolism 982	(MPE) 967	recalcitrance 979
bioinsecticides 988	continuous feed 972	microarray technology 986	reductive dehalogenation 979
biopesticide 988	fermentation 970	microbial transformation 978	regulatory mutant 976
biopolymer 977	forced evolution 969	natural attenuation 985	scaleup 971
bioprospecting 964	gene array 986	natural genetic engineering 969	secondary metabolite 973
bioremediation 982	industrial ecology 990	non-Newtonian broth 971	semisynthetic penicillin 975
biosensor 986	land farming 980	pathway architecture 967	site-directed mutagenesis 966

Questions for Thought and Review

1. What further technological approaches will be required to culture microorganisms that have not yet been grown? Consider the roles of nutrient fluxes, communication molecules, and competition.

2. What makes the area of natural genetic engineering unique? Isn't this simply what has been going on in nature since the time microorganisms were first able to function?

3. What are the advantages of microarrays for the study of gene expression in complex organisms?

4. How is it possible to create a niche or microhabitat for a microorganism? What are the special points of concern in trying to make sure the microbe can find its best place to survive and function?

5. How might the "postgenomic" era differ from the "genomic era"?

6. Most commercial antibiotics are produced by actinomycetes, and only a few are synthesized by fungi and other bacteria. From physiological and environmental viewpoints, how might you attempt to explain this observation?

7. We hear much about the beneficial uses of recombinant DNA technology. What are some of the problems and disadvantages that should be considered when using microorganisms for these applications?

8. Why might *Bacillus thuringiensis* bioinsecticides be of interest in other areas of biotechnology? Consider the molecular aspects of their mode of action.

9. Do you think intrinsic bioremediation can solve all of our environmental pollutant degradation problems? Why or why not?

10. What are some of the possible advantages of biosensors as opposed to more traditional physical and chemical measurement procedures?

11. What are the major types of materials used as nutrients in fermentation media?

12. In what different ways can the term fermentation be used?

13. What parameters can be controlled in a modern industrial fermenter?

14. How do primary and secondary metabolites differ in terms of their synthesis and functions?

Critical Thinking Questions

1. The search for novel plants/microbes and their products can be in direct conflict with the exposure of humans to novel pathogens. Discuss the relative risks and benefits—are there strategies that are more likely to be "win-win"?

2. *Deinococcus radiodurans* is a species of bacteria that is highly resistant to radiation. Can you think of a biotechnological application? How would you test its utility?

3. Discuss the risks of releasing genetically modified microbes or ones that are not natural to the particular environment. What precautions, if any, would you take? What would be your concerns?

4. Why, when a microorganism is removed from a natural environment and grown in the laboratory, will it usually not be able to effectively colonize its original environment if it is grown and added back? Consider the nature of growth media used in the laboratory in comparison to growth conditions in a soil or water when attempting to understand this fundamental problem in microbial ecology.

5. The postgenomic era has been discussed in this and previous chapters of the book. Can you envision the job of a "postgenomicist"?

6. Why is phytoremediation of such current interest for environmental management? Why is it of interest to combine this approach with the use of transgenic plants?

7. The terms biosphere and anthrosphere have been used, together with the term industrial ecology. How does microbial biotechnology relate to these concerns?

For recommended readings on these and related topics, see Appendix V.

Appendix I

A Review of the Chemistry of Biological Molecules

*A*ppendix I provides a brief summary of the chemistry of organic molecules with particular emphasis on the molecules present in microbial cells. Only basic concepts and terminology are presented; introductory textbooks in biology and chemistry should be consulted for a more extensive treatment of these topics.

ATOMS AND MOLECULES

Matter is made of elements that are composed of atoms. An element contains only one kind of atom and cannot be broken down to simpler components by chemical reactions. An atom is the smallest unit characteristic of an element and can exist alone or in combination with other atoms. When atoms combine they form molecules. Molecules are the smallest particles of a substance. They have all the properties of the substance and are composed of two or more atoms.

Although atoms contain many subatomic particles, three directly influence their chemical behavior—protons, neutrons, and electrons. The atom's nucleus is located at its center and contains varying numbers of protons and neutrons (**figure AI.1**). Protons have a positive charge, and neutrons are uncharged. The mass of these particles and the atoms that they compose is given in terms of the atomic mass unit (AMU), which is 1/12 the mass of the most abundant carbon isotope. Often the term dalton (Da) is used to express the mass of molecules. It also is 1/12 the mass of an atom of ^{12}C or 1.661×10^{-24} grams. Both protons and neutrons have a mass of about one dalton. The atomic weight is the actual measured weight of an element and is almost identical to the mass number for the element, the total number of protons and neutrons in its nucleus. The mass number is indicated by a superscripted number preceding the element's symbol (e.g., ^{12}C, ^{16}O, and ^{14}N).

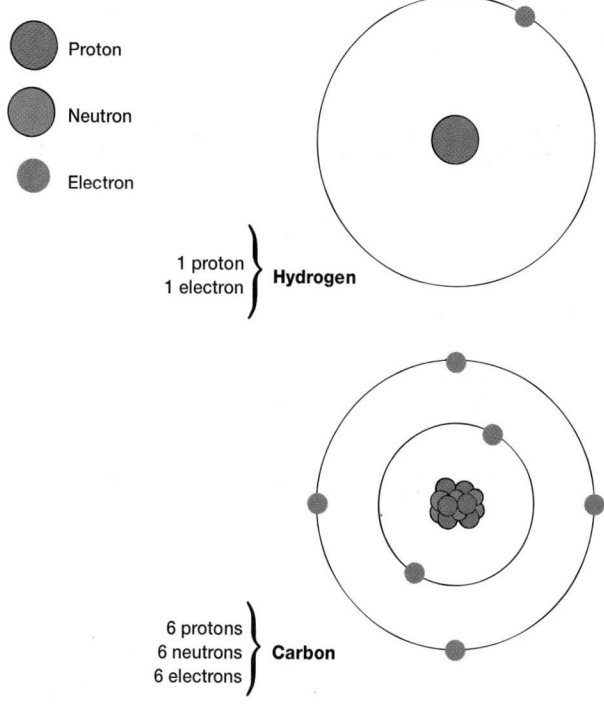

Figure AI.1   Diagrams of Hydrogen and Carbon Atoms. The electron orbitals are represented as concentric circles.

Negatively charged particles called electrons circle the atomic nucleus (figure AI.1). The number of electrons in a neutral atom equals the number of its protons and is given by the atomic number, the number of protons in an atomic nucleus. The atomic number is characteristic of a particular type of atom. For example, carbon has an atomic number of six, hydrogen's number is one, and oxygen's is eight (**table AI.1**).

Table AI.1
Atoms Commonly Present in Organic Molecules

Atom	Symbol	Atomic Number	Atomic Weight	Number of Chemical Bonds
Hydrogen	H	1	1.01	1
Carbon	C	6	12.01	4
Nitrogen	N	7	14.01	3
Oxygen	O	8	16.00	2
Phosphorus	P	15	30.97	5
Sulfur	S	16	32.06	2

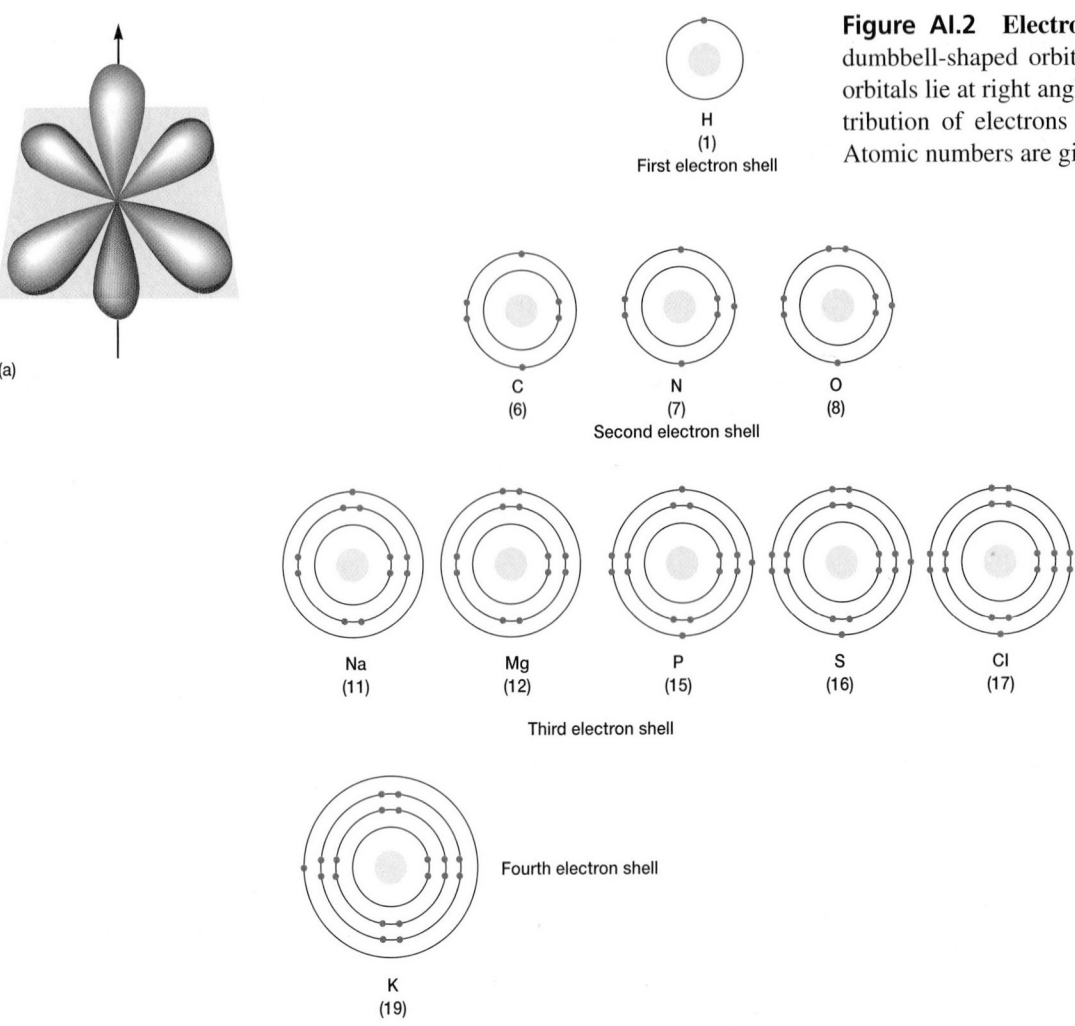

Figure AI.2 Electron Orbitals. (a) The three dumbbell-shaped orbitals of the second shell. The orbitals lie at right angles to each other. (b) The distribution of electrons in some common elements. Atomic numbers are given in parentheses.

The electrons move constantly within a volume of space surrounding the nucleus, even though their precise location in this volume cannot be determined accurately. This volume of space in which an electron is located is called its orbital. Each orbital can contain two electrons. Orbitals are grouped into shells of different energy that surround the nucleus. The first shell is closest to the nucleus and has the lowest energy; it contains only one orbital. The second shell contains four orbitals, one circular and three shaped like dumbbells (**figure AI.2a**). It can contain up to eight electrons. The third shell has even higher energy and holds more than eight electrons. Shells are filled beginning with the innermost and moving outward. For example, carbon has six electrons, two in its first shell and four in the second (figures AI.1 and AI.2b). The electrons in the outermost shell are the ones that participate in chemical reactions. The most stable condition is achieved when the outer shell is filled with electrons. Thus the number of bonds an element can form depends on the number of electrons required to fill the outer shell. Since carbon has four electrons in its outer shell and the shell is filled when it contains eight electrons, it can form four covalent bonds (table AI.1).

CHEMICAL BONDS

Molecules are formed when two or more atoms associate through chemical bonding. Chemical bonds are attractive forces that hold together atoms, ions, or groups of atoms in a molecule or other substance. Many types of chemical bonds are present in organic molecules; three of the most important are covalent bonds, ionic bonds, and hydrogen bonds.

In covalent bonds, atoms are joined together by sharing pairs of electrons (**figure AI.3**). If the electrons are equally shared between identical atoms (e.g., in a carbon-carbon bond), the covalent bond is strong and nonpolar. When two different atoms such as carbon and oxygen share electrons, the covalent bond formed is polar since the electrons are pulled toward the more electronegative atom, the atom that more strongly attracts electrons. A single pair of electrons is shared in a single bond; a double bond is formed when two pairs of electrons are shared.

Atoms often contain either more or fewer electrons than the number of protons in their nuclei. When this is the case, they carry a net negative or positive charge and are called ions. Cations

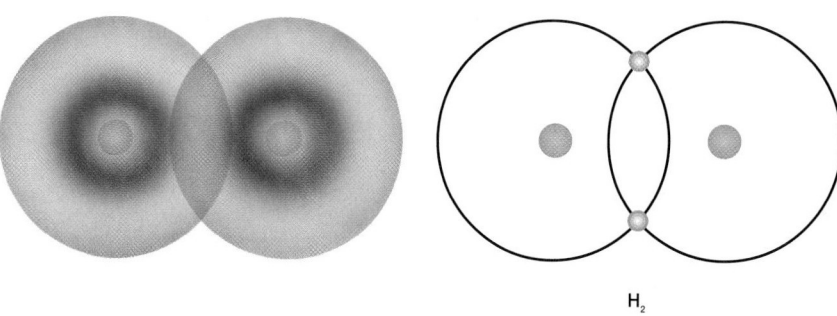

Figure AI.3   The Covalent Bond.   A hydrogen molecule is formed when two hydrogen atoms share electrons.

H_2

$>C=O \cdots H-O-$

$\geqslant N \cdots H-O \overset{O}{\underset{}{\overset{\parallel}{C}}}-$

$>C=O \cdots H-N<$

$-O \overset{H}{\cdots} H-N<$

$\geqslant N \cdots H-N<$

Figure AI.4   Hydrogen Bonds.   Representative examples of hydrogen bonds present in biological molecules.

$$H-\overset{\overset{H}{|}}{\underset{\underset{H}{|}}{C}}-\overset{\overset{H}{|}}{\underset{\underset{H}{|}}{C}}-\overset{\overset{H}{|}}{\underset{\underset{H}{|}}{C}}-\overset{\overset{H}{|}}{\underset{\underset{H}{|}}{C}}-\overset{\overset{H}{|}}{\underset{\underset{H}{|}}{C}}-\overset{\overset{H}{|}}{\underset{\underset{H}{|}}{C}}-H$$
(a) C_6H_{14} (Hexane)

(b) CH_2, CH_2, CH_2, CH_2, CH_2, CH_2 or C_6H_{12} (Cyclohexane)

(c)   benzene structure   or C_6H_6 (Benzene)

Figure AI.5   Hydrocarbons.   Examples of hydrocarbons that are (**a**) linear, (**b**) cyclic, and (**c**) aromatic.

carry positive charges and anions have a net negative charge. When a cation and an anion approach each other, they are attracted by their opposite charges. This ionic attraction that holds two groups together is called an ionic bond. Ionic bonds are much weaker than covalent bonds and are easily disrupted by a polar solvent such as water. For example, the Na^+ cation is strongly attracted to the Cl^- anion in a sodium chloride crystal, but sodium chloride dissociates into separate ions (ionizes) when dissolved in water. Ionic bonds are important in the structure and function of proteins and other biological molecules.

When a hydrogen atom is covalently bonded to a more electronegative atom such as oxygen or nitrogen, the electrons are unequally shared and the hydrogen atom carries a partial positive charge. It will be attracted to an electronegative atom such as oxygen or nitrogen, which carries an unshared pair of electrons; this attraction is called a hydrogen bond (**figure AI.4**). Although an individual hydrogen bond is weak, there are so many hydrogen bonds in proteins and nucleic acids that they play a major role in determining protein and nucleic acid structure.

ORGANIC MOLECULES

Most molecules in cells are organic molecules, molecules that contain carbon. Since carbon has four electrons in its outer shell,

it tends to form four covalent bonds in order to fill its outer shell with eight electrons. This property makes it possible to form chains and rings of carbon atoms that also can bond with hydrogen and other atoms (**figure AI.5**). Although adjacent carbons usually are connected by single bonds, they may be joined by double or triple bonds. Rings that have alternating single and double bonds, like the benzene ring, are called aromatic rings. The hydrocarbon chain or ring provides a chemically inactive skeleton to which more reactive groups of atoms may be attached. These reactive groups with specific properties are known as functional groups. They usually contain atoms of oxygen, nitrogen, phosphorus, or sulfur (**figure AI.6**) and are largely responsible for most characteristic chemical properties of organic molecules.

Organic molecules are often divided into classes based on the nature of their functional groups. Ketones have a carbonyl group within the carbon chain, whereas alcohols have a hydroxyl on the chain. Organic acids have a carboxyl group, and amines have an amino group (**figure AI.7**).

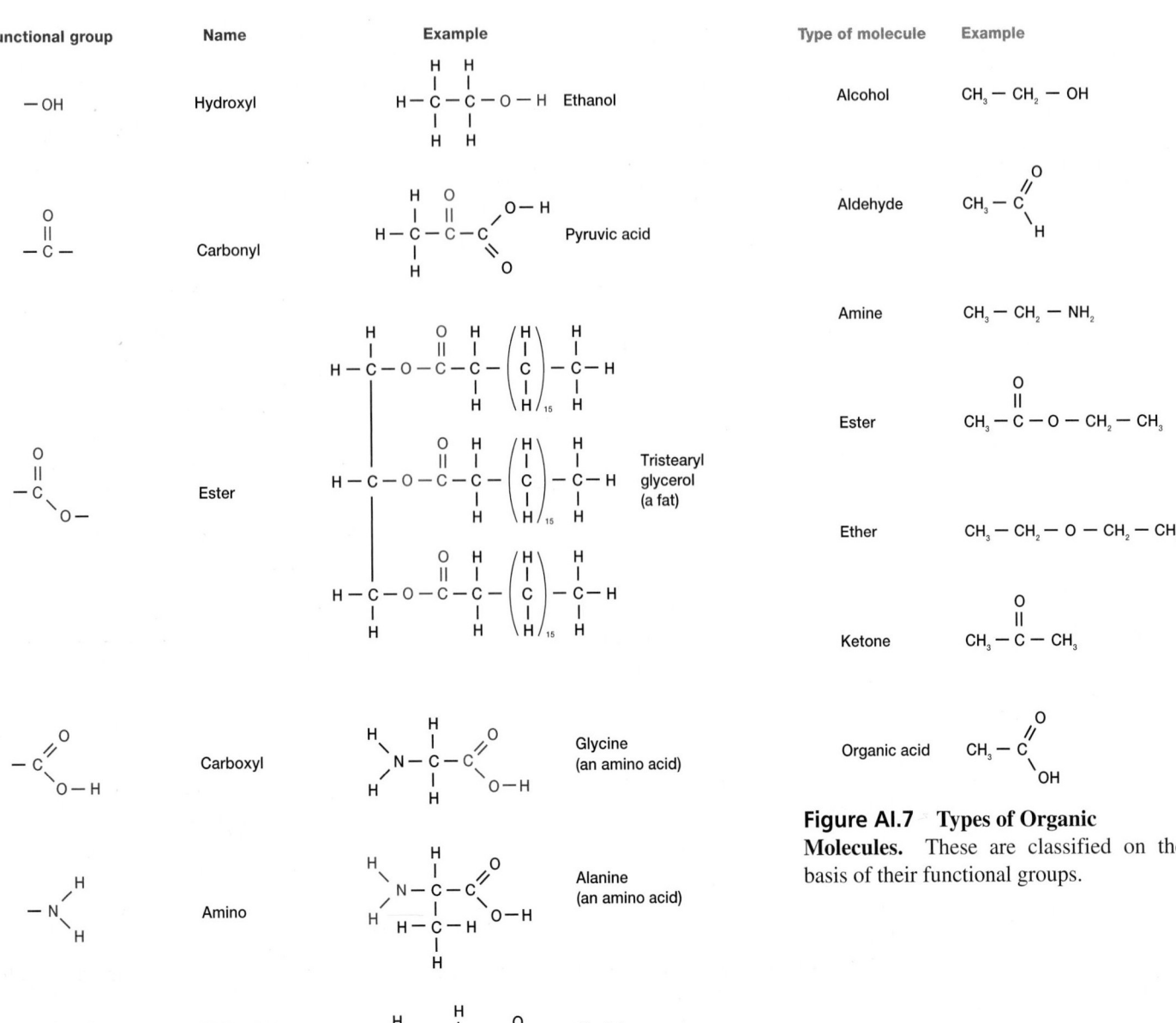

Figure AI.6 Functional Groups. Some common functional groups in organic molecules. The groups are shown in red.

Figure AI.7   Types of Organic Molecules. These are classified on the basis of their functional groups.

Organic molecules may have the same chemical composition and yet differ in their molecular structure and properties. Such molecules are called isomers. One important class of isomers is the stereoisomers. Stereoisomers have the same atoms arranged in the same nucleus-to-nucleus sequence but differ in the spatial arrangement of their atoms. For example, an amino acid such as alanine can form stereoisomers (**figure AI.8**). L-Alanine and other L-amino acids are the stereoisomer forms normally present in proteins.

CARBOHYDRATES

Carbohydrates are aldehyde or ketone derivatives of polyhydroxy alcohols. The smallest and least complex carbohydrates are the simple sugars or monosaccharides. The most common sugars have five or six carbons (**figure AI.9**). A sugar in its ring form has two isomeric structures, the α and β forms, that differ in the orientation of the hydroxyl on the aldehyde or ketone carbon, which is called the anomeric or glycosidic carbon (**figure AI.10**). Mi-

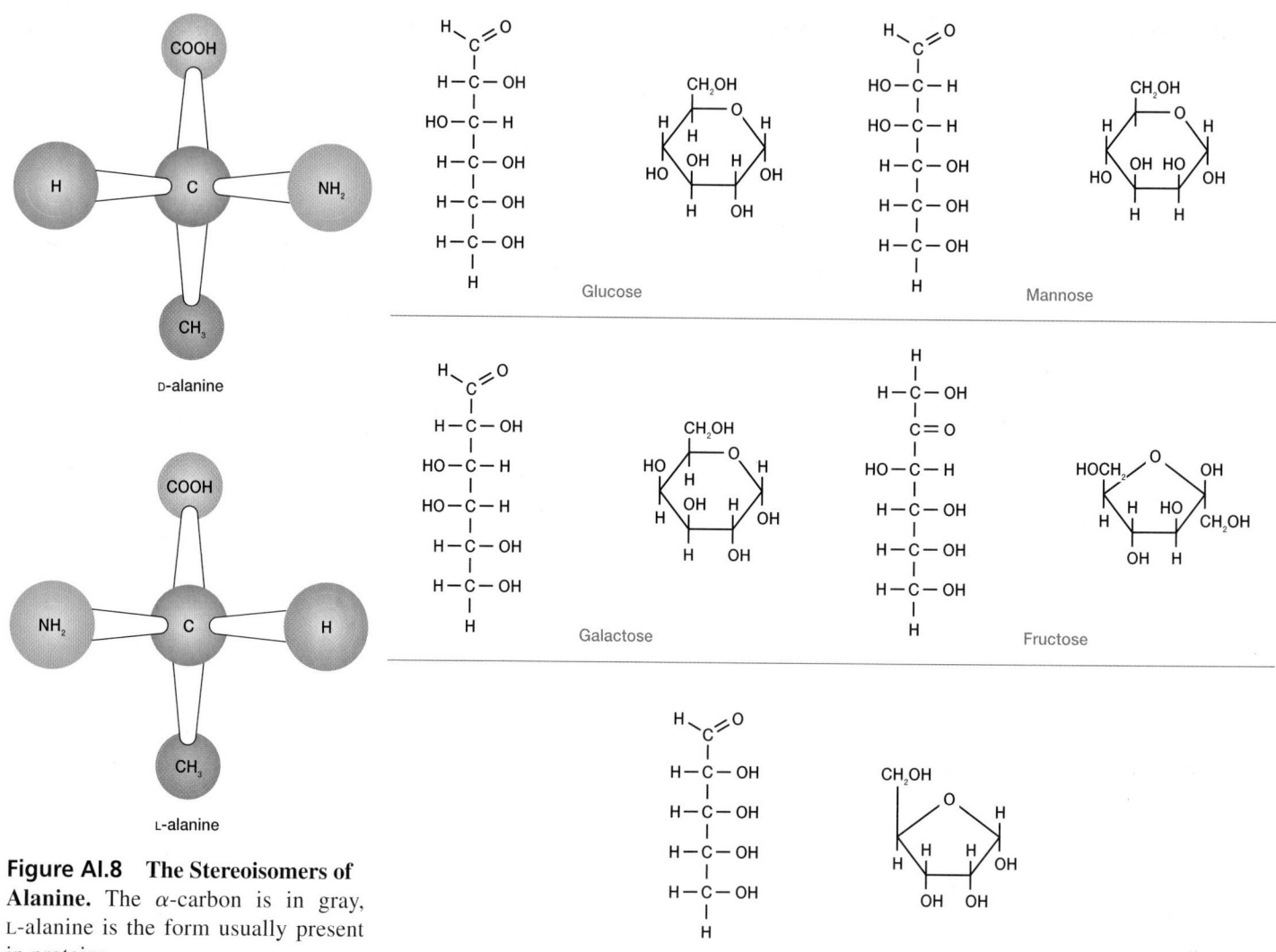

Figure AI.8   The Stereoisomers of Alanine. The α-carbon is in gray, L-alanine is the form usually present in proteins.

Figure AI.9   Common Monosaccharides. Structural formulas for both the open chains and the ring forms are provided.

croorganisms have many sugar derivatives in which a hydroxyl is replaced by an amino group or some other functional group (e.g., glucosamine).

Two monosaccharides can be joined by a bond between the anomeric carbon of one sugar and a hydroxyl or the anomeric carbon of the second (**figure AI.11**). The bond joining sugars is a glycosidic bond and may be either α or β depending on the orientation of the anomeric carbon. Two sugars linked in this way constitute a disaccharide. Some common disaccharides are maltose (two glucose molecules), lactose (glucose and galactose), and sucrose (glucose and fructose). If 10 or more sugars are linked together by glycosidic bonds, a polysaccharide is formed. For example, starch and glycogen are common polymers of glucose that are used as sources of carbon and energy (**figure AI.12**).

LIPIDS

All cells contain a heterogeneous mixture of organic molecules that are relatively insoluble in water but very soluble in nonpolar solvents such as chloroform, ether, and benzene. These molecules are called lipids. Lipids vary greatly in structure and include triacylglycerols, phospholipids, steroids, carotenoids, and many other types. Among other functions, they serve as membrane components, storage forms for carbon and energy, precursors of other cell constituents, and protective barriers against water loss.

Most lipids contain fatty acids, monocarboxylic acids that often are straight chained but may be branched. Saturated fatty acids lack double bonds in their carbon chains, whereas

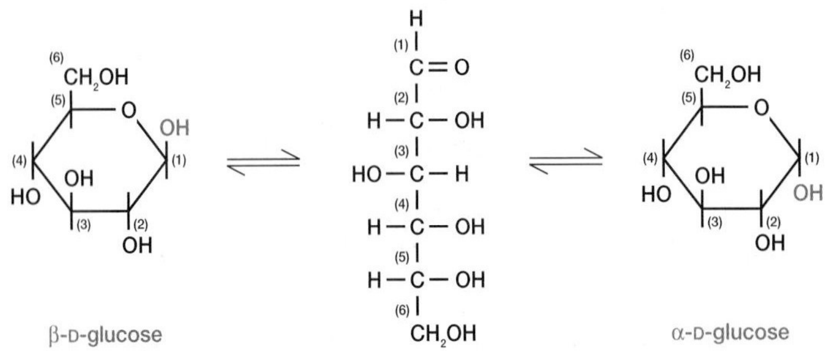

Figure AI.10 **The Interconversion of Monosaccharide Structures.** The open chain form of glucose and other sugars is in equilibrium with closed ring structures (depicted here with Haworth projections). Aldehyde sugars form cyclic hemiacetals, and keto sugars produce cyclic hemiketals. When the hydroxyl on carbon one of cyclic hemiacetals projects above the ring, the form is known as a β form. The α form has a hydroxyl that lies below the plane of the ring. The same convention is used in showing the α and β forms of hemiketals such as those formed by fructose.

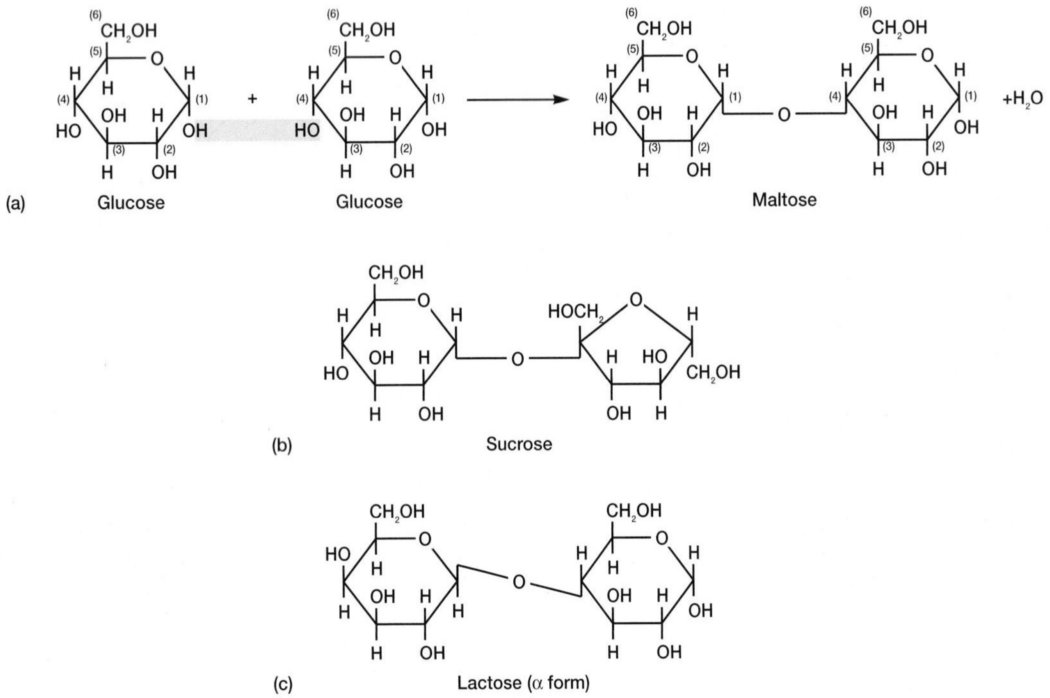

Figure AI.11 **Common Disaccharides.** (**a**) The formation of maltose from two molecules of an α-glucose. The bond connecting the glucose extends between carbons one and four, and involves the α form of the anomeric carbon. Therefore, it is called an α (1 → 4) glycosidic bond. (**b**) Sucrose is composed of a glucose and a fructose joined to each other through their anomeric carbons, and αβ (1 → 2) bond. (**c**) The milk sugar lactose contains galactose and glucose joined by a β (1 → 4) glycosidic bond.

unsaturated fatty acids have double bonds. The most common fatty acids are 16 or 18 carbons long.

Two good examples of common lipids are triacylglycerols and phospholipids. Triacylglycerols are composed of glycerol esterified to three fatty acids (**figure AI.13a**). They are used to store carbon and energy. Phospholipids are lipids that contain at least one phosphate group and often have a nitrogenous constituent as well. Phosphatidyl ethanolamine is an important phospholipid frequently present in bacterial membranes (figure AI.13b). It is composed of two fatty acids esterified to glycerol. The third glycerol hydroxyl is joined with a phosphate group, and ethanolamine is attached to the phosphate. The resulting lipid is very asymmetric with a hydrophobic nonpolar end contributed by the fatty acids and a polar, hydrophilic end. In cell membranes the hydrophobic end is buried in the interior of the membrane, while the polar-charged end is at the membrane surface and exposed to water.

PROTEINS

The basic building blocks of proteins are amino acids. An amino acid contains a carboxyl group and an amino group on its alpha carbon (**figure AI.14**). About 20 amino acids are normally found in proteins; they differ from each other with respect to their side chains (**figure AI.15**). In proteins, amino acids are linked together by peptide bonds between their carboxyls and α-amino groups to form linear polymers called peptides (**figure AI.16**). If a peptide contains more than 10 amino acids, it usually is called a polypeptide. Each protein is composed of one or more polypeptide chains and has a molecular weight greater than about 6,000 to 7,000.

Proteins have three or four levels of structural organization and complexity. The primary structure of a protein is the sequence of the amino acids in its polypeptide chain or chains. The structure of the polypeptide chain backbone is also considered part of the

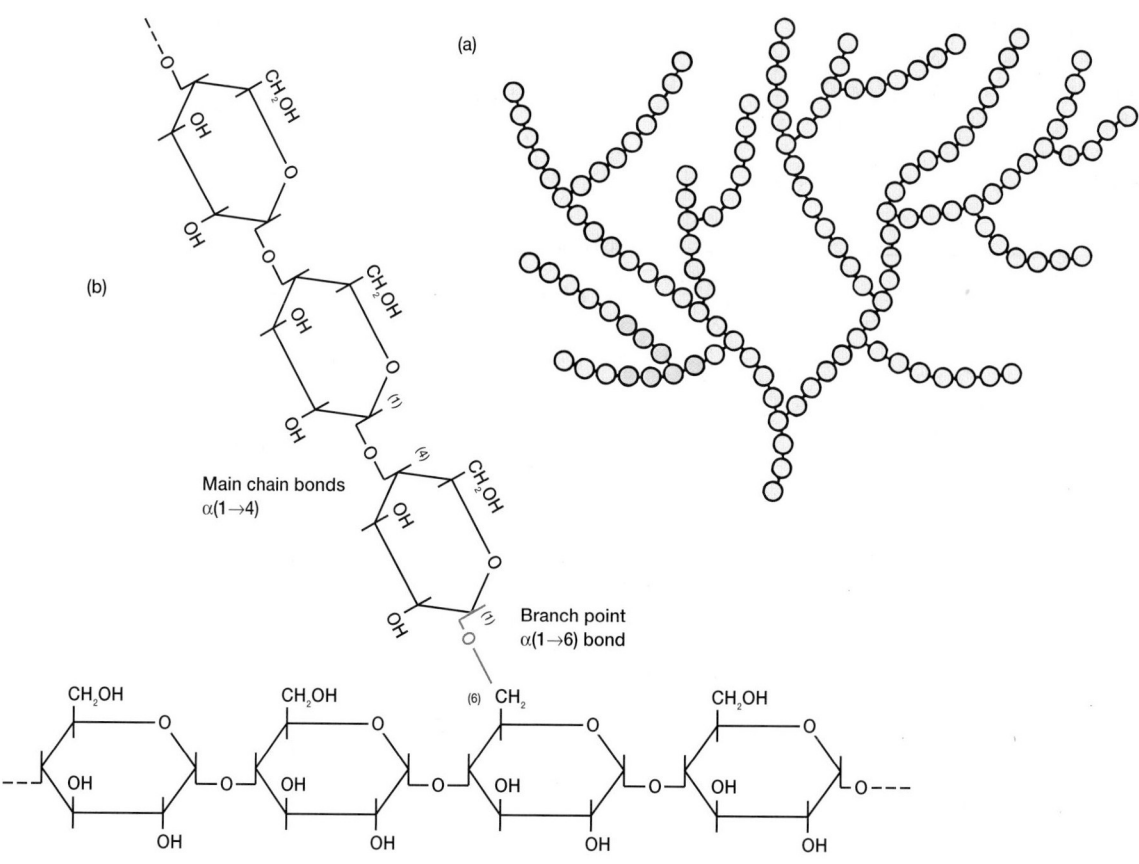

Figure AI.12   Glycogen and Starch Structure.   (a) An overall view of the highly branched structure characteristic of glycogen and most starch. The circles represent glucose residues. (b) A close-up of a small part of the chain (shown in blue in part *a*) revealing a branch point with its α (1 → 6) glycosidic bond, which is colored blue.

(a)

(b)

Figure AI.13   Examples of Common Lipids. (a) A triacyl-glycerol or neutral fat. (b) The phospholipid phosphatidyl ethanolamine. The R groups represent fatty acid side chains.

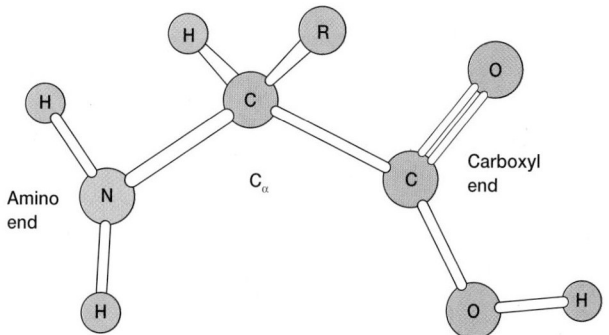

Figure AI.14   L-Amino Acid Structure. The uncharged form is shown.

primary structure. Each different polypeptide has its own amino acid sequence that is a reflection of the nucleotide sequence in the gene that codes for its synthesis. The polypeptide chain can coil along one axis in space into various shapes like the α-helix (**figure AI.17**). This arrangement of the polypeptide in space around a single axis is called the secondary structure. Secondary structure is formed and stabilized by the interactions of amino acids that are fairly close to one another on the polypeptide chain. The

Figure AI.15   The Common Amino Acids.   The structures of the α-amino acids normally found in proteins. Their side chains are shown in blue, and they are grouped together based on the nature of their side chains—nonpolar, polar, negatively charged (acid), or positively charged (basic). Proline is actually an imino acid rather than an amino acid.

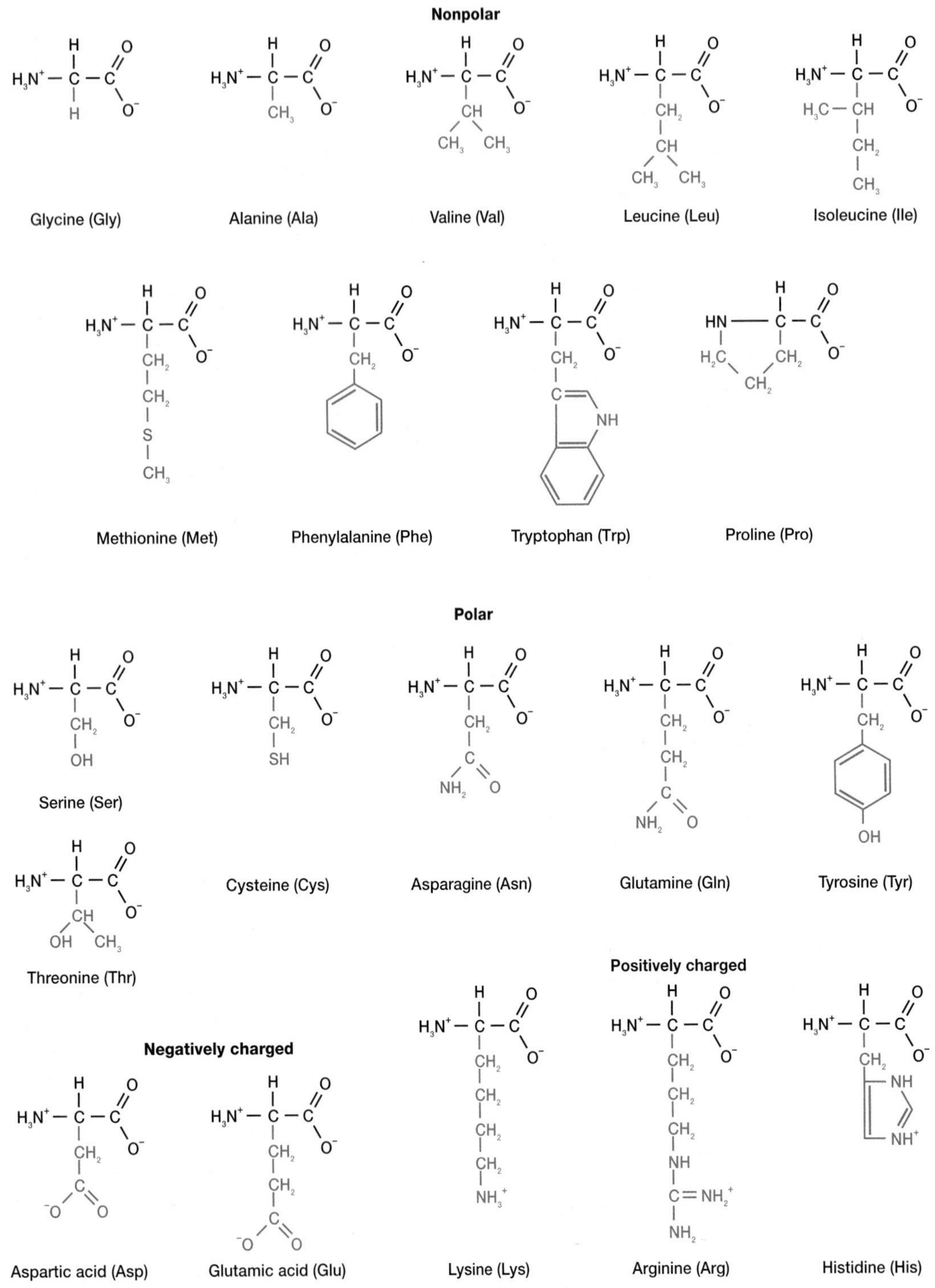

Nonpolar

Glycine (Gly)

Alanine (Ala)

Valine (Val)

Leucine (Leu)

Isoleucine (Ile)

Methionine (Met)

Phenylalanine (Phe)

Tryptophan (Trp)

Proline (Pro)

Polar

Serine (Ser)

Cysteine (Cys)

Asparagine (Asn)

Glutamine (Gln)

Tyrosine (Tyr)

Threonine (Thr)

Negatively charged

Aspartic acid (Asp)

Glutamic acid (Glu)

Positively charged

Lysine (Lys)

Arginine (Arg)

Histidine (His)

Figure AI.16   A Tetrapeptide Chain.   The end of the chain with a free α-amino group is the amino or N terminal. The end with the free α-carboxyl is the carboxyl or C terminal. One peptide bond is shaded in blue.

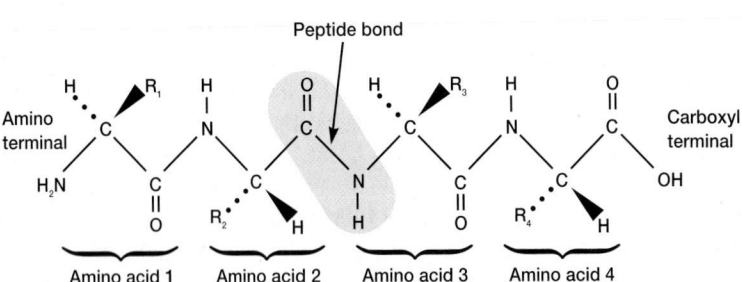

Peptide bond

Amino terminal

Carboxyl terminal

Amino acid 1   Amino acid 2   Amino acid 3   Amino acid 4

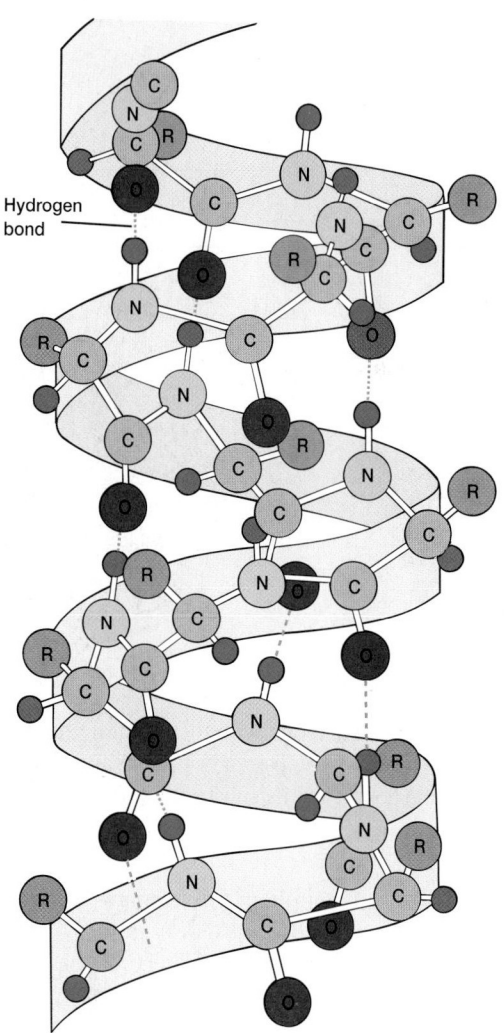

Figure AI.17 The α-Helix. A polypeptide twisted into one type of secondary structure, the α-helix. The helix is stabilized by hydrogen bonds joining peptide bonds that are separated by three amino acids.

polypeptide with its primary and secondary structure can be coiled or organized in space along three axes to form a more complex, three-dimensional shape (**figure AI.18**). This level of organization is the tertiary structure (**figure AI.19**). Amino acids more distant from one another on the polypeptide chain contribute to tertiary structure. Secondary and tertiary structures are examples of conformation, molecular shape that can be changed by bond rotation and without breaking covalent bonds. When a protein contains more than one polypeptide chain, each chain with its own primary, secondary, and tertiary structure associates with the other chains to form the final molecule. The way in which polypeptides associate with each other in space to form the final protein is called the protein's quaternary structure (**figure AI.20**).

The final conformation of a protein is ultimately determined by the amino acid sequence of its polypeptide chains. Under proper conditions a completely unfolded polypeptide will fold into its normal final shape without assistance.

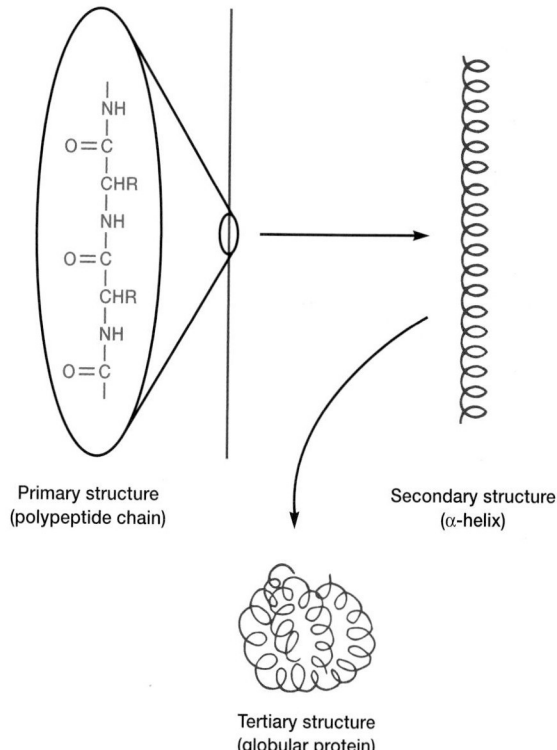

Figure AI.18 Secondary and Tertiary Protein Structures. The formation of secondary and tertiary protein structures by folding a polypeptide chain with its primary structure.

Protein secondary, tertiary, and quaternary structure is largely determined and stabilized by many weak noncovalent forces such as hydrogen bonds and ionic bonds. Because of this, protein shape often is very flexible and easily changed. This flexibility is very important in protein function and in the regulation of enzyme activity. Because of their flexibility, however, proteins readily lose their proper shape and activity when exposed to harsh conditions. The only covalent bond commonly involved in the secondary and tertiary structure of proteins is the disulfide bond. The disulfide bond is formed when two cysteines are linked through their sulfhydryl groups. Disulfide bonds generally strengthen or stabilize protein structure but are not especially important in directly determining protein conformation.

NUCLEIC ACIDS

The nucleic acids, deoxyribonucleic acid (DNA) and ribonucleic acid (RNA), are polymers of deoxyribonucleosides and ribonucleosides joined by phosphate groups. The nucleosides in DNA contain the purines adenine and guanine, and the pyrimidine bases thymine and cytosine. In RNA the pyrimidine uracil is substituted for thymine. Because of their importance for genetics and molecular biology, the chemistry of nucleic acids is introduced earlier in the text. The structure and synthesis of purines and pyrimidines are discussed in chapter 10 (*pp. 211–14*). The structures of DNA and RNA are described in chapter 11 (pp. 224–28).

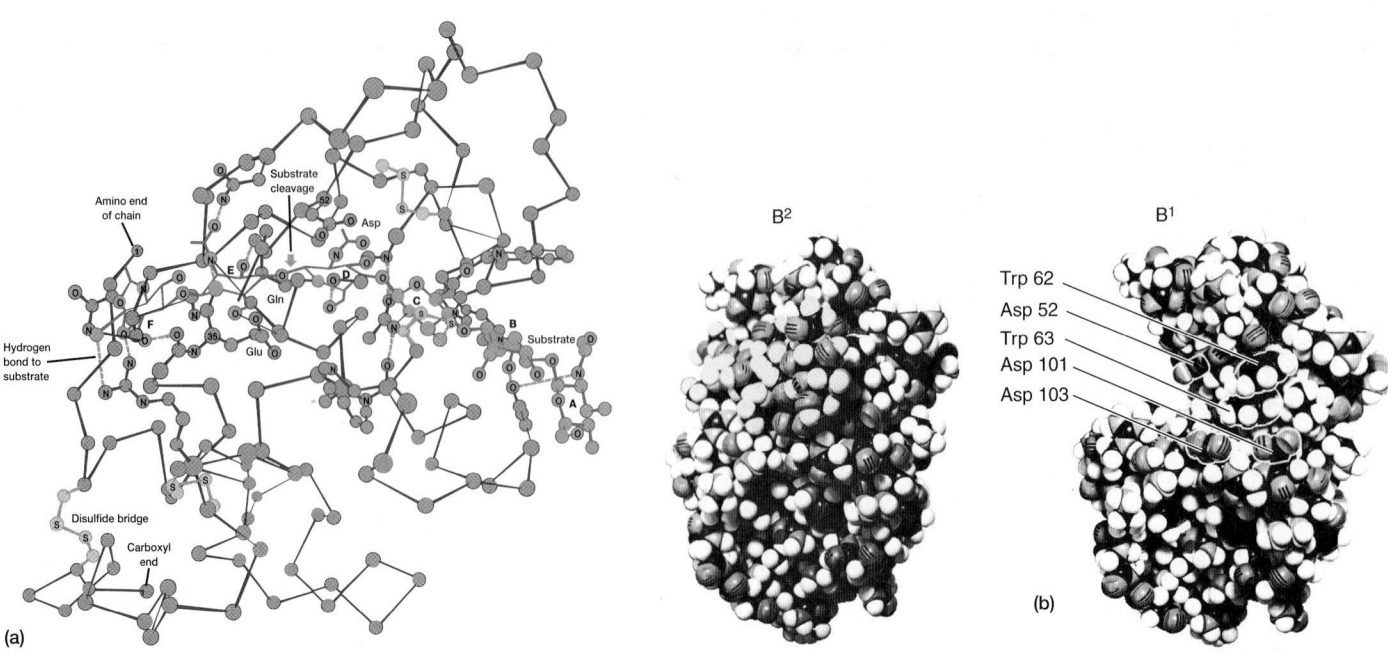

(a)

B²

B¹

Trp 62
Asp 52
Trp 63
Asp 101
Asp 103

(b)

Figure AI.19   Lysozyme.

The tertiary structure of the enzyme lysozyme. (**a**) A diagram of the protein's polypeptide backbone with the substrate hexasaccharide shown in blue. The point of substrate cleavage is indicated. (**b**) A space-filling model of lysozyme. The right figure shows the empty active site with some of its more important amino acids indicated. On the left the enzyme has bound its substrate (in pink).

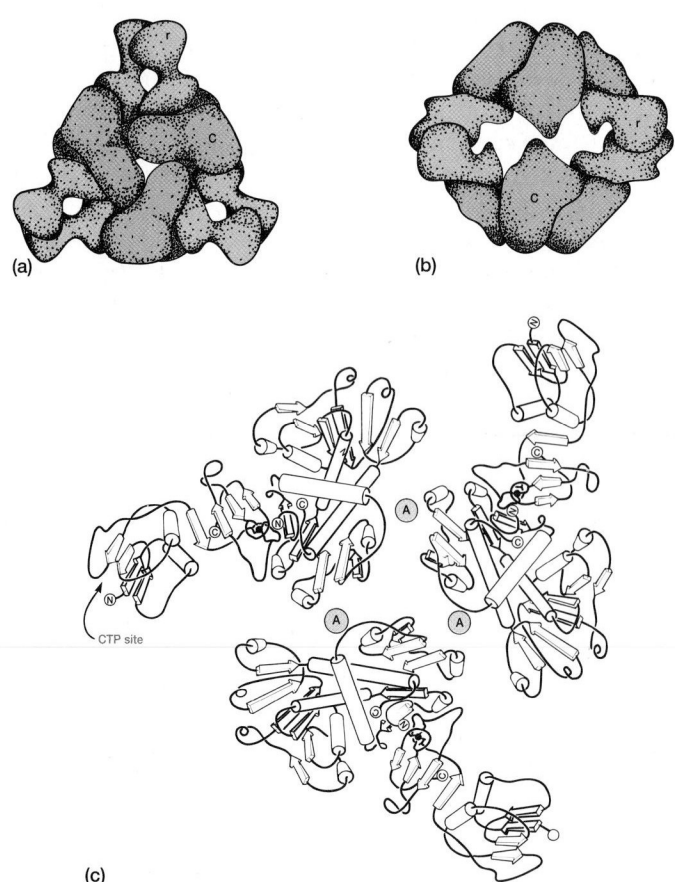

(a)

(b)

(c)

Figure AI.20   An Example of Quaternary Structure.
The enzyme aspartate carbamoyltransferase from *E. coli* has two types of subunits, catalytic and regulatory. The association between the two types of subunits is shown: (**a**) a top view, and (**b**) a side view of the enzyme. The catalytic (C) and regulator (r) subunits are shown in different colors. (**c**) The peptide chains shown when viewed from the top as in (**a**). The active sites of the enzyme are located at the positions indicated by A. (*See pp. 162–63 for more details.*) (**a** and **b**) Adapted from Krause, et al., in *Proceedings of the National Academy of Sciences,* V. 82, 1985, as appeared in *Biochemistry,* 3d edition by Lubert Stryer. Copyright © 1975, 1981, 1988. Reprinted with permission of W. H. Freeman and Company. (**c**) Adapted from Kantrowitz, et al., in *Trends in Biochemical Science,* V. 5, 1980, as appeared in *Biochemistry,* 3d edition by Lubert Stryer. Copyright © 1975, 1981, 1988. Reprinted with permission of W. H. Freeman and Company.

Common Metabolic Pathways

*T*his appendix contains a few of the more important pathways discussed in the text, particularly those involved in carbohydrate catabolism. Enzyme names and final end products are given in color. Consult the text for a description of each pathway and its roles.

Figure AII.1 Glycolysis. The Embden-Meyerhof pathway for the conversion of glucose and other sugars to pyruvate. In some procaryotes glucose is phosphorylated to glucose 6-phosphate during group translocation transport across the plasma membrane.

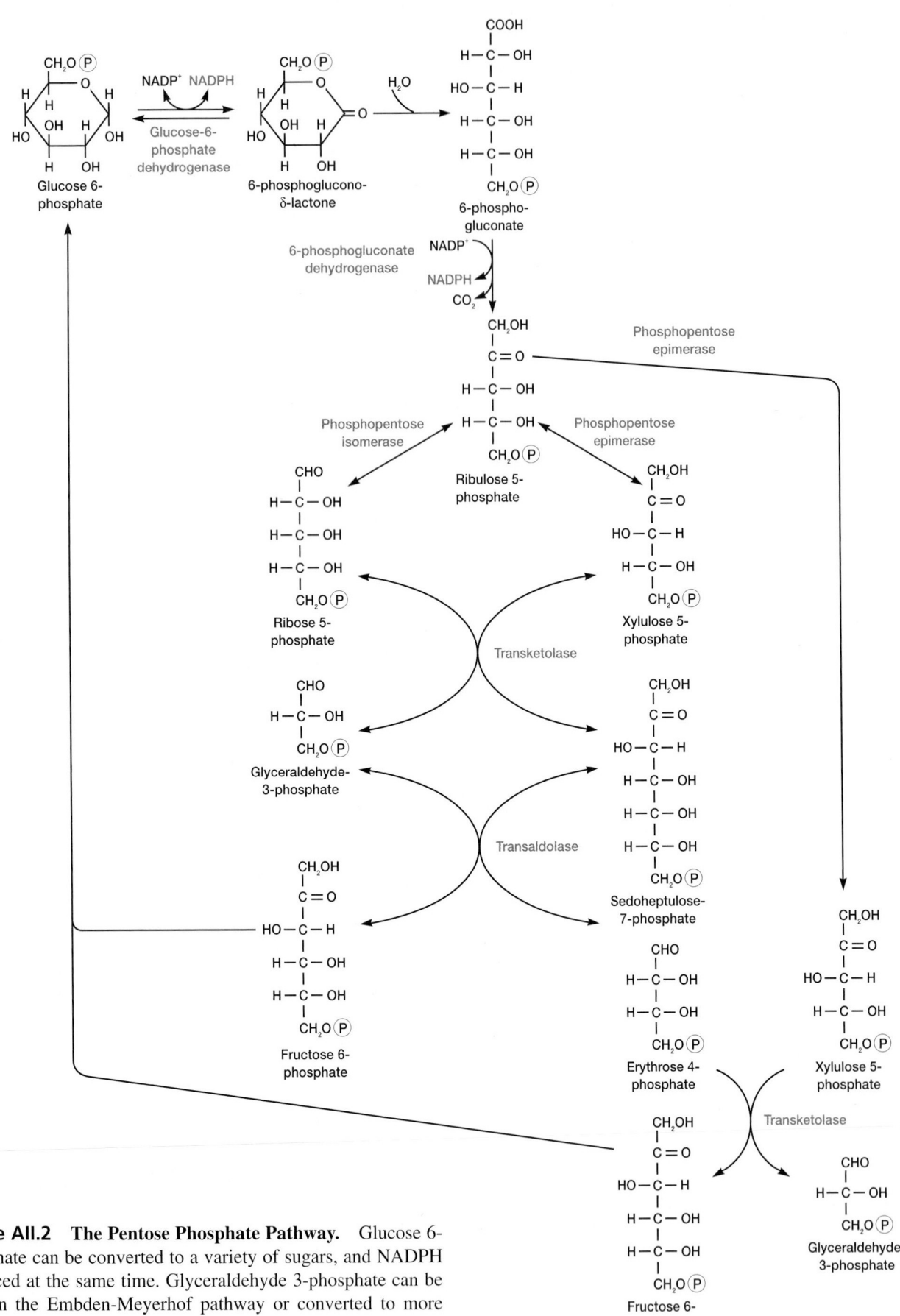

Figure AII.2 The Pentose Phosphate Pathway. Glucose 6-phosphate can be converted to a variety of sugars, and NADPH produced at the same time. Glyceraldehyde 3-phosphate can be used in the Embden-Meyerhof pathway or converted to more fructose 6-phosphate.

Figure AII.3   The Entner-Doudoroff Pathway.

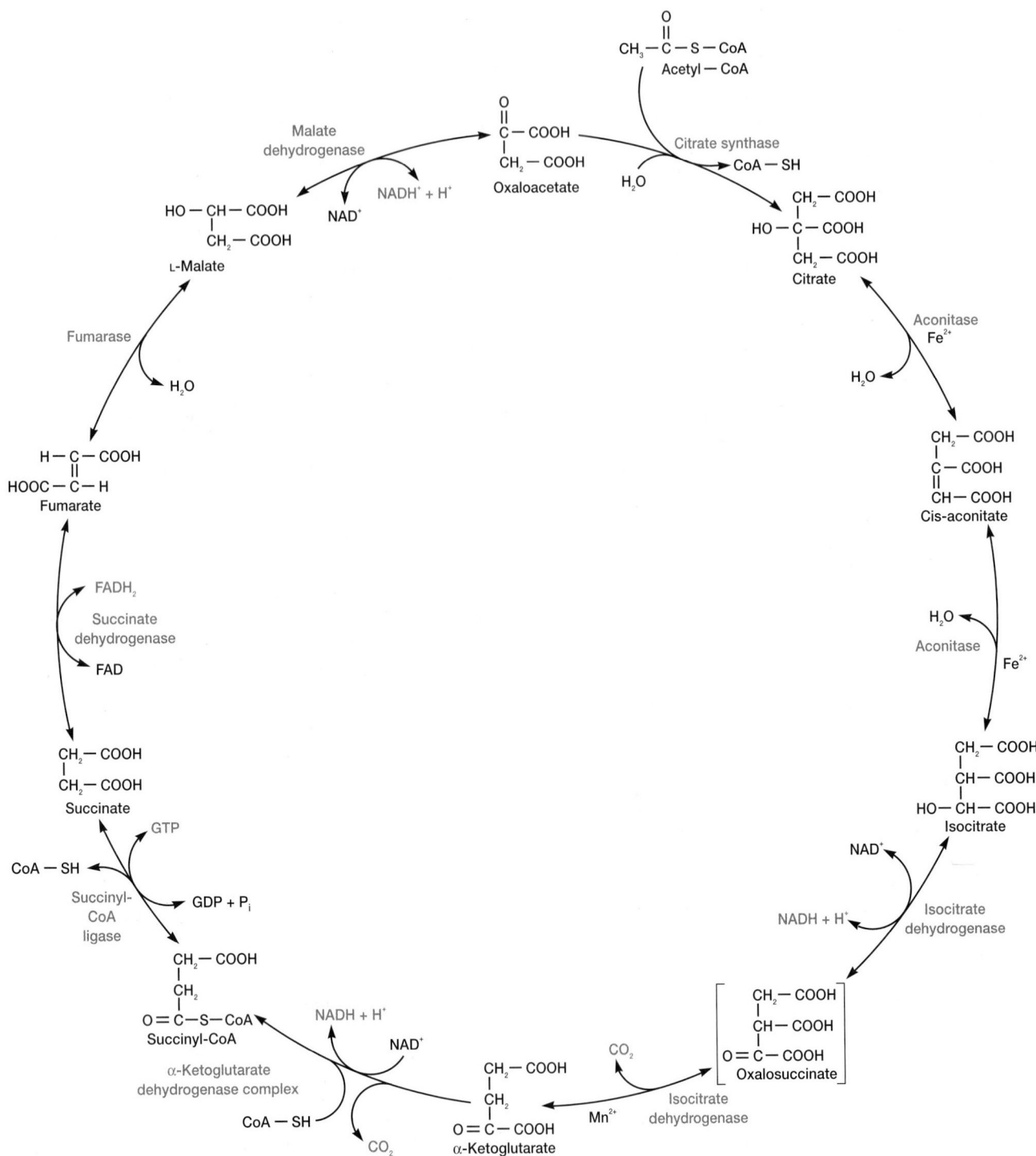

Figure AII.4   The Tricarboxylic Acid Cycle.   Cis-aconitate and oxalosuccinate remain bound to aconitase and isocitrate dehydrogenase. Oxalosuccinate has been placed in brackets because it is so unstable.

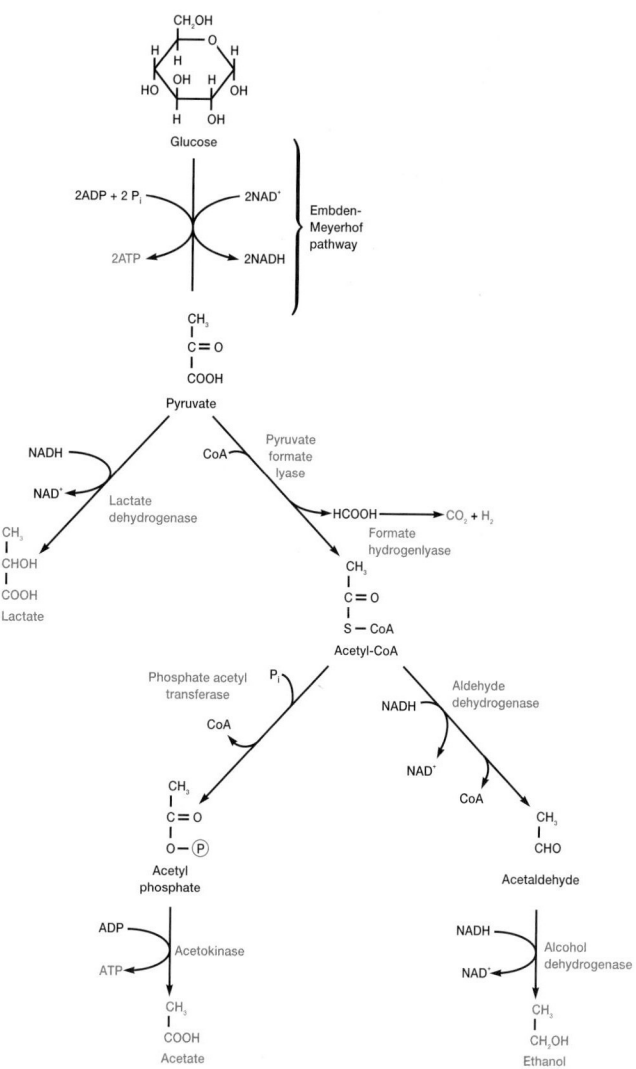

Figure AII.5   The Mixed Acid Fermentation Pathway.   This pathway is characteristic of many members of the *Enterobacteriaceae* such as *E. coli.*

Figure AII.6   The Butanediol Fermentation Pathway.   This pathway is characteristic of members of the *Enterobacteriaceae* such as *Enterobacter.* Other products may also be formed during butanediol fermentation.

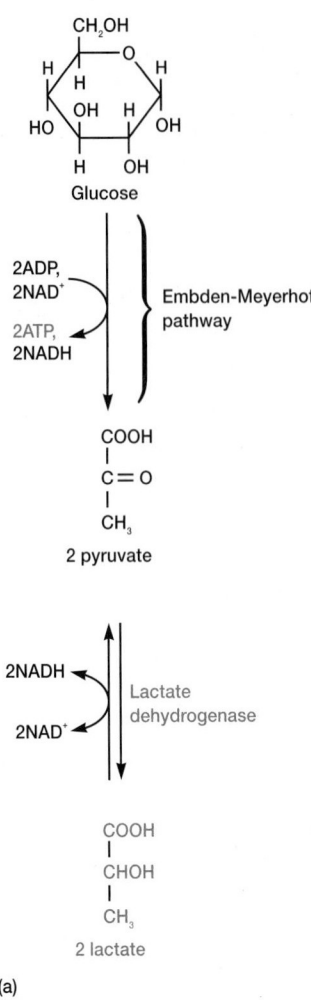

(a)

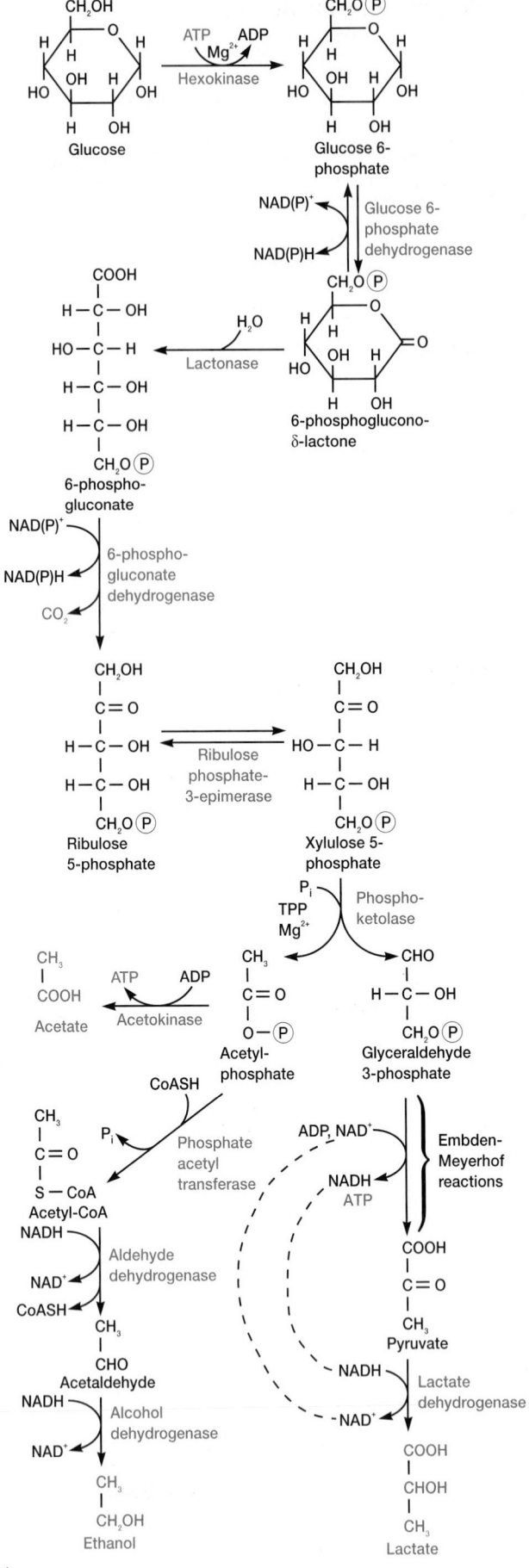

(b)

Figure AII.7 Lactic Acid Fermentations. (a) Homolactic fermentation pathway. (b) Heterolactic fermentation pathway.

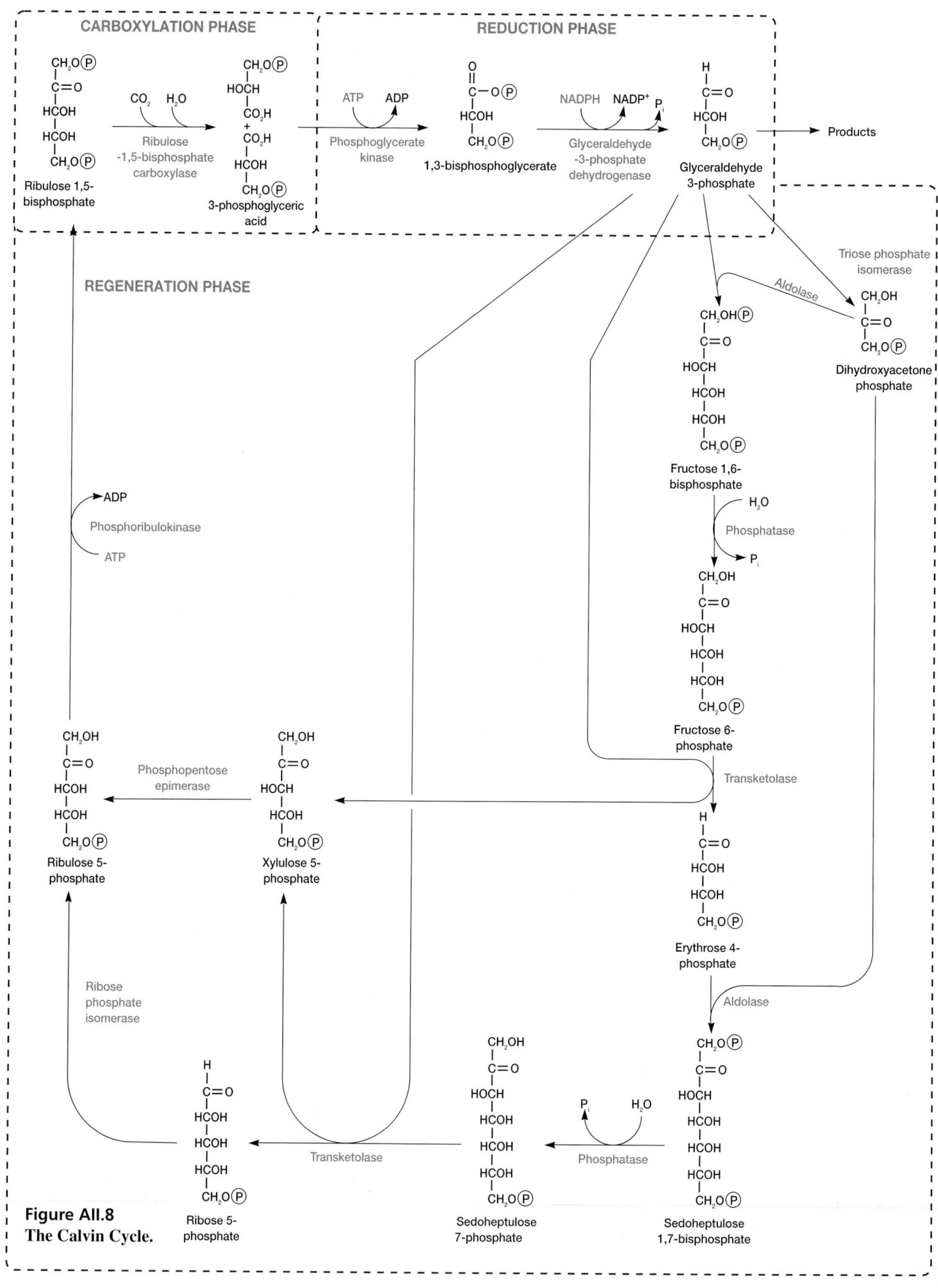

**Figure AII.8
The Calvin Cycle.**

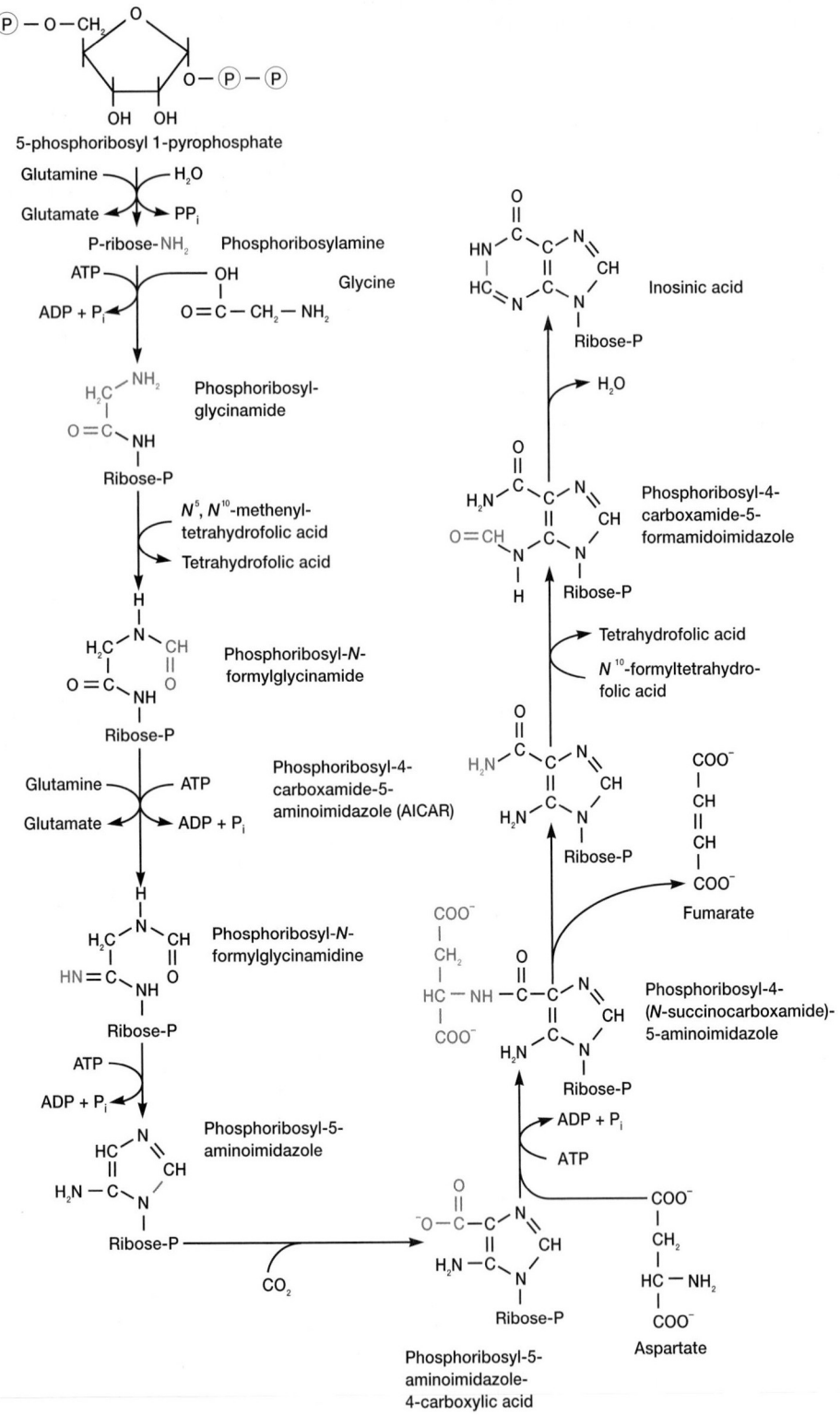

Figure AII.9 The Pathway for Purine Biosynthesis. Inosinic acid is the first purine end product. The purine skeleton is constructed while attached to a ribose phosphate.

Classification of Procaryotes According to the Second Edition of *Bergey's Manual of Systematic Bacteriology*

This appendix summarizes the classification of procaryotes that is being employed by the second edition of *Bergey's Manual.* The second edition is being published in five volumes over a period of several years. Because of advances in taxonomy over the next few years, some of the details in this classification may well change. The overall organization and major taxa should remain constant. The quotation marks around some names indicate that they are not formally approved taxonomic names.

Domain *Archaea*

Phylum AI. *Crenarchaeota*
Class I. *Thermoprotei*
Order I. *Thermoproteales*
Family I. *Thermoproteaceae*
Genus I. *Thermoproteus*
Genus II. *Caldivirga*
Genus III. *Pyrobaculum*
Genus IV. *Thermocladium*
Genus V. *Vulcanisaeta*
Family II. *Thermofilaceae*
Genus I. *Thermofilum*
Order II. *Desulfurococcales*
Family I. *Desulfurococcaceae*
Genus I. *Desulfurococcus*
Genus II. *Acidolobus*
Genus III. *Aeropyrum*
Genus IV. *Ignicoccus*
Genus V. *Staphylothermus*
Genus VI. *Stetteria*
Genus VII. *Sulfophobococcus*
Genus VIII. *Thermodiscus*
Genus IX. *Thermosphaera*
Family II. *Pyrodictiaceae*
Genus I. *Pyrodictium*
Genus II. *Hyperthermus*
Genus III. *Pyrolobus*
Order III. *Sulfolobales*
Family I. *Sulfolobaceae*
Genus I. *Sulfolobus*
Genus II. *Acidianus*
Genus III. *Metallosphaera*
Genus IV. *Stygiolobus*
Genus V. *Sulfurisphaera*
Genus VI. *Sulfurococcus*

Phylum AII. *Euryarchaeota*
Class I. *Methanobacteria*
Order I. *Methanobacteriales*
Family I. *Methanobacteriaceae*
Genus I. *Methanobacterium*
Genus II. *Methanobrevibacter*

Genus III. *Methanosphaera*
Genus IV. *Methanothermobacter*
Family II. *Methanothermaceae*
Genus I. *Methanothermus*

Class II. *Methanococci*
Order I. *Methanococcales*
Family I. *Methanococcaceae*
Genus I. *Methanococcus*
Genus II. *Methanothermococcus*
Family II. *Methanocaldococcaceae*
Genus I. *Methanocaldococcus*
Genus II. *Methanotorris*
Order II. *Methanomicrobiales*
Family I. *Methanomicrobiaceae*
Genus I. *Methanomicrobium*
Genus II. *Methanoculleus*
Genus III. *Methanofollis*
Genus IV. *Methanogenium*
Genus V. *Methanolacinia*
Genus VI. *Methanoplanus*
Family II. *Methanocorpusculaceae*
Genus I. *Methanocorpusculum*
Family III. *Methanospirillaceae*
Genus I. *Methanospirillum*
Genera incertae sedis
Genus I. *Methanocalculus*
Order III. *Methanosarcinales*
Family I. *Methanosarcinaceae*
Genus I. *Methanosarcina*
Genus II. *Methanococcoides*
Genus III. *Methanohalobium*
Genus IV. *Methanohalophilus*
Genus V. *Methanolobus*
Genus VI. *Methanomicrococcus*
Genus VII. *Methanosalsum*
Family II. *Methanosaetaceae*
Genus I. *Methanosaeta*

Class III. *Halobacteria*
Order I. *Halobacteriales*
Family I. *Halobacteriaceae*
Genus I. *Halobacterium*
Genus II. *Haloarcula*

Genus III. *Halobaculum*
Genus IV. *Halococcus*
Genus V. *Haloferax*
Genus VI. *Halogeometricum*
Genus VII. *Halorhabdus*
Genus VIII. *Halorubrum*
Genus IX. *Haloterrigena*
Genus X. *Natrialba*
Genus XI. *Natrinema*
Genus XII. *Natronobacterium*
Genus XIII. *Natronococcus*
Genus XIV. *Natronomonas*
Genus XV. *Natronorubrum*

Class IV. *Thermoplasmata*
Order I. *Thermoplasmatales*
Family I. *Thermoplasmataceae*
Genus I. *Thermoplasma*
Family II. *Picrophilaceae*
Genus I. *Picrophilus*
Family III. *Ferroplasmatacaea*
Genus I. *Ferroplasma*

Class V. *Thermococci*
Order I. *Thermococcales*
Family I. *Thermococcaceae*
Genus I. *Thermococcus*
Genus II. *Palaeococcus*
Genus III. *Pyrococcus*

Class VI. *Archaeoglobi*
Order I. *Archaeoglobales*
Family I. *Archaeoglobaceae*
Genus I. *Archaeoglobus*
Genus II. *Ferroglobus*
Genus III. *Geoglobus*

Class VII. *Methanopyri*
Order I. *Methanopyrales*
Family I. *Methanopyraceae*
Genus I. *Methanopyrus*

Domain *Bacteria*

Phylum BI. *Aquificae*
Class I. *Aquificae*
Order I. *Aquificales*
Family I. *Aquificaceae*
Genus I. *Aquifex*
Genus II. *Calderobacterium*
Genus III. *Hydrogenobaculum*
Genus IV. *Hydrogenobacter*
Genus V. *Hydrogenothermus*
Genus VI. *Persephonella*
Genus VII. *Thermocrinis*
Genera incertae sedis
Genus I. *Desulfurobacterium*

Phylum BII. *Thermotogae*
Class I. *Thermotogae*
Order I. *Thermotogales*
Family I. *Thermotogaceae*
Genus I. *Thermotoga*
Genus II. *Fervidobacterium*
Genus III. *Geotoga*
Genus IV. *Marinitoga*
Genus V. *Petrotoga*
Genus VI. *Thermosipho*

Phylum BIII. *Thermodesulfobacteria*
Class I. *Thermodesulfobacteria*
Order I. *Thermodesulfobacteriales*
Family I. *Thermodesulfobacteriaceae*
Genus I. *Thermodesulfobacterium*

Phylum BIV. "Deinococcus-Thermus"
Class I. *Deinococci*
Order I. *Deinococcales*
Family I. *Deinococcaceae*
Genus I. *Deinococcus*
Order II. *Thermales*
Family I. *Thermaceae*
Genus I. *Thermus*
Genus II. *Meiothermus*

Phylum BV. *Chrysiogenetes*
Class I. *Chrysiogenetes*
Order I. *Chrysiogenales*
Family I. *Chrysiogenaceae*
Genus I. *Chrysiogenes*

Phylum BVI. *Chloroflexi*
Class I. "*Chloroflexi*"
Order I. "*Chloroflexales*"
Family I. "*Chloroflexaceae*"
Genus I. *Chloroflexus*
Genus II. *Chloronema*
Genus III. *Heliothrix*
Genus IV. *Roseiflexus*
Family II. Oscillochloridaceae
Genus I. *Oscillochloris*
Order II. "*Herpetosiphonales*"
Family I. "*Herpetosiphonaceae*"
Genus I. *Herpetosiphon*

Phylum BVII. *Thermomicrobia*
Class I. *Thermomicrobia*
Order I. *Thermomicrobiales*
Family I. *Thermomicrobiaceae*
Genus I. *Thermomicrobium*

Phylum BVIII. *Nitrospira*
Class I. "*Nitrospira*"
Order I. "*Nitrospirales*"
Family I. "*Nitrospiraceae*"
Genus I. *Nitrospira*
Genus II. *Leptospirillum*
Genus III. *Magnetobacterium*
Genus IV. *Thermodesulfovibrio*

Phylum BIX. *Deferribacteres*
Class I. *Deferribacteres*
Order I. *Deferribacterales*
Family I. *Deferribacteraceae*
Genus I. *Deferribacter*
Genus II. *Denitrovibrio*
Genus III. *Flexistipes*
Genus IV. *Geovibrio*
Genera *incertae sedis*
Genus I. *Synergistes*

Phylum BX. *Cyanobacteria*
Class I. "*Cyanobacteria*"
Subsection I.
Family I.
Form genus I. *Chamaesiphon*
Form genus II. *Chroococcus*
Form genus III. *Cyanobacterium*
Form genus IV. *Cyanobium*
Form genus V. *Cyanothece*
Form genus VI. *Dactylococcopsis*
Form genus VII. *Gloeobacter*
Form genus VIII. *Gloeocapsa*
Form genus IX. *Gloeothece*
Form genus X. *Microcystis*
Form genus XI. *Prochlorococcus*
Form genus XII. *Prochloron*
Form genus XIII. *Synechococcus*
Form genus XIV. *Synechocystis*
Subsection II.
Family I.
Form genus I. *Cyanocystis*
Form genus II. *Dermocarpella*
Form genus III. *Stanieria*
Form genus IV. *Xenococcus*
Family II.
Form genus I. *Chroococcidiopsis*
Form genus II. *Myxosarcina*
Form genus III. *Pleurocapsa*
Subsection III.
Family I.
Form genus I. *Arthrospira*
Form genus II. *Borzia*
Form genus III. *Crinalium*
Form genus IV. *Geitlerinema*
Form genus V. *Halospirulina*
Form genus VI. *Leptolyngbya*

Form genus VII. *Limnothrix*
Form genus VIII. *Lyngbya*
Form genus IX. *Microcoleus*
Form genus X. *Oscillatoria*
Form genus XI. *Planktothrix*
Form genus XII. *Prochlorothrix*
Form genus XIII. *Pseudanabaena*
Form genus XIV. *Spirulina*
Form genus XV. *Starria*
Form genus XVI. *Symploca*
Form genus XVII. *Trichodesmium*
Form genus XVIII. *Tychonema*
Subsection IV.
Family I.
Form genus I. *Anabaena*
Form genus II. *Anabaenopsis*
Form genus III. *Aphanizomenon*
Form genus IV. *Cyanospira*
Form genus V. *Cylindrospermopsis*
Form genus VI. *Cylindrospermum*
Form genus VII. *Nodularia*
Form genus VIII. *Nostoc*
Form genus IX. *Scytonema*
Family II.
Form genus I. *Calothrix*
Form genus II. *Rivularia*
Form genus III. *Tolypothrix*
Subsection V.
Family I.
Form genus I. *Chlorogloeopsis*
Form genus II. *Fischerella*
Form genus III. *Geitleria*
Form genus IV. *Iyengariella*
Form genus V. *Nostochopsis*
Form genus VI. *Stigonema*

Phylum BXI. *Chlorobi*
Class I. "*Chlorobia*"
Order I. *Chlorobiales*
Family I. *Chlorobiaceae*
Genus I. *Chlorobium*
Genus II. *Ancalochloris*
Genus III. *Chloroherpeton*
Genus IV. *Pelodictyon*
Genus V. *Prosthecochloris*

Phylum BXII. *Proteobacteria*
Class I. "*Alphaproteobacteria*"
Order I. *Rhodospirillales*
Family I. *Rhodospirillaceae*
Genus I. *Rhodospirillum*
Genus II. *Azospirillum*
Genus III. *Magnetospirillum*
Genus IV. *Phaeospirillum*
Genus V. *Rhodocista*
Genus VI. *Rhodospira*
Genus VII. *Rhodovibrio*
Genus VIII. *Roseospira*
Genus IX. *Skermanella*
Genus X. *Thalassospira*
Family II. *Acetobacteraceae*
Genus I. *Acetobacter*

Genus II. *Acidiphilium*
Genus III. *Acidisphaera*
Genus IV. *Acidocella*
Genus V. *Acidomonas*
Genus VI. *Asaia*
Genus VII. *Craurococcus*
Genus VIII. *Gluconacetobacter*
Genus IX. *Gluconobacter*
Genus X. *Kozakia*
Genus XI. *Paracraurococcus*
Genus XII. *Rhodopila*
Genus XIII. *Roseococcus*
Genus XIV. *Stella*
Genus XV. *Zavarzinia*
Order II. *Rickettsiales*
   Family I. *Rickettsiaceae*
     Genus I. *Rickettsia*
     Genus II. *Orientia*
   Family II. *Anaplasmataceae*
     Genus I. *Anaplasma*
     Genus II. *Aegyptianella*
     Genus III. *Cowdria*
     Genus IV. *Ehrlichia*
     Genus V. *Neorickettsia*
     Genus VI. *Wolbachia*
     Genus VII. *Xenohaliotis*
   Family III. "Holosporaceae"
     Genus I. *Holospora*
     Genus II. *Caedibacter*
     Genus III. *Lyticum*
     Genus IV. *Polynucleobacter*
     Genus V. *Pseudocaedibacter*
     Genus VI. *Symbiotes*
     Genus VII. *Tectibacter*
     Genus VIII. *Odyssella*
Order III. "Rhodobacterales"
   Family I. "Rhodobacteraceae"
     Genus I. *Rhodobacter*
     Genus II. *Ahrensia*
     Genus III. *Amaricoccus*
     Genus IV. *Antarctobacter*
     Genus V. *Gemmobacter*
     Genus VI. *Hirschia*
     Genus VII. *Hyphomonas*
     Genus VIII. *Ketogulonicigenium*
     Genus IX. *Leisingera*
     Genus X. *Maricaulis*
     Genus XI. *Methylarcula*
     Genus XII. *Octadecabacter*
     Genus XIII. *Paracoccus*
     Genus XIV. *Rhodobaca*
     Genus XV. *Rhodothalassium*
     Genus XVI. *Rhodovulum*
     Genus XVII. *Roseibium*
     Genus XVIII. *Roseinatrono-bacter*
     Genus XIX. *Roseivivax*
     Genus XX. *Roseobacter*
     Genus XXI. *Roseovarius*
     Genus XXII. *Rubrimonas*
     Genus XXIII. *Ruegeria*
     Genus XXIV. *Sagittula*
     Genus XXV. *Staleya*
     Genus XXVI. *Stappia*
     Genus XXVII. *Sulfitobacter*

Order IV. "Sphingomonadales"
   Family I. *Sphingomonadaceae*
     Genus I. *Sphingomonas*
     Genus II. *Blastomonas*
     Genus III. *Erythrobacter*
     Genus IV. *Erythromicrobium*
     Genus V. *Erythromonas*
     Genus VI. *Novosphingobium*
     Genus VII. *Porphyrobacter*
     Genus VIII. *Rhizomonas*
     Genus IX. *Sandaracinobacter*
     Genus X. *Sphingobium*
     Genus XI. *Sphingopyxis*
     Genus XII. *Zymomonas*
Order V. *Caulobacterales*
   Family I. *Caulobacteraceae*
     Genus I. *Caulobacter*
     Genus II. *Asticcacaulis*
     Genus III. *Brevundimonas*
     Genus IV. *Phenylobacterium*
Order VI. "Rhizobiales"
   Family I. *Rhizobiaceae*
     Genus I. *Rhizobium*
     Genus II. *Agrobacterium*
     Genus III. *Carbophilus*
     Genus IV. *Chelatobacter*
     Genus V. *Ensifer*
     Genus VI. *Sinorhizobium*
   Family II. *Bartonellaceae*
     Genus I. *Bartonella*
   Family III. *Brucellaceae*
     Genus I. *Brucella*
     Genus II. *Mycoplana*
     Genus III. *Ochrobactrum*
   Family IV. "Phyllobacteriaceae"
     Genus I. *Phyllobacterium*
     Genus II. *Allorhizobium*
     Genus III. *Aminobacter*
     Genus IV. *Aquamicrobium*
     Genus V. *Defluvibacter*
     Genus VI. *Mesorhizobium*
     Genus VII. *Pseudaminobacter*
   Family V. "Methylocystaceae"
     Genus I. *Methylocystis*
     Genus II. *Albibacter*
     Genus III. *Methylopila*
     Genus IV. *Methylosinus*
   Family VI. "Beijerinckiaceae"
     Genus I. *Beijerinckia*
     Genus II. *Chelatococcus*
     Genus III. *Derxia*
     Genus IV. *Methylocapsa*
     Genus V. *Methylocella*
   Family VII. "Bradyrhizobiaceae"
     Genus I. *Bradyrhizobium*
     Genus II. *Afipia*
     Genus III. *Agromonas*
     Genus IV. *Blastobacter*
     Genus V. *Bosea*
     Genus VI. *Nitrobacter*
     Genus VII. *Oligotropha*
     Genus VIII. *Rhodoblastus*
     Genus IX. *Rhodopseudomonas*
   Family VIII. *Hyphomicrobiaceae*
     Genus I. *Hyphomicrobium*

Genus II. *Ancalomicrobium*
Genus III. *Ancylobacter*
Genus IV. *Angulomicrobium*
Genus V. *Aquabacter*
Genus VI. *Azorhizobium*
Genus VII. *Blastochloris*
Genus VIII. *Devosia*
Genus IX. *Dichotomicrobium*
Genus X. *Filomicrobium*
Genus XI. *Gemmiger*
Genus XII. *Labrys*
Genus XIII. *Methylorhabdus*
Genus XIV. *Pedomicrobium*
Genus XV. *Prosthecomicrobium*
Genus XVI. *Rhodomicrobium*
Genus XVII. *Rhodoplanes*
Genus XVIII. *Seliberia*
Genus XIX. *Starkeya*
Genus XX. *Xanthobacter*
   Family IX. "Methylobacteriaceae"
     Genus I. *Methylobacterium*
     Genus II. *Protomonas*
     Genus III. *Roseomonas*
   Family X. "Rhodobiaceae"
     Genus I. *Rhodobium*
     Genus II. *Roseospirillum*

Class II. "Betaproteobacteria"
Order I. "Burkholderiales"
   Family I. "Burkholderiaceae"
     Genus I. *Burkholderia*
     Genus II. *Cupriavidus*
     Genus III. *Lautropia*
     Genus IV. *Limnobacter*
     Genus V. *Pandoraea*
     Genus VI. *Paucimonas*
     Genus VII. *Ralstonia*
     Genus VIII. *Thermothrix*
   Family II. "Oxalobacteraceae"
     Genus I. *Oxalobacter*
     Genus II. *Duganella*
     Genus III. *Herbaspirillum*
     Genus IV. *Janthinobacterium*
     Genus V. *Massilia*
     Genus VI. *Telluria*
   Family III. *Alcaligenaceae*
     Genus I. *Alcaligenes*
     Genus II. *Achromobacter*
     Genus III. *Bordetella*
     Genus IV. *Brackiella*
     Genus V. *Pelistega*
     Genus VI. *Oligella*
     Genus VII. *Pigmentiphaga*
     Genus VIII. *Sutterella*
     Genus IX. *Taylorella*
   Family IV. *Comamonadaceae*
     Genus I. *Comamonas*
     Genus II. *Acidovorax*
     Genus III. *Brachymonas*
     Genus IV. *Caldimonas*
     Genus V. *Delftia*
     Genus VI. *Hydrogenophaga*
     Genus VII. *Macromonas*
     Genus VIII. *Polaromonas*

Genus IX. *Rhodoferax*
Genus X. *Variovorax*
Genus XI. *Xenophilus*
Genus XII. *Xylophilus*
Genera *incertae sedis*
Genus I. *Aquabacterium*
Genus II. *Ideonella*
Genus III. *Leptothrix*
Genus IV. *Roseateles*
Genus V. *Rubrivivax*
Genus VI. *Sphaerotilus*
Genus VII. *Tepidimonas*
Genus VIII. *Thiomonas*
Order II. "Hydrogenophilales"
Family I. "Hydrogenophilaceae"
Genus I. *Hydrogenophilus*
Genus II. *Thiobacillus*
Order III. "Methylophilales"
Family I. "Methylophilaceae"
Genus I. *Methylophilus*
Genus II. *Methylobacillus*
Genus III. *Methylovorus*
Order IV. "Neisseriales"
Family I. *Neisseriaceae*
Genus I. *Neisseria*
Genus II. *Alysiella*
Genus III. *Aquaspirillum*
Genus IV. *Catenococcus*
Genus V. *Chromobacterium*
Genus VI. *Eikenella*
Genus VII. *Formivibrio*
Genus VIII. *Iodobacter*
Genus IX. *Kingella*
Genus X. *Microvirgula*
Genus XI. *Prolinoborus*
Genus XII. *Simonsiella*
Genus XIII. *Vitreoscilla*
Genus XIV. *Vogesella*
Order V. "Nitrosomonadales"
Family I. "Nitrosomonadaceae"
Genus I. *Nitrosomonas*
Genus II. *Nitrosolobus*
Genus III. *Nitrosospira*
Family II. *Spirillaceae*
Genus I. *Spirillum*
Family III. *Gallionellaceae*
Genus I. *Gallionella*
Order VI. "Rhodocyclales"
Family I. *"Rhodocyclaceae"*
Genus I. *Rhodocyclus*
Genus II. *Azoarcus*
Genus III. *Azonexus*
Genus IV. *Azospira*
Genus V. *Azovibrio*
Genus VI. *Dechloromonas*
Genus VII. *Dechlorosoma*
Genus VIII. *Ferribacterium*
Genus IX. *Propionibacter*
Genus X. *Propionivibrio*
Genus XI. *Quadricoccus*
Genus XII. *Thauera*
Genus XIII. *Zoogloea*
Order VII. *"Procabacteriales"*
Family I. *"Procabacteriaceae"*
Genus I. *"Procabacter"*

Class III. "Gammaproteobacteria"
Order I. "Chromatiales"
Family I. *Chromatiaceae*
Genus I. *Chromatium*
Genus II. *Allochromatium*
Genus III. *Amoebobacter*
Genus IV. *Halochromatium*
Genus V. *Halothiobacillus*
Genus VI. *Isochromatium*
Genus VII. *Lamprobacter*
Genus VIII. *Lamprocystis*
Genus IX. *Marichromatium*
Genus X. *Nitrosococcus*
Genus XI. *Pfennigia*
Genus XII. *Rhabdochromatium*
Genus XIII. *Thermochromatium*
Genus XIV. *Thioalkalicoccus*
Genus XV. *Thiobaca*
Genus XVI. *Thiocapsa*
Genus XVII. *Thiococcus*
Genus XVIII. *Thiocystis*
Genus XIX. *Thiodictyon*
Genus XX. *Thioflavicoccus*
Genus XXI. *Thiohalocapsa*
Genus XXII. *Thiolamprovum*
Genus XXIII. *Thiopedia*
Genus XXIV. *Thiorhodococcus*
Genus XXV. *Thiorhodovibrio*
Genus XXVI. *Thiospirillum*
Family II. *Ectothiorhodospiraceae*
Genus I. *Ectothiorhodospira*
Genus II. *Alcalilimnicola*
Genus III. *Alkalispirillum*
Genus IV. *Arhodomonas*
Genus V. *Halorhodospira*
Genus VI. *Nitrococcus*
Genus VII. *Thialkalivibrio*
Genus VIII. *Thiorhodospira*
Order II. *Acidithiobacillales*
Family I. *Acidithiobacillaceae*
Genus I. *Acidithiobacillus*
Family II. *Thermithiobacillaceae*
Genus I. *Thermithiobacillus*
Order III. "Xanthomonadales"
Family I. "Xanthomonadaceae"
Genus I. *Xanthomonas*
Genus II. *Frateuria*
Genus III. *Fulvimonas*
Genus IV. *Luteimonas*
Genus V. *Lysobacter*
Genus VI. *Nevskia*
Genus VII. *Pseudoxanthomonas*
Genus VIII. *Rhodanobacter*
Genus IX. *Schineria*
Genus X. *Stenotrophomonas*
Genus XI. *Thermomonas*
Genus XII. *Xylella*
Order IV. "Cardiobacteriales"
Family I. *Cardiobacteriaceae*
Genus I. *Cardiobacterium*
Genus II. *Dichelobacter*
Genus III. *Suttonella*
Order V. "Thiotrichales"
Family I. "Thiotrichaceae"
Genus I. *Thiothrix*

Genus II. *Achromatium*
Genus III. *Beggiatoa*
Genus IV. *Leucothrix*
Genus V. *Thiobacterium*
Genus VI. *Thiomargarita*
Genus VII. *Thioploca*
Genus VIII. *Thiospira*
Family II. "Piscirickettsiaceae"
Genus I. *Piscirickettsia*
Genus II. *Cycloclasticus*
Genus III. *Hydrogenovibrio*
Genus IV. *Methylophaga*
Genus V. *Thioalkalimicrobium*
Genus VI. *Thiomicrospira*
Family III. "Francisellaceae"
Genus I. *Francisella*
Order VI. "Legionellales"
Family I. *Legionellaceae*
Genus I. *Legionella*
Family II. "Coxiellaceae"
Genus I. *Coxiella*
Genus II. *Rickettsiella*
Order VII. "Methylococcales"
Family I. *Methylococcaceae*
Genus I. *Methylococcus*
Genus II. *Methylobacter*
Genus III. *Methylocaldum*
Genus IV. *Methylomicrobium*
Genus V. *Methylomonas*
Genus VI. *Methylosarcina*
Genus VII. *Methylosphaera*
Order VIII. "Oceanospirillales"
Family I. "Oceanospirillaceae"
Genus I. *Oceanospirillum*
Genus II. *Balneatrix*
Genus III. *Marinomonas*
Genus IV. *Marinospirillum*
Genus V. *Neptunomonas*
Genus VI. *Oceanobacillus*
Genus VII. *Oceanobacter*
Genus VIII. *Pseudospirillum*
Genus IX. *Terasakiella*
Family II. *Alcanivoraxaceae*
Genus I. *Alcanivorax*
Family III. *"Hahellaceae"*
Genus I. *Hahella*
Family IV. *Halomonadaceae*
Genus I. *Halomonas*
Genus II. *Carnimonas*
Genus III. *Chromohalobacter*
Genus IV. *Deleya*
Genus V. *Zymobacter*
Family V. *Oleiphilaceae*
Genus I. *Oleiphilus*
Order IX. *Pseudomonadales*
Family I. *Pseudomonadaceae*
Genus I. *Pseudomonas*
Genus II. *Azomonas*
Genus III. *Azotobacter*
Genus IV. *Cellvibrio*
Genus V. *Chryseomonas*
Genus VI. *Flavimonas*
Genus VII. *Lampropedia*
Genus VIII. *Mesophilobacter*
Genus IX. *Morococcus*

Genus X. *Rhizobacter*
Genus XI. *Rugamonas*
Genus XII. *Serpens*
Genus XIII. *Thermoleophilum*
Family II. *Moraxellaceae*
Genus I. *Moraxella*
Genus II. *Acinetobacter*
Genus III. *Psychrobacter*
Order X. "Alteromonadales"
Family I. *Alteromonadaceae*
Genus I. *Alteromonas*
Genus II. *Alishewanella*
Genus III. *Colwellia*
Genus IV. *Ferrimonas*
Genus V. *Glaciecola*
Genus VI. *Idiomarina*
Genus VII. *Marinobacter*
Genus VIII. *Marinobacterium*
Genus IX. *Microbulbifer*
Genus X. *Moritella*
Genus XI. *Pseudoalteromonas*
Genus XII. *Shewanella*
Genus XIII. *Thalassomonas*
Order XI. "Vibrionales"
Family I. *Vibrionaceae*
Genus I. *Vibrio*
Genus II. *Allomonas*
Genus III. *Enhydrobacter*
Genus IV. *Listonella*
Genus V. *Photobacterium*
Genus VI. *Salinivibrio*
Order XII. "Aeromonadales"
Family I. *Aeromonadaceae*
Genus I. *Aeromonas*
Genus II. *Oceanimonas*
Genus III. *Tolumonas*
Family II. *Succinivibrionaceae*
Genus I. *Succinivibrio*
Genus II. *Anaerobiospirillum*
Genus III. *Ruminobacter*
Genus IV. *Succinimonas*
Order XIII. "Enterobacteriales"
Family I. *Enterobacteriaceae*
Genus I. *Escherichia*
Genus II. *Alterococcus*
Genus III. *Arsenophonus*
Genus IV. *Brenneria*
Genus V. *Buchnera*
Genus VI. *Budvicia*
Genus VII. *Buttiauxella*
Genus VIII. *Calymmatobacterium*
Genus IX. *Cedecea*
Genus X. *Citrobacter*
Genus XI. *Edwardsiella*
Genus XII. *Enterobacter*
Genus XIII. *Erwinia*
Genus XIV. *Ewingella*
Genus XV. *Hafnia*
Genus XVI. *Klebsiella*
Genus XVII. *Kluyvera*
Genus XVIII. *Leclercia*
Genus XIX. *Leminorella*
Genus XX. *Moellerella*
Genus XXI. *Morganella*
Genus XXII. *Obesumbacterium*

Genus XXIII. *Pantoea*
Genus XXIV. *Pectobacterium*
Genus XXV. *"Phlomobacter"*
Genus XXVI. *Photorhabdus*
Genus XXVII. *Plesiomonas*
Genus XXVIII. *Pragia*
Genus XXIX. *Proteus*
Genus XXX. *Providencia*
Genus XXXI. *Rahnella*
Genus XXXII. *Raoultella*
Genus XXXIII. *Saccharobacter*
Genus XXXIV. *Salmonella*
Genus XXXV. *Samsonia*
Genus XXXVI. *Serratia*
Genus XXXVII. *Shigella*
Genus XXXVIII. *Sodalis*
Genus XXXIX. *Tatumella*
Genus XL. *Trabulsiella*
Genus XLI. *Wigglesworthia*
Genus XLII. *Xenorhabdus*
Genus XLIII. *Yersinia*
Genus XLIV. *Yokenella*
Order XIV. "Pasteurellales"
Family I. *Pasteurellaceae*
Genus I. *Pasteurella*
Genus II. *Actinobacillus*
Genus III. *Haemophilus*
Genus IV. *Lonepinella*
Genus V. *Mannheimia*
Genus VI. *Phocoenobacter*

Class IV. "Deltaproteobacteria"
Order I. "Desulfurellales"
Family I. "Desulfurellaceae"
Genus I. *Desulfurella*
Genus II. *Hippea*
Order II. "Desulfovibrionales"
Family I. "Desulfovibrionaceae"
Genus I. *Desulfovibrio*
Genus II. *Bilophila*
Genus III. *Lawsonia*
Family II. "Desulfomicrobiaceae"
Genus I. *"Desulfomicrobium"*
Family III. "Desulfohalobiaceae"
Genus I. *Desulfohalobium*
Genus II. *Desulfomonas*
Genus III. *Desulfonatronovibrio*
Genus IV. *Desulfothermus*
Family IV. "Desulfonatronumaceae"
Genus I. *Desulfonatronum*
Order III. "Desulfobacterales"
Family I. "Desulfobacteraceae"
Genus I. *Desulfobacter*
Genus II. *Desulfobacterium*
Genus III. *Desulfobacula*
Genus IV. *Desulfocella*
Genus V. *Desulfococcus*
Genus VI. *Desulfofaba*
Genus VII. *Desulfofrigus*
Genus VIII. *Desulfomusa*
Genus IX. *Desulfonema*
Genus X. *Desulforegula*
Genus XI. *Desulfosarcina*
Genus XII. *Desulfospira*
Genus XIII. *Desulfotignum*

Family II. *"Desulfobulbaceae"*
Genus I. *Desulfobulbus*
Genus II. *Desulfocapsa*
Genus III. *Desulfofustis*
Genus IV. *Desulforhopalus*
Genus V. *Desulfotalea*
Family III. *"Nitrospinaceae"*
Genus I. *Nitrospina*
Order IV. "Desulfuromonadales"
Family I. "Desulfuromonadaceae"
Genus I. *Desulfuromonas*
Genus II. *Desulfuromusa*
Genus III. *Pelobacter*
Genus IV. *Malonomonas*
Family II. "Geobacteraceae"
Genus I. *Geobacter*
Genus II. *Trichlorobacter*
Order V. "Syntrophobacterales"
Family I. "Syntrophobacteraceae"
Genus I. *Syntrophobacter*
Genus II. *Desulfacinum*
Genus III. *Desulforhabdus*
Genus IV. *Desulfovirga*
Genus V. *Thermodesulforhabdus*
Family II. "Syntrophaceae"
Genus I. *Syntrophus*
Genus II. *Desulfobacca*
Genus III. *Desulfomonile*
Genus IV. *Smithella*
Order VI. "Bdellovibrionales"
Family I. "Bdellovibrionaceae"
Genus I. *Bdellovibrio*
Genus II. *Bacteriovorax*
Genus III. *Micavibrio*
Genus IV. *Vampirovibrio*
Order VII. *Myxococcales*
Suborder I. *Cystobacterinea*
Family I. *Cystobacteraceae*
Genus I. *Cystobacter*
Genus II. *Archangium*
Genus III. *Hyalangium*
Genus IV. *Melittangium*
Genus V. *Stigmatella*
Family II. *Myxococcaceae*
Genus I. *Myxococcus*
Genus II. *Corallococcus*
Genus III. *Pyxicoccus*
Suborder II. *Soranginea*
Family I. *Polyangiaceae*
Genus I. *Polyangium*
Genus II. *Byssophaga*
Genus III. *Chondromyces*
Genus IV. *"Haploangium"*
Genus V. *Jahnia*
Genus VI. *Sorangium*
Suborder III. *Nannocystinea*
Family I. *Nannocystaceae*
Genus I. *Nannocystis*
Family II. *Kofleriaceae*
Genus I. *Kofleria*

Class V. "Epsilonproteobacteria"
Order I. "Campylobacterales"
Family I. *Campylobacteraceae*
Genus I. *Campylobacter*

Genus II. *Arcobacter*
Genus III. *Sulfurospirillum*
Family II. "Helicobacteraceae"
  Genus I. *Helicobacter*
  Genus II. *Thiovulum*
  Genus III. *Wolinella*
Family III. *"Nautilaceae"*
  Genus I. *Nautilia*
  Genus II. *Caminibacter*

Phylum BXIII. *Firmicutes*
Class I. "Clostridia"
Order I. *Clostridiales*
  Family I. *Clostridiaceae*
    Genus I. *Clostridium*
    Genus II. *Acetivibrio*
    Genus III. *Acidaminobacter*
    Genus IV. *Alkaliphilus*
    Genus V. *Anaerobacter*
    Genus VI. *Caloramator*
    Genus VII. *Caloranaerobacter*
    Genus VIII. *Coprobacillus*
    Genus IX. *Dorea*
    Genus X. *Natronincola*
    Genus XI. *Oxobacter*
    Genus XII. *Sarcina*
    Genus XIII. *Sporobacter*
    Genus XIV. *Thermobrachium*
    Genus XV. *Thermohalobacter*
    Genus XVI. *Tindallia*
  Family II. "Lachnospiraceae"
    Genus I. *Lachnospira*
    Genus II. *Acetitomaculum*
    Genus III. *Anaerofilum*
    Genus IV. *Butyrivibrio*
    Genus V. *Catenibacterium*
    Genus VI. *Catonella*
    Genus VII. *Coprococcus*
    Genus VIII. *Johnsonella*
    Genus IX. *Lachnobacterium*
    Genus X. *Pseudobutyrivibrio*
    Genus XI. *Roseburia*
    Genus XII. *Ruminococcus*
    Genux XIII. *Sporobacterium*
  Family III. "Peptostreptococcaceae"
    Genus I. *Peptostreptococcus*
    Genus II. *Anaerococcus*
    Genus III. *Filifactor*
    Genus IV. *Finegoldia*
    Genus V. *Fusibacter*
    Genus VI. *Gallicola*
    Genus VII. *Helococcus*
    Genus VIII. *Micromonas*
    Genus IX. *Peptoniphilus*
    Genus X. *Sedimentibacter*
    Genus XI. *Sporanaerobacter*
    Genus XII. *Tissierella*
  Family IV. "Eubacteriaceae"
    Genus I. *Eubacterium*
    Genus II. *Acetobacterium*
    Genus III. *Anaerovorax*
    Genus IV. *Mogibacterium*
    Genus V. *Pseudoramibacter*
  Family V. *Peptococcaceae*
    Apenus I. *Peptococcus*
    Genus II. *Anaeroarcus*

Genus III. *Anaerosinus*
Genus IV. *Anaerovibrio*
Genus V. *Carboxydothermus*
Genus VI. *Centipeda*
Genus VII. *Dehalobacter*
Genus VIII. *Dendrosporobacter*
Genus IX. *Desulfitobacterium*
Genus X. *Desulfonispora*
Genus XI. *Desulfosporosinus*
Genus XII. *Desulfotomaculum*
Genus XIII. *Mitsuokella*
Genus XIV. *Propionispira*
Genus XV. *Succinispira*
Genus XVI. *Syntrophobotulus*
Genus XVII. *Thermoterrabacterium*
Family VI. "Heliobacteriaceae"
  Genus I. *Heliobacterium*
  Genus II. *Heliobacillus*
  Genus III. *Heliophilum*
  Genus IV. *Heliorestis*
Family VII. "Acidaminococcaceae"
  Genus I. *Acidaminococcus*
  Genus II. *Acetonema*
  Genus III. *Anaeroglobus*
  Genus IV. *Anaeromusa*
  Genus V. *Dialister*
  Genus VI. *Megasphaera*
  Genus VII. *Papillibacter*
  Genus VIII. *Pectinatus*
  Genus IX. *Phascolarctobacterium*
  Genus X. *Propionispora*
  Genus XI. *Quinella*
  Genus XII. *Schwartzia*
  Genus XIII. *Selenomonas*
  Genus XIV. *Sporomusa*
  Genus XV. *Succiniclasticum*
  Genus XVI. *Veillonella*
  Genus XVII. *Zymophilus*
Family VIII. *Syntrophomonadaceae*
  Genus I. *Syntrophomonas*
  Genus II. *Acetogenium*
  Genus III. *Aminobacterium*
  Genus IV. *Aminomonas*
  Genus V. *Anaerobaculum*
  Genus VI. *Anaerobranca*
  Genus VII. *Caldicellulosiruptor*
  Genus VIII. *Dethiosulfovibrio*
  Genus IX. *Pelospora*
  Genus X. *Syntrophospora*
  Genus XI. *Syntrophothermus*
  Genus XII. *Thermaerobacter*
  Genus XIII. *Thermanaerovibrio*
  Genus XIV. *Thermohydrogenium*
  Genus XV. *Thermosyntropha*
Order II. "Thermoanaerobacteriales"
  Family I. "Thermoanaerobacteriaceae"
    Genus I. *Thermoanaerobacterium*
    Genus II. *Ammonifex*
    Genus III. *Carboxydobrachium*
    Genus IV. *Coprothermobacter*
    Genus V. *Gelria*
    Genus VI. *Moorella*
    Genus VII. *Sporotomaculum*
    Genus VIII. *Thermacetogenium*
    Genus IX. *Thermoanaerobacter*
    Genus X. *Thermoanaerobium*

Order III. *Haloanaerobiales*
  Family I. *Haloanaerobiaceae*
    Genus I. *Haloanaerobium*
    Genus II. *Halocella*
    Genus III. *Halothermothrix*
  Family II. *Halobacteroidaceae*
    Genus I. *Halobacteroides*
    Genus II. *Acetohalobium*
    Genus III. *Haloanaerobacter*
    Genus IV. *Halonatronum*
    Genus V. *Natroniella*
    Genus VI. *Orenia*
    Genus VII. *Selenihalanaerobacter*
    Genus VIII. *Sporohalobacter*

Class II. *Mollicutes*
Order I. *Mycoplasmatales*
  Family I. *Mycoplasmataceae*
    Genus I. *Mycoplasma*
    Genus II. *Eperythrozoon*
    Genus III. *Haemobartonella*
    Genus IV. *Ureaplasma*
Order II. *Entomoplasmatales*
  Family I. *Entomoplasmataceae*
    Genus I. *Entomoplasma*
    Genus II. *Mesoplasma*
  Family II. *Spiroplasmataceae*
    Genus I. *Spiroplasma*
Order III. *Acholeplasmatales*
  Family I. *Acholeplasmataceae*
    Genus I. *Acholeplasma*
Order IV. *Anaeroplasmatales*
  Family I. *Anaeroplasmataceae*
    Genus I. *Anaeroplasma*
    Genus II. *Asteroleplasma*
Order V. *Incertae sedis*
  Family I. *"Erysipelotrichaceae"*
    Genus I. *Erysipelothrix*
    Genus II. *Bulleidia*
    Genus III. *Holdemania*
    Genus IV. *Solobacterium*

Class III. "Bacilli"
Order I. *Bacillales*
  Family I. *Bacillaceae*
    Genus I. *Bacillus*
    Genus II. *Amphibacillus*
    Genus III. *Anoxybacillus*
    Genus IV. *Exiguobacterium*
    Genus V. *Filobacillus*
    Genus VI. *Geobacillus*
    Genus VII. *Gracilibacillus*
    Genus VIII. *Halobacillus*
    Genus IX. *Jeotgalibacillus*
    Genus X. *Marinibacillus*
    Genus XI. *Saccharococcus*
    Genus XII. *Salibacillus*
    Genus XIII. *Ureibacillus*
    Genus XIV. *Virgibacillus*
  Family II. *"Alicyclobacillaceae"*
    Genus I. *Alicyclobacillus*
    Genus II. *Pasteuria*
    Genus III. *Sulfobacillus*
  Family III. *Caryophanaceae*
    Genus I. *Caryophanon*
  Family IV. "Listeriaceae"

Genus I. *Listeria*
Genus II. *Brochothrix*
Family V. *"Paenibacillaceae"*
  Genus I. *Paenibacillus*
  Genus II. *Ammoniphilus*
  Genus III. *Aneurinibacillus*
  Genus IV. *Brevibacillus*
  Genus V. *Oxalophagus*
  Genus VI. *Thermicanus*
  Genus VII. *Thermobacillus*
Family VI. *Planococcaceae*
  Genus I. *Planococcus*
  Genus II. *Filibacter*
  Genus III. *Kurthia*
  Genus IV. *Planomicrobium*
  Genus V. *Sporosarcina*
Family VII. *"Sporolactobacillaceae"*
  Genus I. *Sporolactobacillus*
  Genus II. *Marinococcus*
Family VIII. *"Staphylococcaceae"*
  Genus I. *Staphylococcus*
  Genus II. *Gemella*
  Genus III. *Macrococcus*
  Genus IV. *Salinicoccus*
Family IX. *"Thermoactinomycetaceae"*
  Genus I. *Thermoactinomyces*
Family X. *"Turicibacteraceae"*
  Genus I. *Turicibacter*
Order II. *"Lactobacillales"*
Family I. *Lactobacillaceae*
  Genus I. *Lactobacillus*
  Genus II. *Paralactobacillus*
  Genus III. *Pediococcus*
Family II. *"Aerococcaceae"*
  Genus I. *Aerococcus*
  Genus II. *Abiotrophia*
  Genus III. *Dolosicoccus*
  Genus IV. *Eremococcus*
  Genus V. *Facklamia*
  Genus VI. *Globicatella*
  Genus VII. *Ignavigranum*
Family III. *"Carnobacteriaceae"*
  Genus I. *Carnobacterium*
  Genus II. *Agitococcus*
  Genus III. *Alkalibacterium*
  Genus IV. *Alloiococcus*
  Genus V. *Desemzia*
  Genus VI. *Dolosigranulum*
  Genus VII. *Granulicatella*
  Genus VIII. *Isobaculum*
  Genus IX. *Lactosphaera*
  Genus X. *Trichococcus*
Family IV. *"Enterococcaceae"*
  Genus I. *Enterococcus*
  Genus II. *Atopobacter*
  Genus III. *Melissococcus*
  Genus IV. *Tetragenococcus*
  Genus V. *Vagococcus*
Family V. *"Leuconostocaceae"*
  Genus I. *Leuconostoc*
  Genus II. *Oenococcus*
  Genus III. *Weissella*
Family VI. *Streptococcaceae*
  Genus I. *Streptococcus*
  Genus II. *Lactococcus*
Family VII. *Incertae sedis*

Genus I. *Acetoanaerobium*
Genus II. *Oscillospira*
Genus III. *Syntrophococcus*

Phylum BXIV. *Actinobacteria*
Class I. *Actinobacteria*
Subclass I. *Acidimicrobidae*
Order I. *Acidimicrobiales*
Suborder IV. *"Acidimicrobineae"*
Family I. *Acidimicrobiaceae*
  Genus I. *Acidimicrobium*
Subclass II. *Rubrobacteridae*
Order I. *Rubrobacterales*
Suborder V. *"Rubrobacterineae"*
Family I. *Rubrobacteraceae*
  Genus I. *Rubrobacter*
Subclass III. *Coriobacteridae*
Order I. *Coriobacteriales*
Suborder VI. *"Coriobacterineae"*
Family I. *Coriobacteriaceae*
  Genus I. *Coriobacterium*
  Genus II. *Atopobium*
  Genus III. *Collinsella*
  Genus IV. *Crytobacterium*
  Genus V. *Denitrobacterium*
  Genus VI. *Eggerthella*
  Genus VII. *Olsenella*
  Genus VIII. *Slackia*
Subclass IV. *Sphaerobacteridae*
Order I. *Sphaerobacterales*
Suborder VII. *"Sphaerobacterineae"*
Family I. *Sphaerobacteraceae*
  Genus I. *Sphaerobacter*
Subclass V. *Actinobacteridae*
Order I. *Actinomycetales*
Suborder VIII. *Actinomycineae*
Family I. *Actinomycetaceae*
  Genus I. *Actinomyces*
  Genus II. *Actinobaculum*
  Genus III. *Arcanobacterium*
  Genus IV. *Mobiluncus*
Suborder IX. *Micrococcineae*
Family I. *Micrococcaceae*
  Genus I. *Micrococcus*
  Genus II. *Arthrobacter*
  Genus III. *Kocuria*
  Genus IV. *Nesterenkonia*
  Genus V. *Renibacterium*
  Genus VI. *Rothia*
  Genus VII. *Stomatococcus*
Family II. *Bogoriellaceae*
  Genus I. *Bogoriella*
Family III. *Rarobacteraceae*
  Genus I. *Rarobacter*
Family IV. *Sanguibacteraceae*
  Genus I. *Sanguibacter*
Family V. *Brevibacteriaceae*
  Genus I. *Brevibacterium*
Family VI. *Cellulomonadaceae*
  Genus I. *Cellulomonas*
  Genus II. *Oerskovia*
  Genus III. *Tropheryma*
Family VII. *Dermabacteraceae*
  Genus I. *Dermabacter*
  Genus II. *Brachybacterium*
Family VIII. *Dermatophilaceae*

Genus I. *Dermatophilus*
Family IX. *Dermacoccaceae*
  Genus I. *Dermacoccus*
  Genus II. *Demetria*
  Genus III. *Kytococcus*
Family X. *Intrasporangiaceae*
  Genus I. *Intrasporangium*
  Genus II. *Janibacter*
  Genus III. *Knoellia*
  Genus IV. *Ornithinicoccus*
  Genus V. *Ornithinimicrobium*
  Genus VI. *Nostocoidia*
  Genus VII. *Terrabacter*
  Genus VIII. *Terracoccus*
  Genus IX. *Tetrasphaera*
Family XI. *Jonesiaceae*
  Genus I. *Jonesia*
Family XII. *Microbacteriaceae*
  Genus I. *Microbacterium*
  Genus II. *Agreia*
  Genus III. *Agrococcus*
  Genus IV. *Agromyces*
  Genus V. *Aureobacterium*
  Genus VI. *Clavibacter*
  Genus VII. *Cryobacterium*
  Genus VIII. *Curtobacterium*
  Genus IX. *Frigoribacterium*
  Genus X. *Leifsonia*
  Genus XI. *Leucobacter*
  Genus XII. *Mycetocola*
  Genus XIII. *Okibacterium*
  Genus XIV. *Rathayibacter*
  Genus XV. *Subtercola*
Family XIII. *"Beutenbergiaceae"*
  Genus I. *Beutenbergia*
  Genus II. *Georgenia*
  Genus III. *Salana*
Family XIV. *Promicromonosporaceae*
  Genus I. *Promicromonospora*
  Genus II. *Cellulosimicrobium*
Suborder X. *Corynebacterineae*
Family I. *Corynebacteriaceae*
  Genus I. *Corynebacterium*
Family II. *Dietziaceae*
  Genus I. *Dietzia*
Family III. *Gordoniaceae*
  Genus I. *Gordonia*
  Genus II. *Skermania*
Family IV. *Mycobacteriaceae*
  Genus I. *Mycobacterium*
Family V. *Nocardiaceae*
  Genus I. *Nocardia*
  Genus II. *Rhodococcus*
Family VI. *Tsukamurellaceae*
  Genus I. *Tsukamurella*
Family VII. *"Williamsiaceae"*
  Genus I. *Williamsia*
Suborder XI. *Micromonosporineae*
Family I. *Micromonosporaceae*
  Genus I. *Micromonospora*
  Genus II. *Actinoplanes*
  Genus III. *Asanoa*
  Genus IV. *Catellatospora*
  Genus V. *Catenuloplanes*
  Genus VI. *Couchioplanes*

Genus VII. *Dactylosporan-
gium*
Genus VIII. *Pilimelia*
Genus IX. *Spirilliplanes*
Genus X. *Verrucosispora*
Genus XI. *Virgisporangium*
Suborder XII. *Propionibacterineae*
Family I. *Propionibacteriaceae*
Genus I. *Propionibacterium*
Genus II. *Luteococcus*
Genus III. *Microlunatus*
Genus IV. *Propioniferax*
Genus V. *Tessaracoccus*
Family II. *Nocardioidaceae*
Genus I. *Nocardioides*
Genus II. *Aeromicrobium*
Genus III. *Actinopolymorpha*
Genus IV. *Friedmanniella*
Genus V. *Hongia*
Genus VI. *Kribbella*
Genus VII. *Micropruina*
Genus VIII. *Marmoricola*
Suborder XIII. *Pseudonocardineae*
Family I. *Pseudonocardiaceae*
Genus I. *Pseudonocardia*
Genus II. *Actinoalloteichus*
Genus III. *Actinopolyspora*
Genus IV. *Amycolatopsis*
Genus V. *Crossiella*
Genus VI. *Kibdelosporangium*
Genus VII. *Kutzneria*
Genus VIII. *Prauserella*
Genus IX. *Saccharomonospora*
Genus X. *Saccharopolyspora*
Genus XI. *Streptoalloteichus*
Genus XII. *Thermobispora*
Genus XIII. *Thermocrispum*
Family II. *Actinosynnemataceae*
Genus I. *Actinosynnema*
Genus II. *Actinokineospora*
Genus III. *Lechevaliera*
Genus IV. *Lentzea*
Genus V. *Saccharothrix*
Suborder XIV. *Streptomycineae*
Family I. *Streptomycetaceae*
Genus I. *Streptomyces*
Genus II. *Kitasatospora*
Genus III. *Streptoverticillium*
Suborder XV. *Streptosporangineae*
Family I. *Streptosporangiaceae*
Genus I. *Streptosporangium*
Genus II. *Acrocarpospora*
Genus III. *Herbidospora*
Genus IV. *Microbispora*
Genus V. *Microtetraspora*
Genus VI. *Nonomuraea*
Genus VII. *Planobispora*
Genus VIII. *Planomonospora*
Genus IX. *Planopolyspora*
Genus X. *Planotetraspora*
Family II. *Nocardiopsaceae*
Genus I. *Nocardiopsis*
Genus II. *Streptomonospora*
Genus III. *Thermobifida*
Family III. *Thermomono-
sporaceae*

Genus I. *Thermomonospora*
Genus II. *Actinomadura*
Genus III. *Spirillospora*
Suborder XVI. *Frankineae*
Family I. *Frankiaceae*
Genus I. *Frankia*
Family II. *Geodermatophilaceae*
Genus I. *Geodermatophilus*
Genus II. *Blastococcus*
Genus III. *Modestobacter*
Family III. *Microsphaeraceae*
Genus I. *Microsphaera*
Family IV. *Sporichthyaceae*
Genus I. *Sporichthya*
Family V. *Acidothermaceae*
Genus I. *Acidothermus*
Family VI. *"Kineosporiaceae"*
Genus I. *Kineosporia*
Genus II. *Cryptosporangium*
Genus III. *Kineococcus*
Suborder XVII. *Glycomycineae*
Family I. *Glycomycetaceae*
Genus I. *Glycomyces*
Order II. *Bifidobacteriales*
Family I. *Bifidobacteriaceae*
Genus I. *Bifidobacterium*
Genus II. *Falcivibrio*
Genus III. *Gardnerella*
Genus IV. *Parascardovia*
Genus V. *Scardovia*
Family II. Unknown Affiliation
Genus I. *Actinobispora*
Genus II. *Actinocorallia*
Genus III. *Excellospora*
Genus IV. *Pelczaria*
Genus V. *Turicella*

Phylum BXV. *Planctomycetes*
Class I. *"Planctomycetacia"*
Order I. *Planctomycetales*
Family I. *Planctomycetaceae*
Genus I. *Planctomyces*
Genus II. *Gemmata*
Genus III. *Isosphaera*
Genus IV. *Pirellula*

Phylum BXVI. *Chlamydiae*
Class I. *"Chlamydiae"*
Order I. *Chlamydiales*
Family I. *Chlamydiaceae*
Genus I. *Chlamydia*
Genus II. *Chlamydophila*
Family II. *Parachlamydiaceae*
Genus I. *Parachlamydia*
Genus II. *Neochlamydia*
Family III. *Simkaniaceae*
Genus I. *Simkania*
Family IV. *Waddliaceae*
Genus I. *Waddlia*

Phylum BXVII. *Spirochaetes*
Class I. *"Spirochaetes"*
Order I. *Spirochaetales*
Family I. *Spirochaetaceae*
Genus I. *Spirochaeta*

Genus II. *Borrelia*
Genus III. *Brevinema*
Genus IV. *Clevelandina*
Genus V. *Cristispira*
Genus VI. *Diplocalyx*
Genus VII. *Hollandina*
Genus VIII. *Pillotina*
Genus IX. *Treponema*
Family II. *"Serpulinaceae"*
Genus I. *Serpulina*
Genus II. *Brachyspira*
Family III. *Leptospiraceae*
Genus I. *Leptonema*
Genus II. *Leptospira*

Phylum BXVIII. *Fibrobacteres*
Class I. *"Fibrobacteres"*
Order I. *"Fibrobacterales"*
Family I. *"Fibrobacteraceae"*
Genus I. *Fibrobacter*

Phylum BXIX. *Acidobacteria*
Class I. *"Acidobacteria"*
Order I. *"Acidobacteriales"*
Family I. *"Acidobacteriaceae"*
Genus I. *Acidobacterium*
Genus II. *Geothrix*
Genus III. *Holophaga*

Phylum BXX. *Bacteroidetes*
Class I. *"Bacteroides"*
Order I. *"Bacteroidales"*
Family I. *Bacteroidaceae*
Genus I. *Bacteroides*
Genus II. *Acetofilamentum*
Genus III. *Acetomicrobium*
Genus IV. *Acetothermus*
Genus V. *Anaerophaga*
Genus VI. *Anaerorhabdus*
Genus VII. *Megamonas*
Family II. *"Rikenellaceae"*
Genus I. *Rikenella*
Genus II. *Marinilabilia*
Family III. *"Porphyromonadaceae"*
Genus I. *Porphyromonas*
Genus II. *Dysgonomonas*
Genus III. *Tannerella*
Family IV. *"Prevotellaceae"*
Genus I. *Prevotella*

Class II. *"Flavobacteria"*
Order I. *"Flavobacteriales"*
Family I. *Flavobacteriaceae*
Genus I. *Flavobacterium*
Genus II. *Arenibacter*
Genus III. *Bergeyella*
Genus IV. *Capnocytophaga*
Genus V. *Cellulophaga*
Genus VI. *Chryseobacterium*
Genus VII. *Coenonia*
Genus VIII. *Empedobacter*
Genus IX. *Gelidibacter*
Genus X. *Muricauda*
Genus XI. *Ornithobacterium*
Genus XII. *Polaribacter*

Genus XIII. *Psychroflexus*
Genus XIV. *Psychroserpens*
Genus XV. *Riemerella*
Genus XVI. *Saligentibacter*
Genus XVII. *Tenacibaculum*
Genus XVIII. *Weeksella*
Genus XIX. *Zobellia*
Family II. *"Myroidaceae"*
  Genus I. *Myroides*
  Genus II. *Psychromonas*
Family III. *"Blattabacteriaceae"*
  Genus I. *Blattabacterium*

Class III. *"Sphingobacteria"*
  Order I. *"Sphingobacteriales"*
    Family I. *Sphingobacteriaceae*
      Genus I. *Sphingobacterium*
      Genus II. *Pedobacter*
    Family II. *"Saprospiraceae"*
      Genus I. *Saprospira*
      Genus II. *Haliscomenobacter*
      Genus III. *Lewinella*
    Family III. *"Flexibacteraceae"*
      Genus I. *Flexibacter*
      Genus II. *Cyclobacterium*
      Genus III. *Cytophaga*

Genus IV. *Dyadobacter*
Genus V. *Flectobacillus*
Genus VI. *Hymenobacter*
Genus VII. *Meniscus*
Genus VIII. *Microscilla*
Genus IX. *Runella*
Genus X. *Spirosoma*
Genus XI. *Sporocytophaga*
Family IV. *"Flammeovirgaceae"*
  Genus I. *Flammeovirga*
  Genus II. *Flexithrix*
  Genus III. *Persicobacter*
  Genus IV. *Thermonema*
Family V. *Crenotrichaceae*
  Genus I. *Crenothrix*
  Genus II. *Chitinophaga*
  Genus III. *Rhodothermus*
  Genus IV. *Salinibacter*
  Genus V. *Toxothrix*

Phylum BXXI. *Fusobacteria*
Class I. *"Fusobacteria"*
  Order I. *"Fusobacteriales"*
    Family I. *"Fusobacteriaceae"*
      Genus I. *Fusobacterium*
      Genus II. *Ilyobacter*

Genus III. *Leptotrichia*
Genus IV. *Propionigenium*
Genus V. *Sebaldella*
Genus VI. *Streptobacillus*
Genus VII. *Sneathia*
Family II. Incertae sedis
  Genus I. *Cetobacterium*

Phylum BXXII. *Verrucomicrobia*
Class I. *Verrucomicrobiae*
  Order I. *Verrucomicrobiales*
    Family I. *Verrucomicrobiaceae*
      Genus I. *Verrucomicrobium*
      Genus II. *Prosthecobacter*
    Family II. *"Opitutaceae"*
      Genus I. *Opitutus*
    Family III. *"Xiphinematobacteriaceae"*
      Genus I. *Xiphinematobacter*

Phylum BXXIII. *Dictyoglomus*
Class I. *"Dictyoglomi"*
  Order I. *"Dictyoglomales"*
    Family I. *"Dictyoglomaceae"*
      Genus I. *Dictyoglomus*

Appendix IV

Classification of Viruses

In this appendix selected groups of viruses are briefly described and a sketch of each is included. The illustrations are not to scale but should provide a general idea of the morphology of each group. The viruses are separated into five sections based primarily on their host preferences. This material has been adapted with permission from chapter 22 of *Introduction to Modern Virology*, 5th ed. by N. J. Dimmock, A. J. Easton, and K. N. Leppard, copyright © 2001 by Blackwell Scientific Publications, Ltd.

We have not italicized the names of specific viruses in the text. However, it should be noted that in formal taxonomic usage the official name of a virus species—just as with other taxonomic names—is italicized, and the first letter of the first word in the species name is capitalized. Tentative species names and names of strains or isolates are not italicized. In this appendix, we will capitalize and italicize the genus name, but not always the species name.

A. VIRUSES OF VERTEBRATE AND INVERTEBRATE ANIMALS

1. Viruses with double-stranded DNA genomes

a. Family: *Adenoviridae*

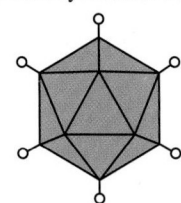

Nonenveloped icosahedral particles of 60 to 90 nm with a fiber protein at each vertex. Contains one molecule of linear double-stranded DNA (dsDNA) of 25,000 to 42,000 base-pairs (bp). A protein is covalently linked to the 5′ end and forms a pseudocircular genome by noncovalently linking to the 3′ end. Replicates and is assembled in the nucleus. Infects vertebrates but one genus infects fungi.

Genera:
Mastadenovirus—infects mammals (e.g., human adenovirus type 2 or type 5)
Aviadenovirus—infects birds (e.g., fowl adenovirus type 1)
Atendovirus—bovine adenoviruses
Siadenovirus (e.g., frog adenovirus 1, turkey adenovirus 3)

b. Family: *Baculoviridae*

Enveloped rod-shaped particles, each with a nucleocapsid of 300 nm × 30 to 60 nm with one molecule of circular dsDNA of 80,000 to 180,000 bp. Virion may be occluded in a protein inclusion body (which makes it very stable). A large and diverse group, common in over 50 species of *Lepidoptera;* also infects *Coleoptera,* etc.

Genera:
Granulovirus—may be occluded in a polyhedron containing one particle (e.g., *Cydia pomonella* granulovirus).
Nucleopolyhedrovirus. Virions may be occluded in a polyhedron containing many particles (e.g., *Autographica californica* multiple nucleopolyhedrosis virus).

c. Family: *Herpesviridae*

Enveloped 120 to 220 nm particle with spikes, enclosing successively a tegument and an icosahedral nucleocapsid of 125 nm. One molecule of linear dsDNA of 120,000 to 240,000 bp. May bud from nuclear membrane. Very large family that infects vertebrates; latency is a common feature of the life cycle. Some are oncogenic.

Subfamily: *Alphaherpesvirinae*

Genera:
Simplexvirus (e.g., human herpes [simplex] virus type 1 and 2—HHV-1 and -2)
Varicellovirus (e.g., HHV-3 or varicella-zoster [chickenpox] virus)
Subfamily: *Betaherpesvirinae*

Genera:
Cytomegalovirus (e.g., HHV-5 or human cytomegalovirus)
Muromegalovirus (e.g., mouse cytomegalovirus)
Roseolavirus (e.g., HHV-6, HHV-7)
Subfamily: *Gammaherpesvirinae* (lymphoproliferative viruses; oncogenic)

Genera:
Lymphocryptovirus (e.g., HHV-4 or Epstein-Barr virus)
Rhadinovirus (e.g., ateline herpesvirus 2, herpesvirus saimiri, HHV-8)
Unclassified: Marek's disease virus, turkey herpesvirus.

d. Family: *Iridoviridae*

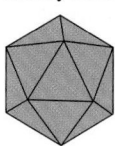

Icosahedral isometric particle of 120 to 200 nm, which may be surrounded by a lipid envelope. Purified virus pellets are iridescent blue or green. One molecule of linear dsDNA of 140,000 to 300,000 bp, but virions may contain one or more copies. Contains several enzymes. Cytoplasmic assembly. Includes viruses of insects, fish (flounder, dab, and goldfish) and many frog species (e.g., frog virus 3).

e. Family: *Papillomaviridae*

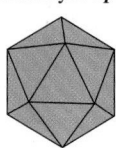

Nonenveloped, 55 nm icosahedral particle with 72 capsomers in a skewed arrangement. One 8,000 bp molecule of double-stranded, covalently closed, circular DNA. Replicates and assembles in the nucleus. Over 60 human papillomaviruses (HPVs) and others from many species. Oncogenic.

f. Family: *Polyomaviridae*

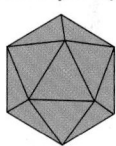

Nonenveloped 40 nm icosahedral particle with 72 capsomers in a skewed arrangement. One 5,000 bp molecule of covalently closed, circular dsDNA. Replicates and assembles in the nucleus. Includes murine polyomavirus and simian virus type 40 (SV40). Oncogenic.

g. Family: *Poxviridae*

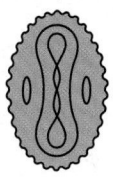

Enveloped brick-shaped virions of 220 to 450 nm × 140 to 260 nm × 140 to 260 nm thick. Complex structure enclosing one or two lateral bodies and a biconcave core, with all enzymes required for mRNA synthesis. One molecule of linear dsDNA of 130,000 to 375,000 bp. Cytoplasmic multiplication. Infects mostly vertebrates but also insects.

Subfamily: *Chondropoxvirinae* (viruses of vertebrates)

Genera:
Orthopoxvirus (e.g. vaccinia and related viruses)
Avipoxvirus (e.g., fowlpox and related viruses)
Capripoxvirus (e.g., sheeppox and related viruses)
Leporipoxvirus (e.g., myxoma of rabbits and related viruses; spread passively by arthropods)
Molluscipoxvirus (e.g., molluscum contagiosum virus)
Parapoxvirus (e.g., orf virus/milker's node virus and related viruses)
Suipoxvirus (e.g., swine pox virus)
Yatapoxvirus (e.g., yaba/tanapox and related viruses of monkeys)

Subfamily: *Entomopoxvirinae* (viruses of insects)

Genera:
Entomopoxvirus A, B, and C

2. **Viruses with ssDNA genomes**

a. Family: *Circoviridae*

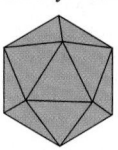

Nonenveloped small icosahedral particles with circular ssDNA. Another genus infects plants.

Genus:
Circovirus—20 nm virion with one molecule of DNA of 1,800 to 2,300 nucleotides (e.g., chicken anemia virus; probably also TT [transfusion-transmitted] virus of humans)

b. Family: *Parvoviridae*

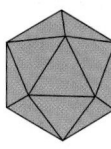

Nonenveloped icosahedral particles of 18 to 26 nm that contain no enzymes. One 5,000 nucleotide linear molecule of *either* negative- *or* positive-sense ssDNA per particle. On extraction, these form an artifactual double strand. Replication is nuclear.

Subfamily: *Parvovirinae* (viruses of vertebrates)

Genera:
Parvovirus—(e.g., canine and feline parvovirus, minute virus of mice)
Erythrovirus—B19 virus, which causes fifth disease of children
Dependovirus—these are satellite viruses
Subfamily: *Densovirinae* (viruses of insects)

3. **Viruses with dsRNA genomes and a virion-associated RNA-dependent RNA polymerase**

a. Family: *Birnaviridae*

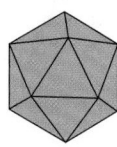

Nonenveloped icosahedral 60 nm particle with a 45 nm core containing two segments of linear dsRNA of 2,800 and 3,200 nucleotides. Has a VPg protein covalently linked to 5′ end of each segment. Cytoplasmic multiplication.

Genera:
Aquabirnavirus (e.g., pancreatic necrosis virus of fish)
Avibirnavirus (e.g., infectious bursal disease of chickens, turkeys, ducks)
Entomobirnavirus (e.g., *Drosophila* X virus)

b. Family: *Reoviridae*

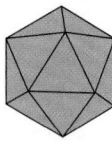

Large family with members found in vertebrates, insects, and *plants*. Nonenveloped 60 to 80 nm icosahedral particle containing an isometric nucleocapsid with 10 to 12 segments of linear dsRNA of 18,000 to 30,000 bp. Cytoplasmic multiplication. Within a genus RNA segments in a mixed infection readily assort to form genetically stable hybrid virus.

Genera:

Orthoreovirus—infects vertebrates (humans, dogs, cattle, birds); 10 dsRNA segments (e.g., reoviruses of humans)

Orbivirus—replicates in vertebrates and their insect vectors; 10 dsRNA segments (e.g., bluetongue virus of sheep)

Rotavirus—name comes from its wheel-like particle with spokes; causes life-threatening diarrhea in very young vertebrates of many species, including humans; 11 dsRNA segments

Aquaerovirus—viruses of fish and *Crustacea;* 11 dsRNA segments

Coltivirus—Colorado tick fever virus of vertebrates; 12 dsRNA segments

Cypovirus—cytoplasmic polyhedrosis viruses of insects (including *Lepidoptera, Hymenoptera,* and *Diptera*); 10 dsRNA segments

4. Viruses with positive-sense ssRNA genomes

a. Family: *Caliciviridae*

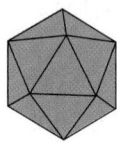

Nonenveloped icosahedral 35 to 40 nm particle with 32 calyxlike (cup-shaped) depressions. Contains one molecule of linear positive-sense ssRNA of 7,500 nucleotides. The 5′ end has a VPg or a cap structure (hepatitis E virus). Cytoplasmic multiplication with subgenomic mRNA. Infects many species (e.g., human calicivirus, Norwalk virus, hepatitis E virus, rabbit hemorrhagic fever virus, swine exanthema virus).

b. Family: *Coronaviridae*

With the *Arteriviridae,* forms the order *Nidovirales.* Enveloped particles of 120 to 160 nm with club-shaped sparse protein spikes. Contains a helical nucleocapsid with one molecule of positive-sense ssRNA of 28,000 to 31,000 nucleotides. One of the largest RNA genomes. Cytoplasmic replication with a nested set of subgenomic mRNAs with a common leader sequence. Buds from Golgi apparatus and endoplasmic reticulum.

Genera:

Coronavirus (e.g., avian infectious bronchitis virus and human coronavirus OC43)

Torovirus—biconcave or toroidal or doughnut-shaped particles (Berne virus of horses and Breda virus of cattle)

c. Family: *Flaviviridae*

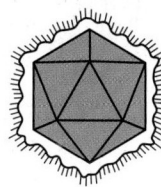

Enveloped 40 to 60 nm particles with an isometric nucleocapsid of 25 to 30 nm and one molecule of linear positive-sense ssRNA of 9,500 to 12,500 nucleotides. Differ from *Alphaviridae* by the presence of a matrix protein, the lack of subgenomic mRNAs, and budding from the endoplasmic reticulum. Cytoplasmic multiplication. Spread by arthropods unless stated.

Genera:

Flavivirus—large group of viruses that multiply in the vertebrate host and insect or tick vector (e.g., yellow fever virus, tick-borne encephalitis virus group, dengue virus group, Japanese encephalitis virus group)

Pestivirus—includes border disease virus (sheep), bovine diarrhea virus, classic swine fever (hog cholera) virus; no vector

Hepacivirus—hepatitis C virus of humans; no vector

d. Family: *Picornaviridae*

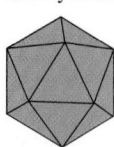

Nonenveloped viruses of mainly vertebrates. One molecule of linear positive-sense ssRNA of 7,000 to 8,500 nucleotides in 30 nm icosahedral particles. Has a 5′ VPg protein. Cytoplasmic multiplication.

Genera:

Enterovirus—acid resistant, primarily of the intestinal tract (e.g., polioviruses, most echoviruses, coxsackieviruses of humans and various nonhuman enteroviruses)

Aphthovirus—economically important foot-and-mouth disease viruses of cattle and other ruminants

Cardiovirus (e.g., encephalomyocarditis [EMC] virus of mice)

Hepatovirus (e.g., human and simian hepatitis A viruses)

Parechovirus (e.g., human echoviruses 22 and 23)

Rhinovirus—acid labile; infects the upper respiratory tract; includes about 100 common cold viruses; also equine rhinoviruses and various unclassified viruses of insects

e. Family: *Tetraviridae*

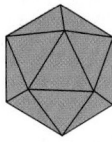

Nonenveloped 40 nm icosahedral particle with one molecule (6,500 nucleotides) or two molecules (2,500 and 5,300 nucleotides) of linear positive-sense ssRNA. No infection of cultured cells. All isolated from *Lepidoptera* (e.g., Naudaurelia β virus)

f. Family: *Togaviridae*

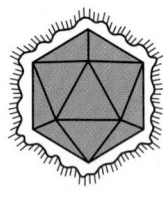

Enveloped 70 nm particles containing an icosahedral nucleocapsid and one molecule of linear positive-sense ssRNA of 9,000 to 12,000 nucleotides. Cytoplasmic multiplication; buds from the plasma membrane. Has an intracellular subgenomic mRNA.

Genera:

Alphavirus—multiplies in vertebrate host and insect vector (e.g., Sindbis virus and Sernliki forest virus)

Rubivirus—rubella virus of humans; no vector

5. Viruses with negative-sense/ambisense ssRNA genomes and a virion-associated RNA-dependent RNA polymerase

a. Family: *Arenaviridae*

Enveloped isometric usually 100 nm particles with club-shaped spikes. Genome is contained in two circular helical nucleocapsids, the larger with one molecule of linear negative-sense ssRNA of 7,500 nucleotides and the smaller, also linear, of 3,500 nucleotides. Both are ambisense. May package more than two nucleocapsids per virion. Virion contains host cell ribosomes for no known function. Cytoplasmic multiplication; buds from plasma membrane. Divided into the LCM/LASV subgroup (Old World arenaviruses: lymphocytic choriomeningitis virus, lassa virus and related viruses) and the Tacaribe complex subgroup (New World arenaviruses: Tacaribe, Junin, Pichinde and related viruses).

b. Family: *Bornaviridae*

Enveloped isometric 90 nm virions with a helical nucleocapsid and one molecule of linear negative-sense ssRNA of 9,000 nucleotides. Nuclear replication. Recently isolated from people with behavioral/neuropsychiatric problems but not known if causal. Known to infect horses since the eighteenth century and several other vertebrate species. Natural host unknown. Borna disease virus.

c. Family: *Bunyaviridae*

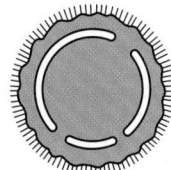

Enveloped isometric 100 nm particle with 10 nm spikes. Contains three helical nucleocapsids, each with one molecule of linear negative-sense ssRNA—large: 6,000 to 12,000 nucleotides, medium: 3,500 to 6,000 nucleotides, and small: 1,000 to 2,000 nucleotides. Cytoplasmic multiplication; buds from the Golgi membranes. Infects vertebrates and spread by arthropods unless stated. Also a genus that infects plants.

Genera:
Bunyavirus—mosquito and gnat vectors (e.g., Bunyamwera and 150 or so other viruses)
Hantavirus—no arthropod vector (e.g., Hantaan virus)
Nairovirus—tick vector (e.g., Nairobi sheep disease virus)
Phlebovirus—S RNA is ambisense; sandfly, tick and gnat vectors (e.g., sandfly fever virus, uukuniemi virus)

d. Family: *Filoviridae*

Genus:

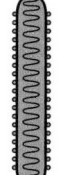

Filovirus—enveloped long filamentous, sometimes branched, particles 800 to 900 (sometimes 14,000) nm × 80 nm with helical nucleocapsid of 50 nm diameter. One molecule of negative-sense ssRNA of 19,000 nucleotides. Buds from plasma membrane. Includes Marburg, Ebola, and Reston viruses. Highly pathogenic for humans. Natural reservoir not known.

e. Family: *Orthomyxoviridae*

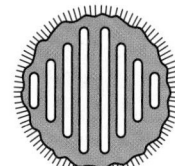

Enveloped pleomorphic 100 nm (sometimes filamentous) particles with a dense layer of protein spikes. Six to eight helical nucleocapsids, 9 nm in diameter with transcriptase activity. Each contains one molecule of linear negative-sense ssRNA, totalling 12,000 to 15,000 nucleotides. All RNA synthesis is nuclear. Within a genus, RNA segments in a mixed infection readily assort to form genetically stable hybrid viruses. Plasma membrane budding.

Genera:
Influenzavirus A—genome comprises eight molecules of RNA ranging from 900 to 2,350 nucleotides. Virions have separate hemagglutinin and neuraminidase spike proteins. Undergoes antigenic shift and drift. Natural reservoir is seashore birds; infects several vertebrate species including humans.
Influenzavirus B—genome comprises eight molecules of RNA ranging from 900 to 2,350 nucleotides. Virions have separate hemagglutinin and neuraminidase spike proteins. Undergoes antigenic drift only. Infects humans.
Influenzavirus C—genome comprises seven molecules of RNA ranging from 1,000 to 2,350 nucleotides. Virions have one hemagglutinin–esterase fusion spike protein. The esterase is a receptor-destroying enzyme. Undergoes minor antigenic variation. Infects humans.
Thogotovirus—carried by ticks and occasionally infects humans. Genome comprises six molecules of RNA. (Thogoto virus and Dhori virus; Asia, Africa, and Europe.)
Possibly also another genus for infectious salmon anemia virus.

f. Family: *Paramyxoviridae*

Enveloped pleomorphic particle usually 150 to 200 nm in diameter. Has a dense layer of fusion protein spikes and attachment protein spikes. Contains one helical nucleocapsid 13 to 18 nm in diameter with one molecule of linear negative-sense ssRNA of 13,000 to 16,000 nucleotides. Filamentous forms of up to 400 nm common. Cytoplasmic multiplication; buds from plasma membrane. Infects vertebrates. Most, but not all, are respiratory viruses.

Subfamily: *Paramyxovirinae*

Genera:

Respirovirus (e.g., human parainfluenzavirus type 1)

Morbillivirus—measles virus, rinderpest virus, and canine distemper virus

Rubulavirus (e.g., mumps virus, Newcastle disease virus of poultry or avian parainfluenzavirus type 1)

Subfamily: *Pneumovirinae*

Genera:

Pneumovirus (e.g., human and bovine respiratory syncytial viruses, pneumonia virus of mice)

Metapneumovirus (e.g., turkey rhinotracheitis virus)

g. Family: *Rhabdoviridae*

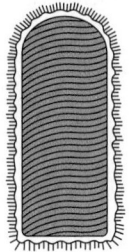

Enveloped, bullet-shaped particles 100 to 430 nm × 45 to 100 nm, containing one molecule of linear negative-sense ssRNA of 11,000 to 15,000 nucleotides. Has 5–10 nm spikes. Inside is a helical nucleocapsid. Cytoplasmic multiplication and buds from plasma membrane. Infects vertebrates. Also genera that infect plants.

Genera:

Vesiculovirus (e.g., vesicular stomatitis virus)

Lyssavirus (e.g., rabies virus, unusual as infects all mammals)

Ephemerovirus (e.g., bovine ephemeral fever virus)

Novirhabdovirus (e.g., infectious hemopoietic virus of fish)

6. Viruses with RNA genomes that replicate through a DNA intermediate

a. Family: *Retroviridae*

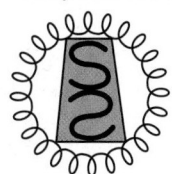

Enveloped 80 to 100 nm particles with spikes. Nucleocapsid can be isometric or a truncated cone; contains two identical copies of a linear positive-sense ssRNA of 7,000 to 11,000 nucleotides. Virions contain reverse transcriptase and integrase enzymes. The DNA provirus is nuclear and integrated with host DNA. Transmission is horizontal or vertical. Associated with many different diseases. Not all viruses are oncogenic.

Genera:

"Mammalian type C retroviruses" (e.g., murine leukemia virus); oncogenic

Alpharetrovirus (e.g., avian leukosis virus, Rous sarcoma virus); oncogenic

Betaretrovirus (e.g., Mason-Pfizer monkey virus)

Gammaretrovirus (e.g., mouse mammary tumour virus); oncogenic

Deltaretrovirus (e.g., bovine leukemia virus, human T-cell lymphotrophic virus types 1–3); oncogenic

Epsilonretrovirus (e.g., walleye dermal sarcoma virus of fish)

Lentivirus—infects primates (human immunodeficiency virus types 1 and 2, simian immunodeficiency viruses), horses, sheep, goats, cattle, and cats; all associated with immunodeficiency diseases

Spumavirus (e.g., human spumaviruses); named after their foamy cytopathogenic effect; no disease known

7. Viruses with a DNA genome that replicate through an RNA intermediate

a. Family: *Hepadnaviridae*

A 40 to 48 nm enveloped particle containing an isometric nucleocapsid with DNA polymerase and protein kinase activities. One partially double-stranded circular DNA molecule that is not covalently closed. This has a complete negative-sense strand of 3,000 nucleotides with a 5'-terminal protein, and a variable length positive-sense strand of 1,700 to 2,800 nucleotides. The circularization overlaps the 3' and 5' termini of the negative-sense DNA. These are "reversiviruses" that have a reverse transcriptase.

Genera:

Orthohepadnavirus (e.g., hepatitis B [HBV] of humans, which is strongly associated with liver cancer, and woodchuck hepatitis virus)

Avihepadnavirus (e.g., duck hepatitis B virus)

B. VIRUSES THAT MULTIPLY IN PLANTS

Knowledge of plant virus multiplication is harder to obtain than that of animal viruses because plant cell culture systems are less manageable. Greater emphasis is given to physical properties and disease characteristics, especially as many are important in agriculture. Differences in virus proteins, translation strategy (which may not be mentioned here), and vector are important criteria in plant virus classification. Some plant viruses also multiply in their animal vector (*Bunyaviridae*, *Marafivirus*, *Reoviridae*, *Rhabdoviridae*), so the plant/animal virus distinction becomes uncertain.

1. Viruses with ssDNA genomes

a. Family: *Geminiviridae*

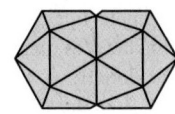

Virions comprise two incomplete icosahedra, 30 nm long × 18 nm, joined as Siamese twins. Nonenveloped. Has closed circular ssDNA. Nuclear replication. Persistent or not in an insect vector but does not multiply in the vector.

Genera:

Curtovirus—genome is one DNA molecule of 3,000 nucleotides. Does not infect monocotyledonous plants (grasses). Whitefly vector (e.g., beet curly top virus).

Mastrevirus—genome is one DNA molecule of 2,600 nucleotides. Infects mainly monocotyledonous plants (grasses). Leafhopper vector (e.g., maize streak virus).

Begomovirus—most genomes are two molecules of DNA each of 2,500 nucleotides. Whitefly vector (e.g., bean golden mosaic virus).

2. Viruses with dsRNA genomes and a virion-associated RNA-dependent RNA polymerase

a. Family: *Reoviridae*

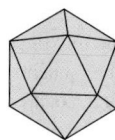

Large and widespread family. Genera also found in vertebrates and insects. Nonenveloped 60 to 80 nm icosahedral particles containing an isometric nucleocapsid with 10 to 12 segments of linear dsRNA of 18,000 to 30,000 bp. Has a transcriptase activity. Cytoplasmic multiplication. Within a genus RNA segments in a mixed infection readily assort to form genetically stable hybrid virus.

Genera:

Fijivirus—10 dsRNA segments. Multiplies in leafhopper insect vector (e.g., Fiji disease virus).

Oryzavirus—10 dsRNA segments. Multiplies in insect vector (e.g., rice ragged stunt virus).

Phytoreovirus—12 dsRNA segments. Multiplies in leafhopper insect vector (e.g., wound tumour virus of clover).

3. Viruses with positive-sense ssRNA genomes
Isometric virions

a. Family: *Comoviridae*

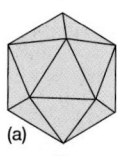
(a)

Two molecules of linear positive-sense ssRNA, each encapsidated separately in a nonenveloped 30 nm icosahedral particle. Both RNAs needed for infectivity. Cytoplasmic multiplication.

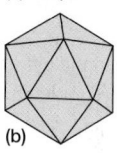
(b)

Genera:

Comovirus—RNAs of <7,000 and 3,981 nucleotides. Nonpersistent in beetle vector (e.g., cowpea mosaic virus).

Fabavirus—RNAs of <7,000 and 4,500 nucleotides. Nonpersistent in aphid vector (e.g., broad-bean wilt virus 1).

Nepovirus—RNAs of >7,000 and 4,000 nucleotides. Does not multiply in nematode vector or not vectored (e.g., tobacco ringspot virus).

b. Family: *Luteoviridae*

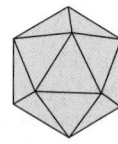

Nonenveloped icosahedral 25 to 30 nm particles with one molecule of linear positive-sense ssRNA of 5,500 to 7,000 nucleotides. Does not multiply in aphid vector, but persists.

Genera:

Enamovirus—no VPg protein (e.g., pea enation mosaic virus)

Luteovirus—no VPg (e.g., barley yellow dwarf virus)

Polerovirus—RNA has a 5′ VPg (e.g., potato leafroll virus)

c. Family: *Tombusviridae*

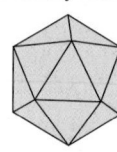

Nonenveloped isometric 32 to 35 nm particle with one or two molecules of linear positive-sense ssRNA totaling 4,000 to 5,000 nucleotides. Cytoplasmic multiplication.

Genera:

Aureusvirus—one RNA. Soil transmitted; no vector (e.g., *Pothos* latent virus).

Avenavirus—one RNA. Transmitted mechanically from plant to plant and possibly also by fungi (e.g., oat chlorotic stunt virus).

Carmovirus—one RNA. Soilborne transmission without vector (e.g., carnation mottle virus).

Dianthovirus—two RNAs (4,000 and 1,500 nucleotides) in one particle. Soilborne transmission without vector (e.g., carnation ringspot virus).

Machlomovirus—one RNA. Infects only grasses. Mechanical or beetle transmission (e.g., maize chlorotic mottle virus).

Necrovirus—one RNA. Fungal vector (e.g., tobacco necrosis virus A).

Panicovirus—one RNA. Transmitted mechanically (e.g., *Panicum* mosaic virus).

Tombusvirus—one RNA. Transmitted mechanically or by grafting (e.g., tomato bushy stunt virus).

Isometric virions and virions that are short rods

d. Family: *Bromoviridae*

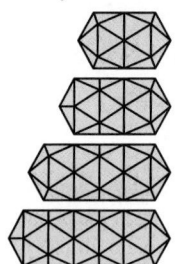

Nonenveloped particles. Linear positive-sense ssRNA. Tripartite genome of 2,000 to 3,000 nucleotides (total 8,000 nucleotides). RNA 4 is coat protein mRNA encoded by RNA 3. Cytoplasmic multiplication.

Genera:

Alfamovirus—four RNAs in four bacilliform particles (short rods) 56, 43, 35, and 30 nm × 18 nm. Particles each with one molecule of RNA 1 (3,644 nucleotides) or RNA 2 (2,593 nucleotides) or RNA 3 (2,037 nucleotides) or two molecules of RNA 4 (881 nucleotides). Infectivity requires RNAs 1 to 3 + coat protein or RNA 4. Cytoplasmic multiplication. Nonpersistent in aphid vector (e.g., alfalfa mosaic virus).

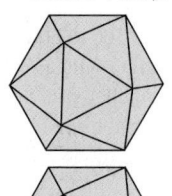
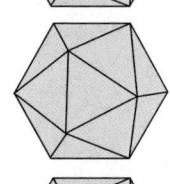
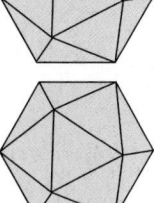

Bromovirus—four RNAs in three isometric 27 nm particles. Particles each with one molecule of RNA 1 (3,234 nucleotides) or RNA 2 (2,865 nucleotides) or RNA 3 (2,117 nucleotides) + RNA 4 (800 nucleotides). Infectivity requires RNAs 1 to 3 + coat protein or RNA 4. Beetle vectors (e.g., brome mosaic virus).

Virions that are rigid rods

These genera have not yet been assigned to a family:

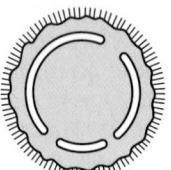

e. *Tobamovirus*—Nonenveloped rod, 300 nm × 18 nm, with helical symmetry. Contains one molecule of linear positive-sense ssRNA of 6,395 nucleotides. Transmitted mechanically or by seed (e.g., tobacco mosaic virus).

f. *Tobravirus*—two nonenveloped rods, 200 nm × 22 nm and 46 to 115 nm × 22 nm, with helical symmetry. Contains, respectively, one molecule of linear positive-sense ssRNA of 7,000 or 2,000 to 4,000 nucleotides. The larger RNA alone is infectious and the smaller one specifies the coat protein. Both are needed for synthesis of new virions. Cytoplasmic multiplication. Nematode vector (e.g., tobacco rattle virus).

Virions that are flexuous rods (i.e., with bends but not necessarily flexible)

g. Family: *Potyviridae*

Nonenveloped flexuous rod(s) with helical symmetry. There are one or two molecules of linear positive-sense ssRNA. Cytoplasmic multiplication. Genera have different vectors.

Genera:

Potyvirus—particle of 700 to 900 nm × 12 nm with one RNA molecule of 10,000 nucleotides. A VPg at the 5′ end. Nonpersistent in aphid vector (e.g., potato virus Y).

Ipomovirus—one particle of 800 to 950 nm × 12 to 16 nm with one RNA molecule of 11,000 nucleotides. Whitefly vector (e.g., sweet potato mild mottle virus).

Macluravirus—nonenveloped flexuous rod, 700 nm × 13 to 16 nm, with one molecule of RNA of 8,000 nucleotides. Nonpersistent in aphid vector (e.g., maclura mosaic virus).

Rymovirus—one particle of 700 nm × 11 to 15 nm with one molecule of RNA of 8,500 to 10,000 nucleotides. Mite vector (e.g., ryegrass mosaic virus).

Tritimovirus—one particle of 700 nm × 13 nm with one molecule of RNA of 9,672 nucleotides. Persistent in mite vector (e.g., wheat streak mosaic virus).

Bymovirus—two particles of 500 to 600 and 250 to 300 nm × 13 nm. The larger with an RNA of 7,632 nucleotides and the smaller with an RNA of 3,585 nucleotides. Fungal vector (e.g., barley yellow mosaic virus).

4. **Viruses with negative-sense/ambisense ssRNA genomes and a virion-associated RNA-dependent RNA polymerase**

a. Family: *Bunyaviridae*

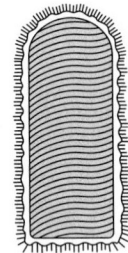

Enveloped isometric 100 nm particle with 10 nm spikes. Each contains three helical nucleocapsids with linear negative-sense ssRNA. The largest nucleocapsid comprises an RNA of 6,000 to 12,000 nucleotides, the medium of 3,500 to 6,000 nucleotides, and the smallest of 1,000 to 2,000 nucleotides. Cytoplasmic multiplication; buds from the Golgi membranes. Spread by arthropods unless stated. Also genera that infect animals.

Genus:

Tospovirus—M and S RNAs are ambisense. Transmitted by insect (thrips) vector. Multiplies in vector (e.g., tomato spotted wilt virus).

b. Family: *Rhabdoviridae*

Enveloped, bullet-shaped particles 100 to 430 nm long and 45 to 100 nm in diameter, containing a helical nucleocapsid with one molecule of linear negative-sense ssRNA of 11,000 to 15,000 nucleotides. Has 5 to 10 nm spikes. Also genera that infect animals.

Genera:

Cytorhabdovirus—plant viruses that mature in the cytoplasm. Spread by (and multiply in) insect vector (e.g., lettuce necrotic yellows virus).

Nucleorhabdovirus—plant viruses that mature in the nucleus. Spread by (and multiply in) insect vector (e.g., potato yellow dwarf virus).

5. **Viruses with DNA genomes that replicate through an RNA intermediate**

a. Family: *Caulimoviridae*

A nonenveloped particle containing one partially double-stranded circular DNA molecule that is not covalently closed (like that of hepadnaviruses), with a complete negative-sense strand of 7,000 to 8,000 nucleotides and an incomplete positive-sense strand with one to three discontinuities. The circularization, overlaps the 3′ and 5′ termini of the negative-sense DNA. These are "reversiviruses" that have a reverse transcriptase. Replication is nuclear and the genome remains episomal and does not integrate.

Genera:

Caulimovirus—icosahedral 50 nm particles. Nonpersistent in aphid vectors (e.g., cauliflower mosaic virus).

Badnavirus—rod-shaped particles 130 nm × 30 nm. Nonpersistent in beetle larvae (e.g., commelina yellow mottle virus).

C.   VIRUSES MULTIPLYING IN ALGAE, FUNGI, AND PROTOZOA

1. Viruses with dsDNA genomes

a. Family: *Adenoviridae*

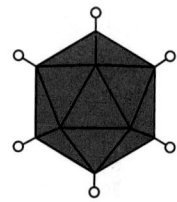

Nonenveloped icosahedral particles of 60 to 90 nm with a fiber protein from each vertex. Contains one molecule of linear ds-DNA of 25,000 to 42,000 bp. Replicate and assembled in the nucleus. Other genera infect animals.

(Putative) genus:

Rhizidiovirus—60 nm virion with one molecule of linear dsDNA of 25,000 bp. Infects fungi (e.g., *Rhizidiomyces* virus).

b. Family: *Phycodnaviridae*

Infects algae. Nonenveloped, isometric, 130 to 190 nm, multilayered particles containing one molecule of linear dsDNA of 160,000 to 380,000 bp. Contain internal lipid. Infects *Paramecium, Chlorella,* and *Hydra* spp. May have a use in controlling algal blooms.

2. Viruses with dsRNA genomes and a virion-associated RNA-dependent RNA polymerase

a. Family: *Hypoviridae*

Genus:

Hypovirus—vesicles of 50 to 80 nm with one molecule of linear dsRNA of 9,000 to 13,000 bp, no structural proteins and polymerase activity. Surrounded by rough endoplasmic reticulum in infected cells. Causes hypovirulence of host chestnut blight fungus (e.g., *Cryphonectria* hypovirus 1).

b. Family: *Partitiviridae*

Two molecules of linear dsRNA of 1,400 to 3,000 bp packaged in one or several nonenveloped isometric particles that lack structural detail. Have a transcriptase. Cytoplasmic multiplication. Also genera that infect plants.

Genera:

Partitivirus—particles 30 to 35 nm. Infects fungi, (e.g., *Gaeumannomyces graminis* virus 019/6A).
Chrysovirus—particles 35 to 40 nm. Infects fungi (e.g., *Penicillium chrysogenum* virus).

c. Family: *Totiviridae*

Nonenveloped particles of 30 to 40 nm with one molecule of linear dsRNA of 5,000 to 7,000 bp. One major capsid protein. Single shell. Cytoplasmic multiplication.

Genera:

Totivirus—some viruses also have other satellite RNA segments that encode killer proteins. Infects the fungus *Saccharomyces cerevisiae.*
Giardiavirus—infects the protozoan *Giardia lamblia.*
Leishmaniavirus—one RNA. Infects protozoa: *Leishmania* spp.

3. Viruses with positive-sense ssRNA genomes

a. Family: *Barnaviridae*

Nonenveloped icosahedral short rod, 50 nm × 19 nm, with one molecule of linear ssRNA of 4,009 nucleotides. Infects fungi (e.g., mushroom bacilliform virus).

b. Family: *Narnaviridae*

One molecule of linear positive-sense ssRNA of 2,500 nucleotides as a ribonucleoprotein with polymerase activity. Infects fungi.

Genera:

Narnavirus (e.g., *Saccharomyces cerevisiae* 20S narnavirus)
Mitovirus (e.g., *Cryphonectria parasitica* mitovirus 1 NB631)

D.   VIRUSES (PHAGES) MULTIPLYING IN *ARCHAEA,* BACTERIA, *MYCOPLASMA,* AND *SPIROPLASMA*

The well-known molecular biology of phages is based on a detailed study of a few representatives, and surprisingly little is known of their comparative biology.

1. Viruses with dsDNA genomes

Viruses that have some head-tail structure, here ordered by decreasing tail length

a. Family: *Siphoviridae*

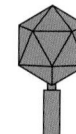

Nonenveloped particle with a long noncontractile tail up to 570 nm × 8 nm. Icosahedral head of 60 nm. Contains one molecule of linear dsDNA of 48,500 bp. Causes no host DNA breakdown. Infects bacteria including enterobacteria, mycobacteria, *Lactococcus, Methanobacterium,* and *Vibrio.* Includes coliphage T1, lambda (λ), chi (χ), and phi (φ) 80.

b. Family: *Myoviridae*

Nonenveloped particle with a complex contractile tail 80 to 455 nm × 16 nm. Contraction requires ATP. Head separated from tail by a neck. Head isometric (as shown) or elongated, 110 nm × 80 nm. Has one molecule of linear dsDNA of 40,000 to 170,000 bp. Infects enterobacteria, *Bacillus,* and *Halobacterium* spp. (*Archaea*). Includes the T-even coliphages T2, T4, and T6, and PBS1, SP8, SP50, P1, P2, 21, 34, and Mu.

c. Family: *Podoviridae*

Nonenveloped particle with a short noncontractile tail of 20 nm × 8 nm. Icosahedral head of 60 nm. One molecule of linear dsDNA of 19,000 to 44,000 bp. Causes host DNA breakdown. Includes the coliphages T3 and T7, enterobacteria phage P22, and bacillus phage phi (ϕ) 29.

Viruses that do not have a head-tail structure

d. Family: *Fuselloviridae*

Enveloped lemon-shaped particle 100 nm × 60 nm with short tail fibers at one pole. One molecule of circular dsDNA of 15,465 bp. Infects *Archaea* such as *Desulfolobus* and *Methanococcus* (e.g. *Sulfolobus* virus 1).

e. Family: *Tectiviridae*

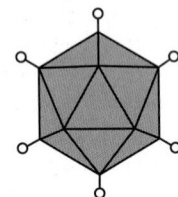

Nonenveloped icosahedral 63 nm particle. Unusual enveloped nucleoprotein surrounded by lipid. One molecule of linear dsDNA of 150,000 bp. On attachment forms a tail-like tube 60 nm × 10 nm. Infects gram-negative bacteria (e.g., enterobacteria phage PRD1). Has similarities with adenoviruses.

f. Family: *Corticoviridae*

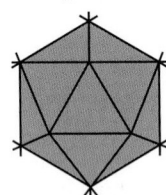

Nonenveloped, nontailed icosahedral 60 nm particle formed of several layers, including one of lipid. Spikes at vertices. One molecule of circular dsDNA of 9,000 bp. Infects *Pseudomonas* (e.g., *Alteromonas* phage PM2).

g. Family: *Plasmaviridae*

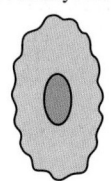

Enveloped pleomorphic 80 nm particle with small dense core; 50 and 125 nm particles also produced. One molecule of circular dsDNA of 12,000 bp. Infects *Mycoplasma* (e.g., *Acholeplasma* phage L2).

h. Family: *Rudiviridae*

Nonenveloped rigid rod of 500 to 780 nm × 23 nm with helical symmetry, with a plug and three tail fibers at each end. One molecule of linear dsDNA of 33,000 bp. Non-lytic. Infects thermophilic *Archaea, Sulfolobus* (e.g., *Sulfolobus* virus SIRV-1).

i. Family: *Lipothrixviridae*

Enveloped, rigid rod of 400 nm × 40 nm with protrusions at both ends that participate in cell attachment. One molecule of linear dsDNA of 16,000 bp. Virions stable at 100°C. Infects *Archaea* (e.g., *Thermoproteus* virus 1).

2. **Viruses with ssDNA genomes**

a. Family: *Inoviridae*

Nonenveloped rod with one molecule of circular positive-sense ssDNA of 5,000 to 10,000 nucleotides. Virion length depends on the length of the DNA and its conformation.

Genera:

Inovirus—flexuous rods of 700 to 2,000 nm × 7 nm with DNA of 4,400 to 8,500 nucleotides. Host bacteria not lysed. Includes enterobacteria phage M13 and enterobacteria phage fd.

Plectrovirus—nearly straight rods of 85 or 280 nm × 10 to 15 nm with DNA of 4,400 to 8,500 nucleotides. Infects *Mycoplasma* (85 nm) and *Spiroplasma* (280 nm) (e.g. *Acholeplasma* phage MV-L51).

b. Family: *Microviridae*

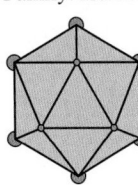

Icosahedral particle with large knobs on the 12 vertices. Minimum and maximum diameters are 22 and 23 nm, respectively. One molecule of circular ssDNA of 4,400 to 5,400 nucleotides.

Genera:

Microvirus—infects bacteria (e.g., enterobacteria phage phi(ϕ)X174)

Spiromicrovirus—infects *Spiroplasma* (e.g., *Spiroplasma* phage 4)

Chlamydiamicrovirus (e.g., *Chlamydia* phage 1)

Bdellomicrovirus (e.g., *Bdellovibrio* phage MAC 1)

3. **Viruses with dsRNA genomes and a virion-associated RNA-dependent RNA polymerase**

a. Family: *Cystoviridae*

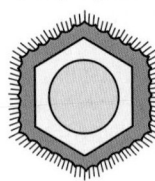

Enveloped icosahedral 85 nm particle with 8 nm spikes. Has a 58 nm nucleocapsid and a 43 nm core. Each particle contains three molecules of linear dsRNA of 6,374, 4,057, and 2,948 bp. Infects *Pseudomonas* (e.g., *Pseudomonas* phage phi[ϕ]6).

4. Viruses with positive-sense ssRNA genomes

a. Family: *Leviviridae*

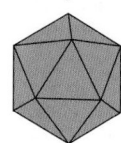

Nonenveloped 26 nm icosahedral particles with one molecule of linear positive-sense ssRNA of 3,400 to 4,300 nucleotides. Include enterobacteria phages R17, MS2, and Qβ.

 E.

SATELLITE VIRUSES AND SATELLITE NUCLEIC ACIDS OF ANIMALS, PLANTS, FUNGI, AND BACTERIA

The replication of satellite viruses and satellite nucleic acids depends upon coinfection of a host cell with a **helper virus.** The satellite genome has no significant homology with that of the helper and hence differs from other types of dependent nucleic acid molecules. Here, we use a broad definition of a satellite virus—one that is incapable of independent production of progeny virus particles, yet may be able to replicate itself. **Satellite viruses** encode their own coat protein, whereas **satellite nucleic acids** do not and use the coat protein of their helper virus. Some satellite viruses/satellite nucleic acids modulate the replication of the helper virus and exacerbate or diminish disease.

1. Satellite nucleic acids with dsDNA genomes

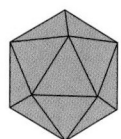

Nonenveloped isometric virions with a *Myoviridae* helper (enterobacteria phage P2). One molecule of linear dsDNA of 11,627 bp. Interesting that the head-tail helper virion protein can form an isometric virion (e.g., enterobacteria P4 satellite).

2. Satellite viruses with ssDNA genomes

a. Family: *Parvoviridae*

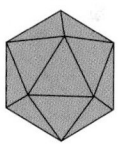

Nonenveloped icosahedral particle of 18 to 26 nm that contains no enzyme. One 5,000 nucleotide linear molecule of *either* negative- *or* positive-sense ssDNA per particle. On extraction these form an artifactual double strand. Nuclear replication. There are also independently replicating (autonomous) parvoviruses of animals.

Subfamily: *Parvovirinae*

Genus:
Dependovirus—dependent on helper adenovirus or herpesvirus for efficient replication, but in some cell cultures replicate independently. Particles package 90% negative-sense DNA. Encode their own coat protein. Adeno-associated viruses. Infects vertebrates.

3. Satellite nucleic acids with ssDNA genomes
A genome of 682 nucleotides with no ORF. Geminiviruses helpers (e.g., tomato leaf curl virus satellite DNA). Infects plants. Particle morphology not yet determined.

4. Satellite nucleic acids with dsRNA genomes

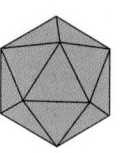

Nonenveloped isometric particles with a *Totiviridae* helper virus. Linear dsRNA of 500 to 1,800 bp inside a virion made of helper virus coat protein (e.g., satellite of *Saccharomyces cerevisiae* M virus). Infects fungi.

5. Satellite viruses with positive-sense ssRNA genomes

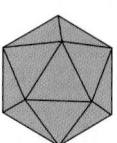

Isometric particles with one linear ssRNA.
Tobacco necrosis virus satellite virus subgroup:
   Particles of 17 nm containing one molecule of RNA of 1,239 nucleotides. Fungal vector. Infects plants.
Chronic bee-paralysis virus-associated satellite virus subgroup:
   Particles of 12 nm containing three molecules of RNA totalling 1,000 nucleotides. Infects animals.

6. Satellite nucleic acids with positive-sense ssRNA genomes

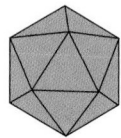

Large linear mRNA satellites:
   Encode a non-structural protein. Rarely modify disease. Linear RNA of 800 to 1,500 nucleotides (e.g., arabis mosaic virus large satellite RNA). Infects plants.

7. Satellite nucleic acids with negative-sense ssRNA genomes

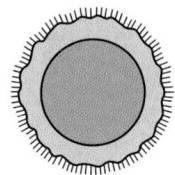

Deltavirus—hepatitis delta virus. Enveloped isometric particle of 36 nm with one circular RNA molecule of 1,700 nucleotides. Dependent on hepatitis B virus, which contributes the envelope, but expresses two delta antigen proteins that form the core. RNA has self-cleaving (ribozyme) activity. May exacerbate hepatitis B virus infection. Infects humans.

8. Satellite nucleic acids with ssRNA genomes (unclassified as make no mRNA)

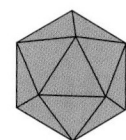

Have no functional ORF.
Small linear RNA satellites:
   Linear RNA of 340 nucleotides (e.g., cucumber mosaic virus satellite RNA). Infects plants.

Circular RNA satellites:
   Covalently closed circular RNA of usually 350 nucleotides. Smallest is 220 nucleotides. All form a double-stranded rodlike viroid RNA. This is encapsidated. Also called virusoids. Have self-cleaving (ribozyme) activity. Five types, all with sobemovirus helpers (e.g., velvet tobacco mottle virus satellite RNA). Infects plants.

Recommended Readings

Chapter 1: The History and Scope of Microbiology

General References

American Society for Microbiology. 1999. Celebrating a century of leadership in microbiology. *ASM News* 65(5).

Beck, R. W. 2000. *A chronology of microbiology in historical context.* Washington, D.C.: ASM Press.

Brock, T. D. 1961. *Milestones in microbiology.* Englewood Cliffs, N.J.: Prentice-Hall.

Bulloch, W. 1979. *The history of bacteriology.* New York: Dover.

Collard, P. 1976. *The development of microbiology.* New York: Cambridge University Press.

de Kruif, P. 1937. *Microbe hunters.* New York: Harcourt, Brace.

Geison, G. L. 1995. *The private science of Louis Pasteur.* Princeton, N.J.: Princeton University Press.

Hellemans, A., and Bunch, B. 1988. *The timetables of science.* New York: Simon and Schuster.

Lechevalier, H. A., and Solotorovsky, M. 1965. *Three centuries of microbiology.* New York: McGraw-Hill.

McNeill, W. H. 1976. *Plagues and peoples.* Garden City, N.Y.: Anchor Press/Doubleday.

Ruestow, E. G. 1996. *The microscope in the Dutch republic: The shaping of discovery.* New York: Cambridge University Press.

Stanier, R. Y. 1978. What is microbiology? In *Essays in microbiology,* J. R. Norris and M. H. Richmond, editors, 1/1–1/32. New York: John Wiley and Sons.

Summers, W. C. 2000. History of microbiology. In *Encyclopedia of microbiology,* vol. 2, J. Lederberg, editor, 677–97. San Diego: Academic Press.

1.1 The Discovery of Microorganisms

Dobell, C. 1960. *Antony van Leeuwenhoek and his "little animals."* New York: Dover.

Ford, B. J. 1981. The Van Leeuwenhoek specimens. *Notes and Records of the Royal Society of London* 36(1):37–59.

Ford, B. J. 1998. The earliest views. *Sci. Am.* 278(4):50–53.

1.2 The Conflict over Spontaneous Generation

Drews, G. 1999. Ferdinand Cohn, a founder of modern microbiology. *ASM News* 65(8):547–53.

Dubos, R. J. 1950. *Louis Pasteur: Free lance of science.* Boston: Little, Brown.

Strick, J. E. 1997. New details add to our understanding of spontaneous generation controversies. *ASM News* 63(4):193–98.

Vallery-Radot, R. 1923. *The life of Pasteur.* New York: Doubleday.

1.3 The Role of Microorganisms in Disease

Brock, T. D. 1988. *Robert Koch: A life in medicine and bacteriology.* Madison, Wis.: Science Tech Publishers.

Fredricks, D. N., and Relman, D. A. 1996. Sequence-based identification of microbial pathogens: A reconsideration of Koch's postulates. *Clin. Microbiol. Rev.* 9(1):18–33.

Hesse, W. 1992. Walther and Angelina Hesse—early contributors to bacteriology. *ASM News* 58(8):425–28.

Hitchens, A. P., and Leikind, M. C. 1939. The introduction of agar-agar into bacteriology. *J. Bacteriol.* 37(5):485–93.

1.4 Industrial Microbiology and Microbial Ecology

Chung, K.-T., and Ferris, D. H. 1996. Martinus Willem Beijerinck (1851–1931): Pioneer of general microbiology. *ASM News* 62(10):539–43.

1.7 The Future of Microbiology

Dixon, B. 1997. Microbiology present and future. *ASM News* 63(3):124–25.

Chapter 2: The Study of Microbial Structure: Microscopy and Specimen Preparation

General References

Rochow, T. G. 1994. *Introduction to microscopy by means of light, electrons, X-rays, or acoustics.* New York: Plenum.

Slayter, E. M. 1992. *Light & electron microscopy.* New York: Cambridge University Press.

2.2 The Light Microscope

Bradbury, S. 1997. *Introduction to light microscopy,* 2d ed. New York: Springer-Verlag.

Perkins, G. A., and Frey, T. G. 2000. Microscopy, optical. In *Encyclopedia of microbiology,* 2d ed., vol. 3, J. Lederberg, editor, 288–306. San Diego: Academic Press.

Stephens, D. J., and Allan, V. J. 2003. Light microscopy techniques for live cell imaging. *Science* 300:82–86.

2.3 Preparation and Staining of Specimens

Clark, G. L., editor. 1973. *Staining procedures used by the Biological Stain Commission,* 3d ed. Baltimore: Williams & Wilkins.

Lillie, R. D. 1969. *H. J. Conn's biological stains,* 8th ed. Baltimore: Williams & Wilkins.

Lippincott-Schwartz, J., and Patterson, G. H. 2003. Development and use of fluorescent protein markers in living cells. *Science* 300:87–91.

Scherrer, Rene. 1984. Gram's staining reaction, Gram types and cell walls of bacteria. *Trends Biochem. Sci.* 9:242–45.

2.4 Electron Microscopy

Koval, S. F., and Beveridge, T. J. 2000. Microscopy, electron. In *Encyclopedia of microbiology,* 2d ed., vol. 3, J. Lederberg, editor, 276–87. San Diego: Academic Press.

Postek, M. T.; Howard, K. S.; Johnson, A. H.; and McMichael, K. L. 1980. *Scanning electron microscopy: A student's handbook.* Burlington, Vt.: Ladd Research Industries.

Wischnitzer, S. 1981. *Introduction to electron microscopy,* 3d ed. New York: Pergamon Press.

2.5 Newer Techniques in Microscopy

Binnig, G., and Rohrer, H. 1985. The scanning tunneling microscope. *Sci. Am.* 253(2):50–56.

Dufrêne, Y. F. 2003. Atomic force microscopy provides a new means for looking at microbial cells. *ASM News* 69(9):438–42.

Kotra, L. P., Amro, N. A., Liu, G.-Y., and Mobashery, S. 2000. Visualizing bacteria at high resolution. *ASM News* 66(11):675–81.

Perkins, G. A., and Frey, T. G. 2000. Microscopy, confocal. In *Encyclopedia of microbiology*, 2d ed., vol. 3, J. Lederberg, editor, 264–75. San Diego: Academic Press.

Weiss, P. 1998. Atom tinkerer's paradise. *Science News* 154:268–70.

Wickramasinghe, H. K. 1989. Scanned-probe microscopes. *Sci. Am.* 261(4):98–105.

Chapter 3: Procaryotic Cell Structure and Function

General References

Goodsell, D. S. 1991. Inside a living cell. In *Trends Biochem. Sci.,* 16:203–6.

Hoppert, M., and Mayer, F. 1999. Prokaryotes. *American Scientist* 87:518–25.

Koch, A. L. 1995. *Bacterial growth and form.* New York: Chapman & Hall.

Koch, A. L. 1996. What size should a bacterium be? A question of scale. *Annu. Rev. Microbiol.* 50:317–48.

Lederberg, J. 2000. *Encyclopedia of microbiology,* 2d ed. San Diego: Academic Press.

Mayer, F. 1986. *Cytology and morphogenesis of bacteria.* Berlin: Gebrüder Borntraeger.

Neidhardt, F. C., editor-in-chief. 1996. *Escherichia coli and Salmonella: Cellular and molecular biology,* 2d ed. Washington, D.C.: ASM Press.

Neidhardt, F. C.; Ingraham, J. L.; and Schaechter, M. 1990. *Physiology of the bacterial cell: A molecular approach.* Sunderland, Mass.: Sinauer Associates.

Schulz, H. N., and Jørgensen, B. B. 2001. Big bacteria. *Annu. Rev. Microbiol.* 55:105–37.

Shapiro, L., and Losick, R. 1997. Protein localization and cell fate in bacteria. *Science* 276:712–18.

3.2 Procaryotic Cell Membranes

Drews, G. 1992. Intracytoplasmic membranes in bacterial cells: Organization, function and biosynthesis. In *Prokaryotic structure and function,* S. Mohan, C. Dow, and J. A. Coles, editors, 249–74. New York: Cambridge University Press.

Ourisson, G.; Albrecht, P.; and Rohmer, M. 1984. The microbial origin of fossil fuels. *Sci. Am.* 251(2):44–51.

3.3 The Cytoplasmic Matrix

Bazylinski, D. A. 1995. Structure and function of the bacterial magnetosome. *ASM News* 61(7):337–43.

Dawes, E. A. 1992. Storage polymers in prokaryotes. In *Prokaryotic structure and function,* S. Mohan, C. Dow, and J. A. Coles, editors, 81–122. New York: Cambridge University Press.

Margolin, W. 1998. A green light for the bacterial cytoskeleton. *Trends Microbiol.* 6(6):233–38.

Stolz, J. F. 1993. Magnetosomes. *J. Gen. Microbiol.* 139:1663–70.

Walsby, A. E. 1977. The gas vacuoles of blue-green algae. *Sci. Am.* 237(2):90–97.

3.4 The Nucleoid

Brock, T. D. 1988. The bacterial nucleus: A history. *Microbiol. Rev.* 52:397–411.

Robinow, C., and Kellenberger, E. 1994. The bacterial nucleoid revisited. *Microbiol. Rev.* 58(2):211–32.

Trun, N. J., and Marko, J. F. 1998. Architecture of a bacterial chromosome. *ASM News* 64(5):276–83.

3.5 The Procaryotic Cell Wall

Beveridge, T. J. 1995. The periplasmic space and the periplasm in gram-positive and gram-negative bacteria. *ASM News* 61(3):125–30.

Ghuysen, J.-M., and Hakenbeck, R., editors. 1994. *Bacterial cell wall.* New York: Elsevier.

Kotra, L. P.; Amro, N. A.; Liu, G.-Y.; and Mobashery, S. 2000. Visualizing bacteria at high resolution. *ASM News* 66(11):675–81.

Navarre, W. W., and Schneewind, O. 1999. Surface proteins of gram-positive bacteria and mechanisms of their targeting to the cell wall envelope. *Microbiol. Mol. Biol. Rev.* 63(1):174–229.

Rietschel, E. T., et al. 1994. Bacterial endotoxin: Molecular relationships of structure to activity and function. *FASEB J.* 8:217–25.

Scherrer, R. 1984. Gram's staining reaction, Gram types and cell walls of bacteria. *Trends Biochem. Sci.* 9:242–45.

3.6 Protein Secretion in Procaryotes

Bernstein, H. D. 2000. The biogenesis and assembly of bacterial membrane proteins. *Curr. Opin. Microbiol.* 3:203–9.

Büttner, D., and Bonas, U. 2002. Port of entry—the type III secretion translocon. *Trends Microbiol.* 10(4):186–92.

Hueck, C. J. 1998. Type III protein secretion systems in bacterial pathogens of animals and plants. *Microbiol. Mol. Biol. Rev.* 62(2):379–433.

Mori, H., and Ito, K. 2001. The Sec protein–translocation pathway. *Trends Microbiol.* 9(10):494–500.

Oliver, D., and Galan, J. 2000. Protein secretion. In *Encyclopedia of microbiology,* 2d ed., vol. 3, J. Lederberg, editor-in-chief, 847–65. San Diego: Academic Press.

3.7 Components External to the Cell Wall

Aizawa, S.-I. 2000. Flagella. In *Encyclopedia of microbiology,* 2d ed., vol. 2, J. Lederberg, editor-in-chief, 380–89. San Diego: Academic Press.

Bayer, M. E., and Bayer, M. H. 1994. Biophysical and structural aspects of the bacterial capsule. *ASM News* 60(4):192–98.

Costerton, J. W.; Geesey, G. G.; and Cheng, K.-J. 1978. How bacteria stick. *Sci. Am.* 238(1):86–95.

DeRosier, D. J. 1998. The turn of the screw: The bacterial flagellar motor. *Cell* 93:17–20.

Harshey, R. M. 2003. Bacterial motility on a surface: Many ways to a common goal. *Annu. Rev. Microbiol.* 57:249–73.

Harshey, R. M., and Toguchi, A. 1996. Spinning tails: homologies among bacterial flagellar systems. *Trends Microbiol.* 4(6):226–31.

Macnab, R. M. 2000. Type III protein pathway exports *Salmonella* flagella. *ASM News* 66(12):738–45.

Mattick, J. S. 2002. Type IV pili and twitching motility. *Annu. Rev. Microbiol.* 56:289–314.

McCarter, L. L. 2001. Polar flagellar motility of the *Vibrionaceae. Microbiol. Mol. Biol. Rev.* 65(3):445–62.

Macnab, R. M. 2003. How bacteria assemble flagella. *Annu. Rev. Microbiol.* 57:77–100.

Mulvey, M. A.; Dodson, K. W.; Soto, G. E.; and Hultgren, S. J. 2000. Fimbriae, Pili. In *Encyclopedia of microbiology,* 2d ed., vol. 2, J. Lederberg, editor-in-chief, 361–79. San Diego: Academic Press.

Sára, M., and Sleytr, U. B. 2000. S-Layer proteins. *J. Bacteriol.* 182(4):859–68.

Sleytr, U. B., and Beveridge, T. J. 1999. Bacterial S-layers. *Trends Microbiol.* 7(6):253–60.

Yonekura, K., et al. 2000. The bacterial flagella cap as the rotary promoter of flagellin self-assembly. *Science* 290:2148–52.

3.8 Chemotaxis

Adler, J. 1976. The sensing of chemicals by bacteria. *Sci. Am.* 234(4):40–47.

Blair, D. F. 1995. How bacteria sense and swim. *Annu. Rev. Microbiol.* 49:489–522.

Manson, M. D.; Armitage, J. P.; Hoch, J. A.; and Macnab, R. M. 1998. Bacterial locomotion and signal transduction. *J. Bacteriol.* 180(5):1009–22.

3.9 The Bacterial Endospore

Driks, A. 2002. Maximum shields: The assembly and function of the bacterial spore coat. *Trends Microbiol.* 10(6):251–54.

Errington, J. 1993. *Bacillus subtilis* sporulation: Regulation of gene expression and control of morphogenesis. *Microbiol. Rev.* 57(1):1–33.

Nicholson, W. L.; Munakata, N.; Horneck, G.; Melosh, H. J.; and Setlow, P. 2000. Resistance of *Bacillus* endospores to extreme terrestrial and extraterrestrial environments. *Microbiol. Mol. Biol. Rev.* 64(3):548–72.

Setlow, P. 1995. Mechanisms for the prevention of damage to DNA in spores of *Bacillus* species. *Annu. Rev. Microbiol.* 49:29–54.

Chapter 4: Eucaryotic Cell Structure and Function

General References

Alberts, B.; Johnson, A.; Lewis, J.; Raff, M.; Roberts, K.; and Walter, P. 2002. *Molecular biology of the cell,* 4th ed. New York: Garland Science.

Becker, W. M.; Kleinsmith, L.; and Hardin, J. 2000. *The world of the cell,* 4th ed. Redwood City, Calif.: Benjamin/Cummings.

Ingber, D. E. 1998. The architecture of life. *Sci. Am.* 278(1):48–57.

Lodish, H.; Baltimore, D.; Berk, A.; Zipursky, S. L.; Matsudaira, P.; and Darnell, J. 1999. *Molecular cell biology,* 4th ed. New York: Scientific American Books.

4.2 The Cytoplasmic Matrix, Microfilaments, Intermediate Filaments and Microtubules

Pumplin, D. W., and Bloch, R. J. 1993. The membrane skeleton. *Trends Cell Biol.* 3:113–17.

Stossel, T. P. 1994. The machinery of cell crawling. *Sci. Am.* 271(3):54–63.

4.3 The Endoplasmic Reticulum

Roy, C. R. 2002. Exploitation of the endoplasmic reticulum by bacterial pathogens. *Trends Microbiol.* 10(9):418–24.

4.4 The Golgi Apparatus

Rothman, J. E. 1985. The compartmental organization of the Golgi apparatus. *Sci. Am.* 253(3):74–89.

Rothman, J. E., and Orci, L. 1996. Budding vesicles in living cells. *Sci. Am.* 274(3):70–75.

4.5 Lysosomes and Endocytosis

Abraham, S. N., and Jeoung-Sook, S. 2001. Caveolae—Not just craters in the cellular landscape. *Science* 293:1447–48.

DeMot, R.; Nagy, I.; Walz, J.; and Baumeister, W. 1999. Proteasomes and other self-compartmentalizing proteases in prokaryotes. *Trends Microbiol.* 7(2):88–92.

Goldberg, A. L., Elledge, S. J., and Harper, J. W. 2001. The cellular chamber of doom. *Sci. Am.* 284(1):68–73.

Mahadevan, L., and Matsudaira, P. 2000. Motility powered by supramolecular springs and ratchets. *Science* 288:95–99.

4.6 Eucaryotic Ribosomes

Craig, E. A.; Gambill, B. D.; and Nelson, R. J. 1993. Heat shock proteins: Molecular chaperones of protein biosynthesis. *Microbiol. Rev.* 57(2):402–14.

Lake, J. A. 1985. Evolving ribosome structure: Domains in archaebacteria, eubacteria, eocytes, and eukaryotes. *Annu. Rev. Biochem.* 54:507–30.

Welch, W. J. 1993. How cells respond to stress. *Sci. Am.* 268(5):56–64.

4.7 Mitochondria

Wallace, D. C. 1997. Mitochondrial DNA in aging and disease. *Sci. Am.* 277(2):40–47.

4.9 The Nucleus and Cell Division

Elledge, S. J. 1996. Cell cycle checkpoints: Preventing an identity crisis. *Science* 274:1664–72.

Glover, D. M.; Gonzalez, C.; and Raff, J. W. 1993. The centrosome. *Sci. Am.* 268(6):62–8.

King, R. W.; Deshaies, R. J.; Peters, J.-M.; and Kirschner, M. W. 1996. How proteolysis drives the cell cycle. *Science* 274:1652–59.

McIntosh, J. R., and McDonald, K. L. 1989. The mitotic spindle. *Sci. Am.* 261(4):48–56.

Murray, A., and Hunt, T. 1993. *The cell cycle: An introduction.* New York: W. H. Freeman.

Stillman, B. 1996. Cell cycle control of DNA replication. *Science* 274:1659–64.

4.11 Cilia and Flagella

Satir, P. 1983. Cilia and related organelles. *Carolina Biology Reader,* no. 123. Burlington, N.C.: Carolina Biological Supply Co.

Chapter 5: Microbial Nutrition

General References

Gottschall, J. C.; Harder, W.; and Prins, R. A. 1992. Principles of enrichment, isolation, cultivation, and preservation of bacteria. In *The prokaryotes,* 2d ed., A. Balows et al., editors, 149–96. New York: Springer-Verlag.

Holt, J. G., and Krieg, N. R. 1994. Enrichment and isolation. In *Methods for general and molecular bacteriology,* 2d ed., P. Gerhardt, editor, 179–215. Washington, D.C.: American Society for Microbiology.

Neidhardt, F. C.; Ingraham, J. L.; and Schaechter, M. 1990. *Physiology of the bacterial cell: A molecular approach.* Sunderland, Mass.: Sinauer.

5.3 Nutritional Types of Microorganisms

Kelly, D. P. 1992. The chemolithotrophic prokaryotes. In *The prokaryotes,* 2d ed., A. Balows et al., editors, 331–43. New York: Springer-Verlag.

5.6 Uptake of Nutrients by the Cell

Abramson, J.; Smirnova, I.; Kasho, V.; Verner, G.; Kaback, H. R.; and Iwata, S. 2003. Structure and mechanism of the lactose permease of *Escherichia coli. Science* 301:610–15.

Chang, G., and Roth, C. B. 2001. Structure of Msba from *E. coli:* A homolog of the multidrug resistance ATP binding cassette (ABC) transporters. *Science* 293:1793–1800.

Dassa, E. 2000. ABC transport. In *Encyclopedia of microbiology,* 2d ed., vol. 1, J. Lederberg, editor-in-chief, 1–12. San Diego: Academic Press.

Doige, C. A., and Ames, G. F.-L. 1993. ATP-dependent transport systems in bacteria and humans: Relevance to cystic fibrosis and multidrug resistance. *Annu. Rev. Microbiol.* 47:291–319.

Earhart, C. F. 2000. Iron metabolism. In *Encyclopedia of microbiology,* 2d ed., vol. 2, J. Lederberg, editor-in-chief, 860–68. San Diego: Academic Press.

Hancock, R. E. W., and Brinkman, F. S. L. 2002. Function of *Pseudomonas* porins in uptake and efflux. *Annu. Rev. Microbiol.* 56:17–38.

Hohmann, S.; Bill, R. M.; Kayingo, G.; and Prior, B. A. 2000. Microbial MIP channels. *Trends Microbiol.* 8(1):33–38.

Locher, K. P.; Lee, A. T.; and Rees, D. C. 2002. The *E. coli* BtuCD structure: A framework for ABC transporter architecture and mechanism. *Science* 296:1091–98.

Neilands, J. B. 1991. Microbial iron compounds. *Annu. Rev. Biochem.* 50:715–31.

Postma, P. W.; Lengeler, J. W.; and Jacobson, G. R. 1993. Phosphoenolpyruvate: carbohydrate phosphotransferase systems of bacteria. *Microbiol. Rev.* 57(3):543–94.

5.7 Culture Media

Atlas, R. M. 1997. *Handbook of microbiological media,* 2d ed. Boca Raton, Fla.: CRC Press.

Cote, R. J., and Gherna, R. L. 1994. Nutrition and media. In *Methods for general and molecular bacteriology,* 2d ed., P. Gerhardt, editor, 155–78. Washington, D.C.: American Society for Microbiology.

Difco Laboratories. 1998. *Difco manual of dehydrated culture media and reagents for microbiology.* 11th ed. Sparks, Md.: BD Bioscience.

Power, D. A., editor. 1988. *Manual of BBL products and laboratory procedures,* 6th ed. Cockeysville, Md.: Becton, Dickinson and Company.

5.8 Isolation of Pure Cultures

Gutnick, D. L., and Ben-Jacob, E. 1999. Complex pattern formation and cooperative organization of bacterial colonies. In *Microbial ecology and infectious disease,* E. Rosenberg, editor, 284–99. Washington, D.C.: ASM Press.

Schindler, J. 1993. Dynamics of Bacillus colony growth. *Trends Microbiol.* 1(9):333–38.

Shapiro, J. A. 1988. Bacteria as multicellular organisms. *Sci. Am.* 258(6):82–89.

Chapter 6: Microbial Growth

General References

Atlas, R. M., and Bartha, R. 1997. *Microbial ecology: Fundamentals and applications,* 4th ed. Menlo Park, Calif.: Benjamin/Cummings.

Cavicchioli, R., and Thomas, T. 2000. Extremophiles. In *Encyclopedia of microbiology,* 2d ed., vol. 2, J. Lederberg, editor-in-chief, 317–37. San Diego: Academic Press.

Koch, A. L. 1995. *Bacterial growth and form.* New York: Chapman & Hall.

Madigan, M. T., and Marrs, B. L. 1997. Extremophiles. *Sci. Am.* 276(4):82–87.

Moat, A. G.; and Foster, J. W.; and Spector, M. P. 2002. *Microbial physiology,* 4th ed. New York: Wiley-Liss.

Neidhardt, F. C.; Ingraham, J. L.; and Schaechter, M. 1990. *Physiology of the bacterial cell: A molecular approach.* Sunderland, Mass.: Sinauer Associates.

Postgate, J. 1994. *The outer reaches of life.* New York: Cambridge University Press.

6.1 The Growth Curve

Frenkiel-Krispin, D., and Minsky, A. 2002. Biocrystallization: a last-resort survival strategy in bacteria. *ASM News* 68(6):277–83.

Kolter, R.; Siegele, D. A.; and Tormo, A. 1993. The stationary phase of the bacterial life cycle. *Annu. Rev. Microbiol.* 47:855–74.

Lazazzera, B. A. 2000. Quorum sensing and starvation: Signals for entry into stationary phase. *Curr. Opin. Microbiol.* 3:177–82.

Marr, A. G. 2000. Growth kinetics, bacterial. In *Encyclopedia of microbiology,* 2d ed., vol. 2, J. Lederberg, editor-in-chief, 584–89. San Diego: Academic Press.

Matin, A. 2000. Starvation, bacterial. In *Encyclopedia of microbiology,* 2d ed., vol. 4, J. Lederberg, editor-in-chief, 394–403. San Diego: Academic Press.

Prosser, J. I., and Tough, A. J. 1991. Growth mechanisms and growth kinetics of filamentous microorganisms. *Critical reviews in biotechnology* 10(4):253–74.

Russell, J. B., and Cook, G. M. 1995. Energetics of bacterial growth: Balance of anabolic and catabolic reactions. *Microbiol. Rev.* 59(1):48–62.

Zambrano, M. M., and Kolter, R. 1996. GASPing for life in stationary phase. *Cell* 86:181–84.

6.4 The Influence of Environmental Factors on Growth

Abe, F.; Kato, C.; and Horikoshi, K. 1999. Pressure-regulated metabolism in microorganisms. *Trends Microbiol.* 7(11):447–53.

Adams, M. W. W. 1993. Enzymes and proteins from organisms that grow near and above 100°C. *Annu. Rev. Microbiol.* 47:627–58.

Bartlett, D. H., and Roberts, M. F. 2000. Osmotic stress. In *Encyclopedia of microbiology,* 2d ed., vol. 3, J. Lederberg, editor-in-chief, 502–15. San Diego: Academic Press.

Blount, P., and Moe, P. C. 1999. Bacterial mechanosensitive channels: integrating physiology, structure and function. *Trends Microbiol.* 7(10):420–24.

Booth, I. R. 1985. Regulation of cytoplasmic pH in bacteria. *Microbiol. Rev.* 49(4):359–78.

Cotter, P.D., and Hill, C. 2003. Surviving the acid test: Responses of gram-positive bacteria to low pH. *Microbiol. Mol. Biol. Rev.* 67 (3):429-53.

Csonka, L. N. 1989. Physiological and genetic responses of bacteria to osmotic stress. *Microbiol. Rev.* 53(1):121–47.

Friedman, S. M. 1992. Thermophilic microorganisms. In *Encyclopedia of microbiology,* 1st ed., vol. 4, J. Lederberg, editor-in-chief, 217–29. San Diego: Academic Press.

Gaill, F. 1993. Aspects of life development at deep sea hydrothermal vents. *FASEB J.* 7:558–65.

Gillis, A. M. 1994. A pressure-filled life. *BioScience* 44(9):584–86.

Gottschall, J. C., and Prins, R. A. 1991. Thermophiles: A life at elevated temperatures. *Trends Ecol. & Evol.* 6(5):157–62.

Inlag, J. A., and Linn, S. 1988. DNA damage and oxygen radical toxicity. *Science* 240:1302–9.

Kashefi, K., and Lovley, D. R. 2003. Extending the upper temperature limit for life. *Science* 301:934.

Krieg, N. R., and Hoffman, P. S. 1986. Microaerophily and oxygen toxicity. *Annu. Rev. Microbiol.* 40:107–30.

Krulwich, T. A., and Guffanti, A. A. 1989. Alkalophilic bacteria. *Annu. Rev. Microbiol.* 43:435–63.

Martinac, B. 2001. Mechanosensitive channels in prokaryotes. *Cell Physiol. Biochem.* 11:61–76.

Morita, R. Y. 2000. Low-temperature environments. In *Encyclopedia of microbiology,* 2d ed., vol. 3, J. Lederberg, editor-in-chief, 93–98. San Diego: Academic Press.

Pivetti, C. D.; Yen, M.-R.; Miller, S.; Busch, W.; Tseng, Y.-H.; Booth, I. R.; and Saier, Jr., M. H. 2003. Two families of mechanosensitive channel proteins. *Microbiol. Mol. Biol. Rev.* 67(1):66–85.

Potts, M. 1994. Desiccation tolerance of prokaryotes. *Microbiol. Rev.* 58(4):755–805.

Stetter, K. O. 1995. Microbial life in hyperthermal environments. *ASM News* 61(6):285–90.

6.5 Microbial Growth in Natural Environments

Dunny, G. M., and Leonard, B. A. B. 1997. Cell-cell communication in gram-positive bacteria. *Annu. Rev. Microbiol.* 51:527–64.

Fuqua, C. 2000. Quorum sensing in gram-negative bacteria. In *Encyclopedia of microbiology,* 2d ed., vol. 4, J. Lederberg, editor-in-chief, 1–13. San Diego: Academic Press.

Losick, R., and Kaiser, D. 1997. Why and how bacteria communicate. *Sci. Am.* 276(2):68–73.

Michiels, J.; Dirix, G.; Vanderleyden, J.; and Xi, C. 2001. Processing and export of peptide pheromones and bacteriocines in Gram-negative bacteria. *Trends Microbiol.* 9(4):164–68.

Miller, M. B., and Bassler, B. L. 2001. Quorum sensing in bacteria. *Annu. Rev. Microbiol.* 55:165–99.

Morita, R. Y. 1997. *Bacteria in oligotrophic environments: Starvation-survival life style.* New York: Chapman & Hall.

Redfield, R. J. 2002. Is quorum sensing a side effect of diffusion sensing? *Trends Microbiol.* 10(8):365–70.

Shapiro, J. A. 1998. Thinking about bacterial populations as multicellular organisms. *Annu. Rev. Microbiol.* 52:81–104.

Watnick, P., and Kolter, R. 2000. Biofilm, city of microbes. *J. Bacteriol.* 182(10):2675–79.

Chapter 7: Control of Microorganisms by Physical and Chemical Agents

General References

Barkley, W. E., and Richardson, J. H. 1994. Laboratory safety. In *Methods for general and molecular bacteriology,* P. Gerhardt, et al., editors, 715–34. Washington, D.C.: American Society for Microbiology.

Block, S. S. 1992. Sterilization. In *Encyclopedia of microbiology,* 1st ed., vol. 4, J. Lederberg, editor-in-chief, 87–103. San Diego: Academic Press.

Block, S. S., editor. 1991. *Disinfection, sterilization and preservation,* 4th ed. Philadelphia: Lea and Febiger.

Centers for Disease Control and National Institutes of Health. 1992. *Biosafety in microbiological and biomedical laboratories,* 3d ed. Washington, D.C.: U.S. Government Printing Office.

Martin, M. A., and Wenzel, R. P. 1995. Sterilization, disinfection, and disposal of infectious waste. In *Principles and practice of infectious diseases,* 4th ed., G. L. Mandell, J. E. Bennett, and R. Dolin editors, 2579–87. New York: Churchill Livingstone.

Perkins, J. J. 1969. *Principles and methods of sterilization in health sciences,* 2d ed. Springfield, Ill.: Charles C. Thomas.

Pike, R. M. 1979. Laboratory-associated infections: Incidence, fatalities, causes, and prevention. *Annu. Rev. Microbiol.* 33:41–66.

Russell, A. D.; Hugo, W. B.; and Ayliffe, G. A. J., editors. 1992. *Principles and practice of disinfection, preservation and sterilization,* 2d ed. Oxford: Blackwell Scientific Publications.

Sewell, D. L. 1995. Laboratory-associated infections and biosafety. *Clin. Microbiol. Rev.* 8(3):389–405.

Sondossi, M. 2000. Biocides. In *Encyclopedia of microbiology,* 2d ed., vol. 1, J. Lederberg, editor-in-chief, 445–60. San Diego: Academic Press.

Strain, B. A., and Gröschel, D. H. M. 1995. Laboratory safety and infectious waste management. In *Manual of clinical microbiology,* 6th ed., P. R. Murray, editor, 75–85. Washington, D.C.: American Society for Microbiology.

Widmer, A. F., and Frei, R. 1999. Decontamination, disinfection, and sterilization. In *Manual of clinical microbiology,* 7th ed., P. R. Murray, et al., editors, 138–64. Washington, D.C.: ASM Press.

7.4 The Use of Physical Methods in Control

Brock, T. D. 1983. *Membrane filtration: A user's guide and reference manual.* Madison, Wis.: Science Tech Publishers.

Sørhaug, T. 1992. Temperature control. In *Encyclopedia of microbiology,* 1st ed., vol. 4, J. Lederberg, editor-in-chief, 201–11. San Diego: Academic Press.

7.5 The Use of Chemical Agents in Control

Gilbert, P., and McBain, A. J. 2003. Potential impact of increased use of biocides in consumer products on prevalence of antibiotic resistance. *Clin. Microbiol. Rev.* 16(2):189–208.

McDonnell, G., and Russell, A. D. 1999. Antiseptics and disinfectants: Activity, action, and resistance. *Clin. Microbiol. Rev.* 12(1):147–79.

Russell, A. D. 1990. Bacterial spores and chemical sporicidal agents. *Clin. Microbiol. Rev.* 3(2):99–119.

Rutala, W. A., and Weber, D. J. 1997. Uses of inorganic hypochlorite (bleach) in health-care facilities. *Clin. Microbiol. Rev.* 10(4):597–610.

Chapter 8: Metabolism: Energy, Enzymes, and Regulation

General References

Becker, W. M.; Kleinsmith, L.; and Hardin, J. 2000. *The world of the cell,* 4th ed. Redwood City, Calif.: Benjamin Cummings.

Garrett, R. H., and Grisham, C. H. 1999. *Biochemistry* 2d ed. New York: Saunders.

Lodish, H.; Baltimore, D.; Berk, A.; Zipursky, S. L.; Matsudaira, P.; and Darnell, J. 1999. *Molecular cell biology,* 4th ed. New York: Scientific American Books.

McKee, T., and McKee, J. R. 2003. *Biochemistry: The molecular basis of life,* 3d ed. New York: McGraw-Hill.

Neidhardt, F. C.; Ingraham, J. L.; and Schaechter, M. 1990. *Physiology of the bacterial cell: A molecular approach.* Sunderland, Mass.: Sinauer Associates.

Nelson, D. L.; and Cox, M. M. 2000. *Lehninger Principles of biochemistry,* 3d ed. New York: Worth Publishers.

Voet, D., and Voet, J. G. 1995. *Biochemistry,* 2d ed. New York: John Wiley and Sons.

8.6 Enzymes

Boyer, Paul D., editor. 1970–1987. *The enzymes.* San Diego: Academic Press.

Branden, C., and Tooze, J. 1991. *Introduction to protein structure.* New York: Garland Publishing.

Fersht, A. 1984. *Enzyme structure and mechanism,* 2d ed. San Francisco: W. H. Freeman.

International Union of Biochemistry and Molecular Biology. 1992. *Enzyme nomenclature.* San Diego: Academic Press.

8.9 Control of Enzyme Activity

Kantrowitz, E. R., and Lipscomb, W. N. 1988. *Escherichia coli* aspartate transcarbamylase: The relation between structure and function. *Science* 241:669–74.

Koshland, D. E., Jr. 1973. Protein shape and biological control. *Sci. Am.* 229(4):52–64.

Saier, M. H., Jr.; Wu, L.-F.; and Reizer, J. 1990. Regulation of bacterial physiological processes by three types of protein phosphorylating systems. *Trends Biochem. Sci.* 15:391–95.

Chapter 9: Metabolism: Energy Release and Conservation

General References

Caldwell, D. R. 2000. *Microbial physiology and metabolism,* 2d ed. Belmont, Calif.: Star Publishing.

Dawes, I. W., and Sutherland, I. W. 1992. *Microbial physiology,* 2d ed. London: Blackwell Scientific Publications.

Ferguson, S. J. 2000. Energy transduction processes: From respiration to photosynthesis. In *Encyclopedia of microbiology,* 2d ed., vol. 2, J. Lederberg, editor-in-chief, 177–86. San Diego: Academic Press.

Garrett, R. H., and Grishman, C. M. 1999. *Biochemistry,* 2d ed. New York: Saunders.

Gottschalk, G. 1986. *Bacterial metabolism,* 2d ed. New York: Springer-Verlag.

Mathews, C. K.; van Holde, K. E.; and Ahern, K.G. 2000. *Biochemistry,* 3d ed. Redwood City, Calif.: Benjamin/Cummings.

McKee, T., and McKee, J. R. 2003. *Biochemistry: The molecular basis of life,* 3d ed. Dubuque, Iowa: McGraw-Hill.

Moat, A. G.; and Foster, J. W.; and Spector, M. P. 2002. *Microbial physiology,* 4th ed. New York: Wiley-Liss.

Neidhardt, F. C.; Ingraham, J. L.; and Schaechter, M. 1990. *Physiology of the bacterial cell: A molecular approach.* Sunderland, Mass.: Sinauer Associates.

Neidhardt, F. C., editor-in-chief. 1996. *Escherichia coli and Salmonella: Cellular and molecular biology,* 2d ed., vol. 1. Washington, D.C.: ASM Press.

Nicholls, D. G., and Ferguson, S. J. 2002. *Bioenergetics,* 3d ed. San Diego: Academic Press.

Voet, D., and Voet, J. G. 1995. *Biochemistry,* 2d ed. New York: John Wiley and Sons.

White, D. 1995. *The physiology and biochemistry of prokaryotes.* New York: Oxford University Press.

9.5 Electron Transport and Oxidative Phosphorylation

Anraku, Y. 1988. Bacterial electron transport chains. *Annu. Rev. Biochem.* 57:101–32.

Baker, S. C.; Ferguson, S. J.; Ludwig, B.; Page, M. D.; Richter, O.-M. H.; and van Spanning, R. J. M. 1998. Molecular genetics of the genus *Paracoccus:* Metabolically versatile bacteria with bioenergetic flexibility. *Microbiol. Mol. Biol. Rev.* 62(4):1046–78.

Boyer, P. D. 1997. The ATP synthase—A splendid molecular machine. *Annu. Rev. Biochem.* 66:717–49.

Deckers-Hebestreit, G., and Altendorf, K. 1996. The F_0F_1-type ATP synthases of bacteria: Structure and function of the F_0 complex. *Annu. Rev. Microbiol.* 50:791–824.

Ingledew, W. J., and Poole, R. K. 1984. The respiratory chains of *Escherichia coli. Microbiol. Rev.* 48(3):222–71.

Kinosita, Jr., K.; Yasuda, R.; Noji, H.; Ishiwata, S.; and Yoshida, M. 1998. F_1-ATPase: A rotary motor made of a single molecule. *Cell* 93:21–4.

Poole, R. K. 2000. Aerobic respiration: Oxidases and globins. In *Encyclopedia of microbiology,* 2d ed., vol. 1, J. Lederberg, editor-in-chief, 53–68. San Diego: Academic Press.

Saraste, M. 1999. Oxidative phosphorylation at the *fin de siècle. Science* 283:1488–93.

Yankovskaya, V., et al. 2003. Architecture of succinate dehydrogenase and reactive oxygen species generation. *Science* 299:700–4.

Zhou, Y.; Duncan, T. M.; and Cross, R. L. 1997. Subunit rotation in *Escherichia coli* F_0F_1-ATP synthase during oxidative phosphorylation. *Proc. Natl. Acad. Sci.* 94:10583–87.

9.6 Anaerobic Respiration

Gunsalus, R. P. 2000. Anaerobic respiration. In *Encyclopedia of microbiology,* 2d ed., vol. 1, J. Lederberg, editor-in-chief, 180–88. San Diego: Academic Press.

Hochstein, L. I., and Tomlinson, G. A. 1988. The enzymes associated with denitrification. *Annu. Rev. Microbiol.* 42:231–61.

Lovley, D. R. 1993. Dissimilatory metal reduction. *Annu. Rev. Microbiol.* 47:263–90.

Zumft, W. G. 1993. The biological role of nitric oxide in bacteria. *Arch. Microbiol.* 160:253–64.

9.7 Catabolism of Carbohydrates and Intracellular Reserve Polymers

Warren, R. A. J. 1996. Microbial hydrolysis of polysaccharides. *Annu. Rev. Microbiol.* 50:183–212.

9.10 Oxidation of Inorganic Molecules

Friedrich, B., and Schwartz, E. 1993. Molecular biology of hydrogen utilization in aerobic chemolithotrophs. *Annu. Rev. Microbiol.* 47:351–83.

Kelly, D. P. 1985. Physiology of the thiobacilli: Elucidating the sulphur oxidation pathway. *Microbiol. Sci.* 2(4):105–9.

9.11 Photosynthesis

Deisenhofer, J.; Michel, H.; and Huber, R. 1985. The structural basis of photosynthetic light reactions in bacteria. *Trends Biochem. Sci.* 10(6):243–48.

Govindjee, and Coleman, W. J. 1990. How plants make oxygen. *Sci. Am.* 262(2):50–58.

Grossman, A. R.; Schaefer, M. R.; Chiang, G. G.; and Collier, J. L. 1993. The phycobilisome, a light-harvesting complex responsive to environmental conditions. *Microbiol. Rev.* 57(3):725–49.

Schlegel, H. G., and Bowien, B., editors. 1989. *Autotrophic bacteria.* Madison, Wis.: Science Tech Publishers.

Youvan, D. C., and Marrs, B. L. 1987. Molecular mechanisms of photosynthesis. *Sci. Am.* 256(6):42–48.

Chapter 10: Metabolism: The Use of Energy in Biosynthesis

General References

Caldwell, D. R. 2000. *Microbial physiology and metabolism* 2d ed. Belmont, Calif.: Star Publishing. Communications, Inc.

Dawes, I. W., and Sutherland, I. W. 1992. *Microbial physiology,* 2d ed. Boston, Mass.: Blackwell Scientific Publications.

Gottschalk, G. 1986. *Bacterial metabolism,* 2d ed. New York: Springer-Verlag.

Mandelstam, J.; McQuillen, K.; and Dawes, I. 1982. *Biochemistry of bacterial growth,* 3d ed. London: Blackwell Scientific Publications.

Mathews, C. K.; van Holde, K. E.; and Ahern, K.G. 2000. *Biochemistry,* 3d ed. Redwood City, Calif.: Benjamin/Cummings.

McKee, T., and McKee, J. R. 2003. *Biochemistry: The molecular basis of life,* 3d ed. Dubuque, Iowa: McGraw-Hill.

Moat, A. G., and Foster, J. W.; and Spector, M. P. 2002. *Microbial physiology,* 4th ed. New York: Wiley-Liss.

Neidhardt, F. C.; Ingraham, J. L.; and Schaechter, M. 1990. *Physiology of the bacterial cell: A molecular approach.* Sunderland, Mass.: Sinauer Associates.

Voet, D., and Voet, J. G. 1995. *Biochemistry,* 2d ed. New York: John Wiley and Sons.

White, D. 1995. *The physiology and biochemistry of procaryotes.* New York: Oxford University Press.

10.2 The Photosynthetic Fixation of CO_2

Schlegel, H. G., and Bowien, B., editors. 1989. *Autotrophic bacteria.* Madison, Wis.: Science Tech Publishers.

Yoon, K.-S.; Hanson, T. E.; Gibson, J. L.; and Tabita, F. R. 2000. Autotrophic CO_2 metabolism. In *Encyclopedia of microbiology,* 2d ed., vol. 1, J. Lederberg, editor-in-chief, 349–58. San Diego: Academic Press.

10.4 The Assimilation of Inorganic Phosphorus, Sulfur, and Nitrogen

Dean, D. R.; Bolin, J. T.; and Zheng, L. 1993. Nitrogenase metalloclusters: Structures, organization, and synthesis. *J. Bacteriol.* 175(21):6737–44.

Knowles, R. 2000. Nitrogen cycle. In *Encyclopedia of microbiology,* 2d ed., vol. 3, J. Lederberg, editor-in-chief, 379–91. San Diego: Academic Press.

Kuykendall, L. D.; Dadson, R. B.; Hashem, F. M.; and Elkan, G. H. 2000. Nitrogen fixation. In *Encyclopedia of microbiology,* 2d ed., vol. 3, J. Lederberg, editor-in-chief, 392–406. San Diego: Academic Press.

Lens, P., and Pol, L. H. 2000. Sulfur cycle. In *Encyclopedia of microbiology,* 2d ed., vol. 4, J. Lederberg, editor-in-chief, 495–505. San Diego: Academic Press.

Luden, P. W. 1991. Energetics of and sources of energy for biological nitrogen fixation. In *Current topics in bioenergetics,* vol. 16, 369–90. San Diego: Academic Press.

Mora, J. 1990. Glutamine metabolism and cycling in *Neurospora crassa. Microbiol. Rev.* 54(3):293–304.

Peters, J. W.; Fisher, K.; and Dean, D. R. 1995. Nitrogenase structure and function: A biochemical-genetic perspective. *Annu. Rev. Microbiol.* 49:335–66.

10.10 Patterns of Cell Wall Formation

Harold, F. M. 1990. To shape a cell: An inquiry into the causes of morphogenesis of microorganisms. *Microbiol. Rev.* 54(4):381–431.

Höltje, J.-V. 1998. Growth of the stress-bearing and shape-maintaining murein sacculus of *Escherichia coli. Microbiol. Mol. Biol. Rev.* 62(1):181–203.

Höltje, J.-V. 2000. Cell walls, bacterial. In *Encyclopedia of microbiology,* 2d ed., vol. 1, J. Lederberg, editor-in-chief, 759–71. San Diego: Academic Press.

Koch, A. L. 1995. *Bacterial growth and form.* New York: Chapman & Hall.

Koch, A. L. 2002. Why are rod-shaped bacteria rod shaped? *Trends Microbiol.* 10(10):452–55.

Chapter 11: Genes: Structure, Replication, and Mutation

General References

Dale, J. W. 1998. *Molecular genetics of bacteria,* 3d ed. New York: John Wiley and Sons.

Griffiths, A. J. F.; Miller, J. H.; Suzuki, D. T.; Lewontin, R. C.; and Gelbart, W. M. 2000. *An introduction to genetic analysis,* 7th ed. New York: W. H. Freeman.

Hartwell, L. H.; Hood, L.; Goldberg, M. L.; Reynolds, A. E.; Silver, L. M.; and Veres, R. C. 2004. *Genetics: From genes to genomes,* 2d ed. New York: McGraw-Hill.

Holloway, B. W. 1993. Genetics for all bacteria. *Ann. Rev. Microbiol.* 47:659–84.

Joset, F., and Guespin-Michel, J. 1993. *Prokaryotic genetics: Genome organization, transfer and plasticity.* Boston: Blackwell.

King, R. C., and Stansfield, W. D. 2002. *A dictionary of genetics,* 6th ed. New York: Oxford University Press.

Lewin, B. 2000. *Genes,* 7th ed. New York: Oxford University Press.

Maloy, S. R.; Cronan, J. E., Jr.; and Freifelder, D. 1994. *Microbial genetics,* 2d ed. Boston: Jones and Bartlett.

Snyder, L., and Champness, W. 1997. *Molecular Genetics of Bacteria.* Washington, D.C.: ASM Press.

Weaver, R. F. 2002. *Molecular biology,* 2d ed., Dubuque, Iowa: McGraw-Hill.

11.2 Nucleic Acid Structure

Bauer, W. R.; Crick, F. H. C.; and White, J. H. 1980. Supercoiled DNA. *Sci. Am.* 243(1):118–33.

Darnell, J. E., Jr. 1985. RNA. *Sci. Am.* 253(4):68–78.

Drlica, K. 2000. Chromosome, Bacterial. In *Encyclopedia of microbiology,* 2d ed., vol. 1, J. Lederberg, editor-in-chief, 808–21. San Diego: Academic Press.

Rich, A., and Kim, S. H. 1978. The three-dimensional structure of transfer RNA. *Sci. Am.* 238(1):52–62.

11.3 DNA Replication

Baker, T. A., and Bell, S. P. 1998. Polymerases and the replisome: Machines within machines. *Cell* 92:295–305.

Cook, P. R. 1999. The organization of replication and transcription. *Science* 284:1790–95.

Hejna, J. A., and Moses, R. E. 2000. DNA replication. In *Encyclopedia of microbiology,* 2d ed., vol. 2, J. Lederberg, editor-in-chief, 82–90. San Diego: Academic Press.

Johnson, K. A. 1993. Conformational coupling in DNA polymerase fidelity. *Ann. Rev. Biochem.* 62:685–713.

Joyce, C. M., and Steitz, T. A. 1994. Function and structure relationships in DNA polymerases. *Ann. Rev. Biochem.* 63:777–822.

Kornberg, A. 1992. *DNA replication,* 2d ed. San Francisco: W. H. Freeman.

Lohman, T. M.; Thorn, K.; and Vale, R. D. 1998. Staying on track: Common features of DNA helicases and microtubule motors. *Cell* 93:9–12.

Losick, R., and Shapiro, L. 1998. Bringing the mountain to Mohammed. *Science* 282:1430–31.

Marians, K. J. 1992. Prokaryotic DNA replication. *Annu. Rev. Biochem.* 61:673–719.

Matson, S. W., and Kaiser-Rogers, K. A. 1990. DNA helicases. *Annu. Rev. Biochem.* 59:289–329.

Meyer, R. R., and Laine, P. S. 1990. The single-stranded DNA-binding protein of *Escherichia coli. Microbiol. Rev.* 54(4):342–80.

Roca, J. 1995. The mechanisms of DNA topoisomerases. *Trends Biochem. Sci.* 20(4):156–60.

Stillman, B. 1994. Smart machines at the DNA replication fork. *Cell* 78:725–28.

Waga, S., and Stillman, B. 1998. The DNA replication fork in eukaryotic cells. *Annu. Rev. Biochem.* 67:721–51.

11.4 The Genetic Code

Andersson, S. G. E., and Kurland, C. G. 1990. Codon preferences in free-living microorganisms. *Microbiol. Rev.* 54(2):198–210.

Osawa, S.; Jukes, T. H.; Watanabe, K.; and Muto, A. 1992. Recent evidence for evolution of the genetic code. *Microbiol. Rev.* 56(1):229–64.

11.5 Gene Structure

Berlyn, M. K. B.; Low, K. B.; and Rudd, K. E. 1996. Linkage map of *Escherichia coli* K-12, edition 9. In *Escherichia coli and Salmonella: Cellular and molecular biology,* 2d ed., vol. 2, F. C. Neidhardt, editor-in-chief, 1715–1902. Washington, D.C.: ASM Press.

Girons, I. S.; Old, I. G.; and Davidson, B. E. 1994. Molecular biology of *Borrelia,* bacteria with linear replicons. *Microbiology* 140:1803–16.

Marinus, M. G. 2000. Methylation of nucleic acids and proteins. In *Encyclopedia of microbiology,* 2d ed., vol. 3, J. Lederberg, editor-in-chief, 240–44. San Diego: Academic Press.

Riley, M. 1993. Functions of the gene products of *Escherichia coli. Microbiol. Rev.* 57(4):862–952.

Weinstock, G. M. 1994. Bacterial genomes: Mapping and stability. *ASM News* 60(2):73–78.

11.6 Mutations and Their Chemical Basis

Bayliss, C. D., and Moxon, E. R. 2002. Hypermutation and bacterial adaptation. *ASM News* 68(11):549–55.

Chicurel, M. 2001. Can organisms speed their own evolution? *Science* 292:1824–27.

Lederberg, J. 1992. Bacterial variation since Pasteur. *ASM News* 58(5):261–65.

Massey, R. C., and Buckling, A. 2002. Environmental regulation of mutation rates at specific sites. *Trends Microbiol.* 10(12):580–84.

Miller J. H. 1996. Spontaneous mutators in bacteria: Insights into pathways of mutagenesis and repair. *Annu. Rev. Microbiol.* 50:625–43.

Singer, B., and Kusmierek, J. T. 1982. Chemical mutagenesis. *Annu. Rev. Biochem.* 52:655–93.

11.7 Detection and Isolation of Mutants

Devoret, R. 1979. Bacterial tests for potential carcinogens. *Sci. Am.* 241(2):40–49.

11.8 DNA Repair

Friedberg, E. C.; Walker, G. C.; and Siede, W. 1995. *DNA repair and mutagenesis.* Herndon, Va.: ASM Press.

Grossman, L. 2000. DNA repair. In *Encyclopedia of microbiology,* 2d ed., vol. 2, J. Lederberg, editor-in-chief, 71–81. San Diego: Academic Press.

Grossman, L., and Thiagalingam, S. 1993. Nucleotide excision repair, a tracking mechanism in search of damage. *J. Biol. Chem.* 268(23):16871–74.

Kuzminov, A. 1999. Recombinational repair of DNA damage in *Escherichia coli* and bacteriophage λ. *Microbiol. Mol. Biol. Rev.* 63(4):751–813.

McCullough, A. K.; Dodson, M. L.; and Lloyd, R. S. 1999. Initiation of base excision repair: Glycosylase mechanisms and structures. *Annu. Rev. Biochem.* 68:255–85.

Miller, R. V. 2000. recA: The gene and its protein product. In *Encyclopedia of microbiology*, 2d ed., vol. 4, J. Lederberg, editor-in-chief, 43–54. San Diego: Academic Press.

Schofield, M. J., and Hsieh, P. 2003. DNA mismatch repair: Molecular mechanisms and biological function. *Annu. Rev. Microbiol.* 57:579–608.

Van Houten, B. 1990. Nucleotide excision repair in *Escherichia coli. Microbiol. Rev.* 54(1):18–51.

Winterling, K. W. 2000. SOS response. In *Encyclopedia of microbiology*, 2d ed., vol. 4, J. Lederberg, editor-in-chief, 336–43. San Diego: Academic Press.

Chapter 12: Genes: Expression and Regulation

General References

Alberts, B.; Johnson, A.; Lewis, J.; Raff, M.; Roberts, K.; and Walter, P. 2002. *Molecular biology of the cell*, 4th ed. New York: Garland Science.

Hartwell, L. H.; Hood, L.; Goldberg, M. L.; Reynolds, A. E.; Silver, L. M.; and Veres, R. C. 2004. *Genetics: From genes to genomes*, 2d ed. New York: McGraw-Hill.

Judson, H. F. 1979. *The eighth day of creation: Makers of the revolution in biology*. London: Jonathan Cape.

Lewin, B. 2000. *Genes*, 7th ed. New York: Oxford University Press.

Lodish, H.; Berk, A.; Zipursky, S. L.; Matsudaira, P.; Baltimore, D.; and Darnell, J. 2000. *Molecular Cell Biology*, 4th ed. New York: W. H. Freeman.

Neidhardt, F. C.; Ingraham, J. L.; and Schaechter, M. 1990. *Physiology of the bacterial cell*. Sunderland, Mass.: Sinauer Associates.

Oltvai, Z. N., and Barabási, A.-L, 2002. Life's complexity pyramid. *Science* 298:763.

Snyder, L., and Champness, W. 1997. *Molecular genetics of bacteria*. Washington, D.C.: ASM Press.

Squires, C. L., and Zaporojets, D. 2000. Proteins shared by the transcription and translation machines. *Annu. Rev. Microbiol.* 54:775–98.

Voet, D., and Voet, J. G. 1995. *Biochemistry*, 2d ed. New York: John Wiley and Sons.

Weaver, R. F. 2002. *Molecular biology*, 2d ed. Dubuque, Iowa: McGraw-Hill.

12.1 DNA Transcription or RNA Synthesis

Ahern, H. 1991. Self-splicing introns: Molecular fossils or selfish DNA? *ASM News* 57(5):258–61.

Cramer, P.; Bushnell, D. A.; and Kornberg, R. D. 2001. Structural basis of transcription: RNA polymerase II at 2.8 angstrom resolution. *Science* 292(5523):1863–76.

Das, A. 1993. Control of transcription termination by RNA-binding proteins. *Annu. Rev. Biochem.* 62:893–930.

Epshtein, V., and Nudler, E. 2003. Cooperation between RNA polymerase molecules in transcription elongation. *Science* 300:801–5.

Gelles, J., and Landick, R. 1998. RNA polymerase as a molecular motor. *Cell* 93:13–16.

Gnatt, A. L.; Cramer, P.; Fu, J.; Bushnell, D. A.; and Kornberg, R. D. 2001. Structural basis of transcription: An RNA polymerase II elongation complex at 3.3 Å resolution. *Science* 292(5523):1876–82.

Koleske, A. J., and Young, R. A. 1995. The RNA polymerase II holoenzyme and its implications for gene regulation. *Trends Biochem. Sci.* 20(3):113–16.

Landick, R. 1997. RNA polymerase slides home: Pause and termination site recognition. *Cell* 88:741–44.

Murakami, K. S.; Masuda, S.; and Darst, S. A. 2002. Structural basis of transcription initiation: RNA polymerase holoenzyme at 4 Å resolution. *Science* 296:1280–84.

Murakami, K. S.; Masuda, S.; Campbell, E. A.; Muzzin, O.; and Darst, S. A. 2002. Structural basis of transcription initiation: An RNA polymerase holoenzyme-DNA complex. *Science* 296:1285–90.

Sarkar, N. 1997. Polyadenylation of mRNA in prokaryotes. *Annu. Rev. Biochem.* 66:173–97.

Staley, J. P., and Guthrie, C. 1998. Mechanical devices of the spliceosome: Motors, clocks, springs, and things. *Cell* 92:315–26.

12.2 Protein Synthesis

Ban, N.; Nissen, P.; Hansen, J.; Moore, P. B.; and Steitz, T. A. 2000. The complete atomic structure of the large ribosomal subunit at 2.4Å resolution. *Science* 289:905–20.

Bukau, B., and Horwich, A. L. 1998. The Hsp70 and Hsp60 chaperone machines. *Cell* 92:351–66.

Cate, J. H.; Yusupov, M. M.; Yusupova, G. Zh.; Earnest, T. N.; and Noller, H. F. 1999. X-ray crystal structures of 70S ribosome functional complexes. *Science* 285:2095–2135.

Cooper, A. A., and Stevens, T. H. 1995. Protein splicing: Self-splicing of genetically mobile elements at the protein level. *Trends Biochem. Sci.* 20:351–56.

Frank, J. 1998. How the ribosome works. *American Scientist* 86:428–39.

Ganoza, M. C.; Kiel, M. C.; and Aoki, H. 2002. Evolutionary conservation of reactions in translation. *Microbiol. Mol. Biol. Rev.* 66(3):460–85.

Gogarten, J. P.; Senejani, A. G.; Zhaxybayeva, O.; Olendzenski, L.; and Hilario, E. 2002. Inteins: Structure, function, and evolution. *Annu. Rev. Microbiol.* 56:263–87.

Green, R., and Noller, H. F. 1997. Ribosomes and translation. *Annu. Rev. Biochem.* 66:679–716.

Hartl, F. U., and Hayer-Hartl, M. 2002. Molecular chaperones in the cytosol: From nascent chain to folded protein. *Science* 295:1852–58.

Jagus, R., and Joshi, B. 2000. Protein biosynthesis. In *Encyclopedia of microbiology*, 2d ed., vol. 3, J. Lederberg, editor-in-chief, 824–46. San Diego: Academic Press.

Narberhaus, F. 2002. α-Crystallin-type heat shock proteins: Socializing minichaperones in the context of a multichaperone network. *Microbiol. Mol. Biol. Rev.* 66(1):64–93.

Sigler, P. B.; Xu, Z.; Rye, H. S.; Burston, S. G.; Fenton, W. A.; and Horwich, A. L. 1998. Structure and function in GroEL-mediated protein folding. *Annu. Rev. Biochem.* 67:581–608.

Yusupov, M. M.; Yusupova, G. Z.; Baucom, A.; Lieberman, K.; Earnest, T. N.; Cate, J. H. D.; and Noller, H. F. 2001. Crystal structure of the ribosome at 5.5 Å resolution. *Science* 292(5518):883–96.

12.3 Regulation of mRNA Synthesis

Amster-Choder, O. 2000. Transcriptional regulation in prokaryotes. In *Encyclopedia of microbiology*, 2d ed., vol. 4, J. Lederberg, editor-in-chief, 610–27. San Diego: Academic Press.

Errington, J. 1993. *Bacillus subtilis* sporulation: Regulation of gene expression and control of morphogenesis. *Microbiol. Rev.* 57(1):1–33.

Ishihama, A. 2000. Functional modulation of *Escherichia coli* RNA polymerase. *Annu. Rev. Microbiol.* 54:499–518.

Losick, R. 1995. Differentiation and cell fate in a simple organism. *BioScience* 45(6):400–5.

Lovett, P. S., and Rogers, E. J. 1996. Ribosome regulation by the nascent peptide. *Microbiol. Rev.* 60(2):366–85.

Ptashne, M. 1992. *A genetic switch*, 2d ed. Cambridge, Mass.: Blackwell Scientific Publications.

Ptashne, M., and Gann, A. 1997. Transcriptional activation by recruitment. *Nature* 386:569–76.

Saier, M. H. 1989. Protein phosphorylation and allosteric control of inducer exclusion and catabolite repression by the bacterial phosphoenolpyruvate:sugar phosphotransferase system. *Microbiol. Rev.* 53(1):109–21.

Severinov, K. 2000. RNA polymerase structure—function: Insights into points of transcriptional regulation. *Curr. Opin. Microbiol.* 3:118–25.

Welch, W. J. 1993. How cells respond to stress. *Sci. Am.* 268(5):56–64.

Werner, M. H., and Burley, S. K. 1997. Architectural transcription factors: Proteins that remodel DNA. *Cell* 88:733–36.

12.4 Attenuation

Landick, R.; Turnbough, C. L., Jr.; and Yanofsky, C. 1996. Transcription attenuation. In *Escherichia coli and Salmonella: Cellular and molecular biology*, 2d ed., vol. 1, F. C. Neidhardt, editor-in-chief, 1263–86. Washington, D.C.: ASM Press.

Yanofsky, C. 1981. Attenuation in the control of expression of bacterial operons. *Nature* 289:751–58.

12.5 Global Regulatory Systems

Neidhardt, F. C., and Savageau, M. A. 1996. Regulation beyond the operon. In *Escherichia coli and Salmonella: Cellular and molecular biology*, 2d ed., vol. 1, F. C. Neidhardt, editor-in-chief, 1310–24. Washington, D.C.: ASM Press.

Nogueira, T., and Springer, M. 2000. Post-transcriptional control by global regulators of gene expression in bacteria. *Curr. Opin. Microbiol.* 3:154–58.

Yura, T.; Nagai, H.; and Mori, H. 1993. Regulation of the heat-shock response in bacteria. *Annu. Rev. Microbiol.* 47:321–50.

12.6 Small RNAs and Regulation

Branch, A. D. 2000. Antisense RNAs. In *Encyclopedia of microbiology,* 2d ed., vol. 1, J. Lederberg, editor-in-chief, 268–85. San Diego: Academic Press.

Feng, X.; Oropeza, R.; Walthers, D.; and Kenney, L. J. 2003. OmpR phosphorylation and its role in signaling and pathogenesis. *ASM News* 69(8), 390–95.

Lau, N. C., and Bartel, D. P. 2003. Censors of the genome. *Sci. Am.* 289(2):34–41.

Moore, S. D.; McGinness, K. E.; and Sauer, R. T. 2003. A glimpse into tmRNA-mediated ribosome rescue. *Science* 300:72–73.

Muto, A.; Ushida, C.; and Himeno, H. 1998. A bacterial RNA that functions as both a tRNA and an mRNA. *Trends Biochem. Sci.* 23:25–29.

Storz, G. 2002. An expanding universe of noncoding RNAs. *Science* 296:1260–63.

Wagner, E. G. H., and Simons, R. W. 1994. Antisense RNA control in bacteria, phages, and plasmids. *Annu. Rev. Microbiol.* 48:713–42.

Wassarman, K. M.; Zhang, A.; and Storz, G. 1999. Small RNAs in *Escherichia coli. Trends Microbiol.* 7(1):37–45.

Weintraub, H. M. 1990. Antisense RNA and DNA. *Sci. Am.* 262(1):40–46.

12.7 Two-Component Phosphorelay Systems

Hoch, J. A. 2000. Two-component and phosphorelay signal transduction. *Curr. Opin. Microbiol.* 3:165–70.

Ninfa, A. J., and Atkinson, M. R. 2000. Two-component systems. In *Encyclopedia of microbiology,* 2d ed., vol. 4, J. Lederberg, editor-in-chief, 742–54. San Diego: Academic Press.

Perego, M. 1998. Kinase-phosphatase competition regulates *Bacillus subtilis* development. *Trends Microbiol.* 6(9):366–70.

Perraud, A.-L.; Weiss, V.; and Gross, R. 1999. Signalling pathways in two-component phosphorelay systems. *Trends Microbiol.* 7(3):115–20.

Piggot, P. J. 2000. Sporulation. In *Encyclopedia of microbiology,* 2d ed., vol. 4, J. Lederberg, editor-in-chief, 377–86. San Diego: Academic Press.

12.8 Control of the Cell Cycle

Donachie, W. D. 1993. The cell cycle of *Escherichia coli. Annu. Rev. Microbiol.* 47:199–230.

Draper, G. C., and Gober, J. W. 2002. Bacterial chromosome segregation. *Annu. Rev. Microbiol.* 56:567–97.

Errington, J.; Daniel, R. A.; and Scheffers, D.-J. 2003. Cytokinesis in bacteria. *Microbiol. Mol. Biol. Rev.* 67(1):52–65.

Gordon, G. S., and Wright, A. 2000. DNA segregation in bacteria. *Annu. Rev. Microbiol.* 54:681–708.

Laub, M. T.; McAdams, H. H.; Feldblyum, T.; Fraser, C. M.; and Shapiro, L. 2000. Global analysis of the genetic network controlling a bacterial cell cycle. *Science* 290:2144–48.

Leonard, A. C., and Grimwade, J. E. 2000. Chromosome replication and segregation. In *Encyclopedia of microbiology,* 2d ed., vol. 1, J. Lederberg, editor-in-chief, 822–33. San Diego: Academic Press.

Lutkenhaus, J., and Mukherjee, A. 1996. Cell division. In *Escherichia coli and Salmonella: Cellular and molecular biology,* 2d ed., vol. 2, F. C. Neidhardt, editor-in-chief, 1615–26. Washington, D.C.: ASM Press.

Lutkenhaus, J., and Addinall, S. G. 1997. Bacterial cell division and the Z ring. *Annu. Rev. Biochem.* 66:93–116.

Marczynski, G. T., and Shapiro, L. 2002. Control of chromosome replication in *Caulobacter crescentus. Annu. Rev. Microbiol.* 56:625–56.

Murray, A., and Hunt, T. 1993. *The cell cycle: An introduction.* New York: W. H. Freeman.

Rothfield, L. I., and Justice, S. S. 1997. Bacterial cell division: The cycle of the ring. *Cell* 88:581–84.

Sherratt, D. J. 2003. Bacterial chromosome dynamics. *Science* 301:780–85.

Chapter 13: Microbial Recombination and Plasmids

Chapters 11 and 12 references also should be consulted, particularly the introductory and advanced textbooks.

General References

Brock, T. D. 1990. *The emergence of bacterial genetics.* Cold Spring Harbor, N.Y.: Cold Spring Harbor Laboratory Press.

Bushman, F. 2002. *Lateral DNA transfer: Mechanisms and consequences.* Woodbury, N.Y.: Cold Spring Harbor Laboratory Press.

Low, K. B. 2000. *Escherichia coli* and *Salmonella,* genetics. In *Encyclopedia of microbiology,* 2d ed., vol. 2, J. Lederberg, editor-in-chief, 270–82. San Diego: Academic Press.

Neidhardt, F. C., editor-in-chief. 1996. *Escherichia coli and Salmonella: Cellular and molecular biology,* 2d ed. Washington, D.C.: ASM Press.

Neidhardt, F. C.; Ingraham, J. L.; and Schaechter, M. 1990. *Physiology of the bacterial cell.* Sunderland, Mass.: Sinauer.

Streips, U. N., and Yasbin, R. E., editors. 1991. *Modern microbial genetics.* New York: Wiley-Liss, Inc.

13.1 Bacterial Recombination: General Principles

Cohan, F. M. 1996. The role of genetic exchange in bacterial evolution. *ASM News* 62(12):631–36.

Kowalczykowski, S. C. 2000. Initiation of genetic recombination and recombination-dependent replication. *Trends Biochem. Sci.* 25:156–65.

Kowalczykowski, S. C.; Dixon, D. A.; Eggleston, A. K.; Lauder, S. D.; and Rehrauer, W. M. 1994. Biochemistry of homologous recombination in *Escherichia coli. Microbiol. Rev.* 58(3):401–65.

Matic, I.; Taddei, F.; and Radman, M. 1996. Genetic barriers among bacteria. *Trends Microbiol.* 4(2):69–73.

Miller, R. V. 1998. Bacterial gene swapping in nature. *Sci. Am.* 278(1):67–71.

Stahl, F. W. 1987. Genetic recombination. *Sci. Am.* 256(2):91–101.

13.2 Bacterial Plasmids

Khan, S. A. 1997. Rolling-circle replication of bacterial plasmids. *Microbiol. Mol. Biol. R.* 61(4):442–55.

Parret, A. H. A., and De Mot, R. 2002. Bacteria killing their own kind: Novel bacteriocins of *Pseudomonas* and other γ-proteobacteria. *Trends Microbiol.* 10(3):107–12.

Rasooly, A., and Rasooly, R. S. 1997. How rolling circle plasmids control their copy number. *Trends Microbiol.* 5(11):40–46.

Summers, D. K. 1996. *The biology of plasmids.* Cambridge: Mass.: Blackwell Science Ltd.

Thomas, C. M. 2000. Plasmids, bacterial. In *Encyclopedia of microbiology,* 2d ed., vol. 3, J. Lederberg, editor-in-chief, 711–29. San Diego: Academic Press.

13.3 Transposable Elements

Bennett, P. M. 2000. Transposable elements. In *Encyclopedia of microbiology,* 2d ed., vol. 4, J. Lederberg, editor-in-chief, 704–24. San Diego: Academic Press.

Davies, D. R.; Goryshin, I. Y.; Reznikoff, W. S.; and Rayment, I. 2000. Three-dimensional structure of the Tn5 synaptic complex transposition intermediate. *Science* 289:77–85.

Haren, L.; Ton-Hoang, B.; and Chandler, M. 1999. Integrating DNA: Transposases and retroviral integrases. *Annu. Rev. Microbiol.* 53:245–81.

Kleckner, N. 1990. Regulation of transposition in bacteria. *Annu. Rev. Cell Biol.* 6:297–327.

Salyers, A. A.; Shoemaker, N. B.; Stevens, A. M.; and Li, L.-Y. 1995. Conjugative transposons: An unusual and diverse set of integrated gene transfer elements. *Microbiol. Rev.* 59(4):579–90.

Scott, J. R., and Churchward, G. G. 1995. Conjugative transposition. *Annu. Rev. Microbiol.* 49:367–97.

13.4 Bacterial Conjugation

Firth, N.; Ippen-Ihler, K.; and Skurray, R. A. 1996. Structure and function of the F factor and mechanism of conjugation. In *Escherichia coli and Salmonella: Cellular and molecular biology,* 2d ed., vol. 2, F. C. Neidhardt, editor-in-chief, 2377–401. Washington, D.C.: ASM Press.

Frost, L. S. 2000. Conjugation, bacterial. In *Encyclopedia of microbiology,* 2d ed., vol. 1, J. Lederberg, editor-in-chief, 847–62. San Diego: Academic Press.

Grohmann, E.; Muth, G.; and Espinosa, M. 2003. Conjugative plasmid transfer in gram-positive bacteria. *Microbiol. Mol. Biol. Rev.* 67(2):277–301.

13.5 DNA Transformation

Claverys, J.-P., and Martin, B. 2003. Bacterial "competence" genes: Signatures of active transformation, or only remnants? *Trends Microbiol.* 11(4):161–65.

Dubnau, D. 1999. DNA uptake in bacteria. *Annu. Rev. Microbiol.* 53:217–44.

McCarty, M. 1985. *The transforming principle: Discovering that genes are made of DNA.* New York: W. W. Norton.

Smeets, L. C.; and Kusters, J. G. 2002. Natural transformation in *Helicobacter pylori:* DNA transport in an unexpected way. *Trends Microbiol.* 10(4):159–62.

Solomon, J. M., and Grossman, A. D. 1996. Who's competent and when: Regulation of natural genetic competence in bacteria. *Trends Genetics* 12(4):150–55.

Wilkins, B. M., and Meacock, P. A. 2000. Transformation, genetic. In *Encyclopedia of microbiology,* 2d ed., vol. 4, J. Lederberg, editor-in-chief, 651–65. San Diego: Academic Press.

13.6 Transduction

Masters, M. 2000. Transduction: Host DNA transfer by bacteriophages. In *Encyclopedia of microbiology,* 2d ed., vol. 4, J. Lederberg, editor-in-chief, 637–50. San Diego: Academic Press.

13.7 Mapping the Genome

Berlyn, M. K. B.; Low, K. B.; and Rudd, K. E. 1996. Linkage map of *Escherichia coli* K-12, edition 9. In *Escherichia coli and Salmonella: Cellular and molecular biology,* 2d ed., vol. 2, F. C. Neidhardt, editor-in-chief, 1715–1902. Washington, D.C.: ASM Press.

Sanderson, K. E.; Hessel, A.; and Rudd, K. E. 1995. Genetic map of *Salmonella typhimurium,* edition VIII. *Microbiol. Rev.* 59(2):241–303.

Chapter 14: Recombinant DNA Technology

General References

Cohen, S. N. 1975. The manipulation of genes. *Sci. Am.* 233(1):25–33.

Falkow, S. 2001. I'll have the chopped liver please, or how I learned to love the clone. *ASM News* 67(11):555–59.

Glick, B. R., and Pasternak, J. J. 2003. *Molecular biotechnology: Principles and applications of recombinant DNA,* 3d ed. Washington, D.C.: ASM Press.

Primrose, S. B.; Twyman, R. M.; and Old, R. W. 2001. *Principles of Gene Manipulation,* 6th ed. Boston: Blackwell Science.

Snyder, L., and Champness, W. 1997. *Molecular genetics of bacteria.* Washington, D.C.: ASM Press.

Watson, J. D.; Gilman, M.; Witkowski, J.; and Zoller, M. 1992. *Recombinant DNA,* 2d ed. San Francisco: W. H. Freeman.

Zyskind, J. W. 2000. Recombinant DNA, basic procedures. In *Encyclopedia of microbiology,* 2d ed., vol. 4, J. Lederberg, editor-in-chief, 55–64. San Diego: Academic Press.

14.1 Historical Perspectives

Bickle, T. A., and Krüger, D. H. 1993. Biology of DNA restriction. *Microbiol. Rev.* 57(2):434–50.

Lear, J. 1978. *Recombinant DNA: The untold story.* New York: Crown Publishers.

Murray, N. E. 2000. DNA restriction and modification. In *Encyclopedia of microbiology,* 2d ed., vol. 2, J. Lederberg, editor-in-chief, 91–105. San Diego: Academic Press.

14.3 The Polymerase Chain Reaction

Arnheim, N., and Levenson, C. H. 1990. Polymerase chain reaction. *Chem Eng. News* 68:36–47.

Cockerill, F. R., and Smith, T. F. 2002. Rapid-cycle real-time PCR: A revolution for clinical microbiology. *ASM News* 68(2):77–83.

Mullis, K. B. 1990. The unusual origin of the polymerase chain reaction. *Sci. Am.* 262(4):56–65.

Palmer, C. J., and Paszko-Kolva, C. 2000. Polymerase chain reaction (PCR). In *Encyclopedia of microbiology,* 2d ed., vol. 3, J. Lederberg, editor-in-chief, 787–91. San Diego: Academic Press.

14.5 Cloning Vectors

Monaco, A. P., and Larin, Z. 1994. YACs, BACs, PACs and MACs: Artificial chromosomes as research tools. *Trends Biotechnol.* 12:280–86.

Shizuya, H., et al. 1992. Cloning and stable maintenance of 300-kilobase-pair fragments of human DNA in *Escherichia coli* using an F-factor-based vector. *Proc. Natl. Acad. Sci.* 89:8794–97.

14.6 Inserting Genes into Eucaryotic Cells

Chilton, M.-D. 1983. A vector for introducing new genes into plants. *Sci. Am.* 248(6):51–59.

Crystal, R. G. 1995. Transfer of genes to humans: Early lessons and obstacles to success. *Science* 270:404–10.

Karcher, S. J. 1994. Getting DNA into a cell: A survey of transformation methods. *Am. Biol. Tech.* 56(1):14–20.

Smith, A. E. 1995. Viral vectors in gene therapy. *Annu. Rev. Microbiol.* 49:807–38.

14.8 Applications of Genetic Engineering

Felgner, P. L. 1997. Nonviral strategies for gene therapy. *Sci. Am.* 276(6):102–6.

Friedmann, T. 1997. Overcoming the obstacles to gene therapy. *Sci. Am.* 276(6):96–101.

Gasser, C. S., and Fraley, R. T. 1992. Transgenic crops. *Sci. Am.* 266(6):62–69.

Gerngross, T. U., and Slater, S. C. 2000. How green are green plastics? *Sci. Am.* 283(2):36–41.

Lillehoj, E. P., and Ford, G. M. 2000. Industrial biotechnology, overview. In *Encyclopedia of microbiology,* 2d ed., vol. 2, J. Lederberg, editor-in-chief, 722–37. San Diego: Academic Press.

Porter, A. G.; Davidson, E. W.; and Liu, J.-W. 1993. Mosquitocidal toxins of bacilli and their genetic manipulation for effective biological control of mosquitoes. *Microbiol. Rev.* 57(4):838–61.

Roessner, C. A., and Scott, A. I. 1996. Genetically engineered synthesis of natural products: From alkaloids to corrins. *Annu. Rev. Microbiol.* 50:467–90.

14.9 Social Impact of Recombinant DNA Technology

Ben-Ari, E. 2002. *Bacillus thuringiensis* as a paradigm for transgenic organisms. *ASM News* 68(12):597–602.

Brown, K. 2001. Seeds of Concern. *Sci. Am.* 284(4):52–57.

Grobstein, C. 1977. The recombinant-DNA debate. *Sci. Am.* 237(1):22–33.

Hopkin, K. 2001. The risks on the table. *Sci. Am.* 284(4):60–61.

Krimsky, S. 1982. *Genetic alchemy: The social history of the recombinant DNA controversy.* Cambridge, Mass.: MIT Press.

Marvier, M. 2001. Ecology of transgenic crops. *American Scientist* 89:160–67.

Poupard, J. A., and Miller, L. A. 2000. Biological warfare. In *Encyclopedia of microbiology,* 2d ed., vol. 1, J. Lederberg, editor-in-chief, 506–19. San Diego: Academic Press.

Teitelman, R. 1989. *Gene dreams: Wall street, academia, and the rise of biotechnology.* New York: Basic Books, Inc.

Tolin, S., and Vidaver, A. 2000. Genetically modified organisms: Guidelines and regulations for research. In *Encyclopedia of microbiology,* 2d ed., vol. 2, J. Lederberg, editor-in-chief, 499–509. San Diego: Academic Press.

Tucker, J. B. Winter 1984–85. Gene wars. *Foreign Policy* 57:58–79.

Wilson, M., and Lindow, S. E. 1993. Release of recombinant microorganisms. *Annu. Rev. Microbiol.* 47:913–44.

Wolfenbarger, L. L., and Phifer, P. R. 2000. The ecological risks and benefits of genetically engineered plants. *Science* 290:2088–93.

Chapter 15: Microbial Genomics

General References

Brown, T. A. 1999. *Genomes.* New York: John Wiley.

Campbell, A. M., and Heyer, L. J. 2003. *Discovering genomics, proteomics, and bioinformatics.* San Francisco, CA: Benjamin Cummings.

Caporale, L. H. 2003. Foresight in genome evolution. *American Scientist* 91:234–41.

Charlebois, R. L., editor 1999. *Organization of the prokaryotic genome.* Washington, D.C.: ASM Press.

Dougherty, B. A. 2000. DNA sequencing and genomics. In *Encyclopedia of microbiology,* 2d ed., vol. 2, J. Lederberg, editor-in-chief, 106–16. San Diego: Academic Press.

Ezzell, C. 2000. Beyond the human genome. *Sci. Am.* 283(1):64–69.

Gibson, G., and Muse, S. V. 2002. *A primer of genome science.* Sunderland, Mass.: Sinauer Associates.

Haseltine, W. A. 1997. Discovering genes for new medicines. *Sci. Am.* 276(3):92–7.

Lander, E. S., and Weinberg, R. A. 2000. Genomics: Journey to the center of biology. *Science* 287:1777–82.

Lawrence, J. G.; Hendrix, R. W.; and Casjens, S. 2001. Where are the pseudogenes in bacterial genomes? *Trends Microbiol.* 9(11):535–40.

Moat, A. G.; Foster, J. W.; and Spector, M. P. 2002. *Microbial physiology,* 4th ed. New York: Wiley-Liss.

Primrose, S. B., and Twyman, R. 2003. *Principles of genome analysis and genomics,* 3d ed. Boston: Blackwell Publishing.

Snyder, M., and Gerstein, M. 2003. Defining genes in the genomics era. *Science* 300:258–60.

15.4 Bioinformatics

Costanzo, M. C., et. al. 2002. Reducing the complexity of the information landscape. *ASM News* 68(1):23–31.

Howard, K. 2000. The bioinformatics gold rush. *Sci. Am.* 283(1):58–63.

Krane, D. E., and Raymer, M. L. 2003. *Fundamental concepts of bioinformatics.* San Francisco, Calif.: Benjamin Cummings.

Saier, Jr., M. H. 2003. Answering fundamental questions in biology with bioinformatics. *ASM News* 69(4):175–81.

15.5 General Characteristics of Microbial Genomes

Andersson, S. G. E., et al. 1998. The genome sequence of *Rickettsia prowazekii* and the origin of mitochondria. *Nature* 396:133–40.

Blattner, F. R., et al. 1997. The complete genome sequence of *Escherichia coli* K-12. *Science* 277:1453–62.

Bult, C. J., et al. 1996. Complete genome sequence of the methanogenic archaeon, *Methanococcus jannaschii. Science* 273:1058–1107.

Cole, S. T., et al. 1998. Deciphering the biology of *Mycobacterium tuberculosis* from the complete genome sequence. *Nature* 393:537–44.

Ferretti, J. J., et al. 2001. Complete genome sequence of an M1 strain of *Streptococcus pyogenes. Proc. Natl. Acad. Sci.* 98(8):4658–63.

Fleischmann, R. D., et al. 1995. Whole-genome random sequencing and assembly of *Haemophilus influenzae* Rd. *Science* 269:496–512.

Fraser, C. M., et al. 1995. The minimal gene complement of *Mycoplasma genitalium. Science* 270:397–403.

Fraser, C. M., et al. 1998. Complete genome sequence of *Treponema pallidum,* the syphilis spirochete. *Science* 281:375–88.

Gaasterland, T. 1999. Archaeal genomics. *Curr. Opin. Microbiol.* 2(5):542–47.

Heidelberg, J. F., et al. 2000. DNA sequence of both chromosomes of the cholera pathogen *Vibrio cholerae. Nature* 406:477–83.

Jain, R.; Rivera, M. C.; and Lake, J. A. 1999. Horizontal gene transfer among genomes: The complexity hypothesis. *Proc. Natl. Acad. Sci.* 96:3801–6.

Kuroda, M., et al. 2001. Whole genome sequencing of meticillin-resistant *Staphylococcus aureus. Lancet* 357:1225–40.

Lawrence, J. G., and Ochman, H. 1998. Molecular archaeology of the *Escherichia coli* genome. *Proc. Natl. Acad. Sci.* 95:9413–17.

Makarova, K. S.; Aravind, L.; Wolf, Y. I.; Tatusov, R. L.; Minton, K. W.; Koonin, E. V.; and Daly, M. J. 2001. Genome of the extremely radiation-resistant bacterium *Deinococcus radiodurans* viewed from the perspective of comparative genomics. *Microbiol. Mol. Biol. Rev.* 65(1):44–79.

Moran, N. A. 2002. Genome evolution in symbiotic bacteria. *ASM News* 68(10):499–505.

Ochman, H.; Lawrence, J. G.; and Groisman, E. A. 2000. Lateral gene transfer and the nature of bacterial innovation. *Nature* 408:299–304.

Pollack, J. D. 1997. *Mycoplasma* genes: A case for reflective annotation. *Trends Microbiol.* 5(10):413–19.

Riley, M., and Serres, M. H. 2000. Interim report on genomics of *Escherichia coli. Annu. Rev. Microbiol.* 54:341–411.

Snel, B.; Bork, P.; and Huynen, M. 2000. Genome evolution: Gene fusion versus gene fission. *Trends Genet.* 16(1):9–11.

Stephens, R. S., et al. 1998. Genome sequence of an obligate intracellular pathogen of humans: *Chlamydia trachomatis. Science* 282:754–59.

Tettelin, H., et al. 2001. Complete genome sequence of a virulent isolate of *Streptococcus pneumoniae. Science* 293:498–506.

White, O., et al. 1999. Genome sequence of the radioresistant bacterium *Deinococcus radiodurans* R1. *Science* 286:1571–77.

15.6 Functional Genomics

Brenner, S. 2000. The end of the beginning. *Science* 287:2173–74.

Eisenberg, D.; Marcotte, E. M.; Xenarios, I.; and Yeates, T. O. 2000. Protein function in the post-genomic era. *Nature* 405:823–826.

Ezzell, C. 2002. Proteins rule. *Sci. Am.* 286(4):40–47.

Friend, S. H., and Stoughton, R. B. 2002. The magic of microarrays. *Sci. Am.* 286(2):38–53.

Gingeras, T. R., and Rosenow, C. 2000. Studying microbial genomes with high-density oligonucleotide arrays. *ASM News* 66(8):463–69.

Graves, P. R., and Haystead, T. A. J. 2002. Molecular biologist's guide to proteomics. *Microbiol. Mol. Biol. Rev.* 66(1):39–63.

Hamadeh, H., and Afshari, C. A. 2000. Gene chips and functional genomics. *American Scientist* 88:508–15.

Huang, S. 2000. The practical problems of post-genomic biology. *Nature Biotechnol.* 18:471–72.

Lockhart, D. J., and Winzeler, E. A. 2000. Genomics, gene expression and DNA arrays. *Nature* 405:827–36.

Machalek, A. Z. 2001. Structural genomics: a slice of the proteomics pie. *ASM News* 67(9):441–7.

Rhodius, V.; Van Dyk, T. K.; Gross, C.; and LaRossa, R. A. 2002. Impact of genomic technologies on studies of bacterial gene expression. *Annu. Rev. Microbiol.* 56:599–624.

Schena, M., editor. 1999. *DNA microarrays: A practical approach.* New York: Oxford University Press.

Van Dyk, T. K. 2002. Lighting the way: Genome-wide, bioluminescent gene expression analysis. *ASM News* 68(5):222–30.

Chapter 16: The Viruses: Introduction and General Characteristics

General References

Cann, A. J. 2001. *Principles of molecular virology,* 3d ed. San Diego: Academic Press.

Dimmock, N. J.; Easton, A. J.; and Leppard, K. N. 2001. *Introduction to modern virology,* 5th ed. London: Blackwell Scientific Publications.

Fields, B. N.; Knipe, D. M.; Chanock, R. M.; Hirsch, M. S.; Melnick, J. L.; Monath, T. P.; and Roizman, B., editors. 1990. *Fields virology,* 2d ed. New York: Raven Press.

Flint, S. J.; Enquist, L. W.; Krug, R. M.; Racaniello, V. R.; and Skalka, A. M. 2000. *Principles of virology: Molecular biology, pathogenesis, and control.* Washington, D.C.: ASM Press.

Hendrix, R. W.; Lawrence, J. G.; Hatfull, G. F.; and Casjens, S. 2000. The origins and ongoing evolution of viruses. *Trends Microbiol.* 8(11):504–8.

Henig, R. M. 1993. *A dancing matrix—Voyages along the viral frontier.* New York: Knopf.

Knight, D. M., and Howley, P. M., editors. 2001. *Fundamental virology,* 4th ed. Philadelphia Lippincott Williams & Wilkins.

Matthews, R. E. F. 1991. *Plant virology,* 3d ed. San Diego: Academic Press

Schlesinger, S., and Schlesinger, M. J. 2000. Viruses. In *Encyclopedia of microbiology,* 2d ed., vol. 4, J. Lederberg, editor-in-chief, 796–810. San Diego: Academic Press.

Tidona, C. A.; Gholamreza, D.; and Buchen-Osmond, C., editors. 2002. *The Springer index of viruses.* New York: Springer-Verlag.

Voyles, B. A. A. 2002. *The biology of viruses,* 2d ed. Chicago: McGraw-Hill.

Webster, R. G., and Granoff, A., editors. 1994. *Encyclopedia of virology.* San Diego: Academic Press.

White, D. O., and Fenner, F. J. 1994. *Medical virology,* 4th ed. San Diego: Academic Press.

16.1 Early Development of Virology

Bos, L. 2000. 100 years of virology: From vitalism via molecular biology to genetic engineering. *Trends Microbiol.* 8(2):82–87.

Eggers, H. J. 1995. Picornaviruses—A historical view. *ASM News* 61(3):121–24.

Fenn, E. A. 2001. *Pox americana: The great smallpox epidemic of 1775–82.* New York: Hill and Wang.

Jennings, F. 1975. *The invasion of America: Indians, colonialism, and the cant of conquest.* Chapel Hill: University of North Carolina Press.

Lechevalier, H. A., and Solotorovsky, M. 1965. *Three centuries of microbiology.* New York: McGraw-Hill.

McNeill, W. H. 1976. *Plagues and peoples.* Garden City, N.Y.: Anchor.

Oldstone, M. B. 1998. *Viruses, plagues & history.* New York: Oxford University Press.

Stearn, E. W., and Stearn, A. E. 1945. *The effect of smallpox on the destiny of the Amerindian.* Boston: Bruce Humphries.

van Helvoort, T. 1996. When did virology start? *ASM News* 62(3):142–45.

Zaitlin, M. 1999. Tobacco mosaic virus and its contributions to virology. *ASM News* 65(10): 675–80.

16.4 Virus Purification and Assays

Henshaw, N. G. 1988. Identification of viruses by methods other than electron microscopy. *ASM News* 54(9):482–85.

Miller, S. E. 1988. Diagnostic virology by electron microscopy. *ASM News* 54(9):475–81.

16.5 The Structure of Viruses

Baker, T. S.; Olson, N. H.; and Fuller, S. D. 1999. Adding the third dimension to virus life cycles: Three-dimensional reconstruction of icosahedral viruses from cryo-electron micrographs. *Microbiol. Mol. Biol. Rev.* 63(4):862–922.

Bresnahan, W. A., and Shenk, T. 2000. A subset of viral transcripts packaged within human cytomegalovirus particles. *Science* 288:2373–76.

Casjens, S. 1985. *Virus structure and assembly.* Boston: Jones and Bartlett.

16.6 Principles of Virus Taxonomy

Eigen, M. 1993. Viral quasispecies. *Sci. Am.* 269(1):42–49.

Van Regenmortel, M. H. V.; Fauquet, C. M.; Bishop, D. H. L.; Carstens, E. B.; Estes, M. K.; Lemon, S. M.; Maniloff, J.; Mayo, M. A.; McGeoch, D. J.; Pringle, C. R.; and Wickner, R. B., editors. 2000. *Virus taxonomy: The classification and nomenclature of viruses. Seventh report of the international committee on taxonomy of viruses.* San Diego: Academic Press.

Chapter 17: The Viruses: Bacteriophages

Chapter 16 references also should be consulted, particularly the introductory and advanced texts.

General References

Ackermann, H.-W. 2000. Bacteriophages. In *Encyclopedia of microbiology,* 2d ed., vol. 1, J. Lederberg, editor-in-chief, 398–411. San Diego: Academic Press.

Banks, D. J.; Beres, S. B.; and Musser, J. M. 2002. The fundamental contribution of phages to GAS evolution, genome diversification and strain emergence. *Trends Microbiol.* 10(11):515–21.

Boyd, E. F.; Davis, B. M.; and Hochhut, B. 2001. Bacteriophage-bacteriophage interactions in the evolution of pathogenic bacteria. *Trends Microbiol.* 9(3):137–44.

Boyd, E. F., and Brüssow, H. 2002. Common themes among bacteriophage-encoded virulence factors and diversity among the bacteriophages involved. *Trends Microbiol.* 10(11):521–29.

Campbell, A. M. 1996. Bacteriophages. In *Escherichia coli and Salmonella: Cellular and molecular biology,* 2d ed., vol. 2, F. C. Neidhardt, editor-in-chief, 2325–38. Washington, D.C.: ASM Press.

Lewin, B. 2000. *Genes,* 7th ed. New York: Oxford University Press.

Stent, G. S., and Calendar, R. 1978. *Molecular genetics: An introductory narrative,* 2d ed. San Francisco: W. H. Freeman.

Stone, R. 2002. Stalin's forgotten cure. *Science* 298:728–31.

17.1 Classification of Bacteriophages

Prangishvili, D.; Stedman, K.; and Zillig, W. 2001. Viruses of the extremely thermophilic archaeon *Sulfolobus. Trends Microbiol.* 9(1):39–43.

Van Regenmortel et al. 2000. *Virus taxonomy: Seventh report of the international committee on taxonomy of viruses.* San Diego: Academic Press.

17.2 Reproduction of Double-Stranded DNA Phages: The Lytic Cycle

Bazinet, C., and King, J. 1985. The DNA translocating vertex of dsDNA bacteriophage. *Annu. Rev. Microbiol.* 39:109–29.

Black, L. W. 1989. DNA packaging in dsDNA bacteriophages. *Annu. Rev. Microbiol.* 43:267–92.

Karam, J. D., editor. 1994. *Molecular biology of bacteriophage T4.* Herndon, Va.: ASM Press.

Lu, M.-J., and Henning, U. 1994. Superinfection exclusion by T-even-type coliphages. *Trends Microbiol.* 2(4):137–39.

Miller, E. S.; Kutter, E.; Mosig, G.; Arisaka, F.; Kunisawa, T.; and Rüger, W. 2003 Bacteriophage T4 genome. *Microbiol. Mol. Biol. Rev.* 67(1):85–156.

Russel, M. 1995. Moving through the membrane with filamentous phases. *Trends Microbiol.* 3(6):223–28.

Rydman, P. S., and Bamford, D. H. 2002. Phage enzymes digest peptidoglycan to deliver DNA. *ASM News* 68(7):330–35.

Wang, I.-N.; Smith, D. L.; and Young, R. 2000. Holins: The protein clocks of bacteriophage infections. *Annu. Rev. Microbiol.* 54:799–825.

Young, R.; Wang, I.-N.; and Roof, W. D. 2000. Phages will out: Strategies of host cell lysis. *Trends Microbiol.* 8(3):120–28.

17.4 Reproduction of RNA Phages

Bernhardt, T. G.; Wang, I.-N.; Struck, D. K.; and Young, R. 2001. A protein antibiotic in the phage Qβ virion: Diversity in lysis targets. *Science* 292:2326–29.

17.5 Temperate Bacteriophages and Lysogeny

Canchaya, C.; Proux, C.; Fournous, G.; Bruttin, A.; and Brüssow, H. 2003. Prophage genomics. *Microbiol. Mol. Biol. Rev.* 67(2):238–76.

Murialdo, H. 1991. Bacteriophage lambda DNA maturation and packaging. *Annu. Rev. Biochem.* 60:125–53.

Ptashne, M. 1992. *A genetic switch,* 2d ed. Cambridge, Mass.: Blackwell Scientific Publications.

Chapter 18: The Viruses: Viruses of Eucaryotes

Chapter 16 references also should be consulted, particularly the introductory and advanced texts.

General References

Evans, A. S. 1997. *Viral infections of humans: Epidemiology & control,* 4th ed. New York: Plenum.

Griffin, D. W.; Donaldson, K. A.; Paul, J. H.; and Rose, J. B. 2003. Pathogenic human viruses in coastal waters. *Clin. Microbiol. Rev.* 16(1):129–43.

Lai, M. M. C. 1992. RNA recombination in animal and plant viruses. *Microbiol. Rev.* 56(1):61–79.

Latchman, D. S. 2000. Transcription, viral. In *Encyclopedia of microbiology,* 2d ed., vol. 4, J. Lederberg, editor-in-chief, 628–36. San Diego: Academic Press.

Morse, S. S. 2000. Viruses, emerging. In *Encyclopedia of microbiology,* 2d ed., vol. 4, J. Lederberg, editor-in-chief, 811–31. San Diego: Academic Press.

Schlesinger, S., and Schlesinger, M. J. 2000. Viruses. In *Encyclopedia of microbiology,* 2d ed., vol. 4, J. Lederberg, editor-in-chief, 796–810. San Diego: Academic Press.

Wommack, K. E., and Colwell, R. R. 2000. Virioplankton: Viruses in aquatic ecosystems. *Micro. Mol. Biol. Rev.* 64(1):69–114.

18.2 Reproduction of Animal Viruses

Basler, C. F., and Palese, P. 2000. Influenza viruses. In *Encyclopedia of microbiology,* 2d ed., vol. 2, J. Lederberg, editor-in-chief, 797–812. San Diego: Academic Press.

Belsham, G. J., and Sonenberg, N. 2000. Picornavirus RNA translation: Roles for cellular proteins. *Trends Microbiol.* 8(7):330–35.

Boehmer, P. E., and Lehman, I. R. 1997. Herpes simplex virus DNA replication. *Annu. Rev. Biochem.* 66:347–84.

Carr, C. M., and Kim, P. S. 1994. Flu virus invasion: Halfway there. *Science* 266:234–36.

Casasnovas, J. M. 2000. The dynamics of receptor recognition by human rhinoviruses. *Trends Microbiol.* 8(6):251–54.

Cello, J.; Paul, A. V.; and Wimmer, E. 2002. Chemical synthesis of poliovirus cDNA: Generation of infectious virus in the absence of natural template. *Science* 297:1016–18.

Chazal, N., and Gerlier, D. 2003. Virus entry, assembly, budding, and membrane rafts. *Microbiol. Mol. Biol. Rev.* 67(2):226–37.

Dornburg, R., and Pomerantz, R. J. 2000. Retroviruses. In *Encyclopedia of microbiology,* 2d ed., vol. 4, J. Lederberg, editor-in-chief, 81–96. San Diego: Academic Press.

Faulkner, G. C.; Krajewski, A. S.; and Crawford, D. H. 2000. The ins and outs of EBV infection. *Trends Microbiol.* 8(4):185–89.

Frankel, A. D., and Young, J. A. T. 1998. HIV-1: Fifteen proteins and an RNA. *Annu. Rev. Biochem.* 67:1–25.

Gallo, R. C., and Montagnier, L. 2002. Prospects for the future. *Science* 298:1730–31.

Garoff, H.; Hewson, R.,; and Opstelten, D.-J. E. 1998. Virus maturation by budding. *Microbiol. Mol. Biol. Rev.* 62(4):1171–90.

Greber, U. F.; Singh, I.; and Helenius, A. 1994. Mechanisms of virus uncoating. *Trends Microbiol.* 2(2):52–56.

Hindmarsh, P., and Leis, J. 1999. Retroviral DNA integration. *Micro. Mol. Biol. Rev.* 63(4):836–43.

Hogle, J. M. 2002. Poliovirus cell entry: Common structural themes in viral cell entry pathways. *Annu. Rev. Microbiol.* 56:677–702.

Katz, R. A., and Skalka, A. M. 1994. The retroviral enzymes. *Annu. Rev. Biochem.* 63:133–73.

Levy, J. A. 1997. *HIV and the pathogenesis of AIDS,* 2d ed. Washington, D.C.: ASM Press.

Melikyan, G. B., and Chernomordik, L. V. 1997. Membrane rearrangements in fusion mediated by viral proteins. *Trends Microbiol.* 5(9):349–55.

Nayak, D. P. 1996. A look at assembly and morphogenesis of orthomyxo- and paramyxoviruses. *ASM News* 62(8):411–14.

Norkin, L. C. 1995. Virus receptors: Implications for pathogenesis and the design of antiviral agents. *Microbiol. Rev.* 8(2):293–315.

Poranen, M. M.; Daugelavičius, R.; and Bamford, D. H. 2002. Common principles of viral entry. *Annu. Rev. Microbiol.* 56:521–38.

Prusiner, S. B. 2002. Discovering the cause of AIDS. *Science* 298:1726–27.

Seeger, C., and Mason, W. S. 2000. Hepatitis B virus biology. *Micro. Mol. Biol. Rev.* 64(1):51–68.

Sodeik, B. 2000. Mechanisms of viral transport in the cytoplasm. *Trends Microbiol.* 8(10):465–72.

Sodeik, B., and Krijnse-Locker, J. 2002. Assembly of vaccinia virus revisited: *de novo* membrane synthesis or acquisition from the host? *Trends Microbiol.* 10(1):15–24.

Stephens, E. B., and Compans, R. W. 1988. Assembly of animal viruses at cellular membranes. *Annu. Rev. Microbiol.* 42:489–516.

Tufaro, R. 1997. Virus entry: Two receptors are better than one. *Trends Microbiol.* 5(7):257–58.

Ugolini, S.; Mondor, I.; and Sattentau, Q. J. 1999. HIV-1 attachment: Another look. *Trends Microbiol.* 7(4):144–49.

Varmus, H. 1988. Retroviruses. *Science* 240:1427–35.

Webster, R. G., and Walker, E. J. 2003. Influenza. *Am. Sci.* 91:122–29.

18.3 Cytocidal Infections and Cell Damage

Buller, R. M. L., and Palumbo, G. J. 1991. Poxvirus pathogenesis. *Microbiol. Rev.* 55(1):80–122.

Oldstone, M. B. A. 1989. Viral alteration of cell function. *Sci. Am.* 261(2):42–48.

18.5 Viruses and Cancer

Burd, E. M. 2003. Human papillomavirus and cervical cancer. *Clin. Microbiol. Rev.* 16(1):1–17.

Dang Do, A. N.; Farrell, L.; Kim, K.; Nguyen, M. L.; and Lambert, P. F. 2000. Oncogenic viruses. In *Encyclopedia of microbiology,* 2d ed., vol. 3, J. Lederberg, editor-in-chief, 456–65. San Diego: Academic Press.

Gibbs, W. W. 2003. Untangling the roots of cancer. *Sci. Am.* 289(1):56–65.

Sherker, A. H., and Marion, P. L. 1991. Hepadnaviruses and hepatocellular carcinoma. *Annu. Rev. Microbiol.* 45:475–508.

18.6 Plant Viruses

Butler, P. J. G., and Klug, A. 1978. The assembly of a virus. *Sci. Am.* 239(5):62–69.

Hull, R. 2000. Plant virology, overview. In *Encyclopedia of microbiology,* 2d ed., vol. 3, J. Lederberg, editor-in-chief, 697–710. San Diego: Academic Press.

Leisner, S. M., and Howell, S. H. 1993. Long-distance movement of viruses in plants. *Trends Microbiol.* 1(8):314–17.

Matthews, R. E. F. 1991. *Plant virology,* 3d ed. New York: Academic Press.

Milner, J. J. 1998. Tobacco mosaic virus: The first century. *Trends Microbiol.* 6(12):466–67.

18.7 Viruses of Fungi, Algae, and Protozoa

La Scola, B., et al. 2003. A giant virus in amoebae. *Science* 299:2033.

Reisser, W. 1993. Viruses and virus-like particles of freshwater and marine eukaryotic algae—a review. *Arch. Protistenkd.* 143:257–65.

18.8 Insect Viruses

Wood, H. A., and Granados, R. R. 1991. Genetically engineered baculoviruses as agents for pest control. *Annu. Rev. Microbiol.* 45:69–87.

Yousten, A. A.; Federici, B.; and Roberts, D. 2000. Insecticides, microbial. In *Encyclopedia of microbiology,* 2d ed., vol. 2, J. Lederberg, editor-in-chief, 813–25. San Diego: Academic Press.

18.9 Viroids and Prions

Diener, T. O. 1993. The viroid: Big punch in a small package. *Trends Microbiol.* 1(8):289–94.

Diener, T. O. 1996. Understanding replication mechanisms in viroids and viroidlike RNAs. *Trends Microbiol.* 4(3):85–87.

Harris, D. A. 1999. Cellular biology of prion diseases. *Clin. Microbiol. Rev.* 12(3):429–44.

Horwich, A. L., and Weissman, J. S. 1997. Deadly conformations—Protein misfolding in prion disease. *Cell* 89:499–510.

Ma, J.; Wollmann, R.; and Lindquist, S. 2002. Neurotoxicity and neurodegeneration when PrP accumulates in the cytosol. *Science* 298:1781–85.

Mestel, R. 1996. Putting prions to the test. *Science* 273:184–89.

Musahl, C., and Aguzzi, A. 2000. Prions. In *Encyclopedia of microbiology,* 2d ed., vol. 3, J. Lederberg, editor-in-chief, 809–23. San Diego: Academic Press.

Prusiner, S. B. 1995. The prion diseases. *Sci. Am.* 272(1):48–57.

Uptain, S. M., and Lindquist, S. 2002. Prions as protein-based genetic elements. *Annu. Rev. Microbiol.* 56:703–41.

Wickner, R. B.; Taylor, K. L.; Edskes, H. K.; Maddelein, M.-L.; Moriyama, H.; and Roberts, B. T. 1999. Prions in *Saccharomyces* and *Podospora* spp.: Protein-based inheritance. *Microbiol. Mol. Biol. Rev.* 63(4):844–61.

Chapter 19: Microbial Taxonomy and Phylogeny

General References

Achenbach, L. A., and Coates, J. D. 2000. Disparity between bacterial phylogeny and physiology. *ASM News* 66(12):714–15.

Balows, A.; Truper, H. G.; Dworkin, M.; Harder, W.; and Schleifer, K-H. 1992. *The prokaryotes,* 2d ed. New York: Springer-Verlag.

Bryant, T. N. 2000. Identification of bacteria, computerized. In *Encyclopedia of microbiology,* 2d ed., vol. 2, J. Lederberg, editor-in-chief, 709–21. San Diego: Academic Press.

Goodfellow, M., and O'Donnell, A. G., editors. 1993. *Handbook of new bacterial systematics.* San Diego, Calif.: Academic Press.

Hillis, D. M.; Moritz, C.; and Mable, B. K., editors. 1996. *Molecular Systematics,* 2d. ed. Sunderland, Mass.: Sinauer Associates.

Hugenholtz, P.; Goebel, B. M.; and Pace, N. R. 1998. Impact of culture-independent studies on the emerging phylogenetic view of bacterial diversity. *J. Bacteriol.* 180(18):4765–74.

Koch, A. L. 2003. Development of the sacculus marked emergence of the bacteria. *ASM News* 69(5):229–33.

Koch, A. L. 2003. Were Gram-positive rods the first bacteria? *Trends Microbiol.* 11(4):166–70.

Logan, N. A. 1994. *Bacterial systematics.* Boston: Blackwell Scientific.

Priest, F., and Austin, B. 1993. *Modern bacterial taxonomy,* 2d ed. New York: Chapman and Hall.

Staley, J. T. 1999. Bacterial biodiversity: A time for place. *ASM News* 65(10):681–87.

Vandamme, P.; Pot, B.; Gillis, M.; De Vos, P.; Kersters, K.; and Swings, J. 1996. Polyphasic taxonomy, a consensus approach to bacterial systematics. *Microbiol. Rev.* 60(2):407–38.

19.2 Microbial Evolution and Diversity

Brown, J. R., and Doolittle, W. F. 1997. *Archaea* and the prokaryote-to-eukaryote transition. *Microbiol. Mol. Biol. Rev.* 61(4): 456–502.

Gogarten, J. P. 1995. The early evolution of cellular life. *Trends Ecol. Evol.* 10(4):147–51.

Gupta, R. S., and Golding, G. B. 1996. The origin of the eukaryotic cell. *Trends Biochem. Sci.* 21:166–71.

Kasting, J. F., and Siefert, J. L. 2002. Life and the evolution of earth's atmosphere. *Science* 296:1066–68.

Kurland, C. G., and Andersson, S. G. E. 2000. Origin and evolution of the mitochondrial proteome. *Micro. Mol. Bio. Rev.* 64(4):786–820.

Saier, M. H., Jr. 2000. Bacterial diversity and the evolution of differentiation. *ASM News* 66(6):337–43.

Simpson, S. 2003. Questioning the oldest signs of life. *Sci. Am.* 288(4):70–77.

Woese, C. R.; Kandler, O.; and Wheelis, M. L. 1990. Towards a natural system of organisms: Proposal for the domains archaea, bacteria, and eucarya. *Proc. Natl. Acad. Sci.* 87:4576–79.

19.3 Taxonomic Ranks

Cohan, F. M. 2002. What are bacterial species? *Annu. Rev. Microbiol.* 56:457–87.

Lan, R., and Reeves, P. R. 2001. When does a clone deserve a name? A perspective on bacterial species based on population genetics. *Trends Microbiol.* 9(9):419–24.

19.4 Classification Systems

Sneath, P. H. A., and Sokal, R. R. 1973. *Numerical taxonomy: The principles and practice of numerical classification.* San Francisco: W. H. Freeman.

Stackebrandt, E. 1992. Unifying phylogeny and phenotypic diversity. In *The prokaryotes,* 2d ed. A. Ballows et al., editors, 19–47. New York: Springer-Verlag.

19.5 Major Characteristics Used in Taxonomy

Johnson, J. L. 1984. Nucleic acids in bacterial classification. In *Bergey's manual of systematic bacteriology,* J. G. Holt, vol. 1, N. R. Krieg, editors, 8–11. Baltimore, Md.: Williams & Wilkins.

Jones, D., and Krieg, N. R. 1984. Serology and chemotaxonomy. In *Bergey's manual of systematic bacteriology,* J. G. Holt, vol. 1, N. R. Krieg, editors, 15–18. Baltimore, Md.: Williams & Wilkins.

19.6 Assessing Microbial Phylogeny

Forterre, P. 1997. Protein versus rRNA: Problems in rooting the universal tree of life. *ASM News* 63(2):89–95.

Gupta, R. S. 2002. Phylogeny of *Bacteria:* Are we now close to understanding it? *ASM News* 68(6):284–91.

Hall, B. G. 2001. *Phylogenetic trees made easy: A how-to manual for molecular biologists.* Sunderland, Mass.: Sinauer Associates.

Ludwig, W., and Schleifer, K.-H. 1999. Phylogeny of *Bacteria* beyond the 16S rRNA standard. *ASM News* 65(11):752–57.

Sneath, P. H. A. 1989. Analysis and interpretation of sequence data for bacterial systematics: The view of a numerical taxonomist. *Syst. Appl. Microbiol.* 12:15–31.

Woese, C. R.; Olsen, G. J.; Ibba, M.; and Söll, D. 2000. Aminoacyl-tRNA synthetases, the genetic code, and the evolutionary process. *Micro. Mol. Biol. Rev.* 64(1):202–36.

19.7 The Major Divisions of Life

Baldauf, S. L.; Roger, A. J.; Wenk-Siefert, I.; Doolittle, W. F. 2000. A kingdom-level phylogeny of eukaryotes based on combined protein data. *Science* 290:972–77.

Cavalier-Smith, T. 1993. Kingdom protozoa and its 18 phyla. *Microbiol. Rev.* 57(4):953–94.

Cavalier-Smith, T. 2002. The neomuran origin of archaebacteria, the negibacterial root of the universal tree and bacterial megaclassification. *Int. J. Syst. Evol. Microbiol.* 52(1):7–76.

Doolittle, W. F. 1999. Phylogenetic classification and the universal tree. *Science* 284:2124–28.

Doolittle, W. F. 2000. Uprooting the tree of life. *Sci. Am.* 282(2):90–95.

Gupta, R. S. 1998. Protein phylogenies and signature sequences: A reappraisal of evolutionary relationships among archaebacteria, eubacteria, and eukaryotes. *Micro. Mol. Biol. Rev.* 62(4):1435–91.

Koonin, E. V.; Makarova, K. S.; and Aravind, L. 2001. Horizontal gene transfer in prokaryotes: Quantification and classification. *Annu. Rev. Microbiol.* 55:709–42.

Lyons, S. L. 2002. Thomas Kuhn is alive and well: the evolutionary relationships of simple life form—a paradigm under siege? *Perspect. Biol. Med.* 45(3):359–76.

Mayr, E. 1998. Two empires or three? *Proc. Natl. Acad. Sci.* 95:9720–23.

Moreira, D., and López-García, P. 2002. The molecular ecology of microbial eukaryotes unveils a hidden world. *Trends Microbiol.* 10(1):31–38.

Pace, N. R. 1997. A molecular view of microbial diversity and the biosphere. *Science* 276:734–40.

Sogin, M. L.; Morrison, H. G.; Hinkle, G.; and Silberman, J. D. 1996. Ancestral relationships of the major eukaryotic lineages. *Microbiologia* 12(1):17–28.

Villarreal, L. P. 2001. Persisting viruses could play role in driving host evolution. *ASM News* 67(10):501–7.

Woese, C. R. 1987. Bacterial evolution. *Microbiol. Rev.* 51(2):221–71.

Woese, C. 1998. The universal ancestor. *Proc. Natl. Acad. Sci.* 95:6854–59.

Zavarzin, G. A.; Stackebrandt, E.; and Murray, R. G. E. 1991. A correlation of phylogenetic diversity in the *Proteobacteria* with the influences of ecological forces. *Can. J. Microbiol.* 37:1–6.

19.8 Bergey's Manual of Systematic Bacteriology

Garrity, G. M., editor-in-chief. 2001. *Bergey's manual of systematic bacteriology,* 2d ed., vol. 1, D. R. Boone and R. W. Castenholz, editors. New York: Springer-Verlag.

Holt, J. G., editor. 1984–1989. *Bergey's manual of systematic bacteriology.* 4 vols. Baltimore, Md.: Williams & Wilkins.

Holt, J. G.; Krieg, N. R.; Sneath, P. H. A.; Staley, J. T.; and Williams, S. T. 1994. *Bergey's manual of determinative bacteriology,* 9th ed. Baltimore, Md.: Williams & Wilkins.

Chapter 20: The *Archaea*

20.1 Introduction to the *Archaea*

Bell, S. D., and Jackson, S. P. 1998. Transcription and translation in Archaea: A mosaic of eukaryal and bacterial features, *Trends Microbiol.* 6(6):222–28.

Bemander, R. 2000. Chromosome replication, nucleoid segregation and cell division in Archaea. *Trends Microbiol.* 8(6):278–83.

Charlebois, R. L. 1999. *Archaea;* Whose sister lineage? In *Organization of the prokaryotic genome.* R. L. Charlebois, editor. 63–76. Washington, D.C.: ASM Press.

Danson, M. J., and Hough, D. W. 1998. Structure, function and stability of enzymes from the Archaea. *Trends Microbiol.* 6(8):307–14.

Eichler, J., and Moll, R. 2001. The signal recognition particle of Archaea. *Trends Microbiol.* 9(3):130–36.

Fuhrman, J. A., and Davis, A. A. 1997. Widespread archaea and novel bacteria from the deep sea as shown by 16S rRNA gene sequences. *Mar. Ecol. Prog. Ser.* 150:275–85.

Garrity, G. M., editor-in-chief. 2001. *Bergey's manual of systematic bacteriology,* 2d. ed., vol. 1, D. R. Boone and R. W. Castenholz, editors. New York: Springer-Verlag.

Grabowski, B., and Kelman, Z. 2003. Archaeal DNA replication: Eukaryal proteins in a bacterial context. *Annu. Rev. Microbiol.* 57:487–516.

Kandler, O., and Zillig, W., editors. 1986. *Archaebacteria '85.* New York: Gustav Fischer Verlag.

Kates, M.; Kushner, D. J.; and Matheson, A. T. 1993. *The biochemistry of Archaea* (Archaebacteria). New York; Elsevier.

Macario, A. J. L.; Lange, M.; Ahring, B. K.; and De Macario, E. C. 1999. Stress genes and proteins in the Archaea. *Micro. Mol. Biol. Rev.* 63(4):923–67.

Myllykallio, H.; Lopez, P.; López-García, P.; Heilig, R.; Saurin, W.; Zivanovic, Y.; Phillipe, H.; and Forterre, P. 2000. Bacterial mode of replication with eukaryotic-like machinery in a hyperthermophilic archaeon. *Science* 288:2212–15.

Olsen, G. J., and Woese, C. R. 1997. Archaeal genomics: An overview. *Cell* 89(7):991–94.

Reeve, J. N.; Sandman, K.; and Daniels, C. J. 1997. Archaeal histones, nucleosomes, and transcription initiation. *Cell* 89(7):999–1002.

Reysenbach, A.-L., and Shock, E. 2002. Merging genomes with geochemistry in hydrothermal ecosystems. *Science* 296:1077–82.

Schäfer, G.; Engelhard, M.; and Müller, V. 1999. Bioenergetics of the Archaea. *Micro. Mol. Biol. Rev.* 63(3):570–620.

Sowers, K. R., and Schreier, H. J. 1999. Gene transfer systems for the Archaea. *Trends Microbiol.* 7(5):212–19.

Vetriani, C., and Reysenbach, A.-L. 2000. *Archaea.* In *Encyclopedia of microbiology.* 2d ed., vol. I. J. Lederberg, editor-in-chief, 319–31. San Diego: Academic Press.

Woese, C. R. 1981. Archaebacteria. *Sci. Am.* 244(6):98–122.

Zillig, W.; Palm, P.; Reiter, W.-D.; Gropp, F.; Pühler, G.; and Klenk, H.-P. 1988. Comparative evaluation of gene expression in archaebacteria. *Eur. J. Biochem.* 173:473–82.

20.2 Phylum *Crenarchaeota*

Burggraf, S.; Huber, H.; and Stetter, K. O. 1997. Reclassification of the crenarchaeal orders and families in accordance with 16S rRNA sequence data. *Int. J. Syst. Bacteriol.* 47(3):657–60.

DeLong, E.F. 2003. Oceans of Archaea. *ASM News* 69(10):503–11.

Pley, U.; Schipka, J.; Gambacorta, A.; Jannasch, H. W.; Fricke, H.; Rachel. R.; and Stetter, K. O. 1991. *Pyrodictium abyssi* sp. nov. represents a novel heterotrophic marine archaeal hyperthermophile growing at 110°C. *System. Appl. Microbiol.* 14:245–53.

20.3 Phylum *Euryarchaeota*

Bult, C. J., et al. 1996. Complete genome sequence of the methanogenic archaeon, *Methanococcus jannaschii. Science* 273:1058–1107.

DiMarco. A. A.; Bobik, T. A.; and Wolfe, R. S. 1990. Unusual coenzymes of methanogenesis. *Ann. Rev. Biochem.* 59:355–94.

Jones, W. J.; Nagle, D. P., Jr.; and Whitman, W. B. 1987. Methanogens and the diversity of archaebacteria. *Microbial. Rev.* 51:135–77.

Luecke, H.; Schobert, B.; Richter, H.-T.; Cartailler, J.-P.; and Lanyi, J. K. 1999. Structural changes in bacteriorhodopsin during ion transport at 2 angstrom resolution. *Science* 286:255–60.

Norton, C. F. 1992. Rediscovering the ecology of halobacteria. *ASM News* 58(7):363–67.

Oren, A. 1999. Bioenergetic aspects of halophilism. *Micro. Mol. Biol. Rev.* 63(2):334–48.

Orphan, V. J.; House, C. H.; Hinrichs, K.-U.; McKeegan, K. D.; and DeLong, E. F. 2001. Methane-consuming archaea revealed by directly coupled isotopic and phylogenetic analysis. *Science* 293:484–87.

Schleper. C.; Puehler, G.; Holz, I.; Gambacorta, A.; Janekovic, D.; Santarias, U.; Klenk, H.-P.; and Zillig, W. 1995. *Picrophilus* gen. nov., fam. nov.; A novel aerobic, heterotrophic, thermoacidophilic genus and family comprising archaea capable of growth around pH O. *J. Bacteriol.* 177(24):7050–59.

Spudich, J. I., 1993. Color sensing in the *Archaea;* a eukaryotic-like receptor coupled to a procaryotic transducer. *J. Bacteriol.* 175(24):7755–61.

Chapter 21: Bacteria: The Deinococci and Nonproteobacteria Gram Negatives

General References

Balows, A.; Trüper, H. G.; Dworkin, M.; Harder, W.; and Schleifer, K.-H. 1992. *The prokaryotes,* 2d ed. New York: Springer-Verlag.

Garrity, G. M., editor-in-chief. 2001. *Bergey's manual of systematic bacteriology,* 2d ed., vol. 1, D. R. Boone and R. W. Castenholz, editors. New York: Springer-Verlag.

Holt, J. G., editor-in-chief. 1984. *Bergey's manual of systematic bacteriology,* vol. 1, N. R. Krieg, editor. Baltimore, Md.: Williams & Wilkins.

Holt, J. G., editor-in-chief. 1989. *Bergey's Manual of Systematic Bacteriology,* vol. 3, J. T. Staley, M. P. Bryant, and N. Pfennig, editors. Baltimore, Md.: Williams & Wilkins.

McBride, M. J. 2001. Bacterial gliding motility: Multiple mechanisms for cell movement over surfaces. *Annu. Rev. Microbiol.* 55:49–75.

21.1 *Aquificae* and *Thermotogae*

Deckert, G., et al. 1998. The complete genome of the hyperthermophilic bacterium *Aquifex aeolicus. Nature* 392:353–58.

Nelson, K. E., et al. 1999. Evidence for lateral gene transfer between Archaea and Bacteria from genome sequence of *Thermotoga maritima. Nature* 399:323–29.

21.2 Deinococcus-Thermus

Battista, J. R. 1997. Against all odds: The survival strategies of *Deinococcus radiodurans. Annu. Rev. Microbiol.* 51:203–24.

Battista, J. R.; Earl, A. M.; and Park, M.-J. 1999. Why is *Deinococcus radiodurans* so resistant to ionizing radiation? *Trends Microbiol.* 7(9):362–65.

Levin-Zaidman, S; Englander, J.; Shimoni, E.; Sharma, A. K.; Minton, K. W.; and Minsky, A. 2003. Ringlike structure of the *Deinococcus radiodurans* genome: A key to radioresistance? *Science* 299:254–56.

Rainey, F. A.; Nobre, M. F.; Schumann, R.; Stackebrandt, E.; and Da Costa, M. S. 1997. Phylogenetic diversity of the deinococci as determined by 16S ribosomal DNA sequence comparison. *Int. J. Syst. Bacteriol.* 47(2):510–14.

21.3 Photosynthetic Bacteria

Adams, D. G. 1992. Multicellularity in cyanobacteria. In *Prokaryotic structure and function,* S. Mohan, C. Dow, and J. A. Coles, editors, 341–84. New York: Cambridge University Press.

Armstrong, G. A. 1994. Eubacteria show their true colors: Genetics of carotenoid pigment biosynthesis from microbes to plants. *J. Bacteriol.* 176(16):4795–802.

Bullerjahn, G. S., and Post, A. F. 1993. The prochlorophytes: Are they more than just chlorophyll a/b-containing cyanobacteria? *Crit. Rev. Microbiol.* 19(1):43–59.

Carmichael, W. W. 1994. The toxins of cyanobacteria. *Sci. Am.* 270(1):78–86.

Garcia-Pichel, F. 2000. Cyanobacteria. In *Encyclopedia of microbiology,* 2d ed., vol. 1, J. Lederberg, editor-in-chief, 907–29. San Diego: Academic Press.

Grossman, A. R.; Schaefer, M. R.; Chiang, G. G.; and Collier, J. L. 1993. The phycobilisome, a light-harvesting complex responsive to environmental conditions. *Microbiol. Rev.* 57(3):725–49.

Partensky, F.; Hess, W. R.; and Vaulot, D. 1999. *Prochlorococcus,* a marine photosynthetic prokaryote of global significance. *Micro. Mol. Biol. Rev.* 63(1):106–27.

Raymond, J; Zhaxybayeva, O; Gogarten, J. P.; Gerdes, S. Y.; and Blankenship, R. E. 2002. Whole-genome analysis of photosynthetic prokaryotes. *Science* 298:1616–20.

Ting, C. S.; Rocap, G.; King, J.; and Chisholm, S. W. 2002. Cyanobacterial photosynthesis in the oceans: The origins and significance of divergent light-harvesting strategies. *Trends Microbiol.* 10(3):134–42.

Wolk, C. P. 1996. Heterocyst formation. *Annu. Rev. Genet.* 30:59–78.

21.4 Phylum *Planctomycetes*

Fuerst, J. A., and Webb, R. I. 1991. Membrane-bounded nucleoid in the eubacterium *Gemmata obscuriglobus. Proc. Natl. Acad. Sci.* 88:8184–88.

21.5 Phylum *Chlamydiae*

Campbell, L. A.; Kuo, C.-C.; and Grayston, J. T. 1998. *Chlamydia pneumoniae* and cardiovascular disease. *Emerg. Infect. Dis.* 4(4):571–79.

Hackstadt, T.; Fischer, E. R.; Scidmore, M. A.; Rockey, D. D.; and Heinzen, R. A. 1997. Origins and functions of the chlamydial inclusion. *Trends Microbiol.* 5(7):288–93.

Hatch, T. P. 1996. Disulfide cross-linked envelope proteins: The functional equivalent of peptidoglycan in chlamydiae? *J. Bacteriol.* 178(1):1–5.

Ngeh, J., and Gupta, S. 2000. *C. pneumoniae* and atherosclerosis: Causal or coincidental link? *ASM News* 66(12):732–37.

Raulston, J. E., and Wyrick, P. B. 2000. Chlamydia. In *Encyclopedia of microbiology,* 2d ed., vol. 1, J. Lederberg, editor-in-chief, 781–88. San Diego: Academic Press.

21.6 Phylum *Spirochaetes*

Harwood, C. S., and Canale-Parola, E. 1984. Ecology of spirochetes. *Annu. Rev. Microbiol.* 38:161–92.

Lilburn, T. G.; Kim, K. S.; Ostrom, N. E.; Byzek, K. R.; Leadbetter, J. R.; and Breznak, J. A. 2001. Nitrogen fixation by symbiotic and free-living spirochetes. *Science* 292:2495–98.

Margulis, L. 2000. Spirochetes. In *Encyclopedia of microbiology,* 2d ed., vol. 4., J. Lederberg, editor-in-chief, 353–63. San Diego: Academic Press.

Radolf, J. D. 1994. Role of outer membrane architecture in immune evasion by *Treponema pallidum* and *Borrelia burgdorferi. Trends Microbiol.* 2(9):307–11.

Radoff J. D.; Steiner, B.; and Shevchenko, D. 1999. *Treponema pallidum:* Doing a remarkable job with what it's got. *Trends Microbiol.* 7(1):7–9.

Saint Girons, I.; Old, I. G.; and Davidson, B. E. 1994. Molecular biology of the *Borrelia,* bacteria with linear replicons. *Microbiology* 140:1803–16.

21.7 Phylum *Bacteroidetes*

McBride, M. J. 2000. Bacterial gliding motility: Mechanisms and mysteries. *ASM News* 66(4):203–10.

Takeuchi, M.; Sakane, T.; Yanagi, M.; Yamasato, K.; Hamana, K.; and Yokota, A. 1995. Taxonomic study of bacteria isolated from plants: Proposal of *Sphingomonas rosa* sp. nov., *Sphingomonas pruni* sp. nov., *Sphingomonas asaccharolytica* sp. nov., and *Sphingomonas mali* sp. nov. *Int. J. Syst. Bacteriol.* 45(2):334–41.

Chapter 22: Bacteria: The Proteobacteria

General References

Balows, A.; Trüper, H. G.; Dworkin, M.; Harder, W.; and Schleifer, K.-H. 1992. *The prokaryotes,* 2d ed. New York: Springer-Verlag.

Brun, Y. V. 2000. Developmental processes in bacteria. In *Encyclopedia of microbiology,* 2d ed., vol. 2. J. Lederberg, editor-in-chief, 15–28. San Diego: Academic Press.

Holt, J. G., editor-in-chief. 1984. *Bergey's Manual of systematic bacteriology,* vol. 1, N. R. Krieg, editor. Baltimore, Md.: Williams & Wilkins.

Holt, J. G., editor-in-chief. 1989. *Bergey's Manual of Systematic Bacteriology,* vol. 3, J. T. Staley, M. P. Bryant, and N. Pfennig, editors. Baltimore, Md.: Williams & Wilkins.

Holt, J. G., editor-in-chief. 1994. *Bergey's Manual of Determinative Bacteriology,* 9th ed. Baltimore, Md.: Williams & Wilkins.

Schlegel, H. G., and Bowien, B. 1989. *Autotrophic bacteria.* Madison, Wis.: Science Tech Publishers.

Shapiro, L.; McAdams, H. H.; and Losick, R. 2002. Generating and exploiting polarity in bacteria. *Science* 298:1942–46.

Zavarzin, G. A.; Stackebrandt, E.; Murray, R. G. E. 1991. A correlation of phylogenetic diversity in the *Proteobacteria* with the influences of ecological forces. *Can. J. Microbiol.* 37:1–6.

22.1 Class *Alphaproteobacteria*

Brun, Y. V.; Marczynski, G.; and Shapiro, L. 1994. The expression of asymmetry during caulobacter cell differentiation. *Annu. Rev. Biochem.* 63:419–50.

Eremeeva, M. E., and Dasch, G. A. 2000. Rickettsiae. In *Encyclopedia of microbiology,* 2d ed., vol. 4, J. Lederberg, editor-in-chief, 140–80. San Diego: Academic Press.

Hooykaas, P. J. J. 2000. Agrobacterium. In *Encyclopedia of microbiology,* 2d ed., vol. 1, J. Lederberg, editor-in-chief, 78–85. San Diego: Academic Press.

Moore, R. L. 1981. The biology of *Hyphomicrobium* and other prosthecate, budding bacteria. *Annu. Rev. Microbiol.* 35:567–94.

Østerås, M., and Jenal, U. 2000. Regulatory circuits in *Caulobacter. Curr. Opin. Microbiol.* 3:171–76.

Winkler, H. H. 1990. Rickettsia species (as organisms). *Annu. Rev. Microbiol.* 44:131–53.

22.2 Class *Betaproteobacteria*

Gillis, M., et al. 1995. Polyphasic taxonomy in the genus *Burkholderia* leading to an emended description of the genus and proposition of *Burkholderia vietnamiensis* sp. nov. for N₂-fixing isolates from rice in Vietnam. *Int. J. Syst. Bacteriol.* 45(2):274–89.

Tettelin, H., et al. 2000. Complete genome sequence of *Neisseria meningitidis* serogroup B strain MC58. *Science* 287(5459):1809–15.

22.3 Class *Gammaproteobacteria*

Dunlap, P. V. 1995. Making a living on cyclic AMP. *ASM News* 61(10):511–16.

Hanson, R. S., and Hanson, T. E. 1996. Methanotrophic bacteria. *Microbiol. Rev.* 60(2):439–71.

Heidelberg, J. F., et al. 2000. DNA sequence of both chromosomes of the cholera pathogen *Vibrio cholerae. Nature* 406:477–83.

Hill, S., and Sawers, G. 2000. Azotobacter. In *Encyclopedia of microbiology,* 2d ed., vol. 1, J. Lederberg, editor-in-chief, 359–71. San Diego: Academic Press.

Kapatral, V.; Zago, A.; Kamath, S.; and Chugani, S. 2000. Pseudomonas. In *Encyclopedia of microbiology,* 2d ed., vol. 3, J. Lederberg, editor-in-chief, 876–92. San Diego: Academic Press.

Larkin, J. M., and Strohl, W. R. 1983. *Beggiatoa, Thiothrix,* and *Thioploca. Annu. Rev. Microbiol.* 37:341–67.

Low, B. K. 2000. *Escherichia coli* and *Salmonella,* Genetics. In *Encyclopedia of microbiology,* 2d ed., vol. 2, J. Lederberg, editor-in-chief, 270–82. San Diego: Academic Press.

Meighen, E. A. 1991. Molecular biology of bacterial bioluminescence. *Microbiol. Rev.* 55(1):123–42.

Neidhardt, F. C., editor-in-chief. 1996. *Escherichia coli* and *Salmonella typhimurium,* 2d ed. Washington, D.C.: ASM Press.

Perna, N. T., et al. 2001. Genome sequence of enterohaemorrhagic *Escherichia coli* O157:H7. *Nature* 409:529–33.

Schaechter, M., and The View from Here Group. 2001. *Escherichia coli* and *Salmonella* 2000: The view from here. *Microbiol. Mol. Biol. Rev.* 65(1):119–30.

Schaechter, M. 2000. *Escherichia coli,* General biology. In *Encyclopedia of microbiology,* 2d ed., vol. 2, J. Lederberg, editor-in-chief, 260–69. San Diego: Academic Press.

Souza, V.; Castillo, A.; and Eguiarte, L. E. 2002. The evolutionary ecology of *Escherichia coli. American Scientist* 90:332–41.

Stover, C. K., et al. 2000. Complete genome sequence of *Pseudomonas aeruginosa* PAO1, an opportunistic pathogen. *Nature* 406:959–64.

22.4 Class *Deltaproteobacteria*

Diedrich, D. L. 1988. Bdellovibrios: Recycling, remodelling and relocalizing components from their prey. *Microbiol. Sci.* 5(4):100–103.

Dworkin, M. 1996. Recent advances in the social and developmental biology of the myxobacteria. *Microbiol. Rev.* 60(1):70–102.

Kearns, D. B., and Shimkets, L. J. 2001. Lipid chemotaxis and signal transduction in *Myxococcus xanthus. Trends Microbiol.* 9(3):126–29.

Spormann, A. M. 1999. Gliding motility in bacteria: Insights from studies of *Myxococcus xanthus. Micro. Mol. Biol. Rev.* 63(3):621–41.

White, D. 2000. Myxobacteria. In *Encyclopedia of microbiology,* 2d ed., vol. 3, J. Lederberg, editor-in-chief, 349–62. San Diego: Academic Press.

22.5 Class *Epsilonproteobacteria*

Dunn, B. E.; Cohen, H.; and Blaser, M. J. 1997. *Helicobacter pylori. Clin. Microbiol. Rev.* 10(4):720–41.

Marais, A.; Mendz, G. L.; Hazell, S. L.; and Mégraud, F. 1999. Metabolism and genetics of *Helicobacter pylori:* The genome era. *Microbiol. Mol. Biol. Rev.* 63(3):642–74.

Parkhill, J., et al. 2000. The genome sequence of the food-borne pathogen *Campylobacter jejuni* reveals hypervariable sequences. *Nature* 403:655–68.

Tomb, J.-F., et al. 1997. The complete genome sequence of the gastric pathogen *Helicobacter pylori. Nature* 388:539–47.

Chapter 23: Bacteria: The Low G + C Gram Positives

General References

Balows, A.; Truper, H. G.; Dworkin, M.; Harder, W.; and Schleifer, K.-H. 1992. *The prokaryotes,* 2d ed. New York: Springer-Verlag.

Holt, J. G., editor-in-chief. 1986. *Bergey's Manual of Systematic Bacteriology,* vol. 2, P. H. A. Sneath, N. S. Mair, and M. E. Sharpe, editors. Baltimore, Md.: Williams & Wilkins.

Holt, J. G., editor-in-chief. 1994. *Bergey's Manual of Determinative Bacteriology,* 9th ed. Baltimore, Md.: Williams & Wilkins.

Hoyle, F., and Wickramasinghe, C. 1981. Where microbes boldly went. *New Scientist* (13 Aug.):412–15.

Ward, J. B. 1981. Teichoic and teichuronic acids: Biosynthesis, assembly, and location. *Microbiol. Rev.* 45(2):211–43.

Weber, P., and Greenberg, J. M. 1985. Can spores survive in interstellar space? *Nature* 316:403–7.

23.2 Class *Mollicutes* (the Mycoplasmas)

Himmelreich, R.; Hilbert, H.; Plagens, H.; Pirkl, E.; Li, B.-C.; and Herrmann, R. 1996. Complete sequence analysis of the genome of the bacterium *Mycoplasma pneumoniae. Nucleic Acids Res.* 24(22):4420–49.

Miles, R. J. 1992. Catabolism in mollicutes. *J. Gen. Microbiol.* 138:1773–83.

Whitcomb, R. F. 1980. The genus *Spiroplasma. Annu. Rev. Microbiol.* 34:677–709.

23.4 Class *Clostridia*

Ahern, H. 1993. A big bacterium—oxymoron of the microbial world. *ASM News* 59(10):519–21.

Amesz, J. 1995. The heliobacteria, a new group of photosynthetic bacteria. *J. Photochem. Photobiol. B* 30:89–96.

Collins, M. D.; Lawson, P. A.; Willems, A.; Cordoba, J. J.; Fernandez-Garayzabal, J.; Garcia, P.; Cai, J.; Hippe, H.; and Farrow, J. A. E. 1994. The phylogeny of the genus *Clostridium:* Proposal of five new genera and eleven new species combinations. *Int. J. Syst. Bacteriol.* 44(4):812–26.

Johnson, E. A. 2000. Clostridia. In *Encyclopedia of microbiology,* 2d ed., vol. 1, J. Lederberg, editor-in-chief, 834–39. San Diego: Academic Press.

Ormerod, J. G.; Kimble, L. K.; Nesbakken, T.; Torgerson, Y. A.; Woese, C. R.; and Madigan, M. T. 1996. *Heliophilum fasciatum* gen. nov. sp. nov. and *Heliobacterium gestii* sp. nov.: Endospore-forming heliobacteria from rice field soils. *Arch. Microbiol.* 165:226–34.

23.5 Class *Bacilli*

Aronson, A. I.; Beckman, W.; and Dunn, P. 1986. *Bacillus thuringiensis* and related insect pathogens. *Microbiol. Rev.* 50(1):1–24.

Ash, C.; Priest, F. G.; and Collins, M. D. 1993. Molecular identification of rRNA group 3 bacilli (Ash, Farrow, Wallbanks, and Collins) using a PCR probe test: Proposal for the creation of a new genus *Paenibacillus. Antonie van Leeuwenhoek* 64:253–60.

Cunningham, M. W. 2000. Pathogenesis of group A streptococcal infections. *Clin. Microbiol. Rev.* 13(3):470–511.

Devine, K. M. 2000. *Bacillus subtilis,* genetics. In *Encyclopedia of microbiology,* 2d ed., vol. 1, J. Lederberg, editor-in-chief, 373–82. San Diego: Academic Press.

Dinges, M. M.; Orwin, P. M.; and Schlievert, P. M. 2000. Exotoxins of *Staphylococcus aureus. Clin. Microbiol. Rev.* 13(1):16–34.

Iandolo, J. J. 2000. *Staphylococcus.* In *Encyclopedia of microbiology,* 2d ed., vol. 4, J. Lederberg, editor-in-chief, 387–93. San Diego: Academic Press.

Kunst, F., et al. 1997. The complete genome sequence of the gram-positive bacterium *Bacillus subtilis. Nature* 390:249–56.

Lambert, B., and Peferoen, M. 1992. Insecticidal promise of *Bacillus thuringiensis:* Facts and mysteries about a successful biopesticide. *BioScience* 42(2):112–22.

Murray, B. E. 1990. The life and times of the enterococcus. *Clin. Microbiol. Rev.* 3(1):46–65.

Nicholson, W. L.; Munakata, N.; Horneck, G.; Melosh, H. J.; and Setlow, P. 2000. Resistance of *Bacillus* endospores to extreme terrestrial and extraterrestrial environments. *Micro. Mol. Biol. Rev.* 64(3):548–72.

Schleifer, K.-H., and Kilpper-Balz, R. 1987. Molecular and chemotaxonomic approaches to the classification of streptococci, enterococci, and lactococci: A review. *Syst. Appl. Microbiol.* 10:1–19.

Setlow, P. 1995. Mechanisms for the prevention of damage to DNA in spores of *Bacillus* species. *Annu. Rev. Microbiol.* 49:29–54.

Somkuti, G. A. 2000. Lactic acid bacteria. In *Encyclopedia of microbiology,* 2d ed., vol. 3, J. Lederberg, editor-in-chief, 1–8. San Diego: Academic Press.

Song, L.; Hobaugh, M. R.; Shustak, C.; Cheley, S.; Bayley, H.; and Gouaux, J. E. 1996. Structure of staphylococcal α-hemolysin, a heptameric transmembrane pore. *Science* 274:1859–66.

Stragier, P., and Losick, R. 1996. Molecular genetics of sporulation in *Bacillus subtilis. Annu. Rev. Genet.* 30:297–341.

Tomasz, A. 2000. *Streptococcus pneumoniae.* In *Encyclopedia of microbiology,* 2d ed., vol. 4, J. Lederberg, editor-in-chief, 444–50. San Diego: Academic Press.

Whiteley, H. R., and Schnepf, H. E. 1986. The molecular biology of parasporal crystal body formation in *Bacillus thuringiensis. Annu. Rev. Microbiol.* 40:549–76.

Chapter 24: Bacteria: The High G + C Gram Positives

24.1 General Properties of the Actinomycetes

Balows, A.; Trüper, H. G.; Dworkin, M.; Harder, W.; and Schleifer, K.-H. 1992. *The prokaryotes,* 2d ed. New York: Springer-Verlag.

Beaman, B. L.; Saubolle, M. A.; and Wallace R. J. 1995. *Nocardia, Rhodococcus, Streptomyces, Oerskovia,* and other aerobic actinomycetes of medical importance. *In Manual of Clinical Microbiology,* 6th ed., P. R. Murray, editor-in-chief, 379–99. Washington, D.C.: American Society for Microbiology.

Dietz, A. 1994. The lives and times of industrial actinomycetes. *ASM News* 60(7): 366–69.

Embley, T. M., and Stackebrandt, E. 1994. The molecular phylogeny and systematics of the actinomycetes. *Annu. Rev. Microbiol.* 48:257–89.

Holt, J. G.; editor-in-chief. 1989. *Bergey's manual of systematic bacteriology,* vol. 4, S. T. Williams and M. E. Sharpe, editors. Baltimore, Md.: Williams & Wilkins.

Holt, J. G., editor-in-chief. 1994. *Bergey's manual of determinative bacteriology,* 9th ed. Baltimore, Md.: Williams & Wilkins.

Krsek, M.; Morris, N.; Egan, S.; and Wellington, E. M. H. 2000. *Actinomycetes.* In *Encyclopedia of microbiology,* 2d ed., vol. 1, J. Lederberg, editor-in-chief, 28–41. San Diego: Academic Press.

Stackebrandt, E.; Rainey, F. A.; and Ward-Rainey, N. L. 1997. Proposal for a new hierarchic classification system, *Actinobacteria* classis nov. *Int. J. Syst. Bacteriol.* 47(2):479–91.

24.5 Suborder *Corynebacterineae*

Belisle, J. T., and Brennan, P. J. 2000. Mycobacteria. In *Encyclopedia of microbiology,* 2d ed., vol. 3. J. Lederberg, editor-in-chief, 312–27. San Diego: Academic Press.

Tortoli, E. 2003. Impact of genotypic studies on mycobacterial taxonomy: The new mycobacteria of the 1990s. *Clin. Microbiol. Rev.* 16(2):319–54.

24.6 Suborder *Micromonosporineae*

Parenti, F., and Coronelli, C. 1979. Members of the genus *Actinoplanes* and their antibiotics. *Annu. Rev. Microbiol.* 33:389–411.

24.8 Suborder *Streptomycineae*

Dyson, P. 2000. *Streptomyces,* genetics. In *Encyclopedia of microbiology,* 2d ed., vol. 4, J. Lederberg, editor-in-chief, 451–66. San Diego: Academic Press.

24.10 Suborder *Frankineae*

Benson, D. R., and Silvester, W. B. 1993. Biology of *Frankia* strains, actinomycete symbionts of actinorhizal plants. *Microbiol. Rev.* 57(2):293–319.

Schwintzer, C. R., and Tjepkema, J. D., editors. 1990. *The biology of Frankia and actinorhizal plants.* San Diego, Calif.: Academic Press.

Chapter 25: The Fungi (Eumycota), Slime Molds, and Water Molds

General Reference

Alexopoulos, C. J.; Mims, C. W.; Blackwell, M. 1996. *Introductory mycology,* 4th ed. New York: John Wiley and Sons.

Barr, D. J. S. 1992. Evolution and kingdoms of organisms from the perspective of a mycologist. *Mycologia* 84:1–11.

Carlile, M. J., and Gooday, G W. 2001. *The fungi,* 2d ed. New York: Academic Press.

Griffin, D. 1993. *Fungal physiology,* 2d ed. New York: Wiley-Liss.

Guarro, J.; Gene, J.; and Stchigel, A. M. 1999. Developments in fungal taxonomy. *Clin. Microbiol. Rev.* 12(3):454–500.

Kurtzman, C. P., and Fell, J. W., editors, 1998. *The Yeasts: A taxonomic study,* 4th ed. New York: Elsevier.

Moore-Landecker, E. 1996. *Fundamentals of fungi,* 4th ed. Englewood Cliffs, N.J.: Prentice-Hall.

Murawski, D. A. 2000. Fungi. *National Geographic* 198(2):59–70.

Schaechter, E. 1997. *In the company of mushrooms.* Boston: Harvard University Press.

25.6 Characteristics of the Fungal Divisions

Bossche, H., editor. 1993. *Dimorphic fungi in biology and medicine.* New York: Plenum Publishing Company.

Gold, M. H., and Alic, M. 1993. Molecular biology of the lignin-degrading basidiomycete *Phanerochaete chrysosporium. Microbiol. Rev.* 57(3):605–22.

Herskowitz, I. 1988. Life cycle of the budding yeast *Saccharomyces cerevisiae. Microbiol. Rev.* 52(4):536–53.

Jackson, S. L., and Heath, I. B. 1993. Roles of calcium ions in hyphal tip growth. *Microbiol. Rev.* 57(2):367–82.

Maresca, B., and Kobayashi, G. S. 1989. Dimorphism in *Histoplasma capsulatum:* A model for the study of cell differentiation in pathogenic fungi. *Microbiol. Rev.* 53(2):186–209.

Marzluf, G. A. 1997. Genetic regulation of nitrogen metabolism in the fungi. *Microbiol. Mol. Biol. Rev.* 61(1):17–32.

Matossian, M. K. 1982. Ergot and the Salem witchcraft affair. *Am. Scientist* 70:355–71.

Newhouse, J. R. 1990. Chestnut blight. *Sci. Am.* 263:106–11.

Ostergaard, S.; Olsson, L.; and Nielsen, J. 2000. Metabolic engineering of *Saccharomyces cerevisiae. Microbiol. Mol. Biol. Rev.* 64(1):34–50.

Ribes, J. A.; Vanover-Sams, C. L.; and Baker, D. J. 2000. Zygomycetes in human disease. *Clin. Microbiol. Rev.* 13(2):236–301.

Strobel, G. A., and Lanier, G. N. 1981. Dutch elm disease. *Sci. Am.* 245:56–66.

25.7 Slime Molds and Water Molds

Gross, J. D. 1994. Developmental decisions in *Dictyostelium discoideum. Microbiol. Rev.* 58(3):330–51.

Loomis, F. W. 1996. Genetic networks that regulate development in *Dictyostelium* cells. *Microbiol. Rev.* 60(1):135–50.

Martin, G. W., and Alexopoulos, C. J. 1969. *The myxomycetes.* Iowa City, University of Iowa Press.

Chapter 26: The Algae

General References

Bold, H. C., and Wynne, M. J. 1985. *Introduction to the algae,* 2d ed. Englewood Cliffs, N.J.: Prentice-Hall.

Darley, W. M. 1982. *Algal biology: A physiological approach.* Boston: Blackwell Scientific Publications.

Lee, R. E. 1989. *Phycology,* 2d ed. New York: Cambridge University Press.

Lembi, C. A., and Waaland, J. R., editors. 1988. *Algae and human affairs.* New York: Cambridge University Press.

Prescott, G. W. 1978. *How to know the freshwater algae,* 3d ed. Dubuque, Iowa: Wm. C. Brown Publishers.

Sze, P. 1997. *A biology of the algae,* 3d ed. Dubuque, Iowa: Wm. C. Brown Publishers.

Van den Hoek, C.; Mann, D.; and Jahns, H. 1995. *An introduction to the algae.* New York: Cambridge University Press.

26.1 Distribution of Algae

Round, F. E. 1984. *The ecology of algae.* New York: Cambridge University Press.

26.2 Classification of Algae

Perasso, R. 1989. Origin of the algae. *Nature* 339:142–44.

26.3 Ultrastructure of the Algal Cell

Grossman, A.; Schaefer, M.; Chiang, G.; and Collier, J. 1993. The phycobilisome, a light-harvesting complex responsive to environmental conditions. *Microbiol. Rev.* 57(3):725–49.
Jacobs, W. P. 1994. *Caulerpa. Sci. Am.* 271(6):100–05.

26.6 Algal Reproduction

Gray, M. W. 1994. One plus one equals one: The making of a crytomonad alga. *ASM News* 60(8):423–27.

26.7 Characteristics of the Algal Divisions

Anderson, D. 1994. Red tides. *Sci. Am.* 271(2):62–69.
Hallegraeff, G. M. 1993. A review of harmful algal blooms and their apparent global increase. *Phycologia* 32(2):79–99.
Lobban, C. S., and Wynne, M. J., editors. 1981. *The biology of seaweeds.* Oxford: Blackwell Scientific Publications.
Saffo, M. B. 1987. New light on seaweeds. *BioScience* 37:170–80.
Van der Meer, J. P. 1983. The domestication of seaweeds. *BioScience* 33:172–76.

Chapter 27: The Protozoa

General References

Jahn, T. L.; Bovee, E. C.; and Jahn, F. F. 1979. *How to know the protozoa.* Dubuque, Iowa: Wm. C. Brown.
Krier, J. P. 1995. *Parasitic protozoa.* New York: Academic Press.
Lee, J. J.; Hunter, S. H.; and Bovee, E. C. 1985. *An illustrated guide to the protozoa.* Lawrence, Kan.: Allen Press, Society of Protozoologists.
Margulis, L.; Corliss, J. O.; and Melkonian, M. 1990. *Handbook of Protoctista.* Boston: Jones and Bartlett.
Sleigh, M. 1992. *Protozoa and other protists.* New York: Cambridge University Press.

27.1 Distribution

Fenchel, T. 1987. *Ecology of protozoa.* New York: Springer-Verlag.

27.3 Morphology

Dunelson, J. E., and Turner, M. J. 1985. How the trypanosome changes its coat. *Sci. Am.* 252(2):44–51.
McFadeen, G.; Gilson, P.; Hofmann, G.; Adcock, G.; and Maier, U.-G. 1994. Evidence that an amoeba acquired a chloroplast by retaining part of an engulfed eukaryotic alga. *Proc. Natl. Acad. Sci.* 91:3690–94.

27.4 Nutrition

Barker, J., and Brown, M. R. W. 1994. Trojan horses of the microbial world: Protozoa and the survival of bacterial pathogens in the environment. *Microbiology* 140:1253–59.
Rudzinska, M. A. 1973. Do suctoria really feed by suction? *BioScience* 23(2):87–94.

27.8 Classification

Cavalier-Smith, T. 1993. Kingdom protozoa and its 18 phyla. *Microbiol. Rev.* 57(4):953–94.
Lee, R. E., and Kugrens, P. 1992. Relationship between the flagellates and the ciliates. *Microbiol. Rev.* 56(4):529–42.

27.9 Representative Types

Adam, R. 1991. The biology of *Giardia spp. Microbiol. Rev.* 55(4):706–32.
Corliss, J. O. 1979. *The ciliated protozoa: Characterization, classification and guide to the literature,* 2d ed. New York: Pergamon Press.
Keeling, P. J., and Fast, N. M. 2002. Microsporidia: Biology and evolution of highly reduced intracellular parasites. *Annu. Rev. Microbiol.* 56:93–116.
Wichterman, R. 1986. *The biology of Paramecium,* 2d ed. New York: Plenum Press.

Chapter 28: Microorganism Interactions and Microbial Ecology

General References

Atlas, R. M., and Bartha, R. 1998. *Microbial ecology: Fundamentals and applications,* 4th ed. Redwood City, Calif.: Benjamin/Cummings.
Brock, T. D. 1978. *Thermophilic microorganisms and life at high temperature.* Springer Verlag: New York.
Pace, N. R. 1999. Microbial ecology and diversity. *ASM News* 65:238–333.
Torsvik, V.; Øvreås, L.; and Thingstad, T. F. 2002. Prokaryotic diversity—magnitude, dynamics, and controlling factors. *Science* 296:1064–66.

28.2 Microbial Interactions

Campbell, B. J., and Cary, S. C. 2000. Characterization of a novel spirochete associated with the hydrothermal vent polychaete annelid, *Alvinella pompejana. Appl. Environ. Microbiol.* 67(1):110–17.
Crespi, B. J. 2001. The evolution of social behavior in microorganisms. *TREE* 16(4):178–83.
Currie, C. R. 2001. A community of ants, fungi, and bacteria: A multilateral approach to studying symbiosis. *Annu. Rev. Microbiol.* 55:357–80.
Hackstein, J. H. P.; Langer, P.; and Rosenberg, J. 1996. Genetic and evolutionary constraints for the symbiosis between animals and methanogenic bacteria. *Env. Mon. Ass.* 42:39–56.
Harb, O. S., and Kwaik, Y. A. 2000. Interaction of *Legionella pneumophila* with protozoa provides lessons. *ASM News* 66(10):609–16.
Hentschel, U., and Steinert, M. 2001. Symbiosis and pathogenesis: common themes, different outcomes. *Trends Microbiol.* 9(12):585.
Keeling, P. J. 2001. Parasites go the full monty. *Nature* 414:401–02.
Nash, T. H., III. 1996. *Lichen biology.* Cambridge, U.K.: Cambridge University Press.
Paracer, S., and Ahmadjian, V. 2000. *Symbiosis.* 2d ed. New York: Oxford University Press.
Potera, C. 2002. Antimicrobial peptides from skin effective in killing pathogens. *ASM News* 68:108–109.
Russell, J. B. 2000. Rumen fermentation. In *Encyclopedia of microbiology,* 2d ed., vol. 4, J. Lederberg, editor-in-chief, 185–94. San Diego: Academic Press.
Segal, G., and Shuman, H. A. 1999. Intracellular multiplication of *Legionella pneumophila* in human and environmental hosts. In *Microbial ecology and infectious disease.* E. Rosenberg, editor, 170–86. Washington, D.C.: ASM Press.
Stouthamer, R.; Breeuwer, J. A. J.; and Hurst, G. D. D. 1999. *Wolbachia pipientis:* Microbial manipulator of arthropod reproduction. *Annu. Rev. Microbiol.* 53:71–102.
Sumner, E. R., and Avery, S. V. 2002. Phenotypic heterogeneity: Differential stress resistance among individual cells of the yeast *Saccharomyces cerevisiae. Microbiology* 148:345–51.
Travis, J. 2001. Human sweat packs a germ-killing punch. *Science News* 160:292.
Velicer, G. J. 2003. Social strife in the microbial world. *Trends Microbial.* 11(7):330–37.
Wiener, P. 2000. Antibiotic production in a spatially structured environment. *Ecol. Letters* 3(2):122–30.

28.3 Nutrient Cycling Interactions

Barkay, T. 2000. Mercury cycle. In *Encyclopedia of microbiology,* 2d ed., vol. 3, J. Lederberg, editor-in-chief, 171–81. San Diego: Academic Press.
Fenchel, T.; King, G. M.; and Blackburn, T. H. 1998. *Bacterial biogeochemistry: The ecophysiology of mineral cycling,* 2d ed. New York: Academic Press.
Lack, J. G.; Chaudhuri, S. K.; Chakraborty, R.; Achenbach, L. A.; and Coates, J. D. 2002. Anaerobic biooxidation of Fe(II) by *Dechlorosoma suillum. Microbiol. Ecol.* 43:424–31.
Lens, P., and Pol, L. H. 2000. Sulfur cycle. In *Encyclopedia of microbiology,* 2d ed., vol. 4, J. Lederberg, editor-in-chief, 495–505. San Diego: Academic Press.

Oremland, R. S., and Stolz, J. F. 2003. The ecology of arsenic. *Science* 300:939–44.

Pray, L. 2002. Microbiologists make discoveries in the sea, in the neighborhood. *The Scientist* 16(19):43.

Stacey, G.; Burris, R. H.; and Evans, H. J. 1991. *Biological nitrogen fixation*. New York: Chapman & Hall.

Vacelet, J.; Boury-Esnault, N.; Flala-Medioni, A.; and Fisher, C. R. 1997. A methanotrophic carnivorous sponge. *Nature* 377:296.

van de Graaf, A. A.; Mulder, A.; de Bruijn, P.; Jetten, M. S. M.; Robertson, L. A.; and Kuenen, J. G. 1995. Anaerobic oxidation of ammonium is a biologically mediated process. *Appl. Environ. Microbiol.* 61:1246–51.

Yrukov, V. V., and Beatty, J. T. 1998. Aerobic anoxygenic phototrophic bacteria. *Microbiol. Mol. Biol. Rev.* 62:695–724.

28.4 The Physical Environment

Brown, M. R. W., and Barker, J. 2000. Unexplored reservoirs of pathogenic bacteria: Protozoa and biofilms. *Trends Microbiol.* 7:46–49.

Edwards, K. A.; Bond, P. L.; Gihring, T. M.; and Banfield, J. F. 2000. An archaeal iron-oxidizing extreme acidophile important in acid mine drainage. *Science* 287:1796–99.

Jeanthon, C. 2000. Molecular ecology of hydrothermal vent microbial communities. *Antonie van Leeuwenhoek.* 77:117–33.

Kashefi, K., and Lovley, D.R. 2003. Extending the upper temperature limit for life. *Science* 301:934.

Kelley, D. S.; Baross, J. A.; and Delany, J. R. 2002. Volcanoes, fluids, and life at mid-ocean ridge spreading centers. *Annu. Rev. Earth Planet. Sci.* 30:385–491.

Levin–Zaidman, S.; Englander, J.; Shimoni, E.; Sharma, A. K.; Minton, K.W.; and Minsky, A. 2003. Ringlike structure of the *Deinococcus radiodurans* genome: A key to radioresistance? *Science* 299:254–56.

O'Toole, G.; Kaplan, H. B.; and Kolter, R. 2000. Biofilm formation as microbial development. *Annu. Rev. Microbiol.* 54:49–79.

Parsek, M. R., and Singh, P.K. 2003. Bacterial biofilms: An emerging link to disease pathogenesis. *Annu. Rev. Microbiol.* 57:677–701.

Pepper, I. L. 2000. Hardy microbe thrives at pH 0. *Science* 287:1731–32.

Rickard, A. H.; Gilbert, P.; High, N. J.; Kolenbrander, P. E.; and Handley, P. S. 2003. Bacterial coaggregation: An integral process in the development of multi-species biofilms. *Trends Microbiol.* 11(2):94–100.

Reysenbach, A.-L., and Cady, S. L. 2001. Microbiology of ancient and modern hydrothermal systems. *Trends Microbiol.* 9(2):79–86.

Thomas, D. N., and Dieckmann, G. S. 2002. Antarctic sea ice—a habitat for extremophiles. *Science* 295:641–44.

Vielle, C., and Zeikus, G. J. 2001. Hyperthermophilic enzymes: Sources, uses, and molecular mechanisms for thermostability. *Microbiol. Mol. Biol. Rev.* 65(1):1–43.

28.5 Microbial Ecology and Its Methods: An Overview

Axtell, C. A., and Beattie, G. A. 2002. Construction and characterization of a proU-gfp transcriptional fusion that measures water availability in a microbial habitat. *Appl. Environ. Microbiol.* 68(9):4604–612.

Boulos, L.; Prévost, M.; Barbeau, B.; Coallier, J.; and Desjardins, R. 1999. LIVE/DEAD bacLight: Application of a new rapid staining method for direct enumeration of viable and total bacteria in drinking water. *J. Microbiol. Meth.* 37:77–86.

Colwell, R. R., and Grimes, D. J. 2000. *Nonculturable microorganisms in the environment*. Washington, D.C.: ASM Press.

Fröhlich, J., and König, H. 2000. New techniques for isolation of single prokaryotic cells. *FEMS Microbiol. Revs.* 24:567–72.

Huber, H.; Hohn, M. J.; Rachel, R.; Fuchs, T.; Wimmer, V. C.; and Stetter, K. O. 2002. A new phylum of archaea represented by a nanosized hyperthermophilic symbiont. *Nature* 417:63–67.

Hurst, C. J.; Knudsen, G. R.; McInerney, M. J.; and Stetzenbach, L. D. 1997. *Manual of environmental microbiology*. Washington, D.C.: American Society for Microbiology.

Liu, J.; Dazzo, F. B.; Glagoleva, O.; Yu, B.; and Jain, A. K. 2001. CMEIAS: A computer-aided system for the image analysis of bacterial morphotypes in microbial communities. *Microbial Ecol.* 41:173–94.

Martin-Laurent, F.; Philoppot, L.; Hallet, S.; Chaussod, R.; Germon, J. C.; Soulas, G.; and Catroux, G. 2001. DNA extraction from soils: Old bias for new microbial diversity analysis methods. *Appl. Environ. Microbiol.* 67(5):2354–59.

Perfilev, B. V., and Gabe, D. R. 1969. *Capillary methods of investigating microorganisms*. Toronto, Ontario: University of Toronto Press.

Postgate, J. R. 1967. Viability measurements and the survival of microbes under minimum stress. *Adv. Microbiol. Phys.* 1:1–23.

Radajewski, S.; Ineson, P.; Parekh, N. R.; and Murrell, J. C. 2000. Stable-isotope probing as a tool in microbial ecology. *Nature* 403:646–49.

Rappé, M. S., and Giovannoni, S. J. 2003. The uncultured microbial majority. *Annu. Rev. Microbiol.* 57:369–94.

Richter, A.; Schwager, C.; Hentze, S.; Ansorge, W.; Hentze, M. W.; and Muckenthaler, M. 2002. Comparison of fluorescent tag DNA labeling methods used for expression analysis by DNA microarrays. *BioTechniques* 33:620–30.

Rochelle, P. A., editor. 2001. *Environmental molecular microbiology, Protocols and applications*. Norfolk, U.K.: Horizon Scientific Press.

Wright, R. T., and Hobbie, J. E. 1966. Use of glucose and acetate by bacteria and algae in aquatic systems. *Ecology* 47:447–64.

Zengler, K.; Toledo, G.; Rappé, M.; Elkins, J.; Mathur, E. J.; Short, J. M.; and Keller, M. 2002. Cultivating the uncultured. *Proc. Natl. Acad. Sci.* 99(24):15681–86.

Chapter 29: Microorganisms in Aquatic Environments

The references provided for chapter 28 also may be consulted for further information.

General References

Kemp, P. F.; Sherr, B. F.; Sherr, E. B.; and Cole, J. J. 1993. *Handbook of methods in aquatic microbial ecology*. Boca Raton: Lewis Publishers.

Overbeck, J., and Chróst, R. J., editors. 1999. *Aquatic microbial ecology, biochemical and molecular approaches*. New York: Springer-Verlag.

29.1 Aquatic Environments and Microorganisms

Crane, S. 2002. Humans encroaching on coastlines. *BioScience* 52(2):191–92.

DeLong, E. F.; Taylor, L. T.; Marsh, T. L., and Preston, C. M. 1999. Visualization and enumeration of marine planktonic archaea and bacteria by using polyribonucleotide probes and fluorescent in situ hybridization. *Appl. Environ. Microbiol.* 65:5554–63.

29.2 The Microbial Community

Bärlocher, F. 1992. *The ecology of aquatic hyphomycetes*. Ecological Studies 94. Berlin: Springer-Verlag.

Beja, O.; Suzuki, M. T.; Heldelberg, J. F.; Nelson, W. C.; Preston, C. M.; Hamada, T.; Elsen, J. A.; Fraser, C. M.; and DeLong, E. F. 2002. Unsuspected diversity among marine aerobic anoxygenic phototrophs. *Nature* 415:630–33.

Berry, J. P.; Reece, K. S.; Rein, K. S.; Baden, D. G.; Haas, L. W.; Riberio, W. L.; Shields, J. D.; Snyder R. V.; Vogelbein, W. K.; and Gawley, R. E. 2002. Are *Pfiesteria* species toxicogenic? Evidence against production of ichthyotoxins by *Pfiesteria shumwayae*. *Proc. Nat. Acad. Sci. USA* 99(17):10970–75.

Feinstein, T. N.; Traslavina, R.; Sun, M. Y.; and Lin, S. J. 2002. Effects of light on photosynthesis, grazing and population dynamics of the heterotropic dinoflagellate *Pfiesteria piscicida* (Dinophyceae). *J. Phycol.* 38(4):659–69.

Hinrichs, K.-U.; Haynes, J. M.; Sylva, S. P., Ewer, P. G., Long, E. F. 1999. Methane-consuming archaebacteria in marine sediments. *Nature* 398:802–805.

Jørgensen, B. B., and Gallardo, V. A. 1999. *Thioploca* spp.: Filamentous sulfur bacteria with nitrate vacuoles. *FEMS Microbiol. Ecol.* 28:301–13.

Schultz, H. N. 2002. *Thiomargarita namibiensis*: Giant microbe holding its breath. *ASM News* 68(3):122–27.

Schulz, H. N.; Brinkhoff, T.; Ferdelman, T. G.; Hernandez Marine, M.; Teske, A.; and Jorgensen, B. B. 1999. Dense populations of a giant sulfur bacterium in Namibian shelf sediments. *Science* 284:493–95.

Schultz, H. N., and Jørgensen, B. B. 2001. Big bacteria. *Annu Rev. Microbiol.* 55:105–37.

29.3 Marine Environments

Anonymous. 1997. The Danube blues. *Discovery* 18:21.

Belgrano, A., and Brown, J. H. 2002. Oceans under the macroscope. *Nature* 419:128–29.

DeLong, E. F.; Taylor, L. T.; Marsh, T. L.; and Preston, C. M. 1999. Visualization and enumeration of marine planktonic archaea and bacteria by using polyribonucleotide probes and fluorescent in situ hybridization. *Appl. Environ. Microbiol.* 65:5554–63.

Forterre, P.; Brochier, C.; and Phillipe, H. 2002. Evolution of the archaea. *Theoret. Pop. Biol.* 61:409–22.

Hinrichs, K.-U.; Haynes, J. M.; Sylva, S. P.; Brewer, P. G.; and DeLong, E. F. 1999. Methane-consuming archaebacteria in marine sediments. *Nature* 398:802–805.

Karl, D. M. 2002. Nutrient dynamics in the deep blue sea. *Trends Microbiol.* 10(9):410–18.

Kirchman, D. L., editor. 2000. *Microbial ecology of the oceans.* New York: Wiley-Liss.

Pahlow, M., and Riebesell, U. 2000. Temporal trends in deep ocean Redfield Ratios. *Science* 287:831–33.

Pratt, S. 2001. Death by dust storm. *Discovery* 22:17.

Raloff, J. 2000. Algae turn fish into a lethal lunch. *Science News* 157:20.

Rayl, A. J. S. 2001. Human viruses at sea. *The Scientist* 15(6):19.

Schulz, H. N.; Brinkhoff, T.; Ferdelman, T. G.; Hernández Marine, M.; Teske, A.; and Jørgensen, B. B. 1999. Dense populations of a giant sulfur bacterium in Namibian shelf sediments. *Science* 284:493–95.

Staley, J. T., and Gosink, J. J. 1999. Poles apart: Biodiversity and biogeography of sea ice bacteria. *Annu. Rev. Microbiol.* 53:189–205.

Thomas, D. N., and Dieckmann, G. S. 2002. Antarctic sea ice—a habitat for extremophiles. *Science* 295:641–44.

Valentine, D. L.; Blanton, D. C.; Reeburgh, W. S. 2000. Hydrogen production by methanogens under low-hydrogen conditions. *Arch. Microbiol.* 174:415–21.

Zhang, C. L.; Li, Y.; Wall, J. D.; Larsen, L.; Sassen, R.; Huang, Y.; Wang, Y.; Peacock, A.; White, D. C.; Horita, J.; and Cole, D. R. 2002. Lipid and carbon isotopic evidence of methane-oxidizing and sulfate-reducing bacteria in association with gas hydrates from the Gulf of Mexico. *Geology* 30(3):239–42.

29.4 Freshwater Environments

Goldman, E. B. 2001. A tale of two lakes. *The Sciences* 41:9–13.

Hart, D. B.; Stone, L.; and Berman, T. 2000. Seasonal dynamics of the lake Kinneret food web: The importance of the microbial loop. *Limnol. Oceanogr.* 45:350–61.

Jouzel, J.; Petit, J. R.; Souchez, R.; Barkov, N. I.; Lipenkov, V. Y.; Raynaud, D.; Stievenard, M.; Vassiliev, N. I.; Verbeke, V., and Veimeux, F. 1999. More than 200 meters of lake ice above subglacial Lake Vostok, Antarctia. *Science* 286:2138–41.

Quayle, W. C.; Peck, L. S.; Peat, H.; Ellis-Evans, J. C.; and Harrigan, P. R. 2002. Extreme responses to climate change in antarctic lakes. *Science* 295:645.

van Etten, J. L.; Meints, R. H. 1999. Giant viruses infecting algae. *Annu. Rev. Microbiol.* 53:447–94.

29.5 Waters and Disease Transmission

Burkhardt, W. I.; Blackstone, G. M.; Skilling, D. E.; and Smith, A. W. 2002. Applied technique for increasing calicivirus detection in shellfish extracts. *J. Appl. Microbiol.* 93:235–40.

Berger, P. S.; Clark, R. M.; and Reasoner, D. J. 2000. Water, drinking. In *Encyclopedia of microbiology,* 2d ed., vol. 4, J. Lederberg, editor-in-chief, 898–913. San Diego: Academic Press.

Clesceri, L. S.; Eaton, A. D.; and Greenberg, A. E. 1998. *Microbiological examination of water and wastewater,* 20th ed. Boca Raton, Fla.: Lewis Publishers.

Grimm, D.; Ludwig, W.; Brandt, B. C.; Michel, R.; Schleifer, K.-H.; Hacker, J.; and Steinert, M. 2002. Development of 18S rRNA-targeted oligonucleotide probes for specific detection of *Hartmannella* and *Naegleria* in *Legionella*-positive environmental samples. *Syst. Appl. Microbiol.* 24:76–82.

Hegarty, J. F.; Dowd, M. T.; and Baker, K. H. 1999. Occurrence of *Helicobacter pylori* in surface water in the United States. *J. Appl. Microbiol.* 87:697–701.

LeClerc, H.; Edberg, S. C.; Pierzo, V.; and Delattire, J. M. 2000. Bacteriophages as indicators of enteric viruses and public health risk in groundwaters. *J. Appl. Microl.* 88:5–21.

LeClerc, H.; Mossel, D. A. A.; Edberg, S. C.; and Struijk, C. B. 2001. Advances in the bacteriology of the coliform group: Their suitability as markers of microbial water safety. *Annu. Rev. Microbiol.* 55:201–34.

Milius, S. 2000. New frog-killing disease may not be so new. *Sci. News* 157:133.

Szewzyk, U.; Szewzyk, R.; Manz, W.; and Schleifer, K.-H. 2000. Microbiological safety of drinking water. *Annu. Rev. Microbiol.* 54:81–127.

Vogelbein, W. K.; Lovko, V. J.; Shields, J. D.; Reece, K. S.; Mason, P. L., Walker, C. C.; and Haas, L. W. 2002. *Pfiesteria shumwayae* kills fish by micropredation not exotoxin secretion. *Nature* 418:967–70.

Yaron, S., and Matthews, K. R. 2002. A reverse transcriptase–polymerase chain reaction assay for detection of viable *Escherichia coli* O157:H7: investigation of specific target genes. *J. Appl. Microbiol.* 92:633–40.

29.6 Wastewater Treatment

Cole, S. 1998. The emergence of treatment wetlands. *Environ. Sci. Technol.* 32:218A–23A.

Falabi, J. A.; Gerba, C. P.; and Karpiscak, M. M. 2002. *Giardia* and *Cryptosporidium* removal from waste-water by a duckweed (*Lemna gibba* L.) covered pond. *Lett. Appl. Microbiol.* 34:384–87.

McKinney, R. E. 2000. Wastewater treatment, municipal. In *Encyclopedia of microbiology,* 2d ed., vol. 4, J. Lederberg, editor-in-chief, 870–83. San Diego: Academic Press.

Rabalais, N. N.; Turner, R. E.; and Scavia, D. 2002. Beyond science into policy: Gulf of Mexico hypoxia and the Mississippi River. *BioScience* 52(2):129–42.

Rocha, C.; Galvao, H.; and Barbosa, A. 2002. Role of transient silicon limitation in the development of cyanobacterial blooms in the Guadiana estuary, south-western Iberia. *Mar. Ecol. Prog. Ser.* 228:35–45.

Toh, S. K.; Webb, R. I.; and Ashbolt, N. J. 2002. Enrichment of autotrophic anaerobic ammonium-oxidizing consortia from various wastewaters. *Microb. Ecol.* 43(1):154–67.

van Dongen, U.; Jetten, M. S. M.; and van Loosdrecht, M. C. M. 2001. The Sharon-Anammox process for treatment of ammonium rich wastewater. *Water Sci. Tech.* 44(1):153–60.

29.7 Groundwater Quality and Home Treatment Systems

Robertson, W. D. 2000. Treatment of wastewater phosphate by reductive dissolution of iron. *J. Environ. Qual.* 29(5):1678–85.

Robertson, W. D.; Blowes, D. W.; Ptacek, C. J.; and Cherry, J. A. 2000. Long-term performance of *in situ* reactive barriers for nitrate remediation. *Ground Water* 38:689–95.

Chapter 30: Microorganisms in Terrestrial Environments

The references provided for chapter 28 also may be consulted for further information.

General References

Alef, K., and Nannipieri, P. 1995. *Methods in applied soil microbiology and biochemistry.* San Diego: Academic Press.

Robertson, G. P.; Coleman, D. C.; Bledsoe, C. S.; and Sollins, P. 1999. *Standard soil methods for long-term ecological research.* New York: Oxford University Press.

Scow, K. M. 2000. Soil microbiology. In *Encyclopedia of microbiology,* 2d ed., vol. 4, J. Lederberg, editor-in-chief, 321–35. San Diego: Academic Press.

Sylvia, D. M.; Fuhrmann, J. J.; Hartel, P. G.; and Zuberer, D. A. 1998. *Principles and applications of soil microbiology.* Upper Saddle River, N.J.: Prentice-Hall.

van Elsas, J. D.; Trevors, J. T.; and Wellington, E. M. H. 1997. *Modern soil microbiology.* New York: Marcel Dekker, Inc.

30.2 Microorganisms in the Soil Environment

Clarholm, M. 1998. The microbial loop in soil. In *Beyond the biomass.* K. Ritz, J. Dighton, and K. E. Giller, editors, 221–30. New York: Wiley-Sayce.

Hedlund, B. P.; Gosink, J. J.; and Staley, J. T. 1997. Verrucomicrobia div. nov., a new division of the Bacteria containing three new species of *Prosthecobacter.* Ant. van Leeuwenhoek, J. Microbiol. 72(1):29–38.

Jurgens, G.; Lindstrom, K.; and Saano, A. 1997. Novel group within the kingdom *Crenarchaeota* from boreal forest soil. *Appl. Environ. Microbiol.* 63:803–5.

Sengeløv, G.; Kowalchuk, G. A.; and Sørensen, S. J. 2000. Influence of fungal-bacterial interactions on bacterial conjugation in the residuesphere. *FEMS Microbiol. Ecol.* 31:39–45.

Vandenkoornhuyse, P.; Baldauf, S. L.; Leyval, C.; Straczek, J.; and Young, J. P. W. 2002. Extensive fungal diversity in plant roots. *Science* 295:2051–61.

Watling, R., and Harper, D. B. 1998. Chloromethane production by wood-rotting fungi and an estimate of the global flux to the atmosphere. *Mycol. Res.* 102(7):769–87.

30.3 Microorganisms and the Formation of Different Soils

Bowker, M. A.; Reed, S. C.; Belnap, J.; and Phillips, S. L. 2002. Temporal variation in community composition, pigmentation and F_vF_m of desert cyanobacterial soil crusts. *Microb. Ecol.* 43:13–25.

Jenny, H. 1980. *The soil resource. Origin and behavior.* New York: Springer-Verlag.

Marquiss, M., and Woudin, S. J. 1997. *Ecology of arctic environments.* Oxford: Blackwell.

Redman, R. S.; Litvintseva, A.; Sheehan, K. B.; Henson, J. M.; and Rodriguez, R. J. 1999. Fungi from geothermal soils in Yellowstone National Park. *Appl. Environ. Microbiol.* 65(12):5193–97.

Snider, C. S.; Hsiang, T.; Zhao, T.; and Griffith, M. 2000. Role of ice nucleation and antifreeze activities in pathogenesis and growth of snow molds. *Phytopathology* 90:354–61.

30.4 Soil Microorganism Associations with Vascular Plants

Allen, M. F. 2000. Mycorrhizae. In *Encyclopedia of microbiology,* 2d ed., vol. 3, J. Lederberg, editor-in-chief, 328–36. San Diego: Academic Press.

Bidartondo, M. I.; Bruns, T. D. 2002. Fine-level mycorrhizal specificity in the monotropoideae (*Ericaceae*): Specificity for fungal species groups. *Molec. Ecol.* 11(3):557–69.

Bidartondo, M. I.; Redecker, D.; Hijri, I.; Wiemken, A.; Bruns, T. D.; Dominguez, L.; Sersic, D. J.; Leake, J. R.; and Read, D. J. 2002. Epiparasitic plants specialized on arbuscular mycorrhizal fungi. *Nature* 419:389–92.

Espinosa-Urgel, M.; Kolter, R.; and Ramos, J.-L. 2002. Root colonization by *Pseudomonas putida:* Love at first sight. *Microbiology* 148:1–3.

Gelvin, S. B. 2003. *Agrobacterium*-mediated plant transformation: the biology behind the "gene-jockeying" tool. *Microbiol. Mol. Biol. Rev.* 67(1):16–37.

Graham, P. H. 2000. Nodule formation in legumes. In *Encyclopedia of microbiology,* 2d ed., vol. 3, J. Lederberg, editor-in-chief, 407–17. San Diego: Academic Press.

Herauart, D.; Baudouin, H. D.; Frendo, E.; Harrison, J.; Santos, R.; Jamet, A.; Van de Sype, G.; Touati, D.; and Puppo, A. 2002. Reactive oxygen species, nitric oxide and glutathione: A key role in the establishment of the legume-*Rhizobium* symbiosis? *Plant Phys. Biochem.* 40(6–8):619–24.

Klein, D. A. 2000. The rhizosphere. In *Encyclopedia of microbiology,* 2d ed., vol. 4, J. Lederberg, editor-in-chief, 117–26. San Diego: Academic Press.

Kuykendall, L. D.; Dadson, R. B.; Hashem, F. M.; and Elkan, G. H. 2000. Nitrogen fixation. In *Encyclopedia of microbiology,* 2d ed., vol. 3, J. Lederberg, editor-in-chief, 392–406. San Diego: Academic Press.

Marie, C.; Broughton, W. J.; and Deakin, W. J. 2001. *Rhizobium* type III secretion systems? legume charmers or alarmers? *Curr. Op. Plant Biol.* 4(4):336–42.

Piagnani, C.; Assante, G.; Scalisi, P.; Zocchi, G.; and Vercesi, A. 2002. Growth and physiological responses of chestnut calli to crude extracts of virulent and hypovirulent strains of *Cryphonectria parasitica. Forest Pathology* 32(1):43–53.

Schmalenberger, A., and Tebbe, C. C. 2002. Bacterial community composition in the rhizosphere of a transgenic, herbicide-resistant maize (*Zea mays*) and comparison to its non-transgenic cultivar Bosphore. *FEMS Microbiol. Ecol.* 40(1):29–37.

Singh, A., and Varma, A. 2000. Orchidaceous mycorrhizal fungi. In *Mycorrhizal biology,* J. Mukerji, B. P. Chamola, and J. Singh, editors, 265–88. New York: Kluwer Academic/Plenum Publishers.

Spaink, H. P. 2000. Root nodulation and infection factors produced by rhizobial bacteria. *Annu. Rev. Microbiol.* 54:257–88.

Sylvia, D. M. 1998. Mycorrhizal symbioses. In *Principles and applications of soil microbiology,* D. M. Sylvia, J. J. Fuhrmann, P. G. Hartel, and D. A. Zuberer, editors, 408–28. Upper Saddle River, N.J.: Prentice-Hall.

Young, B. W.; Massicotte, H. B.; Tackaberry, L. E.; Baldwin, Q. F.; and Egger, K. N. 2002. *Monotropa uniflora:* Morphological and molecular assessment of mycorrhizae retrieved from sites in the sub-boreal spruce biogeoclimatic zone in central British Columbia. *Mycorrhiza* 12(2):75–82.

30.5 Soils, Plants, and Nutrients

Klironomos, J. N. 2002. Feedback with soil biota contributes to plant rarity and invasiveness in communities. *Nature* 417:67–70.

Perakis, S. S., and Hedin, L. O. 2002. Nitrogen loss from unpolluted South American forests mainly via dissolved organic compounds. *Nature* 415:416–19.

Smil, V. 1997. Global population and the nitrogen cycle. *Sci. Am.* 277(1):76–81.

30.6 Soils, Plants, and the Atmosphere

Waibel, A. E.; Peter, T.; Carslaw, K. S.; Oelhaf, H.; Wetzel, G.; Crutzen, P.; Tsias, A.; Reimer, E.; and Fisher, H. 2000. Arctic ozone loss due to denitrification. *Science* 283:2064–69.

Zart, D.; Schmidt, I.; and Bock, E. 2000. Significance of gaseous NO for ammonia oxidation by *Nitrosomonas eutropha. Ant. van Leeuwenhoek.* 77:49–55.

30.8 The Subsurface Biosphere

Amy, P. S., and Haldeman, D. L., editors. 1997. *The microbiology of the terrestrial deep subsurface.* Boca Raton, Fla.: CRC Press.

Chapelle, F. H.; O'Neill, K.; Bradley, P. M.; Methé, B. A.; Clufo, S. A.; Knobel, L. L.; and Lovely, D. R. 2002. A hydrogen-based subsurface microbial community dominated by methanogens. *Nature* 415:312–14.

Ehrlich, H. 2001. *Geomicrobiology,* 4th ed. New York: Marcel Dekker, Inc.

Gold, T. 1999. *The deep hot biosphere.* New York: Springer-Verlag.

Kerr, R. A. 2002. Deep life in the slow, slow lane. *Science* 296:1056–58.

Krumholz, L. R. 2000. Microbial communities in the deep subsurface. *Hydrogeology J.* 8:4–10.

Newman, D. K., and Banfield, J. F. 2002. Geomicrobiology: How molecular-scale interactions underpin biogeochemical systems. *Science* 296:1071–77.

Orphan, V. J.; Taylor, L. T.; Hafenbradl, D.; and DeLong, E. F. 2000. Culture-dependent and culture-independent characterization of microbial assemblages associated with high-temperature petroleum reservoirs. *Appl. Environ. Microbiol.* 66(2):700–11.

30.9 Soil Microorganisms and Human Health

Hamilton, A. 2001. Beware: Toxic mold. *Time* 2 July, 2001, 52–55.

Kuhn, D. M., and Ghannoum, M. A. 2003. Indoor mold, toxigenic fungi, and *Stachybotrys chartarum:* Infectious disease perspective. *Clin. Microbiol. Rev.* 16(1):144–72.

Chapter 31: Normal Microbiota and Nonspecific (Innate) Host Resistance

General References

Darveau, R. P.; McFall-Ngai, M.; Ruby, E.; Miller, S; and Mangam, D. F. 2003. Host tissues may actively respond to beneficial microbes. *ASM News* 69(4): 186–91.

31.1 Gnotobiotic Animals

Gordon, H. A., and Pesti, L. 1971. The gnotobiotic animal as a tool in the study of host microbial relationships. *Bacteriol. Rev.* 35:390–429.

31.2 Normal Microbiota of the Human Body

Drasar, B. S., and Barrow, P. A. 1985. *Intestinal microbiology.* Washington, D.C.: American Society for Microbiology.

Mackowiak, P. A. 1982. The normal microbial flora. *N. Engl. J. Med.* 307:83.

Roth, R., and Jenner, W. 1998. Microbial ecology of the skin. *Annu. Rev. Microbiol.* 42:441–48.

Travis, J. 2003. Gut check. *Science News* 163:344–45.

31.3 Overview of Host Resistance

Medzhitov, R., and Janeway, C. A. 1998. An ancient system of host defense. *Curr. Opin. Immunol.* 10:12–14.

Modlin, R., et al. 1999. The toll of innate immunity on microbial pathogens. *N. Engl. J. Med.* 340(23):1834–35.

31.4 Cells, Tissues, and Organs of the Immune System

Banchereau, J. 2002. The long arm of the immune system. *Sci. Am.* 287(5):52–59.

Golde, D. 1991. The stem cell. *Sci. Am.* 255(6):86–94.

Picker, L., and Siegelman, M. 1999. Lymphoid tissues and organs. In *Fundamental immunology,* 4th ed. Philadelphia: Lippincott-Raven.

31.5 Physical and Chemical Barriers in Nonspecific (Innate) Resistance

Baba, T., and Schneewind, O. 1998. Instruments of microbial warfare: Bacteriocin synthesis, toxicity, and immunity. *Trends Microbiol.* 6(2):66–71.

Edelson, R. L., and Fink, J. M. 1985. The immunologic function of the skin. *Sci. Am.* 256:46–53.

31.6 Inflammation

Baumann, H., and Gauldie, J. 1994. The acute response. *Immunol. Today.* 15:74–80.

31.7 The Complement System

Muller-Eberhard, H. J. 1988. Molecular organization and function of the complement system. *Annu. Rev. Biochem.* 57:321–47.

Walport, Mark J. 2001. Complement: First of two parts. *N. Engl. J. Med.* 344(14):1058–66.

Walport, Mark J. 2001. Complement. *N. Engl. J. Med.* 344(15):1140–44.

31.8 Phagocytosis

Beamer, L. 2002. Human BPI: One protein's journey from laboratory into clinical trials. *ASM News* 68(11):543–48.

Hancock, R. E. W., and Diamond, G. 2000. The role of cationic antimicrobial peptides in innate host defenses. *Trends Microbiol.* 8(9):402–10.

Labro, M.-T. 2000. Interference of antibacterial agents with phagocyte functions: Immunomodulation or "immuno-fairy tales"? *Clin. Microbiol. Rev.* 13(4):615–50.

Lancaster, J. R. 1992. Nitric oxide in cells. *Am. Scientist* 80(3):248–59.

Lehrer, R. 1990. Defensins. Natural peptide antibiotics from neutrophils. *ASM News* 56(6):315–18.

Ofek, I.; Goldhar, J.; Keisari, Y.; and Sharon, N. 1995. Nonopsonic phagocytosis of microorganisms. *Annu. Rev. Microbiol.* 49:239–76.

31.9 Cytokines

Luster, A. 1998. Chemokines—chemotactic cytokines that mediate inflammation. *N. Engl. J. Med.* 338(7):436–45.

Sen, G. C. 2001. Viruses and interferons. *Annu. Rev. Microbiol.* 55:255–81.

31.10 Natural Killer Cells

Berke, G. 1995. Unlocking the secrets of CTL and NK cells. *Immunol. Today* 16:343.

Lanier, L. L. 1998. NK cell receptors. *Annu. Rev. Immunol.* 16:359–70.

Chapter 32: Specific (Adaptive) Immunity

General References

Abbas, A. K.; Lichtman, A. H.; and Pober, J. S. 1997. *Cellular and molecular immunology,* 3d ed. Philadelphia: W. B. Saunders.

Goldsby, R. A.; Kindt, T. J.; and Osborne, B. A. 2000. *Kuby Immunology,* 4th ed. New York: W. H. Freeman.

Janeway, C. A., Jr.; Travers, P.; Walport, M.; and Shlomckik, M. 2001. *Immunobiology: The immune system in health and disease,* 4th ed. New York: Garland Publishing.

Roitt, I. M. 1997. *Essential immunology,* 9th ed. Boston: Blackwell Scientific Publications.

Roitt, I. M.; Brostoff, J.; and Male, D. 1998. *Immunology,* 5th ed. St Louis: C. V. Mosby.

Silverstein, A. M. 1989. *A history of immunology.* San Diego: Academic Press.

32.1 Overview of Specific (Adaptive) Immunity

Science (Special Issue). 1996. Elements of immunity. *Science* 272(5258):50–79.

32.2 Antigens

Engelhard, V. H. 1994. How cells present antigens. *Sci. Am.* 271(2):54–61.

Papageorgiou, A. C., and Acharya, K. R. 2000. Microbial superantigens: From structure to function. *Trends Microbiol.* 8(8):369–75.

32.3 Antibodies

Ada, G. L., and Nossal, G. 1987. The clonal-selection theory. *Sci. Am.* 257(5):62–69.

Collier, R. J., and Koplan, D. A. 1984. Immunotoxins. *Sci. Am.* 251(2):56–64.

Metzger, H. 1990. *Fc receptors and the action of antibodies.* Washington, D.C.: American Society for Microbiology.

Tonegawa, S. 1985. The molecules of the immune system. *Sci. Am.* 252(3):41–57.

32.4 T-Cell Biology

Bahram, S., and Spies, T. 1996. The MIC gene family. *Res. Immunol.* 147:328–40.

Engelhard, V. H. 1994. How cells process antigens. *Sci. Am.* 271(2):54–61.

Gruen, J., and Weissman, S. 1997. Evolving views of the major histocompatibility complex. *Blood* 90:4252–60.

Lanzavecchia, A., and Sallusto, F. 2000. Dynamics of T lymphocyte responses: Intermediates, effectors, and memory cells. *Science* 290:92–97.

Lehner, P., and Trowsdale, J. 1998. Antigen presentation: Coming out gracefully. *Curr. Biol.* 8:R605–10.

Mellman, I., et al. 1998. Antigen processing for amateurs and professionals. *Trends Cell. Biol.* 8:232–30.

Moss, P., et al. 1993. The human T-cell receptor in health and disease. *Annu. Rev. Immunol.* 10:71–85.

Schwartz, R. H. 1993. T cell anergy. *Sci. Am.* 269(2):62–71.

32.5 B-Cell Biology

Benschop, R., and Cambier, J. 1999. B cell development: Signal transduction by antigen receptors and their surrogates. *Curr. Opin. Immunol.* 11:143–50.

Fagarasan, S., and Honjo, T. 2000. T-independent immune response: New aspects of B cell biology. *Science* 290:89–92.

32.6 Action of Antibodies

Keller, M. A., and Stiehm, R. 2000. Passive immunity in prevention and treatment of infectious diseases. *Clin. Microbiol. Rev.* 13(4):602–14.

Rose, N., et al. 1997. *Manual of clinical laboratory immunology,* 5th ed. Washington, D.C.: American Society for Microbiology.

32.7 The Classical Complement Pathway

Caroll, M. C. 1998. The role of complement and complement receptors in induction and regulation of immunity. *Annu. Rev. Immunol.* 16:545–60.

Kinoshita, T. 1991. Biology of complement: The overture. *Immunol. Today* 12:291.

Liszewski, M. K. 1996. Control of the complement system. *Adv. Immunol.* 61:201–31.

32.8 Acquired Immune Tolerance

Ramsdell, F., and Fowlkes, B. 1990. Clonal deletion versus clonal anergy. *Science* 248:342–48.

Rajewsky, K. 1996. Clonal selection and learning in the antibody system. *Nature* 381:751–58.

Chapter 33: Medical Immunology

General References

Kaufmann, S. H.; Sher, A.; and Fafi, A. 2001. *Immunology of infectious diseases.* Washington, D.C.: ASM Press.

Rose, N. R.; Hamilton, R. G.; and Detrick, B. 2002. *Manual of Clinical Laboratory Immunology,* 6th ed. Washington, D.C.: American Society for Microbiology.

Science Special Issue. 1996. Elements of immunity. *Science* 272(5258):50–79.

Scientific American Special Issue. 1993. Life, death, and the immune system. *Sci. Am.* 269(3).

33.1 Vaccines and Immunizations

Ada, G. 2001. Vaccines and vaccination. *N. Engl. J. Med.* 345(14):1042–53.

Arvin, A. M. 2000. Vaccines, viral. In *Encyclopedia of microbiology,* 2d ed., vol. 4, J. Lederberg, editor-in-chief, 779–87. San Diego: Academic Press.

Chattergoon, M., et al. 1997. Genetic immunization: A new era in vaccines and immune therapies. *FASEB J.* 11:754–60.

Dennehy, P. H. 2001. Active immunization in the United States: Developments over the past decade. *Clin Microbiol. Rev.* 14(4):872–908.

Hoiseth, S. K. 2000. Vaccines, bacterial. In *Encyclopedia of microbiology,* 2d ed., vol. 4, J. Lederberg, editor-in-chief, 767–78. San Diego: Academic Press.

Klausner, R. D., et al. 2003. The need for a global HIV vaccine enterprise. *Science* 300:2036–39.

Langridge, W. H. R. 2000. Edible vaccines. *Sci. Am.* 283(3):66–71.

Ogra, P. L.; Faden, H.; and Welliver, R. C. 2001. Vaccination strategies for mucosal immune responses. *Clin. Microbiol. Rev.* 14(2):430–45.

Weiner, D., and Kennedy, R. 1999. Genetic vaccines. *Sci. Am.* 281(1):46–57.

33.2 Immune Disorders

Baggiolini, M., and Dahinden, C. 1994. CC chemokines in allergic inflammation. *Immunol. Today* 15(3):127–33.

Bochner, B. S., and Lichtenstein, L. M. 1991. Anaphylaxis. *N. Engl. J. Med.* 324(25):1785–90.

Buckley, R. H. 1992. Immunodeficiency diseases. *JAMA* 268(20):2797–2806.

Cohen, I. R. 1988. The self, the world, and autoimmunity. *Sci. Am.* 258(4): 52–60.

Cunningham, M. W., and Fujinami, R. S. 2000. *Molecular mimicry, microbes, and autoimmunity.* Herndon, Va.: ASM Press.

Fujinami, R. S. 2001. Viruses and autoimmune disease—two sides of the same coin? *Trends Microbiol.* 9(8):377–81.

Kay, A. B. 2001. Allergy and allergic diseases. *N. Engl. J. Med.* 344(1):30–37.

Lanza R.; Cooper, D.; and Chick, W. 1997. Xenotransplantation. *Sci. Am.* 277(1):54–59.

Platt, J. L., editor. 2000. *Xenotransplantation.* Herndon, Virginia: ASM Press.

Rennie, J. 1990. The body against itself. *Sci. Am.* 263(6):106–15.

Rose, N. R., and Afanasyeva, M. 2003. From infection to autoimmunity: The adjuvant effect. *ASM News* 69(3):132–37.

Steinman, L. 1993. Autoimmune disease. *Sci. Am.* 269(3):106–15.

33.3 Antigen-Antibody Interactions In Vitro

Lopez, M.; Fleisher, T.; and deShazo, R. D. 1992. Use and interpretation of diagnostic immunologic laboratory tests. *JAMA* 268(20):2970–90.

Mahony, J. B., and Chernesky, M. A. 1999. Immunoassays for the diagnosis of infectious diseases. In *Manual of clinical microbiology,* 7th ed., P. R. Murray, editor-in-chief, 202–14. Washington, D.C.: ASM Press.

Chapter 34: Pathogenicity of Microorganisms

General References

Brogden, K. A., and Guthmiller, J. M. 2002. *Polymicrobial diseases,* Washington, D.C.: ASM Press.

Clark, V. 1994. *Bacterial pathogenesis: Identification and regulation of virulence factors.* San Diego, Calif.: Academic Press.

Ewald, P. 1994. *Evolution of infectious diseases.* New York: Oxford University Press.

Falkow, S. 1997. What is a pathogen? *ASM News* 63(7):359–65.

Finlay, B. B., and Falkow, S. 1997. Common themes in microbial pathogenicity revisited. *Microbiol. Mol. Biol. Rev.* 61(2):136–69.

Groisman, E. A., and Ochman, H. 1996. Pathogenicity islands: Bacterial evolution in quantum leaps. *Cell* 87:791–94.

Hacker, J.; Hentschel, U.; and Dobrindt, U. 2003. Prokaryotic chromosomes and disease. *Science* 301:790–93.

Lee, C. 1996. Pathogenicity islands and the evolution of bacterial pathogens. *Infect. Agents Dis.* 5:1–7.

Mims, C.; Nash, A.; and Stephen, J. 2000. *Mims' pathogenesis of infectious disease,* 5th ed. San Diego: Academic Press.

Roth, J.; Bolin, C.; Brogden, K.; Minion, F.; and Wannemuehler, M. 1995. *Virulence mechanisms of bacterial pathogenesis.* Washington, D.C.: ASM Press.

Salyers, A., and Whitt, D. 2002. *Bacterial pathogenesis: A molecular approach,* 2d ed. Washington, D.C.: ASM Press.

Wilson, M.; McNab, R.; and Henderson, B. 2002. *Bacterial disease mechanisms.* Boston, Mass. Cambridge University Press.

34.1 Host-Parasite Relationships

Ewald, P. W. 1993. The evolution of virulence. *Sci. Am.* 268(4):86–93.

Hacker, J., and Kaper, J. B. 2000. Pathogenicity islands and the evolution of microbes. *Annu. Rev. Microbiol.* 54:641–79.

Kaper, J., and Hacker, J. 1999. *Pathogenicity islands and other mobile virulence elements.* Washington, D.C.: ASM Press.

34.2 Pathogenesis of Viral Diseases

Flint, S. J., et al. 1999. *Principles of virology: Molecular biology, pathogenesis, and control.* Washington, D.C.: ASM Press.

Strauss, J., and Strauss, E. G., editors. 2002. *Viruses and human disease.* San Diego: Academic Press.

Tyler, K., and Fields, B. 1996. Pathogenesis of viral infections. In *Fields' virology,* 3d ed., B. N. Fields, et al., editors. Philadelphia: Lippincott-Raven.

Weiss, R. A. 2002. Virulence and pathogenesis. *Trends Microbiol.* 10(7):314–17.

34.3 Pathogenesis of Bacterial Diseases

Brogden, K. A.; Roth, J. A.; Stanton, T. B.; Bolin, C. A.; Minion, F. C.; and Wannemuehler, M. J., editors. 2000. *Virulence mechanisms of bacterial pathogens,* 3rd edition. Washington, D.C.: ASM Press.

Burns, Drusilla L., et al. 2003. *Bacterial protein toxins.* Washington, D.C.: ASM Press.

Gao, L.-Y, and Kwaik, Y. A. 2000. The modulation of host cell apoptosis by intracellular bacterial pathogens. *Trends Microbiol.* 8(7):306–13.

Gaunthier, A., and Finlay, B. B. 2002. Type III secretion system inhibitors are potential antimicrobials. *ASM News* 68(8):383–87.

Gill, D. M. 1982. Bacterial toxins: A table of lethal amounts. *Microbiol. Rev.* 46:86–88.

Groisman, E., editor. 2000. *Principles of Bacterial Pathogenesis.* San Diego, Calif.: Academic Press.

Hueck, C. 1998. Type III protein secretion systems in bacterial pathogens of animals and plants. *Microbiol. Mol. Biol. Rev.* 62(2):379–433.

Johansson, J., and Cossart, P. 2003. RNA-mediated control of virulence gene expression in bacterial pathogens. *Trends Microbiol.* 11(6):280–85.

Lejeune, P. 2003. Contamination of abiotic surfaces: What a colonizing bacterium sees and how to blur it. *Trends Microbiol.* 11(4):179–84.

Mitchell, T. J.; Godfree, A. F.; and Stewart-Tull, D. E. S., editors. 1998. *Toxins,* The Society for Applied Microbiology Symposium Series No. 27. *J. Appl. Microbiol. Symp. Suppl.* 84 (entire issue).

O'Riordan, M., and Portnoy, D. A. 2002. The host cytosol: Front-line or home front? *Trends Microbiol.* 10(8):361–64.

Rietschel, E. T., and Brade, H. 1992. Bacterial endotoxins. *Sci. Am.* 267(2):54–61.

Schantz, E. J., and Johnson, E. A. 1992. Properties and use of botulinum toxin and other microbial neurotoxins in medicine. *Microbiol. Rev.* 56(1):80–99.

Schmitt, C. 1999. Bacterial toxins: Friends or foes? *Emerg. Infect. Dis.* 5(2):224–33.

Sears, C. L., and Kaper, J. B. 1996. Enteric bacterial toxins: Mechanisms of action and linkage to intestinal secretion. *Microbiol. Rev.* 60(1):167–215.

Stanley, P.; Koronakis, V.; and Hughes, C. 1998. Acylation of *Escherichia coli* hemolysin: A unique protein lipidation mechanism underlying toxin function. *Microbiol. Mol. Biol. Rev.* 62(2):309–33.

West, N. P.; Sansonetti, P. J.; Frankel, G.; and Tang, C. M. 2003. Finding your niche: What has been learnt from STM studies on GI colonization. *Trends Microbiol.* 11(7):338–44.

Yoshida, S., and Sasakawa, C. 2003. Exploiting host microtubule dynamics: A new aspect of bacterial invasion. *Trends Microbiol.* 11(3):139–43.

34.4 Microbial Mechanisms for Escaping Host Defenses

Alcami, A., and Koszinowski, U. H. 2000. Viral mechanisms of immune evasion. *Trends Microbiol.* 8(9):410–18.

Stephens, R. 1999. *Chlamydia: Intracellular biology, pathogenesis, and immunity.* Washington, D.C.: ASM Press.

Chapter 35: Antimicrobial Chemotherapy

General References

Brooks, G. F.; Butel, J. S.; and Morse, S. A. 1998. *Jawetz, Melnick & Adelberg's medical microbiology,* 21st ed. Stamford, Conn.: Appleton & Lange.

Bugg, C. E.; Carson, W. M.; and Montgomery, J. A. 1993. Drugs by design. *Sci. Am.* 269(6):92–98.

Butaye, P.; Devriese, L. A.; and Haesebrouck, F. 2003. Antimicrobial growth promoters used in animal feed: Effects of less well known antibiotics on gram-positive bacteria. *Clin. Microbiol. Rev.* 16(2):175–88.

Fernandes, P. B. 1996. Pharmaceutical perspective on the development of drugs to treat infectious diseases. *ASM News* 62(1):21–24.

Kessler, D. A., and Feiden, K. L. 1995. Faster evaluation of vital drugs. *Sci. Am.* 272(3):48–54.

Langer, R. 2003. Where a pill won't reach. *Sci. Am.* 288(4):51–57.

Liu, H., and Reynolds, K. A. 2000. Antibiotic biosynthesis. In *Encyclopedia of microbiology,* 2d ed., vol. 1, J. Lederberg, editor-in-chief, 189–207. San Diego: Academic Press.

Mandell, G. L.; Bennett, J. E.; and Dolin, R. 2000. *Principles and practice of infectious diseases,* 5th ed. New York: Churchill Livingstone.

McDevitt, D., and Rosenberg, M. 2001. Exploiting genomics to discover new antibiotics. *Trends Microbiol.* 9(12):611–17.

Murray, P. R., editor-in-chief. 1999. *Manual of clinical microbiology,* 7th ed. Washington, D.C.: ASM Press.

Physicians' desk reference. Oradell, N.J.: Medical Economics Books (published annually).

Walsh, C. 2003. *Antibiotics: Actions, origins, resistance.* Washington. D.C.: ASM Press.

35.1 The Development of Chemotherapy

Bottcher, H. M. 1964. *Wonder drugs: A history of antibiotics.* Philadelphia: J. B. Lippincott.

Hare, R. 1970. *The birth of penicillin.* Atlantic Highlands, NJ: Allen and Unwin.

35.3 Determining the Level of Antimicrobial Activity

Courvalin, P. 1992. Interpretive reading of antimicrobial susceptibility tests. *ASM News* 58(7):368–75.

Sanders, C. C. 1991. A problem with antimicrobial susceptibility tests. *ASM News* 57(4):187–90.

35.4 Mechanisms of Action of Antimicrobial Agents

Chopra, I., and Roberts, M. 2001. Tetracycline antibiotics: Mode of action, applications, molecular biology, and epidemiology of bacterial resistance. *Microbiol. Mol. Biol. Rev.* 65(2):232–60.

Franklin, T. J. 2001. *Biochemistry and molecular biology of antimicrobial drug action.* New York: Chapman & Hall.

35.6 Antibacterial Drugs

Bayles, K. W. 2000. The bactericidal action of penicillin: New clues to an unsolved mystery. *Trends Microbiol.* 8(6):274–78.

Drlica, K. 1999. Refining the fluoroquinolones. *ASM News* 65(6):410–15.

Drlica, K., and Zhao, X. 1997. DNA gyrase, topoisomerase IV, and the 4-quinolones. *Microbiol. Mol. Biol. Rev.* 61(3):377–92.

Maxwell, A. 1997. DNA gyrase as a drug target. *Trends Microbiol.* 5(3):102–9.

Piepersberg, W. 2000. Aminoglycosides, bioactive bacterial metabolites. In *Encyclopedia of microbiology,* 2d ed., vol. 1, J. Lederberg, editor-in-chief, 162–170. San Diego: Academic Press.

Vakulenko, S. B., and Mobashery, S. 2003. Versatility of aminoglycosides and prospects for their future. *Clin. Microbiol. Rev.* 16(3):430–50.

35.7 Drug Resistance

Amábile-Cuevas, C. F. 2003. New antibiotics and new resistance. *American Scientist* 91:138–49.

Bennett, P. M. 2000. Transposable elements. In *Encyclopedia of microbiology,* 2d ed., vol 4, J. Lederberg, editor-in-chief, 704–24. San Diego: Academic Press.

Borges-Walmsley, M. I., and Walmsley, A. R. 2001. The structure and function of drug pumps. *Trends Microbiol.* 9(2):71–79.

Cetinkaya, Y.; Falk, P.; and Mayhall, C. G. 2000. Vancomycin-resistant enterococci. *Clin. Microbiol. Rev.* 13(4):686–707.

Chambers, H. F. 2003. Solving staphylococcal resistance to β-lactams. *Trends Microbiol.* 11(4):145–48.

Davies, J., and Wright, G. D. 1997. Bacterial resistance to aminoglycoside antibiotics. *Trends Microbiol.* 5(6):234–40.

Davison, H. C.; Low, J. C.; and Woolhouse, M. E. J. 2000. What is antibiotic resistance and how can we measure it? *Trends Microbiol.* 8(12):554–59.

Drlica, K. 2001. A strategy for fighting antibiotic resistance. *ASM News* 67(1):27–33.

Gilbert, P., and McBain, A. J. 2003. Potential impact of increased use of biocides in consumer products on prevalence of antibiotic resistance. *Clin. Microbiol. Rev.* 16(2):189–208.

Gold, H. S., and Moellering, Jr. R. C. 1996. Antimicrobial-drug resistance. *N. Engl. J. Med.* 335(19):1445–53.

Grkovic, S.; Brown, M. H.; and Skurray, R. A. 2002. Regulation of bacterial drug export systems. *Microbiol. Mol. Biol. Rev.* 66(4):671–701.

Grohmann, E.; Muth, G.; and Espinosa, M. 2003. Conjugative plasmid transfer in gram-positive bacteria. *Microbiol. Mol. Biol. Rev.* 67(2):277–301.

Hancock, R. E. W. 1997. The bacterial outer membrane as a drug barrier. *Trends Microbiol.* 5(1):37–42.

Hiramatsu, K.; Cui, L.; Kuroda, M.; and Ito, T. 2001. The emergence and evolution of methicillin-resistant *Staphylococcus aureus. Trends Microbiol.* 9(10):486–93.

Hughes, D., and Andersson, D. I., editors. 2002. *Antibiotic development and resistance.* New York: Taylor & Francis.

Huycke, M. M.; Sahm, D. F.; and Gilmore, M. S. 1998. Multiple-drug resistant enterococci: The nature of the problem and an agenda for the future. *Emerg. Infect. Dis.* 4(2):239–49.

Levy, S. B. 1998. The challenge of antibiotic resistance. *Sci. Am.* 278(3):46–53.

Levy, S. B. 2002. *The Antibiotic Paradox,* 2d ed. Cambridge: Perseus Publishing.

Mah, T.-F. C., and O'Toole, G. A. 2001. Mechanisms of biofilm resistance to antimicrobial agents. *Trends Microbiol.* 9(1):34–9.

Mckeegan, K. S.; Borges-Walmsley, M. I.; and Walmsley, A. R. 2003. The structure and function of drug pumps: An update. *Trends Microbiol.* 11(1):21–29.

Mlot, C. 2001. Antimicrobials in food production: Resistance and alternatives. *ASM News* 67(4):196–200.

Nicolaou, K. C., and Bobby, C. N. C. 2001. Behind enemy lines. *Sci Am.* 284(5):54–61.

Paulsen, I. T., et al. 2003. Role of mobile DNA in the evolution of vancomycin-resistant *Enterococcus faecalis. Science* 299:2071–74.

Rattan, A.; Kalia, A.; and Ahmad, N. 1998. Multidrug-resistant *Mycobacterium tuberculosis:* Molecular perspectives. *Emerg. Infect. Dis.* 4(2):195–209.

Salyers, A.; Shoemaker, N.; Bonheyo, G.; and Frias, J. 1999. Conjugative transposons: Transmissible resistance islands. In *Pathogenicity islands and other mobile virulence elements,* J. B. Kaper and J. Hacker, editors, 331–46. Washington, D.C.: ASM Press.

Walsh, T. R., and Howe, R. A. 2002. The prevalence and mechanisms of vancomycin resistance in *Staphylococcus aureus. Annu. Rev. Microbiol.* 56:657–75.

Wegener, H. C.; Aarestrup, F. M.; Jensen, L. B.; Hammerum, A. M.; and Bager, F. 1999. Use of antimicrobial growth promoters in food animals and *Enterococcus faecium* resistance to therapeutic antimicrobial drugs in Europe. *Emerg. Infect. Dis.* 5(3):329–35.

35.8 Antifungal Drugs

Espinel-Ingroff, A. 2000. Antifungal agents. In *Encyclopedia of microbiology,* 2d ed., vol. 1, J. Lederberg, editor-in-chief, 232–53. San Diego: Academic Press.

Georgopapadakou, N. H., and Tkacz, J. S. 1995. The fungal cell wall as a drug target. *Trends Microbiol.* 3(3):98–104.

Ghannoum, M. A., and Rice, L. B. 1999. Antifungal agents: Mode of action, mechanisms of resistance, and correlation of these mechanisms with bacterial resistance. *Clin. Microbiol. Rev.* 12(4):501–17.

Groll, A. H.; De Lucca, A. J.; and Walsh, T. J. 1998. Emerging targets for the development of novel antifungal therapeutics. *Trends Microbiol.* 6(3):117–24.

Klepser, M. E.; Ernst, E. J.; and Pfaller, M. A. 1997. Update on antifungal resistance. *Trends Microbiol.* 5(9):372–75.

35.9 Antiviral Drugs

Bacon, T. H.; Levin, M. J.; Leary, J. J.; Sarisky, R. T.; and Sutton, D. 2003. Herpes simplex virus resistance to acyclovir and penciclovir after two decades of antiviral therapy. *Clin. Microbiol. Rev.* 16(1):114–28.

Balfour, Jr., H. H. 1999. Antiviral drugs. *N. Engl. J. Med.* 340(16):1255–68.

De Clercq, E. 1997. In search of a selective antiviral chemotherapy. *Clin. Microbiol. Rev.* 10(4):674–93.

Hazeltine, W. A. 2001. Beyond chicken soup. *Sci. Am.* 285(5):56–63.

Whitley, R. J. 2000. Antiviral agents. In *Encyclopedia of microbiology,* 2d ed., vol. 1, J. Lederberg, editor-in-chief, 286–310. San Diego: Academic Press.

Chapter 36: Clinical Microbiology

General References

Alverez-Barrientos, A., et al. 2000. Applications of flow cytometry to clinical microbiology. *Clin. Microbiol. Rev.* 13(2):167–95.

Forbes, B. A.; Sahm, D. F.; and Weissfeld, A. S. 1998. *Bailey and Scott's diagnostic microbiology,* 10th ed. St. Louis: C. V. Mosby.

Garcia, L. S. 1999. *Practical guide to diagnostic parasitology.* Washington, D.C.: ASM Press.

Isenberg, H. D., editor. 1998. *Essential procedures for clinical microbiology.* Washington, D.C.: American Society for Microbiology.

Larone, D. 2002. *Medically important fungi: A guide to identification,* 4th ed. Washington, D.C.: ASM Press.

Murray, P. R., editor-in-chief. 2003. *Manual of clinical microbiology,* 8th. ed. Washington, D.C.: ASM Press.

Murray, P. R., editor-in-chief. 1999. *ASM pocket guide to clinical microbiology.* Washington, D.C.: ASM Press.

Persing, D. H., editor. 1993. *Diagnostic molecular microbiology.* Washington, D.C.: American Society for Microbiology.

Rose, N. R.; Hamilton, R. G.; and Detrick, B., editors. 2002. *Manual of clinical laboratory immunology,* 6th ed. Washington, D.C.: American Society for Microbiology.

Sewell, D. L. 1995. Laboratory-associated infections and biosafety. *Clin. Microbiol. Rev.* 8(3):389–405.

36.1 Specimens

Bartlett, R.; Mazens-Sullivan, M.; Tetreault, J.; Lobel, S.; and Nivard, J. 1994. Evolving approaches to management of quality in clinical microbiology. *Clin. Microbiol. Rev.* 7(1):55–88.

Emori, T., and Gaynes, R. 1993. An overview of nosocomial infections, including the role of the microbiology laboratory. *Clin. Microbiol. Rev.* 6(4):428–42.

Johnson, F. B. 1990. Transport of viral specimens. *Clin. Microbiol. Rev.* 3(2):120–31.

Miller, M. J. 1998. *A guide to specimen management in clinical microbiology.* Washington, D.C.: ASM Press.

36.2 Identification of Microorganisms from Specimens

Arens, M. 1999. Methods for subtyping and molecular comparisons of human viral genomes. *Clin. Microbiol. Rev.* 12(4):612–26.

Belkum, A. 1994. DNA fingerprinting of medically important microorganisms by use of PCR. *Clin. Microbiol. Rev.* 7(2):174–84.

Check, W. 1998. Clinical microbiology eyes nucleic acid—based technologies. *ASM News* 64(2):84–89.

Manafi, M.; Kneifel, W.; and Bascomb, S. 1991. Fluorogenic and chromogenic substrates used in bacterial diagnostics. *Microbiol. Rev.* 55(3):335–48.

Olivo, P. D. 1996. Transgenic cell lines for detection of animal viruses. *Clin. Microbiol. Rev.* 9(3):321–34.

Persing, D. H. 1996. *PCR protocols for emerging infectious diseases.* Washington, D.C.: ASM Press.

Powers, C. 1998. Diagnosis of infectious diseases: A cytopathologists's perspective. *Clin. Microbiol. Rev.* 11(2):341–65.

Stager, C. E., and Davis, J. R. 1993. Automated systems for identification of microorganisms. *Clin. Microbiol. Rev.* 5(3):302–27.

Weiss, J. B. 1995. DNA probes and PCR for diagnosis of parasitic infections. *Clin. Microbiol. Rev.* 8(1):113–30.

Woods, G. L. and Walker, D. H. 1996. Detection of infection or infectious agents by use of cytologic and histologic stains. *Clin. Microbiol. Rev.* 9(3):382–404.

36.3 Susceptibility Testing

Canton, R., et al. 2000. Evaluation of the Wider System, a new computer-assisted image-processing device for bacterial identification and susceptibility testing. *J. Clin. Microbiol.* 38(4):1339–46.

36.4 Computers in Clinical Microbiology

Ryon, K. J., and Peebles, J. E. 1982. On-line computer entry of routine and AutoMicrobic System bacteriology results. In *Rapid methods and automation in microbiology,* R. C. Tilton, editor, 23–27. Washington, D.C.: American Society for Microbiology.

Chapter 37: The Epidemiology of Infectious Disease

General References

Centers for Disease Control and Prevention. *Morbidity and mortality reports.* (A weekly report that discusses infectious diseases.) Atlanta: Centers for Disease Control and Prevention.

Centers for Disease Control and Prevention. *Surveillance.* (Annual summaries of specific infectious diseases.) Atlanta: Centers for Disease Control and Prevention.

Chin, J., editor. 2000. *Control of communicable diseases manual.* 17th edition. Waldorf, Md.: American Public Health Association.

Garrett, L. 2000. *Betrayal of trust: The collapse of global public health.* New York: Hyperion.

Giesecke, J. 1994. *Modern infectious disease epidemiology.* Boston: Little, Brown and Company.

Kiple, K. F., editor. 1993. *The Cambridge world history of human disease.* New York: Cambridge University Press.

Lilienfeld, D. E., and Stolley, P. 1994. *Foundations of epidemiology,* 3d ed. New York: Oxford University Press.

Mack, A., editor. 1991. *In time of plague: The history and social consequences of lethal epidemic disease.* New York: New York University Press.

Mandell, G. L.; Bennett, J. E.; and Dolin, R. 2000. *Principles and practice of infectious diseases,* 5th ed. New York: Churchill Livingstone.

Nesse, R. M., and Williams, G. C. 1998. Evolution and the origins of disease. *Sci. Am.* 279(5):86–93.

Preston, R. 1994. *The hot zone.* New York: Random House.

Rothman, K. J., and Greenland, S. 1998. *Modern epidemiology,* 2d ed. Philadelphia: Lippincott-Raven.

37.1 Epidemiological Terminology

Dorland's Illustrated Medical Dictionary, 28th ed. Philadelphia: W. B. Saunders.
Stedman's medical dictionary, 27th ed. Philadelphia: Lippincott, Williams and Wilkins.

37.2 Measuring Frequency: The Epidemiologist's Tools

The State of World Health. 1997. In The world health report 1996—fighting disease, fostering development. Geneva: World Health Organization, 1–62.

37.3 Infectious Disease Epidemiology

Outbreak of West Nile-like encephalitis—New York. 1999. *Morb. Mortal. Weekly Rep.* 48:845–48.

37.4 Recognition of an Infectious Disease in a Population

Pinner, R. W.; Koo, D.; and Berkelman, R. L. 2000. Surveillance of infectious diseases. In *Encyclopedia of microbiology,* 2d ed., vol. 4, J. Lederberg, editor-in-chief, 506–25. San Diego: Academic Press.

37.5 Recognition of an Epidemic

Mack, A., editor. 1991. *In time of Plague: the history and social consequences of lethal epidemic disease.* New York: New York University Press.

37.6 The Infectious Disease Cycle: Story of a Disease

Anderson, R. M., and May, R. M. 1992. Understanding the AIDS pandemic. *Sci. Am.* 266(5):58–67.

Krauss, H., et al. 2003. *Zoonoses: Infectious diseases transmissible from animals to humans,* 3d ed. Washington, D.C.: ASM Press.

Leavitt, J. 1996. *Typhoid Mary: Captive to the public's health.* Boston, Mass.: Beacon Press.

Moore, P., and Broome, C. 1994. Cerebrospinal meningitis epidemics. *Sci. Am.* 271(5):38–47.

37.7 Virulence and the Mode of Transmission

Ebert, D., and Bull, J. J. 2003. Challenging the trade-off model for the evolution of virulence: Is virulence management feasible? *Trends Microbiol.* 11(1):15–20.

Ewald, P. W. 1993. The evolution of virulence. *Sci. Am.* 268(4):86–93.

Lipsitch, M., and Moxon, E. R. 1997. Virulence and transmissibility of pathogens: What is the relationship? *Trends Microbiol.* 5(1):31–37.

37.8 Emerging and Reemerging Infectious Diseases and Pathogens

Anderson, B. E., and Neuman, M. A. 1997. *Bartonella* spp. as emerging human pathogens. *Clin. Microbiol. Rev.* 10(2):203–19.

Centers for Disease Control and Prevention. *Emerging infectious diseases.* (A journal published since 1995 by the National Center for Infectious Diseases.) Atlanta: Centers for Disease Control and Prevention.

Garrett, L. 1994. *The Coming plague: Newly emerging diseases in a world out of balance.* New York: Farrar, Straus and Giroux.

Krause, R., editor. 1998. *Emerging infectious diseases.* New York: Academic Press.

Levins, R.; Awerbuch, T.; Brinkmann, U.; Eckardt, I.; Epstein, P.; Makhoul, N.; de Possas, C. A.; Puccia, C.; Spielman, A.; and Wilson, M. E. 1994. The emergence of new diseases. *American Scientist* 82:52–60.

Morse, S. S. 2000. Viruses, emerging. In *Encyclopedia of microbiology,* 2d ed., vol. 4, J. Lederberg, editor-in-chief, 811–31. San Diego: Academic Press.

Nathanson, N. 1997. The emergence of infectious diseases: Societal causes and consequences. *ASM News* 63(2):83–88.

Park, S.; Worobo, R. W.; and Durst, R. A. 1999. *Escherichia coli* O157:H7 as an emerging foodborne pathogen: A literature review. *Crit. Rev. Food Sci. Nutr.* 39(6):481–502.

Patz, J. A.; Epstein, P. R.; Burke, T. A.; and Balbus, J. M. 1996. Global climate change and emerging infectious diseases. *JAMA* 275(3):217–23.

Roizman, B. 1995. *Infectious diseases in an age of change: The impact of human ecology and behavior on disease transmission.* Washington, D.C.: National Academy Press.

Sanders, W. E., Jr., and Sanders, C. C. 1997. *Enterobacter* spp.: Pathogens poised to flourish at the turn of the century. *Clin. Microbiol. Rev.* 10(2):220–41.

Scheld, W. M., Craig, W. and Hughes, J. M. 2001. *Emerging infections 5.* Washington, D.C.: ASM Press.

37.9 Control of Epidemics

Jaret, P. 1991. The disease detectives. *Nat. Geographic* 179(1):114–40.

Keller, M. A., and Stiehm, E. R. 2000. Passive immunity in prevention and treatment of infectious diseases. *Clin. Microbiol. Rev.* 13(4):602–14.

37.10 The Emerging Threat of Bioterrorism

Atlas, R. M. 2002. Bioterrorism: From threat to reality. *Annu. Rev. Microbiol.* 56:167–85.

Berche, P. 2001. The threat of smallpox and bioterrorism. *Trends Microbiol.* 9(1):15–18.

Block, S. M. 2001. The growing threat of biological weapons. *American Scientist* 89:28–37.

Casagrande, R. 2002. Technology against terror. *Sci. Am.* 287(4):83–87.

Henderson, D. A.; Inglesby, T. V.; and O'Toole, T. 2002. *Bioterrorism: Guidelines for medical and public health management.* Chicago: AMA Press.

Klietmann, W. F., and Ruoff, K. L. 2001. Bioterrorism: Implications for the clinical microbiologist. *Clin. Microbiol. Rev.* 14(2):364–81.

Milleer, J. et al. 2001. *Germs: Biological weapons and America's secret war.* New York: Simon and Schuster.

Preston, R. 1998. *The cobra event.* New York: Ballantine Books.

Synder, James. 1999. Responding to bioterrorism: The role of the microbiology laboratory. *ASM News:* 65(8):524–25.

Wheelis, M. 2002. Biological warfare at the 1346 siege of Caffa. *Emerg. Infect. Dis.* 8(9):971–75.

37.11 Global Travel and Health Considerations

Ryan, E., and Kain, K. 2000. Health advice and immunizations for travelers. *New Engl. J. Med.* 342(23):1716–25.

37.12 Nosocomial Infections

Aitken, C., and Jeffries, D. J. 2001. Nosocomial spread of viral disease. *Clin. Microbiol. Rev.* 14(3):528–46.

Fridkin, S. K., and Jarvis, W. R. 1996. Epidemiology of nosocomial fungal infections. *Clin. Microbiol. Rev.* 9(4):499–511.

Chapter 38: Human Diseases Caused by Viruses

General References

Dimmock, N. J.; Easton, A. J.; and Leppard, K. N. 2001. *Introduction to modern virology,* 5th ed. Cambridge, Mass.: Blackwell Science, Inc.

Fields, B. N.; Knipe, D. M.; Chanock, R. M.; Hirsch, M. S.; Melnick, J. L.; Monath, T. P.; and Roizman, B., editors. 1995. *Fields' virology,* 3d ed. New York: Raven Press.

Flint, S.; Enquist, L.; Krug, R.; Racaniello, V.; and Skalka, A. 1999. *Principles of virology: Molecular biology, pathogenesis, and control.* Washington, D.C.: ASM Press.

Joklik, W. K.; Willett, H. P.; Amos, D. B.; and Wilfert, C. M. 1992. *Zinsser microbiology,* 20th ed. E. Norwalk, Conn.: Appleton & Lange.

Le Guenno, B. 1995. Emerging viruses. *Sci. Am.* 273(4):56–64.

Levy, J. A. 1997. *HIV and the pathogenesis of AIDS,* 2d ed. Washington, D.C.: ASM Press.

Mandell, G. L.; Douglas, R. G., Jr.; and Bennett, J. E. 2000. *Principles and practice of infectious disease,* 5th ed. New York: John Wiley and Sons.

Murray, P. R., editor-in-chief, 2003. *Manual of clinical microbiology,* 8th ed. Washington, D.C.: ASM Press.

Richman, D.; Whitely, R., and Hayden, F. G. 2002. *Clinical virology,* Washington, D.C.: ASM Press.

Waner, J. L. 1994. Mixed viral infections: Detection and management. *Clin. Microbiol. Rev.* 7(2):143–51.

White, D. O., and Fenner, F. J. 1994. *Medical virology,* 4th ed. San Diego: Academic Press.

38.1 Airborne Diseases

Arvin, A. 1996. Varicella-zoster virus. *Clin. Microbiol. Rev.* 9(3):361–81.

Buller, R. M., and Palumbo, G. 1991. Poxvirus pathogenesis. *Microbiol. Rev.* 55(1):80–122.

Gilden, D., et al. 2000. Neurological complications of the reactivation of varicella-zoster virus. *N. Engl. J. Med.* 342(9):635–45.

Gnann, J. W., and Whitley, R. J. 2002. Herpes zoster. *N. Engl. J. Med.* 347(5):340–46.

Horimoto, T., and Kawaoka, Y. 2001. Pandemic threat posed by influenza A viruses. *Clin. Microbiol. Rev.* 14(1):129–49.

Laver, W.; Bischofberger, N.; and Webster, R. 1999. Disarming flu viruses. *Sci. Am.* 280(1):78–97.

Shaw, M. W.; Arden, H. H.; and Maassab, H. F. 1992. New aspects of influenza virus. *Clin. Microbiol. Rev.* 5(1):74–92.

Sullender, W. 2000. Respiratory syncytial virus genetic and antigenic diversity. *Clin. Microbiol. Rev.* 13(1):1–15.

Taubenberger, J. 1999. Seeking the 1918 Spanish influenza virus. *ASM News* 65(7):472–78.

Vainionpää, R., and Hyypiä, T. 1994. Biology of parainfluenza viruses. *Clin. Microbiol. Rev.* 7(2):265–75.

Webster, R. G., and Walker, E. J. 2003. Influenza. *American Scientist* 91:122–29.

38.2 Arthropod-Borne Diseases

Calisher, C. H. 1994. Medically important arboviruses of the United States and Canada. *Clin. Microbiol. Rev.* 7(1):89–116.

Milius, S. 2003. After West Nile virus: What will it do to the birds and beasts of North America? *Science News* 163:203–5.

Petersen, L. R.; Roehrig, J. T.; and Hughes, J. M. 2002. Outlook: West Nile virus encephalitis. *N. Engl. J. Med.* 347(16):1225–26.

Tsai, T. 1999. Arboviruses. In *Manual of clinical microbiology,* 7th ed., P. Murray, editor-in-chief, 1107–36. Washington, D.C.: ASM Press.

38.3 Direct Contact Diseases

Ablashi, Dharam V. et al. 2002. Spectrum of Kaposi's sarcoma–associated herpesvirus, or human herpes virus 8, disease. *Clin. Microbiol. Rev.* 15(3):439–64.

Ashley, R., and Wald, A. 1999. Genital herpes: Review of the epidemic and potential use of type-specific serology. *Clin. Microbiol. Rev.* 12(1):1–8.

Braun, D. K.; Dominguez, G.; and Pellett, P. E. 1997. Human herpesvirus 6. *Clin. Microbiol. Rev.* 10(3):521–67.

Cuthbert, J. A. 2001. Hepatitis A: Old and new. *Clin. Microbiol. Rev.* 14(1):38–58.

Ezzell, C. 2002. Hope in a vial: Will there be an AIDS vaccine anytime soon? *Sci. Am.* 286(6):38–45.

Falsey, A. R., and Walsh, E. E. 2000. Respiratory syncytial virus infection in adults. *Clin. Microbiol. Rev.* 13(3):371–84.

Gayle, H. D., and Hill, G. L. 2001. Global impact of human immunodeficiency virus and AIDS. *Clin. Microbiol. Rev.* 14(2):327–335.

Hahn, B. H.; Shaw, G. M.; De Cock, K. M.; and Sharp, P. M. 2000. AIDS as zoonosis: Scientific and public health implications. *Science* 287:607–14.

Heegaard, E. D., and Brown, K. E. 2002. Human Parvovirus B19. *Clin. Microbiol. Rev.* 15(3):485–505.

Jones, C. 2003. Herpes simplex virus type 1 and bovine herpesvirus 1 latency. *Clin. Microbiol. Rev.* 16(1):79–95.

Jones, D. S., and Brandt, A. M. 2000. AIDS, Historical. In *Encyclopedia of microbiology,* 2d ed., vol. 1, J. Lederberg, editor-in-chief, 104–15. San Diego: Academic Press.

Lee, J.-Y., and Bowden, D. S. 2000. Rubella virus replication and links to teratogenicity. *Clin. Microbiol. Rev.* 13(4):571–87.

Mahoney, F. 1999. Update on diagnosis, management, and prevention of hepatitis B virus infection. *Clin. Microbiol. Rev.* 12(2):351–66.

Masucci, M. G., and Ernberg, I. 1994. Epstein-Barr virus: Adaptation to a life within the immune system. *Trends in Microbiol.* 2(4)125–30.

Polish, L. B.; Gallagher, M.; Fields, H. A.; Hadler, S. C. 1993. Delta hepatitis: Molecular biology and clinical epidemiological features. *Clin. Microbiol. Rev.* 6(3):211–29.

Rossmann, M. G.; He, Y.; and Kuhn, R. J. 2002. Picornavirus-receptor interactions. *Trends Microbiol.* 10(7):324–31.

Seeger, C., and Mason, W. 2000. Hepatitis B virus biology. *Microbiol. Mol. Biol. Rev.* 64(1):51–68.

Smith, J. S. 1996. New aspects of rabies with emphasis on epidemiology, diagnosis, and prevention of the disease in the United States. *Clin. Microbiol. Rev.* 9(2):166–76.

Steffy, K., and Wong-Staal, F. 1991. Genetic regulation of human immunodeficiency virus. *Microbiol. Rev.* 55(2):193–205.

Tiollais, P., and Buendia, M. A. 1991. Hepatitis B virus. *Sci. Am.* 264(4):116–24.

Vlahov, D., et al. 1998. Prognostic indicators for AIDS and infectious disease death in HIV-infected injection drug users. *JAMA* 279(1):35–40.

Ward, D. E., and Krim, M. 1998. *The AmFAR AIDS handbook: The complete guide to understanding HIV and AIDS.* New York: W. W. Norton.

Winkler, W. G., and Vogel, K. 1992. Control of rabies in wildlife. *Sci. Am.* 266(6):86–93.

Wunner, W. H. 2000. Rabies. In *Encyclopedia of microbiology,* 2d ed., vol. 4, J. Lederberg, editor-in-chief, 15–31. San Diego: Academic Press.

Zein, N. 2000. Clinical significance of hepatitis C virus genotypes. *Clin. Microbiol. Rev.* 13(2):223–35.

38.4 Food-Borne and Waterborne Diseases

Cuthbert, J. A. 1994. Hepatitis C: Progress and problems. *Clin. Microbiol. Rev.* 7(4):505–32.

de Quadros, C. A. 2000. Polio. In *Encyclopedia of microbiology,* 2d ed., vol. 3, J. Lederberg, editor-in-chief, 762–72. San Diego: Academic Press.

Melnick, J. L. 1996. Current status of poliovirus infection. *Clin. Microbiol. Rev.* 9(3):293–300.

Ohka, S., and Nomoto, A. 2001. Recent insights into poliovirus pathogenesis. *Trends Microbiol.* 9(10):501–6.

Xiang, J., and Stapleton, J. 1999. Hepatitis A virus. In *Manual of clinical microbiology,* 7th ed., P. Murray, editor-in-chief, 1014–24. Washington, D.C.: ASM Press.

38.5 Slow Virus and Prion Diseases

Baker, H., and Ridley, R., editors. 1996. *Prion disease.* Totowa, N.J.: Humana Press.

Gambetti, P., and Parchi, P. 1999. Insomnia in prion diseases: Sporadic and familial. *N. Engl. J. Med.* 340(21):1675–77.

Harris, D. 1999. Cellular biology of prion disease. *Clin. Microbiol. Rev.* 12(3):429–44.

Johnson, R., and Gibbs, C. 1998. Creutzfeldt-Jacob disease and related spongiform encephalopathies. *N. Engl. J. Med.* 339(27):1994–2004.

Prusiner, S. B. 2001. Shattuck Lecture: Neurodegenerative diseases and prions. *N. Engl. J. Med.* 344(20):1516–26.

38.6 Other Diseases

Kiviat, N. 1999. Human papillomavirus. In *Manual of clinical microbiology,* 7th ed., P. Murray, editor-in-chief, 1080–88. Washington, D.C.: ASM Press.

Norkin, L. C. 1982. Papoviral persistent infections. *Microbiol. Rev.* 46:384–91.

Roman, A., and Fife, K. 1989. Human papillomaviruses: Are we ready to type? *Clin. Microbiol. Rev.* 2:166–90.

Chapter 39: Human Diseases Caused by Bacteria

General References

Chin, J. 2000. *Control of communicable diseases manual,* 17th ed. Washington, D.C.: American Public Health Association.

Fischetti, V. A., et al. 2000. *Gram-positive pathogens.* Washington, D.C.: ASM Press.

Gulbins, E., and Lang, F. 2001. Pathogens, host-cell invasion and disease. *American Scientist* 89:406–13.

Joklik, W. K.; Willett, H. P.; Amos, D. B.; and Wilfert, C. M. 1992. *Zinsser microbiology,* 20th ed. E. Norwalk, Conn.: Appleton & Lange.

Levy, S. 1998. The challenge of antibiotic resistance. *Sci. Am.* 278(3):46–55.

Mandell, G. L.; Bennett, J. E.; and Dolan, R. 2000. *Principles and practices of infectious diseases.* 5th ed. New York: Churchill Livingston.

Nataro, J.; Blaser, M.; and Cunningham-Rundles, S. 2000. *Persistent bacterial infections.* Washington, D.C.: ASM Press.

Rhen, M.; Eriksson, S.; Clements, M.; Bergström, S.; and Normark, S. J. 2003. The basis of persistent bacterial infections. *Trends Microbiol.* 11(2):80–86.

Salyers, A. A., and Whitt, D. D. 2001. *Bacterial Pathogenesis: A molecular approach,* 2d ed. Washington, D.C.: ASM Press.

Schaechter, M.; Engleberg, N. C.; Eisenstein, B. I.; and Medoff, G. 1998. *Mechanisms of microbial disease,* 3d ed. Philadelphia: Williams & Wilkins.

Stephens, R. 1999. *Chlamydia: Intracellular biology, pathogenesis, and immunity.* Washington, D.C.: ASM Press.

Zinsser, H., 1935. *Rats, lice, and history.* Boston: Little, Brown.

39.1 Airborne Diseases

Brouqui, P., and Raoult, D. 2001. Endocarditis due to rare and fastidious bacteria. *Clin. Microbiol. Rev.* 14(1):177–207.

Clark-Curtiss, J. E., and Haydel, S.E. 2003. Molecular genetics of *Mycobacterium tuberculosis* pathogenesis. *Annu. Rev. Microbiol.* 57:517–49.

Cunninham, M. W. 2000. Pathogenesis of group A streptococcal infections. *Clin. Microbiol. Rev.* 13(3):470–511.

Deuren, M.; Brandzaeg, P.; and van der Meer, J. 2000. Update on meningococcal disease with emphasis on pathogenesis and clinical management. *Clin. Microbiol. Rev.* 13(1):144–66.

Fields, B. S.; Benson, R. F.; and Besser, R. 2002. *Legionella* and Legionnaire's disease: 25 years of investigation. *Clin. Microbiol. Rev.* 15(3):506–26.

Maniloff, J., editor. 1992. *Mycoplasmas: Molecular biology and pathogenesis.* Washington, D.C.: American Society for Microbiology.

Marre, R., et al. 2002. *Legionella.* Washington, D.C.: ASM Press.

Mathema, B., and Kreiswirth, B. N. 2003. Rethinking tuberculosis epidemiology: The utility of molecular methods. *ASM News* 69(2):80–85.

Orme, I. M.; McMurray, D. N.; and Belisle, J. T. 2001. Tuberculosis vaccine development: Recent progress. *Trends Microbiol.* 9(3):115–18.

Paton, J. C. 1996. The contribution of pneumolysin to the pathogenicity of *Streptococcus pneumoniae. Trends Microbiol.* 4(3):103–6.

Peltola, H. 2000. Worldwide *Haemophilus influenzae* type b disease at the beginning of the 21st century: Global analysis of the disease burden 25 years after the use of the polysaccharide vaccine and a decade after the advent of conjugates. *Clin. Microbiol. Rev.* 13(2):302–17.

Smith, I. 2003. *Mycobacterium tuberculosis* pathogenesis and molecular determinants of virulence. *Clin. Microbiol. Rev.* 16(3):463–96.

Todd, J. K. 1988. Toxic shock syndrome. *Clin. Microbiol. Rev.* 1(4):432–46.

Tomasz, A., editor. 2000. *Streptococcus pneumoniae: Molecular biology & mechanisms of disease.* Larchmont, N.Y.: Mary Ann Liebert.

Tunkel, A. R., and Scheld, W. M. 1993. Pathogenesis and pathophysiology of bacterial meningitis. *Clin. Microbiol. Rev.* 6(2):118–36.

Wayne, L. G., and Sramek, H. A. 1992. Agents of newly recognized or infrequently encountered mycobacterial diseases. *Clin. Microbiol. Rev.* 5(1):1–25.

39.2 Arthropod-Borne Diseases

Coburn, J., and Kalish, R. A. 2000. Lyme disease. In *Encyclopedia of microbiology,* 2d ed., vol. 3, J. Lederberg, editor-in-chief, 109–30. San Diego: Academic Press.

Eremeeva, M. E., and Dasch, G. A. 2000. Rickettsiae. In *Encyclopedia of microbiology,* 2d ed., vol. 4, J. Lederberg, editor-in-chief, 140–80. San Diego: Academic Press.

Kantor, F. S. 1994. Disarming lyme disease. *Sci. Am.* 271(3):34–39.

Maurin, M., and Raoult, D. 1999. Q fever. *Clin. Microbiol. Rev.* 12(4):518–53.

Perry, R. D., and Fetherston, J. D. 1997. *Yersinia pestis*—Etiologic agent of plague. *Clin. Microbiol. Rev.* 10(1):35–66.

Reimer, L. G. 1993. Q fever. *Clin. Microbiol. Rev.* 6(3):193–98.

Woodward, T. E. 2000. Typhus fevers and other rickettsial diseases. In *Encyclopedia of microbiology,* 2d ed., vol. 4, J. Lederberg, editor-in-chief, 758–66. San Diego: Academic Press.

39.3 Direct Contact Diseases

Anderson, B. E., and Neuman, M. A. 1997. *Bartonella* spp. as emerging human pathogens. *Clin. Microbiol. Rev.* 10(2):203–19.

Black, C. M. 1997. Current methods of laboratory diagnosis of *Chlamydia trachomatis* infections. *Clin. Microbiol. Rev.* 10(1):160–84.

Blaser, M. J. 1996. The bacteria behind ulcers. *Sci. Am.* 274(2):104–7.

Boman, J., and Hammerschlag, M. R. 2002. *Chlamydia pneumoniae* and atherosclerosis: Assessment of diagnostic methods and relevance to treatment studies. *Clin. Microbiol. Rev.* 15(1):1–20.

Campbell, L.; Kuo, Cho-Chou; and Graystone, J. 1998. *Chlamydia pneumoniae* and cardiovascular disease. *Emerg. Infect. Dis.* 4(4):571–80.

Dunn, B.; Cohen, H.; and Blaser, M. 1997. *Helicobacter pylori. Clin. Microbiol. Rev.* 10(4):720–41.

Ellis, J., et al. 2002. Tularemia. *Clin. Microbiol. Rev.* 15(4):631–46.

Hammerschlag, M. R. 2003. Implications of inappropriate STD testing go beyond pure diagnostics. *ASM News* 69(2):74–79.

Kuo, C-C.; Jackson, L. A.; Campbell, L. A.; Grayson, J. T. 1995. *Chlamydia pneumoniae* (TWAR). *Clin. Microbiol. Rev.* 8(4):451–61.

Lee, A.; Fox, J.; and Hazell, S. 1993. Minireview. Pathogenicity of Helicobacter pylori: A perspective. *Infect. Immun.* 61(5):1601–10.

Lowy, F. 1998. *Staphylococcus aureus* infections. *N. Engl. J. Med.* 339(8):520–32.

Moulder, J. W. 1991. Interaction of chlamydiae and host cells in vitro. *Microbiol. Rev.* 55(1):143–90.

Ngeh, J., and Gupta, S. 2000. *C. pneumoniae* and atherosclerosis: Causal or coincidental link? *ASM News* 66(12):732–37.

Parsek, M. R., and Singh, P. K. 2003. Bacterial biofilms: An emerging link to disease pathogenesis. *Annu. Rev. Microbiol.* 57:677–701.

Saikku, P., et al. 1992. Chronic *Chlamydia pneumoniae* infection as a risk factor for coronary heart disease in the Helsinki Heart Study. *Ann. Intern. Med.* 15(116):273–78.

Singh, A., and Romanowski, B. 1999. Syphilis: Review with emphasis on clinical, epidemiological, and some biologic features. *Clin. Microbiol. Rev.* 12(2):187–209.

Solnick, J. V., and Schauer, D. B. 2001. Emergence of diverse *Helicobacter* species in the pathogenesis of gastric and enterohepatic disease. *Clin. Microbiol. Rev.* 14(1):59–97.

Stephens, R. S. 2003. The cellular paradigm of chlamydial pathogenesis. *Trends Microbiol.* 11(1):44–51.

Suerbaum, S., and Mitchetti, P. 2002. Medical progress: *Helicobacter pylori. N. Engl. J. Med.* 347(15):1175–86.

Swartz, M. N. 2001. Recognition and management of anthrax—an update. *N. Engl. J. Med.* 345(22):162–26.

Wyrick, P. B. 2002. *C. trachomatis:* Infection strategies of the ultimate intracellular pathogen. *ASM News* 68(2):70–76.

39.4 Food-Borne and Waterborne Diseases

Acheson, D. W. K. 2000. Food-borne illnesses. In *Encyclopedia of microbiology,* 2d ed., vol. 2, J. Lederberg, editor-in-chief, 390–411. San Diego: Academic Press.

Darwin, K. H., and Miller, V. 1999. Molecular basis of the interaction of *Salmonella* with the intestinal mucosa. *Clin. Microbiol. Rev.* 12(3):405–28.

Hatheway, C. L. 1990. Toxigenic clostridia. *Clin. Microbiol. Rev.* 3(1):66–98.

Johnson, E. A. 1999. Clostridial toxins as therapeutic agents: Benefits of nature's most toxic proteins. *Annu. Rev. Microbiol.* 53:551–75.

Kopecko, D. J.; Hu, L.; and Zaal, K. J. M. 2001. *Campylobacter jejuni*—microtubule-dependent invasion. *Trends Microbiol.* 9(8):389–96.

Ménard, R.; Dehio, C.; and Sansonetti, P. J. 1996. Bacterial entry into epithelial cells: The paradigm of *Shigella. Trends Microbiol.* 4(6):220–26.

Midura, T. F. 1996. Update: Infant botulism. *Clin. Microbiol. Rev.* 9(2):119–25.

Nachamkin, I., editor. 1992. *Campylobacter jejuni: Current status and future trends.* Washington, D.C.: American Society for Microbiology.

Nataro, J. P., and Kaper, J. B. 1998. Diarrheagenic *Escherichia coli. Clin. Microbiol. Rev.* 11(1):142–201.

Rippey, S. R. 1994. Infectious diseases associated with molluscan shellfish consumption. *Clin. Microbiol. Rev.* 7(4):419–25.

Sanders, W. E., Jr., and Sanders, C. C. 1997. *Enterobacter* spp.: Pathogens poised to flourish at the turn of the century. *Clin. Microbiol. Rev.* 10(2):220–41.

Sansonetti, P. 1999. *Shigella* plays dangerous games. *ASM News* 65(9):611–17.

Vazquez-Boland, J. A., et al. 2001. *Listeria* pathogenesis and molecular virulence determinants. *Clin. Microbiol. Rev.* 14(3):584–640.

Wachsmuth, K.; Blake, P.; and Olsvik, O. 1994. *Vibrio cholerae and cholera: Molecular to global perspectives.* Washington, D.C.: American Society for Microbiology.

39.5 Sepsis and Septic Shock

Bone, R. 1993. Gram-negative sepsis: A dilemma of modern medicine. *Clin. Microbiol. Rev.* 6(1):57–68.

Van Amersfoort, E. S.; Van Berkel, T. J. C.; and Kuiper, J. 2003. Receptors, mediators, and mechanisms involved in bacterial sepsis and septic shock. *Clin. Microbiol. Rev.* 16(3):379–414.

39.6 Dental Infections

Bolstad, A. I.; Jensen, H. B.; and Bakken, V. 1996. Taxonomy, biology, and periodontal aspects of *Fusobacterium nucleatum. Clin. Microbiol. Rev.* 9(1):55–71.

Hamilton, I. R., and Bowden, G. H. 2000. Oral microbiology. In *Encyclopedia of microbiology,* 2d ed., vol. 3, J. Lederberg, editor-in-chief, 466–81. San Diego: Academic Press.

Kolenbrander, P. E.; Ganeshkumar, N.; Cassels, F.; and Hughes, C. 1993. Coaggregation: Specific adherence among human oral plaque bacteria. *The FASEB Journal* 7(5):406–13.

Kolenbrander, P. E., et al. 2002. Communication among oral bacteria. *Microbiol. Mol. Biol. Rev.* 66(3):486–505.

Lamont, R., and Jenkinson, H. 1999. Life below the gum line: Pathogenic mechanisms of *Porphyromonas gingivalis. Microbiol. Mol. Biol. Rev.* 62(4):1244–63.

Li, X.; Kolltveit, K. M.; Tronstad, L.; and Olsen, I. 2000. Systemic diseases caused by oral infection. *Clin. Microbiol. Rev.* 13(4):547–58.

Loesche, W. J., and Grossman, N. S. 2001. Periodontal disease as a specific, albeit chronic, infection: Diagnosis and treatment. *Clin. Microbiol. Rev.* 14(4):727–52.

Marcotte, H., and Lavoie, M. 1998. Oral microbial ecology and the role of salivary immunoglobin A. *Microbiol. Mol. Biol. Rev.* 62(1):71–109.

Chapter 40: Human Diseases Caused by Fungi and Protozoa

General References

Ash, L. R., and Orihel, T. C. 1996. *Atlas of human parasitology.* Washington, D.C.: American Society for Microbiology.

Espinel-Ingroff, A. 1996. History of medical mycology in the United States. *Clin. Microbiol. Rev.* 9(2):235–72.

Fisher, F., and Cook, N. 1998. *Fundamentals of diagnostic mycology.* Philadelphia: W. B. Saunders.

Gutierrez, Y. 1999. *Diagnostic pathology of parasite infections.* Cary, N.Y.: Oxford University Press.

Mandell, G. L.; Bennett, J. E.; and Dolan, R. 2000. *Principles and practice of infectious diseases,* 5th ed. New York: Churchill Livingstone.

Roberts, L. R., and Janovy, J. 2000. *Foundations of parasitology,* 6th ed. Dubuque, Iowa: McGraw-Hill.

Science. 2002. An entire thematic issue devoted to malaria and the *Anopheles gambiae* genome. 298(5591).

40.1 Fungal Diseases

Brakhage, A. A., and Langfelder, K. 2002. Menacing mold: The molecular biology of *Aspergillus fumigatus. Annu. Rev. Microbiol.* 56:433–55.

Calderone, R. A. 2002. *Candida* and candidiasis. Washington, D.C.: ASM Press.

Casadevall, A. 2000. Fungal infections, systemic. In *Encyclopedia of microbiology,* 2d ed., vol. 2, J. Lederberg, editor-in-chief, 460–67. San Diego: Academic Press.

Chandler, F. W., and Watts, J. 1997. *Pathologic diagnosis of fungal infections.* Chicago: ASCP Press.

Channoum, M. 2000. Potential role of phospholipases in virulence and fungal pathogenesis. *Clin. Microbiol. Rev.* 13(1):122–43.

Douglas, L. J. 2003. *Candida* biofilms and their role in infection. *Trends Microbiol.* 11(1):30–36.

Klotz, S. A.; Penn, C. C.; Negvesky, G. J.; and Butrus, S. I. 2000. Fungal and parasitic infections of the eye. *Clin. Microbiol. Rev.* 13(4):662–85.

Larone, D. H. 2002. *Medically important fungi:* A guide to identification, 4th ed. Washington, D.C.: ASM Press.

Latge, J. 1999. *Aspergillus fumigatus* and aspergillosis. *Clin. Microbiol. Rev.* 12(2):310–50.

Sohnle, P. G., and Wagner, D. K. 2000. Fungal infections, cutaneous. In *Encyclopedia of microbiology,* 2d ed., vol. 2, J. Lederberg, editor-in-chief, 451–59. San Diego: Academic Press.

Stringer, J. R. 1996. *Pneumocystis carinii:* What is it, exactly? *Clin. Microbiol. Rev.* 9(4):489–98.

Sundstrom, P. 2003. Fungal pathogens and host response. *ASM News* 69(3):127–31.

40.2 Protozoan Diseases

Adam, R. D. 1991. The biology of *Giardia* spp. *Microbiol. Rev.* 55(4):706–32.

Adam, R. D. 2001. Biology of *Giardia lamblia. Clin. Microbiol. Rev.* 14(3):447–75.

Black, M. W., and Boothroyd, J. C. 2000. Lytic cycle of *Toxoplasma gondii. Microbiol. Mol. Biol. Rev.* 64(3):607–23.

Boothroyd, J. C. 2000. Toxoplasmosis. In *Encyclopedia of microbiology,* 2d ed., vol. 4, J. Lederberg, editor-in-chief, 598–609. San Diego: Academic Press.

Bruckner, D. A. 1992. Amebiasis. *Clin. Microbiol. Rev.* 5(4):356–69.

Carter, R., and Mendis, K. N. 2002. Evolutionary and historical aspects of the burden of malaria. *Clin. Microbiol. Rev.* 15(4):564–94.

Chen, O.; Schlichtherle, M; and Wahlgren, M. 2000. Molecular aspects of severe malaria. *Clin. Microbiol. Rev.* 13(3):439–50.

Chen, X.-M., et al. 2002. Cryptosporidiosis. *N. Engl. J. Med.* 346(22):1723–31.

Clark, D. 1999. New insights into human cryptosporidiosis. *Clin. Microbiol. Rev.* 12(4):554–63.

Dubey, J. P.; Lindsay, D. S.; and Speer, C. A. 1998. Structures of *Toxoplasma gondii* tachyzoites, bradyzoites, and sporozoites and biology and development of tissue cysts. *Clin. Microbiol. Rev.* 11(2):267–99.

Faubert, G. 2000. Immune response to *Giardia duodenalis. Clin. Microbiol. Rev.* 13(1):35–54.

Garcia, L. 1999. *Practical guide to diagnostic parasitology.* Washington, D.C.: ASM Press.

Garcia, L. S., and Bruckner, D. A., editors. 1997. *Diagnostic medical parasitology.* 3d ed. Washington, D.C.: American Society for Microbiology.

Gardner, T. B., and Hill, D. R. 2001. Treatment of giardiasis. *Clin. Microbiol. Rev.* 14(1):114–28.

Handman, E. 2001. Leishmaniasis: Current status of vaccine development. *Clin. Microbiol. Rev.* 14(2):229–43.

Homer, M. J.; Aguilar-Delfin, I.; Telford III, S. R.; Krause, P. J.; and Persing, D. H. 2000. Babesiosis. *Clin. Microbiol. Rev.* 13(3):451–69.

Hyde, J. E. 2000. *Plasmodium.* In *Encyclopedia of microbiology,* 2d ed., vol. 3, J. Lederberg, editor-in-chief, 745–61. San Diego: Academic Press.

Humphreys, M. 2001. *Malaria: Poverty, race and public health in the United States.* Baltimore: Johns Hopkins University Press.

Lindsay, D. S.; Dubey, J. P.; and Blagburn, B. L. 1997. Biology of *Isospora* spp. from humans, nonhuman primates, and domestic animals. *Clin. Microbiol. Rev.* 10(1):19–34.

Long, C. A., and Hoffman, L. 2002. Malaria: From infants to genomic vaccines. *Science* 297(5580):345–47.

Marshall, M. M.; Naumovitz, D.; Ortega, Y.; and Sterling, C. R. 1997. Waterborne protozoan pathogens. *Clin. Microbiol. Rev.* 10(1):67–85.

Mathiopoulos, K. D. 2000. Malaria. In *Encyclopedia of microbiology,* 2d ed., vol. 3, J. Lederberg, editor-in-chief, 131–50. San Diego: Academic Press.

Phillips, R. S. 2001. Current status of malaria and potential for control. *Clin. Microbiol. Rev.* 14(1):208–26.

Que, X., and Reed, S. 2000. Cysteine proteinases and the pathogenesis of amebiasis. *Clin. Microbiol. Rev.* 13(2):196–206.

Seed, R. 2000. Current status of African trypanosomiasis. *ASM News* 66(7):395–402.

Sherman, I. 1995. *Malaria: Parasite biology, pathogenesis, and protection.* Washington, D.C.: ASM Press.

Tanowitz, H. B.; Wittner, M.; Werner, C.; Weiss, L. M.; Kirchhoff, L. V.; and Bacchi, C. 2000. Trypanosomes. In *Encyclopedia of microbiology,* 2d ed., vol. 4, J. Lederberg, editor-in-chief, 725–41. San Diego: Academic Press.

Tanowitz, H. B.; Kirchhoff, D. S.; Morris, S. A.; Weiss, L. M.; and Wittner, M. 1992. Chagas' disease. *Clin. Microbiol. Rev.* 5(4):400–419.

Weber, R.; Bryan, R. T.; Schwartz, D. A.; and Owen, R. L. 1994. Human microsporidial infections. *Clin. Microbiol. Rev.* 7(4):426–61.

Wittner, M. 1999. *The microsporidia and microsporidiosis.* Washington, D.C.: ASM Press.

Wolfe, M. 1992. Giardiasis. *Clin. Microbiol. Rev.* 5(1):93–100.

Chapter 41: Microbiology of Food

General References

Adams, M. R., and Moss, M. O. 2000. *Food microbiology.* Cambridge: Royal Society of Chemistry.

Downes, F. P., and Ito, K., editors. 2001. *Compendium of methods for the microbiological examination of foods,* 4th ed. Washington, D.C.: American Public Health Association.

Doyle, M. P.; Beuchat, L. R.; and Montville, T. J. 2002. *Food microbiology: Fundamentals and frontiers,* 2d ed. Washington, D.C.: ASM Press.

Jay, J. H. 2000. *Modern food microbiology,* 6th ed. Frederick, Md.: Aspen Publishing.

Robinson, R. K.; Batt, C. A.; and Patel, P. D. 2000. *Encyclopedia of food microbiology.* San Diego: Academic Press.

41.1 Microorganism Growth in Foods

Diker, K. S., and Hascelik, G. 1994. The bactericidal activity of tea against *Helicobacter pylori. Let. Appl. Microbiol.* 19:299–300.

Marth, E. H., and Steele, J. L., editors. 1998. *Applied dairy microbiology.* New York: Marcel Dekker, Inc.

Meng, J.; Zhao, S.; Doyle, M. P.; and Kresovich, S. 1997. A multiplex PCR for identifying Shiga-like toxin-producing *Escherichia coli* O157:H7. *Lett. Appl. Microbiol.* 24:172–76.

Navarro, L.; Zarazaga, M.; Sáenz, J.; Ruiz-Larrea, F.; and Torres, C. 2000. Bacteria production by lactic acid bacteria isolated from Rioja red wines. *J. Appl. Microbiol.* 88:44–51.

Ohmomo, S.; Murata, S.; Katayama, N.; Nitisinprasart, M.; Kobayashi, M.; Nakajima, T.; Yajima, M.; and Nakanishi, K. 2000. Purification and some characteristics of enterocin ON-157, a bacteriocin produced by *Enterococcus faecium* NIAI 157. *J. Appl. Microbiol.* 88:81–89.

41.2 Microbial Growth and Food Spoilage

Barrett, J. R. 2000. Mycotoxins: of molds and maladies. *Environ. Health Perspect.* 108:A20–A23.

Henry, S. H.; Bosch, F. X.; Troxell, T. C.; and Bolger, P. M. 2000. Reducing liver cancer—global control of aflatoxin. *Science* 286:2453–54.

Hoogerwerf, S. W.; Kets, E. P. W.; and Dijksterhuis, J. 2002. High-oxygen and high-carbon dioxide containing atmospheres inhibit growth of food associated moulds. *Lett. Appl. Microbiol.* 35:419–22.

Kaper, J. B., and O'Brien, A. D., editors. 1998. *Escherichia coli O157:H7 and other shiga toxin-producing E. coli strains.* Washington, D.C.: American Society for Microbiology.

Potts, S. J.; Slaughter, D. C.; and Thompson, J. F. 2000. A fluorescent lectin test for mold in raw tomato juice. *J. Food Sci.* 65(2):346–50.

van Dolah, F. M. 2000. Marine algal toxins: Origins, health effects, and their increased occurrence. *Environ. Health Perspect.* 108(Suppl. 1):133–41.

41.3 Controlling Food Spoilage

Fung, D. Y. C. 2000. Food spoilage and preservation. In *Encyclopedia of microbiology,* 2d ed., vol. 2, J. Lederberg, editor-in-chief, 412–20. San Diego: Academic Press.

Joerger, R. D.; Hoover, D. G.; Barefoot, S. F.; Harmon, K. M.; Grinstead, D. A.; and Nettles Cutter, C. G. 2000. Bacteriocins. In *Encyclopedia of Microbiology* 2d. ed., vol. 1, J. Lederberg, editor-in-chief, 383–97. San Diego: Academic Press.

Kimura, B.; Yoshiyama, T.; and Fujii, T. 2000. Carbon dioxide inhibition of *Escherichia coli* and *Staphylococcus aureus* on a pH-adjusted surface in a model system. *J. Food Sci.* 64(2):367–70.

41.4 Food-Borne Diseases

Batt, C. 1999. *Listeria* not gone, not forgotten. *Food Microbiol.* 16:103.

Bell, C., and Kyriakides, A. 2000. *Listeria. A practical approach to the organism and its control in foods.* Oxford, U.K.: Blackwell Science.

Burkhardt, W. I., and Calci, K. R. 2000. Pathogen levels in oysters are seasonal. *Appl. Environ. Microbiol.* 66:1375–78.

Castro-Rosas, J., and Escartin, E. F. 2000. Survival and growth of *Vibrio cholerae* 01, *Salmonella typhi,* and *Escherichia coli* O157:H7 in alfalfa sprouts. *J. Food Sci.* 65(1):162–65.

Cox, N. A. 2001. *Campylobacter:* Research advances in sourcing the problem. *Food Safety* 7(5):17–20, 43–44.

D'Aoust, J.-Y. 2001. Foodborne salmonellosis: Current international concerns. *Food Safety* 7(2):10–17.

Detweiler, L. A., and Rubenstein, R. 2000. Bovine spongiform encephalopathy: An overview. *Asaio J.* 46(6):S73–S79.

Doyle, M. P. 2000. Reducing foodborne disease. *FoodTechnology* 54(11):130.

Hui, Y. H.; Pierson, M. D.; and Gorham, J. R. 2001. *Foodborne disease handbook,* 2d ed., vol. 1., *Bacterial pathogens.* New York: Marcel Dekker.

Hui, Y. H.; Sattar, S. A.; Murrell, K. D.; Nip, W.-K.; and Stanfield, P. S. 2001. *Foodborne disease handbook,* 2d ed., vol. 2. *Viruses, parasites, pathogens and HACCP.* New York: Marcel Dekker.

Phillips, C. A. 2001. Arcobacters as emerging human foodborne pathogens. *Food Control* 12(1):1–6.

41.5 Detection of Food-Borne Pathogens

Anonymous. 2002. *Compendium of methods for the microbiological analysis of food and agricultural products.* Gaithersburg, Md.; AOAC International.

Atabay, H. I.; Bang, D. D.; Aydin, F.; Erdogan, H. M.; and Madsen, M. 2002. Discrimination of *Arcobacter butzleri* isolates by polymerase chain reaction-mediated DNA fingerprinting. *Lett. Appl. Microbiol.* 35:141–45.

Cimons, M. 2000. Rapid foodborne pathogen ID system is making a difference. *ASM News* 66(10):617–19.

Lewis, G. D.; Molloy, S. L.; Greening, G. E.; and Dawson, J. 2000. Influence of environmental factors on virus detection by RT-PCR and cell culture. *J. Appl. Microbiol.* 88:633–40.

Momcilovic, D., and Rasooly, A. 2000. Detection and analysis of animal materials in food and feed. *J. Food. Prot.* 63(11):1602–09.

Wan, J. K.; King, K.; Craven, H.; McAuley, C.; Tan, S. E.; and Coventry, M. J. 2000. Probelia™ PCR system for rapid detection of *Salmonella* in milk powder and ricotta cheese. *Lett. Appl. Microbiol.* 30:267–71.

41.6 Microbiology of Fermented Foods

Abraham, A. G., and DeAntoni, G. L. 1999. Characterization of kefir grains grown in cows' milk and in soya milk. *J. Dairy Res.* 66(2):327–33.

Law, B. A., editor. 1997. *Microbiology and biochemistry of cheese and fermented milk,* 2d ed. New York: Chapman & Hall.

Wood, J. B., editor. 1998. *Microbiology of fermented foods.* London: Blackie Academic and Professional.

41.7 Microorganisms as Foods and Food Amendments

Beale, B. 2002. Probiotics: Their tiny worlds are under scrutiny. *The Scientist* 16(15):20–22.

Ehrmann, M. A.; Kurzak, P.; Bauer, J.; and Vogel, R. F. 2002. Characterization of lactobacilli towards their use as probiotic adjuncts in poultry. *J. Appl. Microbiol.* 92:966–75.

Farnworth, E. R. 2003. *Handbook of fermented functional foods.* Boca Raton, FL: CRC Press.

Fritts, C. A.; Kersey, J. H.; Motl, M. A.; Kroger, E. C.; Yan, F. S. J.; Jiang, Q.; Campos, M. M.; Waldroup, A. L.; and Waldroup, P. W. 2000. *Bacillus subtilis* C-3102 (Calsporin) improves live performance and microbiological status of broiler chickens. *J. Appl. Poultry Res.* 9(2):149–55.

Fuller, R., and Perdigon, G., editors 2000. *Probiotics 3 Immunomodulation by the gut microflora and probiotics.* Boston: Kluwer Academic Publishers.

Olando-Martin, E.; Gibson, G. R.; and Rastall, R. A. 2002. Comparison of the in vitro bifidogenic properties of pectins and pectic-oligosaccharides. *J. Appl. Microbiol.* 93:505–11.

Teitelbaum, J. E., and Walker, W. A. 2002. Nutritional impact of pre- and probiotics as protective gastrointestinal organisms. *Annu. Rev. Nutr.* 22:107–38.

Chapter 42: Industrial Microbiology and Biotechnology

General References

Crueger, W., and Crueger, A. 1990. *Biotechnology: A textbook of industrial microbiology.* 2d ed. T. D. Brock, editor. Sunderland, Mass.: Sinauer Associates.

Demain, A. L. 2000. Microbial biotechnology. *Tibtech* 18:26–31.

Demain, A. L., and Davis, J. E., editors. 1999. *Manual of industrial microbiology and biotechnology.* Washington, D.C.: American Society for Microbiology.

Glick, B. R., and Pasternak, J. J. 1998. *Molecular biotechnology: Principles and applications of recombinant DNA,* 2d ed. Washington, D.C.: ASM Press.

Lillehoj, E. P., and Ford, G. M. 2000. Industrial biotechnology, overview. In *Encyclopedia of microbiology*, 2d ed., vol. 2, J. Lederberg, editor-in-chief, 722–37. San Diego: Academic Press.

Rittmann, B. E., and McCarty, P. L. 2001. *Environmental biotechnology: Principles and applications.* New York: McGraw-Hill.

Thayer, A. M., 2002. Biocatalysis. *Chem. Eng. News* (79):28–34.

Waites, M. J.; Morgan, N. L.; Rockey, J. S.; and Higton, G. 2001. *Industrial microbiology. An introduction.* London: Blackwell Science.

Walker, J. M., and Rapley, R. 2001. *Molecular biology and biotechnology.* New York: Springer-Verlag.

42.1 Choosing Microorganisms for Industrial Microbiology and Biotechnology

Bayliss, C. D., and Moxon, E. R. 2002. Hypermutation and bacterial adaption. *ASM News* 68(11):549–55.

Bosma, T.; Damborsky, J.; Stucki, G.; and Janssen, D. B. 2002. Biodegradation of 1,2,-3-trichloropropane through directed evolution and heterologous expression of a haloalkane dehaologenase gene. *Appl. Environ. Microbiol.* 68:3582–87.

Bull, A. T.; Ward, A. C.; and Goodfellow, M. 2000. Search and discovery strategies for biotechnology: The paradigm shift. *Microbiol. Mol. Biol. Rev.* 64(3):573–606.

Ezeronye, O. U., and Okerentugba, P. O. 2001. Optimum conditions for yeast protoplast release and regeneration in *Saccharomyces cerevisiae* and *Candida tropicalis* using gut enzymes of the giant African snail *Achatina achatina. Lett. Appl. Microbiol.* 32:190–93.

Farmer, W. R., and Liao, J. C. 2000. Improving lycopene production in *Escherichia coli* by engineering metabolic control. *Nature Biotechnol.* 18:533–37.

Heesche-Wagner, K.; Schwarz, T.; and Kaufmann, M. 2001. A directed approach to the selection of bacteria with enhanced catabolic activity. *Lett. Appl. Microbiol.* 32:162–165.

Jaeger, K. E.; Reetz, M. T.; and Dijkstra, B. W. 2002. Directed evolution to produce enantioselective biocatalysts. *ASM News* 68:556–62.

Martin, V. J. J.; Smolke, C. D.; and Keasling, J. D. 2002. Redesigning cells for production of complex organic molecules. *ASM News* 68(7):336–43.

Neilsen, J. 2001. *Metabolic engineering.* New York: Springer-Verlag.

Oh, K. H.; Nam, S. H.; and Kim, H. S. 2002. Improvement of oxidative and thermostability of N-carbamyl-D-amino acid amidohydrolase by directed evolution. *Protein Eng.* 15:689–95.

Selifonova, O.; Valle, F.; and Schellenberger, V. 2001. Rapid evolution of novel traits in microorganisms. *Appl. Environ. Microbiol.* 67:3645–49.

Tang, T.-Y.; Went, C.-J.; and Liu, W.-H. 2000. Expression of the creatinase gene from *Pseudomonas putida* RS65 in *Escherichia coli. J. Ind. Microbiol. Biotechnol.* 24:2–6.

Zhang, Y.-X.; Perry, K.; Vinci, V. A.; Powell, K.; Stemmer, W. P. C.; and del Cardayré, S. B. 2002. Genome shuffling leads to rapid phenotypic improvement in bacteria. *Nature* 415:644–46.

42.2 Microorganism Growth in Controlled Environments

Anderson, T. M. 2000. Industrial fermentation processes. In *Encyclopedia of icrobiology*, 2d ed., vol. 2, J. Lederberg, editor-in-chief, 767–81. San Diego: Academic Press.

Regentin, R.; Cadapan, L.; Ou, S.; Zavala, S.; and Licari, P. 2002. Production of a novel FK520 analog in *Streptomyces hygroscopicus:* Improving titer while minimizing impurities. *J. Indust. Microbiol. Biotech.* (28):12–16.

42.3 Major Products of Industrial Microbiology

Gross, R. A., and Kalra, B. 2002. Biodegradable polymers for the environment. *Science* 297:803–807.

Demain, A. L. 2000. Pharmaceutically active secondary metabolites of microorganisms. *Appl. Microbiol. Biotechnol.* 52:455–63.

Holt, S. M.; Al-Sheikh, H.; and Shin, K.-J. 2001. Characterization of dextran-producing *Leuconostoc* strains. *Lett. Appl. Microbiol.* 32:185–89.

Lohner, K. 2001. *Development of novel antimicrobial agents: Emerging strategies.* Norfolk, UK: Horizon Scientific Press.

Strohl, W. R. 1997. *Biotechnology of antibiotics.* New York: Marcel Dekker, Inc.

42.4 Microbial Growth in Complex Natural Environments

Alexander, M. 1999. *Biodegradation and bioremediation,* 2d ed. San Diego, Calif.: Academic Press.

Bollag, W. B.; Dec, J.; and Bollag, J.-M. 2000. Biodegradation. In *Encyclopedia of microbiology,* 2d ed., vol. 1, J. Lederberg, editor-in-chief, 461–71. San Diego: Academic Press.

Hughes, J. B.; Neale, C. N.; and Ward, C. H. 2000. Bioremediation. In *Encyclopedia of microbiology,* 2d ed., vol. 1, J. Lederberg, editor-in-chief, 587–610. San Diego: Academic Press.

Kohler, H.-P. E.; Nickel, K.; and Zipper, C. 2000. Effect of chirality on the microbial degradation and the environmental fate of chiral pollutants. *Adv. Microb. Ecol.* 16:201–31.

Lunsdorf, H.; Erb, R. W.; Abraham, W. R.; and Timmis, K. N. 2000. 'Clay hutches': A novel interaction between bacteria and clay minerals. *Environ. Microbiol.* 2:161–68.

Nishiyama, M.; Senoo, K.; and Matsumoto, S. 1995. Survival of a bacterium in microporous glass in soil. *Soil Biol. Biochem.* 27:1359–61.

Ou, L.-T. 2000. Pesticide biodegradation. In *Encyclopedia of microbiology,* 2d ed., vol. 3, J. Lederberg, editor-in-chief, 594–606. San Diego: Academic Press.

Rawlings, D. E.; Tributsch, H.; and Hansford, G. S. 1999. Reasons why '*Leptospirillum*'-like species rather than *Thiobacillus ferrooxidans* are the dominant iron-oxidizing bacteria in many commercial processes for the biooxidation of pyrite and related ores. *Microbiology* 145:5–13.

Wackett, L. P., and Hershberger, C. D. 2001. *Biocatalysis and biodegradation: Microbial transformation of organic compounds.* Herndon, Virginia: ASM Press.

42.5 Biotechnological Applications

Brush, M. 2001. Making sense of microchip array data. *The Scientist* 15(9):25–28.

Cruse, M.; Telerant, R.; Gallagher, T.; Lee, T.; and Taylor, J. W. 2002. Cryptic species in *Stachybotrys chartarum. Mycologia* 94:814–22.

de Maagd, R. A.; Bravo, A.; and Crickmore, N. 2001. How *Bacillus thuringiensis* has evolved specific toxins to colonize the insect world. *Trends Genet.* 17(4):193–99.

Gabig, M., and Wegrzyn, G. 2001. An introduction to DNA chips: Principles, technology, applications and analysis. *Acta Biochimica Polonica* 48:615–22.

Liao, J. C., and Sabatti, C. 2002. Microanalysis of DNA microarrays. *ASM News* 68(9):432–37.

Palevitz, B. T. 2001. EPA reauthorizes Bt corn. *The Scientist* 15(21):11.

Rajasekaran, K.; Jacks, T. J.; and Finley, J. W. 2002. *Crop Biotechnology.* Washington, D.C.: American Chemical Society.

Rawlings, D. E. 2002. Heavy metal mining using microbes. *Annu. Rev. Microbiol.* 56:65–91.

Savka, M. A.; Dessaux, Y.; Oger, P.; and Rossbach, S. 2002. Engineering bacterial competitiveness and persistence in the phytosphere. *Mol. Plant-Microbe Interact.* 15(9):866–74.

Shelton, A. M.; Zhao, J.-Z.; and Roush, R. T. 2002. Economic, ecological, food safety, and social consequences of Bt transgenic plants. *Annu. Rev. Entomol.* 47:845–81.

Stegmann, R. 2001. *Treatment of contaminated soil: Fundamentals, analysis, applications.* Berlin: Springer-Verlag.

Sun, B.; Griffin, B. M.; Ayala-del-Rio, H. L.; Hashsham, S. A.; and Tiedje, J. M. 2002. Microbial dehalorespiration with 1,1,1-trichloroethane. *Science* 298:1023–25.

Wang, J.-M.; Marlowe, E. M.; Miller-Maier, R. M.; and Brusseau, M. L. 1998. Cyclodextrin-enhanced biodegradation of phenanthrene. *Environ. Sci. Technol.* 32:1907–12.

Wilchek, M., and Bayer, E. A. 1999. Foreword and introduction to the book (strept) avidin-biotin system. *Biomolec. Eng.* 16:1–4.

Yousten, A. A.; Federici, B.; and Roberts, D. 2000. Insecticides, microbial. In *Encyclopedia of microbiology,* 2d ed., vol. 2, J. Lederberg, editor-in-chief, 813–25. San Diego: Academic Press.

42.6 Impacts of Microbial Biotechnology

Florman, S. C. 1996. *The introspective engineer.* New York: St. Martin's Press.

González-Cabera, J.; Herrero, S.; and Ferré, J. 2001. Bt resistance: Complexity belies predictions, defies easy control. *Appl. Environ. Microbiol.* 67:5043–48.

Lifset, R. J. 2000. Full accounting. *The Sciences* 40:32–37.

 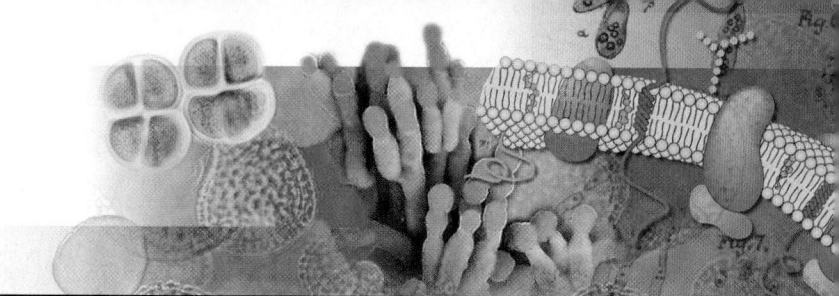

Glossary

Pronunciation Guide

Many of the boldface terms in this glossary are followed by a phonetic spelling in parentheses. These pronunciation aids usually come from *Dorland's Illustrated Medical Dictionary*. The following rules are taken from this dictionary and will help in using its phonetic spelling system.

1. An unmarked vowel ending a syllable (an open syllable) is long; thus *ma* represents the pronunciation of *may; ne*, that of *knee; ri*, of *wry; so*, of *sew; too*, of *two;* and *vu*, of *view*.

2. An unmarked vowel in a syllable ending with a consonant (a closed syllable) is short; thus *kat* represents *cat; bed, bed; hit, hit; not, knot; foot, foot;* and *kusp, cusp.*

3. A long vowel in a closed syllable is indicated by a macron; thus *māt* stands for *mate; sēd*, for *seed; bīl*, for *bile; mōl*, for *mole; fūm*, for *fume;* and *fōōl*, for *fool*.

4. A short vowel that ends or itself constitutes a syllable is indicated by a breve; thus ĕ-*fekt′* for *effect*, ĭ-*mūn′* for *immune*, and ŭ-*klōōd′* for *occlude*.

Primary (′) and secondary (″) accents are shown in polysyllabic words. Unstressed syllables are followed by hyphens.

Some common vowels are pronounced as indicated here.

ə	sof*a*	ē	m*e*t	ŏ	g*o*t
ā	m*a*te	ī	b*i*te	ū	f*ue*l
ă	b*a*t	ĭ	b*i*t	ŭ	b*u*t
ē	b*ea*m	ō	h*o*me		

From Dorland's Illustrated Medical Dictionary. *Copyright © 1988 W. B. Saunders, Philadelphia, Pa. Reprinted by permission.*

A

AB toxins The structure and activity of many exotoxins are based on the AB model. In this model, the B portion of the toxin is responsible for toxin binding to a cell but does not directly harm it. The A portion enters the cell and disrupts its function. (771)

ABC protein secretion pathway Transport systems that use ATP hydrolysis to drive translocation across the plasma membrane. They require an extracytoplasmic substrate-binding protein for proper function. Often called ATP-binding cassette transport systems. (61)

accessory pigments Photosynthetic pigments such as carotenoids and phycobiliproteins that aid chlorophyll in trapping light energy. (191)

acetyl coenzyme A (acetyl-CoA) A combination of acetic acid and coenzyme A that is energy rich; it is produced by many catabolic pathways and is the substrate for the tricarboxylic acid cycle, fatty acid biosynthesis, and other pathways. (178)

acid dyes Dyes that are anionic or have negatively charged groups such as carboxyls. (27)

acid fast Refers to bacteria like the mycobacteria that cannot be easily decolorized with acid alcohol after being stained with dyes such as basic fuchsin. (528)

acid-fast staining A staining procedure that differentiates between bacteria based on their ability to retain a dye when washed with an acid alcohol solution. (28)

acidophile (as′id-o-fīl″) A microorganism that has its growth optimum between about pH 0 and 5.5. (121)

acquired enamel pellicle A membranous layer on the tooth enamel surface formed by selectively adsorbing glycoproteins (mucins) from saliva. This pel-

licle confers a net negative charge to the tooth surface. (911)

acquired immune deficiency syndrome (AIDS) An infectious disease syndrome caused by the human immunodeficiency virus and is characterized by the loss of a normal immune response, followed by increased susceptibility to opportunistic infections and an increased risk of some cancers. (855)

acquired immune tolerance The ability to produce antibodies against nonself antigens while "tolerating" (not producing antibodies against) self-antigens. (734)

acquired immunity Refers to the type of specific (adaptive) immunity that develops after exposure to a suitable antigen or is produced after antibodies are transferred from one individual to another. (706)

actinobacteria (ak″tĭ-no-bak-tēr-e-ah) A group of gram-positive bacteria containing the actinomycetes and their high G + C relatives. (524)

actinomycete (ak″tĭ-no-mi′sēt) An aerobic, grampositive bacterium that forms branching filaments (hyphae) and asexual spores. (522)

actinorhizae Associations between actinomycetes and plant roots. (658)

activated sludge (sluj) Solid matter or sediment composed of actively growing microorganisms that participate in the aerobic portion of a biological sewage treatment process. The microbes readily use dissolved organic substrates and transform them into additional microbial cells and carbon dioxide. (638)

activation energy The energy required to bring reacting molecules together to reach the transition state in a chemical reaction. (157)

active carrier An individual who has an overt clinical case of a disease and who can transmit the infection to others. (828)

active immunization The induction of active immunity by natural exposure to a pathogen or by vaccination. (740)

active site The part of an enzyme that binds the substrate to form an enzyme-substrate complex and catalyze the reaction. Also called the catalytic site. (158)

active transport The transport of solute molecules across a membrane against an electrochemical gradient; it requires a carrier protein and the input of energy. (99)

acute carrier *See* casual carrier. (828)

acute infections Virus infections with a fairly rapid onset that last for a relatively short time. (399)

acute viral gastroenteritis An inflammation of the stomach and intestines, normally caused by Norwalk and Norwalklike viruses, other caliciviruses, rotaviruses, and astroviruses. (868)

acyclovir (a-si′klo-vir) A synthetic purine nucleoside derivative with antiviral activity against herpes simplex virus. (796)

adenine (ad′e-nēn) A purine derivative, 6-aminopurine, found in nucleosides, nucleotides, coenzymes, and nucleic acids. (211)

adenosine diphosphate (ADP) (ah-den′o-sēn) The nucleoside diphosphate usually formed upon the breakdown of ATP when it provides energy for work. (151)

adenosine 5′-triphosphate (ATP) The triphosphate of the nucleoside adenosine, which is a high energy molecule or has high phosphate group transfer potential and serves as the cell's major form of energy currency. (151)

adhesin (ad-he′zin) A molecular component on the surface of a microorganism that is involved in adhesion to a substratum or cell. Adhesion to a specific host tissue usually is a preliminary stage in pathogenesis, and adhesins are important virulence factors. (766)

adjuvant (aj′ə-vənt) Material added to an antigen to increase its immunogenicity. Common examples are alum, killed *Bordetella pertussis,* and an oil emulsion of the antigen, either alone (Freund's incomplete adjuvant) or with killed mycobacteria (Freund's complete adjuvant). (719)

adult T-cell leukemia A type of white blood cell cancer caused by the HTLV-1 virus. (864)

aerobe (a′er-ōb) An organism that grows in the presence of atmospheric oxygen. (125)

aerobic anoxygenic photosynthesis Photosynthetic process in which electron donors such as organic matter or sulfide, which do not result in oxygen evolution, are used under aerobic conditions. (597)

aerobic respiration (res″pī-ra′shun) A metabolic process in which molecules, often organic, are oxidized with oxygen as the final electron acceptor. (150, 168)

aerotolerant anaerobes Microbes that grow equally well whether or not oxygen is present. (125)

aflatoxin (af″lah-tok′sin) A polyketide secondary fungal metabolite that can cause cancer. (942)

agar (ahg′ar) A complex sulfated polysaccharide, usually from red algae, that is used as a solidifying agent in the preparation of culture media. (103)

agglutinates The visible aggregates or clumps formed by an agglutination reaction. (751)

agglutination reaction (ah-gloo″tĭ-na′shun) The formation of an insoluble immune complex by the cross-linking of cells or particles. (732)

agglutinin (ah-gloo″tĭ-nin) The antibody responsible for an agglutination reaction. (732)

AIDS *See* acquired immune deficiency syndrome. (855)

AIDS-related complex (ARC) A collection of symptoms such as lymphadenopathy (swollen lymph glands), fever, malaise, fatigue, loss of appetite, and weight loss. It results from an HIV infection and may progress to frank AIDS. (857)

airborne transmission The type of infectious organism transmission in which the pathogen is truly suspended in the air and travels over a meter or more from the source to the host. (829)

akinetes Specialized, nonmotile, dormant, thick-walled resting cells formed by some cyanobacteria. (459)

alcoholic fermentation A fermentation process that produces ethanol and CO_2 from sugars. (176)

alga (al′gah) A common term for a series of unrelated groups of photosynthetic eucaryotic microorganisms lacking multicellular sex organs (except for the charophytes) and conducting vessels. (554)

algicide (al′jĭ-sīd) An agent that kills algae. (135)

algology (al-gol′o-je) The scientific study of algae. (554)

alkalophile A microorganism that grows best at pHs from about 8.5 to 11.5. (122)

allergen (al′er-jen) A substance capable of inducing allergy or specific susceptibility. (744)

allergic contact dermatitis An allergic reaction caused by haptens that combine with proteins in the skin to form the allergen that produces the immune response. (747)

allergy (al′er-je) *See* hypersensitivity. (744)

allograft (al′o-graft) A transplant between genetically different individuals of the same species. (749)

allosteric enzyme (al′o-ster′ik) An enzyme whose activity is altered by the binding of a small effector or modulator molecule at a regulatory site separate from the catalytic site; effector binding causes a conformational change in the enzyme and its catalytic site, which leads to enzyme activation or inhibition. (161)

allotype Allelic variants of antigenic determinant(s) found on antibody chains of some, but not all, members of a species, which are inherited as simple Mendelian traits. (712)

alpha hemolysis A greenish zone of partial clearing around a bacterial colony growing on blood agar. (516, 772)

alpha-proteobacteria One of the five subgroups of proteobacteria, each with distinctive 16S rRNA sequences. This group contains most of the oligotrophic proteobacteria; some have unusual metabolic modes such as methylotrophy, chemolithotrophy, and nitrogen fixing ability. Many have distinctive morphological features. (474)

alternative complement pathway An antibody-independent pathway of complement activation that includes the C3–C9 components of the classical pathway and several other serum protein factors (e.g., factor B and properdin). (691)

alveolar macrophage A vigorously phagocytic macrophage located on the epithelial surface of the lung alveoli where it ingests inhaled particulate matter and microorganisms. (687)

amantadine (ah-man′tah-den) An antiviral agent used to prevent type A influenza infections. (796)

amebiasis (amebic dysentery) (am″e-bi′ah-sis) An infection with amoebae, often resulting in dysentery; usually it refers to an infection by *Entamoeba histolytica.* (927)

amensalism (a-men′səl-iz-əm) A relationship in which the product of one organism has a negative effect on another organism. (591)

American trypanosomiasis (Chagas' disease) *See* trypanosomiasis. (933)

Ames test A test that uses a special *Salmonella* strain to test chemicals for mutagenicity and potential carcinogenicity. (247)

amino acid activation The initial stage of protein synthesis in which amino acids are attached to transfer RNA molecules. (260)

aminoacyl or acceptor site (A site) The ribosomal site that contains an aminoacyl-tRNA at the beginning of the elongation cycle during protein synthesis; the growing peptide chain is transferred to the aminoacyl-tRNA and lengthens by an amino acid. (265)

aminoglycoside antibiotics (am″ī-no-gli′ko-sīd) A group of antibiotics synthesized by *Streptomyces* and *Micromonospora,* which contain a cyclohexane ring and amino sugars; all aminoglycoside antibiotics

bind to the small ribosomal subunit and inhibit protein synthesis. (791)

amoeboid movement Moving by means of cytoplasmic flow and the formation of pseudopodia (temporary cytoplasmic protrusions). (570)

amphibolic pathways (am″fe-bol′ik) Metabolic pathways that function both catabolically and anabolically. (170)

amphitrichous (am-fit′rĕ-kus) A cell with a single flagellum at each end. (63)

amphotericin B (am″fo-ter′i-sin) An antibiotic from a strain of *Streptomyces nodosus* that is used to treat systemic fungal infections; it also is used topically to treat candidiasis. (796)

anabolism (ah-nab′o-lizm″) The synthesis of complex molecules from simpler molecules with the input of energy. (168)

anaerobe (an-a′er-ōb) An organism that grows in the absence of free oxygen. (125)

anaerobic digestion (an″a-er-o′bik) The microbiological treatment of sewage wastes under anaerobic conditions to produce methane. (639)

anaerobic respiration An energy-yielding process in which the electron transport chain acceptor is an inorganic molecule other than oxygen. (168, 185)

anammox process The coupled use of nitrite as an electron acceptor and ammonium ion as a donor under anaerobic conditions to yield nitrogen gas. (598)

anamnestic response (an″am-nes′tik) The recall, or the remembering, by the immune system of a prior response to a given antigen. (719)

anaphylaxis (an″ah-fĭ-lak′sis) An immediate (type I) hypersensitivity reaction following exposure of a sensitized individual to the appropriate antigen. Mediated by reagin antibodies, chiefly IgE. (744)

anaplerotic reactions (an′ah-plĕ-rot′ik) Reactions that replenish depleted tricarboxylic acid cycle intermediates. (209)

anergy (an′ar-je) A state of unresponsiveness to antigens. Absence of the ability to generate a sensitivity reaction to substances that are expected to be antigenic. (734)

annotation The process of determining the location of specific genes in a genome map after it has been produced by nucleic acid sequencing. (338)

anogenital condylomata (venereal warts) (kon″dĭ-lo″ mah-tah) Warts that are sexually transmitted and caused by types 6, 11, and 42 human papillomavirus. Usually occur around the cervix, vulva, perineum, anus, anal canal, urethra, or glans penis. (871)

anoxic (ā-nok′sik) Without oxygen present. (617)

anoxygenic photosynthesis Photosynthesis that does not oxidize water to produce oxygen; a form of photosynthesis characteristic of purple and green photosynthetic bacteria. (195, 456)

antheridium (an″ther-id′e-um; pl., **antheridia**) A male gamete-producing organ, which may be unicellular or multicellular. (544, 557)

anthrax (an′thraks) An infectious disease of animals caused by ingesting *Bacillus anthracis* spores. Can also occur in humans and is sometimes called woolsorter's disease. (889)

antibiotic (an″tĭ-bi-ot′ik) A microbial product or its derivative that kills susceptible microorganisms or inhibits their growth. (780)

antibody (immunoglobulin) (an′tĭ-bod″e) A glycoprotein produced in response to the introduction of an antigen; it has the ability to combine with the antigen that stimulated its production. Also known as an immunoglobulin (Ig). (708)

antibody-dependent cell-mediated cytotoxicity (ADCC) The killing of antibody-coated target cells by cells with Fc receptors that recognize the Fc region of the bound antibody. Most ADCC is mediated by NK cells that have the Fc receptor or CD16 on their surface. (700)

antibody-mediated immunity *See* humoral immunity. (706)

anticodon triplet The base triplet on a tRNA that is complementary to the triplet codon on mRNA. (260)

antigen (an′tĭ-jen) A foreign (nonself) substance (such as a protein, nucleoprotein, polysaccharide, or sometimes a glycolipid) to which lymphocytes respond; also known as an immunogen because it induces the immune response. (708)

antigen-binding fragment (Fab) "Fragment antigen binding." A monovalent antigen-binding fragment of an immunoglobulin molecule that consists of one light chain and part of one heavy chain, linked by interchain disulfide bonds. (712)

antigenic determinant site (epitope) *See* epitope. (708)

antigenic drift A small change in the antigenic character of an organism that allows it to avoid attack by the immune system. (827, 847)

antigenic shift A major change in the antigenic character of an organism that makes it unrecognized by host immune mechanisms. (827, 847)

antigen-presenting cells Antigen-presenting cells (APCs) are cells that take in protein antigens, process them, and present antigen fragments to B cells and T cells in conjunction with class II MHC molecules so that the cells are activated. Macrophages, B cells, dendritic cells, and Langerhans cells may act as APCs. (721)

antimetabolite (an″tĭ-mĕ-tab′o-līt) A compound that blocks metabolic pathway function by competitively inhibiting a key enzyme's use of a metabolite because it closely resembles the normal enzyme substrate. (785)

antimicrobial agent An agent that kills microorganisms or inhibits their growth. (136)

antisense RNA A single-stranded RNA with a base sequence complementary to a segment of another RNA molecule that can specifically bind to the target RNA and inhibit its activity. (278)

antisepsis (an″tĭ-sep′sis) The prevention of infection or sepsis. (135)

antiseptic (an″tĭ-sep′tik) Chemical agents applied to tissue to prevent infection by killing or inhibiting pathogens. (135)

antiserum (an″tĭ-se′rum) Serum containing induced antibodies. (719)

antitoxin (an″tĭ-tok′sin) An antibody to a microbial toxin, usually a bacterial exotoxin, that combines specifically with the toxin, in vivo and in vitro, neutralizing the toxin. (732, 769)

apical complex (ap′ĭ-kal) A set of organelles characteristic of members of the phylum *Apicomplexa:* polar rings, subpellicular microtubules, conoid, rhoptries, and micronemes. (573)

apicomplexan (a′pĭ-kom-plek′san) A sporozoan protist that lacks special locomotor organelles but has an apical complex and a spore-forming stage. It is either an intra- or extracellular parasite of animals; a member of the phylum *Apicomplexa.* (573)

apoenzyme (ap″o-en′zīm) The protein part of an enzyme that also has a nonprotein component. (156)

apoptosis (ap″o-to′sis) Programmed cell death. The fragmentation of a cell into membrane-bound particles that are eliminated by phagocytosis. Apoptosis is a physiological suicide mechanism that preserves homeostasis and occurs during normal tissue turnover. It causes cell death in pathological circumstances, such as exposure to low concentrations of xenobiotics and infections by HIV and various other viruses. (725, 765, 858)

aporepressor An inactive form of the repressor protein, which becomes the active repressor when the corepressor binds to it. (271)

arbuscular mycorrhizal (AM) fungi The mycorrhizal fungi in a symbiotic fungus-root association that penetrate the outer layer of the root, grow intracellularly, and form characteristic much-branched hyphal structures called arbuscules. (656)

arbuscules Branched, treelike structures formed in cells of plant roots colonized by endotrophic mycorrhizal fungi. (656)

Archaea The domain that contains procaryotes with isoprenoid glycerol diether or diglycerol tetraether lipids in their membranes and archaeal rRNA (among many differences). (411, 438)

artificially acquired active immunity The type of immunity that results from immunizing an animal with a vaccine. The immunized animal now produces its own antibodies and activated lymphocytes. (706)

artificially acquired passive immunity The type of immunity that results from introducing into an animal antibodies that have been produced either in another animal or by in vitro methods. Immunity is only temporary. (707)

ascocarp (as′ko-karp) A multicellular structure in ascomycetes lined with specialized cells called asci in which nuclear fusion and meiosis produce ascospores. An ascocarp can be open or closed and may be referred to as a fruiting body. (545)

ascogenous hypha A specialized hypha that gives rise to one or more asci. (544)

ascogonium (as″ko-go′ne-um; pl., **ascogonia**) The receiving (female) organ in ascomycetous fungi which, after fertilization, gives rise to ascogenous hyphae and later to asci and ascospores. (544)

ascomycetes (as″ko-mi-se′tēz) A division of fungi that form ascospores. (544)

ascospore (as′ko-spor) A spore contained or produced in an ascus. (543)

ascus (as′kus) A specialized cell, characteristic of the ascomycetes, in which two haploid nuclei fuse to produce a zygote, which immediately divides by meiosis; at maturity an ascus will contain ascospores. (544)

aseptic meningitis syndrome *See* meningitis. (878)

aspergillosis (as″per-jil-o′sis) A fungal disease caused by species of *Aspergillus*. (924)

assimilatory reduction The reduction of an inorganic molecule to incorporate it into organic material. No energy is made available during this process. (205, 206, 597)

associative nitrogen fixation Nitrogen fixation by bacteria in the plant root zone (rhizosphere). (651)

athlete's foot *See* tinea pedis. (920)

atomic force microscope A type of scanning probe microscope that images a surface by moving a sharp probe over the surface at a constant distance; a very small amount of force is exerted on the tip and probe movement is followed with a laser. (37)

attenuation (ah-ten″u-a′shun) 1. A mechanism for the regulation of transcription of some bacterial operons by aminoacyl-tRNAs. 2. A procedure that reduces or abolishes the virulence of a pathogen without altering its immunogenicity. (276, 743)

attenuator A rho-independent termination site in the leader sequence that is involved in attenuation. (274)

autoclave (aw′to-klāv) An apparatus for sterilizing objects by the use of steam under pressure. Its development tremendously stimulated the growth of microbiology. (137)

autogenous infection (aw-toj′e-nus) An infection that results from a patient's own microbiota, regardless of whether the infecting organism became part of the patient's microbiota subsequent to admission to a clinical care facility. (841)

autoimmune disease (aw″to-ĭ-mūn′) A disease produced by the immune system attacking self-antigens. Autoimmune disease results from the activation of self-reactive T and B cells that damage tissues after stimulation by genetic or environmental triggers. (748)

autoimmunity (aw″to-ĭ-mun′ĭ-te) Autoimmunity is a condition characterized by the presence of serum autoantibodies and self-reactive lymphocytes. It may be benign or pathogenic. Autoimmunity is a normal consequence of aging; is readily inducible by infectious agents, organisms, or drugs; and is potentially reversible in that it disappears when the offending "agent" is removed or eradicated. (748)

autolysins (aw-tol′ĭ-sins) Enzymes that partially digest peptidoglycan in growing bacteria so that the peptidoglycan can be enlarged. (217)

autotroph (aw′to-trōf) An organism that uses CO_2 as its sole or principal source of carbon. (94)

auxotroph (awk′so-trōf) A mutated prototroph that lacks the ability to synthesize an essential nutrient and therefore must obtain it or a precursor from its surroundings. (239)

axenic (a-zen'ik)   Not contaminated by any foreign organisms; the term is used in reference to pure microbial cultures or to germfree animals. (674)

axial filament   The organ of motility in spirochetes. It is made of axial fibrils or periplasmic flagella that extend from each end of the protoplasmic cylinder and overlap in the middle of the cell. The outer sheath lies outside the axial filament. (66, 467)

B

bacillus (bah-sil'lus)   A rod-shaped bacterium. (41)

bacteremia (bak"ter-e'me-ah)   The presence of viable bacteria in the blood. (767)

Bacteria (bak-te're-a)   The domain that contains procaryotic cells with primarily diacyl glycerol diesters in their membranes and with bacterial rRNA. Bacteria also is a general term for organisms that are composed of procaryotic cells and are not multicellular. (411)

bacterial artificial chromosome (BAC)   A cloning vector constructed from the *E. coli* F-factor plasmid that is used to clone foreign DNA fragments in *E. coli*. (326)

bacterial (septic) meningitis   *See* meningitis. (878)

bacterial vaginosis (bak-te're-əl vaj"ĭ-no'sis)   Bacterial vaginosis is a sexually transmitted disease caused by *Gardnerella vaginalis, Mobiluncus* spp., *Mycoplasma hominis,* and various anaerobic bacteria. Although a mild disease it is a risk factor for obstetric infections and pelvic inflammatory disease. (891)

bactericide (bak-tēr'ĭsid)   An agent that kills bacteria. (135)

bacteriochlorophyll (bak-te"re-o-klo'ro-fil)   A modified chlorophyll that serves as the primary light-trapping pigment in purple and green photosynthetic bacteria. (195)

bacteriocin (bak-te' re-o-sin)   A protein produced by a bacterial strain that kills other closely related strains. (291, 688, 946)

bacteriophage (bak-te're-o-fāj")   A virus that uses bacteria as its host; often called a phage. (354, 372)

bacteriophage (phage) typing   A technique in which strains of bacteria are identified based on their susceptibility to bacteriophages. (815)

bacteriostatic (bak-te"re-o-stat'ik)   Inhibiting the growth and reproduction of bacteria. (135)

bacteroid (bak'tĕ-roid)   A modified, often pleomorphic, bacterial cell within the root nodule cells of legumes; after transformation into a symbiosome it carries out nitrogen fixation. (655)

baeocytes   Small, spherical, reproductive cells produced by pleurocapsalean cyanobacteria through multiple fission. (463)

balanced growth   Microbial growth in which all cellular constituents are synthesized at constant rates relative to each other. (110)

balanitis (bal"ah-ni'tis)   Inflammation of the glans penis usually associated with *Candida* fungi; a sexually transmitted disease. (925)

barophilic (bar"o-fil'ik) or **barophile**   Organisms that prefer or require high pressures for growth and reproduction. (127, 625)

barotolerant   Organisms that can grow and reproduce at high pressures but do not require them. (127, 605)

basal body   The cylindrical structure at the base of procaryotic and eucaryotic flagella that attaches them to the cell. (64, 90)

base analogs   Molecules that resemble normal DNA nucleotides and can substitute for them during DNA replication, leading to mutations. (241)

basic dyes   Dyes that are cationic, or have positively charged groups, and bind to negatively charged cell structures. Usually sold as chloride salts. (27)

basidiocarp (bah-sid'e-o-karp")   The fruiting body of a basidiomycete that contains the basidia. (545)

basidiomycetes (bah-sid"e-o-mi-se'tēz)   A division of fungi in which the spores are borne on club-shaped organs called basidia. (545)

basidiospore (bah-sid'e-ō-spōr)   A spore borne on the outside of a basidium following karyogamy and meiosis. (543)

basidium (bah-sid'e-um; pl., **basidia**)   A structure that bears on its surface a definite number of basidiospores (typically four) that are formed following karyogamy and meiosis. Basidia are found in the basidiomycetes and are usually clubshaped. (545)

basophil (ba'so-fil)   A phagocytic leukocyte whose granules stain bluish-black with a basic dye. It has a segmented nucleus. The granules contain histamine and heparin. (681)

batch culture   A culture of microorganisms produced by inoculating a closed culture vessel containing a single batch of medium. (110)

B cell, also known as a **B lymphocyte**   A type of lymphocyte derived from bone marrow stem cells that matures into an immunologically competent cell under the influence of the bursa of Fabricius in the chicken and bone marrow in nonavian species. Following interaction with antigen, it becomes a plasma cell, which synthesizes and secretes antibody molecules involved in humoral immunity. (681, 728)

B-cell antigen receptor (BCR)   A transmembrane immunoglobulin complex on the surface of a B cell that binds an antigen and stimulates the B cell. It is composed of a membrane-bound immunoglobulin, usually IgD or a modified IgM, complexed with another membrane protein (the Ig-α/Ig-β heterodimer). (728)

benthic (ben'thic)   Pertaining to the bottom of the sea or another body of water. (555)

beta hemolysis   A zone of complete clearing around a bacterial colony growing on blood agar. The zone does not change significantly in color. (516, 772)

β-oxidation pathway   The major pathway of fatty acid oxidation to produce NADH, FADH$_2$, and acetyl coenzyme A. (187)

beta-proteobacteria   One of the five subgroups of proteobacteria, each with distinctive 16S rRNA sequences. Members of this subgroup are similar to the alpha-proteobacteria metabolically, but tend to use substances that diffuse from organic matter decomposition in anaerobic zones. (482)

binal symmetry   The symmetry of some virus capsids (e.g., those of complex phages) that is a combination of icosahedral and helical symmetry. (366)

binary fission   Asexual reproduction in which a cell or an organism separates into two cells. (280, 476, 556, 568)

binomial system   The nomenclature system in which an organism is given two names; the first is the capitalized generic name, and the second is the uncapitalized specific epithet. (413)

bioaugmentation   Addition of pregrown microbial cultures to an environment to perform a specific task. (984)

biochemical oxygen demand (BOD)   The amount of oxygen used by organisms in water under certain standard conditions; it provides an index of the amount of microbially oxidizable organic matter present. (636)

biodegradation (bi"o-deg"rah-da'shun)   The breakdown of a complex chemical through biological processes that can result in minor loss of functional groups, fragmentation into larger constitutents, or complete breakdown to carbon dioxide and minerals. Often the term refers to the undesired microbial-mediated destruction of materials such as paper, paint, and textiles. (979)

biofilms   Organized microbial systems consisting of layers of microbial cells associated with surfaces, often with complex structural and functional characteristics. Biofilms have physical/chemical gradients that influence microbial metabolic processes. They can form on inanimate devices (catheters, medical prosthetic devices) and also cause fouling (e.g., of ships' hulls, water pipes, cooling towers). (602, 897)

biogeochemical cycling   The oxidation and reduction of substances carried out by living organisms and/or abiotic processes that results in the cycling of elements within and between different parts of the ecosystem (the soil, aquatic environment, and atmosphere). (593)

bioinsecticide   A pathogen that is used to kill or disable unwanted insect pests. Bacteria, fungi, or viruses are used, either directly or after manipulation, to control insect populations. (988)

biologic transmission   A type of vector-borne transmission in which a pathogen goes through some morphological or physiological change within the vector. (831)

bioluminescence (bi"o-loo"mĭ-nes'əns)   The production of light by living cells, often through the oxidation of molecules by the enzyme luciferase. (492)

biomagnification   The increase in concentration of a substance in higher-level consumer organisms. (600)

biopesticide   The use of a microorganism or another biological agent to control a specific pest. (988)

bioremediation   The use of biologically mediated processes to remove or degrade pollutants from specific environments. Bioremediation can be carried out by modification of the environment to accelerate biological processes, either with or without the addition of specific microorganisms. (982)

biosensor The coupling of a biological process with production of an electrical signal or light to detect the presence of particular substances. (986)

biosynthesis *See* anabolism. (168)

bioterrorism The intentional or threatened use of viruses, bacteria, fungi, or toxins from living organisms to produce death or disease in humans, animals, and plants. (837)

biotransformation or **microbial transformation** The use of living organisms to modify substances that are not normally used for growth. (978)

black peidra (pe-a′drah) A fungal infection caused by *Piedraia hortae* that forms hard black nodules on the hairs of the scalp. (919)

blastomycosis (blas″to-mi-ko′sis) A systemic fungal infection caused by *Blastomyces dermatitidis* and marked by suppurating tumors in the skin or by lesions in the lungs. (922)

B lymphocyte *See* B cell. (681, 728)

botulism (boch′oo-lizm) A form of food poisoning caused by a neurotoxin (botulin) produced by *Clostridium botulinum* serotypes A–G; sometimes found in improperly canned or preserved food. (905)

bright-field microscope A microscope that illuminates the specimen directly with bright light and forms a dark image on a brighter background. (19)

Bright's disease *See* glomerulonephritis. (881)

broad-spectrum drugs Chemotherapeutic agents that are effective against many different kinds of pathogens. (781)

bronchial-associated lymphoid tissue (BALT) The type of defensive tissue found in the lungs. Part of the nonspecific (innate) immune system. (686)

bronchial asthma An example of an atopic allergy involving the lower respiratory tract. (745)

bubo (bu′bo) A tender, inflamed, enlarged lymph node that results from a variety of infections. (887)

bubonic plague *See* plague. (887)

budding A vegetative outgrowth of yeast and some bacteria as a means of asexual reproduction; the daughter cell is smaller than the parent. (476)

bulking sludge Sludges produced in sewage treatment that do not settle properly, usually due to the development of filamentous microorganisms. (638)

bursa of Fabricius (bər′sə fə-bris′e-əs) Found in birds; the blind saclike structure located on the posterior wall of the cloaca; it performs a thymuslike function. A primary lymphoid organ where B-cell maturation occurs. Bone marrow is the equivalent in mammals. (684)

burst *See* rise period. (373)

burst size The number of phages released by a host cell during the lytic life cycle. (373)

butanediol fermentation A type of fermentation most often found in the family *Enterobacteriaceae* in which 2,3-butanediol is a major product; acetoin is an intermediate in the pathway and may be detected by the Voges-Proskauer test. (176, A-15)

C

Calvin cycle The main pathway for the fixation (or reduction and incorporation) of CO_2 into organic material by photoautotrophs; it also is found in chemolithoautotrophs. (202, A-17)

cancer (kan′ser) A malignant tumor that expands locally by invasion of surrounding tissues and systemically by metastasis. (400)

candidal vaginitis Vaginitis caused by *Candida* species. (925)

candidiasis (kan″dĭ-di′ah-sis) An infection caused by *Candida* species of dimorphic fungi, commonly involving the skin. (924)

capsid (kap′sid) The protein coat or shell that surrounds a virion's nucleic acid. (359)

capsomer (kap′so-mer) The ring-shaped morphological unit of which icosahedral capsids are constructed. (360)

capsule A layer of well-organized material, not easily washed off, lying outside the bacterial cell wall. (61)

carboxysomes Polyhedral inclusion bodies that contain the CO_2 fixation enzyme ribulose 1,5-bisphosphate carboxylase; found in cyanobacteria, nitrifying bacteria, and thiobacilli. (48, 202)

caries (kar′e-ēz) Tooth decay. (914)

carotenoids (kah-rot′e-noids) Pigment molecules, usually yellowish in color, that are often used to aid chlorophyll in trapping light energy during photosynthesis. (191)

carrier An infected individual who is a potential source of infection for others and plays an important role in the epidemiology of a disease. (828)

caseous lesion (ka′se-us) A lesion resembling cheese or curd; cheesy. Most caseous lesions are caused by *M. tuberculosis*. (884)

casual carrier An individual who harbors an infectious organism for only a short period. (828)

catabolism (kah-tab′o-lizm) That part of metabolism in which larger, more complex molecules are broken down into smaller, simpler molecules with the release of energy. (168)

catabolite repression (kah-tab′o-līt) Inhibition of the synthesis of several catabolic enzymes by a metabolite such as glucose. (276)

catalyst (kat′ah-list) A substance that accelerates a reaction without being permanently changed itself. (156)

catalytic site *See* active site. (158)

cathelicidins Antimicrobial peptides that are produced by skin cells and kill bacterial pathogens. They destroy invaders by either punching holes in their membranes or solubilizing membranes through detergent-like action. (677)

catheter (kath′ĕ-ter) A tubular surgical instrument for withdrawing fluids from a cavity of the body, especially one for introduction into the bladder through the urethra for the withdrawal of urine. (800)

caveola (ka-ve-o′lə) A small flask-shaped invagination of the plasma membrane formed during one type of pinocytosis. (80)

CD95 pathway The CD95 receptor is found on many nucleated eucaryotic cells. When the receptor is bound to a specific ligand (CD95L), the CD95-CD95L complex activates several cytoplasmic proteins that initiate a cellular suicide cascade leading to apoptosis. (725)

cell cycle The sequence of events in a cell's growth-division cycle between the end of one division and the end of the next. In eucaryotic cells, it is composed of the G_1 period, the S period in which DNA and histones are synthesized, the G_2 period, and the M period (mitosis). (86, 280)

cell-mediated immunity The type of immunity that results from T cells coming into close contact with foreign cells or infected cells to destroy them; it can be transferred to a nonimmune individual by the transfer of cells. (706)

cellular slime molds Slime molds with a vegetative phase consisting of amoeboid cells that aggregate to form a multicellular pseudoplasmodium; they belong to the division *Acrasiomycota*. (548)

cellulitis (sel″u-li′tis) A diffuse spreading infection of subcutaneous skin tissue caused by streptococci, staphylococci, or other organisms. The tissue is inflamed with edema, redness, pain, and interference with function. (879)

cell wall The strong layer or structure that lies outside the plasma membrane; it supports and protects the membrane and gives the cell shape. (88)

cephalosporin (sef″ah-lo-spōr′in) A group of β-lactam antibiotics derived from the fungus *Cephalosporium*, which share the 7-aminocephalosporanic acid nucleus. (789)

chancre (shang′ker) The primary lesion of syphilis occurring at the site of entry of the infection. (899)

chancroid (shang′kroid) A sexually transmitted disease caused by the gram-negative bacterium *Haemophilus ducreyi*. Worldwide, chancroid is an important cofactor in the transmission of the AIDS virus. Also known as genital ulcer disease due to the painful circumscribed ulcers that form on the penis or entrance to the vagina. (891)

chemical oxygen demand (COD) The amount of chemical oxidation required to convert organic matter in water and wastewater to CO_2. (636)

chemiosmotic hypothesis (kem″e-o-os-mot′ik) The hypothesis that a proton gradient and an electrochemical gradient are generated by electron transport and then used to drive ATP synthesis by oxidative phosphorylation. (182)

chemoheterotroph (ke″mo-het′er-o-trōf″) *See* chemoorganotrophic heterotrophs. (96)

chemolithotroph (ke″mo-lith′o-trōf″) *See* chemolithotrophic autotrophs. (96, 188)

chemolithotropic autotrophs Microorganisms that oxidize reduced inorganic compounds to derive both energy and electrons; CO_2 is their carbon source. Also called chemolithoautotrophs. (96)

chemoorganotrophic heterotrophs Organisms that use organic compounds as sources of energy, hydrogen, electrons, and carbon for biosynthesis. (96)

chemoreceptors Special protein receptors in the plasma membrane or periplasmic space that bind chemicals and trigger the appropriate chemotaxic response. (67)

chemostat (ke′mo-stat) A continuous culture apparatus that feeds medium into the culture vessel at the same rate as medium containing microorganisms is removed; the medium in a chemostat contains one essential nutrient in a limiting quantity. (117)

chemotaxis (ke″mo-tak′sis) The pattern of microbial behavior in which the microorganism moves toward chemical attractants and/or away from repellents. (67)

chemotherapeutic agents (ke″mo-ther-ah-pu′tik) Compounds used in the treatment of disease that destroy pathogens or inhibit their growth at concentrations low enough to avoid doing undesirable damage to the host. (780)

chemotrophs (ke′mo-trōfs) Organisms that obtain energy from the oxidation of chemical compounds. (95)

chickenpox (varicella;) (chik′en-poks) A highly contagious skin disease, usually affecting 2- to 7-year-old children; it is caused by the varicella-zoster virus, which is acquired by droplet inhalation into the respiratory system. (846)

chimera (ki-me′rah) A recombinant plasmid containing foreign DNA, which is used as a cloning vector in genetic engineering. (323)

chiral (ki′rəl) Having handedness: consisting of one or another stereochemical form. (980)

chitin (ki′tin) A tough, resistant, nitrogen-containing polysaccharide forming the walls of certain fungi, the exoskeleton of arthropods, and the epidermal cuticle of other surface structures of certain protists and animals. (539)

chlamydiae (klə-mid′e-e) Members of the genus *Chlamydia*: gram-negative, coccoid cells that reproduce only within the cytoplasmic vesicles of host cells using a life cycle that alternates between elementary bodies and reticulate bodies. (464)

chlamydial pneumonia (klə-mid′e-əl noo-mo′ne-ə) A pneumonia caused by *Chlamydia pneumoniae*. Clinically, infections are mild and 50% of adults have antibodies to the chlamydiae. (891)

chloramphenicol (klo″ram-fen′ĭ-kol) A broad-spectrum antibiotic that is produced by *Streptomyces venzuelae* or synthetically; it binds to the large ribosomal subunit and inhibits the peptidyl transferase reaction. (792)

chlorophyll (klor′o-fil) The green photosynthetic pigment that consists of a large tetrapyrrole ring with a magnesium atom in the center. (191)

chloroplast (klo′ra-plast) A eucaryotic plastid that contains chlorophyll and is the site of photosynthesis. (84)

cholera (kol′er-ah) An acute infectious enteritis, endemic and epidemic in Asia, which periodically spreads to the Middle East, Africa, Southern Europe, and South America; caused by *Vibrio cholerae*. (907)

choleragen (kol′er-ah-gen) The cholera toxin; an extremely potent protein molecule elaborated by strains of *Vibrio cholerae* in the small intestine after ingestion of feces-contaminated water or food. It acts on epithelial cells to cause hypersecretion of chloride and bicarbonate and an outpouring of large quantities of fluid from the mucosal surface. (908)

chromatin (kro′mah-tin) The DNA-containing portion of the eucaryotic nucleus; the DNA is almost always complexed with histones. It can be very condensed (heterochromatin) or more loosely organized and genetically active (euchromatin). (85)

chromoblastomycosis (kro″mo-blas″to-mi-ko′sis) A chronic fungal skin infection, producing wartlike nodules that may ulcerate. It is caused by the black molds *Phialophora verrucosa* or *Fonsecaea pedrosoi*. (921)

chromogen (kro′me-jen) A colorless substrate that is acted on by an enzyme to produce a colored end product. (753)

chromophore group (kro″mo-fōr) A chemical group with double bonds that absorbs visible light and gives a dye its color. (27)

chromosomes (kro′mo-somz) The bodies that have most or all of the cell's DNA and contain most of its genetic information (mitochondria and chloroplasts also contain DNA and genes). (85)

chronic carrier An individual who harbors a pathogen for a long time. (824)

chrysolaminarin The polysaccharide storage product of the chrysophytes and diatoms. (559)

chytrids A term used to describe the *Chytridiomycota*, which are simple terrestrial and aquatic fungi that produce motile zoospores with single, posterior, whiplash flagella. Also considered protists. (546, 622)

cilia (sil′e-ah) Threadlike appendages extending from the surface of some protozoa that beat rhythmically to propel them; cilia are membrane-bound cylinders with a complex internal array of microtubules, usually in a 9 + 2 pattern. (88)

citric acid cycle *See* tricarboxylic acid (TCA) cycle. (178, A-14)

classical complement pathway The antibody-dependent pathway of complement activation; it leads to the lysis of pathogens and stimulates phagocytosis and other host defenses. (734)

classification The arrangement of organisms into groups based on mutual similarity or evolutionary relatedness. (410)

clathrin (klath′rin) A protein that coats the cytoplasmic side of specialized plasma membrane regions and the vesicles they form during receptor-mediated endocytosis. (80)

clone (klōn) A group of genetically identical cells or organisms derived by asexual reproduction from a single parent. (222, 718)

clostridial myonecrosis (klo-strid′e-al mi″o-ne-kro′sis) Death of individual muscle cells caused by clostridia. Also called gas gangrene. (891)

cluster of differentiation molecules (CDs) Functional cell surface proteins or receptors that can be measured in situ from peripheral blood, biopsy samples, or other body fluids. They can be used to identify leukocyte subpopulations. Some examples include interleukin-2 receptor (IL-2R), CD4, CD8, CD25, and intercellular adhesion molecule-1 (ICAM-1). (709)

coaggregation The collection of a variety of bacteria on a surface such as a tooth surface because of cell-to-cell recognition of genetically distinct bacterial types. Many of these interactions appear to be mediated by a lectin on one bacterium that interacts with a complementary carbohydrate receptor on another bacterium. (911)

coagulase (ko-ag′u-las) An enzyme that induces blood clotting; it is characteristically produced by pathogenic staphylococci. (514)

coated vesicles The clathrin coated vesicles formed by receptor-mediated endocytosis. (80)

coccidioidomycosis (kok-sid″e-oi″do-mi-ko′sis) A fungal disease caused by *Coccidioides immitis* that exists in dry, highly alkaline soils. Also known as valley fever, San Joaquin fever, or desert rheumatism. (922)

coccus (kok′us, pl. **cocci**, kok′si) A roughly spherical bacterial cell. (41)

code degeneracy The genetic code is organized in such a way that often there is more than one codon for each amino acid. (234)

codon (ko′don) A sequence of three nucleotides in mRNA that directs the incorporation of an amino acid during protein synthesis or signals the start or stop of translation. (234)

coenocytic (se″no-sit′ik) Refers to a multinucleate cell or hypha formed by repeated nuclear divisions not accompanied by cell divisions. (110, 539)

coenzyme (ko-en′zīm) A loosley bound cofactor that often dissociates from the enzyme active site after product has been formed. (156)

cofactor The nonprotein component of an enzyme; it is required for catalytic activity. (156)

cold sore A lesion caused by the herpes simplex virus; usually occurs on the border of the lips or nares. Also known as a fever blister or herpes labialis. (860)

colicin (kol′ĭ-sin) A plasmid-encoded protein that is produced by enteric bacteria and binds to specific receptors on the cell envelope of sensitive target bacteria, where it may cause lysis or attack specific intracellular sites such as ribosomes. (291, 688)

coliform (ko′lĭ-form) A gram-negative, non-sporing, facultative rod that ferments lactose with gas formation within 48 hours at 35°C. (633)

colonization (kol″ə-nĭ-za′shən) The establishment of a site of microbial reproduction on an inanimate surface or organism without necessarily resulting in tissue invasion or damge. (766)

colony An assemblage of microorganisms growing on a solid surface such as the surface of an agar culture medium; the assemblage often is directly visible, but also may be seen only microscopically. (104)

colony forming units (CFU) The number of microorganisms that form colonies when cultured using spread plates or pour plates, an indication of the number of viable microorganisms in a sample. (115)

Colorado tick fever A disease that occurs in the mountainous regions of the western United States. It is caused by an RNA virus of the genus *Coltivirus* that is spread from ground squirrels, rabbits, and deer to humans by the tick, *Dermacentor andersoni*. Complications are rare. (853)

colorless sulfur bacteria A diverse group of non-photosynthetic proteobacteria that can oxidize reduced sulfur compounds such as hydrogen sulfide. Many are lithotrophs and derive energy from sulfur oxidation. Some are unicellular, whereas others are filamentous gliding bacteria. (483)

combinatorial biology Introduction of genes from one microorganism into another microorganism to synthesize a new product or a modified product, especially in relation to antibiotic synthesis. (966)

comedo (kom′ĕ-do; pl., **comedones**) A plug of dried sebum in an excretory duct of the skin. (677)

cometabolism The modification of a compound not used for growth by a microorganism, which occurs in the presence of another organic material that serves as a carbon and energy source. (982)

commensal (kŏ-men′sal) Living on or within another organism without injuring or benefiting the other organism. (588)

commensalism (kŏ-men′sal-izm″) A type of symbiosis in which one individual gains from the association and the other is neither harmed nor benefited. (588)

common cold An acute, self-limiting, and highly contagious virus infection of the upper respiratory tract that produces inflammation, profuse discharge, and other symptoms. (861)

common-source epidemic An epidemic that is characterized by a sharp rise to a peak and then a rapid, but not as pronounced, decline in the number of individuals infected; it usually involves a single contaminated source from which individuals are infected. (824)

common vehicle transmission The transmission of a pathogen to a host by means of an inanimate medium or vehicle. (831)

communicable disease A disease associated with a pathogen that can be transmitted from one host to another. (827)

community An assemblage of different types of organisms or a mixture of different microbial populations. (578)

competent A bacterial cell that can take up free DNA fragments and incorporate them into its genome during transformation. (299)

competition An interaction between two organisms attempting to use the same resource (nutrients, space, etc.). (592)

competitive exclusion principle Two competing organisms overlap in resource use, which leads to the exclusion of one of the organisms. (592, 960)

complementary DNA (cDNA) A DNA copy of an RNA molecule (e.g., a DNA copy of an mRNA). (313)

complement system A group of plasma proteins that plays a major role in an animal's defensive immune response. (691, 734)

complex medium Culture medium that contains some ingredients of unknown chemical composition. (102)

complex viruses Viruses with capsids having a complex symmetry that is neither icosahedral nor helical. (359)

composting The microbial processing of fresh organic matter under moist, aerobic conditions, resulting in the accumulation of a stable humified product, which is suitable for soil improvement and stimulation of plant growth. (662)

compromised host A host with lowered resistance to infection and disease for any of several reasons. The host may be seriously debilitated (due to malnutrition, cancer, diabetes, leukemia, or another infectious disease), traumatized (from surgery or injury), immunosuppressed, or have an altered microbiota due to prolonged use of antibiotics. (680, 924)

concatemer A long DNA molecule consisting of several genomes linked together in a row. (376)

conditional mutations Mutations that are expressed only under certain environmental conditions. (239)

confocal scanning laser microscope (CSLM) A light microscope in which monochromatic laser-derived light scans across the specimen at a specific level and illuminates one area at a time to form an image. Stray light from other parts of the specimen is blocked out to give an image with excellent contrast and resolution. (35)

congenital (neonatal) herpes An infection of a newborn caused by transmission of the herpesvirus during vaginal delivery. (863)

congenital rubella syndrome A wide array of congenital defects affecting the heart, eyes, and ears of a fetus during the first trimester of pregnancy, and caused by the rubella virus. (851)

congenital syphilis Syphilis that is acquired in utero from the mother. (899)

conidiospore (ko-nid′e-o-spōr) An asexual, thin-walled spore borne on hyphae and not contained within a sporangium; it may be produced singly or in chains. (522, 543)

conidium (ko-nid′e-um; pl., **conidia**) *See* conidiospore. (522)

conjugants (kon′joo-gants) Complementary mating types that participate in a form of protozoan sexual reproduction called conjugation. (568)

conjugation (kon″ju-ga′shun) 1. The form of gene transfer and recombination in bacteria that requires direct cell-to-cell contact. 2. A complex form of sexual reproduction commonly employed by protozoa. (296, 568)

conjugative plasmid A plasmid that carries the genes for sex pili and can transfer copies of itself to other bacteria during conjugation. (288)

conjunctivitis of the newborn *See* ophthalmia neonatorum. (892)

conoid (ko′noid) A hollow cone of spirally coiled filaments in the anterior tip of certain apicomplexan protozoa. (573)

consortium A physical association of two different organisms, usually beneficial to both organisms. (578)

constant region (C$_L$ and C$_H$) The part of an antibody molecule that does not vary greatly in amino acid sequence among molecules of the same class, subclass, or type. (711)

constitutive mutant A strain that produces an inducible enzyme continually, regardless of need, because of a mutation in either the operator or regulator gene. (271)

constructed wetlands Intentional creation of marshland plant communities and their associated microorganisms for environmental restoration or to purify water by the removal of bacteria, organic matter, and chemicals as the water passes through the aquatic plant communities. (p. 641)

consumer An organism that feeds directly on living or dead animals, by ingestion or by phagocytosis. (604)

contact transmission Transmission of the pathogen by contact of the source or reservoir of the pathogen with the host. (830)

continuous culture system A culture system with constant environmental conditions maintained through continual provision of nutrients and removal of wastes. (117)

contractile vacuole (vak′u-ōl) In protists and some animals, a clear fluid-filled cell vacuole that takes up water from within the cell and then contracts, releasing it to the outside through a pore in a cyclical manner. Contractile vacuoles function primarily in osmoregulation and excretion. (567)

convalescent carrier (kon″vah-les′ent) An individual who has recovered from an infectious disease but continues to harbor large numbers of the pathogen. (828)

cooperation A positive but not obligatory interaction between two different organisms. Also called protocooperation. (585)

corepressor (ko″re-pre′sor) A small molecule that inhibits the synthesis of a repressible enzyme. (271)

cortex (kor′teks) The layer of a bacterial endospore that is thought to be particularly important in conferring heat resistance on the endospore. (69)

coryza (kŏ-ri′zah) *See* common cold. (861)

cosmid (koz′mid) A plasmid vector with lambda phage cos sites that can be packaged in a phage capsid; it is useful for cloning large DNA fragments. (326)

cristae (kris′te) Infoldings of the inner mitochondrial membrane. (82)

crossing-over A process in which segments of two adjacent DNA strands are exchanged; breaks occur in both strands, and the exposed ends of each strand join to those of the opposite segment on the other strand. (286)

cryptins Peptides produced by Paneth cells in the intestines. Cryptins are toxic for some bacteria, although their mode of action is not known. (687)

cryptococcosis (krip″to-kok-o′sis) An infection caused by the basidiomycete, *Cryptococcus neoformans*, which may involve the skin, lungs, brain, or meninges. (546, 923)

cryptosporidiosis (krip″to-spo-rid″e-o′sis) Infection with protozoa of the genus *Cryptosporidium*.

The most common symptoms are prolonged diarrhea, weight loss, fever, and abdominal pain. (927)

crystallizable fragment (Fc) The stem of the Y portion of an antibody molecule. Cells such as macrophages bind to the Fc region, and it also is involved in complement activation. (711)

cutaneous anthrax (ku-ta′ne-us an′thraks) A form of anthrax involving the skin. (889)

cutaneous diphtheria (ku-ta′ne-us dif-the′re-ah) A skin disease caused by *Corynebacterium diphtheriae* that infects wound or skin lesions, causing a slow-healing ulceration. (877)

cyanobacteria (si″ah-no-bak-te′re-ah) A large group of bacteria that carry out oxygenic photosynthesis using a system like that present in photosynthetic eucaryotes. (458)

cyclic photophosphorylation (fo″to-fos″for-ĭ-la′shun) The formation of ATP when light energy is used to move electrons cyclically through an electron transport chain during photosynthesis; only photosystem I participates. (193)

cyst (sist) A general term used for a specialized microbial cell enclosed in a wall. Cysts are formed by protozoa and a few bacteria. They may be dormant, resistant structures formed in response to adverse conditions or reproductive cysts that are a normal stage in the life cycle. (568)

cytochromes (si′to-krōms) Heme proteins that carry electrons, usually as members of electron transport chains. (155)

cytokine (si′to-kīn) A general term for nonantibody proteins, released by a cell in response to inducing stimuli, which are mediators that influence other cells. Are produced by lymphocytes, monocytes, macrophages, and other cells. (697)

cytomegalovirus inclusion disease (si″to-meg″ah-lo-vi′rus) An infection caused by the cytomegalovirus and marked by nuclear inclusion bodies in enlarged infected cells. (862)

cytopathic effect (si′to-path′ik) The observable change that occurs in cells as a result of viral replication. Examples include ballooning, binding together, clustering, or even death of the cultured cells. (354, 805)

cytoplasmic matrix (si′to-plaz′mik) The protoplasm of a cell that lies within the plasma membrane and outside any other organelles. In bacteria it is the substance between the cell membrane and the nucleoid. (47, 76)

cytosine (si′to-sēn) A pyrimidine 2-oxy-4-aminopyrimidine found in nucleosides, nucleotides, and nucleic acids. (211)

cytoskeleton (si″to-skel′ĕ-ton) A network of microfilaments, microtubules, intermediate filaments, and other components in the cytoplasm of eucaryotic cells that helps give them shape. (78)

cytostome (si′to-stōm) A permanent site in the protozoan ciliate at which food is ingested. (567)

cytotoxic T (T_C) cell (si″to-tok′sik) A cell that is capable of recognizing virus-infected cells through the major histocompatability molecules and developing into an activated cell that destroys the infected cells. (724)

cytotoxic T lymphocyte (CTL) The activated T cell that can attack and destroy virus-infected cells, tumor cells, and foreign cells. (725)

cytotoxin (si′to-tok′sin) A toxin or antibody that has a specific toxic action upon cells; cytotoxins are named according to the cell for which they are specific (e.g., nephrotoxin). (772)

D

Dane particle A 42 nm spherical particle that is one of three that are seen in hepatitis B virus infections. The Dane particle is the complete virion. (866)

dark-field microscopy Microscopy in which the specimen is brightly illuminated while the background is dark. (21)

dark reactivation The excision and replacement of thymine dimers in DNA that occurs in the absence of light. (128)

deamination (de-am″i-na′shun) The removal of amino groups from amino acids. (188)

death phase The decrease in viable microorganisms that occurs after the completion of growth in a batch culture. (112)

decimal reduction time (D or D value) The time required to kill 90% of the microorganisms or spores in a sample at a specified temperature. (137)

decomposer An organism that breaks down complex materials into simpler ones, including the release of simple inorganic products. Often a decomposer such as an insect or earthworm physically reduces the size of substrate particles. (604)

defensin (de-fens′sin) Specific peptides produced by neutrophils that permeabilize the outer and inner membranes of certain microorganisms, thus killing them. (697)

defined medium Culture medium made with components of known composition. (102)

Delta agent A defective RNA virus that is transmitted as an infectious agent, but cannot cause disease unless the individual is also infected with the hepatitis B virus. *See* hepatitis D. (867)

delta-proteobacteria One of the five subgroups of proteobacteria. Chemoorganotrophic bacteria that usually are either predators on other bacteria or anaerobes that generate sulfide from sulfate and sulfite. (495)

denaturation (de-na″chur-a′shun) A change in enzyme shape that destroys its activity; the term is also applied to changes in nucleic-acid shape. (159)

dendritic cell (den-drit′ik) An antigen-presenting cell that has long membrane extensions resembling the dendrites of neurons. These cells are found in the lymph nodes, spleen, and thymus (interdigitating dendritic cells); skin (Langerhans cells); and other tissues (interstitial dendritic cells). Express MHC class II and B7 costimulatory molecules and present antigens to T-helper cells. (683)

dendrogram A treelike diagram that is used to graphically summarize mutual similarities and relationships between organisms. (415)

denitrification (de-ni″trī-fī-ka′shən) The reduction of nitrate to gaseous products, primarily nitrogen gas, during anaerobic respiration. (185, 598)

dental plaque (plak) A thin film on the surface of teeth consisting of bacteria embedded in a matrix of bacterial polysaccharides, salivary glycoproteins, and other substances. (911)

deoxyribonucleic acid (DNA) (de-ok″se-ri″bo-nu-kle′ik) The nucleic acid that constitutes the genetic material of all cellular organisms. It is a polynucleotide composed of deoxyribonucleotides connected by phosphodiester bonds. (50, 224)

dermatomycosis (der′ma-to-mi-ko′sis) A fungal infection of the skin; the term is a general term that comprises the various forms of tinea, and it is sometimes used to specifically refer to athlete's foot (tinea pedis). (919)

dermatophyte (der′mah-to-fit″) A fungus parasitic on the skin. (919)

desensitization (de-sen″si-ti-za′shun) To make a sensitized or hypersensitive individual insensitive or nonreactive to a sensitizing agent. (746)

desert crust A crust formed by microbial binding of sand grains in the surface zone of desert soil; crust formation primarily involves cyanobacteria. (650)

detergent (de-ter′jent) An organic molecule, other than a soap, that serves as a wetting agent and emulsifier; it is normally used as cleanser, but some may be used as antimicrobial agents. (145)

deuteromycetes (doo″tər-o-mi-se′tēz) In some classification systems, the deuteromycetes or Fungi Imperfecti are a class of fungi. These organisms either lack a sexual stage or it has not yet been discovered. (546)

diatoms (di′ah-toms) Algal protists with siliceous cell walls called frustules. They constitute a substantial subfraction of the phytoplankton. (559)

diauxic growth (di-awk′sik) A biphasic growth pattern or response in which a microorganism, when exposed to two nutrients, initially uses one of them for growth and then alters its metabolism to make use of the second. (276)

differential interference contrast (DIC) microscope A light microscope that employs two beams of plane polarized light. The beams are combined after passing through the specimen and their intereference is used to create the image. (22)

differential media (dif″er-en′shal) Culture media that distinguish between groups of microorganisms based on differences in their growth and metabolic products. (104)

differential staining procedures Staining procedures that divide bacteria into separate groups based on staining properties. (27)

diffusely adhering E. coli (DAEC) DAEC strains of *E. coli* adhere over the entire surface of epithelial cells and usually cause diarrheal disease in immunologically naive and malnourished children. (910)

dikaryotic stage (di-kar-e-ot′ik) In fungi, having pairs of nuclei within cells or compartments. Each cell contains two separate haploid nuclei, one from each parent. (543)

dinoflagellate (di″no-flaj′e-lāt) An algal protist characterized by two flagella used in swimming in a spinning pattern. Many are bioluminescent and an important part of marine phytoplankton, some also are important marine pathogens. (561)

dinospore A motile parasitic stage of a dinoflagellate, such as *Pfiesteria*, that can attach to the surface of a prey fish and cause tissue damage. Dinospores exhibit rapid chemotaxis and micropredatory feeding. Also called swarmers. (627)

diphtheria (dif-the′re-ah) An acute, highly contagious childhood disease that generally affects the membranes of the throat and less frequently the nose. It is caused by *Corynebacterium diphtheriae*. (876)

dipicolinic acid A substance present at high concentrations in the bacterial endospore. It is thought to contribute to the endospore's heat resistance. (69)

diplococcus (dip″lo-kok′us) A pair of cocci. (41)

directed or adaptive mutation A mutation that seems to be chosen so the organism can better adapt to its surroundings. (239)

disease (di-zez) A deviation or interruption of the normal structure or function of any part of the body that is manifested by a characteristic set of symptoms and signs. (822)

disease syndrome (sin′drŏm) A set of signs and symptoms that are characteristic of the disease. (824)

disinfectant (dis″in-fek′tant) An agent, usually chemical, that disinfects; normally, it is employed only with inanimate objects. (135)

disinfection (dis″in-fek′shun) The killing, inhibition, or removal of microorganisms that may cause disease. It usually refers to the treatment of inanimate objects with chemicals. (134)

disinfection by-products (DBPs) Chlorinated organic compounds such as trihalomethanes formed during chlorine use for water disinfection. Many are carcinogens. (631)

dissimilatory nitrate reduction The process in which some bacteria use nitrate as the electron acceptor at the end of their electron transport chain to produce ATP. The nitrate is reduced to nitrite or nitrogen gas. (185)

dissimilatory reduction The use of a substance as an electron acceptor in energy generation. The acceptor (e.g., sulfate or nitrate) is reduced but not incorporated into organic matter during biosynthetic processes. (597)

diurnal oxygen shifts (di-er′nal) The changes in oxygen levels that occur in waters when algae produce and use oxygen on a cyclic basis during day and night. (629)

DNA ligase An enzyme that joins two DNA fragments together through the formation of a new phosphodiester bond. (233)

DNA microarrays (DNA chips) Solid supports that have DNA attached in organized arrays and are used to evaluate gene expression. (345, 986)

DNA polymerase (pol-im′er-ās) An enzyme that synthesizes new DNA using a parental DNA strand as a template. (230)

DNA vaccine A vaccine that contains DNA which encodes antigenic proteins. It is injected directly into the muscle; the DNA is taken up by the muscle cells and encoded protein antigens are synthesized. This produces both humoral and cell-mediated responses. (744)

domains (do-mān′) 1. Compact, self-folding, structurally independent regions of proteins (usually around 100–300 amino acids in length); large proteins may have two or more domains connected by less structured stretches of polypeptide. In the antibody molecule, they are the loops, along with about 25 amino acids on each side, that form compact, globular sections. 2. The primary taxonomic groups above the kingdom level; all living organisms may be placed in one of three domains. (269, 410, 712)

double diffusion agar assay (Ouchterlony technique) An immunodiffusion reaction in which both antibody and antigen diffuse through agar to form stable immune complexes, which can be observed visually. (754)

doubling time *See* generation time. (112)

DPT (diphtheria-pertussis-tetanus) vaccine A vaccine containing three antigens that is used to immunize people against diphtheria, pertussis or whooping cough, and tetanus. (877)

droplet nuclei Small particles (1 to 4 μm in diameter) that represent what is left from the evaporation of larger particles (10 μm or more in diameter) called droplets. (829)

D value *See* decimal reduction time. (137)

E

early mRNA Messenger RNA produced early in a virus infection that codes for proteins needed to take over the host cell and manufacture viral nucleic acids. (375)

Ebola virus hemorrhagic fever (a′bo-lə) An acute infection caused by a virus that produces varying degrees of hemorrhage, shock, and sometimes death. (853)

eclipse period (e-klips′) The initial part of the latent period in which infected host bacteria do not contain any complete virions. (373)

ecosystem (ek″o-sis′tem) A self-regulating biological community and its associated physical and chemical environment. (578)

ectomycorrhizal Referring to a mutualistic association between fungi and plant roots in which the fungus surrounds the root tip with a sheath. (656)

ectoparasite (ek″to-par′ah-sīt) A parasite that lives on the surface of its host. (762)

ectosymbiosis A type of symbiosis in which one organism remains outside of the other organism. (578, 675)

effacing lesion (le′zhən) The type of lesion caused by enteropathogenic strains of *E. coli* (EPEC) when the bacteria destroy the brush border of intestinal epithelial cells. The term AE (attaching-effacing) *E. coli* is now used to designate true EPEC strains that are an important cause of di-

arrhea in children from developing countries and in traveler's diarrhea. (910)

ehrlichiosis (ar-lik″e-o′sis) A tick-borne (*Dermacentor andersoni, Amblyomma americanum*) rickettsial disease caused by *Ehrlichia chaffeensis*. Once inside leukocytes, a nonspecific illness develops that resembles Rocky Mountain spotted fever. (885)

electron acceptor A substance that accepts electrons in an oxidation-reduction reaction. Often called an oxidizing agent or oxidant. (153)

electron donor An electron donor in an oxidation-reduction reaction. Often called a reducing agent or reductant. (153)

electron transport chain A series of electron carriers that operate together to transfer electrons from donors such as NADH and $FADH_2$ to acceptors such as oxygen. (179)

electrophoresis (e-lek″tro-fo-re′sis) A technique that separates substances through differences in their migration rate in an electrical field due to variations in the number and kinds of charged groups they have. (319)

electroporation (e-lek″tro-pə-ra′shən) The application of an electric field to create temporary pores in the plasma membrane in order to insert DNA into the cell and transform it. (327)

elementary body (EB) A small, dormant body that serves as the agent of transmission between host cells in the chlamydial life cycle. (465)

elongation cycle The cycle in protein synthesis that results in the addition of an amino acid to the growing end of a peptide chain. (265)

Embden-Meyerhof pathway (em′den mi′er-hof) A pathway that degrades glucose to pyruvate; the six-carbon stage converts glucose to fructose 1,6-bisphosphate, and the three-carbon stage produces ATP while changing glyceraldehyde 3-phosphate to pyruvate. (171, A-11)

encystation (en-sis-′ta′shen) The formation of a cyst. (568)

endemic disease (en-dem′ik) A disease that is commonly or constantly present in a population, usually at a relatively steady low frequency. (822)

endemic (murine) typhus (mu′rin ti′fus) A form of typhus fever caused by the rickettsia *Rickettsia typhi* that occurs sporadically in individuals who come into contact with rats and their fleas. (885)

endergonic reaction (end″er-gon′ik) A reaction that does not spontaneously go to completion as written; the standard free energy change is positive, and the equilibrium constant is less than one. (152)

endocytosis (en″do-si-to′sis) The process in which a cell takes up solutes or particles by enclosing them in vesicles pinched off from its plasma membrane. (79)

endogenote (en″do-je′nōt) The recipient bacterial cell's own genetic material into which the donor DNA can integrate. (286)

endogenous infection (en-doj′ĕ-nus in-fek′shun) An infection by a member of an individual's own normal body microbiota. (881)

endogenous pyrogen (en-doj′ĕ-nus pi′ro-jen) A substance such as the lymphokine interleukin-1, which is produced by host cells and induces a fever response in the host. It also is called simply a pyrogen. (775)

endomycorrhizal Referring to a mutualistic association of fungi and plant roots in which the fungus penetrates into the root cells and arbuscules and vesicles are formed. (656)

endoparasite (en'do-par'ah-sīt) A parasite that lives inside the body of its host. (762)

endophyte (en'do-fīt) A microorganism living within a plant, but not necessarily parasitic on it. (655)

endoplasmic reticulum (ER) (en"do-plas'mik rĕ-tik'u-lum) A system of membranous tubules and flattened sacs (cisternae) in the cytoplasmic matrix of eucaryotic cells. Rough endoplasmic reticulum (RER) bears ribosomes on its surface; smooth endoplasmic reticulum (SER) lacks them. (78)

endosome (en'do-sōm) A membranous vesicle formed by endocytosis. (79)

endospore (en'do-spōr) An extremely heat- and chemical-resistant, dormant, thick-walled spore that develops within bacteria. (68)

endosymbiont (en"do-sim'be-ont) An organism that lives within the body of another organism in a symbiotic association. (578)

endosymbiosis (en"do-sim"bi-o'sis) A type of symbiosis in which one organism is found within another organism. (578, 675)

endosymbiotic theory or **hypothesis** The theory that eucaryotic organelles such as mitochondria and chloroplasts arose when bacteria established an endosymbiotic relationship with the eucaryotic ancestor and then evolved into organelles. (84, 411)

endotoxin (en"do-tox'sin) The heat-stable lipopolysaccharide in the outer membrane of the cell wall of gram-negative bacteria that is released when the bacterium lyses, or sometimes during growth, and is toxic to the host. (773)

end product inhibition *See* feedback inhibition. (164)

energy The capacity to do work or cause particular changes. (150)

enology The science of wine making. (p. 955)

enteric bacteria (enterobacteria en-ter'ik) Members of the family *Enterobacteriaceae* (gram-negative, peritrichous or nonmotile, facultatively anaerobic, straight rods with simple nutritional requirements); also used for bacteria that live in the intestinal tract. (493)

enterohemorrhagic *E. coli* **(EHEC)** (en"tər-o-hem"ə-raj'ik) EHEC strains of *E. coli* (O157:H7) produce several cytotoxins that provoke fluid secretion in traveler's diarrhea; however, their mode of action is unknown. (910)

enteroinvasive *E. coli* **(EIEC)** (en"tər-o-in-va'siv) EIEC strains of *E. coli* cause traveler's diarrhea by penetrating and binding to the intestinal epithelial cells. EIEC may also produce a cytotoxin and enterotoxin. (910)

enteropathogenic *E. coli* **(EPEC)** (en"tər-o-path-o-jen'ik) EPEC strains of *E. coli* attach to the brush border of intestinal epithelial cells and cause a specific type of cell damage called effacing lesions that lead to traveler's diarrhea. (910)

enterotoxigenic *E. coli* **(ETEC)** (en'tər-o-tok"sī-jen'ik) ETEC strains of *E. coli* produce two plasmid-encoded enterotoxins (which are responsible for traveler's diarrhea) and are distinguished by their heat stability: heat-stable enterotoxin (ST) and heat-labile enterotoxin (LT). (910)

enterotoxin (en"ter-o-tok'sin) A toxin specifically affecting the cells of the intestinal mucosa, causing vomiting and diarrhea. (772, 905)

Entner-Doudoroff pathway A pathway that converts glucose to pyruvate and glyceraldehyde 3-phosphate by producing 6-phosphogluconate and then dehydrating it. (174, A-13)

entropy (en'tro-pe) A measure of the randomness or disorder of a system; a measure of that part of the total energy in a system that is unavailable for useful work. (152)

envelope (en'vě-lōp) 1. All the structures outside the plasma membrane in bacterial cells. 2. In virology it is an outer membranous layer that surrounds the nucleocapsid in some viruses. (53, 359)

enzootic (en"zo-ot'ik) The moderate prevalence of a disease in a given animal population. (823)

enzyme (en'zīm) A protein catalyst with specificity for both the reaction catalyzed and its substrates. (156)

enzyme-linked immunosorbent assay (ELISA) A technique used for detecting and quantifying specific antibodies and antigens. (753)

eosinophil (e"o-sin'o-fil) A polymorphonuclear leukocyte that has a two-lobed nucleus and cytoplasmic granules that stain yellow-red. A mobile phagocyte that is highly antiparasitic. (683)

epidemic (ep"ĭ-dem'ik) A disease that suddenly increases in occurrence above the normal level in a given population. (822)

epidemic (louse-borne) typhus (ep"ĭ-dem'ik tī'fus) A disease caused by *Rickettsia prowazekii* that is transmitted from person to person by the body louse. (885)

epidemiologist (ep"ĭ-de"me-ol'o-jist) A person who specializes in epidemiology. (822)

epidemiology (epi"-de"me-ol'o-je) The study of the factors determining and influencing the frequency and distribution of disease, injury, and other health-related events and their causes in defined human populations. (822)

episome (ep'ĭ-sōm) A plasmid that can exist either independently of the host cell's chromosome or be integrated into it. (288)

epitheca (ep"ĭ-the'kah) The larger of two halves of a diatom frustule (shell). (559)

epitope (ep'i-tōp) An area of the antigen molecule that stimulates the production of, and combines with, specific antibodies; also known as the antigenic determinant site. (708)

epizootic (ep"ĭ-zo-ot'ik) A sudden outbreak of a disease in an animal population. (823)

epizootiology (ep"i-zo-ot"e-ol'o-je) The field of science that deals with factors determining the frequency and distribution of a disease within an animal population. (823)

epsilon-proteobacteria One of the five subgroups of proteobacteria, each with distinctive 16S rRNA sequences. Slender gram-negative rods, some of which are medically important (*Campylobacter* and *Helicobacter*). (500)

equilibrium (e"kwĭ-lib're-um) The state of a system in which no net change is occurring and free energy is at a minimum; in a chemical reaction at equilibrium, the rates in the forward and reverse directions exactly balance each other out. (152)

ergot (er'got) The dried sclerotium of *Claviceps purpurea*. Also, an ascomycete that parasitizes rye and other higher plants causing the disease called ergotism. (544)

ergotism (er'got-izm) The disease or toxic condition caused by eating grain infected with ergot; it is often accompanied by gangrene, psychotic delusions, nervous spasms, abortion, and convulsions in humans and in animals. (544, 942)

erysipelas (er"ĭ-sip'ĕ-las) An acute inflammation of the dermal layer of the skin, occurring primarily in infants and persons over 30 years of age with a history of streptococcal sore throat. (879)

erythema infectiosum (er"ə-the'-mə) A disease in children caused by the parvovirus B19. This disease is common in children between 4 and 11 years of age and is sometimes called fifth disease, since it was the fifth of six erythematous rash diseases in children in an older classification. (864)

erythromycin (ĕ-rith"ro-mi'sin) An intermediate spectrum macrolide antibiotic produced by *Streptomyces erythreus*. (791)

eschar (es'kar) A slough produced on the skin by a thermal burn, gangrene, or the anthrax bacillus. (p. 889)

Eucarya The domain that contains organisms composed of eucaryotic cells with primarily glycerol fatty acyl diesters in their membranes and eucaryotic rRNA. (411)

eucaryotic cells (u"kar-e-ot'ik) Cells that have a membrane-delimited nucleus and differ in many other ways from procaryotic cells; protists, algae, fungi, plants, and animals are all eucaryotic. (11, 90)

euglenoids (u-gle'noids) A group of algae (the division *Euglenophyta*) or protozoa (order *Euglenida*) that normally have chloroplasts containing chlorophyll *a* and *b*. They usually have a stigma and one or two flagella emerging from an anterior gullet. (559)

Eumycota (u"mi-ko'tə) A division of fungi in some classification systems. These are the true fungi consisting of the Zygomycetes, Ascomycetes, Basidiomycetes, and Chytridiomycetes. (538)

eumycotic mycetoma (mi"se-to'mah) *See* maduromycosis. (921)

eutrophic (u-trof'ik) A nutrient-enriched environment. (628)

eutrophication (u"tro-fĭ-ka'shun) The enrichment of an aquatic environment with nutrients. (628)

evolutionary distance A quantitative indication of the number of positions that differ between two aligned macromolecules, and presumably a measure of evolutionary similarity between molecules and organisms. (421)

excystation (ek″sis-ta′shun)   The escape of one or more cells or organisms from a cyst. (568)

exergonic reaction (ek″ser-gon′ik)   A reaction that spontaneously goes to completion as written; the standard free energy change is negative, and the equilibrium constant is greater than one. (152)

exfoliative toxin (eks-fo′le-a″tiv) or **exfoliatin** (eks-fō″le-a′tin)   An exotoxin produced by *Staphylococcus aureus* that causes the separation of epidermal layers and the loss of skin surface layers. It produces the symptoms of the scalded skin syndrome. (899)

exit site (E site)   The location on a ribosome to which an empty tRNA moves from the P site before it finally leaves during protein synthesis. (265)

exoenzymes (ek″so-en′zīms)   Enzymes that are secreted by cells. (53)

exogenote (eks″o-je′nōt)   The piece of donor DNA that enters a bacterial cell during gene exchange and recombination. (286)

exon (eks′on)   The region in a split or interrupted gene that codes for RNA which ends up in the final product (e.g., mRNA). (257)

exotoxin (ek″so-tok′sin)   A heat-labile, toxic protein produced by a bacterium as a result of its normal metabolism or because of the acquisition of a plasmid or prophage. It is usually released into the bacterium's surroundings. (769)

exponential phase (eks″po-nen′shul)   The phase of the growth curve during which the microbial population is growing at a constant and maximum rate, dividing and doubling at regular intervals. (110)

expressed sequence tag (EST)   A partial gene sequence unique to a gene that can be used to identify and position the gene during genomic analysis. (346)

expression vector   A special cloning vector used to express a recombinant gene in host cells; the gene is transcribed and its protein synthesized. (327)

exteins   Polypeptide sequences of precursor self-splicing proteins that are joined together during formation of the final, functional protein. They are separated from one another by intein sequences, which they flank. (270)

extracutaneous sporotrichosis (spo″ro-tri-ko′sis)   An infection by the fungus *Sporothrix schenckii* that spreads throughout the body. (921)

extreme barophilic bacteria   Bacteria that require a high-pressure environment to function. (605)

extreme environment   An environment in which physical factors such as temperature, pH, salinity, and pressure are outside of the normal range for growth of most microorganisms; these conditions allow unique organisms to survive and function. (605)

extremophiles   Microorganisms that grow under harsh or extreme environmental conditions such as very high temperatures or low pHs. (118, 605)

extrinsic factor   An environmental factor such as temperature that influences microbial growth in food. (938)

F

facilitated diffusion   Diffusion across the plasma membrane that is aided by a carrier. (98)

facultative anaerobes (fak′ul-ta″tiv an-a′er-ōbs)   Microorganisms that do not require oxygen for growth, but do grow better in its presence. (125)

facultative psychrophile (fak′ul-ta″tiv si′kro-fīl)   *See* psychrotroph. (123)

fas **gene**   The gene that is active in target cells which are susceptible to killing by cells expressing the Fas ligand, a member of the TNF family of cytokines and cell surface molecules. (725)

fatty acid synthetase (sin′thĕ-tās)   The multienzyme complex that makes fatty acids; the product usually is palmitic acid. (214)

fecal coliform (fe′kal ko′lĭ-form)   Coliforms whose normal habitat is the intestinal tract and that can grow at 44.5°C. (633)

fecal enterococci (fe′kal en″ter-o-kok′si)   Enterococci found in the intestine of humans and other warm-blooded animals. They are used as indicators of the fecal pollution of water. (636)

feedback inhibition   A negative feedback mechanism in which a pathway end product inhibits the activity of an enzyme in the sequence leading to its formation; when the end product accumulates in excess, it inhibits its own synthesis. (164)

fermentation (fer″men-ta′shun)   An energy-yielding process in which an energy substrate is oxidized without an exogenous electron acceptor. Usually organic molecules serve as both electron donors and acceptors. (168, 970)

fever   A complex physiological response to disease mediated by pyrogenic cytokines and characterized by a rise in core body temperature and activation of the immune system. (698)

fever blister   *See* cold sore. (860)

F factor   The fertility factor, a plasmid that carries genes for bacterial conjugation and makes its *E. coli* host the gene donor during conjugation. (288)

fimbria (fim′bre-ah; pl., **fimbriae**)   A fine, hairlike protein appendage on some gram-negative bacteria that helps attach them to surfaces. (62)

final host   The host on/in which a parasite either attains sexual maturity or reproduces. (762)

first law of thermodynamics   Energy can be neither created nor destroyed (even though it can be changed in form or redistributed). (151)

fixation (fik-sa′shun)   The process in which the internal and external structures of cells and organisms are preserved and fixed in position. (26)

flagellin (flaj′ĕ-lin)   The protein used to construct the filament of a bacterial flagellum. (64)

flagellum (flah-jel′um; pl., **flagella**)   A thin, threadlike appendage on many procaryotic and eucaryotic cells that is responsible for their motility. (63, 88)

flat or plane warts   Small, smooth, slightly raised warts. (871)

flavin adenine dinucleotide (FAD) (fla′vin ad′ĕ-nēn)   An electron carrying cofactor often involved in energy production (for example, in the tricarboxylic acid cycle and the β-oxidation pathway). (155)

fluid mosaic model   The currently accepted model of cell membranes in which the membrane is a lipid bilayer with integral proteins buried in the lipid, and peripheral proteins more loosely attached to the membrane surface. (46)

fluid-phase endocytosis   A form of endocytosis in which a small portion of extracellular fluid is pinched off nonselectively and without concentration of the fluid contents. (79)

fluorescence microscope   A microscope that exposes a specimen to light of a specific wavelength and then forms an image from the fluorescent light produced. Usually the specimen is stained with a fluorescent dye or fluorochrome. (25)

fluorescent light (floo″o-res′ent)   The light emitted by a substance when it is irradiated with light of a shorter wavelength. (25)

fomite (fo′mīt; pl., **fomites**)   An object that is not in itself harmful but is able to harbor and transmit pathogenic organisms. Also called fomes. (766, 831)

food-borne infection   Gastrointestinal illness caused by ingestion of microorganisms, followed by their growth within the host. Symptoms arise from tissue invasion and/or toxin production. (905, 946)

food chain   The flow of energy and matter in living organisms through a producer-consumer sequence (*See also* food web). (566)

food intoxication   Food poisoning caused by microbial toxins produced in a food prior to consumption. The presence of living bacteria is not required. (905, 948)

food poisoning   A general term usually referring to a gastrointestinal disease caused by the ingestion of food contaminated by pathogens or their toxins. (905)

food web   A network of many interlinked food chains, encompassing primary producers, consumers, decomposers, and detritivores. (566)

F$_1$ particle   Particle on the inner mitochondrial membrane, which is the site of ATP synthesis by oxidative phosphorylation. (84, 182)

F′ plasmid   An F plasmid that carries some bacterial genes and transmits them to recipient cells when the F′ cell carries out conjugation; the transfer of bacterial genes in this way is often called sexduction. (297)

fragmentation (frag″men-ta′shun)   A type of asexual reproduction in which a thallus breaks into two or more parts, each of which forms a new thallus. (556)

frameshift mutations   Mutations arising from the loss or gain of a base or DNA segment, leading to a change in the codon reading frame and thus a change in the amino acids incorporated into protein. (245)

free energy change   The total energy change in a system that is available to do useful work as the system goes from its initial state to its final state at constant temperature and pressure. (152)

French polio   *See* Guillain-Barré syndrome. (849)

fruiting body A specialized structure that holds sexually or asexually produced spores; found in fungi and in some bacteria. (498, 548)

frustule (frus'tūl) A silicified cell wall in the diatoms. (559)

fungicide (fun'jĭ-sīd) An agent that kills fungi. (135)

fungistatic (fun"jĭ-stat'ik) Inhibiting the growth and reproduction of fungi. (135)

fungus (fung'gus; pl., **fungi**) Achlorophyllous, heterotrophic, spore-bearing eucaryotes with absorptive nutrition; usually, they have a walled thallus. (538)

F value The time in minutes at a specific temperature (usually 250°F) needed to kill a population of cells or spores. (137)

G

gametangium (gam-ĕ-tan'je-um; pl., **gametangia**) A structure that contains gametes or in which gametes are formed. (543)

gamma-proteobacteria One of the five subgroups of proteobacteria, each with distinctive 16S rRNA sequences. This is the largest subgroup and is very diverse physiologically; many important genera are facultatively anaerobic chemoorganotrophs. (485)

gas gangrene (gang'grēn) A type of gangrene that arises from dirty, lacerated wounds infected by anaerobic bacteria, especially species of *Clostridium*. As the bacteria grow, they release toxins and ferment carbohydrates to produce carbon dioxide and hydrogen gas. (891)

gastritis (gas-tri'tis) Inflammation of the stomach. (895)

gastroenteritis (gas"tro-en-ter-i'tis) An acute inflammation of the lining of the stomach and intestines, characterized by anorexia, nausea, diarrhea, abdominal pain, and weakness. It has various causes including food poisoning due to such organisms as *E. coli*, *S. aureus*, *Campylobacter* (camp lobacteriosis), and *Salmonella* species; consumption of irritating food or drink; or psychological factors such as anger, stress, and fear. Also called enterogastritis. (907)

gas vacuole A gas-filled vacuole found in cyanobacteria and some other aquatic bacteria that provides flotation. It is composed of gas vesicles, which are made of protein. (48)

gene (jēn) A DNA segment or sequence that codes for a polypeptide, rRNA, or tRNA. (235)

gene gun A device that uses high-pressure gas or another propellant to shoot a spray of DNA-coated microprojectiles into cells and transform them. Sometimes it is called a biolistic device. (327)

generalized transduction The transfer of any part of a bacterial genome when the DNA fragment is packaged within a phage capsid by mistake. (301)

general recombination Recombination involving a reciprocal exchange of a pair of homologous DNA sequences; it can occur any place on the chromosome. (286)

general secretion pathway (GSP) *See* Sec-dependent pathway. (59)

generation time The time required for a microbial population to double in number. (112)

genetic engineering The deliberate modification of an organism's genetic information by directly changing its nucleic acid genome. (312)

genital herpes (her'pēz) A sexually transmitted disease caused by the herpes simplex virus type 2. (862)

genital ulcer disease *See* chancroid. (891)

genome (je'nōm) The full set of genes present in a cell or virus; all the genetic material in an organism; a haploid set of genes in a cell. (222)

genomic library (je-nom'ik) The collection of clones that contains fragments which represent the complete genome of an organism. (320)

genomic reduction The decrease in genomic information that occurs over evolutionary time as an organism or organelle becomes increasingly dependent on another cell or a host organism. The relationship can be intracellular or extracellular. (591)

genomics The study of the molecular organization of genomes, their information content, and the gene products they encode. (336)

genus (je'nəs) A well-defined group of one or more species that is clearly separate from other genera. (413)

German measles *See* rubella. (851)

germicide (jer'mĭ-sīd) An agent that kills pathogens and many nonpathogens but not necessarily bacterial endospores. (135)

germination (jer"mĭ-na'shun) The stage following bacterial endospore activation in which the endospore breaks its dormant state. Germination is followed by outgrowth. (71)

Ghon complex (gon) The initial focus of parenchymal infection in primary pulmonary tuberculosis. (884)

giardiasis (je"ar-di'ah-sis) A common intestinal disease caused by the parasitic protozoan *Giardia lamblia*. (928)

gingivitis (jin-jĭ-vi'tis) Inflammation of the gingival tissue. (914)

gingivostomatitis (jin"jĭ-vo-sto"mə-ti'tis) Inflammation of the gingiva and other oral mucous membranes. (861)

gliding motility A type of motility in which a microbial cell glides along when in contact with a solid surface. (66, 470)

global regulatory systems Regulatory systems that simultaneously affect many genes and pathways. (276)

glomerulonephritis (glo-mer"u-lo-nĕ-fri'tis) An inflammatory disease of the renal glomeruli. (881)

glucans Polysaccharides composed of glucose units held together by glycosidic linkages. Some types of glucans have $\alpha(1{\rightarrow}3)$ and $\alpha(1{\rightarrow}6)$ linkages and bind bacterial cells together on teeth forming a plaque ecosystem. (912)

gluconeogenesis (gloo"ko-ne"o-jen'e-sis) The synthesis of glucose from noncarbohydrate precursors such as lactate and amino acids. (203)

glycocalyx (gli"ko-kal'iks) A network of polysaccharides extending from the surface of bacteria and other cells. (61)

glycogen (gli'ko-jen) A highly branched polysaccharide containing glucose, which is used to store carbon and energy. (47, A-7)

glycolysis (gli-kol'ĭ-sis) The anaerobic conversion of glucose to lactic acid by use of the Embden-Meyerhof pathway. (171)

glycolytic pathway (gli"ko-lit'ik) *See* Embden-Meyerhof pathway. (171, A-11)

glyoxylate cycle (gli-ok'sĭ-lat) A modified tricarboxylic acid cycle in which the decarboxylation reactions are bypassed by the enzymes isocitrate lyase and malate synthase; it is used to convert acetyl-CoA to succinate and other metabolites. (210)

gnotobiotic (no"to-bi-ot'ik) Animals that are germfree (microorganism free) or live in association with one or more known microorganisms. (674)

Golgi apparatus (gol'je) A membranous eucaryotic organelle composed of stacks of flattened sacs (cisternae), which is involved in packaging and modifying materials for secretion and many other processes. (78)

gonococci (gon'o-kok'si) Bacteria of the species *Neisseria gonorrhoeae*—the organism causing gonorrhea. (892)

gonorrhea (gon"o-re'ah) An acute infectious sexually transmitted disease of the mucous membranes of the genitourinary tract, eye, rectum, and throat. It is caused by *Neisseria gonorrhoeae*. (892)

Gram stain A differential staining procedure that divides bacteria into gram-positive and gram-negative groups based on their ability to retain crystal violet when decolorized with an organic solvent such as ethanol. (27)

grana (gra'nah) A stack of thylakoids in the chloroplast stroma. (85)

granuloma (gran"u-lo'mə) Term applied to nodular inflammatory lesions containing phagocytic cells. (691)

greenhouse gases Gases released from the Earth's surface through chemical and biological processes that interact with the chemicals in the stratosphere to decrease the release of radiation from the Earth. It is believed that this leads to global warming. (665)

griseofulvin (gris"e-o-ful'vin) An antibiotic from *Penicillium griseofulvum* given orally to treat chronic dermatophytic infections of skin and nails. (795)

group translocation A transport process in which a molecule is moved across a membrane by carrier proteins while being chemically altered at the same time. (100)

growth An increase in cellular constituents. (110)

growth factors Organic compounds that must be supplied in the diet for growth because they are essential cell components or precursors of such components and cannot be synthesized. (97)

guanine (gwan'in) A purine derivative, 2-amino-6-oxypurine, found in nucleosides, nucleotides, and nucleic acids. (211)

Guillain-Barré syndrome (ge-yan′bar-ra′) A relatively rare disease affecting the peripheral nervous system, especially the spinal nerves, but also the cranial nerves. The cause is unknown, but it most often occurs after an influenza infection or flu vaccination. Also called French Polio. (849)

gumma (gum′ah) A soft, gummy tumor occurring in tertiary syphilis. (901)

gut-associated lymphoid tissue (GALT) The defensive lymphoid tissue present in the intestines. *See* Peyer's patches. (686)

H

H-2 complex Term for the MHC in the mouse. (722)

halobacteria or **extreme halophiles** A group of archaea that have an absolute dependence on high NaCl concentrations for growth and will not survive at a concentration below about 1.5 M NaCl. (447)

halophile (hal′o-fīl) A microorganism that requires high levels of sodium chloride for growth. (120)

Hansen's disease *See* leprosy. (893)

hantavirus pulmonary syndrome The disease in humans caused by the pulmonary syndrome hantavirus. Deer mice shed the virus in their feces, humans inhale the virus and first develop ordinary flulike aches and pains. Within a few days the hantavirus causes lung damage and capillary leakage. After about a week the infected person enters a crisis phase and may die. (853)

hapten (hap′ten) A molecule not immunogenic by itself that, when coupled to a macromolecular carrier, can elicit antibodies directed against itself. (709)

harborage transmission The mode of transmission in which an infectious organism does not undergo morphological or physiological changes within the vector. (831)

hay fever Allergic rhinitis; a type of atopic allergy involving the upper respiratory tract. (745)

health (helth) A state of optimal physical, mental, and social well-being, and not merely the absence of disease and infirmity. (822)

healthy carrier An individual who harbors a pathogen, but is not ill. (828)

heat-shock proteins Proteins produced when cells are exposed to high temperatures or other stressful conditions. They protect the cells from damage and often aid in the proper folding of proteins. (268)

helical (hel′ĭ-kal) In virology this refers to a virus with a helical capsid surrounding its nucleic acid. (359)

helicases Enzymes that use ATP energy to unwind DNA ahead of the replication fork. (230)

hemadsorption (hem″ad-sorp′shun) The adherence of red blood cells to the surface of something, such as another cell or a virus. (805)

hemagglutination (hem″ah-gloo″tĭ-na′shun) The agglutination of red blood cells by antibodies. (732)

hemagglutinin (hem″ah-gloo′tĭ-nin) The antibody responsible for a hemagglutination reaction. (732)

hemoflagellate (he″mo-flaj′ĕ-lāt) A flagellated protozoan parasite that is found in the bloodstream. (931)

hemolysin (he-mol′ĭ-sin) A substance that causes hemolysis (the lysis of red blood cells). At least some hemolysins are enzymes that destroy the phospholipids in erythrocyte plasma membranes. (772)

hemolysis (he-mol′ĭ-sis) The disruption of red blood cells and release of their hemoglobin. There are several types of hemolysis when bacteria such as streptococci and staphylococci grow on blood agar. In α-hemolysis, a narrow greenish zone of incomplete hemolysis forms around the colony. A clear zone of complete hemolysis without any obvious color change is formed during β-hemolysis. (516, 772)

hemolytic uremic syndrome A kidney disease characterized by blood in the urine and often by kidney failure. It is caused by enterohemorrhagic strains of *Escherichia coli* O157:H7 that produce a Shiga-like toxin, which attacks the kidneys. (910)

hemorrhagic fever A fever usually caused by a specific virus that may lead to hemorrhage, shock, and sometimes death. (853)

hepatitis (hep″ah-ti′tis) Any infection that results in inflammation of the liver. Also refers to liver inflammation as such. (866)

hepatitis A (formerly infectious hepatitis) A type of hepatitis that is transmitted by fecal-oral contamination; it primarily affects children and young adults, especially in environments where there is poor sanitation and overcrowding. It is caused by the hepatitis A virus, a single-stranded RNA virus. (868)

hepatitis B (formerly serum hepatitis) This form of hepatitis is caused by a double-stranded DNA virus (HBV) formerly called the "Dane particle." The virus is transmitted by body fluids. (866)

hepatitis C About 90% of all cases of viral hepatitis can be traced to either HAV or HBV. The remaining 10% is believed to be caused by one and possibly several other types of viruses. At least one of these is hepatitis C (formerly non-A, non-B). (867)

hepatitis D (formerly delta hepatitis) The liver diseases caused by the hepatitis D virus in those individuals already infected with the hepatitis B virus. (867)

hepatitis E (formerly enteric-transmitted NANB hepatitis) The liver disease caused by the hepatitis E virus. Usually, a subclinical, acute infection results; however, there is a high mortality in women in their last trimester of pregnancy. (869)

herd immunity The resistance of a population to infection and spread of an infectious agent due to the immunity of a high percentage of the population. (826)

herpes labialis *See* cold sore. (860)

herpetic keratitis (her-pet′ik ker″ah-ti′tis) An inflammation of the cornea and conjunctiva of the eye resulting from a herpes simplex virus infection. (861)

heterocysts Specialized cells of cyanobacteria that are the sites of nitrogen fixation. (459)

heteroduplex DNA A double-stranded stretch of DNA formed by two slightly different strands that are not completely complementary. (286)

heterogeneous nuclear RNA (hnRNA) The RNA transcript of DNA made by RNA polymerase II; it is then processed to form mRNA. (257)

heterolactic fermenters (het″er-o-lak′tik) Microorganisms that ferment sugars to form lactate, and also other products such as ethanol and CO₂. (176)

heterotroph (het′er-o-trōf″) An organism that uses reduced, preformed organic molecules as its principal carbon source. (94)

heterotrophic nitrification Nitrification carried out by chemoheterotrophic microorganisms. (598)

hexon or **hexamer** A virus capsomer composed of six protomers. (360)

hexose monophosphate pathway (hek′sōs mon″o-fos′fāt) *See* pentose phosphate pathway. (171)

Hfr strain A bacterial strain that donates its genes with high frequency to a recipient cell during conjugation because the F factor is integrated into the bacterial chromosome. (297)

high-energy molecule A molecule whose hydrolysis under standard conditions makes available a large amount of free energy (the standard free energy change is more negative than about −7 kcal/mole); a high-energy molecule readily decomposes and transfers groups such as phosphate to acceptors. (153)

high oxygen diffusion environment A microbial environment in close contact with air and through which oxygen can move at a rapid rate (in comparison with the slow diffusion rate of oxygen through water). (617)

histone (his′tōn) A small basic protein with large amounts of lysine and arginine that is associated with eucaryotic DNA in chromatin. (227)

histoplasmosis (his″to-plaz-mo′sis) A systemic fungal infection caused by *Histoplasma capsulatum var capsulatum*. (923)

hives (hīvz) An eruption of the skin. (745)

holdfast A structure produced by some bacteria and algae that attaches them to a solid object. (478)

holoenzyme A complete enzyme consisting of the apoenzyme plus a cofactor. (156)

holozoic nutrition (hol′o-zo′ik) In this type of nutrition, nutrients (such as bacteria) are acquired by phagocytosis and the subsequent formation of a food vacuole or phagosome. (567)

homolactic fermenters (ho″mo-lak′tik) Organisms that ferment sugars almost completely to lactic acid. (176)

horizontal gene transfer The process in which genes are transferred from one mature, independent organism to another. (286)

hormogonia Small motile fragments produced by fragmentation of filamentous cyanobacteria; used for asexual reproduction and dispersal. (459)

host (hōst) The body of an organism that harbors another organism. It can be viewed as a microenvironment that shelters and supports the growth and multiplication of a parasitic organism. (762)

host restriction The degradation of foreign genetic material by nucleases after the genetic material enters a host cell. (288)

human herpesvirus 6 (HHV-6, type A and B) (hər′pēz)   HHV-6 was initially called the human B-lymphotropic virus. It was later shown to have a marked tropism for CD4⁺ T cells and was renamed HHV-6. HHV-6 causes exanthem subitum (roseola infantum or sixth disease) in infants and has been suspected of involvement in many conditions, including opportunistic infections in immunocompromised patients, hepatitis, lymphoproliferative diseases, synergistic interactions with HIV, lymphadenitis, and chronic fatigue syndrome. (863)

human immunodeficiency virus (HIV)   A lentivirus of the family *Retroviridae* that is associated with the onset of AIDS. (855)

human leukocyte antigen complex (HLA)   An antigen on the surface of cells of human tissues and organs that is recognized by the immune system cells and therefore is important in graft rejection and regulation of the immune response. This is the same as MHC class II. (722)

humoral (antibody-mediated) immunity (hu′moral)   The type of immunity that results from the presence of soluble antibodies in blood and lymph; also known as antibody-mediated immunity. (706)

hybridoma (hi″brĭ-do′mah)   A fast-growing cell line produced by fusing a cancer cell (myeloma) to another cell, such as an antibody-producing cell. (720)

hydrophilic (hi′dro-fil′ik)   A polar substance that has a strong affinity for water (or is readily soluble in water). (44)

hydrophobic (hi′dro-fo′bik)   A nonpolar substance lacking affinity for water (or which is not readily soluble in water). (44)

hyperendemic disease (hi″per-en-dem′ik)   A disease that has a gradual increase in occurrence beyond the endemic level, but not at the epidemic level, in a given population; also may refer to a disease that is equally endemic in all age groups. (822)

hypermutation   A rapid production of multiple mutations in a gene or genes through the activation of special mutator genes. The process may be deliberately used to maximize the possibility of creating desirable mutants. (240)

hypersensitivity (hi″per-sen′si-tiv″i-te)   A condition of increased immune sensitivity in which the body reacts to an antigen with an exaggerated immune response that usually harms the individual. Also termed an allergy. (744)

hyperthermophile (hi″per-ther′mo-fĭl)   A bacterium that has its growth optimum between 80°C and about 113°C. Hyperthermophiles usually do not grow well below 55°C. (124, 607)

hypha (hi′fah; pl., **hyphae**)   The unit of structure of most fungi and some bacteria; a tubular filament. (522, 539)

hypoferremia (hi″po-fĕ-re′me-ah)   Deficiency of iron in the blood. (698)

hypotheca (hi-po-theca)   The smaller half of a diatom frustule. (559)

hypoxic (hi pok′sik)   Having a low oxygen level. (617)

I

icosahedral   In virology this term refers to a virus with an icosahedral capsid, which has the shape of a regular polyhedron having 20 equilateral triangular faces and 12 corners. (359)

identification (i-den″tĭ-fĭ-ka′shun)   The process of determining that a particular isolate or organism belongs to a recognized taxon. (410)

idiotype (id′e-o-tīp′)   A set of one or more unique epitopes in the variable region of an immunoglobulin that distinguishes it from immunoglobulins produced by different plasma cells. (712)

IgA   Immunoglobulin A; the class of immunoglobulins that is present in dimeric form in many body secretions (e.g., saliva, tears, and bronchial and intestinal secretions) and protects body surfaces. IgA also is present in serum. (714)

IgD   Immunoglobulin D; the class of immunoglobulins found on the surface of many B lymphocytes; thought to serve as an antigen receptor in the stimulation of antibody synthesis. (714)

IgE   Immunoglobulin E; the immunoglobulin class that binds to mast cells and basophils, and is responsible for type I or anaphylactic hypersensitivity reactions such as hay fever and asthma. IgE is also involved in resistance to helminth parasites. (714)

IgG   Immunoglobulin G; the predominant immunoglobulin class in serum. Has functions such as neutralizing toxins, opsonizing bacteria, activating complement, and crossing the placenta to protect the fetus and neonate. (713)

IgM   Immunoglobulin M; the class of serum antibody first produced during an infection. It is a large, pentameric molecule that is active in agglutinating pathogens and activating complement. The monomeric form is present on the surface of some B lymphocytes. (713)

immobilization (im-mo″bil-i-za′shun)   The incorporation of a simple, soluble substance into the body of an organism, making it unavailable for use by other organisms. (594)

immune complex (ĭ-mūn′kom′pleks)   The product of an antigen-antibody reaction, which may also contain components of the complement system. (732)

immune surveillance (ĭ-mūn′sur-vāl′ans)   The still somewhat hypothetical process in which lymphocytes such as natural killer (NK) cells recognize and destroy tumor cells; other cells with abnormal surface antigens (e.g., virus-infected cells) also may be destroyed. (736)

immune system   The defensive system in a host consisting of the nonspecific (innate) and specific (adaptive) immune responses. It is composed of widely distributed cells, tissues, and organs that recognize foreign substances and microorganisms and acts to neutralize or destroy them. (680)

immunity (ĭ-mu′nĭ-te)   Refers to the overall general ability of a host to resist a particular disease; the condition of being immune. (680)

immunoblotting   The electrophoretic transfer of proteins from polyacrylamide gels to nitrocellulose sheets to demonstrate the presence of specific proteins through reaction with labeled antibodies. (754)

immunodeficiency (im″u-no-dĕ-fish′en-se; pl., **immunodeficiencies**)   The inability to produce a normal complement of antibodies or immunologically sensitized T cells in response to specific antigens. (750)

immunodiffusion   A technique involving the diffusion of antigen and/or antibody within a semisolid gel to produce a precipitin reaction where they meet in proper proportions. Often both the antibody and antigen diffuse through the gel; sometimes an antigen diffuses through a gel containing antibody. (754)

immunoelectrophoresis (ĭ-mu″no-e-lek″tro-fo-re′sis; pl., **immunoelectrophoreses**)   The electrophoretic separation of protein antigens followed by diffusion and precipitation in gels using antibodies against the separated proteins. (755)

immunofluorescence (im″u-no-floo″o-res′ens)   A technique used to identify particular antigens microscopically in cells or tissues by the binding of a fluorescent antibody conjugate. (755)

immunoglobulin (Ig) (im″u-no-glob′u-lin)   *See* antibody. (710)

immunology (im″u-nol′o-je)   The branch of science that deals with the immune system and attempts to understand the many phenomena that are responsible for both acquired and innate immunity. It also includes the use of antibody-antigen reactions in other laboratory work (serology and immunochemistry). (680)

immunopathology (im″u-no-pə-thol′o-je)   The study of diseases or conditions resulting from immune reactions. (764)

immunoprecipitation (im″u-no-pre-sip″i-ta′shun)   A reaction involving soluble antigens reacting with antibodies to form a large aggregate that precipitates out of solution. (756)

immunotoxin (im′u-no-tok″sin)   A monoclonal antibody that has been attached to a specific toxin or toxic agent (antibody + toxin = immunotoxin) and can kill specific target cells. (721)

impetigo (im″pə-ti′go)   This superficial cutaneous disease, most commonly seen in children, is characterized by crusty lesions, usually located on the face; the lesions typically have vesicles surrounded by a red border. It is the most frequently diagnosed skin infection caused by *S. pyogenes* (impetigo can also be caused by *S. aureus*). (879)

inclusion bodies   Granules of organic or inorganic material lying in the cytoplasmic matrix of bacteria. (47, 399)

inclusion conjunctivitis (in-klu′zhun kon-junk″tĭ-vi′tis)   An infectious disease that occurs worldwide. It is caused by *Chlamydia trachomatis* that infects the eye and causes inflammation and the occurrence of large inclusion bodies. (893)

incubation period   The period after pathogen entry into a host and before signs and symptoms appear. (824)

incubatory carrier   An individual who is incubating a pathogen but is not yet ill. (828)

index case The first disease case in an epidemic within a given population. (823)

indicator organism An organism whose presence indicates the condition of a substance or environment, for example, the potential presence of pathogens. Coliforms are used as indicators of fecal pollution. (633)

inducer (in-dūs′er) A small molecule that stimulates the synthesis of an inducible enzyme. (270)

inducible enzyme An enzyme whose level rises in the presence of a small molecule that stimulates its synthesis. (270)

industrial ecology The study of the ecology of industrial societies with a major focus on material cycling, energy flow, and the ecological impacts of such societies. (990)

infantile paralysis (in′fan-til pah-ral′i-sis) *See* poliomyelitis. (869)

infection (in-fek′shun) The invasion of a host by a microorganism with subsequent establishment and multiplication of the agent. An infection may or may not lead to overt disease. (763)

infection thread A tubular structure formed during the infection of a root by nitrogen-fixing bacteria. The bacteria enter the root by way of the infection thread and stimulate the formation of the root nodule. (655)

infectious disease Any change from a state of health in which part or all of the host's body cannot carry on its normal functions because of the presence of an infectious agent or its products. (763)

infectious disease cycle (chain of infection) The chain or cycle of events that describes how an infectious organism grows, reproduces, and is disseminated. (827)

infectious dose 50 (ID$_{50}$) Refers to the dose or number of organisms that will infect 50% of an experimental group of hosts within a specified time period. (358, 764)

infectious mononucleosis (mono) (mon″o-nu″kle-o′sis) An acute, self-limited infectious disease of the lymphatic system caused by the Epstein-Barr virus and characterized by fever, sore throat, lymph node and spleen swelling, and the proliferation of monocytes and abnormal lymphocytes. (864)

infectivity (in″fek-tiv′i-te) Infectiousness; the state or quality of being infectious or communicable. (764)

inflammation (in″flah-ma′shun) A localized protective response to tissue injury or destruction. Acute inflammation is characterized by pain, heat, swelling, and redness in the injured area. (688)

influenza or flu (in″flu-en′zah) An acute viral infection of the respiratory tract, occurring in isolated cases, epidemics, and pandemics. Influenza is caused by three strains of influenza virus, labeled types A, B, and C, based on capsid antigens. (847)

ingoldian fungi Aquatic hyphomycetes that often have a characteristic tetraradiate hyphal development form and which sporulate under water. Discovered by the British mycologist, C. T. Ingold, in the 1940s. (622)

initial body *See* reticulate body (RB). (465)

innate or **natural immunity** *See* nonspecific resistance. (680)

insertion sequence (in-ser′shun se′kwens) A simple transposon that contains genes only for those enzymes, such as the transposase, that are required for transposition. (292)

integration The incorporation of one DNA segment into a second DNA molecule to form a new hybrid DNA. Integration occurs during such processes as genetic recombination, episome incorporation into host DNA, and prophage insertion into the bacterial chromosome. (384)

integrins (in′tə-grin) A large family of α/β heterodimers. Integrins are cellular adhesion receptors that mediate cell-cell and cell-substratum interactions. Integrins usually recognize linear amino acid sequences on protein ligands. (688)

integron A genetic element with an attachment site for site-specific recombination and an integrase gene. It can capture genes and gene cassettes. (794)

inteins Internal intervening sequences of precursor self-splicing proteins that separate exteins and are removed during formation of the final protein. (270)

intercalating agents Molecules that can be inserted between the stacked bases of a DNA double helix, thereby distorting the DNA and inducing insertion and deletion mutations. (241)

interdigitating dendritic cell Special dendritic cells in the lymph nodes that function as potent antigen-presenting cells and develop from Langerhans cells. (685)

interferon (IFN) (in″tər-fēr′on) A glycoprotein that has nonspecific antiviral activity by stimulating cells to produce antiviral proteins, which inhibit the synthesis of viral RNA and proteins. Interferons also regulate the growth, differentiation, and/or function of a variety of immune system cells. Their production may be stimulated by virus infections, intracellular pathogens (chlamydiae and rickettsias), protozoan parasites, endotoxins, and other agents. (698, 796)

interleukin (in″tər-loo′kin) A glycoprotein produced by macrophages and T cells that regulates growth and differentiation, particularly of lymphocytes. Interleukins promote cellular and humoral immune responses. (697)

intermediate filaments Small protein filaments, about 8 to 10 nm in diameter, in the cytoplasmic matrix of eucaryotic cells that are important in cell structure. (78)

intermediate host The host that serves as a temporary but essential environment for development of a parasite and completion of its life cycle. (762)

interspecies hydrogen transfer The linkage of hydrogen production from organic matter by anaerobic heterotrophic microorganisms to the use of hydrogen by other anaerobes in the reduction of carbon dioxide to methane. This avoids possible hydrogen toxicity. (588)

intertriginous candidiasis A skin infection caused by *Candida* species. Involves those areas of the body,

usually opposed skin surfaces, that are warm and moist (axillae, groin, skin folds). (925)

intoxication (in-tok″si-ka′shun) A disease that results from the entrance of a specific toxin into the body of a host. The toxin can induce the disease in the absence of the toxin-producing organism. (769)

intraepidermal lymphocytes T cells found in the epidermis of the skin that express the γδ T-cell receptor. (685)

intranuclear inclusion body (in″trə-noo′kle-ər) A structure found within cells infected with the cytomegalovirus. (862)

intrinsic factors Food-related factors such as moisture, pH, and available nutrients that influence microbial growth. (938)

intron (in′tron) A noncoding intervening sequence in a split or interrupted gene, which codes for RNA that is missing from the final RNA product. (257)

invasiveness (in-va′siv-nes) The ability of a microorganism to enter a host, grow and reproduce within the host, and spread throughout its body. (764)

ionizing radiation Radiation of very short wavelength or high energy that causes atoms to lose electrons or ionize. (127, 141)

isotype (i′so-tīp) A variant form of an immunoglobulin (e.g., an immunoglobulin class, subclass, or type) that occurs in every normal individual of a particular species. Usually the characteristic antigenic determinant is in the constant region of H and L chains. (712)

J

Jaccard coefficient (S_J) An association coefficient used in numerical taxonomy; it is the proportion of characters that match, excluding those that both organisms lack. (414)

J chain A polypeptide present in polymeric IgM and IgA that links the subunits together. (713)

jock itch *See* tinea cruris. (920)

K

kelp (kelp) A common name for the larger members of the order *Laminariales* of the brown algae. (561)

keratitis (ker″ah-ti′tis) Inflammation of the cornea of the eye. (928)

kinetoplast (ki-ne′to-plast) A special structure in the mitochondrion of kinetoplastid protozoa. It contains the mitochondrial DNA. (570)

Kirby-Bauer method A disk diffusion test to determine the susceptibility of a microorganism to chemotherapeutic agents. (783)

Koch's postulates (koks pos′tu-lāts) A set of rules for proving that a microorganism causes a particular disease. (8)

Koplik's spots (kop′liks) Lesions of the oral cavity caused by the measles (rubeola) virus that are characterized by a bluish white speck in the center of each. (849)

Korean hemorrhagic fever   An acute infection caused by a virus that produces varying degrees of hemorrhage, shock, and sometimes death. (853)

Krebs cycle   *See* tricarboxylic acid (TCA) cycle. (178)

L

lactic acid fermentation (lak'tik)   A fermentation that produces lactic acid as the sole or primary product. (176, A-16)

lager   Pertaining to the process of aging beers to allow flavor development. (956)

lag phase   A period following the introduction of microorganisms into fresh culture medium when there is no increase in cell numbers or mass during batch culture. (110)

laminarin (lam″i-na'rin)   One of the principal storage products of the golden-brown algae; a polymer of glucose. (561)

Lancefield system (group) (lans'feld)   One of the serologically distinguishable groups (as group A, group B) into which streptococci can be divided. (516, 756)

Langerhans cell   Cell found in the skin that internalizes antigen and moves in the lymph to lymph nodes where it differentiates into a dendritic cell. (685)

late mRNA   Messenger RNA produced later in a virus infection, which codes for proteins needed in capsid construction and virus release. (376)

latent period (la'tent)   The initial phase in the one-step growth experiment in which no phages are released. (373)

latent virus infections   Virus infections in which the virus quits reproducing and remains dormant for a period before becoming active again. (399)

leader sequence   A nontranslated sequence at the 5′ end of mRNA that lies between the operator and the initiation codon; it aids in the initiation and regulation of transcription. (237, 254)

lectin complement pathway (lek'tin)   The lectin pathway for complement activation is triggered by the binding of a serum lectin (mannan-binding lectin; MBL) to mannose-containing proteins or to carbohydrates on viruses or bacteria. (693)

legionellosis (le″jə-nel-o'sis)   *See* Legionnaires' disease. (877)

Legionnaires' disease (legionellosis)   A pulmonary form of legionellosis, resulting from infection with *Legionella pneumophila*. (877)

leishmanias (lēsh″ma'ne-ăs)   Zooflagellates, members of the genus *Leishmania,* that cause the disease leishmaniasis. (931)

leishmaniasis (lēsh″mah-ni'ah-sis)   The disease caused by the protozoa called leishmanias. (931)

lepromatous (progressive) leprosy (lep-ro'mah-tus lep′ro-se)   A relentless, progressive form of leprosy in which large numbers of *Mycobacterium leprae* develop in skin cells, killing the skin cells and resulting

in the loss of features. Disfiguring nodules form all over the body. (894)

leprosy (lep′ro-se) or **Hansen's disease**   A severe disfiguring skin disease caused by *Mycobacterium leprae.* (893)

lethal dose 50 (LD$_{50}$)   Refers to the dose or number of organisms that will kill 50% of an experimental group of hosts within a specified time period. (358, 764)

leukemia (loo-ke'me-ah)   A progressive, malignant disease of blood-forming organs, marked by distorted proliferation and development of leukocytes and their precursors in the blood and bone marrow. Certain leukemias are caused by viruses (HTLV-1, HTLV-2). (864)

leukocidin (loo″ko-si'din)   A microbial toxin that can damage or kill leukocytes. (772)

leukocyte (loo′ko-sīt)   Any colorless white blood cell. Can be classified into granular and agranular lymphocytes. (681)

lichen (li'ken)   An organism composed of a fungus and either green algae or cyanobacteria in a symbiotic association. (591)

Liebig's law of the minimum (le'bigz)   Living organisms and populations will grow until lack of a resource begins to limit further growth. (128)

lipopolysaccharide (LPSs) (lip″o-pol″e-sak′ah-rīd)   A molecule containing both lipid and polysaccharide, which is important in the outer membrane of the gram-negative cell wall. (56)

listeriosis (lis-ter″e-o'sis)   A sporadic disease of animals and humans, particularly those who are immunocompromised or pregnant, caused by the bacterium *Listeria monocytogenes.* (908)

lithotroph (lith′o-trōf)   An organism that uses reduced inorganic compounds as its electron source. (95)

log phase   *See* exponential phase. (110)

lophotrichous (lo-fot′rĭ-kus)   A cell with a cluster of flagella at one or both ends. (63)

low oxygen diffusion environment   An aquatic environment in which microorganisms are surrounded by deep water layers that limit oxygen diffusion to the cell surface. In contrast, microorganisms in thin water films have good oxygen transfer from air to the cell surface. (616)

LPS-binding protein   A special plasma protein that binds bacterial lipopolysaccharides and then attaches to receptors on monocytes, macrophages, and other cells. This triggers the release of IL-1 and other cytokines that stimulate the development of fever and additional endotoxin effects. (775)

Lyme disease (LD, Lyme borreliosis) (līm)   A tick-borne disease caused by the spirochete *Borrella burgdorferi.* (886)

lymph node   A small secondary lymphoid organ that contains lymphocytes, macrophages, and dendritic cells. It serves as a site for (1) filtration and removal of foreign antigens and (2) the activation and proliferation of lymphocytes. (684)

lymphocyte (lim′fo-sīt)   A nonphagocytic, mononuclear leukocyte (white blood cell) that is an im-

munologically competent cell, or its precursor. Lymphocytes are present in the blood, lymph, and lymphoid tissues. *See* B cell and T cell. (681)

lymphogranuloma venereum (LGV) (lim″fo-gran″u-lo'mah)   A sexually transmitted disease caused by *Chlamydia trachomatis* serotypes L$_1$–L$_3$, which affect the lymph organs in the genital area. (894)

lymphokine (lim′fo-kin)   A biologically active glycoprotein (e.g., IL-1) secreted by activated lymphocytes, especially sensitized T cells. It acts as an intercellular mediator of the immune response and transmits growth, differentiation, and behavioral signals. (697)

lysis (li'sis)   The rupture or physical disintegration of a cell. (59)

lysogenic (li-so-jen'ik)   *See* lysogens. (301, 380)

lysogens (li'so-jens)   Bacteria that are carrying a viral prophage and can produce bacteriophages under the proper conditions. (301, 380)

lysogeny (li-soj'e-ne)   The state in which a phage genome remains within the bacterial cell after infection and reproduces along with it rather than taking control of the host and destroying it. (301, 380)

lysosome (li'so-sōm)   A spherical membranous eucaryotic organelle that contains hydrolytic enzymes and is responsible for the intracellular digestion of substances. (79)

lysozyme (li'sō-zīm)   An enzyme that degrades peptidoglycan by hydrolyzing the β(1 → 4) bond that joins *N*-acetylmuramic acid and *N*-acetylglucosamine. (59, 685)

lytic cycle (lit'ik)   A virus life cycle that results in the lysis of the host cell. (373)

M

macrolide antibiotic (mak′ro-līd)   An antibiotic containing a macrolide ring, a large lactone ring with multiple keto and hydroxyl groups, linked to one or more sugars. (791)

macromolecule (mak″ro-mol′ĕ-kūl)   A large molecule that is a polymer of smaller units joined together. (200)

macromolecule vaccine   A vaccine made of specific, purified macromolecules derived from pathogenic microorganisms. (743)

macronucleus (mak″ro-nu'kle-us)   The larger of the two nuclei in ciliate protozoa. It is normally polyploid and directs the routine activities of the cell. (567)

macrophage (mak′ro-făj)   The name for a large mononuclear phagocytic cell, present in blood, lymph, and other tissues. Macrophages are derived from monocytes. They phagocytose and destroy pathogens; some macrophages also activate B cells and T cells. (681)

maduromycosis (mah-du′ro-mi-ko'sis)   A subcutaneous fungal infection caused by *Madurella mycetoma;* also termed an eumycotic mycetoma. (921)

madurose   The sugar derivative 3-O-methyl-D-galactose, which is characteristic of several actinomycete genera that are collectively called maduromycetes. (533)

magnetosomes Magnetite particles in magnetotactic bacteria that are tiny magnets and allow the bacteria to orient themselves in magnetic fields. (50)

maintenance energy The energy a cell requires simply to maintain itself or remain alive and functioning properly. It does not include the energy needed for either growth or reproduction. (118)

major histocompatibility complex (MHC) A large set of cell surface molecules in each individual, encoded by a family of genes, that serves as a unique biochemical marker of individual identity. It can trigger T-cell responses that may lead to rejection of transplanted tissues and organs. MHC molecules are also involved in the regulation of the immune response and the interactions between immune cells. (722)

malaria (mah-la're-ah) A serious infectious illness caused by the parasitic protozoan *Plasmodium*. Malaria is characterized by bouts of high chills and fever that occur at regular intervals. (929)

malt (mawlt) Grain soaked in water to soften it, induce germination, and activate its enzymes. The malt is then used in brewing and distilling. (956)

Marburg viral hemorrhagic fever An acute infection caused by a virus that produces varying degrees of hemorrhage, shock, and sometimes death. (853)

mash The soluble materials released from germinated grains and prepared as a microbial growth medium. (956)

mashing The process in which cereals are mixed with water and incubated in order to degrade their complex carbohydrates (e.g., starch) to more readily usable forms such as simple sugars. (955)

mast cell A bone marrow–derived cell present in a variety of tissues that resembles peripheral blood-borne basophils and contains an Fc receptor for IgE. It undergoes IgE-mediated degranulation. (683)

M cell Specialized cell of the intestinal mucosa and other sites, such as the urogenital tract, that delivers the antigen from the apical face of the cell to lymphocytes clustered within the pocket in its basolateral face. (686)

mean growth rate constant (*k*) The rate of microbial population growth expressed in terms of the number of generations per unit time. (113)

measles (rubeola) (me'zelz) A highly contagious skin disease that is endemic throughout the world. It is caused by a morbilli virus in the family *Paramyxoviridae,* which enters the body through the respiratory tract or through the conjunctiva. (849)

medical mycology (mi-kol'o-je) The discipline that deals with the fungi that cause human disease. (918)

meiosis (mi-o'sis) The sexual process in which a diploid cell divides and forms two haploid cells. (86)

melting temperature (*T_m*) The temperature at which double-stranded DNA separates into individual strands; it is dependent on the G + C content of the DNA and is used to compare genetic material in microbial taxonomy. (418)

membrane attack complex (MAC) The complex complement components (C5b–C9) that create a pore

in the plasma membrane of a target cell and leads to cell lysis. C9 probably forms most of the actual pore. (692, 734)

membrane-disrupting exotoxin A type of exotoxin that lyses host cells by disrupting the integrity of the plasma membrane. (772)

membrane filter technique The use of a thin porous filter made from cellulose acetate or some other polymer to collect microorganisms from water, air, and food. (633)

memory B cell A lymphocyte capable of initiating the antibody-mediated immune response upon detection of a specific antigen molecule for which it is genetically programmed. It circulates freely in the blood and lymph and may live for years. (718)

meningitis (men″in-ji'tis) A condition that refers to inflammation of the brain or spinal cord meninges (membranes). The disease can be divided into bacterial (septic) meningitis and aseptic meningitis syndrome (caused by nonbacterial sources). (878)

mesophile (mes'o-fīl) A microorganism with a growth optimum around 20 to 45°C, a minimum of 15 to 20°C, and a maximum about 45°C or lower. (124)

messenger RNA (mRNA) Single-stranded RNA synthesized from a DNA template during transcription that binds to ribosomes and directs the synthesis of protein. (224)

metabolic channeling (mět″ah-bol'ik) The localization of metabolites and enzymes in different parts of a cell. (160)

metabolic control engineering Modification of the controls for biosynthetic pathways without altering the pathways themselves in order to improve process efficiency. (967)

metabolic pathway engineering (MPE) The use of molecular techniques to improve the efficiency of pathways that synthesize specific products. (967)

metabolism (me-tab'o-lizm) The total of all chemical reactions in the cell; almost all are enzyme catalyzed. (168)

metachromatic granules (met″ah-kro-mat'ik) Granules of polyphosphate in the cytoplasm of some bacteria that appear a different color when stained with a blue basic dye. They are storage reservoirs for phosphate. Sometimes called volutin granules. (49)

metastasis (mě-tas'tah-sis) The transfer of a disease like cancer from one organ to another not directly connected with it. (400)

methanogens (meth'ə-no-jens″) Strictly anaerobic archaeons that derive energy by converting CO_2, H_2, formate, acetate, and other compounds to either methane or methane and CO_2. (444)

methylotroph A bacterium that uses reduced one-carbon compounds such as methane and methanol as its sole source of carbon and energy. (477, 488)

Michaelis constant (*K_m*) (mī-ka'lis) A kinetic constant for an enzyme reaction that equals the substrate concentration required for the enzyme to operate at half maximal velocity. (159)

microaerophile (mi″kro-a'er-o-fīl) A microorganism that requires low levels of oxygen for growth,

around 2 to 10%, but is damaged by normal atmospheric oxygen levels. (125)

microbial ecology The study of microorganisms in their natural environments, with a major emphasis on physical conditions, processes, and interactions that occur on the scale of individual microbial cells. (578)

microbial loop The mineralization of organic matter synthesized by photosynthetic phytoplankton through the activity of microorganisms such as bacteria and protozoa. This process "loops" minerals and carbon dioxide back for reuse by the primary producers and makes the organic matter unavailable to higher consumers. (590, 620)

microbial mat A firm structure of layered microorganisms with complementary physiological activities that can develop on surfaces in aquatic environments. (603)

microbial transformation (mi-kro'be-al) *See* bioconversion. (978)

microbiology (mi″kro-bi-ol'o-je) The study of organisms that are usually too small to be seen with the naked eye. Special techniques are required to isolate and grow them. (2)

microbivory The use of microorganisms as a food source by organisms that can ingest or phagocytose them. (648)

microenvironment (mi″kro-en-vi'ron-ment) The immediate environment surrounding a microbial cell or other structure, such as a root. (602)

microfilaments (mi″kro-fil'ah-ments) Protein filaments, about 4 to 7 nm in diameter, that are present in the cytoplasmic matrix of eucaryotic cells and play a role in cell structure and motion. (76)

micronucleus (mi″kro-nu'kle-us) The smaller of the two nuclei in ciliate protozoa. Micronuclei are diploid and involved only in genetic recombination and the regeneration of macronuclei. (567)

micronutrients Nutrients such as zinc, manganese, and copper that are required in very small quantities for growth and reproduction. Also called trace elements. (94)

microorganism (mi″kro-or'gan-izm) An organism that is too small to be seen clearly with the naked eye. (2)

microtubules (mi″kro-tu'buls) Small cylinders, about 25 nm in diameter, made of tubulin proteins and present in the cytoplasmic matrix and flagella of eucaryotic cells; they are involved in cell structure and movement. (77)

miliary tuberculosis (mil'e-a-re) An acute form of tuberculosis in which small tubercles are formed in a number of organs of the body because of dissemination of *M. tuberculosis* throughout the body by the bloodstream. Also known as reactivation tuberculosis. (884)

mineralization The release of inorganic nutrients from organic matter during microbial growth and metabolism. (594)

minimal inhibitory concentration (MIC) The lowest concentration of a drug that will prevent the growth of a particular microorganism. (782)

minimal lethal concentration (MLC) The lowest concentration of a drug that will kill a particular microorganism. (782)

minus, or negative, strand The virus nucleic acid strand that is complementary in base sequence to the viral mRNA. (364)

missense mutation A single base substitution in DNA that changes a codon for one amino acid into a codon for another. (243)

mitochondrion (mi″to-kon′dre-on) The eucaryotic organelle that is the site of electron transport, oxidative phosphorylation, and pathways such as the Krebs cycle; it provides most of a nonphotosynthetic cell's energy under aerobic conditions. It is constructed of an outer membrane and an inner membrane, which contains the electron transport chain. (82)

mitosis (mi-to′sis) A process that takes place in the nucleus of a eucaryotic cell and results in the formation of two new nuclei, each with the same number of chromosomes as the parent. (86)

mixed acid fermentation A type of fermentation carried out by members of the family *Enterobacteriaceae* in which ethanol and a complex mixture of organic acids are produced. (176, A-15)

mixotrophic (mik″so-trof′ik) Refers to microorganisms that combine autotrophic and heterotrophic metabolic processes (they use inorganic electron sources and organic carbon sources). (96)

modified atmosphere packaging (MAP) Addition of gases such as nitrogen and carbon dioxide to packaged foods in order to inhibit the growth of spoilage organisms. (940)

mold Any of a large group of fungi that cause mold or moldiness and that exist as multicellular filamentous colonies; also the deposit or growth caused by such fungi. Molds typically do not produce macroscopic fruiting bodies. (539)

molecular chaperones Proteins that aid in the proper folding of unfolded polypeptides or partly denatured proteins and often also help transport proteins across membranes. (267)

molecular chronometers Nucleic acid and protein sequences that gradually change over time in a random fashion and at a steady rate, and which therefore can be used to determine phylogenetic relationships. (420)

monoclonal antibody (MAb) (mon″o-klōn′al) An antibody of a single type that is produced by a population of genetically identical plasma cells (a clone); a monoclonal antibody is typically produced from a cell culture derived from the fusion product of a cancer cell and an antibody-producing cell (a hybridoma). (719)

monocyte (mon′o-sīt) A mononuclear phagocytic leukocyte that circulates briefly in the bloodstream before migrating to the tissues where it becomes a macrophage. (681)

monocyte-macrophage system The collection of fixed phagocytic cells (including macrophages, monocytes, and specialized endothelial cells) located in the liver, spleen, lymph nodes, and bone marrow. This system is an important component of the host's general nonspecific (innate) defense against pathogens. (681)

monokine (mon′o-kīn) A generic term for a cytokine produced by mononuclear phagocytes (macrophages or monocytes). (697)

monotrichous (mon-ot′rĭ-kus) Having a single flagellum. (63)

morbidity rate (mor-bid′i-te) Measures the number of individuals who become ill as a result of a particular disease within a susceptible population during a specific time period. (823)

mordant (mor′dant) A substance that helps fix dye on or in a cell. (27)

mortality rate (mor-tal′i-te) The ratio of the number of deaths from a given disease to the total number of cases of the disease. (823)

most probable number (MPN) The statistical estimation of the probable population in a liquid by diluting and determining end points for microbial growth. (633)

mucociliary blanket The layer of cilia and mucus that lines certain portions of the respiratory system; it traps microorganisms up to 10 μm in diameter and then transports them by ciliary action away from the lungs. (687)

mucociliary escalator The mechanism by which respiratory ciliated cells move material and microorganisms, trapped in mucus, out of the pharynx, where it is spit out or swallowed. (687)

mucosal-associated lymphoid tissue (MALT) The defensive immune lymphoid tissue located in the intestinal mucosa. (686)

multi-drug-resistant strains of tuberculosis (MDR-TB) A multi-drug-resistant strain is defined as *Mycobacterium tuberculosis* resistant to isoniazid and rifampin, with or without resistance to other drugs. (884)

mumps An acute generalized disease that occurs primarily in school-age children and is caused by a paramyxovirus that is transmitted in saliva and respiratory droplets. The principal manifestation is swelling of the parotid salivary glands. (850)

murein *See* peptidoglycan. (53)

must The juices of fruits, including grapes, that can be fermented for the production of alcohol. (955)

mutagen (mu′tah-jen) A chemical or physical agent that causes mutations. (239)

mutation (mu-ta′shun) A permanent, heritable change in the genetic material. (239)

mutualism (mu′tu-al-izm″) A type of symbiosis in which both partners gain from the association and are unable to survive without it. The mutualist and the host are metabolically dependent on each other. (580)

mutualist (mu′tu-al-ist) An organism associated with another in an obligatory relationship that is beneficial to both. (580)

mycelium (mi-se′le-um) A mass of branching hyphae found in fungi and some bacteria. (41, 539)

mycobiont The fungal partner in a lichen. (591)

mycolic acids Complex 60 to 90 carbon fatty acids with a hydroxyl on the β-carbon and an aliphatic chain on the α-carbon; found in the cell walls of mycobacteria. (528)

mycologist (mi-kol′o-jist) A person specializing in mycology; a student of mycology. (538)

mycology (mi-kol′o-je) The science and study of fungi. (538)

mycoplasma (mi″ko-plaz′mah) Bacteria that are members of the class *Mollicutes* and order *Mycoplasmatales;* they lack cell walls and cannot synthesize peptidoglycan precursors; most require sterols for growth; they are the smallest organisms capable of independent reproduction. (504)

mycoplasmal pneumonia (mi″ko-plaz′mal nu-mo′ne-ah) A type of pneumonia caused by *Mycoplasma pneumoniae*. Spread involves airborne droplets and close contact. (895)

mycorrhizosphere The region around ectomycorrhizal mantles and hyphae in which nutrients released from the fungus increase the microbial population and its activities. (658)

mycosis (mi-ko′sis; pl., **mycoses**) Any disease caused by a fungus. (538, 918)

mycotoxicology (mi-ko′tok″si-kol′o-je) The study of fungal toxins and their effects on various organisms. (538)

myeloma cell (mi″e-lo′mah) A tumor cell that is similar to the cell type found in bone marrow. Also, a malignant, neoplastic plasma cell that produces large quantities of antibodies and can be readily cultivated. (720)

myositis (mi″o-si′tis) Inflammation of a striated or voluntary muscle. (880)

myxamoeba (mik-sah-me′bah; pl., **myxamoebae**) A free-living amoeboid cell that can aggregate with other myxamoebae to form a plasmodium or pseudoplasmodium. Found in cellular slime molds and the myxomycetes. (548)

myxobacteria A group of gram-negative, aerobic soil bacteria characterized by gliding motility, a complex life cycle with the production of fruiting bodies, and the formation of myxospores. (497)

myxospores (mik′so-spōrs) Special dormant spores formed by the myxobacteria. (499)

N

narrow-spectrum drugs Chemotherapeutic agents that are effective only against a limited variety of microorganisms. (781)

natural attenuation The decrease in the level of an enviromental contaminant that results from natural chemical, physical, and biological processes. (969)

natural classification A classification system that arranges organisms into groups whose members share many characteristics and reflect as much as possible the biological nature of organisms. (414)

natural killer (NK) cell A non-T, non-B lymphocyte present in nonimmunized individuals that exhibits MHC-independent cytolytic activity against tumor cells. (700, 736)

naturally acquired active immunity The type of active immunity that develops when an individual's immunologic system comes into contact with an appropriate antigenic stimulus during the course of normal activities; it usually arises as the result of recovering from an infection and lasts a long time. (706)

naturally acquired passive immunity The type of temporary immunity that involves the transfer of antibodies from one individual to another. (706)

necrotizing fasciitis (nek′ro-tīz″ing fas″e-i′tis) A disease that results from a severe invasive group. A streptococcus infection. Necrotizing fasciitis is an infection of the subcutaneous soft tissues, particularly of fibrous tissue, and is most common on the extremities. It begins with skin reddening, swelling, pain, and cellulitis, and proceeds to skin breakdown and gangrene after 3 to 5 days. (880)

negative staining A staining procedure in which a dye is used to make the background dark while the specimen is unstained. (29)

Negri bodies (na′gre) Masses of viruses or unassembled viral subunits found within the brain neurons of rabies-infected animals. (866)

neurotoxin (nu″ro-tok′sin) A toxin that is poisonous to or destroys nerve tissue; especially the toxins secreted by *C. tetani*, *Corynebacterium diphtheriae*, and *Shigella dysenteriae*. (772)

neustonic (nu′ston″ik) The microorganisms that live at the atmospheric interface of a water body. (555)

neutrophil (noo′tro-fil) A mature white blood cell in the granulocyte lineage formed in bone marrow. It has a nucleus with three to five lobes and is very phagocytic. (683)

neutrophile (nu′tro-fīl″) Microorganisms that grow best at a neutral pH range between pH 5.5 and 8.0. (121)

niche (nich) The function of an organism in a complex system, including place of the organism, the resources used in a given location, and the time of use. (602)

nicotinamide adenine dinucleotide (NAD⁺) (nik″o-tin′ah-mīd) An electron-carrying coenzyme; it is particularly important in catabolic processes and usually donates its electrons to the electron transport chain under aerobic conditions. (153)

nicotinamide adenine dinucleotide phosphate (NADP⁺) An electron-carrying coenzyme that most often participates as an electron carrier in biosynthetic metabolism. (154)

nitrification (ni″tri-fi-ka′shun) The oxidation of ammonia to nitrate. (189, 480, 597)

nitrifying bacteria (ni′tri-fi′ing) Chemolithotrophic, gram-negative bacteria that are members of the family *Nitrobacteriaceae* and convert ammonia to nitrate and nitrite to nitrate. (189, 480)

nitrogenase (ni′tro-jen-ās) The enzyme that catalyzes biological nitrogen fixation. (207)

nitrogen fixation The metabolic process in which atmospheric molecular nitrogen is reduced to ammo-

nia; carried out by cyanobacteria, *Rhizobium*, and other nitrogen-fixing procaryotes. (206, 598, 655)

nitrogen oxygen demand (*NOD*) The demand for oxygen in sewage treatment, caused by nitrifying microorganisms. (637)

nitrogen saturation point The point at which mineral nitrogen, when added to an ecosystem, can no longer be incorporated into organic matter through biological processes. (662)

nocardioforms Bacteria that resemble members of the genus *Nocardia;* they develop a substrate mycelium that readily breaks up into rods and coccoid elements (a quality sometimes called fugacity). (528)

nomenclature (no′men-kla″tūr) The branch of taxonomy concerned with the assignment of names to taxonomic groups in agreement with published rules. (410)

noncyclic photophosphorylation (fo″to-fos″for-i-la′shun) The process in which light energy is used to make ATP when electrons are moved from water to NADP⁺ during photosynthesis; both photosystem I and photosystem II are involved. (194)

nondiscrete microorganism A microorganism, best exemplified by a filamentous fungus, that does not have a defined and predictable cell structure or distinct edges and boundaries. The organism can be defined in terms of the cell structure and its cytoplasmic contents. (609)

nongonococcal urethritis (NGU) (u″rə-thri′tis) Any inflammation of the urethra not caused by *Neisseria gonorrhoeae*. (895)

nonsense codon A codon that does not code for an amino acid but is a signal to terminate protein synthesis. (234, 265)

nonsense mutation A mutation that converts a sense codon to a nonsense or stop codon. (244)

nonspecific immune response (innate or natural immunity) *See* nonspecific resistance. (680)

nonspecific resistance Refers to those general defense mechanisms that are inherited as part of the innate structure and function of each animal; also known as nonspecific, innate or natural immunity. (680)

normal microbiota (also **indigenous microbial population, microflora, microbial flora**) mmi″kro-bi-o′tah) The microorganisms normally associated with a particular tissue or structure. (675)

nosocomial infection (nos″o-ko′me-al) An infection that develops within a hospital (or other type of clinical care facility) and is acquired during the stay of the patient. (840)

nuclear envelope (nu′kle-ar) The complex double-membrane structure forming the outer boundary of the eucaryotic nucleus. It is covered by pores through which substances enter and leave the nucleus. (85)

nucleic acid hybridization (nu-kle′ik) The process of forming a hybrid double-stranded DNA molecule using a heated mixture of single-stranded DNAs from two different sources; if the sequences are fairly complementary, stable hybrids will form. (419)

nucleocapsid (nu″kle-o-kap′sid) The nucleic acid and its surrounding protein coat or capsid; the basic unit of virion structure. (359)

nucleoid (nu′kle-oid) An irregularly shaped region in the procaryotic cell that contains its genetic material. (50)

nucleolus (nu-kle′o-lus) The organelle, located within the eucaryotic nucleus and not bounded by a membrane, that is the location of ribosomal RNA synthesis and the assembly of ribosomal subunits. (86)

nucleoside (nu′kle-o-sīd″) A combination of ribose or deoxyribose with a purine or pyrimidine base. (211)

nucleosome (nu′kle-o-sōm″) A complex of histones and DNA found in eucaryotic chromatin; the DNA is wrapped around the surface of the beadlike histone complex. (228)

nucleotide (nu′kle-o-tīd) A combination of ribose or deoxyribose with phosphate and a purine or pyrimidine base; a nucleoside plus one or more phosphates. (50, 211)

nucleus (nu′kle-us) The eucaryotic organelle enclosed by a double-membrane envelope that contains the cell's chromosomes. (85)

numerical aperture The property of a microscope lens that determines how much light can enter and how great a resolution the lens can provide. (29)

numerical taxonomy The grouping by numerical methods of taxonomic units into taxa based on their character states. (414)

nutrient (nu′tre-ent) A substance that supports growth and reproduction. (94)

nystatin (nis′tah-tin) A polyene antibiotic from *Streptomyces noursei* that is used in the treatment of *Candida* infections of the skin, vagina, and alimentary tract. (795)

O

O antigen A polysaccharide antigen extending from the outer membrane of some gram-negative bacterial cell walls; it is part of the lipopolysaccharide. (56)

obligate aerobes Organisms that grow only in the presence of oxygen. (125)

obligate anaerobes Microorganisms that cannot tolerate the presence of oxygen and die when exposed to it. (125)

odontopathogens Dental pathogens. (911)

Okazaki fragments Short stretches of polynucleotides produced during discontinuous DNA replication. (233)

oligotrophic environment (ol″i-go-trof′ik) An environment containing low levels of nutrients, particularly nutrients that support microbial growth. (128, 628)

oncogene (ong″ko-jēn) A gene whose activity is associated with the conversion of normal cells to cancer cells. (400)

one-step growth experiment An experiment used to study the reproduction of lytic phages in which one round of phage reproduction occurs and ends with the lysis of the host bacterial population. (373)

onychomycosis (on″i-ko-mi-ko′sis) A fungal infection of the nail plate producing nails that are opaque, white, thickened, friable, and brittle. Also called ringworm of the nails and tinea unguium. Caused by *Trichophyton* and other fungi such as *C. albicans*. (925)

oocyst (o′o-sist) Cyst formed around a zygote of malaria and related protozoa. (573)

oogonia (o″o-go′ne-a) Mitotically dividing female structures that produce primary oocytes and gametes. (556)

oomycetes (o″o-mi-se′tēz) A collective name for members of the division *Oomycota;* also known as the water molds. (550)

open reading frame (ORF) A reading frame sequence not interrupted by a stop codon; it is usually determined by nucleic acid sequencing studies. (338)

operator The segment of DNA to which the repressor protein binds; it controls the expression of the genes adjacent to it. (271)

operon (op′er-on) The sequence of bases in DNA that contains one or more structural genes together with the operator controlling their expression. (271)

ophthalmia neonatorum (of-thal′me-ah ne″o-nator-um) A gonorrheal eye infection in a newborn, which may lead to blindness. Also called conjunctivitis of the newborn. (892)

opportunistic microorganism or **pathogen** A microorganism that is usually free-living or a part of the host's normal microbiota, but which may become pathogenic under certain circumstances, such as when the immune system is compromised. (680, 764, 924)

opsonization (op″so-ni-za′shun) The action of opsonins in making bacteria and other cells more readily phagocytosed. Antibodies, complement (especially C3b), and fibronectin are potent opsonins. (694, 732)

optical tweezer The use of a focused laser beam to drag and isolate a specific microorganism from a complex microbial mixture. (610)

oral candidiasis *See* thrush. (925)

orchitis (or-ki′tis) Inflammation of the testes. (850)

organelle (or″gah-nel′) A structure within or on a cell that performs specific functions and is related to the cell in a way similar to that of an organ to the body. (75)

organotrophs Organisms that use reduced organic compounds as their electron source. (95)

ornithosis *See* psittacosis. (896)

osmophilic microorganisms (oz″mo-fil′ik) Microorganisms that grow best in or on media of high solute concentration. (939)

osmosis (oz-mo′sis) The movement of water across a selectively permeable membrane from a dilute solution (higher water concentration) to a more concentrated solution. (59)

osmotolerant Organisms that grow over a fairly wide range of water activity or solute concentration. (119)

Ouchterlony technique *See* double diffusion agar assay. (754)

outbreak The sudden, unexpected occurrence of a disease in a given population. (823)

outer membrane A special membrane located outside the peptidoglycan layer in the cell walls of gram-negative bacteria. (53)

oxidation-reduction (redox) reactions Reactions involving electron transfers; the electron donor (reductant) gives electrons to an electron acceptor (oxidant). (153)

oxidative burst The generation of reactive oxygen species, primarily superoxide anion (O_2^-) and hydrogen peroxide (H_2O_2) by a plant or an animal, in response to challenge by a potential bacterial, fungal, or viral pathogen. *See* respiratory burst. (652)

oxidative phosphorylation (fos″for-ĭ-la′-shun) The synthesis of ATP from ADP using energy made available during electron transport. (179)

oxygenic photosynthesis Photosynthesis that oxidizes water to form oxygen; the form of photosynthesis characteristic of algae and cyanobacteria. (195, 456)

P

pacemaker enzyme The enzyme in a metabolic pathway that catalyzes the slowest or rate-limiting reaction; if its rate changes, the pathway's activity changes. (163)

pandemic (pan-dem′ik) An increase in the occurrence of a disease within a large and geographically widespread population (often refers to a worldwide epidemic). (823)

Paneth cell (pah′ net) The granular cell located at the base of glands in the small intestine; it produces the enzyme lysozyme. (687)

pannus (pan′us) A superficial vascularization of the cornea with infiltration of granulation tissue. (903)

panzootic (pan″zo-ot′ik) The wide dissemination of a disease in an animal population. (823)

parasite (par′ah-sīt) An organism that lives on or within another organism (the host) and benefits from the association while harming its host. Often the parasite obtains nutrients from the host. (762)

parasitism (par′ah-si″tizm) A type of symbiosis in which one organism benefits from the other and the host is usually harmed. (590, 762)

parenteral route (pah-ren′ter-al) A route of drug administration that is nonoral (e.g., by injection). (786)

parfocal (par-fo′kal) A microscope that retains proper focus when the objectives are changed. (20)

paronychia (par″o-nik′e-ah) Inflammation involving the folds of tissue surrounding the nail; usually caused by *Candida albicans*. (925)

passive diffusion The process in which molecules move from a region of higher concentration to one of lower concentration as a result of random thermal agitation. (98)

passive immunization The induction of temporary immunity by the transfer of immune products, such as antibodies or sensitized T cells, from an immune vertebrate to a nonimmune one. (741)

Pasteur effect (pas-tur′) The decrease in the rate of sugar catabolism and change to aerobic respiration that occurs when microorganisms are switched from anaerobic to aerobic conditions. (184)

pasteurization (pas″ter-ĭ-za′-shun) The process of heating milk and other liquids to destroy microorganisms that can cause spoilage or disease. (139, 944)

pathogen (path′o-jən) Any virus, bacterium, or other agent that causes disease. (674, 763)

pathogen-associated molecular pattern (PAMP) Conserved molecular structures that occur in patterns on microbial surfaces. The structures and their patterns are unique to particular microorganisms and invariant among members of a given microbial group. (694)

pathogenicity (path″o-je-nis′ĭ-te) The condition or quality of being pathogenic, or the ability to cause disease. (674, 764)

pathogenicity island A large segment of DNA in some pathogens that contains the genes responsible for virulence; often it codes for the type III secretion system that allows the pathogen to secrete virulence proteins and damage host cells. A pathogen may have more than one pathogenicity island. (768)

pathogenic potential The degree that a pathogen causes morbid signs and symptoms. (764)

pathway architecture The analysis, design, and modification of biochemical pathways to increase process efficiency. (967)

pattern recognition receptor A receptor found on macrophages and other phagocytic cells that binds to pathogen-associated molecular patterns on microbial surfaces. (694)

ped A natural soil aggregate, formed partly through bacterial and fungal growth in the soil. (647)

pellicle (pel′ĭ-k′l) A relatively rigid layer of proteinaceous elements just beneath the plasma membrane in many protozoa and algae. The plasma membrane is sometimes considered part of the pellicle. (88, 559, 567)

pelvic inflammatory disease (PID) A severe infection of the female reproductive organs. The disease that results when gonococci and chlamydiae infect the uterine tubes and surrounding tissue. (892, 895)

penicillins (pen″ĭ-sil′ ins) A group of antibiotics containing a β-lactam ring, which are active against gram-positive bacteria. (59, 788)

penton or pentamer A capsomer composed of five protomers. (360)

pentose phosphate pathway (pen′tōs) The pathway that oxidizes glucose 6-phosphate to ribulose 5-phosphate and then converts it to a variety of three to seven carbon sugars; it forms several important products (NADPH for biosynthesis, pentoses, and other sugars) and also can be used to degrade glucose to CO_2. (171, A-12)

peplomer or spike (pep′lo-mer) A protein or protein complex that extends from the virus envelope and often is important in virion attachment to the host cell surface. (364)

peptic ulcer disease A gastritis caused by *Helicobacter pylori*. (895)

peptide interbridge (pep′tīd) A short peptide chain that connects the tetrapeptide chains in some peptidoglycans. (54)

peptidoglycan (pep″tī-do-gli′kan) A large polymer composed of long chains of alternating *N*-acetylglucosamine and *N*-acetylmuramic acid residues. The polysaccharide chains are linked to each other through connections between tetrapeptide chains attached to the *N*-acetylmuramic acids. It provides much of the strength and rigidity possessed by bacterial cell walls. (53, 507)

peptidyl or **donor site (P site)** The site on the ribosome that contains the peptidyl-tRNA at the beginning of the elongation cycle during protein synthesis. (265)

peptidyl transferase The enzyme that catalyzes the transpeptidation reaction in protein synthesis; in this reaction, an amino acid is added to the growing peptide chain. (265)

peptones (pep′tōns) Water-soluble digests or hydrolysates of proteins that are used in the preparation of culture media. (103)

perforin pathway The cytotoxic pathway that uses perforin protein, which polymerizes to form membrane pores that help destroy cells during cell-mediated cytotoxicity. Perforin is produced by cytotoxic T cells and NK cells and stored in granules that are released when a target cell is contacted. (725)

period of infectivity Refers to the time during which the source of an infectious disease is infectious or is disseminating the pathogen. (828)

periodontal disease (per″e-o-don′tal) A disease located around the teeth or in the periodontium. (914)

periodontitis (per″e-o-don-ti′tis) An inflammation of the periodontium. (914)

periodontium (per″e-o-don′she-um) The tissue investing and supporting the teeth, including the cementum, periodontal ligament, alveolar bone, and gingiva. (914)

periodontosis (per″e-o-don-to′sis) A degenerative, noninflammatory condition of the periodontium, which is characterized by destruction of tissue. (914)

periplasm (per′ĭ-plaz-əm) The substance that fills the periplasmic space. (53)

periplasmic flagella The flagella that lie under the outer sheath and extend from both ends of the spirochete cell to overlap in the middle and form the axial filament. Also called axial fibrils and endoflagella. (467)

periplasmic space (per″i-plas′mik) or **periplasm** (per′ĭ-plazm) The space between the plasma membrane and the outer membrane in gram-negative bacteria, and between the plasma membrane and the cell wall in gram-positive bacteria. (53)

peritrichous (pě-rit′rĭ-kus) A cell with flagella evenly distributed over its surface. (63)

permease (per′me-ās) A membrane-bound carrier protein or a system of two or more proteins that transports a substance across the membrane. (98)

pertussis (pər-tus′is) An acute, highly contagious infection of the respiratory tract, most frequently affecting young children, usually caused by *Bordetella pertussis* or *B. parapertussis*. Consists of peculiar paroxysms of coughing, ending in a prolonged crowing or whooping respiration; hence the name whooping cough. (879)

petri dish (pe′tre) A shallow dish consisting of two round, overlapping halves that is used to grow microorganisms on solid culture medium; the top is larger than the bottom of the dish to prevent contamination of the culture. (104)

phage (fāj) *See* bacteriophage. (354)

phagocytic vacuole (fag″o-sit′ik vak′u-ol) A membrane-delimited vacuole produced by cells carrying out phagocytosis. It is formed by the invagination of the plasma membrane and contains solid material. (567)

phagocytosis (fag″o-si-to′sis) The endocytotic process in which a cell encloses large particles in a membrane-delimited phagocytic vacuole or phagosome and engulfs them. (79, 693)

phagolysosome (fag″o-li′so-sōm) The vacuole that results from the fusion of a phagosome with a lysosome. (696)

phagovar (fag′o-var) A specific phage type. (816)

pharyngitis (far″in-ji′tis) Inflammation of the pharynx, often due to a *S. pyogenes* infection. (881)

phase-contrast microscope A microscope that converts slight differences in refractive index and cell density into easily observed differences in light intensity. (21)

phenetic system A classification system that groups organisms together based on the similarity of their observable characteristics. (414)

phenol coefficient test A test to measure the effectiveness of disinfectants by comparing their activity against test bacteria with that of phenol. (146)

phosphatase (fos′fah-tās″) An enzyme that catalyzes the hydrolytic removal of phosphate from molecules. (205)

phosphate group transfer potential A measure of the ability of a phosphorylated molecule such as ATP to transfer its phosphate to water and other acceptors. It is the negative of the $\Delta G°′$ for the hydrolytic removal of phosphate. (153)

photoautotroph (fo′to-aw′to-trōf) *See* photolithotrophic autotrophs. (96)

photolithotrophic autotrophs Organisms that use light energy, an inorganic electron source (e.g., H_2O, H_2, H_2S), and CO_2 as a carbon source. (96)

photoorganotrophic heterotrophs Microorganisms that use light energy and organic electron donors, and also employ simple organic molecules rather than CO_2 as their carbon source. (96)

photoreactivation (fo″to-re-ak″tī-va′shun) The process in which blue light is used by a photoreactivating enzyme to repair thymine dimers in DNA by splitting them apart. (128, 248)

photosynthesis (fo″to-sin′thĕ-sis) The trapping of light energy and its conversion to chemical energy, which is then used to reduce CO_2 and incorporate it into organic form. (150, 190)

photosystem I The photosystem in eucaryotic cells that absorbs longer wavelength light, usually greater than about 680 nm, and transfers the energy to chlorophyll P700 during photosynthesis; it is involved in both cyclic photophosphorylation and noncyclic photophosphorylation. (193)

photosystem II The photosystem in eucaryotic cells that absorbs shorter wavelength light, usually less than 680 nm, and transfers the energy to chlorophyll P680 during photosynthesis; it participates in noncyclic photophosphorylation. (193)

phototrophs Organisms that use light as their energy source. (95)

phycobiliproteins Photosynthetic pigments that are composed of proteins with attached tetrapyrroles; they are often found in cyanobacteria and red algae. (191)

phycobilisomes Special particles on the membranes of cyanobacteria that contain photosynthetic pigments and electron transport chains. (458)

phycobiont (fi″ko-bi′ont) The algal or cyanobacterial partner in a lichen. (591)

phycocyanin (fi″ko-si′an-in) A blue phycobiliprotein pigment used to trap light energy during photosynthesis. (191)

phycoerythrin (fi″ko-er′i-thrin) A red photosynthetic phycobiliprotein pigment used to trap light energy. (191)

phycology (fi-kol′o-je) The study of algae; algology. (554)

phyllosphere The surface of plant leaves. (651)

phylogenetic or **phyletic classification system** (fi″lo-jě-net′ik, fi-let′ik) A classification system based on evolutionary relationships rather than the general similarity of characteristics. (416)

phylogenetic tree A graph made of nodes and branches, much like a tree in shape, that shows phylogenetic relationships between groups of organisms and sometimes also indicates the evolutionary development of groups. (421)

phytoplankton (fi″to-plank′ton) A community of floating photosynthetic organisms, largely composed of algae and cyanobacteria. (555, 619)

phytoremediation The use of plants and their associated microorganisms to remove, contain, or degrade environmental contaminants. (983)

piedra (pe-a′drah) A fungal disease of the hair in which white or black nodules of fungi form on the shafts. (919)

pinocytosis (pi″no-si-to′sis) The endocytotic process in which a cell encloses a small amount of the surrounding liquid and its solutes in tiny pinocytotic vesicles or pinosomes. (79)

pitched Pertaining to inoculation of a nutrient medium with yeast, for example, in beer brewing. (956)

plague (plāg) An acute febrile, infectious disease, caused by the bacillus *Yersinia pestis*, which has a high mortality rate; the two major types are bubonic plague and pneumonic plague. (887)

plankton (plank'ton)   Free-floating, mostly microscopic microorganisms that can be found in almost all waters; a collective name. (555, 566)

planktonic (adj.)   *See* plankton. (555)

plaque (plak)   1. A clear area in a lawn of bacteria or a localized area of cell destruction in a layer of animal cells that results from the lysis of the bacteria by bacteriophages or the destruction of the animal cells by animal viruses. 2. The term also refers to dental plaque, a film of food debris, polysaccharides, and dead cells that cover the teeth. (354, 911)

plasma cell   A mature, differentiated B lymphocyte chiefly occupied with antibody synthesis and secretion; a plasma cell lives for only 5 to 7 days. (685)

plasma membrane   The selectively permeable membrane surrounding the cell's cytoplasm; also called the cell membrane, plasmalemma, or cytoplasmic membrane. (44)

plasmid (plaz'mid)   A double-stranded DNA molecule that can exist and replicate independently of the chromosome or may be integrated with it. A plasmid is stably inherited, but is not required for the host cell's growth and reproduction. (52, 288, 792)

plasmid fingerprinting   A technique used to identify microbial isolates as belonging to the same strain because they contain the same number of plasmids with the identical molecular weights and similar phenotypes. (817)

plasmodial (acellular) slime mold (plaz-mo'de-al)   A member of the division *Myxomycota* that exists as a thin, streaming, multinucleate mass of protoplasm, which creeps along in an amoeboid fashion. (548)

plasmodium (plaz-mo'de-um; pl., **plasmodia**)   A stage in the life cycle of myxomycetes (plasmodial slime molds); a multinucleate mass of protoplasm surrounded by a membrane. Also, a parasite of the genus *Plasmodium*. (548)

plasmolysis (plaz-mol'ĭ-sis)   The process in which water osmotically leaves a cell, which causes the cytoplasm to shrivel up and pull the plasma membrane away from the cell wall. (59)

plastid (plas'tid)   A cytoplasmic organelle of algae and higher plants that contains pigments such as chlorophyll, stores food reserves, and often carries out processes such as photosynthesis. (84)

pleomorphic (ple"o-mor'fik)   Refers to bacteria that are variable in shape and lack a single, characteristic form. (42)

plus strand or **positive strand**   The virus nucleic-acid strand that is equivalent in base sequence to the viral mRNA. (364)

***Pneumocystis* pneumonia (PCP);** (noo"mo-sis-tis)   A type of pneumonia caused by the protist *Pneumocystis jiroveci*. (926)

pneumonic plague   *See* plague. (887)

point mutation   A mutation that affects only a single base pair in a specific location. (242)

polar flagellum   A flagellum located at one end of an elongated cell. (63)

poliomyelitis (po"le-o-mi"e-li'tis)   An acute, contagious viral disease that attacks the central nervous system, injuring or destroying the nerve cells that control the muscles and sometimes causing paralysis; also called polio or infantile paralysis. (869)

poly-β-hydroxybutyrate (PHB) (hi-drok"se-bu'tĭ-rāt)   A linear polymer of β-hydroxybutyrate used as a reserve of carbon and energy by many bacteria. (47)

polymerase chain reaction (PCR)   An in vitro technique used to synthesize large quantities of specific nucleotide sequences from small amounts of DNA. It employs oligonucleotide primers complementary to specific sequences in the target gene and special heat-stable DNA polymerases. (316)

polymorphonuclear leukocyte (PMN) (pol"e-mor"fo-noo'kle-ər)   A leukocyte that has a variety of nuclear forms. (681)

polyphasic taxonomy   An approach in which taxonomic schemes are developed using a wide range of phenotypic and genotypic information. (423)

polyribosome (pol"e-ri'bo-sōm)   A complex of several ribosomes with a messenger RNA; each ribosome is translating the same message. (82, 260)

Pontiac fever   A bacterial disease caused by *Legionella pneumophila* that resembles an allergic disease more than an infection. First described from Pontiac, Michigan. *See* Legionnaires' disease. (877)

population   An assemblage of organisms of the same type. (578)

porin proteins   Proteins that form channels across the outer membrane of gram-negative bacterial cell walls. Small molecules are transported through these channels. (58)

postherpetic neuralgia   The severe pain after a herpes infection. (847)

posttranscriptional modification   The processing of the initial RNA transcript, heterogeneous nuclear RNA, to form mRNA. (257)

potable (po'tah-b'l)   Refers to water suitable for drinking. (633)

pour plate   A petri dish of solid culture medium with isolated microbial colonies growing both on its surface and within the medium, which has been prepared by mixing microorganisms with cooled, still liquid medium and then allowing the medium to harden. (104)

precipitation (or precipitin) reaction (pre-sip"ĭ-ta'shun)   The reaction of an antibody with a soluble antigen to form an insoluble precipitate. (732)

precipitin (pre-sip'ĭ-tin)   The antibody responsible for a precipitation reaction. (732)

prevalence rate   Refers to the total number of individuals infected at any one time in a given population regardless of when the disease began. (823)

Pribnow box   A special base sequence in the promoter that is recognized by the RNA polymerase and is the site of initial polymerase binding. (237, 255)

primary amebic meningoencephalitis   An infection of the meninges of the brain by the free-living amoebae *Naegleria* or *Acanthamoeba*. (928)

primary (frank) pathogen   Any organism that causes a disease in the host by direct interaction with or infection of the host. (764)

primary metabolites   Microbial metabolites produced during the growth phase of an organism. (973)

primary producer   Photoautotrophic and chemoautotrophic organisms that incorporate carbon dioxide into organic carbon and thus provide new biomass for the ecosystem. (604)

primary production   The incorporation of carbon dioxide into organic matter by photosynthetic organisms and chemoautotrophic bacteria. (604)

primary treatment   The first step of sewage treatment, in which physical settling and screening are used to remove particulate materials. (638)

prion (pri'on)   An infectious particle that is the cause of slow diseases like scrapie in sheep and goats; it has a protein component, but no nucleic acid has yet been detected. (406)

probe (prōb)   A short, labeled nucleic acid segment complementary in base sequence to part of another nucleic acid, which is used to identify or isolate the particular nucleic acid from a mixture through its ability to bind specifically with the target nucleic acid. (314, 950)

probiotic   A living organism that may provide health benefits beyond its nutritional value when it is ingested. (679, 959)

procaryotic cells (pro"kar-e-ot'ik)   Cells that lack a true, membrane-enclosed nucleus; bacteria are procaryotic and have their genetic material located in a nucleoid. (11, 90)

procaryotic species   A collection of strains that share many stable properties and differ significantly from other groups of strains. (413)

prodromal stage (pro-dro'məl)   The period during the course of a disease in which there is the appearance of signs and symptoms, but they are not yet distinctive and characteristic enough to make an accurate diagnosis. (824)

proliferative kidney disease (pro-lif'er-a-tiv)   A protozoan disease caused by an unclassified myxozoan in salmonids throughout the world. (574)

promoter   The region on DNA at the start of a gene that the RNA polymerase binds to before beginning transcription. (235, 254)

propagated epidemic   An epidemic that is characterized by a relatively slow and prolonged rise and then a gradual decline in the number of individuals infected. It usually results from the introduction of an infected individual into a susceptible population, and the pathogen is transmitted from person to person. (825)

prophage (pro'fāj)   The latent form of a temperate phage that remains within the lysogen, usually integrated into the host chromosome. (301, 380)

prostheca (pros-the'kah)   An extension of a bacterial cell, including the plasma membrane and cell wall, that is narrower than the mature cell. (476)

prosthetic group (pros-thet'ik)   A tightly bound cofactor that remains at the active site of an enzyme during its catalytic activity. (156)

protease (pro′te-ās) An enzyme that hydrolyzes proteins to their constituent amino acids. Also called a proteinase. (188)

proteasome A large, cylindrical protein complex that degrades ubiquitin-labeled proteins to peptides in an ATP-dependent process. (81)

protein engineering (pro′tēn) The rational design of proteins by constructing specific amino acid sequences through molecular techniques, with the objective of modifying protein characteristics. (966)

protein splicing The post-translational process in which part of a precursor polypeptide is removed before the mature polypeptide folds into its final shape; it is carried out by self-splicing proteins that remove inteins and join the remaining exteins. (269)

proteobacteria (pro″te-o-bak-tēr′e-ah) A large group of bacteria, primarily gram-negative, that 16S rRNA sequence comparisons show to be phylogenetically related; proteobacteria contain the purple photosynthetic bacteria and their relatives and are composed of the α, β, γ, δ, and ε subgroups. (474)

proteome The complete collection of proteins that an organism produces. (346)

protists (pro′tist) Eucaryotes with unicellular organization, either in the form of solitary cells or colonies of cells lacking true tissues. (426)

protomer An individual subunit of a viral capsid; a capsomer is made of protomers. (359)

proton motive force (PMF) The force arising from a gradient of protons and a membrane potential that is thought to power ATP synthesis and other processes. (182)

protoplast (pro′to-plast) A bacterial or fungal cell with its cell wall completely removed. It is spherical in shape and osmotically sensitive. (47, 59)

protoplast fusion The joining of cells that have had their walls weakened or completely removed. (965)

protothecosis (pro″to-the-ko′sis; pl., **protothecoses**) A disease of humans and animals produced by the green alga *Prototheca moriformis.* (557)

prototroph (pro′to-trōf) A microorganism that requires the same nutrients as the majority of naturally occurring members of its species. (239)

protozoan or **protozoon** (pro″to-zo′an, pl. **protozoa**) A microorganism belonging to the *Protozoa* subkingdom. A unicellular or acellular eucaryotic protist whose organelles have the functional role of organs and tissues in more complex forms. Protozoa vary greatly in size, morphology, nutrition, and life cycle. (566)

protozoology (pro″to-zo-ol′o-je) The study of protozoa. (566)

proviral DNA Viral DNA that has been integrated into host cell DNA. In retroviruses it is the double-stranded DNA copy of the RNA genome. (397)

pseudomurein A modified peptidoglycan lacking D-amino acids and containing *N*-acetyltalosaminuronic acid instead of *N*-acetylmuramic acid; found in methanogenic archaea. (439)

pseudoplasmodium (soo″do-plaz-mo′de-um; pl., **pseudoplasmodia**) A sausage-shaped amoeboid structure consisting of many myxamoebae and

behaving as a unit; the result of myxamoebal aggregation in the cellular slime molds; also called a slug. (548)

pseudopodium or **pseudopod** (soo″do-po′de-um) A nonpermanent cytoplasmic extension of the cell body by which amoebae and amoeboid organisms move and feed. (568)

psittacosis (ornithosis; sit″ah-ko′sis) A disease due to a strain of *Chlamydia psittaci,* first seen in parrots and later found in other birds and domestic fowl (in which it is called ornithosis). It is transmissible to humans. (896)

psychrophile (si′kro-fīl) A microorganism that grows well at 0°C and has an optimum growth temperature of 15°C or lower and a temperature maximum around 20°C. (123)

psychrotroph A microorganism that grows at 0°C, but has a growth optimum between 20 and 30°C, and a maximum of about 35°C. (123)

puerperal fever (pu-er′per-al) An acute, febrile condition following childbirth; it is characterized by infection of the uterus and/or adjacent regions and is caused by streptococci. (831)

pulmonary anthrax (pul′mo-ner″e) A form of anthrax involving the lungs. Also known as woolsorter's disease. (889)

pulmonary syndrome hantavirus *See* hantavirus pulmonary syndrome. (853)

pure culture A population of cells that are identical because they arise from a single cell. (104)

purine (pu′rin) A basic, heterocyclic, nitrogen-containing molecule with two joined rings that occurs in nucleic acids and other cell constituents; most purines are oxy or amino derivatives of the purine skeleton. The most important purines are adenine and guanine. (211)

purple membrane An area of the plasma membrane of *Halobacterium* that contains bacteriorhodopsin and is active in photosynthetic light energy trapping. (449)

putrefaction (pu″trĕ-fak′shun) The microbial decomposition of organic matter, especially the anaerobic breakdown of proteins, with the production of foul-smelling compounds such as hydrogen sulfide and amines. (939)

pyrenoid (pi′rĕ-noid) The differentiated region of the chloroplast that is a center of starch formation in green algae and stoneworts. (85, 556)

pyrimidine (pi-rim′i-dēn) A basic, heterocyclic, nitrogen-containing molecule with one ring that occurs in nucleic acids and other cell constituents; pyrimidines are oxy or amino derivatives of the pyrimidine skeleton. The most important pyrimidines are cytosine, thymine, and uracil. (211)

Q

Q fever An acute zoonotic disease caused by the rickettsia *Coxiella burnetii.* (887)

Quellung reaction The increase in visibility or the swelling of the capsule of a microorganism in the

presence of antibodies against capsular antigens. (758)

quorum sensing The process in which bacteria monitor their own population density by sensing the levels of signal molecules that are released by the microorganisms. When these signal molecules reach a threshold concentration, quorum-dependent genes are expressed. (130)

R

rabies (ra′bēz) An acute infectious disease of the central nervous system, which affects all warm-blooded animals (including humans). It is caused by an ssRNA virus belonging to the genus *Lyssavirus* in the family *Rhabdoviridae.* (865)

racking The removal of sediments from wine bottles. (955)

radappertization The use of gamma rays from a cobalt source for control of microorganisms in foods. (946)

radioimmunoassay (RIA) (ra″de-o-im″u-no-as′a) A very sensitive assay technique that uses a purified radioisotope-labeled antigen or antibody to compete for antibody or antigen with unlabeled standard and samples to determine the concentration of a substance in the samples. (757)

reactivation tuberculosis *See* miliary tuberculosis. (884)

reading frame The way in which nucleotides in DNA and mRNA are grouped into codons or groups of three for reading the message contained in the nucleotide sequence. (235)

reagin (re′ah-jin) Antibody that mediates immediate hypersensitivity reactions. IgE is the major reagin in humans. (744)

receptor-mediated endocytosis A type of endocytosis that involves the specific binding of molecules to membrane receptors followed by the formation of coated vesicles. The substances being taken in are concentrated during the process. (80)

recombinant DNA technology The techniques used in carrying out genetic engineering; they involve the identification and isolation of a specific gene, the insertion of the gene into a vector such as a plasmid to form a recombinant molecule, and the production of large quantities of the gene and its products. (312)

recombinant-vector vaccine The type of vaccine that is produced by the introduction of one or more of a pathogen's genes into attenuated viruses or bacteria. The attenuated virus or bacterium serves as a vector, replicating within the vertebrate host and expressing the gene(s) of the pathogen. The pathogen's antigens induce an immune response. (743)

recombination (re″kom-bĭ-na′shun) The process in which a new recombinant chromosome is formed by combining genetic material from two organisms. (286)

recombination repair A DNA repair process that repairs damaged DNA when there is no remaining template; a piece of DNA from a sister molecule is used. (249)

Redfield ratio   The carbon-nitrogen-phosphorus ratio of aquatic microorganisms. This ratio is important for predicting limiting factors for microbial growth. (619)

red tides   Red tides occur frequently in coastal areas and often are associated with population blooms of dinoflagellates. Dinoflagellate pigments are responsible for the red color of the water. Under these conditions, the dinoflagellates often produce saxitoxin, which can lead to paralytic shellfish poisoning. (562)

reductive dehalogenation   The cleavage of carbon-halogen bonds by anaerobic bacteria that creates a strong electron-donating environment. (979)

refraction (re-frak'shun)   The deflection of a light ray from a straight path as it passes from one medium (e.g., glass) to another (e.g., air). (18)

refractive index (re-frak'tiv)   The ratio of the velocity of light in the first of two media to that in the second as it passes from the first to the second. (18)

regulator T cell   Regulator T cells control the development of effector T cells. Two types exist: T-helper cells (CD4$^+$ cells) and T-suppressor cells. There are three subsets of T-helper cells: T_H1, T_H2, and T_H0. T_H1 cells produce IL-2, IFN-γ, and TNF-β. They effect cell-mediated immunity and are responsible for delayed-type hypersensitivity reactions and macrophage activation. T_H2 cells produce IL-4, IL-5, IL-6, IL-10, IL-13. They are helpers for B-cell antibody responses and humoral immunity; they also support IgE responses and eosinophilia. T_H0 cells exhibit an unrestricted cytokine profile. (725)

regulatory mutants   Mutant organisms that have lost the ability to limit synthesis of a product, which normally occurs by regulation of activity of an earlier step in the biosynthetic pathway. (976)

regulon   A collection of genes or operons that is controlled by a common regulatory protein. (276)

replica plating   A technique for isolating mutants from a population by plating cells from each colony growing on a nonselective agar medium onto plates with selective media or environmental conditions, such as the lack of a nutrient or the presence of an antibiotic or a phage; the location of mutants on the original plate can be determined from growth patterns on the replica plates. (245)

replication (rep"li-ka'shun)   The process in which an exact copy of parental DNA or RNA is made with the parental molecule serving as a template. (223)

replication fork   The Y-shaped structure where DNA is replicated. The arms of the Y contain template strand and a newly synthesized DNA copy. (229)

replicative form (RF)   A double-stranded form of nucleic acid that is formed from a single-stranded virus genome and used to synthesize new copies of the genome. (378, 395)

replicon (rep'li-kon)   A unit of the genome that contains an origin for the initiation of replication and in which DNA is replicated. (229, 288)

replisome (rep'li-sōm)   A protein complex or replication factory that copies the DNA double helix to form two daughter chromosomes. (233)

repressible enzyme   An enzyme whose level drops in the presence of a small molecule, usually an end product of its metabolic pathway. (271)

repressor protein (re-pres'or)   A protein coded for by a regulator gene that can bind to the operator and inhibit transcription; it may be active by itself or only when the corepressor is bound to it. (271)

reservoir (rez'er-vwar)   A site, alternate host, or carrier that normally harbors pathogenic organisms and serves as a source from which other individuals can be infected. (765, 828)

reservoir host   An organism other than a human that is infected with a pathogen that can also infect humans. (763)

residuesphere   The region surrounding organic matter such as a seed or plant part in which microbial growth is stimulated by increased organic matter availability. (666)

resolution (rez"o-lu'shun)   The ability of a microscope to separate or distinguish between small objects that are close together. (20)

respiration (res" pĭ-ra' shən)   An energy-yielding process in which the energy substrate is oxidized using an exogenous or externally derived electron acceptor. (168)

respiratory burst   The respiratory burst occurs when an activated phagocytic cell increases its oxygen consumption to support the increased metabolic activity of phagocytosis. The burst generates highly toxic oxygen products such as singlet oxygen, superoxide radical, hydrogen peroxide, hydroxyl radical, and hypochlorite. (696)

respiratory syncytial virus (RSV; sin-sish'al)   A member of the family *Paramyxoviridae* and genus *Pneumovirus;* it is a negative-sense ssRNA virus that causes respiratory infections in children. (851)

restricted transduction   A transduction process in which only a specific set of bacterial genes are carried to another bacterium by a temperate phage; the bacterial genes are acquired because of a mistake in the excision of a prophage during the lysogenic life cycle. (302)

restriction enzymes   Enzymes produced by host cells that cleave virus DNA at specific points and thus protect the cell from virus infection; they are used in carrying out genetic engineering. (312, 375)

reticulate body (RB)   The form in the chlamydial life cycle whose role is growth and reproduction within the host cell. (465)

retroviruses (re"tro-vi'rus-es)   A group of viruses with RNA genomes that carry the enzyme reverse transcriptase and form a DNA copy of their genome during their reproductive cycle. (395)

reverse transcriptase (RT)   An RNA-dependent DNA polymerase that uses a viral RNA genome as a template to form a DNA copy; this is a reverse of the normal flow of genetic information, which proceeds from DNA to RNA. (397, 857)

reversible covalent modification   A mechanism of enzyme regulation in which the enzyme's activity is either increased or decreased by the reversible cova-

lent addition of a group such as phosphate or AMP to the protein. (163)

Reye's syndrome   An acute, potentially fatal disease of childhood that is characterized by severe edema of the brain and increased intracranial pressure, vomiting, hypoglycemia, and liver dysfunction. The cause is unknown but is almost always associated with a previous viral infection (e.g., influenza or varicella-zoster virus infections). (849)

R factors or **R plasmids**   Plasmids bearing one or more drug resistant genes. (289, 793)

rheumatic fever (roo-mat'ik)   An autoimmune disease characterized by inflammatory lesions involving the heart valves, joints, subcutaneous tissues, and central nervous system. The disease is associated with hemolytic streptococci in the body. It is called rheumatic fever because two common symptoms are fever and pain in the joints similar to that of rheumatism. (881)

rhizosphere   A region around the plant root where materials released from the root increase the microbial population and its activities. (651)

rho factor (ro)   The protein that helps RNA polymerase dissociate from the terminator after it has stopped transcription. (257)

rhoptry   Saclike, electron dense structure in the anterior portion of a zoite of a member of the phylum *Apicomplexa;* perhaps involved in the penetration of host cells. (573)

ribonucleic acid (RNA) (ri"bo-nu-kle'ik)   A polynucleotide composed of ribonucleotides joined by phosphodiester bridges. (224)

ribosomal RNA (rRNA)   The RNA present in ribosomes; ribosomes contain several sizes of single-stranded rRNA that contribute to ribosome structure and are also directly involved in the mechanism of protein synthesis. (254)

ribosome (ri'bo-sōm)   The organelle where protein synthesis occurs; the message encoded in mRNA is translated here. (50, 262)

ribotyping   Ribotyping is the use of *E. coli* rRNA to probe chromosomal DNA in Southern blots for typing bacterial strains. This method is based on the fact that rRNA genes are scattered throughout the chromosome of most bacteria and therefore polymorphic restriction endonuclease patterns result when chromosomes are digested and probed with rRNA. (816)

ribulose-1,5-bisphosphate carboxylase (ri'bu-lōs)   The enzyme that catalyzes the incorporation of CO_2 in the Calvin cycle. (202)

ringworm (ring'werm)   The common name for a fungal infection of the skin, even though it is not caused by a worm and is not always ring-shaped in appearance. (919)

rise period or **burst**   The period during the one-step growth experiment when host cells lyse and release phage particles. (373)

RNA polymerase   The enzyme that catalyzes the synthesis of mRNA under the direction of a DNA template. (254)

Rocky Mountain spotted fever A disease caused by *Rickettsia rickettsii*. (888)

rolling-circle mechanism A mode of DNA replication in which the replication fork moves around a circular DNA molecule, displacing a strand to give a tail that is also copied to produce a new double-stranded DNA. (229)

root nodule Gall-like structures on roots that contain endosymbiotic nitrogen-fixing bacteria (e.g., *Rhizobium* or *Bradyrhizobium* is present in legume nodules). (655)

roseola infantum (ro-ze'o-lə) A skin eruption that produces a rose-colored rash in infants. Caused by the human herpesvirus 6. The disease is short-lived and characterized by a high fever of 3 to 4 days' duration. (863)

rubella A moderately contagious skin disease that occurs primarily in children 5 to 9 years of age that is caused by the rubella virus, which is acquired by droplet inhalation into the respiratory system; German measles. (851)

rubeola *See* measles. (849)

rumen (roo-men) The expanded upper portion or first compartment of the stomach of ruminants. (583)

ruminant (roo'mĭ-nant) An herbivorous animal that has a stomach divided into four compartments and chews a cud consisting of regurgitated, partially digested food. (583)

run The straight line movement of a bacterium. (67)

S

salmonellosis (sal"mo-nel-o'sis) An infection with certain species of the genus *Salmonella,* usually caused by ingestion of food containing salmonellae or their products. Also known as *Salmonella* gastroenteritis or *Salmonella* food poisoning. (909)

sanitization (san"ĭ-ti-za'shun) Reduction of the microbial population on an inanimate object to levels judged safe by public health standards; usually, the object is cleaned. (135)

saprophyte (sap'ro-fit) An organism that takes up nonliving organic nutrients in dissolved form and usually grows on decomposing organic matter. (541)

saprozoic nutrition (sap"ro-zo'ik) Having the type of nutrition in which organic nutrients are taken up in dissolved form; normally refers to animals or animal-like organisms. (567)

scaffolding proteins Special proteins that are used to aid procapsid construction during the assembly of a bacteriophage capsid and are removed after the completion of the procapsid. (376)

scale (skāl) A platelike organic structure found on the surface of some cells (chrysophytes). (559)

scanning electron microscope (SEM) An electron microscope that scans a beam of electrons over the surface of a specimen and forms an image of the surface from the electrons that are emitted by it. (34)

scanning probe microscope A microscope used to study surface features by moving a sharp probe over the object's surface (e.g., the scanning tunneling microscope). (35)

scarlatina (skahr"la-te'nah) *See* scarlet fever. (881)

scarlet fever (scarlatina) (skar'let) A disease that results from infection with a strain of *Streptococcus pyogenes* that carries a lysogenic phage with the gene for erythrogenic (rash-inducing) toxin. The toxin causes shedding of the skin. This is a communicable disease spread by respiratory droplets. (881)

schizogony (skĭ-zog'o-ne) Multiple asexual fission. (573)

Sec-dependent pathway System that can transport proteins through the plasma membrane or insert them into the membrane. The bacterial translocon recognizes a signal peptide, and employs SecYEG, SecA, and chaperones such as SecB. (59)

secondary metabolites Products of metabolism that are synthesized after growth has been completed. (973)

secondary treatment The biological degradation of dissolved organic matter in the process of sewage treatment; the organic material is either mineralized or changed to settleable solids. (638)

second law of thermodynamics Physical and chemical processes proceed in such a way that the entropy of the universe (the system and its surroundings) increases to the maximum possible. (152)

secretory IgA (sIgA) The primary immunoglobulin of the secretory immune system. *See* IgA. (714)

secretory vacuole In protists and some animals, these organelles usually contain specific enzymes that perform various functions such as excystation. Their contents are released to the cell exterior during exocytosis. (567)

segmented genome A virus genome that is divided into several parts or fragments, each probably coding for a single polypeptide; segmented genomes are very common among the RNA viruses. (364)

selectins (sə-lek'tins) A family of cell adhesion molecules that are displayed on activated endothelial cells; examples include P-selectin and E-selectin. Selectins mediate leukocyte binding to the vascular endothelium. (688)

selective media Culture media that favor the growth of specific microorganisms; this may be accomplished by inhibiting the growth of undesired microorganisms. (104)

selective toxicity The ability of a chemotherapeutic agent to kill or inhibit a microbial pathogen while damaging the host as little as possible. (781)

self-assembly The spontaneous formation of a complex structure from its component molecules without the aid of special enzymes or factors. (65, 201)

sepsis (sep'sis) Systemic response to infection. This systemic response is manifested by two or more of the following conditions as a result of infection: temperature >38 or <36°C; heart rate >90 beats per min; respiratory rate >20 breaths per min, or pCO$_2$ <32 mm Hg; leukocyte count >12,000 cells per ml^3 or >10% immature (band) forms. Sepsis also has been defined as the presence of pathogens or their toxins in blood and other tissues. (911)

septate (sep'tāt) Divided by a septum or cross wall; also with more or less regular occurring cross walls. (539)

septicemia (sep"tĭ-se'me-ah) A disease associated with the presence in the blood of pathogens or bacterial toxins. (500, 767)

septic shock (sep'tik) Sepsis associated with severe hypotension despite adequate fluid resuscitation, along with the presence of perfusion abnormalities that may include, but are not limited to, lactic acidosis, oliguria, or an acute alteration in mental status. Gram-positive bacteria, fungi, and endotoxin-containing gram-negative bacteria can initiate the pathogenic cascade of sepsis leading to septic shock. (911)

septic tank (sep'tik) A tank used to process domestic sewage. Solid material settles out and is partially degraded by anaerobic bacteria as sewage slowly flows through the tank. The outflow is further treated or dispersed in aerobic soil. (642)

septum (sep'tum; pl., **septa**) A partition or cross-wall that occurs between two cells in a bacterial (e.g., actinomycete) or fungal filament, or which partitions off fungal structures such as spores. Septa also divide parent cells into two daughter cells during bacterial binary fission. (281, 539)

serology (se-rol'o-je) The branch of immunology that is concerned with in vitro reactions involving one or more serum constituents (e.g., antibodies and complement). (751)

serotyping A technique or serological procedure that is used to differentiate between strains (serovars or serotypes) of microorganisms that have differences in the antigenic composition of a structure or product. (757)

serum (se'rum; pl., **serums** or **sera**) The clear, fluid portion of blood lacking both blood cells and fibrinogen. It is the fluid remaining after coagulation of plasma, the noncellular liquid faction of blood. (719)

serum resistance The type of resistance that occurs with bacteria such as *Neisseria gonorrhoeae* because the pathogen interferes with membrane attack complex formation during the complement cascade. (775)

settling basin A basin used during water purification to chemically precipitate out fine particles, microorganisms, and organic material by coagulation or flocculation. (631)

sex pilus (pi'lus) A thin protein appendage required for bacterial mating or conjugation. The cell with sex pili donates DNA to recipient cells. (63, 296)

sheath (shēth) A hollow tubelike structure surrounding a chain of cells and present in several genera of bacteria. (482)

shigellosis (shĭ'gəl-o'sis) The diarrheal disease that arises from an infection with *Shigella* spp. Often called bacillary dysentery. (909)

Shine-Dalgarno sequence A segment in the leader of procaryotic mRNA that binds to a special sequence on the 16S rRNA of the small ribosomal subunit. This helps properly orient the mRNA on the ribosome. (237)

shingles (herpes zoster) (shing'g'lz) A reactivated form of chickenpox caused by the varicella-zoster virus. (847)

siderophore (sid'er-o-for") A small molecule that complexes with ferric iron and supplies it to a cell by aiding in its transport across the plasma membrane. (101)

sigma factor A protein that helps the RNA polymerase core enzyme recognize the promoter at the start of a gene. (254)

sign An objective change in a diseased body that can be directly observed (e.g., a fever or rash). (824)

signal peptide The special amino-terminal sequence on a peptide destined for transport that delays protein folding and is recognized in bacteria by the Sec-dependent pathway machinery. (59)

silage Fermented plant material with increased palatability and nutritional value for animals, which can be stored for extended periods. (958)

silent mutation A mutation that does not result in a change in the organism's proteins or phenotype even though the DNA base sequence has been changed. (242)

simple matching coefficient (S_{SM}) An association coefficient used in numerical taxonomy; the proportion of characters that match regardless of whether or not the attribute is present. (414)

single radial immunodiffusion (RID) assay An immunodiffusion technique that quantitates antigens by following their diffusion through a gel containing antibodies directed against the test antigens. (754)

site-specific recombination Recombination of nonhomologous genetic material with a chromosome at a specific site. (286)

skin-associated lymphoid tissue (SALT) The lymphoid tissue in the skin that forms a first-line defense as a part of nonspecific (innate) immunity. (685)

S-layer A regularly structured layer composed of protein or glycoprotein that lies on the surface of many bacteria. It may protect the bacterium and help give it shape and rigidity. (62)

slime The viscous extracellular glycoproteins or glycolipids produced by staphylococci and *Pseudomonas aeruginosa* bacteria that allows them to adhere to smooth surfaces such as prosthetic medical devices and catheters. More generally, the term often refers to an easily removed, diffuse, unorganized layer of extracellular material that surrounds a bacterial cell. (61, 896)

slime layer A layer of diffuse, unorganized, easily removed material lying outside the bacterial cell wall. (61)

slime mold A common term for members of the divisions *Acrasiomycota* and *Myxomycota*. (548)

slow sand filter A bed of sand through which water slowly flows; the gelatinous microbial layer on the sand grain surface removes waterborne microorganisms, particularly *Giardia*, by adhesion to the gel. This type of filter is used in some water purification plants. (631)

slow virus disease A progressive, pathological process caused by a transmissible agent (virus or prion) that remains clinically silent during a prolonged incubation period of months to years after which progressive clinical disease becomes apparent. (400, 870)

sludge (sluj) A general term for the precipitated solid matter produced during water and sewage treatment; solid particles composed of organic matter and microorganisms that are involved in aerobic sewage treatment (activated sludge). (638)

small RNAs (sRNAs) Special small regulatory RNA molecules that do not function as messenger or ribosomal RNAs. (278)

smallpox (variola) (smawl'poks) Once a highly contagious, often fatal disease caused by a poxvirus. Its most noticeable symptom was the appearance of blisters and pustules on the skin. Vaccination has eradicated smallpox throughout the world. (851)

snapping division A distinctive type of binary fission resulting in an angular or a palisade arrangement of cells, which is characteristic of the genera *Arthrobacter* and *Corynebacterium*. (527)

sorocarp The fruiting structure of the Acrasiomycetes. (550)

sorus A type of fruiting structure composed of a mass of spores or sporangia. (550)

SOS repair A complex, inducible repair process that is used to repair DNA when extensive damage has occurred. (249)

source The location or object from which a pathogen is immediately transmitted to the host, either directly or through an intermediate agent. (828)

Southern blotting technique The procedure used to isolate and identify DNA fragments from a complex mixture. The isolated, denatured fragments are transferred from an agarose electrophoretic gel to a nitrocellulose filter and identified by hybridization with probes. (314)

specialized transduction *See* restricted transduction. (302)

species (spe'shēz) Species of higher organisms are groups of interbreeding or potentially interbreeding natural populations that are reproductively isolated. Bacterial species are collections of strains that have many stable properties in common and differ significantly from other groups of strains. (413)

specific immune response (acquired, adaptive, or specific immunity) *See* acquired immunity. (681)

spheroplast (sfēr'o-plast) A relatively spherical cell formed by the weakening or partial removal of the rigid cell wall component (e.g., by penicillin treatment of gram-negative bacteria). Spheroplasts are usually osmotically sensitive. (59)

spike *See* peplomer. (364)

spirillum (spi-ril'um) A rigid, spiral-shaped bacterium. (41)

spirochete (spi'ro-kēt) A flexible, spiral-shaped bacterium with periplasmic flagella. (41, 466)

spleen (splēn) A secondary lymphoid organ where old erythrocytes are destroyed and blood-borne antigens are trapped and presented to lymphocytes. (684)

split or interrupted gene A structural gene with DNA sequences that code for the final RNA product (expressed sequences or exons) separated by regions coding for RNA absent from the mature RNA (intervening sequences or introns). (257)

spongiform encephalopathies Degenerative central nervous system diseases in which the brain has a spongy appearance; they appear due to prions. (870)

sporadic disease (spo-rad'ik) A disease that occurs occasionally and at random intervals in a population. (822)

sporangiospore (spo-ran'je-o-spōr) A spore born within a sporangium. (522, 543)

sporangium (spo-ran'je-um; pl., **sporangia**) A saclike structure or cell, the contents of which are converted into an indefinite number of spores. It is borne on a special hypha called a sporangiophore. (69, 543)

spore (spōr) A differentiated, specialized form that can be used for dissemination, for survival of adverse conditions because of its heat and dessication resistance, and/or for reproduction. Spores are usually unicellular and may develop into vegetative organisms or gametes. They may be produced asexually or sexually and are of many types. (556)

sporogenesis (spor'o-jen'ĕ-sis) *See* sporulation. (69)

sporotrichosis (spo"ro-tri-ko'sis) A subcutaneous fungal infection caused by the dimorphic fungus *Sporothrix schenckii*. (921)

sporulation (spor"u-la'shun) The process of spore formation. (69)

spread plate A petri dish of solid culture medium with isolated microbial colonies growing on its surface, which has been prepared by spreading a dilute microbial suspension evenly over the agar surface. (104)

sputum (spu'tum) The mucous secretion from the lungs, bronchi, and trachea that is ejected (expectorated) through the mouth. (802)

stalk (stawk) A nonliving bacterial appendage produced by the cell and extending from it. (476)

standard free energy change The free energy change of a reaction at 1 atmosphere pressure when all reactants and products are present in their standard states; usually the temperature is 25°C. (152)

standard reduction potential A measure of the tendency of a reductant to lose electrons in an oxidation-reduction (redox) reaction. The more negative the reduction potential of a compound, the better electron donor it is. (153)

staphylococcal food poisoning (staf"i-lo-kok'al) A type of food poisoning caused by ingestion of improperly stored or cooked food in which *Staphylococcus aureus* has grown. The bacteria produce exotoxins that accumulate in the food. (909)

staphylococcal scalded skin syndrome (SSSS) A disease caused by staphylococci that produce an exfoliative toxin. The skin becomes red (erythema) and sheets of epidermis may separate from the underlying tissue. (899)

starter culture An inoculum, consisting of a mixture of carefully selected microorganisms, used to start a commercial fermentation. (951)

stationary phase (sta′shun-er″e) The phase of microbial growth in a batch culture when population growth ceases and the growth curve levels off. (111)

stem-nodulating rhizobia Rhizobia (members of the genera *Rhizobium, Bradyrhizobium,* and *Azorhizobium*) that produce nitrogen-fixing structures above the soil surface on plant stems. These most often are observed in tropical plants and produced by *Azorhizobium.* (659)

sterilization (ster″ĭ-lĭ-za′shun) The process by which all living cells, viable spores, viruses, and viroids are either destroyed or removed from an object or habitat. (134)

stigma (stig′mah) A light-sensitive eyespot, which is found in some algae and photosynthetic protozoa; it is believed to be involved in phototaxis, at least in some cases. (557)

stoneworts A group of approximately 250 species of algae that have a complex growth pattern, with nodal regions from which whorls of branches arise; they are abundant in fresh to brackish waters. (559)

strain A population of organisms that descends from a single organism or pure culture isolate. (413)

streak plate A petri dish of solid culture medium with isolated microbial colonies growing on its surface, which has been prepared by spreading a microbial mixture over the agar surface, using an inoculating loop. (104)

streptococcal pneumonia An endogenous infection of the lungs caused by *Streptococcus pneumoniae* that occurs in predisposed individuals. (881)

streptococcal sore throat (strep″to-kok′al) One of the most common bacterial infections of humans. It is commonly referred to as "strep throat." The disease is spread by droplets of saliva or nasal secretions and is caused by *Streptococcus* spp. (particularly group A streptococci). (881)

streptolysin-O (SLO) (strep-tol′ĭ-sin) A specific hemolysin produced by *Streptococcus pyogenes* that is inactivated by oxygen (hence the "O" in its name). SLO causes beta-hemolysis of blood cells on agar plates incubated anaerobically. (772)

streptolysin-S (SLS) A product of *Streptococcus pyogenes* that is bound to the bacterial cell but may sometimes be released. SLS causes beta hemolysis on aerobically incubated blood-agar plates and can act as a leukocidin by killing leukocytes that phagocytose the bacterial cell to which it is bound. (772)

streptomycin (strep′to-mi″sin) A bactericidal aminoglycoside antibiotic produced by *Streptomyces griseus.* (791)

strict anaerobes *See* obligate anaerobes. (125)

stroma (stro′mah) The chloroplast matrix that is the location of the photosynthetic carbon dioxide fixation reactions. (84)

stromatolite (stro″mah-to′lit) Dome-like microbial mat communities consisting of filamentous photosynthetic bacteria and occluded sediments (often calcareous or siliceous). They usually have a laminar structure. Many are fossilized, but some modern forms occur. (410)

structural gene A gene that codes for the synthesis of a polypeptide or polynucleotide with a nonregulatory function. (271)

subacute sclerosing panencephalitis Diffuse inflammation of the brain resulting from virus and prion infections. (849)

subgingival plaque (sub-jin′jĭ-val) The plaque that forms at the dentogingival margin and extends down into the gingival tissue. (914)

substrate-level phosphorylation The synthesis of ATP from ADP by phosphorylation coupled with the exergonic breakdown of a high-energy organic substrate molecule. (171)

subsurface biosphere The region below the plant root zone where microbial populations can grow and function. (667)

sulfate reduction (sul′fat) The process of sulfate use as an oxidizing agent, which results in the accumulation of reduced forms of sulfur such as sulfide, or incorporation of sulfur into organic molecules, usually as sulfhydryl groups. (597)

sulfonamide (sul-fon′ah-mīd) A chemotherapeutic agent that has the SO_2-NH_2 group and is a derivative of sulfanilamide. (787)

superantigen Superantigens are bacterial proteins that stimulate the immune system much more extensively than do normal antigens. They stimulate T cells to proliferate nonspecifically through simultaneous interaction with class II MHC proteins on antigen-presenting cells and variable regions on the β chain of the T-cell receptor complex. Examples include streptococcal scarlet fever toxins, staphylococcal toxic shock syndrome toxin-1, and streptococcal M protein. (709)

superinfection (soo″per-in-fek′shun) A new bacterial or fungal infection of a patient that is resistant to the drug(s) being used for treatment. (794)

superoxide dismutase (SOD) (soo″per-ok′sīd dis-mu′tas) An enzyme that protects many microorganisms by catalyzing the destruction of the toxic superoxide radical. (126)

suppressor mutation A mutation that overcomes the effect of another mutation and produces the normal phenotype. (242)

Svedberg unit (sfed′berg) The unit used in expressing the sedimentation coefficient; the greater a particle's Svedberg value, the faster it travels in a centrifuge. (50)

swab (swahb) A wad of absorbent material usually wound around one end of a small stick and used for applying medication or for removing material from an area; also, a dacron-tipped polystyrene applicator. (800)

swarm cell A flagellated cell; the term is usually applied to the motile cells of the *Myxomycota.* (548)

symbiosis (sim″bi-o′sis) The living together or close association of two dissimilar organisms, each of these organisms being known as a symbiont. (578)

symbiosome The final nitrogen-fixing form of *Rhizobium* within root nodule cells. (655)

symptom (simp′təm) A change during a disease that a person subjectively experiences (e.g., pain,

bodily discomfort, fatigue, or loss of appetite). Sometimes the term symptom is used more broadly to include any observed signs. (824)

syndrome *See* disease syndrome. (824)

synthetic medium *See* defined medium. (102)

syntrophism (sin′trŏf-izəm) The association in which the growth of one organism either depends on, or is improved by, the provision of one or more growth factors or nutrients by a neighboring organism. Sometimes both organisms benefit. (585)

syphilis (sif′ĭ-lis) *See* venereal syphilis. (899)

systematic epidemiology The field of epidemiology that focuses on the ecological and social factors that influence the development of emerging and reemerging infectious diseases. (833)

systematics (sis″te-mat′iks) The scientific study of organisms with the ultimate objective being to characterize and arrange them in an orderly manner; often considered synonymous with taxonomy. (410)

systemic lupus erythematosus (loo′pus er″ĭ-them-ah-to′sus) An autoimmune, inflammatory disease that may affect every tissue of the body. (747)

T

taxon (tak′son) A group into which related organisms are classified. (410)

taxonomy (tak-son′o-me) The science of biological classification; it consists of three parts: classification, nomenclature, and identification. (410)

TB skin test Tuberculin hypersensitivity test for a previous or current infection with *Mycobacterium tuberculosis.* (747)

T cell or **T lymphocyte** A type of lymphocyte derived from bone marrow stem cells that matures into an immunologically competent cell under the influence of the thymus. T cells are involved in a variety of cell-mediated immune reactions. (681, 721)

T-cell antigen receptor (TCR) The receptor on the T cell surface consisting of two antigen-binding peptide chains; it is associated with a large number of other glycoproteins. Binding of antigen to the TCR, usually in association with MHC, activates the T cell. (721)

T-dependent antigen An antigen that effectively stimulates B-cell response only with the aid of T-helper cells that produce interleukin-2 and B-cell growth factor. (729)

teichoic acids (ti-ko′ik) Polymers of glycerol or ribitol joined by phosphates; they are found in the cell walls of gram-positive bacteria. (55)

temperate phages Bacteriophages that can infect bacteria and establish a lysogenic relationship rather than immediately lysing their hosts. (301, 380)

template strand (tem′plat) A DNA or RNA strand that specifies the base sequence of a new complementary strand of DNA or RNA. (235)

terminator A sequence that marks the end of a gene and stops transcription. (238, 255)

tertiary treatment (ter′she-er-e) The removal from sewage of inorganic nutrients, heavy metals, viruses, etc., by chemical and biological means after microbes have degraded dissolved organic material during secondary sewage treatment. (640)

test A loose-fitting shell of an amoeba. (571)

tetanolysin (tet″ah-nol′ĭ-sin) A hemolysin that aids in tissue destruction and is produced by *Clostridium tetani*. (903)

tetanospasmin (tet″ah-no-spaz′min) The neurotoxic component of the tetanus toxin, which causes the muscle spasms of tetanus. Tetanospasmin production is controlled by a plasmid gene. (902)

tetanus (tet′ah-nus) An often fatal disease caused by the anaerobic, spore-forming bacillus *Clostridium tetani*, and characterized by muscle spasms and convulsions. (902)

tetracyclines (tet″rah-si′klēns) A family of antibiotics with a common four-ring structure, which are isolated from the genus *Streptomyces* or produced semisynthetically; all are related to chlortetracycline or oxytetracycline. (790)

tetrapartite associations (tet″rah-par′tīt) A symbiotic association of the same plant with three different types of microorganisms. (661)

T$_H$1 cell *See* regulator T cell. (725)

T$_H$2 cell *See* regulator T cell. (725)

T$_H$0 cell *See* regulator T cell. (725)

thallus (thal′us) A type of body that is devoid of root, stem, or leaf; characteristic of some algae, many fungi, and lichens. (522, 539, 556)

T-helper (T$_H$) cell A cell that is needed for T-cell-dependent antigens to be effectively presented to B cells. It also promotes cell-mediated immune responses. (725)

theory A set of principles and concepts that have survived rigorous testing and that provide a systematic account of some aspect of nature. (9)

thermal death time (TDT) The shortest period of time needed to kill all the organisms in a microbial population at a specified temperature and under defined conditions. (137)

thermoacidophiles A group of bacteria that grow best at acid pHs and high temperatures; they are members of the *Archaea*. (444)

thermophile (ther′mo-fīl) A microorganism that can grow at temperatures of 55°C or higher; the minimum is usually around 45°C. (124)

thrush (thrush) Infection of the oral mucous membrane by the fungus *Candida albicans;* also known as oral candidiasis. (925)

thylakoid (thi′lah-koid) A flattened sac in the chloroplast stroma that contains photosynthetic pigments and the photosynthetic electron transport chain; light energy is trapped and used to form ATP and NAD(P)H in the thylakoid membrane. (84)

thymine (thi′min) The pyrimidine 5-methyluracil that is found in nucleosides, nucleotides, and DNA. (211)

thymus (thi′məs) A primary lymphoid organ in the chest that is necessary in early life for the development of immunological functions. T-cell maturation takes place here. (684)

T-independent antigen An antigen that triggers a B cell into immunoglobulin production without T-cell cooperation. (730)

tinea (tin′e-ah) A name applied to many different kinds of superficial fungal infections of the skin, nails, and hair, the specific type (depending on characteristic appearance, etiologic agent, and site) usually designated by a modifying term. (919)

tinea capitis An infection of scalp hair by species of *Trichophyton* or *Microsporum*. (920)

tinea corporis An infection of the smooth parts of the skin by either *Trichophyton rubrum, T. mentagrophytes,* or *Microsporum canis*. (920)

tinea cruris An infection of the groin by either *Epidermophyton floccosum, Trichophyton mentagrophytes,* or *T. rubrum;* also known as jock itch. (920)

tinea pedis A fungal infection of the foot by *Trichophyton rubrum, T. mentagrophytes,* or *E. floccosum;* also known as athlete's foot. (920)

tinea unguium An infection of the nail bed by either *Trichophyton rubrum* or *T. mentagrophytes*. (920)

Ti plasmid A plasmid obtained from *Agrobacterium tumefaciens* that is used to insert genes into plant cells. (330, 660)

titer (ti′ter) Reciprocal of the highest dilution of an antiserum that gives a positive reaction in the test being used. (719)

T lymphocyte *See* T cell. (681)

Toll-like receptor A type of pattern recognition receptor on phagocytes such as macrophages that triggers the proper response to different classes of pathogens. It signals the production of transcription factor NFκB, which stimulates formation of cytokines, chemokines, and other defense molecules. (696)

tonsillitis (ton′si-li′tis) Inflammation of the tonsils, especially the palatine tonsils often due to *S. pyogenes* infection. (881)

toxemia (tok-se′me-ah) The condition caused by toxins in the blood of the host. (769)

toxic shock-like syndrome (TSLS) A disease caused by an invasive group A streptococcus infection that is characterized by a rapid drop in blood pressure, failure of many organs, and a very high fever. It probably results from the release of one or more streptococcal pyrogenic exotoxins. (880)

toxic shock syndrome (TSS) (tok′sik) A staphylococcal disease that most commonly affects females who use certain types of tampons during menstruation. It is associated with the toxic shock syndrome toxin produced by strains of *Staphylococcus aureus*. (898)

toxigenicity (tok″sĭ-jĕ-nis′i-tē) The capacity of an organism to produce a toxin. (764)

toxin (tok′sin) A microbial product or component that injures another cell or organism. Often the term refers to a poisonous protein, but toxins may be lipids and other substances. (769)

toxin neutralization The inactivation of toxins by specific antibodies, called antitoxins, that react with them. (732)

toxoid (tok′soid) A bacterial exotoxin that has been modified so that it is no longer toxic but will still stimulate antitoxin formation when injected into a person or animal. (743, 769)

toxoplasmosis (tok″so-plaz-mo′sis) A disease of animals and humans caused by the parasitic protozoan, *Toxoplasma gondii*. (933)

trachoma (trah-ko′mah) A chronic infectious disease of the conjunctiva and cornea, producing pain, inflammation and sometimes blindness. It is caused by *Chlamydia trachomatis* serotypes A–C. (903)

transamination (trans″am-i-na′shun) The removal of amino acid's amino group by transferring it to an α-keto acid acceptor. (188)

transcriptase (trans-krip′tās) An enzyme that catalyzes transcription; in viruses with RNA genomes, this enzyme is an RNA-dependent RNA polymerase that is used to make RNA copies of the RNA genomes. (394)

transcription (trans-krip′shun) The process in which single-stranded RNA with a base sequence complementary to the template strand of DNA or RNA is synthesized. (224)

transduction (trans-duk′shun) The transfer of genes between bacteria by bacteriophages. (301)

transfer host (trans′fer) A host that is not necessary for the completion of a parasite's life cycle, but is used as a vehicle for reaching a final host. (762)

transfer RNA (tRNA) A small RNA that binds an amino acid and delivers it to the ribosome for incorporation into a polypeptide chain during protein synthesis. (254)

transformation (trans″for-ma′shun) A mode of gene transfer in bacteria in which a piece of free DNA is taken up by a bacterial cell and integrated into the recipient genome. (222, 299)

transgenic animal or **plant** An animal or plant that has gained new genetic information from the insertion of foreign DNA. It may be produced by such techniques as injecting DNA into animal eggs, electroporation of mammalian cells and plant cell protoplasts, or shooting DNA into plant cells with a gene gun. (327)

transient carrier *See* casual carrier. (828)

transition mutations (tran-zish′un) Mutations that involve the substitution of a different purine base for the purine present at the site of the mutation or the substitution of a different pyrimidine for the normal pyrimidine. (241)

translation (trans-la′shun) Protein synthesis; the process by which the genetic message carried by mRNA directs the synthesis of polypeptides with the aid of ribosomes and other cell constituents. (224, 260)

translocon The machinery that transports proteins across membranes. Usually refers to the Sec machinery. (60)

transmission electron microscope (TEM) (trans-mish′un) A microscope in which an image is formed by passing an electron beam through a specimen and focusing the scattered electrons with magnetic lenses. (31)

transovarian passage (trans″o-va′re-an) The passage of a microorganism such as a rickettsia from generation to generation of hosts through tick eggs.

(No humans or other mammals are needed as reservoirs for continued propagation.) (888)

transpeptidation 1. The reaction that forms the peptide cross-links during peptidoglycan synthesis. 2. The reaction that forms a peptide bond during the elongation cycle of protein synthesis. (217, 265)

transposable elements *See* transposon. (291)

transposition (trans″po-zish′un) The movement of a piece of DNA around the chromosome. (291)

transposon (tranz-po′zon) A DNA segment that carries the genes required for transposition and moves about the chromosome; if it contains genes other than those required for transposition, it may be called a composite transposon. Often the name is reserved only for transposable elements that also contain genes unrelated to transposition. (291)

transversion mutations (trans-ver′zhun) Mutations that result from the substitution of a purine base for the normal pyrimidine or a pyrimidine for the normal purine. (241)

traveler's diarrhea A type of diarrhea resulting from ingestion of viruses, bacteria, or protozoa normally absent from the traveler's environment. A major pathogen is enterotoxigenic *Escherichia coli*. (910)

tricarboxylic acid (TCA) cycle The cycle that oxidizes acetyl coenzyme A to CO_2 and generates NADH and $FADH_2$ for oxidation in the electron transport chain; the cycle also supplies carbon skeletons for biosynthesis. (178, A-14)

trichome (tri′kōm) A row or filament of bacterial cells that are in close contact with one another over a large area. (459)

trichomoniasis (trik″o-mo-ni′ah-sis) A sexually transmitted disease caused by the parasitic protozoan *Trichomonas vaginalis*. (934)

trickling filter A bed of rocks covered with a microbial film that aerobically degrades organic waste during secondary sewage treatment. (639)

tripartite associations (tri-par′tīt) A symbiotic association of the same plant with two types of microorganisms. (661)

trophozoite (trof″ o-zo′īt) The active, motile feeding stage of a protozoan organism; in the malarial parasite, the stage of schizogony between the ring stage and the schizont. (568)

tropism (tro′piz-əm) The movement of living organisms toward or away from a focus of heat, light, or other stimulus. (765)

trypanosome (tri-pan′o-sōm) A protozoan of the genus *Trypanosoma*. Trypanosomes are parasitic flagellate protozoa that often live in the blood of humans and other vertebrates and are transmitted by insect bites. (570, 933)

trypanosomiasis (tri-pan″o-so-mi′ah-sis) An infection with trypanosomes that live in the blood and lymph of the infected host. (933)

tubercle (too′ber-k′l) A small, rounded nodular lesion produced by *Mycobacterium tuberculosis*. (884)

tuberculoid (neural) leprosy (too-ber′ku-loid) A mild, nonprogressive form of leprosy that is associated with delayed-type hypersensitivity to antigens

on the surface of *Mycobacterium leprae*. It is characterized by early nerve damage and regions of the skin that have lost sensation and are surrounded by a border of nodules. (893)

tuberculosis (TB) (too-ber″ku-lo′sis) An infectious disease of humans and other animals resulting from an infection by a species of *Mycobacterium* and characterized by the formation of tubercles and tissue necrosis, primarily as a result of host hypersensitivity and inflammation. Infection is usually by inhalation, and the disease commonly affects the lungs (pulmonary tuberculosis), although it may occur in any part of the body. (882)

tuberculous cavity (too-ber′ku-lus) An air-filled cavity that results from a tubercle lesion caused by *M. tuberculosis*. (884)

tularemia (too″lah-re′me-ah) A plaguelike disease of animals caused by the bacterium *Francisella tularensis* subsp. *tularensis* (Jellison type A), which may be transmitted to humans. (903)

tumble Random turning or tumbling movements made by bacteria when they stop moving in a straight line. (67)

tumor (too′mor) A growth of tissue resulting from abnormal new cell growth and reproduction (neoplasia). (400)

turbidostat A continuous culture system equipped with a photocell that adjusts the flow of medium through the culture vessel so as to maintain a constant cell density or turbidity. (118)

twiddle *See* tumble. (67)

two-component phosphorelay system A signal transduction regulatory system that uses the transfer of phosphoryl groups to control gene transcription and protein activity. It has two major components: a sensor kinase and a response regulator. (279)

type I hypersensitivity A form of immediate hypersensitivity arising from the binding of antigen to IgE attached to mast cells, which then release anaphylaxis mediators such as histamine. Examples: hay fever, asthma, and food allergies. (744)

type II hypersensitivity A form of immediate hypersensitivity involving the binding of antibodies to antigens on cell surfaces followed by destruction of the target cells (e.g., through complement attack, phagocytosis, or agglutination). (746)

type III hypersensitivity A form of immediate hypersensitivity resulting from the exposure to excessive amounts of antigens in which antibodies bind to the antigens and produce antibody-antigen complexes. These activate complement and trigger an acute inflammatory response with subsequent tissue damage. Examples: poststreptococcal glomerulonephritis, serum sickness, and farmer's lung disease. (746)

type IV hypersensitivity A delayed hypersensitivity response (it appears 24 to 48 hours after antigen exposure). It results from the binding of antigen to activated T lymphocytes, which then release cytokines and trigger inflammation and macrophage attacks that damage tissue. Type IV hypersensitivity is

seen in contact dermititis from poison ivy, leprosy, and tertiary syphilis. (746)

type I protein secretion pathway *See* ABC protein secretion pathway. (61)

type II protein secretion pathway A system that transports proteins from the periplasm across the outer membrane of gram-negative bacteria. (61)

type III protein secretion pathway A system in gram-negative bacteria that secretes virulence factors and injects them into host cells. (61)

typhoid fever (ti-foid) A bacterial infection transmitted by contaminated food, water, milk, or shellfish. The causative organism is *Salmonella typhi*, which is present in human feces. (910)

U

ultramicrobacteria Bacteria that can exist normally in a miniaturized form or which are capable of miniaturization under low-nutrient conditions. They may be 0.2 μm or smaller in diameter. (621)

ultraviolet (UV) radiation (ul″trah-vi′o-let) Radiation of fairly short wavelength, about 10 to 400 nm, and high energy. (128, 141)

uracil (u′rah-sil) The pyrimidine 2,4-dioxypyrimidine, which is found in nucleosides, nucleotides, and RNA. (211)

V

vaccine (vak′sēn) A preparation of either killed microorganisms; living, weakened (attenuated) microorganisms; or inactivated bacterial toxins (toxoids). It is administered to induce development of the immune response and protect the individual against a pathogen or a toxin. (740)

vaccinomics The application of genomics and bioinformatics to vaccine development. (743)

valence (va′lens) The number of antigenic determinant sites on the surface of an antigen or the number of antigen-binding sites possessed by an antibody molecule. (708)

variable region (V_L and V_H) The region at the N-terminal end of immunoglobulin heavy and light chains whose amino acid sequence varies between antibodies of different specificity. Variable regions form the antigen binding site. (711)

vasculitis (vas″ku-li′tis) Inflammation of a blood vessel. (885)

vector (vek′tor) 1. In genetic engineering, another name for a cloning vector. A DNA molecule that can replicate (a replicon) and transports a piece of inserted foreign DNA, such as a gene, into a recipient cell. It may be a plasmid, phage, cosmid or artificial chromosome. 2. In epidemiology, it is a living organism, usually an arthropod or other animal, that transfers an infective agent between hosts. (314, 765, 829, 831)

vector-borne transmission The transmission of an infectious pathogen between hosts by means of a vector. (831)

vehicle (ve'ĭ-k'l) An inanimate substance or medium that transmits a pathogen. (831)

venereal syphilis (ve-ne're-al sif'ĭ-lis) A contagious, sexually transmitted disease caused by the spirochete *Treponema pallidum*. (899)

venereal warts *See* anogenital condylomata. (871)

verrucae vulgaris (v'ě-roo'se vul-ga'ris; s. **verruca vulgaris**) The common wart; a raised, epidermal lesion with horny surface caused by an infection with a human papillomavirus. (871)

vibrio (vib're-o) A rod-shaped bacterial cell that is curved to form a comma or an incomplete spiral. (41)

viral hemagglutination (vi'ral hem"ah-gloo"tĭ-na'shun) The clumping or agglutination of red blood cells caused by some viruses. (751)

viral neutralization An antibody-mediated process in which IgG, IgM, and IgA antibodies bind to some viruses during their extracellular phase and inactivate or neutralize them. (732)

viremia (vi-re'me-ə) The presence of viruses in the blood stream. (765)

viricide (vir'i-sīd) An agent that inactivates viruses so that they cannot reproduce within host cells. (135)

virion (vi're-on) A complete virus particle that represents the extracellular phase of the virus life cycle; at the simplest, it consists of a protein capsid surrounding a single nucleic acid molecule. (353)

virioplankton Viruses that occur in waters; high levels are found in marine and freshwater environments. (622)

viroid (vi'roid) An infectious agent of plants that is a single-stranded RNA not associated with any protein; the RNA does not code for any proteins and is not translated. (405)

virology (vi-rol'o-je) The branch of microbiology that is concerned with viruses and viral diseases. (352)

virulence (vir'u-lens) The degree or intensity of pathogenicity of an organism as indicated by case fatality rates and/or ability to invade host tissues and cause disease. (764)

virulence factor A bacterial product, usually a protein or carbohydrate, that contributes to virulence or pathogenicity. (766)

virulent bacteriophages (vir'u-lent bak-te're-o-fājs") Bacteriophages that lyse their host cells during the reproductive cycle. (380)

virus (vi'rus) An infectious agent having a simple acellular organization with a protein coat and a single type of nucleic acid, lacking independent metabolism, and reproducing only within living host cells. (353)

vitamin (vi'tah-min) An organic compound required by organisms in minute quantities for growth and reproduction because it cannot be synthesized by the organism; vitamins often serve as enzyme cofactors or parts of cofactors. (97)

volutin granules (vo-lu'tin) *See* metachromatic granules. (49)

W

wart (wort) An epidermal tumor of viral origin. (871)

wastewater treatment The use of physical and biological processes to remove particulate and dissolved material from sewage and to control pathogens. (638)

water activity (a_w) A quantitative measure of water availability in the habitat; the water activity of a solution is one-hundredth its relative humidity. (119)

water mold A common term for a member of the division *Oomycota*. (548)

West Nile fever (encephalitis) A neurological disease that is spread from birds to humans by mosquitoes. It first appeared in the United States in 1999 and subsequently has been reported in almost every state. (854)

white piedra A fungal infection caused by the yeast *Trichosporon beigelii* that forms light-colored nodules on the beard and mustache. (919)

whole-genome shotgun sequencing An approach to genome sequencing in which the complete genome is broken into random fragments, which are then individually sequenced. Finally the fragments are placed in the proper order using sophisticated computer programs. (337)

whole-organism vaccine A vaccine made from complete pathogens, which can be of four types: inactivated viruses; attenuated viruses; killed microorganisms; and live, attenuated microbes. (743)

Widal test (ve-dahl') A test involving agglutination of typhoid bacilli when they are mixed with serum containing typhoid antibodies from an individual having typhoid fever; used to detect the presence of *Salmonella typhi* and *S. paratyphi*. (751)

Winogradsky column A glass column with an anaerobic lower zone and an aerobic upper zone, which allows growth of microorganisms under conditions similar to those found in a nutrient-rich lake. (618)

wort The filtrate of malted grains used as the substrate for the production of beer and ale by fermentation. (955)

X

xenograft (zen"o-graft) A tissue graft between animals of different species. (749)

xerophilic microorganisms (ze"ro-fil'ik) Microorganisms that grow best under low a_w conditions, and may not be able to grow at high a_w values. (939)

Y

yeast (yēst) A unicellular, uninuclear fungus that reproduces either asexually by budding or fission, or sexually through spore formation. (539)

yeast artificial chromosome (YAC) A stretch of DNA that contains all the elements required to propagate a chromosome in yeast and which is used to clone foreign DNA fragments in yeast cells. (326)

yellow fever An acute infectious disease caused by a flavivirus, which is transmitted to humans by mosquitoes. The liver is affected and the skin turns yellow in this disease. (854)

YM shift The change in shape by dimorphic fungi when they shift from the yeast (Y) form in the animal body to the mold or mycelial form (M) in the environment. (541)

Z

zooflagellates (zo"o-flăj'-e-lăts) Flagellate protozoa that do not have chlorophyll and are either holozoic, saprozoic, or symbiotic. (569)

zoonosis (zo"o-no'sis; pl. *zoonoses*) A disease of animals that can be transmitted to humans. (823)

zooplankton (zo"o-plank'ton) A community of floating, aquatic, minute animals and nonphotosynthetic protists. (555)

zoospore (zo'o-spōr) A motile, flagellated spore. (556)

zooxanthella (zo"o-zan-thel'ah) A dinoflagellate found living symbiotically within cnidarians and other invertebrates. (562, 580)

zoster *See* shingles. (847)

z value The increase in temperature required to reduce the decimal reduction time to one-tenth of its initial value. (137)

zygomycetes (zi"go-mi-se'tez) A division of fungi that usually has a coenocytic mycelium with chitinous cell walls. Sexual reproduction normally involves the formation of zygospores. The group lacks motile spores. (544)

zygospore (zi'go-spōr) A thick-walled, sexual, resting spore characteristic of the zygomycetous fungi. (543)

zygote (zi'gōt) The diploid (2n) cell resulting from the fusion of male and female gametes. (557)

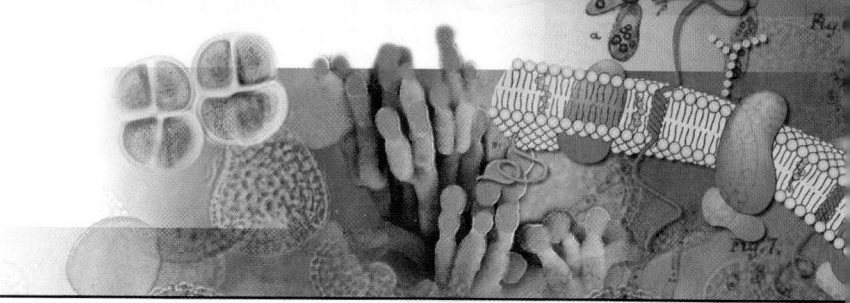

Credits

Photographs

Chapter 1

Opener: © John D. Cunningham/Visuals Unlimited; **Fig. 1.1a:** Corbis; **Figs. 1.2, 1.4:** Corbis; **Fig. 1.5:** American Society for Microbiology; **Fig. 1.6:** North Wind Picture Archives; **Fig. 1.7a:** Rita R. Colwell; **Fig. 1.7b:** Dr. Robert G.E. Murray; **Fig. 1.7c:** American Society for Microbiology Archives Collection; **Fig. 1.7d:** Martha M. Howe; **1.7e:** Frederick C. Neidhardt; **Fig. 1.7f:** Jean E. Brenchley.

Chapter 2

Opener: © George J. Wilder/Visuals Unlimited; **Fig. 2.3:** Courtesy of Leica, Inc.; **Fig. 2.4:** Courtesy of Nikon Inc.; **Fig. 2.8a:** © Arthur M. Siegelman/Visuals Unlimited; **Fig. 2.8b:** © Robert Calentine/Visuals Unlimited; **Fig. 2.8c:** © M. Abbey/Visuals Unlimited; **Fig. 2.8d:** © George J. Wilder/Visuals Unlimited; **Fig. 2.8e:** © M. Abbey/Visuals Unlimited; **Fig. 2.11:** © M. Abbey/Visuals Unlimited; **Fig. 2.13a:** © Arthur M. Siegelman/Visuals Unlimited; **Fig. 2.13b:** Courtesy of Joan Smith Sonneborn, University of Wyoming; **Fig. 2.13c:** Courtesy of Sanofi Diagnostics; **Fig. 2.13d:** Photo contributed by Bruce Roth and Paul Millard, Molecular Probes, Inc.; **Fig. 2.15a:** ©Arthur M. Siegelman/Visuals Unlimted; **Fig. 2.15b:** © Michael A. Gabridge/Visuals Unlimited; **Fig. 2.15c:** Dr. L. Tomalty/Dr. G. Delisle, Queen's University, Kingston, Ontario, Canada; **Fig. 2.15d:** © George J. Wilder/Visuals Unlimited; **Figs. 2.16, 2.17:** © John D. Cunningham/Visuals Unlimited; **Fig. 2.18:** Lansing Prescott; **Fig. 2.19:** © John D. Cunningham/Visuals Unlimited; **Fig. 2.21a:** © George J. Wilder/Visuals Unlimited; **Fig. 2.21b:** © Biology Media/Photo Researchers, Inc.; **Fig. 2.21c:** © KG Murti/Visuals Unlimited; **Fig. 2.22:** © William Ormerod, Jr./Visuals Unlimited; **Fig. 2.24a,b:** © Fred Hossler/Visuals Unlimited; **Fig. 2.26:** Courtesy of E.J. Laishley, University of Calgary; **Fig. 2.28a:** © David M. Phillips/Photo Researches, Inc.; **Fig. 2.28b:** © Paul W. Johnson/Biological Photo Service; **Fig. 2.31a,b:** From L. Tijuis, W.A.J. van Benthum, M.C.M. van Loosdrecht, and J.J. Heinen, "Solids Retention Time in Spherical Biofilms in a Biofilm Airlift Suspension Reactor," *Biotechnology and Bioengineering,* 44:867–879. © 1994 Wiley-Liss, Inc. Reprinted by permission of Wiley Liss, Inc., a subsidiary of John Wiley & Sons, Inc.; **Fig. 2.32:** © Driscoll, Youngquist, and Baldeschwieler, Caltech/SPL/Photo Researchers, Inc.

Chapter 3

Opener: © E.C.S. Chan/Visuals Unlimited; **Fig. 3.1a:** Dr. Leon J. Le Beau; **Fig. 3.1b:** © Arthur M. Siegelman/Visuals Unlimited; **Fig. 3.1c:** © George J. Wilder/Visuals Unlimited; **Fig. 3.1d:** © Thomas Tottleben/Tottleben Scientific Company; **Fig. 3.1e:** Dr. Leon J. Le Beau; **Fig. 3.2a-c:** © David M. Phillips/Visuals Unlimited; **Fig. 3.2d:** Reprinted from the *Shorter Bergey's Manual of Determinative Bacteriology,* 8/e, John G. Holt, editor, 1977, © Bergey's Manual Trust. Published by Williams and Wilkins, Baltimore, MD; **Fig.**

3.2e: From Walther Stoeckenius: *Walsby's Square Bacterium: Fine Structures of an Orthogonal Procaryote*; **Fig. 3.2f:** © Hans Hanert; **Box 3.1a:** Esther Angert, Cornell University; **Box Fig. 3.1b:** Reprinted with permission from Science 284, 16 April 1999, Fig. 1b, page 494. © 1999 American Association for the Advancement of Science. Image courtesy of Heide Schulz; **Fig. 3.8:** © Stanley C. Holt/Biological Photo Service; **Fig. 3.9a:** American Society for Microbiology; **Fig. 3.9b:** Reprinted from the *Shorter Bergey's Manual of Determinative Bacteriology,* 8/e, John G. Holt, editor, 1977, © Bergey's Manual Trust. Published by Williams and Wilkins, Baltimore, MD; **Fig. 3.10:** *American Scientist,* Volume 87, November–December 1999. Image courtesy of David S. Goodsell, the Scripps Research Institute; **Fig. 3.11:** © Ralph A. Slepecky/Visuals Unlimited; **Fig. 3.12a:** American Society for Microbiology; **Fig. 3.12b:** Courtesy of Daniel Branton, Harvard University; **Fig. 3.13a:** National Research Council of Canada; **Fig. 3.13b:** Reprinted from the *Shorter Bergey's Manual of Determinative Bacteriology,* 8/e, John G. Holt, editor, 1977, © Bergey's Manual Trust. Published by Williams and Wilkins, Baltimore, MD ; **Box 3.2a:** D. Balkwill and D. Maratea; **Box 3.2b:** Y. Gorby; **Box 3.2c:** Courtesy of Ralph Wolfe and A. Spormann, University of Illinois at Urbana-Champaign; **Fig. 3.14a,b,c:** American Society for Microbiology; **Fig. 3.15:** © T.J. Beveridge/Biological Photo Service; **Fig. 3.19b:** From H. Formanek et al., "Peptido Glycan Structure" from, *Eur. Journal of Biochemisty* 253, 383–389; **Fig. 3.20:** Courtesy of M.R.J. Salton, NYU Medical Center; **Fig. 3.24:** Reprinted from J.M. Ghuysen and R. Hakenbeck/Bacterial Cell Wall, pp. 263–79. With kind permission from Elsevier Science-NL, Sara Burgerhartstratt 25, 1055 KV Amsterdam, The Netherlands; **Fig. 3.25b:** From M. Kastowsky, T. Gutberlet, and H. Bradaczek, *Journal of Bacteriology,* 774: 4798–4806, 1992; **Fig. 3.28a,b:** © John D. Cunningham/Visuals Unlimited; **Fig. 3.29:** © Georg Musil/Visuals Unlimited; **Fig. 3.30:** Courtesy of R.G. E. Murray, University of Western Ontario; **Fig. 3.31:** © Fred Hossler/Visuals Unlimited; **Fig. 3.32a,b:** © E.C.S. Chan/Visuals Unlimited; **Fig. 3.32c:** © George J. Wilder/Visuals Unlimited; **Figs. 3.33a,b, 3.38, 3.39:** Courtesy of Dr. Julius Adler; **Fig. 3.42:** American Society for Microbiology; **Fig. 3.44a-f:** Academic Press; **Fig. 3.45:** American Society for Microbiology.

Chapter 4

Opener: © Arthur M. Siegelman/Visuals Unlimited; **Fig. 4.1a:** © Eric Grave/Photo Researchers, Inc.; **Fig. 4.1b:** Carolina Biological Supply/Phototake; **Fig. 4.1c:** © Arthur M. Siegelman/Visuals Unlimited; **Fig. 4.1d:** © John D. Cunningham/Visuals Unlimited; **Fig. 4.1e:** © Tom E. Adams/Visuals Unlimited; **Fig. 4.1f:** © John D. Cunningham/Visuals Unlimited; **Fig. 4.2a:** © Richard Rodewald/Biological Photo Service; **Fig. 4.2b:** R.F. Illingworth, A.H. Rose, A. Beckett, "Change in the Lipid Composition and Fine Structure of Saccharomyes Cerevisiae During Ascus Formation," *Journal of Bacteriology* 113:1, 373–386, Fig. 4 on page 381, American Society of Microbiology; **Fig. 4.4:** Reprinted Fig. 3a on page 98, L. Mahadevan & P. Matsudaira, with permis-

sion from *Science,* Volume 288: 95–98, April 7, 2000. © 2000 American Association for the Advancement of Science. Image courtesy of Lewis Tilney; **Figs. 4.6a–4.7b:** © Manfred Schliwa/Visuals Unlimited; **Fig. 4.8:** © B.F. King/Biological Photo Service; **Fig. 4.9a:** © Henry C. Aldrich/Visuals Unlimited; **Fig. 4.10b:** Image courtesy of James Burbach; **Fig. 4.14b:** Academic Press; **Fig. 4.14c:** © Keith Porter/Photo Researchers, Inc.; **Fig. 4.15a:** © Michael J. Dykstra/Visuals Unlimited; **Fig. 4.15b:** © Manfred Schliwa/Visuals Unlimited; **Fig. 4.16a:** Prentice Hall, Upper Saddle River, New Jersey; **Fig. 4.17:** Courtesy of Dr. Garry T. Cole, Univ. of Texas at Austin; **Fig. 4.19:** © Henry C. Aldrich/Visuals Unlimited; **Fig. 4.22:** National Research Council of Canada; **Fig. 4.23:** © Karl Aufderheide/Visuals Unlimited; **Fig. 4.24a:** K.G. Murti/Visuals Unlimited; **Fig. 4.25a:** © Ralph A. Slepecky/Visuals Unlimited; **Fig. 4.25b:** © W.L. Dentler/Biological Photo Service.

Chapter 5

Opener: © Lauritz Jensen/Visuals Unlimited; **Fig. 5.10b 1,2:** Courtesy of Dr. Eshel Ben-Jacob; **Fig. 5.11a–d:** © David M. Phillips/Visuals Unlimited.

Chapter 6

Opener: Courtesy of Nagle Company; **Fig. 6.7a:** Courtesy Nagle Corporation; **Fig. 6.7b:** © B. Otero/Visuals Unlimited; **Fig. 6.7c, d:** Courtesy Nagle Corporation; **Box 6.1:** © Science VU-D. Foster, Woods Hole Oceanographic Institution/Visuals Unlimited; **Fig. 6.16:** Photo Provided by ThermoForma of Marietta, Ohio; **Fig. 6.19a,b:** Courtesy of Jeanne S. Poindexter, Long Island University.

Chapter 7

Opener: © Visuals Unlimited; **Fig. 7.3a:** Courtesy of AMSCO Scientific, Apex, NC; **Fig. 7.4b:** Courtesy of Millipore Corporation; **Fig. 7.5a:** Courtesy of Pall Ultrafine Filter Corporation; **Fig. 7.5b:** © Fred Hossler/Visuals Unlimited; **Fig. 7.6a:** Provided by ThermoForma, Marietta, Ohio.

Chapter 8

Opener: Reprinted by permission W.N. Lipscomb, Harvard University; **Fig. 8.1:** © Artville CD; **Fig. 8.16a,b:** From Voet, *Biochemistry,* 1/e. ©1990 John Wiley & Sons, Inc. Reprinted by permission of John Wiley & Sons, Inc. Image courtesy of Donald Voet.; **Fig. 8.26a,b:** Courtesy of David Eisenberg, UCLA.

Chapter 9

Opener: © The Nobel Foundation 1989; **Fig. 9.28a,b:** © The Nobel Foundation 1989.

Chapter 10

Opener: Fig. 10.15: From A.S. Moffat, "Nitrogenase Structure Revealed," *Science* 250: 1513, December 14, 1990. Photo by M.N. Georgiadis and D.C. Rees, Caltech.

Chapter 11

Fig. 11.7c-1: Illustrations, Irving Geis/Geis Archives Trust. Rights owned by Howard Hughes Medical Institute. Not to be used with out permission; **Fig. 11.7c-2:** From Voet, *Biochemistry,* 1/e. ©1990 John Wiley & Sons, Inc. Reprinted by permission of John Wiley & Sons, Inc. Image courtesy of Donald Voet; **Fig. 11.10a:** Reprinted with permission from Nature 389, 251 Macmillan Magazines Limited. Image courtesy Prof. Dr. Timothy J. Richmond.

Chapter 12

Opener: Lewis, M.. et al. 1996. "Crystal structure of the lactose operon repressor and its complexes with DNA and inducer," *Science* 271:1247–54. Figures 5A (p. 250), 6A (p. 1251), and 1J (p. 153, the left illustration); **Fig. 12.3a:** Reprinted with permission from Science 296, May 17, 2002, Murakami, K.S.; Masuda, S.; and Darst, S.A. © 2002 by the AAAS. Image courtesy of Seth Darst; **Fig. 12.3b:** Reprinted with permission from Science 296, May 17, 2002, Mukarkami, K.S.; Masuda S.; Campbell, E.A.; Muzzin, O.; and Darst, S.A. © 2002 by the AAAS. Image courtesy of Seth Darst; **Fig. 12.3c:** Reprinted with permission from Science 292, June 8, 2002. Gnatt A.L.; Cramer, P; Jianhua, F.; Bushnell, D.A.; and Komberg, R.D. © 2002 by the AAAS. Image courtesy of Dave Bushnell and Dr. Roger Kornberg; **Fig. 12.3d:** Reprinted with permission from Science 292, June 8, 2002 by Aaron Klug. Image courtesy of Dr. Roger Kornberg; **Fig. 12.10:** From R. Rould and T. Steitz, "Structure of E. Coli Glutaminyl-tRNA Synthetase, Complexed with t RNA gln and ATP at 2.8 A Resolution," *Science* 246:1135–1142, Dec. 1, 1989. © 1991 by the AAAS; **Fig. 12.13e,f,g:** Reprinted Figures 2b,f,g from Yusupov, M.M.; Yusupov, G; Baucom, A; Lieberman, K; Earnest, T; Cate, J; and Noller, H. Science 29:88396. May 4, 2001. © 2001 AAAS; **Fig. 12.20a-1, a-2:** Reprinted with permission from *Nature* 388:741–750. Figures 1a and 1b, page 742. Courtesy Paul Sigler; **Fig. 12.20b:** Reprinted with permission from *Nature* 388: 741–750. Figures 1a and 1b, page 742. Courtesy Paul Sigler; **Fig. 12.26:** Lewis, M.. et al. 1996, "Crystal Structure of the Lactose Operon Repressor and its Complexes with DNA and Inducer," *Science* 271: 1247–54. Figures 5A (p. 250), 6A (p. 1251), and 1J (p. 153, the left illustration); **Fig. 12.29b:** From S.C. Schultz, G.C. Shields, and T.A. Steitz, "Crystal Structure of a CAP-DNA Complex: The DNA is Bent by 90 degrees," *Science* 253:1001–1007, Aug. 30, 1991. © 1991 by AAAS.

Chapter 13

Opener: © Oliver Meckes/Photo Researchers, Inc.; **Fig. 13.6:** Courtesy of Charles C. Brinton, Jr. and Judith Carnahan; **Fig. 13.24b,c:** From *Molecular Biology of Bacterial Viruses* by Gunther S. Stent. © 1963 W.H. Freeman and Company. Used with permission.

Chapter 14

Opener: Courtesy of Keith V. Wood; **Fig. 14.1:** Provided by Dr. A.K. Aggarwal (Mount Sinai School of Medicine) from Newman et al. *Science* 269, 656–663(1995). © 1995 by the AAAS; **Fig. 14.9:** Courtesy of the Perkin-Elmer Corporation; **Fig. 14.10b:** Courtesy of Lifetime Technologies, Inc.; **Fig. 14.11 a,b:** Huntington Potter and David Dressler/*Life Magazine* 1980, Time, Inc.; **Box 14.1:** Courtesy of Keith V. Wood; **Box 14.2b:** From D.W. Ow et al., "Transient and Stable Expression of the Firefly Luciferase Gene in Plant Cells and Trasgenic Plants," *Science* 234: 856–859, Nov. 14, 1986. © 1986 by the AAAS.

Chapter 15

Opener: Image courtesy of Affymetrix, Inc. (Santa Clara, CA); **Fig. 15.2a,b:** Reproduced with permission from Invitrogen Corporation; **Fig. 15.5:** Reprinted with permission from Fraser, C.M. et al. "The Minimal Gene Complement of *Mycoplasma Genitalium*," *Science,* 270:397–403, Figure 1, page 398. © 1995 by the AAAS. Photo by The Institute for Genomic Research; **Fig. 15.6:** Reprinted with permission from Fleischman, R.E. et al. 1995. "Whole-genome random sequencing and assembly of *Haemophilus influenzae*," *Science* 269:496–512. Figure 1, page 507. © 1995 by the AAAS. Photo by The Institute for Genomic Research; **Fig. 15.9:** Image courtesy of Affymetrix, Inc. (Santa Clara, CA); **Fig. 15.10:** © Swiss Institute of Bioinformatics, Geneva, Switzerland.

Chapter 16

Opener: Courtesy of Robert C. Liddington and Stephen C. Harrison, Harvard University; **Fig. 16.2:** © Terry C. Hazen/Visuals Unlimited; **Fig. 16.3a,b:** © David M. Phillips/Visuals Unlimited; **Fig. 16.4:** From S.E. Luria, *General Virology,* © 1978 John Wiley & Sons, Inc. Reprinted by permission John Wiley & Sons, Inc.; **Fig. 16.5a:** © Runk/Schoenberger/Grant Heilman Photography, Inc.; **Fig. 16.5b:** © Charles Marden Fitch; **Fig. 16.8:** Courtesy of Janey S. Symington; **Fig. 16.11a:** © Dennis Kunkel/Phototake; **Fig. 16.11c:** Courtesy of Gerald Stubbs and Keiichi Namba, Vanderbilt University; and Donald Caspar, Brandeis University; **Fig. 16.12a:** Courtesy of Michael G. Rossman, Purdue University; **Fig. 16.12b,c:** From J.M. Hogle et al., "Three Dimensional Structure of Poliovirus at 2.9 A resolution," *Science* 229: 1360, Sept. 27, 1985. © 1985 by the AAAS; **Fig. 16.12d:** © George Musil/Visuals Unlimited; **Fig. 16.12e:** Courtesy of Harold Fisher, University of Rhode Island and Robley Williams, University of California at Berkeley; **Fig. 16.12f:** © R. Feldman-Dan McCoy/Rainbow; **Fig. 16.12g:** Courtesy of Harold Fisher, University of Rhode Island and Robley Williams, University of California at Berkeley; **Fig. 16.12h:** Science VU-NIH, R. Feldman/Visuals Unlimited; **Fig. 16.14a,b:** Courtesy of Robert C. Liddington and Stephen C. Harrison, Harvard University; **Figs. 16.15, 16.17a,c:** © K.G. Murti/Visuals Unlimited; **Fig. 16.17d:** © Dr. S. Baum/Peter Arnold, Inc.; **Fig. 16.17e:** © CDC/Science Source/Photo Researchers, Inc.; **Fig. 16.17f:** © R. Feldman-Dan McCoy/Rainbow; **Fig. 16.18b:** Center for Disease Control and Prevention; **Fig. 16.18c:** © K.G. Murti/Visuals Unlimited; **Fig. 16.19b:** © Tomas Broker/Phototake.

Chapter 17

Opener: © Oliver Meckes/Photo Researchers, Inc.; **Fig. 17.4:** © Lee D. Simon/Photo Researchers, Inc.; **Fig. 17.6b-1:** © Fred Hossler/Visuals Unlimited; **Fig. 17.6b-2:** George Chapman, Georgetown University; **Fig. 17.15:** © M. Wurtz/Photo Researchers, Inc.; **Fig. 17.18a:** From F.D. Bushman, C. Shang, and M. Ptashne, "A Single Glutamic Acid Residue Plays a Key Role in the Transcriptional Activation Function of Lambda Repressor," *Cell* 58:1163–1171, September 22, 1989. Cell Press; **Fig. 17.20a:** From A.K. Aggarwal, D.W. Rodgers, M. Drottar, M. Ptashne, and S.C. Harrison, "Recognition of a DNA Operator by the Repressor of Phage 434: A View at High Resolution." *Science* 242:899–907, Nov. 11, 1988. © 1988 by the AAAS.

Chapter 18

Opener: Courtesy of Wayne Hendrickson, Columbia University; **Fig. 18.8:** © K.G. Murti/Visuals Unlimited; **Fig. 18.10a,b:** Center for Disease Control and Prevention; **Fig. 18.11:** Courtesy of J.T. Finch, J.M. Kaper, USDA Agricultural Research Service; **Fig. 18.14a,b:** Courtesy of Russell L. Steere, Advanced Biotechnologies, Inc.; **Fig. 18.15:** © J.R. Adams/Visuals Unlimited.

Chapter 19

Opener: C. Andrew Henley-Larus; **Fig. 19.1a-c:** J. William Schopf. Reprinted with permission from *Science* 260, April 30, Fig. 4a,f,g, p. 643. © 1993 by the American Association of the Advancement of Science; **Fig. 19.2:** C. Andrew Henley-Larus.

Chapter 20

Opener: Lansing Prescott; **Fig. 20.1a,b:** From O. Kandler and H. Konig, "Cell Envelopes of Archaebacteria", *The Bacteria,* Vol. 8, 1985, Fig. 1 (part 1 of 3), page 414, Academic Press; **Fig. 20.8a,b:** Lansing Prescott; **Fig. 20.9a,b:** © Corale L. Brierley/Visuals Unlimited; **Fig. 20.9c:** From J.T. Staley, M.P. Bryant, N. Pfenning, and J.G. Holt (Eds.), *Bergey's Manual of Systematic Bacteriology,* Vol. 3. © 1989 Williams and Wilkins Co., Baltimore. Micrograph courtesy of D. Janekovic and W. Zillig; **Fig. 20.10a:** © Friederich Widdell/Visuals Unlimited; **Fig. 20.1b:** From J.T. Staley, M.P. Bryant, N. Pfenning, and J.G. Hold (Eds.), *Bergey's Manual of Systematic Bacteriology,* Vol. 3 © 1989 Williams and Wilkins Co., Baltimore; **Fig. 20.10c:** © Henry C. Aldrich/Visuals Unlimited; **Fig. 20.10d:** From J.T. Staley, M.P. Bryant, N. Pfenning, and J.G. Holt (Eds.), *Bergey's Manual of Systematic Bacteriology,* Vol. 3. © 1989 Williams and Wilkins Co., Baltimore. R. Robinson, Department of Microbiology, U. of California, Los Angeles; **Box 20.1:** From PNAS, Vol. 99, pages 7663–7668, Fig. 1, Orphan et al, May 2002. © 2002 National Academy of Sciences, U.S.A. Image courtesy of Christopher H. House. **Fig. 20.13a:** From J.T. Staley, M.P. Bryant, N. Pfenning, and J.G. Holt (Eds.), *Bergey's Manual of Systematic Bacteriology,* Vol. 3. © 1989 Williams and Wilkins Co., Baltimore. Prepared by G. Bentzen; photographed by the Laboratory of Clinical Electron Microscopy, U. of Bergen; **Fig. 20.13b:** From J.T. Staley, M.P. Bryant, N. Pfenning, and J.G. Holt (Eds.), *Bergey's Manual of Systematic Bacteriology,* Vol. 3. © 1989 Williams and Wilkins Co., Baltimore. Prepared by A.L. Ustad, Photography by Dept. of Biophysics, Norwegian Institute of Technology; **Fig. 20.14:** From J.T. Staley, M.P. Bryant, N. Pfenning, and J.G. Holt (Eds.), *Bergey's Manual of Systematic Bacteriology,* Vol. 3. © 1989 Williams and Wilkins Co., Baltimore.

Chapter 21

Opener: © Arthur M. Siegelman/Visuals Unlimited; **Fig. 21.1:** R. Huber, H. Konig, K.O. Stetter; **Fig. 21.2a:** From J.G. Holt (Eds.) *Bergey's Manual of Systematic Bacteriology,* Vol. 2, ©1986 Lippincott Williams and Wilkins Co., Baltimore; **Fig. 21.2b:** From Family I. Deinococcaceae. Battista, J.R. and Rainey, F.A. In *Bergey's Manual of Systematic Bacteriology,* 2nd ed., vol. 1, George Garrity © 2001 Springer. Image courtesy John Battista; **Fig. 21.2c:** Image courtesy Prof. Abraham Minsky; **Fig. 21.5a:** From J.T. Staley, M.P. Bryant, N. Pfenning, and J.G. Holt (Eds.), *Bergey's Manual of Systematic Bacteriology,* Vol. 3. © 1989 Williams and Wilkins Co., Baltimore. Micrograph by G. Cohen-Bazire; **Fig. 21.5b,c:** Reprinted from the *Shorter Bergey's Manual of Determinative Bacteriology,* 8/e, John G. Holt, editor, 1977, © Bergey's Manual Trust. Published by Williams and Wilkins, Baltimore, MD; **Fig. 21.6:** © Elizabeth Gentt, Visuals Unlimited; **Fig. 21.7a:** © T.E. Adams/Visuals Unlimited; **Fig. 21.7b:** © Ron Dengler/Visuals Unlimited; **Fig. 21.7c:** © M.I.

Heilman Photography; Box 37.4: Dr. Fred Murphy/Centers for Disease Control and Prevention.

Chapter 38

Opener: Bettmann/Corbis; **Fig. 38.1b:** © John D. Cunningham/Visuals Unlimited; **Fig. 38.2c:** © Carroll H. Weiss/Camera M.D. Studios; **Fig. 38.3a:** Armed Forces Institute of Pathology; **Fig. 38.4:** © /Biophoto Associates/Photo Researchers; **Fig. 38.5:** © Carroll H. Weiss/Camera M.D. Studios; **Fig. 38.6:** Armed Forces Institute of Pathology; **Box 38.2:** © Photo Researchers, Inc.; **Fig. 38.10a:** © Carroll H. Weiss/Camera M.D. Studios; **Fig. 38.10b:** © Science VU/Visuals Unlimited; **Fig. 38.13:** © Carroll H. Weiss/Camera M.D. Studios; **Fig. 38.14a,b:** From N.H. Olson et al., *Proceeding of the National Academy of Sciences,* 90:507. 1993. Photo courtesy of Dr. Michael Rossmann; **Fig. 38.15:** Courtesy of Dan Wiedbrauk, Ph.D, Warde Medical Laboratory, Ann Arbor, Michigan; **Fig. 38.16a:** © CDC/Science Source/Photo Researchers; **Fig. 38.16b:** © Carroll H. Weiss/Camera M.D. Studios; **Fig. 38.17:** © Dr. Brian Eyden/SPL/Photo Researchers; **Fig. 38.18a:** © Tektoff-RM/CNRI/SPL/Custom Medical Stock; **Fig. 38.19:** Courtesy of The National Institute of Health; **Fig. 38.20:** © Eye of Science/Photo Researchers; **Box 38.3:** Corbis; **Fig. 38.21a:** © Kenneth E. Greer/Visuals Unlimited; **Fig. 38.21b:** © Carroll H. Weiss/Camera M.D. Studios; **Fig. 38.21c,d:** © Kenneth E. Greer/Visuals Unlimited.

Chapter 39

Opener: Courtesy of The Royal College of Surgeons Museum, Edinburg, Scotland; **Fig. 39.2:** © Fred Hossler/Visuals Unlimited; **Fig. 39.4:** © Carroll H. Weiss/Camera M.D. Studios; **Fig. 39.5:** Armed Forces Institute of Pathology; **Fig. 39.6:** © M. Abbey/Visuals Unlimited; **Fig. 39.8a:** From *ASM News* 55(2) cover, 1986. American Society for Microbiology; **Fig. 39.8b,c:** © CDC/Peter Arnold, Inc.; **Fig. 39.9:** Centers for Disease Control and Prevention; **Fig. 39.10:** Department of Health & Human Services, courtesy of Dr. W. Burgdorfer; **Fig. 39.11a;** © Science VU/Charles Stratton/Visuals Unlimited; **Fig. 39.12:** © Arthur M. Siegelman/Visuals Unlimited; **Fig. 39.14:** © Science VU-WHO/Visuals Unlimited; **Fig. 39.15:** Armed Forces Institute of Pathology; **Fig. 39.16:** From V. Neman-Simha and F. Megraud, "In Vitro Model for Capylobacter pylori Adherence Properties," *Infection and Immunity,* 56(12):3329–3333, Dec. 1988. American Society for Microbiology; **Fig. 39.17a:** From R. Baselga et al., "Staphylococcus Aureus:Implications in Colonization and Virulence," *Infection and Immunity,* 61(11)L4857–4862, 1993.© American Society for Microbiology; **Fig. 39.17b:** Courtesy of Dr. Dennis G. Maki; **Fig. 39.19a-e:** © Carroll H. Weiss/Camera M.D. Studios; **Fig. 39.19f:** ©Charles Stoer/Camera M.D. Studios; **Fig. 39.20a-c:** © Carroll H. Weiss/Camera M.D. Studios; **Fig. 39.22:** Armed Forces Institute of Pathology; **Fig. 39.23:** From Jacob S. Teppema, "In Vivo Adherence and Colonization of *Vibrio cholerae* Strains That Differ in Hemagglutinating Activity and Motility," *Journal of Infection and Immunity,* 55(9)2093–2102, Sept. 1987. Reprinted by permission of American Society for Microbiology; **Fig. 39.25b:** © Max Listgarten, University of Pennsylvania/Biological Photo Service; **Fig. 39.26:** © E.C.S. Chan/Visuals Unlimited.

Chapter 40

Opener: Lennart Nilsson/Albert Bonniers Forlag AB; **Fig. 40.1:** © Everett S. Beneke/Visuals Unlimited; **Fig. 40.2a,b:** © Everett S. Beneke/Visuals Unlimited; **Figs. 40.3-40.6:** © Carroll H. Weiss/Camera M.D. Studios; **Fig. 40.7:** © Everett S. Beneke/Visuals Unlimited; **Fig.**

40.8: Reprinted by permission of Upjohn Co. from E.S. Beneke et al., 1984 *Human Mycosis in Microbiology;* **Fig. 40.9:** © Everett S. Beneke/Visuals Unlimited; **Fig. 40.10:** Reprinted by permission of Upjohn Co. from E.S. Beneke et al., 1984 *Human Mycosis in Microbiology;* **Fig. 40.11:** © E.C.S. Chan/Visuals Unlimited; **Figs. 40.12, 40.13a:** © Arthur M. Siegelman/Visuals Unlimited; **Fig. 40.13b:** Armed Forces Institute of Pathology; **Fig. 40.14:** © Everett S. Beneke/Visuals Unlimited; **Fig. 40.15a:** David M. Phillips/Visuals Unlimited; **Fig. 40.15b,c:** © Everett S. Beneke/Visuals Unlimited; **Fig. 40.16a:** © Lauritz Jensen/Visuals Unlimited; **Fig. 40.16b:** © Robert Calentine/Visuals Unlimited; **Fig. 40.17a,b:** From M. Schaechter, G. Medoff, and D. Schiessinger (Eds.), *Mechanisms of Microbial Disease,* 1989. Williams and Wilkins; **Fig. 40.19:** Centers for Disease Control and Prevention; **Fig. 40.21a,b:** Armed Forces Institute of Pathology; **Fig. 40.22:** © David M. Phillips/Visuals Unlimited.

Chapter 41

Opener: © Christiana Dittmann/Rainbow; **Fig. 41.2:** Donald A Klein; **Fig. 41.3a:** © Tom E. Adams/Peter Arnold, Inc.; **Fig. 41.3b:** © Martha Powell/Visuals Unlimited; **Fig. 41.6:** © Photo by Mark Seliger, Courtesy of Campbell Soup Company; **Fig. 41.8:** Reprinted from *Applied and Environment Microbiology* (64) 2284–2286, Fig. 1, p. 2284, Sturbaum, G.D., Ortega, Y.R., Gilman, R.H., Sterling, C.R., Caberea, L., and Klein, D.A., "Detection of *Cyclospora cayetanensis* in Wastewater." © 1998 American Society for Microbiology. Image Courtesy of Greg Sturbaum; **Fig. 41.9:** From Peterkin, Idzigk, and Sharpe. "Screening DNA Probes Using the Hydrophobic Probe Grid-Membrane Filter," *Food Microbiology* (6)281–284, 1989. Academic Press, Inc. (London); **Fig. 41.11:** © Elmer Koheman/Visuals Unlimited; **Fig. 41.12:** From D.B. Hughes and D.G. Hover, *Food Technology,* April 1991, Fig. 3, p. 79; **Fig. 41.13a–e:** © John D. Cunningham/Visuals Unlimited; **Fig. 41.14:** © Joe Munroe/Photo Researchers; **Fig. 41.17:** © Vance Henry/Nelson Henry; **Fig. 41.19:** © Stanley Flegler/Visuals Unlimited.

Chapter 42

Opener: Courtesy of General Electric Research and Development Center; **Fig. 42.4a,b:** From B. Atkinson and Daoud, *Advanced Biochemical Engineering* 4:83, Springer Verlag; **Fig. 42.5a:** Society for Industrial Microbiology; **Fig. 42.16:** From J.R. Postgate, *The Sulphate-Reducing Bacteria,* Reprinted with permission of Cambridge University Press.

Appendix

A19b-1, b-2: Illustrations, Irving Geis/Geis Archives Trust. Rights owned by Howard Hughes Medical Institute. Not to be used without permission.

Line Art/Tables/Illustrations

Chapter 2

Fig. 2.23: From William A. Jensen and Roderic B. Park, *Cell Ultrastructure.* Copyright 1967 Wadsworth Publishing Co., Belmont, CA. Reprinted by permission.

Chapter 7

Fig. 7.3b: From John J. Perkins, *Principles and Methods of Sterilization in Health Science,* 2nd edition, 1969.

Courtesy of Charles C. Thomas, Publisher, Springfield, Illinois; **Fig. 7.4a:** Courtesy of Millipore Corporation. Reprinted by permission.

Chapter 10

Table 10.1: "Biosynthesis in *Escherichia coli*" Albert Lehninger, *Bioenergetics,* 2nd edition. Permission Addison Wesley Publishing.

Chapter 11

Table 11.2: Suzuki, Griffiths, Miller, and Lewontin *An Introduction to Genetic Analysis,* 3rd edition, 1986, permission W.H. Freeman.

Chapter 12

Fig. 12.24: "Crystal structure of the lactose operon repressor and its complexes with DNA and inducer" by M. Lewis, et al. from *Science,* 271:I247–54, 1996; **Fig. 12.35:** Adapted from R. Lesick and L. Shapiro, *Science,* 282: p. 1431, 1998.

Chapter 13

Fig. 13.17: "Mechanism of transformation" by J-P Claverys and B Martin from *Trends in Microbiology,* Vol 11, No. 4, April 2003, fig. 1, p.161 Courtesy Elsevier Science, Ltd.

Chapter 15

Fig. 15.7: "Metabolic pathways and transport systems of treponema pallidum" C. M. Fraser, et al. from *Science,* 281: p. 377, July 17, 1998.

Chapter 16

Fig. 16.13: *The Structure of an Icosahedral Capsid* from *Microbiology,* Third Edition by Bernard D. Davis, et al. Copyright 1980 by Harper & Row, Publishers, Inc. Reprinted by permission of HarperCollins Publishers, Inc.

Chapter 17

Fig. 17.1: "Major Bacteriophage Families and Genera" From Van Regenmortel, Fauquet, Bishop, et al., *Virus Taxonomy, 7th Report,* 2000. Academic Press.

Chapter 18

Fig. 18.3: "Diagrammatic Description of Families and Genera of Viruses" From Van Regenmortel, Fauquet, et al., *Virus Taxonomy, 7th Report,* 2000. Academic Press; **Fig. 18.5:** Adapted from S. J. Flint et al., *Principles of Virology,* ASM Press, 2000; **Fig. 18.7:** Adapted from S. J. Flint, et al., *Principles of Virology,* ASM Press, 2000; **Fig. 18.12:** From Van Regenmortel, Fauquet, Bishop, et al., *Virus Taxonomy, 7th Report,* 2000. Academic Press.

Chapter 19

Fig. 19.11: From W. F. Doolittle, *Science* 284: p. 2127. June 25, 1999; **Fig. 19.14:** W.Ludwig & H-P Klenk In Garrity G, ed: *Bergey's Manual of Systematic Bacteriology, Volume I second edition* 2000. Fig. 16. Courtesy Springer-Verlag.

Chapter 20

Fig. 20.7: From H. Huber and K. O. Stettler, *Bergey's Manual of Systematic Bacteriology, Volume I,* Second Edition, George Garrity, Editor-in-Chief. Copyright 2000 Springer-Verlag.

Chapter 21

Table 21.8a: Courtesy Hartwell T. Crim, 1998.

Chapter 22

Fig. 22.16a: From W. Ludwig and H.-P. Klenk, *Bergey's Manual of Systematic Bacteriology, Volume I,* Second Edition, George Garrity, Editor-in-Chief. Copyright 2000 Springer-Verlag; **Table 22.2:** S.W. Watson, et al.: The Family Nitrobacteraceae'. In *The Prokaryotes,* Vol. 1 edited by M.P. Starr et al., 1981. Permission Springer-Verlag Publishing.

Chapter 23

Fig. 23.1a: From W. Ludwig and H.-P. Klenk, *Bergey's Manual of Systematic Bacteriology, Volume I,* Second Edition, George Garrity, Editor-in-Chief. Copyright 2000 Springer-Verlag.

Chapter 24

Fig. 24.3b: From W. Ludwig and H.-P. Klenk, *Bergey's Manual of Systematic Bacteriology, Volume I,* Second Edition, George Garrity, Editor-in-Chief. Copyright 2000 Springer-Verlag.

Chapter 28

Fig. 28.4: From "The Hot Spot" in *Woods Hole Currents* Volume 4, Number 3. Summer 1995; **Fig. 28.8:** From W. W. Mohn and J. M. Tiedje "Microbial Reductive Dehalogenation" in *Microbiological Reviews* 56(3). September 1992; **Fig. 28.13:** R. Guerrero, "Predation as prerequisite to organelle origin: *Daptobacter* as example" in *Symbiosis as a Source of Evolutionary Innovation: Speciation and Morphogenesis,* L. Margulis and R. Fester, eds. MIT, Cambridge MA, 1991; **Fig. 28.16:** M. F. Polz et al., "When Bacteria Hitch a Ride" in *ASM News* 66: 538, 2000; **Fig. 28.34:** J. Fröhlich, J. and H. König "Rapid isolation of single microbial cells from mixed natural and laboratory populations with the aid of a micromanipulator" in *System. Appl. Microbiol.* 2: 253, 1999; **Box 28.2:** "Genetic and evolutionary constraints for the symbiosis between animals and methanogenic bacteria, JHP Hackstein, P. Langer, and J Rosenberg. *Env. Monitoring and Assessment* 42:39-56; **Table 28.2:** Ruby, E. G. Ecolory of benign "infection": Colonization of the squid luminous organ in *Vibrio fisheri.* In E. Rosenberg, *Microbial Ecology and Infectious Disease,* 1999 courtesy ASM.

Chapter 29

Fig. 29.10: E. F. DeLong, et al., "Visualization and enumeration of marine planktonic archaea and bacteria by using polyribonucleotide probes and fluorescent In Situ hybridization" in *Appl. Environ. Microbiol.* 65: 5560,
1999; **Table 29.2:** V.M. Gorlenko et al., *Die Binnengewasser,* Vol XXVIII, Schweizerbart'sche Verlagsbuchhandlung, 1983 and B. Rothe, et al. Archives of Microbiology, 147: 92-99, 1987; **Table 29.3:** A.E. Greenberg et al. Standard Methods for the Examination of Water and Wastewater, 16th edition, 1985, American Public Health Association.

Chapter 30

Fig. 30.19a: Adapted from S. W. Nixon "Coastal marine eutrophication: a definition, social causes, and future concerns" in *Ophelia* 41: 211-212, 1995; **Fig. 30.21b:** Adapted from R. Conrad "Soil microbial processes involved in production and consumption of atmospheric trace gasses." In *Adv. Micro. Ecol.* 14, 1995 with permission Kluwer Academic; **Fig. 30.22:** Adapted from D. K. Lovley "Dissimilatory Fe (III) and Mn (IV) reduction" in *Microbiol. Rev.* 55: 269, 1991; **Fig. 30.23:** From J. M. Hunt *Petroleum Geochemistry and Geology, Second Edition.* Copyright W. H. Freeman and Company, New York NY, 1996; **Table 30.1:** E.W. Russell *Soil Conditions and Plant Growth,* 10th edition, 1973 permission Longman Group LTD; **Table 30.3:** J.W. Woldendorp: "The rhizosphere as part of the plant-soil system." In *Structure and Functioning of Plant Populations* (Amsterdam, Holland; Proceedings, Royal Dutch Academy of Sciences, Natural Sciences Section: 2nd Series, 1978) 70:243; **Table 30.5:** Lengeler, J.W., Drews, G., Schlegel, H.G. 1999 Biology of the Prokaryotes, with permission Blackwell Publishers; **Table 30.6:** Watling, R. and Harper, D.B. 1998 *Mycol, Res.* 102(7): 769-787, with permission Cambridge University Press.

Chapter 32

Fig. 32.27: From Thomas J. Smith, et al., "Structure of a human rhinovirus-bivalently bound antibody complex" in *Proceedings of the National Academy of Science,* Vol. 90, pp. 7015-7018, August 1993. Reprinted by permission.

Chapter 33

Table 33.3: Adapted from Kuby, et al. *Immunology.* Copyright W. H. Freeman and Company, New York NY, 2000.

Chapter 34

Fig. 34.4: Adapted from C. J. Hueck, *Microbiol. and Mol. Biol. Rev.* 62: 379-438. ASM, June 1998.

Chapter 39

Fig. 39.11: Martin Enserink: *Science,* Vol 294, Oct. 2001, pp. 490-491 with permission.

Chapter 41

Fig. 41.4: M.A. Carlson, *Biosensors and Bioelectronics* 14, 2000; **Fig. 41.7:** Adapted from G. D. Lewis et al., "Influence of environmental factors on virus detection by RT-PCR and cell culture" in *J. Appl. Microbiol.* 88: 638. Copyright 2000. with permission Blackwell Publshing; **Fig. 41.10:** Adapted from J. K. Wan, et al., "Probelia ™ PCR system for rapid detection of *Salmonella* in milk powder and ricotta cheese in *Let. Appl. Microbiol.* 30: 269. Copyright 2000; **Table 41.2:** James M. Jay *Modern Food Microbiology,* 6th edition. Permission Chapman & Hall; **Table 41.3:** F.M. van Dolah Environ. Health Persp. 108 (Supplement 1): 133-141; **Table 41.5:** James M. Jay *Modern Food Microbiology,* 6th edition . Permission Chapman & Hall; **Table 41.7:** B. A. Law *Microbiology and Biochemistry of Cheese and Fermented Milk,* 2nd edition, permission Chapman & Hall; **Table 41.9:** James M. Jay *Modern Food Microbiology,* 6th edition. Permission Chapman & Hall.

Chapter 42

Fig. 42.1: Modified from Crueger and Crueger *Biotechnology: A Textbook of Industrial Microbiology, Second Edition.* Copyright Science Tech Publishers, Madison, WI, 1990; **Fig. 42.3:** Adapted from S. S. D. Donadio, et al., "Recent Developments in the genetics of erythromycin formation" in *Industrial Microorganisms: Basic and Applied Molecular Genetics,* R. H. Baltz, et al., eds. ASM, Washington, DC, 1993; **Fig. 42.10:** Modified from Crueger and Crueger *Biotechnology: A Textbook of Industrial Microbiology, Second Edition.* Copyright Science Tech Publishers, Madison, WI, 1990; **Fig. 42.11:** Modified from S. Pedersen, L. Dijkhuizen, B. W. Dijkstra, B.F. Jensen, and S.T. Jorgensen " A better enzyme for cyclodextrins" in *Chemtech,* 25: 19-25. Figure 1, p. 20 permission American Chemical Society; **Box Fig. 42.4:** From M. Wilchek and E. A. Bayer "Introduction to avidin-biotin technology" in *Meth. Enzymology* 184. 1990, and *Biomolecular Engineering* 16: 1-4, 1999; **Table 42.2:** D.A. Cowan: Microbial Genomes – the untapped resource. In *Tibtech* 18: 14-16, 2000. Permission Elsevier Science; **Table 42.3:** Adapted from S. Ostergaard, L. Olsson, and J. Neilson: "Metabolic engineering of *Saccharomyces cerevisiae*" In *Microbiol. Mol. Biol. Rev.* 64(1): 34-50, 2000; **Table 42.4:** Adapted from S. Ostergaard, L. Olsson, and J. Neilson: "Metabolic engineering of *Saccharomyces cerevisiae*" In *Microbiol. Mol. Biol. Rev.* 64(1): 34-50, 2000; **Table 42.5:** J. A. Shapiro: " Natural Genetic Engineering, Adaptive Mutation, and Bacterial Evolution" In *Microbial Ecology of Infectious Disease,* E. Rosenberg, editor, 259-275, 1999 courtesy ASM.

Index